T0179961

Phenomenology of Particle Physics

Written for a two-semester Master's or graduate course, this comprehensive treatise intertwines theory and experiment in an original approach that covers all aspects of modern particle physics. The author uses rigorous step-by-step derivations and provides more than 100 end-of-chapter problems for additional practice to ensure that students will not only understand the material but also be able to apply their knowledge. Featuring up-to-date experimental material, including the discovery of the Higgs boson at CERN and of neutrino oscillations, this monumental volume also serves as a one-stop reference for particle physics researchers of all levels and specialties. Richly illustrated with more than 450 figures, the text guides students through all the intricacies of quantum mechanics and quantum field theory in an intuitive manner that few books achieve.

André Rubbia is a professor in experimental particle physics at ETH Zurich. After obtaining his Ph.D. in particle physics at the Massachusetts Institute of Technology, he worked on the research staff at CERN in Geneva, Switzerland. His research interests include high-energy physics and, in particular, studies of neutrino properties, and he has been a primary contributor to the development of novel particle detectors. He has proposed, developed, and led several international projects in Europe, Asia, and the USA. While continuing his focus on research in particle physics, he has acquired an extended experience in teaching undergraduate and Master's-level courses.

Phenomenology of Particle Physics

André Rubbia

Eidgenössische Technische Hochschule Zürich

CAMBRIDGE
UNIVERSITY PRESS

University Printing House, Cambridge CB2 8BS, United Kingdom

One Liberty Plaza, 20th Floor, New York, NY 10006, USA

477 Williamstown Road, Port Melbourne, VIC 3207, Australia

314–321, 3rd Floor, Plot 3, Splendor Forum, Jasola District Centre, New Delhi – 110025, India

103 Penang Road, #05–06/07, Visioncrest Commercial, Singapore 238467

Cambridge University Press is part of the University of Cambridge.

It furthers the University's mission by disseminating knowledge in the pursuit of education, learning, and research at the highest international levels of excellence.

www.cambridge.org
Information on this title: www.cambridge.org/highereducation/isbn/9781316519349
DOI: 10.1017/9781009023429

First published 2022

Printed in the United Kingdom by TJ Books Limited, Padstow Cornwall 2022

A catalogue record for this publication is available from the British Library.

Library of Congress Cataloging-in-Publication Data
Names: Rubbia, André, 1966– author.
Title: Phenomenology of particle physics / André Rubbia.
Description: Cambridge ; New York, NY : Cambridge University Press, 2022. | Includes bibliographical references and index.
Identifiers: LCCN 2021044753 | ISBN 9781316519349 (hardback)
Subjects: LCSH: Particles (Nuclear physics) | Phenomenological theory (Physics) | Quantum theory. | Quantum field theory. | BISAC: SCIENCE / Physics / Nuclear
Classification: LCC QC793.29 .R83 2022 | DDC 539.7/2–dc23/eng/20211006
LC record available at https://lccn.loc.gov/2021044753

ISBN 978-1-316-51934-9 Hardback

Additional resources for this publication at www.cambridge.org/rubbia

To my family – wife and children

"Life can only be understood backwards; but it must be lived forwards."
Soren Kierkegaard

Contents

Preface

The purpose of this textbook is to teach particle physics, addressing both its phenomenology and its theoretical foundations. **Why bother about the phenomenology of particle physics and quantum field theory?** I would say that one of the main reasons is because the current Standard Model of particle physics can account for (almost) all observed phenomena down to length scales of $\sim 10^{-18}$ m. That's truly impressive! We only know how to compute observables (mostly probabilities for various scattering or decay processes) at those scales using quantum field theories, but we can compute such observables very precisely, and high-precision experiments compare extremely well with those predictions. This is quite a formidable achievement. There are indeed exceptions and the hope is that further theoretical and experimental developments will one day lead to a more complete theory of elementary particles. In the meantime, the Standard Model is the best of what we have, although it is generally accepted that it cannot be the "ultimate theory of everything."

This textbook is primarily addressed to Master's and Doctoral students, as well as young Postdocs. A solid knowledge of classical physics and non-relativistic quantum mechanics coupled to a mastery of mathematical tools is assumed. **Symmetries** and mathematical groups will play an important role throughout the book. We present particle physics according to the so-called **inductive method**, which comes closest to the methodology of real-life progress in physics: starting with some key experiments, the new science is developed. However, this is not the only path. On the contrary, particle physics is filled with examples where theoretical developments guided new experimental discoveries. When this was the case, we present the material in that order. Clearly, students and researchers who want to stay at the forefront of the field need to master both experimental and theoretical aspects. This is one of the reasons I wrote this book, as will be explained below.

The motivation to write this textbook came from my lecturing the course on particle physics at ETH Zurich, since the end of the 1990s. It started off from my hand-written notes to an electronic version that was distributed to the students. After more than two decades of teaching lectures to Master's and Doctoral students at ETHZ, it seemed to me desirable to compile and cleanup the material which formed the basis of the lecture. The substance of the course came from two main sources: on the one hand, I had prepared quite a lot of material based on my own personal experience as a young student and throughout the years as a researcher. On the other hand, I had also accumulated many bits and pieces from different sources, the most relevant ones being listed in the "Textbooks" section at the end of this book. **During the course of my lecturing, this material was subjected to repeated communication with bright ETHZ students, who provided a constant source of constructive questions and criticisms, which triggered on my side the quest for possible improvements**. So, teaching was an opportunity to develop and perfect a vast amount of material under a single coherent form. Based on the very positive feedback I received, I completed and significantly expanded these notes into the present textbook to provide a more coherent and comprehensive version, that would cover both theoretical and experimental aspects of the subject. **While the course was a phenomenological one, my desire was always to remain as mathematically rigorous as possible, for we simply cannot ignore the truly fundamental interplay between theory and experiment in particle physics**. To cite an example that comes to my mind, let us consider one of the brightest physicists of the twentieth century: Enrico Fermi. Within the many successes of his extraordinary career, I was always struck by the fact that he developed the first theory of weak interactions and also ran the first nuclear reactor. What extraordinary achievements in both theory and experiment! These facts were a driving "ideal" to develop the basic structure of this book. The structure of the chapters is graphically shown in Figure 1. The basic idea is to start from classical Newtonian mechanics and

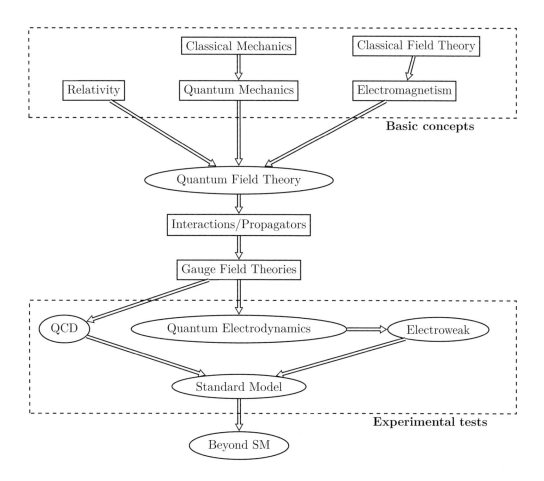

Figure 1 Overview of the material discussed in this book.

Maxwell's electromagnetism, and their failed attempt to explain *all* experimental results, in particular at the atomic level, collected towards the end of the nineteenth century and beginning of the twentieth. On the one hand, it was known that Newtonian mechanics failed at high velocities and needed to be addressed by Einstein's special relativity. On the other hand, it was realized that the description of atomic processes required theories to be "quantized." This means that for **any** given classical theory, one should look for a **quantum theory** that reduces to the classical one in the classical limit! Since Maxwell's theory was known before the actual Quantum Mechanics (QM), the attempts to quantize it actually preceded the developments of QM. But QM was historically developed first as a very successful non-relativistic quantum theory. QM and its **first quantization** introduced new concepts beyond those of classical physics. In particular, QM is "indeterminate," which states that there are physical measurements whose results are not definitely determined by the state of the system prior to the measurement. This understanding represented an incredible step forward in the description of the atomic and subatomic world. As we will discover throughout the chapters, the consequence of combining special relativity with QM implies that energy can quite generally be converted into quanta and particles, and vice versa. Particles are said to be "created" or "annihilated" and conservation of quantum numbers requires the existence of matter as well as antimatter! Theoretically, a relativistic QM theory requires one to abandon the single-particle approach of QM in favor of a multi-particle framework in the context of quantum field theory (QFT). This step involves what is commonly called the **second quantization**. A big success of QFT is its ability to describe interactions and propagators. Gauge field theories introduce the concept that the laws of Nature, to be described within the QFT framework, are constrained by local symmetries. From this concept emerge the three basic descriptions of the electromagnetic, weak, and strong interactions. Inherent to these theories arises the issue of **divergences**, both at large distances (IR divergence) and at short distances (UV divergence). There may be a true physical cutoff at short distances and QFT might not be able to handle physics at infinitely small distances. On the other hand, "effective" quantum field theories with a finite cutoff do make sense and allow us to calculate processes over a very large range of energies. In this context, **the Standard Model (SM) represents the merging of the unified electroweak theory and the strong interaction and is considered the most successful effective theory we have at our disposal today**. But, as the SM cannot be the theory of everything, one seeks a theory "beyond the SM." Whether this theory is just an extended QFT or something completely new is still the subject of debate.

So, this is the program that we would like to pursue. It is a challenge to introduce both experimental and theoretical aspects of a vast field within a single volume. Of course, an introductory course cannot (and should not) cover exhaustively the entire field. Consequently, it has not been possible to be fully comprehensive and much excellent literature exists that treats most of the covered subjects often in more detail. **My hope is that the reader will find this textbook sufficient to get a rock-solid start on the subject**, inviting him/her to look into the list of books and references listed in the "Textbooks" and "References" sections to dive further into this fascinating field. I do believe, however, that a deepening of knowledge beyond the level of this textbook will likely require the young reader to select a specialized "direction," either in experimental or in theoretical particle physics.

I have worked to provide references for every topic that is covered in the book. For the most important topics, I have also included extensive extracts and figures from the original literature, immediately available without the need to access the original publication. Of course, this does not (and should not) preclude the reader also accessing directly the original papers. In this context, it is worth mentioning the existence of InSpire (http://inspirehep.net) and arXiv (http://arxiv.org). InSpire is an online database of essentially every publication relevant to particle physics in history. The arXiv is a preprint server for many fields, including particle physics, where scientists post their completed papers before journal publication. It enables the rapid transmission of ideas, and almost every paper on particle physics written in the past 25 years is available there for free!

Let me conclude with several acknowledgments. For my lectures and for the perfecting of the manuscript, I am greatly indebted to all the students and assistants who actively participated in the course and the exercises over the years and who have carefully read and commented on the entire text in great detail. It is impossible to list them all. However, with my apologies to those I might have inadvertently omitted, I sincerely thank Dr. Andreas Badertscher, Prof. Antonio Bueno, Dr. Laura Molina Bueno, Prof. Mario Campanelli, Tit. Prof. Paolo Crivelli, Dr. Sebastien Murphy, Dr. Balint Radics, Dr. Christian Regenfus, Katharina Lachner, Matthias Schlomberg, and very specially Loris Pedrelli and Alexander Stauffer for their invaluable help and for investing a significant fraction of their time to provide me many corrections, several constructive criticisms, and very useful feedback.

1   Introduction and Notation

A particle (from the Latin *particula*, little part) is a minute portion of matter.

1.1   Subatomic Particles

When we first observe the Universe, it might appear to us as a very complex object. One of the primary goals of the philosophy of Nature (or simply Physics) is to "reduce" ("simplify") this picture in order to find out what the most fundamental constituents of matter (i.e., the atoms from the Greek word *indivisible*) are and to understand the basic forces by which they interact in the otherwise *void* space, along the line of thinking of **Demokritos**[1] who wrote *"Nothing exists except atoms and empty space."*

In this context, **subatomic particles** are physical objects smaller than atoms. In particle physics, particles are objects that are localized in space and that are characterized by intrinsic properties. As we will see later, the set of intrinsic properties used to classify particles is chosen from those which behave in a well-defined way under the action of a transformation. As a matter of fact, we would expect some of these properties not to change at all under particular transformation. For instance, we expect some of its properties to be independent of the velocity of the particle or the direction in which it is traveling. This procedure of classification was actually initiated by **Eugene Wigner**[2] in his seminal paper of 1939 [1]. We will come back to this later.

Elementary particles are particles which, according to current theories, are not made of other particles, or we should rather say, *whose substructure, if any, is unknown*. They are thus considered as **point-like objects**. **Composite particles**, on the other hand, are composed of other particles, in general of elementary particles, and are thus extended in space. All subatomic particles are classified according to their properties and given *common names* such as electrons, muons, taus, protons, neutrons, neutrinos, etc. Particles are indistinguishable in the sense that, for example, one electron is identical to another electron.

Elementary particles can be **massless** or **massive**. Massless particles have a zero **rest mass**, while massive particles have a finite rest mass. Following **Albert Einstein**,[3] energy is equivalent to (inertial) mass and vice versa [2,3]. Consequently, composite particles (i.e., bound states of sub-elementary objects) cannot be massless, as their rest mass should at least account for the binding energy among their constituents.

Particles can be **electrically neutral** or **charged**. An electrically neutral particle has no charge, while charged particles possess an electric charge which is a fraction or multiple of the **elementary unit charge** e. In 2018, the kilogram, Ampère, Kelvin, and mole were redefined in terms of new permanently fixed values of the elementary charge, Planck, Boltzmann, and Avogadro constants (see www.bipm.org). Accordingly, the **exact** value of the elementary charge is given by [4]:

$$e = 1.602\,176\,634 \times 10^{-19} \text{ C} \quad \text{(exact)} \tag{1.1}$$

---

1   Demokritos (c.460–c.370 BC), Ancient Greek philosopher.
2   Eugene Paul "E. P." Wigner (1902–1995), Hungarian–American theoretical physicist, engineer, and mathematician.
3   Albert Einstein (1879–1955), German-born theoretical physicist.

Other than rest mass and charge properties, particles can possess additional **internal degrees of freedom**, such as the **spin**. Such degrees of freedom are characterized by specific integer or half-integer **quantum numbers**. Other internal characteristics such as isospin, strangeness, etc. will be defined as well. Some of these quantities, such as the electric charge, are always conserved, while others, such as strangeness, are violated under some circumstances, as will be discussed later on.

The properties of all of the known particles (e.g., mass, charge, and spin) and the experimental results that measured them are collected in the **Particle Data Group's (PDG) Review of Particle Physics**. The review is published every two years. It can be accessed online at http://pdg.lbl.gov or one can order the book directly from the website. Here, we will most frequently refer to the Particle Data Group's Review published in 2020 [5].

As far as we know, the Universe is composed of 12 elementary (fundamental) spin-1/2 fermions, divided into six **leptons** and six **quarks**. These are listed in Table 1.1. Apart from the differences in their rest masses, the particles of each generation possess the same fundamental interactions. It is natural to ask the origin of this pattern and if there are further generations. This seems not to be the case and consequently the matter content of the Universe of this type can be reduced to these 12 elements. Astronomical observations point

Family	Leptons		Quarks	
	Name	Q	Name	Q
First generation	electron (e)	-1	down	$-1/3$
	electron neutrino (ν_e)	0	up	$+2/3$
Second generation	muon (μ)	-1	strange	$-1/3$
	muon neutrino (ν_μ)	0	charm	$+2/3$
Third generation	tau (τ)	-1	bottom	$-1/3$
	tau neutrino (ν_τ)	0	top	$+2/3$

Table 1.1 A summary of the known fundamental spin-1/2 fermions in Nature.

to the existence of an additional type of matter (called **dark matter**) (see Section 32.5), but the elementary particle content of this latter is unknown as of today.

It is difficult to completely objectively define an elementary particle. We can imagine them as point-like objects with particular characteristics, however, such a picture will never be completely satisfactory. For instance, **Bruno Pontecorvo**[4] stated that a good model to represent some of the neutrino properties (and other spin-1/2 fermions) is with a screw. The helix of the thread of a screw can twist in two possible directions, which is known as **handedness**. See Figure 1.1. Similar considerations apply to the neutrino handedness, and in general to the chiral components of Dirac particles, as will be discussed in Section 8.24! Such macroscopic illustrations can help us visualize the properties of elementary particles.

Identical particles obey certain statistics and are classified accordingly: **fermions** obey Fermi–Dirac statistics while **bosons** obey Bose–Einstein statistics. The **spin-statistics theorem** was first formulated by **Markus Fierz**[5] in 1939 in his Habilitation paper where he noted that, from the very general requirements that the exchange relations of the field quantities should be relativistically invariant and infinitesimal, and that the energy should be positive, it follows that particles with integer spin must always follow Bose statistics, and particles with half-integer spin must follow Fermi statistics.[6] The result was then demonstrated in a more systematic way by **Wolfgang Pauli**[7] in a famous 1940 paper [6]. In other words, in a relativistically invariant wave equation, the requirements of causality and that the energy is positive definite, imply that the intrinsic

---

4   Bruno Pontecorvo (1913–1993), Italian nuclear physicist, early assistant of Enrico Fermi. A convinced communist, he defected to the Soviet Union in 1950.
5   Markus Fierz (1912–2006), Swiss physicist.
6   *"Aus den sehr allgemeinen Forderungen, dass die Vertauschungsrelationen der Feldgrössen relativistisch invariant und infinitesimal sein sollen, und dass die Energie positiv sein soll, folgt nämlich, dass Teilchen mit ganzzahligem Spin stets Bose-Statistik, Teilchen mit halbzahligem Spin Fermi-Statistik haben müssen."* Reprinted from M. Fierz, "Über die relativistische Theorie kräftefreier Teilchen mit beliebigem Spin," *Helv. Phys. Acta*, vol. 12, pp. 3–37, 1939.
7   Wolfgang Ernst Pauli (1900–1958), Austrian-born Swiss and American theoretical physicist.

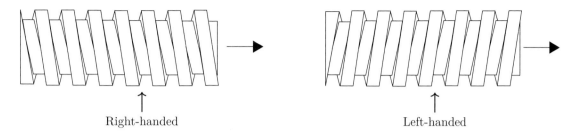

Figure 1.1 Right- and left-handed screw threads. Adapted from Eugene Wycliffe Kerr (1908) *Power and Power Transmission*, 2nd ed., John Wiley & Sons, New York, p. 79, fig. 61 on Google Books, Public Domain, https://commons.wikimedia.org/w/index.php?curid=31218881.

spin of a particle is related to the statistics, and that they obey: *spinless or integer spin particles are bosons while half-integer spin particles are fermions.*

As first postulated by **Louis de Broglie**[8] [7], all particles possess a **dual particle–wave nature** (see Section 4.1). Hence, the rules of **quantum mechanics** apply in the description of subatomic particles and their physical behavior. Actually this has been verified not only for subatomic particles but also for composite objects like atoms and even molecules, and in principle also applies for macroscopic objects. One of the challenges that will be addressed in this book is the need to develop a relativistic formulation of quantum mechanics beyond the "classical" non-relativistic Schrödinger theory.

1.2   Action at Distance

Particles exhibit **interactions at a distance**, described by forces whose strengths are characterized by **coupling constants**. The four **fundamental interactions** – gravitation (related to mass/energy), electromagnetic (related to electric charge), weak (related to the weak isospin and hypercharge), and strong (related to the color charge) – also known as fundamental forces – are the ones that do not appear to be reducible to more basic interactions. As we will see later, interactions among particles are actually *local*. Therefore, forces acting at a distance are thought to be transmitted via **intermediate gauge bosons**. Each force is characterized by a particular set of gauge bosons, with specific properties. The forces, the associated intermediate gauge bosons, and some of their properties are summarized in Table 1.2.

Interaction	Gauge boson			Fermion coupling
	Type	Spin	Mass	
Strong	8 gluons	1	0	$q, \bar{q}$
Electromagnetic (EM)	1 photon	1	0	$\ell^-, \ell^+, q, \bar{q}$
Weak	$W^\pm$ boson	1	80 GeV	$\ell^-, \ell^+, \nu, \bar{\nu}, q, \bar{q}$
	Z^0 boson	1	91 GeV	$\ell^-, \ell^+, \nu, \bar{\nu}, q, \bar{q}$
Gravity	1 graviton (?)	2	0	all

Table 1.2 A summary of the fundamental forces in Nature; ℓ stands for the charged leptons e, μ, or τ.

The strength of a force actually depends on the **energy scale**, which represents the characteristic energy at which the particular interaction is taking place. This fact is illustrated in Figure 1.2, where energy is expressed

---

8   Louis Victor Pierre Raymond de Broglie (1892–1987), French physicist.

in units of **electronvolt**, with 1 GeV = 1 Giga-electronvolt = 10^9 eV and therefore the (exact) value of the unit of energy is given by (see [4]):

$$1 \text{ GeV} = 10^9 \, e \text{ J} = 1.602\,176\,634 \times 10^{-10} \text{ J} \quad \text{(exact)} \tag{1.2}$$

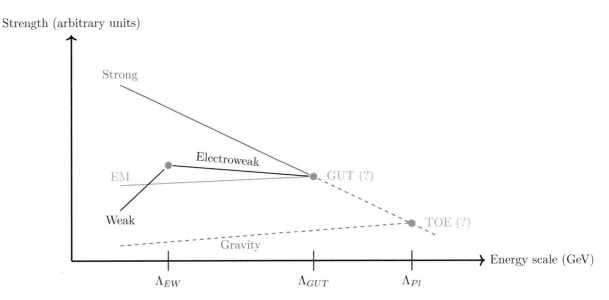

Figure 1.2 Illustration of the strength of the fundamental forces as a function of the energy scale. At the highest energies all forces should unify.

The four forces are believed to be fundamentally related. Special relativity was derived from the attempt to unify Newtonian mechanics with Maxwell's electromagnetism. Later, **Einstein attempted to unify his general theory of relativity with electromagnetism**. The weak and electromagnetic forces have been unified within the **electroweak theory** to be discussed in Chapter 25. These two forces actually behave with a similar strength at the **electroweak scale**, that is determined by the vacuum expectation value (its exact meaning will be discussed in Section 24.5):

$$\Lambda_{EW} \simeq v = \left(\sqrt{2}G_F\right)^{-1/2} \simeq 246 \text{ GeV} \tag{1.3}$$

where G_F is the Fermi coupling constant (see Chapter 21). The hypothesized **Grand Unified Theories (GUT)** unite the electroweak and strong forces at the much higher GUT scale (see Chapter 31), which extrapolating from current experimental data should be around

$$\Lambda_{GUT} \simeq 10^{15\pm1} \text{ GeV} \tag{1.4}$$

(this value will be discussed in Section 31.5). All four forces are believed to unite into a single force at the **Planck**[9] **scale**, characterized by the **Planck mass** M_{Pl}, which corresponds to the simplest combination of the fundamental constants $\hbar$, and c (see Section 1.8), and the Newtonian universal gravitational constant G:

$$\Lambda_{Pl} \simeq M_{Pl} = \sqrt{\frac{\hbar c}{G}} \simeq 2.2 \times 10^{-8} \text{ kg} = 1.2 \times 10^{19} \text{ GeV/c}^2 \tag{1.5}$$

9   Max Karl Ernst Ludwig Planck (1858–1947), German theoretical physicist.

One can also characterize the Planck scale considering the Planck length $l_{Pl} = \sqrt{\hbar G/c^3} \simeq 1.6 \times 10^{-35}$ m or the Planck time $t_{Pl} = l_{Pl}/c \simeq \sqrt{\hbar G/c^5} = 5.4 \times 10^{-44}$ s, which are indeed small numbers compared to our everyday scales! Imagining what happens in those regimes is far from trivial. At present, the development of a unified field theory able to describe all forces in a unified way and up to the Planck scale, the so-called **Theory of Everything** (TOE), remains an open research topic.

1.3   Symmetries and Conservation Laws

Symmetry in physics has been generalized to mean **invariance** – that is, lack of change – under any kind of transformation. Symmetries play an *extremely important role* in particle physics. To illustrate this fact, we for instance quote **Dirac**[10] who wrote on Einstein's theory:

> *"There is one strong reason in support of the theory. It appears as one of the fundamental principles of Nature that the equations expressing basic laws should be invariant under the widest possible group of transformations. The confidence that one feels in Einstein's theory arises because its equations are invariant under a very wide group, the group of transformations of curvilinear coordinates in Riemannian space. Of course even if space were flat, the equations of physics could still be expressed in terms of curvilinear coordinates and would still be invariant under transformations of the coordinates, but there would then exist preferred systems of coordinates, the rectilinear ones, and it would be only the group of transformations of the preferred coordinates that would be physically significant. The wider group of transformations of curvilinear coordinates would then be just a mathematical extension, of no importance for the discussion of physical laws."* [11]

The importance of symmetries is related to the existence of the fundamental **theorem of Noether**[12] (first derived in 1915 and published in 1918 [8], valid for any Lagrangian theory, classical or quantum) which relates such symmetries to *conserved quantities* of the physical system. It states:

> *For every continuous symmetry, there exists a conservation law. For every conservation law, there exists a continuous symmetry.*

Conserved quantities are at the heart of *conservation laws*, which are so fundamental and crucial to describe the behavior of physical systems. Conservation laws highly restrict the possible outcomes of an experiment. In a quantum system, quantum numbers describe quantized properties of the system and are usually subject to constraints. Conserved quantities are the basis to define such good quantum numbers. In particle physics there are many examples of symmetries and their associated conservation laws. Identifying the symmetries of a physical system allows us to express its governing laws in a very elegant and fundamental way. Last but not least, there are also cases where a symmetry is *broken*, and the mechanism has to be understood. The reason why and the mechanism by which a symmetry is broken (or partially broken) can give us insight on the underlying physics. In the case of spontaneous symmetry breaking (see Chapter 24), the underlying theory possesses a fundamental symmetry, however, the dynamical choice of vacuum breaks this symmetry. In all cases, symmetries, broken or not, are fundamental.

Noether's theorem is valid for **continuous transformations**. Specifically, translations in time or in space are examples of such continuous transformations. As a classical case, let us consider two mechanical masses m_1 and

---

10   Paul Adrien Maurice Dirac (1902–1984), English theoretical physicist.
11   Republished with permission of The Royal Society (UK) from P. A. M. Dirac, "Long range forces and broken symmetries,"
      Proc. R. Soc. Lond. A 333: 403–418 (1973); permission conveyed through Copyright Clearance Center, Inc.
12   Amalie Emmy Noether (1882–1935), German mathematician and theoretical physicist.

m_2 and a potential of the system that depends only on the *relative position* of the two masses $V \equiv V(\vec{x}_1 - \vec{x}_2)$. The Newtonian equations of motion are:

$$m_i \frac{\mathrm{d}^2 \vec{x}_i}{\mathrm{d}t^2} = -\frac{\partial}{\partial \vec{x}_i} V(\vec{x}_1 - \vec{x}_2) \qquad (i = 1, 2) \tag{1.6}$$

where $\partial/\partial \vec{x}_i$ is the **3-gradient** of V with respect to the $\vec{x}_i$ coordinates (see Appendix A.9). Transforming with the translation $\vec{x} \to \vec{x}' = \vec{x} + \vec{a}$, where $\vec{a}$ is a given constant vector, transforms

$$\vec{x}_i \to \vec{x}_i' = \vec{x}_i + \vec{a} : \quad \begin{cases} m_i & \to m_i \\[2mm] \dfrac{\mathrm{d}^2 \vec{x}_i}{\mathrm{d}t^2} & \to \dfrac{\mathrm{d}^2 \vec{x}_i'}{\mathrm{d}t^2} = \dfrac{\mathrm{d}^2}{\mathrm{d}t^2}(\vec{x}_i + \vec{a}) = \dfrac{\mathrm{d}^2 \vec{x}_i}{\mathrm{d}t^2} \\[2mm] V(\vec{x}_1 - \vec{x}_2) & \to V(\vec{x}_1' - \vec{x}_2') = V(\vec{x}_1 + \vec{a} - \vec{x}_2 - \vec{a}) = V(\vec{x}_1 - \vec{x}_2) \end{cases}$$

We note that the masses m_i are unchanged quantities during the translation and are hence labeled **scalars**, to define a number quantity which does not change under the transformation. For the gradient we have (dropping the i index and setting $\vec{x} = (x, y, z)$ and $\vec{x}' = (x', y', z')$):

$$\frac{\partial}{\partial \vec{x}} \to \frac{\partial}{\partial \vec{x}'} = \left(\frac{\partial}{\partial x'}, \frac{\partial}{\partial y'}, \frac{\partial}{\partial z'} \right) = \left(\frac{\partial}{\partial x}\frac{\partial x}{\partial x'}, \frac{\partial}{\partial y}\frac{\partial y}{\partial y'}, \frac{\partial}{\partial z}\frac{\partial z}{\partial z'} \right) = \left(\frac{\partial}{\partial x}, \frac{\partial}{\partial y}, \frac{\partial}{\partial z} \right) = \frac{\partial}{\partial \vec{x}} \tag{1.7}$$

The equations of motion in the translated system are then given by:

$$\text{Eq. (1.6)} \quad \longrightarrow \quad m_i \frac{\mathrm{d}^2 \vec{x}_i'}{\mathrm{d}t^2} = -\frac{\partial}{\partial \vec{x}_i'} V(\vec{x}_1' - \vec{x}_2') \qquad (i = 1, 2) \tag{1.8}$$

Consequently, the **equations of motion exhibit the same form in the two systems and are hence considered as translation-invariant**. If we now calculate the total force acting on the masses, we get

$$\vec{F}_{tot} = \vec{F}_1 + \vec{F}_2 = -\frac{\partial}{\partial \vec{x}_1} V(\vec{x}_1 - \vec{x}_2) - \frac{\partial}{\partial \vec{x}_2} V(\vec{x}_1 - \vec{x}_2) = -\frac{\partial}{\partial \vec{x}_1} V(\vec{x}_1 - \vec{x}_2) + \frac{\partial}{\partial \vec{x}_1} V(\vec{x}_1 - \vec{x}_2) = \vec{0} \tag{1.9}$$

hence

$$\frac{\mathrm{d}\vec{P}_{tot}}{\mathrm{d}t} \equiv \vec{F}_{tot} = \vec{0} \tag{1.10}$$

and **the total momentum is constant in time and is hence conserved**.

The above derivation can be generalized to a special class of continuous transformations of space-time, called proper orthochronous Lorentz transformations (see Section 5.4). The **homogeneity in time** corresponds to the invariance under translation in time (this basically implies that there is no absolute time, and that one can define the origin of time $t = 0$ at any time, without changing the physical laws). Similarly, one can define the **homogeneity in space** for the invariance under space translation, and the isotropy of space for the invariance under rotation in space. These transformations lead to the following well-known conserved quantities:

translation in time	$\longrightarrow$	conservation of energy
translation in space	$\longrightarrow$	conservation of momentum
rotation in space	$\longrightarrow$	conservation of angular momentum

We can interpret these results as first principles: it is *natural* (or *aesthetically pleasing*) to *assume* that time and space are homogeneous and that space is isotropic. The conserved quantity corresponding to the time-translation symmetry is defined as energy, the conserved quantity associated with the space-translation symmetry is called momentum, and the conserved quantity related to the rotational symmetry is called angular momentum. **The symmetries of the system guide us in the choice of physical quantities to describe our system**.

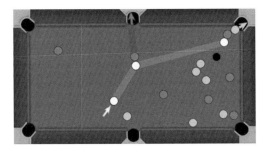 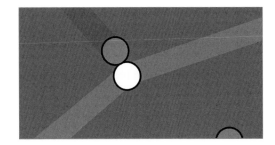

Figure 1.3 Illustration of collisions on a billiard table: (left) view of the entire table; (right) zooming in on one collision.

Transformations of a system are conveniently organized using the mathematical algebra tools of **group theory** (see Appendix B). Continuous transformations form a **Lie group** (see Appendix B.5) and can depend on one or many continuous parameters. For example, translation in time depends on one parameter while translation in space can be represented by three independent parameters corresponding to the three degrees of freedom of space. One can consider in these cases **infinitesimal transformations**, when the parameters are made infinitesimal. For example, the system can undergo an infinitesimal translation in space $\delta\vec{a}$ with $|\delta\vec{a}| \ll 1$ or in time $|\delta t| \ll 1$. A *finite* transformation can then be obtained from *a successive application of infinitesimal transformations*. We say that the finite transformation $T(\alpha)$ can be derived from the product of infinitesimal ones $T(\delta\alpha)$, where $\delta\alpha = \alpha/n$. The net effect of the finite transformation can consequently be written as:

$$T(\alpha) = \underbrace{T(\delta\alpha)\ldots T(\delta\alpha)}_{n} = [T(\delta\alpha)]^n = \left[T\left(\frac{\alpha}{n}\right)\right]^n \qquad (1.11)$$

1.4 Discrete Symmetries

Discrete transformations describe non-continuous changes in a system. They *cannot* be derived from the product of infinitesimal transformations, such as Eq. (1.11). These transformations are associated with *discrete symmetry groups*. There are three important discrete transformations of space and time that we will use throughout the chapters: parity (P), charge conjugation[13] (C), and time reversal (T).

- The **parity transformation** P inverts every spatial coordinate through the origin, while the **time inversion** T (or **time reversal**) means to reverse the direction of time:

$$P : \begin{cases} \vec{x} \to -\vec{x} \\ t \to t \end{cases} ; \qquad T : \begin{cases} \vec{x} \to \vec{x} \\ t \to -t \end{cases} \qquad (1.12)$$

The parity operation is equivalent to a reflection in the $x - y$ plane followed by a rotation about the z-axis.[14] There are a number of ways in which we can consider time reversal. For example, if we look at collisions on a billiard table, as illustrated in Figure 1.3(left), it would clearly violate our sense of how "moving spheres behave" if time were reversed. It is very unlikely that we would have a set of billiard balls moving in just the directions and speeds necessary for them to collect and form a perfect triangle at rest, with the cue ball moving away. However, if we look *at an individual collision* (see Figure 1.3(right)),

13   It will become clearer later why the C transformation is associated with a space-time operation.
14   Often one associates parity with the reflection in a mirror but this is not precisely correct.

reversing time results in a perfectly "normal-looking collision" (if we ignore the small loss in kinetic energy due to inelasticity in the collision). The classical equations of dynamics are invariant under T because they are of second order in time. At the microscopic level, there is time invariance in mechanics. At the macroscopic level, the arrow of time is selected statistically due to entropy. Hence, the former lack of time-reversal invariance has to do with the laws of thermodynamics; in particle physics we are generally interested in individual microscopic processes for which the laws of thermodynamics are not important, and we will consider whether particular interactions between elementary particles are invariant under T or not.

- The **charge conjugation** C is a discrete symmetry that exchanges matter with antimatter and vice versa. Hence, it reverses the sign of the electric charge, color charge, and magnetic moment of a particle (it also reverses the values of the weak isospin and hypercharge charges associated with the weak force).

Figure 1.4 illustrates the actions of P, T, and C on a fundamental (spin-1/2) fermion, for instance an electron or a positron, where the arrow through the sphere corresponds to the direction of the momentum $\vec{p}$ and where the spin $\vec{s}$ (see Section 4.11) is indicated as well as an arrow next to the particle. The parity operator

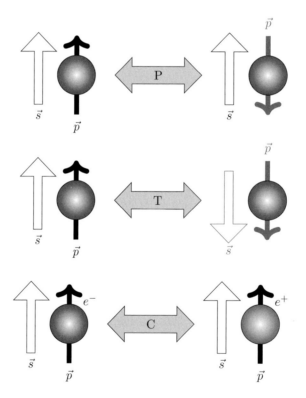

Figure 1.4 Effect of the discrete symmetries P (top), T (middle), and C (bottom) on an electron or positron. Note that the action of the operator applied on the state twice results in the original particle.

changes the direction of motion of a particle, but not the direction of the spin vector. Time reversal changes the sign of both the spin and momentum. Charge conjugation reverses the charge but does not change the direction of the spin nor the momentum of a particle. The action of any operator on a state twice gives back the original particle (up to an arbitrary quantum phase). **Understanding the behavior of a physical system or** its interactions under discrete transformations is an important tool, as described in more detail in the next section and in later chapters.

1.5   Quantities with Well-Defined Transformation Properties

In general, we want to introduce quantities that best describe the properties of the *physical reality*. **To understand the physical properties of a given observable and hence its fundamental meaning, it is important to understand how it behaves under specific transformations of space-time or any other transformation**. A **scalar** is a physical quantity such as a mass or a volume, which only has a magnitude and does not depend on direction. It is not affected by space-time transformations. A **vector** is a physical object that has a magnitude and a direction. A translation in space-time leaves a vector unchanged. A rotation in space is *defined* as a transformation that changes the direction but keeps the magnitude of the vector unchanged. An ordinary vector $\vec{x}$ in space changes sign under parity transformation P:

$$\text{Vector}: \qquad P(\vec{x}) = -\vec{x} \tag{1.13}$$

From any vector $\vec{x}$, one can construct a *scalar* quantity s with the **scalar product** $s \equiv \vec{x} \cdot \vec{x}$. Under parity, we find that s is an invariant scalar, since it remains unchanged:

$$\text{Scalar}: \qquad P(s) = P(\vec{x} \cdot \vec{x}) = P(\vec{x}) \cdot P(\vec{x}) = (-\vec{x}) \cdot (-\vec{x}) = \vec{x} \cdot \vec{x} = +s \tag{1.14}$$

Similarly, with the **cross-product** of two vectors, we can construct the **pseudovector** or **axial vector** $\vec{a} \equiv \vec{x_1} \times \vec{x_2}$, and this new vector does *not* change sign under parity:

$$\text{Pseudovector (or axial vector)}: \qquad P(\vec{a}) = P(\vec{x_1} \times \vec{x_2}) = -\vec{x_1} \times (-\vec{x_2}) = +\vec{a} \tag{1.15}$$

Finally, from an ordinary and pseudovector we can construct the **pseudoscalar** $p \equiv \vec{x} \cdot \vec{a}$, which keeps its magnitude but changes sign under parity. So, ultimately, we have derived quantities that have specific properties under continuous (i.e., translation, rotation) and discrete (parity) transformations of space-time:

$$\begin{array}{rcl} s \text{ scalar} & \Leftrightarrow & P(s) = +s \\ p \text{ pseudoscalar} & \Leftrightarrow & P(p) = -p \\ \vec{x} \text{ vector} & \Leftrightarrow & P(\vec{x}) = -\vec{x} \\ \vec{a} \text{ pseudovector} & \Leftrightarrow & P(\vec{a}) = +\vec{a} \end{array} \tag{1.16}$$

The above expressions can be summarized in the form of an eigenvector equation of the parity with eigenvalues $\lambda_p = \pm 1$:

$$P(c) = \lambda_p c \text{ where } \lambda_p = \begin{cases} +1 \text{ if } c \text{ is scalar/pseudovector} \\ -1 \text{ if } c \text{ is pseudoscalar/vector} \end{cases} \tag{1.17}$$

We can now consider the transformation properties of the common physical properties. A physical position $\vec{x}$ transforms as a vector ($\lambda_p = -1$). So do the velocity $\vec{v} = \mathrm{d}\vec{x}/\mathrm{d}t$ and the linear momentum vector $\vec{p} = m\vec{v}$, $P(\vec{p}) = P(m\vec{v}) = m(-\vec{v}) = -\vec{p}$, $\lambda_p = -1$. The angular momentum vector $\vec{L} = \vec{x} \times \vec{p}$ is an axial vector since it is the cross-product of a position and a momentum vector. Hence, the quantity $\vec{L}$ transforms as $P(\vec{L}) = (-\vec{x}) \times (-\vec{p}) = \vec{L}$, $\lambda_p = +1$, and so does the spin $\vec{s}$ of a particle (see also Figure 1.4).

The transformation property of the classical electric field $\vec{E}$ can be derived applying P on Gauss's law:

$$\frac{\partial}{\partial \vec{x}} \cdot \vec{E}(\vec{x}, t) = \rho(\vec{x}, t) \quad \overset{P}{\to} \quad \frac{\partial}{\partial \vec{x}'} \cdot \vec{E}'(\vec{x}', t) = \rho(\vec{x}', t) \tag{1.18}$$

or

$$-\frac{\partial}{\partial \vec{x}} \cdot \vec{E}'(-\vec{x}, t) = \frac{\partial}{\partial \vec{x}} \cdot \left[-\vec{E}'(-\vec{x}, t) \right] = \rho(-\vec{x}, t) \tag{1.19}$$

which yields the original equation if the electric field transforms as $\vec{E} \to -\vec{E}$ under parity. Similarly, one can show that the magnetic field must transform as $\vec{B} \to +\vec{B}$ under parity.

Other important quantities can be derived by considering the product of two quantities with well-defined transformation properties. For example, we can mention the **electric dipole moment** $\vec{\sigma} \cdot \vec{E}$, the **magnetic dipole moment** $\vec{\sigma} \cdot \vec{B}$, the **longitudinal polarization** $\vec{\sigma} \cdot \vec{p}$, or the **transverse polarization** $\vec{\sigma} \cdot (\vec{p}_1 \times \vec{p}_2)$ (see Section 4.14). Their transformations under parity are summarized in Table 1.3. Similar arguments can also be repeated for the other two discrete transformations T and C, and are also listed in this table. For example, time reversal reverses linear and angular momenta and also spin, and so on. Such quantities are important to understand the invariance properties of fundamental interactions, as will be shown later.

Quantity	Notation	P	C	T
Position	$\vec{x}$	$-\vec{x}$	$+\vec{x}$	$+\vec{x}$
Velocity	$\vec{v} = \mathrm{d}\vec{x}/\mathrm{d}t$	$-\vec{v}$	$+\vec{v}$	$-\vec{v}$
Linear momentum	$\vec{p} = m\vec{v}$	$-\vec{p}$	$+\vec{p}$	$-\vec{p}$
Angular momentum	$\vec{L} = \vec{r} \times \vec{p}$	$+\vec{L}$	$+\vec{L}$	$-\vec{L}$
Spin	$\vec{S}$ or $\vec{\sigma}$	$+\vec{\sigma}$	$+\vec{\sigma}$	$-\vec{\sigma}$
Helicity	$h = \vec{\sigma} \cdot \vec{p}/|p|$	$-h$	$+h$	$+h$
Electric field	$\vec{E}$	$-\vec{E}$	$-\vec{E}$	$+\vec{E}$
Magnetic field	$\vec{B}$	$+\vec{B}$	$-\vec{B}$	$-\vec{B}$
Electric dipole moment	$\vec{\sigma} \cdot \vec{E}$	$-\vec{\sigma} \cdot \vec{E}$	$-\vec{\sigma} \cdot \vec{E}$	$-\vec{\sigma} \cdot \vec{E}$
Magnetic dipole moment	$\vec{\sigma} \cdot \vec{B}$	$+\vec{\sigma} \cdot \vec{B}$	$-\vec{\sigma} \cdot \vec{B}$	$+\vec{\sigma} \cdot \vec{B}$
Longitudinal polarization	$\vec{\sigma} \cdot \vec{p}$	$-\vec{\sigma} \cdot \vec{p}$	$+\vec{\sigma} \cdot \vec{p}$	$+\vec{\sigma} \cdot \vec{p}$
Transverse polarization	$\vec{\sigma} \cdot (\vec{p}_1 \times \vec{p}_2)$	$+\vec{\sigma} \cdot (\vec{p}_1 \times \vec{p}_2)$	$+\vec{\sigma} \cdot (\vec{p}_1 \times \vec{p}_2)$	$-\vec{\sigma} \cdot (\vec{p}_1 \times \vec{p}_2)$

Table 1.3 A summary of the transformations of various physical quantities with well-defined transformation properties under C, P, and T.

1.6 Global and Local Gauge Symmetries

Up to now we have discussed the continuous transformations of space-time (part of the so-called proper orthochronous Lorentz group) and the discrete symmetries of space-time (see Chapter 5). One considers also **internal symmetries**, which do not act on the space-time coordinates but rather on internal degrees of freedom of the quantum systems. **Gauge invariance** is equivalent to the fact that absolute changes in the complex phase of a wave function of a quantum system are unobservable, and only changes in the magnitude of the wave function result in changes to the probabilities (see Section 4.2).

As will be discussed in later chapters, it has become quite evident that the fundamental interactions of Nature can be shown to originate from some sort of *local* gauge symmetry. In this sense local gauge symmetry is considered as a primary concept – a first principle! In a quantum field theory (see Chapter 7) with a **local gauge symmetry** (see Chapter 24), there is an **absolutely** conserved quantity (a charge) and a **long-range force** coupled to the charge. This corresponds for instance to the electric charge Q or e and the long-range electromagnetic interaction, and is the basis of quantum electrodynamics (see Chapter 10). Other examples are the strong (see Chapter 18) and the electroweak (see Chapter 25) interactions.

Global gauge symmetries are also very important, although they seem to lead to approximate laws of Nature. This is, for example, the case for conserved charges such as the lepton number L (see Chapter 21) or the baryon number B (see Chapter 15). **If the baryon number were absolutely conserved from local gauge symmetry, then a long-range force coupled to it should exist**. But there is no evidence for such a force field. However, no experimental evidence for baryon or lepton number violation has been observed, even if not protected by any gauge principle. For practical purposes, one *assumes* that baryon and total lepton numbers are conserved, even if there is no fundamental theoretical reason to suppose that this conservation is absolute. There are in fact very good reasons to believe that they should be violated at some level to explain neutrino masses (see Chapter 31) or the observed baryon–antibaryon asymmetry in the Universe (see Chapter 32).

The simplest gauge transformation depends on one phase and is then represented by the $U(1)$ group (see Appendix B.7). Yang–Mills theories (see Chapter 24) consider bigger groups, such as the $SU(2)$ or $SU(3)$ groups. Invariance under these groups leads to the conserved quantities listed in Table 1.4, which will be discussed throughout the chapters.

$U(1)$ gauge transformation (local)	$\longrightarrow$	electric charge
$U(1)$ gauge transformation (global)	$\longrightarrow$	lepton number (approximate?)
$U(1)$ gauge transformation (global)	$\longrightarrow$	baryon number (approximate?)
$U(1)_Y$ gauge transformation (local)	$\longrightarrow$	weak hypercharge
$SU(2)_L$ gauge transformation (local)	$\longrightarrow$	weak isospin
$SU(3)_c$ gauge transformation (local)	$\longrightarrow$	color
$SU(3)$ gauge transformation (global)	$\longrightarrow$	flavor (approximate)

Table 1.4 A summary of internal (gauge) transformations and their conserved charge.

1.7   Symmetries in Particle Physics and the CPT Theorem

Electromagnetism is C invariant, since Maxwell's equations apply equally to $+$ and $-$ charges. Electromagnetic interactions are invariant under charge conjugation and parity, and conserve C and P quantum numbers. Similar considerations can be extended to the strong and weak forces. A summary of the conservation laws of fundamental forces is shown in Table 1.5. It is experimentally observed that the electromagnetic and strong forces conserve the largest number of quantities, while the weak force does not conserve (or *violates*) several quantities, even some unexpected, such as the parity, the charge conjugation, the time reversal, and even the combination of the two in CP.

- CPT **theorem.** In 1957 **Gerhart Lüders**[15] published a proof [9] that the combined action of C, P, and T, or just CPT, must be conserved in a local Lorentz-invariant quantum field theory (see Chapter 7 for a basic discussion of QFT). The proof was based on the concept of **strong reflection** PT, introduced by Pauli, which consists in:

$$PT : \begin{cases} \vec{x} \to -\vec{x} \\ t \to -t \end{cases} \tag{1.20}$$

The CPT transformation is then composed of the strong reflection and of the matter–antimatter conjugation.

If the Lagrangian of a theory (see Chapter 6) is invariant under CPT, then such CPT conservation has several consequences, for example:

- the rest mass of a particle is identical to the rest mass of its antiparticle ($m = \bar{m}$);

- the lifetime of a particle is identical to the lifetime of its antiparticle ($\tau = \bar{\tau}$);

15   Gerhart Lüders (1920–1995), German theoretical physicist.

Quantity	Type of interaction		
	Strong	Electromagnetic	Weak
Energy and momentum	✓	✓	✓
Electric charge	✓	✓	✓
Baryon number	✓	✓	✓
Lepton number	✓	✓	✓
Isospin	✓	×	×
Strangeness	✓	✓	×
Charm	✓	✓	×
Bottom	✓	✓	×
Parity (P)	✓	✓	×
Charge conjugation (C)	✓	✓	×
Time reversal (T)	✓	✓	×
CP	✓	✓	×
CPT	✓	✓	✓

Table 1.5 A summary of the conservation laws of fundamental forces.

- the magnetic momentum of a particle is the opposite of that of its antiparticle ($\mu = -\bar{\mu}$).

In fact, we will see in subsequent chapters that the marriage between special relativity and quantum mechanics requires the existence of particles *and* antiparticles. In this context, the CPT transformation defines the *correct* relation between a particle and its antiparticle! Experimentally, one can seek deviations in the intrinsic properties of particles and their corresponding antiparticles, and put limits if no deviations are found. For example (values taken from the Particle Data Group's Review [5]):

$$|m_{K^0} - m_{\overline{K^0}}|/m_{\text{average}} < 6 \times 10^{-19} \qquad (90\%\,\text{CL})$$
$$(\tau_{\mu^+} - \tau_{\mu^-})/\tau_{\text{average}} = (2 \pm 8) \times 10^{-5}$$
$$(g_{e^+} - g_{e^-})/g_{\text{average}} = (-0.5 \pm 2.1) \times 10^{-12} \qquad (1.21)$$

where g_{e^-} and g_{e^+} are the g-factors of the electron and the positron, respectively, which are defined in Eq. (4.133) of Chapter 4. Hence, **the profound consequences of CPT invariance have so far been confirmed with stringent experimental precision**. It is therefore accepted that CPT is a true symmetry of Nature.

- **Lorentz violation.** Because CPT invariance in QFT is deeply related to the Lorentz group properties of the fields, CPT violation is believed to be essentially synonymous with a violation of Lorentz invariance. Such violations have never been observed experimentally. However, it is not excluded that they could exist, for example in extreme conditions such as astrophysical events at extreme energies or in any other phenomenon at extremely high energies. If this was the case, it would be equivalent to stating that physics in those regimes deviates from the predictions of special relativity, as was the case with classical mechanics before it.

- CPT **mirror Universe.** CPT conservation implies that if all particles and antiparticles were swapped (C) and the strong reflection was applied (PT), reflecting the space and running time backward, then this new world would look indistinguishable from ours!

1.8   Fundamental Constants and the Natural Units

A quantitative description of the physical observations requires the choice of units and a set of fundamental constants. The **CODATA Task Group on Fundamental Constants** was established to provide a self-consistent set of recommended values of the basic constants and conversion factors based on all the data available at the time of publication [4, 10]. The **Bureau International des Poids et Mesures** (BIPM, founded in 1875) in Paris and the National Institute of Standards and Technology (NIST) in the USA also maintain such databases.[16]

In the MKSA (meter–kilogram–seconds–Ampère) units system, the meter is the unit of length, the kilogram the unit of mass, the second the unit of time, and the Ampère is the unit of current. Most of the time particle physics deals with very small quantities relative to macroscopic units, hence these latter are not practical to describe subatomic particles and one relies on a specific set of units.

In the following we introduce the units commonly used in particle physics. The recommended exact values for the two fundamental constants, the Planck constant and the speed of light in vacuum, are the following [4]:

$$h = 6.626\,070\,15 \times 10^{-34}\,\mathrm{J\,Hz^{-1}}\ (\text{exact}) \quad \text{and} \quad c = 299\,792\,458\ \mathrm{m\,s^{-1}}\ (\text{exact}) \tag{1.22}$$

For the energy, we use the electronvolt (see Eq. (1.2) and corresponding prefixes keV, MeV, GeV, TeV). For the length, the **femtometer** (fm $= 10^{-15}$ m) is often employed. The **reduced Planck constant** $\hbar$ takes the following value:

$$\hbar = \frac{h}{2\pi} = 1.054\,571\,817\ldots \times 10^{-34}\,\mathrm{J\,s} = 6.582\,119\,569\ldots \times 10^{-22}\ \mathrm{MeV\,s} \quad (\text{exact}) \tag{1.23}$$

Hence, if we multiply it by c, we get:

$$\hbar c = 3.161\,526\,77\ldots \times 10^{-26}\ \mathrm{J\,m} = 197.326\,980\,459\ldots\ \mathrm{MeV\,fm} \quad (\text{exact}) \tag{1.24}$$

which can be conveniently approximated to $\hbar c \approx 0.2$ GeV fm ≈ 200 MeV fm. **Our observables can therefore be equally expressed in the MKSA system (m, kg, s, A) or as a function of the derived equivalent set in units of (GeV, $\hbar$, c, e).** The derived units prevent us from having to deal with large exponents. Ad-hoc units can also be defined. For example, it is customary to define areas (e.g., to express cross-sections) in units of **barns**, where

$$1\ \mathrm{barn} = 1\ \mathrm{b} \equiv 10^{-28}\ \mathrm{m^2} = 10^{-24}\ \mathrm{cm^2} \tag{1.25}$$

Typical subatomic cross-sections will be in the range of femtobarns (fb) to millibarns (mb).

• **Natural units:** The constants above are practical to describe the subatomic particles. However, one often goes one step further with the introduction of **natural units**. These latter are employed to make algebraic calculations less cumbersome. There is a choice to set specific constants to unity and different systems will assume different values for these constants. In a "natural" way, we can choose the **Planck unit system**, such that two fundamental constants possess the following specific values:

$$c \equiv 1 \quad \text{and} \quad \hbar \equiv 1 \tag{1.26}$$

This imposes two constraints and therefore leaves us with the electric charge and a free choice for one of the three units of length, time, or mass. The units of choice are usually that of energy and charge. It is then no longer necessary to include the fundamental constants in the equations. We have, for example, for the energy, momentum, and mass relation:

$$E^2 = \vec{p}^2 c^2 + m^2 c^4 \quad \overset{c=1}{\longrightarrow} \quad E^2 = \vec{p}^2 + m^2 \tag{1.27}$$

This equation implies that energy, momentum, and mass possess the same unit. Mass is expressed in energy, and then the rest mass of the electron is $m_e \approx 0.511$ MeV instead of 0.511 MeV c^{-2} (although the electronvolt is not strictly natural). Of course, this reduction does not mean that the factors c (or $\hbar$) have vanished. They must be recovered by dimensional analysis at the end of the calculations, in order to recover a result that can be compared to experiment. This implies reinserting the missing c (and $\hbar$). For example, to recover MKSA units in special relativity problems, we would simply replace t by ct, β by v/c, $\vec{p}$ by $c\vec{p}$, and m by mc^2.

---

16   Up-to-date information is available at www.bipm.org and http://physics.nist.gov.

For example, the units of length and time become inverse energy in natural units:

$$1 \text{ m} = \frac{10^{15} \text{ fm}}{\hbar c} \approx \frac{10^{15} \text{ fm}}{200 \text{ MeV fm}} \approx 5.067\,731\ldots \times 10^{15} \text{ GeV}^{-1} \tag{1.28}$$

and

$$1 \text{ s} = \frac{1 \text{ s}}{\hbar} \approx \frac{1 \text{ s}}{6.582 \times 10^{-22} \text{ MeV s}} \approx 1.519\,267 \times 10^{24} \text{ GeV}^{-1} \tag{1.29}$$

Consequently, if we obtained a result in natural units, a length has the units of an inverse energy $(1/E)$ and would have to be multiplied by $\hbar c$ in order to recover the MKSA units. A time also has the units of an inverse energy $(1/E)$ and would have to be multiplied by $\hbar$ to recover the MKSA units. We can also ask the opposite question. For example, if we have a mass given in natural units of 1 GeV, what is the corresponding value in MKSA units? We should insert c^2 and obtain:

$$m = 1 \text{ GeV}/c^2 = \frac{10^9 e \text{J}}{c^2 \frac{\text{m}^2}{\text{s}^2}} = \frac{1.602177 \times 10^{-10} \text{J}}{(299792458)^2 \frac{\text{m}^2}{\text{s}^2}} = 1.7826619 \times 10^{-27} \text{ kg} \tag{1.30}$$

Similarly found, other conversion factors are presented in Table 1.6.

Quantity u	MKSA	GeV, $\hbar$, c, e	$[u]$	Conversion N.U. to MKSA
Mass	kg	GeV/c^2	1	$1 \text{ GeV} = 1.7826619 \times 10^{-27} \text{ kg}$
Length	m	$(\text{GeV}/\hbar c)^{-1}$	-1	$1 \text{ GeV}^{-1} = 1.9732698 \times 10^{-16} \text{ m}$
Time	s	$(\text{GeV}/\hbar)^{-1}$	-1	$1 \text{ GeV}^{-1} = 6.5821196 \times 10^{-25} \text{ s}$
Energy	$\text{kg m}^2/\text{s}^2$	GeV	1	$1 \text{ GeV} = 1.602176634 \times 10^{-10} \text{ J}$
Momentum	kg m/s	GeV/c	1	$1 \text{ GeV} = 5.3442860 \times 10^{-19} \text{ kg m/s}$
Force	$\text{kg m}/\text{s}^2$	$\text{GeV}^2/\hbar c$	2	$1 \text{ GeV}^2 = 8.1193997 \times 10^5 \text{ N}$
Cross-section (area)	m^2	$(\text{GeV}/\hbar c)^{-2}$	-2	$1 \text{ GeV}^{-2} = 0.3893794 \text{ mb}$
Charge	C = A.s	e	none	$e = 0.303 = 1.602\ldots \times 10^{-19} \text{ C}$
E-field	V/m	$\text{GeV}^2/e\hbar c$	2	$1 \text{ GeV}^2 = 5.0677307 \times 10^{24} \text{ V/m}$
B-field	T = kg / As^2	$\text{GeV}^2/e\hbar c^2$	2	$1 \text{ GeV}^2 = 1.6904130 \times 10^{16} \text{ T}$

Table 1.6 Units, derived units, and conversion for primary physical quantities. The number $[u]$ describes the powers of energy in the unit.

As mentioned before, one needs to perform a dimensional analysis in order to reintroduce the missing constants in order to recover numerical values in MKSA or in derived units. Dimensional analysis is often an art and is learned by practice. We usually write **the energy dimension of a quantity with brackets**, as in $[mass] = 1$, $[length] = -1$, $[time] = -1$, and so on. Some quantities related to energies have positive energy dimensions, while those related to space-time (distances) are negative. More generally, we can write:

$$[E] = [p] = [m] = 1, \ [x] = [t] = -1, \ [\mathrm{d}x] = [\mathrm{d}y] = [\mathrm{d}z] = [\mathrm{d}t] = -1, \ [\mathrm{d}^3\vec{x}] = -3, \ldots \tag{1.31}$$

We give an example – calculation of the lifetime of a muon at rest (see Section 23) in natural units ($[\tau] = -1$) gives:

$$\tau_{n.u.} = \frac{192\pi^3}{G_F^2 m_\mu^5} = 3.43 \times 10^{18} \text{ GeV}^{-1} \tag{1.32}$$

where G_F is the Fermi coupling constant and m_μ the muon rest mass. To recover the lifetime in the MKSA system, this expression should be multiplied by the relevant value in Table 1.6, that is $\tau_{MKS} = \tau_{n.u.}\hbar = \left(3.43 \times 10^{18} \text{ GeV}^{-1}\right)\left(6.58 \times 10^{-25} \text{GeV s}\right) \simeq 2.2 \text{ μs}$.

We give another example: the cross-section σ for a physical process (see Section 2.8) has the units of an area, so dimensionally $[\sigma] = -2$. We might want to express this in millibarns (mb) – see Eq. (1.25):

$$1 \text{ mb} = 10^{-3} \text{ b} \equiv 10^{-31} \text{ m}^2 \tag{1.33}$$

In natural units, the result of our cross-section calculation yields a result in, say, GeV^{-2}. So we need to multiply the result by $(\hbar c)^2$, which has the units of (energy×length)^2, hence we have:

$$\text{GeV}^{-2}(\hbar c)^2 \approx \text{GeV}^{-2}(197.3 \text{ MeV fm})^2 \approx 38937 \times 10^{-36} \text{ m}^2 = 0.38937 \text{ mb} \tag{1.34}$$

• **Heaviside–Lorentz units.** As noted above, the electric charge is a separate physical quantity and can also be redefined by choosing the value of ϵ_0 in Coulomb's law:

$$F = \frac{1}{4\pi\epsilon_0}\frac{q_1 q_2}{r^2}, \quad \text{where } \epsilon_0 = 8.854\,187\,8128 \times 10^{-12} \text{ C}^2 \text{ m}^{-3} \text{ kg}^{-1} \text{ s}^{-2} \tag{1.35}$$

In natural units, the units of charge are defined by setting $\epsilon_0 = 1$. This automatically sets $\mu_0 = 1$ since $c^2 = 1/(\epsilon_0 \mu_0)$, and therefore we reach the intriguing situation where:

$$c = \hbar = \epsilon_0 = \mu_0 = 1 \tag{1.36}$$

Since the units of force and distance are energy squared ($[F] = 2$) and inverse energy ($[r] = -1$) (see Table 1.6), the electric charge is dimensionless ($[q_1] = [q_2] = [e] = 0$)!

The **fine-structure constant** – also being dimensionless – is the same in all units! It has the following form in MKSA units (see Eq. (C.2) and CODATA [4, 10]):

$$\alpha = \frac{e^2}{4\pi\epsilon_0\hbar c} = \frac{1}{137.035999084 \pm 0.000000021} \approx 0.007297353 \tag{1.37}$$

This result can be expressed as:

$$\frac{e^2}{4\pi\epsilon_0} = \alpha\hbar c \simeq 1.45 \text{ MeV fm} \tag{1.38}$$

In natural units, we simply obtain $\alpha = e^2/(4\pi) \approx 1/137$. This clearly shows that the electric charge becomes dimensionless and is equal to:

$$e_{n.u.} = \sqrt{4\pi\alpha} \approx 0.303 \tag{1.39}$$

Hence, **in natural units there is an equivalence between energy, mass, inverse time, and inverse length, and the electric charge is dimensionless.** The units of charge, electric and magnetic field, and their corresponding conversion factors are listed in Table 1.6. We will use the natural units most of the time in the remainder of the text.

Problems

Ex 1.1 Discrete transformation. Derive Table 1.3.

Ex 1.2 Symmetries and conservation laws. Noether's theorem states that for each continuous symmetry, there is a conserved quantity. What is the conserved quantity associated with a Lorentz boost?

Ex 1.3 Local and global gauge symmetries. Gauge symmetries play a fundamental role in particle physics.

    (a) Develop Table 1.4 by expanding on each symmetry and its conserved charge. Discuss the long-range force associated with each local gauge symmetry.

(b) Gravitation is a long-range force. Is there an associated conserved charge? What would be the local gauge symmetry?

Ex 1.4 Fundamental symmetries. The Particle Data Group (PDG) regularly produces a Review of Particle Physics. It contains, among others, a summary of all known properties of subatomic particles and their interactions. It is available online at http://pdg.lbl.gov/. Use the PDG review to describe which symmetry forbids the following elementary reactions or, on the contrary, which interaction governs the decays:

(a) $\Sigma^0 \to \Lambda + \pi^0$ (d) $K^- \to \pi^- + \pi^0$ (g) $\pi^0 \to \gamma + \gamma$ (j) $\mu^- \to e^- + \bar{\nu}_e$

(b) $\Delta^+ \to p + \pi^0$ (e) $\Sigma^0 \to \Lambda + \gamma$ (h) $\eta \to \gamma + \gamma$ (k) $\Xi^- \to \Lambda + \pi^-$

(c) $p \to e^+ + \gamma$ (f) $\Xi^0 \to p + \pi^-$ (i) $\Sigma^- \to n + \pi^-$ (l) $\Sigma^- \to n + e + \bar{\nu}_e$

Ex 1.5 Natural units. Express your age, your mass, your height, and your body surface in natural units.

Ex 1.6 Natural units. Derive Table 1.6.

Ex 1.7 Natural units. What is the typical kinetic energy of a served tennis ball expressed in electronvolt? What is the total energy (including rest mass) of that tennis ball? What is the velocity of a proton with the same kinetic energy? Compare the energy to that of a proton in the Large Hadron Collider (LHC).

2   Basic Concepts

The main drivers of progress in nuclear and particle physics have been theoretical developments and experimental observations. The possibility to perform new observations is critically coupled with the development of new methods in particle acceleration and detection of particles. Historically, the birth of particle physics can be associated with the discovery of the electron in 1897. Before 1897 the modern atomic description of matter did not exist. The absorption and emission spectra of atoms were known, however their description was only based on empirical rules.

2.1   Introduction

In 1766, **Cavendish**[1] wrote a paper called "On factitious airs" [11]. He is attributed the discovery of what he called *inflammable air*, known today as **hydrogen**. Hydrogen was mainly characterized by its physico-chemical properties in order to study in detail its behavior in combustion reactions. In 1855 Ångström,[2] one of the founders of spectroscopy, published the results of his investigations on the line spectrum of hydrogen. Hydrogen then became one of the most important research issues for the physicists of the time. In 1860 **Kirchoff**[3] and **Bunsen**[4] discovered that some substances, when brought into a flame, colored the flame characteristically. They found the existence of sharp spectral lines. They also noted that the spectral lines emitted by a substance are also absorbed by that substance.

The line structures could not be interpreted with the physics of the nineteenth century; the electron was to be discovered by J. J. Thomson only in 1897 (see Section 2.6)! However, this lack of knowledge prompted physicists of the time *to acquire additional spectroscopic data and to improve the measuring apparatus in order to get information.* These spectra have yielded very important information about the structure of chemical elements and the structure of atoms. It was just in this way that the **Zeeman[5] effect** (see Section 4.13) was discovered in 1896 as the fine structure of the spectrum represented by the splitting of the spectral lines in the presence of a strong external magnetic field.

Ångström was the first to point out that the hydrogen spectrum was composed of three lines in the visible band. Later he noted that the violet line was actually formed by two distinct very close lines. In the years that followed, physicists began to notice some regularities of the spectrum, and that some lines were related to others by an empirical equation. In 1885 **Balmer**[6] noticed that a single number had a relation to every line in the hydrogen spectrum, that is in the visible light region. He originally published his empirical formula [12]

---

1   Henry Cavendish (1731–1810), British natural philosopher, scientist, and an important experimental and theoretical chemist and physicist.
2   Anders Jonas Ångström (1814–1874), Swedish physicist.
3   Gustav Robert Kirchhoff (1824–1887), German physicist.
4   Robert Wilhelm Eberhard Bunsen (1811–1899), German chemist.
5   Pieter Zeeman (1865–1943), Dutch physicist.
6   Johann Jakob Balmer (1825–1898), Swiss mathematician and mathematical physicist.

in the form:

$$\lambda = B \frac{m^2}{m^2 - 2^2} \tag{2.1}$$

where λ is the wavelength of the spectral line, B is a constant equal to 3645.6 Å (where 1 Å $= 10^{-10}$ m), and m is an integer greater than 2. In 1888 **Rydberg**[7] generalized the Balmer equation for all transitions of hydrogen:

$$\frac{1}{\lambda} = R_H \left(\frac{1}{2^2} - \frac{1}{n^2} \right) \tag{2.2}$$

where R_H is the **Rydberg constant** for hydrogen and $n = 3, 4, 5, \ldots$ Rydberg's constant was initially empirical but could later be expressed within Bohr's model as a function of other fundamental constants. It is related to the fundamental Rydberg constant corrected by the reduced mass of the hydrogen system:

$$R_H = \frac{R_\infty}{1 + m_e/M_p} \tag{2.3}$$

where m_e and M_p are the electron and proton rest masses, and R_∞ is given by (see Eq. (C.12)):

$$R_\infty \equiv \frac{m_e e^4}{8\epsilon_0^2 h^3 c} = \frac{\alpha^2 m_e c}{2h} \tag{2.4}$$

where h is the Planck constant introduced in Section 1.8 and defined in the next section, and α was defined in Eq. (1.37). Following CODATA 2014 recommendations [10], the Rydberg constant is today one of the most precisely known fundamental constants of nature:

$$R_\infty = (10973731.568160 \pm 0.000021)\,\text{m}^{-1} \tag{2.5}$$

or a relative uncertainty of 1.9×10^{-12}!

2.2   Black-Body Radiation

A **black body** represents an idealized body at a fixed temperature T whose internal structure is constantly emitting and absorbing light. The details of the body are not important. The only assumption is that thermal equilibrium is verified. In 1879 **Stefan**[8] had measured the integrated intensity of a black body and found that it is proportional to the fourth power of the temperature T^4, where T is the absolute temperature of the body. In 1896 **Paschen**[9] published the first measurement of the black-body spectrum and found that it was consistent with the **displacement law** derived by **Wien**[10] that states that the spectrum of the black-body radiation peaks at the wavelength λ_{max} given by:

$$\lambda_{max} \approx \frac{2898\ \mu\text{m K}}{T} \tag{2.6}$$

If we define a metallic, perfectly square box of size L, then classical mechanics allows us to define the boundary conditions for standing electromagnetic waves. The radiation is reflecting back and forth between the walls. If opposing walls are exactly parallel, then the three components of the electric field do not mix and can be treated separately, which we label with the indices x, y, and z corresponding to three perpendicular sides of the box. If we consider the E_x component of the electric field and the wall at $x = 0$ and $x = L$, then the incident and reflected waves must form a standing wave. In addition, since the electric field is transverse to

---

7   Johannes Robert Rydberg (1854–1919), Swedish physicist.
8   Josef Stefan (1835–1893), Austrian physicist.
9   Louis Carl Heinrich Friedrich Paschen (1865–1947), German physicist.
10   Wilhelm Carl Werner Otto Fritz Franz Wien (1864–1928), German physicist.

the direction of propagation of the electromagnetic wave, it must be directed parallel to the wall. However, since the wall is metallic, any electric field parallel to the surface will be zero. Therefore, the electric field component E_x vanishes at both walls $x = 0$ and $x = L$. Obviously, similar conditions apply for the other walls.

• **Rayleigh[11]–Jeans[12] derivation.** Let us consider an electromagnetic wave with wavelength λ and frequency ν. Dimensionally, these two must be related to the speed of light by $c = \lambda\nu$. The wave is assumed to propagate along a direction defined by the three angles α, β, and γ with $\cos^2\alpha + \cos^2\beta + \cos^2\gamma = 1$. This wave must be a standing wave since all three of its components are standing waves. The distance between the nodal planes is given by $\lambda/2$. Decomposing into the three axes, we have:

$$\lambda_x = \frac{\lambda}{\cos\alpha}, \quad \lambda_y = \frac{\lambda}{\cos\beta}, \quad \lambda_z = \frac{\lambda}{\cos\gamma} \tag{2.7}$$

where the spatial dependence of the electric field components is given by

$$E_x \propto \sin\left(\frac{2\pi x}{\lambda_x}\right), \quad E_y \propto \sin\left(\frac{2\pi y}{\lambda_y}\right), \quad E_z \propto \sin\left(\frac{2\pi z}{\lambda_z}\right) \tag{2.8}$$

This form automatically satisfies the boundary condition at the three planes "$x = 0$," "$y = 0$," and "$z = 0$." The conditions on the opposite walls can be written as:

$$\frac{2L}{\lambda_x} = \frac{2L}{\lambda}\cos\alpha = n_x, \quad \frac{2L}{\lambda_y} = \frac{2L}{\lambda}\cos\beta = n_y, \quad \frac{2L}{\lambda_z} = \frac{2L}{\lambda}\cos\gamma = n_z \tag{2.9}$$

where $\vec{n} = (n_x, n_y, n_z)$ is a vector of zero or positive integers ($(0,0,0)$ excluded). Hence, squaring and adding:

$$\left(\frac{2L}{\lambda}\right)^2 \left(\cos^2\alpha + \cos^2\beta + \cos^2\gamma\right) = \left(\frac{2L}{\lambda}\right)^2 = n_x^2 + n_y^2 + n_z^2 = |\vec{n}|^2 \tag{2.10}$$

The allowed frequencies of the standing waves are therefore constrained to be

$$\nu = \frac{c}{\lambda} = \frac{c}{2L}|\vec{n}| \tag{2.11}$$

Now we should count the number of allowed frequencies $N(\nu)\mathrm{d}\nu$ in a given frequency interval $\mathrm{d}\nu$. This is equal to the number of points contained between two concentric shells of radii $r = |\vec{n}|$ and $r + \mathrm{d}r$. Only positive values are allowed for n_x, n_y, and n_z, consequently, we have:

$$N(r)\mathrm{d}r = 4\pi r^2 \mathrm{d}r \times \frac{1}{8} = \frac{\pi}{2}r^2 \mathrm{d}r \tag{2.12}$$

In terms of frequencies, noting $dr = (2L/c)\mathrm{d}\nu$, we obtain:

$$N(\nu)\mathrm{d}\nu = \frac{\pi}{2}\left(\frac{2L}{c}\nu\right)^2 \frac{2L}{c}\mathrm{d}\nu = \frac{\pi}{2}\left(\frac{2L}{c}\right)^3 \nu^2 \mathrm{d}\nu \tag{2.13}$$

This result should actually be multiplied by two in order to take into account the two possible independent polarizations of the electromagnetic radiation. Finally:

$$N_{RJ}(\nu)\mathrm{d}\nu = \frac{8\pi L^3}{c^3}\nu^2 \mathrm{d}\nu \tag{2.14}$$

This result represents the classical Rayleigh–Jeans derivation.

11   John William Strutt, 3rd Baron Rayleigh (1842–1919), British scientist.
12   James Hopwood Jeans (1877–1946), British physicist, astronomer and mathematician.

• **Boltzmann probability distribution.** We now estimate the average energy contained in each standing wave of given frequency ν. Each frequency mode is to be considered as a degree of freedom capable of exchanging energy. Therefore, each mode can store any value of energy $\epsilon(\nu)$ between zero and infinity. However, these modes are assumed to be in thermal equilibrium at a temperature T and able to exchange energy. Their distribution is governed by the **Boltzmann**[13] **probability distribution**

$$P(\epsilon) = Ae^{-\epsilon/k_B T} \tag{2.15}$$

where $k_B = 1.380649 \times 10^{-23}\,\mathrm{J\,K^{-1}}$ is the (exact) **Boltzmann constant** [4]. The average energy of each mode is then

$$\bar{\epsilon} = \frac{\int_0^\infty \epsilon P(\epsilon)\mathrm{d}\epsilon}{\int_0^\infty P(\epsilon)\mathrm{d}\epsilon} = k_B T \tag{2.16}$$

Then, according to the *equipartition theorem of classical physics*, there should be an equal amount of energy in each standing wave given by $\bar{\epsilon}$. The energy per unit volume in the frequency interval ν and $\nu + \mathrm{d}\nu$ of the black body at temperature T is then given by $\bar{\epsilon}(\nu)N(\nu)\mathrm{d}\nu/L^3$, or:

$$\rho_{RJ}(\nu)\mathrm{d}\nu = \frac{8\pi k_B T}{c^3}\nu^2 \mathrm{d}\nu \tag{2.17}$$

As a function of the wavelength $\lambda = c/\nu$, we get:

$$\rho_{RJ}(\lambda)\mathrm{d}\lambda = \rho_{RJ}(\nu)\frac{c}{\lambda^2}\mathrm{d}\lambda = \frac{8\pi k_B T}{c^3}\frac{c}{\lambda^2}\left(\frac{c}{\lambda}\right)^2\mathrm{d}\lambda = 8\pi k_B T\frac{\mathrm{d}\lambda}{\lambda^4} \tag{2.18}$$

This result is called the **Rayleigh–Jeans spectrum distribution of the black body**. Since there are an infinite number of modes, this would imply infinite heat capacity, as well as a non-physical spectrum of emitted radiation that grows without bounds with increasing frequency ν or decreasing wavelength λ, a problem known as the **ultraviolet catastrophe**. Indeed, the total energy predicted is:

$$\int_0^\infty \rho_{RJ}(\lambda)\mathrm{d}\lambda = 8\pi k_B T \lim_{\lambda \to 0}\frac{1}{\lambda^3} \longrightarrow \infty \quad (T > 0) \tag{2.19}$$

This represents the complete failure of classical physics to describe black-body radiation! The theory predicts that the energy goes to infinity as the wavelengths go to zero, a fact which is in total contradiction with experimental data which shows that the energy density always remains finite and tends to zero as the wavelength goes to zero (short wavelengths), as illustrated in Figure 2.1.

• **Planck's theory.** The discrepancy between theory and experiment was resolved by Planck. In order to solve the problem, he introduced a drastically new concept which was in sharp contrast with classical physics. In 1900, he wrote in the proceedings of his contribution to the Deutschen Physikalischen Gesellschaft that the most important point of his calculation was to consider the energy as composed of a certain number of finite equal parts. His postulate may be stated as follows: *Any physical system whose coordinate exhibits a simple harmonic motion with a frequency ν can only possess discrete energies ϵ which satisfy:*

$$\epsilon = nh\nu \quad \text{with} \quad n = 0, 1, 2, 3, \ldots \tag{2.20}$$

where h is the Planck constant (Eq. (1.22)).[14] In order to compute the average energy of a mode, we should therefore replace the integral in Eq. (2.16) by a sum:

$$\bar{\epsilon} = \frac{\sum_{n=0}^{+\infty} \epsilon P(\epsilon)}{\sum_{n=0}^{+\infty} P(\epsilon)} = \frac{\sum_{n=0}^{+\infty} nh\nu Ae^{-nh\nu/k_B T}}{\sum_{n=0}^{+\infty} Ae^{-nh\nu/k_B T}} = \frac{\sum_{n=0}^{+\infty} nh\nu e^{-\beta nh\nu}}{\sum_{n=0}^{+\infty} e^{-\beta nh\nu}} \tag{2.21}$$

13 Ludwig Eduard Boltzmann (1844–1906), Austrian physicist and philosopher.
14 According to Boya [13] Planck wrote h for *Hilfsgrösse* (German), which can be translated as "auxiliary variable."

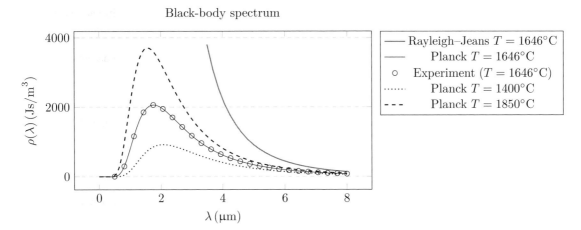

Figure 2.1 A comparison of the experimental black-body data with the Rayleigh–Jeans and the Planck theories. The experimental points illustrate data at a temperature $T = 1646°C$.

where $\beta \equiv 1/k_B T$ and one assumes, as Planck did, that the Boltzmann distribution still applies. Using the series Eq. (A.1), we can simply write:

$$\sum_{n=0}^{+\infty} e^{-\beta n h\nu} = 1 + e^{-\beta h\nu} + (e^{-\beta h\nu})^2 + (e^{-\beta h\nu})^3 + \cdots = \frac{1}{1 - e^{-\beta h\nu}} \tag{2.22}$$

Then, we note that the average energy can be conveniently expressed as:

$$\bar{\epsilon} = -\frac{\mathrm{d}}{\mathrm{d}\beta} \ln \left(\sum_{n=0}^{+\infty} e^{-\beta n h\nu} \right) = -\frac{\mathrm{d}}{\mathrm{d}\beta} \ln \left((1 - e^{-\beta h\nu})^{-1} \right) = \frac{h\nu e^{-\beta h\nu}}{1 - e^{-\beta h\nu}} = \frac{h\nu}{e^{\beta h\nu} - 1} \tag{2.23}$$

Then, evaluating the energy per unit volume in the frequency interval ν and $\nu + \mathrm{d}\nu$ of the black body at temperature T as before now yields:

$$\rho_P(\nu)\mathrm{d}\nu = \frac{8\pi}{c^3}\nu^2 \frac{h\nu}{e^{\beta h\nu} - 1}\mathrm{d}\nu = \frac{8\pi h}{c^3}\frac{\nu^3}{e^{\beta h\nu} - 1}\mathrm{d}\nu \tag{2.24}$$

We can now transform to the wavelength λ to find **Planck's black-body spectrum**:

$$\rho_P(\lambda)\mathrm{d}\lambda = \frac{8\pi hc}{\lambda^5}\frac{1}{e^{hc/\lambda k_B T} - 1}\mathrm{d}\lambda \tag{2.25}$$

When the formula was first derived, the constant h was undetermined and was to be fitted to the experimental data. The success of Planck's derivation lies in the extremely good agreement with data, as illustrated in Figure 2.1, with the value of the constant coming out of the fit. The result was also shown to be in agreement with **Wien's law** and consequently valid for all temperatures T.

Mathematically, Planck's derivation only has the effect of replacing an integral by a sum, with the remarkable consequence for the result. **Planck's postulate can be seen as putting a "cutoff" into the calculation at high frequencies in order to avoid divergences**. Physically, the energy of the standing waves is limited to $\epsilon = 0$, $\epsilon = h\nu$, $\epsilon = 2h\nu$, ... If we now consider Boltzman's distribution factor, then the probability for $\epsilon \neq 0$ becomes negligible when $h\nu \gg k_B T$. Consequently, high frequencies do not contribute to the total energy balance and therefore, unlike in the Rayleigh–Jeans case, the UV behavior is quenched. For low frequencies $h\nu \ll k_B T$, the effect of the limited values $\epsilon = nh\nu$ becomes negligible and we recover the classical result. This is fantastic!

2.3   The Cathode-Ray Tube

The **cathode-ray tube** has played an important role in the development of particle physics. The technical development of the **vacuum pump** at the end of the nineteenth century has permitted the exploration of new phenomena with evacuated glass cylinders. The basic principle is illustrated in Figure 2.2. Two electrodes are located inside a glass cylinder. A **potential difference** is created (at the time with the help of batteries) between the **cathode** and the **anode**. The first cathode-ray tube was built in 1897 by **Karl Ferdinand Braun**.[15] Experimentally one observes: **an electric current through the electrical circuit can be created when the cylinder is slowly filled with gas.**

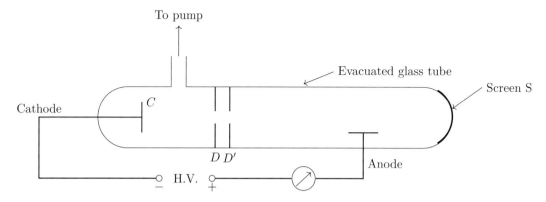

Figure 2.2 The cathode-ray tube (CRT). D and D' are two collimators. The screen S is fluorescent.

A small gas pressure is sufficient to observe an electric current. With the advent of the cathode-ray tube, one could study the properties of the created current as a function of the gas pressure, the type of gas, the electric voltage, etc. One also knew that with specific types of gases one could observe **light** when a fluorescent screen was placed at the end of the cylinder (see S in Figure 2.2).

2.4   Three Types of Rays

In 1895 **X-rays** were discovered "by chance" by **Röntgen**.[16] Such new types of rays could be created in the cathode-ray tube. Röntgen noticed that the new rays were penetrating and invisible, and that they could be detected with the help of **photographic film**, as they blackened the film [14].

In 1896 **Becquerel**[17] discovered **natural radioactivity**, that is, the existence of a natural element (uranium) that spontaneously emits rays, while studying the **phosphorescence** of materials. He wrote:

> *"I would particularly like to emphasize the following fact, which I think is quite important and apart from the phenomena that one might expect to observe (...) Here is how I was brought to make this observation: among the preceding experiences, some of them were prepared on Wednesday February 26th and Thursday February 27th and, as on those days, the sun was only visible intermittently, I had kept the experiments all prepared and tucked the frames in the dark into a drawer of a piece of furniture, leaving in place the slices of salt of uranium. As the Sun did not show up again*

15   Karl Ferdinand Braun (1850–1918), German inventor, physicist.
16   Wilhelm Röntgen (1845–1923), German physicist.
17   Antoine Henri Becquerel (1852–1908), French physicist.

the following days, I developed the photographic plates on March 1st, expecting to find very weak images. The silhouettes, on the contrary, appeared with great intensity."[18]

Becquerel had discovered that, unlike what was expected at the time, uranium emitted rays that blackened photographic film even when the substance was not exposed to the Sun!

In 1898 **Marie Skłodowska Curie**[19] and **Pierre Curie**[20] discovered that several elements are radioactive: **polonium** and **radium**. These elements would also blacken photographic film.

At the end of the nineteenth century, one knew that *three types of radioactive rays* existed in Nature. The rays could be characterized with the help of a magnetic field. The experimental setup is illustrated in Figure 2.3.

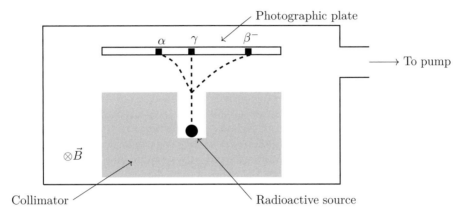

Figure 2.3 An experimental apparatus to study the three types of radioactive rays.

The rays from different radioactive elements were forced in a given flight direction with a *collimator* around the sources. A magnetic field perpendicular to the plane surface was created in an evacuated volume. The rays would be detected with photographic film. One observed experimentally: **three different zones of blackening appear on the photographic film. These three zones correspond to the three types of rays: the so-called alpha (α)-, gamma (γ)-, and beta (β)-rays.**

The rays were observed to penetrate more or less in matter and with the following properties:

1. α**-rays** are electrically charged, as they bend in the magnetic field. They are not visible unless the volume is evacuated. They can only travel a few centimeters in air at normal pressure.

2. β**-rays** are electrically charged and unlike α-rays can bend in positive or negative directions depending on the radioactive element. A few millimeters of material are sufficient to stop β-rays.

3. γ**-rays** are electrically neutral. They share the same properties as X-rays and are the most penetrating type of rays. A few centimeters of lead are necessary to reduce their intensity.

18 *"J'insisterai particulièrement sur le fait suivant, qui me paraît tout à fait important et en dehors des phénomènes que l'on pouvait s'attendre à observer (...) Voici comment j'ai été conduit à faire cette observation: Parmi les experiences qui précèdent, quelques-unes avaient été préparées le mercredi 26 et le jeudi 27 février et, comme ces jours-là, le soleil ne n'est montré que d'une manière intermittente, j'avais conservé les expériences toutes préparées et rentré les châssis à l'obscurité dans le tiroir d'un meuble, en laissait en place les lamelles du sel d'Uranium. Le soleil ne s'étant pas montré de nouveau les jours suivants, j'ai développé les plaques photographiques le 1er mars, en m'attendant à trouver des images très faibles. Les silhouettes, apparurent, au contraire, avec une grande intensité."* Reprinted from H. Becquerel, "Sur les radiations invisibles émises par les corps phosphorescents," *Comptes Rendus,* vol. 122, pp. 501–503, 1896.
19 Marie Skłodowska Curie (1867–1934), Polish physicist and chemist, emigrated to France.
20 Pierre Curie (1859–1906), French physicist.

2.5   The Law of Radioactive Decay

The law of radioactive decay was first experimentally established at the beginning of the twentieth century. In 1900, Rutherford discovered that the radioactivity of the *emanation* of thorium fell relatively fast with time. He could plot his results in a graphical form as shown in Figure 2.4 and introduce the concept of half-time. He noted that the intensity of the radiation given out by the radioactive particles fell off in a geometrical

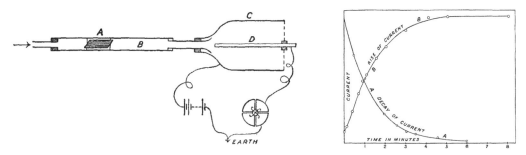

Figure 2.4 (left) Apparatus used by Rutherford in 1900; (right) the observed rise and decay curves of the "emanation." Reprinted from E. Rutherford, "A radio-active substance emitted from thorium compounds," *The London, Edinburgh, and Dublin Philosophical Magazine and Journal of Science*, vol. 49, no. 296, pp. 1–14, 1900, by permission of the publisher (Taylor & Francis Ltd., www.tandfonline.com).

progression with time. The result presented in the figure shows that the intensity of the radiation had fallen to one-half its value after an interval of about one minute. In addition, he could derive the correct formula describing the observed time dependence. The law stated that **the activity of a radioactive source decays exponentially with time**. Compared to this result, all other radioactive substances known at the time seemed to have a perpetually emitting constant activity. Their half-lives were of course later determined to be in the order of thousands of years or more, so the exponential decay was too slow to be easily observed.

The exponential law corresponds to a decay process governed by a constant transition probability per unit time, which is called Γ. The activity of the sample is then directly proportional to the number of radioactive elements in the source as a function of time $N(t)$, which we can write as:

$$N(t) = N(0)e^{-\Gamma t} \tag{2.26}$$

where $N(0)$ is the number of elements at $t = 0$. We see that Γ must have the units of inverse time, which in natural units is just energy (Table 1.6). We say that **the radioactive decay is governed by its decay width Γ**. How does an isotope (or in general a particle) know when to decay? By differentiation with respect to time, one finds:

$$\frac{\mathrm{d}N}{\mathrm{d}t} = -\Gamma N(0)e^{-\Gamma t} = -\Gamma N(t) \quad \Longrightarrow \quad \frac{\mathrm{d}N}{N(t)} = -\Gamma \mathrm{d}t \tag{2.27}$$

Thus, the *mean* fraction of decaying elements in a time $\mathrm{d}t$ is constant and is proportional to Γ. This is true even after many lifetimes, hence, there has been no "aging" in the sense that they would decay faster. Each decaying element decays independently and as long as it exists it has a fixed probability to decay in the next unit of time.

• **Multiple decay modes.** When an element has more than one decay mode, then the total width is the sum of the **partial widths** Γ_i for each individual mode:

$$\Gamma = \sum_j \Gamma_j = \Gamma_1 + \Gamma_2 + \cdots \tag{2.28}$$

The relative frequency of a particular decay mode is referred to as the **branching ratio** or **branching fraction**. The branching ratio $Br(i)$ for a particular decay mode i is given by the partial width Γ_i relative to the total width Γ:

$$Br(i) \equiv \frac{\Gamma_i}{\Gamma} = \frac{\Gamma_i}{\sum_j \Gamma_j} = \frac{\Gamma_i}{\Gamma_1 + \Gamma_2 + \cdots} \tag{2.29}$$

By definition, the sum of the branching ratios of all decay modes must be equal to unity. The branching ratio can be interpreted as a probability. For instance, if the decay has a branching ratio of 10%, this implies that on average that decay will occur 10% of the time.

• **Lifetime.** The **mean lifetime** τ is defined as the inverse of the width, with units of time. It corresponds to the time it takes for the activity of the sample to reduce to a factor $1/e$ of its initial value:

$$\tau \equiv \frac{1}{\Gamma} \tag{2.30}$$

The **half-life** $t_{1/2}$ is defined as the time it takes for the activity to reduce by one-half, so

$$e^{-\Gamma t_{1/2}} = \frac{1}{2} \quad \Longrightarrow \quad t_{1/2} = \frac{\ln 2}{\Gamma} = \tau \ln 2 \approx 0.69315\tau \tag{2.31}$$

• **Activity of a source.** The activity, strength, or decay rate R of a radioactive source is defined as the mean number of its decays per unit time:

$$R \equiv -\frac{\mathrm{d}N}{\mathrm{d}t} = \Gamma N(t) \tag{2.32}$$

The unit is the Becquerel (Bq), defined as:

$$1 \text{ Bq} = 1 \text{ decay/s} \tag{2.33}$$

Historically, the activity was measured in units of Curie (Ci), defined as that of 1 g of pure radium-226. This is in fact a large unit which is equivalent to:

$$1 \text{ Ci} = 3.7 \times 10^{10} \text{ Bq} \tag{2.34}$$

• **Statistical fluctuations in decays.** We consider the case where the rate of decay is relatively small in a given time interval Δt. Then, we can assume $N(t) \approx N(0)$ and the number of decays per unit time R can be considered as constant:

$$R \approx \Gamma N(0) = \text{constant} \tag{2.35}$$

If we perform repeated measurements of the number of decays ΔN in a time interval Δt, then fluctuations will occur due to the statistical nature of the decay process. Each decaying element has a given probability of decaying within Δt, but the exact number of decays at any given time cannot be predicted. The probability of observing k decays in a time interval Δt is given by the **Poisson distribution** (see Appendix G.1):

$$P(k, \mu) = \frac{\mu^k e^{-\mu}}{k!} \tag{2.36}$$

where μ is the average decay rate in the period Δt. A property of the Poisson distribution is that its variance σ^2 is equal to its mean μ, so the standard deviation is (see Eq. (G.14))

$$\sigma = \sqrt{\mu} \tag{2.37}$$

As an example, if a source has an activity $R = 1$ Hz, then the probability of observing zero decays in an interval of 1 s is $P(0, 1) \approx 36.8\%$. For one decay, it is the same since $P(1, 1) \approx 36.8\%$, for two decays it is $P(2, 1) = 18.4\%$, and so on.

• **Radioactive decay chain.** Very often one encounters the situation in which a radioactive isotope decays into a daughter isotope, which itself decays into another unstable nucleus, etc. For example, the thorium series begins with the naturally occurring ^{232}Th and decays into a series of elements including radium, actinium, radon, polonium, bismuth, thallium, and lead:

$$^{232}_{90}\text{Th} \xrightarrow[1.4\times10^{10}\text{y}]{\alpha} {}^{228}_{88}\text{Ra} \xrightarrow[5.7\text{y}]{\beta^-} {}^{228}_{89}\text{Ac} \xrightarrow[6.1\text{h}]{\beta^-} {}^{228}_{90}\text{Th} \xrightarrow[1.9\text{y}]{\alpha} {}^{224}_{88}\text{Ra} \xrightarrow[3.6\text{d}]{\alpha} {}^{220}_{86}\text{Rn} \xrightarrow[55\text{s}]{\alpha} {}^{216}_{84}\text{Po} \xrightarrow[0.14\text{s}]{\alpha} {}^{212}_{82}\text{Pb} \xrightarrow[10.6\text{h}]{\beta^-} {}^{212}_{83}\text{Bi}$$

$$^{212}_{83}\text{Bi} \xrightarrow[3.7\times10^3\text{s}]{\alpha(36\%)} {}^{208}_{81}\text{Tl} \xrightarrow[186\text{s}]{\beta} {}^{208}_{82}\text{Pb(stable)} \quad \text{or} \quad {}^{212}_{83}\text{Bi} \xrightarrow[3.7\times10^3\text{s}]{\beta^-(60\%)} {}^{212}_{84}\text{Po} \xrightarrow[3\times10^{-7}\text{s}]{\alpha} {}^{208}_{82}\text{Pb(stable)} \qquad (2.38)$$

Let us consider the following simple case of a chain with three elements $I_1 \xrightarrow{\tau_1} I_2 \xrightarrow{\tau_2} I_3$(stable). We then have:

$$\frac{\mathrm{d}N_1}{\mathrm{d}t} = -\Gamma_1 N_1, \quad \frac{\mathrm{d}N_2}{\mathrm{d}t} = \Gamma_1 N_1 - \Gamma_2 N_2, \quad \frac{\mathrm{d}N_3}{\mathrm{d}t} = \Gamma_2 N_2 \qquad (2.39)$$

where $\Gamma_i = 1/\tau_i$ are the corresponding decay constants. We always observe the exponential decay law of I_1:

$$N_1(t)/N_1(0) = e^{-\Gamma_1 t} \qquad (2.40)$$

The activity of I_2 and I_3 depends on their initial conditions. If we initially have $N_2(0) = N_3(0) = 0$, then the solution is:

$$N_2(t)/N_1(0) = \frac{\Gamma_1}{\Gamma_2 - \Gamma_1} \left(e^{-\Gamma_1 t} - e^{-\Gamma_2 t} \right)$$

$$N_3(t)/N_1(0) = 1 + \frac{\Gamma_1}{\Gamma_2 - \Gamma_1} e^{-\Gamma_2 t} - \frac{\Gamma_2}{\Gamma_2 - \Gamma_1} e^{-\Gamma_1 t} \qquad (2.41)$$

This is illustrated in Figure 2.5 for the two cases $\tau_2 = \tau_1/2$ ($\Gamma_2 = 2\Gamma_1$) and $\tau_2 = 2\tau_1$ ($\Gamma_2 = \Gamma_1/2$). In both

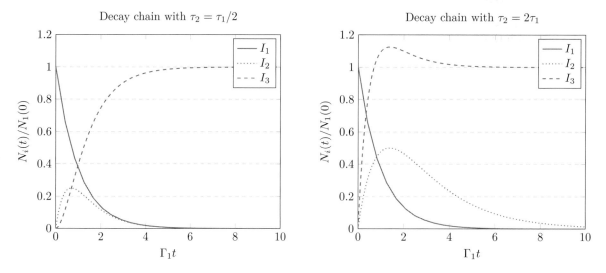

Figure 2.5 The number of decays as a function of time in the three-element decay chain $I_1 \xrightarrow{\tau_1} I_2 \xrightarrow{\tau_2} I_3$(stable); (left) $\tau_2 = \tau_1/2$; (right) $\tau_2 = 2\tau_1$.

cases, the number of decays of isotope I_2 reaches a maximum, which occurs at time t_{max}:

$$\frac{\mathrm{d}N_2(t_{max})}{\mathrm{d}t} = 0 \implies -\Gamma_1 e^{-\Gamma_1 t_{max}} + \Gamma_2 e^{-\Gamma_2 t_{max}} = 0 \implies \frac{\Gamma_1}{\Gamma_2} = e^{-(\Gamma_2 - \Gamma_1)t_{max}} \implies t_{max} = \frac{\ln\left(\frac{\Gamma_2}{\Gamma_1}\right)}{(\Gamma_2 - \Gamma_1)} \qquad (2.42)$$

The ratio of the activities $\mathcal{R}$ of I_2 and I_1 is given by (the activities are given by Eq. (2.32) and not dN_i/dt):

$$\mathcal{R}(t) \equiv \frac{\Gamma_2 N_2(t)}{\Gamma_1 N_1(t)} = \Gamma_2 \frac{\frac{\Gamma_1}{\Gamma_2 - \Gamma_1}\left(e^{-\Gamma_1 t} - e^{-\Gamma_2 t}\right)}{\Gamma_1 e^{-\Gamma_1 t}} = \frac{\Gamma_2}{\Gamma_2 - \Gamma_1}\left(1 - e^{-(\Gamma_2 - \Gamma_1)t}\right) \tag{2.43}$$

We find that at the moment of maximum activity, the activities of I_1 and I_2 are equal (this is called the **ideal equilibrium**):

$$\mathcal{R}(t_{max}) = \frac{\Gamma_2 N_2(t_{max})}{\Gamma_1 N_1(t_{max})} = \frac{\Gamma_2}{\Gamma_2 - \Gamma_1}\left(1 - e^{-(\Gamma_2 - \Gamma_1)\frac{\ln\left(\frac{\Gamma_2}{\Gamma_1}\right)}{(\Gamma_2 - \Gamma_1)}}\right) = \frac{\Gamma_2}{\Gamma_2 - \Gamma_1}\left(1 - \frac{\Gamma_1}{\Gamma_2}\right) = 1 \tag{2.44}$$

In general, one can distinguish different cases depending on the relative lifetimes of the isotopes (see Figure 2.6):

- $\tau_2 \ll \tau_1$ ($\Gamma_2 \gg \Gamma_1$): the ratio of activities rapidly approaches $\simeq 1$, a situation which is called the **secular equilibrium**. Although the isotope I_2 decays rapidly, its activity is dominated by its production rate determined by I_1. For example, in the thorium chain given in Eq. (2.38), this would be the case for all the elements in the chain. Their activity is dominated by the amount of initial thorium which has a lifetime of 1.4×10^{10} years. Knowing the traces of thorium in a given piece of material therefore allows us in principle to compute the activity of all the other elements in the chain under the assumption of secular equilibrium. It should be noted, however, that in real cases, the manufacturing processes can break the equilibrium by separation of the isotopes.

- $\tau_2 < \tau_1$ ($\Gamma_2 > \Gamma_1$): there is an initial buildup of I_2 since its decay lifetime is comparable to that of the parent isotope. The ratio of the activities becomes constant after several lifetimes of the isotopes I_1 at a value $\sim \tau_1/\tau_2 > 1$, a situation called the **transient equilibrium**.

- $\tau_2 > \tau_1$ ($\Gamma_2 < \Gamma_1$): since the I_2 isotope lives longer than I_1, there is buildup as its activity continuously grows with time.

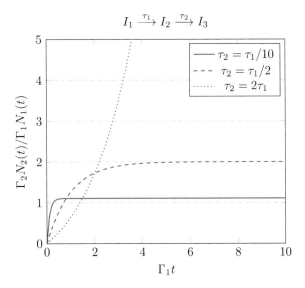

Figure 2.6 The ratio of the activities $\mathcal{R}$ of I_2 and I_1 as a function of time for different scenarios.

2.6   The Discovery of the Electron

In 1897 **J. J. Thomson**[21] discovered and identified the existence of the first subatomic elementary particle – the **electron** – with the help of a modified cathode-ray tube: **with his apparatus he performed the first precise measurement of the e/m ratio of the particle that was emitted by the cathode-ray tube, where e represents the electric charge and m the mass of the particle.**

Figure 2.7 presents the experimental apparatus developed by J. J. Thomson. A fluorescent screen is placed at the far end of the cathode-ray tube to detect the particles and deduce their trajectory after they cross an electric and a magnetic field. The $\vec{E}$ and $\vec{B}$ fields are perpendicular to each other. The bending of the trajectory proved that the particles were electrically negatively charged. The bending $\overline{S_1 S_2}$, located by the spot on the screen and represented by the angle δ, depends on the magnitude of the fields and its study as a function of the fields allows the determination of the ratio e/m.

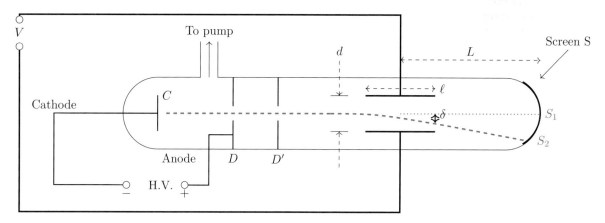

Figure 2.7 The experimental apparatus to measure the e/m ratio of the particles emitted by the cathode-ray tube.

Quantitatively:

• **E field.** Two parallel metallic plates of length l and distance d polarized with an electric potential V give:

$$|\vec{E}| = \frac{V}{d} \quad \longrightarrow \text{Force } |\vec{F_E}| = |-e\vec{E}| = \frac{eV}{d} \tag{2.45}$$

• **B field.** Perpendicular to the electric field and in good approximation perpendicular to the velocity:

$$\text{Force } |\vec{F_B}| = |q\vec{v} \times \vec{B}| = |-e\vec{v} \times \vec{B}| = evB \tag{2.46}$$

where $\vec{v}$ is the velocity of the particle.

The fields are geometrically oriented such that the forces are directed against each other. One selects the magnitude of the fields such that the particles are unbent, that is, the electric and magnetic forces compensate each other:

$$\frac{eV}{d} = evB \quad \longrightarrow \quad v = \frac{V}{dB} \tag{2.47}$$

and in this way one can estimate the velocity of the particle.

---

21   Sir Joseph John Thomson (1856–1940), English physicist.

Then the magnetic field is switched off and the bending $\overline{S_1 S_2}$ of the spot on the screen is measured (see Figure 2.7):

$$\delta(\ell) = \frac{1}{2}at^2 \approx \frac{1}{2}\left(\frac{eV}{md}\right)\left(\frac{\ell}{v}\right)^2 = \frac{1}{2}\left(\frac{e}{m}\right)\left(\frac{V}{d}\right)\left(\frac{\ell}{v}\right)^2 \qquad (2.48)$$

and the bending is equal to:

$$\overline{S_1 S_2} \approx \frac{d\delta(\ell)}{d\ell}L = \frac{1}{2}\left(\frac{e}{m}\right)\left(\frac{V}{d}\right)\left(\frac{1}{v}\right)^2 2\ell L \propto \left(\frac{e}{m}\right)\frac{1}{v^2} \qquad (2.49)$$

where the velocity v was determined previously (see Eq. (2.47)). Therefore the ratio e/m could be determined. The ratio was shown to be independent of the type of gas chosen, hence it represents a property of the particle emitted in the cathode ray.

How could one be sure that the measurements represented the discovery of a new subatomic particle? J. J. Thomson compared the obtained ratio with those known from experiments of **electrolysis**. In this case, the ratio $(e/m)_{ions}$ is obtained with e set to the total electric charge $Q = i\Delta t$ and M the total mass accumulated on the electrode. It was observed that:

$$\left(\frac{e}{m}\right)_{cathode\ rays} \gg \left(\frac{Q}{M}\right) \approx \left(\frac{e}{m}\right)_{ions} \qquad (2.50)$$

and therefore either $e \gg e_{ions}$ or $m \ll m_{ions}$. J. J. Thomson *assumed* that:

$$|e| \simeq |e_{ions}| \quad \text{hence} \quad m \ll m_{ions} \qquad (2.51)$$

that is, the mass of the unknown particle must be much smaller than the mass of an ion. The largest ratio for ions was known to be that of hydrogen, and the results with the cathode-ray tube showed that:

$$\left(\frac{e}{m}\right)_{cathode\ rays} \approx 1836\left(\frac{e}{m}\right)_{H^+} \qquad (2.52)$$

He concluded that the rays that were emitted by the cathode-ray tube corresponded to a new subatomic particle with a very small mass. He had discovered the electron!

2.7   The Structure of Atoms

J. J. Thomson understood that the particle he had discovered was playing a fundamental role in the structure of matter. Nonetheless this triggered at least two unsolved problems: (1) the electron is negatively electrically charged; what compensates this charge in matter? and (2) the electron is very light; where does the mass of the atoms come from? In 1904 he proposed the plum pudding atom model, where *the atoms of the elements consist of a number of negatively electrified corpuscles enclosed in a sphere of uniform positive electrification.*

In 1911 **Geiger**[22] and **Marsden,**[23] and then **Rutherford,**[24] published results on their important **scattering experiment** with α particles. Rutherford had identified the nature of the α particles with a method similar to that used by J. J. Thomson, hence the ratio $(e/m)_\alpha$ was known. The α particle was known to be **a doubly ionized helium atom**. The experimental configuration of the experiment is illustrated in Figure 2.8. A parallel beam of α particles was directed perpendicular to a gold foil with a thickness of 0.4 μm. A **collision** occurs between the α particle and the gold atoms, because of the electromagnetic interaction. During the collision, energy and momentum are exchanged between the two bodies. The α particles are **scattered**!

22   Johannes Hans Wilhelm Geiger (1882–1945), German physicist.
23   Sir Ernest Marsden (1889–1970), English-New Zealand physicist.
24   Ernest Rutherford, 1st Baron Rutherford of Nelson (1871–1937), New Zealand-born British physicist.

Experimentally one observes that the majority of the α particles are **scattered in the forward direction**. Geiger and Marsden were surprised to observe that a small fraction of them is **scattered in the backward direction** [15]. It came as a surprise to them that a finite, although small, fraction of the α particles falling upon a metal plate had their directions changed to such an extent that they emerged again at the side of incidence. Rutherford concluded that collisions with target electrons could not be the origin of the observed

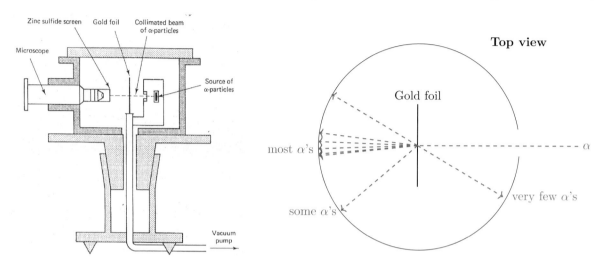

Figure 2.8 Experimental setup for the scattering experiment using α particles on gold foils. (left) Side view of the experiment of Geiger and Marsden. Reprinted from H. Geiger and E. Marsden, "The laws of deflexion of α particles through large angles," *Philosophical Magazine*, vol. 25, no. 148, pp. 604–623, 1913, by permission of the publisher (Taylor & Francis Ltd., www.tandfonline.com); (right) sketch from the top view.

backward scattering, since the known mass of the electrons was much too small to possibly deviate the heavy α particles so strongly. According to the *law of momentum conservation*, the mass of the collision partner must be greater than the mass of the incident α particle. Consequently, Rutherford developed the **atomic model**, by which a heavy and positively charged nucleus is surrounded by a cloud of negatively charged electrons. The experiment was consistent with the existence of the atomic nucleus and the electron cloud. The nucleus of the lightest known element hydrogen was defined by Rutherford as consisting of a **proton**.

These results demonstrated the power of scattering experiments in order to understand the structure of matter. We can state that **the technique of the scattering experiment is one of the most important experimental developments in particle physics.** This model had very serious theoretical weaknesses with the physics known at the time. In particular, it was unstable from an electromagnetic point of view. From Maxwellian electrodynamics, electrons would radiate energy, with the result that the radius of their orbits would decrease until the electrons coalesced with the nucleus. But the experimental data was clear evidence! The results of these experiments will be discussed further in Section 2.11.

2.8   The Differential and Total Cross-Section

We consider an incident beam of parallel particles. We assume that the particles are uniformly distributed on a surface perpendicular to the direction of motion. We define **the flux φ of these incident particles** as:

$$\varphi = \text{number of particles per unit time per unit surface (e.g., particles/m}^2\text{/s)} \tag{2.53}$$

The *interactions* or *collisions* are described in terms of the **cross-section**. This latter quantity essentially gives a measure for the likelihood (or probability) of a given interaction occurring. It may be calculated and predicted from the basic interactions between particles. Let us consider the number of incident particles that cross a surface $d\sigma$ perpendicular to the beam, as shown in Figure 2.9. The surface may be expressed in terms of the **impact parameter** b and the azimuthal angle ϕ as:

$$d\sigma \equiv b\,db\,d\phi \tag{2.54}$$

Consequently, the rate of particles per unit time dF crossing $d\sigma$ is given by:

$$dF \equiv \varphi d\sigma = \varphi b\,db\,d\phi \tag{2.55}$$

We now consider the trajectory shown as the thick black line in Figure 2.9 of the particles crossing the surface

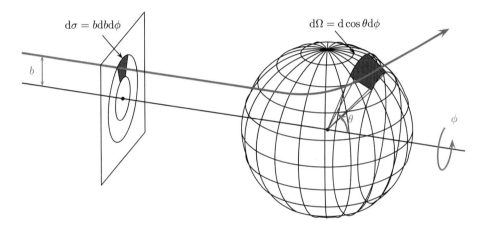

Figure 2.9 An incident particle that crosses the surface $d\sigma$ will be scattered in the solid angle $d\Omega$.

$d\sigma$, which corresponds to a case of repulsive interaction between the incident particle and the scattering center. We note that **the incoming particles crossing the area $d\sigma$ will be scattered in the phase space solid angle $d\Omega$**, with:

$$d\Omega = d\cos\theta d\phi \tag{2.56}$$

where θ is called the **scattering angle** (the angle between incoming and outgoing momentum vectors) and azimuthal symmetry ensures that ϕ is unchanged before and after the interaction. This equality implies that the number of particles per unit time dN, which are scattered in the solid angle $d\Omega$, is equal to:

$$dN(\theta, \phi) \equiv dF = \varphi b\,db\,d\phi = \varphi d\sigma = \varphi\left(\frac{d\sigma}{d\Omega}d\Omega\right) = \varphi\left(\frac{d\sigma}{d\Omega}\right)d\Omega \tag{2.57}$$

and it follows that the **differential cross-section** is proportional to the number of particles per unit time that are scattered in the solid angle $d\Omega$:

$$\left(\frac{d\sigma}{d\Omega}\right) = \frac{1}{\varphi}\frac{dN(\theta, \phi)}{d\Omega} \tag{2.58}$$

Using $d\sigma \equiv b\,db\,d\phi$ and $d\Omega \equiv d\cos\theta d\phi$, the differential cross-section can also be related to the impact parameter as:

$$\left(\frac{d\sigma}{d\Omega}\right) = \left|b\frac{db}{d\cos\theta}\right| = \frac{b}{\sin\theta}\left|\frac{db}{d\theta}\right| \tag{2.59}$$

where the expression $b \equiv b(\theta)$ represents the unique relation between a given scattering angle θ and a given impact parameter b (one takes the absolute value since $db/d\theta$ might be negative).

In addition, the **total cross-section** is defined as the integral over the entire solid angle:

$$\sigma_{tot} = \int \left(\frac{d\sigma}{d\Omega} \right) d\Omega \tag{2.60}$$

• **Hard sphere scattering.** The cross-section has units of area and it can be related to the total geometrical area of the scattering centers. We show this explicitly in the case of a hard scattering sphere of radius R illustrated in Figure 2.10. Geometrically we can see that the condition for a hard collision is:

$$2\alpha + \theta = \pi \quad \text{where} \quad R\sin\alpha = b \tag{2.61}$$

where b is the impact parameter and θ the scattering angle.

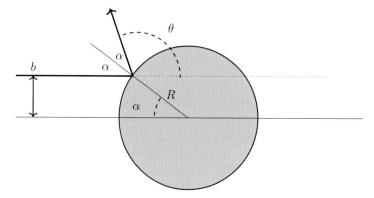

Figure 2.10 Scattering off a hard sphere of radius R: θ is the scattering angle and b the impact parameter.

This implies:

$$b = R\sin\alpha = R\sin\left((\pi - \theta)/2 \right) = R\cos\theta/2 \tag{2.62}$$

We directly compute the differential cross-section for a hard sphere using Eq. (2.59):

$$\left(\frac{d\sigma}{d\Omega} \right) = \frac{b}{\sin\theta} \left| \frac{db}{d\theta} \right| = \frac{b}{\sin\theta} \frac{R}{2} \sin\left(\frac{\theta}{2} \right) = \frac{R^2}{2\sin\theta} \cos\left(\frac{\theta}{2} \right) \sin\left(\frac{\theta}{2} \right) = \frac{R^2}{4} \tag{2.63}$$

The total cross-section therefore, as expected, equals:

$$\sigma_{tot} = \int \left(\frac{d\sigma}{d\Omega} \right) d\Omega = \frac{R^2}{4} \int d\Omega = \pi R^2 \tag{2.64}$$

It represents the area of the cross-section of the sphere. Particles that enter within the area will be scattered; others will be unaffected. It is worth mentioning that the derived association with the physical area is true in the case of the hard sphere, but does not strictly apply to subatomic particles that cannot be represented as simple spheres of matter. However, the concept of cross-section is retained in all cases, for example even when colliding two point-like particles to compute the probability of interactions for a given flux, as discussed in the next section.

2.9   **Interaction Probability**

We note that the cross-section has units of area. We can find its geometrical interpretation in a practical case, where the target is usually a slab of material of some thickness. The scattering centers are assumed to be uniformly distributed in the target slab. The slab has area A and is thin, with thickness $\mathrm{d}x$, as illustrated in Figure 2.11. The black dots in the figure represent the scattering centers with an average density of scatterers per unit volume ρ, and each of them has a scattering area σ. The flux of incoming particles (perpendicular to the surface of the target) is defined as φ. When the incoming particles are also uniformly distributed, then

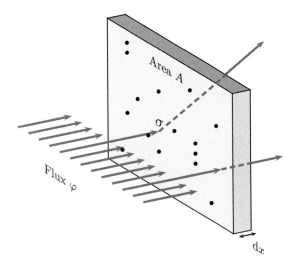

Figure 2.11 Sketch of the scattering of a uniform beam of particles with flux per unit area φ impinging on a thin target of thickness $\mathrm{d}x$ and total area A. The scattering centers are shown as black dots in the target. The cross-section of each scatterer is labeled σ.

the probability that a particle interacts with a scattering center is equal to the ratio of the total area of the scattering centers and the total area of the target:

$$\text{Probability of interaction in } \mathrm{d}x \equiv w = \frac{(\rho A \mathrm{d}x)\sigma}{A} = \rho \sigma \mathrm{d}x \quad (\text{units}: \mathrm{L}^{-3}\mathrm{L}^2\mathrm{L} = 1) \tag{2.65}$$

This is called the **thin target approximation** and is valid when the areas covered by each scattering center do not overlap. The **interaction rate** (number of interactions per unit time) R_{int} is therefore:

$$R_{int} = (\varphi A)w = (A\mathrm{d}x)\rho\varphi\sigma = (V\rho)\varphi\sigma = N_{scatterers}\varphi\sigma \tag{2.66}$$

where V is the total volume of the target, and $\rho V = N_{scatterers}$ the total number of target centers in the target volume. If we consider scattering off *nucleons*, then the rate would be:

$$R_{int} = N_{nucleon}\varphi\sigma = (N_A M)\varphi\sigma \tag{2.67}$$

where $N_A = 6.02214076 \times 10^{23} \text{ mol}^{-1}$ (exact) is the **Avogadro[25] constant** (or Avogadro number) and M is the total mass of the target *in grams*.

• **Thick target.** In order to consider the general case valid for a target of any thickness, we consider the number of surviving particles after a distance x inside the target as $N(x)$. The probability of a particle interacting in

25   Amedeo Carlo Avogadro, Count of Quaregna and Cerreto (1776–1856), Italian scientist.

any thin piece between x and $x + \mathrm{d}x$ is given by w, hence the number of particles interacting within x and $x + \mathrm{d}x$ is:

$$N(x)w = N(x)\rho\sigma\mathrm{d}x \tag{2.68}$$

and the number of particles at a distance $x + \mathrm{d}x$ is equal to their number at x minus those that have interacted, hence:

$$N(x + dx) = N(x) - N(x)\rho\sigma\mathrm{d}x \quad \Longrightarrow \quad \mathrm{d}N(x) = -N(x)\rho\sigma\mathrm{d}x \tag{2.69}$$

The product $\rho\sigma$ has the units of an inverse distance and one defines the **mean free path** λ as the inverse of $\rho\sigma$, which therefore has the units of distance:

$$\lambda \equiv \frac{1}{\rho\sigma} \tag{2.70}$$

Then, we can integrate to find the number of surviving incident particles as a function of the distance in the slab x:

$$\int \frac{\mathrm{d}N}{N} = \ln N = -\int \rho\sigma\mathrm{d}x = -\int \mathrm{d}x/\lambda = -x/\lambda \quad \Longrightarrow \quad N(x) = N(0)e^{-x/\lambda} \tag{2.71}$$

where the initial number of particles is defined as $N(0)$. The probability of an incident particle surviving at a distance x is thus exponentially decreasing with distance and is characterized by the mean free path. After a distance equal to λ, the remaining fraction of particles is $N(\lambda)/N(0) = e^{-1} = 1/e \simeq 0.37$. Similarly, after two or three mean free paths, it is respectively only $N(2\lambda)/N(0) = e^{-2} \simeq 0.14$ or $N(3\lambda)/N(0) = e^{-3} \simeq 0.05$.

2.10   Collinear and Head-On Collisions

Beyond the calculation of the interaction probability on a target, we can further consider the case of collinear and head-on collisions, which is slightly more complicated because it is necessary to account for the movement of two beams of particles. Let us imagine a flux of incoming particles labeled 1 with density n_1 particles per unit volume, traversing a region containing particles labeled 2. If a single particle 1 is traveling with velocity $\vec{\beta}_1$ in a region defined by a cross-sectional area A, which contains n_2 particles 2 per unit volume traveling with a velocity $\vec{\beta}_2$ in the opposite direction of $\vec{\beta}_1$, then in an interval δt, particle 1 crosses a volume containing $n_2 A\delta\ell = n_2 A|\vec{\beta}_1 - \vec{\beta}_2|\delta t$ particles of type 2 (see Figure 2.12). Similar to Eq. (2.65), we can write the probability

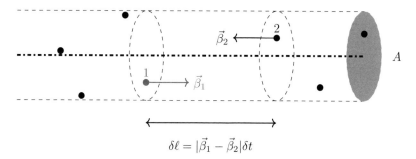

Figure 2.12 Collinear and head-on collisions of particle 1 crossing a region containing particles of type 2.

of interaction of particle 1:

$$\text{Probability of interaction} = \frac{n_2 A|\vec{\beta}_1 - \vec{\beta}_2|}{A}\sigma\delta t = n_2|\vec{\beta}_1 - \vec{\beta}_2|\sigma\delta t \tag{2.72}$$

Consequently, the interaction rate (number of interactions per unit time) of n_1V particles 1 in a volume V is given by:

$$R_{int} = n_1 V n_2 |\vec{\beta}_1 - \vec{\beta}_2| \sigma = n_1 n_2 |\vec{\beta}_1 - \vec{\beta}_2| V \sigma \qquad (2.73)$$

Comparison with Eq. (2.66) allows us to define the relative **flux factor** F as:

$$R_{int} = V(\rho\varphi)\sigma = V F \sigma \qquad \text{where} \qquad F \equiv n_1 n_2 |\vec{\beta}_1 - \vec{\beta}_2| \qquad (2.74)$$

We note that this derivation is valid in non-relativistic cases, which is seldom the case in practical applications. The corresponding relativistic expression will be derived in Chapter 5.

2.11   The Rutherford Scattering Cross-Section

To compare the experimental observation with theory, the **impact parameter** b (before the collision) and the resulting **scattering angle** θ were introduced by Rutherford in a classical calculation of the trajectory of the α particles in the **scattering experiment**. The assumption is that the large-angle deviation of the incoming α particle is due to the collision with a *single* target nucleus, as shown in Figure. 2.13. The motion is similar to the "unbound" Keplerian motion around a planet – the incoming α is approaching from infinity and goes off to infinity. Since the force between the α particle and the nucleus vanishes at large distances, the orbit approaches straight lines at large distances from the scattering center. We are therefore only interested in the deviation of the trajectory, which will be represented by the scattering angle θ.

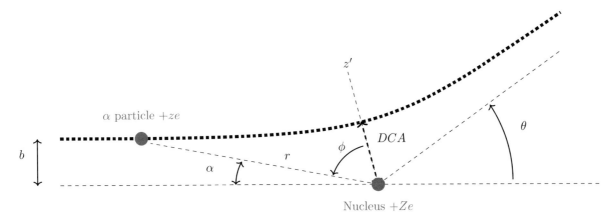

Figure 2.13 Definition of the kinematical quantities relevant for the large-angle scattering experiment of Rutherford. The picture depicts the large-angle deviation due to a single atom. The impact parameter is b (before the collision) and the resulting scattering angle is θ. The position of the α particle along its trajectory is given by (r, α). The distance of closest approach is defined by DCA along an axis z'.

The assumptions are the following:

- The (point-like) nuclei contain protons and are positively charged with charge $+Ze$.

- The α particle is point-like, positively charged with charge $+ze$, and moves at a non-relativistic speed.

- There is a long-range repulsive force between the α particle and the single reacting nucleus in the target.

- Both projectile and target emerge from the collision in their respective ground states (Rutherford scattering); the reaction is hence elastic. The inelastic case where the target nucleus is left in an excited state is called Coulomb scattering and is not considered here.

- If the energy of the α particle is not sufficient to bring it to a distance smaller than the range of the nuclear interaction, there is no other force than the Coulomb force and the trajectories are hyperbolic.

- The trajectories are symmetric with respect to the axis z' defined by the distance of closest approach (DCA). As we will see below, they depend on the impact parameter (see Figure. 2.13).

Then, the interaction between the α particle and the nucleus in the classical description involves the central Coulomb force, which is repulsive between the positively charged α ($+ze$) and the positive target nucleus ($+Ze$):

$$|\vec{F}| = \frac{1}{4\pi\epsilon_0} \frac{zZe^2}{r^2} = \frac{\alpha zZ}{r^2} \tag{2.75}$$

where we used Eq. (1.37). In writing the force, **we have for the moment ignored any magnetic interaction**, which is reasonable in the case of the α particle, which is spinless. But this approximation fails to represent the interaction between particles with spins, as we will see in later chapters.

From a classical point of view (i.e., based on Newton's law, Coulomb's force, and conservation of momentum),[26] the angular momentum of the α particle is conserved and we can assume that the nucleus is much heavier than the incident particle, hence we can neglect its movement. **The fixed nucleus is therefore not recoiling**. From Newton's second law, the change of momentum of the α particle is given by:

$$\vec{p}_f = \vec{p}_i + \Delta\vec{p} = \vec{p}_i + \int \vec{F}\mathrm{d}t \tag{2.76}$$

where $|\vec{p}_f| = |\vec{p}_i|$. The trajectory is symmetric with respect to the z'-axis and the components of the force perpendicular to z' will cancel in the integration. We therefore have:

$$\Delta p = \alpha zZ \int \mathrm{d}t \frac{1}{r^2} \cos\phi \tag{2.77}$$

The angular momentum L and its initial value can be expressed as:

$$L = mr^2\frac{\mathrm{d}\phi}{\mathrm{d}t} \quad \text{and} \quad L = mv_0b \tag{2.78}$$

where b is the impact parameter and v_0 the initial velocity of the α particle. Conservation of angular momentum implies:

$$mr^2\frac{\mathrm{d}\phi}{\mathrm{d}t} = mv_0b \quad \Longrightarrow \quad \Delta p = \alpha zZ \int_{\phi_i}^{\phi_f} \mathrm{d}\phi \frac{1}{v_0b}\cos\phi = \frac{\alpha zZ}{v_0b}(\sin\phi_f - \sin\phi_i) \tag{2.79}$$

By symmetry, $\phi_i = -\phi_f$ and geometrically we have $-\phi_i + \phi_f + \theta = \pi$. Hence:

$$\phi_i = -\frac{1}{2}(\pi - \theta) \quad \text{and} \quad \phi_f = \frac{1}{2}(\pi - \theta) \tag{2.80}$$

So, finally:

$$\Delta p = \frac{\alpha zZ}{v_0b}2\sin\left(\frac{1}{2}(\pi - \theta)\right) = \frac{2\alpha zZ}{v_0b}\cos(\theta/2) \tag{2.81}$$

---

26   We note that these assumptions are not truly applicable to the scattering of elementary particles and we will discuss the quantum-mechanical description of the scattering formula in following chapters. Fortunately, the classical and quantum-mechanical derivation lead to the same result for Rutherford scattering!

At the same time, the change of momentum can be expressed as (see inset to Figure 2.14):

$$\sin(\theta/2) = \frac{\frac{\Delta p}{2}}{p} \implies \Delta p = 2p\sin(\theta/2) \tag{2.82}$$

Combining the two, we can eliminate Δp to find:

$$\Delta p = 2p\sin(\theta/2) = \frac{2\alpha z Z}{v_0 b}\cos(\theta/2) \implies b(\theta) = \frac{\alpha z Z}{v_0 p}\cot\frac{\theta}{2} = \frac{\alpha z Z}{2E_K}\cot\frac{\theta}{2} \tag{2.83}$$

where $E_K = (1/2)mv_0^2$ is the kinetic energy of the α particle. Equation (2.83) defines the relationship between the impact parameter and the scattering angle. Trajectories depend on the impact parameter b, as shown in Figure 2.14.

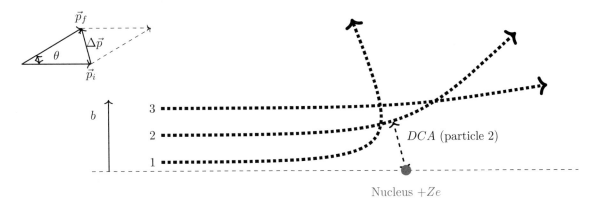

Figure 2.14 Trajectories depend on the impact parameter.

Alternatively, the classical Newtonian solution to this central-force problem yields the following trajectory:

$$\frac{1}{r} = \frac{1}{b}\sin\alpha + \frac{D}{2b^2}(\cos\alpha - 1) \quad \text{where} \quad D \equiv \frac{\alpha z Z}{E_K} \tag{2.84}$$

where the position of the α particle along its trajectory is given in polar coordinates by (r, α). For $r \to \infty$, we find:

$$\sin\alpha + \frac{D}{2b}(\cos\alpha - 1) = 0 \longrightarrow \frac{2b}{D} = \frac{1 - \cos\alpha}{\sin\alpha} = \tan\frac{\alpha}{2} \tag{2.85}$$

With $\theta = \pi - \alpha$ we recover the relation between the scattering angle θ and the impact parameter b of Eq. (2.83):

$$\tan\frac{\pi - \theta}{2} = \cot\frac{\theta}{2} = \frac{2b}{D} \longrightarrow b(\theta) = \frac{\alpha z Z}{2E_K}\cot\frac{\theta}{2} \tag{2.86}$$

We now note that when the incident particle possesses an impact parameter between b and $b+db$, the scattering angle after the collision will be between θ and $\theta + d\theta$. As before, we consider the surface $d\sigma$ and the solid angle $d\Omega$ (see again Figure 2.9): when the incident particle goes through the surface $d\sigma$, it will be scattered through the solid angle $d\Omega$. The surface $d\sigma$ and the solid angle $d\Omega$ can be expressed as a function of the impact parameter and the angles θ and ϕ. Using Eq. (2.59), we get:

$$
\begin{aligned}
\left(\frac{d\sigma}{d\Omega}\right) &\equiv \left|\frac{b\,db}{d\cos\theta}\right| = -\frac{b}{\sin\theta}\frac{db}{d\theta} = -\frac{b}{\sin\theta}\left(\frac{\alpha z Z}{2E_K}\right)\frac{d}{d\theta}\left(\cot\frac{\theta}{2}\right) \\
&= -\left(\frac{\alpha z Z}{2E_K}\right)^2\frac{\cot\frac{\theta}{2}}{\sin\theta}\left(-\frac{1}{\sin^2\frac{\theta}{2}}\right)\frac{1}{2} = \frac{1}{8}\left(\frac{\alpha z Z}{E_K}\right)^2\frac{1}{\sin\theta}\frac{\cos\frac{\theta}{2}}{\sin^3\frac{\theta}{2}} = \frac{1}{16}\left(\frac{\alpha z Z}{E_K}\right)^2\frac{1}{\sin\theta}\frac{2\sin\frac{\theta}{2}\cos\frac{\theta}{2}}{\sin^4\frac{\theta}{2}} \\
&= \left(\frac{\alpha z Z}{4E_K}\right)^2\frac{1}{\sin^4\frac{\theta}{2}}
\end{aligned}
\tag{2.87}
$$

This yields the **classical Rutherford scattering** formula:

$$
\left(\frac{d\sigma}{d\Omega}\right)_{Rutherford} = \frac{(\alpha z Z)^2}{16E_K^2\sin^4\frac{\theta}{2}} = \left(\frac{z Z e^2}{4\pi}\right)^2\frac{1}{16E_K^2\sin^4\frac{\theta}{2}}
\tag{2.88}
$$

where θ is the scattering angle and E_K the kinetic energy of the particle.

• **Experimental confirmation.** A collection of results obtained by Geiger and Marsden is tabulated in Figure 2.15 for silver and gold targets and for scattering angles between 5° and 150° [16]. In this range, the cross-section is expected to change by several orders of magnitude! The number of observed interactions for different angles of deflection are shown in columns III and V. By multiplying by the factor $\sin^4(\theta/2)$, shown in columns IV and VI, one obtains an almost constant number, which is evidence for the validity of Rutherford's formula Eq. (2.88). The cross-section has a sharp dependence as a function of the scattering angle. We speak of a process which is very "**forward-peaked.**"

For small angles, the cross-section becomes very large, and diverges as the angle goes to zero. The total cross-section is therefore also divergent. Excluding the $\theta \to 0$ region where the divergent result is expected to become unphysical, the formula gives an excellent description of the experimental data. The origin of the divergence in the formula is the consequence of the assumed form of the Coulomb force with an infinite range. In reality, we can assume that the atomic electrons shield the charge of the nucleus and limit the effective range of its Coulomb force.

2.12   Classical Thomson Scattering

We illustrate in this section the classical derivation of the interaction of light with free electrons. We want to derive the cross-section for scattering using Maxwell's equations (see Section 10.2), as first performed by J. J. Thomson. We consider a light wave with frequency ω propagating in space incident upon a single electron. The interaction between the light and the electron is due to the interaction of the electron charge with the sinusoidally oscillating electric field. The classical electric field strength can be expressed as:

$$
\vec{E}(t) = \vec{E}_0 e^{-i\omega t}
\tag{2.89}
$$

The fields are transverse with $\hat{k} \perp \vec{E}$, $\hat{k} \perp \vec{B}$, and $\vec{E} \perp \vec{B}$, where $\hat{k}$ is a unit vector in the direction of propagation of the incoming light. The energy flow is given by the Poynting vector (we use MKSA units):

$$
\vec{S} = \frac{1}{\mu_0}\vec{E} \times \vec{B} = \epsilon_0 c^2 \vec{E} \times \vec{B} = \epsilon_0 c|\vec{E}(t)|^2\hat{k}
\tag{2.90}
$$

The time-averaged incoming intensity is then equal to:

$$
I_0 = \langle|\vec{S}|\rangle = \frac{1}{2}\epsilon_0 c E_0^2
\tag{2.91}
$$

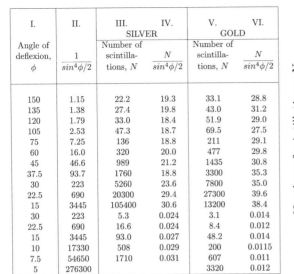
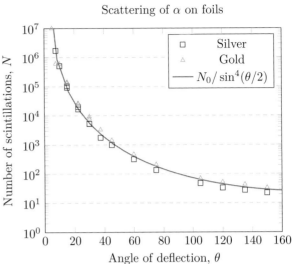

I. Angle of deflexion, ϕ	II. $\dfrac{1}{sin^4\phi/2}$	III. SILVER Number of scintillations, N	IV. $\dfrac{N}{sin^4\phi/2}$	V. GOLD Number of scintillations, N	VI. $\dfrac{N}{sin^4\phi/2}$
150	1.15	22.2	19.3	33.1	28.8
135	1.38	27.4	19.8	43.0	31.2
120	1.79	33.0	18.4	51.9	29.0
105	2.53	47.3	18.7	69.5	27.5
75	7.25	136	18.8	211	29.1
60	16.0	320	20.0	477	29.8
45	46.6	989	21.2	1435	30.8
37.5	93.7	1760	18.8	3300	35.3
30	223	5260	23.6	7800	35.0
22.5	690	20300	29.4	27300	39.6
15	3445	105400	30.6	13200	38.4
30	223	5.3	0.024	3.1	0.014
22.5	690	16.6	0.024	8.4	0.012
15	3445	93.0	0.027	48.2	0.014
10	17330	508	0.029	200	0.0115
7.5	54650	1710	0.031	607	0.011
5	276300			3320	0.012

Figure 2.15 Compilation by Geiger and Marsden of large-angle scattering of α particles by silver and gold foils. Reprinted from H. Geiger and E. Marsden, "The laws of deflexion of α particles through large angles," *Philosophical Magazine*, vol. 25, no. 148, pp. 604–623, 1913, by permission of the publisher (Taylor & Francis Ltd., www.tandfonline.com). In the table, the angle ϕ is the scattering angle. Two sets of data with different sources have been taken for each type of foil and have been rescaled in the plot. The line compares the data to the expected angular dependence.

We now consider the effect of the electric field on an electron. The electric field produces a force and an acceleration on the electron. Under the influence of the acceleration, the electron will oscillate sinusoidally. We assume that the electron mean position is at rest. The equation of motion of the electron (neglecting the magnetic force) is, according to the classical electromagnetic theory, given by (see, e.g., Ref. [17]):

$$m_e \frac{\mathrm{d}^2\vec{x}}{\mathrm{d}t^2} = e\vec{E}_0 e^{-i\omega t} + \frac{2}{3}\frac{e^2}{c^3}\frac{\mathrm{d}^3\vec{x}}{\mathrm{d}t^3} \tag{2.92}$$

where the second term on the r.h.s. depending on the time derivative of the acceleration is the so-called *reaction of the field*. The solution to this equation can be written as an oscillatory term with a damping factor κ:

$$\vec{x}(t) = -\frac{e\vec{E}_0}{m_e\omega^2}e^{-i\omega t}\frac{1}{1+i\kappa} \quad \text{where} \quad \kappa \equiv \frac{2}{3}\frac{e^2}{m_e c^3}\omega \tag{2.93}$$

This means that the electron will oscillate with the same frequency as the incoming light wave with the amplitude determined by the initial wave amplitude damped by the factor κ, and in the direction of the incoming electric field, that is, the polarization direction of the incoming light. It will therefore also *emit a secondary light wave of the same frequency*, which can be approximated by a small emitting dipolar antenna. The intensity of the secondary light wave is given by the magnitude of the Poynting vector as before. In this case, we can write the radially radiated energy from the oscillating electron at a distance R from the electron and in the solid angle $\mathrm{d}\Omega$ as:

$$S_e R^2 \mathrm{d}\Omega = \epsilon_0 c |E_e(t)|^2 R^2 \mathrm{d}\Omega = \epsilon_0 c \left(\frac{\mathrm{d}^2 Z}{\mathrm{d}t^2}\right)^2 \sin^2\vartheta \, \mathrm{d}\Omega \tag{2.94}$$

where ϑ is the angle between the direction of observation (at a distance R) and the direction of polarization of the incident wave $\vec{E}_0$, and $\vec{Z}$ is the Hertzian vector of the oscillating electron:

$$\vec{Z} \equiv \frac{e}{4\pi\epsilon_0 c^2}\vec{x} \tag{2.95}$$

We can now collect the terms to express the radiated wave intensity:

$$|\vec{S}_e| = \frac{\epsilon_0 c}{R^2}\left(\frac{\mathrm{d}^2 Z}{\mathrm{d}t^2}\right)^2 \sin^2\vartheta = \frac{\epsilon_0 c}{R^2}\left(\frac{e}{4\pi\epsilon_0 c^2}\right)^2\left(\frac{\mathrm{d}^2 x}{\mathrm{d}t^2}\right)^2 \sin^2\vartheta \tag{2.96}$$

After time averaging, one finds the radiated intensity:

$$I = \langle S_e \rangle = \frac{1}{2}\frac{\epsilon_0 c}{R^2}\left(\frac{e}{4\pi\epsilon_0 c^2}\right)^2\left(\frac{e^2 E_0^2}{m_e^2}\frac{1}{1+\kappa^2}\right)\sin^2\vartheta = \frac{I_0}{R^2}\underbrace{\left(\frac{1}{4\pi\epsilon_0}\frac{e^2}{m_e c^2}\right)^2\frac{1}{1+\kappa^2}\sin^2\vartheta}_{\equiv\sigma_T} \tag{2.97}$$

where we cast the expression in the form $I = I_0\sigma_T/R^2$ such that the quantity σ_T has units of area. It is called the Thomson differential cross-section:

$$\sigma_T(\vartheta) = \left(\frac{1}{4\pi\epsilon_0}\frac{e^2}{m_e c^2}\right)^2 \sin^2\vartheta\frac{1}{1+\kappa^2} \equiv r_e^2\sin^2\vartheta\frac{1}{1+\kappa^2} \tag{2.98}$$

where r_e is the **classical electron radius**:[27]

$$r_e \equiv \frac{1}{4\pi\epsilon_0}\frac{e^2}{m_e c^2} = 2.8179403227(19)\times 10^{-15}\ \mathrm{m} \tag{2.99}$$

The classical electron radius can be expressed in terms of the fine-structure constant α as (setting $c = 1$ and $\epsilon_0 = 1$):

$$r_e = \frac{\alpha}{m_e} \tag{2.100}$$

If the primary incident light wave is unpolarized, we should average over all directions of polarization, that is, all angles ϑ and we find in this case:

$$\langle\sin^2\vartheta\rangle = \frac{1}{2}(1+\cos^2\theta) \tag{2.101}$$

where θ is the scattering angle. See Figure 2.16. For incoming wave frequencies satisfying $\omega \lesssim 3m_e c^2/2e^2$,

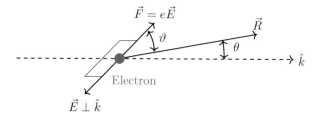

Figure 2.16 Illustration of the relation between the various vectors and the angles (see text).

27   The name derives from the fact that the definition yields $m_e c^2 = e^2/(4\pi\epsilon_0 r_e)$.

which is true up to and including X-rays, the damping term is negligible. Hence, we can simply write the unpolarized Thomson cross-section as:

$$\sigma_T(\theta) = \frac{1}{2}r_e^2(1 + \cos^2\theta) \tag{2.102}$$

The total cross-section can be found by integrating over the solid angle:

$$\sigma_{Thomson}^{tot} = \frac{1}{2}r_e^2 \int d\cos\theta d\phi (1 + \cos^2\theta) = \frac{1}{2}r_e^2 2\pi \frac{8}{3} = \frac{8\pi}{3}r_e^2 \tag{2.103}$$

If we now consider the incoherent interaction with all electrons of an atom with atomic number Z, we get the classical Thomson scattering cross-section per atom:

$$\sigma_{Thomson,atom} = \frac{8\pi}{3}Zr_e^2 \approx 6.65 \times 10^{-25}Z \text{ cm}^2 \tag{2.104}$$

This equation tells us which fraction of the intensity of a beam of light will be scattered by the free electron in an atom due to the Thomson scattering process. Note that light was considered as a wave in this classical derivation. Later we will reconsider this interaction treating light as incoming photons. A corpuscular treatment is supported by the observation of Compton scattering, discussed in the next section.

2.13   The Compton Effect

In 1923 **Arthur Compton**[28] published his analysis of the scattering of X-rays on light elements [18, 19]. He discovered that when a beam of X-rays with well-defined wavelength λ_0 is scattered at an angle θ by a foil, then the scattered rays contain a component with a well-defined wavelength λ_1 with $\lambda_1 > \lambda_0$. This phenomenon is called the **Compton effect**, and is shown in Figure 2.17.

In the figure, the abscissas represent the wavelength while the ordinates are the measured intensity per unit wavelength interval. The spectra were observed to be independent of the material of the scattering foil. Hence, it implies that the phenomenon did not involve the nucleus but rather individual electrons. Compton assumed that these latter can be considered initially free and initially stationary. Some justification in this assumption is that X-rays are an electromagnetic radiation with a much higher frequency than visible or UV light, characteristic of the absorption/emission lines of atoms (see Section 2.1).

Consequently, Compton considered the relativistic collision between an X-ray "quantum" and a free stationary electron. Compton applied the conservation of energy and momentum. If we assign an energy and momentum $(E_\gamma, \vec{p}_\gamma)$ to the incoming X-ray and similarly $(E'_e, \vec{p'}_e)$ and $(E'_\gamma, \vec{p'}_\gamma)$ for the scattered electron and outgoing X-ray, then we have (see Figure 2.18):

$$\vec{p}_\gamma = \vec{p'}_\gamma + \vec{p'}_e \implies \begin{cases} p_\gamma = p'_\gamma \cos\phi + p'_e \cos\theta \\ p'_\gamma \sin\phi = p'_e \sin\theta \end{cases} \tag{2.105}$$

Squaring the two equations, we obtain:

$$(p_\gamma - p'_\gamma \cos\phi)^2 = p'^2_e \cos^2\theta \quad \text{and} \quad p'^2_\gamma \sin^2\phi = p'^2_e \sin^2\theta \tag{2.106}$$

Adding the two, we get:

$$p_\gamma^2 - 2p_\gamma p'_\gamma \cos\phi + p'^2_\gamma = p'^2_e \tag{2.107}$$

---

28   Arthur Holly Compton (1892–1962), American physicist.

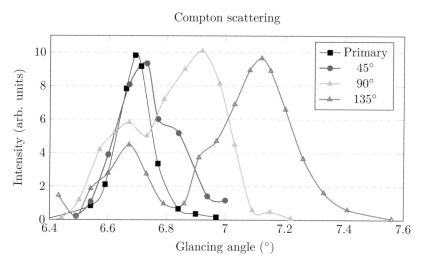

Figure 2.17 Wavelength spectra of X-rays scattered at $45°$, $90°$, and $135°$ by a carbon (graphite) foil, compared to the original $K\alpha$-line. Data taken with permission from A. H. Compton, "The spectrum of scattered X-rays," *Phys. Rev.*, vol. 22, pp. 409–413, Nov 1923. Copyright 1923 by the American Physical Society.

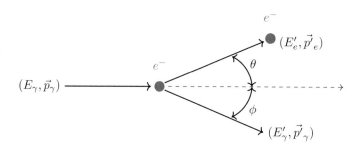

Figure 2.18 Compton effect: a collision between an X-ray quantum and a free stationary electron.

Now moving to the energies, we can express its conservation as:

$$E_\gamma + m_e = E'_\gamma + m_e + T_e \implies E_\gamma - E'_\gamma = T_e \tag{2.108}$$

where T_e is the kinetic energy of the outgoing electron. **We now apply a first assumption on the nature of the X-rays and equate $E_\gamma = p_\gamma$ and $E'_\gamma = p'_\gamma$.** For the electron, we use the relativistic relation between energy and momentum $E_e^2 = p_e^2 + m_e^2$ (see Section 5.5). Thus:

$$E'^2_e = p'^2_e + m_e^2 = (T_e + m_e)^2 = T_e^2 + 2m_e T_e + m_e^2 \tag{2.109}$$

We now combine all the previous results to find:

$$E_\gamma^2 - 2E_\gamma E'_\gamma \cos\phi + E'^2_\gamma = (E_\gamma - E'_\gamma)^2 + 2m_e(E_\gamma - E'_\gamma) \tag{2.110}$$

Expanding the square and regrouping, the relation can be written as:

$$m_e(E_\gamma - E'_\gamma) = E_\gamma E'_\gamma(1 - \cos\phi) \implies \frac{(1 - \cos\phi)}{m_e} = \frac{E_\gamma - E'_\gamma}{E_\gamma E'_\gamma} = \frac{1}{E'_\gamma} - \frac{1}{E_\gamma} \tag{2.111}$$

We now reinsert the speed of light c and the momenta of the X-rays with $E_\gamma = cp_\gamma$ and $E'_\gamma = cp'_\gamma$, so:

$$\frac{(1 - \cos\phi)}{m_e c^2} = \frac{1}{cp'_\gamma} - \frac{1}{cp_\gamma} \tag{2.112}$$

We now apply a second assumption on the nature of the X-rays and equate $p_\gamma = h/\lambda$ **and** $p'_\gamma = h/\lambda'$, where h is the Planck constant and λ the wavelength of the X-rays. Hence, we obtain the main result called the **Compton equation**:

$$\lambda' - \lambda = \lambda_C (1 - \cos\phi) \tag{2.113}$$

where λ_C is the **Compton wavelength** defined as and whose 2018 CODATA recommended value is given by [4]:

$$\lambda_C \equiv \frac{h}{m_e c} = (2.42631023867 \pm 0.00000000073) \times 10^{-12} \text{ m} \tag{2.114}$$

The Compton equation predicts that the increase in wavelength of the scattered X-ray depends only on the scattering angle ϕ and the universal constant λ_C. In particular, it is independent of the wavelength itself! Compton verified his equation experimentally, finding a perfect match. The scattered electron was measured in 1923 by Bothe and by Wilson. These experiments provided strong evidence for the existence of the light quantum or photon obeying the relations:

$$E = cp = h\nu = hc/\lambda \tag{2.115}$$

The Compton experiments show that the photon quantum possesses an energy and a momentum, and that the scattered photon has a lower energy because of energy–momentum conservation in the collision with the electron, and therefore a longer wavelength.

2.14   The Discovery of the Neutron

After the discoveries of the electron and of the proton, the understanding of the atomic structure was still not complete. For instance, the mass and the total spin of isotopes heavier than hydrogen still could not be fully explained. **Rutherford postulated the existence of a new particle that had similar properties to the proton but without elementary electric charge: the so-called neutron.**

In 1930 **Bothe**[29] and his student Becker observed electrically neutral and penetrating rays that were produced when α particles of a strong polonium compound hit light targets, in particular beryllium:

$$\alpha + Be \rightarrow C + \underbrace{X}_{rays} \tag{2.116}$$

They first thought that these were electromagnetic rays (i.e., γ-rays) [20]. A number of elements and compounds were bombarded and in the case of Li, Be, B, F, Mg, and Al, a secondary radiation could be detected.

In 1932 **Irène Curie-Joliot**[30] and **Jean Frédéric Joliot**[31] discovered that the rays could trigger the emission of protons from matter [21]. Thinking that photons of such high energy could perhaps produce some kind of transmutation, they placed thin layers of various substances in contact with the upper wall of their ionization chamber. They noted that on the contrary, the current increases noticeably when screens of hydrogen-containing substances such as paraffin, water, or cellophane are interposed. The most intense effect has been observed with paraffin; the current varies almost twice as much in this case.[32]

---

29   Walther Wilhelm Georg Bothe (1891–1957), German nuclear physicist.
30   Irène Curie-Joliot (1897–1956), French scientist, daughter of Pierre and Marie Curie.
31   Jean Frédéric Joliot-Curie (1900–1958), born Jean Frédéric Joliot, French physicist, husband of Irène Joliot-Curie.
32   *"Au contraire, le courant augmente notablement quand on interpose des écrans de substances contenant de l'hydrogène comme la paraffine, l'eau, la cellophane. L'effet le plus intense a été observé avec la paraffine; le courant varie presque du simple au double dans ce cas."* Reprinted from I. Curie and F. Joliot, "Effet d'absorption de rayons γ de très haute fréquence par projection de noyaux légers," *C. R. Acad. Sci. Paris*, vol. 174, p. 708, 1932.

Joliot and Curie wanted to test if the X-rays behave like ordinary photons and considered that they expel protons in a process similar to that of Compton scattering. Curie and Joliot later realized that the observed intensity was not favoring a Compton-like process and argued that the observed effect could be due to a type of interaction between γ-rays and protons different from that of the Compton effect.

In parallel, **James Chadwick**,[33] triggered by the observations of Curie and Joliot, made measurements which indicated that the radiation could not consist of γ-rays [22]. Chadwick devised an apparatus where he used an **ionization chamber** as a detector, and noted that the secondary penetrating radiation emitted from Be could produce recoil atoms not only from hydrogeneous substances but also from helium, lithium, beryllium, air, and argon (see Figure 2.19).

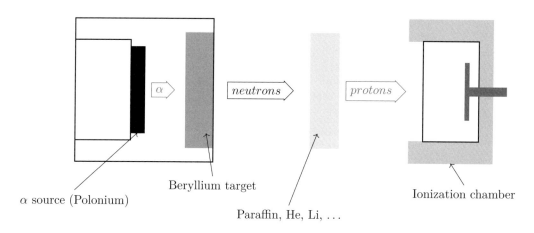

Figure 2.19 Illustration of Chadwick's experimental setup to study the production of secondary particles when α particles hit a beryllium target.

Chadwick could compute the kinematics of the needed γ radiation for both hydrogen and nitrogen. With the help of a **kinematical analysis** assuming a Compton-like process like that suggested by Joliot and Curie, conservation of energy and momentum implies, as illustrated in Figure 2.20:

$$\begin{cases} M_p + E_\gamma = E'_p + E'_\gamma \\ \vec{p}_\gamma = \vec{p}'_p + \vec{p}'_\gamma \end{cases} \qquad (2.117)$$

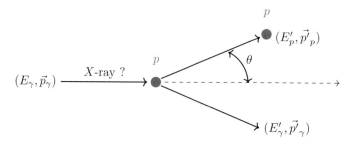

Figure 2.20 Compton-like scattering process to explain the emission of a proton from X-rays.

33   Sir James Chadwick (1891–1974), British physicist.

By introducing the angle of the scattered proton θ, one finds:

$$\vec{p}_\gamma - \vec{p'}_p = \vec{p'}_\gamma \quad \longrightarrow \quad p_\gamma^2 + \left(p'_p\right)^2 - 2p'_p p_\gamma \cos\theta = \left(p'_\gamma\right)^2 \tag{2.118}$$

or

$$E_\gamma^2 + \left(p'_p\right)^2 - 2p'_p E_\gamma \cos\theta = \left(E'_\gamma\right)^2 \tag{2.119}$$

The equation for the energy gives:

$$\left(M_p + E_\gamma - E'_p\right)^2 = \left(E'_\gamma\right)^2 = E_\gamma^2 + \left(p'_p\right)^2 - 2p'_p E_\gamma \cos\theta \tag{2.120}$$

Hence:

$$E_\gamma = \frac{M_p\left(E'_p - M_p\right)}{M_p - E'_p + p'_p \cos\theta} \tag{2.121}$$

The smallest value of the photon energy is obtained with $\cos\theta = 1$. In this case, we find:

$$E_\gamma^{min} = \frac{M_p\left(E'_p - M_p\right)}{M_p - E'_p + p'_p} = \frac{M_p T'}{p'_p - T'} \tag{2.122}$$

where T' is the kinetic energy of the outgoing proton. With the outgoing particles consistent with being protons measured by Chadwick with velocities up to $v'_p \approx 3 \times 10^9$ cm/s $\approx 0.1c$, in this case, $T' = \frac{1}{2} M_p \left(v'_p\right)^2 \approx 5$ MeV and $p'_p c \approx M_p v'_p c \approx 100$ MeV, or:

$$E_\gamma^{min} \approx 50 \text{ MeV} \tag{2.123}$$

Chadwick was struck by the fact that in both cases the high-energy γ radiation was not compatible with the energies available in known radioactive sources. He wrote:

> "*These results, and others I have obtained in the course of the work, are very difficult to explain on the assumption that the radiation from beryllium is a quantum radiation, if energy and momentum are to be conserved in the collisions. The difficulties disappear, however, if it be assumed that the radiation consists of particles of mass 1 and charge 0, or* **neutrons***.*" [34]

Becker and Bothe, who had originally discovered the new type of radiation, also reckoned that they had found a hitherto unknown corpuscular radiation. These were the "neutrons" suspected by Curie-Joliot and Chadwick.[35]

The actual reaction that had been observed was what is called the (α, n) process:

$$\alpha + Be \rightarrow C + n \quad \text{and} \quad n + p \rightarrow n + p \tag{2.124}$$

where the protons were expelled. In this case, the kinematical situation is very different from the Compton-like reaction, because the neutron and the proton have similar masses.

2.15   Interactions of Particles with Matter

We consider the general processes occurring when particles traverse *matter*. Here *matter* can be a gas, a liquid, or a solid of a given material. The material is described by some macroscopic properties, which are supposed to be homogeneous throughout the volume of the considered object. Such quantities are the (1) density ρ, (2) atomic number Z, (3) mass number A, etc. See Figure 2.21. The relevant quantities are summarized in Table 2.1. For compounds, one defines appropriate averages.

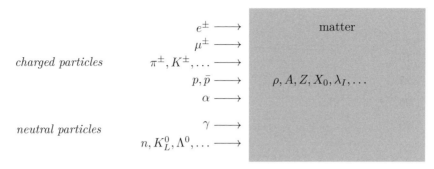

Figure 2.21 Illustration of particles entering a block of homogeneous matter. The matter is characterized by macroscopic quantities. One distinguishes charged and neutral incoming particles.

In general, as the incoming particles enter the block of matter, they will feel the basic building blocks of the target matter that they traverse. Consequently, they will interact with the atomic nuclei and the atomic electrons of the target. These interactions, depending on the type of particle, can be of electromagnetic or strong nature (we neglect neutrinos, which only interact via the very rare weak interactions). The understanding of the interactions between particle and matter is of fundamental importance, since they are at the basis of the detection and measurement of particles. One distinguishes the interactions of charged and neutral particles, as will be discussed in the coming sections.

The interactions (collisions) of the incoming particle with the constituents of the matter being traversed are responsible for the **energy loss** and the **angular deviation** of the particle. Inelastic collisions can also occur. Since each of these processes has a given probability of occurring, one speaks of **stochastic effects**.

2.16   Interactions of Photons with Matter

We distinguish the following domains: (a) very soft photons, which are photons softer than X-rays; (b) hard photons, which include X-rays and γ-rays.

• **Very soft photons.** The dominant processes are coherent effects with many electrons of the atoms. In the coherent process, called Rayleigh (elastic) and Raman (inelastic) scattering, several electrons of the medium participate. When a process is coherent, the cross-section can increase significantly in case of constructive interference. In contrast, we have derived the incoherent process on free electrons in the classical limit called Thomson scattering (see Section 2.12). The calculations of the Rayleigh and Raman scattering cross-sections hence require the definition of the molecular state of the target in order to account for the collective effect. It can be calculated with molecular quantum electrodynamics (see, e.g., Ref. [23]).

For the cases where the energy of the photon $E_\gamma \leq 10$ eV, we enter the **optical domain** and a wave equation

---

34   Reprinted by permission from Springer Nature: J. Chadwick, "Possible existence of a neutron," *Nature*, vol. 129, p. 312, 1932.

35   *"Die Ergebnisse für Bor sind im Einklang mit der Deutung, welche früher der γ-Strahlung des Bors gegeben wurde. Beim Beryllium zeigen die Verhältnisse so viel Ähnlichkeit mit denen beim Bor, daß man auch beim Beryllium **mit einer bisher unbekannten Korpuskularstrahlung** rechnen muß; dies dürften die nach Versuchen von Curie-Joliot und Chadwick vermuteten 'Neutronen' sein."* Reprinted by permission from Springer Nature: H. Becker and W. Bothe, "Die in Bor und Beryllium erregten γ-Strahlen," *Zeitschrift für Physik*, vol. 76, pp. 421–438, Jul 1932.

Material	Z	A (g/mole)	Z/A	ρ (g/cm^3)	X$_0$ (g/cm^2)	X$_0$ (cm)	λ_I (g/cm^2)	Molière radius (cm)	I (eV)
Al	13	26.98	0.48	2.699	24.01	8.90	107.2	4.42	166
Si	14	28.09	0.50	2.329	21.82	9.37	108.4	4.94	173
Fe	26	55.85	0.47	7.874	13.84	1.76	132.1	1.72	286
Cu	29	63.55	0.46	8.960	12.86	1.43	137.3	1.57	322
W	74	183.84	0.40	19.30	6.76	0.35	191.9	0.93	727
Pb	82	207.2	0.40	11.35	6.37	0.56	199.6	1.60	823
Air (1 atm)			0.50	1.2 g/l	36.62	30516	90.1	7330	86
Water (H$_2$O)			0.55	1.00	36.08	36.08	83.3	9.77	80
Concrete (shield)			0.50	2.30	26.57	11.55	97.5	4.90	135
Bismuth germanate (BGO)			0.42	7.13	7.97	1.12	159.1	2.26	534

Table 2.1 A summary of relevant properties of most commonly used materials in particle physics. Data compiled from the Particle Data Group's Atomic and Nuclear Properties of Materials (http://pdg.lbl.gov/AtomicNuclearProperties). See text for definitions.

is used. Following the dispersion relation, one gets:

$$\omega^2 - \frac{k^2 c^2}{\epsilon} = 0 \quad \text{where} \quad E_\gamma = \hbar\omega \tag{2.125}$$

and $\epsilon(\omega)$ is the photon-frequency-dependent dielectric constant of the medium which composes the target matter. When $\epsilon(\omega)$ is real, the photons traverse the transparent medium with a propagation velocity (speed of light in medium) $v = c/n = c/\sqrt{\epsilon}$, where n is the refractive index of the medium. The complex part of $\epsilon(\omega)$ represents the absorption in the medium.

- **X-rays and γ-rays photons.** When the energy of the quantum of light falling on an atom is greater than the ionization energy of the atom, a hit electron will be raised into a state of the continuous spectrum. In this case, all light frequencies can be absorbed and the absorption spectrum is continuous. We consider the various processes as a function of the photon energy, for which the particular process dominates and summarize their basic properties in the following points:

1. The **photoelectric effect** ($E_\gamma \lesssim 100$ keV): this process dominates in this energy range. The photon is absorbed by an atomic electron and a monoenergetic electron from a given shell is emitted:

$$E_e = E_\gamma - \phi \tag{2.126}$$

where ϕ is the binding energy (or work function) of the electron in this shell. The cross-section possesses several maxima which correspond to the various shells.

2. The **scattering process** (100 keV $\lesssim E_\gamma \lesssim 1$ MeV): in this energy range, the incoherent scattering of the photon on one atomic electron dominates. The process can be represented by the Compton $\gamma e \to \gamma e$ scattering on a free electron. The cross-section will be calculated within QED in Section 11.16. The computation leads to the Klein–Nishina formula (see Eq. (11.272)).

3. The **pair creation process** $(E_\gamma \geq 2m_e)$: above the energy required for the creation of an electron–positron pair, this process dominates. The photon converts in such a pair. In order to conserve energy and momentum, the reaction must occur in the presence of a recoiling nucleus. At high energy, the cross-section can be expressed in the following way:

$$\sigma_{pair} = (4\alpha r_e^2) Z^2 \left(\frac{7}{9} \ln \frac{183}{Z^{1/3}} \right) \tag{2.127}$$

where r_e is the classical electron radius (see Eq. (2.99)).

These cross-sections as a function of the photon energy are plotted in Figure 2.22.

• **Photon attenuation length.** Because of their interaction cross-section, the flux of photons will be attenuated in matter. We assume that one photon is absorbed when it interacts, leaving the other photons unaffected. This leads to an exponential attenuation of the photon flux. Consequently, as a function of the traversed path x through matter, the photon intensity can be expressed as:

$$I(x) \equiv I_0 e^{-x/\lambda_{abs}} = I_0 e^{-\mu x} \tag{2.128}$$

where μ is the total absorption coefficient. It is given by:

$$\mu = \left(\frac{\rho N_A}{A} \right) \sigma_{tot} \simeq \left(\frac{\rho N_A}{A} \right) (\sigma_{P.E.} + \sigma_{incoherent} + \sigma_{coherent} + \sigma_{pair}) \tag{2.129}$$

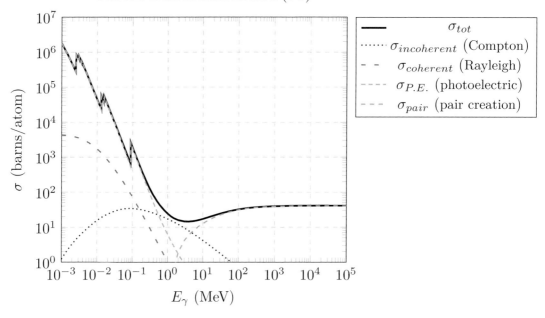

Figure 2.22 The photon cross-section on lead (Pb, $Z = 82$). Data taken from the Physical Measurement Laboratory of the National Institute of Standards and Technology (NIST) (http://physics.nist.gov/).

• **The radiation length** X_0. A very useful quantity is the **radiation length** X_0 defined for a given target material. It is defined as:

$$\frac{1}{X_0} \equiv 4\alpha \left(\frac{\rho N_A}{A} \right) Z^2 r_e^2 \ln \frac{183}{Z^{1/3}} \tag{2.130}$$

where ρ is the density of the target and N_A the Avogadro number. With this definition, one can verify that X_0 has units of length. The *probability* that an incoming photon creates an e^+e^- pair within one radiation length ($1\ X_0$) is then given by:

$$P \simeq \sigma_{pair} \left(\frac{\rho N_A}{A} \right) X_0 \approx \frac{7}{9} \tag{2.131}$$

The mean free path of such a photon is then given by:

$$\lambda_{pair} \simeq \frac{7}{9} X_0 \tag{2.132}$$

The radiation length X_0, as tabulated in Table 2.1, can therefore be used directly to estimate the absorption rate of high-energy photons in a given material. We note that the radiation length is inversely proportional to Z^2. In water it is about 36 cm, while in aluminum, iron, and lead it is respectively 8.9 cm, 1.76 cm, and 0.56 cm. In order to compare materials, one often expresses it in units of g/cm^2 by multiplying it by the density ρ. When given in g/cm^2, one should divide X_0 by the density of the target to recover the length in cm.

2.17   Interactions of Charged Particles with Matter

We want to consider the interactions between a fast, possibly relativistic, charged particle as it enters continuous matter. This case is different from Rutherford's scattering case, where we computed the cross-section for a relatively large deviation of a non-relativistic α particle off a *single* target nucleus.

Here we are interested in the result of **many collisions between the incoming charged particle and the atomic electrons and nuclei of the traversed matter**. If the incident particle is heavier than electrons (in case of electrons or positrons see Section 2.19), the interactions with atomic electrons can transfer a large amount of energy from the incident heavy particle, without causing significant deflection from the latter. Thus, we can assume that in this case the loss of energy is primarily due to the interactions with the atomic electrons. The deflection of the incident particle is, on the other hand, still dominated by collisions with the heavy atomic nuclei like in the case of Rutherford scattering. However, we consider the net result of many collisions confined to relatively small scattering angles, so that overall, the incident particle keeps more or less a straight trajectory. These assumptions are not valid for incident electrons of relatively low energies, where both energy loss and scattering occur dominantly from atomic electrons. Hence their trajectory is much less straight, and electrons will diffuse into the material and can often be back-scattered.

A fast particle of mass M much heavier than electrons and charge ze moves with velocity $v = \beta \approx 1$ and energy $E = \gamma M$ through a medium with atomic electron density n. It passes an atomic electron with impact parameter b, as shown in Figure 2.23. The electron is assumed to initially be at rest relative to the incident particle and to be free. Since the interaction with the electron is very fast, we can furthermore assume that it does not move appreciably during the collision. The energy gained by the electron can be calculated from the change of its momentum, which in this case is simply:

$$\Delta p = \int F_\perp \mathrm{d}t = e \int E_\perp \frac{\mathrm{d}t}{\mathrm{d}x} \mathrm{d}x = \frac{e}{\beta} \int E_\perp \mathrm{d}x \tag{2.133}$$

where only the perpendicular component of the force enters because of the symmetry of the field in the assumed fixed position of the electron during the collision. The electric field felt by the electron is given by the Coulomb force corrected for the Lorentz contraction due to the γ factor of the incoming particle. Although the field is Lorentz contracted, it should be noted that under our assumptions the integral and hence Δp are independent of γ, since it is always symmetric around the electron. The integral can thus be calculated using Gauss's law $\nabla \cdot \vec{E} = \rho/\epsilon_0$ in a cylinder of radius b centered around the incoming particle's trajectory and touching the

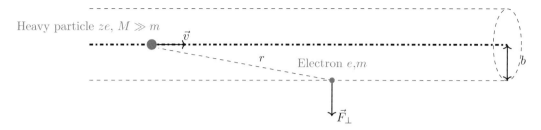

Figure 2.23 Sketch for the calculation of the energy transferred to an atomic electron by a heavy charged particle.

electron. Using the units of Eq. (2.75), we get:

$$\oint_c \vec{E} \cdot \mathrm{d}\vec{A} = \int E_\perp (2\pi b)\mathrm{d}x = ze/\epsilon_0 \tag{2.134}$$

Hence:

$$\Delta p = \frac{e}{\beta} \frac{ze}{2\pi b \epsilon_0} = \frac{2ze^2}{4\pi b \epsilon_0 \beta} = \frac{2\alpha z}{b\beta} \tag{2.135}$$

The energy imparted to the electron during a collision with an impact parameter b is then given by:

$$\Delta E \equiv \frac{\Delta p^2}{2m_e} = \frac{2\alpha^2 z^2}{m_e \beta^2} \frac{1}{b^2} \tag{2.136}$$

The energy gained by the electron is equivalent to the *energy lost* by the incident particle during the collision. We now consider the case when crossing the medium with atomic electron density n. The energy *lost* per unit length $\mathrm{d}x$ to all atomic electrons located at the distance between b and $b + \mathrm{d}b$ from the trajectory of the incident particle is:

$$-\mathrm{d}E = \Delta E(n\mathrm{d}V) = n\Delta E(2\pi b\,\mathrm{d}b\,\mathrm{d}x) = n\frac{4\pi\alpha^2 z^2}{m_e\beta^2}\frac{\mathrm{d}b}{b}\,\mathrm{d}x \tag{2.137}$$

We note that the result is independent of the mass of the incident particle and depends only on its charge and inverse-squared velocity. We should now integrate over the impact parameter b to get the total energy loss, and the question arises which boundaries should be used. **We clearly cannot simply integrate from zero to infinity and therefore need to introduce some physical assumptions.** On the one hand, we note that since $\Delta E \propto 1/b^2$, the expression for the energy lost diverges for $b \to 0$, so there must be a minimum impact parameter. On the other hand, for $b \to \infty$, the approximation used to compute Δp will break down, since the collision cannot be considered as "fast" any more when b is large, since we expect the typical interaction time to be given by $\Delta t \simeq b/(\gamma v)$, where the Lorentz factor accounts for time dilation [17]. Consequently, we write the total energy lost per unit path length as:

$$-\left(\frac{\mathrm{d}E}{\mathrm{d}x}\right) = n\frac{4\pi\alpha^2 z^2}{m_e\beta^2}\int_{b_{min}}^{b_{max}}\frac{\mathrm{d}b}{b} = n\frac{4\pi\alpha^2 z^2}{m_e\beta^2}\ln\left(\frac{b_{max}}{b_{min}}\right) \tag{2.138}$$

where the integral boundaries need to be determined. Several ways to define these boundaries can be found in the literature. Since $\Delta E \propto 1/b^2$, the minimum impact parameter b_{min} can be found by setting the maximum transferable energy corresponding to a "head-on" collision, or $\Delta E_{max} \simeq 2m_e\gamma^2\beta^2$ [17], then:

$$\frac{2\alpha^2 z^2}{m_e\beta^2 b_{min}^2} \simeq 2m_e\gamma^2\beta^2 \quad \Longrightarrow \quad b_{min} \simeq \frac{\alpha z}{m_e\beta^2\gamma} \tag{2.139}$$

To find b_{max}, we require a minimum for ΔE below which our free-electron approximation is not valid anymore. This quantity is called the **mean excitation energy** I and is dependent on the target material. It basically corresponds to the average orbital frequencies of electrons around the atom in a simplified Bohr model (see Appendix C). In order for the atomic electron to be "kicked" by the passing incident particle, the time of the collision Δt defined above should be small compared to the time associated with the orbital frequency. Otherwise the collision can be considered as adiabatic and no energy is transferred. This implies that:

$$\Delta t = \frac{b_{max}}{\gamma\beta} \lesssim \frac{1}{I} \quad\Longrightarrow\quad b_{max} \simeq \frac{\beta\gamma}{I} \tag{2.140}$$

Substituting the boundaries in the integral, we obtain **Bohr's classical stopping power** for heavy charged particles:

$$-\left(\frac{\mathrm{d}E}{\mathrm{d}x}\right) = n\,\frac{4\pi\alpha^2 z^2}{m_e\beta^2}\ln\left(\frac{m_e\beta^3\gamma^2}{\alpha z I}\right) \tag{2.141}$$

This formula, although approximative, contains many of the main effects of the interactions of an incident heavy charged particle with the atomic electrons of a medium. In particular, at low velocities the stopping power is dominated by the $1/\beta^2$ term and increases rapidly as $\beta \to 0$. As the incident particle is slowed down, it will start losing more and more energy and will be slowed down more until it stops in the medium. The heavy particle at rest can then either decay or be absorbed by a nucleus of the medium. At high velocities, the stopping power also increases but only logarithmically as $\ln(\beta^3\gamma^2)$. There is a region around $\beta\gamma \simeq 3\text{--}3.5$, where the incident particle is minimally stopping. It is called a **minimum ionizing particle** (m.i.p.).

- **Bethe–Bloch formula.** The correct prediction of the stopping power of charged particles in matter has been the subject of intense investigations over several years. In particular, a correct quantum-mechanical derivation was performed by several authors, including **Bethe** [24, 25],[36] **Bloch** [26],[37] and **Møller** [27].[38] The passage of particles through matter is discussed extensively in the Particle Data Group's Review [28] in the chapter "Passage of particles through matter," and the reader will find many details there (the chapter can be accessed at https://pdg.lbl.gov/2020/reviews/rpp2020-rev-passage-particles-matter.pdf). The **Bethe–Bloch formula** is commonly used to compute the energy loss (where the speed of light is reintroduced for conventional reasons):

$$-\left(\frac{\mathrm{d}E}{\mathrm{d}x}\right) = K z^2 \frac{Z}{A}\frac{1}{\beta^2}\left[\frac{1}{2}\ln\left(\frac{2m_ec^2\beta^2\gamma^2 W_{max}}{I^2}\right) - \beta^2 - \frac{\delta(\beta\gamma)}{2}\right] \tag{2.142}$$

where K is the coefficient for $\mathrm{d}E/\mathrm{d}x$ given by $K \equiv 4\pi N_A r_e^2 m_e c^2 \simeq 0.31\,\mathrm{MeV}\,\mathrm{mol}^{-1}\,\mathrm{cm}^2$, and $Z(A)$ the atomic number (mass) of the medium. The units of $\mathrm{d}E/\mathrm{d}x$ are given by the constants K/A. For $A = 1$ g per mol, the stopping power is expressed in $\mathrm{MeV}\,\mathrm{g}^{-1}\,\mathrm{cm}^2$. In order to get the actual loss per unit length, one should multiply the stopping power by the density ρ of the medium, and get $\rho\,\mathrm{d}E/\mathrm{d}x$ in units of $\mathrm{MeV}\,\mathrm{cm}^{-1}$. The maximum energy transferred to an electron in a single collision is given by:

$$W_{max} = \frac{2m_ec^2\beta^2\gamma^2}{1 + 2\gamma m_e/M + (m_e/M)^2} \tag{2.143}$$

The last term $\delta(\beta\gamma)$ is a relativistic density correction (see, e.g., the Particle Data Group's Review for details). The Bethe–Bloch formula is valid in the region $0.1 \lesssim \beta\gamma \lesssim 1000$ for heavy particles such as pions or protons, and gives the mean energy lost in most materials with a precision of a few percent.

- **Mean excitation energy.** The mean excitation potential I can be regarded as a tunable parameter of the Bethe–Bloch formula, deduced from actual measurements of $\mathrm{d}E/\mathrm{d}x$ losses in various materials fitted to semi-empirical models. The generally adopted values are those given in *Stopping Powers for Electrons and Positrons*,

---

36   Hans Albrecht Bethe (1906–2005), German-American physicist.
37   Felix Bloch (1905–1983), Swiss physicist.
38   Christian Møller (1904–1980), Danish chemist and physicist.

ICRU Report No. 37 (1984), available at http://physics.nist.gov/PhysRefData/Star/Text/ESTAR.html. A comprehensive list of values can be found on the database "X-Ray Mass Attenuation Coefficients" maintained by the Physical Measurement Laboratory of the National Institute of Standards and Technology (NIST), which can be accessed online at http://physics.nist.gov/PhysRefData/XrayMassCoef/tab1.html.

- **Particle type dependence and particle identification.** The Bethe–Bloch function computed with Eq. (2.142) and neglecting the relativistic density correction, is plotted in Figure 2.24 for muons, pions, kaons, and protons. The curves are quite distinguishable for non-relativistic particles with momenta between 100 MeV and 1 GeV.

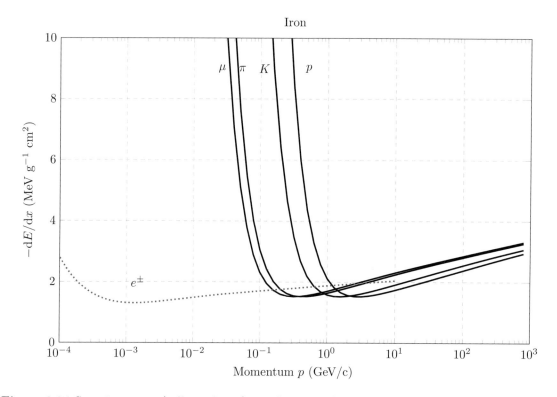

Figure 2.24 Stopping power (collision losses) as a function of momentum for different particles in iron.

In this case, the knowledge of the particle momentum (e.g., by measurement of its curvature in a magnetic field), combined with a dE/dx measurement, allows us to test the particle and therefore provide a mean for particle identification. For completeness, the energy losses due to collisions for electrons and positrons are also shown in the figure. In this case, the approximation that the incoming particle is heavy compared to the target electrons is no longer valid and additional corrections to the Bethe–Bloch formula must be introduced (see Section 2.19). Another significant difference is that above a certain critical energy, typically in the range of a few tens of mega-electronvolts, the loss of energy by radiation is comparable to or greater than the collision loss. In the figure, the curve is extended to 10 GeV for illustration purposes and one should keep in mind that electrons above the critical energy will lose energy primarily by emitting hard photons.

2.18   Molière Multiple Scattering

In the previous section, we have seen that charged particles crossing matter will lose energy due to multiple collisions with the atomic electrons and nuclei of the target material. During these collisions, they will exchange energy and momentum. In a first approximation, we can assume that the nuclei are much more massive than the incident particle. Thus, individual collisions will be governed by the Rutherford scattering cross-section Eq. (2.88). Because of the $1/\sin^4(\theta/2)$ dependence, most collisions will result in a small angular deflection of the incoming particle in the forward direction. We thus expect the particle to follow a random "zig-zag" trajectory as it traverses the target. The cumulative effect is a net deflection from the original particle direction. This process, leading to energy loss and deflection, is illustrated in Figure 2.25.

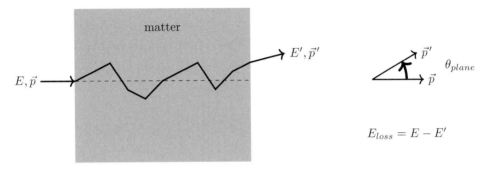

Figure 2.25 Illustration of a muon entering and exiting a block of homogeneous matter after having been subject to energy loss and multiple scattering.

In general, a rigorous calculation of the multiple scattering is complicated. If one ignores the small probability of large-angle scattering, then one can use the Central Limit Theorem to obtain a Gaussian approximation of multiple deviations with a finite angle. The rare large-angle scatterings produce non-Gaussian tails. In this approximation, the scattering distribution is described by the theory of Molière [29]. The central 98% of the projected angular distribution is distributed as Gaussian with RMS given by [28, 30]:

$$\theta_0 = \frac{13.6 \text{ MeV}}{\beta c p} z \sqrt{\frac{x}{X_0}} \left[1 + 0.038 \ln \left(\frac{xz^2}{X_0 \beta^2} \right) \right] \tag{2.144}$$

where

$$\theta_0 = \theta_{plane}^{RMS} = \frac{1}{\sqrt{2}} \theta_{space}^{RMS} \tag{2.145}$$

and p, βc, and z are the momentum, velocity, and charge of the incident particle. The traversed thickness in the scattering medium is given by x, and X_0 is the **radiation length** of the target material, defined in Eq. (2.130). Some values are tabulated in Table 2.1. The choice of normalizing the traversed distance x to the radiation length X_0 is a conventional and practical one. This choice is very clever as it takes into account the nuclear properties of the target, such that the expression of Eq. (2.144) is valid for all targets. It should, however, be remembered that these are approximations and further information on their range of validity can for instance be found in the Particle Data Group's Review [28].

The effect of energy loss and multiple scattering is illustrated in Figure 2.26. The picture represents the result of a GEANT4 Monte-Carlo simulation (see Section 2.22) of 10 positive muons with an initial energy of 1 GeV (tracks entering the front face of the first block) crossing a target composed of 50 slabs of iron, each slab being 2 cm thick. The effect of multiple scattering is visible: each track is deflected in a different direction.

The muons lose energy due to dE/dx and eventually stop in the target. Once stopped, they eventually decay into a positron (visible as short stubs at the end of the muon track) and two neutrinos (not shown). The short tracks visible along the main track correspond to knock-off electrons.

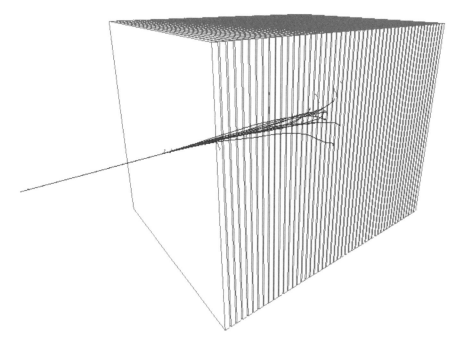

Figure 2.26 GEANT4 simulation (see Section 2.22) of ten 1 GeV positive muons crossing 50 slabs of iron, each 2 cm thick. The effect of multiple scattering is visible. The short tracks visible along the main primary tracks correspond to knock-off electrons.

2.19   Interactions of Electrons and Positrons with Matter

In the case of light, charged particles like electrons or positrons, the energy loss occurs through collisions Eq. (2.142) as well as radiation. The so-called **Bremsstrahlung** (braking radiation in German) occurs because the electrons or positrons are deflected in the field of an atom (or nucleus). The deflection involves an acceleration, therefore there will be emission of radiation in the form of photons. Since the nucleus is heavy compared to the mass of the electron, the nucleus, in a single scatter, can take a large amount of the recoil momentum against the electron and the emitted photon. The average energy loss can be expressed as:

$$\left(\frac{dE}{dx}\right)_{e^+/e^-} = \left(\frac{dE}{dx}\right)_{Bethe-Bloch} + \left(\frac{dE}{dx}\right)_{Bremsstrahlung} \tag{2.146}$$

The cross-section was first computed by **Bethe** and **Heitler**[39] and then corrected and extended for various effects. The original formulation estimates the radiation energy loss of electrons/positrons to be:

$$-\left(\frac{dE}{dx}\right)_{Bremsstrahlung} = 4\alpha N_A \rho \frac{Z^2}{A} r_e^2 E \ln \frac{183}{Z^{1/3}} \simeq \frac{E}{X_0} \tag{2.147}$$

39   Walter Heinrich Heitler (1904–1981), German physicist.

where in the last term we have inserted the already defined radiation length. It looks super simple! Above a few tens of mega-electronvolts, Bremsstrahlung is the dominant process for energy loss. It actually depends on the inverse of the mass of the electron squared (included in the presence of r_e^2). For muons, a similar process occurs, but it becomes important only for muons with energies above a few hundred giga-electronvolts.

If we neglect the collision losses, we note that the energy as a function of the traversed depth will decrease exponentially:

$$-\left(\frac{\mathrm{d}E}{E}\right) \simeq \frac{\mathrm{d}x}{X_0} \quad \Longrightarrow \quad E(x) \simeq E_0 e^{-x/X_0} \tag{2.148}$$

Therefore, the radiation length is equivalent to the distance over which an electron or a positron will have their energy reduced by a factor e. In fact, this last expression can be used to *define* the radiation length. The **critical energy** E_c is defined as the kinetic energy for which collision and radiation losses are identical. Below the critical energy, the dominant loss is via collisions, above the critical energy it is dominated by Bremsstrahlung. Typical values for electrons/positrons are $E_c(Al) = 43$ MeV, $E_c(Fe) = 2$ MeV, and $E_c(Pb) = 7$ MeV (see http://pdg.lbl.gov/2019/AtomicNuclearProperties/).

2.20   Cherenkov Radiation

In 1934 **Cherenkov**[40] discovered that light is emitted when a charged particle traverses a medium at a speed faster than the speed of light in that medium c/n, where n is the refractive index of the medium. Hence, Cherenkov radiation is emitted if the particle moves at velocity v such that (we keep c in this section for clarity):

$$v > c/n \qquad \text{(Cherenkov threshold)} \tag{2.149}$$

The energy carried away by Cherenkov radiation was first computed by **Frank**[41] and **Tamm**:[42]

$$-\frac{\mathrm{d}E}{\mathrm{d}x} = \frac{4\pi e^2}{c^2} \int_{\beta n(\omega) > 1} \mathrm{d}\omega\, \omega \left(1 - \frac{1}{\beta^2 n^2(\omega)}\right) \tag{2.150}$$

where $\beta = v/c$ is the velocity of the particle relative to that of the speed of light in vacuum.

Here $n(\omega)$ is the refractive index of the medium as a function of the emitted light frequency ω. The phenomenon can be assimilated to that of the creation of a shock wave classically created by a body moving faster than the speed of a wave in that medium. In the case of Cherenkov emission, the charged particle passes through the medium, which can be seen as a dielectric. The passage of the charged particle disrupts the electromagnetic field of the medium, which becomes locally polarized. The elementary dipoles of the medium will be displaced. After the disruption has passed, the restoration of the medium creates a coherent wave of photons of a continuous frequency spectrum. The coherently emitted wave front is conical in shape, as illustrated in Figure 2.27.

With respect to the trajectory of the incoming charged particle, the Cherenkov cone has an angle θ defined by:

$$\cos\theta = \frac{ct/n}{\beta ct} = \frac{1}{\beta n} \tag{2.151}$$

The number of emitted Cherenkov photons of a charged particle Z per unit path length x and per wavelength λ can be expressed as:

$$\frac{\mathrm{d}^2 N_\gamma}{\mathrm{d}x \mathrm{d}\lambda} = \frac{2\pi\alpha Z^2}{\lambda^2}\left(1 - \frac{1}{\beta^2 n^2(\lambda)}\right) = \frac{2\pi\alpha Z^2}{\lambda^2}\sin^2\theta \tag{2.152}$$

Cherenkov light is used in so-called Cherenkov counters or detectors, as will be described in Section 3.14.

40   Pavel Alekseyevich Cherenkov (1904–1990), Soviet physicist.
41   Ilya Mikhailovich Frank (1908–1990), Soviet physicist.
42   Igor Yevgenyevich Tamm (1895–1971), Soviet physicist.

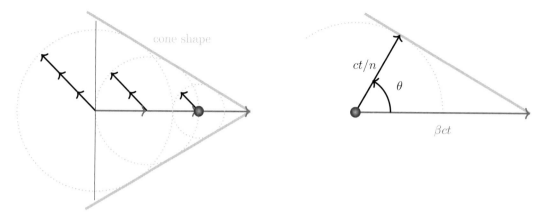

Figure 2.27 (left) Illustration of the Cherenkov cone; (right) definition of the Cherenkov angle.

2.21 Interactions of Neutrons with Matter

The neutron, like the photon, is electrically neutral, so that it is not subject to the Coulomb interactions with the electrons and nuclei in matter. Instead, the main interaction is via the short-range strong force (see Chapter 15) between the neutron and a nucleus. These reactions are much rarer than the Coulomb interactions which act at an extended distance. Neutrons must basically find themselves within the sphere of a nucleus to feel any interaction. Since the scale is the femtometer and matter is mostly "empty" otherwise, then the neutron is said to be a very penetrating particle. However, the lack of electromagnetic interaction also plays in favor of the neutron since compared to a proton, it does not suffer from the very strong Coulomb repulsion by the nucleus.

It is customary to classify neutrons according to their energy because the interaction probability strongly depends on it:

1. Cold or ultra-cold neutrons $E \approx$ meV or μeV.

2. Thermal or slow neutrons $E \simeq kT \approx 0.025$ eV (energy of the thermal agitation energy of matter at room temperature).

3. Epithermal 0.1 eV $\lesssim E \lesssim 100$ keV.

4. Fast 100 keV $\lesssim E \lesssim$ 10s MeV.

5. High energy $E \gtrsim 100$ MeV.

Neutrons may undergo several different nuclear interactions with a given nucleus A.

- **Elastic scattering** $A(n,n)A$: the neutron scatters off a nucleus, bouncing elastically like a billiard ball, resulting in a change of its momentum; this is the dominant reaction for low-energy neutrons.

- **Inelastic scattering** $A(n,n)A^*$: the neutron scatters off inelastically and the target nucleus is left in an excited state, which later decays by emitting a γ-ray or neutrons (spallation). The incoming neutron should have sufficient energy to excite the nucleus to an excited state, which typically happens only for fast neutrons or above.

- **Radiative capture** $n + (Z, A) \rightarrow (Z, A + 1) + \gamma$: the neutron is absorbed by the nucleus, which is left in an excited state; this latter can later decay by emitting a γ-ray. Since there is no Coulomb repulsion, the probability of the neutron being captured by the nucleus is inversely proportional to its velocity (i.e., $\propto 1/v$). Hence, the process occurs mostly for slow neutrons.

- **Other captures** (n, p), (n, d), (n, α), **etc.**: the neutron is absorbed like in the radiative capture but charged debris is emitted by the nucleus instead.

- **Fission** (n, f): the hit nucleus is broken into lighter elements, which later may emit neutrons.

- **High-energy collision**: at high energies, the neutron can be seen to interact dominantly with a single nucleon of the type $n - p$ or $n - n$, the rest of the nucleus acting essentially as a spectator. The hit nucleon can sometimes be broken and a resulting hadron "shower" will be produced.

As an illustration, the total interaction cross-section of neutrons on different targets as a function of energy is plotted in Figure 2.28. The cases of hydrogen, oxygen, and silicon, the relevant components for neutron shielding found in concrete ($\rho = 2.3$ g/cm^3, $\rho_H = 0.051$ g/cm^3, $\rho_O = 1.322$ g/cm^3, $\rho_{Si} = 0.701$ g/cm^3), are shown. In addition, the curve for water is also plotted for comparison. Computations have been carried out with the general-purpose GEANT4 simulation program, which will be described in the next section. The rises at low energy are clearly visible.

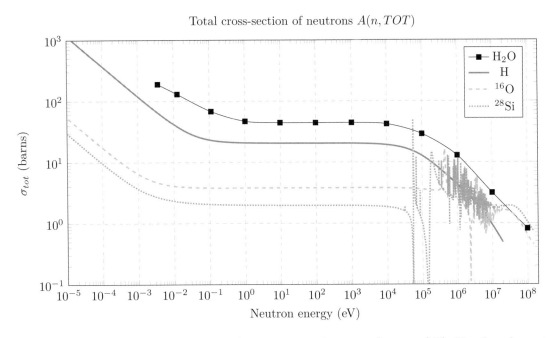

Figure 2.28 Total interaction cross-section of neutrons as a function of energy (eV). The data for water is computed with GEANT4 (see Section 2.22). The rest of the data is taken directly from the ENDF/B-VII.1 database [31].

- **Neutron moderation.** This refers to the process of the slowing down of fast neutrons due to their back-and-forth scattering on the nuclei of matter, both elastically and inelastically, losing energy. The neutron may undergo a nuclear reaction, e.g., produce fission or any other type of reaction listed above, before attaining thermal energies, especially if the cross-section exhibits resonances. If the neutrons reach thermal energies, they will diffuse through matter until they are captured by a nucleus. This process is illustrated in Figure 2.29.

A sample of 100 neutrons was simulated with the help of the GEANT4 simulation package (see next section). Two different configurations are considered: (1) a cubic meter of pure water; (2) a cubic meter of a mixture of water with 10% cadmium (Cd). Cadmium is highly effective in enhancing neutron capture and has the additional feature that γ-rays with a total energy of 9 MeV are emitted when the neutron is captured. This effect can readily be seen when comparing the two configurations. The greatly enhanced gamma emission is clearly visible. We will come back to this feature when we discuss the discovery of the neutrino in Section 21.7. In general, neutron moderation is an important process in nuclear physics and engineering.

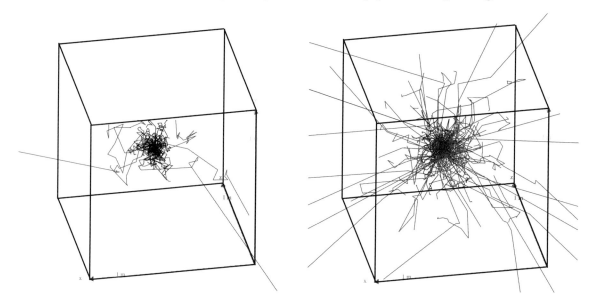

Figure 2.29 GEANT4 simulation (see Section 2.22) of 2 MeV neutrons inside a cubic-meter box filled with (left) pure water; (right) a mixture of water with 10% Cd. Each image show the tracks produced by 100 neutrons. The escaping tracks are γ-rays emitted following a neutron capture.

2.22   Geometry and Tracking (GEANT)

In the early 1970s **René Brun** and collaborators developed the first version of a computer simulation program for particle physics experiments [32] called GEANT, for GEneration of Events ANd Tracks (later also renamed GEometry ANd Tracking). The goal was to create a program to perform simulation software designed to describe the passage of elementary particles through matter, using Monte-Carlo methods (see Appendix H). Developments continued until the version of GEANT3 in the 1980s, which was used for the LEP experiments. Until version 3.21, the code was written in FORTRAN and eventually maintained as part of a large library called CERNLIB.[43] Since about 2000, the last FORTRAN release has been essentially frozen. Its successor, GEANT4, is a complete rewrite in C++ with a modern object-oriented design [33]. GEANT4 was developed at CERN during the period 1994–1998 and is now being maintained and improved by an international collaboration.[44]

GEANT is a library. In order to use it, the user must provide a set of user codes. Basically, the main flow can be summarized as follows: (1) Initialize; (2) Define geometry with materials; (3) Define particles and

43   http://cernlib.web.cern.ch/
44   http://geant4.web.cern.ch

physics processes; (4) Start of a run; (5) Start of an event; (6) Define kinematics of primary particles; (7) Track primary and secondary particles through media; (8) Perform the digitization of sensitive volumes; (9) Loop events; (10) Termination of run. The number of loops over events to accumulate statistics can be specified by the user at runtime. The digitization step transforms the energy deposited in the different volumes representing sensitive detectors into recorded hits that are supposed to correspond to the measurements performed in the actual physical detectors (e.g., an electrical signal). The visualization of the results is also very developed, allowing for several different three-dimensional representations of the setup as well as histograms of various physical quantities (scoring).

This structure is very versatile and powerful. The output has been benchmarked and perfected over many decades. The precision of the simulations today reaches a high degree of accuracy over a wide range of energies. It can be used to optimize a particular setup before construction. After construction and data-taking, it is a fundamental tool to understand and analyze the data produced by particle physics experiments. As an example, the very simple setup of Figure 2.26 and the simulation of the passing muons was performed with GEANT4. On another scale, very complex and detailed GEANT4 descriptions of the LHC experiments (see ATLAS and CMS in Figures 3.35 and 3.36) have been implemented and are routinely used. Originally developed for high-energy physics experiments, GEANT is now used in many other fields, including medical imaging.

Problems

Ex 2.1 Law of radioactive decay. A weak radioactive source is observed for a period of 60 s. In this period of time, 1000 decays are detected. What is the estimation of the decay rate of the source and the error from this measurement?

Ex 2.2 Proton decay. Proton decays are suppressed by the global baryon number symmetry. Check in the Particle Data Group's Review what is the current experimental bound on its lifetime. Assuming its lifetime is given by this bound, how many protons would be needed to obtain on average one decay per year. How many cubic meters of water does that represent (count also the protons in oxygen)?

Ex 2.3 Rutherford scattering. Derive the classical trajectory of the α particle scattered by a point-like non-recoiling nucleus. Show that it can be expressed as:

$$\frac{1}{r} = \frac{1}{b}\sin\alpha + \frac{D}{2b^2}(\cos\alpha - 1) \quad \text{where} \quad D \equiv \frac{\alpha z Z}{E_K} \tag{2.153}$$

where the position of the α particle along its trajectory is given in polar coordinates by (r, α). E_K is the kinetic energy of the incoming particle and b the impact parameter.

Ex 2.4 Distance of closest approach in Rutherford scattering. We consider the classical Rutherford scattering problem (see **Ex 2.3**).

(a) Show that the distance of closest approach d_0 for a head-on collision is given by:

$$d_0 = \frac{zZ\alpha}{E_K} \tag{2.154}$$

(b) Show that the relation between the impact parameter b and the scattering angle simplifies to:

$$b = \frac{d_0}{2}\cot\left(\frac{\theta}{2}\right) \tag{2.155}$$

(c) Now consider the general case, and calculate the distance of closest approach d as a function of the impact parameter b. Show that it can be expressed as:

$$b^2 = d(d - d_0) \tag{2.156}$$

(d) Assume the kinetic energy of the α particle is 130 MeV, impinging on a ^{208}Pb target. What is d_0? Compute d for $b = 1, 4, 7$, and 15 fm.

(e) When the distance of closest approach is in the range of the nuclear force, we expect a deviation from the Rutherford cross-section. Assuming that the ^{208}Pb nucleus has a radius of $R = 12.5$ fm, compute the necessary kinetic energy of the α particles to start seeing such deviations.

Ex 2.5 Origin of β particles. Before the discovery of the neutron, early discussions contemplated the possibility that atomic nuclei are composed of electrons and protons. Using Heisenberg's uncertainty principle, estimate the minimal energy of electrons bound inside a nucleus assumed to have a radius smaller than 10 fm. Is this consistent with an observed β decay spectrum?

Ex 2.6 Order of magnitude of cross-sections. Compare the typical cross-sections for (a) pp interactions, (b) γp interactions, and (c) νp interactions. What can you say about the differences between them?

Ex 2.7 Mean free path of neutrinos in water. A neutrino with an energy of a few mega-electronvolts has an interaction cross-section with free protons of about $\sigma_{\nu p} \approx 10^{-43}$ cm^2. Compute the mean free path of such neutrinos in water.

Ex 2.8 Neutron interactions. Estimate the mean free path of (a) a thermal, (b) a fast, and (c) a high-energy neutron in water.

Ex 2.9 GEANT4 simulation. Using the GEANT4 framework, set up the simulation of muons crossing 50 slabs of iron, each 2 cm thick. Shooting muons of 1 GeV, study the effect of multiple scattering. Do you see δ-rays? What happens if you increase the energy of the muons, to say, 10 GeV, 50 GeV, 200 GeV?

3   Overview of Accelerators and Detectors

It has long been my ambition to have available for study a copious supply of atoms and electrons which have an individual energy far transcending that of the α and β particles from radioactive bodies. I am hopeful that I may yet have my wish fulfilled.

Ernest Rutherford

3.1   Accelerators

Particle accelerators are devices that produce different kinds of energetic, high-intensity beams of stable particles (e, p, ...). They possess many fields of application: nuclear and particle physics, material science, chemistry, biology, medicine, isotope production, medical imaging, medical treatments – just to name a few. Beams of accelerated primary particles can be used to produce beams of secondary metastable particles, such as μ's, π's, K's, etc. Photons (X-rays, γ-rays, visible light) can also be generated from so-called light sources where high-intensity beams of electrons emit synchrotron radiation in undulators or wigglers. Neutrons are generated from high-intensity beams of protons in typical spallation neutron sources.

High-energy accelerators have evidently been a fundamental tool for progress in experimental particle physics. As a matter of fact, particle physics has been one of the main drivers for accelerator developments. Their importance can be understood when one considers them as large microscopes. Because of the wave–particle duality, the spatial "resolution" will improve with energy because of the de Broglie relation $\lambda = h/p$, where λ is the wavelength or the characteristic length of the particle used as the "probe" and p is its momentum (see Section 4.1). We see that the wavelength is inversely proportional to the momentum. Larger momenta correspond to shorter wavelengths and hence access to smaller structures of matter. We have, for example, that $1\text{ GeV}^{-1} \simeq 0.2$ fm, to be compared to the size of a proton of about 1 fm. Hence, GeV is the typical energy scale to study objects of the size of the nucleon. TeV is the scale of today's **Large Hadron Collider** (LHC).

In addition, with the increasing of energy it is possible to reach new kinematical thresholds and produce new particles in collisions thanks to the Einstein relation between energy and mass, $E = mc^2$ (or $E = m$ in natural units). In order to produce new particles, the center-of-mass energy E_{cm} of the collision should satisfy $E_{cm} > m$, or $E_{cm} > 2m$ in the case when the new particles need to be produced in pairs (this depends on the type of collision and of particles, due to the conservation of specific quantum numbers in the interactions). This is the primary reason why particle colliders have been developed (see Section 3.2).

A charged particle is accelerated by an electric field. The creation of high electric fields requires high voltages. One can use a DC (static) potential difference between two conductors to impart kinetic energy to the traversing charged particles. The **Greinacher voltage multiplier** is an electric circuit developed by **Heinrich Greinacher**[1] in 1919. The circuit generates a high DC voltage from a low-voltage AC. The circuit was used for the first time in 1932 by **John Douglas Cockcroft**[2] and **Ernest Thomas Sinton Walton**[3] to power a particle accelerator. This is the reason why the circuit is sometimes called a **Cockcroft–Walton** (CW) generator.

---

1   Heinrich Greinacher (1880–1974), Swiss physicist.
2   Sir John Douglas Cockcroft (1897–1967), British physicist.
3   Ernest Thomas Sinton Walton (1903–1995), Irish physicist.

Cockcroft and Walton achieved a high voltage of about 700 kV and they studied the first nuclear reactions with an accelerator. In their experiment a proton beam of about 400 keV kinetic energy was used to investigate the nuclear reactions $^7\text{Li} + p \rightarrow ^4\text{He} + ^4\text{He}$ and $^7\text{Li} + p \rightarrow ^7\text{Be} + n$.

The **Van de Graaff**[4] **generator** uses a motorized insulating belt (usually made of rubber) to conduct electric charges from a source on one end of the belt to the inside of a metal sphere on the other end. See Figure 3.1. Since electrical charge resides on the outside of the sphere, it accumulates to produce an electrical potential.

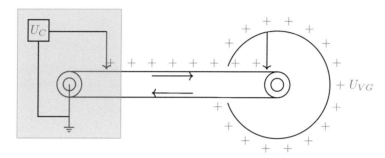

Figure 3.1 Illustration of a Van de Graaff generator.

Practical limitations restrict the potential to about seven million volts. **Tandem Van de Graaff** generators are essentially two generators in series, as illustrated in Figure 3.2, and can produce about 15 million volts. This voltage is then used to accelerate singly charged particles up to an energy of 15 MeV.

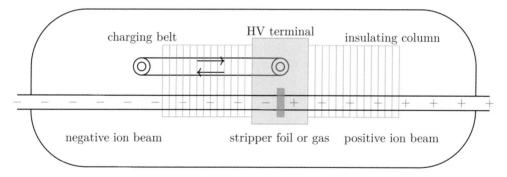

Figure 3.2 A tandem Van de Graaff accelerator.

Electrical breakdown rapidly becomes a fundamental limitation to high voltages. Another solution that avoids very high voltages is to use *time-dependent accelerating fields*. In 1924 **Gustaf Ising**[5] published an accelerator concept with voltage waves propagating from a spark discharge to an array of drift tubes [34]. Voltage pulses arriving sequentially at the drift tubes produce accelerating fields in the sequence of gaps. In 1927 **Rolf Wideröe**,[6] still a graduate student, discovered Ising's 1924 publication in the university library at Aachen. He simplified the concept by replacing the spark gap with an AC oscillator and the idea was experimentally demonstrated by him in 1928. Ising and Wideroe had established the principle of **acceleration by repeated application of time-varying fields**, or **resonance acceleration**: charged particles can gain arbitrarily high kinetic energy from successive traversals through the same accelerating fields with moderate voltages,

---

4   Robert Jemison Van de Graaff (1901–1967), American engineer, physicist.
5   Gustaf (or Gustav) Ising (1883–1960), Swedish accelerator physicist and geophysicist.
6   Rolf Wideröe (1902–1996), Norwegian accelerator physicist.

acquiring a small energy increment with each traversal. There is no "intrinsic" limit to the maximum attainable kinetic energy. The method can be applied to linear accelerators (linac), as shown in Figure 3.3, or to circular accelerators (cyclotron or synchrotron), as discussed in the following.

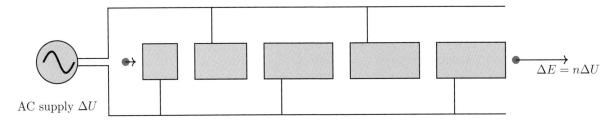

Figure 3.3 Acceleration by repeated application of time-varying fields: the Widerøe structure. The increase in energy of the structure is given by $\Delta E = n\Delta U$, where n is the number of cavities ($n = 5$ in the drawing).

In the **cyclotron** invented by **E. O. Lawrence**[7] in 1934, particles are injected into the center and accelerated with a variable potential while a magnetic field keeps them on spiral trajectories. Finally, particles are extracted. See Figure 3.4. The cyclotron is the first cyclic particle accelerator. It accelerates a beam of

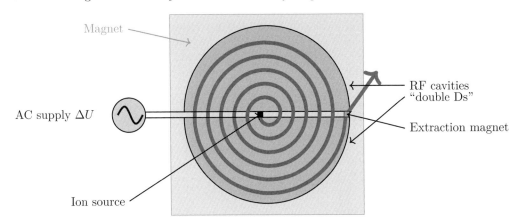

Figure 3.4 The concept of the cyclotron. The spiralling trajectory of a particle is shown as a thick line. The particle is eventually extracted with a magnet at the outer radius of the cyclotron.

charged particles using a high-frequency alternating voltage which is applied between two hollow "D"-shaped sheet metal electrodes. The "D"s are placed face to face with a narrow gap between them, creating a cylindrical space within them for the particles to move. The particles are injected into the center of this space. A large electromagnet applies a static magnetic field perpendicular to the electrode plane. The magnetic field causes the particle's path to bend due to the Lorentz force perpendicular to the direction of motion, with a radius of curvature R given by the **cyclotron condition**:

$$p = \gamma m v = eBR \tag{3.1}$$

This yields the **cyclotron frequency** $\omega = v/R$, where:

$$\omega = \frac{eB}{\gamma m} \tag{3.2}$$

7   Ernest Orlando Lawrence (1901–1958), American nuclear scientist.

As long as the particle is non-relativistic and $\gamma \simeq 1$, the accelerating voltage across the "D"s can have a constant frequency. A **synchrocyclotron** is a cyclotron in which the frequency of the driving radio-frequency (RF) electric field is varied to compensate for relativistic effects. An alternative is the **isochronous cyclotron**, which has a non-uniform magnetic field that increases with the radius.

The largest momentum of a cyclotron for a given type of particle is given by the strength of the magnetic field (or about 2 T for ferromagnetic magnets) and the radius of the "D"s. Very large magnets were constructed for cyclotrons, culminating in Lawrence's 1946 synchrocyclotron, which had pole pieces 4.67 m in diameter. Cyclotrons were the most powerful particle accelerator technology until the 1950s, when they were superseded by the synchrotron, as will be discussed in the next section. As of the time of writing, the largest cyclotron is the 17.1 m multi-magnet TRIUMF accelerator[8] at the University of British Columbia in Vancouver, British Columbia, which can produce 500 MeV protons. The Paul Scherrer Institute (PSI)[9] hosts the most powerful proton **sector cyclotron**, able to provide a continuous operation of 590 MeV protons at a current of 2 mA, corresponding to a beam power of approximately 1 MW! Among other applications, it is used to drive the Swiss Spallation Neutron Source (SINQ). Today, cyclotrons are mostly used worldwide in nuclear medicine for the production of radionuclides.

3.2   Storage Rings and Colliders

In the **synchrotron**, particles travel in a vacuum beam pipe in a closed loop. The trajectory is defined by a set of magnets (dipoles and higher orders for focusing), while high-frequency cavities are used to accelerate the particles. See Figure 3.5. Particles are "injected" into the vacuum pipe at low energy with the magnetic fields in the dipoles at their minimal values. Because the particles traverse acceleration cavities at every

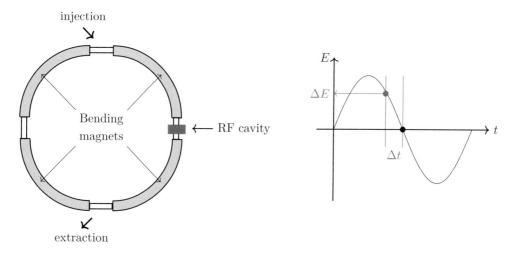

Figure 3.5 (left) Layout of a synchrotron. (right) Illustration of the longitudinal focusing.

turn, the momentum grows accordingly. But with low (MHz) frequencies available before World War II, linacs for high-energy protons and electrons had impractically large gap-to-gap spacings. The developments of the radar during World War II brought higher-frequency microwave sources. The development of pulsed RF power sources at hundreds of megahertz and with megawatts of power removed the technical barrier of having unreasonably long installations for electron and proton linacs, opening the path to the very high-energy storage rings. **L. W. Alvarez**[10] proposed enclosing the gaps between the tubes in a linear accelerator by metallic cavities. The acceleration section would now be composed of a series of tubes forming, together with the outer enclosure, a resonant cavity, and this is still the preferred structure today for the straight section of low-energy accelerators. In the circular ring, the beam has to be kept on the same radius. As the momentum of the particles increases, so does the magnetic field, in a process called "ramping up." The condition for a circular orbit is given by (cf. Eq. (3.1)):

$$p = eBR \quad \Longrightarrow \quad p(\text{GeV/c}) = 0.3\,B(\text{T})\,R(\text{m}) \tag{3.3}$$

for a momentum in giga-electronvolts, a magnetic field in Tesla, and the radius of curvature in meters. Conventional bending magnets reach maximum operation fields of $\simeq 1.5$ T. Higher fields are obtained with superconducting magnets, such as the Fermilab (FNAL)[11] Tevatron that reached 5 T and the CERN[12] LHC whose dipole magnets are designed to reach 10 T. Once the maximal energy is reached, the magnetic fields are kept constant, and the cavities are either switched off, or used to compensate for losses around each turn (see below). In this case the synchrotron is called **a storage ring** in which the energy of the particles is kept constant.

• **Focusing.** A crucially important aspect of a storage ring (also an accelerator) is the type of focusing implemented to keep the particles within the ring. As an illustration, Figure 3.5(right) shows the concept of longitudinal RF focusing. The oscillating field in the RF cavity is adjusted so as to act as a focusing system, such that stored particles which arrive late will receive a bigger kick than the early one and vice versa. Real accelerators also include transverse focusing systems that ensure that particles are always restored towards the ideal trajectory.

• **Energy loss per turn.** A stored charged particle loses energy by **synchrotron radiation** because it is subjected to an acceleration. In a circular ring, the power radiated is proportional to γ^4 and is given by [17]:

$$P_{rad} \simeq \frac{2e^2c\gamma^4}{3R^2} \tag{3.4}$$

Hence, at an equivalent energy, electrons will radiate much more than protons. The radiated energy per turn is:

$$E_{turn} = \frac{4\pi}{3}\frac{e^2\gamma^4}{R} \tag{3.5}$$

The energy radiated, E_{turn}, is lost during each turn and the particles must be re-accelerated by the same amount for the beam to remain at a constant energy. For electron colliders, synchrotron radiation becomes a limiting factor. For example, the CERN LEP collider has a radius $R = 4.3$ km, hence for 100 GeV electrons, we find $\gamma \simeq 2 \times 10^5$ and the loss per turn is $E_{turn} \simeq 2.24$ GeV. In comparison, for the LHC in the same LEP tunnel, protons with an energy of 7 TeV have $\gamma \simeq 7 \times 10^3$ and hence $E_{turn} \simeq 3$ keV, which is negligible.

3.3   High-Energy Colliders

A by-product of the storage ring is the **collider**, in which two beams are accelerated and circulate in opposite directions and are brought into head-on collisions in specific points around the ring. See Figure 3.6. In the case of e^+e^- or $p\bar{p}$ colliders, the particles can circulate in opposite directions within the *same beam pipe*. In a *pp* collider, such as the CERN LHC, the two beams circulate in two separate beam pipes within the same dipole magnet. Most modern high-energy accelerators are composed of a series of linear accelerators and rings in order

10   Luis Walter Alvarez (1911–1988), American experimental physicist and inventor.
11   https://fnal.gov
12   www.cern.ch

to reach the desired final energies. For instance, the CERN accelerator complex is illustrated in Figure 3.7. During LEP time, four experiments – ALEPH, DELPHI, L3, and OPAL – were located in four interaction points. Nowadays the ATLAS, CMS, LHCb, and ALICE experiments are located at the interaction points around the LHC ring to study protons–protons (or ions–ions) collisions. The protons are first accelerated in a linac, then passed to a booster PSB, then sequentially go through the PS, the SPS, and are finally injected as two beams circulating in opposite directions in the LHC ring. Each piece of accelerator is designed to accept and accelerate particles up to a given energy. Extracting the beam from one accelerator and injecting it into the next one is one of the most challenging aspects of an accelerator complex, as it often introduces beam losses.

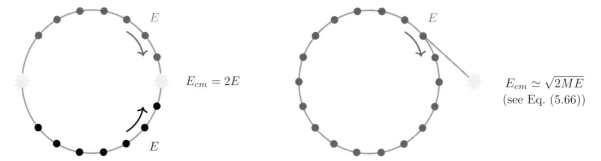

Figure 3.6 Concept of a collider (left), compared to a fixed target configuration (right).

3.4   Lepton Colliders

The electron–positron and electron–electron colliders or lepton colliders that were built are shown in Table 3.1. The first major accelerator facility at the Stanford Linear Accelerator Center (SLAC)[13] was the 2-mile-long linear accelerator, completed in 1966. See Figure 3.8. Beginning operation in 1972, the **SPEAR storage ring** accumulated the high-energy electrons and positrons and allowed them to collide with up to a record center-of-mass energy of 7.4 GeV. A second large electron–positron colliding beam facility called **PEP** began operating in 1980 at center-of-mass energies up to 29 GeV. PEP could host five experiments in different collision points around the ring.

    In Germany, the construction of the first particle accelerator DESY (Deutsches Elektronen Synchrotron, "German Electron Synchrotron")[14] began in 1960. At that time, it was the biggest facility of this kind and was able to accelerate electrons to 7.4 GeV. DORIS (Doppel-Ring-Speicher, "double-ring storage"), built between 1969 and 1974, was DESY's second circular accelerator and its first storage ring with a circumference of nearly 300 m. Constructed as an electron–positron storage ring, one could conduct collision experiments with electrons and their antiparticles at energies of 3.5 GeV per beam. In 1978 the energy of the beams was raised to 5 GeV each. The **PETRA (Positron–Elektron-Tandem-Ring-Anlage, "positron–electron tandem-ring facility")** was built between 1975 and 1978. PETRA covered a large center-of-mass energy region from 12 to 46.8 GeV, thereby extending the energy scale of previous QED tests from SPEAR (SLAC) and DORIS (DESY) by almost an order of magnitude. Much of the data were taken in a scanning mode, where the energy was varied in small step sizes of 20 or 30 MeV.

    Immediately after the PETRA shut down at the end of 1986, the TRISTAN storage ring at KEK, Japan, started its physics program at energies between 50 and 60 GeV. The High Energy Accelerator Research

13   http://slac.stanford.edu
14   www.desy.de

Accelerator	Where	Years	Size	Energy (GeV)		Experiments
		e^-e^+ **accelerators**		e^-	e^+	
AdA	Italy/France	1961–1964	3 m	0.25	0.25	
VEPP-2	USSR	1965–1974	11.5 m	0.7	0.7	OLYA, CMD
ACO	France	1965–1975	22 m	0.55	0.55	
SPEAR	SLAC	1972–1990		3	3	Mark I, Mark II, Mark III
VEPP-2M	USSR	1974–2000	18 m	0.7	0.7	ND, SND, CMD-2
DORIS	DESY	1974–1993	300 m	5	5	ARGUS, Crystal Ball, DASP, PLUTO
PETRA	DESY	1978–1986	2 km	20	20	JADE, MARK-J, CELLO, PLUTO, TASSO
TRISTAN	KEK	1986–1995	700 m	30	30	VENUS, SHIP, TOPAZ, AMY
CESR	Cornell	1979–2002	768 m	6	6	CUSB, CHESS, CLEO, CLEO-2, CLEO-2.5, CLEO-3
PEP	SLAC	1980–1990	700 m	15	15	ASP, DELCO, HRS, MARK II, MAC
SLC	SLAC	1988–1998	linac	45	45	SLD, Mark II
LEP	CERN	1989–2000	27 km	104	104	ALEPH, DELPHI, OPAL, L3
BEPC	China	1989–2004	240 m	2.2	2.2	Beijing Spectrometer (I and II)
VEPP-4M	BINP	**1994-**	366 m	6.0	6.0	KEDR
PEP-II	SLAC	1998–2008	2.2 km	9	3.1	BaBar
KEKB	KEK	**1999-**	3 km	8.0	3.5	Belle
DAΦNE	Italy	**1999-**	98 m	0.7	0.7	KLOE
CESR-c	Cornell	2002–2008	768 m	6	6	CHESS, CLEO-c
VEPP-2000	BINP	**2006-**	24.4 m	1.0	1.0	SND, CMD-3
BEPC II	China	**2008-**	240 m	3.7	3.7	Beijing Spectrometer III
ILC	Japan	?	31 km	250	250	*under study*
CEPC	China	?	50 km	120	120	*under study*
FCC-ee	CERN	?	100 km	180	180	*under study*
CLIC	CERN	?	50 km	1500	1500	*under study*
		e^-e^- **accelerators**		e^-	e^-	
Princeton-Stanford	USA	1962–1967	12 m	0.3	0.3	
VEP-1	USSR	1964–1968	2.7 m	0.13	0.13	
		$\mu^-\mu^+$ **collider**		e^-	e^+	
FCC-$\mu\mu$	CERN	?	?	?	?	*under study*

Table 3.1 List of lepton colliders: BINP = Budker Institute of Nuclear Physics, Novosibirsk, Russia. CERN = European Organization for Nuclear Research, Geneva, Switzerland. DESY = Deutsches Elektronen-Synchrotron, Hamburg, Germany. KEK = Japanese High Energy Accelerator Research Organization. SLAC = SLAC National Accelerator Laboratory, originally named Stanford Linear Accelerator Center, California, USA.

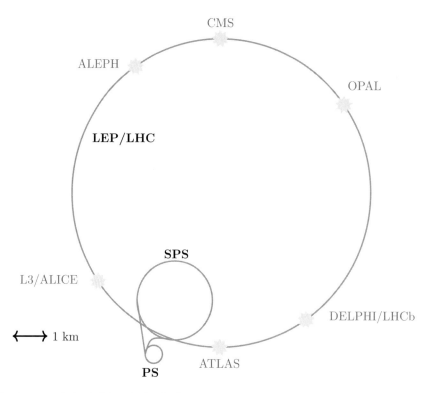

Figure 3.7 The layout of the LEP collider [35] and the current CERN accelerators complex to operate the LHC (PS, SPS, and LHC drawn). During LEP time, four experiments – ALEPH, DELPHI, L3, and OPAL – were located in four interaction points. Currently, the ATLAS, CMS, LHCb, and ALICE experiments are located at the interaction points around the LHC ring.

Organization (KEK – "K Enerug Kasokuki Kenky Kik") is a Japanese organization situated in Tsukuba, Ibaraki prefecture, whose purpose is to operate the largest particle physics laboratory in Japan.[15] Beginning in 1984, the **Transposable Ring Intersecting Storage Accelerator in Nippon (TRISTAN) accumulation ring (AR)** accelerated an electron beam to 6.5 GeV. In 1985 TRISTAN main ring (MR) accelerated both electron and positron beams to 25.5 GeV and then upgraded to 30 GeV with the help of superconducting accelerating cavities. TRISTAN had three detectors: TOPAZ, VENUS, and AMY. Their main purpose was to detect the top quark. TOPAZ was the first experiment ever to confirm vacuum polarization around an electron [36]. By the end of 1988 a maximum energy of 60 GeV was reached and the experiments AMY, TOPAZ, and VENUS have reported results based on an integrated luminosity of about 20 pb^{-1} for each experiment. In 1995 TRISTAN experiments finished and paved the way in 1998 for the first beam at the storage at KEKB (KEK B-factory), ring optimized to provide very high luminosities to produce and study B mesons. KEKB has two rings for accelerating electrons and positrons. The ring for electrons, supporting an energy of 8 GeV, is called the high-energy ring (HER), while the ring for positrons, tuned for an energy of 3.5 GeV, is called the low-energy ring (LER). The HER and LER are constructed side-by-side in the tunnel, which had been excavated for the TRISTAN accelerator.

• **The LEP and SLC colliders.** With the observation of the neutral currents in neutrino interactions (see Section 25.2) in 1973 and the discovery of the W and Z bosons in $p\bar{p}$ collisions in 1983 (see Section 27.4),

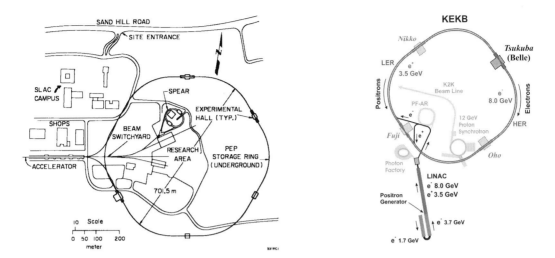

Figure 3.8 (left) The Stanford Linear Accelerator Center (SLAC) areal view. Reprinted figure from J. R. Rees, "The PEP electron-positron ring – PEP stage I*," in *LBL-SLAC Joint Study Group, Berkeley and Stanford, California*; (right) KEK electron–positron accelerators. Reprinted under license CC0 1.0 Universal https://creativecommons.org/publicdomain/zero/1.0/deed.en from https://commons.wikimedia.org/w/index.php?curid=27905978.

the key features of the model of electroweak interactions (see Chapter 25) became established experimentally. Already in 1976, the report from a study group was published by CERN in which the physics potential of a new, very high-energy e^+e^- collider was advocated and approved as the major new infrastructure at CERN. Even after the discovery of the W and Z bosons, the LEP project was continued because it offered the possibility of **precision measurements**. The **Large Electron–Positron Collider (LEP)** is one of the largest particle accelerators ever constructed. Located on the French/Swiss border, it was a circular collider with a circumference of 27 km built in a tunnel roughly 100 m underground. It was built under the leadership of **Picasso**.[16] The collider layout included eight straight sections, with collisions between electron and positron bunches allowed to take place in four of them. The four interaction regions were each instrumented with a multi-purpose detector: L3, ALEPH, OPAL, and DELPHI, as indicated in Figure 3.7. The first collisions were observed in the summer of 1989. LEP was operated in two distinct phases. It was first used from 1989 until 2000 and focused on center-of-mass energies around the Z^0 pole (LEP-I, 90 GeV) and then its center-of-mass energy was increased above the W^+W^- production threshold (LEP-II, 200 GeV). The main bending field was provided by 3280 concrete-loaded dipole magnets, with over 1300 quadrupoles and sextupoles for focusing and correcting the beams in the arcs and in the straight sections. For LEP-I running, the typical energy loss per turn of 125 MeV was compensated by 128 accelerating radio-frequency copper cavities. Synchrotron radiation became a limiting factor to reach energies beyond those of LEP-II (see Eq. (3.5)). Originally four bunches of electrons and four bunches of positrons circulated in the ring, leading to a collision rate of 45 kHz. The dimensions of the bunches were $\sigma_x = 200$ μm, $\sigma_x = 15$ μm, and $\sigma_z = 1.08$ cm. The LEP luminosity was increased in later years by using eight equally spaced bunches, or alternatively four trains of bunches with a spacing of order a hundred meters between bunches in a train. The achieved peak luminosity (see Section 3.7) was 2×10^{31} cm^{-2} s^{-1} above its design luminosity. Under these conditions, each LEP experiment could record approximately 1000 Z^0 decays per hour! The LEP was consequently dubbed a Z^0 **factory**. Much effort was dedicated to the determination of the energy of the colliding beams. A precision of about 2 MeV in the center-

16   Emilio Picasso (1927–2014), Italian physicist.

of-mass energy was achieved, corresponding to a relative uncertainty of about 2×10^{-5} on the absolute energy scale. Details of the energy calibration of LEP can be found in Ref. [37]. This level of accuracy was vital for the precision of the measurements of the mass and width of the Z^0, as described in Chapter 27. To date, LEP is the most powerful accelerator of leptons ever built. It collided electrons with positrons at energies that reached 209 GeV. Around 2001 the LEP accelerator was dismantled to make way for the LHC, which re-used the tunnel.

At SLAC the linac also underwent major upgrades to finally reach 50 GeV in the 1980s. Starting in 1983, extensive modifications and additions to the existing linac enabled it to simultaneously accelerate and collide 50 GeV electron and positron bunches on a pulse-to-pulse basis. This new machine, the **SLAC Linear Collider (SLC)**, separated the beams at the end of the accelerator, guiding them through two opposing magnetic arcs and focusing them on each other, to micron-size spots at just the correct center-of-mass collision energy to produce the Z^0 particle with polarized beams. The SLC linear collider complex is shown in Figure 3.9. The

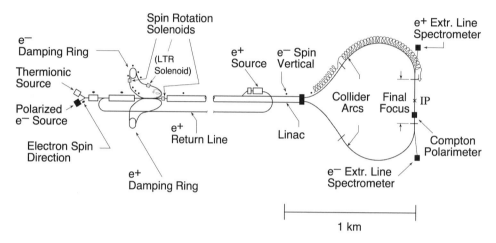

Figure 3.9 The SLC linear collider complex, showing the electron source, the damping rings, the positron source, the 3 km-long linac and arcs, and the final focus. Reprinted figure from M. Woods, "The polarized electron beam for the SLAC linear collider," in *12th International Symposium on High-energy Spin Physics (SPIN 96)*, pp. 623–627, vol. 10, 1996.

electron source, the damping rings, the positron source, the 3 km-long linac and arcs, and the final focus are visible. The helix and arrow superimposed on the upper arc schematically indicate the electron spin precession which occurs during transport. The actual operation was very complex. Two bunches of electrons were initially produced and accelerated. At 30 GeV, the second bunch was diverted to a target, where positrons were created. The positrons were captured, accelerated to 200 MeV, and sent back to the beginning of the linac, where they were then stored in the positron damping ring. The positron bunch was then extracted just before the next two electron bunches, and accelerated. The remaining positron and electron bunches were accelerated to the final energy of ≈ 46.5 GeV and then transported in the arcs to the final focus and interaction point. Approximately 1 GeV was lost in the arcs due to synchrotron radiation, so the center-of-mass energy of the collisions was at the peak of the Z^0 resonance. **The electron spins were manipulated during transport in the arcs, so that the electrons arrived at the interaction point with longitudinal polarization**. The era of high-precision measurements at SLAC started in 1992 with the first longitudinally polarized beams. An electron beam polarization of close to 80% was achieved for the experiment at luminosities up to 8×10^{29} cm^{-2} s^{-1} [38]. To avoid as much as possible any systematic effects, the electron helicity was randomly changed on a pulse-to-pulse basis by changing the circular polarization of the laser. Collisions of polarized electrons with unpolarized positrons allow precise measurements of parity violation in the electroweak interaction and provide a very precise measurement of the weak mixing angle, as will be discussed in Chapter 27.

• **Present and future lepton colliders.** At present, only four lepton colliders are still in operation, in Russia (**VEPP-4M**, **VEPP-2000**[17]), Italy (**DAΦNE**[18]), Japan (**KEKB**), and China (**BEPC II**[19]). New lepton colliders are under study, as listed in Table 3.1 with tentative parameters. Most notably the **International Linear Collider** (ILC) in Japan (see Figure 3.10), the CEPC in China and the FCC-ee at CERN would provide the next-generation electron–positron colliders at center-of-mass energies of a few hundred giga-electronvolts, with the primary goal of precisely studying the properties of the Higgs boson discovered at the LHC (see Chapter 29). In addition, the Compact Linear Collider (CLIC) at CERN and possibly a **muon collider** would allow us to reach multi-tera-electronvolt center-of-mass energies, to continue the exploration of the high-energy frontier.

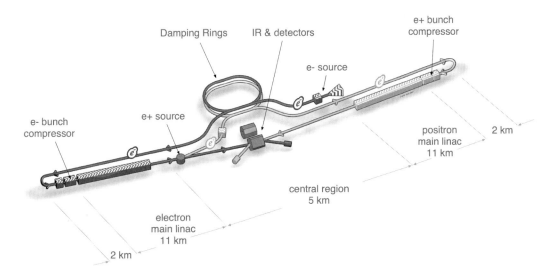

Figure 3.10 Schematic view of the ILC collider (not to scale). Reprinted figure from *"The International Linear Collider Technical Design Report - Volume 1: Executive Summary,"* ILC-REPORT-2013-040 [arXiv:physics.acc-ph/1306.6327 [physics.acc-ph]].

3.5   Hadron Colliders

The list of hadron colliders is shown in Table 3.2. Currently two hadron colliders are still operating worldwide: the **Relativistic Heavy Ion Collider** (RHIC) at BNL[20] and the **Large Hadron Collider** (LHC) at CERN.

The LHC [39] is the world's newest and most powerful accelerator at the energy frontier. It is designed to collide proton beams with a center-of-mass energy of 14 TeV and an unprecedented luminosity of 10^{34} cm^{-2} s^{-1} (see Section 3.7 for the definition of the luminosity). It can also collide heavy (Pb) ions with an energy of 2.8 TeV per nucleon and a peak luminosity of 10^{27} cm^{-2} s^{-1}. The decision to build LHC at CERN was strongly influenced by the cost savings to be made by re-using the LEP tunnel and its injection chain. The tunnel is located 100 m underground and has a circumference of about 27 km (18 miles). The LHC ring is

17   http://v4.inp.nsk.su/
18   www.lnf.infn.it/acceleratori/
19   http://english.ihep.cas.cn
20   www.bnl.gov/rhic/

linked to the CERN accelerators complex, making it possible to re-use the chain up to the SPS as injectors. Hence, the protons have an energy of 450 GeV with a clean phase space as they are transferred into the LHC ring. The approval of the LHC project was given by the CERN Council in December 1994. In 2000, LEP was closed to liberate the tunnel for the LHC.

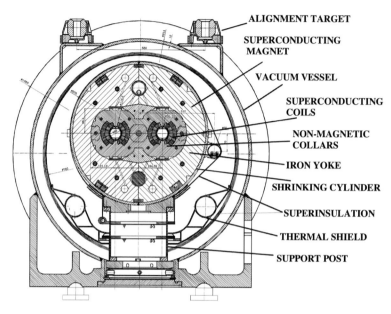

Figure 3.11 Cross-section of the LHC dipole magnets. Reprinted from O. S. Brüning, P. Collier, P. Lebrun, S. Myers, R. Ostojic, J. Poole, and P. Proudlock, *LHC Design Report*. CERN Yellow Reports: Monographs, Geneva: CERN, 2004, under CC BY license https://creativecommons.org/licenses/by/3.0/.

The high beam intensity necessary for the required LHC luminosity of 10^{34} cm^{-2} s^{-1} excludes the use of antiproton beams, since it is too hard to produce these latter in sufficient quantity. Consequently, the particle–antiparticle collider configuration of a common vacuum and magnet system for both circulating beams is excluded. In the LHC accelerator, there are effectively two rings with counter-rotating beams with separate vacuum beam pipes, however within a single cryogenic magnet element. Figure 3.11 shows the cross-section of an LHC dipole magnet. The two visible independent beam pipes are horizontally separated and each surrounded by their own superconducting coils, however with a unique iron-yoke. Each magnet weighs 35 tons, is 15 m long, and the coil current is 11,000 A.

The tunnel geometry was originally designed for the electron–positron machine LEP, and there were eight crossing points flanked by long straight sections for RF cavities that compensated the high synchrotron radiation losses (see Eq. (3.5)). Compared to an electron–positron collider such as LEP, a proton machine does not have the same synchrotron radiation problem and would, ideally, have longer arcs and shorter straight sections for the same circumference, but accepting the existing LEP tunnel was the cost-effective solution. The LHC ring is therefore designed as a proton–proton collider with separate magnet fields and vacuum chambers in the main arcs and with common sections only at the intersection regions where the experimental detectors are located and beams are made to collide. Together with the large number of bunches (2808 for each proton beam), and a nominal bunch spacing of 25 ns, the long common beam pipe implies 34 parasitic collision points at each experimental region. Dedicated crossing-angle orbit bumps separate the two LHC beams left and right from the interaction points, in order to avoid collisions at these parasitic positions. As mentioned above, the LHC accelerator relies on superconducting magnets to reach the required high fields. The peak beam energy depends on the integrated dipole field around the storage ring, which implies a peak dipole field of 8.33 T for 7 TeV proton beams. The main magnets are dipole magnets complemented by several different types of higher-order correcting magnets. The LHC magnets are based on NbTi Rutherford cables. The conductor is cooled to a temperature below 2 K, using superfluid helium, in order to operate at fields above 8 T. A total beam current of 0.5 A corresponds to a stored energy of approximately 360 MJ. In addition, the LHC magnet system has a stored electromagnetic energy of approximately 600 MJ. The beam dumping system and the magnet system must therefore cope with about 1 GJ! The cryogenic system deployed over the circumference of 27 km is also the largest in the world.

New hadron colliders are also under study, for when the LHC reaches the end of its lifespan. The Future Circular Collider (FCC-hh), with a circumference of 100 km, would collide protons with a center-of-mass energy up to 100 TeV. A schematic view showing where the FCC tunnel at CERN is proposed to be located, is shown in Figure 3.12.

Collider	Where	Years	Size (km)	Type	Beam energy	Experiments
ISR	CERN	1971–1984	0.948	$p+p$	31.5 GeV	
Sp$\bar{p}$S	CERN	1981–1984	6.9	$p+\bar{p}$	270–315 GeV	UA1, UA2
Tevatron-I	FNAL	1992–1995	6.3	$p+\bar{p}$	900 GeV	CDF, D0
Tevatron-II	FNAL	2001–2011	6.3	$p+\bar{p}$	980 GeV	CDF, D0
RHIC	BNL	**2001-**	3.8	p	100–255 GeV	PHENIX, STAR
RHIC	BNL	**2001-**	3.8	ions	3.85–100 GeV per nucleon	STAR, PHENIX, BRAHMS, PHOBOS
LHC	CERN	**2008-**	27	$p+p$	7 TeV	ALICE, ATLAS, CMS, LHCb, LHCf, TOTEM
LHC	CERN	**2008-**	27	ions	2.76 TeV per nucleon	ALICE, ATLAS, CMS
FCC-hh	CERN	?	100	$p+p$	50 TeV	*under study*

Table 3.2 List of hadron and ion colliders. BNL = Brookhaven National Laboratory, New York, USA. CERN = European Organization for Nuclear Research, Geneva, Switzerland. FNAL = Fermilab, Illinois, USA.

3.6   Lepton–Hadron Colliders

Only one electron/positron–proton collider was built and operated from 1991 until 2007 at DESY, Hamburg, Germany. Its parameters are summarized in Table 3.3. The Hadron–Electron Ring Accelerator (HERA) electron–proton collider consisted of a ring with 3 km circumference, where electrons and positrons could be accelerated up to 27.5 GeV. The same tunnel hosted a second ring which could accelerate protons up to 820 GeV. Two large experiments called H1 and ZEUS were located at opposite sides of the ring, where the electrons/positrons were made to collide head-on with the protons. See Figure 3.13. A new Large Hadron Electron Collider (LHeC) collider, which would collide electrons/positrons from a new accelerator with protons accelerated from the existing LHC, is under study.

ep-collider	Where	Years	Size (m)	e (GeV)	p (GeV)	Experiments
HERA	DESY	1991–2007	6300	27.5	900	H1, ZEUS, HERMES, HERA-B
LHeC	CERN	?	?	60	7000	*under study*

Table 3.3 List of electron/positron–proton collider(s). DESY = Deutsches Elektronen-Synchrotron, Hamburg, Germany.

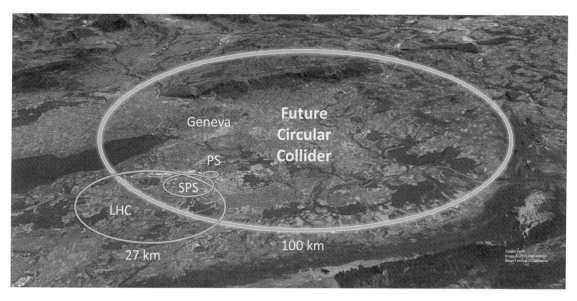

Figure 3.12 A schematic view showing where the FCC tunnel at CERN is proposed to be located. The PS, SPS, and LHC rings are also shown. Reprinted figure by permission from CERN. Copyright 2021 CERN.

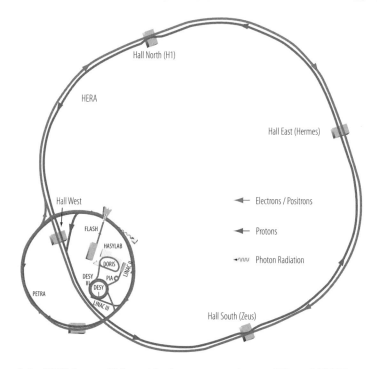

Figure 3.13 Layout of the HERA *ep* collider with the two experiments H1 and ZEUS located in the North and South halls. Reprinted with permission of Springer from P. Schmüser and F. Willeke, "The electron-proton collider HERA." Landolt-Börnstein – Group I Elementary Particles, Nuclei and Atoms 21C (Accelerators and Colliders). Copyright 2013 Springer-Verlag Berlin Heidelberg; permission conveyed through Copyright Clearance Center, Inc.

3.7   The Luminosity of a Collider

In a **collider**, two beams of particles collide head-on. In the LEP collider at CERN, electron–positron collisions were studied. Because particles and antiparticles have opposite electric charges, they can be accelerated in the same ring. Only collisions of stable particles such as ions, proton–antiproton, electron–positron, or proton–proton can be studied. The interaction rate R in a collider is defined as the product of the cross-section with the collider **luminosity** $\mathcal{L}$ (for a review, see Ref. [40]):

$$R \equiv \mathcal{L}\sigma \tag{3.6}$$

If σ is the total cross-section, then the rate R corresponds to the total number of interactions per unit time. If σ is the cross-section of a particular process, then R is the rate of that process. The luminosity contains all the operating parameters of the collider, while the cross-section represents the physics of the interaction. The cross-section has units of area. Hence, the units of the luminosity are those of the inverse area per unit time. The luminosity can be estimated from the physical properties of the colliding beams and the operating parameters of the accelerator, as defined in Figure 3.14 (see, e.g., Ref [40]):

$$\mathcal{L} = \frac{fnN_1N_2}{4\pi\sigma_x\sigma_y} \tag{3.7}$$

where f is the repetition frequency, n is the number of bunches of particles in the ring, N_1, N_2 are the number of particles per bunch for beams 1 and 2, and σ_x, σ_y are the transverse dimensions of the bunches at the collision point (under the assumption that both beams have particles with a Gauss-like distribution in space). Compare with the fixed target situation of Eq. (2.66) and Figure 2.11. As an example, we consider

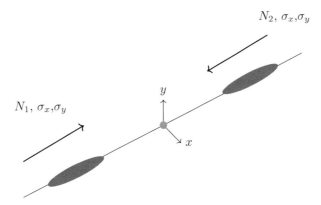

Figure 3.14 Schematic view of colliding bunches with transverse dimensions σ_x and σ_y. N_1, N_2 are the number of particles per bunch.

$N_1 = N_2 = 10^{10}$ particles per bunch, in total $n = 10$ bunches per beam in the ring, a frequency of $f = 1$ MHz, and a beam size of $\simeq 10^{-2}$ cm^2. This gives a luminosity of

$$\mathcal{L} = \frac{10^6 \times 10 \times 10^{20}\,\mathrm{s}^{-1}}{10^{-2}\,\mathrm{cm}^2} \simeq 10^{29}\,\mathrm{cm}^{-2}\,\mathrm{s}^{-1} \tag{3.8}$$

For the *total inelastic cross-section for proton–proton collisions* σ_{pp}, which depends on the total energy of the collider, we have:

$$\sigma_{pp} = \begin{cases} 30\ \mathrm{mb\ for} \sqrt{s} = 10\ \mathrm{GeV} \\ 60\ \mathrm{mb\ for} \sqrt{s} = 7000\ \mathrm{GeV} \end{cases} \tag{3.9}$$

Recalling that **1 millibarn** is

$$1\,\text{mb} = 10^{-3} \times 10^{-24}\,\text{cm}^2 = 10^{-27}\,\text{cm}^2 \tag{3.10}$$

the corresponding **total event rate** at 10 GeV with $\mathcal{L} \simeq 10^{29}\,\text{cm}^{-2}\,\text{s}^{-1}$ is then:

$$R_{pp} = \mathcal{L}\sigma_{pp} = 30 \times 10^{-27}\,\text{cm}^2 \times 10^{29}\,\text{cm}^{-2}\,\text{s}^{-1} \simeq 3000\,\text{Hz} \tag{3.11}$$

and at 7000 GeV for the same luminosity it becomes $R_{pp} \simeq 6$ kHz. Some real examples of colliders and their typical luminosities are listed in Table 3.4. One may notice the very different interaction rate, in particular between hadron colliders and high-energy lepton colliders. This is due to the small total cross-section of e^+e^- interactions compared to the total hadron–hadron cross-section. Furthermore, in the case of PEP and KEKB, the accelerators operate near or on resonances, and the interaction rate varies very strongly with the precise energy, so they are not included in the table. As an example, the luminosity of the CERN LHC collider is in the range of 10^{34} cm^{-2} s^{-1}. The total proton–proton cross-section at a center-of-mass energy of 13 TeV is $\sigma \approx 0.1$ barn $= 10^{-25}$ cm^2. This leads to an interaction rate of $R = 10^9$ per second. The maximum luminosity,

Collider	Energy (GeV)	Typ. $\mathcal{L}$ (cm^{-2} s^{-1})	Rate (s^{-1})	σ_x/σ_y (μm/μm)	Particles per bunch
S$p\bar{p}$S	$315 + 315$	6×10^{30}	4×10^5	60/30	10^{11}
Tevatron $p\bar{p}$	$1000 + 1000$	5×10^{31}	4×10^6	30/30	5×10^{10}
LHC pp	$7000 + 7000$	10^{34}	10^9	17/17	10^{11}
LEP e^+e^-	$105 + 105$	10^{32}	< 1	200/2	5×10^{11}
PEP e^+e^-	$9 + 3$	3×10^{33}	–	150/5	10^{10}
KEKB e^+e^-	$8 + 3.5$	10^{34}	–	77/2	10^{10}

Table 3.4 Compilation of luminosities and machine parameters of recent colliders (the numbers of particles per bunch are indicative).

and therefore the instantaneous number of interactions per second, is very important, but the final figure of merit is the so-called **integrated luminosity**:

$$\mathcal{L}_{integrated} = \int_0^T \mathcal{L}(t)\mathrm{d}t \tag{3.12}$$

Since the units of luminosity are inverse area per unit time, the integrated luminosity has a unit of inverse area, which can be expressed for instance in units of picobarn-inverse (pb^{-1}) or femtobarn-inverse (fb^{-1}). Then, for example, an experiment will observe on average one event of a process with cross-section $\sigma = 1$ fb for each collected integrated luminosity of 1 fb^{-1}.

• **The LHC luminosity.** As a concrete current example, we illustrate the LHC performance during the first 4 years of operation at a center-of-mass energy of $\sqrt{s} = 13$ TeV. Table 3.5 summarizes some of the achieved parameters. The increase in luminosity over the years is truly impressive. The peak luminosity exceeds the design luminosity of 10^{34} cm^{-2} s^{-1}. The integrated luminosity delivered by the LHC to the ATLAS experiment over this period of time amounts to:

$$\text{LHC}: \quad \mathcal{L}_{integrated}(\sqrt{s} = 13\,\text{TeV}) = 156\,\text{fb}^{-1} \tag{3.13}$$

This value corresponds to the raw value provided by the colliders. Not all of it is used for physics analyses. Experiments apply some selections to ensure data of the highest quality. For instance, the integrated cumulative luminosity versus day delivered by the CERN LHC collider to both ATLAS and CMS experiments for high-energy pp collisions is shown in Figure 3.15. These plots show that the experiments indeed collected the collisions provided by the LHC with very high efficiency.

Parameter	2015	2016	2017	2018
Number of colliding bunch pairs	2232	2208	2544/1909	2544
Bunch spacing (ns)	25	25	25	25
Bunch population (10^{11} protons)	1.1	1.1	1.1/1.2	1.2
Peak luminosity (10^{34} cm^{-2} s^{-1})	0.5	1.3	1.6	1.9
Average number of inelastic interactions per crossing	16	41	45/60	55
Total delivered luminosity (fb^{-1})	4.0	38.5	50.2	63.4

Table 3.5 The LHC performance during the first 4 years of operation at a center-of-mass energy of $\sqrt{s} = 13$ TeV. Reprinted from "Luminosity determination in pp collisions at $\sqrt{s} = 13$ TeV using the ATLAS detector at the LHC," Tech. Rep. ATLAS-CONF-2019-021, CERN, Geneva, Jun 2019, under CC BY license http://creativecommons.org/licenses/by/4.0/. Copyright 2019 ATLAS Collaboration.

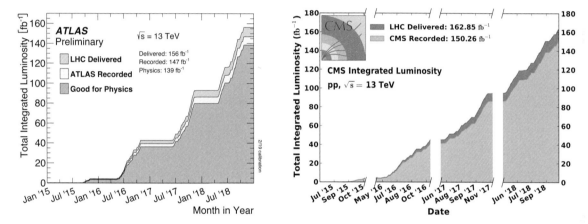

Figure 3.15 The integrated luminosity versus time delivered by the CERN LHC to the ATLAS and CMS experiments for pp collisions. Reprinted with permission from ATLAS and CMS. Copyright 2019 CERN.

- **Factories.** A "factory" is a collider that is specially designed to produce a large number of given particles, generally so that their properties and behavior can be measured with high statistical accuracy. One speaks for example of an experiment at a B-factory – or a beauty factory – to describe a setup of an accelerator coupled to a detector to produce and detect a large number of B mesons (see Chapter 20). Taus and D mesons are also copiously produced at B-factories. The LEP collider is often referred to as a Z^0-factory. Two B-factories were designed and built in the 1990s, and they operated from late 1999 onwards. They are both based on electron–positron colliders with the center-of-mass energy tuned to the $\Upsilon(4S)$ resonance peak, which is just above the threshold for decay into two B mesons (see Section 20.9). The Belle experiment at the KEKB collider in Tsukuba, Japan, and the BaBar experiment at the PEP-II collider at SLAC laboratory in California, USA, completed data collection in 2010 and 2008, respectively. In this context, it is worth noting the **Large Hadron Collider beauty experiment (LHCb)**[21] at the LHC, which studies primarily the physics of the bottom quarks that are produced in proton–proton collisions. It thus could be understood to be a B-factory although the LHC is not used solely for the study of b-quark particles.

21 http://lhcb-public.web.cern.ch

3.8   Ionization Chambers and Geiger Counters

The first electrical devices developed to detect radiation were ionization chambers. In these devices, the passage of particles induces ionization in a gas, which is then detected. A gas is the obvious medium because excited electrons and ions can easily drift over macroscopic distances under the influence of an external electric field. In the regime of interest, one can write a linear relation for the **drift velocity** $v_D = \mu E$, where μ is the mobility and E the external electric field. The electric signal on the electrodes is formed by induction of the moving ions and electrons as they respectively drift towards the cathode and the anode.

In the ionization chamber, the electron–ion pairs produced in the gas where an electric field is present, directly produce a current that is measured by an instrument in an external circuit. Although the current is relatively large, it cannot be generally detected without an external amplifier. Such electronic devices did not exist at the time Rutherford and Geiger made their experiments. Internal amplification in the gas was first introduced by them. In this way, they could achieve **the detection of a single (α) particle**! This relies on the fact that the electric field between a thin central wire and an outer cylinder is inhomogeneous, as it rises steeply near the central wire:

$$E(r) = \frac{CV}{2\pi\epsilon}\frac{1}{r} \quad \text{with} \quad C = \frac{2\pi\epsilon}{\ln(b/a)}; \quad a \leq r \leq b \tag{3.14}$$

where C is the capacitance per unit length, V is the applied voltage, ϵ is the dielectric constant of the gas, r is the radial distance from the inner wire, and a and b are the inner and outer radii.

Multiplication in a gas detector occurs when the primary charge gains sufficient energy from the external electric field to further ionize the medium. This leads to an avalanche, as illustrated in Figure 3.16. Because

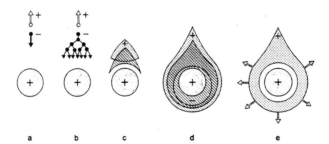

Figure 3.16 Time development of an avalanche in a proportional counter. A single primary electron drifts towards the anode, in regions of increasingly higher fields, starting an avalanche. Due to diffusion, the avalanche has the shape of a drop. The entire process is fast and typically takes a new nanoseconds. Reprinted from F. Sauli, "Principles of operation of multiwire proportional and drift chambers." Copyright CERN 1975–1976, under CC BY license https://creativecommons.org/licenses/by/3.0/.

of the greater mobility of electrons, the avalanche forms a sort of "liquid-drop" shape with electrons grouped near the head and the slower ions trailing behind moving in the opposite direction. The gain is defined by the **first Townsend[22] coefficient** $\alpha(E)$ such that a number $\mathrm{d}n$ of charge carriers in a field E is created by n charges in a path $\mathrm{d}x$:

$$\mathrm{d}n = n\alpha(E)\mathrm{d}x \quad \Longrightarrow \quad n(x) = n_0 e^{\alpha x} \tag{3.15}$$

and the multiplication factor $M = n(x)/n_0 = \exp(\alpha x)$ for constant E. In the case of a non-uniform electric field, the multiplication is expressed as $M = \exp(\int \alpha(x)\mathrm{d}x)$. The units of α are those of an inverse length and $1/\alpha$ is the mean free path for a secondary ionizing collision.

22   John Sealy Edward Townsend (1868–1957), Irish mathematical physicist.

In the cylindrical geometry, electrons drift towards the thin anode wire and, in its vicinity, start creating new electron–ion pairs. Ultraviolet light, which is also produced in the avalanche, can come into play when it reaches the cathode and extracts additional electrons from its surface. These form new avalanches such that the discharge continues even after the original particle has created the primary signal. This autonomous discharge then needs to be "quenched" by some technique. In the **Geiger–Müller[23] counter** the inner wire was covered by a very high-resistivity material such that its charging up would produce a counter-acting field that reduces the original electric field in the medium. On the other hand, if the voltage is kept below a critical value, then the total number of secondary electron–ion pairs is proportional to the primary charge and one speaks in this case of a proportional counter.

3.9   The Wilson Cloud Chamber

Wilson[24] began to make clouds in a **cloud chamber** around the year 1895. A cloud consists of small water droplets. Whether water mixed in air forms clouds depends on its concentration, on temperature, and on pressure. By changing these parameters, vapor can be changed into liquid or vice versa. Vapor is saturated when a tiny change of parameters is sufficient to change its state. Condensation takes place at given points called condensation nuclei. In a super-saturated state, droplets are formed and grow starting from these condensation nuclei. In his first tests, Wilson enclosed saturated vapor in a small chamber and suddenly lowered the pressure by allowing the gas mixture to expand in a larger volume, creating clouds. The cleaner the chamber, the higher the achievable saturation.

After the discovery of X-rays (see Section 2.4), Wilson exposed his chamber to the new rays and found the formation of additional fog. Ten years later Wilson described *"a method of making visible the tracks of ionising particles through a moist gas by condensing water upon the ions immediately after their liberation"* [41]. See Figure 3.17. The number of electron–ion pairs produced per unit length along the track and consequently

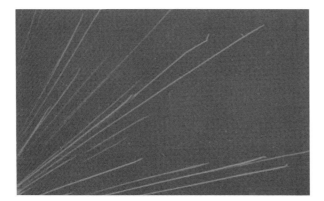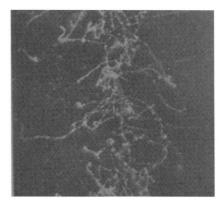

Figure 3.17 Photograph taken by Wilson of tracks produced by (left) α particles; (right) secondary electrons when the chamber is exposed to X-rays. Republished with permission of The Royal Society (UK) from C. T. R. Wilson, "On an expansion apparatus for making visible the tracks of ionising particles in gases and some results obtained by its use," *Proc. Roy. Soc. Lond.*, vol. A87, pp. 277–292, 1912; permission conveyed through Copyright Clearance Center, Inc.

the density of bubbles formed in the cloud chamber, is proportional to the amount of energy released in the

23   Walther Müller (1905–1979), German physicist.
24   Charles Thomson Rees Wilson (1869–1959), Scottish physicist and meteorologist.

medium (see Section 2.17 and Eq. (2.142)). It increases with the charge and is larger at low velocity. Hence, α particles were easier to detect but it was shown that β particles could also be detected.

What is exceptional is the ability to see the track, hence the trajectory, of a single particle and to characterize it. For instance, α and β particles are distinguishable. A slow-moving β particle that ionizes similarly to an α will be much slower and therefore be much more deflected by the heavy atomic nuclei. Such a path is clearly identifiable and different from the path of an α, as illustrated in Figure 3.17(left) and (right).

This Wilson chamber became an extremely valuable instrument to discover or study properties of such particles. If the Wilson chamber is embedded in a magnetic field, the radius of curvature of the curved trajectories allows the determination of the charge and the momentum. Such a setup was for instance used to discover the positron (see Section 8.14) and strange particles (see Section 15.7). In general, interactions with complicated multi-track events can be studied with great accuracy.

In the related **diffusion cloud chamber**, the expansion is replaced by a sudden cooling of the temperature of the medium. Instead of water vapor, alcohol is used because of its lower freezing point. Due to the temperature gradient, a layer of supersaturated vapor is formed above the cold plate. In this region, radiation particles induce condensation and cloud tracks can be readily seen. Such a chamber is continuously sensitive to radiation.

3.10   The Photomultiplier Tube

A photomultiplier tube (PMT) converts light into a measurable electrical signal. They are very sensitive and capable of detecting single photons. Their use has been quite varied in high-energy physics and also in industry. The principle is shown in Figure 3.18. An incident photon hits the photocathode and produces a photoelectron via the photoelectric effect. The probability of producing this is called the quantum efficiency of the PMT. A series of plates called dynodes are held at high voltage by a voltage divider circuit called the base, such that electrons are accelerated from one dynode to the next. At each stage the number of electrons increases due to the collision with the subsequent dynode. The probability of collecting the first photoelectron on the first dynode is called the collection efficiency. After amplification through a series of dynodes, the output signal is seen as a charge pulse on the anode. Typical gains are in the range of 10^6, yielding pC charges. The typical transit time is about 100 ns via a jitter in the range of 1–10 ns. The photocathode is the most delicate part of the PMT. Its performance depends on the material used and on its fabrication. Typical photocathodes are based on bialkali materials. An example of a quantum efficiency curve is shown in Figure 3.19. It has a maximum quantum peak efficiency of about 20% at a wavelength of 370 nm. At the highest frequencies (lower wavelengths) there is a cutoff which is due to the glass window of the PMT. Increased efficiency in the UV can be achieved by replacing the glass with specially made windows.

3.11   Scintillators

Scintillation detectors are most often and widely used to detect particles. They are composed of certain materials that emit light (scintillation) when struck by particles or radiation. This small flash of light is then coupled to an amplifying device such as a photomultiplier to convert light into an electric pulse. The phenomenon is called luminescence. When exposed to certain forms of excitation, luminescent material absorbs and re-emits energy in the form of light, typically visible light. If the re-emission is fast (less than 10 ns) then the process is called fluorescence. If the excited state is metastable, the re-emission can be delayed, and the process is called phosphorescence. Such processes can last up to microseconds or even hours, depending on the material. There are different types of scintillating materials, usually categorized in six main classes: (a) organic crystals, (b) organic liquids, (c) plastics, (d) inorganic crystals, (e) gases, and (f) glasses. The organic scintillators are aromatic hydrocarbon compounds containing linked or condensed benzene-ring structures.

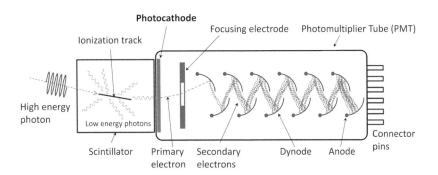

Figure 3.18 Illustration of the concept of the photomultiplier tube. From Qwerty123uiop, CC BY-SA 3.0 <https://creativecommons.org/licenses/by-sa/3.0>, via Wikimedia Commons. https://commons.wikimedia.org/wiki/File:PhotoMultiplierTubeAndScintillator.svg.

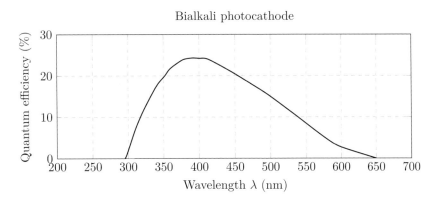

Figure 3.19 Typical PMT quantum efficiency curve as a function of the wavelength. Data taken from www.hamamatsu.com.

Their most distinguishing feature is a very rapid decay time on the order of a few tens of nanoseconds or less. The most common organic crystal is anthracene ($C_{14}H_{10}$), with a decay time of approximately 30 ns. Organic liquids are liquid solutions of one or more organic scintillators in an organic solvent. Like the organic liquids, plastic scintillators are also solutions of organic scintillators but in a solid plastic solvent. The most common and widely used plastic is polystyrene. The inorganic scintillators are mainly crystals, containing a small activator impurity. The most commonly used materials are NaI(Tl) or CsI(Tl), where thallium is the impurity activator. Non-alkali materials can also be used, such as bismuth germanium oxide (BGO) ($Bi_4Ge_3O_{12}$) or lead tungstate ($PbWO_4$). See Table 3.6 for a collection of commonly used inorganic crystals.

Gaseous scintillators rely mostly on noble gases: xenon, krypton, argon, and helium, but nitrogen is also possible. In these scintillators the atoms are individually excited and return to their ground states very rapidly, typically less than 1 ns. Most of the emitted light is generally in the ultraviolet regime, and therefore one method to overcome this is to coat the surface of the detector or the photodetectors themselves with wavelength-shifting molecules such as TPB (tetra-phenyl-butadiene) or POPOP (1,4-bis(5-phenyloxazol-2-yl)benzene) such that the UV light is down-converted to blue, which can be directly detected by the photocathodes of PMTs. Certain glasses can also emit scintillation, and such light is often emitted in conjunction with Cherenkov emission (see Section 2.20).

Parameter units	ρ (g/cm)^3	MP (°C)	X_0 (cm)	R_M (cm)	dE/dx (MeV/cm)	λ_I (cm)	τ_{decay} (ns)	λ_{max} (nm)	n
NaI(Tl)	3.67	651	2.59	4.13	4.8	42.9	245	410	1.85
BGO	7.13	1050	1.12	2.23	9.0	22.8	300	480	2.15
BaF$_2$	4.89	1280	2.03	3.10	6.5	30.7	650	300	1.50
CsI(Tl)	4.51	621	1.86	3.57	5.6	39.3	1 220	550	1.79
CsI(Na)	4.51	621	1.86	3.57	5.6	39.3	690	420	1.84
CsI(pure)	4.51	621	1.86	3.57	5.6	39.3	30	310	1.95
PbWO$_4$	8.30	1123	0.89	2.00	10.1	20.7	30	425	2.20

Table 3.6 Properties of several inorganic crystals. *Source:* Values taken from the Particle Data Group's Review [5].

3.12   Calorimeters (Infinite Mass Detectors)

An ideal calorimeter is a fully sensitive detector with an infinite mass. An incoming particle will interact in the medium and will be fully converted. The energy and position of the incoming particle are measured by **total absorption**.

- **Electromagnetic cascade ("EM-shower").** We consider an incoming $e^\pm$ or γ of high energy. When such a particle enters the calorimeter, it will produce an electromagnetic cascade. Electrons and positrons will emit photons via Bremsstrahlung (see Section 2.19) and photons will convert into electron–positron pairs via pair creation (see Section 2.16). Naively speaking, one can consider that on average half of the energy will be shared among the secondary particles at each step of the electromagnetic shower, hence the number of secondary particles grows as 2^n, while their individual energy is approximately given by $E_0/2^n$, as schematically illustrated in Figure 3.20.

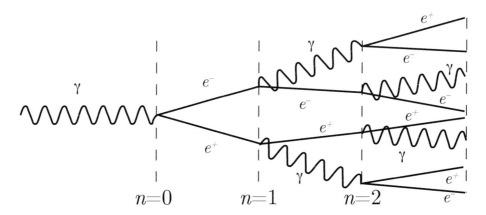

Figure 3.20 Sketch of the development of an electromagnetic shower. At every step, one can assume that the energy is split such that at the n step the energy of the particles is $E_0/2^n$. On the other hand, the number of particles grows as 2^n.

The electromagnetic shower grows until the energy of the particles is smaller than the critical energy E_c, defined in Section 2.19. Once all secondary particles have an energy $E_0/2^n \leq E_c$, no further particles will be created and the existing particles will lose their energy via collisions until they stop. The development of the electromagnetic cascade is a stochastic process. Consequently, there will be fluctuations. These are most practically studied by tracking all particles with Monte-Carlo techniques, as discussed in Section 2.22. An actual simulation of a 50 GeV electron (incoming track at the left) crossing 19 slabs of iron, each 2 cm thick, performed with GEANT4 (see Section 2.22), is shown in Figure 3.21. The formation of an electromagnetic shower is visible. The short tracks correspond to electrons and positrons. Photons are not shown for clarity. Approximately, the total energy of the electron is converted within 20 slabs. We recall that the radiation length of iron (X_0(Fe)) is 1.76 cm, hence each slab crossed transversally corresponds to $1.14 X_0$. In total, 20 slabs amount to $\approx 23 X_0$. The shower profile is distinguishable in the figure.

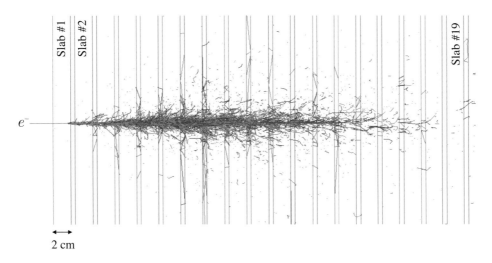

Figure 3.21 Electromagnetic shower: GEANT4 simulation of a 50 GeV electron (incoming track at the left) crossing 19 slabs of iron, each 2 cm thick. The formation of an electromagnetic shower is visible. The short tracks correspond to electrons and positrons. Photons are not drawn for clarity.

• **Longitudinal profile of the electromagnetic shower.** Experimental data confirmed by Monte-Carlo simulations show that the average longitudinal profile of the cascade is determined by incoming particle energy E_0 and the number of radiation lengths X_0 of the calorimeter. In order to describe the profile, one defines the normalized length in units of radiation lengths, i.e., $t \equiv x/X_0$. Longitudinally the number of particles grows exponentially as 2^n until the maximum is reached. The maximum is on average at a depth $t_{max} = x_{max}/X_0$ given by:

$$t_{max} = \ln\left(\frac{E_0}{E_c}\right) + \begin{cases} -0.5 \text{ for electrons/positrons} \\ +0.5 \text{ for photons} \end{cases} \tag{3.16}$$

The average profile can be expressed as [28]:

$$\frac{\mathrm{d}E}{\mathrm{d}t} = E_0 b \frac{(bt)^{a-1} e^{-bt}}{\Gamma(a)} \quad \text{where} \quad a = b t_{max} + 1, \quad b \approx 0.5 \tag{3.17}$$

(b actually depends on the target material and slightly on the energy of the incident particle – see, e.g., Particle Data Group's Review [5] for details).

• **Transverse profile of the electromagnetic shower.** Approximately 90% of the energy of the cascade is deposited within a transverse radius defined by the **Molière radius** R_M; 99% of the energy is deposited within

$3.5R_M$. The Molière radius depends on the properties of the target material as tabulated in Table 2.1 and can be approximated as:

$$R \approx X_0 \frac{21 \text{ MeV}}{E_c} \tag{3.18}$$

• **Other properties of the electromagnetic shower.** We list in the following other approximate properties of the electromagnetic shower:

1. As previously defined, the maximum of the shower occurs by a depth $t = t_{max}$. This depth grows as the logarithm of the incoming particle energy $t_{max} \propto \ln E_0$.

2. At the maximum, the number of secondary particles is approximately proportional to E_0.

3. The total track length is defined as the total path distance traversed by all charged particles. It is proportional to the incoming particle's energy E_0.

4. The fluctuations of the deposited energy are proportional to the fluctuations of the number N of secondary particles produced. As a consequence of the Central Limit Theorem, the fluctuation of the number of particles is proportional to $\sqrt{N}$. Hence, the energy resolution of a calorimeter is roughly proportional to the inverse of the square root of the energy $\sigma_{E_0}/E_0 \propto 1/\sqrt{E_0}$.

• **Sampling calorimeters.** A **sampling calorimeter** consists of a sandwich of an active scintillator packed between a preferably high Z converting material. The concept of sandwich means that this construction can be repeated by adding layers of absorber–scintillator–absorber–scintillator, and so on. The scintillator, which has a density and a thickness smaller than the absorber, detects the particles that are produced in the converter. It accordingly samples the longitudinal development of the electromagnetic shower. The amount of energy measured in a sampling calorimeter can therefore be expressed as:

$$E_{measured} = \int \epsilon \left(\frac{\mathrm{d}E}{\mathrm{d}x} \right) \mathrm{d}x \tag{3.19}$$

where ϵ is the sampling factor (or efficiency) given by:

$$\epsilon \equiv \frac{\left(\frac{\mathrm{d}E}{\mathrm{d}x} \right)_{scint} t_{scint}}{\left(\frac{\mathrm{d}E}{\mathrm{d}x} \right)_{conv} t_{conv} + \left(\frac{\mathrm{d}E}{\mathrm{d}x} \right)_{scint} t_{scint}} \tag{3.20}$$

where t_{conv} and t_{scint} are the corresponding thicknesses in units of radiation length X_0. In a sampling calorimeter, only a fraction of the developing shower will be measured, and hence, the fluctuations will in general be larger than in a fully active calorimeter which measures the entire shower. Therefore, the energy deposited in the active scintillator can fluctuate significantly, in particular if the efficiency is small. Typical electromagnetic sampling calorimeters have energy resolutions of:[25]

$$\text{Electromagnetic calorimeter :} \quad \frac{\sigma_{E_0}}{E_0} \simeq \frac{5\text{–}10\%}{\sqrt{E(\text{GeV})}} \tag{3.21}$$

• **Fully sensitive converters.** There are specific materials with unique properties for building calorimeters. For instance, scintillating crystals such as NaI, CsI, BaF$_2$, or BGO are dense and have high Z, and are in addition transparent to their scintillation. The properties of some of these crystals are given in Table 2.1. One can cite BGO with a density of $\rho_{BGO} = 7.1$ g/cm^3 and a short radiation length of $X_0 = 1.12$ cm. The crystal emits scintillation light when ionizing particles traverse it. Consequently, the total amount of light produced in the crystal is proportional to the energy deposited in the crystal. With several such crystals, one can build large, fully active and segmented calorimeters with excellent energy and spatial resolution properties.

---

25 We note that in a totally active, infinite mass electromagnetic calorimeter _all_ the energy of the incoming particle is converted into ionization, hence, the ideal intrinsic resolution is better than $1\%/\sqrt{E(\text{GeV})}$, driven by very rare photonuclear processes.

- **Hadronic cascade and calorimeters.** A hadronic shower develops when a hadron enters the calorimeter. The hadronic shower is initiated by an **inelastic nuclear collision**, in which several secondary hadrons are generally produced. The majority of the secondary particles are charged pions or neutral pions. They are forward boosted. The secondary charged pions can produce further inelastic collisions, contributing further to the hadronic cascade development. The neutral pion decays very quickly to two photons, which will initiate secondary electromagnetic cascades. The principle is very analogous to that of an electromagnetic shower, however, the development of the hadronic cascade is driven by the nuclear interaction length λ_I, instead of the radiation length X_0. An example of a hadronic shower produced by a 50 GeV positive pion entering the same iron calorimeter configuration as in Figure 3.21, is shown in Figure 3.22. The interaction length of several

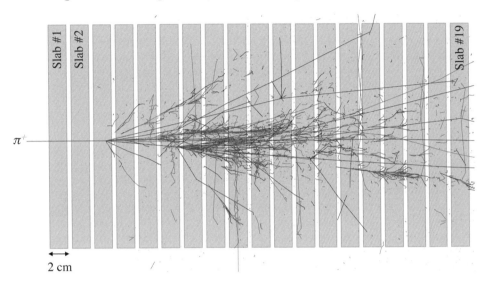

Figure 3.22 Hadronic shower: GEANT4 simulation of a 50 GeV positive pion (incoming track at the left) crossing 19 slabs of iron, each 2 cm thick. The formation of a hadronic cascade is visible. Photons are not drawn for clarity.

materials is tabulated in Table 2.1. We note that:

$$\lambda_I \gg X_0 \tag{3.22}$$

For example in the case of iron, we have $\lambda_I^{Fe} \approx 16$ cm and $X_0^{Fe} \approx 1.76$ cm. Hence, the dimensions of a hadronic shower are larger than that of an electromagnetic one. More (or a denser) absorber is needed to contain a hadronic shower of a given energy, compared to an electromagnetic one of the same energy. Shower fluctuations in hadronic cascades are generally larger than in purely electromagnetic showers. In addition, the process involves nuclear interactions by nature, and these lead to energy losses or gains. Therefore, the typical resolution of hadronic calorimeters is worse than in electromagnetic calorimeters, typically:

$$\text{Hadronic calorimeter}: \quad \frac{\sigma_{E_{had}}}{E_{had}} \simeq \frac{50\text{–}100\%}{\sqrt{E_{had}(\text{GeV})}} \tag{3.23}$$

We have seen that $20\text{–}25X_0$ are needed to fully absorb an electromagnetic shower. In the case of a hadronic shower, a depth of $7\text{–}10\lambda_I$ is generally used. Hence, hadronic calorimeters are always larger than electromagnetic ones.

3.13 Trackers (Massless Detectors)

An ideal tracker is a massless detector that renders the trajectory of a particle without affecting this latter's identity nor its kinematical properties, as opposed to the infinitely massive calorimeter that entirely converts the incoming particle. In practice, one uses a gas as a medium, as in the ionization chamber (see Section 3.8). Inelastic collisions with the atomic electrons of the gas can lead to ionization or excitation of an atom A. For example, for an incident charged pion π:

$$\pi^\pm + A \to \pi^\pm + A^+ + e^-, \qquad \pi^\pm + A \to \pi^\pm + A^* \tag{3.24}$$

The energy necessary to ionize the atom and separate an electron from an atom is called the **ionization energy**. The excitation to atomic levels requires a correct amount of energy to be transferred and is therefore a resonant process. The above reactions are of stochastic nature and two identical particles will not in general produce the same number of electron–ion pairs due to statistical fluctuations. But one can look at the average values. The average energy lost by the incident particle to create an electron–ion pair in the medium is defined as W_i. It is not equal to the ionization energy, since some energy is also expended to excite the atoms. For gaseous argon one measures $W_i = 25$ eV, hence, on average, 40,000 free electrons will be liberated when an incoming particle loses 1 MeV in the medium. For argon gas at STP conditions, this leads to an average of 100 electron–ion pairs produced per centimeter of track. For comparison, in xenon one has on average 340 electron–ion pairs per centimeter, in methane (CH_4) on average 60 electron–ion pairs per centimeter, in ethane (C_2H_6) on average 115 electron–ion pairs per centimeter, and in CO_2 on average 110 electron–ion pairs per centimeter. All these provide good media for tracking detectors and in practice one uses mixtures to improve the stability and performance of the detector. In all cases, the passage of a charged particle leaves a "trace" of electron–ion pairs along its trajectory, which, with the help of appropriate methods, can be used to reconstruct the trajectory of the incoming particle in the medium. Sometimes the collisions can be so hard that the freed electron can be considered as a secondary particle ("hard knock-on electrons"). Such electrons create their own "tracks" in the detector. These are called δ-rays.

• **Multiwire proportional chamber.** A significant breakthrough occurred in 1968 when **Charpak**[26] invented the multiwire proportional chamber (MWPC). Starting from the principle of the proportional counter described in Section 3.8, Charpak showed that an array of many closely spaced anode wires in the same chamber could each act as independent proportional counters. Each wire would have its own electronic amplifier integrated into the chamber to make a practical detector for position sensing. The basic MWPC consists of a plane of equally spaced anode wires centered between two cathode planes, as shown in Figure 3.23. Typically, wires are placed a few millimeters apart, while the anode–cathode gap width is about 1 cm. When a negative voltage is applied to the cathode planes, the electric field is essentially that of a capacitor, with the field lines parallel and constant, except in the regions very close to the anode wires. Indeed, it can be shown that for zero-diameter anode wires, the electric potential is given by (see, e.g., Ref. [42]):

$$V(x, y) = -\frac{CV}{4\pi\varepsilon_0} \ln\left[4\left(\sin^2\frac{\pi x}{s} + \sinh^2\frac{\pi y}{s}\right)\right] \tag{3.25}$$

where V is the applied voltage, s the wire spacing, and C the anode–cathode capacitance. If $\ell \gg s \gg d$, then the capacitance is given by:

$$C = \frac{2\pi\varepsilon_0}{\frac{\pi\ell}{s} - \ln\frac{\pi d}{s}} \tag{3.26}$$

where ℓ is the anode-to-cathode gap distance and d is the anode wire diameter. The electric field has a $1/r$ dependence near the anode, similar to that of the single wire cylindrical proportional chamber. The chamber is filled with appropriate gas. When an incident particle traverses the detectors, electrons and ions are liberated along its path. In the constant-field region, electrons will drift along the field lines towards the nearest anode

26   Georges Charpak (1924–2010), Polish-born French physicist.

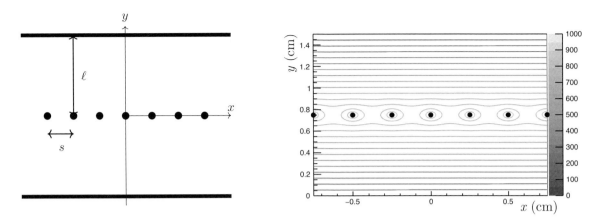

Figure 3.23 (left) Geometry of the multiwire proportional chamber (MWPC). (right) Equipotential lines near the anode wires computed with Garfield++ (http://garfieldpp.web.cern.ch/garfieldpp/).

wire. The ions will instead drift towards the opposing cathode. Upon reaching the high-field region, the electrons will be quickly accelerated to produce an avalanche. The positive ions liberated in the multiplication process then induce a negative signal on the anode wire. The induced signals are amplified and read out with appropriate electronics. The signal of one layer of MWPC gives information on one coordinate of the track trajectory. However, it is possible to stack several MWPC planes oriented in different directions, to then reconstruct offline a two-dimensional position. Several layers are added in practice to avoid ambiguities.

• **The drift chamber.** If an external timestamp (i.e., trigger) is available, then spatial information about a track can be obtained by measuring the **drift time** of the electrons until they reach the anode. One tries to make the drift field as uniform as possible, in order to obtain a linear relationship between position and time, given simply by $x = v_d t$, where v_d is the drift velocity. Such a linear electric field can, for instance, be created by a set of cathode field wires held at appropriate voltages along the drift space. The external trigger can be provided by a thin scintillator, as shown in Figure 3.24.

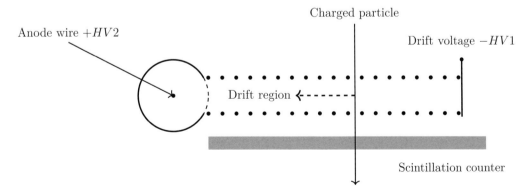

Figure 3.24 Principle of operation of a single-cell drift chamber. Reprinted from F. Sauli, "Principles of operation of multiwire proportional and drift chambers." Copyright CERN 1975–1976, under CC BY license https://creativecommons.org/licenses/by/3.0/.

3.14   Cherenkov Detectors

Cherenkov light is used to detect and identify relativistic particles in so-called **Cherenkov counters**. Such detectors provide accurate determination of the particle velocity threshold. In a Cherenkov counter light is emitted only by particles for which $\beta \geq 1/n$. The velocity parameter $\beta = v/c$ as a function of the momentum for pions, kaons, and protons is plotted in Figure 3.25. If the momentum of the particle is known, then the presence or absence of light signal can be used to identify the type of particle. For example, for a momentum of 1.19 GeV/c, the velocity of protons is $0.78c$, while that of pions is $0.99c$. When the particles traverse a medium with refractive index of $n \simeq 1.26$, corresponding to $1/n \simeq 0.79$, then the presence of Cherenkov light will indicate that the particle is a pion. It should be noted that this is a very powerful way to identify particles in the low-momentum range. Above $\simeq 5$ GeV, most particles become relativistic, as is visible in the figure.

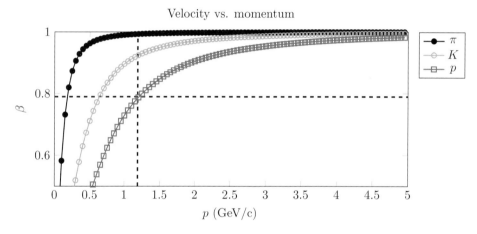

Figure 3.25 The velocity parameter $\beta = v/c$ as a function of momentum for pions, kaons, and protons. The dashed lines correspond to the experimental setup described in Section 15.9.

In **water Cherenkov detectors**, Cherenkov radiation is used to reconstruct an image of the interactions in the medium, for example induced by neutrinos. The refractive index of water as a function of the light wavelength is plotted in Figure 3.26. In the range of sensitivity of typical photomultipliers, i.e., $200 \leq \lambda \leq 600$ nm (see Section 3.10), it has an average value of $n_{H2O} \simeq 1.34$. This corresponds to:

$$\text{Water:}\quad \beta > 1/n_{H2O} \simeq 0.75, \qquad \theta \simeq 42° \tag{3.27}$$

This threshold can be converted into an energy threshold that depends on the type of particle. Using the Lorentz factor (see Chapter 5 and Appendix D), we find $\gamma = 1/\sqrt{1-\beta^2} \simeq 1.5$. Hence, the energy Cherenkov threshold in water for a given particle of mass m corresponds to $E_{threshold} = \gamma mc^2 \simeq 1.5\ mc^2$. These thresholds are illustrated in Figure 3.27 for most common particles. Below the thresholds, particles cannot be observed but become visible as soon as their velocity is above threshold. Integrating Eq. (2.152) over a range of wavelengths $\lambda_1 \leq \lambda \leq \lambda_2$, one obtains simply the number of photons per unit length of the incoming relativistic particle:

$$\frac{\mathrm{d}^2 N_\gamma}{\mathrm{d}x} = 2\pi\alpha Z^2 \sin^2\theta \left(\frac{1}{\lambda_1} - \frac{1}{\lambda_2}\right) \tag{3.28}$$

For the range $300 \leq \lambda \leq 600$ nm in water and for $Z = 1$, this yields about 340 Cherenkov photons per centimeter of track length, which can be detected by standard techniques (we note that pure water is very transparent to such visible light, so photodetectors can be placed at distances relatively far away from the source!).

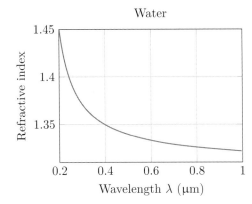

Figure 3.26 Refractive index of water $n_{H2O}(\lambda)$ as a function of the wavelength λ in the range $200 \leq \lambda \leq 1000$ nm. Data taken from Segelstein, D., 1981 M.S. Thesis, University of Missouri–Kansas City.

Particle	Energy threshold (MeV/c^2)
Electrons	0.770
Muons	158
Charged pions	210
Charged kaons	741
Protons	1407

Figure 3.27 Cherenkov thresholds in water.

3.15   Semiconductor Detectors

Semiconductor or solid-state detectors are based on crystalline semiconductor materials, most notably silicon or germanium. The basic operating principle is similar to that of gaseous ionization detectors. Instead of the gas, however, the medium is a solid semiconductor. The passage of ionizing particles liberates electron–hole pairs, while in the gas electron–ion pairs were formed along the particle trajectory. The main advantage of the semiconductor is that the average energy required to create an electron–hole pair is much smaller than in a gas. For instance, it is 3.6 eV for silicon and 2.9 eV for germanium, to be compared to values almost 10 times higher in gases (see Section 3.13). In addition, because of their higher density compared to gases, they have a greater stopping power. They can be manufactured using the constantly developing semiconductor industry. Very high energy and resolution, compact detectors with very fast response time can be built. Years of operation in high-radiation environments have taught the community methods to reduce their sensitivity to radiation damage, which limits their long-term use in high-fluence environments.

• **Pure semiconductors.** Let us first consider pure semiconductors, in which ideally there are no impurities. Solid-state materials exhibit energy band structure. In metals, the valence and conduction bands overlap. In an insulator, there is a "forbidden" gap between the valence and conduction bands. This gap is relatively large, i.e., 6 eV or so. The same structure is present in a semiconductor, but the gap is much smaller, for example it is 1.12 eV in silicon and 0.66 eV in germanium. The energy gap is such that only a few electrons are excited into the conduction band by thermal energy.

Electron–hole pairs are constantly being generated by thermal excitation, which can then be recombined from other electrons. The typical concentrations of free electrons n_i are of the order of 1.5×10^{10}/cm^3 in silicon at $T = 300$ K. This is to be contrasted with the number of atoms, $\approx 10^{22}$/cm^3! So even at room temperature the relative concentration of free electrons is very small.

The excitation of an electron from the valence band to the conduction band leaves a "hole" behind, which can easily be filled by another electron, which itself leaves another hole behind. When an electric field is applied, the observed current through a semiconductor will strongly depend on its temperature, increasing with temperature. The current density is given by the movement of the charge carriers. Electrons drift under the influence of the external electric field. So do holes. This hole movement represents an electric current with positive charge carriers. The total electric current therefore arises from two sources: the movement of electrons in the conduction band and that of holes in the valence band. The relationship between the electric field and the drift velocity is given as before by the mobility. In the case of the semiconductor, one has two independent mobilities, one for the electron and one for the holes, i.e.

$$v_e = \mu_e E \quad \text{and} \quad v_h = \mu_h E \tag{3.29}$$

where μ_e and μ_h are the mobilities of the electrons and holes, respectively. They depend on the actual electric field applied and also on the temperature. For silicon and fields less than 1 kV/cm, the mobilities are constant and the drift velocities are therefore proportional to the applied field. At high electric fields, a saturation effect is reached, and the drift velocity becomes constant with a value of about 10^7 cm/s.

The total current density in the semiconductor is given by:

$$J = en_i(\mu_e + \mu_h)E \tag{3.30}$$

where n_i is the concentration of free electrons. Since $J \equiv \sigma E$, where σ is the conductivity defined as the inverse of the resistivity ρ, one finds that:

$$\rho = \frac{1}{en_i(\mu_e + \mu_h)} \tag{3.31}$$

An electron and a hole might recombine with the emission of a photon. This process is called recombination and is the opposite of electron–hole pair creation. However, due to energy and momentum conservation, the process is very rare and the lifetime of free electrons should theoretically be very long if recombination in pure semiconductors was the only process. In reality, impurities in the crystal create recombination "centers" by adding additional levels in the gap region. These states may capture free electrons, which then recombine with a hole. The lifetime of free carriers in a real semiconductor is therefore a key parameter of the quality of the material. Typical lifetimes at the level of microseconds or more are needed in order to adopt these materials in real-life detectors. A second effect that arises from impurities is trapping. Some impurities are only capable of trapping one type of charge carrier, i.e., either electrons or holes. Such trapping centers simply hold the charge carrier for a given amount of time, after which it is released. If the trapping time is large compared to the charge collection time of the detector, then this trapped charged will seem to be lost. So, the amount of trapping center also defines the quality of the semiconductor. Finally, structural defects may also give additional states in the forbidden gap. These, for example, can be caused by vacancies in the crystal lattice or dislocations. These may occur during the growth of the crystal or created by high doses of radiation. So far, the effects of the impurities described above are detrimental to the operation of the materials as detectors.

• **Doped semiconductors.** The addition of certain elements to a pure semiconductor actually enhances its performance and these are called "doped." In a pure semiconductor, the number of free electrons equals the number of holes. This balance can be changed by doping with atoms having one more or one less valence electron in their outer atomic shell. For example, for silicon which has four valence electrons, it can be doped by "donor" elements of group V such as phosphorous (P), antimony (Sb), or arsenic (As), which have five valence electrons, or by "acceptor" elements of group III such as boron (B), indium (In), gallium (Ga), or aluminum (Al), which have three valence electrons.

The difference between the recombination and trapping impurities and the doping impurities is in the depth of the energy levels created in the energy gap. Doping creates shallow levels very close to either the valence or conduction band. Electrons or holes in those levels are therefore easily excited into the conduction or valence bands, respectively, where they can enhance the conductivity of the semiconductor. In addition, "additional" carriers of one type will fill the other type. For example, in the case of donor elements, the extra electrons also fill up holes, therefore decreasing the hole concentration. Holes will still exist and contribute to the current, however, only as "minority carriers." Doped semiconductors in which the electrons are the majority carriers are called **n-type semiconductors**. Similarly, in the case of doping with acceptor elements, there will be an excess of holes which will decrease the number of free electrons, hence holes will be the majority carriers. Accordingly, doped semiconductors in which holes are the majority carriers are called **p-type semiconductors**. The amount of dopant used is typically in the range of 10^{13} atoms/cm^3, so also here the amount of impurities is very small, yet their effect is very noticeable and dominates the electrical properties. For instance, the resistivity of the doped semiconductors is determined by the donor or acceptor impurity concentrations:

$$\rho \approx \frac{1}{eN_D\mu_e} \qquad \text{n-type} \tag{3.32}$$

and similarly for p-type.

• **Heavily doped semiconductors.** The electrical contact for semiconductors is usually made of heavily doped semiconductors. Impurity concentration in these locations can be as high as 10^{20} atoms/cm^3, so they become highly conductive. A heavily doped p-type semiconductor is labeled p$^+$-type, and a heavily doped n-type is called n$^+$-type.

• **The p–n junction.** Such junctions are known in microelectronics as rectifying diodes and they form the basis of the use of semiconductors for particle detection. Very different geometries can be considered and we will discuss here only the simplest case of the juxtaposition of a p-type and an n-type semiconductor, as shown in Figure 3.28. The formation of a p–n junction creates a zone at the interface of the two materials. Because of

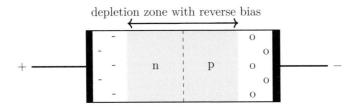

Figure 3.28 Basic principle of a reverse-biased p–n junction, as used in solid-state detectors.

the difference in concentration of carriers, there is an initial diffusion of charges, namely holes move towards the n-region and electrons move towards the p-region. Such movement fills the carriers of the opposite type and leads to the recombination between electrons and holes. This causes an effective charge buildup. The p-region becomes negative and the n-region positive. Consequently, an electric field is created, which will act against the diffusion and an equilibrium is reached. Because of the presence of the electric field, there is now an electric potential difference across the p–n junction (called the "contact potential") and a zone which is depleted of carriers, called the **depletion zone**. This latter has a very special property: **any carrier (electron or hole) created in this zone will be immediately swept away by the electric field**. So, any ionizing particle traversing the depleted zone will liberate electron–hole pairs, which will drift across the junction. If a reverse-bias potential is placed across of the junction, a current signal will be produced, a situation which reminds us of the ionization chamber!

• **Microstrip silicon detectors.** The use of solid-state detectors regained interested in the 1980s when it was demonstrated that spatial resolution in the range of 5 µm could be achieved with a silicon microstrip detector (see, e.g., Ref. [43]), as will be described below. These detectors have played a crucial role in studying the production and decay of short-lived charmed and bottom hadrons at colliders, providing tracking with sufficiently precise spatial resolution near the production point to record the details of their short life and the associated secondary decay vertices and secondary tracks. Such a decay chain will be shown later in Figure 3.33. A typical silicon microstrip detector consists of a wafer of a doped n-type silicon, typically 50–150 µm thick, one side of which is made conducting, and the other side is etched with many thin parallel strips of p-type material, each strip being separated by typically 20–30 µm, as illustrated in Figure 3.29. These detectors are operated in reverse bias with voltages in the range of 100–200 V for full depletion, and act as a series of essentially individually independent p–n junctions. Every electrode is connected to an amplifier. The passage of an ionizing track through the wafer liberates electron–hole pairs, which then drift due to the local electric field. Signals, with a fast rise time ($\lesssim 10$ ns) because of the full depletion and the small size of the detector, are induced on the strips where the track has passed. Charge can typically be shared on several strips (in particular depending on the incoming track angle w.r.t. the wafer plane) and is measured by the individual amplifiers. These signals can then be recorded and combined to obtain a very precise spatial resolution of the order of 5 μm or so.

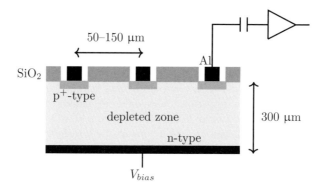

Figure 3.29 Typical layout of a microstrip silicon detector.

• **CMOS pixel detectors.** The ultimate goal of a pixel detector is to directly deliver a 2D "image" of the particles crossing the detector, potentially with spatial, time, and energy information. It is based on the development of modern digital imaging which began in 1969 with the invention of the **charge-coupled device** (CCD) by **Boyle**[27] and **Smith**.[28] Nowadays the development of digital imaging is driven by the possibility of using complementary metal-oxide silicon (CMOS) for imaging sensors. CMOS developments have been the drivers for the imaging camera found in today's consumer electronics. The idea is to have a pixel sensor consisting of millions of active pixels, each containing one amplifier, which convert and amplify the generated charge to a voltage, and which are individually recorded. The current generation of pixel detectors have pixel sizes in the range of 100×100 μm² or even less, down to an impressive size of 20×20 μm². This is a field of continuous progress. A lot of research effort has been put into the optimization of the depletion region, in order to maximize charge collection. Another line of development has been that of hybrid pixel detectors. In this case, each of the readout pixels is put in electrical contact with its absorber layer via a technique called "indium bump bonding" [44]. The development of pixel detectors is an ongoing effort at the time of writing, and the interested reader should refer to the constantly developing literature, in particular that related to upgrades of the inner tracking detectors of the LHC experiments. An example of such pixel detector development for the upgrade of the ALICE tracking detector is shown in Figure 3.30.

3.16   Experiments at Colliders

When two bunches from colliding beams enter in contact with each other, they produce individual interactions called **events**. At the LHC, the intensity of the beam is so high that there is a non-negligible probability that several interactions occur within a single event, in this case, one speaks of pile-up events. This is illustrated in

27   Willard Sterling Boyle (1924–2011), Canadian physicist.
28   George Elwood Smith (born 1930), American scientist and applied physicist.

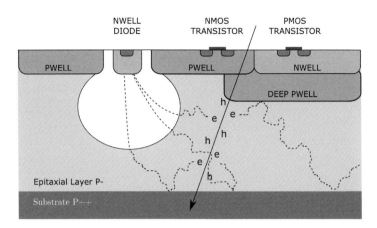

Figure 3.30 Illustration of the ALPIDE pixel sensor developed for the ALICE tracking detector upgrade at the LHC. Reprinted from B. Abelev *et al.*, "Technical design report for the upgrade of the ALICE inner tracking system," *Journal of Physics G: Nuclear and Particle Physics*, vol. 41, p. 087002, Jul 2014, under CC BY license https://creativecommons.org/licenses/by/3.0/.

Figure 3.31, which shows the display of an event recorded by the ATLAS detector at a center-of-mass energy of 13 TeV. The large inset on the left is a cut view in the $(r–\phi)$ plane, perpendicular to the beam. The beams enter in collision at the center of the figure. The curved tracks represent charged particles emanating from collisions and bent by the magnetic field. The curvature allows the measurement of the momentum and the sign of the charge. The energy deposited in calorimeters surrounding the tracking region is also shown as radially directed pillar blocks, whose height represents the energy deposited in that particular cell of the calorimeter. There are actually two sets of calorimeters: first the electromagnetic and then the hadronic calorimeter, as will be discussed below. Surrounding the calorimeters, one can see a set of muon chambers. In the inset bottom-right, the $(r–z)$ plane view parallel to the beams is shown and zoomed around the interaction point. Multiple interactions scattered along the z-axis are visible, representing independent hard interactions. They can be separated by tracing back the origin of the charged tracks and can therefore be studied individually. Detectors at colliders use a wide range of technologies to detect and measure the properties of the particles produced in these collisions. The aim is to be able to retrieve the kinematics of the primary particles produced in the interaction. **Ideally, one would like to reconstruct all the fundamental particles produced in the interaction and be able to identify the basic high-energy (or hard) process that occurred.**

In general, a detector at a collider is an "onion-like" structure with several subdetectors. The center region consists of a cylindrical or polygonal barrel part, with its axis parallel to the z-axis of the colliding beams. The cylindrical structure is closed by two end caps. A detector usually has a solenoid which produces a strong axial magnetic field in the range B = 1–4 T. The inner region of the detector is devoted to the tracking of charged particles, which are bent in the solenoidal magnetic field. The tracking volume is surrounded by an electromagnetic calorimeter (ECAL) and a relatively large-volume hadronic calorimeter (HCAL) for detecting and measuring particles that are not fully stopped in the ECAL. Muon chambers are positioned at the outside to detect high-energy muons produced in the collisions, which are the only particles (apart from neutrinos) that can penetrate through the ECAL+HCAL. The detectors almost cover the complete solid angle, down to the beam pipe, in order to collect all particles produced in the collisions.

The "onion-like" design of a collider experiment is optimized for the identification and energy measurement of the particles produced in high-energy collisions, as indicated in Figure 3.32. The momenta of charged particles are obtained from the curvature of the reconstructed tracks. The energies of neutral particles are obtained from the calorimeters. Particle identification is achieved by comparing the energy deposits in the

Figure 3.31 A display of an event at a center-of-mass energy of 13 TeV, recorded by the ATLAS detector. Reprinted with permission from an image accessed at http://atlas.ch/. Copyright ATLAS Collaboration.

different subdetector systems. Photons and electrons deposit all their energy in the ECAL. Electrons are identified as charged-particle tracks that are associated with an electromagnetic deposit in the ECAL. Charged hadrons are identified as charged-particle tracks associated with a small energy deposit in the ECAL (from ionization energy loss) and a large energy deposit in the HCAL. Neutral hadrons will usually interact in the HCAL. Finally, muons can be identified as charged-particle tracks associated with small energy deposits in both the ECAL and HCAL and signals in the muon detectors on the outside of the detector system.

Neutrinos leave no signals in the detector and their presence can be inferred from the presence of missing energy and momentum. If one measures "all" particles produced in a collision, momentum conservation should give a balanced zero momentum (assuming the collision occurs in the center-of-mass frame). So, the deviation from zero is a measure of the amount of lost particles or "missing" momentum. In practice, one focuses on the "transverse" missing momentum, which can be formally written as:

$$\vec{p}_{T,miss} \equiv -\sum_i \vec{p}_{T,i} \tag{3.33}$$

where the sum extends over the measured transverse momenta of all the observed particles in an event. Significant missing transverse momentum is therefore a sign for the presence of an undetected neutrino. This technique, for instance, is commonly used to study the production and semi-leptonic decay of a W-boson, as will be discussed in Chapter 27:

$$p + p \to W^+ + X \to e^+ + \bar{\nu}_e + X \tag{3.34}$$

• **Isolated leptons.** Isolated electrons, photons, and muons give clear signatures and are easily identified. Tau leptons, which have a very short lifetime, have to be identified from their observed decay products (see Eq. (21.97)). The main tau-lepton decay modes are $\tau^- \to \ell^- \bar{\nu}_\ell \nu_\tau$ ($\ell = e, \mu$), $\tau^- \to \pi^- \left(n\pi^0\right)\nu_\tau$, and $\tau^- \to \pi^- \pi^+ \pi^- \left(n\pi^0\right)\nu_\tau$ ($n = 0, 1, \ldots$). The hadronic decay modes typically lead to final states with one or

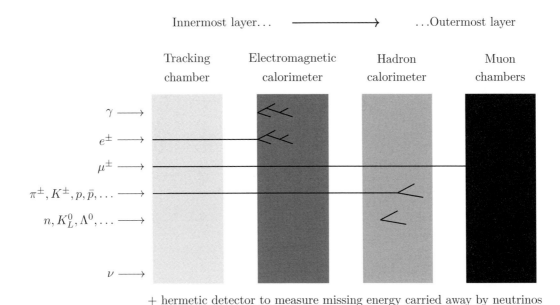

+ hermetic detector to measure missing energy carried away by neutrinos

Figure 3.32 Illustration of high-energy particles being identified by consecutive types of subdetectors in a typical collider experiment. The curvature of the tracks in the magnetic field is not shown for simplicity.

three charged pions and zero, one, or two π^0s which decay to photons $\pi^0 \to \gamma\gamma$. Tau leptons can therefore be identified as narrowly collimated jets of just a few particles and the presence of missing momentum in the event, associated with the tau neutrino in the final state.

• **Reconstruction of jets.** As will be discussed in Chapter 18, quarks are never observed as free particles, but are always found confined within hadrons. However, in high-energy interactions, we are interested in the collisions of the subconstituents, for example the quarks, not the hadrons. In the process $gg \to q\bar{q}$ the two quarks will be produced flying apart at relativistic velocities. The two quarks are converted into further pairs of quarks and antiquarks through a process called hadronization that occurs over a distance scale of 10^{-15} m. Each quark produced in a collision produces a "jet" of hadrons. This will be described in Chapter 19. Hence a quark is observed as an energetic jet of particles. On average, approximately 60% of the energy in a jet is in the form of charged particles (mostly $\pi^\pm$, $K^\pm$, ...), 30% is in the form of photons from $\pi^0 \to \gamma\gamma$ decays, and 10% is composed of neutral hadrons (mostly neutrons, K^0, Λ^0, ...). In high-energy jets, the separation between the individual particles is typically smaller than the segmentation of the calorimeters and not all of the particles in the jet can be resolved. Nevertheless, the energy and momentum of the jet can be determined from the total energy deposited in the calorimeters.

• **Identification of heavy quarks.** In general, it is not possible to tell which flavor of quark was produced, or even whether the jet originated from a quark or a gluon. However, if a heavy quark such as a c- or b-quark is produced (see Chapter 20), the hadronization process will create hadrons which contain such heavy quarks, as the flavor must be conserved. It turns out that some of these hadronic states are relatively long-lived since they can only decay weakly (some of the common decay modes are listed in Chapter 20). When produced in high-energy collisions, these relatively long-lived particles, combined with their Lorentz time-dilation factor, can travel on average a few millimeters before decaying. Consequently, the identification of c- or b-quark jets relies on the ability to resolve the secondary vertices from the primary vertex. Such an example recorded in the

LHCb experiment is shown in Figure 3.33. The event is consistent with the production of a b-jet containing a $\bar{B}^0$ hadron, which undergoes $\bar{B}^0 \to D^{*+}\tau^-\bar{\nu}_\tau$ with $D^{*+} \to D^0\pi^+$ and $D^0 \to K^-\pi^+$, plus a muon from the decay of a very short-lived τ. In practice, such a precise reconstruction is achieved by using dedicated

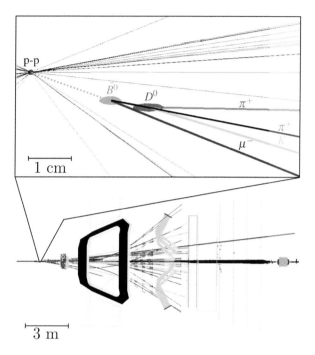

Figure 3.33 Event displays from an event recorded in the LHCb experiment at the LHC, showing clear displaced vertices. The area close to the interaction point is enlarged, showing the tracks of the charged particles produced in the pp interaction, the $\bar{B}^0$ path (dotted), its displaced decay $\bar{B}^0 \to D^{*+}\tau^-\bar{\nu}_\tau$ with $D^{*+} \to D^0\pi^+$ and $D^0 \to K^-\pi^+$, plus the muon from the decay of a very short-lived τ. Figure adapted from F. U. Bernlochner, M. F. Sevilla, D. J. Robinson, and G. Wormser, "Semitauonic b-hadron decays: A lepton flavor universality laboratory," arXiv:2101.08326.

high-precision silicon micro-vertex detectors consisting of several concentric layers of silicon at radii of a few centimeters from the axis of the colliding beams. Such detectors can achieve a single-hit resolution of $O(10\,\mu\mathrm{m})$, sufficient to be able to identify and reconstruct the secondary vertices, even in a dense jet environment. The ability to tag b-quarks has played an important role in a number of recent experiments. This was vital, for instance, in the discovery of the top quark, as discussed in Chapter 29.

3.17　The Detectors at the LHC

Four main experiments are operating at the CERN LHC: ATLAS [45], CMS [46], LHCb [47], and ALICE [48]. The parameters of the ATLAS and CMS detectors are summarized and compared in Table 3.7. As an illustrative example of all the features described in the previous section, Figure 3.34 shows the displays of two candidate events collected by CMS and ATLAS which are consistent with the production of the Higgs boson (see Chapter 25) and its subsequent decay into a pair of muon–antimuon, i.e.

$$p + p \to H^0 + X \to \mu^+\mu^- + X \tag{3.35}$$

The Higgs boson has a very short lifetime and disintegrates (decays) extremely rapidly into the two muons, which appears as the two penetrating tracks in the event displays. The reconstruction of the kinematical properties (energy and directions) of both muons allows the reconstruction of their invariant mass, which shows a value consistent with the Higgs boson rest mass.

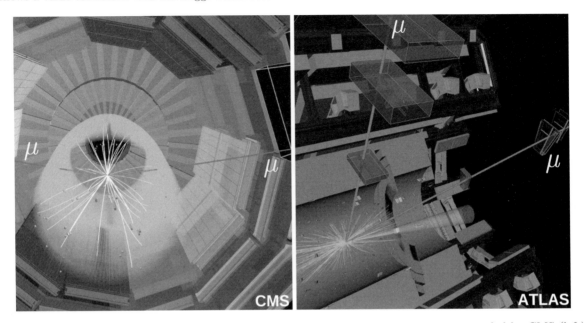

Figure 3.34 Candidate event displays of a Higgs boson decaying into two muons as recorded by CMS (left) and ATLAS (right). Reprinted figures by permission from CERN. Copyright 2021 CERN.

- **ATLAS.** The **ATLAS detector** (a toroidal LHC apparatus) [45] (see Figure 3.35) consists of a series of ever-larger concentric cylinders around the interaction point. It can be divided into four major parts: the inner detector, the calorimeters, the muon spectrometer, and the magnet systems. Each of these is in turn made of multiple layers. The detectors are complementary: the inner detector tracks particles precisely, the calorimeters measure the energy of easily stopped particles, and the muon system makes additional measurements of highly penetrating muons. The two magnet systems bend charged particles in the inner detector and the muon spectrometer, allowing their momenta to be measured.

- **CMS.** The **CMS detector** (compact muon solenoid) (see Figure 3.36) has the following characteristics [46]: the central magnet is a 6 m-diameter, superconducting solenoid providing a 3.8 T magnetic field. Within the magnet, closest to the interaction point, a silicon tracking detector is installed that measures the transverse momenta (p_T) of charged particles with a relative resolution of $\approx 0.7\%$ at 10 GeV up to pseudo-rapidities of $|\eta| < 2.5$ (the pseudo-rapidity will be defined in Section 19.10). Surrounding the tracking volume but still within the solenoid, the calorimetric system covers the region up to $|\eta| < 3$. It consists of a lead–tungstate crystal electromagnetic calorimeter (ECAL) with an energy resolution of $\approx 0.6\%$ for 50 GeV electrons and a brass-scintillator hadronic calorimeter (HCAL) with a resolution of 18% for 50 GeV pions. Steel and quartz-fiber hadron calorimeters extend the coverage to $|\eta| < 5$. Muons are identified in gas-ionization detectors embedded in the steel flux-return yoke of the magnet.

- **LHCb.** The **LHCb experiment** [47] is dedicated to heavy-flavor physics at the LHC. Its main goal is to study CP violation in the quark sector (see Chapter 30) and to search for rare decays of the bottom and charm hadrons to test the Standard Model with very high precision. Unlike the ATLAS and CMS detectors, the LHCb detector is composed of a single-arm spectrometer which covers the forward region from approximately

Parameter	ATLAS	CMS	
Length	46 m	22 m	
Height	22 m	15 m	
Weight	7000 tons	$12,500$ tons	
Solenoid field	2 T	4 T	
Toroid field	≈ 2 T		
ECAL resolution $\left(\frac{\sigma_E}{E}\right)$	$\frac{10\%}{\sqrt{E}} \oplus \frac{30\%}{E} \oplus 1\%$	$\frac{2.7\%}{\sqrt{E}} \oplus \frac{20\%}{E} \oplus 0.55\%$	E (GeV)
HCAL resolution $\left(\frac{\sigma_E}{E}\right)$	$\frac{50\%}{\sqrt{E}} \oplus 0.03$	$\frac{70\%}{\sqrt{E}} \oplus 9.5\%$	E (GeV)
Calo. coverage η	≤ 4.9	≤ 5.0	
Tracking coverage η	≤ 2.5	≤ 2.6	
Nr. track points	$3 + 4 + 36$	12	
Track resolution $\sigma\left(1/p_{\mathrm{T}}\right)^{\eta=0}$	$0.36 \oplus \frac{13}{p_{\mathrm{T}}}$	≈ 0.1	p_T (TeV)
Muon coverage η	≤ 2.7	≤ 2.4	
Muon resolution $(p_{\mathrm{T}} < 0.1$ TeV$)$		9–20%	
$(p_{\mathrm{T}} = 1.0$ TeV$)$	7%	15%–35%	

Table 3.7 Performance parameters of the ATLAS and CMS detectors at the CERN LHC.

10 mrad to 300 mrad in the bending plane. This design choice is justified by the fact that at the LHC energies the bottom and charm hadrons are copiously produced in the forward (and backward) direction, owing to their Lorentz boost in the laboratory system. Because of their large longitudinal Lorentz boost, the decay point of the bottom and charm hadrons will be well displaced from the interaction point due to time dilation, as already illustrated in Figure 3.33. The spectrometer magnet is a warm dipole magnet providing an integral field of 4 Tm. A dedicated system called VELO acts as an efficient vertex locator. Three trackers called TT (for trigger tracker, made of silicon microstrips), IT (for inner tracker, composed as well of silicon microstrip detectors), and OT (consisting of straw tubes) are used to reconstruct the trajectories of charged particles with high efficiency and determine the momenta of the tracks with high resolution. Particle identification, in particular the separation between pions and kaons, is provided by two ring imaging Cherenkov (RICH) counters. A set of electromagnetic and hadronic calorimeters preceded by a preshower complete the energy measurement of charged and neutral particles. Finally, a muon detection system is located behind the calorimeters to perform dedicated identification and measurements of penetrating muons.

• **ALICE.** The **ALICE experiment** [48] is a general-purpose detector optimized to study heavy-ion collisions at the LHC. It focuses on the study of QCD theory of the strong interactions (to be discussed in Chapter 18) in the regime of strongly interacting matter and of the quark–gluon plasma. The detector is composed of several subdetector systems each with its own specific technology. The most stringent challenge is to cope with the extremely large multiplicity of several thousands of particles produced in the collisions of two Pb ions. The central part of ALICE covers the polar region from $45°$ to $135°$ and is embedded in a large magnet. From inside out, the detector consists of an inner tracking system (ITS) with high-resolution silicon pixel detectors, a drift detector (SDD), a strip detector (SSD), and a cylindrical time-projection chamber (TPC). Particle identification is provided by an array of time-of-flight (TOF) detectors, a ring imaging Cherenkov subdetector (HMPID), and transition radiation detectors (TRDs). This system is completed by two electromagnetic

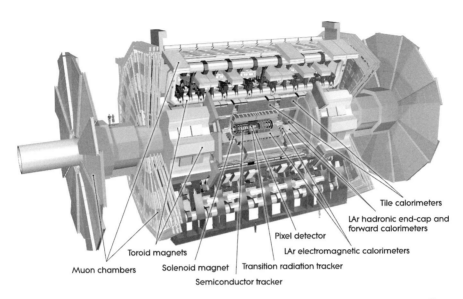

Figure 3.35 Layout of ATLAS detector at the CERN LHC. Reprinted with permission. Copyright ATLAS Collaboration.

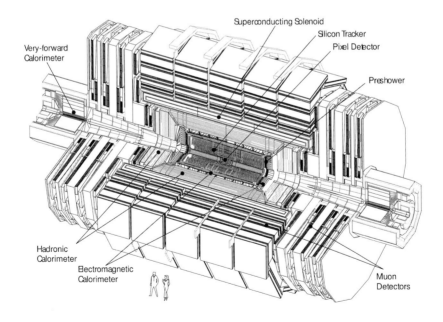

Figure 3.36 Layout of the CMS experiment at the CERN LHC. Reprinted from G. L. Bayatian *et al.*, *CMS Physics: Technical Design Report Volume 1: Detector Performance and Software.* Technical Design Report CMS, Geneva: CERN, 2006, under CC BY license https://creativecommons.org/licenses/by/3.0/. Copyright CMS Collaboration.

calorimeters. There is also a forward muon arm covering the angular region between $2°$ and $9°$.

3.18 Particle Physics Experiments without Accelerators

Progress in particle physics was not only achieved thanks to accelerators of increasing energy, but also with a class of projects that are primarily **non-accelerator experiments**. These include cosmic-ray experiments, astrophysical neutrino detectors, searches for double-beta decay, dark-matter candidates, or magnetic monopoles. The experimental methods are sometimes those familiar with detectors at accelerators but there is also instrumentation either not found at accelerators or applied in a radically different way. Examples are **cosmic-ray experiments** (such as the **Pierre Auger Cosmic Ray Observatory** [49]), massive **water Cherenkov imaging detectors** (e.g., Super-Kamiokande described below and the IceCube Neutrino Observatory [50]), **ultra-cold solid-state detectors** (e.g., CDMS [51]). In addition, radiologically ultra-pure materials and deep underground locations are required for experiments searching for very rare phenomena. There is also a space-based particle physics experiment, the **Alpha Magnetic Spectrometer** (AMS-02) experiment mounted on the International Space Station [52]. To discuss the variety of detectors and the vast domains of science performed with them is beyond the scope of this book. We refer the interested reader to the existing literature.

We want to point out the case of **deep underground detectors**. Large detectors to detect rare processes, located deep underground to suppress backgrounds, tend to be multi-purpose. The physics of interest includes not only solar, reactor, supernova, and atmospheric neutrinos, but also searches for baryon number violation, searches for exotic particles such as dark matter or magnetic monopoles, and neutrino and cosmic-ray astrophysics in different energy regimes. The possibility of detecting such rare events, and the sensitivity to search for new phenomena never observed before, requires an appropriate shielding to reach low or possibly vanishing backgrounds. At the surface level, cosmic rays and their secondary interactions represent a serious source of backgrounds. The rate of cosmic muons at sea level is about $\approx 70\ \mathrm{m^{-2}\,s^{-1}\,sr^{-1}}$ (see Section 15.2). It can be reduced by several orders of magnitude compared to surface level by adopting a deep underground location, such as in a mine or under a mountain. Historically, the depth of an underground facility is measured in kilometers of water-equivalent (labeled km w.e.). A depth of 1 km of w.e. is approximately equal to 370 m of actual rock, if we assume that the average density of the rock is $2.7\ \mathrm{g\,cm^{-3}}$. Figure 3.37(left) shows the flux of cosmic rays as a function of the depth in units equivalent to km w.e. Arrows indicate the depth of some existing facilities. In particular, we see that the Gran Sasso Underground Laboratory and the Homestake mine offer the possibility to perform experiments at a depth greater than 4 km w.e. At this depth, the rate of cosmic muons is at the level of a few $10^{-9}\ \mathrm{cm^{-2}\,s^{-1}\,sr^{-1}}$, several orders of magnitude less than on the surface. The underground detectors may also serve as targets for long-baseline neutrino beams for neutrino oscillation physics studies. A review of the world's **deep underground laboratories**, where these experiments take place, is available in Ref. [53].

• **The Super-Kamiokande detector.** Super-Kamiokande (SK) is a 50 kiloton water Cherenkov imaging detector instrumented with photomultiplier tube light sensors [54]. It is located in the Mozumi mine of the Kamioka Mining and Smelting Company in Japan. The detector cavity lies under the peak of Mt. Ikenoyama, with approximately 1 km of rock or 2700 km w.e. The detector consists of a large tank, 39 m in diameter and 42 m tall, with a total capacity of 50,000 tons of water. A cut-view of the SK detector is shown in Figure 3.37(right). Within the tank, a steel framework supports two separate arrays of photomultipliers (PMTs). One has its PMTs facing inward and the other has them facing outward. The inward-facing array, called ID for inner detector, consists of 11,146 Hamamatsu 50 cm-diameter PMTs. The density of PMTs in the ID is such that about 40% of the surface is covered by photocathodes sensitive to light. The outward-facing array composes the OD (for outer detector). It is optically isolated from the ID and is an array of 1885 20 cm-diameter PMTs. It is used as a veto detector. SK can measure events occurring inside the main ID volume over a wide range of energies, spanning from 4.5 MeV to several hundreds of giga-electronvolts. Events are detected via the Cherenkov light produced by relativistic charged particles with momentum above the Cherenkov threshold in water (see Figure 3.27). This light is emitted in ring patterns which are detected by the light sensors. These rings are used to reconstruct the features (electron/photon or muon, momentum, direction) of the particles which produced them. The SK experiment will be discussed further in Chapter 30 in relation to its crucial role in studying neutrinos.

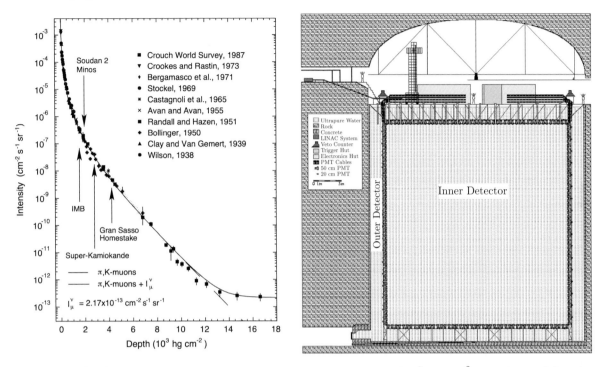

Figure 3.37 (left) Cosmic-ray muon flux as a function of depth in units of 10^3 hg cm^{-2} equivalent to kilometer water-equivalent (km w.e.). (right) Cut-view of the Super-Kamiokande detector. Reprinted from Y. Fukuda *et al.*, "The Super-Kamiokande detector," *Nucl. Instrum. Meth. A*, vol. 501, pp. 418–462, 2003, with permission from Elsevier. Copyright 2003 Elsevier Science B.V.

Problems

Ex 3.1 Synchrotron radiation. A stored charged particle loses energy by synchrotron radiation because it is subjected to an acceleration.

(a) Show that the average power radiated in a circular ring is given by:

$$P_{rad} \simeq \frac{2e^2 c \gamma^4}{3R^2} \tag{3.36}$$

(b) Show that the radiated energy per turn is:

$$E_{turn} = \frac{4\pi}{3} \frac{e^2 \gamma^4}{R} \tag{3.37}$$

(c) Show that at an equivalent energy, electrons will radiate much more than protons.

Ex 3.2 Energy stored in the LHC beams. Verify the claim that for a total current of 0.5 A, the total energy stored in the LHC beams is 360 MJ. In this case, what is the number of protons per bunch?

Ex 3.3 ILC luminosity. The foreseen ILC collider needs to reach a luminosity up to 5×10^{34} cm^{-2} s^{-1}. Estimate the transverse dimensions of the bunches.

Ex 3.4 Probing the Planck scale. The direct study of the quantum nature of the gravitational force would require reaching the Planck scale defined by $\Lambda_{Pl} \simeq \sqrt{\hbar c / G}$. What kind of collider using a reasonable extrapolation of current technologies would be necessary to reach such an energy frontier?

Ex 3.5 Neutrino collider. Is it possible to build a neutrino collider to study $\nu + \bar{\nu} \to Z^0$ production?

4   Non-relativistic Quantum Mechanics

Quantum mechanics was established during the first decades of the twentieth century, thanks to the work of famous physicists like Planck, Bohr, Heisenberg, de Broglie, Compton, Einstein, Schrödinger, Born, von Neumann, Dirac, Fermi, Pauli, von Laue, Dyson, Hilbert, Wien, Bose, Sommerfeld, and others. It became the standard formulation for atomic and subatomic physics and has succeeded spectacularly at describing those phenomena. However, it remained a non-relativistic theory, not adequate to describe particle processes at high energies reached by accelerators or available in cosmic rays. Nonetheless, it was essential to lay the foundation which would later permit its generalization to relativistic quantum mechanics.

4.1   Electron Diffraction

We review in this chapter the foundations of **non-relativistic quantum mechanics** (see also Appendix C), which we like to call "**classical quantum mechanics**" as opposed to the relativistic quantum mechanics.

In 1928 **G. P. Thomson**[1] and **Reid** observed that an electron can behave like a wave in the process called diffraction of electrons, similar to light diffraction. They wrote:

> "*If a fine beam of homogeneous cathode rays is sent nearly normally through a thin celluloid film (of the order 3×10^{-6} cm thick) and then received on a photographic plate 10 cm away and parallel to the film, we find that the central spot formed by the undeflected rays is surrounded by rings, recalling in appearance the haloes formed by mist round the sun. (...) If the density of the plate is measured by a photometer at a number of points along a radius, and the intensity of the rays at these points found by using the characteristic blackening curve of the plate, the rings appear as humps on the intensity-distance curves. In this way rings can be detected which may not be obvious to direct inspection. With rays of about 13,000 volts two rings have been found inside the obvious one.*"[2]

This phenomenon and other observations led to the development of quantum mechanics and the study of the particle–wave nature of subatomic particles.

According to the de Broglie hypothesis, electrons should possess a dual wave and particle nature just as photons do. **Thus all particles can be made to exhibit interference effects and all wave motions have their energy in the form of quanta.** Extending Einstein's postulate for photons $E = h\nu$ and $pc = h/\lambda$ to the electrons yields its frequency ν_e and wavelength λ_e for a given energy E_e and momentum p_e. For a non-relativistic electron with kinetic energy T_e, we have $p = \sqrt{2m_e T_e}$, so its frequency and wavelength are:

$$E_e = m_e + T_e = h\nu_e \quad \text{and} \quad \lambda_e = \frac{h}{p} = \frac{h}{\sqrt{2m_e T_e}} = \frac{hc}{\sqrt{2m_e c^2 T_e}} \tag{4.1}$$

---

1   George Paget Thomson (1892–1975), English physicist.
2   Reprinted by permission from Springer Nature: G. P. Thomson and A. Reid, "Diffraction of cathode rays by a thin film," *Nature*, vol. 119, no. 3007, pp. 890–890, 1927.

An electron at rest accelerated through a potential difference U has after acceleration a kinetic energy $T_e = eU$, hence

$$\lambda_e = \frac{hc}{\sqrt{2m_e c^2 \, eU}} \approx \frac{1.23 \cdot 10^{-9} \text{ m}}{\sqrt{U}} \tag{4.2}$$

For $U = 15$ kV we have $\lambda_e \approx 10^{-9}$ cm $= 0.1$ Å, which is smaller than the size of an atom.

• **Experimental evidence.** The first experimental proof of de Broglie's hypothesis was historically achieved by **Davisson**[3] and **Germer**[4] in 1927 [55]. They created a beam of monoenergetic low-energy electrons by accelerating thermally emitted electrons from a hot filament through a voltage drop. The electron beam was directed normal to the surface of a Ni crystal. They measured the angular distribution of the scattered electrons. They used an accelerating voltage of only $V = 54$ V. Since the energy of the electrons was small, they quickly lost energy in the crystal and it was presumed that the measured electrons must have been scattered by the surface of the crystal. Each atom becomes a scattering center. Their periodicity at the surface of the crystal creates a grating diffraction which they could observe. If we evaluate the de Broglie wavelength of 54 eV electrons, we find $\lambda_e \approx 1.67$ Å. The wavelength of the waves which are interfering constructively can be computed with the **grating equation** of optics $n\lambda = d\sin\theta$, where $n = 1, 2, 3, \ldots$. Setting $n = 1$ and a periodic structure of Ni of $d = 2.15$ Å, one gets $\sin\theta = \lambda/d$ or $\theta \approx 51°$, which is what Davisson and Germer observed, thereby validating de Broglie's hypothesis!

• **Two-slit experiment.** The **two-slit experiment of Young**[5] is the classical interference experiment in optics that directly showed the wave nature of light. Can one repeat the same experiment with electrons? The geometrical setup is illustrated in Figure 4.1. Electrons with momentum $\vec{p}_e$ are incident perpendicular to a plane with two slits. The distance between the two slits is d. A screen is located at a distance $D \gg d$ from the two slits. One measures the intensity at a point P as defined by the angle ϑ. In order to have a maximum of

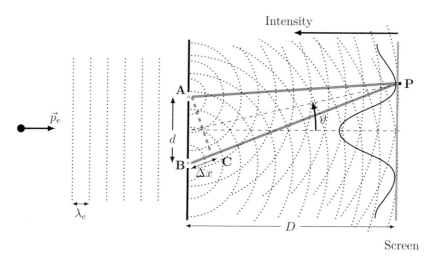

Figure 4.1 Geometrical setup of the double-slit experiment with electrons.

intensity at point P, the two waves coming from the slits should interfere constructively, or, looking at their difference of path $\Delta x \equiv BP - AP = BC$, it should satisfy the well-known relation:

$$\Delta x = d\sin\vartheta = n\lambda_e \qquad n = 0, 1, 2, \ldots \quad \text{(maximum)} \tag{4.3}$$

---

3   Clinton Joseph Davisson (1881–1958), American physicist.
4   Lester Halbert Germer (1896–1971), American physicist
5   Thomas Young (1773–1829), British polymath and physician.

For a minimum, the difference of path Δx should be given by a half-integer multiple of the wavelength:

$$\Delta x = d \sin \vartheta = \left(n + \frac{1}{2} \right) \lambda_e \qquad n = 0, 1, 2, \ldots \qquad \text{(minimum)} \qquad (4.4)$$

The very small wavelength λ_e of the electron has been the reason why several technical challenges had to be overcome before the direct proof of the diffraction of an electron could be achieved.

In 1961 **Jönsson**[6] performed for the first time a version of the double-slit experiment with a single electron [56]. A glass plate was prepared by evaporating a silver film of about 200 Å thickness. This film was irradiated by a line-shaped electron probe in a vacuum of 10^{-4} Torr. This procedure produces a hydrocarbon polymerization film of very low electrical conductivity where the high electron current passed. Then, a copper film is deposited electrolytically, however, the copper does not stick in these zones. In this way, Jönsson could produce grating with slits free of any material 50 μm long and 0.3 μm wide, with a grating constant of 1 μm. He could directly observe the electron diffraction pattern obtained using these slits in an arrangement analogous to Young's light optical interference, as shown in Figure 4.2. With his method, he could test both single and double-slit configurations.

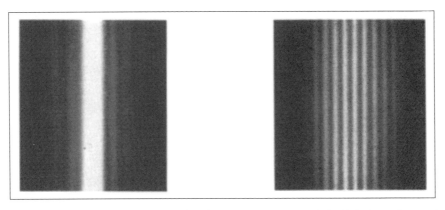

Figure 4.2 Image of the diffraction of electrons (left) through one slit and (right) through two slits. Reprinted from C. Jönsson, "Elektroneninterferenzen an mehreren künstlich hergestellten Feinspalten," *Zeitschrift für Physik*, vol. 161, pp. 454–474, Aug 1961, under permission from Springer Nature. Copyright 1961.

● **Particle–wave duality.** These experiments show the dual nature of elementary particles. We can ask ourselves what would happen in these experiments if one were to separate the incoming electrons in time such that only one electron is in our apparatus at a given moment. The electrons are detected one-by-one on the screen. Figure 4.3 illustrates the image one would obtain on the screen. Each white dot represents one electron. The detection of 10 electrons leads to a pattern which seems somewhat random. We note that all electrons are assumed to have the same initial momentum (same initial condition). The trajectory is no longer classical in the sense that the same initial conditions lead to the observation of the electron in different places on the screen. The electron behaves as a wave when it crosses the two slits and interferes with itself, however, it is detected as a single particle in a given point in space on the screen ("principle of complementarity"). The building up of the interference pattern becomes visible as more electrons are measured, as illustrated in Figure 4.3 for 10, 100, 1000, and 10,000 single electrons. As the statistics increase, the interference pattern becomes visible and tends to the theoretical intensity distribution. It's fantastic!

6   Claus Jönsson (born 1930), German physicist.

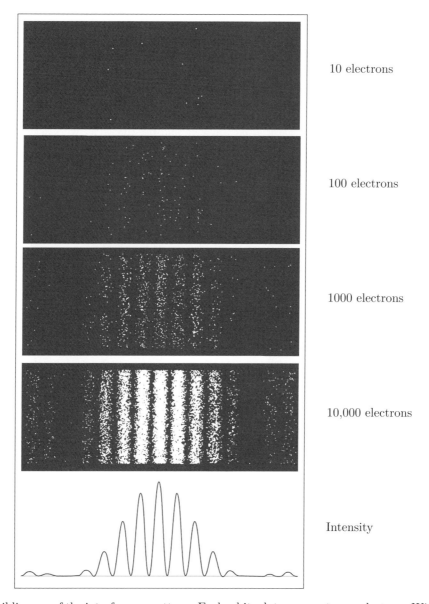

10 electrons

100 electrons

1000 electrons

10,000 electrons

Intensity

Figure 4.3 Building-up of the interference pattern. Each white dot represents one electron. With 10 electrons the pattern is hardly visible. As more electrons are measured, the interference pattern becomes statistically visible and tends to the theoretical intensity distribution.

4.2   The First Quantization

The atomic model developed by **N. Bohr**[7] (see Appendix C.1), which described hydrogen atoms with an electron orbiting the nucleus, was based on classical mechanics with the addition of ad-hoc rules to quantize the allowed orbits to explain the observed atomic line spectra (see Section 2.1). Bohr's correspondence principle demanded that classical and quantum physics give the same answer when the system became macroscopic. **Sommerfeld**[8] added the concept of quantized angular momentum to the system (*"Richtungsquantelung"*) to take into account the observed Zeeman effect. Although successful, the Bohr–Sommerfeld atomic model was not a complete theory of quantum systems. The limitations of the model arise because it does not reflect a fundamental facet of nature, namely the fact that particles possess wave properties.

Particle–wave duality needed to be addressed in a fundamental way. The results from Davisson and Germer, and from other experiments, had shown that this anomalous behavior was not peculiar to light, but was rather quite general. It was understood that all material particles possess wave properties, which can be exhibited under suitable conditions. We have here a very striking and general example of the breakdown of classical mechanics – not merely an inaccuracy in its laws of motion, but an inadequacy of its concepts to supply us with a description of atomic events.

In 1925 **Heisenberg**[9] developed the idea that the quantities appearing in the classical equations had to be reinterpreted but not the equations themselves [57]. He developed his theory based on non-commuting matrices and replaced classical quantities by "ensemble of quantities which are in principle observable." For instance, the position coordinate x is replaced by an ensemble and defined as the product of two such quantities. He developed the theoretical foundation of quantum mechanics, exclusively based on relations between principally observable quantities.

In 1926, **Heisenberg**, **Born**,[10] and **Jordan**[11] published a paper entitled "Zur Quantenmechanik II" [58], where Hamilton's formulation of mechanics with canonical position and momentum is reinterpreted with those quantities being matrices with their commutation rules.

Also in 1926, a major breakthrough towards a theory of quantum mechanics was achieved when Heisenberg's *matrix mechanics* was transformed into *wave mechanics*. This latter was developed by **E. Schrödinger**.[12] The search for a wave equation was apparently initiated by Schrödinger following a remark by Peter Debye[13] at a *Zurich Kolloquium* in 1925 [59]. Where Heisenberg's equations used matrices, the wave equation of Schrödinger introduced differential operators, with the result that the order matters in which the operators are appearing in front of the object upon which they act.

Again in 1926, **Born** first gave the interpretation of Schrödinger's wave which is now generally accepted. He wrote:

"So I would like to give it a try: The guiding field, represented by a scalar function Ψ of the coordinates of all particles involved and the time, propagates according to Schrödinger's differential equation. Momentum and energy, however, are transferred as if corpuscles (electrons) were actually flying around. The trajectories of these particles are only determined as far as the set of energy and momentum limits them; in all other respects, only one probability for a certain path is determined by the value distribution of the function Ψ. You could, somewhat paradoxically, summarize it like this: The motion of the particles follows laws of probability, but the probability itself propagates in accordance with the law of causality." [14]

---

7   Niels Henrik David Bohr (1885–1962), Danish physicist.
8   Arnold Johannes Wilhelm Sommerfeld (1868–1951), German theoretical physicist.
9   Werner Karl Heisenberg (1901–1976), German theoretical physicist.
10   Max Born (1882–1970), German physicist.
11   Ernst Pascual Jordan (1902–1980), German theoretical and mathematical physicist.
12   Erwin Rudolf Josef Alexander Schrödinger (1887–1961), Austrian physicist.
13   Peter Joseph William Debye (1884–1966), Dutch–American physicist and physical chemist.
14   *"Ich möchte also versuchsweise die Vorstellung verfolgen: Das Führungsfeld, dargestellt durch eine skalare Funktion Ψ der Koordinaten aller beteiligten Partikeln und der Zeit, breitet sich nach der Schrödingerschen Differentialgleichung aus.*

Accordingly, a particle (such as the electron) is described by its **wave function** Ψ:

$$\Psi(\vec{x}, t) \equiv \text{complex wave function} \tag{4.5}$$

This function represents the **state of the particle** – all the information – and describes its wave nature (interference, diffraction, two-slit experiment, etc.) as well as its particle properties (momentum, position, etc.). The state can also be represented as a unit vector (called **state vector** $|\Psi\rangle$) residing in a complex Hilbert space (see Section C.2). The common interpretation of the wave function is the following:

$$|\Psi(\vec{x}, t)|^2 \mathrm{d}^3\vec{x} \equiv \quad \text{probability of finding the particle at time } t$$
$$\text{in the volume element } \mathrm{d}^3\vec{x}$$

Its normalization can be expressed as:

$$|\langle\Psi|\Psi\rangle|^2 = \int_V |\Psi(\vec{x}, t)|^2 \mathrm{d}^3\vec{x} = \int_V (\Psi^*\Psi)\, \mathrm{d}^3\vec{x} \equiv 1 \tag{4.6}$$

where we used the bra–ket notation for the inner product of the state $|\Psi\rangle$ (refer to Section C.2 for this notation). The **probability density** at a given point of space-time is defined as:

$$\rho(\vec{x}, t) \equiv \Psi^*(\vec{x}, t)\Psi(\vec{x}, t) = |\Psi(\vec{x}, t)|^2 \tag{4.7}$$

• **Superposition of states.** The quantum state of a system follows the general principle of superposition of quantum mechanics. It requires the assumption that there exists a peculiar relationship between states such that a new state can be expressed as the sum of two or more states in a kind of "superposition." When a state is formed by the superposition of two other states, it will have the properties that are in some sense intermediate between those of the two original states. The state will approach more or less closely that of one state according to the greater or lesser "weight" attached to this state in the superposition.

• **Operators and observables.** The wave function is not itself an "observable," on the other hand it represents the entire information that we have on the particle. The observable quantities (or **observables**) are represented by **Hermitian operators** (A is Hermitian if $A^\dagger = A$) that act on the wave function (see Appendix C.2). This corresponds to a **first quantization** of the problem. For example, the momentum and energy operators are defined in the following way, starting from their classical formulation:

$$\text{Momentum operator}: \quad \vec{p} \longrightarrow -i\hbar\nabla = -i\nabla \tag{4.8}$$

$$\text{Energy operator}: \quad E \longrightarrow +i\hbar\frac{\partial}{\partial t} = i\frac{\partial}{\partial t} \tag{4.9}$$

where we have adopted natural units (see Section 1.8). The operators follow **commutation rules** which are expressed with the help of Poisson brackets (see Appendix C.3). The **canonical commutation rules** of quantum mechanics say that position and momentum operators obey the following commutation rules:

$$[x, p_x] = xp_x - p_x x = i, \quad [y, p_y] = i, \quad \text{and } [z, p_z] = i \tag{4.10}$$

and all other commutations are zero. This can be written in short by setting an index for the coordinates as $x = 1$, $y = 2$ and $z = 3$:

$$[x_i, p_j] = i\delta_{ij} \quad \text{and} \quad [x_i, x_j] = [p_i, p_j] = 0 \quad (i, j = 1, 2, 3) \tag{4.11}$$

Impuls und Energie aber werden so übertragen, als wenn Korpuskeln (Elektronen) tatsächlich herumfliegen. Die Bahnen dieser Korpuskeln sind nur so weir bestimmt, als Energie- und Impulssatz sie einschränken; im übrigen wird für das Einschlagen einer bestimmten Bahn nur eine Wahrscheinlichkeit durch die Werteverteilung der Funktion Ψ bestimmt. Man könnte das, etwas paradox, etwa so zusammenfassen: Die Bewegung der Partikeln folgt Wahrscheinlichkeitsgesetzen, die Wahrscheinlichkeit selbst aber breitet sich im Einklang mit dem Kausalgesetz aus." Reprinted by permission from Springer Nature: M. Born, "Quantenmechanik der Stoßvorgänge," *Zeitschrift für Physik*, vol. 38, pp. 803–827, Nov 1926.

A limitation of non-commuting operators is given by the impossibility to simultaneously assign exact numerical expectation values to both for any particular quantum state. This shows how measurements in quantum mechanics may be incompatible with each other. In the case above of the position and momentum, we can define the (standard) errors σ_x and σ_p with which they can be measured. The **Principle of Uncertainty of Heisenberg** states in this case that the product of these errors must obey:

$$\sigma_{x_i}\sigma_{p_j} \geq \frac{\hbar}{2}\delta_{ij} \tag{4.12}$$

• **Phase advance and interference.** We note that the absolute change by an overall complex phase of the wave function does not alter the value of the probability (Eq. (4.7)) and is therefore unobservable:

$$\Psi(\vec{x},t) \to e^{i\alpha}\Psi(\vec{x},t) : \quad \rho(\vec{x},t) \to |e^{i\alpha}\Psi(\vec{x},t)|^2 = \rho(\vec{x},t) \tag{4.13}$$

A particle with momentum $\vec{p}$ is described by a plane-wave function, such that its spatial dependence is given by:

$$\Psi \propto e^{i\vec{p}\cdot\vec{x}} \quad \Longrightarrow \quad -i\nabla\Psi = (-i)(i\vec{p})\Psi = \vec{p}\,\Psi \tag{4.14}$$

The phase advance along an infinitesimal path $\mathrm{d}\vec{x}$ is $\mathrm{d}\alpha = \vec{p}\cdot\mathrm{d}\vec{x}$, so the phase change along an entire trajectory is obtained by the line integral of $\vec{p}\cdot\mathrm{d}\vec{x}$ along that trajectory:

$$\Delta\alpha = \int_{trajectory} \vec{p}\cdot\mathrm{d}\vec{x} \tag{4.15}$$

In the two-slit experiment described in Section 4.1, the difference of phase between the trajectory AP and BP in Figure 4.1 is:

$$\Delta\alpha_{AP} - \Delta\alpha_{BP} = \int_A^P \vec{p}_1\cdot\mathrm{d}\vec{x} - \int_B^P \vec{p}_2\cdot\mathrm{d}\vec{x} \approx \vec{p}\cdot\left(\int_A^P \mathrm{d}\vec{x} - \int_B^P \mathrm{d}\vec{x}\right) \approx p(BC) = p\Delta x \tag{4.16}$$

where we assumed $\vec{p} = \vec{p}_1 \approx \vec{p}_2$. This is what generates the diffraction pattern observed on the screen.

• **System with many particles.** The wave functions considered up to now describe the state of *one* particle. In order to describe a system of N particles, the wave function must possess $3N$ space coordinates:

$$\Psi(\vec{x}_1, \vec{x}_2, ..., \vec{x}_N, t) \tag{4.17}$$

Nonetheless, the total number of particles in the system must remain constant ($N = $ const.). Consequently, the number of particles in our system described by the wave function cannot change with time. **Hence, this formalism is not adequate to describe the creation or the annihilation of particles.** We will see later that this becomes an important limitation when one searches for a relativistic extension of quantum mechanics. In particular, it was noted early on that the emission or absorption of light quanta would require such a feature. Therefore, since its early developments, the generalization of quantum mechanics to electrodynamics opened up new questions.

4.3   The Schrödinger Equation

The Schrödinger equation is "derived" with the help of the classical relation between energy and momentum. It represents the wave equation (equation of motion) of a single particle of mass m in an external *classical* potential V. The total energy of a particle in classical mechanics is given by:

$$E = T + V \tag{4.18}$$

where $T(\vec{p})$ is the *kinetic* and $V(\vec{x}, t)$ the *potential energy* of the particle. As a function of the momentum, we have:

$$T = \frac{\vec{p}^2}{2m} \quad \text{where } m = \text{ mass of the particle} \tag{4.19}$$

The operator of the total energy (called the **Hamiltonian** operator – see Appendix C.4) is therefore defined as:

$$H(\vec{p}, \vec{x}) = T + V = \frac{\vec{p}^2}{2m} + V = E \tag{4.20}$$

It follows by introducing the wave function Ψ that:

$$H\Psi(\vec{x}, t) = \left(\frac{\vec{p}^2}{2m} + V \right) \Psi(\vec{x}, t) = E\Psi(\vec{x}, t) \tag{4.21}$$

or by using Eqs. (4.8) and (4.9):

$$i\frac{\partial}{\partial t}\Psi(\vec{x}, t) = H\Psi(\vec{x}, t) = \left(-\frac{1}{2m}\nabla^2 + V(\vec{x}, t) \right) \Psi(\vec{x}, t) \tag{4.22}$$

The first quantization occurs when we take the "classical" Hamiltonian $H \equiv H(\vec{p}, \vec{x})$ and use operators acting on the wave function and their implied commutation rules (Eq. (4.11)).

4.4   The Continuity Equation

We consider the time variation of the probability density:

$$\frac{\partial}{\partial t}\rho(\vec{x}, t) \equiv \frac{\partial}{\partial t}\left(\Psi^*(\vec{x}, t)\Psi(\vec{x}, t)\right) = \left(\frac{\partial}{\partial t}\Psi^*(\vec{x}, t) \right) \Psi(\vec{x}, t) + \Psi^*(\vec{x}, t) \left(\frac{\partial}{\partial t}\Psi(\vec{x}, t) \right) \tag{4.23}$$

With the help of the Schrödinger equation, we find:

$$\frac{\partial}{\partial t}\Psi = -i\left(-\frac{1}{2m}\nabla^2 + V \right) \Psi \quad \text{and} \quad \frac{\partial}{\partial t}\Psi^* = +i\left(-\frac{1}{2m}\nabla^2 + V \right) \Psi^* \tag{4.24}$$

By replacing in Eq. (4.23), we get:

$$\begin{aligned}
\frac{\partial\rho}{\partial t} &= \left(+i\left(-\frac{1}{2m}\nabla^2 + V \right) \Psi^* \right) \Psi + \Psi^* \left(-i\left(-\frac{1}{2m}\nabla^2 + V \right) \Psi \right) \\
&= \left(-\frac{i}{2m}\nabla^2\Psi^* \right) \Psi + \Psi^* \left(i\frac{1}{2m}\nabla^2\Psi \right) \\
&= \frac{-i}{2m}\left[\left(\nabla^2\Psi^*\right) \Psi - \left(\nabla^2\Psi\right) \Psi^* \right]
\end{aligned} \tag{4.25}$$

In this equation, the **first time derivative** (i.e., the evolution) of the density is related to the wave function and its **second spatial derivative**. We note that the variation in time of the probability density inside a volume V is related to the net flux leaving the surface A, where $A = \partial V$ is the surface which encompasses the volume V:

$$\frac{\partial}{\partial t}\int_V \rho\, \mathrm{d}^3\vec{x} = -\int_{A=\partial V} \vec{j} \cdot \mathrm{d}\vec{A} \tag{4.26}$$

where $\vec{j}$ is the **probability density current vector**. The minus sign in front of the integral is there to take into account the "outward" flux through the surface. With the help of the **theorem of Gauss**[15] (see Appendix A.11), we get:

$$\frac{\partial}{\partial t} \int_V \rho \, \mathrm{d}^3 \vec{x} = - \int_{A=\partial V} \vec{j} \cdot \mathrm{d}\vec{A} = - \int_V \nabla \cdot \vec{j} \, \mathrm{d}^3 \vec{x} \tag{4.27}$$

Accordingly, we define a vector current that satisfies the continuity equation and describes "*the flow of the probability density*":

$$\vec{j}(\vec{x}, t) \equiv \frac{i}{2m} \left[(\nabla \Psi^*) \, \Psi - (\nabla \Psi) \, \Psi^* \right] \tag{4.28}$$

The concept of current is a very important one that will be used very often! As expected:

$$
\begin{aligned}
\nabla \cdot \vec{j} &= \frac{i}{2m} \left(\nabla \left((\nabla \Psi^*) \Psi \right) - \nabla \left((\nabla \Psi) \Psi^* \right) \right) = \frac{i}{2m} \left(\left(\nabla^2 \Psi^* \right) \Psi + \nabla \Psi^* \nabla \Psi - \nabla \Psi \nabla \Psi^* - \left(\nabla^2 \Psi \right) \Psi^* \right) \\
&= \frac{i}{2m} \left(\left(\nabla^2 \Psi^* \right) \Psi - \left(\nabla^2 \Psi \right) \Psi^* \right) = -\frac{\partial}{\partial t} \rho
\end{aligned}
\tag{4.29}
$$

where in the last term we have used Eq. (4.25). We have therefore found the following **continuity equation**:

$$\frac{\partial \rho}{\partial t} + \nabla \cdot \vec{j} = 0 \tag{4.30}$$

This is also a very important equation!

- **Free particle case.** We consider a free particle, i.e., $V = 0$:

$$V(\vec{x}, t) \equiv 0 \quad \longrightarrow \quad i \frac{\partial}{\partial t} \Psi = -\frac{1}{2m} \nabla^2 \Psi \quad \longrightarrow \quad \Psi(\vec{x}, t) = N e^{-i(Et - \vec{p} \cdot \vec{x})} \tag{4.31}$$

The wave function is a plane wave. It follows that the probability density is a constant in space:

$$\rho = \Psi^* \Psi = |N|^2 \qquad \text{Units}: \ 1/\mathrm{L}^3 \tag{4.32}$$

This means that the particle has the same probability of finding itself in all points in space and the current density describes a constant flux of particles, that moves with a constant speed. The current density is therefore given by the product of the normalization squared and the velocity of the particle:

$$
\begin{aligned}
\vec{j}(\vec{x}, t) &\equiv \frac{i}{2m} \left((\nabla \Psi^*) \, \Psi - (\nabla \Psi) \, \Psi^* \right) = \frac{i}{2m} \left((-i\vec{p}) \Psi^* \Psi - (i\vec{p}) \Psi \Psi^* \right) \\
&= \frac{i}{2m} (-i\vec{p}) 2 |N|^2 = |N|^2 \frac{\vec{p}}{m} = |N|^2 \vec{v}
\end{aligned}
\tag{4.33}
$$

The units of the current density are that of a number per unit time and per unit area:

$$\text{Units}: \ \left[1/L^3 \right] \left[L/T \right] = \left[1/(TL^2) \right] \tag{4.34}$$

It all makes sense!

4.5 The Width of a Metastable State (Breit–Wigner)

A stationary state of energy E takes the form:

$$\Psi(\vec{x}, t) = e^{-iEt} \Psi(\vec{x}) \tag{4.35}$$

---

15   Johann Carl Friedrich Gauss (1777–1855), German mathematician.

All the time dependence is carried by the exponential prefactor. This leads to physical observables of the state that are actually time-independent. For example, the norm of the state is preserved and is given by:

$$|\Psi(\vec{x}, t)|^2 = e^{+iEt}\Psi^*(\vec{x})e^{-iEt}\Psi(\vec{x}) = \Psi^*(\vec{x})\Psi(\vec{x}) = |\Psi(\vec{x})|^2 \tag{4.36}$$

Such a term would, for instance, be appropriate to describe a stable particle. For an unstable particle, we should obtain a time evolution that takes into account the fact that the probability of finding the particle will decrease exponentially with time as $|\Psi(\vec{x}, t)|^2 \propto e^{-t/\tau}$, according to the law of radioactive decay (see Section 2.5). This occurs in the rest frame of the particle, hence the energy is given by its rest mass labeled m_R. The width of the unstable particle is defined as $\Gamma \equiv 1/\tau$ (see Eq. (2.30)), hence, to describe the time evolution of the wave function representing an unstable particle, Eq. (4.35) must be modified to:

$$\Psi(\vec{x}, t) = e^{-im_R t}e^{-(\Gamma/2)t}\Psi(\vec{x}) \tag{4.37}$$

such that the probability density becomes as desired $\propto e^{-t/\tau} = e^{-\Gamma t}$:

$$|\Psi(\vec{x}, t)|^2 = e^{+im_R t}e^{-(\Gamma/2)t}\Psi^*(\vec{x})e^{-im_R t}e^{-(\Gamma/2)t}\Psi(\vec{x}) = e^{-\Gamma t}\Psi^*(\vec{x})\Psi(\vec{x}) = e^{-\Gamma t}|\Psi(\vec{x})|^2 \tag{4.38}$$

The above expression gives us the time evolution of the wave function. In order to transform it into a statement on the energy of the particle, we consider its Fourier transformation, which effectively will translate the function $\Psi(t)$ into $\Psi(E)$, where E is the energy of the state. Using Eq. (A.24) and assuming $t \geq 0$, we find:

$$A(E) = \frac{\Psi(\vec{x})}{\sqrt{2\pi}}\int_0^\infty e^{+i(E-m_R)t}e^{-(\Gamma/2)t}\mathrm{d}t = \frac{\Psi(\vec{x})}{\sqrt{2\pi}}\frac{i}{(E - m_R) + i\Gamma/2} \tag{4.39}$$

where the lower limit on the integral has been set to zero since we are only looking at positive t. The probability density $P(E)$ of finding the state with energy E is proportional to the squared amplitude:

$$P(E) = |A(E)|^2 \propto \frac{|\Psi(\vec{x})|^2}{(E - m_R)^2 + (\Gamma/2)^2} \tag{4.40}$$

Setting the normalization to unity $\int_{-\infty}^{+\infty} P(E)\mathrm{d}E = 1$ leads to the following expression:

$$P(E) = \frac{1}{2\pi}\frac{\Gamma}{(E - m_R)^2 + \Gamma^2/4} \tag{4.41}$$

This function is called the **Breit–Wigner curve**. A graphical representation is shown in Figure 4.4. The probability is half the maximum value at the energies equal to $m_R \pm \Gamma/2$. Hence, the width Γ is called the *full width at half maximum*. The mass of an unstable particle is not *sharp*. The finite lifetime introduces a broadening of the energy. This width of a particle (or in general a quantum state) because of its decay is called the **natural width**, which can be interpreted as a consequence of the Heisenberg uncertainty principle for the time and energy. This phenomenon is well known in atomic spectroscopy, where the **natural line width** is a measure of the lifetime of the corresponding transition. In practice, however, a very narrow line can be affected by the random motion related to the temperature of the source. This effect is called **Doppler broadening** and it can be reduced by using low-temperature sources. Last but not least, there is another effect called **collision broadening**. It typically happens when the observed medium is at relatively high pressure, so that collisions occur with a timescale less than or equal to the lifetime. In this case, it is no longer possible to carry out an undisturbed measurement for a time comparable to, or larger than, the lifetime and consequently, the width of the observed line will be larger than its natural width.

4.6   The Time-Evolution Operator

In quantum mechanics there is no Hermitian operator whose eigenvalues correspond to the time of the system. Unlike position or momentum, time is *not* an observable. **Time is a parameter, not a measurable quantity.**

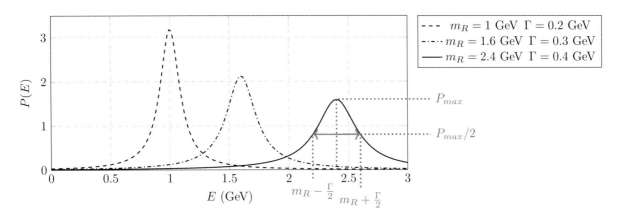

Figure 4.4 The Breit–Wigner probability density function $P(E)$ as a function of energy E. The value Γ represents the full width at half maximum.

This will actually pose a problem when we want to generalize quantum mechanics to a relativistic theory, since this latter teaches us that space and time should be treated on an equal footing. Perhaps surprisingly, we will see in Chapter 6 that the equality of treatment between space and time is addressed in quantum field theories by demoting the position $\vec{x}$ to a parameter level, just as time!

In the meantime, we first consider the non-relativistic case and want to define the Dyson operator $U(t,t_0)$ to describe the "time evolution" of the system from time t_0 to time t. If the system is in a given state at time t_0, there is no reason to expect that the system will remain in this state at a later time t. In particular, we want to understand the evolution of the system due to an external *interaction* (eventually an external *perturbation*). It is customary to use **Heisenberg's representation** for operators (see Appendix C.6):

$$\phi(x^\mu) = e^{iH(t-t_0)}\phi(t_0,\vec{x})e^{-iH(t-t_0)} \tag{4.42}$$

where $H = H_0 + H_{int}$ is the full Hamiltonian of the system. Following Eq. (9.1) we can separate the time evolution of the free system from the interaction part, by defining the field evolution in the **interaction frame** as:

$$\phi_I(x^\mu) = e^{iH_0(t-t_0)}\phi(t_0,\vec{x})e^{-iH_0(t-t_0)} \tag{4.43}$$

By combination of the two above equations, we can replace the term $\phi(t_0,\vec{x})$ on the r.h.s. by $\phi_I(t,\vec{x})$. We have:

$$\begin{aligned}
\phi(t,\vec{x}) &= e^{iH(t-t_0)}\left[e^{-iH_0(t-t_0)}\phi_I(t,\vec{x})e^{iH_0(t-t_0)}\right]e^{-iH(t-t_0)} \\
&= \left[e^{iH(t-t_0)}e^{-iH_0(t-t_0)}\right]\phi_I(t,\vec{x})\left[e^{iH_0(t-t_0)}e^{-iH(t-t_0)}\right] \\
&= U^\dagger(t,t_0)\phi_I(t,\vec{x})U(t,t_0)
\end{aligned} \tag{4.44}$$

where the **Dyson**[16] **time-evolution operator** $U(t,t_0)$ is defined as:

$$U(t,t_0) \equiv e^{iH_0(t-t_0)}e^{-iH(t-t_0)} \tag{4.45}$$

By construction, we have:

$$U^\dagger(t,t_0) = U(t_0,t) \quad \text{and} \quad U(t_2,t_1)U(t_1,t_0) = U(t_2,t_0) \quad \text{for } t_2 \geq t_1 \geq t_0 \tag{4.46}$$

16   Freeman John Dyson (1923–2020), English-born American theoretical physicist.

We note that the time derivative of the operator is equal to:

$$
\begin{aligned}
i\frac{\partial}{\partial t}U(t,t_0) &= e^{iH_0(t-t_0)}(H-H_0)e^{-iH(t-t_0)} = e^{iH_0(t-t_0)}H_{int}e^{-iH(t-t_0)} \\
&= e^{iH_0(t-t_0)}H_{int}e^{-iH_0(t-t_0)}\underbrace{e^{iH_0(t-t_0)}e^{-iH(t-t_0)}}_{=U(t,t_0)}
\end{aligned}
\tag{4.47}
$$

By defining the **time-dependent interaction Hamiltonian operator** $H_{int}(t)$ as:

$$
H_{int}(t) = e^{iH_0(t-t_0)}H_{int}e^{-iH_0(t-t_0)}
\tag{4.48}
$$

we obtain a differential equation for the time-evolution operator:

$$
i\frac{\partial}{\partial t}U(t,t_0) = H_{int}(t)U(t,t_0)
\tag{4.49}
$$

The formal solution with the initial condition that $U(t_0,t_0) = \mathbb{1}$ can be simply expressed as:

$$
U(t,t_0) = \mathbb{1} - i\int_{t_0}^{t}\mathrm{d}t_1\, H_{int}(t_1)U(t_1,t_0)
\tag{4.50}
$$

The solution can be found iteratively by inserting successive replacements of U on the r.h.s. of the equation. It leads to the **Dyson series**:

$$
\begin{aligned}
U(t,t_0) &= \mathbb{1} - i\int_{t_0}^{t}\mathrm{d}t_1\, H_{int}(t_1)U(t_1,t_0) \\
&= \mathbb{1} - i\int_{t_0}^{t}\mathrm{d}t_1\, H_{int}(t_1)\left(\mathbb{1} - i\int_{t_0}^{t_1}\mathrm{d}t_2\, H_{int}(t_2)U(t_2,t_0)\right) \\
&= \mathbb{1} + (-i)\int_{t_0}^{t}\mathrm{d}t_1\, H_{int}(t_1) + (-i)^2\int_{t_0}^{t}\mathrm{d}t_1\int_{t_0}^{t_1}\mathrm{d}t_2\, H_{int}(t_1)H_{int}(t_2) + \cdots \\
&\quad + (-i)^n\int_{t_0}^{t}\mathrm{d}t_1\ldots\int_{t_0}^{t_{n-1}}\mathrm{d}t_n\, H_{int}(t_1)\ldots H_{int}(t_n) + \cdots
\end{aligned}
\tag{4.51}
$$

where we note that by construction $t_0 < \ldots < t_2 < t_1 < t$. We introduce the **time-ordering Dyson operator** T (see Appendix A.7) and transform the time-integration boundaries in each integral to be the same. For example, as illustrated in Figure 4.5, the integral with the condition $t_0 < t_2 < t_1 < t$ can be replaced by two independent time-ordered integrals with $t_0 < t_1 < t$ and $t_0 < t_2 < t$:

$$
\int_{t_0}^{t}\mathrm{d}t_1\int_{t_0}^{t_1}\mathrm{d}t_2\, H_{int}(t_1)H_{int}(t_2) = \frac{1}{2}\int_{t_0}^{t}\mathrm{d}t_1\int_{t_0}^{t}\mathrm{d}t_2\, T\left[H_{int}(t_1)H_{int}(t_2)\right]
\tag{4.52}
$$

If we order all terms with the help of the time-ordering operator, we obtain a Taylor series of time-ordered terms:

$$
U(t,t_0) = \mathbb{1} + \sum_{n=1}^{\infty}\frac{(-i)^n}{n!}\int_{t_0}^{t}\mathrm{d}t_1\ldots\int_{t_0}^{t}\mathrm{d}t_n\, T\left[H_{int}(t_1)\ldots H_{int}(t_n)\right]
\tag{4.53}
$$

It can be formally written with the help of the exponential function:

$$
U(t,t_0) = T\left[e^{-i\int_{t_0}^{t}\mathrm{d}t'\, H_{int}(t')}\right]
\tag{4.54}
$$

We expect the series to converge, and in practice the Dyson operator will be estimated up to a given order in the series. We recall that this expression has been derived in the interaction frame. Even though this result was derived in the context of classical quantum mechanics, we will be able to use it in a relativistic context, as will be shown in Chapter 9!

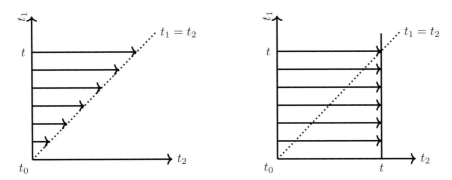

Figure 4.5 (left) The integration in the Dyson series with $t_0 < t_2 < t_1 < t$; (right) the time-ordered integral with $t_0 < t_1 < t$ and $t_0 < t_2 < t$.

4.7   Angular Momentum and Rotations

The space representation of the **angular momentum operator** $\vec{L}$ is given by

$$\vec{L} = (L_x, L_y, L_z) \equiv \vec{x} \times \vec{p} = -i\vec{x} \times \nabla \tag{4.55}$$

where $\vec{x} = (x, y, z)$. See Appendix C.9 for more details. We want to show that **angular momentum is intimately connected with space rotations.** We consider the clockwise space rotation of an angle θ around the z-axis. This rotation is represented by the unitary matrix $R_z(\theta)$ (see Appendix B.6):

$$\begin{pmatrix} x' \\ y' \\ z' \end{pmatrix} = R_z(\theta) \begin{pmatrix} x \\ y \\ z \end{pmatrix} = \begin{pmatrix} \cos\theta & \sin\theta & 0 \\ -\sin\theta & \cos\theta & 0 \\ 0 & 0 & 1 \end{pmatrix} \begin{pmatrix} x \\ y \\ z \end{pmatrix} \tag{4.56}$$

Similarly, one can define matrices for rotations around the x- and y-axis. Rotation matrices form a **non-Abelian Lie algebraic group called** $SO(3)$. The theory of groups is briefly summarized in Appendix B. That $SO(3)$ is non-Abelian is related to the non-commutative nature of rotations.

We consider the wave function Ψ and *assume that it is not directly influenced by a rotation R*. This means that the **wave function possesses a specific symmetry under the rotation** R: everything occurs as if we only rotate our reference system. See Figure 4.6. The rotation (or in general the transformation) we considered is **passive**; it affects the frame of reference but does not act on the physical system itself. The new wave function after the rotation is written as $\Psi'(\vec{x}\,')$. Mathematically:

$$\Psi'(\vec{x}\,') = \Psi(\vec{x}) \quad \Longrightarrow \quad \Psi'(\vec{x}\,') = \Psi(R^{-1}\vec{x}\,') \tag{4.57}$$

Since the latter relation holds for all $\vec{x}\,'$, it can be rewritten as

$$\Psi'(\vec{x}) = \Psi(R^{-1}\vec{x}) \tag{4.58}$$

We note that the relation between Ψ and Ψ' must be represented through **a unitary transformation** $U(\alpha)$ **in the Hilbert space** (see Appendix C). The unitarity ensures the conservation of the normalization $\langle \Psi | \Psi \rangle = \langle \Psi' | \Psi' \rangle$:

$$|\Psi'\rangle = U(\theta)\,|\Psi\rangle \tag{4.59}$$

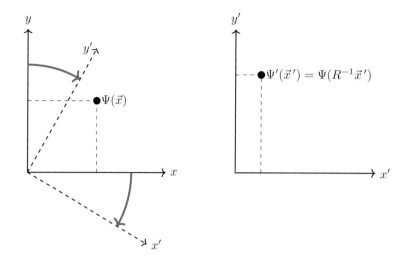

Figure 4.6 Illustration of the transformation of the wave function under the rotation R of the axes (note that we have chosen the clockwise definition of the rotation).

The symbols $R(\theta)$ and $U(\theta)$ are different objects, although they both represent the rotation: R **acts on the space coordinates while the transformation** U **represents the effect on the wave function and acts on the corresponding Hilbert space. They are different representations of the physical space rotation that we have considered.**

For a continuous transformation like a rotation, one can always consider an infinitesimal transformation of angle $\delta\theta$. For example, around the z-axis we have, replacing $\cos\delta\theta \to 1$ and $\sin\delta\theta \to \delta\theta$:

$$R_z(\delta\theta) \simeq \begin{pmatrix} 1 & \delta\theta & 0 \\ -\delta\theta & 1 & 0 \\ 0 & 0 & 1 \end{pmatrix} \implies R_z^{-1}(\delta\theta) = R_z(-\delta\theta) \simeq \begin{pmatrix} 1 & -\delta\theta & 0 \\ \delta\theta & 1 & 0 \\ 0 & 0 & 1 \end{pmatrix} \tag{4.60}$$

For such an infinitesimal rotation, we have, writing out the components:

$$\begin{aligned} \Psi'(\vec{x}) &= \Psi(R_z^{-1}\vec{x}) = \Psi(x - y\delta\theta, y + x\delta\theta, z) \approx \Psi(x,y,z) - y\delta\theta\frac{\partial\Psi}{\partial x} + x\delta\theta\frac{\partial\Psi}{\partial y} \\ &= \Psi(x,y,z) + i\delta\theta L_z\Psi(x,y,z) \end{aligned} \tag{4.61}$$

where we used (see Appendix C.9):

$$L_z = xp_y - yp_x = -i\left(x\frac{\partial}{\partial y} - y\frac{\partial}{\partial x}\right) \tag{4.62}$$

to introduce the angular momentum operator. Hence, the unitary transformation U on the ket is given by:

$$|\Psi'\rangle = U(\delta\theta)|\Psi\rangle = [\mathbb{1} + i\delta\theta L_z]|\Psi\rangle \tag{4.63}$$

where L_z is the angular momentum operator acting on $|\Psi\rangle$.

• **Generator G of a transformation.** This result can be generalized to any rotation around any direction. In the theory of **Lie groups** (see Appendix B.5), one considers the generator G of a transformation. The generator is defined with an infinitesimal transformation with parameter ϵ (do not confuse the $U(\epsilon)$ operator defined here with the Dyson time-evolution operator $U(t, t_0)$ defined in the previous section!):

$$U(\epsilon) \equiv (\mathbb{1} + i\epsilon G) \tag{4.64}$$

The conservation of the normalization of the wave function during the transformation implies:

$$U^\dagger(\epsilon)U(\epsilon) = (\mathbb{1} - i\epsilon G^\dagger)(\mathbb{1} + i\epsilon G) = \mathbb{1} - i\epsilon(G^\dagger - G) + \cdots \tag{4.65}$$

and therefore $U^\dagger(\epsilon)U(\epsilon) = \mathbb{1}$ yields:

$$G^\dagger = G \tag{4.66}$$

Consequently, a group **generator G is a Hermitian operator**, and hence represents a real (i.e., physical) observable. A finite transformation can be obtained by application of successive infinitesimal transformations (see Eq. (B.12)):

$$U(\theta) = \left(U\left(\frac{\theta}{n}\right)\right)^n = \left(\mathbb{1} + i\frac{\theta}{n}G\right)^n \stackrel{n\to\infty}{=} e^{i\theta G} \tag{4.67}$$

In the last term, we have used the exponentiation of an operator A, which is given by:

$$U \equiv \exp(A) = \sum_{k=0}^\infty \frac{A^k}{k!} = \mathbb{1} + A + \frac{A^2}{2} + \frac{A^3}{3!} + \cdots \tag{4.68}$$

Any finite rotation in space can be defined by three parameters (e.g., three Euler angles). To these correspond **three generators of the $SO(3)$ group**. They can, for example, be defined as J_x, J_y, and J_z (see Appendix B.6):

$$J_x = \begin{pmatrix} 0 & 0 & 0 \\ 0 & 0 & -i \\ 0 & i & 0 \end{pmatrix}, \qquad J_y = \begin{pmatrix} 0 & 0 & i \\ 0 & 0 & 0 \\ -i & 0 & 0 \end{pmatrix}, \quad \text{and} \quad J_z = \begin{pmatrix} 0 & -i & 0 \\ i & 0 & 0 \\ 0 & 0 & 0 \end{pmatrix} \tag{4.69}$$

A finite rotation by an angle α around the axis $\vec{n}$ can then be expressed as (see Eq. (B.27)):

$$R_n(\alpha) = e^{i\alpha(n_x J_x + n_y J_y + n_z J_z)} = e^{i\alpha\hat{n}\cdot\vec{J}} \tag{4.70}$$

Its representation on the Hilbert space is then given by the unitary transformation:

$$U(\alpha) = e^{i\alpha\left(\vec{n}\cdot\vec{L}\right)} \tag{4.71}$$

This means that **the angular momentum operator is the generator of spatial rotations in the wave function space!** Going back to the infinitesimal case, we obtain:

$$U(\delta\alpha) = \exp\left(i\delta\alpha\left(\vec{n}\cdot\vec{L}\right)\right) = \sum_{k=0}^\infty \frac{(i\delta\alpha\vec{n}\cdot\vec{L})^k}{k!} \approx 1 + i\delta\alpha\vec{n}\cdot\vec{L} + \cdots \tag{4.72}$$

where the vector $\vec{L}$ is composed of the angular momentum operators for each spatial component. When the Hamiltonian operator H is invariant under the transformation U, the following commutation rule is valid since the order of the operations does not matter:

$$[H,U] = 0 \qquad \Leftrightarrow \qquad HU = UH \tag{4.73}$$

Inserting $U(\delta\alpha)$, we get, keeping only the first term in the expansion:

$$H(1 + i\delta\alpha\vec{n}\cdot\vec{L}) = (1 + i\delta\alpha\vec{n}\cdot\vec{L})H \qquad \Rightarrow \qquad H(\vec{n}\cdot\vec{L}) = (\vec{n}\cdot\vec{L})H \tag{4.74}$$

We can rewrite this using the conventional commutator *for a vector*, which means that the commutation is valid for each component of the vector:

$$H\vec{L} = \vec{L}H \quad \Rightarrow [H,\vec{L}] = 0 \tag{4.75}$$

To summarize: **The angular momentum operator is the generator of the rotation transformation in the Hilbert space. It is Hermitian and is therefore a physical observable. The conservation of angular momentum is a consequence of the invariance of the Hamiltonian under the rotation.** This last feature will be discussed more in Chapter 6.

4.8 Linear Momentum and Translations

Translations in space also form an **Abelian Lie group** (see Appendix B.5). The elements of this group are defined by the vector parameter $\vec{a}$:

$$\vec{x}' = (x', y', z') = \vec{x} + \vec{a} \tag{4.76}$$

In analogy to the space rotation case, the action of the translation on the wave function can be written as:

$$\Psi'(\vec{x}) = \Psi(R^{-1}\vec{x}) = \Psi(\vec{x} - \vec{a}) \tag{4.77}$$

For an infinitesimal translation, we obtain:

$$
\begin{aligned}
\Psi(\vec{x} - \delta\vec{a}) &\approx \Psi(\vec{x}) - \delta a_x \frac{\partial \Psi}{\partial x} - \delta a_y \frac{\partial \Psi}{\partial y} - \delta a_z \frac{\partial \Psi}{\partial z} + \cdots \\
&= \Psi(\vec{x}) - i\delta a_x p_x \Psi(\vec{x}) - i\delta a_y p_y \Psi(\vec{x}) - i\delta a_z p_z \Psi(\vec{x}) + \cdots \\
&= \Psi(\vec{x}) - i\delta\vec{a} \cdot \vec{p}\, \Psi(\vec{x}) + \cdots
\end{aligned}
\tag{4.78}
$$

Hence, for a finite translation, we expect that the transformation $U(\vec{a})$ in Hilbert space will be given by:

$$|\Psi'\rangle \equiv U(\vec{a})|\Psi\rangle = e^{-i\vec{a}\cdot\vec{p}}|\Psi\rangle \tag{4.79}$$

For the probability $\langle\Psi|\Psi\rangle$ to be conserved, the momentum operator must be Hermitian. It represents therefore a physical observable. **The momentum operator is the generator of the space translations.** Finally, we expect that the conservation of linear momentum follows from the invariance of the system under translation, in a similar way as angular momentum conservation followed from invariance under rotations.

4.9 Discrete Transformations

We consider the **parity transformation** applied to the wave function of the quantum system. This gives:

$$P : \begin{cases} \vec{x} \to -\vec{x} \\ t \to t \end{cases} \longrightarrow \text{System } \Psi : \quad \Psi' = P\Psi(\vec{x}) = \Psi(-\vec{x}) \tag{4.80}$$

Two successive space-mirroring operations give the initial wave function, so:

$$\Psi'(\vec{x'}) = P\Psi(\vec{x}) = P^2\Psi'(\vec{x'}) \qquad \text{or} \qquad P^2 = 1 \tag{4.81}$$

Since parity is a Hermitian operator, it also corresponds to an observable physical quantity. The eigenvalues of the transformation are $+1$ or -1. Parity is conserved in a system when $[H, P] = 0$. This is, for example, the case in the central potential problem: $PH(\vec{x}) = H(-\vec{x}) = H(\vec{x})$. A quantum bound system with a radial symmetry has a definite parity. The parity of its wave function $\Psi(\vec{x})$ can be either even, $P = +1$, or odd, $P = -1$ (functionally for example $\Psi(x) = \cos kx$ is even, and $\Psi(x) = \sin kx$ is odd).

As an illustration of a central potential problem, we can mention the **wave functions of the hydrogen atom**, which are a product of a radial function $f(r)$ and the spherical harmonics $Y_{\ell m}(\theta, \phi)$, where ℓ and m are the orbital angular momentum of the state and its projection along an axis (see Appendix C.9):

$$\Psi(r, \theta, \phi) \equiv f(r) Y_{\ell m}(\theta, \phi) \tag{4.82}$$

In spherical polar coordinates the parity operation changes $(r, \vartheta, \varphi) \to (r, \pi - \vartheta, \pi + \varphi)$. From the properties of $Y_{\ell m}$ the wave functions have parity $P = (-1)^\ell$.

Parity is a **multiplicative quantum number**, which means that the parity of a composite system is the product of the parities of the constituents. For a system of particles, the overall parity is given by the product of the intrinsic parities of the particles times the parity of their relative motion defined by their quantum angular momentum ℓ.

From the experimental observation of single photon transitions between atomic states it can be deduced that they obey the **selection rule**:

$$\Delta L = \pm 1 \tag{4.83}$$

• **Intrinsic parity.** In the case of a radiative transition $\ell \to \ell + 1$ of an atom, the initial state has parity $(-1)^\ell$ and the final state has parity $(-1)^{\ell+1} P_\gamma = (-1)^{\ell+1}(-1) = (-1)^\ell$, so parity is conserved in such electromagnetic transitions. Similarly for $\ell \to \ell - 1$. From this it can be deduced that the **intrinsic parity of the photon** is:

$$(-1)^\ell = (-1)^{\ell \pm 1} \times P_\gamma \qquad \text{thus} \qquad P_\gamma = -1 \tag{4.84}$$

Fermions are always created in pairs since they have half-integer spins and angular momentum must be conserved. **The intrinsic parity of an antifermion is therefore opposite to that of a fermion.** By convention, the **intrinsic parity of the proton** and of the **neutron** are taken to be:

$$P(p) = +1, \qquad P(n) = +1 \tag{4.85}$$

For the antinucleons we have $P(\bar{p}) = P(\bar{n}) = -1$. See Table 4.1. The Dirac equation (discussed in Chapter 8)

	$\mathbf{J^{[PC]}}$	**P**	**C**		**P**	**C**
Photon $\lvert\gamma\rangle$	1^{--}	-1	-1			
Several photons $\lvert n\gamma\rangle$		$(-1)^n$	$(-1)^n \lvert\gamma\rangle$			
Electron $\lvert e^-\rangle$	$\frac{1}{2}$	$+1$	$\lvert e^+\rangle$	Positron $\lvert e^+\rangle$	-1	$\lvert e^-\rangle$
Proton $\lvert p\rangle$	$\frac{1}{2}$	$+1$	$\lvert\bar{p}\rangle$	Antiproton $\lvert\bar{p}\rangle$	-1	$\lvert p\rangle$
Neutron $\lvert n\rangle$	$\frac{1}{2}$	$+1$	$\lvert\bar{n}\rangle$	Antineutron $\lvert\bar{n}\rangle$	-1	$\lvert n\rangle$
Pion $\lvert\pi^-\rangle$	0^-	-1	$\lvert\pi^+\rangle$	Pion $\lvert\pi^+\rangle$	$+1$	$\lvert\pi^-\rangle$
Pion $\lvert\pi^0\rangle$	0^{-+}	-1	$\lvert\pi^0\rangle$			
Eta $\lvert\eta\rangle$	0^{-+}	-1	$\lvert\eta\rangle$			
Kaon $\lvert K^0\rangle$	0^-	-1	$\lvert\overline{K^0}\rangle$	Antikaon $\lvert\overline{K^0}\rangle$	$+1$	$\lvert K^0\rangle$

Table 4.1 Example of particle properties under parity and charge conjugation transformations.

also requires that the parities of Dirac particles and antiparticles be opposite of each other. So, for example, the intrinsic parities of the **electron** and of its antiparticle, the **positron**, are:

$$P(e^-) = +1, \qquad P(e^+) = -1 \tag{4.86}$$

Similarly, quarks are assigned parity $+1$ and antiquarks -1. For elementary gauge bosons, the parity of the particle and antiparticle is the same. For other particles (e.g., pions), the parity can be determined by experimental observations. For example, in 1954 **Chinowski** and **Steinberger** showed that the negative pion has a parity $P_{\pi^-} = -1$, hence the pion is a pseudoscalar particle (see Section 15.5).

• **Charge conjugation.** The **charge conjugation** C **reverses the sign of the electric charge and magnetic moment of a particle.** Like the parity operator it satisfies $C^2 = 1$, and has possible eigenvalues $C = \pm 1$. Since the electromagnetic fields change sign under C (see Table 1.3), the photon has:

$$C_\gamma = -1 \tag{4.87}$$

and, consequently, a state $|n\gamma\rangle$ composed of n photons obeys:

$$C|n\gamma\rangle = (-1)^n |\gamma\rangle \tag{4.88}$$

Charge conjugation changes a particle into an antiparticle, so particles are not necessarily eigenstates of C. For example, the charge-conjugated state of the proton is the antiproton (see Section 15.9), and similarly for the charged pion (see Eq. (15.18)):

$$C|p\rangle = |\bar{p}\rangle, \qquad C|\pi^+\rangle = |\pi^-\rangle \tag{4.89}$$

Particles which are their own antiparticles can be expressed as C eigenstates. Examples are the mesons π^0 (see Eq. (15.21)) or η (see Section 15.10); however, not all neutral particles are necessarily C eigenstates, for instance the neutral kaon K^0 (see Eq. (15.62)). We find:

$$C|\pi^0\rangle = |\pi^0\rangle, \qquad C|\eta\rangle = |\eta\rangle, \qquad C|K^0\rangle = |\overline{K^0}\rangle \neq |K^0\rangle \tag{4.90}$$

More generally, we could have defined the charge conjugation operator with an overall "phase" η_C, such that

$$C|\pi^0\rangle \equiv \eta_C |\pi^0\rangle \tag{4.91}$$

but, as mentioned above, two successive applications of the charge conjugation must give back the original state, so $C^2 = 1$ and $\eta_C = \pm 1$.

Furthermore, **electromagnetic and strong interactions conserve both** C **and** P **quantum numbers** (see Table 1.5). We experimentally observe that the π^0 decays into two photons via an electromagnetic interaction, and consequently C must be conserved during the process. Since Eq. (4.87) states that $C_\gamma = -1$, the observation of $\pi^0 \rightarrow \gamma\gamma$ forces the charge conjugation of π^0 to be $\eta_C = (C_\gamma)^2 = +1$, hence $C|\pi^0\rangle = |\pi^0\rangle$. Another consequence of C conservation is that the π^0 decays into two photons but not three. This is very strongly supported by experimental measurements (values taken from the Particle Data Group's Review [5]):

$$\begin{aligned} Br(\pi^0 \rightarrow \gamma\gamma) &= (98.823 \pm 0.034)\% \\ Br(\pi^0 \rightarrow \gamma\gamma\gamma) &< 3.1 \times 10^{-8} \quad (90\% \text{ CL}) \end{aligned} \tag{4.92}$$

Combinations of fermions can also be eigenstates of C; a spin-1/2 fermion–antifermion bound state $\bar{f}f$ will have $C = (-1)^{l+s}$, where l represents the relative angular momentum and s defines their spin configurations. For example, the electron–positron bound state, i.e., **positronium** (see Chapter 14.9), has exactly:

$$C(e^+ e^-) = (-1)^{\ell + s} \tag{4.93}$$

where s is the sum of the spins which can be either 0 or 1. Considering again the π^0 meson as a quark–antiquark bound state $q\bar{q}$, we have in this case $\ell = s = 0$ and hence recover the result $\eta_C = +1$ for the neutral pion. The behavior of our systems under discrete symmetries will be an important aspect for the understanding of their underlying physics, as will be seen in the coming chapters.

4.10   Schrödinger Equation for a Charged Particle in an Electromagnetic Field

As we will discuss in Section 10.2, the magnetic vector potential $\vec{A}(\vec{x}, t)$ is defined such that the magnetic field $\vec{B}$ is given by $\vec{B}(\vec{x}, t) \equiv \nabla \times \vec{A}$. The electric scalar potential $\phi(\vec{x}, t)$ is defined such that the electric field $\vec{E}$

is given by $\vec{E}(\vec{x}, t) \equiv -\nabla \phi(\vec{x}, t) - \frac{\partial \vec{A}(\vec{x}, t)}{\partial t}$. Then, the classical Hamiltonian for a charged particle e in these external electromagnetic fields $(\vec{A}, \phi)$ is given by:

$$H = \frac{1}{2m} \left(\vec{p} - e\vec{A}(\vec{x}, t) \right) \cdot \left(\vec{p} - e\vec{A}(\vec{x}, t) \right) + e\phi(\vec{x}, t) + V(\vec{x}, t) \qquad (4.94)$$

Let us show this result. For simplicity we can set the external potential $V(\vec{r}, t)$ to zero. We also write $\vec{x} = (x_1, x_2, x_3)$ and $\vec{A} = (A_1, A_2, A_3)$. Hamilton and Lagrange formalism will be reviewed in Chapter 6. Anticipating some of the results, we find using Hamilton's equation for the coordinate x_i (see Eq. (6.9)):

$$\frac{\mathrm{d}x_i}{\mathrm{d}t} = \frac{\partial H}{\partial p_i} = \frac{1}{2m}(2p_i - 2eA_i) = \frac{p_i}{m} - \frac{e}{m}A_i \quad \rightarrow \quad m\frac{\mathrm{d}x_i}{\mathrm{d}t} = p_i - eA_i \qquad (4.95)$$

Hence, the time variation of the ith component p_i of the canonical momentum is:

$$\frac{\mathrm{d}p_i}{\mathrm{d}t} = m\frac{\mathrm{d}^2 x_i}{\mathrm{d}t^2} + e\frac{\mathrm{d}A_i}{\mathrm{d}t} = m\frac{\mathrm{d}^2 x_i}{\mathrm{d}t^2} + e\frac{\partial A_i}{\partial t} + e\frac{\partial A_i}{\partial x_j}\frac{\mathrm{d}x_j}{\mathrm{d}t} \qquad (4.96)$$

Also, with Hamilton's equation for p_i (Eq. (6.9)), we get:

$$\frac{\mathrm{d}p_i}{\mathrm{d}t} = -\frac{\partial H}{\partial x_i} = \frac{2e}{2m}(\vec{p} - e\vec{A}) \cdot \frac{\partial \vec{A}}{\partial x_i} - e\frac{\partial \phi}{\partial x_i} = e\frac{\mathrm{d}x_j}{\mathrm{d}t}\nabla_i A_j - e\nabla_i \phi \qquad (4.97)$$

where in the second equality we used $p_i = m\mathrm{d}x_i/\mathrm{d}t + eA_i$. Putting the two results together, we find:

$$
\begin{aligned}
m\frac{\mathrm{d}^2 x_i}{\mathrm{d}t^2} &= -e\left(\frac{\partial A_i}{\partial t} + \frac{\mathrm{d}x_j}{\mathrm{d}t}\nabla_j A_i \right) + e\frac{\mathrm{d}x_j}{\mathrm{d}t}\nabla_i A_j - e\nabla_i \phi \\
&= -e\left(\nabla_j \frac{\mathrm{d}x_j}{\mathrm{d}t}A_i - \nabla_i \frac{\mathrm{d}x_j}{\mathrm{d}t}A_j \right) - e\left(\nabla_i \phi + \frac{\partial A_i}{\partial t} \right) \\
&= -e\left((\frac{\mathrm{d}\vec{x}}{\mathrm{d}t} \cdot \nabla)\vec{A} - \nabla\left(\frac{\mathrm{d}\vec{x}}{\mathrm{d}t}\vec{A} \right) \right)_i - e\left(\nabla\phi + \frac{\partial \vec{A}}{\partial t} \right)_i
\end{aligned} \qquad (4.98)
$$

Vectorially, introducing the velocity and using $\vec{v} \times \vec{B} = \vec{v} \times (\nabla \times \vec{A}) = \nabla(\vec{v} \cdot \vec{A}) - (\vec{v} \cdot \nabla)\vec{A}$, we find:

$$m\frac{\mathrm{d}^2 \vec{x}}{\mathrm{d}t^2} = e\left(-(\vec{v} \cdot \nabla)\vec{A} + \nabla\left(\vec{v} \cdot \vec{A} \right) \right) + e\left(-\nabla\phi - \frac{\partial \vec{A}}{\partial t} \right) = e\vec{v} \times \vec{B} + e\vec{E} \quad \square \qquad (4.99)$$

which indeed corresponds to the Lorentz force! So the ansatz Eq. (4.94) describes the correct motion in the presence of an electromagnetic field. To transfer back the above equations to the quantum-mechanical regime, we must implement the canonical quantization procedure by interpreting momentum and energy as operators acting on the wave function. We now assume that the electromagnetic potential is time-independent, $\vec{A} = \vec{A}(\vec{x})$. The time-independent Schrödinger equation (see Appendix C.5) for a stationary state $|\Psi\rangle$ is then given by (cf. Eq. (C.50)):

$$\frac{1}{2m}\left(\vec{p} - e\vec{A} \right)^2 |\Psi\rangle = (E - e\phi)|\Psi\rangle \qquad (4.100)$$

• **Principle of minimal substitution.** Note that the external fields $\vec{A}(\vec{x})$ and $\phi(\vec{x})$ are still classical and are not quantized at this stage. Comparing this with the time-independent Schrödinger equation for a free particle (cf. Eq. (C.50)), one can derive the **principle of minimal substitution** by noting that the time-independent Schrödinger equation for a charged particle of charge e is obtained by the substitution:

$$\vec{p} \longrightarrow \vec{p} - e\vec{A}, \qquad E \longrightarrow E - e\phi \qquad (4.101)$$

The time-independent Schrödinger equation with the external fields is then:

$$\frac{1}{2m}\left[-i\nabla - e\vec{A}(\vec{x})\right]^2 \Psi(\vec{x}) + e\phi(\vec{x})\Psi(\vec{x}) = E\Psi(\vec{x}) \tag{4.102}$$

This result can be expanded to yield the so-called paramagnetic and diamagnetic terms:

$$-\frac{1}{2m}\nabla^2\Psi + \frac{ie}{2m}\left[\vec{A}\cdot\nabla + \nabla\cdot\vec{A}\right]\Psi + \frac{e^2}{2m}\vec{A}^2\Psi + e\phi\Psi = E\Psi \tag{4.103}$$

or

$$-\frac{1}{2m}\nabla^2\Psi + \underbrace{\frac{ie}{m}\vec{A}\cdot(\nabla\Psi) + \frac{ie}{2m}\Psi\left(\nabla\cdot\vec{A}\right)}_{\text{paramagnetic}} + \underbrace{\frac{e^2}{2m}\vec{A}^2\Psi}_{\text{diamagnetic}} + e\phi\Psi = E\Psi \tag{4.104}$$

In the Coulomb gauge $\nabla\cdot\vec{A} = 0$ (see Section 10.3), the third term vanishes, and we simply get:

$$-\frac{1}{2m}\nabla^2\Psi + \underbrace{\frac{ie}{m}\vec{A}\cdot(\nabla\Psi)}_{\text{paramagnetic}} + \underbrace{\frac{e^2}{2m}\vec{A}^2\Psi}_{\text{diamagnetic}} + e\phi\Psi = E\Psi \tag{4.105}$$

• **A uniform electrostatic field.** In this case, we choose $\vec{A} = 0$ and $\vec{E} = -\nabla\phi$. The stationary Schrödinger equation becomes:

$$-\frac{1}{2m}\nabla^2\Psi(\vec{x}) + e\phi(\vec{x})\Psi(\vec{x}) = E\Psi(\vec{x}) \tag{4.106}$$

which is the expected result for a particle in an electrostatic potential.

• **A uniform magnetic field.** We choose the Coulomb gauge $\nabla\cdot\vec{A} = 0$ and $\phi = 0$. If the magnetic field is uniform, one can define:

$$\vec{A} = -\frac{1}{2}\vec{x}\times\vec{B} \tag{4.107}$$

The Schrödinger equation then becomes:

$$-\frac{1}{2m}\nabla^2\Psi - \frac{ie}{2m}(\vec{x}\times\vec{B})\cdot\nabla\Psi + \frac{e^2}{8m}(\vec{x}\times\vec{B})^2\Psi = E\Psi \tag{4.108}$$

where we used $\nabla\cdot\vec{A} = 0$. This equation can be simplified with $(\vec{x}\times\vec{B})\cdot\nabla\Psi = -\vec{B}\cdot(\vec{x}\times\nabla\Psi)$. With the angular momentum operator $\vec{L} = -\vec{x}\times i\nabla$, the Schrödinger equation can be written:

$$-\frac{1}{2m}\nabla^2\Psi - \underbrace{\frac{e}{2m}(\vec{B}\cdot\vec{L})\Psi}_{\text{paramagnetic}} + \underbrace{\frac{e^2}{8m}(\vec{x}\times\vec{B})^2\Psi}_{\text{diamagnetic}} = E\Psi \tag{4.109}$$

For completeness, we recall that the paramagnetic property of materials is due to the realignment of the electron paths caused by the external magnetic field. Magnetization is not retained in the absence of the external magnetic field. In certain atoms, the paramagnetic terms can cancel and the atoms are said to be diamagnetic. In this case, they are not attracted to a magnetic field, but rather are slightly repelled. We illustrate the use of these equations in the following sections, after we discuss an important additional property of electrons, namely their spin, which will play a crucial role in the magnetic behavior of materials, as will become evident in Eq. (4.132).

4.11   The Discovery of Spin: The Stern–Gerlach Experiment

So far, we have neglected spin. The electron spin is one of the most important discoveries in quantum mechanics and allows us to explain many properties of atoms and molecules. There is no classical analog to the spin. The physicist who contributed most to the theoretical study of the spin has been Pauli who, during the first half of the last century, formulated what today we know as the **Exclusion Principle** [6].

In 1922 **Stern**[17] and **Gerlach**[18] performed a ground-breaking experiment, in which a well-collimated beam of silver atoms passes through a region of inhomogeneous magnetic field before impacting on a photographic plate [60]. An atom of silver has one electron in its outermost shell. The electrons in the inner shells compensate each other, therefore the magnetic moment of the atom is dominated by the effect of the single outermost electron (the nuclear contribution is also negligible). **The idea was to measure the orbital angular momentum of the electron to show that it is quantized.** The magnetic field was directed perpendicular to the beam, and had a strong gradient in the z direction, so that a beam comprised of atoms with a magnetic moment would be bent towards the z- or $-z$-axis. Since a magnetic moment acts as a small magnetic dipole μ, a uniform magnetic field would not exert a resulting force but only a torque, however a non-uniform magnetic field creates a force $\vec{F} = \nabla(\vec{\mu} \cdot \vec{B})$, since the "north" and "south" poles of the tiny magnet created by a single atom are not feeling the same field. For a magnetic moment generated by a non-vanishing angular momentum, the naive expectation would be to observe three distinct directions of quantization for $\ell = 1$ corresponding to $m = -1, 0, +1$.

Curiously, the experiment showed that **the beam of silver atoms exhibited only two quantized directions**. See Figure 4.7. This result implied an angular momentum $\ell = 1/2$ for which there are indeed only two allowed

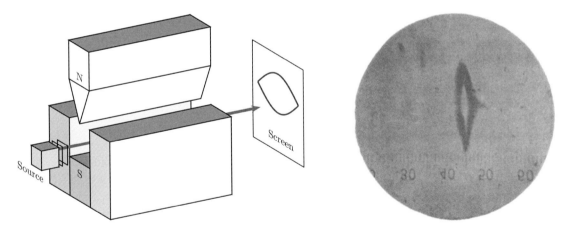

Figure 4.7 (left) Sketch of the Stern–Gerlach experiment. (right) A photograph of the observed beam splitting. Reprinted from W. Gerlach and O. Stern, "Der experimentelle Nachweis der Richtungsquantelung im Magnetfeld," *Zeitschrift für Physik*, vol. 9, pp. 349–352, Dec 1922, under permission from Springer Nature. Copyright 1922.

m values: $m = -1/2, +1/2$. However, such an angular momentum cannot translate into an *orbital* momentum since the z-component of the wave function given by the spherical harmonic Y_{lm} (see Eq. (C.81)) has in this case a factor $e^{\pm i\phi/2}$, and therefore acquires a factor -1 upon rotation through 2π. This yields a **non-single-valued wave function which cannot be interpreted as the solution of the Schrödinger-type problem**. Yet the experiment of Stern–Gerlach was indisputable and it pointed to the existence of a new kind of non-orbital

17   Otto Stern (1888–1969), German American physicist.
18   Walther Gerlach (1889–1979), German physicist.

angular momentum, the so-called **spin**. Although the experiment had been proposed to measure the angular momentum $\vec{L}$ of the outer electron in silver, it had actually discovered the spin $\vec{S}$ of the electron!

• **Spin state.** The spin represents an internal degree of freedom. In the case of the electron, it is a **spin** $1/2$. We have in this case two possible spin states – the "spin-up" and "spin-down" – with spin quantum number s and magnetic spin quantum number s_3:

$$|\uparrow\rangle = |s, s_3\rangle = |\frac{1}{2}, \frac{1}{2}\rangle = \begin{pmatrix} 1 \\ 0 \end{pmatrix} \quad \text{and} \quad |\downarrow\rangle = |\frac{1}{2}, -\frac{1}{2}\rangle = \begin{pmatrix} 0 \\ 1 \end{pmatrix} \tag{4.110}$$

These states are the eigenstates of the diagonal $\vec{S}^2$ and S_3 operators:

$$S_3 = \frac{1}{2} \begin{pmatrix} 1 & 0 \\ 0 & -1 \end{pmatrix} \equiv \frac{1}{2}\sigma_3 \quad \text{and} \quad \vec{S}^2 = S(S+1)\mathbb{1} = \frac{3}{4}\mathbb{1} \tag{4.111}$$

and the two-dimensional ket of a general spin state of a particle is called a **spinor**, represented as:

$$|\chi\rangle \equiv c_1 |\uparrow\rangle + c_2 |\downarrow\rangle = c_1 |\frac{1}{2}, \frac{1}{2}\rangle + c_2 |\frac{1}{2}, -\frac{1}{2}\rangle \tag{4.112}$$

where c_1 and c_2 are complex coefficients, with overall normalization $|c_1|^2 + |c_2|^2 = 1$ and whose squared modulus gives the probability that the particles have respectively a z-component of $+1/2$ or $-1/2$. The **three components of the quantized spin vector** can be defined with the **Pauli matrices** (see Appendix C.12):

$$\vec{S} = (S_1, S_2, S_3) = \frac{1}{2}\vec{\sigma} \tag{4.113}$$

where $\vec{\sigma} = (\sigma_1, \sigma_2, \sigma_3)$ and the **commutation relations** for σ_i are (see Eq. (C.96)):

$$[\sigma_i, \sigma_j] = 2i\epsilon_{ijk}\sigma_k. \tag{4.114}$$

• **Spin operator algebra.** We can check the commutation relations for the component of the spin operator itself:

$$[S_i, S_j] = [\frac{1}{2}\sigma_i, \frac{1}{2}\sigma_j] = \frac{1}{2}\frac{1}{2}[\sigma_i, \sigma_j] = \frac{1}{2}i\epsilon_{ijk}\sigma_k = i\epsilon_{ijk}S_k \tag{4.115}$$

As expected they are the same as those for the angular momentum (see Eq. (C.77)), which proves that the nature of the spin is that of an angular momentum, although it does not correspond to a spatial degree of freedom but is an internal property of the particle.

The **spin raising and lowering operators** $S_{\pm} = S_x \pm iS_y$ with $s = 1/2$ are:

$$S_+ = \begin{pmatrix} 0 & 1 \\ 0 & 0 \end{pmatrix} \quad \text{and} \quad S_- = \begin{pmatrix} 0 & 0 \\ 1 & 0 \end{pmatrix} \tag{4.116}$$

with $S_+ |\downarrow\rangle = |\uparrow\rangle$ and $S_- |\uparrow\rangle = |\downarrow\rangle$.

• **General case.** We can define a unit vector $\hat{n} = (n_x, n_y, n_z)$ in a specific direction. A **given spin state** χ **can be said to point in the** $\hat{n}$ **direction if it is an eigenstate of the operator** $\hat{n} \cdot \vec{\sigma}$:

$$\hat{n} \cdot \vec{\sigma} |\chi\rangle = + |\chi\rangle \implies \begin{pmatrix} n_z & n_x - in_y \\ n_x + in_y & -n_z \end{pmatrix} \begin{pmatrix} c_1 \\ c_2 \end{pmatrix} = \begin{pmatrix} c_1 \\ c_2 \end{pmatrix} \tag{4.117}$$

where we used the property of Eq. (C.101). Using polar coordinates where $\hat{n} = (\sin\theta\cos\phi, \sin\theta\sin\phi, \cos\theta)$, we obtain:

$$\begin{pmatrix} \cos\theta & \sin\theta e^{-i\phi} \\ \sin\theta e^{+i\phi} & -\cos\theta \end{pmatrix} \begin{pmatrix} c_1 \\ c_2 \end{pmatrix} = \begin{pmatrix} c_1 \\ c_2 \end{pmatrix} \tag{4.118}$$

Hence

$$\begin{cases} c_1\cos\theta + c_2\sin\theta e^{-i\phi} = c_1 \\ c_1\sin\theta e^{+i\phi} - c_2\cos\theta = c_2 \end{cases} \implies \begin{cases} c_1(1-\cos\theta) = c_2\sin\theta e^{-i\phi} \\ c_2(1+\cos\theta) = c_1\sin\theta e^{+i\phi} \end{cases} \tag{4.119}$$

So finally:

$$c_2 = \frac{1-\cos\theta}{\sin\theta}e^{i\phi}c_1 = \frac{2\sin^2(\theta/2)}{2\sin(\theta/2)\cos(\theta/2)}e^{i\phi}c_1 = \frac{\sin(\theta/2)}{\cos(\theta/2)}e^{i\phi}c_1 \tag{4.120}$$

Imposing the normalization, we find

$$|c_1|^2 + |c_2|^2 = |c_1|^2 + \left| \frac{\sin(\theta/2)}{\cos(\theta/2)}e^{i\phi}c_1 \right|^2 = 1 \tag{4.121}$$

We can choose

$$c_1 = \cos(\theta/2) \implies c_2 = \sin(\theta/2)e^{i\phi} \tag{4.122}$$

so finally, **the spin eigenstate in a given direction** $\hat{n}$ can be expressed as:

$$|\chi\rangle = \begin{pmatrix} c_1 \\ c_2 \end{pmatrix} = \begin{pmatrix} \cos(\theta/2) \\ \sin(\theta/2)e^{i\phi} \end{pmatrix} \tag{4.123}$$

where $0 \le \theta \le \pi$ and $0 \le \phi < 2\pi$. **For any spinor state, there is a given direction $\hat{n}$ in space along which the spin points "up."**

4.12   The Pauli Equation

The formulation of the Schrödinger equation for an electron, taking into account its spin-1/2, is called the **Pauli equation**, which Pauli first formulated in 1927 [61]. In order to integrate the spin degree of freedom, the wave function must be a **spinor** of the form of Eq. (4.112). Likewise, *the Hamiltonian operator must be a* 2×2 *matrix*. We define first the Hamiltonian of a free spin-1/2 electron to be:

$$H \equiv \frac{\vec{p}^2}{2m_e}\mathbb{1} = \frac{(\sigma \cdot \vec{p})^2}{2m_e} \tag{4.124}$$

where $\mathbb{1}$ is the 2×2 identity matrix, and we used Eq. (C.99) to introduce the Pauli matrices (see Appendix C.12). Hence, we recover the free electron case for each of the spin configurations (the spin is unaffected). In order to obtain the Hamiltonian for an electron in an external electromagnetic field, we perform the minimal substitution Eq. (4.101) and get:

$$H = \frac{1}{2m_e}\vec{\sigma}\cdot\left(\vec{p}-e\vec{A}\right)\vec{\sigma}\cdot\left(\vec{p}-e\vec{A}\right) + e\phi\mathbb{1} \tag{4.125}$$

Again thanks to the properties of the Pauli matrices $(\vec{\sigma}\cdot\vec{a})(\vec{\sigma}\cdot\vec{b}) = (\vec{a}\cdot\vec{b})\mathbb{1} + i\vec{\sigma}\cdot(\vec{a}\times\vec{b})$ (see Eq. (C.98)), we find:

$$H = \frac{1}{2m_e}\left(\vec{p}-e\vec{A}\right)^2\mathbb{1} + \frac{i}{2m_e}\vec{\sigma}\cdot\left(\vec{p}-e\vec{A}\right)\times\left(\vec{p}-e\vec{A}\right) + e\phi\mathbb{1} \tag{4.126}$$

Note that we are dealing with operators, hence:

$$\begin{aligned} \vec{\sigma}\cdot\left(\vec{p}-e\vec{A}\right)\times\left(\vec{p}-e\vec{A}\right) &= \vec{\sigma}\cdot\left(\vec{p}\times\vec{p} - e(\vec{A}\times\vec{p}+\vec{p}\times\vec{A}) + e^2\vec{A}\times\vec{A}\right) \\ &= \epsilon_{ijk}(0 - e(A_i p_j + p_i A_j) + 0)\sigma_k \\ &= -e\epsilon_{ijk}(A_i p_j + p_i A_j)\sigma_k \\ &= -e\epsilon_{ijk}(A_i p_j - p_j A_i)\sigma_k \end{aligned} \tag{4.127}$$

We can make use of the following property:

$$(A_i p_j - p_j A_i)\Psi = A_i(p_j \Psi) - p_j(A_i \Psi) = A_i(p_j \Psi) - (p_j A_i)\Psi - A_i(p_j \Psi) = -(p_j A_i)\Psi \qquad (4.128)$$

Hence

$$\vec{\sigma} \cdot \left(\vec{p} - e\vec{A}\right) \times \left(\vec{p} - e\vec{A}\right) = +e\epsilon_{ijk}(p_j A_i)\sigma_k = -e\epsilon_{ijk}(p_i A_j)\sigma_k = +ie\epsilon_{ijk}(\partial_i A_j)\sigma_k$$
$$= ie\vec{\sigma} \cdot (\nabla \times \vec{A}) = ie\vec{\sigma} \cdot \vec{B} \qquad (4.129)$$

Finally the Hamiltonian is equal to:

$$H = \frac{1}{2m_e}\left(\vec{p} - e\vec{A}\right)^2 \mathbb{1} - \frac{e}{2m_e}\vec{\sigma} \cdot \vec{B} + e\phi\mathbb{1} = \frac{1}{2m_e}\left(\vec{p} - e\vec{A}\right)^2 \mathbb{1} + e\phi\mathbb{1} - \frac{e}{m_e}\left(\frac{\vec{\sigma}}{2}\right) \cdot \vec{B} \qquad (4.130)$$

If H_0 is the spin-independent part of the Hamiltonian, then:

$$H = H_0 - \frac{e}{m_e}\vec{S} \cdot \vec{B} \qquad (4.131)$$

where we have identified the spin-1/2 operator $\vec{S} = (1/2)\vec{\sigma}$. The application of a magnetic field produces the splitting of the two states of spin (**Zeeman effect** – see Section 4.13).

We can apply this result to the case of an electron in a uniform magnetic field. The time-independent Schrödinger equation (Eq. (4.109)) reads in this case:

$$-\frac{1}{2m_e}\nabla^2\Psi - \frac{e}{2m_e}\vec{B} \cdot (\vec{L} + 2\vec{S})\Psi + \frac{e^2}{8m_e}(\vec{x} \times \vec{B})^2\Psi = E\Psi \qquad (4.132)$$

Note the factor 2 between $\vec{L}$ and $\vec{S}$ above.

• **g-factor and Bohr magneton.** For a classical dipole magnet with a magnetic moment $\vec{\mu}$, the magnetic interaction energy is $E_{mag} = -\vec{\mu} \cdot \vec{B}$. This corresponds to the *prediction* that the g-factor of a spin-1/2 electron is equal to 2 (see, e.g., Ref. [62]), since the equation gives a spin magnetic moment equal to:

$$-\vec{\mu}_S = \frac{e}{2m_e}2\vec{S} \equiv g\,\mu_B\vec{S} \qquad (4.133)$$

with the g-factor $g \equiv 2$, and where we introduced **the Bohr magneton** μ_B for an electron:

$$\mu_B \equiv \frac{e}{2m_e} \approx (9.2740100783 \pm 0.0000000028) \times 10^{-24}\,\text{J T}^{-1} \qquad (4.134)$$

In general, the **g-factor** (also called g value or the dimensionless magnetic moment) is the **dimensionless quantity that characterizes the magnetic moment and gyromagnetic ratio**. It is the proportionality constant that relates the observed magnetic moment μ of a particle to its angular momentum quantum number and unit of magnetic moment. This quantity will turn out to be of great importance for the understanding and verification of the modern quantum electrodynamics (QED), as we will see in Chapter 14. The **Landé**[19] g-factor is a particular example of a g-factor, namely for an electron with both spin and orbital angular momenta.

• **Nuclear magneton.** For a proton or a neutron, the magnetic moment is given in units of the **nuclear magneton**

$$\mu_N \equiv \frac{e}{2M_p} \approx (5.0507837461 \pm 0.0000000015) \times 10^{-27}\,\text{J T}^{-1} \qquad (4.135)$$

The experimental values (see Eq. (15.35)) with g-factors different from 2 hints at a complicated inner structure (see Chapter 15 and beyond). We note that the nuclear magneton is slightly more than three orders of magnitude smaller than the Bohr magneton:

$$\mu_N \approx \frac{1}{1836}\mu_B \qquad (4.136)$$

---

19 Alfred Landé (1888–1976), German–American physicist.

4.13   The Zeeman and Stark Effects

The **Zeeman effect** is the splitting of an atomic spectral line into several components in the presence of a static external magnetic field. The total Hamiltonian in a magnetic field is given by:

$$H = H_0 - \vec{\mu} \cdot \vec{B} \tag{4.137}$$

where $\vec{\mu}$ is the magnetic moment of the atom. This latter contains contributions from both electrons and the nucleus. The electronic contribution is dominant, hence, one can write

$$\vec{\mu} \simeq -\frac{e}{2m_e}\left(\vec{L} + g\vec{S}\right) = -\mu_B\left(\vec{L} + g\vec{S}\right) \tag{4.138}$$

where $\vec{L}$ and $\vec{S}$ are the total orbital momentum and spin of the atom (averaged over a state with a given value of the total angular momentum). If the external field is low or moderate, the effect can be treated as a perturbation (see Section 4.16). Under this assumption, the magnetic potential energy of the atom in the applied external magnetic field is just proportional to the magnetic field strength and is given by (see, e.g., Ref. [63]):

$$E - E_0 = \mu_{\rm B}Bm_j\left[1 + (g-1)\frac{j(j+1) - l(l+1) + s(s+1)}{2j(j+1)}\right] \tag{4.139}$$

where j is the total angular momentum, l the angular momentum, and $m_j = -j, -j+1, \ldots, +j$ represents the z-component of the total angular momentum in the quantization axis of the external magnetic field. In the case of a single orbiting electron contributing to the effect (i.e., if all electronic shells below it are filled), one finds:

$$E - E_0 = \mu_{\rm B}Bm_j\left[1 \pm \frac{g-1}{2l+1}\right] \tag{4.140}$$

where the two signs correspond to the two possible electron-spin orientations. The Zeeman effect is very important for applications such as Mössbauer spectroscopy, electron-spin resonance spectroscopy, and nuclear magnetic resonance spectroscopy (MRI).

The **Stark[20] effect** is the analog of the Zeeman effect when the splitting of a spectral line into several components is due to the presence of a static external electric field.

4.14   The Polarization of Electrons

When we consider a beam of particles (e.g., electrons), the term **polarization** indicates the properties of the electrons resulting from a preferred orientation of the spins of the "ensemble" of particles that compose the beam. A **fully (or totally) polarized beam** is characterized by electrons in an eigenstate of the form of Eq. (4.123), where the c_1 and c_2 coefficients of the spin-up and spin-down basis can be calculated directly from the polar angles θ, ϕ of the spin direction. We say that the beam is composed of particles in a "pure" state. In terms of probability, this means that one has 100% probability of finding a particle in the beam in a given configuration. This is hardly achievable in a real experiment.

A **partially polarized beam** is instead represented by a "mixed" state, which is formally described in quantum mechanics with the density matrix ρ:

$$\rho \equiv |\chi\rangle\langle\chi| \tag{4.141}$$

The expectation value of a given operator O is found with the following trace:

$$\langle O \rangle = \langle\chi|O|\chi\rangle = \sum_i \langle\chi|O|i\rangle\langle i|\chi\rangle = \sum_i \langle i|\chi\rangle\langle\chi|O|i\rangle = \text{Tr}\left(|\chi\rangle\langle\chi|O\right) = \text{Tr}\left(\rho O\right) \tag{4.142}$$

20   Johannes Stark (1874–1957), German physicist.

For the case of a beam composed of spin-1/2 particles, the density matrix is represented by the 2×2 matrix of the two fundamental states $|\uparrow\rangle$ and $|\downarrow\rangle$. The matrix ρ is Hermitian and its eigenvalues must be positive or zero and normalized such that $\operatorname{Tr} \rho = 1$. A "pure state, fully polarized beam" is represented by the obvious matrices:

$$\rho_1 = \begin{pmatrix} 1 & 0 \\ 0 & 0 \end{pmatrix}, \quad \text{or} \quad \rho_2 = \begin{pmatrix} 0 & 0 \\ 0 & 1 \end{pmatrix} \tag{4.143}$$

For a fully polarized beam in a given direction $\hat{n} = \hat{n}(\theta, \phi)$ corresponding to the state $|\chi\rangle = c_1 |\uparrow\rangle + c_2 |\downarrow\rangle$, the density matrix is given by:

$$\rho_{\hat{n}} = \begin{pmatrix} |c_1|^2 & c_1 c_2^* \\ c_1^* c_2 & |c_2|^2 \end{pmatrix} = \begin{pmatrix} \cos^2(\theta/2) & \cos(\theta/2)\sin(\theta/2)e^{-i\phi} \\ \cos(\theta/2)\sin(\theta/2)e^{i\phi} & \sin^2(\theta/2) \end{pmatrix} \tag{4.144}$$

where c_1 and c_2 are determined by Eq. (4.123). As expected, the cases $\rho_{1,2}$ are recovered with $\theta = 0$ and π.

A **fully unpolarized beam** is characterized by the density matrix defined by the "coherent sum" of both spin states:

$$\rho_{100\%, unpolarized} = \begin{pmatrix} \frac{1}{2} & 0 \\ 0 & \frac{1}{2} \end{pmatrix} \tag{4.145}$$

In terms of probability, this means that we have an equal 50:50 chance of finding a particle in the beam in spin-up or spin-down configuration. Consequently, a **partially polarized beam** can be represented by the diagonal density matrix:[21]

$$\rho_{partial} = \begin{pmatrix} \rho' & 0 \\ 0 & \rho'' \end{pmatrix} \tag{4.146}$$

with $\rho' + \rho'' = 1$ and where one can always choose $\rho' \geq \rho'' \geq 0$.

The **degree of polarization P** of the beam is then defined as $P = \rho' - \rho''$ such that $0 \leq P \leq 1$ and $1 - P = 1 - \rho' + \rho'' = 2\rho''$. Consequently, we write:

$$\begin{aligned} \rho_{partial} &= \begin{pmatrix} \rho' & 0 \\ 0 & \rho'' \end{pmatrix} = \rho'' \begin{pmatrix} 1 & 0 \\ 0 & 1 \end{pmatrix} + (\rho' - \rho'') \begin{pmatrix} 1 & 0 \\ 0 & 0 \end{pmatrix} \\ &= P \begin{pmatrix} 1 & 0 \\ 0 & 0 \end{pmatrix} + (1 - P) \begin{pmatrix} \frac{1}{2} & 0 \\ 0 & \frac{1}{2} \end{pmatrix} \end{aligned} \tag{4.147}$$

which represents the "incoherent" superposition of the fully polarized beam with weight P and the totally unpolarized beam with weight $1 - P$. One can show that for a pure state one has $\operatorname{Tr} \rho^2 = 1$ while for mixed states $\operatorname{Tr} \rho^2 \leq 1$.

The general density matrix is determined by the two complex numbers c_1 and c_2 constrained by the normalization $\operatorname{Tr} \rho = 1$. This implies that it can be characterized by $4 - 1 = 3$ independent real parameters. One can conveniently introduce the Pauli matrices and write:

$$\rho \equiv \frac{1}{2} \left(\mathbb{1} + \vec{\xi} \cdot \vec{\sigma} \right) \tag{4.148}$$

where $\vec{\xi} = (\xi_1, \xi_2, \xi_3)$ is called the **Bloch vector**.

For a *pure state*, the components are given by:

$$\xi_1 = c_1 c_2^* + c_1^* c_2, \quad \xi_2 = i(c_1 c_2^* - c_1^* c_2), \quad \xi_3 = |c_1|^2 - |c_2|^2 \tag{4.149}$$

---

21 Depending on the choice of basis, the density matrix does not have to be necessarily diagonal, but it can always be diagonalized.

since consequently:

$$
\begin{aligned}
\rho &= \frac{1}{2}\begin{pmatrix} 1 & 0 \\ 0 & 1 \end{pmatrix} + \frac{1}{2}\xi_1\begin{pmatrix} 0 & 1 \\ 1 & 0 \end{pmatrix} + \frac{1}{2}\xi_2\begin{pmatrix} 0 & -i \\ i & 0 \end{pmatrix} + \frac{1}{2}\xi_3\begin{pmatrix} 1 & 0 \\ 0 & -1 \end{pmatrix} \\
&= \frac{1}{2}\begin{pmatrix} 1 & 0 \\ 0 & 1 \end{pmatrix} + \frac{1}{2}\begin{pmatrix} 0 & 2c_1c_2^* \\ 2c_1^*c_2 & 0 \end{pmatrix} + \frac{1}{2}\begin{pmatrix} |c_1|^2 - |c_2|^2 & 0 \\ 0 & -|c_1|^2 + |c_2|^2 \end{pmatrix} \\
&= \frac{1}{2}\begin{pmatrix} 1 + |c_1|^2 - |c_2|^2 & 0 \\ 0 & 1 - |c_1|^2 + |c_2|^2 \end{pmatrix} + \begin{pmatrix} 0 & c_1c_2^* \\ c_1^*c_2 & 0 \end{pmatrix} \\
&= \begin{pmatrix} |c_1|^2 & c_1c_2^* \\ c_1^*c_2 & |c_2|^2 \end{pmatrix} \qquad\qquad \Box
\end{aligned}
\tag{4.150}
$$

where in the last line we used $|c_1|^2 + |c_2|^2 = 1$. From a group-theoretical point of view, we say that a two-dimensional unitary $SU(2)$ transformation of the complex space corresponds to a real three-dimensional orthogonal $SO(3)$ transformation in the space of the vector $\vec{\xi}$ (see Homomorphism in Appendix B.8). In this case, the norm of $\vec{\xi}$ is given by:

$$
\begin{aligned}
|\vec{\xi}|^2 &= \xi_i^2 = (c_1c_2^* + c_1^*c_2)^2 + (i(c_1c_2^* - c_1^*c_2))^2 + (|c_1|^2 - |c_2|^2)^2 \\
&= (c_1c_2^* + c_1^*c_2)^2 - (c_1c_2^* - c_1^*c_2)^2 + (|c_1|^2 - |c_2|^2)^2 = 4c_1c_2^*c_1^*c_2 + (|c_1|^2 - |c_2|^2)^2 \\
&= (|c_1|^2 + |c_2|^2)^2 = 1
\end{aligned}
\tag{4.151}
$$

so for a pure state $\vec{\xi}$ is a unit vector, while for mixed states $P = |\vec{\xi}| < 1$.

In conclusion, to completely define the quantum state of the polarization of a beam (whether fully or partially polarized, pure or mixed) it is sufficient to define *three independent parameters*, which can be represented just by two angles and a magnitude. These can represent the spin vector.

• **Transverse and longitudinal.** As we will see in Chapter 8, the spin is not a covariant quantity in Dirac's relativistic theory. The full quantum state of polarization of electrons will still be defined with three parameters. We will define the **transverse polarization** when the direction of the spin is perpendicular to the momentum, and **longitudinal polarization** when the direction of the spin is parallel or antiparallel to the momentum. As a basis to determine completely the polarization, one can then choose: (1) two states of transverse polarization with opposite spin directions; (2) two other states of transverse polarization turned over by an angle $\pi/2$ relative to (1); (3) two opposite states of longitudinal polarization (parallel and antiparallel).

4.15   The Significance of the Electromagnetic Potentials

In classical mechanics, the force on a charged particle in an electromagnetic field is given by the Lorentz formula $\vec{F} = e\vec{E} + e\vec{v}\times\vec{B}$. Hence, classically the trajectory of the particle is affected by the physical field $\vec{E}$ and $\vec{B}$. The potential fields $\vec{A}$ and ϕ are introduced as mathematical tools, however, one can always reduce the classical problem to a solution as a function of $\vec{E}$ and $\vec{B}$. On the other hand, the principle of minimal substitution, described in the previous sections, introduces the external fields $\vec{A}$ and ϕ as primary fields, so it would appear that the quantum-mechanical solution is affected by these latter and one cannot simply remove them in favor of an expression with only $\vec{E}$ and $\vec{B}$. The apparent paradox is that $\vec{A}$ and ϕ are not unique for a given $\vec{E}$ and $\vec{B}$. How do these statements reconcile each other? Is this an artefact of our mathematical description or can one actually experimentally find evidence that quantum effects can arise from the electromagnetic potentials $\vec{A}$ and ϕ that cannot be described by the Lorentz force?

In 1949 this problem was discussed by **Ehrenberg**[22] and **Siday**[23] in their seminal paper [64] on electron optics. They pointed out that the arbitrariness of the vector potential cannot produce any observable effects.

---

22   Werner Ehrenberg (1901–1975), English physicist.
23   Raymond Eldred Siday (1912–1956), English mathematician.

In particular, any interference effects should be independent of how $\vec{A}$ is chosen. But still the interference between two rays connecting given points depends on their different paths within the $\vec{A}$ field, even in the absence of an absolute fixing of $\vec{A}$. Consequently, we can expect wave phenomena to arise whilst the rays are in electric or magnetic field-free regions. In 1959 **Aharonov**[24] and **Bohm**[25] more specifically analyzed the case and defined a two-slit experiment to test the effect [65], as illustrated in Figure 4.8 (cf. Figure 4.1). Compared to the original two-slit experiment, the minimal substitution introduces an extra phase difference due to the replacement $\vec{p} \to \vec{p} - e\vec{A}$ which is given by:

$$\Delta\delta = e \int_A^P \vec{A} \cdot \mathrm{d}\vec{x} - e \int_B^P \vec{A} \cdot \mathrm{d}\vec{x} = e \int (\nabla \times \vec{A}) \cdot \mathrm{d}\vec{S} = e \int \vec{B} \cdot \mathrm{d}\vec{S} = e\Phi_{solenoid} \qquad (4.152)$$

The net consequence is that the presence of the magnetic field inside the solenoid influences the interference pattern of the electron, although classically the electron's trajectory only ever goes through regions where there is no magnetic field. The effect was first experimentally verified by **Chambers**[26] in 1960 [66].

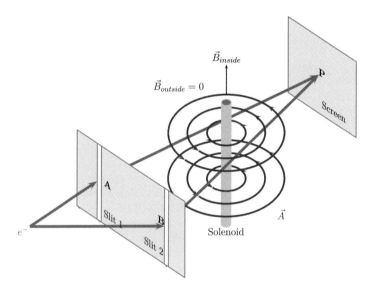

Figure 4.8 Experimental layout to test the Bohm–Aharonov effect: a two-slit experiment with a solenoid placed between the slits.

Today, the **Ehrenberg–Siday–Aharonov–Bohm effect**, also called the **Aharonov–Bohm effect**, is used to express the quantum-mechanical phenomenon in which an electrically charged particle is affected by an electromagnetic potential, despite being confined to a region in which both the magnetic field and the electric field are zero. **It shows that the correct quantum-mechanical description of the physical phenomenon does require the use of the potential fields $\vec{A}$ and ϕ** and supports the relevance of the minimal substitution.

4.16   Non-relativistic Perturbation Theory

Real problems in quantum mechanics are often too complicated to be solved exactly. In this case we use methods of approximation. In **perturbation theory**, we begin with a simple system for which a mathematical

---

24 Yakir Aharonov (born 1932), Israeli physicist.
25 David Joseph Bohm (1917–1992), American scientist.
26 Robert G. Chambers (1924–2016), British physicist.

solution is known, and add an additional "perturbing" Hamiltonian representing a weak disturbance to the system. We consider for instance the following Hamiltonian:

$$H = H_0 + \underbrace{V(\vec{x}, t)}_{\text{perturbation}} \tag{4.153}$$

where H_0 is the unperturbed Hamilton operator and is time-independent. On the other hand, the potential V represents a time-dependent perturbation which is assumed to be small in magnitude compared to H_0. We seek the S **matrix** which generates the final state of the system from the initial system:

$$\Psi(\vec{x}, t = +\infty) = S\Psi(\vec{x}, t = -\infty) \tag{4.154}$$

The idea is that the system is initially characterized by an initial state in the far past, then is subject to a weak perturbation which can be interpreted as acting within a finite time, and is led into a final state in the far future. If there is no interaction, then S reduces to the identity matrix. In the case of the perturbation potential, it is useful to define the **transfer operator** T (or transfer matrix), which generates the change in the state eigenvector produced by the perturbation V:

$$S \equiv \mathbb{1} + T \tag{4.155}$$

We assume that the eigenvalues and eigenstates of the unperturbed system are:

$$H_0 |u_n\rangle = E_n |u_n\rangle \quad \text{hence} \quad |u_n\rangle = \text{ eigenstate} \tag{4.156}$$

The set of eigenstates defines an orthonormal and complete basis of the unperturbed Hamiltonian H_0. Each eigenfunction $u_n(x, t)$ has a well-defined energy E_n. A stationary solution is therefore proportional to:

$$u_n(x, t) = e^{-iE_n t} u_n(\vec{x}) \tag{4.157}$$

because the time dependence introduces a trivial phase $-iE_n t$. The spatial and time-dependent parts can therefore be factorized.

We now consider the solutions of the full Hamiltonian H. If the solutions $|u_n\rangle$ form a complete basis of eigenstates, then we can assume that a solution Ψ of the full Hamiltonian H can be expressed as a **linear combination** of the eigenstates $|u_n\rangle$ of H_0:

$$\Psi(\vec{x}, t) = \sum_n a_n(t) e^{-iE_n t} u_n(\vec{x}) \tag{4.158}$$

From the Schrödinger equation, we can derive that:

$$i\frac{\partial}{\partial t}\Psi = H\Psi \quad \longrightarrow \quad \sum_n \left(i\frac{da_n}{dt} + a_n E_n \right) e^{-iE_n t} u_n = (H_0 + V) \sum_n a_n(t) e^{-iE_n t} u_n \tag{4.159}$$

With $H_0 |u_n\rangle = E_n |u_n\rangle$, we obtain:

$$\sum_n \left(i\frac{da_n}{dt} \right) e^{-iE_n t} u_n = \sum_n a_n(t) V e^{-iE_n t} u_n \tag{4.160}$$

We multiply the equation by a solution u_f^* and perform an integration over space:

$$i \int \sum_n \frac{da_n}{dt} u_f^* e^{-iE_n t} u_n d^3\vec{x} = \int \sum_n a_n(t) u_f^* V e^{-iE_n t} u_n d^3\vec{x} \tag{4.161}$$

or

$$i \sum_n \frac{\mathrm{d}a_n}{\mathrm{d}t} \underbrace{\left(\int \mathrm{d}^3\vec{x}\, u_f^* u_n \right)}_{=\delta_f^n} e^{-iE_n t} = \sum_n a_n(t) \left(\int \mathrm{d}^3\vec{x}\, u_f^* V u_n \right) e^{-iE_n t} \tag{4.162}$$

where we have used the fact that the eigenstates u_n are orthonormal to each other. The time dependence of the coefficient a_f is given by:

$$i \frac{\mathrm{d}a_f}{\mathrm{d}t} e^{-iE_f t} = \sum_n a_n(t) e^{-iE_n t} \underbrace{\left(\int \mathrm{d}^3\vec{x}\, u_f^* V(\vec{x}, t) u_n \right)}_{\equiv V_{fn}(t)} \tag{4.163}$$

or

$$\frac{\mathrm{d}a_f}{\mathrm{d}t} = -i \sum_n a_n(t) e^{i(E_f - E_n)t} V_{fn}(t) \tag{4.164}$$

The function $V_{fn}(t) \equiv \int \mathrm{d}^3\vec{x}\, u_f^*(\vec{x}) V(\vec{x}, t) u_n(\vec{x})$ is defined as the **matrix element**. How can we use this equation? We consider a specific setup: a free particle moves in space. At time $t \simeq 0$ a time-dependent potential V acts on the particle during a short time. We want to find the state of the particle after the action of the potential.

The states of the particle before and after the interaction are defined by its **initial** and **final** states. Since the interaction occurs at $t \simeq 0$, we assume that the initial state is prevalent in the past at time $-T/2$ and the final state in the future at time $+T/2$, where T is a free parameter representing time. At the end of the calculation, we will let $T \to \infty$.

Using Eq. (4.158), the initial and final states are expressed as a linear combination of the unperturbed Hamiltonian $|u_n\rangle$ states. We can actually assume that the initial state corresponds to one actual eigenstate of H_0. Hence, the a_n coefficients at time $t = -T/2$ are set such that the particle is initially in the state $|u_i\rangle$:

$$t = -\frac{T}{2} : \quad \begin{cases} a_i\left(-\frac{T}{2}\right) = 1 \\ a_n\left(-\frac{T}{2}\right) = 0 \quad n \neq i \end{cases} \tag{4.165}$$

The challenge of Eq. (4.164) is that the time variation of the coefficient $a_f(t)$ on the left-hand side is a function of all other a_n on the right-hand side of the equation. We can in principle solve Eq. (4.164) iteratively to retrieve and evolve the a_n coefficients at each time step, relying on numerical methods.

• **Born approximation.** In perturbation theory, however, we will analytically find solutions in successive approximations by an **expansion** in a power series of the calculation. At **first order**, corresponding to the so-called **Born approximation**, we assume that the perturbation is small and that the particle stays in its initial state until the interaction occurs and that the interaction induces a *single transition* between the initial and the final state:

$$\begin{cases} a_i(t < 0) \approx 1 \\ a_n(t < 0) = 0 \quad n \neq i \end{cases} ; \quad \begin{cases} a_f(t > 0) \approx 1 \\ a_n(t > 0) = 0 \quad n \neq f \end{cases} \quad \text{at first order} \tag{4.166}$$

Therefore the summation in Eq. (4.164) can be replaced by a single term, and in addition the coefficient $a_i \approx 1$:

$$\frac{\mathrm{d}a_f}{\mathrm{d}t} \approx -i \underbrace{a_i(t)}_{\approx 1} e^{i(E_f - E_i)t} V_{fi}(t) \tag{4.167}$$

This first-order approximation can easily be integrated over time starting from the initial state $-T/2$ to give:

$$a_f(t) \approx -i \int_{-T/2}^{t} \mathrm{d}t'\, V_{fi}(t') e^{i(E_f - E_i)t'} \tag{4.168}$$

The final state is defined by the single coefficient a_f at time $t = +T/2$ and therefore at first order of approximation:

$$a_f(+\frac{T}{2}) \approx -i \int_{-T/2}^{+T/2} \mathrm{d}t' \int \mathrm{d}^3\vec{x} u_f^*(\vec{x}) V(\vec{x},t') u_i(\vec{x}) e^{i(E_f-E_i)t'} \tag{4.169}$$

4.16.1   The Case of a Time-Independent Perturbation

We consider the case where the perturbing potential is time-independent and we let the free parameter T go to infinity:

$$V(\vec{x},t) = V(\vec{x}) \quad \text{and} \quad T \to \infty \tag{4.170}$$

The transition amplitude from the initial state to the final state is defined as:

$$T_{fi} \equiv -i \int_{-\infty}^{\infty} \mathrm{d}t' e^{i(E_f-E_i)t'} \underbrace{\int \mathrm{d}^3\vec{x} u_f^*(\vec{x}) V(\vec{x}) u_i(\vec{x})}_{\text{Matrix element} \equiv V_{fi}} \tag{4.171}$$

where we note that the time and space integrations are separated. It corresponds to the transfer operator T, defined in Eq. (4.155). The integration over space gives the matrix element between the initial and the final state. The time integration leads to a Dirac δ function. See Appendix A.6. We obtain:

$$T_{fi} = -iV_{fi}(2\pi\delta(E_f - E_i)) \tag{4.172}$$

We note that the amplitude vanishes when the initial and final states have different energies:

$$T_{fi} = 0 \quad \text{if } E_i \neq E_f \tag{4.173}$$

This result implies that **a time-independent perturbation cannot induce a change of energy in the particle.** The **transition probability per unit time** from the initial to the final state is defined as:

$$W_{fi} \equiv \lim_{T\to\infty} \frac{|T_{fi}|^2}{T} \tag{4.174}$$

where we have used the magnitude squared of the amplitude. By replacement we obtain:

$$
\begin{aligned}
W_{fi} &= \lim_{T\to\infty} \frac{|-iV_{fi}(2\pi\delta(E_f - E_i))|^2}{T} \\
&= \lim_{T\to\infty} \frac{1}{T}|V_{fi}|^2(2\pi)^2\delta(E_f - E_i) \int_{-T/2}^{T/2} \frac{\mathrm{d}t}{2\pi} e^{+i(E_f-E_i)t} \\
&= \lim_{T\to\infty} \frac{1}{T}|V_{fi}|^2(2\pi)\delta(E_f - E_i) \underbrace{\int_{-T/2}^{T/2} \mathrm{d}t}_{=T}
\end{aligned} \tag{4.175}
$$

The transition probability per unit time is then given by:

$$W_{fi} = (2\pi)|V_{fi}|^2\delta(E_f - E_i) \tag{4.176}$$

We assume that the particle can transition into several final states of the same energy. The density of these final states $\rho(E)$ is defined in the following way:

$$\rho(E_f)\mathrm{d}E_f = \text{Number of final states in the energy interval } E_f \text{ and } E_f + \mathrm{d}E_f \tag{4.177}$$

The **total transition probability per unit time** W_{fi} from the initial state i to the final states f is found by integration over all final states:

$$W_{fi} = (2\pi) \int dE_f |V_{fi}|^2 \delta(E_f - E_i) \rho(E_f) = (2\pi)|V_{fi}|^2 \rho(E_f) \tag{4.178}$$

where we have used $E_i = E_f$. This result is called **Fermi's**[27] **Golden Rule**.

4.16.2   The Case of a Time-Dependent Perturbation

We now consider the case of a Hamiltonian H of a system which can be split into a time-independent part H_0 and a time-dependent *periodic* perturbation $V(\vec{x}, t) = V(\vec{x})e^{-i\omega t}$:

$$H \equiv H_0 + V(\vec{x})e^{-i\omega t} \tag{4.179}$$

Using an approach similar to that for the time-independent case, we start with the unperturbed eigenstates defined in Eq. (4.156). The matrix element appearing in Eq. (4.164) is simply given by:

$$V_{fn}(t) \equiv \int d^3\vec{x}\, u_f^* \, V(\vec{x}, t)\, u_n = e^{-i\omega t} \int d^3\vec{x}\, u_f^* \, V(\vec{x})\, u_n \tag{4.180}$$

The transition amplitude in the Born approximation from an initial state $|u_i\rangle$ to a final state $|u_f\rangle$ becomes:

$$T_{fi} = -i \int_{-\infty}^{\infty} dt' e^{i(E_f - E_i)t'} e^{-i\omega t'} \int d^3\vec{x}\, u_f^* \, V(\vec{x})\, u_i = -i \int_{-\infty}^{\infty} dt' e^{i(E_f - E_i - \omega)t'} V_{fi} \tag{4.181}$$

where the matrix element V_{fi} was defined in Eq. (4.171). Time integration, similar to the case of Eq. (4.172), yields then the transition amplitude:

$$T_{fi} = -iV_{fi}\left(2\pi\, \delta(E_f - E_i - \omega)\right) \tag{4.182}$$

to be compared with the result for a time-independent perturbation, Eq. (4.172). It follows that the transition is still given by the overlap of the wave functions (matrix element) but is non-vanishing only for $E_f = E_i + \omega$. Consequently, **a time-varying perturbation is needed to increase the energy of the system during the transition from an initial to a final state.**

4.17   The Case of Elastic Scattering

We analyze a scattering experiment with the help of the quantum-mechanical description. We simplify the setup and consider only elastic scattering from a fixed potential.

The potential V is responsible for and induces the scattering of the particle. As a result, the momentum of the particle can change. In the elastic case only the direction of the momentum changes and not its magnitude. We can then write:

$$|\vec{p_i}| = |\vec{p_f}| = p \tag{4.183}$$

The scattering process is described as the transition from an initial state with momentum $\vec{p_i}$ to a final state with momentum $\vec{p_f}$. The kinematical variables (here the three coordinates of each of the momenta) define the states. See Figure 4.9 and compare with the classical picture shown in Figure 2.13. In the quantum-mechanical description we do not consider the trajectory of the particle, but are computing the transition probability from an incoming particle with momentum $\vec{p_i}$ to an outgoing particle with momentum $\vec{p_f}$.

---

27   Enrico Fermi (1901–1954), Italian and naturalized-American physicist.

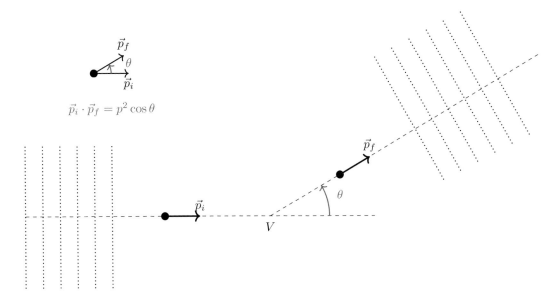

Figure 4.9 Quantum-mechanical description of the scattering experiment. The angle θ represents the scattering angle or the deflection and we assumed an elastic scattering $|\vec{p_i}| = |\vec{p_f}| = p$.

- **Incoming particle.** This is described by an incoming plane wave propagating in the direction given by $\vec{p_i}$ and with a normalization within a box of volume L^3:

$$u_i(\vec{x}) = \frac{1}{L^{3/2}} e^{i\vec{p_i} \cdot \vec{x}} \quad \text{Momentum } \vec{p_i} \tag{4.184}$$

- **Outgoing particle.** Similarly:

$$u_f(\vec{x}) = \frac{1}{L^{3/2}} e^{i\vec{p_f} \cdot \vec{x}} \quad \text{Momentum } \vec{p_f} \tag{4.185}$$

The differential cross-section for the transition between the state $\vec{p_i} \to \vec{p_f}$ is given by:

$$d\sigma_{fi} \equiv \frac{W_{fi}}{j} \quad \text{where } j = \text{flux of the incoming particle} \tag{4.186}$$

Here, W_{fi} has the units of a transition probability per unit time and j is a probability per unit time and unit area. The cross-section has, as expected, units of area.

For the free incoming particle, the corresponding flux is (see Section 4.4):

$$\vec{j} = |N|^2 \frac{\vec{p}}{m} \quad \longrightarrow \quad j = \frac{1}{L^3} \frac{p}{m} \tag{4.187}$$

With the help of Fermi's Golden Rule, Eq. (4.178), we can relate this expression to the matrix element and the density of final states:

$$d\sigma_{fi} \equiv \frac{W_{fi}}{j} = \frac{(2\pi)|V_{fi}|^2 \rho(E_f)}{\frac{1}{L^3} \frac{p}{m}} = (2\pi) \frac{mL^3}{p} |V_{fi}|^2 \rho(E_f) \tag{4.188}$$

How is the density of final states defined? A plane wave (infinitely extended) cannot be enclosed within the assumed box of volume L^3, unless we define boundary conditions, which enforce a periodicity of the wave.

We consider first the **one-dimensional case**:

$$u_f(x) = Ne^{ip_f x} \quad \longrightarrow \quad u_f(x) = u_f(x+L) \quad \longrightarrow \quad p_f = \frac{2\pi n}{L}$$

$$\longrightarrow \quad \mathrm{d}p_f = \frac{2\pi}{L}\mathrm{d}n \quad \longrightarrow \quad \mathrm{d}n = \frac{L}{2\pi}\mathrm{d}p_f \tag{4.189}$$

and in **three dimensions**:

$$\mathrm{d}n = \left(\frac{L}{2\pi}\right)^3 \mathrm{d}p_x \mathrm{d}p_y \mathrm{d}p_z = L^3 \frac{\mathrm{d}^3\vec{p}}{(2\pi)^3} = L^3 \frac{p^2 \mathrm{d}p \mathrm{d}\Omega}{(2\pi)^3} \tag{4.190}$$

where we have introduced the solid angle $\mathrm{d}\Omega$.

The corresponding density of states as a function of the kinetic energy is given by:

$$\rho(E_f) = \frac{\mathrm{d}n}{\mathrm{d}E_f} \quad \text{where} \quad E_f = \frac{p^2}{2m} \longrightarrow \mathrm{d}E_f = \frac{1}{2m}2p\mathrm{d}p = \frac{p\mathrm{d}p}{m} \tag{4.191}$$

or

$$\rho(E_f) = \frac{\mathrm{d}n}{\mathrm{d}E_f} = L^3 \frac{\left(\frac{1}{2\pi}\right)^3 p^2 \mathrm{d}p \mathrm{d}\Omega}{\frac{p\mathrm{d}p}{m}} = L^3 \left(\frac{1}{2\pi}\right)^3 mp\mathrm{d}\Omega \tag{4.192}$$

The differential cross-section becomes (we neglect the indices in the cross-section):

$$\mathrm{d}\sigma = (2\pi)\frac{mL^3}{p}|V_{fi}|^2 L^3 \left(\frac{1}{2\pi}\right)^3 mp\mathrm{d}\Omega = \left(\frac{m}{2\pi}\right)^2 L^6 |V_{fi}|^2 \mathrm{d}\Omega \tag{4.193}$$

or

$$\frac{\mathrm{d}\sigma}{\mathrm{d}\Omega} = \left(\frac{m}{2\pi}\right)^2 L^6 |V_{fi}|^2 \tag{4.194}$$

The matrix element V_{fi} is defined by:

$$V_{fi} = \int \mathrm{d}^3\vec{x}\, u_f^*(\vec{x})V(\vec{x})u_i(\vec{x}) = \frac{1}{L^3}\int \mathrm{d}^3\vec{x}\, e^{-i\vec{p_f}\cdot\vec{x}}V(\vec{x})e^{i\vec{p_i}\cdot\vec{x}} \equiv \frac{1}{L^3}\int \mathrm{d}^3\vec{x}\, V(\vec{x})e^{i\vec{q}\cdot\vec{x}} \tag{4.195}$$

where

$$\vec{q} \equiv \vec{p_i} - \vec{p_f} = \text{change of momentum} \tag{4.196}$$

This last equation represents the **exchange of momentum** during the collision. Finally, the differential cross-section is written as $\mathrm{d}\sigma/\mathrm{d}\Omega$, yielding (the size of the box L disappears):

$$\frac{\mathrm{d}\sigma}{\mathrm{d}\Omega} = \left(\frac{m}{2\pi}\right)^2 \left|\int \mathrm{d}^3\vec{x}\, V(\vec{x})e^{i\vec{q}\cdot\vec{x}}\right|^2 \quad \text{(first order)} \tag{4.197}$$

This equation represents the differential cross-section for the scattering off the potential V. It was derived using the perturbation theory at first order and hence corresponds to the **first-order approximation** or the so-called **Born approximation** of the calculation. It is worth investigating the term in the integral. It corresponds to a Fourier transform of the potential. We can consequently write:

$$\frac{\mathrm{d}\sigma}{\mathrm{d}\Omega} = \left(\frac{m}{2\pi}\right)^2 |\tilde{V}(\vec{q})|^2 \quad \text{(first order)} \tag{4.198}$$

where

$$\tilde{V}(\vec{q}) \equiv \int \mathrm{d}^3\vec{x}\, V(\vec{x})e^{i\vec{q}\cdot\vec{x}} \tag{4.199}$$

is the **Fourier transform** of the potential. The scattering differential cross-section at different $\vec{q}$ therefore informs us about the Fourier transform of the potential. With enough measurements at different values of $\vec{q}$, an inverse Fourier transformation allows us to reconstruct the potential $V(\vec{x})$! As a corollary, the **non-relativistic transition amplitude, Eq. (4.172), in a scattering process** can be expressed quite generally as:

$$T_{fi} = -i\tilde{V}(\vec{q})(2\pi\delta(E_{\vec{p}_f} - E_{\vec{p}_i})) \tag{4.200}$$

where $E_{\vec{p}_i}$ and $E_{\vec{p}_f}$ are the energies of the incoming and outgoing particles with momenta $\vec{p}_i$ and $\vec{p}_f$, and $\vec{q} \equiv \vec{p}_i - \vec{p}_f$.

4.18   Rutherford Scattering Revisited

We consider the setup of the Rutherford scattering and use the previously derived results. In perturbation theory at first order we expect a reasonable result if the acting potential is relatively "weak," hence we cannot only consider the atomic nucleus and its *Coulomb force with infinite range*. We need to assume the potential of a point-like nucleus of charge $+Ze$ surrounded by a cloud of electrons of total charge $-Ze$, represented by the charge distribution $\rho(\vec{x})$. The charge of the scattered particle is equal to ze. The electron cloud will shield the nucleus's charge at large distances, since the atom is electrically neutral. The potential $V(\vec{x})$ can then be expressed as:

$$V(\vec{x}) = \frac{ze^2}{4\pi\epsilon_0}\left(\frac{Z}{|\vec{x}|} - \int\frac{\rho(\vec{x}')}{|\vec{x} - \vec{x}'|}\mathrm{d}^3\vec{x}'\right) \tag{4.201}$$

where

$$\int\rho(\vec{x}')\mathrm{d}^3\vec{x}' = Z \tag{4.202}$$

corresponds to the total charge $-Ze$ of the electrons.

At first order in perturbation theory, we compute the matrix element in the Born approximation:

$$\int\mathrm{d}^3\vec{x}V(\vec{x})e^{i\vec{q}\cdot\vec{x}} = \frac{ze^2}{4\pi\epsilon_0}\int\mathrm{d}^3\vec{x}\left(\frac{Z}{|\vec{x}|} - \int\frac{\rho(\vec{x}')}{|\vec{x} - \vec{x}'|}\mathrm{d}^3\vec{x}'\right)e^{i\vec{q}\cdot\vec{x}} \tag{4.203}$$

We note that:

$$\int\mathrm{d}^3\vec{x}V(\vec{x})\left(\nabla^2 e^{i\vec{q}\cdot\vec{x}}\right) = (iq)^2\int\mathrm{d}^3\vec{x}V(\vec{x})e^{i\vec{q}\cdot\vec{x}} \tag{4.204}$$

Hence:

$$\int\mathrm{d}^3\vec{x}V(\vec{x})e^{i\vec{q}\cdot\vec{x}} = -\frac{1}{q^2}\int\mathrm{d}^3\vec{x}V(\vec{x})\left(\nabla^2 e^{i\vec{q}\cdot\vec{x}}\right) \tag{4.205}$$

We use the **second Green's identity** (see Appendix A.13) and find:

$$\int\mathrm{d}^3\vec{x}V(\vec{x})\left(\nabla^2 e^{i\vec{q}\cdot\vec{x}}\right) = \int\mathrm{d}^3\vec{x}e^{i\vec{q}\cdot\vec{x}}\left(\nabla^2 V(\vec{x})\right) + \int_{A=\partial V}\left(V(\nabla e^{i\vec{q}\cdot\vec{x}}) - e^{i\vec{q}\cdot\vec{x}}(\nabla V)\right)\cdot\mathrm{d}\vec{A} \tag{4.206}$$

When the volume V becomes infinitely large, we then expect that the integral on the surface A, which encompasses V, vanishes, therefore:

$$\begin{cases} V(\vec{x}) \to 0 & \text{when } |\vec{x}| \to \infty \\ \nabla V(\vec{x}) \to 0 & \text{when } |\vec{x}| \to \infty \end{cases} \tag{4.207}$$

and we assume that this happens fast enough so that the surface integral on $A = \partial V$ can be neglected compared to the first integral. In order to estimate the latter, we need to compute the Laplacian of the potential V:

$$\nabla^2 V(\vec{x}) = \frac{ze^2}{4\pi\epsilon_0} \left(Z\nabla^2 \left(\frac{1}{|\vec{x}|} \right) - \int \nabla^2 \left(\frac{1}{|\vec{x} - \vec{x'}|} \right) \rho(\vec{x'}) \mathrm{d}^3\vec{x'} \right) \tag{4.208}$$

We use the result $\nabla^2 \left(1/|\vec{x} - \vec{x'}| \right) = -4\pi\delta(|\vec{x} - \vec{x'}|)$ (see Eq. (A.55)) to find:

$$\nabla^2 V(\vec{x}) = -\frac{ze^2}{\epsilon_0} \left(Z\delta(|\vec{x}|) - \rho(\vec{x}) \right) \tag{4.209}$$

The integration of the potential is therefore equal to:

$$\begin{aligned}
\int \mathrm{d}^3\vec{x} V(\vec{x}) e^{i\vec{q}\cdot\vec{x}} &= -\frac{1}{q^2} \int \mathrm{d}^3\vec{x} V(\vec{x}) \left(\nabla^2 e^{i\vec{q}\cdot\vec{x}} \right) = -\frac{1}{q^2} \int \mathrm{d}^3\vec{x} e^{i\vec{q}\cdot\vec{x}} \left(\nabla^2 V(\vec{x}) \right) \\
&= \frac{ze^2}{\epsilon_0} \frac{1}{q^2} \int \mathrm{d}^3\vec{x} e^{i\vec{q}\cdot\vec{x}} \left(Z\delta(|\vec{x}|) - \rho(\vec{x}) \right)
\end{aligned} \tag{4.210}$$

Following Eq. (4.198), we define the **Fourier transform** of the charge distribution of the electrons as $F(\vec{q})$:

$$F(\vec{q}) \equiv \int \mathrm{d}^3\vec{x} e^{i\vec{q}\cdot\vec{x}} \rho(\vec{x}) \tag{4.211}$$

We can then write:

$$\int \mathrm{d}^3\vec{x} V(\vec{x}) e^{i\vec{q}\cdot\vec{x}} = \frac{ze^2}{\epsilon_0} \frac{1}{q^2} [Z - F(\vec{q})] \tag{4.212}$$

where $\vec{q} \equiv \vec{p_i} - \vec{p_f}$ represents the exchanged momentum. Finally, the differential cross-section is:

$$\frac{\mathrm{d}\sigma}{\mathrm{d}\Omega} = \left(\frac{m}{2\pi} \right)^2 \left(\frac{ze^2}{\epsilon_0} \right)^2 \left(\frac{1}{q^4} \right) [Z - F(\vec{q})]^2 \tag{4.213}$$

We introduce the scattering angle θ in the following way (see Figure 4.9):

$$q^2 \equiv (\vec{q})^2 = (\vec{p_i} - \vec{p_f})^2 = \vec{p_i}^2 + \vec{p_f}^2 - 2\vec{p_i} \cdot \vec{p_f} = 2p^2(1 - \cos\theta) = 4p^2 \sin^2\left(\frac{\theta}{2} \right) \tag{4.214}$$

The maximum momentum transfer is $q_{max}^2 = 4p^2 = (2p)^2$, which occurs for a backscatter at $\theta = 180°$. On the other hand, $q^2 \to 0$ when $\theta \to 0$.

For small scattering angles, the Fourier transformation can be approximated with the first term of its Taylor expansion:

$$e^{i\vec{q}\cdot\vec{x}} \approx 1 + i\vec{q} \cdot \vec{x} + \frac{1}{2}(i\vec{q} \cdot \vec{x})^2 + \cdots \tag{4.215}$$

This means:

$$\begin{aligned}
Z - F(\vec{q}) &= Z - \int \mathrm{d}^3\vec{x} e^{i\vec{q}\cdot\vec{x}} \rho(\vec{x}) \\
&= \underbrace{Z - \int \mathrm{d}^3\vec{x} \rho(\vec{x})}_{=0} \underbrace{-i \int \mathrm{d}^3\vec{x}(\vec{q} \cdot \vec{x})\rho(\vec{x})}_{=0 \text{ when } \rho(\vec{x}) = \rho(-\vec{x})} + \frac{1}{2} \int \mathrm{d}^3\vec{x}(\vec{q} \cdot \vec{x})^2 \rho(\vec{x}) + \cdots
\end{aligned} \tag{4.216}$$

where we have assumed a spherical symmetry for ρ. The result depends on the assumptions on the charge distribution of the electrons. In general, we can define the result as a function of I which depends on the integral of the charge distribution:

$$Z - F(\vec{q}) \approx \frac{1}{2} \int \mathrm{d}^3 \vec{x} (\vec{q} \cdot \vec{x})^2 \rho(\vec{x}) \equiv \frac{q^2 I}{2} \tag{4.217}$$

For small scattering angles, we then get:

$$
\begin{aligned}
\left(\frac{\mathrm{d}\sigma}{\mathrm{d}\Omega}\right)_{\theta \to 0} &= \left(\frac{m}{2\pi}\right)^2 \left(\frac{ze^2}{\epsilon_0}\right)^2 \left(\frac{1}{q^4}\right) [Z - F(\vec{q})]^2 \approx \left(\frac{m}{2\pi}\right)^2 \left(\frac{ze^2}{\epsilon_0}\right)^2 \left(\frac{1}{q^4}\right) \left(\frac{q^2 I}{2}\right)^2 \\
&= \left(\frac{m}{2\pi}\right)^2 \left(\frac{ze^2}{\epsilon_0}\right)^2 \left(\frac{I}{2}\right)^2 \to \text{const.} \tag{4.218}
\end{aligned}
$$

Due to the shielding of the electron cloud, there is no forward divergence as the scattering angle approaches zero (cf. the classical Rutherford scattering formula, Eq. (2.88)).

At the other extreme, we can analyze the situation when the exchanged momentum q becomes large at high energies $p \to \infty$, i.e., $|\vec{q}| \to q_{max}^2 \to \infty$. Because the electron cloud is not a point-like charge, its Fourier transform should decrease and tend to zero for increasing $|\vec{q}|$. In this case, we expect that $F(\vec{q}) \to 0$ when $|\vec{q}| \to \infty$ and we can neglect $F(\vec{q})$ in comparison to Z:

$$
\begin{aligned}
\left(\frac{\mathrm{d}\sigma}{\mathrm{d}\Omega}\right)_{|\vec{q}| \to \infty} &= \left(\frac{m}{2\pi}\right)^2 \left(\frac{ze^2}{\epsilon_0}\right)^2 \left(\frac{1}{q^4}\right) [Z - F(\vec{q})]^2 \approx \left(\frac{m}{2\pi}\right)^2 \left(\frac{zZe^2}{\epsilon_0}\right)^2 \left(\frac{1}{q^4}\right) \\
&= \left(\frac{zZe^2}{4\pi\epsilon_0}\right)^2 \left(\frac{2m}{4p^2 \sin^2(\theta/2)}\right)^2 = \left(\frac{zZe^2}{4\pi\epsilon_0}\right)^2 \left(\frac{1}{4E_K}\right)^2 \frac{1}{\sin^4(\theta/2)} = \left(\frac{\mathrm{d}\sigma}{\mathrm{d}\Omega}\right)_{Rutherford} \tag{4.219}
\end{aligned}
$$

where E_K is the kinetic energy of the particle. Not surprisingly, we recover the classical result – see Eq. (2.88)! Therefore, **the classical Rutherford cross-section represents the Born approximation of the process, when the scattering angle is not too small, and the effect of the electron cloud can be neglected.** However, it remains an astonishing fact that the original derivation of Rutherford using classical mechanics gave the same result as the one derived here with quantum mechanics!

4.19   Higher Orders in Non-relativistic Perturbation

In the Born approximation, we have tackled the challenge of Eq. (4.164) by considering a single transition between the initial and final state. In the so-called first order we retained only one term in the sum to get Eq. (4.167) which we integrated over the time variable to get Eq. (4.168):

$$\frac{\mathrm{d}a_f}{\mathrm{d}t} = -ia_i(t)e^{i(E_f - E_i)t}V_{fi}(t) \implies a_f(t) = -i \int_{-T/2}^{t} \mathrm{d}t' V_{fi}(t')e^{i(E_f - E_i)t'} \tag{4.220}$$

where we have set $a_i(t) = 1$ to perform the integration. We now assume that this equation is a good approximation for all a_n ($n \neq i$) on the right-hand side of the sum and obtain:

$$\frac{\mathrm{d}a_f}{\mathrm{d}t} = -ie^{i(E_f - E_i)t}V_{fi}(t) + (-i)^2 \sum_{n \neq i} \int_{-T/2}^{t} \mathrm{d}t' V_{ni}(t')e^{i(E_n - E_i)t'} e^{i(E_f - E_n)t}V_{fn}(t) \tag{4.221}$$

We assume as before that the potential is time-independent and rewrite accordingly:

$$\frac{\mathrm{d}a_f}{\mathrm{d}t} = \underbrace{-iV_{fi}e^{i(E_f - E_i)t}}_{\text{Born}} + (-i)^2 \sum_{n \neq i} V_{fn}V_{ni} \int_{-T/2}^{t} \mathrm{d}t' e^{i(E_n - E_i)t'} e^{i(E_f - E_n)t} \tag{4.222}$$

We now integrate over time as before, starting from the initial state at $-T/2$ to the final state at $+T/2$ to find:

$$a_f(+T/2) = \underbrace{-iV_{fi} \int_{-T/2}^{+T/2} dt\, e^{i(E_f - E_i)t}}_{\text{Born}} + \underbrace{(-i)^2 \sum_{n \neq i} V_{fn} V_{ni} \int_{-T/2}^{+T/2} dt \int_{-T/2}^{t} dt'\, e^{i(E_n - E_i)t'} e^{i(E_f - E_n)t}}_{\text{First-order correction}} \tag{4.223}$$

In order to find the transition amplitude from the initial to the final state, we need to let $T \to \infty$. This brings us to the complicated integral for $n \neq i$:

$$V_{ni} \int_{-\infty}^{t} dt'\, e^{i(E_n - E_i)t'} = (-i)V_{ni} \left. \frac{e^{i(E_n - E_i)t}}{E_n - E_i} \right|_{-\infty}^{t} \tag{4.224}$$

A trick is introduced here to make the integral possible by noting that by adding a small imaginary term in the exponent $-i\epsilon$ ($\epsilon > 0$), the integrand will tend to zero at $-\infty$. Physically, this is equivalent to simulating the *start* of the interaction, arising from the localizability of the scattered particle and the finite action of the perturbation, by an adiabatic decrease in the interaction strength as $t \to -\infty$ [67]. Indeed:

$$\int_{-\infty}^{t} dt'\, V_{ni} e^{\epsilon t'} e^{i(E_n - E_i)t'} = V_{ni} \int_{-\infty}^{t} dt'\, e^{i(E_n - E_i - i\epsilon)t'} = (-i)V_{ni} \left. \frac{e^{i(E_n - E_i - i\epsilon)t}}{E_n - E_i - i\epsilon} \right|_{-\infty}^{t} = iV_{ni} \frac{e^{i(E_n - E_i - i\epsilon)t}}{E_i - E_n + i\epsilon} \tag{4.225}$$

Similarly, the *cessation* of the interaction at $t \to +\infty$ is obtained with a $+i\epsilon$ term in the exponent. At the end of the calculations, we let $\epsilon \to 0$ so it will disappear from the result. Consequently, we can write:

$$a_f(+T/2) = \underbrace{-iV_{fi} \int_{-T/2}^{+T/2} dt\, e^{i(E_f - E_i)t}}_{\text{First order}} - \underbrace{i \sum_{n \neq i} V_{fn} V_{ni} \int_{-T/2}^{+T/2} dt\, \frac{e^{i(E_n - E_i - i\epsilon)t}}{E_i - E_n \pm i\epsilon} e^{i(E_f - E_n)t}}_{\text{Second order}} \tag{4.226}$$

We can now obtain the Born transition amplitude corrected to the second order by integrating over time and introducing as before the Dirac δ function. The transition amplitude becomes:

$$T_{fi} = T_{fi}^{Born} - (2\pi i) \sum_{n \neq i} \frac{V_{fn} V_{ni}}{E_i - E_n \pm i\epsilon} \delta(E_f - E_i) \tag{4.227}$$

By direct comparison, Eq. (4.172), we see that the obtained second-order result can be expressed as a correction to the matrix element V_{fi} in the Born formula $T_{fi} = -iV_{fi}(2\pi\delta(E_f - E_i))$ with the replacement:

$$V_{fi} \quad \longrightarrow \quad V_{fi} + \sum_{n \neq i} \frac{V_{fn} V_{ni}}{E_i - E_n \pm i\epsilon} \tag{4.228}$$

This result is reminiscent of the **Lippmann[28]–Schwinger[29] equation**, which expresses the relation between the operators as [67]:

$$T = V + V \frac{1}{E - H_0 \pm i\epsilon} T \tag{4.229}$$

with $H = H_0 + V$.

This result can be iterated upon with a sort of recursion to find the correction to the transition amplitude to any order. Indeed, we find by direct replacement:

$$V_{fi} \quad \longrightarrow \quad V_{fi} + \sum_{n \neq i} \frac{V_{fn} V_{ni}}{E_i - E_n \pm i\epsilon} + \sum_{n \neq i} \sum_{m \neq n} \frac{V_{fn} V_{nm} V_{mi}}{(E_m - E_n \pm i\epsilon)(E_i - E_m \pm i\epsilon)} + \cdots \tag{4.230}$$

---

28 Bernard Abram Lippmann (1914–1988), American theoretical physicist.
29 Julian Seymour Schwinger (1918–1994), American theoretical physicist.

This formulation leads to the graphical interpretation shown in Figure 4.10. The transition due to the perturbation can be represented as a series of terms, with increasing order. Each term is characterized by a given number of "vertices." To first order it is just V_{fi}. To second order, there are two interaction vertices characterized by V_{fn} and V_{ni}, where the sum over n represents all the intermediate "virtual" states. These latter are virtual in the sense that energy is not conserved, as $E_n \neq E_i$. Overall, the energy is conserved by the Dirac $\delta(E_f - E_i)$ function.

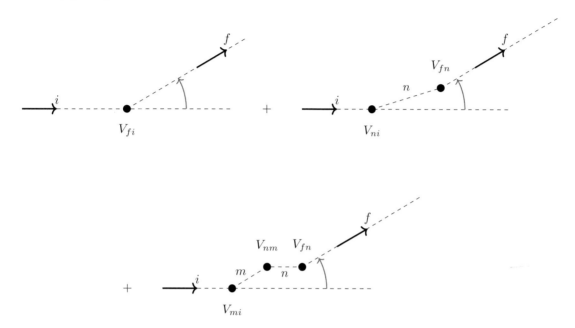

Figure 4.10 Graphical representation of the perturbation series in the $i \to f$ transition to first order, second order, and third order.

Problems

Ex 4.1 Electron polarization. Discuss methods to experimentally determine the polarization of an electron.

Ex 4.2 Photon polarization. Discuss methods to experimentally determine the polarization of a photon.

Ex 4.3 The linear harmonic oscillator. The Hamiltonian operator for the linear harmonic oscillator is given by:

$$H = \frac{p^2}{2m} + \frac{m\omega^2}{2}x^2$$

where the commutator of the operators p and x is $[x, p] = i$. Thus, H is the sum of two squares of operators, and we decompose this sum – analogous to the sum of two positive numbers $\alpha^2 + \beta^2 = (\alpha + i\beta)(\alpha - i\beta)$ – into two Hermitian adjoint operators:

$$a = \sqrt{\frac{m\omega}{2}}x + \frac{i}{\sqrt{2m\omega}}p; \quad a^\dagger = \sqrt{\frac{m\omega}{2}}x - \frac{i}{\sqrt{2m\omega}}p$$

(a) Show that the commutator of the operators a and $a^\dagger$ is $[a, a^\dagger] = 1$.

(b) Show that the Hamiltonian operator can be expressed in the following way through the operators a and $a^\dagger$:

$$H = \omega \left(a^\dagger a + \frac{1}{2} \right) = \omega \left(N + \frac{1}{2} \right) \quad \text{with } N \equiv a^\dagger a = N^\dagger$$

(c) $|\lambda\rangle$ is an eigenstate of the Hermitian operator N with eigenvalue λ. Show that the states $a\,|\lambda\rangle$ and $a^\dagger\,|\lambda\rangle$ are also eigenstates of N with eigenvalues $(\lambda - 1)$ and $(\lambda + 1)$, respectively.

(d) Show that the spectrum of eigenvalues of N corresponds to the integer numbers $n \geq 0$, i.e., the Hamiltonian operator H has the eigenvalue spectrum

$$E_n = \omega \left(n + \frac{1}{2} \right), \quad n = 0, 1, 2, \ldots$$

Ex 4.4 The photoelectric effect. In the electric dipole approximation, the time-dependent potential describing the interaction of an electron, with momentum $\vec{P}$, and an external field $\vec{A}(\vec{r}, t)$ is given by

$$V(\vec{r}, t) = \frac{e}{m} \vec{A}(\vec{r}, t) \cdot \vec{P} \tag{4.231}$$

Let us consider a hydrogen atom in his fundamental state – $(n, l, m) = (1, 0, 0)$, orbital $1s$ – and an external field of the form $\vec{A}(\vec{r}, t) = \vec{A}_0 \cos(\vec{k} \cdot \vec{r} - \omega t)$ impinging on the atom, and compute the probability of the electron being ejected using Fermi's Golden Rule.

(a) Using the time-dependent perturbation theory at first order, show that the potential describing the ejection of an electron can be written as:

$$\tilde{V}(\vec{r}, t) = \frac{e}{2m} e^{i\vec{k} \cdot \vec{r}} \vec{A}_0 \cdot \vec{P} e^{-i\omega t} = \tilde{V}(\vec{r}) e^{-i\omega t} \tag{4.232}$$

Hint: Explicitly develop the integrand of the first-order coefficient (the integration over space is omitted)

$$a_f^{(1)}(t) = -i \int_0^t dt' \, \langle f | \tilde{V}(\vec{r}, t) | i \rangle e^{i(E_f - E_i)t'}$$

(b) Calculate the transition probability using Fermi's Golden Rule. Recall that the hydrogen atom wave function for the $1s$ state is given by $\psi(\vec{r}) = \langle \vec{r} | 1, 0, 0 \rangle = e^{-r/a_0}/\sqrt{\pi a_0^3}$, where a_0 is the Bohr radius, and that the final-state electron vector in momentum space, $|\vec{p}_f\rangle \sim |\vec{k}_f\rangle$, can be expressed in position space as a continuous sum over the vectors $|\vec{r}\rangle$.

Hint: To compute the integral, think about how the term $e^{i\vec{k} \cdot \vec{r}}$ varies on the domain of integration and use the following relations:

$$\int_{\mathbb{R}^3} d^3\vec{r} \, e^{-i\vec{k}_f \cdot \vec{r}} \nabla \left(e^{-r/a_0} \right) = -\int_{\mathbb{R}^3} d^3\vec{r} \, \nabla \left(e^{-i\vec{k}_f \cdot \vec{r}} \right) e^{-r/a_0}$$

$$\int_{\mathbb{R}_{\geq 0}} dr \, r e^{-r/a_0} \sin(k_f r) = 2a_0^3 \frac{k_f}{(1 + a_0^2 k_f^2)^2} \tag{4.233}$$

(c) In which direction is the electron most likely to be ejected?

Ex 4.5 Born approximation in three-dimensional elastic scattering. Compute the differential and total cross-section for a particle scattering off the following potentials in the first-order approximation:

(a) $V(r) = V_0 e^{-r/R}$

(b) $V(r) = V_0 e^{-r^2/R^2}$

Hint: Recall that the total cross-section is given by the integral

$$\sigma = \int d\Omega \, \frac{d\sigma}{d\Omega} \tag{4.234}$$

where $d\Omega = d\Omega(\theta, \phi)$ is the solid-angle element, and think about the change of momentum dependence on the scattering angle θ.

5   Relativistic Formulation and Kinematics

Any theory of a fundamental nature must be consistent with relativity. In order for a physical quantity to be the same for all (inertial) observers, it must be invariant under Lorentz transformations. Physical transformations, either in space-time or internal, of real systems possess the mathematical property of a group and we will study the properties of physical quantities under these transformations. In order to build the theory, we will rely on those physical observables, which have well-defined properties (such as invariance implying symmetries) under the transformations.

5.1   The Poincaré and Lorentz Groups

The Poincaré group and its Lorentz subgroup are of great importance because invariance under the Poincaré group is a fundamental symmetry in particle physics. For example, a relativistic quantum field theory must have a Poincaré-invariant Lagrangian. This means that its fields must transform under representations of the Poincaré group and Poincaré invariance must be implemented. Here we will discuss some properties of the Lorentz and Poincaré groups.

At the basis of the symmetry is the theory of special relativity. Special relativity is based on the principle that one should consider space and time together, and take them to be a four-dimensional flat **Minkowski**[1] space with the **inner product** given by (see Appendix D):

$$x \cdot y \equiv x^0 y^0 - x^1 y^1 - x^2 y^2 - x^3 y^3 = x^0 y^0 - \vec{x} \cdot \vec{y} \tag{5.1}$$

The coordinates x^1, x^2, x^3 are interpreted as spatial coordinates, and the coordinate x^0 is a time coordinate, related to the conventional time coordinate t with respect to chosen units of time and distance given by $x^0 = ct$, where c is the speed of light. In natural units with $c = 1$, the units of time and distance are the same.

The three-vector $\vec{x}$ defines a point in space. The four-vector x^μ defines an **event** in space-time and will be denoted as:

$$x^\mu \equiv (x^0, x^1, x^2, x^3) = (x^0, \vec{x}) \tag{5.2}$$

It is conventional to write formulas using both upper and lower indices and define the inner product by *summing over one upper and one lower index*. This makes use of the Einstein summation convention:

$$x \cdot y = x^\mu y_\mu = x_\mu y^\mu \tag{5.3}$$

where we have introduced the four-vector x_μ with lower index:

$$x_\mu \equiv (x^0, -x^1, -x^2, -x^3) = (x^0, -\vec{x}) \tag{5.4}$$

---

1   Hermann Minkowski (1864–1909), German mathematician.

A four-vector like x^μ with an upper index is a **contravariant vector**, and one like x_μ with a lower index is a **covariant vector** (see Appendix B.2). The relation between a covariant and a contravariant vector is given by introducing the **metric tensor** $g_{\mu\nu}$:

$$x_\mu = g_{\mu\nu}x^\nu \quad \text{and} \quad x^\mu = g^{\mu\nu}x_\nu \tag{5.5}$$

For a flat **Minkowski** space, the metric, in our adopted convention, is simply given by (see Eq. (5.4)):

$$g_{\mu\nu} = g^{\mu\nu} = \begin{pmatrix} 1 & 0 & 0 & 0 \\ 0 & -1 & 0 & 0 \\ 0 & 0 & -1 & 0 \\ 0 & 0 & 0 & -1 \end{pmatrix} \tag{5.6}$$

For Cartesian coordinates, $g_{\mu\nu}$ and $g^{\mu\nu}$ are identical but this does not hold in general for curved space. The metric is used to convert and keep track of covariant and contravariant vectors. The inner product can be expressed with the metric tensor as:

$$x \cdot y = x^\mu y_\mu = g_{\mu\nu}x^\mu y^\nu \tag{5.7}$$

The **Lorentz group** consists of the linear transformations of space-time *conserving the inner product*:

$$x'^\mu = \Lambda^\mu{}_\nu x^\nu \tag{5.8}$$

where Λ is a 4×4 matrix. The components of the matrix are all real and dimensionless. They are independent of x^μ.

The **Poincaré group**, which includes the Lorentz group, consists of the following transformations of space-time:

$$x'^\mu = \Lambda^\mu{}_\nu x^\nu + a^\mu \tag{5.9}$$

The four-vector a^μ defines the relative origin of the coordinate system. Homogeneous transformations of the Poincaré group correspond to the case $a^\mu = 0$.

In order to conserve the inner product, the Λ matrices must satisfy the **relation of orthogonality** (or **Lorentz condition**):

$$g_{\mu\nu}\Lambda^\mu{}_\alpha \Lambda^\nu{}_\beta = g_{\alpha\beta} \tag{5.10}$$

such that **the Lorentz condition is the necessary and sufficient condition to ensure the conservation of the inner product under a homogeneous transformation:**

$$(x \cdot y)' = x' \cdot y' = g_{\mu\nu}x'^\mu y'^\nu = g_{\mu\nu}\Lambda^\mu{}_\alpha \Lambda^\nu{}_\beta x^\alpha y^\beta = g_{\alpha\beta}x^\alpha y^\beta = x \cdot y \tag{5.11}$$

While the **Lorentz transformation (LT)** of a contravariant vector is written as $x'^\mu = \Lambda^\mu{}_\nu x^\nu$, the transformation of a covariant vector is given by $x'_\mu = \Lambda_\mu{}^\nu x_\nu$ (beware of the position of the indices!). Then, the Lorentz condition can also be written with the **Kronecker tensor** (see Appendix D.9) as:

$$x' \cdot y' = \Lambda^\mu{}_\alpha \Lambda_\mu{}^\beta x^\alpha y_\beta = \delta^\beta_\alpha x^\alpha y_\beta = x^\alpha y_\alpha = (x \cdot y) \tag{5.12}$$

hence

$$\Lambda^\mu{}_\alpha \Lambda_\mu{}^\beta = \delta^\beta_\alpha \equiv \begin{cases} 1 \text{ if } \alpha = \beta \\ 0 \text{ if } \alpha \neq \beta \end{cases} \tag{5.13}$$

According to Eq. (5.11), the Lorentz condition implies that the inner product of a contravariant and covariant vector is an **invariant** (called a *scalar* under the transformation). Physically, an invariant observable possesses the same value for each inertial observer.

Starting with $x'^\mu = \Lambda^\mu_{\ \nu} x^\nu$, we note that:

$$\Lambda_\mu^{\ \rho} x'^\mu = \Lambda_\mu^{\ \rho} \Lambda^\mu_{\ \nu} x^\nu = \delta^\rho_\nu x^\nu = x^\rho \tag{5.14}$$

Hence, the **inverse transformation** from the primed system to the unprimed one of a contravariant vector is given by:

$$x^\mu = \Lambda_\nu^{\ \mu} x'^\nu \tag{5.15}$$

and for a covariant vector it is:

$$x_\mu = \Lambda^\nu_{\ \mu} x'_\nu \tag{5.16}$$

How quantities transform under Lorentz transformations is a crucial aspect and this is developed further in the next section.

5.2 Quantities with Well-Defined Lorentz Transformation Properties

We expand the considerations developed in Section 1.5 to the relativistic case. Let us consider a scalar quantity under Lorentz transformation. We know that this observable must have the same value for each inertial observer. As such, we can interpret this observable *as describing a truly fundamental property of the system*, since it is totally independent of the observer! We can, for example, state that the cross-section of a given reaction will be expressed as a function of such scalars.

In general, we are interested in tensors which are objects with well-defined properties under physical transformations, in particular the Lorentz transformation. The example above of the scalar quantity is a particular case of a so-called **tensor of rank 0**. According to Eq. (5.11), a tensor of rank 0 or scalar is an invariant of the Lorentz transformation. As an example, **the invariant space-time interval** (discussed in Section 5.3) is such a scalar physical quantity.

A four-vector is a **tensor of rank 1** with four components that transform under the Lorentz transformation as $x'^\mu = \Lambda^\mu_{\ \nu} x^\nu$. We note that the transformation of a rank 1 tensor has one summation index ν. When any physical observable with four components transforms itself according to $x'^\mu = \Lambda^\mu_{\ \nu} x^\nu$, we define it as a four-vector defined as $a^\mu \equiv (a^0, a^1, a^2, a^3) = (a^0, \vec{a})$:

$$a^\mu \text{ is a four-vector} \quad \Longleftrightarrow \quad a'^\mu = \Lambda^\mu_{\ \nu} a^\nu \tag{5.17}$$

We can consider tensors of higher rank by generalizing the transformation rule according to the number of components and considering an increasing number of summation indices. A **tensor of rank 2** will have $4^2 = 16$ components and two summation indices to describe the Lorentz transformation.

Table 5.1 summarizes the situation in case of Lorentz transformations.

Rank	Name	Components	Property under Lorentz transformation
0	scalar	1	$s' = s$ (invariant)
1	four-vector	4	$a'^\mu = \Lambda^\mu_{\ \nu} a^\nu$
2	tensor 2nd rank	$4^2 = 16$	$t'^{\mu\nu} = \Lambda^\mu_{\ \rho} \Lambda^\nu_{\ \sigma} t^{\rho\sigma}$

Table 5.1 Physical observables classified according to tensors with a given rank.

5.3   The Invariant Space-Time Interval

Let us apply these concepts to a very well-known case. We consider two events in space-time x^μ and y^μ, and the difference

$$\Delta x^\mu \equiv x^\mu - y^\mu = (\Delta x^0, \Delta x^1, \Delta x^2, \Delta x^3) \tag{5.18}$$

The space-time coordinates of such physical events are four-vectors of the type $x^\mu = (ct, x, y, z)$, where we reintroduce t, x, y, z, and c for clarity.

The notion of distance between two points is generalized to the interval between two points in space-time. This interval should be the same for all inertial observers, hence it must be invariant under Lorentz transformation. We hence want to construct the interval as a tensor of rank 0. According to Eq. (5.11), the Lorentz condition implies that the inner product of a contravariant and a covariant vector is such an **invariant**. Hence, noting that Δx^μ is by construction a four-vector, the interval Δs^2 defined as:

$$\Delta s^2 \equiv \Delta x^\mu \Delta x_\mu = (\Delta x^0)^2 - (\Delta x^1)^2 - (\Delta x^2)^2 - (\Delta x^3)^2 \tag{5.19}$$

is invariant. We recognize the **space-time interval of special relativity** (where we again reintroduce t, x, y, z, and c for clarity):

$$\Delta s^2 = (c\Delta t)^2 - (\Delta x)^2 - (\Delta y)^2 - (\Delta z)^2 = (\Delta x)^2 = (x^\mu - y^\mu)^2 \tag{5.20}$$

So, we have started with the space-time coordinates of two physical events x^μ and y^μ, which represent physical observables with the properties of a four-vector. With them, we constructed the new physical observable, the invariant space-time interval $\Delta s^2 \equiv (x^\mu - y^\mu)^2$, using the inner product.

We recall that events which are separated by a **time-like interval** have $\Delta s^2 > 0$; those separated by a **space-like interval** have $\Delta s^2 < 0$; and those separated by a **light-like interval** have $\Delta s^2 = 0$ (see Appendix D.2). So, in general, we classify any four-vector a^μ according to the value of the inner product with itself $a^2 = a^\mu a_\mu$:

$$\begin{cases} a^\mu \text{ is time-like} & \Leftrightarrow & a^2 > 0 \\ a^\mu \text{ is space-like} & \Leftrightarrow & a^2 < 0 \\ a^\mu \text{ is light-like} & \Leftrightarrow & a^2 = 0 \end{cases} \tag{5.21}$$

5.4   Classification of the Poincaré Group

The **Poincaré group**, also sometimes called the **inhomogeneous Lorentz group**, is the **Lie group** that contains all the symmetries of special relativity with translations, rotations, and boosts. Its algebraic structure is described in detail in Appendix B.11. The **Lorentz Lie group** is the subgroup that contains all the homogeneous Poincaré transformations, i.e., rotations and boosts. See Appendix B.10. Physical **rotations** only affect space coordinates and leave the time coordinate unchanged:

$$R : \begin{cases} \vec{x} \to R(\vec{x}) \\ t \to t \end{cases} \tag{5.22}$$

Their matrix representation in four-dimensional Minkowski space is hence given by the matrix (a) in Table 5.2, where R represents the 3×3 space rotation – see Eq. (B.20) in the Appendix.

A **Lorentz boost** affects space and time coordinates. For example, a boost along the x-axis with velocity parameter β_x and Lorentz factor $\gamma^2 = 1/(1 - \beta_x^2)$ is:

$$B_x : \begin{cases} x \to \gamma(x - \beta_x t) \\ y \to y \\ z \to z \\ t \to \gamma(t - \beta_x x) \end{cases} \tag{5.27}$$

Transformation	Matrix	
Proper Lorentz transformation		
(a) Rotation	$\Lambda^\mu{}_\nu(R) = \begin{pmatrix} 1 & 0 & 0 & 0 \\ 0 & & & \\ 0 & & R & \\ 0 & & & \end{pmatrix}$	(5.23)
(b) Boost along x	$\Lambda^\mu{}_\nu(\beta_x) = \begin{pmatrix} \gamma & -\beta_x\gamma & 0 & 0 \\ -\beta_x\gamma & \gamma & 0 & 0 \\ 0 & 0 & 1 & 0 \\ 0 & 0 & 0 & 1 \end{pmatrix}$	(5.24)
Improper Lorentz transformation		
(c) Parity	$\Lambda^\mu{}_\nu(P) = \begin{pmatrix} 1 & 0 & 0 & 0 \\ 0 & -1 & 0 & 0 \\ 0 & 0 & -1 & 0 \\ 0 & 0 & 0 & -1 \end{pmatrix}$	(5.25)
(d) Time reversal	$\Lambda^\mu{}_\nu(T) = \begin{pmatrix} -1 & 0 & 0 & 0 \\ 0 & 1 & 0 & 0 \\ 0 & 0 & 1 & 0 \\ 0 & 0 & 0 & 1 \end{pmatrix}$	(5.26)

Table 5.2 The matrix representations of Lorentz transformations in 4D Minkowski space.

Its matrix representation is given as (b) in Table 5.2. Boosts along other axes are given in Appendix B.10, Eqs. (B.84) and (B.85), and in Appendix D.3, Eq. (D.36).

The **parity transformation** P (also called parity inversion) is a flip in the sign of the spatial coordinates. In three dimensions, it is described by the simultaneous flip in the sign of all three spatial coordinates (a point reflection):

$$P : \begin{cases} \vec{x} \to -\vec{x} \\ t \to t \end{cases}$$ (5.28)

The matrix representation of P in Minskowski space is given as (c) in Table 5.2. Similarly, the **time-reversal** T flips the time coordinate and can be thought of as taking a movie of some process and then playing the movie backwards:

$$T : \begin{cases} \vec{x} \to \vec{x} \\ t \to -t \end{cases}$$ (5.29)

The matrix representation of T in Minskowski space is shown as (d) in Table 5.2.

Note that the matrix of P has its determinant equal to -1, and hence is distinct from a rotation, which has determinant equal to 1. When a transformation has a positive determinant, like a rotation or boost, it is labeled as a **proper Lorentz transformation**. For parity and time-reversal the determinant is -1 and they are defined as an **improper Lorentz transformation**. P and T represent **discrete transformations of space-time**. A Lorentz transformation with $\Lambda^0{}_0 > 0$ is **orthochronous**. Hence, P is an improper orthochronous Lorentz transformation, while T is an improper non-orthochronous Lorentz transformation. In contrast, boosts and rotations are proper orthochronous Lorentz transformations. Infinitesimal proper orthochronous Lorentz transformations can be made infinitesimally close to the identity, while improper Lorentz transformations cannot. See Appendix B.10. The **identity transformation** $\mathbb{1}$ is represented by the identity matrix:

$$\Lambda^{\mu}{}_{\nu}(\mathbb{1}) = \begin{pmatrix} 1 & 0 & 0 & 0 \\ 0 & 1 & 0 & 0 \\ 0 & 0 & 1 & 0 \\ 0 & 0 & 0 & 1 \end{pmatrix} \tag{5.30}$$

5.5   The Energy–Momentum Four-Vector

In special relativity we learn about the phenomenon of time dilation. The **proper time** τ is defined as (for clarity we temporarily reintroduce the speed of light c):

$$\tau \equiv \frac{t}{\gamma} \implies \Delta\tau \equiv \frac{\Delta t}{\gamma} = \frac{\Delta s}{c} \tag{5.31}$$

where γ is the Lorentz factor. The proper velocity of a particle is then defined as:

$$\vec{\eta} \equiv \frac{\mathrm{d}\vec{x}}{\mathrm{d}\tau} = \gamma\frac{\mathrm{d}\vec{x}}{\mathrm{d}t} = \gamma\vec{v} \tag{5.32}$$

- **Velocity four-vector.** It is then easy to extend this quantity to the velocity four-vector or four-velocity:

$$\eta^{\mu} \equiv \frac{\mathrm{d}x^{\mu}}{\mathrm{d}\tau} \tag{5.33}$$

The time component is $\eta^0 = \dfrac{\mathrm{d}x^0}{\mathrm{d}\tau} = \gamma\dfrac{\mathrm{d}(ct)}{\mathrm{d}t} = \gamma c$ and hence the individual components of the four-velocity are:

$$\eta^{\mu} = \gamma(c, \vec{v}) \tag{5.34}$$

The quantity η^2 is certainly an invariant: $\eta^{\mu}\eta_{\mu} = \gamma^2 c^2 - \gamma^2 v^2 = \gamma^2 c^2 (1 - \beta^2) = c^2$.

- **Energy–momentum four-vector.** With the four-velocity we can define the energy–momentum four-vector of a particle as:

$$p^{\mu} \equiv m\eta^{\mu} = m\gamma(c, \vec{v}) = (\gamma mc, \gamma m\vec{v}) = \left(\frac{E}{c}, \vec{p}\right) \tag{5.35}$$

where m is the **rest mass** of the particle (note that we always consider m as the rest mass while historically m_0 was the rest mass and $m = \gamma m_0$ was used to denote the dynamic mass, but we shall not use the historical convention).

The total energy is given by $E = \gamma mc^2$, the kinetic energy is $T = E - mc^2 = (\gamma - 1)mc^2$, and the three-momentum of the particle is $\vec{p} = \gamma m\vec{v}$. We compute:

$$(\vec{p}c)^2 + (mc^2)^2 = (\gamma m\vec{v}c)^2 + m^2 c^4 = m^2 c^4 \left(\gamma^2 v^2/c^2 + 1\right) = m^2 c^4 \gamma^2 = E^2 \tag{5.36}$$

Setting $c = 1$ to recover the natural units, we obtain $E^2 = \vec{p}^2 + m^2$ and:

$$p^{\mu} \equiv (E, \vec{p}), \qquad p_{\mu} \equiv (E, -\vec{p}) \tag{5.37}$$

The inner product of the energy–momentum four-vector is an invariant of the Lorentz transformation, given by:

$$p^2 = p^\mu p_\mu = E^2 - \vec{p}^2 = m^2 \tag{5.38}$$

• **Rest mass.** The **rest mass of a particle is an invariant**, it is the same for all inertial observers. We see the definition of the rest mass from a new standpoint. The rest mass of a particle is the invariant quantity related to the energy–momentum four-vector. It is therefore an important and useful quantity that characterizes the property of the particle and is independent of the observer.

• **Real and virtual particles/on–off shell.** For "real particles" (e.g., in initial and final states of a reaction), the rest mass is fixed by the **on-mass-shell relation**:

$$m^2 = E^2 - \vec{p}^2 \tag{5.39}$$

The particles are said to be "on-shell." For "virtual particles" or "off-shell particles," which can, for example, be exchanged in interactions but never exist as free particles (see Section 1.2), we define the virtual mass m^* with $m^{*2} = E^2 - \vec{p}^2$. This latter can take any value, positive or negative, or zero.

• **Boost of a particle at rest.** Let us consider a particle of mass m at rest in our reference frame. One then has $E = m$ and $\vec{p} = 0$ in this frame (i.e., the rest frame of the particle):

$$p^{\star\,\mu} = (m, \vec{0}) \tag{5.40}$$

If we perform a Lorentz boost of our coordinate system in the $-x$ direction with a velocity β_x, such that relative to this system, the particle moves in the positive x direction with velocity β_x, then we have (see Eq. (5.24)):

$$p^\mu = \Lambda^\mu{}_\nu(-\beta_x)p^{\star\,\nu} = \begin{pmatrix} \gamma & \beta_x\gamma & 0 & 0 \\ \beta_x\gamma & \gamma & 0 & 0 \\ 0 & 0 & 1 & 0 \\ 0 & 0 & 0 & 1 \end{pmatrix} \begin{pmatrix} m \\ 0 \\ 0 \\ 0 \end{pmatrix} = \begin{pmatrix} \gamma m \\ \gamma m\beta_x \\ 0 \\ 0 \end{pmatrix} \tag{5.41}$$

so, as expected, the particle appears to be moving in the positive x direction with a velocity:

$$\beta \equiv \frac{|\vec{p}|}{E} = \frac{\gamma m\beta_x}{\gamma m} = \beta_x \tag{5.42}$$

• **Addition of velocities.** To obtain the law for the composition of two velocities, we can simply apply two Lorentz boosts. In the case of two boosts in the same direction with velocities β_{x1} and β_{x2}, one finds:

$$p^\mu = \Lambda^\mu{}_\alpha(-\beta_{x1})\Lambda^\alpha{}_\nu(-\beta_{x2})p^{\star\,\nu} = \begin{pmatrix} \gamma_1\gamma_2(\beta_{x1}\beta_{x2}+1) & \gamma_1\gamma_2(\beta_{x1}+\beta_{x2}) & 0 & 0 \\ \gamma_1\gamma_2(\beta_{x1}+\beta_{x2}) & \gamma_1\gamma_2(\beta_{x1}\beta_{x2}+1) & 0 & 0 \\ 0 & 0 & 1 & 0 \\ 0 & 0 & 0 & 1 \end{pmatrix} \begin{pmatrix} m \\ 0 \\ 0 \\ 0 \end{pmatrix}$$

$$= \begin{pmatrix} \gamma_1\gamma_2 m(1 + \beta_{x1}\beta_{x2}) \\ \gamma_1\gamma_2 m(\beta_{x1}+\beta_{x2}) \\ 0 \\ 0 \end{pmatrix} \tag{5.43}$$

Hence the final velocity in the x direction is given by:

$$\beta_x = \frac{\beta_{x1}+\beta_{x2}}{1+\beta_{x1}\beta_{x2}} \tag{5.44}$$

which is the standard result of special relativity. We note that **the same result would have been obtained if we had applied the boost β_{x2} first and then β_{x1} second. This means that boosts in the same direction commute.**

• **Numerical boosts.** We illustrate the use of the C++ class available in the PYTHIA Monte-Carlo event generator [68] called Vec4. The Vec4 class gives a simple implementation of four vectors. The code is the following:[2]

```
#include "Pythia8/Pythia.h"
using namespace Pythia8;
int main() {
  Vec4 P(0.,0.,0.,1.);
  cout << "Our first 4-vector : " << P << std::endl;
}
```

The constructor Vec4::Vec4(double x = 0., double y = 0., double z = 0., double t = 0.) creates a four-vector, by default with all components set to 0. In the code above, we use it to create a four-vector momentum P representing a particle at rest with mass $m = 1$ GeV. Running the code yields the following output:

```
Our first 4-vector :     0.000     0.000     0.000     1.000 (    1.000)
```

We now demonstrate the use of the boost class void Vec4::bst(double betaX, double betaY, double betaZ), which boosts the four-momentum by beta = $(\beta_x, \beta_y, \beta_z)$. Let us consider the following operations with $\beta = 0.8$, hence $\gamma = 1/\sqrt{1 - \beta^2} = 1.66\overline{6}$:

```
Vec4 P(0.,0.,0.,1.);
cout << "Our first 4-vector : " << P << std::endl;

// (1) boost along the x-direction forward and then backward
double betax = 0.8;
P.bst(betax,0,0);
cout << "P after boost in x direction: " << P << std::endl;
P.bst(-betax,0,0);
cout << "P after boost in -x direction: " << P << std::endl;

// (2) same along the y-direction forward and then backward
double betay = 0.8;
P.bst(0,betay,0);
cout << "P after boost in y direction: " << P << std::endl;
P.bst(0,-betay,0);
cout << "P after boost in -y direction: " << P << std::endl;

// (3) now combining boost along x and y directions
P.bst(betax,0,0);
P.bst(0,betay,0);
cout << "P after boost in x and y direction: " << P << std::endl;
P.bst(0,-betay,0);
P.bst(-betax,0,0);
cout << "P after boost reverse operations: " << P << std::endl;

// (4) now combining boost along x and y directions but reversing
the order of the backward boosts
P.bst(betax,0,0);
P.bst(0,betay,0);
  cout << "P after boost in x and y direction: " << P << std::endl;
P.bst(-betax,0,0);
```

2   Reprinted code by permission from the PYTHIA author. PYTHIA is licensed under GNU GPL v2 or later. Copyright 2019 Torbjörn Sjöstrand.

```
P.bst(0,-betay,0);
cout << "P after boost reverse operations in opposite order: " << P << std::endl;
```

When we run the code, we obtain the following output:

```
Our first 4-vector :      0.000     0.000     0.000     1.000 (   1.000)
   // (1) boost along the x-direction forward and then backward
P after boost in x direction:     1.333     0.000     0.000     1.667 (   1.000)
P after boost in -x direction:    0.000     0.000     0.000     1.000 (   1.000)

   // (2) same along the y-direction forward and then backward
P after boost in y direction:     0.000     1.333     0.000     1.667 (   1.000)
P after boost in -y direction:    0.000     0.000     0.000     1.000 (   1.000)

   // (3) now combining boost along x and y directions
P after boost in x and y direction:     1.333     2.222     0.000     2.778 (   1.000)
P after boost reverse operations:       0.000     0.000     0.000     1.000 (   1.000)
   // As naively expected, the particle is back in its original momentum state - at rest.

   // (4) now combining boost along x and y directions but reversing the order of the backward boosts
P after boost in x and y direction:     1.333     2.222     0.000     2.778 (   1.000)
P after boost reverse operations in opposite order:    -1.481    -0.099     0.000     1.790 (   1.000)
   // Perhaps surprisingly, the particle is no more at rest after the sequence of boosts!
```

Everything seems fine when we individually perform the boost in the x and y directions. We boost forward, then backward, and we recover the initial particle at rest. So basically, we have numerically demonstrated that (removing the Lorentz indices for clarity):

$$\Lambda(-\beta_x)\Lambda(\beta_x) = \mathbb{1} \quad \text{and} \quad \Lambda(-\beta_y)\Lambda(\beta_y) = \mathbb{1} \tag{5.45}$$

where $\beta_x = \beta_y = \beta = 0.8$ in our example. Now we can also explore what happens when we combine the Lorentz boosts. We note that in step (3) we have done the operation equivalent to the following mathematical formulation:

$$\Lambda(-\beta_x)\Lambda(-\beta_y)\Lambda(\beta_y)\Lambda(\beta_x) = \mathbb{1} \tag{5.46}$$

And perhaps surprisingly, we find in step (4) that we do not recover the original four-vector. The particle initially at rest is not at rest after the sequence of boosts. This is an illustrative example that Lorentz boosts in different directions do not commute:

$$\Lambda(-\beta_y)\Lambda(-\beta_x)\Lambda(\beta_y)\Lambda(\beta_x) \neq \mathbb{1} \quad !! \tag{5.47}$$

This was perhaps to be expected since we can interpret a Lorentz boost as a rotation in the Minkowski space (see Appendix D.3) and we know already that three-dimensional rotations do not commute. Actual matrix multiplication shows that:

$$\Lambda(-\beta_y)\Lambda(-\beta_x)\Lambda(\beta_y)\Lambda(\beta_x) = \begin{pmatrix} \gamma^3\left(\gamma - 2\beta^2\right) & \beta\gamma^3\left(\beta^2 - \gamma + 1\right) & -\beta(\gamma-1)\gamma^2 & 0 \\ \beta(\gamma-1)\gamma^2 & \gamma^2 - \beta^2\gamma^3 & -\beta^2\gamma^2 & 0 \\ \beta\gamma^3\left(-\beta^2 + \gamma - 1\right) & -\beta^2(\gamma-2)\gamma^3 & \gamma^2 - \beta^2\gamma^3 & 0 \\ 0 & 0 & 0 & 1 \end{pmatrix} \tag{5.48}$$

where again we have set $\beta_x = \beta_y = \beta$ and $\gamma = 1/\sqrt{1-\beta^2}$. So

$$\Lambda(-\beta_y)\Lambda(-\beta_x)\Lambda(\beta_y)\Lambda(\beta_x)\begin{pmatrix} m \\ 0 \\ 0 \\ 0 \end{pmatrix} = \begin{pmatrix} m\gamma(\gamma^3 - 2\gamma^2 + 2) \\ m\beta\gamma^2\left(\gamma - 1\right) \\ -m\beta\gamma\left(\gamma^3 - 2\gamma^2 + 1\right) \\ 0 \end{pmatrix} \tag{5.49}$$

Consequently, we have naively boosted the particle along the x, y, $-x$, $-y$ directions. However, instead of ending up with the particle at rest, the particle is still in motion after the sequence of "closed" boosts! This apparent "paradox" of special relativity is discussed in the next section.

5.6   The Thomas–Wigner Rotation

In the previous section, we explored the apparently confusing behavior of the composition of two Lorentz boosts. We will show in the following that from the point of view of group theory, the composition of two non-collinear Lorentz boosts results in a Lorentz transformation that is the composition of a Lorentz boost and a rotation. This rotation is called the **Thomas[3]–Wigner rotation**.

Let us look at this unexpected paradox by again investigating the effect of the two Lorentz boosts on a particle at rest. If we perform the boost in the $+y$ and $+x$ directions, we obtain:

$$\Lambda(\beta_x)\Lambda(\beta_y)p^\star = \begin{pmatrix} \gamma_x\gamma_y & -\beta_x\gamma_x & -\beta_y\gamma_x\gamma_y & 0 \\ -\beta_x\gamma_x\gamma_y & \gamma_x & \beta_x\beta_y\gamma_x\gamma_y & 0 \\ -\beta_y\gamma_y & 0 & \gamma_y & 0 \\ 0 & 0 & 0 & 1 \end{pmatrix} \begin{pmatrix} m \\ 0 \\ 0 \\ 0 \end{pmatrix} = \begin{pmatrix} m\gamma_x\gamma_y \\ -m\gamma_x\gamma_y\beta_x \\ -m\gamma_y\beta_y \\ 0 \end{pmatrix} \tag{5.50}$$

Hence, the velocity of the particle after the boosts is:

$$\vec{\beta}_{x-y} \equiv \frac{\vec{p}}{E} = \left(\frac{-m\gamma_x\gamma_y\beta_x}{m\gamma_x\gamma_y}, \frac{-m\gamma_y\beta_y}{m\gamma_x\gamma_y}, 0 \right) = \left(-\beta_x, -\frac{\beta_y}{\gamma_x}, 0 \right) \tag{5.51}$$

Therefore, the boost in the x direction has been "fully effective," since the first component of the velocity is β_x, however, it has also *reduced* the velocity in the y direction by a factor γ_x. This should actually have been expected since the velocity can never be greater than the speed of light. So, as we perform more boosts, there must be a "mechanism" that ensures this limit. Similarly, swapping the order of the boosts, we find:

$$\Lambda(\beta_y)\Lambda(\beta_x)p^\star = \begin{pmatrix} \gamma_x\gamma_y & -\beta_x\gamma_x\gamma_y & -\beta_y\gamma_y & 0 \\ -\beta_x\gamma_x & \gamma_x & 0 & 0 \\ -\beta_y\gamma_x\gamma_y & \beta_x\beta_y\gamma_x\gamma_y & \gamma_y & 0 \\ 0 & 0 & 0 & 1 \end{pmatrix} \begin{pmatrix} m \\ 0 \\ 0 \\ 0 \end{pmatrix} = \begin{pmatrix} m\gamma_x\gamma_y \\ -m\gamma_x\beta_x \\ -m\gamma_x\gamma_y\beta_y \\ 0 \end{pmatrix} \tag{5.52}$$

and consequently, the velocity is now given by:

$$\vec{\beta}_{y-x} \equiv \frac{\vec{p}}{E} = \left(\frac{-m\gamma_x\beta_x}{m\gamma_x\gamma_y}, \frac{-m\gamma_x\gamma_y\beta_y}{m\gamma_x\gamma_y}, 0 \right) = \left(-\frac{\beta_x}{\gamma_y}, -\beta_y, 0 \right) \tag{5.53}$$

This time the second boost in the y direction has been fully effective and the velocity in the x direction has been reduced. **We note that the sequence of boosts leads to different final velocities**. This seems to be chaos! However, the energy $E = m\gamma_x\gamma_y$ is the same in both cases. We can directly check that the relation $E^2 = p^2 + m^2$ is indeed satisfied by computing, for instance for $\vec{\beta}_{x-y}$, such that:

$$\begin{aligned} E^2 &= p^2 + m^2 = m^2\gamma_x^2\gamma_y^2\beta_x^2 + m^2\gamma_y^2\beta_y^2 + m^2 = m^2\left(\gamma_y^2(\gamma_x^2\beta_x^2 + \beta_y^2) + 1\right) \\ &= m^2\left(\gamma_y^2\left(\gamma_x^2\left(1 - \frac{1}{\gamma_x^2}\right) + 1 - \frac{1}{\gamma_y^2}\right) + 1\right) = m^2\left(\gamma_y^2\left(\gamma_x^2 - 1 + 1 - \frac{1}{\gamma_y^2}\right) + 1\right) \\ &= m^2\left(\gamma_x^2\gamma_y^2 - 1 + 1\right) = m^2\gamma_x^2\gamma_y^2 \qquad \square \end{aligned} \tag{5.54}$$

where we used $\beta_{x/y}^2 = 1 - 1/\gamma_{x/y}^2$. This implies that the magnitudes of the momenta are identical in both cases:

$$p_{x-y}^2 = p_{y-x}^2 = E^2 - m^2 = m^2(\gamma_x^2\gamma_y^2 - 1) \tag{5.55}$$

---

3   Llewellyn Hilleth Thomas (1903–1992), a British physicist and applied mathematician.

Hence, the particle is not gaining energy or momentum, but rather the momentum three-vectors point in different directions! It is straightforward to compute the angle θ_T between the two three-momenta. Indeed:

$$
\begin{aligned}
\cos\theta_T &= \frac{\vec{p}_{x-y} \cdot \vec{p}_{y-x}}{p_{x-y}p_{y-x}} = \frac{m^2\gamma_x^2\gamma_y\beta_x^2 + m^2\gamma_x\gamma_y^2\beta_y^2}{m^2(\gamma_x^2\gamma_y^2 - 1)} = \frac{\gamma_x^2\gamma_y\left(1 - \frac{1}{\gamma_x^2}\right) + \gamma_x\gamma_y^2\left(1 - \frac{1}{\gamma_y^2}\right)}{\gamma_x^2\gamma_y^2 - 1} \\
&= \frac{\gamma_y\left(\gamma_x^2 - 1\right) + \gamma_x\left(\gamma_y^2 - 1\right)}{\gamma_x^2\gamma_y^2 - 1}
\end{aligned}
\tag{5.56}
$$

In the case where $\beta = \beta_x = \beta_y$, we find for $\gamma = \gamma_x = \gamma_y$:

$$
\cos\theta_T = \frac{\gamma\left(\gamma^2 - 1\right) + \gamma\left(\gamma^2 - 1\right)}{\gamma^4 - 1} = \frac{2\gamma\left(\gamma^2 - 1\right)}{(\gamma^2 - 1)(\gamma^2 + 1)} = \frac{2\gamma}{\gamma^2 + 1} = \frac{2\sqrt{1 - \beta^2}}{2 - \beta^2}
\tag{5.57}
$$

The dependence of the angle θ_T as a function of β is plotted in Figure 5.1. The angle goes to zero as $\beta \to 0$, while it tends to 90° for $\beta \to 1$. The effect can be surprising in its magnitude! For relativistic speeds, it leads to a large tangible effect. As such, the Thomas–Wigner rotation of special relativity is as profound and real an effect as for instance time dilation, length contraction, or the relativity of simultaneity. We will see later in Chapter 11 that it has an important consequence when analyzing polarized scattering cross-sections.

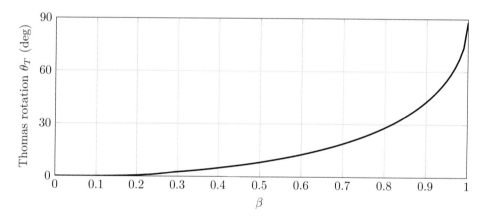

Figure 5.1 The angle of the Thomas–Wigner rotation as a function of the boost parameter β.

We can directly test the validity of the expression with the following PYTHIA code:[4]

```
Vec4 Pxy(0.,0.,0.,1.);
Vec4 Pyx(Pxy);
double beta = 0.8;
Pyx.bst(beta,0,0);
Pyx.bst(0,beta,0);
cout << "P after boost in x then y direction: " << Pyx << std::endl;
Pxy.bst(0,beta,0);
Pxy.bst(beta,0,0);
cout << "P after boost in y then x direction: " << Pxy << std::endl;
double cosT = 2*sqrt(1-beta*beta)/(2-beta*beta);
```

4   Reprinted code by permission from the PYTHIA author. PYTHIA is licensed under GNU GPL v2 or later. Copyright 2019 Torbjörn Sjöstrand.

```
double thetaT = acos(cosT);
Pxy.rotaxis(thetaT, 0,0,1);
cout << "after Thomas rotation : " << Pxy << std::endl;
```

The `Vec4::rotaxis(double theta, double nx, double ny, double nz)` method rotates the three-momentum vector by the azimuthal angle θ around the direction defined by the (n_x, n_y, n_z) three-vector. The output of this program is:

```
P after boost in x then y direction:     1.333    2.222    0.000    2.778 (    1.000)

P after boost in y then x direction:     2.222    1.333    0.000    2.778 (    1.000)

after Thomas rotation :     1.333    2.222    0.000    2.778 (    1.000)
```

This example clearly illustrates that the two momenta are rotated relative to one another by an angle $\theta_T = 28°$, as expected for the chosen value of $\beta = 0.8$. Fantastic!

5.7   Inertial Frames of Reference in Collisions

All inertial systems are equal. We can therefore choose any **inertial reference system** to describe the kinematics of a process. However, it is often worth selecting a reference system in which the equations will be the simplest. The symmetry of the Universe under a translation of space-time implies the conservation of energy and momentum. The reason to define and use these quantities lies in the fact that they are conserved in physical processes, such as collisions or decays of particles. In order to describe the kinematics of a reaction, we use four-vector quantities, in particular the energy–momentum four-vector. Let us consider a two-particle collision process, with two particles A and B with four-momenta $p_A^\mu = (E_A, \vec{p}_A)$ and $p_B^\mu = (E_B, \vec{p}_B)$. For example, the conservation of energy and momentum in the relativistic collision $A + B \rightarrow C + D$ can be expressed as the single equation:

$$p_A^\mu + p_B^\mu = p_C^\mu + p_D^\mu \tag{5.58}$$

where $p_{A,B,C,D}^\mu$ are the energy–momentum four-vectors of the particles A, B, C, and D. The conserved **total energy–momentum four-vector** is

$$P_{tot}^\mu = p_A^\mu + p_B^\mu = p_C^\mu + p_D^\mu \tag{5.59}$$

Different reference frames can be defined by requiring the momentum of the particles to have some specific value. The following are the most frequently used:

- The **laboratory system (LS)** – this is the frame in which the experiment is performed and all the energies and momenta of the initial and final state products are measured.

- The **center-of-mass system (CMS)** – this is the frame in which:

$$\vec{p}_A^\star + \vec{p}_B^\star = 0 \tag{5.60}$$

where the index "$\star$" indicates that the quantities are measured relative to the CMS reference system. In this frame, particles are colliding head-on.

- The **fixed target (FT)** or **target system** – this is the frame in which:

$$\vec{p}_B = 0 \tag{5.61}$$

This corresponds to experiments where a beam of particles A impinges on a stationary target containing particles B, which is at rest in the laboratory system. Experimentally, a beam is shot on a target and secondary particles produced in the collision coming out of the target are measured.

5.8   The Fixed Target vs. Collider Configuration

A particle A collides with a particle B that finds itself at rest relative to the **laboratory system**. We write the energy–momentum four-vectors of the particles relative to the laboratory system as:

$$p_A^\mu = (E_A, \vec{p}_A) \quad \text{and} \quad p_B^\mu = (M_B, 0) \tag{5.62}$$

where B is clearly at rest. See Figure 5.2. We compute the total energy–momentum squared using the

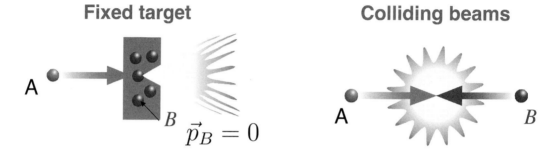

Figure 5.2 The fixed target vs. collider configuration for collisions between two particles.

four-vector algebra:

$$\begin{aligned}
p^2 \equiv (p_A + p_B)^2 &= (E_A + M_B)^2 - (\vec{p}_A + \vec{0})^2 = E_A^2 + 2M_B E_A + M_B^2 - \vec{p}_A^2 \\
&= M_A^2 + M_B^2 + 2M_B E_A
\end{aligned} \tag{5.63}$$

In the center-of-mass system, we have by definition that the total momentum vanishes, $\vec{p}_A^\star + \vec{p}_B^\star = 0$. The total energy–momentum four-vector squared is then equal to:

$$(p^\star)^2 \equiv (p_A^\star + p_B^\star)^2 = (E_A^\star + E_B^\star, 0)^2 = (E^\star)^2 \tag{5.64}$$

where $E^\star = E_A^\star + E_B^\star$ is the total energy in the center-of-mass system.

The inner product of two four-vectors is an *invariant under Lorentz transformation*. Hence:

$$p^2 = (p^\star)^2 \longrightarrow M_A^2 + M_B^2 + 2M_B E_A = (E^\star)^2 \tag{5.65}$$

The total energy in the center-of-mass system $E^\star$ amounts therefore to:

$$E^\star = \sqrt{M_A^2 + M_B^2 + 2M_B E_A} \tag{5.66}$$

This results becomes $E^\star = \sqrt{2M_B E_A}$ when $E_A \gg M_A, M_B$. In the fixed-target configuration, the energy available in the center-of-mass system is smaller than the total energy in the laboratory frame:

$$E^\star = \sqrt{2M_B E_A} < E_A + M_B \tag{5.67}$$

A part of the energy represents the motion of the center-of-mass relative to the laboratory and this energy is not available in the center-of-mass. Experimentally, this result limits the value of the mass of particles that can be created during such collisions. To create a particle of mass M, the energy in the center-of-mass must satisfy $E^\star \geq M$.

In colliding beams, the two beams of energies $E = E_A = E_B$ collide head-on, which means that the center-of-mass system is essentially the laboratory system. In this case, we have

$$E^\star = E_A + E_B = 2E \qquad (5.68)$$

Hence, a collider configuration is clearly much more favorable to produce high energies in the center-of-mass, for instance as needed to produce new heavy particles. This is illustrated in Figure 5.3. For instance, a positron beam of 1 TeV colliding against electrons at rest would only produce collisions with a center-of-mass energy of 1 GeV! With protons colliding on a fixed target consisting of protons (or neutrons), the situation is more favorable, but still much less energy is available in the center-of-mass compared with that in the collider mode. On the other hand, in the case of the fixed target the products of the collisions will be very forward boosted which can sometimes be an experimental advantage.

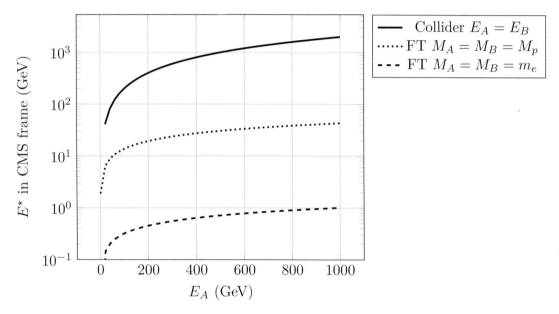

Figure 5.3 Comparison of the energy $E^\star$ in the center-of-mass system (CMS), in collider and fixed target (FT) modes. For the FT, the cases of electrons and protons are assumed.

• **Example Z^0-boson production.** We consider the case of the production of the Z^0-boson with mass $M \approx$ 90 GeV (see Eq. (26.79)), to be created in the process $e^+ e^- \to Z^0$. In the fixed-target configuration, assuming a beam of incident positrons of energy E_A colliding with electrons at rest as the target:

$$E^\star \geq M \implies E_A \geq \frac{M^2}{2M_B} \approx \frac{(90 \text{ GeV})^2}{2(0.511 \times 10^{-3} \text{ GeV})} \approx 8 \times 10^6 \text{ GeV} \gg M \qquad (5.69)$$

In the CERN LEP collider the two beams of energies E_A and E_B could be accelarated up to ≈ 100 GeV. In this case, the $e^+ e^- \to Z^0$ resonance is located at

$$E^\star \approx M \implies E_A + E_B = 2E_A \geq M \implies E_A = E_B \approx \frac{M}{2} \approx 45 \text{ GeV} \qquad (5.70)$$

Hence, a collider configuration is clearly more favorable to produce new heavy particles. In fact, the LEP collider at the highest energies was optimized to study the $e^+ e^- \to W^+ W^-$ reaction (see Section 26.10).

5.9   Transition from Center-of-Mass to Laboratory System

We consider the case of a "daughter" particle D_1 created back-to-back with a second D_2 in the decay of a "parent" particle P. The center-of-mass system is the frame where the parent is at rest. Its velocity relative to the laboratory is given by β_L and its corresponding Lorentz boost factor is $\gamma_L = 1/\sqrt{1 - \beta_L^2}$. In the parent frame, the daughter D_1 has energy $E^\star$ and momentum $p^\star$. We decompose the momentum in transverse $p_\perp^\star$ and parallel $p_\parallel^\star$ components relative to the direction of flight of the parent. The angle of emission $\theta^\star$ with respect to the parent flight direction is then defined by:

$$\cos\theta^\star = \frac{p_\parallel^\star}{p^\star}; \qquad \sin\theta^\star = \frac{p_\perp^\star}{p^\star} \tag{5.71}$$

See Figure 5.4. Although the particle D_1 is emitted in the forward direction in the figure with $p_\parallel^\star > 0$ and $\cos\theta^\star > 0$, it can also be emitted in the backward direction, in which case, $p_\parallel^\star$ would be negative and $\cos\theta^\star < 0$. In the laboratory frame, the energy and momentum are given by E and p. The transverse and

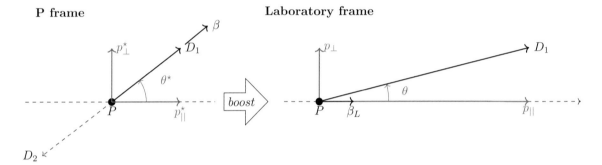

Figure 5.4 Kinematics of a decay in the center-of-mass and laboratory frames.

parallel components are $p_\parallel$ and $p_\perp$. Using the boost, Eq. (5.24), their transformation from the "$\star$" system to the laboratory corresponds to a boost with $-\beta_L$ and can be expressed as:

$$\begin{pmatrix} E \\ p_\parallel \\ p_\perp \end{pmatrix} = \begin{pmatrix} \gamma_L & \gamma_L\beta_L & 0 \\ \gamma_L\beta_L & \gamma_L & 0 \\ 0 & 0 & 1 \end{pmatrix} \begin{pmatrix} E^\star \\ p_\parallel^\star \\ p_\perp^\star \end{pmatrix} \tag{5.72}$$

With the velocity of the daughter D_1 in the parent system being labeled as $\beta \equiv p^\star/E^\star$, it follows:

$$\begin{cases} E = \gamma_L E^\star + \gamma_L \beta_L p_\parallel^\star \\ p_\parallel = \gamma_L(\beta_L E^\star + p_\parallel^\star) = \gamma_L p^\star\left(\beta_L \dfrac{E^\star}{p^\star} + \dfrac{p_\parallel^\star}{p^\star}\right) = \gamma_L p^\star\left(\dfrac{\beta_L}{\beta} + \cos\theta^\star\right) \\ p_\perp = p_\perp^\star \end{cases} \tag{5.73}$$

The decay angle θ in the laboratory frame with respect to the parent direction is given by:

$$\begin{aligned} \tan\theta &= \frac{p_\perp}{p_\parallel} = \frac{p_\perp^\star}{(\gamma_L\beta_L E^\star + \gamma_L p_\parallel^\star)} = \frac{1}{\gamma_L}\left(\frac{p_\perp^\star/p^\star}{\beta_L E^\star/p^\star + p_\parallel^\star/p^\star}\right) \\ &= \frac{1}{\gamma_L}\left(\frac{\sin\theta^\star}{\beta_L/\beta + \cos\theta^\star}\right) \end{aligned} \tag{5.74}$$

The angle θ in the laboratory frame is plotted as a function of $\theta^\star$ in Figure 5.5 for three configurations: (1) $\beta_L = 0.3$, $\beta = 0.5$; (2) $\beta_L = \beta = 0.5$; and (3) $\beta_L = 0.7$, $\beta = 0.5$. If $\beta > \beta_L$, then it is possible that the decay particle flies backwards in the laboratory frame. On the other hand, if $\beta \leq \beta_L$, then the particle is always forward boosted and the angle satisfies $\theta \leq \theta_{max} \leq 90°$, as illustrated in Figure 5.5.

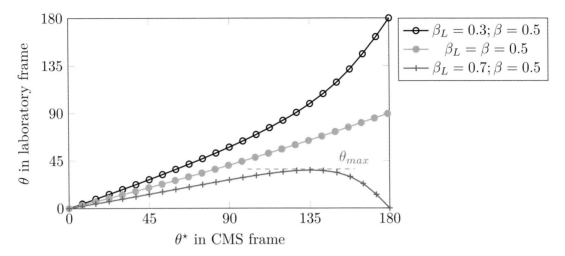

Figure 5.5 Angle in the laboratory frame vs. angle in the center-of-mass for the three cases (a) $\beta_L < \beta$, (b) $\beta_L = \beta$, and (c) $\beta_L > \beta$.

These results can easily be understood if we consider all possible decay angles $\theta^\star$ of the particle in the parent frame. For a fixed momentum $p^\star$ and all directions of emissions, the set of possible longitudinal and transverse momenta can be represented as the **momentum sphere**, as shown in Figure 5.6, where the cylindrical symmetry allows us to just plot the "momentum circle." In the center-of-mass system, this circle has a radius given by (from Eq. (5.71)):

$$p^{\star 2} = p_{||}^{\star 2} + p_{\perp}^{\star 2} = \text{const.} \tag{5.75}$$

We can translate the equation in the laboratory frame using the relations of Eq. (5.73), to obtain:

$$p_{||} = \gamma_L(\beta_L E^\star + p_{||}^\star) \qquad \Longrightarrow \qquad p_{||}^\star = \frac{p_{||}}{\gamma_L} - \beta_L E^\star \tag{5.76}$$

Hence, since $p_T^\star = p_\perp$:

$$p^{\star 2} = p_\perp^2 + \left(\frac{p_{||}}{\gamma_L} - \beta_L E^\star\right)^2 = p_\perp^2 + \frac{1}{\gamma_L^2}\left(p_{||} - \beta_L \gamma_L E^\star\right)^2 \tag{5.77}$$

Consequently, the "momentum sphere" is transformed into an ellipsoid, whose projection in the plane gives an ellipse of the form:

$$\frac{p_\perp^2}{a^2} + \frac{(p_{||} - c)^2}{b^2} = 1 \tag{5.78}$$

where $a = p^\star$, $b = \gamma_L p^\star$, and $c = \beta_L \gamma_L E^\star$. The intercepts of the ellipse with the $p_{||}$-axis are found by boosting the totally forward ($\theta^\star = 0$, $\cos\theta^\star = +1$) and totally backward ($\theta^\star = 180°$, $\cos\theta^\star = -1$) emitted particle in the center-of-mass frame. We find:

$$p_{||} = \gamma_L p^\star\left(\frac{\beta_L}{\beta} \pm 1\right) = \gamma_L \frac{p^\star}{\beta}(\beta_L \pm \beta) = \gamma_L E^\star(\beta_L \pm \beta) \tag{5.79}$$

We recover the cases encountered before, depending on the relative magnitude of β and β_L. The origin of the axis lies inside the ellipse for $\beta_L < \beta$, exactly on the ellipse for $\beta_L = \beta$, and outside for $\beta_L > \beta$. The three cases are illustrated in Figure 5.6. In the first case, the longitudinal momentum $p_\parallel$ can be either positive (corresponding to forward emission) or negative (corresponding to backward emission). In the third case, the longitudinal momentum $p_\parallel$ can only be positive. This classification is natural. In order to change the direction of the momentum from the backward to the forward direction, one needs a boost with a velocity larger than the velocity of the particle. This separation has a practical implication, since for $\beta_L > \beta$, particles will go into the forward hemisphere. The smaller the mass of a particle, the larger its velocity for a given momentum. For photons, with vanishing rest mass, we necessarily have $\beta_L < \beta = 1$, so there is always a fraction of photons that will be emitted backward.

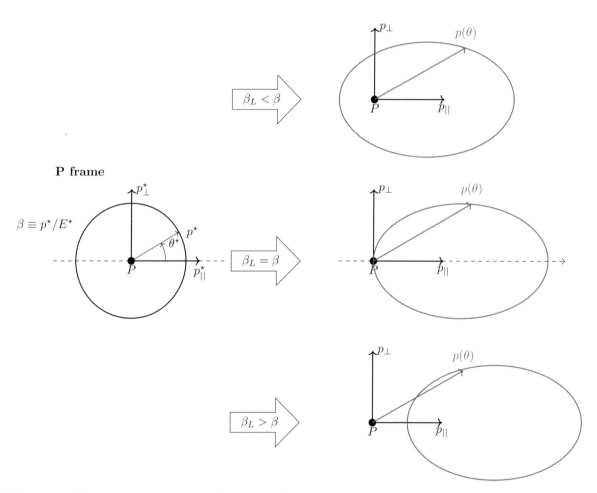

Figure 5.6 The momentum sphere in the center-of-mass and laboratory frame for the three cases $\beta < \beta_L$, $\beta = \beta_L$, and $\beta > \beta_L$.

5.10   A Decay into Two Massless Photons

We consider the decay of the neutral pion π^0 into two photons. We assume that the neutral pion with rest mass $m = m_{\pi^0}$ (see Eq. (15.21) in Section 15.4) moves in the laboratory frame with a velocity β_L ($\gamma = 1/\sqrt{1 - \beta_L^2}$) in the $+x$ axis direction (see Figure 5.7). In the rest frame of the π^0 (the center-of-mass of the neutral pion)

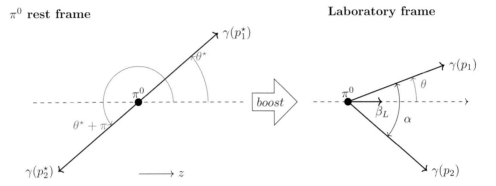

Figure 5.7 Kinematics of $\pi^0 \to \gamma\gamma$ relative to center-of-mass (left) and laboratory frame (right). The **opening angle** α between the two photons in the laboratory frame is also shown.

the two photons are emitted back-to-back, each with an energy $E^\star = m/2$ and momentum $p^\star = E^\star$, since photons are massless.

We label the emission angle of one of the photons in the center-of-mass system as $\theta^\star$. The second photon is emitted along the angle $\theta^\star + \pi$. Then $p_{1,\|}^\star = (m/2)\cos\theta^\star$ and $p_{2,\|}^\star = -(m/2)\cos\theta^\star$. The energies of the photons in the laboratory frame are given by:

$$E_1 = \gamma\frac{m}{2} + \gamma\beta_L p_{1,\|}^\star = \gamma\frac{m}{2} + \gamma\beta_L\frac{m}{2}\cos\theta^\star = \gamma\frac{m}{2}(1 + \beta_L\cos\theta^\star)$$
$$E_2 = \gamma\frac{m}{2} + \gamma\beta_L p_{2,\|}^\star = \gamma\frac{m}{2} - \gamma\beta_L\frac{m}{2}\cos\theta^\star = \gamma\frac{m}{2}(1 - \beta_L\cos\theta^\star) \tag{5.80}$$

and

$$E_1 + E_2 = \gamma\frac{m}{2}(2) = \gamma m \equiv E_{\pi^0} \tag{5.81}$$

is the total energy of the π^0 in the laboratory frame. We derive the components of the momentum:

$$\begin{cases} p_{1,T} = p_{1,T}^\star = \dfrac{m}{2}\sin\theta^\star \\ p_{1,\|} = \gamma(\beta_L E + p_{1,\|}^\star) = \gamma\left(\beta_L\dfrac{m}{2} + \dfrac{m}{2}\cos\theta^\star\right) = \gamma\dfrac{m}{2}(\beta_L + \cos\theta^\star) \end{cases} \tag{5.82}$$

and correspondingly for the second photon. It follows that the relation between the angle in the laboratory frame and the emission angle in the center-of-mass frame is:

$$\tan\theta = \frac{p_{1,T}}{p_{1,\|}} = \frac{\frac{m}{2}\sin\theta^\star}{\gamma\frac{m}{2}(\beta_L + \cos\theta^\star)} = \frac{\sin\theta^\star}{\gamma(\beta_L + \cos\theta^\star)} \tag{5.83}$$

For $\theta^\star \to 0$ we have $\theta \to 0$ and for $\theta^\star \to \pi$ we also have $\theta \to \pi$. In the center-of-mass system the photons that are emitted backwards will also travel in the backward direction in the laboratory frame, even for $\beta_L \to 1$. This is because the photon is massless and always travels with the speed of light. A forward-directed boost cannot change the direction of motion of a backward-emitted photon.

We now consider the product of the photons' energy–momentum four-vectors in the laboratory frame, where we define α as the angle between the two photons (see Figure 5.7):

$$p_1^\mu p_{2\mu} = E_1 E_2 - p_1 p_2 \cos\alpha = E_1 E_2 (1 - \cos\alpha) \tag{5.84}$$

In the center-of-mass frame this angle is $180°$ and therefore:

$$p_1^\mu p_{2\mu} = E_1^\star E_2^\star + p_1^\star p_2^\star = 2\left(\frac{m}{2}\right)^2 \tag{5.85}$$

The product $p_1^\mu p_{2\mu}$ is Lorentz-invariant, and the right-hand terms in the last two equations are equal to each other, hence we obtain:

$$E_1 E_2 (1 - \cos\alpha) = 2\left(\frac{m}{2}\right)^2 \implies 1 - \cos\alpha = \frac{m^2}{2E_1 E_2} \tag{5.86}$$

We express this result as a function of the emission angle in the center-of-mass frame (using Eq. (5.80)):

$$
\begin{aligned}
1 - \cos\alpha &= \frac{m^2}{2E_1 E_2} = \frac{m^2}{2\left(\gamma\frac{m}{2}(1 + \beta_L \cos\theta^\star)\right)\left(\gamma\frac{m}{2}(1 - \beta_L \cos\theta^\star)\right)} \\
&= \frac{2}{\gamma^2(1 - \beta_L^2 \cos^2\theta^\star)}
\end{aligned} \tag{5.87}
$$

We note that the **minimum opening angle** α_m in the laboratory frame is obtained for $\cos\theta^\star = 0$. In this case:

$$1 - \cos\alpha_m = 2\sin^2\left(\frac{\alpha_m}{2}\right) = \frac{2}{\gamma^2} \implies \sin\left(\frac{\alpha_m}{2}\right) = \frac{1}{\gamma} \tag{5.88}$$

The minimum angle depends on the energy of the π^0 in the laboratory frame and decreases as the inverse of the Lorentz factor. For a π^0 energy of about 1 GeV it is roughly $15°$. In the center-of-mass frame, this corresponds to photons being emitted perpendicular to the π^0 flight direction (i.e., $\theta^\star = 90°$).

The **minimum and maximum energies** E_- and E_+ of the photons in the laboratory frame are found with $\theta^\star = 0$ and $\theta^\star = \pi$:

$$E_{1,2} = \frac{\gamma m}{2}(1 \pm \beta_L \cos\theta^\star) = \frac{E_{\pi^0}}{2}(1 \pm \beta_L \cos\theta^\star) \implies \begin{cases} E_+ = E_{\pi^0}\frac{1+\beta_L}{2} \\ E_- = E_{\pi^0}\frac{1-\beta_L}{2} \end{cases} \tag{5.89}$$

For an ultra-relativistic parent π^0 with $\beta_L \to 1$, we have $E_- \simeq 0$ and $E_+ \simeq E_{\pi^0}$. The energy distribution of the photons in the laboratory frame can be found by noting that their emission angle distribution in the center-of-mass frame must be isotropic, since the parent π^0 has spin-0. Isotropy implies that the emission probability in a given solid angle is constant:

$$\frac{dn(\Omega_\star)}{d\Omega_\star} = \frac{d^2 n(\cos\theta^\star, \phi_\star)}{d\cos\theta^\star d\phi_\star} = \text{const.} \implies \frac{dn(\cos\theta^\star)}{d\cos\theta^\star} = \text{const.} \tag{5.90}$$

The energy distribution is obtained with a change of variable (the energy of photon 1 is written as E):

$$\frac{dn(E)}{dE} = \frac{dn(\cos\theta^\star)}{d\cos\theta^\star}\frac{d\cos\theta^\star}{dE} = \text{const.} \times \frac{d\cos\theta^\star}{dE} \tag{5.91}$$

Since

$$E = \frac{E_{\pi^0}}{2}(1 + \beta_L \cos\theta^\star) \implies \cos\theta^\star = \frac{1}{\beta_L}\left(\frac{2E}{E_{\pi^0}} - 1\right) \tag{5.92}$$

the energy distribution of the photons in the laboratory frame is constant between the minimum and the maximum values, as illustrated in Figure 5.8:

$$\frac{dn(E)}{dE} = \text{const.} \times \frac{2}{\beta_L E_{\pi^0}} = \text{const}'. \tag{5.93}$$

In general, the energy distribution of monoenergetic and isotropically emitted particles in the center-of-mass frame is flat in the laboratory frame.

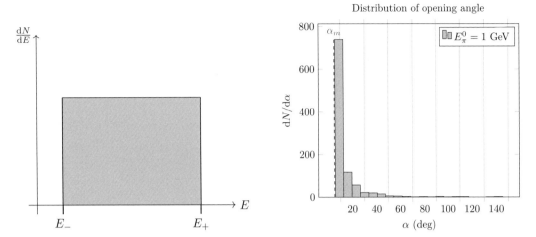

Figure 5.8 (left) Distribution of the energy of the photons in the laboratory frame. In this frame, the parent π^0 of mass m has a velocity β. The maximum and minimum energies are given by $E_\pm = \frac{E_{\pi^0}}{2}(1 \pm \beta)$. For the ultra-relativistic case $\beta \to 1$, we have $E_- \simeq 0$ and $E_+ \simeq E_{\pi^0}$. (right) Distribution of the opening angle between the two photons for $E_\pi^0 = 1$ GeV. The minimum opening angle is $\alpha_m \approx 15°$.

5.11   One-Particle Phase Space: Lorentz Invariance

Assuming a particle is produced when a reaction takes place at $t = 0$, it will subsequently propagate (and eventually decay if unstable) following a trajectory in space fully determined by its three-momentum after the reaction. As illustrated in Fermi's Golden rule, Eq. (4.178), we expect that the probability for any reaction to happen is proportional to the number of states that the final state can reach. This includes all possible momenta that the particle is allowed to have. Within a finite-size box, we have seen that the number of states is given by Eq. (4.190). Anticipating the cancelling out of $L \to \infty$ in our calculations, the integration over all possible momenta can be expressed as before:

$$\int_{-\infty}^{\infty} \frac{\mathrm{d}p_x}{(2\pi)} \frac{\mathrm{d}p_y}{(2\pi)} \frac{\mathrm{d}p_z}{(2\pi)} = \int_{-\infty}^{\infty} \frac{\mathrm{d}^3\vec{p}}{(2\pi)^3} \tag{5.94}$$

Since we integrate over $\mathrm{d}^3\vec{p}$, this integral is not Lorentz-invariant and we should introduce an integration over d^4p. However, final-state particles obey the mass-shell relation (Eq. (5.39)), which can be implemented with a Dirac δ function. In addition, we have chosen that physical particles have positive energies $p^0 > 0$ and this requirement is introduced with the **Heavyside step function** $\Theta(p^0)$ (see Section A.4). Taken all together, the phase space of one particle can hence be written in integral form:

$$\int_{-\infty}^{\infty} \frac{\mathrm{d}^4p}{(2\pi)^4}(2\pi)\delta(p^2 - m^2)\Theta(p^0) \tag{5.95}$$

The integration extends over *all values* of the components of the space p^μ ($\mu = 0, 1, 2, 3$), including p^0 which is a priori unconstrained in the d^4p integration. The step function $\Theta(p^0)$ and Dirac's δ function restore the positive-energy solution with the "on-shell" condition $(p^2 = m^2)$. So this is what we want!

But is this complicated form really Lorentz-invariant? The step function $\Theta(p^0)$ is invariant under the orthochronous Lorentz transformation, since this latter does not change the sign of the time-like component

p^0. The Dirac δ function is also invariant since p^2 and m^2 are Lorentz transformation invariants. The **four-dimensional phase-space element** $\mathrm{d}^4 p$ is also invariant (see Appendix A.8 for the definition of the Jacobi determinant), since:

$$\mathrm{d}^4 p' = \underbrace{\left| \frac{\partial(p'^0, p'^1, p'^2, p'^3)}{\partial(p^0, p^1, p^2, p^3)} \right|}_{\text{Jacobian}} \mathrm{d}^4 p = \underbrace{\left| \det(\Lambda) \right|}_{1} \mathrm{d}^4 p = \mathrm{d}^4 p \qquad (5.96)$$

Now a little magic occurs when realizing that the phase-space element can actually be replaced by:

$$\int \frac{\mathrm{d}^4 p}{(2\pi)^3} \delta(p^2 - m^2)\Theta(p^0) = \int \frac{\mathrm{d}^3 \vec{p}}{(2\pi)^3 2E} \Bigg|_{E=+\sqrt{p^2+m^2}} \qquad (5.97)$$

where the energy E is constrained by energy–momentum conservation for the given particle mass ($E = +\sqrt{\vec{p}^2 + m^2}$) as one integrates over $\mathrm{d}^3 \vec{p}$. We stress that the factor $1/2E$ arises when we integrate over $\mathrm{d}p^0$ under the constraints of the Dirac δ function and the Heaviside step function. On the right-hand side, the integral is taken over $\mathrm{d}^3 \vec{p}$. Each element $\mathrm{d}^3 \vec{p}$ is divided by the factor $2E$.

The equality Eq. (5.97) can be proven by considering the four-dimensional integral of an arbitrary function $f(\vec{p})$:

$$\int \mathrm{d}^4 p\, f(\vec{p})\delta(p^2 - m^2)\Theta(p^0) = \int \mathrm{d}^3 \vec{p}\, f(\vec{p}) \int \mathrm{d}p^0 \delta((p^0)^2 - \vec{p}^{\,2} - m^2)\, \Theta(p^0) \qquad (5.98)$$

We can simplify the Dirac δ function by using its properties summarized in Eq. (A.31), by defining $g(p^0) = (p^0)^2 - \vec{p}^{\,2} - m^2$. The zeros of $g(p^0)$ are found at $p^0 \equiv \pm\sqrt{\vec{p}^{\,2} + m^2}$, and the function also satisfies $\mathrm{d}g(p^0)/\mathrm{d}p^0 = 2p^0$. Then, $\delta(g(p^0))$ can be replaced in the integral to give the following result:

$$\int \mathrm{d}p^0 \left[\frac{\delta(p^0 - \sqrt{\vec{p}^{\,2} + m^2})}{2p^0} + \frac{\delta(p^0 + \sqrt{\vec{p}^{\,2} + m^2})}{2p^0} \right] \Theta(p^0) =$$

$$\int \mathrm{d}p^0 \left[\frac{\delta(p^0 - \sqrt{\vec{p}^{\,2} + m^2})}{2p^0} \right] = \left[\frac{1}{2E} \right] \Bigg|_{E=+\sqrt{p^2+m^2}} \qquad (5.99)$$

It follows that Eq. (5.97) is correct $\square$.

At first glance, the appearance of the energy E in the denominator might be surprising. It is, however, needed for Lorentz invariance, in order to compensate for the non-invariance of the (momentum) volume V, so that the product EV is a Lorentz-invariant quantity. This invariance follows immediately by explicitly noting that a Lorentz boost of each phase-space element $\mathrm{d}^3 \vec{p}/E$ yields:

$$\frac{\mathrm{d}^3 \vec{p}}{E} = \frac{\mathrm{d}p_x \mathrm{d}p_y \mathrm{d}p_z}{E} \implies \frac{\mathrm{d}^3 \vec{p}'}{E'} = \frac{\mathrm{d}p_x' \mathrm{d}p_y' \mathrm{d}p_z'}{E'} = \frac{\mathrm{d}p_x \mathrm{d}p_y \gamma(\mathrm{d}p_z + \beta \mathrm{d}E)}{\gamma(E + \beta p_z)} = \frac{\mathrm{d}^3 \vec{p}}{E} \frac{(1 + \beta \frac{\mathrm{d}E}{\mathrm{d}p_z})}{(1 + \beta \frac{p_z}{E})} \qquad (5.100)$$

Furthermore,

$$\frac{\mathrm{d}E}{\mathrm{d}p_z} = \frac{\mathrm{d}}{\mathrm{d}p_z} \sqrt{\vec{p}^2 + m^2} = \frac{\mathrm{d}}{\mathrm{d}p_z} \sqrt{p_x^2 + p_y^2 + p_z^2 + m^2} = \frac{1}{2} \frac{2p_z}{E} = \frac{p_z}{E} \qquad (5.101)$$

Hence, $\mathrm{d}^3 \vec{p}/E$ is Lorentz-invariant!

Removing the integration, we can now define the **one-particle Lorentz-invariant phase-space (1-Lips) element**:

$$\mathrm{d}Lips(p) \equiv \frac{\mathrm{d}^3 \vec{p}}{(2\pi)^3 2E} \qquad (5.102)$$

with $E^2 = \vec{p}^2 + m^2$.

5.12   **Lorentz-Invariant Phase-Space** $dLips$

So far we have essentially considered the properties of the initial and final particles in a reaction and their Lorentz transformations. Turning now to the complete reaction, we can represent the general case of a collision between two particles A and B, yielding a final state with n particles:

$$A + B \to 1 + 2 + 3 + \cdots + n \tag{5.103}$$

The process can be generically represented as a "**blob**" (representing the interaction) with two **incoming legs** and n **outgoing legs**, as shown in Figure 5.9.

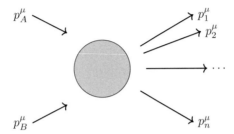

Figure 5.9 Representation of the generic reaction $A + B \to 1 + 2 + \cdots + n$.

The kinematical configuration of the process is defined by the four-vectors of the two incoming particles p_A^μ and p_B^μ and those of the final-state particles p_i^μ ($i = 1, \ldots, n$). In case the particles have additional degrees of freedom, such as spin, these must also be defined, typically with extra variables or indices. The rest masses m_j of the initial and final-state particles are fixed by the following relations:

$$(p_j^\mu)^2 \equiv E_j^2 - \vec{p_j}^{\,2} = m_j^2 \qquad (j = A, B, 1, 2, \ldots, n) \tag{5.104}$$

and consequently do not need to be defined separately. In addition, the final-state momenta cannot vary arbitrarily for a fixed initial state, but four-momentum conservation imposes the obvious extra condition:

$$p_A^\mu + p_B^\mu = \sum_{i=1}^{n} p_i^\mu \tag{5.105}$$

Kinematically the $3n$ coordinates of the final-state vectors $\vec{p_i}$ span an unconstrained momentum space of corresponding dimension $3n$. The Eq. (5.105) condition defines a subspace of $3n - 4$ dimensions, which is called the **phase space**. The total phase space is a space in which all kinematical states of the system are represented. We will be interested in calculating the *probability* of a given final-state kinematical configuration given the initial-state kinematics, namely we will seek the probability of the final-state particles after the collision having their momenta within the infinitesimal phase-space volume $\mathrm{d}^3\vec{p_i}$ ($i = 1, \ldots, n$). The total phase-space volume of the n particles is given by the product:

$$\prod_i \mathrm{d}^3\vec{p_i} \tag{5.106}$$

This phase-space volume is, however, not invariant under Lorentz transformation. We can *define* it in a Lorentz transformation-invariant way as in Section 5.11, by generalizing Eq. (5.102) to the n-**particle Lorentz-invariant phase-space (Lips) element**:

$$\mathrm{d}Lips(p_1, \ldots, p_n) \equiv \prod_i \frac{\mathrm{d}^3\vec{p_i}}{(2\pi)^3 2E_i} \tag{5.107}$$

with $E_i^2 = \vec{p_i}^2 + m_i^2$.

The *n*-**particle phase-space integral** $\Pi_n(s)$ for the collision $A + B$ is defined as:

$$\Pi_n(s) \equiv \int dLips(p_1, \ldots, p_n)(2\pi)^4 \delta^4 \left(p - \sum_i p_i \right) = \int \prod_{i=1}^{n} \frac{d^3\vec{p_i}}{(2\pi)^3 2E_i}(2\pi)^4 \delta^4 \left(p - \sum_i p_i \right) \tag{5.108}$$

where $p^\mu = p_A^\mu + p_B^\mu$, and

$$s \equiv p^2 = (p_A + p_B)^2 \tag{5.109}$$

is a Lorentz transformation-invariant quantity. The invariant s is a very important kinematical variable, being one of the **three Mandelstam variables** (to be defined in Section 11.8). Its value is equal to the square of the center-of-mass energy. The four-dimensional Dirac δ^4 function (see Appendix A.6) accounts for the conservation of the four-momentum (energy and three-momentum), which is a product of four δ functions corresponding to the four components of the momentum four-vector. The factor $(2\pi)^4$ comes from the chosen normalization of the Dirac δ function, Eq. (A.37). The phase-space integral can be simply expressed as:

$$\Pi_n(s) = \frac{1}{(2\pi)^{3n-4}} \int \frac{d^3\vec{p_1}}{2E_1} \cdots \frac{d^3\vec{p_n}}{2E_n} \times \delta^4 \left(p - \sum_i p_i \right) \tag{5.110}$$

with $E_i^2 = \vec{p_i}^2 + m_i^2$ (the m_i dependence is hidden from the $\Pi_n(s)$ term).

5.13   Relativistic Collisions

In Chapter 4 the differential cross-section for the transition between initial and final states has been derived via the non-relativistic perturbation theory. The result is called Fermi's Golden Rule (see Eqs. (4.178) and (4.186)):

$$d\sigma_{fi} \equiv \frac{\text{number of interactions per unit time}}{\text{incoming particles per unit time per unit area}} = \frac{W_{fi}}{j} \tag{5.111}$$

We can rearrange the rule in the following way:

$$d\sigma_{fi} = \frac{W_{fi}}{j} = \underbrace{|V_{fi}|^2}_{\text{dynamics}} \times \underbrace{\frac{1}{j}}_{\text{flux}} \times \underbrace{\rho(E_f)dE_f}_{\text{phase space}} \times \underbrace{(2\pi)\delta(E_f - E_i)}_{\text{energy conservation}} \tag{5.112}$$

where j is the flux of incoming particles. The matrix element V_{fi} contains the dynamical information, that is, the physics of the process. The other factors such as the flux, the phase space, and the condition of energy conservation are needed in order to convert the matrix element to a cross-section that is compared to experiments.

We are seeking a relativistic extension of this result that will allow us to estimate the cross-sections in relativistic processes. We wish to integrate Fermi's Golden Rule over all final states. This can be written using Eq. (4.178):

$$W_{fi} = (2\pi) \int dE_f |V_{fi}|^2 \delta(E_f - E_i) \rho(E_f) = (2\pi) \int |V_{fi}|^2 \delta(E_1 + E_2 - E_i) \frac{d^3\vec{p_1}}{(2\pi)^3} \tag{5.113}$$

where we considered only two particles in the final state and used Eq. (4.190) (see Section 4.17). We can include momentum conservation explicitly by integrating over the momenta of both particles:

$$W_{fi} = (2\pi)^4 \int |V_{fi}|^2 \delta(E_1 + E_2 - E_i) \delta^3(\vec{p_1} + \vec{p_2} - \vec{p_i}) \frac{d^3\vec{p_1}}{(2\pi)^3} \frac{d^3\vec{p_2}}{(2\pi)^3} \tag{5.114}$$

This expression can easily be generalized to n particles in the final state:

$$W_{fi} = \int |V_{fi}|^2 \underbrace{\left(\Pi_j \frac{\mathrm{d}^3 \vec{p}_j}{(2\pi)^3} \right)}_{\text{phase space}} \times \underbrace{(2\pi)^4 \delta^4 \left(p_A + p_B - \sum_j p_j \right)}_{\text{energy–momentum conservation}} \tag{5.115}$$

We are almost there! The δ function enforcing energy–momentum conservation already looks covariant. Following the definition of the Lorentz-invariant phase space, Eq. (5.108), we can make the phase-space factor invariant by dividing it by an extra factor E_j (conventionally $2E_j$). Hence

$$W_{fi} = \underbrace{\int \Pi_j (2E_j) |V_{fi}|^2}_{\equiv |\mathcal{M}_{fi}|^2} \underbrace{\left(\Pi_j \frac{\mathrm{d}^3 \vec{p}_j}{(2\pi)^3 2E_j} \right)}_{\mathrm{d}Lips} \times \underbrace{(2\pi)^4 \delta^4 \left(p_A + p_B - \sum_j p_j \right)}_{\text{energy–momentum conservation}} \tag{5.116}$$

where the extra energy factors have been absorbed into the Lorentz-invariant matrix element $|\mathcal{M}_{fi}|^2$. As a matter of fact, the same goal can be achieved by properly normalizing the particle wave functions in $|V_{fi}|^2$. Consequently, **it is convenient to normalize the wave functions to 2E particles per unit volume**. This choice will be adopted in the remainder of the book (see in particular Section 8.10). The differential cross-section is then expressible as a product of **Lorentz-invariant terms** and a δ function ensuring energy–momentum conservation. The cross-section is therefore manifestly a Lorentz-invariant quantity, since it only depends on Lorentz invariants. Finally, it is written as:

$$\mathrm{d}\sigma = \underbrace{|\mathcal{M}|^2}_{\text{dynamics}} \times \underbrace{\frac{1}{F}}_{\text{flux}} \times \underbrace{S}_{\text{statistics}} \times \underbrace{\left(\Pi_i \frac{\mathrm{d}^3 \vec{p}_i}{(2\pi)^3 2E_i} \right)}_{\text{phase space}} \times \underbrace{(2\pi)^4 \delta^4 \left(p_A + p_B - \sum_i p_i \right)}_{\text{energy–momentum conservation}} \tag{5.117}$$

In the above expression, we find:

- The Lorentz transformation-invariant **matrix element** $|\mathcal{M}(p_A^\mu, p_B^\mu, p_i^\mu)|^2$ contains the dynamics (the physics) of the process and depends on the kinematical configuration.

- The **statistical factor** $S \equiv \frac{1}{j_1!} \frac{1}{j_2!} \dots$ for each group of j_i indistinguishable particles in the final state avoids double counting from identical configurations.

- The Lorentz transformation-invariant **phase space** is proportional to the density of the final states in a volume element $\mathrm{d}^3 \vec{p}_1 \dots \mathrm{d}^3 \vec{p}_n$, defined in Eq. (5.107).

- The **flux factor** F describes the number of incoming particles per unit time and unit area.

To derive the Lorentz transformation-invariant flux factor, we assume that the two particles are collinear and colliding head-on. The non-relativistic expression, Eq. (2.74), is used as a starting point and we write the flux as the product of the particle densities times their relative velocities, here expressed as β_A and β_B:

$$F = n_A n_B |\vec{\beta}_A - \vec{\beta}_B| \tag{5.118}$$

With the chosen normalization, we have $n_A = 2E_A$ particles of type A per unit volume, and $n_B = 2E_B$ particles of type B per unit volume. Since $\beta = p/E$, we obtain:

$$F = (2E_A)(2E_B) \left| \frac{\vec{p}_A}{E_A} - \frac{\vec{p}_B}{E_B} \right| = 4 \left| \vec{p}_A E_B - \vec{p}_B E_A \right| \tag{5.119}$$

hence,

$$
\begin{aligned}
F^2 &= 16\,(\vec{p_A}E_B - \vec{p_B}E_A)^2 = 16\left(\vec{p}_A^2 E_B^2 + \vec{p}_B^2 E_A^2 - 2E_A E_B \vec{p}_A \cdot \vec{p}_B\right) \\
&= 16\left(\vec{p}_A^2(\vec{p}_B^2 + m_B^2) + \vec{p}_B^2(\vec{p}_A^2 + m_A^2) - 2E_A E_B \vec{p}_A \cdot \vec{p}_B\right) \\
&= 16\left(2\vec{p}_A^2\vec{p}_B^2 + \vec{p}_A^2 m_B^2 + \vec{p}_B^2 m_A^2 - 2E_A E_B \vec{p}_A \cdot \vec{p}_B\right)
\end{aligned}
\tag{5.120}
$$

We now introduce the **manifestly Lorentz-invariant flux** F, the so-called Møller flux factor:

$$
F \equiv 4\sqrt{(p_A \cdot p_B)^2 - m_A^2 m_B^2}
\tag{5.121}
$$

The quantity F^2 can be expanded as:

$$
\begin{aligned}
F^2 &= 16\left((E_A E_B - \vec{p}_A \cdot \vec{p}_B)^2 - m_A^2 m_B^2\right) \\
&= 16\left(E_A^2 E_B^2 + (\vec{p}_A \cdot \vec{p}_B)^2 - 2E_A E_B \vec{p}_A \cdot \vec{p}_B - m_A^2 m_B^2\right) \\
&= 16\left((\vec{p}_A^2 + m_A^2)(\vec{p}_B^2 + m_B^2) + \vec{p}_A^2\vec{p}_B^2 - 2E_A E_B \vec{p}_A \cdot \vec{p}_B - m_A^2 m_B^2\right) \\
&= 16\left(2\vec{p}_A^2\vec{p}_B^2 + m_A^2\vec{p}_B^2 + m_B^2\vec{p}_A^2 - 2E_A E_B \vec{p}_A \cdot \vec{p}_B\right)
\end{aligned}
\tag{5.122}
$$

which is equivalent to Eq. (5.120). In the third line we have used the fact that two particles are collinear and colliding head-on. The covariant flux F is manifestly Lorentz-invariant and represents as expected the number of colliding particles per unit time and unit area, in analogy to Eq. (2.74). From now on, we will always use the Lorentz transformation-invariant flux F (Eq. (5.121)).

5.14   Total Phase Space for One Particle

We want to illustrate the integration of d$Lips$. In order to obtain a finite result, we ask how large the one-particle phase space is, when the energy is constrained to be smaller than E_{max}. This integral is best evaluated using spherical coordinates for the three-momenta. The integral over the solid angle dΩ yields the surface of the sphere with unit radius. Therefore, we have:

$$
\begin{aligned}
\int \mathrm{d}Lips(p) &= \int \frac{\mathrm{d}^3\vec{p}}{(2\pi)^3 2E}\bigg|_{E=+\sqrt{p^2+m^2}} = \int \frac{p^2 \mathrm{d}p\,\mathrm{d}\Omega}{(2\pi)^3 2E}\bigg|_{E=+\sqrt{p^2+m^2}} \\
&= \frac{4\pi}{2(2\pi)^3}\int_0^{\sqrt{E_{max}^2-m^2}} \frac{\mathrm{d}p\,p^2}{E}\bigg|_{E=+\sqrt{p^2+m^2}} = \frac{1}{(2\pi)^2}\frac{(\sqrt{E_{max}^2-m^2})^3}{3E_{max}} \\
&= \frac{E_{max}^2}{12\pi^2}\left(1 - \frac{m^2}{E_{\max}^2}\right)^{3/2} \overset{m\to 0}{\longrightarrow} \frac{E_{max}^2}{12\pi^2}
\end{aligned}
\tag{5.123}
$$

The phase space grows quadratically with the maximum energy of the particle, which is understood by the volume of the three-dimensional (momentum) sphere being reduced by the $1/E$ term. The term in brackets is due to the mass of the particle "eating" up the phase space. The heavier the particle, the less momentum phase space is available. For $m^2 \to E_{max}^2$, the phase space vanishes and as expected $m^2 > E_{max}^2$ is unphysical.

5.15   Phase Space for Two-Body Scattering: Adding Four-Momentum Conservation

Since the phase-space integral is a Lorentz-invariant quantity, we can select any frame to compute it. We first consider the two-body reaction $A + B \to 1 + 2$ between particles of masses m_A, m_B, m_1, and m_2. The reaction as seen from the **center-of-mass frame** is shown in Figure 5.10(left). In order to compute the phase-space factor, we use the general result, Eq. (5.117), and simplify it with the evaluation of the integrals over the momenta, constrained by the Dirac δ functions. For two particles in the final state, we have the two-particle phase-space integral Π_2 (see Eq. (5.110)):

$$\Pi_2 \equiv \frac{1}{(2\pi)^2} \int \int \frac{d^3\vec{p_1}}{2E_1} \frac{d^3\vec{p_2}}{2E_2} \delta^4 (p_A + p_B - p_1 - p_2) \tag{5.124}$$

The four-momentum conservation imposes a constraint on the energies and three-momenta, so the two integrals over $d^3\vec{p_1}$ and $d^3\vec{p_2}$ will not be independent. We can decide to first integrate over $\vec{p_2}$ with the following condition, which is specific to the center-of-mass frame (all kinematical quantities are understood in that frame but the "$\star$" index is omitted for clarity):

$$\text{In CMS}: \quad \delta^3 \Big(\underbrace{\vec{p_A} + \vec{p_B}}_{=0} - \vec{p_1} - \vec{p_2} \Big) \quad \Longrightarrow \quad \vec{p_2} = -\vec{p_1} \tag{5.125}$$

The integration over $d^3\vec{p_2}$ disappears, and we are consequently left only with the integral over $d^3\vec{p_1}$ and a set of kinematical constraints. We define the momentum of the initial-state particles and similarly for the final-state particles:

$$\vec{p_i} \equiv \vec{p_A} = -\vec{p_B} \quad \text{and} \quad \vec{p_f} \equiv \vec{p_1} = -\vec{p_2} \tag{5.126}$$

The total energy in the center-of-mass frame $E_\star$ is then just (all quantities are again relative to the center-of-mass frame):

$$\begin{aligned} E_\star &= E_A + E_B = E_1 + E_2 = \sqrt{\vec{p_1}^2 + m_1^2} + \sqrt{\vec{p_2}^2 + m_2^2} \\ &= \sqrt{\vec{p_f}^2 + m_1^2} + \sqrt{\vec{p_f}^2 + m_2^2} \end{aligned} \tag{5.127}$$

where we used $\vec{p_f}^2 = \vec{p_1}^2 = \vec{p_2}^2$. Using Eq. (5.109), we see that the center-of-mass energy is actually just given by the invariant **Mandelstam variable** $\sqrt{s}$:

$$s \equiv p^2 = (p_A + p_B)^2 = (E_A + E_B)^2 - \underbrace{(\vec{p_A} + \vec{p_B})^2}_{=0} = E_\star^2 \quad \Rightarrow \quad E_\star = \sqrt{s} \tag{5.128}$$

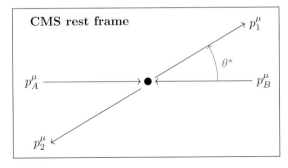
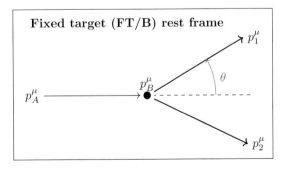

Figure 5.10 Two-body scattering $A + B \to 1 + 2$: (left) in the center-of-mass frame; (right) in the FT/laboratory frame (rest frame of B).

The magnitude $p_f = |\vec{p}_f|$ can be estimated by noting that:

$$E_\star^2 = 2p_f^2 + m_1^2 + m_2^2 + 2\sqrt{p_f^2 + m_1^2}\sqrt{p_f^2 + m_2^2} \quad \Rightarrow \quad p_f^2 + \sqrt{(p_f^2 + m_1^2)(p_f^2 + m_2^2)} = \frac{s - m_1^2 - m_2^2}{2} \tag{5.129}$$

We define $X \equiv p_f^2$, $A \equiv (s - B - C)/2$, $B \equiv m_1^2$, and $C \equiv m_2^2$. We need to solve:

$$X + \sqrt{(X + B)(X + C)} = A \quad \Rightarrow \quad X = \frac{A^2 - BC}{2A + B + C} = \frac{A^2 - BC}{s} \tag{5.130}$$

Then

$$p_f = \frac{\sqrt{A^2 - BC}}{\sqrt{s}} = \frac{1}{2\sqrt{s}}\sqrt{(2A)^2 - 4BC} = \frac{1}{2\sqrt{s}}\sqrt{(s - B - C)^2 - 4BC} \tag{5.131}$$

This result can be neatly expressed introducing the **Källén[5] function** or **triangle function** $\lambda(x, y, z)$:

$$\lambda(x, y, z) \equiv x^2 + y^2 + z^2 - 2xy - 2yz - 2zx = (x - y - z)^2 - 4yz \tag{5.132}$$

to find:

$$|\vec{p}_f| = \frac{\sqrt{(s - B - C)^2 - 4BC}}{2\sqrt{s}} = \frac{\lambda^{1/2}\left(s, m_1^2, m_2^2\right)}{2\sqrt{s}} \tag{5.133}$$

Also, one can derive E_1 by noting:

$$\begin{aligned}
E_1^2 &= p_f^2 + m_1^2 = X + B = \frac{1}{s}\left(A^2 - BC + Bs\right) \\
&= \frac{1}{4s}\left[(s - B - C)^2 - 4BC + 4Bs\right] = \frac{1}{4s}\left(s^2 + B^2 + C^2 - 2BC + 2Bs - 2Cs\right) \\
&= \frac{1}{4s}(s + B - C)^2 \tag{5.134}
\end{aligned}$$

and similarly for E_2, so finally:

$$E_1 = \frac{s + m_1^2 - m_2^2}{2\sqrt{s}}, \qquad E_2 = \frac{s + m_2^2 - m_1^2}{2\sqrt{s}} \tag{5.135}$$

Having derived the conditions arising from the constrained integration on $\mathrm{d}^3\vec{p}_2$, we go back to the phase space and need to perform the integration over $\mathrm{d}^3\vec{p}_1 \equiv \mathrm{d}p_1 \vec{p}_1^2 \mathrm{d}\Omega$:

$$\Pi_2 = \frac{1}{(2\pi)^3}\int \frac{\mathrm{d}^3\vec{p}_1}{2E_1 2E_2}(2\pi)\delta\left(E_\star - E_1 - E_2\right) = \frac{1}{(2\pi)^3}\int \frac{\mathrm{d}p_1 \vec{p}_1^2 \mathrm{d}\Omega}{2E_1 2E_2}(2\pi)\delta\left(E_\star - E_1 - E_2\right) \tag{5.136}$$

The integration over $\mathrm{d}p_1 \vec{p}_1^2$ can be replaced by $\mathrm{d}p_f \vec{p}_f^2$. Since $\mathrm{d}p_f = \mathrm{d}p_1$, we use:

$$\frac{\mathrm{d}E_\star}{\mathrm{d}p_f} = \frac{1}{2}(\vec{p}_f^2 + m_1^2)^{-1/2}2p_f + \frac{1}{2}(\vec{p}_f^2 + m_2^2)^{-1/2}2p_f = p_f\left(\frac{1}{E_1} + \frac{1}{E_2}\right) \tag{5.137}$$

and therefore

$$\mathrm{d}p_f = \mathrm{d}E_\star\left(\frac{p_f}{E_1} + \frac{p_f}{E_2}\right)^{-1} \tag{5.138}$$

---

5   Anders Olof Gunnar Källén (1926–1968), Swedish theoretical physicist.

We can further simplify by integrating over the center-of-mass energy, to keep only the dependence on the solid angle $d\Omega$:

$$
\begin{aligned}
\Pi_2 &= \int d\Omega dE_\star \frac{\vec{p}_f^2}{16\pi^2 E_1 E_2} \left(\frac{p_f}{E_1} + \frac{p_f}{E_2} \right)^{-1} \delta(E_\star - E_1 - E_2) \\
&= \int d\Omega dE_\star \frac{1}{16\pi^2} \frac{p_f}{E_1 E_2} \frac{1}{\left(\frac{1}{E_1} + \frac{1}{E_2} \right)} \delta(E_\star - E_1 - E_2) \\
&= \int d\Omega dE_\star \frac{1}{16\pi^2} \frac{p_f}{E_1 + E_2} \delta(E_\star - E_1 - E_2) \\
&= \int d\Omega \frac{1}{16\pi^2} \frac{p_f}{E_\star}
\end{aligned}
\tag{5.139}
$$

For massless particles, we have $E^\star = 2p_f$ and the integrated phase space simplifies to just this factor:

$$
\Pi_2(m_1 = m_2 = 0) = \frac{4\pi}{32\pi^2} = \frac{1}{8\pi}
\tag{5.140}
$$

We now collect the terms to obtain the differential cross-section:

$$
d\sigma_{CMS} = \frac{1}{(2E_A)(2E_B)|\vec{\beta}_A - \vec{\beta}_B|} \frac{d\Omega}{16\pi^2} \frac{p_f}{E_\star} |\mathcal{M}(AB \to 12)|^2
\tag{5.141}
$$

and get

$$
\left(\frac{d\sigma}{d\Omega} \right)_{CMS} = \frac{1}{(2E_A)(2E_B)|\vec{\beta}_A - \vec{\beta}_B|} \frac{p_f}{(2\pi)^2 4E_\star} |\mathcal{M}(AB \to 12)|^2
\tag{5.142}
$$

We now simplify the flux factor by recalling that $\vec{p}_i = \vec{p}_A = -\vec{p}_B$, so:

$$
(2E_A)(2E_B)|\vec{\beta}_A - \vec{\beta}_B| = 4E_A E_B \left(\frac{p_i}{E_A} + \frac{p_i}{E_B} \right) = 4(p_i E_B + p_i E_A) = 4p_i(E_A + E_B) = 4p_i E_\star
\tag{5.143}
$$

So finally, inserting $s = E_\star^2$, we find:

$$
\left(\frac{d\sigma}{d\Omega} \right)_{CMS} = \frac{1}{64\pi^2 s} \frac{p_f}{p_i} S |\mathcal{M}(AB \to 12)|^2
\tag{5.144}
$$

If all the particles have identical rest masses, or if we consider the ultra-relativistic limit (i.e., $E \gg m_i$ or $m_i \to 0$), we obtain $p_i \sim p_f$ and therefore:

$$
\left(\frac{d\sigma}{d\Omega} \right)_{CMS} = \left(\frac{1}{64\pi^2 s} \right) S |\mathcal{M}(AB \to 12)|^2 \qquad \text{ultra-relativistic}
\tag{5.145}
$$

• **Fixed target frame.** We now derive the corresponding equation in the **FT/laboratory frame** (where B is at rest) – see Figure 5.10(right). Starting from Eq. (5.124) and setting $p_B^\mu = (M_B, \vec{0})$, the momentum δ-function integration fixes $\vec{p}_A + \vec{0} = \vec{p}_1 + \vec{p}_2$ or $\vec{p}_2 = \vec{p}_A - \vec{p}_1$. After integration over $d^3\vec{p}_2$, only the integration over $d^3\vec{p}_1$ remains. Since $p_1 dp_1 = E_1 dE_1$, then $d^3\vec{p}_1 = p_1^2 dp_1 d\Omega = p_1 E_1 dE_1 d\Omega$ and one has:

$$
\Pi_2 = \int \frac{\delta(E_A + M_B - E_1 - E_2) p_1 dE_1 d\Omega}{(2\pi)^2 4E_2}
\tag{5.146}
$$

with the energy E_2 given by:

$$
E_2^2 = m_2^2 + \vec{p}_2^2 = m_2^2 + \vec{p}_A^2 + \vec{p}_1^2 - 2\vec{p}_A \cdot \vec{p}_1 = m_2^2 + \vec{p}_A^2 + \vec{p}_1^2 - 2p_A p_1 \cos\theta
\tag{5.147}
$$

where θ is the scattering angle between the incoming particle A and the outgoing particle 1 (Figure 5.10). The integration is of the form $\int \delta(g(E_1))dE_1$, where:

$$g(E_1) = E_A + M_B - E_1 - \left(m_2^2 + \vec{p}_A^2 + E_1^2 - m_1^2 - 2p_A\sqrt{E_1^2 - m_1^2}\cos\theta \right)^{1/2} \tag{5.148}$$

Using a property of the δ function (see Appendix A.6), we compute:

$$\left| \frac{dg}{dE_1} \right| = \left| -1 - \frac{2E_1 - 2p_A\frac{E_1}{p_1}\cos\theta}{2E_2} \right| = \left| \frac{-E_2 - E_1 + p_A\frac{E_1}{p_1}\cos\theta}{E_2} \right| = \frac{E_A + M_B - p_A\frac{E_1}{p_1}\cos\theta}{E_2} \tag{5.149}$$

where, while removing the absolute value, we used the fact that $E_A + M_B \geq p_A E_1/p_1$. The flux factor (Eq. (5.121)) is $F = 4\sqrt{(E_A M_B)^2 - M_A^2 M_B^2} = 4M_B\sqrt{E_A^2 - M_A^2} = 4p_A M_B$. Collecting the other factors for the cross-section, we get:

$$d\sigma_{lab} = \frac{\Pi_2}{4p_A M_B} S|\mathcal{M}(AB \rightarrow 12)|^2 = \frac{S|\mathcal{M}(AB \rightarrow 12)|^2 E_2 p_1 d\Omega}{4p_A M_B (2\pi)^2 4E_2 \left(E_A + M_B - p_A\frac{E_1}{p_1}\cos\theta \right)} \tag{5.150}$$

Thus the general form for the differential cross-section of $A + B \rightarrow 1 + 2$, where the kinematical quantities are given relative to the laboratory frame (B rest frame), is:

$$\left(\frac{d\sigma}{d\Omega} \right)_{lab} = \frac{p_1}{64\pi^2 p_A M_B \left(E_A + M_B - p_A\frac{E_1}{p_1}\cos\theta \right)} S|\mathcal{M}(AB \rightarrow 12)|^2 \tag{5.151}$$

5.16   Relativistic Decay Rates of Unstable Particles

We consider the decay of an unstable particle of mass M_A and energy E_A into n final-state particles. Such a process follows the law of radioactive decays (see Section 2.5). The **partial width** $d\Gamma$, with a unit of energy, can be written as in the case of scattering, as the product of terms with specific interpretations:

$$d\Gamma = \underbrace{|\mathcal{M}|^2}_{\text{dynamics}} \times \underbrace{\frac{1}{2E_A}}_{\text{normalization}} \times \underbrace{S}_{\text{statistics}} \times \underbrace{\left(\Pi_i \frac{d^3\vec{p}_i}{(2\pi)^3 2E_i} \right)}_{\text{phase-space d}\mathcal{L}ips} \times \underbrace{(2\pi)^4 \delta^4 \left(p_A - \sum_i p_i \right)}_{\text{energy–momentum conservation}} \tag{5.152}$$

The total width is obtained by integrating the partial widths over the phase space of the final state:

$$\Gamma = \int d\Gamma \tag{5.153}$$

When a particle can decay in different channels, the total width is equal to the sum of the widths of each individual channel (also when there are different numbers of particles in the final state):

$$\Gamma = \sum_n \Gamma_n(M_A \rightarrow 1 + 2 + \cdots) \tag{5.154}$$

The width Γ has units of energy, hence in natural units the inverse has units of inverse energy, which corresponds to time. Thus, **the lifetime τ of the particle is simply given by the inverse of the total width** $\tau = 1/\Gamma$ (see Eq. (2.30)).

It should be noted that if the decaying particle has a non-zero spin J and is unpolarized, then the matrix element will be obtained by averaging on the initial spins and summing over the final-state spins (details will be presented in Section 11.9).

5.17   **Phase Space for Two-Body Decay**

We consider the decay of a particle R of rest mass M into two distinguishable daughter particles of masses m_j $(j = 1, 2)$:

$$R \to 1 + 2 \tag{5.155}$$

We compute the two-body phase space in the center-of-mass frame of the parent particle of mass M:

$$
\begin{aligned}
\mathrm{d}^6 R &= \frac{\mathrm{d}^3 \vec{p}_1}{(2\pi)^3 2E_1} \frac{\mathrm{d}^3 \vec{p}_2}{(2\pi)^3 2E_2} (2\pi)^4 \delta^4(p_1 + p_2 - P) \\
&= \frac{\mathrm{d}^3 \vec{p}_1}{(2\pi)^3 2E_1} \frac{\mathrm{d}^3 \vec{p}_2}{(2\pi)^3 2E_2} (2\pi)^4 \delta^3(\vec{p}_1 + \vec{p}_2) \delta(E_1 + E_2 - M)
\end{aligned}
\tag{5.156}
$$

Integration over the momentum of one of the daughter particles taking into account the term $\delta^3(\vec{p}_1 + \vec{p}_2)$ gives the following condition:

$$(\vec{p}_1 + \vec{p}_2) = \vec{0} \quad \Longrightarrow \quad \vec{p} = \vec{p}_1 = -\vec{p}_2 \tag{5.157}$$

and the phase-space factor can be simplified to:

$$\mathrm{d}^3 R = \frac{\mathrm{d}^3 \vec{p}}{(2\pi)^3 2E_1} \frac{1}{(2\pi)^3 2E_2} (2\pi)^4 \delta(E_1 + E_2 - M) = \frac{p^2 \mathrm{d}p \, \mathrm{d}\cos\theta \mathrm{d}\phi}{(2\pi)^2 4 E_1 E_2} \delta(E_1 + E_2 - M) \tag{5.158}$$

We note:

$$E \equiv E_1 + E_2 = \sqrt{\vec{p_1}^2 + m_1^2} + \sqrt{\vec{p_2}^2 + m_2^2} = \sqrt{\vec{p}^2 + m_1^2} + \sqrt{\vec{p}^2 + m_2^2} \tag{5.159}$$

and use a similar calculation as in Eq. (5.137):

$$\mathrm{d}p = \mathrm{d}E \left(\frac{p}{E_1} + \frac{p}{E_2} \right)^{-1} = \mathrm{d}E \left(\frac{pE_1 + pE_2}{E_1 E_2} \right)^{-1} = \frac{E_1 E_2 \mathrm{d}E}{pE} \tag{5.160}$$

Then the phase space becomes:

$$\mathrm{d}^3 R = \frac{p^2 \frac{E_1 E_2 \mathrm{d}E}{pE} \mathrm{d}\cos\theta \mathrm{d}\phi}{(2\pi)^2 4 E_1 E_2} \delta(E - M) = \frac{p \mathrm{d}E \mathrm{d}\cos\theta \mathrm{d}\phi}{4\pi^2 4 E} \delta(E - M) \tag{5.161}$$

One can integrate over the energy to find:

$$\mathrm{d}^2 R = \frac{p}{16\pi^2 M} \mathrm{d}\cos\theta \mathrm{d}\phi \tag{5.162}$$

Combining the factors, the decay width is then equal to:

$$\mathrm{d}\Gamma(M \to 1 + 2) = |\mathcal{M}|^2 \frac{1}{2M} \frac{p}{16\pi^2 M} \mathrm{d}\cos\theta \mathrm{d}\phi = \frac{|\mathcal{M}|^2}{32\pi^2} \frac{p}{M^2} \mathrm{d}\Omega \tag{5.163}$$

where p is given by (cf. Eq. (5.133)):

$$p = \frac{\lambda^{1/2}(M^2, m_1^2, m_2^2)}{2M} \tag{5.164}$$

In the case where the final-state particles are massless or when their rest masses are neglected, we find that $\lambda(M^2, 0, 0) = M^4$, so $p = M/2$ and:

$$\mathrm{d}\Gamma(M \to 1 + 2) = \frac{|\mathcal{M}|^2}{64\pi^2} \frac{1}{M} \mathrm{d}\Omega \qquad \text{(rest frame of } M) \tag{5.165}$$

5.18   Phase Space for Three-Body Decay

We consider the decay of a particle R of rest mass M into three distinguishable daughter particles of masses m_j $(j = 1, 2, 3)$:

$$R \rightarrow 1 + 2 + 3 \tag{5.166}$$

We first estimate the number of independent parameters needed to describe the final state. In the final state with three particles a total of nine coordinates are needed to define the three-momenta (the masses m_j are fixed). The constraints on energy and momentum conservation (four equations) reduce the number of parameters to $9 - 4 = 5$. In addition, we can choose the direction of the coordinate system, hence three more parameters are arbitrary. Finally, two free parameters are sufficient to describe completely the final-state kinematical configuration. We will choose these two relative to the center-of-mass frame of the mother particle R. The momenta of the three final-state particles must lie in a plane. The phase-space factor for three particles is given by:

$$
\begin{aligned}
\mathrm{d}^9 R &= \frac{\mathrm{d}^3 \vec{p}_1}{(2\pi)^3 2E_1} \frac{\mathrm{d}^3 \vec{p}_2}{(2\pi)^3 2E_2} \frac{\mathrm{d}^3 \vec{p}_3}{(2\pi)^3 2E_3} (2\pi)^4 \delta^4(p_1 + p_2 + p_3 - P) \\
&= \frac{1}{(2\pi)^5} \frac{\mathrm{d}^3 \vec{p}_1}{2E_1} \frac{\mathrm{d}^3 \vec{p}_2}{2E_2} \frac{\mathrm{d}^3 \vec{p}_3}{2E_3} \delta^3(\vec{p}_1 + \vec{p}_2 + \vec{p}_3) \delta(E_1 + E_2 + E_3 - M)
\end{aligned} \tag{5.167}
$$

We integrate over the momentum of particle 3:

$$\mathrm{d}^6 R = \frac{1}{8(2\pi)^5} \frac{\mathrm{d}^3 \vec{p}_1}{E_1} \frac{\mathrm{d}^3 \vec{p}_2}{E_2} \frac{1}{E_3} \delta(E_1 + E_2 + E_3 - M) \tag{5.168}$$

with the condition:

$$E_3 = \sqrt{(\vec{p}_1 + \vec{p}_2)^2 + m_3^2} \tag{5.169}$$

Now we choose the coordinate system such that the z-axis is along the direction of the momentum of particle 1. For the momentum of particle 2, it then follows that:

$$\mathrm{d}^3 \vec{p}_2 = p_2^2 \mathrm{d}p_2 \mathrm{d}\cos\theta_{12} \mathrm{d}\phi \tag{5.170}$$

where θ_{12} and ϕ are the polar angles that define the vector $\vec{p}_2$ relative to the coordinate system whose z-axis points in the direction of $\vec{p}_1$. We now employ the relation between E_3 and the momenta $\vec{p}_1$ and $\vec{p}_2$:

$$E_3^2 = (\vec{p}_1 + \vec{p}_2)^2 + m_3^2 = m_3^2 + \vec{p}_1^2 + \vec{p}_2^2 + 2p_1 p_2 \cos\theta_{12} \tag{5.171}$$

Then for fixed vector magnitudes of p_1 and p_2, we have:

$$\frac{\mathrm{d}\cos\theta_{12}}{\mathrm{d}E_3} = \frac{\mathrm{d}}{\mathrm{d}E_3} \frac{(E_3^2 - m_3^2 - \vec{p}_1^2 - \vec{p}_2^2)}{2p_1 p_2} = \frac{E_3}{p_1 p_2} \implies \mathrm{d}\cos\theta_{12} = \frac{E_3 \mathrm{d}E_3}{p_1 p_2} \tag{5.172}$$

We can integrate over the directions of $\vec{p}_2$, keeping its magnitude constant:

$$
\begin{aligned}
\mathrm{d}^4 R &= \frac{1}{8(2\pi)^5} \int \mathrm{d}\phi \int \mathrm{d}\cos\theta_{12} \frac{\mathrm{d}^3 \vec{p}_1}{E_1} \frac{p_2^2 \mathrm{d}p_2}{E_2} \frac{1}{E_3} \delta(E_1 + E_2 + E_3 - M) \\
&= \frac{2\pi}{8(2\pi)^5} \int_{E_{3,min}}^{E_{3,max}} \frac{E_3 \mathrm{d}E_3}{p_1 p_2} \frac{\mathrm{d}^3 \vec{p}_1}{E_1} \frac{p_2^2 \mathrm{d}p_2}{E_2} \frac{1}{E_3} \delta(E_1 + E_2 + E_3 - M) \\
&= \frac{\pi}{4(2\pi)^5} \int_{E_{3,min}}^{E_{3,max}} \mathrm{d}E_3 \frac{\mathrm{d}^3 \vec{p}_1}{p_1 E_1} \frac{p_2 \mathrm{d}p_2}{E_2} \delta(E_1 + E_2 + E_3 - M) \\
&= \frac{\pi}{4(2\pi)^5} \frac{\mathrm{d}^3 \vec{p}_1}{p_1 E_1} \frac{p_2 \mathrm{d}p_2}{E_2} \int_{E_{3,min}}^{E_{3,max}} \mathrm{d}E_3 \delta(E_1 + E_2 + E_3 - M)
\end{aligned} \tag{5.173}
$$

The Dirac δ function can be replaced after integration with the following condition:

$$E_3 = M - E_1 - E_2 \tag{5.174}$$

and the remaining terms in the phase space are:

$$d^4 R = \frac{\pi}{4(2\pi)^5} \frac{d^3 \vec{p}_1}{p_1 E_1} \frac{p_2 dp_2}{E_2} = \frac{\pi}{4(2\pi)^5} \frac{p_1 p_2 dp_1 dp_2}{E_1 E_2} d\phi_z d\cos\theta_z \tag{5.175}$$

where θ_z and ϕ_z define the direction of the z-axis (i.e., the direction of $\vec{p}_1$). Integration over these angles yields the following two-dimensional phase space:

$$d^2 R = 4\pi \frac{\pi}{4(2\pi)^5} \frac{p_1 dp_1}{E_1} \frac{p_2 dp_2}{E_2} = \frac{1}{32\pi^3} \frac{p_1 p_2 dp_1 dp_2}{E_1 E_2} \tag{5.176}$$

With the relation:

$$p_i = \sqrt{E_i^2 - m_i^2} \implies dp_i = \frac{1}{2} \frac{2 E_i dE_i}{\sqrt{E_i^2 - m_i^2}} = \frac{E_i dE_i}{p_i} \tag{5.177}$$

one obtains:

$$d^2 R = \frac{1}{32\pi^3} \frac{p_1 p_2 dp_1 dp_2}{E_1 E_2} = \frac{1}{32\pi^3} dE_1 dE_2 = \frac{1}{32\pi^3} dT_1 dT_2 \tag{5.178}$$

where $T_i = E_i - m_i$ are the kinetic energies of the particles (hence $dE_i = dT_i$). This result implies that the probability that the kinetic energies of the particles 1, respectively 2, are found within T_1 and $T_1 + dT_1$, respectively, T_2 and $T_2 + dT_2$, is constant. In a two-dimensional representation of T_1 vs. T_2, the decays will be uniformly distributed according to phase space in the allowed physical region.

The full phase-space factor can easily be calculated for the case of massless particles. We note that in order to conserve momentum, the energy of a single final-state particle cannot be above $M/2$. Hence, for massless particles, we start by integrating over E_1 between 0 and $M/2$. Once E_1 is chosen, repeating the same argument, the range of E_2 is constrained to be between E_1 and $M/2$. We do not integrate over the last particle since $E_3 = M - E_1 - E_2$. We have:

$$\int d^2 R = \frac{1}{32\pi^3} \int_0^{M/2} dE_1 \int_{E_1}^{M/2} dE_2 = \frac{1}{32\pi^3} \int_0^{M/2} dE_1 (M/2 - E_1) = \frac{M^2}{256\pi^3} \tag{5.179}$$

5.19 Dalitz Plots of Three-Body Decays

The Q-**value** of the reaction is defined as the change in mass. In the center-of-mass system, we obtain:

$$Q = M - m_1 - m_2 - m_3 = T_1 + T_2 + T_3 \tag{5.180}$$

where T_i are the kinetic energies of the particles.

The decay configurations can be graphically plotted in a triangle normalized to Q, as shown in Figure 5.11 for two different cases. The triangle represents a **ternary plot**, which is a barycentric plot of three variables which sum to a constant (here 100%). For illustration, the mass of the parent M was set to 100 GeV. On the left graph, the daughter particles are massless ($m_1 = m_2 = m_3 = 0$). On the right plot, the masses were chosen to be $m_1 = 1$ GeV, $m_2 = 10$ GeV, and $m_3 = 30$ GeV. A pure "phase-space" distribution will populate uniformly the kinematically allowed regions of the triangle, according to Eq. (5.178). For the massless case, this region is an upside-down triangle, where the edges represent the kinematical configuration in which one particle is recoiling against the other two. The case with massive daughter particles cuts the corner of the triangle and reduces the size of the area, since energy is taken away by the mass while momentum conservation implies that heavy particles recoil with larger momentum against lighter ones.

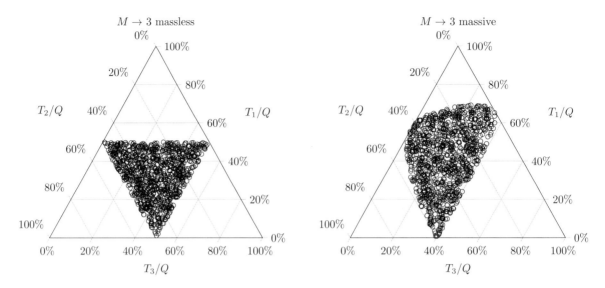

Figure 5.11 Ternary (also Dalitz) plot of a three-body decay. The circles represent 1000 events in the kinematically allowed region. Phase space predicts a uniform population in these regions. For illustration, the mass of the parent particle M is 100 GeV. (left) Massless daughter particles; (right) the masses were chosen to be $m_1 = 1$ GeV, $m_2 = 10$ GeV, and $m_3 = 30$ GeV.

Overall, we can conclude: **phase space predicts that the decay kinematical configuration will be uniformly distributed in the kinematically allowed region of the triangle. A non-uniform distribution is a useful tool to gain information on the dynamics of the process (i.e., the matrix element).**

Such representations of three-body decays are called **Dalitz[6] plots**. Dalitz originally developed this representation technique in order to analyze the decay $K^+ \to \pi^+\pi^+\pi^-$ [69]. Each of the three-body final states produced in an experiment is represented as a dot in the plane defined by the kinetic energies. A prominent example concerns the study of the decay $\omega^0 \to \pi^+\pi^0\pi^-$. The corresponding Dalitz plot is shown in Figure 5.12. The main application was to use it in order to determine spin and parity of the decaying particles.

In their "modern" versions, Dalitz plots are usually drawn from the *invariant masses squared*. Consider the three-particle final state consisting of the three particles (123). Out of these three particles one has three possibilities to form a pair of two particles: (12), (13), and (23). We choose two, e.g.:

$$m_{12}^2 \equiv (p_1 + p_2)^2, \qquad m_{23}^2 \equiv (p_2 + p_3)^2 \tag{5.181}$$

These coordinates are equivalent to the kinetic energies of two of the three decay products in the sense that they are linearly related, e.g.:

$$m_{12}^2 = (p_1 + P - p_1 - p_3)^2 = (P - p_3)^2 = P^2 + p_3^2 - 2P \cdot p_3 = M^2 + m_3^2 - 2M(m_3 + T_3) \tag{5.182}$$

where $P^\mu = p_1^\mu + p_2^\mu + p_3^\mu$ is the momentum of the parent particle and we used $E_3 \equiv m_3 + T_3$.

The basic idea is to plot for example m_{12}^2 vs. m_{23}^2 and to assign each decay event its coordinates with respect to the two axes of the plot. See Figure 5.13 for the same two cases discussed previously. By construction, we have $m_{12}^2 \geq (m_1 + m_2)^2$ and $m_{23}^2 \geq (m_2 + m_3)^2$. The value of m_{12}^2 is constrained to its maximum for $T_3 = 0$, hence, we have:

$$m_{12}^2 \leq M^2 + m_3^2 - 2Mm_3 = (M - m_3)^2 \quad \text{and} \quad m_{23}^2 \leq (M - m_1)^2 \tag{5.183}$$

6 Richard Henry Dalitz (1925–2006), Australian particle physicist.

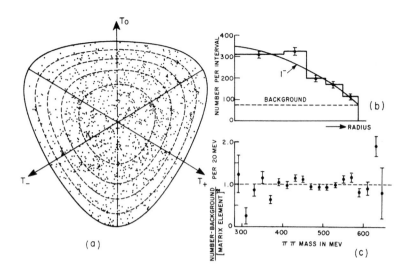

Figure 5.12 Dalitz plot for the decay $\omega^0 \to \pi^+\pi^0\pi^-$. Reprinted with permission from C. Alff *et al.*, "Decays of the omega and eta mesons," *Phys. Rev. Lett.*, vol. 9, pp. 325–327, 1962. Copyright 1962 by the American Physical Society.

For a given value of m_{12}^2, the range of m_{23}^2 is determined by its values when particle 2 is parallel or antiparallel to particle 3. This gives an allowed range $(m_{23}^2)_{min} \leq m_{23}^2 \leq (m_{23}^2)_{max}$ for any given value of m_{12}^2. This is illustrated in Figure 5.13 as a vertical dotted line for a chosen m_{12}^2. Overall, decays are restricted to the regions as shown. For the massless case this is simply half of the $m_{12}^2 - m_{23}^2$ plane, while for massive decay products the corners become inaccessible.

- **Example of resonance production.** Every structure in the density of the plots is due to dynamical characteristics of the reactions and not of kinematical origin. If the reaction proceeds via an almost stable two-body intermediate state:

$$M \to (12)3 \to 123 \tag{5.184}$$

then the three-body final states will be such that the distribution of the invariant mass of the pair (12) is centered around the invariant mass of the two-body intermediate state. This manifests itself in the Dalitz plot as a band of higher than average density of points. With an iso-contour of such two-dimensional histograms, one can obtain a "landscape" where the "mountains" correspond to a lot of events and the "valleys" to very few or no events.

For example, Figure 5.14(left) shows the Dalitz plot for the decay $p\bar{p} \to \pi^+\pi^-\pi^0$. The bands correspond to the existence of the ρ resonance (see Chapter 15). Another example is given in Figure 5.14 (right), which shows $p\bar{p} \to 3\pi^0$. In this case, several resonances called $f_0(980)$, $f_2(1270)$, $f_0(1500)$, and $f_2(1565)$ are clearly observable!

5.20   Three-Body Phase-Space Generation

We consider again the decay of a particle R of rest mass M into three distinguishable daughter particles of masses m_j, as in Section 5.18, but here we want to illustrate the aspects related to **generating the phase space by Monte-Carlo techniques** (see Appendix H). Given the kinematically allowed ranges, it is convenient to

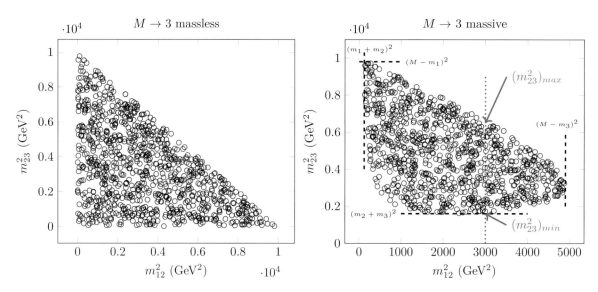

Figure 5.13 A modern Dalitz plot with 1000 events distributed according to phase space. The populated area corresponds to the kinematically allowed regions. For illustration, the mass of the parent particle M is 100 GeV. (left) Massless daughter particles; (right) the masses were chosen to be $m_1 = 1$ GeV, $m_2 = 10$ GeV, and $m_3 = 30$ GeV, as in Figure 5.11.

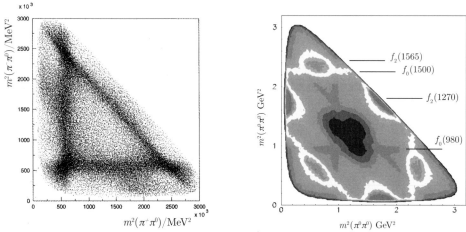

Figure 5.14 Dalitz plot for the decay (left) $p\bar{p} \to \pi^+\pi^-\pi^0$. Reprinted from A. Abele *et al.*, "The rho mass, width and lineshape in $p-\bar{p}$ annihilation at rest into $\pi^+\pi^-\pi^0$," *Phys. Lett. B*, vol. 469, pp. 270–275, 1999, with permission from Elsevier; (right) $p\bar{p} \to 3\pi^0$. Reprinted with permission from C. Amsler, "Proton–antiproton annihilation and meson spectroscopy with the crystal barrel," *Rev. Mod. Phys.*, vol. 70, pp. 1293–1340, 1998. Copyright 1998 by the American Physical Society.

introduce the **energy fractions** $x_i \in [0, 1]$ in the rest mass of R, in the following way:

$$x_i \equiv \frac{2E_i}{M} \quad \text{where} \quad x_1 + x_2 + x_3 = 2 \tag{5.185}$$

(this relation can be used even when we do not consider massless daughter particles, but in this case the range of the x_i will be further constrained). Then, starting from Eq. (5.167) and proceeding along the lines above, we obtain:

$$
\begin{aligned}
d^6 R &= \frac{1}{(2\pi)^5} \frac{dp_1 p_1^2 d\Omega_1}{2E_1} \frac{dp_2 p_2^2 d\cos\theta_{12} d\phi}{2E_2} \frac{\delta(E_1 + E_2 + E_3 - M)}{2E_3} \\
&= \frac{1}{(2\pi)^5} \frac{dE_1 p_1 d\Omega_1}{2} \frac{dE_2 p_2 d\cos\theta_{12} d\phi}{2} \frac{\delta(E_1 + E_2 + E_3 - M)}{2E_3}
\end{aligned}
\tag{5.186}
$$

where Ω_1 is the solid angle subtended by $\vec{p}_1$ and we used the relation derived above, $dp_i = E_i dE_i/p_i$. Then, integrating over the $d\Omega_1$ and $d\phi$ yields:

$$
d^3 R = \frac{8\pi^2}{4(2\pi)^5} dE_1 \sqrt{E_1^2 - m_1^2} dE_2 \sqrt{E_2^2 - m_2^2} d\cos\theta_{12} \frac{\delta(E_1 + E_2 + E_3 - M)}{2E_3}
\tag{5.187}
$$

Thanks to the properties of the Dirac δ function, we can write for massless particles:

$$
\begin{aligned}
\frac{\delta(E_1 + E_2 + E_3 - M)}{2E_3} &= \delta\left((E_1 + E_2 + E_3 - M)^2\right) = \delta\left(\frac{1}{4}(x_1 + x_2 + x_3 - 2)^2 M^2\right) \\
&= \delta\left(\frac{M^2}{4}(x_1^2 + x_2^2 + x_3^2 + 4(1 - x_1 - x_2 - x_3) + 2x_1 x_2 + 2x_2 x_3 + 2x_1 x_3)\right) \\
&= \delta\left(\frac{M^2}{4}(4(1 - x_1 - x_2) + 2x_1 x_2 - 4x_3 + x_1^2 + x_2^2 + x_3^2 + 2x_2 x_3 + 2x_1 x_3)\right) \\
&= \delta\left(M^2(1 - x_1 - x_2 + \frac{1}{2}x_1 x_2) + \frac{M^2}{4}(x_1^2 + x_2^2 - x_3^2))\right) \\
&= \delta\left(M^2(1 - x_1 - x_2 + \frac{1}{2}x_1 x_2) + \frac{M^2}{4}(-2x_1 x_2 \cos\theta_{12})\right)
\end{aligned}
\tag{5.188}
$$

where in the last line we used Eq. (5.171) to obtain $x_3^2 = x_1^2 + x_2^2 + 2x_1 x_2 \cos\theta_{12}$. Finally, we can now integrate over the angle:

$$
\int_{-1}^{1} d\cos\theta_{12} \frac{\delta(E_1 + E_2 + E_3 - M)}{2E_3} = \int_{-1}^{1} d\cos\theta_{12} \delta\left(M^2(1 - x_1 - x_2 + \frac{1}{2}x_1 x_2(1 - \cos\theta_{12}))\right) = \frac{2}{x_1 x_2 M^2}
\tag{5.189}
$$

Hence

$$
d^2 R = \frac{1}{16\pi^3} dx_1 x_1 dx_2 x_2 \frac{2}{x_1 x_2 M^2} \frac{M^4}{16} = \frac{M^2}{128\pi^3} dx_1 dx_2
\tag{5.190}
$$

Now performing the nested integrals, we recover, as anticipated, the result (Eq. (5.179)):

$$
\int d^2 R = \frac{M^2}{128\pi^3} \int_0^1 dx_1 \int_{x_1}^1 dx_2 = \frac{M^2}{256\pi^3}
\tag{5.191}
$$

Although the introduction of the energy fractions x_i is perfectly legitimate and prone to Monte-Carlo (MC) generation, it leads to phase-space factors with energies E_i at the denominator (see Eq. (5.186)). These factors will tend to be very large when the energies are small and are therefore numerically unpractical to handle. Having in mind **importance sampling** (refer to Appendix H), we can select a better choice of integration variable. For instance, let us focus on particle 1 and choose to integrate over $d(p_1^2) = 2p_1 dp_1$. We have:

$$
\frac{dp_1 p_1^2 d\Omega_1}{2E_1} = \frac{d(p_1^2) p_1^2 d\Omega_1}{2p_1 2E_1} = d(p_1^2) d\Omega_1 \frac{p_1}{E_1}
\tag{5.192}
$$

The weight given by the ratio p_1/E_1 is numerically better behaved, and in particular, it becomes $p_1/E_1 \to 1$ for $m_1 \to 0$. So, this is a better choice for a Monte-Carlo generation since it will lead to a reduction of the variance of the weights (see discussion in Appendix H). In addition to defining the energy of the particles, we are now also interested in generating their direction in space. Within the rest frame of M we can then introduce three scaled integration variables x_1, x_2, and x_3 ($x_i \in [0,1]$) that will define the momentum $\vec{p}_1$:

$$p_1^2 = x_1 p_{1,max}^2, \quad \cos\theta_1 = 1 - 2x_2, \quad \phi_1 = 2\pi x_3 \tag{5.193}$$

The maximum momentum of particle 1 is reached when it recoils against the system formed by the (23) particles. This problem can be reduced to a two-body case ((1) against (23)), so we have using the Källén function (see Eq. (5.133)) that:

$$p_{1,max} = \frac{\lambda^{1/2}\left(M^2, m_1^2, (m_2+m_3)^2\right)}{2M} \implies p_{1,max}^2 = \frac{\lambda\left(M^2, m_1^2, (m_2+m_3)^2\right)}{4M^2} \tag{5.194}$$

We can take advantage of having coupled the (23) particles to generate the momentum of each particle in the (23) reference frame. The total energy available in the (23) frame is given by the effective invariant mass of the (23) system, which we call m_{23}. It is related to M and particle 1 in the parent rest frame by:

$$m_{23}^2 = (P - p_1)^2 = M^2 + m_1^2 - 2ME_1 = M^2 + m_1^2 - 2M\sqrt{p_1^2 + m_1^2} \tag{5.195}$$

In the (23) frame, particles 2 and 3 are back-to-back $\vec{p}_2 = -\vec{p}_3$, so we have again a two-body problem and we can use the Källén function (see Eq. (5.133)), this time to get:

$$p_2^2 = p_3^2 = \frac{\lambda\left(m_{23}^2, m_2^2, m_3^2\right)}{4m_{23}^2} \tag{5.196}$$

Two angles are sufficient to define completely the kinematical configuration of the particles within the (23) system. We can choose x_4 and x_5 ($x_i \in [0,1]$) such that:

$$\cos\theta_2 = 1 - 2x_4, \quad \phi_2 = 2\pi x_5 \tag{5.197}$$

We stress again that these momenta are valid in the (23) frame. Once the kinematics are defined in this frame, we must boost them to the M rest frame, but this is straightforward in a numerical generation. The weight of the two-body phase space is given by (cf. Eq. (5.139)):

$$\frac{dp_2 p_2^2 d\Omega_2}{2E_2 2E_3}(2\pi)\delta(m_{23} - E_2 - E_3) = \frac{1}{16\pi^2}d\Omega_2\frac{p_2}{m_{23}}(2\pi)^3 = \frac{\pi}{2}\frac{p_2}{m_{23}}d\cos\theta_2 d\phi_2 = 2\pi^2\frac{p_2}{m_{23}}dx_4 dx_5 \tag{5.198}$$

These results can easily be coded into a program that will sample the integrand and generate events according to the appropriate phase space.

5.21   Monte-Carlo Generation of Multi-particle Phase Space

We have seen in previous sections how to analytically integrate the phase space in cases where the number of final-state particles is relatively small. In practice, as the number of final-state particles increases, one relies on Monte-Carlo methods (see Appendix H) to generate the phase space with many final-state particles and perform the integration of differential cross-sections. One of the first phase-space generators was implemented by **James**[7] [70] as the **GENBOD function (W515 from CERNLIB)**.[8] Another example is the **RAndom Momenta**

---

7   Fred James, CERN physicist, expert in statistics.
8   http://cernlib.web.cern.ch/cernlib/mc/genbod.html

Beautifully Organized (RAMBO) generator based on Ref. [71]. These algorithms have been ported to many programming languages. For example, the official distribution of the PYTHIA Monte-Carlo event generator [68] contains the Rambo class. A version in Python is also available [72]. An interesting implementation for massively parallel platforms in available as MCBooster [73].

Apart from its ease, one advantage of numerically computing phase space is that it provides a natural way to also include decay chains. We illustrate in the following the Python PHASESPACE project,[9] used for example to generate the phase space of the following rare (penguin) decay:

$$B^0 \to K^*(892)\gamma \to K^+\pi^-\gamma \tag{5.199}$$

The primary decay has been observed to be rare with a branching ratio of $Br(B^0 \to K^*(892)\gamma) = (4.18 \pm 0.25) \times 10^{-5}$ (values taken from the Particle Data Group's Review [5]). We neglect for simplicity the intrinsic width of the $K^*(892)$ measured to be approximately 50 MeV. The Python code is the following:

```python
# a Python code to generate phase space for the
# B^0 -> K^*(892)\gamma -> K^+\pi^- \gamma decay chain

from phasespace import GenParticle

# setting up particle masses in MeV (ignore widths, values taken from the PDG)
B0_MASS = 5279.64
KSTARZ_MASS = 895.5
PION_MASS = 139.571
KAON_MASS = 493.677

# defining particles with the decay chain
pion = GenParticle('pi-', PION_MASS)
kaon = GenParticle('K+', KAON_MASS)

kstar = GenParticle('K*', KSTARZ_MASS).set_children(pion, kaon)

gamma = GenParticle('gamma', 0)
bz = GenParticle('B0', B0_MASS).set_children(kstar, gamma)

# generate 5 events
weights, particles = bz.generate(n_events=5)

# print the outputs
print("Parent particle:", bz)
print("Weights:", weights)
print("K*:", particles['K*'])
print("gamma:", particles['gamma'])

print("pi-:", particles['pi-'])
print("K+:", particles['K+'])
```

The code is set to generate five events. The arrays' weights and particles will be filled with the corresponding values. After the arrays are filled with five events, the entries for the parent particle, the event weights, and the four particles $K^*(892)$, γ, π^-, and K^+ are written out. Running the code yields the following output:

---

9   Available at http://pypi.org/project/phasespace/

```
Parent particle: <phasespace.GenParticle: name='B0' mass=5279.64 children=[K*, gamma]>
Weights: tf.Tensor([0.10958271 0.10958271 0.10958271 0.10958271 0.10958271], shape=(5,), dtype=float64)
K*: tf.Tensor(
[[ 1233.57121105    967.98542803 -2028.48795101  2715.76459565]
 [  489.76365112   2512.54784766  -143.84703089  2715.76459565]
 [-2089.66179926    113.61982755  1481.16885888  2715.76459565]
 [ 1959.80170722    -92.82571915 -1650.45985811  2715.76459565]
 [  213.25134347   2553.82852938   -77.0765596   2715.76459565]], shape=(5, 4), dtype=float64)
gamma: tf.Tensor(
[[-1233.57121105   -967.98542803  2028.48795101  2563.87540435]
 [ -489.76365112  -2512.54784766   143.84703089  2563.87540435]
 [ 2089.66179926   -113.61982755 -1481.16885888  2563.87540435]
 [-1959.80170722     92.82571915  1650.45985811  2563.87540435]
 [ -213.25134347  -2553.82852938    77.0765596   2563.87540435]], shape=(5, 4), dtype=float64)
pi-: tf.Tensor(
[[  306.10435503    553.20239127  -979.99899028  1174.56836619]
 [ -132.79030166    674.18111085    54.31240793   703.26619152]
 [ -519.15901901    293.18856804   301.63928216   682.60672721]
 [ 1252.00459374    115.30294938 -1050.47822429  1644.32199907]
 [  170.97671781    195.14526996   -81.20799709   305.59698502]], shape=(5, 4), dtype=float64)
K+: tf.Tensor(
[[  927.46685602    414.78303676 -1048.48896073  1541.19622946]
 [  622.55395278   1838.3667368   -198.15943882  2012.49840413]
 [-1570.50278025   -179.56874049  1179.52957671  2033.15786844]
 [  707.79711348   -208.12866853  -599.98163383  1071.44259659]
 [   42.27462566   2358.68325942     4.13143749  2410.16761063]], shape=(5, 4), dtype=float64)
```

The first three components correspond to the three-momentum vectors, while the fourth entry represents the energy. The phase space is generated in the center-of-mass of the parent B^0 particle. Since the first decay contains only two particles $K^*(892)$ and γ in the final state, these latter are monoenergetic (if we had included the width of the $K^*(892)$ resonance, this would not be the case). Energy and momentum conservation yields $E_{K^*(892)} = 2715.8$ MeV and $E_\gamma = 2563.8$ MeV. These particles are distributed over the 4π solid angle in an isotropic way. The $K^*(892)$ and γ are always back-to-back, as can be observed directly from the components of their three-momentum. For example, in the first event, one has:

```
K*: tf.Tensor(
[ 1233.57121105    967.98542803 -2028.48795101  2715.76459565]
gamma: tf.Tensor(
[-1233.57121105   -967.98542803  2028.48795101  2563.87540435]
```

and similarly for all other events. Each component of the two particles adds to zero. The total energy of the $K^*(892)$ and γ is equal to the mass of the parent B^0 particle. The $K^*(892) \to K^+\pi^-$ is also a two-body decay, so in the rest frame of the $K^*(892)$, both K^+ and π^- are monoenergetic. However, in our case, the $K^*(892)$ is boosted, leading to the given phase-space distribution of the K^+ and π^- energy–momenta. The energy is of course always conserved as one can check by noting that for each event we find $E_{K^*(892)} = E_{\pi^-} + E_{K^+} = 2715.8$ MeV.

Problems

Ex 5.1 Lorentz transformations

(a) Derive the following relations for general Lorentz transformations (which include boosts, rotations, and reflections):

$$\Lambda_\mu{}^\nu = g_{\mu\rho}\,\Lambda^\rho{}_\lambda\,g^{\lambda\nu}$$

$$\Lambda^\mu{}_\nu\,\Lambda_\mu{}^\rho = \Lambda_\mu{}^\nu\,\Lambda^\mu{}_\rho = \delta_\nu^\rho$$

Hint: Use the fact that the inner product of two four-vectors is necessarily Lorentz-invariant.

(b) Prove the following relations for the determinants.

For general Lorentz transformations:

$$\det\left(\Lambda^{\mu}_{\ \nu}\right) \ = \ \pm 1$$

For proper Lorentz transformations (boosts and rotations):

$$\det\left(\Lambda^{\mu}_{\ \nu}\right) \ = \ +1$$

For space reflection and time reversal:

$$\det\left(\Lambda^{\mu}_{\ \nu}\right) \ = \ -1$$

Ex 5.2 Invariant mass. For the decay $R \to 1 + 2$, show that the mass of the particle R can be reconstructed from the daughters as:

$$m_R^2 = m_1^2 + m_2^2 + 2E_1 E_2 (1 - \beta_1 \beta_2 \cos\theta_{12}) \tag{5.200}$$

where β_1 and β_2 are the velocities of particles 1 and 2, E_1 and E_2 their energies, and θ_{12} is the angle between them. In the ultra-relativistic case, one has simply:

$$m_R^2 = 2E_1 E_2 (1 - \cos\theta_{12}) \tag{5.201}$$

Ex 5.3 Antiproton kinematical threshold. Antiprotons can be produced via the following reaction:

$$p + p \to p + p + p + \bar{p} \tag{5.202}$$

Find the minimum proton energy needed in the laboratory frame if one of the initial protons is at rest.

Ex 5.4 General formula for differential cross-section of the scattering process $1 + 2 \to 3 + 4$. Show that the differential cross-section in the center-of-mass system, keeping the rest masses of the four particles, can be expressed as:

$$\left(\frac{d\sigma}{d\Omega}\right)_{CMS} = \frac{1}{64\pi^2 s}\sqrt{\frac{\lambda\left(s, m_3^2, m_4^2\right)}{\lambda\left(s, m_1^2, m_2^2\right)}}|\mathcal{M}|^2 \tag{5.203}$$

where $\lambda(x, y, z)$ is the Källén function.

Ex 5.5 Opening angles. We consider the decay of a particle of mass m and momentum $\vec{p}$ into two daughter particles.

(a) Determine the minimum and maximum opening angle between the two daughter particles in the case they are both massless.

(b) Determine the minimum and maximum opening angle between the two daughter particles in the case they have masses m_1 and m_2, respectively.

(c) Let us assume that a calorimeter is able to separate the electromagnetic showers induced by high-energy photons when they are separated by an angle larger than $10°$. What is the maximum energy of the π^0 which can be reconstructed as a pair of distinct photons in such a calorimeter?

6   The Lagrangian Formalism

If a physical system behaves the same, regardless of how it is oriented in space, the physical laws that govern it are rotationally symmetric; from this symmetry, Noether's theorem shows the angular momentum of the system must be conserved. The physical system itself need not be symmetric; a jagged asteroid tumbling in space conserves angular momentum despite its asymmetry. Rather, the symmetry of the physical laws governing the system is responsible for the conservation law. As another example, if a physical experiment has the same outcome at any place and at any time, then its laws are symmetric under continuous translations in space and time; by Noether's theorem, these symmetries account for the conservation laws of linear momentum and energy within this system, respectively.

Emmy Noether

6.1   The Euler–Lagrange Equations

In quantum mechanics a system is fully described by its Hamiltonian (or Hamilton operator; see Chapter 4). One can also use the **Lagrange**[1] **formalism** as is often the case in classical mechanics. The classical motion of a particle in an external potential V is determined by Newton's law $\vec{F} = m\vec{a}$, which can be formulated as $m\mathrm{d}^2\vec{x}/\mathrm{d}t^2 = -\partial V/\partial \vec{x}$ (see Eq. (1.6)).

In general, a system can be described by a set of n **generalized coordinates** $q_i (i = 1, \ldots, n)$ which depend on time. The spatial position $\vec{r}_i(t)$ of each mass as a function of time in the system is given by a function of the generalized coordinates $\vec{r}_i(t) = \vec{r}_i(q_i(t), t)$ $(i = 1, \ldots, n)$. The time derivatives of the generalized coordinates are called the **generalized velocities** $\mathrm{d}q_i/\mathrm{d}t$ (also written as $\dot{q}_i$). The velocity vectors are estimated as total derivatives:

$$\vec{v_k} \equiv \frac{\mathrm{d}\vec{r}_k}{\mathrm{d}t} = \frac{\partial \vec{r}_k}{\partial q_j}\frac{\mathrm{d}q_j}{\mathrm{d}t} + \frac{\partial \vec{r}_k}{\partial t} \tag{6.1}$$

where the summation $j = 1, \ldots, n$ is implied. Given this, the kinetic energies and the potential of the system depend on the generalized coordinates and generalized velocities. In case of conservative forces only, the **Lagrangian function** is defined as the difference between the kinetic energy T and the potential energy V of the system:

$$L\left(q_i, \frac{\mathrm{d}q_i}{\mathrm{d}t}\right) = T - V \quad (i = 1, ..., n) \tag{6.2}$$

The time integral of the Lagrangian within a time interval given by t_1 and t_2 is called the **action**:

$$S([q]) = \int_{t_1}^{t_2} L\mathrm{d}t \tag{6.3}$$

$S([q])$ is a *functional* because it returns a number from the Lagrangian function integrated over all times between t_1 and t_2. It depends on the chosen trajectory of the system given by the generalized coordinates. It

---

1   Joseph-Louis Lagrange (1736–1813), Italian mathematician and astronomer.

can be computed for the "real" trajectory as well as for any arbitrary one, or more generally virtual variations of the entire trajectories around the actual one. The **principle of least action** states that the system's equation of motion (i.e., the actual path) corresponds to the one for which the action is stationary (at a minimum or maximum): $\delta S = 0$. This condition leads to the **Euler[2]–Lagrange equations**:

$$\frac{\mathrm{d}}{\mathrm{d}t}\left(\frac{\partial L}{\partial\left(\frac{\mathrm{d}q_i}{\mathrm{d}t}\right)}\right) - \frac{\partial L}{\partial q_i} = 0 \qquad i = 1, ..., n \tag{6.4}$$

So, classically, there are n Euler–Lagrange equations for a system with n degrees of freedom. The motion of the system is completely determined with $2n$ initial conditions (at some given time).

In the Hamilton description, the motion is expressed as a set of first-order differential equations. But the number of initial conditions must still remain $2n$, so we consider generalized q_i and we need n additional coordinates. These are chosen to be the **canonical momenta** (or generalized momenta), which are specified by the equation:

$$p_i \equiv \frac{\partial L}{\partial\left(\frac{\mathrm{d}q_i}{\mathrm{d}t}\right)} \tag{6.5}$$

The general correspondence between the Hamiltonian and the Lagrangian functions (including also a possible explicit time dependence) is given by the Legendre transformation changing the variables from $(q_i, \mathrm{d}q_i/\mathrm{d}t, t)$ to (q_i, p_i, t) (with summation implied):

$$H(p_i, q_i, t) = p_i\left(\frac{\mathrm{d}q_i}{\mathrm{d}t}\right) - L\left(q_i, \frac{\mathrm{d}q_i}{\mathrm{d}t}, t\right) \tag{6.6}$$

We find:

$$\begin{aligned}
\mathrm{d}H &= \left(\frac{\mathrm{d}q_i}{\mathrm{d}t}\right)\mathrm{d}p_i + p_i\mathrm{d}\left(\frac{\mathrm{d}q_i}{\mathrm{d}t}\right) - \frac{\partial L}{\partial q_i}\mathrm{d}q_i - \frac{\partial L}{\partial\left(\frac{\mathrm{d}q_i}{\mathrm{d}t}\right)}\mathrm{d}\left(\frac{\mathrm{d}q_i}{\mathrm{d}t}\right) - \frac{\partial L}{\partial t}\mathrm{d}t \\
&= \left(\frac{\mathrm{d}q_i}{\mathrm{d}t}\right)\mathrm{d}p_i - \frac{\partial L}{\partial q_i}\mathrm{d}q_i - \frac{\partial L}{\partial t}\mathrm{d}t = \left(\frac{\mathrm{d}q_i}{\mathrm{d}t}\right)\mathrm{d}p_i - \left(\frac{\mathrm{d}p_i}{\mathrm{d}t}\right)\mathrm{d}q_i - \frac{\partial L}{\partial t}\mathrm{d}t
\end{aligned} \tag{6.7}$$

On the other hand:

$$\mathrm{d}H = \frac{\partial H}{\partial q_i}\mathrm{d}q_i + \frac{\partial H}{\partial p_i}\mathrm{d}p_i + \frac{\partial H}{\partial t}\mathrm{d}t \tag{6.8}$$

So, the equations of motion can then be written as a function of the Hamiltonian:

$$\frac{\mathrm{d}q_i}{\mathrm{d}t} = \frac{\partial H}{\partial p_i}, \qquad \frac{\mathrm{d}p_i}{\mathrm{d}t} = -\frac{\partial H}{\partial q_i}, \qquad \text{and} \qquad -\frac{\partial L}{\partial t} = \frac{\partial H}{\partial t} \tag{6.9}$$

In the following sections, the application of this formalism to our relativistic fields will be introduced.

6.2   Relativistic Tensor Fields

In a relativistic field theory we are interested in the **relativistic equations of motion of fields**. A **field** describes *a continuous system with an infinite number of degrees of freedom*. The state of the system is described by the field ϕ, which depends on the position and time:

$$\phi(\vec{x}, t) = \phi(x^\mu) \tag{6.10}$$

2   Leonhard Euler (1707–1783), Swiss mathematician, physicist, astronomer.

The spatial vector $\vec{x}$ plays the role of an *index* for the infinite number of degrees of freedom. Starting from the (discrete) generalized coordinates defined in the previous section, we replace them with continuous functions:

$$
\begin{cases}
q_i \longrightarrow \phi(x^\mu) \\[2mm]
\dfrac{\mathrm{d}q_i}{\mathrm{d}t} \longrightarrow \dfrac{\partial}{\partial x^\mu}\phi(x) \qquad \mu = 0,1,2,3
\end{cases}
\tag{6.11}
$$

We shall make extensive use of the **four-gradient** (see Appendix D.5):

$$
\partial_\mu \equiv \frac{\partial}{\partial x^\mu} = \left(\frac{\partial}{\partial x^0}, \frac{\partial}{\partial x^1}, \frac{\partial}{\partial x^2}, \frac{\partial}{\partial x^3} \right)
\tag{6.12}
$$

Since the natural units of space and time are inverse energy $[x^\mu] = -1$ (see Eq. (1.31)), the **dimension of the four-gradient is just energy**:

$$
[\partial_\mu] = 1
\tag{6.13}
$$

Please note that in the generalization from the discrete generalized coordinates to the relativistic and continuous field, we have considered both time and spatial derivatives (i.e., the four-gradient) of the field on an equal footing. This comes from the desire to maintain a symmetry between space and time.

In general, the field defines a complex number[3] in each point of space-time, although for simplicity we will often start with real-valued fields. Of course, we can consider the field as observed from different inertial frames. We are in this case interested in the properties of the field under Lorentz transformations. The field must possess specific transformation properties, in the spirit of Section 5.2. Without this, it would be difficult to develop a covariant theory for that field (see Section 6.4).

Under a Lorentz transformation $(x^\mu \to x'^\mu = \Lambda^\mu{}_\nu x^\nu)$, a **scalar field** is defined to transform as (cf. Eq. (4.58) and see Figure 4.6):

$$
\phi(x) \to \phi'(x) = \phi(\Lambda^{-1}x)
\tag{6.14}
$$

This is a general statement which tells us how a scalar field is represented by different inertial observers. An **invariant scalar field** possesses the physical property that it has the same value in all correlated space-time points relative to all observers (note the change of space-time coordinates as each observer sees the field relative to his/her own frame):

$$
\phi'(x') = \phi'(\Lambda^{-1}x') = \phi(x) = \phi(\Lambda^{-1}x)
\tag{6.15}
$$

For example, the following complex field φ is scalar-invariant since its phase $p^\mu x_\mu$ is invariant:[4]

$$
\varphi(x^\mu) = N \underbrace{e^{-ip^\mu x_\mu}}_{\text{invariant}} = N e^{i\vec{p}\cdot\vec{x} - iEt}
\tag{6.16}
$$

More generally, fields can be classified according to how they vary under Lorentz transformations. A **vector field** A^μ defines a four-vector in each point of space-time and must transform as:

$$
A^\mu(x) \to A'^\mu(x) = \Lambda^\mu{}_\nu A^\nu(\Lambda^{-1}x)
\tag{6.17}
$$

Note that here not only the space-time coordinates are affected by the change of observer but also the field itself, unlike the case of a scalar field! Different observers see the four-vector field differently. We can think of this as the generalization of the effect of a rotation in real space of a three-vector field: the transformation affects the orientation of the three-vectors as well as the location of these vectors relative to the rotated observer, as illustrated in Figure 6.1.

A **tensor field** describes a quantity with two Lorentz indices, e.g., $\mu\nu$, and will therefore transform as:

$$
F^{\mu\nu}(x) \to F'^{\mu\nu}(x) = \Lambda^\mu{}_\rho \Lambda^\nu{}_\sigma F^{\rho\sigma}(\Lambda^{-1}x)
\tag{6.18}
$$

---

3   The reason for considering complex numbers is implicit in the fact that one wants to apply the theory to quantum systems.
4   We use the notation $\varphi(x)$ when the field must be complex, to distinguish from the general case $\phi(x)$.

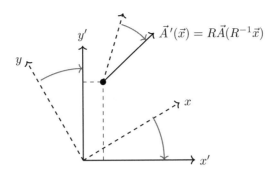

Figure 6.1 Illustration of the transformation of a vector field under the rotation R.

6.3   Lagrangian Density in Field Theory

In the classical case, the Lagrangian function depends on the generalized coordinates and generalized velocities. In the previous section, we have replaced them by a relativistic field and its first derivatives. Now we seek to generalize the Lagrangian, which should lead to the equations of motion of the relativistic field. The Lagrangian function should only depend on the field and its first derivatives at a given $\vec{x}$. Stated differently, in order to be able to transform it from one observer to another, it must be *local*. Consequently, it is introduced as a "density" at each point rather than the absolute quantity L. Let us begin with the case of a real scalar field, and then we will discuss the complex case.

• **Real scalar field.** We define the Lagrangian density $\mathcal{L}$ as a function of the *real field* ϕ and its four-derivatives:

$$\mathcal{L} \equiv \mathcal{L}(\phi, \partial_\mu \phi) \tag{6.19}$$

The action becomes:

$$S \equiv \int L \mathrm{d}t = \int \mathrm{d}t \int \mathrm{d}^3\vec{x}\, \mathcal{L} = \int \mathrm{d}^4 x\, \mathcal{L}(\phi, \partial_\mu \phi) \tag{6.20}$$

where we note that the result should be expressed as a covariant integral. Since the four-dimensional volume element $\mathrm{d}^4 x$ is an invariant of the Lorentz transformation, **we seek Lagrangian densities that are scalar and also invariant under Lorentz transformations.** The invariance of the Lagrange density is going to be a fundamental condition for the covariance of the theory, as will be discussed in the following. In general, other symmetries can be included in the Lagrangian function, for example we will see in later chapters the effect of gauge invariance.

In $\hbar \neq 1$ units, the action has the units of $\hbar$. In natural units, the action should be dimensionless, and all other quantities are measured in units of energy to some power (see Eq. (1.31) in Section 1.8):

$$[S] = \left[\int \mathrm{d}^4 x\, \mathcal{L}(\phi, \partial_\mu \phi) \right] = 0 \tag{6.21}$$

Since p^μ has the units of energy $[p^\mu] = +1$, while x^μ is an inverse energy $[x^\mu] = -1$, the units of the space-time element are $[\mathrm{d}^4 x] = -4$. Then, **the Lagrangian density must have the units of energy to the fourth power in order to yield a dimensionless action**:

$$[\mathcal{L}] = 4 \tag{6.22}$$

The action is the four-dimensional space-time integral of the Lagrangian density $\mathcal{L}(\phi, \partial_\mu \phi)$. The variation δS

depends on the variations of the field $\delta\phi$ and of its derivatives $\delta(\partial_\mu\phi)$:

$$
\begin{aligned}
\delta S &= \delta\phi\frac{\partial}{\partial\phi}\left\{\int \mathrm{d}^4x\mathcal{L}\right\} + \delta(\partial_\mu\phi)\frac{\partial}{\partial(\partial_\mu\phi)}\left\{\int \mathrm{d}^4x\mathcal{L}\right\} = \int \mathrm{d}^4x\left\{\frac{\partial\mathcal{L}}{\partial\phi}\delta\phi + \frac{\partial\mathcal{L}}{\partial(\partial_\mu\phi)}\delta(\partial_\mu\phi)\right\} \\
&= \int \mathrm{d}^4x\left\{\frac{\partial\mathcal{L}}{\partial\phi}\delta\phi + \partial_\mu\left(\frac{\partial\mathcal{L}}{\partial(\partial_\mu\phi)}\delta\phi\right) - \partial_\mu\left(\frac{\partial\mathcal{L}}{\partial(\partial_\mu\phi)}\right)\delta\phi\right\}
\end{aligned}
\tag{6.23}
$$

where we have assumed that the four-gradient of the variation of ϕ is equal to the variation of the four-gradient of ϕ (for a continuous function we can swap the order of differentiations):

$$
\partial_\mu\delta\phi = \delta(\partial_\mu\phi)
\tag{6.24}
$$

We can use the **generalized Stokes theorem** (also Stokes–Cartan theorem) (see Appendix A.12) to rewrite the integral of the derivative over the $n = 4$-dimensional "manifold" Ω as the $(n - 1) = 3$-dimensional integration over the boundary $\partial\Omega$ of the manifold. The four-dimensional integral in Eq. (6.23) is taken over a **space-time "volume"** Ω. We express the integral over the space-time (hyper)volume as an integral over the space-time (hyper)surface $\partial\Omega$:

$$
\int_\Omega \mathrm{d}^4x\partial_\mu\left(\frac{\partial\mathcal{L}}{\partial(\partial_\mu\phi)}\delta\phi\right) = \int_{\partial\Omega}\left(\frac{\partial\mathcal{L}}{\partial(\partial_\mu\phi)}\delta\phi\right)\cdot \mathrm{d}^3S_\mu
\tag{6.25}
$$

We can assume that the field has specific boundary conditions so that it is "fixed" on the hypersurface $\partial\Omega$, hence the variation vanishes over the space-time hypersurface $\partial\Omega$:

$$
\delta\phi|_{\partial\Omega} = 0
\tag{6.26}
$$

and then the integration of the total derivative is zero. Consequently, the second term of Eq. (6.23) vanishes. The principle of least action then implies $\delta S = 0$ and we get:

$$
\delta S = \int \mathrm{d}^4x\left\{\frac{\partial\mathcal{L}}{\partial\phi} - \partial_\mu\left(\frac{\partial\mathcal{L}}{\partial(\partial_\mu\phi)}\right)\right\}\delta\phi = 0
\tag{6.27}
$$

Since this needs to be valid for any variation $\delta\phi$, we can get the **relativistic Euler–Lagrange equation** for the field $\phi(x^\mu)$:

$$
\boxed{\frac{\partial\mathcal{L}}{\partial\phi} - \partial_\mu\left(\frac{\partial\mathcal{L}}{\partial(\partial_\mu\phi)}\right) = 0}
\tag{6.28}
$$

In general, we can consider a theory composed of n (scalar) real fields $\phi_i(x^\mu)$ $(i = 1, \ldots, n)$. In this case we have the Lagrangian $\mathcal{L} \equiv \mathcal{L}(\phi_i, \partial_\mu\phi_i)$ and we obtain a set of Euler–Lagrange equations:

$$
\frac{\partial\mathcal{L}}{\partial\phi_i} - \partial_\mu\left(\frac{\partial\mathcal{L}}{\partial(\partial_\mu\phi_i)}\right) = 0 \qquad (i = 1, \ldots, n)
\tag{6.29}
$$

• **Complex scalar field.** We now consider what happens in the case of a scalar complex field $\varphi(x^\mu)$. Such a field can always be written as the sum of two real fields, e.g., $\varphi(x^\mu) = \phi_1(x^\mu) + i\phi_2(x^\mu)$. So effectively we have the case above with $n = 2$ and two Euler–Lagrange equations. In practice, we can decide to use $\varphi(x^\mu)$ and its complex conjugate $\varphi^*(x^\mu)$ instead of $\phi_1(x^\mu)$ and $\phi_2(x^\mu)$. Then the Lagrangian density can be formally written as:

$$
\mathcal{L} \equiv \mathcal{L}(\varphi, \varphi^*, \partial_\mu\varphi, \partial_\mu\varphi^*)
\tag{6.30}
$$

The two fields are to be treated independently (this is what we would do if we considered ϕ_1 and ϕ_2), and the variation of the action becomes:

$$
\begin{aligned}
\delta S &= \int \mathrm{d}^4x\left\{\frac{\partial\mathcal{L}}{\partial\varphi}\delta\varphi + \frac{\partial\mathcal{L}}{\partial\varphi^*}\delta\varphi^* + \frac{\partial\mathcal{L}}{\partial(\partial_\mu\varphi)}\delta(\partial_\mu\varphi) + \frac{\partial\mathcal{L}}{\partial(\partial_\mu\varphi^*)}\delta(\partial_\mu\varphi^*)\right\} \\
&= \int \mathrm{d}^4x\left\{\frac{\partial\mathcal{L}}{\partial\varphi}\delta\varphi + \frac{\partial\mathcal{L}}{\partial(\partial_\mu\varphi)}\delta(\partial_\mu\varphi) + \mathrm{h.c.}\right\}
\end{aligned}
\tag{6.31}
$$

A treatment similar to the one above leads to the following set of Euler–Lagrange equations for the complex field:

$$\frac{\partial \mathcal{L}}{\partial \varphi} - \partial_\mu \left(\frac{\partial \mathcal{L}}{\partial(\partial_\mu \varphi)} \right) = 0 \quad \text{and} \quad \frac{\partial \mathcal{L}}{\partial \varphi^*} - \partial_\mu \left(\frac{\partial \mathcal{L}}{\partial(\partial_\mu \varphi^*)} \right) = 0 \tag{6.32}$$

6.4   Covariance and Invariance Properties of the Lagrangian

When the equations are made of tensors of the same rank, they keep the same form after a transformation. They are said to be **covariant**. Under the transformation, although the values of both sides of the equations may change, the functional form remains unchanged. In our case, we must take into account that the **principles of relativity require covariance under Lorentz transformations, namely that all equations of Nature be of the same form in any inertial frame**.

Invariance on the other hand means that the value itself does not change under the transformation, which applies to scalar quantities. Invariance under Lorentz transformations and more generally the **specific symmetries of the Lagrangian density** are a *truly fundamental property* of a theory.

We will see that assuming specific symmetries (e.g., gauge symmetries) provides a very powerful tool to "build" the Lagrangian densities of the theories of Nature. Using a Lagrangian formalism, the **theorem of Noether**, already mentioned in Section 1.3, can be stated as: *"if the Lagrangian function describing the motion of a field is invariant under a continuous transformation group, then the field possesses a conserved current associated with this group."* We discuss some specific examples in the following sections.

6.5   Invariance under Pure Space-Time Translation

We define an infinitesimal space-time translation with the help of the fixed four-vector δa^μ. Under the space-time translation, the space-time four-vector changes in the following way:

$$x^\mu \longrightarrow x'^\mu = x^\mu + \delta a^\mu \tag{6.33}$$

How does a scalar field change under an infinitesimal transformation of the space-time coordinates $x \to x'$? We can decompose:

$$\begin{aligned} \delta\phi(x) = \phi'(x') - \phi(x) &= \phi'(x') \underbrace{-\phi'(x) + \phi'(x)}_{=0} -\phi(x) \\ &= \underbrace{\phi'(x') - \phi'(x)}_{\simeq \frac{\partial \phi'}{\partial x^\nu}(x'-x)} + \underbrace{\phi'(x) - \phi(x)}_{\equiv \delta_\phi \phi(x)} \simeq \frac{\partial \phi}{\partial x^\nu}\delta a^\nu \end{aligned} \tag{6.34}$$

where the term $\delta_\phi \phi(x) = \phi'(x) - \phi(x)$ represents the change of the field when x^μ is fixed. Since the translation does not *explicitly* affect the field, but only the coordinates, we have $\delta_\phi \phi(x) = 0$. The above equation states how the field changes (to first order) from one reference frame to the other after the infinitesimal translation of Eq. (6.33). We then have:

$$\delta\phi \simeq \frac{\partial \phi}{\partial x^\nu}\delta a^\nu = (\partial_\nu \phi)\delta a^\nu \tag{6.35}$$

We can write most generally that the Lagrangian density will change due to the variation of the fields and their derivatives under the translation (the fields vary indirectly because of the shift in the reference frame).

For this first derivation, we assume that the Lagrangian *does not explicitly depend* on x^μ, consequently we only consider the functional variation of the field and its derivatives:

$$\delta\mathcal{L} = \left(\frac{\partial\mathcal{L}}{\partial\phi}\right)\delta\phi + \left(\frac{\partial\mathcal{L}}{\partial(\partial_\mu\phi)}\right)\delta(\partial_\mu\phi) = \partial_\mu\left(\frac{\partial\mathcal{L}}{\partial(\partial_\mu\phi)}\right)\delta\phi + \left(\frac{\partial\mathcal{L}}{\partial(\partial_\mu\phi)}\right)\partial_\mu(\delta\phi) = \partial_\mu\left(\frac{\partial\mathcal{L}}{\partial(\partial_\mu\phi)}\delta\phi\right) \quad (6.36)$$

where we used the Euler–Lagrange Eq. (6.28) and swapped the order of the derivatives $\delta(\partial_\mu\phi) = \partial_\mu(\delta\phi)$. Then inserting $\delta\phi$, we have:

$$\delta\mathcal{L} \simeq \partial_\mu\left(\frac{\partial\mathcal{L}}{\partial(\partial_\mu\phi)}(\partial_\nu\phi)\right)\delta a^\nu \quad (6.37)$$

On the other hand, we can formally write the change of the Lagrangian density as the total derivative relative to the space-time coordinates:

$$\delta\mathcal{L} \simeq \frac{\partial\mathcal{L}}{\partial x^\mu}\delta a^\mu = (\partial_\mu\mathcal{L})\delta a^\nu\delta^\mu_\nu = \partial_\mu(\delta^\mu_\nu\mathcal{L})\delta a^\nu \quad (6.38)$$

where we introduced the Kronecker tensor δ^μ_ν (see Appendix D.9). By comparing the last two results, we get:

$$\partial_\mu\left(\frac{\partial\mathcal{L}}{\partial(\partial_\mu\phi)}(\partial_\nu\phi) - \delta^\mu_\nu\mathcal{L}\right)\delta a^\nu = 0 \quad (6.39)$$

Since this result needs to be valid for *any* translation δa^ν, the quantity between brackets must vanish and we obtain the general **four conserved charge/currents** with $\nu = 0, 1, 2, 3$:

$$T_\nu{}^\mu \equiv \frac{\partial\mathcal{L}}{\partial(\partial_\mu\phi)}(\partial_\nu\phi) - \delta^\mu_\nu\mathcal{L} \quad \text{with} \quad \partial_\mu T_\nu{}^\mu = 0 \quad (6.40)$$

6.6   Canonical Energy and Momentum

The translation under space-time led to the definition of $T_\nu{}^\mu$ with four conserved charge/currents ($\nu = 0, \dots, 3$), which must be interpreted as the **canonical energy–momentum quantities of the field**. Hence, $T_\nu{}^\mu$ represents the **energy–momentum tensor**. With this result we can define the **canonical energy and momenta of the theory**, as well as a relation between the Lagrangian and the Hamiltonian, which in this case becomes the **Hamiltonian density** $\mathcal{H}$.

With $\nu = 0$, the $T_0{}^0$ component can be identified as the Hamiltonian density:

$$\mathcal{H} \equiv T_0{}^0 = \frac{\partial\mathcal{L}}{\partial(\partial_0\phi)}(\partial_0\phi) - \mathcal{L} \quad (6.41)$$

while with $\nu = 1, 2, 3$, the $T_i{}^0$ ($i = 1, 2, 3$) correspond to the **momentum densities**. We can define the **canonical momentum operator** Π for the field:

$$\Pi \equiv \frac{\partial\mathcal{L}}{\partial(\partial_0\phi)} \quad (6.42)$$

With these definitions we have an analogy to the classical case (Eq. (6.6)):

$$\mathcal{H} = \Pi(\partial_0\phi) - \mathcal{L} \quad (6.43)$$

The operators are found by integration of the densities over space. The conserved quantity associated with time translations is the Hamiltonian:

$$H = \int d^3\vec{x}\,\mathcal{H} = \int d^3\vec{x}\,T_0{}^0 \quad (6.44)$$

and the conserved quantity associated with space translations is the (three) momentum operator ($i = 1, 2, 3$):

$$P_i \equiv \int \mathrm{d}^3\vec{x}\, T_i{}^0 = \int \mathrm{d}^3\vec{x}\, \frac{\partial \mathcal{L}}{\partial(\partial_0\phi)}(\partial_i\phi) = \int \mathrm{d}^3\vec{x}\, \Pi(\partial_i\phi) \tag{6.45}$$

which corresponds to the total energy and momentum of the system. These results will be used in the upcoming chapters in the context of the second quantization of the fields.

6.7   General Continuity Equation for Charge and Currents

We consider a general infinitesimal transformation of the coordinates and the field defined in the following way:

$$x^\mu \longrightarrow x'^\mu = x^\mu + \delta x^\mu$$
$$\phi(x) \longrightarrow \phi'(x) = \phi(x) + \delta_\phi\phi(x) \tag{6.46}$$

where, as before, the term $\delta_\phi\phi(x) = \phi'(x) - \phi(x)$ represents the change of the field when x^μ is fixed. The **total** change in the field is given by the change of the space-time coordinates $(x \to x')$ and by the explicit change of the field $(\phi \to \phi')$ under the transformation. At first order (cf. Eq. (6.34)):

$$\begin{aligned}\delta\phi(x) &= \phi'(x') - \phi(x) = \underbrace{\phi'(x') - \phi'(x)}_{Eq.\ (6.34)} + \underbrace{\phi'(x) - \phi(x)}_{=\delta_\phi\phi(x)} \\ &\simeq (\partial_\nu\phi)\delta x^\nu + \delta_\phi\phi(x)\end{aligned} \tag{6.47}$$

(note the difference with $\delta_\phi\phi(x) = \phi'(x) - \phi(x)$).
- A *pure* infinitesimal translation $x^\mu \longrightarrow x^\mu + \delta a^\mu$ is given by:

$$\delta x^\mu = \delta a^\mu \quad \text{and} \quad \delta_\phi\phi(x) = 0 \quad \Longrightarrow \quad \delta\phi(x) = (\partial_\mu\phi)\delta a^\mu \tag{6.48}$$

- A *pure* infinitesimal change of the field without explicit change of coordinate system is expressed as:

$$\delta x^\mu = 0 \quad \text{and} \quad \delta\phi(x) = \delta_\phi\phi(x) \neq 0 \tag{6.49}$$

The Lagrangian density can in general also explicitly depend on the four-vector x^μ. Hence, the total variation of $\mathcal{L} = \mathcal{L}(x^\mu, \phi, \partial^\mu\phi)$ under our transformation $x \to x'$ and $\phi(x) \to \phi'(x)$ is given by:

$$\begin{aligned}\delta\mathcal{L} &= \left(\frac{\partial \mathcal{L}}{\partial x^\nu}\right)\delta x^\nu + \left(\frac{\partial \mathcal{L}}{\partial\phi}\right)\delta_\phi\phi + \left(\frac{\partial \mathcal{L}}{\partial(\partial_\mu\phi)}\right)\delta_\phi(\partial_\mu\phi) \\ &= (\partial_\nu\mathcal{L})\delta x^\nu + \left(\frac{\partial \mathcal{L}}{\partial\phi}\right)\delta_\phi\phi + \partial_\mu\left(\frac{\partial \mathcal{L}}{\partial(\partial_\mu\phi)}\delta_\phi\phi\right) - \partial_\mu\left(\frac{\partial \mathcal{L}}{\partial(\partial_\mu\phi)}\right)\delta_\phi\phi\end{aligned} \tag{6.50}$$

where the last term of the first line was rewritten by considering the full derivative ∂_μ and swapping the order of ∂ and $\delta_\phi\phi$. Compare with Eq. (6.23).

Since in general the coordinate system is allowed to change in the transformation, the change of the action is also due to the change of the four-dimensional space-time volume element d^4x and of the Lagrangian density:

$$\delta S = \int \delta(\mathrm{d}^4x)\mathcal{L} + \int \mathrm{d}^4x\,\delta\mathcal{L} \tag{6.51}$$

The change of the volume element d^4x is given by the Jacobian (see Appendix A.8):

$$\begin{aligned}\mathrm{d}^4x' &= \left|\det\left(\frac{\partial x'^\mu}{\partial x^\nu}\right)\right|\mathrm{d}^4x = \left|\det\left(\frac{\partial(x^\mu + \delta x^\mu)}{\partial x^\nu}\right)\right|\mathrm{d}^4x = \left|\det\left(\delta^\mu{}_\nu + \partial_\nu\delta x^\mu\right)\right|\mathrm{d}^4x \\ &\approx (1 + \partial_\nu\delta x^\nu)\mathrm{d}^4x\end{aligned} \tag{6.52}$$

and therefore $\delta(\mathrm{d}^4 x) \approx (\partial_\nu \delta x^\nu)\mathrm{d}^4 x$. Regrouping, the change in action S can be expressed as:

$$
\begin{aligned}
\delta S &= \int \mathrm{d}^4 x (\partial_\nu \delta x^\nu)\mathcal{L} + \int \mathrm{d}^4 x \left[(\partial_\nu \mathcal{L})\delta x^\nu + \left(\frac{\partial \mathcal{L}}{\partial \phi}\right)\delta_\phi \phi + \partial_\mu \left(\frac{\partial \mathcal{L}}{\partial(\partial_\mu \phi)}\delta_\phi \phi\right) - \partial_\mu \left(\frac{\partial \mathcal{L}}{\partial(\partial_\mu \phi)}\right)\delta_\phi \phi \right] \\
&= \int \mathrm{d}^4 x \left[\underbrace{(\partial_\nu \delta x^\nu)\mathcal{L} + (\partial_\nu \mathcal{L})\delta x^\nu}_{=\partial_\nu(\delta x^\nu \mathcal{L})} + \partial_\mu \left(\frac{\partial \mathcal{L}}{\partial(\partial_\mu \phi)}\delta_\phi \phi\right) + \left(\frac{\partial \mathcal{L}}{\partial \phi}\right)\delta_\phi \phi - \partial_\mu \left(\frac{\partial \mathcal{L}}{\partial(\partial_\mu \phi)}\right)\delta_\phi \phi \right] \\
&= \int \mathrm{d}^4 x \left[\partial_\mu \left(\delta^\mu_\nu \delta x^\nu \mathcal{L}\right) + \partial_\mu \left(\frac{\partial \mathcal{L}}{\partial(\partial_\mu \phi)}\delta_\phi \phi\right) \right] + \int \mathrm{d}^4 x \left[\left(\frac{\partial \mathcal{L}}{\partial \phi}\right)\delta_\phi \phi - \partial_\mu \left(\frac{\partial \mathcal{L}}{\partial(\partial_\mu \phi)}\right)\delta_\phi \phi \right] \\
&= \int_\Omega \mathrm{d}^4 x \partial_\mu \left[\left(\delta^\mu_\nu \delta x^\nu \mathcal{L}\right) + \left(\frac{\partial \mathcal{L}}{\partial(\partial_\mu \phi)}\delta_\phi \phi\right) \right] + \int_\Omega \mathrm{d}^4 x \delta_\phi \phi \left[\left(\frac{\partial \mathcal{L}}{\partial \phi}\right) - \partial_\mu \left(\frac{\partial \mathcal{L}}{\partial(\partial_\mu \phi)}\right) \right]
\end{aligned}
\tag{6.53}
$$

where we recall that the four-dimensional integral is over the space-time volume Ω. The first integral above is a total divergence which, with the help of the generalized Stokes theorem, can be rewritten as the integral over the space-time hypersurface $\partial\Omega$ (see Eq. (6.25)):

$$
\delta S = \underbrace{\int_{\partial\Omega} \left[\left(\delta^\mu_\nu \delta x^\nu \mathcal{L}\right) + \left(\frac{\partial \mathcal{L}}{\partial(\partial_\mu \phi)}\delta_\phi \phi\right) \right] \cdot \mathrm{d}^3 S_\mu}_{\equiv \delta S_{\partial\Omega}} + \underbrace{\int_\Omega \mathrm{d}^4 x \delta_\phi \phi \left[\left(\frac{\partial \mathcal{L}}{\partial \phi}\right) - \partial_\mu \left(\frac{\partial \mathcal{L}}{\partial(\partial_\mu \phi)}\right) \right]}_{\equiv \delta S_\Omega}
\tag{6.54}
$$

If we assume, as before, that the integral $\delta S_{\partial\Omega}$ on the hypersurface vanishes, the second integral δS_Ω yields as expected the Euler–Lagrange equation for ϕ when we set $\delta S = 0$. The integral on the hypersurface can be further expanded by adding and subtracting a term:

$$
\begin{aligned}
\delta S_{\partial\Omega} &= \int_{\partial\Omega} \left[\left(\delta^\mu_\nu \delta x^\nu \mathcal{L}\right) + \left(\frac{\partial \mathcal{L}}{\partial(\partial_\mu \phi)}\right)(\partial_\nu \phi)\delta x^\nu - \left(\frac{\partial \mathcal{L}}{\partial(\partial_\mu \phi)}\right)(\partial_\nu \phi)\delta x^\nu + \left(\frac{\partial \mathcal{L}}{\partial(\partial_\mu \phi)}\delta_\phi \phi\right) \right] \cdot \mathrm{d}^3 S_\mu \\
&= \int_{\partial\Omega} \left[\left(\frac{\partial \mathcal{L}}{\partial(\partial_\mu \phi)}\right)\{(\partial_\nu \phi)\delta x^\nu + (\delta_\phi \phi)\} + \left\{(\delta^\mu_\nu \mathcal{L}) - \left(\frac{\partial \mathcal{L}}{\partial(\partial_\mu \phi)}\right)(\partial_\nu \phi)\right\}\delta x^\nu \right] \cdot \mathrm{d}^3 S_\mu \\
&= \int_{\partial\Omega} \left[\left(\frac{\partial \mathcal{L}}{\partial(\partial_\mu \phi)}\right)\delta\phi + \left\{(\delta^\mu_\nu \mathcal{L}) - \left(\frac{\partial \mathcal{L}}{\partial(\partial_\mu \phi)}\right)(\partial_\nu \phi)\right\}\delta x^\nu \right] \cdot \mathrm{d}^3 S_\mu \\
&= \int_{\partial\Omega} \left[\left(\frac{\partial \mathcal{L}}{\partial(\partial_\mu \phi)}\right)\delta\phi - T_\nu{}^\mu \delta x^\nu \right] \cdot \mathrm{d}^3 S_\mu
\end{aligned}
\tag{6.55}
$$

where in the first bracket we recognize the effect of the *total* variation under the transformation. The second term recovers the general conserved charge/current that was found in Eq. (6.40).

Putting all the above results together we obtain the change of the action δS under the general transformation:

$$
\delta S = \int_{\partial\Omega} \left[\left(\frac{\partial \mathcal{L}}{\partial(\partial_\mu \phi)}\right)\delta\phi - T_\nu{}^\mu \delta x^\nu \right] \cdot \mathrm{d}^3 S_\mu + \int_\Omega \mathrm{d}^4 x \delta_\phi \phi \left[\left(\frac{\partial \mathcal{L}}{\partial \phi}\right) - \partial_\mu \left(\frac{\partial \mathcal{L}}{\partial(\partial_\mu \phi)}\right) \right]
\tag{6.56}
$$

Again the second term vanishes if we take ϕ to be a solution of the Euler–Lagrange equation. The first term depends on both $\delta\phi$ and δx. In order to exploit the symmetries of the physical system, we consider the variation under infinitesimal changes of a common parameter given by $\delta\alpha_\nu$ such that:

$$
\delta\phi \equiv \Phi_\rho \delta\alpha^\rho \quad \text{and} \quad \delta x^\nu \equiv X^\nu_\rho \delta\alpha^\rho
\tag{6.57}
$$

where Φ_μ is a set of numbers and X^ν_μ is a matrix of numbers. The first term of the variation of the action then becomes:

$$
\delta S_{\partial\Omega} = \int_{\partial\Omega} \left[\left(\frac{\partial \mathcal{L}}{\partial(\partial_\mu \phi)}\right)\Phi_\rho - T_\nu{}^\mu X^\nu_\rho \right] \cdot \delta\alpha^\rho \mathrm{d}^3 S_\mu
\tag{6.58}
$$

Since $\delta\alpha^\rho$ is arbitrary, the condition $\delta S = 0 = \delta S_{\partial\Omega} + \delta S_\Omega$ allows us to define the general **four-currents** $j^\mu_\rho = (j^0, \vec{j})$ with their **covariant continuity equations** ($\rho = 0, \ldots, 3$):

$$\partial_\mu j^\mu_\rho = 0 \quad \text{where} \quad j^\mu_\rho \equiv \left(\frac{\partial\mathcal{L}}{\partial(\partial_\mu\phi)}\right)\Phi_\rho - T^\mu_\nu X^\nu_\rho \tag{6.59}$$

Integration of the time component of the four-currents j^μ_ρ defines the corresponding total **charges** Q_ρ inside the volume:

$$Q_\rho \equiv \int \mathrm{d}^3\vec{x}\, j^0_\rho \tag{6.60}$$

The relation between the charges Q_ρ and spatial three-currents $\vec{j}_\rho$ can be obtained by noting:

$$\frac{\mathrm{d}Q_\rho}{\mathrm{d}t} = \int \mathrm{d}^3\vec{x}\,\partial_0 j^0_\rho = \int \mathrm{d}^3\vec{x}(\partial_\mu j^\mu_\rho - \vec{\nabla}\cdot\vec{j}_\rho) = -\int \mathrm{d}^3\vec{x}(\vec{\nabla}\cdot\vec{j}_\rho) \tag{6.61}$$

As expected, if the flux of the currents through the surface of the volume vanishes, then the charge is conserved inside the volume.

• **Translation invariance revisited.** We can revisit the case of the invariance under a pure translation in the context of the general expressions derived above. We set in this case $\delta\phi(x) = 0$ and $\delta x^\nu = X^\nu_\rho \delta\alpha^\rho = \delta a^\nu$, so the corresponding four-current tensor is just:

$$\Phi_\rho = 0, X^\nu_\rho = \delta^\nu_\rho \implies j^\mu_\rho = -T^\mu_\nu\delta^\nu_\rho = -T^\mu_\rho \tag{6.62}$$

As expected, the currents j^μ_ρ associated with invariance under pure space-time translation just correspond (up to an overall sign) to the **energy–momentum tensor** T of the field which was derived in Eq. (6.40).

• **Pure gauge (internal phase) transformation.** We now see what happens for a *pure* transformation of a complex field φ given by:

$$\varphi(x) \longrightarrow \varphi'(x) = e^{-i\alpha}\varphi(x) \tag{6.63}$$

where α is a **global phase** (independent of x^μ). We have for the corresponding infinitesimal phase transformation:

$$\delta x^\mu = 0 \quad \text{and} \quad \delta\varphi(x) = \delta_\varphi\varphi(x) = -i\delta\alpha\varphi \tag{6.64}$$

As before, the complex field is written as the sum of two real fields, e.g., $\varphi(x^\mu) = \phi_1(x^\mu) + i\phi_2(x^\mu)$, and we decide to use $\varphi(x^\mu)$ and its complex conjugate $\varphi^*(x^\mu)$ instead of $\phi_1(x^\mu)$ and $\phi_2(x^\mu)$. The two fields are treated independently and a derivation as above gives the variation of the action as a function of the field (as before) plus a Hermitian conjugated expression (see Eq. (6.31)). This propagates to the definition of the currents which becomes:

$$j^\mu_\rho \equiv \left(\frac{\partial\mathcal{L}}{\partial(\partial_\mu\varphi)}\right)\Phi_\rho + \left(\frac{\partial\mathcal{L}}{\partial(\partial_\mu\varphi^*)}\right)\Phi^*_\rho - T^\mu_\nu X^\nu_\rho \tag{6.65}$$

(also the tensor T^μ_ν gets the extra term). For the pure phase transformation ($\delta x^\nu = 0$), we find $\Phi_\rho = -i\varphi$ and $X^\nu_\rho = 0$, so the corresponding four-current is (there is only one):

$$j^\mu = \left(\frac{\partial\mathcal{L}}{\partial(\partial_\mu\varphi)}(-i\varphi) + \text{h.c.}\right) = (-i\varphi)\frac{\partial\mathcal{L}}{\partial(\partial_\mu\varphi)} + (-i\varphi)^*\frac{\partial\mathcal{L}}{\partial(\partial_\mu\varphi^*)} = -i\left(\varphi\frac{\partial\mathcal{L}}{\partial(\partial_\mu\varphi)} - \varphi^*\frac{\partial\mathcal{L}}{\partial(\partial_\mu\varphi^*)}\right) \tag{6.66}$$

We will later use this quantity in interpreting the results of the field equations.

6.8   Proper Lorentz Transformation of a Field

We first consider a Lorentz transformation of a scalar field $\phi(x^\mu)$. Under the transformation, the field undergoes $\phi(x) \rightarrow \phi'(x) = \phi(\Lambda^{-1}x)$ (see Eq. (6.14)). If we write the identity transformation as $\Lambda^\mu_{\ \nu} = \delta^\mu_\nu$, then an **infinitesimal proper Lorentz transformation** derives from it in the following way:

$$x^\mu \rightarrow x'^\mu = \Lambda^\mu_{\ \nu}x^\nu = (\delta^\mu_\nu + \omega^\mu_{\ \nu})x^\nu \tag{6.67}$$

where the $\omega^\mu_{\ \nu}$ define the transformation. As a consequence of the Lorentz orthogonality condition, Eq. (5.10), $\omega_{\alpha\beta}$ is an antisymmetric tensor:

$$\omega_{\alpha\beta} = -\omega_{\beta\alpha} \tag{6.68}$$

(see the derivation of Eq. (B.90) in Appendix B.10). Therefore, it has $4 \times 3/2 = 6$ independent components, which agrees with the number of independent transformations of the Lorentz group: three rotations and three boosts. At first order, the change of the field ϕ is (see Eq. (6.34)):

$$
\begin{aligned}
\delta\phi &= \phi'(x') - \phi(x) = \phi(\Lambda^{-1}x) - \phi(x) = \phi\left((\delta^\mu_\nu - \omega^\mu_{\ \nu})x^\nu\right) - \phi(x) = \phi\left(x^\mu - \omega^\mu_{\ \nu}x^\nu\right) - \phi(x) \\
&\approx -\omega^\mu_{\ \nu}x^\nu \partial_\mu \phi(x)
\end{aligned}
\tag{6.69}
$$

With the help of the antisymmetric condition $\omega_{\alpha\beta} = -\omega_{\beta\alpha}$, we obtain:

$$
\begin{aligned}
\delta\phi &\approx -(\omega_{\mu\nu}x^\nu\partial^\mu\phi(x)) = -\frac{1}{2}(\omega_{\mu\nu}x^\nu\partial^\mu\phi(x) + \omega_{\mu\nu}x^\nu\partial^\mu\phi(x)) = -\frac{1}{2}(\omega_{\mu\nu}x^\nu\partial^\mu\phi(x) - \omega_{\mu\nu}x^\mu\partial^\nu\phi(x)) \\
&= \frac{1}{2}\omega_{\mu\nu}(x^\mu\partial^\nu - x^\nu\partial^\mu)\phi(x) = -\frac{i}{2}\omega_{\mu\nu}\left(x^\mu i\partial^\nu - x^\nu i\partial^\mu\right)\phi(x) \equiv -\frac{i}{2}\omega_{\mu\nu}J^{\mu\nu}\phi(x)
\end{aligned}
\tag{6.70}
$$

In the last line, we have defined six independent quantities as:

$$J^{\mu\nu} \equiv (x^\mu i\partial^\nu - x^\nu i\partial^\mu) = (x^\mu p^\nu - x^\nu p^\mu) \tag{6.71}$$

where we replaced $i\partial^\mu \rightarrow p^\mu$, as will be discussed in Section 7.2. Recalling that Lorentz rotations and boosts are generalizations of the space rotations to space-time rotations (they both keep the inner product constant), **the antisymmetric four-dimension tensor** $J^{\mu\nu}$ is to be understood as the generalization of the three-angular momentum $\vec{L} = \vec{x} \times \vec{p} = \epsilon_{ijk}x^j p^k$. In Section 4.7 we have discussed the relationship between physical space rotations and angular momentum, and have derived the corresponding "generators" and their commutation rules. Here, the $J^{\mu\nu}$ therefore represent six independent generators of the Lorentz group written under their differential form. These six basic generators can be used as a basis for any Lorentz transformation. It's totally beautiful!

With the help of these generators, the **finite Lorentz transformation** is given by (cf. Eq. (4.71)):

$$\Lambda(\Omega_{\mu\nu}) = e^{\frac{i}{2}\Omega_{\mu\nu}J^{\mu\nu}} = \mathbb{1} + \frac{i}{2}\Omega_{\mu\nu}J^{\mu\nu} + \cdots \tag{6.72}$$

where $\Omega_{\mu\nu}$ are numbers. We can derive the commutation relations of the **generators of the Lorentz group** for the field. The **commutation relations of the Lorentz group algebra** can be computed with the definition, Eq. (6.71). The result is the following Lie algebra:

$$[J^{\mu\nu}, J^{\rho\sigma}] = i\left(g^{\nu\rho}J^{\mu\sigma} - g^{\mu\rho}J^{\nu\sigma} - g^{\nu\sigma}J^{\mu\rho} + g^{\mu\sigma}J^{\nu\rho}\right) \tag{6.73}$$

Any complete set of six operators satisfying this algebra will represent generators of the Lorentz group.

In general, the transformation of any field $\Phi(x^\mu)$ of a given rank under Lorentz transformations can be defined by an operator $M \equiv M(\Lambda)$ which depends on the transformation, such that:

$$\Phi(x) \rightarrow \Phi'(x) = M(\Lambda)\Phi(\Lambda^{-1}x) \tag{6.74}$$

Here the $M(\Lambda)$ are said to form **representations of the Lorentz group**, meaning that $M(\Lambda_1\Lambda_2) = M(\Lambda_1)M(\Lambda_2)$, $M(\mathbb{1}) = \mathbb{1}$, and $M(\Lambda^{-1}) = M^{-1}(\Lambda)$. They represent the effect of a given Lorentz transformation on the objects $\Phi(x^\mu)$ of a given rank r, while Λ defines the transformation of a four-vector under the same transformation. The $M(\Lambda)$ for the given rank r will possess $2r$ Lorentz indices. Therefore, although Λ and $M(\Lambda)$ represent the *same physical transformation* (i.e., a given rotation or boost), Λ and $M(\Lambda)$ can be very different as they describe the effect of the physical transformation on different types of quantities.

Problems

Ex 6.1 An oscillating string using Lagrange formalism. As an application on a classical, non-relativistic, continuous system, we derive the equation of motion of an oscillating string (standing wave) from the Euler–Lagrange equation. A string is tensioned with a force F and its mass density (mass per length) is ρ [g/cm]. At $x = 0$ and $x = l$ the string is fixed, and the (small) transverse excitation at the position x and at the time t is $\phi(x,t)$. In classical mechanics the Lagrangian is given by:

$$L = \int_0^l \mathcal{L}\mathrm{d}x = T - V,$$

where $\mathcal{L}$ is the Lagrangian density, T is the kinetic energy of the system, and V its potential energy.

(a) Derive the Lagrangian density $\mathcal{L}$ of the oscillating string.

Hint: The length of the oscillating string is:

$$\int_0^l \mathrm{d}s = \int_0^l \left(\mathrm{d}x^2 + \mathrm{d}\phi^2\right)^{\frac{1}{2}} = \int_0^l \left[1 + \left(\frac{\partial\phi}{\partial x}\right)^2\right]^{\frac{1}{2}} \mathrm{d}x$$

Thus, the expansion (in length) of the string for a small excitation ($\partial\phi/\partial x \ll 1$) from the zero position is $\int_0^l \frac{1}{2}\left(\frac{\partial\phi}{\partial x}\right)^2 \mathrm{d}x$ and, hence, the work necessary to expand the string by this amount against the force F is $\int_0^l \frac{1}{2}F(\frac{\partial\phi}{\partial x})^2\mathrm{d}x$.

(b) According to the principle of least action, the motion of the string between the times t_1 and t_2 is such that the action $S = \int_{t_1}^{t_2} L\mathrm{d}t$ is an extremum ($\delta S = 0$). Derive the Euler–Lagrange equation for the oscillating string. The boundary conditions are $\delta\phi(x,t_1) = \delta\phi(x,t_2) = 0$ for all x and $\delta\phi(0,t) = \delta\phi(l,t) = 0$ for all t.

(c) Derive with the Euler–Lagrange equation the wave equation for the oscillating string (standing wave).

(d) The Hamiltonian density is given by $\mathcal{H} = \Pi \cdot \frac{\partial\phi}{\partial t} - \mathcal{L}$, where $\Pi = \frac{\partial\mathcal{L}}{\left(\frac{\partial\phi}{\partial t}\right)}$ is the canonical momentum density. Show with the Euler–Lagrange equation that the $E = \int_0^l \mathcal{H}\mathrm{d}x$ of the oscillating string is constant.

Ex 6.2 Conserved charge in a Lorentz boost. Show that the conserved charge associated with a Lorentz boost is given by:

$$L^{\mu\nu} = -x^\nu p^\mu + x^\mu p^\nu \tag{6.75}$$

Hint: Consider a Lagrangian of the form $\mathcal{L} = \mathcal{L}(x(\tau), \dot{x}(\tau), \tau)$.

Ex 6.3 Relativistic motion of a charged particle in a uniform magnetic field. We consider a charged particle with momentum $\vec{p}$ and charge e moving in a uniform magnetic field $\vec{B}$.

(a) Show that the relativistic extension of the equation of motion is identical to the classical expression provided that the relativistic definition of the momentum is used:

$$\frac{d\vec{p}}{dt} = e\vec{E} + e\vec{v} \times \vec{B} \tag{6.76}$$

(b) Show that the trajectory of such a particle is a helix.

(c) Find the relation between the radius of curvature of the trajectory and the momentum of the particle.

(d) Show that the transverse momentum p_T expressed in GeV/c of a particle of charge $q = ze$ moving inside a uniform magnetic field $\vec{B}$ whose strength is expressed in Tesla is:

$$p_T = 0.3z|\vec{B}|R \tag{6.77}$$

where R is the radius of curvature expressed in meters.

7 Free Boson Fields

Lorentz symmetry is at the core of modern physics: the kinematical laws of special relativity and Maxwell's field equations in the theory of electromagnetism respect it. The direct relativistic extension of the Schrödinger equation leads to the Klein–Gordon equation, which will be interpreted in the context of the **second quantization**, to describe bosons. In the Standard Model all interactions are induced by intermediate vector gauge boson fields and the Higgs boson is represented by a scalar boson field.

7.1 The Klein–Gordon Equation

The Schrödinger equation is not adequate to describe relativistic processes as it is based on the classical relation between energy and momentum ($E = p^2/2m$). By construction, it relates the first time derivative of the wave function to its second space derivative. It is therefore manifestly not covariant as time and space are treated differently. To derive a relativistic Schrödinger equation we can start with the relativistic relation between energy and momentum $E^2 = \vec{p}^2 + m^2$. We consider a free particle and ignore the presence of an external potential V. In order to "quantize" the problem, we use the canonical ansatz and replace the energy and momentum with their corresponding operators. Namely, we apply the **first quantization** described in Section 4.2:

$$\vec{p} \to -i\nabla \quad \text{and} \quad E \to +i\frac{\partial}{\partial t} \tag{7.1}$$

We introduce the **field** $\phi = \phi(x^\mu) = \phi(\vec{x}, t)$. From the relativistic energy–momentum relation we get the **Klein[1]–Gordon[2] equation**:

$$i^2\frac{\partial^2\phi}{\partial t^2} = i^2\nabla^2\phi + m^2\phi \tag{7.2}$$

or

$$-\frac{\partial^2\phi}{\partial t^2} + \nabla^2\phi = m^2\phi \quad \text{(Klein–Gordon equation)} \tag{7.3}$$

We note that the time and space coordinates are treated on an equal footing in the equation, i.e., the Klein–Gordon equation contains the second time derivative and the second space derivative of the field.

7.2 Covariant Form of the Klein–Gordon Equation

Let us assume that the field is a scalar, invariant under Lorentz transformation. We then want to write the Klein–Gordon equation in a covariant form, such that it is in the same form for all inertial observers. By

---

1   Oskar Benjamin Klein (1894–1977), Swedish theoretical physicist.
2   Walter Gordon (1893–1939), German theoretical physicist.

extension of Eq. (7.1), we can consider the canonical ansatz for the energy–momentum four-vector to obtain the **energy–momentum four-vector operator** $p^\mu = (E, \vec{p})$ (see Appendix D.5):

$$p^\mu \quad \rightarrow \quad i\partial^\mu = i\left(\frac{\partial}{\partial t}, -\nabla\right) \tag{7.4}$$

As it should be, the dimension of $[p^\mu] = [\partial^\mu]$ is that of an energy (see Eq. (6.13)):

$$[p^\mu] = [\partial_\mu] = 1 \tag{7.5}$$

We have:

$$p^2 = (i\partial^\mu)^2 = -\partial^\mu\partial_\mu = -\square^2 = m^2 \tag{7.6}$$

where $\square^2 = \partial^\mu\partial_\mu = \frac{\partial^2}{\partial t^2} - \nabla^2$ is the D'Alembert operator (see Appendix D.6). The **explicitly covariant form of the Klein–Gordon** is then $p^2\phi = -\square^2\phi = m^2\phi$ or:

$$\left(\square^2 + m^2\right)\phi = \left(\partial^\mu\partial_\mu + m^2\right)\phi = 0 \tag{7.7}$$

where $\phi \equiv \phi(x^\mu)$.

The Klein–Gordon (KG) equation can be interpreted as the wave equation of the field. It is consistent with special relativity and is explicitly covariant. In this equation, the D'Alembert operator acting on a scalar field is an invariant scalar. The rest mass squared m^2 is also an invariant scalar. So the entire equation is invariant under Lorentz transformation.

7.3   Current Density of the Klein–Gordon Field

In general, the field can be considered as a **complex scalar field** φ. We consider the complex conjugate of the Klein–Gordon equation:

$$\left(\partial^\mu\partial_\mu + m^2\right)\varphi = 0 \quad \Longrightarrow \quad \left(\partial^\mu\partial_\mu + m^2\right)\varphi^* = 0 \tag{7.8}$$

We multiply the above equations by $-i\varphi^*$ and $i\varphi$ and add them:

$$(-i\varphi^*)\left(\partial^\mu\partial_\mu + m^2\right)\varphi + (i\varphi)\left(\partial^\mu\partial_\mu + m^2\right)\varphi^* = 0$$
$$\Rightarrow \quad (-i\varphi^*\partial^\mu\partial_\mu\varphi) + (i\varphi\partial^\mu\partial_\mu\varphi^*) = 0 \tag{7.9}$$

We now separate the time and spatial dependencies, obtaining the following second-order derivatives:

$$-i\varphi^*\frac{\partial^2\varphi}{\partial t^2} + i\varphi\frac{\partial^2\varphi^*}{\partial t^2} + i\varphi^*\nabla^2\varphi - i\varphi\nabla^2\varphi^* = 0 \tag{7.10}$$

In terms of first-order derivatives:

$$\frac{\partial}{\partial t}\left[i\left(\varphi^*\frac{\partial\varphi}{\partial t} - \varphi\frac{\partial\varphi^*}{\partial t}\right)\right] + \nabla\left[-i\left(\varphi^*\nabla\varphi - \varphi\nabla\varphi^*\right)\right] = 0 \tag{7.11}$$

By analogy with the equation of continuity (see Section. 4.4), we define the density ρ and the current $\vec{j}$:

$$\frac{\partial\rho}{\partial t} + \nabla\cdot\vec{j} = 0 \quad \Longrightarrow \quad \rho \equiv i\left(\varphi^*\frac{\partial\varphi}{\partial t} - \varphi\frac{\partial\varphi^*}{\partial t}\right) \quad \text{and} \quad \vec{j} \equiv -i\left(\varphi^*\nabla\varphi - \varphi\nabla\varphi^*\right) \tag{7.12}$$

We can elegantly write these as a **four-vector current density** $j^\mu \equiv (\rho, \vec{j})$ satisfying the covariant continuity equation $\partial_\mu j^\mu = 0$. The **Klein–Gordon current density** j^μ_{KG} is therefore equal to:

$$j^\mu_{KG} = i\left(\varphi^*\partial^\mu\varphi - \varphi\partial^\mu\varphi^*\right) \tag{7.13}$$

7.4   Lagrangian of the Klein–Gordon Fields

We consider a scalar field $\phi = \phi(x^\mu)$. We seek an invariant Lagrangian which is a function of the field and its derivatives $\mathcal{L} \equiv \mathcal{L}(\phi, \partial_\mu \phi)$. The Lagrangian is defined differently for a purely real field or a complex field.

- For the **real field** ϕ, the Lagrangian is written as follows:

$$\mathcal{L}_{KG} = \frac{1}{2}\left(\partial_\mu \phi\right)\left(\partial^\mu \phi\right) - \frac{1}{2}m^2\phi^2 = \frac{1}{2}\left(\partial_\mu \phi\right)^2 - \frac{1}{2}m^2\phi^2 \tag{7.14}$$

Let us derive the units of the field ϕ. Since the units of $\mathcal{L}$ are energy to the fourth power $[\mathcal{L}] = 4$ (see Eq. (6.22)), $[m^2] = 2$ and $[\partial_\mu] = 1$, then a dimensionless action implies that the **dimensions of the scalar field Klein–Gordon ϕ must be energy**:

$$[\phi] = 1 \tag{7.15}$$

One can verify that the Lagrangian yields the correct answer (Eq. (7.7)) by considering the Euler–Lagrange equations:

$$\frac{\partial \mathcal{L}}{\partial \phi} = -\frac{1}{2}m^2 2\phi = -m^2\phi \tag{7.16}$$

and also:

$$\frac{\partial \mathcal{L}}{\partial(\partial_0 \phi)} = \frac{\partial}{\partial(\partial_0 \phi)}\frac{1}{2}\left[(\partial_0\phi)(\partial_0\phi) - \cdots\right] = \partial_0\phi = \partial^0\phi \tag{7.17}$$

and

$$\frac{\partial \mathcal{L}}{\partial(\partial_1 \phi)} = \frac{\partial}{\partial(\partial_1 \phi)}\frac{1}{2}\left[(\partial_0\phi)(\partial_0\phi) - (\partial_1\phi)(\partial_1\phi)\cdots\right] = -\partial_1\phi = \partial^1\phi \tag{7.18}$$

and similarly for ∂_2 and ∂_3. Hence we can write this result by using the contravariant derivative (note the position of the index):

$$\frac{\partial \mathcal{L}}{\partial(\partial_\mu \phi)} = \frac{\partial}{\partial(\partial_\mu \phi)}\frac{1}{2}(\partial_\mu\phi)^2 = \partial^\mu\phi \tag{7.19}$$

With the Euler–Lagrange equation, we can check that the chosen Lagrangian yields as expected the Klein–Gordon equation:

$$\frac{\partial \mathcal{L}}{\partial \phi} = \partial_\mu\left(\frac{\partial \mathcal{L}}{\partial(\partial_\mu \phi)}\right) \quad \Longrightarrow \quad -m^2\phi = \partial_\mu(\partial^\mu\phi) \quad \Box \tag{7.20}$$

- For the **complex field** φ, we introduce two real fields ϕ_1 and ϕ_2 such that:

$$\varphi = \frac{1}{\sqrt{2}}(\phi_1 + i\phi_2) \quad \text{and} \quad \varphi^* = \frac{1}{\sqrt{2}}(\phi_1 - i\phi_2) \tag{7.21}$$

Both ϕ_1 and ϕ_2 satisfy the Klein–Gordon equation:

$$\left(\partial^\mu\partial_\mu + m^2\right)\phi_1 = 0 \quad \text{and} \quad \left(\partial^\mu\partial_\mu + m^2\right)\phi_2 = 0 \tag{7.22}$$

The Lagrangian is in this case (note the absence of the factor $1/2$ compared to the case where the field is real):

$$\mathcal{L}_{KG} = (\partial_\mu\varphi^*)(\partial^\mu\varphi) - m^2\varphi^*\varphi \tag{7.23}$$

The two real fields ϕ_1 and ϕ_2 are independent from one another, and so are φ and φ^* in the application of the Euler–Lagrange equations. Hence, we can check that each leads to the Klein–Gordon equation:

$$\begin{cases} \partial_\mu\left(\dfrac{\partial \mathcal{L}}{\partial(\partial_\mu \varphi)}\right) - \dfrac{\partial \mathcal{L}}{\partial \varphi} = \partial_\mu(\partial^\mu\varphi^*) + m^2\varphi^* = \partial_\mu\partial^\mu\varphi^* + m^2\varphi^* \\[4mm] \partial_\mu\left(\dfrac{\partial \mathcal{L}}{\partial(\partial_\mu \varphi^*)}\right) - \dfrac{\partial \mathcal{L}}{\partial \varphi^*} = \partial_\mu(\partial^\mu\varphi) + m^2\varphi = \partial_\mu\partial^\mu\varphi + m^2\varphi \end{cases} \tag{7.24}$$

7.5   The Klein–Gordon Density–Current Revisited

We consider the complex Klein–Gordon field and a pure global gauge (internal phase) transformation, as in Section 6.7. In this case, both the field and its complex conjugate will transform as (see Eq. (6.63)):

$$\begin{cases} \varphi(x) \longrightarrow \varphi'(x) = e^{-i\alpha}\varphi(x) \\ \varphi^*(x) \longrightarrow \varphi'^*(x) = e^{+i\alpha}\varphi^*(x) \end{cases} \tag{7.25}$$

In the variation approach we considered that the Lagrangian depends on the two fields independently and hence the total variation of the action is the sum of changes due to each field. Now we have:

$$\frac{\partial \mathcal{L}_{KG}}{\partial(\partial_\mu \varphi)} = \partial^\mu \varphi^* \quad \text{and} \quad \frac{\partial \mathcal{L}_{KG}}{\partial(\partial_\mu \varphi^*)} = \partial^\mu \varphi \tag{7.26}$$

Consequently, the four-current is equal to (see Eq. (6.66)):

$$j^\mu \equiv -i\left(\varphi \frac{\partial \mathcal{L}_{KG}}{\partial(\partial_\mu \varphi)} - \varphi^* \frac{\partial \mathcal{L}_{KG}}{\partial(\partial_\mu \varphi^*)} \right) = i(\varphi^* \partial^\mu \varphi - \varphi \partial^\mu \varphi^*) = j^\mu_{KG} \tag{7.27}$$

As expected, this result coincides with the Klein–Gordon current, Eq. (7.13), however, its derivation is a *consequence of the invariance under the global gauge (internal phase) transformation* of the field. This is a fundamental result which forms the basis of gauge-invariant quantum field theories, as will be discussed later on. We note that a similar argument does not hold for the real scalar field, for which the defined four-vector current would actually vanish.

7.6   Physical Interpretation of the Klein–Gordon Solutions

Can the Klein–Gordon equation be used as a replacement for the Schrödinger equation in the relativistic case and the complex field φ be considered as a single-particle wave function? We consider the free-particle case and attempt to find solutions of the Klein–Gordon equation that describe this state. For a given momentum, the solution is a **plane wave** (which represents a scalar field):

$$\varphi(x^\mu) = N e^{-ipx} = N e^{-ip^\mu x_\mu} = N e^{i\vec{p}\cdot\vec{x} - iEt} \tag{7.28}$$

where $E^2 = \vec{p}^2 + m^2$. Two energies correspond to a given momentum: the **positive** and the **negative** energy. In order to obtain a *complete set*, we must include both types of solutions:

$$\begin{cases} \varphi_+(x) = N e^{i\vec{p}\cdot\vec{x} - iEt} \quad \text{with} \quad E = +\sqrt{p^2 + m^2} > 0 \\ \varphi_-(x) = N e^{i\vec{p}\cdot\vec{x} - iEt} = N e^{i\vec{p}\cdot\vec{x} + i|E|t} \quad \text{with} \quad E = -\sqrt{p^2 + m^2} < 0 \end{cases} \tag{7.29}$$

This situation was expected. In special relativity the energy and the momentum must be considered in a symmetric way (just as the time and space). Since the momentum coordinates extend from $-\infty$ to $+\infty$, so should the energy (the time component of the energy–momentum four-vector). We interpret this as the necessary symmetry between energy and momentum.

We now compute the four-vector current density for the free-particle solution:

$$\begin{aligned} j^\mu_{KG} &= i\left(\varphi^* \partial^\mu \varphi - \varphi \partial^\mu \varphi^* \right) = i|N|^2 \left(e^{+ipx} \partial^\mu e^{-ipx} - e^{-ipx} \partial^\mu e^{+ipx} \right) \\ &= i|N|^2 \left(e^{+ipx}(-ip^\mu)e^{-ipx} - e^{-ipx}(+ip^\mu)e^{+ipx} \right) \\ &= 2i|N|^2(-ip^\mu) = 2|N|^2 p^\mu \end{aligned} \tag{7.30}$$

The Klein–Gordon equation therefore predicts a density $\rho = j^0 = 2|N|^2 E$ and the current $\vec{j} = 2|N|^2 \vec{p}$. In the interpretation of the Schrödinger equation, the quantity $\rho \, d^3 \vec{x}$ represents the probability of finding the particle at the given point in space and time. In the case of the Klein–Gordon equation, **the density is not positive definite!** The negative-energy solutions seem therefore to be problematic. The states with negative energies have a negative density. In addition, how can one define the ground state of the system if energies can be infinitely negative?

Originally developed in 1926, the historical conclusion was to say that the field in the Klein–Gordon equation cannot be interpreted as the wave function of a (relativistic) particle. Because of this, the Klein–Gordon equation was ignored for many years. In 1934 **Pauli** and **Weisskopf** reinstated the Klein–Gordon equation in the context of the second quantization, as discussed in the next section.

7.7   Quantization of the Real Klein–Gordon Field

In quantum mechanics we use operators, however the wave function is still "classical" in the sense that it is a function of the position and time which describes the state of the particle.[3] The system is quantized by the use of the operators with their implied commutation rules. With the derivation of the Klein–Gordon equation we have seen that this prescription is not successful at generalizing the Schrödinger equation to the relativistic case.

In the **quantum field theory (QFT)** the field itself becomes a quantized object (due to the so-called **second quantization**). The field describes a system with an infinite number of continuous degrees of freedom and commutation rules are defined in order to introduce quantization. Hence, in QFT **the field $\phi(x^\mu)$ becomes an operator with commutation rules**, where the variable x^μ plays the role of a continuous "index" for the infinite number of degrees of freedom. QFT can be seen as the framework which allows us to combine special relativity with quantum mechanics. The interpretation of the quantized field is that of a physical system whose excitations describe particles. Particles become the quanta of the field. Such excitations of the quantum field will be described with the use of quantum oscillators (see Appendix C.7).

The QFT formalism is therefore able to develop a "many-particle" theory, where the total number of particles is not a constant, by a mechanism in which particles can be created and annihilated![4] We will see that different types of particles, such as bosons and fermions, will be described, as well as particles and their antiparticles, as is required for a theory consistent with special relativity (where positive and negative energies are mandatory).

To begin with, we consider the real (scalar) field $\phi = \phi(x^\mu)$ and the Klein–Gordon Lagrangian:

$$\mathcal{L}_{KG} = \frac{1}{2} \left(\partial_\mu \phi \right) \left(\partial^\mu \phi \right) - \frac{1}{2} m^2 \phi^2 = \frac{1}{2} \left(\partial_0 \phi \right)^2 - \frac{1}{2} \left(\nabla \phi \right)^2 - \frac{1}{2} m^2 \phi^2 \tag{7.31}$$

The field does not represent the wave function of a particle (in the sense of quantum mechanics) and we should define the canonical quantities of the theory to describe the system. The canonical momentum density of the field is given by (see Eq. (6.42)):

$$\Pi \equiv \frac{\partial \mathcal{L}_{KG}}{\partial (\partial_0 \phi)} = \frac{1}{2} 2 (\partial_0 \phi) = \partial_0 \phi \tag{7.32}$$

and the canonical Hamiltonian (energy) density is (see Eq. (6.43)):

$$\begin{aligned} \mathcal{H} &\equiv \Pi (\partial_0 \phi) - \mathcal{L}_{KG} = (\partial_0 \phi)^2 - \frac{1}{2} (\partial_0 \phi)^2 + \frac{1}{2} (\nabla \phi)^2 + \frac{1}{2} m^2 \phi^2 \\ &= \underbrace{\frac{1}{2} (\partial_0 \phi)^2}_{\geq 0} + \underbrace{\frac{1}{2} (\nabla \phi)^2}_{\geq 0} + \underbrace{\frac{1}{2} m^2 \phi^2}_{\geq 0} \geq 0 \end{aligned} \tag{7.33}$$

---

3   The field is not strictly classical since it is assumed to take real or complex values, having in mind its application to quantum systems.

4   Of course, creation and annihilation will follow some rules, e.g., single fermions cannot simply pop out of the vacuum.

Note that in this way the energy density of the field is always positive!

The real field is expanded as a series of **plane waves**, which correspond to an infinite number of harmonic oscillators continuously indexed with the momentum $\vec{p}$:

$$\phi(x^\mu) \equiv \int \frac{\mathrm{d}^3\vec{p}}{(2\pi)^{3/2}} \frac{1}{\sqrt{2E_p}} \left(a(\vec{p})e^{-ip\cdot x} + a^\dagger(\vec{p})e^{+ip\cdot x}\right) \tag{7.34}$$

where the integral is over all oscillator modes described by the three-vector index $\vec{p}$. Here $p^\mu = (p^0, \vec{p})$ with $p^0 = E_p = +\sqrt{\vec{p}^2 + m^2}$, and the second term $1/\sqrt{2E_p}$ is a normalization factor. The $a(\vec{p})$ and $a^\dagger(\vec{p})$ are the equivalents of the Fourier coefficients. They are **operators**, which are defined by commutation rules. The corresponding equation for the canonical momentum density Π is:

$$\Pi(x^\mu) = \partial_0\phi(x^\mu) = i\int \frac{\mathrm{d}^3\vec{p}}{(2\pi)^{3/2}} \sqrt{\frac{E_p}{2}} \left(-a(\vec{p})e^{-ip\cdot x} + a^\dagger(\vec{p})e^{+ip\cdot x}\right) \tag{7.35}$$

Both $\phi(x^\mu)$ and $\Pi(x^\mu)$ are understood as time-dependent operators in the Heisenberg representation (see Appendix C.6). The prescription to quantize the fields is to introduce the so-called **Equal Time Commutation Rules (ETCR)** for them:

$$\begin{cases} [\phi(\vec{x}, t), \Pi(\vec{x}\,', t)] \equiv i\delta^3(\vec{x} - \vec{x}\,') \\ [\phi(\vec{x}, t), \phi(\vec{x}\,', t)] = [\Pi(\vec{x}, t), \Pi(\vec{x}\,', t)] = 0 \end{cases} \tag{7.36}$$

We can invert the Fourier expansion, Eq. (7.34), to find the commutation rules for $a(\vec{p})$ and $a^\dagger(\vec{p})$. We first consider the field $\phi(x)$, multiply it by a complex exponential and integrate over space at time t:

$$\begin{aligned} \int \mathrm{d}^3\vec{x}\,\phi(x)e^{-ikx} &= \int \frac{\mathrm{d}^3\vec{p}}{(2\pi)^{3/2}} \frac{1}{\sqrt{2E_p}} \left(a(\vec{p})\int \mathrm{d}^3\vec{x}\,e^{-i(p+k)\cdot x} + a^\dagger(\vec{p})\int \mathrm{d}^3\vec{x}\,e^{+i(p-k)\cdot x}\right) \\ &= (2\pi)^{3/2}\int \frac{\mathrm{d}^3\vec{p}}{\sqrt{2E_p}} \left(a(\vec{p})\int \frac{\mathrm{d}^3\vec{x}}{(2\pi)^3}e^{+i(\vec{p}+\vec{k})\cdot\vec{x}}e^{-i(E_p+E_k)t}\right. \\ &\qquad\qquad\qquad\qquad \left. + a^\dagger(\vec{p})\int \frac{\mathrm{d}^3\vec{x}}{(2\pi)^3}e^{-i(\vec{p}-\vec{k})\cdot\vec{x}}e^{+i(E_p-E_k)t}\right) \\ &= (2\pi)^{3/2}\int \frac{\mathrm{d}^3\vec{p}}{\sqrt{2E_p}} \left(a(\vec{p})\delta^3(\vec{p}+\vec{k})e^{-i(E_p+E_k)t} + a^\dagger(\vec{p})\delta^3(\vec{p}-\vec{k})e^{+i(E_p-E_k)t}\right) \\ &= \frac{(2\pi)^{3/2}}{\sqrt{2E_k}} \left(a(-\vec{k})e^{-2iE_k t} + a^\dagger(\vec{k})\right) \end{aligned} \tag{7.37}$$

Similarly for $\Pi(x^\mu)$:

$$\int \mathrm{d}^3\vec{x}\,\Pi(x)e^{-ikx} = i(2\pi)^{3/2}\sqrt{\frac{E_k}{2}} \left(-a(-\vec{k})e^{-2iE_k t} + a^\dagger(\vec{k})\right) \tag{7.38}$$

Putting these together we get:

$$\frac{1}{(2\pi)^{3/2}}\int \mathrm{d}^3\vec{x}\,(E_k\phi(x) - i\Pi(x))\,e^{-ik\cdot x} = \sqrt{\frac{E_k}{2}}\left(2a^\dagger(\vec{k})\right) = \sqrt{2E_k}\,a^\dagger(\vec{k}) \tag{7.39}$$

and its complex conjugate:

$$\frac{1}{(2\pi)^{3/2}}\int \mathrm{d}^3\vec{x}\,(E_k\phi(x) + i\Pi(x))\,e^{+ik\cdot x} = \sqrt{2E_k}\,a(\vec{k}) \tag{7.40}$$

Then, still assuming the same time t for both x^μ and y^μ:

$$
\begin{aligned}
(2\pi)^3 \sqrt{2E_k 2E_p}\, [a(\vec{k}), a^\dagger(\vec{p})] &= \int \mathrm{d}^3\vec{x}\, \mathrm{d}^3\vec{y}\, [E_k \phi(x) + i\Pi(x), E_p \phi(y) - i\Pi(y)]\, e^{+i(k\cdot x - p\cdot y)} \\
&= \int \mathrm{d}^3\vec{x}\, \mathrm{d}^3\vec{y}\, (-iE_k\, [\phi(x), \Pi(y)] + iE_p\, [\Pi(x), \phi(y)])\, e^{+i(k\cdot x - p\cdot y)} \\
&= \int \mathrm{d}^3\vec{x}\, \mathrm{d}^3\vec{y}\, \left(E_k \delta^3(\vec{x} - \vec{y}) + E_p \delta^3(\vec{x} - \vec{y})\right) e^{+i(k\cdot x - p\cdot y)} \\
&= \int \mathrm{d}^3\vec{x}\, (E_k + E_p)\, e^{+i(k-p)\cdot x} \\
&= (2\pi)^3 2E_k\, \delta^3(\vec{k} - \vec{p})
\end{aligned}
\tag{7.41}
$$

where in the last line the δ function allows us to set $E_k = E_p$. The commutator of $a(\vec{k})$ and $a^\dagger(\vec{p})$ then becomes:

$$
[a(\vec{k}), a^\dagger(\vec{p})] = \frac{2E_k}{\sqrt{2E_k 2E_p}}\, \delta^3(\vec{k} - \vec{p}) = \delta^3(\vec{k} - \vec{p})
\tag{7.42}
$$

The other commutators can be shown to vanish. Consequently, we **set the commutators**[5] of the $a(\vec{p})$ and $a^\dagger(\vec{p})$ operators to become the creation and annihilation operators of the harmonic oscillators indexed by $\vec{p}$ and $\vec{p}'$ of the field:

$$
\begin{cases}
[a(\vec{p}), a^\dagger(\vec{p}')] \equiv \delta^3(\vec{p} - \vec{p}') \\
[a(\vec{p}), a(\vec{p}')] = [a^\dagger(\vec{p}), a^\dagger(\vec{p}')] \equiv 0
\end{cases}
\tag{7.43}
$$

In analogy to simple harmonic oscillation (see Appendix C.7), we introduce the **particle number operators**:

$$
N(\vec{p}) \equiv a^\dagger(\vec{p}) a(\vec{p})
\tag{7.44}
$$

where $\vec{p}$ is a continuous index. The operator with the continuous variable "$\vec{p}$" actually represents a particle number density, with a number of particles in a momentum volume $\mathrm{d}^3\vec{p}$. There is an infinite number of independent oscillators, each with its own state. The field represents the cumulative effect of all the independent oscillators.

The **ground state of the field** (which corresponds to **vacuum**) is defined as the state where all oscillators are in their ground state:

$$
\text{Vacuum:} \quad N(\vec{p}) \,|0\rangle = a^\dagger(\vec{p}) a(\vec{p}) \,|0\rangle = 0 \quad \text{for all } \vec{p}
\tag{7.45}
$$

To excite a particular oscillation, we use the corresponding ladder operator of the oscillation, which will correspond to the creation of a particle of momentum $\vec{p}$. For example:

$$
\begin{aligned}
a^\dagger(\vec{p}) \,|0\rangle &\iff |1_{\vec{p}}\rangle \\
a^\dagger(\vec{p}) a^\dagger(\vec{q}) \,|0\rangle &\iff |1_{\vec{p}} 1_{\vec{q}}\rangle \\
a^\dagger(\vec{p}) a^\dagger(\vec{q}) a^\dagger(\vec{q}) \,|0\rangle &\iff |1_{\vec{p}} 2_{\vec{q}}\rangle
\end{aligned}
\tag{7.46}
$$

The $|1_{\vec{p}}\rangle$ state has the oscillator $\vec{p}$ excited in its first state, and so forth. In this way, the full **Hilbert**[6] **space** is spanned by acting on the vacuum with all possible combinations of creator operators. The sum of the n-particle Hilbert spaces for all $n \geq 0$ is called the **Fock**[7] **space**. We also note that because all the creation operators commute among themselves, **a state is symmetric under exchange of two particles**, for example:

$$
a^\dagger(\vec{p}) a^\dagger(\vec{q}) \,|0\rangle = a^\dagger(\vec{q}) a^\dagger(\vec{p}) \,|0\rangle
\tag{7.47}
$$

[5]   Several conventions for Eqs. (7.36) and (7.43) exist in the literature.
[6]   David Hilbert (1862–1943), German mathematician.
[7]   Vladimir Aleksandrovich Fock (1898–1974), Soviet physicist.

7.8   Energy–Momentum Spectrum of the Real Field

The canonical energy and momentum of the field were defined in Section 6.6:

$$H = \int \mathrm{d}^3\vec{x}\,\mathcal{H} = \int \mathrm{d}^3\vec{x}\,T_0^0; \qquad P_i = \int \mathrm{d}^3\vec{x}\,T_i^0 = \int \mathrm{d}^3\vec{x}\,\Pi(\partial_i\phi) \tag{7.48}$$

At the time $t = 0$, the field can be written as:

$$\phi(\vec{x}, t = 0) = \int \frac{\mathrm{d}^3\vec{p}}{(2\pi)^{3/2}} \frac{1}{\sqrt{2E_p}} \left(a(\vec{p}) + a^\dagger(-\vec{p})\right) e^{i\vec{p}\cdot\vec{x}} \tag{7.49}$$

where we used the substitution $\vec{p} \to -\vec{p}$ for the second part. The momentum density at $t = 0$ can also be expressed as (see Eq. (7.35)):

$$\Pi(\vec{x}, t = 0) = i \int \frac{\mathrm{d}^3\vec{p}}{(2\pi)^{3/2}} \sqrt{\frac{E_p}{2}} \left(-a(\vec{p}) + a^\dagger(-\vec{p})\right) e^{i\vec{p}\cdot\vec{x}} = -i \int \frac{\mathrm{d}^3\vec{p}}{(2\pi)^{3/2}} \sqrt{\frac{E_p}{2}} \left(a(\vec{p}) - a^\dagger(-\vec{p})\right) e^{i\vec{p}\cdot\vec{x}} \tag{7.50}$$

The Hamiltonian density of the Klein–Gordon equation gives:

$$\mathcal{H}_{KG} = \frac{1}{2}(\partial_0\phi)^2 + \frac{1}{2}(\nabla\phi)^2 + \frac{1}{2}m^2\phi^2 = \frac{1}{2}(\Pi)^2 + \frac{1}{2}(\nabla\phi)^2 + \frac{1}{2}m^2\phi^2 \tag{7.51}$$

so we need to compute the quantities Π^2, $(\nabla\phi)^2$ and ϕ^2:

$$(\Pi)^2 = (-i)^2 \int\int \frac{\mathrm{d}^3\vec{p}}{(2\pi)^{3/2}} \frac{\mathrm{d}^3\vec{p}'}{(2\pi)^{3/2}} \frac{1}{2}\sqrt{E_p E_{p'}} e^{i(\vec{p}+\vec{p}')\cdot\vec{x}} \times \left(a(\vec{p}) - a^\dagger(-\vec{p})\right)\left(a(\vec{p}') - a^\dagger(-\vec{p}')\right) \tag{7.52}$$

and since the gradient yields $\vec{\nabla}\phi = (i\vec{p})\phi$ we have:

$$(\nabla\phi)^2 = \int\int \frac{\mathrm{d}^3\vec{p}}{(2\pi)^{3/2}} \frac{\mathrm{d}^3\vec{p}'}{(2\pi)^{3/2}} (i)^2 \frac{(\vec{p}\cdot\vec{p}')}{2\sqrt{E_p E_{p'}}} e^{i(\vec{p}+\vec{p}')\cdot\vec{x}} \times \left(a(\vec{p}) + a^\dagger(-\vec{p})\right)\left(a(\vec{p}') + a^\dagger(-\vec{p}')\right) \tag{7.53}$$

And finally:

$$m^2(\phi)^2 = m^2 \int\int \frac{\mathrm{d}^3\vec{p}}{(2\pi)^{3/2}} \frac{\mathrm{d}^3\vec{p}'}{(2\pi)^{3/2}} \frac{1}{2\sqrt{E_p E_{p'}}} e^{i(\vec{p}+\vec{p}')\cdot\vec{x}} \times \left(a(\vec{p}) + a^\dagger(-\vec{p})\right)\left(a(\vec{p}') + a^\dagger(-\vec{p}')\right) \tag{7.54}$$

We then find the expression for the Hamiltonian density by regrouping the terms:

$$H = \int \mathrm{d}^3\vec{x}\,\mathcal{H}_{KG} = \int \mathrm{d}^3\vec{x} \int\int \frac{\mathrm{d}^3\vec{p}\,\mathrm{d}^3\vec{p}'}{(2\pi)^3} \frac{1}{2} e^{i(\vec{p}+\vec{p}')\cdot\vec{x}} [\mathcal{A}(\vec{p}, \vec{p}') + \mathcal{B}(\vec{p}, \vec{p}')] \tag{7.55}$$

where

$$\mathcal{A} = -\frac{1}{2}\sqrt{E_p E_{p'}} \left(a(\vec{p}) - a^\dagger(-\vec{p})\right)\left(a(\vec{p}') - a^\dagger(-\vec{p}')\right) \tag{7.56}$$

and

$$\mathcal{B} = \frac{-(\vec{p}\cdot\vec{p}') + m^2}{2\sqrt{E_p E_{p'}}} \left(a(\vec{p}) + a^\dagger(-\vec{p})\right)\left(a(\vec{p}') + a^\dagger(-\vec{p}')\right) \tag{7.57}$$

The space integration can be replaced by a three-dimensional Dirac δ function (see Eq. (A.41)) on the momenta:

$$H = \frac{1}{2} \int\int \frac{\mathrm{d}^3\vec{p}\,\mathrm{d}^3\vec{p}'}{(2\pi)^3} (2\pi)^3 \delta^3(\vec{p} + \vec{p}') [\mathcal{A}(\vec{p}, \vec{p}') + \mathcal{B}(\vec{p}, \vec{p}')] \tag{7.58}$$

and one of the momentum integrations can be removed by imposing the condition $\vec{p}' = -\vec{p}$:

$$H = \frac{1}{2} \int \mathrm{d}^3\vec{p}\, [\mathcal{A}(\vec{p}, -\vec{p}) + \mathcal{B}(\vec{p}, -\vec{p})] \tag{7.59}$$

where

$$\mathcal{A} = -\frac{1}{2} E_p \left(a(\vec{p}) - a^\dagger(-\vec{p}) \right) \left(a(-\vec{p}) - a^\dagger(\vec{p}) \right) \tag{7.60}$$

and with $E_p^2 = \vec{p}^2 + m^2$, we have:

$$\mathcal{B} = \frac{\vec{p}^2 + m^2}{2E_p} \left(a(\vec{p}) + a^\dagger(-\vec{p}) \right) \left(a(-\vec{p}) + a^\dagger(\vec{p}) \right) = \frac{E_p}{2} \left(a(\vec{p}) + a^\dagger(-\vec{p}) \right) \left(a(-\vec{p}) + a^\dagger(\vec{p}) \right) \tag{7.61}$$

We consider:

$$
\begin{aligned}
\mathcal{A} + \mathcal{B} &= -\frac{E_p}{2} \left[\left(a(\vec{p}) - a^\dagger(-\vec{p}) \right) \left(a(-\vec{p}) - a^\dagger(\vec{p}) \right) - \left(a(\vec{p}) + a^\dagger(-\vec{p}) \right) \left(a(-\vec{p}) + a^\dagger(\vec{p}) \right) \right] \\
&= \frac{E_p}{2} \left[-a(\vec{p})a(-\vec{p}) + a(\vec{p})a^\dagger(\vec{p}) + a^\dagger(-\vec{p})a(-\vec{p}) - a^\dagger(-\vec{p})a^\dagger(\vec{p}) + a(\vec{p})a(-\vec{p}) + a(\vec{p})a^\dagger(\vec{p}) + \right. \\
&\qquad \left. a^\dagger(-\vec{p})a(-\vec{p}) + a^\dagger(-\vec{p})a^\dagger(\vec{p}) \right] \\
&= E_p \left[a(\vec{p})a^\dagger(\vec{p}) + a^\dagger(-\vec{p})a(-\vec{p}) \right]
\end{aligned}
\tag{7.62}
$$

Consequently:

$$H = \frac{1}{2} \int \mathrm{d}^3\vec{p}\, E_p \left[a(\vec{p})a^\dagger(\vec{p}) + a^\dagger(-\vec{p})a(-\vec{p}) \right] = \frac{1}{2} \int \mathrm{d}^3\vec{p}\, E_p \left[a(\vec{p})a^\dagger(\vec{p}) + a^\dagger(\vec{p})a(\vec{p}) \right] \tag{7.63}$$

since $E_p = +\sqrt{\vec{p}^2 + m^2}$. Recalling the commutation rules, Eq. (7.43):

$$a(\vec{p})a^\dagger(\vec{p}') = a^\dagger(\vec{p})a(\vec{p}') + \delta^3(\vec{p} - \vec{p}') \tag{7.64}$$

we find the Hamiltonian is equal to:

$$H = \frac{1}{2} \int \mathrm{d}^3\vec{p}\, E_p \left[2a^\dagger(\vec{p})a(\vec{p}) + \delta^3(\vec{0}) \right] \tag{7.65}$$

which we can then write with the particle number operator as:

$$H = \frac{1}{2} \int \mathrm{d}^3\vec{p}\, E_p \left\{ 2N(\vec{p}) + C \right\} \tag{7.66}$$

This equation describes the total energy of the field as an infinite sum of quantized and independent harmonic oscillators. One problem lies in the so-called "**zero-point energy**" of the field, which is coming from the second term labeled "C." This leads to an apparently divergent energy, which could be expected since we summed over an infinite number of quantum oscillators and the ground state of each of them has non-zero energy. However, we notice that one can never measure an absolute energy. The zero-point term corresponds to an overall constant, and only differences of energy have a physical meaning.[8] Hence, we can redefine the zero point such that the energy of the field is equal to:

$$H = \int \mathrm{d}^3\vec{p}\, E_p\, N(\vec{p}) \tag{7.67}$$

---

8   The Casimir effect is interpreted as a physical appearance of the zero-point energy, which has been observed in the case of the electromagnetic field.

In a similar way one can derive the total momentum of the field to find:

$$\vec{P} = \frac{1}{2}\int \mathrm{d}^3 \vec{p}\,\vec{p}\,\{2N(\vec{p}) + C\} \longrightarrow \vec{P} = \int \mathrm{d}^3\vec{p}\,\{N(\vec{p})\vec{p}\} \tag{7.68}$$

Let us now summarize the results which were obtained thanks to the second quantization of the field.

• **Physical interpretation.** The vacuum state is defined as:

$$N(\vec{p})\,|0\rangle = a^\dagger(\vec{p})a(\vec{p})\,|0\rangle = 0 \quad \text{for all } \vec{p} \tag{7.69}$$

In this case, the total energy and momentum vanish:

$$H\,|0\rangle = \int \mathrm{d}^3\vec{p}\,N(\vec{p})E_p\,|0\rangle = 0 \quad \text{and} \quad \vec{P} = 0 \tag{7.70}$$

We can consider excited states of the field by recalling the effect of the creation operator (see Eq. (7.46)). For example:

$$\begin{aligned}
|1_{\vec{p}}\rangle &= a^\dagger(\vec{p})\,|0\rangle: & E &= E_p = +\sqrt{\vec{p}^2 + m^2}; & \vec{P} &= \vec{p} \\
|1_{\vec{p}}1_{\vec{q}}\rangle &= a^\dagger(\vec{p})a^\dagger(\vec{q})\,|0\rangle: & E &= E_p + E_q; & \vec{P} &= \vec{p} + \vec{q}
\end{aligned} \tag{7.71}$$

It is very natural to associate these states with the corresponding particles with momenta $\vec{p}$ and $\vec{q}$.

• **Negative energies.** When we tried to interpret the Klein–Gordon field as the wave function of a particle, we encountered the so-called negative-energy problem which led to difficulties in its interpretation. In QFT, the field is written as:

$$\phi(x^\mu) \propto \underbrace{a(\vec{p})e^{-ipx}}_{\text{positive frequency}} + \underbrace{a^\dagger(\vec{p})e^{+ipx}}_{\text{negative frequency}} \tag{7.72}$$

where the operator $a(\vec{p})$ **annihilates a particle with positive energy** E_p and the **operator** $a^\dagger(\vec{p})$ **creates a particle with positive energy** E_p. Therefore, such an interpretation solves the issue of the negative energies.

• **On the statistics of the field.** As a direct consequence of the commutation rules of the creation and annihilation operators, we note that the state with two identical particles is symmetric under the exchange of the two particles:

$$a^\dagger(\vec{p})a^\dagger(\vec{q})\,|0\rangle = a^\dagger(\vec{q})a^\dagger(\vec{p})\,|0\rangle \tag{7.73}$$

since $\left[a^\dagger(\vec{p}), a^\dagger(\vec{p}')\right] \equiv 0$. In addition, by successive application of the creation operator, one can obtain a state with several particles in the same state, for example:

$$|3_{\vec{p}}\rangle \propto a^\dagger(\vec{p})a^\dagger(\vec{p})a^\dagger(\vec{p})\,|0\rangle \tag{7.74}$$

Hence, we conclude that the particles associated with the Klein–Gordon field are (scalar) **bosons**. They follow the **Bose–Einstein statistics**.

These results were obtained assuming a real scalar field. More generally, however, the field could be complex, and this will be considered in the next section.

7.9   Quantization of the Complex Klein–Gordon Field

We study the properties of a field $\varphi(x^\mu)$ which is complex. We can in this case separate the field into its real and imaginary parts:

$$\varphi = \frac{1}{\sqrt{2}}(\phi_1 + i\phi_2) \quad \text{and} \quad \varphi^\dagger = \frac{1}{\sqrt{2}}(\phi_1 - i\phi_2) \tag{7.75}$$

where ϕ_1 and ϕ_2 are real fields. Each of these fields must satisfy the Klein–Gordon equation:

$$\left(\partial^\mu \partial_\mu + m^2\right)\phi_1 = 0 \quad \text{and} \quad \left(\partial^\mu \partial_\mu + m^2\right)\phi_2 = 0 \tag{7.76}$$

Each of these real fields can be expanded as previously done with the creation and annihilation operators $a_i(\vec{p})$ $(i = 1, 2)$:

$$\begin{cases} \phi_1(x) \equiv \displaystyle\int \frac{\mathrm{d}^3\vec{p}}{(2\pi)^{3/2}} \frac{1}{\sqrt{2E_p}} \left(a_1(\vec{p})e^{-ip\cdot x} + a_1^\dagger(\vec{p})e^{+ip\cdot x}\right) & (7.77) \\[3mm] \phi_2(x) \equiv \displaystyle\int \frac{\mathrm{d}^3\vec{p}}{(2\pi)^{3/2}} \frac{1}{\sqrt{2E_p}} \left(a_2(\vec{p})e^{-ip\cdot x} + a_2^\dagger(\vec{p})e^{+ip\cdot x}\right) & (7.78) \end{cases}$$

As in the case of the real field, the four-vector $p^\mu = (p^0, \vec{p})$ is computed with $p^0 = E_p = +\sqrt{\vec{p}^2 + m^2}$, and the term $1/\sqrt{2E_p}$ is a normalization factor. We introduce the new operators $a(\vec{p})$ and $b(\vec{p})$ as follows:

$$\begin{cases} a(\vec{p}) = \dfrac{1}{\sqrt{2}}\left(a_1(\vec{p}) + ia_2(\vec{p})\right) \\[3mm] b^\dagger(\vec{p}) = \dfrac{1}{\sqrt{2}}\left(a_1^\dagger(\vec{p}) + ia_2^\dagger(\vec{p})\right) \end{cases} \tag{7.79}$$

and rewrite the field φ and its conjugate $\varphi^\dagger$ accordingly:

$$\begin{cases} \varphi(x) \equiv \displaystyle\int \frac{\mathrm{d}^3\vec{p}}{(2\pi)^{3/2}} \frac{1}{\sqrt{2E_p}} \left(a(\vec{p})e^{-ip\cdot x} + b^\dagger(\vec{p})e^{+ip\cdot x}\right) \\[3mm] \varphi^\dagger(x) \equiv \displaystyle\int \frac{\mathrm{d}^3\vec{p}}{(2\pi)^{3/2}} \frac{1}{\sqrt{2E_p}} \left(a^\dagger(\vec{p})e^{+ip\cdot x} + b(\vec{p})e^{-ip\cdot x}\right) \end{cases} \tag{7.80}$$

We note that the field φ is associated with the annihilation operator a and the creation operator $b^\dagger$. For the conjugate field we have the opposite situation. Hence, the field operators contain the joint creation-annihilation process. How shall we interpret this? In the complex field, we have two "species" of particles, each one associated with a particular set of creation/annihilation operators. We note that the new operators obey the following commutation rules:

$$\begin{aligned} [a(\vec{p}), a^\dagger(\vec{p}')] &= \frac{1}{2}[(a_1 + ia_2)(a_1^\dagger - ia_2^\dagger) - (a_1^\dagger - ia_2^\dagger)(a_1 + ia_2)] \\ &= \frac{1}{2}[a_1a_1^\dagger - ia_1a_2^\dagger + ia_2a_1^\dagger + a_2a_2^\dagger - (a_1^\dagger a_1 + ia_1^\dagger a_2 - ia_2^\dagger a_1 + a_2^\dagger a_2)] \\ &= \frac{1}{2}\left([a_1, a_1^\dagger] - i\underbrace{[a_1, a_2^\dagger]}_{=0} + i\underbrace{[a_2, a_1^\dagger]}_{=0} + [a_2, a_2^\dagger]\right) \\ &= \delta^3(\vec{p} - \vec{p}') \end{aligned} \tag{7.81}$$

and analogously $[b(\vec{p}), b^\dagger(\vec{p}')] = \delta^3(\vec{p} - \vec{p}')$. All other commutators vanish, e.g.:

$$\begin{aligned} [a(\vec{p}), b^\dagger(\vec{p}')] &= \frac{1}{2}[(a_1 + ia_2)(a_1^\dagger + ia_2^\dagger) - (a_1^\dagger + ia_2^\dagger)(a_1 + ia_2)] \\ &= \frac{1}{2}[a_1a_1^\dagger + ia_1a_2^\dagger + ia_2a_1^\dagger - a_2a_2^\dagger - (a_1^\dagger a_1 + ia_1^\dagger a_2 + ia_2^\dagger a_1 - a_2^\dagger a_2)] \\ &= \frac{1}{2}\left([a_1, a_1^\dagger] + i\underbrace{[a_1, a_2^\dagger]}_{=0} + i\underbrace{[a_2, a_1^\dagger]}_{=0} - [a_2, a_2^\dagger]\right) \\ &= 0 \end{aligned} \tag{7.82}$$

Thus it follows that (annihilation operator at $\vec{p}$, creator operator at $\vec{p}\,'$):

$$\left[a(\vec{p}), a^\dagger(\vec{p}\,')\right] = \left[b(\vec{p}), b^\dagger(\vec{p}\,')\right] = \delta^3(\vec{p} - \vec{p}\,')$$
$$\text{all others } = 0 \tag{7.83}$$

hence they possess an algebra similar to that of the a_i operators. We can define correspondingly two types of **particle number operators** $N^+(\vec{p})$ and $N^-(\vec{p})$:

$$N^+(\vec{p}) \equiv a^\dagger(\vec{p})a(\vec{p}) \quad \text{and} \quad N^-(\vec{p}) \equiv b^\dagger(\vec{p})b(\vec{p}) \tag{7.84}$$

The quantized complex fields $\varphi(x^\mu)$ and $\varphi^\dagger(x^\mu)$ obey the following Klein–Gordon Lagrangian:

$$\mathcal{L}_{KG} = \left(\partial_\mu \varphi^\dagger\right)\left(\partial^\mu \varphi\right) - m^2 \varphi^\dagger \varphi = \left(\partial_0 \varphi^\dagger\right)\left(\partial_0 \varphi\right) - \left(\nabla \varphi^\dagger\right)\left(\nabla \varphi\right) - m^2 \varphi^\dagger \varphi \tag{7.85}$$

Each of the two fields φ and $\varphi^\dagger$ has its own canonical momentum defined as $\Pi = \frac{\partial \mathcal{L}_{KG}}{\partial(\partial_0 \varphi)}$ and $\frac{\partial \mathcal{L}_{KG}}{\partial(\partial_0 \varphi^\dagger)}$, which are conjugates of one another (cf. Eq. (7.32)):

$$\Pi \equiv \frac{\partial \mathcal{L}_{KG}}{\partial(\partial_0 \varphi)} = \partial_0 \varphi^\dagger \quad \text{and} \quad \frac{\partial \mathcal{L}_{KG}}{\partial(\partial_0 \varphi^\dagger)} = \partial_0 \varphi = \Pi^\dagger \tag{7.86}$$

The Hamiltonian density is given by (cf. Eq. (6.41)):

$$\begin{aligned}
\mathcal{H}_{KG} &= \frac{\partial \mathcal{L}_{KG}}{\partial(\partial_0 \varphi)}(\partial_0 \varphi) + \frac{\partial \mathcal{L}_{KG}}{\partial(\partial_0 \varphi^\dagger)}(\partial_0 \varphi^\dagger) - \mathcal{L}_{KG} = \Pi\Pi^\dagger + \Pi^\dagger \Pi - \left(\Pi\Pi^\dagger - \left(\nabla \varphi^\dagger\right)\left(\nabla \varphi\right) - m^2 \varphi^\dagger \varphi\right) \\
&= \Pi^\dagger \Pi + \left(\nabla \varphi^\dagger\right)\left(\nabla \varphi\right) + m^2 \varphi^\dagger \varphi
\end{aligned} \tag{7.87}$$

Since as before H is constant in time, we compute H at the "easiest" point in time, $t = 0$. We have:

$$\varphi(\vec{x}, 0) = \int \frac{d^3\vec{p}}{(2\pi)^{\frac{3}{2}}} \frac{1}{\sqrt{2E_p}}(a(\vec{p})e^{i\vec{p}\cdot\vec{x}} + b^\dagger(\vec{p})e^{-i\vec{p}\cdot\vec{x}}) = \int \frac{d^3\vec{p}}{(2\pi)^{\frac{3}{2}}} \frac{1}{\sqrt{2E_p}}[a(\vec{p}) + b^\dagger(-\vec{p})]e^{i\vec{p}\cdot\vec{x}} \tag{7.88}$$

$$\varphi^\dagger(\vec{x}, 0) = \int \frac{d^3\vec{p}}{(2\pi)^{\frac{3}{2}}} \frac{1}{\sqrt{2E_p}}(a^\dagger(\vec{p})e^{-i\vec{p}\cdot\vec{x}} + b(\vec{p})e^{i\vec{p}\cdot\vec{x}}) = \int \frac{d^3\vec{p}}{(2\pi)^{\frac{3}{2}}} \frac{1}{\sqrt{2E_p}}[a^\dagger(\vec{p}) + b(-\vec{p})]e^{-i\vec{p}\cdot\vec{x}} \tag{7.89}$$

where since the integrals run over the entire $\vec{p}$-space, we reorganized the sum and substituted $\vec{p} \leftrightarrow -\vec{p}$ for the b terms. Similarly:

$$\begin{aligned}
\Pi(\vec{x}, 0) &= \int \frac{d^3\vec{p}}{(2\pi)^{\frac{3}{2}}} i\sqrt{\frac{E_p}{2}}(a^\dagger(\vec{p})e^{-i\vec{p}\cdot\vec{x}} - b(\vec{p})e^{i\vec{p}\cdot\vec{x}}) = \int \frac{d^3\vec{p}}{(2\pi)^{\frac{3}{2}}} i\sqrt{\frac{E_p}{2}}[a^\dagger(-\vec{p}) - b(\vec{p})]e^{i\vec{p}\cdot\vec{x}} \\
\Pi^\dagger(\vec{x}, 0) &= \int \frac{d^3\vec{p}}{(2\pi)^{\frac{3}{2}}} (-i)\sqrt{\frac{E_p}{2}}(a(\vec{p})e^{i\vec{p}\cdot\vec{x}} - b^\dagger(\vec{p})e^{-i\vec{p}\cdot\vec{x}}) = \int \frac{d^3\vec{p}}{(2\pi)^{\frac{3}{2}}} (-i)\sqrt{\frac{E_p}{2}}[a(-\vec{p}) - b^\dagger(\vec{p})]e^{-i\vec{p}\cdot\vec{x}}
\end{aligned}$$
$$\tag{7.90}$$

and finally:

$$\begin{aligned}
\vec{\nabla}\varphi(\vec{x}, 0) &= \int \frac{d^3\vec{p}}{(2\pi)^{\frac{3}{2}}} \frac{i\vec{p}}{\sqrt{2E_p}}(a(\vec{p})e^{i\vec{p}\cdot\vec{x}} - b^\dagger(\vec{p})e^{-i\vec{p}\cdot\vec{x}}) = \int \frac{d^3\vec{p}}{(2\pi)^{\frac{3}{2}}} \frac{i\vec{p}}{\sqrt{2E_p}}[a(\vec{p}) + b^\dagger(-\vec{p})]e^{i\vec{p}\cdot\vec{x}} \\
\vec{\nabla}\varphi^\dagger(\vec{x}, 0) &= \int \frac{d^3\vec{p}}{(2\pi)^{\frac{3}{2}}} \frac{-i\vec{p}}{\sqrt{2E_p}}(a^\dagger(\vec{p})e^{-i\vec{p}\cdot\vec{x}} - b(\vec{p})e^{i\vec{p}\cdot\vec{x}}) = \int \frac{d^3\vec{p}}{(2\pi)^{\frac{3}{2}}} \frac{-i\vec{p}}{\sqrt{2E_p}}[a^\dagger(\vec{p}) + b(-\vec{p})]e^{-i\vec{p}\cdot\vec{x}}
\end{aligned}$$
$$\tag{7.91}$$

We must now consider the products of field operators found in the Hamiltonian. This introduces the additional integration of the space coordinate $\vec{x}$:

$$
\int d^3x \, \Pi^\dagger \Pi = \int d^3x \frac{d^3\vec{p}\,d^3\vec{p}'}{(2\pi)^3} \frac{\sqrt{E_p E_{p'}}}{2} e^{-i(\vec{p}-\vec{p}')\cdot\vec{x}}[a(-\vec{p}) - b^\dagger(\vec{p})][a^\dagger(-\vec{p}') - b(\vec{p}')]
$$

$$
\int d^3x \, \vec{\nabla}\phi^\dagger \vec{\nabla}\phi = \int d^3x \frac{d^3\vec{p}\,d^3\vec{p}'}{(2\pi)^3} \frac{\vec{p}\cdot\vec{p}'}{2\sqrt{E_p E_{p'}}} e^{-i(\vec{p}-\vec{p}')\cdot\vec{x}}[a^\dagger(\vec{p}) + b(-\vec{p})][a(\vec{p}') + b^\dagger(-\vec{p}')]
$$

$$
m^2 \int d^3x \, \phi^\dagger \phi = \int d^3x \frac{d^3\vec{p}\,d^3\vec{p}'}{(2\pi)^3} \frac{m^2}{2\sqrt{E_p E_{p'}}} e^{-i(\vec{p}-\vec{p}')\cdot\vec{x}}[a^\dagger(\vec{p}) + b(-\vec{p})][a(\vec{p}') + b^\dagger(-\vec{p}')] \tag{7.92}
$$

Performing the space integration along with the $\exp(-i(\vec{p}-\vec{p}')\cdot\vec{x})$ terms yields a Dirac δ^3 function $(2\pi)^3\delta^3(\vec{p}-\vec{p}')$ (see Eq. (A.41)), which after integrating over $d^3\vec{p}'$ enforces $\vec{p} = \vec{p}'$ and $\sqrt{E_p E_{p'}} = E_p$. Consequently, the Hamiltonian can be expressed with a single integral over $d^3\vec{p}$, regrouping the three pieces:

$$
\begin{aligned}
H &= \frac{1}{2}\int d^3\vec{p}\,\Big\{ E_p[a(-\vec{p})a^\dagger(-\vec{p}) - a(-\vec{p})b(\vec{p}) - b^\dagger(\vec{p})a^\dagger(-\vec{p}) + b^\dagger(\vec{p})b(\vec{p})] \\
&\quad + \frac{\vec{p}^2}{E_p}[a^\dagger(\vec{p})a(\vec{p}) + a^\dagger(\vec{p})b^\dagger(-\vec{p}) + b(-\vec{p})a(\vec{p}) + b(-\vec{p})b^\dagger(-\vec{p})] \\
&\quad + \frac{m^2}{E_p}[a^\dagger(\vec{p})a(\vec{p}) + a^\dagger(\vec{p})b^\dagger(-\vec{p}) + b(-\vec{p})a(\vec{p}) + b(-\vec{p})b^\dagger(-\vec{p})] \Big\} \\
&= \frac{1}{2}\int d^3\vec{p}\,\Big\{ E_p[a(-\vec{p})a^\dagger(-\vec{p}) - a(-\vec{p})b(\vec{p}) - b^\dagger(\vec{p})a^\dagger(-\vec{p}) + b^\dagger(\vec{p})b(\vec{p})] \\
&\quad + \frac{\vec{p}^2 + m^2}{E_p}[a^\dagger(\vec{p})a(\vec{p}) + a^\dagger(\vec{p})b^\dagger(-\vec{p}) + b(-\vec{p})a(\vec{p}) + b(-\vec{p})b^\dagger(-\vec{p})] \Big\} \\
&= \frac{1}{2}\int d^3\vec{p}\, E_p[a(\vec{p})a^\dagger(\vec{p}) + a^\dagger(\vec{p})a(\vec{p}) + b^\dagger(\vec{p})b(\vec{p}) + b(\vec{p})b^\dagger(\vec{p})] \tag{7.93}
\end{aligned}
$$

Since $a(\vec{p})a^\dagger(\vec{p}') = a^\dagger(\vec{p}')a(\vec{p}) + \delta^3(\vec{p}-\vec{p}')$ and $b(\vec{p})b^\dagger(\vec{p}') = b^\dagger(\vec{p}')b(\vec{p}) + \delta^3(\vec{p}-\vec{p}')$, the above expression just becomes, as expected:

$$
H = \frac{1}{2}\int d^3\vec{p}\, E_p[2a^\dagger(\vec{p})a(\vec{p}) + 2b^\dagger(\vec{p})b(\vec{p}) + 2\delta^3(\vec{p}-\vec{p})] \tag{7.94}
$$

After subtraction of the (infinite) zero-point energy as in Eq. (7.67) and inserting the particle number operators N^+ and N^-, one is left with:

$$
H = \int d^3\vec{p}\, E_p[a(\vec{p})a^\dagger(\vec{p}) + b^\dagger(\vec{p})b(\vec{p})] = \int d^3\vec{p}\, E_p(N^+(\vec{p}) + N^-(\vec{p})) \tag{7.95}
$$

One can similarly derive the canonical momentum $\vec{P}$. Finally, the **energy–momentum four-vector operator** becomes:

$$
P_\mu = \int d^3\vec{p}\,(p_\mu)\left[N^+(\vec{p}) + N^-(\vec{p})\right] \tag{7.96}
$$

Hence, **the complex Klein–Gordon (scalar) field describes a system which a priori contains two different types of particles (although of the same rest mass). The number of particles of each type is given by the particle number operators N^+ and N^-.**

Let us consider the four-current density $j^\mu = i(\varphi^*\partial^\mu\varphi - \varphi\partial^\mu\varphi^*)$ (see Eq. (7.13)). We have seen in Section 7.5 that this current can be interpreted as a consequence of the gauge symmetry of the Lagrangian of the *complex* scalar field. The time component corresponds to the total charge:

$$
Q \equiv \int d^3\vec{x}\,j^0 = i\int d^3\vec{x}\,(\varphi^\dagger\partial^0\varphi - \varphi\partial^0\varphi^\dagger) = i\int d^3\vec{x}\,(\varphi^\dagger\Pi^\dagger - \varphi\Pi) \tag{7.97}
$$

We compute the operator Q of the conserved charge at $t = 0$:

$$Q(t=0) = i \int \mathrm{d}^3\vec{x} \int \frac{\mathrm{d}^3\vec{p}\,\mathrm{d}^3\vec{p}'}{(2\pi)^3} e^{-i(\vec{p}+\vec{p}')\cdot\vec{x}} \times \left\{ \frac{-i}{2}\sqrt{\frac{E_{p'}}{E_p}}[a^\dagger(\vec{p}) + b(-\vec{p})][a(-\vec{p}') - b^\dagger(\vec{p}')] \right.$$
$$\left. -\frac{i}{2}\sqrt{\frac{E_{p'}}{E_p}}[a(\vec{p}) + b^\dagger(-\vec{p})][a^\dagger(-\vec{p}') - b(\vec{p}')] \right\} \qquad (7.98)$$

By integration over $\mathrm{d}^3\vec{p}'$ we get, as before, $\vec{p}' = -\vec{p}$ and $E_{p'} = E_p$. Hence:

$$Q = i\left(-\frac{i}{2}\right)\int \mathrm{d}^3\vec{p}\,\{[a^\dagger(\vec{p}) + b(-\vec{p})][a(\vec{p}) - b^\dagger(-\vec{p})] + [a(\vec{p}) + b^\dagger(-\vec{p})][a^\dagger(\vec{p}) - b(-\vec{p})]\} \qquad (7.99)$$

Brute force distribution yields:

$$\begin{aligned}
Q &= \frac{1}{2}\int \mathrm{d}^3\vec{p}\,\{a^\dagger(\vec{p})a(\vec{p}) - a^\dagger(\vec{p})b^\dagger(-\vec{p}) + b(-\vec{p})a(\vec{p}) - b(-\vec{p})b^\dagger(-\vec{p}) \\
&\qquad\qquad a(\vec{p})a^\dagger(\vec{p}) - a(\vec{p})b(-\vec{p}) + b^\dagger(-\vec{p})a^\dagger(\vec{p}) - b^\dagger(-\vec{p})b(-\vec{p})\} \\
&= \frac{1}{2}\int \mathrm{d}^3\vec{p}\,\Big\{a^\dagger(\vec{p})a(\vec{p}) + a(\vec{p})a^\dagger(\vec{p}) - b(-\vec{p})b^\dagger(-\vec{p}) - b^\dagger(-\vec{p})b(-\vec{p}) - \underbrace{[a^\dagger(\vec{p}), b^\dagger(\vec{p})]}_{=0} + \underbrace{[b(-\vec{p}), a(\vec{p})]}_{=0}\Big\}
\end{aligned}$$
$$(7.100)$$

We can swap $-\vec{p} \to \vec{p}$ and with $a(\vec{p})a^\dagger(\vec{p}') = a^\dagger(\vec{p}')a(\vec{p}) + \delta^3(\vec{p} - \vec{p}')$ and $b(\vec{p})b^\dagger(\vec{p}') = b^\dagger(\vec{p}')b(\vec{p}) + \delta^3(\vec{p} - \vec{p}')$, the charge Q is given by:

$$\begin{aligned}
Q &= \frac{1}{2}\int \mathrm{d}^3\vec{p}\,\{a^\dagger(\vec{p})a(\vec{p}) + a^\dagger(\vec{p})a(\vec{p}) + \delta^3(\vec{p}-\vec{p}') - b^\dagger(\vec{p})b(\vec{p}) - b^\dagger(\vec{p})b(\vec{p}) - \delta^3(\vec{p}-\vec{p}')\} \\
&= \frac{1}{2}\int \mathrm{d}^3\vec{p}\,\{2a^\dagger(\vec{p})a(\vec{p}) - 2b^\dagger(\vec{p})b(\vec{p})\} = \int \mathrm{d}^3\vec{p}\,\{a^\dagger(\vec{p})a(\vec{p}) - b^\dagger(\vec{p})b(\vec{p})\} \qquad (7.101)
\end{aligned}$$

So finally:

$$Q = \int \mathrm{d}^3\vec{p}\,[N^+(\vec{p}) - N^-(\vec{p})] \qquad (7.102)$$

Because of the negative sign between the operators N^+ and N^-, **the complex Klein–Gordon field describes two types of particles with the same rest mass but opposite charges**. The $a^\dagger$ operator creates a particle of charge $+1$ and the $b^\dagger$ operator a particle with charge -1, and correspondingly for the annihilation operators a and b. The natural assumption is to say that the complex field describes a set of particle–antiparticle states, with opposite charges!

Problems

Ex 7.1 Transformation of scalar field Lagrangian density. Consider the Lagrangian density

$$\mathcal{L} = \frac{1}{2}\sum_{i=1,2}\left[(\partial_\mu\phi_i)(\partial^\mu\phi_i) - m^2\phi_i^2\right]$$

where ϕ_1, ϕ_2 are real scalar fields.

(a) Consider the field transformation given by:

$$\phi_1 \to \phi_1' = \phi_1 \cos\alpha - \phi_2 \sin\alpha \quad \text{and} \quad \phi_2 \to \phi_2' = \phi_1 \sin\alpha + \phi_2 \cos\alpha \tag{7.103}$$

You may recognize this as a rotation by angle α in the two-dimensional space with ϕ_1 along the horizontal axis and ϕ_2 along the vertical axis.

Show that the Lagrangian density is invariant under this transformation.

(b) Find the infinitesimal form of the transformation given by Eq. (7.103).

(c) Show that

$$j^\mu = \phi_1 \partial^\mu \phi_2 - \phi_2 \partial^\mu \phi_1$$

Ex 7.2 Discrete transformation of the free scalar field. How does a complex scalar field transform under a discrete P, C, and T transformation? (*Hint:* consider what happens on the annihilation–creation operators.) What about CP and CPT? What happens in the case of the real scalar field?

8   Free Fermion Dirac Fields

The equation, developed by Dirac as the union of quantum mechanics and relativity, historically led to the prediction of the existence of a new form of matter – **the antimatter** – previously unsuspected and unobserved and which was experimentally confirmed several years later with the discovery of the positron. The equation also entailed the explanation of spin. Altogether it represented one of the great triumphs of theoretical physics. In the context of quantum field theory, the Dirac equation is reinterpreted to describe quantum fields corresponding to spin-1/2 particles. In the Standard Model all fundamental building blocks of matter – the quarks and leptons – are represented with such Dirac fields.

8.1   Original Derivation of the Dirac Equation

Before the Dirac equation, the Klein–Gordon equation was the only known relativistic extension of the Schrödinger equation. As discussed in the previous chapter, its interpretation as the equation of motion of the wave function of a particle was problematic. Before the reinterpretation of the Klein–Gordon equation in the context of quantum field theory, one searched for alternative solutions to the problem, with an equation of motion of a wave function that is consistent with special relativity. Dirac developed such an equation. The so-called Dirac equation can therefore be interpreted as the equation of motion of a wave function that describes a particle. We will see that already in this context, the equation predicts the **existence of particles and antiparticles**.

In 1927 **Dirac** notes that one limitation of the Klein–Gordon equation is the second time derivative of the wave function, which leads to the negative probabilities. An alternative is to develop an equation which contains the time derivative to first order (as in the Schrödinger equation) to describe the wave function Ψ of a single particle:

$$\text{Ansatz}: \quad i\frac{\partial \Psi}{\partial t} = H_{Dirac}\Psi = E\Psi \tag{8.1}$$

The covariance enforces a symmetry of the equation between space and time. The Dirac Hamilton operator must therefore also contain a first-order derivative with respect to space coordinates:

$$\text{Dirac's postulate}: \quad H_{Dirac} \equiv -i\vec{\alpha}\cdot\nabla + \beta m = \vec{\alpha}\cdot\vec{p} + \beta m \tag{8.2}$$

where the $\vec{\alpha} = (\alpha_1, \alpha_2, \alpha_3)$ and β parameters must be determined. For a free particle, special relativity implies that the total energy of the particle is $E^2 = \vec{p}^2 + m^2$, hence:

$$E^2 = (\vec{\alpha}\cdot\vec{p} + \beta m)^2 = (\alpha_i p_i + \beta m)^2 = (\alpha_i p_i + \beta m)(\alpha_j p_j + \beta m) \tag{8.3}$$

where we sum over i and j. It follows:

$$E^2 = \left(\sum_i \alpha_i^2 p_i^2 + \sum_{i\neq j} \alpha_i \alpha_j p_i p_j + m\sum_i (\alpha_i\beta + \beta\alpha_i)p_i + \beta^2 m^2 \right) \tag{8.4}$$

212

Comparing the above result with:

$$E^2 = \vec{p}^2 + m^2 = \sum_i p_i^2 + m^2 \tag{8.5}$$

we obtain the following requirements on $\vec{\alpha}$ and β:

$$\begin{cases} \alpha_i^2 = 1 \\ \alpha_i \alpha_j + \alpha_j \alpha_i = 0 \quad (i \neq j) \\ \alpha_i \beta + \beta \alpha_i = 0 \\ \beta^2 = 1 \end{cases} \tag{8.6}$$

To obtain a compact form, we can use the **anticommutator** $\{a, b\} = ab + ba$:

$$\alpha_i^2 = \beta^2 = 1; \quad \{\alpha_i, \alpha_j\} = 2\delta_j^i; \quad \{\alpha_i, \beta\} = 0 \tag{8.7}$$

Dirac proposed that in order to satisfy these conditions, $\vec{\alpha}$ and β must be taken as **matrices** [74]. In order to obtain real energies, the Hamilton operator must be Hermitian, hence **the $\vec{\alpha}$ and β matrices must also be Hermitian**. Then:

$$\alpha_i^\dagger = \alpha_i \quad \text{and} \quad \beta^\dagger = \beta \tag{8.8}$$

In addition, their eigenvalues must be $+1$ or -1 since $\alpha_i^2 = \beta^2 = \mathbb{1}$, since from here on the α_i and β are matrices. Finally, the anticommutation between α_i and β implies:

$$\alpha_i \beta + \beta \alpha_i = 0 \quad \implies \quad \alpha_i \beta \beta = -\beta \alpha_i \beta$$
$$\implies \quad \alpha_i = -\beta \alpha_i \beta \tag{8.9}$$

Considering the trace, we can find (using $\text{Tr}(AB) = \text{Tr}(BA)$):

$$\text{Tr}(\alpha_i) = \text{Tr}(-\beta \alpha_i \beta) = \text{Tr}(-\beta \beta \alpha_i) = \text{Tr}(-\alpha_i) = -\text{Tr}(\alpha_i) \tag{8.10}$$

Hence, the matrices α_i are **traceless**, and similarly for β. Hence:

$$\text{Tr}(\alpha_i) = \text{Tr}(\beta) = 0 \tag{8.11}$$

Therefore, one must find **four** traceless matrices that satisfy the anticommutation rules of Eq. (8.7). The 2×2 **Pauli matrices** σ_i are traceless and anticommuting but there are only *three* (see Appendix C.12), so it is not possible to find a fourth anticommuting one. Likewise, 3×3 matrices cannot furnish a set of four anticommuting matrices. Dirac wrote:

> "We make use of the matrices which Pauli introduced to describe the three components of spin angular momentum. These matrices have just the properties $\sigma_i^2 = 1$, $\sigma_i \sigma_j + \sigma_j \sigma_i = 0$ for $i \neq j$ that we require for our α's. We cannot, however, just take the σ's to be three of our α's, because then it would not be possible to find the fourth. We must extend the σ's in a diagonal manner to bring in two more rows and columns, so that we can introduce three more matrices ρ_1, ρ_2, ρ_3 of the same form as σ_1, σ_2, σ_3, but referring to different rows and columns." [1]

Hence, the simplest representation (i.e., the representation with the smallest dimension) of these matrices is a 4×4 representation:

$$\alpha_i = \begin{pmatrix} \cdot & \cdot & \cdot & \cdot \\ \cdot & \cdot & \cdot & \cdot \\ \cdot & \cdot & \cdot & \cdot \\ \cdot & \cdot & \cdot & \cdot \end{pmatrix}; \qquad \rho_i = \begin{pmatrix} \cdot & \cdot & \cdot & \cdot \\ \cdot & \cdot & \cdot & \cdot \\ \cdot & \cdot & \cdot & \cdot \\ \cdot & \cdot & \cdot & \cdot \end{pmatrix} \tag{8.12}$$

---

1   Republished with permission of The Royal Society (UK) from P. A. M. Dirac, "The quantum theory of the electron," *Proc. Roy. Soc. Lond.*, vol. A117, pp. 610–624, 1928; permission conveyed through Copyright Clearance Center, Inc.

Dirac effectively proposed building up a complete set as a direct product of the 2×2 unit matrix and Pauli matrices:

$$\sigma_{i,Dirac} = \mathbb{1} \otimes \sigma_{i,Pauli} \quad \text{and} \quad \rho_{i,Dirac} = \sigma_{i,Pauli} \otimes \mathbb{1} \tag{8.13}$$

See Table 8.1. Dirac then originally chose:

$$\alpha_1 = \rho_1 \sigma_{1,Dirac}, \quad \alpha_2 = \rho_1 \sigma_{2,Dirac}, \quad \alpha_3 = \rho_1 \sigma_{3,Dirac}, \quad \text{and} \quad \beta = \rho_3 \tag{8.14}$$

Dirac matrices

$\cdot$	$\mathbb{1}$	$\sigma_{1,Dirac}$	$\sigma_{2,Dirac}$	$\sigma_{3,Dirac}$
$\mathbb{1}$	$\begin{pmatrix} 1 & 0 & 0 & 0 \\ 0 & 1 & 0 & 0 \\ 0 & 0 & 1 & 0 \\ 0 & 0 & 0 & 1 \end{pmatrix}$	$\begin{pmatrix} 0 & 1 & 0 & 0 \\ 1 & 0 & 0 & 0 \\ 0 & 0 & 0 & 1 \\ 0 & 0 & 1 & 0 \end{pmatrix}$	$\begin{pmatrix} 0 & -i & 0 & 0 \\ i & 0 & 0 & 0 \\ 0 & 0 & 0 & -i \\ 0 & 0 & i & 0 \end{pmatrix}$	$\begin{pmatrix} 1 & 0 & 0 & 0 \\ 0 & -1 & 0 & 0 \\ 0 & 0 & 1 & 0 \\ 0 & 0 & 0 & -1 \end{pmatrix}$
ρ_1	$\begin{pmatrix} 0 & 0 & 1 & 0 \\ 0 & 0 & 0 & 1 \\ 1 & 0 & 0 & 0 \\ 0 & 1 & 0 & 0 \end{pmatrix}$	$\alpha_1 = \begin{pmatrix} 0 & 0 & 0 & 1 \\ 0 & 0 & 1 & 0 \\ 0 & 1 & 0 & 0 \\ 1 & 0 & 0 & 0 \end{pmatrix}$	$\alpha_2 = \begin{pmatrix} 0 & 0 & 0 & -i \\ 0 & 0 & i & 0 \\ 0 & -i & 0 & 0 \\ i & 0 & 0 & 0 \end{pmatrix}$	$\alpha_3 = \begin{pmatrix} 0 & 0 & 1 & 0 \\ 0 & 0 & 0 & -1 \\ 1 & 0 & 0 & 0 \\ 0 & -1 & 0 & 0 \end{pmatrix}$
ρ_2	$\begin{pmatrix} 0 & 0 & -i & 0 \\ 0 & 0 & 0 & -i \\ i & 0 & 0 & 0 \\ 0 & i & 0 & 0 \end{pmatrix}$	$\begin{pmatrix} 0 & 0 & 0 & -i \\ 0 & 0 & -i & 0 \\ 0 & i & 0 & 0 \\ i & 0 & 0 & 0 \end{pmatrix}$	$\begin{pmatrix} 0 & 0 & 0 & -1 \\ 0 & 0 & 1 & 0 \\ 0 & 1 & 0 & 0 \\ -1 & 0 & 0 & 0 \end{pmatrix}$	$\begin{pmatrix} 0 & 0 & -i & 0 \\ 0 & 0 & 0 & i \\ i & 0 & 0 & 0 \\ 0 & -i & 0 & 0 \end{pmatrix}$
ρ_3	$\beta = \begin{pmatrix} 1 & 0 & 0 & 0 \\ 0 & 1 & 0 & 0 \\ 0 & 0 & -1 & 0 \\ 0 & 0 & 0 & -1 \end{pmatrix}$	$\begin{pmatrix} 0 & 1 & 0 & 0 \\ 1 & 0 & 0 & 0 \\ 0 & 0 & 0 & -1 \\ 0 & 0 & -1 & 0 \end{pmatrix}$	$\begin{pmatrix} 0 & -i & 0 & 0 \\ i & 0 & 0 & 0 \\ 0 & 0 & 0 & i \\ 0 & 0 & -i & 0 \end{pmatrix}$	$\begin{pmatrix} 1 & 0 & 0 & 0 \\ 0 & -1 & 0 & 0 \\ 0 & 0 & -1 & 0 \\ 0 & 0 & 0 & 1 \end{pmatrix}$

Table 8.1 The Dirac matrices σ_i and ρ_i and their products.

However, any set of four 4×4 matrices that obey the algebra of Eq. (8.7) will work. Today the two most common representations are called the **Pauli–Dirac representation** and the **Weyl (or chiral) representation**.

• **The Pauli–Dirac representation.** In this case, the β matrix is diagonal and corresponds to Dirac's original choice, and it follows that:

$$\beta = \rho_3 = \begin{pmatrix} 1 & 0 & 0 & 0 \\ 0 & 1 & 0 & 0 \\ 0 & 0 & -1 & 0 \\ 0 & 0 & 0 & -1 \end{pmatrix} = \begin{pmatrix} \mathbb{1} & 0 \\ 0 & -\mathbb{1} \end{pmatrix}, \quad \alpha_i = \rho_1 \sigma_{i,Dirac} = \begin{pmatrix} 0 & \sigma_i \\ \sigma_i & 0 \end{pmatrix} \tag{8.15}$$

where we have used the 2×2 block compact notation.

• **Chiral or Weyl representation.** In this case, the α_i matrices are chosen diagonal in the 2×2 reduced form, and it follows that:

$$\alpha_i = -\rho_3 \sigma_{i,Dirac} = \begin{pmatrix} -\sigma_i & 0 \\ 0 & \sigma_i \end{pmatrix}, \quad \beta = \rho_1 = \begin{pmatrix} 0 & \mathbb{1} \\ \mathbb{1} & 0 \end{pmatrix} \tag{8.16}$$

• **Transformation to/from Pauli–Dirac from/to Weyl representation.** The two representations can be related by a transformation $\mathcal{R}$, such that for example for β, we have:

$$\beta_{PD} = \mathcal{R} \beta_{chiral} \mathcal{R}^{-1} \quad \text{where} \quad \mathcal{R} \equiv \frac{1}{\sqrt{2}} \begin{pmatrix} \mathbb{1} & \mathbb{1} \\ -\mathbb{1} & \mathbb{1} \end{pmatrix} \quad \to \mathcal{R}^{-1} = \frac{1}{\sqrt{2}} \begin{pmatrix} \mathbb{1} & -\mathbb{1} \\ \mathbb{1} & \mathbb{1} \end{pmatrix} \tag{8.17}$$

Indeed:

$$
\beta_{PD} \overset{?}{=} \frac{1}{2}\begin{pmatrix} \mathbb{1} & \mathbb{1} \\ -\mathbb{1} & \mathbb{1} \end{pmatrix}\begin{pmatrix} 0 & \mathbb{1} \\ \mathbb{1} & 0 \end{pmatrix}\begin{pmatrix} \mathbb{1} & -\mathbb{1} \\ \mathbb{1} & \mathbb{1} \end{pmatrix} = \frac{1}{2}\begin{pmatrix} \mathbb{1} & \mathbb{1} \\ -\mathbb{1} & \mathbb{1} \end{pmatrix}\begin{pmatrix} \mathbb{1} & \mathbb{1} \\ \mathbb{1} & -\mathbb{1} \end{pmatrix}
$$

$$
= \begin{pmatrix} \mathbb{1} & 0 \\ 0 & -\mathbb{1} \end{pmatrix} \quad \Box \tag{8.18}
$$

Naturally the physical results will not depend on the particular chosen representation. One will select the most natural representation in order to find solutions to the Dirac equation, depending on which properties are of most interest. In particular, whether we are interested in the diagonality of the α_i or β, which correspond, as shown later, to particular particle properties.

Summarizing, the consequence of requiring consistency with special relativity is that the wave function is forced to have four degrees of freedom. Having a massless particle (i.e., $m = 0$) would imply that there would be no β term in the Dirac equation and the α_i matrices could be reduced to the 2×2 Pauli matrices. We therefore note that massless Dirac particles can be described by a two-component object (the **bi-spinor** or **Weyl**[2] **spinor**) and that it is the particles' massive nature that imposes the four degrees of freedom.

● **Why the Pauli matrices?** At first the reader might be surprised by Dirac's choice to extend the Pauli matrices. Within group theory, we can motivate this choice by the structure of the Lorentz group (see Appendix B.10). In the appendix, we show that the generators $\vec{J}$ and $\vec{K}$ of spatial rotations and boosts can be recast more simply into a new set of generators $\vec{L}$ and $\vec{R}$ (see Eq. (B.111)), with the commutation rules of two independent $SO(3)$ or $SU(2)$ algebras. Hence, the representations of the Lorentz group can be classified as $SU(2) \times SU(2)$ representations! Dirac's equation corresponds to the two independent Lorentz sub-representations called $(\frac{1}{2}, 0)$ and $(0, \frac{1}{2})$ describing left-handed and right-handed spinors. The Dirac spinors will generally be introduced in the next section.

8.2   The Dirac Spinors

The Dirac equation can be represented as a system of four coupled equations as a consequence of the dimensionality of the α_i and β matrices. The wave function Ψ must therefore consequently also possess four elements. We consider Eq. (8.2) and write it as:

$$
\left(i\frac{\partial}{\partial t} + i\vec{\alpha} \cdot \nabla - \beta m \right)\Psi = 0 \qquad \text{(Dirac equation)} \tag{8.19}
$$

The wave function Ψ is called a **Dirac spinor**, **four-spinor**, or **bi-spinor** and is represented by a column vector with four complex functions ψ_i $(i = 1, ..., 4)$:

$$
\Psi(x^\mu) = \begin{pmatrix} \cdot \\ \cdot \\ \cdot \\ \cdot \end{pmatrix} = \begin{pmatrix} \psi_1(x^\mu) \\ \psi_2(x^\mu) \\ \psi_3(x^\mu) \\ \psi_4(x^\mu) \end{pmatrix} \tag{8.20}
$$

We note that although the Dirac spinor has four components, it is **not** a four-vector. It does not transform as $x'^\mu = \Lambda^\mu_{\ \nu} x^\nu$. The physical interpretation of the four components will be discussed later. However, we can already note that a consequence of introducing an equation that is first order in time/space derivatives is that the wave function has new degrees of freedom!

---

2   Hermann Klaus Hugo Weyl (1885–1955), German theoretical physicist.

8.3   Covariant Form of the Dirac Equation

In order to express the Dirac equation in a covariant form, one introduces the γ matrices in order to handle the $\vec{\alpha}$ (spatial) and β (time) components in a symmetric way. The Dirac equation will be written as a function of the newly defined γ matrices and the covariance of the Dirac equation will provide a prescription for the transformation of the spinors under the Lorentz group transformations.

The **Dirac γ matrices** are defined as:

$$\gamma^0 \equiv \beta \quad \text{and} \quad \gamma^k \equiv \beta\alpha_k \quad (k = 1, 2, 3) \tag{8.21}$$

or as a compact form (note that the matrices do **not** represent a four-vector and the index μ is just to identify the matrix):

$$\gamma^\mu \equiv (\beta, \beta\alpha_k) = \begin{pmatrix} \cdot & \cdot & \cdot & \cdot \\ \cdot & \cdot & \cdot & \cdot \\ \cdot & \cdot & \cdot & \cdot \\ \cdot & \cdot & \cdot & \cdot \end{pmatrix} \tag{8.22}$$

The properties of the γ matrices follow directly from those of the α and β matrices. We note that:

$$\begin{aligned} (\gamma^0)^2 &= \beta^2 = \mathbb{1} \\ (\gamma^k)^2 &= \beta\alpha_k\beta\alpha_k = -\alpha_k\beta\beta\alpha_k = -(\alpha_k)^2 = -\mathbb{1} \\ \gamma^\mu\gamma^\nu &= -\gamma^\nu\gamma^\mu \quad \text{for } \mu \neq \nu \end{aligned} \tag{8.23}$$

All these relations can be summarized into the following **anticommutation rule,** called the **Clifford[3] algebra:**

$$\gamma^\mu\gamma^\nu + \gamma^\nu\gamma^\mu = \{\gamma^\mu, \gamma^\nu\} = (2g^{\mu\nu}) \cdot \mathbb{1} \tag{8.24}$$

Are the γ matrices Hermitian? The **conjugate transpose of the matrices** are:

$$\begin{aligned} (\gamma^0)^\dagger &= (\beta)^\dagger = \beta = \gamma^0 \\ (\gamma^k)^\dagger &= (\beta\alpha_k)^\dagger = \alpha_k^\dagger\beta^\dagger = \alpha_k\beta = -\beta\alpha_k = -\gamma^k \quad (k = 1, 2, 3) \end{aligned} \tag{8.25}$$

So γ^0 is Hermitian, and the γ^k are anti-Hermitian. This can be summarized in the expression:

$$(\gamma^\mu)^\dagger = \gamma^0\gamma^\mu\gamma^0 \tag{8.26}$$

Finally, the Dirac equation can be written more elegantly by using the four Dirac gamma matrices. Multiplying the Dirac equation by β, we obtain:

$$\left(i\beta\frac{\partial}{\partial t} + i\beta\vec{\alpha} \cdot \nabla - \beta\beta m\right)\Psi = \left(i\gamma^0\frac{\partial}{\partial t} + i\vec{\gamma} \cdot \nabla - \mathbb{1}m\right)\Psi = 0 \tag{8.27}$$

The **covariant form of the Dirac equation** can hence be expressed with the γ matrices as:

$$(i\gamma^\mu\partial_\mu - m)\Psi = 0 \quad \text{or} \quad i\gamma^\mu\partial_\mu\Psi = m\Psi \tag{8.28}$$

It can even more elegantly be written as

$$\gamma^\mu p_\mu\Psi = m\Psi \tag{8.29}$$

3   William Kingdon Clifford (1845–1879), English mathematician and philosopher.

8.4   The Dirac Equation in Matrix Form

The Dirac equation actually represents a set of coupled differential equations that relate the four components of the spinor:

$$\sum_{k=1}^{4}\left[\sum_{\mu} i\,(\gamma^{\mu})_{jk}\,\partial_{\mu} - m\delta_{jk}\right]\Psi_k = 0 \quad (j=1,2,3,4) \tag{8.30}$$

The anticommutation rule, Eq. (8.24), fully defines the algebra of the γ matrices. It can be used to perform calculations without explicitly writing down the 4×4 matrices. Nevertheless, it is often useful to choose a particular representation and perform calculations for each component.

For example, in the **Pauli–Dirac representation** the γ matrices are (in 2×2 blocks and 4×4):

$$\gamma^0_{PD} = \begin{pmatrix} \mathbb{1} & 0 \\ 0 & -\mathbb{1} \end{pmatrix} = \begin{pmatrix} 1 & 0 & 0 & 0 \\ 0 & 1 & 0 & 0 \\ 0 & 0 & -1 & 0 \\ 0 & 0 & 0 & -1 \end{pmatrix}; \quad \gamma^1_{PD} = \begin{pmatrix} 0 & \sigma_1 \\ -\sigma_1 & 0 \end{pmatrix} = \begin{pmatrix} 0 & 0 & 0 & 1 \\ 0 & 0 & 1 & 0 \\ 0 & -1 & 0 & 0 \\ -1 & 0 & 0 & 0 \end{pmatrix}$$

$$\gamma^2_{PD} = \begin{pmatrix} 0 & \sigma_2 \\ -\sigma_2 & 0 \end{pmatrix} = \begin{pmatrix} 0 & 0 & 0 & -i \\ 0 & 0 & i & 0 \\ 0 & i & 0 & 0 \\ -i & 0 & 0 & 0 \end{pmatrix}; \quad \gamma^3_{PD} = \begin{pmatrix} 0 & \sigma_3 \\ -\sigma_3 & 0 \end{pmatrix} = \begin{pmatrix} 0 & 0 & 1 & 0 \\ 0 & 0 & 0 & -1 \\ -1 & 0 & 0 & 0 \\ 0 & 1 & 0 & 0 \end{pmatrix}$$

$$\tag{8.31}$$

The Dirac equation in the Pauli–Dirac representation becomes (in 2×2 blocks):

$$\left(i\underbrace{\begin{pmatrix} \mathbb{1} & 0 \\ 0 & -\mathbb{1} \end{pmatrix}}_{=\gamma^0}\partial_t + i\underbrace{\begin{pmatrix} 0 & \sigma_k \\ -\sigma_k & 0 \end{pmatrix}}_{=\gamma^k}\partial_k - m\underbrace{\begin{pmatrix} \mathbb{1} & 0 \\ 0 & \mathbb{1} \end{pmatrix}}_{=\mathbb{1}} \right) \begin{pmatrix} \Psi_1 \\ \Psi_2 \\ \Psi_3 \\ \Psi_4 \end{pmatrix} = 0 \tag{8.32}$$

• **Relation to the Klein–Gordon equation.** The Dirac equation mixes up different components of Ψ through the matrices γ^{μ}. However, **each individual component of the spinor field satisfies the Klein–Gordon equation.** To see this, we multiply the Dirac equation by $(i\gamma^{\nu}\partial_{\nu} + m)$:

$$(i\gamma^{\nu}\partial_{\nu} + m)(i\gamma^{\mu}\partial_{\mu} - m)\Psi = 0 = -(\gamma^{\mu}\gamma^{\nu}\partial_{\mu}\partial_{\nu} + m^2)\Psi \tag{8.33}$$

We can use a trick, noting that the derivatives commute, to write:

$$(\gamma^{\mu}\gamma^{\nu}\partial_{\mu}\partial_{\nu}) = \frac{1}{2}(\gamma^{\mu}\gamma^{\nu}\partial_{\mu}\partial_{\nu} + \gamma^{\mu}\gamma^{\nu}\partial_{\nu}\partial_{\mu}) = \frac{1}{2}(\gamma^{\mu}\gamma^{\nu}\partial_{\mu}\partial_{\nu} + \gamma^{\nu}\gamma^{\mu}\partial_{\mu}\partial_{\nu}) = \frac{1}{2}\{\gamma^{\mu},\gamma^{\nu}\}\partial_{\mu}\partial_{\nu} \tag{8.34}$$

With the Clifford algebra, we have $\frac{1}{2}\{\gamma^{\mu},\gamma^{\nu}\}\partial_{\mu}\partial_{\nu} = \frac{1}{2}(2g^{\mu\nu})\partial_{\mu}\partial_{\nu} = \partial^{\mu}\partial_{\mu}$, hence:

$$-(\partial^{\mu}\partial_{\mu} + m^2)\Psi = 0 \tag{8.35}$$

which has no matrices and hence is independently valid for each of the four components of Ψ!

8.5   The Slash Notation

We now introduce a very useful notation called the **"slash" notation** . For any four-vector a^{μ} we write *"a-slash"* to mean its product with the γ matrices:

$$\slashed{a} \equiv \gamma^{\mu}a_{\mu} = \gamma^0 a_0 - \vec{\gamma}\cdot\vec{a} \tag{8.36}$$

With this notation, the covariant Dirac equation has the very compact form:

$$(\not{\partial} - m)\Psi = 0 \quad \text{or} \quad \not{\partial}\Psi = m\Psi \tag{8.37}$$

We will rely heavily on the slash notation when computing interaction processes.

8.6 The Adjoint Equation and the Current Density

We would like to define a four-vector current density for the Dirac equation similar to the case of the Schrödinger and Klein–Gordon equations. This is where the perceived problems with the Klein–Gordon equation arose when interpreting the field as a particle wave function. Starting from the Dirac equation, the **adjoint** is found with the dagger operator:

$$\left(-i\frac{\partial \Psi^\dagger}{\partial t}\underbrace{(\gamma^0)^\dagger}_{=\gamma^0} - i\frac{\partial \Psi^\dagger}{\partial x^k}\underbrace{(\gamma^k)^\dagger}_{=-\gamma^k} - m\Psi^\dagger \right) = 0 \tag{8.38}$$

or

$$\left(-i\frac{\partial \Psi^\dagger}{\partial t}\gamma^0 + i\frac{\partial \Psi^\dagger}{\partial x^k}\gamma^k - m\Psi^\dagger \right) = 0 \tag{8.39}$$

This equation is similar to the original Dirac equation, except for the relative sign between the time and space coordinates. We multiply the equation by γ^0 from the right:

$$\left(-i\frac{\partial \Psi^\dagger}{\partial t}\gamma^0\gamma^0 + i\frac{\partial \Psi^\dagger}{\partial x^k}\underbrace{\gamma^k\gamma^0}_{=-\gamma^0\gamma^k} - m\Psi^\dagger\gamma^0 \right) = 0 \tag{8.40}$$

or

$$\left(-i\left(\frac{\partial \Psi^\dagger}{\partial t}\gamma^0\right)\gamma^0 - i\left(\frac{\partial \Psi^\dagger}{\partial x^k}\gamma^0\right)\gamma^k - m\Psi^\dagger\gamma^0 \right) = 0 \tag{8.41}$$

By defining the **adjoint spinor** $\overline{\Psi} \equiv \Psi^\dagger\gamma^0$ (a row vector):

$$\overline{\Psi} \equiv \left(\bar{\Psi}_1, \bar{\Psi}_2, \bar{\Psi}_3, \bar{\Psi}_4\right) \tag{8.42}$$

the **adjoint (covariant) Dirac equation** is:

$$\left(i\partial_\mu\overline{\Psi}\gamma^\mu + m\overline{\Psi}\right) = 0 \quad \text{or} \quad i\partial_\mu\overline{\Psi}\gamma^\mu = -m\overline{\Psi} \tag{8.43}$$

In the Pauli–Dirac representation, the adjoint spinor is simply:

$$\overline{\Psi} = \Psi^\dagger\gamma^0 = (\Psi^*)^T\gamma^0 = (\Psi_1^*, \Psi_2^*, \Psi_3^*, \Psi_4^*)\begin{pmatrix} 1 & 0 & 0 & 0 \\ 0 & 1 & 0 & 0 \\ 0 & 0 & -1 & 0 \\ 0 & 0 & 0 & -1 \end{pmatrix} = (\Psi_1^*, \Psi_2^*, -\Psi_3^*, -\Psi_4^*) \tag{8.44}$$

Using the Dirac equation and its adjoint, one can construct the current density. We multiply the Dirac equation on the left by $\overline{\Psi}$ and the adjoint Dirac equation on the right by Ψ and add them:

$$\left(i\overline{\Psi}\gamma^\mu\partial_\mu\Psi - m\overline{\Psi}\Psi\right) + \left(i(\partial_\mu\overline{\Psi})\gamma^\mu\Psi + m\overline{\Psi}\Psi\right) = 0$$
$$\implies i\overline{\Psi}\gamma^\mu\partial_\mu\Psi + i(\partial_\mu\overline{\Psi})\gamma^\mu\Psi = 0 \implies \partial_\mu(\overline{\Psi}\gamma^\mu\Psi) = 0 \tag{8.45}$$

Noting the continuity equation $\partial_\mu j^\mu$, we identify the current-density four-vector of the Dirac equation:

$$j^\mu \equiv \overline{\Psi}\gamma^\mu\Psi \qquad \text{Current-density four-vector} \tag{8.46}$$

Its time coordinate corresponds to the **probability density**:

$$\rho = j^0 = \overline{\Psi}\gamma^0\Psi = \Psi^\dagger \underbrace{\gamma^0\gamma^0}_{=\mathbb{1}}\Psi = \Psi^\dagger\Psi = \begin{pmatrix} \Psi_1^* & \Psi_2^* & \Psi_3^* & \Psi_4^* \end{pmatrix}\begin{pmatrix} \Psi_1 \\ \Psi_2 \\ \Psi_3 \\ \Psi_4 \end{pmatrix} = \sum_{i=1}^4 \Psi_i^*\Psi_i = \sum_{i=1}^4 |\Psi_i|^2 \tag{8.47}$$

and it follows that it is positive definite $\rho \geq 0$. **The Dirac equation therefore can be interpreted as the equation of motion of the wave function Ψ of a particle!**

8.7   Dirac Equation in an External Electromagnetic Field

In analogy to what was successfully derived in the case of the Schrödinger equation in Section 4.10, the Dirac equation for a particle of charge e in an external electromagnetic field can be obtained by applying **the principle of minimal substitution** $\vec{p} \longrightarrow \vec{p} - e\vec{A}, E \longrightarrow E - e\phi$ (see Eq. (4.101)) on the free Dirac equation to find:

$$(\vec{\alpha}\cdot(\vec{p} - e\vec{A}) + \beta m + e\phi\mathbb{1})\Psi = E\Psi = i\frac{\partial\Psi}{\partial t} \tag{8.48}$$

Hence, by defining the four-vector potential $A^\mu = (A^0, \vec{A}) = (\phi, \vec{A})$ (see Section 10.2), we can write:

$$i\frac{\partial\Psi}{\partial t} = (\vec{\alpha}\cdot(\vec{p} - e\vec{A}) + \beta m + eA^0\mathbb{1})\Psi \tag{8.49}$$

which after multiplication by β on the left becomes:

$$\begin{aligned}
\left[i\beta\frac{\partial}{\partial t} - \left(\beta\vec{\alpha}\cdot(\vec{p} - e\vec{A}) + \beta^2 m + e\beta A^0\right)\right]\Psi &= \left[i\beta\frac{\partial}{\partial t} - \left(\beta\vec{\alpha}\cdot\vec{p} - e\beta\vec{\alpha}\cdot\vec{A} + e\beta A^0\right) - m\mathbb{1}\right]\Psi \\
&= \left[i\beta\frac{\partial}{\partial t} + i\beta\vec{\alpha}\cdot\nabla - e\beta A^0 + e\beta\vec{\alpha}\cdot\vec{A} - m\mathbb{1}\right]\Psi \\
&= \left[i\gamma^0\frac{\partial}{\partial t} + i\vec{\gamma}\cdot\nabla - e\gamma A^0 + e\vec{\gamma}\cdot\vec{A} - m\mathbb{1}\right]\Psi \\
&= 0
\end{aligned} \tag{8.50}$$

or in covariant form:

$$[\gamma^\mu(i\partial_\mu - eA_\mu) - m]\Psi = 0 \tag{8.51}$$

This fundamental equation will be very handy later (see, e.g., Section 8.22). But first, in the next sections, we will focus on the free Dirac particle to derive its properties.

8.8   Solutions of the Free Dirac Equation

In general, we seek a solution for the free Dirac equation. As already mentioned, the solution is called a spinor, which is a quantity with four components. We expect that spinors depend on space-time and the four-momentum of the particle. Hence:

$$\Psi = \Psi(x^\mu, p^\mu) \tag{8.52}$$

When searching for a complete set of solutions, the relativistic relation between energy and momentum needs to be taken into account and therefore both positive and negative energy solutions will coexist:

$$E = \pm \left| \sqrt{\vec{p}^2 + m^2} \right| \quad \Longrightarrow \quad E > 0 \quad \text{or} \quad E < 0 \tag{8.53}$$

where m is the rest mass of the particle.[4] We can assume that for the free particle the space-time dependence and the momentum dependence factorize and that the space-time dependence is trivial and simply given by the phase similar to the case of the plane wave, while for a given rest mass the three-momentum defines a quantity $u(E, \vec{p})$ (called **the spinor** u) with four components. Accordingly:

$$\Psi(x^\mu) \equiv u(E, \vec{p}) e^{-ip \cdot x} = \begin{pmatrix} \cdot \\ \cdot \\ \cdot \\ \cdot \end{pmatrix} e^{-ip \cdot x} \tag{8.54}$$

8.8.1   Solution for a Particle at Rest

We seek the plane-wave solution for a Dirac particle of mass m at rest $\vec{p} = 0$, $E = \pm m$:

$$\Psi(x^\mu) = u(\pm m, \vec{p} = 0) e^{-ip^\mu x_\mu} = u e^{\mp imt} \tag{8.55}$$

The wave function is in this case independent of the space coordinates and the Dirac equation simplifies to:

$$(i\gamma^\mu \partial_\mu \Psi - m\Psi) = 0 \quad \Longrightarrow \quad (i\gamma^0 \partial_0 \Psi - m\Psi) = 0 \tag{8.56}$$

We implement the Pauli–Dirac representation in which γ^0 is diagonal:

$$(i\gamma^0 \partial_0 \Psi - m\Psi) = 0 \quad \text{where} \quad \gamma^0 = \gamma^0_{PD} = \begin{pmatrix} \mathbb{1} & 0 \\ 0 & -\mathbb{1} \end{pmatrix} \tag{8.57}$$

By inspection it seems natural to split the spinor Ψ into two two-component spinors labeled Ψ_A and Ψ_B:

$$\Psi \equiv \begin{pmatrix} \Psi_A \\ \Psi_B \end{pmatrix} = \begin{pmatrix} \begin{pmatrix} \cdot \\ \cdot \end{pmatrix} \\ \begin{pmatrix} \cdot \\ \cdot \end{pmatrix} \end{pmatrix} \tag{8.58}$$

Then:

$$i \begin{pmatrix} \mathbb{1} & 0 \\ 0 & -\mathbb{1} \end{pmatrix} \partial_0 \begin{pmatrix} \Psi_A \\ \Psi_B \end{pmatrix} = m \begin{pmatrix} \Psi_A \\ \Psi_B \end{pmatrix} \tag{8.59}$$

One obtains two independent equations:

$$i\partial_0 \Psi_A = m\Psi_A \quad \text{and} \quad i\partial_0 \Psi_B = -m\Psi_B \tag{8.60}$$

The solutions are written as a function of the **two-component spinors** u_A **and** u_B:

$$\begin{cases} \Psi_A(t) = u_A e^{-imt} & \Longrightarrow \quad E > 0 \quad \text{positive-energy solutions} \\ \Psi_B(t) = u_B e^{+imt} & \Longrightarrow \quad E < 0 \quad \text{negative-energy solutions} \end{cases} \tag{8.61}$$

Going back to the 4×4 Pauli–Dirac representation of Eq. (8.59), we find:

$$\begin{pmatrix} m\mathbb{1} & 0 \\ 0 & m\mathbb{1} \end{pmatrix} \begin{pmatrix} u_A \\ u_B \end{pmatrix} = m \begin{pmatrix} u_A \\ u_B \end{pmatrix} \tag{8.62}$$

---

4   Actually the rest mass is $|m|$ and the m parameter in the Dirac equation could be positive or negative.

Since the first matrix on the left-hand side is diagonal, we can find decoupled solutions for u_A and u_B, noting that the set u_A corresponds to the positive-energy solutions and u_B to the negative ones. As a set of eigenvectors, we can finally choose:

$$u_A = \begin{pmatrix} 1 \\ 0 \end{pmatrix} \quad \text{or} \quad u_A = \begin{pmatrix} 0 \\ 1 \end{pmatrix} \quad \text{with} \quad E = +m \tag{8.63}$$

and similarly

$$u_B = \begin{pmatrix} 1 \\ 0 \end{pmatrix} \quad \text{or} \quad u_B = \begin{pmatrix} 0 \\ 1 \end{pmatrix} \quad \text{for} \quad E = -m \tag{8.64}$$

Putting these results together, we have found a complete set of four orthogonal solutions $\Psi_0^{(s)}(\vec{p} = 0)$ ($s = 1, ..., 4$) of the Dirac equation ($\Psi_0^{(1)}$ and $\Psi_0^{(2)}$ with positive energy, and $\Psi_0^{(3)}$ and $\Psi_0^{(4)}$ with negative energy) in the case of a particle of mass m at rest:

$$\begin{cases} \Psi_0^{(1)} = \mathcal{N} \begin{pmatrix} 1 \\ 0 \\ 0 \\ 0 \end{pmatrix} e^{-imt}; \quad \Psi_0^{(2)} = \mathcal{N} \begin{pmatrix} 0 \\ 1 \\ 0 \\ 0 \end{pmatrix} e^{-imt} \quad \text{with positive energy} \\[2em] \Psi_0^{(3)} = \mathcal{N} \begin{pmatrix} 0 \\ 0 \\ 1 \\ 0 \end{pmatrix} e^{+imt}; \quad \Psi_0^{(4)} = \mathcal{N} \begin{pmatrix} 0 \\ 0 \\ 0 \\ 1 \end{pmatrix} e^{+imt} \quad \text{with negative energy} \end{cases} \tag{8.65}$$

where $\mathcal{N}$ is a normalization factor to be defined. While the positive and negative solutions are formally expected for consistency with special relativity, their physical interpretation must be explained. In addition, there is a twofold degeneracy for each solution which also needs to be interpreted. This will be done after the derivation of the solutions for a moving Dirac particle.

8.8.2   Solution for a Moving Particle

We seek the solutions for a particle with four-momentum $p^\mu = (E, \vec{p})$:

$$(\gamma^\mu p_\mu - m)\Psi = 0 \quad \Longrightarrow \quad \gamma^\mu p_\mu = \gamma^0 p_0 - \vec{\gamma} \cdot \vec{p} = \gamma^0 E - \vec{\gamma} \cdot \vec{p} \tag{8.66}$$

We note that by doing so, we have effectively replaced the operator p^μ in the Dirac equation with its four-momentum counterpart with real time-like and space-like components given by E and $\vec{p}$. In the following, we will indistinguishably use p^μ in equations to indicate the four-momentum or the operator, since we only consider the plane-wave solutions for particles of a well-determined momentum p^μ. Choosing again the Pauli–Dirac representation, we obtain:

$$\gamma^\mu p_\mu = \begin{pmatrix} E\mathbb{1} & 0 \\ 0 & -E\mathbb{1} \end{pmatrix} - \begin{pmatrix} 0 & \vec{\sigma} \cdot \vec{p} \\ -\vec{\sigma} \cdot \vec{p} & 0 \end{pmatrix} = \begin{pmatrix} E\mathbb{1} & -\vec{\sigma} \cdot \vec{p} \\ \vec{\sigma} \cdot \vec{p} & -E\mathbb{1} \end{pmatrix} \tag{8.67}$$

The situation is no longer diagonal as in the case of the particle at rest. Had we chosen the Weyl representation, the situation would be similar. While in the case of the particle at rest the solutions could be decoupled into two 2×2 problems, in the general case we are faced with the two coupled 2×2 system (or equivalently a system of four coupled equations). We retain the two 2×2 solutions and write in analogy to the particle at rest case:

$$u(\vec{p}) = \begin{pmatrix} u_A(\vec{p}) \\ u_B(\vec{p}) \end{pmatrix} = \begin{pmatrix} \begin{pmatrix} \cdot \\ \cdot \end{pmatrix} \\ \begin{pmatrix} \cdot \\ \cdot \end{pmatrix} \end{pmatrix} \tag{8.68}$$

where the bi-spinors u_A and u_B now depend on the three-momentum $\vec{p}$. In this case:

$$(\gamma^\mu p_\mu - m) \begin{pmatrix} u_A \\ u_B \end{pmatrix} = \begin{pmatrix} (E-m)\mathbb{1} & -\vec{\sigma}\cdot\vec{p} \\ \vec{\sigma}\cdot\vec{p} & -(E+m)\mathbb{1} \end{pmatrix} \begin{pmatrix} u_A \\ u_B \end{pmatrix} = 0 \tag{8.69}$$

Hence, we obtain two coupled equations:

$$\begin{cases} (E-m)u_A - (\vec{\sigma}\cdot\vec{p})u_B = 0 \\ (\vec{\sigma}\cdot\vec{p})u_A - (E+m)u_B = 0 \end{cases} \implies \begin{cases} u_A = \left(\dfrac{\vec{\sigma}\cdot\vec{p}}{E-m} \right) u_B \\ u_B = \left(\dfrac{\vec{\sigma}\cdot\vec{p}}{E+m} \right) u_A \end{cases} \tag{8.70}$$

We can, for example, eliminate u_B to keep u_A:

$$u_A = \left(\frac{(\vec{\sigma}\cdot\vec{p})^2}{(E-m)(E+m)} \right) u_A \tag{8.71}$$

The product of the three-momentum with the Pauli matrices gives $(\vec{\sigma}\cdot\vec{p})^2 = \vec{p}^{\,2}\mathbb{1}$ (see Appendix C.12). Hence:

$$u_A = \left(\frac{\vec{p}^{\,2}}{(E^2 - m^2)} \right) u_A \implies (E^2 - m^2)u_A = \vec{p}^{\,2}\mathbb{1}u_A \tag{8.72}$$

which is expected since $E^2 - m^2 = \vec{p}^{\,2}$. We must find two orthogonal basis vectors for u_A. We take the solutions for the particle at rest:

$$\begin{cases} u_A = \begin{pmatrix} 1 \\ 0 \end{pmatrix} \implies u_B = \left(\dfrac{\vec{\sigma}\cdot\vec{p}}{E+m} \right) \begin{pmatrix} 1 \\ 0 \end{pmatrix} = \dfrac{1}{E+m} \begin{pmatrix} p_z \\ p_x + ip_y \end{pmatrix} \\ u_A = \begin{pmatrix} 0 \\ 1 \end{pmatrix} \implies u_B = \left(\dfrac{\vec{\sigma}\cdot\vec{p}}{E+m} \right) \begin{pmatrix} 0 \\ 1 \end{pmatrix} = \dfrac{1}{E+m} \begin{pmatrix} p_x - ip_y \\ -p_z \end{pmatrix} \end{cases} \tag{8.73}$$

where we have used the explicit 2×2 matrix for $\vec{\sigma}\cdot\vec{p}$ (see Appendix C.12). Repeating the same procedure for u_B we obtain the corresponding solutions:

$$\begin{cases} u_B = \begin{pmatrix} 1 \\ 0 \end{pmatrix} \implies u_A = \left(\dfrac{\vec{\sigma}\cdot\vec{p}}{E-m} \right) \begin{pmatrix} 1 \\ 0 \end{pmatrix} = \dfrac{1}{E-m} \begin{pmatrix} p_z \\ p_x + ip_y \end{pmatrix} \\ u_B = \begin{pmatrix} 0 \\ 1 \end{pmatrix} \implies u_A = \left(\dfrac{\vec{\sigma}\cdot\vec{p}}{E-m} \right) \begin{pmatrix} 0 \\ 1 \end{pmatrix} = \dfrac{1}{E-m} \begin{pmatrix} p_x - ip_y \\ -p_z \end{pmatrix} \end{cases} \tag{8.74}$$

With the above results, we can now write the four orthogonal plane-wave solutions of the free Dirac particle $\Psi^{(s)}$ ($s = 1, ..., 4$) as follows:

$$\begin{cases} E > 0: \quad \Psi^{(1)} = \mathcal{N} \begin{pmatrix} 1 \\ 0 \\ \dfrac{p_z}{E+m} \\ \dfrac{p_x + ip_y}{E+m} \end{pmatrix} e^{-ip\cdot x}; \quad \Psi^{(2)} = \mathcal{N} \begin{pmatrix} 0 \\ 1 \\ \dfrac{p_x - ip_y}{E+m} \\ \dfrac{-p_z}{E+m} \end{pmatrix} e^{-ip\cdot x} \\ \\ E < 0: \quad \Psi^{(3)} = \mathcal{N} \begin{pmatrix} \dfrac{p_z}{E-m} \\ \dfrac{p_x + ip_y}{E-m} \\ 1 \\ 0 \end{pmatrix} e^{-ip\cdot x}; \quad \Psi^{(4)} = \mathcal{N} \begin{pmatrix} \dfrac{p_x - ip_y}{E-m} \\ \dfrac{-p_z}{E-m} \\ 0 \\ 1 \end{pmatrix} e^{-ip\cdot x} \end{cases} \tag{8.75}$$

where $\mathcal{N}$ is the normalization to be defined. For $\vec{p} = 0$ giving $E = \pm m$, we recover the two solutions in Eq. (8.65). Hence, the two solutions $\Psi^{(1)}$ and $\Psi^{(2)}$ correspond to the positive energy while the two solutions $\Psi^{(3)}$ and $\Psi^{(4)}$ correspond to the negative energy. As already mentioned, the solutions have a twofold energy degeneracy.

8.9   Dirac Four-Component and Two-Component Spinors

Let us examine the structure of the Dirac spinors. The free solutions derived above represent plane waves. Let us consider $\Psi^{(1)}$ and $\Psi^{(2)}$ which correspond to positive energies. By direct comparison with the solutions of Eq. (8.75), we can write for $s = 1, 2$:

$$\Psi^{(1,2)} = u^{(1,2)}(\vec{p})e^{-ip\cdot x} = \mathcal{N}\begin{pmatrix} u_A^{(1,2)} \\ \dfrac{\vec{\sigma}\cdot\vec{p}}{E+m}u_A^{(1,2)} \end{pmatrix}e^{-ip\cdot x} \tag{8.76}$$

where we have used the four-component spinor $u(\vec{p})$ and the two-component bi-spinor $u_A^{(s)}$ as given in Eq. (8.63):

$$u_A^{(1)} = \begin{pmatrix} 1 \\ 0 \end{pmatrix} \quad \text{or} \quad u_A^{(2)} = \begin{pmatrix} 0 \\ 1 \end{pmatrix} \tag{8.77}$$

The product $(\vec{\sigma}\cdot\vec{p})$ is a 2×2 matrix given in Eq. (C.100). Similarly, we can write down the spinors for the negative-energy solutions ($E < 0$):

$$\Psi^{(3,4)} = u^{(3,4)}(\vec{p})e^{-ip\cdot x} = \mathcal{N}\begin{pmatrix} \dfrac{\vec{\sigma}\cdot\vec{p}}{E-m}u_A^{(1,2)} \\ u_A^{(1,2)} \end{pmatrix}e^{-ip\cdot x} \tag{8.78}$$

Without loss of generality, we can consider the case when the **particle travels in the z direction** $\vec{p} = (0, 0, p)$ (either in the z direction or opposite with $p > 0$ or $p < 0$). In this case the solutions of Eq. (8.75) are equal to $\Psi^{(1,2)} = u_z^{(1,2)}\exp(-ip\cdot x)$ (positive energy) and $\Psi^{(3,4)} = u_z^{(3,4)}\exp(-ip\cdot x)$ (negative energy), where:

$$u_z^{(1)} = \mathcal{N}\begin{pmatrix} 1 \\ 0 \\ \dfrac{p}{E+m} \\ 0 \end{pmatrix}; \quad u_z^{(2)} = \mathcal{N}\begin{pmatrix} 0 \\ 1 \\ 0 \\ \dfrac{-p}{E+m} \end{pmatrix}; \quad u_z^{(3)} = \mathcal{N}\begin{pmatrix} \dfrac{p}{E-m} \\ 0 \\ 1 \\ 0 \end{pmatrix}; \quad u_z^{(4)} = \mathcal{N}\begin{pmatrix} 0 \\ \dfrac{-p}{E-m} \\ 0 \\ 1 \end{pmatrix} \tag{8.79}$$

In the **non-relativistic limit** $p \simeq m\beta$ and $E \simeq m + \frac{1}{2}m\beta^2$, hence

$$\frac{p}{E+m} \simeq \frac{\beta}{2 + \frac{\beta^2}{2}} = \left(\frac{\beta}{2}\right)\frac{1}{1 + (\beta/2)^2} \to 0 \quad \text{when} \quad \beta \to 0 \tag{8.80}$$

In this particular case, let us consider the positive-energy solutions $u_z^{(1)}$ and $u_z^{(2)}$. The upper two components remain $\mathcal{O}(1)$ while the lower two components vanish as $\beta \to 0$. A similar argument holds for the negative solutions:

$$\text{Non-relativistic: } u_z^{(1)} \simeq \mathcal{N}\begin{pmatrix} 1 \\ 0 \\ \cdot \\ \cdot \end{pmatrix}; \quad u_z^{(2)} \simeq \mathcal{N}\begin{pmatrix} 0 \\ 1 \\ \cdot \\ \cdot \end{pmatrix}; \quad u_z^{(3)} \simeq \mathcal{N}\begin{pmatrix} \cdot \\ \cdot \\ 1 \\ 0 \end{pmatrix}; \quad u_z^{(4)} \simeq \mathcal{N}\begin{pmatrix} \cdot \\ \cdot \\ 0 \\ 1 \end{pmatrix} \tag{8.81}$$

Due to this, a four-component Dirac spinor can be split into two two-components, called the "**large components**" $\mathcal{O}(1)$ and the "**small components**," represented with dots which are negligible in the low-energy limit. This point will play an important role when one expands the Dirac equation to the non-relativistic case. The four components effectively reduce to a two-component equation, equivalent to the Pauli theory of spin (Eq. (4.124)). Such a non-relativistic expansion of the Dirac equation will be discussed in Chapter 14.

8.10   Normalization, Orthogonality, and Completeness

As discussed in Sections 5.13 and 5.16, the convention is to normalize wave functions in order to have a convenient formulation of cross-sections and rates as products of invariant quantities of the Lorentz transformation. Since the energies can be positive or negative, this implies that we normalize to $2|E|$ particles per unit volume, where the factor two is conventional. From Eq. (8.47) this implies for $\Psi = \Psi^{(1)}$ or $\Psi = \Psi^{(2)}$:

$$\rho = \overline{\Psi}\gamma^0\Psi = \Psi^\dagger\Psi = \sum_{i=1}^{4}|\Psi_i|^2 = |\mathcal{N}|^2\left(1 + \frac{p_x^2 + p_y^2 + p_z^2}{(E+m)^2}\right) = |\mathcal{N}|^2\frac{2E}{E+m} \tag{8.82}$$

Similarly for $\Psi = \Psi^{(3)}$ or $\Psi = \Psi^{(4)}$. Hence the normalization of $2|E|$ particles per unit volume implies:

$$\mathcal{N} = \sqrt{|E| + m} \tag{8.83}$$

For the plane-wave solutions $\Psi = u(E, \vec{p})\exp(\pm ip \cdot x)$, we write the normalization condition as:

$$\Psi^\dagger\Psi = u^\dagger u = \sum_i |u_i|^2 = 2|E| \tag{8.84}$$

It is straightforward to verify that the spinors are orthogonal, $u^{(r)\dagger}u^{(s)} = 0$ for $r \neq s$. We can summarize these orthonormality conditions into the following single equation:

$$u^{(r)\dagger}u^{(s)} = 2|E|\delta_{r,s} \qquad r,s = 1,\ldots,4 \tag{8.85}$$

A related result is the term $\bar{u}^{(r)}u^{(s)}$. Starting from the Dirac equation, Eq. (8.37):

$$\slashed{p}u^{(s)} = mu^{(s)} \implies \bar{u}^{(r)}\gamma^0\gamma^\mu p_\mu u^{(s)} = m\bar{u}^{(r)}\gamma^0 u^{(s)} = mu^{(r)\dagger}u^{(s)} \tag{8.86}$$

Yet,

$$\bar{u}^{(r)}\gamma^0(\gamma^0 p_0 - \gamma^k p_k)u^{(s)} = \bar{u}^{(r)}(E\mathbb{1} - \gamma^0\gamma^k p_k)u^{(s)} = \bar{u}^{(r)}(E\mathbb{1} + \gamma^k\gamma^0 p_k)u^{(s)} \tag{8.87}$$

since $\gamma^0\gamma^k = -\gamma^k\gamma^0$. Similarly, the Hermitian of the compact Dirac equation, Eq. (8.37), gives us $[(\slashed{p} - m)u]^\dagger = u^\dagger(\gamma^\mu)^\dagger p_\mu - mu^\dagger = u^\dagger\gamma^0\gamma^\mu\gamma^0 p_\mu - mu^\dagger = 0$. If we multiply by γ^0 from the right, we obtain $u^\dagger\gamma^0\gamma^\mu\gamma^0 p_\mu\gamma^0 - mu^\dagger\gamma^0 = \bar{u}\gamma^\mu p_\mu - m\bar{u} = 0$, or:

$$\bar{u}(\gamma^\mu p_\mu - m) = \bar{u}(\slashed{p} - m) = 0 \tag{8.88}$$

Hence

$$\bar{u}^{(r)}\slashed{p} = m\bar{u}^{(r)} \implies \bar{u}^{(r)}\gamma^\mu p_\mu\gamma^0 u^{(s)} = m\bar{u}^{(r)}\gamma^0 u^{(s)} = mu^{(r)\dagger}u^{(s)} \tag{8.89}$$

and so:

$$\bar{u}^{(r)}(\gamma^0 p_0 - \gamma^k p_k)\gamma^0 u^{(s)} = \bar{u}^{(r)}(E\mathbb{1} - \gamma^k\gamma^0 p_k)u^{(s)} \tag{8.90}$$

We can add the last two results each separately, equaling $mu^{(r)\dagger}u^{(s)}$ to find:

$$2\bar{u}^{(r)}(E\mathbb{1})u^{(s)} = 2mu^{(r)\dagger}u^{(s)} \implies \bar{u}^{(r)}u^{(s)} = \frac{m}{E}u^{(r)\dagger}u^{(s)} = \frac{|E|}{E}2m\delta_{r,s} \tag{8.91}$$

Another important relation that will be used several times in calculations is the completeness relation of the spinors. We consider the positive-energy solutions $s = 1, 2$ and note that:

$$u^{(s)} = \mathcal{N}\begin{pmatrix} u_A^{(s)} \\ \dfrac{\vec{\sigma}\cdot\vec{p}}{E+m}u_A^{(s)} \end{pmatrix} \quad \text{hence} \quad \bar{u}^{(s)} = u^{(s)\dagger}\gamma^0 = \mathcal{N}\begin{pmatrix} u_A^{(s)\dagger}, & -u_A^{(s)\dagger}\dfrac{\vec{\sigma}\cdot\vec{p}}{E+m} \end{pmatrix} \tag{8.92}$$

The product $u^{(s)}\bar{u}^{(s)}$ is equal to:

$$u^{(s)}\bar{u}^{(s)} = \mathcal{N}^2 \begin{pmatrix} u_A^{(s)} u_A^{(s)\dagger} & \dfrac{-\vec{\sigma}\cdot\vec{p}}{E+m} u_A^{(s)} u_A^{(s)\dagger} \\ \dfrac{\vec{\sigma}\cdot\vec{p}}{E+m} u_A^{(s)} u_A^{(s)\dagger} & -\left(\dfrac{\vec{\sigma}\cdot\vec{p}}{E+m}\right)^2 u_A^{(s)} u_A^{(s)\dagger} \end{pmatrix} \tag{8.93}$$

By considering the sum over the two possible solutions, we obtain:

$$\sum_{s=1,2} u_A^{(s)} u_A^{(s)\dagger} = \begin{pmatrix} 1 \\ 0 \end{pmatrix}\begin{pmatrix} 1 & 0 \end{pmatrix} + \begin{pmatrix} 0 \\ 1 \end{pmatrix}\begin{pmatrix} 0 & 1 \end{pmatrix} = \begin{pmatrix} 1 & 0 \\ 0 & 1 \end{pmatrix} = \mathbb{1} \tag{8.94}$$

Hence the completeness relation is obtained by summing the product of the spinor and its adjoint over the two solutions:

$$\sum_{s=1,2} u^{(s)}\bar{u}^{(s)} = (E+m)\begin{pmatrix} \mathbb{1} & \dfrac{-\vec{\sigma}\cdot\vec{p}}{E+m} \\ \dfrac{\vec{\sigma}\cdot\vec{p}}{E+m} & -\left(\dfrac{\vec{\sigma}\cdot\vec{p}}{E+m}\right)^2 \end{pmatrix} = \begin{pmatrix} E+m & -\vec{\sigma}\cdot\vec{p} \\ \vec{\sigma}\cdot\vec{p} & -\dfrac{\vec{p}^2}{E+m} \end{pmatrix} = \begin{pmatrix} E+m & -\vec{\sigma}\cdot\vec{p} \\ \vec{\sigma}\cdot\vec{p} & -(E-m) \end{pmatrix}$$
$$\tag{8.95}$$

where we used $(\vec{\sigma}\cdot\vec{p})^2 = \vec{p}^2\mathbb{1}$ (see Appendix C.12) and $\vec{p}^2 = E^2 - m^2 = (E+m)(E-m)$. We can compare this result to the following sum (a 4×4 matrix):

$$\gamma^\mu p_\mu + m\mathbb{1} = \gamma^0 E - \gamma^k p_k + m\mathbb{1} = E\begin{pmatrix} \mathbb{1} & 0 \\ 0 & -\mathbb{1} \end{pmatrix} - \begin{pmatrix} 0 & \vec{\sigma}\cdot\vec{p} \\ -\vec{\sigma}\cdot\vec{p} & 0 \end{pmatrix} + m\begin{pmatrix} \mathbb{1} & 0 \\ 0 & \mathbb{1} \end{pmatrix} \tag{8.96}$$

Finally, we derived the **completeness relation of the Dirac spinors**, which states:

$$\sum_{s=1,2} u^{(s)}(p)\bar{u}^{(s)}(p) = \gamma^\mu p_\mu + m\mathbb{1} = \slashed{p} + m\mathbb{1} \tag{8.97}$$

8.11   Energy States Projection Operators

We have seen that the Dirac equation admits four solutions, two with positive energy and two with negative energy. Given any spinor, we can define projection operators Λ_+ (resp. Λ_-) that project over positive or negative energy states:

$$\Lambda_+ \equiv \frac{\slashed{p} + m\mathbb{1}}{2m} \quad \text{and} \quad \Lambda_- \equiv \frac{-\slashed{p} + m\mathbb{1}}{2m} \tag{8.98}$$

An arbitrary spinor can be written as a linear combination of our four solutions:

$$u = \sum_{s=1,4} c_s u^{(s)} \tag{8.99}$$

Then, for instance, the positive-energy projection gives:

$$\Lambda_+ u = \Lambda_+ \left(\sum_{s=1,4} c_s u^{(s)}\right) = \frac{\slashed{p} + m\mathbb{1}}{2m}\left(\sum_{s=1,4} c_s u^{(s)}\right) = \frac{1}{2m}\left(\sum_{r=1,2} u^{(r)}\bar{u}^{(r)}\right)\left(\sum_{s=1,4} c_s u^{(s)}\right) \tag{8.100}$$

where we have used Eq. (8.97). This result can be rewritten using Eq. (8.91):

$$\Lambda_+ u = \frac{1}{2m}\sum_{r=1,2} u^{(r)}\sum_{s=1,4} c_s \bar{u}^{(r)} u^{(s)} = \frac{1}{2m}\sum_{r=1,2} u^{(r)}\sum_{s=1,4} c_s 2m\delta_{r,s} = \sum_{r=1,2} c_r u^{(r)} \tag{8.101}$$

and similarly for Λ_-. So, as expected, the operators project two out of the four solutions. We also verify that the operators have the property of projections by checking that successive projections do not alter the result:

$$(\Lambda_+)^2 = \frac{(\not{p} + m\mathbb{1})^2}{4m^2} = \frac{1}{4m^2}\left(\not{p}\not{p} + 2m\not{p} + m^2\mathbb{1}\right) \tag{8.102}$$

However, using the Clifford algebra (Eq. (8.24)), we get:

$$\not{p}\not{p} = \gamma^\nu\gamma^\mu p_\nu p_\mu = (-\gamma^\mu\gamma^\nu + 2g^{\mu\nu}\mathbb{1})p_\nu p_\mu = -\not{p}\not{p} + 2p^2\mathbb{1} \tag{8.103}$$

So:

$$(\Lambda_+)^2 = \frac{1}{4m^2}\left((p^2 + m^2)\mathbb{1} + 2m\not{p}\right) = \frac{2m}{4m^2}(m\mathbb{1} + \not{p}) = \Lambda_+ \qquad \square \tag{8.104}$$

A similar derivation can be performed for Λ_-.

8.12 Spin of a Dirac Particle

The free-particle solutions of the Dirac equation have a twofold energy degeneracy. In order to interpret this result, we will first consider the orbital angular momentum operator $\vec{L} \equiv \vec{x} \times \vec{p} = -i\vec{x} \times \nabla$ (see Appendix C.9) and check if it commutes with the Dirac Hamiltonian:

$$
\begin{aligned}
[\vec{L}, H_{Dirac}]\Psi &= [\vec{x} \times \vec{p}, (\vec{\alpha} \cdot \vec{p} + \beta m)]\Psi = [\vec{x} \times \vec{p}, \vec{\alpha} \cdot \vec{p}]\Psi + \underbrace{[\vec{x} \times \vec{p}, \beta m]\Psi}_{=0} \\
&= (\vec{x} \times \vec{p})(\vec{\alpha} \cdot \vec{p})\Psi - (\vec{\alpha} \cdot \vec{p})(\vec{x} \times \vec{p})\Psi \\
&= (\vec{x}(\vec{\alpha} \cdot \vec{p}) \times \vec{p})\Psi - ((\vec{\alpha} \cdot \vec{p})\vec{x}) \times \vec{p})\,\Psi \\
&= ([\vec{x}, \vec{\alpha} \cdot \vec{p}] \times \vec{p})\,\Psi
\end{aligned} \tag{8.105}
$$

where we treated $\vec{\alpha} \cdot \vec{p}$ as a scalar quantity with respect to the vector product. We note that because of the commutation rules of each component of the space and momentum coordinates $[x_i, p_j] = i\delta_{ij}$, one has that the commutator is just a vector formed by the three α matrices up to a factor i:

$$
\begin{aligned}
[\vec{x}, \vec{\alpha} \cdot \vec{p}] &= \left[(x_1, x_2, x_3), \sum_j \alpha_j p_j\right] = \left(\left[x_1, \sum_j \alpha_j p_j\right], \left[x_2, \sum_j \alpha_j p_j\right], \left[x_3, \sum_j \alpha_j p_j\right]\right) \\
&= i(\alpha_1, \alpha_2, \alpha_3)
\end{aligned} \tag{8.106}
$$

So finally, we find that the angular momentum **does not commute** with the Dirac Hamiltonian:

$$[\vec{L}, H_{Dirac}] = i\vec{\alpha} \times \vec{p} \neq 0 \tag{8.107}$$

As a consequence, the orbital angular momentum is not a conserved quantity of the quantum system.

We now define the 4×4 **operator** Σ as an extension of the Pauli spin operator using the 2×2 Pauli matrices:

$$\vec{\Sigma} \text{ operator}: \quad \vec{\Sigma} = (\Sigma_1, \Sigma_2, \Sigma_3) \equiv \begin{pmatrix} \vec{\sigma} & 0 \\ 0 & \vec{\sigma} \end{pmatrix} \tag{8.108}$$

where:

$$\Sigma_1 = \begin{pmatrix} 0 & 1 & 0 & 0 \\ 1 & 0 & 0 & 0 \\ 0 & 0 & 0 & 1 \\ 0 & 0 & 1 & 0 \end{pmatrix}; \quad \Sigma_2 = \begin{pmatrix} 0 & -i & 0 & 0 \\ i & 0 & 0 & 0 \\ 0 & 0 & 0 & -i \\ 0 & 0 & i & 0 \end{pmatrix}; \quad \Sigma_3 = \begin{pmatrix} 1 & 0 & 0 & 0 \\ 0 & -1 & 0 & 0 \\ 0 & 0 & 1 & 0 \\ 0 & 0 & 0 & -1 \end{pmatrix} \tag{8.109}$$

We compute the commutator of the Σ operator with the Hamiltonian as in the case of the orbital momentum. Using the Pauli–Dirac representation, we note that:

$$[\vec{\Sigma}, H_{Dirac}] = [\vec{\Sigma}, (\vec{\alpha} \cdot \vec{p} + \beta m)] = \left[\begin{pmatrix} \vec{\sigma} & 0 \\ 0 & \vec{\sigma} \end{pmatrix}, \begin{pmatrix} 0 & \vec{\sigma} \cdot \vec{p} \\ \vec{\sigma} \cdot \vec{p} & 0 \end{pmatrix} + \begin{pmatrix} \mathbb{1} & 0 \\ 0 & -\mathbb{1} \end{pmatrix} m\right] \tag{8.110}$$

The first term is:

$$\left[\begin{pmatrix} \vec{\sigma} & 0 \\ 0 & \vec{\sigma} \end{pmatrix}, \begin{pmatrix} 0 & \vec{\sigma} \cdot \vec{p} \\ \vec{\sigma} \cdot \vec{p} & 0 \end{pmatrix}\right] = \begin{pmatrix} 0 & [\vec{\sigma}, \vec{\sigma} \cdot \vec{p}] \\ [\vec{\sigma}, \vec{\sigma} \cdot \vec{p}] & 0 \end{pmatrix} = -2i \begin{pmatrix} 0 & \vec{\sigma} \\ \vec{\sigma} & 0 \end{pmatrix} \times \vec{p} \tag{8.111}$$

because

$$\begin{aligned}
[\vec{\sigma}, \vec{\sigma} \cdot \vec{p}] &= [(\sigma_1, \sigma_2, \sigma_3), \sigma_j] p_j = ([\sigma_1, \sigma_j] p_j, [\sigma_2, \sigma_j] p_j, [\sigma_3, \sigma_j] p_j) \\
&= (2i\epsilon_{1jk}\sigma_k p_j, 2i\epsilon_{2jk}\sigma_k p_j, 2i\epsilon_{3jk}\sigma_k p_j) \\
&= -2i (\epsilon_{1jk}, \epsilon_{2jk}, \epsilon_{3jk}) \sigma_j p_k = -2i(\vec{\sigma} \times \vec{p})
\end{aligned} \tag{8.112}$$

The second term is:

$$\left[\begin{pmatrix} \vec{\sigma} & 0 \\ 0 & \vec{\sigma} \end{pmatrix}, \begin{pmatrix} \mathbb{1} & 0 \\ 0 & -\mathbb{1} \end{pmatrix}\right] = \begin{pmatrix} \vec{\sigma} & 0 \\ 0 & \vec{\sigma} \end{pmatrix}\begin{pmatrix} \mathbb{1} & 0 \\ 0 & -\mathbb{1} \end{pmatrix} - \begin{pmatrix} \mathbb{1} & 0 \\ 0 & -\mathbb{1} \end{pmatrix}\begin{pmatrix} \vec{\sigma} & 0 \\ 0 & \vec{\sigma} \end{pmatrix} = 0 \tag{8.113}$$

so we finally obtain:

$$[\vec{\Sigma}, H_{Dirac}] = -2i \begin{pmatrix} 0 & \vec{\sigma} \\ \vec{\sigma} & 0 \end{pmatrix} \times \vec{p} = -2i(\vec{\alpha} \times \vec{p}) \tag{8.114}$$

By comparing this last result with that of the orbital momentum, Eq. (8.107), we define the **total angular momentum operator** $\vec{J}$ as the sum of the orbital momentum and the **spin operator** $\vec{S} = (1/2)\Sigma$:

$$\vec{J} \equiv \vec{L} + \frac{1}{2}\vec{\Sigma} = \vec{L} + \vec{S} \tag{8.115}$$

such that its commutator with the Hamiltonian vanishes:

$$[\vec{J}, H_{Dirac}] = [\vec{L} + \vec{S}, H_{Dirac}] = 0 \tag{8.116}$$

We can conclude that the Dirac equation describes a relativistic particle with spin-$1/2$. The 4×4 matrix spin operator S is:

$$\vec{S} = (S_1, S_2, S_3) = \frac{1}{2}\vec{\Sigma} = \frac{1}{2}\begin{pmatrix} \vec{\sigma} & 0 \\ 0 & \vec{\sigma} \end{pmatrix} \tag{8.117}$$

The components of $\vec{\Sigma}$ have the same commutation relations as the Pauli matrices (see Eq. (B.42) in Appendix B.8 and also Appendix C.12), hence the $\vec{S}$ components have the same commutation relations as those of the orbital angular momentum (see Eq. (B.30) in Appendix B.6):

$$[\Sigma_i, \Sigma_j] = 2i\epsilon_{ijk}\Sigma_k \implies [\frac{1}{2}\Sigma_i, \frac{1}{2}\Sigma_j] = i\epsilon_{ijk}\frac{1}{2}\Sigma_k \implies [S_i, S_j] = i\epsilon_{ijk}S_k \tag{8.118}$$

The spin magnitude of the Dirac particle is given by $\vec{S}^2\Psi = s(s+1)\Psi$, where:

$$\vec{S}^2 = \frac{1}{4}(\Sigma_1^2 + \Sigma_2^2 + \Sigma_3^2) = \frac{3}{4}\begin{pmatrix} 1 & & & \\ & 1 & & \\ & & 1 & \\ & & & 1 \end{pmatrix} \tag{8.119}$$

Let us consider the spinors for the particle at rest $\Psi_0^{(i)}$ $(i = 1, ..., 4)$ – see Eq. (8.65). They are clearly eigenstates of the diagonal operator S_3:

$$S_3 = \frac{1}{2}\Sigma_3 = \frac{1}{2}\begin{pmatrix} \sigma_3 & 0 \\ 0 & \sigma_3 \end{pmatrix} = \frac{1}{2}\begin{pmatrix} 1 & 0 & 0 & 0 \\ 0 & -1 & 0 & 0 \\ 0 & 0 & 1 & 0 \\ 0 & 0 & 0 & -1 \end{pmatrix} \tag{8.120}$$

and therefore, represent spin-up $|\uparrow\rangle$ and spin-down $|\downarrow\rangle$ eigenstates.

Let us consider again the special case when the **particle travels in the z direction** $\vec{p} = (0, 0, \pm p)$. The solutions of Eq. (8.75) are equal to $\Psi^{(1,2)} = u_z^{(1,2)}\exp(-ip\cdot x)$ (positive energy) and $\Psi^{(3,4)} = u_z^{(3,4)}\exp(-ip\cdot x)$ (negative energy), where the spinors $u_z^{(i)}$ $(i = 1,\ldots,4)$ were given in Eq. (8.79). We can immediately deduce that:

$$S_3\Psi^{(1)} = +\frac{1}{2}\Psi^{(1)}; \quad S_3\Psi^{(2)} = -\frac{1}{2}\Psi^{(2)}; \quad S_3\Psi^{(3)} = +\frac{1}{2}\Psi^{(3)}; \quad S_3\Psi^{(4)} = -\frac{1}{2}\Psi^{(4)} \tag{8.121}$$

Hence, for particles traveling along the z direction, the spinors $\Psi^{(1)}$ and $\Psi^{(3)}$ represent spin-up, while the spinors $\Psi^{(2)}$ and $\Psi^{(4)}$ represent spin-down:

$$\vec{p} = (0, 0, \pm p) \quad \Longrightarrow \quad \begin{cases} \Psi^{(1)}, \Psi^{(3)} & \text{spin-up} \quad |\uparrow\rangle \\ \Psi^{(2)}, \Psi^{(4)} & \text{spin-down} \quad |\downarrow\rangle \end{cases} \tag{8.122}$$

In general, solutions of the Dirac equation **cannot be eigenstates of the spin, since** $[\vec{\Sigma}, H_{Dirac}] \neq 0$. Nevertheless, we will use the spinors even though their spin interpretation will not be directly possible. What matters is that the four solutions represent a complete set of solutions, with which any solution (including spin eigenstates) can be written.

8.13   Dirac Hole Theory – One-Particle Quantum Mechanics

Until this moment in scientific history the negative energy solutions have been a riddle. Since the Dirac equation is found to describe a spin-1/2 particle, the natural candidate for a Dirac particle is the **electron**. The question is how to interpret the electrons with negative energies? Also, why wouldn't electrons of a given energy not "spontaneously" decay into more negative energy states?

In order to answer this problem, Dirac postulated the existence of an **infinite sea** in the vacuum state, which is populated with negative-energy electrons. As a consequence, the negative-energy states are filled, and the Pauli exclusion principle was involved in order to explain why electrons (with positive energy) will not fall into states of negative energy. Furthermore, in the Dirac sea picture, an electron with negative energy $-E < -m$ could be excited by adding energy $> 2m$ to a state with positive energy $E' > m$, leaving **a hole in the sea** as shown in Figure 8.1. A hole in the sea represents a positive charge relative to the fully occupied phase space. Dirac originally wished to identify these "holes" with protons, but this had to be abandoned when he found that the holes necessarily have the same mass as negative electrons. The Dirac hole in the sea therefore predicts the existence of positively charged "electrons." In other words, **the absence of a negative-energy electron in the sea (a hole) represents the presence of a positive-energy positron** with opposite electric charge. This picture would explain the process of e^+e^- pair creation as well as the annihilation of an e^+e^- pair, as will be discussed later.

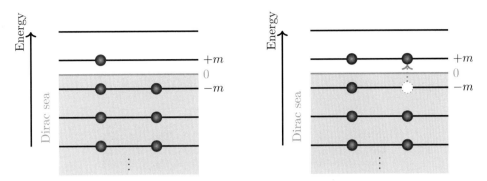

Figure 8.1 Dirac hole theory: an electron with negative energy "$-E$" is excited in a state with positive energy $E' > 0$, leaving a hole in the sea.

8.14   Discovery of the Positron

Around 1911, **Victor Hess**[5] discovered **cosmic rays** with experiments on ionizing radiation at high altitude using hot air balloons. Cosmic rays provided a natural source of particles at the highest energies (see Section 15.2). Until the end of the 1920s their composition was still unclear and investigations were ongoing to try to characterize their nature. In particular, cosmic rays were thought to be a potential source of particles other than the "conventional" electrons, protons, and neutrons that constituted ordinary matter. If the primary cosmic rays were composed of high-energy photons, then the observed flux of particles could be in the majority composed of electrons produced in the Compton process. However, it was found that positive and negative tracks were about equally frequent. At this stage, it was believed that the positive tracks must have been protons.

However, **in 1933 Anderson**[6] observed positive tracks in cosmic rays which were too long to be interpreted as protons. **Anderson** and **Millikan**[7] used a **Wilson cloud chamber** (see Section 3.9) to detect tracks produced by ionizing particles. The cloud chamber uses supersaturated vapor. Little bubbles are formed along the trajectory of the ionizing tracks, which can easily be photographed (see Section 3.9). Anderson had succeeded in building a cloud chamber that could be fitted in a large and powerful magnet. The momentum and charge could be measured from the curvature of the tracks in the magnetic field with a strength up to 2.5 T. The apparatus also contained a 6 mm-thick lead plate in order to identify the direction of the particle.

With this setup Anderson found 15 tracks out of 1300 photographs which were positively charged and not compatible with being produced by protons. An example is shown in Figure 8.2. In order to conclude on the existence of positive electrons, Anderson relied on techniques of "**particle identification**" based on the **track range** and the **energy loss**. Anderson published his results in the famous paper entitled "The positive electron" [75], where he introduced the name **positron**.

Precisely during the same period **Blackett**[8] and **Occhialini**[9] published a paper [76] were they show many photographs of tracks originating from cosmic rays. Their study of interactions induced by cosmic-ray particles is extensive. Compared to Anderson, their apparatus was more advanced as it was able to **self-trigger when a particle crossed the apparatus**, significantly increasing the efficiency of the experiment. They wrote:

"It is clear that there are several distinct processes giving rise to the complex tracks (...). In these,

5   Victor Franz Hess (1883–1964), Austrian–American physicist.
6   Carl David Anderson (1905–1991), American physicist.
7   Robert Andrews Millikan (1868–1953), American experimental physicist.
8   Patrick Maynard Stuart Blackett, Baron Blackett (1897–1974), British experimental physicist.
9   Giuseppe Paolo Stanislao "Beppo" Occhialini (1907–1993), Italian physicist.

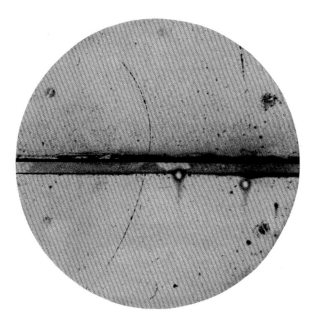

Figure 8.2 An example of a track observed by Anderson consistent with being produced by a positive electron or positron. Reprinted with permission from C. D. Anderson, "The positive electron," *Phys. Rev.*, vol. 43, pp. 491–494, Mar 1933. Copyright 1933 by the American Physical Society.

> *the typical process seems to be the simultaneous ejection of a number of particles of high energy (...). There are three possible hypotheses that can be made about the origin of these particles. They may have existed previously in the struck nucleus, or they may have existed in the incident particle, or they may have been created during the process of collision. Failing any independent evidence that they existed as separate particles previously, it is reasonable to adopt the last hypothesis. (...) One would then describe these showers as involving a creation of particles. If, however, the conservation of electric charge is to be fulfilled, then positive and negative electrons must be produced in equal numbers (...). In this way one can imagine that negative and positive electrons may be born in pairs."*

In the same paper, they also raised the question of the lifetime of the positron. They wrote:

> *"The existence of positive electrons in these showers raises immediately the question of why they have hitherto eluded observation. It is clear that they can have only a limited life as free particles since they do not appear to be associated with matter under normal conditions. It is conceivable that they can enter into combination with other particles to form stable nuclei and so cease to be free, but it seems more likely that they disappear by reacting with a negative electron to form two or more quanta."* [10]

Actually the paper of Blackett and Occhialini was received on February 7th, 1933, while Anderson's paper mentioned above was received a few weeks later on February 28th, 1933! However, Anderson had already

---

10 Republished with permission of The Royal Society (UK) from P. Blackett and G. Occhialini, "Some photographs of the tracks of penetrating radiation," *Proc. Roy. Soc. Lond. A*, vol. A139, no. 839, pp. 699–720, 1933; permission conveyed through Copyright Clearance Center, Inc.

published first evidence for the positive electron in an earlier report, entitled *The Apparent Existence of Easily Deflectable Positives*, where he claimed:

> *"The interpretation of these tracks as due to protons, or other heavier nuclei, is ruled out on the basis of range and curvature. (...) For the interpretation of these effects it seems necessary to call upon a positively charged particle having a mass comparable with that of an electron, or else admit the chance occurrence of independent tracks on the same photograph so placed as to indicate a common point of origin of two particles. The latter possibility on a probability basis is exceedingly unlikely."* [11]

For this reason, Anderson is recognized as the discoverer of the positron.

It interesting to note that both Anderson's and Blackett–Occhialini's original motivation was to study cosmic rays, not discover Dirac's positive electron. However, these observations provided strong experimental evidence in support of Dirac's model. In fact, Blackett and Occhialini were in close contact with Dirac and gave supportive arguments to Dirac's hole theory. For example, they wrote in the same paper as before:

> *"On Dirac's theory the positive electrons should only have a short life, since it is easy for a negative electron to jump down into an unoccupied state, so filling up a hole and leading to the simultaneous annihilation of a positive and negative electron, the energy being radiated as two quanta."*

Nonetheless, the hole theory was confronted with a number of conceptual problems, and presently it is considered for its historical value, as will be discussed in the next section.

8.15   Stückelberg–Feynman Interpretation and Antiparticles

In a 1934 paper **Stückelberg**[12] published a fully covariant perturbative treatment of Compton scattering, Bremsstrahlung, as well as annihilation and creation of particle–antiparticle pairs involving Dirac electrons [77]. Manifest covariance was kept by the use of the four-dimensional Fourier transform of the wave function, which leads to the **energy–momentum representation**, avoiding thereby the use of space and time variables. In the calculation of processes, Stückelberg's formalism includes the contributions from positive and negative energies into a single propagation function, which corresponds to what will later be called the **Feynman propagator**.

In 1949 **Feynman**[13] published in the **Theory of Positrons** a solution to replace the hole theory [78]. In his paper he writes:

> *"In this solution, the 'negative energy states' appear in a form which may be pictured (as by Stückelberg) in space-time as waves traveling away from the external potential backwards in time. Experimentally, such a wave corresponds to a positron approaching the potential and annihilating the electron. A particle moving forward in time (electron) in a potential may be scattered forward in time (ordinary scattering) or backward (pair annihilation). When moving backward (positron) it may be scattered backward in time (positron scattering) or forward (pair production). For such a particle the amplitude for transition from an initial to a final state is analyzed to any order in the potential by considering it to undergo a sequence of such scatterings."* [14]

11   From C. D. Anderson, "The apparent existence of easily deflectable positives," *Science*, vol. 76, pp. 238–239, 1932. Reprinted with permission from the American Association for the Advancement of Science (AAAS).

12   Ernst Carl Gerlach Stückelberg (1905–1984), Swiss mathematician and physicist (full name after 1911: Baron Ernst Carl Gerlach Stueckelberg von Breidenbach zu Breidenstein und Melsbach).

13   Richard Phillips Feynman (1918–1988), American theoretical physicist.

14   Reprinted excerpt with permission from R. P. Feynman, "The theory of positrons," *Phys. Rev.*, vol. 76, pp. 749–759, Sep 1949. Copyright 1949 by the American Physical Society.

Hence, the modern interpretation is that a negative-energy solution is interpreted as a particle which propagates backward in time. Physically it corresponds to an antiparticle propagating forward in time. See Figure 8.3. This can be naively understood by noting that the phase of the plane-wave function is unchanged under the $E \to -E$ and $t \to -t$ transformation:

$$e^{-i(-E)(+t)} = e^{-i(+E)(-t)} \tag{8.123}$$

In principle it is possible to use the negative-energy solutions and perform calculations using the above Stückelberg–Feynman interpretation. However, in order to simplify their use, it is convenient to define spinors for antiparticles in terms of their physical energy ($E > 0$) and momentum.

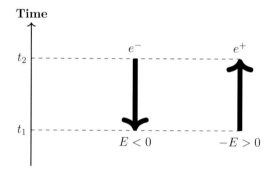

Figure 8.3 A negative-energy electron E propagating backward in time is interpreted as a positron with positive energy $-E$ moving forward in time.

We define the new solutions Φ of the Dirac equation by reversing the signs of E and $\vec{p}$ (cf. Eq. (8.54)) to give:

$$\Phi^{(i)}(x^\mu) \equiv v^{(i)}(E, \vec{p})e^{+ip\cdot x} = u^{(j)}(-E, -\vec{p})e^{-i(-p)\cdot x} \tag{8.124}$$

with the spinors $v^{(i)}(E, \vec{p})$ and their correspondence to the spinors $u^{(j)}(-E, -\vec{p})$ still needing to be defined. There is an important point related to the use of the spinors for the antiparticles. The positive $+ip \cdot x$ sign in the exponent of the function implies that the action of the energy and momentum operators will not yield the physical quantities (e.g., positive energy) since:

$$i\frac{\partial \Phi^{(i)}}{\partial t} = i(iE)\Phi^{(i)} = -E\Phi^{(i)} \tag{8.125}$$

As expected, even if $E > 0$, the wave functions $\Phi^{(i)}$ still represent negative-energy solutions of the Dirac equation. The spinor for the antiparticle must obey a modified Dirac equation to reflect the physical energy and momentum. Substituting Φ into the Dirac equation $(i\gamma^\mu \partial_\mu - m)\Phi = 0$ yields the Dirac equation for antiparticles in momentum space:

$$(\gamma^\mu p_\mu + m)v = 0 \tag{8.126}$$

To find the solutions for $v(E, \vec{p})$, we repeat the derivations of Section 8.8 and finally find (cf. Eq. (8.75)):

$$
\begin{cases}
E > 0: \quad \Phi^{(1)} = \mathcal{N} \begin{pmatrix} \dfrac{p_x - ip_y}{E + m} \\ \dfrac{-p_z}{E + m} \\ 0 \\ 1 \end{pmatrix} e^{+ip \cdot x}; \quad
\Phi^{(2)} = \mathcal{N} \begin{pmatrix} \dfrac{p_z}{E + m} \\ \dfrac{p_x + ip_y}{E + m} \\ 1 \\ 0 \end{pmatrix} e^{+ip \cdot x} \\[2em]
E < 0: \quad \Phi^{(3)} = \mathcal{N} \begin{pmatrix} 1 \\ 0 \\ \dfrac{p_z}{E - m} \\ \dfrac{p_x + ip_y}{E - m} \end{pmatrix} e^{+ip \cdot x}; \quad
\Phi^{(4)} = \mathcal{N} \begin{pmatrix} 0 \\ 1 \\ \dfrac{p_x - ip_y}{E - m} \\ \dfrac{-p_z}{E - m} \end{pmatrix} e^{+ip \cdot x}
\end{cases}
\tag{8.127}
$$

where $\Phi^{(1)}$ and $\Phi^{(2)}$ correspond to the positive-energy solutions, and $\Phi^{(3)}$ and $\Phi^{(4)}$ to the negative ones. Hence, combining the solutions Ψ and Φ we have collected eight wave functions, however, only four are independent from one another. We can use both sets of Ψs and Φs but it is natural **to use the positive-energy solutions $\Psi^{(1)}$ and $\Psi^{(2)}$ for particles (electrons) and the positive-energy solutions $\Phi^{(1)}$ and $\Phi^{(2)}$ for antiparticles (positrons).**

We note:

$$
\Phi^{(1)}(p) = \mathcal{N} \begin{pmatrix} \dfrac{p_x - ip_y}{E + m} \\ \dfrac{-p_z}{E + m} \\ 0 \\ 1 \end{pmatrix} e^{+ip \cdot x} = \mathcal{N} \begin{pmatrix} \dfrac{-(p_x - ip_y)}{-E - m} \\ \dfrac{p_z}{-E - m} \\ 0 \\ 1 \end{pmatrix} e^{-i(-p) \cdot x} = u^{(4)}(-p) e^{-i(-p) \cdot x} = \Psi^{(4)}(-p) \tag{8.128}
$$

and

$$
\Phi^{(2)}(p) = \mathcal{N} \begin{pmatrix} \dfrac{p_z}{E + m} \\ \dfrac{p_x + ip_y}{E + m} \\ 1 \\ 0 \end{pmatrix} e^{+ip \cdot x} = \mathcal{N} \begin{pmatrix} \dfrac{-p_z}{-E - m} \\ \dfrac{-(p_x + ip_y)}{-E - m} \\ 1 \\ 0 \end{pmatrix} e^{-i(-p) \cdot x} = u^{(3)}(-p) e^{-i(-p) \cdot x} = \Psi^{(3)}(-p) \tag{8.129}
$$

Hence, we have the following correspondence:

$$
v^{(1,2)}(p) \Longleftrightarrow u^{(4,3)}(-p) \tag{8.130}
$$

This correspondence should come as no surprise. The replacement $p^\mu \longrightarrow -p^\mu$ affects also the orbital angular momentum $\vec{L} = \vec{r} \times \vec{p} \longrightarrow -\vec{L}$. In order for the total momentum to be conserved, the commutator $[\vec{J}, H_{Dirac}] = [\vec{L} + \vec{S}, H_{Dirac}]$ must remain zero (see Eq. (8.116)), hence the spin operator should also change sign $\vec{S} \longrightarrow -\vec{S}$. In terms of the Dirac sea hole theory, we can interpret the situation as the absence of a negative-energy electron with spin-up representing the presence of a positive-energy positron with spin-down, and vice versa.

To summarize, we will use solutions in terms of the *physical positive energies* $E > 0$. For the **Dirac particles** (i.e., electrons), the *two* solutions are $\Psi^{(i)} = u^{(i)} \exp(-ip \cdot x)$ $(i = 1, 2)$ with:

$$
E > 0: \quad u^{(1)}(E, \vec{p}) = \sqrt{E + m} \begin{pmatrix} 1 \\ 0 \\ \dfrac{p_z}{E + m} \\ \dfrac{p_x + ip_y}{E + m} \end{pmatrix}; \quad
u^{(2)}(E, \vec{p}) = \sqrt{E + m} \begin{pmatrix} 0 \\ 1 \\ \dfrac{p_x - ip_y}{E + m} \\ \dfrac{-p_z}{E + m} \end{pmatrix} \tag{8.131}
$$

and for the **Dirac antiparticles** (i.e., positrons) the *two* solutions are $\Phi^{(i)} = v^{(i)} \exp(+ip \cdot x)$ $(i = 1, 2)$ with

$$E > 0: \quad v^{(1)}(E, \vec{p}) = \sqrt{E+m} \begin{pmatrix} \dfrac{p_x - ip_y}{E+m} \\ \dfrac{-p_z}{E+m} \\ 0 \\ 1 \end{pmatrix}; \quad v^{(2)}(E, \vec{p}) = \sqrt{E+m} \begin{pmatrix} \dfrac{p_z}{E+m} \\ \dfrac{p_x + ip_y}{E+m} \\ 1 \\ 0 \end{pmatrix} \tag{8.132}$$

where in all cases the conventional normalization is $\mathcal{N} = \sqrt{E+m}$. In compact form, the four solutions can be simply and conveniently expressed as a function of two-component spinors, as in Eq. (8.76):

$$u^{(1,2)}(E, \vec{p}) = \sqrt{E+m} \begin{pmatrix} u_A^{(1,2)} \\ \left(\frac{\vec{\sigma} \cdot \vec{p}}{E+m}\right) u_A^{(1,2)} \end{pmatrix}; \quad v^{(1,2)}(E, \vec{p}) = \sqrt{E+m} \begin{pmatrix} \left(\frac{\vec{\sigma} \cdot \vec{p}}{E+m}\right) u_A^{(2,1)} \\ u_A^{(2,1)} \end{pmatrix} \tag{8.133}$$

where the two-component bases $u_A^{(1,2)}$ are given in Eq. (8.77).

There is a subtle difference between the u and v solutions, however. We note that the action of the energy and momentum operators on the antiparticle solutions for positive energy will still give a negative energy since $i\partial_0 \Phi(E, \vec{p}) = -E\Phi(E, \vec{p})$. This is because they are indeed still the negative-energy solutions of the Dirac equation (as they should be). If we want the same p^μ for both particle and antiparticle, the consequence is that the spinors $u(E, \vec{p})$ and $v(E, \vec{p})$ will satisfy two apparently "different" Dirac equations in momentum space. We can express this as:

$$(\gamma^\mu p_\mu - m)u = 0 \quad \text{and} \quad (\gamma^\mu p_\mu + m)v = 0 \tag{8.134}$$

We can also consider the adjoint spinors $\bar{u}(E, \vec{p})$ and $\bar{v}(E, \vec{p})$. We recall that Eq. (8.88) was $\bar{u}(\not{p} - m) = 0$ and this can also be extended to $\bar{v}$. Finally, we summarize these results as:

$$(\not{p} - m)u = 0 \quad \text{and} \quad \bar{u}(\not{p} - m) = 0$$
$$(\not{p} + m)v = 0 \quad \text{and} \quad \bar{v}(\not{p} + m) = 0 \tag{8.135}$$

The **orthonormality relations** can be summarized as (for $E > 0$) – see Eq. (8.91):

$$\begin{cases} \bar{u}^{(r)} u^{(s)} = 2m\delta_{r,s} \quad (u^{\dagger(r)} u^{(s)} = 2E\delta_{r,s}) \\ \bar{v}^{(r)} v^{(s)} = -2m\delta_{r,s} \end{cases} \tag{8.136}$$

The completeness relation for u is Eq. (8.97), while for v it is:

$$\sum_{s=1,2} v^{(s)}(p) \bar{v}^{(s)}(p) = \gamma^\mu p_\mu - m\mathbb{1} = \not{p} - m\mathbb{1} \tag{8.137}$$

The Stückelberg–Feynman interpretation will have important consequences for the analysis of interactions in our relativistic theory. Let us consider the interaction illustrated in Figure 8.4. The time and space axes are shown on the left. In the left diagram an incoming electron emits a photon and then absorbs another one. So, the actual reaction depicted is $e^-\gamma \to e^-\gamma$. In reality, we only ever see the final-state particles, so positive and negative energy states must always be included. So, for the same reaction $e^-\gamma \to e^-\gamma$ we should also consider the possibility that the electron emits the photon later in time then it absorbs the incoming photon. This is illustrated on the right of Figure 8.4. The intermediate electron flowing backward in time will be interpreted as a positron flowing forward in time. Hence, the reaction should be interpreted as the incoming photon creating an e^-e^+ pair. The positron at a later time annihilates with the incoming electron to produce a photon.

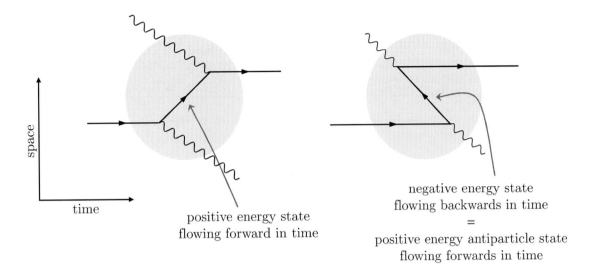

Figure 8.4 The Stückelberg–Feynman interpretation has important consequences for the analysis of interactions. The case of the $e^-\gamma \to e^-\gamma$ reaction is illustrated.

8.16   Helicity of a Dirac Particle

We have seen that the mathematical constraints to develop a quantum-mechanical wave equation consistent with special relativity led to the Dirac equation with its positive and negative-energy solutions and its twofold spin degeneracy. We have, however, seen that the commonly used spinors for particle $u(E, \vec{p})$ and antiparticles $v(E, \vec{p})$ are only spin eigenstates in the specific case where the particles travel along the z direction. Furthermore, we have seen that the spin operator $\vec{S} = (1/2)\vec{\Sigma}$ (Eq. (8.114)) does not commute with the Hamiltonian, hence, it is impossible to find a common set of eigenstates of spin and energy.

A solution to this problem is to define and use the **helicity operator**, which represents the **normalized projection of the spin along the flight direction** of the particle:[15]

$$h \equiv \frac{\vec{\Sigma} \cdot \vec{p}}{|\vec{p}|} = \begin{pmatrix} \dfrac{\vec{\sigma} \cdot \vec{p}}{|\vec{p}|} & 0 \\ 0 & \dfrac{\vec{\sigma} \cdot \vec{p}}{|\vec{p}|} \end{pmatrix} \tag{8.138}$$

We can verify that the helicity operator commutes with the Dirac Hamilton for a free particle, hence it is possible to define spinors that are simultaneous eigenstates of the Hamiltonian and the helicity operator (cf. Eqs. (8.105) and (8.114)):

$$[\vec{\Sigma} \cdot \vec{p}, H_{Dirac}]\Psi = \left([\vec{\Sigma}, H_{Dirac}] \cdot \vec{p}\right)\Psi = (-2i(\vec{\alpha} \times \vec{p}) \cdot \vec{p})\Psi = 0 \tag{8.139}$$

We note that (with $p = |\vec{p}|$):

$$h^2 = \frac{1}{p^2}\begin{pmatrix} (\vec{\sigma} \cdot \vec{p})^2 & 0 \\ 0 & (\vec{\sigma} \cdot \vec{p})^2 \end{pmatrix} = \frac{1}{p^2}\begin{pmatrix} (\vec{p})^2\mathbb{1} & 0 \\ 0 & (\vec{p})^2\mathbb{1} \end{pmatrix} = \mathbb{1} \tag{8.140}$$

15   We define the helicity with the $\vec{\Sigma}$ operator which leads to the eigenvalues ± 1 for spin-1/2 particles. It is also possible to define the helicity with the spin operator $\vec{S}$, in which case the eigenvalues of the helicity would be $\pm 1/2$.

Consequently, $h = \pm 1$ and for a spin-1/2 particle, the spin is quantized to be either "up" or "down" and this shows the helicity to be a good quantum number with eigenvalues respectively +1 or −1. These states are respectively called **positive** or **right-handed** and **negative** or **left-handed** helicity states, as illustrated in Figure 8.5. The large arrows symbolize the spin orientations relative to the momentum $\vec{p}$. Hence, the spin of the particle is parallel to its momentum for $h = +1$, and antiparallel for $h = -1$.

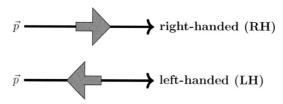

Figure 8.5 The positive $h = +1$ (i.e., right-handed) and negative $h = -1$ (i.e., left-handed) helicity states of a Dirac particle. The large arrows symbolize the spin orientations relative to $\vec{p}$.

It should be noted that although it is conserved for a free Dirac particle, **helicity is not a Lorentz-invariant quantity**. Indeed, a massive particle always has a velocity less than the speed of light, hence it is possible to find a boosted inertial frame relative to which the particle momentum appears reversed (this is not true for a massless particle traveling at the speed of light; this case is mostly relevant for the special case of neutrinos, although these latter are not strictly massless). Relative to the boosted observer, the helicity of the particle will appear reversed, hence helicity is invariant under Lorentz transformation only for massless particles.

The **helicity eigenstates** with $h = +1$ and $h = -1$ will be respectively denoted with the $\uparrow$ and $\downarrow$ signs. Starting with particle spinors, this implies:

$$hu_\uparrow = +u_\uparrow \qquad \text{and} \qquad hu_\downarrow = -u_\downarrow \tag{8.141}$$

We rewrite the equation with the two two-component bi-spinors u_A and u_B (see Eq. (8.68)) and consider the eigenvalue of the helicity to be λ, where $\lambda = \pm 1$:

$$h \begin{pmatrix} u_A \\ u_B \end{pmatrix} = \frac{1}{p} \begin{pmatrix} \vec{\sigma} \cdot \vec{p} & 0 \\ 0 & \vec{\sigma} \cdot \vec{p} \end{pmatrix} \begin{pmatrix} u_A \\ u_B \end{pmatrix} = \frac{1}{p} \begin{pmatrix} (\vec{\sigma} \cdot \vec{p})u_A \\ (\vec{\sigma} \cdot \vec{p})u_B \end{pmatrix} = \lambda \begin{pmatrix} u_A \\ u_B \end{pmatrix} \tag{8.142}$$

We can solve for u_A or u_B and will actually find the same results since the two individual equations for u_A or u_B are identical. However, the final solution should obey the Dirac equation, hence u_A and u_B are not independent. We can choose to solve for u_A and then deduce u_B to be coherent as before (see Eqs. (8.67), (8.69), and (8.70)). Therefore:

$$\frac{1}{p}(\vec{\sigma} \cdot \vec{p})u_A = \lambda u_A \qquad \text{with} \quad u_A = \begin{pmatrix} A_1 \\ A_2 \end{pmatrix} \tag{8.143}$$

Introducing the momentum in cylindrical coordinates $\vec{p} = (p_x, p_y, p_z) = (p \sin\theta \cos\phi, p \sin\theta \sin\phi, p \cos\theta)$, one obtains:

$$\frac{1}{p} \begin{pmatrix} p_z & p_x - ip_y \\ p_x + ip_y & -p_z \end{pmatrix} \begin{pmatrix} A_1 \\ A_2 \end{pmatrix} = \begin{pmatrix} \cos\theta & \sin\theta e^{-i\phi} \\ \sin\theta e^{i\phi} & -\cos\theta \end{pmatrix} \begin{pmatrix} A_1 \\ A_2 \end{pmatrix} = \lambda \begin{pmatrix} A_1 \\ A_2 \end{pmatrix} \tag{8.144}$$

or (cf. Eq. (4.119)):

$$\begin{cases} A_1 \cos\theta + A_2 \sin\theta e^{-i\phi} = \lambda A_1 \\ A_1 \sin\theta e^{i\phi} - A_2 \cos\theta = \lambda A_2 \end{cases} \implies \begin{cases} (\lambda - \cos\theta)A_1 = \sin\theta e^{-i\phi} A_2 \\ (\lambda + \cos\theta)A_2 = \sin\theta e^{i\phi} A_1 \end{cases} \tag{8.145}$$

From the first equation and for $\lambda = +1$, we get:

$$A_2 = \frac{1 - \cos\theta}{\sin\theta} e^{i\phi} A_1 = \frac{2\sin^2(\theta/2)}{2\sin(\theta/2)\cos(\theta/2)} e^{i\phi} A_1 = \frac{\sin(\theta/2)}{\cos(\theta/2)} e^{i\phi} A_1 \qquad (8.146)$$

and from the second equation and for $\lambda = -1$, we get:

$$A_2 = -\frac{\sin\theta}{(1 - \cos\theta)} e^{i\phi} A_1 = -\frac{\cos(\theta/2)}{\sin(\theta/2)} e^{i\phi} A_1 \qquad (8.147)$$

Choosing $A_1 = \cos(\theta/2)$ for $\lambda = +1$ and $A_1 = -\sin(\theta/2)$ for $\lambda = -1$, and making use of the relation Eq. (8.70), we can define the **positive/right-handed and negative/left-handed helicity eigenstate spinors for particles** as $\Psi_{\uparrow,\downarrow}(x^\mu) = u_{\uparrow,\downarrow}(p)\exp(-ip\cdot x)$ where:

$$u_\uparrow(E, p, \theta, \phi) = \mathcal{N} \begin{pmatrix} \cos(\theta/2) \\ \sin(\theta/2)e^{i\phi} \\ \frac{p}{E+m}\cos(\theta/2) \\ \frac{p}{E+m}\sin(\theta/2)e^{i\phi} \end{pmatrix} \quad \text{and} \quad u_\downarrow(E, p, \theta, \phi) = \mathcal{N} \begin{pmatrix} -\sin(\theta/2) \\ \cos(\theta/2)e^{i\phi} \\ \frac{p}{E+m}\sin(\theta/2) \\ -\frac{p}{E+m}\cos(\theta/2)e^{i\phi} \end{pmatrix} \qquad (8.148)$$

(cf. Eq. (4.123)). With the conventional normalization for the spinors, we should have:

$$u_{\uparrow\downarrow}^\dagger u_{\uparrow\downarrow} = \mathcal{N}^2 \left(u_A^\dagger u_A + \frac{p^2}{(E+m)^2} u_A^\dagger u_A \right) = \mathcal{N}^2 \left(1 + \frac{p^2}{(E+m)^2} \right) u_A^\dagger u_A = 2E \qquad (8.149)$$

For example, for $u_\uparrow$, we find:

$$u_A^\dagger u_A = \begin{pmatrix} \cos(\theta/2) & \sin(\theta/2)e^{-i\phi} \end{pmatrix} \begin{pmatrix} \cos(\theta/2) \\ \sin(\theta/2)e^{i\phi} \end{pmatrix} = \cos^2(\theta/2) + \sin^2(\theta/2) = 1 \qquad (8.150)$$

and similarly for the solution of $u_\downarrow$. Consequently:

$$u_{\uparrow\downarrow}^\dagger u_{\uparrow\downarrow} = \mathcal{N}^2 \left(\frac{(E+m)^2 + p^2}{(E+m)^2} \right) = \mathcal{N}^2 \frac{2E}{(E+m)} = 2E \qquad (8.151)$$

therefore, the corresponding conventional spinor normalization is $\mathcal{N} = \sqrt{E+m}$.

In order to find the corresponding antiparticle states $v_\uparrow$ and $v_\downarrow$, we recall the association Eq. (8.130), hence, in the correspondence between a negative-energy particle and a positive-energy antiparticle, the spin flips. We can therefore interpret the replacement as $(E, \vec{p}, \vec{S}) \to (-E, -\vec{p}, -\vec{S})$. The helicity $h \propto \vec{S}\cdot\vec{p} \to (-\vec{S})\cdot(-\vec{p}) = \vec{S}\cdot\vec{p}$ remains unchanged.

Finally, the **positive/right-handed and negative/left-handed helicity eigenstate spinors for antiparticles** are defined as $\Phi_{\uparrow,\downarrow}(x^\mu) = v_{\uparrow,\downarrow}(p)\exp(+ip\cdot x)$ where:

$$v_\uparrow(E, p, \theta, \phi) = \mathcal{N} \begin{pmatrix} \frac{p}{E+m}\sin(\theta/2) \\ -\frac{p}{E+m}\cos(\theta/2)e^{i\phi} \\ -\sin(\theta/2) \\ \cos(\theta/2)e^{i\phi} \end{pmatrix} \quad \text{and} \quad v_\downarrow(E, p, \theta, \phi) = \mathcal{N} \begin{pmatrix} \frac{p}{E+m}\cos(\theta/2) \\ \frac{p}{E+m}\sin(\theta/2)e^{i\phi} \\ \cos(\theta/2) \\ \sin(\theta/2)e^{i\phi} \end{pmatrix} \qquad (8.152)$$

and $\mathcal{N} = \sqrt{E+m}$.

8.17   Covariance of the Dirac Equation

In Section 8.3 we have stated the covariance of the Dirac equation. To establish covariance we must consider Dirac's equation in two different reference frames x and x' related to each other by a Lorentz transformation. Replacing $x \to x'$, $p \to p'$ (or $i\partial \to i\partial'$) and $\Psi \to \Psi'$ in Eq. (8.28), we have:

$$\left(i\gamma^\mu \partial'_\mu - m \right) \Psi'(x') = 0 \quad \text{where} \quad \Psi'(x') = u(p') \underbrace{e^{-ip' \cdot x'}}_{\text{invariant}} = u(p')e^{-ip \cdot x} \tag{8.153}$$

We have seen that under a Lorentz transformation $(x^\mu \to x'^\mu = \Lambda^\mu{}_\nu x^\nu)$, the effect on any physical quantity can be written according to Eq. (6.74). If we consider Dirac spinors, we can connect each Lorentz transformation Λ with an operator $U(\Lambda)$ acting on the (Hilbert) space of the Ψ solutions, which satisfies:

$$U(\Lambda)\Psi(x)U^{-1}(\Lambda) = S(\Lambda)\Psi(\Lambda^{-1}x) \tag{8.154}$$

Here we are interested in finding the transformation of a four-component Dirac spinor under Lorentz transformation. Hence, we seek the 4×4 **representation of the Lorentz group** $S(\Lambda)$ such that:

$$\Psi'(x) = S(\Lambda)\Psi(\Lambda^{-1}x) \tag{8.155}$$

In terms of components, it actually represents the four coupled relations ($\alpha = 0, ..., 3$) and $S(\Lambda)$ is a 4×4 matrix:

$$\Psi'^\alpha(x) = S(\Lambda)^\alpha{}_\beta \Psi^\beta(\Lambda^{-1}x). \tag{8.156}$$

We note that although both four-vectors and spinors possess four components, and therefore both $\Lambda^\mu{}_\nu$ and $S(\Lambda)^\alpha{}_\beta$ can be represented with 4×4 matrices, in general they will be totally different (see Section 6.8)!

In Eq. (8.153), we clearly see that $S(\Lambda)$ must act only on the spinor $u(p)$, since the exponential phase is invariant. It is also independent of x^μ. After substitution and multiplying on the left with S^{-1}, we get:

$$S^{-1}\left(i\gamma^\mu \frac{\partial}{\partial x^{\mu'}} - m \right) S\Psi = S^{-1}\left(i\gamma^\mu \frac{\partial}{\partial x^\nu}\frac{\partial x^\nu}{\partial x^{\mu'}} - m \right) S\Psi = \left(i\left(S^{-1}\gamma^\mu \frac{\partial x^\nu}{\partial x^{\mu'}}S \right)\partial_\nu - m \right)\Psi = 0 \tag{8.157}$$

which coincides with Dirac's equation if:

$$S^{-1}\gamma^\mu \frac{\partial x^\nu}{\partial x^{\mu'}}S = \gamma^\nu \quad \Longrightarrow \quad S^{-1}\gamma^\mu S = \frac{\partial x^{\mu'}}{\partial x^\nu}\gamma^\nu = \frac{\partial}{\partial x^\nu}\Lambda^\mu{}_\alpha x^\alpha \gamma^\nu = \Lambda^\mu{}_\nu \gamma^\nu \tag{8.158}$$

Hence, the Dirac equation is indeed covariant as long as the spinor changes as shown above with $S^{-1}\gamma^\mu S = \Lambda^\mu{}_\nu \gamma^\nu$. For each transformation Λ, one can always find such an $S(\Lambda)$. For instance, for a proper *infinitesimal* Lorentz transformation, we can write $x'^\mu = \Lambda^\mu{}_\nu x^\nu = (\delta^\mu_\nu + \omega^\mu{}_\nu)x^\nu$ (see Eq. (6.67)) where the $\omega^\mu{}_\nu$'s define the transformation. We consequently seek $S(\Lambda)$ near the identity:

$$S(\Lambda) \simeq \mathbb{1} + \mathcal{S} \quad \Rightarrow \quad S^{-1}(\Lambda) \simeq \mathbb{1} - \mathcal{S} \tag{8.159}$$

where $\mathcal{S}$ is a 4×4 matrix with infinitesimally small elements. We have:

$$(\mathbb{1} - \mathcal{S})\gamma^\mu(\mathbb{1} + \mathcal{S}) = \gamma^\mu - \mathcal{S}\gamma^\mu + \gamma^\mu\mathcal{S} + \cdots \overset{\text{Eq. (8.158)}}{=} (\delta^\mu_\nu + \omega^\mu{}_\nu)\gamma^\nu = \gamma^\mu + \omega^\mu{}_\nu\gamma^\nu \tag{8.160}$$

After removal of γ^μ on both sides, we find:

$$[\gamma^\mu, \mathcal{S}] = -\mathcal{S}\gamma^\mu + \gamma^\mu\mathcal{S} = \omega^\mu{}_\nu\gamma^\nu \tag{8.161}$$

Since $\mathcal{S}$ is a 4×4 matrix with infinitesimally small elements, it must depend on $\omega^\mu{}_\nu$. Also, because $\omega^\mu{}_\nu$ is antisymmetric, an educated ansatz (as suggested by Dirac) is to seek a solution proportional to the commutator of 4×4 matrices γ^μ, since these latter already contain the Pauli matrices. Consequently, we write:

$$\mathcal{S} = C\left[\gamma^\alpha, \gamma^\beta \right]\omega_{\alpha\beta} \quad \text{where} \quad C = \text{a number} \tag{8.162}$$

From Eq. (8.161):

$$[\gamma^\mu, \mathcal{S}] = -C\left[\gamma^\alpha, \gamma^\beta\right]\omega_{\alpha\beta}\gamma^\mu + \gamma^\mu C\left[\gamma^\alpha, \gamma^\beta\right]\omega_{\alpha\beta} = \left(-\gamma^\alpha\gamma^\beta\gamma^\mu + \gamma^\beta\gamma^\alpha\gamma^\mu + \gamma^\mu\gamma^\alpha\gamma^\beta - \gamma^\mu\gamma^\beta\gamma^\alpha\right)C\omega_{\alpha\beta} \quad (8.163)$$

Since ω is antisymmetric, we can rewrite the previous expression swapping the sign of a term and reversing the order α and β:

$$
\begin{aligned}
[\gamma^\mu, \mathcal{S}] &= \left(\gamma^\beta\gamma^\alpha\gamma^\mu + \gamma^\beta\gamma^\alpha\gamma^\mu + \gamma^\mu\gamma^\alpha\gamma^\beta + \gamma^\mu\gamma^\alpha\gamma^\beta\right)C\omega_{\alpha\beta} = 2C\left(\gamma^\beta\gamma^\alpha\gamma^\mu + \gamma^\mu\gamma^\alpha\gamma^\beta\right)\omega_{\alpha\beta} \\
&= 2C\left(\gamma^\beta(2g^{\alpha\mu} - \gamma^\mu\gamma^\alpha) + (2g^{\mu\alpha} - \gamma^\alpha\gamma^\mu)\gamma^\beta\right)\omega_{\alpha\beta} \\
&= 2C\left(2g^{\alpha\mu}\gamma^\beta - \gamma^\beta\gamma^\mu\gamma^\alpha + 2g^{\mu\alpha}\gamma^\beta - \gamma^\alpha\gamma^\mu\gamma^\beta\right)\omega_{\alpha\beta} = 4C\left(g^{\alpha\mu}\gamma^\beta + g^{\mu\alpha}\gamma^\beta\right)\omega_{\alpha\beta} \\
&= 8C\left(g^{\alpha\mu}\gamma^\beta\right)\omega_{\alpha\beta} \quad\quad (8.164)
\end{aligned}
$$

We obtain:

$$8C\omega^\mu{}_\beta\gamma^\beta \equiv \omega^\mu{}_\nu\gamma^\nu \implies C = \frac{1}{8} \quad (8.165)$$

Finally, we found that the infinitesimal Lorentz transformation of the spinor can be expressed in terms of the γ^μ matrices and the $\omega^\mu{}_\nu$:

$$S(\Lambda) \simeq \mathbb{1} + \frac{1}{8}\left[\gamma^\mu, \gamma^\nu\right]\omega_{\mu\nu} \quad (8.166)$$

where the $\omega^\mu{}_\nu$ are determined by $x'^\mu = \Lambda^\mu{}_\nu x^\nu = (\delta^\mu_\nu + \omega^\mu{}_\nu)x^\nu$. This is a direct demonstration that $\Lambda^\mu{}_\nu$ and $S(\Lambda)^\alpha_\beta$ are different 4×4 matrices! In the next section, we analyze the case of a finite Lorentz transformation.

8.18   Proper Lorentz Transformation of a Spinor

The general expression to transform a spinor with $S(\Lambda)$ for a given Lorentz transformation $x^\mu \to x'^\mu = \Lambda^\mu{}_\nu x^\nu$ (not necessarily infinitesimal) can be written as above where $S(\Lambda)$ is given by (cf. Eq. (6.72)):

$$\Psi'(x) = S(\Lambda)\Psi(\Lambda^{-1}x) \quad \text{where} \quad S(\Lambda) = \exp(\frac{i}{2}\Omega_{\mu\nu}S^{\mu\nu}) \quad (8.167)$$

The $\Omega_{\mu\nu}$ are exactly the same numbers as in Eq. (6.72); this ensures that we are doing the same Lorentz transformation on x and Ψ!

We need to find the **generators** $S^{\mu\nu}$ of $S(\Lambda)$ (although similar in notation, be aware of the difference between $S^{\mu\nu}$ and $S(\Lambda)^\alpha_\beta$!). How can we connect the Dirac equation and the Clifford algebra of the γ matrices with the representations of the Lorentz group? As before, we should use a combination of the Dirac γ^μ matrices to define them. We should recover the case of an infinitesimal transformation, hence:

$$S(\Lambda) \simeq \mathbb{1} + \frac{i}{2}\Omega_{\mu\nu}S^{\mu\nu} + \cdots = \mathbb{1} + \frac{1}{8}\left[\gamma^\mu, \gamma^\nu\right]\omega_{\mu\nu} = \mathbb{1} + \frac{i}{2}\left(-\omega_{\mu\nu}\right)\frac{i}{4}\left[\gamma^\mu, \gamma^\nu\right] \quad (8.168)$$

So, setting $\Omega_{\mu\nu} = -\omega_{\mu\nu}$, it turns out that **the generators of the Lorentz transformation for the spinors** can be expressed as:

$$S^{\mu\nu} = \frac{i}{4}\left[\gamma^\mu, \gamma^\nu\right] = \frac{i}{4}\left(\gamma^\mu\gamma^\nu - \gamma^\nu\gamma^\mu\right) = \frac{i}{2}\left(\gamma^\mu\gamma^\nu - g^{\mu\nu}\right) \quad (8.169)$$

Selecting the Pauli–Dirac representation, the three boosts are given by S^{0i}/S^{i0} ($i = 1, 2, 3$):

$$S^{0i}_{PD} = -S^{i0}_{PD} = \frac{i}{2}\left(\gamma^0\gamma^i\right) = \frac{i}{2}\begin{pmatrix} \mathbb{1} & 0 \\ 0 & -\mathbb{1} \end{pmatrix}\begin{pmatrix} 0 & \sigma_i \\ -\sigma_i & 0 \end{pmatrix} = \frac{i}{2}\begin{pmatrix} 0 & \sigma_i \\ \sigma_i & 0 \end{pmatrix} \quad (8.170)$$

and the rotations are represented by S^{ij} $(i \neq j \neq 0)$:

$$
\begin{aligned}
S_{PD}^{ij} &= \frac{i}{4}\left[\gamma^i, \gamma^j\right] = \frac{i}{4}\left\{\gamma^i\gamma^j - \gamma^j\gamma^i\right\} = \frac{i}{4}\left\{\begin{pmatrix} 0 & \sigma_i \\ -\sigma_i & 0 \end{pmatrix}\begin{pmatrix} 0 & \sigma_j \\ -\sigma_j & 0 \end{pmatrix} - (i \leftrightarrow j)\right\} \\
&= -\frac{i}{4}\left\{\begin{pmatrix} \sigma_i\sigma_j & 0 \\ 0 & \sigma_i\sigma_j \end{pmatrix} - (i \leftrightarrow j)\right\} \\
&= -\frac{i}{4}\left\{\begin{pmatrix} \sigma_j\sigma_i + 2i\epsilon_{ijk}\sigma_k & 0 \\ 0 & \sigma_j\sigma_i + 2i\epsilon_{ijk}\sigma_k \end{pmatrix} - \begin{pmatrix} \sigma_j\sigma_i & 0 \\ 0 & \sigma_j\sigma_i \end{pmatrix}\right\} \\
&= \frac{1}{2}\epsilon_{ijk}\begin{pmatrix} \sigma_k & 0 \\ 0 & \sigma_k \end{pmatrix}
\end{aligned}
\tag{8.171}
$$

From these, we can extract six generators: three for the rotations which we cast into a vector $\vec{J}$ and three for the boosts into $\vec{K}$ (see Appendix B.10) by selecting:

$$
\vec{J} = i(S^{23}, S^{31}, S^{12}) = \frac{i}{2}\begin{pmatrix} \vec{\sigma} & 0 \\ 0 & \vec{\sigma} \end{pmatrix} \quad \text{and} \quad \vec{K} = i(S^{10}, S^{20}, S^{30}) = -\frac{1}{2}\begin{pmatrix} 0 & \vec{\sigma} \\ \vec{\sigma} & 0 \end{pmatrix}
\tag{8.172}
$$

We could now use the generators to compute the representation matrices for finite boosts and rotation transformations. For the boosts, they turn out to be not completely straightforward in the Pauli–Dirac representation, because they are not diagonal, as obvious from the above calculation. In the following, the Weyl or chiral representation will be used to gain insight on the spinor properties under such transformations, where indeed the generators of the boosts will be diagonal!

8.19 Weyl Spinors and Chirality

In the previous section, the generators of the Lorentz boosts and rotations for spinors were derived. We have however noticed that in the Pauli–Dirac representation, their matrix representations are not diagonal. Another possibility for the γ matrices is the **Weyl or chiral representation**, where in 2×2 blocks, the γ matrices are represented as (see Eq. (8.17)):

$$
\gamma_{chiral}^0 = \mathcal{R}^{-1}\gamma_{PD}^0\mathcal{R} = \begin{pmatrix} 0 & \mathbb{1} \\ \mathbb{1} & 0 \end{pmatrix}; \quad \gamma_{chiral}^k = \mathcal{R}^{-1}\gamma_{PD}^k\mathcal{R} = \begin{pmatrix} 0 & \sigma_k \\ -\sigma_k & 0 \end{pmatrix} = \gamma_{PD}^k
\tag{8.173}
$$

We note that although the α_i and β matrices are different than in the Pauli–Dirac representation, the γ^k matrices are the same and γ^0 is different (see Eq. (8.31)). It is now easy to compute the boost and rotation generators plugging into Eq. (8.169) and note that indeed **they are diagonal in the Weyl representation**:

$$
S^{0i} = \frac{i}{2}\left(\gamma^0\gamma^i\right) = \frac{i}{2}\begin{pmatrix} 0 & \mathbb{1} \\ \mathbb{1} & 0 \end{pmatrix}\begin{pmatrix} 0 & \sigma_i \\ -\sigma_i & 0 \end{pmatrix} = \frac{i}{2}\begin{pmatrix} -\sigma_i & 0 \\ 0 & \sigma_i \end{pmatrix} \quad (i = 1, 2, 3)
\tag{8.174}
$$

and

$$
S^{ij} = \frac{i}{4}\left[\gamma^i, \gamma^j\right] = \frac{1}{2}\epsilon_{ijk}\begin{pmatrix} \sigma_k & 0 \\ 0 & \sigma_k \end{pmatrix} \quad (i \neq j \neq 0)
\tag{8.175}
$$

With the diagonal generators, the representation of the finite transformation of boosts and rotations will also be 2×2 block diagonal:

$$
S(\Lambda)_{chiral} = \exp\left(\frac{i}{2}\Omega_{\mu\nu}S^{\mu\nu}\right) = \begin{pmatrix} e^{s_1} & 0 \\ 0 & e^{s_2} \end{pmatrix}
\tag{8.176}
$$

where s_1 and s_2 can be directly computed from the definition of $S^{\mu\nu}$. As done previously, we write the Dirac spinor as a function of two two-component bi-spinors (see Eq. (8.68)), which we call Ψ_L and Ψ_R for reasons evident below:

$$\Psi_{chiral} = \begin{pmatrix} \Psi_L \\ \Psi_R \end{pmatrix} \tag{8.177}$$

We say that the Dirac spinor representation of the Lorentz group is *reducible* into two representations acting on the two-component spinors Ψ_L and Ψ_R. In other words, a **Lorentz rotation or a boost transforms the chiral bi-spinor Ψ_L into itself, and Ψ_R into itself as well**:

$$\Psi_L \xrightarrow{proper\,L.T.} \Psi'_L = e^{s_1}\Psi_L \qquad \text{and} \qquad \Psi_R \xrightarrow{proper\,L.T.} \Psi'_R = e^{s_2}\Psi_R \tag{8.178}$$

Ψ_L and Ψ_R never mix under such transformations. On the other hand, the Dirac equation of motion in the Weyl representation is (cf. Eq. (8.32) for the Pauli–Dirac representation):

$$\begin{aligned} (i\gamma^\mu\partial_\mu - m)\Psi &= \left(i\begin{pmatrix} 0 & \mathbb{1} \\ \mathbb{1} & 0 \end{pmatrix}\partial_t + i\begin{pmatrix} 0 & \sigma_k \\ -\sigma_k & 0 \end{pmatrix}\partial_k - m\begin{pmatrix} \mathbb{1} & 0 \\ 0 & \mathbb{1} \end{pmatrix} \right)\begin{pmatrix} \Psi_L \\ \Psi_R \end{pmatrix} \\ &= \begin{pmatrix} -m & i(\partial_t + \sigma_k\partial_k) \\ i(\partial_t - \sigma_k\partial_k) & -m \end{pmatrix}\begin{pmatrix} \Psi_L \\ \Psi_R \end{pmatrix} = 0 \end{aligned} \tag{8.179}$$

The two components Ψ_L and Ψ_R are mixed by the mass term in the Dirac equation! But for the case $m = 0$, they are not, and the equations of motion for Ψ_L and Ψ_R decouple into what is called the **Weyl equations of motion**:

$$i(\partial_t - \vec{\sigma}\cdot\nabla)\Psi_L = 0 \quad \text{and} \quad i(\partial_t + \vec{\sigma}\cdot\nabla)\Psi_R = 0 \tag{8.180}$$

These equations are especially important when treating **massless neutrinos**.

So far, the representations of the Lorentz rotations and boosts came out diagonal because of the particular Weyl or chiral representation chosen. In fact, this is why the representation is called chiral – it is because of the decomposition of the spinor in Eq. (8.177) and the corresponding two 2×2 irreducible representations of the Lorentz group. Is there an invariant way to define the chirality of a Dirac particle, that does not depend on the chosen representation?

• **Chirality and γ^5.** One introduces the **"fifth" gamma matrix** γ^5:

$$\gamma^5 \equiv i\gamma^0\gamma^1\gamma^2\gamma^3 \tag{8.181}$$

One can check that γ^5 satisfies the following relations (see Appendix E.1):

$$\{\gamma^5, \gamma^\mu\} = 0; \qquad (\gamma^5)^2 = 1; \qquad (\gamma^5)^\dagger = \gamma^5; \qquad [S^{\mu\nu}, \gamma^5] = 0 \tag{8.182}$$

The last relation implies that γ^5 is an invariant under rotations or boosts. It is easy to verify that γ^5 has the following forms in the chiral and Pauli–Dirac representations with $\gamma^5_{PD} = \mathcal{R}\gamma^5_{chiral}\mathcal{R}^{-1}$ (see Eq. (8.17)):

$$\gamma^5_{chiral} = \begin{pmatrix} -\mathbb{1} & 0 \\ 0 & \mathbb{1} \end{pmatrix} \quad \rightarrow \quad \gamma^5_{PD} = \mathcal{R}\gamma^5_{chiral}\mathcal{R}^{-1} = \begin{pmatrix} 0 & \mathbb{1} \\ \mathbb{1} & 0 \end{pmatrix} \tag{8.183}$$

Independently of the chosen representation, we define **the chiral projection operators P_L and P_R** as:

$$P_L = \frac{1}{2}\left(1 - \gamma^5\right) \xrightarrow[\text{chiral repr.}]{} \begin{pmatrix} \mathbb{1} & 0 \\ 0 & 0 \end{pmatrix}; \qquad \xrightarrow[\text{PD repr.}]{} \frac{1}{2}\begin{pmatrix} \mathbb{1} & -\mathbb{1} \\ -\mathbb{1} & \mathbb{1} \end{pmatrix}$$

$$P_R = \frac{1}{2}\left(1 + \gamma^5\right) \xrightarrow[\text{chiral repr.}]{} \begin{pmatrix} 0 & 0 \\ 0 & \mathbb{1} \end{pmatrix}; \qquad \xrightarrow[\text{PD repr.}]{} \frac{1}{2}\begin{pmatrix} \mathbb{1} & \mathbb{1} \\ \mathbb{1} & \mathbb{1} \end{pmatrix} \tag{8.184}$$

where we see that they are in block *diagonal* form in the *chiral* representation. We note that they project out the Weyl spinors:

$$P_L \begin{pmatrix} \Psi_L \\ \Psi_R \end{pmatrix} = \begin{pmatrix} \Psi_L \\ 0 \end{pmatrix}; \qquad P_R \begin{pmatrix} \Psi_L \\ \Psi_R \end{pmatrix} = \begin{pmatrix} 0 \\ \Psi_R \end{pmatrix} \tag{8.185}$$

Also:

$$\gamma^5 \begin{pmatrix} \Psi_L \\ 0 \end{pmatrix} = \begin{pmatrix} -\mathbb{1} & 0 \\ 0 & \mathbb{1} \end{pmatrix} \begin{pmatrix} \Psi_L \\ 0 \end{pmatrix} = -\begin{pmatrix} \Psi_L \\ 0 \end{pmatrix}; \qquad \gamma^5 \begin{pmatrix} 0 \\ \Psi_R \end{pmatrix} = +\begin{pmatrix} 0 \\ \Psi_R \end{pmatrix} \tag{8.186}$$

So, **a Dirac spinor with only a left-handed chiral component is an eigenstate of γ^5 with eigenvalue -1, and respectively with a right-handed chiral component is an eigenstate of γ^5 with eigenvalue $+1$.** Indeed, we saw that these spinors transform under Lorentz transformation without mixing.

We can further verify that the chiral operators satisfy the requirements for a complete set of **projectors**, since $P_L^2 = P_L$, $P_R^2 = P_R$, and that:

$$P_L P_R = \frac{1}{2}\left(1 - \gamma^5\right)\frac{1}{2}\left(1 + \gamma^5\right) = \frac{1}{4}\left(1 + \gamma^5 - \gamma^5 - (\gamma^5)^2\right) = 0 \tag{8.187}$$

and

$$P_L + P_R = \frac{1}{2}\left(1 - \gamma^5\right) + \frac{1}{2}\left(1 + \gamma^5\right) = \mathbb{1} \tag{8.188}$$

Hence, any spinor Ψ can be decomposed into its chiral left-handed and right-handed components:

$$\Psi \equiv \Psi_L + \Psi_R \tag{8.189}$$

where $\Psi_L = P_L\Psi$ and $\Psi_R = P_R\Psi$ are each four-component spinors (not to be confused with the bi-spinor defined above; be careful with the notation!).

• **Lorentz boost in the Pauli–Dirac representation.** We start again from the $S(\Lambda)$ in the chiral representation and express a Lorentz boost as (recalling $\Omega_{i0} = -\omega_{i0}$):

$$S(\Lambda(\vec{\phi}))_{chiral} = e^{\frac{i}{2}(\Omega_{i0}S^{i0} + \Omega_{0i}S^{0i})} = e^{-i\phi_i S^{0i}} = \begin{pmatrix} e^{-\frac{1}{2}\vec{\sigma}\cdot\vec{\phi}} & 0 \\ 0 & e^{\frac{1}{2}\vec{\sigma}\cdot\vec{\phi}} \end{pmatrix} \tag{8.190}$$

where $\vec{\phi} = (\phi_x, \phi_y, \phi_z)$ represent the parameters of the boost. So as before the transformation of the chiral spinor is (cf. Eq. (8.178)):

$$\Psi_{chiral} \xrightarrow{proper\,L.T.} \Psi'_{chiral} = \begin{pmatrix} \Psi'_L \\ \Psi'_R \end{pmatrix} = S(\Lambda)\begin{pmatrix} \Psi_L \\ \Psi_R \end{pmatrix} = \begin{pmatrix} e^{-\frac{1}{2}\vec{\sigma}\cdot\vec{\phi}} & 0 \\ 0 & e^{\frac{1}{2}\vec{\sigma}\cdot\vec{\phi}} \end{pmatrix}\begin{pmatrix} \Psi_L \\ \Psi_R \end{pmatrix} \tag{8.191}$$

In the Pauli–Dirac representation:

$$\Psi_{PD} = \mathcal{R}\Psi_{chiral} = \frac{1}{\sqrt{2}}\begin{pmatrix} \mathbb{1} & \mathbb{1} \\ -\mathbb{1} & \mathbb{1} \end{pmatrix}\begin{pmatrix} \Psi_L \\ \Psi_R \end{pmatrix} = \frac{1}{\sqrt{2}}\begin{pmatrix} \Psi_R + \Psi_L \\ \Psi_R - \Psi_L \end{pmatrix} \tag{8.192}$$

and

$$\begin{aligned} S(\Lambda)_{PD} &= \mathcal{R}S(\Lambda)_{chiral}\mathcal{R}^{-1} = \frac{1}{2}\begin{pmatrix} \mathbb{1} & \mathbb{1} \\ -\mathbb{1} & \mathbb{1} \end{pmatrix}\begin{pmatrix} e^{-\frac{1}{2}\vec{\sigma}\cdot\vec{\phi}} & 0 \\ 0 & e^{\frac{1}{2}\vec{\sigma}\cdot\vec{\phi}} \end{pmatrix}\begin{pmatrix} \mathbb{1} & -\mathbb{1} \\ \mathbb{1} & \mathbb{1} \end{pmatrix} \\ &= \frac{1}{2}\begin{pmatrix} e^{\frac{1}{2}\vec{\sigma}\cdot\vec{\phi}} + e^{-\frac{1}{2}\vec{\sigma}\cdot\vec{\phi}} & e^{\frac{1}{2}\vec{\sigma}\cdot\vec{\phi}} - e^{-\frac{1}{2}\vec{\sigma}\cdot\vec{\phi}} \\ e^{\frac{1}{2}\vec{\sigma}\cdot\vec{\phi}} - e^{-\frac{1}{2}\vec{\sigma}\cdot\vec{\phi}} & e^{\frac{1}{2}\vec{\sigma}\cdot\vec{\phi}} + e^{-\frac{1}{2}\vec{\sigma}\cdot\vec{\phi}} \end{pmatrix} = \begin{pmatrix} \cosh\left(\frac{1}{2}\vec{\sigma}\cdot\vec{\phi}\right) & \sinh\left(\frac{1}{2}\vec{\sigma}\cdot\vec{\phi}\right) \\ \sinh\left(\frac{1}{2}\vec{\sigma}\cdot\vec{\phi}\right) & \cosh\left(\frac{1}{2}\vec{\sigma}\cdot\vec{\phi}\right) \end{pmatrix} \end{aligned} \tag{8.193}$$

We can set $\vec{\phi} \equiv \phi \hat{n}$ where $\hat{n}$ is a unit vector, then using Eqs. (A.5) and (A.6), we find:

$$
\begin{cases}
\cosh\left(\frac{1}{2}\phi\vec{\sigma}\cdot\hat{n}\right) = \mathbb{1} + \frac{1}{2}(\phi/2)^2(\vec{\sigma}\cdot\hat{n})^2 + \frac{1}{24}(\phi/2)^4(\vec{\sigma}\cdot\hat{n})^4 + \cdots = \cosh(\phi/2)\mathbb{1} & (8.194) \\[2mm]
\sinh\left(\frac{1}{2}\vec{\sigma}\cdot\vec{\phi}\right) = (\phi/2)(\vec{\sigma}\cdot\hat{n}) + \frac{1}{6}(\phi/2)^3(\vec{\sigma}\cdot\hat{n})^3 + \frac{1}{120}(\phi/2)^5(\vec{\sigma}\cdot\hat{n})^5 + \cdots \\[2mm]
\qquad = (\vec{\sigma}\cdot\hat{n})\left((\phi/2) + \frac{1}{6}(\phi/2)^3(\vec{\sigma}\cdot\hat{n})^2 + \frac{1}{120}(\phi/2)^5(\vec{\sigma}\cdot\hat{n})^4 + \cdots\right) \\[2mm]
\qquad = \sinh(\phi/2)(\vec{\sigma}\cdot\hat{n}) & (8.195)
\end{cases}
$$

where we used Eq. (C.99). Since $\cosh\phi = \gamma$ and $\sinh\phi = \beta\gamma$ (see Eq. (D.27)), we have:

$$
\begin{aligned}
\cosh\left(\frac{\phi}{2}\right) &= \sqrt{\frac{\cosh\phi + 1}{2}} = \sqrt{\frac{\gamma + 1}{2}} = \sqrt{\frac{E+m}{2m}} \\[2mm]
\sinh\left(\frac{\phi}{2}\right) &= \frac{\sinh(\phi)}{\sqrt{2(\cosh(\phi)+1)}} = \sqrt{\frac{p^2}{2m(E+m)}} = \sqrt{\frac{(E+m)(E-m)}{2m(E+m)}} = \sqrt{\frac{E-m}{2m}} \\[2mm]
\tanh\left(\frac{\phi}{2}\right) &= \sqrt{\frac{E-m}{E+m}} = \frac{p}{E+m}
\end{aligned}
\qquad (8.196)
$$

where we used $E = \gamma m$ and $p = \beta E$. Regrouping all the terms, we obtain:

$$
S(\Lambda)_{PD} = \begin{pmatrix} \cosh(\phi/2)\mathbb{1} & \sinh(\phi/2)(\vec{\sigma}\cdot\hat{n}) \\ \sinh(\phi/2)(\vec{\sigma}\cdot\hat{n}) & \cosh(\phi/2)\mathbb{1} \end{pmatrix} = \sqrt{\frac{E+m}{2m}}\begin{pmatrix} \mathbb{1} & \frac{\vec{\sigma}\cdot\vec{p}}{E+m} \\ \frac{\vec{\sigma}\cdot\vec{p}}{E+m} & \mathbb{1} \end{pmatrix} \qquad (8.197)
$$

To cross-check this result, we apply this Lorentz transformation to the spinors at rest, Eq. (8.65):

$$
S(\Lambda)_{PD}\, u(\vec{p}=\vec{0}) = \sqrt{\frac{E+m}{2m}}\begin{pmatrix} \mathbb{1} & \frac{\vec{\sigma}\cdot\vec{p}}{E+m} \\ \frac{\vec{\sigma}\cdot\vec{p}}{E+m} & \mathbb{1} \end{pmatrix} \underbrace{\sqrt{2m}\begin{pmatrix} u_A \\ 0 \end{pmatrix}}_{=u(\vec{p}=0)} = \sqrt{E+m}\begin{pmatrix} u_A \\ \frac{\vec{\sigma}\cdot\vec{p}}{E+m}u_A \end{pmatrix} \qquad (8.198)
$$

which match, as expected, the results of Eq. (8.133) found by explicit solutions of Dirac's equation! A similar derivation can be done for all four solutions of the Dirac equations. Hence, as should be, one can construct the Dirac spinor for any momentum $\vec{p}$ starting from the solutions for a particle at rest and boosting them appropriately with $S(\Lambda)$.

8.20 Improper Lorentz Transformations – Dirac's Equation under Parity

So far, we have only considered proper transformations Λ which are continuously connected to the identity; these are the ones which have an infinitesimal form. However, we have seen that there are also two discrete symmetries which are part of the Lorentz group (see Section B.11):

$$
\begin{aligned}
\text{Parity P} &: \quad x^0 \to x^0; \quad x^i \to -x^i \\
\text{Time reversal T} &: \quad x^0 \to -x^0; \quad x^i \to x^i \quad (i=1,2,3)
\end{aligned}
\qquad (8.199)
$$

We now look for the representation $S(P)$ and $S(T)$ of these discrete transformations on a Dirac spinor, so that (writing for clarity $x^0 = t$ and $\vec{x} = (x^1, x^2, x^3)$):

$$
\Psi(t,\vec{x}) \to S(P)\Psi(t,-\vec{x}) \qquad \text{and} \qquad \Psi(t,\vec{x}) \to S(T)\Psi(-t,\vec{x}) \qquad (8.200)
$$

with corresponding operators:

$$P : P\Psi(x)P^{-1} = S(P)\Psi(t, -\vec{x}); \quad T : T\Psi(x)T^{-1} = S(T)\Psi(-t, \vec{x}) \tag{8.201}$$

Just as for rotations and boosts, the discrete transformations P and T should be representable by a 4×4 matrix. We use the γ matrices to construct them. In this section, we discuss the parity operation, and the time-reversal operator will be derived in the next section.

We first note that **the operator P will reverse the momentum $\vec{p}$ of a particle but should retain its spin (no spin flip but the helicity will change!)**. Also, the parity operation must satisfy $P^{-1} = P$ and $P^2 = \mathbb{1}$. In order to handle spin states in a straightforward way, we consider the case where the particle travels in the z direction. The corresponding spinors are shown in Eq. (8.79). We saw that $u_z^{(1)}$ and $u_z^{(3)}$ represent "spin-up" and respectively $u_z^{(2)}$ and $u_z^{(4)}$ "spin-down" configurations. However, since $u_z^{(1)}$ and $u_z^{(3)}$ correspond to positive and negative-energy solutions, we do not want to match them. Finally, up to an overall phase change, the parity should connect $u_z^{(i)}(E, \vec{p}) \leftrightarrow u_z^{(i)}(E, -\vec{p})$. We have for $\vec{p} = (0, 0, p)$:

$$S(P)u_z^{(1)}(E, -p) = \mathcal{N}\begin{pmatrix} 1 & 0 & 0 & 0 \\ 0 & ? & 0 & 0 \\ 0 & 0 & -1 & 0 \\ 0 & 0 & 0 & ? \end{pmatrix}\begin{pmatrix} 1 \\ 0 \\ \frac{-p}{E+m} \\ 0 \end{pmatrix} = \mathcal{N}\begin{pmatrix} 1 \\ 0 \\ \frac{p}{E+m} \\ 0 \end{pmatrix} = u_z^{(1)}(E, p) \tag{8.202}$$

where the entries with "?" cannot yet be defined, and

$$S(P)u_z^{(2)}(E, -p) = \mathcal{N}\begin{pmatrix} 1 & 0 & 0 & 0 \\ 0 & 1 & 0 & 0 \\ 0 & 0 & -1 & 0 \\ 0 & 0 & 0 & -1 \end{pmatrix}\begin{pmatrix} 0 \\ 1 \\ 0 \\ \frac{-p}{E+m} \end{pmatrix} = \mathcal{N}\begin{pmatrix} 0 \\ 1 \\ 0 \\ \frac{p}{E+m} \end{pmatrix} = u_z^{(2)}(E, p) \tag{8.203}$$

where now the 4×4 matrix is fully constrained. So, we realize that the representation of the parity operator is simply γ^0. The spinor representation of the parity is hence:

$$\boxed{P : \Psi \to S(P)\Psi = \eta_P \gamma^0 \Psi} \tag{8.204}$$

where η_P is an unobservable global phase. It is instructive to check the effect of the parity operator on the chiral states. By using the 2×2 block form of γ^0 in the Weyl representation (see Eq. (8.173)), we immediately see that:

$$S(P)\Psi_{chiral} = \begin{pmatrix} 0 & \mathbb{1} \\ \mathbb{1} & 0 \end{pmatrix}\begin{pmatrix} \Psi_L \\ \Psi_R \end{pmatrix} = \begin{pmatrix} \Psi_R \\ \Psi_L \end{pmatrix} \tag{8.205}$$

which shows that **parity exchanges right-handed and left-handed chiral spinors**. Therefore, while the chirality of a Dirac particle is invariant under a proper Lorentz transformation, it flips its sign under parity. Finally, the parity transformation will be expressed as:

$$\boxed{\text{Parity}: \quad P\Psi(t, \vec{x})P = \eta_P \gamma^0 \Psi(t, -\vec{x})} \tag{8.206}$$

Applying the spatial inversion to the Dirac equation, Eq. (8.27), gives:

$$\left(i\gamma^0\frac{\partial}{\partial t} - i\vec{\gamma} \cdot \nabla - m\right)\Psi(t, -\vec{x}) = 0 \tag{8.207}$$

This is not the same as the Dirac equation, Eq. (8.27), because there is a change of sign of the first derivative in the spatial coordinates. If we multiply from the left by γ^0 and use the relation $\gamma^0\gamma^i + \gamma^i\gamma^0 = 0$, we get back a valid Dirac equation:

$$\left(i\gamma^0\frac{\partial}{\partial t} + i\vec{\gamma} \cdot \nabla - m\right)\gamma^0\Psi(t, -\vec{x}) = 0 \tag{8.208}$$

which is consistent with Eq. (8.206).

● **Intrinsic parity of Dirac fermions.** We use the parity operator, Eq. (8.204). Directly applying it to the Dirac spinors, we get $P(u^{(1,2)}) = +u^{(1,2)}$ and $P(v^{(1,2)}) = -v^{(1,2)}$. Hence:

- The intrinsic parity of Dirac fermions is $P = +1$ (even).

- The intrinsic parity of Dirac antifermions is $P = -1$ (odd).

Parity is a *multiplicative quantum number*, so the parity of a many-particle system is equal to the product of the intrinsic parities of the particles times the parity of the spatial wave function, which is $(-1)^\ell$ (see Eq. (C.85)). As an example, positronium is an $e^+ e^-$ atom with:

$$P(e^+ e^-) = P(e^-)P(e^+)(-1)^\ell = (-1)^{\ell+1} \qquad (8.209)$$

where ℓ is the relative orbital angular momentum between the e^+ and e^-.

8.21   Dirac's Equation under Time Reversal

We recall that the operation is simply:

$$\text{Time reversal T} \quad : \quad x^0 \to -x^0; \quad x^i \to x^i \quad (i = 1, 2, 3) \qquad (8.210)$$

The **time-reversal operator** should flip the momentum $\vec{p}$ and also the spin. In analogy to the parity transformation case, we seek the operator T according to Wigner's definition: *"First we have to examine, which operation K transforms a wavefunction ϕ into the wavefunction $K\phi$ of that state, whose time runs inversely to that of ϕ, whose future thus is identical with the past and whose past is identical with the future of ϕ."* [16] Wigner's definition of time reversal can be understood by considering for simplicity the Schrödinger equation. If we take its complex conjugate and also $t \to -t$, we obtain:

$$-i\frac{\partial \psi^*(-t, \vec{x})}{\partial(-t)} = i\frac{\partial \psi^*(-t, \vec{x})}{\partial t} = H\psi^*(-t, \vec{x}) \qquad (8.211)$$

This equation actually just means that $\psi^*(-t, \vec{x})$ **is** a solution of the Schrödinger equation with the same energy. The degeneracy between $\psi(t, \vec{x})$ and $\psi^*(-t, \vec{x})$ is called **Kramers'[17] degeneracy.** The trick in Wigner's T is to take the complex conjugation of the equation and then complex conjugate the wave function. We note that Wigner's T operator satisfies $T^{-1} = -T$ and $T^2 = -\mathbb{1}$ (T is anti-unitary; see Ref. [79] for details). In the case of Dirac's spinor, we have:

$$\text{Time-reversal} : \Psi(t, \vec{x}) \to S(T)\Psi^*(-t, \vec{x}) \qquad (8.212)$$

In order to derive the form of $S(T)$, we start with Dirac's equation explicitly written out (see Eq. (8.27)) and applying $t \to -t$ in the equation, we seek a solution for:

$$\left(-i\gamma^0 \frac{\partial}{\partial t} + i\vec{\gamma} \cdot \nabla - \mathbb{1}m\right)\Psi(-t, \vec{x}) = 0 \qquad (8.213)$$

---

16   *"Zunächst müssen wir untersuchen, welche Operation K eine Wellenfunktion ϕ in die Wellenfunktion $K\phi$ desjenigen Zustandes überführt, dessen Zeit umgekehrt der von ϕ läuft, dessen Zukunft also identisch mit der Vergangenheit und dessen Vergangenheit identisch mit der Zukunft von ϕ ist."* Reprinted with permission from Springer Nature: Wigner E.P. (1993) Über die Operation der Zeitumkehr in der Quantenmechanik. In: Wightman A.S. (eds) The Collected Works of Eugene Paul Wigner. The Collected Works of Eugene Paul Wigner (Part A: The Scientific Papers. Part B: Historical, Philosophical, and Socio-Political Papers), vol A / 1. Springer, Berlin, Heidelberg.

17   Hendrik Anthony "Hans" Kramers (1894–1952), Dutch physicist.

According to Wigner, we apply the complex conjugation, multiply on the left by $S(T)$, and insert $S^{-1}(T)S(T) = \mathbb{1}$ at the right, to find:

$$S(T)\left(i\left(\gamma^0\right)^*\frac{\partial}{\partial t} - i\left(\vec{\gamma}\right)^*\cdot\nabla - \mathbb{1}m\right)S^{-1}(T)\underbrace{S(T)\Psi^*(-t,\vec{x})}_{\equiv\Psi_T} = 0 \tag{8.214}$$

where we define the time-reversed spinor as $\Psi_T = S(T)\Psi^*$. This leads to the following equation:

$$\left(iS(T)\left(\gamma^0\right)^*S^{-1}(T)\frac{\partial}{\partial t} - iS(T)\left(\vec{\gamma}\right)^*S^{-1}(T)\cdot\nabla - \mathbb{1}m\right)\Psi_T = 0 \tag{8.215}$$

In order to recover Dirac's equation for Ψ_T, we must have:

$$S(T)\left(\gamma^0\right)^*S^{-1}(T) = \gamma^0 \quad\text{and}\quad S(T)\left(\gamma^k\right)^*S^{-1}(T) = -\gamma^k \quad (k = 1,2,3) \tag{8.216}$$

Now we note that:

$$\left(\gamma^0\right)^* = \gamma^0, \quad \left(\gamma^1\right)^* = \gamma^1, \quad\text{and}\left(\gamma^3\right)^* = \gamma^3 \tag{8.217}$$

and

$$\left(\gamma^2\right)^* = \begin{pmatrix} 0 & 0 & 0 & -i \\ 0 & 0 & i & 0 \\ 0 & i & 0 & 0 \\ -i & 0 & 0 & 0 \end{pmatrix}^* = \begin{pmatrix} 0 & 0 & 0 & +i \\ 0 & 0 & -i & 0 \\ 0 & -i & 0 & 0 \\ +i & 0 & 0 & 0 \end{pmatrix} = -\gamma^2 \tag{8.218}$$

therefore $S(T)$ must satisfy the following conditions:

$$\begin{cases} S(T)\gamma^0 S^{-1}(T) = +\gamma^0, & S(T)\gamma^1 S^{-1}(T) = -\gamma^1, \\ S(T)\gamma^2 S^{-1}(T) = +\gamma^2, & S(T)\gamma^3 S^{-1}(T) = -\gamma^3 \end{cases} \tag{8.219}$$

For these equations to be satisfied, we choose the product $S(T) \propto \gamma^1\gamma^3$. The spinor representation of the time reversal is hence:

$$\boxed{T : \Psi \to \Psi_T = S(T)\Psi^* = \eta_T\gamma^1\gamma^3\Psi^*} \tag{8.220}$$

To illustrate the effect on the spinor $u(E,\vec{p})$, recalling that the γ^k are the same in the Pauli–Dirac and in the Weyl representation, and noting that $\sigma_1\sigma_3 = i\epsilon_{132}\sigma_2 = -i\sigma_2$ (see Appendix C.12), we can express the 4×4 $S(T)$ as the diagonal of two 2×2 block matrices:

$$S(T) \equiv \eta_T\gamma^1\gamma^3 = \eta_T\begin{pmatrix} 0 & \sigma_1 \\ -\sigma_1 & 0 \end{pmatrix}\begin{pmatrix} 0 & \sigma_3 \\ -\sigma_3 & 0 \end{pmatrix} = \eta_T\begin{pmatrix} -\sigma_1\sigma_3 & 0 \\ 0 & -\sigma_1\sigma_3 \end{pmatrix} = i\eta_T\begin{pmatrix} \sigma_2 & 0 \\ 0 & \sigma_2 \end{pmatrix} \tag{8.221}$$

where η_T is an unobservable global phase. We consider as in the case of parity the special case of a particle traveling in the z direction for which the identification of the spin is straightforward. We have for $\vec{p} = (0,0,p)$ (see Eq. (8.79)), writing down the 4×4 elements:

$$S(T)u_z^{(1)}(E,p) = i\eta_T\mathcal{N}\begin{pmatrix} 0 & -i & 0 & 0 \\ i & 0 & 0 & 0 \\ 0 & 0 & 0 & -i \\ 0 & 0 & i & 0 \end{pmatrix}\begin{pmatrix} 1 \\ 0 \\ \frac{p}{E+m} \\ 0 \end{pmatrix} = -\eta_T\mathcal{N}\begin{pmatrix} 0 \\ 1 \\ 0 \\ \frac{p}{E+m} \end{pmatrix} = -\eta_T u_z^{(2)}(E,p) \tag{8.222}$$

and

$$S(T)u_z^{(2)}(E,p) = i\eta_T\mathcal{N}\begin{pmatrix} 0 & -i & 0 & 0 \\ i & 0 & 0 & 0 \\ 0 & 0 & 0 & -i \\ 0 & 0 & i & 0 \end{pmatrix}\begin{pmatrix} 0 \\ 1 \\ 0 \\ \frac{p}{E+m} \end{pmatrix} = \eta_T\mathcal{N}\begin{pmatrix} 1 \\ 0 \\ \frac{p}{E+m} \\ 0 \end{pmatrix} = \eta_T u_z^{(1)}(E,p) \tag{8.223}$$

which is exactly what we expect. So finally, the time-reversal transformation will be expressed as (replacing $T^{-1} = -T$):

$$\text{Time reversal}: \quad T\Psi(t,\vec{x})T = -\eta_T \gamma^1 \gamma^3 \Psi(-t,\vec{x}) \tag{8.224}$$

8.22 Dirac's Equation under Charge Conjugation

The operation that **replaces each particle by its antiparticle and vice versa** is called the **charge conjugation operator** C keeping the spin unchanged (the operator is used for both charged particles such as electrons and positrons, as well as neutral particles such as neutrinos!).

We define the **charge conjugated spinor** Ψ_C as follows:[18]

$$\text{Charge conjugation}: \quad \Psi \rightarrow \Psi_C = C\Psi^* = C\Psi^{\dagger T} \tag{8.225}$$

where T denotes the transpose matrix. We consider the Dirac equation for a particle of charge e in an external electromagnetic field given by Eq. (8.51). The complex conjugate of this equation is just:

$$[(\gamma^\mu)^*(-i\partial_\mu - eA_\mu) - m]\Psi^* = 0 \tag{8.226}$$

On the other hand, for the charged conjugated particle (of charge $-e$), the Dirac equation should be:

$$[\gamma^\mu(i\partial_\mu + eA_\mu) - m]\Psi_C \overset{?}{=} 0 \tag{8.227}$$

We want the spinor Ψ_C to satisfy the above equation and Ψ_C should be uniquely related to Ψ. Taking the complex-conjugated equation and multiplying on the left by the searched-for operator C, we get:

$$\begin{aligned}
C[(\gamma^\mu)^*(-i\partial_\mu - eA_\mu) - m]\underbrace{C^{-1}C}_{=\mathbb{1}}\Psi^* &= [C(\gamma^\mu)^*(-i\partial_\mu - eA_\mu)C^{-1} - mCC^{-1}]C\Psi^* \\
&= [-C(\gamma^\mu)^*C^{-1}(i\partial_\mu + eA_\mu) - m]\Psi_C = 0 \tag{8.228}
\end{aligned}$$

In order to satisfy Eq. (8.227) with $\Psi_C = C\Psi^*$, we require that there exists an operator C such that:

$$-C(\gamma^\mu)^*C^{-1} = \gamma^\mu \quad \text{or} \quad -C(\gamma^\mu)^* = \gamma^\mu C \tag{8.229}$$

In our representations (see, e.g., Eq. (8.31)), only γ^2 is imaginary, and it turns out that the above condition is satisfied with $C \equiv -i\gamma^2$. For this choice, we have for $\mu = 0, 1, 3$, that:

$$-C(\gamma^\mu)^* = -(-i\gamma^2)\gamma^\mu = i\gamma^2\gamma^\mu = -i\gamma^\mu\gamma^2 = \gamma^\mu(-i\gamma^2) = \gamma^\mu C \quad \square \quad (\mu \neq 2) \tag{8.230}$$

For the case $\mu = 2$, we note that $(\gamma^2)^* = -\gamma^2$ (see Eq. (8.218)), hence:

$$-C(\gamma^2)^* = +(-i\gamma^2)\gamma^2 = -i\gamma^2\gamma^2 = \gamma^2(-i\gamma^2) = \gamma^2 C \quad \square \tag{8.231}$$

So finally, with $\Psi_C = C\Psi^*$, the charge-conjugation operation can be expressed as:

$$\text{Charge conjugation}: \quad \Psi \rightarrow \Psi_C = -i\eta_C \gamma^2 \Psi^* \tag{8.232}$$

where η_C is an unobservable global phase.

---

18   We note that sometimes in the literature the charge-conjugation operator C is defined slightly differently as $\Psi_C = C\gamma^0\Psi^* = C\overline{\Psi}^T$.

The C operator does not change the spin or the momentum, and hence not the helicity. We verify directly the effect of the charge conjugation on our specific particle and antiparticle solutions using the spinors $u^{(i)}$ and $v^{(i)}$ of the Dirac equation. In the Pauli–Dirac representation, we have (ignoring the global phase):

$$Cv^{(1)}(E, \vec{p}) = i\gamma^2 v^{(1)*} = i \begin{pmatrix} 0 & 0 & 0 & -i \\ 0 & 0 & i & 0 \\ 0 & i & 0 & 0 \\ -i & 0 & 0 & 0 \end{pmatrix} \mathcal{N} \begin{pmatrix} \dfrac{p_x + ip_y}{E + m} \\ \dfrac{-p_z}{E + m} \\ 0 \\ 1 \end{pmatrix} = \mathcal{N} \begin{pmatrix} 1 \\ 0 \\ \dfrac{p_z}{E + m} \\ \dfrac{p_x + ip_y}{E + m} \end{pmatrix} = u^{(1)}(E, \vec{p}) \quad (8.233)$$

and

$$Cv^{(2)}(E, \vec{p}) = i\gamma^2 v^{(2)*} = i \begin{pmatrix} 0 & 0 & 0 & -i \\ 0 & 0 & i & 0 \\ 0 & i & 0 & 0 \\ -i & 0 & 0 & 0 \end{pmatrix} \mathcal{N} \begin{pmatrix} \dfrac{p_z}{E + m} \\ \dfrac{p_x + ip_y}{E + m} \\ 1 \\ 0 \end{pmatrix} = -\mathcal{N} \begin{pmatrix} 0 \\ 1 \\ \dfrac{p_x - ip_y}{E + m} \\ \dfrac{-p_z}{E + m} \end{pmatrix} = -u^{(2)}(E, \vec{p})$$

$$(8.234)$$

Therefore the effect of the charge-conjugation operator on the antiparticle spinors $v^{(1)}$ and $v^{(2)}$ is to transform them into $u^{(1)}$ and $u^{(2)}$ (up to an unobservable global complex phase).

8.23 Dirac's Equation under CPT Transformation

As we discussed in Section 1.7, a further discrete transformation of fundamental importance is the combination of charge conjugation, parity, and time reversal, or CPT. The transformation of the spinor under each individual discrete Lorentz transformation is summarized in Table 8.2. Combining these properties, a spinor will transform under CPT according to:

$$S(CPT)\Psi(x) = -i\eta_C\gamma^2\eta_P\gamma^0\eta_T\gamma^1\gamma^3\Psi^*(-x) = -i\eta_{CPT}\gamma^0\gamma^1\gamma^2\gamma^3\Psi^*(-x) = -\eta_{CPT}\gamma^5\Psi^*(-x) \quad (8.235)$$

Now starting from the Dirac equation for a particle in an external magnetic field, Eq. (8.51), and applying

Space-time transformation	Spinor transformation
P	$\Psi(x) \to \eta_P\gamma^0\Psi(t, -\vec{x})$
T	$\Psi(x) \to \eta_T\gamma^1\gamma^3\Psi(-t, \vec{x})$
C	$\Psi(x) \to -i\eta_C\gamma^2\Psi^*$
CPT	$\Psi(x) \to -\eta_{CPT}\gamma^5\Psi^*(-x)$

Table 8.2 Summary of the spinors' discrete Lorentz transformations.

CPT, we can write:

$$[\gamma^\mu(i\partial_\mu - eA_\mu) - m]\Psi = 0 \xrightarrow{CPT} [\gamma^\mu(-i\partial_\mu + eA_\mu) - m]\Psi_{CPT} = 0 \quad (8.236)$$

This result is now quite equivalent to Dirac's equation. However, if we change the sign of the energy and the three-momentum, namely $p^\mu \to -p^\mu$, thus $-i\partial_\mu \to i\partial_\mu$, we obtain the correct form for a Dirac particle with opposite electric charge:

$$[\gamma^\mu(i\partial_\mu + eA_\mu) - m]\,\Psi_{CPT} = 0 \tag{8.237}$$

This result supports the Stückelberg–Feynman interpretation that a particle of mass m is equivalent to an antiparticle of mass m traveling backward in space-time. As mentioned in Section 1.7, the CPT transformation is crucial since the CPT-theorem (see, e.g., Ref. [9]) states that all relativistic quantum field theories must be invariant under this transformation. Also, that particles and their corresponding antiparticles have the same rest mass is a sign of CPT conservation.

8.24 Chirality and Helicity

It is important to realize the difference between chirality and helicity. The chiral states are defined as the eigenstates of γ^5 which can be obtained from any spinor with the left-handed and right-handed chiral projections P_L and P_R, whereas the helicity is defined as the projection of the spin operator onto the direction of the momentum of the particle. Because of the properties of quantization of the spin, the eigenvalues of the helicity are ± 1, which we called positive or right-handed and negative or left-handed helicity states. Note therefore that the **handedness** (left or right) is used to characterize *both* the chirality and the helicity, although they correspond to different physical properties (for the helicity we clearly have a physical interpretation of positive or negative helicity eigenvalues). **The reader might prefer to adopt the distinction left/right-handed chiral and positive/negative helicity states** – but this is usually not the case in the literature. **The situation is further complicated by the fact that in the so-called ultra-relativistic case $E \gg m$ where rest masses can be neglected, or for actually massless particles, chiral and helicity eigenstates coincide but are opposite for particle and antiparticles!** We derive these results in the following paragraphs.

The relationship between chiral and helicity eigenstates can be found by decomposing the helicity eigenstates which we derived in Eq. (8.148) by studying the effect of the chiral projection operators. Representing for convention:

$$c = \cos(\theta/2), \quad s = \sin(\theta/2), \quad \text{and} \quad \alpha = \frac{p}{E+m} \tag{8.238}$$

we have in the Pauli–Dirac representation for particles:

$$u_\uparrow(E,p,\theta,\phi) = \mathcal{N} \begin{pmatrix} c \\ se^{i\phi} \\ \alpha c \\ \alpha s e^{i\phi} \end{pmatrix} \quad \text{and} \quad u_\downarrow(E,p,\theta,\phi) = \mathcal{N} \begin{pmatrix} -s \\ ce^{i\phi} \\ \alpha s \\ -\alpha c e^{i\phi} \end{pmatrix} \tag{8.239}$$

and for antiparticles:

$$v_\uparrow(E,p,\theta,\phi) = \mathcal{N} \begin{pmatrix} \alpha s \\ -\alpha c e^{i\phi} \\ -s \\ ce^{i\phi} \end{pmatrix} \quad \text{and} \quad v_\downarrow(E,p,\theta,\phi) = \mathcal{N} \begin{pmatrix} \alpha c \\ \alpha s e^{i\phi} \\ c \\ se^{i\phi} \end{pmatrix} \tag{8.240}$$

In the same representation, we have (see Eq. (8.184)):

$$P_L = \frac{1}{2}\begin{pmatrix} 1 & 0 & -1 & 0 \\ 0 & 1 & 0 & -1 \\ -1 & 0 & 1 & 0 \\ 0 & -1 & 0 & 1 \end{pmatrix} \quad \text{and} \quad P_R = \frac{1}{2}\begin{pmatrix} 1 & 0 & 1 & 0 \\ 0 & 1 & 0 & 1 \\ 1 & 0 & 1 & 0 \\ 0 & 1 & 0 & 1 \end{pmatrix} \tag{8.241}$$

We first analyze the state $u_\uparrow$. Since any spinor can be decomposed into its chiral left-handed and right-handed components, we compute:

$$P_L u_\uparrow = \frac{\mathcal{N}}{2} \begin{pmatrix} 1 & 0 & -1 & 0 \\ 0 & 1 & 0 & -1 \\ -1 & 0 & 1 & 0 \\ 0 & -1 & 0 & 1 \end{pmatrix} \begin{pmatrix} c \\ se^{i\phi} \\ \alpha c \\ \alpha se^{i\phi} \end{pmatrix} = \frac{1}{2}(1-\alpha)\mathcal{N} \begin{pmatrix} c \\ se^{i\phi} \\ -c \\ -se^{i\phi} \end{pmatrix} \tag{8.242}$$

and

$$P_R u_\uparrow = \frac{\mathcal{N}}{2} \begin{pmatrix} 1 & 0 & 1 & 0 \\ 0 & 1 & 0 & 1 \\ 1 & 0 & 1 & 0 \\ 0 & 1 & 0 & 1 \end{pmatrix} \begin{pmatrix} c \\ se^{i\phi} \\ \alpha c \\ \alpha se^{i\phi} \end{pmatrix} = \frac{1}{2}(1+\alpha)\mathcal{N} \begin{pmatrix} c \\ se^{i\phi} \\ c \\ se^{i\phi} \end{pmatrix} \tag{8.243}$$

Remembering that pure left or right chiral states are eigenstates of γ^5 (see Eq. (8.186)), we can define the left and right chiral eigenspinors for particles u_L, u_R and antiparticles v_L, v_R, as:

$$\gamma^5 u_L = -u_L \quad \text{and} \quad \gamma^5 u_R = +u_R; \qquad \gamma^5 v_R = -v_R \quad \text{and} \quad \gamma^5 v_L = +v_L \tag{8.244}$$

The apparently opposite convention for particle and antiparticle will become clear below. As expected, we have for the chiral left-handed projections:

$$P_L u_L = \frac{1}{2}\left(u_L - \gamma^5 u_L\right) = \frac{1}{2}\left(u_L + u_L\right) = u_L; \quad P_L u_R = 0; \quad P_L v_R = v_R; \quad P_L v_L = 0 \tag{8.245}$$

and similarly for the chiral right-handed projections:

$$P_R u_L = 0; \quad P_R u_R = u_R; \quad P_R v_R = 0; \quad P_R v_L = v_L \tag{8.246}$$

Combining the above results, the positive-helicity eigenstate $u_\uparrow$ can be expressed in terms of chiral components as:

$$
\begin{aligned}
u_\uparrow &= P_L u_\uparrow + P_R u_\uparrow = \frac{1}{2}(1-\alpha)\mathcal{N} \begin{pmatrix} c \\ se^{i\phi} \\ -c \\ -se^{i\phi} \end{pmatrix} + \frac{1}{2}(1+\alpha)\mathcal{N} \begin{pmatrix} c \\ se^{i\phi} \\ c \\ se^{i\phi} \end{pmatrix} \\
&\equiv A_L^\uparrow u_L + A_R^\uparrow u_R
\end{aligned}
\tag{8.247}
$$

where $A_L^\uparrow$ and $A_R^\uparrow$ are complex coefficients. From the above equality, we can conclude that the left and right-handed coefficients are proportional to:

$$A_L^\uparrow \propto \frac{1}{2}(1-\alpha) \quad \text{and} \quad A_R^\uparrow \propto \frac{1}{2}(1+\alpha) \tag{8.248}$$

Similarly, we would find, starting from the negative-helicity state, that $u_\downarrow = A_L^\downarrow u_L + A_R^\downarrow u_R$, where:

$$A_L^\downarrow \propto \frac{1}{2}(1+\alpha) \quad \text{and} \quad A_R^\downarrow \propto \frac{1}{2}(1-\alpha) \tag{8.249}$$

For the **ultra-relativistic limit** $E \gg m$ where rest masses can be neglected, or for actually massless particles, we have $m \to 0$ or $\alpha \to 1$. Hence, in this case, the helicity eigenstates are equivalent to the chiral eigenstates, hence **for particles in the limit $E \gg m$ positive helicity is identical to right-handed chirality, and negative helicity is identical to left-handed chirality**:

$$
\begin{aligned}
A_L^\uparrow \to 0; & \quad \text{and} \quad A_R^\uparrow \to 1 \quad \text{so} \quad u_\uparrow \leftrightarrow u_R \\
A_L^\downarrow \to 1; & \quad \text{and} \quad A_R^\downarrow \to 0 \quad \text{so} \quad u_\downarrow \leftrightarrow u_L
\end{aligned}
\tag{8.250}
$$

The convention of Eq. (8.244) leads to **negative-helicity antiparticles being identical to their right-handed chiral state, and positive-helicity being equal to their left-handed chiral state**:

$$v_\uparrow \leftrightarrow v_L \qquad \text{and} \qquad v_\downarrow \leftrightarrow v_R \tag{8.251}$$

This conclusion can also be stated as follows: **in the ultra-relativistic limit, the left and right-handed chiral solutions of the Dirac equation (i.e., which are eigenstates of γ^5) are identical to helicity eigenstates. Inversely, states with a defined helicity (either positive or negative) correspond to pure left-handed or right-handed chiral states.** This is a fundamental result to analyze the so-called chiral interactions, as discussed later.

8.25   Spin Projection Operator for a Dirac Particle

Although we will commonly use the helicity and/or chirality eigenstates, it is still useful to define spin-projection operators in order to obtain eigenstates with given polarization. The spin operator $\vec{\Sigma}$ was defined in Eq. (8.108) as a function of the 2×2 Pauli matrices. We now seek a covariant form for this operator that can be used in any reference frame. We start by considering the product of two γ matrices. In the Pauli–Dirac representation, one can explicitly show that for $i, j = 1, 2, 3$:

$$\gamma^i \gamma^j = \begin{pmatrix} 0 & \sigma_i \\ -\sigma_i & 0 \end{pmatrix} \begin{pmatrix} 0 & \sigma_j \\ -\sigma_j & 0 \end{pmatrix} = -\begin{pmatrix} \sigma_i \sigma_j & 0 \\ 0 & \sigma_i \sigma_j \end{pmatrix} \tag{8.252}$$

Consequently, for $i \neq j$, the product is:

$$
\begin{aligned}
i\gamma^i \gamma^j &= \frac{i}{2}\left(\gamma^i \gamma^j + \gamma^i \gamma^j\right) = \frac{i}{2}\left(\gamma^i \gamma^j - \gamma^j \gamma^i\right) = -\frac{i}{2}\begin{pmatrix} \sigma_i \sigma_j - \sigma_j \sigma_i & 0 \\ 0 & \sigma_i \sigma_j - \sigma_j \sigma_i \end{pmatrix} \\
&= -\frac{(2i)i}{2}\begin{pmatrix} \epsilon_{ijk}\sigma_k & 0 \\ 0 & \epsilon_{ijk}\sigma_k \end{pmatrix} = \epsilon_{ijk}\begin{pmatrix} \sigma_k & 0 \\ 0 & \sigma_k \end{pmatrix}
\end{aligned}
\tag{8.253}
$$

The spin operator $\vec{\Sigma}$ (see Eq. (8.108)) can then be written simply as:

$$\vec{\Sigma} = i\left(\gamma^2 \gamma^3, \gamma^3 \gamma^1, \gamma^1 \gamma^2\right) \tag{8.254}$$

For example, the diagonal spin operator S_3 becomes:

$$S_3 = \frac{1}{2}\Sigma_3 = \frac{i}{2}\gamma^1 \gamma^2 = -\frac{i}{2}\underbrace{\gamma^0 \gamma^0}_{=\mathbb{1}}\gamma^1 \gamma^2 \underbrace{\gamma^3 \gamma^3}_{=-\mathbb{1}} = -\frac{1}{2}\gamma^0 \gamma^5 \gamma^3 = -\frac{1}{2}\gamma^5 \gamma^3 \gamma^0 \tag{8.255}$$

We have seen in Section 8.12 that the eigenvectors of S_3 correspond to the spin-up and spin-down configurations for a particle traveling along the z direction. In general, the **polarization vector $\vec{s}$ is defined in the rest frame of the particle, as the unit vector ($\vec{s}^2 = 1$) that points in the direction of the spin**. This definition can be easily extended to the spin-polarization four-vector s^μ. In the rest frame, we have:

$$s^\mu = (0, \vec{s}), \qquad s^2 = -1 \tag{8.256}$$

In the rest frame, we also have $p^\mu = (m, \vec{0})$, hence $s^\mu \cdot p_\mu = 0$. We seek the spin operator for any spin direction. The following operator:

$$S(s_\mu) \equiv \frac{1}{2}\gamma^5 \gamma^\mu s_\mu \gamma^0 = \frac{1}{2}\gamma^5 \slashed{s}\gamma^0 \tag{8.257}$$

reduces to S_3 in Eq. (8.255) for the specific case $s^\mu = (0,0,0,1)$. The operator $S(s_\mu)$ actually is of the correct form for any polarization vector $\vec{s}$ in the particle rest frame. Since $\not{p}u$ reduces to $m\gamma^0 u$ in the rest frame, we can replace γ^0 by $\not{p}/m$ to get the covariant spin operator:

$$S(s_\mu) \equiv \frac{1}{2m}\gamma^5 \not{s}\not{p} \tag{8.258}$$

This expression can be used for any frame, by boosting the momentum four-vector from the rest frame to that of the moving particle, and applying the same boost to the spin four-vector:

$$p^\mu = \Lambda^\mu_{\;\nu}\left[(m,\vec{0})\right]^\nu \quad \longrightarrow \quad s^\mu = \Lambda^\mu_{\;\nu}\left[(0,\vec{s})\right]^\nu \tag{8.259}$$

In order to find the projection operator, similar to the one that projects energy states derived in Section 8.11, we note that for a particle traveling along the z direction, we found (see Eq. (8.121)):

$$S_3 u_z^{(1)}(p) = +\frac{1}{2}u_z^{(1)}(p), \qquad S_3 u_z^{(2)}(p) = -\frac{1}{2}u_z^{(2)}(p) \tag{8.260}$$

Defining $u_z^{(+1)}(p) = u_z^{(1)}(p)$ and $u_z^{(-1)}(p) = u_z^{(2)}(p)$, the two above relations can be expressed in a very compact form as with $s = \pm 1$:

$$\frac{1}{2}(1 + 2sS_3)u_z^{(s)} = \begin{cases} \frac{1}{2}(1 + 2(1)\frac{1}{2})u_z^{(1)} = +u_z^{(1)} \\ \frac{1}{2}(1 + 2(-1)\left(-\frac{1}{2}\right))u_z^{(2)} = u_z^{(2)} \end{cases} = u_z^{(s)} \tag{8.261}$$

and

$$\frac{1}{2}(1 + 2sS_3)u_z^{(-s)} = \begin{cases} \frac{1}{2}(1 + 2(-1)\frac{1}{2})u_z^{(1)} = 0 \\ \frac{1}{2}(1 + 2(+1)\left(-\frac{1}{2}\right))u_z^{(2)} = 0 \end{cases} \tag{8.262}$$

so finally for the particle along z with momentum $\vec{p}$:

$$\frac{1}{2}(1 + 2sS_3)u_z^{(s')}(\vec{p}) = \delta_{s'}^s u_z^{(s')}(\vec{p}) \tag{8.263}$$

Let us consider the case $s = +1$, where the spin points in the direction of the polarization vector. Because of the covariant form of this expression, we may apply it to any polarization vector s^μ with $s^\mu p_\mu = 0$ and get for any spinor $u(\vec{p})$:

$$\frac{1}{2}(1 + 2\frac{1}{2m}\gamma^5 \not{s}\not{p})u^{(s')}(\vec{p}) = \frac{1}{2}(1 + \gamma^5 \not{s}\frac{\not{p}}{m})u^{(s')}(\vec{p}) = \frac{1}{2}(1 + \gamma^5 \not{s})u^{(s')}(\vec{p}) = \delta_{s'}^{+1}u^{(s')}(\vec{p}) \tag{8.264}$$

where we used $\not{p}u = mu$. This equation is of the form:

$$\Sigma(s)u^{(s')}(\vec{p}) = \delta_{s'}^{+1}u^{(s')}(\vec{p}) \tag{8.265}$$

where

$$\Sigma(s) \equiv \frac{1}{2}(1 + \gamma^5 \not{s}) \tag{8.266}$$

is the spin-projection operator that projects out the state with spin corresponding to the spin four-vector s^μ. We can cast this into:

$$\Sigma(s)u(\vec{p}, s) = u(\vec{p}, s) \quad \text{and} \quad \Sigma(-s)u(\vec{p}, s) = 0 \tag{8.267}$$

Similar arguments can be repeated for the v spinors to find the same result, so finally the spin projector is applicable to both u and v spinors.

8.26   Dirac Lagrangian – Searching for the Action

In order to derive the action associated with the Dirac wave equation, we should construct the Dirac Lagrangian which is an invariant under Lorentz transformation. We therefore seek to build an invariant out of a spinor. A real number is formed from a spinor using the complex transpose. As a first attempt, we consider the quantity (see Eq. (8.47)):

$$\Psi^\dagger \Psi = (\Psi^*)^T \Psi = \begin{pmatrix} \Psi_1^* & \Psi_2^* & \Psi_3^* & \Psi_4^* \end{pmatrix} \begin{pmatrix} \Psi_1 \\ \Psi_2 \\ \Psi_3 \\ \Psi_4 \end{pmatrix} = \sum_{i=1}^4 |\Psi_i|^2 \tag{8.268}$$

How does it transform under Lorentz transformation? Since $\Psi(x) \to S(\Lambda)\Psi(\Lambda^{-1}x)$ and hence $\Psi(x)^\dagger \to \Psi^\dagger(\Lambda^{-1}x)S^\dagger(\Lambda)$ we have:

$$\Psi^\dagger \Psi \to \Psi^\dagger(\Lambda^{-1}x)S^\dagger(\Lambda)S(\Lambda)\Psi(\Lambda^{-1}x) \tag{8.269}$$

We are therefore interested in the properties of $S^\dagger(\Lambda)S(\Lambda)$. With the proper Lorentz transformation representation given by $S(\Lambda) = \exp(-\frac{1}{2}\Omega_{\mu\nu}S^{\mu\nu})$ (see Eq. (8.167)), we have $S^\dagger(\Lambda) = \exp(+\frac{i}{2}\Omega_{\mu\nu}(S^{\mu\nu})^\dagger)$, where using Eq. (8.169) we find:

$$(S^{\mu\nu})^\dagger = \left[\frac{i}{4}[\gamma^\mu, \gamma^\nu]\right]^\dagger = \frac{i}{4}\left[(\gamma^\mu)^\dagger, (\gamma^\nu)^\dagger\right] = \frac{i}{4}\left[\gamma^0\gamma^\mu\gamma^0, \gamma^0\gamma^\nu\gamma^0\right] = \gamma^0 S^{\mu\nu}\gamma^0 \tag{8.270}$$

since $(\gamma^\mu)^\dagger = \gamma^0\gamma^\mu\gamma^0$ (see Eq. (8.26)). Finally:

$$S^\dagger(\Lambda) = \exp\left(-\frac{i}{2}(-\Omega_{\mu\nu})(\gamma^0 S^{\mu\nu}\gamma^0)\right) = \gamma^0 \exp\left(-\frac{i}{2}(-\Omega_{\mu\nu})S^{\mu\nu}\right)\gamma^0 = \gamma^0 S(\Lambda)^{-1}\gamma^0 \tag{8.271}$$

where we used the fact that the inverse of the Lorentz transformation can be found by replacing $\Omega_{\mu\nu}$ by $-\Omega_{\mu\nu}$. So, in general (mathematically S is not unitary, i.e., $S^\dagger \neq S^{-1}$ because its generators are not Hermitian):

$$S^\dagger(\Lambda)S(\Lambda) = \gamma^0 S(\Lambda)^{-1}\gamma^0 S(\Lambda) \neq \mathbb{1} \tag{8.272}$$

Therefore, the "naive" product $\Psi^\dagger\Psi$ is not the sought-for invariant. We must use the adjoint spinor $\overline{\Psi}$ (see Eq. (8.42)) introduced to develop the covariant Dirac equation. We have (assuming for example the Pauli–Dirac representation):

$$\begin{aligned}
\overline{\Psi}\Psi &= \Psi^\dagger\gamma^0\Psi = \begin{pmatrix} \Psi_1^* & \Psi_2^* & \Psi_3^* & \Psi_4^* \end{pmatrix} \begin{pmatrix} 1 & 0 & 0 & 0 \\ 0 & 1 & 0 & 0 \\ 0 & 0 & -1 & 0 \\ 0 & 0 & 0 & -1 \end{pmatrix} \begin{pmatrix} \Psi_1 \\ \Psi_2 \\ \Psi_3 \\ \Psi_4 \end{pmatrix} \\
&= |\Psi_1|^2 + |\Psi_2|^2 - |\Psi_3|^2 - |\Psi_4|^2
\end{aligned} \tag{8.273}$$

Its transformation under proper Lorentz transformation is:

$$\begin{aligned}
\overline{\Psi}(x)\Psi(x) = \Psi^\dagger(x)\gamma^0\Psi(x) &\to \Psi^\dagger(\Lambda^{-1}x)S^\dagger(\Lambda)\gamma^0 S(\Lambda)\Psi(\Lambda^{-1}x) \\
&= \Psi^\dagger(\Lambda^{-1}x)\gamma^0 S(\Lambda)^{-1}\gamma^0\gamma^0 S(\Lambda)\Psi(\Lambda^{-1}x) \\
&= \Psi^\dagger(\Lambda^{-1}x)\gamma^0\Psi(\Lambda^{-1}x) \\
&= \overline{\Psi}(\Lambda^{-1}x)\Psi(\Lambda^{-1}x) \qquad \square
\end{aligned} \tag{8.274}$$

Finally, **the quantity $\overline{\Psi}\Psi$ constructed from the spinor and its adjoint is a scalar invariant under Lorentz transformation**.

What about the **current quantity** $j^\mu \equiv \overline{\Psi}\gamma^\mu\Psi$ constructed from spinors and one gamma matrix? For a given index μ, the product is of the form:

$$
\overline{\Psi}\gamma^\mu\Psi = \begin{pmatrix} \Psi_1^* & \Psi_2^* & \Psi_3^* & \Psi_4^* \end{pmatrix} \gamma^0 \begin{pmatrix} \cdot & \cdot & \cdot & \cdot \\ \cdot & \cdot & \cdot & \cdot \\ \cdot & \cdot & \cdot & \cdot \\ \cdot & \cdot & \cdot & \cdot \end{pmatrix}^\mu \begin{pmatrix} \Psi_1 \\ \Psi_2 \\ \Psi_3 \\ \Psi_4 \end{pmatrix} = [.]^\mu \qquad (\mu = 0, 1, 2, 3)
$$

$$
= \begin{pmatrix} . & . & . & . \end{pmatrix} \tag{8.275}
$$

What are its properties? We first note that the divergence of the current vanishes:

$$
\partial_\mu j^\mu = (\partial_\mu \overline{\Psi})\gamma^\mu\Psi + \overline{\Psi}\gamma^\mu\partial_\mu\Psi = (im\overline{\Psi})\Psi + \overline{\Psi}(-im\Psi) \overset{!}{=} 0 \tag{8.276}
$$

Hence the current j^μ is conserved if the spinor Ψ satisfies the Dirac equation. Under the proper Lorentz transformation, the current will be transformed as (suppressing the space-time coordinates x^μ):

$$
\overline{\Psi}\gamma^\mu\Psi \to \overline{\Psi}S(\Lambda)^{-1}\gamma^\mu S(\Lambda)\Psi \overset{?}{=} \Lambda^\mu{}_\nu (\overline{\Psi}\gamma^\nu\Psi) \tag{8.277}
$$

Note that it is the quantity $\overline{\Psi}\gamma^\nu\Psi$ that is transforming, not the γ^ν matrix! In order for $\overline{\Psi}\gamma^\mu\Psi$ to transform as a vector, we must satisfy:

$$
S(\Lambda)^{-1}\gamma^\mu S(\Lambda) = \Lambda^\mu{}_\nu \gamma^\nu \tag{8.278}
$$

In order to show that this is the case, one can consider the infinitesimal transformations and note that (see Eq. (6.72)):

$$
(\mathbb{1} + \frac{i}{2}\Omega_{\alpha\beta}S^{\alpha\beta} + \cdots)\gamma^\mu(\mathbb{1} - \frac{i}{2}\Omega_{\alpha\beta}S^{\alpha\beta} + \cdots) = (\mathbb{1} - \frac{i}{2}\Omega_{\alpha\beta}J^{\alpha\beta} + \cdots)^\mu{}_\nu \gamma^\nu \tag{8.279}
$$

which leads to the requirement (at first order):

$$
\frac{i}{2}S^{\alpha\beta}\gamma^\mu - \gamma^\mu\frac{i}{2}S^{\alpha\beta} = -\frac{i}{2}(J^{\alpha\beta})^\mu{}_\nu \gamma^\nu \quad \text{or} \quad -\left[S^{\alpha\beta}, \gamma^\mu\right] = (J^{\alpha\beta})^\mu{}_\nu \gamma^\nu \tag{8.280}
$$

On the one hand (see Eq. (8.169)) a calculation shows:

$$
\left[S^{\alpha\beta}, \gamma^\mu\right] = \left[\frac{i}{2}\left(\gamma^\alpha\gamma^\beta - g^{\alpha\beta}\right), \gamma^\mu\right] = \frac{i}{2}\left[\gamma^\alpha\gamma^\beta, \gamma^\mu\right] = i\left(\gamma^\alpha g^{\beta\mu} - \gamma^\beta g^{\mu\alpha}\right) \tag{8.281}
$$

On the other hand, one can write the generators as:

$$
(J^{\alpha\beta})^{\mu\nu} = i\left(g^{\alpha\mu}g^{\beta\nu} - g^{\beta\mu}g^{\alpha\nu}\right) \quad \text{or} \quad (J^{\alpha\beta})^\mu{}_\nu = i\left(g^{\alpha\mu}\delta^\beta_\nu - g^{\beta\mu}\delta^\alpha_\nu\right) \tag{8.282}
$$

hence

$$
(J^{\alpha\beta})^\mu{}_\nu \gamma^\nu = g^{\alpha\mu}\gamma^\beta - g^{\beta\mu}\gamma^\alpha \quad \square \tag{8.283}
$$

Finally, the **current quantity $\overline{\Psi}\gamma^\mu\Psi$ constructed from the spinor, its adjoint, and one gamma matrix transforms as a four-vector under the Lorentz transformation.**

Having all the necessary building blocks which transform covariantly under Lorentz transformation, we can now construct the desired function. The **Lagrangian density of the Dirac equation** is the following, where Ψ is the spinor (or later the spinor field) and $\overline{\Psi}$ its adjoint, and it can be used to generate both the Dirac equation and its adjoint:

$$
\mathcal{L}_{Dirac} = i\overline{\Psi}\gamma^\mu\partial_\mu\Psi - m\overline{\Psi}\Psi \tag{8.284}
$$

In order to verify this result, we use the relativistic Euler–Lagrange equations, Eq. (6.28), varying Ψ and $\overline{\Psi}$ independently. We can obtain Eq. (8.28) with:

$$
\frac{\partial\mathcal{L}}{\partial(\partial_\mu\overline{\Psi})} = 0 \quad \text{and} \quad \frac{\partial\mathcal{L}}{\partial\overline{\Psi}} = i\gamma^\mu\partial_\mu\Psi - m\Psi \implies i\gamma^\mu\partial_\mu\Psi - m\Psi = 0 \quad \square \tag{8.285}
$$

similarly to obtain the adjoint Eq. (8.43):

$$\frac{\partial \mathcal{L}}{\partial (\partial_\mu \Psi)} = i\overline{\Psi}\gamma^\mu \quad \text{and} \quad \frac{\partial \mathcal{L}}{\partial \Psi} = -m\overline{\Psi} \quad \Longrightarrow \quad i\partial_\mu (\overline{\Psi}\gamma^\mu) + m\overline{\Psi} = 0 \quad \square \qquad (8.286)$$

It is completely gorgeous! Note again that it is because of the covariant nature of $\overline{\Psi}\Psi$ and $\overline{\Psi}\gamma^\mu \Psi$ (thanks to the "magic" of the γ matrices) that we could build the invariant Dirac Lagrangian.

8.27   Second Quantization of the Dirac Field

In analogy to the quantization of the real or complex Klein–Gordon fields, we want to understand how to treat the Dirac spinor field $\Psi(x^\mu)$ as a *quantum field*. The prescription is similar to that of the Klein–Gordon scalar field where one considers a series of plane waves which can be interpreted as an infinite number of harmonic oscillators continuously indexed with a momentum $\vec{p}$. We have introduced creation and annihilation $a^\dagger$, a operators to excite or de-excite these quantum oscillators. In the case of the complex field, we had to introduce two types of operators $a^\dagger$, a, $b^\dagger$, b which were interpreted as two types of particles with the same rest mass but opposite charge.

We have seen that a Dirac field Ψ has four components describing spin-1/2 particles and antiparticles. To develop the plane waves, we use the four spinors $u^{(1)}$, $u^{(2)}$, $v^{(1)}$, and $v^{(2)}$. In the quantized Dirac field, one can create or annihilate particles $u^{(s)}(E, \vec{p})$ or antiparticles $v^{(s)}(E, \vec{p})$ of momentum $\vec{p}$ and spin-state s ($s = 1, 2$).

Finally, the Dirac field operators Ψ and adjoint $\overline{\Psi}$ are defined with two pairs of creation–annihilation operators $a_s(\vec{p})$, $a_s^\dagger(\vec{p})$, $b_s(\vec{p})$, $b_s^\dagger(\vec{p})$ ($s = 1, 2$) as:

$$\begin{cases} \Psi \equiv \displaystyle\int \frac{\mathrm{d}^3\vec{p}}{(2\pi)^{3/2}} \frac{1}{\sqrt{2E_p}} \sum_{s=1,2} \left(a_s(\vec{p}) u^{(s)}(\vec{p}) e^{-ip\cdot x} + b_s^\dagger(\vec{p}) v^{(s)}(\vec{p}) e^{+ip\cdot x} \right) \\ \overline{\Psi} \equiv \displaystyle\int \frac{\mathrm{d}^3\vec{p}}{(2\pi)^{3/2}} \frac{1}{\sqrt{2E_p}} \sum_{s=1,2} \left(a_s^\dagger(\vec{p}) \bar{u}^{(s)}(\vec{p}) e^{+ip\cdot x} + b_s(\vec{p}) \bar{v}^{(s)}(\vec{p}) e^{-ip\cdot x} \right) \end{cases} \qquad (8.287)$$

where the integrals are over all oscillator modes described by the momentum index $\vec{p}$ and the sum over the spin index s, and the second term is a normalization with $E_p = +\sqrt{\vec{p}^2 + m^2}$.

As expected, the quantized Dirac field describes two types of particles with same rest mass, which could for example be the electron and the positron. In this case:

- $a^\dagger(\vec{p})$ (resp. $a(\vec{p})$) creates (resp. annihilates) an electron with energy $+E_p > 0$, momentum $\vec{p}$ and spin state s.

- $b^\dagger(\vec{p})$ (resp. $b(\vec{p})$) creates (resp. annihilates) a positron with energy $+E_p > 0$, momentum $\vec{p}$ and spin state s.

What shall be the **commutation rules** of these operators? If we use the commutation as in the case of the Klein–Gordon equation, this will lead to *bosons*. On the other hand, Dirac particles with half-integer spin should be *fermions*, according to the theorem of spin statistics [6]. Hence one should adopt **anticommutation rules** for the operators. For a quantum oscillator whose ladder operator obeys an anticommutation rule (see Appendix C.8), there are only two possible states: the ground state and the first excited state. In quantum field theory, this leads to the following interpretation: the particles are fermions and hence obey **Pauli's exclusion principle**. More than one particle cannot be present in a given specific quantum state. This means that a particular quantum oscillator representing a specific momentum $\vec{p}$ and spin state s can only be in the $|0\rangle$ or $|1\rangle$ state. No more than one particle in the system can occupy that state.

The anticommutation rules of the creation–annihilation operators of the Dirac quantum field are accordingly chosen to be the following:

$$
\begin{cases}
\left\{a_r(\vec{p}), a_s^\dagger(\vec{p}')\right\} = \left\{b_r(\vec{p}), b_s^\dagger(\vec{p}')\right\} = \delta^3(\vec{p} - \vec{p}')\delta_{r,s} \\
\left\{a_r(\vec{p}), a_s(\vec{p}')\right\} = \left\{a_r^\dagger(\vec{p}), a_s^\dagger(\vec{p}')\right\} \equiv 0
\end{cases}
\tag{8.288}
$$

Following a calculation similar to that of the quantum Klein–Gordon field case, one can find the energy and momentum operators of the Dirac field. The Hamiltonian is:

$$
H = \int \mathrm{d}^3\vec{p} \sum_{s=1,2} E_p \left(\underbrace{a_s^\dagger(\vec{p})a_s(\vec{p})}_{=N_s^a(\vec{p})} + \underbrace{b_s^\dagger(\vec{p})b_s(\vec{p})}_{=N_s^b(\vec{p})} \right)
\tag{8.289}
$$

and the total momentum is equal to:

$$
\vec{P} = \int \mathrm{d}^3\vec{p} \sum_{s=1,2} \vec{p} \left(\underbrace{a_s^\dagger(\vec{p})a_s(\vec{p})}_{=N_s^a(\vec{p})} + \underbrace{b_s^\dagger(\vec{p})b_s(\vec{p})}_{=N_s^b(\vec{p})} \right)
\tag{8.290}
$$

where we recognize the number operators $N_s^a(\vec{p})$ and $N_s^b(\vec{p})$ for each type of particle, given momentum $\vec{p}$ and spin state s, which correspond to their numbers per unit momentum volume $\mathrm{d}^3\vec{p}$.

Starting from the vacuum ground state, the state with a single electron of momentum $\vec{p}$ and spin state s is defined as (neglecting normalization for the moment):

$$
|e^-(\vec{p}, s)\rangle \equiv a_s^\dagger(\vec{p}) |0\rangle
\tag{8.291}
$$

Similarly, a single positron of momentum $\vec{p}$ and spin state s is defined as:

$$
|e^+(\vec{p}, s)\rangle \equiv b_s^\dagger(\vec{p}) |0\rangle
\tag{8.292}
$$

A state with two particles is found by successive application of the creation operator, e.g., $a^\dagger(\vec{p})a^\dagger(\vec{q}) |0\rangle$. As a direct consequence of the anticommutation rules, we have in this case:

$$
a^\dagger(\vec{p})a^\dagger(\vec{q}) |0\rangle = -a^\dagger(\vec{q})a^\dagger(\vec{p}) |0\rangle
\tag{8.293}
$$

which shows that the **state is antisymmetric under the exchange of the two particles**, which is what is expected for a system of particles following the **Fermi–Dirac statistics**.

All the discussion so far was in the momentum space. The field operators Ψ and $\overline{\Psi}$ depend on the space-time coordinate x^μ. From the development of the field as a function of the plane waves and the corresponding anticommutation rules for the creation and annihilation operators, one can show that the fields will also obey the following **equal-time anticommutation rules**:

$$
\begin{cases}
\left\{\Psi_i(\vec{x}, t), \Psi_j^\dagger(\vec{x}', t)\right\} = \delta^3(\vec{x} - \vec{x}')\delta_{i,j} \\
\left\{\Psi_i(\vec{x}, t), \Psi_j(\vec{x}', t)\right\} = 0 \\
\left\{\Psi_i^\dagger(\vec{x}, t), \Psi_j^\dagger(\vec{x}', t)\right\} = 0
\end{cases}
\tag{8.294}
$$

where $i, j = 1, 2, 3, 4$ are the indices of the four components of the spinor fields.

8.28   Chiral Interactions – Bilinear Covariants

We have seen in Section 8.26 that the covariants of the type $\overline{\Psi}\Psi$ and $\overline{\Psi}\gamma^\mu\Psi$ played a fundamental role in constructing the Lagrangian of the free Dirac particles. One can generalize to the so-called **bilinear covariants** by considering the product of the spinor and its adjoint with a 4×4 constant matrix Γ:

$$\overline{\Psi}\Gamma\Psi \qquad (8.295)$$

Each 4×4 Γ matrix possesses 16 independent components, hence it can be expressed in a basis of 16 independent 4×4 matrices, which can be conveniently classified according to their transformation properties under Lorentz transformation. Such a basis can be the following:

$$\Gamma_C(C = 1,\ldots,16): \quad \mathbb{1}, \underbrace{\gamma^\mu}_{4 \text{ matrices}}, \underbrace{i\sigma^{\mu\nu} = 2iS^{\mu\nu}}_{6 \text{ matrices}}, \underbrace{i\gamma^\mu\gamma^5}_{4 \text{ matrices}}, \gamma^5 \qquad (8.296)$$

where the factor-two difference with $S^{\mu\nu}$ (see Eq. (8.169)) in the totally antisymmetric tensor $\sigma^{\mu\nu}$ is a convention:

$$\sigma^{\mu\nu} \equiv \frac{i}{2}[\gamma^\mu, \gamma^\nu] \qquad (8.297)$$

The six independent non-vanishing matrices are σ^{01}, σ^{02}, σ^{03}, σ^{12}, σ^{13}, and σ^{23}. Hence, **any 4×4 constant matrix can be decomposed into terms that have well-defined transformation properties under the Lorentz group!** This is a very important result that allows us to understand the physical properties of the $\overline{\Psi}\Gamma\Psi$ quantities. In addition, it will be important for the quantities $\overline{\Psi}\Gamma\Psi$ to be Hermitian, such that their observables will be real.

We note that one of the basis matrices is γ^5. We consider the quantity $J^5 \equiv \overline{\Psi}\gamma^5\Psi$, which is an invariant under proper Lorentz transformation and we compute its transformation under the **parity transformation**, noting that $P(\Psi) = \gamma^0\Psi$ implies $P(\overline{\Psi}) = \overline{\Psi}\gamma^0$ and since $\gamma^\mu\gamma^5 = -\gamma^5\gamma^\mu$:

$$P(\overline{\Psi}\gamma^5\Psi) = \overline{\Psi}\gamma^0\gamma^5\gamma^0\Psi = -\overline{\Psi}\gamma^5\gamma^0\gamma^0\Psi = -(\overline{\Psi}\gamma^5\Psi) \qquad (8.298)$$

(another way of writing this is to note that $S(P)\gamma^5 = -\gamma^5 S(P)$ – see Eq. (8.204)). Hence, the quantity $\overline{\Psi}\gamma^5\Psi$ is defined as a **pseudoscalar** because it is invariant under a proper Lorentz transformation, however, it changes sign under parity. In contrast, the quantity $J \equiv \overline{\Psi}\Psi$ remains unchanged under parity and proper Lorentz transformation, and hence is a pure **scalar**:

$$P(\overline{\Psi}\Psi) = \overline{\Psi}\gamma^0\gamma^0\Psi = \overline{\Psi}\Psi \qquad (8.299)$$

Furthermore, we have seen in the previous section that $J^\mu \equiv \overline{\Psi}\gamma^\mu\Psi$ transforms as a four-vector under proper Lorentz transformation. What happens to it under parity? We look at the time and space components independently:

$$P(\overline{\Psi}\gamma^\mu\Psi) = \overline{\Psi}\gamma^0\gamma^\mu\gamma^0\Psi = \begin{cases} \overline{\Psi}\gamma^0\gamma^0\gamma^0\Psi = +\overline{\Psi}\gamma^0\Psi \\ \overline{\Psi}\gamma^0\gamma^k\gamma^0\Psi = -\overline{\Psi}\gamma^k\Psi \end{cases} \qquad (8.300)$$

This can be written in compact form as:

$$P(\overline{\Psi}\gamma^\mu\Psi) = \eta[\mu]\overline{\Psi}\gamma^\mu\Psi \qquad (8.301)$$

where we introduced the $\eta[\mu]$ **symbol**:

$$\eta[\mu] \equiv \begin{cases} +1 \text{ if } \mu = 0 \\ -1 \text{ if } \mu = 1,2,3 \end{cases} \qquad (8.302)$$

Hence, the vector acquires the same negative sign as the space component of the ordinary space-time vector x^μ. Therefore $\overline{\Psi}\gamma^\mu\Psi$ is a **pure vector** quantity. On the other hand, by inserting the γ^5 we note that $J^{5\mu} \equiv i\overline{\Psi}\gamma^\mu\gamma^5\Psi$ transforms as (ignoring the i for clarity):

$$P(\overline{\Psi}\gamma^\mu\gamma^5\Psi) = \overline{\Psi}\gamma^0\gamma^\mu\gamma^5\gamma^0\Psi = \begin{cases} \overline{\Psi}\gamma^0\gamma^0\gamma^5\gamma^0\Psi = -\overline{\Psi}\gamma^0\gamma^5\Psi \\ \overline{\Psi}\gamma^0\gamma^k\gamma^5\gamma^0\Psi = +\overline{\Psi}\gamma^k\gamma^5\Psi \end{cases} \tag{8.303}$$

which in compact form is just:

$$P(\overline{\Psi}\gamma^\mu\gamma^5\Psi) = -\eta[\mu]\overline{\Psi}\gamma^\mu\gamma^5\Psi \tag{8.304}$$

Hence, the quantity $i\overline{\Psi}\gamma^\mu\gamma^5\Psi$ is labeled an **axial vector or pseudovector** in order to distinguish its behavior under parity from that of a vector. One could repeat the same considerations for the quantity built with the tensor $\sigma^{\mu\nu}$.

That's all! The 16 bilinear covariants are summarized in Table 8.3. Any quantity of the form $\overline{\Psi}\Gamma\Psi$ can be expressed as a linear combination of the bilinear covariants. We are now armed with new terms involving γ^μs and γ^5 that we can start adding to our Lagrangian to construct a theory of interactions between particles. When a theory treats Ψ_L and Ψ_R on an equal footing, it is called a vector-like theory. A combination of scalar or vector terms with pseudoscalar or axial-vector terms will break parity invariance. In such a theory, called a **chiral theory**, the chiral components Ψ_L and Ψ_R will be treated differently. This will be the case for the "Standard Model" theory.

Quantity	Property	How many	Parity P
$J \equiv \overline{\Psi}\Psi$	scalar	1	$+1$
$J^\mu \equiv \overline{\Psi}\gamma^\mu\Psi$	vector	4	$\eta[\mu]$
$J^{\mu\nu} \equiv i\overline{\Psi}\sigma^{\mu\nu}\Psi$	tensor	6	$\eta[\mu]\eta[\nu]$
$J^{5\mu} \equiv i\overline{\Psi}\gamma^\mu\gamma^5\Psi$	axial vector	4	$-\eta[\mu]$
$J^5 \equiv \overline{\Psi}\gamma^5\Psi$	pseudoscalar	1	-1

Table 8.3 List of 16 bilinear covariants, where $\eta[\mu] = +1$ if $\mu = 0$ and $\eta[\mu] = +1$ if $\mu = 1, 2, 3$.

Problems

Ex 8.1 Gamma matrices. The 4×4 γ-matrices are given (in the Pauli–Dirac representation) by

$$\gamma^0 = \begin{pmatrix} \mathbf{1} & 0 \\ 0 & -\mathbf{1} \end{pmatrix} \qquad \gamma^i = \begin{pmatrix} 0 & \sigma^i \\ -\sigma^i & 0 \end{pmatrix}$$

Prove the following relations:

(a) $(\gamma^0)^2 = \mathbf{1}$, $(\gamma^k)^2 = -\mathbf{1}$ $k = 1, 2, 3$.

(b) $\gamma^{0\dagger} = \gamma^0$, $(\gamma^k)^\dagger = -\gamma^k$.

(c) $\{\gamma^\mu, \gamma^\nu\} = 2\, g^{\mu\nu}\, \mathbf{1}$.

Ex 8.2 Bi-linear covariants. The Lorentz transformation of spinors $\psi(x)$ between two inertial frames can be written as:

$$\psi'(x') = S(\Lambda)\psi(x); \quad x' = \Lambda\, x$$

where S is a 4×4 matrix, which depends on the Lorentz transformation between the two inertial frames.

(a) For a boost in the x direction, S is given by:

$$S = \begin{pmatrix} a_+ & a_-\sigma^1 \\ a_-\sigma^1 & a_+ \end{pmatrix}, \text{ with } a_\pm = \pm\sqrt{\frac{1}{2}(\gamma\pm1)} \; ; \; \gamma = (1-\beta^2)^{-1/2}$$

Show that: $S^\dagger S = \gamma\begin{pmatrix} 1 & -\beta\sigma^1 \\ -\beta\sigma^1 & 1 \end{pmatrix}, S^\dagger\gamma^0 S = \gamma^0.$

(b) Show that $\bar{\psi}\gamma^5\psi$ is invariant under the Lorentz transformation S for a boost in the x direction and that it changes sign under the parity transformation $\psi' = \gamma^0\psi$, i.e., it is a pseudoscalar.

(c) Show that $v^\mu \equiv \bar{\psi}\gamma^\mu\psi$ in the Dirac representation is a polar vector, i.e., under a Lorentz boost in the x direction the components transform like a four-vector, and under the parity transformation as:

$$(v^0)' = v^0$$
$$\vec{v}' = -\vec{v}$$

Ex 8.3 Algebra of the Lorentz group for the spinors. Given the anticommutation relation of the γ matrices:

$$\{\gamma^\mu, \gamma^\nu\} = 2g^{\mu\nu}\mathbb{1}$$

prove that the matrices

$$S^{\mu\nu} = \frac{i}{4}[\gamma^\mu, \gamma^\nu]$$

satisfy the generator algebra of the Lorentz group:

$$[S^{\mu\nu}, S^{\rho\sigma}] = i\left(g^{\mu\sigma}S^{\nu\rho} + g^{\nu\rho}S^{\mu\sigma} - g^{\mu\rho}S^{\nu\sigma} - g^{\nu\sigma}S^{\mu\rho}\right)$$

Ex 8.4 Chirality and helicity

(a) Show that the chirality is not a good quantum number for a massive fermion by checking the commutator $[H, \gamma_5]$.

(b) Show that helicity is conserved although it depends on the choice of the coordinate system.

9 Interacting Fields and Propagator Theory

In the approximation of classical relativistic theory the creation of an electron pair (electron A, positron B) might be represented by the start of two world lines from the point of creation, 1. The world lines of the positron will then continue until it annihilates another electron, C, at a world point 2. Between the times t_1 and t_2 there are then three world lines, before and after only one. However, the world lines of C, B, and A together form one continuous line albeit the "positron part" B of this continuous line is directed backwards in time. Following the charge rather than the particles corresponds to considering this continuous world line as a whole rather than breaking it up into its pieces. It is as though a bombardier flying low over a road suddenly sees three roads and it is only when two of them come together and disappear again that he realizes that he has simply passed over a long switchback in a single road. (...) This view is quite different from that of the Hamiltonian method which considers the future as developing continuously from out of the past. Here we imagine the entire space-time history laid out, and that we just become aware of increasing portions of it successively.

R. Feynman[1]

9.1 Direct or Non-derivative Interaction Terms

Until now we have only considered *free* Klein–Gordon and Dirac particles. How can we introduce interactions among such particles? In general, we can split the Lagrangian of the system into two terms: the first $\mathcal{L}_0$ which describes free particles (i.e., their kinetic term and their rest mass) and a second $\mathcal{L}_{int}$ which describes their interactions:

$$\mathcal{L} \equiv \mathcal{L}_0 + \mathcal{L}_{int} \tag{9.1}$$

As a consequence, we can think of the Hamiltonian operator as well as the sum of two terms $H = H_0 + H_{int}$, where H_{int} represents the interaction part. In this separation, we implicitly assume that the interaction is "weak" enough to be considered as a *perturbation* to the free Hamiltonian theory of the particles.

Let us consider the theory of a quantized scalar field ϕ. The Lagrangian depends in general on the field ϕ, its derivatives $\partial\phi$, and also possibly explicitly from x^μ. In the case of the Klein–Gordon and Dirac equations, we saw that they depend on ϕ and its derivatives $\partial_\mu \phi$, where the derivative term corresponds to the *kinetic term* of the field. To be concrete, we can write out the case of the real scalar field:

$$\mathcal{L}_0 = \frac{1}{2}(\partial\phi)^2 - \frac{1}{2}m^2\phi^2 = \frac{1}{2}(\partial\phi)^2 - V_0(\phi) \tag{9.2}$$

Here we have identified the (bare) mass term as due to the potential $V_0(\phi)$. We have seen in previous chapters that this theory (of free particles) is exactly solvable. The eigenstates, or better yet the Hilbert space, is fully known and can be generated by applying the creation operators on the vacuum $|0\rangle$. However, there are no interactions between these particles.

---

1 Reprinted excerpt with permission from R. P. Feynman, "The theory of positrons," *Phys. Rev.*, vol. 76, pp. 749–759, Sep 1949. Copyright 1949 by the American Physical Society.

So, what about an interaction term of the Lagrangian? We can think of changing the potential V_0 by adding higher-order terms. In our example of the real scalar field, we would for instance expand the potential in this way:

$$V_0(\phi) = \frac{1}{2}m^2\phi^2 \quad \longrightarrow \quad V(\phi) = \underbrace{\frac{1}{2}m^2\phi^2}_{V_0} + \underbrace{\lambda\phi^3 + g\phi^4 + \cdots}_{V_{int}} \tag{9.3}$$

where λ and g need to be determined and have a dimension that ensures that each term in the sum has the correct unit of energy to the fourth power (see Eq. (6.22)). In the case of **direct or non-derivative coupling** as above, the interaction terms **depend only on the field**, and not on its kinetics, i.e., its derivatives. Using Eq. (6.41) which relates the Lagrangian and the Hamiltonian, we find for the interaction part, under the assumption of non-derivative coupling, that:

$$\mathcal{H}_{int} = \frac{\partial \mathcal{L}_{int}}{\partial(\partial^0\phi)}(\partial^0\phi) - \mathcal{L}_{int} = -\mathcal{L}_{int} \tag{9.4}$$

Now that we have a Lagrangian, how shall we handle interactions? Introducing interactions leads to important changes to the solutions. The first major change is that the particles now interact; hence, the **state of the system can change with time**. In the previous chapters, we have several times insisted on the fact that the theory must be covariant and have shown the direct consequences. In particular, covariance imposed requirements on the role of *time* with respect to space coordinates. Shall the role of time be again singled out?

In the classical approach to quantum mechanics, we have used perturbation theory (see Section 4.16) to compute transition amplitudes between different states of the system. We have considered the transition between the initial past $-T/2$ and the final future $+T/2$, where at the end we let $T \to \infty$. See for example the computation of the Rutherford scattering cross-section in Chapter 4. In this way, the time variable finally disappears from the problem, although it is needed in order to perform the computation. We shall adopt a similar approach here to describe scattering between particles.

Another consequence of interactions is that the spectrum of particles may contain bound states of the elementary particle fields. These are harder to handle than scattering, since the approach of perturbation theory generally fails. The dream would be to find the *exact* solution of an interacting theory, computing all bound states (exactly) and all interactions (exactly). This has not been possible for most practical theories so far, and one assumes that the interaction terms can be viewed as small perturbations. Bounds states are then factorized out of the problem.

We will see that another consequence of interactions is that the mass term m that we have put in our Lagrangian will no longer be equal to the rest mass of the particles, but this latter will be modified by radiative corrections (as will be discussed in Section 9.12 and Chapter 12). In general, the (bare) terms in our Lagrangian will be affected by radiative corrections and will need to be connected to physically measurable quantities. In practice, this will lead to the requirement of *renormalizing* them.

Finally, the fact that the Hilbert space of states is different in the interacting state as in the case of a free theory is also true for the vacuum itself! Whereas the vacuum of the free theory was defined as $|0\rangle$, the ground state of the interacting theory will in general be different and is customarily labeled $|\Omega\rangle$. We will actually initially neglect this fact, labeling the ground state as $|0\rangle$ for simplicity, and will come back to it in Section 9.11.

9.2 Transition Amplitudes: S Matrix, $\mathcal{T}$, and $\mathcal{M}$

We consider the scattering of an initial (or incoming) state $|i\rangle$ to the final (or outgoing) state $|f\rangle$ with the aim of computing the transition amplitude. In analogy to the non-relativistic case (see Section 4.16), the process is formulated in terms of the theory of asymptotic in- and out-states. Given the locality of the interaction which we assume, the initial and final states are free states with well-defined kinematical configurations. Near $t \simeq 0$, the interaction occurs, and the state of the system is changed from the initial to the final state. Consequently, the S **matrix** is defined as the transition amplitude between an initial (past) state defined at time $-T/2$ to a final (future) state at time $+T/2$, in the limit $T \to \infty$. As in the case of Fermi's Golden Rule, a transition probability per unit time can then be derived. Formally then, the S matrix is given by:

$$S \equiv \lim_{T \to \infty} U\left(-\frac{T}{2}, +\frac{T}{2}\right) = U(-\infty, +\infty) \tag{9.5}$$

where we introduced the Dyson time-evolution operator described in Section 4.6. Using the result, Eq. (4.53), it can be expressed as:

$$S = \mathbb{1} + \sum_{n=1}^{\infty} \frac{(-i)^n}{n!} \int_{-\infty}^{\infty} \mathrm{d}t_1 \dots \int_{-\infty}^{\infty} \mathrm{d}t_n \, T\left[H_{int}(t_1) \dots H_{int}(t_n)\right] \tag{9.6}$$

The use of the interaction Hamiltonian density $\mathcal{H}_{int}$ allows us to write a covariant expression, where the integration is over the entire space-time:

$$S = \mathbb{1} + \sum_{n=1}^{\infty} \frac{(-i)^n}{n!} \int \mathrm{d}^4 x_1 \dots \int \mathrm{d}^4 x_n \, T\left[\mathcal{H}_{int}(x_1) \dots \mathcal{H}_{int}(x_n)\right] \tag{9.7}$$

The S matrix can be represented as a series of S_n terms corresponding to the nth order of the calculation:

$$S \equiv \mathbb{1} + S_1 + S_2 + \cdots \tag{9.8}$$

where the nth term is given by:

$$S_n = \frac{(-i)^n}{n!} \int \mathrm{d}^4 x_1 \dots \int \mathrm{d}^4 x_n \, T\left[\mathcal{H}_{int}(x_1) \dots \mathcal{H}_{int}(x_n)\right] \tag{9.9}$$

The terms S_n can in principle be individually computed up to a given order n and the total amplitude is then given by their sum up to the given order.

• **Transfer matrix and amplitudes.** As before (see Eq. (4.155)), we write $S \equiv 1 + i\mathcal{T}$, where $\mathcal{T}$ is the transfer matrix. The factor i is conventional. In a free theory without interactions, the transfer matrix vanishes, and the S matrix is the identity. Since the S matrix should also vanish unless the initial and final states have the same total four-momentum, it is useful to factor an overall momentum-conserving Dirac δ function, in the spirit of the relativistic kinematical expressions derived previously (see Section 5.12). So finally, we can define:

$$S \equiv \mathbb{1} + i\mathcal{T} = \mathbb{1} + i(2\pi)^4 \delta^4\left(p - \sum_i p_i\right)\mathcal{M} \tag{9.10}$$

where the momenta are defined as in Eq. (5.108) and $\mathcal{M}$ is the **transition amplitude** describing the situation where some energy–momentum-conserving scattering took place. In general, we will be interested in the **scattering amplitudes** between an initial $|i\rangle$ and final $|f\rangle$ state of given kinematical configuration, formally:

$$\langle f|\mathcal{M}|i\rangle = \frac{\langle f|S - \mathbb{1}|i\rangle}{i(2\pi)^4 \delta^4\left(p - \sum_i p_i\right)} \tag{9.11}$$

which allows us to focus on the "interesting part" of the transition matrix, removing the trivial identity and the overall energy–momentum conservation.

• **Optical theorem.** The total scattering matrix S should be unitary because the total probability (for any outcome of a scattering process) is one, i.e., $S^\dagger S = \mathbb{1}$. Writing as before the scattering amplitude as $S = \mathbb{1} + i\mathcal{T}$, we obtain:

$$\mathbb{1} = S^\dagger S = \mathbb{1} + i(\mathcal{T} - \mathcal{T}^\dagger) + \mathcal{T}^\dagger \mathcal{T} = \mathbb{1} + i(2i\Im m(\mathcal{T})) + \mathcal{T}^\dagger \mathcal{T} \tag{9.12}$$

which yields the **optical theorem**, i.e., $\mathcal{T}^\dagger \mathcal{T} = 2\Im m(\mathcal{T})$. The scattering cross-section is determined by the imaginary part of the amplitude.

9.3   A Toy Model of Local Interaction

We build the hypothetical toy model with a real scalar field $\sigma(x)$ of rest mass M and three real scalar fields $\phi_a(x)$ with rest masses m. The toy model is built to illustrate the methods and does not need to represent a physical reality.

The field $\sigma(x)$ represents a spinless neutral particle which interacts with the particles described by the other fields. The fields $\phi_1(x)$ and $\phi_2(x)$ can be combined to form a complex scalar field, which we label φ, where:

$$\varphi \equiv \frac{1}{\sqrt{2}}\left(\phi_1 + i\phi_2\right) \tag{9.13}$$

The complex field φ describes charged particles of opposite charge, which can for example be labeled π^+ and π^-. The real field $\phi(x) = \phi_3(x)$ describes an associated neutral particle, labeled π^0. In the absence of interactions, the free system corresponds to the following Lagrangian density:

$$\begin{aligned}
\mathcal{L}_0 &= \frac{1}{2}\left(\partial_\mu\sigma\right)^2 - \frac{1}{2}M^2\sigma^2 + \sum_a\left(\frac{1}{2}\left(\partial_\mu\phi_a\right)\left(\partial^\mu\phi_a\right) - \frac{1}{2}m^2\phi_a^2\right) \\
&= \frac{1}{2}\left(\partial_\mu\sigma\right)^2 - \frac{1}{2}M^2\sigma^2 + \frac{1}{2}\left(\partial_\mu\phi\right)\left(\partial^\mu\phi\right) - \frac{1}{2}m^2\phi^2 + \left(\partial_\mu\varphi^\dagger\right)\left(\partial^\mu\varphi\right) - m^2\varphi^\dagger\varphi
\end{aligned} \tag{9.14}$$

The Lagrangian density $\mathcal{L}_0$ describes the free system and we will add interaction terms $\mathcal{L}_{int}$ which will act as perturbations that can trigger state transitions. The full theory is given by $\mathcal{L} = \mathcal{L}_0 + \mathcal{L}_{int}$.

In our toy model, we illustrate two possible reactions: the neutral decay $\sigma \to \pi^0\pi^0$ and the charged decay $\sigma \to \pi^+\pi^-$.

9.3.1   Neutral Decay $\sigma \to \pi^0\pi^0$ in the Toy Model

In the case of direct or non-derivative interactions, we assume that the interaction terms of the Lagrangian only contain the fields and not their derivatives. In addition, the simplest form of interaction is *local*, i.e., the terms of the interaction are taken in a single point of space-time.

In our toy model, the reaction $\sigma \to \pi^0\pi^0$ is driven by a local, direct coupling among the particles, which is expressed as the **product of fields**. We search for Lorentz-invariant terms and we consider for reasons that will become clear later, the following products of fields:

$$\phi^4, \sigma\phi^3, \sigma^2\phi^2, \sigma^3\phi, \sigma^4, \phi^3, \sigma\phi^2, \sigma^2\phi, \text{ or } \sigma^3 \tag{9.15}$$

The strength of the coupling is given by a **coupling constant**. A dimensional analysis shows that the terms like ϕ^4 have a dimensionless coupling constant, while the terms like ϕ^3 must contain a coupling constant with the dimension of energy (mass). Finally, we can write these coupling constants as gs, or λs and obtain the following direct coupling terms:

$$g\phi^4, g'\sigma\phi^3, g''\sigma^2\phi^2, \ldots, \lambda\phi^3, \lambda'\sigma\phi^2, \lambda''\sigma^2\phi, \ldots \tag{9.16}$$

At this stage, we can consider and *impose further symmetries on the Lagrangian*. For example, we could assume that the field $\sigma(x)$ is a scalar under the discrete parity transformation, while the field $\phi(x)$ is pseudoscalar. These are just *assumptions we adopt to explore our toy model*. **In a real theory of Nature, specific assumptions on the underlying symmetry of natural laws and their fields will be confronted with experimental data.** We can write:

$$P: \begin{cases} \sigma(t,\vec{x}) \to \sigma(t,-\vec{x}) = +\sigma(t,\vec{x}) \\ \phi(t,\vec{x}) \to \phi(t,-\vec{x}) = -\phi(t,\vec{x}) \end{cases} \tag{9.17}$$

We note that the free Lagrangian $\mathcal{L}_0$ is invariant under the parity transformation. If we want the interacting system to conserve this property under the assumptions above for the scalar σ and pseudoscalar ϕ fields, then the coupling terms can only contain an *even* number of ϕ. For σ there are no special constraints. This requirement reduces the number of possible terms in the interaction Lagrangian of our toy model to be:

$$\mathcal{L}_{int} = -g\phi^4 - g'\sigma^2\phi^2 - g''\sigma^4 - \lambda\sigma\phi^2 - \lambda'\sigma^3 \tag{9.18}$$

Because we have only direct couplings, the Hamiltonian and Lagrangian interaction densities are trivially related by $\mathcal{H}_{int} = -\mathcal{L}_{int}$. In order to compute transition amplitudes, the fields must be expressed in the *interaction frame*, as the time-evolution operator and the S-matrix were derived in that context (see Section 9.2). So we write:

$$\mathcal{H}_{int} = g\phi_I^4 + g'\sigma_I^2\phi_I^2 + g''\sigma_I^4 + \lambda\sigma_I\phi_I^2 + \lambda'\sigma_I^3 \tag{9.19}$$

From now on, we drop for clarity the subscript "I" from the fields, so we write ϕ **for** ϕ_I **and** σ **for** σ_I. These two scalar fields $\sigma(x)$ and $\phi(x)$ are quantized according to the prescription for free bosons, outlined in Chapter 7. We introduce two sets of creation–annihilation operators, i.e., $a(\vec{p})$ and $a^\dagger(\vec{p})$ for the field ϕ and $c(\vec{k})$ and $c^\dagger(\vec{k})$ for the field σ:

$$
\begin{aligned}
\phi(\vec{x},t) &\equiv \int \frac{d^3\vec{p}}{(2\pi)^{3/2}} \frac{1}{\sqrt{2E_p}} \left(a(\vec{p})e^{-ip\cdot x} + a^\dagger(\vec{p})e^{+ip\cdot x} \right) \\
\sigma(\vec{x},t) &\equiv \int \frac{d^3\vec{k}}{(2\pi)^{3/2}} \frac{1}{\sqrt{2E_k}} \left(c(\vec{k})e^{-ik\cdot x} + c^\dagger(\vec{k})e^{+ik\cdot x} \right)
\end{aligned}
\tag{9.20}
$$

These operators obey the following commutation rules:

$$
\begin{cases}
[a(\vec{p}), a^\dagger(\vec{p}')] \equiv \delta^3(\vec{p} - \vec{p}') & [a(\vec{p}), a(\vec{p}')] = [a^\dagger(\vec{p}), a^\dagger(\vec{p}')] \equiv 0 \\
[c(\vec{p}), c^\dagger(\vec{p}')] \equiv \delta^3(\vec{p} - \vec{p}') & [c(\vec{p}), c(\vec{p}')] = [c^\dagger(\vec{p}), c^\dagger(\vec{p}')] \equiv 0 \\
[a(\vec{p}), c(\vec{p}')] = [a^\dagger(\vec{p}), c^\dagger(\vec{p}')] = [a(\vec{p}), c^\dagger(\vec{p}')] = [a^\dagger(\vec{p}), c(\vec{p}')] = 0
\end{cases}
\tag{9.21}
$$

The *initial* and *final states* will be defined by their **kinematical configuration**. For scalar particles (i.e., spinless), we just need to define the four-momenta. We call them k^μ, q_1^μ, and q_2^μ, such that:

$$\sigma(k^\mu) \to \pi^0(q_1^\mu) + \pi^0(q_2^\mu) \tag{9.22}$$

Energy–momentum conservation imposes of course that $k^\mu = q_1^\mu + q_2^\mu$.

In quantum field theories, the states in the Fock space of the theory are defined starting from the **vacuum** $|0\rangle$ and applying the creation operators (see Eq. (7.46)). Accordingly, the initial state with a single particle σ with four-momentum k^μ in our decay process is defined as:

$$|k\rangle \equiv \underbrace{\sqrt{(2\pi)^3 2E_k}}_{\text{normalization}} c^\dagger(\vec{k})|0\rangle = \mathcal{N}_i c^\dagger(\vec{k})|0\rangle \tag{9.23}$$

where a conventional normalization factor was introduced. Similarly, the final state $|q_1 q_2\rangle$ is defined as:

$$|q_1 q_2\rangle \equiv \sqrt{(2\pi)^3 2E_{q_1}} \sqrt{(2\pi)^3 2E_{q_2}} a^\dagger(\vec{q}_1)a^\dagger(\vec{q}_2)|0\rangle = \mathcal{N}_f a^\dagger(\vec{q}_1)a^\dagger(\vec{q}_2)|0\rangle \tag{9.24}$$

or better considering its *bra* $\langle q_1 q_2|$:

$$\langle q_1 q_2| \equiv \mathcal{N}_f \langle 0| a(\vec{q}_1)a(\vec{q}_2) \tag{9.25}$$

So finally, the ket $|k\rangle$ creates the particle σ out of the vacuum, and the bra $\langle q_1 q_2|$ annihilates the two π^0s back into the vacuum, in agreement with the asymptotic picture described in Section 9.2. Consequently, we seek an amplitude of the form $\langle 0|a(\vec{q}_1)a(\vec{q}_2)\dots c^\dagger(\vec{k})|0\rangle$. We now need the interaction term that will connect the two states. Which terms in the Hamiltonian Eq. (9.19) should be considered? We note that the coupling term $\lambda\sigma\phi^2(x)$ connects the three fields in the desired way. At first order the corresponding S matrix term is simply given by:

$$\langle q_1 q_2| S_1 |k\rangle = (-i)\langle q_1 q_2| \int d^4x_1 \lambda\sigma(x_1)\phi^2(x_1)|k\rangle \tag{9.26}$$

where we note that as advertised, "local" means that we consider the interaction at the single space-time point x_1^μ. Replacing the initial and final states with the creation and annihilation operators, we find the amplitude starting from the vacuum $|0\rangle$ and ending in the vacuum $\langle 0|$:

$$\langle q_1 q_2|\, S_1\, |k\rangle = (-i\lambda)\mathcal{N}_i\mathcal{N}_f \,\langle 0|\, a(\vec{q}_1)a(\vec{q}_2) \int \mathrm{d}^4 x_1 \sigma(x_1)\phi^2(x_1)c^\dagger(\vec{k})\,|0\rangle \equiv (-i\lambda)\mathcal{N}_i\mathcal{N}_f\mathcal{S} \qquad (9.27)$$

We consider the amplitude $\mathcal{S} = \langle 0|\ldots|0\rangle$ and compute the integral over the single space-time point, which we relabel x^μ. We begin with the replacement of the $\sigma(x)$ field by its quantized plane wave expansion with modes labeled $\vec{k}'$:

$$
\begin{aligned}
\mathcal{S} &= \langle 0|\, a(\vec{q}_1)a(\vec{q}_2) \int \mathrm{d}^4 x\, \sigma(x)\phi^2(x)c^\dagger(\vec{k})\,|0\rangle \\
&= \langle 0|\, a(\vec{q}_1)a(\vec{q}_2) \int \mathrm{d}^4 x\, \phi^2(x) \int \frac{\mathrm{d}^3\vec{k}'}{(2\pi)^{3/2}}\frac{1}{\sqrt{2E_{k'}}}\left(c(\vec{k}')e^{-ik'\cdot x} + c^\dagger(\vec{k}')e^{+ik'\cdot x}\right)c^\dagger(\vec{k})\,|0\rangle \\
&= \langle 0|\, a(\vec{q}_1)a(\vec{q}_2) \int \mathrm{d}^4 x\, \phi^2(x) \int \frac{\mathrm{d}^3\vec{k}'}{(2\pi)^{3/2}}\frac{1}{\sqrt{2E_{k'}}}e^{-ik'\cdot x}c(\vec{k}')c^\dagger(\vec{k})\,|0\rangle \qquad (9.28)
\end{aligned}
$$

where we used the fact that the bracket $\langle 0|c^\dagger(\vec{k}')c^\dagger(\vec{k})|0\rangle$ vanishes. The commutation rules for the c, $c^\dagger$ operators can be used to swap the order of the latter, as follows:

$$\langle 0|\, c(\vec{k}')c^\dagger(\vec{k})\,|0\rangle = \langle 0|\, \delta^3(\vec{k}'-\vec{k})\,|0\rangle + \underbrace{\langle 0|\, c^\dagger(\vec{k})c(\vec{k}')\,|0\rangle}_{=0} = \langle 0|\, \delta^3(\vec{k}'-\vec{k})\,|0\rangle \qquad (9.29)$$

The annihilation and creation operators are gone and only a Dirac δ function is left. This allows us to perform the integral over the modes $\vec{k}'$ to obtain:

$$
\begin{aligned}
\mathcal{N}_i\mathcal{S} &= \mathcal{N}_i\,\langle 0|\, a(\vec{q}_1)a(\vec{q}_2) \int \mathrm{d}^4 x\, \phi^2(x) \int \frac{\mathrm{d}^3\vec{k}'}{(2\pi)^{3/2}}\frac{1}{\sqrt{2E_{k'}}}e^{-ik'\cdot x}\delta^3(\vec{k}'-\vec{k})\,|0\rangle \\
&= \langle 0|\, a(\vec{q}_1)a(\vec{q}_2) \int \mathrm{d}^4 x\, \phi^2(x)e^{-ik\cdot x}\,|0\rangle \qquad (9.30)
\end{aligned}
$$

In a similar way, we replace the field ϕ by its quantized plane-wave expansion in the previous result:

$$\mathcal{N}_i\mathcal{S} = \langle 0|\, a(\vec{q}_1)a(\vec{q}_2) \int \mathrm{d}^4 x\left(\int \frac{\mathrm{d}^3\vec{p}}{(2\pi)^{3/2}}\frac{1}{\sqrt{2E_p}}\left(a(\vec{p})e^{-ip\cdot x} + a^\dagger(\vec{p})e^{+ip\cdot x}\right)\right)^2 e^{-ik\cdot x}\,|0\rangle \qquad (9.31)$$

We note, however, that this time the field enters as ϕ^2, hence we will obtain two integrals over two modes, which we label $\vec{p}_1$ and $\vec{p}_2$. We see that the following combination of operators gives:

$$
\begin{aligned}
\langle 0|\, a(\vec{q}_1)a(\vec{q}_2)a^\dagger(\vec{p}_1)a^\dagger(\vec{p}_2)\,|0\rangle &= \langle 0|\, a(\vec{q}_1)\left(\delta^3(\vec{q}_2-\vec{p}_1) + a^\dagger(\vec{p}_1)a(\vec{q}_2)\right)a^\dagger(\vec{p}_2)\,|0\rangle \\
&= \langle 0|\, \delta^3(\vec{q}_2-\vec{p}_1)a(\vec{q}_1)a^\dagger(\vec{p}_2)\,|0\rangle + \langle 0|\, a(\vec{q}_1)a^\dagger(\vec{p}_1)a(\vec{q}_2)a^\dagger(\vec{p}_2)\,|0\rangle \\
&= \delta^3(\vec{q}_1-\vec{p}_2)\delta^3(\vec{q}_2-\vec{p}_1) + \delta^3(\vec{q}_1-\vec{p}_1)\delta^3(\vec{q}_2-\vec{p}_2) \qquad (9.32)
\end{aligned}
$$

while all other combinations vanish:

$$\langle 0|\, a(\vec{q}_1)a(\vec{q}_2)a(\vec{p}_1)a(\vec{p}_2)\,|0\rangle = \langle 0|\, a(\vec{q}_1)a(\vec{q}_2)a(\vec{p}_1)a^\dagger(\vec{p}_2)\,|0\rangle = \langle 0|\, a(\vec{q}_1)a(\vec{q}_2)a^\dagger(\vec{p}_1)a(\vec{p}_2)\,|0\rangle = 0 \qquad (9.33)$$

So we obtain:

$$
\begin{aligned}
\mathcal{N}_i\mathcal{S} &= \langle 0|\, a(\vec{q}_1)a(\vec{q}_2) \int \mathrm{d}^4 x \int\int \frac{\mathrm{d}^3\vec{p}_1}{(2\pi)^{3/2}\sqrt{2E_{p_1}}}\frac{\mathrm{d}^3\vec{p}_2}{(2\pi)^{3/2}\sqrt{2E_{p_2}}}a^\dagger(\vec{p}_1)e^{+ip_1\cdot x}a^\dagger(\vec{p}_2)e^{+ip_2\cdot x}e^{-ik\cdot x}\,|0\rangle \\
&= \int \mathrm{d}^4 x\, e^{-ik\cdot x} \int \frac{\mathrm{d}^3\vec{p}_1}{(2\pi)^{3/2}} \int \frac{\mathrm{d}^3\vec{p}_2}{(2\pi)^{3/2}}\frac{1}{\sqrt{2E_{p_1}}}\frac{1}{\sqrt{2E_{p_2}}}e^{+ip_1\cdot x}e^{+ip_2\cdot x}\, \times \\
&\qquad \left(\delta^3(\vec{q}_1-\vec{p}_2)\delta^3(\vec{q}_2-\vec{p}_1) + \delta^3(\vec{q}_1-\vec{p}_1)\delta^3(\vec{q}_2-\vec{p}_2)\right) \qquad (9.34)
\end{aligned}
$$

So finally, the S matrix at first order is computed by performing the integrals over the two modes $\vec{p}_1$ and $\vec{p}_2$, and also x^μ:

$$
\begin{aligned}
\langle q_1 q_2 | S_1 | k \rangle &= (-i\lambda) \mathcal{N}_i \mathcal{N}_f S = (-i\lambda) \sqrt{(2\pi)^3 2 E_{q_1}} \sqrt{(2\pi)^3 2 E_{q_2}} \times \\
&\quad \int \mathrm{d}^4 x\, e^{-ik\cdot x} \int \frac{\mathrm{d}^3 \vec{p}_1}{(2\pi)^{3/2}} \int \frac{\mathrm{d}^3 \vec{p}_2}{(2\pi)^{3/2}} \frac{1}{\sqrt{2E_{p_1}}} \frac{1}{\sqrt{2E_{p_2}}} e^{+ip_1\cdot x} e^{+ip_2\cdot x} \times \\
&\quad \left(\delta^3(\vec{q}_1 - \vec{p}_2) \delta^3(\vec{q}_2 - \vec{p}_1) + \delta^3(\vec{q}_1 - \vec{p}_1) \delta^3(\vec{q}_2 - \vec{p}_2) \right) \\
&= (-i\lambda) \sqrt{2E_{q_1}} \sqrt{2E_{q_2}} \int \mathrm{d}^4 x \int \mathrm{d}^3 \vec{p}_1 \int \mathrm{d}^3 \vec{p}_2 \frac{1}{\sqrt{2E_{p_1}}} \frac{1}{\sqrt{2E_{p_2}}} e^{+i(p_1+p_2-k)\cdot x} \times \\
&\quad \left(\delta^3(\vec{q}_1 - \vec{p}_2) \delta^3(\vec{q}_2 - \vec{p}_1) + \delta^3(\vec{q}_1 - \vec{p}_1) \delta^3(\vec{q}_2 - \vec{p}_2) \right) \\
&= (-i\lambda) \sqrt{2E_{q_1}} \sqrt{2E_{q_2}} \int \mathrm{d}^4 x \frac{1}{\sqrt{2E_{q_1}}} \frac{1}{\sqrt{2E_{q_2}}} \left(e^{+i(q_2+q_1-k)\cdot x} + e^{+i(q_1+q_2-k)\cdot x} \right) \\
&= (-i\lambda) \int \mathrm{d}^4 x \left(e^{+i(q_2+q_1-k)\cdot x} + e^{+i(q_1+q_2-k)\cdot x} \right) \\
&= (-2i\lambda)(2\pi)^4 \delta^4(q_1 + q_2 - k) \quad\quad (9.35)
\end{aligned}
$$

We have seen in Chapter 5 that cross-sections (see Eq. (5.117)) and decay widths (see Eq. (5.152)) are practically expressed as a product of **Lorentz-invariant terms** and a δ function ensuring energy–momentum conservation. We identify such terms in the S matrix result above and indeed recognize the δ function which is equivalent to the expected condition $k^\mu = q_1^\mu + q_2^\mu$, and also see that the amplitude $\mathcal{M}$ (see Eq. (9.10)) for the decay in our toy model is given at first order by:

$$
i\mathcal{M}(\sigma \to \pi^0 \pi^0) = -2i\lambda \quad\quad (9.36)
$$

This result can be graphically represented as a **Feynman diagram**, as illustrated in Figure 9.1. The diagram contains (1) a vertex and (2) three external legs, which represent the σ and the two π^0 particles, with their given kinematical configuration labeled with their corresponding four-vectors. The vertex contains a factor $-i\lambda$ which symbolizes the strength of the coupling. The apparently missing factor 2 compared to $\mathcal{M}$ comes from the symmetrization of the final state: the two π^0 are identical bosons and there are two ways to couple them to the vertex. **This symmetry factor is not represented in the Feynman diagrams.** The partial decay rate

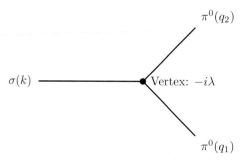

Figure 9.1 Feynman diagram of the decay $\sigma \to \pi^0 \pi^0$ in our toy model at first order.

of the σ particle must be computed with the proper kinematical factors. Similar to Eq. (5.163) and making use of the six-dimensional phase space (Eq. (5.156)) and a statistical factor $S = 1/2$ for two identical particles in the final state, we obtain:

$$
\begin{aligned}
\mathrm{d}\Gamma(M \to 1+2) &= \frac{|\mathcal{M}|^2}{2E_\sigma} S \frac{\mathrm{d}^3\vec{q_1}}{(2\pi)^3 2E_{q_1}} \frac{\mathrm{d}^3\vec{q_2}}{(2\pi)^3 2E_{q_2}} (2\pi)^4 \delta^4(k - q_1 - q_2) \\
&= \frac{|2i\lambda|^2}{32\pi^2} \frac{1}{2} \frac{p}{M^2} \mathrm{d}\Omega \quad \text{where} \quad p = \frac{1}{2}\sqrt{M^2 - 4m^2} \\
&= \frac{4\lambda^2}{32\pi^2} \frac{\sqrt{M^2 - 4m^2}}{4M^2} \mathrm{d}\Omega = \frac{\lambda^2}{32\pi^2 M} \sqrt{1 - \left(\frac{2m}{M}\right)^2} \, \mathrm{d}\Omega
\end{aligned}
\tag{9.37}
$$

which obviously constrains $M \geq 2m$. The integration over the solid angle $\mathrm{d}\Omega$ yields a multiplicative factor 4π, and we obtain the full width of the decay:

$$
\Gamma(\sigma \to \pi^0\pi^0) = \frac{\lambda^2}{8\pi M}\sqrt{1 - \left(\frac{2m}{M}\right)^2} \overset{m \ll M}{\longrightarrow} \frac{\lambda^2}{8\pi M}
\tag{9.38}
$$

As expected, the amplitude for the decay process is proportional to λ, and therefore the partial decay width is proportional to λ^2, consistent with the quantum-mechanical prescription for probabilities. Since the coupling constant λ has units of energy (see Eq. (9.16)), the width also has units of energy, and it corresponds to the uncertainty on the mass of the particle given its finite lifetime. The lifetime τ is given by:

$$
\tau = \Gamma^{-1} = \frac{8\pi M}{\lambda^2}\left(1 - \left(\frac{2m}{M}\right)^2\right)^{-1/2}
\tag{9.39}
$$

As expected, the larger the coupling strength λ^2, the shorter is the lifetime. Also, for a given mass M, the larger the mass of the decay products m, the smaller is the width, since the decay is suppressed by phase space, and hence the longer the lifetime.

9.3.2   Charged Decay in the Toy Model

We repeat the derivation in the case of the decay into two charged particles. The kinematical configuration is written as:

$$
\sigma(k^\mu) \to \pi^+(q_1^\mu) + \pi^-(q_2^\mu)
\tag{9.40}
$$

We can assume a similar coupling as in the case of the decay into the two neutral particles. The term, which describes the direct coupling, is hence:

$$
\mathcal{L}_{int} = -\lambda\sigma(x)\left(\phi_1^2(x) + \phi_2^2(x)\right)\ldots = -2\lambda\sigma(x)\left(\varphi^\dagger(x)\varphi(x)\right)\ldots
\tag{9.41}
$$

The field φ is scalar and complex and is quantized according to Eq. (7.80):

$$
\begin{cases}
\varphi(x) \equiv \displaystyle\int \frac{\mathrm{d}^3\vec{p}}{(2\pi)^{3/2}} \frac{1}{\sqrt{2E_p}}\left(a(\vec{p})e^{-ip\cdot x} + b^\dagger(\vec{p})e^{+ip\cdot x}\right) \\
\varphi^\dagger(x) \equiv \displaystyle\int \frac{\mathrm{d}^3\vec{p}}{(2\pi)^{3/2}} \frac{1}{\sqrt{2E_p}}\left(a^\dagger(\vec{p})e^{+ip\cdot x} + b(\vec{p})e^{-ip\cdot x}\right)
\end{cases}
\tag{9.42}
$$

As in the case of the neutral decay, we compute the matrix element to first order using the matrix element $\langle q_1 q_2 | S_1 | k \rangle$ to find:

$$
\begin{aligned}
\langle q_1 q_2 | S_1 | k \rangle &= (-i)\langle q_1 q_2 | \int \mathrm{d}^4 x_1 \, 2\lambda\sigma(x_1)\varphi^\dagger(x_1)\varphi(x_1) | k \rangle \\
&= (-2i\lambda)\langle 0 | b(\vec{q_2})a(\vec{q_1}) \int \mathrm{d}^4 x \int \mathrm{d}^3\vec{p_1} a^\dagger(\vec{p_1})e^{+ip_1\cdot x} \int \mathrm{d}^3\vec{p_2}b^\dagger(\vec{p_2})e^{+ip_2\cdot x} \int \mathrm{d}^3\vec{k'}c(\vec{k'})e^{-ik'\cdot x}c^\dagger(\vec{k})|0\rangle \\
&= (-2i\lambda)(2\pi)^4\delta^4(q_1 + q_2 - k)
\end{aligned}
\tag{9.43}
$$

The amplitude of the charged decay is therefore equal to that of the neutral pion decay mode:

$$iM(\sigma \to \pi^+\pi^-) = -2i\lambda \tag{9.44}$$

The corresponding Feynman diagram is shown in Figure. 9.2. In this case, there is no need to symmetrize because the final state is composed of two distinguishable particles because of their opposite charges. The vertex factor on the diagram is hence $2i\lambda$.

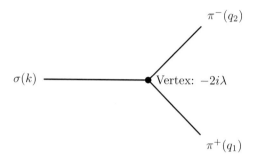

Figure 9.2 Feynman diagram of the decay $\sigma \to \pi^+\pi^-$ in our toy model at first order.

The full decay width can easily be computed as in the case of the decay to neutral particles. The phase space can be constructed as it was in the case of the neutral decay, but now with the statistical factor $S = 1$ we have:

$$\Gamma(\sigma \to \pi^+\pi^-) = 2\Gamma(\sigma \to \pi^0\pi^0) = \frac{\lambda^2}{4\pi}\sqrt{1 - \left(\frac{2m}{M}\right)^2} \tag{9.45}$$

9.4   The Feynman Propagator of the Real Scalar Field

In the previous section, we have seen how local interactions were introduced in the theory and used to describe, for example, the decay of a scalar particle. This form of direct coupling terms represents a fundamental type of interaction. At the same time, we need to find a prescription to describe the interaction of particles that do not find themselves in the same point of space-time. Since all interactions are *local*, this kind of **interaction at distance** is described by an **exchange of a field quantum, that carries energy and momentum between the two distant particles. This kind of exchange process is characterized by the amplitude of the creation of the field quantum in a given point in space-time and its annihilation in a different point of space-time.** This amplitude is called the **propagator**.

The second quantization of a field involves its expansion as a function of harmonic oscillators associated with plane waves indexed by the momentum $\vec{p}$. A particle with given momentum $\vec{p}$ is defined as such:

$$|p\rangle \equiv N a^\dagger(\vec{p}) |0\rangle \tag{9.46}$$

where the normalization is $N = \sqrt{(2\pi)^3 2E_p}$. This corresponds to a given state in the momentum space. In order to define the propagator from one point of space-time to another, we need to define a state corresponding to a particle located in a given point of space-time x^μ. We consider the application of the real-field operator on the vacuum and note:

$$\phi(x^\mu) |0\rangle = \int \frac{d^3\vec{p}}{(2\pi)^{3/2}} \frac{1}{\sqrt{2E_p}} \left(\underbrace{a(\vec{p}) e^{-ip\cdot x} |0\rangle}_{\to 0} + a^\dagger(\vec{p}) e^{+ip\cdot x} |0\rangle \right) \propto \int d^3\vec{p} \, e^{+ip\cdot x} |\vec{p}\rangle \tag{9.47}$$

where the four-vector $p^\mu = (p^0, \vec{p})$ with $p^0 = E_p = +\sqrt{\vec{p}^2 + m^2}$. This equation represents the linear combination of eigenstates with given momentum $\vec{p}$ and resembles the expression in quantum mechanics of a wave function for a perfectly localized particle (**Heisenberg's uncertainty principle**, see Appendix C). We can therefore say that the application of the field operator on the vacuum represents the state of a particle perfectly localized at a point of space-time, in other words, **the field operator $\phi(x^\mu)$ creates a particle at the space-time coordinate** x^μ.

Similarly, we can consider the conjugate expression:

$$\langle 0 | \phi(x^\mu) = \int \frac{d^3 p}{(2\pi)^{3/2}} \frac{1}{\sqrt{2E_p}} \langle 0 | \left(a(\vec{p}) e^{-ip\cdot x} + \underbrace{a^\dagger(\vec{p}) e^{+ip\cdot x}}_{\to 0} \right) \propto \int d^3 p\, e^{-ip\cdot x} \langle \vec{p} | \tag{9.48}$$

and conclude that the field operator applied to its left **annihilates a particle at the space-time coordinate** x^μ.

As a consequence, **the amplitude for the propagation of a particle from the space-time point y^μ to x^μ corresponds to its creation at y^μ and annihilation at x^μ, hence, it can be expressed as** $D(x, y) = \langle 0 | \phi(x) \phi(y) | 0 \rangle$. Because of translation invariance, we expect the propagator to depend only on the **space-time displacement** $x - y$, so finally:

$$\text{Propagator:} \quad D(x - y) \equiv \langle 0 | \phi(x) \phi(y) | 0 \rangle \tag{9.49}$$

It is easy to show that:

$$\begin{aligned}
\phi(x)\phi(y) &= \int \frac{d^3 p}{(2\pi)^{3/2}} \int \frac{d^3 q}{(2\pi)^{3/2}} \frac{1}{\sqrt{4E_p E_q}} \left(a(\vec{p}) e^{-ip\cdot x} + a^\dagger(\vec{p}) e^{+ip\cdot x} \right) \left(a(\vec{q}) e^{-iq\cdot y} + a^\dagger(\vec{q}) e^{+iq\cdot y} \right) \\
&\propto \left((a(\vec{p}) a(\vec{q}) \ldots) + (a(\vec{p}) a^\dagger(\vec{q}) \ldots) + (a^\dagger(\vec{p}) a(\vec{q}) \ldots) + (a^\dagger(\vec{p}) a^\dagger(\vec{q}) \ldots) \right)
\end{aligned} \tag{9.50}$$

where the several "..." represent the exponential terms, which are omitted for clarity. From all the possible combinations of operators, we have $\langle 0 | a(\vec{p}) a^\dagger(\vec{q}) | 0 \rangle = \delta^3(\vec{p} - \vec{q})$, while all others vanish (e.g., $\langle 0 | a(\vec{p}) a(\vec{q}) | 0 \rangle = 0$, etc.). Hence, we can perform one integral over $d^3 q$ for example and obtain:

$$D(x - y) = \int \frac{d^3 p}{(2\pi)^3} \frac{1}{2E_p} e^{-ip\cdot(x-y)} \tag{9.51}$$

where again $p^\mu = (p^0, \vec{p})$ and $p^0 = E_p = +\sqrt{\vec{p}^2 + m^2} > 0$. This effectively means that the integral is running over the three-momentum with the energy p^0 of the particle fixed on the mass shell by the relation $p^2 - m^2 = E_p^2 - \vec{p}^2 - m^2 = 0$. We further note:

- $D(x - y)$ is the quantum amplitude for a particle to travel from $y \to x$;

- then $D(y - x)$ is the quantum amplitude for a particle to travel from $x \to y$.

We find a very important property of the propagator $D(x - y)$ by noting that $D(x - y)$ and $D(y - x)$ are in general not identical but are however related:

$$D(y - x) = \int \frac{d^3 p}{(2\pi)^3} \frac{1}{2E_p} e^{-ip\cdot(y-x)} = \int \frac{d^3 p}{(2\pi)^3} \frac{1}{2E_p} e^{-i(-p)\cdot(x-y)} \tag{9.52}$$

Generally, **the propagation amplitude from $y^\mu \to x^\mu$ is identical to the propagation amplitude from $x^\mu \to y^\mu$** if we swapped $p^\mu \to -p^\mu$ in the exponent. And this reminds us of the Stückelberg–Feynman interpretation of the Dirac equation (see Section 8.15). However, in the present case, we cannot simply consider $p^\mu \to -p^\mu$ because that would imply $p^0 \to -p^0$, while we had only positive-energy solutions in Eq. (9.51).

We note that so far we have not specified which of the events x^μ or y^μ comes first in *time*, i.e., whether the time components of the four-vectors satisfy $y^0 > x^0$ or $y^0 < x^0$. When it comes to understanding the physical interpretation of the propagator, recalling that $D(x - y)$ corresponds to creation at y^μ and annihilation at x^μ, we then distinguish two cases:

- $x^0 > y^0$: forward propagation in time.

- $x^0 < y^0$: backward propagation in time.

The Feynman prescription is to distinguish the two cases and use the following form of propagator, called the **Feynman propagator**:

$$D_F(x-y) \equiv \begin{cases} D(x-y) & \text{if } x^0 > y^0 \\ D(y-x) & \text{if } x^0 < y^0 \end{cases} \tag{9.53}$$

We can rewrite the Feynman propagator with the Heaviside step function (see Appendix A.4). Using Eq. (9.51):

$$
\begin{aligned}
D_F(x-y) &= \Theta(x^0-y^0)D(x-y) + \Theta(y^0-x^0)D(y-x) \\
&= \Theta(x^0-y^0)\langle 0| \phi(x)\phi(y) |0\rangle + \Theta(y^0-x^0)\langle 0| \phi(y)\phi(x) |0\rangle \\
&= \Theta(x^0-y^0)\int \frac{d^3\vec{p}}{(2\pi)^3}\frac{1}{2E_p}e^{-ip\cdot(x-y)} + \Theta(y^0-x^0)\int \frac{d^3\vec{p}}{(2\pi)^3}\frac{1}{2E_p}e^{-ip\cdot(y-x)}
\end{aligned} \tag{9.54}
$$

where $p^\mu = (E_p, \vec{p})$ and $E_p = +\sqrt{\vec{p}^2 + m^2} > 0$. The Heaviside step functions can conveniently be expressed as integrals in the complex plane (see Appendix A.4). For example, for the first term:

$$
\begin{aligned}
\Theta(x^0-y^0)\int \frac{d^3\vec{p}}{(2\pi)^3}\frac{1}{2E_p}e^{-ip\cdot(x-y)} &= \lim_{\epsilon\to 0^+}\frac{-1}{2\pi i}\int_{-\infty}^{+\infty}dz \frac{e^{-iz(x^0-y^0)}}{z+i\epsilon}\int \frac{d^3\vec{p}}{(2\pi)^3}\frac{1}{2E_p}e^{-iE_p(x^0-y^0)}e^{+i\vec{p}\cdot(\vec{x}-\vec{y})} \\
&= -\lim_{\epsilon\to 0^+}\int \frac{d^3\vec{p}}{(2\pi)^3}\int_{-\infty}^{+\infty}\frac{dz}{2\pi i}\frac{e^{-i(z+E_p)(x^0-y^0)}}{z+i\epsilon}\frac{1}{2E_p}e^{+i\vec{p}\cdot(\vec{x}-\vec{y})}
\end{aligned} \tag{9.55}
$$

The integral over dz has a pole at $z = -i\epsilon$ where ϵ is infinitesimal and taken in the limit $\epsilon \to 0^+$. This has the effect of slightly shifting the pole off the real axis, which allows us to carry out the integral (**to regularize it**) and thereafter we can take the limit. With the change of variable $p^0 \equiv z + E_p$, we have (note that in this case p^0 is a complex number and we label it as such, even with a risk of confusion with the time component of p^μ, for reasons that will become clear later):

$$
\Theta(x^0-y^0)\int \frac{d^3\vec{p}}{(2\pi)^3}\frac{1}{2E_p}e^{-ip\cdot(x-y)} = -\lim_{\epsilon\to 0^+}\int \frac{d^3\vec{p}}{(2\pi)^3}\frac{e^{+i\vec{p}\cdot(\vec{x}-\vec{y})}}{2E_p}\int_{-\infty}^{+\infty}\frac{dp^0}{2\pi i}\frac{e^{-ip^0(x^0-y^0)}}{(p^0-E_p+i\epsilon)} \tag{9.56}
$$

This last result should be understood as follows: we integrate on dp^0 before we integrate on $d^3\vec{p}$. The integral $\int dp^0$ has a pole at $+E_p$, which is regularized by shifting the pole off the real axis with $i\epsilon$, and finally we take the limit $\epsilon \to 0$. Then the integral over $d^3\vec{p}$ should be evaluable without any singularities. During the two separate integrations, p^0 and $\vec{p}$ are independent and do not need to satisfy the on-shell mass relation. Finally, we obtain:

$$
\Theta(x^0-y^0)\int \frac{d^3\vec{p}}{(2\pi)^3}\frac{1}{2E_p}e^{-ip\cdot(x-y)} = i\lim_{\epsilon\to 0^+}\int \frac{d^4p}{(2\pi)^4}\frac{1}{2E_p}\frac{e^{-ip\cdot(x-y)}}{(p^0-E_p+i\epsilon)} \tag{9.57}
$$

We note that the integration is over the four-momentum space d^4p! The second term can be found by exchanging $x \leftrightarrow y$ in the previous result:

$$
\begin{aligned}
\Theta(y^0-x^0)\int \frac{d^3\vec{p}}{(2\pi)^3}\frac{1}{2E_p}e^{-ip\cdot(y-x)} &= -\lim_{\epsilon\to 0^+}\int \frac{d^3\vec{p}}{(2\pi)^3}\frac{e^{+i\vec{p}\cdot(\vec{y}-\vec{x})}}{2E_p}\int_{-\infty}^{+\infty}\frac{dp^0}{2\pi i}\frac{e^{-ip^0(y^0-x^0)}}{(p^0-E_p+i\epsilon)} \\
&= -\lim_{\epsilon\to 0^+}\int \frac{d^3\vec{p}}{(2\pi)^3}\frac{e^{+i\vec{p}\cdot(\vec{x}-\vec{y})}}{2E_p}\int_{-\infty}^{+\infty}\frac{dp^0}{2\pi i}\frac{e^{-ip^0(x^0-y^0)}}{(-p^0-E_p+i\epsilon)}
\end{aligned} \tag{9.58}
$$

where, in between the first and second line, we have flipped the sign of $\vec{p}$. This is valid since we integrate over the entire $d^3\vec{p}$ space while E_p depends only on $\vec{p}^2$. We also flipped the sign of p^0. Hence:

$$
\Theta(y^0-x^0)\int \frac{d^3\vec{p}}{(2\pi)^3}\frac{1}{2E_p}e^{-ip\cdot(y-x)} = i\lim_{\epsilon\to 0^+}\int \frac{d^4p}{(2\pi)^4}\frac{1}{2E_p}\frac{e^{-ip\cdot(x-y)}}{(-p^0-E_p+i\epsilon)} \tag{9.59}
$$

We can combine the two results to rewrite the Feynman propagator in the following form:

$$
\begin{aligned}
D_F(x-y) &= i \lim_{\epsilon \to 0^+} \int \frac{\mathrm{d}^4 p}{(2\pi)^4} \frac{1}{2E_p} \left[\frac{e^{-ip \cdot (x-y)}}{(p^0 - E_p + i\epsilon)} + \frac{e^{-ip \cdot (x-y)}}{(-p^0 - E_p + i\epsilon)} \right] \\
&= i \lim_{\epsilon \to 0^+} \int \frac{\mathrm{d}^4 p}{(2\pi)^4} \frac{e^{-ip \cdot (x-y)}}{2E_p} \left[\frac{2E_p}{(p^0 - E_p + i\epsilon)(p^0 + E_p - i\epsilon)} \right] + \mathcal{O}(\epsilon) \\
&= i \lim_{\epsilon' \to 0^+} \int \frac{\mathrm{d}^4 p}{(2\pi)^4} \left[\frac{e^{-ip \cdot (x-y)}}{((p^0)^2 - E_p^2 + i\epsilon')} \right] + \mathcal{O}(\epsilon')
\end{aligned}
\tag{9.60}
$$

In the second line, the term proportional to ϵ in the numerator can be neglected. In the last line, all the terms proportional to $i\epsilon$ in the denominator have been collected into a single term $i\epsilon'$ and the higher-order terms have again been neglected under the assumption of the limit. The denominator can be expressed as follows, where m is the rest mass of the particle:

$$
(p^0)^2 - E_p^2 + i\epsilon' = (p^0)^2 - \vec{p}^2 - m^2 + i\epsilon' = p^2 - m^2 + i\epsilon'
\tag{9.61}
$$

We see that the integrals were regularized using the so-called "$i\epsilon$" **prescription of Feynman**. In his original paper, Feynman wrote:

> "*The integrals are not yet completely defined for there are poles in the integrand when $p^2 - m^2 = 0$. We can define how these poles are to be evaluated by the rule that m is considered to have an infinitesimal negative imaginary part. That is m is replaced by $m - i\delta$ and the limit taken as $\delta \to 0$ from above. If we call $E_p = +\sqrt{m^2 + \vec{p}^2}$ then the integrals which involve p^0 essentially is $\int \exp(-ip^0(t_2 - t_1)) dp^0 ((p^0)^2 - E_p^2)^{-1}$ which has poles at $p^0 = +E_p$ and $p^0 = -E_p$. The replacement of m by $m - i\delta$ means that E_p has a small negative imaginary part; the first pole is below and the second above the real axis. Now if $t_2 - t_1 > 0$ the contour can be completed around the semicircle below the real axis thus giving a residue from the $p^0 = +E_p$ pole, or $-(2E_p)^{-1} \exp(-iE_p(t_2 - t_1))$. If $t_2 - t_1 < 0$ the upper semicircle must be used, and $p^0 = -E_p$ at the pole, so that the function varies in each case as required (...).*" [2]

See Figure 9.3.

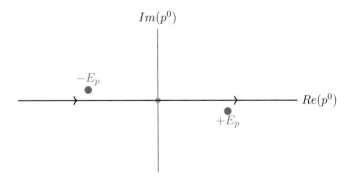

Figure 9.3 Illustration of the regularization of the propagator integral, following the $i\epsilon$ prescription of Feynman.

2   Reprinted excerpt with permission from R. P. Feynman, "The theory of positrons," *Phys. Rev.*, vol. 76, pp. 749–759, Sep 1949. Copyright 1949 by the American Physical Society.

Finally, we can express the **Feynman propagator of the real Klein–Gordon field** as the following integral of four-momentum:

$$D_F(x - y) \equiv i \int \frac{\mathrm{d}^4 p}{(2\pi)^4} \frac{e^{-ip\cdot(x-y)}}{(p^2 - m^2 + i\epsilon)} \tag{9.62}$$

where, with the $i\epsilon$ term, the limit $\epsilon \to 0^+$ is implicitly assumed. The expression looks manifestly covariant! We note that although the result looks very compact, it was actually obtained as the sum of two terms, one valid for forward propagation, and the other for backward propagation in time. In the Feynman propagator, both cases are simultaneously taken into account. It is by summing the two Heaviside functions, which respectively took care of the $x^0 > y^0$ and $x^0 < y^0$ cases, that we could obtain a covariant integral over the four-momentum $\mathrm{d}^4 p$ starting from the propagator of the quantized field defined at a given time t and expanded as superpositions of plane waves summed over $\mathrm{d}^3\vec{p}$!

In fact, the propagator in this form looks like a Fourier integral. Accordingly, we define the **Feynman propagator for the real scalar field in momentum space** as:

$$\tilde{D}_F(p) \equiv \frac{i}{(p^2 - m^2 + i\epsilon)} \tag{9.63}$$

where

$$D_F(x - y) = \int \frac{\mathrm{d}^4 p}{(2\pi)^4} e^{-ip\cdot(x-y)} \tilde{D}_F(p). \tag{9.64}$$

It was first derived by Feynman in 1949, where he argued that computations are often simplified by working with momentum and energy variables rather than space and time, and is at the basis of the theory of the Feynman diagrams. The propagator defined above will be attached to internal lines of the diagrams, representing the propagation of virtual particles.

9.5 The Generalized Breit–Wigner

In Section 4.5, we have seen how quantum mechanics introduces the width of a metastable state. We'd like to generalize this concept in the context of our relativistic propagator theory. The introduction of the exponential decay term in Eq. (4.37) can also be formally achieved by replacing the bare mass with a complex mass term:

$$m_R \longrightarrow m_R - i\Gamma/2 \tag{9.65}$$

This result suggests that the Feynman propagator Eq. (9.63) of a particle with finite lifetime can be found by the replacement:

$$\tilde{D}_F(p) \propto \frac{1}{(p^2 - m_R^2 + i\epsilon)} \longrightarrow \tilde{D}_F(p, \Gamma) \propto \frac{1}{(p^2 - (m_R - i\Gamma/2)^2)} = \frac{1}{(p^2 - (m_R^2 - im_R\Gamma - \Gamma^2/4))} \tag{9.66}$$

For a width sufficiently smaller than the mass $\Gamma \ll m_R$, we can absorb the (real) last term into the resonance mass $m_R^2 - \Gamma^2/4 \to m_R^2$, and finally **the propagator of a resonance of mass m_R and width Γ is given by**:

$$\tilde{D}_F(p, m_R, \Gamma) \propto \frac{1}{p^2 - m_R^2 + im_R\Gamma} \tag{9.67}$$

We notice that the effect of the finite width is to displace the pole of the propagator into the complex plane, and hence it effectively replaces the $i\epsilon$ prescription. Since the propagator enters into the amplitude of the process, the cross-section in an s-channel in the region of the resonance becomes proportional to:

$$\sigma(s) \propto \left| \frac{1}{s - m_R^2 + im_R\Gamma} \right|^2 = \frac{1}{(s - m_R^2)^2 + m_R^2\Gamma^2} \tag{9.68}$$

This form is called the **relativistic Breit[3]–Wigner distribution**. It can also be interpreted as a probability density function. In general, Γ can also be a function of s; this dependence is typically only important when the width is not small compared to the resonance mass and the phase-space dependence of the width needs to be taken into account, for example when some decay modes are kinematically forbidden or suppressed.

9.6   The Feynman Propagator and Time Ordering

Starting from Eq. (9.54) we can conveniently write the Feynman propagator with the **time-ordering operator** T:

$$
\begin{aligned}
D_F(x-y) &= \Theta(x^0 - y^0) \langle 0| \phi(x)\phi(y) |0\rangle + \Theta(y^0 - x^0) \langle 0| \phi(y)\phi(x) |0\rangle \\
&= \langle 0| T[\phi(x)\phi(y)] |0\rangle
\end{aligned}
\tag{9.69}
$$

The time ordering T implies that the operators are to be ordered by time, with the latest located at the left. This notation is very practical and will often be used. We can easily verify that the propagator actually represents the **Green's function** of the field equation. Using the Klein–Gordon operator (see Eq. (7.7)), we have:

$$
\begin{aligned}
\left(\partial^\mu \partial_\mu + m^2\right) \langle 0| T[\phi(x)\phi(y)] |0\rangle &= i\left(\partial^\mu \partial_\mu + m^2\right) \int \frac{\mathrm{d}^4 p}{(2\pi)^4} \frac{e^{-ip\cdot(x-y)}}{(p^2 - m^2)} \\
&= i \int \frac{\mathrm{d}^4 p}{(2\pi)^4} \left(\partial^\mu \partial_\mu + m^2\right) \frac{e^{-ip\cdot(x-y)}}{(p^2 - m^2)} \\
&= i \int \frac{\mathrm{d}^4 p}{(2\pi)^4} \left(-p^2 + m^2\right) \frac{e^{-ip\cdot(x-y)}}{(p^2 - m^2)} \\
&= -i \int \frac{\mathrm{d}^4 p}{(2\pi)^4} e^{-ip\cdot(x-y)} = -i\delta^4(x-y)
\end{aligned}
\tag{9.70}
$$

This fact will be useful when deriving the propagators for other fields, as shown below. In Section 9.10, we will also see that there is a very powerful graphical method that can be used to compute time-ordered products of field operators, making use of the so-called Wick's theorem.

9.7   The Feynman Propagator of the Complex Scalar Field

We consider the Feynman propagator of a quantized complex scalar field $\varphi(x)$. Recalling Eq. (7.80), we must consider the field and its conjugate:

$$
\varphi(x) \equiv \int \frac{\mathrm{d}^3 \vec{p}}{(2\pi)^{3/2}} \frac{1}{\sqrt{2E_p}} \left(a(\vec{p})e^{-ip\cdot x} + b^\dagger(\vec{p})e^{+ip\cdot x}\right); \quad \varphi^\dagger(x) \equiv \int \frac{\mathrm{d}^3 \vec{p}}{(2\pi)^{3/2}} \frac{1}{\sqrt{2E_p}} \left(a^\dagger(\vec{p})e^{+ip\cdot x} + b(\vec{p})e^{-ip\cdot x}\right)
\tag{9.71}
$$

We remember that this complex field is adequate to describe a system composed of two sets of particles with opposite charges and identical rest masses. The field φ annihilates a particle of type a and creates a particle of type b, while the conjugate $\varphi^\dagger$ annihilates a particle of type b and creates a particle of type a.

In analogy to Eq. (9.49), the **amplitude for the propagation from the space-time point y^μ to x^μ corresponds to the amplitude for the complex field**, which we can label as $D^q(x - y)$:

$$
D^q(x-y) = \langle 0| \varphi(x)\varphi^\dagger(y) |0\rangle .
\tag{9.72}
$$

---

3   Gregory Breit (1899–1981), Russian-born American physicist.

In order for this to make sense, every scalar particle a of the theory has a corresponding scalar "antiparticle" b with opposite charge. We note that the propagator is actually composed of the amplitude for the particle of type a to be created in y^μ and annihilated in point x^μ and of the amplitude for the particle of type b to be created in x^μ and annihilated in y^μ. Both types of particles are included in the amplitude and they are propagating in opposite directions. This is consistent with the Stückelberg–Feynman interpretation that both situations are equivalent, when we interpret one particle as the antiparticle of the other.

The corresponding **Feynman propagator** $D_F^q(x-y)$ and its **representation in four-momentum space** $\tilde{D}_F^q(p)$ can be written from $D^q(x-y)$ by considering the time-ordering operator T:

$$D_F^q(x-y) \equiv \langle 0 | T(\varphi(x)\varphi^\dagger(y)) | 0 \rangle = \int \frac{\mathrm{d}^4 p}{(2\pi)^4} e^{-ip\cdot(x-y)} \tilde{D}_F^q(p) \tag{9.73}$$

The Feynman propagator is a Green's function of the Klein–Gordon equation for the complex field; it must satisfy:

$$\left(\partial^\mu \partial_\mu + m^2\right) \langle 0 | T(\varphi(x)\varphi^\dagger(y)) | 0 \rangle = -i\delta^4(x-y) \tag{9.74}$$

In the four-momentum space, it becomes (see Eq. (9.70)):

$$(-p^2 + m^2)\tilde{D}_F^q(p) = -i \qquad \text{or} \qquad \tilde{D}_F^q(p) = \frac{i}{p^2 - m^2} \tag{9.75}$$

which is the same result obtained for the real Klein–Gordon field! Hence, taking into account the regularization with the $i\epsilon$ prescription, we found that the Feynman propagator $D_F^q(x-y)$ of the complex Klein–Gordon field equals:

$$D_F^q(x-y) = i \int \frac{\mathrm{d}^4 p}{(2\pi)^4} \frac{e^{-ip\cdot(x-y)}}{(p^2 - m^2 + i\epsilon)} \quad \text{and} \quad \tilde{D}_F^q(p) \equiv \frac{i}{(p^2 - m^2 + i\epsilon)} \tag{9.76}$$

It is the same as the one for the real field (cf. Eqs. (9.62) and (9.63)). One can express this fact by saying that in the case of a real field, the particle and its antiparticle are the same (not distinguishable by a charge), and hence both types do not appear explicitly, although they are embedded in the result.

9.8 Charged Scalar Particle Scattering in our Toy Model

We go back to our toy model and illustrate how interaction terms and also propagators can enter in the description of a scattering process. In the toy model, the scalar particles are charged and are described by three real scalar fields $\phi_i(x)$ $(i=1,2,3)$. Starting from our toy Lagrangian, Eq. (9.18), we now retain the **direct** σ–ϕ **local interaction** terms $\lambda\sigma\phi_i^2$ and the **self-coupling terms** $g\phi_i^4$. The relevant part of the interaction Lagrangian is then composed of:

$$\begin{aligned}
\mathcal{L}_{int} &= -g(\phi_1^2 + \phi_2^2 + \phi_3^2)^2 - \lambda\sigma(\phi_1^2 + \phi_2^2 + \phi_3^2) + \cdots \\
&= -g\phi^4 - \lambda\sigma\phi^2 - 4g(\varphi^\dagger\varphi)^2 - 4g\varphi^\dagger\varphi\phi^2 - 2\lambda\sigma(\varphi^\dagger\varphi) + \cdots
\end{aligned} \tag{9.77}$$

where in the second line we have introduced the fields φ and $\varphi^\dagger$, as in Eq. (9.41). We consider the scattering between four scalar particles and define its kinematical configuration with the four-vectors q_1, q_2, q_3, and q_4:

$$\pi^+(q_1) + \pi^-(q_2) \to \pi^+(q_3) + \pi^-(q_4) \tag{9.78}$$

There are several ways to obtain this reaction.

• **Four-point interaction.** At first order there is a contribution from the **four-point vertex**, as illustrated in Figure 9.4, which can be obtained with the $4g(\varphi^\dagger\varphi)^2$ interaction term. The corresponding amplitude, inserting

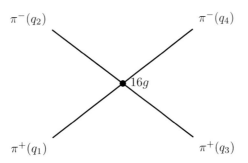

Figure 9.4 Four-point vertex for pion scattering $\pi^+(q_1) + \pi^-(q_2) \to \pi^+(q_3) + \pi^-(q_4)$.

initial and final states and integrating over one point in space-time x_1, is equal to:

$$\langle q_3 q_4 | S_1 | q_1 q_2 \rangle = (-i) \langle q_3 q_4 | \int \mathrm{d}^4 x_1 \, 4g (\varphi^\dagger(x_1)\varphi(x_1))^2 | q_1 q_2 \rangle \tag{9.79}$$

By inserting the common expressions with plane waves for the quantized fields for φ and $\varphi^\dagger$, the term $(\varphi^\dagger \varphi)^2$ yields momentum integrals over four modes, which we label $\vec{p}_1, \ldots, \vec{p}_4$:

$$
\begin{aligned}
\langle q_3 q_4 | S_1 | q_1 q_2 \rangle &= (-4ig) \mathcal{N}_i \mathcal{N}_f \langle 0 | b(\vec{q}_4) a(\vec{q}_3) \int \mathrm{d}^4 x_1 (\varphi^\dagger(x_1)\varphi(x_1))^2 b^\dagger(\vec{q}_2) a^\dagger(\vec{q}_1) | 0 \rangle \\
&= (-4ig) \langle 0 | b(\vec{q}_4) a(\vec{q}_3) \int \mathrm{d}^4 x_1 \int \mathrm{d}^3 \vec{p}_1 \left(a^\dagger(\vec{p}_1) e^{+ip_1 \cdot x_1} + b(\vec{p}_1) e^{-ip_1 \cdot x_1} \right) \times \\
&\quad \int \mathrm{d}^3 \vec{p}_2 \left(a(\vec{p}_2) e^{-ip_2 \cdot x_1} + b^\dagger(\vec{p}_2) e^{+ip_2 \cdot x_1} \right) \times \int \mathrm{d}^3 \vec{p}_3 \left(a^\dagger(\vec{p}_3) e^{+ip_3 \cdot x_1} + b(\vec{p}_3) e^{-ip_3 \cdot x_1} \right) \times \\
&\quad \int \mathrm{d}^3 \vec{p}_4 \left(a(\vec{p}_4) e^{-ip_4 \cdot x_1} + b^\dagger(\vec{p}_4) e^{+ip_4 \cdot x_1} \right) b^\dagger(\vec{q}_2) a^\dagger(\vec{q}_1) | 0 \rangle
\end{aligned}
\tag{9.80}
$$

Here $\mathcal{N}_i$ and $\mathcal{N}_f$ are the initial and final state normalizations (see Eqs. (9.23) and (9.25)). In the second line, we have omitted the normalization factors early on for readability purposes. We will do the same also in the following, unless otherwise noted. We can simplify this expression by defining $a_i \equiv a(\vec{p}_i) e^{-ip_i \cdot x_1}$ and $b_i \equiv b(\vec{p}_i) e^{-ip_i \cdot x_1}$. Hence:

$$
\begin{aligned}
\langle q_3 q_4 | S_1 | q_1 q_2 \rangle &= (-4ig) \langle 0 | b(\vec{q}_4) a(\vec{q}_3) \int \mathrm{d}^4 x_1 \int \mathrm{d}^3 \vec{p}_1 \int \mathrm{d}^3 \vec{p}_2 \int \mathrm{d}^3 \vec{p}_3 \int \mathrm{d}^3 \vec{p}_4 \\
&\quad \left(a_1^\dagger + b_1 \right) \times \left(a_2 + b_2^\dagger \right) \times \left(a_3^\dagger + b_3 \right) \times \left(a_4 + b_4^\dagger \right) b^\dagger(\vec{q}_2) a^\dagger(\vec{q}_1) | 0 \rangle
\end{aligned}
\tag{9.81}
$$

Multiplying by brute force the four products in parentheses gives 16 terms but not all of them are relevant. We note that we need to consider the products with contain *exactly* two a-type and two b-type operators, in total with two creation and two annihilation operators in order to connect to the initial and final states. Let us start with the product $a_1^\dagger b_2^\dagger b_3 a_4$ that resembles Eq. (9.32) and noting that a and b freely commute, we get:

$$
\begin{aligned}
\langle 0 | b(\vec{q}_4) a(\vec{q}_3) a_1^\dagger b_2^\dagger b_3 a_4 b^\dagger(\vec{q}_2) a^\dagger(\vec{q}_1) | 0 \rangle &= \langle 0 | b(\vec{q}_4) b_2^\dagger b_3 b^\dagger(\vec{q}_2) a(\vec{q}_3) a_1^\dagger a_4 a^\dagger(\vec{q}_1) | 0 \rangle \\
&= \langle 0 | b(\vec{q}_4) b_2^\dagger b_3 b^\dagger(\vec{q}_2) a(\vec{q}_3) a_1^\dagger \left(\delta^3(\vec{p}_4 - \vec{q}_1) e^{-ip_4 \cdot x_1} + a^\dagger(\vec{q}_1) a_4 \right) | 0 \rangle \\
&= \delta^3(\vec{p}_4 - \vec{q}_1) e^{-ip_4 \cdot x_1} \langle 0 | b(\vec{q}_4) b_2^\dagger b_3 b^\dagger(\vec{q}_2) a(\vec{q}_3) a_1^\dagger | 0 \rangle \\
&= \delta^3(\vec{p}_4 - \vec{q}_1) \delta^3(\vec{q}_3 - \vec{p}_1) e^{-i(p_4 - p_1) \cdot x_1} \times \\
&\quad \delta^3(\vec{p}_3 - \vec{q}_2) \delta^3(\vec{q}_4 - \vec{p}_2) e^{-i(p_3 - p_2) \cdot x_1}
\end{aligned}
\tag{9.82}
$$

Starting from the above term $a_1^\dagger b_2^\dagger b_3 a_4$ and keeping constant the number of a-type, b-type operators and daggers, we obtain the three additional products:

$$a_1^\dagger b_2^\dagger b_3 a_4 \implies b_1 b_2^\dagger a_3^\dagger a_4, \quad a_1^\dagger a_2 b_3 b_4^\dagger, \quad b_1 a_2 a_3^\dagger b_4^\dagger \tag{9.83}$$

Since a and b commute, we can write the four terms as:

$$b_2^\dagger b_3 a_1^\dagger a_4, \quad b_1 b_2^\dagger a_3^\dagger a_4, \quad b_3 b_4^\dagger a_1^\dagger a_2, \quad b_1 b_4^\dagger a_2 a_3^\dagger \tag{9.84}$$

The b terms in the second product yield:

$$
\begin{aligned}
\langle 0| b(\vec{q}_4) b_1 b_2^\dagger b^\dagger(\vec{q}_2) |0\rangle &= \langle 0| b(\vec{q}_4) \left(\delta^3(\vec{p}_1 - \vec{p}_2) e^{-i(p_1 - p_2)\cdot x_1} + b_2^\dagger b_1 \right) b^\dagger(\vec{q}_2) |0\rangle \\
&= \langle 0| \delta^3(\vec{p}_1 - \vec{p}_2) e^{-i(p_1 - p_2)\cdot x_1} b(\vec{q}_4) b^\dagger(\vec{q}_2) + b(\vec{q}_4) b_2^\dagger b_1 b^\dagger(\vec{q}_2) |0\rangle \\
&= \delta^3(\vec{p}_1 - \vec{p}_2) \delta^3(\vec{q}_4 - \vec{q}_2) e^{-i(p_1 - p_2)\cdot x_1} + \delta^3(\vec{q}_4 - \vec{p}_2) \delta^3(\vec{p}_1 - \vec{q}_2) e^{-i(p_1 - p_2)\cdot x_1} \\
&\to \delta^3(\vec{q}_4 - \vec{p}_2) \delta^3(\vec{p}_1 - \vec{q}_2) e^{-i(p_1 - p_2)\cdot x_1}
\end{aligned}
\tag{9.85}
$$

and the a terms give:

$$\langle 0| a(\vec{q}_3) a_3^\dagger a_4 a^\dagger(\vec{q}_1) |0\rangle \to \delta^3(\vec{q}_3 - \vec{p}_3) \delta^3(\vec{p}_4 - \vec{q}_1) e^{-i(p_4 - p_3)\cdot x_1} \tag{9.86}$$

and similarly for the two last terms. Finally, the result is:

$$
\begin{aligned}
\langle q_3 q_4 | S_1 | q_1 q_2 \rangle &= (-4ig) \int d^4 x_1 \int d^3\vec{p}_1 \int d^3\vec{p}_2 \int d^3\vec{p}_3 \int d^3\vec{p}_4 \times \\
&\quad \left[\delta^3(\vec{p}_4 - \vec{q}_1) \delta^3(\vec{q}_3 - \vec{p}_1) \delta^3(\vec{p}_3 - \vec{q}_2) \delta^3(\vec{q}_4 - \vec{p}_2) e^{-i(p_4 + p_3 - p_2 - p_1)\cdot x_1} + \right. \\
&\quad \left. \delta^3(\vec{q}_4 - \vec{p}_2) \delta^3(\vec{p}_1 - \vec{q}_2) \delta^3(\vec{q}_3 - \vec{p}_3) \delta^3(\vec{p}_4 - \vec{q}_1) e^{-i(p_4 + p_1 - p_2 - p_3)\cdot x_1} + \right. \\
&\quad \left. \ldots \right] \\
&= (-4gi)\, 4 \int d^4 x_1 e^{-i(q_1 + q_2 - q_3 - q_4)\cdot x_1} \\
&= (-16ig)(2\pi)^4 \delta^4(q_1 + q_2 - q_3 - q_4)
\end{aligned}
\tag{9.87}
$$

Recognizing the terms representing energy–momentum conservation, we can extract the matrix element amplitude of the four-point vertex diagram, which we label $\mathcal{M}_1$:

$$\boxed{i\mathcal{M}_1(\pi^+ \pi^- \to \pi^+ \pi^-) = -16ig} \tag{9.88}$$

The calculation gives us the vertex coupling factor of $16g$, where g is the assigned coupling constant.

• **Scattering process.** At next order we can consider the term $-2\lambda \sigma(\varphi^\dagger \varphi)$. Including the initial and final states and integrating over two space-time points x_1 and x_2, the amplitude is:

$$\langle q_3 q_4 | S_2 | q_1 q_2 \rangle = \frac{(-2\lambda i)^2}{2!} \langle q_3 q_4 | \int d^4 x_1 \int d^4 x_2 T \left[\sigma(x_1)(\varphi^\dagger(x_1)\varphi(x_1))\sigma(x_2)(\varphi^\dagger(x_2)\varphi(x_2)) \right] | q_1 q_2 \rangle \tag{9.89}$$

Direct application of the quantized field operators φ and $\varphi^\dagger$ on the initial $|q_1 q_2\rangle$ and final $\langle q_3 q_4|$ states, followed by some algebra similar to the previous case, gives the following result, where we recognize exponential functions of all possible combinations of the four-momenta vectors:

$$
\begin{aligned}
\langle q_3 q_4 | \varphi^\dagger(x_1)\varphi(x_1)\varphi^\dagger(x_2)\varphi(x_2) | q_1 q_2 \rangle &= e^{ix_1(q_3 - q_1)} e^{ix_2(q_4 - q_2)} + e^{ix_1(q_3 + q_4)} e^{-ix_2(q_1 + q_2)} + \\
&\quad e^{-ix_1(q_1 + q_2)} e^{ix_2(q_3 + q_4)} + e^{ix_1(q_4 - q_2)} e^{ix_2(q_3 - q_1)} \\
&= 2 \left[e^{ix_1(q_3 + q_4)} e^{-ix_2(q_1 + q_2)} + e^{ix_1(q_4 - q_2)} e^{-ix_2(q_1 - q_3)} \right]
\end{aligned}
\tag{9.90}
$$

First we note that a combinatorial factor 2 has appeared in front of the brackets. **Such a term will always cancel the $1/n!$ term in the Taylor series of the S matrix – see Eq. (9.89). So we will implicitly keep this in mind in the S_i expansion.** The terms with the field σ do not act on external four-momenta of the initial or final states. They deliver internal lines corresponding to the Feynman propagator of virtual particles. Recalling that the σ particle has a rest mass M, we have:

$$\langle 0 | T [\sigma(x_1)\sigma(x_2)] | 0 \rangle = D_F(x_1 - x_2) = i \int \frac{\mathrm{d}^4 p}{(2\pi)^4} \frac{e^{-ip\cdot(x_1 - x_2)}}{(p^2 - M^2 + i\epsilon)} \tag{9.91}$$

Finally, we obtain the following S-matrix element:

$$
\begin{aligned}
& \langle q_3 q_4 | S_2 | q_1 q_2 \rangle \\
&= (-2\lambda i)^2 \int \mathrm{d}^4 x_1 \mathrm{d}^4 x_2 \langle 0 | T [\sigma(x_1)\sigma(x_2)] | 0 \rangle \left\{ e^{ix_1(q_3+q_4)} e^{-ix_2(q_1+q_2)} + e^{ix_1(q_4-q_2)} e^{-ix_2(q_1-q_3)} \right\} \\
&= (-2\lambda i)^2 i \int \mathrm{d}^4 x_1 \mathrm{d}^4 x_2 \frac{\mathrm{d}^4 p}{(2\pi)^4} \frac{e^{-ip\cdot(x_1-x_2)}}{(p^2 - M^2 + i\epsilon)} \left\{ e^{ix_1(q_3+q_4)} e^{-ix_2(q_1+q_2)} + e^{ix_1(q_4-q_2)} e^{-ix_2(q_1-q_3)} \right\} \\
&= (-2\lambda i)^2 i \int \mathrm{d}^4 x_1 \mathrm{d}^4 x_2 \frac{\mathrm{d}^4 p}{(2\pi)^4} \left\{ \frac{e^{ix_1(q_3+q_4-p)} e^{-ix_2(q_1+q_2-p)}}{(p^2 - M^2 + i\epsilon)} + \frac{e^{ix_1(q_4-q_2-p)} e^{-ix_2(q_1-q_3-p)}}{(p^2 - M^2 + i\epsilon)} \right\} \\
&= (-2\lambda i)^2 (2\pi)^4 i \int \mathrm{d}^4 p \left\{ \frac{\delta^4(q_3+q_4-p)\delta^4(q_1+q_2-p)}{(p^2 - M^2 + i\epsilon)} + \frac{\delta^4(q_4-q_2-p)\delta^4(q_1-q_3-p)}{(p^2 - M^2 + i\epsilon)} \right\} \\
&= (-2\lambda i)^2 (2\pi)^4 \delta^4(q_3+q_4-q_1-q_2) \left\{ \frac{i}{((q_1+q_2)^2 - M^2 + i\epsilon)} + \frac{i}{((q_1-q_3)^2 - M^2 + i\epsilon)} \right\}
\end{aligned}
\tag{9.92}
$$

The term $(2\pi)^4 \delta^4(q_3 + q_4 - q_1 - q_2)$ corresponds once again to the overall energy–momentum conservation between initial and final states. We also recognize Feynman propagators for the σ particle in the momentum space of the form $\tilde{D}_F(p)$ with well-defined kinematical configurations given by $q_1 + q_2$ and $q_1 - q_3$. The second-order amplitude is therefore equal to:

$$i\mathcal{M}_2(\pi^+\pi^- \to \pi^+\pi^-) = (-2\lambda i)^2 \left\{ \tilde{D}_F(q_1 + q_2) + \tilde{D}_F(q_1 - q_3) \right\} \tag{9.93}$$

We can interpret this result by looking at the corresponding diagrams in Figure 9.5. The first diagram (left) represents the annihilation of the $\pi^+\pi^-$ pair into a virtual σ followed by its decay into a $\pi^+\pi^-$ pair. For reasons that will become clear later, this is called the **s-channel**. The four-momentum of the σ is constrained by energy–momentum conservation at the vertices to be equal to $q_1 + q_2$, as shown on the diagram. The second diagram (right) represents the scattering of the $\pi^+\pi^-$ pair via the **exchange of a virtual σ particle**, whose energy–momentum is equal to $q_1 - q_3$.

• **Total amplitude.** Finally, the total amplitude $\mathcal{M}$ up to second order is given by the sum of the two amplitudes:

$$i\mathcal{M} = i\mathcal{M}_1 + i\mathcal{M}_2 \tag{9.94}$$

We stress that these are amplitudes (complex numbers) and not probabilities which are proportional to $|\mathcal{M}|^2 = |\mathcal{M}_1|^2 + |\mathcal{M}_2|^2 + \mathcal{M}_1^*\mathcal{M}_2 + \mathcal{M}_1\mathcal{M}_2^*$. Hence the sum of such amplitudes can lead to interference effects among the three diagrams considered: (1) four-point interaction; (2) σ-annihilation; (3) σ-exchange. In order to calculate the cross-section for the reaction, the common phase-space factors need to be included, as was shown in Chapter 5. Similar arguments can be considered for the scattering of neutral scalar particles. This is left to the exercises (see **Ex 9.1**).

9.9   Causality and Propagators

The causal properties of Minkowski space-time are encoded in its light-cone structure, which requires, consistent with relativity, that $v \leq c$ for all signals (see Appendix D). An event at the origin of a light cone may influence

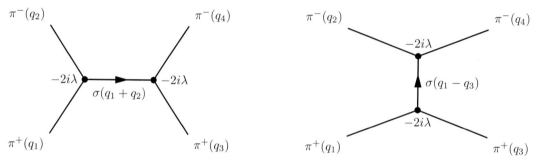

Figure 9.5 The diagrams for the $\pi^+\pi^-$-scattering process in our toy model: (left) σ-annihilation diagram; (right) σ-exchange diagram.

any event in its forward light cone (the "future"). The event at the origin of the light cone may have been influenced by events in its backward light cone (the "past"). Those outside the light cone termed **space-like** could be connected only by signals with $v > c$, which would violate causality, so such events must be disconnected.

This condition can be considered for the case of the scalar field ϕ at the two space-time points x and y. If the two points are to be causally disconnected, then the two operators must commute. So, it is crucial that $[\phi(x), \phi(y)] = 0$ when x and y are space-like separated, namely when $(x - y)^2 < 0$. In order to verify this statement, we start with the equal time commutation rules given in Eq. (7.36), which are defined for $x^0 = y^0$. Under this condition, we have $x^0 - y^0 = 0$, hence the separation is always space-like:

$$(x - y)^2 = (x^0 - y^0)^2 - (\vec{x} - \vec{y})^2 = -(\vec{x} - \vec{y})^2 < 0 \tag{9.95}$$

so $[\phi(x), \phi(y)] = 0$ in the case of equal times. We now note that $(x - y)^2$ remains invariant under Lorentz transformations, hence space-like separations remain, as expected, space-like in all frames. The commutator $[\phi(x), \phi(y)] = 0$ is also Lorentz-invariant, since the fields are made of products of Lorentz-invariant terms! Therefore, if the commutator vanishes in one frame in which the separation is space-like, then it vanishes in all frames. Hence

$$[\phi(x), \phi(y)] = 0 \quad \text{when} \quad (x - y)^2 < 0 \quad \square \tag{9.96}$$

A similar argument holds for $[\phi(x), \Pi(y)]$, so causality is preserved quite generally.

What about the propagator? Should the amplitude for the propagation from space-like separated x and y also naively vanish? We note that by definition:

$$\begin{aligned}
\langle 0| [\phi(x), \phi(y)] |0\rangle &= \langle 0|\phi(x)\phi(y) - \phi(y)\phi(x)|0\rangle \\
&= D(x - y) - D(y - x)
\end{aligned} \tag{9.97}$$

Considering the first term for instance, we have for $x^0 = y^0$ and $\vec{x} - \vec{y} = \vec{r}$:

$$\begin{aligned}
\langle 0|\phi(x)\phi(y)|0\rangle &= D(x - y) = \int \frac{\mathrm{d}^3\vec{p}}{(2\pi)^3} \frac{1}{2E_p} e^{-ip\cdot(x-y)} = \int \frac{\mathrm{d}^3\vec{p}}{(2\pi)^3} \frac{1}{2E_p} e^{-i\vec{p}\cdot\vec{r}} \\
&= \frac{1}{(2\pi)^3} \int \mathrm{d}p\, \mathrm{d}\cos\theta \mathrm{d}\phi \frac{p^2}{2E_p} e^{-ipr\cos\theta} = \frac{2\pi}{(2\pi)^3} \int_0^\infty \mathrm{d}p\, \frac{p^2}{2E_p} \frac{e^{ipr} - e^{-ipr}}{ipr} \\
&= \frac{1}{2(2\pi)^2 r} \int_0^\infty (-ip)\mathrm{d}p\, \frac{e^{ipr} - e^{-ipr}}{\sqrt{p^2 + m^2}} = \frac{1}{8\pi^2 r} \int_{-\infty}^\infty (-ip)\mathrm{d}p\, \frac{e^{ipr}}{\sqrt{p^2 + m^2}}
\end{aligned} \tag{9.98}$$

One can now set $\rho \equiv -ip$ to obtain:

$$D(x - y) \propto \int_m^\infty \rho\mathrm{d}\rho\, \frac{e^{-\rho r}}{\sqrt{\rho^2 - m^2}} \quad \xrightarrow{\text{when } r \to \infty} e^{-mr} = e^{-m|\vec{x} - \vec{y}|} \tag{9.99}$$

So, the amplitude is non-zero even for space-like separations, i.e., outside the light cone! This result gives us a clue on how we should handle the plain propagator. We cannot simply consider $D(x-y)$ alone, because there is no reason to choose x before or after y. Hence, physically both terms add $D(x-y) - D(y-x)$, and the total amplitude cancels as it should for space-like separations. For a complex scalar field or for a Dirac field, the amplitude for a particle traveling from x to y is always to be considered together with that of its antiparticle traveling from y to x (in the case of the real scalar field, the particle is *de facto* its own antiparticle). This is a direct requirement from the covariant formulation. *As a consequence, particles and antiparticles must exist!*

9.10   Normal Ordering and Wick's Theorem

9.10.1   The Case of Two Real Scalar Fields

Using the Dyson series for the S matrix, Eq. (9.9), we can in principle compute all matrix elements to any order n in the series, however, as we have already noticed, calculations can be rather cumbersome. As the order increases, the calculation implies evaluation of an increasing number of time-ordered products of field operators squeezed between asymptotic initial and final states. For example, considering only real scalar fields for the moment, we have an expression of the type:

$$\langle f | \, T \left(\phi(x_1) \dots \phi(x_n) \right) |i\rangle = \langle 0 | \, a..a \, T \left(\phi(x_1) \dots \phi(x_n) \right) a^\dagger ..a^\dagger \, |0\rangle \qquad (9.100)$$

where the initial (final) states are constructed by creation from (annihilation into) the vacuum. We note that the field operators themselves yield products of creation and annihilation operators. Computing such matrix elements often involves a complex "bookkeeping" of all such time-ordered operators.

Wick's[4] theorem [80] provides an extremely useful tool to achieve our computational goal. It basically connects a **time-ordered product** T into a **normal-ordered product** N (also called Wick's ordering) of field operators. **Normal ordering means that all annihilation operators are moved to the right of the expression, while the creation operators are moved to the left.** For example:

$$N \left(a(\vec{q}_1) a^\dagger(\vec{q}_2) a(\vec{q}_3) \right) = a^\dagger(\vec{q}_2) a(\vec{q}_1) a(\vec{q}_3) \qquad (9.101)$$

Normal ordering is powerful since **the vacuum expectation value of any normal product vanishes**:

$$\langle 0 | \, N(\text{any operator}) \, |0\rangle = 0 \qquad (9.102)$$

This will greatly simplify the calculations.

We begin with the simplest case of the time-ordered product of just two field operators $T[\phi(x)\phi(y)]$, noting that its vacuum expectation gives just the Feynman propagator – see Eq. (9.69):

$$\langle 0 | \, T[\phi(x)\phi(y)] \, |0\rangle = D_F(x-y) \qquad (9.103)$$

We can always decompose the real scalar field $\phi(x)$ into so-called "positive" and "negative" parts:

$$\phi(x) \equiv \phi^+(x) + \phi^-(x) = \int \frac{\mathrm{d}^3\vec{p}}{(2\pi)^{3/2}\sqrt{2E_p}} \left(a(\vec{p}) e^{-ip\cdot x} \right) + \int \frac{\mathrm{d}^3\vec{p}}{(2\pi)^{3/2}\sqrt{2E_p}} \left(a^\dagger(\vec{p}) e^{+ip\cdot x} \right) \qquad (9.104)$$

which is useful because:

$$\phi^+(x) |0\rangle \propto a(\vec{p}) |0\rangle = 0 \qquad \text{and} \qquad \langle 0| \phi^-(x) \propto \langle 0| a^\dagger(\vec{p}) = 0 \qquad (9.105)$$

---

4   Gian Carlo Wick (1909–1992), Italian theoretical physicist.

If we assume $x^0 > y^0$, then:

$$
\begin{aligned}
T[\phi(x)\phi(y)] \ \overset{x^0 \geq y^0}{=} \ & \phi(x)\phi(y) = \left(\phi^+(x) + \phi^-(x)\right)\left(\phi^+(y) + \phi^-(y)\right) \\
= \ & \phi^+(x)\phi^+(y) + \phi^+(x)\phi^-(y) + \phi^-(x)\phi^+(y) + \phi^-(x)\phi^-(y)
\end{aligned}
\tag{9.106}
$$

By inspection we note that the "++" (resp. "−−") terms have only $a(\vec{p})$ (resp. $a^\dagger(\vec{p})$) operators. The "−+" has all annihilation operators to the right of the creation operators. Therefore, all terms except the "+−" are normal ordered! We get rid of the "+−" by considering the commutator:

$$
\left[\phi^+(x), \phi^-(y)\right] = \phi^+(x)\phi^-(y) - \phi^-(y)\phi^+(x)
\tag{9.107}
$$

We obtain an expression where *all terms are normal ordered except the commutator*:

$$
\begin{aligned}
T[\phi(x)\phi(y)] \ \overset{x^0 \geq y^0}{=} \ & \phi^+(x)\phi^+(y) + \left[\phi^+(x), \phi^-(y)\right] + \phi^-(y)\phi^+(x) + \phi^-(x)\phi^+(y) + \phi^-(x)\phi^-(y) \\
= \ & N\left(\phi(x)\phi(y)\right) + \left[\phi^+(x), \phi^-(y)\right]
\end{aligned}
\tag{9.108}
$$

If we assume $x^0 < y^0$, then we get:

$$
\begin{aligned}
T[\phi(x)\phi(y)] \ \overset{x^0 \leq y^0}{=} \ & \phi^+(y)\phi^+(x) + \left[\phi^+(y), \phi^-(x)\right] + \phi^-(x)\phi^+(y) + \phi^-(y)\phi^+(x) + \phi^-(y)\phi^-(x) \\
= \ & \phi^+(x)\phi^+(y) + \left[\phi^+(y), \phi^-(x)\right] + \phi^-(x)\phi^+(y) + \phi^-(y)\phi^+(x) + \phi^-(x)\phi^-(y) \\
= \ & N\left(\phi(x)\phi(y)\right) + \left[\phi^+(y), \phi^-(x)\right]
\end{aligned}
\tag{9.109}
$$

In order to combine the two cases $x^0 > y^0$ and $x^0 < y^0$, one defines the **contraction of the two real scalar fields**:

$$
\overparen{\phi(x)\phi(y)} = \begin{cases} \left[\phi^+(x), \phi^-(y)\right] & \text{if} \quad x^0 > y^0 \\ \left[\phi^+(y), \phi^-(x)\right] & \text{if} \quad x^0 < y^0 \end{cases}
\tag{9.110}
$$

We note that the vacuum expectation value of the contraction of the two fields is just its Feynman propagator:

$$
\langle 0| \overparen{\phi(x)\phi(y)} |0\rangle = D_F(x-y)
\tag{9.111}
$$

So, finally, Wick's theorem for the case of two fields relates their time ordering with their normal ordering as follows:

$$
T(\phi(x)\phi(y)) = N(\phi(x)\phi(y)) + \overparen{\phi(x)\phi(y)} = N(\phi(x)\phi(y) + \overparen{\phi(x)\phi(y)})
\tag{9.112}
$$

where for convenience we have inserted the fully contracted pair of fields inside the normal-ordering operator. The beauty here is that the expression is valid for both cases $x^0 > y^0$ and $x^0 < y^0$!

9.10.2 The Case of Any Number of Real Scalar Fields

This result can be generalized to a time-ordered product of any number of fields. **Wick's theorem states that the time-ordered product of n fields is equivalent to the normal-ordered sum of the product of the fields and all their possible distinct contractions.** It is expressed as:

$$
T(\phi(x_1)\dots\phi(x_n)) = N(\phi(x_1)\dots\phi(x_n) + \text{all possible contractions})
\tag{9.113}
$$

For $n = 2$ we recover the previous result. For the proof at higher multiplicities, we assume that it is valid for $n - 1$ and use induction. First, we note that time-ordered products of fields are invariant under interchange of fields by definition, since regardless of labels, time ordering always prevails. Since normal-ordered products and the contractions are also invariant under such changes of order, we may choose $t_i < t_{i+1}$ for $i \in \{1, 2, \dots, n-1\}$. Assuming Wick's theorem is valid for $n - 1$ fields, defining $\phi_a = \phi(x_a)$, we can write:

$$
T(\phi_1\phi_2\dots\phi_n) = \phi_1 T(\phi_2\dots\phi_n) = \phi_1 N(\phi_2\dots\phi_n + \text{all possible contractions } \phi_2\dots\phi_n)
\tag{9.114}
$$

Let us begin with the term that has no contractions $\phi_1 N(\phi_2 \ldots \phi_n)$. We need to show that

$$\phi_1 N(\phi_2 \ldots \phi_n) = \underbrace{N(\phi_1 \phi_2 \ldots \phi_n)}_{=W_1} + \underbrace{\sum_{j=2}^{n} N\left(\overbrace{\phi_1 \ldots \phi_j} \ldots\right)}_{=W_2} \tag{9.115}$$

as this gives us the extra term with ϕ_1 under the normal ordering and also the extra terms where ϕ_1 is contracted. We split ϕ_1 into positive and negative parts and compute:

$$(\phi_1^+ + \phi_1^-) N(\phi_2 \ldots \phi_n) \tag{9.116}$$

Recalling Eq. (9.105), we see that the term ϕ_1^- can readily be put on the left inside the normal-ordering bracket, since it is already normally ordered, hence we can write:

$$W_- = \phi_1^- N(\phi_2 \ldots \phi_n) = N(\phi_1^- \phi_2 \ldots \phi_n) \tag{9.117}$$

To handle the term with ϕ_1^+ and move it inside the normal product, one considers the commutator:

$$\begin{aligned} W_+ &= \phi_1^+ N(\phi_2 \ldots \phi_n) - N(\phi_2 \ldots \phi_n)\phi_1^+ + N(\phi_2 \ldots \phi_n)\phi_1^+ \\ &= \left[\phi_1^+, N(\phi_2 \ldots \phi_n)\right] + N(\phi_2 \ldots \phi_n)\phi_1^+ \end{aligned} \tag{9.118}$$

Since the normal ordering brings the fields with positive frequencies to the right, we can rewrite the second normal product including ϕ_1^+, i.e., $N(\phi_2 \ldots \phi_n \phi_1^+)$. This term can actually be written as $N(\phi_1^+ \phi_2 \ldots \phi_n)$. Then combining it with W_- we get:

$$N(\phi_1^+ \phi_2 \ldots \phi_n) + W_- = N(\phi_1 \phi_2 \ldots \phi_n) = W_1 \tag{9.119}$$

which is exactly the first term W_1, i.e., the term of Wick's theorem with no contractions. So we have found that:

$$W^+ + W^- = W_1 + \left[\phi_1^+, N(\phi_2 \ldots \phi_n)\right] \tag{9.120}$$

Then, starting from $[\phi_1^+, \phi_2 \phi_3] = [\phi_1^+, \phi_2]\phi_3 + \phi_2[\phi_1^+, \phi_3]$ and iterating, we get that:

$$\left[\phi_1^+, N(\phi_2 \ldots \phi_n)\right] = \sum_{j=2}^{n} N\left(\ldots [\phi_1^+, \phi_j] \ldots\right) = \sum_{j=2}^{n} N\left(\ldots [\phi_1^+, \phi_j^-] \ldots\right) = \sum_{j=2}^{n} N\left(\ldots \overbrace{\phi_1 \phi_j} \ldots\right) \tag{9.121}$$

which is exactly W_2. This proves the case where we had no contractions in the normal product $N(\phi_2 \ldots \phi_n)$. A similar situation happens when we start with a normal product with one contraction. For example, we find (in the sum the values $j = k$ and $j = l$ are excluded):

$$\phi_1 N(\phi_2 \ldots \overbrace{\phi_k \ldots \phi_l} \ldots \phi_n) = N(\phi_1 \phi_2 \ldots \overbrace{\phi_k \ldots \phi_l} \ldots \phi_n) + \sum_{j=2,\, j \neq k, j \neq l}^{n} N\left(\overbrace{\phi_1 \ldots \phi_j} \ldots \overbrace{\phi_k \ldots \phi_l} \ldots\right) \tag{9.122}$$

Therefore, also in this case, we find all the terms that we need when adding ϕ_1 to the normal product of fields with one contraction. And so on, repeating this with all possible contractions, one finds all the additional terms that include ϕ_1. So, finally, all the terms of Wick's theorem that need to be added when going from $n - 1$ to n have therefore been added. This proves Wick's theorem $\square$.

9.10.3   The Use of Feynman Diagrams

We can consider one illuminating example of the application of Wick's theorem. For instance, for the time-ordered product of four fields ($n = 4$), we have:

$$T(\phi_1\phi_2\phi_3\phi_4) = N(\phi_1\phi_2\phi_3\phi_4 +$$
$$\phi_1\phi_2\phi_3\phi_4 + \phi_1\phi_2\phi_3\phi_4 + \phi_1\phi_2\phi_3\phi_4 + \phi_1\phi_2\phi_3\phi_4 + \phi_1\phi_2\phi_3\phi_4 + \phi_1\phi_2\phi_3\phi_4 +$$
$$\phi_1\phi_2\phi_3\phi_4 + \phi_1\phi_2\phi_3\phi_4 + \phi_1\phi_2\phi_3\phi_4) \tag{9.123}$$

The contraction of two fields in a string of products replaces the two fields by their contraction while leaving the others untouched. Recalling the identity with the Feynman propagator, we have for instance:

$$N\left(\phi_1\phi_2\phi_3\phi_4\right) = D_F(x_1 - x_2) \cdot N(\phi_3\phi_4) \tag{9.124}$$

and so on. We now note that **the vacuum expectation value of any normal-ordered terms that are not fully contracted vanishes**. Finally, most terms go away and we are left with *only the following three terms*:

$$\langle 0|\, T(\phi_1\phi_2\phi_3\phi_4)\,|0\rangle = \langle 0|\,\phi_1\phi_2\phi_3\phi_4\,|0\rangle + \langle 0|\,\phi_1\phi_2\phi_3\phi_4\,|0\rangle + \langle 0|\,\phi_1\phi_2\phi_3\phi_4\,|0\rangle \tag{9.125}$$

In terms of Feynman propagators, we therefore simply obtain:

$$\langle 0|\, T(\phi_1\phi_2\phi_3\phi_4)\,|0\rangle = D_F(x_1 - x_2)D_F(x_3 - x_4) + D_F(x_1 - x_3)D_F(x_2 - x_4) + D_F(x_1 - x_4)D_F(x_2 - x_3) \tag{9.126}$$

Wick's theorem has transformed the vacuum expectation value of the time-ordered product of field operators into a sum of products of Feynman propagators! These can be graphically represented as Feynman diagrams, as illustrated in Figure 9.6. We interpret this result as follows: (1) two particles are created at two space-

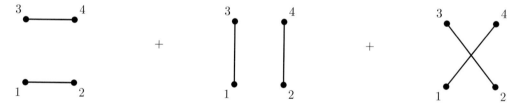

Figure 9.6 The vacuum expectation of the time-ordered product of four real scalar fields $\langle 0|\, T(\phi_1\phi_2\phi_3\phi_4)\,|0\rangle$ yields the three amplitudes, which are represented graphically by their Feynman diagrams.

time points; (2) they propagate until they annihilate into two other space-time points. This can happen in three ways, which correspond to the number of ways we can connect four points in pairs of two. The total amplitude $\langle 0|\, T(\phi_1\phi_2\phi_3\phi_4)\,|0\rangle$ is the sum of the three diagrams, and there is no way to distinguish which path the particles have actually taken.

To summarize, Wick's theorem allows us to arrange systematically the computation of processes in an organized and graphical way. This method will be used to compute the amplitudes of physical reactions in a given theory. We can, for example, refer the reader to the direct derivations of scalar particle decays (see Section 9.3) and scattering (see Section 9.8). Following our new approach, we would first have derived the Feynman diagrams responsible for the desired reaction, using a set of **predefined Feynman rules for a given theory**. Once all diagrams are considered to a given order in the Dyson expansion, we use field contractions to compute the propagators and eventually the amplitudes. This will be done later for specific theories.

9.10.4   The Case of the Complex Scalar Field

The expression for the Feynman propagator of the complex scalar field was derived in Section 9.7. It was shown that the proper propagation amplitude to consider was the product of the field and its conjugate. The formalism of Wick's contraction can therefore be straightforwardly extended to the complex scalar field. We illustrate here the case for $n = 2$:

$$T(\varphi(x)\varphi^\dagger(y)) = N(\varphi(x)\varphi^\dagger(y) + \overline{\varphi(x)\varphi^\dagger(y)}) \tag{9.127}$$

where **the contraction of two complex scalar fields** is:

$$\langle 0| \overline{\varphi(x)\varphi^\dagger(y)} |0\rangle = D_F^q(x - y) \tag{9.128}$$

In Feynman diagrams, the real and complex scalar fields are represented in the same way. Their propagator is simply a line joining the two vertices.

9.11   So What About the Ground State?

We have seen the importance of the time-ordered two-point correlation $\langle 0|T(\phi_1\phi_2)|0\rangle$, which represents the Feynman propagator of free-particle states. In general, we care about n-point correlations of n time-ordered fields $\langle 0|T(\phi_1 \ldots \phi_n)|0\rangle$. In scattering theory, the total amplitude contains such time-ordered products of fields squeezed between the asymptotic initial and final states.

We have already mentioned in Section 9.1 that **the ground state $|\Omega\rangle$ of the interacting theory is different from the vacuum of the free theory** $|0\rangle$. Indeed, in an interacting theory, the vacuum state itself will evolve with time and we need to consider the n-point correlation squeezed between the initial and final states defined by the new ground state $\langle \Omega|T(\phi_1 \ldots \phi_n)|\Omega\rangle$.

Since the perturbative approach assumes a localized interaction, as discussed in the first part of this chapter (see also Section 4.16), we expect that the ground state $|\Omega\rangle$ will be the vacuum of the free theory $|0\rangle$ in the asymptotic regimes (i.e., at $t = -T/2$ and $t = +T/2$ for $T \to \infty$). Hence, in the interaction picture, we can write for the ground state at a given reference time t_0 such that $+T/2 \geq t_0 \geq -T/2$:

$$|\Omega\rangle|_{t_0} = \eta \lim_{T\to\infty} U(t_0, -T/2)|0\rangle \quad \text{and} \quad {}_{t_0}|\langle\Omega| = \eta^* \lim_{T\to\infty} \langle 0| U(+T/2, t_0) \tag{9.129}$$

These definitions state that the ground state of the interacting theory will evolve from and into the free vacuum. The factor η is some global normalization to be determined. It should be set such that:

$$\langle\Omega|\Omega\rangle = 1 \quad \longrightarrow \quad |\eta|^2 \lim_{T\to\infty} \langle 0|U(+T/2, t_0)U(t_0, -T/2)|0\rangle = |\eta|^2 \lim_{T\to\infty} \langle 0|U(+T/2, -T/2)|0\rangle = 1 \tag{9.130}$$

or

$$|\eta|^2 = \lim_{T\to\infty} \frac{1}{\langle 0|U(+T/2, -T/2)|0\rangle} \tag{9.131}$$

We now consider a space-time point y^μ with $y_0 \geq t_0 \geq -T/2$. We relate the Heisenberg picture and the interaction picture using Eq. (4.44) and compute $\phi(y)|\Omega\rangle$:

$$\phi(y)|\Omega\rangle|_{t_0} = U^\dagger(y^0, t_0)\phi_I(y)U(y^0, t_0)|\Omega\rangle|_{t_0} = \eta \lim_{T\to\infty} U^\dagger(y^0, t_0)\phi_I(y)U(y^0, t_0)U(t_0, -T/2)|0\rangle \tag{9.132}$$

where $-T/2 \leq t_0 \leq y_0$. Likewise:

$${}_{t_0}|\langle\Omega| \phi(x) = \eta^* \lim_{T\to\infty} \langle 0| U(+T/2, t_0)U^\dagger(x^0, t_0)\phi_I(x)U(x^0, t_0) \tag{9.133}$$

where $+T/2 \geq x_0 \geq t_0$. We can now compute in our interacting theory the relevant two-point correlation between two space-time points x^μ and y^μ assuming $+T/2 \geq x_0 \geq y_0 \geq t_0 \geq -T/2$:

$$
\begin{aligned}
\langle\Omega|\phi(x)\phi(y)|\Omega\rangle &= |\eta|^2 \lim_{T\to\infty} \langle 0|U(+T/2,t_0)U^\dagger(x^0,t_0)\phi_I(x)U(x^0,t_0)U^\dagger(y^0,t_0)\phi_I(y)U(y^0,t_0)U(t_0,-T/2)|0\rangle \\
&= |\eta|^2 \lim_{T\to\infty} \langle 0|U(+T/2,x^0)\phi_I(x)U(x^0,y^0)\phi_I(y)U(y^0,-T/2)|0\rangle
\end{aligned}
\tag{9.134}
$$

Finally, we arrive at:

$$
\langle\Omega|\phi(x)\phi(y)|\Omega\rangle = \lim_{T\to\infty} \frac{\langle 0|U(+T/2,x^0)\phi_I(x)U(x^0,y^0)\phi_I(y)U(y^0,-T/2)|0\rangle}{\langle 0|U(+T/2,-T/2)|0\rangle}
\tag{9.135}
$$

We will actually be interested in the time-ordered product of the fields. We can insert the Dyson operator and this latter actually allows us to reorder the fields inside the bracket. So:

$$
\begin{aligned}
\langle\Omega|\phi(x)\phi(y)|\Omega\rangle &= \lim_{T\to\infty} \frac{\langle 0|T\left[\phi_I(x)\phi_I(y)U(+T/2,x^0)U(x^0,y^0)U(y^0,-T/2)\right]|0\rangle}{\langle 0|U(+T/2,-T/2)|0\rangle} \\
&= \lim_{T\to\infty} \frac{\langle 0|T\left[\phi_I(x)\phi_I(y)U(+T/2,-T/2)\right]|0\rangle}{\langle 0|U(+T/2,-T/2)|0\rangle}
\end{aligned}
\tag{9.136}
$$

We can remove the requirement that $x_0 \geq y_0$ by inserting the time-ordering operator on the left-hand side. Consequently, we finally obtain:

$$
\langle\Omega|T(\phi(x)\phi(y))|\Omega\rangle = \lim_{T\to\infty} \frac{\langle 0|T\left(\phi_I(x)\phi_I(y)U(+T/2,-T/2)\right)|0\rangle}{\langle 0|U(+T/2,-T/2)|0\rangle}
\tag{9.137}
$$

We use Eq. (4.54) to express this result as a function of the interaction Hamiltonian:

$$
\langle\Omega|T(\phi(x)\phi(y))|\Omega\rangle = \lim_{T\to\infty} \frac{\langle 0|T\left(\phi_I(x)\phi_I(y)e^{-i\int_{-T/2}^{+T/2}dt' H_{int}(t')}\right)|0\rangle}{\langle 0|e^{-i\int_{-T/2}^{+T/2}dt' H_{int}(t')}|0\rangle}
\tag{9.138}
$$

We now introduce the interaction Hamiltonian density $\mathcal{H}_{int}$ to write a covariant-looking expression, setting $T\to\infty$, and generalize to the notation that will be used for more than two fields:

$$
\langle\Omega|T(\phi_1\phi_2)|\Omega\rangle = \frac{\langle 0|T\left(\phi_I(x_1)\phi_I(x_2)e^{-i\int dt'\int d^3\vec{z}\,\mathcal{H}_{int}}\right)|0\rangle}{\langle 0|T\left(e^{-i\int dt'\int d^3\vec{z}\,\mathcal{H}_{int}}\right)|0\rangle} = \frac{\langle 0|T\left(\phi_I(x_1)\phi_I(x_2)e^{-i\int d^4 z\,\mathcal{H}_{int}}\right)|0\rangle}{\langle 0|T\left(e^{-i\int d^4 z\,\mathcal{H}_{int}}\right)|0\rangle}
\tag{9.139}
$$

This last expression can then easily be expanded to the n-point correlation for more than two fields to find:

$$
\langle\Omega|T(\phi_1\ldots\phi_n)|\Omega\rangle = \frac{\langle 0|T\left(\phi_I(x_1)\ldots\phi_I(x_n)e^{-i\int d^4 z\,\mathcal{H}_{int}}\right)|0\rangle}{\langle 0|T\left(e^{-i\int d^4 z\,\mathcal{H}_{int}}\right)|0\rangle}
\tag{9.140}
$$

We see that the numerator contains exactly the time-ordered products of fields squeezed between the free vacuum ground state. Gorgeous! The additional effect comes from the denominator that we must understand. In order to do so, we will take the concrete case of the ϕ^4 theory in the next section.

9.12   A Specific Theory: The ϕ^4 Interaction

Let us consider the real scalar field ϕ. The so-called ϕ^4 theory is constructed by adding the **quartic interaction term** $\phi^4(x)$ to the free Klein–Gordon Lagrangian with rest mass m, in a similar fashion as for our toy model:

$$\mathcal{L} = \mathcal{L}_{KG} + \mathcal{L}_{int} = \frac{1}{2}(\partial_\mu \phi)^2 - \frac{1}{2}m^2\phi^2 - \frac{g}{4!}\phi^4 \tag{9.141}$$

The constant g is a free parameter of the model. The extra factor 4! in the denominator of the ϕ^4 term is conventionally included in order to reduce the symmetry factor, as we will see in the following.

In this theory, we consequently have $\mathcal{H}_{int} = g\phi^4/4!$. We analyse the two-point correlation (dropping the subscript "T"):

$$\langle\Omega|T(\phi_1\phi_2)|\Omega\rangle = \frac{\langle 0|T\left(\phi_1\phi_2 e^{-i\int d^4z \frac{g\phi^4(z)}{4!}}\right)|0\rangle}{\langle 0|T\left(e^{-i\int d^4z \frac{g\phi^4(z)}{4!}}\right)|0\rangle} \tag{9.142}$$

We expand the numerator up to the first order in g, setting $\phi(z) = \phi_3$, to find:

$$\langle 0|T\left(\phi_1\phi_2 e^{-i\int d^4z \frac{g\phi_3^4}{4!}}\right)|0\rangle = \underbrace{\langle 0|T(\phi_1\phi_2)|0\rangle}_{=D_F(x_1-x_2)} + \left(\frac{-ig}{4!}\right)\langle 0|T\left(\int d^4z\, \phi_1\phi_2\phi_3\phi_3\phi_3\phi_3\right)|0\rangle + \mathcal{O}(g^2) \tag{9.143}$$

The first term in the expansion corresponds to the Feynman propagator $D_F(x_1-x_2)$ between x_1 to x_2, which is graphically represented by just a line between the two space-time points. We apply Wick's theorem to expand the term proportional to g. We need to consider all possible fully contracted contractions of $\phi_1\phi_2\phi_3\phi_3\phi_3\phi_3$:

$$
\phi_1\phi_2\phi_3\phi_3\phi_3\phi_3 + \phi_1\phi_2\phi_3\phi_3\phi_3\phi_3 + \phi_1\phi_2\phi_3\phi_3\phi_3\phi_3 +
$$
$$
\phi_1\phi_2\phi_3\phi_3\phi_3\phi_3 + \phi_1\phi_2\phi_3\phi_3\phi_3\phi_3 + \phi_1\phi_2\phi_3\phi_3\phi_3\phi_3 +
$$
$$
\phi_1\phi_2\phi_3\phi_3\phi_3\phi_3 + \phi_1\phi_2\phi_3\phi_3\phi_3\phi_3 + \phi_1\phi_2\phi_3\phi_3\phi_3\phi_3 +
$$
$$
\phi_1\phi_2\phi_3\phi_3\phi_3\phi_3 + \phi_1\phi_2\phi_3\phi_3\phi_3\phi_3 + \phi_1\phi_2\phi_3\phi_3\phi_3\phi_3 +
$$
$$
\phi_1\phi_2\phi_3\phi_3\phi_3\phi_3 + \phi_1\phi_2\phi_3\phi_3\phi_3\phi_3 + \phi_1\phi_2\phi_3\phi_3\phi_3\phi_3 \tag{9.144}
$$

The first three terms contain the contraction $\phi_1\phi_2$, which can be written in terms of the Feynman propagator $D_F(x_1-x_2)$. Their combinatorial factor is equivalent to the ways we can contract the four ϕ_3 fields, two by two, the two being indistinguishable, or $\binom{4}{2}\frac{1}{2} = 4!/(2!2!) \times 1/2 = 3$. The rest of the terms are proportional to $D_F(x_1-z)D_F(x_2-z)$. Their number is given by four ways to contract ϕ_1 with any of the ϕ_3 times three ways to contract ϕ_2 with the remaining ϕ_3, hence $4 \times 3 = 12$. We can summarize the full result as:

$$\frac{-ig}{4!}\left[\binom{4}{2}\frac{1}{2}\int d^4z\, D_F(x_1-x_2)D_F(z-z)D_F(z-z)\right.$$
$$\left. +(4\times 3)\int d^4z\, D_F(x_1-z)D_F(x_2-z)D_F(z-z)\right] =$$
$$\frac{-ig}{8}\int d^4z\, D_F(x_1-x_2)D_F(z-z)D_F(z-z) + \frac{-ig}{2}\int d^4z\, D_F(x_1-z)D_F(x_2-z)D_F(z-z) \tag{9.145}$$

Graphically, these convert into the two diagrams illustrated in Figure 9.7. We still have two external points but note that the internal space-time point defined by z needs to be integrated upon. The factors $1/8$ and $1/2$

Figure 9.7 The two Feynman diagrams corresponding to the first order of the ϕ^4 two-point correlation point.

are called **symmetry factors**. In order to understand them, we define the components of a diagram as either the two ends of a line starting and ending in the same point, entire lines between points, or internal vertices. The symmetry factor of a diagram is given by the number of ways one can exchange its components without changing the diagram itself. In the first diagram, the double loop is symmetric under the swap of the in- and out-going ends of both loops. In addition, it is symmetric under the interchange of the two loops. Therefore, the symmetric factor is $2 \times 2 \times 2 = 8$. In the second diagram, the diagram is symmetric under the swap of the in- and out-going end of the loop, so the symmetry factor is simply 2.

The space-time point defined by z is a vertex or internal point. In the case of the ϕ^4 theory, there are always four lines meeting at such a vertex. One must integrate over it. The space-time points x_1 and x_2 associated with the $\phi_1\phi_2$ correlation refer to external lines and are fixed (no integration).

- **Disconnected diagrams (vacuum bubbles).** The first diagram that we have discussed above contains a loop which is "disconnected" from the two space-time points x_1 and x_2. In general, diagrams will contain such sub-diagrams which are not connected to any of the external points. A disconnected piece contains only internal space-time points, which need to be integrated upon. In contrast, a sub-diagram that is connected to at least one external point is called a partially connected diagram.

The disconnected diagrams will add up to an overall factor in our calculations. In general, a diagram can contain zero, one, two, and so on disconnected diagrams. If a disconnected diagram $\mathcal{D}$ appears n times, we must include a symmetry factor $n!$ to account for the exchange of two such identical sub-diagrams. The sum of the terms will be of the form:

$$1 + \mathcal{D} + \frac{1}{2!}\mathcal{D}\mathcal{D} + \frac{1}{3!}\mathcal{D}\mathcal{D}\mathcal{D} + \cdots = \sum_n \frac{1}{n!}\mathcal{D}^n \tag{9.146}$$

Let us now assume that there are different types of disconnected diagrams $\mathcal{D}_i$. Then, the sum over all disconnected diagrams will yield an exponentiation of the pieces:

$$\prod_i \sum_n \frac{1}{n!}\mathcal{D}_i^n = \prod_i e^{\mathcal{D}_i} = e^{\sum_i \mathcal{D}_i} \tag{9.147}$$

Now the denominator in Eq. (9.142) contains no external points as well and therefore yields **precisely** the sum of disconnected diagrams we just computed. Hence, the disconnected diagrams cancel the denominator! We found the incredible result that:

$$\langle\Omega|T(\phi(x_1)\phi(x_2))|\Omega\rangle = \sum \text{all partially connected diagrams with two external legs} \tag{9.148}$$

This expression can be generalized to the n-point correlation:

$$\langle\Omega|T(\phi(x_1)\ldots\phi(x_n))|\Omega\rangle = \sum \text{all partially connected diagrams with } n \text{ external legs} \tag{9.149}$$

In practice, this means that **we will never need to worry about disconnected diagrams, and we will only ever need to focus on the connected ones!**

• **Free and dressed propagator (one-particle-irreducible diagram).** We have seen that in the interacting theory, the following two Feynman propagators are different since they refer to two different ground states:

$$D_F(x_1 - x_2) = \langle 0|T(\phi_1\phi_2)|0\rangle \quad \text{and} \quad D_F^d(x_1 - x_2) = \langle \Omega|T(\phi_1\phi_2)|\Omega\rangle \tag{9.150}$$

The propagator of the free theory is $D_F(x_1 - x_2)$, while $D_F^d(x_1 - x_2)$ is called the "dressed" propagator. This latter can be calculated perturbatively as before, considering all partially connected diagrams. We are now interested in seeing what happens at higher orders in g. These are called the **radiative corrections to the free Feynman propagator**.

The corresponding Feynman diagrams are shown in Figure 9.8. The last diagram in the figure is *not irreducible*. This means that we can cut a single line and we still obtain a valid diagram. In this case, just cut the line in the loop. The other four diagrams are on the contrary called **one-particle-irreducible (1PI)** because we cannot produce two separate non-trivial diagrams from them by cutting a line. In order to organize higher-order loop diagrams, it is therefore useful to define the list of all non-trivial 1PI diagrams. These are "amputated" by not attaching any external points. They only contain the internal z structure. The sum of these diagrams is defined as $-i\Sigma(p)$.

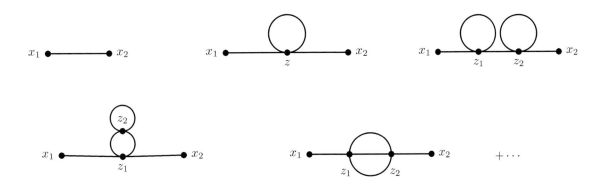

Figure 9.8 The connected Feynman diagrams corresponding to the higher order of the ϕ^4 two-point correlation point.

We can now write down the dressed propagator $\tilde{D}_F^d(p)$ in terms of the free propagator $\tilde{D}_F(p)$ and the 1PI sum $-i\Sigma(p)$ as the following series:

$$\begin{aligned}
\tilde{D}_F^d(p) &= \tilde{D}_F(p) + \tilde{D}_F(p)\left(-i\Sigma(p)\right)\tilde{D}_F(p) + \cdots \\
&= \frac{i}{p^2 - m^2 + i\epsilon} + \frac{i}{p^2 - m^2 + i\epsilon}\left(-i\Sigma(p)\right)\frac{i}{p^2 - m^2 + i\epsilon} + \cdots \\
&= \frac{i}{p^2 - m^2 + i\epsilon}\left[1 + \frac{\Sigma(p)}{p^2 - m^2 + i\epsilon} + \left(\frac{\Sigma(p)}{p^2 - m^2 + i\epsilon}\right)^2 + \cdots\right]
\end{aligned} \tag{9.151}$$

This sum is graphically represented in Figure 9.9. We can use the "resumming" technique by assuming that the higher-order terms in the bracket of the above expression will be of the type of the geometric series (see Eq. (A.1)). We then obtain the **Dyson resummation of the propagator**:

$$\tilde{D}_F^d(p) = \frac{i}{p^2 - m^2 + i\epsilon}\left[1 - \frac{\Sigma(p)}{p^2 - m^2 + i\epsilon}\right]^{-1} = \frac{i}{p^2 - m^2 - \Sigma(p) + i\epsilon} \tag{9.152}$$

Figure 9.9 Higher-order corrections to the propagator: (left) "plain" (or naked); (middle) "dressed" with one 1PI; (right) two 1PI.

The function $\Sigma(p)$ is a very complicated function and it needs to be calculated perturbatively. We will perform this explicitly for the case of QED in Chapter 12. This is a beautiful result that explicitly shows that the particle states of the interacting theory are different from those of the free theory! We can interpret this as an effective shift of the mass, which now introduces a pole at $p^2 = m^2 + \Sigma$ instead of $p^2 = m^2$. We can say that in our perturbative approach to scattering, we have "switched off" the interaction in the asymptotic regimes, however, the self-interaction of the incoming and outgoing particles cannot be removed. Thus, the incoming and outgoing states need to be radiatively resummed.

9.13   Scattering in the ϕ^4 Theory

We continue our discussion of a real scalar field within the ϕ^4 theory and this time we want to revisit the scattering process in more general terms. As before, we consider the scattering between four identical scalar particles, labeled σ, and define the kinematical configuration with the four-vectors q_1, q_2, q_3, and q_4:

$$\sigma(q_1) + \sigma(q_2) \to \sigma(q_3) + \sigma(q_4) \tag{9.153}$$

At the lowest order we should compute:

$$
\begin{aligned}
\langle q_3 q_4 | S_1 | q_1 q_2 \rangle &= (-i) \langle q_3 q_4 | T \left(\int \mathrm{d}^4 x_1 \frac{g}{4!} \phi^4(x_1) \right) | q_1 q_2 \rangle \\
&= \frac{-ig}{4!} \langle q_3 q_4 | T \left(\int \mathrm{d}^4 x_1 \phi_1 \phi_1 \phi_1 \phi_1 \right) | q_1 q_2 \rangle
\end{aligned} \tag{9.154}
$$

where the time-ordering operator has been left in to remind us that we want to use Wick's theorem. The time ordering can now be replaced by the normal-ordered sum:

$$
\begin{aligned}
T(\phi_1 \phi_1 \phi_1 \phi_1) &= N(\phi_1 \phi_1 \phi_1 \phi_1 + \\
&\quad \phi_1\phi_1\phi_1\phi_1 + \phi_1\phi_1\phi_1\phi_1 + \phi_1\phi_1\phi_1\phi_1 + \phi_1\phi_1\phi_1\phi_1 + \phi_1\phi_1\phi_1\phi_1 + \phi_1\phi_1\phi_1\phi_1 + \\
&\quad \phi_1\phi_1\phi_1\phi_1 + \phi_1\phi_1\phi_1\phi_1 + \phi_1\phi_1\phi_1\phi_1) \\
&= \phi\phi\phi\phi + \phi\phi\phi\phi + \phi\phi\phi\phi
\end{aligned} \tag{9.155}
$$

where in the last line we adopted a shorthand notation for all the possible contractions, in this particular case the normal-ordered products with zero, one, and two contractions. We analyze them one by one. The first term, omitting for clarity the factor $-ig/4!$ and the initial state and final normalization factors, is:

$$\langle q_3 q_4 | \int \mathrm{d}^4 x_1 \phi\phi\phi\phi |q_1 q_2\rangle \propto \langle 0| a(\vec{q}_4)a(\vec{q}_3) \int \mathrm{d}^4 x_1 \phi\phi\phi\phi a^\dagger(\vec{q}_2)a^\dagger(\vec{q}_1) |0\rangle$$

$$\propto \langle 0| a(\vec{q}_4)a(\vec{q}_3) \int \mathrm{d}^4 x_1 \int \mathrm{d}^3 \vec{p}_1 \left(a^\dagger(\vec{p}_1)e^{+ip_1\cdot x_1} + a(\vec{p}_1)e^{-ip_1\cdot x_1}\right) \times$$

$$\int \mathrm{d}^3 \vec{p}_2 \left(a(\vec{p}_2)e^{-ip_2\cdot x_1} + a^\dagger(\vec{p}_2)e^{+ip_2\cdot x_1}\right) \times \int \mathrm{d}^3 \vec{p}_3 \left(a^\dagger(\vec{p}_3)e^{+ip_3\cdot x_1} + a(\vec{p}_3)e^{-ip_3\cdot x_1}\right) \times$$

$$\int \mathrm{d}^3 \vec{p}_4 \left(a(\vec{p}_4)e^{-ip_4\cdot x_1} + a^\dagger(\vec{p}_4)e^{+ip_4\cdot x_1}\right) a^\dagger(\vec{q}_2)a^\dagger(\vec{q}_1) |0\rangle \qquad (9.156)$$

where we introduced the plane waves for the quantized field for ϕ, with momentum integrals over four modes, which we label $\vec{p}_1, \ldots, \vec{p}_4$. We can simplify this expression by defining $a_i \equiv a(\vec{p}_i)e^{-ip_i\cdot x_1}$. Hence:

$$\langle q_3 q_4 | \int \mathrm{d}^4 x_1 \phi\phi\phi\phi |q_1 q_2\rangle \propto \langle 0| a(\vec{q}_4)a(\vec{q}_3) \int \mathrm{d}^4 x_1 \int \mathrm{d}^3 \vec{p}_1 \int \mathrm{d}^3 \vec{p}_2 \int \mathrm{d}^3 \vec{p}_3 \int \mathrm{d}^3 \vec{p}_4$$

$$\left(a_1^\dagger + a_1\right) \times \left(a_2 + a_2^\dagger\right) \times \left(a_3^\dagger + a_3\right) \times \left(a_4 + a_4^\dagger\right) a^\dagger(\vec{q}_2)a^\dagger(\vec{q}_1) |0\rangle \qquad (9.157)$$

After a lengthy algebra (see Eqs. (9.80)–(9.87)), we finally get:

$$\frac{-ig}{4!} \langle q_3 q_4 | \int \mathrm{d}^4 x_1 \phi\phi\phi\phi |q_1 q_2\rangle = \frac{-ig}{4!}(4!) \int \mathrm{d}^4 x_1 e^{-i(q_3+q_4-q_1-q_2)\cdot x}$$

$$= -ig(2\pi)^4 \delta^4(q_3 + q_4 - q_1 - q_2) \qquad (9.158)$$

This kind of diagram is illustrated in Figure 9.10.

Figure 9.10 Scattering in the ϕ^4 theory: four-point interaction term.

A term of the second type, with one field contraction, is given by:

$$\frac{-ig}{4!} \langle q_3 q_4 | \int \mathrm{d}^4 x_1 \phi\phi\phi\phi |q_1 q_2\rangle \propto \langle 0| a(\vec{q}_4)a(\vec{q}_3) \int \mathrm{d}^4 x_1 \int \mathrm{d}^3 \vec{p}_1 \int \mathrm{d}^3 \vec{p}_2$$

$$D_F(x_1 - x_1) \left(a_1^\dagger + a_1\right) \times \left(a_2 + a_2^\dagger\right) a^\dagger(\vec{q}_2)a^\dagger(\vec{q}_1) |0\rangle \qquad (9.159)$$

Only the terms with equal numbers of a and $a^\dagger$ will survive. Then, we obtain:

$$\langle 0| a(\vec{q}_4)a(\vec{q}_3)D_F(x_1 - x_1) \left(a_1^\dagger a_2 + a_1 a_2^\dagger\right) a^\dagger(\vec{q}_2)a^\dagger(\vec{q}_1) |0\rangle \qquad (9.160)$$

This kind of diagram is illustrated in Figure 9.11.

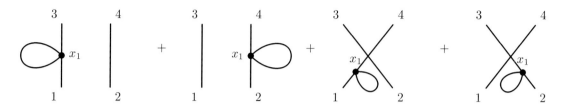

Figure 9.11 Scattering in the ϕ^4 theory: trivial diagrams with one loop.

The third type of term is simply:

$$\langle q_3 q_4 | \int \mathrm{d}^4 x_1 \overset{\frown\frown}{\phi\phi\phi\phi} |q_1 q_2\rangle \quad \propto \quad \langle 0 | a(\vec{q}_4) a(\vec{q}_3) \int \mathrm{d}^4 x_1 D_F(x_1 - x_1) D_F(x_1 - x_1) a^\dagger(\vec{q}_2) a^\dagger(\vec{q}_1) | 0\rangle$$

$$= \quad \left(\delta(\vec{q}_3 - \vec{q}_1)\delta(\vec{q}_4 - \vec{q}_2) + \delta(\vec{q}_3 - \vec{q}_2)\delta(\vec{q}_4 - \vec{q}_1)\right) \int \mathrm{d}^4 x_1 D_F(x_1 - x_1) D_F(x_1 - x_1)$$

$$(9.161)$$

This kind of term leads to a trivial propagation of the particles (i.e., with no scattering) multiplied by a vacuum bubble. The disconnected vacuum is to be neglected, so finally this term is a trivial contribution to the S matrix and does not contribute to the scattering amplitude. Diagrammatically we can trace this to the fact that the fields are fully contracted and the outer legs are then just trivially connected. This is illustrated in Figure 9.12.

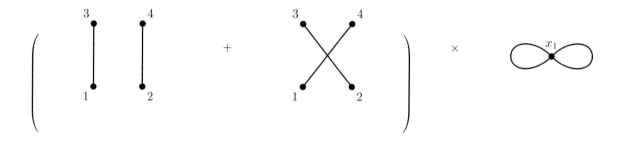

Figure 9.12 Scattering in the ϕ^4 theory: trivial diagrams with a disconnected loop.

9.14 Feynman Rules of the ϕ^4 Theory

Following the Feynman diagram approach, we can define the list of rules that define our ϕ^4 theory. Through calculations of momentum integrals similar to those performed in the previous sections, we would arrive at the following rules in momentum space for the scattering process of the type $\langle q_n \ldots q_{m+1} | S | q_1 \ldots q_m \rangle$:

- Draw the connected Feynman diagrams with n external legs to a given order in g (= the number of internal vertices).

- Assign a vertex factor to each vertex (joining four lines):

ϕ^4 vertex: $= -ig$

- Assign incoming four-momenta p_is and outgoing four-momenta q_is to each external leg. Apply the factor:

External scalar leg: ——————▶ $= 1$

- Label the four-momenta of internal legs and apply the factor for the scalar propagator for each given four-momentum (e.g., for p) of the type:

Scalar propagator: ——————▶ $= \dfrac{i}{p^2 - m^2 + i\epsilon}$

- Apply energy–momentum conservation at each vertex. This defines the momentum of each scalar propagator.

- Integrate over all internal momenta with a factor $\int \frac{\mathrm{d}^4 p}{(2\pi)^4}$.

- Divide by the symmetry factor.

- Sum up all diagrams and apply an overall conservation of energy–momentum between initial and final states.

We note that the four-point vertex factor is conveniently just given by $-ig$ thanks to our $g/4!$ term in the Lagrangian. This is a general result which states that for a scalar to the power n, the corresponding vertex factor is going to be $-ign!$. For the product of r fields interacting via $g\phi_1^{n_1} \dots \phi_r^{n_r}$, the vertex factor is $-ign_1! \dots n_r!$. Hence, the quartic interaction of a real field $g\phi^4/4!$ gives us the vertex factor $-ig$.

• **Complex field.** For the case of a complex field, we should remember to treat the field and its conjugate as independent, so the term $g(\phi^\dagger \phi)^2$ is of the form $g\phi_1^2\phi_2^2$ hence the vertex factor is $-ign_1!n_2! = -ig2!2! = -4ig$. This result is consistent with Eq. (9.88), where the full derivation starting from the interaction term $4g(\phi^\dagger\phi)^2$ gave us the vertex term $2!2!i4g = 16ig$.

• **Scalar pions scattering.** Applying these Feynman rules to our toy model Lagrangian, Eq. (9.77), we would immediately find the following amplitudes for the scattering processes via the four-point interaction:

$$
\begin{aligned}
i\mathcal{M}(\pi^+\pi^- \to \pi^+\pi^-) &= -i(2!2!)4g = -16ig \\
i\mathcal{M}(\pi^0\pi^0 \to \pi^0\pi^0) &= -i(4!)g = -24ig
\end{aligned}
\tag{9.162}
$$

which is consistent with what we had previously (tediously) derived. Similarly, with an additional coupling between the neutral and charged fields given by g', we would obtain:

$$
i\mathcal{M}(\pi^+\pi^- \to \pi^0\pi^0) = i\mathcal{M}(\pi^0\pi^0 \to \pi^+\pi^-) = -i(2!2!)g' = -4ig'
\tag{9.163}
$$

The calculations of the cross-sections are left to the exercises (see **Ex 9.2**).

9.15   The Feynman Propagator of the Dirac Field

The concept of the Feynman propagator can naturally be extended to non-scalar fields, such as Dirac fields. In all cases, the Feynman propagator is the **time-ordered correlation function of two free fields in the vacuum state**, and it must be **a Green's function of the equation of motion of the Dirac field**.

Recalling the results on the quantization of the Dirac field obtained in Section 8.27, we know that a Dirac field Ψ has four components describing spin-1/2 particles and antiparticles. To develop the plane waves, we used the four spinors $u^{(1)}$, $u^{(2)}$, $v^{(1)}$, and $v^{(2)}$, and one can create or annihilate particles $u^{(s)}(\vec{p})$ or antiparticles $v^{(s)}(\vec{p})$ of momentum $\vec{p}$ and spin state s ($s = 1, 2$). The Dirac field operators Ψ and adjoint $\overline{\Psi}$ were defined with two pairs of creation–annihilation operators $a_s(\vec{p})$, $a_s^\dagger(\vec{p})$, $b_s(\vec{p})$, $b_s^\dagger(\vec{p})$ ($s = 1, 2$) as:

$$\Psi \equiv \int \frac{\mathrm{d}^3\vec{p}}{(2\pi)^{3/2}\sqrt{2E_p}} \sum_{s=1,2} \left(a_s u^{(s)}(\vec{p}) e^{-ip\cdot x} + b_s^\dagger v^{(s)}(\vec{p}) e^{+ip\cdot x} \right)$$

$$\overline{\Psi} \equiv \int \frac{\mathrm{d}^3\vec{p}}{(2\pi)^{3/2}\sqrt{2E_p}} \sum_{s=1,2} \left(a_s^\dagger \bar{u}^{(s)}(\vec{p}) e^{+ip\cdot x} + b_s \bar{v}^{(s)}(\vec{p}) e^{-ip\cdot x} \right) \qquad (9.164)$$

where $E_p = +\sqrt{\vec{p}^2 + m^2}$. By inspection we see that just as in the case of the complex scalar Klein–Gordon field, the Dirac propagator involves the field and its adjoint. The fields are, however, spinors with four components and one should consider the propagation of each combination! The **amplitude for the propagation from the space-time point y^μ to x^μ corresponds to the amplitude for the Dirac field** and is hence:

$$S_{\alpha\beta}(x - y) = \langle 0 | \Psi_\alpha(x) \overline{\Psi}_\beta(y) | 0 \rangle \qquad (9.165)$$

Just as in the case of the complex field, this makes sense only because every Dirac particle (i.e., electron) has a corresponding "antiparticle" (i.e., the positron). Alternatively, we can also omit the indices and write $S(x - y)$, but in this case one should remember that it is actually a 4×4 matrix!

We can directly extend this definition to the **Feynman propagator of the Dirac field** $S_F(x - y)$ (a 4×4 matrix!) by using the time-ordered product of spinor fields. One has to take into account that amplitudes for fermions should be antisymmetric under interchange (Fermi–Dirac statistics) and therefore one must take the difference between the two time-ordered contributions:

$$\begin{aligned} S_F(x - y) &= \theta(x^0 - y^0) \langle 0 | \Psi(x) \overline{\Psi}(y) | 0 \rangle - \theta(y^0 - x^0) \langle 0 | \overline{\Psi}(y) \Psi(x) | 0 \rangle \\ &\equiv \langle 0 | T \left(\Psi(x) \overline{\Psi}(y) \right) | 0 \rangle \end{aligned} \qquad (9.166)$$

Note that one uses the same symbol T for both Klein–Gordon and Dirac fields, so one should remember that in the case of the Dirac field there is a minus sign! Its **representation in four-momentum space** $\tilde{S}_F(p)$ is given by the Fourier transform:

$$S_F(x - y) = \int \frac{\mathrm{d}^4 p}{(2\pi)^4} e^{-ip(x-y)} \tilde{S}_F(p) \qquad (9.167)$$

This propagator must be a Green's function of the 4×4 Dirac equation:

$$(i\gamma^\mu \partial_\mu - m\mathbb{1}) S_F(x - y) = i\delta^4(x - y) \cdot \mathbb{1} \qquad (9.168)$$

Writing the indices, it must satisfy the following set of equations:

$$(i\gamma^\mu \partial_\mu - m\mathbb{1})_{\alpha\beta} S_F^{\beta\rho}(x - y) = i\delta_\alpha^\rho \delta^4(x - y) \qquad (9.169)$$

Writing the propagator in terms of its four-momentum space representation and applying the Dirac operator we get:

$$(i\gamma^\mu \partial_\mu - m\mathbb{1})_{\alpha\beta} \int \frac{\mathrm{d}^4 p}{(2\pi)^4} e^{-ip(x-y)} \tilde{S}_F^{\beta\rho}(p) = \int \frac{\mathrm{d}^4 p}{(2\pi)^4} e^{-ip(x-y)} (\gamma^\mu p_\mu - m\mathbb{1})_{\alpha\beta} \tilde{S}_F^{\beta\rho}(p) = i\delta_\alpha^\rho \delta^4(x - y) \quad (9.170)$$

and hence:

$$\tilde{S}_F^{\beta\rho}(p) = \frac{i\delta^\rho_\alpha}{(\gamma^\mu p_\mu - m\mathbb{1})_{\alpha\beta}} = \frac{(i\delta^\rho_\alpha)\,(\gamma^\mu p_\mu + m\mathbb{1})^{\alpha\beta}}{p^2 - m^2} \tag{9.171}$$

Going back to the matrix form and using the slash notation (see Eq. (8.36)), we obtain:

$$\tilde{S}_F(p) = \frac{i\,(\slashed{p} + m\mathbb{1})}{p^2 - m^2} \tag{9.172}$$

Like for the scalar field propagators, this expression must be regularized, thus adding the $i\epsilon$ prescription, we obtain the **Feynman propagator for the Dirac field (a 4 × 4 matrix)**:

$$S_F(x - y) = i \int \frac{\mathrm{d}^4 p}{(2\pi)^4} e^{-ip(x-y)} \frac{\slashed{p} + m\mathbb{1}}{(p^2 - m^2 + i\epsilon)} \tag{9.173}$$

Again, the propagator looks manifestly covariant! This is the beauty of the Feynman propagators.

We recall its physical interpretation. In the non-covariant form of the propagation amplitude, we are forced to distinguish both the forward and backward propagation in time. We have considered the two cases: $x^0 > y^0$ and $x^0 < y^0$. This is illustrated in Figure 9.13. The arrows indicate the direction of propagation of the

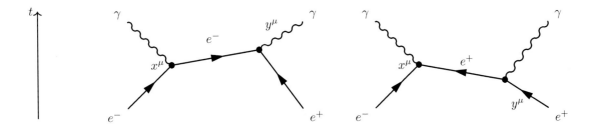

Figure 9.13 Non-covariant formulation of the annihilation of an electron–positron pair into two photons. The arrows indicate the direction of propagation of the particles. (left) $x^0 < y^0$; (right) $x^0 > y^0$. Time flows in the vertical direction.

particles, and time flows in the vertical direction in the graphs. In the case $x^0 < y^0$, we say that an electron propagates forward in time, emits a photon at time x^0, and then continues to propagate until it annihilates with a positron at time y^0, yielding the photon. In the case $x^0 > y^0$, a positron is propagating forward in time and emits a photon at time y^0, then continues to travel until it annihilates with an electron at time x^0, yielding a photon. In both cases, both the electron and positron are virtual in the intermediate path between the two vertices x^μ and y^μ (a real electron and positron pair cannot annihilate into a single photon).

In the covariant prescription of Feynman, the two cases are represented by the single diagram shown in Figure 9.14. We consider two incoming lines representing the electron and the positron. There is a single Feynman propagator that connects the two incoming lines at each vertex. A photon is emitted at each vertex. The Feynman propagator includes both physical cases $x^0 < y^0$ and $x^0 > y^0$ at once, and hence there is no need to use the time coordinate. The direction of the arrow is conventional, since the propagator contains both the electron and positron terms. In the case of the external leg, the direction of the arrow actually represents the *incoming* positron. Having abandoned the space-time coordinates x^μ and y^μ (they disappeared from the diagram), this formalism also strongly suggests the use of the four-momentum space-time representation, and to rely on a graphical representation of the diagrams – a method that can clearly greatly simplify computations.

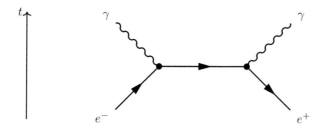

Figure 9.14 Feynman covariant formulation of a diagram contributing to the annihilation of an electron–positron pair into two photons. The direction of the arrows in Feynman diagrams is conventional.

We can say more about the physical interpretation of the propagator of a particle, by using the spinors completeness relation Eq. (8.97) and noting that:

$$\tilde{S}_F(p) = \frac{i\,(\not{p} + m\mathbb{1})}{p^2 - m^2} = \frac{i}{p^2 - m^2}\left(\sum_{s=1,2} u^{(s)}\bar{u}^{(s)}\right) \tag{9.174}$$

This result is actually true in general. **The form of the propagator of a virtual particle includes an amplitude term $i/(p^2 - m^2)$ which represents the pole that we encountered also in the case of scalar particles. However, there is an extra term in the numerator which represents the sum over all possible spin states of the propagating particle!**

9.16   Contraction of the Dirac Field

In the previous section we have shown that the Feynman propagator for a Dirac spinor is represented as a 4×4 matrix – see Eq. (9.173). The definition of the Feynman propagator took into account that the time-ordering operator for fermions is defined with an opposite sign. This implies that in order to recover this definition of the propagator, the Wick's contraction of two Dirac fields must be defined as an anticommutator and with a negative sign:

$$\overbrace{\Psi(x)\overline{\Psi}(y)} = \left\{ \begin{array}{ll} \left\{\Psi^+(x), \overline{\Psi}^-(y)\right\} & \text{if } x^0 > y^0 \\ -\left\{\overline{\Psi}^+(y), \Psi^-(x)\right\} & \text{if } x^0 < y^0 \end{array} \right\} = S_F(x - y) \tag{9.175}$$

Since the Feynman propagator is in this case a 4×4 matrix, so is the contraction of two Dirac spinors. It is also easy to note that:

$$\overbrace{\Psi(x)\Psi(y)} = \overbrace{\overline{\Psi}(x)\overline{\Psi}(y)} = 0 \tag{9.176}$$

To generalize Wick's theorem to spinor fields, we must in addition assume that the time- and normal-ordering operators change sign for each swap of operators that is necessary to bring them into the proper order! With these conventions, **Wick's theorem for a product of any number of Dirac fields** is the following:

$$T\left(\Psi(x_1)\overline{\Psi}(x_2)\Psi(x_3)\ldots\right) = N\left(\Psi(x_1)\overline{\Psi}(x_2)\Psi(x_3)\ldots + \text{all possible contractions}\right) \tag{9.177}$$

The negative sign convention for the contractions is illustrated in the following example:

$$\begin{aligned} N(\overbrace{\Psi(x_1)\Psi(x_2)\overline{\Psi}(x_3)}\overline{\Psi}(x_4)) &= -\overbrace{\Psi(x_1)\overline{\Psi}(x_3)}N(\Psi(x_2)\overline{\Psi}(x_4)) \\ &= -S_F(x_1 - x_3)N(\Psi(x_2)\overline{\Psi}(x_4)) \end{aligned} \tag{9.178}$$

9.17   The Yukawa Interaction

Before we move to the discussion of QED in the next chapter, we discuss a simplified model which involves scalar and Dirac particles interacting with each other. The interaction between the real scalar field, labeled ϕ, with the fermion fields, labeled Ψ and $\overline{\Psi}$, is described by the so-called **Yukawa**[5] **interaction**. In order to conserve electric charge, the neutral scalar particle must couple to a fermion–antifermion pair $f\bar{f}$. The simplest expression we can think of, is just the equivalent of the term $\propto \lambda(\varphi^\dagger\varphi)$ that we considered in our toy model Eq. (9.77), by replacing the complex scalar field φ with the Dirac field Ψ. Hence, we obtain:

$$\mathcal{L}_{int} = -g\phi(\overline{\Psi}\Psi) \tag{9.179}$$

where g is the conventional coupling constant. The full Lagrangian of the Yukawa theory is then simply given by:

$$\mathcal{L}_{Yukawa} = \mathcal{L}_{KG} + \mathcal{L}_{Dirac} + \mathcal{L}_{int} = \frac{1}{2}(\partial_\mu\phi)^2 - \frac{1}{2}m_\phi^2\phi^2 + \overline{\Psi}\left(i\gamma^\mu\partial_\mu - m\right)\Psi - g\phi(\overline{\Psi}\Psi) \tag{9.180}$$

where m_ϕ is the rest mass of the neutral scalar particle ϕ, m is the rest mass of the fermion f, and g is the coupling constant of the Yukawa theory.

As in the case of the toy model, we consider the decay process $\phi \to f + \bar{f}$ and the scattering process $f\bar{f} \to f\bar{f}$ via the exchange of the scalar particle ϕ.

• **The decay** $\phi \to f + \bar{f}$. The kinematical configuration is expressed as in Eq. (9.40), however now taking into account also the spins of the fermions:

$$\phi(k^\mu) \to f(q_1^\mu, s_1) + \bar{f}(q_2^\mu, s_2) \tag{9.181}$$

The matrix element to first order is given by an expression similar to Eq. (9.43), i.e.:

$$\langle q_1 s_1, q_2 s_2| S_1 |k\rangle = (-i)\langle q_1 s_1, q_2 s_2| \int \mathrm{d}^4 x_1 g\phi(x_1)\overline{\Psi}(x_1)\Psi(x_1) |k\rangle \tag{9.182}$$

where the field expansions for Ψ and $\overline{\Psi}$ are given as a function of the a and b operators in Eq. (9.164), and the field expansion for ϕ is defined with the c operator:

$$\phi(\vec{x}, t) \equiv \int \frac{\mathrm{d}^3\vec{p}}{(2\pi)^{3/2}} \frac{1}{\sqrt{2E_p}} \left(c(\vec{p})e^{-ip\cdot x} + c^\dagger(\vec{p})e^{+ip\cdot x}\right) \tag{9.183}$$

The initial state will be defined exactly as in the case of the real scalar field, applying the creation operator on the vacuum. This is just given by Eq. (9.23). For the final state, we should take into account the spins of the fermions and that we have a fermion–antifermion pair:

$$\langle q_1 s_1, q_2 s_2| = \sqrt{(2\pi)^3 2E_{q_1}} \sqrt{(2\pi)^3 2E_{q_2}} \langle 0| a_{s_1}(\vec{q}_1)b_{s_2}(\vec{q}_2) \tag{9.184}$$

Hence, we obtain:

$$\begin{aligned}
\langle q_1 s_1, q_2 s_2| S_1 |k\rangle &= (-ig)\sqrt{(2\pi)^3 2E_{q_1}} \sqrt{(2\pi)^3 2E_{q_2}} \sqrt{(2\pi)^3 2E_k} \times \\
&\quad \langle 0| a_{s_1}(\vec{q}_1)b_{s_2}(\vec{q}_2) \int \mathrm{d}^4 x \phi(x)\overline{\Psi}(x)\Psi(x)c^\dagger(\vec{k}) |0\rangle
\end{aligned} \tag{9.185}$$

As before, we must replace the field operators by their quantized plane-wave expansion. This is going to again be equivalent to the result derived in Eq. (9.43) however with the extra spinor factor that takes into account

5   Hideki Yukawa (1907–1981), Japanese theoretical physicist.

the spin of the final-state fermions. Making use of the anticommutation rules defined in Eq. (8.288), one obtains:

$$
\begin{aligned}
\langle q_1 s_1, q_2 s_2 | S_1 | k \rangle &= (-ig) \langle 0 | b_{s_2}(\vec{q}_2) a_{s_1}(\vec{q}_1) \int \mathrm{d}^4 x \times \\
&\quad \sum_{r,s} \int \mathrm{d}^3\vec{p}_1 a_r^\dagger(\vec{p}_1) \bar{u}^{(r)}(p_1) e^{+ip_1 \cdot x} \int \mathrm{d}^3\vec{p}_2 b_s^\dagger(\vec{p}_2) v^{(s)}(p_2) e^{+ip_2 \cdot x} \int \mathrm{d}^3\vec{k}' c(\vec{k}') e^{-ik' \cdot x} c^\dagger(\vec{k}) | 0 \rangle \\
&= (-ig) \bar{u}^{(s_1)}(q_1) v^{(s_2)}(q_2) (2\pi)^4 \delta^4(q_1 + q_2 - k)
\end{aligned}
\tag{9.186}
$$

In the above derivation, **we should pay attention to the fact that the boson fields commute while the fermionic fields anticommute.** Hence, we have (see Eq. (9.29)):

$$
\langle 0 | c(\vec{k}') c^\dagger(\vec{k}) | 0 \rangle = \langle 0 | \delta^3(\vec{k}' - \vec{k}) | 0 \rangle + \langle 0 | c^\dagger(\vec{k}) c(\vec{k}') | 0 \rangle = \langle 0 | \delta^3(\vec{k}' - \vec{k}) | 0 \rangle
\tag{9.187}
$$

while

$$
\langle 0 | a_r(\vec{p}') a_s^\dagger(\vec{p}) | 0 \rangle = \langle 0 | \delta^3(\vec{p}' - \vec{p}) \delta_{r,s} | 0 \rangle - \langle 0 | a_s^\dagger(\vec{p}) a_r(\vec{p}') | 0 \rangle = \langle 0 | \delta^3(\vec{p}' - \vec{p}) \delta_{r,s} | 0 \rangle
\tag{9.188}
$$

and similarly for the b, $b^\dagger$ operators. **The difference of sign actually plays no role in the above case, but in general it will, as will be shown below.** Finally, the amplitude of the decay of a scalar into a fermion–antifermion pair is therefore given by:

$$
i\mathcal{M}(\sigma \to f\bar{f}) = (-ig) \bar{u}^{(s)}(q_1) v^{(s)}(q_2)
\tag{9.189}
$$

It is interesting to note that the above result is, as expected, analogous to the result of Eq. (9.44), apart from the additional term $\bar{u}^{(s)}(q_1) v^{(s)}(q_2)$ which describes the spin dependence of the amplitude!

• **Contraction of external legs.** While straightforward, the calculation above is a bit tedious and it can be simplified by noting that the structure of the result can be derived more systematically. First, we realize that the spinor $\bar{u}$ is associated with the outgoing fermion, while the spinor v is coming from the outgoing antifermion. A similar derivation would easily show that a spinor u should be associated with an incoming fermion, and a spinor $\bar{v}$ to an incoming antifermion. We also stress that the term in the amplitude is always given by the product of an adjoint spinor term with a spinor, i.e., $\overline{\Psi}\Psi$, which guarantees that it leads to an invariant scalar (see Section 8.28).

In order to calculate more complex or higher-order diagrams, we will rely on Wick's theorem to transform time-ordered products into normal products of field contractions. We can elegantly extend the concept of contraction to the case of initial and final states. These are called **contractions of external legs** and are defined below (see also Section 10.11):

- Incoming scalar field: $\phi | p^\mu \rangle \to 1$ Outgoing: $\langle p^\mu | \phi \to 1$

- Incoming Dirac fermion: $\Psi | p^\mu, s \rangle \to u^{(s)}(\vec{p})$ Outgoing: $\langle p^\mu, s | \overline{\Psi} \to \bar{u}^{(s)}(\vec{p})$

- Incoming Dirac antifermion: $\overline{\Psi} | p^\mu, s \rangle \to \bar{v}^{(s)}(\vec{p})$ Outgoing: $\langle p^\mu s | \Psi \to v^{(s)}(\vec{p})$

To illustrate the contraction of external legs, we can consider once again the decay of the scalar particle into a fermion–antifermion pair $\phi(k) \to f(q_1) + \bar{f}(q_2)$, which was calculated above. Using the rules, the computation of the amplitude can easily be reduced to the following single contraction:

$$
\begin{aligned}
\langle q_1 s_1, q_2 s_2 | S_1 | k \rangle &= (-i) \langle q_1 s_1, q_2 s_2 | \int \mathrm{d}^4 x_1 g \phi(x_1) \overline{\Psi}(x_1) \Psi(x_1) | k \rangle \\
&= (-ig) \bar{u}^{(s_1)}(q_1) v^{(s_2)}(q_2) (2\pi)^4 \delta^4(q_1 + q_2 - k)
\end{aligned}
\tag{9.190}
$$

a result which exactly coincides with what we obtained in Eq. (9.186).

9.18   Scattering and Feynman Rules for the Yukawa Theory

The Yukawa theory predicts the interaction between pairs of fermions (or antifermions), as well as interaction between fermion–antifermion pairs. The non-relativistic limit has interesting implications. We consider these in the following.

• **Scattering amplitude for two fermions.** We now consider the scattering between two identical fermions, labeled f, and define the kinematical configuration with the four-vectors q_1, q_2, q_3, and q_4:

$$f(q_1) + f(q_2) \to f(q_3) + f(q_4) \tag{9.191}$$

At the lowest order we should compute:

$$\langle q_3 q_4 | S_2 | q_1 q_2 \rangle = \frac{(-ig)^2}{2!} \langle q_3 q_4 | T \left(\int \mathrm{d}^4 x_1 \int \mathrm{d}^4 x_2 \phi(x_1) \overline{\Psi}(x_1) \Psi(x_1) \phi(x_2) \overline{\Psi}(x_2) \Psi(x_2) \right) | q_1 q_2 \rangle \tag{9.192}$$

The time ordering can now be replaced by the normal-ordered sum using Wick's theorem. We should consider all possible contractions. There are no external scalar particles, so we can only contract the two ϕ fields at x_1 and x_2 together, yielding a scalar particle propagator. The expression is clearly identical if we swap $x_1 \leftrightarrow x_2$. Hence, for each contraction, there are two possibilities, and this cancels the 2! term in front of the expression. This is always true also at higher orders and therefore **the $n!$ factor in front of the S matrix Taylor series will always cancel the sum of the time-ordered fields with n space-time points x_i, since there will always be $n!$ ways of interchanging the x_i ($i = 1, \ldots, n$) vertices within the integrals and obtaining the same result.** Hence, taking a particular contraction, we will have:

$$-g^2 \langle q_3, s_3; q_4, s_4 | \int \mathrm{d}^4 x_1 \int \mathrm{d}^4 x_2 \phi(x_1) \overline{\Psi}(x_1) \Psi(x_1) \phi(x_2) \overline{\Psi}(x_2) \Psi(x_2) | q_1, s_1; q_2, s_2 \rangle$$

$$= -g^2 \bar{u}^{(s_4)}(q_4) u^{(s_2)}(q_2) \bar{u}^{(s_3)}(q_3) u^{(s_1)}(q_1) \times$$

$$\int \frac{\mathrm{d}^4 q}{(2\pi)^4} \frac{i}{q^2 - m_\phi^2 + i\epsilon} (2\pi)^4 \delta^4(q - q_1 + q_3)(2\pi)^4 \delta^4(q_4 - q_2 - q)$$

$$= -ig^2 \left[\bar{u}^{(s_4)}(q_4) u^{(s_2)}(q_2) \frac{1}{q^2 - m_\phi^2 + i\epsilon} \bar{u}^{(s_3)}(q_3) u^{(s_1)}(q_1) \right] (2\pi)^4 \delta^4(q_1 + q_2 - q_3 - q_4) \tag{9.193}$$

with the momentum of the scalar propagator constrained to be equal to $q^\mu = q_1^\mu - q_3^\mu = q_4^\mu - q_2^\mu$. Because we are dealing with fermions, we should pay attention to the signs that can appear due to the anticommutation of the creation–annihilation operators. In fact, in the above contraction, focusing only on the fermion fields, we have a term of the type:

$$\langle 0 | a_{s_3}(q_3) a_{s_4}(q_4) \underbrace{\overline{\Psi}(x_1)}_{a^\dagger + b} \underbrace{\Psi(x_1)}_{a + b^\dagger} \underbrace{\overline{\Psi}(x_2)}_{a^\dagger + b} \underbrace{\Psi(x_2)}_{a + b^\dagger} a_{s_1}^\dagger(q_1) a_{s_2}^\dagger(q_2) | 0 \rangle \tag{9.194}$$

In shorthand, we can express this term, keeping only relevant products, as:

$$a_3 a_4 a_{x_1}^\dagger a_{x_1} a_{x_2}^\dagger a_{x_2} a_1^\dagger a_2^\dagger \quad \rightsquigarrow \quad -a_3 a_{x_1}^\dagger a_4 a_{x_1} a_{x_2}^\dagger a_{x_2} a_1^\dagger a_2^\dagger$$

$$\rightsquigarrow +a_3 a_{x_1}^\dagger a_4 a_{x_2}^\dagger a_{x_1} a_{x_2} a_1^\dagger a_2^\dagger \quad \rightsquigarrow \quad -a_3 a_{x_1}^\dagger a_4 a_{x_2}^\dagger a_{x_1} a_1^\dagger a_{x_2} a_2^\dagger \tag{9.195}$$

Another possible and different contraction is obtained by exchanging one of the field contractions, say of the adjoint spinors (we already took into account the swapping of both normal and adjoint spinor fields when removing the 2! factor):

$$-g^2 \langle q_3, s_3; q_4, s_4 | \int d^4x_1 \int d^4x_2 \phi(x_1)\overline{\Psi}(x_1)\Psi(x_1)\phi(x_2)\overline{\Psi}(x_2)\Psi(x_2) | q_1, s_1; q_2, s_2 \rangle$$

$$= +ig^2 \left[\bar{u}^{(s_3)}(q_3)u^{(s_2)}(q_2)\frac{1}{q^2 - m_\phi^2 + i\epsilon}\bar{u}^{(s_4)}(q_4)u^{(s_1)}(q_1) \right](2\pi)^4\delta^4(q_1 + q_2 - q_3 - q_4) \qquad (9.196)$$

with the momentum of the scalar propagator constrained to be equal to $q^\mu = q_1^\mu - q_4^\mu = q_3^\mu - q_2^\mu$. We notice that a "mysterious" plus sign has appeared in front of the amplitude compared to the minus sign in the previous amplitude. In this case, the order of the fields is given by (using as before the shorthand notation):

$$a_3 a_4 a_{x_1}^\dagger a_{x_1} a_{x_2}^\dagger a_{x_2} a_1^\dagger a_2^\dagger \quad \leadsto \quad -a_3 a_4 a_{x_1}^\dagger a_{x_2}^\dagger a_{x_1} a_{x_2} a_1^\dagger a_2^\dagger \quad \leadsto \quad +a_3 a_4 a_{x_1}^\dagger a_{x_2}^\dagger a_{x_1} a_1^\dagger a_{x_2} a_2^\dagger$$

$$\leadsto -a_3 a_{x_1}^\dagger a_4 a_{x_2}^\dagger a_{x_1} a_1^\dagger a_{x_2} a_2^\dagger \quad \leadsto \quad +a_3 a_{x_1}^\dagger a_{x_2}^\dagger a_4 a_{x_1} a_1^\dagger a_{x_2} a_2^\dagger \qquad (9.197)$$

So, the operators can be untangled with just two swaps, and this leads to a sign opposite to the one obtained in the first contraction! We note that the two amplitudes (up to the sign) could simply be obtained from one another under the exchange $3 \leftrightarrow 4$! This was expected since the final states $f(q_3)$ and $f(q_4)$ are identical particles, hence we cannot distinguish the cases where these are swapped, and therefore both possibilities must appear in the amplitude. However, we can say that dealing with fermions, the swap introduces the extra minus sign.

Therefore, the **total amplitude for the process can now be found by adding the two amplitudes. Because of the fermion statistics, there will be a negative sign between the two terms corresponding to the two diagrams. Conventionally, we say that the two diagrams must be "subtracted," rather than "added."** Finally, we have:

$$\langle q_3 q_4 | S_2 | q_1 q_2 \rangle = (i\mathcal{M})(2\pi)^4\delta^4(q_1 + q_2 - q_3 - q_4) \qquad (9.198)$$

where

$$i\mathcal{M}(ff \to ff) = -ig^2 \left[\bar{u}^{(s_4)}(q_4)u^{(s_2)}(q_2)\frac{1}{(q_1 - q_3)^2 - m_\phi^2 + i\epsilon}\bar{u}^{(s_3)}(q_3)u^{(s_1)}(q_1) - \right.$$

$$\left. \bar{u}^{(s_3)}(q_3)u^{(s_2)}(q_2)\frac{1}{(q_1 - q_4)^2 - m_\phi^2 + i\epsilon}\bar{u}^{(s_4)}(q_4)u^{(s_1)}(q_1) \right] \qquad (9.199)$$

The two amplitudes are graphically represented as Feynman diagrams in Figure 9.15. In this example, we have derived an important rule which will be relevant in the following chapters. It can be summarized in the following way: **there is a minus sign between the amplitude of two diagrams that differ only by the permutation of two fermion operators.** We will come back to this in the next chapter.

• **Feynman rules for the Yukawa theory.** We can use the previous example to derive a set of Feynman rules particular to the Yukawa theory, that will allow us to quickly write down the amplitude of a given diagram without the need for extended calculations. The rules are organized in tabular form with (1) item, (2) diagram representation, and (3) amplitude term:

• [YUK1] **External scalar leg:** $p^\mu \rightarrow$ $= 1$

• [YUK2] **External fermion leg:** $p^\mu \rightarrow$ $= \begin{cases} u^{(s)}(\vec{p})\text{-incoming} \\ \bar{u}^{(s)}(\vec{p})\text{-outgoing} \end{cases}$

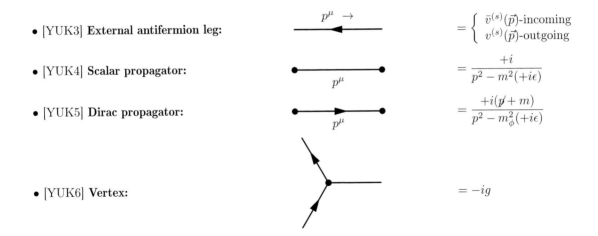

- [YUK3] **External antifermion leg:**

$$= \begin{cases} \bar{v}^{(s)}(\vec{p})\text{-incoming} \\ v^{(s)}(\vec{p})\text{-outgoing} \end{cases}$$

- [YUK4] **Scalar propagator:**

$$= \frac{+i}{p^2 - m^2(+i\epsilon)}$$

- [YUK5] **Dirac propagator:**

$$= \frac{+i(\not{p} + m)}{p^2 - m_\phi^2(+i\epsilon)}$$

- [YUK6] **Vertex:**

$$= -ig$$

The careful reader might have been puzzled by the direction of the arrow for the leg of an antifermion. We illustrate this point in the next paragraph. It should also be noted that the derivation of the amplitudes with the rules does not explicitly take into account the overall sign and that the relative sign between diagrams will have to be made "by hand," as will be shown next.

- **Scattering amplitude for fermion–antifermion pair.** We now consider the scattering between two fermion–antifermion particles, labeled f and $\bar{f}$, and define the kinematical configuration with the four-vectors q_1, q_2, q_3, and q_4:

$$f(q_1) + \bar{f}(q_2) \rightarrow f(q_3) + \bar{f}(q_4) \tag{9.200}$$

Rather than deriving the amplitudes from the calculations, we decide to first draw the Feynman diagrams and then read off the amplitudes using the rules defined above! The diagrams can be inferred from the one shown in Figure 9.15. They are shown in Figure 9.16. **Note the convention adopted for the arrow in the case of antiparticles opposite to the direction of the momentum**, as discussed in Section 9.15. The first diagram is the same as in the case of the scattering of two fermions, apart from the direction of the arrows. The second is new and corresponds to the annihilation of two fermions into our scalar particle.

Direct application of the Feynman rule gives us the amplitudes. Starting with the diagram on the left of Figure 9.16, we find:

$$i\mathcal{M}_1(f\bar{f} \rightarrow f\bar{f}) = \underbrace{\bar{u}^{(s_3)}(q_3)u^{(s_1)}(q_1)}_{\text{external legs}} \underbrace{(-ig)}_{\text{vertex}} \underbrace{\frac{+i}{q^2 - m_\phi^2 + i\epsilon}}_{\text{propagator}} \underbrace{(-ig)}_{\text{vertex}} \underbrace{\bar{v}^{(s_2)}(q_2)v^{(s_4)}(q_4)}_{\text{external legs}}$$

$$= -ig^2 \bar{u}^{(s_3)}(q_3)u^{(s_1)}(q_1) \frac{1}{(q_1 - q_3)^2 - m_\phi^2 + i\epsilon} \bar{v}^{(s_2)}(q_2)v^{(s_4)}(q_4) \tag{9.201}$$

When writing the propagator, we have used energy–momentum conservation at each vertex, which implies that $q^\mu = q_1^\mu - q_3^\mu$. The amplitude of the diagram on the right can be derived in a similar way to find:

$$i\mathcal{M}_2(f\bar{f} \rightarrow f\bar{f}) = -ig^2 \bar{u}^{(s_3)}(q_3)v^{(s_4)}(q_4) \frac{1}{(q_1 + q_2)^2 - m_\phi^2 + i\epsilon} \bar{v}^{(s_2)}(q_2)u^{(s_1)}(q_1) \tag{9.202}$$

We note that momentum conservation this time implies that the four-momentum of the scalar particle is given by $q^\mu = q_1^\mu + q_2^\mu$. Detailed calculations as in the case of the fermion–fermion scattering would show that here also the two amplitudes would have opposite signs due to Fermi statistics. Consequently, the two amplitudes

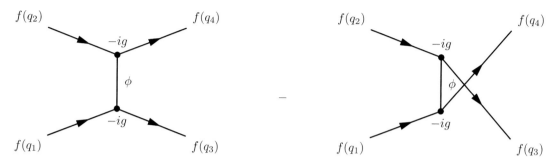

Figure 9.15 The diagrams for the scattering process of two fermions in the Yukawa theory. The amplitudes of the two diagrams must be subtracted from one another.

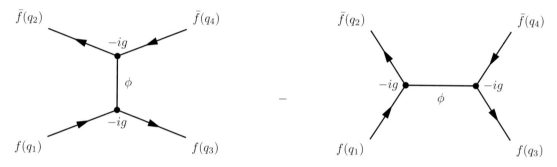

Figure 9.16 The diagrams for the scattering process of a fermion–antifermion pair in the Yukawa theory. The amplitudes of the two diagrams must be subtracted from one another.

must be subtracted from another. We found a corollary to the "minus-sign rule," which can be expressed as follows: **there is a a minus sign between the amplitude of two diagrams that differ only by the interchange of one initial (resp. final) fermion with a final (resp. initial) antifermion.** So, finally, the result is simply $i\mathcal{M} = i\mathcal{M}_1 - i\mathcal{M}_2$, or:

$$
\begin{aligned}
i\mathcal{M}(f\bar{f} \to f\bar{f}) &= -ig^2 \Big[\bar{u}^{(s_3)}(q_3)u^{(s_1)}(q_1) \frac{1}{(q_1 - q_3)^2 - m_\phi^2 + i\epsilon} \bar{v}^{(s_2)}(q_2)v^{(s_4)}(q_4) - \\
&\quad \bar{u}^{(s_3)}(q_3)v^{(s_4)}(q_4) \frac{1}{(q_1 + q_2)^2 - m_\phi^2 + i\epsilon} \bar{v}^{(s_2)}(q_2)u^{(s_1)}(q_1) \Big]
\end{aligned} \tag{9.203}
$$

The amplitude could also have been derived from Eq. (9.199) by noting that an incoming antifermion is represented by the spinor $\bar{v}$ and an outgoing antifermion is described by v.

Problems

Ex 9.1 Toy model with interacting scalar fields. In Section 9.8, we have considered a simple toy model with real scalar and complex scalar fields. We have computed the scattering amplitude for the process $\pi^+\pi^- \to \pi^+\pi^-$.

(a) In a similar fashion, compute the amplitude for the $\pi^0\pi^0 \to \pi^0\pi^0$ process.

(b) Assuming an additional coupling between the neutral and charged fields given by g', compute the amplitudes for $\pi^+\pi^- \to \pi^0\pi^0$ and $\pi^0\pi^0 \to \pi^+\pi^-$.

Ex 9.2 Scattering in the ϕ^4 theory. Compute the amplitudes and the differential cross-sections for the following scattering processes within the ϕ^4 theory:

(a) $\pi^+\pi^- \to \pi^+\pi^-$.

(b) $\pi^0\pi^0 \to \pi^0\pi^0$.

(c) $\pi^+\pi^- \to \pi^+\pi^-$.

Ex 9.3 Yukawa potential. The full Lagrangian of the Yukawa theory was introduced to be given by:

$$\mathcal{L}_{Yukawa} = \mathcal{L}_{KG} + \mathcal{L}_{Dirac} + \mathcal{L}_{int} = \frac{1}{2}(\partial_\mu \phi)^2 - \frac{1}{2}m_\phi^2\phi^2 + i\overline{\Psi}\left(\gamma^\mu\partial_\mu - m\right)\Psi + g\phi(\overline{\Psi}\Psi) \qquad (9.204)$$

where m_ϕ is the rest mass of the neutral scalar particle, m is the rest mass of the Dirac fermion f, and g is the coupling constant of the Yukawa theory. Consider the scattering of distinguishable fermions in the non-relativistic limit.

(a) Sketch the corresponding Feynman diagram at tree level.

(b) In the non-relativistic limit, one can write the four-vectors of the incoming and outgoing particles as $p = (m, \vec{p})$, $p' = (m, \vec{p}')$, $k = (m, \vec{k})$, and $k' = (m, \vec{k}')$. Show that

$$(p' - p)^2 \simeq -|\vec{p}' - \vec{p}|^2 \quad \text{and} \quad u^{(s)}(p) \simeq \sqrt{m}\begin{pmatrix} u_A^s \\ u_A^s \end{pmatrix} \qquad (9.205)$$

where u_A^s is an orthonormal spin-dependent bi-spinor.

(c) Show that
$$\bar{u}^{(s')}(p')u^{(s)}(p) = 2m\delta^{s,s'} \qquad (9.206)$$
and similarly for $\bar{u}^{(r')}(k')u^{(r)}(k)$.

(d) Show that the scattering amplitude then becomes:

$$i\mathcal{M} = \frac{ig^2}{|\vec{q}|^2 + m_\phi^2}2m\delta^{s,s'}2m\delta^{r,r'} \qquad (9.207)$$

where $\vec{q} \equiv \vec{p}' - \vec{p}$.

(e) What is the consequence for the spins?

(f) Convince yourself that the amplitude shows that the interaction can be expressed as due to a potential, whose Fourier transform is given by:

$$\tilde{V}(\vec{q}) = \frac{-g^2}{|\vec{q}|^2 + m_\phi^2} \qquad (9.208)$$

(g) Invert the Fourier transform to find the Yukawa potential:

$$V(r) = -\frac{g^2}{4\pi}\frac{1}{r}e^{-m_\phi r} \qquad (9.209)$$

This result shows that the Yukawa interaction between two scalar fermions is attractive, with a limited range given by the mass of the exchanged particle ϕ.

(h) Now consider the interaction between a fermion and an antifermion, and show that in this case also, the Yukawa interaction is attractive.

10   Quantum Electrodynamics

Though not a fundamentally new theory, like general relativity or quantum mechanics, quantum electrodynamics, as an effort to obtain a consistent relativistic treatment of the interaction of matter with radiation, gave birth to most fertile concepts of modern particle physics such as quantum fields, gauge symmetries, and renormalization techniques.

10.1   Historical Note on the Development of Quantum Electrodynamics

Historically, the behavior of light (or more generally electromagnetic radiation) represented a failure of classical mechanics known at the beginning of the twentieth century. On the one hand, the phenomena of interference and diffraction could be explained only on the basis of a wave theory. On the other hand, phenomena such as photoelectric emission and scattering by free electrons showed that light is composed of small particles. These particles, which were called photons, each had a definite energy and momentum, depending on the frequency of the light, and appeared to have just as real an existence as electrons, or any other particles known in physics. A fraction of a photon was never observed.

Quantum electrodynamics (QED) is the **quantum field theory** that describes the electromagnetic interactions between electrons, positrons, and photons. It intimately combines the Dirac theory of spin-1/2 fermions with that of the quantized electromagnetic field, derived from Maxwell's classical theory. In a form that we will discuss later, QED is the prototype for **gauge theories**. QED hence represents the foundation of the modern Standard Model theory of electroweak and strong interactions. The basic building blocks of QED are the electrons, the positrons, and the photons. In previous chapters, we saw that the electron was discovered by J. J. Thomson and the positron by C. D. Anderson. **Who discovered the photon?**

The story begins with Planck in 1900 and his famous attempts to describe the law of the electromagnetic radiation emitted by black bodies, as discussed in Section 2.2. Planck assumed that the necessary energy quantization occurred during emission or absorption of light, but this latter did not necessarily need to be an intrinsic property of electromagnetic radiation.

Einstein, on the other hand, argued that quantization was an **intrinsic property of electromagnetic radiation**. In his famous 1905 paper, he wrote:

> *"Indeed, it seems to me that the observations of 'black radiation', photoluminescence, the generation of cathode rays by ultraviolet light and other phenomena concerning the generation or transformation of light seem to be better understood under the assumption that the energy of light is discontinuously distributed in space. According to the assumption to be considered here, when a ray of light is propagated from a point, the energy is not continuously distributed over larger and larger spaces, but consists of a* **finite number of energy quanta localized in space points, which move without subdivision, and can only be absorbed and generated as a whole.**" [1]

---

1   *"Es scheint mir nun in der Tat, dass die Beobachtungen über die „schwarze Strahlung", Photolumineszenz, die Erzeugung von Kathodenstrahlen durch ultraviolettes Licht und andere die Erzeugung bez. Verwandlung des Lichtes betreffende Er-*

Accordingly, light is a quantum system!

In the 1920s the main focus indeed moved toward creating a quantum-mechanical theory of the electromagnetic field. In 1923 **Compton** published his analysis of the scattering of X-rays on light elements wherein he derived his scattering equation (see Section 2.13). He wrote:

> *"The present theory depends essentially upon the assumption that each electron which is effective in the scattering scatters a complete quantum. It involves also the hypothesis that the quanta of radiation are received from definite directions and are scattered in definite directions.* **The experimental support of the theory indicates very convincingly that a radiation quantum carries with it directed momentum as well as energy."** [2]

His last sentence clearly supports the idea that the light quanta is an elementary particle, hence implies the existence of the massless **photon**.

In 1924, **de Broglie** introduced the general idea of the dual description of elementary systems [7]. Guided by the idea of a general relationship between the notions of frequency and energy, he considered the existence of a periodic phenomenon of a nature yet to be specified, which would be linked to any isolated piece of energy and which would depend on its own mass by the Planck–Einstein equation. This would then lead to associating the uniform movement of any material point with the propagation of a certain wave.

In 1926, **Heisenberg**, **Born**[3], and **Jordan**[4] developed the necessary tools to quantum-mechanically describe a system with an infinite number of degrees of freedom, and in particular the mathematical tools that led to the **canonical formalism** and the Hamiltonian formulation (see Section 4.2).

In 1927, **Dirac** publishes the first reasonably complete theory of quantum electrodynamics, which included both the electromagnetic field and matter as quantum-mechanical objects. He wrote:

> *"It will also be shown that the Hamiltonian which describes the interaction of the atom and the electromagnetic waves can be made identical with the Hamiltonian for the problem of interactions of the atom with an assembly of particles moving with the velocity of light and satisfying the Einstein–Bose statistics, by a suitable choice of the interaction energy for the particles. The number of particles having any specified direction of motion and energy (...) is equal to the number of quanta of energy in the corresponding wave in the Hamiltonian for the waves."* [5]

The theory was shown to be able to model important processes such as the emission of a photon by an electron dropping into a quantum state of lower energy, **a process in which the number of particles is not constant**: one atom in the initial state becomes an atom plus a photon in the final state. The ability to describe creation and annihilation of particles is one of the most important features of quantum field theory.

Finally, in 1934, in a step unrelated to electrodynamics, **Fermi** published his essay on the theory of weak β-decay. He wrote:

scheinungsgruppen besser verständlich erscheinen unter der Annahme, dass die Energie des Lichtes diskontinuierlich im Raume verteilt sei. Nach der hier ins Auge zu fassenden Annahme ist bei Ausbreitung eines von einem Punkte ausgehenden Lichtstrahles die Energie nicht kontinuierlich auf grösser und grösser werdende Räume verteilt, sondern es besteht dieselbe aus einer endlichen Zahl von in Raumpunkten lokalisierten Energiequanten, welche sich bewegen, ohne sich zu teilen und nur als Ganze absorbiert und erzeugt werden können." Republished with permission of John Wiley & Sons – Books from A. Einstein, "Über einen die Erzeugung und Verwandlung des Lichtes betreffenden heuristischen Gesichtspunkt," *Annalen der Physik*, vol. 322, no. 6, pp. 132–148, 1905; permission conveyed through Copyright Clearance Center, Inc.

2   Reprinted excerpt with permission from A. H. Compton, "A quantum theory of the scattering of X-rays by light elements," *Phys. Rev.*, vol. 21, pp. 483–502, May 1923. Copyright 1923 by the American Physical Society.

3   Max Born (1882–1970), German physicist.

4   Ernst Pascual Jordan (1902–1980), German theoretical and mathematical physicist.

5   Republished with permission of The Royal Society (UK) from P. A. M. Dirac, "The quantum theory of the emission and absorption of radiation," *Proc. Roy. Soc. Lond.*, vol. A114, no. 767, pp. 243–265, 1927; permission conveyed through Copyright Clearance Center, Inc.

"It therefore seems more appropriate to follow Heisenberg to assume that a nucleus consists only of heavy particles, protons and neutrons. In order to understand the possibility of β emission, we will try to build a theory of light particle emission from a nucleus in analogy to the theory of light quantum emission from an excited atom in the ordinary radiation process. In radiation theory, the total number of light quanta is not a constant: light quanta are created when they are emitted by an atom and disappear when they are absorbed." [6]

His theory led to the evidence that **not only the quanta of the electromagnetic field but also other particles might emerge and disappear as a result of their interaction with other particles**. The creation and annihilation operators of quantum field theory were seen as fundamental to describe such phenomena.

Despite its early successes, QED was plagued by several serious theoretical difficulties. Basic physical quantities, such as the self-energy of the electron, the energy shift of electron states due to the presence of the electromagnetic field, gave infinite, divergent contributions – a nonsensical result – when computed using the perturbative techniques. This triggered a QED crisis. The "divergence problems" were solved in the late 1940s through a procedure known as **renormalization**. Radiative corrections and renormalization will be discussed in Chapter 12. Meanwhile, let us dive into QED by first recalling the classical Maxwell theory.

10.2   Classical Maxwell Theory

The starting point for the quantization of the electromagnetic field is the classical field equations of **Maxwell**,[7] which, as we know, represent a classical theory already consistent with special relativity. The electric field is described by the three-vector field $\vec{E}(\vec{x}, t)$ and the magnetic field by the three-vector field $\vec{B}(\vec{x}, t)$. The *sources* are the electric charge density $\rho(\vec{x}, t)$ and the current density $\vec{j}(\vec{x}, t)$. In natural units (see Section 1.8), we have $c = \epsilon_0 = \mu_0 = 1$. The four Maxwell equations are then *in vacuum*:

$$\left. \begin{aligned} \nabla \cdot \vec{E} &= \rho \\ \nabla \times \vec{B} - \frac{\partial \vec{E}}{\partial t} &= \vec{j} \end{aligned} \right\} \quad \textbf{inhomogeneous equations}$$

$$\left. \begin{aligned} \nabla \cdot \vec{B} &= 0 \\ \nabla \times \vec{E} + \frac{\partial \vec{B}}{\partial t} &= 0 \end{aligned} \right\} \quad \textbf{homogeneous equations} \tag{10.1}$$

Vector fields, such as the electric and magnetic fields, are associated with forces. In order to obtain the energy, which is the relevant quantity to build the Hamiltonian and also the Lagrangian, we should consider **potentials**. The divergence equation for $\vec{B}$ suggests that we can still introduce a vector potential. Consequently, the magnetic vector potential $\vec{A}$ is defined such that the magnetic field $\vec{B}$ is given by $\vec{B} \equiv \nabla \times \vec{A}$:

$$\vec{B} \equiv \nabla \times \vec{A} \quad \Longrightarrow \quad \nabla \cdot \vec{B} = \nabla \cdot (\nabla \times \vec{A}) = 0 \quad \square \tag{10.2}$$

Any conservative vector field $\vec{F}$ (that means $\nabla \times \vec{F} = 0$) can be derived from a scalar potential ϕ as $\vec{F} \equiv -\nabla \phi$. We hence find:

$$\nabla \times \vec{E} + \frac{\partial \vec{B}}{\partial t} = 0 \quad \Longrightarrow \quad \nabla \times \left(\vec{E} + \frac{\partial \vec{A}}{\partial t} \right) = 0 \quad \Longrightarrow \quad \vec{E} + \frac{\partial \vec{A}}{\partial t} \equiv -\nabla \phi \tag{10.3}$$

---

6   *"Sembra per conseguenza più appropriato ammettere con Heisenberg che tutti i nuclei consistano soltanto di particelle pesanti, protoni e neutroni. Per comprendere tuttavia la possibilità dell'emissione dei raggi β, noi tenteremo di costruire una teoria dell'emissione delle particelle leggere da un nucleo in analogia alla teoria dell'emissione di un quanto di luce da un atomo eccitato nell'ordinario processo della irradiazione. Nella teoria dell'irradiazione, il numero totale dei quanti di luce non è costante; i quanti vengono creati all'atto della loro emissione da un atomo eccitato, e spariscono invece quando sono assorbiti."* Reprinted by permission from Springer Nature: E. Fermi, "Tentativo di una teoria dei raggi β," *Il Nuovo Cimento (1924–1942)*, vol. 11, p. 1. Copyright 2008.

7   James Clerk Maxwell (1831–1879), Scottish physicist.

Consequently, the electric scalar potential ϕ is defined such that the electric field $\vec{E}$ satisfies Eq. (10.3). Finally, we have:

$$\vec{B} \equiv \nabla \times \vec{A} \quad \text{and} \quad \vec{E} \equiv -\nabla \phi - \frac{\partial \vec{A}}{\partial t} \tag{10.4}$$

Classically it is possible to argue that only the electromagnetic fields are physical, while the electromagnetic potentials are purely mathematical constructs (in addition, not even unique because of the gauge transformation discussed below!). However, as was discussed in Section 4.15, the **Ehrenberg–Siday–Aharonov–Bohm effect** is a quantum-mechanical phenomenon in which an electrically charged particle is affected by the electromagnetic potential, despite being confined to a region in which both the electric and magnetic fields are zero. It is generally argued that the effect illustrates the physicality of electromagnetic potentials in quantum mechanics.

The equations for the potentials ϕ and $\vec{A}$ are determined by eliminating the fields $\vec{E}$ and $\vec{B}$ from Maxwell's equations. For the electric field, this is just:

$$\nabla \cdot \vec{E} = \rho \quad \Longrightarrow \quad -\frac{\partial}{\partial t}(\nabla \cdot \vec{A}) - \nabla^2 \phi = \rho \tag{10.5}$$

For the magnetic field, we get:

$$\nabla \times (\nabla \times \vec{A}) = \vec{j} - \frac{\partial^2 \vec{A}}{\partial t^2} - \nabla \frac{\partial \phi}{\partial t} \tag{10.6}$$

Since $\nabla \times (\nabla \times \vec{A}) = \nabla(\nabla \cdot \vec{A}) - \nabla^2 \vec{A}$, we can rearrange to find:

$$\nabla \left(\nabla \cdot \vec{A} + \frac{\partial \phi}{\partial t} \right) - \nabla^2 \vec{A} + \frac{\partial^2 \vec{A}}{\partial t^2} = \vec{j} \tag{10.7}$$

In the **covariant formulation of classical electromagnetism**, Maxwell's equations are expressed in a form that is manifestly invariant under Lorentz transformations. These expressions both make it simple to prove that the laws of classical electromagnetism take the same form in any inertial coordinate system, and also provide a way to translate the fields from one frame to another. In this formulation, the **principal physical object is the antisymmetric electromagnetic field tensor $F^{\mu\nu}$**, which is the combination of the electric and magnetic fields into a covariant term. It is given by:

$$F^{\mu\nu} \equiv \partial^\mu A^\nu - \partial^\nu A^\mu \quad \text{where} \quad A^\mu(x) \equiv (\phi(\vec{x}, t), \vec{A}(\vec{x}, t)) \tag{10.8}$$

Its antisymmetric nature is seen immediately, since $F^{\mu\nu} = -F^{\nu\mu}$ and $F^{\mu\mu} = 0$. The four-vector quantity A^μ is called the **electromagnetic four-potential**. The 4×4 antisymmetric electromagnetic tensor transforms itself under a Lorentz transformation as expected according to (see Section 5.2):

$$F^{\mu\nu} \to \Lambda^\mu{}_\alpha \Lambda^\nu{}_\beta F^{\alpha\beta} \tag{10.9}$$

It is straightforward to show that the tensor entries are directly given by the component of the electric and magnetic fields $\vec{E} = (E_x, E_y, E_z)$ and $\vec{B} = (B_x, B_y, B_z)$:

$$F^{\mu\nu} = \begin{pmatrix} 0 & -E_x & -E_y & -E_z \\ E_x & 0 & -B_z & B_y \\ E_y & B_z & 0 & -B_x \\ E_z & -B_y & B_x & 0 \end{pmatrix} \tag{10.10}$$

One can verify that the Lorentz transformation of $F^{\mu\nu}$ leads to the correct transformation for the fields $\vec{E}$ and $\vec{B}$. Since the tensor contains both electric and magnetic fields, we understand why in general a Lorentz transformation will mix the $\vec{E}$ and $\vec{B}$ fields. In this form, the common origin of the electric and magnetic phenomena, deriving from the existence of the electric charge under the requirements of special relativity, is clear!

In order to include the sources of the fields, we define the **four-vector electric charge–current density** $J^\mu(x)$. With this definition, the two inhomogeneous Maxwell equations can be simply expressed as:

$$\partial_\mu F^{\mu\nu} = J^\nu \quad \text{where} \quad J^\nu(x) \equiv (\rho(\vec{x}, t), \vec{j}(\vec{x}, t)) \tag{10.11}$$

The two homogeneous Maxwell equations are automatically verified. In this form, the entire Maxwell theory can be summarized as one equation whose covariance is very explicit!

In order to develop the QED theory, we need the Lagrangian function. For the free electromagnetic field ($J^\nu = 0$), it is given by the **Proca[8]–Lagrange function**:

$$\mathcal{L}_{free} = -\frac{1}{4} F_{\mu\nu} F^{\mu\nu} \tag{10.12}$$

By construction, this Lagrangian density is invariant under Lorentz transformations (it should also be gauge-invariant). We verify that it yields the Maxwell equations, Eq. (10.11) for a free field. We vary it relative to the A^μ potential:

$$
\begin{aligned}
\frac{\partial \mathcal{L}_{free}}{\partial(\partial^\mu A^\nu)} &= \frac{\partial}{\partial(\partial^\mu A^\nu)} \left[-\frac{1}{4} (\partial_\alpha A_\beta - \partial_\beta A_\alpha)(\partial^\alpha A^\beta - \partial^\beta A^\alpha) \right] \\
&= -\frac{1}{4} \frac{\partial}{\partial(\partial^\mu A^\nu)} \left[(2\partial_\alpha A_\beta \partial^\alpha A^\beta - 2\partial_\alpha A_\beta \partial^\beta A^\alpha) \right] \\
&= -\frac{1}{4} 2 \left[\frac{\partial}{\partial(\partial^\mu A^\nu)} (\partial_\alpha A_\beta \partial^\alpha A^\beta) - \frac{\partial}{\partial(\partial^\mu A^\nu)} (\partial_\alpha A_\beta \partial^\beta A^\alpha) \right] \\
&= -\frac{1}{4} 2 \left[2(\partial_\mu A_\nu - \partial_\nu A_\mu) \right] = -(\partial_\mu A_\nu - \partial_\nu A_\mu)
\end{aligned}
\tag{10.13}
$$

Therefore:

$$\frac{\partial \mathcal{L}_{free}}{\partial(\partial^\mu A^\nu)} = -F_{\mu\nu} \quad \text{and} \quad \frac{\partial \mathcal{L}_{free}}{\partial A^\nu} = 0 \tag{10.14}$$

Hence, using the Euler–Lagrange equation, we get:

$$\frac{\partial \mathcal{L}_{free}}{\partial A^\nu} - \partial^\mu \left(\frac{\partial \mathcal{L}_{free}}{\partial(\partial^\mu A^\nu)} \right) = 0 \quad \Longrightarrow \quad \partial^\mu F_{\mu\nu} = 0 \quad \square \tag{10.15}$$

We now consider a term that represents the case where the electromagnetic field is created by the sources J^ν. We add a new term with a direct coupling between the field and the sources of the form $J_\mu A^\mu$ to the free Lagrangian:

$$\mathcal{L} = \mathcal{L}_{free} + \mathcal{L}_{sources} = -\frac{1}{4} F_{\mu\nu} F^{\mu\nu} - J_\mu A^\mu \tag{10.16}$$

We can verify that it has the correct form since:

$$\frac{\partial \mathcal{L}_{sources}}{\partial(\partial^\mu A^\nu)} = 0 \quad \text{and} \quad \frac{\partial \mathcal{L}_{sources}}{\partial A^\nu} = -J_\nu \quad \xrightarrow{\mathcal{L} = \mathcal{L}_{free} + \mathcal{L}_{sources}} \quad \partial^\mu F_{\mu\nu} = J_\nu \quad \square \tag{10.17}$$

From this last result, we can also notice that the continuity equation relating charges and currents $\partial^\nu J_\nu = 0$ is automatically satisfied, since $F_{\mu\nu}$ is antisymmetric:

$$\partial^\nu J_\nu = \underbrace{\partial^\nu \partial^\mu}_{\text{symmetric}} F_{\mu\nu} = 0 \quad \square \tag{10.18}$$

---

8   Alexandru Proca (1897–1955), Romanian physicist.

So, finally, **the entire classical Maxwell theory of electromagnetism can be summarized into the following single Lorentz-invariant Lagrangian**:

$$\mathcal{L}_{Maxwell} = -\frac{1}{4}F_{\mu\nu}F^{\mu\nu} - J_\mu A^\mu \tag{10.19}$$

It is beautiful!

10.3   Gauge Freedom

We have argued that the electromagnetic four-potential A^μ – from where the electric and magnetic fields can be derived according to the definitions given above – needs to be taken as the fundamental physical quantity in quantum mechanics. A famous complication is, however, known to be given by the so-called **freedom in the choice of the gauge**. Indeed, the four-potential A^μ is not uniquely defined in terms of the electric and magnetic fields. Consider the **gauge transformation** of the form:

$$A^\mu \to A^\mu + \partial^\mu \lambda \tag{10.20}$$

where $\lambda = \lambda(x^\mu)$ is any scalar function of space-time. Under the gauge change, the electric and magnetic fields remain unchanged, since:

$$F^{\mu\nu} \to \partial^\mu(A^\nu + \partial^\nu \lambda) - \partial^\nu(A^\mu + \partial^\mu \lambda) = \partial^\mu A^\nu - \partial^\nu A^\mu = F^{\mu\nu} \tag{10.21}$$

The classical electric and magnetic fields contain only "physical" degrees of freedom, in the sense that every mathematical degree of freedom in an electromagnetic field configuration has a separately measurable effect on the motions of test charges in the vicinity. In contrast, there is more freedom to choose A^μ than there are real physical degrees of freedom. The gauge freedom is to be viewed as a "redundancy" in our mathematical description. Ultimately the **physical observables must be independent of the chosen gauge**.

There are many ways to **fix the gauge**, and the practical choice often depends on the problem to be solved. The gauge freedom must be understood by the fact that we are using the potential A^μ with a priori four independent components, whereas the actual components are not totally independent from one another. The gauge is fixed by imposing an additional condition on the four-potential A^μ, which effectively imposes a relation among the four components (this is similar to the case of the four-energy–momentum vector where the relation $E^2 = \vec{p}^2 + m^2$ implies that the four quantities are not fully independent).

- **Lorenz gauge.** In the **Lorenz gauge**[9] or **Landau gauge**, the freedom fixing condition is expressed as:

$$\text{Lorenz gauge:} \quad \partial_\mu A^\mu(x) = 0 \tag{10.22}$$

In order to satisfy this gauge, one should find the gauge functions λ which satisfy the condition $\partial_\mu \partial^\mu \lambda = 0$. In this case, the covariant expression of the Maxwell theory, Eq. (10.11), becomes even simpler:

$$\partial_\mu F^{\mu\nu} = \partial_\mu \partial^\mu A^\nu - \underbrace{\partial_\mu \partial^\nu A^\mu}_{=0} = J^\nu \xrightarrow{\text{Lorenz gauge}} \partial_\mu \partial^\mu A^\nu = J^\nu \tag{10.23}$$

The general solution in vacuum (i.e., a free electromagnetic field in vacuum with no sources $J^\nu = 0$) is a plane-wave solution with four degrees of freedom, as will be considered in Eq. (10.29). The Lorenz gauge reduces the number of degrees of freedom by one, with the requirement $\partial_\mu A^\mu = 0$ or the condition on the gauge fixing function $\partial_\mu \partial^\mu \lambda = 0$. However, there are still several ways to find a function λ that satisfies $\partial_\mu \partial^\mu \lambda = 0$. This can be resolved by adding a new condition, for example the Coulomb gauge-fixing condition, as below.

9   This gauge is named after the Danish physicist Ludvig Lorenz and not after Hendrik Lorentz!

• **Coulomb gauge.** The **Coulomb gauge-fixing condition (or transverse gauge)** imposes the requirement that:

$$\text{Coulomb gauge:} \quad \nabla \cdot \vec{A}(x) \equiv 0 \qquad (10.24)$$

Together with the Lorenz gauge condition, this leads to:

$$\partial_\mu A^\mu = \partial_0 A^0 + \nabla \cdot \vec{A} = \partial_0 A^0 + 0 = 0 \qquad (10.25)$$

So $A^0(\vec{x})$ is time-independent. It is fully determined by $\vec{A}$ and $\partial_t \vec{A}$, as a consequence of Maxwell's equation, Eq. (10.5):

$$\nabla^2 A^0 + \nabla \cdot \frac{\partial \vec{A}}{\partial t} = 0 \quad \Longrightarrow \quad A^0(\vec{x}) = \int \mathrm{d}^3 \vec{y} \, \frac{(\nabla \cdot \partial \vec{A}/\partial t)\,(\vec{y})}{4\pi\,|\vec{x} - \vec{y}|} = 0 \qquad (10.26)$$

So, in cases where there are no sources (i.e., $J^\nu(x) = 0$), we find the following **Coulomb gauge-fixing condition (or transverse gauge) in vacuum:**

$$\text{Coulomb gauge in vacuum:} \quad A^0(x) \equiv 0 \quad \text{and} \quad \nabla \cdot \vec{A}(x) \equiv 0 \qquad (10.27)$$

The Coulomb gauge is, however, not Lorentz-covariant. A further gauge transformation has to be made to retain the Coulomb gauge condition after a Lorentz transformation. Because of this, the Coulomb gauge is not practical in covariant perturbation theories, such as QED. However, one advantage of the Coulomb gauge is that the polarization of the photon takes a simple form, as will be shown below.

10.4   The Photon Polarization States

It is known experimentally that when polarized light is used for ejecting electrons via the photoelectric effect, there is a preferential direction for the electron emission. Hence, the polarization properties of light must be closely connected with its corpuscular properties and one must ascribe a polarization to the photon. Hence, in a beam of light consisting of photons, every photon is in a certain state of polarization.

In addition, since gauge invariance represents a redundancy of the system, we might be tempted to formulate the photons purely in terms of the local, physical, gauge-invariant electric and magnetic fields. However, we have seen that this is not satisfactory to explain all phenomena, in particular, the Ehrenberg–Siday–Aharonov–Bohm effect requires the use of the four-potential field A^μ.

As a consequence, the four-potential field A^μ is a four-component object and we can express it with **the polarization four-vector** ϵ^μ. It has been known for a long time that the photon carries spin angular momentum that does not depend on its frequency. **Bose,**[10] in his 1924 paper on the hypothesis of the light quantum, already noted:

> "In any case, the total number of cells must be regarded as the number of possible arrangements of a quantum in the given volume. It seems, however, appropriate to multiply this number once again by 2 in order to take into account the fact of the polarization (...)." [11]

Bose envisaged the possibility of the light quantum possessing, besides energy $h\nu$ and linear momentum $h\nu/c$, also an intrinsic spin or angular momentum $\pm h/2\pi$ around an axis parallel to the direction of its motion. The

---

10   Satyendra Nath Bose (1894–1974), Indian physicist.
11   *"Indessen muss die Gesamtzahl der Zellen als die Zahl der möglichen Anordnungen eines Quants in dem gegebenen Volumen angesehen werden. Um der Tatsache der Polarisation Rechnung zu tragen, erscheint es dagegen geboten, diese Zahl noch mit 2 zu multiplizieren (...)."* Reprinted by permission from Springer Nature: S. N. Bose, "Plancks Gesetz und Lichtquantenhypothese," *Zeitschrift für Physik*, vol. 26, pp. 178–181, Dec 1924.

weight factor 2 thus arises from the possibility of the spin of the light quantum being either right-handed or left-handed, corresponding to the two alternative signs of the angular momentum.

Following Dirac (see, e.g., Ref. [81]), we must regard a linearly polarized light as carried by a quantum which has an even chance of having plus or minus one Bohr unit as its angular momentum. Elliptically polarized light would be similarly regarded as characterized by unequal chances of possessing spins with the two alternative signs. This is allowed by the fundamental principle of superposition of quantum states.

The spin-1 nature of the photon was supported by the selection rules for the emission or absorption of radiation from atoms as a direct consequence of conservation of angular momentum in those processes. In addition, experimental studies on the scattering of light by gases by **Raman**[12] and **Bhagavantam**[13] [82] provided direct evidence that the light quantum carries spin angular momentum that does not depend on its frequency, consistent with the theoretical interpretations above. Consequently, the "real" photon emitted or absorbed in quantum-electrodynamic processes is identified as a **spin-1 particle with only two possible polarizations (or helicities)**.

In quantum mechanics, however, a spin-1 boson has three spin projections corresponding to three possible helicity states $s = +1, 0, -1$. The case $s = 0$ is known as longitudinal polarization, and the $s = \pm 1$ are transverse polarizations (or left and right-handed circular polarizations). If the real photon has only transverse polarizations, then the $s = 0$ state does not exist for it, or generally, for massless particles.

Nonetheless, it is still not obvious how the massless photon with *two* spin states can be described by the four-potential field A^μ that has *four* degrees of freedom. The answer to these complications lies in the properties of electromagnetism and in the gauge freedom.

It is obvious that the field potential A^μ must satisfy the Maxwell equation Eq. (10.11). In the Lorenz gauge and for a free field ($J^\nu = 0$), we note that **this latter is equivalent to the Klein–Gordon equation for a massless particle** (see Eq. (10.23)). Each one of the four components of the $A^\mu(x)$ potential satisfies the Klein–Gordon equation for a massless particle:

$$\partial_\mu \partial^\mu A^\nu(x) = 0 \quad \text{where} \quad \nu = 0, 1, 2, 3 \quad \Leftrightarrow \quad \underbrace{(\partial_\mu \partial^\mu + m^2)\phi(x)}_{\text{Klein–Gordon}} \overset{m=0}{=} \partial_\mu \partial^\mu \phi(x) = 0 \tag{10.28}$$

The equation has a plane-wave solution, of the form:

$$A^\mu(x) = \epsilon^\mu(k)e^{-ik\cdot x} \tag{10.29}$$

where $\epsilon^\mu(k)$ is a four-vector describing the **polarization of the electromagnetic field** and k^μ being the four-momentum of the quanta of the field, satisfying $k^2 = m^2 = 0$. The polarization four-vector has a priori four degrees of freedom and fixing the gauge will provide constraints on its components, reducing its effective number of degrees of freedom. With the Lorenz gauge, we have for instance:

$$\partial_\mu A^\mu = (-ik_\mu)e^{-ik\cdot x}\epsilon^\mu(k) = 0 \quad \Longrightarrow \quad k_\mu \epsilon^\mu = 0 \tag{10.30}$$

This equation reduces the number of independent components of the polarization to three. We saw that the single Lorenz gauge is still not sufficient to uniquely define the four-potential. We can still apply a further gauge transformation (Eq. (10.20)) with the condition $\partial_\mu \partial^\mu \lambda = 0$. Consider $\lambda(x) = -iae^{-ik\cdot x}$ with a constant a. The function satisfies $\partial_\mu \partial^\mu \lambda = 0$, since $k^2 = 0$ for the real photon. For our plane wave, this gives:

$$A^\mu \to A^\mu - \partial^\mu \lambda = \epsilon^\mu(k)e^{-ik\cdot x} + ia(-ik^\mu)e^{-ik\cdot x} = \left(\epsilon^\mu(k) + ak^\mu\right)e^{-ik\cdot x} \tag{10.31}$$

This implies that the physical electromagnetic field is unchanged under the following change of the polarization vector:

$$\epsilon^\mu(k) \to \epsilon^\mu(k) + ak^\mu \tag{10.32}$$

---

12   Chandrasekhara Venkata Raman (1888–1970), Indian physicist.
13   Suri Bhagavantam (1909–1989), Indian scientist.

Any polarization that differs by a multiple of the four-momentum of the photon corresponds to the same physical photon polarization. Here we encounter again the gauge effect. There are more degrees of freedom in our formalism than encountered in the physical system. So the use of the covariant (four-vector) formalism comes at the cost of gauge complications. This is the price to pay!

If we take the Coulomb gauge-fixing condition, A^0 is constant in time (see Eq. (10.25)) and we can set $\epsilon^0 = 0$ and finally Eq. (10.30) becomes:

$$\epsilon^0 = 0 \quad \text{and} \quad \vec{\epsilon} \cdot \vec{k} = 0 \tag{10.33}$$

The polarization vector is forced to be *perpendicular to the direction of propagation* or *transverse* (that is the reason why the Coulomb gauge is also called the transverse gauge). The total number of independent degrees of freedom for the polarization has been reduced from four to two, and these correspond to the two independent helicity states of our real photon, consistent with the theoretical interpretation and experimental evidence.

The condition $\vec{\epsilon} \cdot \vec{k} = 0$ means that we can chose two independent three-vectors orthogonal to the direction of motion of the particle. If we set the z-axis as the direction of motion, two independent basis vectors $\vec{\epsilon}_1$ and $\vec{\epsilon}_2$ can, for example, be chosen as follows:

$$\epsilon_1^\mu = (0,1,0,0) \equiv (0,\vec{\epsilon}_1) \quad \text{and} \quad \epsilon_2^\mu = (0,0,1,0) \equiv (0,\vec{\epsilon}_2) \tag{10.34}$$

We also have the invariant property $\epsilon^2 = \epsilon^\mu \epsilon_\mu = -1$ (cf. Eq. (8.256)). Since $\vec{B} = \nabla \times \vec{A} \propto \vec{k} \times \vec{\epsilon}$, the two basis vectors ϵ_1^μ and ϵ_2^μ actually correspond to the **transverse linear polarizations of the electromagnetic waves**. Right- or left-handed *circularly polarized* waves can be obtained by the overlap of two linearly polarized waves appropriately shifted by 90° in phase. They could be written as (the choice is not unique):

$$\epsilon_R^\mu = -\frac{1}{\sqrt{2}}(0,1,i,0) \quad \text{and} \quad \epsilon_L^\mu = \frac{1}{\sqrt{2}}(0,-1,i,0) \tag{10.35}$$

Such circular polarizations are associated with the two degrees of freedom of the helicities ($h = \pm 1$) of the real photon. And we recover the classical result of electromagnetism, namely that electromagnetic fields are purely transverse.

In general, we note that the polarization vectors are orthonormal and that they obey (note the complex conjugation):

$$\vec{\epsilon}_\lambda^{\,*} \cdot \vec{\epsilon}_{\lambda'} = \delta_{\lambda'}^\lambda \tag{10.36}$$

The same equation considering the four-vector polarizations ϵ_λ^μ instead of the three-vectors reads:

$$\epsilon_\lambda^{*\mu} \epsilon_{\mu,\lambda'} = -\delta_{\lambda'}^\lambda \tag{10.37}$$

This is again for $\lambda = 1, 2$ (or R, L in Eq. (10.35)). Only for the real photon can we get rid of two of the four degrees of freedom of $A^\mu(x)$ through our gauge-fixing conditions (here the Coulomb gauge).

10.5   The Second Quantization of the Electromagnetic Field

By analogy with the real scalar field, we consider both "positive" and "negative" frequencies for a given momentum $\vec{k}$:

$$A_+^\mu(x) \propto e^{-ik\cdot x}\epsilon^\mu(k); \qquad A_-^\mu(x) \propto e^{+ik\cdot x}(\epsilon^\mu)^*(k) \tag{10.38}$$

where we assumed ϵ^μ and $(\epsilon^\mu)^*$ in anticipation that the electromagnetic potential $A^\mu(x) = A_+^\mu(x) + A_-^\mu(x)$ is real. In the case of the real photon, the energy is just $\omega_k = k^0 = |\vec{k}|$.

In order to quantize the electromagnetic field in the *canonical* way, we should now proceed as in the case of the Klein–Gordon or Dirac fields, i.e., by describing the infinite number of continuous degrees of freedom with a set of continuously indexed quantum harmonic oscillators. This will yield the set of creation and annihilation operators with specific commutation rules. However, the gauge fixing poses an extra problem, since the field depends on it, and hence so will the quantized operators.[14] **The canonical quantization of the electromagnetic field therefore imposes that the field A^μ is quantized within a gauge-fixing condition.** The actual choice will affect the form of the field operators with their own creation and annihilation operators with corresponding Fock spaces. At the end, of course, all physical observables should be independent of the gauge, and it turns out that this will be a powerful cross-check of the calculations. All quantum field theory calculations for the same problem performed under different gauges should lead to the same physical observables, although their actual form and computations can largely differ from one another.

Accordingly, we write the **quantized electromagnetic field potential A^μ, to be combined with a gauge-fixing condition**, as:

$$A^\mu(x) \equiv A^\mu_+(x) + A^\mu_-(x) = \int \frac{\mathrm{d}^3\vec{k}}{(2\pi)^{3/2}} \frac{1}{\sqrt{2\omega_k}} \sum_{\lambda=1}^{2} \left(\epsilon^\mu_\lambda(\vec{k}) a_\lambda(\vec{k}) e^{-ik\cdot x} + \epsilon^{\mu*}_\lambda(\vec{k}) a^\dagger_\lambda(\vec{k}) e^{+ik\cdot x} \right) \qquad (10.39)$$

Only *two* polarization states have been included in the sum over λ, and this clearly only makes sense because it is assumed that the choice of gauge will bring the number of degrees of freedom for A^μ to the desired two! In other words, for each mode $\vec{k}$, the waves $A^\mu_+(x)|_{\vec{k}}$ and $A^\mu_-(x)|_{\vec{k}}$ obey Maxwell's equation as well as the gauge-fixing condition. For instance, in the Coulomb gauge, the field could be expanded with the set of polarization vectors defined in Eq. (10.34).

The creation and annihilation operators $a^\dagger_\lambda(\vec{k})$ and $a_\lambda(\vec{k})$ create or destroy a photon with energy $\omega_k = |\vec{k}|$, momentum $\vec{k}$, and polarization state λ. Their commutation rules are defined as follows, consistent with the fact that photons are bosons:

$$\begin{cases} \left[a_\lambda(\vec{k}), a^\dagger_{\lambda'}(\vec{k}') \right] = \delta^\lambda_{\lambda'} \delta^3(\vec{k} - \vec{k}') \\ \left[a_\lambda(\vec{k}), a_\lambda(\vec{k}') \right] = \left[a^\dagger_\lambda(\vec{k}), a^\dagger_\lambda(\vec{k}') \right] = 0 \end{cases} \qquad (10.40)$$

One can show that the free Hamiltonian is then given by:

$$H = \int \mathrm{d}^3\vec{k} \, \omega_k \sum_{\lambda=1}^{2} \left(a^\dagger_\lambda(\vec{k}) a_\lambda(\vec{k}) \right) + \text{const.} \qquad (10.41)$$

where the constant is the (neglected) zero-point energy of the field. There is a corresponding equation for the total momentum $\vec{K}$ of the field, which is the analog of that derived for the boson field in Eq. (7.68).

10.6 The Photon Propagator in the Coulomb Gauge

We have shown the fundamental role played by the propagators of the fields in developing interactions. Anticipating that QED will deal with the interactions between photons and particles (i.e., light and matter but also virtual photons), this will have an impact on the degrees of freedom of the field. Thus, we need to compute the photon propagator. Since the electromagnetic field has an index μ, the photon propagator will actually have two indices, e.g., μ and ν.

Specifically, the **Feynman photon propagator $G^{\mu\nu}_F(x-y)$** is defined as:

$$G^{\mu\nu}_F(x-y) \equiv \langle 0| T\left(A^\mu(x) A^\nu(y) \right) |0\rangle \qquad (10.42)$$

---

14   We note that indeed the form of the photon propagator operator depends on the gauge!

Its exact form depends on the choice of the gauge-fixing condition. The propagator is computed as the Green's function of the field equation, in our case the Maxwell equation:

$$\partial_\mu F^{\mu\nu} = \partial_\mu(\partial^\mu A^\nu - \partial^\nu A^\mu) = \partial_\mu \partial^\mu A^\nu - \partial^\nu \partial_\mu A^\mu = J^\nu \rightarrow \mathcal{D}^\nu_\alpha A^\alpha(x) = J^\nu(x) \tag{10.43}$$

where $\mathcal{D}^\nu_\alpha \equiv [\partial_\mu \partial^\mu \delta^\nu_\alpha - \partial^\nu \partial_\alpha] = [\Box^2 \delta^\nu_\alpha - \partial^\nu \partial_\alpha]$ (we have used the D'Alembert operator, Eq. (D.42)). The Green's function should satisfy:

$$\mathcal{D}^\mu_\alpha G^{\alpha\nu}_F(x-y) = ig^{\mu\nu}\delta^4(x-y) \tag{10.44}$$

Integration over $(-i)\mathrm{d}^4 y J_\nu(y)$ yields the expected result:

$$(-i)\int \mathrm{d}^4 y \mathcal{D}^\mu_\alpha G^{\alpha\nu}_F(x-y)J^\nu(y) = \mathcal{D}^\mu_\alpha\left[(-i)\int \mathrm{d}^4 y G^{\alpha\nu}_F(x-y)J_\nu(y)\right] = g^{\mu\nu}J_\nu(x) = J^\mu(x) \tag{10.45}$$

Comparing Eqs. (10.43) and (10.45) we get:

$$A^\mu(x) = -i\int \mathrm{d}^4 y G^{\mu\nu}_F(x-y)J_\nu(y) \tag{10.46}$$

Let us decide to use the *Coulomb gauge* $\nabla \cdot \vec{A} \equiv 0$ which implies that A^0 is time-independent (see Eq. (10.25)) and with $\partial_\mu A^\mu = \partial_0 A^0 = 0$, we get:

$$\begin{cases} \mathcal{D}^0_\alpha A^\alpha(x) = J^0(x) & \rightarrow \quad \partial_\mu \partial^\mu A^0 - \partial^0(\partial_0 A^0) = (\partial_0 \partial^0 - \nabla^2)A^0 = -\nabla^2 A^0 = J^0 \\ \\ \mathcal{D}^k_\alpha A^\alpha(x) = J^k(x) & \rightarrow \quad \partial_\mu \partial^\mu \vec{A} + \nabla(\partial_0 A^0) = \vec{J} \end{cases} \tag{10.47}$$

Noting that the charge density is $J^0 = \rho$ and that the electric scalar potential is $A^0 = \phi$, we get $\nabla^2\phi = -\rho$ which is Poisson's equation for the classical electrostatic field. The term A^0 is then just given by the Coulomb potential created by the electric charge distribution ρ! Now we understand why this is called the Coulomb gauge. This implies (see Eq. (10.26)):

$$A^0(\vec{x}) = \phi(\vec{x}) = \int \mathrm{d}^3\vec{y}\,\frac{\rho(\vec{y})}{4\pi|\vec{x}-\vec{y}|} = \int \mathrm{d}^3\vec{y}\,\frac{J_0(\vec{y})}{4\pi|\vec{x}-\vec{y}|} = \int \mathrm{d}^4 y\,\frac{\delta(x^0-y^0)}{4\pi|\vec{x}-\vec{y}|}J_0(y) \tag{10.48}$$

Comparing with Eq. (10.46) yields:

$$G^{00}_F(x-y) = \frac{i\delta(x^0-y^0)}{4\pi|\vec{x}-\vec{y}|} \quad \text{and} \quad G^{0i}_F(x-y) = 0 \quad (i=1,2,3) \tag{10.49}$$

Having found the time component G^{00}_F, we now need to find a way to constrain the space components G^{ij}_F. We do this by noting that the continuity equation $\partial_\mu J^\mu = 0$ provides a link between the time and space components. Formally, $-\nabla^2 A^0 = J^0$ means that:

$$\partial_0 A^0 = -(\partial_0/\nabla^2)J^0 = +(1/\nabla^2)\nabla \cdot \vec{J} = (1/\nabla^2)\partial^j J^j \tag{10.50}$$

Therefore, we have for each component of the three-vector $\vec{A}$:

$$\partial_\mu \partial^\mu A^i + \partial^i \frac{1}{\nabla^2}\partial^j J^j = J^i \tag{10.51}$$

We have already noted that the representation in momentum space is usually more practical for computations, and this was actually one of the important outcomes of Feynman's covariant theory. In this context we want to rewrite the results for A^0 and $\vec{A}$ for a given frequency mode $k^\mu = (k^0, \vec{k})$, and hence:

$$A^\mu(x) = \int \frac{\mathrm{d}^4 k}{(2\pi)^4} e^{-ik\cdot x}\tilde{A}^\mu(k) \quad \text{and} \quad J^\mu(x) = \int \frac{\mathrm{d}^4 k}{(2\pi)^4} e^{-ik\cdot x}\tilde{J}^\mu(k) \tag{10.52}$$

Considering then the Fourier transform of Eq. (10.51), we get:

$$-((k^0)^2 - \vec{k}^2)\tilde{A}^i(\vec{k}) = \tilde{J}^i(\vec{k}) - \frac{k^i k^j}{\vec{k}^2}\tilde{J}^j(\vec{k}) = \left(\delta^{ij} - \frac{k^i k^j}{\vec{k}^2}\right)\tilde{J}^j(\vec{k}) \tag{10.53}$$

or with $k^2 = (k^0)^2 - \vec{k}^2$:

$$\tilde{A}^i(\vec{k}) = -\frac{1}{((k^0)^2 - \vec{k}^2)}\left(\delta^{ij} - \frac{k^i k^j}{\vec{k}^2}\right)\tilde{J}^j(\vec{k}) = -\frac{1}{k^2}\left(\delta^{ij} - \frac{k^i k^j}{\vec{k}^2}\right)\tilde{J}^j(\vec{k}) \tag{10.54}$$

or in terms of the Fourier transform of the Green's function:

$$\tilde{G}_F^{i0}(k) = 0 \quad \text{and} \quad \tilde{G}_F^{ij}(k) = \frac{i}{k^2}\left(\delta^{ij} - \frac{k^i k^j}{\vec{k}^2}\right) \tag{10.55}$$

From Eq. (10.49) we can also find:

$$\tilde{G}_F^{00}(k) = \frac{i}{(\vec{k})^2} = \frac{i}{k^2}\frac{k^2}{(\vec{k})^2} = \frac{i}{k^2}\left(\frac{(k^0)^2 - (\vec{k})^2}{(\vec{k})^2}\right) = \frac{i}{k^2}\left(\frac{(k^0)^2}{(\vec{k})^2} - 1\right) \quad \text{and} \quad \tilde{G}_F^{0i}(k) = 0 \tag{10.56}$$

So, altogether we obtain the **Feynman propagator for the photon in momentum space in the Coulomb gauge**:

$$\tilde{G}_F^{\mu\nu}(k) \equiv \frac{i}{k^2}C^{\mu\nu}(k) \quad \text{where} \quad C^{00} = -1 + \frac{(k^0)^2}{(\vec{k})^2}, \; C^{0i} = C^{i0} = 0, \; C^{ij} = \left(\delta^{ij} - \frac{k^i k^j}{\vec{k}^2}\right) \tag{10.57}$$

A convenient way to summarize this result is to write a single equation for $C^{\mu\nu}$:

$$C^{\mu\nu}(k) \equiv -g^{\mu\nu} + k^\mu c^\nu(k) + k^\nu c^\mu(k) \tag{10.58}$$

where, *for the Coulomb condition*, we have:

$$c^\mu(k) = \frac{(k^0, -\vec{k})}{2(\vec{k})^2} \tag{10.59}$$

We can express the photon propagator in the space-time coordinates integrating over $d^4 k = dk^0 d^3\vec{k}$ and taking care of the regularization of the poles at $k^0 = \pm|\vec{k}|$, as in the case of the other propagators. We use the usual "$i\epsilon$" prescription and get the **Feynman photon propagator** $G_F^{\mu\nu}(x - y)$:

$$G_F^{\mu\nu}(x - y) = i\int \frac{d^4 k}{(2\pi)^4} e^{-ik\cdot(x-y)}\frac{(-g^{\mu\nu} + k^\mu c^\nu(k) + k^\nu c^\mu(k))}{k^2 + i\epsilon} \quad \text{(Coulomb gauge)} \tag{10.60}$$

where, in the Coulomb gauge, $c^\mu(k)$ is given by Eq. (10.59). As expected, this form is not Lorentz-covariant because of the definition of c^μ. However, it is instructive to note some features of this result. The photon propagator's dependence $1/k^2$ is identical to that of the Feynman propagator of a real scalar field (see Eq. (9.62)) for a boson of zero rest mass ($1/(p^2 - m^2) \to 1/k^2$). **The difference between the photon and the scalar boson propagators is contained in the structure tensor** $(-g^{\mu\nu} + k^\mu c^\nu(k) + k^\nu c^\mu(k))$ **in the numerator of Eq. (10.60).**

10.7   The Photon Propagator under Any Gauge-Fixing Condition

Although we will not prove it here, it can be shown that the Feynman photon propagator under any gauge condition can be written in the form of Eq. (10.60), with the $c^\mu(k)$ **depending on the chosen gauge**. In practice, one chooses a gauge condition which can a priori simplify the calculations at hand. As already mentioned, any physical observable will be gauge-independent, so it should not depend on the choice of $c^\mu(k)$.

In the case of QED, instead of fixing the gauge by constraining the gauge field via an auxiliary equation such as the Lorenz or the Coulomb gauge discussed previously, which leads to the non-Lorentz-covariant definition in Eq. (10.59), one can directly impose a constraint on the photon propagator form, for example by fixing $c^\nu(k)$. In a useful class of gauges, the Lorentz-invariance condition can be imposed by choosing an expression similar to Eq. (10.59), but manifestly covariant, by taking the c^μ parallel to k^μ:

$$c^\mu(k) = \xi \frac{k^\mu}{2k^2} \quad \Longrightarrow \quad C^{\mu\nu}(k) \equiv -g^{\mu\nu} + \xi \frac{k^\mu k^\nu}{k^2} \tag{10.61}$$

where ξ is a real number which, as we will show below, is "gauge fixing." Under this assumption, the **Feynman photon propagator** would become:

$$G_F^{\mu\nu}(x - y) = i \int \frac{d^4 k}{(2\pi)^4} \frac{e^{-ik \cdot (x-y)}}{k^2 + i\epsilon} \left(-g^{\mu\nu} + \xi \frac{k^\mu k^\nu}{k^2 + i\epsilon} \right) \tag{10.62}$$

A way of deriving this result is to return to the original Lagrangian of electromagnetism and modify the part related to Maxwell's equation in order to fix the gauge by adding a term. Starting from Eq. (10.19) and setting $J^\nu = 0$, we write:

$$\mathcal{L}_{Maxwell} = \mathcal{L}_{Proca} + \mathcal{L}_{GF} = -\frac{1}{4} F_{\mu\nu} F^{\mu\nu} - \frac{1}{2(1-\xi)} (\partial_\mu A^\mu)^2 \tag{10.63}$$

where $\mathcal{L}_{GF}$ is the "gauge-fixing" term. We immediately see that the extra term is exactly what we need since:

$$\frac{\partial \mathcal{L}_{GF}}{\partial A^\nu} = 0 \quad \text{and} \quad \frac{\partial \mathcal{L}_{GF}}{\partial(\partial^\mu A^\nu)} = -\frac{1}{2(1-\xi)} \frac{\partial}{\partial(\partial^\mu A^\nu)} (\partial_\alpha A_\beta)(\partial^\alpha A^\beta) = -\frac{1}{(1-\xi)} \partial_\mu A_\nu \tag{10.64}$$

which yields the following electrodynamics (in vacuum without sources):

$$\partial_\mu F^{\mu\nu} = \partial_\mu \partial^\mu A^\nu - \partial_\mu \partial^\nu A^\mu + \frac{1}{(1-\xi)} \partial_\mu \partial^\nu A^\mu = 0 \tag{10.65}$$

For $\xi \to 1$ the last term diverges, and the theory is only consistent if we impose $\partial_\mu A^\mu = 0$, a.k.a. the Lorenz/Landau gauge! The other interesting case, called the **Feynman gauge**, is $\xi \to 0$. In this case the last term cancels exactly the second term and the theory is just given by $\partial_\mu F^{\mu\nu} = \partial_\mu \partial^\mu A^\nu = 0$. Summarizing:

$$\begin{cases} \xi \to 0 : \text{Feynman gauge} \\ \xi \to 1 : \text{Lorenz/Landau gauge} \end{cases} \tag{10.66}$$

The form of the photon propagator depends on the chosen gauge! We can show that Eq. (10.62) is the right expression by showing that it is the Green's function of our wave equation. Starting from Eq. (10.65), we can simply express it as four equations $\mathcal{D}^\nu_\alpha A^\alpha = 0$ with $\nu = 0, 1, 2, 3$ and the operator (cf. Eq. (10.43)):

$$\mathcal{D}^\mu_\alpha(\xi) \equiv \left[\partial_\lambda \partial^\lambda \delta^\mu_\alpha - \partial^\mu \partial_\alpha + \frac{1}{1-\xi} \partial^\mu \partial_\alpha \right] \tag{10.67}$$

The Green's function should satisfy as before Eq. (10.44). Hence we need to show that:

$$i \mathcal{D}^\mu_\alpha \int \frac{d^4 k}{(2\pi)^4} \frac{e^{-ik \cdot (x-y)}}{k^2} \left(-g^{\alpha\nu} + \xi \frac{k^\alpha k^\nu}{k^2} \right) \overset{?}{=} i g^{\mu\nu} \int \frac{d^4 k}{(2\pi)^4} e^{-ik \cdot (x-y)} \tag{10.68}$$

where we used Eq. (D.45) for the Dirac δ function. This is equivalent to:

$$i \left[\partial_\lambda \partial^\lambda \delta^\mu_\alpha - \partial^\mu \partial_\alpha + \frac{1}{1-\xi} \partial^\mu \partial_\alpha \right] \frac{e^{-ik \cdot (x-y)}}{k^2} \left(-g^{\alpha\nu} + \xi \frac{k^\alpha k^\nu}{k^2} \right) \overset{?}{=} i g^{\mu\nu} e^{-ik \cdot (x-y)} \tag{10.69}$$

or

$$\frac{i}{k^2}\left[(-ik_\lambda)(-ik^\lambda)\delta^\mu_\alpha - (-ik^\mu)(-ik_\alpha) + \frac{1}{1-\xi}(-ik^\mu)(-ik_\alpha)\right]\left(-g^{\alpha\nu} + \xi\frac{k^\alpha k^\nu}{k^2}\right) \overset{?}{=} ig^{\mu\nu} \qquad (10.70)$$

so

$$\frac{1}{k^2}\left[-k^2\delta^\mu_\alpha + k^\mu k_\alpha - \frac{1}{1-\xi}k^\mu k_\alpha\right]\left(-g^{\alpha\nu} + \xi\frac{k^\alpha k^\nu}{k^2}\right) \overset{?}{=} g^{\mu\nu} \qquad (10.71)$$

And finally

$$\frac{1}{k^2}\left[\underbrace{k^2 g^{\mu\nu} - k^\mu k^\nu + \frac{1}{1-\xi}k^\mu k^\nu - k^2\xi\frac{k^\mu k^\nu}{k^2} + k^\mu\xi\frac{k^2 k^\nu}{k^2} - \frac{\xi}{1-\xi}k^\mu k^\nu}_{=0}\right] \overset{?}{=} g^{\mu\nu} \qquad \square \qquad (10.72)$$

Hence, Eq. (10.62) is the correct general form of the photon propagator.

It is worth noting that the Feynman gauge makes the photon Feynman propagator become compact and practical, since with $\xi = 0$, one just has:

$$G_F^{\mu\nu}(x-y) = \int\frac{\mathrm{d}^4 k}{(2\pi)^4}e^{-ik\cdot(x-y)}\left(\frac{-ig^{\mu\nu}}{k^2 + i\epsilon}\right) \qquad (10.73)$$

or

$$\tilde{G}_F^{\mu\nu}(x-y) = \left(\frac{-ig^{\mu\nu}}{k^2 + i\epsilon}\right) \qquad \text{(Feynman gauge)} \qquad (10.74)$$

This form is most often used in tree-level calculations.

Regardless of the chosen gauge, the photon propagator always retains a specific form. This latter is equivalent to that of a boson with a structure contained in the gauge-dependent numerator, and it has two Lorentz indices. In analogy to the general result obtained for the propagator of the Dirac particle (see Eq. (9.174)), we expect the photon propagator to represent the sum over the polarization states of the photon at the numerator and the usual propagation amplitude at the denominator. Indeed, ignoring the pole at the denominator, we have:

$$\tilde{G}_F^{\mu\nu}(k) = \frac{i}{k^2}\left(-g^{\mu\nu} + \xi\frac{k^\mu k^\nu}{k^2}\right) \leftrightarrow \frac{i}{k^2}\left(\sum_\lambda \epsilon^{\mu,*}_{(\lambda)}\epsilon^\nu_{(\lambda)}\right) \qquad (10.75)$$

where in the last term we sum over all polarization states. Hence, we expect a condition of **completeness on the sum of the photon polarization states** that should be of the form (recall the case of the Dirac spinor where Eq. (9.174) has the sum of the term $\bar{u}u$):

$$\sum_\lambda \epsilon^{\mu,*}_{(\lambda)}\epsilon^\nu_{(\lambda)} \leftrightarrow -g^{\mu\nu} + \xi\frac{k^\mu k^\nu}{k^2} \qquad (10.76)$$

We have used the left–right arrows to indicate that the last statements are not really "equality" statements but rather should be understood as a rule, stating that the sum over the photon polarization states can be replaced by the r.h.s. term with $g^{\mu\nu}$, but it is still to be determined in which cases it can be used. This will be discussed in Section 10.14.

Still, complications arise here. First of all, we see that the relation apparently depends on the gauge via ξ, but physical observables should not depend on the gauge. Secondly, which and how many polarization states λ should be included in the sum? The two transverse and the longitudinal one? How do we reconcile this result with the two *transverse* states of polarization of a real photon? Furthermore, the photon propagator describes a "virtual" photon for which the "on-shell" mass condition $k^2 = 0$ need not be verified. How shall we take that into account? We note that for such **virtual photons**, or any **spin-1 massive vector field** B^μ, the field is obtained from Maxwell's equation with the substitution $\partial_\mu \partial^\mu$ by $\partial_\mu \partial^\mu + M^2 = \Box^2 + M^2$ (see Eq. (D.42)), where M is the mass of the spin-1 particle, to give (see Eq. (10.23)):

$$\left(\Box^2 + M^2\right) B^\nu - \partial^\nu \partial_\mu B^\mu = J^\nu \qquad (\nu = 0, \ldots, 3) \tag{10.77}$$

If we attempt to include a "gauge transformation" of the type

$$B^\nu \to B^\nu + \partial^\nu \lambda \tag{10.78}$$

then the following equation is obtained:

$$\left(\Box^2 + M^2\right) B^\nu - \partial^\nu \partial_\mu B^\mu + M^2 \partial^\nu \lambda = J^\nu \tag{10.79}$$

This equation is manifestly different from Eq. (10.77), and is only equal for the case $M = 0$. Hence, for massive spin-1 vector fields there is no "additional" constraint that reduces the number of polarization states of the real photon to two. **The allowed polarization states for a massive spin-1 particle are therefore three: the two transverse with $s = \pm 1$ and the longitudinal one** $s = 0$. Consequently, we are noting that the longitudinal state is unphysical for a real (massless) photon. However, for a virtual photon ($k^2 \neq 0$), the propagator includes three polarization states as expected for a spin-1 massive particle. We will come back to this point again later in Sections 10.13 and 10.14.

10.8 The Propagator of a Massive Spin-1 Particle

The propagator for a massive spin-1 particle can be directly derived in a similar fashion as for the massless photon. The starting point is the definition of the propagator, where this time we use the time-ordered correlation function of a **massive vector field** B^μ **of mass** M computed at two space-time points, sandwiched within the vacuum states (see Eq. (10.42)):

$$G_F^{\mu\nu}(x - y) \equiv \langle 0| \, T \left(B^\mu(x) B^\nu(y) \right) |0\rangle \tag{10.80}$$

We now use the general property that the propagator is the Green's function of the equations of motion. For the free massive vector field, the equations of motion are given by Eq. (10.77), consequently we can write (see Eq. (10.44)):

$$\mathcal{D}_\alpha^\mu G_F^{\alpha\nu}(x - y) = \left[(\Box^2 + M^2)\delta_\alpha^\mu - \partial^\mu \partial_\alpha \right] G_F^{\alpha\nu}(x - y) = ig^{\mu\nu}\delta^4(x - y) \tag{10.81}$$

As before, we must formally invert the operator $\mathcal{D}_\alpha^\mu$ to find $G_F^{\alpha\nu}(x - y)$. Applying the Fourier transform of the propagator yields (using Eq. (D.45) for the Dirac δ function):

$$\left[(\Box^2 + M^2)\delta_\alpha^\mu - \partial^\mu \partial_\alpha \right] \int \frac{\mathrm{d}^4 k}{(2\pi)^4} e^{-ik\cdot(x-y)} \tilde{G}_F^{\alpha\nu}(k) = ig^{\mu\nu} \int \frac{\mathrm{d}^4 k}{(2\pi)^4} e^{-ik\cdot(x-y)} \tag{10.82}$$

This can be written as:

$$\left[(-k^2 + M^2)\delta_\alpha^\mu + k^\mu k_\alpha \right] \tilde{G}_F^{\alpha\nu}(k) = ig^{\mu\nu} \tag{10.83}$$

We can multiply this equation by the four-vector k_μ to obtain:

$$\left(-k_\alpha k^2 + k_\alpha M^2 + k^2 k_\alpha \right) \tilde{G}_F^{\alpha\nu}(k) = M^2 k_\alpha \tilde{G}_F^{\alpha\nu}(k) = ik_\mu g^{\mu\nu} \quad \to \quad k_\alpha \tilde{G}_F^{\alpha\nu}(k) = \frac{ik^\nu}{M^2} \tag{10.84}$$

By reinserting this result in the second term of Eq. (10.83) we get:

$$(-k^2 + M^2)\delta_\alpha^\mu \tilde{G}_F^{\alpha\nu}(k) + k^\mu \frac{ik^\nu}{M^2} = ig^{\mu\nu} \tag{10.85}$$

or

$$(-k^2 + M^2)\tilde{G}_F^{\mu\nu}(k) = ig^{\mu\nu} - i\frac{k^\mu k^\nu}{M^2} = i\left(g^{\mu\nu} - \frac{k^\mu k^\nu}{M^2}\right) \tag{10.86}$$

Finally, the **Feynman propagator in momentum space for a massive spin-1 boson of mass** M is given by:

$$\tilde{G}_F^{\mu\nu}(k) = \frac{i\left(-g^{\mu\nu} + \frac{k^\mu k^\nu}{M^2}\right)}{k^2 - M^2 + i\epsilon} \tag{10.87}$$

We note that in the limit where k^2 is small compared to the mass of the particle, we have $k^2 \ll M^2$ and the propagator simplifies to:

$$\tilde{G}_F^{\mu\nu}(k) \simeq \frac{-ig^{\mu\nu}}{M^2} \qquad (k^2 \ll M^2) \tag{10.88}$$

i.e the interaction appears point-like (no k^2 dependence).

10.9 The QED Lagrangian – Coupling Photons to Fermions

To start, we are looking for the theory of interactions between photons and electrons/positrons. While we mention electrons/positrons, we can extend the theory to include all charged leptons, i.e., (anti)muons, (anti)taus, and also (anti)quarks. Later we will also include coupling to other types of particles, such as charged scalar bosons.

The QED Lagrangian will therefore be composed of three distinct parts: (1) the term describing the free Dirac fields; (2) the term describing the free (quantized) electromagnetic field; and (3) the interaction or coupling between the two types of fields:

$$\mathcal{L}_{QED} = \mathcal{L}_{free\ Dirac} + \mathcal{L}_{free\ Maxwell} + \mathcal{L}_{coupling} \tag{10.89}$$

The Lagrangian for free Dirac particles $\mathcal{L}_{Dirac}$ was given in Eq. (8.284). In Eq. (10.19) the entire Maxwell theory was summarized in the Lagrangian $\mathcal{L}_{Maxwell}$, including also the source term $J_\mu A^\mu$. We can hence write:

$$\mathcal{L}_{QED} = \overline{\Psi}\left[i\gamma^\mu \partial_\mu - m\right]\Psi - \frac{1}{4}F_{\mu\nu}F^{\mu\nu} - J^\mu A_\mu \tag{10.90}$$

We need to define J^μ to correspond to the four-current of a Dirac particle coupling to a photon. The four-current–density of the free Dirac field was derived in Eq. (8.46) to be $j^\mu \equiv \overline{\Psi}\gamma^\mu\Psi$. We have shown in Eq. (8.276) that it is conserved if the spinor Ψ satisfies the Dirac equation. In addition, Eq. (8.283) proves that it transforms as a four-vector under a Lorentz transformation. Hence, the four-vector current $j_V^\mu = \overline{\Psi}\gamma^\mu\Psi$ is an excellent candidate to couple to A_μ.

In order to finalize the coupling, we go back to classical electrodynamics and recall that the motion of an electrically charged particle e in an electromagnetic potential can be obtained via the canonical minimal substitution of the momentum $\vec{p} \to \vec{p} - e\vec{A}$ (see Eq. (4.101)). We generalize the substitution to:

$$p^\mu \to p^\mu - eA^\mu \quad \text{or} \quad i\partial^\mu \to i\partial^\mu - eA^\mu \tag{10.91}$$

Inserting the canonical substitution into the free Dirac Lagrangian yields:

$$\mathcal{L}_{Dirac+canonical} = \overline{\Psi}\left[\gamma^\mu(i\partial_\mu - eA_\mu) - m\right]\Psi = \overline{\Psi}\left(i\gamma^\mu \partial_\mu - m\right)\Psi - e\overline{\Psi}\gamma^\mu\Psi A_\mu \tag{10.92}$$

where we obtain the current $j^\mu = e\overline{\Psi}\gamma^\mu\Psi$ with the desired properties, and where e is the elementary electric charge.

Finally, the **complete QED Lagrangian coupling electromagnetic and Dirac fields** in the absence of external sources ($J^\nu = 0$) can be expressed as:

$$\mathcal{L}_{QED} \equiv \overline{\Psi}\left[\gamma^\mu i\partial_\mu - m\right]\Psi - \frac{1}{4}F_{\mu\nu}F^{\mu\nu} - e\overline{\Psi}\gamma^\mu\Psi A_\mu \tag{10.93}$$

It looks simple and clean!

We note that this Lagrangian predicts that the interaction between a Dirac field and the electromagnetic field is the direct and local product of the charge e and the bilinear covariant of the vector form $\overline{\Psi}\gamma^\mu\Psi$ (see Eq. (8.295)) with the four-vector potential A_μ. In analogy to Eq. (9.16), the charge e is the coupling constant which represents the strength of the interaction. In natural units, $e = \sqrt{4\pi\alpha}$, where $\alpha \approx 1/137$ is the fine-structure constant (see Eq. (1.37)).

We note that the **QED Lagrangian is invariant under a Lorentz transformation**. The first two terms are manifestly covariant and the interaction term is the product of two four-vectors, the bilinear covariant four-vector and the electromagnetic potential A^μ, and is hence also invariant.

● **What about the gauge?** For the free Maxwell theory, we have seen that the existence of a gauge freedom and its fixing was crucial in order to cut down the physical degrees of freedom to the requisite two for the electromagnetic field. Does our interacting theory above still have a gauge freedom? In other words, is the Lagrangian invariant under the following gauge transformation:

$$A_\mu(x) \to A_\mu(x) + \partial_\mu\lambda(x) \tag{10.94}$$

where $\lambda(x)$ is an arbitrary function? The interaction term alone is *not* invariant, since it becomes:

$$-e\overline{\Psi}\gamma^\mu\Psi(A_\mu + \partial_\mu\lambda) = \mathcal{L}_{coupling} - e\overline{\Psi}\gamma^\mu\Psi\partial_\mu\lambda \tag{10.95}$$

The trick to restore gauge invariance is to consider its representation on the spinor field. If we impose that the spinor will change under the gauge transformation as follows:

$$\Psi \to \Psi e^{-ie\lambda(x)} \quad \text{so} \quad \overline{\Psi} \to \overline{\Psi}e^{+ie\lambda(x)} \tag{10.96}$$

then we have (ignoring the mass term):

$$\begin{aligned}
\overline{\Psi}e^{+ie\lambda(x)}\gamma^\mu i\partial_\mu\left(\Psi e^{-ie\lambda(x)}\right) &= \left(\overline{\Psi}e^{+ie\lambda(x)}\right)i\gamma^\mu\left((\partial_\mu\Psi)e^{-ie\lambda(x)} + \Psi e^{-ie\lambda(x)}(-ie\partial_\mu\lambda(x))\right) \\
&= \overline{\Psi}i\gamma^\mu\partial_\mu\Psi + e\overline{\Psi}\gamma^\mu\Psi\partial_\mu\lambda(x)
\end{aligned} \tag{10.97}$$

where the extra term coming from the derivative of $\lambda(x)$ *cancels exactly* the extra term in the coupling! Wow!

● **The covariant derivative D_μ.** An elegant way to rewrite this result is to introduce the **covariant derivative** D_μ defined as:

$$D_\mu \equiv \partial_\mu + ieA_\mu \tag{10.98}$$

and the QED Lagrangian can be simply expressed as:

$$\mathcal{L}_{QED} \equiv \overline{\Psi}\left[\gamma^\mu iD_\mu - m\right]\Psi - \frac{1}{4}F_{\mu\nu}F^{\mu\nu} \tag{10.99}$$

where **the coupling term is automatically taken into account in the covariant derivative**. Under gauge transformation, we have:

$$D_\mu\Psi \quad \to \quad \partial_\mu(\Psi e^{-ie\lambda(x)}) + ie(A_\mu + \partial_\mu\lambda)\Psi e^{-ie\lambda(x)} = e^{-ie\lambda(x)}D_\mu\Psi \tag{10.100}$$

hence the QED Lagrangian is invariant under gauge transformation:

$$\mathcal{L}_{QED} \quad \to \quad \overline{\Psi}e^{+ie\lambda(x)}e^{-ie\lambda(x)}\gamma^\mu iD_\mu\Psi - m\overline{\Psi}\Psi - \frac{1}{4}F_{\mu\nu}F^{\mu\nu} = \mathcal{L}_{QED} \quad \square \tag{10.101}$$

This completely fundamental observation will be discussed further in Chapter 24.

10.10   *S*-Matrix Amplitudes for QED

We now want to work out the rules for the Feynman diagrams of the QED theory. Once we have found the rules, it will be easy to graphically consider all possible contributing diagrams to a given QED process to a given order in the expansion and compute the corresponding amplitudes.

We start by considering the Dyson *S*-matrix series Eq. (9.9), noting that the Hamiltonian density of the QED coupling is simply $\mathcal{H}_{int} = (e\bar{\Psi}\gamma^\mu\Psi)A_\mu$. The *S*-matrix is therefore a series of terms, where the nth term is proportional to e^n:

$$S \equiv 1 + \frac{-ie}{1!}(\cdots) + \frac{(-ie)^2}{2!}(\cdots) \tag{10.102}$$

where $e \approx 0.303$ in natural units (see Eq. (1.37)). Given its value, we expect the series of the expansion of the *S*-matrix to be reasonably rapidly converging. Explicitly, the first term S_1 is equal to:

$$S_1 = \frac{(-ie)}{1!} \int d^4x_1 \bar{\Psi}(x_1)\gamma^\mu\Psi(x_1)A_\mu(x_1) \tag{10.103}$$

and the second term S_2 is the time-ordered product:

$$S_2 = \frac{(-ie)^2}{2!} \int\int d^4x_1 d^4x_2 T\left[\bar{\Psi}(x_1)\gamma^\mu\Psi(x_1)A_\mu(x_1)\bar{\Psi}(x_2)\gamma^\nu\Psi(x_2)A_\nu(x_2)\right] \tag{10.104}$$

We will rely on **Wick's theorem** (see Section 9.10) to handle the time-ordered product of the fields. Also, as discussed in detail in Chapters 7 and 8, the Fock space of the particles is generated with the creation operators, starting from the vacuum. For QED with electrons/positrons and photons, we have the following cases:

- **Electron** (spin-1/2 Dirac fermion) of four-momentum p^μ and spin state s:
  Initial state: $|p^\mu s\rangle = \sqrt{(2\pi)^3 2E_p} a_s^\dagger(\vec{p})|0\rangle$
  Final state: $\langle p^\mu s| = \langle 0|\sqrt{(2\pi)^3 2E_p} a_s(\vec{p})$
  where $E_p = +\sqrt{\vec{p}^2 + m^2}$.

- **Positron** (spin-1/2 Dirac antifermion) of four-momentum p^μ and spin state s:
  Initial state: $|p^\mu s\rangle = \sqrt{(2\pi)^3 2E_p} b_s^\dagger(\vec{p})|0\rangle$
  Final state: $\langle p^\mu s| = \langle 0|\sqrt{(2\pi)^3 2E_p} b_s(\vec{p})$
  where $E_p = +\sqrt{\vec{p}^2 + m^2}$.

- **Photon** (spin-1 boson) of four-momentum k^μ and polarization ϵ^μ (recall that the polarization vector is constrained to two degrees of freedom by the choice of gauge):
  Initial state: $|k^\mu \epsilon^\mu\rangle = \sqrt{(2\pi)^3 2\omega_k} a_\epsilon^\dagger(\vec{k})|0\rangle$
  Final state: $\langle k^\mu \epsilon^\mu| = \langle 0|\sqrt{(2\pi)^3 2\omega_k} a_\epsilon(\vec{k})$
  where $\omega_k = +\sqrt{\vec{k}^2}$.

For a given QED process with m incoming initial particles $(A + B + C + \cdots)$ and n outgoing final state particles $(1 + 2 + 3 + \cdots)$, the *S*-matrix is computed as follows:

$$\langle p_1^\mu s_1, p_2^\mu s_2, p_3^\mu s_3, \ldots | S | p_A^\mu s_A, p_B^\mu s_B, p_C^\mu s_C, \ldots \rangle \tag{10.105}$$

where each particle is defined by its four-momentum and (if any) its internal degrees of freedom (such as the spin state). We will use these results to derive the Feynman rules for QED. As already noted in Chapter 9, the $n!$ terms in front of the S_n term of order n of the Taylor series will always cancel out, since there will always be $n!$ ways of interchanging the x_i $(i = 1, \ldots, n)$ vertices and obtaining the same result.

10.11   Contractions of External Legs

We consider an incoming (initial-state) electron. Its state is defined as $|e^-\rangle = |p_1^\mu s_1\rangle \propto a_{s_1}^\dagger(\vec{p}_1)|0\rangle$. The S_1 matrix is given in Eq. (10.103). In order to compute it, we will be led to consider the action of the field $\Psi(x)$ on the electron state. If we consider only the term with the $a^\dagger$ creation operator:

$$\Psi(x)|e^-\rangle = \Psi(x)|p_1^\mu s_1\rangle \propto \Psi(x)a_{s_1}^\dagger(\vec{p}_1)|0\rangle \tag{10.106}$$

Including the conventional normalization factors, we obtain from Eq. (8.287) for $|e^-\rangle = |p_1^\mu s_1\rangle$:

$$
\begin{aligned}
\Psi(x)|e^-\rangle &= \int \frac{d^3\vec{p}}{(2\pi)^{3/2}} \frac{1}{\sqrt{2E_p}} \sum_{s=1,2} \left(a_s(\vec{p})u^{(s)}(\vec{p})e^{-ip\cdot x} + b_s^\dagger(\vec{p})v^{(s)}(\vec{p})e^{+ip\cdot x} \right) \sqrt{(2\pi)^3 2E_1}\, a_{s_1}^\dagger(\vec{p}_1)|0\rangle \\
&= \int d^3\vec{p}\, \sqrt{\frac{E_1}{E_p}} \sum_{s=1,2} \left(u^{(s)}(\vec{p})e^{-ip\cdot x} \right) a_s(\vec{p}) a_{s_1}^\dagger(\vec{p}_1)|0\rangle + \cdots \\
&= \int d^3\vec{p}\, \sqrt{\frac{E_1}{E_p}} \sum_{s=1,2} \left(u^{(s)}(\vec{p})e^{-ip\cdot x} \right) \delta^3(\vec{p}-\vec{p}_1)\delta_{s,s_1}|0\rangle + \cdots \\
&\rightsquigarrow e^{-ip_1\cdot x} u^{(s_1)}(\vec{p}_1)|0\rangle
\end{aligned}
\tag{10.107}
$$

where the $\cdots$ term is not relevant here since it contains a $b^\dagger$ operator and we are only interested in an electron state. In a similar way, we can convince ourselves that for $\langle e^-| = \langle p_2^\mu s_2|$, we would get:

$$\langle p_2^\mu, s_2|\overline{\Psi} \rightsquigarrow \langle 0|\, e^{+ip_2\cdot x} \bar{u}^{(s_2)}(\vec{p}_2) \tag{10.108}$$

For external positrons $\langle e^+|$ and $|e^+\rangle$, similar expressions containing the spinor $v^{(s)}(\vec{p})$ can be derived. And finally for an external initial- or final-state photon $|\gamma\rangle = |k,\epsilon\rangle$, we find:

$$A_\mu|k,\epsilon\rangle \rightsquigarrow \epsilon_\mu(k)e^{-ik\cdot x}|0\rangle \qquad \text{and} \qquad \langle k,\epsilon|A_\mu \rightsquigarrow \langle 0|\, \epsilon_\mu^*(k)e^{+ik\cdot x} \tag{10.109}$$

In Section 9.10 we have seen the power of Wick's theorem and normal ordering, and the concept of field contraction was introduced to conveniently keep track of all possible diagrams (i.e., contractions). In Section 9.17, we have extended the concept of contraction to the case of initial and final states and called them **contractions of external legs**. We do the same here (note that the states below labeled by the four-momentum and spin correspond to different particles):

- Incoming electron: $\overline{\Psi|p^\mu, s\rangle} \rightarrow u^{(s)}(\vec{p})$ Outgoing electron: $\overline{\langle p^\mu, s|\overline{\Psi}} \rightarrow \bar{u}^{(s)}(\vec{p})$

- Incoming positron: $\overline{\overline{\Psi}|p^\mu, s\rangle} \rightarrow \bar{v}^{(s)}(\vec{p})$ Outgoing positron: $\overline{\langle p^\mu s|\Psi} \rightarrow v^{(s)}(\vec{p})$

- Incoming photon: $\overline{A_\mu|k,\epsilon\rangle} \rightarrow \epsilon_\mu(k)$ Outgoing photon: $\overline{\langle k^\mu, \epsilon|A_\mu} \rightarrow \epsilon_\mu^*(k)$

The Dirac field external lines carry plane-wave spinors $u(p,s)$, $v(p,s)$, $\bar{u}(p,s)$, or $\bar{v}(p,s)$, depending on the charge of the fermion and whether it is incoming or outgoing. For the photon, we just have the polarization four-vector. In these definitions, we removed the plane waves represented by the exponential terms $\exp(\pm ip\cdot x)$, because they will lead to trivial energy–momentum-conservation relations when we integrate over d^4x. The use of external legs will be further illustrated in the next sections.

10.12   Vertices – the Vertex Factor

Let us, for instance, consider a hypothetical process where an electron absorbs a photon and thereby changes its four-momentum. The initial state is given by $|e^-\gamma\rangle = |p_1^\mu s_1, k^\mu \epsilon^\mu\rangle$ and the final state by $|e^-\rangle = |p_2^\mu s_2\rangle$. We want to compute the first-order S-matrix term:

$$\langle e^- | S_1 | e^- \gamma \rangle = (-ie) \int d^4 x \, \langle p_2^\mu s_2 | \overline{\Psi}(x) \gamma^\mu \Psi(x) A_\mu(x) | p_1^\mu s_1, k^\mu \epsilon^\mu \rangle \tag{10.110}$$

We first contract the external legs to get:

$$\langle p_2^\mu s_2 | \overline{\Psi}(x) \gamma^\mu \Psi(x) A_\mu(x) | p_1^\mu s_1, k^\mu \epsilon^\mu \rangle \tag{10.111}$$

Since there are *no more possible contractions left* in the above expression, we are "done" and we can **immediately** write the amplitude for S_1 by replacing the contracted fields by their coefficients and reintroducing the corresponding exponential terms $\exp(\pm ip \cdot x)$ with the proper signs, to get:

$$\langle e^- | S_1 | e^- \gamma \rangle = (-ie) \int d^4 x \, e^{-i(k + p_1 - p_2) \cdot x} \langle 0 | \bar{u}^{(s_2)}(\vec{p}_2) \gamma^\mu u^{(s_1)}(\vec{p}_1) \epsilon_\mu(\vec{k}) | 0 \rangle \tag{10.112}$$

This method is simple and effective! The integration over $d^4 x$ yields the following result:

$$\langle e^- | S_1 | e^- \gamma \rangle = (2\pi)^4 \delta^4 (k + p_1 - p_2) \left(\bar{u}^{(s_2)}(\vec{p}_2)(-ie\gamma^\mu) u^{(s_1)}(\vec{p}_1) \epsilon_\mu(\vec{k}) \right) \tag{10.113}$$

As previously anticipated, the integration over $d^4 x$ provides **the Dirac δ function which ensures the conservation of energy–momentum at the vertex**, in other words $p_2^\mu = p_1^\mu + k^\mu$. Factoring out the terms, the *amplitude* of the process is given by:

$$i\mathcal{M}^{e^-}(p_1, s_1, k, \epsilon, p_2, s_2) = \bar{u}^{(s_2)}(\vec{p}_2)(-ie\gamma^\mu) u^{(s_1)}(\vec{p}_1) \epsilon_\mu(\vec{k}) \tag{10.114}$$

The μ index is summed over the γ matrices and the four components of the polarization. Although the vertex has a priori four independent components $-ie\gamma^\mu$ ($\mu = 0, 1, 2, 3$), only two are independent once they get multiplied by the polarization vector whose degrees of freedom are fixed by the gauge. The Feynman diagram corresponding to this amplitude is shown in Figure 10.1(left).

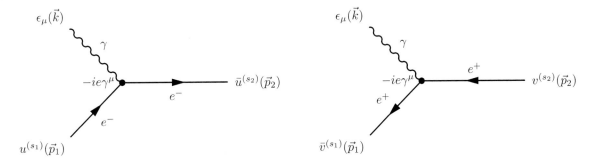

Figure 10.1 Feynman diagram of the hypothetical photon absorption process: (left) by an electron; (right) by a positron. The direction of the arrows in Feynman diagrams is conventional (as explained in the text).

We can write directly the amplitude for the same process involving a positron instead of an electron by looking at the corresponding diagram, by considering an incoming positron with spinor $\bar{v}$, a vertex with factor $-ie\gamma^\mu$, and an outgoing positron with spinor v. See Figure 10.1(right).

$$iM^{e^+}(p_1, s_1, k, \epsilon, p_2, s_2) = \bar{v}^{(s_1)}(\vec{p}_1)(-ie\gamma^\mu)v^{(s_2)}(\vec{p}_2)\epsilon_\mu(\vec{k}) \qquad (10.115)$$

As already noted in Chapter 9, **the direction of the arrows in Feynman diagrams is conventional!** For the incoming and outgoing electrons, the arrow on the external line has the same direction as the particle – incoming for an incoming e^- and outgoing for an outgoing e^-, but for the positrons, the arrow points in the opposite direction of the particle and its momentum. An incoming e^+ has an outgoing line (but in-flowing momentum) while an outgoing e^+ has an incoming line (but out-flowing momentum). In general, **the arrows in the fermion lines follow the flow of the electric charge (in units of $-e$), hence opposite directions for the electrons and the positrons.**

In conclusion, we have found that **the vertex factor** for QED, which couples the photon to the electron or to the positron, is proportional to the elementary electric charge (coupling strength) and has the vector form determined by γ^μ and is equal to:

$$\text{QED vertex factor:} \quad -ie\gamma^\mu \qquad (10.116)$$

10.13   Rules for Propagators

Let us consider the second-order term in the S-matrix expansion (Eq. (10.104)). In order to see how to handle these cases, we begin by focusing on a typical $2 \to 2$ scattering process with two fermions in the initial state and two in the final state, hence:

$$\begin{aligned}
\langle e^+ e^- | S_2 | e^+ e^- \rangle &= \frac{(-ie)^2}{2!} \int\int d^4x_1 d^4x_2 \times \\
&\quad \langle p_2, s_2; k_2, t_2 | T\left(\overline{\Psi}(x_1)\gamma^\mu\Psi(x_1)A_\mu(x_1)\overline{\Psi}(x_2)\gamma^\nu\Psi(x_2)A_\nu(x_2)\right) | p_1, s_1; k_1, t_1 \rangle
\end{aligned} \qquad (10.117)$$

We use **Wick's theorem** (see Eq. (9.113)) to replace the time-ordered product of the fields by a sum of normal-ordered products, where all possible contractions must be added. To see how this works, we consider one of the many possible contractions:

$$\langle p_2, s_2; k_2, t_2 | \overline{\Psi}\gamma^\mu\Psi A_\mu \overline{\Psi}\gamma^\nu\Psi A_\nu | p_1, s_1; k_1, t_1 \rangle \qquad (10.118)$$

We realize that something is not quite right in the above expression since it is not fully contracted, unless we also contract the two remaining potentials A_μ and A_ν. However, since we have not considered any external photon legs, we must consider the contraction between them too. We recall that this is the actual definition of the photon propagator (see Eq. (10.42))! Hence:

$$\overline{A^\mu(x)A^\nu}(y) = \langle 0| T\left(A^\mu(x)A^\nu(y)\right) |0\rangle = G_F^{\mu\nu}(x-y) \qquad (10.119)$$

So finally the S-matrix element becomes:

$$\frac{(-ie)^2}{2!}\int\int d^4x_1 d^4x_2 \langle p_2, s_2; k_2, t_2 | (\overline{\Psi}\gamma^\mu\Psi)(x_1)A_\mu(x_1)(\overline{\Psi}\gamma^\nu\Psi)(x_2)A_\nu(x_2) | p_1, s_1; k_1, t_1 \rangle \qquad (10.120)$$

This contraction corresponds to the Feynman diagram shown in Figure 10.2(left). We note that it leads to the exchange of a photon between the fermions! We can read off the factors directly from the Feynman diagram

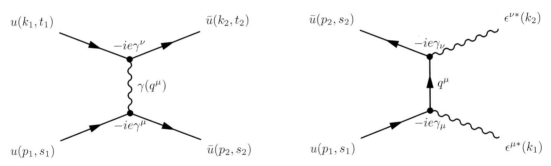

Figure 10.2 (left) One example of a Feynman diagram contributing to the $2 \to 2$ scattering process, involving the photon propagator. (right) Example of a diagram contributing to the annihilation of two fermions into two photons.

and construct the second term of the scattering matrix. Remember that the energy–momentum conservation is to be applied at each of the two vertices (we note that the combinatorial 2! factor cancels out for the same reasons given in Section 9.8):

$$(\bar{u}(p_2)(-ie\gamma_\mu)u(p_1))\, \tilde{G}_F^{\mu\nu}(q^2)\,(\bar{u}(k_2)(-ie\gamma_\nu)u(k_1))\,\delta^4(p_1 - p_2 - q)\delta^4(k_1 + q - k_2) \tag{10.121}$$

where we can choose the **covariant photon propagator in the Feynman gauge** ($\xi = 0$), Eq. (10.74):

$$\tilde{G}_F^{\mu\nu}(q^2) = \frac{-ig^{\mu\nu}}{q^2 + i\epsilon} \tag{10.122}$$

We notice that the indices μ and ν of the γ matrices get exactly fully contracted with the $\mu\nu$ indices of the photon propagator. The four-momentum of the photon q^μ is constrained to be $q^\mu = p_1^\mu - p_2^\mu = k_2^\mu - k_1^\mu$. Finally, the amplitude for the considered $2 \to 2$ scattering diagram can be expressed with a single μ index (noting the positions of the μ indices in the first and last brackets), as:

$$i\mathcal{M}(p_1, s_1; k_1, t_1; p_2, s_2; k_2, t_2) = ie^2\,(\bar{u}(p_2)\gamma^\mu u(p_1))\left(\frac{1}{q^2 + i\epsilon}\right)(\bar{u}(k_2)\gamma_\mu u(k_1)) \tag{10.123}$$

Something "magical" happened with the photon propagator. We note that all four components of the γ^μ and γ^ν matrices at each vertex contribute to the amplitude. Basically, at each of the two vertices the photon couples to the fermion current via the four components, or in other words, apparently to four polarization states!

How do we reconcile this with the fact that the real photon polarization has to be transverse and with only two degrees of freedom? The gauge fixing will have the effect of reducing the number of degrees of freedom in the photon propagator from four to three. As already discussed in Section 10.7, the kinematics of the photon q^μ is constrained by energy–momentum conservation at the vertices, however this does *not* imply that the photon will necessarily be "on-shell." In fact, the propagator represents an **"off-shell" virtual photon.** Therefore, **the propagating photon possesses a "virtual mass" and hence three polarization degrees of freedom – two transverse and one longitudinal – as expected from a massive spin-1 particle.** It's marvelous!

Let us see what happens if we replace the final state with two outgoing photons and the inital state with an electron (fermion–antifermion) pair. In this case, we have (see Figure 10.2(right)):

$$\langle\gamma\gamma|S_2|e^+e^-\rangle = \frac{(-ie)^2}{2!}\int\int \mathrm{d}^4x_1\mathrm{d}^4x_2 \times$$
$$\langle k_1, \epsilon_1; k_2, \epsilon_2|\,T\left(\overline{\Psi}(x_1)\gamma^\mu\Psi(x_1)A_\mu(x_1)\overline{\Psi}(x_2)\gamma^\nu\Psi(x_2)A_\nu(x_2)\right)|p_1, s_1; p_2, s_2\rangle \tag{10.124}$$

We contract the fields to obtain:

$$\langle k_1, \epsilon_1; k_2, \epsilon_2 | \overline{\Psi}(x_1)\gamma^\mu \Psi(x_1) A_\mu(x_1) \overline{\Psi}(x_2)\gamma^\nu \Psi(x_2) A_\nu(x_2) | p_1, s_1; p_2, s_2 \rangle \tag{10.125}$$

where we see that it is not fully contracted until we consider also the contraction of the leftover Dirac field giving the propagator of a Dirac particle:

$$\overline{\Psi(x_1)\overline{\Psi}(x_2)} \longrightarrow \frac{i(\slashed{q} + m)}{q^2 - m^2(+i\epsilon)} \tag{10.126}$$

where $q^\mu = p_1^\mu + p_2^\mu = k_1^\mu + k_2^\mu$. Finally, the amplitude can be written down directly off the series of contractions. It is equal to:

$$i\mathcal{M}(p_1, s_1; p_2, s_2; k_1, \epsilon_1; k_2, \epsilon_2) = -ie^2 \bar{u}(p_2)\gamma^\mu \frac{(\slashed{q} + m)}{q^2 - m^2(+i\epsilon)} \gamma^\nu u(p_1) \epsilon_\mu^*(k_1) \epsilon_\nu^*(k_2) \tag{10.127}$$

We recall that in the case of real photons, the gauge fixing reduces the number of independent degrees of freedom of the polarization to *two*. For example, in the case of the Coulomb gauge, we would have: $\vec{\epsilon}(k_1) \cdot \vec{k}_1 = \vec{\epsilon}(k_2) \cdot \vec{k}_2 = 0$ and $\epsilon^0(k_1) = \epsilon^0(k_2) = 0$. **So, in the case of an external photon leg representing a real physical photon, the number of polarization states λ to consider will be just two.**

To summarize, in this section we have seen how propagators for photons and Dirac fermions appear in the amplitude once all the external legs have been contracted. We also understood the questions of the spin degrees of freedom. For Dirac particles, the situation is simply related to the two independent spin orientations. Real (massless) photons in external legs will have two independent polarizations, while virtual (massive) photons will behave as spin-1 vector particles with three independent polarization states.

10.14   Real, Virtual Photons and the Ward Identities

We consider a QED scattering process in which a (real) photon is emitted:

$$X \to Y + \gamma(k^\mu) \tag{10.128}$$

where X and Y are some combinations of electrons and positrons, never mind the details. Any Feynman diagram contributing to such a process has an outgoing photon line, where the corresponding amplitude can always be generically expressed as:

$$\mathcal{M}(X \to Y + \gamma) = \epsilon_\mu^*(k)\mathcal{M}^\mu \tag{10.129}$$

where $\mathcal{M}^\mu$ comprises the rest of the diagram – the vertices, the internal lines, and the external lines for the electrons and positrons. The net $\mathcal{M}^\mu$ factor depends on the momenta of all the particles and on the spin states of the electrons and positrons, but it does not depend on the outgoing photon's polarization – that dependence is carried by the $\epsilon_\mu^*(k)$ factor.

Likewise, a process in which a photon is absorbed:

$$X + \gamma(k^\mu) \to Y \tag{10.130}$$

has an amplitude of the form:

$$\mathcal{M}(X + \gamma(k^\mu) \to Y) = \epsilon_\mu(k)\mathcal{M}^\mu \tag{10.131}$$

where $\mathcal{M}^\mu$ depends on all the momenta and also electron/positron spins, but not on the incoming photon's polarization.

We now recall what we had derived in Eq. (10.32). Any change of gauge that affects polarization in the manner of $\epsilon^\mu(k) \to \epsilon^\mu(k) + ak^\mu$ leaves the physical photon unchanged. Therefore, the above amplitudes, which should not depend on the gauge, should be unchanged by such a transformation. Consequently, this implies that:

$$(\epsilon^*_\mu(k) + a^* k_\mu)\mathcal{M}^\mu = \epsilon^*_\mu(k)\mathcal{M}^\mu + a^* k_\mu \mathcal{M}^\mu \stackrel{gauge}{\equiv} \epsilon^*_\mu(k)\mathcal{M}^\mu \tag{10.132}$$

and similarly:

$$(\epsilon_\mu(k) + a k_\mu)\mathcal{M}^\mu = \epsilon_\mu(k)\mathcal{M}^\mu + a k_\mu \mathcal{M}^\mu \stackrel{gauge}{\equiv} \epsilon_\mu(k)\mathcal{M}^\mu \tag{10.133}$$

Hence, we *must* have:

$$\boxed{k_\mu \mathcal{M}^\mu \equiv 0} \tag{10.134}$$

This equation – and similar formulae for amplitudes involving multiple photons – are called the **Ward identities**.[15] The Ward identity $k_\mu \mathcal{M}^\mu = 0$ provides for the gauge invariance of the amplitudes $\epsilon_\mu(k)\mathcal{M}^\mu$ and $\epsilon^*_\mu(k)\mathcal{M}^\mu$. **It must hold by Lorenz invariance and the fact that the (real) massless photon has two physical polarization states**.

Let us not be confused by the Ward identity. In general, an individual Feynman diagram is not always gauge-independent. However, when one sums over all diagrams contributing to some scattering process at some order, the sum will always be gauge-invariant and satisfy the Ward identities. Hence, the Ward identities represent a very powerful tool to check the coherence of our amplitudes, for instance to make sure that we have correctly accounted for and computed all diagrams contributing to a given order.

In the arguments above, we have as always considered a covariant formulation of the polarization of the photon, in extending the two physical transverse polarizations to the four polarization states. Considering again an amplitude of the form $\epsilon_\mu(k)\mathcal{M}^\mu$ (the same conclusions hold for $\epsilon^*_\mu(k)\mathcal{M}^\mu$), we find the probability for the process to be proportional to:

$$|\epsilon_\mu(k)\mathcal{M}^\mu|^2 = \epsilon^{*,\alpha}(k)\epsilon^\beta(k)\mathcal{M}^*_\alpha \mathcal{M}_\beta \tag{10.135}$$

Without loss of generality, we can take a Lorentz frame in which the photon four-momentum $k^\mu = (k^0, \vec{k})$ is:

$$k^\mu = (\omega, 0, 0, |\vec{k}|) \tag{10.136}$$

that is we choose the z-axis to be along the photon $\vec{k}$. The situation for photons is always complicated by the freedom of the choice of gauge. In the Coulomb gauge, the two transverse physical polarizations of the photon are, Eq. (10.34):

$$\epsilon^\mu_1 = (0, 1, 0, 0) \quad \text{and} \quad \epsilon^\mu_2 = (0, 0, 1, 0) \tag{10.137}$$

The probability summed over the two photon-transverse polarization states is then:

$$\sum_{\lambda=1}^{2} \epsilon^{*,\alpha}_\lambda(k)\epsilon^\beta_\lambda(k)\mathcal{M}^*_\alpha \mathcal{M}_\beta = |\mathcal{M}_1|^2 + |\mathcal{M}_2|^2 \tag{10.138}$$

We can express this result, extending it to *four* polarization states, introducing corresponding amplitudes $\mathcal{M}_\lambda$ ($\lambda = 0, \ldots, 3$):

$$|\mathcal{M}_1|^2 + |\mathcal{M}_2|^2 = -g^{\alpha\beta}\mathcal{M}^*_\alpha \mathcal{M}_\beta + |\mathcal{M}_0|^2 - |\mathcal{M}_3|^2 \tag{10.139}$$

where the last two terms correspond to the time-like and the longitudinal polarizations, respectively. However, these two are unphysical for real photons and this can readily be verified with the Ward identity. Since $k_\mu \mathcal{M}^\mu = 0$, we have in our case $k^\mu = (\omega, 0, 0, |\vec{k}|)$:

$$\omega \mathcal{M}_0 - |\vec{k}|\mathcal{M}_3 = 0 \tag{10.140}$$

15   John Clive Ward (1924–2000), British–Australian physicist.

This guarantees that for a real photon for which $\omega = |\vec{k}|$, the two amplitudes cancel:

$$\omega = |\vec{k}| \quad \Leftrightarrow \quad \mathcal{M}_0 = \mathcal{M}_3 \tag{10.141}$$

Hence, **for a real photon, the sum over the two polarizations can be expressed in a covariant form with** $g^{\mu\nu}$ **since the amplitudes of the two unphysical polarization states will cancel each other**:

$$\sum_{\lambda=1}^{2} \epsilon_\lambda^{*,\alpha}(k)\epsilon_\lambda^\beta(k)\mathcal{M}_\alpha^*\mathcal{M}_\beta = |\mathcal{M}_1|^2 + |\mathcal{M}_2|^2 = -g^{\alpha\beta}\mathcal{M}_\alpha^*\mathcal{M}_\beta \tag{10.142}$$

and equivalently **the completeness relation says that for a real photon in the initial or final state of a process, we can make the following replacement** (cf. Eq. (10.76)):

$$\sum_{\lambda=1}^{2} \epsilon_\lambda^{*,\mu}(k)\epsilon_\lambda^\nu(k) \leftrightarrow -g^{\mu\nu} \qquad \text{for } k^2 = 0 \text{ (real photon)} \tag{10.143}$$

This result is gauge-independent.

In the above discussion, a real photon was defined with $k^2 = 0$ and hence had only two physical transverse polarization states. However, in some sense, it is difficult to say that a photon is "real," and all photons can be considered as virtual since they will eventually be detected somewhere (e.g., in a detector). Hence, the strict separation between real and virtual photons is somewhat artificial and we can admit that all photons will have some degree of virtuality, albeit it can be "very very small." How can one reconcile these two descriptions? This is discussed below.

• **Tree-level scattering.** Let us consider a tree-level QED $2 \to 2$ scattering process with a virtual photon exchange represented by a photon propagator (see Figure 10.2(left) and Eq. (10.123)). We can write the amplitude as:

$$i\mathcal{M} = -ie^2 \left(\bar{u}(p_2)\gamma_\mu u(p_1)\right) \tilde{G}_F^{\mu\nu}(k) \left(\bar{u}(k_2)\gamma_\nu u(k_1)\right) = -ie^2 j_\mu(p_1, p_2)\tilde{G}_F^{\mu\nu}(k)j_\nu(k_1, k_2) \tag{10.144}$$

where the $j^\mu = \bar{u}\gamma^\mu u$ are the four-vector currents associated with the Dirac spinors. Since energy–momentum conservation at the vertices imposes $k = p_1 - p_2 = k_2 - k_1$, we have:

$$k^\mu j_\mu(p_1, p_2) = \bar{u}(p_2)k^\mu\gamma_\mu u(p_1) = \bar{u}(p_2)\left(p_1^\mu - p_2^\mu\right)\gamma_\mu u(p_1) \tag{10.145}$$

Recalling that $u(p_1)$ satisfies the Dirac equation, Eq. (8.135) we get:

$$k^\mu j_\mu(p_1, p_2) = \bar{u}(p_2)\left(m - m\right)u(p_1) = 0 \tag{10.146}$$

and similarly $k^\mu j_\mu(k_1, k_2) = 0$. This is a spectacular result. Using this fact we can compute the amplitude with the covariant photon propagator:

$$\begin{aligned}
i\mathcal{M} &= -ie^2 j_\mu(p_1, p_2)\frac{i}{k^2}\left(-g^{\mu\nu} + \xi\frac{k^\mu k^\nu}{k^2}\right)j_\nu(k_1, k_2) \\
&= e^2 j_\mu(p_1, p_2)\left(\frac{-g^{\mu\nu}}{k^2}\right)j_\nu(k_1, k_2) + e^2 j_\mu(p_1, p_2)\frac{1}{k^2}\underbrace{\left(\xi\frac{k^\mu k^\nu}{k^2}\right)}_{=0}j_\nu(k_1, k_2) \\
&= e^2 j_\mu(p_1, p_2)\left(\frac{-g^{\mu\nu}}{k^2}\right)j_\nu(k_1, k_2)
\end{aligned} \tag{10.147}$$

So, this shows that the $\xi k^\mu k^\nu / k^2$ term cancels in the amplitude! Here we used a pedestrian approach and showed that we can do this for a typical tree-level Feynman diagram with one photon exchange. There is a general proof using current conservation. However, one must be careful here. We have already mentioned that gauge invariance does not imply that every diagram is gauge-invariant but only that the sum of diagrams must be gauge-invariant! In the tree-level diagram that we considered, the matrix element alone must be gauge-invariant because we assumed that there are no other diagrams at this order (which is the case if the particles are distinguishable).

It can be concluded that the term $k^\mu k^\nu / p^2$ in the covariant photon propagator does not contribute to the tree-level matrix element and can hence be discarded. In practice, for tree-level diagrams we can define the photon propagator as in the case of the Feynman gauge:

$$\tilde{G}_F^{\mu\nu}(k) \equiv \frac{-ig^{\mu\nu}}{k^2} \qquad \text{(tree-level diagrams)} \tag{10.148}$$

In higher-order diagrams, only the sum of the amplitudes for all contributing diagrams is gauge-invariant. In this case, the matrix element for a particular diagram may not be gauge-invariant and photon propagators must be written in full with the gauge-dependent terms $\xi k^\mu k^\nu / k^2$, although we still have the freedom to choose the Feynman gauge where $\xi = 0$. In general, the choice of gauge is critical for computations with higher-order diagrams.

• **Interpretation of the polarization states in the covariant photon propagator.** Let us now return to the question of the polarization states in the covariant photon propagator. The $g^{\mu\nu}$ term implies that we must consider a basis with four independent polarization vectors (instead of the two polarization three-vectors we used for real photons in the Lorenz+Coulomb gauge). The polarization vectors depend on the photon four-momentum k^μ. Without loss of generality, we can assume again that the photon flight direction is along the z-axis, then $k^\mu = (\omega, 0, 0, |\vec{k}|)$. We want to retain the transverse polarizations as before, hence we keep the two four-vectors:

$$\underbrace{\epsilon_1^\mu = (0, 1, 0, 0); \quad \epsilon_2^\mu = (0, 0, 1, 0)}_{\text{transverse}} \tag{10.149}$$

which satisfy $\epsilon_1^\mu \cdot k_\mu = \epsilon_2^\mu \cdot k_\mu = 0$. For other four-momenta k^μ, the polarization vectors are the appropriate Lorentz transformations of these vectors $\epsilon^\mu \to \Lambda^\mu{}_\nu \epsilon^\nu$, since $\epsilon_1^\mu \cdot k_\mu$ and $\epsilon_2^\mu \cdot k_\mu$ are Lorentz-invariant.

The third vector ϵ_3^μ should represent the longitudinal polarization. In the "rest frame of the virtual photon of mass $m^2 = k^2$," we would pick $\epsilon_3^\mu = (0, 0, 0, 1)$. Applying the Lorentz boost of the photon, the polarization vector becomes $\epsilon_3^\mu \to (\beta\gamma, 0, 0, \gamma)$, or:

$$\underbrace{\epsilon_3^\mu = \frac{1}{m}(|\vec{k}|, 0, 0, \omega)}_{\text{longitudinal}} \tag{10.150}$$

Note that the three polarization vectors ϵ_i^μ ($i = 1, 2, 3$) satisfy, as they should, $\epsilon_i^2 = -1$ and $\epsilon_i^\mu \cdot k_\mu = 0$. They are space-like, so the last polarization vector ϵ_0^μ is taken to be time-like:

$$\underbrace{\epsilon_0^\mu = (1, 0, 0, 0)}_{\text{time-like}} \tag{10.151}$$

With this basis, we can now compute the numerator of the covariant photon propagator assuming it corresponds to the **sum over the three physical polarization states** of the virtual photon, which we can write in tensor form

$\mathcal{E}^{\mu\nu}$:

$$
\begin{aligned}
\mathcal{E}^{\mu\nu} &= \sum_{\lambda=1}^{3} \epsilon_\lambda^{\mu,*} \epsilon_\lambda^{\nu} = \begin{pmatrix} 0 & 0 & 0 & 0 \\ 0 & 1 & 0 & 0 \\ 0 & 0 & 1 & 0 \\ 0 & 0 & 0 & 0 \end{pmatrix} + \frac{1}{m^2} \begin{pmatrix} |\vec{k}|^2 & 0 & 0 & \omega|\vec{k}| \\ 0 & 0 & 0 & 0 \\ 0 & 0 & 0 & 0 \\ \omega|\vec{k}| & 0 & 0 & \omega^2 \end{pmatrix} \\
&= \begin{pmatrix} 0 & 0 & 0 & 0 \\ 0 & 1 & 0 & 0 \\ 0 & 0 & 1 & 0 \\ 0 & 0 & 0 & 0 \end{pmatrix} + \frac{1}{m^2} \begin{pmatrix} \omega^2 - m^2 & 0 & 0 & \omega|\vec{k}| \\ 0 & 0 & 0 & 0 \\ 0 & 0 & 0 & 0 \\ \omega|\vec{k}| & 0 & 0 & |\vec{k}|^2 + m^2 \end{pmatrix} \\
&= \begin{pmatrix} -1 & 0 & 0 & 0 \\ 0 & 1 & 0 & 0 \\ 0 & 0 & 1 & 0 \\ 0 & 0 & 0 & 1 \end{pmatrix} + \frac{1}{m^2} \begin{pmatrix} \omega^2 & 0 & 0 & \omega|\vec{k}| \\ 0 & 0 & 0 & 0 \\ 0 & 0 & 0 & 0 \\ \omega|\vec{k}| & 0 & 0 & |\vec{k}|^2 \end{pmatrix}
\end{aligned} \tag{10.152}
$$

This impressive result can be written recalling the four-momentum of the photon as $k^\mu = (k^0, k^1, k^2, k^3)$ and $k^2 = m^2$:

$$
\sum_{\lambda=1}^{3} \epsilon_\lambda^{\mu,*} \epsilon_\lambda^{\nu} = -g^{\mu\nu} + \frac{1}{k^2} \begin{pmatrix} k^0 k^0 & 0 & 0 & k^0 k^3 \\ 0 & 0 & 0 & 0 \\ 0 & 0 & 0 & 0 \\ k^3 k^0 & 0 & 0 & k^3 k^3 \end{pmatrix} = -g^{\mu\nu} + \frac{k^\mu k^\nu}{k^2} \tag{10.153}
$$

In summary, we have addressed the difficulties introduced by the Lorentz covariant description of the photon propagator. The apparent issues are resolved since we have seen that for a virtual (massive) spin-1 photon, the propagator actually describes the sum of the three physical polarization states. For a real (massless) photon, we have found that the longitudinal and time-like terms cancel each other (magically thanks to the minus sign in the metric!). So, the covariant form is adequate in all cases. We noted that for practical purposes it can be reduced to the simple form $-ig^{\mu\nu}/k^2$, since the extra term must cancel at the end due to gauge invariance. Impressive!

- **Longitudinal polarization in massive gauge bosons.** At this stage, we can add that massive spin-1 fields are not only encountered in virtual photons, but fundamental massive spin-1 gauge bosons exist in Nature. As we will detail in Chapter 23, the carriers of the weak interactions, the $W^\pm$ and Z bosons, are such massive spin-1 bosons. A feature is that at high energy $E \gg M$, the longitudinal polarization vector of massive gauge vector bosons of mass M becomes (replacing $\omega \to E$ and $m \to M$ in Eq. (10.150)):

$$
\epsilon_3^\mu \to \frac{E}{M}(1, 0, 0, 1) \tag{10.154}
$$

with $E/M \to \infty$ as $E \to \infty$! In processes involving these modes, the cross-sections will tend to behave like $\sigma \sim \epsilon_{(3)}^2 \sim g^2 E^2/M^2$, where g is a coupling constant. So, no matter how small the coupling constant g is, the cross-section will grow at high energy and eventually reach the unitarity limit. One can say that perturbation theory will break down in these regimes. Therefore, in general, longitudinal polarizations of massive spin-1 propagators pose problems.

Naively, for $M \simeq 100$ GeV and $g \simeq 0.1$, the unitarity bound is violated at $E \simeq 1$ TeV. In the electroweak theory, the longitudinal degrees of freedom are the result of spontaneous symmetry breaking and are acquired from the Goldstone bosons of the Higgs field (see Chapter 25). The divergent behavior of W and Z bosons at high energy is tamed by the appearance of diagrams with the Higgs boson. Cancellation occurs when the energy is higher than the Higgs boson mass. For a light Higgs boson, as discovered experimentally ($m_h \simeq 125$ GeV – this will be discussed in Chapter 29), it turns out that the longitudinal scattering remains in the weak domain and is calculable within the SM. It is wonderful!

10.15   Internal Vertices – Closed Loops and Self-energies

So far, we have considered processes at the lowest orders connecting initial- to final-state fields. Without changing the initial and final states, i.e., for the same set of external legs, we can look at higher-order diagrams with additional internal vertices (each vertex adds a factor e, so the contribution corresponds to a higher-order term in the Dyson series, with fields to be contracted internally rather than to external legs). Such additional diagrams form a class with "internal vertices." They can be computed separately.

• **Self-energy of the photon/closed loops.** As an example, we can consider, for each photon propagator, a higher-order diagram where a virtual e^+e^- pair is created and then annihilated, forming a "closed loop." See Figure 10.3(left). This diagram is going to be important to compute the "self-energy of the photon" (see Section 12.3). The fermions in the loop are virtual and hence not constrained to be on-shell. Their four-momenta is constrained by energy–momentum conservation at each vertex to be q^μ and $k^\mu + q^\mu$, where k^μ is the four-momentum of the original photon. By itself q^μ is not fixed and is not measurable. It must hence be integrated upon over all possible values. For every closed loop, we will therefore have to include an integration factor of the type:

$$\int \frac{\mathrm{d}^4 q}{(2\pi)^4} \tag{10.155}$$

for each unfixed internal four-momentum.

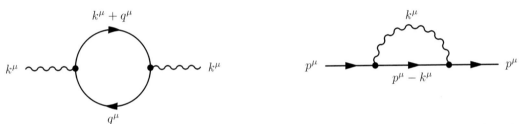

Figure 10.3 (left) Diagram of a closed loop with a virtual e^+e^- pair creation–annihilation. (right) Contribution to Dirac fermion self-energy.

In order to illustrate the Feynman rule for a loop, we start with the photon propagator in the Feynman gauge with four-momentum k^μ and realize that the loop will modify it accordingly:

$$\tilde{G}_0^{\mu\nu}(k^2) = \frac{-ig^{\mu\nu}}{k^2} \quad \rightarrow \quad \tilde{G}_{1\text{-loop}}^{\mu\nu}(k^2) = \frac{-ig^{\mu\mu'}}{k^2} \mathcal{G}_{\mu'\nu'} \frac{-ig^{\nu'\nu}}{k^2} \tag{10.156}$$

where we have the original two indices $\mu\nu$ of the naked photon propagator $\tilde{G}_0^{\mu\nu}(k^2)$, and two new indices $\mu'\nu'$ for the loop. So, in any diagram, we could replace the $\tilde{G}_0^{\mu\nu}(k^2)$ propagator and "plug in" the one-loop propagator $\tilde{G}_{1\text{-loop}}^{\mu\nu}(k^2)$ instead. Using the building blocks derived in the previous sections, we can directly write down the expression for $\mathcal{G}^{\mu'\nu'}$. The result for a given pair of indices $\mu'\nu'$ should be a number and following the arrow direction in Figure 10.3(left), it should contain from left to right: a vertex factor, a fermion propagator, a vertex factor, and a second fermion propagator that links back to the first vertex. Each of these is represented by a 4×4 matrix which, when multiplied for a given pair of indices $\mu'\nu'$, should be a number. We write down the indices of each 4×4 matrix and we ignore the poles of the propagators for clarity purposes:

$$\mathcal{G}^{\mu'\nu'}(k^2) = \int \frac{\mathrm{d}^4 q}{(2\pi)^4} \left[(-ie\gamma^{\mu'})_{\alpha\beta} \frac{i(\not{k}+\not{q}+m)_{\beta\delta}}{(k+q)^2 - m^2} (-ie\gamma^{\nu'})_{\delta\epsilon} \frac{i(\not{q}+m)_{\epsilon\alpha}}{q^2 - m^2} \right] \tag{10.157}$$

where we note that the order of matrix indices is $\alpha\beta \rightarrow \beta\delta \rightarrow \delta\epsilon \rightarrow \epsilon\alpha$. We have used the same index α for the first and last index. This is because we want to describe a loop and hence, as we follow the direction of the arrows, we start from the first vertex, go to the second, and come back to the first!

In summary, **closed loops will always result in the trace of a product of Dirac γ matrices and it will be necessary to integrate for every unfixed momentum "around the loop" with a term of the type** $\mathrm{d}^4 q/(2\pi)^4$. In our example, it would be (removing the internal indices $\alpha\beta\delta\epsilon$):

$$\mathcal{G}^{\mu'\nu'}(k^2) = \int \frac{\mathrm{d}^4 q}{(2\pi)^4} \,\mathrm{Tr}\left[(-ie\gamma^{\mu'})\frac{i(\not{k}+\not{q}+m)}{(k+q)^2-m^2}(-ie\gamma^{\nu'})\frac{i(\not{q}+m)}{(q^2-m^2)}\right] \tag{10.158}$$

Note that a trace of a matrix product depends only on the cyclic order of the matrices. Thus, one may start the product with any vertex or propagator. As long as the matrices are in the correct cyclic order, the trace will be the same.

Finally, one should include a factor of (-1) for a loop in a diagram. This means that for **a diagram containing n fermion loops, one should include an overall factor** $(-1)^n$ in front of the diagram.

- **Self-energy of fermions.** Another interesting diagram is that of a contribution to the so-called *fermion self-energy*, where the propagator of a fermion with four-momentum p^μ is altered by the emission and reabsorption of a virtual photon with four-momentum k^μ (see Figure 10.3(right)). Between the two photon vertices, the fermion is constrained to have a four-momentum $p^\mu - k^\mu$. In this case, k^μ is unconstrained and we must hence integrate over $\mathrm{d}^4 k$ with all possible momenta.

Without going into details, we note that in both cases the integrals will need to be regularized to avoid poles of the propagators. The poles correspond to the virtual particles becoming "on-shell," i.e., real. Furthermore, these kinds of higher-order contributions – also called **radiative corrections** – lead to infinities. They can be classified into **vertex corrections**, **external leg corrections**, and (as above) **propagator corrections**. This divergence problem seriously plagued the development of QED for many years. This issue was finally solved with the development of the technique of **renormalization** (see Chapter 12). Without it, QED could not have been so successful and become the template to describe all elementary forces (except gravity).

We are now ready to move to the next section, where we will summarize all the rules to write down the amplitudes of QED processes associated with Feynman diagrams.

10.16 Practical Rules for Building QED Amplitudes

In the previous section we saw how to explicitly compute the amplitude of a particular Feynman diagram from the QED Lagrangian, starting from the S-matrix using the field contraction methods. These led to practical rules for each of the "building blocks" such as the vertex, the propagator, and the external leg. The usage of simple rules can greatly reduce the otherwise rather tedious and repetitive computations of field contractions.

We hence collect and summarize 14 diagram rules for these individual pieces obtained so far, with the idea that in the future we will directly draw the Feynman diagrams contributing to a particular reaction of interest, and then derive from them the corresponding amplitudes, without explicit calculations. In the next chapter, we will discuss how to compute cross-sections or decay rates starting from the amplitudes. For completeness, we recall that the amplitudes are representations of the physics in the energy–momentum space, whereas the vertices in the Feynman diagrams represent vertices in space-time coordinates.

QED rules are usually extended to include other charged fermions besides electrons and positrons. The straightforward extension includes muon and tau leptons, and quarks, which behave exactly like the electrons, except for different masses, and the fractional charges for quarks. In terms of the Feynman rules, the muons and the taus have exactly the same vertices, propagators, or external line factors as the electrons, except for a different mass m in the propagators. For the quarks, in addition to the mass, the vertex factor is also modified to $(iQe\gamma^\mu)$, where Q is the quark charge ($Q = (+2/3)$ for u, c, t and $Q = (-1/3)$ for d, s, b). To distinguish between the fermion species, one should label the solid lines with e, μ, τ or the corresponding quark symbol. Different species do not mix, so a label belongs to the entire continuous fermionic line; for an external line, the species must agree with the incoming/outgoing particles at the ends of the line; for a closed loop, one should sum over the species.

The rules are organized in tabular form with (1) item, (2) diagram representation, and (3) amplitude term:

- [R1] **Dirac propagator:** $\quad\qquad\qquad$ p^μ $\qquad\qquad = \dfrac{i(\not{p}+m)}{p^2-m^2(+i\epsilon)}$

- [R2] **Photon propagator:** $\quad\mu \quad q^2 \quad \nu \qquad = \dfrac{-ig^{\mu\nu}}{q^2(+i\epsilon)}$ (Feynman gauge)

- [R3] **External fermion leg:** $\qquad p^\mu \;\rightarrow \qquad = \begin{cases} u^{(s)}(\vec{p})\text{-incoming} \\ \bar{u}^{(s)}(\vec{p})\text{-outgoing} \end{cases}$

- [R4] **External antifermion leg:** $\qquad p^\mu \;\rightarrow \qquad = \begin{cases} \bar{v}^{(s)}(\vec{p})\text{-incoming} \\ v^{(s)}(\vec{p})\text{-outgoing} \end{cases}$

- [R5] **Arrows for fermions:** The arrows in the fermion/antifermion lines follow the flow of the electric charge (in units of $-e$).

- [R6] **External photon leg:** $\quad k^\mu \;\rightarrow \qquad = \begin{cases} \epsilon_\mu(\vec{k})\text{-incoming} \\ \epsilon^*_\mu(\vec{k})\text{-outgoing} \end{cases}$

- [R7] **QED vertex:** 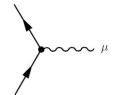 $\mu \qquad = -iQe\gamma^\mu$

- [R8] **At each vertex:** Include a Dirac $\delta^4(\sum p_i^\mu)$ function to impose total energy–momentum conservation at each vertex.

- [R9] **Internal propagators:** From the above rule, compute and assign the four-momenta of all internal propagators. For internal photons, choose a direction for the momentum flow beforehand.

- [R10] **Internal loops:** For closed internal loops, integrate with $\mathrm{d}^4q/(2\pi)^4$ for every unfixed momentum. Include a factor of (-1) for each fermion loop. A closed loop always gives the trace of a product of Dirac γ matrices.

- [R11] **Overall:** Impose an overall energy–momentum conservation between initial and final state with the factor: $(2\pi)^4\delta^4\left(\sum_{initial} p_i - \sum_{final} p_f\right)$.

Given the initial and final states, the total amplitude is given by the sum of the ensemble of all possible diagrams to a given order in the Dyson series. There are additional rules, which we will not derive explicitly here, that are needed when adding diagrams corresponding to the same amplitude, and are listed below:

- [R12] **Add all diagrams:** Draw and sum over the amplitudes of *all* topologically distinct diagrams with the proper relative sign (see [R13] bullet) and to a given order in the Dyson series to compute the total amplitude of the process.

- [R13] **Relative sign:** Add a minus sign between the amplitude of two diagrams that differ only by the permutation of two fermion operators to ensure that the overall amplitude is antisymmetric under the interchange of two initial or two final-state fermions, or of one initial (resp. final) fermion with a final (resp. initial) antifermion, as required by the Fermi–Dirac statistics.

- [R14] **Flavor:** The flavor of the fermion is conserved at the QED vertex, e.g., $ee\gamma$, $\mu\mu\gamma$, but the vertex $e\mu\gamma$ is **forbidden**.

The use of these last rules will be illustrated in the next chapter.

So – we did it! We are now ready to move on and see how we can build and compute diagrams associated with real physical QED processes and then compute their cross-sections or decay widths, or any other physical observable that characterizes them. Before we do that, we first point out an important result on the "chiral" structure of QED and then discuss the electrodynamics of scalar particles and the Gordon decomposition in the next sections. Computations of processes involving Dirac fermions and photons within QED will be illustrated in Chapter 11.

10.17 Chiral Structure of QED

We have seen that the fundamental QED interaction between a photon and an electrically charged spin-1/2 fermion is given by the vertex factor of the vector form $-iQe\overline{\Psi}\gamma^\mu\Psi A_\mu$. We can write this as the following term $-iQej^\mu A_\mu$, with the current for any spinor Ψ:

$$j^\mu = \overline{\Psi}\gamma^\mu\Psi \tag{10.159}$$

We have seen that any spinor can always be decomposed into its *chiral* left-handed and right-handed components (see Eq. (8.189)):

$$\Psi = \Psi_L + \Psi_R \quad \text{where} \quad \Psi_L = P_L\Psi = \frac{1}{2}\left(1 - \gamma^5\right)\Psi \quad \text{and} \quad \Psi_R = P_R\Psi = \frac{1}{2}\left(1 + \gamma^5\right)\Psi. \tag{10.160}$$

Hence we obtain four possible combinations labeled RR, RL, LR, LL:

$$j^\mu = \overline{\Psi}\gamma^\mu\Psi = \overline{(\Psi_R + \Psi_L)}\gamma^\mu(\Psi_R + \Psi_L) = \overline{\Psi_R}\gamma^\mu\Psi_R + \overline{\Psi_R}\gamma^\mu\Psi_L + \overline{\Psi_L}\gamma^\mu\Psi_R + \overline{\Psi_L}\gamma^\mu\Psi_L \tag{10.161}$$

We note that *only two of the four* are non-vanishing, since the RL and LR combinations give zero:

$$\overline{\Psi_R}\gamma^\mu\Psi_L = \Psi_R^\dagger\gamma^0\gamma^\mu\Psi_L = \Psi^\dagger P_R^\dagger\gamma^0\gamma^\mu\Psi_L = \Psi^\dagger P_R\gamma^0\gamma^\mu\Psi_L = \Psi^\dagger\gamma^0 P_L\gamma^\mu\Psi_L = \overline{\Psi}P_L\gamma^\mu\Psi_L \tag{10.162}$$

where we used $P_R^\dagger = \left(\frac{1+\gamma^5}{2}\right)^\dagger = P_R$ because $\gamma^{5\dagger} = \gamma^5$, and $P_R\gamma^0 = \gamma^0 P_L$ because $\gamma^5\gamma^\mu = -\gamma^\mu\gamma^5$. Finally:

$$\overline{\Psi_R}\gamma^\mu\Psi_L = \overline{\Psi}P_L\gamma^\mu P_L\Psi = \overline{\Psi}\gamma^\mu P_R P_L\Psi = 0 \tag{10.163}$$

and similarly:

$$\overline{\Psi_L}\gamma^\mu\Psi_R = \overline{\Psi}\gamma^\mu P_L P_R\Psi = 0 \tag{10.164}$$

We found that only the two combinations RR and LL contribute with equal weights:

$$j^\mu \equiv \overline{\Psi}\gamma^\mu\Psi = \overline{\Psi_R}\gamma^\mu\Psi_R + \overline{\Psi_L}\gamma^\mu\Psi_L \tag{10.165}$$

One must be very careful with how one interprets this statement. The result implies that **chirality is conserved at the QED interaction vertex**. What it does not say is that there are left-handed chiral electrons and right-handed chiral electrons which are distinct particles. Though it is useful when describing particle interactions, chirality is not conserved in the propagation of a free particle. In fact, the chiral states Ψ_L and Ψ_R do not even satisfy the Dirac equation. Since chirality is not a good quantum number, it can evolve with time. A massive particle starting off as a completely left-handed chiral state will soon evolve a right-handed chiral component. By contrast, helicity is a conserved quantity during free particle propagation. Only in the case of massless particles, for which helicity and chirality are identical and are conserved in free-particle propagation, can left-handed and right-handed particles be considered distinct. For neutrinos this mostly holds since they are extremely light and therefore can be approximated as massless in several cases. Neutrino masses will be discussed in later chapters.

The intimate although complex relationship between chirality and helicity was discussed in Section 8.24. In the ultra-relativistic limit, where helicity and chirality eigenstates simply match, chirality conservation imposes a constraint on the contributing helicity states at the vertex, with the caveat that particles and antiparticles behave in opposite ways: the $\overline{\Psi_R}\gamma^\mu\Psi_R$ current corresponds to positive helicity fermions and negative helicity antifermions, while the $\overline{\Psi_L}\gamma^\mu\Psi_L$ current is responsible for the vertex of negative helicity fermions and positive helicity antifermions. In the annihilation $f\bar{f}$ the helicities must be opposite to one another. See Figure 10.4.

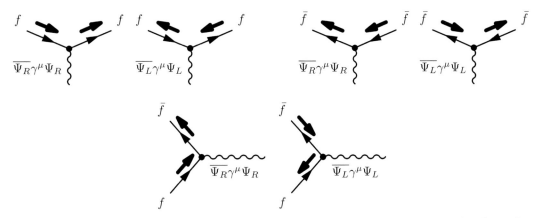

Figure 10.4 Non-vanishing chiral configurations and correspondence to helicity states in the ultra-relativistic limit for: (up-left) ff, (up-right) $\bar{f}\bar{f}$, and (bottom) $f\bar{f}$.

Since the parity transformation exchanges right-handed and left-handed chiral spinors $\Psi_R \leftrightarrow \Psi_L$ (see Eq. (8.205)), the QED current remains invariant:

$$P : j^\mu \to \overline{\Psi_L}\gamma^\mu\Psi_L + \overline{\Psi_R}\gamma^\mu\Psi_R = j^\mu \tag{10.166}$$

We say that the QED interactions (or matrix elements) are *parity invariant*, or that:

Parity is conserved in QED

10.18 Electrodynamics of Scalar Fields

In the previous sections, we have developed the theory of interactions between the photon and fundamental spin-1/2 Dirac particles, such as electrons or positrons (in general leptons and quarks). In Nature we also encounter charged particles that are scalar, for example, the pions π^+, π^0, and π^- are spin-0 mesons (see Section 15.4). The quantum field associated with the pion could be described by the Klein–Gordon equation. We stress, however, that the pion is not an elementary particle (instead it is composed of elementary electrically charged spin-1/2 quarks; see Chapter 17) and therefore our description is not completely satisfactory. Nonetheless, it is worthwhile to investigate the hypothetical electrodynamics of fundamental scalar fields, although elementary charged scalar particles have never been seen in Nature.

Accordingly, we consider the real scalar Klein–Gordon field ϕ for the "neutral pion" π^0, while the complex scalar Klein–Gordon field φ (see Eq. (9.13)) is adequate for the "charged pions" $\pi^\pm$. The real scalar field has no associated charge, while we have seen that the complex scalar field naturally accommodates a set of particles with opposite charges (see Eq. (7.102)). The charged pion will therefore interact electromagnetically via a vertex with the photon.

We can construct a gauge-invariant Lagrangian for the electrodynamics of scalar particles with the covariant derivative Eq. (10.98) (cf. Eq. (10.99)):

$$\mathcal{L} = (D_\mu \varphi)^\dagger D^\mu \varphi - m^2 \varphi^\dagger \varphi - \frac{1}{4} F_{\mu\nu} F^{\mu\nu} \tag{10.167}$$

Expanding the first term in the Lagrangian leads to the following (non-direct) interaction terms:

$$\mathcal{L}_{int} = ieA_\mu \left[(\partial^\mu \varphi^\dagger)\varphi - \varphi^\dagger \partial^\mu \varphi \right] + e^2 A^\mu A_\mu \varphi^\dagger \varphi \tag{10.168}$$

Using the methods described in the previous chapters, one can then develop a set of Feynman rules to compute the electromagnetic interactions for charged scalars. Introducing as before for the initial state $\langle 0|\varphi(x)|\vec{p}_1\rangle \to e^{+ip_1 \cdot x}$ and $\langle \vec{p}_2|\varphi^\dagger(x)|0\rangle \to e^{-ip_2 \cdot x}$ for the final state, the derivative terms in the brackets lead to the vertex factor:

$$-i(ie)\left[(-ip_2^\mu) - (+ip_1^\mu)\right] = -ie(p_1 + p_2)^\mu \tag{10.169}$$

The term proportional to e^2 gives a four-point coupling of a scalar with two photons proportional to $2ie^2$ (the factor 2 comes from the symmetry of the two photons). The rules are given below, organized in tabular form with (1) item, (2) diagram representation, and (3) amplitude term:

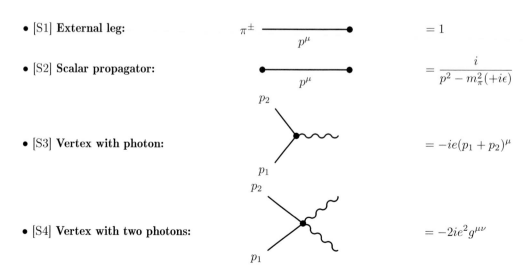

- [S1] **External leg:** $\pi^\pm$ $\quad p^\mu \quad$ $= 1$

- [S2] **Scalar propagator:** $\quad p^\mu \quad$ $= \dfrac{i}{p^2 - m_\pi^2 (+i\epsilon)}$

- [S3] **Vertex with photon:** $\quad p_2,\ p_1 \quad$ $= -ie(p_1 + p_2)^\mu$

- [S4] **Vertex with two photons:** $\quad p_2,\ p_1 \quad$ $= -2ie^2 g^{\mu\nu}$

As a further example, we consider the scattering of two charged pions $\pi^\pm + \pi^\pm \to \pi^\pm + \pi^\pm$ via the electromagnetic exchange of a photon. At the tree level we have two diagrams illustrated in Figure 10.5. The two diagrams must be added, since for bosons the result must be symmetric under the interchange of two particles. The tree-level amplitude is simply:

$$\mathcal{M} = (ie)^2 \left[\frac{(p_1 + p_3)_\mu (p_2 + p_4)^\mu}{(p_1 - p_3)^2} + \frac{(p_1 + p_4)_\mu (p_2 + p_3)^\mu}{(p_1 - p_4)^2} \right] \tag{10.170}$$

The derivation of the cross-section is left to the exercises (see **Ex 10.3**).

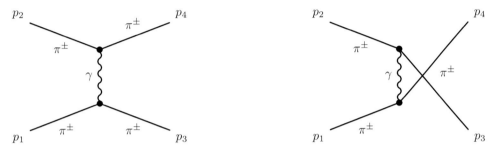

Figure 10.5 Tree-level diagram for the process of two charged pions $\pi^\pm + \pi^\pm \to \pi^\pm + \pi^\pm$ via electromagnetic scattering.

10.19   The Gordon Decomposition

The **Gordon identity** first derived in 1928 states that any vector current can be expressed in the following way:

$$\bar{u}(p')\gamma^\mu u(p) = \frac{1}{2m}\bar{u}(p')\left[(p' + p)^\mu + i\sigma^{\mu\nu}(p' - p)_\nu\right]u(p) \tag{10.171}$$

This result is always true and gives the separation of the current into the so-called conduction and the polarization.[16] The decomposition can be proven by considering the term with the definition $\sigma^{\mu\nu} \equiv (i/2)\left[\gamma^\mu, \gamma^\nu\right]$:

$$
\begin{aligned}
\bar{u}(p')i\sigma^{\mu\nu}(p'_\nu - p_\nu)u(p) &= -\frac{1}{2}\bar{u}(p')\left[\gamma^\mu, \gamma^\nu\right](p'_\nu - p_\nu)u(p) \\
&= -\bar{u}(p')\left[(-\gamma^\nu\gamma^\mu + g^{\mu\nu}\mathbb{1})\,p'_\nu - (\gamma^\mu\gamma^\nu - g^{\mu\nu}\mathbb{1})\,p_\nu\right]u(p) \\
&= -\bar{u}(p')\left[(-p\!\!\!/\,'\gamma^\mu + p'^\mu) - (\gamma^\mu p\!\!\!/ - p^\mu)\right]u(p) \\
&= -\bar{u}(p')\left[(p + p')^\mu - \gamma^\mu p\!\!\!/ - p\!\!\!/\,'\gamma^\mu\right]u(p) \\
&= \bar{u}(p')\left[-(p + p')^\mu + 2m\gamma^\mu\right]u(p) \tag{10.172}
\end{aligned}
$$

where we used the following results (see Appendix E.1):

$$\left[\gamma^\mu, \gamma^\nu\right] = \gamma^\mu\gamma^\nu - \gamma^\nu\gamma^\mu = 2\gamma^\mu\gamma^\nu - 2g^{\mu\nu}\mathbb{1} = 2\left(-\gamma^\nu\gamma^\mu + 2g^{\mu\nu}\mathbb{1}\right) - 2g^{\mu\nu}\mathbb{1} = -2\gamma^\nu\gamma^\mu + 2g^{\mu\nu}\mathbb{1} \tag{10.173}$$

and also the Dirac equation $p\!\!\!/\,u(p) = mu(p)$ and $\bar{u}(p')p\!\!\!/\,' = m\bar{u}(p')$.

---

16   *"Der Strom der Diracschen Elektronentheorie wird in Leitungs- und Polarisationsstrom zerlegt."* Reprinted by permission from Springer Nature: W. Gordon, *"Der Strom der Diracschen Elektronentheorie,"* *Zeitschrift für Physik*, vol. 50, pp. 630–632, Sep 1928.

The Gordon identity allows us to decompose the point-like QED current into two parts. The first term is the coupling of the form $(p' + p)$, which is the same as Eq. (10.169) that we found in the previous section for the case of a charged scalar particle. This first term therefore represents the purely electric interaction, hence is called the **electric term**. The second part $i\sigma^{\mu\nu}q_\nu$ is not present for charged scalar particles and is related to the spin of the Dirac particle, stemming from their magnetic moments. The $i\sigma^{\mu\nu}q_\nu$ term is therefore called the **magnetic term**. In summary, the QED vector interaction current of a point-like Dirac particle of the form $\overline{\Psi}\gamma^\mu\Psi$ contains the electric and magnetic contributions to the interaction. The Gordon identity exhibits that a Dirac particle interacts via both its charge and its magnetic moment.

The Gordon decomposition can be viewed as a specific case of decomposition of a four-current. If we consider the general form $\bar{u}(p')\Gamma^\mu u(p)$, then Lorentz covariance considerably narrows down the possible forms. If Γ^μ is to transform as a four-vector, then it must be a linear combination of matrices γ^μ and the available momenta p^μ and p'^μ. Consequently we can write, using as before $p + p'$, and hence also $p - p'$:

$$\bar{u}(p')\Gamma^\mu u(p) = \bar{u}(p')\left(A_1 \cdot (p + p')^\mu + A_2 \cdot (p - p')^\mu + A_3\gamma^\mu\right)u(p) \tag{10.174}$$

where A_1, A_2, and A_3 can depend on p and p' and in principle they themselves involve γ matrices. However, using as before the Dirac equation, one could always simplify terms of the type $\not{p}u(p) = mu(p)$ and $\bar{u}(p')\not{p}' = m\bar{u}(p')$. Therefore, the coefficients should non-trivially depend on scalars such as $(p + p')^2 = 2p \cdot p' + 2m^2$ or $(p - p')^2 = -2p \cdot p' + 2m^2$. Without loss of generality, it is nicer to unify these into the single non-trivial scalar $q^2 = (p - p')^2 = -2p \cdot p' + 2m^2$. So finally, we have found that the most general form of Γ^μ which is consistent with Lorentz covariance is given by:

$$\bar{u}(p')\Gamma^\mu u(p) = \bar{u}(p')\left(A_1(q^2, m^2) \cdot (p + p')^\mu + A_2(q^2, m^2) \cdot (p - p')^\mu + A_3(q^2, m^2)\gamma^\mu\right)u(p) \tag{10.175}$$

We will use this result recurrently.

Problems

Ex 10.1 Classical Coulomb potential. Derive Coulomb's law using classical field theory. Start from the Lagrangian

$$\mathcal{L} = -\frac{1}{4}F_{\mu\nu}F^{\mu\nu} - J_\mu A^\mu \tag{10.176}$$

and represent a single charge e at the origin as:

$$J_\mu = \left(e\delta^3(\vec{x}), \vec{0}\right) \tag{10.177}$$

In the Lorenz gauge, the solution is $\Box A_\nu(x) = J_\nu(x)$. Since $\delta^3(\vec{x})$ is time-independent, show that the solution to find is:

$$\Box A_0(x) = e\delta^3(\vec{x}) \tag{10.178}$$

Derive the known result:

$$A_0(x) = \frac{e}{4\pi}\frac{1}{r} \tag{10.179}$$

where $r = |\vec{x}|$.

Ex 10.2 The Coulomb force in QED. Refer to **Ex 9.3** and compute the non-relativistic limit of the interaction between two fermions in QED at tree level via the exchange of a photon.

(a) Sketch the corresponding Feynman diagram at tree level.

(b) Comparing to the Yukawa case, show that the QED potential is a repulsive potential with an infinite range:

$$V(r) = \frac{e^2}{4\pi r} \tag{10.180}$$

(c) Show that, on the other hand, for fermion–antifermion scattering, the potential is attractive. Hence, the exchange of a vector boson makes the interaction repulsive for like fermions (i.e., equal charge) and attractive for fermion–antifermion (i.e., opposite charges), as expected from the classical Coulomb force!

Ex 10.3 Scattering of two charged scalar particles via photon exchange. Consider the scattering of two charged pions $\pi^\pm + \pi^\pm \to \pi^\pm + \pi^\pm$ via the electromagnetic exchange of a photon. Compute the differential cross-section in the center-of-mass system.

Ex 10.4 Annihilation of two charged scalar particles into two other charged scalar particles via photon exchange. Consider the following hypothetical reaction within electrodynamics (see Section 10.18) of scalar particles $\phi^+\phi^- \to \gamma^* \to \psi^+\psi^-$, where $\phi^\pm$ and $\psi^\pm$ are point-like spinless scalar particles with rest masses m and M.

(a) Write down the amplitude at tree level assuming the coupling given by the electromagnetism of scalar particles.

(b) Show that the amplitude can be written as:

$$\mathcal{M} = e^2 \frac{u - t}{s} \tag{10.181}$$

where s, t, u are the Mandelstam variables.

(c) Use the general phase space result derived in **Ex 5.4** to show that this leads to the following differential cross-section in the center-of-mass system:

$$\left(\frac{d\sigma}{d\Omega}\right)_{CMS} = \frac{\alpha^2}{4s}\sqrt{\frac{s - 4M^2}{s - 4m^2}}\left(1 - \frac{4m^2}{s}\right)\left(1 - \frac{4M^2}{s}\right)\cos^2\theta \tag{10.182}$$

(d) Compute the total cross-section and show that it is given by:

$$\sigma = \frac{\pi\alpha^2}{3s}\sqrt{\frac{s - 4M^2}{s - 4m^2}}\left\{1 - \frac{4\left(m^2 + M^2\right)}{s} + \frac{16m^2 M^2}{s^2}\right\} \tag{10.183}$$

(e) Take the high-energy limit $s \gg m^2, M^2$ and show that the results become:

$$\left(\frac{d\sigma}{d\Omega}\right)_{CMS} = \frac{\alpha^2}{4s}\cos^2\theta \quad \text{and} \quad \sigma = \frac{\pi\alpha^2}{3s} \tag{10.184}$$

where α is the fine-structure constant.

(f) Can you motivate these results in terms of the spin of the photon?

(g) Compare the results obtained to that for $e^+e^- \to \mu^+\mu^-$ derived in Chapter 11.

11   Computations in QED

The underlying physical laws necessary for the mathematical theory of a large part of physics and the whole of chemistry are thus completely known, and the difficulty is only that the exact application of these laws leads to equations much too complicated to be soluble. It therefore becomes desirable that approximate practical methods of applying quantum mechanics should be developed, which can lead to an explanation of the main features of complex atomic systems without too much computation.

P. A. M. Dirac[1]

11.1   Introduction

In principle and with the help of Feynman diagrams, it could be possible to *compute any reaction at any desired precision*. In practice, calculations become increasingly complex endeavors as the precision increases, with the number of diagrams to compute growing rapidly with the increasing order under consideration. The **tree-level calculation** corresponds to the lowest possible order. The first, second, ... order radiative corrections are added when adding additional vertices, with either external legs and/or internal lines or loops.

In all cases, the starting point is the elaboration of the Feynman diagrams for a given process. One considers all topologically non-equivalent diagrams. Once the diagrams are laid down, their amplitudes are derived using the Feynman rules. The invariant amplitude will ultimately be written as the product of **a matrix element** $\mathcal{M}$ and a Dirac δ function to impose energy–momentum conservation between the initial and the final states:

$$(-i\mathcal{M}) \times (2\pi)^4 \delta^4 \left(\sum_{initial} p_i - \sum_{final} p_f \right) \tag{11.1}$$

The amplitudes of all topologically non-equivalent diagrams are added (they are complex numbers) with the proper relative sign defined by the Feynman rule **[R13]** (see Section 10.16). We can express this as follows:

$$\mathcal{M} = \mathcal{M}_1 + (\pm 1)\mathcal{M}_2 + \cdots (\pm 1)\mathcal{M}_n \tag{11.2}$$

The cross-section of a reaction or the decay width of a particle depends on the *probability* of the process, which is proportional to the square of the absolute value of the amplitude, i.e., the product of the amplitude with its complex conjugate. For instance:

$$d\sigma \propto |\mathcal{M}|^2 = M^* M = [\mathcal{M}_1 + (\pm 1)\mathcal{M}_2 + \cdots (\pm 1)\mathcal{M}_n]^* [\mathcal{M}_1 + (\pm 1)\mathcal{M}_2 + \cdots (\pm 1)\mathcal{M}_n] \tag{11.3}$$

For example, for the case of just two diagrams with amplitudes $\mathcal{M}_1$ and $\mathcal{M}_2$, the total matrix element $\mathcal{M} = \mathcal{M}_1 \pm \mathcal{M}_2$ contains contributions from the sum of three terms which may be calculated separately: the

---

1   Republished with permission of The Royal Society (UK) from P. A. M. Dirac, "Quantum mechanics of many-electron systems," *Proc. Roy. Soc. Lond.*, vol. A123, pp. 714–733, 1929; permission conveyed through Copyright Clearance Center, Inc.

two diagrams $|\mathcal{M}_1|^2$ and $|\mathcal{M}_2|^2$ which are simply the square of each amplitude and a third which is the sum of the "interference terms" $\mathcal{M}_1^* \mathcal{M}_2$ and $\mathcal{M}_1 \mathcal{M}_2^*$ between the diagrams:

$$
\begin{aligned}
|\mathcal{M}|^2 &= (\mathcal{M}_1 \pm \mathcal{M}_2)^* (\mathcal{M}_1 \pm \mathcal{M}_2) = |\mathcal{M}_1|^2 + |\mathcal{M}_2|^2 \pm \mathcal{M}_1^* \mathcal{M}_2 \pm \mathcal{M}_1 \mathcal{M}_2^* \\
&= |\mathcal{M}_1|^2 + |\mathcal{M}_2|^2 \pm 2\mathcal{R}e(\mathcal{M}_1 \mathcal{M}_2^*)
\end{aligned}
\tag{11.4}
$$

where we used the fact that for any two complex numbers a and b, one has $|a \pm b|^2 = |a|^2 + |b|^2 \pm 2\mathcal{R}e(ab^*)$. There are different approaches to solve this problem:

- Plug Dirac spinors into amplitudes $\mathcal{M}_i$ and compute $\sum_i \mathcal{M}_i$ and then $|\mathcal{M}|^2$.

- Trace method and theorems to compute $|\mathcal{M}|^2$ (**Casimir's[2] trick**).

- Computer algebra system (CAS) to compute $\mathcal{M}_i$ and/or $|\mathcal{M}|^2$.

The results depend on the kinematics and on the spin configurations of the particles. Experimentally one defines the **spin polarization** of a particular beam, as discussed in Section 4.14. The polarization defines the **degree to which the spin of the particles of the beam are aligned with a given direction. Zero polarization corresponds to an unpolarized beam.**

Many experiments are performed with unpolarized beams. For an electron or a positron beam, the helicity can be either positive or negative. To compute the unpolarized beam cross-section, **we average over the initial-state spin configurations**. In the final state, particles can be produced with different spin configurations. In many experiments the polarization of the final-state particles is not measured. To compare these results with theory, **we must add over the final-state spin configurations**.

For example for a $2 \to 2$ scattering with incoming spins r, s and outgoing spins r', s', the squared matrix element averaged over initial spins and summed over final-state spins $\langle |\mathcal{M}|^2 \rangle$ is equal to the **spin-averaged matrix element**:

$$
\langle |\mathcal{M}|^2 \rangle \equiv \frac{1}{2} \sum_{r=1}^{2} \frac{1}{2} \sum_{s=1}^{2} \sum_{r'=1}^{2} \sum_{s'=1}^{2} |\mathcal{M}(r, s \to r', s')|^2
\tag{11.5}
$$

The sums are over the two spin states, where the $1/2$ factors account for the averaging over the two initial-state spin configurations.

The obtained matrix elements can be integrated over the phase space to obtain total cross-sections or full widths. It can, however, happen that the results will diverge in given kinematical regions. For example, in the classical Rutherford scattering formula, there is a divergence in the forward direction. In QED, such divergences related to external legs will also occur, and are interpreted as due to the truncation of the Dyson series to a given order. If all diagrams could be calculated to an infinite order, cross-sections would remain finite. There is another class of divergences that are related to internal lines and/or loops, and these will be resolved with the process of **renormalization**.

In modern experiments, differential cross-sections are implemented in **event generators**, which integrate the processes using Monte-Carlo techniques (see Appendix H). These codes actually have the ability to *generate final-state kinematical configurations distributed according to the calculated differential cross-sections*. Such **generated events** are then passed through complicated detector simulations (see Section 2.22) and compared with detector data.

11.2   A First Example: Mott Scattering

11.2.1   Deriving the Amplitude

As a first example we consider the scattering of an electron off a static electromagnetic field given by the classical Coulomb potential. We start with the Feynman diagram at first order as shown in Figure 11.1. The

---

2   Hendrik Brugt Gerhard Casimir (1909–2000), Dutch physicist.

incoming (initial-state) electron with momentum p and spin state s scatters off the potential A^μ and becomes the outgoing (final-state) electron with momentum p' and spin state s'. The classic Coulomb potential for the point-like charge Ze is simply:

$$A^\mu = A_\mu = \left(\frac{Ze}{4\pi\,|\vec{x}|}, 0, 0, 0 \right) \qquad \text{or} \qquad \gamma^\mu A_\mu = \gamma^0 \frac{Ze}{4\pi\,|\vec{x}|} \tag{11.6}$$

The incoming and outgoing electrons give the following factors when contracted with their corresponding Dirac field operators:

$$\Psi\,|p^\mu, s\rangle \simeq e^{-ip\cdot x} u^{(s)}(p)\,|0\rangle \qquad \text{and} \qquad \langle p'^\mu, s'|\,\overline{\Psi} \simeq \langle 0|\,e^{+ip'\cdot x} \bar{u}^{(s')}(p') \tag{11.7}$$

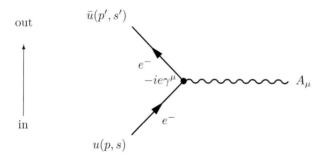

Figure 11.1 Feynman diagram representing an electron scattering off a classical electromagnetic potential A^μ.

- **Scattering amplitude.** With Eq. (10.103) we get the following first-order amplitude:

$$\langle e^- |S_1| e^-\rangle = (-ie) \int \mathrm{d}^4 x\, \langle p'^\mu s'|\,\overline{\Psi}(x)\gamma^0 \Psi(x) A_0(x)\,|p^\mu s\rangle \tag{11.8}$$

Omitting the bracket, we find:

$$(-ie)\frac{Ze}{4\pi} \int \mathrm{d}^4 x\, \bar{u}^{(s')}(p')\gamma^0 u^{(s)}(p)\frac{e^{-i(p-p')\cdot x}}{|\vec{x}|} = -i\frac{Ze^2}{4\pi} \int \mathrm{d}^3\vec{x}\, \bar{u}^{(s')}(p')\gamma^0 u^{(s)}(p)\frac{e^{+i(\vec{p}-\vec{p}')\cdot\vec{x}}}{|\vec{x}|} \int \mathrm{d}t\, e^{-i(E-E')t}$$

Integration over time yields a Dirac function for energy conservation between initial and final states:

$$\langle e^- |S_1| e^-\rangle = -i\frac{Ze^2}{4\pi}\bar{u}^{(s')}(p')\gamma^0 u^{(s)}(p) \int \mathrm{d}^3\vec{x}\, \frac{e^{+i(\vec{p}-\vec{p}')\cdot\vec{x}}}{|\vec{x}|} 2\pi\delta(E - E') \tag{11.9}$$

Defining the momentum transfer $\vec{q} = \vec{p} - \vec{p}'$ and its spherical coordinates relative to $\vec{x}$, the spatial integration is of the type:

$$\begin{aligned}
\int \mathrm{d}^3\vec{x}\, \frac{e^{+i\vec{q}\cdot\vec{x}}}{|\vec{x}|} &= \int r\sin\theta \mathrm{d}\phi r\mathrm{d}\theta \mathrm{d}r \frac{e^{+iqr\cos\theta}}{r} = \int_0^{2\pi} \mathrm{d}\phi \int_0^\infty r\mathrm{d}r \int_{-1}^1 \mathrm{d}\cos\theta\, e^{+iqr\cos\theta} \\
&= 2\pi \int_0^\infty r\mathrm{d}r \frac{\left(e^{iqr} - e^{-iqr}\right)}{iqr} = 2\pi \int_0^\infty \mathrm{d}r \frac{2i\sin(qr)}{iq} \\
&= \frac{4\pi}{q} \int_0^\infty \mathrm{d}r\, \sin(qr)
\end{aligned} \tag{11.10}$$

In order to perform the integral, let us consider the modified integral that corresponds to the **Yukawa**[3] or **screened Coulomb potential** ($a > 0$):

$$\int_0^\infty \mathrm{d}r \, \sin(qr) e^{-ar} = -\frac{e^{-a\cdot r}}{q^2 + a^2} \left(q\cos(qr) + a\sin(qr) \right) \Big|_0^\infty = \frac{q}{q^2 + a^2} \tag{11.11}$$

Hence, we solve the spatial integration corresponding to the Fourier transform of the Coulomb potential:

$$\int \mathrm{d}^3\vec{x} \, \frac{e^{+i\vec{q}\cdot\vec{x}}}{|\vec{x}|} = \lim_{a\to 0} \frac{4\pi}{q} \int_0^\infty \mathrm{d}r \, \sin(qr) e^{-ar} = \frac{4\pi}{q} \lim_{a\to 0} \frac{q}{q^2 + a^2} = \frac{4\pi}{q^2} \tag{11.12}$$

Finally, the amplitude can be expressed as:

$$\langle e^- |S_1| e^- \rangle = -i\mathcal{M} 2\pi\delta(E - E') \quad \text{where} \quad \mathcal{M} \equiv \frac{Ze^2}{q^2} \bar{u}^{(s')}(p')\gamma^0 u^{(s)}(p) \tag{11.13}$$

This expression can be used to calculate the amplitude for a given incoming and outgoing kinematical configuration and spin configuration. In the following sections, we treat the problem of computing the resulting differential cross-sections using different methods.

• **A word about Lorentz invariance.** The amplitude, on which the cross-section depends, must be a Lorentz invariant, since different inertial observers cannot conclude on different probabilities for reactions to happen. On the other hand, our expression for $\mathcal{M}$ does not look invariant! What happened? The answer is that we implicitly calculated the amplitude in the reference frame of the field, as we defined in Eq. (11.6). In order to make the covariance explicit, we note that our resulting amplitude Eq. (11.13) can be expressed as:

$$\mathcal{M} = A_0(q^2)\bar{u}^{(s')}(p')\gamma^0 u^{(s)}(p) = A_\mu(q^2)\bar{u}^{(s')}(p')\gamma^\mu u^{(s)}(p) \tag{11.14}$$

where the Fourier transform of the electromagnetic potential in its reference frame is:

$$A^\mu(q^2) \equiv \frac{Ze^2}{q^2}\epsilon^\mu \quad \text{with} \quad \epsilon^\mu = (1, 0, 0, 0) \tag{11.15}$$

Under a Lorentz transformation, the electromagnetic potential transforms as a four-vector (we note that q^2 is a scalar!):

$$A^\mu(q^2) \to \Lambda^\mu_{\ \nu} A^\nu(q^2) = \frac{Ze^2}{q^2} \Lambda^\mu_{\ \nu}\epsilon^\nu \tag{11.16}$$

and we already showed in Section 8.26 that the quantity of the type $\bar{u}\gamma^\mu u$ transforms as a four-vector as well. Hence, finally, our amplitude Eq. (11.13) is indeed invariant since it is the product of two four-vectors!

11.2.2   Computing Helicity Amplitudes with Spinors

We illustrate the spinor method. We use the positive ($\uparrow$) and negative ($\downarrow$) helicity eigenstates (Eq. (8.148)) derived in Section 8.16) in spherical coordinates for a particle with energy E, momentum p, and direction given by the two angles θ and ϕ:

$$u_\uparrow(p) = \mathcal{N} \begin{pmatrix} c \\ se^{i\phi} \\ \alpha c \\ \alpha s e^{i\phi} \end{pmatrix} \quad \text{and} \quad u_\downarrow(p) = \mathcal{N} \begin{pmatrix} -s \\ ce^{i\phi} \\ \alpha s \\ -\alpha c e^{i\phi} \end{pmatrix} \tag{11.17}$$

---

3   Hideki Yukawa (1907–1981), Japanese theoretical physicist.

where

$$\mathcal{N} = \sqrt{E + m_e}, \quad c = \cos\left(\frac{\theta}{2}\right), \quad s = \sin\left(\frac{\theta}{2}\right), \quad \text{and} \quad \alpha = \frac{p}{E + m_e} \tag{11.18}$$

In the previously derived amplitude Eq. (11.13), the spinor dependence is included in the last term. We note that for any two spinors $u(p)$ and $u(p')$, we have:

$$
\begin{aligned}
\bar{u}(p')\gamma^0 u(p) &= u^\dagger(p')\gamma^0\gamma^0 u(p) = u^\dagger(p')u(p) = (u_1^*(p'), u_2^*(p'), u_3^*(p'), u_4^*(p')) \begin{pmatrix} u_1(p) \\ u_2(p) \\ u_3(p) \\ u_4(p) \end{pmatrix} \\
&= u_1^*(p')u_1(p) + u_2^*(p')u_2(p) + u_3^*(p')u_3(p) + u_4^*(p')u_4(p)
\end{aligned}
\tag{11.19}
$$

So we need to consider the spinors of the incoming and outgoing electrons, and for each, we have two helicity states: up and down. This gives us four amplitudes which depend on the incoming and outgoing electron momenta. Without loss of generality, we can fix the orientation of the momentum of the incoming electron in the positive z direction, and assume it scatters in the x–z plane, such that we can ignore the trivial ϕ dependence, setting $\phi = 0$. The δ function in the amplitude Eq. (11.13) enforces the incoming and outgoing electron to have equal energies $E = E'$. Accordingly, we set:

$$p^\mu = (E, 0, 0, p) \quad \text{and} \quad p'^\mu = (E, p\sin\theta, 0, p\cos\theta) \tag{11.20}$$

where θ is the electron scattering angle (with our choice of axes, the scattering angle coincides with the θ angle of the spherical coordinate system!). We plug the spinors into the amplitude, for example for the case where both incoming and outgoing electrons have a positive helicity, and we obtain:

$$
\begin{aligned}
\mathcal{M}_{\uparrow\uparrow} &\propto u_1^*(p')u_1(p) + u_2^*(p')u_2(p) + u_3^*(p')u_3(p) + u_4^*(p')u_4(p) = |\mathcal{N}|^2 (c, s, \alpha c, \alpha s) \begin{pmatrix} 1 \\ 0 \\ \alpha \\ 0 \end{pmatrix} \\
&= |\mathcal{N}|^2 c(1 + \alpha^2)
\end{aligned}
\tag{11.21}
$$

Similarly, considering the other combinations yields:

$$
\begin{aligned}
\mathcal{M}_{\uparrow\uparrow} &\propto \bar{u}_\uparrow(p')\gamma^0 u_\uparrow(p) = |\mathcal{N}|^2 c(1 + \alpha^2), & \mathcal{M}_{\downarrow\downarrow} &\propto \bar{u}_\downarrow(p')\gamma^0 u_\downarrow(p) = |\mathcal{N}|^2 c(1 + \alpha^2) \\
\mathcal{M}_{\uparrow\downarrow} &\propto \bar{u}_\uparrow(p')\gamma^0 u_\downarrow(p) = |\mathcal{N}|^2 s(1 - \alpha^2), & \mathcal{M}_{\downarrow\uparrow} &\propto \bar{u}_\downarrow(p')\gamma^0 u_\uparrow(p) = -|\mathcal{N}|^2 s(1 - \alpha^2)
\end{aligned}
\tag{11.22}
$$

The scattering of the electron is expressed in terms of its transition probability from one initial state to its final state. The transition probability is proportional to the squared magnitude of the amplitude. Hence, we need to compute $|\mathcal{M}|^2$ for each particular helicity configuration. This is just given by:

$$|\mathcal{M}_{\uparrow\uparrow}|^2 \propto c^2(1+\alpha^2)^2, \quad |\mathcal{M}_{\uparrow\downarrow}|^2 \propto s^2(1-\alpha^2)^2, \quad |\mathcal{M}_{\downarrow\uparrow}|^2 \propto s^2(1-\alpha^2)^2, \quad |\mathcal{M}_{\downarrow\downarrow}|^2 \propto c^2(1+\alpha^2)^2 \tag{11.23}$$

We note that $|\mathcal{M}_{\uparrow\uparrow}|^2 = |\mathcal{M}_{\downarrow\downarrow}|^2$ and $|\mathcal{M}_{\uparrow\downarrow}|^2 = |\mathcal{M}_{\downarrow\uparrow}|^2$, as should be. This is a direct consequence of **parity P conservation** in QED (see Section 10.17)! Also, as expected, the result satisfies **helicity conservation in the ultra-relativistic limit**, since in this case $\alpha \to 1$, hence $|\mathcal{M}_{\uparrow\downarrow}|^2 = |\mathcal{M}_{\downarrow\uparrow}|^2 \to 0$.

Up to this point, the expression retains freely specified spin and polarization states for the electrons. However, if we are not interested in the spin state of the electrons (this is many times the case since experimentally polarizations are difficult to control and measure), then we want to find the spin-averaged matrix element (see Eq. (11.5)) by averaging over the initial-state helicities and summing over the final-state helicities:

$$
\begin{aligned}
\langle |\mathcal{M}|^2 \rangle &\equiv \frac{1}{2}\left(\underbrace{|\mathcal{M}_{\uparrow\uparrow}|^2 + |\mathcal{M}_{\downarrow\downarrow}|^2}_{\text{no helicity flip}} + \underbrace{|\mathcal{M}_{\uparrow\downarrow}|^2 + |\mathcal{M}_{\downarrow\uparrow}|^2}_{\text{helicity flip}} \right) = \frac{Z^2 e^4}{q^4} \frac{|\mathcal{N}|^4}{2} \left(2c^2(1+\alpha^2)^2 + 2s^2(1-\alpha^2)^2 \right) \\
&= \frac{Z^2 e^4}{q^4} |\mathcal{N}|^4 (1 + \alpha^4) + 2\frac{Z^2 e^4}{q^4} |\mathcal{N}|^4 \alpha^2 (c^2 - s^2)
\end{aligned}
\tag{11.24}
$$

We now plug in the values of $\mathcal{N}$ and α to get:

$$
\begin{aligned}
\langle |\mathcal{M}|^2 \rangle /(\frac{Z^2 e^4}{q^4}) &= (E + m_e)^2 \left(\frac{(E + m_e)^4 + p^4}{(E + m_e)^4} \right) + 2(E + m_e)^2 \frac{p^2}{(E + m_e)^2}(c^2 - s^2) \\
&= (E + m_e)^2 + (E - m_e)^2 + 2p^2(c^2 - s^2) = 2E^2 + 2m_e^2 + 2p^2 \cos\theta \\
&= 4E^2 - 2p^2(1 - \cos\theta) = 4E^2 - 4p^2 \sin^2(\theta/2)
\end{aligned}
\tag{11.25}
$$

Finally, the matrix element averaged over initial helicities and summed over final helicities is just:

$$
\langle |\mathcal{M}|^2 \rangle = 4 \frac{Z^2 e^4}{q^4} E^2 (1 - \beta^2 \sin^2(\theta/2))
\tag{11.26}
$$

where the electron velocity is $\beta = p/E$. We have detailed this calculation to show that this method does provide the largest flexibility, although sometimes it can actually be algebraically tedious to work with the spinors.

11.2.3   Unpolarized Scattering Probability – Trace Technique

We want to compute $|\mathcal{M}|^2$ for the unpolarized case without having to plug in explicit spinors. The most complicated part is the product of the spinors and the γ^0 matrix, which we express as the product of the term times its complex conjugate:

$$
\left| \bar{u}^{(s')}(p') \gamma^0 u^{(s)}(p) \right|^2 = \left[\bar{u}^{(s')}(p') \gamma^0 u^{(s)}(p) \right] \left[\bar{u}^{(s')}(p') \gamma^0 u^{(s)}(p) \right]^*
\tag{11.27}
$$

We note that the complex conjugate term can be expressed as:

$$
\left[u^{(s'),\dagger}(p') \gamma^{0} \gamma^0 u^{(s)}(p) \right]^\dagger = \left[u^{(s),\dagger}(p) \gamma^{0,\dagger} \gamma^{0,\dagger} u^{(s')}(p') \right] = \left[\bar{u}^{(s)}(p) \gamma^0 u^{(s')}(p') \right]
\tag{11.28}
$$

where we used $\gamma^{0,\dagger} = \gamma^0$. Hence

$$
|\mathcal{M}|^2 \propto \left[\bar{u}^{(s')}(p') \gamma^0 u^{(s)}(p) \right] \left[\bar{u}^{(s)}(p) \gamma^0 u^{(s')}(p') \right]
\tag{11.29}
$$

In the unpolarized case, we define as before the spin-averaged matrix element (see Eq. (11.5)) by averaging over the initial-state spin and summing over the final-state spin:

$$
\langle |\mathcal{M}|^2 \rangle = \frac{1}{2} \sum_{s=1}^{2} \sum_{s'=1}^{2} |\mathcal{M}|^2 = \frac{1}{2} \frac{Z^2 e^4}{q^4} \sum_{s=1}^{2} \sum_{s'=1}^{2} \left[\bar{u}^{(s')}(p') \gamma^0 u^{(s)}(p) \right] \left[\bar{u}^{(s)}(p) \gamma^0 u^{(s')}(p') \right]
\tag{11.30}
$$

We introduce a powerful method called "**Casimir's trick**" to compute the spin-averaged squared matrix element, which will be developed further in Section 11.9. We make use of the **completeness relations** valid for any spinor (see Eq. (8.97)) to remove the sum over the spins:

$$
\langle |\mathcal{M}|^2 \rangle \propto \frac{1}{2} \sum_{s'=1}^{2} \bar{u}^{(s')}(p') \gamma^0 \sum_{s=1}^{2} \left[u^{(s)}(p) \bar{u}^{(s)}(p) \right] \gamma^0 u^{(s')}(p') = \frac{1}{2} \sum_{s'=1}^{2} \bar{u}^{(s')}(p') \gamma^0 \left(\not{p} + m_e \mathbb{1} \right) \gamma^0 u^{(s')}(p')
\tag{11.31}
$$

If we make the spinor and matrices indices explicit, we are free to rearrange this. Let's define the matrix:

$$
\mathcal{G}^{\alpha\beta} = \left[\gamma^0 \left(\not{p} + m_e \mathbb{1} \right) \gamma^0 \right]^{\alpha\beta}
\tag{11.32}
$$

then we can write:

$$
\langle |\mathcal{M}|^2 \rangle \propto \frac{1}{2} \sum_{s'=1}^{2} \sum_{\alpha,\beta} \left[\bar{u}^{(s')}(p') \right]^\alpha \mathcal{G}^{\alpha\beta} \left[u^{(s')}(p') \right]^\beta
\tag{11.33}
$$

Note that in the last expression the indices are all at the same position since we are computing the sum of all products. By rearrangement we get:

$$\sum_{s'=1}^{2}\sum_{\alpha,\beta}\left[\bar{u}^{(s')}(p')\right]^{\alpha}\mathcal{G}^{\alpha\beta}\left[u^{(s')}(p')\right]^{\beta} = \sum_{\alpha,\beta}\mathcal{G}^{\alpha\beta}\sum_{s'=1}^{2}\left[u^{(s')}(p')\right]^{\beta}\left[\bar{u}^{(s')}(p')\right]^{\alpha} = \sum_{\alpha,\beta}\mathcal{G}^{\alpha\beta}\left[\slashed{p}'+m_e\mathbb{1}\right]^{\beta\alpha}$$

$$= \sum_{\alpha}\left[\mathcal{G}\left(\slashed{p}'+m_e\mathbb{1}\right)\right]^{\alpha\alpha} \tag{11.34}$$

where we again made use of the spinor completeness relation. We can multiply the two matrices and find:

$$\langle|\mathcal{M}|^2\rangle \propto \mathrm{Tr}\left[\gamma^0\left(\slashed{p}+m_e\mathbb{1}\right)\gamma^0\left(\slashed{p}'+m_e\mathbb{1}\right)\right] \tag{11.35}$$

We have found a result that is called the Casimir trick. Namely, the unpolarized square matrix element can be found by the trace of the product of the γ matrices and replacing the spinors by their momenta slashed plus their rest masses! The trick has translated the problem of spin averaging to the computation of a trace of γ matrices. Indeed, let us analyse the trace. If for the moment we ignore the momentum four-vectors and consider only the γ matrices, we have a term of the type:

$$\mathrm{Tr}\left[\gamma^0\left(\gamma^\mu+m_e\mathbb{1}\right)\gamma^0\left(\gamma^\nu+m_e\mathbb{1}\right)\right] \tag{11.36}$$

where the μ index is related to the $\slashed{p}$ term and the ν index to the $\slashed{p}'$ term. At this stage we consider and expand the products of γ matrices:

$$\mathrm{Tr}\left[\gamma^0\left(\gamma^\mu+m_e\mathbb{1}\right)\gamma^0\left(\gamma^\nu+m_e\mathbb{1}\right)\right] = \mathrm{Tr}\left[(\gamma^0\gamma^\mu+m_e\gamma^0)\gamma^0\left(\gamma^\nu+m_e\mathbb{1}\right)\right]$$

$$= \mathrm{Tr}\left[\gamma^0\gamma^\mu\gamma^0\gamma^\nu+m_e\gamma^0\gamma^0\gamma^\nu+m_e\gamma^0\gamma^\mu\gamma^0+m_e^2\gamma^0\gamma^0\right]$$

$$= \mathrm{Tr}\left[\gamma^0\gamma^\mu\gamma^0\gamma^\nu\right]+m_e\,\mathrm{Tr}\left[\gamma^0\gamma^0\gamma^\nu\right]+m_e\,\mathrm{Tr}\left[\gamma^0\gamma^\mu\gamma^0\right]+m_e^2\,\mathrm{Tr}\left[\mathbb{1}\right] \tag{11.37}$$

We have now reached a very practical situation where all we need to solve are traces of products of γ matrices. These kinds of terms will come back very often in our calculations of QED (and beyond). This stage of the calculation is generally called the step where we make use of the trace techniques or "traceology." There is a list of results (called **trace theorems**) which can help us. They are illustrated in Appendix E.2.

For example, an extremely useful theorem to reduce the number of terms is **[T3]**: *The trace of any product of an odd number of γ^μ is zero.* Hence two terms in our above expression vanish (recall also $\mathrm{Tr}(ABC) = \mathrm{Tr}(CAB) = \mathrm{Tr}(BCA)$ (cyclic)):

$$\mathrm{Tr}\left[\gamma^0\gamma^0\gamma^\nu\right] = \mathrm{Tr}\left[\gamma^0\gamma^\mu\gamma^0\right] = 0 \tag{11.38}$$

Cool!

Also, quite trivially $\mathrm{Tr}\left[\mathbb{1}\right] = 4$ so the only term that needs work is the first with the product of four matrices. In Appendix E.2, we also find **[T5]**, the proof that: $\mathrm{Tr}(\gamma^\mu\gamma^\nu\gamma^\rho\gamma^\sigma) = 4(g^{\mu\nu}g^{\rho\sigma}-g^{\mu\rho}g^{\nu\sigma}+g^{\mu\sigma}g^{\nu\rho})$. Hence (keeping track of the indices!):

$$\mathrm{Tr}\left[\gamma^0\gamma^\mu\gamma^0\gamma^\nu\right] = 4(g^{0\mu}g^{0\nu}-g^{00}g^{\mu\nu}+g^{0\nu}g^{\mu0}) \tag{11.39}$$

Putting back the four-vectors associated with the γ^μ and γ^ν matrices, and taking $p^\mu = (E,\vec{p})$ and similarly for $p'^\nu = (E',\vec{p'})$, we get:

$$p_\mu p'_\nu\,\mathrm{Tr}\left[\gamma^0\gamma^\mu\gamma^0\gamma^\nu\right] = 4(EE'-p\cdot p'+E'E) = 4(2EE'-p\cdot p') \tag{11.40}$$

Finally

$$\langle|\mathcal{M}|^2\rangle = \frac{1}{2}\sum_{s,s'=1}^{2}|\mathcal{M}|^2 = \frac{Z^2e^4}{q^4}2(2EE'-p\cdot p'+m_e^2) \tag{11.41}$$

We can now plug in the kinematics. As before, we fix $p^\mu = (E, p, 0, 0)$ and with the δ function imposing $E = E'$ we get $|\vec{p}| = |\vec{p}'| = p$ and $p^\mu = (E, 0, 0, p)$, $p'^\mu = (E, p\sin\theta, 0, p\cos\theta)$, where θ is the electron scattering angle (see Eq. (11.20)). We have with $E^2 = p^2 + m_e^2$:

$$p \cdot p' = E^2 - p^2 \cos\theta = m_e^2 + p^2(1 - \cos\theta) = m_e^2 + 2p^2 \sin^2(\theta/2) \tag{11.42}$$

so

$$4(2EE' - p \cdot p' + m_e^2) = 4(2E^2 - m_e^2 - 2p^2 \sin^2(\theta/2) + m_e^2) = 8(E^2 - p^2 \sin^2(\theta/2)) \tag{11.43}$$

Finally,

$$\langle |\mathcal{M}|^2 \rangle = 4 \frac{Z^2 e^4}{q^4} E^2 (1 - \beta^2 \sin^2(\theta/2)) \tag{11.44}$$

where again the electron velocity is $\beta = p/E$. This compares, as it should, exactly with the result Eq. (11.26) obtained with the spinors!

The electron scattering differential cross-section $d\sigma/d\Omega$ can therefore be expressed, making use of the phase-space factor for $e^- + \mathcal{A} \rightarrow e^- + \mathcal{A}$ (see Eq. (5.151)):

$$\left(\frac{d\sigma}{d\Omega} \right) = \frac{1}{64\pi^2 M_\mathcal{A} (E + M_\mathcal{A} - p\cos\theta)} \times \langle |\mathcal{M}|^2 \rangle \times (2M_\mathcal{A})^2 \tag{11.45}$$

where the last term takes into account the relativistic normalization of the target $\mathcal{A}$. Since we assume that the scattering is off a static potential, there is no recoiling target, and we can achieve this by letting $M_\mathcal{A} \rightarrow \infty$. This yields:

$$\left(\frac{d\sigma}{d\Omega} \right) = \frac{1}{64\pi^2} \times \langle |\mathcal{M}|^2 \rangle \times 4 = \frac{1}{16\pi^2} \times \underbrace{\frac{Z^2 e^4}{q^4} 4E^2 (1 - \beta^2 \sin^2(\theta/2))}_{= \langle |\mathcal{M}|^2 \rangle} \tag{11.46}$$

Noting that:

$$q^2 = (\vec{p} - \vec{p}')^2 = p^2 - 2p^2 \cos\theta + p^2 = 2p^2(1 - \cos\theta) = 4p^2 \sin^2(\theta/2) \tag{11.47}$$

we get:

$$\frac{4E^2}{q^4} = \frac{E^2}{4p^4 \sin^4(\theta/2)} \tag{11.48}$$

Consequently, the differential cross-section is given by:

$$\left(\frac{d\sigma}{d\Omega} \right) = \frac{1}{16\pi^2} \times \frac{(4\pi\alpha Z)^2 E^2}{4p^4 \sin^4(\theta/2)} \left(1 - \beta^2 \sin^2(\theta/2) \right) = \frac{(\alpha Z)^2 E^2}{4p^4 \sin^4(\theta/2)} \left(1 - \beta^2 \sin^2(\theta/2) \right) \tag{11.49}$$

where $\alpha = e^2/(4\pi)$ is the fine-structure constant (see Eq. (1.37)).

11.2.4 Analyzing the Result

The derived differential cross-section is valid when we scatter off a static potential and the scattered electron (or particle) is a spin-1/2 Dirac particle. This can be seen as scattering off a target particle which does not recoil. We now analyze the result in two kinematical regimes: (1) when the electron is non-relativistic $\beta \rightarrow 0$; (2) when the electron is relativistic $\beta \rightarrow 1$.

• **Non-relativistic limit.** We can replace the energy by the rest mass, $E \rightarrow m_e$, and the kinetic energy is $E_K = p^2/2m_e$. Hence

$$\left(\frac{d\sigma}{d\Omega} \right) = \frac{(\alpha Z)^2 m_e^2}{4p^4 \sin^4(\theta/2)} \left(1 - \beta^2 \sin^2(\theta/2) \right) = \underbrace{\frac{(\alpha Z)^2}{16 E_K^2 \sin^4(\theta/2)}}_{\text{Rutherford}} \underbrace{\left(1 - \beta^2 \sin^2(\theta/2) \right)}_{\text{spin correction}} \tag{11.50}$$

Compare with Eq. (2.88).

● **Relativistic case.** For $E \approx p$, $\beta \to 1$, we find what is called the **Mott scattering formula** first derived by Mott[4] in 1929 [83]:

$$\left(\frac{\mathrm{d}\sigma}{\mathrm{d}\Omega}\right)_{Mott} = \left(\frac{\mathrm{d}\sigma}{\mathrm{d}\Omega}\right)_{Rutherford} \times \cos^2\left(\frac{\theta}{2}\right) = \frac{(\alpha Z)^2 \cos^2\left(\frac{\theta}{2}\right)}{4p^2 \sin^4\left(\frac{\theta}{2}\right)} \tag{11.51}$$

The Mott cross-section is valid in the limit where the target recoil is neglected and the scattered electron (or particle) is relativistic with spin-1/2. In the case of the Rutherford scattering, Eq. (2.88), we were in the limit where the target recoil was neglected and the scattered particle was spinless and non-relativistic. In both cases, the target is always considered point-like, which is surely an approximation of a real target nucleus, but the goal here is to estimate the effect of the electron spin on the differential cross-section.

The graphical comparisons of the Rutherford and Mott scattering cross-sections as a function of the scattering angle θ and $\cos\theta$ are shown in Figure 11.2. The extra $\cos^2(\theta/2)$ term in the Mott cross-section is the correction due to the electron spin. With the help of the double-angle trigonometry relations, $\cos^2(\theta/2) = (1 + \cos\theta)/2$ and $\sin^2(\theta/2) = (1 - \cos\theta)/2$, we find:

$$\left(\frac{\mathrm{d}\sigma}{\mathrm{d}\Omega}\right)_{Mott} \propto \frac{1 + \cos\theta}{(1 - \cos\theta)^2} \tag{11.52}$$

In the limit $\theta \to 180°$ (or $\cos\theta \to -1$), the Mott cross-section vanishes because the electron cannot be fully back scattered while conserving its helicity. This configuration would correspond to the 180° reversal of the direction of the electron spin, which is not allowed by total angular momentum conservation.

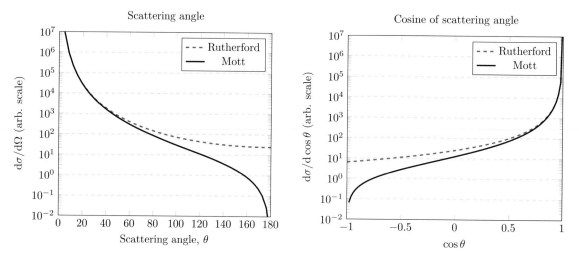

Figure 11.2 Angular dependence of the Mott scattering cross-section compared to the Rutherford formula: (left) as a function of the scattering angle θ; (right) as a function of $\cos\theta$.

---

4   Sir Nevill Francis Mott (1905–1996), British physicist.

11.3   Spin-Flip Paradox and Thomas Precession

We discuss in this section an apparent paradox related to the spin-dependent differential cross-sections. We take the example of Mott scattering, but the conclusions apply to any QED process. In order to simplify the discussion, we take the concrete case where the incoming electron has a velocity $\beta = 2\sqrt{2}/3 \approx 0.94$ which implies a Lorentz factor $\gamma = 3$, and the outgoing electron is scattered at a right angle from the incoming electron, i.e., the scattering angle in this frame is $\theta^\star = 90°$. We have $E^\star = \gamma m_e = 3m_e$ and $p^\star = \gamma m_e \beta = 2\sqrt{2}m_e$. Consequently, the kinematics in this frame are simply given by:

$$p^{\star\mu} = \begin{pmatrix} 3m_e \\ 0 \\ 0 \\ 2\sqrt{2}m_e \end{pmatrix} \quad \text{and} \quad p^{\star'\mu} = \begin{pmatrix} 3m_e \\ 2\sqrt{2}m_e \\ 0 \\ 0 \end{pmatrix} \tag{11.53}$$

We need to define our spinor basis in order to compute the spin-dependent matrix element corresponding to this kinematical configuration. This time we decide to use the two positive-energy Dirac solutions from Eq. (8.75) in order to compute our spinor. We have seen in Section 8.12 that these solutions represent spin-up and spin-down eigenstates when the particle travels along the z-axis. This is the case of the initial-state electron. For the scattered electron, this choice simply means we are keeping the z-axis as our quantization axis for its polarization. Insertion of our kinematics in the relevant expressions yields:

$$u_\uparrow(p^\star) = \sqrt{m_e} \begin{pmatrix} 2 \\ 0 \\ \sqrt{2} \\ 0 \end{pmatrix}, \quad u_\downarrow(p^\star) = \sqrt{m_e} \begin{pmatrix} 0 \\ 2 \\ 0 \\ -\sqrt{2} \end{pmatrix}, \quad u_\uparrow(p'^\star) = \sqrt{m_e} \begin{pmatrix} 2 \\ 0 \\ 0 \\ \sqrt{2} \end{pmatrix}, \quad u_\downarrow(p'^\star) = \sqrt{m_e} \begin{pmatrix} 0 \\ 2 \\ \sqrt{2} \\ 0 \end{pmatrix} \tag{11.54}$$

A piece of cake! We can then find the corresponding amplitudes simply by using Eq. (11.19):

$$\mathcal{M}_{\uparrow\uparrow} = \left(\frac{Ze^2}{q^2}\right) \bar{u}_\uparrow(p'^\star)\gamma^0 u_\uparrow(p^\star) = \left(\frac{Ze^2}{q^2}\right) m_e \left(2,0,0,\sqrt{2}\right) \begin{pmatrix} 2 \\ 0 \\ \sqrt{2} \\ 0 \end{pmatrix} = \left(\frac{Ze^2}{q^2}\right) 4m_e \tag{11.55}$$

and similarly:

$$\mathcal{M}_{\uparrow\downarrow} = \left(\frac{Ze^2}{q^2}\right) \bar{u}_\uparrow(p'^\star)\gamma^0 u_\downarrow(p^\star) = \left(\frac{Ze^2}{q^2}\right) m_e \left(2,0,0,\sqrt{2}\right) \begin{pmatrix} 0 \\ 2 \\ 0 \\ -\sqrt{2} \end{pmatrix} = -\left(\frac{Ze^2}{q^2}\right) 2m_e \tag{11.56}$$

Therefore, using parity conservation, we find:

$$|\mathcal{M}_{\uparrow\uparrow}|^2 = |\mathcal{M}_{\downarrow\downarrow}|^2 = 16m_e^2 \left(\frac{Ze^2}{q^2}\right)^2, \quad \text{and} \quad |\mathcal{M}_{\uparrow\downarrow}|^2 = |\mathcal{M}_{\downarrow\uparrow}|^2 = 4m_e^2 \left(\frac{Ze^2}{q^2}\right)^2 \tag{11.57}$$

where $q^2 = (p^\star - p'^\star)^2 = -16m_e^2$. Then, using Eq. (11.46), we derive the differential cross-section:

$$\begin{aligned} \left(\frac{d\sigma}{d\Omega}\right) &= \frac{\langle|\mathcal{M}|^2\rangle}{16\pi^2} = \frac{1}{16\pi^2}\frac{1}{2}\left(2|\mathcal{M}_{\uparrow\uparrow}|^2 + 2|\mathcal{M}_{\uparrow\downarrow}|^2\right) = \frac{1}{16\pi^2}\frac{Z^2 e^4}{q^4}m_e^2(16+4) \\ &= \frac{Z^2\alpha^2}{64m_e^2}(\underbrace{4}_{\text{no spin flip}} + \underbrace{1}_{\text{spin flip}}) \end{aligned} \tag{11.58}$$

where the factor $1/2$ in the first line comes from the averaging over the initial-state spin. Let us compare this result to the one obtained by direct application of the (spin-averaged) Mott formula with $\sin^2(\theta^\star/2) = 1/2$ and check that we indeed get the same result:

$$\left(\frac{d\sigma}{d\Omega}\right)_{Mott} = \frac{Z^2\alpha^2 E^{\star 2}}{4p^{\star 4}\sin^4(\theta^\star/2)}(1 - \beta^2\sin^2(\theta^\star/2)) = \frac{Z^2\alpha^2(3m_e)^2}{4(2\sqrt{2}m_e)^4\left(\frac{1}{2}\right)^2}\left(1 - \frac{8}{9}\frac{1}{2}\right) = \frac{9Z^2\alpha^2}{4\times 16\times 4m_e^2\times\frac{1}{4}}\left(\frac{5}{9}\right)$$

$$= \frac{Z^2\alpha^2}{64m_e^2}(5) \qquad \square \tag{11.59}$$

So far, so good. Now let us analyze the same scattering process from the perspective of a different inertial observer. We consider a Lorentz boost of our coordinate system in the $-x$ direction with a velocity $\beta_x = 2\sqrt{2}/3$ (we note that this velocity coincides with the incoming velocity of the electron along the z direction in the previous frame) with $\gamma_x = 3$, such that relative to this system, we have (see Eqs. (5.24) and (11.20)):

$$p^\mu = \underbrace{\begin{pmatrix} \gamma_x & \beta_x\gamma_x & 0 & 0 \\ \beta_x\gamma_x & \gamma_x & 0 & 0 \\ 0 & 0 & 1 & 0 \\ 0 & 0 & 0 & 1 \end{pmatrix}}_{=\Lambda^\mu{}_\nu(-\beta_x)}\begin{pmatrix} E^\star \\ 0 \\ 0 \\ p^\star \end{pmatrix} = \begin{pmatrix} \gamma_x E^\star \\ \gamma_x\beta_x E^\star \\ 0 \\ p^\star \end{pmatrix} = \begin{pmatrix} 9m_e \\ 6\sqrt{2}m_e \\ 0 \\ 2\sqrt{2}m_e \end{pmatrix} \tag{11.60}$$

where the $\star$ quantities refer to the unboosted reference frame, and:

$$p'^\mu = \begin{pmatrix} \gamma_x & \beta_x\gamma_x & 0 & 0 \\ \beta_x\gamma_x & \gamma_x & 0 & 0 \\ 0 & 0 & 1 & 0 \\ 0 & 0 & 0 & 1 \end{pmatrix}\begin{pmatrix} E^\star \\ p^\star\sin\theta^\star \\ 0 \\ p^\star\cos\theta^\star \end{pmatrix} = \begin{pmatrix} \gamma_x(E^\star + \beta_x p^\star\sin\theta^\star) \\ \gamma_x(\beta_x E^\star + p^\star\sin\theta^\star) \\ 0 \\ p^\star\cos\theta^\star \end{pmatrix} = \begin{pmatrix} 17m_e \\ 12\sqrt{2}m_e \\ 0 \\ 0 \end{pmatrix} \tag{11.61}$$

Our spinors then become:

$$u_\uparrow(p) = \sqrt{m_e}\begin{pmatrix} \sqrt{10} \\ 0 \\ 2/\sqrt{5} \\ 6/\sqrt{5} \end{pmatrix}, \quad u_\downarrow(p) = \sqrt{m_e}\begin{pmatrix} 0 \\ \sqrt{10} \\ 6/\sqrt{5} \\ -2/\sqrt{5} \end{pmatrix}, \quad u_\uparrow(p') = \sqrt{m_e}\begin{pmatrix} 3\sqrt{2} \\ 0 \\ 0 \\ 4 \end{pmatrix}, \quad u_\downarrow(p') = \sqrt{m_e}\begin{pmatrix} 0 \\ 3\sqrt{2} \\ 4 \\ 0 \end{pmatrix} \tag{11.62}$$

We must also boost the electromagnetic potential as described in Eq. (11.16), and this yields:

$$A^\mu(q^2) = \frac{Ze^2}{q^2}\begin{pmatrix} \gamma_x & \beta_x\gamma_x & 0 & 0 \\ \beta_x\gamma_x & \gamma_x & 0 & 0 \\ 0 & 0 & 1 & 0 \\ 0 & 0 & 0 & 1 \end{pmatrix}\begin{pmatrix} 1 \\ 0 \\ 0 \\ 0 \end{pmatrix} = \frac{Ze^2}{q^2}\begin{pmatrix} \gamma_x \\ \beta_x\gamma_x \\ 0 \\ 0 \end{pmatrix} \tag{11.63}$$

Then, according to Eq. (11.14), the amplitude of the scattering process in the boosted frame is given by:

$$\mathcal{M} = A_\mu(q^2)\bar{u}(p')\gamma^\mu u(p) = \gamma_x\frac{Ze^2}{q^2}\left[\bar{u}(p')\gamma^0 u(p) - \beta_x\bar{u}(p')\gamma^1 u(p)\right] \tag{11.64}$$

where the minus sign comes from the metric when computing $A_\mu\gamma^\mu$. The first term in the bracket with γ^0 was already computed in Eq. (11.19). Similarly, we see that:

$$\bar{u}(p')\gamma^1 u(p) = u^\dagger(p')\gamma^0\gamma^1 u(p)$$

$$= (u_1^*(p'), u_2^*(p'), u_3^*(p'), u_4^*(p'))\begin{pmatrix} 1 & 0 & 0 & 0 \\ 0 & 1 & 0 & 0 \\ 0 & 0 & -1 & 0 \\ 0 & 0 & 0 & -1 \end{pmatrix}\begin{pmatrix} 0 & 0 & 0 & 1 \\ 0 & 0 & 1 & 0 \\ 0 & -1 & 0 & 0 \\ -1 & 0 & 0 & 0 \end{pmatrix}\begin{pmatrix} u_1(p) \\ u_2(p) \\ u_3(p) \\ u_4(p) \end{pmatrix}$$

$$= u_1^*(p')u_4(p) + u_2^*(p')u_3(p) + u_3^*(p')u_2(p) + u_4^*(p')u_1(p) \tag{11.65}$$

These lead to the following matrix element:

$$
\mathcal{M}'_{\uparrow\uparrow} = \frac{\gamma_x Z e^2}{q^2} m_e \left(\begin{pmatrix} 3\sqrt{2} \\ 0 \\ 0 \\ 4 \end{pmatrix}^T \begin{pmatrix} \sqrt{10} \\ 0 \\ 2/\sqrt{5} \\ 6/\sqrt{5} \end{pmatrix} - \frac{2\sqrt{2}}{3} \begin{pmatrix} 3\sqrt{2} \\ 0 \\ 0 \\ 4 \end{pmatrix}^T \begin{pmatrix} 6/\sqrt{5} \\ 2/\sqrt{5} \\ 0 \\ \sqrt{10} \end{pmatrix} \right) = \frac{3Z e^2}{q^2} m_e \left(\frac{54}{\sqrt{5}} - \frac{152}{3\sqrt{5}} \right)
$$

$$
= \frac{Z e^2}{q^2} m_e \left(2\sqrt{5} \right) \tag{11.66}
$$

and similarly:

$$
\mathcal{M}'_{\uparrow\downarrow} = \frac{\gamma_x Z e^2}{q^2} m_e \left(\begin{pmatrix} 0 \\ 3\sqrt{2} \\ 4 \\ 0 \end{pmatrix}^T \begin{pmatrix} \sqrt{10} \\ 0 \\ 2/\sqrt{5} \\ 6/\sqrt{5} \end{pmatrix} - \frac{2\sqrt{2}}{3} \begin{pmatrix} 0 \\ 3\sqrt{2} \\ 4 \\ 0 \end{pmatrix}^T \begin{pmatrix} 6/\sqrt{5} \\ 2/\sqrt{5} \\ 0 \\ \sqrt{10} \end{pmatrix} \right) = \frac{3Z e^2}{q^2} m_e \left(\frac{8}{\sqrt{5}} - \frac{8}{\sqrt{5}} \right)
$$

$$
= 0 \tag{11.67}
$$

Consequently, using parity conservation, we find:

$$
|\mathcal{M}'_{\uparrow\uparrow}|^2 = |\mathcal{M}'_{\downarrow\downarrow}|^2 = 20 m_e^2 \left(\frac{Z e^2}{q^2} \right)^2, \quad \text{and} \quad |\mathcal{M}'_{\uparrow\downarrow}|^2 = |\mathcal{M}'_{\downarrow\uparrow}|^2 = 0 \tag{11.68}
$$

We are baffled by the fact that these expressions do not look like Eq. (11.57) obtained in the first reference frame. Of course, the factor $Z e^2/q^2 m_e$ is an invariant as expected, but **we found that different inertial observers observe different probabilities of spin-conserving and spin-flipping processes! Is this possible?** This result leads to what is sometimes called the **spin-flip paradox.**

Indeed, in the frame at rest, we computed that the ratio of spin conserving to spin flipping reactions was $16 : 4$, while we find in the boosted frame that this ratio is $20 : 0$. The spin flip is prohibited in the boosted frame (given our choice of parameters). The reader is perhaps not surprised to note that the *unpolarized* cross-section is however identical in both frames since:

$$
\left(\frac{d\sigma}{d\Omega} \right) = \frac{Z^2 \alpha^2}{64 m_e^2} (\underbrace{4}_{\text{no spin flip}} + \underbrace{1}_{\text{spin flip}}) \quad \text{and} \quad \left(\frac{d\sigma}{d\Omega} \right)' = \frac{Z^2 \alpha^2}{64 m_e^2} (\underbrace{5}_{\text{no spin flip}} + \underbrace{0}_{\text{spin flip}}) \tag{11.69}
$$

Thus, even though we obtained two apparently inconsistent sets of polarized cross-sections, the unpolarized cross-sections agree (and in fact agree with the Mott scattering formula!). The key to understanding the paradox is to realize that the Lorentz boost that we are applying is not parallel to the direction of flight of the incoming electron! In order to move from the observer's reference frame to the electron rest frame (i.e., the frame in which the spin is defined), one should perform two *non-collinear* boosts. Hence, the electron's rest frame is *Thomas–Wigner rotated* with respect to the observer's rest frame (see Section 5.6). Hence, the electron spin points in a rotated direction compared to a boosted observer. These conclusions remain valid for any other kinematical parameter. Hence, we conclude that different inertial observers will measure different polarized cross-sections. Another way of stating this is to say that the polarization of a particle depends on the frame from which it is observed. A particle, let us say fully polarized in the center-of-mass frame of a reaction, will in general not be fully polarized when observed in the laboratory frame if this latter is a different frame!

11.4   Electron–Muon/Tau Scattering at Tree Level

As a further example for the application of the QED Feynman rules, consider the elastic scattering of an electron and a muon or a tau lepton at **tree level** (i.e., the lowest order in the Dyson series). The reaction is

$e^-\ell^- \to e^-\ell^-$, where $\ell = \mu$ or τ, and we define the initial and final-state kinematics with four-momenta and spin states (p^μ, s), (k^μ, r), (p'^μ, s'), and (k'^μ, r'). Then we have:

$$e^-(p,s) + \ell^-(k,r) \to e^-(p',s') + \ell^-(k',r') \tag{11.70}$$

with the corresponding tree-level Feynman diagram shown in Figure 11.3. By using the Feynman rules, we can directly construct the amplitude, reading the factors off the diagram, considering the two fermion currents – electron at the bottom and other charged lepton at the top of the diagram – linked together via the photon propagator. **At tree level the interaction between the electron and the lepton ℓ occurs via the exchange of a virtual photon with four-momentum q^μ.**

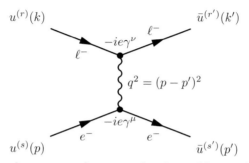

Figure 11.3 Coulomb scattering between an electron and a charged lepton ℓ: tree-level Feynman diagram.

Energy–momentum conservation at the two vertices imposes $p = q + p'$ and $k + q = k'$ or $q = p - p' = k' - k$ (alternatively we could have chosen $q = p' - p = k - k'$ without any effect on observables). The overall energy–momentum conservation yields two Dirac δ functions $\delta^4(q - p + p')$ and $\delta^4(q - k' + k)$. Finally, the amplitude for Coulomb e-ℓ scattering at first order $\langle e^-\ell^-|S_2|e^-\ell^-\rangle$ is proportional to:

$$\left(\bar{u}^{(s')}(p')(-ie\gamma^\mu)u^{(s)}(p)\right)\frac{-ig_{\mu\nu}}{q^2}\left(\bar{u}^{(r')}(k')(-ie\gamma^\nu)u^{(r)}(k)\right)(2\pi)^4\delta^4(q-p+p')\delta^4(q-k'+k)$$

$$= \frac{ie^2}{(p-p')^2}\left(\bar{u}(p')\gamma^\mu u(p)\right)\left(\bar{u}(k')\gamma_\mu u(k)\right)(2\pi)^4\delta^4(p+k-p'-k') \tag{11.71}$$

where in the last line we removed the spin indices for readability, so **we need to remember to associate again the spins s, s', r, r' for specific calculations with the spinors.** Finally, the corresponding amplitude $\mathcal{M}$ is equal to:

$$-i\mathcal{M}_{e\ell \to e\ell} = \frac{ie^2}{(p-p')^2}\left(\bar{u}(p')\gamma^\mu u(p)\right)\left(\bar{u}(k')\gamma_\mu u(k)\right) \tag{11.72}$$

It is practical to split this amplitude into the four-currents corresponding to the electron and the lepton ℓ, as follows:

$$-i\mathcal{M}_{e\ell \to e\ell} = \frac{ie^2}{(p-p')^2}j^\mu(p,p')j_\mu(k,k') \tag{11.73}$$

where the electron and the muon/tau currents have exactly the same form and are simply the product of the adjoint spinor, the gamma matrices γ^μ, and the spinor of the corresponding fermion:

$$j^\mu(p,p') = \bar{u}(p')\gamma^\mu u(p) \quad \text{and} \quad j^\mu(k,k') = \bar{u}(k')\gamma^\mu u(k) \tag{11.74}$$

11.5   Heavy Lepton Pair Creation

An electron and a positron can annihilate each other into a single virtual photon, and if the center-of-mass energy is sufficient, the photon can create a heavier fermion–antifermion pair $\ell^+\ell^-$, where $\ell = \mu$ or τ. The kinematics of the event is defined as follows:

$$e^-(p,s) + e^+(k,r) \rightarrow \ell^-(p',s') + \ell^+(k',r') \qquad \ell = \mu, \tau \tag{11.75}$$

The tree-level Feynman diagram is shown in Figure 11.4. The amplitude of this diagram can be directly

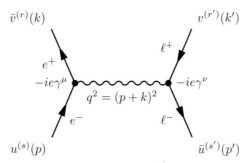

Figure 11.4 Feynman diagram for heavy lepton $\ell^+\ell^-$ pair creation, where $\ell = \mu$ or τ.

expressed as:

$$-i\mathcal{M}_{e^+e^- \rightarrow \ell^+\ell^-} = \frac{ie^2}{(p+k)^2} \left(\bar{v}(k)\gamma^\mu u(p) \right) \left(\bar{u}(p')\gamma_\mu v(k') \right) \tag{11.76}$$

where $\ell = \mu$ or τ. Note that we have used the u spinors for the particles and v spinors for the antiparticles. The order of the spinors in the amplitude follows that of the corresponding arrows in the Feynman diagram.

11.6   Møller Scattering

The **Møller**[5] reaction $e^-e^- \rightarrow e^-e^-$ is similar to $e^-\ell^- \rightarrow e^-\ell^-$, but involves only electrons. We define the initial and final-state kinematics with four-momenta and spin states (p^μ, s), (k^μ, r), (p'^μ, s'), and (k'^μ, r'). We have now:

$$e^-(p,s) + e^-(k,r) \rightarrow e^-(p',s') + e^-(k',r') \tag{11.77}$$

We note that unlike the $e^-\ell^- \rightarrow e^-\ell^-$ case, the two electrons in the final state are *indistinguishable* and we cannot tell from which vertex they have originated. Hence, we cannot distinguish the case (1) where the final-state electron (p'^μ, s') comes from vertex 1 and the final-state electron (k'^μ, r') comes from vertex 2 from the opposite situation (2) where (p'^μ, s') comes from vertex 2 and (k'^μ, r') comes from vertex 1. **As these two cases are indistinguishable, we must draw both diagrams and add their amplitudes.** At tree level we obtain the two diagrams shown in Figure 11.5.

The amplitude of the diagram on the left, which we call $\mathcal{M}_t$ (for reasons that will be explained later) is the same as the one for $e^-\ell^- \rightarrow e^-\ell^-$. The amplitude called $\mathcal{M}_u$ for the diagram on the right is obtained by swapping $(p',s') \leftrightarrow (k',r')$ in $\mathcal{M}_t$, and **adding an overall minus sign since the interchange of two indistinguishable fermions is antisymmetric.** This is a direct consequence of rule **[R13]**: the minus sign between the amplitude of two diagrams ensures that the overall amplitude is antisymmetric under the interchange of the two initial and two final-state electrons. Hence the antisymmetrization yields (ignoring the Dirac function of the four-momentum conservation):

$$
\begin{aligned}
-i\mathcal{M}_{e^-e^- \rightarrow e^-e^-} &= \mathcal{M}_t - \mathcal{M}_u \\
&= \frac{ie^2}{(p-p')^2} \left(\bar{u}(p')\gamma^\mu u(p) \right) \left(\bar{u}(k')\gamma_\mu u(k) \right) - \frac{ie^2}{(p-k')^2} \left(\bar{u}(k')\gamma^\mu u(p) \right) \left(\bar{u}(p')\gamma_\mu u(k) \right)
\end{aligned}
\tag{11.78}
$$

5   Christian Møller (1904–1980), Danish chemist and physicist.

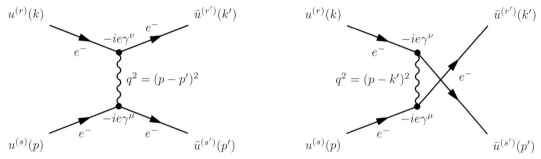

Figure 11.5 Møller scattering: two Feynman diagrams contributing at tree level. (left) The amplitude $\mathcal{M}_t$; (right) the amplitude $\mathcal{M}_u$.

11.7   Bhabha Scattering

The Bhabha scattering reaction, named after **H. J. Bhabha**,[6] corresponds to electron–positron scattering. The process is $e^-e^+ \to e^-e^+$. The kinematics are defined as:

$$e^-(p,s) + e^+(k,r) \to e^-(p',s') + e^+(k',r') \tag{11.79}$$

There are two diagrams contributing at tree level: photon exchange and annihilation. See Figure 11.6.

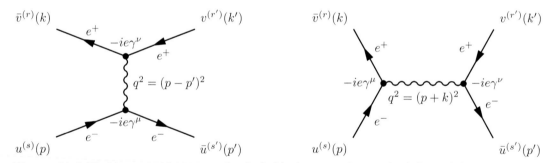

Figure 11.6 Bhabha scattering at tree level: (left) photon exchange; (right) annihilation process.

The total amplitude becomes:

$$
\begin{aligned}
-i\mathcal{M}_{e^+e^- \to e^+e^-} &= \mathcal{M}_t - \mathcal{M}_s \\
&= \frac{ie^2}{(p-p')^2}\left(\bar{u}(p')\gamma^\mu u(p)\right)\left(\bar{v}(k)\gamma_\mu v(k')\right) - \frac{ie^2}{(p+k)^2}\left(\bar{v}(k)\gamma^\mu u(p)\right)\left(\bar{u}(p')\gamma_\mu v(k')\right)
\end{aligned}
\tag{11.80}
$$

There is again a relative minus sign between the two terms of the amplitude. This is due to rule **[R13]**, which states in this case that the overall amplitude must be antisymmetric under the interchange of one initial (resp. final) electron with a final (resp. initial) positron.

6   Homi Jehangir Bhabha (1909–1966), Indian nuclear physicist, founding director, and professor of physics at the Tata Institute of Fundamental Research.

This minus sign is also consistent with the fact that the Bhabha scattering amplitude is related to that of the Møller scattering via crossing symmetry, as discussed in the next section. This enforces the same relative sign between the two diagrams of each reaction.

11.8   Mandelstam Variables and Crossing Symmetry

In 1958, **Mandelstam**[7] introduced three **relativistic scalar-invariant variables**, now named after him. For a generic $2 \to 2$ reaction with particles f_i of rest mass m_i, like:

$$f_1 + f_2 \to f_3 + f_4 \tag{11.81}$$

the **Mandelstam variables** are defined as the scalar products of pairs of four-momentum vectors of the following combinations of the particles:

$$\begin{aligned}
s &= (f_1 + f_2)^2 = (f_3 + f_4)^2 \\
t &= (f_1 - f_3)^2 = (f_4 - f_2)^2 \\
u &= (f_1 - f_4)^2 = (f_3 - f_2)^2
\end{aligned} \tag{11.82}$$

Because of energy–momentum conservation, each Mandelstam variable can be defined in the two identical ways, as shown above. The letters s, t, u are used in the terms *s-channel* (or **space-channel**), *t-channel* (or **time-channel**), and *u-channel* (like *t*-channel but with the final-state particles swapped). In terms of our kinematics defined by p, k, p', and k' we would have:

$$f_1(p) + f_2(k) \to f_3(p') + f_4(k') \tag{11.83}$$

with

$$s = (p + k)^2 = (p' + k')^2; \qquad t = (p - p')^2 = (k' - k)^2; \qquad u = (p - k')^2 = (p' - k)^2 \tag{11.84}$$

The center-of-mass frame (CMS) corresponds *per definition* to the condition $\vec{p}^{\star} + \vec{k}^{\star} = 0$, where $\star$ denotes quantities in the CMS. The center-of-mass energy is then:

$$E_\star = E_p^{\star} + E_k^{\star} \quad \text{with} \quad \vec{p}^{\star} + \vec{k}^{\star} = 0 \tag{11.85}$$

As noted earlier (see Eq. (5.128)), this implies that the invariant s is equal to the **square of the center-of-mass energy**:

$$s = (p + k)^2 = (E_p + E_k)^2 - (\vec{p} + \vec{k})^2 = (E_p^{\star} + E_k^{\star})^2 - 0 = E_\star^2 > 0 \tag{11.86}$$

or

$$E_\star = \sqrt{s} \tag{11.87}$$

Summing the three Mandelstam variables, we obtain:

$$\begin{aligned}
s + t + u &= (p + k)^2 + (p - p')^2 + (p - k')^2 \\
&= p^2 + k^2 + 2p \cdot k + p^2 + (p')^2 - 2p \cdot p' + p^2 + (k')^2 - 2p \cdot k' \\
&= m_1^2 + m_2^2 + m_3^2 + m_4^2 + 2p^2 + 2p \cdot k - 2p \cdot p' - 2p \cdot k' \\
&= m_1^2 + m_2^2 + m_3^2 + m_4^2 + 2p \cdot (p + k - p' - k') \\
&= \sum_{i=1}^{4} m_i^2
\end{aligned} \tag{11.88}$$

7   Stanley Mandelstam (1928–2016), South Africa-born American theoretical physicist.

where in the last line we used the overall energy–momentum conservation. **Note that t and u can be negative since s is positive.**

In the so-called **ultra-relativistic limit** defined as $E_i \gg m_i$ or $m_i \to 0$, the rest masses are negligible in comparison to the energies of the particles. In this case, the Mandelstam variables simplify to products of two four-vectors and the sum $s + t + u$ vanishes:

$$
\textbf{Ultra-relativistic limit}: \begin{cases} s = m_1^2 + m_2^2 + 2p \cdot k \simeq 2p \cdot k \simeq 2p' \cdot k' \\[2mm] t = m_1^2 + m_3^2 - 2p \cdot p' \simeq -2p \cdot p' \simeq -2k' \cdot k \\[2mm] u = m_1^2 + m_4^2 - 2p \cdot k' \simeq -2p \cdot k' \simeq -2p' \cdot k \\[2mm] \qquad\qquad\qquad\qquad s + t + u \simeq 0 \end{cases} \tag{11.89}
$$

The Mandelstam variables can be used to classify the amplitudes of the processes that we have discussed in the previous sections. Each process is characterized by the exchange of a virtual photon whose kinematics depend on the diagram configuration. They are summarized in Table 11.1.

Reaction	Channel	Diagram	Amplitude $-i\mathcal{M}/ie^2$
$e^- \ell^- \to e^- \ell^-$	t	Fig. 11.3	$\dfrac{(\bar{u}(p')\gamma^\mu u(p))\,(\bar{u}(k')\gamma_\mu u(k))}{t}$
$e^+ e^- \to \ell^+ \ell^-$	s	Fig. 11.4	$\dfrac{(\bar{v}(k)\gamma^\mu u(p))\,(\bar{u}(p')\gamma_\mu v(k'))}{s}$
$e^- e^- \to e^- e^-$ (Møller)	$t+u$	Fig. 11.5	$\dfrac{(\bar{u}(p')\gamma^\mu u(p))\,(\bar{u}(k')\gamma_\mu u(k))}{t} - \dfrac{(\bar{u}(k')\gamma^\mu u(p))\,(\bar{u}(p')\gamma_\mu u(k))}{u}$
$e^+ e^- \to e^+ e^-$ (Bhabha)	$s+t$	Fig. 11.6	$\dfrac{(\bar{u}(p')\gamma^\mu u(p))\,(\bar{v}(k)\gamma_\mu v(k'))}{t} - \dfrac{(\bar{v}(k)\gamma^\mu u(p))\,(\bar{u}(p')\gamma_\mu v(k'))}{s}$

Table 11.1 Classification of the basic QED processes as a function of the Mandelstam channel for the kinematics of the photon propagator. In the table, ℓ stands for μ or τ.

Let us investigate the annihilation (pair creation) process $e^+ e^- \to \ell^+ \ell^-$ and the scattering process $e^- \ell^- \to e^- \ell^-$. According to the definitions of the Mandelstam variables, the annihilation is an s-channel process while scattering is a t-channel, as shown in Figure 11.7(left) and (middle). The electron (p, p') and lepton ℓ (k, k') currents for the annihilation are equal to:

$$
j^\mu(p, k) = \bar{v}(k)\gamma^\mu u(p) \quad \text{and} \quad j^\mu(p', k') = \bar{u}(p')\gamma^\mu v(k') \tag{11.90}
$$

For the scattering process they are:

$$
j^\mu(p, p') = \bar{u}(p')\gamma^\mu u(p) \quad \text{and} \quad j^\mu(k, k') = \bar{u}(k')\gamma^\mu u(k) \tag{11.91}
$$

The scattering amplitude can be derived from the annihilation amplitude by making the following replacement:

$$
k \to -p', \quad p \to p, \quad p' \to k', \quad k' \to -k \tag{11.92}
$$

where the negative sign accounts for swapping the spinors $u \leftrightarrow v$. Hence, the negative sign arises when we replace an incoming fermion (resp. antifermion) in one diagram by an outgoing antifermion (resp. fermion)

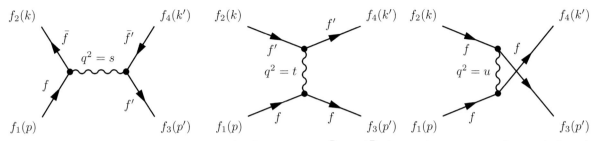

Figure 11.7 Mandelstam classification: (left) s-channel $f\bar{f} \to f'\bar{f}'$; (middle) t-channel $ff' \to ff'$; (right) u-channel $ff \to ff$.

in the other diagram. We recover the Stückelberg–Feynman interpretation that a particle which is traveling forward in time is equivalent to an antiparticle which is traveling backward in time (and vice versa). **This property of Feynman diagrams is called crossing symmetry.**

• **Crossing symmetry.** In summary, *crossing symmetry* implies that the amplitude for the scattering process $e^- \ell^- \to e^- \ell^-$ can be derived from that of the annihilation process $e^+ e^- \to \ell^+ \ell^-$ with the following substitution:

$$\mathcal{M}_{e^- \ell^- \to e^- \ell^-}(p, k, p', k') = \mathcal{M}_{e^+ e^- \to \ell^+ \ell^-}(p, -p', k', -k) \tag{11.93}$$

Furthermore, the Mandelstam variables are modified in this way:

$$e^- \ell^- \to e^- \ell^- \quad \begin{cases} s = (p - p')^2 \\ t = (p - k')^2 \\ u = (p + k)^2 \end{cases} \quad \leftrightarrow \quad \begin{cases} s = (p + k)^2 \\ t = (p - p')^2 \\ u = (p - k')^2 \end{cases} \quad e^+ e^- \to \ell^+ \ell^- \tag{11.94}$$

so, s **becomes** t, t **becomes** u, **and** u **becomes** s.

• **Møller and Bhabha reactions revisited.** We can go a step further and consider the Møller $e^- e^- \to e^- e^-$ and Bhabha $e^+ e^- \to e^+ e^-$ reactions. In the Møller case, the indistinguishable electrons in the final state imply that the tree-level amplitude derived in Eq. (11.78) has a t-channel as well as a u-channel. The amplitude $\mathcal{M}_t$ in Figure 11.5(left) can be directly constructed from the $e^- \ell^- \to e^- \ell^-$ amplitude, while the amplitude $\mathcal{M}_u$ in Figure 11.5(right) is obtained with the swap $(p', s') \leftrightarrow (k', r')$. Hence, $t = (p - p')^2$ and $u = (p - k')^2$ swap to finally give:

$$\mathcal{M}_{e^- e^- \to e^- e^-}(p, k, p', k') = \frac{(\bar{u}(p')\gamma^\mu u(p))\,(\bar{u}(k')\gamma_\mu u(k))}{t} - \frac{(\bar{u}(k')\gamma^\mu u(p))\,(\bar{u}(p')\gamma_\mu u(k))}{u} \tag{11.95}$$

as listed in Table 11.1.

The amplitude for the Bhabha reaction can be derived from the Møller one by crossing symmetry, swapping one incoming and one outgoing electron with one outgoing and one incoming positron. This implies for $e^-(p, s) + e^+(k, r) \to e^-(p', s') + e^+(k', r')$ that $k \to -k'$ and $k' \to -k$ (see Figure 11.6):

$$\mathcal{M}_{e^- e^+ \to e^- e^+}(p, k, p', k') = \mathcal{M}_{e^- e^- \to e^- e^-}(p, -k', p', -k) \tag{11.96}$$

and this implies that $t = (p - p')^2$ remains t, and $u = (p - k')^2$ becomes $s = (p + k)^2$. The Bhabha amplitude is then:

$$\mathcal{M}_{e^- e^+ \to e^- e^+}(p, k, p', k') = \frac{(\bar{u}(p')\gamma^\mu u(p))\,(\bar{v}(k)\gamma_\mu v(k'))}{t} - \frac{(\bar{v}(k)\gamma^\mu u(p))\,(\bar{u}(p')\gamma_\mu v(k'))}{s} \tag{11.97}$$

as can be cross-checked from Table 11.1.

The properties of crossing symmetries are very useful to reduce the number of calculations, or simply as a consistency cross-check.

11.9   Trace Technique – Casimir's Trick

A typical QED building block has an amplitude that might look something like this:

$$\bar{u}(p_1)\gamma^\mu u(p_2) \equiv \bar{u}_1 \Gamma u_2 \tag{11.98}$$

where we simplified the spinor notation and assume a generic bilinear covariant Γ. This is where the Feynman rules take us, but to get a number we need to substitute explicit forms for the external particles. When we do not care about the polarizations of the particles then we need to:

- Average over the polarizations of the initial-state particles;

- Sum over the polarizations of the final-state particles in the squared amplitude $|\mathcal{M}|^2$.

This is what we called the **spin-averaged matrix element** and we denoted it by $\langle|\mathcal{M}|^2\rangle$. Note that the averaging over initial-state polarizations involves summing over all polarizations and then dividing by the number of independent polarizations, so $\langle|\mathcal{M}|^2\rangle$ involves a sum over the polarizations of all external particles.

Considering our basic amplitude building block, we note that:

$$|\mathcal{M}|^2 \propto (\bar{u}_1 \Gamma u_2)^2 = [\bar{u}_1 \Gamma u_2][\bar{u}_1 \Gamma u_2]^* \tag{11.99}$$

We can compute the complex conjugate of the second term (see Eq. (11.28)):

$$[\bar{u}_1 \Gamma u_2]^* = [u_2^\dagger \Gamma^\dagger \gamma^{0\dagger} u_1] = [u_2^\dagger \gamma^0 \gamma^0 \Gamma^\dagger \gamma^0 u_1] = [\bar{u}_2 \bar{\Gamma} u_1] \tag{11.100}$$

where $\bar{\Gamma} \equiv \gamma^0 \Gamma^\dagger \gamma^0$. For the QED vertex, we have $\Gamma = \gamma^\mu$ and

$$\overline{\gamma^\mu} = \gamma^0 \gamma^{\mu\dagger} \gamma^0 = \begin{cases} (\gamma^0 \gamma^{0\dagger} \gamma^0) = \gamma^0 \\ (\gamma^0 \gamma^{k\dagger} \gamma^0) = (-\gamma^{k\dagger} \gamma^0 \gamma^0) = \gamma^k \end{cases} \qquad k = 1, 2, 3$$

so in the case that $\bar{\Gamma} = \Gamma$, we have:

$$|\mathcal{M}|^2 \propto (\bar{u}_1 \Gamma u_2)^2 = [\bar{u}_1 \Gamma u_2][\bar{u}_2 \Gamma u_1] \tag{11.101}$$

We reintroduce the spin indices for the spinors and consider the sum over the spins:

$$\langle|\mathcal{M}|^2\rangle \propto \sum_{s_1} \sum_{s_2} [\bar{u}_1^{(s_1)} \Gamma u_2^{(s_2)}][\bar{u}_2^{(s_2)} \Gamma u_1^{(s_1)}] \tag{11.102}$$

We can simplify and get rid of the sums by using the **completeness relations** valid for any spinor, see Eqs. (8.97) and (8.137):

$$\sum_s \left(u^{(s)}(p)\bar{u}^{(s)}(p) \right) = \gamma^\mu p_\mu + m\mathbb{1} = \not{p} + m\mathbb{1}$$

$$\sum_s \left(v^{(s)}(p)\bar{v}^{(s)}(p) \right) = \not{p} - m\mathbb{1} \tag{11.103}$$

Then for example for u_2:

$$\sum_{s_2} [\bar{u}_1^{(s_1)} \Gamma u_2^{(s_2)}][\bar{u}_2^{(s_2)} \Gamma u_1^{(s_1)}] = \bar{u}_1^{(s_1)} \Gamma \left(\not{p}_2 + m_2\right) \Gamma u_1^{(s_1)} = \bar{u}_1^{(s_1)} \mathcal{G} u_1^{(s_1)} \tag{11.104}$$

We consider the indices of each term in the product, rearranging the terms:

$$\bar{u}_1^{(s_1)} \mathcal{G} u_1^{(s_1)} = \sum_{\alpha,\beta} (\bar{u}_1^{(s_1)})_\alpha \mathcal{G}_{\alpha\beta} (u_1^{(s_1)})_\beta = \sum_{\alpha,\beta} \mathcal{G}_{\alpha\beta} (u_1^{(s_1)})_\beta (\bar{u}_1^{(s_1)})_\alpha = \sum_{\alpha,\beta} \mathcal{G}_{\alpha\beta} (u_1^{(s_1)} \bar{u}_1^{(s_1)})_{\beta\alpha}$$

$$= \sum_\alpha (\mathcal{G} u_1^{(s_1)} \bar{u}_1^{(s_1)})_{\alpha\alpha} = \text{Tr}(\mathcal{G} u_1^{(s_1)} \bar{u}_1^{(s_1)}) \tag{11.105}$$

We can now perform the sum over the spins of u_1:

$$\sum_{s_1} \text{Tr}(\mathcal{G}u_1^{(s_1)}\bar{u}_1^{(s_1)}) = \text{Tr}\left(\mathcal{G}\left(\not{p}_1 + m_1\right)\right) \tag{11.106}$$

Finally, we have found again **Casimir's trick** to compute the sums over spins:

$$\langle|\mathcal{M}|^2\rangle \propto \sum_{s_1}\sum_{s_2}[\bar{u}_1^{(s_1)}\Gamma u_2^{(s_2)}][\bar{u}_2^{(s_2)}\Gamma u_1^{(s_1)}] = \text{Tr}\left(\Gamma\left(\not{p}_2 + m_2\right)\Gamma\left(\not{p}_1 + m_1\right)\right) \tag{11.107}$$

This result is valid for fermions. In the case of antifermions, we replace u by v and use the corresponding completeness relation, for example for v_2 we would get:

$$\langle|\mathcal{M}|^2\rangle \propto \sum_{s_1}\sum_{s_2}[\bar{u}_1^{(s_1)}\Gamma v_2^{(s_2)}][\bar{v}_2^{(s_2)}\Gamma u_1^{(s_1)}] = \text{Tr}\left(\Gamma\left(\not{p}_2 - m_2\right)\Gamma\left(\not{p}_1 + m_1\right)\right) \tag{11.108}$$

where the sign in front of m_2 was flipped. With Casimir's trick it is that simple! Useful results on traces of products of γ matrices are summarized in the so-called **trace theorems**. See Appendix E.2.

11.10   Unpolarized Scattering Cross-Sections

Let us consider the heavy lepton pair creation $e^-(p, s) + e^+(k, r) \to \ell^-(p', s') + \ell^+(k', r')$, where $\ell = \mu, \tau$. The amplitude $\mathcal{M}(p, k, p', k')$ is given in Eq. (11.76). The complex conjugate amplitude is (we use the index ν in place of μ):

$$\mathcal{M}^* = \frac{-ie^2}{(p+k)^2}\left(\bar{v}(k')\gamma^\nu u(p')\right)\left(\bar{u}(p)\gamma_\nu v(k)\right) \tag{11.109}$$

The square of the matrix element $|\mathcal{M}|^2$ is then directly computed as:

$$\begin{aligned}|\mathcal{M}|^2 &= \mathcal{M}\mathcal{M}^* = \frac{e^4}{(p+k)^4}\left(\bar{v}(k)\gamma^\mu u(p)\right)\left(\bar{u}(p')\gamma_\mu v(k')\right)\left(\bar{v}(k')\gamma^\nu u(p')\right)\left(\bar{u}(p)\gamma_\nu v(k)\right) \\ &= \frac{e^4}{s^2}\left(\bar{v}(k)\gamma^\mu u(p)\right)\left(\bar{u}(p)\gamma^\nu v(k)\right)\left(\bar{u}(p')\gamma_\mu v(k')\right)\left(\bar{v}(k')\gamma_\nu u(p')\right)\end{aligned} \tag{11.110}$$

Each of the four terms is a current of the type $\overline{\Psi}\gamma^\mu\Psi$ – see Eq. (8.275), so:

$$|\mathcal{M}|^2 = \frac{e^4}{s^2}[.]^\mu[.]^\nu[.]_\mu[.]_\nu = [.]^{\mu\nu}[.]_{\mu\nu} \tag{11.111}$$

which we can conveniently write as:

$$|\mathcal{M}|^2 = \frac{e^4}{s^2}L_e^{\mu\nu}(p,k)L_{\mu\nu}^\ell(p',k') \tag{11.112}$$

where the parts involving the electron and positron spinors have been regrouped in the electron tensor $L_e^{\mu\nu}$ and similarly the parts with the spinors of the heavy leptons have been regrouped in the tensor $L_{\mu\nu}^\ell$. We recognize the s-channel propagator, which yields a factor $1/s^2$ in the matrix element. The unpolarized (spin-averaged) matrix element is simply given by:

$$\langle|\mathcal{M}|^2\rangle \equiv \frac{e^4}{s^2}\langle L_e^{\mu\nu}(p,k)\rangle\langle L_{\mu\nu}^\ell(p',k')\rangle \tag{11.113}$$

When *averaging* over initial-state spins and using Casimir's trick (see Eq. (11.107)), we get:

$$\langle L_e^{\mu\nu}\rangle = \frac{1}{2}\sum_s \frac{1}{2}\sum_r \left(\bar{v}(k)\gamma^\mu u(p)\right)\left(\bar{u}(p)\gamma^\nu v(k)\right) = \frac{1}{4}\,\mathrm{Tr}\left((\slashed{k}-m_e)\gamma^\mu(\slashed{p}+m_e)\gamma^\nu\right) \tag{11.114}$$

and *summing* over final-state spins yields:

$$\langle L_{\mu\nu}^{\ell}\rangle = \sum_{s'}\sum_{r'} \left(\bar{u}(p')\gamma_\mu v(k')\right)\left(\bar{v}(k')\gamma_\nu u(p')\right) = \mathrm{Tr}\left((\slashed{p}'+m_\ell)\gamma_\mu(\slashed{k}'-m_\ell)\gamma_\nu\right) \tag{11.115}$$

where m_ℓ is the rest mass of the heavy lepton. We note that all spinors have disappeared from the tensors! The traces are just numbers. It's fantastic!

We can now rely on the trace theorems to simplify the traces:

$$\mathrm{Tr}\left((\slashed{k}-m_e)\gamma^\mu(\slashed{p}+m_e)\gamma^\nu\right) = \mathrm{Tr}\left(\slashed{k}\gamma^\mu\slashed{p}\gamma^\nu + \underbrace{(-m_e)\gamma^\mu\slashed{p}\gamma^\nu}_{=0} + \underbrace{(m_e)\slashed{k}\gamma^\mu\gamma^\nu}_{=0} - m_e^2\gamma^\mu\gamma^\nu\right)$$
$$\text{odd number of } \gamma \text{ matrices}$$
$$= \mathrm{Tr}\left(\slashed{k}\gamma^\mu\slashed{p}\gamma^\nu\right) - m_e^2\underbrace{\mathrm{Tr}\left(\gamma^\mu\gamma^\nu\right)}_{=4g^{\mu\nu}} = k_\alpha p_\beta\,\mathrm{Tr}\left(\gamma^\alpha\gamma^\mu\gamma^\beta\gamma^\nu\right) - m_e^2 4g^{\mu\nu} \tag{11.116}$$

The trace of four γ-matrices is equal to:

$$k_\alpha p_\beta\,\mathrm{Tr}\left(\gamma^\alpha\gamma^\mu\gamma^\beta\gamma^\nu\right) = 4k_\alpha p_\beta\left(g^{\alpha\mu}g^{\beta\nu} - g^{\alpha\beta}g^{\mu\nu} + g^{\alpha\nu}g^{\mu\beta}\right) = 4\left(k^\mu p^\nu - (k\cdot p)g^{\mu\nu} + k^\nu p^\mu\right) \tag{11.117}$$

Finally:

$$\langle L_e^{\mu\nu}\rangle = \frac{1}{4}4\left(k^\mu p^\nu + k^\nu p^\mu - g^{\mu\nu}\left(k\cdot p + m_e^2\right)\right) \tag{11.118}$$

Direct comparison of Eq. (11.114) with Eq. (11.115) shows that we can readily write down the heavy lepton tensor by the replacements $k\to -p'$, $p\to -k'$, and $m_e\to m_\ell$, giving (note that there is no factor 1/4 here):

$$\langle L_{\mu\nu}^{\ell}\rangle = 4\left(p'_\mu k'_\nu + p'_\nu k'_\mu - g_{\mu\nu}\left(p'\cdot k' + m_\ell^2\right)\right) \tag{11.119}$$

We decide to neglect the electron mass relative to the heavy lepton mass, in order to simplify the analytic expressions. In this case, the spin-averaged matrix element is:

$$\langle|\mathcal{M}|^2\rangle = \frac{16}{4}\frac{e^4}{s^2}\left(k^\mu p^\nu + k^\nu p^\mu - g^{\mu\nu}(k\cdot p)\right)\left(p'_\mu k'_\nu + p'_\nu k'_\mu - g_{\mu\nu}\left(p'\cdot k' + m_\ell^2\right)\right)$$
$$= \frac{4e^4}{s^2}\left\{2\left((k\cdot p')(p\cdot k') + (p\cdot p')(k\cdot k')\right) - 2(p'\cdot k')(k\cdot p) + \underbrace{g^{\mu\nu}g_{\mu\nu}}_{=4}(k\cdot p)(p'\cdot k' + m_\ell^2)\right.$$
$$\left. -2(k\cdot p)(p'\cdot k' + m_\ell^2)\right\}$$
$$= 2\left(\frac{4e^4}{s^2}\right)\left\{(kp')(pk') + (pp')(kk') - (p'k')(kp) + 2(kp)(p'k') - (kp)(p'k') + m_\ell^2(kp)\right\}$$
$$= \left(\frac{8e^4}{s^2}\right)\left\{(kp')(pk') + (pp')(kk') + m_\ell^2(kp)\right\} \tag{11.120}$$

Finally, the unpolarized spin-averaged invariant squared matrix element is given by:

$$\langle|\mathcal{M}_{e^+e^-\to\ell^+\ell^-}|^2\rangle = \left(\frac{8e^4}{s^2}\right)\left\{(p\cdot p')(k\cdot k') + (k\cdot p')(p\cdot k') + m_\ell^2(k\cdot p)\right\} \tag{11.121}$$

It looks so explicitly covariant! It's really nice.

In the ultra-relativistic limit, the Mandelstam variables reduce to products of the kinematical four-vector of the types found in the matrix element. See Eq. (11.89). We find:

$$(p \cdot p')(k \cdot k') \approx \frac{t}{2}\frac{t}{2} = \frac{t^2}{4} \qquad \text{and} \qquad (p \cdot k')(p' \cdot k) \approx \frac{u^2}{4} \tag{11.122}$$

We can then rewrite the matrix element of the heavy lepton pair creation in terms of s, t, and u:

$$\langle |\mathcal{M}_{e^+e^- \to \ell^+\ell^-}|^2 \rangle \simeq \frac{8e^4}{s^2}\left\{ \frac{t^2}{4} + \frac{u^2}{4} \right\} \tag{11.123}$$

so finally for the ultra-relativistic limit we have:

$$\langle |\mathcal{M}(e^+e^- \to \ell^+\ell^-)|^2 \rangle \simeq 2e^4\left\{ \frac{t^2 + u^2}{s^2} \right\} \tag{11.124}$$

The s-channel nature of the interaction is visible in the matrix element, as s appears in the denominator. This result is plotted as a function of t/s in Figure 11.8. We recall that the denominator corresponds to the photon propagator, and as expected, carries the energy and momentum of the center-of-mass.

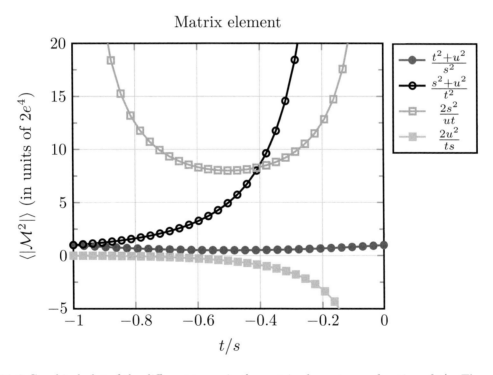

Figure 11.8 Graphical plot of the different terms in the matrix elements as a function of t/s. The u variable is given by $u/s = -(1 + t/s)$. The interference terms are also shown separately.

We can now illustrate the power of the crossing symmetry in order to derive the matrix element for the process $e^-\ell^- \to e^-\ell^-$ from the annihilation process $e^+e^- \to \ell^+\ell^-$. We have seen (see Eq. (11.93)) that:

$$\mathcal{M}_{e^-\ell^- \to e^-\ell^-}(p, k, p', k') = \mathcal{M}_{e^+e^- \to \ell^+\ell^-}(p, -p', k', -k) \tag{11.125}$$

so

$$\langle|\mathcal{M}_{e^-\ell^-\to e^-\ell^-}|^2\rangle = \left(\frac{8e^4}{t^2}\right)\left\{(p\cdot k')(p'\cdot k)+(p'\cdot k')(p\cdot k)-m_\ell^2(p'\cdot p)\right\} \tag{11.126}$$

Also $s\to t$, $t\to u$, and $u\to s$. So neglecting fermion masses, this yields:

$$\langle\mathcal{M}(e^-\ell^-\to e^-\ell^-)|^2\rangle \simeq (2e^4)\left\{\frac{s^2+u^2}{t^2}\right\} \tag{11.127}$$

We have so far seen two cases where only a single diagram contributes at tree level. We can consider the heavy lepton pair creation as the prototype for an s-channel reaction, and the electron–muon scattering as a prototype t-channel. We can go a step further and consider the **Møller reaction**. Its amplitude $\mathcal{M}$ is given in Eq. (11.78) and we first need to compute $|\mathcal{M}|^2$:

$$|\mathcal{M}_{e^-e^-\to e^-e^-}|^2 = |\mathcal{M}_t|^2 + |\mathcal{M}_u|^2 - \mathcal{M}_t^*\mathcal{M}_u - \mathcal{M}_t\mathcal{M}_u^* \tag{11.128}$$

The matrix elements $|\mathcal{M}_t|^2$ and $|\mathcal{M}_u|^2$ can be read off directly from Eqs. (11.124) and (11.127), invoking cross-symmetry. We consider the interference term:

$$\begin{aligned}
\mathcal{M}_t^*\mathcal{M}_u &= \frac{-ie^2}{t}\left[(\bar{u}(p')\gamma^\mu u(p))\,(\bar{u}(k')\gamma_\mu u(k))\right]^*\frac{ie^2}{u}\,(\bar{u}(k')\gamma^\nu u(p))\,(\bar{u}(p')\gamma_\nu u(k)) \\
&= \frac{e^4}{ut}\,(\bar{u}(k)\gamma^\mu u(k'))\,(\bar{u}(p)\gamma_\mu u(p'))\,(\bar{u}(k')\gamma^\nu u(p))\,(\bar{u}(p')\gamma_\nu u(k)) \\
&= \frac{e^4}{ut}\,(\bar{u}(k)\gamma^\mu u(k'))\,(\bar{u}(k')\gamma^\nu u(p))\,(\bar{u}(p)\gamma_\mu u(p'))\,(\bar{u}(p')\gamma_\nu u(k))
\end{aligned} \tag{11.129}$$

We use Casimir's trick in order to compute the unpolarized matrix element, translating the products of spinors into traces of products of γ matrices:

$$\langle\mathcal{M}_t^*\mathcal{M}_u\rangle = \frac{1}{4}\frac{e^4}{ut}\,\mathrm{Tr}\left((\slashed{k}+m_e)\gamma^\mu(\slashed{k}'+m_e)\gamma^\nu(\slashed{p}+m_e)\gamma_\mu(\slashed{p}'+m_e)\gamma_\nu\right) \tag{11.130}$$

We decide to neglect the electron masses and get the approximate expression valid in the ultra-relativistic case, which is still the product of eight γ matrices:

$$\langle\mathcal{M}_t^*\mathcal{M}_u\rangle \simeq \frac{e^4}{4ut}\,\mathrm{Tr}\left(\slashed{k}\gamma^\mu\slashed{k}'\gamma^\nu\slashed{p}\gamma_\mu\slashed{p}'\gamma_\nu\right) \tag{11.131}$$

We note that:

$$\gamma^\nu\slashed{p}\gamma_\mu\slashed{p}'\gamma_\nu = p_\alpha p'_\beta\gamma^\nu\gamma^\alpha\gamma_\mu\gamma^\beta\gamma_\nu = -2p_\alpha p'_\beta\gamma^\beta\gamma_\mu\gamma^\alpha = -2\,\slashed{p}'\gamma_\mu\slashed{p} \tag{11.132}$$

where we have used the property $\gamma^\nu\gamma^\alpha\gamma^\mu\gamma^\beta\gamma_\nu = -2\gamma^\beta\gamma^\mu\gamma^\alpha$ (see Appendix E.1). Then again:

$$\slashed{k}\gamma^\mu\slashed{k}'\gamma^\nu\slashed{p}\gamma_\mu\slashed{p}'\gamma_\nu = -2\,\slashed{k}\gamma^\mu\slashed{k}'\slashed{p}'\gamma_\mu\slashed{p} = -2\,\slashed{k}k'_\alpha p'_\beta\gamma^\mu\gamma^\alpha\gamma^\beta\gamma_\mu\slashed{p} = -2\,\slashed{k}k'_\alpha p'_\beta 4g^{\alpha\beta}\slashed{p} = -8(k'\cdot p')\,\slashed{k}\slashed{p} \tag{11.133}$$

because $\gamma^\mu\gamma^\alpha\gamma^\beta\gamma_\mu = 4g^{\alpha\beta}\mathbb{1}$. Finally:

$$\langle\mathcal{M}_t^*\mathcal{M}_u\rangle \simeq \frac{-8e^4}{4ut}(k'\cdot p')Tr\,(\slashed{k}\slashed{p}) = \frac{-2e^4}{ut}(k'\cdot p')4(k\cdot p) \simeq \frac{-2e^4}{ut}4\left(\frac{s}{2}\right)\left(\frac{s}{2}\right) \tag{11.134}$$

where we have used the trace theorem **[T4]** $Tr(\gamma^\mu\gamma^\nu) = 4g^{\mu\nu}$ (see Appendix E.2). The total spin-averaged interference term is finally equal to:

$$\langle\mathcal{M}_t^*\mathcal{M}_u\rangle + \langle\mathcal{M}_t\mathcal{M}_u^*\rangle \simeq 2(-2e^4)\frac{s^2}{ut} = -(2e^4)\frac{2s^2}{ut} \tag{11.135}$$

Adding the various pieces together, the spin-averaged matrix element for the **Møller reaction** is equal to:

$$\langle |\mathcal{M}_{e^-e^-\to e^-e^-}|^2 \rangle \simeq (2e^4) \left(\frac{s^2+u^2}{t^2} + \frac{2s^2}{ut} + \frac{s^2+t^2}{u^2} \right) \tag{11.136}$$

We can now use crossing symmetry to directly derive the matrix element for the **Bhabha scattering** reaction $e^-(p) + e^+(k) \to e^-(p') + e^+(k')$. The replacement is $k \to -k'$ and $k' \to -k$ (see Eq. (11.96)), $t = (p - p')^2$ remains t, and $u = (p - k')^2$ swaps with $s = (p + k)^2$:

$$\mathcal{M}_{e^-e^+\to e^-e^+}(p, k, p', k') = \mathcal{M}_{e^-e^-\to e^-e^-}(p, -k', p', -k) \tag{11.137}$$

and this implies:

$$\langle |\mathcal{M}_{e^-e^+\to e^-e^+}|^2 \rangle \simeq (2e^4) \left(\frac{s^2+u^2}{t^2} + \frac{2u^2}{ts} + \frac{u^2+t^2}{s^2} \right) \tag{11.138}$$

The different terms entering in the matrix elements are graphically summarized in Figure 11.8.

11.11   Angular Dependence and Forward–Backward Peaks

Our four main reactions and their matrix elements are summarized in Table 11.2. We observe that in all cases the denominators (originating from the photon propagator) are functions of specific Mandelstam variables. While s is positive and different from zero (see Eq. (11.88)), t and u can vanish. Since $t = (p - p')^2 = (k' - k)^2$, $t \to 0$ in the limit $p \to p'$. This corresponds to a **forward-peaked** scattering, the case when the incoming electron is almost unaffected since before and after it has the same four-momentum. Also since $u = (p - k')^2 = (p' - k)^2$, $u \to 0$ in the limit $p \to k'$. This corresponds to a **backward-peaked** scattering. In these cases, the matrix elements with t or u in the denominator will diverge at tree level. As shown in Table 11.2, $e^-\ell^- \to e^-\ell^-$ ($\ell = \mu, \tau$) has a forward peak, $e^-e^- \to e^-e^-$ has forward and backward peaks (the two electrons are indistinguishable), and $e^-e^+ \to e^-e^+$ has a forward peak. The s-channels have no divergences.

Reaction	Forward peak	Backward peak	$\langle |\mathcal{M}|^2 \rangle / 2e^4$ ultra-relativistic
$e^-\ell^- \to e^-\ell^-$	yes		$\dfrac{s^2+u^2}{t^2}$
$e^-e^+ \to \ell^-\ell^+$			$\dfrac{t^2+u^2}{s^2}$
$e^-e^- \to e^-e^-$	yes	yes	$\dfrac{s^2+u^2}{t^2} + \dfrac{2s^2}{ut} + \dfrac{s^2+t^2}{u^2}$
$e^-e^+ \to e^-e^+$	yes		$\dfrac{s^2+u^2}{t^2} + \dfrac{2u^2}{ts} + \dfrac{u^2+t^2}{s^2}$

Table 11.2 Summary of the four QED reactions and their matrix elements at tree level in the ultra-relativistic limit.

These "divergent" behaviors can be understood in physical terms, by noticing the properties of the photon propagator. The divergence comes about when the mass of the exchanged virtual photon $m_\gamma^* = q^2$ tends to zero, i.e., the virtual photon becomes "on mass-shell" or "real." In the case of the s-channel, the mass of the virtual photon is never real since $m_\gamma^* = \sqrt{s} = E_\star$, and we can represent the s-channel scattering as:

$$e^+ e^- \to \gamma^* \to \ell^+ \ell^- \tag{11.139}$$

So divergences occur in the case of the t- and u-channels. Note that although the tree-level cross-section leads to a mathematical infinity, the actual cross-section is strongly peaked but remains finite. Mathematically, the tree-level cross-section becomes an inadequate approximation in that region of phase space.

11.12   Scattering Angle in the Center-of-Mass System

Up until now we have expressed the matrix elements as a function of invariant quantities such as the Mandelstam variables or other products of kinematical four-vectors.

We now choose the special center-of-mass reference frame (CMS) and compute angular scattering variables. The example of electron–positron annihilation into two muons is drawn in Figure 11.9. In the center-of-mass, the total three-momentum vanishes and the two incoming particles collide head-on coming from opposite directions. The outgoing particles are back-to-back at an angle θ relative to the incoming particles. Note that the scattering angle θ is defined as the deflection of the incoming electron towards the outgoing muon, hence by choosing the incoming and outgoing negatively charged particles (we could also have chosen the two positively charged particles; however, one usually does not mix charges). With this definition of θ, forward scattering corresponds to the limit $\theta \to 0$ (or $\cos\theta \to 1$), and backward scattering is the case $\theta \to \pi$ (or $\cos\theta \to -1$).

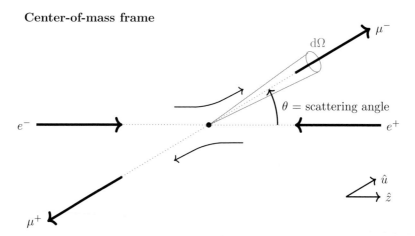

Figure 11.9 The $e^+ e^- \to \mu^+ \mu^-$ scattering in the center-of-mass frame. The cone symbolizes the solid angle $\mathrm{d}\Omega = \mathrm{d}\cos\theta \mathrm{d}\phi$.

In the center-of-mass frame, the four-vectors of the incoming and outgoing particles for $e^-(p) + e^+(k) \to \mu^-(p') + \mu^+(k')$ are given by:

$$p^\mu = (E_p^\star, \vec{p}^{\,\star}) \qquad k^\mu = (E_k^\star, -\vec{p}^{\,\star}) \qquad p'^\mu = (E_{p'}^\star, \vec{p}^{\,\prime\,\star}) \qquad k'^\mu = (E_{k'}^\star, -\vec{p}^{\,\prime\,\star}) \tag{11.140}$$

where

$$E_p^\star = E_k^\star = \sqrt{(\vec{p}^{\,\star})^2 + m_e^2}, \quad E_{p'}^\star = E_{k'}^\star = \sqrt{(\vec{p}^{\,\prime\,\star})^2 + m_\mu^2} \quad \text{and} \quad \cos\theta^\star = \hat{p}^{\,\prime\,\star} \cdot \hat{p}^{\,\star} \tag{11.141}$$

We can directly compute the following useful invariant products of our four-vectors, as a function of the kinematical variables of the center-of-mass frame. We start with the Mandelstam variables. The center-of-mass energy $E_\star$ with $E_p^\star = E_k^\star = E^\star$ is:[8]

$$s = (p + k)^2 = (E^\star + E^\star)^2 - 0 = (2E^\star)^2 \quad \longrightarrow \quad E_\star \equiv \sqrt{s} = 2E^\star \tag{11.142}$$

In the ultra-relativistic limit, we neglect the rest masses of the particles and can express the kinematical four-vectors as follows:

$$p^\mu = (E^\star, E^\star \hat{z}), \qquad k^\mu = (E^\star, -E^\star \hat{z}); \qquad p'^\mu = (E^\star, E^\star \hat{u}), \qquad k'^\mu = (E^\star, -E^\star \hat{u}) \tag{11.143}$$

where $\cos\theta = \hat{u} \cdot \hat{z}$. In this case, we have:

$$
\begin{aligned}
p \cdot p' &= E^\star E^\star - E^\star \hat{z} \cdot E^\star \hat{u} = E^\star E^\star (1 - \cos\theta) \\
k \cdot k' &= E^\star E^\star - (-E^\star \hat{z}) \cdot (-E^\star \hat{u}) = E^\star E^\star (1 - \cos\theta) \\
p \cdot k' &= E^\star E^\star - E^\star \hat{z} \cdot (-E^\star \hat{u}) = E^\star E^\star (1 + \cos\theta) \\
p' \cdot k &= E^\star E^\star - E^\star \hat{u} \cdot (-E^\star \hat{z}) = E^\star E^\star (1 + \cos\theta)
\end{aligned}
\tag{11.144}
$$

The t and u variables can then be computed directly as a function of the scattering angle θ:

$$t = (p - p')^2 \simeq -2p \cdot p' = -2E^\star E^\star (1 - \cos\theta) = -\frac{s}{2}(1 - \cos\theta) \tag{11.145}$$

and similarly:

$$u = (p - k')^2 \simeq -2p \cdot k' = -2E^\star E^\star (1 + \cos\theta) = -\frac{s}{2}(1 + \cos\theta) \tag{11.146}$$

We recover the kinematical limits for $\theta \to 0$ (or $\cos\theta \to 1$) and $\theta \to \pi$ (or $\cos\theta \to -1$), which are:

$$-s \leq t \leq 0 \quad \text{and} \quad -s \leq u = -(s + t) \leq 0 \tag{11.147}$$

Inspection of Table 11.2 leads us to calculating quantities of the type:

$$\frac{t^2 + u^2}{s^2} = \frac{\frac{s^2}{4}(1 - \cos\theta)^2 + \frac{s^2}{4}(1 + \cos\theta)^2}{s^2} = \frac{1}{2}\left(1 + \cos^2\theta\right) \tag{11.148}$$

and

$$\frac{s^2 + u^2}{t^2} = \frac{s^2 + \frac{s^2}{4}(1 + \cos\theta)^2}{\frac{s^2}{4}(1 - \cos\theta)^2} = \frac{4 + (1 + \cos\theta)^2}{(1 - \cos\theta)^2} \tag{11.149}$$

Finally

$$\frac{s^2 + t^2}{u^2} = \frac{s^2 + \frac{s^2}{4}(1 - \cos\theta)^2}{\frac{s^2}{4}(1 + \cos\theta)^2} = \frac{4 + (1 - \cos\theta)^2}{(1 + \cos\theta)^2} \tag{11.150}$$

With these results we can easily compute the spin-averaged matrix element squared for our four common processes. We want to eventually compute the differential cross-section $\mathrm{d}\sigma/\mathrm{d}\Omega = \mathrm{d}^2\sigma/\mathrm{d}\cos\theta\,\mathrm{d}\phi$. We refer to Eq. (5.145) and write:

$$\left(\frac{\mathrm{d}\sigma}{\mathrm{d}\Omega}\right)_{CMS} = \left(\frac{1}{64\pi^2}\right)\left(\frac{1}{E_\star^2}\right)|\mathcal{M}|^2 = \left(\frac{1}{64\pi^2 s}\right)|\mathcal{M}|^2 \tag{11.151}$$

It is customary to write these cross-sections as a function of the fine-structure constant α, recalling that $e = \sqrt{4\pi\alpha}$ and hence $2e^4 = 32\pi^2\alpha^2$. For the **electron–lepton scattering**, where the lepton is either a muon or a tau, we find:

$$\left(\frac{\mathrm{d}\sigma(e^-\ell^- \to e^-\ell^-)}{\mathrm{d}\Omega}\right)_{CMS} = \left(\frac{2e^4}{64\pi^2 s}\right)\frac{4 + (1 + \cos\theta)^2}{(1 - \cos\theta)^2} = \left(\frac{\alpha^2}{2s}\right)\frac{4 + (1 + \cos\theta)^2}{(1 - \cos\theta)^2} \tag{11.152}$$

---

8   Note that $E_\star$ denotes the center-of-mass energy, while $E^\star$ is the energy of each incoming particle in the center-of-mass frame.

and for **the heavy lepton pair creation**:

$$\left(\frac{\mathrm{d}\sigma(e^+e^- \to \ell^+\ell^-)}{\mathrm{d}\Omega}\right)_{CMS} = \left(\frac{2e^4}{64\pi^2 s}\right)\frac{1}{2}\left(1+\cos^2\theta\right) = \left(\frac{\alpha^2}{4s}\right)\left(1+\cos^2\theta\right) \qquad (11.153)$$

One can also compute the center-of-mass differential cross-sections for the Møller and Bhabha processes using a similar method. We note that:

$$\frac{2s^2}{ut} = \frac{2s^2}{\frac{s^2}{4}(1+\cos\theta)(1-\cos\theta)} = \frac{8}{1-\cos^2\theta} = \frac{8}{\sin^2\theta} \qquad (11.154)$$

We then obtain **the ultra-relativistic Møller formula for the center-of-mass scattering angle**:

$$
\begin{aligned}
\left(\frac{\mathrm{d}\sigma}{\mathrm{d}\Omega}\right)_{CMS} &= \left(\frac{2e^4}{64\pi^2 s}\right)\left\{\frac{4+(1+\cos\theta)^2}{(1-\cos\theta)^2} + \frac{8}{\sin^2\theta} + \frac{4+(1-\cos\theta)^2}{(1+\cos\theta)^2}\right\} \\
&= \left(\frac{\alpha^2}{2s}\right)\left\{\frac{\left(4+(1+\cos\theta)^2\right)(1+\cos\theta)^2 + \left(4+(1-\cos\theta)^2\right)(1-\cos\theta)^2}{(1-\cos\theta)^2(1+\cos\theta)^2} + \frac{8}{\sin^2\theta}\right\} \\
&= \left(\frac{\alpha^2}{2s}\right)\left\{\frac{2\left(\cos^4\theta + 10\cos^2\theta + 5\right)}{\sin^4\theta} + \frac{8}{\sin^2\theta}\right\}
\end{aligned}
\qquad (11.155)
$$

or

$$\left(\frac{\mathrm{d}\sigma(e^-e^- \to e^-e^-)}{\mathrm{d}\Omega}\right)_{CMS} = \left(\frac{\alpha^2}{2s}\right)\frac{(7+\cos 2\theta)^2}{2\sin^4\theta} \qquad (11.156)$$

Similarly, we have:

$$\frac{2u^2}{ts} = \frac{2\frac{s^2}{4}(1+\cos\theta)^2}{-\frac{s^2}{2}(1-\cos\theta)} = -\frac{(1+\cos\theta)^2}{(1-\cos\theta)} \qquad (11.157)$$

Hence, the **ultra-relativistic Bhabha formula for the center-of-mass scattering angle** is:

$$\left(\frac{\mathrm{d}\sigma(e^-e^+ \to e^-e^+)}{\mathrm{d}\Omega}\right)_{CMS} = \left(\frac{\alpha^2}{2s}\right)\frac{\left(3+\cos^2\theta\right)^2}{2(\cos\theta - 1)^2} \qquad (11.158)$$

These angular dependencies are summarized in Figure 11.10.

11.13   Automatic Computation of Feynman Diagrams

Several tools exist in order to perform algebraic computations of Feynman diagrams. Historically, **Veltman**[9] developed the program SCHOONSCHIP in 1967 to cross-check the evaluation of traces. Some other codes that were subsequently developed are, for example, MACSYMA, REDUCE, and FORM. Mathematica,[10] today's highly developed algebraic system, was actually developed in the 1980s by the particle physicist **Wolfram**.[11] These programs became increasingly important as the automation of Feynman diagram calculations is essentially mandatory for higher-order perturbative calculations.

9   Martinus Justinus Godefriedus "Tini" Veltman (1931–2021), Dutch theoretical physicist.
10   www.wolfram.com/mathematica/
11   Stephen Wolfram (born 1959), a British–American computer scientist, physicist, and businessman.

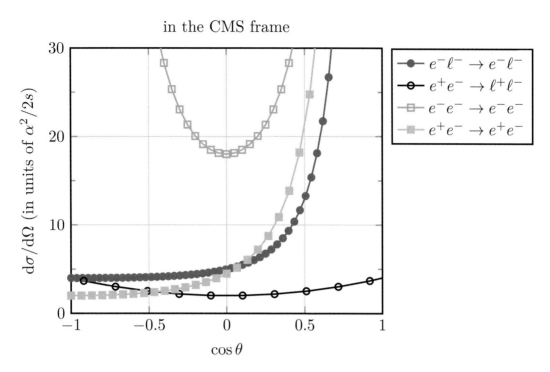

Figure 11.10 Angular dependencies in the center-of-mass frame of the differential cross-sections for the main QED processes. The label ℓ stands for either muon or tau.

Here we briefly illustrate lowest-order calculations with FeynCalc[12] and FeynArts.[13] FeynCalc is a Mathematica package for symbolic evaluation of Feynman diagrams and algebraic calculations [84, 85] It is a very comprehensive package and we refer the interested reader to the website for updated information. FeynArts [86] is used to generate Feynman diagrams, given a number of loops and external particles.

Here we want to study the $e^+e^- \to \ell^+\ell^-$ process within QED. One begins with the functions to generate Feynman diagrams. The function `CreateTopologies[l,ni -> no]` finds all distinct ways of connecting n_i incoming and n_o outgoing lines with l loops. With the created topologies *top*, the function `InsertFields[top,i1, i2, ... -> o1,o2,...]` determines all allowed combinations of field diagrams. The incoming fields are `i1, i2, ...` and the outgoing fields are `o1, o2, ...`. The output is a set of diagrams.

```
diags = InsertFields[ CreateTopologies[0, 2 -> 2],
               {F[2, {1}], -F[2, {1}]} ->{F[2, {2}], -F[2, {2}]},
               InsertionLevel -> {Classes}, Restrictions -> QEDOnly];
```

Here `F` is a generic fermion. `F[n]` is a fermion of a given class and $n = 2$ stands for leptons. `F[n, i, ...]` is a member of the fermions of a given class, so `F[2, 1]` is the electron and `F[2, 2]` is the muon. The negative fields correspond to antifermions.

The function `CreateFeynAmp` takes as input the diagrams and applies the Feynman rules in order to compute the amplitude. The function `FCFAConvert` converts the diagrams output from FeynArts into a

12   FeynCalc is covered by the GNU General Public License 3. Copyright 1990–2021 Rolf Mertig, Copyright 1997–2021 Frederik Orellana, Copyright 2014–2021 Vladyslav Shtabovenko. Reprinted with permission.
13   Feynarts is distributed under the GNU Lesser General Public License. Copyright 1998–2021 Thomas Hahn.

suitable input for FeynCalc:

```
amp[0] = FCFAConvert[CreateFeynAmp[diags], IncomingMomenta -> {p, k},
        OutgoingMomenta -> {pp, kp}, UndoChiralSplittings -> True,
        ChangeDimension -> 4, List -> False, SMP -> True, Contract -> True]
```

The result of the amplitude will be given as functions of scalar products of four-vectors. There is a way to preset these latter for ease of use. The function `FCClearScalarProducts` removes all user-performed specific settings. Then the function `SetMandelstam[s, t, u, p1, p2, p3, p4, m1, m2, m3, m4]` defines scalar products in terms of Mandelstam variables and puts them on-shell:

```
FCClearScalarProducts[];
SetMandelstam[s, t, u, p, k, -k1, -k2,
                    SMP["m_e"], SMP["m_e"], SMP["m_l"], SMP["m_l"]];
```

where the SMP return the symbols for the model parameter corresponding to the mass of the electron and lepton. The amplitude squared is directly computed as `amp[0] (ComplexConjugate[amp[0]])`. The function `FeynAmpDenominatorExplicit` writes out the propagator explicitly as a fraction. `FermionSpinSum` constructs traces out of squared amplitudes. `DiracSimplify` contracts all Lorentz indices and attempts to simplify the result. Putting these all together gives:

```
ampSquared[0] = (amp[0] (ComplexConjugate[amp[0]])) //
        FeynAmpDenominatorExplicit //
        FermionSpinSum[#, ExtraFactor -> 1/2^2] & //
        DiracSimplify // Simplify
```

The computation yields directly the following result:

$$\langle|\mathcal{M}(e^+e^- \to \ell^+\ell^-)|^2\rangle = 2e^4 \frac{2m_e^2\left(2m_\ell^2 + s - t - u\right) + 2m_e^4 + 2m_\ell^4 + 2m_\ell^2(s - t - u) + t^2 + u^2}{s^2} \quad (11.159)$$

which, as expected, reduces to our expression Eq. (11.124) for $m_e, m_\ell \to 0$. Not bad!

Similarly, one could repeat the same exercise for the process $e^-\ell^- \to e^-\ell^-$ to obtain:

$$\langle|\mathcal{M}(e^-\ell^- \to e^-\ell^-)|^2\rangle = 2e^4 \frac{-2m_e^2\left(-2m_\ell^2 + s - t + u\right) + 2m_e^4 + 2m_\ell^4 - 2m_\ell^2(s - t + u) + s^2 + u^2}{t^2} \quad (11.160)$$

to be exactly compared to Eq. (11.127) in the case $m_e, m_\ell \to 0$.

What is the effect of retaining the mass terms in these matrix elements? In order to get to the differential cross-sections, one easily introduces the phase-space factors in the following way, for example for the $e^+e^- \to \ell^+\ell^-$ process:

```
prefac1 = 1/(64 Pi^2 s);
integral1 = (Factor[ampSquared[0] /. {t -> -s/2 (1 - Cos[Th]), u -> -s/2 (1 + Cos[Th]),
    SMP["e"]^4 -> (4 Pi SMP["alpha_fs"])^2}])
    diffXSection1 = prefac1 integral1
```

where we insert the kinematical factor $1/64\pi^2 s$ and introduce the fine-structure constant α and the scattering angle θ in the CMS frame by setting the Mandelstam variables t and u. This immediately gives us:

$$\left(\frac{d\sigma(e^+e^- \to \ell^+\ell^-)}{d\Omega}\right)_{CMS} = \left(\frac{\alpha^2}{4s}\right)\left[1 + \cos^2\theta + \frac{8m_e^2 m_\ell^2 + 8sm_e^2 + 4m_e^4 + 4m_\ell^4 + 8sm_\ell^2}{s^2}\right] \quad (11.161)$$

Compare with Eq. (11.153). Kinematical constraints enforce the center-of-mass energy to be larger than twice the final state lepton rest mass, so $\sqrt{s} \geq 2m_\ell$, hence the terms proportional to the electron mass are negligible for $m_\ell \gg m_e$, which is the case for muon and tau final states. The differential cross-section can then be simplified by setting $m_e = 0$ to get:

$$\left(\frac{\mathrm{d}\sigma(e^+e^- \to \ell^+\ell^-)}{\mathrm{d}\Omega} \right)_{CMS} = \left(\frac{\alpha^2}{4s} \right) \left[1 + \cos^2\theta + \frac{(4m_\ell^4 + 8sm_\ell^2)}{s^2} \right] \qquad (11.162)$$

We conclude that taking into account the lepton rest masses did not alter the dependence on the scattering angle, which is only given by $\cos^2\theta$. It only modified the total cross-section. This latter can be computed by integration over the solid angle. This can be done with the following command:

```
2 Pi Integrate[% Sin[Th], {Th, 0, Pi}]
```

where "%" means the previous result. The computation yields the following expression:

$$\sigma_{tot}(e^+e^- \to \ell^+\ell^-) = \frac{4\pi\alpha^2}{3s} \left(1 + \frac{3m_\ell^4 + 6sm_\ell^2}{s^2} \right) \overset{s \gg 4m_\ell^2}{\Longrightarrow} \frac{4\pi\alpha^2}{3s} \qquad (11.163)$$

where we see that the term depending on m_ℓ becomes negligible compared to the first term when the center-of-mass energy is well above threshold, namely $\sqrt{s} \gg 2m_\ell$.

11.14 Helicity Conservation at High Energies

We have until now focused on unpolarized cross-sections which are generally easier to measure in experiments. However, a more comprehensive understanding of the dynamics is achieved when we study the spin dependence of the reactions. As an example, we illustrate the case of the heavy lepton pair creation $e^-e^+ \to \ell^-\ell^+$. The differential cross-section in the center-of-mass was derived in Eq. (11.153). Where does its $(1 + \cos\theta)^2$ angular dependence come from?

We can answer this question by analyzing the cross-section for each set of spin configurations in the initial and final states separately. To do this, we must first choose a basis of polarization states for the initial and final-state spinors. We could decide to use chirality, however this basis is not practical for experimental comparison, where the polarization and not the chirality of the particles can be fixed or measured. Hence, the natural choice of basis for a comparison with experimental data is the helicity basis. On the other hand, the discussion of Section 10.17 shows that from a theoretical point of view the chiral basis is really the most practical one!

In order to simplify the problem and yet reach our desired level of understanding, throughout this section we are interested in the behavior of the spins "in the high-energy limit" when all masses can be effectively neglected ($E \gg m$). We have seen in Section 8.24 that in this case helicity and chiral states match exactly.

Under this assumption, we consider the reaction $e^-e^+ \to \ell^-\ell^+$ and focus on the initial-state electron and positron. We can express their helicity or chirality state as e_R^-, e_L^-, e_R^+, and e_L^+. Following the result that chirality is always conserved at the QED vertex (see Eq. (10.165)), the helicity is also conserved in the high-energy limit. Hence, a QED vertex couples e_R^- with e_R^- and e_L^- with e_L^-. Because particles and antiparticles of the same chiralities possess opposite helicities (see Eq. (8.244)) this implies that electrons and positrons will possess opposite helicities at the vertex, hence e_R^+ couples with e_L^- and e_L^+ with e_R^-. See Figure 11.11.

Therefore we only need to consider the following two initial-state configuration spins:

$$\frac{\mathrm{d}\sigma}{\mathrm{d}\Omega}(e_L^+e_R^-) \quad \text{and} \quad \frac{\mathrm{d}\sigma}{\mathrm{d}\Omega}(e_R^+e_L^-) \qquad \text{since} \quad \frac{\mathrm{d}\sigma}{\mathrm{d}\Omega}(e_L^+e_L^-) = \frac{\mathrm{d}\sigma}{\mathrm{d}\Omega}(e_R^+e_R^-) = 0 \qquad (11.164)$$

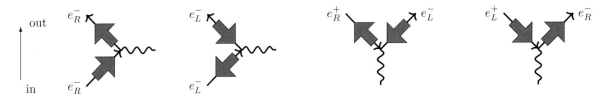

Figure 11.11 Possible helicity configurations at a QED vertex: (a) $e_R^- e_R^-$, (b) $e_L^- e_L^-$, (c) $e_R^+ e_L^-$, and (d) $e_L^+ e_R^-$.

Repeating the same arguments, we would find that the final-state leptons follow a similar rule. Hence, the helicities of the final-state lepton–antilepton pair must be opposite. In summary, **there are only four non-vanishing helicity configurations in the** $e^+ e^- \to \ell^+ \ell^-$ **reaction**: $e_R^+ e_L^- \to \ell_R^+ \ell_L^-$, $e_R^+ e_L^- \to \ell_L^+ \ell_R^-$, $e_L^+ e_R^- \to \ell_R^+ \ell_L^-$, and $e_L^+ e_R^- \to \ell_L^+ \ell_R^-$. These four processes build the entire cross-section and we should average over the initial-state spins and sum over the final-state spins. Hence:

$$\frac{d\sigma}{d\Omega}(e^+ e^- \to \ell^+ \ell^-) = \frac{1}{4}\left[\frac{d\sigma}{d\Omega}\left(e_R^+ e_L^- \to \ell_R^+ \ell_L^-\right) + \frac{d\sigma}{d\Omega}\left(e_R^+ e_L^- \to \ell_L^+ \ell_R^-\right) + \right.$$
$$\left. \frac{d\sigma}{d\Omega}\left(e_L^+ e_R^- \to \ell_R^+ \ell_L^-\right) + \frac{d\sigma}{d\Omega}\left(e_L^+ e_R^- \to \ell_L^+ \ell_R^-\right)\right] \qquad (11.165)$$

We now want to compute the polarized cross-section for each of the four configurations above. At tree level, our starting point is the amplitude Eq. (11.76) with $e^-(p, s) + e^+(k, r) \to \ell^-(p', s') + \ell^+(k', r')$ – see Figure 11.4. In order to benefit from the trace techniques, even though we now want to consider only one set of polarizations at a time, we note that in our high-energy limit (or for massless fermions), the chiral projection operators P_R and P_L (see Eq. (8.184)) will project out our desired helicity states. Thus, if in our amplitude we make the replacement (we omit the spin indices in the spinors):

$$\bar{v}(k)\gamma^\mu u(p) \longrightarrow \bar{v}(k)\gamma^\mu P_R u(p) = \bar{v}(k)\gamma^\mu \left(\frac{1 + \gamma^5}{2}\right) u(p) \qquad (11.166)$$

the amplitude for a right-handed electron e_R^- is unchanged while that of a left-handed electron e_L^- becomes zero! Note that the previous substitution also gives:

$$v^\dagger(k)\gamma^0 \gamma^\mu \left(\frac{1 + \gamma^5}{2}\right) u(p) = v^\dagger(k)\gamma^0 \left(\frac{1 - \gamma^5}{2}\right)\gamma^\mu u(p) = v^\dagger(k)\left(\frac{1 + \gamma^5}{2}\right)\gamma^0 \gamma^\mu u(p) \qquad (11.167)$$

hence the same replacement imposes a requirement on the spinor v but this corresponds to the opposite-helicity positron, hence to the expected e_L^+. So the projection P_R yields the initial-state combination $e_L^+ e_R^-$. Equivalent arguments apply to the final-state lepton pair. Hence, the initial and final polarized states cross-section can be found by inserting the desired chiral projection operators in the two currents. The polarized amplitude for $e^-(p, s) + e^+(k, r) \to \ell^-(p', s') + \ell^+(k', r')$ becomes (valid in the high-energy limit):

$$-i\mathcal{M} = \frac{ie^2}{(p+k)^2}\left(\bar{v}(k)\gamma^\mu P_{L,R} u(p)\right)\left(\bar{u}(p')\gamma_\mu P_{L,R} v(k')\right)$$
$$= \frac{ie^2}{(p+k)^2}\left(\bar{v}(k)\gamma^\mu \left(\frac{1 \mp \gamma^5}{2}\right) u(p)\right)\left(\bar{u}(p')\gamma_\mu \left(\frac{1 \mp \gamma^5}{2}\right) v(k')\right) \qquad (11.168)$$

where the four non-vanishing amplitudes are chosen with the projection operator combinations LL, LR, RL, and RR (or equivalently "$--$," "$-+$," "$+-$," and "$++$") for respectively $e_R^+ e_L^- \to \ell_R^+ \ell_L^-$, $e_R^+ e_L^- \to \ell_L^+ \ell_R^-$, $e_L^+ e_R^- \to \ell_R^+ \ell_L^-$, and $e_L^+ e_R^- \to \ell_L^+ \ell_R^-$.

Let us see what happens for the electron–positron current in the $e_L^+ e_R^-$ case. The sum over the spins of the electron and positron in the squared amplitude is then:

$$\sum_{s,r} \left| \bar{v}(k) \gamma^\mu \left(\frac{1+\gamma^5}{2} \right) u(p) \right|^2 = \sum_{s,r} \left[\bar{v}(k) \gamma^\mu \left(\frac{1+\gamma^5}{2} \right) u(p) \right] \left[\bar{v}(k) \gamma^\nu \left(\frac{1+\gamma^5}{2} \right) u(p) \right]^\dagger$$

$$= \sum_{s,r} \left[\bar{v}(k) \gamma^\mu \left(\frac{1+\gamma^5}{2} \right) u(p) \bar{u}(p) \gamma^\nu \left(\frac{1+\gamma^5}{2} \right) v(k) \right] \qquad (11.169)$$

where we used

$$\left[v(k)^\dagger \gamma^0 \gamma^\mu \left(\frac{1+\gamma^5}{2} \right) u(p) \right]^\dagger = u(p)^\dagger \left(\frac{1+\gamma^5}{2} \right)^\dagger (\gamma^\mu)^\dagger (\gamma^0)^\dagger v(k) = u(p)^\dagger \left(\frac{1+\gamma^5}{2} \right) (\gamma^0 \gamma^\mu \gamma^0) \gamma^0 v(k)$$

$$= u(p)^\dagger \gamma^0 \left(\frac{1-\gamma^5}{2} \right) \gamma^\mu v(k)$$

$$= \bar{u}(p) \gamma^\mu \left(\frac{1+\gamma^5}{2} \right) v(k) \qquad (11.170)$$

Hence

$$\sum_{s,r} \left| \bar{v}(k) \gamma^\mu \left(\frac{1+\gamma^5}{2} \right) u(p) \right|^2 = \mathrm{Tr} \left[\slashed{k} \gamma^\mu \left(\frac{1+\gamma^5}{2} \right) \slashed{p} \gamma^\nu \left(\frac{1+\gamma^5}{2} \right) \right] = \mathrm{Tr} \left[\slashed{k} \gamma^\mu \slashed{p} \left(\frac{1-\gamma^5}{2} \right) \gamma^\nu \left(\frac{1+\gamma^5}{2} \right) \right]$$

$$= \mathrm{Tr} \left[\slashed{k} \gamma^\mu \slashed{p} \gamma^\nu \left(\frac{1+\gamma^5}{2} \right)^2 \right] = \mathrm{Tr} \left[\slashed{k} \gamma^\mu \slashed{p} \gamma^\nu \left(\frac{1+\gamma^5}{2} \right) \right] \qquad (11.171)$$

We now make use of the two trace theorems (see Appendix E.2): the product of four γ matrices is given by **[T5]** $Tr(\gamma^\mu \gamma^\nu \gamma^\rho \gamma^\sigma) = 4(g^{\mu\nu} g^{\rho\sigma} - g^{\mu\rho} g^{\nu\sigma} + g^{\mu\sigma} g^{\nu\rho})$ and that of four γs plus γ^5 by **[T9]** $Tr(\gamma^\mu \gamma^\nu \gamma^\rho \gamma^\sigma \gamma^5) = -4i\epsilon^{\mu\nu\rho\sigma}$. We get:

$$\frac{1}{2} k_\alpha p_\beta \, \mathrm{Tr} \left[\gamma^\alpha \gamma^\mu \gamma^\beta \gamma^\nu (1+\gamma^5) \right] = 2 \left(k^\mu p^\nu + k^\nu p^\mu - g^{\mu\nu} (k \cdot p) \right) - 2i\epsilon^{\alpha\mu\beta\nu} k_\alpha p_\beta \qquad (11.172)$$

So

$$\sum_{s,r} \left| \bar{v}(k) \gamma^\mu \left(\frac{1+\gamma^5}{2} \right) u(p) \right|^2 = 2 \left(k^\mu p^\nu + k^\nu p^\mu - g^{\mu\nu} (k \cdot p) - i\epsilon^{\alpha\mu\beta\nu} k_\alpha p_\beta \right) \qquad (11.173)$$

An identical calculation for the final-state lepton pair yields:

$$\sum_{s',r'} \left| \bar{u}(p') \gamma_\mu \left(\frac{1+\gamma^5}{2} \right) v(k') \right|^2 = 2 \left(p'_\mu k'_\nu + p'_\nu k'_\mu - g_{\mu\nu} (p' \cdot k') - i\epsilon_{\rho\mu\sigma\nu} p'^\rho k'^\sigma \right) \qquad (11.174)$$

The amplitude squared is proportional to the product of the two terms. We get:

$$\left(k^\mu p^\nu + k^\nu p^\mu - g^{\mu\nu} (k \cdot p) - i\epsilon^{\alpha\mu\beta\nu} k_\alpha p_\beta \right) \left(p'_\mu k'_\nu + p'_\nu k'_\mu - g_{\mu\nu} (p' \cdot k') - i\epsilon_{\rho\mu\sigma\nu} p'^\rho k'^\sigma \right)$$

$$= \qquad \left(k^\mu p^\nu + k^\nu p^\mu \right) \left(p'_\mu k'_\nu + p'_\nu k'_\mu - g_{\mu\nu} (p' \cdot k') - i\epsilon_{\rho\mu\sigma\nu} p'^\rho k'^\sigma \right)$$

$$- g^{\mu\nu} (k \cdot p) \left(p'_\mu k'_\nu + p'_\nu k'_\mu - g_{\mu\nu} (p' \cdot k') - i\epsilon_{\rho\mu\sigma\nu} p'^\rho k'^\sigma \right)$$

$$- i\epsilon^{\alpha\mu\beta\nu} k_\alpha p_\beta \left(p'_\mu k'_\nu + p'_\nu k'_\mu - g_{\mu\nu} (p' \cdot k') - i\epsilon_{\rho\mu\sigma\nu} p'^\rho k'^\sigma \right) \qquad (11.175)$$

The Levi-Civita tensor is totally antisymmetric, its products vanish when they are multiplied by terms that are symmetric under the interchange $\mu \leftrightarrow \nu$:

$$-i\epsilon^{\alpha\mu\beta\nu} k_\alpha p_\beta \left(p'_\mu k'_\nu + p'_\nu k'_\mu - g_{\mu\nu} (p' \cdot k') \right) = 0 \qquad (11.176)$$

and the same happens for the term involving $-i\epsilon_{\rho\mu\sigma\nu}$. Hence the following terms remain:

$$
\begin{aligned}
(\ldots) &= (k^\mu p^\nu + k^\nu p^\mu)\left(p'_\mu k'_\nu + p'_\nu k'_\mu - g_{\mu\nu}(p' \cdot k')\right) - g^{\mu\nu}(k \cdot p)\left(p'_\mu k'_\nu + p'_\nu k'_\mu - g_{\mu\nu}(p' \cdot k')\right) \\
&\quad - \epsilon^{\alpha\mu\beta\nu} k_\alpha p_\beta \epsilon_{\rho\mu\sigma\nu} p'^\rho k'^\sigma \\
&= (k \cdot p')(p \cdot k') + (k \cdot k')(p \cdot p') - (k \cdot p)(p' \cdot k') + (k \cdot k')(p \cdot p') + (k \cdot p')(p \cdot k') - (k \cdot p)(p' \cdot k') \\
&\quad - 2(k \cdot p)(p' \cdot k') + g^{\mu\nu} g_{\mu\nu}(k \cdot p)(p' \cdot k') - \epsilon^{\alpha\mu\beta\nu} k_\alpha p_\beta \epsilon_{\rho\mu\sigma\nu} p'^\rho k'^\sigma \\
&= 2(k \cdot p')(p \cdot k') + 2(k \cdot k')(p \cdot p') - 4(k \cdot p)(p' \cdot k') + 4(k \cdot p)(p' \cdot k') - \epsilon^{\alpha\mu\beta\nu} k_\alpha p_\beta \epsilon_{\rho\mu\sigma\nu} p'^\rho k'^\sigma \\
&= 2(k \cdot p')(p \cdot k') + 2(k \cdot k')(p \cdot p') - \epsilon^{\alpha\mu\beta\nu} k_\alpha p_\beta \epsilon_{\rho\mu\sigma\nu} p'^\rho k'^\sigma \qquad (11.177)
\end{aligned}
$$

The last term with the Levi-Civita tensor can be simplified using the contraction formulae in Appendix D.9. We have:

$$
\begin{aligned}
k_\alpha p_\beta p'^\rho k'^\sigma \epsilon^{\alpha\mu\beta\nu} \epsilon_{\rho\mu\sigma\nu} &= k_\alpha p_\beta p'^\rho k'^\sigma (-\epsilon^{\mu\nu\alpha\beta})(-\epsilon_{\mu\nu\rho\sigma}) = -2 k_\alpha p_\beta p'^\rho k'^\sigma \left(\delta^\alpha_\rho \delta^\beta_\sigma - \delta^\alpha_\sigma \delta^\beta_\rho\right) \\
&= -2 p'^\rho k'^\sigma (k_\rho p_\sigma - k_\sigma p_\rho) = -2(k \cdot p')(k' \cdot p) + 2(k \cdot k')(p \cdot p') \qquad (11.178)
\end{aligned}
$$

Finally, we collect all the terms:

$$
\begin{aligned}
|\mathcal{M}|^2 &= \frac{4e^4}{(p+k)^4}\left(2(k \cdot p')(p \cdot k') + 2(k \cdot k')(p \cdot p') + 2(k \cdot p')(k' \cdot p) - 2(k \cdot k')(p \cdot p')\right) \\
&= \frac{16e^4}{(p+k)^4}(k \cdot p')(p \cdot k') \qquad (11.179)
\end{aligned}
$$

In the center-of-mass frame we can use the kinematics in the ultra-relativistic case from Eq. (11.143) and find $k \cdot p' = p \cdot k' = E^* E^* (1 + \cos\theta)$ and $(p+k)^4 \simeq (2p \cdot k)^2 = (2E^*)^4$, then:

$$
|\mathcal{M}|^2 = \frac{16e^4}{(2E^*)^4}(E^*)^4(1 + \cos\theta)^2 = e^4(1 + \cos\theta)^2 \qquad (11.180)
$$

We now put this result into the phase-space factor Eq. (5.145) to get our final result valid in the center-of-mass frame:

$$
\frac{d\sigma}{d\Omega}\left(e^+_L e^-_R \to \ell^+_L \ell^-_R\right) = \left(\frac{e^4}{64\pi^2 E^2_\star}\right)(1 + \cos\theta)^2 = \frac{\alpha^2}{4s}(1 + \cos\theta)^2 \qquad (11.181)
$$

where we have used $e^4/(64\pi^2) = \alpha^2/4$ and the Mandelstam variable $s = E^2_\star$.

In order to compute the other three non-vanishing differential cross-sections, we do not need to repeat the entire calculation. To see what happens when we swap the spins of the incoming electron–positron pair, we simply need to replace $1 + \gamma^5$ by $1 - \gamma^5$ in Eq. (11.173), which implies that the term $i\epsilon^{\alpha\mu\beta\nu}$ becomes $-i\epsilon^{\alpha\mu\beta\nu}$. Hence,

$$
\begin{aligned}
|\mathcal{M}|^2 &= \frac{4e^4}{(p+k)^4}\left(2(k \cdot p')(p \cdot k') + 2(k \cdot k')(p \cdot p') - 2\left((k \cdot p')(k' \cdot p) - (k \cdot k')(p \cdot p')\right)\right) \\
&= \frac{16e^4}{(p+k)^4}(k \cdot k')(p \cdot p') \qquad (11.182)
\end{aligned}
$$

and

$$
\frac{d\sigma}{d\Omega}\left(e^+_R e^-_L \to \ell^+_L \ell^-_R\right) = \frac{\alpha^2}{4s}(1 - \cos\theta)^2 \qquad (11.183)
$$

Having computed the two cases $LR \to LR$ and $RL \to LR$ to find respectively $(1 + \cos\theta)^2$ and $(1 - \cos\theta)^2$, we could compute the other two cases $RL \to RL$ and $LR \to RL$. However, we note that these last two can be obtained by a parity transformation from the first two, since the transformation flips the momenta but not the spins! We saw previously that QED is parity-invariant (see Eq. (10.166)), hence the corresponding differential cross-section must be the same. Indeed:

$$\frac{d\sigma}{d\Omega}\left(e_R^+ e_L^- \to \ell_R^+ \ell_L^-\right) = \frac{d\sigma}{d\Omega}\left(e_L^+ e_R^- \to \ell_L^+ \ell_R^-\right) = \frac{\alpha^2}{4s}(1+\cos\theta)^2 \tag{11.184}$$

and

$$\frac{d\sigma}{d\Omega}\left(e_L^+ e_R^- \to \ell_R^+ \ell_L^-\right) = \frac{d\sigma}{d\Omega}\left(e_R^+ e_L^- \to \ell_L^+ \ell_R^-\right) = \frac{\alpha^2}{4s}(1-\cos\theta)^2 \tag{11.185}$$

If we do not measure the polarizations of final-state leptons, we should add their corresponding cross-sections, for example the contributions $LR \to LR$ and $LR \to RL$. We get:

$$\begin{aligned}\frac{d\sigma}{d\Omega}\left(e_L^+ e_R^- \to \ell_L^+ \ell_R^-\right) + \frac{d\sigma}{d\Omega}\left(e_L^+ e_R^- \to \ell_R^+ \ell_L^-\right) &= \frac{\alpha^2}{2s}\left[\frac{(1+\cos\theta)^2}{2} + \frac{(1-\cos\theta)^2}{2}\right] \\ &= \frac{\alpha^2}{2s}(1+\cos^2\theta)\end{aligned} \tag{11.186}$$

We thus obtain a $1+\cos^2\theta$ dependence from the sum of the individual $(1+\cos\theta)^2$ and $(1-\cos\theta)^2$ contributions, as is graphically illustrated in Figure 11.12. The same is true for the other two helicity configurations $RL \to LR$ and $RL \to RL$.

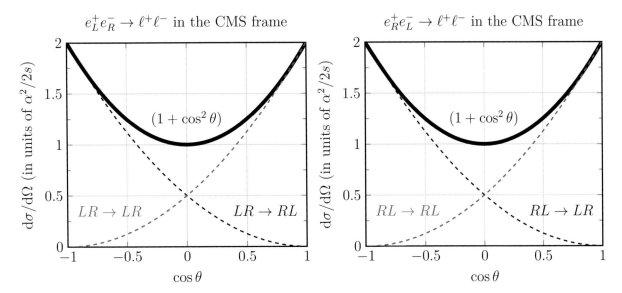

Figure 11.12 Angular dependence in the center-of-mass frame of the differential cross-section of the lepton pair creation process (left) $e_L^+ e_R^- \to \ell^+ \ell^-$ and (right) $e_R^+ e_L^- \to \ell^+ \ell^-$.

The unpolarized cross-section, Eq. (11.153), is recovered by adding all four contributions and dividing by four to average over the initial-state electron–positron spins (see Eq. (11.165)):

$$\frac{d\sigma}{d\Omega}(e^+e^- \to \ell^+\ell^-) = \frac{1}{4}\left[\underbrace{\frac{\alpha^2}{2s}(1+\cos^2\theta)}_{LR\to LR,\,LR\to RL} + \underbrace{\frac{\alpha^2}{2s}(1+\cos^2\theta)}_{RL\to LR,\,RL\to RL}\right] = \underbrace{\frac{\alpha^2}{4s}(1+\cos^2\theta)}_{\text{unpolarized}} \tag{11.187}$$

We recall that these explicit equations are only valid in the ultra-relativistic limit.

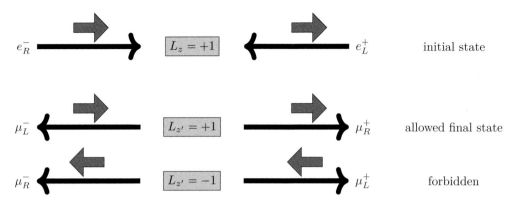

Figure 11.13 Conservation of angular momentum in the backward-scattering case at $180°$.

We note that these results on the angular dependence can readily be attributed to conservation of the total angular momentum. Let us illustrate this in the LR case. We note that:

$$\frac{d\sigma}{d\Omega}\left(e_L^+ e_R^- \to \ell_R^+ \ell_L^-\right)\Big|_{\cos\theta=-1} = \frac{\alpha^2}{4s}\left(1-(-1)\right)^2 = \frac{\alpha^2}{s}$$

$$\frac{d\sigma}{d\Omega}\left(e_L^+ e_R^- \to \ell_L^+ \ell_R^-\right)\Big|_{\cos\theta=-1} = \frac{\alpha^2}{4s}(1+(-1))^2 = 0 \tag{11.188}$$

This is explained by the fact that for our choice of LR in the initial state, the z-component of the total angular momentum is $L_z = +1$. For the combination $LR \to RL$, the z-component of the final state is also $L_{z'} = +1$ and in fact the overlap is full and the cross-section maximal, see Figure 11.13. Meanwhile for the opposite helicity of the final state we have $L_{z'} = -1$ and this configuration is not allowed from the point of view of angular momentum conservation. In between these two extreme cases, the cross-section is determined by the Clebsch–Gordan coefficient (see Appendix C.11).

11.15   Pair Annihilation into Two Photons $e^+e^- \to \gamma\gamma$

We now discuss processes with external photons. We begin with the annihilation of an electron–positron pair into two photons. The kinematics is given by:

$$e^-(p,s) + e^+(k,r) \to \gamma(k_1,\epsilon_1) + \gamma(k_2,\epsilon_2) \tag{11.189}$$

The corresponding Feynman diagrams at tree level are shown in Figure 11.14. These are examples of diagrams in which there is a fermion propagator between the two vertices between a fermion and a photon. The fermion propagator has four-momentum $q^\mu = p^\mu - k_1^\mu = k_2^\mu - k^\mu$. We recall that the direction of the arrow in the fermion propagator is conventional and that the Feynman propagator includes both contributions from an electron and a positron (i.e., forward and backward) into a single term. Starting at the end of the positron and working backwards, the contribution from the first diagram using the Feynman rules is:

$$i\mathcal{M}_1 = \epsilon_2^{\mu*}(k_2)\epsilon_1^{\nu*}(k_1)\bar{v}(k,r)(-ie\gamma_\mu)\frac{i(\slashed{q}+m_e)}{q^2-m_e^2}(-ie\gamma_\nu)u(p,s) \tag{11.190}$$

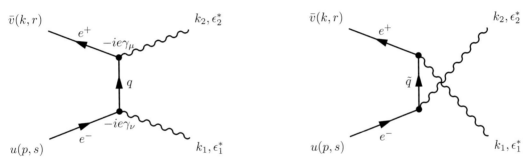

Figure 11.14 Feynman diagrams for $e^+e^- \to \gamma\gamma$ at tree level.

with $q = p - k_1 = k_2 - k$. Likewise, for the crossed diagram we swap the two photons by exchanging the indices on the vertices, and obtain:

$$i\mathcal{M}_2 = \epsilon_2^{\mu*}(k_2)\epsilon_1^{\nu*}(k_1)\bar{v}(k,r)(-ie\gamma_\nu)\frac{i(\tilde{\slashed{q}} + m_e)}{\tilde{q}^2 - m_e^2}(-ie\gamma_\mu)u(p,s) \tag{11.191}$$

with $\tilde{q} = p - k_2 = k_1 - k$. The net amplitude of the two diagrams can be expressed by factorizing the outgoing photon polarization vectors:

$$i\mathcal{M} = \epsilon_{2\mu}^*(k_2)\epsilon_{1\nu}^*(k_1)i\mathcal{M}^{\mu\nu} = \epsilon_{2\mu}^*(k_2)\epsilon_{1\nu}^*(k_1)\left(i\mathcal{M}_1^{\mu\nu} + i\mathcal{M}_2^{\mu\nu}\right) \tag{11.192}$$

We use the Mandelstam variable $t = (p - k_1)^2$ and rewrite $i\mathcal{M}_1^{\mu\nu}$:

$$i\mathcal{M}_1^{\mu\nu} = \frac{(-ie^2)}{t - m_e^2}\bar{v}(k,r)\gamma^\mu(\slashed{p} - \slashed{k}_1 + m_e)\gamma^\nu u(p,s) \tag{11.193}$$

and similarly with $u = (p - k_2)^2$, we find for $i\mathcal{M}_2^{\mu\nu}$:

$$i\mathcal{M}_2^{\mu\nu} = \frac{(-ie^2)}{u - m_e^2}\bar{v}(k,r)\gamma^\nu(\slashed{p} - \slashed{k}_2 + m_e)\gamma^\mu u(p,s) \tag{11.194}$$

Note the opposite orders of the γ^μ and γ^ν vertices in the $\mathcal{M}_1^{\mu\nu}$ and $\mathcal{M}_2^{\mu\nu}$ amplitudes. At this stage it is instructive to check the Ward identities. For the first photon we should have $k_{1\nu}\mathcal{M}^{\mu\nu} = 0$, and for the second photon $k_{2\mu}\mathcal{M}^{\mu\nu} = 0$.

We start with the first photon and the first diagram. We have (ignoring the first term of the amplitude):

$$\begin{aligned}
k_{1\nu}\bar{v}\gamma^\mu(\slashed{p} - \slashed{k}_1 + m_e)\gamma^\nu u &= \bar{v}\gamma^\mu(\slashed{p} - \slashed{k}_1 + m_e)\slashed{k}_1 u \\
&= \bar{v}\gamma^\mu(\slashed{p} + m_e)\slashed{k}_1 u \qquad \text{since } \slashed{k}_1\slashed{k}_1 = k_1^2 = 0 \\
&= \bar{v}\gamma^\mu\left(2(p \cdot k_1) - \slashed{k}_1\slashed{p} + \slashed{k}_1 m_e\right)u \qquad \text{since } \slashed{p}\slashed{k}_1 = 2(p \cdot k_1) - \slashed{k}_1\slashed{p} \\
&= \bar{v}\gamma^\mu\left(2(p \cdot k_1) - \slashed{k}_1(\slashed{p} - m_e)\right)u \\
&= 2(p \cdot k_1)\bar{v}\gamma^\mu u \qquad \text{since } (\slashed{p} - m_e)u = 0 \\
&= -(t - m_e^2)\bar{v}\gamma^\mu u \tag{11.195}
\end{aligned}$$

where in the last line we have used $2(p \cdot k_1) = -(p - k_1)^2 + p^2 + k_1^2 = -t + m_e^2$, and consequently:

$$k_{1\nu}\mathcal{M}_1^{\mu\nu} = e^2\bar{v}\gamma^\mu u \tag{11.196}$$

This means that for the first diagram alone, the Ward identity is not satisfied. Similarly, for the second diagram, we get:

$$k_{1\nu}\mathcal{M}_2^{\mu\nu} = -e^2\bar{v}\gamma^\mu u \tag{11.197}$$

and again the Ward identity is not satisfied. **But together both diagrams satisfy the Ward identity**:

$$k_{1\nu}\mathcal{M}^{\mu\nu} = k_{1\nu}\mathcal{M}_1^{\mu\nu} + k_{1\nu}\mathcal{M}_2^{\mu\nu} = 0 \tag{11.198}$$

As mentioned previously, the Ward identity does not necessarily apply diagram-by-diagram but only for entire amplitudes. The same arguments and calculations can be repeated for the second photon to get:

$$k_{2\mu}\mathcal{M}^{\mu\nu} = k_{2\mu}\mathcal{M}_1^{\mu\nu} + k_{2\mu}\mathcal{M}_2^{\mu\nu} = 0 \tag{11.199}$$

Combining the two diagrams, we will get the following matrix element squared:

$$|\mathcal{M}|^2 = |\mathcal{M}_1|^2 + |\mathcal{M}_2|^2 + 2\Re e(\mathcal{M}_1\mathcal{M}_2^*) \tag{11.200}$$

since for any two complex numbers a and b we have $|a+b|^2 = (a+b)(a^*+b^*) = |a|^2 + |b|^2 + ab^* + a^*b = |a|^2 + |b|^2 + 2\Re e(ab^*)$. Let us now focus first on the square of the matrix element of $\mathcal{M}_1$:

$$|\mathcal{M}_1|^2 = \frac{e^4}{(t-m_e^2)^2} \bar{v}(k,r)\not{\epsilon}_2^*(k_2)(\not{p}-\not{k}_1+m_e)\not{\epsilon}_1^*(k_1)u(p,s)\bar{u}(p,s)\not{\epsilon}_1(k_1)(\not{p}-\not{k}_1+m_e)\not{\epsilon}_2(k_2)v(k,r) \tag{11.201}$$

In order to get the unpolarized matrix element, we *average* over the incoming fermion helicities and find:

$$\frac{1}{4}\sum_{s,r}|\mathcal{M}_1|^2 = \frac{e^4}{4(t-m_e^2)^2}\,\mathrm{Tr}\left((\not{k}-m_e)\not{\epsilon}_2^*(k_2)(\not{p}-\not{k}_1+m_e)\not{\epsilon}_1^*(k_1)(\not{p}+m_e)\not{\epsilon}_1(k_1)(\not{p}-\not{k}_1+m_e)\not{\epsilon}_2(k_2)\right) \tag{11.202}$$

Now we also sum over the *two* helicities of each of the final-state photons and use the completeness relation for real photons, Eq. (10.143):

$$\sum_{\lambda=1}^{2}\epsilon_\lambda^\mu\epsilon_\lambda^{*\nu} = -g^{\mu\nu} \implies \sum_{\lambda=1}^{2}\not{\epsilon}(\lambda)\not{\epsilon}^*(\lambda) = \sum_{\lambda=1}^{2}\gamma^\mu\epsilon_\mu(\lambda)\gamma^\nu\epsilon_\nu^*(\lambda) = -\gamma^\mu\gamma^\nu g_{\mu\nu} = -\gamma^\mu\gamma_\mu \tag{11.203}$$

to find:

$$\frac{1}{4}\sum_{s,r,\lambda_1,\lambda_2}|\mathcal{M}_1|^2 = \frac{e^4}{4(t-m_e^2)^2}\,\mathrm{Tr}\left((\not{k}-m_e)\gamma^\mu(\not{p}-\not{k}_1+m_e)\gamma^\nu(\not{p}+m_e)\gamma_\nu(\not{p}-\not{k}_1+m_e)\gamma_\mu\right) \tag{11.204}$$

We note that the trace is cyclic and focus on the term (see Appendix E.1):

$$\gamma_\mu(\not{k}-m_e)\gamma^\mu = \gamma_\mu k_\alpha\gamma^\alpha\gamma^\mu - m_e\gamma_\mu\gamma^\mu = -2k_\alpha\gamma^\alpha - 4m_e\mathbb{1} = -2\not{k} - 4m_e\mathbb{1} \tag{11.205}$$

and similarly:

$$\gamma^\nu(\not{p}+m_e)\gamma_\nu = -2\not{p} + 4m_e\mathbb{1} \tag{11.206}$$

We can recompute the spin-averaged matrix element:

$$\langle|\mathcal{M}_1|^2\rangle = \frac{e^4}{4(t-m_e^2)^2}\,\mathrm{Tr}\left((-2\not{k}-4m_e\mathbb{1})(\not{p}-\not{k}_1+m_e\mathbb{1})(-2\not{p}+4m_e\mathbb{1})(\not{p}-\not{k}_1+m_e\mathbb{1})\right) \tag{11.207}$$

Expanding the trace in 16 factors and using the fact that the trace of a product of an odd number of γ matrices vanishes, we obtain the following terms, noting $\not{p}\not{p} = p\cdot p = m_e^2$ and for the photons $\not{k}_1\not{k}_1 = k_1\cdot k_1 = 0$:

1. $(-2\not{k})(\not{p}-\not{k}_1)(-2\not{p})(\not{p}-\not{k}_1) = +4\not{k}(\not{p}-\not{k}_1)\not{p}(\not{p}-\not{k}_1)$

2. $(-2\not{k})(\not{p}-\not{k}_1)(4m_e\mathbb{1})(m_e\mathbb{1}) = -8m_e^2\not{k}(\not{p}-\not{k}_1) = -8m_e^2(\not{k}\not{p}-\not{k}\not{k}_1)$

3. $(-2\not{k})(m_e\mathbb{1})(-2\not{p})(m_e\mathbb{1}) = +4m_e^2\not{k}\not{p}$

4. $(-2\not k)(m_e\mathbb{1})(4m_e\mathbb{1})(\not p - \not k_1) = -8m_e^2\not k(\not p - \not k_1) = -8m_e^2(\not k\not p - \not k\not k_1)$

5. $(-4m_e\mathbb{1})(\not p - \not k_1)(-2\not p)(m_e\mathbb{1}) = +8m_e^2(\not p - \not k_1)\not p = +8m_e^2(\not p\not p - \not k_1\not p) = +8m_e^4\mathbb{1} - 8m_e^2\not k_1\not p$

6. $(-4m_e\mathbb{1})(\not p - \not k_1)(4m_e\mathbb{1})(\not p - \not k_1) = -16m_e^2(\not p - \not k_1)(\not p - \not k_1) = -16m_e^4\mathbb{1} + 32m_e^2\not p\not k_1$

7. $(-4m_e\mathbb{1})(m_e\mathbb{1})(-2\not p)(\not p - \not k_1) = 8m_e^4\mathbb{1} - 8m_e^2\not p\not k_1$

8. $(-4m_e\mathbb{1})(m_e\mathbb{1})(4m_e\mathbb{1})(m_e\mathbb{1}) = -16m_e^4\mathbb{1}$

We insert these results and do more "traceology" to find:

$$
\begin{aligned}
&\operatorname{Tr}\left\{4\not k(\not p - \not k_1)\not p(\not p - \not k_1) - 16m_e^2(\not k\not p - \not k\not k_1) + 4m_e^2\not k\not p - 8m_e^2\not k_1\not p + 24m_e^2\not p\not k_1 - 16m_e^4\mathbb{1}\right\} \\
={}& 4\operatorname{Tr}\left\{\not k(\not p - \not k_1)\not p(\not p - \not k_1)\right\} - 4m_e^2\operatorname{Tr}\left\{4\not k\not p - 4\not k\not k_1 - \not k\not p + 2\not k_1\not p - 6\not p\not k_1\right\} - 16m_e^4 Tr\left\{\mathbb{1}\right\} \\
={}& 4\operatorname{Tr}\left\{\not k(\not p - \not k_1)\not p(\not p - \not k_1)\right\} - 4m_e^2\operatorname{Tr}\left\{3\not k\not p - 4\not k\not k_1 + 2\not k_1\not p - 6\not p\not k_1\right\} - 64m_e^4
\end{aligned}
\tag{11.208}
$$

The terms in the first trace give:

$$
\begin{aligned}
\not k(\not p - \not k_1)\not p(\not p - \not k_1) &= \not k(\not p - \not k_1)(\not p\not p - \not p\not k_1) = \not k(\not p\not p\not p - \not p\not p\not k_1 - \not k_1\not p\not p + \not k_1\not p\not k_1) = \\
&= \not k(m_e^2\not p - m_e^2\not k_1 - \not k_1 m_e^2 + \not k_1\not p\not k_1) \\
&= m_e^2\not k\not p - 2m_e^2\not k\not k_1 + \not k\not k_1\not p\not k_1
\end{aligned}
\tag{11.209}
$$

Consequently, the trace becomes:

$$
\begin{aligned}
&4\operatorname{Tr}\left\{m_e^2\not k\not p - 2m_e^2\not k\not k_1 + \not k\not k_1\not p\not k_1\right\} - 4m_e^2\operatorname{Tr}\left\{3\not k\not p - 4\not k\not k_1 + 2\not k_1\not p - 6\not p\not k_1\right\} - 64m_e^4 = \\
&4m_e^2\operatorname{Tr}\left\{\not k\not p - 2\not k\not k_1\right\} + 4\operatorname{Tr}\left\{\not k\not k_1\not p\not k_1\right\} - 4m_e^2\operatorname{Tr}\left\{3\not k\not p - 4\not k\not k_1 + 2\not k_1\not p - 6\not p\not k_1\right\} - 64m_e^4 = \\
&4\left\{\operatorname{Tr}\left(\not k\not k_1\not p\not k_1\right) + m_e^2\operatorname{Tr}\left(-2\not k\not p + 2\not k\not k_1 - 2\not k_1\not p + 6\not p\not k_1\right) - 16m_e^4\right\}
\end{aligned}
\tag{11.210}
$$

Finally, we insert this result in the initial spin-averaged matrix element:

$$
\langle|\mathcal{M}_1|^2\rangle = \frac{e^4}{(t - m_e^2)^2}\left\{\operatorname{Tr}\left(\not k\not k_1\not p\not k_1\right) + m_e^2\operatorname{Tr}\left(-2\not k\not p + 2\not k\not k_1 - 2\not k_1\not p + 6\not p\not k_1\right) - 16m_e^4\right\}
\tag{11.211}
$$

We arrive at the kinematics using:

$$
\not k_1\not p\not k_1 = k_1^\alpha p^\beta k_1^\delta \gamma_\alpha\gamma_\beta\gamma_\delta = k_1^\alpha p^\beta k_1^\delta\gamma_\alpha(2g_{\beta\delta} - \gamma_\delta\gamma_\beta) = 2(p\cdot k_1)\not k_1 - \not k_1\not k_1\not p = 2(p\cdot k_1)\not k_1
\tag{11.212}
$$

and for any two four-vectors:

$$
\operatorname{Tr}(\not a\not b) = a_\alpha b_\beta\operatorname{Tr}(\gamma^\alpha\gamma^\beta) = 4a_\alpha b_\beta g^{\alpha\beta} = 4a\cdot b
\tag{11.213}
$$

The spin-averaged matrix element corresponding to the first diagram is ultimately:

$$
\begin{aligned}
\langle|\mathcal{M}_1|^2\rangle &= \frac{e^4}{(t - m_e^2)^2}\left\{8(p\cdot k_1)(k\cdot k_1) + 4m_e^2(-2k\cdot p + 2k\cdot k_1 - 2k_1\cdot p + 6p\cdot k_1) - 16m_e^4\right\} \\
&= \frac{e^4}{(t - m_e^2)^2}\left\{8(p\cdot k_1)(k\cdot k_1) + 4m_e^2(-2k\cdot p + 2k\cdot k_1 + 4p\cdot k_1) - 16m_e^4\right\}
\end{aligned}
\tag{11.214}
$$

We may further simplify this formula by expressing all the momenta products in terms of the Mandelstam's variables s, t, and u. Recalling $s + t + u = 2m_e^2$, we have:

$$
s = (p + k)^2 = p^2 + k^2 + 2(p\cdot k) = 2m_e^2 + 2(p\cdot k) \quad\to\quad p\cdot k = \frac{1}{2}(s - 2m_e^2) = \frac{1}{2}(-t - u)
\tag{11.215}
$$

also
$$s = (k_1 + k_2)^2 = k_1^2 + k_2^2 + 2(k_1 \cdot k_2) = 2(k_1 \cdot k_2) \quad \rightarrow \quad k_1 \cdot k_2 = \frac{s}{2} \tag{11.216}$$

and
$$t = (p - k_1)^2 = p^2 + k_1^2 - 2(p \cdot k_1) = m_e^2 - 2(p \cdot k_1) \quad \rightarrow \quad p \cdot k_1 = -\frac{1}{2}(t - m_e^2) \tag{11.217}$$

Further
$$u = (p - k_2)^2 = p^2 + k_2^2 - 2(p \cdot k_2) = m_e^2 - 2(p \cdot k_2) \quad \rightarrow \quad p \cdot k_2 = -\frac{1}{2}(u - m_e^2) \tag{11.218}$$

and also
$$k \cdot k_1 = (k_2 + k_1 - p) \cdot k_1 = k_2 \cdot k_1 + k_1^2 - p \cdot k_1 = \frac{1}{2}(s + t - m_e^2) = \frac{1}{2}(m_e^2 - u) \tag{11.219}$$

and finally
$$k \cdot k_2 = (k_2 + k_1 - p) \cdot k_2 = k_2^2 + k_1 \cdot k_2 - p \cdot k_2 = \frac{1}{2}(s + u - m_e^2) = \frac{1}{2}(m_e^2 - t) \tag{11.220}$$

Then we have:
$$\begin{aligned}
\langle |\mathcal{M}_1|^2 \rangle &= \frac{4e^4}{(t - m_e^2)^2} \left\{ \frac{1}{2}(m_e^2 - t)(m_e^2 - u) + m_e^2 \left(-(-t - u) + (m_e^2 - u) - 2(t - m_e^2) \right) - 4m_e^4 \right\} \\
&= \frac{2e^4}{(t - m_e^2)^2} \left(tu - m_e^2(3t + u) - m_e^4 \right) \\
&= \frac{2e^4}{(t - m_e^2)^2} \left((t - m_e^2)(u - 3m_e^2) - 4m_e^4 \right)
\end{aligned} \tag{11.221}$$

The matrix element for the crossed diagram can be found by crossing symmetry, swapping k_1 and k_2, and hence $t \leftrightarrow u$:

$$\langle |\mathcal{M}_2|^2 \rangle = \frac{2e^4}{(u - m_e^2)^2} \left((u - m_e^2)(t - 3m_e^2) - 4m_e^4 \right) \tag{11.222}$$

A similar procedure must be applied to compute the interference term between the two graphs. For the interference term we have averaging as before on the initial-state spins and summing over the helicities of the final-state photons:

$$\begin{aligned}
\frac{1}{4} \sum_{s,r} 2\Re e(\mathcal{M}_1 \mathcal{M}_2^*) &= \frac{1}{2} \sum_{s,r,\lambda_1,\lambda_2} \Re e \left(\epsilon_2^{\mu*}(k_2)\epsilon_1^{\nu*}(k_1)\mathcal{M}_{1\mu\nu}\epsilon_2^{\rho}(k_2)\epsilon_1^{\sigma}(k_1)\mathcal{M}_{2\rho\sigma}^* \right) \\
&= \frac{1}{2} \sum_{s,r} \Re e \left(g^{\mu\rho} g^{\nu\sigma} \mathcal{M}_{1\mu\nu} \mathcal{M}_{2\rho\sigma}^* \right) = \frac{1}{2} \sum_{s,r} \Re e \left(\mathcal{M}_{1\mu\nu} \mathcal{M}_2^{*\mu\nu} \right) \\
&= \frac{2e^4}{4(t - m_e^2)(u - m_e^2)} \sum_{s,r} \Re e \left(\bar{v}(k)\gamma^\mu(\slashed{q} + m_e)\gamma^\nu u(p)\bar{u}(p)\gamma_\mu(\slashed{\tilde{q}} + m_e)\gamma_\nu v(k) \right)
\end{aligned} \tag{11.223}$$

The sum over the spins gives us the following trace:

$$\mathrm{Tr} \left((\slashed{k} - m_e)\gamma^\mu(\slashed{q} + m_e)\gamma^\nu(\slashed{p} + m_e)\gamma_\mu(\slashed{\tilde{q}} + m_e)\gamma_\nu \right) \tag{11.224}$$

which expands again in 16 factors. Using the fact that the trace of a product of an odd number of γ matrices vanishes, the number of terms reduces to eight. We can further use the properties in Appendices E.1 and E.2 to remove the γ^νs:

$$\gamma^\nu \gamma_\nu = 4\mathbb{1}, \quad \gamma^\nu \gamma^\mu \gamma_\nu = -2\gamma^\mu, \quad \gamma^\nu \slashed{a} \slashed{b} \gamma_\nu = 4(a \cdot b)\mathbb{1}, \quad \text{and} \quad \gamma^\nu \slashed{a} \gamma^\mu \slashed{b} \gamma_\nu = -2\slashed{b}\gamma^\mu \slashed{a} \tag{11.225}$$

We can do the same with the γ^μ's. We also use $\mathrm{Tr}(\slashed{a}\slashed{b}) = 4(a \cdot b)$ and we go again:

1. $\mathrm{Tr}\left(\not{k}\gamma^\mu \not{q}\underbrace{\gamma^\nu \not{p}\gamma_\mu \tilde{\not{q}}\gamma_\nu}_{=-2\tilde{\not{q}}\gamma_\mu \not{p}}\right) = -2\,\mathrm{Tr}\left(\not{k}\underbrace{\gamma^\mu \not{q}\tilde{\not{q}}\gamma_\mu}_{=4(q\cdot\tilde{q})}\not{p}\right) = -8(q\cdot\tilde{q})\,\mathrm{Tr}(\not{k}\not{p}) = -32(q\cdot\tilde{q})(k\cdot p)$

2. $m_e^2\,\mathrm{Tr}\left(\not{k}\gamma^\mu \not{q}\underbrace{\gamma^\nu \gamma_\mu \gamma_\nu}_{=-2\gamma_\mu}\right) = -2m_e^2\,\mathrm{Tr}\left(\not{k}\underbrace{\gamma^\mu \not{q}\gamma_\mu}_{=-2\not{q}}\right) = 4m_e^2\,\mathrm{Tr}(\not{k}\not{q}) = 16m_e^2(k\cdot q)$

3. $m_e^2\,\mathrm{Tr}\left(\not{k}\underbrace{\gamma^\mu \gamma^\nu \not{p}\gamma_\mu \gamma_\nu}_{=4\not{p}}\right) = 4m_e^2\,\mathrm{Tr}(\not{k}\not{p}) = 16m_e^2(k\cdot p)$

4. $m_e^2\,\mathrm{Tr}\left(\not{k}\underbrace{\gamma^\mu \gamma^\nu \gamma_\mu}_{=-2\gamma^\nu} \tilde{\not{q}}\gamma_\nu\right) = -2m_e^2\,\mathrm{Tr}\left(\not{k}\underbrace{\gamma^\nu \tilde{\not{q}}\gamma_\nu}_{=-2\tilde{\not{q}}}\right) = 4m_e^2\,\mathrm{Tr}(\not{k}\tilde{\not{q}}) = 16m_e^2(k\cdot\tilde{q})$

5. $-m_e^2\,\mathrm{Tr}\left(\underbrace{\gamma^\mu \not{q}\gamma^\nu \not{p}\gamma_\mu}_{=-2\not{p}\gamma^\nu \not{q}}\gamma_\nu\right) = 2m_e^2\,\mathrm{Tr}\left(\not{p}\underbrace{\gamma^\nu \not{q}\gamma_\nu}_{=-2\not{q}}\right) = -4m_e^2\,\mathrm{Tr}(\not{p}\not{q}) = -16m_e^2(p\cdot q)$

6. $-m_e^2\,\mathrm{Tr}\left(\gamma^\mu \not{q}\gamma^\nu \gamma_\mu \tilde{\not{q}}\gamma_\nu\right) = -m_e^2\,\mathrm{Tr}\left(\underbrace{\gamma_\nu \gamma^\mu \not{q}\gamma^\nu \gamma_\mu}_{=4\not{q}}\tilde{\not{q}}\right) = -4m_e^2\,\mathrm{Tr}(\not{q}\tilde{\not{q}}) = -16m_e^2(q\cdot\tilde{q})$

7. $-m_e^2\,\mathrm{Tr}\left(\gamma^\mu \underbrace{\gamma^\nu \not{p}\gamma_\mu \tilde{\not{q}}\gamma_\nu}_{=-2\tilde{\not{q}}\gamma_\mu \not{p}}\right) = 2m_e^2\,\mathrm{Tr}\left(\underbrace{\gamma^\mu \tilde{\not{q}}\gamma_\mu}_{=-2\tilde{\not{q}}}\not{p}\right) = -4m_e^2\,\mathrm{Tr}(\tilde{\not{q}}\not{p}) = -16m_e^2(\tilde{q}\cdot p)$

8. $-m_e^4\,\mathrm{Tr}\left(\gamma^\mu \underbrace{\gamma^\nu \gamma_\mu \gamma_\nu}_{=-2\gamma_\mu}\right) = 2m_e^4\,\underbrace{\mathrm{Tr}\left(\gamma^\mu \gamma_\mu\right)}_{=4\,\mathrm{Tr}(\mathbb{1})} = 32m_e^4$

Hence the 16 factors of the trace reduce to the following eight terms:

$$\mathrm{Tr}(\ldots) = -32(q\cdot\tilde{q})(k\cdot p) + 16m_e^2\left[(k\cdot q) + (k\cdot p) + (k\cdot\tilde{q}) - (p\cdot q) - (q\cdot\tilde{q}) - (\tilde{q}\cdot p)\right] + 32m_e^4 \qquad (11.226)$$

The useful products of four-vectors are:

$$
\begin{aligned}
k\cdot q &= k\cdot(k_2 - k) = k\cdot k_2 - m_e^2 = \tfrac{1}{2}(m_e^2 - t) - m_e^2 = -\tfrac{1}{2}(t + m_e^2) \\
k\cdot\tilde{q} &= k\cdot(k_1 - k) = k\cdot k_1 - m_e^2 = \tfrac{1}{2}(m_e^2 - u) - m_e^2 = -\tfrac{1}{2}(u + m_e^2) \\
p\cdot q &= p\cdot(p - k_1) = m_e^2 - p\cdot k_1 = m_e^2 + \tfrac{1}{2}(t - m_e^2) = \tfrac{1}{2}(t + m_e^2) \\
p\cdot\tilde{q} &= p\cdot(p - k_2) = m_e^2 - p\cdot k_2 = m_e^2 + \tfrac{1}{2}(u - m_e^2) = \tfrac{1}{2}(u + m_e^2) \\
q\cdot\tilde{q} &= (p - k_1)\cdot(p - k_2) = m_e^2 - p\cdot k_1 - p\cdot k_2 + k_1\cdot k_2 = m_e^2 + \tfrac{1}{2}(t - m_e^2) + \tfrac{1}{2}(u - m_e^2) + \tfrac{s}{2} \\
&= \tfrac{1}{2}(s + t + u) = m_e^2 \qquad\qquad\qquad\qquad\qquad\qquad\qquad\qquad (11.227)
\end{aligned}
$$

and, as before, $k \cdot p = \frac{1}{2}(s - 2m_e^2)$. By insertion of the four products into the trace terms, the result becomes:

$$
\begin{aligned}
\mathrm{Tr}(\ldots) &= -32m_e^2 \frac{1}{2}(s - 2m_e^2) \\
&\quad +16m_e^2\left[-\frac{1}{2}(t + m_e^2) + \frac{1}{2}(s - 2m_e^2) - \frac{1}{2}(u + m_e^2) - \frac{1}{2}(t + m_e^2) - m_e^2 - \frac{1}{2}(u + m_e^2)\right] + 32m_e^4 \\
&= -16m_e^2(s - 2m_e^2) + 8m_e^2\left[-t - m_e^2 + s - 2m_e^2 - u - m_e^2 - t - m_e^2 - 2m_e^2 - u - m_e^2\right] + 32m_e^4 \\
&= -16m_e^2 s + 32m_e^4 + 8m_e^2\left[-8m_e^2 + s - 2t - 2u\right] + 32m_e^4 \\
&= -8m_e^2 s - 16m_e^2(u + t) = -8m_e^2 s - 16m_e^2(2m_e^2 - s) = 8m_e^2(s - 4m_e^2) \qquad (11.228)
\end{aligned}
$$

We can now sum the two matrix elements $\langle|\mathcal{M}_1|^2\rangle$ and $\langle|\mathcal{M}_2|^2\rangle$, and the interference terms together. We get:

$$
\begin{aligned}
\langle|\mathcal{M}|^2\rangle &= \langle|\mathcal{M}_1|^2\rangle + \langle|\mathcal{M}_2|^2\rangle + 2\langle\mathcal{M}_1\mathcal{M}_2^*\rangle \\
&= \frac{2e^4}{(t - m_e^2)^2}\left((t - m_e^2)(u - 3m_e^2) - 4m_e^4\right) + \frac{2e^4}{(u - m_e^2)^2}\left((u - m_e^2)(t - 3m_e^2) - 4m_e^4\right) \\
&\quad + \frac{2e^4}{4(t - m_e^2)(u - m_e^2)}\left(8m_e^2(s - 4m_e^2)\right) \\
&= 2e^4\left[\frac{\left((t - m_e^2)(u - 3m_e^2) - 4m_e^4\right)}{(t - m_e^2)^2} + \frac{\left((u - m_e^2)(t - 3m_e^2) - 4m_e^4\right)}{(u - m_e^2)^2} + \frac{2m_e^2(s - 4m_e^2)}{(t - m_e^2)(u - m_e^2)}\right] \\
&= 2e^4\left[\frac{u - m_e^2}{t - m_e^2} + \frac{t - m_e^2}{u - m_e^2} + \mathcal{A}\right] \qquad (11.229)
\end{aligned}
$$

where

$$
\mathcal{A} \equiv -\frac{2m_e^2}{t - m_e^2} - \frac{2m_e^2}{u - m_e^2} - \frac{4m_e^4}{(t - m_e^2)^2} - \frac{4m_e^4}{(u - m_e^2)^2} + \frac{2m_e^2 s}{(t - m_e^2)(u - m_e^2)} - \frac{8m_e^4}{(t - m_e^2)(u - m_e^2)} \qquad (11.230)
$$

At this stage, it is worth noting that:

$$
\begin{aligned}
1 - \left(1 + \frac{2m_e^2}{t - m_e^2} + \frac{2m_e^2}{u - m_e^2}\right)^2 &= -\frac{4m_e^2}{t - m_e^2} - \frac{4m_e^2}{u - m_e^2} - \frac{8m_e^4}{(t - m_e^2)(u - m_e^2)} - \frac{4m_e^4}{(t - m_e^2)^2} - \frac{4m_e^4}{(u - m_e^2)^2} \\
&= \mathcal{A} - \frac{2m_e^2}{t - m_e^2} - \frac{2m_e^2}{u - m_e^2} - \frac{2m_e^2 s}{(t - m_e^2)(u - m_e^2)} \\
&= \mathcal{A} - 2m_e^2\left[\frac{1}{t - m_e^2} + \frac{1}{u - m_e^2} + \frac{s}{(t - m_e^2)(u - m_e^2)}\right] \\
&= \mathcal{A} - 2m_e^2 \underbrace{\left[\frac{u - m_e^2 + t - m_e^2 + s}{(t - m_e^2)(u - m_e^2)}\right]}_{=0} \qquad (11.231)
\end{aligned}
$$

Consequently, we can further shrink the matrix element to the common form:

$$
\langle|\mathcal{M}|^2\rangle\,(e^+e^- \to \gamma\gamma) = 2e^4\left[\frac{u - m_e^2}{t - m_e^2} + \frac{t - m_e^2}{u - m_e^2} + 1 - \left(1 + \frac{2m_e^2}{t - m_e^2} + \frac{2m_e^2}{u - m_e^2}\right)^2\right] \qquad (11.232)
$$

Note that if we were to neglect the electron rest mass, the result would turn out to be beautifully simple and just given by:

$$
\langle|\mathcal{M}|^2\rangle\,(e^+e^- \to \gamma\gamma) \approx 2e^4\left[\frac{u}{t} + \frac{t}{u}\right] \qquad (m_e \longrightarrow 0) \qquad (11.233)
$$

We now define the kinematics in the center-of-mass frame where $p = (E, 0, 0, p)$ and $k = (E, 0, 0, -p)$ for the initial state and $k_{1,2} = (\omega, \pm \vec{k})$ for the final state. Note that $\omega = \left|\vec{k}\right| = E$. Consequently, the Mandelstam variables are equal to:

$$
\begin{aligned}
s &= (p + k)^2 = (2E)^2 = 4E^2 \\
t - m_e^2 &= -2(p \cdot k_1) = -2(E^2 - \vec{p} \cdot \vec{k}_1) = -2(E^2 - pE \cos\theta) = -2E(E - p\cos\theta) \\
u - m_e^2 &= -2(p \cdot k_2) = -2E(E + p\cos\theta)
\end{aligned}
\tag{11.234}
$$

where θ is the scattering angle, see Figure 11.15. Then one can prove that:

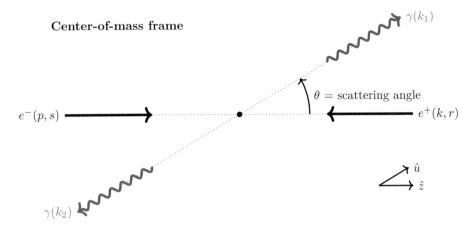

Figure 11.15 The kinematics for the pair annihilation into two photons in the center-of-mass frame.

$$
\begin{aligned}
\frac{u - m_e^2}{t - m_e^2} + \frac{t - m_e^2}{u - m_e^2} + 1 &= \frac{(E + p\cos\theta)}{(E - p\cos\theta)} + \frac{(E - p\cos\theta)}{(E + p\cos\theta)} + 1 \\
&= \frac{(E + p\cos\theta)^2 + (E - p\cos\theta)^2 + (E^2 - p^2\cos^2\theta)}{E^2 - p^2\cos^2\theta} \\
&= \frac{3E^2 + p^2\cos^2\theta}{p^2 + m_e^2 - p^2\cos^2\theta} = \frac{3m_e^2 + 3p^2 + p^2\cos^2\theta}{m_e^2 + p^2\sin^2\theta}
\end{aligned}
\tag{11.235}
$$

and

$$
1 + \frac{2m_e^2}{t - m_e^2} + \frac{2m_e^2}{u - m_e^2} = 1 - \frac{m_e^2}{E(E - p\cos\theta)} - \frac{m_e^2}{E(E + p\cos\theta)} = \frac{p^2\sin^2\theta - m_e^2}{p^2\sin^2\theta + m_e^2}
\tag{11.236}
$$

Thus

$$
\langle|\mathcal{M}|^2\rangle = 2e^4 \left[\frac{3m_e^2 + p^2(3 + \cos^2\theta)}{m_e^2 + p^2\sin^2\theta} - \left(\frac{p^2\sin^2\theta - m_e^2}{p^2\sin^2\theta + m_e^2} \right)^2 \right]
\tag{11.237}
$$

Finally, we put this result in the phase-space element using Eq. (5.145) to find:

$$
\begin{aligned}
\left(\frac{d\sigma(e^+ e^- \rightarrow \gamma\gamma)}{d\Omega} \right)_{\text{CMS}} &= \frac{\langle|\mathcal{M}|^2\rangle}{2! 64\pi^2 s} = \frac{2e^4}{2! 64\pi^2 (4E^2)} [\ldots] = \frac{2(4\pi\alpha)^2}{2! 64\pi^2 (4E^2)} [\ldots] \\
&= \frac{\alpha^2}{16E^2} \left[\frac{3m_e^2 + p^2(3 + \cos^2\theta)}{m_e^2 + p^2\sin^2\theta} - \left(\frac{p^2\sin^2\theta - m_e^2}{p^2\sin^2\theta + m_e^2} \right)^2 \right]
\end{aligned}
\tag{11.238}
$$

where we have introduced a symmetry factor $S = 1/2!$ to account for the identical particles in the final state. In the high-energy limit, we can simplify the result by setting $m_e \to 0$ and $p \to E$ to find:

$$\left(\frac{d\sigma}{d\Omega}\right)_{\text{CMS}} (e^+ e^- \to \gamma\gamma) \simeq \frac{\alpha^2}{16E^2} \left[\frac{(3 + \cos^2\theta)}{\sin^2\theta} - 1\right] = \frac{\alpha^2}{2s}\left(\frac{1 + \cos^2\theta}{\sin^2\theta}\right) \tag{11.239}$$

The cross-section is strongly peaked in the forward direction, as shown in Figure 11.16. In the opposite limit of

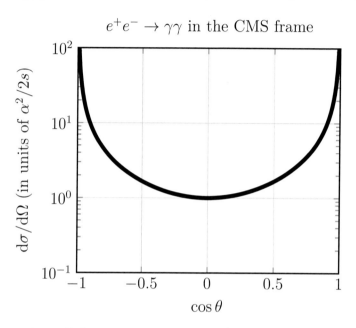

Figure 11.16 Differential cross-section for $e^+ e^- \to \gamma\gamma$ at high energies.

slow electron–positron pairs, we get $p \ll m_e$ and the expression in the square brackets becomes $3 - (-1)^2 = 2$, hence the angular distribution of the photons is isotropic:

$$\left(\frac{d\sigma}{d\Omega}\right)_{\text{CMS,non rel.}} (e^+ e^- \to \gamma\gamma) \simeq \frac{\alpha^2}{8m_e p} \tag{11.240}$$

11.16   Compton Scattering $e^-\gamma \to e^-\gamma$

Compton scattering of an electron and a photon $e^-\gamma \to e^-\gamma$ is related to the $e^+ e^- \to \gamma\gamma$ annihilation by crossing symmetry. Indeed, at the tree level there are two diagrams, shown in Figure 11.17, which are obviously related to the annihilation diagrams of Figure 11.14 by $s \leftrightarrow t$ crossing.

The amplitudes of the two tree-level Feynman diagrams are then (see Eqs. (11.190) and (11.191)):

$$\begin{aligned}
i\mathcal{M}_1 &= \epsilon_2^{\nu*}(k_2)\bar{u}(p',s')(-ie\gamma_\nu)\frac{i(\slashed{p} + \slashed{k}_1 + m_e)}{(p+k_1)^2 - m_e^2}(-ie\gamma_\mu)\epsilon_1^\mu(k_1)u(p,s) \\
&= -ie^2\bar{u}(p')\slashed{\epsilon}_2^*\frac{\slashed{p} + \slashed{k}_1 + m_e}{(p+k_1)^2 - m_e^2}\slashed{\epsilon}_1 u(p)
\end{aligned} \tag{11.241}$$

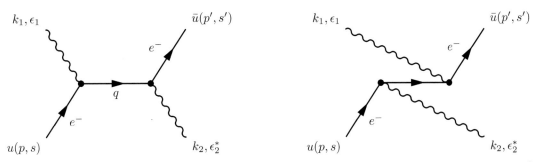

Figure 11.17 Feynman diagrams for Compton scattering at tree level. In the left diagram, $q = p + k_1 = p' + k_2$. On the right, $\tilde{q} = p - k_2 = p' - k_1$.

and

$$
\begin{aligned}
i\mathcal{M}_2 &= \epsilon_2^{\mu*}(k_2)\bar{u}(p',s')(-ie\gamma_\nu)\frac{i(\slashed{\tilde{q}}+m_e)}{(p-k_2)^2-m_e^2}(-ie\gamma_\mu)\epsilon_1^\nu(k_1)u(p,s) \\
&= -ie^2\bar{u}(p')\slashed{\epsilon}_1\frac{\slashed{\tilde{q}}+m_e}{(p-k_2)^2-m_e^2}\slashed{\epsilon}_2^*u(p)
\end{aligned}
\tag{11.242}
$$

For this case, we illustrate the explicit calculation with spinors and photon polarizations. The laboratory coordinate system is chosen, such that the initial electron is at rest and the incoming photon is propagating along the z-axis. The four-momenta before and after scattering are expressed as:

$$
\begin{aligned}
k_1^\mu &= (k, \vec{k}) = (k,\ 0,\ 0,\ k) \\
p^\mu &= (m_e,\ 0,\ 0,\ 0) \\
k_2^\mu &= (k', \vec{k}') = (k',\ 0,\ k'\sin\theta,\ k'\cos\theta) \\
p'^\mu &= (E', \vec{p}'_e) = (E',\ 0,\ p'\sin\alpha,\ p'\cos\alpha),\quad p' \equiv |\vec{p}'_e|
\end{aligned}
\tag{11.243}
$$

See Figure 11.18. We select the Coulomb gauge where the photons are transverse. In the laboratory frame

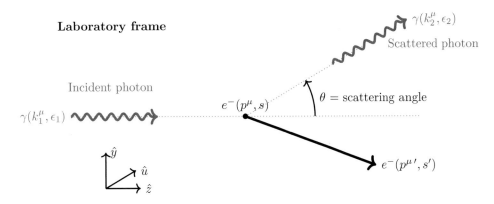

Figure 11.18 The kinematics of the Compton process in the laboratory frame.

and given the selected kinematics, we can choose the following linear transverse vectors:

$$
\begin{aligned}
\epsilon_1(k_1) &\equiv (0,\ \vec{\epsilon_1}) & \epsilon_1 \cdot k_1 &= -\vec{\epsilon_1}\cdot\vec{k} = 0 & (11.244) \\
\epsilon_2(k_2) &= (0,\ \vec{\epsilon_2}) & \epsilon_2 \cdot k_2 &= -\vec{\epsilon_2}\cdot\vec{k}' = 0 & (11.245)
\end{aligned}
$$

We note that they are real so we can ignore the complex conjugation on the polarization vectors. We also assume a normalization $\epsilon_1^2 = \epsilon_2^2 = -1$. The total amplitude can be expressed as:

$$i\mathcal{M} = -ie^2 \bar{u}(p') \left[\not{\epsilon}_2 \frac{\not{p} + \not{k}_1 + m_e}{(p+k_1)^2 - m_e^2} \not{\epsilon}_1 + \not{\epsilon}_1 \frac{\not{p} - \not{k}_2 + m_e}{(p-k_2)^2 - m_e^2} \not{\epsilon}_2 \right] u(p,s) \tag{11.246}$$

The denominators simplify, noting again that:

$$(p+k_1)^2 = p^2 + k_1^2 + 2p \cdot k_1 = m_e^2 + 2p \cdot k_1 \tag{11.247}$$

and similarly $(p-k_2)^2 = m_e^2 - 2p \cdot k_2$. For the numerator, we apply the Dirac equation $(\not{p} - m_e)u = 0$ to get:

$$(\not{p} + m_e)\not{\epsilon}_1 u = 2(p \cdot \epsilon_1)u - \not{\epsilon}_1(\not{p} - m_e)u = 0 \tag{11.248}$$

since $\epsilon_1 \cdot p = 0$. And similarly $(\not{p} + m_e)\not{\epsilon}_2 u = 0$ since $\epsilon_2 \cdot p = 0$. Then:

$$i\mathcal{M} = -ie^2 \bar{u}(p') \left[\not{\epsilon}_2 \frac{\not{k}_1}{2p \cdot k_1} \not{\epsilon}_1 + \not{\epsilon}_1 \frac{-\not{k}_2}{-2p \cdot k_2} \not{\epsilon}_2 \right] u(p) \tag{11.249}$$

For the transverse polarization vectors, we always have that $\not{k}_1\not{\epsilon}_1 = -\not{\epsilon}_1\not{k}_1$ and $\not{k}_2\not{\epsilon}_2 = -\not{\epsilon}_2\not{k}_2$ since $\epsilon_1 \cdot k_1 = \epsilon_2 \cdot k_2 = 0$. Finally, we obtain:

$$i\mathcal{M} = ie^2 \bar{u}(p') \left[\frac{\not{\epsilon}_2\not{\epsilon}_1\not{k}_1}{2p \cdot k_1} + \frac{\not{\epsilon}_1\not{\epsilon}_2\not{k}_2}{2p \cdot k_2} \right] u(p) \tag{11.250}$$

We now use Casimir's trick to compute the average over initial-state spins and sum over the final states. The amplitude squared is then simply:

$$\langle |\mathcal{M}|^2 \rangle = \frac{e^4}{2} \text{Tr} \left[(\not{p}' + m) \left(\frac{\not{\epsilon}_2\not{\epsilon}_1\not{k}_1}{2p \cdot k_1} + \frac{\not{\epsilon}_1\not{\epsilon}_2\not{k}_2}{2p \cdot k_2} \right) (\not{p} + m) \left(\frac{\not{k}_1\not{\epsilon}_1\not{\epsilon}_2}{2p \cdot k_1} + \frac{\not{k}_2\not{\epsilon}_2\not{\epsilon}_1}{2p \cdot k_2} \right) \right] \tag{11.251}$$

where the factor $1/2$ takes into account the initial-state electron spin. With the trace theorems, the expression can be reduced to:

$$\langle |\mathcal{M}|^2 \rangle = e^4 \left[\frac{k_2 \cdot p}{k_1 \cdot p} + \frac{k_1 \cdot p}{k_2 \cdot p} + 4(\epsilon_1 \cdot \epsilon_2)^2 - 2 \right] \tag{11.252}$$

In order to include the phase-space factors in the laboratory frame, we use the general form of Eq. (5.151) to get:

$$\left(\frac{d\sigma}{d\Omega} \right)_{lab} = \frac{k'}{64\pi^2 k m_e \, (k + m_e - k\cos\theta)} \langle |\mathcal{M}|^2 \rangle \tag{11.253}$$

Energy conservation simply gives $k + m_e = k' + E'$, hence:

$$(k + m_e)^2 = (k' + E')^2 \longrightarrow k^2 = k'^2 + 2k'E' + E'^2 - 2m_e k - m_e^2 \tag{11.254}$$

At the same time, momentum conservation yields:

$$\begin{aligned} E'^2 - m_e^2 &= \vec{p}_e'^2 = (\vec{k} - \vec{k}')^2 = k^2 + k'^2 - 2kk'\cos\theta \\ &= 2k'^2 + 2k'E' + E'^2 - 2m_e k - m_e^2 - 2kk'\cos\theta \end{aligned} \tag{11.255}$$

hence

$$2k'^2 + 2k'E' - 2m_e k - 2kk'\cos\theta = 0 \tag{11.256}$$

This can be written as

$$k'(k' + E' - k\cos\theta) = k'(k + m_e - k\cos\theta) = m_e k \tag{11.257}$$

so finally

$$\frac{1}{k + m_e - k\cos\theta} = \frac{k'}{m_e k} \tag{11.258}$$

and the differential cross-section in the laboratory frame becomes:

$$\left(\frac{d\sigma}{d\Omega}\right)_{\text{Lab.}} (e^-\gamma \to e^-\gamma) = \frac{\langle|\mathcal{M}|^2\rangle}{64\pi^2 m_e^2} \left(\frac{k'}{k}\right)^2 \tag{11.259}$$

and we find with $e^2 = 4\pi\alpha$:

$$\left(\frac{d\sigma}{d\Omega}\right)_{\text{Lab.}} (e^-\gamma \to e^-\gamma) = \frac{\alpha^2}{4m_e^2}\left(\frac{k'}{k}\right)^2\left[\frac{k_2 \cdot p}{k_1 \cdot p} + \frac{k_1 \cdot p}{k_2 \cdot p} + 4(\epsilon_1 \cdot \epsilon_2)^2 - 2\right] \tag{11.260}$$

In order to find the unpolarized cross-section, we should average over the initial and sum over the final-state photon polarizations. This gives, with $k_1 \cdot p = m_e k$ and $k_2 \cdot p = m_e k'$:

$$\left(\frac{d\sigma}{d\Omega}\right)_{\text{Lab.,unpolarized}} (e^-\gamma \to e^-\gamma) = \frac{1}{2}\sum_{\lambda_1,\lambda_2}\frac{\alpha^2}{4m_e^2}\left(\frac{k'}{k}\right)^2\left[\frac{k'}{k} + \frac{k}{k'} + 4(\epsilon_1 \cdot \epsilon_2)^2 - 2\right] \tag{11.261}$$

We can explicitly define the photon polarization four-vectors ϵ_1 and ϵ_2 for each possible transverse linear polarization state. For the initial-state photon, we label them as follows with $\lambda = 1, 2$ (see Eq. (10.34)):

$$\epsilon_1^\mu(\lambda = 1) = (0,\ 1,\ 0,\ 0), \qquad \epsilon_1^\mu(\lambda = 2) = (0,\ 0,\ 1,\ 0) \tag{11.262}$$

For the final-state photon, we must rotate the vectors around the z-axis in order to keep the transversality with the final-state momentum. Hence:

$$\epsilon_2^\mu(\lambda = 1) = (0,\ 1,\ 0,\ 0), \qquad \epsilon_2^\mu(\lambda = 2) = (0,\ 0,\ \cos\theta,\ -\sin\theta) \tag{11.263}$$

To obtain the unpolarized cross-section, we compute the summation of the four products $(\epsilon_1 \cdot \epsilon_2)^2$ to find:

$$\sum_{\lambda_1,\lambda_2=1}^{2} (\epsilon_1(\lambda_1) \cdot \epsilon_2(\lambda_2))^2 = 1 + 0 + 0 + \cos^2\theta \tag{11.264}$$

Note that we could equally have chosen the two helicity states ϵ_R^μ and ϵ_L^μ (see Eq. (10.35)). We would have in this case:

$$\epsilon_{1R}^\mu = -\frac{1}{\sqrt{2}}(0, 1, i, 0) \quad \text{and} \quad \epsilon_{1L}^\mu = \frac{1}{\sqrt{2}}(0, -1, i, 0) \tag{11.265}$$

and for the final-state photon, after rotation, we get:

$$\epsilon_{2R}^\mu = -\frac{1}{\sqrt{2}}\begin{pmatrix} 1 & 0 & 0 & 0 \\ 0 & 1 & 0 & 0 \\ 0 & 0 & \cos\theta & \sin\theta \\ 0 & 0 & -\sin\theta & \cos\theta \end{pmatrix}\begin{pmatrix} 0 \\ 1 \\ i \\ 0 \end{pmatrix} = -\frac{1}{\sqrt{2}}\begin{pmatrix} 0 \\ 1 \\ i\cos\theta \\ -i\sin\theta \end{pmatrix} \quad \text{and} \quad \epsilon_{2L}^\mu = \frac{1}{\sqrt{2}}\begin{pmatrix} 0 \\ -1 \\ i\cos\theta \\ -i\sin\theta \end{pmatrix} \tag{11.266}$$

Consequently, the sum over the four helicity states yields:

$$\sum_{\lambda_1,\lambda_2=L,R} (\epsilon_1(\lambda_1)\cdot\epsilon_2(\lambda_2))^2 = (\epsilon_{1L}\cdot\epsilon_{2L})^2 + (\epsilon_{1L}\cdot\epsilon_{2R})^2 + (\epsilon_{1R}\cdot\epsilon_{2L})^2 + (\epsilon_{1R}\cdot\epsilon_{2R})^2$$

$$= \left(\frac{1}{2}(0,-1,i,0)\cdot(0,-1,i\cos\theta,-i\sin\theta)\right)^2$$

$$+ \left(-\frac{1}{2}(0,-1,i,0)\cdot(0,1,i\cos\theta,-i\sin\theta)\right)^2$$

$$+ \left(-\frac{1}{2}(0,1,i,0)\cdot(0,-1,i\cos\theta,-i\sin\theta)\right)^2$$

$$+ \left(\frac{1}{2}(0,1,i,0)\cdot(0,1,i\cos\theta,-i\sin\theta)\right)^2$$

$$= \frac{1}{4}(1-\cos\theta)^2 + \frac{1}{4}(1+\cos\theta)^2 + \frac{1}{4}(1+\cos\theta)^2 + \frac{1}{4}(1-\cos\theta)^2$$

$$= \frac{1}{2}(1-\cos\theta)^2 + \frac{1}{2}(1+\cos\theta)^2$$

$$= 1+\cos^2\theta \tag{11.267}$$

Then, needless to say, either way, the term in brackets of Eq. (11.261) becomes:

$$\sum_{\lambda_1,\lambda_2}\left[\frac{k'}{k}+\frac{k}{k'}+4(\epsilon_1\cdot\epsilon_2)^2-2\right] = 4\left[\frac{k'}{k}+\frac{k}{k'}\right]+4(1+\cos^2\theta)-4\times 2 = 4\left[\frac{k'}{k}+\frac{k}{k'}+(\cos^2\theta-1)\right] \tag{11.268}$$

This result leads to the **Klein–Nishina**[14] equation for unpolarized photons scattered from a free electron in lowest order, derived for the first time in 1929 [87]:

$$\left(\frac{d\sigma}{d\Omega}\right)_{\text{Lab.,unpolarized}} (e^-\gamma\to e^-\gamma) = \frac{r_e^2}{2}\left(\frac{k'}{k}\right)^2\left[\frac{k'}{k}+\frac{k}{k'}-\sin^2\theta\right] \tag{11.269}$$

where r_e is the **classical electron radius** given in Eq. (2.100). Using Eq. (11.257), we can express the energy of the scattered (outgoing) photon as:

$$k' = \frac{m_e k}{(k+m_e-k\cos\theta)} = \frac{k}{1+\alpha(1-\cos\theta)} \tag{11.270}$$

where $\alpha\equiv k/m_e$ (not to be confused with the fine-structure constant!). Then

$$\left[\frac{k'}{k}+\frac{k}{k'}-\sin^2\theta\right] = \frac{k'^2+k^2}{k\,k'}-\sin^2\theta = \frac{1+(1+\alpha(1-\cos\theta))^2}{1+\alpha(1-\cos\theta)}-\sin^2\theta$$

$$= \frac{2+2\alpha(1-\cos\theta)+\alpha^2(1-\cos\theta)^2}{1+\alpha(1-\cos\theta)}-\sin^2\theta$$

$$= 2+\frac{\alpha^2(1-\cos\theta)^2}{1+\alpha(1-\cos\theta)}-\sin^2\theta = 1+\cos^2\theta+\frac{\alpha^2(1-\cos\theta)^2}{1+\alpha(1-\cos\theta)} \tag{11.271}$$

Replacing this into the differential cross-section obtained above and integrating over $d\phi$ leads to the common form of the **Compton scattering formula**:

$$\left(\frac{d\sigma}{d\cos\theta}\right) = \pi r_e^2 \frac{1}{(1+\alpha(1-\cos\theta))^2}\left[1+\cos^2\theta+\frac{\alpha^2(1-\cos\theta)^2}{1+\alpha(1-\cos\theta)}\right] \tag{11.272}$$

14 Yoshio Nishina (1890–1951), Japanese physicist.

This angular distribution of the scattered photon is plotted in Figure 11.19. At low incoming photon energies, the photon is scattered forward and backward. As the energy of the incoming photon increases, the probability of forward scattering increases and the differential cross-section becomes strongly forward peaked.

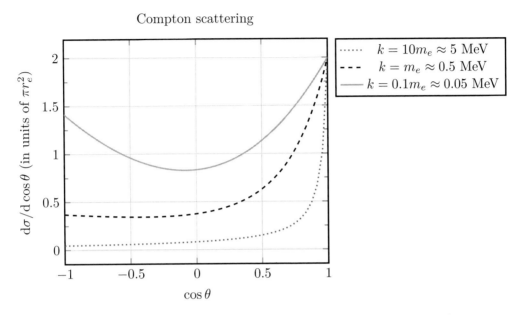

Figure 11.19 Compton scattering formula: angular distribution of the scattered photon for various incoming energies.

It is customary to express it as a function of the **kinetic energy of the scattered electron** T. By energy conservation, we find:

$$T = k - k' = k \left(1 - \frac{1}{1 + \alpha(1 - \cos\theta)} \right) = k \frac{\alpha(1 - \cos\theta)}{1 + \alpha(1 - \cos\theta)} \tag{11.273}$$

and

$$\left| \frac{dT}{d\cos\theta} \right| = \frac{k\alpha}{(1 + \alpha(1 - \cos\theta))^2} \tag{11.274}$$

Defining $x \equiv T/k$, we find:

$$\frac{x}{\alpha} = \frac{1 - \cos\theta}{1 + \alpha(1 - \cos\theta)} \implies 1 - \cos\theta = \frac{x}{\alpha}(1 + \alpha(1 - \cos\theta)) \tag{11.275}$$

hence

$$\alpha(1 - \cos\theta) = \frac{x}{1 - x}, \quad \cos\theta = 1 - \frac{x}{\alpha(1 - x)}, \quad \text{and} \quad \frac{\alpha^2(1 - \cos\theta)^2}{1 + \alpha(1 - \cos\theta)} = \frac{x^2}{1 - x} \tag{11.276}$$

Consequently

$$
\begin{aligned}
\left(\frac{\mathrm{d}\sigma}{\mathrm{d}T}\right) &= \frac{\pi r_e^2}{k\alpha}\left[1+\left(1-\frac{x}{\alpha(1-x)}\right)^2+\frac{x^2}{1-x}\right] \\
&= \frac{\pi r_e^2}{k\alpha}\left[1+\frac{\alpha^2(1-x)^2-2\alpha x(1-x)+x^2}{\alpha^2(1-x)^2}+\frac{x^2}{1-x}\right] \\
&= \frac{\pi r_e^2}{k\alpha}\left[2-\frac{2x}{\alpha(1-x)}+\frac{x^2}{\alpha^2(1-x)^2}+\frac{x^2}{1-x}\right]
\end{aligned}
\tag{11.277}
$$

Finally, writing also $k=\alpha m_e$, the Compton differential cross-section for a photon scattering off a single (free) electron is given by:

$$
\left(\frac{\mathrm{d}\sigma}{\mathrm{d}T}\right)=\frac{\pi r_e^2}{m_e\alpha^2}\left[2+\frac{x^2}{\alpha^2(1-x)^2}+\frac{x}{1-x}\left(x-\frac{2}{\alpha}\right)\right]
\tag{11.278}
$$

with $x=T/\alpha m_e$. Figure 11.20 shows the distribution of this differential cross-section for several incident photon energies. The kinematics of the scattered electron enforces the energy to be in the range $0\leq T\leq T_{max}$,

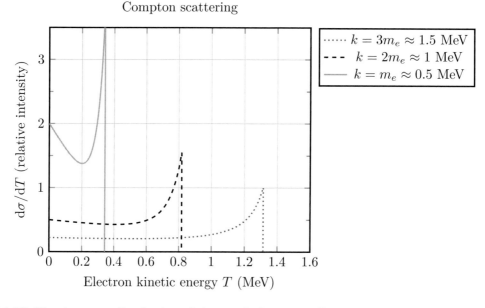

Figure 11.20 Kinetic-energy distribution of the recoil electron in Compton scattering for several incident photon energies.

where the maximum electron kinetic energy corresponds to the configuration where the photon is back-scattered with $\theta=180°$ or $\cos\theta=-1$. This gives the so-called **Compton edge**:

$$
T_{max}=k\frac{2\alpha}{1+2\alpha}
\tag{11.279}
$$

which is visible in the distributions of the differential cross-section as a sharp drop at the maximum recoil energy.

The Compton effect was derived assuming that the initial electron is free and at rest, and the kinematics is given by energy–momentum conservation of two-body scattering. This is adequate to describe scattering off quasi-free atomic electrons, where each atomic electron acts as an independent target. Therefore, the Compton effect is also called **incoherent scattering** and the cross-section *per atom* is equal to Z times the cross-section per electron. The **total cross-section per atom** is then given by:

$$\sigma_{tot}(k) = Z \int_{T_{min}}^{T_{max}} \left(\frac{d\sigma}{dT}\right) dT = \frac{Z\pi r_e^2}{\alpha} \int_0^{2\alpha/(1+2\alpha)} dx \left[2 + \frac{x^2}{\alpha^2(1-x)^2} + \frac{x}{1-x}\left(x - \frac{2}{\alpha}\right)\right] \tag{11.280}$$

where we used $dx = dT/\alpha m_e$. A direct indefinite integration gives:

$$\int dx [\cdots] = 2x - \frac{(\alpha^2 - 2\alpha - 2)}{\alpha^2} \ln(x-1) + \frac{x - \frac{1}{x-1}}{\alpha^2} - \frac{\alpha x^2 + (2\alpha - 4)x}{2\alpha} \tag{11.281}$$

The boundaries of the definite integral for the log term yield (recalling $x_{min} = 0$):

$$\ln\left(\frac{x_{max}-1}{x_{min}-1}\right) = \ln(1 - x_{max}) = \ln\left(1 - \frac{2\alpha}{1+2\alpha}\right) = \ln\left(\frac{1}{1+2\alpha}\right) = -\ln(1+2\alpha) \tag{11.282}$$

and the definite integral for the remaining terms is:

$$2x + \frac{x - \frac{1}{x-1}}{\alpha^2} - \frac{\alpha x^2 + (2\alpha - 4)x}{2\alpha} \Bigg|_0^{2\alpha/(1+2\alpha)} = \frac{2\alpha^3 + 18\alpha^2 + 16\alpha + 4}{4\alpha^3 + 4\alpha^2 + \alpha} \tag{11.283}$$

So finally the total Compton cross-section per atom is simply:

$$\sigma_{tot}(k) = 2\pi Z r_e^2 \left[\frac{(\alpha^2 - 2\alpha - 2)}{2\alpha^3} \ln(1+2\alpha) + \frac{\alpha^3 + 9\alpha^2 + 8\alpha + 2}{4\alpha^4 + 4\alpha^3 + \alpha^2}\right] \tag{11.284}$$

In the very high-energy limit, as $k \to \infty$ so $\alpha = k/m_e \gg 1$, the cross-section goes to zero as:

$$\sigma_{tot}(k \to \infty) \sim \pi Z r_e^2 \frac{\ln(2\alpha)}{\alpha} \tag{11.285}$$

One can also show that the limit as α approaches zero is given by:

$$\lim_{\alpha \to 0} \left[\frac{(\alpha^2 - 2\alpha - 2)}{2\alpha^3} \ln(1+2\alpha) + \frac{\alpha^3 + 9\alpha^2 + 8\alpha + 2}{4\alpha^4 + 4\alpha^3 + \alpha^2}\right] = \frac{4}{3} \tag{11.286}$$

so at low energies the Compton coherent cross-section reduces to the classical Thomson result (see Section 2.12):

$$\sigma_{tot}(k \to 0) \sim \frac{8}{3}\pi Z r_e^2 \tag{11.287}$$

Thomson scattering is therefore the incoherent scattering of photons by free electrons in the classical limit. On the other hand, Rayleigh scattering is the coherent scattering of photons by the entire atom. All the electrons in the atom participate in a coherent manner and it is therefore called coherent scattering.

In fact, however, for incoherent scattering when $k \lesssim 100$ keV, the binding energy of the atomic electron must be taken into account by a corrective factor to the Klein–Nishina cross-section, called the scattering function $S(k, k')$:

$$\left(\frac{d\sigma}{dT}\right) = \left(\frac{d\sigma}{dT}\right)_{K.N.} \times S(k, k') \tag{11.288}$$

One effect is that at very low energy, the cross-section vanishes instead of being constant, unlike the prediction of the classical Thomson formula.

Incoherent and coherent scattering processes are characterized by the fact that no energy is transferred to the atoms of the medium. The atoms are neither excited nor ionized. Only the direction of the photon is changed with the emission of an electron (the electron itself can, however, further excite or ionize the medium). Such scattering processes become very small at high energies.

Problems

Ex 11.1 Electron–muon scattering amplitudes with helicity spinors. We consider the following QED process $e^- + \mu^- \to e^- + \mu^-$.

(a) Derive the amplitude at first order choosing p_1 and p_2 as the initial four-momenta for the e^- and the μ^-, and p_3, p_4 the corresponding final momenta.

(b) Show that in the center-of-mass frame with incoming and outgoing electron given by $p_1^\mu = (E_1, 0, 0, p)$ and $p_3^\mu = (E_1, p\sin\theta, 0, p\cos\theta)$, one has:

$$
\begin{aligned}
(\bar{u}_\downarrow(p_3)\gamma^\mu u_\downarrow(p_1)) &= 2\left(E_1\cos\frac{\theta}{2}, p\sin\frac{\theta}{2}, -ip\sin\frac{\theta}{2}, p\cos\frac{\theta}{2}\right) \\
(\bar{u}_\uparrow(p_3)\gamma^\mu u_\downarrow(p_1)) &= 2\left(m_e\sin\frac{\theta}{2}, 0, 0, 0\right) \\
(\bar{u}_\uparrow(p_3)\gamma^\mu u_\uparrow(p_1)) &= 2\left(E_1\cos\frac{\theta}{2}, p\sin\frac{\theta}{2}, ip\sin\frac{\theta}{2}, p\cos\frac{\theta}{2}\right) \\
(\bar{u}_\downarrow(p_3)\gamma^\mu u_\uparrow(p_1)) &= -2\left(m_e\sin\frac{\theta}{2}, 0, 0, 0\right)
\end{aligned}
\tag{11.289}
$$

where the arrows correspond to each helicity state of the spinor.

(c) Write down the incoming and outgoing muon four-momenta p_2 and p_4, and the helicity eigenstate spinors $u_\downarrow(p_2)$, $u_\downarrow(p_4)$, $u_\uparrow(p_2)$, $u_\uparrow(p_4)$ (take the muon mass as M and the muon energy to be E_2). By comparing the forms of the muon and electron spinors, explain how the muon currents can be written down without any further calculation.

$$
\begin{aligned}
(\bar{u}_\downarrow(p_4)\gamma^\mu u_\downarrow(p_2)) &= 2\left(E_2\cos\frac{\theta}{2}, -p\sin\frac{\theta}{2}, -ip\sin\frac{\theta}{2}, -p\cos\frac{\theta}{2}\right) \\
(\bar{u}_\uparrow(p_4)\gamma^\mu u_\downarrow(p_2)) &= 2\left(M\sin\frac{\theta}{2}, 0, 0, 0\right) \\
(\bar{u}_\uparrow(p_4)\gamma^\mu u_\uparrow(p_2)) &= 2\left(E_2\cos\frac{\theta}{2}, -p\sin\frac{\theta}{2}, ip\sin\frac{\theta}{2}, -p\cos\frac{\theta}{2}\right) \\
(\bar{u}_\downarrow(p_4)\gamma^\mu u_\uparrow(p_2)) &= -2\left(M\sin\frac{\theta}{2}, 0, 0, 0\right)
\end{aligned}
\tag{11.290}
$$

(d) Explain why some of the above currents vanish in the relativistic limit where the electron and muon mass can be neglected. Sketch the spin configurations which are allowed in this limit.

(e) Show that in the relativistic limit, the matrix element squared $|\mathcal{M}_{LL}|^2$ for the case where the incoming electron and incoming muon are both left-handed is given by:

$$
|\mathcal{M}_{LL}|^2 = \frac{4e^4 s^2}{(p_1 - p_3)^4}
\tag{11.291}
$$

where $s = (p_1 + p_2)^2$. Why is the result independent of θ?

(f) Find a similar expression for the matrix element $|\mathcal{M}_{RL}|^2$ for a right-handed incoming electron and a left-handed incoming muon, and explain why it vanishes when $\theta = \pi$. Write down the corresponding results for $|\mathcal{M}_{LL}|^2$ and $|\mathcal{M}_{LR}|^2$.

(g) Show that, in the relativistic limit, the differential cross-section for unpolarized electron–muon scattering in the center-of-mass frame is:

$$\frac{d\sigma}{d\Omega} = \frac{2\alpha^2}{s} \frac{1 + \frac{1}{4}(1 + \cos\theta)^2}{(1 - \cos\theta)^2} \tag{11.292}$$

(h) Show that the spin-averaged matrix element squared for unpolarized electron–muon scattering can be written as:

$$|\mathcal{M}|^2 = \frac{1}{4} \frac{e^4}{(p_1 - p_3)^4} L^{\mu\nu} W_{\mu\nu} \tag{11.293}$$

where the electron and muon tensors are given by:

$$L^{\mu\nu} = \sum_{spins} (\bar{u}(p_3)\gamma^\mu u(p_1))(\bar{u}(p_3)\gamma^\nu u(p_1))^*$$

$$W_{\mu\nu} = \sum_{spins} (\bar{u}(p_4)\gamma_\mu u(p_2))(\bar{u}(p_4)\gamma_\nu u(p_2))^* \tag{11.294}$$

(i) Using the currents from part (b), show that the components of electron tensor $L^{\mu\nu}$ are:

$$\begin{Bmatrix} L_{00} & L_{01} & L_{02} & L_{03} \\ L_{10} & L_{11} & L_{12} & L_{13} \\ L_{20} & L_{21} & L_{22} & L_{23} \\ L_{30} & L_{31} & L_{32} & L_{33} \end{Bmatrix} = 8 \begin{Bmatrix} E_1^2 c^2 + m_e^2 s^2 & E_1 psc & 0 & E_1 pc^2 \\ E_1 psc & p^2 s^2 & 0 & p^2 sc \\ 0 & 0 & p^2 s^2 & 0 \\ E_1 pc^2 & p^2 sc & 0 & p^2 c^2 \end{Bmatrix}$$

where $s = \sin\frac{\theta}{2}$ and $c = \cos\frac{\theta}{2}$.

(j) Verify that $L^{\mu\nu} = 4(p_1^\mu p_3^\nu + p_3^\mu p_1^\nu + g^{\mu\nu}(m^2 - p_1 \cdot p_3))$.

Ex 11.2 Mott scattering in a different way. Show that the Mott scattering differential cross-section can be derived from the $e^- + \mu^- \to e^- + \mu^-$ process relative to the muon rest frame and letting the mass of the muon go to infinity.

Ex 11.3 Pair creation. Compute the amplitude and the cross-section for the process $\gamma\gamma \to e^+ e^-$.

Ex 11.4 Electron–positron annihilation into two charged scalars. Consider the reaction $e^+(p)e^-(p') \to \gamma^* \to \phi^+(k_+)\phi^-(k_-)$, where $\phi^\pm$ are (spinless) scalar particles.

(a) Show that the amplitude at tree level can be written as:

$$i\mathcal{M} = ie^2 \bar{v}(p')\gamma_\mu u(p) \frac{1}{q^2}(k_- - k_-)^\mu \tag{11.295}$$

where $q^\mu = k_+ - k_-$.

(b) Compute the differential cross-section in the center-of-mass system.

(c) Compute the total cross-section.

(d) What differences do you expect between this process $e^+(p)e^-(p') \to \gamma^* \to \phi^+(k_+)\phi^-(k_-)$ and the muon–antimuon pair production $e^+(p)e^-(p') \to \gamma^* \to \mu^+(k_+)\mu^-(k_-)$?

12 QED Radiative Corrections

Horror vacui ("Nature abhors vacuum.")

Aristotle

12.1 The Naive Expectation and the Occurrence of Divergences

Quantum field theory is complex. Ask ourselves why we bother with all these quantum fields in the first place? QED is a field theory of well-defined perturbation expansion and in principle any physical prediction can be calculated with **practically infinite precision**. So, in this chapter we explore the techniques associated with computing "higher-order" or "purely quantum" effects of electromagnetism. The S-matrix was written as the Dyson expansion (see Section 9.2), where the factor in the expansion is the electric elementary charge e. We have seen that a basic $2 \to 2$ scattering process (such as $e^- \ell^- \to e^- \ell^-$) has two QED vertices at tree level, hence the amplitude $\mathcal{M}$ is proportional to e^2 and the cross-section is $|\mathcal{M}|^2 \propto e^4$. It is customary to express this in terms of the fine-structure constant α related to $e = \sqrt{4\pi\alpha}$. Hence $\alpha = e^2/4\pi \simeq 1/137$, yielding in the $2 \to 2$ scattering case:

$$\mathcal{M} \propto e^2 \propto \alpha \qquad \text{then} \qquad |\mathcal{M}|^2 \propto e^4 \propto \alpha^2 \simeq \left(\frac{1}{137}\right)^2 \tag{12.1}$$

Adding one vertex, the order of the amplitude $\mathcal{M}$ is one more $e \propto \sqrt{\alpha}$ and the order of the cross-section gains a factor $e^2 \propto \alpha$. There exists a standard way of calculating the errors on QED predictions at a given order n. The idea is to calculate predictions for a given observable P at different orders of perturbation expansion: for example, for the scattering process the tree level is $P(\alpha^2)$, at the first-order correction $P(\alpha^3)$, etc., and the calculation should be continued until the difference $|P(\alpha^{(n-1)}) - P(\alpha^n)|$ is smaller than the desired error for the observable P. There are serious, although not fundamental, difficulties in applying the above scheme. Tree-level calculations are relatively quick but higher-order calculations become very complex since the number of diagrams increases very rapidly.

As an illustrative example, we show in Figure 12.1 all the diagrams at one-loop level entering in the amplitude of the $e^-\mu^- \to e^-\mu^-$ process. In the diagrams, e_m stands for the leptons e, μ, and τ; u_m for the up-type quarks u, c, and t; and d_m for the down-type quarks d, s, and b. The diagrams contain two more vertices compared to those at tree level. Hence they are of order:

$$\mathcal{M}_{\text{1-loop}} \propto e^4 \propto \alpha^2 \qquad \text{then} \qquad |\mathcal{M}_{\text{1-loop}}|^2 \propto e^8 \propto \alpha^4 \simeq \left(\frac{1}{137}\right)^4 \tag{12.2}$$

In order to tackle the problem, they are categorized as follows:

- Diagrams (1) and (2) correspond to **vertex corrections**.

- Diagrams (3) and (4) correspond to **box diagrams**.

- Diagrams (5)–(7) correspond to **photon self-energy (or vacuum polarization)**.

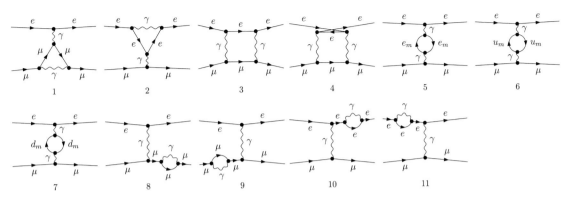

Figure 12.1 Feynman diagrams entering the one-loop level calculation of the $e^- \mu^- \to e^- \mu^-$ process, where e_m stands for the leptons e, μ, and τ; u_m for the up-type quarks u, c, and t; and d_m for the down-type quarks d, s, and b (the diagrams have been created with FeynArts [86]).

- Diagrams (8)–(11) correspond to **fermion self-energy**.

In order to compute their effect, one has to calculate the squared matrix element of each of the 11 individual diagrams and also take into account all possible interference terms, of which there are 55 in total.

Similarly, one can consider the diagrams contributing to the **Bremsstrahlung emission** of a photon leading to the final state $e^- \mu^- \to e^- \mu^- \gamma$. The corresponding diagrams are shown in Figure 12.2. The diagrams

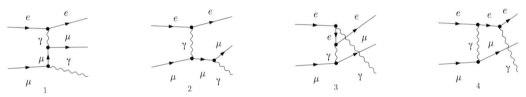

Figure 12.2 Feynman diagrams entering the one-loop level calculation of the $e^- \mu^- \to e^- \mu^- \gamma$ process (the diagrams have been created with FeynArts [86]).

contain one more vertex compared to those at tree level. Hence they are of order:

$$\mathcal{M}_{e^- \mu^- \gamma} \propto e^3 \propto \alpha^{3/2} \qquad \text{then} \qquad |\mathcal{M}_{e^- \mu^- \gamma}|^2 \propto e^6 \propto \alpha^3 \simeq \left(\frac{1}{137}\right)^3 \tag{12.3}$$

Naively one could think that since α is rather small (of the order of 1%), the one-loop diagrams will be suppressed compared to tree level by typically 10^{-4} and therefore negligible. Similarly, the photon radiation is suppressed by 10^{-2} and therefore also very rare. However, corrections are larger than naively expected due to various enhancement factors present in the QED dynamics. The source of these enhancements is well understood and governed by the structure of QED singularities called **ultraviolet**, **infrared**, and **collinear**. A multitude of techniques were developed to control these enhanced terms. Many of these theoretical developments were connected to the calculation of the anomalous magnetic moment, to be discussed in more detail in Chapter 14. In 1947, Schwinger was able to develop a non-divergent formulation of QED that permitted him to show that the lowest-order radiative correction to the electron anomaly was equal to $\alpha/2\pi$ ($= 0.00116$). Schwinger's application of the concepts of **mass and charge renormalization** (first suggested by Kramers) provided the key to eliminating the divergences. Although the renormalization aspects of the theory were considered by many to be aesthetically unsatisfying, the practical success of the theory was undeniable. In 1949, Dyson showed that Schwinger's theory could be extended to permit the calculation of higher-order corrections to the properties of quantum systems. By showing the equivalence of the Schwinger–Tomonaga and Feynman formulations of the theory, Dyson was able to simplify the calculational procedure, devise an unambiguous program for obtaining the nth-order contribution to quantities which can be calculated using QED, and show that these contributions would remain finite to arbitrary order in α. Such methods were fundamental to establish QED as the theory of electromagnetic interactions, with which processes and related quantities could be calculated with extremely high accuracy. We illustrate some of these calculations in the following sections.

12.2   Soft Photons

Let us start with the amplitude where Bremsstrahlung occurs from an incoming electron. We compare the tree-level diagram of the electron scattering off a "blob" with the radiative correction in which the electron emits a photon before it scatters off the blob – the corresponding Feynman diagrams are shown in Figure 12.3.

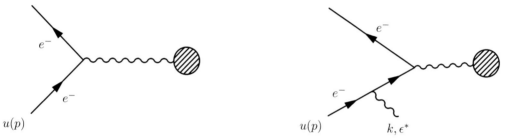

Figure 12.3 Photon Bremsstrahlung: (left) tree-level interaction; (right) with one real photon emission from the incoming fermion.

If we call $\mathcal{M}_0(p)u(p,s)$ the tree-level amplitude for scattering of the electron with momentum p^μ off the blob, then we can express the diagram with the photon emission by adding the photon leg with momentum k^μ and the electron propagator after the photon emission taking into account that the electron now interacts with the blob with a momentum $(p-k)^\mu$:

$$
\begin{aligned}
\mathcal{M}(p,k) &= i\mathcal{M}_0(p-k)\frac{\not{p}-\not{k}+m_e}{(p-k)^2-m_e^2}(-ie\gamma^\mu)u(p,s)\epsilon_\mu^*(k) \\
&= e\mathcal{M}_0(p-k)\frac{\not{p}-\not{k}+m_e}{(p-k)^2-m_e^2}\not{\epsilon}^*(k)u(p,s)
\end{aligned}
\tag{12.4}
$$

Since $(p-k)^2 = p^2 + k^2 - 2p\cdot k = m_e^2 - 2p\cdot k$, we have:

$$
\mathcal{M}(p,k) = e\mathcal{M}_0(p-k)\frac{\not{p}-\not{k}+m_e}{-2p\cdot k}\not{\epsilon}^*(k)u(p,s)
\tag{12.5}
$$

We now take the **soft-photon limit**, which means that the energy of the photon is small (or soft) compared to the momentum p of the electron. Hence, we can neglect $\not{k}$ in the numerator:

$$
\mathcal{M}(p,k) \simeq e\mathcal{M}_0(p)\frac{\not{p}+m_e}{-2p\cdot k}\not{\epsilon}^*(k)u(p,s) \simeq \frac{-e}{2p\cdot k}\mathcal{M}_0(p)(\not{p}+m_e)\not{\epsilon}^*(k)u(p,s)
\tag{12.6}
$$

Now we use $\gamma^\mu \gamma^\nu = -\gamma^\nu \gamma^\mu + 2g^{\mu\nu}\mathbb{1}$ (see Appendix E.1), so $\not{p}\not{\epsilon}^*(k) = -\not{\epsilon}^*(k)\not{p} + 2p \cdot \epsilon^*$ to find:

$$
\begin{aligned}
\frac{-e}{2p \cdot k} \mathcal{M}_0(p)(\not{p} + m_e)\not{\epsilon}^*(k)u(p,s) &= \frac{-e}{2p \cdot k}\mathcal{M}_0(p)\left(2p \cdot \epsilon^* + \not{\epsilon}^*(k)(-\not{p} + m_e)\right)u(p,s) \\
&= \frac{-e}{2p \cdot k}\mathcal{M}_0(p)(2p \cdot \epsilon^*)u(p,s) \\
&= -e\left(\frac{p \cdot \epsilon^*}{p \cdot k}\right)[\mathcal{M}_0(p)u(p,s)]
\end{aligned}
\tag{12.7}
$$

If we assume that for soft photons $\mathcal{M}_0(p-k)u(p-k,s) \approx \mathcal{M}_0(p)u(p,s)$, then **in the soft photon limit the matrix element can be represented as a multiplicative factor of the lower-order matrix element times the soft photon factor, which turns out to be independent from the properties of the particular tree-level process under consideration.**

Let us look at the soft photon factor. Defining $p^\mu = (E_p, \vec{p})$ and for the photon $k^\mu = (E_k = |\vec{k}|, \vec{k})$, we find:

$$
p \cdot k = (E_p E_k - \vec{p} \cdot \vec{k})
\begin{cases}
|\vec{k}| \to 0 \implies p \cdot k \to 0 \\[2mm]
\vec{k} \parallel \vec{p} \implies p \cdot k = E_p|\vec{k}| - |\vec{p}||\vec{k}| = |\vec{k}|(E_p - |\vec{p}|) \to 0 \text{ if } E_p \gg m_e
\end{cases}
\tag{12.8}
$$

We encounter the first type of enhancements in the dynamics of QED. When the energy of the photon becomes smaller, the denominator $p \cdot k$ decreases and the soft photon factor increases. The factor diverges in the **infrared limit**, i.e., when the energy of the photon tends to zero ($|\vec{k}| \to 0$). Similarly, when the photon and the fermion are close to parallel ($\vec{k} \parallel \vec{p}$), we have an enhancement, leading to the **collinear divergence** in the high-energy limit.

To address the infrared divergence, we need to ask which question we want to answer. If the goal is to compare theory to actual experimental data, then we must assume that photons will only be detected by the apparatus in question if their energy is above a certain threshold, which we call k_{min}. If soft photons with energy below k_{min} are not visible, then any states with additional extremely soft photons are indistinguishable, by any method, from the ones in which they are absent. We can say that they are physically identical and that, as a principle, our calculations do not need to necessarily bring meaningful answers to non-physical questions. One then divides the photons into two groups, of energy larger and smaller than k_{min}. The latter are omitted from the kinematical consideration and "summed" over, as an overall **soft-photons correction** to the cross-section. The real photons, above k_{min}, can then be computed without divergences.

In a similar fashion, we can handle the collinear divergence by kinematically considering only photons emitted with a minimum angle with respect to the parent electron. The very forward enhancement is then handled inclusively, assuming that in a real experimental calorimeter, for instance, the electromagnetic showers produced by the overlapping energy and photon will be experimentally indistinguishable.

Although we have taken as an example the case of Bremsstrahlung from an electron, similar rules apply of course to the positron and all other electrically charged fermions such as muons, taus, and quarks.

12.3   Self-energy of the Photon

Let us consider the photon propagator with momentum k^μ. See Figure 12.4(left). At one-loop correction we interrupt the photon propagator in its middle and add a fermion–antifermion loop of mass m, as shown in Figure 12.4(middle). These correspond to diagrams (5)–(7) in Figure 12.1. According to the Feynman rules, the loop introduces the trace of the associated γ matrices, an integration over the unmeasurable momentum, and an overall (-1) sign, so the photon propagator with one fermion loop $\tilde{G}^{\mu\nu}_{F,1L}(k)$ can be expressed as:

$$
\begin{aligned}
\tilde{G}^{\mu\nu}_{F,1L}(k) &= (-1)\left(\frac{-ig^{\mu\alpha}}{k^2}\right)\int\frac{\mathrm{d}^4p}{(2\pi)^4}\,\mathrm{Tr}\left\{(ie\gamma_\alpha)\left(\frac{-i(\not{k}+\not{p}+m)}{(k+p)^2-m^2}\right)(ie\gamma_\beta)\left(\frac{-i(\not{p}+m)}{p^2-m^2}\right)\right\}\left(\frac{-ig^{\beta\nu}}{k^2}\right) \\
&= (-1)(ie)^2\tilde{G}^{\mu\alpha}_F(k)\int\frac{\mathrm{d}^4p}{(2\pi)^4}\,\mathrm{Tr}\left\{\gamma_\alpha\left(\frac{-i(\not{k}+\not{p}+m)}{(k+p)^2-m^2}\right)\gamma_\beta\left(\frac{-i(\not{p}+m)}{p^2-m^2}\right)\right\}\tilde{G}^{\beta\nu}_F(k) \\
&= \tilde{G}^{\mu\alpha}_F(k)\cdot ie^2\Pi_{\alpha\beta}(k)\cdot\tilde{G}^{\beta\nu}_F(k)
\end{aligned}
\tag{12.9}
$$

where $\tilde{G}^{\mu\alpha}_F(k)$ is the "plain" photon Feynman propagator (see Eq. (10.74)) and the one-loop fermion diagram *amputated* from its legs is:

$$
\Pi^{\alpha\beta}_{1L}(k) = -i\int\frac{\mathrm{d}^4p}{(2\pi)^4}\,\mathrm{Tr}\left\{\gamma^\alpha\left(\frac{-i(\not{k}+\not{p}+m)}{(k+p)^2-m^2}\right)\gamma^\beta\left(\frac{-i(\not{p}+m)}{p^2-m^2}\right)\right\}
\tag{12.10}
$$

In order to compute the effect of the first-order photon loop correction to a particular tree-level diagram, we can simply replace the "plain" photon propagator by that of Eq. (12.9), which includes one fermion loop, so basically replacing $\tilde{G}^{\mu\nu}_F(k) \to \tilde{G}^{\mu\nu}_{F,1L}(k)$ in our tree-level amplitudes.

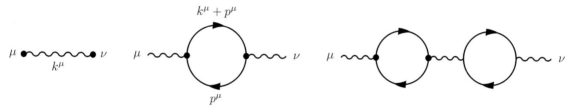

Figure 12.4 One-fermion-loop corrections to the photon propagator: (left) "plain" or naked; (middle) "dressed" with one fermion–antifermion loop; (right) two loops.

We can further continue this process and for example express the photon propagator with two fermion loops (see Figure 12.4) as a function of Π^{1L}:

$$
\begin{aligned}
\tilde{G}^{\mu\nu}_{F,2L}(k) &= \tilde{G}^{\mu\alpha}_F(k)\cdot ie^2\Pi^{1L}_{\alpha\beta}(k)\cdot\tilde{G}^{\beta\gamma}_F(k)\cdot ie^2\Pi^{1L}_{\gamma\delta}(k)\cdot\tilde{G}^{\delta\nu}_F(k) \\
&= (-e^4)\tilde{G}^{\mu\alpha}_F(k)\cdot\Pi^{1L}_{\alpha\beta}(k)\cdot\tilde{G}^{\beta\gamma}_F(k)\cdot\Pi^{1L}_{\gamma\delta}(k)\cdot\tilde{G}^{\delta\nu}_F(k) \\
&= (-e^4)\tilde{G}^{\mu\alpha}_F(k)\cdot\Pi^{2L}_{\alpha\beta}(k)\cdot\tilde{G}^{\beta\nu}_F(k)
\end{aligned}
\tag{12.11}
$$

We can generalize the procedure and consider all higher-order corrections to the photon as a "blob" – see Figure 12.5. The blobs can contain any diagram that cannot be split into two by removing a single line (we rediscover the so-called **one-particle irreducible (1PI) diagrams** – see Section 9.12). The fully "dressed" photon can then be expressed as:

$$
\tilde{G}^{\mu\nu}(k) \equiv \tilde{G}^{\mu\nu}_F(k) + ie^2\tilde{G}^{\mu\alpha}_F(k)\Pi_{\alpha\beta}(k)\tilde{G}^{\beta\nu}_F(k) - e^4\tilde{G}^{\mu\alpha}_F(k)\Pi_{\alpha\beta}(k)\tilde{G}^{\beta\gamma}_F(k)\Pi_{\gamma\delta}(k)\tilde{G}^{\delta\nu}_F(k) + \cdots
\tag{12.12}
$$

Once the full photon propagator is computed, it can in principle be inserted into any diagram. However, Feynman diagrams with loops can diverge. For example, the one-loop fermion correction of Eq. (12.10) diverges quadratically with energy:

$$
\Pi^{\alpha\beta}_{1L}(k) \propto \int\mathrm{d}^4p\frac{p^2}{p^4} \propto \int\mathrm{d}^4p\frac{1}{p^2} \propto \int p^3\mathrm{d}p\frac{1}{p^2} \propto p^2 \to \infty
\tag{12.13}
$$

This is the **ultraviolet divergence of QED**, since the integral diverges due to its high-energy behavior.

Figure 12.5 Higher-order corrections to the photon propagator: (left) "plain" or naked; (middle) "dressed" with one blob; (right) two blobs.

The first thing to do is to introduce some **regularization**, as we already did in the case of the "plain" Feynman propagator. *Regularized diagrams must converge.* When the regularization parameter tends to some value, the original theory must be restored. There exist many regularization methods, and its choice is crucial for any computation. If possible, the regularization should preserve as much of the symmetry of the theory as possible.

A **cutoff regularization** is a technique to remove the divergence with a cutoff at some energy scale Λ to simulate the effect of unknown physics. This essentially means that **QED is a theory only valid up to the energy Λ** and hence it does not make sense to integrate up to infinity.[1] This is done by introducing the new function $\Pi^{\alpha\beta}(k,\Lambda)$, such that:

$$\Pi^{\alpha\beta}(k,\Lambda) \to \Pi^{\alpha\beta}(k) \quad \text{when} \quad \Lambda \to \infty. \tag{12.14}$$

For example, one can define (within the integral of Eq. (12.10)):

$$\Pi_{1L}^{\alpha\beta}(k,\Lambda) = -i \int \frac{\mathrm{d}^4 p}{(2\pi)^4} \, \mathrm{Tr}\{\cdots\} \times \left(\frac{\Lambda^4}{(\Lambda^2 - p^2)^2}\right) \tag{12.15}$$

At finite Λ, this is now convergent since $\Pi_{1L}^{\alpha\beta}(k,\Lambda) \propto \int \mathrm{d}p/p^3$. One can perform the calculations and then analyze the results as a function of Λ.

What actually matters is the Λ-dependence of the result, the way the result diverges as $\Lambda \to \infty$. One can often separate divergent parts from finite contributions. Let us explore Eq. (12.10) further. The function $\Pi^{\alpha\beta}(k,\Lambda)$ can be simplified to:

$$\Pi_{1L}^{\alpha\beta}(k,\Lambda) = i \int \frac{\mathrm{d}^4 p}{(2\pi)^4 \mathcal{D}_0 \mathcal{D}_1} \, \mathrm{Tr}\left\{\gamma^\alpha (\not{k} + \not{p} + m)\gamma^\beta (\not{p} + m)\right\} \left(\frac{\Lambda^4}{(\Lambda^2 - p^2)^2}\right) \tag{12.16}$$

where

$$\mathcal{D}_0 = p^2 - m^2 \quad \text{and} \quad \mathcal{D}_1 = (k+p)^2 - m^2 \tag{12.17}$$

Let us consider each term (see Appendices E.2 and E.3):

- $\mathrm{Tr}\left(\gamma^\alpha (\not{k} + \not{p})\gamma^\beta \not{p}\right) = 4\left[p^\alpha (k+p)^\beta + p^\beta (k+p)^\alpha - (p \cdot (k+p))g^{\alpha\beta}\right]$

- $m\,\mathrm{Tr}\left(\gamma^\alpha (\not{k} + \not{p})\gamma^\beta\right) = 0$

- $m\,\mathrm{Tr}\left(\gamma^\alpha \gamma^\beta \not{p}\right) = 0$

- $m^2\,\mathrm{Tr}\left(\gamma^\alpha \gamma^\beta\right) = 4m^2 g^{\alpha\beta}$

---

1   Similarly, one can consider the introduction of new very massive particles at a very high energy scale which "modulates" the divergences of the propagator.

therefore:

$$\Pi_{1L}^{\alpha\beta}(k,\Lambda) = 4i \int \frac{d^4p}{(2\pi)^4 \mathcal{D}_0 \mathcal{D}_1} \left\{ p^\alpha(k+p)^\beta + p^\beta(k+p)^\alpha - (p\cdot(k+p) - m^2)g^{\alpha\beta} \right\} \left(\frac{\Lambda^4}{(\Lambda^2 - p^2)^2} \right) \qquad (12.18)$$

The integration must lead to a covariant result and we can admit that its general form will be:

$$\Pi_{1L}^{\alpha\beta}(k,\Lambda) = g^{\alpha\beta} B_{00}(k^2,\Lambda) + k^\alpha k^\beta B_{11}(k^2,\Lambda) \qquad (12.19)$$

where $B_{00}(k^2,\Lambda)$ and $B_{11}(k^2,\Lambda)$ are **scalar functions** to be determined. This basically means that any covariant function of k^μ with two indices μ,ν can always be written as a sum of a term depending on k^2 and a second on $k^\mu k^\nu$. The term B_{11} does not contribute to a typical QED vector current, since:

$$
\begin{aligned}
k_\mu \bar{u}(p')\gamma^\mu u(p) &= \bar{u}(p')\slashed{k}u(p) = \bar{u}(p')(\slashed{p} - \slashed{p}')u(p) = \bar{u}(p')(\slashed{p}u(p)) - (\bar{u}(p')\slashed{p}')u(p) \\
&= m\bar{u}(p')u(p) - m\bar{u}(p')u(p) = 0
\end{aligned}
\qquad (12.20)
$$

The same is evidently true for terms with v spinors instead of u spinors, so this is a general result for QED vertices. Considering just B_{00}, we can decide to expand it in k^2 as follows:

$$B_{00}(k^2,\Lambda) = A_0(\Lambda) + k^2 A_1(\Lambda) + k^4 A_2(k^2,\Lambda) \qquad (12.21)$$

where A_0 and A_1 are independent of k^2. In the expansion, each successive term introduces an extra power of k^2 and this determines the dimensionality of the A_1 and A_2 functions. When we integrate, we find (e.g., cf. Eq. (12.13)):

$$A_0 \propto \int_0^\Lambda p\,dp \propto p^2 \to \Lambda^2, \quad A_1 \propto \int_0^\Lambda dp/p \propto \ln p \to \ln\Lambda, \quad A_2(k^2) \propto \int_0^\Lambda dp/p^3 \to \text{ finite} \qquad (12.22)$$

Finally, the radiative corrections to the photon propagator (called photon self-energy for historical reasons) will have three parts: (1) a divergent "Λ^2" term; (b) a softly divergent "$\ln\Lambda$" term; and (3) a calculable finite term. One includes the finite corrections in the **photon structure function of k^2** and deals separately with the divergent and softly divergent terms. One can find in the literature several ways to solve the photon self-energy. We illustrate in the following one of these methods.

• **Computing the photon self-energy.** Let us start from Eq. (12.18) and define the integral $\Pi^{\alpha\beta}(k)$ corresponding to the loop with amputated external legs and without cutoff, reinserting the $+i\epsilon$ terms in the denominators. The integral will diverge:

$$\Pi^{\alpha\beta}(k,m) = 4i \int \frac{d^4p}{(2\pi)^4} \frac{p^\alpha(k+p)^\beta + p^\beta(k+p)^\alpha - (p\cdot(k+p) - m^2)g^{\alpha\beta}}{(p^2 - m^2 + i\epsilon)((k+p)^2 - m^2 + i\epsilon)} \qquad (12.23)$$

In order to handle the denominators, we note that they can be written with exponential factors (see Appendix F.2):

$$(-i)\int_0^\infty dz\, e^{iz(t+i\epsilon)} = \frac{1}{t+i\epsilon} \quad \text{when} \quad \epsilon > 0 \qquad (12.24)$$

Then,

$$
\begin{aligned}
\Pi^{\alpha\beta}(k,m) = &-4i \int_0^\infty dz_1 \int_0^\infty dz_2 \int \frac{d^4p}{(2\pi)^4} \left(p^\alpha(k+p)^\beta + p^\beta(k+p)^\alpha - (p\cdot(k+p) - m^2)g^{\alpha\beta} \right) \\
&\times e^{iz_1(p^2 - m^2 + i\epsilon)} e^{iz_2((k+p)^2 - m^2 + i\epsilon)}
\end{aligned}
\qquad (12.25)
$$

where the order of integrals has been changed. We can now change the integration variable by defining:

$$\ell^\mu \equiv p^\mu + \frac{z_2}{z_1 + z_2}k^\mu \quad \Longrightarrow \quad p^\mu = \ell^\mu - \frac{z_2}{z_1 + z_2}k^\mu \qquad (12.26)$$

hence

$$k^\mu + p^\mu = \ell^\mu + k^\mu - \frac{z_2}{z_1 + z_2} k^\mu = \ell^\mu + \frac{z_1}{z_1 + z_2} k^\mu \qquad (12.27)$$

We note that $d\ell^4 = dp^4$ and need to consider the following terms in $\Pi^{\alpha\beta}(k)$:

- $p^\alpha(k+p)^\beta = \ell^\alpha \ell^\beta - \frac{z_2}{z_1+z_2} k^\alpha \ell^\beta + \frac{z_1}{z_1+z_2} \ell^\alpha k^\beta - \frac{z_1 z_2}{(z_1+z_2)^2} k^\alpha k^\beta$

- $p^\beta(k+p)^\alpha = \ell^\alpha \ell^\beta - \frac{z_2}{z_1+z_2} k^\beta \ell^\alpha + \frac{z_1}{z_1+z_2} \ell^\beta k^\alpha - \frac{z_1 z_2}{(z_1+z_2)^2} k^\alpha k^\beta$

- $p^2 = \ell^2 - 2\frac{z_2}{z_1+z_2} \ell \cdot k + \left(\frac{z_2}{z_1+z_2}\right)^2 k^2$

- $(p+k)^2 = \ell^2 + 2\frac{z_1}{z_1+z_2} \ell \cdot k + \left(\frac{z_1}{z_1+z_2}\right)^2 k^2$

The exponential term of Eq. (12.25) is then:

$$
\begin{aligned}
i\Big(z_1 p^2 + z_2(k+p)^2 - (z_1 + z_2)m^2\Big) &= i\Big(z_1(\ell^2 - 2\frac{z_2}{z_1+z_2}\ell \cdot k + \left(\frac{z_2}{z_1+z_2}\right)^2 k^2) \\
&\quad + z_2(\ell^2 + 2\frac{z_1}{z_1+z_2}\ell \cdot k + \left(\frac{z_1}{z_1+z_2}\right)^2 k^2) - (z_1+z_2)m^2\Big) \\
&= i(z_1+z_2)\ell^2 + i\left(\frac{z_1 z_2}{z_1+z_2}k^2 - (z_1+z_2)m^2\right) \qquad (12.28)
\end{aligned}
$$

Consequently, setting $Z \equiv z_1 + z_2$, we find:

$$
\begin{aligned}
\Pi^{\alpha\beta}(k,m) &= -4i \int_0^\infty dz_1 \int_0^\infty dz_2 \int \frac{d^4\ell}{(2\pi)^4} \left(\ell^\alpha \ell^\beta - \frac{z_2}{Z}k^\alpha \ell^\beta + \frac{z_1}{Z}\ell^\alpha k^\beta - \frac{z_1 z_2}{Z^2}k^\alpha k^\beta \right. \\
&\quad + \ell^\alpha \ell^\beta - \frac{z_2}{Z}k^\beta \ell^\alpha + \frac{z_1}{Z}\ell^\beta k^\alpha - \frac{z_1 z_2}{Z^2}k^\alpha k^\beta \\
&\quad \left. - ((\ell^\mu - \frac{z_2}{Z}k^\mu) \cdot (\ell_\mu + \frac{z_1}{Z}k_\mu) - m^2)g^{\alpha\beta}\right) \times e^{iZ\ell^2} e^{i\left(\frac{z_1 z_2}{Z}k^2 - Zm^2\right)} \\
&= -4i \int_0^\infty dz_1 \int_0^\infty dz_2 \int \frac{d^4\ell}{(2\pi)^4} \left(2\ell^\alpha \ell^\beta - \frac{z_2}{Z}k^\alpha \ell^\beta + \frac{z_1}{Z}\ell^\alpha k^\beta - 2\frac{z_1 z_2}{Z^2}k^\alpha k^\beta \right. \\
&\quad \left. - \frac{z_2}{Z}k^\beta \ell^\alpha + \frac{z_1}{Z}\ell^\beta k^\alpha - (\ell^2 - \frac{z_2}{Z}k \cdot \ell + \frac{z_1}{Z}k \cdot \ell - \frac{z_1 z_2}{Z^2}k^2 - m^2)g^{\alpha\beta}\right) \\
&\quad \times e^{iZ\ell^2} e^{i\left(\frac{z_1 z_2}{Z}k^2 - Zm^2\right)} \qquad (12.29)
\end{aligned}
$$

We can now use the identity **[FI6]** from Appendix F.5 to remove several terms which will vanish during the integration over ℓ. Consequently, we are left with the following terms:

$$
\begin{aligned}
\Pi^{\alpha\beta}(k,m) &= -4i \int_0^\infty dz_1 \int_0^\infty dz_2 \int \frac{d^4\ell}{(2\pi)^4} \left(2\ell^\alpha \ell^\beta - 2\frac{z_1 z_2}{Z^2}k^\alpha k^\beta - (\ell^2 - \frac{z_1 z_2}{Z^2}k^2 - m^2)g^{\alpha\beta}\right) \\
&\quad \times e^{iZ\ell^2} e^{i\left(\frac{z_1 z_2}{Z}k^2 - Zm^2\right)} \qquad (12.30)
\end{aligned}
$$

Further, we use equalities **[FI5]**, **[FI7]**, and **[FI8]** from Appendix F.5 to perform the integration over ℓ, and get for $\Pi^{\alpha\beta}(k,m)$:

$$
-4i \int_0^\infty dz_1 \int_0^\infty dz_2 \left(\frac{2}{32\pi^2 i Z^2} \frac{ig^{\alpha\beta}}{Z} - \frac{1}{16\pi^2 i Z^2} 2 \frac{z_1 z_2}{Z^2} k^\alpha k^\beta + \frac{1}{16\pi^2 i Z^2} \left(-\frac{2i}{Z} + \frac{z_1 z_2}{Z^2} k^2 + m^2 \right) g^{\alpha\beta} \right)
$$
$$
\times e^{i\left(\frac{z_1 z_2}{Z} k^2 - Z m^2 \right)}
$$
$$
= -4i \int_0^\infty dz_1 \int_0^\infty dz_2 \frac{1}{16\pi^2 i Z^2} \left(\frac{ig^{\alpha\beta}}{Z} - 2\frac{z_1 z_2}{Z^2} k^\alpha k^\beta + \left(-\frac{2i}{Z} + \frac{z_1 z_2}{Z^2} k^2 + m^2 \right) g^{\alpha\beta} \right) e^{i\left(\frac{z_1 z_2}{Z} k^2 - Z m^2 \right)}
$$
$$
= -\frac{1}{4\pi^2} \int_0^\infty dz_1 \int_0^\infty \frac{dz_2}{Z^2} \left((-k^\alpha k^\beta) \frac{2 z_1 z_2}{Z^2} + \left(\frac{-i}{Z} + \frac{z_1 z_2}{Z^2} k^2 + m^2 \right) g^{\alpha\beta} \right) e^{i\left(\frac{z_1 z_2}{Z} k^2 - Z(m^2 - i\epsilon) \right)}
$$

$$ \tag{12.31} $$

where we have reintroduced $i\epsilon$ in the exponential term. This looks complicated! Let us consider the integral of the three terms proportional to $g^{\alpha\beta}$. We note that it can be written as:

$$
-\frac{1}{4\pi^2} \int_0^\infty \int_0^\infty \frac{dz_1 dz_2}{Z^2} \left(2\frac{z_1 z_2}{Z^2} k^2 + \frac{-i}{Z} - \frac{z_1 z_2}{Z^2} k^2 + m^2 \right) g^{\alpha\beta} e^{i\left(\frac{z_1 z_2}{Z} k^2 - Z(m^2 - i\epsilon) \right)}
$$
$$
\equiv -\frac{1}{2\pi^2} \int_0^\infty \int_0^\infty \frac{dz_1 dz_2}{Z^2} \left(\frac{z_1 z_2}{Z^2} k^2 \right) g^{\alpha\beta} e^{i\left(\frac{z_1 z_2}{Z} k^2 - Z(m^2 - i\epsilon) \right)} - \frac{1}{4\pi^2} \mathcal{A} g^{\alpha\beta} \tag{12.32}
$$

We now want to show that $\mathcal{A}$ vanishes. We replace $z_1 \to \lambda z_1$ and $z_2 \to \lambda z_2$, i.e., $Z \to \lambda Z$. This means that we are *scaling* the z_1 and z_2 variables. Consequently, $\mathcal{A}$ is expressed as:

$$
\mathcal{A} \equiv \int_0^\infty \int_0^\infty \frac{dz_1 dz_2}{Z^2} \left(\frac{-i}{\lambda Z} - \frac{z_1 z_2}{Z^2} k^2 + m^2 \right) e^{i\lambda\left(\frac{z_1 z_2}{Z} k^2 - Z(m^2 - i\epsilon) \right)} \tag{12.33}
$$

Now we use a "trick" by observing that:

$$
i\lambda \frac{\partial}{\partial \lambda} \left[\int_0^\infty \int_0^\infty \frac{dz_1 dz_2}{\lambda Z^3} e^{i\lambda\left(\frac{z_1 z_2}{Z} k^2 - Z(m^2 - i\epsilon) \right)} \right] = i\lambda \int_0^\infty \int_0^\infty \frac{dz_1 dz_2}{Z^3} \left[\frac{i(\cdots) e^{i\lambda(\cdots)} \lambda - e^{i\lambda(\cdots)}}{\lambda^2} \right]
$$
$$
= i\lambda \int_0^\infty \int_0^\infty \frac{dz_1 dz_2}{Z^3} \left[\frac{i(\cdots)\lambda - 1}{\lambda^2} \right] e^{i\lambda(\cdots)}
$$
$$
= \int_0^\infty \int_0^\infty \frac{dz_1 dz_2}{Z^2} \left[-\frac{z_1 z_2}{Z^2} k^2 + (m^2 - i\epsilon) - \frac{i}{Z\lambda} \right] e^{i\lambda(\cdots)}
$$
$$
= \mathcal{A} \tag{12.34}
$$

No kidding! So, finally:

$$
\mathcal{A} = i\lambda \frac{\partial}{\partial \lambda} \underbrace{\left[\int_0^\infty \int_0^\infty dz_1 dz_2 (\ldots) \right]}_{\text{must be independent of } \lambda} = 0 \tag{12.35}
$$

where we used the fact that the result of the integration must be independent of λ since we are integrating from zero to infinity. Therefore, we have found that $\mathcal{A}$ is identically equal to zero, even for $\lambda = 1$. We also ignore the terms proportional to $k^\alpha k^\beta$ as they vanish by current conservation, as in Eq. (10.147). Consequently, we now only need to estimate the remaining terms:

$$
\Pi^{\alpha\beta}(k,m) = -\frac{1}{2\pi^2} g^{\alpha\beta} k^2 \int_0^\infty \int_0^\infty dz_1 dz_2 \frac{z_1 z_2}{Z^4} e^{i\left(\frac{z_1 z_2}{Z} k^2 - Z(m^2 - i\epsilon) \right)} \tag{12.36}
$$

We can use another trick with the following identity:

$$
\int_0^\infty \frac{d\lambda}{\lambda} \delta\left(1 - \frac{Z}{\lambda} \right) = \int_0^\infty \frac{d\lambda}{\lambda} \frac{\delta(Z - \lambda)}{\left| \frac{Z}{\lambda^2} \right|} = \int_0^\infty d\lambda \frac{\lambda \delta(Z - \lambda)}{|Z|} = \frac{Z}{|Z|} = 1 \qquad (Z > 0) \tag{12.37}
$$

to get:

$$\Pi^{\alpha\beta}(k,m) = -\frac{1}{2\pi^2}g^{\alpha\beta}k^2 \int_0^\infty \int_0^\infty \int_0^\infty \frac{d\lambda}{\lambda}dz_1 dz_2 \frac{z_1 z_2}{Z^4}\delta\left(1-\frac{Z}{\lambda}\right)e^{i\left(k^2\frac{z_1 z_2}{Z}-(m^2-i\epsilon)Z\right)} \quad (12.38)$$

We can apply again the scaling trick $Z \to \lambda Z$, in order to rewrite the result as:

$$
\begin{aligned}
\Pi^{\alpha\beta}(k,m) &= -\frac{1}{2\pi^2}g^{\alpha\beta}k^2 \int_0^\infty \int_0^\infty dz_1 dz_2 z_1 z_2 \delta\left(1-z_1-z_2\right)\int_0^\infty \frac{d\lambda}{\lambda}e^{i\lambda\left(k^2 z_1 z_2 - m^2 + i\epsilon\right)} \\
&= -\frac{1}{2\pi^2}g^{\alpha\beta}k^2 \int_0^1 dz_1 z_1(1-z_1)\int_0^\infty \frac{d\lambda}{\lambda}e^{i\lambda\left(k^2 z_1(1-z_1)-m^2+i\epsilon\right)}
\end{aligned}
\quad (12.39)
$$

In this form, the logarithmic divergence has been absorbed in the integration over λ. Some regularization is needed to proceed further.

• **Pauli–Villars[2] regularization [88].** In order to control the infinities, we need to regularize the loop integral in one way or another. We adopt here the so-called Pauli–Villars regularization. The divergence is ultraviolet in the sense that it is coming from integrating arbitrarily high momenta in the loop integral. The Pauli–Villars regularization subtracts off the same loop integral with a much larger mass. The subtracted piece is regarded as a contribution of another field (the Pauli–Villars field) with the same quantum numbers as the original field, but with the opposite statistics. As long as we take its mass Λ very large and deal with physics at energy scales much lower than Λ, this method does not pose any problems. We hence regularize the integral as follows:

$$\tilde{\Pi}^{\alpha\beta}(k) \equiv \Pi^{\alpha\beta}(k,m) - \Pi^{\alpha\beta}(k,\Lambda) \quad (12.40)$$

In doing so, we can extract the remaining k^2 dependence of the result out of the infinities. Replacing $z_1 \to z$, we find:

$$
\begin{aligned}
\tilde{\Pi}^{\alpha\beta}(k) &\simeq -\frac{1}{2\pi^2}g^{\alpha\beta}k^2 \int_0^1 dz\, z(1-z)\ln\left(\frac{\Lambda^2}{m^2-k^2 z(1-z)}\right) \\
&= -\frac{1}{2\pi^2}g^{\alpha\beta}k^2 \int_0^1 dz\, z(1-z)\left[\ln\left(\frac{\Lambda^2}{m^2}\right)+\ln\left(\frac{m^2}{m^2-k^2 z(1-z)}\right)\right]
\end{aligned}
\quad (12.41)
$$

where, in the first line, we have used the properties of the **exponential integrals** (see Appendix F.2). Finally, with $\int_0^1 dz\, z(1-z) = 1/6$, the term can be rewritten as:

$$\tilde{\Pi}^{\alpha\beta}(k) = -\frac{1}{12\pi^2}g^{\alpha\beta}k^2\left(\ln\left(\frac{\Lambda^2}{m^2}\right)-6\int_0^1 dz\, z(1-z)\ln\left(1-\frac{k^2}{m^2}z(1-z)\right)\right) \quad (12.42)$$

Going back to Eq. (12.9) and introducing the fine-structure constant $\alpha = e^2/4\pi$ on the right-hand side yields the final expression:

$$ie^2\tilde{\Pi}^{\alpha\beta}(k) = -ig^{\alpha\beta}k^2\left[\left(\frac{\alpha}{3\pi}\right)\left(\ln\left(\frac{\Lambda^2}{m^2}\right)-f\left(-\frac{k^2}{m^2}\right)\right)\right] \quad (12.43)$$

The most important term is the $\alpha/3\pi$! The result is logarithmically divergent with the cutoff Λ^2. The function $f(x)$ is on the other hand finite for finite x:

$$f(x) = 6\int_0^1 dz\, z(1-z)\ln(1+xz(1-z)) \quad \to \quad \begin{cases} \ln x & \text{if } x \gg 1 \\ 6x\int_0^1 dz\, z^2(1-z)^2 = \frac{x}{5} & \text{if } x \ll 1 \end{cases} \quad (12.44)$$

---

2   Felix Villars (1921–2002), Swiss-born American physicist.

The dressed photon propagator to order $\mathcal{O}(\alpha)$, according to Eq. (12.12), can then be expressed as:

$$
\begin{aligned}
\tilde{G}^{\mu\nu}(k, \Lambda) &= \tilde{G}_F^{\mu\nu}(k) + \tilde{G}_F^{\mu\alpha}(k)ie^2\tilde{\Pi}_{\alpha\beta}(k)\tilde{G}_F^{\beta\nu}(k) \\
&= \frac{-ig^{\mu\nu}}{k^2} - i\left(\frac{-ig^{\mu\alpha}}{k^2}\right)g_{\alpha\beta}k^2\left[\left(\frac{\alpha}{3\pi}\right)\left(\ln\left(\frac{\Lambda^2}{m^2}\right) - f\left(-\frac{k^2}{m^2}\right)\right)\right]\left(\frac{-ig^{\beta\nu}}{k^2}\right) \\
&= \frac{-ig^{\mu\nu}}{k^2} - \left(\frac{-ig^{\mu\nu}}{k^2}\right)\left[\left(\frac{\alpha}{3\pi}\right)\left(\ln\left(\frac{\Lambda^2}{m^2}\right) - f\left(-\frac{k^2}{m^2}\right)\right)\right]
\end{aligned}
\tag{12.45}
$$

So finally:

$$
\tilde{G}^{\mu\nu}(k, \Lambda) = \left(\frac{-ig^{\mu\nu}}{k^2}\right)\left[1 - \frac{\alpha}{3\pi}\left(\ln\left(\frac{\Lambda^2}{m^2}\right) - f\left(-\frac{k^2}{m^2}\right)\right)\right]
\tag{12.46}
$$

The minus sign in $f(-k^2/m^2)$ should not trouble us, when we realize that in a scattering process $k^2 < 0$. The physics of the corrections is included in this four-momentum-dependent term, and the divergence with the cutoff Λ will be handled with the concept of renormalization discussed in the next sections.

12.4   Renormalization

An extremely important concept is **renormalization**. A given theory cannot predict *everything* and it must rely on a set of fundamental input parameters, such as the elementary electric charge or the mass of the electron. Based on this set of parameters, the theory is used to **predict observables in given experimental conditions**. A theory should predict finite observables and we can question the validity of a theory if it predicts infinite values.

In the case of QED, serious mathematical issues with divergences are a priori encountered when computing higher-order diagrams. In QFT, finite results can be obtained through a process of renormalization. So, the question on the validity of a theory is whether it **can or cannot be renormalized**. QED is renormalizable. Quantum chromodynamics (QCD – see Chapter 18) and the electroweak theory (see Chapter 25) are renormalizable. So, it does not matter if a theory appears to be giving infinite results **if such divergences can be cured through a well-defined process**.

Although the actual renormalization scheme is not unique, a renormalizable theory can be cured from its a priori divergences. The most widely used renormalization scheme is called the **minimal subtraction scheme** or $\overline{\text{MS}}$. It depends on a single parameter – the **renormalization scale**. Physical results and observables will not depend on this artificial quantity. The **on-shell renormalization scheme** is also widely used, especially in QED at low energies or in QCD for heavy quarks. It turns out that renormalization also provides a deep physical insight on Nature, as we will illustrate below.

Renormalization can be viewed in several ways. From one perspective it is a mathematical manipulation, which allows us to calculate finite, testable observables. From a more physical perspective, we start by noting that the key divergences come from the high-energy limits of the momentum integration. However, experimentally, we do not know what kind of physics describes high-energy processes. For instance, there may be new heavy yet undiscovered particles which contribute to the loop diagrams (as virtual particles in the loop) and would change the divergence of the integral. From this perspective, renormalization is a procedure which allows us to sensibly calculate the effects of the **low-energy physics**, independent of how it behaves at very high energies.

Let us consider the basic QED Lagrangian, Eq. (10.93):

$$
\mathcal{L}_{bare} \equiv \overline{\Psi}\left[\gamma^\mu i\partial_\mu - m_0\right]\Psi - \frac{1}{4}F_{\mu\nu}F^{\mu\nu} - e_0\overline{\Psi}\gamma^\mu\Psi A_\mu
\tag{12.47}
$$

where we have explicitly replaced the mass and the elementary charges by what we call the **"bare electron mass"** m_0 and the **"bare elementary charge"** e_0. These are the "fundamental" input parameters of the theory, with which we are going to compute actual physical observables in given experimental conditions.

We note that the elementary electric charge e is an observable, and as such, it will be predicted by the theory as a function of the input parameters e_0, i.e., $e = f_{QED}(e_0)$. Since an observable depends on the given experimental conditions, it is possible that our measured electric charge e will depend on these conditions. On the other hand, the input parameter e_0 is constant. **Hence, e and e_0 are not the same.** As a matter of fact, e_0 is simply an input parameter of the theory that needs to be tuned in order to reproduce experimental observables, and by itself e_0 is *not* a physical observable.

12.5   Electric Charge Renormalization

We illustrate the basic principle of **charge renormalization** using the self-energy corrections to the photon propagator. We consider a photon with two QED vertices. We found that the tree-level propagator is modified by the quantity in the square brackets of Eq. (12.46), where we replace m by m_e to consider for concreteness the electron–positron loop:

$$(-ie_0\gamma_\mu)\cdots\left(\frac{-ig^{\mu\nu}}{k^2}\right)\left[1-\frac{e_0^2}{12\pi^2}\left(\ln\left(\frac{\Lambda^2}{m_e^2}\right)-f\left(-\frac{k^2}{m_e^2}\right)\right)+\mathcal{O}(e_0^4)\right]\cdots(-ie_0\gamma_\nu) \qquad (12.48)$$

where we have used the electric charge $e_0^2 = 4\pi\alpha$. Should the input parameter to the theory be e_0 or the experimentally measured electric charge e? We remind ourselves that the correction has two parts: a divergent part handled by the cutoff Λ and the computable correction $f(k^2)$, which depends on the kinematical configuration as k^2 (see Eq. (12.43)). The idea behind renormalization is to **include the divergence into a redefinition of the electric charge** e, such that the photon propagator is calculable and expressed as:

$$(-ie\gamma_\mu)\cdots\left(\frac{-ig^{\mu\nu}}{k^2}\right)\left[1+\frac{e^2}{12\pi^2}f\left(-\frac{k^2}{m_e^2}\right)+\mathcal{O}(e^4)\right]\cdots(-ie\gamma_\nu) \qquad (12.49)$$

and where the divergence is absorbed in the **renormalized charge** e, as follows:

$$e \equiv e_0\sqrt{1-\frac{e_0^2}{12\pi^2}\ln\left(\frac{\Lambda^2}{m_e^2}\right)} \qquad (12.50)$$

Since e is to be finite, the original input parameter of the theory e_0 will become singular, but it does not matter since we are not interested in the theory at very high energy (in the ultraviolet regime). The important point is that the above procedure allows us to compute the k^2-dependent radiative correction to the photon propagator in a regularized and finite way. What is important is that the photon propagator now contains this finite k^2-dependent correction term, whose effect can be directly tested in experiments.

This procedure can be pushed further and the remaining k^2 dependence of the photon propagator can also be included in the scale-dependent electric charge $e(k^2)$! Then, we have simply:

$$(-ie(k^2)\gamma_\mu)\cdots\left(\frac{-ig^{\mu\nu}}{k^2}\right)\cdots(-ie(k^2)\gamma_\nu) \qquad (12.51)$$

where the **"running"** electric charge is:

$$e(k^2) \equiv e(0)\sqrt{1+\frac{e^2(0)}{12\pi^2}f\left(-\frac{k^2}{m_e^2}\right)+\mathcal{O}(e^4(0))} \qquad (12.52)$$

"Running" is the term used to express the fact that the electric charge which we "see" when we probe an electron with a photon, is varying with the four-momentum k^2. The magnitude of k^2 is to be interpreted as the "resolution" of the probing photon, in the sense of the particle–wave duality of de Broglie ($\lambda \propto 1/p$). We interpret this result as the existence of a **charge screening effect** around a point charge. The naked electric charge is surrounded by a cloud of virtual e^+e^- pairs (those which we computed as the fermion–antifermion loop in the photon propagator).

The phenomenon of **vacuum polarization** is illustrated in Figure 12.6. At relatively "small" $|k^2|$, the photon probes the shielded charge. At "larger" $|k^2|$, the photon probes more of the naked charge, hence the electric charge that we see increases with $|k^2|$. It is often expressed as the **"running of the coupling constant α."** Using

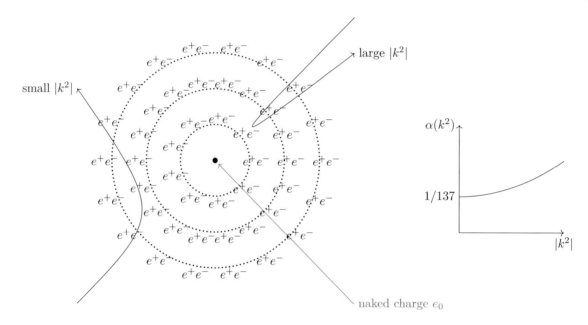

Figure 12.6 Basic illustration of the vacuum polarization effect inducing the "running" of the coupling constant α.

$e^2 = 4\pi\alpha$, we directly get:

$$\alpha(k^2) \equiv \alpha(0)\left[1 + \frac{\alpha(0)}{3\pi}f\left(-\frac{k^2}{m_e^2}\right) + \mathcal{O}(\alpha^2)\right] \tag{12.53}$$

Rather than relying on $\alpha(0)$, we experimentally measure the physical elementary charge at **some momentum scale** $\mu^2 > 0$. Then the coupling constant $\alpha(\mu^2)$ can be computed at this scale using the above equation replacing $-k^2 \to \mu^2$. We find:

$$\begin{aligned}
\alpha(k^2) &= \alpha(\mu^2)\frac{\alpha(k^2)}{\alpha(\mu^2)} = \alpha(\mu^2)\frac{\left[1 + \frac{\alpha(0)}{3\pi}f\left(-\frac{k^2}{m_e^2}\right) + \cdots\right]}{\left[1 + \frac{\alpha(0)}{3\pi}f\left(\frac{\mu^2}{m_e^2}\right) + \cdots\right]} \\
&\simeq \alpha(\mu^2)\left[1 + \frac{\alpha(0)}{3\pi}f\left(-\frac{k^2}{m_e^2}\right) + \cdots\right]\left[1 - \frac{\alpha(0)}{3\pi}f\left(\frac{\mu^2}{m_e^2}\right) + \cdots\right] \\
&\simeq \alpha(\mu^2)\left[1 + \frac{\alpha(0)}{3\pi}\left(f\left(-\frac{k^2}{m_e^2}\right) - f\left(\frac{\mu^2}{m_e^2}\right)\right) + \cdots\right]
\end{aligned} \tag{12.54}$$

where we used the geometric power series, Eq. (A.1). Since $f(x) \to \ln x$ when $x \gg 1$, we can write:

$$\alpha(k^2) = \alpha(\mu^2) \left[1 + \frac{\alpha(\mu^2)}{3\pi} \ln\left(\frac{-k^2}{\mu^2} \right) + \mathcal{O}(\alpha^2) \right] \tag{12.55}$$

where inside the brackets we have also set $\alpha(\mu^2)$, since the difference between $\alpha(\mu^2)$ and $\alpha(0)$ in the correction is a higher-order effect. Recalling that $k^2 < 0$ in a scattering process, the momentum squared scale is defined as the positive quantity $Q^2 \equiv -k^2 > 0$. Then:

$$\alpha(Q^2) = \alpha(\mu^2) \left[1 + \frac{\alpha(\mu^2)}{3\pi} \ln\left(\frac{Q^2}{\mu^2} \right) + \mathcal{O}(\alpha^2) \right] \tag{12.56}$$

We can do better by "resumming" the result by assuming that the higher-order terms will be of the type of the geometric power series $1 + x + x^2 + x^3 + \cdots = 1/(1 - x)$ for $|x| \ll 1$ (see Eq. (A.1)). Hence, the **resummed running of the coupling constant** is finally:

$$\alpha(Q^2) = \frac{\alpha(\mu^2)}{1 - \frac{\alpha(\mu^2)}{3\pi} \ln\left(\frac{Q^2}{\mu^2} \right)} \tag{12.57}$$

The minus sign implies that the QED coupling constant α increases with Q^2, as illustrated in the sketch of Figure 12.6. This result is consistent with the physical interpretation of the vacuum polarization (i.e., the incorporation of virtual fermion loops in the photon propagator), inducing a charge screening. Hence, the electromagnetic coupling constant or the apparent elementary charge of a fermion increases as Q^2 increases. To really define what we have done, we can say that **we have fixed the parameters of the theory by doing some particular experiment at some scale μ^2 and used the QED higher-order calculations to extrapolate observables to another experimental condition at a higher scale Q^2.**

12.6 Confronting Reality – the Running of α

No one can actually confront the concept of renormalization for the first time without a bit of skepticism. The running of the electromagnetic coupling constant α as a function of the momentum transfer is an immediate consequence of the renormalization of the photon propagator and its experimental demonstration, therefore it represents a clear "smoking gun" for the validity of the theoretical concept developed in this chapter. Measurements at low energy can determine the fine-structure constant at $Q^2 \simeq 0$. The recommended world average value from the CODATA group [10] is:

$$\alpha_{\text{CODATA}} = \frac{e^2}{4\pi\epsilon_0 \hbar c} = (7.2973525664 \pm 0.0000000017) \times 10^{-3} \tag{12.58}$$

with an impressive relative uncertainty of 2.3×10^{-10}. This gives a reciprocal value of:

$$\alpha(Q^2 \simeq 0)_{\text{CODATA}} = \frac{1}{137.035999138 \pm 0.000000032} \tag{12.59}$$

The coupling has been measured at high energy in e^+e^- collisions at colliders of increasing energy. Table 12.1 is a compilation of results from data collected with experiments at DORIS, PEP, PETRA, TRISTAN, and LEP colliders (the experiments will be discussed in more detail in Chapter 13). At high energies, the value of the electromagnetic constant $\alpha_{\text{SM}}(Q^2)$ is extracted from a fit of experimental measurements within the Standard

Model theory (see Chapter 29). The measurements at the highest energies are from the OPAL experiment, using data at center-of-mass energies up to 209 GeV. Their fitted value at an average Q of 193 GeV is:

$$\alpha_{\mathrm{SM}}(Q = 193 \text{ GeV}) = \frac{1}{126.7^{+2.4}_{-2.3}} \tag{12.60}$$

This value is significantly lower than the value at $Q^2 \simeq 0$ and represents a direct experimental observation of the running of the QED coupling constant!

$\sqrt{s}$ (GeV)	$1/\alpha$	Experiments (Collider) [Ref.]
10.4	131.9 ± 3.0	CRYSTAL BALL(DORIS) [89]
29.0	130.1 ± 1.3	HRS/MAC/MARKII(PEP) [89]
34.7	130.9 ± 1.5	CELLO/JADE/MARKJ/PLUTO/TASSO (PETRA) [89]
43.7	$130.9^{+2.5}_{-2.4}$	CELLO/JADE/MARKJ/PLUTO/TASSO (PETRA) [89]
57.8	125.0 ± 2.0	AMY/TOPAZ/VENUS(TRISTAN) [89]
57.8	128.5 ± 1.9	TOPAZ(TRISTAN) [90]
193	$126.7^{+2.4}_{-2.3}$	OPAL(LEP) [91]
$Q \simeq 0$	$137.035999138 \pm 0.000000032$	CODATA [10]

Table 12.1 A summary of the measurements of α^{-1} at high energies compared to the CODATA value at $Q \simeq 0$.

Figure 12.7 graphically represents these results. The solid line shows the theoretical expectation. An accurate theoretical evaluation of $\alpha_{SM}(s)$ valid at high energies must take into account several effects. One can conveniently split the different contributions as:

$$\alpha_{SM}(s) = \frac{\alpha(0)}{1 - \Delta\alpha_{leptons}(s) - \Delta\alpha_{hadrons}(s) - \Delta\alpha_{top}(s) - \Delta\alpha_{new}(s)} \tag{12.61}$$

where $\Delta\alpha_{leptons}$ is the correction introduced by the leptonic loops (electron, muon, and tau), $\Delta\alpha_{hadrons}$ is the correction introduced by hadronic corrections corresponding to loops of quarks, excluding the top quark, since it is much heavier than the other quarks. $\Delta\alpha_{top}$ is the contribution of the top quark loop. Finally, $\Delta\alpha_{new}(s)$ could contain the effect of "new" virtual particles in the higher-order diagrams. The agreement between theory and experiment is clearly visible, strongly supporting the running of the fine-structure constant predicted by QED and the absence of a significant contribution from new physics at the attained energies.

Due to the logarithmic nature of the effects, a precise determination of α_{SM} is usually referred to a fixed point at high energy, for instance conveniently taken to be the mass M_Z of the Z^0 boson (see Chapters 25 and 26). A major contribution originates from the vacuum polarization of leptons. The dominant term is given by [92]:

$$\begin{aligned}
\Delta\alpha_{\mathrm{leptons}}(M_Z^2) &= \frac{\alpha(0)}{\pi} \sum_{i=e,\mu,\tau} \left(-\frac{5}{9} + \frac{1}{3}\ln\frac{M_Z^2}{m_i^2} - 2\frac{m_i^2}{M_Z^2} + \mathcal{O}\left(\frac{m_i^4}{M_Z^4}\right) \right) + \Delta\alpha_{\mathrm{lep,2l}}(M_Z^2) + \mathcal{O}\left(\alpha^3\right) \\
&\approx 314.19 \times 10^{-4} + \Delta\alpha_{\mathrm{lep,2l}}(M_Z^2) + \mathcal{O}\left(\alpha^3\right)
\end{aligned} \tag{12.62}$$

where the two-loop correction is small, $\Delta\alpha_{\mathrm{lep,2l}}(M_Z^2) \approx 0.78 \times 10^{-4}$ [93], but for example not negligible compared to the top quark contribution. The contribution from hadronic vacuum polarization is nearly as

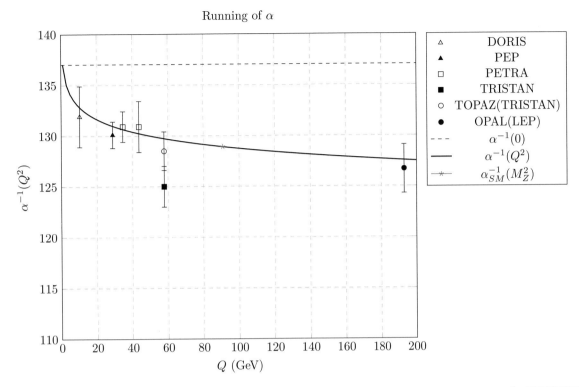

Figure 12.7 Measured values of the running coupling constant α at the DORIS, PEP, PETRA, TRISTAN, and LEP e^+e^- colliders. The solid line shows the theoretical expectation.

large as the leptonic contribution, but its theoretical calculation is complicated by the fact that the quark–antiquark loops will create hadronic resonances (see Section 20.11). However, it can be related through dispersion relations to the cross-section for hadron production in electron–positron annihilation defined as the $R(s)$ ratio as a function of the Mandelstam variable s (see Section 19.2). It can be expressed as [92]:

$$\Delta\alpha_{hadrons}(M_Z^2) \;=\; -\frac{\alpha(0)M_Z^2}{3\pi}\,\mathrm{Re}\int_{4m_\pi^2}^{\infty}\mathrm{d}s\,\frac{R(s)}{s\,(s - M_Z^2 - i\epsilon)} \tag{12.63}$$

The integrand can be obtained from experimental data. Several authors have calculated this contribution and a precise theoretical prediction yields (see Ref. [92]):

$$\Delta\alpha_{hadrons}(M_Z^2) = (277.45 \pm 1.68) \times 10^{-4} \tag{12.64}$$

The contribution of heavy quarks is known to three-loop QCD corrections [94]. In the case of the top quark, this leads to $\Delta\alpha_{top}(M_Z^2) = (-0.70 \pm 0.05) \times 10^{-4}$ for the strong coupling constant $\alpha_s(M_Z^2) = 0.118$ (see Section 18.15) and $M_t = 175.6 \pm 5.5$ GeV (see Section 29.2). Putting these results together, one obtains the theoretical prediction:

$$\alpha_{SM}(M_Z^2) = \frac{1}{128.927 \pm 0.023} \tag{12.65}$$

This result is quite precise compared to the direct experimental measurements illustrated above. There is a lot of physics entering in the running of such an a priori simple quantity as the fine-structure constant, and the agreement between data and theory with such precision is a striking and very important achievement!

12.7   Electron Self-energy

As in the case of the photon, let us define the sum of all 1PI diagrams as $-i\Sigma(p)$. We then write the fully resummed electron propagator as (see Section 9.12):

$$
\begin{aligned}
\tilde{S}_F(p) &= \frac{i}{\not{p}-m_0} + \frac{i}{\not{p}-m_0}\left(-i\Sigma(p)\right)\frac{i}{\not{p}-m_0} + \frac{i}{\not{p}-m_0}\left(-i\Sigma(p)\right)\frac{i}{\not{p}-m_0}\left(-i\Sigma(p)\right)\frac{i}{\not{p}-m_0} + \cdots \\
&= \frac{i}{\not{p}-m_0}\left[1 + \Sigma(p)\frac{1}{\not{p}-m_0} + \Sigma(p)\frac{1}{\not{p}-m_0}\Sigma(p)\frac{1}{\not{p}-m_0} + \cdots\right]
\end{aligned}
\tag{12.66}
$$

We can use the "resumming" of the propagator as in Section 9.12. Consequently

$$
\tilde{S}_F(p) = \frac{i}{\not{p}-m_0}\left[1 - \Sigma(p)\frac{1}{\not{p}-m_0}\right]^{-1} = \frac{i}{\not{p}-m_0}\left[\frac{\not{p}-m_0-\Sigma(p)}{\not{p}-m_0}\right]^{-1} = i\left[\not{p}-m_0-\Sigma(p)\right]^{-1}
\tag{12.67}
$$

Then, the fully dressed fermion propagator $\tilde{S}_F(p)$ for a particle of bare mass m goes simply as follows:

$$
\tilde{S}_F(p) = \frac{i}{\not{p}-m_0-\Sigma(p)}
\tag{12.68}
$$

What is m_0? In this case, we should reinterpret the "bare mass" m_0 as an input parameter to the theory and not the actual physical mass of the fermion. Hence, we replaced m by m_0. We need to calculate $\Sigma(p)$ perturbatively. This calculation is analogous to the photon self-energy. With the contributions from tree- and loop-level terms (see Figure 12.8), we have:

$$
\begin{aligned}
\tilde{S}_F(p) &= \frac{i\left(\not{p}+m_0\mathbb{1}\right)}{p^2-m_0^2} + \frac{i\left(\not{p}+m_0\mathbb{1}\right)}{p^2-m_0^2}\times\int\frac{\mathrm{d}^4k}{(2\pi)^4}\left[\left(-ie_0\gamma^\alpha\right)\frac{i\left(\not{p}-\not{k}+m_0\mathbb{1}\right)}{(p-k)^2-m_0^2}\left(-ie_0\gamma^\beta\right)\frac{-ig_{\alpha\beta}}{k^2}\right]\times\frac{i\left(\not{p}+m_0\mathbb{1}\right)}{p^2-m_0^2} \\
&\equiv \frac{i\left(\not{p}+m_0\mathbb{1}\right)}{p^2-m_0^2} + \frac{i\left(\not{p}+m_0\mathbb{1}\right)}{p^2-m_0^2}\times ie_0^2\Sigma_2(p)\times\frac{i\left(\not{p}+m_0\mathbb{1}\right)}{p^2-m_0^2}
\end{aligned}
\tag{12.69}
$$

where we introduced the amputated two-point term $\Sigma_2(p)$:

$$
\Sigma_2(p) \equiv i\int\frac{\mathrm{d}^4k}{(2\pi)^4}\gamma^\alpha\frac{\not{p}-\not{k}+m_0\mathbb{1}}{(p-k)^2-m_0^2+i\epsilon}\gamma_\alpha\frac{1}{k^2+i\epsilon}
\tag{12.70}
$$

This function corresponds to the diagram in Figure 12.8(right) without the incoming and outgoing fermion

Figure 12.8 Electron self-energy: (left) bare fermion propagator; (right) first-order photon loop.

legs. As we might have expected, the integral is ultraviolet divergent and must be regularized in the very high-energy behavior. We define the new function $\Sigma_2(p,\Lambda)$ with the high-energy cutoff Λ, where $\Sigma_2(p,\Lambda)\to\Sigma_2(p)$ when $\Lambda\to\infty$. For example (assuming that k^2 means $k^2+i\epsilon$):

$$
\Sigma_2(p,\Lambda) = i\int\frac{\mathrm{d}^4k}{(2\pi)^4}\gamma^\alpha\frac{\not{p}-\not{k}+m_0\mathbb{1}}{(p-k)^2-m_0^2}\gamma_\alpha\frac{1}{k^2}\left[\frac{\Lambda^2}{k^2+\Lambda^2}\right]
\tag{12.71}
$$

This is a highly non-trivial integration over all possible four-momenta k^μ of the virtual photon! In addition, we are in the presence of products of γ matrices. On Lorentz-covariance grounds, we can expand $\Sigma(p, \Lambda)$ (and also $\Sigma_2(p, \Lambda)$) as a linear combination of the identity $\mathbb{1}$ and powers of $(\not{p})^n$ matrices – compare with Eq. (12.21). We can decide to expand around the point $\not{p} - m_0$:

$$\Sigma(p, \Lambda) = A_0(\Lambda) + (\not{p} - m_0)A_1(\Lambda) + (\not{p} - m_0)^2 A_2(p, \Lambda) \tag{12.72}$$

where A_0, A_1, and A_2 do not contain γ matrices. In the sandwich of $\Sigma(p, \Lambda)$ with the free-particle spinors $\bar{u}(p)$ and $u(p)$, we get, recognizing the Dirac equation:

$$\bar{u}(p)\Sigma(p, \Lambda)u(p) = \bar{u}(p)A_0(\Lambda)u(p) + 0 + 0 \tag{12.73}$$

We found that the self-energy is equivalent to adding a mass term correction, since the mass in the Dirac equation has the form $m\overline{\Psi}\Psi$. We can therefore obtain the **physical interpretation that the bare mass m_0 will be affected by the electromagnetic interactions** surrounding the charged fermion:

$$m = m_0 + \delta m = m_0 + \bar{u}(p)A_0(\Lambda)u(p) \tag{12.74}$$

This result was first obtained by **V. F. Weisskopf**[3] in 1939 (using old-fashioned perturbation theory). He showed that the self-energy of the free Dirac fermion is positive and logarithmically dependent on the cutoff Λ [95]. We outline in the following the derivation of this result. First, we simplify the numerator of $\Sigma_2(p, \Lambda)$, sandwiched between the spinors $\bar{u}(p)$ and $u(p)$:

$$
\begin{aligned}
\bar{u}(p)\gamma^\alpha \left(\gamma^\beta(p - k)_\beta + m_0\mathbb{1}\right)\gamma_\alpha u(p) &= \bar{u}(p)\left(\gamma^\alpha\gamma^\beta(p - k)_\beta\gamma_\alpha + \gamma^\alpha m_0\gamma_\alpha\right)u(p) \\
&= \bar{u}(p)\left[(2g^{\alpha\beta} - \gamma^\beta\gamma^\alpha)(p - k)_\beta\gamma_\alpha + 4m_0\mathbb{1}\right]u(p) \\
&= \bar{u}(p)\left[2(p - k)_\beta\gamma^\beta - \gamma^\beta\gamma^\alpha(p - k)_\beta\gamma_\alpha + 4m_0\mathbb{1}\right]u(p) \\
&= \bar{u}(p)\left[-2(\not{p} - \not{k}) + 4m_0\mathbb{1}\right]u(p) \\
&= \bar{u}(p)\left[2\not{k} + 2m_0\mathbb{1}\right]u(p) \tag{12.75}
\end{aligned}
$$

where in the last line we have used Dirac's equation to replace $\not{p}$ with m_0. The denominator is simply:

$$(p - k)^2 - m_0^2 = p^2 - 2p \cdot k + k^2 - m_0^2 = -2p \cdot k + k^2 \tag{12.76}$$

and finally, we can express the Λ cutoff as an integral by noting:

$$\int_0^{\Lambda^2} \mathrm{d}t \frac{1}{(k^2 + t)^2} = \frac{1}{k^2}\left[\frac{\Lambda^2}{k^2 + \Lambda^2}\right] \tag{12.77}$$

Putting all the bits together, we obtain:

$$(ie_0^2)\bar{u}(p)\Sigma_2(p, \Lambda)u(p) = -2e_0^2 \int \frac{\mathrm{d}^4k}{(2\pi)^4} \int_0^{\Lambda^2} \mathrm{d}t \frac{\not{k} + m_0\mathbb{1}}{(-2p \cdot k + k^2)((k^2 + t)^2)} \tag{12.78}$$

We now make use of the trick known as the **Feynman parametrization method** to evaluate the integral (see Appendix F.4). The useful relation is Eq. (F.21) in Appendix F. Setting $a = k^2 + t$ and $b = (-2p \cdot k + k^2)$ we can write:

$$
\begin{aligned}
(ie_0^2)\bar{u}(p)\Sigma_2(p, \Lambda)u(p) &= -2e_0^2\bar{u}(p) \int \frac{\mathrm{d}^4k}{(2\pi)^4} \int_0^{\Lambda^2} \mathrm{d}t \int_0^1 \mathrm{d}x \frac{2x(\not{k} + m_0\mathbb{1})}{((k^2 + t)x + (-2p \cdot k + k^2)(1 - x))^3} u(p) \\
&= -e_0^2\bar{u}(p) \int_0^1 \mathrm{d}x \int_0^{\Lambda^2} \mathrm{d}t \int \frac{\mathrm{d}^4k}{(2\pi)^4} \frac{4x(\not{k} + m_0\mathbb{1})}{(tx - 2p \cdot k(1 - x) + k^2 + i\epsilon)^3} u(p) \tag{12.79}
\end{aligned}
$$

---

3   Victor Frederick "Viki" Weisskopf (1908–2002), Austrian-born American theoretical physicist.

where in the last line we recalled that k^2 means $k^2 + i\epsilon$. The **Feynman integrals** found in Appendix F are now readily applicable. Making use of Eqs. (F.26) and (F.27) to perform the integration over d^4k, we immediately have:

$$
\begin{aligned}
(ie_0^2)\bar{u}(p)\Sigma_2(p,\Lambda)u(p) &= -e_0^2\bar{u}(p)\int_0^1 dx \int_0^{\Lambda^2} dt\, \frac{4x(i\not{p}(1-x)+im_0\mathbb{1})}{32\pi^2(tx-p^2(1-x)^2)}u(p) \\
&= -(ie_0^2)\bar{u}(p)\int_0^1 dx \int_0^{\Lambda^2} dt\, \frac{4x(m_0(1-x)+m_0)}{32\pi^2(tx+m_0^2(1-x)^2)}u(p) \\
&= -\frac{ie_0^2 m_0}{8\pi^2}\int_0^1 dx \int_0^{\Lambda^2} dt\, \frac{x(2-x)}{tx+m_0^2(1-x)^2}\bar{u}(p)u(p)
\end{aligned}
\tag{12.80}
$$

Now, we should integrate over t to find:

$$
\begin{aligned}
\int_0^{\Lambda^2} dt\, \frac{x(2-x)}{tx+m_0^2(1-x)^2} &= (2-x)\ln\left(\frac{\Lambda^2 x+m_0^2(1-x)^2}{m_0^2(1-x)^2}\right) \approx (2-x)\ln\left(\frac{\Lambda^2 x}{m_0^2(1-x)^2}\right) \\
&= (2-x)\left[\ln\left(\frac{\Lambda^2}{m_0^2}\right)+\ln\left(\frac{x}{(1-x)^2}\right)\right]
\end{aligned}
\tag{12.81}
$$

where in the second approximation we assume $\Lambda^2 \gg m_0^2$. The integration over x yields:

$$
\int_0^1 dx(2-x)\left[\ln\left(\frac{\Lambda^2}{m^2}\right)+\ln\left(\frac{x}{(1-x)^2}\right)\right] = \frac{3}{4}\left(2\ln\left(\frac{\Lambda^2}{m_0^2}\right)+1\right)
\tag{12.82}
$$

In order to estimate the divergent electron self-energy contribution, we only keep the Λ^2 dependence. So, finally, we have found that:

$$
\delta m = e_0^2 \Sigma(\Lambda) \simeq \frac{e_0^2}{8\pi^2}\frac{3m_0}{2}\ln\left(\frac{\Lambda^2}{m_0^2}\right) = \left(\frac{3\alpha}{2\pi}\right)m_0\ln\left(\frac{\Lambda}{m_0}\right)
\tag{12.83}
$$

Since the correction is ultraviolet-divergent, it needs to be renormalized with a procedure similar to that of the bare elementary electric charge e_0. The important term is the factor $3\alpha/2\pi$.

12.8 QED Vertex Corrections

We consider the full QED vertex with contributions from tree- and loop-level terms. The correction diagrams are shown in Figure 12.9. This case is rather more complicated than the previous ones, since more diagrams will contribute. Note that the leftmost and rightmost diagrams correspond to self-energy corrections to the fermion mass, so we have already included them in the renormalization of the bare mass. The middle diagram is the one that we need to consider, which connects the incoming with the outgoing fermion legs (before and after the vertex). It is clear that this will change the kinematics of the fermion at the photon vertex, hence is a real correction. The QED full vertex can be expressed as:

$$
\bar{u}(p')\Gamma^\mu u(p) = \bar{u}(p')\gamma^\mu u(p) + \bar{u}(p')\delta\Gamma^\mu u(p)
\tag{12.84}
$$

where

$$
\bar{u}(p')\delta\Gamma^\mu u(p) = \int \frac{d^4k}{(2\pi)^4}\frac{-ig_{\nu\rho}}{(k-p)^2}\bar{u}(p')(-ie\gamma^\nu)\frac{i(\not{k}'+m)}{(k'^2-m^2)}\gamma^\mu \frac{i(\not{k}+m)}{(k^2-m^2)}(-ie\gamma^\rho)u(p)
\tag{12.85}
$$

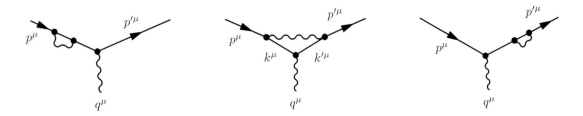

Figure 12.9 QED vertex corrections: (left) and (right) these contributions are already included in the electron self-energy corrections; (middle) vertex correction to $\mathcal{O}(e_0^3)$. In this case, we note that $q = p' - p = k' - k$

The general form of $\delta\Gamma^\mu$ is constrained to be of the type of Eq. (10.175). Now the **Gordon identity**, Eq. (10.171), lets us write the tree-level vector current in the following way:

$$(ie)\bar{u}(p')\gamma^\mu u(p) = \frac{ie}{2m}\bar{u}(p')\left[(p'+p)^\mu + i\sigma^{\mu\nu}(p'-p)_\nu\right]u(p) \tag{12.86}$$

The electron vertex, now including higher orders, can be conveniently expressed in a similar fashion:

$$(ie)\bar{u}(p')\Gamma^\mu u(p) \equiv (ie)\bar{u}(p')\left[\gamma^\mu F_1(q^2) + \frac{i}{2m}\sigma^{\mu\nu}q_\nu F_2(q^2)\right]u(p) \tag{12.87}$$

where $q = p' - p = k' - k$ and the $F_1(q^2)$ and $F_2(q^2)$ functions describe the structure of the electron. For tree level only, we would set $F_1(q^2) = 1$ and $F_2(q^2) = 0$ to recover the vertex factor.

We can rearrange Eq. (12.85) and use $k' = q + k$ to get:

$$\bar{u}(p')\delta\Gamma^\mu u(p) = 2ie^2 \int \frac{\mathrm{d}^4 k}{(2\pi)^4} \frac{\bar{u}(p')\left[\slashed{k}\gamma^\mu(\slashed{q}+\slashed{k}) + m^2\gamma^\mu - 2m(2k+q)^\mu\right]u(p)}{(k-p)^2((q+k)^2-m^2)(k^2-m^2)} \tag{12.88}$$

where we have used $\gamma^\nu\gamma^\mu\gamma_\nu = -2\gamma^\mu$ (Appendix E.1). Here we are not interested in the modifications to F_1, which have already been discussed in the context of the charge renormalization, rather we focus on the vertex diagrams which modify F_2. The numerator from the integrand is reworked in the following way to get the relevant term:

$$\begin{aligned}
\slashed{k}\gamma^\mu(\slashed{q}+\slashed{k}) + m^2\gamma^\mu - 2m(2k+q)^\mu &= \slashed{k}\gamma^\mu\slashed{q} - 2mq^\mu + \slashed{k}\gamma^\mu\slashed{k} + m^2\gamma^\mu - 4mk^\mu \\
&= 2\slashed{k}q^\mu - \slashed{k}\slashed{q}\gamma^\mu - 2mq^\mu + 2\slashed{k}k^\mu - \slashed{k}\slashed{k}\gamma^\mu + m^2\gamma^\mu - 4mk^\mu
\end{aligned} \tag{12.89}$$

The denominator is the product of three terms. We must employ a trick in order to combine it into a single term. We use Eq. (F.23) from Appendix F, setting $a = (k-p+i\epsilon)^2$, $b = (q+k+i\epsilon)^2 - m^2$, and $c = k^2 - m^2 + i\epsilon$:

$$\frac{1}{(k-p+i\epsilon)^2((q+k)^2-m^2+i\epsilon)(k^2-m^2+i\epsilon)} = \int_0^1 \mathrm{d}x_1\mathrm{d}x_2\mathrm{d}x_3\delta(x_1+x_2+x_3-1)\frac{2}{\Delta^3} \tag{12.90}$$

where with $x_1 + x_2 + x_3 = 1$, we have:

$$\begin{aligned}
\Delta = x_1 a + x_2 b + x_3 c + i\epsilon &= x_1(k-p)^2 + x_2(q^2 + 2q\cdot k + k^2 - m^2) + x_3(k^2 - m^2) + i\epsilon \\
&= k^2 - 2k\cdot(x_1 p - x_2 q) + x_1 p^2 + x_2 q^2 - (x_2 + x_3)m^2 + i\epsilon
\end{aligned} \tag{12.91}$$

The integral over k is now shifted by completing the square in k by defining $\ell^\mu \equiv k^\mu - x_1 p^\mu + x_2 q^\mu$, so that (see Ref. [96]):

$$
\begin{aligned}
\Delta &= \ell^2 - (x_1 p - x_2 q)^2 + x_1 p^2 + x_2 q^2 - (x_2 + x_3) m^2 + i\epsilon \\
&= \ell^2 - (x_1^2 m^2 + x_2^2 q^2 - 2 x_1 x_2 p \cdot q) + x_1 m^2 + x_2 q^2 - (1 - x_1) m^2 + i\epsilon \\
&= \ell^2 - x_2^2 q^2 + 2 x_1 x_2 (p' - q) \cdot q + x_2 q^2 - (1 - 2 x_1 + x_1^2) m^2 + i\epsilon \\
&= \ell^2 + x_2 (-x_1 + x_3) q^2 - (1 - x_1)^2 m^2 + 2 x_1 x_2 p' \cdot q + i\epsilon \\
&= \ell^2 + x_2 x_3 q^2 - (x_2 + x_3)^2 m^2 + x_1 x_2 \underbrace{(2 p \cdot q + q^2)}_{=(p+q)^2 - p^2 = p'^2 - p^2 = 0} + i\epsilon \\
&\equiv \ell^2 - M^2 + i\epsilon
\end{aligned}
\tag{12.92}
$$

where $M^2 \equiv -x_2 x_3 q^2 + (x_2 + x_3)^2 m^2$ is positive if q^2 is negative and $|q^2| > m^2$. **Therefore, we have achieved the goal to shrink the three terms in the denominator** $(k - p)^2 ((q + k)^2 - m^2)(k^2 - m^2)$ **into a single term** $\ell^2 - M^2$. This term behaves like a pole in the propagator in the ℓ^2 momentum and can hence be interpreted as a mass term. With the shifted momentum, the vertex factor becomes (see Ref. [96] for details):

$$
\begin{aligned}
\bar{u}(p') \delta \Gamma^\mu u(p) &= 2 i e^2 \int \frac{\mathrm{d}^4 \ell}{(2\pi)^4} \int_0^1 \mathrm{d}x_1 \mathrm{d}x_2 \mathrm{d}x_3 \delta(x_1 + x_2 + x_3 - 1) \times \frac{2}{(\ell^2 - M^2 + i\epsilon)^3} \\
&\quad \bar{u}(p') \left[\gamma^\mu \left(-\frac{1}{2} \ell^2 + (1 - x_2)(1 - x_3) q^2 + (1 - 2 x_1 - x_1^2) m^2 \right) + (p + p')^\mu m x_1 (x_1 - 1) \right. \\
&\quad \left. + q^\mu m (x_1 - 2)(x_3 - x_2) \right] u(p)
\end{aligned}
\tag{12.93}
$$

We can see that the integral of the third term proportional to q^μ is antisymmetric under the interchange of x_2 and x_3, while the denominator Δ is symmetric under $x_2 \leftrightarrow x_3$. Hence the integral of this term must vanish. The other two terms can be rewritten with the help of the Gordon identity in order to obtain the terms proportional to γ^μ and $\sigma^{\mu\nu}$. One finally finds [96]:

$$
\begin{aligned}
\bar{u}(p') \delta \Gamma^\mu u(p) &= 2 i e^2 \int \frac{\mathrm{d}^4 \ell}{(2\pi)^4} \int_0^1 \mathrm{d}x_1 \mathrm{d}x_2 \mathrm{d}x_3 \delta(x_1 + x_2 + x_3 - 1) \times \frac{2}{(\ell^2 - M^2 + i\epsilon)^3} \\
&\quad \bar{u}(p') \left[\gamma^\mu \left(-\frac{1}{2} \ell^2 + (1 - x_2)(1 - x_3) q^2 + (1 - 4 x_1 + x_1^2) m^2 \right) \right. \\
&\quad \left. + \frac{i \sigma^{\mu\nu} q_\nu}{2m} \left(2 m^2 x_1 (1 - x_1) \right) \right] u(p)
\end{aligned}
\tag{12.94}
$$

Comparing with Eq. (12.87) we now have explicit expressions for $F_1(q^2)$ and $F_2(q^2)$. For F_2 we obtain:

$$
F_2(q^2) = 8 i e^2 m^2 \int \frac{\mathrm{d}^4 \ell}{(2\pi)^4} \int_0^1 \mathrm{d}x_1 \mathrm{d}x_2 \mathrm{d}x_3 \delta(x_1 + x_2 + x_3 - 1) \frac{x_1 (1 - x_1)}{(\ell^2 - M^2 + i\epsilon)^3}
\tag{12.95}
$$

This integral over ℓ is finite – there is no UV divergence so it can be evaluated directly without the need for a cutoff! Its value is found in Appendix F by setting $\Delta = M^2$ in Eq. (F.24), so we get:

$$
\begin{aligned}
F_2(q^2) &= 8 e^2 m^2 \int_0^1 \mathrm{d}x_1 \mathrm{d}x_2 \mathrm{d}x_3 \delta(x_1 + x_2 + x_3 - 1) \frac{x_1 (1 - x_1)}{32 \pi^2 M^2} \\
&= \frac{e^2 m^2}{4 \pi^2} \int_0^1 \mathrm{d}x_1 \mathrm{d}x_2 \mathrm{d}x_3 \delta(x_1 + x_2 + x_3 - 1) \frac{x_1 (1 - x_1)}{-x_2 x_3 q^2 + (1 - x_1)^2 m^2}
\end{aligned}
\tag{12.96}
$$

We can now use the constraint that $x_1 + x_2 + x_3 = 1$ to remove the x_2 integral and introduce the fine-structure constant:

$$
F_2(q^2) = \frac{\alpha m^2}{\pi} \int_0^1 \mathrm{d}x_1 \int_0^{1 - x_1} \mathrm{d}x_3 \frac{x_1 (1 - x_1)}{-(1 - x_1 - x_3) x_3 q^2 + (1 - x_1)^2 m^2}
\tag{12.97}
$$

For $q^2 \to 0$ we get the correction to the static magnetic moment of the electron, as will be discussed in Chapter 14:

$$F_2(q^2 \to 0) \approx \frac{\alpha}{\pi} \int_0^1 \mathrm{d}x_1 \int_0^{1-x_1} \mathrm{d}x_3 \frac{x_1}{(1-x_1)} = \frac{\alpha}{\pi} \int_0^1 \mathrm{d}x_1 x_1 = \frac{\alpha}{2\pi} \tag{12.98}$$

This result was first derived by **Schwinger** in 1948 [97]. Finally, we have found that the QED vertex corrections of Eq. (12.88) can be embedded in the structure functions F_1 and F_2 in the following way:

$$F_1(q^2) = 1 + \mathcal{O}(\alpha, q^2) \quad \text{and} \quad F_2(q^2) = \frac{\alpha}{2\pi} + \mathcal{O}(\alpha, q^2) \tag{12.99}$$

The first-order radiatively corrected vertex can be expressed as:

$$(ie)\bar{u}(p')\left[\gamma^\mu(1 + \mathcal{O}(\alpha, q^2)) + \frac{i}{2m}\frac{\alpha}{2\pi}\sigma^{\mu\nu}q_\nu\right]u(p) \tag{12.100}$$

which, using the Gordon identity, leads to the following result:

$$(ie)\bar{u}(p')\left[\frac{1}{2m}(p'+p)^\mu + \frac{i}{2m}\left(1 + \frac{\alpha}{2\pi}\right)\sigma^{\mu\nu}q_\nu + \gamma^\mu\mathcal{O}(\alpha, q^2))\right]u(p) \tag{12.101}$$

We will discuss further the implications of the second term in the bracket in Chapter 14.

12.9   Dimensional Regularization

When we derived higher-order QED diagrams, we encountered divergences for instance in the loop diagrams and we had to find an "ad-hoc" way to treat these problems. In general, momentum integration of higher-order terms is not convergent in four-dimensional space-time. Dimensional regularization, first introduced by **'t Hooft**[4] and **Veltman** [98], is to replace four-dimensional loop integrals by "D-dimensional" ones in order to make the divergence explicit. The advantage of dimensional regularization over the cutoff scheme discussed in the previous sections is that it preserves gauge invariance. The idea is that by lowering the dimensionality, one can effectively reduce the divergence, since:

$$\underbrace{\int \mathrm{d}^4 p \frac{1}{p^2} \propto \int p^3 \mathrm{d}p \frac{1}{p^2} \propto p^2}_{\text{quadratic divergence}}, \quad \underbrace{\int \mathrm{d}^3 p \frac{1}{p^2} \propto p}_{\text{linear divergence}}, \quad \underbrace{\int \mathrm{d}^2 p \frac{1}{p^2} \propto \ln p}_{\text{log divergence}}, \quad \underbrace{\int \mathrm{d}p \frac{1}{p^2} \propto \frac{1}{p}}_{\text{convergence}} \tag{12.102}$$

It is rather hard to imagine D-dimensional space; however, we can construct a formal calculus in D-dimensional Minkoswki space-time (one dimension is time and $D-1$ are Euclidian space dimensions). The metric $g^{\mu\nu}$ becomes D-dimensional:

$$\text{metric } g^{\mu\nu}: \quad \mu, \nu = 0, 1, \ldots, D-1 \quad \text{and} \quad g^{\mu\nu}g_{\mu\nu} = D \tag{12.103}$$

The Dirac matrices are also D-dimensional and must obey the Clifford algebra. The following relations are maintained:

$$\{\gamma^\mu, \gamma^\nu\} = 2g^{\mu\nu}\mathbb{1}, \quad \text{Tr } \mathbb{1} = D \tag{12.104}$$

and the indices are D-dimensional. Then, for instance:

$$\gamma^\mu\gamma_\mu = D\mathbb{1}, \quad \gamma^\mu\gamma_\nu\gamma_\mu = 2g_{\mu\nu}\gamma^\mu - \gamma^\mu\gamma_\mu\gamma_\nu = (2-D)\gamma_\nu \tag{12.105}$$

4   Gerard 't Hooft (born 1946), Dutch theoretical physicist.

and similarly:

$$\gamma^\mu \slashed{a} \gamma_\mu = (2-D)\slashed{a}, \quad \gamma^\mu \slashed{a}\slashed{b}\gamma_\mu = 4(a \cdot b)\mathbb{1} + (D-4)\slashed{a}\slashed{b}, \quad \gamma^\mu \slashed{a}\slashed{b}\slashed{c}\gamma_\mu = -2\slashed{c}\slashed{b}\slashed{a} - (D-4)\slashed{a}\slashed{b}\slashed{c} \tag{12.106}$$

The D-dimensional integral can be written as:

$$\int d^D p (\cdots) \tag{12.107}$$

All algebraic manipulations, including loops, are those of convergent integrals when $D < 4$, and only after the integrations, one takes the limit $D \to 4$. The divergences appear then as poles in $1/(D-4)$. One often introduces the parameter ϵ such that:[5]

$$D \equiv 4 - 2\epsilon \quad \Longrightarrow \quad \epsilon \equiv (4-D)/2 \tag{12.108}$$

Like in any other regularization scheme, dimensional regularization introduces an arbitrary energy (or mass) scale, usually denoted μ. If we consider, for instance, the one-loop integral, one multiplies any quantity by an appropriate D-dependent power of μ such that the energy dimension of the result is independent of D! The result of the integration will be μ-dependent and one will have to rely, as usual, on a renormalization scheme in order to remove this dependence. In general, the artificial energy scale μ can be expressed as a modification of the coupling constant with a D-dependent dimension.

12.10   Photon Self-energy Revisited

In this section we revisit the calculation of the photon self-energy and illustrate the use of the dimensional regularization. We start from Eq. (12.23) and note that the denominator is composed of the product of two terms. As in the case of the vertex correction, we introduce the Feynman parameter x and shift the integrating momentum to $\ell^\mu \equiv p^\mu + x k^\mu$. We can then express:

$$(p^2 - m^2)((k+p)^2 - m^2) = (\ell^2 - \Delta)^2 \tag{12.109}$$

where

$$\Delta \equiv m^2 - x(1-x)k^2 \tag{12.110}$$

Now the numerator $p^\alpha(k+p)^\beta + p^\beta(k+p)^\alpha - (p \cdot (k+p) - m^2)g^{\alpha\beta}$ should be rewritten as a function of ℓ^μ to find:

$$\Pi^{\alpha\beta}(k,m) = i \int_0^1 dx \int \frac{d^4\ell}{(2\pi)^4} \frac{2\ell^\alpha \ell^\beta - g^{\alpha\beta}\ell^2 - 2x(1-x)k^\alpha k^\beta + g^{\alpha\beta}(m^2 + x(1-x)k^2)}{(\ell^2 - \Delta)^2} \tag{12.111}$$

where the terms proportional to ℓ^μ have been removed since they give a vanishing integral. This integral will diverge in the UV regime! We can cure this divergence by generalizing it to $D = 4 - 2\epsilon$ dimensions:

$$e_0^2 \Pi^{\alpha\beta}(k,m) = ie_0^2 \mu^{2\epsilon} \int_0^1 dx \int \frac{d^D\ell}{(2\pi)^D} \frac{2\ell^\alpha \ell^\beta - g^{\alpha\beta}\ell^2 - 2x(1-x)k^\alpha k^\beta + g^{\alpha\beta}(m^2 + x(1-x)k^2)}{(\ell^2 - \Delta)^2} \tag{12.112}$$

where the $\mu^{2\epsilon}$ term has been added to make the units of the result independent of D. The first two terms of the numerator depend on ℓ^μ while the last two, do not. We treat them separately. We have:

$$\begin{aligned}
\mathcal{A} &= \int \frac{d^D\ell}{(2\pi)^D} \frac{2\ell^\alpha \ell^\beta - g^{\alpha\beta}\ell^2}{(\ell^2 - \Delta)^2} &&= \int \frac{d^D\ell}{(2\pi)^D} \frac{\frac{2}{D}\ell^2 g^{\alpha\beta} - g^{\alpha\beta}\ell^2}{(\ell^2 - \Delta)^2} = \int \frac{d^D\ell}{(2\pi)^D} \frac{\left(\frac{2}{D} - 1\right)\ell^2 g^{\alpha\beta}}{(\ell^2 - \Delta)^2} \\[2mm]
&= i\frac{(-1)^{2-1}}{2^D \pi^{D/2}} \frac{D}{2} \frac{\Gamma(2 - D/2 - 1)}{\Gamma(2)} \frac{\left(\frac{2}{D} - 1\right)g^{\alpha\beta}}{\Delta^{2-D/2-1}} \\[2mm]
&= \frac{i}{16\pi^2} \frac{\Gamma(2 - D/2)}{1 - D/2} \frac{g^{\alpha\beta}\Delta}{\Delta^{2-D/2}} = \frac{-i}{16\pi^2} \frac{\Gamma(2-D/2)}{\Delta^{2-D/2}} g^{\alpha\beta}\Delta \tag{12.113}
\end{aligned}$$

5   One also finds $D \equiv 4 - \epsilon$ in the literature.

where in the second line we used the result Eq. (F.38) and also $\Gamma(1-D/2) = \Gamma(2-D/2)/(1-D/2)$. For the last two terms we simply use Eq. (F.34) and get:

$$\mathcal{B} = \int \frac{d^D\ell}{(2\pi)^D} \frac{1}{(\ell^2 - \Delta)^2} = \frac{i(-1)^2}{2^D \pi^{D/2}} \frac{\Gamma(2-D/2)}{\Gamma(2)} \frac{1}{\Delta^{2-D/2}} = \frac{i}{16\pi^2} \frac{\Gamma(2-D/2)}{\Delta^{2-D/2}} \tag{12.114}$$

Then

$$\mathcal{A} + \mathcal{B}\left(-2x(1-x)k^\alpha k^\beta + g^{\alpha\beta}(m^2 + x(1-x)k^2)\right)$$

$$= \frac{-i}{16\pi^2} \frac{\Gamma(2-D/2)}{\Delta^{2-D/2}} \times \left(g^{\alpha\beta}(m^2 - x(1-x)k^2) + 2x(1-x)k^\alpha k^\beta - g^{\alpha\beta}(m^2 + x(1-x)k^2)\right)$$

$$= \frac{-i}{16\pi^2} \frac{\Gamma(2-D/2)}{\Delta^{2-D/2}} \times \left(2x(1-x)k^\alpha k^\beta - 2g^{\alpha\beta}(x(1-x)k^2)\right)$$

$$= \frac{i}{16\pi^2} \frac{\Gamma(2-D/2)}{\Delta^{2-D/2}} \times 2x(1-x)\left(k^2 g^{\alpha\beta} - k^\alpha k^\beta\right) \tag{12.115}$$

Consequently

$$e^2 \Pi^{\alpha\beta}(k,\mu) = ie^2 \mu^{2\epsilon} \int_0^1 dx \frac{i}{16\pi^2} \frac{\Gamma(2-D/2)}{\Delta^{2-D/2}} \times 2x(1-x)\left(k^2 g^{\alpha\beta} - k^\alpha k^\beta\right) \tag{12.116}$$

We want to estimate the effect on the bare charge and consider only the $k^2 g^{\alpha\beta}$ term (we ignore the $k^\alpha k^\beta$ term which, as before, vanishes by current conservation). Note that Δ depends on x. The divergence of the result is embedded in the Gamma function and its denominator. We use Eq. (F.35) to expand the behavior near $D = 4$:

$$e^2 \Pi^{\alpha\beta}(k,\mu) \approx -\frac{e^2}{2\pi^2} \mu^{2\epsilon} k^2 g^{\alpha\beta} \int_0^1 dx\, x(1-x)\left(\frac{1}{\epsilon} - \ln\Delta - \gamma_E + \ln(4\pi)\right) \tag{12.117}$$

Finally, we recover the fact that our correction to the bare charge e diverges as $1/\epsilon$! But this is not what we want. After renormalization, we are interested in the k^2 dependence of our result. Hence, we consider:

$$\tilde{\Pi}^{\alpha\beta}(k) \equiv \Pi^{\alpha\beta}(k) - \Pi^{\alpha\beta}(0) \overset{\epsilon\to 0}{=} -\frac{1}{2\pi^2} k^2 g^{\alpha\beta} \int_0^1 dx\, x(1-x) \ln\left(\frac{m^2}{m^2 - x(1-x)k^2}\right) \tag{12.118}$$

where we recovered the result derived in Eq. (12.41) with the Pauli–Villars regularization ☐. Mind-blowing!

Problems

Ex 12.1 Running coupling. The coupling constant given in Eq. (12.57) is found to be scale-dependent.

(a) Prove that for any measurement, the parameter μ in this equation is arbitrary. To achieve this, assume two experiments using different values μ_a and μ_b and that Eq. (12.57) holds for one of them. Then show that it will necessarily hold for the other as well.

(b) Usually the running of the coupling is described using α in the low-energy limit (since the value is just given by the well-known fine-structure constant in this regime). Which value is usually chosen for μ^2 in this case? Explain where this choice originates from.

(c) The running coupling has a singularity at a certain energy, which is called the Landau pole. Determine the energy scale at which this happens. Does this result carry physical meaning? Is the scale at which the Landau pole appears accessible with current accelerator technology?

(d) Calculate the energy scales at which you expect the coupling to be 1%, 10%, and 100% stronger than $\alpha(0) \approx 1/137$, respectively. Are these scales accessible by current accelerator technology?

Ex 12.2 Light scattering by light. Compute the amplitude and cross-section for the process of light scattering by light.

13 Tests of QED at High Energy

At a lepton collider, every event is a signal. At a hadron collider, every event is background.

Samuel C. C. Ting

13.1 Studying QED at Lepton Colliders

Electron–positron colliders and their associated high-performance detectors have provided excellent environments to test QED at increasingly higher center-of-mass energies. Table 3.1 presented a comprehensive list of e^+e^- colliders. Each new generation of collider, with their corresponding experiments, has provided excellent environments to study the properties of QED at high energy. A subset, most relevant for our present discussion, is reported in Table 13.1.

Accelerator	Where	Years	CMS energy $\sqrt{s}$ (GeV)	Experiments
SPEAR	SLAC	1972–1990	3–8	Mark I, Mark II, Mark III
DORIS	DESY	1974–1993	10	ARGUS, Crystal Ball, DASP, PLUTO
CESR	Cornell	1979–2003	12	CLEO
PETRA	DESY	1978–1986	12–46	JADE, MARK-J, CELLO, PLUTO, TASSO
PEP	SLAC	1980–1990	29	DELCO, HRS, MAC, MARK-2, TPC
TRISTAN	KEK	1986–1995	52–64	VENUS, SHIP, TOPAZ, AMY
SLC	SLAC	1988–1998	91	SLD
LEP	CERN	1989–2000	88–208	ALEPH, DELPHI, OPAL, L3

Table 13.1 List of electron–positron colliders and the corresponding experiments that recorded e^+e^- collisions at high energies.

A major experimental advantage of e^+e^- colliders over other types of accelerators is that they provide events in the center-of-mass frame. Events produced in e^+e^- collisions are also genuinely clean compared to, say, those in hadron collisions at proton–proton or proton–antiproton colliders. At an e^+e^- collider, essentially all events can be precisely identified, reconstructed and studied with high efficiency and high purity. In addition, the charge of the final-state particles can be identified and the angular dependence of the scattering cross-sections in $2 \to 2$ processes can be measured in a straightforward way. To illustrate some of these features, an example $e^+e^- \to$ hadrons candidate event recorded in the TASSO experiment at PETRA is shown in Figure 13.1.

Most experiments at e^+e^- colliders have been built as general multi-purpose detectors in order to exploit the large variety of physics questions. They had to be able to detect the different topologies of low-multiplicity

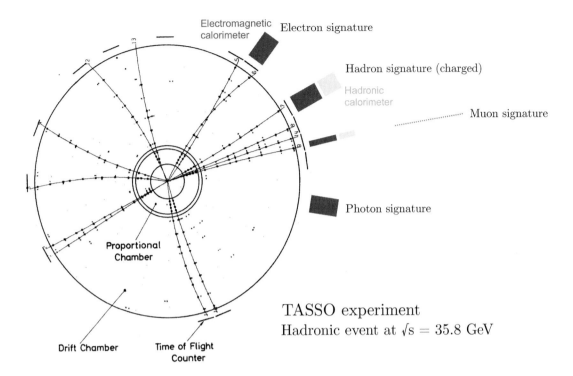

Figure 13.1 An $e^+e^- \to$ hadrons candidate event recorded in the TASSO experiment at the PETRA storage ring.

(leptons, photons) and high-multiplicity (hadrons) events and be ready for unexpected exotic reactions. This required hermetically closed "4π detectors" with momentum analysis of charged particles through bending in a magnetic field and particle identification of electrons, muons, and photons. The functions of the e^+e^- detectors were relatively clear, and for the most part are shared by all the detectors at the different colliders. We summarize below some of their main features in more detail:

- **Tracking.** The trackers provided a way of measuring points along the trajectory of charged particles. This permitted one to count the number of charged particles, to find the point of origin of the tracks, called vertices, to measure the direction of travel of charged particles, and to measure their momentum by the curvature of the trajectory in a magnetic field. Usually, tracking was provided over a large solid angle ($> 70\%$ of 4π). Some detectors placed more emphasis on the quality of the tracking than others. Most detectors gave special attention to precision reconstruction of production vertices of the particles.

- **Electromagnetic calorimetry.** While tracking afforded a measurement of charged particles, some means was needed to measure the kinematical properties of photons. The implementation of this function meant building a calorimeter which forced a conversion of the photon (or electron) into an electromagnetic cascade, so that the total energy of the particle could be measured (see Section 3.12).

- **Hadronic calorimetry.** In addition to the electromagnetic calorimeter, a hadronic calorimeter was needed to estimate the total energy of all the particles (except neutrinos). More material was used to reach several interaction lengths, so that all hadrons would cascade into lower-energy particles, eventually depositing all their energy.

- **Muon detectors.** Typically placed behind or around all the previous subdetectors, and hence, only being reached by very penetrating particles like muons, the muon detectors could either be simply used for identification or provide some kind of measurement if combined with a magnetic field.

- **Particle identification.** Several experiments had placed emphasis on identifying the charged particles. Knowing the identity helped greatly in trying to reconstruct decaying particles. Generally speaking, identifying charged, long-lived particles involves measuring both the momentum and velocity of the particle. Measuring the velocity was done with the use of Cherenkov counters, or specific ionization (dE/dx), or by the time of flight over a known distance.

A typical example is the JADE detector at PETRA, shown in Figure 13.2. Similarly, the three-dimensional view of L3, one of the four detectors at LEP, is shown later in Figure 13.10. The L3 detector is described in more detail in Section 13.7. The other LEP detectors are shown in Figure 27.14.

Figure 27.15 shows real events collected near the Z^0 pole at LEP, classified as $e^+e^- \to q\bar{q}$, $e^+e^- \to e^+e^-$, $e^+e^- \to \mu^+\mu^-$, and $e^+e^- \to \tau^+\tau^-$ final states, visualized with the event displays of the OPAL, DELPHI, L3, and ALEPH experiments. The front view of an event classified as $Z^0 \to b\bar{b}$ collected by the SLD detector at the SLC collider is shown in Figure 27.16. Due to the b-quark finite lifetime (see Section 20.9), a secondary decay point is to be expected. Indeed, a displaced secondary vertex, reconstructed thanks to the silicon vertex detector, is visible in the expanded side view (r–z view) of the beam interaction point. All these examples do show the relative "cleanliness" of the events recorded at e^+e^- colliders.

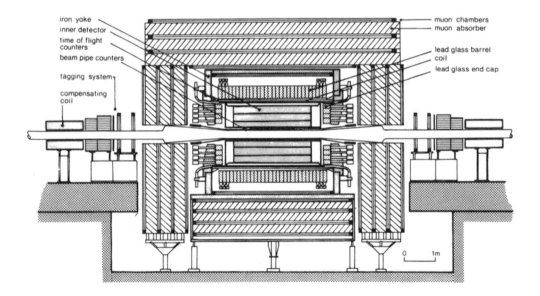

Figure 13.2 The JADE detector at the PETRA e^+e^- storage ring. Reprinted figure from W. Bartel *et al.*, "Total cross-section for hadron production by e^+e^- annihilation at PETRA energies," *Phys. Lett. B*, vol. 88, pp. 171–176. Copyright 1979, with permission from Elsevier.

13.2   Deviations from QED and Electroweak Effects before LEP/SLC

In the following, we will focus on the data collected by the experiments at the SPEAR, PEP, DORIS, PETRA, and TRISTAN colliders, which means before the advent of LEP and SLC. The discussion of the electroweak studies at LEP and SLC will be the main subject of Chapter 27. Here we are focusing on QED and are therefore considering the case of reactions in which the electroweak effects are expected to be small compared to the dominant electromagnetic interaction. Below is a list of the most important processes that were recorded with high statistics in this energy domain:

- **Bhabha scattering**: $e^+e^- \to e^+e^-$

- **Lepton pair production**: $e^+e^- \to \mu^+\mu^-, \tau^+\tau^-$ (or any new lepton pair, i.e., from a fourth generation)

- **Photon pair production**: $e^+e^- \to \gamma\gamma$

- **Hadrons through quark pair formation**: $e^+e^- \to q\bar{q} \to$ hadrons (see Section 19.2)

- **Fermion pair production through photon–photon scattering**: $e^+e^- \to e^+e^- f\bar{f}$

QED remains an internally consistent theory within the electroweak model, provided that its invariant quantities (the center-of-mass energy squared, the space-like momentum squared) are small relative to values at which the weak interaction becomes important. For e^+e^- collisions, this happens around the energy of the Z^0 mass. Hence, **at the PETRA, PEP and TRISTAN e^+e^- colliders, even at their highest energies, the conditions are such that QED is the dominating force, the purely weak contributions being at the percent level.** This criterion was not satisfied at LEP or SLC, which were built on purpose to test the **electroweak theory** (discussed in Chapter 25). An exception is the reaction $e^+e^- \to \gamma\gamma$, for which electroweak effects remained negligible even at the highest energies of LEP, as will be discussed in Section 13.8.

The unpolarized cross-sections for all basic QED processes involving e^+e^- initial states have been derived in Chapter 11. Since all the cross-sections were found to have a $1/s$ dependence, we can consider the normalized cross-sections by s, valid at any center-of-mass energy. The tree-level QED differential cross-section in the center-of-mass frame for $e^+e^- \to \mu^+\mu^-$ and $e^+e^- \to \tau^+\tau^-$ can be expressed as $s\frac{\mathrm{d}\sigma}{\mathrm{d}\cos\theta}$ (see Eq. (11.153) and noticing that the integration over $\mathrm{d}\phi$ gives a trivial 2π factor):

$$\text{QED:} \quad s\frac{\mathrm{d}\sigma}{\mathrm{d}\cos\theta}(e^+e^- \to \ell^+\ell^-) = \frac{\pi\alpha^2}{2}\left(1+\cos^2\theta\right) \tag{13.1}$$

Similarly, for the Bhabha cross-section, we have the following prediction (see Eq. (11.158)):

$$\text{QED:} \quad s\frac{\mathrm{d}\sigma}{\mathrm{d}\cos\theta}(e^+e^- \to e^+e^-) = \frac{\pi\alpha^2}{2}\frac{\left(3+\cos^2\theta\right)^2}{(1-\cos\theta)^2} \tag{13.2}$$

Finally, the $e^+e^- \to \gamma\gamma$ reaction has the following predicted differential cross-section (see Eq. (11.239)):

$$\text{QED:} \quad s\frac{\mathrm{d}\sigma}{\mathrm{d}\cos\theta}(e^+e^- \to \gamma\gamma) = \pi\alpha^2\frac{1+\cos^2\theta}{\sin^2\theta} = \pi\alpha^2\frac{1+\cos^2\theta}{1-\cos^2\theta} \tag{13.3}$$

These tree-level cross-sections are plotted in Figure 13.3. By definition, the scattering angle θ is measured between the direction of the positron beam and the direction of the scattered positively charged lepton. In the case of $e^+e^- \to \gamma\gamma$, the photons cannot be distinguished so the range is taken as $0 \leq \cos\theta \leq 1$. We see that the Bhabha $e^+e^- \to e^+e^-$ cross-section is always larger than the cross-section of the other reactions and the angular distribution increases steeply in the forward direction. The cross-section at $\cos\theta = +0.8$ is, for example, almost two orders of magnitude larger than at $\cos\theta = -0.8$.

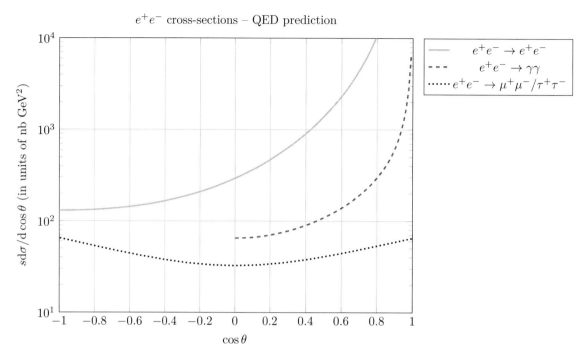

Figure 13.3 The center-of-mass energy squared normalized differential cross-sections $s\frac{\mathrm{d}\sigma}{\mathrm{d}\cos\theta}$ for $e^+e^- \to e^+e^-$, $e^+e^- \to \mu^-\mu^+$ or $\tau^-\tau^+$, and $e^+e^- \to \gamma\gamma$. Note the logarithmic vertical scale.

In general, measured cross-sections can be compared to the tree-level (Born) QED cross-section modified by radiative corrections and potential electroweak effects. These effects can be formally introduced as corrections to the QED cross-section, as follows:

$$\left(\frac{\mathrm{d}\sigma}{\mathrm{d}\Omega}\right)_{measured} = \left(\frac{\mathrm{d}\sigma}{\mathrm{d}\Omega}\right)_{QED,Born} (1 + \delta_{radcorr} + \delta_{electroweak}) \tag{13.4}$$

The radiative corrections $\delta_{radcorr}$ are calculable order by order. Typically, complete calculations for the e^+e^- processes have been performed up to $\mathcal{O}(\alpha^3)$, or even in some special cases up to $\mathcal{O}(\alpha^4)$, to be compared to experimental results. A convenient way to compare measurements with the lowest-order theory is to unfold the data by using the calculated radiative corrections, thus arriving at "lowest-order data." This method to present results was applied by almost all the experiments at PETRA, PEP, and TRISTAN. Below LEP and SLC energies, the electroweak corrections $\delta_{electroweak}$ were at percent level, unless one focused on the forward–backward (or charge) asymmetry, where the interference between the photon and the neutral weak intermediate boson Z^0 (whose existence was at the time not experimentally established) created an asymmetry at the level of 10%. This result will be discussed later in Section 13.5.

To summarize, lepton colliders provided very clean environments to perform measurements with very high precision (often limited by statistics). The precise study of these fundamental reactions provided a sensitive means to test for unexpected departures from QED at small distances. No significant differences from QED have been reported, aside from electroweak effects that eventually superseded the measurements at the LEP energies, as will be shown in the coming sections.

13.3   Bhabha Scattering ($e^+e^- \rightarrow e^+e^-$)

The Bhabha differential cross-section is dominated by large t-channel contributions and shows therefore a strong forward peaking. Bhabha scattering is thus extremely useful to determine with high rates the luminosity in e^+e^- experiments at very small angles ($\theta \simeq 30$ mrad). Since the effects of weak interactions vanish at small angles, it is possible to measure the luminosity independently of a precise knowledge of the values of the weak interaction parameters. The precision of the luminosity measurement is mainly limited by systematic errors created by the steepness of the angular distribution and by uncertainties in the knowledge of the angular acceptance of the detectors. The resulting systematic error of the cross-section, and thus of the luminosity, are at the level of 2–4% depending on the experiment.

The reaction $e^+e^- \rightarrow e^+e^-$ can be separated from other reactions such as $e^+e^- \rightarrow$ hadrons because electrons deposit their total energy in the electromagnetic calorimeter and by the low charged multiplicity of the event. The two-photon reaction $e^+e^- \rightarrow e^+e^-e^+e^-$ and the tau events with $e^+e^- \rightarrow \tau^-\tau^+ \rightarrow e^+e^- + \nu's$, can be removed from the data sample by acollinearity and momentum cuts, since both reactions show missing momentum from the undetected e^+e^- or neutrinos, respectively. The most difficult background for Bhabha scattering is the reaction $e^+e^- \rightarrow \gamma\gamma$. In principle, photons can be separated from electrons or positrons because they are neutral and do not leave a track in the drift chambers of a detector. However, photons can convert into an e^+e^- pair by interacting with the beam pipe or with additional material in front of the inner drift chambers and simulate an electron or a positron track. All these effects can be taken into account and finally handled on a statistical basis.

Figure 13.4 shows the angular distribution measured by MARK J at $\sqrt{s} = 34.6$ GeV. The angular acceptance of this detector covers the range from $12°$ to $168°$ and thus extends to relatively small angles. Because no attempt was made to distinguish electrons from positrons, the angular distribution is folded around $90°$ and therefore the cross-section is expressed as a function of $|\cos\theta|$. Note the very steep rise of the cross-section at small angles. The measured points agree very well with the QED prediction. In order to better display the

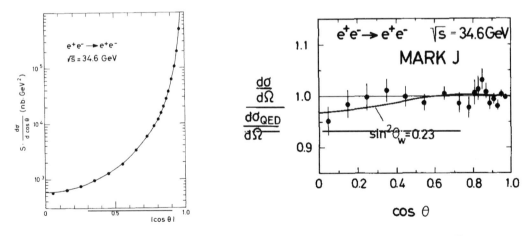

Figure 13.4 Bhabha cross-section measurements from MARK J at PETRA taken at $\sqrt{s} = 34.6$ GeV (left) differential cross-section in the central and forward regions; (right) ratio between measurements and QED predictions. Reprinted from B. Adeva *et al.*, "A summary of recent experimental results from Mark-J: High-energy e^+e^- collisions at PETRA," *Phys. Rept.*, vol. 109, p. 131, 1984, with permission from Elsevier.

details of the measurements, the ratio of the measured cross-section is divided by the QED cross-section. This is shown in Figure 13.4(right). The horizontal line corresponds by construction to the pure QED prediction, while the thick black line includes the electroweak corrections. From this data alone, measurements cannot demonstrate the presence of weak interaction because the effect is too small. The measurements are consistent with the pure QED prediction. The measurements of Bhabha scattering are a very important test of QED (and in principle of the Standard Model at the energies of, for example, LEP or SLC). Since the measurements follow precisely the theoretical prediction, they can be used to exclude new physics beyond QED. In particular, they exclude that the electron or the positron are "composite" objects and support their point-like Dirac behavior up to a scale of 250 GeV, or equivalently to the size of 10^{-16} cm, as will be discussed in Section 13.6.

13.4 The Total $e^+e^- \to \mu^+\mu^-$ Cross-Section

The total cross-sections are computed by integrating over their angular scattering angle. In case of processes without divergence, such as the heavy lepton pair creation, the total cross-section remains finite. Using Eq. (11.153), we find:

$$\sigma_{tot} = \int d\Omega \left(\frac{d\sigma}{d\Omega}\right)_{CMS} = \left(\frac{\alpha^2}{4s}\right) \int_{-1}^{1} d\cos\theta \int d\phi \left(1 + \cos^2\theta\right) = \left(\frac{\alpha^2}{4s}\right)(2\pi)\left(\frac{8}{3}\right) \tag{13.5}$$

The **total cross-section for muon pair creation** is then:

$$\sigma_{tot}(e^+e^- \to \mu^+\mu^-) \equiv \sigma_{\mu\mu}^{QED} = \frac{4\pi\alpha^2}{3s} \tag{13.6}$$

where we note the inverse proportional with the center-of-mass energy squared s. The total cross-section decreases sharply with energy. This is due to the phase-space factors and is generally true for any process involving point-like initial-state particles like electrons or positrons. This is one of the reasons why the luminosity of colliders must increase with the center-of-mass energy.

The expression of the total cross-section is in natural units and the $1/s$ dependence yields a GeV^{-2} unit. We can transform this in a more practical unit, such as the barn, by recalling that 1 barn $\equiv 10^{-28}$ m^2 and 1 m$^2 \approx 25.7 \times 10^{30}$ GeV^{-2} (see Section 1.8). We find:

$$\sigma_{tot}(e^+e^- \to \mu^+\mu^-) \simeq \frac{86.8}{E_\star^2(\text{GeV}^2)} \text{ nb} \tag{13.7}$$

where $E_\star = 2E_e$ is the center-of-mass energy of the collider with beams of energy E_e.

The total cross-section is usually presented in terms of $R_{\mu\mu}$, the cross-section normalized by the point-like QED cross-section:

$$R_{\mu\mu}(s) \equiv \frac{\sigma_{\mu\mu}(s)}{\sigma_{\mu\mu}^{QED}(s)} \tag{13.8}$$

As will be described in more detail in Section 26.6, the electroweak model predicts a small deviation from the QED prediction even at low energies due to the interference between the γ and Z^0 exchange diagrams shown in Figure 26.5, and the Z^0 exchange diagram itself (although it is an even smaller effect). In the electroweak model, the heavy lepton pair production cross-section is given by:

$$\sigma_{\ell\ell}(s) = \frac{4\pi\alpha^2}{3s}\left\{1 - 2c_V^e c_V^\ell \chi(s) + \left(\left[(c_V^e)^2 + (c_A^e)^2\right]\left[(c_V^\ell)^2 + (c_A^\ell)^2\right]\chi^2(s)\right)\right\} \tag{13.9}$$

where the vector and axial-vector coupling constants for the given charged leptons are given in Table 26.2, and where:

$$\chi(s) = \frac{1}{4\sin^2\theta_W \cos\theta_W}\frac{s}{M_Z^2 - s} = \frac{G_F M_Z^2}{2\sqrt{2}\pi\alpha}\frac{s}{M_Z^2 - s} \simeq \frac{G_F s}{2\sqrt{2}\pi\alpha} \quad \text{for } s \ll M_Z^2 \tag{13.10}$$

(here the width of the Z^0 boson was neglected – cf. Eq. (26.81)). Hence, in the case of the muon pairs, we expect:

$$R_{\mu\mu} = 1 - 2c_V^e c_V^\mu \chi + \cdots \approx 1 - c_V^e c_V^\mu \frac{G_F s}{\sqrt{2}\pi\alpha} + \cdots \qquad (13.11)$$

The effect is small at energies below the Z^0 pole: for instance, numerically we find $R_{\mu\mu} = 1.0015$ for $\sqrt{s} = 30$ GeV and $R_{\mu\mu} = 1.0248$ for $\sqrt{s} = 50$ GeV. Its value as a function of s is plotted as the thick line labeled "Electroweak" in Figure 13.5.

The ratio $R_{\mu\mu}$ was measured by experiments at the PETRA collider (CELLO, JADE, MARK J, PLUTO, TASSO), at the PEP collider (HRS, MARK II, MAC), and at the TRISTAN collider (AMY). Figure 13.5 shows, for example, the results from CELLO, MARK J, MAC, and AMY. Other experiments find consistent results with these. The AMY results at TRISTAN reach the highest energies, corresponding to $s = 3250$ GeV^2, so they a priori offer the highest sensitivity to high-energy deviations. The total cross-section measurement

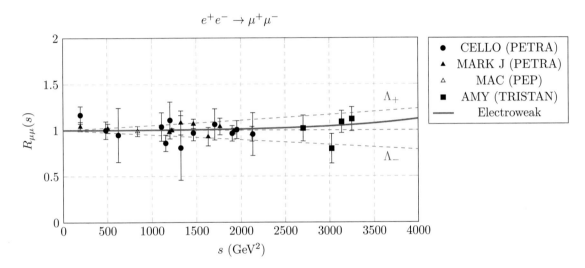

Figure 13.5 Measurements of $R_{\mu\mu}$ from CELLO, MARK J [99] at PETRA, MAC at PEP, and AMY [100] at TRISTAN as a function of the center-of-mass energy squared s. The error bars include statistical and systematic errors. The solid line labeled "Electroweak" shows the electroweak prediction. The horizontal dashed line is the QED prediction. The dashed curves labeled $\Lambda_\pm$ show the deviation from QED for cutoff parameters $\Lambda_\pm = 200$ GeV (see Section 13.6).

for the reaction $e^+e^- \to \mu^+\mu^-$ is well described by QED within a few percent up to energies $\sqrt{s} = 57$ GeV. The horizontal line represents the expectation within QED. QED remains a perfect description even at those high energies. Possible deviations from QED will be discussed in Section 13.6.

13.5 The Forward–Backward Charge Asymmetry

The measurement of the forward–backward asymmetry in $e^+e^- \to f^+f^-$ has been one of the main physics goals at PETRA and PEP because it showed the existence of the interference between the photon and the Z^0 exchange and thereby helped determine the weak couplings (this will be discussed in Section 26.7). In the center-of-mass system, QED predicts an unpolarized angular dependence of the type $(1 + \cos^2\theta)$. To study potential deviations, we can express the differential cross-section of $e^+e^- \to \mu^+\mu^-$ as:

$$\frac{\mathrm{d}\sigma}{\mathrm{d}\cos\theta} = \frac{\pi\alpha^2}{2s} \times \left[R_{\mu\mu}(1+\cos^2\theta) + B\cos\theta\right] \tag{13.12}$$

where $R_{\mu\mu}$ is given by Eq. (13.8). In pure QED, we have $B \equiv 0$ at tree level, and an experimental verification of the predicted differential cross-section is an important test of the theory. Experimentally, it requires the electric charge of the final-state particles to be measured. See Figure 13.6.

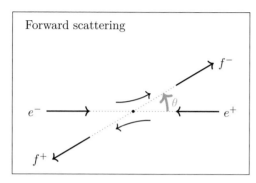

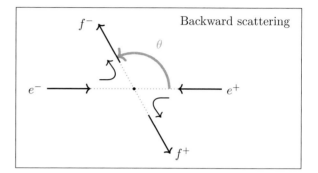

Figure 13.6 Forward $(\cos\theta > 0)$ and backward $(\cos\theta < 0)$ scattering.

The measurements of the angular distribution of $e^+e^- \to \mu^+\mu^-$ at $\sqrt{s} = 34.5$ GeV made by MARK J and AMY experiments are shown in Figure 13.7. They show clearly the negative forward–backward asymmetry predicted by the electroweak theory. The data are corrected for QED radiative effects. The dashed lines indicate the lowest-order QED prediction of $(1+\cos^2\theta)$ and the solid lines are fits to the data using the form $A(1+\cos^2\theta) + B\cos\theta$. The significant evidence on a forward–backward charge asymmetry in the angular distributions was first announced by PETRA experiments in 1981. After that, extensive measurements were performed at PETRA, PEP, and TRISTAN. All experiments measured a clear deviation from QED with a statistical significance of several standard deviations. **The asymmetry could be well interpreted within the electroweak theory.** Therefore, we note that while the total cross-section was not able to find significant deviation from QED, the measurement of the forward–backward asymmetry on the other hand was able to collect strong support for the predictions of the electroweak theory.

It is customary to define two regions of scattering: (a) the forward region $\cos\theta > 0$ and (b) the backward region $\cos\theta < 0$. See Figure 13.6. The **forward–backward asymmetry** is defined as:

$$A_{FB} \equiv \frac{\int_{\cos\theta>0} \mathrm{d}\Omega \frac{\mathrm{d}\sigma}{\mathrm{d}\Omega} - \int_{\cos\theta<0} \mathrm{d}\Omega \frac{\mathrm{d}\sigma}{\mathrm{d}\Omega}}{\int \mathrm{d}\Omega \frac{\mathrm{d}\sigma}{\mathrm{d}\Omega}} \tag{13.13}$$

Direct integration of the assumed differential cross-section yields:

$$A_{FB} = \frac{\int_0^1 \left(R_{\mu\mu}(1+x^2) + Bx\right)\mathrm{d}x - \int_{-1}^0 \left(R_{\mu\mu}(1+x^2) + Bx\right)\mathrm{d}x}{\int_{-1}^1 \left(R_{\mu\mu}(1+x^2) + Bx\right)\mathrm{d}x} = \frac{\frac{4}{3}R_{\mu\mu} + \frac{B}{2} - \frac{4}{3}R_{\mu\mu} + \frac{B}{2}}{\frac{8}{3}R_{\mu\mu}} = \frac{3}{8}\frac{B}{R_{\mu\mu}} \tag{13.14}$$

In the electroweak theory, the asymmetry, just as in the case of $R_{\mu\mu}$, is influenced by the Z^0 exchange diagram and the interference between this latter and the one with the gamma exchange. It can be calculated to be equal to (the Z^0 exchange diagram will be explicitly calculated in Section 26.6):

$$B_{electroweak} = -4c_A^e c_A^\mu \chi + 8c_V^e c_V^\mu c_A^e c_A^\mu \chi^2 = -4c_A^e c_A^\mu \chi(1 - 2c_V^e c_V^\mu \chi) \simeq -4c_A^e c_A^\mu \chi \tag{13.15}$$

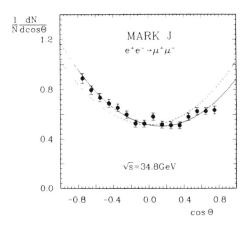
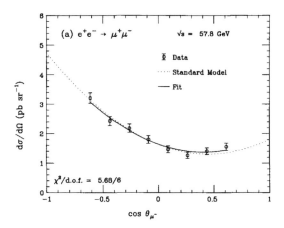

Figure 13.7 (left) The measurements of the angular distribution of $e^+e^- \to \mu^+\mu^-$ at $\sqrt{s} = 34.8$ GeV made by the PETRA experiment MARK J. Reprinted with permission from B. Adeva *et al.*, "Electroweak studies in e^+e^- collisions: $12 < \sqrt{s} < 46.78$ GeV," *Phys. Rev. D*, vol. 38, pp. 2665–2678, 1988. Copyright 1988 by the American Physical Society. (right) The same at $\sqrt{s} = 57.8$ GeV made by the TRISTAN experiment AMY. Reprinted from, C. Velissaris *et al.*, "Measurements of cross-section and charge asymmetry for $e^+e^- \to \mu^+\mu^-$ and $e^+e^- \to \tau^+\tau^-$ at $\sqrt{s} = 57.8$ GeV," *Phys. Lett. B*, vol. 331, pp. 227–235, 1994, with permission from Elsevier. Both plots show clearly the negative forward–backward asymmetry predicted by the electroweak theory. The data are corrected for QED radiative effects. The dashed lines indicate the lowest-order QED prediction of $(1 + \cos^2\theta)$ and the solid lines are fits to the data using the form $A(1 + \cos^2\theta) + B\cos\theta$.

since $c_V^e = c_V^\mu \approx -0.04$ for $\sin^2\theta_W = 0.23$ (see Table 26.2). Hence, the asymmetry practically depends only on the axial-vector couplings c_A^e and c_A^μ. Numerically, the effect is noticeable even at energies below the Z^0 pole: for instance, we find $A_{\mu\mu} = -0.068$ for $\sqrt{s} = 30$ GeV and $A_{\mu\mu} = -0.238$ for $\sqrt{s} = 50$ GeV! Its value as a function of s is plotted as the thick line labeled "Electroweak" in Figure 13.8.

Experimentally, its measurement is relatively simple and therefore represents an important test of the theory. The measured forward–backward asymmetry is just given by the ratio of number of events falling in each of the two categories:

$$A_{\ell\ell} = \frac{N(\theta < 90°) - N(\theta > 90°)}{N(\theta < 90°) + N(\theta > 90°)} \tag{13.16}$$

where N represents the observed number of events in that region. This measurement is the cleanest, with muons and taus in the final state. Hence, it is a simple matter of counting events and taking the ratio of the difference over the total. In this ratio, several systematic errors will cancel. The measurements of the forward–backward asymmetry $A_{\mu\mu}^{FB}$ in the reaction $e^+e^- \to \mu^+\mu^-$ as a function of the center-of-mass energy squared s are shown in Figure 13.8. The deviation from zero is very visible and the trend with energy can be used to predict some of the parameters of the electroweak theory. Because of the low energy accessible compared to the mass of the weak scale, one cannot measure the two main electroweak parameters that drive $A_{\mu\mu}$ and $R_{\mu\mu}$ independently. Basically, one cannot disentangle the cases of a higher Z^0 boson mass and a smaller mixing angle from a lower Z^0 boson mass and a higher mixing angle. However, if one of the two parameters is fixed, then the other can be evaluated. We report here the results from Mark-J [99]:

$$\sin^2\theta_W = 0.21^{+0.04}_{-0.02} \pm 0.01 \quad \text{for } m_Z = 91.9 \pm 2 \text{ GeV} \tag{13.17}$$

or

$$m_Z = 89^{+4}_{-3} \pm 1 \text{ GeV} \quad \text{for } \sin^2\theta_W = 0.230 \pm 0.005 \tag{13.18}$$

As we will see later in Chapter 27, each of these is a quite precise prediction!

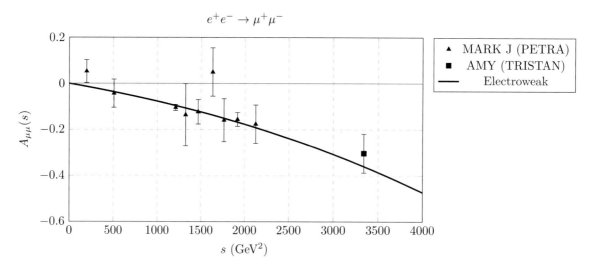

Figure 13.8 Measurements of the forward–backward asymmetry $A_{\mu\mu}$ in the reaction $e^+e^- \to \mu^+\mu^-$ from Mark-J [99] and AMY [101], plotted as a function of the center-of-mass energy squared s. The thick line labeled "Electroweak" corresponds to the prediction of the electroweak theory, while the horizontal line at zero would be the case with only QED.

13.6   Deviations from Point-Like Interaction at the Highest Energies

QED describes the interaction of point-like particles (e.g., Dirac fermions) with the electromagnetic field. In general, the testing of QED breakdown is based on the hypothesis that QED is embedded in a more general theory characterized by an as of yet inaccessible energy region. Historically, deviations from the theory were expressed in terms of a fictitious "cutoff" Λ up to which the theory is holding. Deviation corresponds to a regime in which new phenomena would become explicitly visible, i.e., at energies close to Λ. The breakdown of QED can be tested in $e^+e^- \to \mu^+\mu^-$ reactions by searching for deviations from the $1/s$ prediction by a cutoff parameter. This is performed by introducing the cutoff parameters $\Lambda_\pm^2$, which modify the total QED cross-section by a multiplicative term, which according to Ref. [102] is of the type:

$$\sigma_\Lambda(e^+e^- \to \mu^+\mu^-) = \sigma_{\mu\mu}^{QED}(s) \times \left(1 \mp \frac{s}{s - \Lambda_\pm^2}\right)^2 \tag{13.19}$$

As already mentioned, the data show no significant deviation from QED. This allows one to place a limit on the cutoff parameters. For example, the dashed curve in Figure 13.5 shows the deviation from QED for cutoff parameters $\Lambda_\pm = 200$ GeV. Lower limits on the lepton QED cutoff parameters $\Lambda_\pm$ exceed values of 250 GeV. Similar results have been obtained for the $e^+e^- \to \tau^+\tau^-$ final state. Agreement with QED is also obtained, just as in the case of the muon final states. These results suggest that the electron, the muon, and the tau behave as point-like particles down to distances smaller than 10^{-16} cm. This is remarkable.

In the context of a possible substructure of the quarks and leptons (**compositeness** discussed in Section 32.4), this would be the energy region in which the binding interactions of new constituents would start to manifest themselves. The **excited electron model** is a specific example of a possible QED breakdown mechanism. In this model the QED Lagrangian is left intact but a new effective Lagrangian is added corresponding to the possible existence of an excited electron vertex $e^*e\gamma$ and proportional to the lowest possible power of $1/\Lambda$. In e^+e^- collisions, this opens the possibility of a diagram in which a virtual excited electron is exchanged, as shown in Figure 13.9. First attempts to formalize possible deviations from QED in a gauge-invariant way

Figure 13.9 Feynman diagrams for $e^+e^- \to \gamma\gamma$ via the coupling to an excited electron e^*.

led to the possibility of a new interaction of the electron to a new heavy excited state first proposed by **Low**[1] to be of the magnetic form [103]:

$$\mathcal{L}_{int,e^*} = \frac{\lambda e}{2m_{e^*}}\overline{\Psi_{e^*}}\sigma^{\mu\nu}\Psi_e F_{\mu\nu} + \text{h.c.} = \frac{e}{2\Lambda}\overline{\Psi_{e^*}}\sigma^{\mu\nu}\Psi_e F_{\mu\nu} + \text{h.c.} \tag{13.20}$$

where we have introduced the notation "+h.c." to represent the Hermitian conjugated term of the first term. The cutoff Λ is in units of energy and can be expressed as:

$$\frac{1}{\Lambda} \equiv \frac{\lambda}{m_{e^*}} \tag{13.21}$$

where the coupling constant λ is unitless and m_{e^*} is the mass of the excited electron. As noted in Ref. [103], such a Lagrangian is not renormalizable and should be considered as an effective Lagrangian which describes the low-energy manifestation of a more complete theory of composite fermions.

This new interaction would, for example, modify the $e^+e^- \to \gamma\gamma$ cross-section. To lowest order it would be given by [102]:

$$\frac{d\sigma}{d\Omega}(e^+e^- \to \gamma\gamma) = \left(\frac{d\sigma}{d\Omega}\right)_{QED} \times \left[1 + \frac{s^2}{2\Lambda^4}\sin^2\theta\right] \tag{13.22}$$

where the term in brackets contains the deviation from QED. As expected, it tends to unity for $\sqrt{s} \ll \Lambda$. Nonetheless, if one could for instance measure the term in brackets with a precision of $\approx 1\%$, it would allow us to be sensitive to a scale $\Lambda \approx 2.5\sqrt{s}$. In general, and without necessarily connecting it to the excited electron, one can parameterize deviations from QED in the $e^+e^- \to \gamma\gamma$ reaction with two cutoff parameters $\Lambda_\pm$ expressing the differential cross-section as:

$$\frac{d\sigma}{d\Omega}(e^+e^- \to \gamma\gamma) = \left(\frac{d\sigma}{d\Omega}\right)_{QED} \times \left[1 \pm \frac{s^2}{2\Lambda_\pm^4}\sin^2\theta\right] = \left(\frac{d\sigma}{d\Omega}\right)_{QED} \times \left[1 \pm \frac{s^2}{2\Lambda_\pm^4}\frac{\sin^4\theta}{1-\cos^2\theta}\right] \tag{13.23}$$

where the Λ_- is introduced for symmetry purposes. Similar modifications of the differential cross-sections can be introduced for other reactions such as $e^+e^- \to e^+e^-$, $e^+e^- \to \mu^+\mu^-$, and $e^+e^- \to \tau^+\tau^-$. As no significant deviations were found, limits on the Λ parameters could be set at the level of a few hundred giga-electronvolts. The measurements of $e^+e^- \to \gamma\gamma$ are summarized in Section 13.8.

---

1   Francis Eugene Low (1921–2007), American theoretical physicist.

13.7   The L3 Detector at LEP

As a specific example of an experiment that studied e^+e^- collisions at the highest energies, we discuss in this section the L3 detector at LEP. The OPAL, DELPHI, and ALEPH experiments had similar features and their layouts are shown in Figure 27.14. The layout of the L3 detector and of its subdetectors is shown in Figure 13.10. The overall parameters of the detector are summarized in Table 13.2. We outline in the following the main features of each subsystem that composed the L3 detector.

• **Tracking chamber.** The central tracking chamber (TEC) was composed of two concentric cylindrical drift chambers on common endplates. There were two cylindrical proportional chambers with cathode strip readout, the Z detector, and a plastic scintillating fiber system. The TEC was conceived to optimally function in the limited space available within the electromagnetic calorimeter (ECAL, see below) and was designed with the following goals: (1) to measure precisely the r–ϕ path of charged particles with a space resolution of 50 µm; (2) to determine the sign of charged particles up to energies of 50 GeV; (3) to provide track multiplicity and (4) to reconstruct the interaction point and secondary vertices. The TEC was composed of three parts: the inner, outer, and Z chambers. The inner and outer chambers contained signal wires to measure precisely the r–ϕ coordinates. Charge division wires determined the z coordinate by measuring the asymmetry of the charge collected at both ends of the wire. Given the resistivity of the wire, this asymmetry can be used to determine the coordinate along the wire. The Z chambers were added to specifically improve the z-coordinate measurements. The PSF (plastic scintillating fibers) surrounding the outer TEC was composed of fibers running parallel to the beam pipe. These fibers yielded information on whether a minimum ionizing particle traversed the TEC and provided an independent measurement for calibration.

• **Electromagnetic calorimeter (ECAL).** The electromagnetic calorimeter located between the tracking chamber and the hadron calorimeter was composed of nearly 11,000 bismuth germanate (BGO) crystals pointing to the interaction region. Each crystal was 24 cm long (about 22 radiation lengths) and was 2×2 cm^2 at the inner end and 3×3 cm^2 at the outer end. BGO crystals were chosen for the following reasons (see also Table 3.6):

  • to provide excellent energy resolution for electrons and photons over the energy range between 100 MeV and 50 GeV;

  • for their short radiation length (1.12 cm), short Molière radius (2.3 cm), large nuclear interaction length (22 cm), high density $\left(7.13 \text{ g/cm}^3\right)$, and high refractive index (2.15);

  • for their radiation hardness and non-hygroscopicity.

The overall resolution achieved was 1.6% at 2 GeV and better than 1% at 50 GeV. The ECAL also achieved an excellent position resolution of 0.5 mm using the center-of-gravity method to localize the transverse peak of the electromagnetic shower.

• **Hadron calorimeter (HCAL).** The HCAL measured the energy of hadrons via total absorption calorimetry with a uranium hadron calorimeter and the BGO crystals. It was a fine-sampling calorimeter made of depleted uranium absorber plates interspersed with proportional wire chambers. Uranium was used because it has the following properties:

  • short absorption length (the BGO and HCAL corresponded to a total of seven to eight absorption lengths);

  • HCAL acted as a filter so that only non-showering particles reached the muon detector;

  • uranium radioactivity provided a gamma source to aid in calibration of the wire chamber.

The chambers were planes of brass tubes oriented alternatively perpendicular to each other for determination of the z and ϕ coordinate. The polar angle was measured by stretching the endcap wires azimuthally. The signal

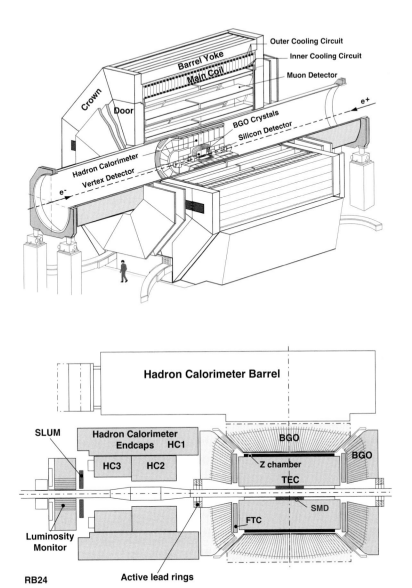

Figure 13.10 The L3 detector at LEP: (top) perspective view; (bottom) side view. Courtesy L3 Collaboration.

Parameter	L3
Length	14 m
Height	8 m
Magnetic field	0.5 T solenoid
Tracking (TEC)	1 m (length)$\times$ 0.5 m (radial)
coverage	$|\cos\theta| < 0.8$
inner chambers	12 segments r–ϕ measurement (8 signal and 8 charge division wires each)
outer chambers	24 segments r–ϕ measurement (54 wires each)
gas mixture	80% CO_2 and 20% C_4H_{10}
Z chambers	two layers of drift chambers
gas mixture	80% argon and 20% CO_2
PSF	143 plastic scintillating fibers per outer TEC segment
ECAL barrel	7680 BGO crystals
ECAL endcaps	2×1536 BGO crystals
ECAL coverage	$12°$ to $160°$
ECAL resolution	1.6% at 2 GeV, < 1% at 50 GeV
HCAL barrel	4.725 m long composed of nine rings (16 modules each)
HCAL barrel coverage	$35°$ to $145°$
HCAL readout	7968 chambers and 371,764 wires and 3960 readout towers
HCAL proportional chambers	gas mixture 80% argon and 20% CO_2
HCAL endcaps	composed of three rings
HCAL endcaps coverage	$5.5°$ to $35°$ on either side
HCAL resolution $\left(\frac{\sigma}{E}\right)_{had}$	$\approx \left(\frac{55}{\sqrt{E}} \oplus 6\right)$%
Muon chambers coverage	$44° < \theta < 136°$ (barrel)
Muon chambers	eight octants, each with five P chambers (one MI, two MM, and two MO)
Single wire resolution	less than 200 μm in the P chambers and 500 μm in the Z chambers
Gas mixture	61.5% Ar, 38.5% C_2H_6 (P chambers) 91.5% Ar, 8.5% CH_4 (Z chambers)
Muon resolution $\Delta p_\mu/p_\mu$	2% at $p_\mu = 45$ GeV/c
Scintillator counter system	30 plastic counters between the ECAL and HCAL
coverage	$|\cos\theta| \leq 0.83$ and 93% of ϕ
time resolution	$\sigma \approx 0.3 \pm 0.1$ ns

Table 13.2 Parameters of the L3 detector at the CERN LEP collider.

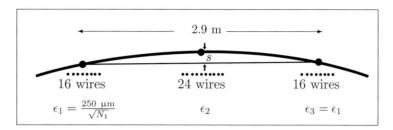

Figure 13.11 Measurement of the sagitta s of a muon track in the L3 detector at LEP.

wires were grouped into readout towers. The hadronic energy resolution was $\approx \left(55/\sqrt{E} \oplus 6\right)\,\%$. Outside the hadron calorimeter was a muon filter designed to reduce punch-throughs necessary for muon identification. It consisted of eight octants each of which had six brass absorber plates interleaved with five layers of proportional chambers and a 1.5 cm-thick absorber plate. In total, it added one absorption length to the hadron calorimeter.

- **Muon chambers.** The design goal of the muon spectrometer was to precisely measure muon momenta with a resolution $\Delta p/p \approx 2\%$ at $p_\mu = 45$ GeV/c. Three layers of high-precision drift chambers measured the curvature of the muon track between the support tube and the magnet coil. Each octant had a special mechanical structure which supported five drift chambers (called P chambers), which measured the muon transverse momentum. The five chambers were subdivided into one inner (MI), two middle (MM), and two outer (MO). The MO, MI, and MM chambers had 21, 19, and 15 cells, respectively. Each MO and MI cell contained 16 sense wires and the MM cell contained 24 sense wires. The Z chambers were located on the tops and bottoms of the inner and outer chambers. They measured the z coordinate of the muon track using two layers of drift chambers with wires perpendicular to the beam direction. The five P chambers were located to provide the best **sagitta** measurement (see Figure 13.11):

$$s = X_{MM} - \frac{X_{MI} - X_{MO}}{2} \tag{13.24}$$

where X_i was the measured position in the chamber i. For a 45 GeV muon, the sagitta was $s = 3.7$ mm. In order to achieve a momentum resolution of 2%, the necessary accuracy in the sagitta measurement was $\Delta s = 2\% \times 3700\,\mu\text{m} = 74\,\mu\text{m}$. There were three main sources of systematic errors in the sagitta measurement:

- intrinsic drift chamber resolution, which was dependent on the single wire resolution (200 μm) and on the number of wires used for reconstruction of the track. This contributed to a sagitta uncertainty of $\approx 54\,\mu\text{m}$;

- multiple scattering inside chambers added 31 μm of uncertainty;

- accuracy of alignment of the three different layers and knowledge of the wire positions. This added 33 μm to the uncertainty.

Adding these errors in quadrature yielded the desired $\Delta s = 69.8\,\mu\text{m}$!

The determination of the sagitta to high precision required a critical alignment between chambers in a given octant. An alignment system consisting of LEDs, lenses, and quadrant photodiodes defined the octant central line and the positions of the wires were measured relative to it. This method yielded an impressive accuracy of 10 μm. To assure two octant center lines were parallel to each other, a laser beacon was used to measure the angle between the two octant center lines with a precision of 25 μm, corresponding to an error in the sagitta of less than 10 μm. The long-term stability was monitored and could be demonstrated over a decade.

● **Scintillation counters.** The scintillator counter system consisted of 30 plastic counters situated between the electromagnetic and hadronic calorimeters. They had two primary purposes: (1) measurement of the time-of-flight between two opposite counters in order to reject cosmic events; (2) measurement of the hit multiplicity used as a trigger for hadronic events. The counters were divided into 32 ϕ sectors positioned along the beam line following the shape of the hadron calorimeter. Photomultiplier tubes at both ends of each scintillator were read out to provide very precise timing resolution at the sub-nanosecond level.

● **Luminosity monitor.** The luminosity monitors were located in the very forward and backward positions relative to the vertex point (see Figure 13.10) and consisted of two cylindrical BGO detectors and charged particle-tracking chambers with good position resolution. They measured the integrated luminosity by counting the low-angle Bhabha events and comparing the obtained rate with the theoretical Bhabha cross-section within the appropriate angular region. As already pointed out in Section 13.3, Bhabha scattering is extremely useful to measure the luminosity of the collider, since the differential cross-section is dominated in the forward and backward direction by the large t-channel contribution. Hence, the event rate is very large and high statistics can be rapidly accumulated. In the case of L3, the coverage in the forward angular region was 24.7 mrad $< \theta < 69.3$ mrad with 100% efficiency, which corresponded to an effective Bhabha cross-section of ≈ 100 nb. The reached luminosity error was about 0.5% for a year of running.

13.8 Photon Pair Production ($e^+e^- \rightarrow \gamma\gamma$)

The reaction $e^+e^- \rightarrow \gamma\gamma$ provides a clean QED test up to the highest energies. This process has been studied by many experiments at e^+e^- colliders at PEP, PETRA, TRISTAN, and LEP. Figure 13.12 shows the differential cross-section distributions from the JADE experiment at four energies with $\sqrt{s} = 14, 22, 34.5$ and 43.1 GeV. The agreement with QED is excellent. Limits on the different cutoff scale parameters $\Lambda_\pm$ have been set by experiments at PEP, PETRA, and TRISTAN. With the advent of LEP, the situation changed. The LEP era at CERN has been described in Section 3.4. Four large experiments called ALEPH, DELPHI, L3, and OPAL collected data at LEP. The ALEPH, DELPHI, and OPAL detectors will be described in Section 27.6. The L3 detector, with its specific focus on measuring electrons, muons, and photons, was described in detail in Section 13.7. The $e^+e^- \rightarrow \gamma\gamma$ reaction provided a means to test QED up to several hundreds of giga-electronvolts.

One unique aspect of the photon pair production is that it is not much influenced by electroweak diagrams, compared to the other channels such as $e^+e^- \rightarrow e^+e^-$ or $e^+e^- \rightarrow \ell^+\ell^-$. Hence, the QED contribution remains the dominant one even at much higher energies, such as the Z^0 resonance and above. For this reason, QED could be tested at the energies of LEP-I and LEP-II. Since the sensitivity to possible deviations from QED increases rapidly with the center-of-mass energy, the results obtained at the lower energies reached with PEP, PETRA, and TRISTAN were superseded by LEP data. At LEP energies, however, photon Bremsstrahlung becomes important and studies have been extended to the reactions $e^+e^- \rightarrow \gamma\gamma(n\gamma)$.

Referring ourselves to the derivation performed in Section 11.15, we set the kinematics to be:

$$e^-(p) + e^+(k) \rightarrow \gamma(k_1) + \gamma(k_2) \tag{13.25}$$

The corresponding Feynman diagrams at tree level were shown in Figure 11.14. The Born-level cross-section was computed and given in Eq. (11.239) (also Eq. (13.3)) in the ultra-relativistic limit. If one keeps the mass of the electron in the denominator, one gets for the differential cross-section:

$$\left(\frac{d\sigma}{dc}\right)_{Born}(e^+e^- \rightarrow \gamma\gamma) = \frac{\pi\alpha^2}{s}\frac{1+c^2}{e^2-c^2} \tag{13.26}$$

where $e \equiv E_k/|\vec{k}|$ and $c \equiv \cos\theta$. The QED virtual and soft photons (i.e., below detection threshold) have been

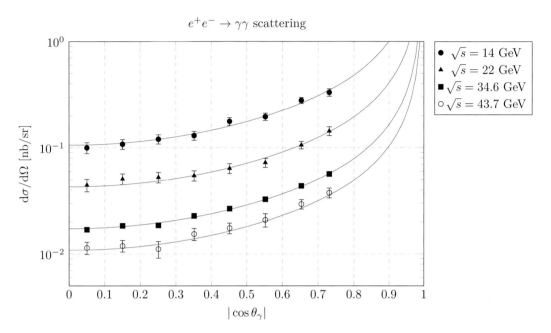

$e^+e^- \to \gamma\gamma$ scattering

Figure 13.12 Cross-section distributions for the reaction $e^+e^- \to \gamma\gamma$ measured by JADE [104] at PETRA: differential cross-sections at $\sqrt{s} = 14, 22, 34.6$ and 43.7 GeV compared to QED predictions (straight lines).

computed (see Ref. [105]) and modify the cross-section as follows:

$$\left(\frac{\mathrm{d}\sigma}{\mathrm{d}c}\right)(e^+e^- \to \gamma\gamma) = \left(\frac{\mathrm{d}\sigma}{\mathrm{d}c}\right)_{Born} \times \left\{1 + \delta_{radcorr}\right\} \tag{13.27}$$

where

$$\delta_{radcorr} = -\frac{\alpha}{\pi}\left\{2(1-2v)(\ln k_{cut} + v) + \frac{3}{2} - \frac{1}{3}\pi^2 + \frac{1}{2(1+c^2)}\left[-4v^2\left(3-c^2\right) - 8vc^2 + 4uv\left(5 + 2c + c^2\right)\right.\right.$$
$$\left.\left. + 4wv\left(5 - 2c + c^2\right) - u\left(5 - 6c + c^2\right) - w\left(5 + 6c + c^2\right) - 2u^2\left(5 + 2c + c^2\right) - 2w^2\left(5 - 2c + c^2\right)\right]\right\} \tag{13.28}$$

where $k_{cut} = E_\gamma^{soft}/|\vec{k}|$ is the soft photon cutoff and:

$$m = m_e/\left|\vec{k}\right|, \quad v = \frac{1}{2}\ln\frac{4}{m^2}, \quad u = \frac{1}{2}\ln\frac{2(e+c)}{m^2}, \quad w = \frac{1}{2}\ln\frac{2(e-c)}{m^2} \tag{13.29}$$

This expression can be computed numerically. The angular dependence and the associated small forward–backward asymmetry given by the terms containing c^2 and c turns out to be small. Overall, the factor is determined by the term with the logarithmic dependence on k_{cut}. For instance, for a center-of-mass energy $\sqrt{s} = 189$ GeV and $k_{cut} = 0.1(0.01)$, one finds a correction of approximately $-0.17(-0.43)$. We note that the radiative correction is negative, i.e., it reduces the cross-section and the correction becomes more significant for smaller values of the cutoff k_{cut}. We understand this fact by recalling that the term takes into account the radiative corrections due to virtual photons and all emitted photons with energies $E_\gamma \leq E_\gamma^{soft}$. On the other hand, the cross-section for the emission of a "real" photon defined by an energy $E_\gamma > E_\gamma^{soft}$ must be considered exclusively. The expression for the explicit three-photon final state has been reported in Ref. [105]. Its amplitude is related to that of the so-called double Compton effect $\gamma e^- \to e^- \gamma\gamma$ first computed by Mandl and Skyrme in the 1950s [106]. In this case the kinematics is a bit more complicated and one writes:

$$e^-(p) + e^+(k) \to \gamma(k_1) + \gamma(k_2) + \gamma(k_3) \tag{13.30}$$

with the reduced variables:

$$x_i \equiv k_i^0/|\vec{k}| \quad \text{and} \quad z_2 \equiv \cos\alpha \tag{13.31}$$

where α is the angle between the first and the third photon. Hence, the x_i represent the fractional energies of the final-state gammas. Since we have three photons in the final state, the number of degrees of freedom (d.o.f.) of the final-state kinematics is equal to $3 \times 3 - 4 = 5$. We can choose the solid angles of the first (2 d.o.f.) and third photon (2 d.o.f.), and the fractional energy of the third photon x_3 (1 d.o.f.) to express the differential cross-section, which then becomes [105]:

$$\frac{d^5\sigma}{d\Omega_1 d\Omega_2 dx_3} = \frac{\alpha^3}{8\pi^2 s} \frac{x_1 x_3}{2e - x_3 + x_3 z_2} \times \sum_{\text{permutations}} \left[-2m^2 \frac{\kappa_2'}{\kappa_3^2 \kappa_1'} - 2m^2 \frac{\kappa_2}{\kappa_3'^2 \kappa_1} + \frac{2}{\kappa_3 \kappa_3'} \left(\frac{\kappa_2^2 + \kappa_2'^2}{\kappa_1 \kappa_1'} \right) \right] \tag{13.32}$$

where "permutations" denotes all the permutations of $(1,2,3)$ in the indices of the quantities in the bracket, and:

$$\kappa_i \equiv x_i(e - c) \quad \text{and} \quad \kappa_i' = x_i(e + c) \tag{13.33}$$

Finally we note that also the explicit cross-section for the process with four final-state photons $e^-(p) + e^+(k) \to \gamma(k_1) + \gamma(k_2) + \gamma(k_3) + \gamma(k_4)$ has been calculated in Ref. [107]. The same authors developed the GGG Monte-Carlo generator that correctly deals with the proper combination of the virtual and soft photon correction, the exclusive three-photon and four-photon final state. We refer the interested reader to the given literature for more information.

These cross-sections can be used to compare QED predictions to events observed at high energy. For example, the study performed by the L3 experiment [108] is reported here. The total cross-section of $e^+e^- \to \gamma\gamma(n\gamma)$ as a function of the center-of-mass energy is presented in Figure 13.13. The angular acceptance for the gammas is $16° < \theta_\gamma < 164°$, which corresponds to $|\cos\theta_\gamma| < 0.96$. The total cross-section corresponds to the integration over this angular acceptance. In order to deal with radiative corrections, the observed data are corrected (unfolded) using the theoretical predictions of the GGG Monte-Carlo generator discussed above. Hence, the presented results can be directly compared to the Born-level cross-section and are independent of experimental threshold effects. The measurements extend up to a center-of-mass energy of 200 GeV and show excellent agreement with the QED prediction. The differential cross-section as a function of $|\cos\theta|$ presented at a given $\sqrt{s} = 189$ GeV is shown as well and exhibits good agreement with the solid line, corresponding to the lowest-order QED prediction.

To illustrate the reality of hard Bremsstrahlung, a spectacular $e^+e^- \to \gamma\gamma\gamma\gamma$ recorded by the L3 experiment is shown in Figure 13.14. The four photons have energies of 58.1, 48.5, 11.1, and 7.1 GeV. Two of the photons are interpreted as Bremsstrahlung from the initial-state particles.

These data show that QED and its radiative corrections are valid up to high energies. In order to quantify the possible deviations from QED, a maximum likelihood fit to the differential cross-sections was performed at each center-of-mass energy, using Eq. (13.23). At the 95% CL the following lower limits are obtained: $\Lambda_+ > 321$ GeV and $\Lambda_- > 282$ GeV. The dashed and dotted lines in Figure 13.13 represent the deviations from QED obtained for these parameters $\Lambda_+ = 321$ GeV and $\Lambda_- = 282$ GeV. For the coupling to an excited electron of the form of Eq. (13.21), this implies a lower limit on its mass for a unit coupling of $m_{e^*} > 283$ GeV at the 95% CL. The most up-to-date lower limit on the mass of the excited lepton from $e^+e^- \to \gamma\gamma$ measurements at high energies is given by (values taken from the Particle Data Group's Review [5]):

$$m_{e^*} > 356 \text{ GeV} \quad (95\% \text{ CL}) \tag{13.34}$$

where one assumed $\lambda_\gamma = 1$.

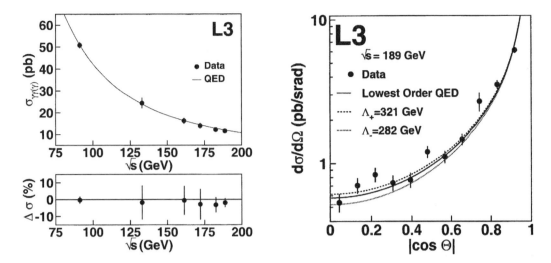

Figure 13.13 (left) Measured cross-section as a function of the center-of-mass energy compared with the Born-level QED prediction. The bottom part of the figure presents the relative deviation of the measurements with respect to the QED expectations. (right) Differential cross-section for the process $e^+e^- \to \gamma\gamma(n\gamma)$ measured by the L3 experiment at LEP. The points show the measurements corrected for efficiency and additional photons. The solid line corresponds to the lowest-order QED prediction. The dashed and dotted lines represent the limits obtained for deviations from QED. Reprinted from M. Acciarri *et al.*, "Hard-photon production and tests of QED at LEP," *Phys. Lett. B*, vol. 475, pp. 198–205. Copyright 2000, with permission from Elsevier.

13.9 Direct Search for Excited Leptons

In the previous section, the search for small departures from the QED prediction was used to set a constraint on the existence of new couplings, for instance in the context of an excited electron. The study of radiative processes, which increase with energy, also provides a means to study QED predictions and opens the path to a direct search for compositeness. Charged (e^*, μ^*, and τ^*) excited leptons but possibly also excited neutrinos (ν_e^*, ν_μ^*, ν_τ^*) are predicted by composite models (see Section 32.4). The underlying theoretical background generally used to model the interactions of excited leptons is discussed in Section 32.4. In this section, we discuss the phenomenology and the strategies for experimental searches of excited states.

When searching for excited leptons, one should remember that even if the compositeness scale is very large, the mass of the excited states could be in a much lower-energy domain, since the dynamics at the subconstituent level is completely unspecified. This motivates the searches for excited states within all the kinematical range accessible at colliders. With the availability of higher center-of-mass energies, the prospects for finding the lowest excited states of the spectrum increase.

Searches for excited quarks and leptons have been performed over the last decades in experiments at the LEP, HERA, Tevatron, and LHC colliders. In particular, electron–positron collisions provide a very clean environment to search for leptonic final states with extra photons. For instance, if they exist in a kinematically accessible domain at the LEP collider, these excited leptons should be produced through pair production:

$$e^+e^- \to \ell^*\ell^* \to \ell\ell\gamma\gamma \quad \text{and} \quad e^+e^- \to \nu^*\nu^* \to \nu\nu\gamma\gamma \tag{13.35}$$

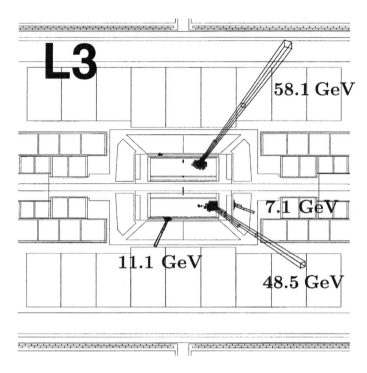

Figure 13.14 Display of an event with four photons detected by the L3 experiment at $\sqrt{s} = 183$ GeV. Reprinted from M. Acciarri *et al.*, "Hard-photon production and tests of QED at LEP," *Phys. Lett. B*, vol. 475, pp. 198–205. Copyright 2000, with permission from Elsevier.

or possibly via single production:

$$e^+e^- \to \ell\ell^* \to \ell\ell\gamma \quad \text{and} \quad e^+e^- \to \nu\nu^* \to \nu\nu\gamma \tag{13.36}$$

where we have assumed that the excited lepton decays radiatively, $\ell^* \to \ell\gamma$. Within the theoretical model for compositeness outlined in Section 32.4, the radiative decay width for $\ell^* \to \ell\gamma$ is:

$$\Gamma(\ell^* \to \ell\gamma) = \frac{\alpha}{4} \left(\frac{\lambda}{m_{\ell^*}}\right)^2 m_{\ell^*}^3 \tag{13.37}$$

The details of the derivation of this result are left to **Ex 13.1**. Excited leptons, if sufficiently massive, could in principle also decay into light ordinary leptons and a massive virtual or real gauge boson. The heavy gauge boson subsequently decays into a pair of fermions yielding the final states $\ell^* \to \ell Z^{0*} \to \ell f\bar{f}$ (neutral-current decays) or $\ell^* \to \nu W^* \to \nu f f'$ (charged-current decays). If the excited lepton is more massive than the gauge boson, then it can decay into a real gauge boson. The decay width is in this case given by [109]:

$$\Gamma(\ell^* \to \ell V) = \frac{\alpha}{4} \left(\frac{\lambda_V}{m_{\ell^*}}\right)^2 m_{\ell^*}^3 \left(1 - \frac{M_V^2}{m_{\ell^*}^2}\right)^2 \left(1 + \frac{M_V^2}{2m_{\ell^*}^2}\right) \tag{13.38}$$

Assuming that $\lambda \approx \lambda_W \approx \lambda_Z$, the radiative decay mode is dominant for masses of the excited leptons below M_V and decreases above M_V due to the channels $\ell^* \to \nu W^*$ and $\ell^* \to \ell Z^{0*}$. In e^+e^- collisions, pair production searches are sensitive to excited leptons of masses up to values close to the kinematical limit $\sqrt{s}/2$, while the single production mechanism extends the search potential up to masses close to $\sqrt{s}$. Since not much is a priori known on these new particles, the experimental searches generally consider both single and pair production. Within the assumed theoretical model and neglecting contributions from anomalous magnetic moments, the cross-section for pair production of excited leptons derived from the Lagrangian Eq. (32.23) depends only on their mass and on $\sqrt{s}$. The single production cross-section for excited leptons derived from Eq. (32.24) depends also on the unknown parameters f/λ and f'/λ.

Several selection strategies are put in place to select the various possible final-state signatures to include radiative decay modes shown in Eqs. (13.35) and (13.36), as well as neutral and charged decay modes of the type:

$$e^+e^- \to \ell^*\ell^* \to \nu\nu WW \quad \text{and} \quad e^+e^- \to \nu^*\nu^* \to \ell\ell WW \tag{13.39}$$

or possibly via single production:

$$e^+e^- \to \ell\ell^* \to \ell\nu W \quad \text{and} \quad e^+e^- \to \nu\nu^* \to \nu\ell W \tag{13.40}$$

The searches therefore combine a mixture of requirements on electrons, muons, photons, jets, and missing energy. A beautiful event collected by L3 compatible with a $\mu\mu\gamma\gamma$ final state is shown in Figure 13.15. The LEP experiments did not find evidence for the existence of excited leptons and set lower limits on their masses under specific theoretical assumptions. These limits at the 95% CL are summarized in the following (values taken from the Particle Data Group's Review [5]):

$$
\begin{aligned}
m_{e^*} &> 103.2 \text{ GeV} & e^+e^- &\to e^*e^* \\
m_{\mu^*} &> 103.2 \text{ GeV} & e^+e^- &\to \mu^*\mu^* \\
m_{\tau^*} &> 103.2 \text{ GeV} & e^+e^- &\to \tau^*\tau^* \\
m_{\nu^*} &> 102.6 \text{ GeV} & e^+e^- &\to \nu^*\nu^*
\end{aligned}
\tag{13.41}
$$

Searches in the single production channel can reach twice these values but are dependent on the assumed value for the couplings. Typically, upper limits on f/Λ and f'/Λ ranging from 10^{-1} to 10^{-4} GeV^{-1} are set in the mass range from 100 to 200 GeV.

The largest kinematical constraints on the existence of excited leptons actually come from the ATLAS and CMS experiments at the LHC, owing to their high reach in center-of-mass energies, reaching several tera-electronvolts. From the searches for the single production via the contact interaction $q\bar{q} \to e^*e$ and the decay via a neutral or charged processes:

$$pp \to ee^* \to eeq\bar{q} \quad \text{or} \quad pp \to ee^* \to e\nu q_1\bar{q}_2 \tag{13.42}$$

they find no excess in data. The neutral reaction results in two energetic electrons and at least two hadronic jets. The charged reaction yields exactly one energetic electron, at least two hadronic jets produced by two collimated quarks, and missing transverse momentum. Assuming $f = f' = 1$ and $m_{e^*} = \Lambda$, the absence of signal gives a lower bound $m_{e^*} > 4.8$ TeV at the 95% CL In the case of the radiative decay $\ell^* \to \ell\gamma$ of a single excited lepton produced in pp collisions, one searches for an energetic photon accompanied by a charged lepton. No excess in data excludes electron and muon masses up to 3.9 and 3.8 TeV, respectively, assuming as before $\Lambda = m_{\ell^*}$. For the excited tau leptons, similar conclusions can be drawn with the resulting lower bound $m_{\tau^*} > 2.5$ TeV at the 95% CL From the search of the pair production of excited neutrinos and under the same assumptions, they derive a lower bound on the mass $m_{\nu^*} > 1.6$ TeV. Overall, it looks like leptons do not get so easily excited!

Problems

Ex 13.1 Lifetime of the ℓ^*. We consider the decay $\ell^* \to \ell + \gamma$. Compute the partial decay width. In the process $e^+e^- \to \ell\ell^* \to \ell\ell\gamma$ at the Z^0 pole, estimate the lifetime and the mean free path of the ℓ^*.

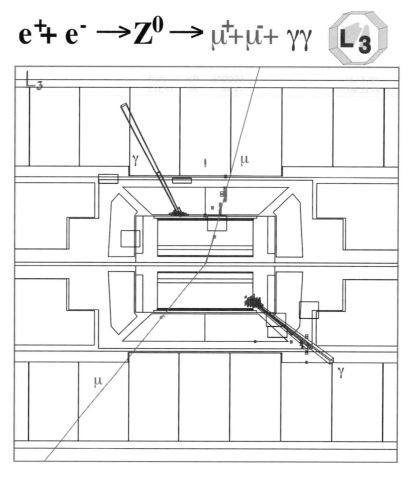

Figure 13.15 Display of an event with two oppositely charged muons and two photons detected by the L3 experiment at a center-of-mass energy close to the Z^0 pole.

(a) Starting with the Lagrangian Eq. (32.20) with $|a_\gamma| \approx |b_\gamma| \approx 1$, show that the matrix element can be expressed as:

$$\mathcal{M} = \frac{e\lambda}{2m_{\ell^*}} \left[\bar{u}(p_2) \sigma^{\mu\nu}(1 - \gamma^5) q_\nu \epsilon_\mu^* u(p_1) \right] \tag{13.43}$$

(b) Show that the corresponding squared matrix element averaged over incoming spins and summed over outgoing spins is given by:

$$\frac{1}{2} \sum_{spins} \mathcal{M}\mathcal{M}^* = -\frac{1}{2} \left(\frac{e\lambda}{4m_{\ell^*}} \right)^2 q_\alpha q_\mu \epsilon_\mu^* \epsilon_\beta \, \mathrm{Tr} \left[\not{p}_2 (\gamma^\mu \gamma^\nu - \gamma^\nu \gamma^\mu)(1 - \gamma^5) \not{p}_1 (\gamma^\alpha \gamma^\beta - \gamma^\beta \gamma^\alpha)(1 - \gamma^5) \right] \tag{13.44}$$

(c) Use the photon completeness relation to find that:

$$\langle |\mathcal{M}|^2 \rangle = \left(\frac{e\lambda}{4m_{\ell^*}} \right)^2 [32(p_1 \cdot q)(p_2 \cdot q)] \tag{13.45}$$

(d) Choose the rest frame of the ℓ^*, thus assign $p_1 = (m_{\ell^*}, \vec{0})$ and $p_2 = (E_2, \vec{p}_2)$. Obtain the partial width of the decay:

$$d\Gamma = \frac{1}{8m_{l^*}} \frac{1}{(2\pi)^2} \left(\frac{e\lambda}{4m_{l^*}}\right)^2 32 \left[m_{l^*}^3 E_2 - m_{l^*}^2 E_2^2\right] \frac{1}{E_2 E_3} \delta\left(m_{l^*} - E_2 - E_3\right) \delta^3\left(-\vec{p}_2 - \vec{p}_3\right) \quad (13.46)$$

(e) Integrate, neglecting the mass of the ordinary lepton, to find:

$$\Gamma(\ell^* \to \ell\gamma) = \frac{\alpha}{4} \left(\frac{\lambda}{m_{\ell^*}}\right)^2 m_{\ell^*}^3 \qquad (13.47)$$

(f) Estimate numerically the lifetime and the mean free path of a 45 GeV ℓ^* for a coupling $\gamma/m_{\ell^*} \approx 10^{-4}$ GeV^{-1}, produced in the reaction $e^+e^- \to Z^0 \to \ell\ell^*$.

Ex 13.2 Dark photon production in e^+e^- collisions. In addition to gravity, dark matter could interact feebly to the ordinary particles through some mediators. We shall study the production of a hypothetical massive spin-1 A' boson (dark photon) in electron–positron collisions and its corresponding decay properties for the following effective Lagrangian:

$$\mathcal{L} \supset -\frac{1}{4}\tilde{F}_{\mu\nu}\tilde{F}^{\mu\nu} + \frac{1}{2}m_{A'}^2 X_\mu X^\mu - \epsilon e Q_f X_\mu \bar{\psi}\gamma^\mu\psi + (D_\mu\phi)^* D^\mu\phi - \mu^2\phi^*\phi - \frac{1}{2}\rho^2(\phi\phi + \phi^*\phi^*) \quad (13.48)$$

where $\tilde{F}^{\mu\nu} = \partial^\mu X^\nu - \partial^\nu X^\mu$ is the A' boson field strength tensor, $D_\mu = \partial_\mu + ig_D X_\mu$ with g_D the dark coupling parameter, and $\phi = 1/\sqrt{2}(\phi_1 + i\phi_2)$ the complex dark-matter scalar field; ϵ is the mixing parameter between the $U(1)_Y$ and $U(1)_{A'}$ gauge fields.

(a) Draw the lowest-order Feynman diagrams corresponding to the process $e^+e^- \to A'\gamma$ and compute the differential cross-section $d\sigma/d\cos\theta$ where θ is the angle between the incoming electron and the outgoing A' boson. How does it depend on ϵ and $m_{A'}$?

(b) Compute the total cross-section for the above process. What is the center-of-mass energy dependence? To which collider energy would the experiment be the most sensitive?

(c) Given the above Lagrangian, determine what are the possible decay channels of the A' boson.

(d) Compute the spin-average decay width for each of the above final states. How do the branching ratios depend on the parameters ϵ, g_D and the dark photon masses?

(e) How would you detect the final states in a fixed-target experiment and thus infer the presence of the A' boson?

14   Tests of QED at Low Energy

Historically, the earliest approach to calculation of a lepton g factor was to treat the electron as a charged rigid spinning object of finite dimensions. The spinning charge gives rise to equivalent current loops, so that if the charge distribution and spin are known, the magnetic moment can be calculated from elementary considerations. For a body with charge density $\rho_e(r)$ and mass density $\rho_m(r)$, the g factor is determined by the functional dependence of the ratio ρ_e/ρ_m. If this ratio is constant throughout the body, then $g = 1$ regardless of the geometrical shape of the object. In 1925 Uhlenbeck and Goudsmit were able to make more quantitative speculations about the intrinsic properties of the electron. By postulating that the electron had a spin of $1/2$ and $g = 2$, they were able to use the Bohr–Sommerfeld quantum theory to explain atomic fine structures (...) From 1947 to the present, the precision of both theoretical calculation and experimental measurement has improved (...) One of the most striking features of this improvement has been an almost exact correspondence between the accuracy of theory and experiment. (...) In spite of the unsatisfactory nature of the renormalization aspects of the theory, there has been no significant evidence for breakdown of QED.

Rich and Wesley[1]

14.1   The General Electromagnetic Form Factors

The Gordon identity, introduced in Section 10.19, allows us to decompose the point-like QED current into two parts (see Eq. (10.171)):

$$\bar{u}(p')\gamma^\mu u(p) = \frac{1}{2m}\bar{u}(p')\left[(p'+p)^\mu + i\sigma^{\mu\nu}(p'-p)_\nu\right]u(p) \tag{14.1}$$

The first term, called the **electric term**, is the coupling of the form $(p'+p)$, which is the same as Eq. (10.169) and represents the purely electric interaction. The second part $i\sigma^{\mu\nu}q_\nu$, called the **magnetic term**, is the interaction associated with the magnetic moment of the Dirac particle. As already pointed out, the Gordon identity shows that a Dirac particle interacts via both its charge and its magnetic moment.

In Section 12.8, we have demonstrated that radiative corrections, such as the vertex correction, can be interpreted as a modification of the electric and magnetic terms with the introduction of two functions $F_1(q^2)$ and $F_2(q^2)$ that describe the radiatively corrected electron. This led us to the following expression, Eq. (12.87):

$$(ie)\bar{u}(p')\Gamma^\mu u(p) \equiv (ie)\bar{u}(p')\left[\underbrace{\gamma^\mu F_1(q^2)}_{\text{electric+magnetic}} + \underbrace{\frac{i}{2m}\sigma^{\mu\nu}q_\nu F_2(q^2)}_{\text{anomalous magnetic}}\right]u(p) \tag{14.2}$$

where $q = p' - p = k' - k$ and $F_1(q^2)$ and $F_2(q^2)$ are the structure functions of the electron. We would set $F_1(q^2) = 1$ and $F_2(q^2) = 0$ to recover the QED vertex factor at tree level. We already saw that in the static

1   Reprinted excerpt with permission from A. Rich and J. C. Wesley, "The current status of the lepton g factors," *Rev. Mod. Phys.*, vol. 44, pp. 250–283, Apr 1972. Copyright 1972 by the American Physical Society.

limit $q^2 \to 0$, the form factor is just the electric charge:

$$F_1(q^2 = 0) = 1 \tag{14.3}$$

Similarly, as we already mentioned in Section 4.12, Pauli's equation predicts that the magnetic moment associated with the spin yields an interaction term in the Hamiltonian of the type:

$$-g_{Pauli}\mu_B(\vec{S} \cdot \vec{B}) = -2\frac{e}{2m}(\vec{S} \cdot \vec{B}) = -2\frac{eF_1(q^2 = 0)}{2m}(\vec{S} \cdot \vec{B}) \tag{14.4}$$

where $g_{Pauli} = 2$ and $\vec{S} = (1/2)\vec{\sigma}$ is the spin operator. We will show in the next section that this result remains true also for a point-like Dirac particle, i.e., $g_{Dirac} = g_{Pauli} = 2$. Recall, it was shown in Chapter 12 that QED radiative corrections give a non-zero value for $F_2(q^2)$. In the static limit given by $q^2 \to 0$, this leads to an additional magnetic moment term. Hence, the magnetic momentum is now given by the sum of the magnetic moment of a point-like Dirac particle *plus* a so-called **anomalous magnetic moment**. This can be written as:

$$-g\mu_B(\vec{S} \cdot \vec{B}) = (-2)\frac{e}{2m}\left[F_1(0) + F_2(0)\right](\vec{S} \cdot \vec{B}) \tag{14.5}$$

Consequently, in general, g will differ from $g_{Dirac} = 2$! **Starting from the point-like bare electron, we observe that radiative corrections alter its nature. Higher-order corrections in perturbation theory produce the "structure." A physical picture behind this is that the "bare" electron is always accompanied with a cloud of virtual particle–antiparticle pairs, as discussed in Chapter 12, which makes the electron practically look like an "extended object." The electron does not appear as simply point-like any longer!** The anomalous magnetic moment a defines the deviation from $g = 2$. It can be defined as:

$$a \equiv \frac{g}{2} - 1 = \frac{g - 2}{2} = F_2(0) \tag{14.6}$$

Experimental results will be discussed and compared to theory in the following sections of this chapter. We will also see in Chapter 16 that this method will also be applicable in order to describe extended objects such as protons or neutrons!

• **Terms with discrete symmetry violation.** The method outlined above is a general procedure. Now we need to ask ourselves a related question. Can the form of the current be even more general than Eq. (14.2)? Or rather, what is the most general form of the electron current that can be considered? The most naive requirement is simply that our current retains its four-vector nature. However, while in the case of derived terms starting from QED as above, where the structure is "calculable" from higher-order corrections, discrete transformation such as P, T, or CP are conserved, since this is intrinsic to the theory. On the other hand, when seeking the most general form of the current, there is no a priori reason to believe that this must be the case (recall the discussion on symmetries in Section 1.7). Accepting violation of the discrete symmetries opens the door for new possible terms. The general current, still consistent with Lorentz covariance, can then be expressed as a function of four structure functions or form factors $F_i(q^2)$ (see Ref. [110]):

$$\bar{u}(p')\left[\underbrace{\gamma^\mu F_1(q^2)}_{\text{electric+magnetic}} + \underbrace{\frac{i}{2m}\sigma^{\mu\nu}q_\nu F_2(q^2)}_{\text{anomalous magnetic}} + \underbrace{i\epsilon^{\mu\nu\alpha\beta}\frac{\sigma_{\alpha\beta}}{4m}q_\nu F_3(q^2)}_{\text{electric dipole}} + \underbrace{\frac{1}{2m}\left(q^\mu - \frac{q^2}{2m}\gamma^\mu\right)\gamma^5 F_4(q^2)}_{\text{anapole}}\right]u(p) \tag{14.7}$$

The functions $F_i(q^2)$ are real. The first two terms, proportional to F_1 and F_2, are the same as before. There are two new terms proportional to the so-called F_3 and F_4. What is the meaning of the remaining form factors? The term with F_3 is related to the electric dipole moment of the particle. It can be shown that it leads to an electric interaction of the form:

$$-d\vec{\sigma} \cdot \vec{E} = \frac{1}{2m}F_3(0)\left(\vec{\sigma} \cdot \vec{E}\right) \tag{14.8}$$

where d is the **electric dipole moment of the particle**. The last term involving F_4 is called the **Zeldovich anapole moment**. It can be shown to lead to the following peculiar-looking interaction:

$$F_4(0)\vec{\sigma} \cdot \left[\nabla \times \vec{B} - \frac{\partial \vec{E}}{\partial t} \right] \tag{14.9}$$

Up until now no one has measured a non-vanishing electric dipole moment at any value of q^2 for any particle. This is believed to be related to the fact that the electric dipole moment breaks the time-reversal symmetry. Indeed, recalling Table 1.3, we see that under T we have: $\vec{\sigma} \to -\vec{\sigma}$ and $\vec{E} \to +\vec{E}$. This does not mean that it should be strictly zero, as T violation is not excluded by a gauge principle, but does lead to the suspicion that the electric dipole moment, if non-zero, must at least be very small. As is the case with F_3, there is no direct experimental evidence for F_4. The anapole moment can be shown to violate parity.

14.2 The Magnetic Moment of a Point-Like Dirac Particle

In Sections 4.10 and 4.11, we have derived the behavior of a particle in an external magnetic field within a non-relativistic approach. We want to verify that the prediction of the Dirac equation reduces to the Schrödinger case in the non-relativistic limit. Starting from Dirac's ansatz (see Eq. (8.2)), we apply the **principle of minimal substitution**, Eq. (4.101), and find:

$$i\frac{\partial \Psi}{\partial t} = \left(\vec{\alpha} \cdot (\vec{p} - e\vec{A}) + \beta m + e\phi \mathbb{1} \right) \Psi = (\vec{\alpha} \cdot \vec{\pi} + \beta m + e\phi \mathbb{1}) \Psi \tag{14.10}$$

where in the last term we introduced:

$$\vec{\pi} \equiv \vec{p} - e\vec{A} \tag{14.11}$$

At low momentum (or slow velocities) this equation should reduce to the Pauli theory. We decompose the spinor Ψ into two parts and further split off the rest mass factor:

$$\Psi(x) = e^{-imx^0} \begin{pmatrix} \Psi_A(x) \\ \Psi_B(x) \end{pmatrix} \tag{14.12}$$

It follows that:

$$\begin{aligned} H\Psi &= i\partial_0 \Psi = i\partial_0 e^{-imx^0} \begin{pmatrix} \Psi_A \\ \Psi_B \end{pmatrix} = m e^{-imx^0} \begin{pmatrix} \Psi_A \\ \Psi_B \end{pmatrix} + e^{-imx^0} i\partial_0 \begin{pmatrix} \Psi_A \\ \Psi_B \end{pmatrix} \\ &= m\Psi + e^{-imx^0} i\partial_0 \begin{pmatrix} \Psi_A \\ \Psi_B \end{pmatrix} \end{aligned} \tag{14.13}$$

Hence

$$i\partial_0 \begin{pmatrix} \Psi_A(x) \\ \Psi_B(x) \end{pmatrix} = (H - m\mathbb{1}) \begin{pmatrix} \Psi_A(x) \\ \Psi_B(x) \end{pmatrix} \tag{14.14}$$

We express this result in terms of 2×2 blocks:

$$H - m\mathbb{1} = \begin{pmatrix} 0 & \vec{\sigma} \cdot \vec{\pi} \\ \vec{\sigma} \cdot \vec{\pi} & 0 \end{pmatrix} + \begin{pmatrix} 0 & 0 \\ 0 & -2m \end{pmatrix} + \begin{pmatrix} e\phi & 0 \\ 0 & e\phi \end{pmatrix} \tag{14.15}$$

This leads to the system of two coupled equations:

$$\begin{cases} i\partial_0 \Psi_A(x) = (\vec{\sigma} \cdot \vec{\pi})\Psi_B(x) + e\phi\Psi_A(x) \\ i\partial_0 \Psi_B(x) = (\vec{\sigma} \cdot \vec{\pi})\Psi_A(x) - (2m - e\phi)\Psi_B(x) \end{cases} \tag{14.16}$$

It is important to note that the rest mass term appears only in one equation. If the energy of the particle and the field interaction are small compared to the rest mass, we can write using the second equation (since $i\partial_0\Psi/m \ll 1$, which implies that $i\partial_0\Psi_B(x)$ also vanishes):

$$(\vec{\sigma}\cdot\vec{\pi})\Psi_A(x) - (2m - e\phi)\Psi_B(x) \simeq 0 \tag{14.17}$$

Hence

$$\Psi_B(x) = \frac{(\vec{\sigma}\cdot\vec{\pi})}{2m - e\phi}\Psi_A(x) = \left(\frac{1}{2m}\right)\frac{(\vec{\sigma}\cdot\vec{\pi})}{1 - \frac{e}{2m}\phi}\Psi_A(x) \simeq \frac{1}{2m}\left(1 + \frac{e}{2m}\phi\right)(\vec{\sigma}\cdot\vec{\pi})\Psi_A(x) \tag{14.18}$$

Finally we find the following expression valid in the non-relativistic and weak potentials case:

$$i\partial_0\Psi_A(x) \simeq \frac{1}{2m}\left(1 + \frac{e}{2m}\phi\right)(\vec{\sigma}\cdot\vec{\pi})^2\Psi_A(x) + e\phi\Psi_A(x) \tag{14.19}$$

Recalling the algebra of the Pauli matrices (see Appendix C.12), we get:

$$(\vec{\sigma}\cdot\vec{\pi})^2 = \vec{\pi}^2 + i\vec{\sigma}\cdot(\vec{\pi}\times\vec{\pi}) \tag{14.20}$$

The cross-product of the second term does not vanish since we are dealing with operators acting on a bi-spinor ψ, and it can readily be computed:

$$
\begin{aligned}
(\vec{\pi}\times\vec{\pi})\psi &= ((\vec{p}-e\vec{A})\times(\vec{p}-e\vec{A}))\psi = -\frac{e}{i}\left[\nabla\times\vec{A} + \vec{A}\times\nabla\right]\psi \\
&= -\frac{e}{i}\left[(\nabla\times\vec{A})\psi + (\nabla\psi)\times\vec{A} + \vec{A}\times\nabla\psi\right] \\
&= -\frac{e}{i}(\nabla\times\vec{A})\psi = -\frac{e}{i}\vec{B}\psi
\end{aligned} \tag{14.21}
$$

where in the last term we have identified the magnetic field $\vec{B} \equiv \nabla\times\vec{A}$. Thus, for simplicity in absence of an external potential ϕ, we have found that the non-relativistic limit of the Dirac equation leads to the following expression:

$$i\partial_0\Psi_A(x) \simeq \frac{1}{2m}\left((\vec{p}-e\vec{A})^2 - e\vec{\sigma}\cdot\vec{B}\right)\Psi_A(x) = \left(\frac{(\vec{p}-e\vec{A})^2}{2m} - \frac{e}{2m}\vec{\sigma}\cdot\vec{B}\right)\Psi_A(x) \tag{14.22}$$

This is exactly equivalent to the result obtained starting from the Schrödinger equation – see Eq. (4.130)! This represents another success of the Dirac theory, manifesting itself in its non-relativistic approximation yielding back the classical Schrödinger result. In this context, the g-**factor** (also g value, Landé factor, or dimensionless magnetic moment) was defined (see Section 4.11). Making use of the **Bohr magneton** μ_B defined in Eq. (4.134), the interaction of the Dirac particle with the magnetic field can then be expressed as a function of the spin operator $\vec{S} = (1/2)\vec{\sigma}$:

$$-\frac{e}{2m}\vec{\sigma}\cdot\vec{B} = -2\frac{e}{2m}\vec{S}\cdot\vec{B} = -g_{Dirac}\mu_B(\vec{S}\cdot\vec{B}) \equiv \vec{\mu}\cdot\vec{B} \tag{14.23}$$

where the g-factor of a Dirac particle is *predicted by the theory* to be:

$$g_{Dirac} = 2 \tag{14.24}$$

and hence the magnetic moment of a Dirac electron is:

$$-\vec{\mu}_{Dirac} = g_{Dirac}\mu_B\vec{S} \tag{14.25}$$

Thus, the interaction of the magnetic moment with the external magnetic field $\vec{B}$ with $g = 2$ is a consequence of the Dirac equation. It's a big success! However, as discussed in the introduction to this chapter and as will be developed in the next section, QED radiative corrections alter this result, which is true only for a "bare" point-like Dirac particle.

14.3   Electron Magnetic Moment

Is the electron a point-like Dirac particle? Sure, and following its introduction, the Dirac theory enjoyed great success in predicting many phenomena of atomic physics. In 1947, however, Nafe, Nelson, and Rabi obtained precision measurements of the hyperfine-structure intervals in hydrogen and deuterium [111]. They noted that their precise measurements showed *a 0.2% discrepancy from predictions of the hyperfine-structure interval based on g = 2*. These experiments indicated that the Dirac theory of the electron was no longer completely satisfactory, and thus set the stage for the introduction of the current theory of quantum electrodynamics.

Two distinct experimental techniques have been developed to permit precision measurements of the *g*-factor: (1) "**precession experiments**" – the direct observation of spin precession of polarized electrons (or muons) in a constant magnetic field; (2) "**resonance experiments**" – where an oscillating electromagnetic field induces transitions between energy levels of the electron interacting with a static magnetic field. The two techniques are presented in the next paragraphs.

(1) Precession experiments. A particle of rest mass m and charge e moves with velocity $\vec{v}$ in a constant magnetic field $\vec{B}$. The orbital motion is a uniform rotation at the **cyclotron frequency** ω_c:

$$\omega_c \equiv \underbrace{\frac{\omega_0}{\gamma}}_{\text{cyclotron}} \qquad \text{where} \qquad \omega_0 \equiv \frac{eB}{m} \tag{14.26}$$

where the relativistic correction due to γ has been taken into account. The spin motion, as viewed from the laboratory frame, is a uniform **Larmor**[2] **precession** at the frequency:

$$\omega_s = \underbrace{\frac{g}{2}\gamma\omega_c}_{=\frac{g}{2}\omega_0} + \underbrace{(1-\gamma)\omega_c}_{\text{Thomas}} \tag{14.27}$$

The first term is equivalent to the precession for a particle at rest and the second is the **Thomas precession** frequency due to the acceleration of the circular motion (see Section 5.6). The **relative precession** of the spin relative to the velocity occurs at the frequency ω_D:

$$\omega_D \equiv \omega_s - \omega_c = \frac{g}{2}\gamma\omega_c + (1-\gamma)\omega_c - \omega_c = \left(\frac{g}{2}-1\right)\gamma\omega_c = \left(\frac{g}{2}-1\right)\gamma\frac{eB}{\gamma m} \equiv a\omega_0 \tag{14.28}$$

where $a \equiv (g/2-1) = (g-2)/2$ is the **anomalous magnetic moment** defined in Eq. (14.6). For a point-like Dirac particle we have $g = 2$ and therefore $a_{Dirac} = 0$! This is already great but what makes this result even more fantastic is that the relative precession frequency is independent of γ. This fortunate circumstance allows us to measure a without a first-order correction to the velocity.

The schematic principle of a precession experiment is shown in Figure 14.1. One uses a polarized source of electrons, which are stored in a constant magnetic field for a time T after which they are analyzed by a **polarimeter**, an apparatus which measures the amount of polarization $\hat{S} \cdot \hat{n}$ relative to a fixed direction $\hat{n}$. The output of the polarimeter will oscillate as a function of the storage time T with a behavior proportional to $\cos\omega_D T$. The output of the polarimeter as a function of T can be fitted precisely to extract ω_D. The accuracy with which it can be determined will increase in direct proportion to T, all other factors being equal, so that the most obvious way to improve the precision of the measurement is to increase the time the particle spends in the magnetic field. In a real experiment, the magnetic field will never be "exactly constant" and also typically electric fields will be necessary to guide the particles in the desired direction. Thus the fields in the storage region will consist of a nearly homogeneous magnetic field $B(r)$, as well as a weak, but not necessarily homogeneous, electric field $E(r)$. The problem of computing the general trajectory in these electric

2   Sir Joseph Larmor (1857–1942), a Northern Irish physicist and mathematician.

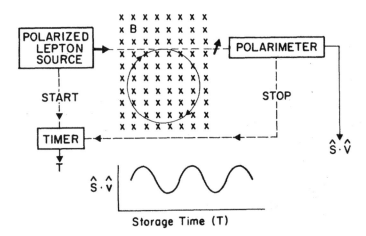

Figure 14.1 Schematic of a precession experiment. Reprinted with permission from A. Rich and J. C. Wesley, "The current status of the lepton g factors," *Rev. Mod. Phys.*, vol. 44, pp. 250–283, Apr 1972. Copyright 1972 by the American Physical Society.

and magnetic fields determined by the Lorentz force plus a spin precession is given by the **Bargmann–Michel–Telegdi**[3] equation (BMT) [112].

(2) Resonance experiments. The storage method employed in resonance experiments is based on the confinement of low-energy (0.01–10 eV) electrons in a **Penning trap**, see Figure 14.2. The trap was named after **Penning**[4] by **Dehmelt**[5] who built the first trap. It consists of a uniform axial magnetic field $\vec{B} \equiv B_0 \hat{z}$ and a superimposed electric quadrupole field generated by a pair of hyperbolic electrodes that surround the storage region. The magnetic field confines the electrons radially, while the electric field confines them axially. With a potential V_0 across the electrodes, the electrostatic potential in the trap is:

$$V(r, z) = \left(\frac{V_0}{r_0^2} \right) \left(\frac{r^2}{2} - z^2 \right) \tag{14.29}$$

where $r^2 = x^2 + y^2$ and r_0 are defined in Figure 14.2.

• **Motion in a Penning trap.** The motion in the trap is composed of three decoupled periodic oscillations, as shown in Figure 14.3. Their derivation is left to **Ex 14.1**. The three motions are listed in the following, with their approximate frequencies for $B = 0.4$ T, $V_0 = 10$ V, and $r_0 = 0.8$ cm:

1. A cyclotron motion at frequency $\omega_B = \omega_0 - \omega_{EB}$ ($\nu_B \simeq 12$ GHz).

2. An axial oscillation at frequency ($\nu_E \simeq 40$ MHz):

$$\omega_E = \sqrt{\frac{2eV_0}{mr_0^2}} \tag{14.30}$$

3   Valentine Louis Telegdi (1922–2006), Hungarian-born American physicist.
4   Frans Michel Penning (1894–1953), Dutch experimental physicist.
5   Hans Georg Dehmelt (1922–2017), German and American physicist.

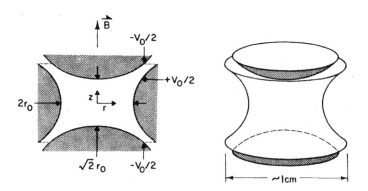

Figure 14.2 Schematic of a Penning trap for the resonance experiment. Reprinted with permission from A. Rich and J. C. Wesley, "The current status of the lepton g factors," *Rev. Mod. Phys.*, vol. 44, pp. 250–283, Apr 1972. Copyright 1972 by the American Physical Society.

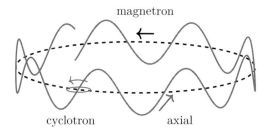

Figure 14.3 Schematic of the three decoupled periodic motions in a Penning trap called magnetron (dashed), axial (wavy), and cyclotron (fast circular).

3. A slow drift of the cyclotron orbit center around z at the magneton frequency ($\nu_{EB} \simeq 70$ kHz):

$$\omega_{EB} = \frac{\omega_0}{2} - \sqrt{\frac{\omega_0^2}{4} - \frac{\omega_E^2}{2}} = \frac{\omega_0}{2} - \frac{\omega_0}{2}\sqrt{1 - \frac{4\omega_E^2}{2\omega_0^2}} \approx \frac{\omega_0}{2}\left(1 - \left(1 - \frac{1}{2}\frac{2\omega_E^2}{\omega_0^2}\right)\right) = \frac{\omega_E^2}{2\omega_0} \quad (14.31)$$

where in the last term we have used the fact that $\omega_E \ll \omega_B$ for the considered experimental parameters.

The energy eigenvalues of an electron in this configuration can be calculated exactly and are given as a function of three quantum state numbers n_B, n_E, and n_{EB} and the spin $s_z = \pm 1/2$:

$$E(n_B, n_E, n_{EB}, s_z) = \left(n_B + \frac{1}{2}\right)\omega_B + \left(n_E + \frac{1}{2}\right)\omega_E - \left(n_{EB} + \frac{1}{2}\right)\omega_{EB} - (1 + a)\omega_0 s_z \quad (14.32)$$

Transitions between eigenstates are induced by appropriate excitation of oscillating electromagnetic fields. The resonance curves can be measured to precisely determine the corresponding frequencies. There are three transitions of particular interest:

1. $\Delta n_B = \pm 1$ – cyclotron orbit transitions with energy $\Delta E = \omega_B$.

2. $\Delta s_z = \pm 1$ – spin-flip transition with energy $\Delta E = \omega_S = (1 + a)\omega_0$.

3. $\Delta n_B = \pm 1, \Delta s_z = \mp 1$ – the so-called direct $g - 2$ transition $\Delta E = \omega_S - \omega_B$.

In order to measure the anomaly a, one measures the direct $g - 2$ transition frequency $\omega_S - \omega_B$ and either ω_S or ω_0.

- **Historical developments.** After the measurements of Nafe, Nelson, and Rabi in 1947 [111], Breit suggested that the hyperfine-structure discrepancy may be due to $g_e \neq 2$ [113]. Kusch and Foley performed the first quantitative measurement of a_e using an atomic beam technique to measure Zeeman splittings. They determined that [114]:

$$a_e(\exp) = 0.00119 \pm 0.00005 \tag{14.33}$$

In 1948, **Schwinger** used the mass renormalization concept suggested by Kramers (see Section 12.4) to cure divergences in QED and calculated the second-order correction to the electron magnetic moment [97], which is basically related to the QED vertex correction Eq. (12.101):

$$a_e(\text{theory}) = \frac{\alpha}{2\pi} \simeq 0.00116 \tag{14.34}$$

This result is in agreement with experimental data! In 1949 Dyson showed that lepton anomalies (as well as all QED observables) can in principle be calculated to an arbitrary order in α. He gave a precise "program" for computing the nth-order contribution [115]:

$$a_e(\text{theory}) = A_e\left(\frac{\alpha}{\pi}\right) + B_e\left(\frac{\alpha}{\pi}\right)^2 + C_e\left(\frac{\alpha}{\pi}\right)^3 + D_e\left(\frac{\alpha}{\pi}\right)^4 + E_e\left(\frac{\alpha}{\pi}\right)^5 + \cdots \tag{14.35}$$

Schwinger's result therefore gives $A_e = 1/2$. It seemed that the deviation from $g_e = 2$ was therefore well understood within QED. Electrons we encounter in Nature do not behave as point-like Dirac particles. An extended appearance, in the vacuum around them, is engendered when virtual particles are being constantly created and annihilated from electromagnetic interactions. These advances triggered a "race" between theory – to compute higher-order QED corrections – and experimentalists – to invent new ingenious methods to improve measurement accuracy.

In 1963 Wilkinson and Crane refined a measurement using a Mott double-scattering technique with a magnetic field between scatterings to find [116]:

$$a_e(\exp) = 0.001159622 \pm 0.000000027 \tag{14.36}$$

In 1966 Rich and Crane extended the technique to positrons and measured for the first time the **positron anomalous magnetic moment** [117]:

$$a_{e^+}(\exp) = 0.001168 \pm 0.000011 \tag{14.37}$$

The result agrees at 10 ppm (parts per million) with the one obtained for the electron, as expected by CPT invariance. In 1968, the direct observation of spin and cyclotron resonances of free thermal electrons in a Penning configuration ion trap was performed by Gräff et al. [118], finding $a_e(\exp) = 0.001159 \pm 0.000002$. In 1969, Gräff, Klempt, and Werth reported the first evidence of $g - 2$ transitions by RF absorption and measured $a_e(\exp) = 0.001159660 \pm 0.000000030$ [119].

- **CODATA value.** The current a_e value is known with extremely high precision and is compiled by the CODATA [10] by NIST to be:

$$a_e(\exp) = 0.00115965218091 \pm 0.00000000000026 \tag{14.38}$$

or a precision of about 200 parts per trillion. This result is based largely on the most precise measurement done by the Harvard group in 2008 using a resonance measurement in a Penning trap with a single electron (one-electron quantum cyclotron) [120].

- **Latest theoretical developments.** In order to test QED to such a precision, it is necessary to have a theoretical value up to the fifth term in the Dyson expansion since $(\alpha/\pi)^5 \simeq 0.07 \times 10^{-12}$. In the Standard Model and at the level of this precision, the theoretical value has contributions from three types of interactions: electromagnetic, hadronic, and electroweak. The last two are very small but cannot be ignored at this level of precision. The first three terms are known analytically. The fourth term has contributions from 891 Feynman diagrams and the fifth term is the result of 12,672 diagrams, which were calculated thanks to a gigantic numerical analysis [121]. In order to obtain the theoretical prediction of a_e, however, one needs the input parameter α. QED itself cannot determine what the fine-structure constant α is. Its value can only be derived from measurements.

• **An independent measurement of α.** While the measurement of the anomalous magnetic moment can be used to determine the fine-structure constant α, a precise and independent measurement of α tests the overall coherence of the theory. In 2018, Parker *et al.* published a value of α with an accuracy of 0.2 parts per billion [122]. They used matter–wave interferometry with a cloud of cesium-133 atoms to make the most accurate measurement to date. Starting from Eq. (2.4), their measurement is based on the fact that the fine structure can be expressed as:

$$R_\infty = \frac{\alpha^2 m_e c}{2h} \quad \Longrightarrow \quad \alpha^2 = \frac{2R_\infty h}{m_e c} = \frac{2R_\infty}{c}\frac{h}{m_e} = \underbrace{\left(\frac{2R_\infty}{c}\right)}_{\text{H spectroscopy}} \times \underbrace{\left(\frac{m_{Cs}}{m_e}\right)}_{\text{Penning traps}} \times \underbrace{\left(\frac{2\pi\hbar}{m_{Cs}}\right)}_{\text{interferometry [122]}} \tag{14.39}$$

The idea to determine the last term is to measure the kinetic energy of the cesium-133 atom that recoils from scattering a photon. The kinetic energy is in this case:

$$E_{kin} = \frac{\hbar^2 k^2}{2m_{Cs}} \quad \text{with} \quad k = 2\pi/\lambda \tag{14.40}$$

where λ is the wavelength of the laser. The other terms are known with very high precision: 0.006 ppb for the Rydberg constant and the ratio of atom-to-electron is known to better than 0.1 ppb for many atoms, including cesium-133. Hence, the measurement of the recoil kinetic energy combined with the two other ratios provides an independent measurement of α. Parker *et al.* found:

$$\alpha^{-1}(^{133}Cs) = 137.035999046 \pm 0.000000027 \quad (0.2 \text{ ppb}) \tag{14.41}$$

• **Comparison between experiment and theory.** This is still an evolving field. At the time of writing, we can report on the recent conclusion from Aoyama *et al.* in Ref. [123]. With the cesium-133 measurement of α, the theoretical computation up to the fifth order yields:

$$a_e(\text{theory}) = 0.001159652181606(229)_\alpha(11)_{\text{theory}}(12)_{\text{hadronic}} \tag{14.42}$$

where uncertainties are due to the fine-structure constant, the numerical evaluation of the fifth order, and the hadronic contribution in this order. The dominant error still comes from the knowledge of the α coupling constant! The difference between experiment and theory then becomes:

$$\delta a_e = a_e(\text{exp}) - a_e(\text{theory}) = -(0.696 \pm 0.316) \times 10^{-12} \tag{14.43}$$

This value could be interpreted as a slight tension between theory and experiment, since it shows a $\sim 2\sigma$ deviation from zero. Maybe with more precision some deviations from QED will hint at new physics at very high energy. But within the precisions obtained so far, it is not possible to conclude that this is the case.

14.4   Muon Magnetic Moment

The (known) interactions of the muons are identical to those of the electrons, yet the rest mass of the muon is about 200 times that of the electron. In 1957 Suura and Wichman [124] and independently Petermann [125] calculated the difference between the electron and the muon anomalous magnetic moment. The difference between a_μ and a_e comes from the dependence of the vacuum polarization terms on the mass of the fermion and on hadronic corrections. It is approximately 0.6% of a.

In 1957 Coffin *et al.* [126] obtained a first measurement by measuring the spin precession frequency of stopped muons in a magnetic field. In order to extract g_μ from this measurement, one needs to know the muon mass m_μ, which at the time limited the accuracy of the comparison with a_e. A direct determination of a_μ avoids the necessity of correcting the results for the ratio m_μ/m_e. In 1961 the first direct determination of a_μ was performed at CERN by Charpak *et al.* [127] with an accuracy of 2%, which was then improved to 0.5% in 1965. As for the case of the electron, the anomalous magnetic moment of the muon was measured by comparing the relative precession of the muon spin with respect to the velocity in a nearly constant magnetic field.

In further refinements of this method by Farley *et al.* (1966) [128] and Bailey *et al.* (1968) [129], the magnet was replaced by a **muon storage ring** in order to increase the total $g-2$ precession angle of the trapped muons. The concept of such an experiment is illustrated in Figure 14.4. The 1968 measurement reached an accuracy of 265 ppm of a_μ. They also performed a stringent CPT test by comparing g_μ for μ^+ and μ^-. Central to

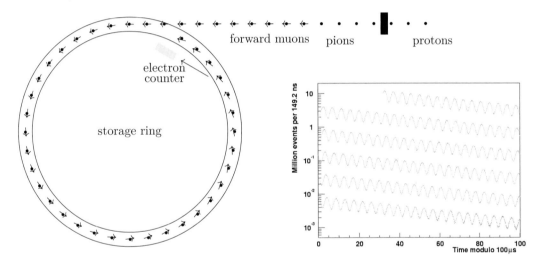

Figure 14.4 Schematic of the $g-2$ muon storage ring experiment. High-energy protons hit a target, producing secondary pions. These latter shortly decay into muons (and neutrinos). The muons are polarized. The arrows symbolize the muon spin direction. Muons of a given momentum are trapped in the storage ring until they decay. The decay electron is analyzed to determine the muon polarization. At every turn, the angle between the momentum and the spin of the muon increases due to the anomalous $g-2$ term, as schematically illustrated. **Inset:** The distribution of electron counts versus time for 3.6×10^9 muon decays from the BNL experiment, reprinted with permission from G. Bennett *et al.*, "Final report of the muon E821 anomalous magnetic moment measurement at BNL," *Phys. Rev. D*, vol. 73, p. 072003, 2006. Copyright 2006 by the American Physical Society.

the experiment is the production of polarized muons in the weak decays $\pi \to \mu\nu$ of pions produced by high-energy protons hitting a target. The experiment hence starts with a proton accelerator from which high-energy protons are extracted and sent on a target. Secondary pions are created in the target. The weak decays produce particles with polarization depending on their momentum, due to parity violation in the weak process (to be introduced in Chapter 22). By selecting forward muons with the highest or lowest momentum, one obtains a highly polarized muon beam. The muons are stored in the ring which selects particles of a given momentum. The time-dilated lifetime of the muons $\gamma\tau_\mu \simeq E_\mu/m_\mu \times (2.2~\mu\text{s})$ allows them to circle several times in the ring, while precession occurs. When muons decay $\mu \to e\nu\bar{\nu}$ again via a weak process, the energy spectrum of the electrons contains information about the muon polarization (to be discussed in Chapter 23). The number of electrons detected above a certain energy threshold will oscillate with time, as the spin precesses.

● **The Brookhaven era.** The measurements of g-2 of the muon performed over the years are summarized in Table 14.1. The currently most precise results were obtained by the E821 Muon (g-2) experiment at

±	Measurement	σ_{a_μ}/a_μ	Sensitivity	Reference
μ^+	$g = 2.00 \pm 0.10$		$g = 2$	Garwin *et al.* [130], Nevis (1957)
μ^+	$0.001\,13^{+0.00016}_{-0.00012}$	12.4%	$\frac{\alpha}{\pi}$	Garwin *et al.* [131], Nevis (1959)
μ^+	$0.001\,145(22)$	1.9%	$\frac{\alpha}{\pi}$	Charpak *et al.* [132] CERN 1 (SC) (1961)
μ^+	$0.001\,162(5)$	0.43%	$\left(\frac{\alpha}{\pi}\right)^2$	Charpak *et al.* [133] CERN 1 (SC) (1962)
$\mu^\pm$	$0.001\,166\,16(31)$	265 ppm	$\left(\frac{\alpha}{\pi}\right)^3$	Bailey *et al.* [129] CERN 2 (PS) (1968)
$\mu^\pm$	$0.001\,165\,895(27)$	23 ppm	$\left(\frac{\alpha}{\pi}\right)^3$ + Hadronic	Bailey *et al.* [134] CERN 3 (PS) (1975)
$\mu^\pm$	$0.001\,165\,911(11)$	7.3 ppm	$\left(\frac{\alpha}{\pi}\right)^3$ + Hadronic	Bailey *et al.* [135] CERN 3 (PS) (1979)
μ^+	$0.001\,165\,919\,1(59)$	5 ppm	$\left(\frac{\alpha}{\pi}\right)^3$ + Hadronic	Brown *et al.* [136] BNL (2000)
μ^+	$0.001\,165\,920\,2(16)$	1.3 ppm	$\left(\frac{\alpha}{\pi}\right)^4$ + Weak	Brown *et al.* [137] BNL (2001)
μ^+	$0.001\,165\,920\,3(8)$	0.7 ppm	$\left(\frac{\alpha}{\pi}\right)^4$ + Weak + ?	Bennett *et al.* [138] BNL (2002)
μ^-	$0.001\,165\,921\,4(8)(3)$	0.7 ppm	$\left(\frac{\alpha}{\pi}\right)^4$ + Weak + ?	Bennett *et al.* [139] BNL (2004)
$\mu^\pm$	$0.001\,165\,920\,80(63)$	0.54 ppm	$\left(\frac{\alpha}{\pi}\right)^4$ + Weak + ?	Bennett *et al.* [139, 140] BNL WA (2004)

Table 14.1 Historical measurements of the muon anomalous magnetic moment.

the Brookhaven National Laboratory with a precision below 1 ppm. Recalling the BMT equation [112], the equation of motion of the muon spin vector $\vec{\omega}_s$ is:

$$\begin{aligned}
\vec{\omega}_s &= \frac{e}{m_\mu}\left[\left(\frac{g_\mu}{2} - 1 + \frac{1}{\gamma}\right)\vec{B} - \left(\frac{g_\mu}{2} - 1\right)(\vec{\beta}\cdot\vec{B})\vec{\beta} - \left(\frac{g_\mu}{2} - \frac{\gamma}{\gamma+1}\right)\left(\vec{\beta}\times\vec{E}\right)\right] \\
&= \frac{e}{m_\mu}\left[\left(a_\mu + \frac{1}{\gamma}\right)\vec{B} - a_\mu(\vec{\beta}\cdot\vec{B})\vec{\beta} - \left(a_\mu + \frac{\gamma-1}{\gamma^2-1}\right)\left(\vec{\beta}\times\vec{E}\right)\right]
\end{aligned} \tag{14.44}$$

The trajectory is defined by the Lorentz force $\mathrm{d}\vec{p}/\mathrm{d}t = \vec{\omega}_c \times \vec{p} = e(\vec{E} + \vec{\beta}\times\vec{B})$, where the cyclotron frequency $\vec{\omega}_c$ is:

$$\vec{\omega}_c = \frac{e}{m_\mu}\left[\frac{1}{\gamma}\vec{B} - \frac{\gamma}{\gamma^2-1}\left(\vec{\beta}\times\vec{E}\right)\right] \tag{14.45}$$

The difference $\vec{\omega}_a = \vec{\omega}_s - \vec{\omega}_c$ is given by:

$$\vec{\omega}_a = \frac{e}{m_\mu}\left[a_\mu\vec{B} - a_\mu(\vec{\beta}\cdot\vec{B})\vec{\beta} - \left(a_\mu - \frac{1}{\gamma^2-1}\right)\left(\vec{\beta}\times\vec{E}\right)\right] \tag{14.46}$$

where $a_\mu = (g_\mu - 2)/2$. If we assume that the motion of the muon is always perpendicular to the magnetic field in the storage ring, then $(\vec{\beta}\cdot\vec{B}) \simeq 0$. Furthermore, for the "magic momentum" defined as:

$$a_\mu - \frac{1}{\gamma^2-1} = 0 \quad \to \quad \gamma = \sqrt{1 + \frac{1}{a_\mu}} \simeq 29.26 \tag{14.47}$$

the term proportional to $\vec{\beta} \times \vec{E}$ also vanishes. Hence, for a storage ring capturing muons of momentum $\simeq 3.1$ GeV/c, the spin difference frequency is directly proportional to a_μ:

$$\vec{\omega}_a|_{magic} = a_\mu \frac{e\vec{B}}{m_\mu} \tag{14.48}$$

The E821 results for negative and positive muons are [140]:

$$a_{\mu^-} = 11659214(8)(3) \times 10^{-10} \quad \text{and} \quad a_{\mu^+} = 11659204(7)(5) \times 10^{-10} \tag{14.49}$$

They are consistent with each other and can be combined to give:

$$a_\mu(\exp) = 11659208(6) \times 10^{-10} \quad (0.5 \text{ ppm}) \tag{14.50}$$

The Standard Model prediction consists of QED, hadronic, and weak contributions. Its theoretical uncertainty is dominated by the knowledge on the lowest-order hadronic vacuum polarization. This contribution can be determined directly from the annihilation of e^+e^- to hadrons through a dispersion integral [141]. The indirect determination using data from hadronic τ decays, the conserved vector current hypothesis, plus the appropriate isospin corrections, could in principle improve the precision. However, discrepancies between the results exist. The two data sets do not give consistent results. Using e^+e^- annihilation data, the corresponding theoretical value is:

$$a_\mu(\text{theory}) = 11659181(8) \times 10^{-10} \quad (0.7 \text{ ppm}) \tag{14.51}$$

The value deduced from τ decay is larger by 15×10^{-10} and has a stated uncertainty of 7×10^{-10} (0.6 ppm). The difference between the experimental determination and the theory using the e^+e^- or τ data is 2.7σ and 1.4σ, respectively. This is clearly a very impressive achievement. However, it does also point to a limitation of this technique. In case of significant discrepancy between experiment and theory, the interpretation requires making assumptions on the source of the discrepancy, which can be either theoretical or experimental in nature.

14.5   Bound States – Hydrogenic Atoms

The study of **hydrogenic atoms** has been a parallel line of very fruitful investigation and precision tests of QED. Examples of such bound systems with one electron are the hydrogen (the e^-p bound state) and the "**exotic atoms**," such as the **positronium** (the e^-e^+ bound state), the **muonium** (the $e^-\mu^+$ bound state), the **pionic hydrogen** (the π^-p bound state), or the **muonic hydrogen** (the μ^-p bound state). Recently, the measurement of the Lamb shift in muonic hydrogen has raised a lot of excitement [142].

We recall that **atomic states** are specified using a long-established notation, $n^{2s+1}X_j$, where n is the principal quantum number, s is the spin quantum number, X denotes the orbital angular momentum number (S for $l = 0$, P for $l = 1$, D for $l = 2$, F for $l = 3$), and j denotes the total angular momentum quantum number. A state with $n = 3$, $s = 1/2$, $l = 2$, and $j = 3/2$ is denoted as $3^2D_{3/2}$. For atomic hydrogen s is always $1/2$ and so the $2s + 1$ term is often dropped from the specification, $3^2D_{3/2}$ becoming $3D_{3/2}$, for example. This is the notation we will use in the following.

The atomic structure of hydrogen is summarized in Figure 14.5 with an increasing level of sophistication. In the leftmost column, we have the naive Bohr levels, where the energy only depends on the principal quantum number n. Their energy is simply given by (see Eq. (C.11)):

$$E_n = -\frac{1}{2}\frac{\alpha^2 m_e c^2}{n^2} = -\frac{Ry}{n^2}, \qquad Ry \approx 13.605 \text{ eV} \tag{14.52}$$

A first (non-relativistic) correction that takes into account the finite mass of the proton in the hydrogen atom is to replace the mass of the electron by the reduced mass of the system μ, where

$$\mu \equiv \frac{m_e M_p}{M_p + m_e} \tag{14.53}$$

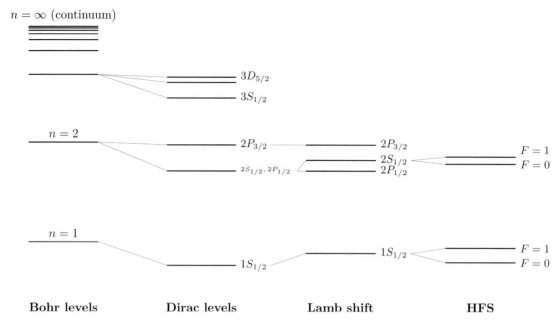

Figure 14.5 Atomic energy levels of the hydrogen atom. For the Dirac levels, the fine-structure constant has been multiplied by a factor 50 in order to make the difference more visible. The scales for the Lamb levels and the hyperfine-structure (HFS) levels have also been expanded for visibility.

which is a correction at the level of $1/1836$. Additional corrections are listed in the following:

- **Fine structure.** Relativistic corrections from the Dirac equation, discussed in the next section.

- **Quantum electrodynamic corrections.** The **Lamb shift**, discussed in Section 14.7, can be explained by QED by its effect on $g_e \neq 2$ and also vacuum polarization, causing the splitting between the $2S_{1/2}$ and $2P_{1/2}$ states.

- **Hyperfine structure.** Corrections due to the interaction between the nuclear and the electron spins, which leads to a nuclear magnetic moment interacting with the orbital and spin degrees of freedom of the electron. This effect is discussed in Section 14.8.

- **Recoil corrections.** Takes into account the nuclear motion.

- **Proton radius.** Nuclear structure corrections (due to the nuclear charge distribution).

These sometimes small shifts might seem like tiny insignificant effects, but they provided considerable insight into QED! In the next sections, we will do an overview of these different effects on the atoms.

14.6   Dirac Solution of the Hydrogen Atom – Fine Structure

We aim to find the energy levels of the hydrogen atom using a relativistic theory. The Dirac formulation accounts for energy shifts due to relativistic corrections and magnetic interactions between the orbital and spin magnetic moments. We have to solve the Dirac equation, Eq. (8.2) with a central field potential $V(r) =$

$-Ze^2/(4\pi r) = -Z\alpha/r$, where $Z = 1$ for hydrogen, so the Dirac equation for such a problem leads to the following Dirac Hamiltonian H_{DC} for a central force:

$$H_{DC} = \vec{\alpha} \cdot \vec{p} + \beta m_e \mathbb{1} + V(r)\mathbb{1} \tag{14.54}$$

In general, we can make the ansatz that the solution spinor $\Psi \equiv (\varphi, \chi)$ is composed of two two-component functions. The Dirac equation becomes:

$$H_{DC} \begin{pmatrix} \varphi \\ \chi \end{pmatrix} = \left[\begin{pmatrix} 0 & \vec{\sigma} \\ \vec{\sigma} & 0 \end{pmatrix} \cdot \vec{p} + \begin{pmatrix} 1 & 0 \\ 0 & -1 \end{pmatrix} m_e + \begin{pmatrix} 1 & 0 \\ 0 & 1 \end{pmatrix} V \right] \begin{pmatrix} \varphi \\ \chi \end{pmatrix} = i\frac{\partial}{\partial t} \begin{pmatrix} \varphi \\ \chi \end{pmatrix} \tag{14.55}$$

This leads to the following eigenvalues equation for a stationary solution:

$$(H_{DC} - E)\Psi = \begin{pmatrix} V - E + m_e & \vec{\sigma} \cdot \vec{p} \\ \vec{\sigma} \cdot \vec{p} & V - E - m_e \end{pmatrix} \begin{pmatrix} \varphi \\ \chi \end{pmatrix} = 0 \tag{14.56}$$

which can be written as two coupled equations:

$$\begin{cases} (V - E + m)\varphi + (\vec{\sigma} \cdot \vec{p})\chi = 0 \\ \\ (V - E - m)\chi + (\vec{\sigma} \cdot \vec{p})\varphi = 0 \end{cases} \tag{14.57}$$

where φ and χ are column vectors with two components. However, the potential is central and the problem is spherically symmetric, and hence invariant under parity P. It follows that the Hamiltonian obeys the following commutations:

$$[H_{DC}, \vec{J}] = 0 \quad \text{and} \quad [H_{DC}, P] = 0 \tag{14.58}$$

where $\vec{J} = \vec{L} + \vec{S}$ is the total angular momentum operator, $\vec{L}$ the orbital angular momentum, and $\vec{S}$ the spin. We start by seeking solutions that are simultaneous eigenfunctions of energy E, J^2, and J_z. The angular part of the eigenfunctions can be constructed by generalizing the concept of spherical harmonics to what is called **spherical spinors**, which must combine the orbital angular momentum dependence and the spin dependence. The spherical spinors are denoted $\Omega_{jlm}(\theta, \phi)$ with:

$$\Omega_{jlm}(\theta, \phi) \equiv \sum_{\mu=\pm 1/2} C(l, 1/2, j; m - \mu, \mu, m)Y_{l,m-\mu}(\theta, \phi)\chi_\mu \tag{14.59}$$

where the spherical harmonics $Y_{l,m}(\theta, \phi)$ associated with the orbital angular momentum are given in Appendix C.9 and the spin part of the wave function is described by the two-component spinors:

$$\chi_{+1/2} = \begin{pmatrix} 1 \\ 0 \end{pmatrix} \quad \text{and} \quad \chi_{-1/2} = \begin{pmatrix} 0 \\ 1 \end{pmatrix} \tag{14.60}$$

The coefficient $C(j_1, j_2, j_3; m_1, m_2, m_3)$ is the Clebsch–Gordan coefficient connecting the angular momentum and the spin with the total angular momentum $\vec{J}$. For each quantum number l we obtain the following two solutions $j = l \pm 1/2$ corresponding to the two possible spin orientations:

$$\Omega_{l+1/2,lm}(\theta, \phi) = \begin{pmatrix} \sqrt{\frac{l+m+1/2}{2l+1}}Y_{l,m-1/2}(\theta, \phi) \\ \sqrt{\frac{l-m+1/2}{2l+1}}Y_{l,m+1/2}(\theta, \phi) \end{pmatrix}; \quad \Omega_{l-1/2,lm}(\theta, \phi) = \begin{pmatrix} -\sqrt{\frac{l-m+1/2}{2l+1}}Y_{l,m-1/2}(\theta, \phi) \\ \sqrt{\frac{l+m+1/2}{2l+1}}Y_{l,m+1/2}(\theta, \phi) \end{pmatrix} \tag{14.61}$$

Recalling that the Hamiltonian commutes with the parity operator, we can also consider parity as a good quantum number. Let us define the operator K as follows:

$$K \equiv -\vec{\sigma} \cdot \vec{L} - \mathbb{1} \tag{14.62}$$

Spherical functions are eigenvectors of $\vec{\sigma} \cdot \vec{L}$ hence also of K, so we can replace $\Omega_{jlm}(\theta, \phi)$ with $\Omega_{\kappa m}(\theta, \phi)$ to obtain the following notation:

$$K\Omega_{\kappa m}(\theta, \phi) = \kappa\Omega_{\kappa m}(\theta, \phi) \tag{14.63}$$

where $\kappa = -l - 1$ for $j = l + 1/2$ and $\kappa = l$ for $j = l - 1/2$. The quantum number κ is hence a non-zero integer number. Its value determines both j and l:

$$j = l + 1/2 \quad : \quad \kappa = -l - 1 = -(j - 1/2) - 1 = -(j + 1/2)$$
$$j = l - 1/2 \quad : \quad \kappa = l = (j + 1/2) \tag{14.64}$$

or for a given j, we have both positive and negative κ solutions:

$$\kappa = \mp(j + 1/2) \qquad \text{for} \quad l = j \pm 1/2 \tag{14.65}$$

We recall that the parity of the spherical harmonics $Y_{l,m}$ is $(-1)^l$ (see Eq. (C.85)). Hence, the two spinors $\Omega_{\kappa m}$ and $\Omega_{-\kappa m}$ have opposite parities since they have the same j but $l = j \pm 1/2$ differing by one unit.

We now need to solve the coupled differential equations. Since a central problem is conveniently expressed in spherical coordinates, we need to write the Dirac equation using those variables. In the Hamiltonian H_{DC} the first term $\vec{\alpha} \cdot \vec{p}$ and the last term $V(r)$ depend on the coordinates. We introduce $\hat{x} = \vec{x}/r$ and in order to obtain another expression for $\vec{\sigma} \cdot \vec{p}$, we calculate with the use of Eq. (C.98):

$$(\vec{\sigma} \cdot \hat{x})(\vec{\sigma} \cdot \vec{p}) = \hat{x} \cdot \vec{p} + i\sigma \cdot (\hat{x} \times \vec{p}) = \hat{x} \cdot \vec{p} + i\frac{\vec{\sigma} \cdot \vec{L}}{r} \tag{14.66}$$

By multiplying this last result by $\vec{\sigma} \cdot \hat{x}$ and noting that $(\vec{\sigma} \cdot \hat{x})^2 = \mathbb{1}$, we obtain:

$$\vec{\sigma} \cdot \vec{p} = \frac{\vec{\sigma} \cdot \hat{x}}{r}\left(r\hat{x} \cdot \vec{p} + i\vec{\sigma} \cdot \vec{L}\right) = \frac{\vec{\sigma} \cdot \hat{x}}{r}\left(-ir\frac{\partial}{\partial r} + i\vec{\sigma} \cdot \vec{L}\right) = -i(\vec{\sigma} \cdot \hat{x})\left(\frac{\partial}{\partial r} + \frac{-\vec{\sigma} \cdot \vec{L}}{r}\right) \tag{14.67}$$

where we used $\vec{p} = -i\nabla = -i\left(\hat{r}\partial/\partial r + \hat{\theta}r^{-1}\partial/\partial\theta + \hat{\phi}(r\sin\theta)^{-1}\partial/\partial\phi\right)$. The operator $(\vec{\sigma} \cdot \hat{x}) = \vec{\sigma} \cdot \vec{x}/r$ is a pseudoscalar under parity, thus:

$$P\frac{\vec{\sigma} \cdot \vec{x}}{r}\Omega_{\kappa m} = -\frac{\vec{\sigma} \cdot \vec{x}}{r}P(\Omega_{\kappa m}) = -\frac{\vec{\sigma} \cdot \vec{x}}{r}(-1)^\ell\Omega_{\kappa m} = (-1)^{\ell-1}\frac{\vec{\sigma} \cdot \vec{x}}{r}\Omega_{\kappa m} \tag{14.68}$$

It follows that $(\vec{\sigma} \cdot \hat{x})\Omega_{\kappa m}$ is an eigenfunction of the parity with eigenvalue $(-1)^{\ell-1} = (-1)^{j\pm1/2-1} = (-1)^{j\mp1/2}$. Hence, the operator $\vec{\sigma} \cdot \hat{x}$ swaps the parity of the spherical spinors:

$$(\vec{\sigma} \cdot \hat{x})\Omega_{\kappa m} = \eta\Omega_{-\kappa m} \tag{14.69}$$

where η is a complex phase which can be conventionally set to $\eta = -1$.

Solutions to the Dirac equation in the spherically symmetric potential for a state with angular momentum quantum numbers (κ, m) can be constructed using both spin orientations, hence using $\Omega_{\kappa m}$ and $\Omega_{-\kappa m}$ (of opposite parities) with the following ansatz:

$$\Psi_{\kappa,m}(\vec{x}) \equiv \begin{pmatrix} \varphi \\ \chi \end{pmatrix} = \begin{pmatrix} if_k(r)\Omega_{\kappa m}(\theta, \phi) \\ g_k(r)\Omega_{-\kappa m}(\theta, \phi) \end{pmatrix} \tag{14.70}$$

where $\vec{x} = (r, \theta, \phi)$. The wave functions $\Psi_{\kappa,m}$ are understood to describe possible states of the electron satisfying the one-electron Dirac equation. The radial part and the angular parts have been factorized and the (arbitrary) functions $f_k(r)$ and $g_k(r)$ describe the radial dependence of the solution. We insert this ansatz into the two coupled equations, Eq. (14.57), and find:

$$\begin{cases} (V - E + m_e)if_k(r)\Omega_{\kappa m}(\theta, \phi) + (\vec{\sigma} \cdot \vec{p})g_k(r)\Omega_{-\kappa m}(\theta, \phi) = 0 \\ \\ (V - E - m_e)g_k(r)\Omega_{-\kappa m}(\theta, \phi) + (\vec{\sigma} \cdot \vec{p})if_k(r)\Omega_{\kappa m}(\theta, \phi) = 0 \end{cases} \tag{14.71}$$

We use Eq. (14.67) and note that $-\vec{\sigma} \cdot \vec{L} = K + \mathbb{1}$ so $-(\vec{\sigma} \cdot \vec{L})\Omega_{\kappa m} = (\kappa + 1)\Omega_{\kappa m}$, thus:

$$\begin{cases} (V - E + m_e)if_k(r)\Omega_{\kappa m} - i\left(\vec{\sigma} \cdot \hat{x}\left(\dfrac{\partial}{\partial r} + \dfrac{-\kappa + 1}{r}\right)\right)g_k(r)\Omega_{-\kappa m} = 0 \\[3mm] (V - E - m_e)g_k(r)\Omega_{-\kappa m} - i\left(\vec{\sigma} \cdot \hat{x}\left(\dfrac{\partial}{\partial r} + \dfrac{\kappa + 1}{r}\right)\right)if_k(r)\Omega_{\kappa m} = 0 \end{cases} \qquad (14.72)$$

and using Eq. (14.69) to remove the angular dependence, we find the *radial* coupled equations:

$$\begin{cases} (V - E + m_e)f_k(r) + \left(\dfrac{d}{dr} + \dfrac{-\kappa + 1}{r}\right)g_k(r) = 0 \\[3mm] (V - E - m_e)g_k(r) - \left(\dfrac{d}{dr} + \dfrac{\kappa + 1}{r}\right)f_k(r) = 0 \end{cases} \qquad (14.73)$$

We can somewhat simplify the expressions with the substitutions $F = rf$ and $G = rg$:

$$\begin{cases} (V - E + m_e)\dfrac{F}{r} = -\left(\dfrac{d}{dr} + \dfrac{-\kappa + 1}{r}\right)\dfrac{G}{r} = -\dfrac{1}{r}\dfrac{dG}{dr} + \dfrac{G}{r^2} - \dfrac{(-\kappa + 1)G}{r^2} = -\dfrac{1}{r}\dfrac{dG}{dr} + \dfrac{\kappa G}{r^2} \\[3mm] (V - E - m_e)\dfrac{G}{r} = \left(\dfrac{d}{dr} + \dfrac{\kappa + 1}{r}\right)\dfrac{F}{r} = \dfrac{1}{r}\dfrac{dF}{dr} - \dfrac{F}{r^2} + \dfrac{(\kappa + 1)F}{r^2} = \dfrac{1}{r}\dfrac{dF}{dr} + \dfrac{\kappa F}{r^2} \end{cases} \qquad (14.74)$$

which finally become:

$$\begin{cases} (V - E + m_e)F = -\dfrac{dG}{dr} + \dfrac{\kappa G}{r} \\[3mm] (V - E - m_e)G = \dfrac{dF}{dr} + \dfrac{\kappa F}{r} \end{cases} \qquad (14.75)$$

This result actually holds for any central potential. We now introduce Coulomb's potential $V = -Z\alpha/r$ and define $k_1 = m_e + E$, $k_2 = m_e - E$ and the dimensionless $\rho = \sqrt{k_1 k_2}\, r$:

$$\begin{cases} -\dfrac{dG}{d\rho} + \dfrac{\kappa G}{\rho} = \dfrac{1}{\sqrt{k_1 k_2}}\left(-\dfrac{Z\alpha}{r} + k_2\right)F = \left(-\dfrac{Z\alpha}{\rho} + \sqrt{\dfrac{k_2}{k_1}}\right)F \\[3mm] \dfrac{dF}{d\rho} + \dfrac{\kappa F}{\rho} = \dfrac{1}{\sqrt{k_1 k_2}}\left(-\dfrac{Z\alpha}{r} - k_1\right)G = \left(-\dfrac{Z\alpha}{\rho} - \sqrt{\dfrac{k_1}{k_2}}\right)G \end{cases} \qquad (14.76)$$

As a first step, we examine these equations at large values of ρ and take a power series with an exponential decay as ansatz:

$$\begin{cases} F(\rho) \equiv \rho^s e^{-\rho} \displaystyle\sum_{m=0}^{\infty} a_m \rho^m = e^{-\rho} \sum_{m=0}^{\infty} a_m \rho^{s+m} \\[3mm] G(\rho) \equiv \rho^s e^{-\rho} \displaystyle\sum_{m=0}^{\infty} b_m \rho^m = e^{-\rho} \sum_{m=0}^{\infty} b_m \rho^{s+m} \end{cases} \qquad (14.77)$$

We assume that the exponential will make the functions go to zero as $\rho \to \infty$. Taking the derivatives with respect to ρ leads to:

$$\dfrac{dF}{d\rho} = \rho^s e^{-\rho}\left[-\sum_{m=0}^{\infty} a_m \rho^m + \sum_{m=0}^{\infty} a_m(s + m)\rho^{m-1}\right] = \rho^s e^{-\rho}\left[\sum_{m=0}^{\infty}\left(a_m(s + m)\rho^{-1} - a_m\right)\rho^m\right] \qquad (14.78)$$

or finally:

$$\begin{cases} \dfrac{\mathrm{d}F}{\mathrm{d}\rho} = \rho^s e^{-\rho}\left[s\,a_0\,\rho^{-1} + \displaystyle\sum_{m=0}^{\infty}((s+m+1)a_{m+1} - a_m)\rho^m\right] \\[3mm] \dfrac{\mathrm{d}G}{\mathrm{d}\rho} = \rho^s e^{-\rho}\left[s\,b_0\,\rho^{-1} + \displaystyle\sum_{m=0}^{\infty}((s+m+1)b_{m+1} - b_m)\rho^m\right] \end{cases} \tag{14.79}$$

and dividing by ρ this can be expressed as:

$$\begin{cases} \dfrac{F}{\rho} = \rho^s e^{-\rho}\left[a_0\,\rho^{-1} + \displaystyle\sum_{m=0}^{\infty} a_{m+1}\rho^m\right] \\[3mm] \dfrac{G}{\rho} = \rho^s e^{-\rho}\left[b_0\,\rho^{-1} + \displaystyle\sum_{m=0}^{\infty} b_{m+1}\rho^m\right] \end{cases} \tag{14.80}$$

We now substitute these results into Eqs. (14.76) to obtain recursion relations:

$$(\kappa - s)\,b_0\,\rho^{-1} - \sum_{m=0}^{\infty}((s+m+1-\kappa)b_{m+1} - b_m)\rho^m = -Z\alpha\left[a_0\,\rho^{-1} + \sum_{m=0}^{\infty} a_{m+1}\rho^m\right] + \sqrt{\frac{k_2}{k_1}}\sum_{m=0}^{\infty} a_m\rho^m \tag{14.81}$$

and

$$(\kappa + s)\,a_0\,\rho^{-1} + \sum_{m=0}^{\infty}((s+m+1+\kappa)a_{m+1} - a_m)\rho^m = -Z\alpha\left[b_0\,\rho^{-1} + \sum_{m=0}^{\infty} b_{m+1}\rho^m\right] - \sqrt{\frac{k_1}{k_2}}\sum_{m=0}^{\infty} b_m\rho^m \tag{14.82}$$

The coefficients of ρ^{-1} and ρ^m must vanish term by term. This gives us four recursion relations:

$$\begin{cases} (\kappa - s)\,b_0 + Z\alpha a_0 = 0 \\[2mm] (\kappa + s)\,a_0 + Z\alpha b_0 = 0 \\[2mm] -(s+m+1-\kappa)b_{m+1} + b_m + Z\alpha a_{m+1} - \sqrt{\dfrac{k_2}{k_1}}a_m = 0 \\[3mm] (s+m+1+\kappa)a_{m+1} - a_m + Z\alpha b_{m+1} + \sqrt{\dfrac{k_1}{k_2}}b_m = 0 \end{cases} \tag{14.83}$$

The first two equations can be written as:

$$\begin{pmatrix} Z\alpha & \kappa - s \\ \kappa + s & Z\alpha \end{pmatrix}\begin{pmatrix} a_0 \\ b_0 \end{pmatrix} = 0 \implies \kappa^2 - s^2 - Z^2\alpha^2 = 0 \implies s^2 = \kappa^2 - Z^2\alpha^2 \tag{14.84}$$

In order for the solutions to behave as $\rho \to 0$, we take the positive solution $s = +\sqrt{\kappa^2 - Z^2\alpha^2}$. We multiply the third expression by $\sqrt{k_1/k_2}$ to get:

$$\begin{cases} -\sqrt{\dfrac{k_1}{k_2}}(s+m+1-\kappa)b_{m+1} + \sqrt{\dfrac{k_1}{k_2}}b_m + \sqrt{\dfrac{k_1}{k_2}}Z\alpha a_{m+1} - a_m = 0 \\[3mm] (s+m+1+\kappa)a_{m+1} - a_m + Z\alpha b_{m+1} + \sqrt{\dfrac{k_1}{k_2}}b_m = 0 \end{cases} \tag{14.85}$$

We subtract the two equations from one another and obtain:

$$-\sqrt{\frac{k_1}{k_2}}(s+m+1-\kappa)b_{m+1} + \sqrt{\frac{k_1}{k_2}}Z\alpha a_{m+1} - (s+m+1+\kappa)a_{m+1} - Z\alpha b_{m+1} = 0 \tag{14.86}$$

or

$$-\sqrt{\frac{k_1}{k_2}}\left(s+m+1-\kappa+\sqrt{\frac{k_2}{k_1}}Z\alpha\right)b_{m+1}-\left(-\sqrt{\frac{k_1}{k_2}}Z\alpha+s+m+1+\kappa\right)a_{m+1}=0 \qquad (14.87)$$

The expansions for F and G are finite only if the series terminate, i.e., if there exists $t=m+1$ with $a_t\neq 0$, $b_t\neq 0$, and $a_{t+1}=b_{t+1}=0$ (we assume that both series terminate at the same index). From this we note that $t=0,1,2,\ldots$ and from the last equations of Eq. (14.83), this implies that:

$$b_t=\sqrt{\frac{k_2}{k_1}}a_t \qquad (14.88)$$

and thus, the condition for terminating the series is:

$$-\sqrt{\frac{k_1}{k_2}}\left(s+t-\kappa+\sqrt{\frac{k_2}{k_1}}Z\alpha\right)b_t-\left(-\sqrt{\frac{k_1}{k_2}}Z\alpha+s+t+\kappa\right)a_t=0 \qquad (14.89)$$

or

$$2\sqrt{k_1 k_2}(s+t)+(k_2-k_1)Z\alpha=0 \qquad (14.90)$$

We now use this expression to yield a constraint on the energy by substituting back k_1 and k_2:

$$2\sqrt{(m_e+E)(m_e-E)}(s+t)-2EZ\alpha=0 \quad\Longrightarrow\quad \sqrt{m_e^2-E^2}(s+t)=Z\alpha E \qquad (14.91)$$

By squaring and collecting the energy terms we find:

$$(s+t)^2 m_e^2=\left(Z^2\alpha^2+(s+t)^2\right)E^2 \quad\Longrightarrow\quad E^2=\frac{m_e^2}{(Z^2\alpha^2+(s+t)^2)/(s+t)^2} \qquad (14.92)$$

Finally:

$$E=m_e\left[\frac{Z^2\alpha^2+(s+t)^2}{(s+t)^2}\right]^{-\frac{1}{2}}=m_e\left[1+\frac{Z^2\alpha^2}{(s+t)^2}\right]^{-\frac{1}{2}} \qquad (14.93)$$

The form $m(1+\cdots)^{-1/2}$ comes from the fact that the relativistic energy includes also the rest mass m_e of the electron. We can identify the index $t=0,1,2,\ldots$ as $t=n-(j+1/2)$, where n is the principal quantum number of the electron. We have $0\leq j+1/2\leq n$. We also recall that $s=\sqrt{\kappa^2-Z^2\alpha^2}$, which depends on the quantum number $\kappa=\mp(j+1/2)$ defining angular momentum and spin. We can hence write the energy E_{nj} for the quantum numbers n and j as:

$$E_{nj}=m_e\left[1+\frac{Z^2\alpha^2}{\left(n-(j+\frac{1}{2})+\sqrt{(j+\frac{1}{2})^2-Z^2\alpha^2}\right)^2}\right]^{-\frac{1}{2}} \qquad (14.94)$$

This result is **the exact relativistic Dirac solution** for hydrogen-like atoms. Impressive! **This equation tells us that the energy levels depend only on the principal quantum number n and the total angular momentum j, i.e., states (for a given n) with the same value of j will have the same energy.** Compared to the classical quantum-mechanical result, the states of the principle quantum number n are not completely degenerate any more, as shown in Figure 14.5.

However, the agreement is not perfect. The Dirac equation predicts the states $^2S_{1/2}$ and $^2P_{1/2}$ to be still degenerate. They have the same quantum numbers n and J, but a different parity. Experimentally they were found to still be split – a shift called the Lamb shift and described in the next section, which could only be calculated in the context of QED!

Nonetheless, the exact Dirac prediction for hydrogen-like atoms was a great advancement compared to the non-relativistic result based on the Schrödinger equation. By noting that the **relativistic terms should be**

treatable as corrections to the classical Schrödinger result, one can write Dirac's result, for instance for $Z = 1$ and reintroducing the constant c, in an expansion series as:

$$
\begin{aligned}
E_{nj} &\simeq m_e c^2 \left[1 - \frac{\alpha^2}{2n^2} - \frac{(\alpha^2)^2}{2n^4} \left(\frac{n}{j + \frac{1}{2}} - \frac{3}{4} \right) + \cdots \right] \\
&= m_e c^2 - \frac{\alpha^2 m_e c^2}{2n^2} - \frac{(\alpha^2)^2 m_e c^2}{2n^4} \left(\frac{n}{j + \frac{1}{2}} - \frac{3}{4} \right) + \cdots
\end{aligned}
\tag{14.95}
$$

This result is called the **Sommerfeld fine-structure expression**. The first term corresponds to the rest mass of the electron. The second term gives exactly the classical Schrödinger energy levels, which only depend on the principal quantum number n (see Eq. (14.52)). The next term is the first relativistic correction and the fine-structure splitting. It removes the degeneracy of the levels with respect to n. For a given n, the energy increases with j. This corresponds to the $L - S$ coupling which distinguishes $j = l \pm 1/2$. As mentioned above, the next correction would be the Lamb shift which removes the degeneracy for fixed n and j.

To summarize, the **Dirac theory provides the fully relativistic treatment of the Coulomb potential** which is the dominant term, however, the other corrections such as the Lamb shift (see next section) represent the "radiative corrections" which are to be treated within QED. The various levels were summarized in Figure 14.5.

14.7 The Lamb Shift

Historically, the observation of the **Lamb shift** in the Balmer α transition of atomic hydrogen has been a major milestone. As previously mentioned, the effect could not be explained by the Schrödinger or Dirac formulations of quantum mechanics and needed the power of QED to include higher-order effects due to vacuum polarization.

We start by recalling Eq. (12.49) which describes the correction to the photon propagator in the renormalized QED theory:

$$
(-ie\gamma_\mu) \cdots \left(\frac{-ig^{\mu\nu}}{k^2} \right) \left[1 + \frac{e^2}{12\pi^2} f\left(-\frac{k^2}{m_e^2} \right) + \mathcal{O}(e^4) \right] \cdots (-ie\gamma_\nu)
\tag{14.96}
$$

where e is the renormalized charge. In the very low $|k^2| \ll m_e^2$ regime relevant here, the correction function $f(x) \approx x/5$ (see Eq. (12.44)). Hence, the term in brackets becomes:

$$
\left[1 - \frac{e^2}{60\pi^2 m_e^2} k^2 + \cdots \right]
\tag{14.97}
$$

The unity factor represents the pure Coulomb potential between the nucleus and the electron in the atom, which we already encountered in the Rutherford scattering derivation (see Eq. (4.201)):

$$
V_0(\vec{x}) = -\frac{e^2}{4\pi\epsilon_0} \left(\frac{Z}{|\vec{x}|} \right) = -\frac{Ze^2}{\epsilon_0} \int \frac{\mathrm{d}^3\vec{k}}{(2\pi)^3} \frac{e^{i(\vec{k}\cdot\vec{x})}}{|\vec{k}|^2}
\tag{14.98}
$$

The second term in brackets leads to a correction of the Coulomb potential. The modified potential can be written as:

$$
\begin{aligned}
V(\vec{x}) &= -\frac{Ze^2}{\epsilon_0} \int \frac{\mathrm{d}^3\vec{k}}{(2\pi)^3} \frac{e^{i(\vec{k}\cdot\vec{x})}}{|\vec{k}|^2} \left[1 - \frac{e^2}{60\pi^2 m_e^2} k^2 + \cdots \right] = -\frac{Ze^2}{\epsilon_0} \int \frac{\mathrm{d}^3\vec{k}}{(2\pi)^3} \frac{1}{|\vec{k}|^2} \left[1 - \frac{e^2}{60\pi^2 m_e^2} \nabla^2 + \cdots \right] e^{i(\vec{k}\cdot\vec{x})} \\
&= -\frac{Ze^2}{4\pi\epsilon_0} \left(1 - \frac{e^2}{60\pi^2 m_e^2} \nabla^2 \right) \frac{1}{|\vec{x}|} \\
&= -\frac{Ze^2}{4\pi\epsilon_0} \frac{1}{|\vec{x}|} - \frac{Ze^4}{60\pi^2 m_e^2 \epsilon_0} \delta(\vec{x})
\end{aligned}
\tag{14.99}
$$

where in the last line we used $\nabla^2 \left(1/|\vec{x} - \vec{x'}|\right) = -4\pi\delta(|\vec{x} - \vec{x'}|)$ (see Eq. (A.55)). Consequently, the addition of the QED correction to the photon propagator at the one-loop level adds the effect of the virtual e^+e^- pairs and represents the correction to the Coulomb potential due to the k^2 dependence of the charge renormalization (see Section 12.5). Indeed, the extra term has a negative sign in front of it! It leads to an additional *attractive* interaction between the electron and the nucleus. This is understood in the sense that as the electron "penetrates" the cloud of virtual e^+e^- pairs surrounding the nucleus, it will probe a larger nucleus charge. Alternatively, we can say that the actual nucleus charge is "screened" at large distances by the e^+e^- cloud surrounding it.

The extra term in the Coulomb potential will alter the energy level of the atoms, for instance the hydrogen atom. Its magnitude can be computed treating the second term as a perturbation in the Schrödinger equation. It will be proportional to the spatial overlap of the electron wave function and the δ function, leading to the energy-level correction for primary quantum number n and ℓ:

$$\Delta E_{n\ell} = -\frac{e^4}{60\pi^2 m_e^2 \epsilon_0}|\Psi_{n\ell}(0)|^2 \tag{14.100}$$

where $\Psi_{nl}(0)$ represents the wave function of the electron at the origin, where the nucleus lies. Hence, **the correction only affects those levels which have a non-vanishing wave function at their origin, i.e., only those with $\ell = 0$** (see Appendix. C.10). Finally, the so-called "Lamb shift" correction to the hydrogen atom can be expressed as a function of the Rydberg unit of energy, Eq. (2.4) and Eq. (C.12) as:

$$\Delta E_{n\ell} = -\frac{8\alpha^3}{15\pi n^3}Ry\delta_{\ell 0} \quad \text{(Lamb shift)} \tag{14.101}$$

where the $\delta_{\ell 0}$ takes into account that only $\ell = 0$ states are affected. What is amazing is that this originally virtual additional interaction leads to a real observable effect which can be detected experimentally, as will be shown in the following. As already mentioned, the observation of this effect has been a tremendously important milestone to give support to the existence of virtual corrections and to the validity of the renormalization procedure, as described in Chapter 12.

- **Experimental observation of the Lamb shift.** Energy levels are studied experimentally by observing the transitions between energy levels. Selection rules, which govern how much the angular momentum can change in a transition, prohibit transitions between certain sub-levels. The observed lines correspond to allowed transitions. Experimentally, Doppler broadening occurs due to the thermal motion of the atoms. So nearby lines can overlap in experiments and hence cannot be distinguished. In 1938, Williams [143] exploited both of these factors by measuring the Balmer α spectrum from a deuterium discharge lamp cooled to 100 K. He succeeded in resolving for the first time a third Balmer α component, the component corresponding to the $3P_{1/2} \to 2S_{1/2}$ transition. The measurements showed this separation to be larger than predicted by the Dirac theory.

In 1947 **W. E. Lamb**[6] and R. C. Retherford performed a brilliant experiment in which they directly measured the energy difference between the $2S_{1/2}$ and $2P_{1/2}$ states [144]. A schematic of the experimental layout is shown in Figure 14.6. A beam of hydrogen atoms in the $1S_{1/2}$ ground state is transported in a region where it is bombarded with electrons. As a result, some of the atoms are excited to the $2S_{1/2}$ state. Optical transitions from this state to the $1S_{1/2}$ ground state are prohibited by selection rules. Hence, the lifetime of the $2S_{1/2}$ is very long ($\simeq 0.1$ s). The beam then passes through a region where it is exposed to **electromagnetic radiation with energy equal to the energy difference between the $2S_{1/2}$ and $2P_{1/2}$ states to induce such transitions.** Once an atom is in the $2P_{1/2}$ state, it decays to the ground state with a lifetime of $\simeq 10^{-9}$ s. Finally, the atoms strike a tungsten foil. Upon striking the foil, atoms still in the $2S_{1/2}$ state decay to the ground state and in doing so liberate electrons from the foil by **Auger emission**. By measuring the emission current from the foil with and without the electromagnetic radiation, Lamb and Retherford were

---

6   Willis Eugene Lamb Jr. (1913–2008), American physicist.

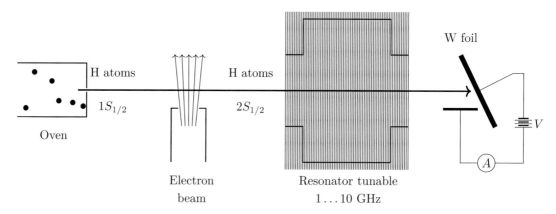

Figure 14.6 Schematic layout of the Lamb–Retherford experiment.

able to determine the energy difference between the $2S_{1/2}$ and $2P_{1/2}$. Their initial measurements indicated this difference to be about 1000 MHz.

14.8   Hyperfine Splitting and Finite Size Effect

The **hyperfine structure** was discovered in 1891 by Michelson using an interferometer. Such an interferometer was used to measure the spectroscopy of a typical atom (see Section 2.1). Many spectral lines were found to consist of several components. These very closely spaced lines implied the existence of a hyperfine splitting of the energy levels of an atom. The effect was first explained by Pauli in 1924. The hyperfine splitting is interpreted as the interaction between the magnetic field produced by the moving electron and the inherent magnetic field of the nucleus. Since nuclear magnetic moments are three orders of magnitude smaller than the atomic magnetic moments (see Section 4.12), the hyperfine splitting is smaller than the spin–orbit splitting by the same order.

If the nuclear spin angular momentum is defined as $\vec{I}$ and the total angular momentum of the electron as $\vec{J}$, then it is customary to define the resultant **grand total angular momentum** as $\vec{F} \equiv \vec{I} + \vec{J}$. The magnitude of this latter is given by $F = \sqrt{f(f+1)}$ and its z-component by m_f, where $-f \leq m_f \leq f$. The relation between the nuclear magnetic moment and the nuclear spin can be expressed similarly to Eq. (4.133):

$$\vec{\mu}_I \equiv g_i \mu_N \vec{I} \tag{14.102}$$

where g_i depends on the details of the nuclear structure, and μ_N is the nuclear magneton (see Eq. (4.135)). In general, the nuclear magnetic moment is difficult to calculate for complex nuclei. It is of the order of unity, and can be both positive or negative depending on the nucleus. The simplest case is clearly hydrogen. Finally, the term that gives rise to the hyperfine splitting is proportional to:

$$\Delta E_{\vec{I} \cdot \vec{J}} \propto f(f+1) - j(j+1) - i(i+1) \tag{14.103}$$

where j and i are the quantum numbers associated with $\vec{J}$ and $\vec{I}$. This term leads to the further splitting of the lines which depend on f, as illustrated in Figure 14.5.

The nuclear charge distribution (the so-called finite-size effect) affects the S states in leading order (the ones that have a non-zero probability distribution at the origin).

14.9   Positronium: The Lightest Known Atom

Positronium (Ps), the positron–electron bound state, is the lightest known atom, which is bounded and self-annihilates through the same electromagnetic interaction. Its existence was postulated in 1934 by Stepan Mohorovii [145], only one year after the discovery of the positron by Anderson [75]. Triggered by the Nobel lecture of Dirac, Mohorovii wrote at the time:

> *"But now the question arises, what will happen, if a positron and an electron meet in close proximity. For it is clear that – because of the electrical attraction – the two elementary particles will move around their common center of gravity, thus forming a system that would act outwardly as a neutral system. Such a system is nothing else than an atom, which would have all similarities with a hydrogen atom, only it would be 920.5 times lighter, and such an 'abaric' element we will call 'Electrum', because it would be built of only two polar electric elementary particles."* [7]

The bound state was renamed *positronium* by Ruark in 1945 [146]. Ruark already pointed out the great interest of such a hydrogen-like atom and gave a qualitative discussion of its spectroscopic structure. Indeed, the energy levels of positronium may be calculated in a first approximation using non-relativistic quantum mechanics, similar to the hydrogen atom. For such calculations, positronium can be considered like the hydrogen atom, but with two particles of equal mass orbiting a common center. In classical mechanics, this problem is solved as a one-body problem with the reduced mass μ of the system (see Eq. (14.52)):

$$\mu = \frac{m_1 m_2}{m_1 + m_2} = \frac{m_e}{2} \qquad (14.104)$$

This implies that the Bohr levels of positronium can be derived from those of hydrogen with the replacement $m_e \to m_e/2$. The principal energy levels are then:

$$E_n^{Ps} = \frac{1}{2} E_n^{\text{Hydrogen}} = -\frac{\alpha^2 m_e c^2}{4n^2} \qquad (14.105)$$

The ground state of Ps has a binding energy of $E_1 \simeq 13.605 \text{ eV}/2 = 6.803 \text{ eV}$ (see Eq. (14.52)). Classically, one can imagine positronium as an electron and a positron orbiting around their center of mass and separated by twice the hydrogen radius (i.e., ~ 1 for $n = 1$). In classical quantum mechanics, the Schrödinger wave functions are the same as for hydrogen atoms (see Appendix C.10), except that the equivalent Bohr radius a_0 for the positronium is twice the value for hydrogen:

$$a_0^{Ps} = 2a_0 \simeq 1.06 \times 10^{-8} \text{ cm} \qquad (14.106)$$

- **First experimental evidence.** Positronium was observed experimentally for the first time in 1951 by Deutsch [147]. This opened a new field whose interest rapidly grew. Being free of finite-size effects, positronium has proven to be an ideal and clean system for testing the accuracy of bound-state QED calculations.

- **Properties of the system.** The positronium properties are greatly dependent on the relative orientation of the positron and electron spins, namely parallel (the triplet ground state $^3S_1$, called **ortho-positronium** or

---

[7] *"Es erhebt sich aber jetzt von selbst die Frage, was geschehen wird, wenn sich ein Positron und ein Elektron in grosser Nähe begegnen. Es ist nämlich klar, dass sich – wegen der elektrischen Anziehung – die beiden Elementarteilchen um den gemeinsamen Schwerpunkt bewegen werden und so ein System bilden, das nach aussen als ein neutrales System wirken würde. Ein solches System ist nichts anderes als ein Atom, welches alle Ähnlichkeiten mit einem Wasserstoffatom hätte, nur ware es 920.5 mal leichter und ein solches 'abarisches' Element werden wir 'Electrum' nennen, da es nur aus zwei polaren elektrischen Elementarteilchen gebaut wäre."* Republished excerpt with permission of John Wiley & Sons – books from S. Mohorovii, "Möglichkeit neuer Elemente und ihre Bedeutung für die Astrophysik," *Astronomische Nachrichten*, vol. 253, no. 4, pp. 93–108. Copyright 1934 Wiley-VCH Verlag GmbH & Co. KGaA, Weinheim.

$o - Ps$) or antiparallel (the singlet ground state $^1S_0$, called **para-positronium** or $p - Ps$). These correspond to the following spin states:

$$\begin{cases} S = 1: \quad \begin{cases} |1,1\rangle = |\uparrow,\uparrow\rangle \\ |1,0\rangle = \frac{1}{\sqrt{2}}(|\uparrow,\downarrow\rangle + |\downarrow,\uparrow\rangle) \\ |1,-1\rangle = |\downarrow,\downarrow\rangle \end{cases} \\ S = 0: \quad |0,0\rangle = \frac{1}{\sqrt{2}}(|\uparrow,\downarrow\rangle - |\downarrow,\uparrow\rangle) \end{cases}$$

(14.107)

The energy levels of the two spin configurations split when the spin–orbit $V \sim \vec{L} \cdot \vec{S}$ and spin–spin $V \sim \vec{\mu}_1 \cdot \vec{\mu}_2 = \vec{S}_1 \cdot \vec{S}_2$ couplings are considered. Therefore the Hamiltonian reads $H = H_0 + H_1 + H_2 + H_3 + H_4 + H_5$ where H_0 is the Coulomb term, H_1 is the first-order relativistic correction, H_2 is the Darwin term, H_3 is the contribution of the spin–orbit, H_4 is the spin–spin coupling, and H_5 is the virtual annihilation, i.e., the emission and reabsorption of a virtual photon by the positronium atom, which only occurs for $o - Ps$ due to angular momentum conservation.

The energy shift for para-positronium is given by contributions of mainly H_1 and H_4:

$$\Delta E_{p-Ps} = -\frac{1}{64}m_e \alpha^4 - \frac{1}{4}m_e \alpha^4$$

(14.108)

For the ortho-positronium case, the virtual annihilation contribution H_5 is also considered:

$$\Delta E_{o-Ps} = -\frac{1}{64}m_e \alpha^4 + \frac{1}{4}m_e \alpha^4 + \frac{1}{12}m_e \alpha^4$$

(14.109)

Hence the hyperfine splitting for the positronium ground state is:

$$\Delta E_{HFS} = \Delta E_{p-Ps} - \Delta E_{o-Ps} = -\frac{7}{12}m_e \alpha^4 \simeq -8.45 \times 10^{-4} \text{ eV}$$

(14.110)

• **Positronium lifetime calculation.** The positronium is a short-lived system. It decays electromagnetically into photons. The number of photons can be deduced from *selection rules*. From the relative intrinsic parities of electrons and positrons, we obtain the parity $P = (-1)^{l+1}$ (see Eq. (8.209)). The charge conjugation can be seen as an exchange of the electron and positron, hence it behaves like the parity transformation and the exchange of the spin labels. Hence, $C = (-1)^{l+s}$ (see Eq. (4.93)) and consequently $CP = (-1)^{s+1}$. The $p - Ps$ and the $o - Ps$ are respectively CP eigenstates with eigenvalues of -1 and $+1$. Recalling that the photon charge conjugation eigenvalue for a system of n photons is just $C(n\gamma) = (-1)^n$ (see Eq. (4.87)), we find the following decay selection rule:

$$(-1)^{l+s} = (-1)^n$$

(14.111)

Hence, electromagnetic interactions cause $p - Ps$ to only decay into an even number of photons, the two-photon decay being its main annihilation channel, while the $o - Ps$ decays dominantly into three photons. The corresponding Feynman graphs are shown in Figure 14.7. The selection rules and the dominant decay modes are listed in Table 14.2.

The decay rate of the $p - Ps$ mode was first calculated in 1946 by **Wheeler**[8] [148] and **Pirenne**[9] [149], as illustrated in Figure 14.8. Some kind of factorization is needed to separate the bound state from the relativistic decay process. Their result is embedded in the **Wheeler–Pirenne formula**:

$$\Gamma(Ps \to n\gamma) = \frac{1}{2J+1}|\phi(0)|^2 \left(4v_{rel}\sigma(e^+e^- \to n\gamma)\right)$$

(14.112)

[8] John Archibald Wheeler (1911–2008), American theoretical physicist.
[9] Jean Pirenne (1913–2004), Belgian theoretical physicist.

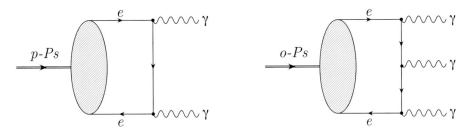

Figure 14.7 Feynman graphs for the main decay process of $p - Ps$ and $o - Ps$.

$^{2S+1}L_J$	J^{PC}	$\#\gamma$	Dominant decay mode
$^1S_0$	0^{-+}	even	$\gamma\gamma$
$^3S_1$	1^{--}	odd	$\gamma\gamma\gamma$
$^1P_1$	1^{+-}	odd	$\gamma\gamma\gamma$
$^3P_0$	0^{++}	even	$\gamma\gamma$
$^3P_1$	1^{++}	even	$\gamma\gamma\gamma\gamma$
$^3P_2$	2^{++}	even	$\gamma\gamma$
$^1D_2$	2^{-+}	even	$\gamma\gamma$
$^3D_1$	1^{--}	odd	$\gamma\gamma\gamma$
$^3D_2$	2^{--}	odd	$\gamma\gamma\gamma$

Table 14.2 Table of photon selection rules and dominant decay modes of positronium as a function of its momentum state.

The bound-state physics is encoded in the wave function (spherical harmonics), and the hard process is given by the non-relativistic approximation of the QED calculation (see, e.g., Eq. (11.240)). The factorization basically implies that the Coulomb binding in the positronium atom has a negligible effect upon the decay probability. The cross-section $\sigma(e^+e^- \to n\gamma)$ and flux factor correspond to the collision of an electron and a positron moving relative to each other. The result for para-positronium is:

$$\Gamma^{\text{th}}(p - Ps \to \gamma\gamma) \equiv \Gamma^0 = \frac{4\pi\alpha^2}{m_e^2}|\phi(0)|^2 = \frac{\alpha^5 m_e}{2} \simeq 8032.5\,\mu\text{s}^{-1} \tag{14.113}$$

In contrast to the singlet state, ortho-positronium can only decay into an odd number of photons. The main decay mode is thus $o - Ps \to 3\gamma$, whose decay rate was first calculated in 1949 by Ore and Powel [150] to be:

$$\Gamma^{\text{th}}(o - Ps \to \gamma\gamma\gamma) = \frac{16\alpha^3}{9m_e^2}\left(\pi^2 - 9\right)|\phi(0)|^2 = \frac{2}{9}\left(\pi^2 - 9\right)\frac{\alpha^6 m_e}{\pi} \simeq 7.2111\,\mu\text{s}^{-1} \tag{14.114}$$

Hence, the lifetime of the $o - Ps$ is approximately a thousand times longer than that of $p - Ps$.

Since those first calculations were performed, significant efforts have been engaged to compute higher-order corrections. Positronium provides indeed a fertile ground for developing QED calculations with bound states. The present theoretical knowledge of the widths of the ground states to three and two photons may be

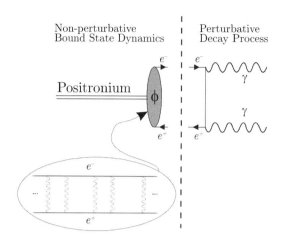

Figure 14.8 Schematic representation of the factorization of the non-perturbative and perturbative parts of the decay rate calculation.

summarized as follows [151]:

$$
\begin{aligned}
\Gamma^{\mathrm{th}}(p-Ps \to \gamma\gamma) &= \frac{\alpha^5 m_e}{2}\left\{1 + \frac{\alpha}{\pi}\left(\frac{\pi^2}{4} - 5\right) + \left(\frac{\alpha}{\pi}\right)^2\left[-2\pi^2\ln\alpha + 5.1243(33)\right]\right. \\
&\left. + \frac{\alpha^3}{\pi}\left[-\frac{3}{2}\ln^2\alpha + \left(\frac{533}{90} - \frac{\pi^2}{2} + 10\ln 2\right)\ln\alpha + \frac{D_p}{\pi^2}\right]\right\}
\end{aligned}
\tag{14.115}
$$

$$
\begin{aligned}
\Gamma^{\mathrm{th}}(o-Ps \to \gamma\gamma\gamma) &= \frac{2(\pi^2 - 9)\alpha^6 m_e}{9\pi}\left\{1 - \frac{\alpha}{\pi}10.286\,606(10) + \left(\frac{\alpha}{\pi}\right)^2\left[\frac{\pi^2}{3}\ln\alpha + 44.87(26)\right]\right. \\
&\left. + \frac{\alpha^3}{\pi}\left[-\frac{3}{2}\ln^2\alpha + \left(3.428\,869(3) - \frac{229}{30} - 8\ln 2\right)\ln\alpha + \frac{D_o}{\pi^2}\right]\right\}
\end{aligned}
\tag{14.116}
$$

The coefficients $D_{o,p}$ parameterize the unknown non-logarithmic $O(\alpha^3)$ terms. The $o - Ps$ decays into five photons and the $p - Ps$ decays into four photons, leading to an increase of the numerical coefficients in front of $(\alpha/\pi)^2$ by 0.187(11) and 0.274290(8), respectively (see, e.g., Ref. [151]). Including all the terms known so far, the best theoretical expectations for the $p - Ps$ and $o - Ps$ total decay rates are:

$$
\begin{aligned}
\Gamma^{\mathrm{th}}(p - Ps) &= 7989.6178(2) \ \mu\mathrm{s}^{-1} \tag{14.117} \\
\Gamma^{\mathrm{th}}(o - Ps) &= 7.039979(11) \ \mu\mathrm{s}^{-1} \tag{14.118}
\end{aligned}
$$

where the given errors stem only from the uncertainty in the numerical values of the perturbative coefficients.

• **Positronium lifetime puzzle.** The most recent experimental data on the $p - Ps$ decay rate [152] yields:

$$
\Gamma^{\mathrm{exp}}(p - Ps) = 7990.9(1.7) \ \mu\mathrm{s}^{-1} \tag{14.119}
$$

is consistent with theoretical expectations within error bars. The experimental determination of the $o - Ps$ decay rate, however, has a colorful history of inconsistent results and poor agreement with theoretical predictions. Over many years, significant discrepancies were found between the theoretical rate and experimental results, resulting in the so-called ortho-positronium lifetime puzzle (see, e.g., Ref. [153]). This triggered a large interest in improving the accuracy of both experiments and theory. The measurements of the $o - Ps$ lifetime performed after 1987 are summarized in Table 14.3. The most recent independent measurements are the SiO_2 powder [154] and vacuum [155] experiments:

$$\Gamma^{\exp}(o-Ps) = 7.0396(12)_{\text{stat.}}(11)_{\text{syst.}} \ \mu s^{-1} \quad (SiO_2 \text{ powder}) \qquad (14.120)$$
$$\Gamma^{\exp}(o-Ps) = 7.0404(10)_{\text{stat.}}(8)_{\text{syst.}} \ \mu s^{-1} \quad (\text{vacuum}) \qquad (14.121)$$

which are, however, in very good agreement with the theoretical expectations. At the moment of writing, we can state that the agreement between QED predictions and experiments is excellent. Future improvements in precision will be needed to further investigate positronium.

Year	Experiment	Rate (μs^{-1})	Errors (ppm)	Technique	Exp.−theory	Exp.−theory (sigma)
1987	Ann Arbor	7.0516	180	Gas	0.01162	9.2
1989	Ann Arbor [156]	7.0514	200	Gas	0.01142	8.1
1990	Ann Arbor [157]	7.0482	230	Vacuum	0.00822	5.1
1995	Tokyo [158]	7.0398	412	Powder	−0.00018	−0.06
2003	Tokyo [154]	7.0396	227	Powder	−0.00038	−0.24
2003	Ann Arbor [155]	7.0404	185	Vacuum	0.00042	0.32

Table 14.3 Experimental results on ortho-positronium lifetime. The last column shows the deviation between experiment and theory, expressed in standard deviations.

Problems

Ex 14.1 Motion in a Penning trap. We consider a Penning trap within a uniform magnetic field given by $\vec{B} = B_0 \hat{z}$ and an electric potential given by Eq. (14.29). The equation of motion of a charged particle with charge e is given by:

$$m\ddot{\vec{x}} = e\left(\vec{E} + \vec{v} \times \vec{B}\right) \qquad (14.122)$$

(a) Show that the electric field is given by:

$$\vec{E} = \frac{V_0}{r_0^2}(x, y, -2z) \qquad (14.123)$$

(b) Show that the equation of motion setting $\vec{x} = (x, y, z)$ then becomes:

$$\begin{pmatrix} \ddot{x} \\ \ddot{y} \\ \ddot{z} \end{pmatrix} = \frac{eV_0}{mr_0^2}\begin{pmatrix} x \\ y \\ -2z \end{pmatrix} + \omega_0 \begin{pmatrix} \dot{y} \\ -\dot{x} \\ 0 \end{pmatrix} \qquad (14.124)$$

where $\omega_0 = eB_0/m$ is the non-relativistic cyclotron frequency (see Eq. (14.26)).

(c) Note that in the above expression the z coordinate is decoupled from the radial ones. The movement along the z coordinate leads to the **axial oscillation**. Use the following ansatz:

$$z(t) = z_0 e^{i\omega_E t} \qquad (14.125)$$

to prove Eq. (14.30).

(d) We now need to solve the radial motion. Show that it is determined by the equation:

$$\begin{pmatrix} \ddot{x} \\ \ddot{y} \end{pmatrix} = \frac{\omega_E^2}{2} \begin{pmatrix} x \\ y \end{pmatrix} + \omega_0 \begin{pmatrix} \dot{y} \\ -\dot{x} \end{pmatrix} \qquad (14.126)$$

Introduce the complex function $u(t) = x(t) + iy(t)$ and show that it follows the following equation:

$$\ddot{u} = \frac{\omega_E^2}{2} u - i\omega_0 \dot{u} \qquad (14.127)$$

(e) Make the following ansatz $u(t) = u_0 e^{-i\omega t}$ and find the two solutions $\omega_\pm$:

$$\omega_\pm = \frac{1}{2}\left(\omega_0 \pm \sqrt{\omega_0^2 - 2\omega_E^2}\right) \qquad (14.128)$$

(f) Show that the slow frequency defined as the **magneton frequency** ω_{EB} in Eq. (14.31) can be identified with ω_-. While the fast **cyclotron frequency** is given by:

$$\omega_B \equiv \omega_+ = \omega_0 - \omega_{EB} \qquad (14.129)$$

(g) Compute the three frequencies for an electron in a Penning trap with the following parameters: $B = 0.4$ T, $V_0 = 10$ V, and $r_0 = 0.8$ cm.

Ex 14.2 True muonium decay. True muonium is the bound state comprising a muon (μ^-) and its antiparticle (μ^+)[10]. Since both muons are fermions, there are two spin configurations the bound state can have. One is called para-true-muonium (henceforth pMu) and has total spin 0, while the total spin-1 configuration is called ortho-true-muonium (henceforth oMu).

(a) C-parity for the bound state is given by $(-1)^{L+S}$, where L is the orbital angular momentum and S the spin angular momentum of the bound state. Assuming the atom is in the s state (spectroscopic notation), how many photons need to be created in the decay for C-parity to be conserved?

(b) Draw the lowest-order Feynman diagrams for pMu and oMu to decay which respect C-parity conservation.

   Hint: It can also decay to lighter fermions!

(c) Calculate the total cross-section at tree level for the decay of pMu and oMu.

   Hint: Calculate the unpolarized total cross-section in the low-energy limit of the respective graphs of part (b).

(d) Using the Pirenne–Wheeler factorization formula, Eq. (14.112):

$$\Gamma \approx \frac{1}{2J+1} |\psi(0)|^2 \times (4v_{rel}\sigma_{LO})$$

where Γ is the decay rate, $\psi(0)$ is the hydrogen wave function (given by the solution of the Schrödinger equation for $n = 1$ and $l = m = 0$ and using the appropriate reduced mass of the true muonium atom) at the origin, and v_{rel} is the relative velocity of the constituents, calculate the lifetime of oMu and pMu, respectively.

(e) Calculate the ratio of the pMu to the oMu lifetimes. Argue why this value is so different from the case of positronium, which is on the order of 10^{-3}.

---

10 For historical reasons, the bound state of an electron and a positive muon is called Muonium, even if "-onium" atoms actually describe atoms of particles and their corresponding antiparticles, hence the qualifier "true."

15 Hadrons

If I could remember the names of all these particles, I'd be a botanist.

Enrico Fermi

15.1 The Strong Force

The successful development of QED represented a great achievement: the theory was very useful, it handled matter and antimatter (electrons and positrons), it introduced the technique of renormalization, and it proved to be extremely useful and precise (for example in computing the anomalous magnetic moments). Nonetheless, QED could not simply explain even the existence of the nucleus of atoms! Indeed, what holds the nucleus together? After all, the positively charged protons should repel one another violently, packed together as they are in such close proximity. Evidently there must be some other force, more powerful than the force of electrical repulsion, that binds the protons (and neutrons) together. Likely due to a lack of a better name, the new force was labeled **the strong force**. How should the quantum field theory of the strong force be introduced?

In 1935 Yukawa makes a **first attempt at a quantized theory of the strong force** to address the type of interaction between the neutron and the proton that would explain the nuclear structure. In a paper entitled "On the Interaction of Elementary Particles. I," Yukawa proposed the existence of a new field of nuclear forces since the electromagnetic and weak (from Fermi's theory – see Section 21.3) could not account for nuclear forces, and thereby predicted the existence of a new quanta, the **meson**. He wrote (translated from Japanese):

> "To remove this defect, it seems natural to modify the theory of Heisenberg and Fermi in the following way. The transition of a heavy particle from neutron state to proton state is not always accompanied by the emission of light particles, i.e., a neutrino and an electron, but the energy liberated by the transition is taken up sometimes by another heavy particle, which in turn will be transformed from a proton state into a neutron state. (...) The potential of force between the neutron and proton should, however, not be of the Coulomb type, but decrease more rapidly with distance." [1]

Yukawa postulates the existence of the potential U in analogy to the electromagnetic potential to explain the *action at distance* of the strong force. The proton and the neutron will interact strongly, and in particular, stick together in the compact nucleus of atoms. Considering the "nature of the quanta accompanying the field," Yukawa stated in the same paper:

> "The U-field should be quantized according to the general method of the quantum theory. Since the neutron and the proton both obey Fermi's statistics, the quanta accompanying the U-field should

---

[1] Reprinted excerpt with permission of Oxford University Press from H. Yukawa, "On the interaction of elementary particles. I," *Proceedings of the Physico-Mathematical Society of Japan, 3rd Series*, vol. 17, pp. 48–57, 1935. Copyright 1935 The Physical Society of Japan and The Mathematical Society of Japan.

obey Bose's statistics and the quantization can be carried out the line similar to that of the electromagnetic field. The law of conservation of the electric charge demands that the quantum should have charge either $+e$ or $-e$. The field quantity U corresponds to the operator which increases the number of negatively charged quanta and decreases the number of positively charged quanta by one respectively."

In modern language, the strong interaction between the proton and the neutron can be graphically represented in analogy to the QED tree-level diagram as shown in Figure 15.1. The diagram on the left illustrates the electromagnetic interaction via the exchange of the spin-1 photon (the virtual gauge boson). In analogy, the strong interaction between a proton and a neutron is constructed via the exchange of a *scalar charged meson* called the **pion** and labeled $\pi^{\pm}$. The strength of the force is represented by the to-be-determined **strong coupling constant** e_H. Since the pion must be a boson, its spin must be integer, and is taken to be zero for simplicity. For reasons that are explained below, the mass of the pion was predicted to be "middle-weight," hence the name *meson*.

Figure 15.1 (left) Electromagnetic interaction via the exchange of a spin-1 gauge boson (photon). (right) The strong interaction between a proton and a neutron via the exchange of a scalar charged boson (meson).

But if there exists such a potent force in nature, why don't we notice it in everyday life? In order to explain such observations (e.g., that the strong force is confined to the nucleus and not observed elsewhere), Yukawa postulated that *its range of action was to be finite*, in contrast to the electromagnetic interaction. The electrostatic potential satisfies the Laplace equation in vacuum $\nabla^2 U(\vec{x}) = 0$. The well-known solution is the Coulomb potential inversely proportional to the distance $U(\vec{x}) \propto 1/|\vec{x}| = 1/r$ for the point charge at $\vec{x} = 0$. Yukawa modified the Laplace equation to $(\nabla^2 - \lambda^2)U(\vec{x}) = 0$ and introduced its solution, the **screened Coulomb potential** (which we already encountered in Eq. (11.11)) [159]:

$$U(|\vec{x}|) = U(r) \propto g^2 \frac{e^{-\lambda r}}{r}, \tag{15.1}$$

where g is the coupling constant with the dimension of an electric charge, and λ with the dimension of inverse distance. Compared to the Coulomb potential which has an infinite range, the Yukawa potential corresponds to a force with a range determined by the inverse of the λ parameter. See Figure 15.2, which illustrates the Coulomb potential (dashed curve) and the Yukawa potentials (lines) with the arbitrarily chosen values $\lambda = 1$, 2, and 4 fm^{-1}. The larger the λ, the shorter the corresponding range.

For the strong force to be confined to the nucleus, the constant λ must be approximately [159]:

$$\lambda \simeq 5 \times 10^{12} \text{ cm}^{-1} = \frac{1}{2 \times 10^{-15} \text{ m}} = \frac{1}{2 \text{ fm}} \tag{15.2}$$

This is possible by assuming that the exchanged virtual boson responsible for the strong force is massive. In the case of electromagnetism, the range is not finite since the real photon is massless. Yukawa succeeded

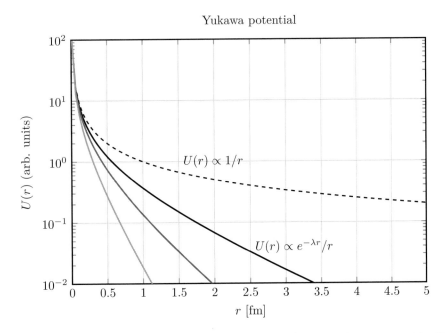

Figure 15.2 Illustration of the Coulomb potential (dashed curve) and Yukawa potentials (lines) with the arbitrarily chosen values $\lambda = 1$, 2, and 4 fm^{-1}. The larger the λ, the shorter the corresponding range.

in estimating the mass of the scalar meson. We can use simple arguments invoking **Heisenberg's uncertainty principle** for energy and time, and stating that the exchanged particle will travel with the speed of light as far as permitted during the time interval Δt before it is absorbed again. Thus (we reintroduce $\hbar$ and c for clarity):

$$c\Delta t \simeq \lambda^{-1} \quad \text{where} \quad \Delta E \Delta t \simeq \hbar \tag{15.3}$$

The energy ΔE corresponds to the rest mass of the exchanged boson, hence we obtain:

$$\Delta E \Delta t = (mc^2)(\lambda^{-1}/c) \simeq \hbar \tag{15.4}$$

or that *the range of the force λ^{-1} is of the order of* (see Eq. (1.24)):

$$\lambda^{-1} \simeq \frac{\hbar c}{mc^2} \simeq \frac{200}{m \,[\text{MeV}]} \,\text{fm} \tag{15.5}$$

With the value $\lambda^{-1} \simeq 2$ fm one obtains $m \simeq 100$ MeV or, as Yukawa put it *"a value 2×10^2 times as large as the electron mass. Such a quantum with large mass and positive or negative charge has never been found by the experiment, the above theory seems to be on a wrong line"* [2]!

Indeed, an elementary particle with the mass postulated by Yukawa was not known in 1934. Such a particle π, if it existed, was going to be a "**meson**" with a "*middle weight*" mass, much larger than the one of the electron (light-weight or "**lepton**"), yet smaller than that of the proton or neutron (heavy-weight or "**baryons**"), and of spin 0. Mathematically:

$$m_e \lesssim m_{\pi^\pm} \lesssim M_p, M_n \tag{15.6}$$

---

2   Reprinted excerpt with permission of Oxford University Press from H. Yukawa, "On the interaction of elementary particles. I," *Proceedings of the Physico-Mathematical Society of Japan, 3rd Series*, vol. 17, pp. 48–57, 1935. Copyright 1935 The Physical Society of Japan and The Mathematical Society of Japan.

It is natural to extend the argument and postulate the existence of the electrically neutral spin-0 pion, labeled π^0. The neutral boson can explain the strong interaction without change of electric charge, see Figure 15.3. Finally we should seek at least three mesons π^+, π^0, and π^-.

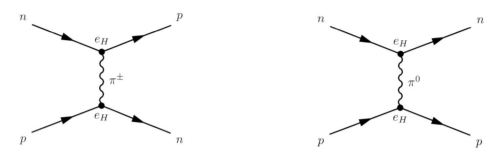

Figure 15.3 The strong interaction between a proton and a neutron via the exchange of a spin-0 (left) charged pion, (right) neutral pion.

15.2   The Study of Cosmic Rays

How shall Yukawa's postulated particle be detected? It was reasonable to think that such particles would be present in high-energy **cosmic rays**. Primary cosmic rays, predominately composed of high-energy protons, continuously hit the atmosphere and interact in its upper layers. Above a few giga-electronvolts, the differential energy spectrum of primary protons follows a **power law**, which means that the flux can be written as:

$$\phi_p(E) \propto E^{-\gamma_p} \tag{15.7}$$

where it is experimentally determined that $\gamma_p \approx 2.7$. High-energy interactions lead to the production of secondary particles. The energy spectra of secondary particles are also well described by a power law. Pions of all possible electric charges are produced for example via the following nucleon–nucleon reactions:

$$\begin{cases} pp \to pp\pi^0 & pn \to pp\pi^- \\ pp \to pn\pi^+ & pn \to pn\pi^0 \\ & pn \to nn\pi^+ \end{cases} \tag{15.8}$$

For higher incident proton energies, the number of pions produced in a single reaction can increase significantly, which can be expressed as:

$$NN \to NN + n\pi's \quad \text{with} \quad n \geq 1 \tag{15.9}$$

where N represents a nucleon. In addition, although less abundant, other mesons such as kaons, or baryons such as hyperons or nucleon–antinucleon pairs, can emerge from the strong interactions of the energetic primary cosmic rays. If sufficiently energetic, the secondary hadrons will themselves initiate new hadronic interactions, building up what is called a **hadronic shower cascade** (see Section 3.12). This process is graphically illustrated in Figure 15.4(left). Secondary neutral pions will almost immediately decay into photons with a lifetime of $\tau \approx 10^{-16}$ s (see Eq. (15.21)), leading to secondary electrons and positrons:

$$\pi^0 \to \gamma + \gamma \to e^-, e^+, \ldots \tag{15.10}$$

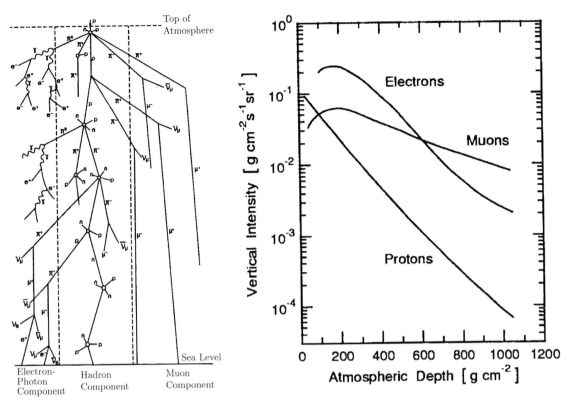

Figure 15.4 (left) Schematic drawing of the development of the shower cascade in the atmosphere after the interaction of a primary cosmic ray proton. (right) Representation of the composition of cosmic rays as a function of the altitude (depth in atmosphere). Reprinted from P. K. Grieder, "Chapter 1 – cosmic ray properties, relations and definitions," in *Cosmic Rays at Earth* (P. K. Grieder, ed.), pp. 1–53, Amsterdam: Elsevier. Copyright 2001, with permission from Elsevier.

The charged pions will also decay in the upper atmosphere, giving muons. The muons themselves can decay before reaching Earth's ground, giving the complete chain:

$$\pi^+ \rightarrow \quad \mu^+ \quad +\nu_\mu \tag{15.11}$$
$$\quad\quad\quad \hookrightarrow \quad e^+ + \nu_e + \bar{\nu}_\mu$$
$$\pi^- \rightarrow \quad \mu^- \quad +\bar{\nu}_\mu \tag{15.12}$$
$$\quad\quad\quad \hookrightarrow \quad e^- + \bar{\nu}_e + \nu_\mu$$

The composition of secondary cosmic rays as a function of the atmospheric depth is illustrated in Figure 15.4(right). The depth is related to the altitude and represents the amount of material a secondary particle has traversed at a given altitude. The total depth of the atmosphere at the level of the ground is 1033 g/cm^2. It is expressed in units of g/cm^2, which are naturally used for the stopping power (see Eq. (2.142)) of charged particles traversing a medium.

Cosmic rays at the Earth's surface are predominantly composed of muons. The probability of muons decaying before reaching the ground decreases with their energy. A muon typically loses 2 GeV from production point to the surface [28]. A 2.4 GeV muon has a decay length of $\gamma c\tau \approx 15$ km. Because of energy losses, it is reduced to about 9 km. It turns out that this length is about the same as the height of the atmosphere. Hence, the muons that we observe on the surface of the Earth are those which have not decayed within the atmosphere. The integral intensity of atmospheric muons above 1 GeV/c at sea level is $\approx 70 \text{ m}^{-2} \text{ s}^{-1} \text{ sr}^{-1}$ [28].

15.3 The Existence of Exotic Atoms

Initial experiments to study cosmic rays relied on thick absorbers surrounded by active detectors. Charged particles lose their energy according to the Bethe–Bloch formula (see Eq. (2.142)) and can either interact or stop. When a positively charged particle is stopped, it will generally stay and eventually decay. If we only consider the electromagnetic interactions, we predict that negatively charged particles form exotic atoms. An **exotic atom** is an otherwise normal atom in which an electron has been replaced by another negative particle.

However, mesons, like pions, have a special behavior when they are stopped in matter, due to their strong force. Being bosons, they could interact and be absorbed by a nucleus. Yukawa estimated that the majority of slow mesons would indeed be captured by atomic nuclei after having been stopped by losing their energy through ionization in dense materials.

In 1940 Tomonaga and Araki considered the effect of the "nuclear Coulomb field" [160] and noted that **positive pions would be repelled by the nucleus, while negative mesons would be attracted by it. Hence, they estimated that the capture probability of positive and negative pions was to be very different**. In particular, in the case of positive pions, the lifetime would be long enough that there is a possibility that the positive pion *decays* before it is captured! Tomonaga and Araki predicted that *"practically all positive mesons, which come to rest, should therefore be necessarily accompanied by a disintegration electron at the end of their range."* [3] Therefore (see Figure 15.5):

- Negative pions tend to form **pionic atoms**, with the pion becoming bound due to the Coulomb attraction to the nucleus. Produced in an excited state, the pion can release energy and fall into lower bound states. Because of its mass being much larger than that of the electron, the pion is located at small radii and will interact with the nucleus with high probability as soon as its spatial distribution comes within range of the strong force.

- Positive pions do not form bound atomic states. They stop and will eventually decay, emitting positrons.

The predictions of Tomonaga/Araki were eventually experimentally verified, and the difference in lifetime between positive and negative pions was the subject of many experimental investigations. These latter indeed led to the discovery of the pion. Historically, the muon was first discovered, with a puzzling mass in the range of that predicted by Yukawa! This will be discussed in the next section.

A pionic atom is a special kind of exotic atom. The case of **hydrogenic atoms** will be discussed in Section 14.5. In a **muonic atom**, a muon replaces an electron. Since the muon is not sensitive to the strong force, muonic atoms are governed by the electromagnetic interaction. The Bohr orbits of the muon are much closer to the nucleus than in an ordinary atom, since one must replace m_e by m_μ in Eq. (C.9). For an atom with atomic number Z, we have:

$$r(Z) = \left(\frac{m_e}{m_\mu}\right)\frac{a_0 n^2}{Z} \approx \frac{1}{200}\frac{a_0 n^2}{Z} \ll a_0 \qquad \text{for } n \ll 14\sqrt{Z} \tag{15.13}$$

In the pionic (resp. kaonic) atoms a negative pion (resp. kaon) replaces an electron. Since hadrons can interact via the strong force, unlike the case of the muon, the orbitals can be influenced by the nuclear force between the hadron and the nucleus. Since the strong force has a short range, these effects are the strongest for levels

---

3   Reprinted excerpt with permission from S. Tomonaga and G. Araki, "Effect of the nuclear Coulomb field on the capture of slow mesons," *Phys. Rev.*, vol. 58, pp. 90–91, Jul 1940. Copyright 1940 by the American Physical Society.

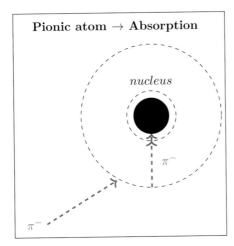
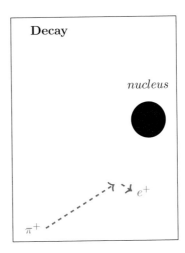

Figure 15.5 Illustration of the difference in behavior of stopping negative and positive pions in matter, leading to absorption and decay.

close to the nucleus. The lowest energy levels are absent since the hadron has a very high chance of being absorbed by the nucleus.

When an exotic atom is formed, it needn't be in the ground state. In fact, it is usually not in the ground state, since naively the "large radius" at which the particle is trapped by the atom translates into a high quantum state n (for instance for the muonic atom the relation between r and n is given by Eq. (15.13)). An excited exotic atom can easily cascade into its ground state $1s$ by Auger emission (typically at the higher levels) and by X-ray emissions at the lower n levels.

As it reaches the ground state, the particle has a higher and higher probability of being closer to the nucleus. In the case of the muon, the capture can eventually occur, in which the muon is absorbed by a proton in the nucleus via the following two weak interaction reactions:

$$\mu^- + p \rightarrow n + \nu_\mu \quad \text{or} \quad \mu^- + p \rightarrow n + \nu_\mu + \gamma \tag{15.14}$$

In order to detect the capture, one should either identify the neutron or the associated photon. The total lifetime of the negative muon in an exotic atom is going to be the result of two competing processes: (1) the muon decay process $\mu \rightarrow e\nu\bar{\nu}$ and (2) the muon capture process, and is given by:

$$\tau_\mu^{-1} = \tau_{decay}^{-1} + \tau_{capture}^{-1} \tag{15.15}$$

Some examples of muon lifetimes in typical materials are listed in Table 15.1. A theory of muon capture was developed by **Primakoff**[4] [161]. Although the capture rate varies significantly by several orders of magnitude over the range of elements considered, it can be phenomenologically reasonably described by the simple formula for an atom of atomic number Z and atomic mass A [162]:

$$\tau_{capture}^{-1}(A, Z) = Z_{eff}^4 X_1 \left[1 - X_2 \left(\frac{A - Z}{2A} \right) \right] \tag{15.16}$$

where $X_1 = 170 \text{ s}^{-1}$ and $X_2 = 3.125$. Z_{eff} is the effective atomic number of the nucleus. For light nuclei, the nucleus can be considered as point-like. Since the Bohr radius of the muon decreases as Z (see Eq. (15.13)), the probability of finding the muon within the nucleus increases proportionally to Z^3. Since there are Z protons,

---

4   Henry Primakoff (1914–1983), theoretical physicist.

Material	A	Z	Z_{eff}	Free lifetime (μs)	Capture lifetime (μs)	μ^- lifetime (μs)	μ^+ lifetime (μs)
Free decay	–	–	–	2.2	–	2.2	2.2
Hydrogen	1	1	1.0	2.2	2400	2.2	2.2
Carbon	12	6	5.7	2.2	25	2.0	2.2
Aluminum	27	13	11	2.2	1.5	0.88	2.2
Iron	56	26	20	2.2	0.22	0.20	2.2
Lead	207	82	33	2.2	0.09	0.08	2.2

Table 15.1 A compilation of muon lifetimes in various materials. The lifetime of μ^+ is always 2.2 μs.

the total capture rate should naively scale like Z^4. However, for heavier nuclei, the point-like approximation becomes less valid and an effective Z_{eff} is empirically introduced to take into account this effect. Its values are listed in Table 15.1. The term in brackets take into account Pauli blocking. When the muon captures, a proton is transformed into a neutron but this latter can only populate a free state. Hence, this factor reduces the capture for heavier nuclei, for which the relative number of neutrons is larger.

15.4   The Discovery of the Pion and the Muon

Both the muon and the pion were discovered in the study of such cosmic rays, more than 12 years after Yukawa had postulated the existence of his particle. As already mentioned, the muon was discovered chronologically first with terrestrial experiments (since most cosmic pions do not reach the surface!) and initially confused with the Yukawa particle. The real Yukawa particle was discovered afterwards. This confusion led to both being historically called **mesons**. Today the pion is still labeled as a meson, while the muon is a charged lepton.

In 1937 the two independent groups of **Anderson, Neddermeyer** and **Street, Stevenston** detected the "mesotron," a type of particle with the apparent characteristics of the meson. The measurements of the energy loss of particles found in cosmic rays as they traverse thick slabs of material led to the existence of penetrating particles which are hence unlike electrons that develop into an electromagnetic shower. The penetrating particles were compatible with being lighter than the proton. However, they could not study in all detail the properties of the newly discovered particles.

In 1939 **Rossi**[5] and collaborators showed that the mesotron is unstable. They performed the first electronic experiment that can measure the lifetime at rest [163]. They found $\tau_\mu = (2.4 \pm 0.3)$ μs, in agreement with current values, and provided the first experimental confirmation of the time dilation predicted by special relativity!

In 1946 **Conversi**, **Pancini**, and **Piccioni** proved that the particle discovered by Anderson, Neddermeyer, Street, and Stevenston could *not* be the particle of Yukawa! Its interaction with matter was too weak [164]. Instead of the strongly interacting Yukawa particle, the newly discovered partile was the muon! The properties of the muon (μ^- and μ^+) are the same as those of the electron (e^- and e^+) except for their rest mass, which is $m_\mu \simeq 105$ MeV and is thus much heavier than the electron. Muons (weakly) decay with a relatively long lifetime $\tau \simeq 2.2$ μs and the radiative decay $\mu \to e\gamma$ is not seen. The measured properties of the muon are

---

5   Bruno Benedetto Rossi (1905–1993), Italian experimental physicist.

(values taken from the Particle Data Group's Review [5]):

$$\boxed{\mu^+,\mu^-}: \quad m_{\mu^\pm} = (105.6583745 \pm 0.0000024) \text{ MeV}$$
$$\tau = (2.1969811 \pm 0.0000022) \times 10^{-6} \text{ s}$$
$$Br(\mu^- \to e^- + \bar{\nu}_e + \nu_\mu) \simeq 100\%$$
$$Br(\mu^- \to e^- + \nu_e + \bar{\nu}_\mu) < 1.2\%$$
$$Br(\mu^- \to e^- + \gamma) < 4.2 \times 10^{-13} \quad (90\% \, \text{CL})$$
$$Br(\mu^- \to e^- + e^- + e^+) < 1.0 \times 10^{-12} \quad (90\% \, \text{CL})$$

(15.17)

The layout of the Conversi *et al.* experiment is shown in Figure 15.6. Charged cosmic rays are focalized by magnetic fields and slowed down and a fraction is stopped by an absorber. The decay electrons are detected. The table presented in Figure 15.6 compares the number of electrons observed for positive and negative polarities, for iron and carbon targets. The lines (d) and (e) show that the observed positive and negative mesons had similar probabilities of decaying (their measured rates were $36(+)$ vs. $27(-)$ per 100 h, hence the *observed particle was not compatible with being the strongly interacting Yukawa particle*).

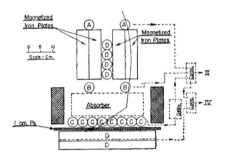

TABLE I. Results of measurements on β-decay rates for positive and negative mesons.

Sign	Absorber	III	IV	Hours	M/100 hours
(a) +	5 cm Fe	213	106	155.00′	67 ± 6.5
(b) −	5 cm Fe	172	158	206.00′	3
(c) −	none	71	69	107.45′	−1
(d) +	4 cm C	170	101	179.20′	36 ± 4.5
(e) −	4 cm C+5 cm Fe	218	146	243.00′	27 ± 3.5
(f) −	6.2 cm Fe	128	120	240.00′	0

Figure 15.6 Layout of the Conversi, Pancini, and Piccioni experiment. Reprinted with permission from M. Conversi, E. Pancini, and O. Piccioni, "On the disintegration of negative mesons," *Phys. Rev.*, vol. 71, pp. 209–210, 1947. Copyright 1947 by the American Physical Society.

In 1947 the groups of **Powell, Fowler, Perkins**[6] at Imperial College, London and of **Lattes, Occhialini, Powell** (alias Bristol group) succeeded in studying tracks obtained in the exposition of photographic emulsions at high altitudes, in airplanes or mountains. Photographic emulsions provide a direct record of cosmic-ray events with extremely fine resolution. The emulsion consists of crystals of silver halide embedded in gelatin. Exposition to light changes the structure of the crystals and creates a latent image. X-rays, as found by Röntgen (see Section 2.4), also create such a latent image. Charged particles traversing the emulsion produce ionization that also modifies the halide grains. These modified halides turn into small grains of silver with a diameter of about 1 μm in the development process. Silver appears black and the developed emulsions can be observed with an optical microscope. Perkins was able to take advantage of the advances in the technology of emulsions produced by the Ilford company and flew them in an aircraft at 30,000 feet [165]. He recorded tracks of charged pions present at high altitudes in the atmosphere. He was able to detect "stars" following the capture of the meson: after the cosmic ray stopped and was absorbed, the nucleus was blasted apart and fragments were observed in the emulsion (forming the "stars"). One of the events recorded is shown in Figure 15.7. This showed the behavior predicted by Tomonaga and Araki, contrary to the results of the Conversi *et al.* group.

In the exposition of emulsions by the Bristol group at the Pic du Midi at an altitude of 2800 m and at the Bolivian Andes at 5500 m, it was observed that while some mesons interacted, others decayed [166]. See Figure 15.8. They had succeeded in reconstructing the full decay chain of the stopping cosmic meson: the track of a charged particle stops and decays into a secondary charged particle whose track also stops. Their

---

6   Donald Hill Perkins (born 1925), British physicist.

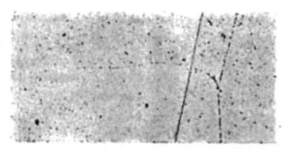

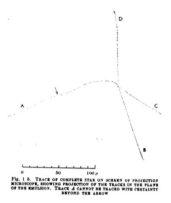

Fig. 1 *a*. PHOTOMICROGRAPH OF CENTRE OF STAR, SHOWING TRACK OF MESON PRODUCING DISINTEGRATION. (LEITZ 2 MM. OIL-IMMERSION OBJECTIVE. × 500)

Fig. 1 *b*. TRACK OF COMPLETE STAR ON SCREEN OF PROJECTION MICROSCOPE, SHOWING PROJECTION OF THE TRACKS IN THE PLANE OF THE EMULSION. TRACK *A* CANNOT BE TRACED WITH CERTAINTY BEYOND THE ARROW

Figure 15.7 A "star" event recorded in the emulsions flown at high altitude in an airplane. Reprinted with permission from Springer Nature: D. H. Perkins, "Nuclear disintegration by meson capture," *Nature*, vol. 159, pp. 126–127. Copyright 1947.

work established that there were indeed two different particles, one of which decayed into the other, and could establish the mass difference $m_\pi > m_\mu$.

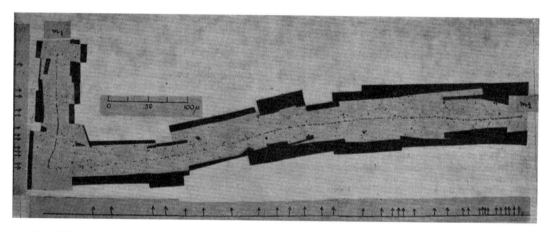

Figure 15.8 The first observation of the π decay into μ. Reprinted with permission from Springer Nature: C. M. G. Lattes, H. Muirhead, G. P. S. Occhialini, and C. F. Powell, "Processes involving charged mesons," *Nature*, vol. 159, pp. 694–697. Copyright 1947.

Today we know that when a charged pion decays into a muon and a neutrino, the muon acquires a fixed kinetic energy of 4.1 MeV due to the small difference of mass between the pion and the muon (and the mass of the neutrino is negligible with respect to this). The lifetime of the charged pion is approximately 26 ns, which is long-lived in comparison to a typical interaction rate of the strong force. The pion, being the lightest meson, cannot decay hadronically. Pions are then produced in strong interactions and decay through the weak interaction with the production of a lepton and a neutrino. When one observes such a decay, the neutrino generally goes undetected. The measured properties of the charged pion are (values taken from the Particle

Data Group's Review [5]):

$$\boxed{\pi^+, \pi^-}: \quad m_{\pi^\pm} = (139.57061 \pm 0.00024) \text{ MeV}$$
$$\tau = (2.6033 \pm 0.0005) \times 10^{-8} \text{ s}$$
$$Br(\pi^+ \to \mu^+ + \nu_\mu) = (99.98770 \pm 0.00004)\% \qquad T_\mu \simeq 4.1 \text{ MeV} \tag{15.18}$$
$$Br(\pi^+ \to e^+ + \nu_e) = (1.230 \pm 0.004) \times 10^{-4}$$

As discussed in Section 15.2, pions are copiously produced in the upper atmosphere. However, because of their short lifetime, most of them actually do not reach the surface of the Earth before decaying. This is the reason why the much longer-lived muons were detected first. The full decay chain of a stopping muon is hence:

$$\pi^+ \to \quad \mu^+ \quad + \nu_\mu$$
$$\qquad\qquad \hookrightarrow \quad e^+ + \nu_e + \bar\nu_\mu \tag{15.19}$$

where the neutrinos escape detection.

In 1950 **Bjorklund** *et al.* studied hadronic reactions at the then high energies produced by a synchrocyclotron. They observed the reactions as a function of the energy of the accelerated particles. They discovered an energy threshold for the incident protons of kinetic energy T_p to produce photons in the interaction of protons on carbon or beryllium targets. The best explanation for the existence of a threshold for photons is from the creation of **neutral pions** labeled π^0 ("pi-zero" or "pi-naught"), which subsequently very quickly decay into two photons:

$$\begin{cases} p + C \to \ldots + \pi^0 \qquad \text{when } T_p \geq 200 \text{ MeV} \\ \qquad\qquad\quad \hookrightarrow \gamma\gamma \\ p + Be \to \ldots + \pi^0 \\ \qquad\qquad\quad \hookrightarrow \gamma\gamma \end{cases} \tag{15.20}$$

As for the charged pion, the neutral pion cannot decay into other mesons. But unlike the charged pion which decays via the weak interaction, the neutral pion decays rather rapidly with a lifetime of about 10^{-16} s via the electromagnetic interaction. In the observations of Bjorklund *et al.*, incident protons with kinetic energies above approximately 200 MeV made the creation of pions kinematically possible. The produced neutral pions were not directly observed due to their short lifetime but led to a sharp increase of observed photon pairs as the energy of the incident protons was increased above threshold. The measured properties of the neutral pion are (values taken from the Particle Data Group's Review [5]):

$$\boxed{\pi^0 \to \gamma\gamma} \quad Br(\pi^0 \to \gamma\gamma) = (98.823 \pm 0.034\%)$$
$$m_{\pi^0} = (134.9770 \pm 0.0005) \text{ MeV}$$
$$\tau = (8.52 \pm 0.18) \times 10^{-17} \text{ s} \tag{15.21}$$
$$m_{\pi^\pm} - m_{\pi^0} = 4.5936 \pm 0.0005 \text{ MeV}$$

Summarizing some properties, charged and neutral pions will be copiously produced in strong interactions. However, being the lightest hadrons, they will not decay hadronically via the strong interaction. Instead, neutral pions decay electromagnetically into two photons and charged pions weakly and dominantly into a muon and a neutrino. The lifetime of the muon, which is also governed by weak interactions, is even longer. Thus we have the lifetime inequality:

$$\underbrace{\tau_{\pi^0}(10^{-16} \text{ s})}_{\text{electromagnetic}} \ll \underbrace{\tau_{\pi^\pm}(10^{-8} \text{ s}) \ll \tau_{\mu^\pm}(10^{-6} \text{ s})}_{\text{weak}} \tag{15.22}$$

15.5 The Spin and Parity of the Pion

Let us consider the **deuteron** d, which is a nuclear bound state composed of a proton (p) and a neutron (n). Its properties are rather well known (see, e.g., Ref. [167]):

$$\boxed{\text{Deuteron } d \text{ (stable)}} : \quad J^P = 1^+, I = 0$$
$$m_d = (1875.612928 \pm 0.000012) \text{ MeV}$$
$$\text{Binding energy } \epsilon = (2.22456612 \pm 0.00000048) \text{ MeV} \tag{15.23}$$
$$\mu_d = (0.8574382284 \pm 0.0000000094) \, \mu_B$$

It is remarkable that the deuteron is a stable system: if the neutron in the deuteron were to decay to form a proton, electron, and antineutrino, the combined mass energies of these particles would be $2M_p + m_e \simeq 1877.05$ MeV $> m_d$! Hence, a neutron in a deuteron cannot decay (note that most neutrons in nuclei behave in a similar way; what is surprising is that the deuteron is such a simple stable system). The total spin and parity are experimentally known to be $J^P = 1^+$. The wave function of the deuteron contains three parts: (1) space, (2) spin, and (3) isospin (see Section 15.6 for the definition of isospin). The total spin of the deuteron comes from the total spin of the two nucleons $S = s_p + s_n$ and their relative angular momentum L. The positive parity enforces the angular momentum L to be even, since $(-1)^L$, so $L = 0, 2, \ldots$. Furthermore, $S = 0$ or $S = 1$ since the proton and neutron are spin-1/2. Then only two configurations yield $J = 1$ of the deuteron: $S = 1$, $L = 0$ or $S = 1$, $L = 2$. The total wave function of the deuteron must be antisymmetric with respect to the interchange of the proton and the neutron, hence the isospin wave function has to be antisymmetric. Since isospin follows the same algebra as that of the spin, the antisymmetric isospin combination corresponds to the deuteron isospin state $I_d \equiv |I = 0, I_3 = 0\rangle = \frac{1}{\sqrt{2}} \left(|pn\rangle - |np\rangle \right)$.

The binding energy is $\epsilon \simeq 2.2$ MeV or just 1.1 MeV per nucleon, which is less than the typical binding energy of nucleons in ordinary nuclei of about 8 MeV per nucleon. As a consequence, excited states of the deuteron do not exist, and the proton and neutron can easily be dissociated. Such a low binding energy makes the deuteron an ideal target of (quasi)free neutrons (there is no free neutron target!). The deuteron can for example be dissociated into two neutrons or two protons via the reactions:

$$\pi^- + d \to n + n \quad \text{and} \quad \pi^+ + d \to p + p \tag{15.24}$$

In 1951 **W. B. Cheston** showed that a comparison of the total cross-sections for the forward (pion production) and backward (pion absorption) rates of the reaction:

$$p + p \leftrightarrows \pi^+ + d \tag{15.25}$$

at appropriate energies allows a unique determination of the spin of the charged pion, independent of the (at the time unknown) details of the weak interaction [168]. The unpolarized cross-sections of the two reactions can be shown to differ only by phase space and sums over spin final states, so their ratio is given by:

$$\frac{\sigma(p + p \to \pi^+ + d)}{\sigma(\pi^+ + d \to p + p)} = \frac{(2s_\pi + 1)(2J_d + 1)}{(2s_p + 1)^2} \frac{p_\pi^2}{p_p^2} = \frac{3}{4}(2s_\pi + 1)\frac{p_\pi^2}{p_p^2} \tag{15.26}$$

where p_π (resp. p_p) is the momentum of the pion (resp. proton) in the center-of-mass system of the considered reaction, J_d is the spin of the deuteron, and s_π (resp. s_p) the spin of the pion (resp. proton). In the last term we inserted $J_d = 1$ and $s_p = 1/2$.

In the same year **Cartwright, Richman, Whitehead, and Wilcos** measured the forward cross-section, corresponding to a pion and a deuteron produced in pp collisions [169]. They used 340 MeV protons hitting a hydrogen target. Also the same year **Durbin, Loar, and Steinberger** measured the backward reaction [170] by comparing the rate of production of protons by pions on ordinary water (H_2O) and heavy water (D_2O). The pions were produced by 380 MeV protons hitting a beryllium target. The secondary particles were selected magnetically to have a momentum $p = 75 \pm 4$ MeV. The composition of the beam was 90% π^+ and 10% μ^+. Combination of the two results showed that the spin of the pion is zero.

- **Intrinsic parity of the pion.** The deuteron reactions can also be used to determine the intrinsic parity of the pion. In the initial state the angular momentum is zero; since the pion has no spin and the deuteron has spin 1, the total angular momentum must be $J = L + S = 1$. Therefore, also in the final state we must have $J = 1$. Let us consider for instance $\pi^- + d \to n + n$. The symmetry of the final state wave function (under interchange of the two neutrons) is given by:

$$(-1)^{S+1}(-1)^L = (-1)^{L+S+1} \tag{15.27}$$

Since we have two identical fermions, the state must be antisymmetric under interchange of the two neutrons, which implies that $L + S$ is even. Probing options under the condition $J = 1$, we find the following cases: $L = 0$ $S = 1 \to$ forbidden; $L = 1$ $S = 0 \to$ forbidden; $L = 2$ $S = 1 \to$ forbidden; so finally we are left with the only possible solution with $L = 1$ and $S = 1$. Therefore the parity of the final state is $P = (-1)^L = -1$. Since the parity of the deuteron is $P_d = +1$, we obtain for the intrinsic parity of the pion $P_\pi = -1$. In summary, the **charged pion is a pseudoscalar meson**:

$$\boxed{\pi^+, \pi^-}: \quad J^P = 0^- \tag{15.28}$$

In 1954 **Chinowsky and Steinberger** detected the $\pi^- + d \to n + n$ reaction [171]. The requirement that the two neutrons be back-to-back allowed them to separate the reaction from, e.g., $\pi^- + d \to n + n + \gamma$, and also note the absence of $\pi^- + d \to n + n + \pi^0$. In $\pi^- + d \to n + n$, the Pauli principle enforces the final state to be antisymmetric under the exchange of the neutrons, which corresponds also to the parity swap. Therefore, the parity of the final state is $(-1)^\ell(+1)^2 = -1$, where $\ell = 1, 3, \ldots$ is the angular momentum between the two neutrons. Angular momentum conservation enforces $\ell = 1$ since the pion is spinless and the deuteron has spin 1, assigning a zero angular momentum in the capture process. Hence, the observation of $\pi^- + d \to n + n$ provided the equation $P(\pi)P(d)(-1)^0 = P(n)^2(-1)^\ell = -1$. Consequently, the pion was indeed a pseudoscalar $P(\pi) = -1$. The non-observation of $\pi^- + d \to n + n + \pi^0$ indicated that the neutral pion is also a pseudoscalar.

In 1959 **Plano, Prodell, Samios, Schwartz and Steinberger** published *direct* evidence that the π^0 is indeed a pseudoscalar particle. They performed an experiment based on an idea by **C. N. Yang** [172], using the decay $\pi^0 \to \gamma\gamma$. Yang demonstrated that angular momentum and parity conservation require a perpendicular correlation of the polarization of the two photons for a pseudoscalar particle. For a scalar particle, the correlation must be parallel. But the measurement requires the determination of the polarization of the two photons. Since the polarization of the photons is difficult to measure directly, one relies on the double Dalitz decay, which can be seen as the internal conversion of the two photons. Plano *et al.* pointed out that the double Dalitz decay of the π^0 would allow the determination of these polarizations:

$$\pi^0 \to \gamma_1 + \gamma_2 \to (e^+e^-)(e^+e^-) \tag{15.29}$$

The planes of the pairs "remember" the polarization of the virtual intermediate photons. The correlation of the planes in the double Dalitz decay was computed by Kroll and Wada [173]. In the rest frame of the π^0, we can reconstruct two planes subtended by each electron–positron pair, and compute the momentum $\vec{k}$ and the two polarization vectors $\vec{\epsilon}_1$ and $\vec{\epsilon}_2$, as shown in Figure 15.9. Since the pion has no spin, it must be either a scalar or a pseudoscalar particle (if it had a spin-1 it could not decay into two photons, since these latter can only have $L = 0$ or $L = 2$ corresponding to antiparallel and parallel spins). The wave function of the two-photon state must be a scalar or pseudoscalar function f of the three vectors $\vec{k}, \vec{\epsilon}_1, \vec{\epsilon}_2$. Since we have that $\vec{k} \cdot \vec{\epsilon}_1 = 0$ and $\vec{k} \cdot \vec{\epsilon}_2 = 0$, we can only construct two appropriately behaved quantities:

$$P = 1 : f = f(\vec{\epsilon}_1 \cdot \vec{\epsilon}_2) \propto \cos\phi \quad \text{and} \quad P = -1 : f = f(\vec{\epsilon}_1 \times \vec{\epsilon}_2) \cdot \vec{k} \propto \sin\phi \tag{15.30}$$

where ϕ is defined in Figure 15.9. Consequently, the matrix element for a final state with given parity is going to be proportional to the two cases:

$$|\mathcal{M}_+| \propto \cos^2\phi = \frac{1 + \cos 2\phi}{2} \quad \text{and} \quad |\mathcal{M}_-| \propto \sin^2\phi = \frac{1 - \cos 2\phi}{2} \tag{15.31}$$

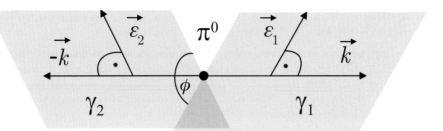

Figure 15.9 Spin–Parity of the pion: definition of the angle ϕ.

The actual calculation of Wada and Kroll leads to the theoretical correlation:

$$|\mathcal{M}_\pm| \propto 1 \pm \alpha(\phi, x_1, x_2, x_3) \cos 2\phi \tag{15.32}$$

where x_1, x_2, and x_3 are the energy fractions of three of the final-state particles. The function α is universal and the parity dependence is only in the sign. Experimentally, the π^0 were produced at rest in the capture reaction $\pi^- p \to \pi^0 n$ in a hydrogen bubble chamber of 30 cm diameter and 15 cm depth in a magnetic field of 0.55 T. A total of 700,000 pictures were taken. From these, a sample of 103 pictures were selected which exhibited the desired topology. While the charge of the particles is measured, there are still two ways to match two electrons and two positrons. They found that in most cases the angle between the pair is small ($\sim 10°$), so they can be naturally matched. After quality cuts, 64 events remained from which the angle between the planes produced by the electron–positron pairs is plotted, weighted by the correlation coefficient α, as shown in Figure 15.10. The observed distribution is then fitted with the $1 - \alpha \cos 2\phi$ form. It is clearly consistent with the data, hence it favors the pseudoscalar hypothesis over the scalar one.

More recently, in 2008, the measurement was repeated by the KTeV experiment [174]. The in-flight $K_L \to \pi^0 \pi^0 \pi^0$ decays were studied, leading to significant statistics of the Dalitz double decay. With 30,511 candidates, they confirmed the negative parity of the neutral pion and measured:

$$Br(\pi^0 \to (e^+ e^-)(e^+ e^-)) = (3.26 \pm 0.18) \times 10^{-5} \tag{15.33}$$

Their reconstructed angle ϕ is shown in Figure 15.11, very consistent with expectations that the neutral pion is pseudoscalar. In summary, like the charged pion, the **neutral pion is a pseudoscalar meson**:

$$\boxed{\pi^0}: \quad J^P = 0^- \tag{15.34}$$

15.6 The Isospin Symmetry

Let us consider the two lightest baryons: the **proton** (p) and the **neutron** (n). We note that these two particles possess similar properties (for example their rest masses differ only by a permil) except for their electromagnetic properties. Their electric charges differ by one unit and their magnetic moments are totally different (values

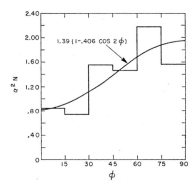

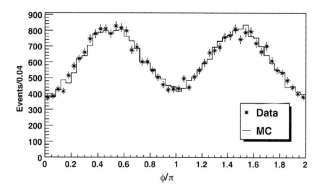

Figure 15.10 Plot of the weighted frequency distribution of the angle between planes of polarization. Reprinted with permission from N. P. Samios, R. Plano, A. Prodell, M. Schwartz, and J. Steinberger, "Parity of the neutral pion and the decay $\pi^0 \to 2e^+ + 2e^-$," *Phys. Rev.*, vol. 126, pp. 1844–1849, 1962. Copyright 1962 by the American Physical Society.

Figure 15.11 Plot of the weighted frequency distribution of the angle between planes of polarization measured by KTeV. Reprinted with permission from E. Abouzaid *et al.*, "Determination of the parity of the neutral pion via the four-electron decay," *Phys. Rev. Lett.*, vol. 100, p. 182001, 2008. Copyright 2008 by the American Physical Society.

taken from the Particle Data Group's Review [5]):

> Proton p (stable):
> $$M_p = (938.272081 \pm 0.000006)\text{ MeV}$$
> $$Q_p = +e \qquad (|Q_p + q_e|/e) < 1 \times 10^{-21}$$
> $$\mu_p = (2.792847351 \pm 0.000000009)\,\mu_N$$
>
> Neutron $n \to p + e^- + \bar{\nu}_e$:
> $$M_n = (939.565413 \pm 0.000006)\text{ MeV}$$
> $$\tau_n = (880.2 \pm 1.0)\text{ s}$$
> $$Q_n = (-2 \pm 8) \times 10^{-22}e,$$
> $$\mu_n = (-1.9130427 \pm 0.0000005)\,\mu_N$$

Their rest mass difference is indeed at the level of one permil, with the neutron being slightly heavier:

$$M_n - M_p = 1.2933321 \pm 0.0000005\text{ MeV} \tag{15.35}$$

In 1932 **Heisenberg** introduced the concept of **isospin** (also called **isotopic spin**) while building a model to explain the construction of the nucleus in atoms. At the time, the possibility that the neutron (just discovered by Chadwick based on the works of Curie and Joliot – see Section 2.14) be "composed" of a proton and an electron was still considered. This fact would explain the "decay" of a neutron into a proton and electron. Heisenberg actually found it much more sensible to assume that the neutron is an independent elementary particle. Heisenberg considered the short-range strong interaction between protons and neutrons. He assumed that the range of the strong interaction was of the order of 10^{-12} cm = 10 fm. He then described the dynamics of the system with the introduction of a new quantum number to distinguish neutrons from protons. He wrote:

> *"Rather, the neutron should be considered as an independent fundamental building block, however, which, under suitable circumstances, is assumed to be capable of splitting into a proton and an electron, whereby the conservation laws for energy and momentum are presumably no longer applicable. (...) About the action of the forces of the elementary nuclear building blocks on each other,*

we first consider the action between neutron and proton. If the neutron and proton are brought into a distance comparable to the dimensions of a nucleus, (...) an exchange of negative charge will occur. (...) Similarly, the interaction of two neutrons will be described by an interaction energy, whereby one can (...) assume that this energy leads to an attractive force between the neutrons. (...) About the functions (...) only some very general statements can be made. It is assumed that in regions of the order 10^{-12} cm they quickly sink to zero as r increases. (...) To write down the Hamilton function of the atomic nucleus, the following variables are useful: each particle in the nucleus is characterized by five quantities, the three spatial coordinates, the spin and by a fifth number ρ^ζ, which corresponds to values +1 and −1. $\rho^\zeta = +1$ means that the particle is a neutron, $\rho^\zeta = -1$ means that the particle is a proton. Since the Hamilton function also contains transition elements from $\rho^\zeta = +1$ to $\rho^\zeta = -1$ due to the change of place, it is useful to also introduce the matrices

$$\rho^\xi = \begin{pmatrix} 0 & 1 \\ 1 & 0 \end{pmatrix}, \quad \rho^\eta = \begin{pmatrix} 0 & -i \\ i & 0 \end{pmatrix}, \quad \rho^\zeta = \begin{pmatrix} 1 & 0 \\ 0 & -1 \end{pmatrix}, \tag{15.36}$$

The space of ξ,η,ζ has of course nothing to do with the real space." [7]

In modern language, we can say that in the *absence of electromagnetic interaction*,[8] the proton and neutron would be considered as a single unit called the **nucleon** with an internal degree of freedom in an internal space of isospin. The electromagnetic interaction lifts this degeneracy. This situation could, for example, be interpreted in analogy to the Zeemann effect, where the spin degeneracy in the atom is lifted by the external magnetic field.

A given nucleon (proton or neutron) is characterized by its state of isospin and its associated **(flavor) isospin quantum number**. Since we have two possible states, in analogy to the spin, we can assign an isospin $I = 1/2$ to the nucleon. In the modern assignment, the proton is given an isospin $I_3 = +1/2$ and the neutron $I_3 = -1/2$. Hence, we can write the **isospin state spinor** χ as:

$$\chi = \begin{pmatrix} p \\ n \end{pmatrix} \quad \text{where} \quad p \equiv \begin{pmatrix} 1 \\ 0 \end{pmatrix} \quad \text{and} \quad n \equiv \begin{pmatrix} 0 \\ 1 \end{pmatrix} \tag{15.37}$$

The p and n isospin eigenstates are basically degenerate in the absence of electromagnetic interaction. The strong interaction is insensitive to the third component of isospin and does not distinguish protons from neutrons (apart from their tiny difference in mass). The degeneracy is lifted by the electromagnetic interaction.

We note that there is a complete correspondence between the algebraic properties of spin S and isospin I. They are both representable as an $SU(2)$ group. The isospin spinor and the nucleon isospin eigenstates $|I, I_3\rangle$ can be further written as "isospin up" and "isospin-down":

$$I = \frac{1}{2}: \quad p \equiv |\frac{1}{2}, +\frac{1}{2}\rangle, \quad n \equiv |\frac{1}{2}, -\frac{1}{2}\rangle \tag{15.38}$$

---

7   *"Vielmehr soll das Neutron als selbständiger Fundamentalbestandteil betrachtet werden, von dem allerdings angenommen wird, dass er unter geeigneten Umständen in Proton und Elektron aufspalten kann, wobei vermutlich die Erhaltungssätze für Energie und Impuls nicht mehr anwendbar sind. (...) Von den Kraftwirkungen der elementaren Kernbausteine aufeinander betrachten wir zunächst die zwischen Neutron und Proton. Bringt man Neutron und Proton in einen mit Kerndimensionen vergleichbaren Abstand, so wird (...) ein Platzwechsel der negativen Ladung eintreten. (...) Ähnlich wird die Wechselwirkung zweier Neutronen durch eine Wechselwirkungsenergie beschrieben werden, wobei man (...) annehmen kann, dass diese Energie zu einer Anziehungskraft zwischen den Neutronen führt. (...) Über die Funktionen (...) lassen sich nur einige ganz allgemeine Aussagen machen. Man wird vermuten, dass sie in Bereichen der Ordnung 10^{-12} cm mit wachsendem r rasch nach Null absinken. (...) Um nun die Hamiltonfunktion des Atomkerns aufzuschreiben, erweisen sich folgende Variablen als zweckmässig: Jedes Teilchen im Kern wird charakterisiert durch fünf Grössen, die drei Ortskoordinaten, den Spin und durch eine fünfte Zahl ρ^ζ, die der beiden Werte +1 und −1 fähig ist. $\rho^\zeta = +1$ soll bedeuten, das Teilchen sei ein Neutron, $\rho^\zeta = -1$ bedeutet, das Teilchen sei ein Proton. Da in der Hamiltonfunktion wegen des Platzwechsels auch Übergangselemente von $\rho^\zeta = +1$ nach $\rho^\zeta = -1$ vorkommen, erweist es sich als zweckmässig, auch die Matrizen $\rho^\xi = ...$, $\rho^\eta = ...$, $\rho^\zeta = ...$ einzuführen. Der Raum der ξ,η,ζ hat aber natürlich nichts mit dem wirklichen Raum zu tun."* Reprinted by permission of Springer Nature: W. Heisenberg, "Über den Bau der Atomkerne. I," *Zeitschrift für Physik*, vol. 77, pp. 1–11, Jan 1932. Copyright 1932.

8   That is to say, in an hypothetical condition where the electromagnetic force could be switched off!

We label the corresponding "Pauli matrices of the isospin" as τ_1, τ_2 and the diagonal τ_3. The **isospin raising and lowering operators** $\tau_\pm$ are defined as:

$$\tau_\pm \equiv \frac{1}{2}\left(\tau_1 \pm i\tau_2\right) \tag{15.39}$$

such that

$$\tau_+ \left|n\right\rangle = \left|p\right\rangle \quad \text{and} \quad \tau_- \left|p\right\rangle = \left|n\right\rangle \tag{15.40}$$

We can now consider an **isospin transformation** as an equivalent to a *rotation in isospin space*. This transformation does form an $SU(2)$ group. **The nucleon forms an $SU(2)$-doublet in isospin space.** This means that under the isospin transformation, the nucleon remains the nucleon although the proton and neutron are exchanged (or mixed). This is graphically represented in Figure 15.12.

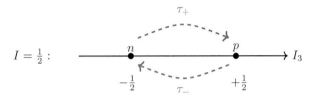

Figure 15.12 Representation of the nucleon as a doublet in the isospin space.

Heisenberg's assumption that the proton and neutron are equivalent from the point of view of the strong force can be expressed as an invariance property. We say that **the strong interaction is invariant under the $SU(2)$ isospin transformation.** This is called the **isospin symmetry** of the strong force.[9] If there is *symmetry under the isospin transformation* from the point of view of the strong force, then Noether's theorem tells us that **isospin is a conserved quantity in strong interactions!**

• **Pion multiplet.** What about Yukawa's meson, the so-called pion? In this case, we have three particles $\pi^\pm$ and π^0. We have already noted that their rest masses are almost degenerate – see Eqs. (15.18) and (15.21), at the few percent level, so slightly less degenerate than the nucleons, but nonetheless. It is hence natural to assign the pions to an **isospin multiplet with $I = 1$:**

$$I = 1: \quad \pi^+ \equiv \left|1, +1\right\rangle, \quad \pi^0 \equiv \left|1, 0\right\rangle, \quad \pi^- \equiv \left|1, -1\right\rangle \tag{15.41}$$

Evidently $\tau_+ \left|\pi^-\right\rangle = \left|\pi^0\right\rangle$, $\tau_+ \left|\pi^0\right\rangle = \left|\pi^+\right\rangle$, ... and similarly for the τ_- operator. The isospin transformation mixes the three pions of different electric charge among themselves, which are otherwise similar from the point of view of the strong force.

• **Isospin properties.** We can derive a lesson from the above arguments. Isospin is a good quantum number which is conserved in strong interactions as a consequence of the symmetry properties under the isospin transformation. This allows us to classify and organize the elementary particles according to *their multiplet representation in the associated group.* From a list of a priori unrelated elementary particles p, n, $\pi^\pm$, and π^0, we obtained a clear and organized picture and a prescription for classification. Nucleons and pions acquire a distinctive place within the existing class of hadrons.

9   We note that today's interpretation is not exactly as such. The proton and the neutron are no longer considered as elementary particles but are known to be composed of the elementary quarks. In this picture, the proton is composed of two u-quarks and one d-quark. The neutron is made of one u-quark and two d-quarks. In the context of QCD, it is known that the strong force does not strictly obey isospin symmetry because the rest masses of the u and d-quarks are slightly different. We will ignore this "breaking" of the isospin symmetry by the strong force.

• **Baryon number.** Since nucleons and pions belong to separate multiplets, it also means that they will not transform into each other under the isospin transformation. Nucleons and pions then clearly belong to different classes of particles, which split the hadrons into the class of "baryons" and the class of "mesons." In order to distinguish the two classes, one introduces the **baryon quantum number** B. Baryons have a baryon number of $B = +1$, mesons have a baryon number of $B = 0$, and antibaryons have a baryon number of $B = -1$. We can then summarize the classification of our elementary hadrons in the following way:

$$
\begin{aligned}
&B = 0, I(J^P) = 1(0^-)\,\text{multiplet}: & \pi^+ \equiv |1, +1\rangle, \quad \pi^0 \equiv |1, 0\rangle, \quad \pi^- \equiv |1, -1\rangle \\
&B = 1, I(J^P) = 1/2(1/2^+)\,\text{doublet}: & p \equiv \left|\frac{1}{2}, +\frac{1}{2}\right\rangle, \quad n \equiv \left|\frac{1}{2}, -\frac{1}{2}\right\rangle
\end{aligned}
\tag{15.42}
$$

where J is the **spin**, P the **intrinsic parity**, I the **isospin**, and B the **baryonic quantum number**.

We already noted that the electromagnetic interaction breaks the isospin symmetry. We hence expect the electric charge to be uniquely related to the quantum numbers defined above. Indeed, we see that:

$$
Q = \frac{B}{2} + I_3
\tag{15.43}
$$

This relation will be further extended in the coming chapters (see, e.g., Chapter 17) into what is referred to as the Gell-Mann/Nishijima formula.

• **Dynamical consequences.** We discuss in the following some of the dynamical consequences of isospin invariance in the strong interaction. In general the amplitude of a strong process from an initial state of isospin I, I_3 to a final state of total isospin I', I_3' can be expressed as:

$$
\mathcal{M} = \langle I', I_3'|A|I, I_3\rangle
\tag{15.44}
$$

where A is some operator. Under the assumption that the strong force is completely invariant under isospin rotation, we can conclude that:

- the total isospin must be conserved in strong interactions, so $I' = I$;

- the amplitude of the strong interaction must not depend on I_3.

The amplitude of a strong process will hence be decomposed in terms that depend only on I:

$$
\mathcal{M}_I = \langle I|A|I\rangle
\tag{15.45}
$$

Let us consider a *system composed of two nucleons*. The isospin state of this system will follow rules similar to those of the addition of spins. For example, we recall that the addition of two spin-1/2 leads to a *spin singlet* state and a *spin triplet* state: $2 \otimes 2 = 3 \oplus 1$. In analogy, we can for example write the isospin states for the combination of a proton and a neutron, one with total isospin 1 and the other with total isospin 0. We hence obtain the symmetric total isospin $I = 1$ iso-triplet:

$$
\begin{cases}
|I = 1, I_3 = +1\rangle = pp \\
|I = 1, I_3 = 0\rangle = (pn + np)/\sqrt{2} \\
|I = 1, I_3 = -1\rangle = nn
\end{cases}
\tag{15.46}
$$

and the antisymmetric total isospin $I = 0$ iso-singlet:

$$
|I = 0, I_3 = 0\rangle = (pn - np)/\sqrt{2}
\tag{15.47}
$$

From a strong force point of view, this system should be described by two main amplitudes corresponding to the isospin 0 and 1 cases: $\mathcal{M}_0$ and $\mathcal{M}_1$.

As a second example, let us consider the scattering of a pion with a nucleon. We can for example look at the three reactions:

$$
\begin{aligned}
(1) &\quad \pi^+ + p \to \pi^+ + p \\
(2) &\quad \pi^- + p \to \pi^- + p \\
(3) &\quad \pi^- + p \to \pi^0 + n
\end{aligned}
\tag{15.48}
$$

Since the pion has isospin $I = 1$ and the nucleon an isospin $I = 1/2$, the total isospin can be decomposed in a $I = 3/2$ and $I = 1/2$ state. We introduce the two corresponding amplitudes:

$$
\mathcal{M}_3 = \langle \tfrac{3}{2} | A | \tfrac{3}{2} \rangle \quad \text{and} \quad \mathcal{M}_1 = \langle \tfrac{1}{2} | A | \tfrac{1}{2} \rangle
\tag{15.49}
$$

The existence of an isospin symmetry allows us to derive relations among the cross-sections of the various reactions. The Clebsch–Gordan coefficients for the various isospin configurations are:

$$
\begin{aligned}
\pi^+ + p: &\quad |1,1\rangle \otimes |\tfrac{1}{2}, \tfrac{1}{2}\rangle = |\tfrac{3}{2}, \tfrac{3}{2}\rangle \\
\pi^- + p: &\quad |1,-1\rangle \otimes |\tfrac{1}{2}, \tfrac{1}{2}\rangle = \frac{1}{\sqrt{3}} |\tfrac{3}{2}, -\tfrac{1}{2}\rangle - \sqrt{\frac{2}{3}} |\tfrac{1}{2}, -\tfrac{1}{2}\rangle \\
\pi^0 + n: &\quad |1,0\rangle \otimes |\tfrac{1}{2}, -\tfrac{1}{2}\rangle = \sqrt{\frac{2}{3}} |\tfrac{3}{2}, -\tfrac{1}{2}\rangle + \frac{1}{\sqrt{3}} |\tfrac{1}{2}, -\tfrac{1}{2}\rangle
\end{aligned}
\tag{15.50}
$$

By considering the brackets for the initial and final states factorized by their total isospin values, we derive the relations between the amplitudes of the three processes:

$$
\begin{cases}
\mathcal{M}(\pi^+ + p \to \pi^+ + p) = \mathcal{M}_3 \\
\mathcal{M}(\pi^- + p \to \pi^- + p) = \dfrac{1}{3}\mathcal{M}_3 + \dfrac{2}{3}\mathcal{M}_1 \\
\mathcal{M}(\pi^- + p \to \pi^0 + n) = \dfrac{\sqrt{2}}{3}\mathcal{M}_3 - \dfrac{\sqrt{2}}{3}\mathcal{M}_1
\end{cases}
\tag{15.51}
$$

This allows us to get ratios among the cross-sections:

$$
\sigma_1 : \sigma_2 : \sigma_3 = |\mathcal{M}_3|^2 : \frac{1}{9}|\mathcal{M}_3 + 2\mathcal{M}_1|^2 : \frac{2}{9}|\mathcal{M}_3 - \mathcal{M}_1|^2
\tag{15.52}
$$

For illustration, if we assume that the amplitude of the $I = 3/2$ process is much larger than the $I = 1/2$ one, then one would expect that the ratio of the cross-sections is simply given by:

$$
\sigma_1 : \sigma_2 : \sigma_3 \simeq 9 : 1 : 2 \quad \text{for} \quad |\mathcal{M}_3| \gg |\mathcal{M}_1|
\tag{15.53}
$$

and hence:

$$
\frac{\sigma(\pi^+ p)}{\sigma(\pi^- p)} \simeq \frac{9}{1 + 2} = 3
\tag{15.54}
$$

This relation is actually experimentally validated in the region of the so-called $\Delta(1232)$ resonance with isospin $I = 3/2$ [28]!

15.7 The Discovery of Strangeness

In 1943 **Leprince-Ringuet**[10] and **L'héritier** from l'Ecole Polytechnique of Paris reported on the first evidence of the existence of a particle with a mass of about 990 times the mass of the electron in cosmic rays. They wrote:

> *"In 1943, in the laboratory of Largentière (Hautes-Alpes) located at an altitude of 1000 m, we took a series of 10000 pictures of cosmic trajectories controlled by counters. The rays, filtered by 10 cm of lead, passed through a 75 cm high Wilson chamber, placed in a H magnetic field of about 2500 gauss. We placed ourselves in the most favorable experimental conditions to make the best use of the pictures of collisions between penetrating particles and electrons in the gas of the chamber, in order to determine the rest mass of the incident particle. We obtained about ten interesting pictures."* [11]

The technique used to determine the mass of the incident particle was to reconstruct the kinematics of secondary electrons (δ-rays) produced in the collisions of the cosmic rays with the electrons in a Wilson chamber [175]. One picture reported in Figure 15.13 was particularly impressive. It shows a gas collision for which excellent conditions of reconstruction are achieved, since the secondary track makes a large angle and its projected radius of curvature, as well as the deflection by which it deviates from the primary, are precisely measurable.

Let us define the four-momenta in the laboratory frame before and after the collision of the incoming particle of mass M with p^μ and p'^μ and of the electron with k^μ and k'^μ. We have:

$$p^\mu = \left(\sqrt{\vec{p}^2 + M^2}, \vec{p}\right), \quad k^\mu = \left(m_e, \vec{0}\right) \quad \text{and} \quad p'^\mu = \left(\sqrt{\vec{p}'^2 + M^2}, \vec{p}'\right), \quad k'^\mu = \left(E'_k, \vec{k}'\right) \tag{15.55}$$

Energy–momentum conservation yields $p + k = p' + k'$, or squaring:

$$(p+k)^2 = p^2 + 2p \cdot k + k^2 = (p'+k')^2 = M^2 + 2p' \cdot k' + k'^2 \tag{15.56}$$

Since $p^2 = p'^2 = M^2$ and $k^2 = k'^2 = m_e^2$, we simply have:

$$p \cdot k = p' \cdot k' \implies m_e \sqrt{\vec{p}^2 + M^2} = E'_k \sqrt{\vec{p}'^2 + M^2} - \vec{k}' \cdot \vec{p}' \tag{15.57}$$

As is obvious from Figure 15.13, we can safely assume that the incoming particle is undeflected, hence $\vec{p} \approx \vec{p}'$, consequently:

$$\sqrt{\vec{p}^2 + M^2}\,(E'_k - m_e) \approx \vec{k}' \cdot \vec{p} \implies M^2 = \vec{p}^2 \left(\frac{(\vec{k}' \cdot \vec{p})^2}{\vec{p}^2 (E'_k - m_e)^2} - 1\right) \tag{15.58}$$

We note that:

$$\frac{(\vec{k}' \cdot \vec{p})^2}{\vec{p}^2 (E'_k - m_e)^2} = \frac{\vec{k}'^2 \vec{p}^2 \cos^2 \zeta}{\vec{p}^2 (E'_k - m_e)^2} = \frac{(E_k'^2 - m_e^2)\cos^2 \zeta}{(E'_k - m_e)^2} = \frac{(E'_k + m_e)\cos^2 \zeta}{(E'_k - m_e)} \tag{15.59}$$

---

10   Louis Leprince-Ringuet (1901–2000), French physicist.

11   *"Nous avons pris, au cours de l'année 1943, dans le laboratoire de Largentière (Hautes-Alpes) situé à 1000m d'altitude, une série de 10000 clichés de trajectoires cosmiques commandées par compteurs. Les rayons, filtrés par 10 cm de plomb, traversaient une chambre de Wilson de 75 cm de hauteur, placée dans un champ magnétique H de 2500 gauss environ. Nous nous sommes placés dans les conditions expérimentales les plus favorables pour profiter au mieux des clichés de collision entre particules pénétrantes et électrons du gaz de la chambre, dans le but de déterminer la masse au repos de la particule incidente. Nous avons obtenu une dizaine de clichés intéressants."* Reprinted from L. Leprince-Ringuet and M. L'héritier, "Existence probable d'une particle de masse 990 m_0 dans le rayonnement cosmique," *Comptes Rendus Hebd. Séances Acad. Sci.*, vol. 219, 1944.

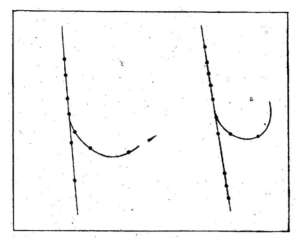

Dessin stéréoscopique de la collision.

Figure 15.13 Tracks obtained by Leprince-Ringuet interpreted as an incoming positive cosmic-ray particle with mass 900 m_e kicking an energetic secondary δ-ray measured with a Wilson tracking chamber. Reprinted from L. Leprince-Ringuet and M. L'héritier, "Existence probable d'une particle de masse 990 m_0 dans le rayonnement cosmique," *Comptes Rendus Hebd. Séances Acad. Sci.*, vol. 219, 1944.

where ζ is the measured deflection angle between the incoming particle and the δ ray. Finally:

$$M \approx p \left(\frac{E'_k + m_e}{E'_k - m_e} \cos^2 \zeta - 1 \right)^{1/2} \tag{15.60}$$

Using their values reported in Ref. [175], Leprince-Ringuet and L'héritier compute that $M \simeq 990\, m_e$. They were very surprised by this value! This new particle with an apparent rest mass of $M \simeq 500 \pm 60$ MeV could not be interpreted as a muon but was not unanimously accepted as a heavy meson; it was argued that the upper limit on the mass could reach the proton mass if all the errors were doubled and added together. Despite the low probability of the proton hypothesis, the evidence was not deemed decisive [176].

In 1947, **Rochester and Butler**, working at the University of Manchester, observed two photographs where cosmic rays apparently produced peculiar V-shaped tracks or "forks." They used a cloud chamber with a magnetic field to study the products of the interactions of high-energy cosmic rays in a block of lead. One of the two photographs is reported in Figure 15.14 and was consistent with the decay of a neutral particle into two charged particles. The second showed the decay of a charged particle into another charged and at least one neutral particle. These events reflected the decay of unknown particles with masses roughly 1,000 times that of an electron. They write:

> "*We conclude from all the evidence that Photograph 1 represents the decay of a neutral particle, the mass of which is unlikely to be less than* $770m_e$ *or greater than* $1,600m_e$, *into the two observed charged particles. (...) It may be noted that no neutral particle of mass* $1000m_e$ *has yet been observed; a charged particle of mass* $990m_e \pm 12$ *per cent has, however, been observed by Leprince-Ringuet and L'héritier.*"[12]

---

12   Reprinted by permission from Springer Nature: G. D. Rochester and C. C. Butler, "Evidence for the existence of new unstable elementary particles," *Nature*, vol. 160, pp. 855–857, 1947. Copyright 1947.

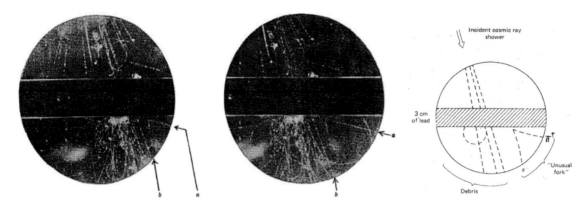

Figure 15.14 Photograph from the Rochester cloud chamber showing an usual "fork." Reprinted by permission from Springer Nature: G. D. Rochester and C. C. Butler, "Evidence for the existence of new unstable elementary particles," *Nature*, vol. 160, pp. 855–857, 1947.

In 1949, **Brown, Camerini, Fowler, Muirhead, Powell, and Ritson** offered more extensive corroborating evidence for these "V-particles" [177] using a different technique based on photographic emulsions.

The experiments described above led to the first observations of strange particles – now known as **kaons**. They come as charged and neutral particles and have no spin. The charged kaon properties are the following (values taken from the Particle Data Group's Review [5]):

$$
\boxed{K^+, K^-} \quad
\begin{aligned}
&m_{K^\pm} = (493.677 \pm 0.016) \text{ MeV} \\
&\tau = (1.2380 \pm 0.0020) \times 10^{-8} \text{ s} \qquad (c\tau \approx 371 \text{ cm}) \\
&Br(K^+ \to \mu^+\nu_\mu) = (63.56 \pm 0.11)\% \\
&Br(K^+ \to \pi^+\pi^0) = (20.67 \pm 0.08)\% \quad (K\pi_2) \\
&Br(K^+ \to \pi^+\pi^+\pi^-) = (5.583 \pm 0.024)\% \quad (K\pi_3) \\
&Br(K^+ \to e^+\pi^0\nu_e) = (5.07 \pm 0.04)\% \quad (Ke_3) \\
&Br(K^+ \to \mu^+\pi^0\nu_\mu) = (3.352 \pm 0.033)\% \quad (K\mu_3) \\
&Br(K^+ \to \pi^+\pi^0\pi^0) = (1.760 \pm 0.023)\%
\end{aligned}
\tag{15.61}
$$

This implies that $c\tau \simeq 3.7$ m, so the mean free flight path is $l = \beta\gamma c\tau \simeq \gamma c\tau$. Hence, the kaon is relatively long-lived. This is similar to the charged pion which is $\tau \simeq 26$ ns or $c\tau \simeq 7.8$ m. This comes from the fact that the decay of the kaons is mediated by the weak force.

The neutral kaons have a shorter lifetime and two types exist: K^0 and $\bar{K}^0$. The neutral kaon properties reported are the following (values taken from the Particle Data Group's Review [5]):

$$
\boxed{K^0, \bar{K}^0} : \quad
\begin{aligned}
&m_{K^0} = (497.611 \pm 0.013) \text{ MeV} \\
&m_{K^0} - m_{K^\pm} = (3.934 \pm 0.020) \text{ MeV}
\end{aligned}
\tag{15.62}
$$

The physical states are mixed and come in one short-lived and one long-lived particle: the "K-short" labeled

K_S^0 and "K-long" written K_L^0. Their properties are (values taken from the Particle Data Group's Review [5]):

$$\boxed{K_S^0}: \quad \tau(K_S^0) = (8.954 \pm 0.004) \times 10^{-11} \text{ s} \quad (c\tau \approx 2.68 \text{ cm})$$
$$Br(K_S^0 \to \pi^+\pi^-) = (69.20 \pm 0.05)\%$$
$$Br(K_S^0 \to \pi^0\pi^0) = (30.69 \pm 0.05)\%$$
$$Br(K_S^0 \to \pi^+\pi^-\pi^0) = (3.5^{+1.1}_{-0.9}) \times 10^{-7}$$

$$\boxed{K_L^0}: \quad \tau(K_L^0) = (5.116 \pm 0.021) \times 10^{-8} \text{ s} \quad (c\tau \approx 1530 \text{ cm})$$
$$Br(K_L^0 \to \pi^\pm e^\mp \nu_e) = (40.55 \pm 0.11)\%$$
$$Br(K_L^0 \to \pi^\pm \mu^\mp \nu_\mu) = (27.04 \pm 0.07)\%$$ \hfill (15.63)
$$Br(K_L^0 \to 3\pi^0) = (19.52 \pm 0.12)\%$$
$$Br(K_L^0 \to \pi^+\pi^-\pi^0) = (12.54 \pm 0.05)\%$$
$$Br(K_L^0 \to \pi^+\pi^-) = (1.967 \pm 0.010) \times 10^{-3}$$
$$Br(K_L^0 \to \pi^0\pi^0) = (8.64 \pm 0.06) \times 10^{-4}$$

In 1951, **Armenteros, Barker, Butler, Cachon, and Chapman** published an analysis that showed that actually were at least two different kinds of strange particles which produced protons and pions when they decayed. In their paper, entitled "Decay of V-particles" [178], they concluded that if one assumed that only two particles were produced by the neutral decays, then two different decay modes were needed to explain the photographs: $V^0 \to p^+ + \pi^-$, with the V^0 mass in the range $(2000\text{–}2500)\,m_e$ and $V^0 \to \pi^+ + \pi^-$, with the V^0 mass about $1000\,m_e$. Hence, they had found strong evidence for the existence of two types of V^0 particle. They had discovered heavier particles, which also had the strange behavior, the Λ hyperon, whose properties are described below.

A **hyperon** is any baryon containing *strangeness* (but no charm, bottom, or top as will be seen later), such as Λ, Σ, Ξ, and Ω. The decay of a hyperon is also weak and different from the case of kaons, since there is a nucleon (a proton or a neutron) in the decay products. Their properties and main decay modes are (values taken from the Particle Data Group's Review [5]):

$$\boxed{\Lambda^0}: \quad m_\Lambda = (1115.683 \pm 0.006) \text{ MeV}$$
$$\tau(\Lambda) = (2.632 \pm 0.020) \times 10^{-10} \text{ s} \quad (c\tau \approx 7.89 \text{ cm})$$
$$\mu_\Lambda = (-0.613 \pm 0.004)\mu_N$$ \hfill (15.64)
$$Br(\Lambda \to p\pi^-) = (63.9 \pm 0.5)\%$$
$$Br(\Lambda \to n\pi^0) = (35.8 \pm 0.5)\%$$
$$Br(\Lambda \to n\gamma) = (1.75 \pm 0.15) \times 10^{-3}$$

$$\boxed{\Sigma^+}: \quad m_{\Sigma^+} = (1189.37 \pm 0.07) \text{ MeV}$$
$$\tau(\Sigma^+) = (8.018 \pm 0.026) \times 10^{-11} \text{ s} \quad (c\tau \approx 2.40 \text{ cm})$$
$$\mu_{\Sigma^+} = (-0.613 \pm 0.004)\mu_N$$ \hfill (15.65)
$$Br(\Sigma^+ \to p\pi^0) = (51.57 \pm 0.30)\%$$
$$Br(\Sigma^+ \to n\pi^+) = (48.31 \pm 0.30)\%$$
$$Br(\Sigma^+ \to p\gamma) = (1.23 \pm 0.05) \times 10^{-3}$$

$$\boxed{\Sigma^0}: \quad m_{\Sigma^0} = (1192.642 \pm 0.024) \text{ MeV}$$
$$\tau(\Sigma^0) = (74 \pm 7) \times 10^{-21} \text{ s} \quad \text{(short lived)}$$ \hfill (15.66)
$$Br(\Sigma^0 \to \Lambda\gamma) = (100)\%$$

$$\boxed{\Sigma^-}: \quad m_{\Sigma^-} = (1197.449 \pm 0.030) \text{ MeV}$$
$$m_{\Sigma^-} - m_{\Sigma^+} = (8.08 \pm 0.08) \text{ MeV}$$
$$\tau(\Sigma^-) = (1.479 \pm 0.011) \times 10^{-10} \text{ s} \quad (c\tau \approx 4.43 \text{ cm})$$ \hfill (15.67)
$$Br(\Sigma^- \to n\pi^-) = (99.848 \pm 0.005)\%$$

15.8   The Strangeness Quantum Number

The discovery of this new type of particle in cosmic rays came as a surprise, hence the name "strange." Such strange particles were heavy enough to decay rapidly into lighter particles. Yet the observation was indicating that the strange particles were rather long-lived.

In 1952 **Pais**[13] was the first to propose that different mechanisms were responsible for the creation and decay of strange particles [179]. Indeed, strange particles could be copiously produced in strong interactions. If the same strong interaction was responsible for their decays, then their lifetime would be extremely short in the range of $\tau \approx 10^{-23}$ s and certainly much shorter than those observed at $\tau \approx 10^{-9}$ s! Hence, the decay via the strong force was to be forbidden. To explain the long lifetime, Pais invoked a **very weak coupling so far only considered in neutrino processes, such as** β-, μ-, π-**decay**.

In 1953 **Kazuhiko Nishijima**[14] and **Tadao Nakano**[15] developed arguments that led to the concept of strangeness [180]. Nishijima originally called it the "η-charge," following the nomenclature of the time which assigned it to the η meson [181]. Nishijima derived the formula:

$$Q = I_3 + \frac{B}{2} + \frac{\eta}{2} \tag{15.68}$$

In 1956 **Gell-Mann**[16] published the same concept independently [182], based on work he had performed earlier. He labeled strangeness S:

$$Q = I_3 + \frac{B}{2} + \frac{S}{2} \tag{15.69}$$

The **strangeness** S is a new quantity to be considered as an intrinsic property of elementary particles, in the same way as electric charge and isospin are. Antiparticles are assigned the opposite strangeness $-S$. Elementary particles such as the proton, the neutron, or the pion have no strangeness. The kaons, lambdas, and sigmas possess strangeness:

$$S = 0 : \pi, p, n \tag{15.70}$$

$$S = +1 : K^+, K^0 \tag{15.71}$$

$$S = -1 : K^-, \bar{K}^0, \Lambda, \Sigma^+, \Sigma^-, \Sigma^0 \tag{15.72}$$

Note that Σ^- is *not* the antiparticle of the Σ^+, and that they both possess $S = -1$.

• **Strangeness conservation.** Strangeness is *additive* and not multiplicative, unlike for example the parity which is multiplicative. Since electric charge Q is always conserved as well as the baryon number B, strangeness follows isospin. As a consequence, S **is conserved in strong interactions, whereas it is not conserved in weak interactions**. S is also conserved in electromagnetic interactions, as a consequence of the conservation of I_3.

The above rules imply that strange particles are produced in pairs in collisions of non-strange particles, such as for example the following pion–nucleon reactions:

$$\left.\begin{array}{c|ccc} & \pi^- p \to K^+ \Sigma^- & \pi^- p \to K^0 \Sigma^0 & \pi^- p \to K^0 \Lambda^0 \\ S & 0+0 = +1-1 & 0+0 = +1-1 & 0+0 = +1-1 \end{array}\right. \tag{15.73}$$

Strangeness conservation in the strong interaction implies that the following processes are **forbidden**, although they conserve other quantities such as the electric charge:

$$\left.\begin{array}{c|ccc} & \pi^- p \nrightarrow \pi^+ \Sigma^- & \pi^- p \nrightarrow \pi^0 \Sigma^0 & \pi^- p \nrightarrow K^- \Sigma^+ \\ S & 0+0 \neq 0-1 & 0+0 \neq 0-1 & 0+0 \neq -1-1 = -2 \end{array}\right. \tag{15.74}$$

---

13 Abraham Pais (1918–2000), Dutch-born American physicist and science historian.
14 Kazuhiko Nishijima (1926–2009), Japanese physicist.
15 Tadao Nakano (1926–2004), Japanese physicist.
16 Murray Gell-Mann (1929–2019), American physicist.

15.9   The Baryon Quantum Number and Antibaryons

We recall that the existence of antimatter was first discovered by Anderson in 1932 when he detected the positron (e^+). Theoretically the existence of the positron was a prediction of the Dirac theory. In 1955 it was not clear if antibaryons such as antiprotons, antineutrons, antilambdas, antisigmas, etc., existed or not. One issue was the observation of their magnetic moments, which clearly indicated that they were not simple point-like Dirac particles:

$$\mu_p = +2.79\mu_N \quad \text{and} \quad \mu_n = -1.91\mu_N \tag{15.75}$$

where $\mu_N = e/(2M_p)$ (see Eq. (4.135)). If antiprotons existed, how could they be produced?

In 1938 **Stückelberg** originally introduced the "heavy charge" (today's baryon number) to explain the absence of decay of heavier particles into lighter particles. He wrote:

> *"Except the law of conservation of electric charge caused by Maxwell's theory, there is obviously still another conservation theorem: In all observed transformations of matter, no transformations from heavy particles (neutron and proton) to light particles (electron and neutrino) have been observed. We therefore want to establish a conservation law of heavy charge."* [17]

Under the assumption that *"everything that is not forbidden will happen,"* [18] the radiative decay of the proton could induce a positron. Hence, the observed stability of the proton arises from the conservation of the baryon number:

$$p \nrightarrow e^+\gamma \qquad \text{forbidden by } B \text{ conservation} \tag{15.76}$$

If the baryon quantum number B is conserved in all interactions, then antibaryons must be produced in pairs. Hence, the conservation of the baryonic number implies that only proton–antiproton pairs will be produced, for instance in the reaction:

$$\underbrace{p + p}_{B=2} \rightarrow \underbrace{pp\bar{p}p}_{B=2} \tag{15.77}$$

If we assume an experimental configuration where high-energy protons hit a fixed target, then using Eq. (5.66) we see that the above reaction requires that the energy $E_A = T_p + M_p$ of the incoming proton is:

$$E^* = \sqrt{2M_p^2 + 2M_pE_A} > 4M_p \quad \rightarrow \quad 2M_p + 2E_A > 16M_p \quad \rightarrow \quad E_A > 7M_p$$
$$\rightarrow \quad T_p > 6M_p \tag{15.78}$$

Consequently, the kinetic energy of the proton beam must be at least $T_p \approx 5.6$ GeV.

• **Discovery of the antiproton.** In 1955 **Chamberlain, Segrè, Wiegand, and Ypsilantis** discovered the antiproton [183] at the Bevatron accelerator at the Lawrence Berkeley National Laboratory in California. In their experiment high-energy protons were shot on the copper target:

$$p + Cu \rightarrow p + X \tag{15.79}$$

If the target nucleon is in a nucleus and has some momentum, the threshold is lowered. Assuming a Fermi energy of 25 MeV, one may calculate that the threshold for the formation of a proton–antiproton pair is

---

17   *"Ausser diesem, durch die Maxwell'sche Theorie bedingten, Erhaltungssatz der elektrischen Ladung, gibt es aber offenbar noch einen weiteren Erhaltungssatz: Bei allen beobachteten Umwandlungen der Materie, wurden noch keine Umwandlungen von schweren Partikeln (Neutron und Proton) in leichte Partikel (Elektron und Neutrino) beobachtet. Wir wollen daher einen Erhaltungssatz der schweren Ladung fordern."* Reprinted from E. C. G. Stueckelberg, "Die Wechselwirkungskräfte in der Elektrodynamik und in der Feldtheorie der Kernkräfte. Teil II und III," *Helva. Phys. Acta.*, vol. 11, p. 299, 1939.

18   *"Alles, das nicht verboten ist, wird geschehen."*

approximately 4.3 GeV [183]. The challenge was to detect the production of antiprotons in a large background of pions:

$$p + Cu \to p + \pi's + X \tag{15.80}$$

The expected rate of antiprotons was 1 in $\approx 60{,}000$ pions. The experimental requirement was hence to identify the antiproton. This was achieved with a simultaneous measurement of the momentum p and the velocity β of particles produced in the pCu interactions. The mass of the particle was inferred from the relation $p = \gamma m\beta$. The setup was prepared to select only those secondary particles with negative electric charge $q = -1$ and a momentum $p = 1.19$ GeV/c (see Figure 15.15). For that chosen momentum, the velocity of a proton is very different from that of a pion (see Figure 3.25):

$$\beta_p \simeq 0.78 \qquad \text{and} \qquad \beta_\pi \simeq 0.99 \tag{15.81}$$

The velocity was estimated with two independent methods:

- Cherenkov counters (see Section 3.14) – tuned to produce light when $\beta > 0.79$. The light signal was used in anti-coincidence and produced a "pion-veto."

- Time-of-flight measurement.

In their famous experiment, a total of about 60 antiprotons were successfully identified and detected.

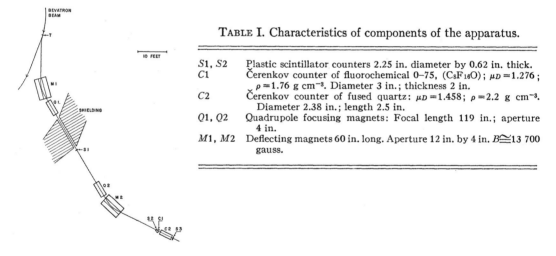

TABLE I. Characteristics of components of the apparatus.

$S1$, $S2$	Plastic scintillator counters 2.25 in. diameter by 0.62 in. thick.
$C1$	Čerenkov counter of fluorochemical 0–75, (C₈F₁₆O); $\mu_D = 1.276$; $\rho = 1.76$ g cm⁻³. Diameter 3 in.; thickness 2 in.
$C2$	Čerenkov counter of fused quartz: $\mu_D = 1.458$; $\rho = 2.2$ g cm⁻³. Diameter 2.38 in.; length 2.5 in.
$Q1$, $Q2$	Quadrupole focusing magnets: Focal length 119 in.; aperture 4 in.
$M1$, $M2$	Deflecting magnets 60 in. long. Aperture 12 in. by 4 in. $B \cong 13\,700$ gauss.

Figure 15.15 Layout of the antiproton discovery experiment at the Bevatron. Reprinted with permission from O. Chamberlain, E. Segrè, C. Wiegand, and T. Ypsilantis, "Observation of antiprotons," *Phys. Rev.*, vol. 100, pp. 947–950, Nov 1955. Copyright 1955 by the American Physical Society.

Cork, Lambertson, Piccioni, and Wenzel subsequently discovered the antineutron at the Bevatron accelerator [184]. They studied the secondary interactions of antiprotons produced in a similar setup to Chamberlain *et al.*, using a beryllium target (instead of copper) and protons of 6.2 GeV and selecting negatively charged particles of 1.4 GeV. They could reach the production rate of 300–600 $\bar{p}$ per hour. Antineutrons were expected to be produced via the "charge-exchange" process:

$$p + \bar{p} \to n + \bar{n} \tag{15.82}$$

The antineutrons were observed via their annihilation coupled to a charged veto to suppress antiprotons. See Figure 15.16. X is the charge-exchange scintillator; S_1, S_2 are scintillation counters and C is a lead-glass Cerenkov counter. The pulse-height spectrum in the lead glass counter for neutral events is shown in

Figure 15.16(right). The solid histogram is for 54 antineutron events identified with an energy loss in the charge-exchange scintillator less than 100 MeV. The dashed histogram is for 20 other neutral events. The smooth solid curve is for antiprotons.

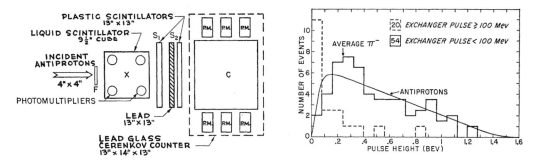

Figure 15.16 (left) Layout of the antineutron discovery experiment; (right) pulse-height spectrum in lead glass counter (see text). Reprinted with permission from B. Cork, G. R. Lambertson, O. Piccioni, and W. A. Wenzel, "Antineutrons produced from antiprotons in charge-exchange collisions," *Phys. Rev.*, vol. 104, pp. 1193–1197, Nov 1956. Copyright 1956 by the American Physical Society.

• **Discovery of $\bar{\Lambda}^0$.** In 1958 **Prowse and Baldo-Ceolin** reported evidence for the antilambda at the meson beam of the Bevatron. The antilambda would be pair-created and subsequently decay into an antiproton and a π^+ [185]:

$$\pi^- + p \to \Lambda^0 + \bar{\Lambda}^0 + n$$
$$ \bar{p} \;+\; \pi^+ \qquad\qquad (15.83)$$
$$\phantom{\pi^- + p \to \Lambda^0 + \bar{p}} \text{star} \quad \mu^+ \to e^+$$

The threshold for $\bar{\Lambda}^0$ production with a free nucleon is 4.73 GeV and extends down to ≈ 4.3 GeV with a bound nucleon. The experiment used photographic emulsion stacks inserted into the pion beam. The emulsions were scanned for stopping mesons and these were followed back to their origins if it appeared that they were produced in the region of high meson flux. One event was found with a V^0 decay compatible with an anti lambda (see Figure 15.17). The star was identified as an antiproton interaction in flight, with an estimated energy of 230^{+22}_{-7} MeV. The opening angle of the V is $64 \pm 1°$ and the π-meson energy was 32 MeV, deduced from its range of 1.70 cm in the emulsion. From these values the Q value in the decay was calculated to be in excellent agreement with that for the normal Λ^0, and hence made the observation of an antilambda the most plausible explanation.

15.10   A Jungle of Resonances

Around 1950, the number of known particles with strong interactions (**hadrons**) was already quite large: p, n, π, K, Λ, Σ, ... The systematic study of scattering processes at accelerators of increasing energy revealed the existence of a "jungle" of elementary particles in the form of **resonances**. Total scattering cross-sections were measured as a function of the center-of-mass energy.

In 1952 **Anderson, Fermi, Long, and Nagle**, working at the Chicago Cyclotron, were the first to discover such a resonance [186]. They noticed the striking fact that the pion–proton cross-section becomes very different

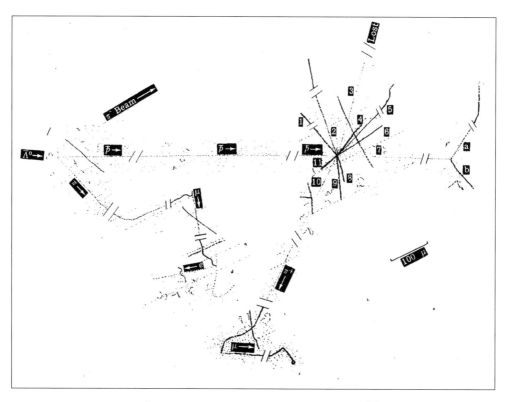

Figure 15.17 Event showing a V^0 decay compatible with an antilambda $\bar{\Lambda}^0 \to \bar{p} + \pi^+$ observed in a stack of photographic emulsions. Reprinted with permission from D. J. Prowse and M. Baldo-Ceolin, "Anti-lambda hyperon," *Phys. Rev. Lett.*, vol. 1, pp. 179–180, Sep 1958. Copyright 1958 by the American Physical Society.

for positive and negative pions when the pion energy is around 150 MeV:

$$\sigma(\pi^+ p) \gg \sigma(\pi^- p) \qquad \text{if } E_\pi > 150 \text{ MeV} \tag{15.84}$$

Data are shown in Figure 15.18. A spectacular feature was that such particles possessed a broad width typically in the range of $\Gamma \simeq 100$ MeV, hence the name resonance. The explanation suggested by **Brueckner** was that the process proceeded through a very short-lived **isobar resonance** [187]. If a particle is unstable with a lifetime in the range of $\tau \approx 10^{-23}$ s, then the energy uncertainty from Heisenberg's principle would be reflected in the determination of its mass, hence providing a mechanism to explain the width:

$$\Delta E \Delta t \simeq \hbar \Longrightarrow \Gamma \simeq \Delta E \simeq \frac{\hbar}{\tau} = \frac{6,6 \times 10^{-22} \text{ MeV} \cdot \text{s}}{10^{-23} \text{ s}} = \mathcal{O}(100 \text{ MeV}) \tag{15.85}$$

Brueckner concluded:

> "It is found that the steep variation of cross-section with energy, the broad plateau suggesting a resonance peak of the π^+ scattering at about 200 MeV, and the anomalous ratio of π^+ to π^- scattering can be interpreted to be an indication of the existence of an excited nucleon isobaric state. Assignment of the isobar to the state with total angular momentum $J=3/2$ and total isotopic angular momentum $I=3/2$ gives predictions in excellent agreement with experiment." [19]

19 Reprinted with permission from K. A. Brueckner, "Meson–nucleon scattering and nucleon isobars," *Phys. Rev.*, vol. 86, pp. 106–109, Apr 1952. Copyright 1952 by the American Physical Society.

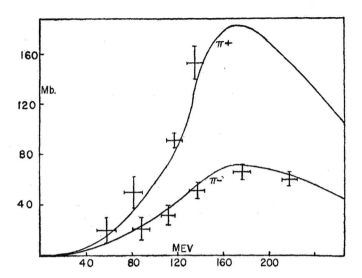

Figure 15.18 Total measured cross-section for scattering of π^+ and π^- mesons in hydrogen. Reprinted with permission from K. A. Brueckner, "Meson–nucleon scattering and nucleon isobars," *Phys. Rev.*, vol. 86, pp. 106–109, Apr 1952. Copyright 1952 by the American Physical Society.

• $\Delta(1232)$ **resonance.** The resonance discovered in $\sigma(\pi^+ p)$ by Anderson *et al.* is the particle called $\Delta^{++}(1232)$'s resonance with a total spin $J = 3/2$ and isospin $I = 3/2$:

$$\pi^+ + p \to \Delta^{++}(1232) \to \pi^+ + p \qquad (15.86)$$

Each resonance will be uniquely identified by its rest mass and its quantum numbers, such as spin, isospin, etc. The spin of $\Delta^{++}(1232)$ was determined from the angular distribution of the decay products in its main decay $\Delta^{++}(1232) \to \pi^+ p$. The isospin $I = 3/2$ is derived from the dynamical analysis. The observation that $\sigma(\pi^+ p) \simeq 3\sigma(\pi^- p)$ at the resonance strongly supports the $I = 3/2$ hypothesis (see Eq. (15.52)).

The resonance such as Δ^{++} was first found by studying the behavior of the scattering cross-section of two colliding particles (e.g., proton–pion) as a function of the center-of-mass energy. In this case, the cross-section is enhanced by the effect of the propagator in the s-channel. This is illustrated in Figure 15.19(left).

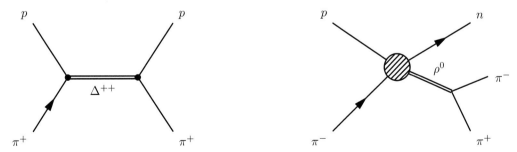

Figure 15.19 (left) Cross-section enhancement via s-channel resonance production; (right) example of resonance production in a scattering experiment.

• **Invariant mass method to discover new resonances.** In a resonance production experiment, one collides high-energy particles and searches for the production of new particles. The presence of a resonance is inferred when it is found that outgoing particles prefer to emerge with a particular value of combined invariant mass. For example, see Figure 15.19(right), we obtain for $p_1^\mu = (E_1, \vec{p}_1)$ and $p_2^\mu = (E_2, \vec{p}_2)$ the invariant mass:

$$m^2 = (p_1 + p_2)^2 = p_1^2 + p_2^2 + 2p_1 \cdot p_2 = m_1^2 + m_2^2 + 2(E_1 E_2 - \vec{p}_1 \cdot \vec{p}_2) \simeq 2E_1 E_2 (1 - \cos \alpha) \qquad (15.87)$$

where α is the opening angle between the two particles and we have neglected the outgoing particle rest masses in the last expression.

Finding a resonance in this fashion is more difficult because it is necessary to look at all possible combinations of outgoing particles which might have arisen from a resonance. All the combinations are plotted, and one observes if there is a "preferred" value, usually above some background. One advantage of the production method is that it does not require that the resonance be directly produced by the incoming particles. Some examples of such mass distributions are shown in Figure 15.20. The data are from a study of final-state particles in $\pi^+ + p$ interactions at the BNL-AGS [188]. The final states were observed thanks to the BNL 20-foot bubble chamber, which permitted exclusive final states to be reconstructed. The histograms show clear resonances in various channels: $\pi^+ \pi^0$ pairs, $\pi^+ \pi^-$ pairs, and also $\pi^+ \pi^0 \pi^-$ triplets. The pairs correspond to the ρ and the triplets to the ω^0 and η^0. In high-energy collisions, typically as the center-of-mass energy

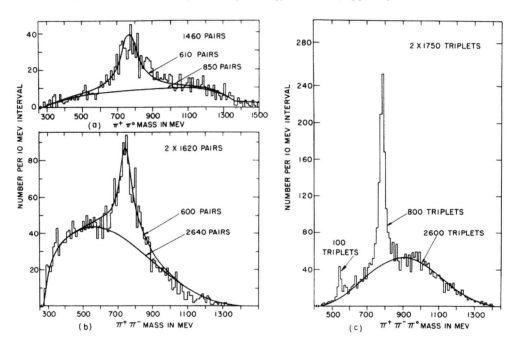

Figure 15.20 Mass distributions for (a) for $\pi^+ \pi^0$ pairs from $\pi^+ + p \to \pi^+ + p + \pi^0$ events; (b) $\pi^+ \pi^-$ pairs from $\pi^+ + p \to \pi^+ + p + \pi^+ \pi^-$ events; and (c) $\pi^+ \pi^0 \pi^-$ triplets from $\pi^+ + p \to \pi^+ + p + \pi^+ \pi^0 \pi^-$ events. Reprinted with permission from C. Alff, D. Berley, D. Colley, N. Gelfand, U. Nauenberg, D. Miller *et al.*, "Production of pion resonances in $\pi^+ p$ interactions," *Phys. Rev. Lett.*, vol. 9, pp. 322–324, Oct 1962. Copyright 1962 by the American Physical Society.

increases, any number of new particles may emerge from the hard interactions and it is possible to estimate if they have originated from some resonance by using the invariant mass technique. Note that the masses of the decay productions (above m_1 and m_2) need to be generally included, hence final-state particle identification is important. Then, resonances decaying into pions, kaons, etc., can be produced and studied, while it would not be possible to produce interacting pion–kaon beams.

This method, pushed by the particle accelerator development in the 1950s, led to the discovery of literally hundreds of **hadrons** (baryons and mesons), all of which may have been legitimately regarded as "elementary," as the proton or the neutron. Some of these particles are listed in Tables 15.2 and 15.3. This surely generated a great deal of interest, especially when regularities in the properties of the resonances started to become apparent. This will be discussed further in Chapter 17.

	Mass (MeV)	Q	$I(J^P)$	B	S	$\tau(s)$ or Γ(MeV)	Typical decay
π^0	135.0	0	$1(0^-)$	0	0	8.5×10^{-17}	$\gamma\gamma$
π^+	139.6	+1	$1(0^-)$	0	0	2.6×10^{-8}	$\mu^+\nu_\mu$
π^-	139.6	−1	$1(0^-)$	0	0	2.6×10^{-8}	$\mu^-\bar{\nu}_\mu$
K^+	493.7	+1	$1/2(0^-)$	0	+1	1.2×10^{-8}	$\mu^+\nu_\mu$
K^-	493.7	−1	$1/2(0^-)$	0	−1	1.2×10^{-8}	$\mu^-\bar{\nu}_\mu$
$K^0, \bar{K}^0$	497.6	0	$1/2(0^-)$	0	± 1	$K_S^0 : 9.0 \times 10^{-11}$	$\pi\pi$
						$K_L^0 : 5.1 \times 10^{-8}$	$\pi\pi\pi$
η^0	547.3	0	$0(0^-)$	0	0	$\approx 10^{-18}$	$\gamma\gamma$
$\rho(770)$	769.0	$0, \pm 1$	$1(1^-)$	0	0	150 MeV	$\pi\pi$
$\omega(782)$	782.7	0	$0(1^-)$	0	0	8.49 MeV	$\pi^+\pi^-\pi^0$

Table 15.2 Illustration of the jungle of mesons.

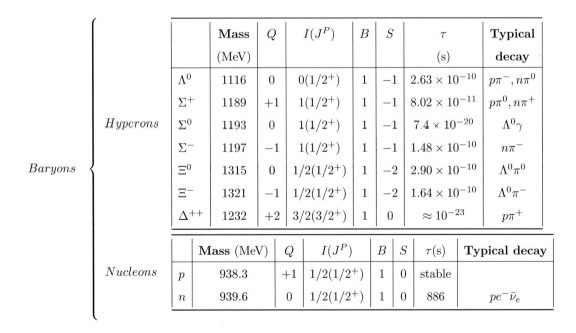

		Mass (MeV)	Q	$I(J^P)$	B	S	τ (s)	Typical decay
	Λ^0	1116	0	$0(1/2^+)$	1	−1	2.63×10^{-10}	$p\pi^-, n\pi^0$
	Σ^+	1189	+1	$1(1/2^+)$	1	−1	8.02×10^{-11}	$p\pi^0, n\pi^+$
Hyperons	Σ^0	1193	0	$1(1/2^+)$	1	−1	7.4×10^{-20}	$\Lambda^0\gamma$
	Σ^-	1197	−1	$1(1/2^+)$	1	−1	1.48×10^{-10}	$n\pi^-$
	Ξ^0	1315	0	$1/2(1/2^+)$	1	−2	2.90×10^{-10}	$\Lambda^0\pi^0$
	Ξ^-	1321	−1	$1/2(1/2^+)$	1	−2	1.64×10^{-10}	$\Lambda^0\pi^-$
	Δ^{++}	1232	+2	$3/2(3/2^+)$	1	0	$\approx 10^{-23}$	$p\pi^+$

		Mass (MeV)	Q	$I(J^P)$	B	S	$\tau(s)$	Typical decay
Nucleons	p	938.3	+1	$1/2(1/2^+)$	1	0	stable	
	n	939.6	0	$1/2(1/2^+)$	1	0	886	$pe^-\bar{\nu}_e$

Baryons {

Table 15.3 Illustration of the jungle of baryons.

15.11 Decays of Hadrons

There are selection rules to decide if a hadronic decay is weak, electromagnetic, or strong. Based on the type of decay, different quantities are conserved:

- For strong interactions the total isospin I, and the third component I_3, are conserved. Strong interactions always lead to hadronic final states.

- For electromagnetic decays the total isospin I is not conserved, but I_3 is. There are often photons or charged lepton–antilepton pairs in the final state.

- For weak decays I and I_3 are not conserved. They are the only ones that can change flavor, e.g., $\Delta S = 1$. Weak decays can lead to hadronic, semi-leptonic, or leptonic final states. A neutrino in the final state is a clear signature of a weak interaction.

The lightest hadrons cannot decay to other hadrons via strong interactions. The π^0, η, and σ_0 decay electromagnetically. The $\pi^\pm$, $K^\pm$, K^0, n, Λ, $\Sigma^\pm$, Ξ, and Ω^- decay via flavor-changing weak interactions, with long lifetimes.

Problems

Ex 15.1 π^0 lifetime. Estimate the $\pi^0 \to \gamma\gamma$ decay width and corresponding lifetime.

Ex 15.2 Deuteron production. The deuteron d is a bound state composed of a neutron and a proton (see Section 15.5). Its isospin is equal to $I = 0$. Compute the ratio of the cross-sections of the following processes:

$$p + p \to d + \pi^+, \qquad p + n \to d + \pi^0, \qquad n + n \to d + \pi^- \tag{15.88}$$

Ex 15.3 The α particle. The bound state of two protons and two neutrons, in other words the nucleus of ^{4}He, is called an α particle.

(a) We know that the isotopes of hydrogen or lithium with an atomic weight of four (^{4}H, ^{4}Li) do not exist. What can we conclude on the isospin of the α particle?

(b) The reaction $d + d \to \alpha + \pi^0$ was never observed. Explain why.

(c) Would you expect that the isotope ^{4}Be exists?

(d) What about the bound state of four neutrons?

Ex 15.4 ^{3}H and ^{3}He production in the pd scattering. The isospin I of a nucleus must follow the rule $|I_3| \le I \le \frac{1}{2} A$. In the ground state of the nucleus, the isospin must always take the smallest possible value. Compute the following ratio of the cross-sections:

$$\sigma(p + d \to \pi^+ + {}^3\text{H}) \, / \, \sigma(p + d \to \pi^0 + {}^3\text{He}) \tag{15.89}$$

Ex 15.5 Pion nucleus scattering. We consider the six following elastic πN scattering processes:

(a) $\pi^+ + p \to \pi^+ + p$ (b) $\pi^0 + p \to \pi^0 + p$

(c) $\pi^- + p \to \pi^- + p$ (d) $\pi^+ + n \to \pi^+ + n$

(e) $\pi^0 + n \to \pi^0 + n$ (f) $\pi^- + n \to \pi^- + n$

and the four following charge-exchange processes:

(g) $\pi^+ + n \to \pi^0 + p$ (h) $\pi^0 + p \to \pi^+ + n$

(i) $\pi^0 + n \to \pi^- + p$ (j) $\pi^- + p \to \pi^0 + n$

Since a πN state can have either a total isospin $I = 3/2$ or $I = 1/2$, the 10 amplitudes for the processes above can be described as a function of two independent amplitudes $\mathcal{M}_3$ ($I = 3/2$) and $\mathcal{M}_1$ ($I = 1/2$).

(a) Express the 10 amplitudes $\mathcal{M}_a, \ldots, \mathcal{M}_j$ as a function of $\mathcal{M}_1$ and $\mathcal{M}_3$.

(b) The ratios of the cross-sections σ_a, σ_c, and σ_j are given by:

$$\sigma_a : \sigma_c : c_j = 9|\mathcal{M}_3|^2 : |\mathcal{M}_3 + 2\mathcal{M}_1|^2 : 2|\mathcal{M}_3 - \mathcal{M}_1|^2$$

Compute the ratios of all 10 cross-sections $\sigma_a : \sigma_b : \ldots : \sigma_j$.

(c) The measured cross-sections summarized by the Particle Data Group's Review can be found at http://pdg.lbl.gov/current/xsect/. Search the tables corresponding to πN scattering. At a center-of-mass energy of 1232 MeV the πN-scattering cross-sections exhibit a pronounced peak, which corresponds to the excitation of the Δ resonance with isospin $I = 3/2$. At the resonance, one expects $\mathcal{M}_3 \gg \mathcal{M}_1$, hence $\sigma_a : \sigma_c : \sigma_j = 9 : 1 : 2$. Compute the ratios of all 10 cross-sections at the resonance and compare with experimental data.

16   Electron–Proton Scattering

> Information about the internal structure of individual nucleons is contained in the results of a variety of experiments performed in recent years. Those experiments in which the interaction with the nucleon is electromagnetic are susceptible of a considerably more precise and unambiguous interpretation than those involving meson interactions.
>
> *Yennie, Lévy, Ravenhall[1]*

16.1   Relativistic Electron–Proton Elastic Scattering

Just in the same way as the scattering of α particles on nuclei has been fundamental to develop the atomic model, our understanding about **the internal structure of the proton** has enormously benefited from scattering studies using electrons as probes. In this kind of experiment, we bombard target protons with electrons of generally fixed energy and study the scattered electron properties (its energy and angle) to infer the structure properties of the proton.

Scattering of electrons and protons is primarily an electromagnetic interaction and should be described by QED. Electron beams have been used to probe the structure of the proton (and neutron) since the 1960s, with the most recent results coming from the high-energy HERA electron–proton collider at DESY in Hamburg, where electroweak corrections became visible. These experiments provided direct evidence for the composite nature of protons and neutrons, and measured the distributions of the quarks and gluons inside the nucleon.

Let us begin with a hypothetical *point-like proton*. In Section 11.2 we have derived the Mott differential cross-section, Eq. (11.51). The result was valid in the limit where the target recoil was neglected and the scattered electron was relativistic with spin-1/2. This was compared to the case of the Rutherford scattering, Eq. (2.88), where the target recoil was neglected, and the scattered electron was spinless and non-relativistic. It was inferred that the Mott scattering included a term that followed from angular momentum conservation due to the spin-1/2 of the electron. This approximation was for a fixed target. If the target is a *recoiling proton which also has a spin-1/2*, then magnetic spin–spin interactions are also expected.

We now want to obtain the relativistic differential cross-section for the elastic process $e^- p \to e^- p$, including the effect of the proton recoil. The tree-level Feynman diagram is shown in Figure 16.1, where the "blob" at the proton vertex illustrates the fact that the proton is not going to be a point-like Dirac particle. However, if the proton *were* a point-like Dirac particle, we could directly infer the results obtained for the electron–muon scattering process $e^- \mu^- \to e^- \mu^-$, replacing the muon with a proton, and applying the same calculations. Neglecting the electron mass and *treating the proton as a point-like Dirac particle*, the matrix element for unpolarized elastic electron–proton scattering $e^-(p) + p(k) \to e^-(p') + p(k')$ is hence (see Eq. (11.126)):

$$\langle |\mathcal{M}_{e^- p \to e^- p}|^2 \rangle = \left(\frac{8e^4}{(p - p')^4} \right) \left[(p \cdot k')(p' \cdot k) + (p' \cdot k')(p \cdot k) - M_p^2(p' \cdot p) \right] \tag{16.1}$$

---

1   Reprinted excerpt with permission from D. R. Yennie, M. M. Lévy, and D. G. Ravenhall, "Electromagnetic structure of nucleons," *Rev. Mod. Phys.*, vol. 29, pp. 144–157, Jan 1957. Copyright 1957 by the American Physical Society.

where M_p is the rest mass of the proton. Historically, most electron–proton elastic scattering experiments did not observe the recoiling proton and consequently the cross-section was computed in terms of the electron observables, which are the incoming electron energy E_1, the outgoing electron energy E_3, and the scattering angle θ.

In the **non-relativistic limit** $E_1 \ll M_p$, we can write $E_1 \simeq E_3 \equiv E$ and the energy of the outgoing proton is essentially its rest mass M_p. Then, $p \cdot k' = p' \cdot k = p' \cdot k' = p \cdot k = E M_p$, and thus the matrix element is in this case:

$$\langle |\mathcal{M}_{e^- p \to e^- p}|^2 \rangle = \left(\frac{8e^4}{q^4} \right) \left[2E^2 M_p^2 - M_p^2(p' \cdot p) \right] \tag{16.2}$$

where $q^2 = (p - p')^2$. Hence, as expected, the matrix element is of the form of Eq. (11.41) and we recover the Mott scattering formula in the non-relativistic limit.

In the **ultra-relativistic limit** $E_1 \gg m_e$, and given the kinematical configuration of the proton rest frame defined in Figure 16.1, we have (neglecting the electron rest mass):

$$p^\mu = (E_1, 0, 0, E_1); \quad k^\mu = (M_p, 0, 0, 0); \quad p'^\mu = (E_3, 0, E_3 \sin\theta, E_3 \cos\theta); \quad k'^\mu = (E_{k'}, \vec{k}') \tag{16.3}$$

Obviously energy–momentum conservation implies that $k' = p + k - p'$. Hence, from these definitions we can compute the four-vector scalar products present in the matrix element:

$$\begin{aligned}
p \cdot k &= E_1 M_p; \qquad p' \cdot k = E_3 M_p; \qquad p \cdot p' = E_1 E_3(1 - \cos\theta) \\
p \cdot k' &= p \cdot (p + k - p') = m_e^2 + p \cdot k - p \cdot p' \simeq E_1 \left(M_p - E_3(1 - \cos\theta) \right) \\
p' \cdot k' &= p' \cdot (p + k - p') = p' \cdot p + p' \cdot k - m_e^2 \simeq E_3 \left(E_1(1 - \cos\theta) + M_p \right)
\end{aligned} \tag{16.4}$$

where the terms depending on the electron rest mass are neglected. Consequently, the matrix element as

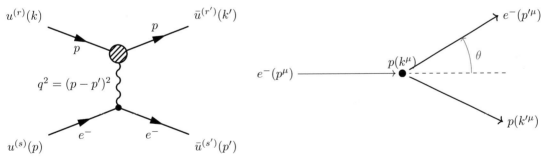

Figure 16.1 Electron–proton elastic scattering: (left) tree-level Feynman diagram, the blob illustrates the fact that the proton is not a point-like Dirac particle; (right) kinematics in the proton rest frame.

a function of the incoming electron energy, the outgoing electron energy, and its scattering angle, can be expressed as:

$$\begin{aligned}
\langle |\mathcal{M}_{e^- p \to e^- p}|^2 \rangle &= \left(\frac{8e^4 M_p E_1 E_3}{q^4} \right) \left[2M_p - E_3(1 - \cos\theta) + E_1(1 - \cos\theta) - M_p(1 - \cos\theta) \right] \\
&= \left(\frac{8e^4 M_p E_1 E_3}{q^4} \right) \left[(E_1 - E_3)(1 - \cos\theta) + M_p(1 + \cos\theta) \right]
\end{aligned} \tag{16.5}$$

which can be rewritten in terms of the half-angle:

$$\langle |\mathcal{M}_{e^- p \to e^- p}|^2 \rangle = \left(\frac{16e^4 M_p E_1 E_3}{q^4} \right) \left[(E_1 - E_3) \sin^2 \frac{\theta}{2} + M_p \cos^2 \frac{\theta}{2} \right] \tag{16.6}$$

The four-momentum squared of the virtual photon $q^2 = (p - p')^2$ can also be expressed as a function of the electron energy and the scattering angle in the laboratory frame:

$$q^2 = (p - p')^2 = p^2 + p'^2 - 2p \cdot p' \simeq -2E_1 E_3 (1 - \cos\theta) = -4E_1 E_3 \sin^2 \frac{\theta}{2} \tag{16.7}$$

where in the last two terms we have neglected the rest mass of the electrons. We note that:

$$(E_1 - E_3)M_p = (p \cdot k) - (p' \cdot k) = (p - p') \cdot k = q \cdot k \tag{16.8}$$

Considering the proton, we can write $(k')^2 = M_p^2$ and also:

$$(k')^2 = (k + q)^2 = k^2 + q^2 + 2k \cdot q = M_p^2 + q^2 + 2(E_1 - E_3)M_p \tag{16.9}$$

Hence the **energy transferred** ν from the electron to the target proton in the case of elastic scattering is:

$$\nu \equiv E_1 - E_3 = -\frac{q^2}{2M_p} \tag{16.10}$$

Since $q^2 \simeq -4E_1 E_3 \sin^2(\theta/2)$ is always negative (q^2 is *space-like*), the scattered electron always has less energy than the incoming electron, i.e., $E_1 - E_3 > 0$. So finally:

$$\langle |\mathcal{M}_{e^- p \to e^- p}|^2 \rangle = \left(\frac{e^4 M_p^2}{E_1 E_3 \sin^4 \frac{\theta}{2}} \right) \left[\cos^2 \frac{\theta}{2} - \frac{q^2}{2M_p^2} \sin^2 \frac{\theta}{2} \right] \tag{16.11}$$

Although the differential cross-section is expressed as a function of θ and q^2, it actually depends on a *single parameter* for a fixed incident electron energy E_1. Indeed, for a given scattering angle θ, we obtain:

$$q^2 = -2E_1 E_3 (1 - \cos\theta) = -2M_p(E_1 - E_3) \tag{16.12}$$

then $M_p E_1 = M_p E_3 + E_1 E_3 (1 - \cos\theta)$, or

$$\frac{M_p E_1}{E_3} = M_p + E_1 (1 - \cos\theta) \tag{16.13}$$

Substituting back and defining a positive **momentum transfer squared** $Q^2 \equiv -q^2$, we then directly relate Q^2 and the **energy transferred** $E_1 - E_3$ to θ for a fixed E_1, in the case of elastic scattering, by:

$$Q^2(\theta) = \frac{2M_p E_1^2 (1 - \cos\theta)}{M_p + E_1(1 - \cos\theta)} \quad \text{and} \quad \nu \equiv E_1 - E_3 = \frac{Q^2}{2M_p} \tag{16.14}$$

The differential cross-section can be obtained with the phase-space expression Eq. (5.151) and using again that $M_p E_1/E_3 = M_p + E_1(1 - \cos\theta)$, to obtain:

$$\left(\frac{d\sigma}{d\Omega} \right) = \frac{E_3}{64\pi^2 E_1 M_p (E_1 + M_p - E_1 \cos\theta)} |\mathcal{M}|^2 = \frac{E_3^2}{64\pi^2 E_1^2 M_p^2} |\mathcal{M}|^2 \tag{16.15}$$

so the differential cross-section for the scattering of relativistic electrons from a Dirac-like proton initially at rest and including its recoil, is:

$$\left(\frac{d\sigma}{d\Omega} \right)_{Dirac} = \frac{\alpha^2}{4E_1^2 \sin^4(\theta/2)} \frac{E_3}{E_1} \left[\cos^2 \frac{\theta}{2} + \frac{Q^2}{2M_p^2} \sin^2 \frac{\theta}{2} \right] \tag{16.16}$$

The term E_3/E_1 is due to the proton recoil, and the new term $\propto \sin^2(\theta/2)$ is a purely **magnetic spin–spin interaction between the spin of the electron and the spin of the proton**. Consequently, the QED differential cross-section can be seen as composed of the following terms:

$$\left(\frac{d\sigma}{d\Omega}\right)_{Dirac} = \underbrace{\frac{\alpha^2}{4E_1^2 \sin^4(\theta/2)}}_{Rutherford} \underbrace{\frac{E_3}{E_1}}_{proton\ recoil} \left[\underbrace{\cos^2\frac{\theta}{2}}_{overlap\ spin\ 1/2} + \underbrace{\frac{Q^2}{2M_p^2}\sin^2\frac{\theta}{2}}_{spin–spin\ magnetic}\right] \quad (16.17)$$

In the limit where $Q^2/M_p^2 \ll 1$, the second term in the bracket becomes negligible compared to the cosine term, and the expression reduces to that of a "recoil-corrected Mott scattering cross-section," which is valid in the limit where the scattered electron is relativistic (cf. Eq. (11.51)) and has spin-1/2:

$$\left(\frac{d\sigma}{d\Omega}\right)_{Mott,recoil} = \frac{\alpha^2}{4E_1^2 \sin^4(\theta/2)} \frac{E_3}{E_1}\left[\cos^2\left(\frac{\theta}{2}\right)\right] \quad (16.18)$$

In summary, we have the following line of approximations:

$$\left(\frac{d\sigma}{d\Omega}\right)_{Dirac} \xrightarrow{Q^2/M_p^2 \ll 1} \left(\frac{d\sigma}{d\Omega}\right)_{Mott,recoil} \xrightarrow{no\ recoil} \left(\frac{d\sigma}{d\Omega}\right)_{Mott} \xrightarrow{spinless} \left(\frac{d\sigma}{d\Omega}\right)_{Rutherford} \quad (16.19)$$

16.2 Form Factors

Quite generally, a **form factor** is introduced in scattering problems to account for the spatial extent of the scatterer. The probability amplitude for a point-like scatterer is modified by a form factor, which takes into account the spatial extent and shape of the target.

For the elastic scattering of relativistic electrons from a point-like Dirac proton, we have found that the cross-section can be factorized as the product of specific terms in Eq. (16.17). In order to account for the extended nature of the target, we begin with basic considerations of an electron scattering off a static potential due to an extended electric charge Ze distributed in space according to the density function $\rho(\vec{x})$. The electric potential at the position $\vec{x}$ from the center is given by (e.g., cf. Eq. (4.201)):

$$V(\vec{x}) = (Ze)\int \frac{\rho(\vec{x}')}{4\pi|\vec{x}-\vec{x}'|}d^3\vec{x}' \quad \text{with} \quad \int \rho(\vec{x})d^3\vec{x} \equiv 1 \quad (16.20)$$

In first-order perturbation theory the amplitude is proportional to:

$$\begin{aligned}
\mathcal{M} &\propto \langle \Psi_f|V(\vec{x})|\Psi_i\rangle = \int e^{-i\vec{p}_f \cdot \vec{x}}V(\vec{x})e^{+i\vec{p}_i \cdot \vec{x}}d^3\vec{x} = (Ze)\int e^{-i(\vec{p}_f - \vec{p}_i)\cdot\vec{x}}\int \frac{\rho(\vec{x}')}{4\pi|\vec{x}-\vec{x}'|}d^3\vec{x}'d^3\vec{x} \\
&= (Ze)\int\int d^3\vec{x}d^3\vec{x}' e^{-i\vec{q}\cdot\vec{x}}\frac{\rho(\vec{x}')}{4\pi|\vec{x}-\vec{x}'|} = \int\int d^3\vec{x}d^3\vec{x}' e^{-i\vec{q}\cdot(\vec{x}-\vec{x}')}e^{-i\vec{q}\cdot\vec{x}'}\frac{Ze\rho(\vec{x}')}{4\pi|\vec{x}-\vec{x}'|} \quad (16.21)
\end{aligned}$$

where $\vec{q} = \vec{p}_f - \vec{p}_i$. By introducing $\vec{X} \equiv \vec{x} - \vec{x}'$ and substituting in one of the integrals, we see that the resulting amplitude is equivalent to that of scattering from a point source multiplied by a new term called the **form factor** F:

$$\mathcal{M} \propto (Ze)\underbrace{\int d^3\vec{X}\frac{e^{-i\vec{q}\cdot\vec{X}}}{4\pi|\vec{X}|}}_{\mathcal{M}_{point-like}}\underbrace{\int d^3\vec{x}' e^{-i\vec{q}\cdot\vec{x}'}\rho(\vec{x}')}_{F(\vec{q})} \quad (16.22)$$

We note that due to the integration over the entire space, the resulting form factor does not depend on the direction of $\vec{q}$, hence it can be written as:

$$F(\vec{q}^{\,2}) \equiv \int d^3\vec{x}\, e^{-i\vec{q}\cdot\vec{x}} \rho(\vec{x}) \tag{16.23}$$

At this stage we can refer to the QED calculation of the Mott scattering derived in Section 11.2 for the scattering of an electron off a static electromagnetic field given by the classical Coulomb potential of a point-like charge. The calculation for the point source gave us the Mott scattering cross-section. Here we can replace the potential A^μ (Eq. (11.6)) by the following:

$$A^\mu = \left((Ze) \int \frac{\rho(\vec{x}^{\,\prime})}{4\pi|\vec{x}-\vec{x}^{\,\prime}|} d^3\vec{x}^{\,\prime}, 0, 0, 0 \right) \tag{16.24}$$

Finally the result for the QED differential cross-section scattering off a static extended charge distribution is readily derived from Eq. (11.51) by adding the form factor amplitude squared:

$$\left(\frac{d\sigma}{d\Omega}\right)_\rho = \frac{Z^2\alpha^2}{4E^2\sin^4(\theta/2)} \cos^2\frac{\theta}{2} |F(\vec{q}^{\,2})|^2 \tag{16.25}$$

The form factor F represents the Fourier transform of the charge distribution ρ. It is similar to diffraction of plane waves in optics, and we have already encountered it in Eq. (4.212). The finite size of the scattering centers introduces a phase difference between plane waves. If the wavelength is long compared to the size, all waves are effectively in phase and hence we would recover the Mott scattering of a point-like particle, Eq. (11.51), implying $F(\vec{q}^{\,2}) \to 1$. This is clearly also the case for a *point-like target*. We can describe such a charge distribution by a Dirac δ function, hence the form factor is simply:

$$F(\vec{q}^{\,2}) = \int d^3\vec{x}\, e^{-i\vec{q}\cdot\vec{x}} \delta(\vec{x}) = e^{-i\vec{q}\cdot\vec{x}}\big|_{\vec{x}=0} = 1 \qquad \text{point-like} \tag{16.26}$$

As expected, the form factor is flat with a constant value equal to one, $F(\vec{q}^{\,2}) = 1$, for all values of q^2. When the wavelength is very small compared to the spatial extension of the charge distribution, the phases from different regions of the charge will tend to cancel each other and $F(\vec{q}^{\,2} \to \infty) = 0$. Hence, **for any finite extension of the target, the form factor will tend to zero for high values of q^2.** This is illustrated in Figure 16.2.

The computation of the form factor simplifies in the case of distributions with spherical symmetry. By taking a coordinate system with an axis along the direction of $\vec{q}$ and defining θ as the angle between $\vec{q}$ and $\vec{x}$, we get:

$$
\begin{aligned}
F(\vec{q}^{\,2}) &\equiv \int d^3\vec{x}\, e^{-i\vec{q}\cdot\vec{x}} \rho(\vec{x}) = \int dr\, r^2 \sin\theta d\theta d\phi\, e^{-iqr\cos\theta} \rho(r) \\
&= 2\pi \int dr\, r^2 \rho(r) \int dy\, e^{-iqry} = \frac{4\pi}{q} \int_0^\infty dr \rho(r) r \sin(qr)
\end{aligned}
\tag{16.27}
$$

where the integration over θ was performed with the substitution $y = \cos\theta$. As a first application, we imagine a **spherical target with a constant charge density** with radius a:

$$\rho_S(r) = \begin{cases} \frac{1}{4\pi a^3/3} & \text{for } r < a \\ 0 & \text{elsewhere} \end{cases}$$

Then, performing integration by parts, the form factor is given by ($q > 0$):

$$
\begin{aligned}
\left|F_S(\vec{q}^{\,2})\right| &= \left| \frac{3}{qa^3} \int_0^a dr\, r \sin(qr) \right| = \frac{3}{qa^3} \left[\frac{r\cos qr}{q} \Big|_0^a - \int_0^a \frac{\cos qr}{q} dr \right] \\
&= \frac{3}{qa^3} \left[\frac{a\cos qa}{q} - \frac{\sin qa}{q^2} \right] = \frac{3\cos qa}{(qa)^3} [qa - \tan qa]
\end{aligned}
\tag{16.28}
$$

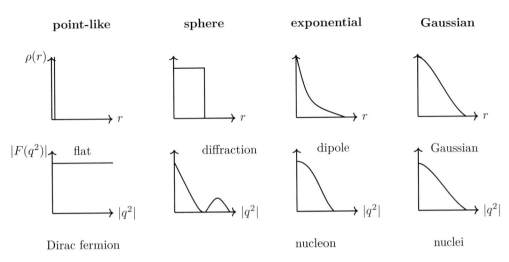

Figure 16.2 Illustrations of spherically symmetric charge distributions and their corresponding form factors.

The form factor is zero when the expression in square brackets is zero, i.e., when $\tan(qa) = qa$, which is true for various values of q. These correspond to various scattering angles and we would expect a diffraction-like pattern from electron scattering off this target. This is also illustrated in Figure 16.2.

Some light nuclei do not have enough protons to build up a constant charge density in their interior and it just falls off from the center. This can sometimes be expressed as an exponential where b dictates the fall-off:

$$\rho_D(r) = \frac{b^3}{8\pi} e^{-br} \tag{16.29}$$

The corresponding form factor is given by the **dipole form**:

$$F_D(q^2) = \left(1 + \frac{q^2}{b^2}\right)^{-2} = \frac{1}{(1 + q^2/b^2)^2} \tag{16.30}$$

We will see in Section 16.5 that experimental data on nucleons fit this distribution rather well.

Finally, some nuclei can also be described with a **Gaussian distribution**:

$$\rho_G(r) = \left(\frac{b^2}{2\pi}\right)^{3/2} e^{-b^2 r^2/2} \tag{16.31}$$

The corresponding form factor is also Gaussian:

$$F_G(q^2) = e^{-q^2/2b^2} \tag{16.32}$$

The inclusion of the form factor alters the angular distribution of the scattering. Measurements of the deviation of the angular distribution from Mott scattering can then be used to extract information about the spatial extent and shape of the charge distribution.

16.3   The Rosenbluth Formula

In general, the finite size of a particle can be accounted for by introducing structure functions. For the case of the proton (or the neutron), two functions are considered: one related to the charge distribution, and the other related to the distribution of the magnetic moment.

The QED current for a point-like Dirac particle is of the form $j^\mu = \bar{\Psi}\gamma^\mu\Psi \propto \bar{u}\gamma^\mu u$. The most general form consistent with current conservation can be derived with the help of the Gordon identity, Eq. (10.171):

$$2m\bar{u}(p')\gamma^\mu u(p) = \bar{u}(p')\left[(p'+p)^\mu + i\sigma^{\mu\nu}(p'-p)_\nu\right]u(p) = \bar{u}(p')\left[(p'+p)^\mu + i\sigma^{\mu\nu}q_\nu\right]u(p) \qquad (16.33)$$

where $q = p' - p$. As stated in Section 10.19, the Gordon decomposition of the point-like QED current contains two parts. The first term is the coupling of the form $(p'+p)$ which is the same as that of a scalar particle (see Eq. (10.169)). This represents an electric term. The second part $i\sigma^{\mu\nu}q_\nu$ is related to the spins and represents a magnetic term stemming from the magnetic moments of the Dirac particles.

To generalize the form of the current, we can multiply each term, i.e., the electric and magnetic ones, by an arbitrary function. To ensure covariance, the arbitrary functions must depend on any scalar quantity which can be constructed. One candidate is $p^2 = M^2$, but it is trivial. The only non-trivial quantity to describe the interaction is q^2. The two functions are called **structure functions** and are generally defined as $F_1(q^2)$ and $F_2(q^2)/2M$, where the factor $2M$ is introduced such that $F_2(0)$ is directly related to the anomalous magnetic moment of the particle. $F_1(q^2)$ and $F_2(q^2)$ are respectively called the **Dirac and Pauli form factors**.

With the two Dirac and Pauli form factors, the general form of a conserving current between the initial-state proton of momentum p^μ and the final-state proton of momentum p'^μ is (cf. Eqs. (12.87) and (14.2)):

$$\langle p(p')|J^\mu(q^2)|p(p)\rangle \equiv \bar{u}(p')\left[\gamma^\mu F_1(q^2) + \frac{i}{2M}\sigma^{\mu\nu}q_\nu F_2(q^2)\right]u(p) \qquad (16.34)$$

It is the most general form of the hadronic current for the spin 1/2-nucleon, satisfying relativistic invariance and current conservation, and including an internal structure for the target proton (or neutron).

In order to derive the matrix element for the $e^-p \to e^-p$ process shown in Figure 16.1 to take into account the structure of the proton, we replace the proton current $\bar{u}(k')\gamma_\mu u(k)$ in the matrix element Eq. (11.72), while keeping the current $\bar{u}(p')\gamma_\mu u(p)$ associated with the point-like Dirac electron unchanged, yielding:

$$
\begin{aligned}
-i\mathcal{M}_{ep\to ep} &= \frac{ie^2}{(p-p')^2}\left(\bar{u}(p')\gamma_\mu u(p)\right)\cdot\langle p(k')|J^\mu(q^2)|p(k)\rangle \\
&= \frac{ie^2}{(p-p')^2}\left(\bar{u}(p')\gamma_\mu u(p)\right)\cdot\bar{u}(k')\left[\gamma^\mu F_1(q^2) + \frac{i}{2M_p}\sigma^{\mu\nu}q_\nu F_2(q^2)\right]u(k) \qquad (16.35)
\end{aligned}
$$

recalling that $q = p - p' = k' - k$ and M_p is the proton rest mass. This matrix element can be conveniently rewritten, using Gordon's identity, as:

$$-i\mathcal{M} = \frac{ie^2}{q^2}\left(\bar{u}(p')\gamma_\mu u(p)\right)\cdot\bar{u}(k')\left[(F_1+F_2)\gamma^\mu - \frac{F_2}{2M_p}(k+k')^\mu\right]u(k) \qquad (16.36)$$

The structure functions $F_1(q^2)$ and $F_2(q^2)$ of the proton are also called "proton form factors." Recall that for a point-like Dirac proton we would have $F_1(q^2) = 1$ and $F_2(q^2) = 0$ to get $\bar{u}\gamma^\mu u$, which would still include the electric and magnetic interaction of a Dirac particle, so F_2 really describes **an anomalous magnetic term**, as discussed in Section 14.1.

In order to calculate the cross-section, we need to evaluate the matrix element squared, which is of the form of a product of an electron (lepton) and proton (hadronic) tensor – compare this form with that of Eq. (11.112):

$$|\mathcal{M}|^2 = \frac{e^4}{q^4}L^{\mu\nu}(p,p')W_{\mu\nu}(k,k') \qquad (16.37)$$

where the **lepton tensor** is simply

$$L^{\mu\nu} = \bar{u}(p')\gamma^\mu u(p)\bar{u}(p)\gamma^\nu u(p') \qquad (16.38)$$

while the **hadronic tensor** is

$$W_{\mu\nu} = \bar{u}(k')\left[(F_1 + F_2)\gamma^\mu - \frac{F_2}{2M_p}K^\mu\right]u(k)\bar{u}(k)\left[(F_1 + F_2)\gamma^\nu - \frac{F_2}{2M_p}K^\nu\right]u(k') \tag{16.39}$$

with $K^\mu \equiv k^\mu + k'^\mu$.

We consider the unpolarized cross-section, hence our calculations will involve the usual averaging and summing over spins. The evaluation of the lepton tensor $L^{\mu\nu}$ gives the same result as for a point-like Dirac case (cf. Eqs. (11.114) and (11.118)):

$$\langle L^{\mu\nu}\rangle = \frac{1}{2}\operatorname{Tr}\left((\slashed{p}' + m_e)\gamma^\mu(\slashed{p} + m_e)\gamma^\nu\right) = \frac{1}{2}4\left(p'^\mu p^\nu + p'^\nu p^\mu - g^{\mu\nu}\left(p'\cdot p - m_e^2\right)\right) \tag{16.40}$$

The hadronic tensor $\langle W^{\mu\nu}\rangle$ can be decomposed by distribution into the four terms:

$$
\begin{aligned}
\frac{1}{2}(F_1 + F_2)^2\operatorname{Tr}\left((\slashed{k}' + M_p)\gamma^\mu(\slashed{k} + M_p)\gamma^\nu\right) &= \frac{1}{2}(F_1 + F_2)^2\operatorname{Tr}\left(\slashed{k}'\gamma^\mu\slashed{k}\gamma^\nu + M_p\gamma^\mu\slashed{k}\gamma^\nu + M_p\slashed{k}'\gamma^\mu\gamma^\nu + M_p^2\gamma^\mu\gamma^\nu\right) \\
&= \frac{1}{2}(F_1 + F_2)^2\operatorname{Tr}\left(\slashed{k}'\gamma^\mu\slashed{k}\gamma^\nu + M_p^2\gamma^\mu\gamma^\nu\right) \\
&= \frac{1}{2}(F_1 + F_2)^2\left(4(k'^\mu k^\nu + k'^\nu k^\mu - (k'\cdot k)g^{\mu\nu}) + M_p^2 4g^{\mu\nu}\right) \\
&= 2(F_1 + F_2)^2\left(k'^\mu k^\nu + k'^\nu k^\mu - g^{\mu\nu}(k'\cdot k - M_p^2)\right) \tag{16.41}
\end{aligned}
$$

and the cross-product term:

$$
\begin{aligned}
\frac{1}{2}\left(\frac{-F_2}{2M_p}\right)(F_1 + F_2)\operatorname{Tr}\left((\slashed{k}' + M_p)\gamma^\mu(\slashed{k} + M_p)\right)K^\nu &= \frac{1}{2}\left(\frac{-F_2}{2M_p}\right)(F_1 + F_2)\operatorname{Tr}\left(M_p\gamma^\mu\slashed{k} + M_p\slashed{k}'\gamma^\mu\right)K^\nu \\
&= \left(\frac{-F_2}{4}\right)(F_1 + F_2)4(k^\mu + k'^\mu)K^\nu \\
&= -F_2(F_1 + F_2)K^\mu K^\nu \tag{16.42}
\end{aligned}
$$

and similarly:

$$\frac{1}{2}\left(\frac{-F_2}{2M_p}\right)(F_1 + F_2)K^\mu\operatorname{Tr}\left((\slashed{k}' + M_p)(\slashed{k} + M_p)\gamma^\nu\right) = -F_2(F_1 + F_2)K^\mu K^\nu \tag{16.43}$$

and the last term:

$$
\begin{aligned}
\frac{1}{2}\frac{F_2^2}{4M_p^2}\operatorname{Tr}\left((\slashed{k}' + M_p)K^\mu(\slashed{k} + M_p)K^\nu\right) &= \frac{1}{2}\frac{F_2^2}{4M_p^2}K^\mu K^\nu\operatorname{Tr}\left(\slashed{k}'\slashed{k} + M_p^2\mathbb{1}\right) \\
&= \frac{F_2^2}{2M_p^2}(k'\cdot k + M_p^2)K^\mu K^\nu = \frac{F_2^2}{4M_p^2}K^2 K^\mu K^\nu \tag{16.44}
\end{aligned}
$$

From now on, we neglect the electron mass relative to that of the proton. We can compute the matrix element by direct multiplication:

$$
\begin{aligned}
\langle|\mathcal{M}|^2\rangle &= \frac{4e^4}{q^4}\left(p'_\mu p_\nu + p'_\nu p_\mu - g_{\mu\nu}(p'\cdot p)\right) \\
&\quad\times\left[(F_1 + F_2)^2\left(k'^\mu k^\nu + k'^\nu k^\mu - g^{\mu\nu}(k\cdot k' - M_p^2)\right) + K^\mu K^\nu\left(\frac{F_2^2}{8M_p^2}K^2 - F_2(F_1 + F_2)\right)\right] \\
&= \frac{4e^4}{q^4}\left[(F_1 + F_2)^2\left(2(p'\cdot k')(p\cdot k) + 2(p'\cdot k)(p\cdot k') - 2M_p^2(p'\cdot p)\right)\right. \\
&\quad\left. + \left(\frac{F_2^2}{8M_p^2}K^2 - F_2(F_1 + F_2)\right)\left(2(p'\cdot K)(p\cdot K) - K^2(p'\cdot p)\right)\right] \tag{16.45}
\end{aligned}
$$

One can check that indeed one recovers Eq. (16.1) for $F_1 = 1$ and $F_2 = 0$. We now note that:

$$
\begin{aligned}
t &= (k' - k)^2 = 2M_p^2 - 2k' \cdot k \quad \longrightarrow \quad k \cdot k' = \frac{1}{2}(2M_p^2 - t) \\
K^2 &= (k + k')^2 = 2M_p^2 + 2k \cdot k' = 4M_p^2 - t
\end{aligned}
\tag{16.46}
$$

Hence,

$$
\begin{aligned}
\left(\frac{F_2^2}{8M_p^2} K^2 - F_2(F_1 + F_2) \right) &= \left(\frac{F_2^2}{2} - \frac{t}{8M_p^2} F_2^2 - F_2 F_1 - F_2^2 \right) = \frac{1}{2}\left(-F_2^2 - \frac{t}{4M_p^2} F_2^2 - 2F_2 F_1 - F_1^2 + F_1^2 \right) \\
&= \frac{1}{2}\left(-(F_1 + F_2)^2 + F_1^2 - \frac{t}{4M_p^2} F_2^2 \right)
\end{aligned}
\tag{16.47}
$$

Consequently, the matrix element is given by:

$$
\langle |\mathcal{M}|^2 \rangle = \frac{4e^4}{q^4} \left[2B \left((p' \cdot k')(p \cdot k) + (p' \cdot k)(p \cdot k') - M_p^2(p' \cdot p) \right) + (A - B) \left((p' \cdot K)(p \cdot K) - \frac{K^2}{2}(p' \cdot p) \right) \right]
\tag{16.48}
$$

where

$$
A \equiv F_1^2 - \frac{t}{4M_p^2} F_2^2 \quad \text{and} \quad B \equiv (F_1 + F_2)^2
\tag{16.49}
$$

As before, we can further note that:

$$
\begin{aligned}
s &= (p + k)^2 = (p' + k')^2 = M_p^2 + 2p \cdot k = M_p^2 + 2p' \cdot k' \quad \longrightarrow \quad p \cdot k = p' \cdot k' = \frac{1}{2}(s - M_p^2) \\
t &= (p - p')^2 = 2m_e^2 - 2p \cdot p' \approx -2p \cdot p' \quad \longrightarrow \quad p \cdot p' \approx -\frac{t}{2} \\
u &= (k - p')^2 = (p - k')^2 \approx M_p^2 - 2k \cdot p' = M_p^2 - 2k' \cdot p \quad \longrightarrow \quad p \cdot k' = p' \cdot k = \frac{1}{2}(M_p^2 - u) \\
p \cdot K &= p \cdot k + p \cdot k' = \frac{1}{2}(s - u) \\
p' \cdot K &= p' \cdot k + p' \cdot k' = \frac{1}{2}(s - u)
\end{aligned}
\tag{16.50}
$$

Putting the products into the matrix element, we obtain:

$$
\langle |\mathcal{M}|^2 \rangle = \frac{4e^4}{q^4} \left[2B \left(\frac{1}{4}(s - M_p^2)^2 + \frac{1}{4}(M_p^2 - u)^2 + \frac{M_p^2 t}{2} \right) + \left(\frac{A - B}{4} \right) \left((s - u)^2 + 4t M_p^2 - t^2 \right) \right]
\tag{16.51}
$$

Collecting the terms proportional to A and using $s + t + u = 2M_p^2$, we find:

$$
\frac{A}{4} \left((s - u)^2 + 4t M_p^2 - t^2 \right) = \frac{A}{4} \left((2s - 2M_p^2 + t)^2 + 4t M_p^2 - t^2 \right) = \frac{A}{4} \left(4(s - M_p^2)^2 + 4st \right)
\tag{16.52}
$$

With a bit more algebra, we can show that the terms proportional to B simply reduce to:

$$
\begin{aligned}
-\frac{1}{2}(s - M_p^2)^2 + \frac{1}{2}(M_p^2 - u)^2 + M_p^2 t - st &= -\frac{1}{2}(s - M_p^2)^2 + \frac{1}{2}((s - M_p^2) + t)^2 + M_p^2 t - st \\
&= \frac{1}{2}\left(2(s - M_p^2)t + t^2 \right) + M_p^2 t - st = \frac{t^2}{2}
\end{aligned}
\tag{16.53}
$$

Finally, we have the matrix element as a function of the Mandelstam variables in a very compact form:

$$
\langle |\mathcal{M}|^2 \rangle = \frac{4e^4}{t^2} \left[A \left((s - M_p^2)^2 + st \right) + B \frac{t^2}{2} \right]
\tag{16.54}
$$

where A and B are given in Eq. (16.49). In the chosen laboratory frame where the target proton is at rest, and neglecting the electron mass as before, we find $s = (p+k)^2 \simeq M_p^2 + 2M_pE_1$. From Eqs. (16.7) and (16.12), we have $t = q^2 = -4E_1E_3\sin^2(\theta/2) = -2M_p(E_1 - E_3)$, where θ is the scattering angle. So

$$
\begin{aligned}
(s - M_p^2)^2 + st &= (2M_pE_1)^2 - 2M_p(E_1 - E_3)(M_p^2 + 2M_pE_1) = 2M_p^2(2E_1E_3 - M_p(E_1 - E_3)) \\
&= 2M_p^2(2E_1E_3 - 2E_1E_3\sin^2(\theta/2)) \\
&= 4M_p^2 E_1 E_3 \cos^2(\theta/2)
\end{aligned}
\tag{16.55}
$$

and

$$
B\frac{t^2}{2} = -\frac{B}{2}q^2 4E_1E_3\sin^2(\theta/2) \tag{16.56}
$$

The spin-averaged matrix element squared expressed in terms of the laboratory system variables is then:

$$
\begin{aligned}
\langle|\mathcal{M}|^2\rangle &= \frac{16e^4}{16E_1^2E_3^2\sin^4(\theta/2)}M_p^2 E_1 E_3 \left[A\cos^2(\theta/2) - B\frac{q^2}{2M_p^2}\sin^2(\theta/2)\right] \\
&= \frac{M_p^2 e^4}{E_1E_3\sin^4(\theta/2)}\left[A\cos^2(\theta/2) - B\frac{q^2}{2M_p^2}\sin^2(\theta/2)\right]
\end{aligned}
\tag{16.57}
$$

Finally, we use Eq. (16.15) to find the differential cross-section:

$$
\left(\frac{d\sigma}{d\Omega}\right) = \frac{E_3^2}{64\pi^2 E_1^2 M_p^2}|\mathcal{M}|^2 = \frac{e^4}{64\pi^2 E_1^2\sin^4(\theta/2)}\frac{E_3}{E_1}\left[A\cos^2(\theta/2) - B\frac{q^2}{2M_p^2}\sin^2(\theta/2)\right] \tag{16.58}
$$

This equation is known as the **Rosenbluth[2] formula** [189] for the **elastic** $e^-p \to e^-p$ **scattering** expressed in terms of the incident electron energy E_1 and scattering angle θ. Reintroducing $Q^2 = -q^2$, it can be written as:

$$
\left(\frac{d\sigma}{d\Omega}\right)_{Rosenbluth} = \frac{\alpha^2}{4E_1^2\sin^4(\theta/2)}\frac{E_3}{E_1}\left[A\cos^2\frac{\theta}{2} + B\frac{Q^2}{2M_p^2}\sin^2\frac{\theta}{2}\right] \tag{16.59}
$$

where Q^2 and E_3 are uniquely determined by the energy E_1 and the angle θ via Eq. (16.14), and where:

$$
A(Q^2) \equiv F_1^2(Q^2) + \frac{Q^2}{4M_p^2}F_2^2(Q^2) \quad \text{and} \quad B(Q^2) \equiv (F_1(Q^2) + F_2(Q^2))^2 \tag{16.60}
$$

For the case of a *point-like Dirac proton*, we obviously have $A(Q^2) = B(Q^2) = 1$, and, as should be, we recover Eq. (16.16), which was the cross-section for the scattering of a spin-1/2 electron on a point-like Dirac target.

We can write the Rosenbluth cross-section even more explicitly and compactly as a correction to the recoil-corrected Mott scattering formula, Eq. (16.18). Introducing the commonly used **dimensionless and Lorentz-invariant quantity** τ:

$$
\tau \equiv \frac{Q^2}{4M_p^2} \tag{16.61}
$$

we have:

$$
\left(\frac{d\sigma}{d\Omega}\right)_{Rosenbluth} = \left(\frac{d\sigma}{d\Omega}\right)_{Mott,recoil} \times \left[A(Q^2) + 2\tau B(Q^2)\tan^2\frac{\theta}{2}\right] \tag{16.62}
$$

Rosenbluth's formula shows that there is a contribution to the elastic scattering from both the Dirac and the Pauli components of the proton's magnetic moment. In this expression the pure QED dependence is included in the Mott cross-section, while the effect of the target structure is contained in the bracketed term.

2   Marshall Nicholas Rosenbluth (1927–2003), American plasma physicist.

16.4   The Finite Size of the Proton

In 1954 the first elastic electron scattering experiments were undertaken by **R. W. McAllister** and **R. Hofstadter**[3] and collaborators. They were using electron beams of the High Energy Physics Laboratory (HEPL) at Stanford in California. The layout of the experiment is shown in Figure 16.3(left).

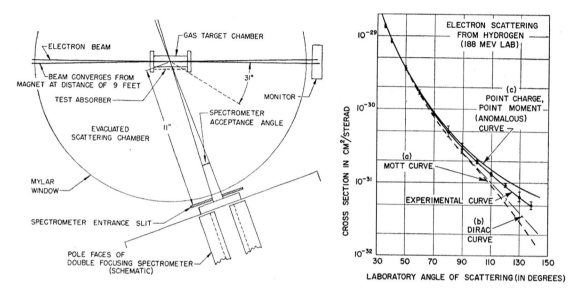

Figure 16.3 (left) Layout of the electron scattering experiment. (right) Results for electrons with incident energy of 188 MeV impinging on hydrogen. Reprinted figures with permission from R. W. McAllister and R. Hofstadter, "Elastic scattering of 188-MeV electrons from the proton and the alpha particle," *Phys. Rev.*, vol. 102, pp. 851–856, May 1956. Copyright 1956 by the American Physical Society.

Their primary aim was to discover the finite size of *nuclei*, and even that of the *proton*. In 1956, they wrote:

> "*The proton, deuteron, and alpha particle are most interesting to study because they are among the simplest nuclear structures. Furthermore, nuclei are built up out of protons and neutrons and it is fascinating to think of what the proton itself is built.*"[4]

They performed pioneering studies of electron scattering in atomic nuclei and thereby achieved discoveries concerning the structure of the nucleons. The form factors of the proton were measured for the first time. The experimental cross-section could not be fit if the proton was treated as a point-like Dirac particle (the "Mott" cross-section), nor with the addition of an anomalous magnetic moment alone. More structure was required, thereby showing that the nucleon is likely composite.

The results for electrons with incident energy of 188 MeV impinging on hydrogen are shown in Figure 16.3(right). Curve (a) shows the theoretical Mott curve for a spinless point proton. Curve (b) shows the theoretical curve for a point proton with a Dirac magnetic moment (Landé factor $g = 2$). Curve (c) shows

3   Robert Hofstadter (1915–1990), American physicist.
4   Reprinted excerpt with permission from R. W. McAllister and R. Hofstadter, "Elastic scattering of 188-MeV electrons from the proton and the alpha particle," *Phys. Rev.*, vol. 102, pp. 851–856, May 1956. Copyright 1956 by the American Physical Society.

the theoretical curve for a point proton having the anomalous contribution in addition to the Dirac value of magnetic moment (Landé factor $g = 2\mu_p \simeq 2 \times 2.79 = 5.58$).

The experimental curve falls between curves (b) and (c), so none of the predictions for the point-like proton fit the data, not even with the anomalous magnetic moment. The deviation from the theoretical curves represents the effect of a form factor for the proton and indicates structure within the proton. **The experiment clearly indicated that the proton cannot be described as a point-like particle** [190].

The data collected by Hofstadter was spectacular and allowed one to test hypotheses on the charge distribution of the proton. The elastic scattering data are consistent with that of an exponential charge distribution [191]. It shows that the root-mean-square radius of the proton charge distribution is consistent with being 0.80 fm (see Figure 16.4).

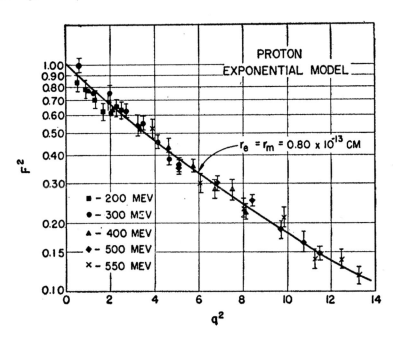

Figure 16.4 The square of the form factor plotted against Q^2. Q^2 is given in units of $10^{-26}\,\mathrm{cm}^2$. The solid line is calculated for the exponential model with the root-mean-square radius 0.80×10^{-13} cm. Reprinted figure from R. Hofstadter, "Electron scattering and nuclear structure," *Rev. Mod. Phys.*, vol. 28, pp. 214–254, Jul 1956. Copyright 1956 by the American Physical Society.

16.5   The Electric and Magnetic Form Factors

In 1957 **Yennie, Lévy, and Ravenhall** derived the expressions for the e-nucleon cross-sections in terms of the Dirac and Pauli form factors $F_{1(p,n)}$ and $F_{2(p,n)}$ for the proton and the neutron [192]. Their interpretation for these form factors is based on the assumption that "physical" nucleons (proton or neutron) are made of "bare" nucleons surrounded by a cloud of pions interacting in a charge-symmetrical manner.

In 1960 **Sachs**[5] *et al.* introduced the **electric** G_E and the **magnetic** G_M **form factors** and conjectured that their new choice of form factors is more physically meaningful than F_1 and F_2. The relations between $G_E(Q^2)$

5   Robert G. Sachs (1916–1999), American theoretical physicist, founder and director of the Argonne National Laboratory.

and $G_M(Q^2)$ and $F_1(Q^2)$ and $F_2(Q^2)$ for the proton and neutron are simply given by:

$$\begin{cases} G_{E(p,n)} \equiv F_{1(p,n)} - \tau F_{2(p,n)} \\ G_{M(p,n)} \equiv F_{1(p,n)} + F_{2(p,n)} \end{cases} \tag{16.63}$$

With the Sachs form factors, the **Rosenbluth elastic differential cross-section** becomes:

$$\left(\frac{d\sigma}{d\Omega} \right)_{Rosenbluth} = \frac{\alpha^2}{4E_1^2 \sin^4(\theta/2)} \frac{E_3}{E_1} \underbrace{\left[\frac{G_E^2 + \tau G_M^2}{1 + \tau} \cos^2 \frac{\theta}{2} + 2\tau G_M^2 \sin^2 \frac{\theta}{2} \right]}_{\text{Sachs form factors}} \tag{16.64}$$

The functions G_E and G_M take account of the extended size of the target. In the limit $Q^2 \to 0$ the electric and magnetic form factors yield the static properties of the nucleon. Thus, the electric form factor in the limit $Q^2 \to 0$ yields the charge of the proton and neutron:

$$\text{Proton}: G_E^p(0) = 1 \qquad \text{Neutron}: G_E^n(0) = 0 \tag{16.65}$$

Similarly, the magnetic form factor corresponds to the total magnetic moment of the nucleon, that is:

$$\text{Proton}: G_M^p(0) = \mu_p = 1 + \mu_p^a \qquad \text{Neutron}: G_M^n(0) = \mu_n = \mu_n^a \tag{16.66}$$

where we recall that the anomalous magnetic moments for the proton and neutron are $\mu_p^a \simeq 1.793$ and $\mu_n^a \simeq -1.913$. The non-vanishing magnetic moments immediately point to the fact that the nucleon has a structure (whereas the electron is a Dirac point-like particle).

The form factors *cannot* be calculated from first principles. Since the first measurements in the 1950s up to the present time, they remain empirical functions, extracted from measured differential cross-sections. The earliest experiments at relatively small Q^2 found that protons and neutrons can be approximately described by:

$$G_E^p(Q^2) \simeq \frac{G_M^p(Q^2)}{\mu_p} \simeq \frac{G_M^n(Q^2)}{\mu_n} \simeq G_D(Q^2) \tag{16.67}$$

where $G_D(Q^2)$ is the **dipole form factor** [193] given in Eq. (16.30):

$$G_D(Q^2) \equiv \frac{1}{(1 + Q^2/Q_0^2)^2} \qquad \text{where} \quad Q_0^2 = 0.71 \text{ GeV}^2 \tag{16.68}$$

The elastic scattering cross-section can be written in a simpler form, without an interference term between G_E and G_M by noting:

$$\begin{aligned} \left(\frac{d\sigma}{d\Omega} \right)_{Rosenbluth} &= \left(\frac{d\sigma}{d\Omega} \right)_{Mott} \left(\frac{E_3}{E_1} \right) \left[\frac{1}{1+\tau} \right] \left[G_E^2 + \tau G_M^2 \left(1 + 2(1+\tau) \tan^2 \frac{\theta}{2} \right) \right] \\ &= \left(\frac{d\sigma}{d\Omega} \right)_{Mott} \left(\frac{E_3}{E_1} \right) \left[\frac{1}{1+\tau} \right] \left[G_E^2 + \frac{\tau}{\epsilon} G_M^2 \right] \end{aligned} \tag{16.69}$$

where

$$\epsilon \equiv \frac{1}{\left(1 + 2(1+\tau) \tan^2 \frac{\theta}{2} \right)} \tag{16.70}$$

The electric and magnetic form factors of the nucleon can be extracted from measurements of cross-sections at a constant Q^2 and different beam energies. This is known as the Rosenbluth separation technique. This latter takes advantage of the linear dependence in ϵ in the reduced cross-section σ_R, as follows:

$$\sigma_R \equiv \epsilon(1+\tau) \left(\frac{E_1}{E_3} \right) \left(\frac{d\sigma}{d\Omega} \right)_{Rosenbluth} \bigg/ \left(\frac{d\sigma}{d\Omega} \right)_{Mott} = \epsilon G_E^2 + \tau G_M^2 \tag{16.71}$$

showing that σ_R is expected to have a linear dependence on ϵ, with the slope proportional to G_E^2 and the intercept equal to τG_M^2. These measurements led to the following conclusions:

- Form factors fall rapidly with Q^2 so the proton is definitely not point-like. The data are reasonably well accommodated by the dipole form factor.

- Taking the Fourier transform, the spatial charge and magnetic moment distributions are consistent with $\rho(r) \simeq \rho_0 e^{-r/r_0}$, where $r_0 \simeq 0.24$ fm.

- The RMS charge radius of the proton is $r_p \simeq 0.88$ fm.

After the first investigation during the 1950s, the field has continued to make progress towards measurements of the nucleon form factor with increasing precision. The range of Q^2 was also extended to reach high values up to 30 GeV^2 and beyond. A compilation of data and best fit from Ref. [194] for the electric G_{Ep} and magnetic G_{Mp} and G_{Mn} form factors for the proton and the neutron, normalized to the ideal dipole form factor G_D, are shown in Figure 16.5. From this, we see that the dipole form is a good approximation for the form factors but at the same time high-precision data shows significant deviations from it in particular at high Q^2. Understanding the exact nature of the nucleon is subject to intense theoretical investigations. Experimentally, new techniques relying on the double polarization methods have been developed. With the continued experimental, theoretical, and computing (lattice) progress, the goal is to one day be able to compute the form factors of the nucleon in terms of the strong interactions (QCD) between their basic constituents (quarks and gluons).

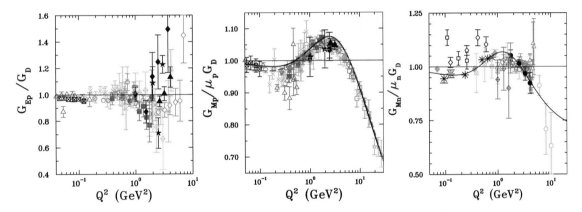

Figure 16.5 Form factors extracted from cross-section measurements vs. Q^2: (left) proton G_{Ep}/G_D; (middle) proton $G_{Mp}/\mu_p G_D$; (right) neutron $G_{Mn}/\mu_n G_D$. Reprinted by permission from Springer Nature: V. Punjabi, C. F. Perdrisat, M. K. Jones, E. J. Brash, and C. E. Carlson, "The structure of the nucleon: Elastic electromagnetic form factors," *The European Physical Journal A*, vol. 51, p. 79, Jul 2015. https://doi.org/10.1140/epja/i2015-15079-x. Copyright Società Italiana di Fisica / Springer-Verlag 2015.

- **Proton charge radius.** Prior to 2010, the most precise measurements of the proton's charge radius came from electron–proton scattering and hydrogen spectroscopy [142]. When combined, these two methods yielded a radius of 0.8751 ± 0.0061 fm. In 2010, high-precision measurements of the proton radius from laser spectroscopy of muonic hydrogen found values that were 6σ smaller than that value. This was known as the **proton radius puzzle**. Following this, new electron–proton scattering and hydrogen spectroscopy experiments were performed and found results consistent with that obtained with muonic hydrogen, albeit with larger errors. The latest recommended 2018 CODATA value for the proton charge radius is:

$$r_p = (0.8414 \pm 0.0019) \times 10^{-15} \text{ m} \tag{16.72}$$

dominated by the measurement on muonic hydrogen.

16.6   Deep Inelastic Electron–Proton Scattering

The above results formed the prejudice that the proton was a soft ("mushy") extended object, possibly with a hard core surrounded by a cloud of mesons, mainly pions.

Continuum inelastic electron scattering was first carried out in the famous SLAC-MIT program headed by **Friedman**,[6] **Kendall**,[7] and **Taylor**.[8] A photograph of the SLAC-MIT experiment is shown in Figure 16.6. The principal goals of the experiment were first elastic scattering, then electro-production of resonances, and

Figure 16.6 A photograph of the SLAC-MIT experiment that was located in the End Station A at the 2-mile Stanford Linear Accelerator, which was at the time able to accelerate electrons up to 20 GeV.

finally continuum electro-production, for which most people at the time had few expectations. The SLAC-MIT team saw its objective in searching for the hard core of the proton, but it was not obvious for everyone at the time that the most important discovery would actually come from the deep inelastic regime. This region corresponds to the excitation of the continuum well beyond the nucleon resonances. Too little was known theoretically. This breakthrough could be done by exploiting the higher energy of the electron beam that became available with the commissioning of the 2-mile linac with $E = 20$ GeV.

The reaction equation for the inclusive **deep inelastic scattering (DIS regime)** is written:

$$e^-(p^\mu) + p(k^\mu) \rightarrow e^-(p^{'\mu}) + X(k^{'\mu}) \tag{16.73}$$

where X is a system of outgoing hadrons (mostly pions). See Figure 16.7. In this regime, the reaction becomes inelastic because the proton breaks up in the collision and a new hadronic system X composed of one or more hadrons is formed. During inelastic scattering the proton breaks up into its constituents, which then form the hadronic jet composed of several hadrons.

6   Jerome Isaac Friedman (born 1930), American physicist.
7   Henry Way Kendall (1926–1999), American particle physicist.
8   Richard Edward Taylor (1929–2018), Canadian physicist.

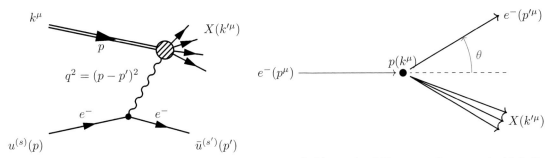

Figure 16.7 Electron–proton deep inelastic scattering: (left) tree-level Feynman diagram, the blob illustrates the fact that the proton is not a point-like Dirac particle and breaks up; (right) kinematics in the proton rest frame.

Experimentally, only the scattered electron is observed in the first round of experiments. The kinematical ranges covered are summarized in Table 16.1. The experimental layout is illustrated in Figure 16.8.   The

Angle θ	Range E (GeV)	Range E' (GeV)	Range Q^2 (GeV^2)	Max. W (GeV)
6°	7–16	3.25–13.2	0.25–2.31	4.8
10°	7–17.7	3.0–12.46	0.80–6.7	5.2
18°	4.5–17	2.0–8.0	1.0–14.0	5.1
26°	4.5–18	1.75–5.5	1.5–20.0	5.0
34°	4.5–15	1.2–3.25	2.0–16.7	4.3

Table 16.1 The kinematical range of the SLAC-MIT experiment [195, 196].

unobserved hadronic system X is the missing mass. The energy of the incident electron beam is accurately known. The proton is the target particle. In the SLAC experiments (and many later experiments at CERN) the target is at rest in the laboratory, which defines the laboratory frame. Hence in this frame, as in Eq. (16.3), we have:

$$p^\mu = (E_1, 0, 0, E_1); \quad k^\mu = (M_p, 0, 0, 0); \quad p'^\mu = (E_3, 0, E_3 \sin\theta, E_3 \cos\theta); \quad q = (p - p') \qquad (16.74)$$

In the elastic case, we found the relation $Q^2 = 2M_p\nu$ (Eq. (16.14)). This relation no longer holds in the inelastic case. Instead we define **the invariant mass of the hadronic system** W:

$$W^2 \equiv (k')^2 = (k + q)^2 = M_p^2 + 2k \cdot q + q^2 \qquad (16.75)$$

W represents the center-of-mass energy of the outgoing hadrons. In the laboratory frame, since $k \cdot q = M_p(E_1 - E_3) = M_p\nu$, we find:

$$W^2 = M_p^2 + 2M_p\nu - Q^2 \qquad (16.76)$$

We recall that ν is the energy transferred via the virtual photon from the incoming electron to the target proton. This relation has an important consequence: when we discussed the elastic case, we effectively considered only the reactions where $W^2 = M_p^2$, since the proton retained its identity. The elastic case can then be seen as the

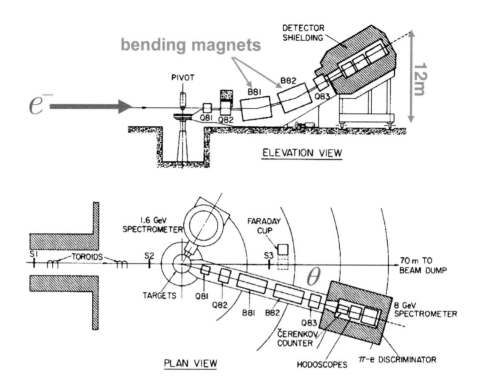

Figure 16.8 Sketch of the SLAC-MIT experiment. Reprinted figure from P. N. Kirk *et al.*, "Elastic electron–proton scattering at large four-momentum transfer," *Phys. Rev. D*, vol. 8, pp. 63–91, Jul 1973. Copyright 1973 by the American Physical Society.

"special" case of inelastic scattering where the outgoing hadronic system is simply composed of a proton. We then have $W = M_p$ or:

$$W^2_{elastic} = M^2_p = M^2_p + 2M_p\nu - Q^2 \quad \longrightarrow \quad Q^2 = 2M_p\nu \quad \text{elastic scattering} \tag{16.77}$$

In the elastic scattering, the kinematics and the differential cross-section depended only on *one parameter* for a fixed incoming electron energy. We chose it to be the scattering angle θ but we could as well have conveniently used Q^2, since:

$$Q^2 \simeq 4E_1E_3\sin^2\left(\frac{\theta}{2}\right) \quad \text{then} \quad \frac{\mathrm{d}Q^2}{\mathrm{d}\theta} = 4E_1E_3 2\sin\left(\frac{\theta}{2}\right)\cos\left(\frac{\theta}{2}\right)\frac{1}{2} = 2E_1E_3\sin(\theta) \tag{16.78}$$

which leads to:

$$\mathrm{d}Q^2 = 2E_1E_3\sin\theta\mathrm{d}\theta = 2E_1E_3\mathrm{d}\cos\theta = \frac{E_1E_3}{\pi}\mathrm{d}\Omega \tag{16.79}$$

where we used $\mathrm{d}\Omega = 2\pi\,\mathrm{d}\cos\theta$ in case of ϕ symmetry. We hence obtain the following relation for the elastic scattering:

$$\frac{\mathrm{d}\sigma}{\mathrm{d}Q^2} = \frac{\pi}{E_1E_3}\frac{\mathrm{d}\sigma}{\mathrm{d}\Omega} \quad \text{where} \quad Q^2 = 2M_p\nu \tag{16.80}$$

This implies that the Rosenbluth elastic scattering formula, Eq. (16.59), becomes:

$$
\left(\frac{d\sigma}{dQ^2}\right)_{Rosenbluth} = \frac{\pi\alpha^2}{4E_1^4 \sin^4(\theta/2)}\left[A\cos^2\frac{\theta}{2} + B\frac{Q^2}{2M_p^2}\sin^2\frac{\theta}{2}\right] \simeq \frac{4\pi\alpha^2}{Q^4}\left(\frac{E_3}{E_1}\right)^2\left[A\cos^2\frac{\theta}{2} + B\frac{Q^2}{2M_p^2}\sin^2\frac{\theta}{2}\right]
$$
$$(16.81)$$

We note that the Rosenbluth differential cross-section depends only on the single parameter Q^2, as expected.

In the inelastic case, the relation $Q^2 = 2M_p\nu$ is generally no longer valid and we hence expect the inelastic double differential cross-section to depend on two parameters, which we can choose to be Q^2 and ν:

$$
\left(\frac{d\sigma}{dQ^2}\right)_{elastic} \quad \longrightarrow \quad \left(\frac{d^2\sigma}{dQ^2 d\nu}\right)_{inelastic}
$$
$$(16.82)$$

We must satisfy the condition that we recover the elastic cross-section for the cases where $Q^2 = 2M_p\nu$, when integrating over ν, so using the property of the Dirac δ function, Eq. (A.31):

$$
\left(\frac{d\sigma}{dQ^2}\right) \equiv \int d\nu \left(\frac{d^2\sigma}{dQ^2 d\nu}\right)\delta\left(\nu - \frac{Q^2}{2M_p}\right) = \frac{1}{\left|\frac{dg}{d\nu}\right|}\left(\frac{d^2\sigma}{dQ^2 d\nu}\right) = \frac{E_3}{E_1}\left(\frac{d^2\sigma}{dQ^2 d\nu}\right)
$$
$$(16.83)$$

since

$$
\begin{aligned}
\frac{dg}{d\nu} &= 1 - \frac{1}{2M_p}\frac{dQ^2}{d\nu} = 1 - \frac{1}{2M_p}\frac{d}{d\nu}\left[4E_1(E_1 - \nu)\sin^2(\theta/2)\right] = 1 + \frac{4E_1 E_3 \sin^2(\theta/2)}{2M_p E_3} = 1 + \frac{Q^2}{2M_p E_3}\\
&= \frac{2M_p E_3 + Q^2}{2M_p E_3} = \frac{E_3 + \nu}{E_3} = \frac{E_1}{E_3}
\end{aligned}
$$
$$(16.84)$$

The Rosenbluth cross-section can then also be written as a double differential cross-section in the following way:

$$
\left(\frac{d^2\sigma}{dQ^2 d\nu}\right)_{Rosenbluth} = \frac{E_1}{E_3}\left(\frac{d\sigma}{dQ^2}\right) = \frac{4\pi\alpha^2}{Q^4}\left(\frac{E_3}{E_1}\right)\left[A\cos^2\frac{\theta}{2} + B\frac{Q^2}{2M_p^2}\sin^2\frac{\theta}{2}\right]\delta\left(\nu - \frac{Q^2}{2M_p}\right)
$$
$$(16.85)$$

where the Dirac δ function ensures that the expression is valid for the elastic case. For the inelastic case, Q^2 and ν are independent variables while for the elastic case, they are constrained by $Q^2 = 2M_p\nu$. Another difference is that the form factors $A(Q^2)$ and $B(Q^2)$ are only going to be valid for the elastic case.

In 1964 **Sidney Drell**[9] and **J. D. Walecka**[10] published a comprehensive theoretical study on electromagnetic processes with nuclear targets such as nucleons [197]. They considered the inelastic regime, and based on previous work by **J. D. Bjorken**,[11] Von Gehlen, Gourdin, and Hand, defined on general grounds the **two structure functions** W_i $(i = 1, 2)$ which depend on *two* independent kinematical variables. The resulting inelastic scattering cross-section is defined as:

$$
\left(\frac{d^2\sigma}{dQ^2 d\nu}\right)_{inelastic} = \frac{4\pi\alpha^2}{Q^4}\left(\frac{E_3}{E_1}\right)\left[W_2(Q^2, \nu)\cos^2\frac{\theta}{2} + 2W_1(Q^2, \nu)\sin^2\frac{\theta}{2}\right]
$$
$$(16.86)$$

where the structure functions $W_i = W_i(Q^2, \nu)$ depend on the two independent variables (the factor two in front of W_1 is conventional). The two functions depend on the properties of the target nucleon.

Drell and Walecka pointed out that the same two general inelastic form factors appear also in the pair production or Bremsstrahlung events since they characterize the target. The structure functions need to be measured experimentally since they cannot be constructed from theory in the absence of a complete theory of the nucleon (this is still mostly the case today).

---

9   Sidney David Drell (1926–2016), American theoretical physicist.
10   John Dirk Walecka (born 1932), American physicist.
11   James Daniel "BJ" Bjorken (born 1934), American theoretical physicist.

Based on the general discussion of the form factors for a target of finite size (see Section 16.2), the inelastic structure functions were expected to drop with increasing Q^2 as rapidly as the elastic form factors, since this would correspond to the situation of the so-called "mushy proton." **But the experimental results were surprisingly very different!** Inelastic collisions indicated a lot of events at large scattering angles.

In 1969 the SLAC-MIT experiment published the measured cross-sections for inelastic scattering of electrons from hydrogen for incident energies from 7 to 17 GeV at scattering angles of 6 to 10° covering a range of squared four-momentum transfers up to 7.4 GeV² [195]. Some of this data is reported in Figure 16.9.

$$ep \to eX \text{ scattering}$$

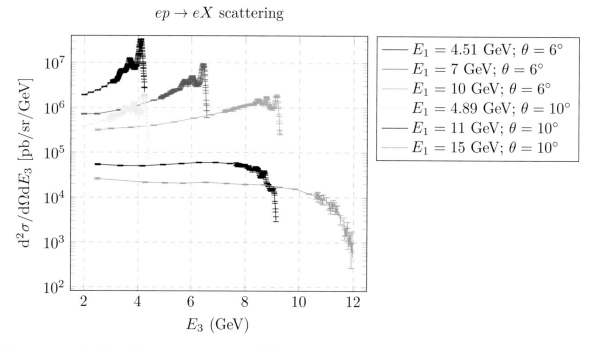

Figure 16.9 Double differential cross-section $\mathrm{d}^2\sigma/\mathrm{d}\Omega\mathrm{d}E_3$ in ep scattering. Data reprinted with permission from E. D. Bloom *et al.*, "High-energy inelastic $e - p$ scattering at 6 and 10 deg," *Phys. Rev. Lett.*, vol. 23, pp. 930–934, Oct 1969. Copyright 1969 by the American Physical Society.

Each measured cross-section as a function of E_3 can be mapped directly to a single measurement for a given hadronic invariant mass W, since with Eq. (16.76) we obtain:

$$W^2(E_1, E_3, \theta) = M_p^2 + 2M_p\nu - Q^2 \approx M_p^2 + 2M_p(E_1 - E_3) + 4E_1E_3\sin^2(\theta/2) \tag{16.87}$$

Consequently, the measured differential cross-section $\mathrm{d}^2\sigma/\mathrm{d}\Omega\mathrm{d}E_3$ plotted versus the hadronic invariant mass W can be readily plotted, as shown in Figure 16.10, and data collected at various initial electron energies E_1 and scattering angles θ can be directly compared. A very important outcome can be seen: up to about a hadronic invariant mass $W = 1.8$ GeV, there is a structure in the cross-section corresponding to the production of the known hadronic resonances. **Above 1.8 GeV there is no structure, and this corresponds to the deep inelastic regime.**

In the inelastic continuum, the behavior of the differential cross-section was found to be very surprising. Although the elastic cross-section falls very fast compared to scattering from a point charge, **the inelastic form factors stay large, indicating that there are point-like objects inside the nucleon!** See Figure 16.11. Such objects were called "partons" and will be discussed in Section 16.8 and in the next chapters.

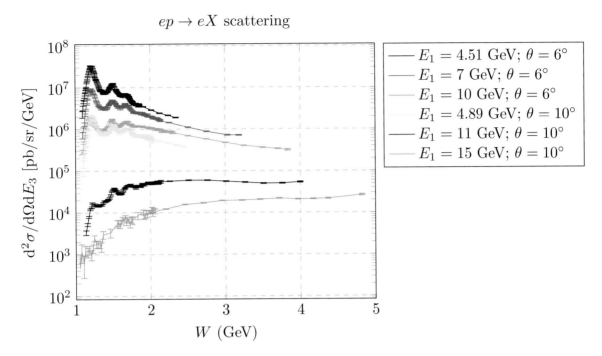

Figure 16.10 Double differential cross-section $\mathrm{d}^2\sigma/\mathrm{d}\Omega\mathrm{d}E_3$ in ep scattering plotted as a function of the hadronic invariant mass W. Data reprinted with permission from E. D. Bloom *et al.*, "High-energy inelastic $e-p$ scattering at 6 and 10 deg," *Phys. Rev. Lett.*, vol. 23, pp. 930–934, Oct 1969. Copyright 1969 by the American Physical Society.

16.7   Lorentz-Invariant Kinematics – the Bjorken x and y Variables

In this section, we introduce the Lorentz-invariant definition of the kinematical variables for the scattering processes. In the elastic $e-p$ scattering case, the final state is composed of the scattered electron and the final state proton. For inelastic scattering, the mass of the final-state hadronic system is no longer the proton mass M_p. The final state must however always contain at least one baryon to conserve the total baryon number, which implies that the final-state invariant mass W satisfies, using Eq. (16.76):

$$W^2 \geq M_p^2 \Rightarrow -Q^2 + 2M_p\nu \geq 0 \Rightarrow Q^2 \leq 2M_p\nu \tag{16.88}$$

We have $W^2 = (k')^2 = E_X^2 - |\vec{p}_X|^2$, where E_X is the energy of the hadronic system. We define the following **kinematical regimes**, which correspond to the physical observations:

$$\begin{cases} W = M_p \longrightarrow \text{elastic} \\ 1 \lessapprox W \lessapprox 2 \text{ GeV} \longrightarrow \text{inelastic isobar resonance excitation} \\ W \gtrapprox 2 \text{ GeV} \longrightarrow \text{deep inelastic} \end{cases} \tag{16.89}$$

For inelastic scattering we have seen that we need two independent variables in order to completely describe the differential cross-section. We have previously used ν and Q^2, which have the units of energy and (energy)^2. All the scattering processes considered so far can be expressed in a similar fashion as a double differential

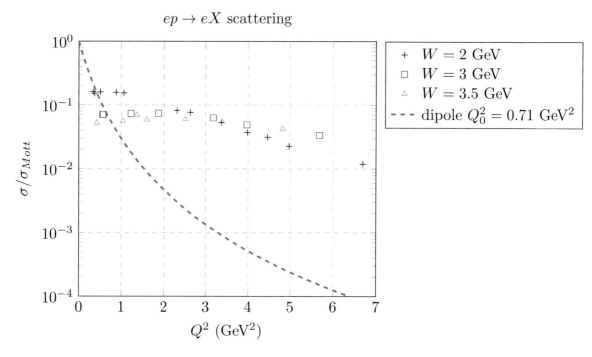

Figure 16.11 Measured total cross-section σ normalized to the Mott cross-section vs. Q^2. There is a striking difference between the data at the chosen W values and the elastic cross-section illustrated by the dipole form. No errors on the data are displayed. Data reprinted with permission from M. Breidenbach *et al.*, "Observed behavior of highly inelastic electron–proton scattering," *Phys. Rev. Lett.*, vol. 23, pp. 935–939, Oct 1969. Copyright 1969 by the American Physical Society.

cross-section of the type:

$$\left(\frac{\mathrm{d}^2\sigma}{\mathrm{d}Q^2\mathrm{d}\nu}\right) = \frac{4\pi\alpha^2}{Q^4}\left(\frac{E_3}{E_1}\right)\left[\cdots\right] \tag{16.90}$$

where for the elastic scattering with a point-like Dirac particle, which we take here to be for instance the muon, we have:

$$\left[\cdots\right]_{e\mu\to e\mu} = \left[\cos^2\frac{\theta}{2} + \frac{Q^2}{2m_\mu^2}\sin^2\frac{\theta}{2}\right]\delta\left(\nu - \frac{Q^2}{2m_\mu}\right) \tag{16.91}$$

while for the elastic scattering off a proton we would write:

$$\left[\cdots\right]_{ep\to ep} = \left[\frac{G_E^2 + \tau G_M^2}{1 + \tau}\cos^2\frac{\theta}{2} + 2\tau G_M^2\sin^2\frac{\theta}{2}\right]\delta\left(\nu - \frac{Q^2}{2M_p}\right) \tag{16.92}$$

where the electric $G_E(Q^2)$ and magnetic $G_M(Q^2)$ form factors depend on Q^2. And finally, for the inelastic case, we would have:

$$\left[\cdots\right]_{ep\to eX} = \left[W_2(Q^2,\nu)\cos^2\frac{\theta}{2} + 2W_1(Q^2,\nu)\sin^2\frac{\theta}{2}\right] \tag{16.93}$$

where the two structure functions $W_1(Q^2,\nu)$ and $W_2(Q^2,\nu)$ in general depend on both kinematical variables.

It is also convenient to introduce the two **dimensionless Bjorken scaling variables** x and y. They are defined as a function of the four-vectors as:

$$x \equiv \frac{-q^2}{2k \cdot q} \qquad \text{and} \qquad y \equiv \frac{k \cdot q}{p \cdot k} \tag{16.94}$$

and are hence Lorentz invariants. Both x and y are kinematically bounded within the range $[0, 1]$. Indeed, $W^2 = (k + q)^2 = M_p^2 + 2k \cdot q + q^2$, so:

$$Q^2 = M_p^2 + 2k \cdot q - W^2 \longrightarrow Q^2 \leq 2k \cdot q \longrightarrow x \leq 1 \tag{16.95}$$

For $x = 1$ the reaction is *elastic* and for $0 \leq x < 1$ it is *inelastic*. Similarly, since y is the fractional energy loss of the incoming electron, we have:

$$0 \leq y \leq 1 \tag{16.96}$$

We note also that the variables ν and Q^2 can be defined as invariants, since:

$$\nu = \frac{k \cdot q}{M_p} \qquad \text{and} \qquad Q^2 = -q^2 = -(p - p')^2 \tag{16.97}$$

Table 16.2 summarizes the expressions for the four kinematical variables in their Lorentz-invariant form and as a function of the variables in the laboratory frame.

Quantity	Lorentz invariant	Laboratory frame ($k^\mu = (M_p, \vec{0})$)
Energy transferred ν	$\dfrac{k \cdot q}{M_p}$	$E_1 - E_3$
Four-momentum transfer Q^2	$-q^2 = -(p - p')^2$	$\simeq 4E_1 E_3 \sin^2 \theta/2$
Bjorken x (x in $[0,1]$)	$\dfrac{-q^2}{2k \cdot q}$	$\dfrac{Q^2}{2M_p \nu}$
Bjorken y (y in $[0,1]$)	$\dfrac{k \cdot q}{p \cdot k}$	$\dfrac{\nu}{E_1} = 1 - \dfrac{E_3}{E_1}$

Table 16.2 Commonly used Lorentz-invariant and laboratory frame kinematical variables.

We note that:

$$x = \frac{Q^2}{2M_p \nu} = \frac{Q^2}{2M_p y E_1} \longrightarrow Q^2 = 2M_p E_1 xy \tag{16.98}$$

Useful quantities are found depending on the ways we express the product xy as shown below, and then the variable y can be conveniently expressed as:

$$xy = \frac{Q^2}{2M_p \nu} \frac{\nu}{E_1} = \frac{Q^2}{2M_p E_1} = \frac{Q^2}{s - M_p^2} \qquad \text{and} \qquad y = \frac{Q^2}{x(s - M_p^2)} \tag{16.99}$$

where we have neglected the electron rest mass. We hence find for $s \gg M_p^2$:

$$xy \approx \frac{4E_1 E_3 \sin^2 \frac{\theta}{2}}{2M_p E_1} = \frac{2E_3}{M_p} \sin^2 \frac{\theta}{2} \qquad \text{and} \qquad y \approx \frac{Q^2}{xs} \tag{16.100}$$

The Jacobian for the transformation from the (ν, Q^2) to the (x, y) variables is:

$$\frac{\mathrm{d}^2\sigma}{\mathrm{d}x\,\mathrm{d}y} = \frac{\mathrm{d}^2\sigma}{\mathrm{d}Q^2\,\mathrm{d}\nu} \cdot \left| \frac{\partial(Q^2, \nu)}{\partial(x, y)} \right| \tag{16.101}$$

where on the r.h.s. we have the absolute value of the Jacobi determinant. The **Jacobi determinant** is equal to:

$$\frac{\partial(Q^2, \nu)}{\partial(x, y)} = \begin{vmatrix} \dfrac{\partial Q^2}{\partial x} & \dfrac{\partial Q^2}{\partial y} \\[2mm] \dfrac{\partial \nu}{\partial x} & \dfrac{\partial \nu}{\partial y} \end{vmatrix} = \begin{vmatrix} 2M_p E_1 y & 2M_p E_1 x \\ 0 & E_1 \end{vmatrix} = 2M_p E_1^2 y \tag{16.102}$$

Thus

$$\mathrm{d}Q^2\mathrm{d}\nu \longrightarrow \left(2M_p E_1^2 y\right)\mathrm{d}x\mathrm{d}y \tag{16.103}$$

The differential cross-section, Eq. (16.86), can then be rearranged as follows:

$$\left(\frac{\mathrm{d}^2\sigma}{\mathrm{d}x\mathrm{d}y} \right)_{inelastic} = \frac{\pi\alpha^2 M_p y}{2E_1 E_3 \sin^4(\theta/2)} \left[W_2 \cos^2 \frac{\theta}{2} + 2W_1 \sin^2 \frac{\theta}{2} \right] \tag{16.104}$$

where the structure functions W_i are functions of any two independent kinematical variables, e.g., (Q^2, ν) or (x, y).

16.8 Bjorken Scaling (1969)

The obvious way to determine if the proton is built up out of point-like particles is to compare the way the cross-section behaves, since it depends heavily on the proton structure functions.

In the case of elastic scattering, the two structure functions can be directly found by comparing Eq. (16.86) with Eq. (16.85):

$$\begin{cases} 2W_1^{elastic}(Q^2, \nu) = B\dfrac{Q^2}{2M_p^2}\delta\left(\nu - \dfrac{Q^2}{2M_p}\right) = 2\tau G_M^2 \delta\left(\nu - \dfrac{Q^2}{2M_p}\right) \\[4mm] W_2^{elastic}(Q^2, \nu) = A\delta\left(\nu - \dfrac{Q^2}{2M_p}\right) = \left(\dfrac{G_E^2 + \tau G_M^2}{1 + \tau}\right)\delta\left(\nu - \dfrac{Q^2}{2M_p}\right) \end{cases} \tag{16.105}$$

where the form factors G_E and G_M are evaluated at $\nu = Q^2/2M_p$.

The structure functions for scattering off a Dirac "point-like" proton are recovered by setting $A(Q^2) = B(Q^2) = 1$, or

$$\begin{cases} 2M_p W_1^{Dirac\ pointlike}(Q^2, \nu) = \dfrac{Q^2}{2M_p}\delta\left(1 - \dfrac{Q^2}{2M_p\nu}\right) = \nu\delta\left(1 - \dfrac{Q^2}{2M_p\nu}\right) \\[4mm] \nu W_2^{Dirac\ pointlike}(Q^2, \nu) = \nu\delta\left(1 - \dfrac{Q^2}{2M_p\nu}\right) \end{cases} \tag{16.106}$$

Hence for point-like scattering, the two structure functions do not depend on ν and Q^2 separately but only on their ratio Q^2/ν. The fact that the structure functions at very high energies become less dependent on Q^2 and only dependent on a single variable, was predicted by Bjorken in the late 1960s. He reasoned that, if large high-energy photons (leading to high q^2 events) resolved point-like constituents in the proton, the structure functions would scale with the dimensionless variable ω defined as:

$$\omega \equiv \frac{2k \cdot q}{-q^2} = \frac{2M_p\nu}{Q^2} = \frac{1}{x} \tag{16.107}$$

Experimentally one can consider:

$$\omega' \equiv 1 + \frac{W^2}{Q^2} = \frac{Q^2 + W^2}{Q^2} = \frac{M_p^2 + 2M_p\nu}{Q^2} \approx \omega \tag{16.108}$$

in the region $\nu \gg M_p/2$. Bjorken conjectured that in the limit of Q^2 and ν approaching infinity, with the ratio ω held fixed, the two quantities νW_2 and W_1 become functions of ω only. That is:

$$\begin{cases} M_p W_1(Q^2, \nu) \to F_1(\omega) \\ \nu W_2(Q^2, \nu) \to F_2(\omega) \end{cases} \quad \text{when} \quad \begin{array}{c} \nu \to \infty \\ Q^2 \to \infty \\ \omega \text{ fixed} \end{array} \tag{16.109}$$

The F_1 and F_2 structure functions[12] depend only on the dimensionless variable and hence no energy scale is present. They are independent of Q^2 for fixed values of ω.

Starting from Eq. (16.104) and noting $\nu = yE_1$, we can rewrite the differential electron–proton scattering cross-section using Bjorken's structure functions F_1^{ep} and F_2^{ep}:

$$\left(\frac{\mathrm{d}^2\sigma}{\mathrm{d}x\mathrm{d}y}\right)_{ep} = \frac{\pi\alpha^2}{2E_1E_3\sin^4(\theta/2)}\left[\underbrace{F_2^{ep}\frac{M_p}{E_1}\cos^2\frac{\theta}{2}}_{e^- \text{ spin overlap}} + \underbrace{2yF_1^{ep}\sin^2\frac{\theta}{2}}_{\text{spin–spin}}\right] \tag{16.110}$$

and since $Q^2 = 4E_1E_3\sin^2\theta/2$, we have:

$$\sin^2\left(\frac{\theta}{2}\right) = \frac{Q^2}{4E_1E_3} = \frac{2M_pE_1xy}{4E_1E_3} = \frac{M_pxy}{2E_3} \tag{16.111}$$

and also $y = 1 - E_3/E_1$ thus $E_3/E_1 = 1 - y$, so finally:

$$\begin{aligned} \left(\frac{\mathrm{d}^2\sigma}{\mathrm{d}x\mathrm{d}y}\right)_{ep} &= \frac{\pi\alpha^2}{2E_1E_3\sin^4(\theta/2)}\left[F_2^{ep}\frac{M_p}{E_1}(1 - \frac{M_pxy}{2E_3}) + 2yF_1^{ep}\frac{M_pxy}{2E_3}\right] \\ &= \frac{8\pi\alpha^2 M_pE_1}{16E_1^2E_3^2\sin^4(\theta/2)}\left[(1 - y - \frac{M_p}{2E_1}xy)F_2^{ep} + \frac{y^2}{2}2xF_1^{ep}\right] \\ &= \frac{8\pi\alpha^2 M_pE_1}{Q^4}\left[(1 - y - \frac{M_p}{2E_1}xy)F_2^{ep} + \frac{y^2}{2}2xF_1^{ep}\right] \end{aligned} \tag{16.112}$$

which can be written, replacing $\mathrm{d}x\mathrm{d}y$ by $(2M_pE_1x)^{-1}\mathrm{d}x\mathrm{d}Q^2$, as:

$$\left(\frac{\mathrm{d}^2\sigma}{\mathrm{d}x\mathrm{d}Q^2}\right)_{ep} = \frac{4\pi\alpha^2}{Q^4}\frac{1}{x}\left[\underbrace{(1 - y - \frac{M_p}{2E_1}xy)F_2^{ep}}_{e^- \text{ spin overlap}} + \underbrace{\frac{y^2}{2}2xF_1^{ep}}_{\text{spin–spin}}\right] \tag{16.113}$$

The reason for writing the spin–spin term as a function of $y^2/2$ will become apparent later in the quark–parton model (see Chapter 17). At high energy where $E_1 \gg M_p$, the expression simply becomes:

$$\left(\frac{\mathrm{d}^2\sigma}{\mathrm{d}x\mathrm{d}Q^2}\right)_{ep} = \frac{4\pi\alpha^2}{Q^4}\frac{1}{x}\left[(1 - y)F_2^{ep}(x, Q^2) + \frac{y^2}{2}2xF_1^{ep}(x, Q^2)\right] \tag{16.114}$$

---

12   Unfortunately, historically labeled in the same way as the Dirac and Pauli form factors.

where we reintroduced dependence of the structure functions on the kinematics with the variables (x, Q^2). Although we did derive explicit expressions for F_1^{ep} or F_2^{ep}, the obtained result for $\mathrm{d}^2\sigma/\mathrm{d}x\mathrm{d}Q^2$ makes a striking prediction, since it factorizes the kinematics and the photon propagator dependence in the cross-section, while leaving the terms in brackets to study the properties of F_1^{ep} and F_2^{ep} as a function of x and Q^2. It basically allowed us to factorize the effect of the probe (the photon) as predicted by QED from the target (the proton), which is a dynamical object described by the structure functions.

Scaling behavior, which implies a dependence on x only and not on Q^2, was first investigated at high energy in the SLAC-MIT experiment using hydrogen and deuterium targets. Scaling was not expected in the kinematical regime where resonance production dominates. Nor is scaling expected at low Q^2. In Figure 16.12, the experimental values for the quantity $F_2^{ep} = \nu W_2$ are plotted vs. Q^2 for a fixed value of $\omega = 4$ obtained at different scattering angles $\theta = 6°$, $10°$, $18°$, and $26°$.

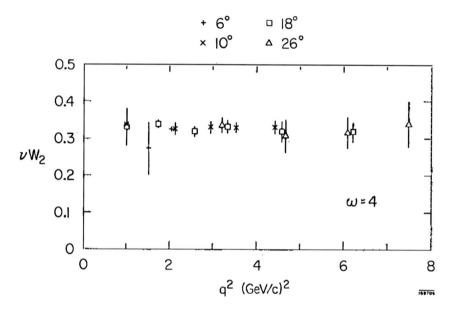

Figure 16.12 The quantity νW_2 vs. Q^2 for the proton. Reprinted figure from E. D. Bloom *et al.*, "Recent results in inelastic electron scattering," in *15th International Conference on High Energy Physics*, 9 (1970).

In conclusion, if scattering in some kinematical regime was caused by point-like constituents, the structure functions for fixed ω would be independent of Q^2 in that region. This is the analog to the behavior of the momentum transfer in the Rutherford scattering, which led to the identification of the atomic nuclei. This was the main argument that the photon is interacting with point-like "partons." The SLAC-MIT data provided first evidence for the existence of a point-like substructure inside the proton. However, a good part of the community at that time did not believe that such partons (quarks) represented real physical objects.

Problems

Ex 16.1 The right electron energy. The proton has a size of about 1 fm. Discuss what is the "optimal" energy of an electron beam needed to study a system of such dimensions.

Ex 16.2 The most general form factor. In the lowest-order QED the electron–proton scattering $e^-(p)p(k) \to$

$e^-(p')p(p')$, the spin-averaged amplitude is given by:

$$\langle |\mathcal{M}|^2 \rangle = \frac{e^4}{q^4} L^{\mu\nu}_{electron} W_{\mu\nu,proton} \tag{16.115}$$

where $W_{\mu\nu,proton}$ is an unknown function describing the photon–proton vertex, and

$$L^{\mu\nu}_{electron} = 2\left(p^\mu \cdot p'^\nu + p^\nu \cdot p'^\mu + g^{\mu\nu}[m^2 - (p \cdot p')] \right) \tag{16.116}$$

The most general form of the $W_{\mu\nu}$ tensor is given by (neglecting the W_3 term which enters into neutrino–proton scattering):

$$W^{\mu\nu} = -W_1 g^{\mu\nu} + \frac{W_2}{M_p^2} k^\mu k^\nu + \frac{W_4}{M_p^2} q^\mu q^\nu + \frac{W_5}{M_p^2} (k^\mu q^\nu + k^\nu q^\mu) \tag{16.117}$$

(a) Prove that $q_\mu W^{\mu\nu} = 0$ and $q_\mu L^{\mu\nu} = 0$.

(b) Prove that it follows from (a) that the following are true:

$$W_4 = \frac{M_p^2}{q^2} W_1 + \frac{1}{4} W_2 \qquad \text{and} \qquad W_5 = \frac{1}{2} W_2 \tag{16.118}$$

that is, $W^{\mu\nu}$ can be expressed in terms of just two functions, $W_1(q^2)$ and $W_2(q^2)$.

(c) Derive the spin-averaged amplitude by contracting Eq. (16.115), substituting the obtained expression for $W^{\mu\nu}$.

Ex 16.3 Form factors for a Dirac proton. Consider a Dirac proton: starting from the assumption that $W^{\mu\nu}_{Dirac\,p}$ has the same form as $L^{\mu\nu}_{electron}$, calculate the form factors $W_{1,Dirac\,p}(q^2)$ and $W_{2,Dirac\,p}(q^2)$.

17   Partons

I have difficulty in writing this note because it is not in the nature of a deductive paper, but is the result of an induction. I am more sure of the conclusions than of any single argument which suggested them to me for they have an internal consistency which surprises me and exceeds the consistency of my deductive arguments which hinted at their existence.

Richard Feynman[1]

17.1   The Eightfold Way (1962)

In 1961 M. Gell-Mann published a report entitled *The Eightfold way: a theory of strong interactions symmetry.* He wrote:

> *"It has seemed likely for many years that the strongly interacting particles, grouped as they are into isotopic multiplets, would show traces of a higher symmetry that is somehow broken. Under the higher symmetry, the eight familiar baryons would be degenerate and form a super-multiplet. As the higher symmetry is broken, the Ξ, Λ, Σ, and N would split apart, leaving inviolate only the conservation of isotopic spin, of strangeness, and of baryons. Of these three, the first is partially broken by electromagnetism and the second is broken by the weak interactions. Only the conservation of baryons and of electric charge are absolute."[2]*

The assumption is hence that as a consequence of an internal symmetry of the strong force, we expect that the eight known spin-1/2 baryons Ξ, Λ, Σ, n, and p would be degenerate. This is to be understood as a generalization of the concept that the proton and the neutron are degenerate and form the nucleon under the isospin symmetry. A natural classification of particles (quantum systems) that obey symmetries, is to put them into **irreducible representations of the corresponding symmetry groups** or **multiplets**. It is to be assumed that in the context of the hadrons these multiplets are organized according to the symmetries of isospin, strangeness, and baryon number.

- $SU(2)$ **isospin** $I = 1/2$. The symmetry group for isospin was found to be $SU(2)$. To describe the nucleon system, we used the two-dimensional representation of $SU(2)$ in analogy to the spin-1/2 case. A **nucleon state** χ was then represented in the following basis:

$$\chi \equiv \begin{pmatrix} p \\ n \end{pmatrix} \quad \text{with} \quad |p\rangle = \begin{pmatrix} 1 \\ 0 \end{pmatrix} \quad \text{and} \quad |n\rangle = \begin{pmatrix} 0 \\ 1 \end{pmatrix} \tag{17.1}$$

---

1   Reprinted excerpt with permission, from R. P. Feynman, "Very high-energy collisions of hadrons," *Phys. Rev. Lett.*, vol. 23, pp. 1415–1417, Dec 1969. Copyright 1969 by the American Physical Society.

2   Reprinted from M. Gell-Mann, *The eightfold way: A theory of strong interaction symmetry*, Report CTSL-20, California Institute of Technology, March 1961.

The isospin transformation (i.e., the iso-rotation) transforms χ into χ' as:

$$\chi' \equiv \begin{pmatrix} p' \\ n' \end{pmatrix} = U \begin{pmatrix} p \\ n \end{pmatrix} = \begin{pmatrix} U_{11} & U_{12} \\ U_{21} & U_{22} \end{pmatrix} \begin{pmatrix} p \\ n \end{pmatrix} \tag{17.2}$$

where U is a 2×2 unitary matrix of the $SU(2)$ group (see Appendix B.8) of unitary matrices with determinant equal to unity. The unitary condition ensures that the normalization of the quantum state is preserved during the corresponding transformation. Since U_{ij} can be complex numbers, the $n \times n$ complex matrix U can in general be described by $2n^2$ independent real parameters. The unitarity condition $U^\dagger U = 1$ reduces the number to $2n^2 - n^2 = n^2$ real parameters. In addition, the condition $\det U = 1$ gives one more constraint, hence a $SU(n)$-matrix has $n^2 - 1$ real parameters. For the case $n = 2$, this implies that an $SU(2)$ unitary matrix U can be simply written as:

$$U = \begin{pmatrix} U_{11} & U_{12} \\ -U_{12}^* & U_{11}^* \end{pmatrix} \quad \text{where} \quad \det U = |U_{11}|^2 + |U_{12}|^2 = 1 \tag{17.3}$$

fully constrained by $2^2 - 1 = 3$ independent real parameters. To these correspond the three Hermitian generators of the group which can be defined by their Lie algebra (see Appendix B.8):

$$[\tau_i, \tau_j] = 2i\epsilon_{ijk}\tau_k \quad \text{or} \quad \left[\frac{1}{2}\tau_i, \frac{1}{2}\tau_j\right] = i\epsilon_{ijk}\frac{1}{2}\tau_k \tag{17.4}$$

and any $SU(2)$ transformation can be written as:

$$\begin{pmatrix} p' \\ n' \end{pmatrix} = U(\vec{\alpha}) \begin{pmatrix} p \\ n \end{pmatrix} = e^{i\vec{\alpha}\cdot\vec{\tau}/2} \begin{pmatrix} p \\ n \end{pmatrix} \tag{17.5}$$

where $\vec{\alpha} = (\alpha_1, \alpha_2, \alpha_3)$. For the fundamental generators τ_i of the two-dimensional representation of $SU(2)$, we can choose the Pauli matrices:

$$\tau_1 = \begin{pmatrix} 0 & 1 \\ 1 & 0 \end{pmatrix}, \quad \tau_2 = \begin{pmatrix} 0 & -i \\ i & 0 \end{pmatrix}, \quad \tau_3 = \begin{pmatrix} 1 & 0 \\ 0 & -1 \end{pmatrix} \tag{17.6}$$

Each generator corresponds to an observable, but since they do not commute among each other, they define a set of incompatible variables. Hence, the states are defined in terms of the third component of the isospin and of the total isospin. If we define the physical isospin operators as $I_i = \tau_i/2$, then the isomorphism with the $SO(3)$ group is explicit (note the absence of factor two in the commutation rules):

$$I_i \equiv \frac{\tau_i}{2}, \quad I^2 \equiv I_1^2 + I_2^2 + I_3^2 \quad \text{with} \quad [I_i, I_j] = i\epsilon_{ijk}I_k \tag{17.7}$$

The graphical representation of the nucleon doublet was shown in Figure 15.12.

We define the corresponding isospin raising and lowering operators $I_+ \equiv (\tau_1 + i\tau_2)/2$ and $I_- \equiv (\tau_1 - i\tau_2)/2$, also called **ladder operators**, and then express an infinitesimal $SU(2)$ transformation as:

$$U = \mathbb{1} + \sum \frac{i}{2}\alpha_i\tau_i = \mathbb{1} + i\left(I_+\frac{\alpha_1 - i\alpha_2}{2} + I_-\frac{\alpha_1 + i\alpha_2}{2} + I_3\alpha_3\right) \tag{17.8}$$

We can therefore express and compute any $SU(2)$ transformation with the help of the isospin raising and lowering operators and the I_3 operator. In matrix form, we have:

$$I_+ = \begin{pmatrix} 0 & 1 \\ 0 & 0 \end{pmatrix}; \quad I_- = \begin{pmatrix} 0 & 0 \\ 1 & 0 \end{pmatrix}; \quad I_3 = \frac{1}{2}\begin{pmatrix} 1 & 0 \\ 0 & -1 \end{pmatrix} \tag{17.9}$$

- $SU(2)$ **isospin** $I = 1$. In order to describe the $I = 1$ isospin (for example for the pion), we look for a **three-dimensional representation of** $SU(2)$, hence for a set of 3×3 matrices that follow the $SU(2)$ Lie algebra. We seek in this case:

$$[S_i, S_j] = i\epsilon_{ijk}S_k \quad \text{where} \quad S_{i,j,k} = 3 \times 3 \text{ matrices} \tag{17.10}$$

A possible set is given by the following 3×3 traceless matrices corresponding to the generators of the three-dimensional representation of the $SU(2)$ group:

$$S_1 \equiv \frac{1}{\sqrt{2}} \begin{pmatrix} 0 & -1 & 0 \\ -1 & 0 & 1 \\ 0 & 1 & 0 \end{pmatrix}; \quad S_2 \equiv \frac{i}{\sqrt{2}} \begin{pmatrix} 0 & 1 & 0 \\ -1 & 0 & -1 \\ 0 & 1 & 0 \end{pmatrix}; \quad S_3 \equiv \begin{pmatrix} 1 & 0 & 0 \\ 0 & 0 & 0 \\ 0 & 0 & -1 \end{pmatrix} \tag{17.11}$$

Just as in the case of the two-dimensional representation of $SU(2)$, we still only have one diagonal matrix S_3 with the following eigenvalues: $-1, 0, 1$. The three-dimensional representation of $SU(2)$ is for example adequate to describe the $SU(2)$ isospin $I = 1$ multiplet of the pion:

$$\pi \equiv \begin{pmatrix} \pi^+ \\ \pi^0 \\ \pi^- \end{pmatrix} \quad \text{pion multiplet of the SU(2) isospin group} \tag{17.12}$$

A graphical representation of the multiplet is shown in Figure 17.1, where the effect of the isospin ladder operators $I_+ \equiv I_1 + iI_2$ and $I_- \equiv I_1 - iI_2$ has been represented with the dashed lines.

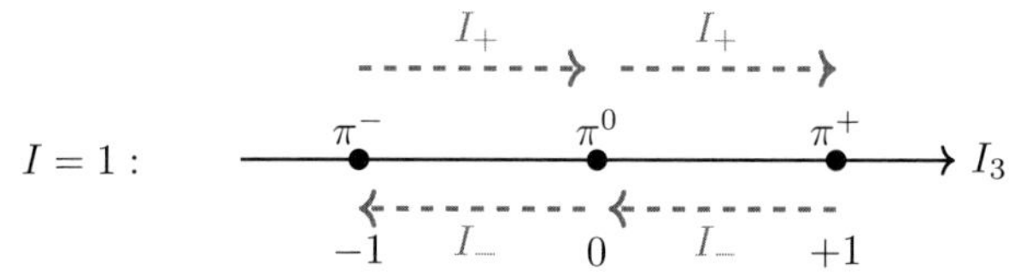

Figure 17.1 Representation of the pion as an isospin $I = 1$ multiplet in isospin space.

- $SU(3)$ **flavor; including strangeness.** In order to describe all hadrons discussed in the previous chapters, we need to extend the symmetry group to include isospin as well as strangeness and baryonic number. The Gell-Mann/Nishijima relation can be written as:

$$Q = I_3 + \frac{B}{2} + \frac{S}{2} = I_3 + \frac{B + S}{2} \tag{17.13}$$

It is practical to define a new derived quantity called the **hypercharge** Y:

$$Y \equiv B + S \tag{17.14}$$

If we assume that the baryonic number B is strictly conserved, then the conservation of the hypercharge follows from that of strangeness, so both are equivalent. With this definition, the Gell-Mann/Nishijima relation can be simply expressed as:

$$Q = I_3 + \frac{Y}{2} \quad \text{or} \quad Y = 2(Q - I_3) \tag{17.15}$$

We can then derive the following quantum numbers for the spin-1/2 baryons:

$$\begin{cases} \text{Nucleon}: n, p & I_3 = \pm\frac{1}{2}; Q = 0, 1 \longrightarrow Y = 1 \\ \text{Strange-baryons}: \Sigma^+, \Sigma^0, \Sigma^- & I_3 = \pm 1, 0; Q = \pm 1, 0 \longrightarrow Y = 0 \\ \Lambda^0: & I_3 = 0; Q = 0 \longrightarrow Y = 0 \\ \Xi^-, \Xi^0: & I_3 = \pm\frac{1}{2}; Q = -1, 0 \longrightarrow Y = -1 \end{cases} \tag{17.16}$$

Hence, the strong force symmetry is expressed with the isospin and the hypercharge. **The isospin group is $SU(2)$ and adding hypercharge requires the extension to the $SU(3)$ group** (see Appendix B.9). The fundamental representation of $SU(3)$ is three-dimensional and we can express it with a set of 3×3 unitary matrices (with determinant unity). Such a set has $3^2 - 1 = 8$ generators, of which two independent ones are diagonal generators. In analogy to $SU(2)$, we write:

$$U(\alpha_a) \equiv \exp\left(i \sum_{a=1}^{8} \alpha_a \lambda_a / 2 \right) \qquad (17.17)$$

where α_a $(a = 1, \ldots, 8)$ are eight real parameters. As before, we define the charge raising and lowering operators $I_+ \equiv (\lambda_1 + i\lambda_2)/2$ and $I_- \equiv (\lambda_1 - i\lambda_2)/2$ and one can directly check that:

$$I_+ \begin{pmatrix} 0 \\ 1 \\ 0 \end{pmatrix} = \begin{pmatrix} 1 \\ 0 \\ 0 \end{pmatrix} \implies I_+ = \begin{pmatrix} 0 & 1 & 0 \\ 0 & 0 & 0 \\ 0 & 0 & 0 \end{pmatrix} \qquad (17.18)$$

and

$$I_- \begin{pmatrix} 1 \\ 0 \\ 0 \end{pmatrix} = \begin{pmatrix} 0 \\ 1 \\ 0 \end{pmatrix} \implies I_- = \begin{pmatrix} 0 & 0 & 0 \\ 1 & 0 & 0 \\ 0 & 0 & 0 \end{pmatrix} \qquad (17.19)$$

as well as:

$$I_+ \begin{pmatrix} 1 \\ 0 \\ 0 \end{pmatrix} = 0; \quad I_+ \begin{pmatrix} 0 \\ 0 \\ 1 \end{pmatrix} = 0 \quad \text{and} \quad I_- \begin{pmatrix} 0 \\ 1 \\ 0 \end{pmatrix} = 0; \quad I_- \begin{pmatrix} 0 \\ 0 \\ 1 \end{pmatrix} = 0 \qquad (17.20)$$

The third component is then given by $I_3 = \frac{1}{2}\lambda_3$, so with:

$$I_3 \begin{pmatrix} 1 \\ 0 \\ 0 \end{pmatrix} = +\frac{1}{2} \begin{pmatrix} 1 \\ 0 \\ 0 \end{pmatrix}; \quad I_3 \begin{pmatrix} 0 \\ 1 \\ 0 \end{pmatrix} = -\frac{1}{2} \begin{pmatrix} 0 \\ 1 \\ 0 \end{pmatrix}; \quad \text{and} \quad I_3 \begin{pmatrix} 0 \\ 0 \\ 1 \end{pmatrix} = 0 \qquad (17.21)$$

we obtain:

$$\lambda_1 = I_+ + I_- = \begin{pmatrix} 0 & 1 & 0 \\ 1 & 0 & 0 \\ 0 & 0 & 0 \end{pmatrix}; \quad \lambda_2 = -i(I_+ - I_-) = \begin{pmatrix} 0 & -i & 0 \\ i & 0 & 0 \\ 0 & 0 & 0 \end{pmatrix}; \quad \lambda_3 = \begin{pmatrix} 1 & 0 & 0 \\ 0 & -1 & 0 \\ 0 & 0 & 0 \end{pmatrix} \qquad (17.22)$$

The matrices λ_1, λ_2, and λ_3 are therefore equal to the Pauli matrices extended by one null row and column. In a similar way, one wants to define four other matrices to define two additional sets of ladder operators, so in total we get the following $SU(3)$ **ladder operators**:

$$I_\pm = \frac{1}{2}(\lambda_1 \pm i\lambda_2); \quad U_\pm = \frac{1}{2}(\lambda_6 \pm i\lambda_7); \quad V_\pm = \frac{1}{2}(\lambda_4 \pm i\lambda_5) \qquad (17.23)$$

The $\lambda_4, \ldots, \lambda_7$ can be obtained by swapping rows and columns of the λ_1 and λ_2 matrices. Their interpretation can be checked in a straightforward way by direct application on a state-vector. Last but not least, one should define λ_8 as a diagonal matrix. Putting these all together, the corresponding traceless matrices λ_i are called

the Gell-Mann matrices λ_a $(a = 1, \ldots, 8)$ – see also Eq. (B.61) in the appendix:

$$
\lambda_1 = \begin{pmatrix} 0 & 1 & 0 \\ 1 & 0 & 0 \\ 0 & 0 & 0 \end{pmatrix}; \quad
\lambda_2 = \begin{pmatrix} 0 & -i & 0 \\ i & 0 & 0 \\ 0 & 0 & 0 \end{pmatrix}; \quad
\lambda_3 = \begin{pmatrix} 1 & 0 & 0 \\ 0 & -1 & 0 \\ 0 & 0 & 0 \end{pmatrix}
$$

$$
\lambda_4 = \begin{pmatrix} 0 & 0 & 1 \\ 0 & 0 & 0 \\ 1 & 0 & 0 \end{pmatrix}; \quad
\lambda_5 = \begin{pmatrix} 0 & 0 & -i \\ 0 & 0 & 0 \\ i & 0 & 0 \end{pmatrix}
$$

$$
\lambda_6 = \begin{pmatrix} 0 & 0 & 0 \\ 0 & 0 & 1 \\ 0 & 1 & 0 \end{pmatrix}; \quad
\lambda_7 = \begin{pmatrix} 0 & 0 & 0 \\ 0 & 0 & -i \\ 0 & i & 0 \end{pmatrix}; \quad
\lambda_8 = \frac{1}{\sqrt{3}} \begin{pmatrix} 1 & 0 & 0 \\ 0 & 1 & 0 \\ 0 & 0 & -2 \end{pmatrix} \tag{17.24}
$$

with the generators $T^a \equiv \lambda_a/2$ $(a = 1, \ldots, 8)$ satisfying the commutation relations:

$$
\left[T^a, T^b \right] = i f^{abc} T^c \quad \text{or} \quad \left[\frac{\lambda^a}{2}, \frac{\lambda^b}{2} \right] = i f^{abc} \frac{\lambda^c}{2} \tag{17.25}
$$

where **the** $SU(3)$ **structure constants** f^{abc} are real and totally antisymmetric. The only non-zero (up to permutations) f^{abc} constants are (see Appendix B.9):

$$
f^{123} = 1, \; f^{147} = -f^{156} = f^{246} = f^{257} = f^{345} = -f^{367} = \frac{1}{2}, \; f^{458} = f^{678} = \frac{\sqrt{3}}{2} \tag{17.26}
$$

The three-dimensional generators of $SU(3)$ operate on three-dimensional column vectors. Clearly, there are three linearly independent vectors, which we may denote by u_i $(i = 1, 2, 3)$. We note that both λ_3 and λ_8 are diagonal. They commute and therefore describe a compatible set of the eigenstates with the following eigenvalues:

$$
\begin{pmatrix} 1 \\ 0 \\ 0 \end{pmatrix} : (\lambda_3, \lambda_8) = \left(1, \frac{1}{\sqrt{3}} \right); \quad
\begin{pmatrix} 0 \\ 1 \\ 0 \end{pmatrix} : (\lambda_3, \lambda_8) = \left(-1, \frac{1}{\sqrt{3}} \right); \quad
\begin{pmatrix} 0 \\ 0 \\ 1 \end{pmatrix} : (\lambda_3, \lambda_8) = \left(0, -\frac{2}{\sqrt{3}} \right) \tag{17.27}
$$

The graphical representation of the three pairs of eigenvalues is shown in the λ_3, λ_8 plane in Figure 17.2. The

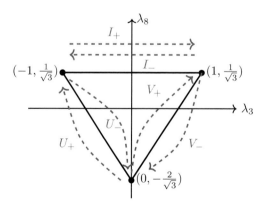

Figure 17.2 Fundamental representation of the $SU(3)$ group.

Casimir operator of $SU(3)$ (defined equivalently to the total isospin of $SU(2)$) is written as:

$$T^2 \equiv \sum_{a=1}^{8} T_a^2 = \frac{1}{4}\sum_{a=1}^{8} \lambda_a^2 = \frac{4}{3}\begin{pmatrix} 1 & 0 & 0 \\ 0 & 1 & 0 \\ 0 & 0 & 1 \end{pmatrix} \tag{17.28}$$

• **Physical interpretation.** Summarizing, the basic assumption of the eightfold way is that there is an underlying $SU(3)$ symmetry of isospin and hypercharge, that can be considered as a generalization of the $SU(2)$ isospin symmetry, postulated with the assignments:

$$I_3 \equiv \frac{\lambda_3}{2} \quad \text{and} \quad Y \equiv \frac{\lambda_8}{\sqrt{3}} \tag{17.29}$$

The eight known **spin-1/2 baryons** can be naturally accommodated in an eight-dimensional irreducible representation (octet) of the group. If the underlying symmetries were exact, then these particles would be degenerate in mass and indistinguishable. The $SU(3)$ transformation converts one state into another within the octet. The electromagnetic interaction breaks the symmetry and allows us to distinguish the states, and so does their rest mass. In particular, while we had already stressed that the masses of the proton and of the neutron are similar at the per-mil level, the differences within the octet are as large as 400 MeV! The model is nonetheless predictive as it allows us to naturally classify the observed spin-1/2 baryons. The **spin-1/2 baryon octet** is represented graphically in Figure 17.3. The masses of the particles (in MeV) are given in brackets.

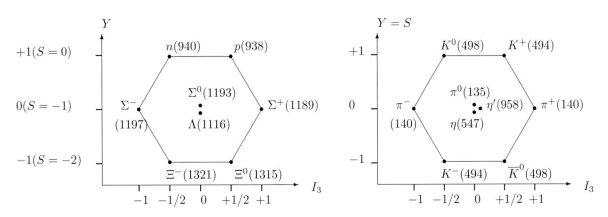

Figure 17.3 The spin-1/2 baryon octet.   **Figure 17.4** The spin-0 meson multiplet.

Gell-Mann also promptly noted that the pion and strange mesons fit into a similar set of eight particles, as shown in Figure 17.4. The **spin-0 mesons (pseudoscalar mesons $J = 0$, $P = -1$)** including the pions and the kaons, as well as the η are assigned to the octet. The existence of the η was actually predicted by the eightfold way.

We note that the π^+ and π^- form a set of particle–antiparticle, similarly for the K^+ and K^- pair. For the π^0 and η, particle and antiparticle are the same and represented as a single entry in the octet. This is possible when the particles are electrically neutral. Note however that this is not always the case. For the neutral kaon, the octet clearly distinguishes the K^0 and $\bar{K}^0$, which are hence not identical although electrically neutral! They are differentiated by their isospin and hypercharge quantum numbers, which are respectively $(I_3, Y) = (-1/2, +1)$ and $(+1/2, -1)$. Such discrimination is not possible for the π^0 and η, which have both $I = Y = 0$.

The predictive power of the eightfold way was further confirmed when it was noted that the known spin-1 vector mesons and spin-3/2 baryons could also be organized respectively in an octet plus singlet (see

Figure 17.5) and a decuplet (see Figure 17.6). When the spin-3/2 Δ, Σ^*, Ξ^* resonances were arranged in the decuplet (see Figure 17.6), one member was still missing, but predicted by the eightfold way model. Gell-Mann called it Ω^-, predicted to be negatively charged with a strangeness $S = -3$ and that the spin-parity should be the same as those of the other members of the decuplet.

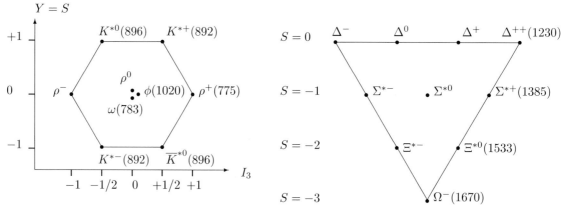

Figure 17.5 The spin-1 vector meson multiplet.

Figure 17.6 Spin-3/2 baryon decuplet.

• **Discovery of the Ω^-.** In 1964 an event compatible with the production of Ω^- was observed in the 80-inch hydrogen bubble chamber at the Brookhaven National Laboratory (BNL, USA) [198]. The bubble chamber photograph and the corresponding line diagram are shown in Figure 17.7. The interpretation of the event is:

$$K^- + p \rightarrow \Omega^- + K^+ + K^0$$
$$\hookrightarrow \Xi^0 \ + \ \pi^-$$
$$\hookrightarrow \Lambda^0 \ + \ \pi^0$$
$$\hookrightarrow p + \pi^- \quad \hookrightarrow \gamma \ + \ \gamma$$
$$\hookrightarrow e^+ e^- \quad \hookrightarrow e^+ e^- \tag{17.30}$$

The prediction of the existence of the Ω^- represented an important milestone towards the acceptance of the eightfold model.

17.2   The Quark Model

The predictions of the eightfold way were very successful and powerful. The spectrum of the known hadrons could be explained as a consequence of the $SU(3)$ symmetry. The experimental look into the jungle of hadronic particles revealed the richness and variety of the hadronic spectrum. Nonetheless, a basic question remained unanswered: why isn't a particle multiplet found which corresponds to the fundamental representation of $SU(3)$ and only octets and decuplets are realized in Nature? From a group theory point of view, such multiplets can be constructed from the $SU(3)$ fundamental and conjugate representations, for example in the following ways

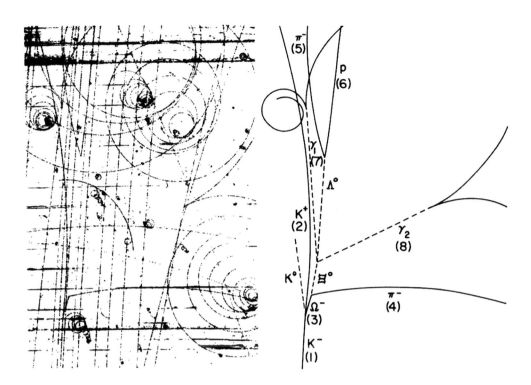

Figure 17.7 Bubble chamber photograph and corresponding line diagram of an event consistent with the production of Ω^-. Reprinted figure from V. E. Barnes *et al.*, "Observation of a hyperon with strangeness minus three," *Phys. Rev. Lett.*, vol. 12, pp. 204–206, Feb 1964. Copyright 1964 by the American Physical Society.

(see Appendix B.12):

$$
\begin{aligned}
3 \otimes 3 &= 6 \oplus \bar{3} \\
3 \otimes \bar{3} &= 8 \oplus 1 \\
3 \otimes 3 \otimes 3 &= 10 \oplus 8 \oplus 8 \oplus 1
\end{aligned}
\tag{17.31}
$$

How can one reconcile this result with the physical reality?

• **Introducing quarks as the fundamental representation of** $SU(3)$. In 1964 **Gell-Mann and Zweig**[3] postulated that hadrons must be composed of sub-elementary particles [199]. Gell-Mann called them "**quarks**." **Quarks are spin-**$1/2$ **point-like Dirac particles** with a baryon number of $B = 1/3$. Initially, three types of quarks with three flavors called upness U, downness D, and strangeness S were postulated:

$$
|u\rangle = \begin{pmatrix} 1 \\ 0 \\ 0 \end{pmatrix}; \quad |d\rangle = \begin{pmatrix} 0 \\ 1 \\ 0 \end{pmatrix}; \quad |s\rangle = \begin{pmatrix} 0 \\ 0 \\ 1 \end{pmatrix}
\tag{17.32}
$$

with the quantum numbers summarized in Table 17.1.

3   George Zweig (born 1937), Russian–American physicist.

Quark flavor	u	d	s	$\bar{u}$	$\bar{d}$	$\bar{s}$
Mass (MeV)	$2.2^{+0.6}_{-0.4}$	$4.7^{+0.5}_{-0.4}$	96^{+8}_{-4}			
Baryonic B	$+1/3$	$+1/3$	$+1/3$	$-1/3$	$-1/3$	$-1/3$
Upness U	$+1$	0	0	-1	0	0
Downness D	0	-1	0	0	$+1$	0
Strangeness S	0	0	-1	0	0	$+1$
Charge Q	$+2/3$	$-1/3$	$-1/3$	$-2/3$	$+1/3$	$+1/3$
Isospin I_3	$+1/2$	$-1/2$	0	$-1/2$	$+1/2$	0
Hypercharge Y	$+1/3$	$+1/3$	$-2/3$	$-1/3$	$-1/3$	$+2/3$
	$SU(2)$		$SU(2)$			
	$SU(3)$			$SU(3)$		

Table 17.1 The quantum numbers of the u, d, and s quarks and corresponding antiquarks.

Choosing the baryon number $1/3$ implies that baryons are composed of three quarks. The relation to isospin is given by:

$$I_3 = \frac{U + D}{2} \tag{17.33}$$

and the Gell-Mann/Nishijima relation can be expressed as a function of the quark quantum numbers as:

$$Q = \frac{U + D + S}{2} + \frac{B}{2} \tag{17.34}$$

We note that:

$$B = \frac{U - D - S}{3} \quad \text{and} \quad Y = S + B = \frac{3S + U - D - S}{3} = \frac{U - D + 2S}{3} \tag{17.35}$$

- **Antiquarks and the conjugate states.** The antiquarks are defined as the following conjugate states:

$$|\bar{u}\rangle = \overline{\begin{pmatrix} 1 \\ 0 \\ 0 \end{pmatrix}}; \quad |\bar{d}\rangle = \overline{\begin{pmatrix} 0 \\ 1 \\ 0 \end{pmatrix}}; \quad |\bar{s}\rangle = \overline{\begin{pmatrix} 0 \\ 0 \\ 1 \end{pmatrix}} \tag{17.36}$$

with the quantum numbers summarized in Table 17.1. We note that all additive quantum numbers should have the opposite sign for antiquarks, e.g., the electric charge, the isospin, the strangeness, hypercharge, etc. From $\bar{I}_3 = -I_3$ and $\bar{Y} = -Y$, we obtain:

$$\bar{\lambda}_3 = -\lambda_3 \quad \text{and} \quad \bar{\lambda}_8 = -\lambda_8 \tag{17.37}$$

For the complex matrices, we should in addition apply complex conjugation, hence, in general, we have:

$$\bar{\lambda}_a = -\lambda_a^* \quad (a = 1, \ldots, 8) \tag{17.38}$$

The $\bar{\lambda}_a$ matrices follow similar commutation rules as the λ_as. One can construct the corresponding ladder operators. For instance, we have:

$$\bar{I}_+ = \frac{1}{2}(\bar{\lambda}_1 + i\bar{\lambda}_2) = \begin{pmatrix} 0 & 0 & 0 \\ -1 & 0 & 0 \\ 0 & 0 & 0 \end{pmatrix} \tag{17.39}$$

Consequently:

$$\bar{I}_+\bar{u} = -\bar{d} \qquad (17.40)$$

with a negative sign.

• **The 3 and $\bar{3}$ representations.** The isospin symmetry has been enlarged to a threefold $SU(3)$ flavor symmetry between u, d, and s quarks. The **fundamental representation of the quark** is called 3, while for an antiquark it is represented by $\bar{3}$ (called the **conjugate representation**). The difference between 3 and $\bar{3}$ can be appreciated from the illustration in Figure 17.8. It can be verified directly from the matrix representation of the ladder

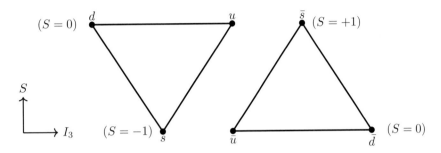

Figure 17.8 Fundamental representations of the $SU(3)$ flavor of (left) quarks and (right) conjugate for antiquarks.

operators, Eq. (17.23), that for the quarks we have (see Figure 17.2):

$$I_+d = u, \quad I_-u = d, \quad U_-d = s, \quad U_+s = d, \quad V_-u = s, \quad \text{and} \quad V_+s = u \qquad (17.41)$$

with all other combinations giving zero (e.g., $I_+u = 0$, $I_-d = 0$, ...). For the antiquarks, we find (with a negative sign):

$$I_+\bar{u} = -\bar{d}, \quad I_-\bar{d} = -\bar{u}, \quad U_+\bar{d} = -\bar{s}, \quad U_-\bar{s} = -\bar{d}, \quad V_+\bar{u} = -\bar{s}, \quad \text{and} \quad V_-\bar{s} = -\bar{u} \qquad (17.42)$$

The $SU(3)$ flavor symmetry is broken by the strange quark mass, which is much larger than the up and down quark masses. However, the strong interactions are almost invariant under this $SU(3)$ symmetry. In this way, we assume that the strong force treats all quark flavors equally. Because the rest masses of the up and down quarks are approximately the same, the strong interaction possesses an approximate isospin symmetry. This can be seen by noting that from the point of view of the strong force, nothing changes in the reaction if all up quarks are replaced by down quarks and vice versa.

• **Hadrons are bound states.** Hadrons are to be considered as bound states of quarks. One of the difficulties of the quark model was to find the reason why free quarks did not seem to be observed in nature. As a matter of fact, elementary particles with fractional electric charge have never been directly observed. Nonetheless, the quark model was successful in predicting the classification of hadrons in octets and decuplets. Today the absence of free quarks is understood in the context of "**color confinement,**" which can be shown to be a consequence of quantum chromodynamics (QCD), as will be described in Section 18.13.

Since quarks carry a 1/3 baryon number, a baryon is a bound state of three quarks $B \equiv qqq$, while an antibaryon is a state of three antiquarks $\bar{B} \equiv \bar{q}\bar{q}\bar{q}$. Mesons have a vanishing baryon number and are therefore bound states of a quark–antiquark pair $M \equiv q\bar{q}$. These are by far the dominant and naturally arising combinations in nature. In principle, other bound states could be possible as long as they are "colorless," as will be described later. The existence of a pentaquark state was predicted. A **pentaquark** is a particle state consisting of four quarks and one antiquark bound together, and experimental evidence for pentaquarks has been collected in the recent years.

17.2.1 Quark Model for Light Mesons States

We can build a meson by combining the fundamental three-dimensional representation of $SU(3)$ for a quark and its conjugate for an antiquark. Group theory tells us that the combination of 3 and $\bar{3}$ yields an octet and a singlet (see Appendix B.12), and in its formalism, this reads simply:

$$3 \otimes \bar{3} = 8 \oplus 1 \tag{17.43}$$

This operation is graphically shown in Figure 17.9. Under the $SU(3)$ transformation, all members of the octet will transform into each other. The singlet state remains unchanged under the transformation (it is an invariant of $SU(3)$).

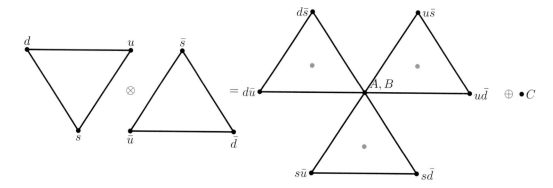

Figure 17.9 Illustration of the group operation $3 \otimes \bar{3} = 8 \oplus 1$ giving an $SU(3)$ octet plus a singlet.

The quark contents for the octet are uniquely defined for the six states to be $d\bar{s}$, $u\bar{s}$, $d\bar{u}$, $u\bar{d}$, $s\bar{u}$, and $s\bar{d}$, while the three states A, B, and C have the same quantum numbers $Y = I_3 = 0$ and can be realized in Nature as "mixed" states. The C state is a singlet; hence it must be symmetric under an $SU(3)$ transformation. A possible symmetric and normalized state which transforms into itself is given by:

$$C \equiv \frac{1}{\sqrt{3}} \left(u\bar{u} + d\bar{d} + s\bar{s} \right) \tag{17.44}$$

By inspection, we also can define A as the state belonging to the $SU(2)$ isospin triplet and B as the isospin singlet. In analogy to $SU(2)$ spin, we associate the spin-up with the quark state u and the spin-down with the d quark. For antiquarks, the association becomes $-\bar{d}$ and $\bar{u}$ (where the negative sign is a conventional phase). Accordingly, the $SU(2)$ isospin triplet can be written as:

$$\text{isospin } I = 1 \text{ triplet}: \quad \begin{cases} |1, +1\rangle = - u\bar{d} \\ |1, 0\rangle = \frac{1}{\sqrt{2}} \left(u\bar{u} - d\bar{d} \right) \\ |1, -1\rangle = d\bar{u} \end{cases} \tag{17.45}$$

which can also be directly verified with the ladder operators, since:

$$I_+(d\bar{u}) = (I_+ d)\bar{u} + d(I_+ \bar{u}) = u\bar{u} - d\bar{d} \quad \text{and} \quad I_-(-u\bar{d}) = -(I_- u)\bar{d} - u(I_- \bar{d}) = -d\bar{d} + u\bar{u} \tag{17.46}$$

We can then identify A as the $I_3 = 0$ member of the isospin $I = 1$ triplet:

$$A \equiv |1, 0\rangle = \frac{1}{\sqrt{2}} \left(u\bar{u} - d\bar{d} \right) \tag{17.47}$$

and B can finally be defined in an orthogonal way to both A and C. We can write in general the state composed of the quark–antiquark pairs with three factors α, β, and γ to be determined:

$$B = \alpha u\bar{u} + \beta d\bar{d} + \gamma s\bar{s} \tag{17.48}$$

The condition of orthogonality with the state A gives:

$$\langle A|B\rangle \propto \alpha \langle u\bar{u}|u\bar{u}\rangle - \beta \langle d\bar{d}|d\bar{d}\rangle = 0 \quad \Longrightarrow \quad \alpha = \beta \tag{17.49}$$

Similarly, orthogonality with the state C gives:

$$\langle B|C\rangle \propto \alpha \langle u\bar{u}|u\bar{u}\rangle + \beta \langle d\bar{d}|d\bar{d}\rangle + \gamma \langle s\bar{s}|s\bar{s}\rangle = 0 \quad \Longrightarrow \quad \gamma = -(\alpha + \beta) = -2\alpha \tag{17.50}$$

where in the last term we set $\alpha = \beta$. So, finally, normalizing the state gives:

$$B \equiv \frac{1}{\sqrt{6}} \left(u\bar{u} + d\bar{d} - 2s\bar{s} \right) \tag{17.51}$$

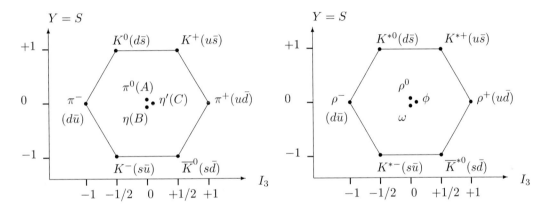

Figure 17.10 The spin-0 meson multiplet.   **Figure 17.11** The spin-1 meson multiplet.

We are now ready to map the quark–antiquark predicted bound states with the pseudoscalar mesons observed in Nature. See Figure 17.10. We compare these states with the classification in Figure 17.4 and note that because the $SU(3)$ flavor symmetry is only approximate, the physical states can be mixtures of the octet and singlet states. Empirically one finds:

$$|\pi^0\rangle = A = \frac{1}{\sqrt{2}} \left(u\bar{u} - d\bar{d} \right), \quad |\eta\rangle \approx B = \frac{1}{\sqrt{6}} \left(u\bar{u} + d\bar{d} - 2s\bar{s} \right), \quad |\eta'\rangle \approx C = \frac{1}{\sqrt{3}} \left(u\bar{u} + d\bar{d} + s\bar{s} \right) \tag{17.52}$$

so, the physical particle η' represents essentially the singlet.

• **Additional degrees of freedom.** Mesons behave like a bound system with a quark and an antiquark, which can possess additional internal degrees of freedom. Such a quantum system possesses discrete energy levels. In particular, the two quarks can have a relative orbital momentum L between them. In what follows, J represents the total angular momentum of a state. $J = L + S$, where L is the orbital angular momentum and S is the spin angular momentum. The spin and mass of the meson depend on the orientation of the spins of the quarks and of their relative orbital angular momentum L. In order to compute the total spin of the quark–antiquark pair, we add the two quarks' spin-1/2. Using the rules of spin addition, we obtain the triplet states with $S=1$ and the singlet state with $S=0$. We can represent the two configurations in the following way:

$$\text{Triplet}: \quad (q \uparrow \bar{q} \uparrow), (q \uparrow \bar{q} \downarrow + q \downarrow \bar{q} \uparrow), (q \downarrow \bar{q} \downarrow), \quad \text{Singlet}: \quad q \uparrow \bar{q} \downarrow - q \downarrow \bar{q} \uparrow \tag{17.53}$$

With the triplet spin states, we can explain the $J^P = 1^-$ vector mesons. As expected, they have the same quark content as the pseudoscalar mesons, as shown in Figure 17.11, but the triplet spin state explains the total vector spin of the bound state. By comparison with the octet in Figure 17.5, we can associate them with the corresponding particles. Unlike the pseudoscalar mesons which decay weakly, the vector mesons can decay strongly. Empirically, one finds that:

$$|\rho^0\rangle = \frac{1}{\sqrt{2}} \left(u\bar{u} - d\bar{d}\right), \quad |\omega\rangle = \frac{1}{\sqrt{2}} \left(u\bar{u} + d\bar{d}\right), \quad |\phi\rangle \approx s\bar{s} \tag{17.54}$$

This is consistent with the fact that we observe predominantly $\omega \to 3\pi$ and $\phi \to K\bar{K}$.

Both pseudoscalar and vector mesons have the same $SU(3)$ quark content. However, the rest masses are very different, compare for instance $m_{\pi^0} = 135$ MeV and $m_{\rho^0} = 775$ MeV. The $q\bar{q}$ bound states have internal degrees of freedom, resulting in discrete energy levels. The observed mass difference points to the spin–spin interactions, which can be expressed as:

$$m(q_1 q_2) = m_1 + m_2 + \frac{A}{m_1 m_2} \vec{S}_1 \cdot \vec{S}_2 \tag{17.55}$$

where the parameters m_1, m_2 correspond to the bare quark masses, and the interaction term A is adjusted to reproduce the observed pseudoscalar and vector meson states.

17.2.2   Quark Model for Baryonic States

Baryons are more complicated. We should consider three quarks $q_1 q_2 q_3$, which gives us $3^3 = 27$ possible quark flavor configurations. Of course, we should not just take all combinations of three quarks, since for instance uud and udu correspond both to the same particle state. One rather combines states that possess a particular symmetry under the interchange of the quarks. If we combine two quarks of $SU(3)$ flavors, we obtain $3^2 = 9$ combinations. Among them the uu, dd, and ss; however ud and du are identical and should be written as $ud + du$ (symmetric) or $ud - du$ antisymmetric. If we consider the normalization, we have the following six combinations symmetric under the quark interchange $1 \leftrightarrow 2$:

$$\boxed{1\,1}, \boxed{2\,2}, \boxed{3\,3}, \quad \boxed{1\,2}, \quad \boxed{1\,3}, \quad \boxed{2\,3}$$

$$6: \quad uu, \quad dd, \quad ss, \frac{1}{\sqrt{2}}(ud + du), \frac{1}{\sqrt{2}}(us + su), \frac{1}{\sqrt{2}}(ds + sd) \tag{17.56}$$

and the corresponding three antisymmetric combinations are:

$$\boxed{\begin{smallmatrix}1\\2\end{smallmatrix}}, \quad \boxed{\begin{smallmatrix}1\\3\end{smallmatrix}}, \quad \boxed{\begin{smallmatrix}2\\3\end{smallmatrix}}$$

$$\bar{3}: \quad \frac{1}{\sqrt{2}}(ud - du), \frac{1}{\sqrt{2}}(us - su), \frac{1}{\sqrt{2}}(ds - sd) \tag{17.57}$$

In $SU(3)$ group theory, the above result is expressed as the product $3 \otimes 3 = 6 \oplus \bar{3}$. For the combination of three $SU(3)$ objects, we need to add one quark to the bi-quark states found previously. Indeed, group theory tells us that (see Appendix B.12):

$$\text{Flavor}: 3 \otimes 3 \otimes 3 = (6 \oplus \bar{3}) \otimes 3 = \underbrace{10}_{S} \oplus \underbrace{8}_{MS} \oplus \underbrace{8}_{MA} \oplus \underbrace{1}_{A} \tag{17.58}$$

where S, MS, MA, and A stand respectively for symmetric, mixed symmetric, mixed antisymmetric and antisymmetric. The antisymmetric and mixed state correspond to the situation under the interchange of the first two quarks $1 \leftrightarrow 2$. Consequently, the 10 symmetric flavor states are given by:

$$\phi_S \quad : \quad ddd, \frac{1}{\sqrt{3}}(udd + ddu + dud), \frac{1}{\sqrt{3}}(uud + udu + duu), uuu$$

$$\frac{1}{\sqrt{3}}(dds + dsd + sdd), \frac{1}{\sqrt{6}}(uds + dsu + sud + usd + sdu + dus), \frac{1}{\sqrt{3}}(uus + usu + suu)$$

$$\frac{1}{\sqrt{3}}(dss + ssd + sds), \frac{1}{\sqrt{3}}(uss + ssu + sus), sss \tag{17.59}$$

We compare these combinations with the spin $3/2$ baryon decuplet (see Figure 17.6). The first four flavor combinations form the four isospin $I = 3/2$, $S = 0$ states of the Δ, which in short can be written as ddd, ddu, duu, and uuu. The next three are the isospin $I = 1$, $S = -1$ states of the Σ or in short dds, dus, and uus. The next two are the isospin $I = 1/2$, $S = -2$ states of the Ξ or in short dss and uss. Finally, the last state corresponds to the $I = 0$, $S = -3$ state of the Ω^- which is sss.

The mixed symmetric and antisymmetric states can be found by considering the symmetry under the interchange of the quarks $1 \leftrightarrow 2$ while adding a third. As can be seen from Figure 17.12, the eight mixed antisymmetric states after normalization are given by:

$$\phi_{MA} \quad : \quad \frac{1}{\sqrt{2}}(udd - dud), \frac{1}{\sqrt{2}}(udu - duu)$$

$$\frac{1}{\sqrt{2}}(dsd - sdd), \frac{1}{2}(usd - sud + dsu - sdu), \frac{1}{\sqrt{2}}(usu - suu)$$

$$\frac{1}{\sqrt{2}}(dss - sds), \frac{1}{\sqrt{2}}(uss - sus)$$

$$\frac{1}{2\sqrt{3}}(sdu - sud + usd - dsu - 2dus + 2uds) \tag{17.60}$$

where $A \propto (us - su)d + (ds - sd)u = usd - sud + dsu - sdu$ has been defined to be the $Y = I_3 = 0$ member of the isospin triplet and $B \propto (us - su)d - (ds - sd)u + 2(ud - du)s = sdu - sud + usd - dsu - 2dus + 2uds$ is an orthogonal state.

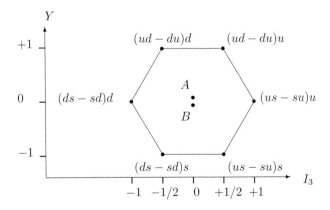

Figure 17.12 The baryon octet of the mixed antisymmetric states under interchange $1 \leftrightarrow 2$.

Of course, it is possible to define antisymmetric states under the interchange $1 \leftrightarrow 3$ and $2 \leftrightarrow 3$. Rather, the eight mixed symmetric states ϕ_{MS}, orthogonal to the corresponding ones in ϕ_S and ϕ_{MA}, can be derived to be:

$$\phi_{MS} \quad : \quad -\frac{1}{\sqrt{6}}\left(udd + dud - 2ddu\right), \frac{1}{\sqrt{6}}\left(udu + duu - 2uud\right)$$

$$\frac{1}{\sqrt{6}}\left(dsd + sdd - 2dds\right), \frac{1}{2\sqrt{3}}\left(sdu + sud + dsu + usd - 2(dus + uds)\right), \frac{1}{\sqrt{6}}\left(usu + suu - 2uus\right)$$

$$\frac{1}{2}\left(dsu - usd + sdu - sud\right)$$

$$-\frac{1}{\sqrt{6}}\left(dss + sds - 2ssd\right), -\frac{1}{\sqrt{6}}\left(uss + sus - 2ssu\right) \tag{17.61}$$

The first two correspond to the nucleon $S = 0$ states, in short udd (neutron-like) and uud (proton-like). The next three form the Σ isospin triplet $S = -1$ states, in short dds, uds, and uus. The next isospin singlet is the Λ. And the last two are the isospin doublet Ξ states ($S = -2$), with, in short, dss and uss. Finally, the totally antisymmetric state is given by:

$$\phi_A : \frac{1}{\sqrt{6}}\left(uds - dus + dsu - sdu + sud - usd\right) \tag{17.62}$$

In addition, the spin of each quark can be either in the up or in the down state. This gives another $2^3 = 8$ possible spin states. Finally, we have two independent relative orbital momenta L and L'. In the present discussion, we consider the case where $L = L' = 0$. The mixture of flavor and spin requires the combination $SU(3)_{flavor} \otimes SU(2)_{spin}$. Here we already recognize the prediction for the decuplet and the octets. For the spins, we get:

$$\text{Spin} : 2 \otimes 2 \otimes 2 = (3 \oplus 1) \otimes 2 = \underbrace{4}_{\text{spin } 3/2} \oplus \underbrace{2}_{\text{spin } 1/2} \oplus \underbrace{2}_{\text{spin } 1/2} \tag{17.63}$$

and we recognize the $J = 1/2$ and $J = 3/2$ states.

Hence, the flavor $3 \otimes 3 \otimes 3$ combination gives us the prediction for the decuplet and octet states, while the spin $2 \otimes 2 \otimes 2$ predicts the existence of $J = 1/2$ and $J = 3/2$ states. Still assuming no additional angular momentum, so $L = L' = 0$, we can reconstruct the known spin-1/2 octuplet (obtained by a combination of mixed symmetry flavor and mixed symmetry spin) and the spin-3/2 decuplet (obtained by combination of symmetric flavor states) in terms of their quark content. These are shown in Figures 17.13 and 17.14 (cf. Figures 17.3 and 17.6).

17.2.3 Hadrons Flavor Classification

The existence of a deeper level of elementary constituents of matter can clearly explain the large number of known mesonic and baryonic states. What appeared to be a messy hadronic world became understood in terms of a few constituent spin-$\frac{1}{2}$ quarks of given flavors. Today we know that there are in total six quarks of different flavors. We will discuss later the existence of the c and b quarks in Chapter 20 and of the t quark in Chapter 29. This requires the flavor symmetry group to be extended to $SU(6)_{flavor}$. Although within this group, the symmetry becomes more and more broken with the addition of heavier and heavier quarks, one can still nicely classify the entire hadronic spectrum. For example, we can identify the following known hadrons with the following quark contents:

Light and strange mesons:	$\pi^+ = u\bar{d},$	$K^+ = u\bar{s},$	$K^0 = d\bar{s},$	$\pi^0 = (u\bar{u} - d\bar{d})/\sqrt{2} \dots$
Charmed mesons:	$D^+ = c\bar{d},$	$D^0 = c\bar{u},$	$D_s^+ = c\bar{s}$	$\dots$
Bottom mesons:	$B^+ = u\bar{b},$	$B^0 = d\bar{b},$	$B_s^0 = s\bar{b},$	$B_c^+ = c\bar{b} \dots$
Light and strange baryons:	$p = uud,$	$n = udd,$	$\Sigma^+ = uus,$	$\Sigma^0 = uds \dots$
Charmed baryons:	$\Sigma_c^+ = udc,$	$\Sigma_c^{++} = uuc,$	$\Xi_c^+ = usc,$	$\Xi_c^0 = dsc \dots$
	$\Xi_{cc}^+ = dcc,$	$\Xi_{cc}^{++} = ucc,$	$\Omega_{cc}^+ = scc$	$\dots$

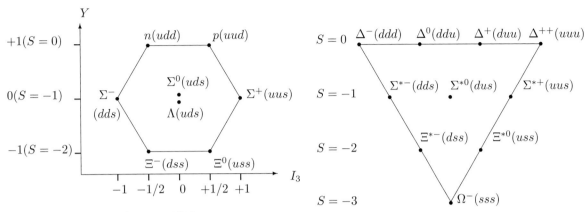

Figure 17.13 The spin-1/2 baryon octet.

Figure 17.14 The spin-3/2 baryon decuplet.

The $SU(4)_{flavor}$ weight diagrams for scalar and vector mesons and for baryons are illustrated in Figure 20.12.

17.3   The Evidence of Color

There is a series of experimental verifications that support the fact that **a new degree of freedom must be introduced for each species of quark.** The new property is defined as the "color" [200]. The minimal solution is $N_C = 3$ colors, which are labeled R, G, and B. The color state can then be represented as the vector:

$$c = \begin{pmatrix} R \\ G \\ B \end{pmatrix} \tag{17.64}$$

Every quark thus carries an internal color degree of freedom and we need to consider the following quark states:

$$u_R, u_G, u_B$$
$$d_R, d_G, d_B$$
$$s_R, s_G, s_B \tag{17.65}$$

We realize that the color also follows an $SU(3)$ symmetry! It is called the $SU(3)_{color}$ or $SU(3)_C$ symmetry, which transforms one color into another. The transformation represents a rotation in the internal color space. Unlike the $SU(3)_{flavor}$ symmetry which was not exact, the $SU(3)_C$ symmetry is an *exact* symmetry of Nature. We will come back to this point in the next chapter.

• **Color confinement.** In order to avoid the existence of extra states with non-zero color which are not observed in Nature, one needs to further postulate that **all asymptotic states are colorless, i.e., singlets under rotations in color space.** This means that all hadrons (all bound states) are color-singlets. This assumption is known as the confinement hypothesis, because it implies the non-observability of free quarks: since quarks carry color they have to be confined within color-singlet bound states. For baryons, this means that the quarks are always R, G, and B. For mesons, the quark–antiquark pairs are always color-singlets, since antiquarks carry anticolor.

We now illustrate some of the key evidence for the existence of color.

• **The Δ^{++} resonance.** There is a one-to-one correspondence between the observed hadrons and the states predicted by the simple meson and baryon classification; thus, the quark model provides a sort of very useful "Periodic Table" for hadrons. However, the quark picture faces a problem concerning the Fermi–Dirac statistics of the constituents. Let us consider the Δ^{++} baryon, which possesses a spin-3/2. We note that the spin and quark configuration is symmetric under the exchange of two quarks:

$$|\Delta^{++}(J = 3/2), J_3 = +3/2\rangle = |u \uparrow\rangle |u \uparrow\rangle |u \uparrow\rangle \tag{17.66}$$

Such a state is in contradiction with Pauli's exclusion principle, since the three quark fermions cannot be in the same quantum state. In order to accommodate the Δ^{++} there must be at least three independent colors. We can now define the state of the Δ^{++} baryon as:

$$|\Delta^{++}(J = 3/2), J_3 = +3/2\rangle = |u_R \uparrow\rangle |u_G \uparrow\rangle |u_B \uparrow\rangle \quad + \text{permutations} \tag{17.67}$$

• **The $e^+e^- \to hadrons$ cross-section.** the experimental data support a factor $N_C = 3$ for each quark flavor, see Section 19.2.

• **The $\pi^0 \to \gamma\gamma$ decay width.** The decay of the neutral pion occurs at first order via a quark loop, as shown in Figure 17.15. The decay rate can be expressed as:

$$\Gamma(\pi^0 \to \gamma\gamma) = \left(\frac{\alpha}{2\pi}\right)^2 N_C^2 (Q_u^2 - Q_d^2) \frac{m_\pi^3}{8\pi f_\pi} \tag{17.68}$$

where $f_\pi \approx m_\pi$ is the pion decay constant derived from the charged pion decay (see Section 23.10). The

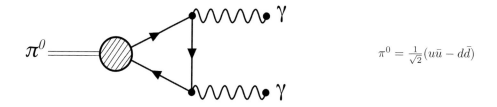

$$\pi^0 = \frac{1}{\sqrt{2}}(u\bar{u} - d\bar{d})$$

Figure 17.15 The tree-level diagram for the $\pi^0 \to \gamma\gamma$ decay.

predicted values are consequently

$$\Gamma(\pi^0 \to \gamma\gamma) = \begin{cases} 0.86 \text{ eV} & \text{for } N_C = 1 \\ 7.75 \text{ eV} & \text{for } N_C = 3 \end{cases} \tag{17.69}$$

to be compared to the experimental value of 7.73 ± 0.16 eV (see Eq. (15.21)).

• **The $\tau \to e\nu\nu$ branching ratio.** The tau lepton decays via the weak process, which couples universally to an electron and a neutrino, to a muon and a neutrino, or to an "up–down" quark pair (this will be discussed in detail in Chapter 23). Kinematically, we therefore expect the three branching ratios $\tau \to e\nu_e\nu_\tau$, $\tau \to \mu\nu_\mu\nu_\tau$, $\tau \to ud\nu_\tau$ to be of the same order, so 1/3 each without color. With three possible colors for the quarks, the branching ratio to electron and muon would, on the other hand, each be of the order of 1/5. This is very close to the measured value $Br(\tau \to e\nu_e\nu_\tau) = (17.82 \pm 0.04)\%$ (see Eq. (21.97)).

17.4   **Quark Flux Diagrams**

Since hadrons are formed by bound states of quarks, their strong interactions are to be understood as the result of a more basic strong interaction among their constituents. Such interactions are represented with quark flux diagrams, where each quark is considered as an elementary particle. We show in the following a few examples.

- The scattering process via a resonance $p + \pi^+ \to \Delta^{++} \to p + \pi^+$: the diagram for resonance production can be drawn in reduced form showing just the quark lines, which are continuous and do not change flavor. See Figure 17.16.

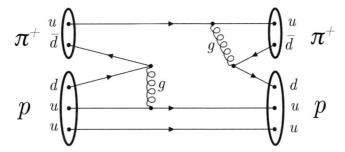

Figure 17.16 The quark flux diagram for the reaction $p + \pi^+ \to \Delta^{++} \to p + \pi^+$.

- Decay of $\phi \to K^+ K^-$: see Figure 17.17. The reason for assigning an $s\bar{s}$ state to the ϕ is discussed in **Ex 17.3**.

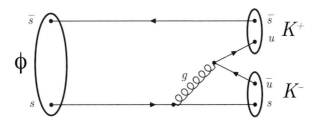

Figure 17.17 The quark flux diagram for the reaction $\phi \to K^+ K^-$.

- Decay of a meson into a fermion pair: see Figure 17.18.

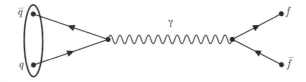

Figure 17.18 The quark flux diagram for the decay of a meson into a fermion–antifermion pair.

17.5   Electron–Parton Scattering

We want to describe the scattering off *one* parton (=quark) constituting the proton. The requirements for our assumptions to be justified are that during the time the photon interacts with the parton, we can neglect other parton–parton (strong) interactions. The assumption is then that **during the time the photon interacts the parton can be regarded as quasi-free**. As we will see later, the reason for this assumption stems from the asymptotic freedom to be understood within QCD. The strong force between the quarks becomes small at the Q^2 characteristic of the photon interaction. It is as if before and after the photon interaction, the strong force strongly binds together the quarks in hadrons, while during the hard Q^2 interaction the strong force is comparatively negligible, and the quarks are behaving as if they were free.

Let us then assume that a (quasi-free) quark of mass m_q inside the proton carries a fraction ξ of the proton's four-momentum. See Figure 17.19. When we write the four-momentum of the parton as ξk^μ, we are neglecting the transverse parton momenta. The corresponding parton mass is simply $m_q = \sqrt{(\xi k)^2} = \xi M_p$. The frame of reference where this happens is called the **infinite momentum frame** or **Breit frame**. See Table 17.2.

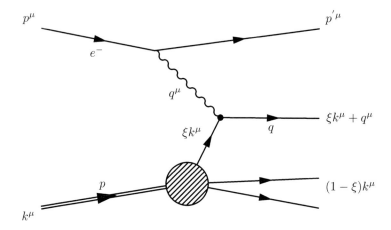

Figure 17.19 The tree-level Feynman diagram of the electron–parton scattering; the blob illustrates the fact that the proton is composed of partons and the photon interacts with one parton q, initially carrying a fraction ξ of the proton momentum.

Quantity	Proton	Parton
Four-momentum	P^μ	xP^μ
Energy	E	xE
Longitudinal momentum	P_L	$p_L = xP_L$
Transverse momentum	$P_T = 0$	$p_T = 0$
Mass	$M_p = \sqrt{P^2}$	$m_q = \sqrt{(xP)^2} = xM$

Table 17.2 Kinematics in the Breit frame with $\xi = x$.

After the scattering, the struck quark's four-momentum is $\xi k + q$, where $q = p - p'$. Then:

$$(\xi k + q)^2 = m_q^2 \simeq 0 \longrightarrow \xi^2 k^2 + q^2 + 2\xi k \cdot q \simeq q^2 + 2\xi k \cdot q = 0 \tag{17.70}$$

Hence, we get:

$$\xi = \frac{Q^2}{2k \cdot q} \quad \longrightarrow \quad \xi = x \tag{17.71}$$

Then **Bjorken's scaling variable x can be identified as the fraction of the proton momentum carried by the struck quark** (in a frame where the proton has very high energy).

In terms of the electron–proton kinematical variables, we have:

$$s = (k+p)^2 \simeq 2k \cdot p; \quad y = \frac{k \cdot q}{p \cdot k}; \quad x = \frac{Q^2}{2k \cdot q} \tag{17.72}$$

But for the underlying quark interaction, we note that the kinematical quantities are (we use the hat notation for them):

$$\begin{cases} \hat{s} = (xk + p)^2 \simeq 2x(k \cdot p) = xs \\[2mm] \hat{y} = \dfrac{xk \cdot q}{xp \cdot k} = y \\[2mm] \hat{x} = 1 \quad \text{elastic scattering} \end{cases} \tag{17.73}$$

Since quarks are assumed to be point-like Dirac particles, we can use the results obtained for the $e^-\mu^- \to e^-\mu^-$ scattering (see Eq. (11.152)) expressed as a function of the scattering angle θ^* in the **center-of-mass frame** of the reaction, rewriting it using the kinematical quantities in the underlying electron–parton (here muon) system:

$$\left(\frac{d\sigma}{d\Omega}\right)_{e\mu} = \left(\frac{\alpha^2}{2\hat{s}}\right)\frac{4 + (1 + \cos\theta^*)^2}{(1 - \cos\theta^*)^2} = \frac{e^4}{8\pi^2\hat{s}}\frac{1 + \frac{1}{4}(1 + \cos\theta^*)^2}{(1 - \cos\theta^*)^2} \tag{17.74}$$

In the center-of-mass of the underlying electron–muon system, we can write when neglecting the rest masses:

$$p^\mu = (E, 0, 0, E); \quad k^\mu = (E, 0, 0, -E); \quad p'^\mu = (E, E\sin\theta^*, 0, E\cos\theta^*) \tag{17.75}$$

where θ^* is the scattering angle of the electron. Then

$$\hat{y} = \frac{k \cdot q}{p \cdot k} = \frac{k \cdot (p - p')}{2E^2} = \frac{E^2(1 - \cos\theta^*)}{2E^2} = \frac{1}{2}(1 - \cos\theta^*) \tag{17.76}$$

So the inelasticity variable $\hat{y}$ is uniquely related to the scattering angle in the center-of-mass frame. It vanishes when the scattering angle is zero, and it is equal to one when the electron backscatters at $180°$. This relation is usually written as:

$$1 - \hat{y} = \frac{1}{2}(1 + \cos\theta^*) \tag{17.77}$$

It is convenient to express the $e^-\mu^- \to e^-\mu^-$ differential cross-section as a differential of the inelasticity:

$$\left(\frac{d\sigma}{d\hat{y}}\right)_{e\mu} = \left(\frac{d\sigma}{d\Omega}\right)_{e\mu}\left|\frac{2\pi d\cos\theta^*}{d\hat{y}}\right| = \frac{e^4}{8\pi^2\hat{s}}\frac{(1 + (1 - \hat{y})^2)}{4\hat{y}^2}4\pi = \frac{e^4}{8\pi\hat{s}}\frac{(1 + (1 - \hat{y})^2)}{\hat{y}^2} \tag{17.78}$$

We want to change the variables once again and express the differential cross-section as a function of q^2. Since:

$$q^2 = (p - p')^2 \approx -2p \cdot p' = -2E^2(1 - \cos\theta^*) = -\hat{s}\hat{y} \tag{17.79}$$

we can write:

$$\left(\frac{d\sigma}{dq^2}\right)_{e\mu} = \left(\frac{d\sigma}{d\hat{y}}\right)_{e\mu}\left|\frac{d\hat{y}}{dq^2}\right| = \frac{e^4}{8\pi\hat{s}}\frac{(1 + (1 - \hat{y})^2)}{q^4/\hat{s}^2}\frac{1}{\hat{s}} = \underbrace{\frac{e^4}{8\pi q^4}}_{\text{propagator}}\underbrace{\left(1 + (1 - \hat{y})^2\right)}_{\text{angular dependence}} \tag{17.80}$$

This form is great because the photon propagator term, proportional to $1/Q^4$, and the angular dependence in the center-of-mass, have been factorized. Note, however, that because of Eq. (17.79), q^2 and $\hat{y}$ are not independent variables and the differential cross-section does depend on only one variable. Indeed, replacing $\hat{y}$ in our previous expression yields:

$$\left(\frac{d\sigma}{dq^2}\right)_{e\mu} = \frac{e^4}{8\pi q^4}\left(1 + \left(1 + \frac{q^2}{\hat{s}}\right)^2\right) \tag{17.81}$$

However, we keep the first form with the q^2 and $\hat{y}$ explicit. Since quarks are assumed to be point-like fermions like muons, we can directly apply this result to the $e^- q \to e^- q$ scattering by simply introducing the fractional charge of the quark, given by e_q, at one vertex. This simply leads to an additional factor $e_q^2 = (2/3)^2$ for up-type quarks and $e_q^2 = (1/3)^2$ for down-type quarks in the differential cross-section:

$$\left(\frac{d\sigma}{dq^2}\right)_{eq} = \frac{e^4 e_q^2}{8\pi q^4}\left(1 + (1 - \hat{y})^2\right) \tag{17.82}$$

So, finally the differential cross-section for the electron–quark scattering is given by:

$$\left(\frac{d\sigma}{dq^2}\right)_{eq} = \frac{2\pi\alpha^2 e_q^2}{q^4}(1 + (1 - \hat{y})^2) = \frac{4\pi\alpha^2 e_q^2}{q^4}\left((1 - \hat{y}) + \frac{\hat{y}^2}{2}\right) \tag{17.83}$$

We discuss in the next section how to embed this result in the nucleon in order to compare it with experimental data.

17.6 The Naive Quark–Parton Model

The parton model was proposed by Feynman in 1969, to describe deep inelastic scattering in terms of point-like constituents inside the nucleon known as partons with an effective mass $m < M_p$. We start with a rough model of the proton, assuming that it consists of some number of Dirac point-like spin-1/2 partons, each one carrying a given fraction of the proton momentum. Let us ignore any strong (i.e., QCD) interactions and let us assume that the nucleon (either proton or neutron) constituents are free spin-$\frac{1}{2}$ partons. This is called the **free parton model**. Within this model, the nucleons have three point-like constituents ($p = u_v u_v d_v$, $n = u_v d_v d_v$), which are called **valence quarks**.

We will see in Chapter 18 that the carriers of the strong force in QCD are called **gluons** (the analog of the photon in QED). Gluons are of course present in the nucleon; however, they do not interact directly with the photon probe. The photon–gluon interaction only occurs through the virtual q-$\bar{q}$ pairs coupled to the gluon constituents. Thus, instead of gluons, the photon feels a **sea** of q-$\bar{q}$ partons within the nucleon.

The **parton distribution functions** $q^p(x)$ define the probability of finding a quasi-free parton q *in the proton* which carries an energy fraction between x and $x + dx$. Hence, $u^p(x)$, $\bar{u}^p(x)$, $d^p(x)$, $\bar{d}^p(x)$, $s^p(x)$, $\bar{s}^p(x)$, ... denote the probability distributions for u, $\bar{u}$, d, $\bar{d}$, s, $\bar{s}$, ... quarks with momentum fraction x in the proton. Here we assumed the scaling of the structure functions, as first observed by the SLAC-MIT experiment, hence, the parton distribution functions are also written to scale, at least approximately, on the dimensionless scaling variable x:

$$q^p(x, Q^2) \to q^p(x) \tag{17.84}$$

The effect is expected in the naive quark–parton model from the elastic scattering on dimensionless, i.e., point-like scattering centers inside the nucleon. Even in this case, scaling would become exact in the Bjorken limit where $Q^2 \to \infty$ at constant x, such that the transverse momentum of partons in the infinite momentum frame of the proton becomes negligible.

Scattering experiments can be used to measure the structure functions $q^p(x)$. The parton model restores the *elastic scattering* relationship between q^2 and ν (Eq. (16.10)), with the parton mass m_q replacing M_p:

$$\nu = -\frac{q^2}{2m_q} \qquad \text{(elastic parton level)} \tag{17.85}$$

In the quasi-free parton model, the cross-section for scattering from a particular quark within the proton with a momentum fraction in the range x and $x + \mathrm{d}x$ is:

$$\left(\frac{\mathrm{d}\sigma}{\mathrm{d}Q^2}\right)_{eq} = \frac{4\pi\alpha^2}{Q^4}\left((1-y) + \frac{y^2}{2}\right)e_q^2 q^p(x)\mathrm{d}x \tag{17.86}$$

where e_q is the electric charge of the parton ($+2/3$ and $-1/3$ for quarks). Summing over all types of quarks within the proton should give the expression for the electron–proton scattering cross-section. We find:

$$\left(\frac{\mathrm{d}^2\sigma}{\mathrm{d}x\mathrm{d}Q^2}\right)_{ep} = \frac{4\pi\alpha^2}{Q^4}\left((1-y) + \frac{y^2}{2}\right)\underbrace{\sum_q e_q^2 q^p(x)}_{\text{sum over all quarks}} = \frac{4\pi\alpha^2}{Q^4}\frac{1}{x}\left((1-y) + \frac{y^2}{2}\right)x\sum_q e_q^2 q^p(x) \tag{17.87}$$

By comparing with Eq. (16.114), we obtain the free parton model prediction for the structure functions as a function of the underlying quark distribution functions:

$$F_2^{ep}(x, Q^2) = 2xF_1^{ep}(x, Q^2) = x\sum_q e_q^2 q^p(x) \tag{17.88}$$

and

$$\left(\frac{\mathrm{d}^2\sigma}{\mathrm{d}x\mathrm{d}Q^2}\right)_{ep} = \frac{4\pi\alpha^2}{Q^4}\sum_q e_q^2 q^p(x)\left[(1-y) + \frac{y^2}{2}\right] = \frac{2\pi\alpha^2}{Q^4}\sum_q e_q^2 q^p(x)\left[1 + (1-y)^2\right] \tag{17.89}$$

At high energy $y \approx Q^2/xs$, consequently, this expression can also be written as:

$$\left(\frac{\mathrm{d}^2\sigma}{\mathrm{d}x\mathrm{d}Q^2}\right)_{ep} = \frac{2\pi\alpha^2}{Q^4}\sum_q e_q^2 q^p(x)\left[1 + \left(1 - \frac{Q^2}{xs}\right)^2\right] \tag{17.90}$$

The sum is over all the different flavors of partons in the proton. The structure function just describes the parton content of the proton. This latter depends on the details of the proton, a complicated object indeed! It is not possible to compute it from first principles using perturbation theory. But still our previous expression makes a striking prediction: in the deep inelastic regime, the complicated structure of the proton embedded in the $q^p(x)$ functions, can be factorized from the QED predicted cross-section by dividing by the factor:

$$f = \frac{1 + \left(1 - \dfrac{Q^2}{xs}\right)^2}{Q^4} \tag{17.91}$$

If $q^p(x)$ is only indeed a function of x, then $\mathrm{d}^2\sigma/\mathrm{d}\Omega\mathrm{d}E_3/f$ should be independent of Q^2. This provides a direct cross-check of the Bjorken scaling hypothesis. Indeed, the data from the SLAC-MIT data for $Q^2 > 1$ GeV^2 is reported in Figure 17.20. It clearly *does* exhibit scaling! This was not as obvious from Figures 16.9 and 16.10.

We note that F_1 and F_2 are not at all independent! Writing out explicitly the quark flavors and including the respective electric charges ($+2/3$ and $-1/3$), dropping the p index in the quark–parton distribution functions, yields:

$$F_2^{ep}(x, Q^2) = 2xF_1^{ep}(x, Q^2) = x\left(\frac{4}{9}u(x) + \frac{1}{9}d(x) + \frac{4}{9}\bar{u}(x) + \frac{1}{9}\bar{d}(x)\right) \tag{17.92}$$

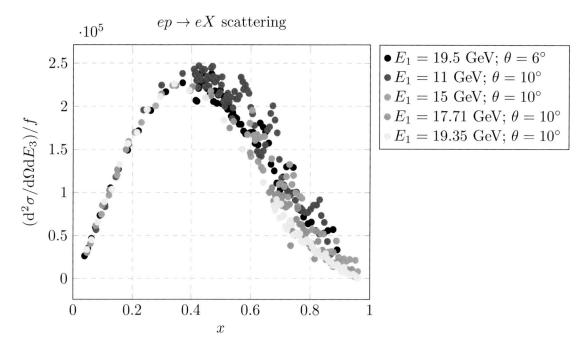

Figure 17.20 Double differential cross-section divided by the factor f (see text) as a function of Bjorken x in ep scattering for data with $Q^2 > 1$ GeV^2. Compare with Figures 16.9 and 16.10. Adapted with permission from E. D. Bloom *et al.*, "High-energy inelastic $e - p$ scattering at 6 and 10 deg," *Phys. Rev. Lett.*, vol. 23, pp. 930–934, Oct 1969. Copyright 1969 by the American Physical Society.

Assuming isospin symmetry, we can expect the parton distribution functions to be symmetric under the interchange $n \leftrightarrow p$ and $u \leftrightarrow d$. Hence, the neutron ($= ddu$) should be the same as the proton (uud) with the interchange up–down. We can relate the up and down quark distributions in a neutron to the ones in a proton:

$$u^n(x) = d^p(x) \equiv d(x) \ ; \qquad d^n(x) = u^p(x) \equiv u(x) \tag{17.93}$$

the remaining parton distributions for the sea of quarks and antiquarks are assumed to be the same for protons and neutrons. Thus, combining data from different deep inelastic processes, it is possible to obtain separate information on the individual parton distribution functions. Therefore, the structure functions for the neutron are:

$$F_2^{en}(x, Q^2) = 2x F_1^{en}(x, Q^2) = x \left(\frac{4}{9} d(x) + \frac{1}{9} u(x) + \frac{4}{9} \bar{d}(x) + \frac{1}{9} \bar{u}(x) \right) \tag{17.94}$$

where $u(x)$, $d(x)$, ... are the parton distribution functions of the proton.

We can define the total fraction of proton momentum carried by the q and $\bar{q}$ quark with the following integral:

$$f_q = \int_0^1 \left[x q(x) + x \bar{q}(x) \right] \mathrm{d}x \tag{17.95}$$

Then we have:

$$\int_0^1 F_2^{ep}(x) \mathrm{d}x = \frac{4}{9} f_u + \frac{1}{9} f_d \quad \text{and} \quad \int_0^1 F_2^{en}(x) \mathrm{d}x = \frac{4}{9} f_d + \frac{1}{9} f_u \tag{17.96}$$

The quark distributions must satisfy some constraints. Since both the proton and the neutron have zero strangeness:

$$\int_0^1 dx\, [s(x) - \bar{s}(x)] = 0 \tag{17.97}$$

Similar relations follow for the heavier flavors $(c, \bar{c}, b, \bar{b}, \ldots)$. The proton and neutron electric charges imply two additional sum rules:

$$\int_0^1 dx\, [u(x) - \bar{u}(x)] = 2 \quad \text{and} \quad \int_0^1 dx\, [d(x) - \bar{d}(x)] = 1 \tag{17.98}$$

which just give the excess of u and d quarks over antiquarks.

The quark-model concept of valence quarks gives further insight into the nucleon structure. We can decompose the u and d distribution functions into the sum of valence and sea contributions, and take the remaining parton distributions to be pure sea. Since gluons are flavor singlets, one expects the sea to be flavor-independent. In this way, the number of independent distributions is reduced to three:

$$\begin{aligned} u(x) &= u_v(x) + q_s(x) \quad \text{and} \quad d(x) = d_v(x) + q_s(x) \\ \bar{u}(x) &= \bar{d}(x) = s(x) = \bar{s}(x) = \cdots = q_s(x) \end{aligned} \tag{17.99}$$

Within this model, the strangeness sum rule Eq. (17.97) is automatically satisfied, while Eqs. (17.98) imply constraints on the valence-quark distributions alone.

17.7   The Proton and Neutron Structure Functions

Following the successes of electron–proton scattering, the SLAC-MIT experiment was augmented to measure electron–deuteron scattering. Some of this data is shown in Figure 17.21. The data gathered permitted a detailed comparison of the inelastic electron–proton (ep) and electron–neutron (en) scattering cross-sections, and hence to extract the **neutron structure function** $F_2^{en}(x, Q^2)$.

The measured values of the proton structure function $F_2^{ep}(x, Q^2)$ are shown in Figure 17.22(left) as a function of x, for many different values of Q^2 between 2 and 18 GeV^2; the concentration of data points along a curve indicates that Bjorken scaling is correct, to quite a good approximation.

The naive three-quark model for the nucleon would suggest the existence of a peak at $x = 1/3$ in the proton and neutron structure functions. However, the distribution shown in Figure 17.22(left) does not show such behavior. The difference can easily be understood as originating from the parton–sea contributions. Taking the difference between the proton and neutron structure functions, where the contribution from the sea cancels, the data exhibits indeed a broad peak around $x = 1/3$, as illustrated in Figure 17.22(right).

A crucial prediction of the naive parton model is the **Callan[4]–Gross[5] relation**

$$F_2(x) = 2x F_1(x) \tag{17.100}$$

which is a consequence of our assumption of spin-$\frac{1}{2}$ partons. For Dirac particles the magnetic moment is directly related to the charge. Hence, the two form factors, reminiscent of the electric and magnetic terms, are fixed with respect to each other. It is also easy to check that spin-0 partons would have led to $F_1(x) = 0$. Figure 17.23 shows that the Callan–Gross relation is also reasonably well satisfied by the data, supporting the spin-$\frac{1}{2}$ assignment for the partons. **Quarks behave like spin-1/2 Dirac particles.**

---

[4]   Curtis Gove Callan Jr. (born 1942), American theoretical physicist.
[5]   David Jonathan Gross (born 1941), American theoretical physicist and string theorist.

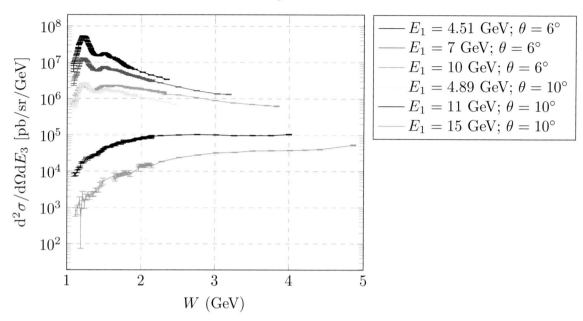

Figure 17.21 Double differential cross-section $\mathrm{d}^2\sigma/\mathrm{d}\Omega\mathrm{d}E_3$ in eD scattering plotted as a function of the hadronic invariant mass W. Adapted with permission from J. S. Poucher *et al.*, "High-energy single-arm inelastic $e - p$ and $e - d$ scattering at 6-degrees and 10-degrees," *Phys. Rev. Lett.*, vol. 32, p. 118, 1974. Copyright 1974 by the American Physical Society.

One can also integrate the experimentally measured structure functions to find:

$$\int_0^1 F_2^{ep}(x)\mathrm{d}x \simeq 0.18 \quad \text{and} \quad \int_0^1 F_2^{en}(x)\mathrm{d}x \simeq 0.12 \tag{17.101}$$

With the definitions for f_u and f_d (Eq. (17.95)) one gets the striking result that:

$$f_u \simeq 0.36 \quad \text{and} \quad f_d \simeq 0.18 \tag{17.102}$$

so, as expected $f_u \approx 2f_d$ since the two up quarks in the proton should carry twice the momentum of the down quarks. However, this result also shows that **all quarks carry just over $f_u + f_d \approx 50\%$ of the total proton momentum**. The rest is carried by *gluons*, which do not contribute to electron–nucleon scattering, since they are electrically neutral (as will be discussed in Chapter 18).

Structure functions are often characterized by so-called "sum-rules." For example, the **Gottfried sum rule** [201] is given by:

$$\begin{aligned}
G_s &\equiv \int_0^1 \frac{\mathrm{d}x}{x} \left[F_2^{ep}(x) - F_2^{en}(x)\right] = \int_0^1 \mathrm{d}x \left[\frac{4}{9}(u + \bar{u}) + \frac{1}{9}(d + \bar{d}) - \frac{1}{9}(u + \bar{u}) - \frac{4}{9}(d + \bar{d})\right] \\
&= \frac{1}{3} \int_0^1 \mathrm{d}x \left[u - d + \bar{u} - \bar{d}\right] = \frac{1}{3} \int_0^1 \mathrm{d}x \left[u_v - d_v\right] + \frac{1}{3} \int_0^1 \mathrm{d}x \left[2\bar{u} - 2\bar{d}\right] \\
&= \frac{1}{3} + \frac{2}{3} \int_0^1 \mathrm{d}x \left[\bar{u} - \bar{d}\right]
\end{aligned} \tag{17.103}$$

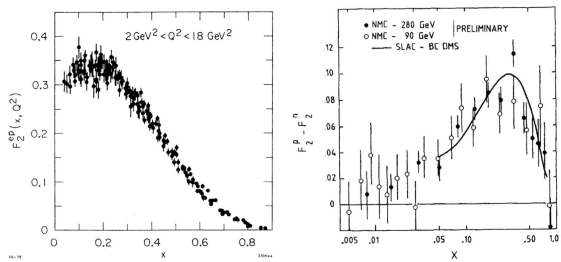

Figure 17.22 (left) Experimental data on F_2^{ep} as a function of x, for different values of Q^2. Taken from A. Mosher (ed.), *Quantum Chromodynamics: Proceedings, 7th SLAC Summer Institute on Particle Physics (SSI 79), Stanford, Calif., 9–20 Jul 1979*, 1980. (right) Measured difference $F_2^{ep} - F_2^{en}$ as a function of x. Reprinted figure from S. Rock, "Proton and neutron structure functions," *Nucl. Phys.*, vol. A527, pp. 553C–557C, 1991, with permission from Elsevier.

where we used that $u = u_v + \bar{u}$ and $d = d_v + \bar{d}$. The parton model implies that the sea is isospin symmetric, hence $G_s = 1/3$, which turns out not to be so well described by experimental measurements. Current data are in favor of a value around 0.2, which implies a light-quark flavor asymmetry in the sea of the proton.

Another interesting quantity is the ratio:

$$\frac{F_2^{en}(x)}{F_2^{ep}(x)} = \frac{4d_v(x) + u_v(x) + \Sigma_{sea}}{4u_v(x) + d_v(x) + \Sigma_{sea}} \tag{17.104}$$

where Σ_{sea} is the total sea contribution. Since all probability distributions must be positive-definite, this ratio should satisfy the bounds $\frac{1}{4} \leq F_2^{en}(x)/F_2^{ep}(x) \leq 4$, which are consistent with the data (see Figure 17.24). The measured ratio appears to tend to 1 at small x, indicating that the sea contributions dominate in that region, while for $x \to 1$ the ratio tends to $1/4$. Naively, if the u and d valence quarks had the same shape as $x \to 1$, then we would expect $u_v(x) = 2d_v(x)$ in this region, and therefore the ratio $F_2^{en}(x)/F_2^{ep}(x)$ would tend to $2/3$. Data shows that $F_2^{en}(x)/F_2^{ep}(x) \to 1/4$ when $x \to 1$. Therefore in reality we have rather $d_v(x)/u_v(x) \to 0$ for $x \to 1$. This is not fully understood and is a clear sign that some features of the nucleon structure functions remain mysterious.

One of the big challenges is that, at present, parton distribution functions cannot be calculated from basic principles of quantum chromodynamics. The primary reason is that they represent the strong binding and dynamics of the quarks inside the nucleon, and perturbation theory cannot be used in this regime. Hence, their investigation relies primarily on experimental data.

The perhaps surprising fact concerning the relatively successful predictions of the naive parton model, is that we have assumed the existence of quasi-free independent point-like quarks inside the proton, in spite of the fact that quarks are supposed to be confined by very strong color forces. Bjorken scaling suggests that the strong interactions must have the property of **asymptotic freedom**: they should become weaker at short distances, so that quarks behave as free particles for $Q^2 \to \infty$. We will see in Section 18.12 that this feature can be explained by the properties of the strong force.

Thus, the interaction between a $q\bar{q}$ pair looks like some kind of rubber band. If we try to separate the

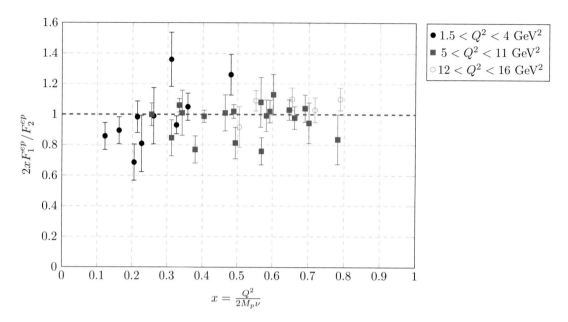

Figure 17.23 The ratio $2xF_1^{ep}/F_2^{ep}$ vs. x, for different Q^2 values (1.5 GeV^2 < Q^2 < 16 GeV^2). The data is from the SLAC-MIT experiment described in Section 16.6.

quark from the antiquark, the force joining them increases. At some point, the energy on the elastic band is bigger than $2m_q$, so that it becomes energetically favorable to create an additional $q\bar{q}$ pair; then the band breaks down into two mesonic systems, $q\bar{q}$ and $q\bar{q}$, each one with its corresponding half-band joining the quark pair. Increasing more and more the energy, we can only produce more and more mesons, but quarks remain always confined within color-singlet bound states. Conversely, if one tries to confine two quark constituents into a very short-distance region, the elastic band loses the energy and becomes very soft; quarks behave then as free particles! This a priori peculiar behavior will be discussed further in Section 18.13.

Problems

Ex 17.1 Mass formula of Gell-Mann–Okubo. The so-called Gell-Mann–Okubo mass formula yields a relation between the masses of the members of the baryon octet. It is expressed as:

$$2(m_N + m_\Xi) = 3m_\Lambda + m_\Sigma \qquad (17.105)$$

Use the known values of m_N, m_Ξ, and m_Σ to estimate the rest mass of the Λ. How large is the deviation from the experimentally observed value?

Ex 17.2 The ρ meson. The ρ meson has an isospin $I = 1$ and comes in three charge eigenstates: ρ^+, ρ^0 and ρ^-. It decays almost solely into two pions $\rho \to \pi\pi$. Show that the decays $\rho^\pm \to \pi^\pm\pi^0$ and $\rho^0 \to \pi^+\pi^-$ are allowed, while the decay $\rho^0 \to \pi^0\pi^0$ is forbidden.

Ex 17.3 The light vector mesons. The spin-triplet combination ($\uparrow\uparrow$) of a $q\bar{q}$ pair with $\ell = 0$ yields the vector mesons ($J = 1$). In order to explain the experimental observations, one must assume that the octet

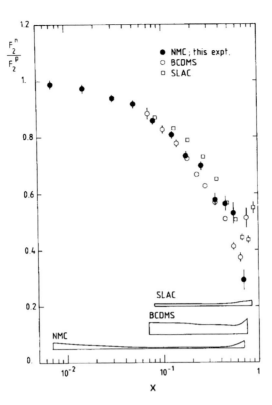

Figure 17.24 The ratio F_2^{en}/F_2^{ep} vs. x. Reprinted from D. Allasia *et al.*, "Measurement of the neutron and the proton F_2 structure function ratio," *Phys. Lett. B*, vol. 249, no. 2, pp. 366–372, 1990, with permission from Elsevier.

member with $I = S = 0$ and the corresponding singlet state are mixed. This can be expressed as:

$$\omega = \phi_0 \cos\theta + \phi_8 \sin\theta \quad \text{and} \quad \phi = \phi_0 \sin\theta - \phi_8 \cos\theta \tag{17.106}$$

where ϕ_0 and ϕ_8 are the singlet and octet states with $I = S = 0$. The ϕ and ω represent the physical states and θ is the mixing parameter. According to the $SU(3)$ symmetry, the ϕ_0 and ϕ_8 are classified as:

$$\phi_0 = \frac{1}{\sqrt{3}}(u\bar{u} + d\bar{d} + s\bar{s}) \quad \text{and} \quad \phi_8 = \frac{1}{\sqrt{6}}(u\bar{u} + d\bar{d} - 2s\bar{s}) \tag{17.107}$$

(a) It is assumed that the expectation value of the energy operator between two eigenstates is proportional to the mass squared of the mesons. Show that the mixing parameter θ is given by:

$$\tan^2\theta = \frac{M_\phi^2 - M_8^2}{M_8^2 - M_\omega^2} \tag{17.108}$$

(b) Use the empirical formula $M_8^2 = \frac{1}{3}\left(4M_{K^*}^2 - M_\rho^2\right)$ to estimate the value of θ.

(c) Show that for $\theta = 1/\sqrt{3}$, one has simply:

$$\phi = s\bar{s} \quad \text{and} \quad \omega = \frac{1}{\sqrt{2}}\left(u\bar{u} + d\bar{d}\right) \tag{17.109}$$

(d) Draw the corresponding quark flux diagrams (see Section 17.4) and explain qualitatively the branching ratios of the following decays:

$$\phi(1020) \quad \left. \begin{array}{l} \to K^+ K^- \\ \to K^0 \bar{K}^0 \end{array} \right\} \quad 84\% \\ \to \pi^+ \pi^- \pi^0 \quad 15\% \qquad (17.110)$$

and

$$\omega(783) \quad \begin{array}{l} \to \pi^+ \pi^- \pi^0 \quad 90\% \\ \left. \begin{array}{l} \to \pi^+ \pi^- \\ \to \pi^0 \gamma \end{array} \right\} \quad 10\% \end{array} \qquad (17.111)$$

Ex 17.4 Electron–neutron experiments. Electron–neutron experiments are harder to do than electron–proton experiments because you cannot make a target of free neutrons. Nevertheless, the essential data can be inferred from electron–deuteron scattering, and it is found that:

$$\int_0^1 \mathrm{d}x \, F_2^{en} = 0.12 \qquad (17.112)$$

Use this together with the proton result:

$$\int_0^1 \mathrm{d}x \, F_2^{ep} = 0.18 \qquad (17.113)$$

to confirm $\int_0^1 \mathrm{d}x x u(x) = 2 \int_0^1 \mathrm{d}x x d(x)$. (*Hint:* How do you suppose that $u^n(x)$ and $d^n(x)$ are related to the corresponding functions of the proton?)

Ex 17.5 Sum rules for the proton. From the known flavor content of the proton, find the value of $\int_0^1 \mathrm{d}x (u(x) - \bar{u}(x))$. State the corresponding *sum rules* for d and s.

18   Quantum Chromodynamics

> What is especially striking and remarkable is that in fundamental physics, a beautiful or elegant theory is more likely to be right than a theory that is inelegant. A theory appears to be beautiful or elegant (or simple, if you prefer) when it can be expressed concisely in terms of mathematics we already have. Symmetry exhibits the simplicity.

Murray Gell-Mann

18.1   Quarks and Gluons

Quantum chromodynamics (QCD), the theory of the strong interaction, was formulated in the early 1970s as a **non-Abelian gauge field theory** of **quarks that interact by the exchange of spin-1 massless gluons**. Today QCD is part of the Standard Model of particle physics and is one of the pillars on which our understanding of Nature rests.

The underlying concepts of QCD are in many aspects similar to those of QED. The QED vertex couples a spin-1 photon with zero rest mass (a real photon is massless) to an electrically charged fermion and action-at-distance is mediated by a photon. In QCD, the strong force is mediated by eight (massless) spin-1 gluons. The electron carries one unit of electric charge $-e$, while the positron has an opposite charge $+e$. Quarks, in addition to their electric charge, carry the "**color charge**," which comes in three types called R, G, and B (see Section 17.3). Antiquarks carry the anticolors $\overline{R}$, $\overline{G}$, and $\overline{B}$.

The strong interaction is invariant under a transformation in color space, e.g.:

$$R \to G, \quad G \to B, \quad B \to R \tag{18.1}$$

This can be expressed as the existence of an $SU(3)$ color symmetry, i.e., the strong force is the same for all three colors. Unlike the *uds* flavor symmetry which we discussed in the previous chapter, the $SU(3)$ color is an **exact** symmetry of Nature.

Under this light, QCD appears *at first sight* to be a sort of generalization of QED with eight gluons replacing the photon. But because the **gluons themselves carry color**, there is a fundamental difference between QED and QCD. The photon does not carry electric charge and being electrically neutral, a QED vertex factor between photons does not exist. The QED vertex strictly couples a photon to an electrically charged fermion. On the other hand, since gluons are colored and therefore carry a strong charge, gluons interact with other gluons or "self-couple" via QCD vertices involving three or four gluons. This self-coupling property is of fundamental importance and leads to asymptotic freedom, a key ingredient to understanding color confinement. Yang and Mills have shown that **all gauge field theories based on non-Abelian symmetries generate gauge bosons with self-couplings**. This will be discussed in Chapter 24.

18.2   The Colored Quark Dirac Fields

A quark is a spin-1/2 Dirac fermion. In addition, a quark has an internal color degree-of-freedom. Hence, a quark spinor must also include a color part. Let us denote q_f^i a quark Dirac field of color i ($i = 1, 2, 3$) and flavor f ($f = u, d, s, \ldots$). We can effectively write:

$$q_f^i \equiv c_i \Psi_f(p) \tag{18.2}$$

where Ψ_f is the four-component Dirac spinor for the quark of momentum p^μ, flavor f and the c_i, a three-element column vector, representing one of the possible *color states*:

$$c_i \equiv \begin{pmatrix} \delta_{1i} \\ \delta_{2i} \\ \delta_{3i} \end{pmatrix} : \quad c_1 = R = \begin{pmatrix} 1 \\ 0 \\ 0 \end{pmatrix}; \quad c_2 = G = \begin{pmatrix} 0 \\ 1 \\ 0 \end{pmatrix}; \quad c_3 = B = \begin{pmatrix} 0 \\ 0 \\ 1 \end{pmatrix} \tag{18.3}$$

Let us adopt a *vector notation in color space* (distinguished by the absence of i index):

$$q_f \equiv \begin{pmatrix} q_f^1 \\ q_f^2 \\ q_f^3 \end{pmatrix} = \begin{pmatrix} q_f^R \\ q_f^G \\ q_f^B \end{pmatrix} \quad \text{and} \quad \bar{q}_f = \begin{pmatrix} \bar{q}_f^1 & \bar{q}_f^2 & \bar{q}_f^3 \end{pmatrix} = \begin{pmatrix} \bar{q}_f^R & \bar{q}_f^G & \bar{q}_f^B \end{pmatrix} \tag{18.4}$$

We will use interchangeably both notations for the color index: $i = 1, 2, 3$ and $i = R, G, B$. The Lagrangian for the *free* Dirac colored quark fields can then simply be written in compact form as:

$$\mathcal{L}_{Dirac,color} = \sum_f \bar{q}_f \left(i\gamma^\mu \partial_\mu - m_f \mathbb{1} \right) q_f \tag{18.5}$$

In terms of the vector in color space, the Lagrangian actually represents the three-color sum, summed over all flavors f:

$$\mathcal{L}_{Dirac,color} = \sum_f \begin{pmatrix} \bar{q}_f^1 & \bar{q}_f^2 & \bar{q}_f^3 \end{pmatrix} \left(i\gamma^\mu \partial_\mu - m_f \mathbb{1} \right) \begin{pmatrix} q_f^1 \\ q_f^2 \\ q_f^3 \end{pmatrix} \tag{18.6}$$

It is also useful to express it with explicit indices for the color space, noting that:

$$\mathcal{L}_{Dirac,color} = \sum_f \sum_i \bar{q}_f^i \left(i\gamma^\mu \partial_\mu - m_f \mathbb{1} \right) q_f^i = \sum_f \sum_{ij} \bar{q}_f^i \left(i\gamma^\mu \partial_\mu - m_f \mathbb{1} \right) \delta_{ij} q_f^j \tag{18.7}$$

This Lagrangian is, as it should be, invariant under an arbitrary *global* $SU(3)_C$ transformations in color space.

18.3   Color Interactions via Gluons

The strong force is mediated by eight massless gluons corresponding to the eight generators of the $SU(3)_C$ symmetry. The three color charges (R,G,B) are conserved. Just like in QED where antiparticles have opposite electric charge, the antiquarks carry the opposite color charge, labeled $\bar{R}$, $\bar{G}$, and $\bar{B}$. Only particles that have non-zero color charge couple to the gluons. For instance leptons, which have no color, do not feel the strong force.

We have already mentioned that $SU(3)_C$ is an exact (gauge) symmetry of Nature (this is unlike the $SU(3)$ flavor which was shown to be approximate). Since QCD is invariant under the $SU(3)_C$ transformation, the strength of the gluon coupling is independent of the actual color of the quark (or anticolor of the antiquark). It is defined with a single constant **strong coupling constant** g_s.

How shall the gluons be constructed? Let us assume the presence of an incoming and an outgoing quark, which for the sake of argument each carry a different color, for example R and B. The interaction with the gluon occurs via the exchange of color whereby the incoming quark changes color. **In order to conserve color at the vertex, the gluon must carry a combination of color–anticolor.** In our example, the gluon must have an $R\bar{B}$ color state (see Figure 18.1).

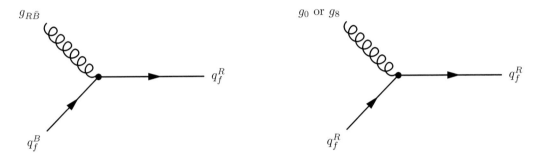

Figure 18.1 Feynman diagram of a quark–gluon vertex with a net flow of color. (left) Scattering of B–R quarks; (right) scattering of R–R quarks. See text for the definitions of g_0 and g_8.

From this first example, we conclude that **gluons carry color–anticolor pairs**. With three colors and three anticolors, we would naively expect to have $3 \times 3 = 9$ gluons! In order to get the correct answer, we need to recall the $SU(3)$ group properties that $3 \otimes \bar{3} = 8 \oplus 1$. The gluon color state can therefore be derived directly from the picture of the color of a quark–antiquark state, such as the one describing mesons (see Section 17.2 and Figure 18.2).

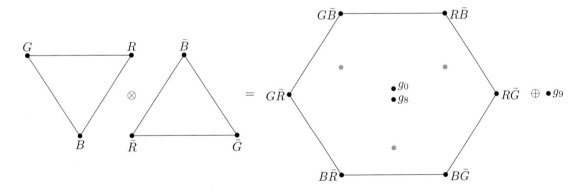

Figure 18.2 The $SU(3)$ color symmetry for gluons composed of a color–anticolor pair leads to $3 \otimes \bar{3} = 8 \oplus 1$ giving an $SU(3)$ octet plus a singlet.

Accordingly, the **color-singlet gluon**, namely:

$$g_9 \equiv g_{(R\bar{R}+G\bar{G}+B\bar{B})/\sqrt{3}} \tag{18.8}$$

does not carry any net color, and is postulated not to carry any force, since such a singlet would give rise to long-range strong forces since the gluons are massless! On the contrary, we empirically know that the strong force is short-ranged and therefore the physical gluons are confined. The color singlet state must not exist in Nature! This argument is consistent with the strong interaction arising from the fundamental $SU(3)$ symmetry. The gluons arise from the generators of the symmetry group (the Gell-Mann matrices). There are eight such matrices and hence eight gluons. For example, a hypothetical $U(3)$ symmetry would imply nine gluons, with the additional gluon being the color singlet state and QCD would be an unconfined long-range force! This leaves us with the **eight physical gluons as the members of the colored $SU(3)$ octet members** represented in Figure 18.2:

$$g_{G\overline{R}}, \ g_{G\overline{B}}, \ g_{R\overline{B}}, \ g_{R\overline{G}}, \ g_{B\overline{G}}, \ g_{B\overline{R}}, \quad g_0 \equiv g_{(R\overline{R}-G\overline{G})/\sqrt{2}}, \quad \text{and} \quad g_8 \equiv g_{(R\overline{R}+G\overline{G}-2B\overline{B})/\sqrt{6}} \tag{18.9}$$

Note that the last two labeled g_0 and g_8 appear "colorless" but are not color singlets since they are actually members of the octet. Consequently, to describe the color state of the gluon, we need an **eight-element column vector** C^a $(a = 1, \ldots, 8)$, where:

$$C^a \equiv \begin{pmatrix} \delta_{1a} \\ \delta_{2a} \\ \delta_{3a} \\ \delta_{4a} \\ \delta_{5a} \\ \delta_{6a} \\ \delta_{7a} \\ \delta_{8a} \end{pmatrix} : \quad C^1 = \begin{pmatrix} 1 \\ 0 \\ 0 \\ 0 \\ 0 \\ 0 \\ 0 \\ 0 \end{pmatrix}; \quad C^2 = \begin{pmatrix} 0 \\ 1 \\ 0 \\ 0 \\ 0 \\ 0 \\ 0 \\ 0 \end{pmatrix}; \quad \ldots \quad C^8 = \begin{pmatrix} 0 \\ 0 \\ 0 \\ 0 \\ 0 \\ 0 \\ 0 \\ 1 \end{pmatrix} \tag{18.10}$$

In a second example, we consider what happens if the two quarks have the same color, say R and R? In this case, there are two possibilities as illustrated in Figure 18.1(right). Thus, the color factor involved at the vertex for scattering two red quarks is $g_s/\sqrt{2}$ and $g_s/\sqrt{6}$ for the gluons $(R\overline{R} - G\overline{G})/\sqrt{2}$ and $(R\overline{R} + G\overline{G} - 2B\overline{B})/\sqrt{6}$, respectively.

Turning to scattering, we consider two quarks or two antiquarks mediated by a t-channel exchange of a gluon. In QED with the exchange of a photon, the strength of the amplitude would be proportional to $(ee_1)(ee_2)$, where e_i are the unit charges of the quarks. Similarly, in QCD, the strength of the coupling is given by the product of the color charges, say $(g_s c_1)(g_s c_2)$. We need to take into account that the possible gluons which can be exchanged in the scattering of two quarks depend on the colors of these latter. We have (see Figure 18.3):

- $RR \to RR$. The possible gluons mediating are g_0 and g_8. The color factors involved at each vertex are $c_1 = c_2 = 1/\sqrt{2}$ for the g_0 exchange and $c_1 = c_2 = 1/\sqrt{6}$ for g_8, respectively. The strength of the scattering amplitude is therefore given by $g_s^2 c_1 c_2 = g_s^2(1/\sqrt{2})(1/\sqrt{2}) = g_s^2/2$ and $g_s^2(1/\sqrt{6})(1/\sqrt{6}) = g_s^2/6$, respectively. The total color factor is the *sum* of these latter, so $g_s^2/2 + g_s^2/6 = (2/3)g_s^2$ for the case of two red quarks. The **color factor C_F is defined as half the value of this product**, where the factor one-half is due to the conventional normalization of the Gell-Mann matrices ($T^a = \lambda^a/2$ – see Eq. (18.21)):

$$C_F = \frac{1}{2} c_1 c_2 = \frac{1}{2} \frac{2}{3} = \frac{1}{3} \tag{18.11}$$

- $RB \to RB$. The possible gluon mediating is g_8 with a color factor:

$$C_F = \frac{1}{2} \frac{1}{\sqrt{6}} \frac{-2}{\sqrt{6}} = -\frac{1}{6} \tag{18.12}$$

- $RB \to BR$. This is the case swapped from the previous one and the possible gluon mediating is $R\overline{B}$ with a color factor:

$$C_F = \frac{1}{2}(1) = \frac{1}{2} \tag{18.13}$$

- All other combinations give the same results as they should because of color symmetry.

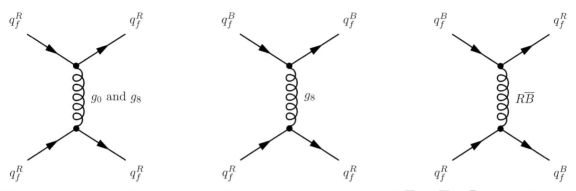

Figure 18.3 Feynman diagrams for quark–quark scattering. g_0 is the $(R\overline{R} - G\overline{G})/\sqrt{2}$ gluon and g_8 is the $(R\overline{R} + G\overline{G} - 2B\overline{B})/\sqrt{6}$ gluon.

Note that we have implicitly assumed that the color factors had to be summed. Actually, one should be careful with indistinguishable amplitudes since we are dealing with fermions. For instance, the indistinguishable $RB \to RB$ and $RB \to BR$ should be added (or subtracted) if the overall color state is symmetric (or antisymmetric), hence:

$$C_F = \frac{1}{2} - \frac{1}{6} = \frac{1}{3} \quad \text{or} \quad C_F^A = \frac{1}{2} + \frac{1}{6} = \frac{2}{3} \tag{18.14}$$

The color factors in t-channel interactions between a quark and an antiquark can be derived in a similar way. Therefore:

- $R\overline{R} \to R\overline{R}$. The red–antired combination of the mediated gluon can be g_0 or g_8 with strength $g_s^2/2$ and $g_s^2/6$, giving:

$$C_F = \frac{1}{2}\left(\frac{1}{2} + \frac{1}{6}\right) = \frac{1}{3} \tag{18.15}$$

- $R\overline{R} \to B\overline{B}$ or $R\overline{R} \to G\overline{G}$. The gluon carries $R\overline{B}$ or $R\overline{G}$ and the color factor is:

$$C_F = \frac{1}{2} \tag{18.16}$$

- $R\overline{B} \to R\overline{B}$. For the case quark–antiquark of different color–anticolor, the gluon is g_8 and the color factor is:

$$C_F = \frac{1}{2}\frac{1}{\sqrt{6}}\frac{-2}{\sqrt{6}} = -\frac{1}{6} \tag{18.17}$$

Getting color factors right is crucial to QCD calculations! In the upcoming sections, however, we will put them aside for a moment and come back to them in Section 18.7. There, they will be directly determined by the Gell-Mann matrices.

18.4 The Gluonic Field and the Quark–Gluon Vertex

The gluonic fields $G_a^\mu(x)$ $(a = 1, \ldots, 8)$ are defined in analogy to the electromagnetic field potential $A^\mu(x)$. In QED, the photon is the quantum of the field potential $A^\mu(x)$ and the electromagnetic field tensor $F^{\mu\nu}$ is given by Eq. (10.8):

$$F^{\mu\nu}(x) = \partial^\mu A^\nu(x) - \partial^\nu A^\mu(x) \tag{18.18}$$

Similarly, the eight gluons are the quanta of the gluonic fields $G_a^\mu(x)$. We now seek the gluonic field tensors, which conventionally are called $G_a^{\mu\nu}$. The careful reader might notice that both the gluonic field potentials $G_a^\mu(x)$ and the gluonic field tensors $G_a^{\mu\nu}$ are abbreviated with the letter G. They must therefore be distinguished by their number of Lorentz indices! We could try to generalize an expression for the eight gluonic field tensors of the form $G_a^{\mu\nu} = \partial^\mu G_a^\nu - \partial^\nu G_a^\mu$, however, this would *not* be $SU(3)$ gauge-invariant (see Chapter 24 for details). It turns out that the correct generalization has an additional term, which is a consequence of the $SU(3)$ generators being non-commutative (generally said to be non-Abelian). Because of this, QCD is defined as a non-Abelian gauge field theory. The gauge-invariant expression for the **eight gluonic field tensors** $G_a^{\mu\nu}(x)$ $(a = 1, \ldots, 8)$ is then (see Eq. (24.58)):

$$G_a^{\mu\nu}(x) \equiv \partial^\mu G_a^\nu(x) - \partial^\nu G_a^\mu(x) - g_s f_{abc} G_b^\mu(x) G_c^\nu(x) \tag{18.19}$$

where the f_{abc} are the group structure constants of $SU(3)$ defined in Eq. (17.26). We note that the extra term has the very peculiar feature of containing the strong coupling constant g_s. This term gives rise to gluon self-interactions which are hence totally defined as a consequence of the non-Abelian $SU(3)$ symmetry!

Let us consider the coupling between a quark and a gluon. Starting from the QED coupling between an electrically charged fermion and a photon, we recall the interacting term leading to the conserved vector current at the vertex has the form (see Eq. (10.93)):

$$\mathcal{L}_{int,QED} = -e\overline{\Psi}(x)\gamma^\mu\Psi(x)A_\mu(x) \tag{18.20}$$

The generalization to the strong color interaction corresponds to a Lagrangian term for a given flavor f:

$$\mathcal{L}_{f,int,QCD} = -g_s\bar{q}_f(x)(T^a)\gamma^\mu q_f(x)G_\mu^a(x) = -g_s\bar{q}_f(x)\left(\frac{\lambda^a}{2}\right)\gamma^\mu q_f(x)G_\mu^a(x) \tag{18.21}$$

where T_a are the $SU(3)$ generators and λ_a the Gell-Mann matrices. In this expression, recall that the quark field q_f is the three-dimensional color vector of spinors given in Eq. (18.4) and the λ^a are 3×3 matrices. The index a implies a sum over eight gluons $a = 1, \ldots, 8$. More explicitly and considering all flavors, this is equal to:

$$\mathcal{L}_{int,QCD} = -g_s \sum_f \sum_{a=1}^{8} \begin{pmatrix} \bar{q}_f^1 & \bar{q}_f^2 & \bar{q}_f^3 \end{pmatrix} \left(\frac{\lambda^a}{2}\right) \begin{pmatrix} \gamma^\mu q_f^1 \\ \gamma^\mu q_f^2 \\ \gamma^\mu q_f^3 \end{pmatrix} G_\mu^a(x) \tag{18.22}$$

The QCD interaction term can be compared to that of QED and the Feynman rules can be derived similarly. The quark–gluon vertex is analogous to the QED electron–photon case. The associated QED current was found to be of the type $j_{QED}^\mu = -ie\bar{u}(p')\gamma^\mu u(p)$. Consequently, introducing the conventional Dirac quark spinors u_f and antiquark spinors v_f (neglecting spin indices for clarity) for a given flavor f, the vector current associated with a given qqg vertex can be written as (see Figure 18.4):

$$j_f^\mu(p, i, p', j, a) \equiv \bar{u}_f(p')c_j^\dagger \left(-ig_s\frac{\lambda^a}{2}\gamma^\mu\right) c_i u_f(p) \tag{18.23}$$

where the c_i and c_j are the color column vectors defined in Eq. (18.3).

The 3×3 **Gell-Mann matrices** λ^a **act on the three-dimensional color vectors, whereas the** 4×4 **Dirac matrices act on the four components of the spinors.** Therefore, each matrix product factorizes, allowing us to write:

$$j_f^\mu = \left(-\frac{i}{2}g_s\right)\left[(\cdot \quad \cdot \quad \cdot)\begin{pmatrix} \cdot & \cdot & \cdot \\ \cdot & \cdot & \cdot \\ \cdot & \cdot & \cdot \end{pmatrix}^a \begin{pmatrix} \cdot \\ \cdot \\ \cdot \end{pmatrix}\right]\left[(\cdot \quad \cdot \quad \cdot \quad \cdot)\begin{pmatrix} \cdot & \cdot & \cdot & \cdot \\ \cdot & \cdot & \cdot & \cdot \\ \cdot & \cdot & \cdot & \cdot \\ \cdot & \cdot & \cdot & \cdot \end{pmatrix}^\mu \begin{pmatrix} \cdot \\ \cdot \\ \cdot \\ \cdot \end{pmatrix}\right]$$

$$= -\frac{i}{2}g_s\left[c_j^\dagger\lambda^a c_i\right][\bar{u}_f(p')\gamma^\mu u_f(p)] \tag{18.24}$$

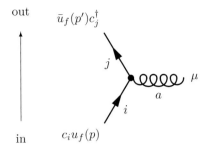

Figure 18.4 Feynman diagram for the quark–gluon vertex. The quarks have flavor f and color indices i and j. The gluon state is defined by the index a. Compare with the QED case in Figure 11.1.

where $c^\dagger$ is a row vector. The first bracket defines the **color factor** λ_{ij}^a, which is just a number, as:

$$\lambda_{ji}^a \equiv c_j^\dagger \lambda^a c_i = \begin{pmatrix} \delta_{1j} & \delta_{2j} & \delta_{3j} \end{pmatrix} \begin{pmatrix} \cdot & \cdot & \cdot \\ \cdot & \cdot & \cdot \\ \cdot & \cdot & \cdot \end{pmatrix}^a \begin{pmatrix} \delta_{1i} \\ \delta_{2i} \\ \delta_{3i} \end{pmatrix} = \left(\lambda^a\right)_{ji} \tag{18.25}$$

which, given the definition of the c_is, is just the (j, i) element of the λ^a matrix. In this manner, the **Feynman rule for the** qqg **vertex** for quark spinors $u_f(p)$ with color index i and $\bar{u}_f(p')$ with color index j can be summarized to the factor:

$$-\frac{i}{2}g_s\left(\lambda^a\right)_{ji}\gamma^\mu \tag{18.26}$$

18.5   The Gluonic Kinetic Term and Gluon Self-coupling

The kinetic term for the gluon fields is similar to the corresponding term for QED. Each of the eight gluon fields has its own kinetic term and the total Lagrangian is given by the sum of the eight terms. This can be written as:

$$\mathcal{L}_{glue} \equiv -\frac{1}{4}\sum_a (G_a)_{\mu\nu}(G_a)^{\mu\nu} = -\frac{1}{4}\vec{G}_{\mu\nu} \cdot \vec{G}^{\mu\nu} \tag{18.27}$$

A mass term for the gluon would be of the type $\frac{1}{2}m^2(G_a)^\mu(G_a)_\mu$ but since the gluon is massless it is not needed. We will see in Chapter 24 that this mass term is not gauge-invariant, therefore, in the context of the gauge field theory the gluon is predicted to be massless.

The equation of motion for the free gluon field corresponds to a generalization of the covariant Maxwell's equation, Eq. (10.11), with $J_\nu = 0$ in vacuum:

$$\partial^\mu(G_a)_{\mu\nu}(x) + gf_{abc}G_b^\mu(x)(G_c)_{\mu\nu}(x) = 0 \tag{18.28}$$

This expression represents a set of equations for $\nu = 0, \ldots, 3$ and $a = 1, \ldots, 8$. The **gluon Feynman propagator** is defined as in the case of the photon to be:

$$iD_{ab}^{\mu\nu}(x - y) = \langle 0|T\left(G_a^\mu(x)G_b^\nu(y)\right)|0\rangle \tag{18.29}$$

Its computation leads to the following expression:

$$iD_{ab}^{\mu\nu}(x - y) = \delta_{ab}\int\frac{\mathrm{d}^4q}{(2\pi)^4}\frac{-g^{\mu\nu} + (1 - \xi)\frac{q^\mu q^\nu}{q^2}}{q^2 + i\epsilon} \tag{18.30}$$

where ξ represents the **gluons field gauge**. The corresponding Feynman diagram is shown in Figure 18.5(a). The kinetic Lagrangian, Eq. (18.27), combined with the gluonic field tensor, Eq. (18.19), generates *non-linear*

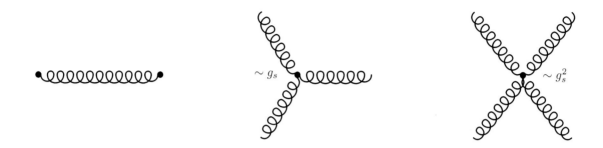

Figure 18.5 Feynman diagrams for (a) the gluon propagator; (b) the triple gluon vertex coupling; (c) the quadruple gluon vertex coupling.

gluon interactions, which have no analogs in QED. The **three-gluon vertex** is proportional to g_s and can be computed to be:

$$\mathcal{L}^{(3)} = -g_s f_{abc} \left(\partial^\mu (G_a)^\nu - \partial^\nu (G_a)^\mu \right) (G_b)_\mu (G_c)_\nu \tag{18.31}$$

The **four-gluon vertex** is proportional to g_s^2 and is equal to:

$$\mathcal{L}^{(4)} = \frac{g_s^2}{4} f_{abe} f_{cde} (G_a)_\mu (G_b)_\nu (G_c)^\mu (G_d)^\nu \tag{18.32}$$

The corresponding Feynman diagrams are shown in Figures 18.5(b) and (c).

In QED, the basic current led us to consider several tree-level $2 \to 2$ scattering processes, such as $f_1 + f_2 \to f_1 + f_2$, $f + \bar{f} \to f + \bar{f}$, $\gamma + f \to \gamma + f$, $f + \bar{f} \to \gamma + \gamma$, $\gamma + \gamma \to f + \bar{f}$, however, there is no $\gamma + \gamma \to \gamma + \gamma$ process (this last is called the scattering of light by light). In QCD, in addition to the scattering of quarks with gluons, the gluon self-coupling forces us to consider also the process involving only gluons. Hence, replacing photons by gluons and adding the additional process, we find the following basic $2 \to 2$ QCD processes (see Section 18.9):

$$q_1 + q_2 \to q_1 + q_2,\, q + \bar{q} \to q + \bar{q},\, g + q \to g + q,\, q + \bar{q} \to g + g,\, g + g \to q + \bar{q}, \quad \text{and} \quad g + g \to g + g \tag{18.33}$$

18.6   The QCD Lagrangian and Feynman Rules

We now have all the pieces to write down the full QCD Lagrangian by adding the three terms for (a) the free Dirac colored field Eq. (18.7), (b) the gluonic field kinetic term Eq. (18.27) and (c) the interaction term between the colored Dirac and the gluonic fields Eq. (18.21):

$$\mathcal{L}_{QCD} = \mathcal{L}_{Dirac,color} + \mathcal{L}_{glue} + \mathcal{L}_{int,QCD} \tag{18.34}$$

We can write the Lagrangian more elegantly by noting, for a given flavor f and colors i, j:

$$
\begin{aligned}
\mathcal{L}_{Dirac,color} + \mathcal{L}_{int,QCD} &= \sum_{ij} \left[\bar{q}_f^i \left(i\gamma^\mu \partial_\mu - m_f \mathbb{1} \right) \delta_{ij} q_f^j - g_s \bar{q}_f^i \left(T^a \right)_{ij} \gamma^\mu q_f^j G_\mu^a \right] \\
&= \sum_{ij} \bar{q}_f^i \left[\left(i\gamma^\mu \partial_\mu - m_f \mathbb{1} \right) \delta_{ij} - g_s \left(T^a \right)_{ij} \gamma^\mu G_\mu^a \right] q_f^j \\
&= \sum_{ij} \bar{q}_f^i \left[i\gamma^\mu \left(\partial_\mu \delta_{ij} + ig_s \left(T^a \right)_{ij} G_\mu^a \right) - \left(m_f \mathbb{1} \right) \delta_{ij} \right] q_f^j
\end{aligned}
\tag{18.35}
$$

We can hence introduce the covariant derivative D_{ij}^μ:

$$
D_{ij}^\mu \equiv \partial^\mu \delta_{ij} + ig_s \left(T^a \right)_{ij} \left(G^a \right)^\mu
\tag{18.36}
$$

thus for all flavors:

$$
\mathcal{L}_{Dirac,color} + \mathcal{L}_{int,QCD} = \sum_f \sum_{ij} \bar{q}_f^i \left[i\gamma_\mu D_{ij}^\mu - \left(m_f \mathbb{1} \right) \delta_{ij} \right] q_f^j
\tag{18.37}
$$

Finally, the full QCD Lagrangian in matrix form is simply, removing the sum over the color indices i, j:

$$
\mathcal{L}_{QCD} = \sum_f \bar{q}_f \left(i\gamma^\mu D_\mu - m_f \mathbb{1} \right) q_f - \frac{1}{4} \vec{G}^{\mu\nu} \cdot \vec{G}_{\mu\nu}
\tag{18.38}
$$

with

$$
D^\mu \equiv \partial^\mu \mathbb{1} + ig_s \vec{T} \cdot \vec{G}^\mu
\tag{18.39}
$$

In spite of the rich physics contained in it, the Lagrangian looks very simple, because of its color-symmetry properties. All interactions are contained in the covariant derivative and given in terms of a single universal strong coupling constant g_s. The existence of self-interactions among the gauge fields is a new feature that was not present in the QED case; these gauge self-interactions explain properties like asymptotic freedom and confinement, which do not appear in QED.

In QED, an important step in the quantization of the photon field was the proper handling of the gauge. In an analogous way, in order to properly quantize the QCD Lagrangian, one needs to add the so-called **gauge-fixing Faddeev–Popov terms** [202]. Since this is a rather complex technical issue, it is not discussed here.

We will consider QCD in its perturbative regime (**perturbative QCD**), which means that we can perform an expansion of observables in terms of the constant $\alpha_s \lesssim 1$, where:

$$
\alpha_s \equiv \frac{g_s^2}{4\pi}
\tag{18.40}
$$

Under the assumption of applicability of perturbative QCD, we move to the corresponding **QCD Feynman rules** that follow from the Lagrangian. The indices a, b $(a, b = 1, \dots, 8)$ correspond to the gluons color octet, and the indices $i, j = 1, 2, 3$ are quark colors:

- [R1] **Quark propagator:**

  $i \bullet\!\!\longrightarrow\!\!\bullet j$
  p^μ
  $= \dfrac{i(\slashed{p} + m)}{p^2 - m_f^2 (+i\epsilon)} \delta_{ij}$

- [R2] **Gluon propagator:**

  $a, \mu \bullet\!\!\text{0000000}\!\!\bullet b, \nu$
  q^2
  $= \dfrac{-i\delta_{ab}}{q^2 (+i\epsilon)} g^{\mu\nu}$

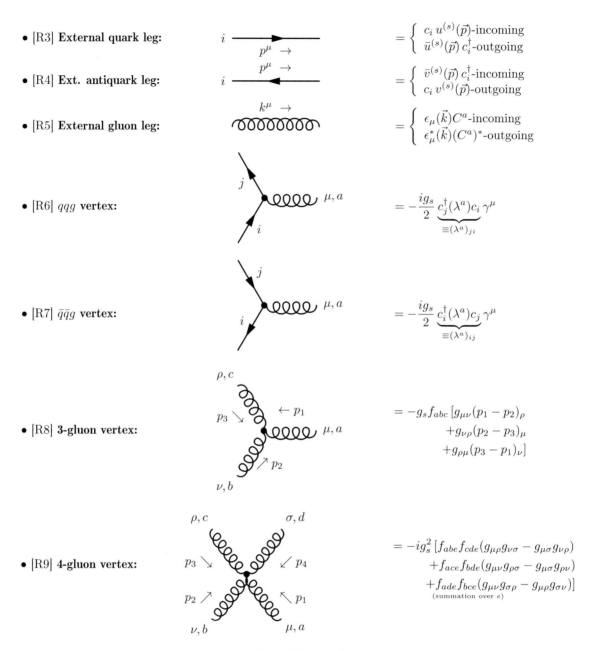

- [R3] **External quark leg:** $\quad i \xrightarrow{\quad\quad}_{p^\mu \to} \quad = \begin{cases} c_i\, u^{(s)}(\vec{p})\text{-incoming} \\ \bar{u}^{(s)}(\vec{p})\, c_i^\dagger\text{-outgoing} \end{cases}$

- [R4] **Ext. antiquark leg:** $\quad i \xleftarrow[\;\;]{p^\mu \to} \quad = \begin{cases} \bar{v}^{(s)}(\vec{p})\, c_i^\dagger\text{-incoming} \\ c_i\, v^{(s)}(\vec{p})\text{-outgoing} \end{cases}$

- [R5] **External gluon leg:** $\quad = \begin{cases} \epsilon_\mu(\vec{k})\, C^a\text{-incoming} \\ \epsilon_\mu^*(\vec{k})\,(C^a)^*\text{-outgoing} \end{cases}$

- [R6] qqg **vertex:** $\quad = -\dfrac{ig_s}{2}\, \underbrace{c_j^\dagger(\lambda^a)c_i}_{\equiv(\lambda^a)_{ji}}\, \gamma^\mu$

- [R7] $\bar{q}\bar{q}g$ **vertex:** $\quad = -\dfrac{ig_s}{2}\, \underbrace{c_i^\dagger(\lambda^a)c_j}_{\equiv(\lambda^a)_{ij}}\, \gamma^\mu$

- [R8] **3-gluon vertex:** $\quad \begin{aligned} = -g_s f_{abc} [&g_{\mu\nu}(p_1 - p_2)_\rho \\ &+ g_{\nu\rho}(p_2 - p_3)_\mu \\ &+ g_{\rho\mu}(p_3 - p_1)_\nu] \end{aligned}$

- [R9] **4-gluon vertex:** $\quad \begin{aligned} = -ig_s^2 [&f_{abe}f_{cde}(g_{\mu\rho}g_{\nu\sigma} - g_{\mu\sigma}g_{\nu\rho}) \\ &+ f_{ace}f_{bde}(g_{\mu\nu}g_{\rho\sigma} - g_{\mu\sigma}g_{\rho\nu}) \\ &+ f_{ade}f_{bce}(g_{\mu\nu}g_{\sigma\rho} - g_{\mu\rho}g_{\sigma\nu})] \\ &\text{(summation over } e) \end{aligned}$

- [R10] Useful relations (see Appendix B.9): $\left[\lambda^a, \lambda^b\right] = 2if^{abc}\lambda^c$

$$\text{Tr}\left(\lambda^a\lambda^b\right) = 2\delta^{ab}, \quad \sum_a \text{Tr}\left(\lambda^a\lambda^b\lambda^a\lambda^c\right) = -(4/3)\delta^{bc}, \quad \sum_a \text{Tr}\left(\lambda^a\lambda^a\lambda^b\lambda^c\right) = (32/3)\delta^{bc}$$

The rest of the rules are similar to those for QED. Energy and momentum must be conserved at each vertex to determine the internal four momenta, and so forth.

18.7   From QED to QCD Diagrams

We have already mentioned the similarities between QED and QCD. We can take advantage of this in order to help us compute QCD processes. We recall that:

- Quarks carry electric charge and color. There are three colors; since color is not directly measurable, cross-sections must be averaged over initial-state colors and summed over final-state colors (just as spins are handled in unpolarized cross-sections).

- Color is exchanged by eight spin-1 massless colored gluons while electric charge couples to the electrically neutral photon.

- Color interactions are vector-like and the QCD amplitudes can be computed similarly to those of QED (see example below).

- The vertex factor in QCD is given by the product of the strong coupling constant and the color factor C_F, while in QED it is just given by the elementary electric charge. Consequently, in QCD, there is some freedom in defining the coupling constant and the color factor, since only the product of the two enters the cross-sections and is physically meaningful. The adopted convention is just:

$$(g_s)^2 C_F \tag{18.41}$$

- QCD perturbative calculations only make sense if the coupling is small enough, i.e., at short distances or $Q^2 \gg 1$ GeV^2 (see Section 18.15).

We illustrate the QCD Feynman rules with the computation of the quark–quark scattering process. Using the rules defined above, we can directly derive the tree-level amplitude corresponding to the diagram shown in Figure 18.6. The initial-state quarks have color indices i and m, and the final-state quarks j and n. The

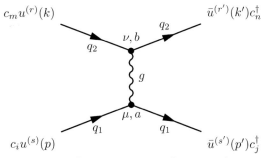

Figure 18.6 Strong scattering $q_1 q_2 \rightarrow q_1 q_2$ between two quarks via exchange of a gluon: tree-level Feynman diagram. Compare with the QED diagram in Figure 11.3.

amplitude is then given by:

$$
\begin{aligned}
-i\mathcal{M}_{q_1 q_2 \rightarrow q_1 q_2} &= \left[\bar{u}^{(r')}(k')c_n^\dagger \left(-ig_s \gamma_\nu \frac{\lambda^b}{2} \right) c_m u^{(r)}(k) \right] \left[\frac{-i\delta_{ab} g^{\mu\nu}}{(p-p')^2} \right] \left[\bar{u}^{(s')}(p')c_j^\dagger \left(-ig_s \gamma_\mu \frac{\lambda^a}{2} \right) c_i u^{(s)}(p) \right] \\
&= ig_s^2 \left[\left(c_n^\dagger \frac{\lambda^a}{2} c_m \right) \bar{u}^{(r')}(k')\gamma_\nu u^{(r)}(k) \right] \left[\frac{g^{\mu\nu}}{(p-p')^2} \right] \left[\left(c_j^\dagger \frac{\lambda^a}{2} c_i \right) \bar{u}^{(s')}(p')\gamma_\mu u^{(s)}(p) \right] \\
&= \frac{ig_s^2}{(p-p')^2} \underbrace{\frac{1}{4} \left(\lambda_{nm}^a \lambda_{ji}^a \right)}_{\text{color factor } C_F} \left[\bar{u}^{(r')}(k')\gamma^\mu u^{(r)}(k) \right] \left[\bar{u}^{(s')}(p')\gamma_\mu u^{(s)}(p) \right]
\end{aligned}
\tag{18.42}
$$

Note that the sum over the index a ($a = 1, \ldots, 8$) is implicit. It corresponds to summing over all possible gluon states that can be exchanged at once. Compare this result with the QED amplitude in Eq. (11.72) for $e - \mu$ scattering. The comparison reveals that the QCD $2 \to 2$ amplitude is indeed identical to the QED one **after the replacement of the coupling constants** $e^2 \leftrightarrow g_s^2$ or equivalently (with the convention of Eq. (18.41)):

$$\alpha = \frac{e^2}{4\pi} \quad \leftrightarrow \quad \alpha_s = \frac{g_s^2 C_F}{4\pi} \tag{18.43}$$

with **the inclusion of the "color factor"** given by:

$$C_F(im \to jn) \equiv \frac{1}{4} \sum_{a=1}^{8} \left(\lambda_{nm}^a \lambda_{ji}^a \right) \tag{18.44}$$

The amplitudes have the same spinor structure of type $\bar{u}\gamma^\mu u$ in their currents at the vertices. In general, the transcription between QED and QCD is straightforward, however, one should be careful that the transition is not always so simple, as there are cases of QCD diagrams (e.g., involving gluon–gluon couplings) which do not exist in QED! This will be discussed in Section 18.9.

The color factor expressed as Eq. (18.44) can readily be applied to recover the results derived in Section 18.3. For example, for the $2 \to 2$ quark scattering of the same color, say RR, we find:

$$C_F(RR \to RR) = \frac{1}{4} \sum_{a=1}^{8} \left(\lambda_{11}^a \lambda_{11}^a \right) = \frac{1}{4} \left(\lambda_{11}^3 \lambda_{11}^3 + \lambda_{11}^8 \lambda_{11}^8 \right) = \frac{1}{4} \left(1 + \frac{1}{3} \right) = \frac{1}{3} \tag{18.45}$$

Because of $SU(3)_C$ symmetry, we have:

$$C_F(RR \to RR) = C_F(GG \to GG) = C_F(BB \to BB) = \frac{1}{3} \tag{18.46}$$

In the case of two different colors, for example $RB \to RB$, we keep only the λ^a matrices that have non-zero entries in positions 11 and 33, hence:

$$C_F(RB \to RB) = \frac{1}{4} \sum_{a=1}^{8} \left(\lambda_{11}^a \lambda_{33}^a \right) = \frac{1}{4} \left(\lambda_{11}^8 \lambda_{33}^8 \right) = \frac{1}{4} \left(\frac{1}{\sqrt{3}} \cdot \frac{-2}{\sqrt{3}} \right) = -\frac{1}{6} \tag{18.47}$$

and again by symmetry, all exchanges of a color give:

$$C_F(RB \to RB) = C_F(RG \to RG) = \cdots = -\frac{1}{6} \tag{18.48}$$

And finally the swapped color case yields, for example for $RB \to BR$, a color factor equal to:

$$C_F(RB \to BR) = \frac{1}{4} \sum_{a=1}^{8} \left(\lambda_{12}^a \lambda_{21}^a \right) = \frac{1}{4} \left(\lambda_{12}^1 \lambda_{21}^1 + \lambda_{12}^2 \lambda_{21}^2 \right) = \frac{1}{4} \left(1 - i^2 \right) = \frac{2}{4} = \frac{1}{2} \tag{18.49}$$

By symmetry:

$$C_F(RB \to BR) = C_F(RG \to GR) = \cdots = \frac{1}{2} \tag{18.50}$$

These values coincide with the color factors we had derived in a heuristic way in Section 18.3.

What happens with antiquarks and their anticolors? Since ingoing and outgoing are swapped for antiparticles with respect to particles, the indices of the colors also swap. For example, the color factors for the t- and s-channels of the $q\bar{q} \to q\bar{q}$ scattering process (see Figure 18.7) are given by:

$$C_F^t(i\bar{m} \to j\bar{n}) \equiv \frac{1}{4} \sum_{a=1}^{8} \left(\lambda_{mn}^a \lambda_{ji}^a \right) \quad \text{and} \quad C_F^s(i\bar{m} \to j\bar{n}) \equiv \frac{1}{4} \sum_{a=1}^{8} \left(\lambda_{mi}^a \lambda_{jn}^a \right) \tag{18.51}$$

For example, one can compute:

$$C_F^t(R\overline{R} \to R\overline{R}) = \frac{1}{4}\left(\lambda_{11}^3\lambda_{11}^3 + \lambda_{11}^8\lambda_{11}^8\right) = \frac{1}{3}, \quad C_F^t(R\overline{G} \to R\overline{G}) = -\frac{1}{6}, \quad C_F^t(R\overline{R} \to G\overline{G}) = \frac{1}{2} \quad (18.52)$$

As expected, these coincide with the color factors derived in Section 18.3. Also for the s-channel, one for instance finds:

$$C_F^s(R\overline{R} \to R\overline{R}) = \frac{1}{3}, \quad C_F^s(R\overline{G} \to R\overline{G}) = \frac{1}{2}, \quad C_F^s(R\overline{R} \to G\overline{G}) = -\frac{1}{6} \quad (18.53)$$

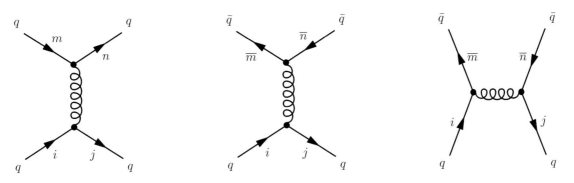

Figure 18.7 Feynman diagrams for $qq \to qq$ and t- and s-channels for $q\bar{q} \to q\bar{q}$ scattering.

18.8 Summing Initial and Final-State Color Factors

When dealing with QCD diagrams, it will always be necessary to consider products of the type $\lambda_{nm}^a\lambda_{ji}^a$ when computing color factors. A very useful theorem following directly from the structure of $SU(3)$ is the following:

$$C_F(im \to jn) = \frac{1}{4}\sum_{a=1}^{8}\left(\lambda_{nm}^a\lambda_{ji}^a\right) = \frac{1}{2}\delta_{ni}\delta_{mj} - \frac{1}{6}\delta_{nm}\delta_{ji} \quad (18.54)$$

The theorem can be verified directly by application of the Gell-Mann matrices as before. Indeed:

(a) With $i = m = j = n = 1$, we get:

$$C_F(11 \to 11) = \frac{1}{2}\delta_{11}\delta_{11} - \frac{1}{6}\delta_{11}\delta_{11} = \frac{1}{2} - \frac{1}{6} = \frac{1}{3} \quad (18.55)$$

which is exactly Eq. (18.45).

(b) With $n = m = 1$; $j = i = 2$, we get:

$$C_F(21 \to 21) = \frac{1}{2}\delta_{12}\delta_{12} - \frac{1}{6}\delta_{11}\delta_{11} = -\frac{1}{6} \quad (18.56)$$

which is exactly Eq. (18.47).

(c) With $n = i = 1$; $m = j = 2$, we get:

$$C_F(12 \to 21) = \frac{1}{2}\delta_{11}\delta_{22} - \frac{1}{6}\delta_{12}\delta_{12} = \frac{1}{2} \qquad (18.57)$$

which is exactly Eq. (18.49).

We can now use this theorem to show that:

$$
\begin{aligned}
\sum_a \text{Tr}(\lambda^a \lambda^a) &= \sum_a \text{Tr}\left(\sum_j \lambda_{ij}^a \lambda_{jk}^a\right) = \sum_a \sum_{i,k} \delta_{ik} \sum_j \lambda_{ij}^a \lambda_{jk}^a = \sum_a \sum_{i,j} \lambda_{ij}^a \lambda_{ji}^a = \sum_{i,j}\left(2\delta_{ii}\delta_{jj} - \frac{2}{3}\delta_{ij}\delta_{ji}\right) \\
&= 2\sum_{i=1}^{3}1\sum_{j=1}^{3}1 - \frac{2}{3}\sum_{i=1}^{3}\delta_{ii} = 2\times3\times3 - \frac{2}{3}\times3 = 18 - 2 = 16 \qquad (18.58)
\end{aligned}
$$

This is consistent with the result obtained by brute force in Appendix B.9 (Eq. (B.66)). We now reach this very important $SU(3)$ property (without the summation over a and b !):

$$\text{Tr}\left(\lambda^a \lambda^b\right) = 2\,\delta^{ab} \quad \Leftrightarrow \quad \text{Tr}\left(T^a T^b\right) = \frac{1}{2}\,\delta^{ab} \qquad (18.59)$$

We recognize Eq. (B.65). This result turns out to be very useful when computing the averaged color factors, as shown below.

18.9 Elementary $2 \to 2$ Quark Processes at Leading Order

We now study the elementary $2 \to 2$ QCD processes, in a similar way that we derived basic properties of the elementary $2 \to 2$ QED processes in Chapter 11. Starting from the results summarized in Table 11.2 and using the expressions derived in Section 11.12, we note that:

$$\frac{\mathrm{d}\sigma}{\mathrm{d}t} = \frac{2}{s}\frac{\mathrm{d}\sigma}{\mathrm{d}\cos\theta^\star} = \frac{4\pi}{s}\frac{\mathrm{d}\sigma}{\mathrm{d}\Omega^\star} = \frac{4\pi}{s}\frac{1}{64\pi^2 s}|\mathcal{M}|^2 = \frac{1}{16\pi s^2}|\mathcal{M}|^2 \qquad (18.60)$$

Consequently, we have for example:

$$\frac{\mathrm{d}\sigma}{\mathrm{d}t}(e^-\mu^- \to e^-\mu^-)_{QED} = \frac{1}{16\pi s^2}(2e^4)\left(\frac{s^2+u^2}{t^2}\right) = 2\frac{(4\pi\alpha)^2}{16\pi s^2}\left(\frac{s^2+u^2}{t^2}\right) = \frac{2\pi\alpha^2}{s^2}\left(\frac{s^2+u^2}{t^2}\right) \qquad (18.61)$$

To convert this expression to the scattering of two quarks of different flavors q_1 and q_2 in QCD, we need to replace the fine-structure constant α by the strong coupling α_s and add the corresponding color factor C_F. Hence

$$\frac{\mathrm{d}\sigma}{\mathrm{d}t}(q_1 q_2 \to q_1 q_2) = \frac{2\pi\alpha_s^2}{s^2}\langle C_F^2\rangle\left(\frac{s^2+u^2}{t^2}\right) \qquad (18.62)$$

where we have taken into account an averaging of the color factor. Indeed, as was the case for the spin, the fact that we do not measure color implies that our cross-sections will be averaged over initial colors i,m and summed over final colors j,n. Using the property of Eq. (18.59), we get:

$$
\begin{aligned}
\langle C_F^2\rangle &= \underbrace{\frac{1}{3}\frac{1}{3}\sum_{i,m}}_{\text{average initial colors}} \times \underbrace{\sum_{j,n}}_{\text{sum final colors}} \times C_F^2(im \to jn) = \frac{1}{9}\sum_{i,j,m,n}\sum_{a,b}\frac{1}{4}\left(\lambda_{nm}^a \lambda_{ji}^a\right)\frac{1}{4}\left(\lambda_{ji}^{*b}\lambda_{nm}^{*b}\right) \\
&= \frac{1}{9}\frac{1}{16}\sum_{a,b}\sum_{i,j,m,n}\lambda_{nm}^a\lambda_{mn}^b\lambda_{ji}^a\lambda_{ij}^b = \frac{1}{9}\sum_{a,b}\frac{1}{16}\text{Tr}(\lambda^a\lambda^b)\,\text{Tr}(\lambda^a\lambda^b) = \frac{1}{9}\sum_{a,b}\frac{1}{16}(2\delta^{ab})(2\delta^{ab}) \\
&= \frac{1}{9}\times\frac{1}{4}\times 8 = \frac{2}{9} \qquad (18.63)
\end{aligned}
$$

Thus

$$\frac{d\sigma}{dt}(q_1 q_2 \to q_1 q_2) = \frac{\pi \alpha_s^2}{s^2} \left(\frac{4}{9} \frac{s^2 + u^2}{t^2} \right) \tag{18.64}$$

We can apply the crossing symmetry from the t-channel to the s-channel to find the cross-section for the annihilation of two quarks into two quarks of different flavors (cf. $e^+ e^- \to \ell^+ \ell^-$):

$$\frac{d\sigma}{dt}(q_1 \bar{q}_1 \to q_2 \bar{q}_2) = \frac{\pi \alpha_s^2}{s^2} \left(\frac{4}{9} \frac{t^2 + u^2}{s^2} \right) \tag{18.65}$$

and also for the process of annihilation of a quark and antiquark of different flavors, which remains a t-channel:

$$\frac{d\sigma}{dt}(q_1 \bar{q}_2 \to q_1 \bar{q}_2) = \frac{\pi \alpha_s^2}{s^2} \left(\frac{4}{9} \frac{s^2 + u^2}{t^2} \right) \tag{18.66}$$

These results are summarized in Table 19.1 and plotted in Figure 18.8 as a function of the ratio t/s. We remind the reader that, from the discussion of the scattering kinematics in Section 11.12, we found that this ratio is related to the scattering angle via:

$$t = -\frac{s}{2}(1 - \cos\theta) \implies \frac{t}{s} = -\frac{1}{2} + \frac{\cos\theta}{2} \tag{18.67}$$

In order to compute the scattering of quarks of the same flavor, we should start from the Bhabha scattering cross-section (see Section 11.10):

$$\frac{d\sigma}{dt}(e^- e^+ \to e^- e^+)_{QED} = \frac{1}{16\pi s^2}(2e^4)\left(\frac{s^2 + u^2}{t^2} + \frac{2u^2}{ts} + \frac{u^2 + t^2}{s^2} \right) \tag{18.68}$$

The first and last term in parentheses correspond to the t- and s-channels, while the middle term is the interference term. We can apply the same color factor as before, except for the interference term, which has a color factor given by (see rule [R10]):

$$\frac{1}{3}\frac{1}{3}\frac{1}{16} \sum_{a,b} \text{Tr}\left(\lambda^a \lambda^b \lambda^a \lambda^b \right) = \frac{1}{9}\frac{1}{16} \times \left(-\frac{4}{3} \right) \times 8 = -\frac{2}{27} \tag{18.69}$$

So finally, we have:

$$\frac{d\sigma}{dt}(q_1 \bar{q}_1 \to q_1 \bar{q}_1) = \frac{2\pi \alpha_s^2}{s^2} \left(\frac{2}{9} \frac{s^2 + u^2}{t^2} - \frac{2}{27}\frac{2u^2}{ts} + \frac{2}{9}\frac{u^2 + t^2}{s^2} \right) \tag{18.70}$$

or

$$\frac{d\sigma}{dt}(q_1 \bar{q}_1 \to q_1 \bar{q}_1) = \frac{\pi \alpha_s^2}{s^2} \left(\frac{4}{9}\left(\frac{s^2 + u^2}{t^2} + \frac{u^2 + t^2}{s^2} \right) - \frac{8}{27}\frac{u^2}{ts} \right) \tag{18.71}$$

By crossing symmetry, one immediately obtains:

$$\frac{d\sigma}{dt}(q_1 q_1 \to q_1 q_1) = \frac{\pi \alpha_s^2}{s^2} \left(\frac{4}{9}\left(\frac{s^2 + u^2}{t^2} + \frac{s^2 + t^2}{u^2} \right) - \frac{8}{27}\frac{s^2}{ut} \right) \tag{18.72}$$

These results are also summarized in Table 19.1 and plotted in Figure 18.8.

18.10   The Pair-Annihilation Process $q\bar{q} \to gg$

We consider the process mediating the pair annihilation of quarks in QCD $q\bar{q} \to gg$ at tree level. This is analogous to the QED pair annihilation into two photons $e^+e^- \to \gamma\gamma$, described in Section 11.15. The kinematics is given by:

$$q(p,s) + \bar{q}(k,r) \to g(k_1, \epsilon_1) + g(k_2, \epsilon_2) \tag{18.73}$$

The corresponding Feynman diagrams at tree level are shown in Figure 18.9. The first two diagrams have their exact analogs in QED, as shown in Figure 11.14. The third diagram, with the triple gluon vertex, is only present in QCD.

We can use the results derived in QED to compute the first two diagrams. For these, the amplitudes can be found from their QED counterparts replacing the coupling e by g_s and adding the color factors. We have, using the Feynman rules **[R1]** and **[R3]** to **[R5]** (cf. Eq. (11.190)):

$$
\begin{aligned}
i\mathcal{M}_1 &= \epsilon_2^{\mu*}(k_2) C^{a*} \epsilon_1^{\nu*}(k_1) C^{b*} \bar{v}(k,r) c_2^\dagger (-i\frac{g_s}{2}\lambda^a \gamma_\mu) \frac{i(\slashed{q}+m_q)}{q^2 - m_q^2} (-i\frac{g_s}{2}\lambda^b \gamma_\nu) c_1 u(p,s) \\
&= \left(\frac{-g_s^2}{4}\right) \underbrace{\left[C^{a*} C^{b*} c_2^\dagger \lambda^a \lambda^b c_1 \right]}_{\text{color factor}} \underbrace{\epsilon_2^{\mu*}(k_2)\epsilon_1^{\nu*}(k_1) \bar{v}(k,r) \gamma_\mu \frac{i(\slashed{q}+m_q)}{q^2 - m_q^2} \gamma_\nu u(p,s)}_{\equiv \mathcal{M}_1^{QED}/e^2}
\end{aligned} \tag{18.74}
$$

where we recognize the QED amplitude (divided by the coupling e^2) by replacing the electron rest mass m_e with the quark m_q. The final-state gluons carry indices ν and b, and μ and a, respectively. Likewise, we can get the amplitude of the second diagram, where the two gluons are crossing, by swapping the μ and ν indices and obtain:

$$
i\mathcal{M}_2 = \left(\frac{-g_s^2}{4}\right) \underbrace{\left[C^{a*} C^{b*} c_2^\dagger \lambda^b \lambda^a c_1 \right]}_{\text{color factor}} \underbrace{\epsilon_2^{\mu*}(k_2)\epsilon_1^{\nu*}(k_1) \bar{v}(k,r) \gamma_\nu \frac{i(\slashed{\tilde{q}}+m_q)}{\tilde{q}^2 - m_q^2} \gamma_\mu u(p,s)}_{\equiv \mathcal{M}_2^{QED}/e^2} \tag{18.75}
$$

This is the analog of the QED amplitude, Eq. (11.191). Note that the order of the Gell-Mann (indices a and b) and the Gamma matrices (indices μ and ν) are inverted relative to each other in the amplitudes $\mathcal{M}_1$ and $\mathcal{M}_2$, but the final-state gluons always carry the indices ν and b, and μ and a, respectively.

For the third diagram which is specific to QCD, we use the Feynman rule **[R8]** for the triple gluon vertex and find:

$$
\begin{aligned}
i\mathcal{M}_3 &= \underbrace{\epsilon_2^{\mu*}(k_2) C^{a*} \epsilon_1^{\nu*}(k_1) C^{b*}}_{\text{final-state gluons}} \underbrace{\bar{v}(k,r) c_2^\dagger (-i\frac{g_s}{2}\lambda^d \gamma_\delta) c_1 u(p,s)}_{q\bar{q}g \text{ vertex}} \underbrace{\left(\frac{-ig^{\delta\lambda}\delta^{dc}}{q^2}\right)}_{\text{gluon propagator}} \\
&\times \underbrace{\left\{ -g_s f_{abc} \left[g_{\mu\nu} (-k_2 + k_1)_\lambda + g_{\nu\lambda}(-k_1 - q)_\mu + g_{\lambda\mu}(q + k_2)_\nu \right] \right\}}_{\text{triple gluon vertex}}
\end{aligned} \tag{18.76}
$$

A tricky point here is to keep track of all those indices! Moving from the left to the right of the diagram, the $q\bar{q}g$ vertex carries the Lorentz and color indices δ and d, respectively. The gluon propagator with the factor $g^{\delta\lambda}\delta^{dc}$ and momentum q connects this vertex to one leg of the triple gluon vertex with indices λ and c. The triple gluon vertex has color indices a, b, and c and Lorentz indices λ, μ, and ν. The external gluons carry indices ν and b, and μ and a, respectively, as was the case for the amplitudes $\mathcal{M}_1$ and $\mathcal{M}_2$. To construct the triple gluon vertex, we first take the structure constant f_{abc} of the $SU(3)$ group associated with the three color indices a, b, c. Then the Lorentz structure is constructed with three terms of the type $g_{\mu\nu}(\ldots)_\lambda$ + permutations. Since we chose the first term to be proportional to $g_{\mu\nu}$, the corresponding momenta are the gluon momenta k_2 and k_1. These latter are outgoing. Hence, the first term is $g_{\mu\nu}(-k_2 + k_1)_\lambda$. Then we permute, recalling that q^μ is incoming, to get $g_{\nu\lambda}(-k_1 - q)_\mu$. And finally, $g_{\lambda\mu}(q + k_2)_\nu$.

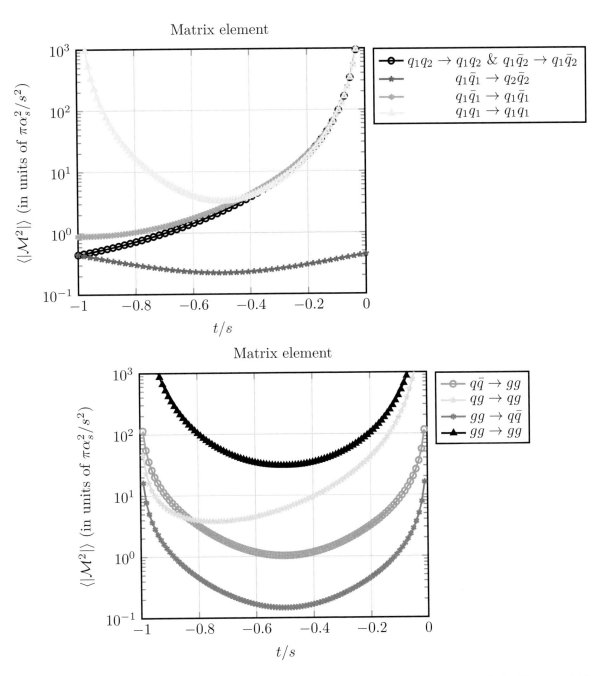

Figure 18.8 Graphical plots of the different terms in the matrix elements as a function of t/s. The u variable is given by $u/s = -(1 + t/s)$.

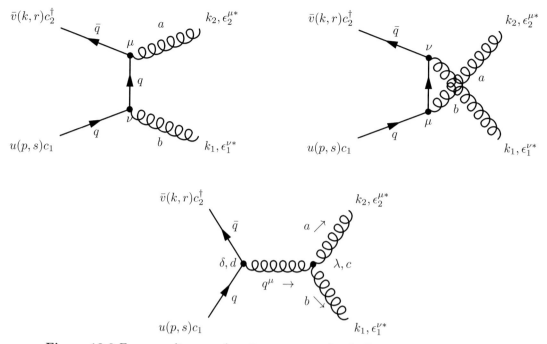

Figure 18.9 Feynman diagrams for $q\bar{q} \to gg$ at tree level. Compare with Figure 11.14.

- **Gauge invariance / Ward identities in QCD.** We have already discussed the importance of dealing with gauge invariance in QED. We now address this issue in the context of QCD, where the polarization degrees of freedom of the photons are now replaced by those of the gluons. Both photons and gluons should behave identically in the sense that they are massless spin-1 bosons.

From now on, we neglect the quark rest mass $m_q \to 0$. As in the QED case, we factorize the polarization vectors of the outgoing gluons and write:

$$i\mathcal{M} = \epsilon^*_{2\mu}(k_2)\epsilon^*_{1\nu}(k_1)C^{a*}C^{b*}i\mathcal{M}^{\mu\nu} = \epsilon^*_{2\mu}(k_2)\epsilon^*_{1\nu}(k_1)C^{a*}C^{b*}\left(i\mathcal{M}^{\mu\nu}_1 + i\mathcal{M}^{\mu\nu}_2 + i\mathcal{M}^{\mu\nu}_3\right) \tag{18.77}$$

where

$$i\mathcal{M}^{\mu\nu}_1 = \frac{-ig_s^2}{4t}\left[c_2^\dagger \lambda^a \lambda^b c_1\right]\bar{v}(k,r)\gamma^\mu(\not{p} - \not{k}_1)\gamma^\nu u(p,s) \tag{18.78}$$

and

$$i\mathcal{M}^{\mu\nu}_2 = \frac{-ig_s^2}{4u}\left[c_2^\dagger \lambda^b \lambda^a c_1\right]\bar{v}(k,r)\gamma^\nu(\not{p} - \not{k}_2)\gamma^\mu u(p,s) \tag{18.79}$$

Compare with Eqs. (11.193) and (11.194). Finally

$$
\begin{aligned}
i\mathcal{M}^{\mu\nu}_3 &= \frac{g_s^2}{2s}\left[c_2^\dagger \lambda^d c_1\right]\bar{v}(k,r)\gamma^\lambda u(p,s)\left\{f_{abd}\left[g^{\mu\nu}\left(-k_2 + k_1\right)_\lambda + g^\nu_\lambda\left(-k_1 - q\right)^\mu + g^\mu_\lambda\left(q + k_2\right)^\nu\right]\right\} \\
&= \frac{g_s^2 f_{abd}}{2s}\left[c_2^\dagger \lambda^d c_1\right]\bar{v}\left\{\left[g^{\mu\nu}\left(-\not{k}_2 + \not{k}_1\right) + \gamma^\nu\left(-k_1 - q\right)^\mu + \gamma^\mu\left(q + k_2\right)^\nu\right]\right\}u
\end{aligned} \tag{18.80}
$$

We saw that the Ward identity was satisfied in QED when both diagrams $\mathcal{M}_1$ and $\mathcal{M}_2$ were added together. It is instructive to see what happens in the case of QCD. We have:

$$k_{1\nu}\mathcal{M}^{\mu\nu}_1 = \frac{g_s^2}{4}\left[c_2^\dagger \lambda^a \lambda^b c_1\right]\bar{v}\gamma^\mu u \quad \text{and} \quad k_{1\nu}\mathcal{M}^{\mu\nu}_2 = -\frac{g_s^2}{4}\left[c_2^\dagger \lambda^b \lambda^a c_1\right]\bar{v}\gamma^\mu u \tag{18.81}$$

therefore

$$k_{1\nu}\left(\mathcal{M}_1^{\mu\nu} + \mathcal{M}_2^{\mu\nu}\right) = \frac{g_s^2}{4}\left(c_2^\dagger\left[\lambda^a, \lambda^b\right]c_1\right)\bar{v}\gamma^\mu u \tag{18.82}$$

hence, the Ward identity is *not* satisfied since the Gell-Mann matrices do not generally commute and $\left[\lambda^a, \lambda^b\right] = 2if^{abc}\lambda^c$ (see rule [**R10**]). So

$$k_{1\nu}\left(\mathcal{M}_1^{\mu\nu} + \mathcal{M}_2^{\mu\nu}\right) = \frac{ig_s^2 f^{abd}}{2}\left(c_2^\dagger\lambda^d c_1\right)\bar{v}\gamma^\mu u \tag{18.83}$$

The gauge cancellation that took place in QED between the two diagrams is spoiled by the non-Abelian nature of the coupling of quarks to gluons! The only possible way out is to add additional diagrams. We should consider the third diagram with the triple gluon vertex. We have:

$$\begin{aligned}
k_{1\nu}\mathcal{M}_3^{\mu\nu} &= \frac{1}{i}\frac{g_s^2 f_{abd}}{2s}\left[c_2^\dagger\lambda^d c_1\right]\bar{v}\left\{\left[k_1^\mu\left(-\not{k}_2 + \not{k}_1\right) + \not{k}_1\left(-2k_1 - k_2\right)^\mu + \gamma^\mu k_1\cdot(k_1 + 2k_2)\right]\right\}u \\
&= \frac{-ig_s^2 f_{abd}}{2s}\left[c_2^\dagger\lambda^d c_1\right]\bar{v}\left\{\left[-k_1^\mu\not{k}_2 + k_1^\mu\not{k}_1 - 2k_1^\mu\not{k}_1 - k_2^\mu\not{k}_1 + \gamma^\mu k_1\cdot(k_1 + 2k_2)\right]\right\}u \\
&= \frac{-ig_s^2 f_{abd}}{2s}\left[c_2^\dagger\lambda^d c_1\right]\bar{v}\left[-k_1^\mu\not{k}_2 - k_1^\mu\not{k}_1 - k_2^\mu\not{k}_1 + \gamma^\mu\underbrace{k_1\cdot k_1}_{=0} + \gamma^\mu\underbrace{(2k_1\cdot k_2)}_{=s}\right]u \\
&= \frac{-ig_s^2 f_{abd}}{2s}\left[c_2^\dagger\lambda^d c_1\right]\bar{v}\left[-k_1^\mu(\not{k}_2 + \not{k}_1) - k_2^\mu\not{k}_1 + s\gamma^\mu\right]u
\end{aligned} \tag{18.84}$$

However, we can once again use Dirac's equation:

$$\bar{v}(\not{k}_1 + \not{k}_2)u = \bar{v}(\not{p}' + \not{p})u = \bar{v}(-m_q + m_q)u = 0 \tag{18.85}$$

Thus

$$k_{1\nu}\mathcal{M}_3^{\mu\nu} = -i\frac{g_s^2 f_{abd}}{2}\left[c_2^\dagger\lambda^d c_1\right]\bar{v}\gamma^\mu u + i\frac{g_s^2 f_{abd}}{2s}\left[c_2^\dagger\lambda^d c_1\right]k_2^\mu\bar{v}\not{k}_1 u \tag{18.86}$$

and we are left with:

$$k_{1\nu}k_{2\mu}\left(\mathcal{M}_1^{\mu\nu} + \mathcal{M}_2^{\mu\nu}\right) + k_{1\nu}k_{2\mu}\mathcal{M}_3^{\mu\nu} = 0 \tag{18.87}$$

since $k_2^2 = 0$ for a real gluon. Finally, we therefore find that **the combined amplitude of all three diagrams is indeed gauge-invariant for physical transverse on-shell gluons!** That's great!

• **Matrix element.** We can now proceed to compute the squared matrix element averaged over initial and summed over final-state spins and colors. Since we have three diagrams $\mathcal{M} = \mathcal{M}_1 + \mathcal{M}_2 + \mathcal{M}_3$, we must consider six different pieces, but we can actually conveniently regroup them as follows:

$$|\mathcal{M}|^2 = |\mathcal{M}_{12}|^2 + |\mathcal{M}_3|^2 + 2\Re e(\mathcal{M}_{12}\mathcal{M}_3^*) \tag{18.88}$$

where $\mathcal{M}_{12} = \mathcal{M}_1 + \mathcal{M}_2$ and

$$|\mathcal{M}_{12}|^2 = |\mathcal{M}_1|^2 + |\mathcal{M}_2|^2 + 2\Re e(\mathcal{M}_1\mathcal{M}_2^*) \tag{18.89}$$

Compare with Eq. (11.200). We can get a hint on how to compute the $|\mathcal{M}_1|^2$, $|\mathcal{M}_2|^2$ and interference $2\Re e(\mathcal{M}_1\mathcal{M}_2^*)$ averaged over initial and summed over final helicities by using the results obtained in QED, neglecting the fermion rest mass (see Eqs. (11.221) and (11.222)):

$$\langle|\mathcal{M}_1|^2\rangle_{QED} = 2e^4\frac{u}{t} \quad \text{and} \quad \langle|\mathcal{M}_2|^2\rangle_{QED} = 2e^4\frac{t}{u} \tag{18.90}$$

These expressions led to the forward and backward peaks of the cross-section as $t \to 0$ or $u \to 0$, respectively. The interference term vanishes when we neglect the fermion rest mass (see Eq. (11.232)):

$$2 \langle \Re e(\mathcal{M}_1 \mathcal{M}_2^*) \rangle_{QED} = 0 \tag{18.91}$$

We should now evaluate the color factors for $|\mathcal{M}_1|^2$ and $|\mathcal{M}_2|^2$, averaged over initial quark colors i, m and summed over final state gluon colors. This gives (see rule **[R10]**):

$$\frac{1}{3}\frac{1}{3}\frac{1}{16} \sum_{a,b} \mathrm{Tr}\left(\lambda^a \lambda^b \lambda^b \lambda^a\right) = \frac{1}{3}\frac{1}{3}\frac{1}{16} \sum_{a,b} \mathrm{Tr}\left(\lambda^a \lambda^a \lambda^b \lambda^b\right) = \frac{1}{9}\frac{1}{16}\frac{32}{3} \times 8 = \frac{16}{27} \tag{18.92}$$

So we expect the dominant part of the amplitudes to be of that form, namely $\langle|\mathcal{M}_1|^2\rangle \simeq 2g_s^4 \frac{16}{27}\frac{u}{t}$ and $\langle|\mathcal{M}_2|^2\rangle \simeq 2g_s^4 \frac{16}{27}\frac{t}{u}$. Are these matrix elements then also applicable to the QCD case? Let us consider the calculation of $|\mathcal{M}_1|^2$. We can follow the steps performed for the QED case, as described in Section 11.15. In QCD, we would get (cf. Eq. (11.201)):

$$|\mathcal{M}_1|^2 = \frac{g_s^4}{16t^2}\left[\cdots\right]^*\left[\cdots\right]\bar{v}(k,r)\slashed{\epsilon}_2^*(k_2)(\slashed{p}-\slashed{k}_1)\slashed{\epsilon}_1^*(k_1)u(p,s)\bar{u}(p,s)\slashed{\epsilon}_1(k_1)(\slashed{p}-\slashed{k}_1)\slashed{\epsilon}_2(k_2)v(k,r) \tag{18.93}$$

where the dots in the big square brackets represent the color factors. In order to get the unpolarized matrix element, we *average* over the incoming quark helicities and find (cf. Eq. (11.202)):

$$\frac{1}{4}\sum_{s,r}|\mathcal{M}_1|^2 = \frac{g_s^4}{4 \times 16t^2}\mathrm{Tr}\left((\slashed{k})\slashed{\epsilon}_2^*(k_2)(\slashed{p}-\slashed{k}_1)\slashed{\epsilon}_1^*(k_1)(\slashed{p})\slashed{\epsilon}_1(k_1)(\slashed{p}-\slashed{k}_1)\slashed{\epsilon}_2(k_2)\right) \tag{18.94}$$

The next step is to sum over the gluon polarization states. One would be tempted to replace the gluon polarization using the completeness relation used for real physical photons (see Eq. (10.143)). But this is where it gets messier than in the QED case. We have already pointed out in Section 10.14 that the validity of this expression is related to the gauge problem and the Ward identities. In fact, we saw that the amplitudes for the production of the time-like and longitudinal components of the photon always cancelled each other in QED (see Eq. (10.141)). This is no longer true in the case of QCD and the amplitudes for unphysical gluons do not necessary cancel. In fact, in the case of $q\bar{q} \to gg$, we saw that the Ward identity was not satisfied for the two diagrams $\mathcal{M}_1$ and $\mathcal{M}_2$, and that we needed to add the third diagram with the triple gluon vertex in order to restore its validity. This observation already hinted at the fact that one cannot directly transcribe QED to QCD. Consequently, for the gluons, we should use the completeness relation given in Eq. (10.153), *de facto* including unphysical contributions:

$$\sum_{\lambda} \epsilon_{(\lambda)}^{\mu,*}\epsilon_{(\lambda)}^{\nu} \leftrightarrow -g^{\mu\nu} + \frac{k^{\mu}k^{\nu}}{k^2} \tag{18.95}$$

After lengthy calculations, one indeed obtains a result which contains an extra factor compared to the naive translation from QED:

$$\langle|\mathcal{M}_1|^2\rangle = 2g_s^4 \frac{16}{27}\frac{u}{t}\left(\frac{3t^2+u}{s^2}\right) \tag{18.96}$$

The $\langle|\mathcal{M}_2|^2\rangle$ is found by swapping $u \leftrightarrow t$, hence:

$$\langle|\mathcal{M}_2|^2\rangle = 2g_s^4 \frac{16}{27}\frac{t}{u}\left(\frac{3u^2+t}{s^2}\right) \tag{18.97}$$

The interference term, while vanishing in the QED case, can be computed:

$$2 \langle \Re e(\mathcal{M}_1 \mathcal{M}_2^*) \rangle = -\frac{8}{27}\frac{(t-u)^2}{s^2} \tag{18.98}$$

Consequently

$$\langle |\mathcal{M}_{12}|^2 \rangle = \frac{32g_s^4}{27} \left(\frac{u}{t} \left(\frac{3t^2 + u}{s^2} \right) + \frac{t}{u} \left(\frac{3u^2 + t}{s^2} \right) - \frac{1}{4} \frac{(t-u)^2}{s^2} \right) \tag{18.99}$$

Shall we worry about the extra contributions from unphysical gluons? Well, the short answer is no. Of course, they only appear when we consider diagrams individually for which the Ward identities are not independently satisfied. We have seen that in our present case the inclusion of the missing third diagram containing the triple gluon does lead, as expected, to the Ward identities being satisfied. Hence, we expect that the unphysical contributions will be "cured" when we eventually take all diagrams contributing to our process into account. So, we need not worry if these terms appear when we consider each diagram individually (or a subset of diagrams), which do not satisfy the Ward identities. More generally, however, the presence of unphysical states of gluons can be worrisome and is intrinsically embedded in the non-Abelian nature of QCD. This was not present in QED. As was already pointed out in the case of QED, we do want to stick to a covariant quantization. The other clearly disfavored solution was to adopt a non-covariant quantization where only the physical transverse polarizations do exist. This would be rather inconvenient and break the beauty of the construction. A more clever solution consists in additional unphysical scalar fields, the so-called *ghosts*, whose inclusion will exactly cancel the longitudinal gluon polarization. In practice, this magic trick is achieved by adding the so-called Faddeev–Popov term to the Lagrangian [202]. The exact mechanism giving rise to these terms can only be understood in the broader context of path integrals, which are beyond the scope of the present book. So we won't elaborate further, except to say that, this procedure is once again related to the gauge-fixing problem and it allows us to develop a covariant formalism, and hence the set of Feynman rules, to perform our calculations!

Let us go back to our calculation and now focus on $|\mathcal{M}_3|^2$. We begin by rewriting $\mathcal{M}_3$ by replacing q^2 with s and collapsing the indices c and δ to get:

$$
\begin{aligned}
i\mathcal{M}_3 &= \frac{g_s^2 f_{abd}}{2s} \epsilon_2^{\mu*}(k_2) C^{a*} \epsilon_1^{\nu*}(k_1) C^{b*} \bar{v}(k,r) c_2^\dagger (\lambda^d \gamma^\lambda) c_1 u(p,s) \\
&\quad \times \left\{ g_{\mu\nu} (-k_2 + k_1)_\lambda + g_{\nu\lambda} (-2k_1 - k_2)_\mu + g_{\lambda\mu} (k_1 + 2k_2)_\nu \right\} \\
&= \frac{g_s^2 f_{abd}}{2s} \left[C^{a*} C^{b*} c_2^\dagger (\lambda^d) c_1 \right] \bar{v} \gamma^\lambda \left\{ (\epsilon_1^* \cdot \epsilon_2^*)(-k_2 + k_1)_\lambda + \epsilon_{1\lambda}^* \epsilon_2^* \cdot (-2k_1 - k_2) + \epsilon_{2\lambda}^* \epsilon_1^* \cdot (k_1 + 2k_2) \right\} u \\
&= \frac{g_s^2 f_{abd}}{2s} \left[C^{a*} C^{b*} c_2^\dagger (\lambda^d) c_1 \right] \bar{v} \left\{ (\epsilon_1^* \cdot \epsilon_2^*)(-\slashed{k}_2 + \slashed{k}_1) + \slashed{\epsilon}_1^* \epsilon_2^* \cdot (-2k_1 - k_2) + \slashed{\epsilon}_2^* \epsilon_1^* \cdot (k_1 + 2k_2) \right\} u \\
&= \frac{g_s^2 f_{abd}}{s} \left[C^{a*} C^{b*} c_2^\dagger (\lambda^d) c_1 \right] \bar{v} \left\{ \frac{(\epsilon_1^* \cdot \epsilon_2^*)}{2} (-\slashed{k}_2 + \slashed{k}_1) - (\epsilon_2^* \cdot k_1) \slashed{\epsilon}_1^* + (\epsilon_1^* \cdot k_2) \slashed{\epsilon}_2^* \right\} u \tag{18.100}
\end{aligned}
$$

since $\epsilon_1 \cdot k_1 = 0$ and $\epsilon_2 \cdot k_2 = 0$. A very tedious calculation yields:

$$\langle |\mathcal{M}_3|^2 \rangle = \frac{16g_s^4}{3} \left(\frac{ut}{s^2} \right) \tag{18.101}$$

A similarly lengthy calculation gives:

$$2 \langle \Re e(\mathcal{M}_{12} \mathcal{M}_3^*) \rangle = -\frac{32g_s^4}{3} \left(\frac{ut}{s^2} \right) \tag{18.102}$$

Collecting all the pieces we get:

$$
\begin{aligned}
\langle|\mathcal{M}|^2\rangle &= \langle|\mathcal{M}_{12}|^2\rangle + \langle|\mathcal{M}_3|^2\rangle + 2\langle\Re e(\mathcal{M}_{12}\mathcal{M}_3^*)\rangle \\
&= \frac{32g_s^4}{27}\left(\frac{u}{t}\left(\frac{3t^2+u}{s^2}\right)+\frac{t}{u}\left(\frac{3u^2+t}{s^2}\right)-\frac{1}{4}\frac{(t-u)^2}{s^2}\right)+\frac{16g_s^4}{3}\left(\frac{ut}{s^2}\right)-\frac{32g_s^4}{3}\left(\frac{ut}{s^2}\right) \\
&= \frac{32g_s^4}{27s^2}\left[\frac{u}{t}(3t^2+u^2)+\frac{t}{u}(3u^2+t^2)-\frac{(t-u)^2}{4}-\frac{9}{2}ut\right] \\
&= \frac{32g_s^4}{27s^2ut}\frac{1}{4}\left[4u^2(3t^2+u^2)+4t^2(3u^2+t^2)-ut(t-u)^2-18u^2t^2\right] \\
&= \frac{32g_s^4}{27s^2ut}\frac{1}{4}\left[4u^4-tu^3+8t^2u^2-ut^3+4t^4\right] \\
&= \frac{32g_s^4}{27s^2ut}\left[(u^2+t^2)(t^2-\frac{tu}{4}+u^2)\right]=\frac{32g_s^4}{27s^2ut}(u^2+t^2)\left[\underbrace{(t+u)^2}_{\approx s^2}-\frac{9}{4}tu\right] \\
&= \frac{32g_s^4}{27}\left[\frac{u^2+t^2}{ut}-\frac{9}{4}\frac{u^2+t^2}{s^2}\right] \tag{18.103}
\end{aligned}
$$

So, ultimately:

$$
\langle|\mathcal{M}|^2\rangle = g_s^4\left[\frac{32}{27}\frac{u^2+t^2}{ut}-\frac{8}{3}\frac{u^2+t^2}{s^2}\right] \tag{18.104}
$$

Taking into account the phase-space factor Eq. (18.60), one finally obtains:

$$
\frac{d\sigma}{dt}=\frac{1}{16\pi s^2}\langle|\mathcal{M}|^2\rangle=\frac{(4\pi\alpha_s)^2}{16\pi s^2}\left\{\frac{32}{27}\left(\frac{u^2+t^2}{ut}\right)-\frac{8}{3}\left(\frac{u^2+t^2}{s^2}\right)\right\} \tag{18.105}
$$

or

$$
\frac{d\sigma}{dt}(q\bar{q}\to gg)=\frac{\pi\alpha_s^2}{s^2}\left(\frac{32}{27}\frac{u^2+t^2}{ut}-\frac{8}{3}\frac{u^2+t^2}{s^2}\right) \tag{18.106}
$$

By crossing symmetry, we can obtain the cross-section for the $qg\to qg$ process. The difference is that one must average over one initial-state gluon color instead of a quark color. Since we previously averaged over three initial-state colors of quarks and summed over the eight colors of two gluons, the new color factor after crossing one quark and one gluon will be $(3/8)$ times the previous color factor. Hence, one finds:

$$
\frac{d\sigma}{dt}(qg\to qg)=\frac{\pi\alpha_s^2}{s^2}\left(-\frac{4}{9}\frac{u^2+s^2}{us}+\frac{u^2+s^2}{t^2}\right) \tag{18.107}
$$

By further crossing symmetry, we can obtain the cross-section for the $gg\to q\bar{q}$ process. The difference is that this time one must average over initial-state gluon colors instead of quark colors. Consequently, the crossing of two quarks and two colors implies rescaling the color factor by $(3/8)^2$, thus one finds:

$$
\frac{d\sigma}{dt}(gg\to q\bar{q})=\frac{\pi\alpha_s^2}{s^2}\left(\frac{1}{6}\frac{u^2+t^2}{ut}-\frac{3}{8}\frac{u^2+t^2}{s^2}\right) \tag{18.108}
$$

These results are summarized in Table 19.1 and plotted in Figure 18.8.

18.11 The Purely Gluonic Process $gg\to gg$

We now consider the $2\to2$ scattering with only gluons, or gluon–gluon scattering:

$$
g(k_1,\epsilon_1)+g(k_2,\epsilon_2)\to g(k_3,\epsilon_3)+g(k_4,\epsilon_4) \tag{18.109}
$$

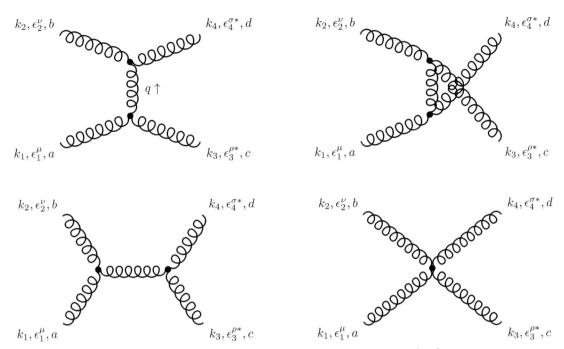

Figure 18.10 Feynman diagrams for $gg \rightarrow gg$ at tree level.

This is the most notorious and difficult case, with no analog in QED, since there is no tree-level diagram for $\gamma\gamma \rightarrow \gamma\gamma$ (scattering of light by light). The four lowest-order diagrams representing the interaction of two gluons are shown in Figure 18.10. The external gluons have incoming momenta k_1 and k_2, and outgoing momenta k_3 and k_4, with their respective polarization vectors ϵ_i. We assign their polarization four-vector Lorentz indices as μ, ν and ρ, σ, respectively. Finally, their color indices are given by a, b and c, d. The first two diagrams are t- and u-channel exchanges. We start with $\mathcal{M}_1$ and write the amplitude with the help of two triple-gluon vertices. The momentum of the exchanged gluon is $q = k_1 - k_3 = k_4 - k_2$. We assign a Lorentz index λ and a color index e. We therefore have (note that for simplicity we have already swallowed the indices on the gluon propagator):

$$
\begin{aligned}
i\mathcal{M}_1 &= \epsilon_{1\mu} C^a \epsilon_{2\nu} C^b \underbrace{(-g_s) f^{ace} \left[g^{\mu\rho}(k_1 + k_3)^\lambda + g^{\rho\lambda}(-k_3 + q)^\mu + g^{\lambda\mu}(-q - k_1)^\rho \right]}_{\text{triple gluon}} \underbrace{\frac{-i}{(k_1 - k_3)^2}}_{\text{gluon propagator}} \\
&\quad \times \underbrace{(-g_s) f^{bde} \left[g^{\nu\sigma}(k_2 + k_4)_\lambda + g^\sigma_\lambda(-k_4 - q)^\nu + g^\nu_\lambda(q - k_2)^\sigma \right]}_{\text{triple gluon}} \epsilon^*_{3\rho} C^{c*} \epsilon^*_{4\sigma} C^{d*} \\
&= \frac{-ig_s^2 f^{ace} f^{bde}}{t} \left[g^{\mu\rho}(k_1 + k_3)^\lambda - g^{\rho\lambda}(2k_3 - k_1)^\mu - g^{\lambda\mu}(2k_1 - k_3)^\rho \right] \\
&\quad \times \left[g^{\nu\sigma}(k_2 + k_4)_\lambda - g^\sigma_\lambda(2k_4 - k_2)^\nu - g^\nu_\lambda(2k_2 - k_4)^\sigma \right] \left[\epsilon_{1\mu} \epsilon_{2\nu} \epsilon^*_{3\rho} \epsilon^*_{4\sigma} C^a C^b C^{c*} C^{d*} \right] \quad (18.110)
\end{aligned}
$$

We recognize the t-channel nature of the amplitude. In using rule **[R8]**, we again paid attention to the signs of the momenta, recalling that the rule is defined assuming that all momenta are incoming. In the gluon triple vertex involving k_1 and k_3, we have that k_1 is incoming and k_3 is outgoing, and the definition of q makes it outgoing as well relative to this vertex. Hence, the first term is written $g^{\mu\rho}(k_1 + k_3)^\lambda$. Then, following the same reasoning, the second term is $g^{\rho\lambda}(-k_3 + q)^\mu$ and finally the third term is $g^{\lambda\mu}(-q - k_1)^\rho$. Similar arguments define the signs inside the square bracket in the second line of the expression, noting that this time k_2 and q

are incoming, while k_4 is outgoing. We now drop the C^a factors and further simplify the amplitude by making use of the fact that $\epsilon_i \cdot k_i = 0$:

$$
\begin{aligned}
i\mathcal{M}_1 &= \frac{-ig_s^2 f^{ace} f^{bde}}{t} \left[(\epsilon_1 \cdot \epsilon_3^*)(k_1 + k_3)^\lambda - \epsilon_3^{*\lambda}(2k_3 - k_1) \cdot \epsilon_1 - \epsilon_1^\lambda (2k_1 - k_3) \cdot \epsilon_3^* \right] \\
&\quad \times \left[(\epsilon_2 \cdot \epsilon_4^*)(k_2 + k_4)_\lambda - \epsilon_{4\lambda}^*(2k_4 - k_2) \cdot \epsilon_2 - \epsilon_{2\lambda}(2k_2 - k_4) \cdot \epsilon_4^* \right] \\
&= \frac{-ig_s^2 f^{ace} f^{bde}}{t} \left[(\epsilon_1 \cdot \epsilon_3^*)(k_1 + k_3)^\lambda - 2\epsilon_3^{*\lambda}(k_3 \cdot \epsilon_1) - 2\epsilon_1^\lambda(k_1 \cdot \epsilon_3^*) \right] \\
&\quad \times \left[(\epsilon_2 \cdot \epsilon_4^*)(k_2 + k_4)_\lambda - 2\epsilon_{4\lambda}^*(k_4 \cdot \epsilon_2) - 2\epsilon_{2\lambda}(k_2 \cdot \epsilon_4^*) \right] \\
&= \frac{-ig_s^2 f^{ace} f^{bde}}{t} \times \left[(\epsilon_1 \cdot \epsilon_3^*)(\epsilon_2 \cdot \epsilon_4^*)(k_1 + k_3) \cdot (k_2 + k_4) \right. \\
&\quad - 2\epsilon_1 \cdot (k_2 + k_4)(k_1 \cdot \epsilon_3^*)(\epsilon_2 \cdot \epsilon_4^*) - 2\epsilon_2 \cdot (k_1 + k_3)(k_2 \cdot \epsilon_4^*)(\epsilon_1 \cdot \epsilon_3^*) \\
&\quad - 2\epsilon_3^* \cdot (k_2 + k_4)(k_3 \cdot \epsilon_1)(\epsilon_2 \cdot \epsilon_4^*) - 2\epsilon_4^* \cdot (k_1 + k_3)(k_4 \cdot \epsilon_2)(\epsilon_1 \cdot \epsilon_3^*) \\
&\quad + 4(\epsilon_1 \cdot \epsilon_2)(k_1 \cdot \epsilon_3^*)(k_2 \cdot \epsilon_4^*) + 4(\epsilon_1 \cdot \epsilon_4^*)(k_1 \cdot \epsilon_3^*)(k_4 \cdot \epsilon_2) \\
&\quad \left. + 4(\epsilon_2 \cdot \epsilon_3^*)(k_3 \cdot \epsilon_1)(k_2 \cdot \epsilon_4^*) + 4(\epsilon_3^* \cdot \epsilon_4^*)(k_3 \cdot \epsilon_1)(k_4 \cdot \epsilon_2) \right]
\end{aligned}
\tag{18.111}
$$

Although this result looks rather unmanageable, there is some sense of symmetry in it that comforts us. True scholars would have guessed it! We can now evaluate the squared matrix elements averaged over incoming polarizations and summed over final-state polarizations. By realizing that there are two initial-state polarizations per gluon and eight possible colors to average on, one then gets:

$$
\langle |\mathcal{M}_1|^2 \rangle = \frac{1}{2^2} \sum_{\lambda_1, \lambda_2, \lambda_3, \lambda_4} \frac{1}{8^2} \sum_{colors} |\mathcal{M}_1|^2
\tag{18.112}
$$

These are very tedious calculations and we only show the results here. After a long journey, one finally finds:

$$
\langle |\mathcal{M}_1|^2 \rangle = \frac{9}{8} g_s^4 \frac{\left(t^6 + 4t^5 u + 38t^4 u^2 + 72t^3 u^3 + 57t^2 u^4 + 20tu^5 + 4u^6 \right)}{s^4 t^2}
\tag{18.113}
$$

For $\langle |\mathcal{M}_2|^2 \rangle$, we can luckily (no calculation involved!) just swap $t \leftrightarrow u$ in the previous result:

$$
\langle |\mathcal{M}_2|^2 \rangle = \frac{9}{8} g_s^4 \frac{\left(u^6 + 4u^5 t + 38u^4 t^2 + 72u^3 t^3 + 57u^2 t^4 + 20ut^5 + 4t^6 \right)}{s^4 u^2}
\tag{18.114}
$$

We then have the amplitude for the s-channel, which can be obtained in a similar way, setting $q = k_1 + k_2 = k_3 + k_4$:

$$
\begin{aligned}
i\mathcal{M}_3 &= \epsilon_{1\mu} C^a \epsilon_{2\nu} C^b (-g_s) f^{abe} \left[g^{\mu\nu}(k_1 - k_2)^\lambda + g^{\nu\lambda}(k_2 + q)^\mu + g^{\lambda\mu}(-q - k_1)^\nu \right] \left(\frac{-i}{(k_1 + k_2)^2} \right) \\
&\quad \times (-g_s) f^{cde} \left[g^{\rho\sigma}(-k_3 + k_4)_\lambda + g_\lambda^\sigma(-k_4 - q)^\rho + g_\lambda^\rho(q + k_3)^\sigma \right] \epsilon_{3\rho}^* C^{c*} \epsilon_{4\sigma}^* C^{d*} \\
&= \frac{-ig_s^2 f^{abe} f^{cde}}{s} \left[g^{\mu\nu}(k_1 - k_2)^\lambda + g^{\nu\lambda}(k_1 + 2k_2)^\mu - g^{\lambda\mu}(2k_1 + k_2)^\nu \right] \\
&\quad \times \left[-g^{\rho\sigma}(k_3 - k_4)_\lambda - g_\lambda^\sigma(k_3 + 2k_4)^\rho + g_\lambda^\rho(2k_3 + k_4)^\sigma \right] \left[\epsilon_{1\mu} \epsilon_{2\nu} \epsilon_{3\rho}^* \epsilon_{4\sigma}^* C^a C^b C^{c*} C^{d*} \right] \\
&= \frac{-ig_s^2 f^{abe} f^{cde}}{s} \left[(\epsilon_1 \cdot \epsilon_2)(k_1 - k_2)^\lambda + 2\epsilon_2^\lambda(k_2 \cdot \epsilon_1) - 2\epsilon_1^\lambda(k_1 \cdot \epsilon_2) \right] \\
&\quad \times \left[-(\epsilon_3^* \cdot \epsilon_4^*)(k_3 - k_4)_\lambda - 2\epsilon_{4\lambda}^*(k_4 \cdot \epsilon_3^*) + 2\epsilon_{3\lambda}^*(k_3 \cdot \epsilon_4^*) \right]
\end{aligned}
\tag{18.115}
$$

A similar approach as to the t-channel leads to the following s-channel squared matrix element:

$$\langle |\mathcal{M}_3|^2 \rangle = \frac{9}{8} g_s^4 \frac{(t^2 - u^2)^2}{s^4} \tag{18.116}$$

Finally, we write the $\mathcal{M}_4$ amplitude with the help of the four-gluon vertex:

$$
\begin{aligned}
i\mathcal{M}_4 \;=\; -ig_s^2 \; \Big[& f^{abe} f^{cde} \big((\epsilon_1 \cdot \epsilon_3^*)(\epsilon_2 \cdot \epsilon_4^*) - (\epsilon_1 \cdot \epsilon_4^*)(\epsilon_2 \cdot \epsilon_3^*) \big) \\
+ & f^{ace} f^{bde} \big((\epsilon_1 \cdot \epsilon_2)(\epsilon_3^* \cdot \epsilon_4^*) - (\epsilon_1 \cdot \epsilon_4^*)(\epsilon_2 \cdot \epsilon_3^*) \big) \\
+ & f^{ade} f^{bce} \big((\epsilon_1 \cdot \epsilon_2)(\epsilon_3^* \cdot \epsilon_4^*) - (\epsilon_1 \cdot \epsilon_3^*)(\epsilon_2 \cdot \epsilon_4^*) \big) \Big] \times C^a C^b C^{c*} C^{d*}
\end{aligned} \tag{18.117}
$$

It is worth noting that this amplitude does not explicitly contain any dependence on the four-momenta of the gluons. A dependence on them will however arise via the polarization vectors. Taking the matrix element multiplied by its complex conjugate, and performing the color contraction, one obtains:

$$
\begin{aligned}
\frac{1}{8^2} \sum_{colors} |\mathcal{M}_4|^2 \;=\; \frac{27}{16} g_s^4 \Big(& (\epsilon_1^* \cdot \epsilon_4)(\epsilon_2^* \cdot \epsilon_3) \Big[2(\epsilon_1 \cdot \epsilon_4^*)(\epsilon_2 \cdot \epsilon_3^*) - (\epsilon_1 \cdot \epsilon_3^*)(\epsilon_2 \cdot \epsilon_4^*) - (\epsilon_1 \cdot \epsilon_2)(\epsilon_3^* \cdot \epsilon_4^*) \Big] \\
& + (\epsilon_1^* \cdot \epsilon_3)(\epsilon_2^* \cdot \epsilon_4) \times \Big[2(\epsilon_1 \cdot \epsilon_3^*)(\epsilon_2 \cdot \epsilon_4^*) - (\epsilon_1 \cdot \epsilon_4^*)(\epsilon_2 \cdot \epsilon_3^*) - (\epsilon_1 \cdot \epsilon_2)(\epsilon_3^* \cdot \epsilon_4^*) \Big] \\
& + (\epsilon_1^* \cdot \epsilon_2)(\epsilon_3 \cdot \epsilon_4) \times \Big[2(\epsilon_1 \cdot \epsilon_2)(\epsilon_3^* \cdot \epsilon_4^*) - (\epsilon_1 \cdot \epsilon_3^*)(\epsilon_2 \cdot \epsilon_4^*) - (\epsilon_1 \cdot \epsilon_4^*)(\epsilon_2 \cdot \epsilon_3^*) \Big] \Big)
\end{aligned} \tag{18.118}
$$

where in the last two terms in brackets one has the same terms as in the first bracket but with the shown replacement of polarization vectors. By averaging over the two transverse polarizations of the initial-state gluons and summing over those of the final states, we then get:

$$\langle |\mathcal{M}_4|^2 \rangle = \left(\frac{27}{16} \right) g_s^4 \frac{(3t^4 + 4t^3 u + 10t^2 u^2 + 4tu^3 + 3u^4)}{s^4} \tag{18.119}$$

The derivation of the interference terms is left to the reader. They turn out to be:

$$
\begin{aligned}
\langle |\mathcal{M}_{12}|^2 \rangle &= \frac{9g_s^4 \left(2t^6 + 21t^5 u + 94t^4 u^2 + 158t^3 u^3 + 94t^2 u^4 + 21tu^5 + 2u^6 \right)}{32 s^4 tu} \\
\langle |\mathcal{M}_{13}|^2 \rangle &= \frac{9g_s^4 \left(t^2 - u^2 \right) \left(t^3 + 10t^2 u + 9tu^2 + 2u^3 \right)}{32 s^4 t} \\
\langle |\mathcal{M}_{14}|^2 \rangle &= -\frac{27 u g_s^4 \left(t^3 + 12t^2 u + 13tu^2 + 6u^3 \right)}{16 s^4} \\
\langle |\mathcal{M}_{23}|^2 \rangle &= \frac{9g_s^4 \left(u^2 - t^2 \right) \left(2t^3 + 9t^2 u + 10tu^2 + u^3 \right)}{32 s^4 u} \\
\langle |\mathcal{M}_{24}|^2 \rangle &= -\frac{27 t g_s^4 \left(6t^3 + 13t^2 u + 12tu^2 + u^3 \right)}{16 s^4} \\
\langle |\mathcal{M}_{34}|^2 \rangle &= 0
\end{aligned} \tag{18.120}
$$

Putting all the four matrix elements and their six independent interferences together and using the fact that $s + t + u = 0$ gives the long-sought-for answer:

$$
\begin{aligned}
\langle |\mathcal{M}|^2 \rangle \;=\; & \langle |\mathcal{M}_1|^2 \rangle + \langle |\mathcal{M}_2|^2 \rangle + \langle |\mathcal{M}_3|^2 \rangle + \langle |\mathcal{M}_4|^2 \rangle \\
& + 2 \left(\langle |\mathcal{M}_{12}|^2 \rangle + \langle |\mathcal{M}_{13}|^2 \rangle + \langle |\mathcal{M}_{14}|^2 \rangle + \langle |\mathcal{M}_{23}|^2 \rangle + \langle |\mathcal{M}_{24}|^2 \rangle + \langle |\mathcal{M}_{34}|^2 \rangle \right) \\
=\; & \frac{9g_s^4}{2} \frac{(t^2 + tu + u^2)^3}{s^2 t^2 u^2}
\end{aligned} \tag{18.121}
$$

It is impressively compact and symmetric! Finally, adding the phase-space factor gives:

$$\frac{d\sigma}{dt} = \frac{9}{2}\frac{\pi\alpha_s^2}{s^2}\left(\frac{(t^2+tu+u^2)^3}{s^2t^2u^2}\right) \tag{18.122}$$

This expression can be shown to be equivalent to the standard form of the differential cross-section, given by:

$$\frac{d\sigma}{dt}(gg \to gg) = \frac{9\pi\alpha_s^2}{2s^2}\left(3 - \frac{ut}{s^2} - \frac{us}{t^2} - \frac{st}{u^2}\right) \tag{18.123}$$

This result is summarized in Table 19.1 and plotted in Figure 18.8.

18.12   Running of α_s and Asymptotic Freedom

In the context of QED, we have learned that the evaluation of diagrams that contain loops or higher-order diagrams leads to divergent amplitudes. In the context of renormalization, divergent integrals are regularized (for example with an additional ultraviolet cutoff Λ). In this way, the divergences of the perturbation expansion can be treated by absorbing them into the definition of the bare parameters of the theory. Renormalization introduces a new scale μ and the higher-order corrections to the observables are computed for a given scale Q^2 relative to μ. Ultimately, the running of the QED coupling constant α as a function of the scale Q^2 is given by Eq. (12.57):

$$\alpha(Q^2) = \frac{\alpha(\mu^2)}{1 - \frac{\alpha(\mu^2)}{3\pi}\ln\left(\frac{Q^2}{\mu^2}\right)} \tag{18.124}$$

The treatment of renormalization in QCD is similar to that of QED. However, there is a very important difference. Considering for example vacuum polarization in QED, we found that the photon propagator was modified by loop diagrams containing a fermion–antifermion (e.g., e^+e^-) pair. In QCD, because of the existence of gluon self-coupling, there are additional diagrams that contribute. Higher-order corrections to the gluon propagator are illustrated in Figure 18.11. We learned that amplitudes with fermionic loops enter with a

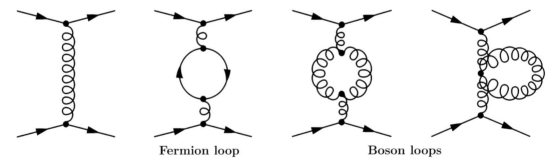

Fermion loop   Boson loops

Figure 18.11 Illustration of the diagrams contributing to the gluon propagator at higher orders. The loops involving fermions and bosons contribute with an opposite sign.

(-1) sign. On the other hand, the additional QCD diagrams with gluon loops are bosonic loops which do not carry the (-1) sign! The net effect is that while the fermionic loops acted as a "screening" effect, the bosonic loops will effectively do the opposite, hence leading to an "antiscreening" effect. The running of the strong coupling constant α_s is hence driven by two opposing effects: the bosonic loops entering in the expression with an opposite sign to the fermionic loops. The actual expression, analog to Eq. (12.57), is given by [203]:

$$\alpha_s(Q^2) = \frac{\alpha_s(\mu^2)}{1 + \alpha_s(\mu^2)\frac{\beta_0}{4\pi}\ln\left(\frac{Q^2}{\mu^2}\right)} \tag{18.125}$$

where β_0 depends on the number of colors N_C and quark flavors N_f in the following way:

$$\beta_0 = \frac{11N_C - 2N_f}{3} \tag{18.126}$$

For $N_C = 3$ and $N_f \leq 16$ quarks, β_0 is greater than zero and hence α_s *decreases* with increasing Q^2! This is a totally amazing result!

We now have a justification for the asymptotic freedom of QCD. The strong coupling constant decreases rapidly with Q^2, much faster than α increases in QED. The value of α_s varies from $\alpha_s(Q^2 \sim 1\text{ GeV}^2) \simeq 0.5$ to $\alpha_s(Q^2 \sim 100\text{ GeV}^2) \simeq 0.15$. And $\alpha_s(Q^2) \to 0$ as $Q^2 \to \infty$. This has many implications. It is, for example, the reason why quarks within the nucleon could be treated as quasi-free particles in the discussion of the deep inelastic electron–proton scattering, rather than being strongly bound inside the nucleon.

Because of asymptotic freedom, the strong interaction physics can now be calculated in perturbation theory when the momentum transfer is large. In particular, the tree-level one-gluon exchange diagram becomes a good approximation for quark interaction as $Q^2 \gg 1\text{ GeV}^2$. At present, the most accomplished result of QCD research is in the perturbative region, where many experimental data have been explained well by **perturbative QCD** (pQCD).

18.13 Color Confinement

A free quark would be detectable as a fractionally electrically charged particle with $|Q| = 2/3$ or $1/3$. Despite many attempts to find free quarks, none have been successful so far. The hypothesis of color confinement can explain the absence of observation of free quarks. It states that all naturally occurring particles are color-singlets, or equivalently, no colored particles can propagate in vacuum freely. As already mentioned above, a color-singlet gluon, being massless, would mediate an infinite-range force like the Coulomb force. This is in total contradiction with the observation that the strong force has an effective short range and is limited to the range of the nucleus (recall Yukawa's conjecture in Section 15.1). Color confinement also ensures that all hadrons are color-singlets; thus hadrons, although they are actual quark bound states, cannot exert a long-range strong force.

Color confinement is believed to be a consequence of the gluon self-interaction terms, which arise due to the non-Abelian nature of QCD, implying that gluons carry color and thereby interact with each other. There is not yet a formal proof of this![1] A qualitative understanding is reached if one considers two free quarks, e.g., a quark and an antiquark, being pulled apart, as shown in Figure 18.12. This would happen if one wanted to hypothetically free them from each other. When "pulling" the quarks apart, QCD tells us that the force between them increases roughly proportionally to their separation. The potential energy between the quarks grows like $V(r) \simeq \kappa r$, where $\kappa \simeq 1\text{ GeV/fm}$. As they move apart, the energy stored between them grows and before they can be separated it appears that there is sufficient energy to create a new quark–antiquark pair. Such pairs created out of vacuum are colorless, e.g., of the type $R\bar{R}$, or $G\bar{G}$, ... Each of the newly formed quarks in the pair can combine with one of the original quarks to form color-singlets, i.e., hadrons! Thus, any attempt to isolate a colored quark merely results in the production of colorless hadrons.

---

1   See www.claymath.org/millennium-problems/yang-mills-and-mass-gap. If you find a solution, you could win 1M$.

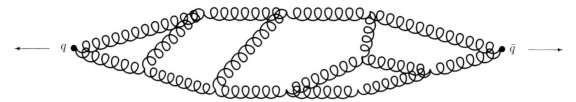

Figure 18.12 Graphical illustration of the formation of a "tube" of gluons stretching between the two quarks as they become more and more apart.

18.14 Hadronization

Colored quarks and gluons can be considered as quasi-free particles during a "high-Q^2" hard interaction. But subsequently, as a consequence of color confinement, they will rapidly reorganize into color-singlet hadrons. This process is called hadronization or fragmentation. It involves the creation of additional quark–antiquark pairs created out of the color field. See Figure 18.13. Although we expect it to be governed by QCD, this process is difficult to calculate from first principles because it is dominated by soft non-perturbative effects. Semi-empirical methods of fragmentation have therefore been developed. They often rely on computer Monte-Carlo simulations. The basic assumption is that each fast parton produced in the hard process will fragment independently of the other, at least in first approximation.

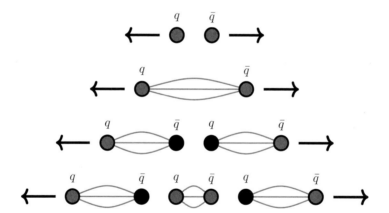

Figure 18.13 Graphical illustration of the formation of color-singlet hadrons during the hadronization process.

For example, the hard process $e^+e^- \to q\bar{q}$ will produce two fast flying-apart quarks (one quark and one antiquark). Each of them will fragment into a set of hadrons, leading to what is commonly called "two jets." In the process, energy and momentum must be conserved, so we expect all hadrons in the jet to be boosted in the direction of the original parton. The total energy of all particles in the jet corresponds to the initial energy of the quark. The same happens with hard gluons.

In general, let us call E_p the energy of the initial parton (quark, antiquark, or gluon). The parton will fragment into a jet composed of several hadrons. Let us consider one hadron h in the jet with energy fraction z, where:

$$z \equiv \frac{E_h}{E_p} \quad \text{with} \quad 0 \le z \le 1 \tag{18.127}$$

The p-to-h **fragmentation function** $D_p^h(z)$ is defined such that $D_p^h(z)\mathrm{d}z$ equals the probability of finding h in the jet with a fractional energy in the range z and $z+\mathrm{d}z$. In this approximation, one assumes that the variable z is the only significant variable, and that masses and transverse momenta can be neglected. It is assumed that the fragmentation function is sort of universal and does not depend on the details of the underlying hard scattering process that produced the partons. In this sense, we really factorize the hard process which can be typically computed within perturbative QCD and the hadronization process which leads to hadrons in the final state which is empirically studied and universally tuned (this is conceptually identical to what is done when defining parton distribution functions to, for instance, compute deep inelastic scattering).

With the help of the fragmentation function $D_p^h(z)$, the inclusive cross-section for the production of the particle h is then related to the cross-section for producing the parent parton p in this way:

$$
\begin{aligned}
\frac{\mathrm{d}\sigma}{\mathrm{d}E_h}(AB \to hX) &= \sum_p \int \frac{\mathrm{d}\sigma}{\mathrm{d}E_p}(AB \to pX)D_p^h(z)\mathrm{d}z\frac{\mathrm{d}E_p}{\mathrm{d}z}\frac{\mathrm{d}z}{\mathrm{d}E_h} \\
&= \sum_p \int \frac{\mathrm{d}\sigma}{\mathrm{d}E_p}(AB \to pX)D_p^h\left(\frac{E_h}{E_p}\right)\frac{\mathrm{d}E_p}{E_p}
\end{aligned}
\tag{18.128}
$$

since $\mathrm{d}z = \mathrm{d}E_h/E_p$.

Integrating the fragmentation function $D_p^h(z)$ over a range z_{min} to z_{max} gives the probability of finding a hadron h with an energy fraction within that range. The integration over the full kinematical range corresponds to the average number of hadrons h in the jet arising from the parton p. This number is called the mean multiplicity of h:

$$
n_p^h \equiv \int_{z_{min}}^1 \mathrm{d}z D_p^h(z)
\tag{18.129}
$$

where $z_{min} \simeq m_h/E_p$.

• **Light quarks.** The fragmentation of light quarks (u, d, s) is empirically found to be well parameterized within the form:

$$
D(z) = \frac{f(1-z)^n}{z}
\tag{18.130}
$$

where f is a constant. The term $(1-z)^n$ determines the behavior at large z while the $1/z$ term dominates at low z. It leads to a mean multiplicity of hadrons that increases logarithmically with the energy:

$$
\bar{n}_p^h \simeq \int_{z_{min}}^1 \mathrm{d}z \frac{f(1-z)^n}{z} \simeq f\ln(E_p/m_h)
\tag{18.131}
$$

where we assumed $z_{min} \ll 1$. These predictions are supported by data. For example, Figure 18.14(left) shows the measured fragmentation of hadrons vs. $x_p = p/p_{beam} \approx z$ compared to several Monte-Carlo hadronization models. These data are used to tune the parameters in the models. The measured average number of charged hadrons in e^+e^- collisions as a function of the center-of-mass energy is shown in Figure 18.14(right). It is consistent with a logarithmic rise.

• **Heavy quarks.** When the initial fragmenting parton is a heavy quark, the fragmentation function is seen to be harder, concentrated at large values of z. This effect increases with increasing mass of the initial quark. The first reason is that heavy quarks will radiate less gluons than light quarks. The second reason is associated with the process of fragmentation itself. Naively we expect partons to combine into hadrons when their velocities are comparable. A heavy quark can be seen to be slow even at higher energies, given its large mass, hence it is more likely to combine with other softer quarks created in the fragmentation in the early stages of the shower, thereby retaining most of its energy. The **Peterson fragmentation model** captures these features. It predicts that:

$$
D_Q(z) \propto \frac{1}{z}\left(1 - \frac{1}{z} - \frac{\epsilon_Q}{1-z}\right)^{-2}
\tag{18.132}
$$

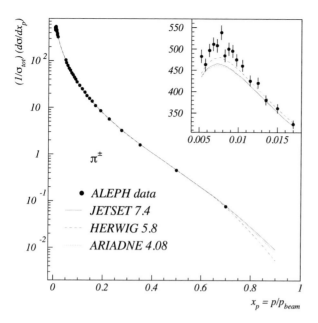
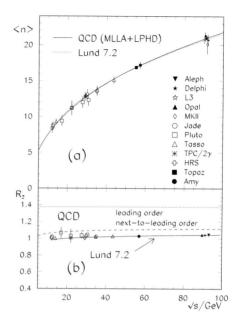

Figure 18.14 (left) Measurement fragmentation of hadrons in ALEPH vs. $x_p \approx z$ compared to several Monte-Carlo hadronization models. (right) Charged hadron multiplicities measured at several center-of-mass energies. Reprinted from R. Barate *et al.*, "Studies of quantum chromodynamics with the ALEPH detector," *Phys. Rept.*, vol. 294, pp. 1–165, 1998, with permission from Elsevier.

where ϵ_Q is a heavy quark-dependent parameter, expected to be proportional to m_Q^{-2}. This effect can be directly observed in data, as is visible in Figure 18.15, which shows the measured fragmentation cross-section as a function of $x_p \approx z$ for D^0, D^*, and inclusive B hadrons. This is compatible with $\epsilon_c \approx 0.2$ and $\epsilon_b \approx 0.02$, with a ratio $10:1$ suggested by the predicted m_Q^{-2} dependence.

18.15   Determination of the Strong Coupling Constant

In the massless quark limit, QCD has only one free parameter: the strong coupling constant α_s. Thus, all strong interaction phenomena should be described in terms of this single input. The coupling constant in itself is not a physical observable, but rather a quantity defined in the context of perturbation theory, which enters in the calculations of the predictions for experimentally measurable observables. The measurements of α_s in different processes and at different mass scales provide then a crucial test of QCD: if QCD is the right theory of the strong interactions, all measured observables should lead to the same coupling.

The measured values of α_s at different scales ranging over approximately three orders of magnitude from ≈ 1 GeV and ≈ 10000 GeV are shown in Figure 18.16. This compilation [28] relies on several different processes which are sensitive to α_s. From the lowest Q to the highest:

- **Hadronic decays of τ leptons**, which give a precise value at the scale $Q = m_\tau \simeq 1777$ MeV.

- **Deep inelastic lepton–nucleon scattering**, including analyzing structure functions in next-to-next-to leading order QCD (called NNLO), and a combination of precision measurements at HERA (see Section 19.8).

- **Analysis of radiative Υ decays**, which determines α_s at the scale $Q = m_\Upsilon \simeq 9460$ MeV.

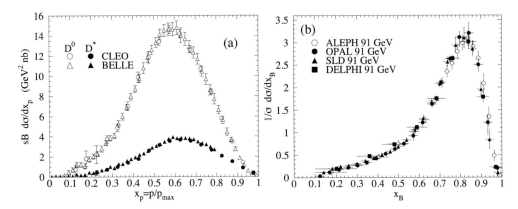

Figure 18.15 Measured fragmentation cross-section as a function of $x_p \approx z$ for (left) D^0, D^* and (right) inclusive B hadrons. Reprinted with permission from P. A. Zyla *et al.* (Particle Data Group), *Prog. Theor. Exp. Phys.* **2020**, 083C01 (2020).

- **Hadronic final states of e^+e^- annihilations**, studying event shapes measured around the Z^0 peak and at LEP2 center-of-mass energies up to 209 GeV.

- **Hadron collider results**, resulting from studies of inclusive jet cross-sections and from jet angular correlations; the determinations use the ratio of inclusive 3-jet to 2-jet cross-sections, from inclusive jet production and from the 3-jet differential cross-sections; the $t\bar{t}$ cross-section is also used to determine the strong coupling.

- **Electroweak precision fits.** global fits of many observables can extract all parameters within the Standard Model theory. Here, the hadronic width of the Z^0 gauge boson is a parameter that can be precisely fitted and used to extract the strong coupling in next-to-NNLO precision.

All the different measurements are combined and the "world" average value of the strong coupling constant is given by convention at the scale $Q = M_Z$. The current world average has a precision of $\sim 1\%$ and is:

$$\alpha_s(M_Z^2) = 0.1179 \pm 0.0010 \qquad (18.133)$$

18.16　　The Q^2 Evolution of the Parton Distribution Functions

As a consequence of QCD, quarks inside hadrons can be depicted in a dynamic state, constantly interacting with each other via exchange of gluons. We can use this observation to gain further insight on the parton distribution functions. We first note that for a point proton that would simply consist of a single quark, $q^p(x)$ would just be a δ function at $x = 1$, since all the momentum would be carried by one object. For a nucleon consisting of three quarks at rest (in the nucleon rest frame), each quark would carry 1/3 of the momentum in the infinite-momentum frame, and $q^p(x)$ would then be a δ function at $x = 1/3$. For three interacting quarks, we expect a smeared-out distribution peaked in the region $x \approx 1/3$. Finally, for the case of three interacting quarks plus a "sea" of quark–antiquark pairs we expect that the low-x region would become populated by soft pairs. When a quark emits a quark–antiquark pair, all three resulting particles have lower x than the original quark. These behaviors are depicted in Figure 18.17.

We have seen in the discussion of the deep inelastic scattering of electrons on protons, that the existence of subconstituents (partons) led to the scaling of the nucleon structure functions. This experimental observation

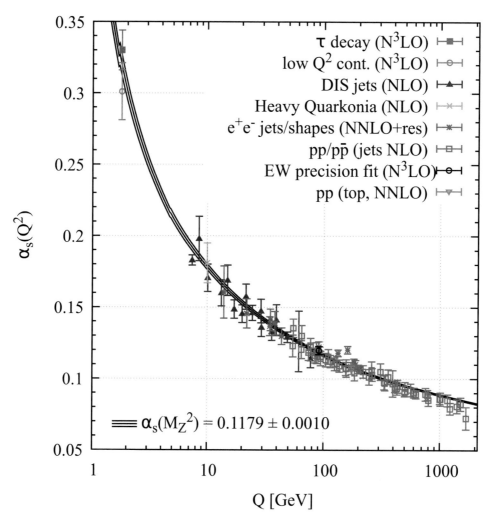

Figure 18.16 Summary of measurements of α_s as a function of the energy scale Q. The respective degree of QCD perturbation theory used in the extraction of α_s is indicated in brackets (NLO: next-to-leading order; NNLO: next-to-next-to leading order; NNLO+res: NNLO matched with resummed next-to-leading logs; N^{3}LO: next-to-NNLO). Reprinted with permission from P. A. Zyla *et al.* (Particle Data Group), *Prog. Theor. Exp. Phys.* **2020**, 083C01 (2020).

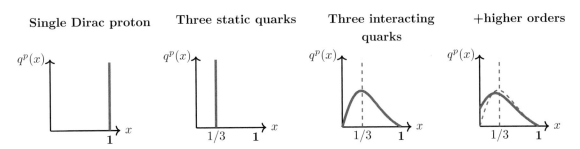

Figure 18.17 Parton distribution functions assuming different models for the proton.

was clearly a critical milestone in the understanding of the parton model. In the context of QCD, **scaling violations** occur naturally since, at large parton momenta x and increasing Q^2, the structure functions are increasingly depleted by hard gluon radiation from quarks; at small x, they are enriched by gluon conversion into low-momentum quark–antiquark pairs.

Increasing further the momentum transfer, the photon probe has a greater sensitivity to smaller distances, and it is able to resolve the scattered quark into a quark and a gluon. Thus, a parton with momentum fraction x can be resolved into a parton and a gluon of smaller momentum fractions, $x' < x$ and $x - x'$, respectively. In a similar way, a gluon with momentum fraction x can be resolved into a quark and an antiquark with momentum fractions $x' < x$ and $x - x'$. These anticipated variations are graphically sketched in Figure 18.18. Thus, one

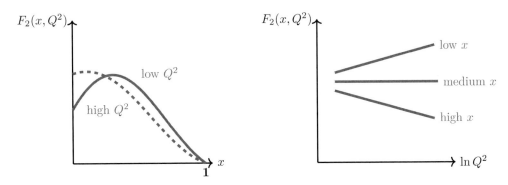

Figure 18.18 Sketch of the expected scaling violation in F_2 as a function of x and Q^2.

can expect to find a definite Q^2 dependence in the parton distributions, i.e., **Bjorken scaling violations** due to the underlying QCD interactions. And this is indeed the case! The structure functions have an explicit dependence on Q^2 that comes from either **higher twist effects** or from **higher-order perturbative QCD effects**. The former are not calculable unambiguously; the latter are predictable by the method of the **Dokshitzer–Gribov–Lipatov–Altarelli–Parisi (DGLAP) evolution equations**, as outlined in the following.

We start with the case of electron–proton scattering in the deep inelastic regime. At first radiative order of α_s, we need to consider the emission of a gluon from the initial or final-state quark. The corresponding Feynman diagrams are shown in Figure 18.19. It is convenient to forget about the incoming and outgoing electrons and see the interactions as the interaction of the virtual photon γ^* with a quark. The underlying process can therefore be expressed as follows:

$$\gamma^*(k_1, \epsilon_1) + q(p, s) \rightarrow q(k, r) + g(k_2, \epsilon_2) \tag{18.134}$$

In the first diagram, a quark with momentum fraction y radiates a gluon, reducing its momentum to a fraction

x and then interacts with the photon. By construction, $y > x$. In the second diagram, the quark interacts with the photon and then radiates a gluon in the final state.

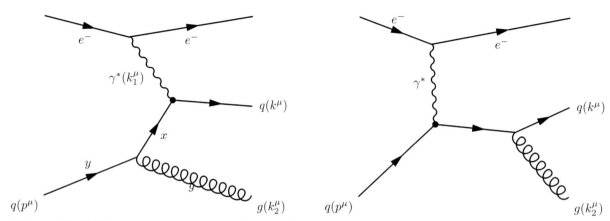

Figure 18.19 Electron–parton scattering: higher-order gluon Bremsstrahlung in electron deep inelastic scattering.

In order to compute the amplitudes, we can start once again from the QED counterpart, i.e., Compton scattering. Since we are interested in the unpolarized cross-section, we can alternatively consider our derivation for the $e^+e^- \to \gamma\gamma$ process (see Section 11.15 and Eq. (11.232)) and use crossing symmetry to find (neglecting the rest masses) the matrix element for the Compton process:

$$\langle |\mathcal{M}(\gamma e^- \to \gamma e^-)|^2\rangle_{QED} = (2e^4)\left(-\frac{u}{s} - \frac{s}{u}\right) \tag{18.135}$$

We should, however, modify this result to take into account that in the deep inelastic process that we are interested in, the incident photon γ^* is virtual with a mass $k_1^2 = -Q^2$. One can show that this leads to:

$$\langle |\mathcal{M}(\gamma^* e^- \to \gamma e^-)|^2\rangle_{QED} = (2e^4)\left(-\frac{u}{s} - \frac{s}{u} + \frac{2tQ^2}{su}\right) \tag{18.136}$$

We can now infer the result for the desired process $\gamma^* q \to qg$ by replacing e^4 by $e_q^2 e^2 g_s^2$ and by adding the color factor $C_F = 4/3$, where e_q is the unitless quark charge. Taking into account the conventionally chosen order of the outgoing particles yields:

$$\langle |\mathcal{M}(\gamma^* q \to qg)|^2\rangle = (32\pi^2 e_q^2 \alpha\alpha_s)C_F \left(\underbrace{\frac{-t}{s}}_{\text{final-state radiation}} + \underbrace{\frac{s}{-t}}_{\text{initial-state radiation}} + \underbrace{\frac{2uQ^2}{st}}_{\text{interference}} \right) \tag{18.137}$$

(recall that t is negative). The amplitude is strongly forward peaked as $|t| \to 0$ and we can neglect the first term which corresponds to the diagram of final-state radiation, to write:

$$\langle |\mathcal{M}(\gamma^* q \to qg)|^2\rangle \approx \frac{32\pi^2 e_q^2 \alpha\alpha_s}{(-t)}C_F\left(s - \frac{2uQ^2}{s}\right) \tag{18.138}$$

We focus on the small t region where we can approximate $u \approx -Q^2 - s$. We then find:

$$
\begin{aligned}
\langle |\mathcal{M}(\gamma^* q \to qg)|^2\rangle &\approx \frac{32\pi^2 e_q^2 \alpha\alpha_s}{(-t)}C_F\left(s - \frac{2(-Q^2 - s)Q^2}{s}\right) = \frac{32\pi^2 e_q^2 \alpha\alpha_s}{(-t)}C_F\left(\frac{s^2 + 2(Q^2 + s)Q^2}{s}\right) \\
&= \frac{32\pi^2 e_q^2 \alpha\alpha_s}{(-t)}C_F\left(\frac{(s + Q^2)^2 + Q^4}{s}\right)
\end{aligned} \tag{18.139}
$$

In analogy to Bjorken $x \equiv Q^2/(2P \cdot q)$, we define:

$$z \equiv \frac{Q^2}{2p \cdot k_1} = \frac{Q^2}{(p+k_1)^2 - k_1^2} = \frac{Q^2}{s + Q^2} \tag{18.140}$$

We note that:

$$
\begin{aligned}
\frac{1+z^2}{1-z} &= \left(1 + \frac{Q^4}{(s+Q^2)^2}\right)\left(1 - \frac{Q^2}{s+Q^2}\right)^{-1} = \left(\frac{(s+Q^2)^2 + Q^4}{(s+Q^2)^2}\right)\left(\frac{s}{s+Q^2}\right)^{-1} \\
&= \left(\frac{(s+Q^2)^2 + Q^4}{(s+Q^2)^2}\right)\left(\frac{s+Q^2}{s}\right) = \left(\frac{(s+Q^2)^2 + Q^4}{s(s+Q^2)}\right)
\end{aligned}
\tag{18.141}
$$

Hence, we may write:

$$\langle |\mathcal{M}(\gamma^* q \to qg)|^2 \rangle = 32\pi^2 e_q^2 \alpha \alpha_s \left(\frac{s+Q^2}{-t}\right) P_{qq}(z) \tag{18.142}$$

where

$$P_{qq}(z) \equiv C_F \left(\frac{1+z^2}{1-z}\right) \tag{18.143}$$

The variable z represents the fraction by which the relative momentum of the initial-state quark is changed by the gluon emission. The function $P_{qq}(z)$ gives the probability of a quark having a reduced momentum by a fraction z due to gluon emission. The singularity at $z \to 1$ is an infrared divergence which comes from the emission of soft gluons. This divergence reflects the fact that we have only computed a single gluon emission. Higher-order diagrams should be included to cure this divergence.

We now include the phase-space factor to find the differential cross-section valid for small $|t|$:

$$\frac{d\sigma}{dt}(\gamma^* q \to qg) = \frac{1}{16\pi s^2} 32\pi^2 e_q^2 \alpha \alpha_s \left(\frac{s+Q^2}{-t}\right) P_{qq}(z) \approx \underbrace{\left(\frac{4\pi^2 \alpha e_q^2}{s}\right)}_{\equiv \sigma_0} \cdot \left(\frac{\alpha_s}{2\pi}\right)\left(\frac{1}{-t}\right) P_{qq}(z) \tag{18.144}$$

We can now integrate over t from an infrared cutoff μ (to avoid the $1/t$ divergence) and its maximum given by Q^2. We consequently get:

$$\sigma(\gamma^* q \to qg) = \sigma_0 \cdot \left(\frac{\alpha_s}{2\pi}\right) \ln\left(\frac{Q^2}{\mu^2}\right) P_{qq}(z) \quad \text{where} \quad \sigma_0 \equiv \left(\frac{4\pi^2 \alpha e_q^2}{s}\right) \tag{18.145}$$

At the Born level without gluon emission we would have $z = 1$ and the cross-section could be expressed as:

$$\sigma(\gamma^* q \to q) = \sigma_0 \cdot \delta(1-z) \tag{18.146}$$

In general, we can then separate the hard interaction with the photon and assign the initial-state gluon emission to a change of the parton distribution function of the target! A quark with initial momentum y inside the target has a probability of radiating a gluon, reducing its momentum to a fraction $x = zy$. Let us define the parton distribution function of the quark q by $q(x)$. The cross-section at a given x will then be given by the sum over all possible cross-sections with reduced fraction z in the following way (recall $y > x$ by construction):

$$\sigma(x) = \int_x^1 dy\, q(y) \int_0^1 dz\, \delta(x - zy)\sigma(z) = \int_x^1 dy q(y) \frac{\sigma(x/y)}{y} \tag{18.147}$$

We can add the contributions from initial-state radiation by modifying the parton distribution function as follows:

$$q(x, Q^2) = \int_x^1 \frac{dy}{y} q(y) \left(\delta\left(1 - \frac{x}{y}\right) + \frac{\alpha_s}{2\pi} P_{qq}\left(\frac{x}{y}\right) \ln\left(\frac{Q^2}{\mu^2}\right)\right) \tag{18.148}$$

It is customary to express the "change" in the parton distribution function as a formal derivative. We finally reach the first **Altarelli[2]–Parisi[3] equation**:

$$\frac{dq(x, Q^2)}{d \ln Q^2} = \left(\frac{\alpha_s}{2\pi}\right) \int_x^1 \frac{dy}{y} P_{qq}\left(\frac{x}{y}\right) q(y) \tag{18.149}$$

The gluon emission introduces a Q^2 dependence of the parton distribution function $q(x)$. One speaks of the **evolution of the parton distribution function** with Q^2. The function $P_{qq}(x/y)$ is called the **splitting funtion** and represents the probability of finding another quark with momentum fraction x starting with a quark of momentum y.

A second type of diagram also influences the evolution of the parton distribution function, as shown in Figure 18.20. This is due to the process called gluon splitting, where a gluon splits into a quark–antiquark pair $g \to q\bar{q}$, with the quark or the antiquark subsequently interacting with the photon. The QED analog to this process is the pair creation $\gamma^*\gamma \to e^+e^-$. A similar derivation as above would yield the following differential cross-section:

$$\frac{d\sigma}{dt}(\gamma^* g \to q\bar{q}) = \sigma_0 \frac{\alpha_s}{2\pi} \frac{1}{(-t)} P_{qg}(z) \tag{18.150}$$

where

$$P_{qg}(z) = C_F \left(z^2 + (1-z)^2\right) \tag{18.151}$$

The color factor is $C_F = 1/2$. The **splitting function** $P_{qg}(z)$ gives the probability of finding a quark with relative momentum fraction z inside a gluon.

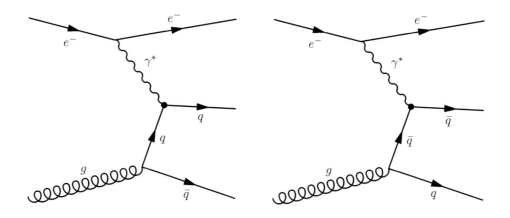

Figure 18.20 Electron–parton scattering: higher-order gluon splitting diagrams in electron deep inelastic scattering.

Putting the two contributions together, we obtain the DGLAP evolution equations for the parton distribution functions $q_i(x, Q^2)$ of the six quarks at leading order:

$$\frac{dq_i(x, Q^2)}{d \ln Q^2} = \left(\frac{\alpha_s(Q^2)}{2\pi}\right) \int_x^1 \frac{dy}{y} \left(P_{qq}\left(\frac{x}{y}\right) q_i(y, Q^2) + P_{qg}\left(\frac{x}{y}\right) g(y, Q^2)\right) \tag{18.152}$$

2   Guido Altarelli (1941–2015), Italian theoretical physicist.
3   Giorgio Parisi (born 1948), Italian theoretical physicist.

where $g(x, Q^2)$ is the parton distribution function of the gluons. These latter will also evolve. One can show that they will be given in an analogous way by:

$$\frac{dg(x, Q^2)}{d \ln Q^2} = \left(\frac{\alpha_s(Q^2)}{2\pi}\right) \int_x^1 \frac{dy}{y} \left(\sum_i P_{gq}\left(\frac{x}{y}\right) q_i(y, Q^2) + P_{gg}\left(\frac{x}{y}\right) g(y, Q^2)\right) \qquad (18.153)$$

where the sum is over all flavors of quarks and antiquarks (so $i = 1, \ldots, 2N_f$), and where:

$$P_{gq}(z) = \frac{4}{3}\left(\frac{1 + (1-z)^2}{z}\right) \quad \text{and} \quad P_{gg}(z) = 6\left(\frac{z}{1-z} + \frac{1-z}{z} + z(1-z)\right) \qquad (18.154)$$

To conclude, we stress once again that the DGLAP evolution equations determine the Q^2 dependence of the parton distribution functions. The x dependence, on the other hand, has to be obtained from fitting experimental results, such as the deep inelastic and other hard-scattering data.

Problems

Ex 18.1 Color factors. Calculate the color factor for the quark–antiquark octet using the state:

    (a) $b\bar{q}$

    (b) $\frac{r\bar{r} - b\bar{b}}{\sqrt{2}}$

    (c) $\frac{r\bar{r} + b\bar{b} - 2g\bar{g}}{\sqrt{6}}$

Ex 18.2 Scale independence. The strong coupling constant (just as the coupling in QED) is found to be scale-dependent (see Eq. (18.125)). Prove that for any measurement, the parameter μ in this equation is arbitrary. To achieve this, assume two experiments use different values μ_a and μ_b and that the formula above holds for one of them. Then show that it will necessarily hold for the other as well.

Ex 18.3 The Landau pole. Both for QED and QCD, we find that the coupling has a singularity at a certain energy scale, the so-called Landau pole.

    (a) In QED we find the running coupling to be given by Eq. (12.57). Determine the energy $\sqrt{Q^2}$ at which the Landau pole occurs. Do you expect this energy level to be reached with current accelerator technology?

    (b) Determine at what energy scale you would expect a 1% difference from $\alpha(0)$. Is that energy scale accessible with current technology?

    (c) In QCD the scale-dependent coupling is given by Eq. (18.125). We can define a scale:

$$\ln \Lambda^2 = \ln \mu^2 - \frac{4\pi}{\beta_0 \alpha_s(\mu^2)}$$

    and express $\alpha_s(Q^2)$ only in terms of this single parameter (and Q^2, of course). Derive this expression.

    (d) Experimentally there is a large error associated with determining this parameter Λ. The value is measured to be on the order of 100 MeV to 1 GeV. Determine $\alpha_s(Q^2)$ for both 10 GeV and 100 GeV at three points within the range of Λ. Compare the relative change in the coupling strength between QED and QCD in the accessible energy range.

Ex 18.4 $qg \to qg$ **scattering**

(a) Draw the tree-level diagrams intervening in the quark–gluon scattering process, $qg \to qg$, and use the QCD Feynman rules to write down the corresponding amplitudes.

(b) Show whether or not the amplitudes or their combinations are gauge-invariant.

Ex 18.5 $gg \to gg$ **scattering**

(a) Draw the four lowest-order diagrams representing the process $gg \to gg$.

(b) Compute the resulting amplitudes.

(c) Go to the center-of-mass frame. Compute all the amplitudes as a function of the kinematics of the four gluons and for each their polarization state (do not attempt to sum over the gluon polarization states).

(d) Add the amplitudes to get the total $\mathcal{M}$.

(e) Find the differential scattering cross-section. The result should be equivalent to Eq. (18.123).

(f) Determine whether the force is attractive or repulsive; if it is the former, this may allow pure gluonic states (glueballs).

(g) What would be the quantum numbers of the lightest glueball state?

19 Experimental Tests of QCD

Theorists have wonderful ideas which take years and years to be verified.

David Gross

19.1 QCD Measurements at e^+e^- Colliders

Hadrons are copiously produced at high-energy electron–positron or hadron–hadron colliders and provide a well-suited environment to study QCD. Electron–positron colliders are particularly well suited, since the tree-level process is the s-channel annihilation into a virtual photon and a quark–antiquark pair in the final state. The QED part of the process is perfectly understood. Compared to deep inelastic scattering or hadron–hadron collisions, there is no need to know parton structure functions, and experimentally e^+e^- collisions are very clear with no target proton remnants with a laboratory frame coinciding with the center-of-mass frame. Detectors with high-resolution tracking and calorimeters with 4π coverage surrounded by muon sub-detectors are ideal to study reactions in an exclusive way, often permitting the identification and measurement of individual hadrons and reconstruction of jets. The principal e^+e^- colliders, also most relevant for QCD studies, were shown in Table 13.1.

The first observation of jets and the first direct observation of the gluon were achieved at e^+e^- colliders. In addition, many unique and classic (and often still best) studies which helped to establish QCD, were done at e^+e^- as well as hadron colliders. We outline the most relevant studies in the following list:

1. Event shape variables (sphericity, thrust, ...).

2. Spins of the quark and gluon.

3. Precise, consistent measurements of α_s.

4. The running of α_s.

5. The triple-gluon vertex and other non-Abelian characteristics.

6. Precise measurements of the color factors and establishment of SU(3) gauge structure.

7. Differences between gluon and quark jets.

8. First experimental demonstration of coherence effects in QCD (string effect).

This is just to name a few. We illustrate a few of these measurements in the following sections.

19.2 e^+e^- **Annihilation into Hadrons**

The development of accelerators has permitted the study of the e^+e^- annihilation cross-section into hadrons at an ever increasing center-of-mass energy. The tree-level process is the s-channel annihilation into a virtual photon and a quark–antiquark pair in the final state. Since quarks are confined, the probability of hadronizing is just one; therefore, the sum over all possible quarks in the final state gives the total inclusive cross-section into hadrons.

In QED we computed the differential cross-section for the $e^+e^- \to \mu^+\mu^-$ reaction. The result was simply (see Eq. (11.153)):

$$\left(\frac{\mathrm{d}\sigma}{\mathrm{d}\Omega}\right)_{\mu\mu} = \left(\frac{\alpha^2}{4s}\right)(1+\cos^2\theta) \quad \text{and} \quad \sigma_{tot} = \frac{4\pi\alpha^2}{3s} \tag{19.1}$$

The (QED) differential cross-section for a given quark q is:

$$\left(\frac{\mathrm{d}\sigma}{\mathrm{d}\Omega}\right)_{q\bar{q}} = N_C\left(\frac{\alpha^2 e_q^2}{4s}\right)(1+\cos^2\theta) \quad \text{and} \quad \sigma_{tot} = N_C\frac{4\pi\alpha^2 e_q^2}{3s} \tag{19.2}$$

The factor $N_C = 3$ arises because **there is a diagram for each quark color and the cross-sections need to be added**. At the lowest energies, the $q\bar{q}$ will collapse into a bound state. At the highest energies, the two quarks fly apart and produce jets (as will be discussed in the next section). To obtain the cross-section for producing all types of hadrons, all quark flavors $q = u, d, s, \ldots$ for which $\sqrt{s} \geq 2m_q$ must be added. Therefore:

$$\sigma_{tot}(e^+e^- \to \text{hadrons}) = \sum_q \sigma_{tot}(e^+e^- \to q\bar{q}) = N_C\sum_q e_q^2\sigma_{tot}(e^+e^- \to \mu^+\mu^-) \tag{19.3}$$

This simple result leads to the dramatic observation that a *direct counting* of the color quantum number and the number of quark flavors can be obtained from the ratio:

$$R_{e^+e^-} \equiv \frac{\sigma(e^+e^- \to \text{hadrons})}{\sigma(e^+e^- \to \mu^+\mu^-)} = N_C\sum_q e_q^2 \tag{19.4}$$

For $N_C = 3$ we have:

$$\begin{aligned}
R_{e^+e^-} &= 3\left[\left(\frac{2}{3}\right)^2 + \left(\frac{1}{3}\right)^2 + \left(\frac{1}{3}\right)^2\right] = 2 \quad \text{for } u, d, s \\
&= 2 + 3\left(\frac{2}{3}\right)^2 = \frac{10}{3} \quad \text{for } u, d, s, c \\
&= \frac{10}{3} + 3\left(\frac{1}{3}\right)^2 = \frac{11}{3} \quad \text{for } u, d, s, c, b \\
&= \frac{11}{3} + 3\left(\frac{2}{3}\right)^2 = 5 \quad \text{for } u, d, s, c, b, t
\end{aligned} \tag{19.5}$$

where we added the heavier charm c, the bottom b, and the top t quarks. These predictions are compared to the measured ratio in Figure 19.1 (see also Figure 20.21). These results are amazing! Although the simple formula of Eq. (19.4) cannot explain the complicated structure around the different quark thresholds, it gives the right average value of the ratio (away from the thresholds), provided that N_C is taken to be three. **The sharp peaks correspond to the production of narrow resonances near the quark flavor production thresholds.** Notice that strong interactions have not been taken into account; only the confinement hypothesis has been used.

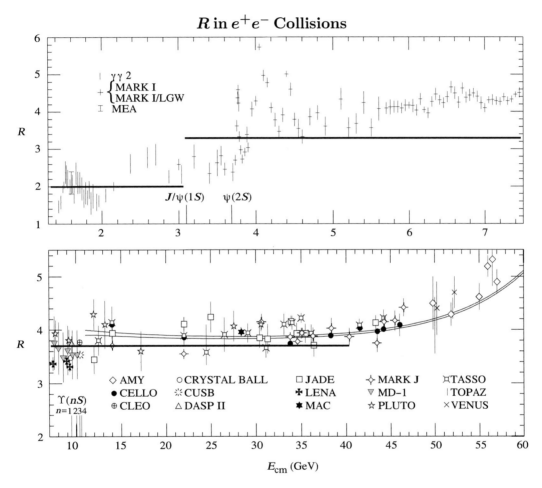

Figure 19.1 Measurements of $R_{e^+e^-}$ shown as a function of the center-of-mass energy (GeV). The straight lines correspond to the naive expectations for a given number of quark flavors, respectively $R = 2$, $10/3$, and $11/3$. Two smooth curves are overlaid for $E_{cm} > 11$ GeV, showing the theoretical prediction for R, including higher-order QCD and electroweak corrections. Reprinted with permission from R. M. Barnett *et al.*, "Review of particle physics. Particle Data Group," *Phys. Rev. D*, vol. 54, pp. 1–720, 1996. Copyright 1996 by the American Physical Society.

19.3   The First Observation of Jets

In 1970, following up on work by Bjorken and Brodsky [204], and Drell and collaborators [205], **Cabibbo**,[1] **Parisi**, and **Testa** [206] started discussing "jets" as a possible signature of quarks. For many years the existence of such jets had remained doubtful, because the available energies and transverse momenta in collision experiments were not large enough to clearly identify the emerging hadrons as fragmentation products of single quarks. As Bjorken and Brodsky stated:

> *"In the next few years, electron–positron storage rings will be developed capable of producing hadron systems of total mass $\sqrt{s}$ up to 6 GeV or higher. Aside from predictions for the energy dependence of the total annihilation cross-section into hadrons, there has been little discussion concerning the composition, multiplicity, and other properties expected for the multibody hadron final states. It is not so clear what to expect, even qualitatively. The process $e^+e^- \to$ hadrons at high energy differs from almost all other hadron processes inasmuch as the hadrons are produced via the decay of an arbitrarily heavy virtual photon. One picture of such a decay would be that the virtual photon decays into an intermediate state consisting of a virtual pair of 'bare' constituent partons (such as a bare quark–antiquark pair) which subsequently decay in some way into hadrons – mainly pions. If this were the case, one could anticipate anisotropy and the existence of an axis in the distribution of hadron products; in other words, the hadrons 'remember' the direction along which the bare constituents were emitted. Under these circumstances, the transverse momenta $p_\perp$, of the secondaries relative to the axis for a given event would be no more than a few hundred MeV, while the longitudinal momenta could be much larger."* [2]

The underlying processes are illustrated in Figure 19.2. At the partonic level, the tree-level cross-section for a

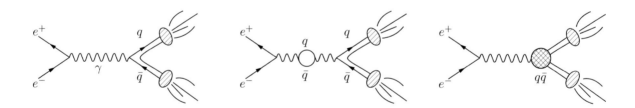

Figure 19.2 Feynmans diagram for $e^+e^- \to q\bar{q} \to$ jets: (left) continuum; (middle) vacuum polarization enhanced by a nearby resonance; (right) direct resonance production. The shaded areas represent the nonperturbative hadronization process, which lead to the formation of final-state hadrons, regrouped (in this case) in two jets.

given quark flavor is just $e^+e^- \to q\bar{q}$ with a cross-section given by (see Eq. (19.2)):

$$\sigma(e^+e^- \to q\bar{q}) = 3e_q^2 \left(\frac{4\pi\alpha^2}{3s} \right) \tag{19.6}$$

The two final-state quarks will then fragment and hadronize to produce hadrons, as described in Section 18.14. Experimentally a **jet** can be defined as a collimated spray of stable particles arising from the fragmentation

---

1   Nicola Cabibbo (1935–2010), Italian physicist.
2   Reprinted excerpt with permission from J. D. Bjorken and S. J. Brodsky, "Statistical model for electron–positron annihilation into hadrons," *Phys. Rev. D*, vol. 1, pp. 1416–1420, Mar 1970. Copyright 1970 by the American Physical Society.

and hadronization of a parton (quark or gluon) after a collision. Jet reconstruction algorithms are used to combine the calorimetry and tracking information coming from the detectors to define jets. **Ideally the jets would provide a link between the observed colorless stable particles and the underlying physics at the partonic level.**

For the first time in 1975, in a ground-breaking analysis of the data recorded at the electron–positron collider SPEAR at SLAC [207], the process:

$$e^+ e^- \to q\bar{q} \to \text{jet} + \text{jet} \tag{19.7}$$

was observed at an energy in the center-of-mass system of 7.4 GeV. The jets were not directly visible to the unaided eye (i.e., as narrow bundles of particles in the detector). However, using the sphericity tensor to analyse the final states of hadrons and checking the angular distribution of the jet axis with respect to the incident beam yielded convincing evidence for the existence of the quark jets.

To search for jets, they found for each event the direction which minimized the sum of squares of transverse momenta of the detected particles. **The sphericity tensor is the inertia tensor of classical mechanics with coordinate vectors replaced by momentum vectors as a measure of the shape of a multi-particle state in momentum space**, and thereby of its "jettiness." The **sphericity** S is defined as:

$$S = \frac{3}{2} \min \left(\frac{\sum_i p_{\perp,i}^2}{\sum_i p_i^2} \right) \tag{19.8}$$

with respect to the event axis (to be defined through minimization). It is a measure of the summed $p_\perp^2$ with respect to the axis. The minimum value corresponds to the axis being the event axis. It can be shown that $0 \le S \le 1$. **A jet will have S tending towards zero while an isotropic phase-space distribution will have larger values of S.** The data collected at SPEAR is illustrated in Figure 19.3 for the center-of-mass energies of $\sqrt{s} = 3$ GeV and 7.4 GeV. The observation is clearly consistent with jets being produced at the higher energies. The hadrons become more clustered around an axis (corresponding to the parent quark's direction) than predicted by isotropic phase space. Jets were observed!

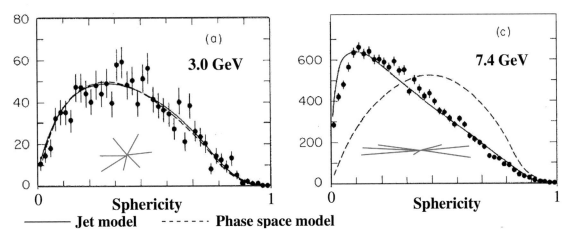

Figure 19.3 Sphericity distribution observed at center-of-mass energies of $\sqrt{s} = 3$ GeV and 7.4 GeV. Reprinted with permission from G. Hanson *et al.*, "Evidence for jet structure in hadron production by e^+e^- annihilation," *Phys. Rev. Lett.*, vol. 35, pp. 1609–1612, Dec 1975. Copyright 1975 by the American Physical Society.

19.4   Event Jet Variables

Clustering algorithms have become an indispensable tool in the study of multi-hadronic events. They take the large number of particles produced in high-energy scatterings and cluster them into a small number of "jets." Such a simplified characterization of the event helps focus on the main properties of the underlying dynamics. In particular, the reconstructed jets should reflect the partons, and thus allow a separation of perturbative and non-perturbative QCD physics aspects. We have seen in the previous section that such a jet algorithm was crucial in the understanding of $e^+e^- \rightarrow$ hadrons events. In this context, we discuss here another method of finding jets in an event called the **"thrust" axis method**. This jet algorithm is a computationally more demanding optimization which requires one to find the direction vector $\vec{n}$ that maximizes the variable:

$$T \equiv \max_{|\vec{n}|=1} \left(\frac{\sum_i |\vec{p}_i \cdot \vec{n}|}{\sum_i |\vec{p}_i|} \right) \tag{19.9}$$

where $\vec{p}_i$ is the reconstructed momentum of particle i in the event. The vector $\vec{n}$ that maximizes T is called the thrust axis. The thrust is constrained to be $1/2 \leq T \leq 1$, with $T = 1$ for two back-to-back pencil jets and $T = 1/2$ for a spherically symmetric distribution. Once the thrust axis is found, one can reiterate to find the "major" and "minor" axes:

$$\text{Major} = \max_{|\vec{n}'|=1, \vec{n}' \cdot \vec{n}=0} \left(\frac{\sum_i |\vec{p}_i \cdot \vec{n}'|}{\sum_i |\vec{p}_i|} \right) \tag{19.10}$$

and

$$\text{Minor} = \left(\frac{\sum_i |\vec{p}_i \cdot \vec{n}''|}{\sum_i |\vec{p}_i|} \right) \quad \text{with} \quad \vec{n}'' \cdot \vec{n} = \vec{n}'' \cdot \vec{n}' = 0 \tag{19.11}$$

Finally, the "oblateness" is equal to the difference between major and minor. The various algorithms are optimized within each experimental condition to best reproduce the parton characteristics.

The SPEAR data also allowed the study of the scattering angle distribution in the center-of-mass frame, for which we expect a $d\sigma/d\Omega \propto 1 + A\cos^2 \theta$ distribution from QED. Indeed, the SPEAR data showed not only jets but that their angular distribution was consistent with originating from spin-1/2 partons. Such studies were also confirmed at the higher energies provided by the electron–positron collider PETRA of DESY, Hamburg. The measurements of the angular distribution of the sphericity axis by the CELLO experiment [208] is shown in Figure 19.4. At these energies, the event is characterized by two jets and it is not possible to distinguish which came from the quark and which came from the antiquark. Hence, one is limited to the study of $|\cos \theta|$. The data collected at center-of-mass energies between 38.8 and 46.8 GeV are wonderfully consistent with a $1 + \cos^2 \theta$ distribution, predicted by QED for two spin-1/2 Dirac quarks.

At higher energies, we expect the jets to become thinner, as particles will be more longitudinally boosted. This is also observed, as is illustrated in Figure 19.5, where the average thrust values measured in different experiments are plotted against the center-of-mass energies. This behavior is well reproduced by the theoretical models, shown as the curves.

19.5   The Discovery of the Gluon

The gluons do not scatter leptons directly as quarks do. Being colored, gluons interact among each other and with the quarks; they are permanently confined like the quarks, hence they cannot be detected as free particles. A first indication of the existence of gluons in the nucleon came from the data on lepton–nucleon scattering: the momentum sum rule of the nucleon structure functions suggested that about half of a fast nucleon's momentum must be carried by flavorless parton constituents, presumably gluons. Scaling violations compatible with QCD observed in deep inelastic scattering were also supporting evidence for the existence of

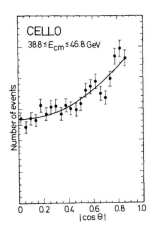

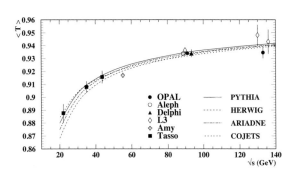

Figure 19.4 Measurement of the angular distribution of the sphericity axis. Reprinted from H.-J. Behrend *et al.*, "Determination of α_s and $\sin^2\theta_W$ from measurements of the total hadronic cross-section in e^+e^- annihilation," *Phys. Lett. B*, vol. 183, no. 3, pp. 400–411, 1987, with permission from Elsevier.

Figure 19.5 Average thrust in e^+e^- events as a function of center-of-mass energy. Reprinted by permission from Springer Nature: G. Alexander *et al.*, "QCD studies with e^+e^- annihilation data at 130-GeV and 136-GeV," *Z. Phys. C – Particles and Fields*, **72**, pp. 191–206 (1996). https://doi.org/10.1007/s002880050237. Copyright Springer-Verlag 1996.

gluons. The effect remained mysterious, however, and it was thought that Nature was just "behaving" as if gluons existed, partons being merely a mathematical construct mirroring an underlying symmetry. The crucial step was taken in 1979 by experiments at PETRA of DESY, Hamburg. It is worth noting that the accelerator had been primarily motivated by the search for the "top" quark in e^+e^- collisions.

In 1976 **John Ellis**,[3] **Mary K. Gaillard**, and **Graham Ross**[4] suggested the existence of the emission of hard gluon Bremsstrahlung by quarks, in analogy with the radiation of electromagnetic Bremsstrahlung by electrically charged particles. An emitted hard gluon would subsequently generate a jet similar to a quark jet. They wrote:

> "*Motivated by the approximate validity of the naive parton model and by asymptotic freedom, we suggest that hard gluon Bremsstrahlung may be the dominant source of hadrons with large momenta transverse to the main jet axis. This process should give rise to three-jet final states.*"[5]

This is illustrated in Figure 19.6. This kind of final state would lead to what they called "Mercedes-Benz" events, where the three jets should be in a plane.

At the partonic level, the kinematics can be defined by introducing the following scalars:

$$s = (q + \bar{q})^2, \quad \text{and} \quad t = (q + g)^2, \quad \text{and} \quad u = (\bar{q} + g)^2 \qquad (19.12)$$

The differential cross-section at lowest order is then given by:

$$\frac{\mathrm{d}^2\sigma(e^+e^- \to q\bar{q}g)}{\mathrm{d}t\mathrm{d}u} = \left(\frac{4}{3}\right)\left(\frac{\alpha_s}{2\pi}\right)\left(3e_q^2\left(\frac{4\pi\alpha^2}{3s}\right)\right)\left(\frac{t^2 + u^2 + 2sQ^2}{tuQ^4}\right) \qquad (19.13)$$

---

3   Jonathan Richard Ellis (born 1946), British theoretical physicist.
4   Graham Garland Ross (1944–2021), British theoretical physicist.
5   Reprinted from J. Ellis, M. K. Gaillard, and G. G. Ross, "Search for gluons in e^+e^- annihilation," *Nuclear Physics B*, vol. 111, no. 2, pp. 253–271, 1976, with permission from Elsevier. Copyright 1976 North-Holland Publishing Company.

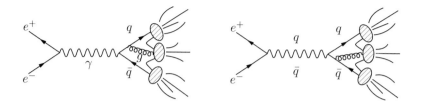

Figure 19.6 Tree-level Feynman diagram for $e^+e^- \to q\bar{q}g \to 3$ jets. The shaded areas represent the non-perturbative hadronization process, which lead to the formation of final-state hadrons, regrouped (in this case) in three jets.

where $Q^2 = s + t + u$. At this level, the expression diverges for t or u going to zero. This happens whenever the gluon is collinear with the quark or the antiquark (collinear divergence) or when its energy goes to zero (infrared divergence). We have already encountered similar divergences when we discussed photon radiation in QED (see Section 12.2). These unphysical results need to be taken care of using usual methods, such as restricting the kinematical phase space and summing higher-order diagrams.

The momentum vectors of the three elementary partons (quark, antiquark, and gluon), produced by the annihilation of an electron–positron pair, span a plane, as illustrated in Figure 19.7. Consequently, the three jets generated by the hadronization of the partons are forming an (approximately) "planar event." This approximate planarity of the array of momentum vectors of the hadrons is simply a consequence of momentum conservation and the limited transverse momenta within a jet.

Following the successful method of the sphericity, a "tri-jettiness" variable was devised [209]. The idea was that once the invariant mass of each pair of jets in a three-jet event was at least about 7.4 GeV (the mass at which the two-jet states had first been identified at SPEAR), a three-jet state would be identified by the method. This led to the estimate that three-jet events could be detected once PETRA reached a center-of-mass energy of more than ≈ 22 GeV. In 1979 the PETRA accelerator was operated with each beam up to an energy of 13.7 GeV, yielding $\sqrt{s} = 27.4$ GeV.

The four detectors MARK-J, PLUTO, TASSO, and later JADE recorded data and all found evidence for planar three-jet events [210–213]. One beautiful candidate event from PLUTO is shown in Figure 19.8, a clear indication of the process:

$$e^+e^- \to q\bar{q}g \to \text{jet} + \text{jet} + \text{jet} \tag{19.14}$$

19.6 Monte-Carlo Parton Shower Simulations

Figure 19.9 illustrates hadronic final states collected at the high energies of LEP by the OPAL experiment. One clearly identifies different topologies, which can be classified in two and three final-state jets. At the partonic level, these would naively translate to $e^+e^- \to q\bar{q}$ and $e^+e^- \to q\bar{q}g$.

Leading-order (LO) calculations give qualitative understanding of observables but often not accurately enough. In particular, they cannot describe these multiple-jet events. We have seen in the previous sections that exact matrix elements for $2 \to 2$ and $2 \to 3$ scattering processes can be calculated in a straightforward way. However, it is also clear that the computations become increasingly complex as the number of final-state partons increases. In any case, exact calculations do suffer from divergences and are always at the partonic level, so they need to be cured and then matched to experimental data which on the other hand collects events with final-state hadrons.

In this context, a Monte-Carlo method, called **parton shower simulations**, based on leading logarithm

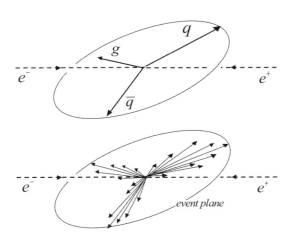

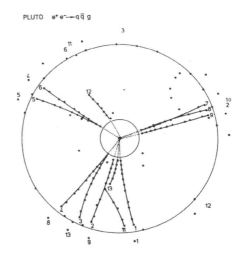

Figure 19.7 The momentum vectors of the three elementary partons (quark, antiquark, and gluon), produced by annihilation of an electron–positron pair, span a plane (upper figure). Consequently, the three jets generated by the hadronization of the partons are forming an (approximately) "planar event."

Figure 19.8 An evident 3-jet event from PLUTO at PETRA: 13 charged tracks and 5 neutral energy deposits were reconstructed (the numbers are just labels). From Ch. Berger, "Results from the PLUTO experiment on e^+e^- reactions at high energies," in *Proc. of the 1979 International Symposium on Lepton and Photon Interactions at High Energies*, Fermilab (USA).

approximations has been developed. Compared to LO calculations, the addition of the parton showers LO+PS, as illustrated in Figure 19.10, gives a better description of the data. They simulate the large number of soft gluons which are emitted by the final-state partons before they fragment into hadrons. Parton shower codes can describe most of the features of the common events (they are "tuned" to do so!). But they usually rely on soft and collinear approximations, so they do not necessarily reproduce correctly the hard, large-angle gluon radiation. But it is often the rare, hard multi-jet configurations that are interesting.

These latter are correctly described by NLO calculations. This is sketched in Figure 19.10 as the NLO diagram with the emission of a single hard gluon by one of the final-state quarks. NLO calculations are necessary to have a reliable estimate of the normalization and shape of complex events by taking into account effects of extra radiation (in technical terms, they reduce the QCD μ-scale uncertainty).

In practice, one merges the two approaches by combining explicit LO/NLO or even NNLO calculations of matrix elements with parton shower developments. This is represented graphically in Figure 19.10 as the NLO+PS case, where a final-state configuration with two quarks and a hard gluon emitted at a large angle is shown. Each of the three partons (the quark, the antiquark, and the gluon) is then subjected to parton showers, a process which describes the emission of additional soft gluons. The proper matching between the "hard" and "soft" gluons needs attention to avoid double counting and the methods are well developed today. As an illustration, the rates of three-jet events observed in different experiments at different center-of-mass energies are shown in Figure 19.11. They are compared to the theoretical predictions which exhibit a very good match. Overall, the process of hadronic event generation is rather complex, but many well-proven simulation models exist today to properly reproduce experimental data in a very large kinematical domain.

We illustrate in the following the actual generation of an e^+e^- event with the PYTHIA Monte-Carlo event generator [68]. The (simplified) main code, written in C++, is shown below. It basically sets up the colliding beams of electrons against positrons, giving each beam an energy of 10 GeV. The program has an event loop,

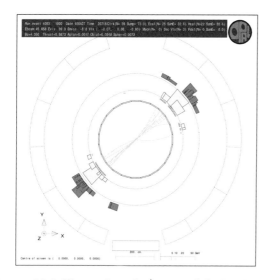

Figure 19.9 Illustration of e^+e^- annihilation into two- and three-jet final states in the OPAL detector at LEP. Copyright CERN.

which repeatedly generates new event configurations based on the known differential cross-sections. Each event is printed in tabular form. This is repeated 10,000 times. Then the program ends. The source code is the following:[6]

```cpp
// This is a simple test program.
#include "Pythia8/Pythia.h"
using namespace Pythia8;

int main() {
  // Generator.
  Pythia pythia;

  // Process selection.
  pythia.readString("Beams:frameType = 2");
  pythia.readString("Beams:idA =  11");
  pythia.readString("Beams:idB = -11");
  pythia.settings.parm("Beams:eA", 10.);
  pythia.settings.parm("Beams:eB", 10.);
    // Allow no substructure in e+- beams as we are neglecting
    // initial state radiation at low energy
  pythia.readString("PDF:lepton = off");
  pythia.init();

  // Begin event loop. Generate event. Skip if error. List first few.
  for (int iEvent = 0; iEvent < 10000; ++iEvent) {
    if (!pythia.next()) continue;
```

---

6   Reprinted code by permission from the PYTHIA author. PYTHIA is licensed under GNU GPL v2 or later. Copyright 2019 Torbjörn Sjöstrand.

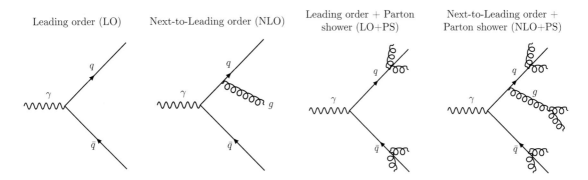

Figure 19.10 Graphical illustration of the matrix elements at leading order (LO), next-to-leading order with the explicit emission of a hard gluon (NLO), leading order with parton shower simulating many soft gluons (LO+PS), and next-to-leading order with an explicit emission of a hard gluon and with a parton shower simulating the emission of many soft gluons (NLO+PS).

```
    pythia.event.list();

    // End of event loop. Statistics. Output histograms.
    }
    pythia.stat();
    return 0;
}
```

A typical event output by the PYTHIA code above is shown below. The event is presented in tabular form containing the number of the line, an identifier, the name of the object, its status, its mothers and daughters, which are indices that allow us to reconstruct the origin and the descendants of the object, the colors, and the kinematical momenta (p_x, p_y, p_z), energy and rest mass in units of GeV. The truncated output is the following:

```
-------- PYTHIA Event Listing  (truncated) --------------------------------------------------

no      id  name         status    mothers   daughters    colours       p_x       p_y       p_z        e         m
0       90  (system)      -11      0    0     0    0      0      0      0.000     0.000     0.000    20.000    20.000
1       11  (e-)          -12      0    0     3    0      0      0      0.000     0.000    10.000    10.000     0.001
2      -11  (e+)          -12      0    0     4    0      0      0      0.000     0.000   -10.000    10.000     0.001
3       11  (e-)          -21      1    0     5    6      0      0      0.000     0.000    10.000    10.000     0.000
4      -11  (e+)          -21      2    0     5    6      0      0      0.000     0.000   -10.000    10.000     0.000
5        2  (u)           -23      3    4     9    9    101      0      5.708    -7.330    -3.685    10.000     0.330
6       -2  (ubar)        -23      3    4     7    8      0    101     -5.708     7.330     3.685    10.000     0.330
                        Charge sum:  0.000         Momentum sum:      0.000    -0.000    -0.000    20.000    20.000

-------- End PYTHIA Event Listing  ------------------------------------------------------------
```

Lines 0–4 describe the initial state. We recognize the electron and positron each with an energy of 10 GeV and moving in opposite p_z directions. Lines 5 and 6 list the final-state partons, in this case we had $e^+e^- \to u\bar{u}$. As a result of the scattering, the quark–antiquark pairs have non-vanishing p_x and p_z momenta, i.e., they have, as expected, gained transverse momenta. Still they are back-to-back, balanced as one can see by adding their respective p_x and p_y.

The next step is the parton shower:

```
-------- PYTHIA Event Listing  (truncated) --------------------------------------------------
no      id  name         status    mothers   daughters    colours       p_x       p_y       p_z        e         m
7       -2  (ubar)        -51      6    0    16   16      0    105     -4.440     7.090     2.419     8.715     0.330
8       21  (g)           -51      6    0    10   11    105    101     -1.094     0.017     1.153     1.590     0.000
9        2  (u)           -52      5    5    12   12    101      0      5.534    -7.107    -3.572     9.696     0.330
```

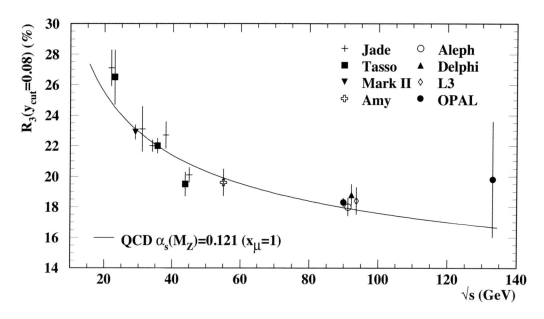

Figure 19.11 Rate of three-jet events in e^+e^- events as a function of the center-of-mass energy. Reprinted by permission from Springer Nature: G. Alexander *et al.*, "QCD studies with e^+e^- annihilation data at 130-GeV and 136-GeV," *Z. Phys. C - Particles and Fields*, **72**, pp. 191–206 (1996). https://doi.org/10.1007/s002880050237. Copyright Springer-Verlag 1996.

10	21	(g)	-51	8	0	15	15	105	106	-0.096	0.227	0.957	0.988	0.000
11	21	(g)	-51	8	0	14	14	106	101	-0.810	-0.452	0.075	0.931	0.000
12	2	(u)	-52	9	9	13	13	101	0	5.346	-6.865	-3.451	9.366	0.330

The lines with status equal to -51 correspond to outgoing objects produced by parton branching, while the -52 are the outgoing copy of recoiler with changed momentum to conserve energy momentum. In the case above, the $\bar{u}$ quark has radiated a gluon of 1.590 GeV, leaving a recoiler quark u in line 9. The gluon has further branched into two gluons.

Finally, we obtain the following partons ready to hadronize:

```
-------- PYTHIA Event Listing  (truncated) -------------------------------------------
no    id  name        status    mothers   daughters    colours    p_x      p_y      p_z       e        m
13     2  (u)          -71    12   12    17   20    101    0     5.346   -6.865   -3.451    9.366    0.330
14    21  (g)          -71    11   11    17   20    106  101    -0.810   -0.452    0.075    0.931    0.000
15    21  (g)          -71    10   10    17   20    105  106    -0.096    0.227    0.957    0.988    0.000
16    -2  (ubar)       -71     7    7    17   20      0  105    -4.440    7.090    2.419    8.715    0.330
```

And finally we have the hadronization and decay of metastable particles:

```
-------- PYTHIA Event Listing  (truncated) -------------------------------------------
17   323  (K*+)        -83    13   16    21   22      0    0     5.147   -6.576   -3.723    9.184    0.866
18  3122  (Lambda0)    -83    13   16    25   26      0    0    -0.985   -0.663    0.729    1.785    1.116
19  -211  pi-           84    13   16     0    0      0    0    -1.062    1.823    0.482    2.169    0.140
20 -2114  (Deltabar0)  -84    13   16    23   24      0    0    -3.100    5.416    2.512    6.862    1.353
21   321  K+            91    17    0     0    0      0    0     3.826   -5.026   -2.548    6.829    0.494
22   111  (pi0)        -91    17    0    27   28      0    0     1.321   -1.550   -1.175    2.355    0.135
23 -2112  nbar0         91    20    0     0    0      0    0    -2.686    4.042    1.991    5.329    0.940
24   111  (pi0)        -91    20    0    29   30      0    0    -0.414    1.374    0.521    1.533    0.135
25  2212  p+            91    18    0     0    0      0    0    -0.921   -0.667    0.683    1.625    0.938
26  -211  pi-           91    18    0     0    0      0    0    -0.064    0.003    0.046    0.160    0.140
27    22  gamma         91    22    0     0    0      0    0     0.982   -1.087   -0.894    1.716    0.000
28    22  gamma         91    22    0     0    0      0    0     0.339   -0.463   -0.281    0.639    0.000
29    22  gamma         91    24    0     0    0      0    0    -0.422    1.307    0.490    1.459    0.000
30    22  gamma         91    24    0     0    0      0    0     0.008    0.067    0.030    0.074    0.000
```

The event can therefore be expressed in terms of its final-state particles as:

$$e^+ e^- \to u + \bar{u} \quad \to \quad K^{*+} + \pi^- + \Lambda^0 + \quad \bar{\Delta}^0$$

(19.15)

The event jet analysis leads to the following result for this event:

```
-------- PYTHIA Thrust Listing ------------
        value      e_x       e_y       e_z
Thr   0.93707   -0.56296   0.71922   0.40718
Maj   0.15832   -0.51792  -0.69093   0.50435
Min   0.04800    0.64407   0.07304   0.76147
-------- End PYTHIA Thrust Listing --------
```

The thrust value is $T = 0.937$, the major is equal to 0.16, and the minor is 0.05. Two different jet algorithms, the LUND P_T and the JADE algorithms (see, e.g., Ref. [214]), can also be applied and give the following answers:

```
-------- PYTHIA ClusterJet Listing,   Lund pT =  2.000 GeV ---
no   mult     p_x       p_y       p_z        e         m
 0    6     -5.147     6.576     3.723     10.816     5.778
 1    3      5.147    -6.576    -3.723      9.184     0.866
-------- End PYTHIA ClusterJet Listing ----------------------
```

The LUND P_T with a jet parameter set at 2 GeV finds two jets matching well the initial direction and energies of the final-state partons. The gluons are not resolved as separate jets. On the other hand, the JADE algorithm with a parameter $m = 2$ GeV finds three jets, the third one with an energy of about 1.8 GeV, which reasonably matches the emission of the gluon.

```
-------- PYTHIA ClusterJet Listing,   JADE m =  2.000 GeV ---
no   mult     p_x       p_y       p_z        e         m
 0    3      5.147    -6.576    -3.723      9.184     0.866
 1    4     -4.162     7.239     2.993      9.031     1.691
 2    2     -0.985    -0.663     0.729      1.785     1.116
-------- End PYTHIA ClusterJet Listing ----------------------
```

This example shows that there is no absolute way of reconstructing the energy flow from hadronic events. The number of jets actually reconstructed depends on the parameters chosen for the algorithms. This is not a problem *per se*, but requires then that both experimental data and theoretical models, generated via Monte-Carlo samples of events, are passed through the same jet algorithms and their output compared. Of course, the jet reconstruction can never be perfect, since there will always be smearing effects that cannot be compensated, and since there is not even a well-defined transition from perturbative to non-perturbative QCD.

19.7   The Polarized Deep Inelastic Muon Scattering

The CERN M2 muon beam line was first commissioned in 1978 and is still in operation today, with only minor modifications. It is a high-energy muon beam, combining a wide range of momenta up to 300 GeV with high intensities and minimal halo background. The beam has a natural longitudinal polarization that can be tuned by varying the momentum ratio of decay muons to parent pions, and can reach values up to $\approx 80\%$. Longitudinal polarization occurs naturally due to parity violation in the $\pi, K \to \mu\nu$ decay. A high beam polarization is an essential prerequisite for the measurement of spin-dependent structure functions.

The beam design is based on two independent sections: (1) a hadron acceptance and decay channel 600 m in length (on the order of 10% of the pion decay length at 100 GeV/c) to obtain an appreciable muon yield; (2) followed by a muon transport section that selects and cleans the muon beam produced upstream and transports and focuses it to the experiment. The muons are emitted along the decay channel, by pions and

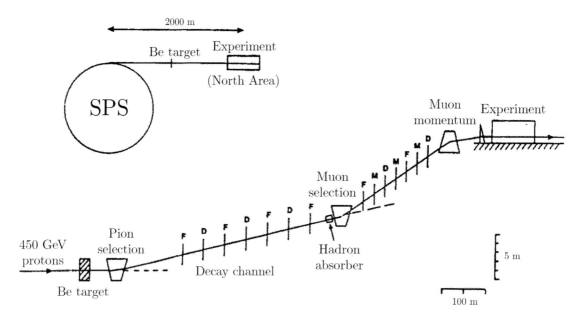

Figure 19.12 Layout of the CERN muon beam, where F and D denote focusing and defocusing magnets and M are magnetic collimators. The beam is bent in the vertical direction by means of four dipole magnets. Note the very different horizontal and vertical scales. Reprinted from S. Kullander, "Highlights of the European Muon Collaboration," *Nucl. Phys.*, vol. A518, pp. 262–296, 1990, with permission from Elsevier. Copyright 1990 Elsevier Science Publishers B.V. (North Holland).

kaons which are produced by the interaction with beryllium of 450 GeV protons from the CERN Super Proton Synchrotron. The layout is shown in Figure 19.12.

The CERN muon beam was built to serve the European Muon Collaboration (EMC), and the Bologna–CERN–Dubna–Munich–Saclay (BCDMS) Collaboration, which took data from 1978 until 1985, both with liquid hydrogen, liquid deuterium, and solid nuclear targets. The beam was also used by the subsequent New Muon Collaboration (NMC, 1986–1989) and Spin Muon Collaboration (SMC, 1992–1996) experiments. Since 2002, the tradition is continued by the COMPASS experiment, which represents the most comprehensive rebuild of the EMC/NMC/SMC spectrometer. The main goal of the early experiments was to perform (a) a precision measurement of scaling violations for tests of perturbative QCD and measurements of the strong coupling constant, and (b) the measurement of spin-dependent structure functions.

The deep inelastic differential cross-section discussed in the previous chapter was valid for unpolarized electrons interacting on an unpolarized nuclear target. In the CERN muon beam the incoming muons will be naturally polarized. If the incoming lepton is assumed to be longitudinally polarized with helicity $\lambda = \pm 1$, then we can replace its current $(\bar{u}(p')\gamma_\mu u(p))$ in Eq. (16.35) with $(\bar{u}(p')\gamma_\mu (1 + \lambda\gamma^5) u(p))$ to get

$$-i\mathcal{M}^\lambda_{\ell p \to \ell p} = \frac{ie^2}{(p - p')^2} \left(\bar{u}(p')\gamma_\mu \left(1 + \lambda\gamma^5 \right) u(p) \right) \cdot \bar{u}(k') \left[\gamma^\mu F_1(q^2) + \frac{i}{2M_p}\sigma^{\mu\nu}q_\nu F_2(q^2) \right] u(k) \qquad (19.16)$$

The resulting cross-section is, however, unchanged because of parity conservation in QED. The same conclusion would be reached if one assumed that the target but not the lepton is polarized. However, if both the lepton and the target are polarized, then the cross-section will depend on the internal spin structure of the target, which can then be studied.

Since both quarks and gluons possess spin and the forces between them are spin-dependent, one expects important information on these forces and on nucleon structure to be obtained through the study of the spin-dependent aspects of the nucleon wave function. Deep inelastic polarization experiments thus probe the spin distribution of quark constituents inside the nucleon, which should be explainable within QCD. As expected, the calculation of the quark spin distributions inside the nucleon is a non-perturbative calculation and it is therefore difficult to calculate spin-dependent structure functions from the basics of QCD. However, models of nucleon structure have been developed, which can be compared to experimental data. Establishing the relationship between the simple spin-1/2 of the proton and the neutron, and their internal structure consisting of quarks and gluons, is today the primary motivation for polarized deep inelastic experiments.

Experimentally, the measured asymmetry A compares the cross-sections for the case of parallel and antiparallel spins of the colliding particles and is given by [215]:

$$A \equiv \frac{\mathrm{d}\sigma^{\uparrow\downarrow} - \mathrm{d}\sigma^{\uparrow\uparrow}}{\mathrm{d}\sigma^{\uparrow\downarrow} + \mathrm{d}\sigma^{\uparrow\uparrow}} \tag{19.17}$$

where $\sigma^{\uparrow\uparrow}$ is the cross-section when the spins of lepton and proton are parallel and along the direction of motion of the incident lepton; $\sigma^{\uparrow\downarrow}$ is the cross-section for antiparallel spins.

In order to compute the differential cross-section, the hadronic tensor $W^{\mu\nu}$ of Eq. (16.39) is decomposed into the sum of a symmetric spin-independent part and an antisymmetric spin-dependent part:

$$W^{\mu\nu} \equiv W_S^{\mu\nu} + W_A^{\mu\nu} \tag{19.18}$$

The spin-independent symmetric part is defined with the common spin-independent structure (functions $W_1(\nu, Q^2)$ and $W_2(\nu, Q^2)$):

$$W_S^{\mu\nu} = \left(-g^{\mu\nu} + \frac{q^\mu q^\nu}{q^2}\right) W_1 + \frac{1}{M_p^2} \left(k^\mu - \frac{k \cdot q}{q^2} q^\mu\right) \left(k^\nu - \frac{k \cdot q}{q^2} q^\nu\right) W_2 \tag{19.19}$$

where the kinematical variables are the usual ones, as defined in Figure 16.1. The antisymmetric tensor amplitude defines the **spin-dependent structure functions** $G_1(\nu, Q^2)$ and $G_2(\nu, Q^2)$ [215]:

$$W_A^{\mu\nu} = M_p \epsilon^{\mu\nu\alpha\beta} q_\alpha s_\beta G_1 + \frac{1}{M_p} \epsilon^{\mu\nu\alpha\beta} q_\alpha \left[(k \cdot q)s_\beta - (s \cdot q)k_\beta\right] G_2 \tag{19.20}$$

where s^μ is the **spin polarization four-vector of the lepton** (see Eq. (8.256)). It can be shown that the difference of the differential cross-sections in terms of the spins gives directly G_1 and G_2, as:

$$\frac{\mathrm{d}\sigma^{\uparrow\downarrow}}{\mathrm{d}Q^2\mathrm{d}\nu} - \frac{\mathrm{d}\sigma^{\uparrow\uparrow}}{\mathrm{d}Q^2\mathrm{d}\nu} = \frac{4\pi\alpha^2}{E_1^2 Q^2} \left[M_P G_1(E_1 + E_3 \cos\theta) - Q^2 G_2\right] \tag{19.21}$$

In the Bjorken scaling limit, we replace the two spin-dependent structure functions by two **scaling parton distribution functions** $g_1(x)$ **and** $g_2(x)$ (cf. Eq. (16.109)):

$$M_p^2 \nu G_1(\nu, Q^2) \to g_1(x) \quad \text{and} \quad M_p \nu^2 G_2(\nu, Q^2) \to g_2(x) \tag{19.22}$$

The scaling functions $g_1(x)$ and $g_2(x)$ have direct physical interpretations in terms of quark partons. For a proton, the spin structure $g_1^p(x)$ can be expressed as:

$$g_1^p(x) = \frac{1}{2} \sum e_i^2 \left[q_i^+(x) - q_i^-(x)\right] \equiv \frac{1}{2} \sum e_i^2 \left[\Delta q_i(x)\right] \tag{19.23}$$

where i denotes the quark flavor, e_i is the quark charge, and q_i^+ and q_i^- are the probability distributions of quarks with spin parallel, respectively, antiparallel with the proton spin. The functions $\Delta q_i(x)$ are the shorthand notation for the **differences of probabilities for each spin configuration**.

The spin-dependent proton structure function expressed in terms of quarks can then be written as:

$$2g_1^p(x) = \frac{4}{9}\Delta u(x) + \frac{4}{9}\Delta \bar{u}(x) + \frac{1}{9}\Delta d(x) + \frac{1}{9}\Delta \bar{d}(x) + \frac{1}{9}\Delta s(x) + \frac{1}{9}\Delta \bar{s}(x) \tag{19.24}$$

where the factors 4/9 and 1/9 come from the electromagnetic nature of the interaction. Only the x dependence is written – however, one should bear in mind that scaling violation will introduce a Q^2 dependence from QCD higher orders.

The integral of $\Delta u(x) + \Delta \bar{u}(x)$ gives the fraction of the proton's spin in the quark–parton model, which is carried by the up and anti-up quarks. Analogous contributions can be written for down and strange quarks, giving the **equivalent of Bjorken's sum rule for the spins**:

$$\int_0^1 g_1^p(x)\mathrm{d}x = \frac{1}{2}\left[\frac{4}{9}\Delta u + \frac{1}{9}\Delta d + \frac{1}{9}\Delta s\right] \tag{19.25}$$

where $\Delta q = \int_0^1 (\Delta q(x) + \Delta \bar{q}(x))\mathrm{d}x$. The **proton spin** J_p can then be decomposed as:

$$J_p \equiv \sum_q (\Delta q + L_q) + \Delta G + L_g \tag{19.26}$$

where Δq is the intrinsic contribution of the quarks of given flavor, ΔG is the gluon contribution, and L_q and L_g are the quark and gluon angular momentum contributions, respectively. In order to explain the spin-1/2 of the proton, the sum of each contribution has to exactly add up to $J_p = 1/2$.

No theoretical predictions exist for the exact x dependence of spin-dependent structure functions; this is similar to the situation in unpolarized scattering. Separate sum rules for the proton and the neutron were first derived by Ellis and Jaffe [216]. A key assumption in these studies is that isospin symmetry is valid, such that an up quark in the proton is equivalent to a down quark in the neutron. If this is true, then the total quark contribution will be the same for a proton and a neutron. Hence, from isospin invariance, the spin-dependent neutron structure function is:

$$\int_0^1 g_1^n(x)\mathrm{d}x = \frac{1}{2}\left[\frac{1}{9}\Delta u + \frac{4}{9}\Delta d + \frac{1}{9}\Delta s\right] \tag{19.27}$$

where the Δq are for the proton specifically.

The total contribution $\Delta\Sigma$ of the quarks to the nucleon spin is given by:

$$\Delta\Sigma \equiv \Delta u + \Delta d + \Delta s \tag{19.28}$$

In 1973, Ellis and Jaffe made the assumption that the strange quarks in the nucleon are not polarized and hence $\Delta s = 0$. This leads to the Ellis–Jaffe sum rule, which when combined with beta-decay data, gives the following predictions:

$$\int_0^1 g_1^p(x)\mathrm{d}x\bigg|_{EJ} = 0.176 \pm 0.006, \quad \text{and} \quad \int_0^1 g_1^n(x)\mathrm{d}x\bigg|_{EJ} = -0.002 \pm 0.005 \tag{19.29}$$

at $Q^2 \approx 10$ GeV^2.

In 1987 the EMC Collaboration reported its first measurement. Their data combined with others gave the unexpected result:

$$\int_0^1 g_1^p(x)\mathrm{d}x\bigg|_{EMC} = 0.126 \pm 0.018 \tag{19.30}$$

in sharp disagreement with the Ellis–Jaffe sum rule. This result was also confirmed by the SMC Collaboration. This situation, dubbed the "**proton spin crisis**," implied that the spin structure of the proton cannot be explained by the polarized valence quarks only. So, while in a static quark model where the quarks should carry essentially all of the nucleon spin, actual experimental data seem to indicate that it is actually less than a third. A compilation of data on g_1 is presented in Figure 19.13. Recent COMPASS results [217] indicate a somewhat larger quark spin contribution of $\Delta\Sigma = 0.33 \pm 0.03(\text{stat.}) \pm 0.05(\text{syst.})$ and a strange quark contribution $\Delta s + \Delta\bar{s} = -0.08 \pm 0.01(\text{stat.}) \pm 0.02(\text{syst.})$ different from zero. COMPASS data are consistent with the EMC conclusion that the quark spins do not account for most of the proton spin. In conclusion, there is today clear evidence that the spin structure of the nucleon is more complicated than the prediction of the naive quark–parton model.

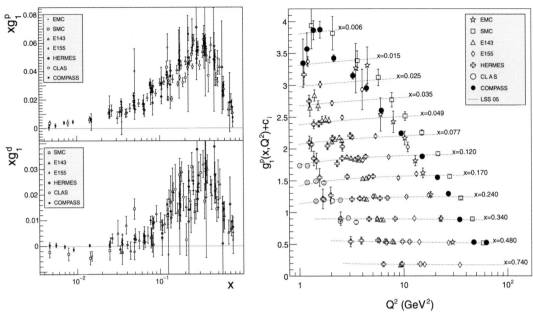

Figure 19.13 (left) $xg_1(x, Q^2)$ as a function of x for $Q^2 > 1$ GeV^2 for the proton and the deuteron; (right) proton data for $g_1(x, Q^2)$ as a function of x and Q^2 with $W > 2.5$ GeV. For clarity, the g_1 data for the ith x bin (starting from $i = 0$) were offset by $c_i = 0.28(11.6 - i)$. Reprinted from G. K. Mallot and R. Voss, "Deep inelastic scattering with the SPS muon beam," *Adv. Ser. Direct. High Energy Phys.*, vol. 23, pp. 287–311, 2015. *Int. J. Mod. Phys.* A30, 1530052 (2015), under CC BY license https://creativecommons.org/licenses/by/3.0/

19.8   Electron (Positron)–Proton Scattering at Very High Energies with HERA

The HERA electron–proton collider was able to accelerate electrons or positrons up to 27.5 GeV and collide them with protons up to an energy of 820 GeV (see Section 3.3 and Figure 3.13). Two independent experiments, H1 and ZEUS, studied these collisions.

- **H1 at HERA.** The **H1 detector** [218] (see Figure 19.14(left)) focused on the clean identification of electrons and their energy measurement, produced in $e^{\pm}p$ collisions. It consisted of a central and a forward tracking system, each containing layers of drift chambers and trigger proportional chambers. Precise measurement of electron showers was provided by a liquid argon calorimeter. A liquid argon cryostat surrounded the trackers; it housed the lead absorber plates and readout gaps of the electromagnetic section. It was followed by the steel plates acting as the hadronic section of the calorimeter. A superconducting cylindrical coil with a diameter of 6 m and a length of 5.75 m provided a track-bending field of 1.15 T. The iron return yoke of the magnet was instrumented with streamer tubes. Muon tracks were registered there. Muon identification further benefited from additional chambers inside and outside the iron. Stiff muon tracks in the forward direction were analyzed in a supplementary toroidal magnet sandwiched between drift chambers.

- **ZEUS at HERA.** The **ZEUS detector** (see Figure 19.14(right)) also focused the precise study of $e^{\pm}p$ collisions. Its dimensions were 12 m × 10 m × 19 m and its total weight was 3600 tons. The heart of the ZEUS detector was the uranium scintillator calorimeter (CAL) which measured energies and directions of particles and particle jets with high precision. The CAL hermetically enclosed the tracking detectors which measured the tracks of charged particles using wire chambers and which consisted of a vertex detector (VXD), a central drift chamber (CTD), forward (FTD) and backward (RTD) drift chambers, and in the forward direction a transition radiation detector (TRD) to identify high-energy electrons. These chambers were surrounded by a thin superconducting

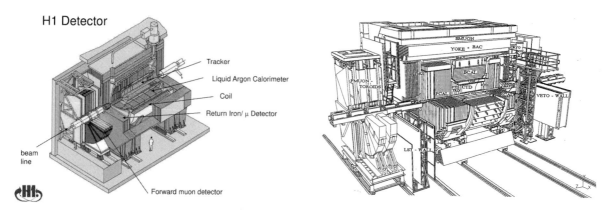

Figure 19.14 (left) Side view of the H1 detector at HERA. Reprinted from image accessed at www-h1.desy.de/h1det/. (right) Side view of the ZEUS detector at HERA. Reprinted from image accessed at www-zeus.desy.de/.

solenoid coil producing an axial magnetic field of 1.8 T for determining the momenta of charged particles from track curvature. Energy not fully absorbed in the uranium calorimeter was measured in the backing calorimeter (BAC) which used the 7.3 cm-thick iron plates of the return yoke as absorber and proportional tube chambers for observing penetrating particles. Particles which were not absorbed in the substantial material of the uranium scintillator and backing calorimeter were identified as muons. Their tracks were measured before and after the iron yoke by limited streamer tube chambers (BMUON and FMUON). The muon momenta were determined by the deflections of their paths by the solenoid and by the iron yoke which was magnetized toroidally up to 1.6 T by copper coils. In the forward direction, magnetized iron toroids instrumented with limited streamer tube and drift chambers measured very energetic muons (up to 150 GeV/c). An iron wall equipped with two layers of scintillation counters (VETO-WALL) was placed near the tunnel exit for detection of background particles produced upstream by the proton beam. In the very forward direction the leading proton spectrometer was installed in the beam line to measure forward scattered protons. In the direction of the electron beam, photons and electrons were detected in the luminosity monitor.

• **Data collection.** each experiment collected over one million $e^{\pm} - p$ inelastic neutral-current[7] events:

$$e^{\pm} + p \rightarrow e^{\pm} + X \qquad \text{(neutral current)} \tag{19.31}$$

The highest center-of-mass energy of the collisions was $\sqrt{s} = 318$ GeV. The data from deep inelastic scattering collected by H1 and ZEUS probed the proton in kinematical domains never previously reached. It enabled the study of the structure functions of the proton in the range $200 < Q^2 < 30,000$ GeV^2 and x down to $\simeq 10^{-4}$! An example of a positron–proton collision recorded by H1 at very high Q^2 is shown in Figure 19.15. The scattered positron energy and scattering angle were precisely measured in the tracker and electromagnetic calorimeter. These two measurements were sufficient to completely determine the Q^2 and ν of the event. The Bjorken x and y were also fully determined for each event. The proton was broken up and led to a final-state hadronic system visible as a jet of particles, balanced against the outgoing lepton (the total transverse momentum must be equal to zero). The energy and direction of the hadronic jet provided redundant although less precise information on the kinematics.

The QED contribution to $e^{\pm} - p$ scattering, derived in Eq. (16.114) under the assumption $E_e \gg M_p$, is the starting point to describe the neutral-current interactions at HERA. Making use of the Callan–Gross relation

7   A neutral current is defined as a current in which the exchanged boson is electrically neutral.

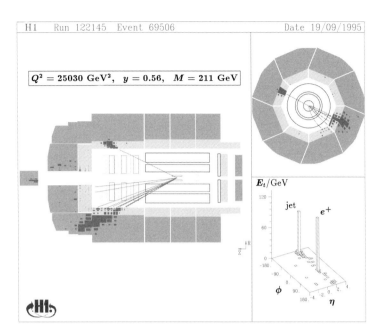

Figure 19.15 A high-energy positron–proton collision recorded by the H1 detector at a measured $Q^2 \simeq$ $25,000$ GeV^2. The event kinematics is determined from the electron angle and its energy. The measurement of the hadronic system gives some redundancy, although it is not as precise as the measurement based on the electron kinematics. Reprinted from image accessed at www-h1.desy.de/pictures/.

Eq. (17.100), the unpolarized double differential cross-section can be written as:

$$
\left(\frac{\mathrm{d}^2\sigma}{\mathrm{d}x\mathrm{d}Q^2}\right)_{QED} = \left(\frac{4\pi\alpha^2}{Q^4}\right)\frac{1}{x}\left[(1-y)F_2^{ep} + \frac{y^2}{2}2xF_1^{ep}\right] = \left(\frac{2\pi\alpha^2}{Q^4}\right)\frac{1}{x}\left[2(1-y)F_2^{ep} + y^2 F_2^{ep}\right]
$$

$$
= \left(\frac{2\pi\alpha^2}{Q^4}\right)\frac{1}{x}\left[1 + (1-y)^2\right]F_2^{ep} \tag{19.32}
$$

At the very high Q^2 achievable at HERA, the *weak* neutral-current contribution should also be taken into account, in addition to the electromagnetic current. At tree level, its Feynman diagram is similar to the QED one, however where the exchange of a photon is replaced by the exchange of a Z^0 gauge boson. The Feynman rules for computing its amplitude are defined within the electroweak theory (see Chapter 25) and will be introduced in later chapters discussing the weak interaction. The additional interaction also adds an interference term between the diagrams with the photon and the Z^0. To some extent, we already encountered such an effect when we discussed the tests of QED at high energies in Chapter 13. There, we saw that the QED cross-section was to be corrected by electroweak effects. Consequently, for describing electron–proton scattering at HERA, the QED $e - p$ cross-section was modified to account for the weak interaction, this latter having the peculiarity of introducing parity violation. This requires the introduction of a new structure function called F_3, as will become fully evident when we discuss neutrino–nucleon interactions in Chapter 28. A physical consequence is that the $e^- p$ and $e^+ p$ cross-sections are no longer equivalent (although they were within QED). Including also the so-called "longitudinal term," the differential cross-section for the neutral-current process is finally written as:

$$
\frac{\mathrm{d}^2\sigma_{\mathrm{NC}}^{e^{\pm}p}}{\mathrm{d}x\mathrm{d}Q^2} = \left(\frac{2\pi\alpha^2}{Q^4}\right)\frac{1}{x}\left\{(1+(1-y)^2)\tilde{F}_2(x,Q^2) \mp (1-(1-y)^2)x\tilde{F}_3(x,Q^2) - y^2\tilde{F}_{\mathrm{L}(x,Q^2)}\right\} \tag{19.33}
$$

where $\tilde{F}_2 = F_2^{ep} +$ weak and interference terms, and $\tilde{F}_3$ has a purely weak origin. The longitudinal structure function $\tilde{F}_L$ is important at low Q^2 (cf. Eq. (28.66)).

In order to take into account various effects, the experiments prefer to present their data as a function of the reduced neutral-current differential cross-sections $\sigma_{r,\mathrm{NC}}^{\pm}$, defined as [219]:

$$\sigma_{r,\mathrm{NC}}^{\pm} \equiv \frac{\mathrm{d}^2\sigma_{\mathrm{NC}}^{e^{\pm}p}}{\mathrm{d}x\mathrm{d}Q^2} \cdot \frac{Q^4 x}{2\pi\alpha^2 Y_+} = \tilde{F}_2 \mp \frac{Y_-}{Y_+}x\tilde{F}_3 - \frac{y^2}{Y_+}\tilde{F}_L \qquad (19.34)$$

where $Y_{\pm} = 1 \pm (1 - y^2)$. The measured combined HERA data for the inclusive neutral current e^+p and e^-p reduced cross-sections [219] together with fixed-target data [220, 221] and the predictions of HERAPDF2.0 NNLO (available at www.desy.de/h1zeus/herapdf20/) are shown in Figure 19.16(left). The lines indicate the predictions of the theoretical fits to the data. For clarity, the data at different values of x are shifted vertically.

At the intermediate and high values of x, the measured cross-sections are almost independent of Q^2 and hence are very consistent with Bjorken scaling. This is true up to the highest values of $Q^2 \approx 10^4 - 10^5$ GeV^2, a fact that can be interpreted as the quark being point-like up to at least those values of Q^2. A sudden deviation from Bjorken scaling at very high Q^2 could have signaled the presence of a substructure inside the quark! The consistency with Bjorken scaling can be translated into a limit on the "radius" of the quark, which must be smaller than $\sim 10^{-18}$ m. At low x the data shows very strong violations of scaling, as predicted by QCD. In the plot, the points are fitted to next-to-next-to-leading order QCD (labeled HERAPDF NNLO) and results are presented as lines on top of the data points. The older data from the deep inelastic muon scattering data collected by BCDMS and NMC are also shown in the plot as squares (fixed target). They extend the intermediate x region at low Q^2. Overall, the QCD predictions describe very well the entire set of data from the different experiments, both at HERA and at muon fixed-target facilities, and hence very strongly support the proton model over a very wide kinematical range.

At these very high energies, the weak charged-current contribution[8] also becomes non-negligible and is observed via the presence of a final-state neutrino (missing charged lepton + missing energy):

$$e^{\pm} + p \rightarrow \nu + X \qquad \text{(charged current)} \qquad (19.35)$$

Being a pure weak process, it has no analog in QED and is much rarer than the neutral current. It is analyzed in a similar fashion as the neutral-current reaction. The double differential cross-section is in this case given by:

$$\frac{\sigma_{\mathrm{CC}}^{e^{\pm}p}}{\mathrm{d}x\mathrm{d}Q^2} = \frac{G_F^2}{2\pi}\left[\frac{M_W^2}{M_W^2 + Q^2}\right]^2 \frac{1}{x}\left(\frac{Y_+}{2}W_2^{\pm} \mp \frac{Y_-}{2}xW_3^{\pm} - \frac{y^2}{2}W_L^{\pm}\right) \qquad (19.36)$$

where G_F is the weak Fermi coupling constant, to be introduced in Chapter 21, and the term in square brackets takes into account the propagation of the massive $W^{\pm}$ weak boson. The functions $W_2^{\pm}(x, Q^2)$, $W_3^{\pm}$, and $W_L^{\pm}$ are the charged-current structure functions and play an equivalent role to $W_2^{\pm}(x, Q^2)$, $W_3^{\pm}(x, Q^2)$, and $W_L^{\pm}(x, Q^2)$ in the neutral-current reaction. From this expression, the reduced cross-section used in the experimental analysis is defined as [219]:

$$\sigma_{r,\mathrm{CC}}^{\pm} = \frac{2\pi x}{G_F^2}\left[\frac{M_W^2 + Q^2}{M_W^2}\right]^2 \frac{\mathrm{d}^2\sigma_{\mathrm{CC}}^{e^{\pm}p}}{\mathrm{d}x\mathrm{d}Q^2} \qquad (19.37)$$

The data collected at HERA are shown in Figure 19.16(right). Although the charged-current data is less abundant, it provides a picture of the proton structure totally consistent with that obtained with neutral-current data.

---

8   The weak charged current will be introduced in detail in Chapter 23.

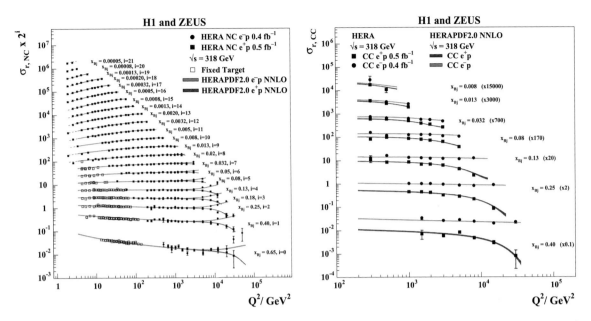

Figure 19.16 The combined HERA data for the inclusive reduced cross-sections for (left) neutral current e^+p and e^-p and (right) charged current e^+p and $\nu+X$. The fixed-target data [220,221] and the predictions of HERAPDF2.0 NNLO are overlaid. The lines indicate the predictions of the theoretical fits to the data. For clarity, the data at different values of x are shifted vertically. Reprinted from H. Abramowicz *et al.*, "Combination of measurements of inclusive deep inelastic $e^\pm p$ scattering cross-sections and QCD analysis of HERA data," *Eur. Phys. J. C*, vol. 75, no. 12, p. 580, 2015, under CC BY license http://creativecommons.org/licenses/by/4.0/.

19.9   Global Fits of the Parton Distribution Functions

A global fit of different types of measurements, including deep inelastic scattering of electron–proton, muon–proton, as well as neutrino deep inelastic scattering, allows us to extract the most precise parton distribution functions. This work is performed by several groups around the world. It is an art! As an example, the results of the LHAPDF5 parton distribution functions for the proton are shown in Figure 19.17. The curves represent $xq(x)$ for u, d, $\bar{u}$, $\bar{d}$, s, and gluons. They have been compiled with the software "A PDF Evolution Library" (APFEL) [222,223]. Because of scaling violations, we expect the parton distribution functions to depend on Q^2. This is indeed observed, as can be readily seen from the two plots which correspond to $Q^2 = 1$ GeV^2 and $Q^2 = 100$ GeV^2.

The data are indeed consistent with the parton model and clearly confirm that with increasing Q^2, the parton distributions change significantly, consistent with the DGLAP evolution equations, and in the following way:

- Valence quarks will lose their momentum in favor of gluons; i.e., in favor of the sea.

- Gluon Bremsstrahlung will shift the valence and sea distributions to smaller x values.

- The splitting of a gluon into a quark–antiquark pair will increase the amount of sea (mostly at small x).

Persistent experimental and theoretical work over the last 40 years has significantly enlarged the kinematical domain of the parton distribution functions and their resulting precision. Figure 19.18 shows the kinematical region in the x vs. Q^2 plane covered by experimental data. It is vast and extends to very low x values down to the range of 10^{-6}.

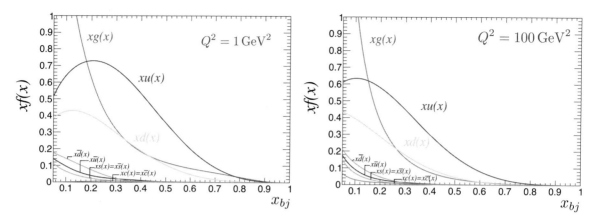

Figure 19.17 LHAPDF5 parton distribution functions for the proton. The curves represent $xq(x)$ for u, d, $\bar{u}$, $\bar{d}$, s, and gluons at $Q^2 = 1$ GeV^2 and $Q^2 = 100$ GeV^2. Compiled with the software APFEL (available at www.hepforge.org and https://apfel.mi.infn.it/).

The global parton distribution functions are plotted in log scale in Figure 19.19 for $Q^2 = 100$ GeV^2. Generally speaking, we can separate the valence quarks from the sea:

$$q(x) = q_v(x) + q_s(x) \tag{19.38}$$

The parton content of the proton (or in general of a hadron) is defined by its valence content. This is the result of the classification coming out of the $SU(3)$ flavor analysis. Naively, we found for instance the expectations that:

$$\int_0^1 dx(u(x) - \bar{u}(x)) = 2; \quad \int_0^1 dx(d(x) - \bar{d}(x)) = 1; \quad \int_0^1 dx(s(x) - \bar{s}(x)) = 0 \tag{19.39}$$

This is supported by the hard quarks at high x. However, the small-x region of soft partons is a fascinating world by itself. The pdf can be parameterized reasonably well by the following expressions:

$$\text{Valence quarks}: \quad \begin{cases} xq_v(x \to 1) \simeq (1-x)^3 \\ xq_v(x \to 0) \simeq x^{0.5} \end{cases}$$

$$\text{Sea quarks}: \quad \begin{cases} xq_s(x \to 1) \simeq (1-x)^7 \\ xq_s(x \to 0) \simeq x^{-0.25} \end{cases} \tag{19.40}$$

We observe an apparent divergence in the pdf as $x \to 0$ and

$$xq(x) \simeq x\bar{q}(x) \simeq x^{-0.25} \implies \int_0^1 dx(q(x) + \bar{q}(x)) \to \infty \tag{19.41}$$

How can there be an infinite number of partons in a proton? Obviously, the number of extra quark–antiquark pairs and gluons can be infinite, as long as the total momentum they carry is finite. This is the case, since the divergence is at $x \to 0$. The proton is indeed a very complex object!

19.10   Hadron–Hadron Collisions

The bulk of events at hadron–hadron colliders (see Section 3.3) arises from multiple "soft" collisions between partons. Such processes cannot be computed within perturbative QCD, which requires the presence of at least

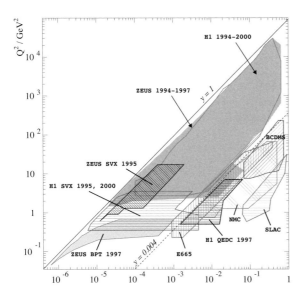

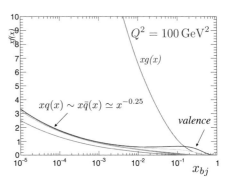

Figure 19.18 Kinematical region in the x vs. Q^2 plane covered by experimental data. Note the log–log scale! Reprinted from figure accessed at www-zeus.desy.de/physics/sfew/PUBLIC/sfew_res ults/preliminary/moriond04/zeush1.php.

Figure 19.19 The parton distribution functions plotted in log scale. Compiled with the software APFEL (available at www.hepforge.org and https://apfel.mi.infn.it/).

one high Q^2 scale. At high energies, a small fraction of the interactions, however, do involve hard collisions. Studies of hadron–hadron collisions at increasing energies have shown that the transverse momentum (p_T) of jets relative to the incident beam direction extended to very large values. This was first observed in the 1970s at the CERN ISR accelerator where pp collisions up to $\sqrt{s} \simeq 60$ GeV were achieved. The maximum p_T was also shown to increase with increasing s. This was contrary to the early observation in cosmic rays, where it was thought that the transverse momentum of hadrons was always limited in hadron–hadron collisions.

The observations at the CERN ISR were compatible with the occurrence of hard parton–parton collisions among sub-constituents of the hadrons: $qq \to qq$, $qg \to qg$, $gg \to gg$, etc. Such hard collisions, when they occurred, were responsible for the large p_T events. Large p_T events were subsequently studied at the CERN $p\bar{p}$ collider with $\sqrt{s} \simeq 600$ GeV, at the FNAL Tevatron with up to $\sqrt{s} \simeq 1.8$ TeV, and finally at the CERN LHC up to $\sqrt{s} \simeq 13$ TeV.

At hadron–hadron colliders, such as for example a proton–proton or proton–antiproton collider, two beams of protons (or protons–antiprotons) are brought into head-on collisions. When the two colliding beams have the same energy E_p, the collisions occur in the center-of-mass frame of the pp or $p\bar{p}$ system. We hence define the accelerator center-of-mass energy squared as:

$$s = (P_1 + P_2)^2 \simeq 2P_1 \cdot P_2 = 4E_p^2 \tag{19.42}$$

where P_1 and P_2 are the four-momenta of the colliding particles.

However, the hard scattering occurs in the parton–parton frame. It was first pointed out by **Drell** and **Yan** that **parton model ideas developed for deep inelastic scattering could be extended to certain processes in hadron–hadron collisions** (see Section 20.2 and Ref. [224]). The key point was that *all* logarithms appearing in the Drell–Yan corrections could be factored into the parton distributions, and this showed that it was a general feature of hard scattering processes (factorization theorem).

Let us define x_1, x_2 as the fraction of momentum carried by the partons i and j relative to their parents P_1 and P_2. In the center-of-mass frame of the two hadrons, the components of momenta of the incoming partons may then be written as:

$$p_1^\mu = x_1 P_1 = \frac{\sqrt{s}}{2}(x_1, 0, 0, x_1)$$

$$p_2^\mu = x_2 P_2 = \frac{\sqrt{s}}{2}(x_2, 0, 0, -x_2) \qquad (19.43)$$

The center-of-mass energy squared in the parton–parton system is:

$$\hat{s} \equiv (p_1 + p_2)^2 = (x_1 P_1 + x_2 P_2)^2 \simeq 2x_1 x_2 P_1 \cdot P_2 = x_1 x_2 s \qquad (19.44)$$

The cross-sections for the parton–parton subprocess must hence be combined with the relevant quark, antiquark and gluon density parton distribution functions in order to obtain their contribution to the total cross-section (see Figure 19.20). This can be expressed as:

$$d\sigma(P_1 P_2 \to X)(s) = \sum_{i,j} \int \int dx_1 dx_2 f_{q_i/P_1}(x_1) f_{q_j/P_2}(x_2) \hat{\sigma}_{ij \to X}(x_1 x_2 s) \qquad (19.45)$$

where the parton–parton center-of-mass energy is $\hat{s} = x_1 x_2 s$ and x_1, x_2 are the fractions of momentum carried by the partons i and j relative to their parents P_1 and P_2.

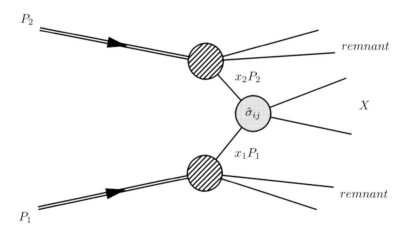

Figure 19.20 Feynman diagram for a hard collision $P_1 P_2 \to X$.

In actual hadron–hadron collisions, the momentum fractions x_1 and x_2 of the two interacting partons are unknown. The event kinematics must then be described by three variables, for example, x_1, x_2, and Q^2. This is more complicated than the lepton–proton deep inelastic scattering case, where the kinematics was fully constrained by the outgoing lepton. The three independent variables x_1, x_2, and Q^2 in hadron–hadron scattering need to be related to three experimentally well-measured quantities. The scattered partons are observed as jets. In a process such as $pp \to 2$ jets $+ X$, the angles of the two jets relative to the beam axis and their energies are well measured. Consequently, the momenta of the jets in the transverse plane, referred to as the transverse momenta p_T, are used to define the kinematics.

The net longitudinal momentum of the colliding parton is simply given by:

$$p_L = \frac{\sqrt{s}}{2}(x_1 - x_2) = (x_1 - x_2)E_p \qquad (19.46)$$

Therefore, the final state produced in the parton–parton collision is boosted along the beam direction, either in the forward or backward direction (i.e., if $x_1 > x_2$ or $x_1 < x_2$). The transverse components are not affected, and the total transverse momentum should balance to zero (when we neglect the transverse momenta of the initial partons within the protons).

The **pseudo-rapidity** η and the **azimuthal angle** ϕ in the plane transverse to the beam are commonly used variables to define the spatial coordinates describing the direction of a particle relative to the beam axis. If θ is the angle between the particle three-momentum and the positive direction of the beam axis, then the pseudo-rapidity η of the particle is equal to:

$$\eta \equiv -\ln\left[\tan\left(\frac{\theta}{2}\right)\right] = \frac{1}{2}\ln\left(\frac{|p| + p_z}{|p| - p_z}\right) \simeq \frac{1}{2}\ln\left(\frac{E + p_z}{E - p_z}\right) \tag{19.47}$$

where p_z is the component of the momentum along the beam axis (i.e., the longitudinal momentum), and $E \simeq |p|$ is the energy or momentum of the particle when the rest masses are negligible. The pseudo-rapidity is used to define a measure of angular separation between particles commonly used in particle physics:

$$\Delta R \equiv \sqrt{(\Delta \eta)^2 + (\Delta \phi)^2} \tag{19.48}$$

The difference in azimuthal angle, $\Delta \phi$, is invariant under Lorentz boosts along the beam line (z-axis) because it is measured in a plane (i.e., the "transverse" x–y plane) orthogonal to the beam line. The difference of rapidities $\Delta \eta$ is also Lorentz-invariant if the involved particles are massless (or ultra-relativistic such that the masses can be neglected compared to their energy). This can be seen by considering an actual boost along the beam z-axis. The boosted pseudo-rapidity is:

$$\eta' = \frac{1}{2}\ln\left(\frac{E' + p_z'}{E' - p_z'}\right) = \frac{1}{2}\ln\left(\frac{\gamma(E - \beta p_z) + \gamma(p_z - \beta E)}{\gamma(E - \beta p_z) - \gamma(p_z - \beta E)}\right) = \frac{1}{2}\ln\left(\frac{(1 - \beta)(E + p_z)}{(1 + \beta)(E - p_z)}\right) = \eta + \frac{1}{2}\ln\left(\frac{1 - \beta}{1 + \beta}\right) \tag{19.49}$$

hence $\Delta \eta = \Delta \eta'$. Therefore, the unknown boost along the beam direction does not affect the difference of the pseudo-rapidities, and this latter can hence be used to construct jet quantities related to the partonic system. The total energy flow in the (η, ϕ) plane can be used to reconstruct jets and other final-state particles in hadron–hadron collisions. This is, for example, illustrated in Figure 19.21 which shows a high-energy proton–antiproton collision collected by the CDF experiment at FNAL. The event is consistent with the production of a top quark. The LEGO view represents the transverse energy deposited in the (η, ϕ) grid. Four jets are identified and a single high-p_T positron.

Jet **cone algorithms** assume that particles in jets will show up in conical regions and thus they cluster based on (η, ϕ) space, resulting in jets with rigid circular boundaries. To measure the "separation" between particles within a jet one uses the separation ΔR (Eq. (19.48)) in the (η, ϕ) plane. All particles within a cone $\Delta R < \Delta R^{cut}$ are basically put together to form a jet. In practice, there are technical issues with jet algorithms called "collinear safety" and "infrared safety." This basically means that ideally two hard partons flying within the cone should lead to two separate jets. The infrared safety implies that soft gluon emissions should not alter jets. An accurate jet algorithm should also be able to calculate the correct amount of missing energy in the detector. Jets are an essential tool for a variety of measurements at hadron colliders and are still the subject of intense investigations and improvements. See for example the software package for jet finding in pp and e^+e^- collisions, FastJet (available at http://fastjet.fr).

19.11   High-Energy Proton–(Anti)proton Collisions

Figure 19.22 shows the predictions for some important cross-sections at $p\bar{p}$ and pp colliders. The total cross-section σ_{tot} is very large and orders of magnitude above any hard scattering process. The second dominant process, labeled σ_b, is the inclusive production of b quarks. Jet production cross-sections are labeled σ_{jet} and are shown for particular requirements on the transverse energy of the jet. In particular, $\sigma_{jet}(E_T^{jet}) > 100$ GeV corresponds to the cross-section for producing at least one jet with a transverse energy above 100 GeV.

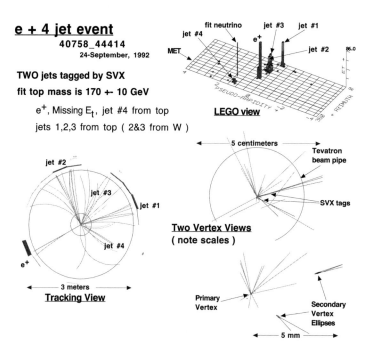

Figure 19.21 A high-energy proton–antiproton collision collected by the CDF experiment at FNAL. The event is consistent with the production of a top quark (see Section 29.4). Figure accessed online from www.hep.uiuc.edu/CDF/events.html.

Since gluons carry a sizeable fraction of the proton's momentum, important contributions from hard scatterings involving gluons are expected. It turns out that quark–gluon and gluon–gluon elastic scatterings are the most important processes besides quark–quark scattering. All possible $2 \to 2$ parton scattering processes were computed at tree level in Chapter 18. The partonic collision is specified by its center-of-mass energy squared $\hat{s} = x_1 x_2 s$ and the momentum transfer squared $\hat{t}$. The corresponding differential cross-section in the parton frame is written as [225]:

$$\frac{\mathrm{d}\sigma}{\mathrm{d}\hat{t}}(p_1 p_2 \to p_3 p_4) = \frac{\pi \alpha_s^2}{\hat{s}^2} \Sigma \tag{19.50}$$

where $\hat{s} = (p_1 + p_2)^2$, $\hat{t} = (p_1 - p_3)^2$, and the expressions for Σ are given in Table 19.1. Their relative strength for an arbitrary scattering angle of $\theta = \pi/2$ is also shown.

Starting from the partonic four-momenta $p_1^\mu = (\sqrt{s}/2)(x_1, 0, 0, x_1)$ and $p_2^\mu = (\sqrt{s}/2)(x_2, 0, 0, -x_2)$ (see Eq. (19.43)), the total four-momentum of the partonic collision is given by:

$$p_{tot}^\mu = p_1^\mu + p_2^\mu = \frac{\sqrt{s}}{2}(x_1 + x_2, 0, 0, x_1 - x_2) \tag{19.51}$$

and thus the rapidity y of the partonic system is:

$$y = \frac{1}{2} \ln \left(\frac{E + p_z}{E - p_z} \right) = \frac{1}{2} \ln \left(\frac{x_1}{x_2} \right) \tag{19.52}$$

Defining the invariant mass of the final state $M^2 = p_{tot}^2 \sim \hat{s} = x_1 x_2 s$, we can write:

$$x_1 = \frac{M}{\sqrt{s}} e^y \qquad \text{and} \qquad x_2 = \frac{M}{\sqrt{s}} e^{-y} \tag{19.53}$$

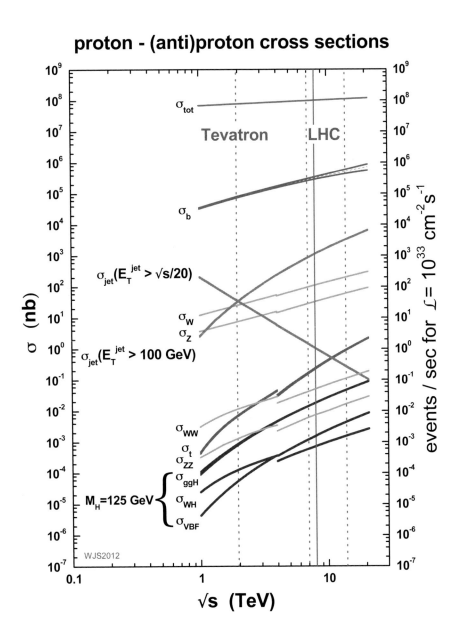

Figure 19.22 Hard proton–(anti)proton cross-sections predicted by QCD as a function of $\sqrt{s}$. For $\sqrt{s} < 4$ TeV, the cross-sections are computed for $p\bar{p}$ collisions. Above 4 TeV, pp collisions are assumed. The center-of-mass energies of the Tevatron and LHC colliders are also indicated. Republished with permission of IOP Publishing Ltd, from J. M. Campbell, J. W. Huston, and W. J. Stirling, "Hard interactions of quarks and gluons: A primer for LHC physics," *Rept. Prog. Phys.*, vol. 70, p. 89, 2007, permission conveyed through Copyright Clearance Center, Inc. Copyright 2007 IOP Publishing Ltd.

Process	Σ	Σ for $\theta = \pi/2$
$q_1 q_2 \to q_1 q_2$, $q_1 \bar{q}_2 \to q_1 \bar{q}_2$	$\dfrac{4}{9} \dfrac{s^2 + u^2}{t^2}$	2.2
$q_1 q_1 \to q_1 q_1$	$\dfrac{4}{9} \left(\dfrac{s^2 + u^2}{t^2} + \dfrac{s^2 + t^2}{u^2} \right) - \dfrac{8}{27} \dfrac{s^2}{ut}$	3.3
$q_1 \bar{q}_1 \to q_2 \bar{q}_2$	$\dfrac{4}{9} \dfrac{t^2 + u^2}{s^2}$	0.2
$q_1 \bar{q}_1 \to q_1 \bar{q}_1$	$\dfrac{4}{9} \left(\dfrac{s^2 + u^2}{t^2} + \dfrac{t^2 + u^2}{s^2} \right) - \dfrac{8}{27} \dfrac{u^2}{st}$	2.6
$q\bar{q} \to gg$	$\dfrac{32}{27} \dfrac{u^2 + t^2}{ut} - \dfrac{8}{3} \dfrac{u^2 + t^2}{s^2}$	1.0
$gg \to q\bar{q}$	$\dfrac{1}{6} \dfrac{u^2 + t^2}{ut} - \dfrac{3}{8} \dfrac{u^2 + t^2}{s^2}$	0.2
$qg \to qg$	$-\dfrac{4}{9} \dfrac{u^2 + s^2}{us} + \dfrac{u^2 + s^2}{t^2}$	6.1
$gg \to gg$	$\dfrac{9}{2} \left(3 - \dfrac{ut}{s^2} - \dfrac{us}{t^2} - \dfrac{st}{u^2} \right)$	30

Table 19.1 Hard-scattering subprocesses in QCD and the associated differential cross-sections in lowest order. Σ is defined by Eq. (19.50). The initial (final) colors and spins have been averaged (summed). q and g denote quark and gluon, respectively. Subscripts 1, 2 denote distinct flavors; s, t, u are the Mandelstam variables of the subprocess, where the hats have been omitted for clarity. Reprinted from B. Combridge, J. Kripfganz, and J. Ranft, "Hadron production at large transverse momentum and QCD," *Phys. Lett. B*, vol. 70, no. 2, pp. 234–238, 1977, with permission from Elsevier.

Thus, different values of M and y probe different values of the parton x of the colliding beams. The formulae relating x_1 and x_2 to M and y also apply to the production of any final state with this invariant mass and rapidity, for example, when producing a particle of mass M. Assuming the factorization scale Q is equal to M, the relationship between the parton (x, Q^2) values and the kinematical variables M and y can be computed.

For the LHC collision energy $\sqrt{s} = 13$ TeV, this relationship is illustrated Figure 19.23. For a given rapidity y there are two (dashed) lines, corresponding to the values of x_1 and x_2. For $y = 0$, $x_1 = x_2 = M/\sqrt{s}$.

Computing the jet transverse momentum distribution requires as input, in addition to the expressions for the various hard-scattering differential cross-sections $d\sigma/d\hat{t}$, the structure functions $F(x, Q^2)$ and the fragmentation functions $D(z, Q^2)$, where it is assumed that the following parts will factorize:

1. **Structure of the colliding hadrons:** the information on the hadron structure is encapsulated in the universal parton distribution functions.

2. **The hard-scattering process** is computed in perturbation theory in the parton (hat) frame.

3. **Parton shower**, where one or more gluons can be emitted and additional quark–antiquark pairs can be created.

4. **Underlying event:** the multiple-parton "soft" scattering of the other partons is approximated with a few tunable parameters to reproduce the "underlying" activity.

5. **Hadronization into jets** of the partons is computed via the process of fragmentation.

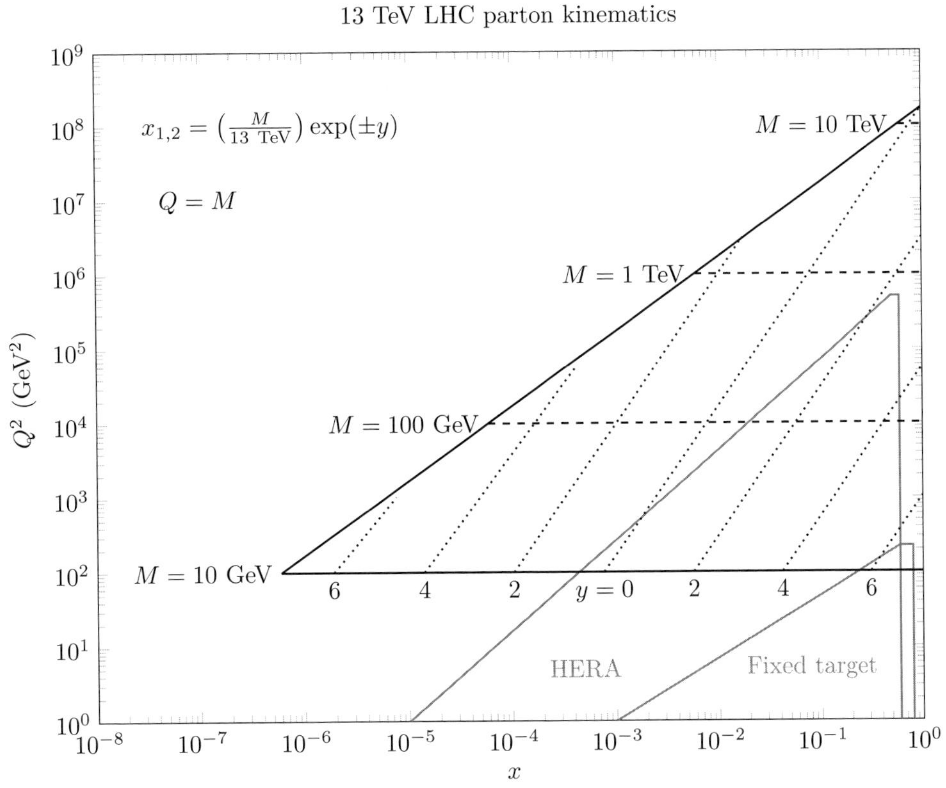

Figure 19.23 Graphical representation of the relationship between parton (x, Q^2) variables and the kinematical variables corresponding to a final state of invariant mass M produced with rapidity y at the LHC collider with $\sqrt{s} = 13$ TeV.

6. **Unstable particle decays:** all metastable particles are propagated and decayed according to their decay branching ratios and lifetimes.

This complicated sequence of events is graphically illustrated in Figure 19.24. In order to take into account all of the above features, Monte-Carlo (MC) event generators are very widely used to make predictions for collider experiments and to compare predictions with experimental data. For a review see, e.g., Ref. [226]. The most commonly used programs at present are PYTHIA [68] (already discussed in previous sections), SHERPA [227], and HERWIG [228].

In Section 19.6, we have illustrated the generation of an e^+e^- collision with the PYTHIA generator. In order to compare directly the difference between lepton and hadron collisions, we use in the following the same generator to generate a proton–proton collision at the LHC center-of-mass energy of $\sqrt{s} = 13$ TeV. The difference will be quite remarkable, as will become evident. We use the very simple program below. This latter basically instructs the generator about the center-of-mass energy, the type of process (here hard QCD), and we set a minimum transverse momentum in the partonic system $\hat{p}_T > 20$ GeV. This is to avoid the numerical divergence at low transverse momenta and also reflect approximately the threshold set by the LHC experiments to record the events: given the crossing rate at the LHC (see Section 3.7), it would be impossible to read all collisions and experiments must make decisions on which events to register (a process called "triggering"). The source code is the following:[9]

9   Reprinted code by permission from the PYTHIA author. PYTHIA is licensed under GNU GPL v2 or later. Copyright 2019

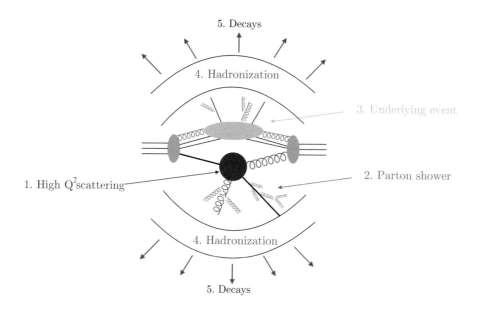

Figure 19.24 Illustration of the basic elements of a hadron–hadron collision at high energy.

```
// This is a simple test program.
// It studies the charged multiplicity distribution at the LHC.

#include "Pythia8/Pythia.h"
using namespace Pythia8;
int main() {
  // Generator. Process selection. LHC initialization. Histogram.
  Pythia pythia;
  pythia.readString("Beams:eCM = 13000.");          // set the center-of-mass energy
  pythia.readString("HardQCD:all = on");      // we want to look at jets
  pythia.readString("PhaseSpace:pTHatMin = 20."); // only those with transverse p > 20 GeV
  pythia.init();
  Hist mult("charged multiplicity", 100, -0.5, 799.5);
  // Begin event loop. Generate event. Skip if error. List first one.
  for (int iEvent = 0; iEvent < 100; ++iEvent) {
    if (!pythia.next()) continue;
    // Find number of all final charged particles and fill histogram.
    int nCharged = 0;
    for (int i = 0; i < pythia.event.size(); ++i)
      if (pythia.event[i].isFinal() && pythia.event[i].isCharged())
        ++nCharged;
    mult.fill( nCharged );
  // End of event loop. Statistics. Histogram. Done.
  }
  pythia.stat();
```

Torbjörn Sjöstrand.

```
    cout << mult;
    return 0;
}
```

A typical event output by the above PYTHIA code is shown below. The event is presented in tabular form as for the case of the e^+e^- collision. The chosen "hard" event is a $gg \to u\bar{u}$ process. It is rather asymmetric since the longitudinal momenta of the initial-state partons are respectively 1.679 and 647.645 GeV. The event is expected to be very forward boosted. The two final-state quarks u and $\bar{u}$ are produced back-to-back in the transverse plane and the longitudinal momenta are respectively -229.088 and -416.878 GeV, consistent with a strong boost along the beam axis. The transverse momentum of the quarks is $p_T = \sqrt{30.419^2 + 8.439^2} = 31.568$ GeV.

```
-------- PYTHIA Event Listing  (hard process)  ---------------------------------------------------------------------

no      id   name         status   mothers   daughters    colours      p_x       p_y        p_z         e         m
0       90   (system)       -11    0    0    0    0      0    0       0.000     0.000      0.000    13000.000  13000.000
1     2212   (p+)           -12    0    0    3    0      0    0       0.000     0.000   6500.000     6500.000      0.938
2     2212   (p+)           -12    0    0    4    0      0    0       0.000     0.000  -6500.000     6500.000      0.938
3       21   (g)            -21    1    0    5    6    101  102       0.000     0.000      1.679        1.679      0.000
4       21   (g)            -21    2    0    5    6    102  103       0.000     0.000   -647.645      647.645      0.000
5        2   u               23    3    4    0    0    101    0     -30.419     8.439   -229.088      231.253      0.330
6       -2   ubar            23    3    4    0    0      0  103      30.419    -8.439   -416.878      418.072      0.330
                    Charge sum:  0.000           Momentum sum:      0.000     0.000   -645.966      649.324     65.959

-------- End PYTHIA Event Listing  ---------------------------------------------------------------------------------
```

The parton shower generates a very large number of secondary gluons and quarks, more than 300!! We list some of them below:

```
-------- PYTHIA Event Listing  (hard process)  ---------------------------------------------------------------------
  no      id  name         status   mothers   daughters    colours      p_x       p_y        p_z         e         m
   9       2  (u)            -44    5    5   54   55    101    0     -31.080    10.691   -247.601      249.773      0.330
  10      -2  (ubar)         -44    6    6   60   61      0  103      30.055    -7.198   -398.219      399.416      0.330
  11      21  (g)            -43    7    0   77   77    102  104       1.025    -3.493     23.454       23.735      0.000
  12      21  (g)            -31   16    0   14   15    106  105       0.000     0.000      1.488        1.488      0.000
  13       2  (u)            -31   17   17   14   14    105    0       0.000     0.000   -297.173      297.173      0.000
  14      21  (g)            -33   12   13   18   18    106  107      -3.496    -1.012     -0.764        3.718      0.000
  15       2  (u)            -33   12   13   19   19    107    0       3.496     1.012   -294.920      294.943      0.330
  16      21  (g)            -41   33    0   20   12    108  105      -0.000     0.000      1.848        1.848      0.000
  17       2  (u)            -42   34   34   13   13    105    0      -0.000    -0.000   -297.173      297.173      0.000
  18      21  (g)            -44   14   14   32   32    106  107      -0.306    -1.090      1.319        1.739      0.000
  19       2  (u)            -44   15   15   30   31    107    0       3.520     1.011   -288.532      288.555      0.330
  20      21  (g)            -43   16    0   37   37    106  108      -3.214    -0.079     -8.113        8.727      0.000
  21      21  (g)            -31   25   25   23   24    109  110       0.000     0.000      5.358        5.358      0.000
  22      21  (g)            -31   26    0   23   24    111  112       0.000     0.000     -8.997        8.997      0.000
  23      21  (g)            -33   21   22   27   27    109  112       0.693     2.656      4.773        5.506      0.000
  24      21  (g)            -33   21   22   28   28    111  110      -0.693    -2.656     -8.413        8.849      0.000
  25      21  (g)            -42   40    0   21   21    109  110       0.000    -0.000      5.358        5.358      0.000
 ...
 333      21  (g)            -73  305  321  433  433    104  117       0.166    -0.155     61.833       61.834      0.282
 334      21  (g)            -73  313  330  418  418    121  161       0.501    -0.016     -0.789        0.990      0.326
 335      21  (g)            -73  295  311  424  424    110  106       0.130     1.517     -2.630        3.062      0.372
 336      21  (g)            -73  299  303  428  428    159  146      -0.442    -1.120      0.995        1.609      0.385
 337      21  (g)            -73  285  312  423  423    106  140       0.397    -1.657      3.897        4.275      0.427
 338      21  (g)            -73  318  319  383  383    154  156       0.137     1.242      0.886        1.572      0.355
 339       1  (d)            -71  327  327  341  349    163    0       0.741     0.906      2.611        2.880      0.330
 340    2203  (uu_1)         -71  326  326  341  349      0  163       1.331     1.788   5825.536     5825.536      0.771
-------- End PYTHIA Event Listing  ---------------------------------------------------------------------------------
```

After the process of parton showers, the partons are regrouped into stable and unstable hadrons during the hadronization process. Finally, all unstable particles are decayed into stable final states. This again generates several hundreds of final-state particles, some of which are listed below:

```
-------- PYTHIA Event Listing  (hard process)  ---------------------------------------------------------------------
no      id   name         status   mothers   daughters    colours      p_x       p_y        p_z         e         m
341    -211   pi-             83  339  340    0    0      0    0       0.021     0.124      0.316        0.368      0.140
342     211   pi+             83  339  340    0    0      0    0       0.555     1.025      4.193        4.354      0.140
343     311   (K0)           -83  339  340  512  512      0    0      -0.303    -0.649     11.223       11.257      0.498
```

344	3312	(Xi-)	-83	339	340	593	594	0	0	0.280	0.201	57.357	57.373	1.322
345	113	(rho0)	-84	339	340	513	514	0	0	0.182	0.268	252.052	252.053	0.728
346	-3114	(Sigma*bar+)	-84	339	340	515	516	0	0	0.372	0.283	247.006	247.010	1.391
347	113	(rho0)	-84	339	340	517	518	0	0	-0.456	-0.175	1362.113	1362.113	0.929
348	2114	(Delta0)	-84	339	340	519	520	0	0	1.476	1.502	3605.762	3605.763	1.290
349	211	pi+	84	339	340	0	0	0	0	-0.055	0.116	288.125	288.125	0.140
...														
729	111	(pi0)	-91	572	0	805	806	0	0	0.382	0.023	-0.044	0.408	0.135
730	111	(pi0)	-91	572	0	807	808	0	0	-0.090	0.004	-0.087	0.184	0.135
731	211	pi+	91	573	0	0	0	0	0	-0.287	-0.215	0.962	1.036	0.140
732	-211	pi-	91	573	0	0	0	0	0	0.068	0.066	0.264	0.314	0.140
733	22	gamma	91	575	0	0	0	0	0	-0.060	0.090	-0.107	0.152	0.000
734	22	gamma	91	575	0	0	0	0	0	-0.077	0.552	-0.325	0.645	0.000
735	211	pi+	91	581	0	0	0	0	0	-0.219	-0.551	3.044	3.104	0.140
736	-211	pi-	91	581	0	0	0	0	0	-0.363	-1.003	3.228	3.402	0.140
...														
789	-2212	pbar-	91	674	0	0	0	0	0	0.124	-0.875	2.356	2.685	0.938
790	211	pi+	91	674	0	0	0	0	0	-0.070	-0.166	0.593	0.635	0.140
791	22	gamma	91	677	0	0	0	0	0	-0.048	0.014	0.143	0.152	0.000
792	22	gamma	91	677	0	0	0	0	0	-0.054	-0.040	-0.007	0.067	0.000
793	22	gamma	91	679	0	0	0	0	0	-0.009	0.012	-0.785	0.785	0.000
794	22	gamma	91	679	0	0	0	0	0	-0.019	-0.159	-1.408	1.418	0.000
795	22	gamma	91	682	0	0	0	0	0	0.122	-0.149	-7.791	7.794	0.000
796	22	gamma	91	682	0	0	0	0	0	0.023	-0.081	-8.706	8.706	0.000
797	22	gamma	91	685	0	0	0	0	0	0.033	-0.085	-10.002	10.003	0.000
798	22	gamma	91	685	0	0	0	0	0	0.140	0.007	-12.097	12.098	0.000
799	22	gamma	91	687	0	0	0	0	0	0.112	0.117	40.258	40.259	0.000
800	22	gamma	91	687	0	0	0	0	0	0.003	-0.014	0.530	0.531	0.000
801	22	gamma	91	698	0	0	0	0	0	1.135	-0.310	-15.075	15.120	0.000
802	22	gamma	91	698	0	0	0	0	0	6.106	-1.373	-81.746	81.985	0.000
803	22	gamma	91	699	0	0	0	0	0	2.162	-0.515	-26.193	26.287	0.000
804	22	gamma	91	699	0	0	0	0	0	1.041	-0.310	-13.680	13.723	0.000
805	22	gamma	91	729	0	0	0	0	0	0.274	-0.013	0.024	0.275	0.000
806	22	gamma	91	729	0	0	0	0	0	0.108	0.036	-0.068	0.133	0.000
807	22	gamma	91	730	0	0	0	0	0	-0.035	-0.050	0.008	0.062	0.000
808	22	gamma	91	730	0	0	0	0	0	-0.055	0.054	-0.095	0.122	0.000
809	22	gamma	91	739	0	0	0	0	0	-0.064	0.293	-248.841	248.842	0.000
810	22	gamma	91	739	0	0	0	0	0	0.053	0.276	-293.317	293.317	0.000
811	22	gamma	91	740	0	0	0	0	0	-0.214	0.071	-146.007	146.007	0.000
812	22	gamma	91	740	0	0	0	0	0	-0.050	0.106	-98.974	98.974	0.000

```
                    Charge sum:   2.000            Momentum sum:  -0.000     0.000     0.000  13000.000  13000.000
-------- End PYTHIA Event Listing ------------------------------------------------------------------------------
```

After the generation of 100 events, the generator prints out some statistics. We learn that 63% of the hard processes were $gg \to gg$, and 28% were $qg \to qg$ and the rest involved quarks. As has been mentioned before, we do expect such a ratio and this result confirms the fact that proton–proton colliders are predominantly producing gluon–gluon interactions. Further information is that the average number of charged hadrons in the final state was 205. LHC events are indeed complex and very sophisticated detectors are needed in order to reconstruct precisely the interesting features of these events. Actual jet events collected by the ATLAS and CMS experiments are presented in the next section.

19.12 The Observation of Jets at the CERN LHC

At the LHC collider, the role of QCD becomes fundamental. QCD is always present! The complex hadronic environment is very rich in theory, which deserves exploration and understanding. QCD processes have enormous cross-sections which can hide many possible signals of new physics. Hadron–hadron collisions become an important background for the search of the Higgs boson or other new physics searches. QCD introduces uncertainties on other measurements. The uncertainties on the parton distribution functions affect the measurements (for example of the Higgs boson properties), in particular in the low-x region, which is very important at the LHC.

A di-jet event in proton–proton collisions at $\sqrt{s} = 7$ TeV recorded by the CMS experiment is shown in Figure 19.25. This is the event with a di-jet invariant mass $M = 2.130$ TeV, the highest ever recorded at those energies. The versatility of the LHC experiments can be exploited to identify and reconstruct individually

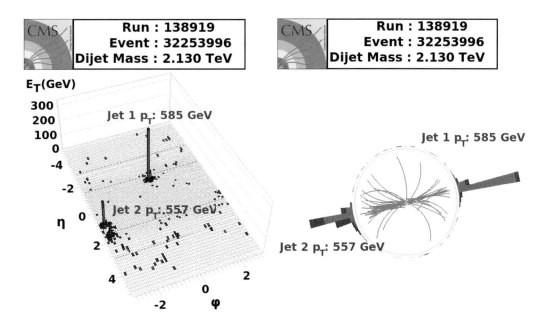

Figure 19.25 A di-jet event in proton–proton collisions at $\sqrt{s} = 7$ TeV recorded by the CMS experiment. LEGO display of calorimeter energies in (η, ϕ) coordinates (left) and in the transverse (r, ϕ) plane (right) of the highest mass di-jet event. Figure reprinted with permission from https://twiki.cern.ch/twiki/bin/view/CMSPublic/PhysicsResultsJME#Events.

each particle arising from the collisions. This method relies on the combination of the information from all sub-detectors. It aims at reconstructing and identifying all stable particles in the event (i.e., electrons, muons, photons, charged hadrons, and neutral hadrons) towards an optimal determination of their direction, energy, and type. The resulting particle-flow event reconstruction leads to an improved expected performance for jets, taus, and missing transverse energy. The list of individually reconstructed particles is used to build jets (from which the quark and gluon energies and directions are inferred), to determine the missing transverse energy (which gives an estimate of the direction and energy of the neutrinos and other invisible particles). Let us illustrate this method with the CMS detector (see Section 3.17 for details on its subdetectors). With its large silicon tracker immersed in a uniform magnetic field, charged-particle tracks can be reconstructed with large efficiency down to a momentum transverse to the beam of 150 MeV/c, for pseudo-rapidities within $|\eta| \leq 2.6$. Photons are reconstructed with an excellent energy resolution by the electromagnetic calorimeter (ECAL) surrounding the tracker within pseudo-rapidities of ±3.0. Together with the large magnetic field, the ECAL granularity is a key element in the feasibility of a particle-flow event reconstruction, as it generally permits photons to be separated from charged-particle energy deposits. Charged and neutral hadrons deposit their energy in the hadron calorimeter (HCAL) surrounding the ECAL, with a similar pseudo-rapidity coverage. The energy of charged hadrons is determined from a combination of the track momentum and the corresponding ECAL and HCAL energies. Electrons are reconstructed by a combination of a track and several energy deposits in the ECAL, from the electron itself and from possible Bremsstrahlung photons radiated by the electron in the tracker material on its way to the ECAL. Muons are reconstructed and identified, in isolation as well as in jets, with very large efficiency and purity from a combination of the tracker and muon chamber information.

Jets are reconstructed starting from the above objects. Two key parameters of the jet algorithm are the jet energy scale (JES) and jet energy resolution (JER). These two performance parameters are crucial for precise physics studies. Jets need to be calibrated in order to have the correct energy scale. The reconstruction

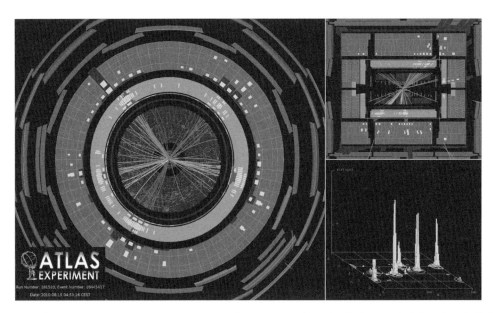

Figure 19.26 A six-jet event in proton–proton collisions at $\sqrt{s} = 7$ TeV recorded by the ATLAS experiment. The lower-right inset shows the LEGO plot of the measured transverse energy in the (η, ϕ) plane. Reprinted from G. Aad *et al.*, "Measurement of multi-jet cross-sections in proton–proton collisions at a 7 TeV center-of-mass energy," *Eur. Phys. J.*, vol. C71, p. 1763, 2011, under CC BY-NC-SA 4.0 license http://creativecommons.org/licenses/by-nc-sa/4.0/.

implements an algorithm to mitigate the effect of pile-up events. The basic idea of the jet energy scale is to exploit the transverse momentum balance between the jet to be calibrated and a reference object. An improper jet energy scale would tend to generate imbalance at the reconstructed level. The typical systematic error on the achieved jet energy scale is 2–4%. The typical jet energy resolutions are 15–20% at 30 GeV, about 10% at 100 GeV, and 5% at 1 TeV [229].

The precise study of jets plays a fundamental role at the LHC. Their multiplicity and their kinematical features provide important physics results and vital information for any other analysis. The LHC provides an ideal playground to study pQCD and parton showers. A large kinematical range is accessible to the LHC experiments, typically from tens of giga-electronvolts up to several tera-electronvolts, within a rapidity range $|\eta| < 2.5$. A six-jet event in proton–proton collisions at $\sqrt{s} = 7$ TeV recorded by the ATLAS experiment is shown in Figure 19.26. The number of jets in proton–proton collisions at $\sqrt{s} = 7$ TeV and the transverse momentum of the leading jet (jet with highest p_T) are shown in Figure 19.27. The data is compared to event simulators ALPGEN+HERWIG, PYTHIA, and SHERPA. An agreement between theory and experiment at the level of 10–20% is observed.

Problems

Ex 19.1 Monte-Carlo generation of high-energy e^+e^- collisions

(a) Download and install PYTHIA (version 8) on your personal computer (you can access the relevant information and download the code at http://home.thep.lu.se/Pythia/).

(b) Generate 10,000 e^+e^- events at LEP energies (choose the center-of-mass energy $\sqrt{s} = M_Z \simeq 91.2$ GeV).

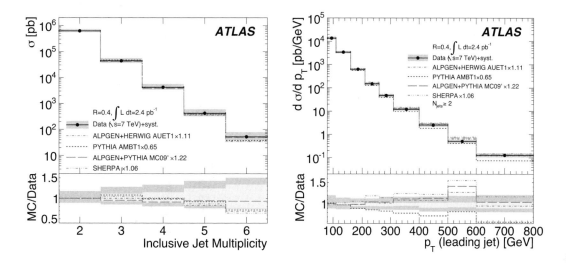

Figure 19.27 (left) Measurement of the number of jets in proton–proton collisions at $\sqrt{s} = 7$ TeV. (right) Transverse momentum of the leading jet (jet with highest p_T). The data are compared to event simulators. Reprinted from G. Aad *et al.*, "Measurement of multi-jet cross-sections in proton–proton collisions at a 7 TeV center-of-mass energy," *Eur. Phys. J.*, vol. C71, p. 1763, 2011, under CC BY-NC-SA 4.0 license http://creativecommons.org/licenses/by-nc-sa/4.0/.

    (c) Estimate the number of events detected in the OPAL detector with two, three, and four jets (see Section 19.6 and Figure 19.9).

    (d) Compare the rate of events to three jets with the result shown in Figure 19.11.

Ex 19.2 Monte-Carlo generation of ep collisions

    (a) Download and install PYTHIA (version 8) on your personal computer (see **Ex 19.1**).

    (b) Generate 100,000 electron–proton collisions at HERA energies (see Section 19.8).

    (c) Plot the kinematical distribution of the events, i.e., Q^2, W, x, y.

    (d) Compare with the kinematical regions shown in Figure 19.18.

Ex 19.3 Monte-Carlo generation of pp collisions

    (a) Download and install PYTHIA (version 8) on your personal computer (see **Ex 19.1**).

    (b) Generate 100,000 proton–proton collisions at $\sqrt{s} = 13$ TeV (i.e., the LHC configuration, see Section 19.11). Focus on the hard QCD process with a minimum $p_T = 20$ GeV.

    (c) Study the features of the events, such as charged multiplicity, jet numbers, etc.

    (d) What is the average center-of-mass energy $\sqrt{\hat{s}}$ in the hard-scattering partonic frame?

    (e) Estimate the parameters of an e^+e^- collider with center-of-mass energy equal to the average center-of-mass energy $\sqrt{\hat{s}}$ above. Discuss the advantages/disadvantages of the pp vs. e^+e^- options.

20   Heavy Quarks: Charm and Bottom

We called the new quark the "charmed quark" because we were pleased, and fascinated by the symmetry it brought to the subnuclear world.

Sheldon Lee Glashow[1]

20.1   The Existence of a Fourth Quark

The first speculations about "charm" were made in the mid-1960s [230], and full attention to it was given in the 1970s with the advent of the Cabibbo–GIM mechanism, as discussed in Section 23.13. In 1970 Drell and Yan discussed the production of massive lepton pairs in hadron–hadron collisions. In the early 1970s a group led by Lederman and Zavattini studied the production of muon pairs produced in high-energy proton collisions at the AGS at Brookhaven [231, 232]. They concluded that muon pairs with effective masses 1 GeV and 6.5 GeV had been observed in the collisions of 30 GeV protons with a uranium target. The production cross-section was seen to vary smoothly with mass exhibiting no resonant structure. In 1974, during what is called the "November revolution," two independent groups discovered a new resonance identified as a bound state composed of a charm–anticharm pair $c\bar{c}$, as will be described in Section 20.5.

20.2   The Drell–Yan Process

In Section 19.2 we have seen that the QED process $e^+e^- \to \gamma^* \to q\bar{q}$ leads to hadrons in the final state. The time-reversed process is called the **Drell–Yan[2] process** (DY) and leads to the production of lepton pairs, $q\bar{q} \to \gamma^* \to \ell^+\ell^-$. Such hard collisions can occur in hadron–hadron interactions, leading to the following process (see Figure 20.1):

$$P_1 P_2 \to \ell^+\ell^- + \text{anything} \tag{20.1}$$

Drell and Yan proposed this electromagnetic process to account for the production of the **continuum of lepton pairs** from hadron–hadron collisions [233]. In proton–proton collisions (pp), the antiquarks can only arise from the sea. Consequently, the Drell–Yan process has a higher cross-section in proton–antiproton ($p\bar{p}$) collisions or $\pi^{\pm}$-p or $K^{\pm}$-p collisions, since the antiquarks can come from the valence in antiprotons, while mesons are directly $q\bar{q}$ bound states. In the CMS frame of the two colliding hadrons P_1 and P_2, we can readily write, neglecting hadron masses:

$$P_1 = (E, 0, 0, E) \quad \text{and} \quad P_2 = (E, 0, 0, -E) \tag{20.2}$$

with the center-of-mass energy squared given by $s = 4E^2$. Neglecting further the parton masses and their transverse momenta in the parent hadrons, we have that the four-momenta of the quark and the antiquark

---

1   From M. Riordan, *The Hunting of the Quark: A True Story of Modern Physics* (1987), Simon & Schuster. p. 210. ISBN 978-0671504663.
2   Tung-Mow Yan (born 1937), Taiwanese-born American physicist.

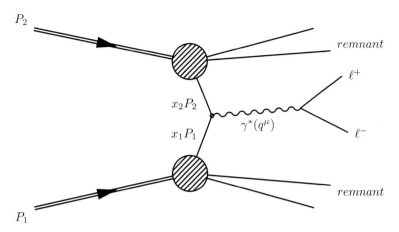

Figure 20.1 Tree-level Feynman diagram for the Drell–Yan process $P_1 P_2 \to \ell^+ \ell^- + X$.

are given by respectively $x_1 P_1$ and $x_2 P_2$ (see Eq. (19.43)). Subsequently, the four-momentum of the virtual photon and therefore of the **lepton pair** is given by:

$$q^\mu \equiv (q_0, q_1, q_2, q_3) = \left((x_1 + x_2)\frac{\sqrt{s}}{2}, 0, 0, (x_1 - x_2)\frac{\sqrt{s}}{2} \right) \quad \Longrightarrow \quad q^2 = x_1 x_2 s \tag{20.3}$$

In the Drell–Yan process, the virtual photon is "time-like" because $q^2 > 0$, while we recall that in the deep-inelastic scattering the photon was "space-like" since we had $q^2 < 0$. However, according to the parton model, both Drell–Yan and deep inelastic scattering probe the same parton distribution functions in the hadron and provide different ways to study these functions. The **invariant mass of the lepton pair** $m_{\ell^+\ell^-}$ is given by:

$$m_{\ell^+\ell^-} = \sqrt{q^2} = \sqrt{x_1 x_2 s} \tag{20.4}$$

The longitudinal momentum of the pair is $q_3 = (x_1 - x_2)E$ and it is customary to define the **"Feynman-x"** variable x_F as:

$$x_F \equiv \frac{2q_3}{\sqrt{s}} = \frac{2(x_1 - x_2)E}{2E} = x_1 - x_2 \tag{20.5}$$

with $-1 \le x_F \le 1$. The kinematics of the process are fully defined by the lepton-pair invariant mass and its Feynman x_F. It is often convenient to introduce the **dimensionless variable** τ:

$$\tau \equiv \frac{q^2}{s} = \frac{m_{\ell^+\ell^-}^2}{s} = x_1 x_2 \quad \Longrightarrow \quad m_{\ell^+\ell^-}^2 = \tau s \tag{20.6}$$

Then:

$$x_1 = \frac{1}{2}(x_1 + x_2 + x_1 - x_2) = \frac{1}{2}\left((x_1^2 + 2\tau + x_2^2)^{1/2} + x_F \right) = \frac{1}{2}\left((x_F^2 + 4\tau)^{1/2} + x_F \right) \tag{20.7}$$

and a similar calculation for x_2 finally gives the two expressions:

$$\begin{cases} x_1 = \frac{1}{2}\left((x_F^2 + 4\tau)^{1/2} + x_F \right) \\ x_2 = \frac{1}{2}\left((x_F^2 + 4\tau)^{1/2} - x_F \right) \end{cases} \tag{20.8}$$

The relation between the parton kinematical variables x_1, x_2 and the lepton-pair variables τ and x_F is plotted in the plane (x_1, x_2) in Figure 20.2. Both x_1 and x_2 are physically limited in the range $0 < x_i < 1$. Lines

Drell–Yan process kinematics

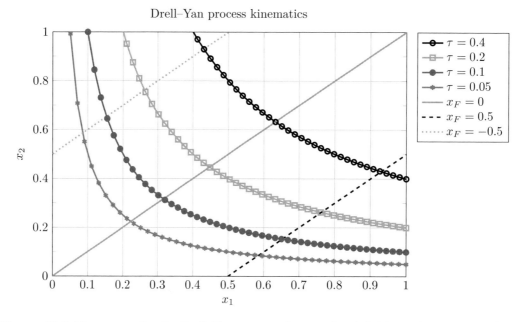

Figure 20.2 Phase space for the Drell–Yan process. See text for definitions of x_1, x_2, τ, and x_F.

of constant lepton-pair mass and τ are rectangular hyperbolae, and lines of constant x_F are diagonals. It is obvious from the plot that a set of parameters x_F and τ defines a unique point in phase space in the plane (x_1, x_2), which completely defines the kinematics of an event.

Let us now derive the Drell–Yan differential cross-section. Starting from the $e^+ e^- \to \mu^+ \mu^-$ process, the cross-section $q\bar{q} \to \ell^+ \ell^-$ at the parton level is given by:

$$\frac{\mathrm{d}\hat{\sigma}}{\mathrm{d}q^2}\left(q_i \bar{q}_i \to \ell^+ \ell^-\right) = \frac{4\pi \alpha^2}{3q^2} e_i^2 \delta(q^2 - \tau s) \tag{20.9}$$

We note that for this process to function, the quark–antiquark pair ($q\bar{q}$) must necessarily be colorless, otherwise it will not annihilate into the colorless photon. Among the N_C^2 (nine) possible color combinations, only N_C (three) are colorless, hence one must include a factor $1/N_C$ in a quark–parton model calculation. The cross-section for $P_1 P_2 \to \ell^+ \ell^- + X$ is hence expressed as:

$$\frac{\mathrm{d}\sigma}{\mathrm{d}q^2}\left(P_1 P_2 \to \ell^+ \ell^- + X\right) = \frac{1}{N_C} \sum_i \int \mathrm{d}x_1 \mathrm{d}x_2 \left[q_i(x_1)\bar{q}_i(x_2) + \bar{q}_i(x_1)q_i(x_2)\right] \frac{\mathrm{d}\hat{\sigma}}{\mathrm{d}q^2}\left(q_i \bar{q}_i \to \ell^+ \ell^-\right) \tag{20.10}$$

where $q_i(x_1)$ (resp. $q_i(x_2)$) and $\bar{q}_i(x_1)$ (resp. $\bar{q}_i(x_2)$) are the parton distribution functions of the quark q_i and antiquark $\bar{q}_i$ in the hadrons P_1 (resp. P_2). We can define the function $\mathcal{F}$ which depends on the colliding hadrons P_1, P_2:

$$\mathcal{F}^{P_1, P_2}(x_1, x_2) = \sum_i e_i^2 \left[q_i(x_1)\bar{q}_i(x_2) + \bar{q}_i(x_1)q_i(x_2)\right] \tag{20.11}$$

then the Drell–Yan cross-section is given by:

$$\frac{\mathrm{d}\sigma}{\mathrm{d}q^2}\left(P_1 P_2 \to \ell^+ \ell^- + X\right) = \frac{4\pi \alpha^2}{3N_C q^2} \int \mathrm{d}x_1 \mathrm{d}x_2 \mathcal{F}^{P_1, P_2}(x_1, x_2) \delta(q^2 - x_1 x_2 s) \tag{20.12}$$

Hence, the double-differential Drell–Yan cross-section (setting $N_C = 3$) can be expressed as:

$$\frac{\mathrm{d}^2\sigma}{\mathrm{d}x_1 \mathrm{d}x_2}\left(P_1 P_2 \to \ell^+ \ell^- + X\right) = \frac{4\pi \alpha^2}{9x_1 x_2 s} \mathcal{F}^{P_1, P_2}(x_1, x_2) \tag{20.13}$$

We can re-express the result in terms of measurable parameters of the lepton pair. With the help of the Jacobian, we find:

$$
dm_{\ell^+\ell^-}^2 \, dx_F = \begin{vmatrix} \dfrac{\partial x_F}{\partial x_1} & \dfrac{\partial x_F}{\partial x_2} \\[2mm] \dfrac{\partial m_{\ell^+\ell^-}^2}{\partial x_1} & \dfrac{\partial m_{\ell^+\ell^-}^2}{\partial x_2} \end{vmatrix} dx_1 dx_2 = (x_1 + x_2) s \, dx_1 dx_2
\tag{20.14}
$$

then we have:

$$
\frac{d^2\sigma}{dm_{\ell^+\ell^-}^2 \, dx_F} \left(P_1 P_2 \to \ell^+\ell^- + X \right) = \frac{4\pi\alpha^2}{9 m_{\ell^+\ell^-}^2} \frac{1}{(x_1+x_2)s} \mathcal{F}^{P_1,P_2}(x_1, x_2) = \frac{4\pi\alpha^2}{9 m_{\ell^+\ell^-}^4} \frac{x_1 x_2}{(x_1+x_2)} \mathcal{F}^{P_1,P_2}(x_1, x_2)
\tag{20.15}
$$

where x_1 and x_2 are given by Eq. (20.8). We have seen in Eq. (19.52) that the rapidity of the parton system is given by $y = \frac{1}{2}\ln\left(\frac{x_1}{x_2}\right)$. In the case of Drell–Yan, this corresponds to the rapidity of the lepton pair system. Using Eq. (19.53), we can write:

$$
x_1 = \frac{m_{\ell^+\ell^-}}{\sqrt{s}} e^y \qquad \text{and} \qquad x_2 = \frac{m_{\ell^+\ell^-}}{\sqrt{s}} e^{-y}
\tag{20.16}
$$

and the integration in terms of $dx_1 dx_2$ can be expressed in terms of the rapidity and the invariant mass of the lepton pair, again using the corresponding Jacobian:

$$
dm_{\ell^+\ell^-} \, dy = \begin{vmatrix} \dfrac{\partial y}{\partial x_1} & \dfrac{\partial y}{\partial x_2} \\[2mm] \dfrac{\partial m_{\ell^+\ell^-}}{\partial x_1} & \dfrac{\partial m_{\ell^+\ell^-}}{\partial x_2} \end{vmatrix} dx_1 dx_2 = \frac{s}{2 m_{\ell^+\ell^-}} dx_1 dx_2
\tag{20.17}
$$

Hence, the double-differential Drell–Yan cross-section written in terms of the invariant mass and the rapidity of the lepton pair is just given by:

$$
\frac{d^2\sigma}{dm_{\ell^+\ell^-} \, dy} \left(P_1 P_2 \to \ell^+\ell^- + X \right) = \frac{8\pi\alpha^2}{9 s m_{\ell^+\ell^-}} \mathcal{F}^{P_1,P_2}\left(\frac{m_{\ell^+\ell^-}}{\sqrt{s}} e^y, \frac{m_{\ell^+\ell^-}}{\sqrt{s}} e^{-y} \right)
\tag{20.18}
$$

The above expressions are adequate to describe the **continuum of the Drell–Yan process**. If we assume scaling at the parton level, then the function $\mathcal{F}^{P_1,P_2}$ scales and then the product:

$$
m_{\ell^+\ell^-}^3 \frac{d^2\sigma}{dm_{\ell^+\ell^-} \, dy} = \frac{8\pi\alpha^2 m_{\ell^+\ell^-}^2}{9s} \mathcal{F}^{P_1,P_2} = \frac{8\pi\alpha^2}{9} x_1 x_2 \mathcal{F}^{P_1,P_2}
\tag{20.19}
$$

has lost its energy dependence. Since x_1 and x_2 both depend on $m_{\ell^+\ell^-}$ and y, integrating over y gives a differential cross-section that depends only on $m_{\ell^+\ell^-}$, which can be expressed as a scaling as a function of $\tau = m_{\ell^+\ell^-}^2 / s$. This simple prediction of Drell–Yan is to be tested experimentally.

Modifications to the tree-level cross-section due to higher-order QCD corrections caused by gluon emission and absorption by the quark or antiquark have been computed. In addition, in the above treatment, we only considered the virtual photon exchange within QED. However, there are also important corrections from the electroweak theory (discussed in Chapter 25), which predicts in addition the exchange of a virtual Z^0. The inclusion of the two diagrams $q\bar{q} \to \gamma^*/Z^{0*} \to \ell^+\ell^-$ leads to a modification of the cross-section and also to a spectacularly strong enhancement of the cross-section for $m_{\ell^+\ell^-} \simeq m_{Z^0} \approx 91$ GeV (see, e.g., Ref [234] for details).

- **QCD corrections and introduction of a K-factor.** To the lowest order, the Drell–Yan hard process is the purely electromagnetic annihilation of a $q\bar{q}$ pair into a virtual photon with mass $q^2 = m_{\ell^+\ell^-}^2$. In addition, we should consider first-order QCD corrections proportional to α_s. These latter include gluon radiation in the final state, scattering with an initial-state gluon, as well as gluon vertex corrections. The corresponding diagrams are illustrated in Figure 20.3. These corrections can be calculated and generally introduce quite non-trivial modifications to the cross-section. At first approximation, one can consider that these effects give an *overall correction*, i.e., they simply modify the total cross-section, typically increasing it by some factor, which is historically called the *K*-**factor**. The QCD-corrected cross-section, Eq. (20.19), is then expressed as:

$$\frac{\mathrm{d}^2\sigma}{\mathrm{d}m_{\ell^+\ell^-}\mathrm{d}y} = \frac{8\pi\alpha^2 K}{9m_{\ell^+\ell^-}^3}x_1 x_2 \mathcal{F}^{P_1,P_2} \tag{20.20}$$

with (see Ref. [235]):

$$K = 1 + \frac{\alpha_s(Q^2)}{2\pi}\frac{4}{3}\left(1 + \frac{4}{3}\pi^2\right) \approx 1 + 3\alpha_s(Q^2) \approx 1.6 \tag{20.21}$$

where we assumed $\alpha_s(10~\text{GeV}) \approx 0.2$ (see Figure 18.16). The correction is large, at the level of 60% with the chosen values, and cannot generally be neglected. Now, physicists use full higher-order QCD-corrected differential cross-sections to the Drell–Yan process in order to compare with precise experimental data.

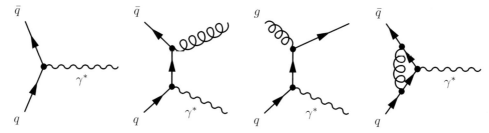

Figure 20.3 Tree-level (left) and some higher-order QCD corrections to the Drell–Yan process.

20.3   Observation of Muon Pairs in High-Energy Proton–Nucleus Collisions

The simple Drell–Yan model makes statements about the lepton pairs produced in hadronic collisions and not about the accompanying hadrons, so it has been sufficient to detect solely leptons in order to make many tests of the model. **The appearance of the propagator of the virtual photon in the amplitude leads to the rapid fall of the cross-section with increasing lepton pair mass.**

Experiments have initially been performed with fixed targets, which allow the use of a range of beam particles ($\pi^{\pm}$, $K^{\pm}$, p, $\bar{p}$), however the Drell–Yan reaction has also been observed at colliders. These kinds of experiments are rather difficult because of the necessity to find the few electron or muon pairs in the high background of hadrons. Muons can be distinguished from the copiously produced hadrons by their highly penetrating character; electrons, by their electromagnetic showering properties. In fixed targets, the detection of muons is usually preferred rather than the detection of electrons, because of the relative simplicity. The main background in a muon experiment is muons from the decay of pions and kaons produced in the target. To suppress this it is necessary to place material immediately downstream of the target to absorb these particles before they can decay. By placing a wall of heavy material, several interaction lengths thick, immediately downstream of the target, the majority of π and K mesons can be caused to interact before decaying. Muons that emerge from the absorber can be detected and form part of the trigger. Preferably a magnetic spectrometer downstream of the absorber measures the muon momenta. The search for Drell–Yan events relies on the study of unlike-sign pairs (e^+e^- or $\mu^+\mu^-$) while the *background* can be estimated from like-sign pairs ($\ell^+\ell^+$ or $\ell^-\ell^-$), which occur from the accidental production of leptons.

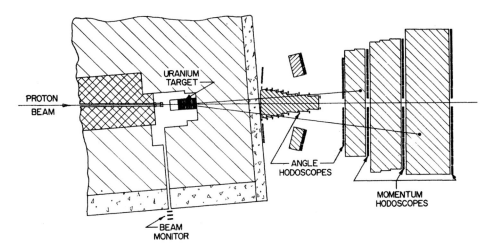

Figure 20.4 Layout of the Lederman *et al.* experiment to study muon pair production in high-energy proton–nucleus collisions. Reprinted with permission from J. H. Christenson, G. S. Hicks, L. M. Lederman, P. J. Limon, B. G. Pope, and E. Zavattini, "Observation of muon pairs in high-energy hadron collisions," *Phys. Rev.*, vol. D8, pp. 2016–2034, 1972. Copyright 1973 by the American Physical Society.

The first fundamental step to study scattering with "time-like" photons in hadron collisions was performed at BNL by a Columbia University team led by **Leon Lederman**[3] in the period 1968–1969 [231, 232]. One of the primary goals was to search for intermediate vector bosons (IVBs). They observed muon pairs produced in high-energy collisions of protons with a uranium target. The apparatus and experimental layout are sketched in Figure 20.4.

About 2.5×10^{11} protons were accelerated by the AGS of BNL at each cycle and "slow-extracted" from the ring during a 300 ms long spill onto the uranium target. Immediately downstream of the target, a 10-feet thick steel shield followed by a 2-feet thick concrete wall was acting as a filter for hadrons. Only muons with momentum greater than 6 GeV could penetrate through that amount of material and thus the shield was suppressing most of the soft muons originating from decays of soft hadrons (mostly pions and kaons). Additional absorber was placed after the concrete wall. The flux of muons during the spill was about 10^6 per second.

The invariant mass of the identified muon pairs was reconstructed from their estimated momenta p_1 and p_2 and their production angles θ_1 and θ_2 with:

$$m_{\mu\mu}^2 = 2p_1p_2(1 - \cos(\theta_1 + \theta_2)) \qquad (20.22)$$

Multiple scattering (see Section 2.18) in the absorber was a limiting factor in determining the production angles of the muon at the target. This angular uncertainty limited the resolution with which the invariant mass could be reconstructed with Eq. (20.22). The momentum of the muons was estimated using their range in the steel. The muon pairs could be measured in the mass range $1 < m_{\mu\mu} < 6.7$ GeV and the reconstructed mass resolution was estimated to be $\pm 15\%$ at 2 GeV and $\pm 8\%$ at 5 GeV.

A real challenge for the time was to acquire and filter the data coming from the experiment at this high rate. A hardware trigger logic was implemented and if it accepted an event based on criteria of consistency between counters compatible with the production of muons, the data was sent to a computer for analysis and storage. In the course of the experiment, more than 300 million events were recorded and analyzed online by a computer. After offline correction for detector efficiencies, the differential cross-sections for the continuum

---

3   Leon Max Lederman (1922–2018), American experimental physicist.

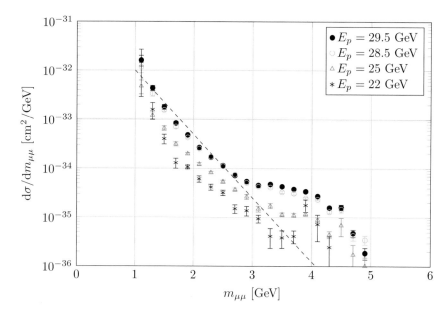

Figure 20.5 Measured invariant mass of the muon pair in proton–nucleus interactions for incident proton energies of 22, 25, 28.5, and 29.5 GeV. Data taken with permission from J. H. Christenson, G. S. Hicks, L. M. Lederman, P. J. Limon, B. G. Pope, and E. Zavattini, "Observation of muon pairs in high-energy hadron collisions," *Phys. Rev.*, vol. D8, pp. 2016–2034, 1972. Copyright 1973 by the American Physical Society. The dashed line illustrates an exponentially decaying spectrum.

muon pair production were calculated. The results as a function of the invariant mass $m_{\mu\mu}$ are shown in Figure 20.5 for incident proton energies of 22, 25, 28.5, and 29.5 GeV. The data was found consistent with a very rapid falloff $\mathrm{d}\sigma/\mathrm{d}m_{\mu\mu} \simeq \sigma_0/m_{\mu\mu}^5$, decreasing by about seven orders of magnitude between 1 and 6 GeV.

The experiment was built to explore the nature of the continuum and search for possible resonances. The authors concluded that *"the cross-section varies smoothly (...) and exhibits no resonant structure"* [231]. However, the "shoulder" above ~ 3 GeV was very striking but difficulties due to the small signal-to-noise ratio, the modest mass resolution, and the data being sensitive to background subtraction led to the above *negative* conclusion about the possible existence of a new effect not explained by a continuum.

20.4   Scaling Properties of the Drell–Yan Process

The properties of the Drell–Yan process have been further experimentally tested for example at CERN and FNAL studying dilepton pairs produced in fixed target proton–nucleus $p + A$ collisions at proton energies of 400 and 800 GeV. Beams composed of positive and negative pions have also been used.

The FNAL E772 experiment could report the angular distribution of a lepton in the virtual photon rest frame, which is a direct test of the QED coupling, analogous to the tests for example in $e^+e^- \to \ell^+\ell^-$ (see Section 11.12). Helicity conservation at the vertices (we are in the unpolarized case) determines the common $1 + \cos^2\theta$ dependence. E772 tested the hypothesis by fitting their data with the function:

$$\frac{\mathrm{d}\sigma}{\mathrm{d}\Omega} = \sigma_0 \left(1 + \lambda \cos^2\theta\right) \tag{20.23}$$

The result is shown in Figure 20.6. The dashed curve represents a fit to the data with this form, yielding the

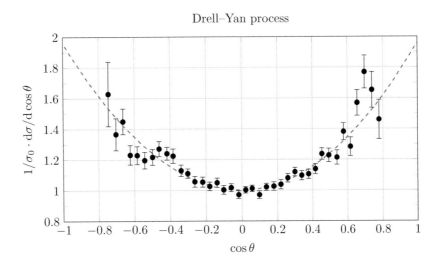

Figure 20.6 Angular distribution of dilepton pairs in $p + Cu$ collisions at 800 GeV measured by FNAL E772. The solid curve is a fit to $1 + \lambda \cos^2 \theta$. Data taken from P. McGaughey, "Recent measurements of quarkonia and Drell–Yan production in proton nucleus collisions," *Nucl. Phys. A*, vol. 610, pp. 394C–403C, 1996, with permission from Elsevier.

best-fit value of $\lambda = 0.96 \pm 0.04(\text{stat.}) \pm 0.06(\text{syst.})$, nicely consistent with theoretical expectations.

The scaling properties of the Drell–Yan process predicted by the parton model (see Eq. (20.19)) have been thoroughly tested. The compilation of the results from the CERN NA3 [236] and FNAL E605 [237] and E772 [238] experiments is shown in Figure 20.7. The data for each x_F bin is scaled by powers of ten for visibility. The graph reports the measured scaled cross-section $m_{\ell\ell}^3 \mathrm{d}^2\sigma/\mathrm{d}x_F \mathrm{d}m_{\ell\ell}$ as a function of the dimensionless scaling variable $\sqrt{\tau} = m_{\ell\ell}/\sqrt{s}$ (see Eq. (20.6)). This data is clearly consistent with scaling between the 400 and 800 GeV data over most of the explored kinematical region and is strong supporting evidence for the parton model prediction for the Drell–Yan process.

20.5   The Discovery of the J/Ψ

The discovery of a new, very massive (3.1 GeV) particle was announced in November 1974 simultaneously by two different laboratories, and consequently kept the "double name" J/Ψ. The short period of time over which these two simultaneous discoveries were made, was called the "November 1974 revolution."

One of the two projects was the MIT experiment by **Samuel C. C. Ting**[4] and collaborators who were aiming at a similar objective as Lederman's experiment, namely to search for new vector particles which decay into lepton pairs. Unlike Lederman's experiment, Ting aimed at very high precision, and devised a very high-resolution double-arm spectrometer, each arm including magnets and 8000 wire proportional chambers. See Figure 20.8(left). The high-energy protons were shot on a beryllium target in order to minimize multiple scattering. In addition, Ting focused on electron–positron pairs instead of unlike-sign muon pairs. Unlike the muons that are subject to large multiple scattering in the absorber shield, the electrons would be measured directly – however, facing the very difficult experimental challenge to detect and measure such rare pairs in a very large hadronic background. Hence, he relied on Cerenkov counters and lead-glass calorimeters to separate electrons from hadrons with a suppression factor of more than 10^8. The detector was able to reconstruct the

4   Samuel Chao Chung Ting (born 1936), American physicist.

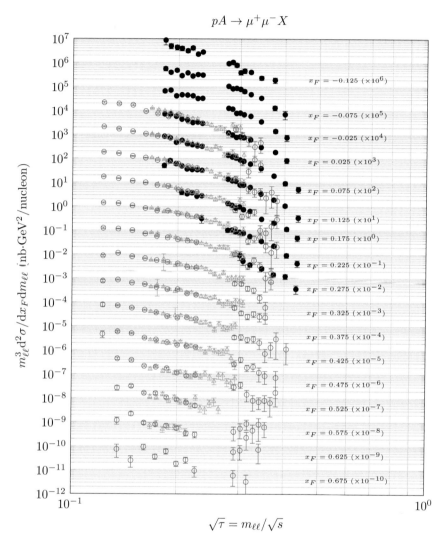

Figure 20.7 Dimuon pair measurements in proton–nucleus collisions in the CERN NA3 experiment at 400 GeV (open triangles) [236] and FNAL E605 (filled circles – data reproduced with permission from G. Moreno *et al.*, "Dimuon production in proton–copper collisions at $\sqrt{s}$ = 38.8-GeV," *Phys. Rev. D*, vol. 43, pp. 2815–2836, 1991. Copyright 1991 by the American Physical Society) and E772 at 800 GeV (open circles – data reproduced with permission from P. L. McGaughey *et al.*, "Erratum: Cross sections for the production of high-mass muon pairs from 800 GeV proton bombardment of ^{2}H [Phys. Rev. D 50, 3038 (1994)]," *Phys. Rev. D*, vol. 60, p. 119903, Oct 1999. Copyright 1999 by the American Physical Society). Data for each x_F bin are scaled by powers of 10 for visibility.

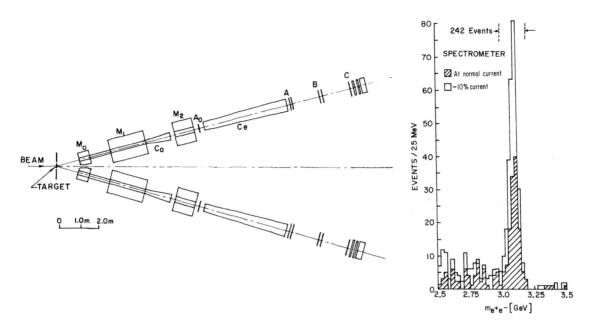

Figure 20.8 (left) Layout of the MIT experiment. Reprinted from J. J. Aubert *et al.*, "Discovery of the new particle J," *Nucl. Phys. B*, vol. 89, no. 1, pp. 1–18, 1975, with permission from Elsevier. (right) Reconstructed invariant mass of e^+e^- pairs. Reprinted with permission from J. J. Aubert *et al.*, "Experimental observation of a heavy particle J," *Phys. Rev. Lett.*, vol. 33, pp. 1404–1406, Dec 1974. Copyright 1974 by the American Physical Society.

invariant mass of the pairs with an incredible resolution of 20 MeV (a precision of better than 1%)! The entire setup consisted in addition to 10 kiloton of shielding, 100 tons of lead and 5 tons of uranium. One of the main challenges for the team had apparently been to convince agencies to build such a complicated device with such high resolution, and someone on a committee once remarked that such a resolution was not needed since no narrow resonances were known to exist!

The data collected with the AGS protons at 30 GeV showed a very clear and narrow peak in the reconstructed e^+e^- pairs, see Figure 20.8(right). From the data, the width of the resonance is consistent with zero since the measured width is compatible with that expected from the experimental resolution. Many checks were performed to ensure that this peak was really produced by a new narrow resonance and was not instrumental. For instance, the magnet currents were changed but the peak position remained fixed. Also, half of the data was taken with a reversed polarity of the spectrometer. All these and other tests convinced the team that they had observed a new very narrow and massive resonance at 3.1 GeV via the reaction:

$$p + Be \rightarrow J + X \rightarrow e^+e^- + X \tag{20.24}$$

In the meantime at Stanford, the SLAC group led by **Burton Richter**[5] and **Martin Perl**, and the LBL group led by **Maurice Goldhaber** had built a huge magnetic detector called Mark I, which completely surrounded the interaction point of the electron–positron beams of the SPEAR storage ring. The Mark I detector (see Figure 20.9(left)) had a solenoidal field coaxial with the beam of about 0.4 T in a volume of 20 m^3. Particles moving radially outwards from the interaction point were curved by the magnetic field. Their passage was detected and measured by a set of subdetectors, including trigger counters, 16 concentric cylindrical wire chambers, a cylindrical array of 24 lead-scintillator shower counters, and an additional array of spark chambers

---

5   Burton Richter (1931–2018), American physicist.

dedicated to muons. The Mark I detector with its 4π coverage and sophisticated triggers represented a first model for many future detectors at colliders. The detector was optimal to study e^+e^- annihilation into hadrons and the idea was to study the cross-section as a function of the center-of-mass energy $\sqrt{s}$ and extend to higher energies the knowledge of the ratio R (Eq. (19.4)) discussed in Section 19.2. In fact, measurements made at the Cambridge Electron Accelerator seemed to indicate a rise of R above 3 GeV. The Mark I experiment performed energy scans, initially in steps of 200 MeV. The size of these steps was thought to be sufficient to study the behavior of R as a function of $\sqrt{s}$, and seemed to indicate an increase of R. It was then decided to take more data and scan the region between 3.1 and 3.3 GeV in very small steps of ≈ 2 MeV. This very fine energy scan revealed a large and very narrow resonance, as shown in Figure 20.9(right). This resonance was also visible in the e^+e^-, as well as $\mu^+\mu^-$ final states. They had observed the Ψ via the reaction:

$$e^+e^- \to \Psi \to \text{hadrons}, e^+e^-, \dots \tag{20.25}$$

Both experiments using very different reactions had detected the extremely narrow charmonium bound $1S$ state of the charm quark $c\bar{c}$. The properties of the $J/\Psi(1S)$ are the following (values taken from the Particle Data Group's Review [5]):

$$\boxed{J/\Psi(1S)}: \quad J^{PC} = 1^{--}, \quad m_{J/\Psi} = (3096.900 \pm 0.006)\text{ MeV}$$
$$\Gamma = (92.9 \pm 2.8)\text{ keV}$$
$$Br(J/\Psi \to \text{hadrons}) = (87.7 \pm 0.5)\% \tag{20.26}$$
$$Br(J/\Psi \to e^+e^-) = (5.971 \pm 0.032)\%$$
$$Br(J/\Psi \to \mu^+\mu^-) = (5.961 \pm 0.033)\%$$

So, the $J/\Psi(1S)$ is indeed extremely narrow, consequently long-lived (!) and yet so massive! It was not possible to explain this striking observation in terms of the known quarks u, d, and s and the need to introduce the new charm flavor, whose existence had been predicted earlier in the context of the Cabibbo–GIM mechanism (Section 23.13), was evident. The mass of the J/Ψ also immediately indicated that the c quark is much heavier than the other known flavors $m_c \gg m_u, m_d, m_s$. The value was in the required mass range [239] to offer a consistent explanation of the strong suppression of strangeness (flavor)-changing neutral currents (FCNC) with the GIM mechanism (see Section 23.13).

20.6   The Existence of Excited States – Charmonium

Very shortly after the discovery of the $J/\Psi(3096)$, the SLAC group found another narrow state at a slightly higher center-of-mass energy, called Ψ' or $\Psi(3685)$. Since both $J/\Psi(3096)$ and $\Psi(3685)$ are produced via the annihilation $e^+e^- \to \Psi$, they must possess $J^{PC} = 1^{--}$. The $\Psi(3685)$ is identified as the $\Psi(2S)$ state of charmonium. The properties of the $\Psi(3685)$ are the following (values taken from the Particle Data Group's Review [5]):

$$\boxed{\Psi(3685)}: \quad J^{PC} = 1^{--}, \quad m_{\Psi(2S)} = (3686.097 \pm 0.025)\text{ MeV}$$
$$\Gamma = (296 \pm 8)\text{ keV}$$
$$Br(\Psi(3685) \to \text{hadrons}) = (97.85 \pm 0.13)\%$$
$$Br(\Psi(3685) \to e^+e^-) = (7.89 \pm 0.17) \times 10^{-3} \tag{20.27}$$
$$Br(\Psi(3685) \to \mu^+\mu^-) = (7.9 \pm 0.9) \times 10^{-3}$$
$$Br(\Psi(3685) \to \tau^+\tau^-) = (3.1 \pm 0.4) \times 10^{-3}$$

Charmonium is sometimes considered as the "positronium of QCD" because the $c\bar{c}$ bound state is a powerful tool for understanding the strong interaction. The high mass of the c quark makes it possible to perform a description of the $c\bar{c}$ dynamical properties through a non-relativistic potential model. The state of the bound system can be described with the spectroscopic notation $n^{2S+1}L_J$, where S is the sum of the spins and

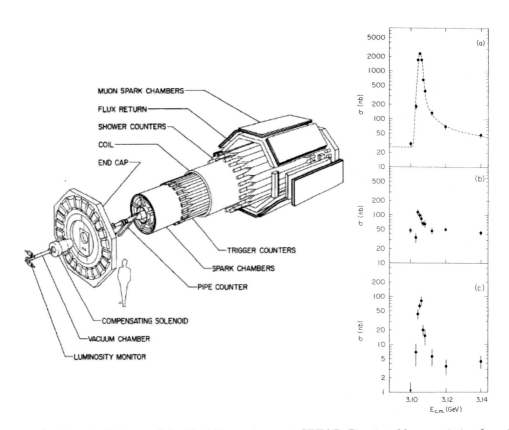

Figure 20.9 (left) Exploded view of the Mark I experiment at SPEAR. Reprinted by permission from Springer Nature: H. Lynch *et al.*, "Recent results for e^+e^- annihilation at SPEAR." In: Perlmutter A., Widmayer S.M. (eds) *Theories and Experiments in High-Energy Physics*. Studies in the Natural Sciences. Springer, Boston, MA. https://doi.org/10.1007/978-1-4613-4464-3_1. (right) Measured cross-section vs. $\sqrt{s}$ for (a) $e^+e^- \rightarrow$ hadrons, (b) $e^+e^- \rightarrow e^+e^-$, (c) $e^+e^- \rightarrow \mu^+\mu^-$, $\pi^+\pi^-$, and K^+K^-. Reprinted with permission from J. E. Augustin *et al.*, "Discovery of a narrow resonance in e^+e^- annihilation," *Phys. Rev. Lett.*, vol. 33, pp. 1406–1408, Dec 1974. Copyright 1974 by the American Physical Society.

$\vec{J} = \vec{L} + \vec{S}$ is the total angular momentum. The intrinsic parity is $P = (-1)^{L+1}$ and the charge conjugation $C = (-1)^{L+S}$. The masses and the J^{PC} assignments of the observed $c\bar{c}$ bound states are shown in Figure 20.10. As already mentioned, the $J^{PC} = 1^{--}$ states can be directly produced in $e^{+}e^{-}$ annihilation. Other states may be accessed through radiative decays or in decays of heavier particles, such as, for instance, bottom hadrons (see Section 20.9).

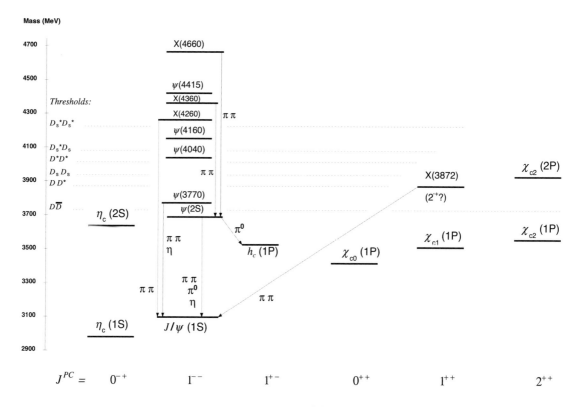

Figure 20.10 Charmonium spectroscopy: masses and J^{PC} assignments of the observed $c\bar{c}$ bound states. Reprinted with permission from C. Patrignani *et al.*, "Review of particle physics," *Chin. Phys.*, vol. C40, no. 10, p. 100001, 2016.

A description of the dynamical properties of charmonium in terms of non-relativistic potential models can be obtained with a functional form of the potential, chosen to reproduce the known asymptotic properties of the strong interaction. The interaction between the quark and the antiquark is expressed with a potential of the form $V_{q\bar{q}}(r)$, where r represents the "distance" between the two quarks. At large distance, confinement gives rise to a potential linearly growing with r:

$$V_{q\bar{q}}(r) \xrightarrow{r \to \infty} \kappa r \qquad (20.28)$$

where adjustment to data leads to $\kappa \simeq 1$ GeV/fm (see Section 18.13). At short distances, asymptotic freedom effectively weakens the strong force. The situation becomes similar to the electromagnetic force and positronium $e^{+}e^{-}$. The non-relativistic limit of QED leads to the common Coulomb potential, which is attractive in the case of positronium, so $V_{e^{+}e^{-}} \propto -\alpha/r$. Similarly, the short-distance non-relativistic potential of the strong force between the quark–antiquark pair can be expressed as:

$$V_{q\bar{q}}(r) \xrightarrow{r \to 0} -\frac{C_{F}\alpha_{S}(r)}{r} \qquad (20.29)$$

where in addition to replacing α with α_S, the color factor C_F has been added as an overall multiplicative term. The color wave function of the quarkonium bound state must be the color singlet state. It therefore contains equal contributions from the $R\overline{R}$, $G\overline{G}$, and $B\overline{B}$ combinations. It can be written $(R\overline{R} + G\overline{G} + B\overline{B})/\sqrt{3}$. It is then sufficient to compute the term for $R\overline{R}$ and multiply by three:

$$3 \times \frac{1}{\sqrt{3}}(R\overline{R}) \times \frac{1}{\sqrt{3}}(R\overline{R} + G\overline{G} + B\overline{B}) \tag{20.30}$$

The color factors for $R\overline{R} \to R\overline{R}$, $R\overline{R} \to G\overline{G}$, and $R\overline{R} \to B\overline{B}$ are respectively $C_F = 1/3$, $1/2$, and $1/2$ – see Eqs. (18.15) and (18.16). The total color factor is found by adding them. It is quite satisfying to get an overall color factor which is positive (corresponding to an attractive potential) of:

$$C_F = 3 \times \frac{1}{3} \times \left(\frac{1}{3} + \frac{1}{2} + \frac{1}{2} \right) = \frac{4}{3} \tag{20.31}$$

Thus, the potential between the quark and antiquark color singlet in the non-relativistic limit can be approximated by the sum of a short-range Coulomb-like attractive potential and a long-range confining term:

$$V_{q\bar{q}}(r) \simeq -\frac{4}{3}\frac{\alpha_S \hbar c}{r} + \kappa r \tag{20.32}$$

where $\hbar c \simeq 0.2$ GeV/fm (see Eq. (1.24)), $\kappa \simeq 1$ GeV/fm, and r is given in femtometers. This potential for $\alpha_S = 0.2$ is shown in Figure 20.11. From the curve of the potential, we can conclude that for the chosen

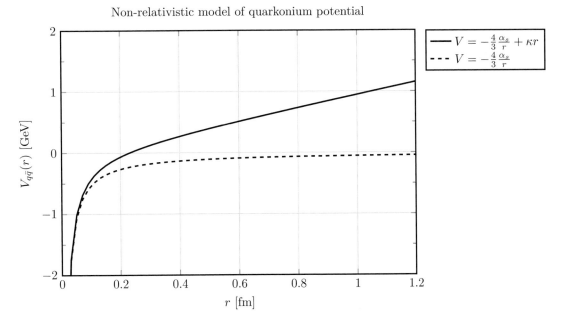

Figure 20.11 Non-relativistic potential $V_{q\bar{q}}(r)$ between the quark and antiquark in the quarkonium system for $\alpha_S = 0.2$ and $\kappa = 1$ GeV/fm.

parameters it becomes positive around 0.2 fm, beyond which the linear term representing the confinement effect becomes dominant.

As in the positronium system, $q\bar{q}$ bound states possess radial and orbital excitations. In order to reproduce the structure of charmonium, one must also take into account "spin–orbit" and "spin–spin" interactions (and also tensor currents). This leads to a family of charmonium states with definite masses and spins, which depend on the levels of excitement and momentum state of the $c\bar{c}$ pair. Such additional states lead to the spectroscopy with $J = 0, 1, 2$, with particles labeled $\eta_c(2S)$, χ_{c0}, χ_{c1}, χ_{c2}, Ψ^*s, and Xs, as illustrated in Figure 20.10(left). The conventional nomenclature is that the singlet S states with spin 0 are called η_c; the triplet S states with spin-1 are called Ψs (and the J/Ψ); the triplet P states with spins 0, 1, and 2 are called χ_{cJ}s.

20.7 Charmed Hadrons – $SU(4)$ Flavor Symmetry

We have seen that the $SU(3)_{flavor}$ symmetry, although approximate, is very powerful in predicting the spectrum of hadrons. Just like the case of isospin and strangeness, the charm flavor corresponds to an additive quantum number C, whose value is $C = +1$ for the c quark and $C = -1$ for the $\bar{c}$ quark. For the lighter quarks u, d, and s we have naturally $C = 0$. The c quark carries an electric charge $Q = +2/3$, isospin $I = 0$, and strangeness $S = 0$, so the Gell-Mann/Nishijima relation of Eq. (17.34) can be updated accordingly:

$$Q = I_3 + \frac{Y}{2} = \frac{U + D + S + C}{2} + \frac{B}{2} \tag{20.33}$$

where B is the baryon number, as defined in Eq. (15.42). With four quarks at hand, we must extend the $SU(3)_{flavor}$ to an $SU(4)_{flavor}$ symmetry. The resulting multiplets can be built to represent all possible hadrons that can be constructed with the u, d, s, and c quarks. **The $SU(4)$ group symmetry predicts the existence of many possible charmed hadrons.**

The fundamental $SU(4)$ representation of a quark is a tetrahedron with the u, d, s in the base with the isospin I_3 and hypercharge Y axis, and c at the top of the pyramid with the vertical C axis. For the antiquark, the representation is an upside-down pyramid with its bottom located at $C = -1$. The $4 \otimes \bar{4}$ combination of the two representations gives us 16 $q\bar{q}$ mesons. The non-charmed meson octets are shown in Figure 17.4 for spin-0 and Figure 17.5 for spin-1. Their extensions to $SU(4)$ are shown in Figures 20.12(left) (a) and (b). Each meson multiplet contains a $c\bar{c}$ bound state with a **"hidden charm"**: the spin-0 is the $\eta_c(2983)$ and the spin-1 is the original J/Ψ. These are called *hidden*, since the total charm quantum number of a $C = +1$ quark and a $C = -1$ antiquark makes them appear "charmless." The **"open charm"** states correspond to charmed mesons composed of a charmed quark and a non-charmed antiquark (resp. charmed antiquark and non-charmed quark). These are (see Figure 20.12):

$$\begin{cases} \text{Spin-0}: D^0(c\bar{u}), D^+(c\bar{d}), \overline{D^0}(u\bar{c}), D^-(d\bar{c}), D_s^+(c\bar{s}), D_s^-(s\bar{c}) \\[2mm] \text{Spin-1}: D^{0*}(c\bar{u}), D^{*+}(c\bar{d}), \overline{D^{*0}}(u\bar{c}), D^{*-}(d\bar{c}), D_s^{*+}(c\bar{s}), D_s^{*-}(s\bar{c}) \end{cases} \tag{20.34}$$

20.8 Charmed Hadrons Properties

The charmed mesons composed of a c and a non-c quark are discussed below. Their measured properties are the following (values taken from the Particle Data Group's Review [5]):

$$\boxed{D^\pm}: \quad \begin{aligned} &I(J^P) = 1/2(0^-), \quad m_{D^+} = (1869.59 \pm 0.09) \text{ MeV} \\ &\tau = (1.040 \pm 0.007) \times 10^{-12} \text{ s} \\ &Br(D^+ \to e^+ + X) = (16.07 \pm 0.30)\% \\ &Br(D^+ \to \mu^+ + X) = (17.6 \pm 3.2)\% \\ &Br(D^+ \to K^- + X) = (25.7 \pm 1.4)\% \\ &Br(D^+ \to K^0 + X, \bar{K}^0 + X) = (61 \pm 5)\% \end{aligned} \tag{20.35}$$

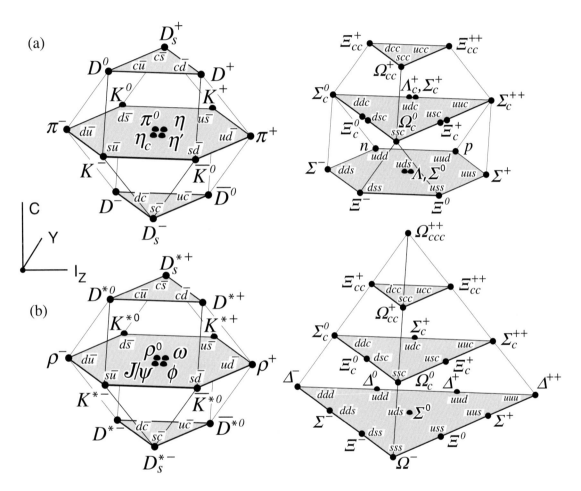

Figure 20.12 (left) The $SU(4)_F$ weight diagrams for scalar and vector mesons. (right) The $SU(4)_F$ weight diagrams for baryons. Reprinted with permission from P. A. Zyla *et al.* (Particle Data Group), *Prog. Theor. Exp. Phys.* **2020**, 083C01 (2020).

and

$$\boxed{D^0}: \quad I(J^P) = 1/2(0^-), \quad m_{D^0} = (1864.83 \pm 0.05) \text{ MeV}$$
$$\tau = (4.101 \pm 0.015) \times 10^{-13} \text{ s}$$
$$Br(D^0 \to e^+ + X) = (6.49 \pm 0.11)\%$$
$$Br(D^0 \to \mu^+ + X) = (6.7 \pm 0.6)\%$$
$$Br(D^0 \to K^- + X) = (54.7 \pm 2.8)\%$$
$$Br(D^0 \to K^0 + X, \bar{K}^0 + X) = (47 \pm 4)\%$$

(20.36)

Similar considerations lead to the charmed spin-1/2 and spin-3/2 baryons Λ_c^+, Σ_c, Ξ_c, Ω_c, including the triple charmed state Ω_{ccc}^{++}, listed in Figure 20.12(right). Their measured properties are the following (values taken from the Particle Data Group's Review [5]):

$$\boxed{\Lambda_c^+}: \quad I(J^P) = 0(1/2^+), \quad m_{\Lambda_c^+(udc)} = (2286.46 \pm 0.14) \text{ MeV}$$
$$\tau = (2.00 \pm 0.06) \times 10^{-13} \text{ s}$$
$$Br(\Lambda_c^+ \to p + X) = (50 \pm 16)\%$$
$$Br(\Lambda_c^+ \to n + X) = (50 \pm 16)\%$$
$$Br(\Lambda_c^+ \to \Lambda + X) = (38.2^{+2.9}_{-2.4})\%$$
$$Br(\Lambda_c^+ \to 3prongs) = (24 \pm 8)\%$$

(20.37)

$$\boxed{\Sigma_c(2455)}: \quad I(J^P) = 1(1/2^+)$$
$$m_{\Sigma_c^{++}(uuc)} = (2453.97 \pm 0.14) \text{ MeV}$$
$$m_{\Sigma_c^+(udc)} = (2452.9 \pm 0.4) \text{ MeV}$$
$$m_{\Sigma_c^0(ddc)} = (2453.75 \pm 0.14) \text{ MeV}$$
$$Br(\Sigma_c(2455) \to \Lambda_c^+ \pi) = 100\%$$

(20.38)

$$\boxed{\Xi_c^{+/0}}: \quad I(J^P) = 1/2(1/2^+)$$
$$m_{\Xi_c^+(usc)} = (2467.93 \pm 0.18) \text{ MeV}$$
$$\tau_{\Xi_c^+(usc)} = (4.42 \pm 0.26) \times 10^{-13} \text{ s}$$
$$m_{\Xi_c^0(dsc)} = (2470.91 \pm 0.25) \text{ MeV}$$
$$\tau_{\Xi_c^0(usc)} = (1.12 \pm 0.13) \times 10^{-13} \text{ s}$$
$$Br(\Xi_c \to \Sigma + K + X) = 100\%$$

(20.39)

$$\boxed{\Omega_c^0}: \quad I(J^P) = 0(1/2^+)$$
$$m_{\Omega_c^0(ssc)} = (2695.2 \pm 1.7) \text{ MeV}$$
$$\tau_{\Omega_c^0} = (2.68 \pm 0.26) \times 10^{-13} \text{ s}$$

(20.40)

• **c Quark mass.** From the masses of the observed charmed hadrons, one can conclude that the c quark is much heavier than the u, d, s quarks. Since quarks are never observed as free particles, the definition of their masses is somewhat more involved than for other particles. The pole mass (which corresponds to the mass one uses in the quark propagator) of the c quark is found to be [28]:

$$m_c = (1.67 \pm 0.07) \text{ GeV}$$

(20.41)

• **Lifetimes of charmed hadrons.** Charged D mesons have many possible decay modes. Why are they long-lived? They are the lightest mesons containing a single charm quark. Hence, they must change the charm quark into a quark of another type to decay. The Feynman diagrams of two typical decay modes of D^0 are shown in Figure 20.13. An important consideration is that the flavor change to a lighter quark implies a charged weak decay with the exchange of a $W^\pm$ boson, as will be discussed in Chapter 23. **This explains the long lifetime of light charmed mesons.**

• **Fragmentation.** The hadronization of quarks has been introduced in Section 18.14. Since the c quark is a heavy quark, the fragmentation function is seen to be harder, concentrated at large values of z and is well

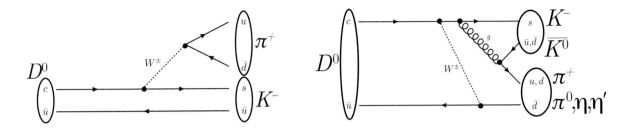

Figure 20.13 Decay mode (left) $D^0 \to K^-\pi^+$; (right) higher-order $D^0 \to K^-\pi^+$ or $D^0 \to \overline{K^0}(\pi^0, \eta, \eta')$.

described by the Peterson fragmentation function with parameter $\epsilon_c = 0.2$ (see Eq. (18.132)). This is clearly visible in Figure 18.15(left).

• **Spectator approximation.** In the diagrams of Figure 20.13, one has considered the weak decay of the heavy charm quark, consequently, it turns into a lighter quark. Since the energy (or the scale Q^2) involved is much larger than the typical quark binding energy, it is reasonable to assume that the heavy quark is "free" during its decay, the other quark(s) and constituents acting as **passive spectators**. A direct consequence of the spectator model is that **all hadrons containing one heavy quark of a particular flavor should have approximately the same lifetime**. From the values of the Particle Data Group's Review, this can be seen to be not precisely true:

$$\tau(D^\pm) \approx 10 \times 10^{-13} \text{ s} > \tau(D^0) \approx 4 \times 10^{-13} \text{ s} > \tau(\Lambda_c^+) \approx 2 \times 10^{-13} \text{ s} \tag{20.42}$$

Nonetheless, they are of the same order. In the case of the charm decay, the energy release is at the level of giga-electronvolts, so we could have expected deviations from the pure simplistic spectator model. We will see in the following that the spectator approximation becomes better the heavier the quark.

• **Okubo–Zweig–Iizuka rule.** Why are some charmed states so narrow? For instance, the width of the J/Ψ is about 90 keV. In fact, all charmonium states with a mass $\lesssim 3700$ MeV are narrow. Indeed, the charmonium states with $n = 1$ and the $n = 2$ η_c and $\Psi(2S)$ have a mass below the $D\overline{D}$ threshold, consequently, they are below the open charmed threshold. However, since the charm is hidden and hence the states appear "charmless," the conservation of the charm flavor is not responsible for their very narrow width. Instead, their long lifetime is understood within the **Okubo–Zweig–Iizuka (OZI) rule** that explains why certain decay modes appear less frequently than otherwise might be expected. The OZI rule was independently proposed by Susumu Okubo, George Zweig, and Jugoro Iizuka. The OZI rule is a consequence of QCD: the decrease of the coupling constant with increasing Q^2. For the OZI-suppressed channels, the gluons must have high Q^2 so the coupling constant will appear small to these gluons. **For these states, the conservation of the charm flavor is not responsible for their very narrow widths**.

The $J/\Psi(1S)$ and $\Psi(2S)$ decays to hadrons (e.g., $\pi^+\pi^0\pi^-$) are OZI suppressed, since the disconnected quark lines require three gluons, as shown in Figure 20.14(left). The OZI suppression for the J/Ψ explains its long lifetime and its width of $\Gamma(J\Psi) = 5$ keV, which is much smaller than the width of the $\rho(770) \approx 150$ MeV and $\omega(784) \approx 8.5$ MeV. The width of the $\Psi(2S)$ is still very narrow, $\Gamma(\Psi(2S)) = 296 \pm 8$ keV. All charmonium states with a mass smaller than ≈ 3700 MeV possess a narrow width. States with $n \geq 3$ have a greater mass and possess a much larger width in the range of tens of mega-electronvolts. These states lie above the threshold for the OZI-allowed strong decay to two charmed D mesons. These states have a much shorter lifetime and are called quasi-bound states. The OZI-allowed decay $\Psi(3770) \to D^+D^-$ is illustrated in Figure 20.14(right).

• **Charmonium spectroscopy.** According to the picture of charmonium from the spectroscopy shown in Figure 20.10, the $J/\Psi(1S)$ is a $c\bar{c}$ bound state with relative angular momentum $\ell = 0$ and parallel quark spins $S = 1$ in the triplet-state configuration. This state is analogous to the e^+e^- ortho-positronium state described in Section 14.9. Selection rules force the $o - Ps$ to decay into an odd number of photons. Its decay width into

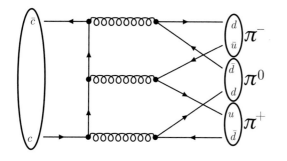 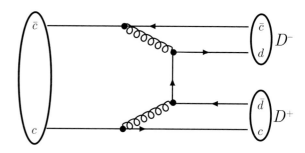

Figure 20.14 (left) OZI-suppressed decay J/Ψ or $\Psi(2S) \to \pi^+\pi^0\pi^-$; (right) OZI-allowed decay $\Psi(3770) \to D^+D^-$ above threshold.

three photons was given in Eq. (14.114). In analogy, the decay of the $J/\Psi(1S)$ into three gluons is found by replacing the fine-structure constant by the strong coupling times an appropriate color factor:

$$\Gamma(c\bar{c} \to ggg \to \text{hadrons}) \simeq \underbrace{\frac{5}{18}}_{\text{color factor}} \times \frac{16\alpha_s^3}{9m_{c\bar{c}}^2} \left(\pi^2 - 9\right) |\phi(0)|^2 \tag{20.43}$$

In comparison, the decays into leptons $J/\Psi(1S) \to e^+e^-$ and $J/\Psi(1S) \to \mu^+\mu^-$ occur via the electromagnetic interaction. In a non-relativistic approach, these latter decay rates can be estimated with the **Van Royen–Weisskopf** formula [240]:

$$\Gamma(c\bar{c} \to \gamma^* \to \ell^+\ell^-) \simeq \frac{16\pi\alpha^2 e_c^2}{3} |\phi(0)|^2 \frac{1}{m_{c\bar{c}}^2} \left(1 - \frac{4m_\ell^2}{m_{c\bar{c}}^2}\right)^{1/2} \left(1 + \frac{2m_\ell^2}{m_{c\bar{c}}^2}\right) \approx \frac{16\pi\alpha^2 e_c^2}{3m_{c\bar{c}}^2} |\phi(0)|^2 \tag{20.44}$$

where $\ell = e, \mu$ and $e_c = +2/3$ is the c quark elementary charge. Consequently, one can write:

$$\begin{aligned}
\Gamma(c\bar{c} \to ggg \to \text{hadrons}) &\simeq \frac{5}{18} \frac{\pi^2 - 9}{3\pi} \frac{\alpha_s^3}{e_c^2\alpha^2} \Gamma(c\bar{c} \to e^+e^-) \\
&\approx 5.76 \times 10^{-2} \left(\frac{\alpha_s^3}{\alpha^2}\right) \Gamma(c\bar{c} \to e^+e^-) \\
&\approx 19 \times \Gamma(c\bar{c} \to e^+e^-)
\end{aligned} \tag{20.45}$$

where the value of the strong coupling constant at $Q^2 \simeq m_{J/\Psi}^2$ is approximately $\alpha_s(m_{J/\Psi}^2) \approx 0.26$ (see Figure 18.16). We should still include the purely electromagnetic decay into hadrons:

$$\begin{aligned}
\Gamma(c\bar{c} \to \gamma^* \to q\bar{q} \to \text{hadrons}) &= \underbrace{3}_{\text{colors}} \left(\sum_{q=u,d,s} e_q^2\right) \Gamma(c\bar{c} \to e^+e^-) = 3 \left[\left(\frac{2}{3}\right)^2 + 2\left(\frac{1}{3}\right)^2\right] \Gamma(c\bar{c} \to e^+e^-) \\
&= 3 \times \frac{6}{9} \Gamma(c\bar{c} \to e^+e^-) = 2 \times \Gamma(c\bar{c} \to e^+e^-)
\end{aligned} \tag{20.46}$$

Hence, finally, the width into hadrons is given by:

$$\Gamma(c\bar{c} \to \text{hadrons}) = \Gamma(c\bar{c} \to ggg \to \text{hadrons}) + \Gamma(c\bar{c} \to \gamma^* \to q\bar{q} \to \text{hadrons}) \approx (19+2)\Gamma(c\bar{c} \to e^+e^-) \tag{20.47}$$

Combining these results, we can estimate the ratio of the hadronic decay width to the leptonic ones. We find:

$$\Gamma(c\bar{c} \to \text{hadrons}) : \Gamma(c\bar{c} \to e^+e^-) : \Gamma(c\bar{c} \to \mu^+\mu^-) \approx 21 : 1 : 1 \simeq 91\% : 4\% : 4\% \tag{20.48}$$

which, given the naive treatment, is really pretty close to the measured branching ratios of Eq. (20.26): $Br(J/\Psi \rightarrow \text{hadrons}) = (87.7 \pm 0.5)\%$, $Br(J/\Psi \rightarrow e^+e^-) = (5.971 \pm 0.032)\%$, and $Br(J/\Psi \rightarrow \mu^+\mu^-) = (5.961 \pm 0.033)\%$. We conclude that the non-relativistic model of a $c\bar{c}$ bound state can make good predictions of the charmonium! Relativistic and higher-order QCD corrections to this model have been implemented and can be found in the literature.

20.9   The Bottom Quark

The discovery of the charm quark triggered speculations about the existence of a third family. It seemed quite reasonable that the first family (u, d, e, ν_e) and the second family (c, s, μ, ν_μ) should be complemented by a third family composed of the bottom (or beauty) and the top quark (or truth), hence forming the family (t, b, τ, ν_τ). The regular pattern of the three lepton and quark families is surely an intriguing puzzle of particle physics.

In 1973 **Makoto Kobayashi**[6] and **Toshihide Maskawa**[7] pointed out the necessity of the third family in their model of the "CKM" quark mixing to explain CP violation [241] (to be discussed in Chapter 30). The bottom quark was discovered in 1977 by a FNAL experiment led by Lederman. After the missed opportunity to discover the charm quark, the team pursued the search for new peaks in the Drell–Yan $\mu^+\mu^-$ pairs with protons reaching 400 GeV impinging on nuclear targets. Their apparatus was a double-arm spectrometer set to measure the di-muon pairs with an excellent resolution of about 2% for invariant masses above 5 GeV. The data collected shows a clear, statistically significant peak in the muon pair mass at 9.5 GeV. A more detailed analysis indicated the presence of at least two peaks, one at 9.44 GeV and the second at 10.17 GeV, which were given the name Υ. The measured properties of $\Upsilon(1S)$ and $\Upsilon(2S)$ are the following (values taken from the Particle Data Group's Review [5]):

$$\boxed{\Upsilon(1S)}: \quad \begin{aligned} &J^{PC} = 1^{--}, \quad m_{\Upsilon(1S)} = (9460.30 \pm 0.26) \text{ MeV} \\ &\Gamma = (54.02 \pm 1.25) \text{ keV} \\ &Br(\Upsilon(1S) \rightarrow \text{hadrons}) = (81.7 \pm 0.7)\% \\ &Br(\Upsilon(1S) \rightarrow e^+e^-) = (2.38 \pm 0.11)\% \\ &Br(\Upsilon(1S) \rightarrow \mu^+\mu^-) = (2.48 \pm 0.05)\% \\ &Br(\Upsilon(1S) \rightarrow \tau^+\tau^-) = (2.60 \pm 0.10)\% \end{aligned} \tag{20.49}$$

and

$$\boxed{\Upsilon(2S)}: \quad \begin{aligned} &J^{PC} = 1^{--}, \quad m_{\Upsilon(2S)} = (10023.26 \pm 0.31) \text{ MeV} \\ &\Gamma = (31.98 \pm 2.63) \text{ keV} \\ &Br(\Upsilon(2S) \rightarrow \Upsilon(1S) + \pi^+ + \pi^-) = (17.85 \pm 0.26)\% \\ &Br(\Upsilon(2S) \rightarrow \Upsilon(1S) + \pi^0 + \pi^0) = (8.6 \pm 0.4)\% \\ &Br(\Upsilon(2S) \rightarrow e^+e^-) = (1.91 \pm 0.16)\%\% \\ &Br(\Upsilon(2S) \rightarrow \mu^+\mu^-) = (1.93 \pm 0.17)\% \\ &Br(\Upsilon(2S) \rightarrow \tau^+\tau^-) = (2.00 \pm 0.21)\% \end{aligned} \tag{20.50}$$

Owing to its large mass, the bottomonium has even more bound states than the charmonium, as illustrated in Figure 20.15. The open bottom states are the B mesons referred to as the $B^\pm$, B_0, B_s, and B_c mesons. The bottom baryons are called the Λ_b^0, Σ_b, Ξ_b, and Ω_b. **According to the current definition, a meson having a b quark is considered as a $\bar{B}$ meson, while a meson with an antiquark $\bar{b}$ is a B meson.** Note that this is in opposition to what is used for charmed hadrons, where the correspondence is $D = c\bar{q}$ and $\overline{D} = \bar{c}q$. See Table 20.1. For the bottom baryons, on the other hand, we have $N_b = bqq$ and $\bar{N}_b = \bar{b}\bar{q}\bar{q}$.

---

6   Makoto Kobayashi (born 1944), Japanese physicist.
7   Toshihide Maskawa (born 1940), Japanese physicist.

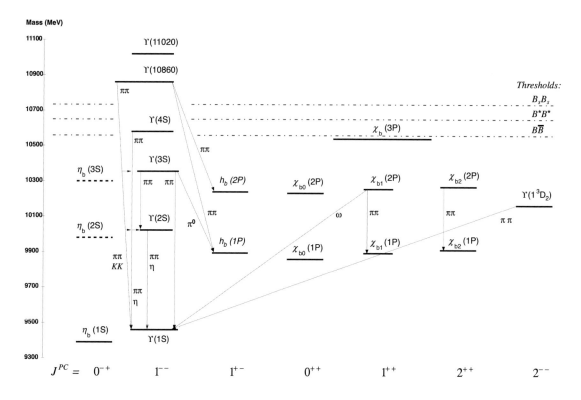

Figure 20.15 Bottomonium spectroscopy: masses and J^{PC} assignments of the observed $b\bar{b}$ bound states. Reprinted with permission from C. Patrignani *et al.*, "Review of particle physics," *Chin. Phys.*, vol. C40, no. 10, p. 100001, 2016.

The measured properties of the B mesons are given below (values taken from the Particle Data Group's Review [5]):

$$\boxed{B^{\pm}}: \quad \begin{aligned} &I(J^P) = 1/2(0^-), \quad m_{B^{\pm}} = (5279.33 \pm 0.13)\text{ MeV} \\ &\tau = (1638 \pm 4) \times 10^{-15}\text{ s} \\ &Br(B^+ \to \ell^+ \nu_{\ell} + X) = (10.99 \pm 0.28)\% \\ &Br(B^+ \to \overline{D}^0 + X) = (79 \pm 4)\% \\ &Br(B^+ \to D^- + X) = (9.9 \pm 1.2)\% \end{aligned} \quad (20.51)$$

$$\boxed{B^0}: \quad \begin{aligned} &I(J^P) = 1/2(0^-), \quad m_{B^0} = (5279.64 \pm 0.13)\text{ MeV} \\ &\tau = (1519 \pm 4) \times 10^{-15}\text{ s} \\ &Br(B^0 \to \ell^+ \nu_{\ell} + X) = (10.33 \pm 0.28)\% \\ &Br(B^0 \to K^{\pm} + X) = (78 \pm 8)\% \\ &Br(B^0 \to \overline{D}^0 + X) = (47.4 \pm 2.8)\% \end{aligned} \quad (20.52)$$

$$\boxed{B_s^0}: \quad \begin{aligned} &I(J^P) = 0(0^-), \quad m_{B_s^0} = (5366.88 \pm 0.17)\text{ MeV} \\ &\Gamma = (66.24 \pm 18) \times 10^{10}\text{ s}^{-1} \\ &Br(B_s^0 \to D_s^- + X) = (93 \pm 25)\% \\ &Br(B_s^0 \to \ell^+ \nu_{\ell} + X) = (9.6 \pm 0.8)\% \end{aligned} \quad (20.53)$$

$$
\overline{B} \Rightarrow
\begin{array}{|l|}
\hline
B^- \equiv b\bar{u} \\
\overline{B}_d^0 \equiv b\bar{d} \\
\overline{B}_s^0 \equiv b\bar{s} \\
\overline{B}_c^0 \equiv b\bar{c} \\
\hline
N_b \Rightarrow \quad bqq
\end{array}
\qquad
B \Rightarrow
\begin{array}{|l|}
\hline
B^+ \equiv \bar{b}u \\
B_d^0 \equiv \bar{b}d \\
B_s^0 \equiv \bar{b}s \\
B_c^0 \equiv \bar{b}c \\
\hline
\overline{N}_b \Rightarrow \quad \bar{b}\bar{q}\bar{q}
\end{array}
$$

Table 20.1 Quark content of some bottom mesons and baryons. Note the way mesons are defined is different than for baryons.

$$
\boxed{B_c^+}: \quad
\begin{aligned}
&I(J^P) = 0(0^-), \quad m_{B_c^+} = (6274.9 \pm 0.8) \text{ MeV} \\
&\tau = (0.510 \pm 0.009) \times 10^{-12} \text{ s} \\
&Br(B_c^0 \to J/\Psi(1S)\ell^+\nu_\ell) = (8.1 \pm 1.2) \times 10^{-5}
\end{aligned}
\tag{20.54}
$$

• b **Quark mass.** From the masses of the observed bottom hadrons, one can conclude that the b quark is much heavier than the u, d, s, and c quarks. The pole mass of the b quark is found to be [28]:

$$
m_b = (4.78 \pm 0.06) \text{ GeV}
\tag{20.55}
$$

• **Fragmentation.** Since the b quark is a very heavy quark, the fragmentation function is harder, concentrated at large values of z, and well described by the Peterson fragmentation function with parameter $\epsilon_b = 0.02$ (see Eq. (18.132)). This is clearly visible in Figure 18.15(right).

• **Decay modes.** Since the b quark is the lighter element of the third-generation quark doublet, the decays of b-flavored hadrons must occur via generation-changing processes through the CKM mixing matrix (see Section 23.14). Only weak Cabibbo suppressed flavor transitions are possible at tree level (see Chapter 23). Some typical decay modes of B mesons are illustrated in Figure 20.16. At higher levels, the virtual top quark can enter in a loop, as shown on the right side.

20.10   Leptonic B-factories: BaBar and Belle

B-factories exploit the $\Upsilon(4S)$ resonance as a clean source of $B^0\overline{B}^0$ and B^+B^- pairs. As shown in Figure 20.15, the $\Upsilon(4S)$ is a bottomonium state with $J^{PC} = 1^{--}$, allowing its exclusive production in e^+e^- collisions, and with a mass just above the production thresholds for $B^0\overline{B}^0$ and B^+B^- pairs. The $\Upsilon(4S)$ decays with a probability of about 52% to B^+B^- and with a probability of 48% to $B^0\overline{B}^0$ [28].

The absolute $e^+e^- \to b\bar{b}$ production cross-section at the $\Upsilon(4S)$ resonance is about 1 nb and the ratio of the $b\bar{b}$ production cross-section to the total inelastic e^+e^- cross-section is about 25%. Taking into account the $\Upsilon(4S)$ branching fraction into $B^0\overline{B}^0$ of about 50% and assuming a combined duty factor of accelerator and experiment of 30%, the production of $10^7 B^0\overline{B}^0$ pairs per year requires an instantaneous luminosity of about 2×10^{33} cm^{-2} s^{-1}. The design luminosity of the PEP-II and KEKB colliders (see Table 3.1) hosting the BaBar and Belle experiments was 3×10^{33} cm^{-2} s^{-1} and the record peak luminosity reached at KEKB was 2×10^{34} cm^{-2} s^{-1} (see Table 3.4).

The BaBar experiment operated from 1999 until April 2008 and collected an integrated luminosity of 433 fb^{-1} at the $\Upsilon(4S)$ resonance, corresponding to 467×10^6 produced $B\overline{B}$ pairs (including $B^0\overline{B}^0$ and B^+B^-); the Belle experiment operated from 1999 until November 2010 and collected an integrated luminosity of 711 fb^{-1} at the $\Upsilon(4S)$ resonance, corresponding to 772×10^6 produced $B\overline{B}$ pairs. Both BaBar and Belle

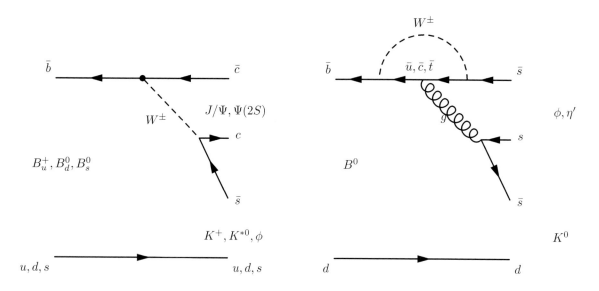

Figure 20.16 Some typical decay modes of B mesons. (left) Spectator diagram of the Cabibbo suppressed tree-level decays $B_u^+ \to J/\Psi K^+$, $B_d^+ \to J/\Psi K^{*0}$, $B_s^+ \to J/\Psi \phi$. (right) Spectator diagram for $B^0 \to K^0 \phi$ or $B^0 \to K^0 \eta'$, the virtual top quark can enter in the process.

collected smaller event samples also at other $\Upsilon(nS)$ resonances and at center-of-mass energies in between the resonances. The Japanese program has now been upgraded with the Super KEKB program,[8] with an impressive target luminosity of 8×10^{35} cm^{-2} s^{-1}.

PEP-II and KEKB are asymmetric B-factories, since they collide electrons and positrons with unequal beam energies. The center-of-mass frame of the collision is then Lorentz boosted along the beam axis and this extra Lorentz boost gives the B mesons a mean decay length in the laboratory system that is large enough to be measured with sufficient precision. At the PEP-II accelerator, 9 GeV electrons were collided with 3.1 GeV positrons, resulting in a Lorentz boost of $\beta\gamma = 0.56$ and a mean decay length of the B mesons of $\beta\gamma c\tau_{B^0} \approx 255$ μm, while 8 GeV electrons were collided with 3.5 GeV positrons in KEKB, resulting in a Lorentz boost of $\beta\gamma = 0.425$ and a mean decay length of the B mesons of $\beta\gamma c\tau_{B^0} \approx 195$ μm.

A side view of the Belle detector is presented in Figure 20.17. Detailed technical descriptions of the BaBar and Belle detectors can be found in Refs. [242] and [243]. Both detectors had the traditional barrel/endcap layout of collider experiments, but with a forward–backward asymmetric design reflecting the asymmetric beam energies. Both detectors comprised a silicon-microstrip vertex detector (see Section 3.15) mounted closely around the beam pipe, surrounded by a drift chamber embedded in a solenoidal magnetic field to reconstruct the trajectories and momenta of charged particles, followed by detectors for kaon/pion identification, an electromagnetic calorimeter for electron/positron identification, photon and π^0 reconstruction, and finally a detector for the identification of muons and long-lived neutral hadrons such as K_L^0 mesons. The main difference between the BaBar and Belle detectors was in the respective approaches toward kaon/pion identification. The BaBar experiment employed a special type of ring-imaging Cherenkov detector for this purpose, while kaon/pion separation in Belle relied on a combination of threshold Cherenkov counters (see Section 3.14), a time-of-flight system, and dE/dx measurements in the tracking drift chamber.

• **Secondary vertex reconstruction.** The precise reconstruction of the decay vertices of the two B mesons is particularly important for many physics analyses, as will be discussed in Chapter 30. Given the lifetimes

---

8   Up-to-date information can be found at www-superkekb.kek.jp.

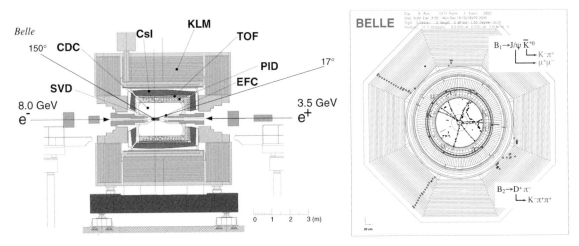

Figure 20.17 (left) Side view of the Belle detector. Reprinted from A. Abashian *et al.*, "The Belle detector," *Nucl. Instrum. Meth. A*, vol. 479, pp. 117–232, 2002, with permission from Elsevier. (right) Fully reconstructed $\Upsilon(4S) \rightarrow B\overline{B}$ recorded by the Belle detector. Reprinted from image accessed at https://belle.kek.jp/belle/events/.

mentioned above, the resolution must be close to or better than 100 μm. To achieve this, both BaBar and Belle relied on their silicon-microstrip detectors located near the interaction point. In BaBar, the achieved vertex resolution for fully reconstructed decays of B mesons was about 55 μm in the direction along the beam axis and about 65 μm in the plane orthogonal to the beam axis [242]. An example of a secondary vertex produced in proton–proton collisions at the LHC and recorded by the LHCb detector is shown in Figure 3.33. The area close to the interaction point is enlarged, showing the tracks of the charged particles produced in the pp interaction, the $\overline{B}^0$ path (dotted), its displaced decay $\overline{B}^0 \rightarrow D^{*+}\tau^-\bar{\nu}_\tau$ with $D^{*+} \rightarrow D^0\pi^+$ and $D^0 \rightarrow K^-\pi^+$, plus the muon from the decay of a very short-lived τ. Another example is presented in Figure 27.16, where the front view of an event classified as $Z^0 \rightarrow b\bar{b}$ collected by the SLD detector at the SLC collider is shown. A displaced secondary vertex reconstructed with the silicon vertex detector, is visible in the expanded side view (r–z view) of the beam interaction point.

• **Lepton flavor tagging.** Lepton flavor-tagging algorithms rely mostly on the charge of high-momentum leptons from the semi-leptonic decays $\bar{b} \rightarrow \bar{c}\ell^+\nu_\ell$ and $b \rightarrow c\ell^-\bar{\nu}_\ell$: a positively charged lepton indicates a B^0 decay, a negatively charged lepton indicates a $\overline{B}^0$ decay. Leptons of opposite charge are produced via cascade decays $\bar{b} \rightarrow \bar{c} \rightarrow \bar{s}\ell^-\bar{\nu}_\ell$ and $b \rightarrow c \rightarrow s\ell^+\nu_\ell$.

• **Charged kaon tagging.** The dominant source of charged kaons is cascade decays $\bar{b} \rightarrow \bar{c} \rightarrow \bar{s}$ and $b \rightarrow c \rightarrow s$, where a positive kaon indicates the decay of a B^0 meson, while a negative kaon indicates a $\overline{B}^0$ decay.

• **Λ baryon tagging.** The strange quarks and antiquarks produced in cascade decays $\bar{b} \rightarrow \bar{c} \rightarrow \bar{s}$ and $b \rightarrow \bar{c} \rightarrow \bar{s}$ can also hadronize into $\bar{\Lambda}^0$ and Λ^0 baryons, where a $\bar{\Lambda}^0$ baryon indicates a B^0 decay while a Λ^0 baryon indicates a $\overline{B}^0$ decay. Λ^0 and $\bar{\Lambda}^0$ candidates are formed from a track that has been identified as a proton or antiproton and a second track of opposite charge, which comes from a common displaced decay vertex and yields an invariant mass compatible with the known Λ^0 mass.

• **Slow pion tagging.** If the charmed quark or antiquark from the decay $\bar{b} \rightarrow \bar{c}W^+$ or $b \rightarrow cW^-$ hadronizes into a $D^{*\pm}$ meson, tagging information is provided by the charge of the low-momentum pion, $\pi^\pm_{\mathrm{sl}}$ from the subsequent decay $D^{*+} \rightarrow \overline{D}^0\pi^+_{\mathrm{sl}}$ or $D^{*-} \rightarrow D^0\pi^-_{\mathrm{sl}}$. A π^-_{sl} indicates a B^0 decay while a π^+_{sl} indicates a $\overline{B}^0$ decay.

20.11   Vector Meson Dominance

The theory of vector meson dominance (VMD) arose much earlier than the quark model, but it is still compatible with it. It was originally developed by **J. J. Sakurai**[9] as a theory of strong interactions. It is based on the observation that **the vector mesons have the same quantum numbers as the photon** $J^{PC} = 1^{--}$. The idea of the VMD consists of assuming that the interaction of a photon with hadrons is dominantly due to conversion of the γ into a hadronic vector meson $J^{PC} = 1^{--}$, which then interacts strongly with the hadronic target, as in any other strong collision, as illustrated in Figure 20.18. The restricted VMD corresponds to the use of the low-lying vector meson states (ρ, ω, ϕ); the so-called extended (or generalized) VMD uses a complete series of excited vector meson states (ρ, ρ', ω, ω', etc).

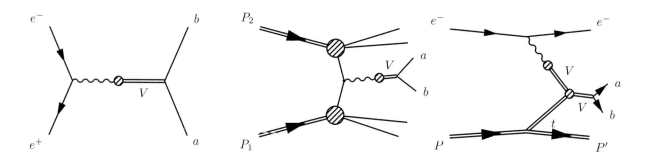

Figure 20.18 Vector meson dominance: (left) e^+e^- collisions; (middle) hadron–hadron collisions; (right) electron–hadron inelastic scattering.

To explain the VDM in quark language, we restrict ourselves to the low-energy domain, where the momentum scale Q^2 is of the order of hadronic masses. Because the typical distance of interaction is of the same order of magnitude as the confinement radius, the $q\bar{q}$ pair converted from the photon cannot be free. The confinement effects will drive the $q\bar{q}$ pair to the nearby bound state with the same quantum numbers as the photon, therefore to vector mesons (see Figure 20.19). The process in which the hadronic bound state is formed requires a low Q^2, since at high Q^2 asymptotic freedom comes into play. **The VDM effect is therefore relevant at low Q^2 and decreases rapidly for increasing Q^2 in favor of quark flux lines.**

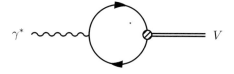

Figure 20.19 Vector meson dominance in quark language: the photon structure function comes from the creation of a quark–antiquark pair which merges into a bound state.

Once the resonance V is produced, it has a short lifetime which determines its natural width $\Gamma_V \equiv 1/\tau_V$. It will typically decay into various decay modes. The **total width** can be expressed as the sum of the partial decay widths. The **branching ratio** $Br(i)$ gives the relative probability of decaying via a given mode i and is

9   Jun John Sakurai (1933–1982), Japanese–American particle physicist and theorist.

defined in Eq. (2.29) as a function of Γ_i. Hence:

$$\Gamma = \sum_j \Gamma_j \quad \text{and} \quad Br(i) = \frac{\Gamma_i}{\Gamma} \qquad (20.56)$$

Consequently, the measurement of the branching fractions can be used to reconstruct the partial widths. For example, the partial widths associated with the exclusive decays $\rho(770) \to e^+e^-$, $\omega(782) \to e^+e^-$, and $\phi(1020) \to e^+e^-$ are measured to be (values taken from the Particle Data Group's Review [5]):

$$
\begin{aligned}
\Gamma_{ee}(\rho(770)) &= (7.04 \pm 0.06) \text{ keV}, & \Gamma_{ee}(\omega(782)) = (0.60 \pm 0.02) \text{ keV}, \\
\Gamma_{ee}(\phi(1020)) &= (1.27 \pm 0.04) \text{ keV}
\end{aligned}
\qquad (20.57)
$$

These widths are rather small compared to the total width of those resonances (see Table 15.2). However, when a resonance decays into $V \to e^+e^-$ and $V \to a+b$, then the reaction $e^+e^- \to a+b$ can proceed via the resonance V. Hence, the process $e^+e^- \to V \to a+b$ becomes possible, as shown in Figure 20.18(left).

This effect can explain the following observation: when the e^+e^- center-of-mass energy is varied, one measures a series of peaks in the total cross-section for some specific channels $e^+e^- \to$ hadrons. These peaks are due to the formation in the s-channel of vector meson resonances which we interpret as quark–antiquark bound states but can also be seen as a consequence of the VMD effect. The general features of e^+e^- annihilation are naively given by a cross-section expression calculated as a product of the usual $1/s$ dependence of the quark–antiquark pair production, the width of the resonance V to e^+e^-, and the Breit–Wigner resonance shape (see Eq. (4.41)) for the vector meson:

$$\sigma(e^+e^- \to V \to \text{all})\big|_V \propto \sigma(e^+e^- \to \text{hadrons}) \times \Gamma_{ee} \times BW(\sqrt{s}, m_V, \Gamma_V) \qquad (20.58)$$

For a vector resonance, we can also safely write $\sigma(e^+e^- \to V \to \text{hadrons}) \simeq \sigma(e^+e^- \to V \to \text{all})$. So finally, the cross-section is expected to behave near the resonance as:

$$\sigma(e^+e^- \to V \to \text{hadrons})\big|_V \propto \frac{\Gamma_{ee}\Gamma_V}{(\sqrt{s} - m_V)^2 + \Gamma_V^2/4} \qquad (20.59)$$

where Γ_{ee} is the width of $V \to e^+e^-$ and $\Gamma_V(s)$ is the total (mostly hadronic) energy-dependent width of the vector meson. This naive approach cannot be directly compared to experiment, because it neglects for example the interference terms. These occur if multiple resonances are produced within the same energy range (e.g., the ρ^0 and the ω have very similar masses compared to their widths). In addition, one usually has to consider a non-resonant background. Detailed calculations are needed to predict the cross-section with the required precision. A review of low-energy e^+e^- hadronic annihilation data and its application can be found in Ref. [244].

Nonetheless, within the naive approach, we can understand the qualitative picture obtained experimentally. The compilation of data on $\sigma(e^+e^- \to \text{hadrons})$ as a function of the center-of-mass energy $\sqrt{s}$ is shown in Figure 20.20. The regions around the light, charm, and bottom quarks are shown in Figure 20.21. The cross-section shows clear peaks corresponding to the production of the vector meson resonances with the $J^{PC} = 1^{--}$ of the photon. At very high energy the very large Z^0 resonance is visible. This latter is explained within the electroweak theory, to be discussed in Chapter 25 and following. For quite a while after the successful discovery of the charm and bottom quark, it was hoped that the top quark would also be discovered in an analogous way. However, its mass turned out to be much heavier (see Section 29.2) than the other quarks and even the Z^0, and incidentally **too short-lived to produce the $t\bar{t}$ toponium bound state** [245].

Similar considerations also apply to deep inelastic scattering, as illustrated in Figure 20.18(right). In this case, once the virtual meson is formed, it has a relatively large cross-section with the target nucleon. The interaction is mostly of **diffractive nature**, which implies that the cross-section will strongly depend on the variable $t = Q^2$, typically with a decreasing exponential dependence, leading to a very rapid fall-off with t:

$$\frac{\text{d}\sigma}{\text{d}t} \simeq e^{-b|t|} \qquad (20.60)$$

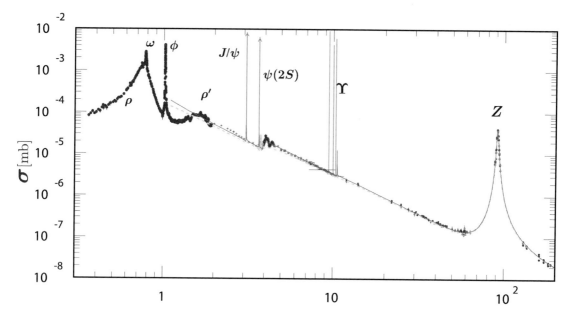

Figure 20.20 Compilation of data $\sigma(e^+e^- \to \text{hadrons})$ as a function of $\sqrt{s}$ in giga-electronvolts. Reprinted figure from V. V. Ezhela, S. B. Lugovsky, and O. V. Zenin, "Hadronic part of the muon $g-2$ estimated on the $\sigma(tot)(e^+e^- \to \text{hadrons})$ evaluated data compilation," [arXiv:hep-ph/0312114 [hep-ph]].

where b is a parameter that depends on the "sizes" of the vector meson and the nucleon. The diffractive processes have recently been measured in $e^{\pm}p$ collisions at HERA, exploiting the fact that the electron/positron can be seen to emit the photon before hitting the proton. Hence, one is effectively searching for the reaction $\gamma^*p \to Vp$. The following processes $\gamma^*p \to \rho^0 p$, $\gamma^*p \to \omega p$, $\gamma^*p \to \phi p$, and $\gamma^*p \to J/\Psi p$ were measured by the ZEUS and HERA experiments, leading to $b \approx 8 \text{ GeV}^{-2}$ for ρ^0 and ω, $\approx 7 \text{ GeV}^{-2}$ for ϕ, and $\approx 4 \text{ GeV}^{-2}$ for J/Ψ (see, e.g., Ref. [246]). These values indicate the needed scale $Q^2 \gg b^{-1}$ above which the deep inelastic scattering process is reasonably dominated by point-like scattering.

In diffractive hadron–hadron scattering, the energy transfer between the two interacting hadrons remains small, however one or both hadrons can break up into multi-particle final states, however preserving the quantum numbers of the associated initial hadrons. These are called single or double-dissociation diffractive events. In contrast, in elastic scattering, the two hadrons emerge preserved and no other particles are produced. Finally, all other configurations would correspond to the inelastic scattering. Early attempts to describe diffraction led to the postulate of the exchange of the **Pomeron** [247], which possesses the quantum numbers of vacuum. In the language of QCD, similar properties can be obtained with the exchange of two gluons [248].

Problems

Ex 20.1 Lepton isolation. Because the CKM matrix is almost diagonal (see Eqs. (23.232) and (30.16)), the most favored decay path for a heavy quark in charged currents is either the same generation, or if this is kinematically prohibited, to the nearest generation.

   (a) Show that as a result we have the following main modes (including also the top quark to be discussed in Section 29.2): $c \to s$ and $b \to c \to s$ and $t \to b \to c \to s$ with a virtual W boson at each step.

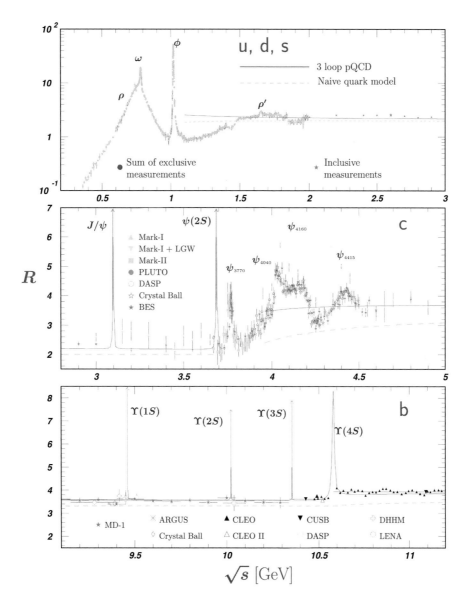

Figure 20.21 Figure 20.20 rescaled to the hadronic R ratio, zoomed in region of quark thresholds. Reprinted figure from V. V. Ezhela, S. B. Lugovsky, and O. V. Zenin, "Hadronic part of the muon $g-2$ estimated on the $\sigma(tot)(e^+e^- \to \text{hadrons})$ evaluated data compilation," [arXiv:hep-ph/0312114 [hep-ph]].

(b) Considering that the W bosons could decay leptonically, show that there is a possibility of producing $\ell^-\bar{\nu}_\ell$ or $\ell^+\nu_\ell$ pairs at each step.

(c) What is the maximum number of charged leptons that can be produced in charm, bottom, or top quark decays?

(d) Considering the kinematics of the decays and in particular that the parent quark is heavy with mass m_Q, explain why the method of lepton isolation is a practical method to identify heavy quarks (use the fact that the lepton can have a transverse momentum relative to the quark flight direction up to $p_T^\ell \simeq (1/2)m_Q$).

Ex 20.2 Direct photons. The production of high-p_T photons is similar to the Drell–Yan process and can be used to test QCD predictions. It is also a source of irreducible background at the CERN LHC for the search of the Higgs boson decaying into two photons.

(a) Draw the Feynman diagrams of the lowest-order processes at the partonic level $q\bar{q} \rightarrow \gamma g$ and $gq(\bar{q}) \rightarrow \gamma q(\bar{q})$.

(b) Show that the corresponding differential cross-section can be expressed as:

$$\frac{d\sigma(q\bar{q} \rightarrow \gamma g)}{d\hat{t}} = \frac{8\pi\alpha\alpha_S e_q^2}{9}\left(\frac{\hat{t}^2 + \hat{u}^2}{\hat{s}^2\hat{t}\hat{u}}\right) \tag{20.61}$$

and

$$\frac{d\sigma(gq \rightarrow \gamma q)}{d\hat{t}} = -\frac{\pi\alpha\alpha_S e_q^2}{3}\left(\frac{\hat{s}^2 + \hat{u}^2}{\hat{s}^3\hat{u}}\right) \tag{20.62}$$

21   Neutrinos and the Three Lepton Families

Fermi gave the first account of this theory to several of his Roman friends while we were spending the Christmas vacation of 1933 in the Alps. It was in the evening after a full day of skiing; we were all sitting on one bed in a hotel room, and I could hardly keep still in that position, bruised as I was after several falls on icy snow. Fermi was fully aware of the importance of his accomplishment and said that he would be remembered for this paper, his best so far. He sent a letter to Nature advancing his theory, but the editor refused it because he thought it contained speculations that were too remote from physical reality; and instead the paper (Tentative Theory of Beta Rays) was published in Italian and in the Zeitschrift für Physik.

Emilio Segré

21.1   Are Neutrinos Special?

Neutrinos can be considered as special particles under a certain point of view. They are elementary fermions which are exceptionally elusive and penetrating (e.g., capable of traversing the Earth unaffected, although at energies above tera-electronvolts, the Earth starts to become less transparent). This puzzling property comes from the fact that neutrinos do not carry color and are electrically neutral, hence do not feel the strong nor the electromagnetic force and interact only via the *weak force* (not considering gravity). Neutrinos are also very light, practically massless, and so far, no one has succeeded in measuring their rest (absolute) masses.

The neutrino was originally interpreted as an "electrically neutral electron," although with a much smaller rest mass. The existence of the "neutral electron" (and correspondingly "neutral muon" and "neutral tau") represents a chance for the particle physicist: it allows us to directly test the weak force! This comes at a price: it is impossible to directly detect a neutrino since it does not ionize when it crosses matter, nor does it deposit its energy in a calorimeter. We can detect neutrinos only indirectly when they rarely interact with matter and produce secondary particles which can be detected and measured.

The neutrinos and their charged lepton partners (electron, muon, and tau) compose in total **the six known leptons**, organized in **three lepton families** identified by the **leptonic charges** L_e, L_μ, L_τ (to be discussed in Section 21.9):

$$\begin{array}{l} Q=0 \\ Q=-1 \end{array} \rightarrow \left(\begin{array}{c} \nu_e \\ e^- \end{array} \right), \qquad \left(\begin{array}{c} \nu_\mu \\ \mu^- \end{array} \right), \qquad \left(\begin{array}{c} \nu_\tau \\ \tau^- \end{array} \right) \tag{21.1}$$

$$L_e = +1, \qquad L_\mu = +1, \qquad L_\tau = +1 \tag{21.2}$$

To these correspond the three families of antiparticles with opposite leptonic charges:

$$\begin{array}{l} Q=0 \\ Q=+1 \end{array} \rightarrow \left(\begin{array}{c} \bar{\nu}_e \\ e^+ \end{array} \right), \qquad \left(\begin{array}{c} \bar{\nu}_\mu \\ \mu^+ \end{array} \right), \qquad \left(\begin{array}{c} \bar{\nu}_\tau \\ \tau^+ \end{array} \right) \tag{21.3}$$

$$L_e = -1, \qquad L_\mu = -1, \qquad L_\tau = -1 \tag{21.4}$$

Note that we have here *assumed a distinction between neutrinos and antineutrinos from their leptonic charge*, as it naturally happens if the neutrino is a Dirac particle with the properties of a "neutral electron." This is however still an open experimental question, whose consequences will be discussed in Chapter 31.

21.2   Pauli's Postulate (1930)

We recall that radioactivity was discovered by Becquerel in 1896 (see Chapter 2). By 1907, it was clear that there were different types of rays (α, β, and γ): the α-rays are composed of helium nuclei and they are emitted with specific energies from the decay of radioactive isotopes. On the other hand, the β-rays, which appeared to be composed of electrons, seemed to consist of "lines" over a continuous background. While the stopping range of α particles was well defined, making them appear monochromatic, the range of β particles did not seem to exhibit such property.

In 1907 **Hahn**[1] and **Meitner**[2] studied the absorption of β particles in thin foils and (erroneously) concluded that the β spectrum is monochromatic. In 1909 Rutherford's group in Manchester performed related measurements. Williams succeeded in obtaining a monochromatic electron beam via scattering of electrons in a magnetic field and showed that the energy absorption of electrons in matter is not exponential. This observation contradicted the conclusion of Hahn and Meitner. In 1910, these latter employed a 180° magnetic spectrometer coupled to a photographic setup. The blackening of the photographic emulsion showed regions where monochromatic electrons group. This was a proof that radioactive isotopes emit β-rays of specific energy. However, they withdrew their claim in 1911, after they realized that the blackening of an emulsion is not linear and that the measured spectrum was effectively inhomogeneous. They therefore embarked on trying to find a reason for this observed inhomogeneity, even if the mechanism of production created monoenergetic rays! Rutherford's discovery of the nucleus supports the possibility that β-rays are monoenergetic.

In 1914 **James Chadwick**, at the time a student of Rutherford, worked with Hans Geiger in Berlin. He replaced the photographic plane in the 180° magnetic spectrometer by a slit aperture and a counter. His data seemed to indicate a continuous spectrum. In 1921 **Charles Ellis**,[3] another co-worker of Rutherford, showed that the lines of the β-spectrum are electrons which originate from the internal conversion of nuclear γs. The atom was at the time already represented with Bohr's model and the energy levels were known. For the first time, Ellis observed emissions of γs from nuclear transitions.

In 1922 Meitner published new data compared to a model of monochromatic emission of β particles coupled to a mechanism of energy loss, which could lead to the observed continuous spectrum. In parallel, Ellis and Chadwick repeated their experiment adding a brass absorption plate and looking at backward scattering. They concluded that the continuous spectrum was a correct reality and that any explanation of it as due to secondary sources was untenable. Compton scattering, discovered in 1923, supported Meitner's view that even a monochromatic source can lead to the detection of a continuous spectrum.

In 1925 Ellis and Wooster invented the **calorimetric technique** and performed a decisive experiment. The calorimeter was supposed to collect the total energy of the β-ray, regardless of the actual mechanism of detection. In a specific measurement, they observed that the endpoint energy of ^{210}Bi is 1.05 MeV, however the measured average calorimetric energy was only 0.34 MeV. So, they showed that one could do nothing to trap the lost energy. In 1927 they claimed that the long controversy about the origin of the continuous spectrum appeared to be settled.

In parallel with experimental progress, theoretical developments were also very important. By that time, quantization in quantum mechanics was being understood as the origin for the discrete atomic energy levels, and the continuous spectrum of β-rays represented a real puzzle. Actually, the "energy spectrum" crisis was only part of the problem, and several observations on the nuclear spin, statistics, and magnetic moments were

1   Otto Hahn (1879–1968), German chemist, expert in radioactivity and radiochemistry.
2   Lise Meitner (1878–1968), Austrian–Swedish physicist.
3   Sir Charles Drummond Ellis (1895–1980), English physicist.

also not understandable. The debate amplified and different thoughts emerged: **Niels Bohr** expressed the opinion that one should be ready to abandon energy–momentum conservation in β emission.

On December 4th, 1930, the solution to the crisis came in a famous open letter written by Pauli while he was at the Physikalisches Institut der Eidgenössischen Technischen Hochschule (Zürich) to a group of colleagues for a reunion in Tübingen (Switzerland). Therein, he postulated the **existence of a new particle to explain the experimentally observed continuous β spectrum**. He wrote:

> *"Dear Radioactive Ladies and Gentlemen,*
> *As the bearer of these lines, to whom I graciously ask you to listen, will explain to you in more detail, how because of the 'wrong' statistics of the N and ^6Li nuclei and the continuous beta spectrum, I have hit upon a desperate remedy to save the 'exchange theorem' of statistics[4] and the law of conservation of energy. Namely, the possibility that there could exist electrically neutral particles in the nuclei, that I wish to call neutrons, which have spin 1/2 and obey the exclusion principle and which further differ from light quanta in that they do not travel with the velocity of light. The mass of the neutrons should be of the same order of magnitude as the electron mass and in any event not larger than 0.01 proton masses. The continuous beta spectrum would then become understandable with the assumption that in beta decay a neutron is emitted in addition to the electron such that the sum of the energies of the neutron and the electron is constant. I agree that my remedy could seem incredible because one should have seen these neutrons much earlier if they really exist. But only the one who dares can win and the difficult situation, due to the continuous structure of the beta spectrum, is highlighted by a remark of my honoured predecessor, Mr Debye, who told me recently in Bruxelles: 'Oh, it's well better not to think about this at all, like new taxes'. Therefore, every solution to the issue must be discussed. Thus, dear radioactive people, examine and judge. Unfortunately, I cannot appear in Tubingen personally since I am indispensable here in Zurich because of a ball on the night of 6/7 December. With my best regards to you, and also to Mr Back. Your humble servant, W. Pauli"* [5]

4   This means: Exclusion principle (Fermi statistics) and half-integer spin by odd number of particles; Bose–Einstein statistics and integer spin by even number of particles (see Section 1.1).

5   Translated from *"Liebe Radioaktive Damen und Herren, wie der Überbringer dieser Zeilen, den ich huldvollst anzuhören bitte, Ihnen des näheren auseinandersetzen wird, bin ich angesichts der „falschen" Statistik der N - und Li 6-Kerne, sowie des kontinuierlichen β-Spektrums auf einen verzweifelten Ausweg verfallen, um den „Wechselsatz" der Statistik und den Energiesatz zu retten. Nämlich die Möglichkeit, es könnten elektrisch neutrale Teilchen, die ich Neutronen nennen will, in den Kernen existieren, welche den Spin 1/2 haben und das Ausschliessungsprinzip befolgen und sich von den Lichtquanten ausserdem noch dadurch unterscheiden, dass sie nicht mit Lichtgeschwindigkeit laufen. Die Masse der Neutronen müsste von derselben Grössenordnung wie die Elektronmasse sein, und jedenfalls nicht grösser als 0,01 Protonenmassen. - Das kontinuierliche β-Spektrum wäre dann verständlich unter der Annahme, dass beim β-Zerfall mit dem Elektron jeweils noch ein Neutron emittiert wird, derart, dass die Summe der Energien von Neutron und Elektron konstant ist. Nun handelt es sich weiter darum, welche Kräfte auf die Neutronen wirken. Das wahrscheinlichste Modell für das Neutron scheint mir aus wellenmechanischen Gründen (näheres weiss der Überbringer dieser Zeilen) dieses zu sein, dass das ruhende Neutron ein magnetischer Dipol von einem gewissen Moment μ ist. Die Experimente verlangen wohl, dass die ionisierende Strahlung eines solchen Neutrons nicht grösser sein kann als die eines Gammastrahls, und dann darf μ wohl nicht grösser sein als e · 10^{-13} cm. Ich traue mich aber vorläufig nicht, etwas über diese Idee zu publizieren, und wende mich erst vertrauensvoll an euch, liebe Radioaktive, mit der Frage, wie es um den experimentellen Nachweis eines solchen Neutrons stände, wenn dieses ein ebensolches oder etwa 10mal grösseres Durchdringungsvermögen besitzen würde wie ein γ-Strahl. Ich gebe zu, dass mein Ausweg vielleicht von vornherein wenig wahrscheinlich erscheinen mag, weil man die Neutronen, wenn sie existieren, wohl längst gesehen hätte. Aber nur wer wagt, gewinnt, und der Ernst der Situation beim kontinuierlichen β-Spektrum wird nur durch den Ausspruch meines verehrten Vorgängers im Amte, Herrn Debye, beleuchtet, der mir kürzlich in Brüssel gesagt hat: „O, daran soll man am besten gar nicht denken, so wie an die neuen Steuern". Darum soll man jeden Weg zur Rettung ernstlich diskutieren. - Also, liebe Radioaktive, prüfet, und richtet. - Leider kann ich nicht persönlich in Tübingen erscheinen, da ich infolge eines in der Nacht vom 6. zum 7. Dezember in Zürich stattfindenden Balles hier unabkömmlich bin. - Mit vielen Grüssen an Euch, sowie auch an Herrn Back, Euer untertänigster Diener –W. Pauli."* Reprinted excerpt from the Pauli letter collection: letter to Lise Meitner, https://cds.cern.ch/record/83282/files/meitner_0393.pdf. Copyright CERN.

There are various decay modes that are associated with β decays (we write ν for ν_e):

$$\beta^- \text{ decay}: \quad P(Z,A) \to D(Z+1,A) + e^- + \bar{\nu}$$
$$\beta^+ \text{ decay}: \quad P(Z,A) \to D(Z-1,A) + e^+ + \nu$$
$$\epsilon \text{ electron capture (EC)}: \quad P(Z,A) \to D(Z-1,A) + \nu \tag{21.5}$$

where P is the parent and D the daughter atom. The parameters of the β decays of some common isotopes are listed in Table 21.1.

Decay	Type	Q-value	Half-life $t_{1/2} = \tau \ln 2$	$ft_{1/2}$ (s)
$n \to p + e^- + \bar{\nu}_e$	β^-	7823 keV	609.6 ± 0.4 s	1100
$^6\text{He} \to {}^6\text{Li} + e^- + \bar{\nu}_e$	β^-	3508 keV	806.7 ms	810
$^{10}\text{C} \to {}^{10}\text{B}^* + e^+ + \nu_e$	β^+	3647 keV	19.2 s	3100
$^{14}\text{O} \to {}^{14}\text{N}^* + e^+ + \nu_e$	β^+	5143 keV	70.6 s	3076
$^{60}\text{Co} \to {}^{60}\text{Ni}^* + e^- + \bar{\nu}_e$	β^-	2824 keV	5.2 yr	3×10^7
$^{152}\text{Eu} + e^- \to {}^{152}\text{Sm}^* + \nu_e$	EC	1874 keV	13.5 yr	10^{10}
$^{22}\text{Na} \to {}^{22}\text{Ne} + e^+ + \nu_e$	β^+	2842 keV	2.60 yr	6×10^{13}

Table 21.1 Beta decays of the free neutron and some common isotopes. The definition of $ft_{1/2}$ is given in Eq. (21.42).

If we assume that the parent nucleus is at rest, then **energy conservation in β decays** brings clearly testable signatures. It is traditional to define the Q-value of atomic reactions referring to the masses of the *atoms*, while ignoring electron binding energies. Accordingly, the maximum available energy in β^- decays, corresponding to the emission of an electron, is:

$$E_{max}(\beta^-) = M_{Z,A} - M_{Z+1,A} \tag{21.6}$$

while the maximum energy available in β^+ decays, corresponding to the emission of a positron, is:

$$E_{max}(\beta^+) = M_{Z,A} - M_{Z-1,A} - 2m_e \tag{21.7}$$

For the electron capture case ϵ, we have:

$$E_{max}(\epsilon) = M_{Z,A} - M_{Z-1,A} \tag{21.8}$$

The three equations are written in terms of the mass of the atom $M_{Z,A}$ purely conventionally and because it is convenient, and they properly take into account the rest mass energies of the electrons involved in the process.

Now comes the main question of what happens to the available energy E_{max}. Let us consider, for example, β^-. In the absence of a neutrino, the process is a two-body reaction and there is only one way to share this energy between the electron and the recoiling nucleus. Since the nucleus is much more massive than the electron, its recoil can be neglected, and the electron will carry most of the energy away in the form of kinetic energy:

$$P(Z,A) \to D(Z+1,A) + e^- : \quad E_e = E_{max} \tag{21.9}$$

Given the assumed form for the β reaction, which involves on the other hand the electron and the neutrino in the final state, the electron energy distribution is expected to be continuous:

$$P(Z,A) \to D(Z+1,A) + e^- + \bar{\nu} : \quad E_e + E_\nu = E_{max} \tag{21.10}$$

The neutrino hypothesis was not a trivial one in 1930. Pauli had actually initially thought of the neutrino as a building block of the nucleus. In 1933, Chadwick discovered the actual neutron (see Section 2.14). Fermi decided to relabel Pauli's neutron as "**neutrino**," for "small neutron" in Italian. Pauli abandons the idea of the neutrino as the neutral building block of the nucleus. Nonetheless, the idea that a light neutral particle be emitted in beta decays in conjunction with the observed electron remained and indeed was included in the first attempt to develop a quantitative theory of the weak interaction by Fermi, as discussed in the next chapter.

Following his now famous letter, Pauli continued to perfect his thinking on the neutrino. The laws of conservation of energy and momentum had to apply and the emission of β particles happened together with very penetrating radiation from neutral particles, which had not been observed so far. The sum of the energies of the β particle and the neutral particle (or particles) emitted by the nucleus in one process was equal to the energy corresponding to the upper limit of the β spectrum. It was obvious that also conservation of momentum, angular momentum, and type of statistics in all elementary processes was to be considered. It became evident that the mass of neutrinos could not be much greater than that of the electron. It was even possible that the neutrino mass was zero, so that it would move at the speed of light like the photon. Nevertheless, its penetrating force would be much greater than that of a photon of the same energy. It seemed permissible that neutrinos had spin-1/2, and that they followed the Fermi statistics, although experiments did not provide us with proof of these hypotheses. Hence, by the mid-1930s, it had become most likely that β decays were accompanied by neutrinos with the following properties.

- **Conservation laws:** energy, momentum, and angular momentum are conserved in nuclear processes.

- **Neutral:** a neutral particle is emitted in β disintegrations in addition to the observed charged β-ray.

- **Mass:** the neutrino could be massless, like the photon, in which case, it travels with the speed of light; should it be considered as a "matter" particle or as "radiation"?

- **Spin:** it must have a spin-1/2 and follow Fermi–Dirac statistics (it is a fermion like the electron).

- **Magnetic properties:** it could possess a magnetic moment but not necessarily.

- **Weakly interacting:** the neutrino must be extremely penetrating to have escaped detection.

21.3   Nuclear β Decays: Fermi's Four-Point Interaction (1934)

In 1934 **Enrico Fermi** presented his theory of β decays in analogy to the electromagnetic interaction [249, 250] (the paper was published in English in 1968 [251]). He developed it as a first *attempt* of a quantitative theory of the weak interaction to explain the decay of an isotope. The various β decay and electron-capture modes in Eq. (21.5) correspond to the underlying reactions:

$$\beta^- : n \rightarrow p \qquad \text{and} \qquad \beta^+, \epsilon : p \rightarrow n \tag{21.11}$$

Inspired by the successful developments of the quantum field theory of electromagnetism, Fermi proposed the existence of a four-fermion contact term for the conversion of a neutron into a proton and vice versa, including Pauli's postulated neutrino. This is illustrated in Figure 21.1. With his theory, Fermi was able to successfully derive the essential features of β decays. One assumption was that the interaction responsible for the decay is weak compared to the strong interaction that forms the nucleus, which is certainly valid for decays with lifetimes greater than typical nuclear interaction times ($\simeq 10^{-23}$ s). Consequently, the decay can be treated as a weak perturbation and its rate can be estimated with Fermi's Golden Rule if we limit ourselves to a non-relativistic description (see Eq. (4.178)):

$$W_{fi} = (2\pi)|V_{fi}|^2 \rho(E_f) = (2\pi) \left| \int \Psi_f^* V \Psi_i \mathrm{d}^3 \vec{x} \right|^2 \rho(E_f) \tag{21.12}$$

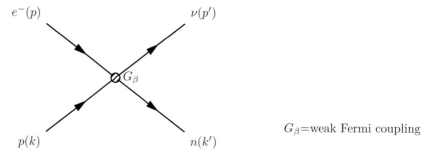

$$G_\beta = \text{weak Fermi coupling}$$

Figure 21.1 Basic weak current–current diagram for the four-fermion contact term postulated by Fermi.

where we recognize the matrix element V_{fi} given by the space integral of the interaction potential V between the initial and final (stationary) nuclear states of the system. The final-state wave function must not only include the nucleus but also the electron and the neutrino which are produced in the process. Fermi could not derive the mathematical form of the potential V. Instead he considered all of the five possible covariant terms, whose transformation properties have been derived in the context of the bilinear covariants of Dirac's theory (see Section 8.28). Consequently, Fermi defined the matrix element as:

$$V_{fi} \equiv G_\beta \sum_i \int \left(\Psi_f^* \phi_e^* \phi_\nu^* O_i \Psi_i \right) \mathrm{d}^3 \vec{x} \tag{21.13}$$

where the **five forms of operators** O_i with $i = V, A, S, P, T$ correspond to the **vector** (V), **axial vector** (A), **scalar** (S), **pseudoscalar** (P), or **tensor** (T) terms (see also Table 8.3). The final-state electron and neutrino are represented by their wave functions ϕ_e and ϕ_ν, while Ψ_f and Ψ_i are the nuclear final and initial states.

The value of the **weak Fermi coupling constant** G_β determines the strength of the β interaction and must be determined experimentally. Because of the four-point nature of the interaction, G_β has units of inverse energy squared. Today the weak coupling strength is expressed in terms of G_F, which is related to $G_\beta \equiv G_F |V_{ud}|$ (see Chapter 23).[6] The Fermi constant G_F is very precisely determined experimentally via the muon decay rate (see Section 23.6) with a relative error of 0.5×10^{-6} to be (value taken from CODATA [10]):

$$G_F = (1.1663787 \pm 0.0000006) \times 10^{-5} \text{ GeV}^{-2} \tag{21.14}$$

At this stage, we pause and compare the above expression with that derived in quantum electrodynamics. The invariant amplitude corresponding to the four-fermion contact term is inspired from and is a generalization of the electromagnetic vector current. Using the formalism of spinors, it is given by the product of the nucleon current ($p \to n$) and of the lepton current ($e^- \to \nu_e$). Using modern language, the vector part of the weak interaction in Fermi's theory would be expressed as:

$$\mathcal{M}_{weak}^V = G_\beta \left[\bar{u}_n(k') \gamma^\mu u_p(k) \right] \times \left[\bar{u}_\nu(p') \gamma_\mu u_e(p) \right] \tag{21.15}$$

where u_e is the spinor of the electron and u_ν is the spinor of the neutrino. We take note of some points here.

- **Neutrino nature:** the neutrino enters as an electrically neutral spin-1/2 Dirac particle in the weak current, indeed similar to a neutral electron.

- **Charge changing:** both weak currents change the electric charge by one unit $|\Delta Q| = 1$. The nucleon current between the proton and the neutron reduces the electric charge by one unit, while the leptonic current increases it by one unit.

6    $G_\beta \approx G_F$ to within 5%.

- **Universality:** the strength of all weak interactions is given by a single value, the **Fermi coupling constant**, an assumption that needs to be experimentally verified.

- **Four-point interaction:** there is no propagator between the two currents.

We now turn back to the quantitative discussion of β decays. The advantages of a quantitative theory were clearly that precise quantitative predictions could be made for many observables in β decays, testable in experiments! Because of the very nature of the escaping neutrino, however, these tests were limited to the observation of the **electron energy spectrum**, **direction of emission**, or more challenging measurements of the recoiling daughter nucleus.

Here, we follow the non-relativistic formalism developed in Chapter 4. The electron and neutrino wave functions ϕ_e and ϕ_ν are given by free-particle plane waves normalized within an arbitrary box of volume $V = L^3$ (see Eq. (4.184)):

$$\phi_e(\vec{x}) = \frac{1}{L^{3/2}} e^{i\vec{p_e}\cdot\vec{x}}, \qquad \phi_\nu(\vec{x}) = \frac{1}{L^{3/2}} e^{i\vec{p_\nu}\cdot\vec{x}} \tag{21.16}$$

where $\vec{p_e}$ and $\vec{p_\nu}$ are the outgoing momenta. The typical energy of a β electron is given by the nuclear scale Q, in the range of mega-electronvolts (see Eq. (21.10)).

We now perform an approximation which is valid for so-called **allowed transitions**: the momentum of an electron with kinetic energy $T = 1$ MeV is $p_e = \sqrt{(T + m_e)^2 - m_e^2} \approx 1.4$ MeV/c. The corresponding de Broglie wavelength is $\lambda_e = h/p_e c \gg 1$ fm. And similarly for the neutrino. Hence, in the volume integral over the nuclear volume Eq. (21.13) we can approximate the exponentials of the wave functions of the electron and neutrino to $\exp(i\vec{p}\cdot\vec{x}) = 1 + (i\vec{p}\cdot\vec{x}) + \cdots \approx 1$ to get:

$$V_{fi} = \frac{G_\beta}{L^3} \sum_i \int \left(\Psi_f^* e^{-i\vec{p_e}\cdot\vec{x}} e^{-i\vec{p_\nu}\cdot\vec{x}} O_i \Psi_i\right) \mathrm{d}^3\vec{x} \approx \frac{G_\beta}{L^3} \sum_i \int \left(\Psi_f^* O_i \Psi_i\right) \mathrm{d}^3\vec{x} \equiv \frac{G_\beta}{L^3} M_{fi} \tag{21.17}$$

where M_{fi} is the **nuclear matrix element**.

Therefore, under the allowed transition approximation, the only factors in the decay rate that depend on the electron or neutrino energy come from the density of states factor $\rho(E_f)$. The partial decay rate for electrons and neutrinos is derived from Eq. (21.12):

$$W_{fi} = \frac{2\pi G_\beta^2}{L^6} |M_{fi}|^2 \rho(E_f) \tag{21.18}$$

where the final energy $E_f = E_e + E_\nu \simeq E_e + p_\nu$ if $m_\nu \ll E_\nu$. We arrive at the conclusion that **the energy shape of the so-called allowed beta transitions is determined (to lowest order) only by the density of states $\rho(E_f)$**. To find the density of states, we focus on the energies, and ignore the direction of emission of the electron and the neutrino. Using Eq. (4.190), we get the density of states:

$$\mathrm{d}n = \left(\frac{L}{2\pi}\right)^6 p_e^2 \mathrm{d}p_e \mathrm{d}\Omega_e \, p_\nu^2 \mathrm{d}p_\nu \mathrm{d}\Omega_\nu = (4\pi)^2 \left(\frac{L}{2\pi}\right)^6 p_e^2 \mathrm{d}p_e p_\nu^2 \mathrm{d}p_\nu \tag{21.19}$$

We obtain:

$$\rho(E_f) = \frac{\mathrm{d}n}{\mathrm{d}E_f} = (4\pi)^2 \left(\frac{L}{2\pi}\right)^6 p_e^2 \mathrm{d}p_e p_\nu^2 \frac{\mathrm{d}p_\nu}{\mathrm{d}E_f} \propto L^6 p_e^2 \mathrm{d}p_e p_\nu^2 \tag{21.20}$$

since $\mathrm{d}p_\nu/\mathrm{d}E_f = 1$ at a fixed electron energy E_e. The electron energy spectrum $N(E_e)$ can be expressed as:

$$\frac{\mathrm{d}N(E_e)}{\mathrm{d}E_e} = \frac{\mathrm{d}W}{\mathrm{d}E_e} = \frac{\mathrm{d}W}{\mathrm{d}p_e}\frac{\mathrm{d}p_e}{\mathrm{d}E_e} \propto \frac{2\pi G_\beta^2}{L^6} |M_{fi}|^2 \, L^6 p_e^2 p_\nu^2 \frac{\mathrm{d}p_e}{\mathrm{d}E_e} \propto p_e^2 p_\nu^2 \frac{E_e}{p_e} \propto p_e E_e p_\nu^2 \tag{21.21}$$

since $\mathrm{d}p_e/\mathrm{d}E_e = E_e/p_e$. As it should, the size of the box L disappeared from our solution since the L^6 power at the numerator of the density cancelled the corresponding term in the transition rate at the denominator.

Energy conservation, Eq. (21.10), applies to calculate the kinematics. Since the minimum energy of the neutrino is m_ν, the upper limit of the electron energy spectrum determines the **endpoint E_0 of the spectrum**. One can conveniently write the neutrino energy as a function of the endpoint energy $E_\nu = (E_0 - E_e) + m_\nu$, then:

$$p_\nu^2 = E_\nu^2 - m_\nu^2 = (E_0 - E_e)^2 + 2m_\nu(E_0 - E_e) = (E_0 - E_e)(E_0 - E_e + 2m_\nu) \tag{21.22}$$

The electron spectrum is thus proportional to:

$$N(E_e)\mathrm{d}E_e \propto p_e E_e (E_0 - E_e)(E_0 - E_e + 2m_\nu)\mathrm{d}E_e \tag{21.23}$$

This form clearly exhibits the sensitivity of the electron spectrum to the neutrino mass. In particular, it is close to the endpoint, when $E_e \approx E_0$, that the sensitivity is the greatest. In the case that the neutrino rest mass is zero, we obtain the expression:

$$N(E_e)\mathrm{d}E_e \propto p_e E_e (E_0 - E_e)^2 \mathrm{d}E_e \qquad \text{for } m_\nu = 0 \tag{21.24}$$

From this, it is customary to plot the **Kurie function** as a function of the reduced dimensionless variable $0 \le x \le 1$ as:

$$K(x) \equiv \sqrt{\frac{N(E_e)}{p_e E_e}} \qquad \text{where } E_e \equiv x E_0 \tag{21.25}$$

and where $N(E_e)$ represents the rate of emitted electrons. For a vanishing neutrino mass, the theory of allowed transitions gives:

$$K(x) \propto \sqrt{\frac{p_e E_e (E_0 - E_e)^2}{p_e E_e}} = E_0 - E_e = E_0(1 - x) \tag{21.26}$$

A Kurie plot for **tritium (or hydrogen-3)** is shown in Figure 21.2. The beta decay reaction is:

$$^3_1 H \to {}^3_2 He + e^- + \bar\nu \tag{21.27}$$

Circles are the data points and curves are predictions for various assumptions on neutrino rest masses. A clear success was the ability to derive a stringent upper limit on the rest mass of the neutrino by careful measurement of the beta-energy spectrum in tritium decay near its end point [252]. In order to get a constraint on the neutrino rest mass, it was indeed not necessary to have a complete theory of the weak interactions, as long as one restricted oneself to the so-called allowed transitions. The behavior of the Kurie plot near the endpoint led to the common belief that the neutrino rest mass was exactly zero.

The derivation so far can be applied to the neutron decay or to other light nuclei without problems. For heavier nuclei, one observes systematic differences which are attributed to the Coulomb interaction between the β particle and the daughter nucleus. Naively we expect the electron to feel a repulsive force while the positron will feel an attractive potential. These effects will modify the spectrum of the emitted particles. This modification is taken into account by introducing an additional factor called the **Fermi function** $F(Z', E_e)$, where Z' is the atomic number of the daughter nucleus. The corresponding Kurie function is accordingly modified to include the Fermi function and becomes:

$$K(x) \equiv \sqrt{\frac{N(E_e)}{F(Z', E_e) p_e E_e}} \tag{21.28}$$

Deviations from the simple straight line would in this case indicate a dependence on the weak matrix element (or of course a non-zero neutrino mass).

Finally, the assumption that the effect of the nuclear matrix element is negligible on the spectrum is not totally correct. There are known cases where it introduces an additional momentum dependence. Cases where this happens are called, for historical reasons, **forbidden decays**. These are not absolutely forbidden, but

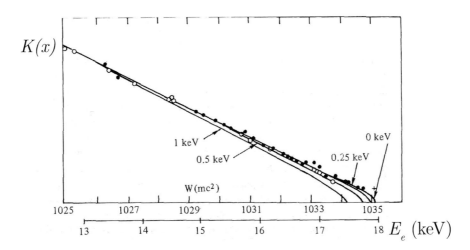

Figure 21.2 The Kurie plot for tritium beta decay. Circles are the data points and curves are predictions for various assumptions on neutrino rest masses. Reprinted with permission from L. M. Langer and R. J. D. Moffat, "The beta-spectrum of tritium and the mass of the neutrino," *Phys. Rev.*, vol. 88, pp. 689–694, Nov 1952. Copyright 1952 by the American Physical Society.

are less likely to occur than allowed decays (hence have longer lifetimes). The complete β spectrum can be expressed as the product of these three factors:

$$N(E_e)\mathrm{d}E_e \propto p_e E_e (E_0 - E_e)^2 F(Z', E_e) |M_{fi}|^2 S(p_e, p_\nu)\mathrm{d}E_e \tag{21.29}$$

where the additional electron and neutrino momentum dependence from forbidden terms is parameterized by the function $S(p_e, p_\nu)$.

The matrix element M_{fi} can be written as a function of the nuclear wave functions of the initial and final states nuclei as follows:

$$M_{fi} \equiv \int \mathrm{d}^3\vec{x}\, \Psi_{Z\pm1,A}\Psi_{Z,A} \tag{21.30}$$

We notice that this definition will inevitably lead to selection rules. In fact, the overlap M_{fi} can vanish (or be very near zero), and hence $|M_{fi}|^2$ too, if the wave functions are orthogonal or of opposite parity. The first condition is for example true for nuclei of different total momentum J. The second condition arises from the fact that the integration is over the entire space and if the integrand is odd under parity, the integral will necessarily vanish. These two restrictions compose the **Fermi selection rules for allowed transitions**:

$$\text{Fermi}: \quad \Delta J = 0, \quad \text{Parity}: + \to + \quad \text{or} \quad - \to - \tag{21.31}$$

Albeit very successful, Fermi's theory had certain problems. Bohr was impressed by Fermi's theory of β decays with neutrinos and considered it sufficient evidence to accept that all conservation laws are obeyed in such decays. It turns out that this was not entirely correct, since it was later discovered that parity and other quantum numbers are violated in weak decays. In addition, the four-point interaction predicts cross-sections that rise linearly with energy. This leads to unitarity problems at high energy and clearly shows that the Fermi model can only be an effective approximation at low energies and is bound to break down at some high energy scale. This apparent limitation would later be resolved with the introduction of the weak boson propagator, to be introduced in Section 23.2.

21.4   Total Decay Rate of Beta Decays

The total decay rate Γ_β is given by the integration over the full phase space. One can, for instance, take the β spectrum and integrate over the electron energy to find:

$$\Gamma_\beta \equiv \int \frac{dW}{dE_e} dE_e = \frac{G_\beta^2}{(2\pi)^3} \int_{m_e}^{E_0} |M_{fi}|^2 p_e E_e (E_0 - E_e)^2 F(Z', E_e) S(p_e, p_\nu) dE_e \tag{21.32}$$

which, for a matrix element independent of the electron momentum, can be written as:

$$\Gamma_\beta = \frac{G_\beta^2 m_e^5}{(2\pi)^3} |M_{fi}|^2 f(Z', E_0) \tag{21.33}$$

where

$$f(Z', E_0) = \frac{1}{m_e^5} \int_{m_e}^{E_0} p_e E_e (E_0 - E_e)^2 F(Z', E_e) S(p_e, p_\nu) dE_e \tag{21.34}$$

The function f is called the **Fermi integral** and is tabulated for values of Z' and E_0. The factor m_e^5 has been introduced in order to make f dimensionless. Consequently, the half-life is given by (see Eq. (2.31)):

$$t_{1/2} = \tau \ln 2 = \frac{\ln 2}{\Gamma_\beta} = \frac{(2\pi)^3 \ln 2}{G_\beta^2 m_e^5 |M_{fi}|^2 f(Z', E_0)} \tag{21.35}$$

As illustrated in Table 21.1, Nature provides an enormous range of values for the half-life. It is interesting to see how these variations are due to the matrix element $|M_{fi}|^2$. Neglecting the Coulomb effect and the forbidden terms allows us to integrate the expression analytically:

$$m_e^5 f(0, E_0) = \int_{m_e}^{E_0} p_e E_e (E_0 - E_e)^2 dE_e = \int_{m_e}^{E_0} E_e \sqrt{E_e^2 - m_e^2} \left(E_0^2 - 2 E_0 E_e + E_e^2 \right) dE_e \tag{21.36}$$

The first term is simply:

$$E_0^2 \int_{m_e}^{E_0} E_e \sqrt{E_e^2 - m_e^2} \, dE_e = E_0^2 \frac{1}{3} (E_e^2 - m_e^2)^{3/2} \Big|_{m_e}^{E_0} = \frac{E_0^2}{3} (E_0^2 - m_e^2)^{3/2} \approx \frac{E_0^5}{3} \tag{21.37}$$

where in the last part we have assumed $E_0 \gg m_e$. This is totally reasonable, since most endpoint energies are much greater than the electron rest mass. The second term yields, also for $E_0 \gg m_e$:

$$-2E_0 \int_{m_e}^{E_0} E_e^2 \sqrt{E_e^2 - m_e^2} dE_e \approx (-2E_0) \frac{1}{8} \sqrt{E_e^2 - m_e^2} (2E_e^3) \Big|_{m_e}^{E_0} \approx -\frac{E_0^5}{2} \tag{21.38}$$

The third term is straightforward:

$$\int_{m_e}^{E_0} E_e \sqrt{E_e^2 - m_e^2} \left(E_e^2 \right) dE_e = \left(\frac{2m_e^2}{15} + \frac{E_e^2}{5} \right) (E_e^2 - m_e^2)^{3/2} \Big|_{m_e}^{E_0} = \frac{1}{5} \left(\frac{2m_e^2}{3} + E_0^2 \right) (E_0^2 - m_e^2)^{3/2}$$

$$\approx \frac{E_0^5}{5} \tag{21.39}$$

Altogether, we find:

$$m_e^5 f(0, E_0) \approx \frac{E_0^5}{3} - \frac{E_0^5}{2} + \frac{E_0^5}{5} = \frac{E_0^5}{30} \qquad \text{for } E_0 \gg m_e \tag{21.40}$$

Hence, we expect the Fermi integral to have a fifth-power dependence on the Q value of the reaction. **The decay rate rises very rapidly with the energy released**. It can, for example, be seen in the difference between the short lifetime of the muon (see Eq. (15.17)) compared to that of the neutron (see Eq. (15.35)). Kinematically, we expect:

$$\left(\frac{\tau_n}{\tau_\mu}\right)^{kin} \simeq \left(\frac{m_\mu}{M_n - M_p}\right)^5 \approx \left(\frac{105.6 \text{ MeV}}{1.293 \text{ MeV}}\right)^5 \approx 3.6 \times 10^9 \tag{21.41}$$

to be compared to the experimental ratio $(\frac{\tau_n}{\tau_\mu})^{exp} \approx 4 \times 10^9$. It makes sense! This kinematical factor can be compensated for by defining the **comparative half-life** $ft_{1/2}$:

$$ft_{1/2} = \frac{(2\pi)^3 \ln 2}{G_\beta^2 m_e^5 |M_{fi}|^2} \tag{21.42}$$

which provides a direct link to the product of the weak mixing constant and the matrix element $G_\beta^2 |M_{fi}|^2$. The $ft_{1/2}$ values of the β decays of some common isotopes are compiled in Table 21.1. We see that the values still vary over a very wide range of values, from 10^3 up to 10^{14} s. One finds isotopes in Nature with values up to 10^{23} s. The assumption that the interaction vertex is universal and given by G_β^2 points towards a large range of values of $|M_{fi}|^2$. We will return to this point in Section 22.1.

21.5 The Inverse Beta Decay Reaction

The free-neutron lifetime of $\tau = 885.7 \pm 0.8$ s allows us to determine the Fermi coupling constant. By crossing symmetry where ingoing (outgoing) particles are replaced by outgoing (ingoing) antiparticles, the free-neutron decay implies the existence of two other related processes, the electron capture (see Eq. (21.5)) and the inverse beta decay (see Figure 21.3):

$$n \to p + e^- + \bar{\nu} \quad \Longleftrightarrow \quad \begin{cases} e^- + p \to n + \nu & (\text{EC}) \\ \bar{\nu} + p \to n + e^+ & (\text{IBD}) \\ \nu + n \to e^- + p & (\text{QE}) \end{cases} \tag{21.43}$$

The EC corresponds to the capture of an electron from one of the inner atomic shells by a nuclear proton, with the emission of a monoenergetic neutrino. We also note that crossing symmetry requires, after having associated the emission of an "antineutrino" $\bar{\nu}$ with the β^- (electron) in the neutron decay, that a neutrino ν is produced in EC, and that an antineutrino $\bar{\nu}$ is involved in the inverse beta decay (IBD) process, while the interaction of a ν on a neutron would give an electron and a proton in the final state. This latter is called quasi-elastic (or QE), as will be discussed in Section 21.9. This assumption on the type of neutrino produced in β disintegrations can be summarized as:

$$\beta^- : n \to p + e^- + \bar{\nu} \quad \text{and} \quad \beta^+ : p \to n + e^+ + \nu \tag{21.44}$$

However, being electrically neutral, the non-equality of the neutrino ν and antineutrino $\bar{\nu}$ requires the existence of a **lepton number**. The question of identity ($\nu = \bar{\nu}$) or non-identity ($\nu \neq \bar{\nu}$) is still an open question of particle physics, as will be discussed in Chapter 31, but it has no consequence to the single β decays.

The inverse beta decay reaction has played a crucial role in many experiments. When a neutrino interacts with a proton, a positron and a neutron will emerge. Both particles can be experimentally detected, providing a unique signature for the event:

$$\bar{\nu} + p \to e^+ + n \tag{21.45}$$

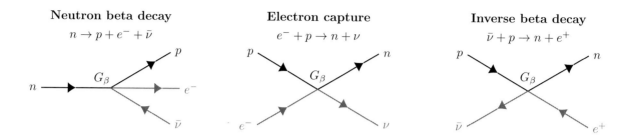

Figure 21.3 Crossing symmetry relates beta decay, electron capture, and inverse beta decay.

The kinematics of the reaction is illustrated in Figure 21.4. Neglecting the neutron recoil, we expect:

$$E_{\bar{\nu}} \simeq E_{e^+} + (M_n - M_p) \approx T_{e^+} + m_e + 1.293 \text{ MeV} \approx T_{e^+} + 1.804 \text{ MeV} \tag{21.46}$$

where T_{e^+} is the kinetic energy of the outgoing positron. Since the kinetic energy T_{e^+} of the positron must be positive, the kinematical threshold for the reaction is that the incoming antineutrino must have an energy $E_{\bar{\nu}} \gtrsim 1.804$ MeV. A precise calculation, including the recoiling neutron, leads to the energy threshold condition:

$$E_{\bar{\nu}} > 1.806 \text{ MeV} \tag{21.47}$$

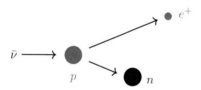

Figure 21.4 Kinematics of the inverse beta decay used to detect reactor neutrinos.

In 1934 **Bethe** and **Peierls** estimated the cross-section for the inverse beta decay process. Quoting them:

"The view has recently been put forward that a neutral particle of about electronic mass, and spin $1/2\hbar$ exists, and that this 'neutrino' is emitted together with an electron in β-decay. This assumption allows the conservation laws for energy and angular momentum to hold in nuclear physics. Both the emitted electron and neutrino could be described either (a) as having existed before in the nucleus or (b) as being created at the time of emission. In a recent paper Fermi has proposed a model of β-disintegration using (b) which seems to be confirmed by experiment using assumptions of the Fermi theory. (...) The possibility of creating neutrinos necessarily implies the existence of annihilation processes. The most interesting amongst them would be the following: a neutrino hits a nucleus and a positive or negative electron is created while the neutrino disappears and the charge of the nucleus changes by 1. The cross-section σ for such processes for a neutrino of given energy may be estimated from the lifetime t of β-radiating nuclei giving neutrinos of the same energy (this estimate is in accord with Fermi's model but is more general). Dimensionally, the connection will be $\sigma = A/t$ where A has the dimensions $cm^2 s$. The longest length and time which can possibly be involved are $\hbar/mc$ and $\hbar/mc^2$. Therefore, $\sigma < \hbar^3/(m^3 c^4 t)$. For an energy of $2 \cdot 3 \times 10^6$ volts, t is 3 minutes and therefore $\sigma < 10^{-44}$ cm^2. It is therefore absolutely impossible to observe processes of this kind with the neutrinos created in nuclear transformations." [7]

7   Reprinted excerpt by permission from Springer Nature: H. Bethe and R. Peierls, "The neutrino," *Nature*, vol. 133, pp. 532. Copyright 1934 Nature Publishing Group.

The IBD cross-section can be accurately computed, relating it to other measured quantities in β decays. Detailed calculations have been performed (see, e.g., Ref. [253]) and a naive approximation valid at mega-electronvolt energies is:

$$\sigma(\bar{\nu} + p \rightarrow e^+ + n) \simeq \frac{2\pi^2}{\tau_n m_e^5 f} E_e p_e \qquad (21.48)$$

where E_e and p_e are respectively the energy and momentum of the outgoing positron, and f is the free neutron phase-space factor equal to 1.715. The cross-section is given as a function of the neutron lifetime $\tau_n = 885.7 \pm 0.8$ s. The measurable e^+ energy is strongly correlated with the incoming neutrino energy. Numerically, the cross-section is:

$$\sigma(\bar{\nu} + p \rightarrow e^+ + n) \simeq 10^{-43} \left(\frac{E_e p_e}{\text{MeV}^2} \right) \times \left(\frac{\tau}{886 \text{ s}} \right) \text{ cm}^2 \qquad (21.49)$$

It grows quadratically with the energy. The total cross-section is very small, on the order of 10^{-43} cm^2. Bethe and Peierls indeed further noted: *"With increasing energy, σ increases (in Fermi's model for large energies as $(E/mc^2)^2$) but even if one assumes a very steep increase, it seems highly improbable that, even for cosmic ray energies, σ becomes large enough to allow the process to be observed. If, therefore, the neutrino has no interaction with other particles besides the processes of creation and annihilation mentioned (...) one can conclude that there is no practically possible way of observing the neutrino."* In order to get a feel for how small the cross-section is, we can compute the hypothetical mean free path of a neutrino of that energy traveling through water. The mean free path λ is the average distance traveled between successive interactions. It is given by (see Eq. (2.70)):

$$\lambda = \frac{1}{n\sigma} \qquad (21.50)$$

where n is the density of the target (cm^{-3}) and the cross-section is given in square centimeters. Numerically, we can assume water molecules as the target, where there are two free protons per H_2O molecule. Since 1 g or equivalently 1 cm^3 of water corresponds to 1/18 mol of H_2O, the density n of free protons in water is:

$$n = 2 \times \left(\frac{1}{18} \frac{\text{mol}}{\text{cm}^3} \right) \times (6 \times 10^{23} \text{mol}^{-1}) \simeq 6.7 \times 10^{22} \text{ cm}^{-3} \qquad (21.51)$$

The mean free path λ of the neutrinos in water is hence:

$$\lambda \simeq \frac{1}{(6.7 \times 10^{22} \text{ cm}^{-3})(10^{-43} \text{ cm}^2)} = 1.5 \times 10^{20} \text{ cm} \qquad (21.52)$$

or about 150 light-years!! Bethe and Peierls were surely right to express their concerns!

In 1946 **Pontecorvo** proposed proving the existence of the neutrino by its direct detection. He suggested detecting neutrinos from nuclear reactors via the inverse beta decay process:

$$\nu + {}^{37}\text{Cl} \rightarrow \beta^- + {}^{37}\text{Ar} \qquad (21.53)$$

He estimated backgrounds and concluded that the experiment was feasible. However, it turns out that reactor "neutrinos" are *antineutrinos* and Pontecorvo's idea was later used by **Davis** to detect solar neutrinos.

In 1956, more than 20 years after Pauli's postulate, neutrinos were first successfully detected in the experiment of **Cowan, Reines, Harrison, and Kruse**. Although the cross-sections for neutrino interactions are indeed extremely small, technical advances during the second half of the twentieth century led to the realization of very powerful neutrino sources, which eventually led to an observable interaction rate! This is the subject of the next section.

21.6 Reactor Neutrinos

Nuclear reactors are intense, pure, and controllable sources of low-energy electron antineutrinos. Fission product yields by mass for thermal neutron fission of ^{235}U, ^{239}Pu (and ^{233}U used in the thorium cycle) are illustrated in Figure 21.5. The fission of heavy elements produces light elements which must shed neutrons to

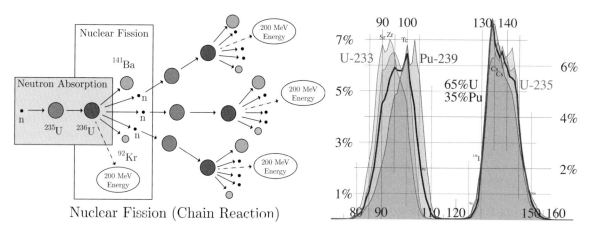

Figure 21.5 (left) Schematic view of thermal fission of ^{235}U. (right) Fission product yields by mass for thermal neutron fission of ^{235}U, ^{239}Pu, and ^{233}U used in the thorium cycle. From JWB at en.wikipedia [CC BY 3.0 (http://creativecommons.org/licenses/by/3.0) or GFDL (www.gnu.org/copyleft/fdl.html)], via Wikimedia Commons.

approach the line of stability. Reactor neutrinos are emitted in the subsequent β decays of unstable fission fragments. Only antineutrinos are produced in this way; the neutrino flux is negligible. In modern reactors, the uranium fuel is enriched to a few percent in ^{235}U, but there are also significant contributions to the neutrino flux from the fissioning of ^{238}U, ^{239}Pu, and ^{241}Pu. The four fuel isotopes, ^{235}U, ^{238}U, ^{239}Pu, and ^{241}Pu, generate more than 99% of the thermal power and reactor antineutrinos, and other sources are usually neglected. The average energy released is ≈ 200 MeV per fission [254], and six antineutrinos are emitted on average. **The average thermal energy released in each fission for the ith isotope is called e_i.** The corresponding effective energies per fission are determined from the energy released in fission, minus the energy carried off by the antineutrinos, plus the energy produced by neutron captures on the reactor materials. They are given in Table 21.2.

Nucleus	Energy from mass excess	Without ν	Including n-captures
^{235}U	202.7±0.1	192.9±0.5	201.7±0.6
^{238}U	205.9±0.3	193.9±0.8	205.0±0.9
^{239}Pu	207.2±0.3	198.5±0.8	210.0±0.9
^{241}Pu	210.6±0.3	200.3±0.8	212.4±1.0

Table 21.2 Transforming the thermal power into the fission rate (all energies in MeV/fission). Reprinted from M. F. James, "Energy released in fission," *J. Nucl. Energy*, vol. 23, no. 9, pp. 517–536, 1969, with permission from Elsevier.

The reactor neutrino flux as a function of the antineutrino energy $E_{\bar{\nu}}$ can be predicted as:

$$\phi^0(E_{\bar{\nu}}) = \frac{W_{\text{thermal}}}{\sum_i f_i e_i} \cdot \sum_i f_i \cdot S_i(E_{\bar{\nu}}) \tag{21.54}$$

where f_i and $S_i(E_{\bar{\nu}})$ are the fission fraction and the neutrino flux per fission for the ith isotope, respectively. The resulting antineutrino spectrum for ^{235}U fission is illustrated in Figure 21.6. The units are arbitrary. The spectrum is represented with a black line. It is rapidly falling with energy. Such a prediction was expected to carry an uncertainty of about 2–3% [255]. However, recently, reactor neutrino experiments [256–258] found a large discrepancy between the predicted and measured spectra in the 4–6 MeV region. Hence, an *ab-initio* calculation of the reactor neutrino flux with a percent precision is still lacking.

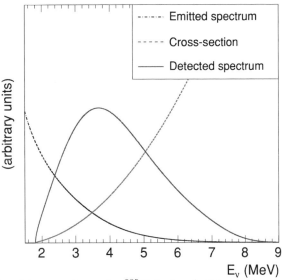

Figure 21.6 Detected antineutrino spectrum for ^{235}U fission (solid curve). Units are arbitrary. Reprinted with permission from T. A. Mueller *et al.*, "Improved predictions of reactor antineutrino spectra," *Phys. Rev. C*, vol. 83, p. 054615, May 2011. Copyright 2011 by the American Physical Society.

The reactor neutrino flux is isotropic, large, and proportional to the thermal power of the core W_{thermal}. The total flux integrated over energy and full solid angle is evaluated to be:

$$\phi_{\bar{\nu}}^0 \simeq (2 \times 10^{20}) \times \left(\frac{W_{\text{thermal}}}{\text{GW}}\right) \ \bar{\nu}/\text{s} \tag{21.55}$$

Since the flux is isotropic, it reduces as a function of the distance L between the source (the core of the nuclear reactor) and the detector as:

$$\phi_\nu = \frac{\phi_{\bar{\nu}}^0}{4\pi L^2} \simeq 1.6 \times 10^{15} \left(\frac{1}{L\,(\text{m})}\right)^2 \times \left(\frac{W_{\text{thermal}}}{\text{GW}}\right) \ \bar{\nu}/\text{cm}^2/\text{s} \tag{21.56}$$

where we have expressed the distance L in meters.

Reactor neutrinos are most practically detected via the inverse beta decay reaction (see Eq. (21.45)). The event spectrum can be computed with the product of the flux times the cross-section: $N(E_{\bar{\nu}}) = \phi(E_{\bar{\nu}}) \times \sigma(E_{\bar{\nu}})$. The detected rate rises from the threshold value at about 1.8 MeV, reaches a maximum around 4 MeV, and vanishes after 8 MeV. This shape, shown in Figure 21.6 as the bell-shaped curve, is the result of folding the emitted spectrum (dashed–dotted curve), and the inverse beta-decay cross-section (dashed curve).

To compute the total event rate R of reactor neutrinos in a detector, we use the thin target approximation:

$$R \equiv N_T \times \phi_{\bar{\nu}} \times \sigma \qquad (21.57)$$

where N_T is the number of target elements (free protons), ϕ is the neutrino flux ($\nu/\text{cm}^2/\text{s}$), and σ the cross-section (cm^2). If one assumes a 1-ton water (H_2O) detector at a distance L (m) from the core and considers interactions on free protons, the number of target protons per ton is given by:

$$N_T = (10^6\,\text{g}) \times 2 \times \left(\frac{1}{18}\,\frac{\text{mol}}{\text{g}}\right) \times (6 \times 10^{23}\,\text{mol}^{-1}) \simeq 6.7 \times 10^{28}\,\text{protons/ton} \qquad (21.58)$$

Hence the rate of IBD events R in a detector of mass m_{detector} in tons located at a distance L from a reactor with power W_{thermal} is simply:

$$
\begin{aligned}
R &= (6.7 \times 10^{28}\,\text{ton}^{-1})(m_{\text{detector}}) \times \left(1.6 \times 10^{15} \left(\frac{1}{L\,(\text{m})}\right)^2 \left(\frac{W_{\text{thermal}}}{\text{GW}}\right)\text{cm}^{-2}\text{s}^{-1}\right) \times 10^{-43}(\text{cm}^2) \\
&\approx 11 \left(\frac{W_{\text{thermal}}}{\text{GW}}\right)\left(\frac{m_{\text{detector}}}{\text{ton}}\right)\left(\frac{1}{L^2}\right)\text{s}^{-1} \qquad (21.59)
\end{aligned}
$$

For a distance $L = 100$ m, a thermal power $W_{\text{thermal}} = 10$ GW, and a detector target mass of $m = 1$ ton, the rate of inverse beta decay events is $R \approx 0.01$ s^{-1} or about 900 events per day. This is not a very large rate but is not negligible either, owing to the very strong source of neutrinos provided by the nuclear reactors. This simple calculation also points to an important fact: neutrino detectors need to be massive! The problem, then, is to identify and separate the few neutrino interactions with reasonable efficiency against other sources of backgrounds of reactor neutrons and gamma rays, natural radioactivity, and cosmic rays, all of which are abundant at the involved energies of a few mega-electronvolts.

21.7 First Direct Detection of Neutrinos: Cowan and Reines (1958)

The *direct* detection of the neutrino required specific knowledge on its interaction properties. This was the case after the advent of the Fermi theory and its successful prediction of the β-disintegration properties. The weak interaction followed the principle of the electromagnetic vector current with the difference that the electric charge changes by one unit in the weak interaction. Weak interactions hence involved charged currents. With this information, the detection of "the neutrino" postulated by Pauli had a clear path: produce an interaction with matter via its transformation into a positron.

During the first 20 years after Pauli's neutrino hypothesis, several discussions took place on the possibility of experimentally detecting a neutrino. It became clear that the IBD reaction was going to be the most promising reaction to detect neutrinos, since:

1. The cross-section could be accurately computed within the Fermi theory.

2. The small cross-section can be compensated by a high-flux neutrino source and massive neutrino detectors (see Eq. (21.59)).

3. The threshold was low enough to efficiently detect low-energy neutrinos coming from β decays, i.e., from nuclear reactors (see Figure 21.6).

4. Materials rich in free protons are cheap (e.g., water, hydrocarbon, liquid scintillators) and this permits one to build large detectors [259].

5. In scintillators, it is possible to tag both the e^+ and the neutron, reducing the backgrounds [259].

6. The measurable e^+ energy is strongly correlated with the neutrino energy.

Some experimental proposals were more practical than others, and the key issue was to produce the source of neutrinos strong enough to yield detectable interaction rates. A sufficiently powerful neutrino source would be needed in order to obtain a significant count rate above background. It is striking to recall that a proposal approved by the Los Alamos authorities consisted in exploding an atomic bomb in the air and detecting the neutrinos produced in a nearby detector. See Figure 21.7(left). More than 10^{24} neutrinos were expected to be produced during the explosion.[8]

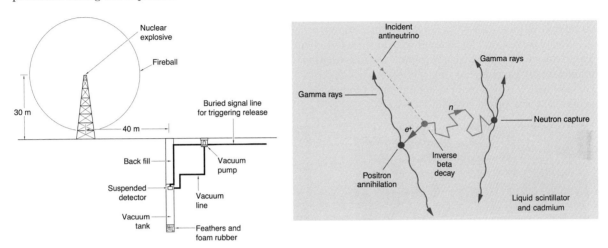

Figure 21.7 (left) Sketch of a proposal to detect neutrinos nearby an exploding nuclear atomic bomb. (right) Improved experimental signature proposed by Reines and Cowan to detect neutrinos at the nuclear power reactor. Reprinted from F. Reines and C. L. Cowan, *Reines-Cowan Experiments: Detecting the Poltergeist*, LA-UR-97-2534-02.

The atomic bomb proposal was soon superseded by the rapid developments of nuclear reactors, and by the idea to fully exploit the correlation between the positron and the neutron in the final state of the IBD reaction to strongly suppress non-neutrino-induced backgrounds coming from the environmental conditions. Fermi had been informed of the early attempts to detect free neutrinos with an atomic bomb. When he was told about the possibility of using a nuclear reactor, Fermi replied in a letter to Reines with the following comment: *"Certainly your new method should be much simpler to carry out and have the great advantage that the measurement can be repeated any number of times."*

In late 1951, **Reines**[9] **and Cowan**[10] began a project called "Project Poltergeist," the first experiment in neutrino physics [259]. The idea for detecting the neutrino was to detect both products of inverse beta decay [260]. See Figure 21.7(right). The incident antineutrino (dashed line) interacts with a proton with the emission of a positron and a neutron. The reaction takes place in a liquid scintillator. The short, solid arrow indicates that, shortly after it has been created, the positron annihilates with an electron, creating a back-to-back pair of gammas. This will cause the liquid scintillator to produce a flash of visible light. In the meantime, the neutron wanders about following a random path (longer, solid arrow) until it is captured by a nucleus. To enhance the capture signal, cadmium chloride is added to the liquid scintillator. Cadmium is a highly effective neutron absorber and gives off a gamma ray when it absorbs a neutron (see Section 2.21):

$$n + \mathrm{Cd} \rightarrow \ ^m\mathrm{Cd} \rightarrow \mathrm{Cd} + \gamma \ (9 \ \mathrm{MeV}) \tag{21.60}$$

---

8   The nuclear bombs used in World War II were equivalent to about 15 kton of TNT or about 60 TJ of energy. Modern nuclear weapons can carry up to 84000 TJ.

9   Frederick Reines (1918–1998), American physicist.

10   Clyde Lorrain Cowan Jr. (1919–1974), American physicist.

The arrangement was such that the neutron capture occurred with a lifetime of 5 μs after the signal from the positron annihilation. The neutron capture on Cd releases about 9 MeV of energy in gamma rays that will again cause the liquid to produce a flash of light. This sequence of two flashes of light separated by a few microseconds is the double signature of inverse beta decay.

In 1953 their first attempt at the **Hanford nuclear reactor** found first evidence for events consistent with inverse beta decays [261]. The Hanford nuclear site located in the state of Washington (USA) had been established in 1943 as part of the Manhattan Project. A photograph of the first detector built is shown in Figure 21.8. It had been called "Herr Auge" and with its 300 L capacity was at the time the largest apparatus ever built. The inner volume had a 10 ft^3 vat for the liquid scintillator doped with cadmium and 90 photomultiplier tubes, each with a 2-inch-diameter face and a thin, photosensitive surface.

The expected rate for delayed coincidences from neutrino-induced events was 0.1 to 0.3 counts per minute. The delayed-coincidence background, present whether or not the reactor was on, was about 5 counts per minute, many times higher than the expected signal rate. Cosmic rays with their neutron secondaries generated backgrounds 10 times more abundant than the neutrino signals. Although it saw hints of a signal, the Hanford experiment was inconclusive!

Figure 21.8 (top-left) "Herr Auge:" the first large liquid scintillator neutrino detector ever built; (top-right) view of the Hanford setup. Reprinted from F. Reines and C. L. Cowan, *Reines-Cowan Experiments: Detecting the Poltergeist* LA-UR-97-2534-02.

Three years later Reines and Cowan had redesigned and rebuilt an improved experiment [262]. The new experiment, based on the lessons learned at Hanford, was designed to further reduce sources of background. See Figure 21.9. The entire detector was surrounded by a lead-paraffin shield and located underground, nearby a reactor at the **Savannah River Plant**. The new detector was segmented in multiple layers (which they called "club-sandwich") where the functions of target and energy detection via scintillator were separated. The tanks, labeled Detector I, II, and III in Figure 21.9, contained 1400 L of liquid triethylbenzol (TEB), to which p-terphenyl and POPOP-wavelength shifter (see Section 3.11) were added to give the scintillation properties. The target tanks labeled A and B, which are located between tanks I, II, and III, contained each 200 L of water which acted as the proton target, and which contained about 40 kg of dissolved CdCl$_2$ to capture the neutrons. The electronic system was improved to record coincidences and graphically record the traces on oscilloscopes. The expected signatures of the prompt annihilation and the delayed neutron signal are illustrated in Figure 21.9. One event recorded as compatible with the neutrino signature is also shown.

With the new setup, the following results were obtained:

- **Rate vs. nuclear power.** The signal is shown to depend on the reactor power, and is in relatively good

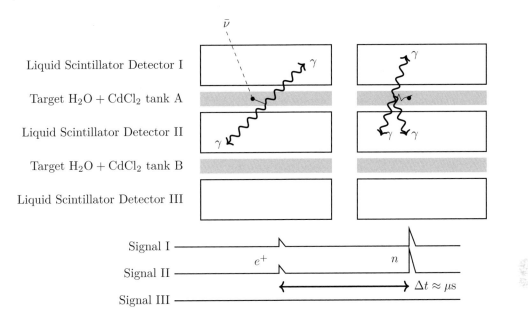

Figure 21.9 Improved setup at the Savannah River power plant with a separation between target (H_2O + $CdCl_2$) and detector consisting of liquid scintillator: (left) the two gammas from the annihilation of the positron are detected in the external tanks filled with liquid scintillator (I and II in the picture); (right) after diffusion in the target, the neutron is captured on Cd, which releases gammas that are also detected in the detectors.

agreement with the predicted IBD cross-section [263]; the signal-to-background from the reactor was 20/1; the signal-to-background from the environment was 3/1.

- **Number of target protons.** The dilution of the target water (H_2O) with heavy water (D_2O) halves the observed rate as expected.

- **Detection efficiency of prompt signal.** By dissolving β^+ radioactive sources in the target, the prompt signal was shown to be compatible with that of positron annihilation.

- **Neutron signal.** By removing the $CdCl_2$, the signal from delayed neutrons was totally suppressed.

- **External backgrounds.** The effectiveness of the external shield was tested with an external strong americium–beryllium source producing neutrons, and the signal was shown not to be caused by neutrons and gamma rays emitted from the reactor.

The reality of the neutrino had been proven!

21.8   Evidence for μ-Meson Decays to an Electron Plus Two Neutrals

In 1946 **Jack Steinberger**,[11] at the time a graduate student of Enrico Fermi, measured for the first time the spectrum of electrons emitted in muon decays. His PhD results showed that this was a three-body decay and implied the participation of two neutral particles in the decay rather than one [264]. As discussed in Section 15.4,

11 Hans Jakob "Jack" Steinberger (1921–2020), German–American physicist.

the difference between the muon and the pion was just being understood, and Steinberger interprets his observations as the consequence of the charged meson decaying into a neutral meson as: $\text{meson}^{\pm} \to \text{meson}^{0} + e^{\pm} + \nu$ ("the three-particle decay hypothesis"). Note that this measurement was performed before the direct detection of the neutrino by Reines and Cowan. In 1949 **Tionmo** and **Wheeler** calculated the electron spectrum treating all the particles involved in the decay process as spin-1/2 Dirac particles (in contrast to the Yukawa π-meson which had to be a spin-zero boson), and this led to the conclusion that the two neutral particles were two neutrinos participating in the reaction [265]:

$$\mu^{\pm} \to e^{\pm} + \nu + \bar{\nu} \tag{21.61}$$

21.9   Studies of Weak Interactions at High Energies

The direct detection of the neutrino by Reines and Cowan represented a major milestone for Fermi's theory of weak interactions, but it did not solve all neutrino puzzles, and in fact raised several questions.

The question of the **identity of the neutrino** was related to the fact that the neutrino is the only elementary fermion which is electrically neutral. There is therefore a priori no way to directly distinguish the neutrino from the antineutrino (in contrast to say the electron and the positron, which have opposite electric charges). In 1955 **Davis**[12] attempted to look for the following reaction at a reactor in Brookhaven [266] and later also at Savannah River:

$$\nu + {}^{37}\text{Cl} \overset{?}{\to} {}^{37}\text{Ar} + e^{-} \qquad \text{where } \nu \text{ from reactor} \tag{21.62}$$

hoping to detect a capture on neutrons. He did not find any evidence for such a reaction and set $\sigma < 0.9 \times 10^{-45}$ cm^2.

On the theoretical side it was known that the Fermi theory **violates unitarity at high energy**. Since the interaction was a four-fermion interaction, it diverged quickly at energy scales of the order of $(G_F)^{-1/2} \simeq 300$ GeV. A solution to this problem was to propose a $W^{\pm}$ **boson-mediated interaction**. This hypothesis should be tested with neutrinos of increasing energies, by searching for the production of a real $W^{\pm}$.

In the Fermi four-point theory, the amplitude of the loop shown in the top of Figure 21.10 is zero. However, **Feinberg**[13] showed that if the weak interaction is mediated by an intermediate boson $W^{\pm}$, then the process $\mu \to e + \gamma$ becomes possible under the assumption $\nu = \nu_e = \nu_\mu$ via the diagrams shown in the lower part of Figure 21.10, with a rate:

$$Br(\mu \to e\gamma) \approx 10^{-4} \tag{21.63}$$

The rate computed by Feinberg [267] is much larger than the experimental limit known at the time, set on electrons in muon capture [268]. However, this would not happen if neutrinos associated with muons are different than neutrinos associated with electrons.

T. D. Lee[14] and **C. N. Yang**[15] underlined the importance of experiments with high-energy neutrinos to solve these questions [269]. Several processes involving neutrinos were considered and the issue was to experimentally verify which reactions involved which type of neutrino or antineutrino:

$$
\begin{aligned}
\text{pion decay}: \quad & \pi^{+} \to \mu^{+} + \nu_{\mu} \\
\text{muon capture}: \quad & \mu^{-} + p \to n + \nu_{\mu} \\
\beta^{+} \text{ decay}: \quad & Z \to (Z-1) + e^{+} + \nu_{e} \\
\beta^{-} \text{ decay}: \quad & Z \to (Z+1) + e^{-} + \bar{\nu}_{e}
\end{aligned}
\tag{21.64}
$$

The study of the positron spectrum in $\mu^{+} \to e^{+}\nu\bar{\nu}$ favored a value for the Michel parameter ρ close to the 0.75 predicted for the case where the two neutrinos were of a different nature. On the other hand, Lee and Yang

---

12   Raymond (Ray) Davis Jr. (1914–2006), American chemist and physicist.
13   Gerald Feinberg (1933–1992), American physicist.
14   Tsung-Dao Lee (born 1926), Chinese–American physicist.
15   Chen-Ning Frank Yang (born 1922), Chinese physicist.

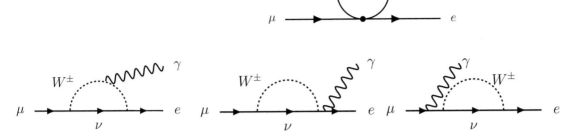

Figure 21.10 (upper) Feynman diagrams for $\mu - e$ coupling with the Fermi four-point contact term and (lower) diagrams contributing to $\mu \to e\gamma$ with an intermediate vector boson $W^\pm$ under the assumption of $\nu = \nu_e = \nu_\mu$.

stated in their paper that *"It is simplest to assume that they are also the same particle [$\nu_e = \nu_\mu$]. However, a test of this assumption is clearly desirable. To obtain such a test it is necessary to do some kind of capture experiment on the neutrinos or antineutrinos."* [16]

• **Conservation of lepton number.** B. Pontecorvo [270] also raised the question about the nature of the neutrinos produced in pairs with an electron or a muon. Are they the same or different particles? He proposed experiments to solve this question. He introduced the notation ν_e and ν_μ for the neutrinos (with corresponding antineutrinos $\bar{\nu}_e$ and $\bar{\nu}_\mu$) and assigned to each an **electron leptonic charge** L_e. If neutrinos and antineutrinos are different particles, then there must be a conserved charge (**the lepton number**) associated with these states:

$$(e^-, \nu_e) : L_e = +1; \qquad (e^+, \bar{\nu}_e) : L_e = -1 \qquad (21.65)$$

This charge corresponds to an internal degree of freedom of the neutrino. The individual sums must be conserved in all possible reactions. For instance, to the β^+ decay corresponds the inverse reaction:

$$\bar{\nu}_e + p \;\to\; e^+ + n$$
$$L_e = -1 + 0 \;\; = \;\; -1 + 0 \qquad (21.66)$$

This is the particle that was discovered by Reines and Cowan, which was conventionally assigned a negative leptonic charge and is labeled as an antineutrino. The charge-conjugated state has an opposite leptonic charge and is involved in the β^- decay:

$$\nu_e + n \;\to\; e^- + p$$
$$L_e = +1 + 0 \;\; = \;\; +1 + 0 \qquad (21.67)$$

The existence of a **leptonic charge** L_μ is associated with the muon neutrinos as follows:

$$(\mu^-, \nu_\mu) : L_\mu = +1; \qquad (\mu^+, \bar{\nu}_\mu) : L_\mu = -1 \qquad (21.68)$$

which can be checked by comparing the following reactions:

$$\bar{\nu}_\mu + p \;\to\; \mu^+ + n \qquad \text{and} \qquad \nu_\mu + p \;\overset{?}{\to}\; \mu^+ + n$$
$$L_\mu = -1 + 0 \;\; = \;\; -1 + 0 \qquad L_\mu = +1 + 0 \neq -1 + 0 \qquad (21.69)$$

16   Reprinted excerpt with permission from T. D. Lee and C. N. Yang, "Theoretical discussions on possible high-energy neutrino experiments," *Phys. Rev. Lett.*, vol. 4, pp. 307–311, Mar 1960. Copyright 1960 by the American Physical Society.

The following reactions are also never observed experimentally:

$$\begin{aligned}
\mu + p &\not\to e + p \\
\mu &\not\to e + e + e \\
\mu &\not\to e + \gamma
\end{aligned} \tag{21.70}$$

These strongly suppressed reactions did not seem to violate any conservation law known at the time. Following the principle stating that everything that is not prohibited will eventually happen, their non-observation implies a conservation law that needs to be experimentally tested, and supports the hypothesis of the conserved leptonic charges.

- **Existence of the neutral current.** The possible existence of a **neutral weak current** was pointed out. By using high-energy neutrinos, it would be possible to study whether such currents exist, for example, via the search for the reactions:

$$\nu + p \overset{?}{\to} \nu + p \quad \text{and} \quad \nu + n \overset{?}{\to} \nu + n \tag{21.71}$$

- **Universality of weak interactions involving $e^\pm$ and $\mu^\pm$.** The electron and the muon have identical electromagnetic interactions and seem to differ only by their mass. The strong suppression of the "β decay of the pion"

$$\pi^+ \overset{?}{\to} e^+ + \nu_e \tag{21.72}$$

compared to the rate of $\pi^+ \to \mu^+ + \nu_\mu$ was still a great puzzle at the time,[17] since the weak currents (e, ν) and (μ, ν) were believed to be identical and universal. The dominance of the muonic versus the electronic decay of the pion is described in Section 23.10. The decay modes of the charged pion have been summarized in Eq. (15.18).

- **Test of the four-point structure of the Fermi interaction.** The assumption that the lepton current acts only at a single space-time point introduces restrictions on the forms of the cross-sections for all neutrino and antineutrino reactions. In contrast, we have seen that the helicity conservation at the QED vertex had a strong impact on the cross-section. It is hence important to study precisely the kinematics of neutrino interactions. The Fermi theory which was known to work well at low energies implied that the cross-section increases with phase space, where all the details of the matrix element can be approximated by a constant. Dimensionally, this leads to:

$$\sigma \simeq \kappa G_F^2 s \tag{21.73}$$

where κ is a unitless constant that represents the point-like four-fermion interaction. In the laboratory frame, we can write the cross-section for a neutrino $p_1 = (E_\nu, 0, 0, E_\nu)$ hitting a target nucleon at rest $p_2 = (M_N, \vec{0})$, as:

$$\sigma \simeq \kappa G_F^2 (p_1 + p_2)^2 = \kappa G_F^2 \left(p_1^2 + p_2^2 + 2 p_1 \cdot p_2 \right) \approx 2\kappa G_F^2 (p_1 \cdot p_2) = 2\kappa G_F^2 M_N E_\nu \tag{21.74}$$

where E_ν is the incoming neutrino energy and $M_N = (M_p + M_n)/2$ is the target nucleon mass. **Consequently, for the point-like Fermi interaction, the cross-section increases linearly with the laboratory neutrino energy, thereby raising an issue of unitarity at high energy.** Inserting the value of the Fermi coupling constant

---

17 *"I think that everybody will agree that the absence of hyperon β-decay and the absence of pion β-decay are two great mysteries."* (Gell-Mann, 1958). During the same year, Feynman gave a lecture called "Forbidding of $\pi - \beta$ decay."

Eq. (21.14), we find:

$$
\begin{aligned}
\sigma \;\simeq\;& 2\kappa(1.167 \times 10^{-5}\ \mathrm{GeV}^{-2})^2(1\ \mathrm{GeV})E_\nu(\mathrm{GeV}) \approx 3\kappa \times 10^{-10}\left(\frac{E_\nu}{1\ \mathrm{GeV}}\right)\ \mathrm{GeV}^{-2} \\
=\;& \kappa \times 10^{-10}\left(\frac{E_\nu}{1\ \mathrm{GeV}}\right)\ \mathrm{mb} \qquad (1\ \mathrm{GeV}^{-2} = 0.389\ \mathrm{mb}\ \text{from Table 1.6}) \\
=\;& \kappa \times 10^{-37}\left(\frac{E_\nu}{1\ \mathrm{GeV}}\right)\ \mathrm{cm}^2 \qquad (1\ \mathrm{mb} = 10^{-27}\ \mathrm{cm}^2) \\
=\;& 10^{-38}\left(\frac{E_\nu}{1\ \mathrm{GeV}}\right)\left(\frac{\kappa}{0.1}\right)\ \mathrm{cm}^2
\end{aligned}
\tag{21.75}
$$

where in the last term, we have used the educated guess that $\kappa \simeq (2\pi)^{-1} \simeq 0.1$. The cross-section for neutrinos in the range of 1 GeV naively predicted by Fermi's theory should consequently be in the range $\mathcal{O}\left(10^{-38}\right)$ cm^2. Owing to the increased neutrino energy, this is much larger than the cross-section for MV neutrinos (e.g., from reactors), which was $\mathcal{O}\left(10^{-43}\right)$ cm^2 (see Eq. (21.49)). As a general rule, it was discovered that high-energy neutrinos are "easier" to detect since the cross-section increases linearly with neutrino energy.

• **Conserved vector current with the electromagnetic current (CVC).** In 1968 **Feynman** and **Gell-Mann** published a paper [271] where they proposed that the weak current possesses a "universal" $V - A$ nature. Furthermore, it is suggested that the vector part V of the current must be equal to the corresponding part of the electromagnetic current (times a constant). Hence, one can extract the nucleon form factor measured in the **electromagnetic elastic scattering** process:

$$
e^- + p \to e^- + p
\tag{21.76}
$$

to be used in the vector part of the weak process. Under this assumption, one can calculate the cross-section for the corresponding **weak quasi-elastic processes**:

$$
\nu + n \to p + e^- \qquad \text{and} \qquad \bar\nu + p \to n + e^+
\tag{21.77}
$$

They are called *quasi*-elastic processes or QE, since under the assumption that the difference of mass between the proton and the neutron can be neglected, they appear to be "elastic" (the nucleon remains itself). The result of the calculation is shown in Figure 21.11 as a function of the incoming neutrino energy. A striking observation is that the cross-section is:

$$
\sigma(\nu + n \to p + e^-) \approx \sigma(\bar\nu + p \to n + e^+) \simeq \mathcal{O}(10^{-38})\ \mathrm{cm}^2 \qquad \text{for } E_\nu \simeq 1\ \mathrm{GeV}
\tag{21.78}
$$

This value is several orders of magnitude larger than the IBD cross-section for MeV neutrinos (see Eq. (21.49)), hence, giga-electronvolt neutrinos interact much more than mega-electronvolt ones. A precise study of these reactions would allow one to compare them with existing data on electron–nucleon scattering at the same momentum transfer to the nucleus, thereby testing the nature of the weak interaction and the conserved vector current (CVC) hypothesis.

21.10   Accelerator Neutrinos

In 1959 **Schwartz**[18] from Columbia (NY, USA) published an experimental concept to use "high-energy neutrinos" to study weak interactions and address the crisis of unitarity of the cross-section [272]. The goal

---

18   Melvin Schwartz (1932–2006), American physicist.

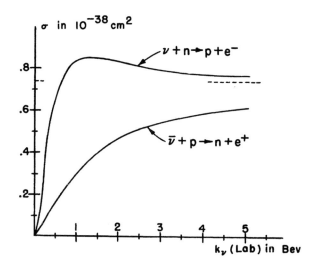

Figure 21.11 Elastic cross-sections under the hypothesis of $V - A$ coupling and CVC. "BeV" is the old nomenclature which is today replaced by "GeV." Reprinted with permission from T. D. Lee and C. N. Yang, "Theoretical discussions on possible high-energy neutrino experiments," *Phys. Rev. Lett.*, vol. 4, pp. 307–311, Mar 1960. Copyright 1960 by the American Physical Society.

was to reach neutrinos with energies in the giga-electronvolt range. The natural source of such neutrinos are high-energy positive pions, undergoing the decay:

$$\pi^+ \to \mu^+ \nu_\mu \tag{21.79}$$

and in accordance with the "absence of the β decay" of the pion $\pi^+ \to e^+ \nu_e$ (see Eq. (15.18)). Hence, high-energy protons were sent on a target where high-energy pions were produced and allowed to decay in a large volume. This produced forward-boosted neutrinos, which we call an "accelerator neutrino beam" (see Figure 21.12). Schwartz, intrigued by the first computation of the neutrino cross-section at giga-electronvolt energies, proposed this method to study the neutrino–nucleon scattering cross-section at high energies.

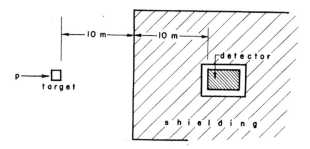

Figure 21.12 Proposed experimental arrangement for a high-energy neutrino experiment. Reprinted with permission from M. Schwartz, "Feasibility of using high-energy neutrinos to study the weak interactions," *Phys. Rev. Lett.*, vol. 4, pp. 306–307, Mar 1960. Copyright 1960 by the American Physical Society.

Following Schwartz's original derivation, we define as I_p the rate per second of high-energy protons ($E_p = 3$ GeV) impinging on the target. Let us assume that the rate of high-energy pions (above 2 GeV) is roughly

1/10 that of the protons, hence $I_\pi = I_p/10$, emitted in a forward cone of $\theta = 45°$ or a solid angle $\Omega_\pi = 2\pi(1 - \cos\theta) \approx 2$. The probability of pions decaying before hitting a shield at an assumed distance of $L = 10$ m is $\epsilon_{decay} \approx 10\%$. The detector is located at a distance $d = 20$ m. The solid angle in 1 cm^2 is then $\Omega_{det} = 1/d^2 \approx 0.25 \times 10^{-6}$ ster. The resulting neutrino flux per square centimeter per second at the detector is:

$$\phi \simeq I_\pi\, \epsilon_{decay} \frac{\Omega_{det}}{\Omega_\pi} \approx \left(\frac{I_p}{10}\right) \times (0.1) \times \left(\frac{0.25 \times 10^{-6}}{2}\right) \approx 10^{-9} I_p \; \nu/\text{cm}^2/\text{s} \tag{21.80}$$

where I_p is given in protons per second. The event rate in a detector of mass $m_{detector}$ given in tons is equal to:

$$R = (10^{-9} I_p) \times \left(\frac{m_{\text{detector}}}{1 \text{ ton}}\right) \times (10^6 \times 6 \times 10^{23}) \times \sigma \approx 6 \times 10^{20} \times \left(\frac{m_{\text{detector}}}{1 \text{ ton}}\right) \times I_p \times \sigma \; \text{s}^{-1} \tag{21.81}$$

where the cross-section for neutrino interactions should be given in square centimeters. For a detector of 10 tons and a cross-section $\sigma \simeq 10^{-38}$ cm^2 (Eq. (21.75)), we get:

$$R \approx \left(6 \times 10^{-17} \times I_p\right) \; \text{s}^{-1} \quad\Longrightarrow\quad I_p = \frac{R\,(\text{s}^{-1})}{6 \times 10^{-17}} \; \text{p/s} \tag{21.82}$$

In order to reach a rate of 1 neutrino event per hour in 10 tons, the rate of high-energy protons must be:

$$I_p = \frac{3 \times 10^{-4}\text{s}^{-1}}{6 \times 10^{-17}} = 5 \times 10^{12} \; \text{p/s} \tag{21.83}$$

In 1959 **Pontecorvo** independently proposed a similar neutrino experiment to check if the electron and muon neutrino are distinguishable particles [270]. He recalled that we already knew that the neutrino $\bar{\nu}$ produced in β^- disintegrations induces:

$$\bar{\nu} + p \rightarrow e^+ + n \tag{21.84}$$

So the question was:

$$\bar{\nu}_\mu + p \overset{?}{\rightarrow} \mu^+ + n \tag{21.85}$$

$$\bar{\nu}_\mu + p \overset{?}{\rightarrow} e^+ + n \tag{21.86}$$

where he underlined the fact that the $\bar{\nu}_\mu$ represents the neutral particle that is produced in association with the negative muon in the decay of the negative pion:

$$\pi^- \rightarrow \mu^- + \bar{\nu}_\mu \tag{21.87}$$

(Similarly $\pi^+ \rightarrow \mu^+ + \nu_\mu$ and $\nu_\mu + n \overset{?}{\rightarrow} \mu^- + p$.) Pontecorvo envisioned a stopped pion beam, although he mentioned briefly that neutrinos from decay-in-flight of pions have higher energy and hence a higher cross-section. Neutrinos produced by stopped pions are isotropic and in order to have enough flux, Pontecorvo concluded that a 1 GeV proton beam with an intensity of 10^{15} s^{-1} would have been necessary (such a source would only be available 30 years later!).

21.11   The Discovery of the Muon Neutrino (1962)

These ideas were actually implemented in the famous BNL-Columbia experiment at the Brookhaven AGS led by **Lederman, Schwartz,** and **Steinberger**. In 1962 the results from the first experiment at a high-energy neutrino beam produced at an accelerator were published, showing that $\nu_e \neq \nu_\mu$ [273]. The experimental setup is shown in Figure 21.13.

The basic principle of the experiment was the following:

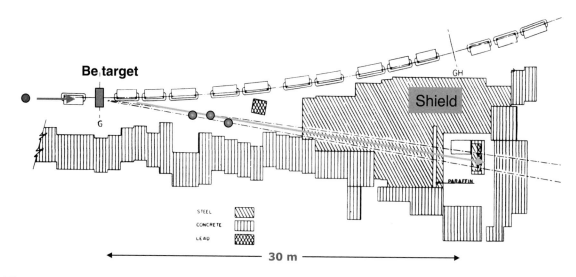

Figure 21.13 The experimental setup of the BNL-Columbia experiment. Reprinted with permission from G. Danby, J.-M. Gaillard, K. Goulianos, L. M. Lederman, N. Mistry, M. Schwartz, and J. Steinberger, "Observation of high-energy neutrino reactions and the existence of two kinds of neutrinos," *Phys. Rev. Lett.*, vol. 9, pp. 36–44, Jul 1962. Copyright 1962 by the American Physical Society.

- **High-energy protons.** The AGS accelerator provided protons with an energy of 15 GeV and an intensity of $(2-4) \times 10^{11}$ protons per pulse with a repetition rate of 3000 pulses per hour (repetition period 1.2 s). This yielded an average of $\approx (1\text{–}2) \times 10^{11}$ protons per second.

- **Proton extraction.** At the end of acceleration, a rapid beam deflector drove the protons onto an 8 cm-thick beryllium target over a period of 20–30 μs.

- **Decay of secondaries.** The flight path of the secondaries to the neutrino detector made an angle of 7.5° relative to the incoming direction of the protons.

- **Neutrino flux.** The expected neutrino flux from $\pi^{\pm} \to \mu^{\pm} + \nu(\bar{\nu})$ is shown in Figure 21.14. A smaller flux from $K^{\pm} \to \mu^{\pm} + \nu(\bar{\nu})$ is also expected at high energy. Most of the flux is coming as expected from pions and is below 1 GeV. Above 1 GeV there is a component coming from kaons which dominates at the highest end of the spectrum. Neutrinos from kaons are more energetic and are less frequent due to the kinematical suppression at the production vertex in the interaction of the protons on the target.

- **Hadron stop/shield.** The experimental challenge was to detect the neutrinos among the very high flux of other particles produced in the high-energy collisions. The high flux of secondary particles produced in the proton–Be interaction flew in the direction of the neutrino detector and reached a 13.5 m-thick iron shield after a distance of 21 m from the target. The mean free path of hadrons in the shield was 0.25 m. The total shield was designed to suppress hadrons by a factor 10^{24}. This shield was enough to range out muons up to an energy of 17 GeV: the proton energy was chosen to reduce as much as possible the rate of muons crossing the detector.

- **Neutrino detector.** A 10-ton neutrino detector located behind the shield would detect neutrino interactions.

Once the proton intensity seemed to be achievable, the next experimental challenge was to construct a detector of 10 tons that would be able to distinguish the interactions of electron neutrinos from muon neutrinos.

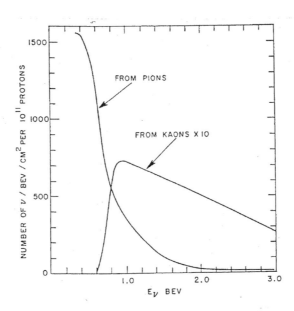

Figure 21.14 Expected high-energy neutrino flux from 15 GeV protons in the BNL experiment. Reprinted with permission from G. Danby, J.-M. Gaillard, K. Goulianos, L. M. Lederman, N. Mistry, M. Schwartz, and J. Steinberger, "Observation of high-energy neutrino reactions and the existence of two kinds of neutrinos," *Phys. Rev. Lett.*, vol. 9, pp. 36–44, Jul 1962. Copyright 1962 by the American Physical Society.

This condition is a general feature of neutrino detection, and is still an issue today in modern experiments. In general, we can say that the type of neutrino (the neutrino flavor) will be tagged via a charged-current interaction, which produces a charged lepton of the corresponding flavor: electron neutrinos produce leading electrons while muon neutrinos produce leading muons. A neutrino detector must hence be a massive device in order to collect a sufficiently large number of events but at the same time provide enough granularity to distinguish final-state particles. The neutrino detector plays two roles: (1) the neutrino target and (2) the detection device. In the BNL experiment, the detector (which was also the neutrino target) was made of individual modules of spark chambers separated by 9 mm-thick plexiglas plates. Each module weighed one ton and carried nine aluminum plates of $112 \times 112 \times 2.5$ cm^3. In total, the detector contained 10 modules or equivalently a target of 10 tons (see Figure 21.15).

In between each aluminum plate, a high-pressure spark gap was created such that a spark would be produced after the passage of a charged particle through the aluminum plate. When the passage of a particle was detected in a counter, a high voltage, producing a field of 1 kV/mm in the gap, was applied across two neighboring plates. This resulted in sparks being produced in the regions where particles had ionized the gas. The trigger was also used to take a photographic picture of the sparks. At the top and bottom, different counters were also acting as anti-coincidence shields.

A typical neutrino interaction is shown in Figure 21.16. In this picture we can observe the features of the events. A long straight track (a muon candidate) is recoiling against a few tracks (the remnant of the nucleon). The radiation length X_0 of aluminum is 8.9 cm, hence every aluminum plate corresponded to $0.3X_0$. While muons are penetrating, one expects electrons to produce a rather short shower. Hence, the rather good granularity of the detector (which allowed one to see individual hadrons produced in the event) provided the identification of the event.

The BNL experiment ran for an exposure of 3.48×10^{17} protons on target (300 h running). Under normal operation the event rate was 10 s^{-1}. The events were recorded on photographic film. A total of 113 triggers

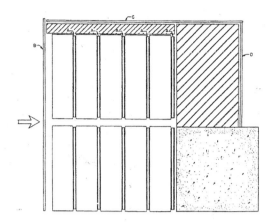 

Figure 21.15 Neutrino detector with spark chambers and counters. (left) Schematic view; A is the readout counters and B, C, D the anti-coincidence counters. Reprinted with permission from G. Danby, J.-M. Gaillard, K. Goulianos, L. M. Lederman, N. Mistry, M. Schwartz, and J. Steinberger, "Observation of high-energy neutrino reactions and the existence of two kinds of neutrinos," *Phys. Rev. Lett.*, vol. 9, pp. 36–44, Jul 1962. Copyright 1962 by the American Physical Society. (right) A photograph of the experimental setup (J.-M. Gaillard, private communication).

satisfied the selection criteria for neutrino events occurring within the fiducial volume of the spark chamber (i.e., the neutrino interacted with the Al plates). The events had to satisfy the following criteria to be accepted: (1) the origin of the event had to be within the detector; (2) the first two chambers were used as veto and did not have to detect any track; (3) if only one track was observed, it had to point back to the neutrino source; (4) the secondary tracks had to have an angle less than 60° with respect to the neutrino flight direction. After further rejection of low-energy events not easily identifiable (49 events with a short track), the remaining events were classified as such:

- 34 single-track events, consistent with a muon of momentum greater than 300 MeV – the tracks were identified as muons if their range was larger than the typical interaction length of hadrons in aluminum, which is about 40 cm. Hence, muons are penetrating tracks (see Figure 21.17);

- 22 multi-track events (see Figure 21.18);

- 8 showers, consistent with electromagnetic showers.

It was checked whether these events were really neutrino interactions. In particular that:

1. **They were not induced by neutrons.** The event vertices were uniformly distributed over the detector volume, while the interaction length in aluminum is 40 cm and for electromagnetic particles only 8 cm. No change of rate when 1.2 m of the shielding was removed. If the events were due to escaping neutrons, then the observed rate of events would have increased dramatically by the removal of shielding. Neutron interactions are expected to produce charged and neutral pions in the ratio of approximately 2 : 1, hence shower-like events would have been observed more frequently.

2. **They were not cosmic rays.** Data taken with the AGS accelerator off led to an expectation of 5 ± 1 cosmic-ray induced events after cuts and could not explain all observed events.

3. **They came from decays of hadrons in the decay tunnel.** Adding 1.2 m of shielding close to the target to absorb secondary hadrons before they decayed, reduced the rate of observed events, as expected. The

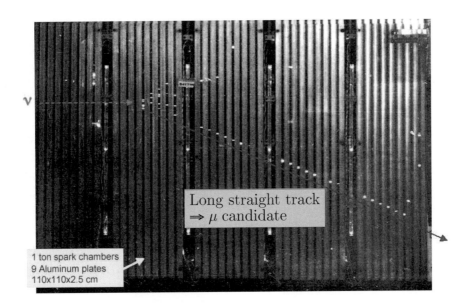

Figure 21.16 A typical neutrino interaction at BNL with a long penetrating track (J.-M. Gaillard, private communication).

pion decay probability was reduced by a factor of eight and the observed event rate was reduced from 1.46 ± 0.2 to 0.3 ± 0.2 per 10^{16} incident protons.

The events classified as showers were compared to the electromagnetic showers produced by 400 MeV electrons in a test beam. Only two of the neutrino events resembled the electrons from the test beam. This number of events was consistent with the rate expected from the decay $K^+ \rightarrow e^+ + \nu_e + \pi^0$ and $K^0 \rightarrow e^- + \bar{\nu}_e + \pi^+$ or $K^0 \rightarrow e^+ + \nu_e + \pi^-$.

If $\nu_e = \nu_\mu$, there should have been of the order of 30 electron showers consistent with the 34 single muon events. On the contrary, the few showers not consistent with electrons, except maybe two, do not support this hypothesis. Hence, the authors concluded that the most plausible explanation for the absence of the electron showers is that the electron and muon neutrinos are different particles $\nu_e \neq \nu_\mu$. They had discovered and detected the muon neutrino!

21.12   The Discovery of the Tau Lepton (1975)

The accelerator experiment at BNL had demonstrated that neutrinos come with an intrinsic property which we call **neutrino flavor**: $\nu_e \neq \nu_\mu$. In 1976 **Perl**[19] and collaborators discovered the third family of charged leptons, the tau, through the study of e^+e^- collisions at the SPEAR accelerator. They used the Mark I detector (see Figure 20.9(left) and Figure 21.19(a)) with augmented muon towers. The center-of-mass energy ranged from 3 to 8 GeV and an "anomalous lepton production" was observed above a certain threshold [274]:

$$e^+e^- \rightarrow l^+l'^- \quad + \quad \text{missing energy} \tag{21.88}$$

An $e - \mu$ event is shown in Figure 21.19. These events were interpreted as the pair production of the heavy

19   Martin Lewis Perl (1927–2014), American physicist.

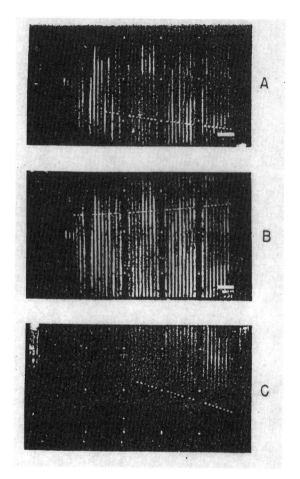

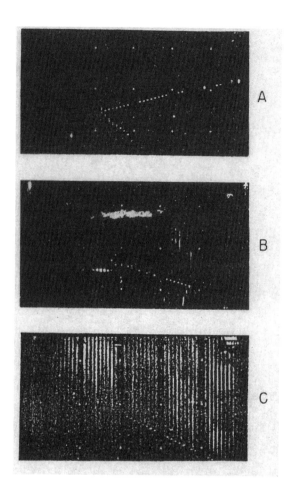

Figure 21.17 Single muon events: (A) $p_\mu >$ 540 MeV/c (lower track) and δ-ray (upper track) that indicated the direction of the track (neutrinos from left); (B) $p_\mu > 700$ MeV/c; (C) $p_\mu > 440$ MeV/c with δ-ray (J.-M. Gaillard, private communication).

Figure 21.18 Events with vertex: (A) single muon $p_\mu > 500$ MeV/c and recoiling activity; (B) two tracks leaving the detector; (C) four tracks with a long track with $p_\mu > 600$ MeV/c (J.-M. Gaillard, private communication).

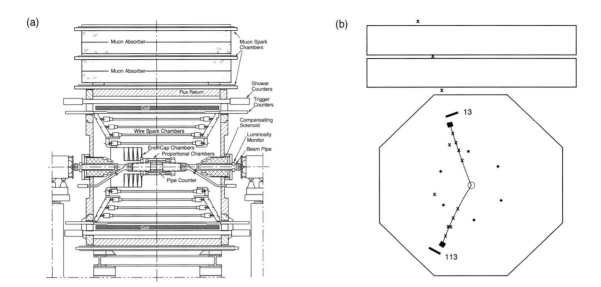

Figure 21.19 (left) side-view of the Mark I detector with muon towers; (right) an $e - \mu$ events observed in e^+e^- collisions. Reprinted by permission from Springer Nature: M. L. Perl, "The Discovery of the tau lepton and the changes in elementary particle physics in forty years," *Phys. Perspect.*, vol. 6, pp. 401–427, 2004. Copyright 2004 Birkhäuser Verlag, Basel.

lepton τ. The letter τ is from Greek "$\tau\rho\iota\tau o\nu$" for thirdly – the third charged lepton. With sufficiently high energy of the beams, the tau leptons were pair produced via the s-channel single-photon exchange reaction:

$$e^+e^- \to \gamma^* \to \tau^+\tau^- \tag{21.89}$$

The cross-section for the process can readily be derived from the equivalent $e^+e^- \to \gamma^* \to \mu^+\mu^-$. The spin-averaged matrix element for $e^-(p,s) + e^+(k,r) \to \ell^-(p',s') + \ell^+(k',r')$ was given in Eq. (11.121), so setting $\ell = \tau$, we find:

$$\langle |\mathcal{M}_{e^+e^- \to \tau^+\tau^-}|^2 \rangle = \frac{8(4\pi\alpha)^2}{s^2} \left\{ (p \cdot p')(k \cdot k') + (k \cdot p')(p \cdot k') + m_\tau^2 (k \cdot p) \right\} \tag{21.90}$$

We are interested in the differential cross-section in the center-of-mass system. Neglecting the rest mass of the electrons, but keeping those of the taus, we find (see Eq. (11.143)):

$$p \cdot p' = k \cdot k' = E^2 \left(1 - \beta \cos\theta\right)$$
$$p \cdot k' = p' \cdot k = E^2 \left(1 + \beta \cos\theta\right)$$
$$k \cdot p = 2E^2 \tag{21.91}$$

where E is the energy and β the velocity parameter of the τ^- or τ^+, and θ is the scattering angle. The differential cross-section is then (the first β factor is reminiscent of the p_f/p_i kinematical term):

$$
\begin{aligned}
\mathrm{d}\sigma &= \frac{\beta}{64\pi^2 s} \frac{8(4\pi\alpha)^2}{s^2} \left\{ E^4 \left(1 - \beta\cos\theta\right)^2 + E^4 \left(1 + \beta\cos\theta\right)^2 + 2m_\tau^2 E^2 \right\} \mathrm{d}\Omega \\
&= \frac{\beta}{64\pi^2 s} \frac{8(4\pi\alpha)^2}{s^2} 2E^2 \left\{ E^2 \left(1 + \beta^2 \cos^2\theta\right) + m_\tau^2 \right\} \mathrm{d}\Omega
\end{aligned} \tag{21.92}
$$

We can write:

$$E^2\left(1 + \beta^2\cos^2\theta\right) + m_\tau^2 = E^2 + E^2\beta^2(1 - \sin^2\theta) + m_\tau^2 = E^2 - E^2\beta^2\sin^2\theta + \underbrace{E^2\beta^2 + m_\tau^2}_{=E^2} = E^2(2 - \beta^2\sin^2\theta)$$

(21.93)

Hence, recalling that $s = 4E^2$, we find:

$$d\sigma = \frac{\beta}{64\pi^2 s}\frac{16(4\pi\alpha)^2}{s^2}\frac{s^2}{16}\left\{2 - \beta^2\sin^2\theta\right\}d\Omega = \frac{\alpha^2}{4s}\beta\left\{2 - \beta^2\sin^2\theta\right\}d\Omega$$

(21.94)

Integrating over the solid angle yields the total cross-section:

$$\sigma(e^+e^- \to \tau^+\tau^-) = \left(\frac{4\pi\alpha^2}{3s}\right)\beta\left\{\frac{3 - \beta^2}{2}\right\} = \beta\left\{\frac{3 - \beta^2}{2}\right\} \times \sigma(e^+e^- \to \mu^+\mu^-)$$

(21.95)

In the last term we compare the tau cross-section to that for muon production, Eq. (13.6) (when the energy is much larger than the muon mass). One defines the useful ratio R_τ:

$$R_{\tau\tau} \equiv \frac{\sigma(e^+e^- \to \tau^+\tau^-)}{\sigma_{\mu\mu}^{QED}} = \beta\left\{\frac{3 - \beta^2}{2}\right\} \longrightarrow 1 \qquad \text{when } \beta \to 1$$

(21.96)

The value $R_{\tau\tau}$ near the kinematical threshold can be used to derive the spin of the tau. Figure 21.20 shows the experimental measurements compared to theoretical curves under the assumption of different spins. The dashed curve with $s = 1/2$ fits perfectly the data, providing direct evidence that the tau is a spin $1/2$ Dirac fermion, analogous to the electron and muon.

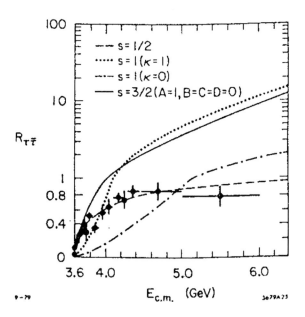

Figure 21.20 Measurements of the $R_{\tau\tau}$ ratio measured in $e^+e^- \to \tau^+\tau^-$ as a function of the center-of-mass energy. See text for definition. Reprinted from J. Kirkby, "Review of e^+e^- reactions in the energy range 3-GeV to 9-GeV," in *Proceedings: International Symposium on Lepton and Photon Interactions at High Energies, Batavia, Ill., Aug 23–29, 1979.*

• **Tau lepton properties and leptonic charge.** The tau lepton is very short-lived and rapidly decays in different modes. The measured properties of the tau are (values taken from the Particle Data Group's Review [5]):

$$\boxed{\tau^+, \tau^-} : \quad \begin{aligned} &m_{\tau^\pm} = (1776.86 \pm 0.12) \text{ MeV} \\ &\tau = (2.903 \pm 0.05) \times 10^{-13} \text{ s} \qquad (c\tau \simeq 86.93 \text{ } \mu m) \\ &Br(\tau^- \to \mu^- + \bar{\nu}_\mu + \nu_\tau) = (17.39 \pm 0.04)\% \\ &Br(\tau^- \to e^- + \bar{\nu}_e + \nu_\tau) = (17.82 \pm 0.04)\% \\ &Br(\tau^- \to h^- + \nu_\tau) = (11.51 \pm 0.05)\% \\ &Br(\tau^- \to h^- + \geq 1h^0 + \nu_\tau) = (37.00 \pm 0.09)\% \\ &Br(\tau^- \to h^- h^- h^+ + \geq 0h^0 + \nu_\tau) = (15.21 \pm 0.06)\% \end{aligned}$$

(21.97)

where h^- (h^+) stands for a negatively (positively) charged hadron. Here, a third independent **leptonic charge** L_τ **associated with the tau** was introduced as follows:

$$(\tau^-, \nu_\tau) : L_\tau = +1; \qquad (\tau^+, \bar{\nu}_\tau) : L_\tau = -1$$

(21.98)

Consequently, there are three types of neutrinos:

$$\nu_e \neq \nu_\mu \neq \nu_\tau$$

(21.99)

For a long-time, the ν_τ was known to exist through indirect evidence: (1) from the overall consistency of the theory and (2) because of the kinematical properties of the tau decays. The best indirect evidence came from the precision measurements of the Z^0-lineshape at LEP which yielded the number of ordinary light neutrinos $N_\nu = 3$ with high precision (see Section 27.10). As discussed in the next section, the direct detection of the tau neutrino has now been achieved.

• **Tau lifetime.** The weak tau decay width at tree level will be computed in Section 23.11. The measurement of the tau lifetime was best performed in high-energy $e^+ + e^- \to \tau^+ + \tau^-$ events at LEP and SLC. The produced taus are nearly mono-energetic due to small radiation losses and the back-to-back event topologies of the taus can be reconstructed with high efficiency and low background. The decay length, even with the boost, is still relatively short, $\beta\gamma c\tau \approx 2.2$ mm (it is smaller than the radius of the beampipe!). The measurement of the tau lifetime with a precision at the percent level was achieved due to the following factors that played favorably. (a) The beam spot was very small: at LEP $\sigma_x \approx 150$ μm and $\sigma_y \approx 10$ μm; at SLC $\sigma_x \approx \sigma_y \approx 2.5$ μm. So, the production point of the tau was well defined. (b) Experiments at LEP and SLC have very precise solid-state detectors just around the beam pipe, which provided tracking with very high spatial resolution in the region of interest, typically with precisions at the level of 10–25 μm. (c) The number of taus produced was large, about 100,000 $\tau^+\tau^-$ for each LEP experiment. Many methods are available to measure the τ lifetime. Initially, only three-prong vertexing was used (see Figure 21.21), with its direct access to the flight distance from the interaction point. Later, it became useful to consider the more numerous one-prong decays through the impact parameter δ approach (see Figure 21.21). Finally, a three-dimensional impact parameter (3D-IP) method was used. For illustration purposes, the results from the L3 experiment at LEP are shown in Figure 21.22. For the impact parameter method, only one-prong tau decays were used, while only three-prong tau decays were selected for the decay length method. The hatched areas represent the distributions of background events carrying no lifetime information. Its width is determined by the spatial resolution of the track reconstruction. In both cases, one sees that the data has a clear peak at zero with an asymmetric tail towards larger impact parameters and longer decay lengths, which allow the extraction of the tau lifetime. The other LEP experiments found similar distributions. More recently, the most precise tau lifetime measurement has been obtained by Belle (see Section 20.10), based on ≈ 1.1 million events. Based on similar vertexing techniques as at LEP, but owing to the very large event sample, they find [275]:

$$\tau_\tau = (290.17 \pm 0.53(\text{stat.}) \pm 0.33(\text{syst.})) \times 10^{-15} \text{ s}$$

(21.100)

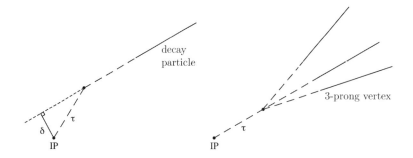

Figure 21.21 Measurement of the τ lifetime through two different approaches: (left) the impact parameter and (right) the decay length.

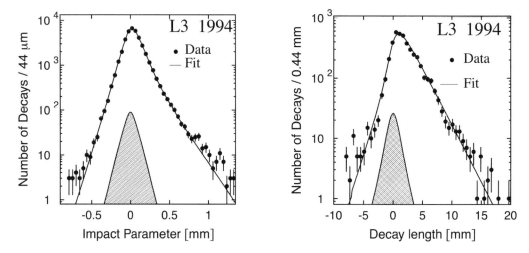

Figure 21.22 Impact parameter and decay length distributions obtained by the L3 experiment at LEP. Reprinted from M. Acciarri *et al.*, "Measurement of the lifetime of the τ lepton," *Phys. Lett. B*, vol. 479, pp. 67–78, 2000, with permission from Elsevier.

21.13   The First Direct Detection of the Tau Neutrino

The direct interaction of ν_τ was first achieved by the **DONUT (Direct Observation of NU Tau) experiment** in 2000 [276]. DONUT was designed to produce and identify tau neutrinos and to search for non-Standard Model interactions. A high-energy neutrino beam containing tau neutrinos was directed onto a neutrino interaction target in which the interactions of tau neutrinos and nuclei were recorded. The experimental layout is shown in Figure 21.23.

The identification of these interactions was based on the detection of the tau lepton and the high-momentum hadrons produced in the charged-current interaction of a neutrino and a nucleon (see Chapter 28):

$$CC: \quad \nu_\tau + \mathcal{N} \longrightarrow \quad \tau^- + X, \qquad \bar{\nu}_\tau + \mathcal{N} \longrightarrow \quad \tau^+ + X$$
$$\phantom{CC: \quad \nu_\tau + \mathcal{N} \longrightarrow \quad} \hookrightarrow Y + \nu_\tau \qquad\qquad\qquad \hookrightarrow Y + \bar{\nu}_\tau \qquad (21.101)$$

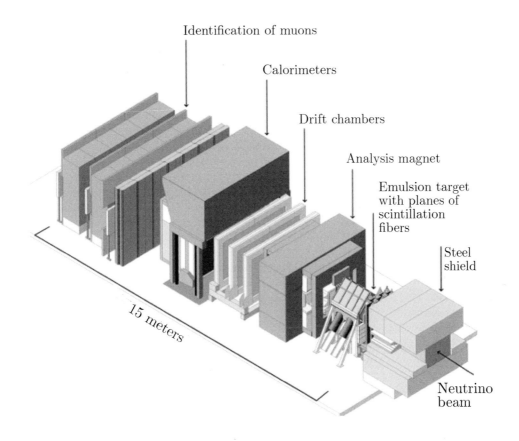

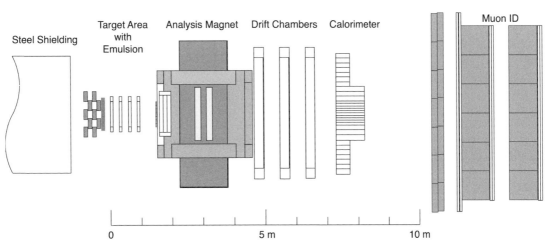

Figure 21.23 Schematic view of the DONUT experiment: (top) 3D view of the DONUT experimental setup; (bottom) side view. Reprinted with permission from K. Kodama *et al.*, "Final tau neutrino results from the DONUT experiment," *Phys. Rev.*, vol. D78, p. 052002, 2008. Copyright 2008 by the American Physical Society.

Since the tau leptons produced in the charged-current interactions at the typical energy of 50 GeV of DONUT had path lengths of approximately 2.5 mm, a high-resolution detector was required. These short tau tracks were recorded with sheets of photographic emulsion interleaved with thin steel plates. Photographic emulsions were employed as an integrating detector similarly to photographic film. We have already discussed how photographic emulsions were successfully employed in the study of light and strange mesons (see Chapter 15). Charged particles passing through an emulsion sheet initiate a chemical process that modifies the molecular structure. This modification becomes visible after the emulsion is developed.

About 80% of the taus decay to one charged particle and several neutral particles, hence the tau itself and its decay can be identified in the emulsion target as a short track and a kink produced in its decay. Hadrons and other leptons produced in neutrino–nucleon interactions were recorded with a conventional spectrometer, which was used to identify the neutrino interaction type, measure event parameters, and determine the location of the neutrino interaction vertex in the emulsion.

The primary source of high-energy particles was the Fermilab Tevatron, a superconducting synchrotron that accelerated protons to a maximum energy of 800 GeV. To produce the neutrino beam for the experiment, 800 GeV protons extracted from the Tevatron were steered onto a tungsten target, where the protons collided with target nucleons and generated many different final-state particles. Neutrinos from charm meson decays formed the prompt component of the neutrino beam. Most light long-lived particles such as pions and kaons interacted and lost energy in the target material before decaying. The decays of these particles produced a background of low-momentum non-prompt neutrinos. It was expected that 10^{17} protons on the target should give about 400 neutrino charged-current interactions in the target, of which about 5% should be charged-current interactions of taus.

Since the neutrino interaction targets contained emulsion, which records every charged particle track passing through it, the total number of charged particles (mostly muons) in the neutrino target region had to be kept below about $10^5/\text{cm}^2$. Active shielding consisting of dipole magnets was used to deflect the high-momentum muons away from the neutrino target. The first magnet downstream of the proton beam target was a 7 m-long aperture-less dipole operating in saturation at a field of 3.0 T. The second magnet was a 5 m-long toroidal magnet operating at a field of 2.1 T. Together, the two magnets divided the beam of high-energy muons into two "plumes" separated by about 2 m at the emulsion target.

The DONUT target emulsions were placed on 0.5 m by 0.5 m large plastic sheets to provide mechanical strength and improve the ease of handling. Two types of emulsion configurations were explored (see Figure 21.24):

- **Bulk.** The first type had a 90 µm-thick plastic base coated with 330 µm-thick emulsion. Many sheets were stacked to form 70 mm-thick emulsion modules as shown in Figure 21.24. Nuclear emulsion constituted 95% of the mass of a bulk module. Emulsion used in this configuration is a volume tracking detector; the vertex and all of the charged particle tracks from a neutrino interaction are recorded in the emulsion.

- **ECC (emulsion cloud chamber).** The second type had 1 mm-thick iron (or steel) plates interleaved with emulsion sheets. An ECC module is shown on the right-hand side of Figure 21.24. The emulsion sheet had a 200 µm-thick plastic base with coatings of 100 µm emulsion on each side. Since emulsion contributes only 5% to the mass of a module, an ECC module is a sampling detector with very high resolution perpendicular to the beam direction. The vertex of a neutrino interaction will most likely not be visible in the emulsion, 95% of the time it will occur in the iron. The charged particle tracks are visible only as track segments in each emulsion layer.

A total of 3.6×10^{17} protons were sent on target. From 6.6 million triggers, about 1000 candidate events in the emulsions were selected for further analysis. A total of 578 neutrino interactions were located in the emulsions and reconstructed. Out of those, nine were found compatible with a ν_τ CC interaction. The resulting cross-section was found to be:

$$\sigma(\nu_\tau)/E_\nu = (0.72 \pm 0.24(\text{stat.}) \pm 0.36(\text{syst.})) \times 10^{-38} \text{ cm}^2 \text{ GeV}^{-1} \tag{21.102}$$

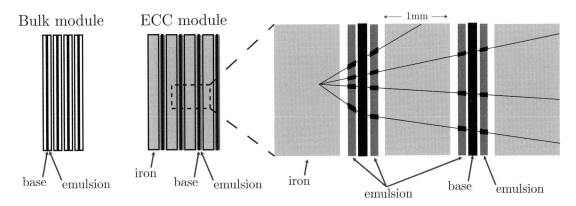

Figure 21.24 Types of emulsions used in DONUT: (left) bulk and ECC (right) details of the ECC configuration. The charged particle tracks are visible only as track segments in each emulsion layer.

in agreement with theoretical expectations within the experimental uncertainties [277]. **This result can be considered as the discovery of the ν_τ neutrino.** Two tau candidate events that survived all cuts are reported in Figure 21.25. We note that the antineutrino-tau ($\bar{\nu}_\tau$) is the only particle of the Standard Model that has not been directly detected, but whose existence is inferred from the detection of the ν_τ!

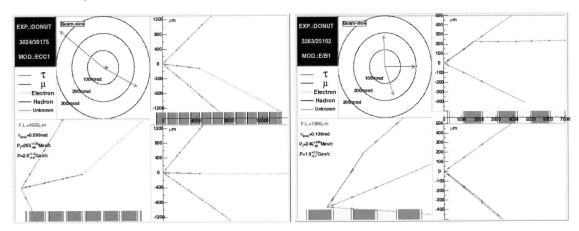

Figure 21.25 Two events compatible with ν_τ charged-current interactions found in the emulsions of DONUT.

Problems

Ex 21.1 Inverse β decay and the detection of neutrinos. Consider the decay of a free neutron at rest (mean life $\tau_n \approx 880\,\mathrm{s}$).

(a) What is the maximal energy of the antineutrino?

(b) Why does a free proton not decay by weak interaction?

The inverse reaction was used in 1956 by Cowan and Reines to detect antineutrinos from a nuclear reactor for the first time.

(c) Write down the reaction and explain how the final-state particles were detected.

(d) What is the energy threshold for $\bar{\nu}_e$ to undergo "inverse β-decay" with a proton at rest?

(e) Following the original reasoning of Bethe and Peierls, *Nature* **133**, 689–690 (5 May 1934), make an order-of-magnitude estimate of the interaction cross-section for this process, using the neutron lifetime.

Ex 21.2 Event rate in a 1 GeV ν_μ beam. The muon neutrino ν_μ was discovered in 1962 at the Brookhaven National Laboratory. A 15 GeV proton beam was hitting a Be target, producing mainly high-energy pions (and some kaons). From the decay $\pi^+ \to \mu^+ + \nu_\mu$, a high-energy ν_μ beam was obtained. With a detector consisting of 10 modules of 1 ton each built from aluminum plates acting as spark chambers, it was shown that in charged-current (CC) neutrino–nucleon interactions muons were produced, but not electrons:

$$\nu_\mu + N \to \mu^- + X \tag{21.103}$$

where N is a nucleon and X represents the hadronic remnants (for instance a proton in the case of quasi-elastic interactions on neutrons). The neutrino flux at the detector (perpendicular to the detector of cross-section A) was $\phi_\nu = 5000/(\text{cm}^2\,\text{s})$, the cross-section for neutrino–nucleon scattering in the relevant energy range can be approximated to $\sigma_{\nu N} \approx 10^{-38}$ cm^2, and the Al density is $\rho_{Al} = 2.7$ g/cm^3.

Calculate the event rate R in the Al detector for the quasi-elastic νN interactions.

Ex 21.3 Decay distribution of tau decay products. Tau leptons are produced in the process $e^+e^- \to \tau^+\tau^-$ at a center-of-mass energy of 91.2 GeV. The angular distribution of the π^- from the decay $\tau^- \to \pi^-\nu_\tau$ is:

$$\frac{\mathrm{d}N}{\mathrm{d}\cos\theta^\star} \propto 1 + \cos\theta^\star \tag{21.104}$$

where $\theta^\star$ is the polar angle of the π^- in the tau lepton rest frame, relative to the direction defined by the tau spin. Determine the laboratory frame energy distribution of the π^- for the following cases: (a) the tau spin is aligned with its flight direction; (b) the tau spin is opposite its flight direction.

22 Parity Violation in Weak Interactions

What is very striking, from the presented data on asymmetry and polarization effects, is the remarkable kindness of Nature to the experimental physicists: all these effects have their maximum possible value. To be more precise and technical, it seems that the violation of C and P is maximum.

Louis Michel [1]

22.1 Parity Violation in Weak Decays

During the 1940s and 1950s, the studies continued on the β decays. It was found that not all β decays occur between nuclear states with identical angular momenta, so the **Fermi allowed transitions** defined in Section 21.3, which represent a $\Delta J = 0$ operator (see Eq. (21.31)), could *not* be a complete description. For instance, a known $\Delta J = 0$ decay is:

$$0^+ \to 0^+ : {}^{14}\mathrm{O} \to {}^{14}\mathrm{N}^* + \beta^+ + \nu_\mathrm{e} \tag{22.1}$$

while a known $\Delta J = 1$ transition is:

$$0^+ \to 1^+ : {}^6\mathrm{He} \to {}^6\mathrm{Li} + \beta^- + \bar{\nu}_\mathrm{e} \tag{22.2}$$

Note that both conserve the parity of the initial and final states.

In the approximation of the allowed transitions, the integral of the wave functions of the electron and neutrino were replaced by a constant (see Section 21.3), and we derived a result which depended only on phase space. On the other hand, the change of the angular momentum in the nucleus can result from the spins of the electrons and the neutrino, each of which carries $s = 1/2$. The $e - \nu$ pair can be in the singlet state with a total $S = 0$ (antiparallel spins), or in the triplet state with $S = 1$ (parallel spins). In addition, the pair can also be emitted with its own angular momentum ℓ. Then the **Fermi allowed transitions** correspond to the $e - \nu$ pair in an $\ell = 0$, $S = 0$ state. As a result, there can be no change in the angular momentum of the nucleus, consistent with the Fermi selection rule of Eq. (21.31). The transitions in which the $e - \nu$ pair is emitted in an $S = 1$ state are called **Gamow**[2]**–Teller**[3] **transitions**. In the allowed approximation, the pair has no relative angular momentum $\ell = 0$, hence the parity of the initial and final states must be unchanged, since a state with an orbital angular momentum ℓ is $(-1)^\ell$ (see Section 4.9). This leads to the **Gamow–Teller (GT) selection rule for allowed β decays:**

$$\text{Gamow–Teller:} \quad \Delta J = 0, 1 \,(\text{not}\, 0^+ \to 0^+) \quad \text{parity} : + \to + \quad \text{or} \quad - \to - \tag{22.3}$$

Decays can be either a Fermi or a Gamov–Teller transition or a mixture of both. For example, we note that the decay of a free neutron leads to the $1/2^+ \to 1/2^+$ transition, hence $\Delta J = 0$, which satisfies both Fermi

---

1   Reprinted from "Weak interactions: leptonic modes – Theoretical I," in *8th Annual International Conference on High Energy Physics, CERN, Geneva, Switzerland,* 30 Jun–5 Jul, 1958, pp.251–255.
2   George Gamow (1904–1968), born Georgiy Antonovich Gamow, Russian–American theoretical physicist and cosmologist.
3   Edward Teller (1908–2003), Hungarian–American theoretical physicist.

and GT selection rules. This is an example of "mixed" F and GT transitions, in which the exact proportions of F and GT are to be determined experimentally or calculated from the matrix element of the initial and final wave functions. The forbidden decays are classified according to the angular momentum carried off by the $e - \nu$ pair. In an nth forbidden decay, the $e - \nu$ pair takes away n units of angular momentum ℓ. For $\ell = 1, 3, 5, \ldots$ the parity of the nuclear state changes. When $\ell = 2, 4, 6, \ldots$ the parity does not change, but the initial and final nuclear spins can differ by more than one. A summary of the classification of nuclear β decay processes is shown in Table 22.1.

Type of decay	Selection rule	ℓ	ΔP	$ft_{1/2}$ (s)
super-allowed	$\Delta J = 0, \pm 1$	0	no	$10^3 - 10^4$
allowed	$\Delta J = 0, \pm 1$	0	no	$2 \times 10^3 - 10^6$
1st forbidden	$\Delta J = 0, \pm 1$	1	yes	$10^6 - 10^8$
unique 1st forbidden	$\Delta J = \pm 2$	1	yes	$10^8 - 10^9$
2nd forbidden	$\Delta J = \pm 1, \pm 2$	2	no	$2 \times 10^{10} - 2 \times 10^{13}$
unique 2nd forbidden	$\Delta J = \pm 3$	2	no	10^{12}
3rd forbidden	$\Delta J = \pm 2, \pm 3$	3	yes	10^{18}
unique 3rd forbidden	$\Delta J = \pm 4$	3	yes	4×10^{15}
4th forbidden	$\Delta J = \pm 3, \pm 4$	4	no	10^{23}
unique 4th forbidden	$\Delta J = \pm 5$	4	no	10^{19}

Table 22.1 A summary of the classification of nuclear β decay processes. The ΔP column indicates if the nuclear parity is changed by the decay. The last column displays the comparative lifetime parameter $ft_{1/2}$ (see Eq. (21.42)).

To generalize the Fermi theory, one had to consider the possible forms of the currents acting on single particles (i.e., proton, neutron, electron, neutrino). These currents can be conveniently organized with the classification of the **bilinear covariants** (see Table 8.3). **Before 1956, it was presumed that parity was conserved in weak interactions.**[4] The assumption that parity is conserved in weak interactions imposes a constraint on the possible forms of the current. For example, terms S (scalar), V (vector), and T (tensor) are allowed, but terms mixing V (vector) and A (axial vector), or terms with P (pseudoscalar) are forbidden. Several experimental tests were proposed in order to clarify the actual nature of the weak current.

In 1956, **Lee** and **Yang** were trying to solve a very puzzling problem called **the $\tau - \theta$ problem**. Two strange mesons, called the τ and the θ, appeared to be identical in every respect: mass, spin, charge, etc. The problem was that the τ^+ was observed to decay into three pions $\pi^+\pi^+\pi^-$ or $\pi^+\pi^0\pi^0$. The other one, the θ^+, decayed into two pions $\pi^+\pi^0$. Both are spin-zero particles of strangeness one. The analysis of the final state showed that the τ^+ decayed into a parity odd state, $P(\tau^+) = -1$, while the θ^+ decayed into a parity even state, $P(\theta^+) = +1$. This seemed impossible if the two particles were the same.

Lee and Yang, after studying this, pointed out in 1956 that maybe these two particles could be the same particle with two distinctive modes of decay. Of course, this would be possible only if the parity is violated in these decays [278].

$$\theta^+\text{-decay}: \qquad K^+ \to \pi^+\pi^0 \qquad\qquad P(\pi\pi) = (-1)(-1) = +1 \qquad\qquad (22.4)$$

$$\tau^+\text{-decay}: \quad K^+ \to \pi^+\pi^+\pi^-, \pi^+\pi^0\pi^0 \quad P(\pi\pi\pi) = (-1)^3 = -1 \qquad\qquad (22.5)$$

---

4 Parity conservation was considered as a sort of dogma. Still today the origin of its violation is not explained (e.g., as a consequence of a more fundamental theory).

They examined carefully the available evidence for parity conservation, and concluded that there was a great deal of evidence for parity conservation in the strong and the electromagnetic interactions, while there was none in the weak interaction. The weak decay of the kaon is just one case, among many, where the parity is not conserved. They further proposed various ways the parity non-conservation could be tested experimentally in the weak interaction. This proposal came as a big surprise and was received with great skepticism since one strongly believed in the conservation of parity in all fundamental interactions.

To test for possible parity violation it is necessary to observe a dependence of a decay rate or of a cross-section on a term that changes under the parity operation. To distinguish a process from its "parity transformed" one, we for example identify a quantity O that changes under parity transformation, i.e., under P: $O \to O' \neq O$. Then, if in a given process the probability $W(O)$ of the quantity O is found to have the following property:

$$W(O) \neq W(O') \tag{22.6}$$

then the process violates parity. A particular class is given by quantities which are *pseudoscalars*, namely that just change sign under parity P: $O \to -O$. Then parity violation implies:

$$W(O) \neq W(-O) \tag{22.7}$$

An important example for such a quantity is given by the angular correlation between the momentum $\vec{p}$ of a particle and its spin orientation $\vec{s}$:

$$O \equiv \cos\theta = \frac{\vec{s} \cdot \vec{p}}{|\vec{s}||\vec{p}|} \propto \vec{s} \cdot \vec{p} \tag{22.8}$$

Parity reverses the momentum $\vec{p} \to -\vec{p}$, but not the spin $\vec{s} \to \vec{s}$ (in general angular momentum) of a particle. Hence the angle between the momentum and the spin undergoes $\theta \to \pi - \theta$ under parity. See Figure 22.1.

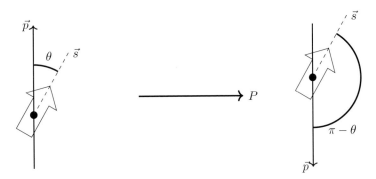

Figure 22.1 Angle between momentum $\vec{p}$ of a particle and the orientation of its spin $\vec{s}$ before and after parity transformation.

So parity violation implies that the **probabilities are different for processes that are identical to each other expect for a spin flip:**

$$W(\vec{s} \cdot \vec{p}) \neq W(-\vec{s} \cdot \vec{p}) = W((-\vec{s}) \cdot \vec{p}) \tag{22.9}$$

In other words, **an asymmetry in the observed distribution of the angle between momentum and spin is an indication for parity violation.** A concrete example is given by the function:

$$W(\cos\theta) \propto 1 + \alpha\cos\theta \quad \text{with} \quad \alpha \neq 0 \tag{22.10}$$

Parity also reverses helicities $h \propto \vec{s} \cdot \vec{p}$, so we expect parity violation to also appear in observables related to the helicities of the initial and final-state particles.

22.2   Wu Experiment on Parity Violation (1957)

In 1957 the non-conservation of parity in weak interactions was experimentally demonstrated by **Wu**[5] in the study of β disintegrations of polarized ^{60}Co nuclei [279]. The observed reaction is:

$$^{60}\text{Co} \rightarrow ^{60}\text{Ni}^* + e^- + \bar{\nu}_e \qquad \beta^- \ (99.88\%), \quad E_0 = 317 \text{ keV} \qquad (22.11)$$

where E_0 is the maximum kinetic energy of the electron. The decay occurs via a Gamov–Teller transition ($\Delta J = 1$), with a lifetime $\tau = 5.2$ yr $= 1925$ days. Two photons of energy 1.17 and 1.33 MeV are emitted after each β-decay from the de-excitation of the ^{60}Ni$^*(4^+) \rightarrow ^{60}Ni^*(2^+) \rightarrow ^{60}Ni(0^+)$. See Figure 22.2.

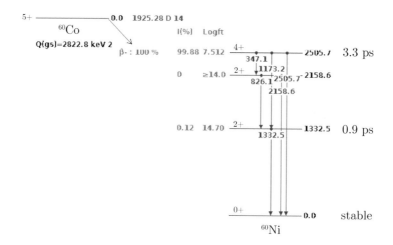

Figure 22.2 The nuclear levels in the β disintegration of ^{60}Co $\rightarrow ^{60}$Ni$^* + e^- + \bar{\nu}_e$, with subsequent de-excitation of the ^{60}Ni* via gamma emission. Image from the National Nuclear Data Center (NNDC) at BNL accessed via www.nndc.bnl.gov.

It was known that a high degree of polarization could be reached via paramagnetic demagnetization of crystals such as cerium–magnesium (CeMg) nitrate. The radioactive source of ^{60}Co used by Mrs. Wu was grown on such a crystal and after cooling at a temperature of $T \approx 0.01$ K, a large number of these nuclei would align with the external magnetic field. Low temperatures were needed for thermal motion not to destroy the polarization. In this experimental condition, the spins of the ^{60}Co nuclei aligned in the direction of the B field. The amount of polarization of the nuclei could be monitored with the spatial asymmetry of the emitted photons. The β transition induced a $\Delta J = 1$ change. Since the ^{60}Co nuclei have a spin $J = 5$ and the final state ^{60}Ni* nuclei has a spin $J = 4$, the electron and the neutrino form a system that is emitted with a given angular momentum state $S = 1$. Hence, the spin of the electron and of the neutrino must point in the direction of the ^{60}Co spin (i.e., the direction of the magnetic field). See Figure 22.3. In the three-body decay ^{60}Co $\rightarrow ^{60}$Ni$^* + e^- + \bar{\nu}_e$, particles can be emitted in different directions with the condition:

$$\vec{P}_{^{60}\text{Co}} = \vec{P}_{^{60}\text{Ni}^*} + \vec{p}_e + \vec{p}_\nu \qquad (22.12)$$

In particular, we can consider the two extreme cases where the electron is emitted in the direction or in the opposite direction of the magnetic field (i.e., the direction of the ^{60}Co spin)! The main goal of Mrs. Wu's experiment was to compare the rate of these two configurations.

The basic concept of the experiment is illustrated in Figure 22.4. The electrons from ^{60}Co decays were detected by a thin anthracene scintillation crystal placed above the radioactive source. Scintillation light

5   Chien Shiung Wu (1912–1997), Chinese–American particle and experimental physicist, known as Mrs. Wu.

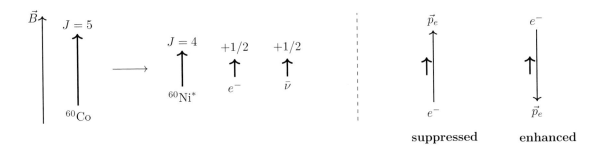

Figure 22.3 β disintegration of $^{60}\text{Co} \rightarrow ^{60}\text{Ni}^* + e^- + \bar{\nu}_e$: spin orientation (in thick lines) and flight direction (thinner lines on the right-hand side). The $\vec{B}$-field defines the axis of quantization. The electron can be emitted in the direction or opposite to the direction of the magnetic field.

is transmitted to the photomultiplier tube (PMT) on top of the cryostat. The anisotropy of gamma-rays from the $^{60}\text{Ni}^*$ de-excitation was measured with two NaI crystals. This is a measure of the degree of the polarization of the ^{60}Co nuclei. The actual experimental setup of Mrs. Wu and the asymmetry results are shown in Figure 22.5.

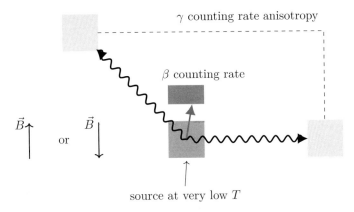

Figure 22.4 Illustration of Mrs. Wu's experimental concept to study the behavior of beta decays under the parity transformation.

The results obtained were spectacular! The top and middle graph in Figure 22.5(right) represent the γ anisotropy (difference in counting rate between two NaI crystals). This graph is a control of polarization. It is independent of the orientation (parity conservation) and is just a measure of the amount of polarization. The bottom graph in Figure 22.5(right) shows the β asymmetry defined as the counting rate in the anthracene crystal relative to the rate without polarization (after the set up was warmed up) for two orientations of the magnetic field. The β asymmetry rate was different for the two magnetic field orientations. The asymmetry disappeared when the crystal was warmed up (the magnetic field was still present). Thus, it is indeed connected to the spins orientation and is not just due to the presence of the magnetic field.

The spin of an emitted electron is therefore correlated with the direction of the spin of the nucleus. It points in the direction of the magnetic field. Mrs. Wu measured that the detected electrons were not emitted isotropically. Because the asymmetry of the photons is independent of the direction of the magnetic field (e.g., the asymmetry is a measure of the degree of polarization), the asymmetry of the electrons had to depend

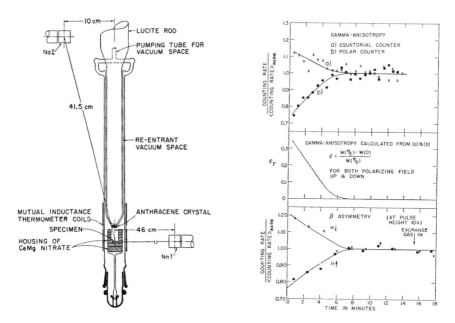

Figure 22.5 (left) Apparatus used in Mrs. Wu's experiment: (right) the photon and electron counting asymmetries as a function of time. Reprinted with permission from C. S. Wu, E. Ambler, R. W. Hayward, D. D. Hoppes, and R. P. Hudson, "Experimental test of parity conservation in beta decay," *Phys. Rev.*, vol. 105, pp. 1413–1415, Feb 1957. Copyright 1957 by the American Physical Society.

on the direction of the field. There were fewer electrons emitted in the direction along the nuclei spins than in the opposite direction, as illustrated in Figure 22.3. In conclusion, Mrs. Wu's experiment experimentally demonstrated that β-**rays from oriented** ^{60}**Co nuclei are preferentially emitted in the direction opposite to that of the** ^{60}**Co spin**. This implies a violation of parity in nuclear β decay!

22.3   Helicity in Weak Interactions – $V - A$ Structure

The first attempt at a theory of the weak interactions by Fermi was inspired by QED and did not violate parity. The leptonic part of the vector current for β decays was written as (see Eq. (21.15)):

$$[\bar{u}_\nu(p')\gamma_\mu u_e(p)] \equiv \bar{\nu}\gamma^\mu e \tag{22.13}$$

The **chiral structure of QED** was discussed in Section 10.17. It was shown that chirality is conserved at the QED interaction vertex and that *the V (vector) form of the QED current enforces parity conservation* (Eq. (10.166)). In general, the electron and the neutrino spinors can be decomposed into their left-handed and right-handed *chiral* components. The chiral projections of the electron e_R and e_L can be obtained with the left-handed and right-handed chiral projections P_L and P_R (Eq. (8.184)). When we apply this concept to the QED electromagnetic current of the electron, we find:

$$\bar{e}\gamma^\mu e = (\overline{e_L} + \overline{e_R})\gamma^\mu(e_L + e_R) = \overline{e_L}\gamma^\mu e_L + \underbrace{\overline{e_R}\gamma^\mu e_L}_{=0} + \underbrace{\overline{e_L}\gamma^\mu e_R}_{=0} + \overline{e_R}\gamma^\mu e_R \tag{22.14}$$

since $P_L + P_R = \frac{1}{2}\left[\mathbb{1} - \gamma^5 + \mathbb{1} + \gamma^5\right] = \mathbb{1}$.

In Eq. (8.205) we have seen that the parity transformation P exchanges right-handed and left-handed chiral spinors. Therefore, while the chirality of a Dirac particle is invariant under a proper Lorentz transformation (Eq. (8.182)), it flips under parity. Thus, the **QED electromagnetic current conserves parity** since the term

$$\bar{e}\gamma^\mu e = \overline{e_L}\gamma^\mu e_L + \overline{e_R}\gamma^\mu e_R \tag{22.15}$$

remains unchanged under the interchange $R \leftrightarrow L$. The QCD vertex factor of Eq. (18.23) has the same form and similarly the **QCD strong current conserves parity**.

The observation of parity violation in weak interactions is an indication that the current cannot be of the simple vector form, such as the QED one. In order for the weak current to be covariant, its chiral current is expressed in terms of the bilinear covariants of the form of Eq. (8.295). Before 1958 the correct form of β interactions was thought to be tensor and scalar (with the consequence that the neutrino was right-handed, and the antineutrino left-handed). Following Mrs. Wu's pioneering measurements, many experiments were performed over several years on the study of the weak processes and β decays.

In 1958 **George Sudarshan**[6] and **Robert Marshak**[7] [280], and independently **Feynman and Gell-Mann** [271], determined the **correct tensor structure $V - A$ (vector minus axial vector)** of the weak charged-current interaction (more in Chapter 23). The vector term is proportional to γ^μ while the axial term is given by $\gamma^\mu\gamma^5$. Starting from Eq. (22.13), we thus modify its form to be $V - A$:

$$\underbrace{\frac{1}{2}\bar{\nu}\gamma^\mu e}_{V} - \underbrace{\frac{1}{2}\bar{\nu}\gamma^\mu\gamma^5 e}_{A} = \bar{\nu}\gamma^\mu\left(\frac{1-\gamma^5}{2}\right)e = \bar{\nu}\gamma^\mu P_L e \tag{22.16}$$

We note that exactly the same result can be obtained with the current $\overline{\nu_L}\gamma^\mu e_L$ since:

$$\begin{aligned}
\overline{\nu_L}\gamma^\mu e_L &= \bar{\nu}\left(\frac{1+\gamma^5}{2}\right)\gamma^\mu\left(\frac{1-\gamma^5}{2}\right)e = \bar{\nu}\gamma^\mu\left(\frac{1-\gamma^5}{2}\right)\left(\frac{1-\gamma^5}{2}\right)e = \bar{\nu}\gamma^\mu P_L^2 e \\
&= \bar{\nu}\gamma^\mu P_L e
\end{aligned} \tag{22.17}$$

where, as in Eq. (10.162), we found the general result:

$$\underbrace{\overline{\Psi_L}\gamma^\mu\Psi_L}_{V} = \overline{\Psi}P_R\gamma^\mu P_L\Psi = \overline{\Psi}\gamma^\mu P_L P_L\Psi = \overline{\Psi}\gamma^\mu P_L\Psi = \underbrace{\overline{\Psi}\gamma^\mu\left(\frac{1-\gamma^5}{2}\right)\Psi}_{V-A} \tag{22.18}$$

Hence, **the weak charged current can be seen as a V (vector) current (i.e., like the electromagnetic interaction) that only couples to the left-handed chiral projection** of the spinor! The weak interaction is a chiral theory that is not symmetric with respect to the left-handed and right-handed chiral projections. It does not treat Ψ_L and Ψ_R on an equal footing and as a result, it violates parity.

Note that a purely vector or purely axial form of the current will not break parity. It is the combination of both terms that introduces parity violation. The $V + A$ (vector plus axial vector) combination would also violate parity but would not be compatible with the direction of asymmetry observed in experiments (e.g., Mrs. Wu's experiment).

We now consider the weak processes, where X and Y decay into a charged lepton $\ell^\pm$ and the corresponding neutrino $\nu_\ell(\bar{\nu}_\ell)$:

$$X \to \ell^- + \bar{\nu}_\ell \quad \text{or} \quad Y \to \ell^+ + \nu_\ell \tag{22.19}$$

As was mentioned before, the helicity operator corresponds to the projection of the spin operator onto the direction of the momentum of the particle (see Eq. (8.138)):

$$h \equiv \frac{\vec{\sigma} \cdot \vec{p}}{|\vec{p}|} \tag{22.20}$$

---

6   Ennackal Chandy George Sudarshan (1931–2018), Indian theoretical physicist.
7   Robert Eugene Marshak (1916–1992), American physicist.

According to the $V - A$ current, the **weak current only couples to the left-handed chiral projection**. As discussed in Section 8.24, the chirality and the helicity are related yet distinguishable properties of a Dirac particle. In Eq. (8.247), we had derived the positive helicity eigenstate $u_\uparrow$ in terms of chiral components as:

$$u_\uparrow = A_L^\uparrow u_L + A_R^\uparrow u_R \propto \frac{1}{2}(1 - \alpha)u_L + \frac{1}{2}(1 + \alpha)u_R \qquad \text{where} \qquad \alpha = \frac{p}{E + m} \tag{22.21}$$

and

$$u_\downarrow = A_L^\downarrow u_L + A_R^\downarrow u_R \propto \frac{1}{2}(1 + \alpha)u_L + \frac{1}{2}(1 - \alpha)u_R \tag{22.22}$$

Since the weak current only couples to left-handed chiral projection, the final state is a combination of positive and negative helicity states as given by the factors in front of u_L in the expressions above. Hence, the probability of finding the final-state lepton with a helicity $h = \pm 1$ is:

$$W(h = \pm 1) = \frac{(\langle u_{\uparrow, \downarrow}|A|u\rangle)^2}{(\langle u|A|u\rangle)^2} = \frac{1}{2}(1 \mp \alpha)^2 \tag{22.23}$$

The **longitudinal polarization** P_L of a particle is proportional to the difference of probabilities to find the particle in the helicity $+1$ and -1 states:

$$P_L = \frac{W(+1) - W(-1)}{W(+1) + W(-1)} \tag{22.24}$$

Zero longitudinal polarization means that the particle has an equal probability to be in the $+1$ and -1 helicity states.

Particles produced in weak decays are expected to be polarized as a consequence of the $V - A$ current. We have for the lepton ℓ^-:

$$\begin{aligned}
P_L(\ell^-) &= \frac{(1 - \alpha)^2 - (1 + \alpha)^2}{(1 - \alpha)^2 + (1 + \alpha)^2} = \frac{-2\alpha}{1 + \alpha^2} = -2\left(\frac{p}{E + m}\right)\frac{1}{1 + \left(\frac{p}{E+m}\right)^2} \\
&= -\frac{2p(E + m)}{E^2 + 2mE + m^2 + p^2} = -\frac{p}{E} = -\beta
\end{aligned} \tag{22.25}$$

where β is simply **the velocity of the lepton**! The amount of polarization is just given by the velocity of the particle. The more relativistic the particle is, the more it is polarized. For the antilepton, the weak coupling still couples to the left-handed chiral states, however, recall that the *left-handed chiral state* maps to the *positive* helicity for antiparticles (see Eq. (8.244)). Hence, the longitudinal polarization of the antilepton ℓ^+ is:

$$P_L(\ell^+) = +\beta \tag{22.26}$$

These two results for the lepton and antilepton can be combined into a single expression:

$$\text{Lepton polarization in weak decay}: P_L(\ell) = \mathcal{H}\beta \quad \text{where} \quad \mathcal{H} = \begin{cases} -1 \text{ for } \ell^- \\ +1 \text{ for } \ell^+ \end{cases} \tag{22.27}$$

Massless particles always travel at the speed of light, hence we have $\beta \equiv 1$, and they are expected to be always fully polarized in weak decays. Neutrinos are special in this case. They are known to *only* couple to the weak interactions, hence under that assumption would *only* appear in Nature in a fully polarized and hence unique helicity configuration. When considered massless, neutrinos have therefore the following longitudinal polarization:

$$\text{Massless neutrinos}: P_L = \begin{cases} -1 \text{ for } \nu \\ +1 \text{ for } \overline{\nu} \end{cases} \tag{22.28}$$

The massless neutrino is fully polarized with a negative helicity, and a massless antineutrino is fully polarized with a positive helicity. In this scenario, neutrinos with a positive helicity and antineutrinos with a negative helicity do not appear in the theory.

22.4   Two-Component Theory of Massless Weyl Neutrinos

Neutrinos were postulated as spin-1/2 particles in order to explain the statistics (i.e., angular momentum conservation in beta decays). Unlike the electron, we recall that the neutrino is electrically neutral and hence the distinction between neutrino and antineutrino requires the introduction of the leptonic charge (see Eqs. (21.66), (21.67), and (21.69)). Another striking difference between neutrino and electron (and muon) is that the Kurie plots obtained in the study of β disintegrations were indicating a neutrino rest mass consistent with zero. **Is the neutrino a Dirac particle?**

It turns out that this question is still unresolved today (this will be discussed in Chapter 31). Regardless, in our theory we use a four-component Dirac spinor $\Psi(x)$ for all charged leptons (electron, muon, and tau) and all quarks, and want to do the same for neutrinos. The four independent components of the Dirac spinor represent the four possibilities of "particle" and "antiparticle," each with a spin $+1/2$ and $-1/2$. Let us use the Weyl representation (see Section 8.19) and write any spinor as a two two-component spinor as in Eq. (8.177):

$$\Psi(x) = u(\vec{p})e^{-ip\cdot x} = \begin{pmatrix} \cdot \\ \cdot \\ \cdot \\ \cdot \end{pmatrix} e^{-ip\cdot x} \qquad \text{where} \qquad u(\vec{p}) = \begin{pmatrix} u_L \\ u_R \end{pmatrix} = \begin{pmatrix} \left\{ \begin{matrix} \cdot \\ \cdot \end{matrix} \right. \\ \left\{ \begin{matrix} \cdot \\ \cdot \end{matrix} \right. \end{pmatrix} \tag{22.29}$$

We have seen in Eq. (8.179) that the spinor obeys the Dirac equation expressed as two coupled equations. The two-component spinors u_L and u_R are mixed by the mass term in the Dirac equation. But for the case $m = 0$, they do not mix, and the equations of motion for Ψ_L and Ψ_R decouple into what are called the **Weyl equations of motion**:

$$\begin{cases} (p_0 \mathbb{1} - \vec{\sigma} \cdot \vec{p})u_R = \begin{pmatrix} \cdot & \cdot \\ \cdot & \cdot \end{pmatrix}\begin{pmatrix} \cdot \\ \cdot \end{pmatrix} = 0 \\ (p_0 \mathbb{1} + \vec{\sigma} \cdot \vec{p})u_L = \begin{pmatrix} \cdot & \cdot \\ \cdot & \cdot \end{pmatrix}\begin{pmatrix} \cdot \\ \cdot \end{pmatrix} = 0 \end{cases} \implies \begin{cases} u_R = \left(+\dfrac{\vec{\sigma}\cdot\vec{p}}{p_0}\right)u_R \\ u_L = \left(-\dfrac{\vec{\sigma}\cdot\vec{p}}{p_0}\right)u_L \end{cases} \tag{22.30}$$

with $p_0 = E = \pm|\vec{p}|$ and $p_0^2 = E^2 = \vec{p}^2$. In both cases we need to consider two solutions for the energy. Let us take u_L. For the positive energy solution $p_0 = +|\vec{p}|$, we find a negative helicity:

$$u_L = \left(-\frac{\vec{\sigma}\cdot\vec{p}}{p_0}\right)u_L = -\underbrace{\left(\frac{\vec{\sigma}\cdot\vec{p}}{|\vec{p}|}\right)}_{\text{helicity}} u_L \tag{22.31}$$

Such a state hence corresponds to a massless neutrino with negative helicity $\nu_\downarrow$ and is often labeled as ν_L. For the negative energy solution $p_0 = -|\vec{p}|$, we find the positive helicity:

$$u_L = \left(-\frac{\vec{\sigma}\cdot\vec{p}}{p_0}\right)u_L = +\underbrace{\left(\frac{\vec{\sigma}\cdot\vec{p}}{|\vec{p}|}\right)}_{\text{helicity}} u_L \tag{22.32}$$

Such a state corresponds to a positive helicity massless antineutrino $\bar{\nu}_\uparrow$ and is often labeled $\bar{\nu}_R$. In other words, the two-component spinor u_L describes a neutrino $\nu_\downarrow = \nu_L$ and an antineutrino $\bar{\nu}_\uparrow = \bar{\nu}_R$. In a similar way, we find that the two-component spinor u_R describes a neutrino $\nu_\uparrow = \nu_R$ and an antineutrino $\bar{\nu}_\downarrow = \bar{\nu}_L$.

The decoupling of the Dirac equation in the massless case has important physical consequences. For the neutrinos, we obtain two types of states of different chirality, which are totally decoupled. We recall that chirality is a Lorentz-invariant and that left-handed and right-handed components are never mixed in a Lorentz transformation. If neutrinos are massless, we obtain two independent types: (1) a "left-handed" neutrino and a "right-handed" antineutrino; (2) a "right-handed" neutrino and a "left-handed" antineutrino. This is called the two-components theory of massless neutrinos.

Note that since the weak charged current couples only to the left-handed chiral projection of the spinor (i.e., $\bar{\nu}_L \gamma^\mu e_L$) the condition of the two components is automatically verified in the sense that the four-components spinor will then be exactly equivalent to the two-components u_L, since, e.g., in the Weyl representation, we obtain:

$$\frac{1}{2}\left(1 - \gamma^5\right) u = \begin{pmatrix} 1 & 0 \\ 0 & 0 \end{pmatrix} \begin{pmatrix} u_L \\ u_R \end{pmatrix} = \begin{pmatrix} u_L \\ 0 \end{pmatrix} \tag{22.33}$$

As expected, only the two-component u_L survives and hence the weak $V - A$ interaction involves only the left-handed neutrino ν_L and the right-handed antineutrino $\bar{\nu}_R$.

Since the only known interaction of neutrinos is the weak force (we neglect gravitation!), only left-handed neutrinos and right-handed antineutrinos are observable in our theory. The parity transformation would take a physical left-handed neutrino into an unobservable right-handed neutrino. This is clearly in violation of the weak interaction and is another sign of parity violation. In fact, **the left-handed neutrino ν_L and the right-handed antineutrino $\bar{\nu}_R$ are related by the CP transformation**, since P will flip the helicity and C will change particle–antiparticle. So, indeed the neutrino states observed in Nature are related by the CP transformation:

$$CP\,|\nu_L\rangle = \eta\,|\bar{\nu}_R\rangle \qquad \text{and} \qquad CP\,|\bar{\nu}_R\rangle = \eta'\,|\nu_L\rangle \tag{22.34}$$

where η, η' represent phases. This connection is illustrated in Figure 22.6. In the 1950s, this relation was perceived as reassuring news after the "shock" of parity violation, since it appeared as if CP was a good symmetry of Nature. Soon after though, CP symmetry was also shown to be violated (see Chapter 30)!

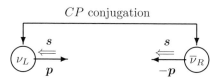

Figure 22.6 The left-handed neutrino and the right-handed antineutrino observed in Nature are related by the CP transformation.

22.5   The Helicity of the Neutrino (1958)

The $V - A$ theory makes very firm predictions on the helicity of fermions in weak interactions, and in particular for that of neutrinos. In order to measure the final-state polarizations, one needs detectors that are sensitive to the spins of the incoming particles. A statistical measurement over an ensemble of particles leads to a measurement of the degree of polarization in a given reaction. The measurement of the polarization of neutrinos is exceptionally challenging and requires the use of some tricks. In 1968 **Goldhaber**,[8] **Grodzins**,[9] and **Sunyar**[10] succeeded in determining the helicity of neutrinos [281]. They concluded their paper stating that the neutrino was compatible with a left-handedness of 100%!

8   Maurice Goldhaber (1911–2011), Austrian-born American physicist.
9   Lee Grodzins (born 1926), American physicist.
10   Andrew William Sunyar (1920–1986), American physicist.

The layout of the experiment is shown in Figure 22.7. The main goal was to study the electron capture of europium, ^{152}Eu:

$$^{152}\text{Eu}(0^-) + e^- \quad \rightarrow \quad ^{152}\text{Sm}^*(1^-) + \nu_e \quad \rightarrow \quad ^{152}\text{Sm}(0^+) + \gamma + \nu_e \tag{22.35}$$

Naturally occurring Eu is composed of two isotopes, ^{151}Eu and ^{153}Eu. The ^{152}Eu(0^-) is synthesized and has a half-life of 9.3 h. The probability of electron capture is $\sim 27\%$. The neutrino is emitted recoiling against the ^{152}Sm atom. The clever idea behind this scheme is that one can select those γs from the decay of the Sm excited state which travel oppositely to the direction of the electron-capture ν_es (i.e., in the direction of the nuclear recoil) by having them resonantly scatter from an Sm ring target. By angular momentum conservation the helicity of the downward-going γ is the same as that of the upward-traveling ν_e.

The angular momenta and the spin correlations are shown in detail in Figure 22.8 for the two a priori possible spin states of the neutrinos labeled ν_L and ν_R. Because of angular momentum conservation, the spin J of the recoiling nucleus ^{152}Sm* must be directed in the opposite direction of the spin of the neutrino. Therefore, the recoiling nucleus possesses the same helicity as the neutrino!

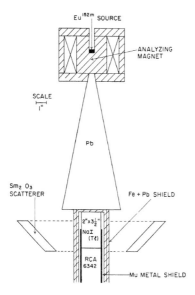

$$^{152}\text{Eu} + e^- \rightarrow {}^{152}\text{Sm}^* + \nu$$
$$J: \quad 0 \ \oplus \ \tfrac{1}{2} \rightarrow \quad 1 \ \oplus \ \tfrac{1}{2}$$

$$^{152}\text{Sm}^* \rightarrow {}^{152}\text{Sm} + \gamma$$

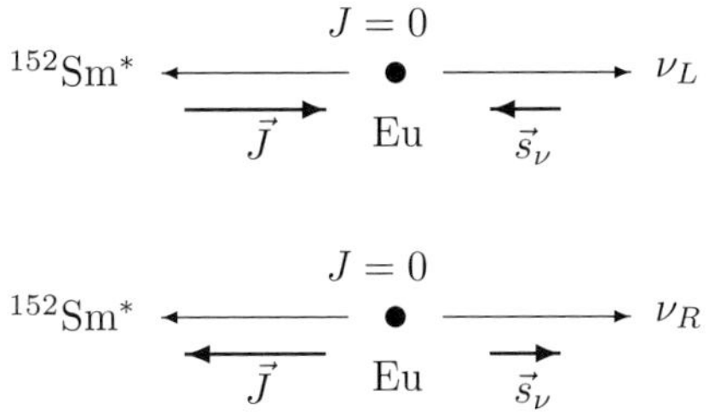

Figure 22.7 Sketch of the experimental apparatus of Goldhaber *et al.* to determine the helicity of the neutrino. Reprinted from M. Goldhaber, "Weak interactions: leptonic modes–experimental," in *Proceedings of the 1958 Annual International Conference on High Energy Physics at CERN, Geneva*, July 1958.

Figure 22.8 Electron capture of ^{152}Eu and spin correlations.

The excited state ^{152}Sm* promptly decays ($\tau \approx 10$ fs) with the emission of a single photon. The transition is $J = 1 \rightarrow J' = 0$ with a photon energy of 961 keV in the rest frame of the excited nucleus. We note that the photon takes over the angular momentum of its parent nucleus. We can now consider two cases: the photons which are emitted forward (resp. backward) relative to the flight direction of the recoiling excited ^{152}Sm*. We note that the *forward-emitted photons* have the *same helicity as the neutrinos*, while the backward-emitted photons have an opposite helicity. See Figure 22.9.

In the case of forward emission, the energy of the emitted photon is slightly larger than 961 keV because of the recoil against the neutrino. These slightly more energetic photons possess exactly the right energy to excite the following resonant reaction:

$$^{152}\text{Sm} + \gamma \rightarrow {}^{152}\text{Sm}^* \rightarrow {}^{152}\text{Sm} + \gamma \tag{22.36}$$

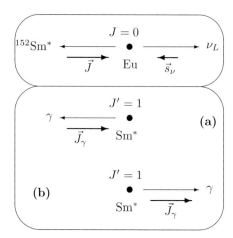

Figure 22.9 Correlation among spins of the emitted particles in the sequence $^{152}\text{Eu}(0^-) + e^- \to {}^{152}\text{Sm}^*(1^-)$ $+\nu_e \to {}^{152}\text{Sm}(0^+) + \gamma + \nu_e$. (a) Forward-emitted photon; (b) backward-emitted photon.

The slightly higher energy is needed to conserve energy and momentum in the resulting recoil of the excited nucleus. In the case of backward-emitted photons, their energy is slightly less than 961 keV and is therefore not sufficient to excite the resonance. In the experiment, the resonant scattering was obtained with a ring composed of Sm_2O_3. An NaI counter, shielded by lead, was placed at a large angle (see Figure 22.7) and detected only those photons that underwent resonant scattering. These events corresponded to the forward-emitted photons, which carried the same helicity as the neutrinos!

Finally, having selected the photons that carry the information of the neutrino helicity, one must still determine the polarization of the photons in order to derive the polarization of the neutrinos. We note that the target was embedded in a large block of magnetized iron. In this configuration, the detected photons had to travel through the magnetized iron. As they traveled the iron, they could Compton scatter with electrons in the medium. These electrons were polarized in the magnetized iron. As a result, the transmission probabilities of photons for left-circular and right-circular polarizations were different due to the electron–photon spin–spin dependence of the scattering cross-section (see Section 11.16).

After taking into account all possible effects that would reduce the observed effect, the obtained results were showing a neutrino polarization consistent with 100% negative helicity. The error on the helicity measurement was difficult to estimate, however, it can be estimated to have been around 30%. The result hence clearly disfavoured the existence of neutrinos with positive helicity.

22.6   Helicity in the Weak Decays of the π-Meson (Pion)

We have previously mentioned in Section 21.9 that the apparent absence of β decay of the π-meson was a great puzzle of weak interactions. More sensitive measurements led to the observation that the β decay of the π-meson is possible; however, its rate is strongly suppressed relative to the decay into muon (value taken from the Particle Data Group's Review [5]):

$$\frac{\Gamma(\pi \to e\nu_e)}{\Gamma(\pi \to \text{all})} = (1.230 \pm 0.004) \times 10^{-4} \tag{22.37}$$

This is a priori a very surprising result since in apparent contradiction with the universality of the weak couplings (e, ν) and (μ, ν) and the phase-space factor for the electronic decay being larger than for the muonic decay. Using the result of Eq. (5.163), we find:

$$\Gamma(\pi^\pm \to \ell^\pm + \nu_\ell) = \int \frac{|\mathcal{M}|^2}{32\pi^2} \frac{p_\ell}{m_\pi^2} d\Omega = \frac{|\mathcal{M}|^2}{8\pi} \frac{p_\ell}{m_\pi^2} \qquad (22.38)$$

where $p_\ell = |\vec{p}_\ell| = |\vec{p}_\nu|$ is the magnitude of the momentum of the lepton ℓ or the neutrino in the parent pion rest frame. In the expression above, we must remember that the matrix element $|\mathcal{M}|^2$ carries units of energy squared since the resulting width is in units of energy (see Section 5.16). From energy and momentum conservation, we have:

$$m_\pi = E_\ell + E_\nu = \sqrt{p_\ell^2 + m_\ell^2} + p_\ell \quad \implies \quad p_\ell = \frac{m_\pi^2 - m_\ell^2}{2m_\pi} \qquad (22.39)$$

Also:

$$E_\ell^2 = p^2 + m_\ell^2 = \left(\frac{m_\pi^2 - m_\ell^2}{2m_\pi}\right)^2 + m_\ell^2 = \frac{\left(m_\pi^2 - m_\ell^2\right)^2 + 4m_\pi^2 m_\ell^2}{4m_\pi^2} = \frac{\left(m_\pi^2 + m_\ell^2\right)^2}{4m_\pi^2} \qquad (22.40)$$

hence

$$E_\ell = \frac{m_\pi^2 + m_\ell^2}{2m_\pi} \quad \text{and} \quad \beta_\ell = \frac{p_\ell}{E_\ell} = \frac{m_\pi^2 - m_\ell^2}{m_\pi^2 + m_\ell^2} \qquad (22.41)$$

The matrix element $|\mathcal{M}|^2$ contains the spinors of the lepton and of the neutrino, and as a consequence, its normalization is proportional to:

$$|\mathcal{M}|^2 \propto E_\ell E_\nu = E_\ell |\vec{p}_\nu| = E_\ell p_\ell \qquad (22.42)$$

The phase-space factor Γ_{PS} in the total width $\Gamma(\pi^\pm \to \ell^\pm + \nu_\ell)$ will then contribute as:

$$\Gamma_{PS} \propto \frac{E_\ell p_\ell^2}{m_\pi^2} = \frac{\left(m_\pi^2 + m_\ell^2\right)}{2m_\pi} \times \frac{\left(m_\pi^2 - m_\ell^2\right)^2}{4m_\pi^2} \times \frac{1}{m_\pi^2} = \frac{\left(m_\pi^2 + m_\ell^2\right)\left(m_\pi^2 - m_\ell^2\right)^2}{8m_\pi^5} \qquad (22.43)$$

Hence the ratio R_{PS} of the space-phase factors for the decay into electrons relative to muons is:

$$R_{PS} = \frac{\Gamma_{PS}(\pi^\pm \to e^\pm + \nu_e)}{\Gamma_{PS}(\pi^\pm \to \mu^\pm + \nu_\mu)} = \frac{\left(m_\pi^2 + m_e^2\right)\left(m_\pi^2 - m_e^2\right)^2}{\left(m_\pi^2 + m_\mu^2\right)\left(m_\pi^2 - m_\mu^2\right)^2} \approx 3.5 \qquad (22.44)$$

where we used $m_\pi \approx 139$ MeV, $m_\mu \approx 105$ MeV, and $m_e \approx 0.511$ MeV. Based on the phase space-factor, we would expect decays to electrons to be *more probable* than decays to muons, and dominate the decay branching fraction, a fact which is in strong contraction with data!

As anticipatable, the parity-violating dynamics of the weak interaction dominate over phase space! In Feynman's own words:

> *"Assume, then, that the universal theory is right. We can deduce that the rate of $\pi \to \mu + \nu$ and $\pi \to e + \nu$ is equal to the ratio of the phase space available, which is better for the $e + \nu$ than for the $\mu + \nu$, however, by one additional factor, which was originally pointed out many years ago by Ruderman for axial vector coupling, and that is this: in the universal theory this is an antineutrino and that is an electron, then all particles in the relativistic limit come out spinning to the left, antiparticles to the right. So if, for the moment, we take the relativistic limit, this particle should come out spinning to the left and this particle spinning to the right, and they just go in opposite directions, one spinning to the left and the other to the right, and the result is that there is a total angular momentum, but the π has no angular momentum, so the result is impossible. It cannot decay at all, except that when the particle is not relativistic the probability of finding its spinning counter-clockwise, that is oppositely to the rule about the left, is $(1 - v/c)$, so this is an additional factor for the two processes. But since the electron is so much lighter than the meson, its v is so*

much closer to c that it makes the rate of the electron disintegration much less theoretically than the rate of the other reaction. It turns out then that the amplitude is proportional to the mass of the electron, the probability is proportional to the square of the mass, and when you work it out numerically for the known masses, this ratio should be 13.6×10^{-5}. Therefore, this should safely predict no electron decays." [11]

Let us repeat the argument, where for concreteness we focus on $\pi^+ \to \ell^+ + \nu$ decays. The orientations of the spins in these decays are illustrated in Figure 22.10(a). The neutrino is very highly relativistic and can

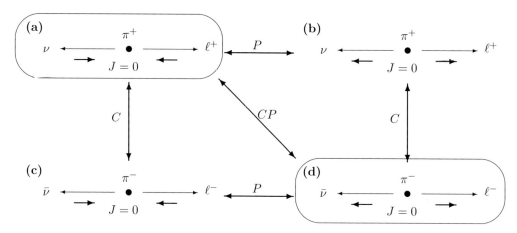

Figure 22.10 Momenta and spin orientations in the π meson decay into a lepton and a neutrino, and related decays via P, C, and CP transformation. Only the two reactions in boxes are observed in Nature (assuming the neutrino is ultra-relativistic and hence 100% polarized).

be assumed to be 100% polarized with negative helicity, i.e., with a spin pointing in the opposite direction to the neutrino momentum. The antilepton ℓ^+ and the neutrino ν travel back-to-back in opposite directions. The spin of the antilepton must point in the opposite direction to that of the neutrino to conserve angular momentum of the spin-0 parent. Hence, the antilepton must have a negative helicity! The probability of this happening is suppressed by its "natural" positive longitudinal polarization, Eq. (22.26), equal to $P_L(\ell^+) = +\beta$. Naively the wrong helicity is given for the electron by a factor $1 - \beta_e \approx 2.7 \times 10^{-5}$, while for the muon case it is $1 - \beta_\mu \approx 0.72$. They are very different because the muon is so much less relativistic than the electron owing to its mass m_μ comparable to the pion mass m_π.

According to Eq. (22.23), the fraction of antileptons with negative helicity is given by (flipping the signs in front of α since we are considering an antiparticle):

$$\frac{W(h=-1)}{W(h=+1)+W(h=-1)} = \frac{\frac{1}{2}(1-\alpha)^2}{\frac{1}{2}(1+\alpha)^2 + \frac{1}{2}(1-\alpha)^2} = \frac{1}{2}\frac{(1-\alpha)^2}{1+\alpha^2} = \frac{1}{2}\left(1 - \frac{2\alpha}{1+\alpha^2}\right) \tag{22.45}$$

where $\alpha = p_\ell/(E_\ell + m_\ell)$. We note:

$$\alpha^2 = \frac{p_\ell^2}{(E_\ell + m_\ell)^2} = \frac{E_\ell^2 - m_\ell^2}{(E_\ell + m_\ell)^2} = \frac{E_\ell - m_\ell}{E_\ell + m_\ell} = \frac{(m_\pi - m_\ell)^2}{(m_\pi + m_\ell)^2} \tag{22.46}$$

so

$$\frac{2\alpha}{1+\alpha^2} = \frac{2\frac{(m_\pi - m_\ell)}{(m_\pi + m_\ell)}}{1 + \frac{(m_\pi - m_\ell)^2}{(m_\pi + m_\ell)^2}} = \frac{2(m_\pi - m_\ell)(m_\pi + m_\ell)}{(m_\pi + m_\ell)^2 + (m_\pi - m_\ell)^2} = \frac{2(m_\pi^2 - m_\ell^2)}{2(m_\pi^2 + m_\ell^2)} \tag{22.47}$$

11   Reprinted from R. P. Feynman, "Forbidding of $\pi - \beta$ decay," in *Proceedings of the 1958 Annual International Conference on High Energy Physics at CERN, Geneva*, July 1958.

Hence, the helicity contribution to the width $\Gamma_\mathcal{H}$ is proportional to:

$$\Gamma_\mathcal{H} \propto \frac{1}{2}\left(1 - \frac{(m_\pi^2 - m_\ell^2)}{(m_\pi^2 + m_\ell^2)}\right) = \frac{1}{2}\left(\frac{2m_\ell^2}{m_\pi^2 + m_\ell^2}\right) \tag{22.48}$$

The total width is proportional to the product of the helicity factor and the phase-space factor (the exact result will be given in Eq. (23.183)):

$$\Gamma \propto \Gamma_\mathcal{H}\Gamma_{PS} = \left(\frac{m_\ell^2}{m_\pi^2 + m_\ell^2}\right)\frac{\left(m_\pi^2 + m_\ell^2\right)\left(m_\pi^2 - m_\ell^2\right)^2}{8m_\pi^5} = \frac{1}{8m_\pi^3}\left(\frac{m_\ell}{m_\pi}\right)^2\left(m_\pi^2 - m_\ell^2\right)^2 \tag{22.49}$$

The desired ratio of the pion decay to electron and to muon including the helicity factor is then:

$$R \equiv \frac{\Gamma(\pi^+ \to e^+ + \nu_e)}{\Gamma(\pi^+ \to \mu^+ + \nu_\mu)} = \left(\frac{m_e}{m_\mu}\right)^2\left(\frac{m_\pi^2 - m_e^2}{m_\pi^2 - m_\mu^2}\right)^2 \approx 1.28 \times 10^{-4} \tag{22.50}$$

This value is precisely of the order of the experimental result, Eq. (22.37)! The small difference at the level of $\approx 5\%$ can be accounted for by our tree-level approximation and can be fixed by applying radiative corrections (see Section 23.10). Hence, the strong suppression of the β decay of the π meson, $\pi \to e + \nu$, can be entirely attributed to parity violation in weak decays!

Similar arguments can be made for the π^- decays. Figure 22.10 summarizes the decay modes of the positive and negative mesons. The various reactions are related by the P, C, and CP transformations. Only the two reactions (a) and (d) represented inside of the boxes are observed in Nature, and are related by CP. The absence of (b) and (c) can be understood by the requirement on the neutrinos, which are fully polarized. The fact that the parity transformation on, for example, reaction (a) leads to reaction (b), which is absent in Nature, is a clear sign of parity violation!

22.7 Helicity of the Electron and Positron in Muon Decays (1958)

In 1958 the helicity of the electron and positron from muon decays was measured by **Macq, Crowe, and Haddock** [282]. The main goal of the experiment was to test the relation between the sign of the polarization of the muon and of the β particle, and confirm the emission of two neutrinos in such decays. At the high-energy end of the beta spectrum the neutrino and antineutrino are emitted together in the opposite direction from the beta particle. Because the neutrino and antineutrino have spins in opposite directions their net spin is zero, therefore the electron has the same spin as the muon. See Figure 22.11.

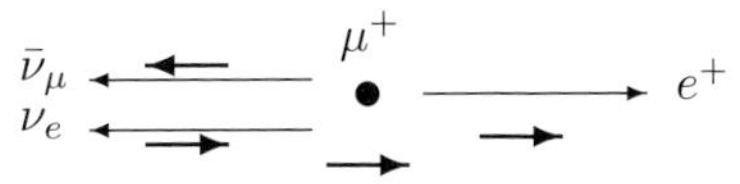

Figure 22.11 Muon (μ-meson) decay into a lepton and two neutrinos. At the high-energy end of the beta spectrum the electron has the same spin as the muon.

From the $V - A$ in the case of $\mu^+ \to e^+ + \nu + \bar{\nu}$, it is expected that those states in which the positron momentum is antiparallel to the μ^+ momentum in the pion rest frame, are strongly favored. If the positron has positive helicity then the muon has negative helicity, and vice versa. This result can be shown to hold for all electron energies. Then it follows that the μ^+ has negative helicity, and similarly μ^- has positive helicity and the e^- has negative helicity.

Experimentally the helicity was measured by determining the sense of circular polarization of their Bremsstrahlung by the method of absorption in iron magnetized against or along the direction of motion of the particles. The results were found to be consistent with the predictions from the $V - A$ form of the weak interaction [282].

In 1967 an actual measurement of the electron helicity was reported by **D. Schwartz** [283], using the spin dependence of electron–electron (Møller) scattering to determine the longitudinal polarization. The scattering of an unpolarized beam by an unpolarized target leaves the beam unpolarized. Here beam polarization is understood in the sense described in Section 4.14. Also, in the scattering of a polarized beam by an unpolarized target (or an unpolarized beam by a polarized target), no azimuthal asymmetry appears. This is a consequence of parity conservation in electromagnetic interactions. Hence, both beam and target must be polarized in order to study the polarization. The relativistic unpolarized differential $e^- e^- \to e^- e^-$ scattering cross-section is given in Eq. (11.156). The polarization effects in the scattering of polarized electrons (or positrons) on a polarized target were first calculated by **Kresnin** and **Rosentsveig** and can be introduced as a multiplicative factor to the unpolarized Møller cross-section [284]:

$$\left(\frac{d\sigma(e^- e^- \to e^- e^-)}{d\Omega}\right)_{polarized} = \left(\frac{d\sigma}{d\Omega}\right)_{unpolarized} \times \left[1 - P^b_\parallel P^t_\parallel A_\parallel(\theta) - P^b_\perp P^t_\perp A_\perp(\theta) \cos(2\phi - \phi_1 - \phi_2)\right]$$

(22.51)

where the analyzing power is defined by the transverse and longitudinal asymmetry functions $A_{\perp,\parallel}$:

$$A_\parallel = \frac{(7 + \cos^2\theta)\sin^2\theta}{(3 + \cos^2\theta)^2} \leq \frac{7}{9} \quad \text{and} \quad A_\perp = \frac{\sin^4\theta}{(3 + \cos^2\theta)^2} \leq \frac{1}{9}$$

(22.52)

$P^{b,t}_{\perp,\parallel}$ are the beam and target transverse/longitudinal polarizations, θ, ϕ are the electron scattering angles, and ϕ_1, ϕ_2 are the azimuths of the transverse polarization vectors. This method was shown to be more precise than the Bremsstrahlung method and led to a measured value $h_{e^-} = -0.89 \pm 0.28$, consistent with the $V - A$ prediction [283].

22.8   Helicity of the Muon Neutrino (1986)

Strictly considered, there is no theoretical argument that would constrain the weak charged current to be *purely* $V - A$, and it is an experimental goal to try to quantify any possible deviations from this fact. In particular, the question of whether $V + A$ currents, even suppressed, do exist in Nature, is a very important one and motivated by certain models (see Chapter 31). The existence of $V + A$ currents would affect the polarization of neutrinos. A stringent test on the muon-neutrino helicity was obtained by the Berkeley–Northwestern–TRIUMF Collaboration at TRIUMF in the 1980s [285, 286]. The following chain reaction was precisely studied:

$$\pi^+_{stopped} \to \mu^+ + \nu_\mu \quad \text{and} \quad \mu^+_{surface} \to \overline{\nu}_\mu + e^+ + \nu_e$$

(22.53)

They measured the decay asymmetry of the emitted positrons near the endpoint of the muon decay spectrum. We describe in the following the main features of the experiment.

• **Surface muon beam.** A source of highly polarized muons was obtained with the so-called "surface" muons. These were muons that come from the decay of pions which were created at the surface of the primary target after this latter was shot with protons. At TRIUMF, the accelerator provided a high-intensity source of protons up to an energy of 500 MeV (see Section 3.1). The layout of the so-called M13 beamline used in the experiment is shown in Figure 22.12. The protons from the beam line labeled 1A were shot on a 2 or 10 mm-thick carbon target. The secondary particles were collected at an angle of 135° with respect to the primary proton beam and were transported in vacuum through a system of two bending magnets, B1 and B2, which allowed for a momentum selection with $\Delta p/p \simeq 0.5\%$. At a typical current of 100 μA from the proton accelerator, the beamline delivered approximately 15,000 μ^+ per second. The flux of positive particles as a function of the beamline momentum selected by the appropriate settings of the fields in B1 and B2, is shown in Figure 22.13.

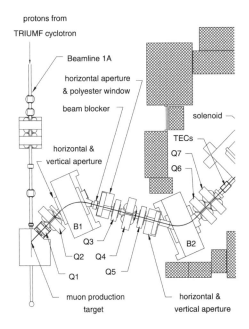

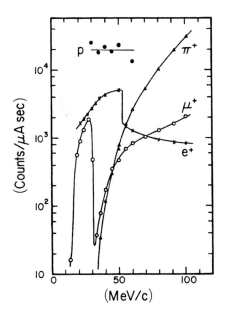

Figure 22.12 Illustration of the proton and M13 beam lines at TRIUMF. Surface muons were selected from the muon production target. Reprinted from J. F. Bueno *et al.*, "Precise measurement of parity violation in polarized muon decay," *Phys. Rev.*, vol. D84, p. 032005, 2011, under CC BY license https://creativecommons.org/licenses/by/3.0/.

Figure 22.13 Positive secondary particle flux as a function of the momentum selection in the dipole bending magnets. Reprinted with permission from A. Jodidio *et al.*, "Search for right-handed currents in muon decay," *Phys. Rev.*, vol. D34, p. 1967 (1986). [Erratum: *Phys. Rev.*, vol. D37, p. 237 (1988)]. Copyright 1986 by the American Physical Society.

The time structure of the primary beam consisted of a proton burst lasting 2–5 ns, repeating every 43 ns. The times of arrival of the secondary particles helped identify their origin. Below 29.8 MeV/c the muon flux was dominated by muons produced by pions decaying at rest near the surface of the production target. By selecting a momentum of 29.5 MeV/c, these surface muons constituted over 98% of the muon flux. Their arrival rate was measured to be exponentially distributed according to the lifetime of the pion, as expected.

As shown in Figure 22.10, neglecting neutrino masses and in the case of a pure $V - A$ weak charged current, one expects positive muons to be fully backward polarized. If the muons cross material, they will suffer from multiple scattering which can alter their state of polarization. That was the reason for selecting only "surface" muons.

• **Michel electron spectrum.** The setup of the experiment is shown in Figure 22.14(left). P1–P3 are proportional chambers; S1–S3 are scintillators; and D1–D4 are drift chambers. The momentum-selected surface muons were stopped in a pure metal foil target (e.g., Al, Cu, Ag, and Au). The target was located within a 1.1 T longitudinal field which "held" the spins. The interaction in the target could, however, alter the polarization of the muons (see below). This was a reason for repeating the measurement with different targets and choosing pure metal foils.

After the decays of the stopped positive muons, one precisely measured the momentum spectrum of the positrons. The combination of the proportional and drift chambers determined the trajectories of the incoming muons and outgoing positrons. The scintillators defined the precise timing of the events. The decay positrons emitted near the beam direction were focused into the spectrometer, where their momentum was analyzed by a cylindrical dipole magnet of 1 m diameter.

We define x as the reduced positron energy given by $x = E_e/W_e$, with W_e the maximum possible electron energy ($\approx m_\mu/2$, see Eq. (23.72)) and θ as the angle between the muon spin and the electron momentum. Since the muons are expected to be backward polarized, the positrons emitted near the beam direction correspond to an angle $\theta \approx 180°$. In addition, only the electrons emitted with a momentum near the endpoint are analyzed, so with $x \to 1$. As we will see in Section 23.8, the differential distribution of the decay positron can be expressed as (see Eq. (23.141)):

$$\frac{1}{\Gamma}\frac{\mathrm{d}^2\Gamma}{\mathrm{d}x\,\mathrm{d}\cos\theta}(x \to 1) \propto \left[\frac{2\rho}{3} + P_\mu\xi\left(\frac{2\delta}{3}\right)\cos\theta\right] = \frac{2\rho}{3}\left[1 + P_\mu\xi\left(\frac{\delta}{\rho}\right)\cos\theta\right] \qquad (22.54)$$

where P_μ is the muon polarization. The parameters ρ, δ, and ξ are called the Michel parameters. In the pure $V - A$ weak charged current, we have $\rho = \delta = 3/4$ and $\xi = 1$ (see Table 23.3), therefore **one expects that the decay rate of the positron completely vanishes at $x = 1$ for $P_\mu = 1$ and $\cos\theta = -1$.**

Experimentally, the endpoint value of the spectrum was obtained by fitting the spectrum within some angular acceptance and some positron momentum range. One clever idea to determine the endpoint (and thereby calibrate the spectrometer) is to take data with a target field configuration such that the muons will precess inside the target until they decay. In this way, muons will appear unpolarized since their spins will point in all directions after averaging over many decays. An example of data taken in the two configurations (labeled "spin precessed" and "spin held") is shown in Figure 22.14(right). The striking and huge effect of the spin on the electron momentum is clearly visible!

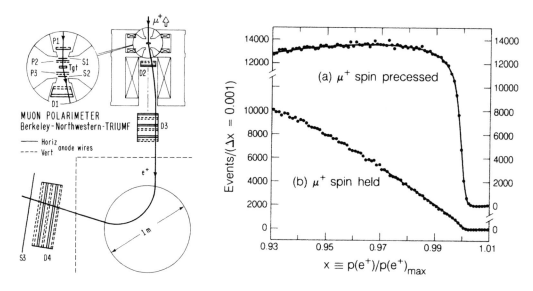

Figure 22.14 (left) Plan view of the Berkeley–Northwestern–TRIUMF experiment. (right) The measured positron spectra for (a) spin-precession mode and (b) spin-held mode. Reprinted with permission from A. Jodidio *et al.*, "Search for right-handed currents in muon decay," *Phys. Rev.*, vol. D34, p. 1967, 1986. [Erratum: *Phys. Rev.*, vol. D37, p. 237 (1988)]. Copyright 1986 by the American Physical Society.

The measurement of the rate at the endpoint was repeated for different targets and the deviation from zero was determined. This translated into measured values for the product $P_\mu(\xi\delta/\rho)$. To finalize the measurement, one should also account for possible sources of systematic errors. One of them is the possibility that the surface muons were depolarized in the target. Table 22.2 shows the results obtained for each metal individually, and

Material	$P_\mu(\xi\delta/\rho)$
Ag	$0.9983^{+0.0022}_{-0.0023}$
Al	$0.9985^{+0.00056}_{-0.00056}$
Au	$0.9986^{+0.0011}_{-0.0011}$
Cu	$1.0007^{+0.0015}_{-0.0015}$
He	$0.9979^{+0.0022}_{-0.0023}$
All targets	$0.9986^{+0.00046}_{-0.00046}$

Table 22.2 Measured $P_\mu(\xi\delta/\rho)$ for each metal, liquid He, and all targets combined. The errors are statistical only. From Ref. [285].

liquid He. Results are consistent. Since the results with the various target material do not show discrepancies, they can be combined. Finally, the combined result with all different targets and including systematic errors was estimated to be:

$$P_\mu(\xi\delta/\rho) = 0.99863 \pm 0.00046(\text{stat.}) \pm 0.00075(\text{syst.}) \tag{22.55}$$

which is consistent with the $V - A$ prediction of unity.

It was actually difficult to prove that all possible sources of muon depolarization were taken into account. Still, it was possible to interpret the result as a conservative lower limit on the possible values of $P_\mu(\xi\delta/\rho)$. The probability distribution function (pdf) of $P_\mu(\xi\delta/\rho) \times 10^3$ is plotted in Figure 22.15. It is given by a Gaussian distribution of mean $\mu = 998.63$ with a standard deviation given by $\sigma = \sqrt{46^2 + 75^2}$ (the statistical and systematic errors have been added in quadrature). The 90% CL lower limit on $P_\mu(\xi\delta/\rho)$ corresponds to the value at which the shaded integral is equal to 10%; i.e., only 10% of the experiments will yield a lower value than this. It can be computed to be equal to:

$$P_\mu(\xi\delta/\rho) > 0.99747 \quad (90\% \text{ CL}) \tag{22.56}$$

Assuming the Michel parameters, this represents a very stringent constraint on the polarization of the muon. These measurements have been gradually improved over the years and in particular new results have been obtained in 2011 by the TWIST Collaboration using the same surface beam at TRIUMF. The current world average is (value taken from the Particle Data Group's Review [5]):

$$\xi P_\mu = 1.0009^{+0.0016}_{-0.0007} \tag{22.57}$$

in very good agreement with the pure $V - A$ hypothesis.

• **Helicity of the muon neutrino.** As commented by **Fetscher** in Ref. [287], the result (Eq. (22.56)) can be used to derive a limit on the helicity of the muon neutrino. We first note that the product $(\xi\delta/\rho)$ must be less than or equal to one in order for the differential cross-section to be always positive, i.e., $(\xi\delta/\rho) \leq 1$. The maximal polarization is also 100%. Consequently:

$$1 \geq |P_\mu| > \frac{0.99747}{\xi\delta/\rho} > 0.99747 \tag{22.58}$$

On the other hand, the measured polarization is either smaller or equal to the helicity of the muon at the moment of the pion decay, since the muon can only be depolarized in the stopping target and when crossing materials. Thus, one gets a lower limit for the muon neutrino since, up to a sign, the helicity of the neutrino is identical to that of the muon. Therefore, we can write:

$$|P_{\nu_\mu}| > 0.99747 \tag{22.59}$$

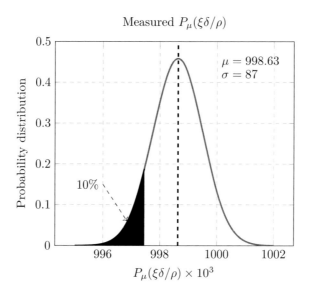

Figure 22.15 The probability distribution of the measurement (see text).

Problems

Ex 22.1 Charged pion decay – parity non-conservation. Charged pions decay weakly to muons and neutrinos. For charged pions decaying at rest the positive muon spin is aligned along its momentum, and therefore can form a polarized beam along this direction.

(a) Calculate the expected degree of polarization, P_L, for this system.

(b) Check if your result agrees with the measurement of Friedman and Telegdi, *Phys. Rev.* **106**, 1290 (1957) and if not, explain the reasons for the difference.

Ex 22.2 Neutrino polarization. It is experimentally known that neutrinos have non-vanishing mass, although they are tiny compared to other fundamental fermions. Let us assume that the rest mass of a neutrino is 1 eV. Estimate its degree of polarization in (a) pion decays, (b) kaon decays, and (c) at an energy of 1 GeV.

Ex 22.3 Kaon decay. Estimate the ratio:

$$R \equiv \frac{\Gamma(K^+ \to e^+ + \nu_e)}{\Gamma(K^+ \to \mu^+ + \nu_\mu)} \tag{22.60}$$

and compare this with the experimentally observed value.

Ex 22.4 Where is left? Imagine that you would like to explain to "someone" in a galaxy far, far away, that the human heart is on our left side. How could you describe this uniquely, without sending a left-handed or right-handed object (the polarization of the beam of light could be destroyed during travel)? As far as one knows, the far galaxy could be composed of antimatter.

23   The Weak Charged-Current Interaction

> If weak forces are assumed to have been mediated by intermediate bosons (W), the boson mass would have to equal $137M_p$ (...). In the sequel we assume just this. For the outrageous mass value itself ($M_W \sim 137M_p$) we can offer no explanation.
>
> *A. Salam and J.C. Ward*[1]

23.1   The Universal Weak Force

The weak force acts among all elementary fermions, that is, all leptons (e, μ, τ, ν) and all quarks (u, d, s, ...). That is to say, all elementary fermions carry a "**weak charge g**." Although all particles feel the weak interaction, their weak interactions are often hidden because of their "weakness" relative to the strong or electromagnetic interactions, which can occur with several orders of magnitude higher probability. The relative strengths of the forces can be appreciated when comparing typical lifetimes of particles: a hadronic resonance decays strongly while the neutral pion decays electromagnetically:

$$\begin{aligned} \rho &\to \pi\pi & \text{with} & & \tau \approx 10^{-23}\ \text{s strong} \\ \pi^0 &\to \gamma\gamma & \text{with} & & \tau \approx 10^{-18}\ \text{s electromagnetic} \\ \Sigma^0(1192) &\to \Lambda\gamma & \text{with} & & \tau \approx 10^{-19}\ \text{s electromagnetic} \end{aligned} \tag{23.1}$$

For comparison, typical weak decays have lifetimes of:

$$\begin{aligned} \pi^{\pm} &\to \mu^{\pm} + \nu & \text{with} & & \tau \approx 2.6 \times 10^{-8}\ \text{s weak} \\ \mu &\to e\nu\nu & \text{with} & & \tau \approx 2.2 \times 10^{-6}\ \text{s weak} \\ \Lambda &\to p\pi & \text{with} & & \tau \approx 2.6 \times 10^{-10}\ \text{s weak} \\ \Sigma^+(1189) &\to p\pi^0 & \text{with} & & \tau \approx 10^{-10}\ \text{s weak} \end{aligned} \tag{23.2}$$

Hence, weak interactions can be practically studied when other types of interactions are strongly suppressed or forbidden. Namely, in the cases where:

1. neutrinos are involved in the reaction, since they do not possess a strong or an electromagnetic coupling (e.g., $\pi \to \mu\nu$, $\mu \to e\nu\nu$, ...);

2. electromagnetic interactions are forbidden, e.g., $\mu \to e\gamma$ violates conservation of the lepton flavor;

3. strong interactions are forbidden by conservation of quantum numbers such as isospin, strangeness, baryon number, ..., for example $\Lambda \to p\pi$, $\Sigma^+(1189) \to p\pi^0$, ...

---

1   Reprinted with permission from A. Salam and J. C. Ward, "Electromagnetic and weak interactions," *Phys. Lett.*, vol. 13, pp. 168–171, 1964. Copyright 1964. Published by Elsevier B.V.

Looking into weak decays in more detail, we note that there are strangeness conserving $|\Delta S| = 0$ and changing $|\Delta S| = 1$ processes:

$$n \to p e^- \bar{\nu}_e \qquad |\Delta S| = 0 \tag{23.3}$$

$$\Sigma^- \to n e^- \bar{\nu}_e \qquad |\Delta S| = 1 \tag{23.4}$$

Experimentally one observes that the dynamics of the $\Delta S = 1$ reactions are suppressed with respect to the $\Delta S = 0$.

By the early 1960s, all experiments were confirming the parity-violating $V - A$ Lorentz form of the lepton and hadron currents participating in weak interactions. It was natural to expect that **all weak processes are described by $V - A$ charged currents of the type** (n, p), (Λ, p), (e, ν_e), (μ, ν_μ) **with a universal coupling** G_F. But this fact had to be proven experimentally.

Starting from the original product of Fermi's theory, Eq. (21.15), the amplitude for the $\beta^\pm$ decays including the $V - A$ form are given by:

$$\mathcal{M}_{V-A}(n \to p e^- \bar{\nu}_e) = \frac{G_\beta}{\sqrt{2}} \left[\bar{u}_p(k') \gamma^\mu (1 - \gamma^5) u_n(k) \right] \times \left[\bar{u}_e(p') \gamma_\mu (1 - \gamma^5) v_{\bar{\nu}_e}(p) \right] \tag{23.5}$$

and

$$\mathcal{M}_{V-A}(p \to n e^+ \nu_e) = \frac{G_\beta}{\sqrt{2}} \left[\bar{u}_n(k') \gamma^\mu (1 - \gamma^5) u_p(k) \right] \times \left[\bar{u}_{\nu_e}(p') \gamma_\mu (1 - \gamma^5) v_e(p) \right] \tag{23.6}$$

The factor $(\sqrt{2})^{-1}$ and the absence of a factor $1/2$ in front of the $(1 - \gamma^5)$ terms are conventional. The label G_β is to take into account potential differences between nuclear processes and leptonic decays (see Section 23.3). The amplitude for the muon decay is at tree level given by:

$$\mathcal{M}_{V-A}(\mu^- \to e^- \nu_\mu \bar{\nu}_e) = \frac{G_F}{\sqrt{2}} \left[\bar{u}_{\nu_\mu} \gamma^\mu (1 - \gamma^5) u_{\mu^-} \right] \times \left[\bar{u}_e \gamma_\mu (1 - \gamma^5) v_{\bar{\nu}_e} \right] \tag{23.7}$$

The diagrams corresponding to the weak reactions $n \to p e \nu$, $\mu^- \to e^- \nu_\mu \bar{\nu}_e$ and the strangeness changing decay $\Lambda \to p e \nu$ are shown in Figure 23.1. The coupling constant for a strangeness changing weak process has been labeled G.

23.2 The IVB-Mediated Weak Force

The fast $\mu \to e\gamma$ decay (see Section 21.9) can be avoided with the existence of more than one neutrino and the conservation of the lepton flavor (see Section 21.11). This and other observations opened the path towards **the universal $V - A$ charged current mediated by the electrically charged intermediate vector bosons (IVB) called $W^\pm$.** The $W^\pm$ bosons have a spin-1 and a weak coupling g, in analogy to the photon with coupling e for the electromagnetic force and the gluon with coupling g_S for the strong force. The $W^\pm$ **boson propagator** is the one for a massive spin-1 field (Eq. (10.87)) and is hence given in momentum space by:

$$\tilde{G}_W^{\mu\nu}(q^2) \equiv \frac{i \left(-g^{\mu\nu} + \frac{q^\mu q^\nu}{M_W^2} \right)}{q^2 - M_W^2 + i M_W \Gamma_W} \tag{23.8}$$

where M_W is the mass and Γ_W the width of the boson (see Section 26.4). Compare to the case of the massless photon $\tilde{G}_\gamma^{\mu\nu}(q^2) \equiv -i g^{\mu\nu}/q^2$.

The IVB couples to leptons of a specific flavor family, namely to the following "doublets" (the reason for this notation will become clearer in Chapter 25):

$$W^\pm : \qquad \begin{pmatrix} \nu_e \\ e^- \end{pmatrix}; \quad \begin{pmatrix} \nu_\mu \\ \mu^- \end{pmatrix}; \quad \begin{pmatrix} \nu_\tau \\ \tau^- \end{pmatrix} \tag{23.9}$$

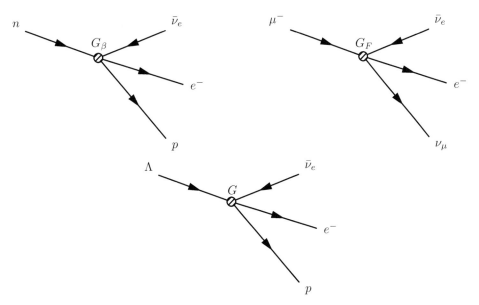

Figure 23.1 Fermi's four-point contact theory: (upper-left) β^- decay of neutron $n \to pe\nu$ (strangeness conserving); (right) muon decay; (lower) $\Lambda \to pe\nu$ (strangeness changing).

The universal coupling weak charged vertex factor to leptons is given by the $V - A$ Lorentz current:

$$W^\pm \text{ −lepton vertex factor}: \qquad \frac{-ig}{\sqrt{2}}\gamma^\mu\left(\frac{1-\gamma^5}{2}\right) = \frac{-ig}{\sqrt{2}}\gamma^\mu P_L \qquad (23.10)$$

The factor $(\sqrt{2})^{-1}$ is also conventional and in this case one keeps the factor $1/2$ in front of the $(1 - \gamma^5)$ terms. The corresponding Feynman diagrams are illustrated in Figure 23.2.

The Fermi current between hadrons translates into a weak coupling between the $W^\pm$ and the quarks. We note that the existence of $\Delta S = 0$ and $\Delta S = 1$ weak reactions (both have $\Delta Q = 1$) implies that the weak $W^\pm$ couples to all quarks. In analogy to the lepton case, (e, ν_e) and (μ, ν_μ), we seek a coupling of the $W^\pm$ to pairs of quarks. If we start with the three quarks u, d, and s, conservation of electric charge and $\Delta Q = 1$ yields the two possible choices:

$$W^\pm: \qquad \begin{pmatrix} u \\ d \end{pmatrix}; \quad \begin{pmatrix} u \\ s \end{pmatrix} \qquad (23.11)$$

The (u, d) current induces the $\Delta S = 0$ reactions and the (u, s) current the $\Delta S = 1$ reactions.

In 1963 **N. Cabibbo** explains the decay patterns in the weak interactions by postulating that these two apparently different transitions should be regarded as a single current coupling both $\Delta S = 0$ and $\Delta S = 1$ transitions [288]. Initially the mixing had only to do with the d and s quarks but it will be extended to all quarks in the CKM matrix (see Section 23.14 and Chapter 30). Considering only strangeness, the total weak current is then:

$$j^\mu = aj^\mu(\Delta S = 0) + bj^\mu(\Delta S = 1) \qquad (23.12)$$

where a and b are complex numbers to be determined. Cabibbo conjectured that "j^μ has unit length," hence $a^2 + b^2 = 1$ and therefore the two currents $\Delta S = 0$ and $\Delta S = 1$ are related and determined by only one free parameter, the so-called **Cabibbo angle** θ_C:

$$j^\mu = \cos\theta_C j^\mu(\Delta S = 0) + \sin\theta_C j^\mu(\Delta S = 1) \qquad (23.13)$$

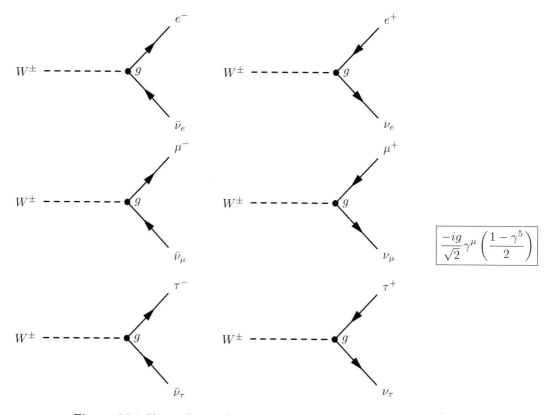

Figure 23.2 Vertex factors for the weak coupling of the charged $W^\pm$ to leptons.

Because it turns out that $\cos^2\theta_C > \sin^2\theta_C$, the $\Delta S = 0$ reaction is called *Cabibbo favored* while the $\Delta S = 1$ reaction is *Cabibbo suppressed*. The corresponding vertex factors for the Cabibbo favored and suppressed coupling to the charged $W^\pm$ are shown in Figure 23.3.

The Cabibbo current implies that quarks are "mixed." This means that the weak interaction basis is different from the strong interaction (flavor) basis. The weak current couples to **weak eigenstates that are not pure quark eigenstates but are composed of mixtures of mass (flavor) eigenstates**. Defining the weak eigenstate d' as:

$$d' \equiv d\cos\theta_C + s\sin\theta_C \tag{23.14}$$

where d and s are the quark mass (flavor) eigenstates, we get that the weak current couples to the quark doublet:

$$\begin{pmatrix} u \\ d' \end{pmatrix} = \begin{pmatrix} u \\ d\cos\theta_C + s\sin\theta_C \end{pmatrix} \tag{23.15}$$

We say that **the weak and mass (flavor) eigenstates of the quarks are mixed**. The universal weak $W^\pm$ vertex factor then becomes:

$$W^\pm \text{ vertex factor}: \quad \frac{-ig}{\sqrt{2}}\gamma^\mu\left(\frac{1-\gamma^5}{2}\right) = \frac{-ig}{\sqrt{2}}\gamma^\mu P_L \tag{23.16}$$

and couples to the following doublets of fermions:

$$W^\pm: \quad \begin{pmatrix} \nu_e \\ e^- \end{pmatrix} ; \begin{pmatrix} \nu_\mu \\ \mu^- \end{pmatrix} ; \begin{pmatrix} \nu_\tau \\ \tau^- \end{pmatrix} ; \begin{pmatrix} u \\ d' \end{pmatrix} \tag{23.17}$$

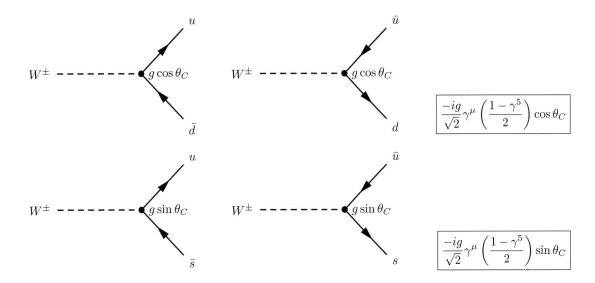

Figure 23.3 Vertex factors for the Cabibbo favored $u \to d$ and suppressed $u \to s$ weak coupling of the charged $W^{\pm}$ to quarks.

The Cabibbo angle can be determined by comparing the rates of processes that distinguish themselves only by $\Delta S = 0$ and $\Delta S = 1$. For instance, the decay width of the processes $\pi^+ \to \mu^+ \nu$ ($\Delta S = 0$) and $K^+ \to \mu^+ \nu$ ($\Delta S = 1$) give (see Eq. (22.49) and the result of the full calculation at tree level is given in Eq. (23.183)):

$$\Gamma(\pi^+ \to \mu^+ \nu) = \frac{G_F^2 \cos^2 \theta_C}{8\pi} \frac{f_\pi^2}{m_\pi} \left(\frac{m_\mu}{m_\pi}\right)^2 \left(m_\pi^2 - m_\mu^2\right)^2 \tag{23.18}$$

where f_π is **the decay constant of the pion** with units of mass. The width for the kaon decay $\Gamma(K^+ \to \mu^+ \nu)$ can be found by replacing m_π with m_K, f_π with f_K, and $\cos^2 \theta_C$ by $\sin^2 \theta_C$ in the expression above. The ratio of the widths is given by:

$$\frac{\Gamma(K^+ \to \mu^+ \nu)}{\Gamma(\pi^+ \to \mu^+ \nu)} \simeq \frac{f_K^2}{f_\pi^2} \frac{m_\pi^3}{m_K^3} \frac{\left(m_K^2 - m_\mu^2\right)^2}{\left(m_\pi^2 - m_\mu^2\right)^2} \frac{\sin^2 \theta_C}{\cos^2 \theta_C} \approx \left(\frac{f_K^2}{f_\pi^2}\right) \times 18 \times \tan^2 \theta_C \tag{23.19}$$

where we used $m_K \simeq 494$ MeV, $m_\pi \simeq 139$ MeV, and $m_\mu \simeq 105$ MeV. Assuming $f_\pi \approx f_K$ and in absence of the Cabibbo factor, one would naively predict that the decay width $\Gamma(K^+ \to \mu^+ \nu) \approx 18\Gamma(\pi^+ \to \mu^+ \nu)$. On the other hand, with the measured values $\tau_\pi \approx 26$ ns, $Br(\pi^+ \to \mu^+ \nu) \approx 100\%$, $\tau_K \approx 12.4$ ns, and $Br(K^+ \to \mu^+ \nu) \approx 63.5\%$, the experimentally observed ratio is much smaller:

$$\frac{\Gamma(K^+ \to \mu^+ \nu)}{\Gamma(\pi^+ \to \mu^+ \nu)} = \frac{\Gamma(K^+ \to \text{all})Br(K^+ \to \mu^+ \nu)}{\Gamma(\pi^+ \to \mu^+ \nu)} = \frac{\tau_\pi}{\tau_K} Br(K^+ \to \mu^+ \nu) \approx 1.33$$

The degree of suppression of $\Delta S = 1$ vs. $\Delta S = 0$ is consistent with a rather *small mixing between the d and s quarks*, namely:

$$18 \tan^2 \theta_C \approx 1.33 \quad \Longrightarrow \tan^2 \theta_C \approx 0.07 \quad \Longrightarrow \theta_C \approx 15° \tag{23.20}$$

A more precise determination gives a value:

$$\theta_C \approx 13° \quad \Longrightarrow \quad \cos^2 \theta_C \approx 0.95, \quad \sin^2 \theta_C \approx 0.05 \tag{23.21}$$

The Cabibbo theory also predicts the leptonic decays of hyperons, such as the rate of $\Lambda \to p + e^- + \bar{\nu}$, $\Sigma^- \to n + e^- + \bar{\nu}$, $\Xi^- \to \Lambda + e^- + \bar{\nu}$, $\Xi^- \to \Sigma^0 + e^- + \bar{\nu}$, and $\Xi^0 \to \Sigma^+ + e^- + \bar{\nu}$. The CKM matrix formalism will be introduced in detail in Chapter 30.

23.3   The Weak Coupling Constant and the Mass of the IVB

Let us consider again the two decays $\Sigma^0(1192) \to \Lambda\gamma$ and $\Sigma^+(1189) \to p\pi^0$. The two particles have very similar masses and naively we expect them to have similar phase-space factors. However, their lifetimes differ by nine orders of magnitude! Does this imply that the weak coupling g is very small compared to the electromagnetic coupling e? It would appear so from the measured small value of the Fermi coupling constant (see Eq. (21.14)).

However, in the $W^\pm$ there are two parameters: the coupling g and the mass of the boson M_W. At low energy, we have $q^2 \ll M_W^2$, the IVB propagator, Eq. (23.8), becomes a constant independent of the value of q^2:

$$\tilde{G}_W^{\mu\nu}(q^2) \xrightarrow{q^2 \ll M_W^2} \frac{-ig^{\mu\nu}}{M_W^2} \tag{23.22}$$

The physical interpretation is that at low energy $q^2 \ll M_W^2$ the effect of the propagator is not noticeable. The IVB propagator and weak vertices must in this case reduce to the four-point contact term of Fermi's theory.

The tree-level amplitude for the muon decay $\mu^-(k) \to e^-(p)\nu_\mu(k')\bar{\nu}_e(p')$ mediated by the $W^\pm$ is of the form (see Figure 23.4):

$$
\begin{aligned}
i\mathcal{M}_\mu(\mu^- \to e^- \nu_\mu \bar{\nu}_e) &= \frac{-ig}{\sqrt{2}} \left(\bar{u}(k')\gamma_\mu P_L u(k)\right) \tilde{G}_W^{\mu\nu}(q^2) \frac{-ig}{\sqrt{2}} \left(\bar{u}(p)\gamma_\nu P_L v(p')\right) \\
&= \frac{-ig}{\sqrt{2}} \left(\bar{u}(k')\gamma_\mu \left(\frac{1-\gamma^5}{2}\right) u(k)\right) \left(\frac{-ig^{\mu\nu}}{M_W^2}\right) \frac{-ig}{\sqrt{2}} \left(\bar{u}(p)\gamma_\nu \left(\frac{1-\gamma^5}{2}\right) v(p')\right) \\
&\simeq \frac{ig^2}{8M_W^2} \left(\bar{u}(k')\gamma^\mu \left(1-\gamma^5\right) u(k)\right) \left(\bar{u}(p)\gamma_\mu \left(1-\gamma^5\right) v(p')\right)
\end{aligned}
\tag{23.23}
$$

where we assumed $q^2 = m_\mu^2 \ll M_W^2$. Comparison with Eq. (23.7) yields:

$$\mathcal{M}_{V-A}(\mu^- \to e^- \nu_\mu \bar{\nu}_e) = \frac{G_F}{\sqrt{2}} \left(\bar{u}(k')\gamma^\mu \left(1-\gamma^5\right) u(k)\right) \left(\bar{u}(p)\gamma_\mu \left(1-\gamma^5\right) v(p')\right) \tag{23.24}$$

which gives the fundamental relation between the Fermi coupling constant G_F, the weak coupling constant g and the mass of the $W^\pm$ boson M_W:

$$\frac{G_F}{\sqrt{2}} \equiv \frac{g^2}{8M_W^2} \tag{23.25}$$

We will see later that this relation is, however, modified by radiative corrections (see Eq. (27.8)). As per the definition, the term g^2 is dimensionless (cf. $e^2 = 4\pi\alpha$) and accordingly we recover the units of G_F of $[E]^{-2}$. The relation, however, does not allow one to predict the weak coupling constant g and the mass of the $W^\pm$ independently.

During the period 1964–1968, **Abdus Salam**[2] and **John Ward**[3] [289, 290] and independently **Steven Weinberg**[4] [291] developed the electroweak unified theory based on the observations that electromagnetic and weak interactions share the same characteristics:

1. both forces affect equally all forms of leptons and hadrons;

---

2   Mohammad Abdus Salam (1926–1996), Pakistani theoretical physicist.
3   John Clive Ward (1924–2000), British–Australian physicist.
4   Steven Weinberg (1933–2021), American theoretical physicist.

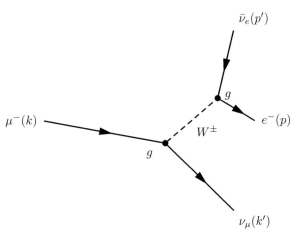

Figure 23.4 Tree-level Feynman diagram for the muon decay $\mu^-(k) \to e^-(p)\nu_\mu(k')\bar{\nu}_e(p')$.

2. both are vector V in character (and the weak interaction has in addition an axial component);

3. both (individually) possess universal coupling strength.

These facts strongly supported the gauge nature of the weak interaction (just like QED is a gauge QFT) and pointed to the existence of a massive intermediate vector boson to mediate the weak force. In the unified electromagnetic-weak (gauge) theory, the weak coupling constant must have a value similar to that of the electromagnetic coupling constant:

$$g \approx e \tag{23.26}$$

The "weakness" of the weak force at low energies is mainly attributed to the *"outrageous massiveness"* [289] of the $W^\pm$ boson. Indeed, under $g \approx e$:

$$M_W^2 = \frac{\sqrt{2}g^2}{8G_F} \approx \frac{\sqrt{2}e^2}{8G_F} = \frac{\pi\alpha}{\sqrt{2}G_F} \tag{23.27}$$

Inserting the values for G_F and α, one gets:

$$M_W \approx 37.3 \text{ GeV} \tag{23.28}$$

The $W^\pm$ boson was discovered at the CERN proton–antiproton collider by the UA1 [292] and UA2 [293] experiments with a mass of $M_W \approx 80$ GeV (see Section 27.4)! The difference between the naive expectation and the experimental value shows that the assumption that $g = e$ was not completely correct, as will be described in Chapter 25. Interestingly, if we use the experimental value for M_W, we find:

$$g^2 \equiv 4\pi\alpha_W = \frac{8G_F M_W^2}{\sqrt{2}} \approx 0.42 \tag{23.29}$$

which is larger than $e^2 = 4\pi\alpha \approx 0.09$. So, the weak coupling is in reality actually stronger than the electromagnetic coupling!

23.4 Casimir's Trick, Traces, and the Fierz Transformation

In Section 11.9 we have seen how Casimir's trick and trace theorems can be used to compute the spin-averaged matrix element squared for QED (and QCD) processes. In this section, we derive the rules for using Casimir's trick for the $V - A$ theory. For a general product of spinors with a bilinear covariant, we found that:

$$|\mathcal{M}|^2 \propto (\bar{u}_1 \Gamma u_2)^2 = [\bar{u}_1 \Gamma u_2][\bar{u}_1 \Gamma u_2]^* = [\bar{u}_1 \Gamma u_2][\bar{u}_2 \bar{\Gamma} u_1] \tag{23.30}$$

where $\bar{\Gamma} \equiv \gamma^0 \Gamma^\dagger \gamma^0$. The average over initial spins and sum of final-state spins leads to the following trace:

$$\langle |\mathcal{M}|^2 \rangle \propto \mathrm{Tr} \left[\Gamma(\not{p}_1 + m_1)\bar{\Gamma}(\not{p}_2 + m_2) \right] \tag{23.31}$$

The bilinear covariants of interest in weak processes are γ^μ, γ^5, and $\gamma^\mu \gamma^5$. We have:

- If Γ is a product of matrices, i.e., $\Gamma = \gamma^a \gamma^b$, then $\bar{\Gamma} = \overline{\gamma^b \gamma^a}$.

- $\overline{\gamma^\mu} = \gamma^\mu$.

- $\overline{\gamma^5} = \gamma^0 \gamma^{5,\dagger} \gamma^0 = \gamma^0 \gamma^5 \gamma^0 = -\gamma^0 \gamma^0 \gamma^5 = -\gamma^5$.

- $\overline{\gamma^\mu \gamma^5} = \overline{\gamma^5} \overline{\gamma^\mu} = -\gamma^5 \gamma^\mu = \gamma^\mu \gamma^5$.

- $\overline{\gamma^\mu (1 - \gamma^5)} = \overline{\gamma^\mu} - \overline{\gamma^\mu \gamma^5} = \gamma^\mu - \gamma^\mu \gamma^5 = \gamma^\mu \left(1 - \gamma^5\right)$.

Hence, the rules derived for QED can be readily applied to weak $V - A$ processes. We will discuss a few examples in the next sections. However, we first introduce the **Fierz transformation** [294], which is a handy operation to change the order of fermion fields in a four-fermion contact term. We can rewrite the tree-level amplitude for the muon decay (Eq. (23.7)) for a general $V - A$ interaction between four spin-1/2 particles (replacing $\mu \to 1$, $e \to 2$, $\nu_\mu \to 3$, and $\bar{\nu}_e \to \bar{4}$):

$$\mathcal{M}_{V-A}(1 \to 23\bar{4}) = \frac{G_F}{\sqrt{2}} \left[\bar{u}_3 \gamma^\mu (1 - \gamma^5) u_1 \right] \times \left[\bar{u}_2 \gamma_\mu (1 - \gamma^5) v_4 \right] \tag{23.32}$$

In the above equation, the assumption that the interaction is a pure $V - A$ has been put by hand. In order to test this fundamental hypothesis (and possible deviations from it), it is appropriate to write a general amplitude just as in the case of β decays (see Eq. (21.13)). Four fermion interactions can be constructed out of any of the covariant bilinears, accordingly we can consider the general interaction between four spin-1/2 Dirac spinors, invariant under proper Lorentz transformation, and assuming lepton number conservation (hence the combination of u and v spinors). The general amplitude is then given by:

$$\mathcal{M}(1 \to 23\bar{4}) = \frac{G_F}{\sqrt{2}} \sum_{i=S,V,T,A,P} \left[\bar{u}_3 \Gamma_i u_1 \right] \times \left[\bar{u}_2 \Gamma^i (A_i + A'_i \gamma^5) v_4 \right] \tag{23.33}$$

where the Γ_i are our usual products of Dirac matrices, introduced in the context of the bilinear covariants, corresponding to the S, V, T, A, and P bilinears (see Table 8.3):

$$\Gamma^i = \mathbb{1}, \gamma^\mu, i\sigma^{\mu\nu}, i\gamma^\mu \gamma^5, \gamma^5 \tag{23.34}$$

The coefficients A_i and A'_i are 10 complex numbers to be determined. Taking into account an overall arbitrary phase, one can define 19 real independent parameters to be determined experimentally. In an explicit way, keeping for instance the V and A terms, we find:

$$
\begin{aligned}
\frac{\sqrt{2}}{G_F} \mathcal{M}(1 \to 23\bar{4}) &= A_V \left(\bar{u}_3 \gamma_\mu u_1\right) \left(\bar{u}_2 \gamma^\mu v_4\right) + A'_V \left(\bar{u}_3 \gamma_\mu u_1\right) \left(\bar{u}_2 \gamma^\mu \gamma^5 v_4\right) \\
&\quad - A_A \left(\bar{u}_3 \gamma_\mu \gamma^5 u_1\right) \left(\bar{u}_2 \gamma^\mu \gamma^5 v_4\right) - A'_A \left(\bar{u}_3 \gamma_\mu \gamma^5 u_1\right) \left(\bar{u}_2 \gamma^\mu v_4\right) \\
&= \left[\bar{u}_3 \gamma_\mu (A_V - A'_A \gamma^5) u_1\right] \left(\bar{u}_2 \gamma^\mu v_4\right) + \left[\bar{u}_3 \gamma_\mu (A'_V - A_A \gamma^5) u_1\right] \left(\bar{u}_2 \gamma^\mu \gamma^5 v_4\right) \tag{23.35}
\end{aligned}
$$

The pure $V - A$ case (Eq. (23.32)) corresponds to the interaction with $A_V = 1$, $A'_V = -1$, $A_A = -1$, $A'_A = 1$, and all other A_is $= A'_i$s $= 0$.

The choice of the location of the A_i and $A'_i\gamma^5$ can appear arbitrary, and could have been equally placed between the $\bar{u}_3$ and u_1 spinors. Further, since we have a direct four-point interaction, the amplitude does not seem unique because we could satisfy the general assumptions of a Lorentz-invariant amplitude by writing spinors in the so-called **charge retention order**. Namely, by swapping the position of the u_2 and u_3 spinors (still retaining $\mu \to 1$, $e \to 2$, $\nu_\mu \to 3$, and $\bar{\nu}_e \to 4$):

$$\mathcal{M}(1 \to 32\bar{4}) = \frac{G_F}{\sqrt{2}} \sum_{i=S,V,T,A,P} [\bar{u}_2\Gamma_i u_1] \times [\bar{u}_3\Gamma^i(B_i + B'_i\gamma^5)v_4] \tag{23.36}$$

Fortunately, it was shown by Fierz that both forms are equivalent. The complex coefficients A_i, A'_i can be expressed in terms of linear combinations of the B_i, B'_i, so that the two forms are related by a Fierz transformation. Let us first consider the term without γ^5. We want:

$$\sum_i A_i [\bar{u}_3\Gamma_i u_1][\bar{u}_2\Gamma^i v_4] = -\sum_j B_j [\bar{u}_2\Gamma_j u_1][\bar{u}_3\Gamma^j v_4] \tag{23.37}$$

where the minus sign in front is conventionally inserted to follow the anticommutation rules of the Dirac fields. By inserting the explicit indices on the spinors, the expression becomes:

$$\sum_i A_i [\bar{u}_{3\gamma}(\Gamma_i)_{\gamma\alpha} u_{1\alpha}][\bar{u}_{2\beta}(\Gamma^i)_{\beta\delta} v_{4\delta}] = -\sum_j B_j [\bar{u}_{2\beta}(\Gamma_j)_{\beta\alpha} u_{1\alpha}][\bar{u}_{3\gamma}(\Gamma^j)_{\gamma\delta} v_{4\delta}] \tag{23.38}$$

or for any spinors and after multiplying by $(\Gamma_k)_{\delta\beta}$:

$$\sum_i A_i (\Gamma_i)_{\gamma\alpha} \underbrace{(\Gamma^i)_{\beta\delta}(\Gamma_k)_{\delta\beta}}_{=\mathrm{Tr}(\Gamma^i\Gamma_k)} = -\sum_j B_j (\Gamma_j)_{\beta\alpha}(\Gamma^j)_{\gamma\delta}(\Gamma_k)_{\delta\beta} = -\sum_j B_j (\Gamma^j)_{\gamma\delta}(\Gamma_k)_{\delta\beta}(\Gamma_j)_{\beta\alpha} \tag{23.39}$$

Noting the property that for any Γ_i ($i = S, V, T, A, P$) (see Appendix E.4):

$$\mathrm{Tr}\left(\Gamma^i\Gamma_k\right) = 4\delta^i_k \tag{23.40}$$

we obtain the desired relation between the coefficients:

$$4A_k(\Gamma_k)_{\gamma\alpha} = -\sum_j B_j (\Gamma^j)_{\gamma\delta}(\Gamma_k)_{\delta\beta}(\Gamma_j)_{\beta\alpha} = -\sum_j B_j (\Gamma^j\Gamma_k\Gamma_j)_{\gamma\alpha} \tag{23.41}$$

We can rewrite this equation in matrix form to get:

$$A_k\Gamma_k = -\frac{1}{4}\sum_j B_j (\Gamma^j\Gamma_k\Gamma_j) \tag{23.42}$$

The coefficients can now be evaluated explicitly. For example, we take $\Gamma_k = \mathbb{1}$ and find:

$$A_S\mathbb{1} = -\frac{1}{4}\sum_j B_j (\Gamma^j\mathbb{1}\Gamma_j) = -\frac{1}{4}\sum_j B_j (\Gamma^j\Gamma_j) = -\frac{1}{4}\sum_j B_j\mathbb{1} \quad\Longrightarrow\quad A_S = -\frac{1}{4}\sum_j B_j \tag{23.43}$$

The other coefficients can be computed as well, and can be shown to be [294]:

$$\begin{cases} 4A_S = -\sum_j B_j \\ 4A_V = -4B_S + 2B_V - 2B_A + 4B_P \\ 4A_T = -6B_S + 2B_T - 6B_P \\ 4A_A = -4B_S - 2B_V + 2B_A + 4B_P \\ 4A_P = -B_S + B_V - B_T + B_A - B_P \end{cases} \tag{23.44}$$

We note that there is no connection between the tensor T and the vector V and axial vector A. On the other hand, if there is a V or A in one form, the other form will contain in general S, V, A, and P. For the term containing γ^5, we should consider:

$$\sum_i A_i' \left[\bar{u}_3 \Gamma_i u_1\right] \left[\bar{u}_2 \Gamma^i \gamma^5 v_4\right] = -\sum_j B_j' \left[\bar{u}_2 \Gamma_j u_1\right] \left[\bar{u}_3 \Gamma^j \gamma^5 v_4\right] \tag{23.45}$$

We note that the position of γ^5 is not affected by the reordering. An explicit calculation shows that the relationships between the coefficients A_i' and B_j' are identical to those for A_i and B_j. Consequently, the relationship between all coefficients can be expressed in the form:

$$A_i^{(l)} = -\frac{1}{4} \sum_{j=S,V,T,A,P} F_{ij} B_j^{(l)} \tag{23.46}$$

where the elements F_{ij} are listed in Table 23.1.

i/j	S	V	T	A	P
S	1	1	1	1	1
V	4	-2	0	2	-4
T	6	0	-2	0	6
A	4	2	0	-2	-4
P	1	-1	1	-1	1

Table 23.1 Coefficients F_{ij} of the Fierz transformation.

In order to perform computations, one can employ either the traditional or the "charge retention ordered" forms. For example, an often-used result is the reordering of the basic $V - A$ interaction between the four Dirac spinors; Eq. (23.32) is then just very simply:

$$\mathcal{M}(1 \to 23\bar{4}) \propto \left[\bar{u}_3 \gamma^\mu (1-\gamma^5) u_1\right] \left[\bar{u}_2 \gamma_\mu (1-\gamma^5) v_4\right] = -\left[\bar{u}_2 \gamma^\mu (1-\gamma^5) u_1\right] \left[\bar{u}_3 \gamma_\mu (1-\gamma^5) v_4\right] \tag{23.47}$$

or similarly for a two-body scattering process:

$$\mathcal{M}(12 \to 34) \propto \left[\bar{u}_3 \gamma^\mu (1-\gamma^5) u_1\right] \left[\bar{u}_4 \gamma_\mu (1-\gamma^5) u_2\right] = -\left[\bar{u}_4 \gamma^\mu (1-\gamma^5) u_1\right] \left[\bar{u}_3 \gamma_\mu (1-\gamma^5) u_2\right] \tag{23.48}$$

23.5 The Muon Decay Rate

In this section we apply standard techniques to compute the lifetime of the muon starting from its tree-level Feynman diagram. We define the kinematics of the muon and of the outgoing particles with the following four-vectors (see Figure 23.4):

$$\mu^-(k) \to e^-(p) + \nu_\mu(k') + \bar{\nu}_e(p') \tag{23.49}$$

The energy–momentum conservation at each vertex yields:

$$k^\mu = q^\mu + k'^\mu = p^\mu + p'^\mu + k'^\mu \tag{23.50}$$

In the rest frame of the muon, we write the quantities:

$$k^\mu = (m_\mu, \vec{0}), \quad p^\mu = (E_e, \vec{p}), \quad k'^\mu = (E_{\nu_\mu}, \vec{k'}), \quad p'^\mu = (E_{\bar{\nu}_e}, \vec{p'}) \tag{23.51}$$

so, neglecting the neutrino masses, the momentum squared of the $W^{\pm}$ boson is equal to:

$$q^2 = (k^\mu - k'^\mu)^2 = m_\mu^2 - 2k \cdot k' = m_\mu^2 - 2m_\mu E_{\nu_\mu} \leq m_\mu^2 \tag{23.52}$$

So clearly $q^2 \leq m_\mu^2 \ll M_W^2$, since $m_\mu/M_W \approx 10^{-3}$, and the virtual boson mediating the decay is very far from its mass shell. This explains the observed, relatively long lifetime of the muon. We can safely use the Fermi current approximation and write the amplitude for the decay $\mu^- \to e^- \nu_\mu \bar{\nu}_e$ in the following way:

$$\mathcal{M}(\mu^- \to e^- \nu_\mu \bar{\nu}_e) = \frac{G_F}{\sqrt{2}} \left(\bar{u}(k')\gamma^\mu(1 - \gamma^5)u(k) \right) \left(\bar{u}(p)\gamma_\mu(1 - \gamma^5)v(p') \right) \tag{23.53}$$

Note the use of the spinor $u(k)$ for the decaying muon, $\bar{u}(k')$ for the outgoing ν_μ, $\bar{u}(p)$ for the outgoing electron, and $v(p')$ for the outgoing antineutrino $\bar{\nu}_e$. To compute the amplitude for the charge conjugate decay $\mu^+ \to e^+ \bar{\nu}_\mu \nu_e$, one needs of course to use the proper combination of spinors. The square of the matrix element $|\mathcal{M}|^2$ becomes (cf. Eq. (11.110)):

$$|\mathcal{M}|^2 = \mathcal{M}\mathcal{M}^* = \frac{G_F^2}{2} \left(\bar{u}(k')\gamma^\mu(1 - \gamma^5)u(k) \right) \left(\bar{u}(k)\gamma^\nu(1 - \gamma^5)u(k') \right)$$
$$\times \left(\bar{u}(p)\gamma_\mu(1 - \gamma^5)v(p') \right) \left(\bar{v}(p')\gamma_\nu(1 - \gamma^5)u(p) \right) \tag{23.54}$$

As for the electromagnetic case, the parts involving the electron and electron-neutrino spinors are regrouped in the electron tensor $L_e^{\mu\nu}$ and similarly the part with the muon spinors are regrouped in the tensor $L_{\mu\nu}^\mu$. We find (cf. Eq. (11.112) and note the absence of a propagator):

$$|\mathcal{M}|^2 = \frac{G_F^2}{2} L_\mu^{\mu\nu}(k, k') L_{\mu\nu}^e(p, p') \tag{23.55}$$

We now assume that the muon is unpolarized. Making use of the results derived in Section 23.4, the averaged matrix element squared becomes the product of two traces:

$$\langle |\mathcal{M}|^2 \rangle = \frac{1}{2}\frac{G_F^2}{2} \text{Tr}\left(\gamma^\mu(1 - \gamma^5)(\slashed{k} + m_\mu)\gamma^\nu(1 - \gamma^5)\slashed{k}' \right) \text{Tr}\left(\gamma_\mu(1 - \gamma^5)\slashed{p}'\gamma_\nu(1 - \gamma^5)(\slashed{p} + m_e) \right) \tag{23.56}$$

Since the mass of the electron is much smaller than the muon mass ($m_e \approx m_\mu/200$) and in practice also than any momenta involved, **we neglect the terms proportional to the electron rest mass**, setting $m_e \to 0$. The averaged matrix element becomes:

$$\langle |\mathcal{M}|^2 \rangle \simeq \frac{G_F^2}{4} \text{Tr}\left(\gamma^\mu(1 - \gamma^5)(\slashed{k} + m_\mu)\gamma^\nu(1 - \gamma^5)\slashed{k}' \right) \text{Tr}\left(\gamma_\mu(1 - \gamma^5)\slashed{p}'\gamma_\nu(1 - \gamma^5)\slashed{p} \right) \tag{23.57}$$

We now recall that the trace of an odd product of matrices vanishes as well as the trace of a product of γ^5 with an odd product of γ matrices, hence, we consider separately the terms proportional to $\slashed{k}$ and m_μ since they will lead to an even/odd number of matrices. The term proportional to $\slashed{k}$ is readily computed using the relation **[TS8]** from Appendix E.3:

$$\text{Tr}\left(\gamma^\mu(1 - \gamma^5)\slashed{k}\gamma^\nu(1 - \gamma^5)\slashed{k}' \right) \text{Tr}\left(\gamma_\mu(1 - \gamma^5)\slashed{p}'\gamma_\nu(1 - \gamma^5)\slashed{p} \right) = 256(k \cdot p')(k' \cdot p) \tag{23.58}$$

The term proportional to the muon mass vanishes:

$$m_\mu \underbrace{\text{Tr}\left(\gamma^\mu(1 - \gamma^5)\gamma^\nu(1 - \gamma^5)\slashed{k}' \right)}_{0} \text{Tr}\left(\gamma_\mu(1 - \gamma^5)\slashed{p}'\gamma_\nu(1 - \gamma^5)\slashed{p} \right) = 0 \tag{23.59}$$

since the first trace contains an odd number of γ matrices. The averaged matrix element squared (having neglected the electron rest mass) is then just simply:

$$\langle |\mathcal{M}|^2 \rangle \simeq 64G_F^2(k \cdot p')(k' \cdot p) \tag{23.60}$$

Using Eq. (23.51), we find:

$$k \cdot p' = m_\mu E_{\bar{\nu}_e} \qquad \text{and} \qquad 2(k' \cdot p) \approx (k' + p)^2 = (k - p')^2 \simeq m_\mu^2 - 2m_\mu E_{\bar{\nu}_e} \tag{23.61}$$

which is quite an amazing result since it means that the dynamics of the decay can be completely determined only by fixing the energy of the antineutrino $\bar{\nu}_e$:

$$\langle |\mathcal{M}|^2 \rangle \simeq 32 G_F^2 m_\mu^2 E_{\bar{\nu}_e}(m_\mu - 2E_{\bar{\nu}_e}) \tag{23.62}$$

In particular, it vanishes for the two extreme cases where the electron neutrino has zero energy or the maximum possible energy (below are the exact boundaries in our approximation where we neglect the electron mass):

$$\langle |\mathcal{M}|^2 \rangle = 0 \qquad \Leftrightarrow \qquad E_{\bar{\nu}_e} = 0 \quad \text{or} \quad E_{\bar{\nu}_e} = \frac{m_\mu}{2} \tag{23.63}$$

The highest bound corresponds to the limiting kinematical configuration shown in Figure 23.5, where the electron antineutrino is back-to-back with the electron and muon neutrino pair with energies $E_e = E_{\nu_\mu} = E_{\bar{\nu}_e}/2$. This is the only possible way to balance momentum. In this configuration, angular momentum conservation forces the electron (or equivalently the muon neutrino) to have the "wrong helicity." In our approximation where $m_e \to 0$ this is prohibited and hence our matrix element vanishes. This is a direct consequence of the $V - A$ parity-violating nature of the weak process. Another way of expressing this fact is to assume that the electron has the same helicity as the muon neutrino (since both are particles). In this configuration the spins of the three decay products add to the state of total angular momentum of $3/2$, which is not compatible with the initial state spin-$1/2$ of the muon. In conclusion, the electron antineutrino cannot reach its kinematical limit, although this would be allowed from pure point of view of kinematics. The $V - A$ structure of the weak interaction suppresses this.

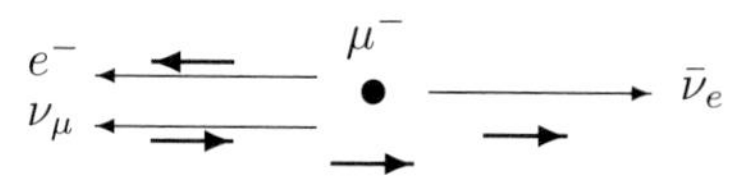

Figure 23.5 Muon (μ-meson) decay into a lepton and two neutrinos. In the kinematical configuration where the electron antineutrino recoils against the electron and the muon neutrino, these latter must have opposite helicity which forces one of them to have the "wrong helicity."

We now compute the partial decay width of the process $\mu^-(k) \to e^-(p)\nu_\mu(k')\bar{\nu}_e(p')$ by considering the three-body phase-space factor (see Section 5.18):

$$d^9\Gamma = \frac{1}{2m_\mu} \langle |\mathcal{M}|^2 \rangle \frac{d^3\vec{p}}{(2\pi)^3 2E_e} \frac{d^3\vec{k'}}{(2\pi)^3 2E_{\nu_\mu}} \frac{d^3\vec{p'}}{(2\pi)^3 2E_{\bar{\nu}_e}} (2\pi)^4 \delta^4(k - p - k' - p') \tag{23.64}$$

Following the method described in Section 5.18, the kinematics in our chosen rest frame of the parent muon is totally defined by two independent parameters. We can decide to integrate out the muon neutrino with momentum $\vec{k'}$ and can choose as variables the energy of the electron antineutrino $E_{\bar{\nu}_e}$ already appearing in the matrix element and its opening angle θ with the electron. We get:

$$d^9\Gamma = \frac{\langle |\mathcal{M}|^2 \rangle}{8(2\pi)^5 m_\mu} \frac{d^3\vec{p}}{E_e} \frac{d^3\vec{k'}}{2E_{\nu_\mu}} \frac{d^3\vec{p'}}{E_{\bar{\nu}_e}} \delta^4(k - p - k' - p') \tag{23.65}$$

In order to rewrite the integral over $d^3\vec{k'}$, we use the general expression (see Section 5.12):

$$\int \frac{d^3\vec{k'}}{2E_{\nu_\mu}} = \int d^4k' \theta(E_{\nu_\mu}) \delta((k')^2) = \int d^4k' \theta(m_\mu - E_e - E_{\bar{\nu}_e}) \delta((k - p - p')^2) \tag{23.66}$$

which, after integration over d^4k', leaves the partial width over the outgoing electron and electron neutrino:

$$d^6\Gamma = \frac{\langle|\mathcal{M}|^2\rangle}{8(2\pi)^5 m_\mu} \frac{d^3\vec{p}\,d^3\vec{p}'}{E_e\,E_{\bar{\nu}_e}} \theta(m_\mu - E_e - E_{\bar{\nu}_e})\delta((k-p-p')^2) \tag{23.67}$$

The Dirac δ function can be simplified by noting:

$$(k-p-p')^2 = k^2 + p^2 + p'^2 - 2(k\cdot p) - (2k\cdot p') + 2(p\cdot p') = m_\mu^2 - 2m_\mu E_e - 2m_\mu E_{\bar{\nu}_e} + 2E_e E_{\bar{\nu}_e}(1 - \cos\theta) \tag{23.68}$$

The phase-space factor yields, always neglecting neutrino masses:

$$\frac{d^3\vec{p}\,d^3\vec{p}'}{E_e\,E_{\bar{\nu}_e}} = \frac{p^2 dp d\Omega}{E_e} \frac{E_{\bar{\nu}_e}^2 dE_{\bar{\nu}_e}(2\pi)d\cos\theta}{E_{\bar{\nu}_e}} \simeq (8\pi^2)E_e E_{\bar{\nu}_e} dE_e dE_{\bar{\nu}_e} d\cos\theta \tag{23.69}$$

where we have integrated over the solid angle and in the last term we have neglected the electron mass. The partial width simplifies further to:

$$d^3\Gamma = \frac{\langle|\mathcal{M}|^2\rangle}{32\pi^3 m_\mu} E_e E_{\bar{\nu}_e} \delta(m_\mu^2 - 2m_\mu E_e - 2m_\mu E_{\bar{\nu}_e} + 2E_e E_{\bar{\nu}_e}(1 - \cos\theta))dE_e dE_{\bar{\nu}_e} d\cos\theta \tag{23.70}$$

The Dirac δ function gives us the following kinematical constraint:

$$\frac{m_\mu}{2} - E_e - E_{\bar{\nu}_e} + \frac{E_e E_{\bar{\nu}_e}}{m_\mu}(1 - \cos\theta) = 0 \tag{23.71}$$

The maximum energy of the electron E_e is equal to (this corresponds to a kinematical configuration equivalent to that shown in Figure 22.11):

$$W_e = \max(E_e) = \frac{m_\mu^2 + m_e^2}{2m_\mu} \approx \frac{m_\mu}{2} \tag{23.72}$$

where in the last term, as before, we neglect the electron mass relative to the muon mass. Hence:

$$E_e + E_{\bar{\nu}_e} - \frac{E_e E_{\bar{\nu}_e}}{m_\mu}(1 - \cos\theta) = W_e \quad\Longrightarrow\quad E_e + E_{\bar{\nu}_e} = W_e + \frac{E_e E_{\bar{\nu}_e}}{2W_e}(1 - \cos\theta) \tag{23.73}$$

We can write this as a constraint on the energy carried away by the electron and the electron antineutrino, taking into account that $0 \leq (1 - \cos\theta) \leq 2$:

$$W_e \leq E_e + E_{\bar{\nu}_e} \leq W_e + \frac{E_e E_{\bar{\nu}_e}}{W_e} \tag{23.74}$$

From the first inequality we can write $W_e - E_e \leq E_{\bar{\nu}_e}$. From the second follows:

$$E_{\bar{\nu}_e} - \frac{E_e E_{\bar{\nu}_e}}{W_e} \leq W_e - E_e \quad\rightarrow\quad E_{\bar{\nu}_e}\left(1 - \frac{E_e}{W_e}\right) \leq W_e\left(1 - \frac{E_e}{W_e}\right) \quad\rightarrow\quad E_{\bar{\nu}_e} \leq W_e \tag{23.75}$$

Then we decide to integrate over all possible electron-antineutrino energies $E_{\bar{\nu}_e}$ for a given electron energy E_e:

$$W_e - E_e \leq E_{\bar{\nu}_e} \leq W_e \quad \text{with} \quad 0 \leq E_e \leq W_e \tag{23.76}$$

We must still integrate over the opening angle θ between the electron and the electron antineutrino. We note that this gives simply:

$$\int \delta(\cdots + 2E_e E_{\bar{\nu}_e}(1 - \cos\theta))d\cos\theta = \frac{1}{2E_e E_{\bar{\nu}_e}} \tag{23.77}$$

With this we can get the partial width by integrating over the opening angle, an expression which looks very clean indeed:

$$d^2\Gamma = \frac{\langle|\mathcal{M}|^2\rangle}{64\pi^3 m_\mu}dE_e dE_{\bar{\nu}_e} \tag{23.78}$$

We can derive the electron spectrum by integrating over the electron-antineutrino energy:

$$\frac{d\Gamma}{dE_e} = \frac{1}{64\pi^3 m_\mu}\int_{W_e-E_e}^{W_e}dE_{\bar{\nu}_e}\langle|\mathcal{M}|^2\rangle = \frac{G_F^2 m_\mu}{2\pi^3}\int_{W_e-E_e}^{W_e}E_{\bar{\nu}_e}(m_\mu - 2E_{\bar{\nu}_e})dE_{\bar{\nu}_e} \tag{23.79}$$

Integration over the electron-neutrino energy yields:

$$\begin{aligned}
\int_{W_e-E_e}^{W_e}E_{\bar{\nu}_e}(m_\mu - 2E_{\bar{\nu}_e})dE_{\bar{\nu}_e} &= \left.\frac{m_\mu E_{\bar{\nu}_e}^2}{6}\left(3-\frac{4E_{\bar{\nu}_e}}{m_\mu}\right)\right|_{W_e-E_e}^{W_e} \\
&= \frac{m_\mu}{6}\left[W_e^2\left(3-\frac{4W_e}{m_\mu}\right)-(W_e-E_e)^2\left(3-\frac{4W_e}{m_\mu}+\frac{4E_e}{m_\mu}\right)\right]
\end{aligned} \tag{23.80}$$

Using $W_e \approx m_\mu/2$, we find:

$$3-\frac{4W_e}{m_\mu} \approx 3-\frac{4m_\mu}{2m_\mu} = 1 \tag{23.81}$$

Hence

$$\begin{aligned}
\int_{W_e-E_e}^{W_e}E_{\bar{\nu}_e}(m_\mu - 2E_{\bar{\nu}_e})dE_{\bar{\nu}_e} &= \frac{m_\mu}{6}\left[W_e^2-(W_e-E_e)^2\left(1+\frac{4E_e}{m_\mu}\right)\right] \\
&= \frac{m_\mu}{6}\left[2W_e E_e - E_e^2 - (W_e-E_e)^2\left(\frac{4E_e}{m_\mu}\right)\right] \\
&\approx \frac{m_\mu E_e}{6}\left[m_\mu - E_e - \frac{4}{m_\mu}\left(\frac{m_\mu^2}{4}-m_\mu E_e + E_e^2\right)\right] \\
&= \frac{m_\mu E_e^2}{6}\left[3-\frac{4E_e}{m_\mu}\right]
\end{aligned} \tag{23.82}$$

$$\frac{d\Gamma}{dE_e} = \frac{G_F^2 m_\mu^2 E_e^2}{12\pi^3}\left[3-\frac{4E_e}{m_\mu}\right] = \frac{G_F^2 m_\mu^4}{12\pi^3}\frac{1}{4}\left(\frac{2E_e}{m_\mu}\right)^2\left[3-2\left(\frac{2E_e}{m_\mu}\right)\right] \tag{23.83}$$

- **Michel electron spectrum.** Finally, the **spectrum of the electron in unpolarized muon decays** introducing the **electron reduced energy** $x \equiv E_e/W_e = 2E_e/m_\mu$ ($0 \le x \le 1$), is given by:

$$\left(\frac{d\Gamma}{dx}\right)_e = \frac{G_F^2 m_\mu^5}{12\pi^3}\frac{x^2}{8}(3-2x) = \frac{G_F^2 m_\mu^5}{12\pi^3}\frac{1}{16}2x^2(3-2x) \tag{23.84}$$

or

$$\left(\frac{d\Gamma}{dx}\right)_e = \frac{G_F^2 m_\mu^5}{192\pi^3}[n(x)] \quad \text{where} \quad n(x) = 2x^2(3-2x) \tag{23.85}$$

The function $n(x)$ gives the electron reduced-energy spectrum (neglecting the electron rest mass). It is plotted in Figure 23.6.

- **Muon lifetime.** One can integrate over the electron reduced-energy in the full interval $x \in [0,1]$ to get:

$$\Gamma = \int dx \frac{d\Gamma}{dx} = \frac{G_F^2 m_\mu^5}{192\pi^3}\underbrace{\int_0^1 2x^2(3-2x)\,dx}_{=1} = \frac{G_F^2 m_\mu^5}{192\pi^3} \tag{23.86}$$

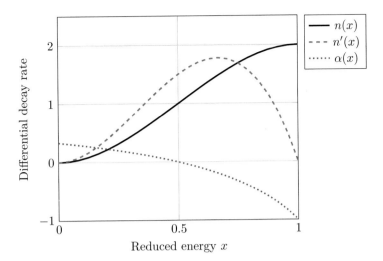

Figure 23.6 Decay energy distribution for $\mu^- \to e^- \bar{\nu}_e \nu_\mu$. $n(x)$ is the energy distribution for the electron (neglecting its rest mass) and $n'(x)$ is the energy distribution of the electron antineutrino. $\alpha(x)$ is the asymmetry parameter of the muon (see Section 23.7).

The **total width of the muon** Γ and its lifetime τ_μ are then equal to:

$$\Gamma(\mu^- \to e^- \nu_\mu \bar{\nu}_e) = \frac{1}{\tau_\mu} = \frac{G_F^2 m_\mu^5}{192\pi^3} \tag{23.87}$$

• **Electron-antineutrino spectrum.** Let us go back to Eq. (23.78) and derive the electron-antineutrino spectrum by integrating over the electron energy:

$$\mathrm{d}^2\Gamma = \frac{32 G_F^2 m_\mu^2 E_{\bar{\nu}_e}(m_\mu - 2E_{\bar{\nu}_e})}{64\pi^3 m_\mu}\mathrm{d}E_e\mathrm{d}E_{\bar{\nu}_e} = \frac{G_F^2 m_\mu}{2\pi^3}E_{\bar{\nu}_e}(m_\mu - 2E_{\bar{\nu}_e})\mathrm{d}E_{\bar{\nu}_e}\mathrm{d}E_e \tag{23.88}$$

Since the matrix element does not depend on E_e, we simply have to integrate:

$$\int_{W_e - E_{\bar{\nu}_e}}^{W_e} \mathrm{d}E_e = E_e\Big|_{W_e - E_{\bar{\nu}_e}}^{W_e} = W_e - W_e + E_{\bar{\nu}_e} = E_{\bar{\nu}_e} \tag{23.89}$$

Hence

$$\mathrm{d}\Gamma = \frac{G_F^2 m_\mu}{2\pi^3}E_{\bar{\nu}_e}^2(m_\mu - 2E_{\bar{\nu}_e})\mathrm{d}E_{\bar{\nu}_e} \tag{23.90}$$

Introducing again the reduced energy, this time for the electron antineutrino, i.e., $x = E_{\bar{\nu}_e}/W_e = 2E_{\bar{\nu}_e}/m_\mu$ (neglecting the electron rest mass, the electron and the neutrinos have the same maximum energy), we get:

$$\mathrm{d}\Gamma = \frac{G_F^2 m_\mu}{2\pi^3}W_e^2 x^2(m_\mu - 2xW_e)W_e\mathrm{d}x = \frac{G_F^2 m_\mu}{2\pi^3}\left(\frac{m_\mu}{2}\right)^3 x^2(m_\mu - 2x\frac{m_\mu}{2})\mathrm{d}x = \frac{G_F^2 m_\mu^5}{16\pi^3}x^2(1-x)\mathrm{d}x \tag{23.91}$$

or

$$\left(\frac{\mathrm{d}\Gamma}{dx}\right)_{\bar{\nu}_e} = \frac{G_F^2 m_\mu^5}{192\pi^3}\left[n'(x)\right] \quad \text{where} \quad n'(x) = 12x^2(1-x) \tag{23.92}$$

The function $n'(x)$ gives the electron reduced-energy spectrum (neglecting the electron rest mass). It is normalized such that its integral is $\int_0^1 n'(x)\mathrm{d}x = 1$. It is plotted in Figure 23.6.

• **Discussion.** There is a lot of physics embedded in the functions $n(x)$ and $n'(x)$ which follows directly from the $V - A$ nature of the weak interaction and the conservation of angular momentum. We have already noted in Eq. (23.63) that the partial decay rate vanishes for the two extreme cases where the electron neutrino has zero energy or the maximum possible energy (below are the exact boundaries in our approximation where we neglect the electron mass). Hence, we expect:

$$\left(\frac{d\Gamma}{dx}\right)_{\bar{\nu}_e} = 0 \quad \Leftrightarrow \quad n'(x = 0) = 0 \quad \text{or} \quad n'(x = 1) = 0 \tag{23.93}$$

The corresponding spin configuration for $x = 1$ is shown in Figure 23.5. For the same reason, angular momentum conservation prohibits the electron to have zero energy, hence

$$\left(\frac{d\Gamma}{dx}\right)_e = 0 \quad \Leftrightarrow \quad n(x = 0) = 0 \tag{23.94}$$

Finally, the decay rate is *maximum* when the electron carries away the maximum energy since in this case it nicely recoils against the two neutrinos whose spins compensate one another. We have already noted in Section 22.7 that in this case the electron has the same spin as the muon (see Figure 22.11). Hence:

$$\left(\frac{d\Gamma}{dx}\right)_e \text{ is maximum} \quad \Leftrightarrow \quad n(x = 1) = 2 \tag{23.95}$$

Lovely!

23.6 Electron Mass Term and Radiative Corrections to the Muon Lifetime

The total muon decay rate provides a very accurate determination of the Fermi coupling constant G_F and therefore a relationship between the weak coupling constant g and the mass of the IVB boson M_W via Eq. (23.25). It is therefore fundamental to reach the highest possible precision in its determination. Theoretically, one has to improve our previous result, Eq. (23.87), by including mass terms and also QED radiative and electroweak corrections. Following **Rittbergen** and **Stuart**, the result can be expressed as [295]:

$$\Gamma(\mu^- \to e^-(\gamma)\nu_\mu\bar{\nu}_e) = \underbrace{\left(\frac{G_F^2 m_\mu^5}{192\pi^3}\right)}_{\text{Born}} \underbrace{\left(1 + \delta f\left(\frac{m_e^2}{m_\mu^2}, \frac{m_{\nu_\mu}^2}{m_\mu^2}\right)\right)}_{\text{finite mass}} \underbrace{(1 + \delta_{\text{QED}})}_{\text{QED correction}} \underbrace{\left(1 + \frac{3m_\mu^2}{5M_W^2}\right)}_{\text{electroweak correction}} \tag{23.96}$$

The mass of the electron neutrino has been neglected in the space-phase factor. Under this assumption, one finds [295]:

$$\delta f(x, y) = -8x + 8x^3 - x^4 - 12x^2 \ln x - 8y + \cdots \tag{23.97}$$

Note that as a consequence of the Fierz theorem (see Eq. (23.47)), the electron and the muon neutrino spinors can be exchanged and as a result both terms lead to the same pre-factor in δf. The mass of the muon neutrino can be safely neglected, but computing this theoretical correction requires the electron-to-muon mass ratio m_e/m_μ. The recommended value by CODATA is [10]:

$$m_\mu/m_e = 206.7682830(46) \quad \Longrightarrow \quad m_e/m_\mu = 4.83633169(11) \times 10^{-3} \tag{23.98}$$

With this value, the phase-space factor becomes:

$$\delta f = -187.051 \times 10^{-6} \tag{23.99}$$

The one-loop QED contributions to the muon lifetime were first calculated by **Kinoshita** and **Sirlin** [296] and by **Berman** [297]. Rittbergen and Stuart have also calculated the QED corrections to order α^2 [298] and found:

$$\delta_{\text{QED}} = \underbrace{\left(\frac{\alpha(m_\mu)}{\pi}\right)\left(\frac{25}{8} - \frac{\pi^2}{2}\right)}_{1\text{ loop}} + \underbrace{\left(\frac{\alpha(m_\mu)}{\pi}\right)^2 6.743}_{2\text{ loops}} \tag{23.100}$$

where the fine-structure constant $\alpha(m_\mu)$ at the relevant momentum transfer $Q^2 = m_\mu^2$ (see Section 12.6) is $\alpha(m_\mu) = 1/135.902660087(44)$ computed starting from the CODATA value at zero momentum transfer $\alpha(0) = 1/137.035999074(44)$. One then finds:

$$\delta_{\text{QED}} = -4238.9 \times 10^{-6} + 37.0 \times 10^{-6} = -4201.9 \times 10^{-6} \tag{23.101}$$

Finally, the electroweak correction is negligible in comparison and is about 1×10^{-6}. We can write:

$$\Gamma^{theo}(\mu^- \to e^-(\gamma)\nu_\mu\bar{\nu}_e) = \left(\frac{G_F^2 m_\mu^5}{192\pi^3}\right)(1 + \delta f)(1 + \delta_{QED})(1 + \delta_{EW}) \equiv \left(\frac{G_F^2 m_\mu^5}{192\pi^3}\right)\left(1 + \delta\Gamma^{corr}\right) \tag{23.102}$$

The theoretical estimations were subsequently further improved by **Pak** and **Czarnecki** to include electron mass corrections to the original two-loop calculations [299]. Consequently, the total theoretical error is estimated to be $\Delta\Gamma^{theo}/\Gamma^{theo} = 0.14 \times 10^{-6}$.

• **Precise determination of the Fermi coupling constant.** The previous result can be inverted and expressed as a function of the muon lifetime $\tau_\mu = 1/\Gamma(\mu^- \to e^-(\gamma)\nu_\mu\bar{\nu}_e)$ to find:

$$G_F = \sqrt{\frac{192\pi^3}{\tau_\mu m_\mu^5}\frac{1}{1 + \delta\Gamma^{corr}}} \tag{23.103}$$

The error in the Fermi constant G_F is then:

$$\frac{\Delta G_{\text{F}}}{G_{\text{F}}} = \frac{1}{2}\sqrt{\left(\frac{\Delta\tau_\mu}{\tau_\mu}\right)^2 + \left(5\frac{\Delta m_\mu}{m_\mu}\right)^2 + \left(\frac{\Delta\Gamma^{theo}}{\Gamma^{theo}}\right)^2} \tag{23.104}$$

The theoretical determination of G_F in natural units (GeV^{-2}) from the measurement of τ_μ in units of time requires a unit conversion via Planck's constant, $\hbar$, which is an exact number. See Eq. (1.23) and Table 1.6.

Recent experimental measurements of the muon lifetime sorted according to their publication year are shown in Table 23.2. The most precise muon lifetime is measured by the MuLan experiment at the Paul Scherrer Institute [300]:

$$\tau_\mu(\text{MuLan}) = 2196980.3 \pm 2.1(\text{stat.}) \pm 0.7(\text{syst.}) \text{ ps} \tag{23.105}$$

The experimental error is $\Delta\tau_\mu/\tau_\mu = 1 \times 10^{-6}$ or one part per million (ppm)! This result dominates the value recommended by the Particle Data Group's Review – see Eq. (15.17).

The uncertainty in the CODATA value of the muon mass is $\Delta m_\mu/m_\mu = 22 \times 10^{-9}$, and its contribution to the error in G_F is five times larger, but still negligible at 0.1 ppm. The theoretical error is 0.14 ppm. The error in G_F is dominated by the experimental error of 1 ppm, but reaches nonetheless the amazing precision of:

$$\frac{\Delta G_{\text{F}}}{G_{\text{F}}} = \frac{1}{2}\sqrt{(1 \text{ ppm})^2 + (0.1 \text{ ppm})^2 + (0.14 \text{ ppm})^2} = 0.5 \text{ ppm} \tag{23.106}$$

The CODATA recommended value for G_F is given in Eq. (21.14).

Measured muon lifetime (μs)	Reference	Publication year
2.1973 ± 0.0003	Duclos *et al.* [301]	1973
2.19711 ± 0.00008	Balandin *et al.* [302]	1974
2.19695 ± 0.00006	Giovanetti *et al.* [303]	1984
2.197078 ± 0.000073	Bardin *et al.* [304]	1984
2.197013 ± 0.000024	Chitwood *et al.* [305]	2007
2.197083 ± 0.000035	Barczyk *et al.* [306]	2008
2.1969803 ± 0.0000022	V. Tishchenko *et al.* [300]	2013

Table 23.2 Experimental measurements of the muon lifetime sorted according to their publication year.

23.7   Michel Electron Spectrum (Polarized Case)

We now focus on the momentum spectrum of the electron in the rest frame of the decaying muon. In Section 22.8 we have seen that the spectrum is influenced by the polarization of the muon, and that the chain $\pi \to \mu$ provides a natural way to experimentally study polarized muons.

Therefore, we consider again Eq. (23.55) and this time **we assume that the muon is polarized.** Accordingly, we introduce the spins of the muon and of the electron with the polarization four-vectors s_1^μ and s_2^μ:

$$\mu^-(k, s_1^\mu) \to e^-(p, s_2^\mu) + \nu_\mu(k') + \bar{\nu}_e(p') \tag{23.107}$$

Neglecting their rest masses, the neutrinos are fully polarized. The spin projection operator $\Sigma(s) = \frac{1}{2}(1 + \gamma^5 \not{s})$ for a Dirac particle was introduced in Section 8.25. As we did for the electromagnetic interaction (see Section 11.14), the polarized decay rate is computed using the common Casimir trick for averaging and summing over the spin degrees of freedom, however, inserting "by hand" the spin projection operator. Consequently, making use of Eq. (8.266), we can write the muon tensor $L_{\mu\nu}^\mu(k, k')$ as (cf. Eq. (23.56)):

$$L_\mu^{\mu\nu}(k, k', s_1) = \mathrm{Tr}\left[\gamma^\mu(1 - \gamma^5)(\not{k} + m_\mu)\underbrace{\frac{1}{2}(1 + \gamma^5 \not{s}_1)}_{\text{spin projection}}\gamma^\nu(1 - \gamma^5)\not{k}'\right] \tag{23.108}$$

We now use the trace theorems of Appendix E.2 to simplify this expression. We first note that we have four separate terms:

$$\begin{aligned}
L_\mu^{\mu\nu}(k, k', s_1) &= \frac{1}{2}\mathrm{Tr}\left[\gamma^\mu(1 - \gamma^5)\not{k}\gamma^\nu(1 - \gamma^5)\not{k}'\right] + \underbrace{\frac{1}{2}\mathrm{Tr}\left[\gamma^\mu(1 - \gamma^5)\not{k}\gamma^5\not{s}_1\gamma^\nu(1 - \gamma^5)\not{k}'\right]}_{=0} \\
&\quad + \underbrace{\frac{m_\mu}{2}\mathrm{Tr}\left[\gamma^\mu(1 - \gamma^5)\gamma^\nu(1 - \gamma^5)\not{k}'\right]}_{=0} + \frac{m_\mu}{2}\mathrm{Tr}\left[\gamma^\mu(1 - \gamma^5)\gamma^5\not{s}_1\gamma^\nu(1 - \gamma^5)\not{k}'\right] \\
&= \frac{1}{2}\mathrm{Tr}\left[\gamma^\mu(1 - \gamma^5)(\not{k} + m_\mu\gamma^5\not{s}_1)\gamma^\nu(1 - \gamma^5)\not{k}'\right]
\end{aligned} \tag{23.109}$$

where we used the fact that the trace of an *odd* number of gamma matrices vanishes. **We note that the spin term is proportional to m_μ, so it only contributes to the decay amplitude if we do not neglect the muon rest mass.** Using the property **[G53]** $\gamma^5\gamma^\mu = -\gamma^\mu\gamma^5$ from Appendix E.1, we further obtain:

$$
\begin{aligned}
L_\mu^{\mu\nu}(k, k', s_1) &= \frac{1}{2}\operatorname{Tr}\left[\gamma^\mu(\slashed{k} + m_\mu\gamma^5\slashed{s}_1)\gamma^\nu(1-\gamma^5)(1-\gamma^5)\slashed{k}'\right] \\
&= \operatorname{Tr}\left[\gamma^\mu(\slashed{k} + m_\mu\gamma^5\slashed{s}_1)\gamma^\nu(1-\gamma^5)\slashed{k}'\right]
\end{aligned}
\tag{23.110}
$$

since $(1-\gamma^5)(1-\gamma^5) = 1 - 2\gamma^5 + (\gamma^5)^2 = 2(1-\gamma^5)$. The spin-dependent term can be further simplified by bringing the γ^5 matrix to the right:

$$
\begin{aligned}
m_\mu\operatorname{Tr}\left[\gamma^\mu\gamma^5\slashed{s}_1\gamma^\nu(1-\gamma^5)\slashed{k}'\right] &= -m_\mu\operatorname{Tr}\left[\gamma^\mu\slashed{s}_1\gamma^5\gamma^\nu(1-\gamma^5)\slashed{k}'\right] = m_\mu\operatorname{Tr}\left[\gamma^\mu\slashed{s}_1\gamma^\nu\gamma^5(1-\gamma^5)\slashed{k}'\right] \\
&= -m_\mu\operatorname{Tr}\left[\gamma^\mu\slashed{s}_1\gamma^\nu(1-\gamma^5)\slashed{k}'\right]
\end{aligned}
\tag{23.111}
$$

Finally, we find:

$$
L_\mu^{\mu\nu}(k, k', s_1) = \operatorname{Tr}\left[\gamma^\mu(\slashed{k} - m_\mu\slashed{s}_1)\gamma^\nu(1-\gamma^5)\slashed{k}'\right] = (k - m_\mu s_1)_\alpha k'_\beta \operatorname{Tr}\left[\gamma^\mu\gamma^\alpha\gamma^\nu(1-\gamma^5)\gamma^\beta\right]
\tag{23.112}
$$

Now similarly, for the electron tensor:

$$
L_e^{\mu\nu}(p, p', s_2) = (p - m_e s_2)_\delta p'_\rho \operatorname{Tr}\left[\gamma^\mu\gamma^\delta\gamma^\nu(1-\gamma^5)\gamma^\rho\right]
\tag{23.113}
$$

The matrix element squared is found with the product of the two tensors which can be calculated using the general property **[T10]** found in Appendix E.2. This simply yields (cf. Eq. (23.60)):

$$
|\mathcal{M}|^2 = \frac{G_F^2}{2}(k - m_\mu s_1)_\alpha k'_\beta(p - m_e s_2)_\delta p'_\rho\left(64\delta_\rho^\alpha\delta_\delta^\beta\right) = 32G_F^2\left((k - m_\mu s_1)\cdot p'\right)(k'\cdot(p - m_e s_2))
\tag{23.114}
$$

The next step is to integrate over the phase space of the neutrinos. This integral was facilitated for the unpolarized case since we saw that the matrix element in Eq. (23.62) could be cast in the muon rest frame to depend only on the energy of the outgoing electron neutrino. We employ here a trick by rewriting the matrix element as:

$$
|\mathcal{M}|^2 = 32G_F^2\left((k - m_\mu s_1)^\alpha(p - m_e s_2)^\beta\right)p'_\alpha k'_\beta
\tag{23.115}
$$

where p' and k' are the four-vectors of the neutrinos. We thus obtain (cf. Eq. (23.65)):

$$
d^9\Gamma = \frac{2G_F^2}{(2\pi)^5 m_\mu}\left((k - m_\mu s_1)^\alpha(p - m_e s_2)^\beta\right)p'_\alpha k'_\beta\frac{d^3\vec{p}}{E_e}\frac{d^3\vec{k}'}{E_{\nu_\mu}}\frac{d^3\vec{p}'}{E_{\bar{\nu}_e}}\delta^4(k - p - k' - p')
\tag{23.116}
$$

In order to integrate over the neutrinos, we separate the terms that depend on them and write:

$$
d^3\Gamma = \frac{2G_F^2}{(2\pi)^5 m_\mu}\frac{d^3\vec{p}}{E_e}\left((k - m_\mu s_1)^\alpha(p - m_e s_2)^\beta\right)I_{\alpha\beta}(k - p)
\tag{23.117}
$$

where

$$
I_{\alpha\beta}(P^\mu) \equiv \int p'_\alpha k'_\beta\frac{d^3\vec{k}'}{E_{\nu_\mu}}\frac{d^3\vec{p}'}{E_{\bar{\nu}_e}}\delta^4(P - k' - p')
\tag{23.118}
$$

The integral depends on invariant quantities and on the four-vector with indices α and β. It is therefore **covariant!** In order to solve it, we can then use the following trick: after integration, it can only depend on P^μ and the Lorentz covariance imposes its form to be, in general:

$$
I_{\alpha\beta}(P^\mu) = A(P^2)g_{\alpha\beta} + B(P^2)P_\alpha P_\beta
\tag{23.119}
$$

where A and B need to be determined. In order to do that, we need two equations. The first one can be found by multiplying $I_{\alpha\beta}$ with our metric:

$$
\begin{aligned}
g^{\alpha\beta} I_{\alpha\beta} &= \int g^{\alpha\beta} p'_\alpha k'_\beta \frac{\mathrm{d}^3\vec{k'}}{E_{\nu_\mu}} \frac{\mathrm{d}^3\vec{p'}}{E_{\bar{\nu}_e}} \delta^4(P - k' - p') = \int (p' \cdot k') \frac{\mathrm{d}^3\vec{k'}}{E_{\nu_\mu}} \frac{\mathrm{d}^3\vec{p'}}{E_{\bar{\nu}_e}} \delta^4(P - k' - p') \\
&= \frac{P^2}{2} \int \frac{\mathrm{d}^3\vec{k'}}{E_{\nu_\mu}} \frac{\mathrm{d}^3\vec{p'}}{E_{\bar{\nu}_e}} \delta^4(P - k' - p')
\end{aligned}
\tag{23.120}
$$

where we used $P^2 = (k' + p')^2 = k'^2 + (2k' \cdot p') + p'^2 = 2k' \cdot p'$ if we assume the neutrinos to be massless. By further defining the integral:

$$
I \equiv \int \frac{\mathrm{d}^3\vec{k'}}{E_{\nu_\mu}} \frac{\mathrm{d}^3\vec{p'}}{E_{\bar{\nu}_e}} \delta^4(P - k' - p')
\tag{23.121}
$$

we obtain

$$
g^{\alpha\beta} I_{\alpha\beta} = \frac{P^2}{2} I = A(P^2) g_{\alpha\beta} g^{\alpha\beta} + B(P^2) P_\alpha P_\beta g^{\alpha\beta} = 4A(P^2) + P^2 B(P^2)
\tag{23.122}
$$

A second equation can be found considering the product $P^\alpha P^\beta I_{\alpha\beta}$:

$$
\begin{aligned}
P^\alpha P^\beta I_{\alpha\beta}(P^\mu) &= (P \cdot p')(P \cdot k')I = ((k' + p') \cdot p')((k' + p') \cdot k')I = (k' \cdot p' + p'^2)(k'^2 + k' \cdot p')I \\
&= (k' \cdot p')^2 I = \frac{P^4}{4} I
\end{aligned}
\tag{23.123}
$$

where we again used $P^2 = 2k' \cdot p'$ and assumed massless neutrinos. This leads to:

$$
P^\alpha P^\beta I_{\alpha\beta}(P^\mu) = \frac{P^4}{4} I = (P^2)A(P^2) + P^4 B(P^2)
\tag{23.124}
$$

The system of equation can easily be solved:

$$
\begin{cases} 4A + P^2 B = \frac{P^2}{2} I \\ A + P^2 B = \frac{P^2}{4} I \end{cases} \longrightarrow \begin{cases} A = \frac{P^2}{12} I \\ B = \frac{1}{6} I \end{cases}
\tag{23.125}
$$

so

$$
I_{\alpha\beta} = \frac{1}{12} \left(P^2 g_{\alpha\beta} + 2P_\alpha P_\beta \right) I
\tag{23.126}
$$

The missing integral I is invariant! We can compute it in any inertial frame we like. We decide to evaluate it in the rest frame of the two neutrinos. In this case, they must be back-to-back with the same momenta magnitudes, so we can write $|\vec{k'}| = |\vec{p'}| = E_{\nu_\mu} = E_{\bar{\nu}_e}$. Hence, defining $P^\mu = (E, \vec{P})$, we get:

$$
\begin{aligned}
I &= \int \frac{\mathrm{d}^3\vec{k'}}{E_{\nu_\mu}} \frac{\mathrm{d}^3\vec{p'}}{E_{\bar{\nu}_e}} \delta^4(P - k' - p') = \int \frac{\mathrm{d}^3\vec{k'}}{E_{\nu_\mu}^2} \delta(E - 2E_{\nu_\mu}) = (4\pi) \int \frac{k'^2 \mathrm{d}k'}{E_{\nu_\mu}^2} \delta(E - 2E_{\nu_\mu}) \\
&= (4\pi) \int \mathrm{d}E_{\nu_\mu} \delta(E - 2E_{\nu_\mu}) = (2\pi)
\end{aligned}
\tag{23.127}
$$

where we used $\mathrm{d}k' = \mathrm{d}E_{\nu_\mu}$. Therefore

$$
I_{\alpha\beta} = \frac{\pi}{6} \left(P^2 g_{\alpha\beta} + 2P_\alpha P_\beta \right)
\tag{23.128}
$$

Collecting all the bits and pieces yields the impressive result:

$$
\mathrm{d}^3\Gamma = \frac{G_F^2}{6(2\pi)^4 m_\mu} \frac{\mathrm{d}^3\vec{p}}{E_e} \left((k - m_\mu s_1)^\alpha (p - m_e s_2)^\beta \right) \left((k - p)^2 g_{\alpha\beta} + 2(k - p)_\alpha (k - p)_\beta \right)
\tag{23.129}
$$

Note that this expression is valid when the polarization of both the muon and the electron is defined. At this point it is reasonable to neglect the electron rest mass relative to the muon one (and we are not interested here in the dependence of the electron polarization) and express quantities in the muon rest frame. The kinematics is then given by Eq. (23.51). We need in addition to define the muon spin and the angle θ between the electron momentum and the muon spin:

$$s_1^\mu = (0, \hat{s}_\mu) \quad \text{and} \quad \cos\theta \equiv \frac{\hat{s}_\mu \cdot \vec{p}}{E_e} \tag{23.130}$$

See Figure 23.7. Consequently

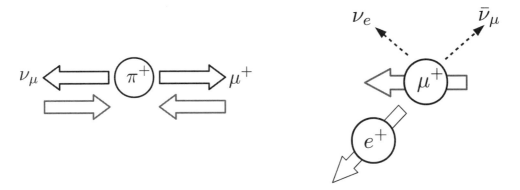

Figure 23.7 Momenta and spin orientations in the decay chain $\pi \to \mu\nu$; $\mu \to e\nu\nu$. The angle between the electron momentum and the muon spin is defined as θ.

$$
\begin{aligned}
\left((k - m_\mu s_1)^\alpha p^\beta\right) & \left((k-p)^2 g_{\alpha\beta} + 2(k-p)_\alpha (k-p)_\beta\right) \\
= & \quad (k-p)^2 \left((k - m_\mu s_1) \cdot p\right) + 2\left((k - m_\mu s_1) \cdot (k-p)\right)\left(p \cdot (k-p)\right) \\
= & \quad (k-p)^2 (k \cdot p) - m_\mu (k-p)^2 (s_1 \cdot p) + 2\left(k^2 - m_\mu \underbrace{s_1 \cdot k}_{=0} - k \cdot p + m_\mu s_1 \cdot p\right)\left(p \cdot k - \underbrace{p^2}_{=0}\right) \\
= & \quad (k-p)^2 (k \cdot p) - m_\mu (k-p)^2 (s_1 \cdot p) + 2k^2 (p \cdot k) - 2(k \cdot p)^2 + 2m_\mu (s_1 \cdot p)(p \cdot k) \\
= & \quad (k-p)^2 (k \cdot p) + 2k^2 (p \cdot k) - 2(k \cdot p)^2 - m_\mu (k-p)^2 (s_1 \cdot p) + 2m_\mu (s_1 \cdot p)(p \cdot k) \\
= & \quad (p \cdot k)\left((k-p)^2 + 2k^2 - 2(k \cdot p)\right) - m_\mu (s_1 \cdot p)\left((k-p)^2 - 2(p \cdot k)\right) \\
= & \quad (p \cdot k)\left(3k^2 - 4k \cdot p\right) - m_\mu (s_1 \cdot p)(k^2 - 4k \cdot p) \tag{23.131}
\end{aligned}
$$

where we used $(k-p)^2 = k^2 - 2k \cdot p$ when we neglect the electron rest mass. Also, kinematically, we have $k \cdot p = m_\mu E_e$ and $k^2 = m_\mu^2$ so:

$$
\begin{aligned}
(k \cdot p)\left(3k^2 - 4k \cdot p\right) - m_\mu (k^2 - 4k \cdot p)(s_1 \cdot p) & = m_\mu E_e \left(3m_\mu^2 - 4m_\mu E_e\right) - m_\mu (m_\mu^2 - 4m_\mu E_e)(-E_e \cos\theta) \\
& = m_\mu^3 E_e \left(3 - 4\frac{E_e}{m_\mu}\right) + m_\mu^3 (1 - 4\frac{E_e}{m_\mu})(E_e \cos\theta) \\
& = \frac{m_\mu^4}{2} x (3 - 2x) + \frac{m_\mu^4}{2} x(1 - 2x)\cos\theta \\
& = \frac{m_\mu^4}{2} x (3 - 2x - (2x - 1)\cos\theta) \tag{23.132}
\end{aligned}
$$

where we reintroduced the electron reduced energy $x = E_e/W_e \simeq 2E_e/m_\mu$. Hence, setting $\mathrm{d}^3\vec{p} = p^2 \mathrm{d}p\, \mathrm{d}\cos\theta \mathrm{d}\phi$, the differential width is:

$$
\begin{aligned}
\mathrm{d}^3\Gamma &= 2 \times \frac{G_F^2}{6(2\pi)^4 m_\mu} \frac{p^2 \mathrm{d}p \, \mathrm{d}\cos\theta \mathrm{d}\phi}{E_e} \frac{m_\mu^4}{2} x \left(3 - 2x - (2x - 1)\cos\theta\right) \\
&= \frac{G_F^2 m_\mu^5}{24(2\pi)^4} x^2 \left(3 - 2x - (2x - 1)\cos\theta\right) \mathrm{d}x \mathrm{d}\cos\theta \mathrm{d}\phi
\end{aligned}
\tag{23.133}
$$

where the direction of the electron momentum has been defined with respect to the muon spin direction and we have again neglected the electron rest mass. The extra factor two in front of the expression comes from the sum over the two polarization states of the final-state electron. The integration over ϕ is trivial and we find:

$$
\mathrm{d}^2\Gamma = \frac{G_F^2 m_\mu^5}{24(2\pi)^3} x^2 \left(3 - 2x - (2x - 1)\cos\theta\right) \mathrm{d}x \, \mathrm{d}\cos\theta
\tag{23.134}
$$

We want to put this expression in the form of Eq. (23.85). We note:

$$
x^2 \left(3 - 2x - (2x - 1)\cos\theta\right) = \frac{1}{2} \underbrace{2x^2(3 - 2x)}_{=n(x)} \left(1 - \frac{2x - 1}{3 - 2x}\cos\theta\right)
\tag{23.135}
$$

Accordingly, we write:

$$
\frac{\mathrm{d}^2\Gamma}{\mathrm{d}x \, \mathrm{d}\cos\theta} = \frac{G_F^2 m_\mu^5}{192\pi^3} [n(x)] \frac{1}{2} \left(1 - \frac{2x - 1}{3 - 2x}\cos\theta\right)
\tag{23.136}
$$

This derivation is valid for the decay of a negative muon. In the case of the positive muon, a similar derivation can be performed. The amplitude will be similar except that the signs in front of the m_μ and m_e terms will change when we pass from a particle to an antiparticle (and vice versa). These modifications lead to the change of sign in front of the $\cos\theta$ term. Hence, we can write the differential decay rate for negative (μ^-) and positive (μ^+) muons in compact form, as:

$$
\frac{\mathrm{d}^2\Gamma(\mu^\mp \to e^\mp \nu\bar{\nu})}{\mathrm{d}x \, \mathrm{d}\cos\theta} = \frac{G_F^2 m_\mu^5}{192\pi^3} [n(x)] \frac{1}{2} \left(1 \pm \alpha(x)\cos\theta\right)
\tag{23.137}
$$

where the **asymmetry parameter of the muon** $\alpha(x)$ is:

$$
\alpha(x) = \frac{1 - 2x}{3 - 2x}
\tag{23.138}
$$

and the sign of α is sort of conventional. The asymmetry is plotted in Figure 23.6. When the asymmetry is maximal at $x = 1$, the effect of the angle θ on the decay rate is maximal. The asymmetry changes sign in the middle of the spectrum at $x = 1/2$.

• **Experimental confirmation.** A first experimental confirmation of the electron spectrum from muon decay with high statistics was achieved at the Nevis Laboratory of Columbia University in the 1960s [307]. The experimental setup is shown in Figure 23.8(left). Positive pions were stopped in a plastic scintillator target. About 80% of the decay muons stopped and subsequently decayed within the same target. Due to angular acceptance, about 5% of the decay positrons entered the vacuum chamber of the electron counter, where their momentum is analyzed by a homogeneous magnetic field. Two trajectories with 35 and 52.5 MeV (endpoint) are illustrated as dashed lines in Figure 23.8. The positron trajectory is inferred from the positions reconstructed by a setup of sonic spark chambers, which provide via microphones located in the corners of the plane, the reconstruction of the x–y position of the spark within the plane. A total of 13×10^6 events were collected, of which about 1 million were used for the analysis of the spectrum, resulting in the distribution shown in Figure 23.8(right). The data are clearly totally consistent with the prediction of the $V - A$ theory (cf. Figure 23.6).

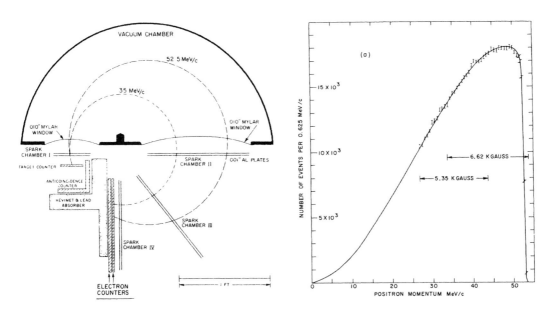

Figure 23.8 (left) Apparatus to measure the momentum spectrum of the electron in unpolarized muon decays. (right) The measured spectrum compared to $V - A$ expectations. Reprinted with permission from M. Bardon, P. Norton, J. Peoples, A. M. Sachs, and J. Lee-Franzini, "Measurement of the momentum spectrum of positrons from muon decay," *Phys. Rev. Lett.*, vol. 14, pp. 449–453, Mar 1965. Copyright 1965 by the American Physical Society.

23.8　　Weak Structure in Muon Decays – Michel Parameters

The decay of the muon is a purely leptonic process occurring via the weak charged current. It offers the possibility to test the Lorentz structure of the weak interaction with high precision and without complications from the hadronic interactions. From the lack of the $\mu \to e\gamma$ process, it is reasonable to require the conservation of the leptonic number. But the question of the existence of $V + A$ currents in addition to the dominant $V - A$ one is very important. In addition, generally speaking, the interaction could contain other couplings than vector or axial vector. Therefore, the muon decay can be described in the most general lepton-number-conserving four-fermion point interaction, as shown in the 1950s by **Michel**[5] in Refs. [308, 309], allowing for scalar (S), vector (V), tensor (T), axial-vector (A), and pseudoscalar (P) couplings. As shown in Eq. (23.33), it can be expressed as:

$$\mathcal{M}(\mu^- \to e^- \nu_\mu \bar{\nu}_e) = \frac{G_F}{\sqrt{2}} \sum_{i=S,V,T,A,P} \left[\bar{u}(k')\Gamma_i u(k) \right] \times \left[\bar{u}(p)\Gamma^i (A_i + A'_i \gamma^5) v(p') \right] \tag{23.139}$$

The order of the spinors above is in the so-called **charge exchange order**, but we have seen in Section 23.4, that the amplitude can be rewritten in the so-called **charge retention order** by using the Fierz transformation. Therefore, we get (see Eq. (23.36)):

$$\mathcal{M}(\mu^- \to e^- \nu_\mu \bar{\nu}_e) = \frac{G_F}{\sqrt{2}} \sum_{i=S,V,T,A,P} \left[\bar{u}(p)\Gamma_i u(k) \right] \times \left[\bar{u}(k')\Gamma^i (B_i + B'_i \gamma^5) v(p') \right] \tag{23.140}$$

where the B coefficients are related to the As via the Fierz transformation (see Table 23.1).

5　Louis Michel (1923–1999), French mathematical physicist.

A lengthy calculation using the above amplitude, similar to that performed in the previous sections, yields the differential decay rate as a function of the electron energy and direction. Neglecting the electron mass and for the muon at rest with a degree of polarization $P_\mu = |\vec{P}_\mu|$, **Kinoshita**[6] and **Sirlin**[7] first obtained [310]:

$$\frac{d^2\Gamma(\mu^\mp \to e^\mp \nu \bar{\nu})}{dx d\cos\theta} = \frac{G_F^2 m_\mu^5}{192\pi^3} \frac{A}{8} x^2 \left\{ 3(1-x) + \frac{2\rho}{3}(4x-3) \mp P_\mu \xi \left[(1-x) + \frac{2\delta}{3}(4x-3)\right] \cos\theta \right\} \quad (23.141)$$

where the so-called **Michel parameters** ρ, ξ, **and** δ are given by [310]:

$$\begin{aligned} \rho &= (3b + 6c)/A \\ \delta &= (3b' - 6c')/(-3a' + 4b' - 14c') \\ \xi &= -(3a' - 4b' + 14c')/A \end{aligned} \quad (23.142)$$

where

$$\begin{aligned} A &= a + 4b + 6c \\ a &= |B_S|^2 + |B_S'|^2 + |B_P|^2 + |B_P'|^2, & a' &= 2\,\mathrm{Re}\,(B_S B_P^* + B_P B_S'^*) \\ b &= |B_V|^2 + |B_V'|^2 + |B_A|^2 + |B_A'|^2, & b' &= 2\,\mathrm{Re}\,(B_V B_A'^* + B_A B_V'^*) \\ c &= |B_T|^2 + |B_T'|^2, & c' &= 2\,\mathrm{Re}\,(B_T B_T'^*) \end{aligned} \quad (23.143)$$

and $x = E_e/W_e = 2E_e/m_\mu$ (see Eq. (23.72)) is the electron reduced energy and θ is the angle between the muon spin and the electron momentum (see Eq. (23.130)).

It was first pointed out that $\rho = \delta = 3/4$ for the muon decay into a neutrino–antineutrino pair ($\mu \to e + \nu + \bar{\nu}$, i.e., the two-component theory, see Section 22.4), while it would be zero if the muon decayed into two neutrinos ($\mu \to e + 2\nu$) or two antineutrinos ($\mu \to e + 2\bar{\nu}$) [308]. Later it was understood that **precision measurements of the Michel parameters can quantify deviations from the pure $V - A$ interaction**. We can consider the following cases:

- Pure $V - A$ (valid for both charge exchange and charge retention). In this case, we have $A_V = 1$, $A_V' = -1$, $A_A = 1$, $A_A' = -1$, and all other A_is $= 0$ (see Section 23.4). As a consequence, after the Fierz transformation, we have: $B_V = 1$, $B_V' = -1$, $B_A = -1$, $B_A' = 1$, and all other B_is $= 0$. Hence, $a = a' = 0$, $b = 4$, $b' = 4$, and $c = c' = 0$. Then $A = 16$, and:

$$V - A: \quad \rho = 3 \cdot 4/16 = 3/4, \quad \delta = 12/16 = 3/4, \quad \xi = 16/16 = 1 \quad (23.144)$$

As predictable, we find in this case:

$$\begin{aligned} \frac{d^2\Gamma_\mp}{dx d\cos\theta} &\propto 2x^2 \left\{ 3(1-x) + \frac{2}{3}\frac{3}{4}(4x-3) \mp P_\mu \left[(1-x) + \frac{2}{3}\frac{3}{4}(4x-3)\right] \cos\theta \right\} \\ &= 2x^2 \left\{ 3(1-x) + \left(2x - \frac{3}{2}\right) \mp P_\mu \left[(1-x) + \left(2x - \frac{3}{2}\right)\right] \cos\theta \right\} \\ &= x^2 \left\{ 3 - 2x \mp P_\mu(2x - 1)\cos\theta \right\} \end{aligned} \quad (23.145)$$

so we recover Eq. (23.137) with $P_\mu = 1$.

- Pure S, P (charge retention). For example, $B_S = B_S' = 1$ and all others vanishing, we have $a = 2$ and $a' = b = b' = c = c' = 0$, with $A = 2$, hence:

$$S, P: \quad \rho = 0, \quad \delta = 0, \quad \xi = 0 \quad (23.146)$$

and the decay is totally insensitive to the spin correlations! Similar arguments hold for pure P.

6   Tichir Kinoshita (born 1925), Japanese–American theoretical physicist.
7   Alberto Sirlin (born 1930), Argentine theoretical physicist.

- <u>Pure T (charge retention)</u>. For $B_T = B'_T = 1$ and all others vanishing, we have $a = a' = 0$, $b = b' = 0$, and $c = c' = 2$, with $A = 12$, hence:

$$T: \quad \rho = 1, \quad \delta = -6(2)/(-14 \cdot 2) = 3/7, \quad \xi = -14(2)/12 = -7/3 \tag{23.147}$$

Hence, **the precise measurement of the Michel parameters provides a stringent test of the nature of the charged weak interaction**. Experimentally, the ρ parameter can be measured from precision measurements of the decay spectrum, as illustrated in Figure 23.9. The δ and ξ parameters can be determined from precision measurements of polarized muon decays, as illustrated in Figure 23.10. Such experiments were discussed in Sections 22.8 and 23.7.

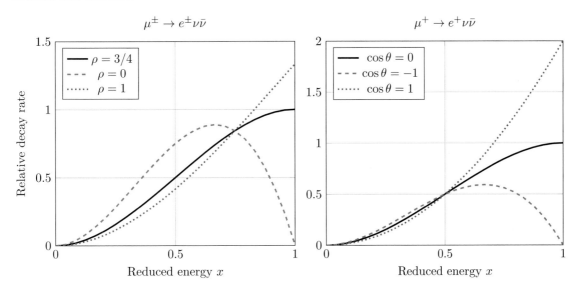

Figure 23.9 Decay electron/positron energy distribution for different values of the Michel parameter ρ.

Figure 23.10 Decay positron energy distribution for different values of the emission angle w.r.t. the muon spin orientation.

- **Correction for the electron mass and electron polarization term.** The distribution of the decay electron (positron) computed above neglected the electron mass. The effect of the finite electron mass and inclusion of its polarization should be taken into account for higher precision and further testing of the $V - A$ theory. The differential decay probability (neglecting radiative corrections) is in this case [311]:

$$\frac{\mathrm{d}^2\Gamma(\mu^\mp \to e^\mp \nu\bar{\nu})}{\mathrm{d}x \mathrm{d}\cos\vartheta} = \frac{G_F^2 m_\mu W_e^4}{4\pi^3} \sqrt{x^2 - x_0^2} \left(F_{IS}(x) \mp P_\mu \cos\theta F_{AS}(x) \right) \left(1 + \vec{P}_e(x,\theta) \cdot \hat{\zeta} \right) \tag{23.148}$$

where

$$
\begin{aligned}
F_{IS}(x) &= x(1-x) + \frac{2\rho}{9}\left(4x^2 - 3x - x_0^2\right) + \eta x_0(1-x) \\
F_{AS}(x) &= \frac{1}{3}\xi\sqrt{x^2 - x_0^2}\left[1 - x + \frac{2\delta}{3}(4x - 3 + (\sqrt{1 - x_0^2} - 1))\right]
\end{aligned}
\tag{23.149}
$$

and $W_e = m_\mu^2 + m_e^2/2m_\mu$ (see Eq. (23.72)) and $x_0 \equiv m_e/W_e \approx 9.67 \times 10^{-3}$. We note the appearance of a term proportional to an additional Michel parameter η. The other quantities P_μ, x, θ, and the Michel parameters ρ, ξ, δ are as before. Of course, setting $m_e = 0$ so $x_0 = 0$, we recover Eq. (23.141).

The electron polarization is given by $\vec{P}_e$ and $\widehat{\zeta}$ is the direction in which the polarization detector is most sensitive (i.e., such that the effect is maximal when the polarization of the electron is parallel to that direction). The polarization $\vec{P}_e$ can be decomposed along three natural axes: $\hat{z}$ is along the electron momentum, $\hat{y}$ is transverse to the electron momentum and perpendicular to the decay plane formed by the end products, $\hat{x}$ is transverse to the electron momentum and in the decay plane. Then the polarization vector can be expressed as a function of the longitudinal polarization P_L and two transverse ones P_{T_1} and P_{T_2}:

$$\vec{P}_e(x,\theta) = P_{T_1}(x,\theta)\widehat{x} + P_{T_2}(x,\theta)\widehat{y} + P_L(x,\theta)\widehat{z} \tag{23.150}$$

where

$$
\begin{aligned}
P_{T_1}(x,\theta) &= P_\mu \sin\theta F_{T_1}(x)/\left(F_{IS}(x) \pm P_\mu \cos\theta F_{AS}(x)\right) \\
P_{T_2}(x,\theta) &= P_\mu \sin\theta F_{T_2}(x)/\left(F_{IS}(x) \pm P_\mu \cos\theta F_{AS}(x)\right) \\
P_L(x,\theta) &= \pm F_{IP}(x) + P_\mu \cos\theta F_{AP}(x)/\left(F_{IS}(x) \pm P_\mu \cos\theta F_{AS}(x)\right)
\end{aligned}
\tag{23.151}
$$

and the functions $F_{T_1}(x)$, $F_{T_2}(x)$, $F_{IP}(x)$, and $F_{AP}(x)$ are given in Ref. [311]. They are expressible as functions of the four Michel parameters ρ, ξ, δ, η as well as seven additional ones called ξ', ξ'', η', α/A, β/A, α'/A, and β'/A. We will not discuss all of them here. However, we note that ξ' is a measure of the longitudinal polarization of the electron, which is expected to be equal to unity in the pure $V - A$ theory.

Experimentally, the measurement of the electron momentum and three polarizations is a challenge but has been achieved. These experiments have been combined and provide a very stringent test of the $V - A$ nature of the charged weak interaction. Some of these measurements compared to their expected values are compiled in Table 23.3, together with other important measurements related to the muon and its decay properties. From these results we can only conclude "Live long and prosper $V - A$"!

Parameter	Value	
Muon mass m_μ	$105.6583745 \pm 0.0000024$ MeV	
m_μ/m_e	$206.7682826 \pm 0.0000046$	
Muon neutrino m_{ν_μ}	< 0.19 MeV 90% CL	
Max electron energy W_e $(m_\nu = 0)$	52.828 MeV	
Lifetime τ	$(2.1969811 \pm 0.0000022) \times 10^{-6}$ s	
Michel parameters	Value	Theory ($V - A$)
ρ	0.74979 ± 0.00026	$3/4$
δ	0.75047 ± 0.00034	$3/4$
$P_\mu\xi$	$1.0009^{+0.0016}_{-0.0007}$	1
η	0.057 ± 0.034	0
ξ' (longitudinal polarization of e)	1.00 ± 0.04	1

Table 23.3 Properties of the muon and its decay parameters (from P. A. Zyla *et al.* (Particle Data Group), *Prog. Theor. Exp. Phys.* **2020**, 083C01 (2020)).

23.9   The Inverse Muon Decay

Consider the **inverse muon decay (IMD) process** (compare with IBD in Section 21.5):

$$\nu_\mu(k,r) + e^-(p,s) \to \mu^-(k',r') + \nu_e(p',s') \tag{23.152}$$

The amplitude of this process is related to the decay $\mu^- \to e^- \bar{\nu}_e \nu_\mu$ by the replacement of ingoing and outgoing particles while keeping an outgoing ν_e. The tree-level Feynman diagram is represented in Figure 23.11 and its corresponding amplitude is:

$$\mathcal{M}(\nu_\mu + e^- \to \mu^- + \nu_e) = \frac{G_F}{\sqrt{2}} \left(\bar{u}(k') \gamma^\mu (1-\gamma^5) u(k) \right) \left(\bar{u}(p') \gamma_\mu (1-\gamma^5) u(p) \right) \tag{23.153}$$

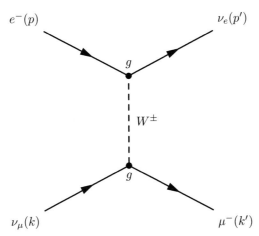

Figure 23.11 Tree-level Feynman diagram for the inverse muon decay $\nu_\mu + e^- \to \mu^- + \nu_e$.

Applying Casimir's trick one finds, assuming massless neutrinos (note that we sum over all spins, although for massless neutrino spinors we have seen that only two components survive the $V - A$ factor):

$$\sum_{s,s',r,r'} |\mathcal{M}|^2 = \frac{G_F^2}{2} \text{Tr} \left(\gamma^\mu (1-\gamma^5)(\slashed{p}+m_e)\gamma^\nu(1-\gamma^5)\slashed{p}' \right) \text{Tr} \left(\gamma_\mu(1-\gamma^5)\slashed{k}\gamma_\nu(1-\gamma^5)(\slashed{k}'+m_\mu) \right) \tag{23.154}$$

Since we have neglected the masses of the neutrinos, the traces proportional to m_e or m_μ vanish, making use of the trace theorems **[T3]** and **[T8]** of Appendix E.2. We consider, for instance, the following trace:

$$m_e \text{Tr} \left(\gamma^\mu (1-\gamma^5)\gamma^\nu(1-\gamma^5)\slashed{p}' \right) = 0 \tag{23.155}$$

which has an odd number of γ^μ matrices, and the same happens for the other combinations. Consequently, we are left with just:

$$\sum_{s,s',r,r'} |\mathcal{M}|^2 = \frac{G_F^2}{2} \text{Tr} \left(\gamma^\mu (1-\gamma^5)\slashed{p}\gamma^\nu(1-\gamma^5)\slashed{p}' \right) \text{Tr} \left(\gamma_\mu(1-\gamma^5)\slashed{k}\gamma_\nu(1-\gamma^5)\slashed{k}' \right) \tag{23.156}$$

The trace gives, using **[TS8]** of Appendix E.3:

$$\sum_{s,s',r,r'} |\mathcal{M}|^2 = \frac{G_F^2}{2} 256 \left[(p \cdot k)(p' \cdot k') \right] = 128 G_F^2 (p \cdot k)(p' \cdot k') \tag{23.157}$$

We need to average over initial spins so we include a factor $1/2$ for the incoming electron. On the other hand, the incoming neutrino has only one helicity (it is fully polarized), then the factor is 1. So, finally:

$$\langle |\mathcal{M}|^2 \rangle = \left(\frac{1}{2} \right) (1) \sum_{s,s',r,r'} |\mathcal{M}|^2 = 64 G_F^2 (p \cdot k)(p' \cdot k') \tag{23.158}$$

We now consider the reaction in the laboratory frame where an incident neutrino ν_μ scatters off a target electron e^- at rest. We have:

$$\text{Laboratory frame: } k^\mu = (E_\nu, \vec{p}_\nu), \quad p^\mu = (m_e, \vec{0}) \tag{23.159}$$

In order to produce a muon in the final state, the energy in the center-of-mass frame should satisfy:

$$s = (k+p)^2 \simeq 2k \cdot p \simeq 2m_e E_\nu > m_\mu^2 \tag{23.160}$$

hence the energy threshold for the reaction in the laboratory frame is:

$$E_\nu > \frac{m_\mu^2}{2m_e} \approx 11 \text{ GeV} \tag{23.161}$$

and in an experimental apparatus the reaction will produce a single energetic and forward-boosted muon. The differential cross-section in the center-of-mass frame, neglecting the electron mass, is simply given by noting:

$$\text{CMS frame: } k^\mu = (E^\star, \vec{p}), \quad p^\mu = (E^\star, -\vec{p}), k'^\mu = (E_\mu^\star, \vec{k}), \quad p'^\mu = (E_\nu^\star, -\vec{k}) \tag{23.162}$$

Hence

$$(p \cdot k) = E^{\star 2} + |\vec{p}|^2 \simeq 2E^{\star 2}, \quad (p' \cdot k') = E_\mu^\star E_\nu^\star + |\vec{k}|^2 \tag{23.163}$$

but since $E_\mu^\star + E_\nu^\star = 2E^\star$ then $E_\mu^{\star 2} + E_\nu^{\star 2} + 2E_\mu^\star E_\nu^\star = 4E^{\star 2}$, and we can write:

$$2(p' \cdot k') = 4E^{\star 2} - E_\mu^{\star 2} - E_\nu^{\star 2} + 2|\vec{k}|^2 = 4E^{\star 2} - (E_\mu^{\star 2} - |\vec{k}|^2) - (E_\nu^{\star 2} - |\vec{k}|^2) = 4E^{\star 2} - m_\mu^2 \tag{23.164}$$

The averaged squared matrix element becomes (having neglected the electron mass):

$$\langle |\mathcal{M}|^2 \rangle \simeq 32G_F^2(2E^{\star 2})(4E^{\star 2} - m_\mu^2) = 256G_F^2 E^{\star 4}\left(1 - \left(\frac{m_\mu}{2E^\star}\right)^2\right) \tag{23.165}$$

We can use this result in the phase-space formula of Eq. (5.145) to obtain:

$$\left(\frac{d\sigma}{d\Omega}\right)_{CMS} = \frac{256G_F^2 E^{\star 4}}{64\pi^2 s}\left(1 - \left(\frac{m_\mu}{2E^\star}\right)^2\right) = \frac{G_F^2 E^{\star 2}}{\pi^2}\left(1 - \left(\frac{m_\mu}{2E^\star}\right)^2\right) \simeq \frac{G_F^2 s}{4\pi^2} \tag{23.166}$$

where we have used $s = (2E^\star)^2$ and assumed $E^\star \gg m_\mu$. The differential cross-section is constant, therefore isotropic; this means that all scattering solid angles $d\Omega$ in the center-of-mass system have equal probability. This is explained by the spin configurations, as shown in Figure 23.12. The electron (assumed massless) and the neutrino have negative helicities and hence their total angular momentum adds to a $J = 0$ state. As a consequence, the differential cross-section is going to be isotropic in the center-of-mass frame.

Figure 23.12 Helicities in the inverse muon decay (mass of the electron is neglected).

We can exploit the isotropy in the center-of-mass to directly calculate the total cross-section. This latter is given by:

$$\sigma(\nu_\mu + e^- \to \mu^- + \nu_e) \simeq \int d\Omega \frac{G_F^2 s}{4\pi^2} = \frac{G_F^2 s}{\pi} \tag{23.167}$$

In the laboratory frame, where $s \simeq 2m_e E_\nu$, the cross-section rises linearly and can be expressed as:

$$\sigma(\nu_\mu + e^- \to \mu^- + \nu_e) = \frac{2G_F^2 m_e E_\nu}{\pi} \approx 1.720 \times 10^{-41}\left(\frac{E_\nu}{\text{GeV}}\right) \text{ cm}^2 \tag{23.168}$$

to be compared to the neutrino–nucleon cross-section $\sigma(\nu_\mu + n \to \mu^- + p) \approx 10^{-38}$ cm^2 (see Eq. (21.78))! These cross-sections are plotted in Figure 28.13. An experimental observation of the process is described in Section 27.3.

23.10   Semi-leptonic Decay Rate of the Pion and the Kaon

We consider the **semi-leptonic decay of the pion** $\pi^- \to \mu^- + \bar{\nu}_\mu$ (also called $\pi_{\mu 2}$) and similarly of the kaon $K^- \to \mu^- + \bar{\nu}_\mu$ ($K_{\mu 2}$). The kinematics is defined as follows:

$$\text{Semi-leptonic decays:} \quad \pi^-(q^\mu) \to \mu^-(p^\mu) + \bar{\nu}_\mu(k^\mu), \quad K^-(q^\mu) \to \mu^-(p^\mu) + \bar{\nu}_\mu(k^\mu) \tag{23.169}$$

where $q^\mu \equiv p^\mu + k^\mu$. See Figure 23.13. Let us begin with the pion semi-leptonic decay (the semi-leptonic kaon decay is similar). The amplitude contains two terms: (1) the hadronic current and (2) the leptonic current. We can define the leptonic current with the $W^\pm$ coupling to the leptons. For the pion, the system is composed of a $q\bar{q}$ bound state and this hadronic current cannot be expressed in terms of free Dirac spinors. When we discussed the lifetime of the positronium, we saw the Wheeler–Pirenne equation, where the bound-state physics is encoded in the wave function at the origin (see Eq. (14.112)). In the covariant formalism, we consider the hadronic four-current. We do not know a priori the exact form of this hadronic current. If the quarks were

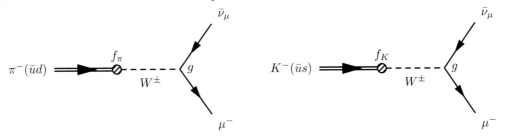

Figure 23.13 (left) The $\pi^- \to \mu^- \bar{\nu}_\mu$ decay ($\Delta S = 0$). (right) The $K^- \to \mu^- \bar{\nu}_\mu$ decay ($\Delta S = 1$).

free Dirac fermions, we would write the Cabibbo favored transition as follows:

$$\mathcal{M} = \frac{G_F}{\sqrt{2}} \underbrace{\left(\bar{u}(p)\gamma_\mu(1-\gamma^5)v(k)\right)}_{\text{lepton current}} \cos\theta_C \underbrace{\left(\bar{v}_u\gamma^\mu\left(1-\gamma^5\right)u_d\right)}_{\text{hadronic current}} \tag{23.170}$$

where v_u and u_d represent the spinors of the $\bar{u}$ and d quarks. Because of our lack of knowledge on the quark bound state it represents, we cannot compute the hadronic current from first principles, however, we can parametrize it as coupling to the $W^\pm$ in a general way. Thus, the amplitude can be generically expressed as:

$$\mathcal{M} = \frac{G_F}{\sqrt{2}} \underbrace{\left(\bar{u}(p)\gamma_\mu(1-\gamma^5)v(k)\right)}_{\text{lepton current}} \underbrace{\langle 0|J^\mu|\pi^-\rangle}_{\text{hadronic current}} \tag{23.171}$$

We search a Lorentz-invariant matrix element, hence, the hadronic current can have a vector or an axial vector form. Since the pion is spinless, the only degree of freedom of the pion is its four-momentum q^μ. As an ansatz, we assume that the current has the form:

$$\text{ansatz}: \langle 0|J^\mu|\pi^-\rangle \equiv \cos\theta_C \, q^\mu f(q^2) \approx \cos\theta_C \, q^\mu f(q^2 \approx 0) \approx \cos\theta_C \, q^\mu f_\pi \tag{23.172}$$

where f is the pion structure function and f_π is the pion decay constant.

• **Pion semi-leptonic decay** $\pi_{\mu 2}$**.** The amplitude, which is Cabibbo favored for the pion, then becomes:

$$\begin{aligned}
\mathcal{M}_\pi &= \frac{G_F \cos\theta_C}{\sqrt{2}} f_\pi (p^\mu + k^\mu) \left(\bar{u}(p)\gamma_\mu(1-\gamma^5)v(k)\right) \\
&= \left(\frac{G_F \cos\theta_C}{\sqrt{2}} f_\pi\right) \left[\bar{u}(p)\slashed{p}(1-\gamma^5)v(k) + \bar{u}(p)\slashed{k}(1-\gamma^5)v(k)\right]
\end{aligned} \tag{23.173}$$

With the help of the Dirac equation, Eq. (8.135), we note that:

$$\bar{u}(p)\not{p} = m_\mu \bar{u}(p) \quad \text{and} \quad (\not{k} + m_\nu)v(k) \approx \not{k}v(k) = 0 \tag{23.174}$$

where we have set the neutrino mass to zero. Then the amplitude simply becomes:

$$\mathcal{M}_\pi = \frac{G_F \cos\theta_C}{\sqrt{2}} f_\pi m_\mu \left[\bar{u}(p)(1-\gamma^5)v(k)\right] \tag{23.175}$$

With the help of the rules for Casimir's trick derived in Section 23.4, we can compute the matrix element squared. Since the pion is (pseudo)scalar (spinless), there is no initial state averaging. From Section 23.4, we recall $\overline{\Gamma} = \overline{1-\gamma^5} = 1+\gamma^5$, and hence:

$$\begin{aligned}
\langle|\mathcal{M}_\pi|^2\rangle &= \frac{G_F^2 \cos^2\theta_C}{2} f_\pi^2 m_\mu^2 \operatorname{Tr}\left((\not{p}+m_\mu)(1-\gamma^5)\not{k}(1+\gamma^5)\right) \\
&= \frac{G_F^2 \cos^2\theta_C}{2} f_\pi^2 m_\mu^2 \left[\operatorname{Tr}\left(\not{p}(1-\gamma^5)\not{k}(1+\gamma^5)\right) + m_\mu \underbrace{\operatorname{Tr}\left((1-\gamma^5)\not{k}(1+\gamma^5)\right)}_{0}\right]
\end{aligned} \tag{23.176}$$

Using Appendix E.3, the first trace is equal to:

$$\begin{aligned}
\operatorname{Tr}\left(\not{p}(1-\gamma^5)\not{k}(1+\gamma^5)\right) &= \operatorname{Tr}(\not{p}\not{k}) - \underbrace{\operatorname{Tr}(\not{p}\gamma^5\not{k})}_{0} + \underbrace{\operatorname{Tr}(\not{p}\not{k}\gamma^5)}_{0} - \operatorname{Tr}(\not{p}\gamma^5\not{k}\gamma^5) = \operatorname{Tr}(\not{p}\not{k}) + \operatorname{Tr}\left(\not{p}\not{k}\underbrace{\gamma^5\gamma^5}_{\mathbb{1}}\right) \\
&= 2\operatorname{Tr}(\not{p}\not{k}) = 8(p\cdot k)
\end{aligned} \tag{23.177}$$

So finally:

$$\langle|\mathcal{M}_\pi|^2\rangle = 4G_F^2 \cos^2\theta_C f_\pi^2 m_\mu^2 (p\cdot k) \tag{23.178}$$

In order to perform the phase-space integration, we must select a reference frame. In the rest frame of the pion, we simply write:

$$q^\mu = (m_\pi, \vec{0}), \quad p^\mu = (E, \vec{p}) \quad \text{and} \quad k^\mu = (\omega, -\vec{p}) \tag{23.179}$$

and

$$q^2 = m_\pi^2 = (p+k)^2 = p^2 + k^2 + 2p\cdot k \implies 2p\cdot k = m_\pi^2 - m_\mu^2 \tag{23.180}$$

so the matrix element squared is just a constant:

$$\langle|\mathcal{M}_\pi|^2\rangle = 2G_F^2 \cos^2\theta_C f_\pi^2 m_\mu^2 (m_\pi^2 - m_\mu^2) \tag{23.181}$$

We make use of Eq. (22.38) with the momentum of the muon given by Eq. (22.39):

$$\Gamma(\pi^- \to \mu^- + \nu_\mu) = \frac{|\mathcal{M}_\pi|^2}{8\pi} \frac{p_\ell}{m_\pi^2} = \frac{2G_F^2 \cos^2\theta_C f_\pi^2 m_\mu^2 (m_\pi^2 - m_\mu^2)}{8\pi} \frac{m_\pi^2 - m_\mu^2}{2m_\pi^3} \tag{23.182}$$

so finally, one gets:

$$\Gamma(\pi \to \mu\nu) = \frac{G_F^2 \cos^2\theta_C}{8\pi} \frac{f_\pi^2}{m_\pi} \left(\frac{m_\mu}{m_\pi}\right)^2 (m_\pi^2 - m_\mu^2)^2 \tag{23.183}$$

This result contains as expected the helicity properties discussed in Section 22.6. As mentioned previously, it correctly explains the strong suppression of the electronic decay mode of the pion versus the muonic decay, which was empirically derived in Eq. (22.50):

$$\frac{\Gamma(\pi \to e\nu_e)}{\Gamma(\pi \to \mu\nu_\mu)} = \left(\frac{m_e}{m_\mu}\right)^2 \left(\frac{m_\pi^2 - m_e^2}{m_\pi^2 - m_\mu^2}\right)^2 \approx 1.28 \times 10^{-4} \tag{23.184}$$

Amazing!

• **Universality in $\pi_{\mu2}$ vs. π_{e2}.** A precise comparison must include radiative corrections due to virtual exchange and real photon emission and include in addition the mass term corrections. Hence, we write:

$$R_{e/\mu} \equiv \frac{\Gamma(\pi \to e\nu_e)}{\Gamma(\pi \to \mu\nu_\mu)} = \left(\frac{m_e}{m_\mu}\right)^2 \left(\frac{m_\pi^2 - m_e^2}{m_\pi^2 - m_\mu^2}\right)^2 \left(1 + \delta R_{e/\mu}\right) \tag{23.185}$$

The dominant correction is the mass correction:

$$\delta R_{e/\mu} = \frac{3\alpha}{\pi} \ln\left(\frac{m_e}{m_\mu}\right) \approx -3.72\% \tag{23.186}$$

The refined calculations from **Marciano** and **Sirlin** yield [312]:

$$\delta R_{e/\mu} = -(3.76 \pm 0.04)\% \tag{23.187}$$

Hence

$$R_{e/\mu}^{theory} = (1.2356 \pm 0.0001) \times 10^{-4} \tag{23.188}$$

These corrections bring the theoretical expectation very close to the experimental value given in Eq. (22.37). This result can be seen as another stringent test of universality.

• **Kaon semi-leptonic decay.** We can obtain a similar expression for the kaon decay replacing the mass m_π with m_K, the decay constant f_π with f_K (to be determined), and taking into account the Cabibbo suppression substituting $\cos\theta_C$ with $\sin\theta_C$, see Figure 23.13(right):

$$\Gamma(K^+ \to \mu^+\nu) = \frac{G_F^2 \sin^2\theta_C}{8\pi} \frac{f_K^2}{m_K} \left(\frac{m_\mu}{m_K}\right)^2 \left(m_K^2 - m_\mu^2\right)^2 \tag{23.189}$$

The observed decay time of the pion is $\tau_\pi \approx 26$ ns, and this corresponds to the total width of:

$$\Gamma(\pi) = \tau_\pi^{-1} = \frac{6.6 \times 10^{-25} \text{ s} \cdot \text{GeV}}{2.6 \times 10^{-8} \text{ s}} \approx 2.5 \times 10^{-17} \text{ GeV} \tag{23.190}$$

The π decay constant f_π can be expressed as a function of the decay width, the Fermi coupling constant G_F, the Cabibbo angle, and the masses of the pion and the muon:

$$f_\pi = \frac{\sqrt{8\pi} m_\pi^{3/2} \Gamma^{1/2}}{G_F \cos\theta_C m_\mu (m_\pi^2 - m_\mu^2)} \simeq (2.6 \times 10^7)\sqrt{2.5 \times 10^{-17}} \text{ GeV} \approx 0.130 \text{ GeV} \approx m_\pi \tag{23.191}$$

Hence, the observed lifetime of the pion is consistent with $f_\pi \approx m_\pi$. The actual value recommended by the Particle Data Group's Review after a treatment of radiative corrections is:

$$f_\pi |V_{ud}| = (127.13 \pm 0.02(\text{exp.}) \pm 0.13(\text{rad. corr.})) \text{ MeV} \tag{23.192}$$

where the errors are from the experimental rate measurement and the radiative correction factor, respectively (V_{ud} corresponds to the generalization of the Cabibbo angle and is defined in Section 23.14). In Eq. (23.20) the further assumption $f_K \approx f_\pi$ was used in order to determine the Cabibbo angle from the ratio of the observed lifetimes of the pion and kaon. Under this assumption, the similar lifetime between the kaon and the pion dominantly comes from the Cabibbo suppression of the kaon decay, which would otherwise have a much shorter lifetime based on its kinematical phase space. See Eq. (23.19). Similarly, as before, the value for the kaon decay constant recommended by the Particle Data Group's Review is:

$$f_K |V_{us}| = (35.09 \pm 0.04(\text{exp.}) \pm 0.04(\text{rad. corr.})) \text{ MeV} \tag{23.193}$$

which is consistent with the assumption $f_K \approx 150$ MeV $\approx 1.2 f_\pi$.

• **General semi-leptonic decays.** Charged mesons formed from a quark and an antiquark can always decay to a lepton and a neutrino, via the weak charged current. We have seen the case of the pion and kaon, but these apply equally to heavier states such as charmed or bottomed mesons. Let us denote with P any of these pseudoscalar mesons composed of the $q_1 \bar{q}_2$ quark pair. The decay width is then:

$$\Gamma(P^+ \to \ell^+ \nu_\ell) = \frac{G_F^2}{8\pi} |V_{q_1 q_2}|^2 \frac{f_P^2}{m_P} \left(\frac{m_\ell}{m_P}\right)^2 \left(m_P^2 - m_\ell^2\right)^2 \tag{23.194}$$

where M_P is the mass of the pseudoscalar meson and $V_{q_1 q_2}$ is the Cabibbo–Kobayashi–Maskawa matrix element between the quarks q_1 and $\bar{q}_2$ defined in Section 23.14. As before, the decay constant f_P is given by the matrix element between the meson and the vacuum.

23.11   Tau Decay Modes and Rates

The charged tau lepton is heavy enough that it can also decay into a muon or to mesons composed of light quarks, see Figure 23.14. The decay width of the tau lepton to electrons or muons is then:

$$\Gamma(\tau^- \to e^- \nu_\tau \bar{\nu}_e) = \Gamma(\tau^- \to \mu^- \nu_\tau \bar{\nu}_\mu) = \frac{G_F^2 m_\tau^5}{192\pi^3} \tag{23.195}$$

The total width of the tau is given by the sum of all its partial decay widths:

$$\begin{aligned} \Gamma(\tau \to \text{all}) &= \frac{1}{\tau_\tau} = \sum_i \Gamma_i \simeq \Gamma(\tau^- \to e^- \nu_\tau \bar{\nu}_e) + \Gamma(\tau^- \to \mu^- \nu_\tau \bar{\nu}_\mu) + 3\Gamma(\tau^- \to \bar{u}d'\nu_\tau) \\ &\approx 5 \times \Gamma(\tau^- \to e^- \nu_\tau \bar{\nu}_e) \end{aligned} \tag{23.196}$$

where Γ_i is the partial width of the decay mode i. The factor three in front of the quark decay mode accounts for color.

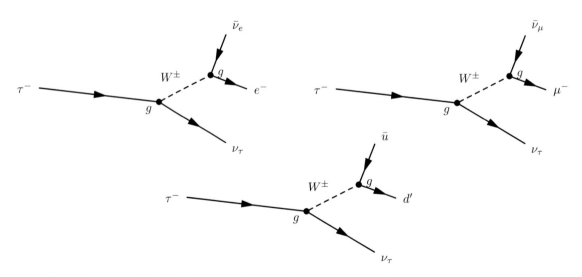

Figure 23.14 Tree-level Feynman diagrams for the tau decays into $e\nu\bar{\nu}$, $\mu\nu\bar{\nu}$, and hadrons.

Many exclusive decay modes of the tau can be calculated analytically using the previously developed methods. The leptonic decay modes can be directly rescaled from the muon decay to electrons. The hadronic decay modes can be derived using the techniques used, for example, the $\pi_{\ell 2}$ decay.

- $\tau^- \to \pi^- \nu_\tau$ **decay.** The method is very similar to the $\pi_{\ell 2}$ decay with the hadronic current taking the form $\langle 0|J^\mu|\pi^-\rangle \approx \cos\theta_C q^\mu f_\pi$ (cf. Eq. (23.171)). The result is:

$$\Gamma\left(\tau^- \to \pi^- \nu_\tau\right) = \frac{G_F^2 \cos^2\theta_C f_\pi^2 m_\tau^3}{16\pi}\left(1 - \frac{m_\pi^2}{m_\tau^2}\right)^2 \tag{23.197}$$

- $\tau^- \to K^- \nu_\tau$ **decay.** The result is:

$$\Gamma\left(\tau^- \to K^- \nu_\tau\right) = \frac{G_F^2 \sin^2\theta_C f_K^2 m_\tau^3}{16\pi}\left(1 - \frac{m_K^2}{m_\tau^2}\right)^2 \tag{23.198}$$

- **General case** $\tau^- \to h^- \nu_\tau$ **decay.** We consider the general case of a hadronic decay into a system h^- with total four-momentum p^μ. The amplitude of the reaction $\tau^-(k) \to h^-(p) + \nu_\tau(k')$ can be written as:

$$\mathcal{M}(\tau^- \to h^- + \nu_\tau) = \frac{G_F}{\sqrt{2}}\bar{u}(k')\gamma_\mu\left(1 - \gamma^5\right)u(k)\langle h^-|J^\mu|0\rangle \tag{23.199}$$

In order to calculate the cross-section, we need to evaluate the matrix element squared, which is of the form of a product of a leptonic and a hadronic tensor $|\mathcal{M}|^2 \propto L^{\mu\nu}(k,k')W_{\mu\nu}(p)$ – compare this form with those of Eqs. (16.37) and (23.55). The leptonic part can be computed as usual and the hadronic tensor (or **spectral function**) needs to be written in a general form with some form factors to be determined experimentally. We write:

$$W^{\mu\nu}(p) = \left(\frac{p^\mu p^\nu}{p^2} - g^{\mu\nu}\right)\rho\left(p^2\right) + \frac{p^\mu p^\nu}{p^2}\rho'\left(p^2\right) \tag{23.200}$$

where $\rho(p^2)$ and $\rho'(p^2)$ parameterize our ignorance on the details of the hadronic current. The leptonic tensor of the tau current is simply:

$$L^{\mu\nu}(k,k') = k_\alpha k'_\beta \text{Tr}\left(\gamma^\mu(1-\gamma^5)\gamma^\alpha\gamma^\nu(1-\gamma^5)\gamma^\beta\right) = 8\left(k^\mu k'^\nu + k^\nu k'^\mu - (k\cdot k')g^{\mu\nu}\right) + \underbrace{8i\epsilon^{\mu\alpha\nu\beta}k_\alpha k'_\beta}_{\text{antisymmetric}} \tag{23.201}$$

where we used the trace theorem **[TS4]** from Appendix E.3. The second term of $L_{\mu\nu}$ is antisymmetric under the interchange $\mu \leftrightarrow \nu$, while $W^{\mu\nu}$ is symmetric. Hence, their product vanishes. We also note that:

$$\frac{1}{8}L_{\mu\nu}\frac{p^\mu p^\nu}{p^2} = \left(\frac{2(k\cdot p)(k'\cdot p)}{p^2} - (k\cdot k')\right) \quad \text{and} \quad -\frac{1}{8}L_{\mu\nu}g^{\mu\nu} = -2(k\cdot k') + (k\cdot k')\underbrace{g_{\mu\nu}g^{\mu\nu}}_{=4} = 2(k\cdot k') \tag{23.202}$$

Hence, collecting the terms to compute the contraction $L_{\mu\nu}W^{\mu\nu}$ averaging over the initial spin, one finds:

$$\langle|\mathcal{M}|^2\rangle = \frac{1}{2}\frac{G_F^2}{2}L_{\mu\nu}W^{\mu\nu} = 2G_F^2\left[\left(\frac{2(k\cdot p)(k'\cdot p)}{p^2} + k\cdot k'\right)\rho(p^2) + \left(\frac{2(k\cdot p)(k'\cdot p)}{p^2} - k\cdot k'\right)\rho'(p^2)\right] \tag{23.203}$$

The computation of the scalar product is best done in the tau rest frame where we have $k^\mu = (m_\tau, \vec{0})$ and $p = (E_p, \vec{p})$. Using the kinematical constraint $k^\mu = p^\mu + k'^\mu$, we then easily find that:

$$k^2 = (p + k')^2 = p^2 + 2p\cdot k' = m_\tau^2 \quad \to \quad (p\cdot k') = \frac{m_\tau^2 - p^2}{2} \tag{23.204}$$

but also

$$(p\cdot k') = p\cdot(k - p) = (p\cdot k) - p^2 = m_\tau E_p - p^2 \tag{23.205}$$

hence

$$\frac{m_\tau^2 - p^2}{2} = m_\tau E_p - p^2 \quad \rightarrow \quad E_p = \frac{m_\tau^2 + p^2}{2m_\tau} \tag{23.206}$$

Finally

$$(k \cdot p) = m_\tau E_p = \frac{1}{2}\left(m_\tau^2 + p^2\right) \quad \text{and} \quad (k \cdot k') = k \cdot (k - p) = k^2 - k \cdot p = \frac{1}{2}\left(m_\tau^2 - p^2\right) \tag{23.207}$$

Using the scalar products, the averaged matrix element can then be expressed as:

$$\langle |\mathcal{M}|^2 \rangle \;=\; G_F^2 m_\tau^2 \left(\frac{1-x}{x}\right) [(1 + 2x)\rho(x) + \rho'(x)] \tag{23.208}$$

where $x \equiv p^2/m_\tau^2$. We use Eq. (5.163) to include the phase space:

$$\mathrm{d}\Gamma = \frac{\langle |\mathcal{M}|^2 \rangle}{32\pi^2} \frac{|\vec{p}|}{m_\tau^2} \mathrm{d}\Omega = \frac{G_F^2 m_\tau^2}{32\pi^2}\left(\frac{1-x}{x}\right)\left[\cdots\right]\frac{m_\tau}{2}(1-x)\frac{1}{m_\tau^2}(4\pi) \tag{23.209}$$

where we used $|\vec{p}| = \sqrt{E_p^2 - p^2} = m_\tau(1 - x)/2$ and integrated over the solid angle. The total decay rate is found by integration over all possible masses of the hadronic system, which is equivalent to integrating over x between 0 and 1. Consequently

$$\Gamma\left(\tau^- \to h^- \nu_\tau\right) = \frac{G_F^2 m_\tau}{16\pi} \int_0^1 \mathrm{d}x \frac{(1-x)^2}{x} [(1 + 2x)\rho(x) + \rho'(x)] \tag{23.210}$$

If we consider the case $h^- = \pi^-$ and compare with the result of Eq. (23.197), we note that:

$$\rho(x) = 0 \quad \text{and} \quad \rho'(x) = \left(\cos^2 \theta_C f_\pi^2 m_\tau^2\right) x\, \delta\left(x - \frac{m_\pi^2}{m_\tau^2}\right) \tag{23.211}$$

These functions describe the decay into a pseudoscalar meson.

• $\tau^- \to \pi^- \pi^0 \nu_\tau$ **decay.** Now we consider the decay of the tau to a vector meson ρ^-. Since the ρ is unstable and broad, this mode actually dominates the $\pi^- \pi^0$ final state:

$$\tau^- \to \rho^- \nu_\tau \to \pi^- \pi^0 \nu_\tau \tag{23.212}$$

We first assume that the ρ^- is stable and that its polarization is given by the four-vector ϵ^μ. The hadronic current has the form $\langle \rho^-(p, \epsilon^\nu)|J^\mu|0\rangle \simeq \cos\theta_C \epsilon^\mu f_\rho$. Summing over the polarization states of the massive vector particle (i.e., three states), one would get:

$$\rho(x) = \left(\cos^2 \theta_C f_\rho^2 m_\tau^2\right) x\, \delta\left(x - \frac{m_\rho^2}{m_\tau^2}\right) \quad \text{and} \quad \rho'(x) = 0 \tag{23.213}$$

Hence

$$\begin{aligned}
\Gamma\left(\tau^- \to \rho^- \nu_\tau\right) &= \frac{G_F^2 m_\tau}{16\pi} \int_0^1 \mathrm{d}x \frac{(1-x)^2}{x}\left[(1 + 2x)\left(\cos^2 \theta_C f_\rho^2 m_\tau^2\right) x\, \delta\left(x - \frac{m_\rho^2}{m_\tau^2}\right)\right] \\
&= \frac{G_F^2 \cos^2 \theta_C f_\rho^2 m_\tau^3}{16\pi m_\rho^2}\left(1 - \frac{m_\rho^2}{m_\tau^2}\right)^2\left(1 + \frac{2m_\rho^2}{m_\tau^2}\right)
\end{aligned} \tag{23.214}$$

The parameter f_ρ needs to be determined experimentally. Invoking CVC (see Section 21.9), its magnitude can be related to the vector current that can be measured in the $e^+ e^- \to \gamma^* \to \rho$ reaction. In fact, this is the method used to account for the finite width of the ρ meson. A sophisticated approach is to use cross-sections measured in e^+e^- collisions to derive the spectral functions to be used in tau decays. For instance, the $e^+e^- \to \pi^+\pi^-$ in the relevant center-of-mass energy range $s \simeq m_\rho^2$, gives:

$$\Gamma\left(\tau^- \to \rho^- \nu_\tau\right) = \frac{G_F^2 \cos^2\theta_C m_\tau^7}{128\pi^4\alpha^2} \int_{\sim 0}^1 x(1-x)^2(1+2x)\sigma(xm_\tau^2)\mathrm{d}x \tag{23.215}$$

- **Monte-Carlo generation of τ decays.** The necessity to accurately model tau decays, including the spin effects, requires the implementation of sophisticated Monte-Carlo generation codes. Historically, dedicated packages such as TAUOLA [313, 314] were developed to handle these. More recently, fully modeled hadronic currents with spin correlations, based on prior modelling work in TAUOLA, were implemented in PYTHIA 8 [315]. All known τ lepton decays with a branching fraction greater than 0.04% are modeled. These are summarized in Table 23.4.

Multiplicity	Model		Decay Products
2-body	single hadron		π^-, K^-
3-body	leptonic		$e^-\bar{\nu}_e$, $\mu^-\bar{\nu}_\mu$
	Kühn and Santamaria	[316]	$\pi^0\pi^-$, K^0K^-, ηK^-
	Finkemeier and Mirkes	[317]	$\pi^-\bar{K}^0$, $\pi^0 K^-$
4-body	CLEO	[318]	$\pi^0\pi^0\pi^-$, $\pi^-\pi^-\pi^+$
	Finkemeier and Mirkes	[319]	$K^-\pi^-K^+$, $K^0\pi^-\bar{K}^0$, $K_S^0\pi^-K_S^0$, $K_L^0\pi^-K_L^0$, $K_S^0\pi^-K_L^0$, $K^-\pi^0K^0$, $\pi^0\pi^0K^-$, $K^-\pi^-\pi^+$, $\pi^-\bar{K}^0\pi^0$
	Decker $et\ al.$	[320]	$\pi^0\pi^0\pi^+$, $\pi^-\pi^-\pi^+$, $K^-\pi^-K^+$, $K^0\pi^-\bar{K}^0$, $K^-\pi^0K^0$, $\pi^0\pi^0K^-$, $K^-\pi^-\pi^+$, $\pi^-\bar{K}^0\pi^0$, $\pi^-\pi^0\eta$
	Jadach $et\ al.$	[313]	$\gamma\pi^0\pi^-$
5-body	Novosibirsk	[321]	$\pi^0\pi^-\pi^-\pi^+$, $\pi^0\pi^0\pi^0\pi^-$
6-body	Kühn and Wąs	[322]	$\pi^0\pi^0\pi^-\pi^-\pi^+$, $\pi^0\pi^0\pi^0\pi^0\pi^-$, $\pi^-\pi^-\pi^-\pi^+\pi^+$

Table 23.4 Summary of available τ lepton decay models in PYTHIA 8 sorted by multiplicity. For each model the decays available and the corresponding reference are given. The implicit ν_τ is omitted for brevity. Reprinted from P. Ilten, "Tau decays in Pythia 8," *Nucl. Phys. Proc. Suppl.*, vol. 253–255, pp. 77–80, 2014, with permission from Elsevier.

23.12   Test of Lepton Universality

From many experimental processes it is found that the strength of the weak interaction is the same for all lepton flavors. The $W^\pm$ couples to the three families of leptons with the same coupling g:

$$\begin{pmatrix} \nu_e \\ e^- \end{pmatrix}; \quad \begin{pmatrix} \nu_\mu \\ \mu^- \end{pmatrix}; \quad \begin{pmatrix} \nu_\tau \\ \tau^- \end{pmatrix} \tag{23.216}$$

This is called the universality of the lepton coupling. Muon and tau decays can be used to test this hypothesis. The muon decay, whose tree-level Feynman diagram is shown in Figure 23.4, involves two weak vertices (μ, ν_μ) and (e, ν_e). Leaving the possibility that the two couplings G_F^e to electrons and G_F^μ to muons are different from one another, the muon decay rate Eq. (23.87) can be expressed as:

$$\Gamma(\mu^- \to e^-\nu_\mu\bar{\nu}_e) = \frac{1}{\tau_\mu} = \frac{G_F^e G_F^\mu m_\mu^5}{192\pi^3} \tag{23.217}$$

Similarly, the total decay widths of the tau lepton to electrons or muons are then:

$$\Gamma(\tau^- \to e^- \nu_\tau \bar{\nu}_e) = \frac{G_F^e G_F^\tau m_\tau^5}{192\pi^3} \quad \text{and} \quad \Gamma(\tau^- \to \mu^- \nu_\tau \bar{\nu}_\mu) = \frac{G_F^\mu G_F^\tau m_\tau^5}{192\pi^3} \tag{23.218}$$

The ratio of a partial width to the total width is the branching ratio:

$$Br(\tau^- \to e^- \bar{\nu}_e \nu_\tau) = \frac{\Gamma(\tau^- \to e^- \bar{\nu}_e \nu_\tau)}{\Gamma(\tau \to \text{all})} = \tau_\tau \times \Gamma(\tau^- \to e^- \bar{\nu}_e \nu_\tau) \tag{23.219}$$

Then the tau lifetime can be expressed as:

$$\tau_\tau = \frac{Br(\tau^- \to e^- \bar{\nu}_e \nu_\tau)}{\Gamma(\tau^- \to e^- \bar{\nu}_e \nu_\tau)} = \frac{192\pi^3}{G_F^e G_F^\tau m_\tau^5} Br(\tau^- \to e^- \bar{\nu}_e \nu_\tau) \tag{23.220}$$

and similarly we have for the muon:

$$\tau_\mu = \frac{Br(\mu^- \to e^- \bar{\nu}_e \nu_\mu)}{\Gamma(\mu^- \to e^- \bar{\nu}_e \nu_\mu)} = \frac{192\pi^3}{G_F^e G_F^\mu m_\mu^5} Br(\mu^- \to e^- \bar{\nu}_e \nu_\mu) \tag{23.221}$$

So the ratio of the lifetimes gives:

$$\frac{\tau_\tau}{\tau_\mu} = \frac{G_F^\mu m_\mu^5}{G_F^\tau m_\tau^5} \frac{Br(\tau^- \to e^- \bar{\nu}_e \nu_\tau)}{Br(\mu^- \to e^- \bar{\nu}_e \nu_\mu)} = \frac{(g^\mu)^2 m_\mu^5}{(g^\tau)^2 m_\tau^5} \frac{Br(\tau^- \to e^- \bar{\nu}_e \nu_\tau)}{Br(\mu^- \to e^- \bar{\nu}_e \nu_\mu)} \tag{23.222}$$

where the g^ℓ are the weak couplings corresponding to the $W^\pm \leftrightarrow \ell \nu_\ell$ vertex. By precisely measuring the branching fractions, the rest masses and lifetimes of the muon and tau, the ratio of the muon and the tau weak charged coupling is found to be [28]:

$$\frac{g_\tau}{g_\mu} = 1.0006 \pm 0.0022 \tag{23.223}$$

Similarly, by comparing the electronic and muonic decay of the tau, the ratio of the coupling is determined to be:

$$\frac{g_\mu}{g_e} = 1.0000 \pm 0.0020 \tag{23.224}$$

Within the experimental accuracy, there is clear evidence for universality and it can be concluded that $g^e = g^\mu = g^\tau$. The Michel decay parameters of the τ have also been measured experimentally. They are summarized in Table 23.5 and compared to those of the muon. The parameters are consistent within experimental errors between muon and tau and fully support the $V - A$ theory.

We can conclude that there is no experimental evidence that the $W^\pm$ does not couple to all three families of leptons identically via a pure $V - A$ current.

23.13   The Glashow–Illiopoulos–Maiani Mechanism

In 1963 Cabibbo suggested that the $d \to u + W^-$ vertex carries the factor $\cos\theta_C$ (Cabibbo favored), whereas the $s \to u + W^-$ vertex has a factor $\sin\theta_C$ (Cabibbo suppressed). The Cabibbo theory is very successful in correlating a multitude of $\Delta S = 1$ weak decay rates relative to $\Delta S = 0$ reactions. However, historically, there remained a problem with the observed branching ratio of the $K_L^0 \to \mu^+\mu^-$ decay. This kind of process is called **flavor-changing neutral currents (FCNC)**. The experimentally measured value is (value taken from the Particle Data Group's Review [5]):

$$\frac{\Gamma(K_L^0 \to \mu^+\mu^-)}{\Gamma(K_L^0 \to \text{all})} \simeq (6.84 \pm 0.11) \times 10^{-9} \tag{23.225}$$

Michel parameter	Value μ Table 23.3	Value τ	Theory $(V - A)$
ρ	0.74979 ± 0.00026	0.745 ± 0.008	$3/4$
δ	0.75047 ± 0.00034	0.746 ± 0.021	$3/4$
ξ	$1.0009^{+0.0016}_{-0.0007}$	0.985 ± 0.030	1
η	0.057 ± 0.034	0.013 ± 0.020	0

Table 23.5 Properties of the tau decay parameters compared to muon and theory (from P. A. Zyla *et al.* (Particle Data Group), *Prog. Theor. Exp. Phys.* **2020**, 083C01 (2020)).

The decay into two muons was predicted by the Cabibbo theory to occur at tree level via the so-called **box diagram** with the exchange of two $W^{\pm}$ bosons. See Figure 23.15. Because the diagram contains both the $d \to u$ ($\Delta S = 0$) and $s \to u$ ($\Delta S = 1$) vertices, the decay rate is proportional to the product $(\cos\theta_C \sin\theta_C)^2 \approx 0.05$! This was not sufficient to explain the observed strong suppression of $\sim 10^{-8}$! In general, one could show that any type of FCNC process was highly suppressed.

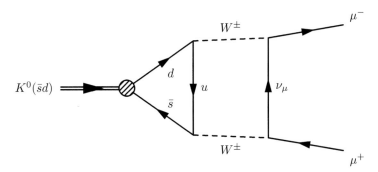

Figure 23.15 The tree-level box diagram for the $K_L^0 \to \mu^+\mu^-$ decay. The quark propagator in the box is the u quark.

In 1970 **Glashow**,[8] **Iliopoulos**,[9] and **Maiani**[10] introduced the existence of a massive fourth-flavor **charmed quark** (c) [323], in addition to the u, d, and s, and explained the suppression of the decay by introducing the so-called **GIM mechanism**. The c quark carries an electric charge of $+2/3$, and is therefore a sort of "heavy" u quark. The c quark has weak couplings to the s and d quarks that carry factors of $-\sin\theta_C$ (the minus sign is extremely important!) and $\cos\theta_C$. See Figure 23.16.

The existence of the additional charm quark changes the prediction for the tree-level amplitude of the K^0 decay. We must now consider the *two* tree-level diagrams shown in Figure 23.17. The box with the u quark has an amplitude proportional to $\sin\theta_C \cos\theta_C$, while the one with the c quark has an amplitude with a factor $-\sin\theta_C \cos\theta_C$. The resulting amplitude would exactly vanish if the two diagrams were identical. But this also requires that the masses of the u and c quarks are identical:

$$\mathcal{M}_u + \mathcal{M}_c = 0 \quad \text{when} \quad m_c \equiv m_u \tag{23.226}$$

But the two diagrams differ if the masses of the u and c quarks are not identical! This was known to be the case at the time, since the c quark had never been observed! So, for $m_c \gg m_u$, the amplitudes do not cancel

8   Sheldon Lee Glashow (born 1932), American theoretical physicist.
9   John Iliopoulos (born 1940), a Greek physicist.
10   Luciano Maiani (born 1941), San Marino citizen physicist.

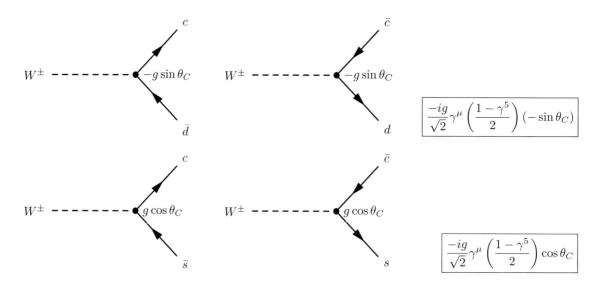

Figure 23.16 Vertex factors for the Cabibbo suppressed $c \to d$ and favored $c \to s$ weak coupling of the charged $W^{\pm}$ to quarks. Compare with Figure 23.3.

exactly, however, the process will be strongly suppressed, as is the case for the $K^0 \to \mu^+ + \mu^-$ decay. This compensation is called the **GIM mechanism**.

The GIM mechanism introduced a very striking symmetry between quarks and leptons. The weak charged current simply couples to the three leptons $W \to e\nu_e$, $W \to \mu\nu_\mu$, $W \to \tau\nu_\tau$ and the two quarks $W \to ud'$, $W \to cs'$ (known at the time). These can be organized in the following "doublets":

$$\text{Leptons:} \begin{pmatrix} \nu_e \\ e^- \end{pmatrix} ; \begin{pmatrix} \nu_\mu \\ \mu^- \end{pmatrix} ; \begin{pmatrix} \nu_\tau \\ \tau^- \end{pmatrix} \quad \text{Quarks:} \begin{pmatrix} u \\ d' \end{pmatrix} ; \begin{pmatrix} c \\ s' \end{pmatrix} \tag{23.227}$$

The weak eigenstates d' and s' are given as functions of the mass (flavor) eigenstates via the Cabibbo angle θ_C as a single parameter:

$$d' = d\cos\theta_C + s\sin\theta_C \quad \text{and} \quad s' = -d\sin\theta_C + s\cos\theta_C \tag{23.228}$$

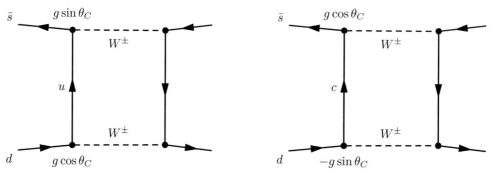

Figure 23.17 The amplitudes $\mathcal{M}_u$ (left) and $\mathcal{M}_c$ (right) at higher-order corrections illustrating the GIM mechanism.

This "mixing" can be conveniently represented as a rotation in the space of the weak and mass (flavor) eigenstates:

$$
\underbrace{\begin{pmatrix} d' \\ s' \end{pmatrix}}_{\text{weak eigenstates}} = \underbrace{\begin{pmatrix} \cos\theta_C & \sin\theta_C \\ -\sin\theta_C & \cos\theta_C \end{pmatrix}}_{\text{unitary mixing matrix}} \underbrace{\begin{pmatrix} d \\ s \end{pmatrix}}_{\text{flavor eigenstates}}
\tag{23.229}
$$

The mixing matrix is a special 2×2 unitary matrix.

We note that it is by convention that the rotation acts on the down-type quarks d and s. The same result could be accomplished by rotating the up-type quarks and by introducing u' and c'.

23.14 Weak Charged Coupling to Quarks and Leptons

We have seen the significance of the quark mixing theory developed by Cabibbo and the importance of the GIM mechanism in weak interactions. The Cabibbo–GIM mechanism was generalized by **Kobayashi** and **Maskawa** to handle three generations of quarks to explain CP violation [241], as will be discussed in Chapter 30. The theory naturally accommodates mixing among all types of fermions, namely quarks as well as leptons. The $W^\pm$ boson couples to the following doublets of weak eigenstate fermion fields in a universal way (i.e., with the same coupling g):[11]

$$
W^\pm: \qquad \begin{pmatrix} \nu_e \\ e^- \end{pmatrix}; \begin{pmatrix} \nu_\mu \\ \mu^- \end{pmatrix}; \begin{pmatrix} \nu_\tau \\ \tau^- \end{pmatrix}; \begin{pmatrix} u \\ d' \end{pmatrix}; \begin{pmatrix} c \\ s' \end{pmatrix}; \begin{pmatrix} t \\ b' \end{pmatrix}
\tag{23.230}
$$

Both share very similar properties and can therefore be discussed collectively.

We first discuss the quark case. The term in the Lagrangian of our theory that describes the weak charged-current interaction that reproduces the wanted vertex factors shown in Figures 23.3 and 23.16 will be of the form:

$$
\begin{aligned}
\mathcal{L}_{quarks}^{CC} &= -\frac{ig}{\sqrt{2}} \left[\bar{u}\gamma^\mu \frac{1-\gamma^5}{2} d' + \bar{c}\gamma^\mu \frac{1-\gamma^5}{2} s' + \bar{t}\gamma^\mu \frac{1-\gamma^5}{2} b' \right] + \text{h.c.} \\
&= -\frac{ig}{\sqrt{2}} \begin{pmatrix} \bar{u} & \bar{c} & \bar{t} \end{pmatrix} \gamma^\mu \frac{1}{2}(1-\gamma^5) \begin{pmatrix} d' \\ s' \\ b' \end{pmatrix} + \text{h.c.}
\end{aligned}
\tag{23.231}
$$

The mixing matrix that governs the quark sector mixing is called the **Cabibbo–Kobayashi–Maskawa (CKM)** matrix. In this case, the weak eigenstates are related to the mass (flavor) eigenstates by:

$$
\begin{pmatrix} d' \\ s' \\ b' \end{pmatrix} = \begin{pmatrix} V_{ud} & V_{us} & V_{ub} \\ V_{cd} & V_{cs} & V_{cb} \\ V_{td} & V_{ts} & V_{tb} \end{pmatrix} \begin{pmatrix} d \\ s \\ b \end{pmatrix} = V_{\text{CKM}} \begin{pmatrix} d \\ s \\ b \end{pmatrix}
\tag{23.232}
$$

In terms of the mass eigenstates, the weak charged current will have the form:

$$
\mathcal{L}_{quarks}^{CC} = -\frac{g}{\sqrt{2}} \begin{pmatrix} \bar{u} & \bar{c} & \bar{t} \end{pmatrix} \gamma^\mu \frac{1}{2}(1-\gamma^5) V_{\text{CKM}} \begin{pmatrix} d \\ s \\ b \end{pmatrix} + \text{h.c.}
\tag{23.233}
$$

The corresponding Feynman diagrams are shown in Figure 23.18. Note the complex conjugation.

---

11   We note that $m_t \gg m_W$ so the tb' coupling only occurs for virtual $W^\pm$s or ts.

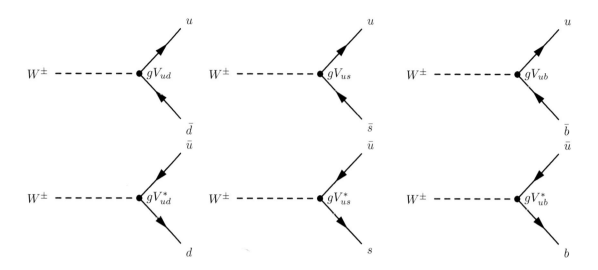

Figure 23.18 Vertex factors for the u quark weak coupling to the charged $W^\pm$. Similar diagrams can be derived for the other up-type quarks with their corresponding CKM element $V_{qq'}$. Note the complex conjugation for antiquarks.

What about the leptons? Why did we not consider mixing? It is natural to assume that the weak charged currents between the six quarks and the six leptons have a similar structure. Indeed, the answer is that the leptons are also mixed but this was experimentally verified much later when **neutrino flavor oscillations** were discovered (see Section 30.6). The term in the Lagrangian that corresponds to the vertex factors listed in Figure 23.2 can be expressed as:

$$
\begin{aligned}
\mathcal{L}^{CC}_{leptons} &= -\frac{ig}{\sqrt{2}} \left[\bar{e}\gamma^\mu \frac{1-\gamma^5}{2} \nu_e + \bar{\mu}\gamma^\mu \frac{1-\gamma^5}{2} \nu_\mu + \bar{\tau}\gamma^\mu \frac{1-\gamma^5}{2} \nu_\tau \right] + \text{h.c.} \\
&= -\frac{ig}{\sqrt{2}} \begin{pmatrix} \bar{e} & \bar{\mu} & \bar{\tau} \end{pmatrix} \gamma^\mu \frac{1}{2}(1-\gamma^5) \begin{pmatrix} \nu_e \\ \nu_\mu \\ \nu_\tau \end{pmatrix} + \text{h.c.}
\end{aligned}
\tag{23.234}
$$

Hence, the weak charged-current interaction will couple a lepton to a neutrino of the same flavor. However, neutrinos propagate in space as mass eigenstates. We hence consider their mixing. For the lepton sector the mixing matrix is called the **Pontecorvo–Maki[12]–Nakagawa[13]–Sakata[14] (PMNS)** matrix (see Section 30.6). The weak eigenstates ν_α ($\alpha = e, \mu, \tau$) (neutrino flavor eigenstates) are related to the mass eigenstates ν_i ($i = 1, 2, 3$) of masses m_i ($i = 1, 2, 3$) in the following way:

$$
\begin{pmatrix} \nu_e \\ \nu_\mu \\ \nu_\tau \end{pmatrix} \equiv \begin{pmatrix} U_{e1} & U_{e2} & U_{e3} \\ U_{\mu1} & U_{\mu2} & U_{\mu3} \\ U_{\tau1} & U_{\tau2} & U_{\tau3} \end{pmatrix} \begin{pmatrix} \nu_1 \\ \nu_2 \\ \nu_3 \end{pmatrix} = U_{PMNS} \begin{pmatrix} \nu_1 \\ \nu_2 \\ \nu_3 \end{pmatrix}
\tag{23.235}
$$

---

12   Ziro Maki, Japanese physicist.
13   Masami Nakagawa, Japanese physicist.
14   Shoichi Sakata (1911–1970), Japanese physicist.

Hence, the charged-current interaction of leptons can be written in terms of their mass eigenstates as:

$$
\mathcal{L}_{leptons}^{CC} = -\frac{g}{\sqrt{2}} \left(\bar{e} \quad \bar{\mu} \quad \bar{\tau} \right) \gamma^\mu \frac{1}{2}(1 - \gamma^5) U_{PMNS} \begin{pmatrix} \nu_1 \\ \nu_2 \\ \nu_3 \end{pmatrix} + \text{h.c.} \tag{23.236}
$$

In the case of massless neutrinos, the masses are degenerate and $m_1 = m_2 = m_3 = 0$. Hence, one can always redefine the mass eigenstates such that the mixing matrix is the unity $U_{PMNS} = \mathbb{1}$. In this case, it is unnecessary to distinguish weak and mass eigenstates and we recover the trivial case:

$$
\mathcal{L}_{leptons}^{CC} = -\frac{g}{\sqrt{2}} \left(\bar{e} \, \bar{\mu} \, \bar{\tau} \right) \gamma^\mu \frac{1}{2}(1 - \gamma^5) \mathbb{1} \begin{pmatrix} \nu_1 \\ \nu_2 \\ \nu_3 \end{pmatrix} + \text{h.c.} = -\frac{g}{\sqrt{2}} \left(\bar{e} \, \bar{\mu} \, \bar{\tau} \right) \gamma^\mu \frac{1}{2}(1 - \gamma^5) \begin{pmatrix} \nu_e \\ \nu_\mu \\ \nu_\tau \end{pmatrix} + \text{h.c.} \tag{23.237}
$$

On the other hand, if the neutrino masses are not degenerate, then we should consider the mixing matrix and understand the phenomenological consequences that can lead to the measurement of the mixing matrix elements.

The physical consequences of the quark and lepton mixing will be discussed in Chapter 30. It is worth noting that the motivation for introducing the CKM matrix was not merely an extension from two to three generations. Indeed, Kobayashi and Maskawa proposed it in 1973 before the discovery of the charm quark. Their motivation was to introduce a complex phase to explain CP violation, as will be explained in Chapter 30.

23.15   Weak Decays of Charmed or Bottom Hadrons

The quark "spectator" approximation was introduced in Chapter 20 when discussing the decay of charmed or bottom hadrons. We can now use this approximation to derive the decay widths and corresponding decay lifetimes using the weak charged current. Consequently, we consider the weak decay of a hadron containing one heavy quark where, via the weak charged current, this heavy quark turns into a lighter one. We recall that we expect this approximation to improve the higher the mass of the parent hadron, since the higher the Q^2, the more "free" are the quarks during the decay process (see running of α_s in Section 18.15).

- **Semi-leptonic decay modes.** The most probable mode for the decay of a charmed meson is via a $c \to W^+ s$ transition, where the W^+ subsequently decays into $u\bar{d}$, $e^+\nu_e$, or $\mu^+\nu_\mu$. This gives us five $(3 + 1 + 1 = 5)$ possibilities in total, when accounting for the three quark colors. The tree-level diagram is very similar to that of the muon decay shown in Figure 23.4 and the total decay rate can be derived from Eq. (23.96):

$$
\Gamma(c \to \text{all}) \simeq 5 \times \left(\frac{G_F^2 m_c^5}{192\pi^3} \right) \left(1 + \delta f \left(\frac{m_s^2}{m_c^2}, 0 \right) \right) \tag{23.238}
$$

where the correction for the finite mass is given by Eq. (23.97). Under this simple model, the semi-leptonic branching ratio is simply $1/5$:

$$
Br(c \to se\nu) = Br(c \to s\mu\nu) = \frac{\Gamma(c \to se\nu)}{\Gamma(c \to \text{all})} = \frac{1}{5} \tag{23.239}
$$

This prediction can be compared with actually measured values as listed in Eqs. (20.35) and (20.36). For the $D^\pm$, the branchings $Br(D^+ \to e^+ + X) = (16.07 \pm 0.30)\%$ and $Br(D^+ \to \mu^+ + X) = (17.6 \pm 3.2)\%$ are relatively compatible with the simple prediction, while the branchings $Br(D^0 \to e^+ + X) = (6.49 \pm 0.11)\%$ and $Br(D^0 \to \mu^+ + X) = (6.7 \pm 0.6)\%$ indicate an enhancement of the non-leptonic modes for the D^0. Nonetheless, the spectator model provides a qualitative picture of the decay of the charmed mesons.

• **Estimation of the decay lifetime.** By direct comparison, the charmed meson lifetimes can be expressed in terms of the muon lifetime as follows:

$$\tau_c \simeq \frac{\tau_\mu}{5} \left(\frac{m_\mu}{m_c}\right)^5 \left(1 + \delta f\left(\frac{m_s^2}{m_c^2}, 0\right)\right)^{-1} \tag{23.240}$$

The prediction is highly sensitive to the choice of the m_c and to some extent m_s (we have neglected m_u and m_d). We have already seen in Section 18.13 that quarks cannot be encountered as free particles, hence one cannot directly determine the rest mass of a bare quark (in fact, it doesn't make much sense to do so). Nonetheless, we can make educated guesses and check if these yield the correct order of magnitude. Taking $m_c = (1/2)m_{J/\Psi} \approx 1.55$ GeV and $m_s = (1/2)m_\phi \approx 0.50$ GeV yields a large mass correction factor $\delta f \approx -0.53$ (compare with the muon lifetime case, Eq. (23.99)). The estimated lifetime is then:

$$\tau_c \approx \frac{2.2~\mu s}{5} \left(\frac{0.105~\text{GeV}}{1.55~\text{GeV}}\right)^5 (1 - 0.53)^{-1} \approx 1.34 \times 10^{-12}~\text{s} \tag{23.241}$$

which is amazingly good when compared to the measured value $\tau(D^\pm) = (1.040 \pm 0.007) \times 10^{-12}$ s (see Eq. (20.35)).

• **Cabibbo suppressed decays.** Let us consider as an example the following decays of the charmed D^0 meson (the experimentally measured branching ratios are shown in brackets):

$$
\begin{aligned}
D^0 &\rightarrow K^-\pi^+ \quad ((3.950 \pm 0.031)\%) \\
&\rightarrow \pi^-\pi^+ \quad ((1.455 \pm 0.024) \times 10^{-3}) \\
&\rightarrow K^+\pi^- \quad ((1.50 \pm 0.07) \times 10^{-4})
\end{aligned} \tag{23.242}
$$

Can one understand their relative decay branching ratios? The decay tree-level diagrams are shown in Figure 23.19. In the spectator approximation, we consider the decay c quark as free, and we use the Cabibbo

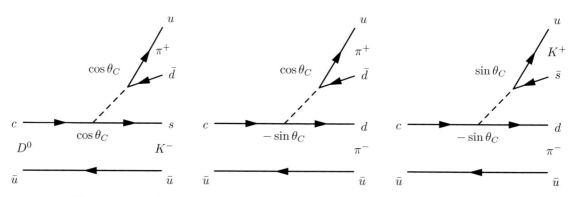

Figure 23.19 Cabibbo enhanced and suppressed decays of the charmed D^0 meson.

theory to estimate the relative rates of the three decay modes. We can assume that the phase-space factor is similar in all cases. Then we have simply:

$$\mathcal{M}(D^0 \rightarrow K^-\pi^+) \propto \cos^2\theta_C, \mathcal{M}(D^0 \rightarrow \pi^-\pi^+) \propto -\sin\theta_C\cos\theta_C, \mathcal{M}(D^0 \rightarrow K^+\pi^-) \propto -\sin^2\theta_C \tag{23.243}$$

thus

$$
\begin{aligned}
\Gamma(D^0 \rightarrow K^-\pi^+) : \Gamma(D^0 \rightarrow \pi^-\pi^+) : \Gamma(D^0 \rightarrow K^+\pi^-) &\approx \cos^4\theta_C : \sin^2\theta_C\cos^2\theta_C : \sin^4\theta_C \\
&\approx 1 : \tan^2\theta_C : \tan^4\theta_C \\
&\approx 1 : 0.05 : 0.003
\end{aligned} \tag{23.244}
$$

which is experimentally relatively well verified. This confirms that the spectator model is a good approximation and that the ratio of the decay modes is clearly driven by the Cabibbo suppression mechanism!

• **Bottom meson decays.** It is expected that the spectator model should be a better approximation in the case of the b quark than for the c quark, because of its larger mass. At the b vertex, only the transitions $b \to c$ and $b \to u$ are allowed, since the $b \to t$ is kinematically prohibited (the top quark will be discussed in Section 29.2). As will be shown in Chapter 30, the transition $b \to c$ is the dominant one. Hence, there are $1+1+1+3+3 = 9$ possible decay modes:

$$b \to c\,e\,\nu_e, c\,\mu\,\nu_\mu, c\,\tau\,\nu_\tau, c\,d\,\bar{u}, c\,s\,\bar{c} \tag{23.245}$$

where, as before, we have accounted for the color factor. One cannot neglect the masses in case of the $\tau\nu_\tau$ and the $s\bar{c}$ and we should use a correction $\delta f(x,y) = (m_c^2/m_b^2, m_\tau^2/m_b^2)$ and $\delta f(x,y) = (m_c^2/m_b^2, m_c^2/m_b^2)$ (see Eq. (23.97)) for these modes. These phase-space factors suppress those two decay modes by about a factor five, consequently, we have instead of the factor nine, simply $1+1+\frac{1}{5}+3+\frac{3}{5} \simeq 5.8$. The b-quark lifetime can then be expressed as a function of the muon lifetime as (cf. Eq. (23.240)):

$$\tau_b \simeq \frac{\tau_\mu}{5.8|V_{cb}|^2} \left(\frac{m_\mu}{m_b}\right)^5 \left(1 + \delta f\left(\frac{m_c^2}{m_b^2}, 0\right)\right)^{-1} \tag{23.246}$$

where we have introduced the CKM matrix element V_{cb} to take into account the Cabibbo suppression of the $b \to c$ transition (cf. Figure 23.18).

This model predicts a semi-leptonic branching ratio equal to:

$$Br(b \to ce\nu) \simeq \frac{1}{5.8} \approx 17\% \tag{23.247}$$

to be compared with the experimental values given in Eqs. (20.51) and (20.52): $Br(B^+ \to \ell^+\nu_\ell + X) = (10.99 \pm 0.28)\%$ and $Br(B^0 \to \ell^+\nu_\ell + X) = (10.33 \pm 0.28)\%$. It is clear that the agreement is not perfect and here too the non-semi-leptonic modes are enhanced compared to the naive tree-level spectator model.

With the same word of caution concerning the assignment of the quark masses as in the case of the charmed meson, one can use the lifetime of the bottom meson to estimate the magnitude of the CKM element V_{cb}. Indeed, setting $m_b = (1/2)m_\Upsilon \approx 4.73$ GeV and $m_c = (1/2)m_{J/\Psi} \approx 1.55$ GeV, one finds a mass correction factor $\delta f \approx -0.54$ (which is numerically very similar to the case of the charmed meson):

$$\tau_b \simeq \frac{2.2 \ \mu s}{5.8|V_{cb}|^2} \left(\frac{0.105 \ \text{GeV}}{4.73 \ \text{GeV}}\right)^5 (1 - 0.54)^{-1} \approx 1.78 \times 10^{-12} \times \left(\frac{0.05}{|V_{cb}|}\right)^2 \ \text{s} \tag{23.248}$$

When comparing this result with the experimentally measured value $\tau_{B^\pm} = (1638 \pm 4) \times 10^{-15}$ s (see Eq. (20.51)), one can conclude that the magnitude of the CKM matrix element V_{cb} is rather small, i.e., in the range of 5%, which is amazingly close to the square of the Cabibbo suppression $|V_{us}|^2 \simeq \sin^2 \theta_C \approx 0.05$ (see Eq. (23.21)). This fact will be readdressed in a broader context in Section 30.2.

23.16   The Neutron Lifetime

The classical way to study the weak interaction is of course via β decay. In this context, the simplest and most easily studied system is the free neutron decay into a proton, an electron, and an antineutrino. We define the kinematics for $n \to pe^-\bar{\nu}_e$ as:

$$n(p^\mu) \to p(p'^\mu) + e^-(k^\mu) + \bar{\nu}_e(k'^\mu) \tag{23.249}$$

The CODATA recommended value for the neutron–proton mass difference is [10] (see also Eq. (15.35)):

$$\Delta \equiv M_n - M_p = (1.29333236 \pm 0.00000046) \ \text{MeV} \tag{23.250}$$

so we cannot neglect the mass of the electron in the calculations. As in the case of weak decays involving hadrons (e.g., pion, kaon decays, tau decays into hadrons, etc.), we need to take into account the proton and neutron structures in order to compute the process. The amplitude can be expressed as (cf. Eq. (23.171)):

$$\mathcal{M} = \frac{G_F}{\sqrt{2}} \underbrace{\left(\bar{u}(k)\gamma_\mu(1-\gamma^5)v(k')\right)}_{\text{lepton current}} \underbrace{\langle p|J^\mu|n\rangle}_{\text{hadronic current}} \qquad (23.251)$$

When we discussed the electromagnetic process $e^- p \to e^- p$ in Chapter 16, we expressed the conserved hadronic current in its general Lorentz-covariant form with the help of the Dirac and Pauli form factors $F_1(Q^2)$ and $F_2(Q^2)$ (see Eq. (16.34)). In the case of the weak interaction, the proton and the neutron are to be seen as a nucleon (since the weak force does not distinguish the electric charge), hence members of an isospin doublet. Therefore, the weak current should be expressed as an isospin changing current between two nucleons. In addition, the momentum transfer in the decay is very low so essentially we care about the form factors at $Q^2 = 0$, i.e., their static limits. The current should also violate parity, so a mixture of scalar, pseudoscalar, vector, axial vector or tensor should be considered. Fortunately, it has been shown that only vector (V) and axial vector (A) contribute significantly [324]. So, finally, we can write the hadronic current as a function of just two constants, historically labeled g_V and g_A:[15]

$$< N(p')\left|J_\mu^+\right|N(p) >= V_{ud}\,\bar{u}(p')\gamma_\mu\left(g_V + g_A\gamma_5\right)u(p) \qquad (23.252)$$

Hence

$$\mathcal{M}(n \to pe^-\bar{\nu}) = \frac{G_F V_{ud}}{\sqrt{2}}\left(\bar{u}(k)\gamma^\mu(1-\gamma^5)v(k')\right)\left(\bar{u}(p')\gamma_\mu\left(g_V + g_A\gamma_5\right)u(p)\right) \qquad (23.253)$$

The decay rate is then computed using standard trace and phase-space integration techniques. The electron spectrum is given by:

$$\frac{d\Gamma_n}{dE_e} \propto E_e\sqrt{E_e^2 - m_e^2}\left[\Delta - E_e\right]^2 \qquad (23.254)$$

Integration over the electron spectrum yields the total width of the neutron decay:

$$\Gamma_n = \frac{G_F^2|V_{ud}|^2 m_e^5}{16\pi^3} f^R(r)\left(g_V^2 + 3g_A^2\right) \qquad (23.255)$$

where $r = \Delta/m_e$ and the phase-space factor is:

$$f^R(r) = \left[\frac{1}{15}\left(2r^4 - 9r^2 - 8\right)\sqrt{r^2 - 1} + r\ln(r + \sqrt{r^2 - 1})\right] \qquad (23.256)$$

• **Conserved vector current hypothesis.** We have already mentioned CVC in Section 21.9. It says that the vector part of the current in weak and electromagnetic interactions should be identical. This leads to $g_V = 1$. Any deviation of g_V from unity would indicate a breakdown of CVC. Assuming CVC, and with the measured neutron lifetime $\tau_n = (880.2\pm1.0)$ s and the CKM matrix element $V_{ud} \simeq \cos\theta_C \approx 0.97$, we find $g_V^2 + 3g_A^2 \simeq 6.1$ or:

$$|g_A| \approx 1.3 \qquad (23.257)$$

• **Correlation measurements.** The most accurate measurement of the axial coupling of the neutron comes from correlation measurements with polarized neutrons. The decay of polarized neutrons can be written as [324]:

$$\frac{d^3\Gamma}{dE_e d\Omega} = \Gamma(E_e)\left(1 + P_n A(E_e)\beta\cos\theta\right) \qquad (23.258)$$

where $\Gamma(E_e)$ is the unpolarized neutron differential decay rate, P_n is the polarization of the neutron, β is the velocity of the electron, θ is the angle between the neutron spin direction and the electron momentum

---

15   We use the convention of the Particle Data Group's Review with the "+" leading to a negative value for g_A.

direction, and $A(E_e)$ is the asymmetry parameter, which in general depends on the decay electron energy. If one neglects corrections from the neutron recoil and also from radiative corrections, the asymmetry is just a constant given by [324]:

$$A_0 = \frac{-2(\lambda^2 - |\lambda|)}{1 + 3\lambda^2} \qquad (23.259)$$

where $\lambda \equiv g_A/g_V$.

• **Neutron decay parameter.** The latest experimental results from the UCNA experiment [325] at the Los Alamos Neutron Science Center are displayed in Figure 23.20, compiled with other experimental results. They find:

$$A_0 = -0.12015 \pm 0.00034(\text{stat.}) \pm 0.00063(\text{syst.}) \implies \lambda = \frac{g_A}{g_V} = -1.2772 \pm 0.0020 \qquad (23.260)$$

Historically there has been some discussion on the actual value of λ, also related to differing measurements of the neutron lifetime. Recent measurements (marked as post-2002) indicate a higher value for λ, consistent

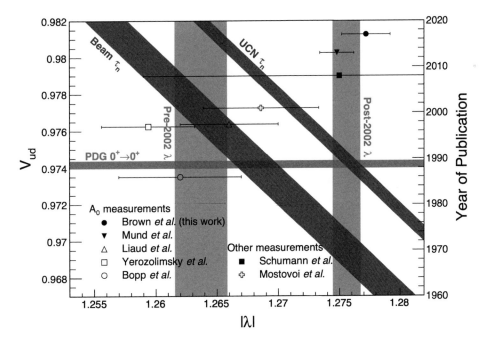

Figure 23.20 Compilation of neutron decay parameter results in the V_{ud} vs. $|\lambda|$ plane. Recent measurements (marked as post-2002) indicate a higher value for λ. Reprinted with permission from M. A. P. Brown *et al.*, "New result for the neutron β-asymmetry parameter A_0 from UCNA," *Phys. Rev.*, vol. C97, no. 3, p. 035505, 2018. Copyright 2018 by the American Physical Society.

with the independently determined V_{ud} (horizontal band labeled PDG $0^+ \to 0^+$) and the neutron lifetime τ_n measured with ultra-cold neutrons (diagonal band labeled UCN τ_n). Its value is (value taken from the Particle Data Group's Review [5]):

$$\lambda \equiv \frac{g_A}{g_V} = -1.2732 \pm 0.0023 \qquad (23.261)$$

If the nucleon had been composed of free quarks, then the constants would be $g_V = 1$ and $g_A = -1$. Hence, the departure of the ratio λ from unity is a manifestation of the strong interaction between the partons of the nucleon.

Problems

Ex 23.1 Muon decay width computed with FeynCalc. We consider the decay of the muon and want to calculate its decay width using the FeynCalc package (see Section 11.13).

(a) Calculate the partial width of the process $\mu \to e + \nu_\mu + \bar{\nu}_e$ at tree level, neglecting the electron mass.

(b) What is the result in the case where the mass of the electron is not neglected ?

Ex 23.2 Decays of charged and neutral pions. The neutral pion has a very short lifetime compared to the charged pion.

(a) Discuss quantitatively why this is the case.

(b) Check in the Particle Data Group's Review the experimental methods used to measure the π^0 lifetime and discuss the methods.

Ex 23.3 Bottom meson decays. The charged-current decay $B^+ \to \mu^+ \nu_\mu$ is analogous to the $\pi^+ \to \mu^+ \nu_\mu$ decay.

(a) Estimate its partial width $\Gamma(B^+ \to \mu^+ \nu_\mu)$. Use the decay constant $f_B \approx 190$ MeV (this value can be found in the Particle Data Group's Review).

(b) Compare this branching to the semi-leptonic decay $B \to D\ell\nu_\ell$. What can you say?

(c) Compute the branching ratio for $B^+ \to \mu^+ \nu_\mu$ and compare with experimental results.

Ex 23.4 Bottom meson decays. The decay $B^0 \to \mu^+ \mu^-$ is a very rare decay. Explain why you expect its branching ratio to be very small. Check and elaborate on the latest experimental results in the Particle Data Group's Review.

Ex 23.5 General $V \pm A$ decay rate. Consider the general weak decay of a heavy quark a into three particles $a(p_1) \to b(p_2)c(p_3)\bar{d}(p_4)$ driven by an effective Lagrangian with both $V - A$ and $V + A$ contributions. We can generally write this in terms of left-handed and right-handed couplings:

$$\mathcal{L} = \left(g_L \bar{b}_L \gamma^\mu a_L + g_R \bar{b}_R \gamma^\mu a_R\right)\left(g'_L \bar{c}_L \gamma_\mu d_L + g'_R \bar{c}_R \gamma_\mu d_R\right) \tag{23.262}$$

where g_L, g_R, g'_L, and g'_R are the left-handed and right-handed couplings.

(a) Show that the squared matrix element, summed over spins and averaged over the initial one, is given by:

$$\begin{aligned}
\frac{1}{2}\sum |\mathcal{M}|^2 &= 8\Big[\left(g_L^2 g'^2_L + g_R^2 g'^2_R\right)(p_1 \cdot p_4)(p_2 \cdot p_3) + \left(g_L^2 g'^2_R + g_R^2 g'^2_L\right)(p_1 \cdot p_3)(p_2 \cdot p_4) \\
&\quad - m_a m_b\, g_L g_R \left(g'^2_L + g'^2_R\right)(p_3 \cdot p_4) - m_c m_d\left(g_L^2 + g_R^2\right)g'_L g'_R(p_1 \cdot p_2) \\
&\quad - 4 m_a m_b m_c m_d\, g_L g_R g'_L g'_R\Big]
\end{aligned} \tag{23.263}$$

(b) We now assume that the masses of c and d are negligible compared to those of a and b. Show that the decay width is given by:

$$\Gamma = \frac{m_a^5}{128\pi^3}\left[(g_L^2 + g_R^2)(g'^2_L + g'^2_R)I_1 + g_L g_R(g'^2_L + g'^2_R)I_2\right] \tag{23.264}$$

where

$$I_1 = \int_0^{(1-x_b)^2} \mathrm{d}z\, z(1 + x_b^2 - z)\lambda^{1/2}(1, z, x_b^2) \quad \text{and} \quad I_2 = -2\int_0^{(1-x_b)^2} \mathrm{d}z\, z x_b \lambda^{1/2}(1, z, x_b^2) \tag{23.265}$$

with $x_b \equiv m_b/m_a$ and λ the Källén function given in Eq. (5.132).

(c) Now consider the pure $V - A$ case and compare with the discussion in Section 23.15.

24   Gauge Field Theories and Spontaneous Symmetry Breaking

A gauge theory is a type of field theory in which the Lagrangian is invariant under a continuous group of local transformations. A Yang–Mills theory is a gauge theory based on the $SU(N)$ group. They are at the core of the unification of the electromagnetic and weak forces as well as quantum chromodynamics.

Wikipedia

24.1   Gauge Invariance Principle

In 1921 **Weyl** studied the invariance properties of quantum mechanics. In particular, he noted that the absolute phase of the wave function is not observable. In a formal and generic way, we consider a **global transformation of the phase** of a Dirac spinor. This can be expressed with the following **unitary global gauge transformation** $U(\alpha)$:

$$\text{Global gauge:} \quad \Psi(x) \longrightarrow \Psi'(x) \equiv U(\alpha)\Psi(x) = e^{iQ\alpha}\,\Psi(x) \tag{24.1}$$

where α is an arbitrary real phase constant and Q is a **charge characteristic of the spinor** Ψ. The phase of $\Psi(x)$ is a pure convention-dependent quantity without physical meaning. The set of gauge transformations forms a continuous group of transformations $U(\alpha)$, which depend on the single continuous parameter α. This group is the unitary group $U(1)$. We note that **the $U(1)$ group is an Abelian group**, which means that two transformations commute:

$$U(1) \text{ Abelian:} \quad U(\alpha_1)U(\alpha_2) = U(\alpha_1 + \alpha_2) = U(\alpha_2)U(\alpha_1) \tag{24.2}$$

According to Weyl, this transformation leaves *all physical observables unchanged*. Hence, the physical laws must be invariant under a global $U(1)$ transformation. Let us apply the gauge transformation to the *free* Dirac equation:

$$\mathcal{L}_{Dirac} = i\,\overline{\Psi}(x)\gamma^\mu\partial_\mu\Psi(x) - m\,\overline{\Psi}(x)\Psi(x) \tag{24.3}$$

We note that the gauge transformation will have the following effect:

$$\begin{cases} \Psi(x) \longrightarrow e^{iQ\alpha}\,\Psi(x) \\ \overline{\Psi}(x) \longrightarrow e^{-iQ\alpha}\,\overline{\Psi}(x) \\ \partial_\mu\Psi(x) \longrightarrow e^{iQ\alpha}\,\partial_\mu\Psi(x) \end{cases} \tag{24.4}$$

In the last line, we have explicitly made use of the fact that the phase α is *global*, hence is the same in all points of space-time, and consequently does not depend on x^μ. Under these assumptions, we see that the *free* Dirac equation is indeed invariant under the global gauge transformation:

$$\mathcal{L}_{Dirac} \longrightarrow i\,e^{-iQ\alpha}\overline{\Psi}(x)\gamma^\mu e^{iQ\alpha}\partial_\mu\Psi(x) - m\,e^{-iQ\alpha}e^{iQ\alpha}\overline{\Psi}(x)\Psi(x) = \mathcal{L}_{Dirac} \tag{24.5}$$

For any given Lagrangian $\mathcal{L}$ we can consider its invariance properties under an **infinitesimal $U(1)$ gauge transformation**:

$$U(\delta\alpha)\Psi(x) \simeq (1 + iQ\delta\alpha)\Psi(x) = \Psi(x) + iQ\delta\alpha\Psi(x) \quad \text{with } |\delta\alpha| \to 0 \tag{24.6}$$

The variation of the Lagrangian $\mathcal{L}(\Psi(x), \partial_\mu\Psi(x))$ under this infinitesimal transformation can be expressed as:

$$
\begin{aligned}
\delta\mathcal{L} &= \frac{\partial\mathcal{L}}{\partial\Psi}\delta\Psi + \frac{\partial\mathcal{L}}{\partial(\partial_\mu\Psi)}\delta(\partial_\mu\Psi) - \left[\Psi \to \overline{\Psi}\right] \\
&= \frac{\partial\mathcal{L}}{\partial\Psi}iQ\delta\alpha\Psi(x) + \frac{\partial\mathcal{L}}{\partial(\partial_\mu\Psi)}(iQ\delta\alpha\partial_\mu\Psi) - \left[\Psi \to \overline{\Psi}\right] \\
&= iQ\delta\alpha\partial_\mu\left(\frac{\partial\mathcal{L}}{\partial(\partial_\mu\Psi)}\Psi\right) + \underbrace{iQ\delta\alpha\left(\frac{\partial\mathcal{L}}{\partial\Psi}\right)\Psi - iQ\delta\alpha\left(\partial_\mu\frac{\partial\mathcal{L}}{\partial(\partial_\mu\Psi)}\right)\Psi}_{=0} - \left[\Psi \to \overline{\Psi}\right]
\end{aligned}
\tag{24.7}
$$

where we have used the fact that the field Ψ is a solution of the Euler–Lagrange equation. Hence, if the Lagrangian is invariant under the global $U(1)$ gauge transformation, then:

$$\delta\mathcal{L} = iQ\delta\alpha\partial_\mu\left(\frac{\partial\mathcal{L}}{\partial(\partial_\mu\Psi)}\Psi - \overline{\Psi}\frac{\partial\mathcal{L}}{\partial(\partial_\mu\overline{\Psi})}\right) = 0 \quad \longrightarrow \quad \partial_\mu j^\mu = 0 \tag{24.8}$$

where the conserved gauge four-current j^μ associated with the gauge-invariant Lagrangian is defined as:

$$j^\mu \equiv iQ\left(\frac{\partial\mathcal{L}}{\partial(\partial_\mu\Psi)}\Psi - \overline{\Psi}\frac{\partial\mathcal{L}}{\partial(\partial_\mu\overline{\Psi})}\right) \tag{24.9}$$

Considering again the *free* Dirac Lagrangian and setting $Q = e = \sqrt{4\pi\alpha}$, we "magically" obtain the current of the well-known Lorentz-vector form:

$$j^\mu_{Dirac} \equiv ie\left(i\overline{\Psi}\gamma^\mu\Psi - 0\right) = -e\overline{\Psi}\gamma^\mu\Psi \tag{24.10}$$

It is impressive! We obtain a direct relation between the global $U(1)$ invariance and the electric charge of QED. As a consequence of the Noether theorem, a conserved quantity was to be expected as a result of an invariance. Here we interpret the electric charge of the Dirac spinor as the conserved quantity under the global $U(1)$ invariance.

24.2   QED as a $U(1)$ Local Gauge Field Theory

Until now we have analyzed the consequences of a global phase invariance. What happens if the phase is a "local" gauge, in other words, if the phase parameter α becomes a function of the space-time coordinates? Under a *local phase change* we have $\alpha \equiv \alpha(x)$ and the local gauge transformation of the field is then:

$$\text{Local gauge:} \quad \Psi(x) \xrightarrow{\text{Local } U(1)} \Psi'(x) \equiv U(\alpha)\Psi(x) = e^{iQ\alpha(x)}\Psi(x) \tag{24.11}$$

Let us consider again the QED Lagrangian describing a free Dirac fermion. However, the free Lagrangian is no longer invariant if one allows the phase transformation to depend on the space-time coordinate, i.e., under a local phase change $\alpha \equiv \alpha(x)$, because:

$$\partial_\mu\Psi(x) \xrightarrow{\text{Local } U(1)} e^{iQ\alpha}\partial_\mu\Psi(x) + iQe^{iQ\alpha}\left(\partial_\mu\alpha(x)\right)\Psi(x) \tag{24.12}$$

The second term which depends on the derivatives of the phase breaks the invariance of the Lagrangian.

The **gauge field invariance principle** is the requirement that the $U(1)$ phase invariance should hold *locally*, i.e., with space-time-dependent transformations $\alpha \equiv \alpha(x^\mu)$. This is only possible if one adds some additional term to the Lagrangian, transforming in such a way as to cancel the $\partial_\mu \alpha$ term in Eq. (24.12). The needed modification is completely fixed by introducing the **gauge-covariant derivative**:

$$D_\mu \Psi(x) \equiv \left[\partial_\mu + iQA_\mu(x)\right] \Psi(x) \tag{24.13}$$

where **the gauge vector field** $A_\mu(x)$ is a spin-1 field (since ∂_μ has a Lorentz index). We can recover gauge invariance (see Eq. (24.5)):

$$D_\mu \Psi(x) \xrightarrow{U(1)} (D_\mu \Psi)'(x) \equiv e^{iQ\alpha(x)} D_\mu \Psi(x) \tag{24.14}$$

if the gauge vector field transforms as:

$$A_\mu(x) \xrightarrow{U(1)} A'_\mu(x) \equiv A_\mu(x) - \partial_\mu \alpha(x) \tag{24.15}$$

since

$$
\begin{aligned}
(D_\mu \Psi)'(x) &= \left[\partial_\mu + iQA'_\mu(x)\right] e^{iQ\alpha(x)} \Psi(x) \\
&= e^{iQ\alpha} \partial_\mu \Psi(x) + iQ e^{iQ\alpha} (\partial_\mu \alpha(x)) \Psi(x) + iQA'_\mu(x) e^{iQ\alpha} \Psi(x) \\
&= e^{iQ\alpha} \partial_\mu \Psi(x) + iQ e^{iQ\alpha} (\partial_\mu \alpha(x)) \Psi(x) + iQ(A_\mu(x) - \partial_\mu \alpha(x)) e^{iQ\alpha} \Psi(x) \\
&= e^{iQ\alpha} (\partial_\mu + iQA_\mu(x)) \Psi(x)
\end{aligned}
\tag{24.16}
$$

At this stage the gauge field $A_\mu(x)$ is an arbitrary function of space-time that compensates for the phase of the local gauge transformation, such that the equation is finally gauge-invariant. Its physical interpretation is not yet explicit. By inserting the gauge-covariant derivative into the free Dirac Lagrangian, one obtains the following Lagrangian **invariant under local $U(1)$ transformations**:

$$
\begin{aligned}
\mathcal{L}_{U(1)} &\equiv i\overline{\Psi}(x)\gamma^\mu D_\mu \Psi(x) - m\overline{\Psi}(x)\Psi(x) \\
&= i\overline{\Psi}(x)\gamma^\mu \partial_\mu \Psi(x) - m\overline{\Psi}(x)\Psi(x) - QA_\mu(x)\overline{\Psi}(x)\gamma^\mu \Psi(x) \\
&= \mathcal{L}_{Dirac} - QA_\mu(x)\overline{\Psi}(x)\gamma^\mu \Psi(x)
\end{aligned}
\tag{24.17}
$$

The gauge principle has generated an interaction between the Dirac spinor and the gauge field A_μ, which is nothing else than the familiar QED vertex if we *interpret* the charge $Q = e$.

The covariant derivative can be used to construct other covariant quantities. For example, the antisymmetric product of two covariant derivatives on the field spinor Ψ is written as:

$$[D_\mu, D_\nu] \Psi \equiv D_\mu(D_\nu \Psi) - D_\nu(D_\mu \Psi) \tag{24.18}$$

Explicit calculation of one term, e.g., $D_\mu(D_\nu \Psi)$, gives:

$$
\begin{aligned}
D_\mu(D_\nu \Psi) &= \left[\partial_\mu + iQA_\mu\right] (\partial_\nu + iQA_\nu) \Psi \\
&= \partial_\mu \partial_\nu \Psi + iQ\partial_\mu(A_\nu \Psi) + iQA_\mu \partial_\nu \Psi - Q^2 A_\mu A_\nu \Psi \\
&= \partial_\mu \partial_\nu \Psi + iQ(\partial_\mu A_\nu)\Psi + iQA_\nu \partial_\mu \Psi + iQA_\mu \partial_\nu \Psi - Q^2 A_\mu A_\nu \Psi
\end{aligned}
\tag{24.19}
$$

and similarly, $D_\nu(D_\mu \Psi)$ is found by swapping $\mu \leftrightarrow \nu$. Hence, all symmetric terms in $\mu\nu$ cancel in the commutation (note also $[A_\mu, A_\nu] = 0$) and we simply find:

$$
\begin{aligned}
[D_\mu, D_\nu] \Psi &= iQ\left[(\partial_\mu A_\nu)\Psi + A_\nu \partial_\mu \Psi + A_\mu \partial_\nu \Psi - (\partial_\nu A_\mu)\Psi - A_\mu \partial_\nu \Psi - A_\nu \partial_\mu \Psi\right] \\
&= iQ(\partial_\mu A_\nu - \partial_\nu A_\mu)\Psi \equiv iQF_{\mu\nu}\Psi
\end{aligned}
\tag{24.20}
$$

where $F_{\mu\nu} \equiv \partial_\mu A_\nu - \partial_\nu A_\mu$ is the **usual electromagnetic field strength** (see Eq. (10.8))! Hence, we found the amazing result that **the field strength associated with the gauge field can be found with the antisymmetric product of two covariant derivatives of the theory!**

Further, if one wants A_μ to be a true propagating field, one needs to add a gauge-invariant kinetic term of the form of Eq. (10.12) to the Lagrangian:

$$\mathcal{L}_{\text{kin}} \equiv -\frac{1}{4} F_{\mu\nu} F^{\mu\nu} \tag{24.21}$$

Hence the QED Lagrangian, Eq. (10.93), can be simply constructed with the local gauge invariance principle:

$$\boxed{\mathcal{L}_{QED} \equiv \mathcal{L}_{U(1)} + \mathcal{L}_{\text{kin}}} \tag{24.22}$$

and the gauge field $A_\mu(x)$ can be interpreted as the photon field. Note that the gauge transformation, Eq. (24.15), is always allowed by the degree of freedom already embedded into the Maxwell equations (see Eq. (10.20))!

The field equation for a massive boson was given in Eq. (10.77) starting from Maxwell's equation and substituting $\partial_\mu \partial^\mu$ by $\partial_\mu \partial^\mu + M^2$. A possible mass term for the gauge field in the Lagrangian is then $\frac{1}{2} M^2 A^\mu A_\mu$. It is, however, forbidden because it would violate gauge invariance, since:

$$\frac{1}{2} M^2 A'^\mu A'_\mu = \frac{1}{2} M^2 \left(A_\mu(x) - \partial_\mu \alpha(x)\right) \left(A^\mu(x) - \partial^\mu \alpha(x)\right) \neq \frac{1}{2} M^2 A^\mu A_\mu \tag{24.23}$$

Therefore, **the gauge photon field is predicted to be massless ($M = 0$).** The experimental limit on the photon rest mass is:

$$m_\gamma < 1 \times 10^{-18} \text{ eV} \tag{24.24}$$

In conclusion, from the "simple $U(1)$ gauge-symmetry requirement," one could deduce precisely and uniquely the correct QED Lagrangian and that its gauge boson, the photon, must be massless if local invariance is to hold. That's fantastic!

24.3 Yang–Mills Gauge Field Theories

We want to discuss local gauge field theories that are invariant under local $SU(N)$ transformations. Let us first consider a field spinor Ψ in the case $N = 2$ [326].[1] Such a spinor field can be represented with a "doublet" structure:

$$\Psi \equiv \begin{pmatrix} \Psi_1 \\ \Psi_2 \end{pmatrix} \quad \text{and} \quad \overline{\Psi} = \begin{pmatrix} \overline{\Psi}_1 & \overline{\Psi}_2 \end{pmatrix} \tag{24.25}$$

Its free Lagrangian can be expressed as:

$$\mathcal{L}_{Dirac} = i\, \overline{\Psi}(x) \gamma^\mu \partial_\mu \Psi(x) - \overline{\Psi}(x) M \Psi(x) \tag{24.26}$$

where M is the mass matrix:

$$M \equiv \begin{pmatrix} m_1 & 0 \\ 0 & m_2 \end{pmatrix} \tag{24.27}$$

We can consider a **global unitary transformation of the field doublet**:

$$\Psi \to U\Psi \tag{24.28}$$

---

[1]   Yang and Mills (Robert Laurence Mills (1927–1999), theoretical physicist) constructed the gauge theory based not on the one-dimensional group $U(1)$ of electrodynamics, but on the three-dimensional group $SU(2)$ of isospin conservation, in the hope that this would become a theory of the strong interactions.

where U is an element of $SU(2)$ that can be represented with a 2×2 matrix. Since $\overline{\Psi} \to \overline{\Psi} U^\dagger$, then unitarity $U^\dagger U = \mathbb{1}$ implies:

$$\overline{\Psi}\Psi \to \overline{\Psi} U^\dagger U \Psi = \overline{\Psi}\Psi \tag{24.29}$$

and $\overline{\Psi}\Psi$ is an invariant of the $SU(2)$ transformation. If the symmetry is exact, then the two particles must also possess the same rest mass, and we have $m_1 = m_2$.

Any unitary U matrix of $SU(2)$ can be expressed as a function of the (three) generators of $SU(2)$, where I (for "isospin") plays the role of the charge Q in the $U(1)$ case. Then we have the transformation:

$$\Psi \to U(\vec{\alpha})\Psi = e^{iI\alpha_i \tau_i}\Psi = e^{iI\vec{\alpha}\cdot\vec{\tau}}\Psi \tag{24.30}$$

where the vector $\vec{\alpha} = (\alpha_1, \alpha_2, \alpha_3)$ is determined by three real numbers, and τ_i are the Pauli matrices.

A **local $SU(2)$ unitary transformation** can be introduced by allowing the phase vector $\vec{\alpha} = (\alpha_1, \alpha_2, \alpha_3)$ to depend on x^μ:

$$\Psi(x) \overset{\text{Local } SU(2)}{\longrightarrow} U(\vec{\alpha}(x))\Psi(x) = e^{iI\alpha_i(x)\tau_i}\Psi = e^{iI\vec{\alpha}(x)\cdot\vec{\tau}}\Psi \tag{24.31}$$

and in this case:

$$\partial_\mu \Psi(x) \to U\left(\partial_\mu \Psi(x)\right) + \left(\partial_\mu U\right)\Psi(x) \tag{24.32}$$

As in the case of the local Abelian $U(1)$ transformation, a local $SU(2)$ transformation will *not* keep the free Lagrangian invariant. In order to ensure gauge invariance, the generalized covariant derivative should be used:

$$D_\mu \Psi(x) \equiv \left[\mathbb{1}\partial_\mu + i\frac{g}{2}\tau_i W^i_\mu(x)\right]\Psi(x) = \left[\mathbb{1}\partial_\mu + i\frac{g}{2}\vec{\tau}\cdot\vec{W}_\mu(x)\right]\Psi(x) \tag{24.33}$$

where $g(= 2I)$ is a "coupling constant" and **the three gauge vector fields** $W^i_\mu(x)$ $(i = 1, \ldots, 3)$ are a set of spin-1 gauge fields. In the case of a group like $SU(2)$, one has three generators and hence needs three independent functions of space-time to compensate for the local phase changes:

$$\vec{W}_\mu(x) \equiv (W^1_\mu(x), W^2_\mu(x), W^3_\mu(x)) \tag{24.34}$$

For clarity one generally adopts the convention that the Lorentz index and the field index are placed symmetrically, hence, we will write W^i_μ and $W_{i,\mu}$ synonymously. The sufficient condition for the invariance of the Lagrangian is to have the covariant derivative of the spinor satisfy:

$$D_\mu \Psi \overset{SU(2)}{\to} (D_\mu \Psi)' = U(D_\mu \Psi) \tag{24.35}$$

We note that on the one hand:

$$(D_\mu \Psi)' = \left[\mathbb{1}\partial_\mu + i\frac{g}{2}\vec{\tau}\cdot\vec{W}'_\mu\right]\Psi' = U(\partial_\mu\Psi) + (\partial_\mu U)\Psi + i\frac{g}{2}\left(\vec{\tau}\cdot\vec{W}'_\mu\right)U\Psi \tag{24.36}$$

On the other hand:

$$U(D_\mu \Psi) = U\left[\mathbb{1}\partial_\mu + i\frac{g}{2}\vec{\tau}\cdot\vec{W}_\mu\right]\Psi = U(\partial_\mu\Psi) + i\frac{g}{2}U(\vec{\tau}\cdot\vec{W}_\mu\Psi) \tag{24.37}$$

This leads to the equality:

$$i\frac{g}{2}\left(\vec{\tau}\cdot\vec{W}'_\mu\right)U\Psi = i\frac{g}{2}U(\vec{\tau}\cdot\vec{W}_\mu\Psi) - (\partial_\mu U)\Psi = i\frac{g}{2}U(\vec{\tau}\cdot\vec{W}_\mu)\Psi - (\partial_\mu U)\Psi \tag{24.38}$$

Since this should be valid for any field, including the field $U^{-1}\Psi$, we obtain:

$$\vec{\tau}\cdot\vec{W}'_\mu = U(\vec{\tau}\cdot\vec{W}_\mu)U^{-1} + \frac{2i}{g}(\partial_\mu U)U^{-1} \tag{24.39}$$

The first term represents the $SU(2)$ rotation while the second term is called the gradient term. We note that in the case of $U(1)$ where $U \equiv \exp(iQ\alpha(x))$, then $Q \leftrightarrow g/2$, and we would get:

$$A'_\mu = U(A_\mu)U^{-1} + \frac{i}{Q}(\partial_\mu U)U^{-1} = A'_\mu = A_\mu + \frac{i}{Q}(iQ(\partial_\mu\alpha)U)U^{-1} = A_\mu - \partial_\mu\alpha \tag{24.40}$$

which is what is expected (see Eq. (24.15)).

If we now consider an infinitesimal $SU(2)$ gauge transformation of the form:

$$U\Psi = e^{ig\vec{\alpha}(x)\cdot\vec{\tau}/2}\Psi \simeq \left[\mathbb{1} + \frac{ig}{2}\vec{\tau}\cdot\vec{\alpha}(x) + \cdots\right]\Psi \quad \text{with} \quad |\alpha_i| \to 0 \tag{24.41}$$

then the transformation of the gauge field is in first approximation:

$$
\begin{aligned}
\vec{\tau}\cdot\vec{W}'_\mu &\simeq \left(\mathbb{1} + \frac{ig}{2}\vec{\tau}\cdot\vec{\alpha}\right)(\vec{\tau}\cdot\vec{W}_\mu)\left(\mathbb{1} - \frac{ig}{2}\vec{\tau}\cdot\vec{\alpha}\right) + \frac{2i}{g}(\frac{ig}{2}\vec{\tau}\cdot(\partial_\mu\vec{\alpha}))\left(\mathbb{1} - \frac{ig}{2}\vec{\tau}\cdot\vec{\alpha}\right) \\
&\simeq (\vec{\tau}\cdot\vec{W}_\mu) + (\vec{\tau}\cdot\vec{W}_\mu)\left(-\frac{ig}{2}\vec{\tau}\cdot\vec{\alpha}\right) + \left(\frac{ig}{2}\vec{\tau}\cdot\vec{\alpha}\right)(\vec{\tau}\cdot\vec{W}_\mu) - (\vec{\tau}\cdot(\partial_\mu\vec{\alpha})) \\
&= (\vec{\tau}\cdot\vec{W}_\mu) - \frac{ig}{2}\left[(\vec{\tau}\cdot\vec{W}_\mu)(\vec{\tau}\cdot\vec{\alpha}) - (\vec{\tau}\cdot\vec{\alpha})(\vec{\tau}\cdot\vec{W}_\mu)\right] - (\vec{\tau}\cdot(\partial_\mu\vec{\alpha})) \\
&= (\vec{\tau}\cdot\vec{W}_\mu) - \frac{ig}{2}\left[2i\vec{\tau}\cdot\left(\vec{W}_\mu\times\vec{\alpha}\right)\right] - (\vec{\tau}\cdot(\partial_\mu\vec{\alpha})) \\
&= (\vec{\tau}\cdot\vec{W}_\mu) - g\vec{\tau}\cdot\left(\vec{\alpha}\times\vec{W}_\mu\right) - (\vec{\tau}\cdot(\partial_\mu\vec{\alpha}))
\end{aligned}
\tag{24.42}
$$

where we used the properties of the Pauli matrices, Eq. (C.98). The expression above can be reduced to the following rule for the transformation of the $SU(2)$ gauge fields:

$$\vec{W}'_\mu = \vec{W}_\mu - g\vec{\alpha}\times\vec{W}_\mu - \partial_\mu\vec{\alpha} \tag{24.43}$$

This is the $SU(2)$ analog to the transformation $A'_\mu = A_\mu - \partial_\mu\alpha$ in $U(1)$, where the additional vector product comes from the non-Abelian nature of $SU(2)$. The appearance of this term with a vector product might appear surprising at first. Its origin can indeed be traced back to the non-Abelian structure of $SU(2)$ by considering each gauge field separately and noting that by using Eq. (A.19), we get:

$$W'_{i,\mu} = W_{i,\mu} - g\epsilon_{ijk}\alpha^j W^k_\mu - \partial_\mu\alpha_i \equiv W_{i,\mu} - gf_{ijk}\alpha^j W^k_\mu - \partial_\mu\alpha_i \tag{24.44}$$

where we recognize the structure constants $[t_i, t_j] = if_{ijk}t_k$ of the t_i generators of our group $SU(2)$ (see Eq. (B.43)). This last equation represents the transformation of the gauge field under an infinitesimal $SU(2)$ transformation.

The gauge principle based on the $U(1)$ Abelian symmetry applied to the free Dirac equation led us to the derivation of the QED Lagrangian from first principles. Similarly, we can now build the gauge field theory based on the non-Abelian $SU(2)$ symmetry by considering the effect of the covariant derivative on the free Dirac Lagrangian. We find:

$$
\begin{aligned}
\mathcal{L}_{SU(2)} &= i\overline{\Psi}(x)\left(\gamma^\mu D_\mu - M\right)\Psi(x) = \mathcal{L}_{free} - \frac{g}{2}\overline{\Psi}(x)\gamma^\mu\left(\vec{\tau}\cdot\vec{W}_\mu\right)\Psi \\
&= \mathcal{L}_{free} - g\overline{\Psi}(x)\gamma^\mu\left(\vec{t}\cdot\vec{W}_\mu\right)\Psi
\end{aligned}
\tag{24.45}
$$

where $\vec{t} \equiv (t_1, t_2, t_3)$ are the $SU(2)$ generators corresponding to our structure constants f_{ijk}. We now understand the origin of the $1/2$ factor in the $g/2$ term in our original Lagrangian, Eq. (24.33).

We note that **the form of the interaction term is completely and uniquely defined by the gauge principle and the structure of the symmetry group**. In order to build a theory and promote the gauge fields to real propagative boson fields, we should still consider the kinetic terms associated with the $\vec{W}_\mu$ fields. We can write:

$$\mathcal{L}_{kin} = -\frac{1}{4} \vec{W}_{\mu\nu} \cdot \vec{W}^{\mu\nu} \tag{24.46}$$

where the tensor field strengths $\vec{W}_{\mu\nu} = \left(W_{\mu\nu}^1, W_{\mu\nu}^2, W_{\mu\nu}^3 \right)$ are defined to ensure the gauge invariance of the kinetic term. As in the case of the $U(1)$ symmetry, we can **use the commutator of two covariant derivatives, Eq. (24.20), to obtain the field strength tensors**. For instance:

$$\begin{aligned}
D_\mu(D_\nu \Psi) &= \left(\partial_\mu + ig\, t_i W_\mu^i \right) \left[\partial_\nu \Psi + ig\, t_j W_\nu^j \Psi \right] \\
&= \partial_\mu \partial_\nu \Psi + ig\, t_i W_\mu^i (\partial_\nu \Psi) + ig\, \partial_\mu (t_j W_\nu^j \Psi) - g^2 (t_i W_\mu^i)(t_j W_\nu^j \Psi) \\
&= \partial_\mu \partial_\nu \Psi + ig\, t_i W_\mu^i (\partial_\nu \Psi) + ig\, (\partial_\mu t_j W_\nu^j)\Psi + ig\, (t_j W_\nu^j)(\partial_\mu \Psi) - g^2 (t_i W_\mu^i)(t_j W_\nu^j \Psi)
\end{aligned} \tag{24.47}$$

Thus, analogous to the $U(1)$ case, we get:

$$\begin{aligned}
[D_\mu, D_\nu]\,\Psi &= D_\mu(D_\nu \Psi) - D_\nu(D_\mu \Psi) \\
&= igt_i \left[(\partial_\mu W_\nu^i) - (\partial_\nu W_\mu^i) \right] \Psi - g^2 \left[(t_j W_\mu^j)(t_k W_\nu^k) - (t_j W_\nu^j)(t_k W_\mu^k) \right] \Psi \\
&= igt_i \left[(\partial_\mu W_\nu^i) - (\partial_\nu W_\mu^i) \right] \Psi - g^2 \left[t_j t_k - t_k t_j \right] W_\mu^j W_\nu^k \Psi \\
&= igt_i \left[(\partial_\mu W_\nu^i) - (\partial_\nu W_\mu^i) - g f_{jk}^i W_\mu^j W_\nu^k \right] \Psi \\
&\equiv igt_i W_{\mu\nu}^i \Psi
\end{aligned} \tag{24.48}$$

where we have introduced our field strength tensors. Consequently, these latter turn out to be given by:

$$W_{\mu\nu}^i \equiv \partial_\mu W_\nu^i - \partial_\nu W_\mu^i - g f_{jk}^i W_\mu^j W_\nu^k \tag{24.49}$$

So, using the structure constants of $SU(2)$, we can replace $f_{ijk} = \epsilon_{ijk}$, and hence obtain the vectorial equation:

$$\vec{W}_{\mu\nu} = \partial_\mu \vec{W}_\nu - \partial_\nu \vec{W}_\mu - g \vec{W}_\mu \times \vec{W}_\nu \tag{24.50}$$

The extra cross-product term, which was not present in the Abelian $U(1)$ case, represents a boson self-coupling term, like the one we already encountered in QCD. From a phenomenological point of view, the gauge bosons carry the charge g of the interaction. This is a direct consequence of the non-Abelian nature of the $SU(2)$ group. Such a term is not present in QED where the photon is electrically neutral. **Consequently, in non-Abelian gauge theories, the gauge bosons will self-interact**.

24.4 QCD as an $SU(3)$ Gauge Field Theory

The Yang–Mills theory based on the local $SU(2)$ gauge symmetry can be generalized to a local $SU(N)$ by considering the field N-multiplet:

$$\Psi = \begin{pmatrix} \Psi_1 \\ \dots \\ \Psi_N \end{pmatrix} \tag{24.51}$$

Then we have the local $SU(N)$ transformation defined as (see Eq. (24.11)):

$$\text{Local gauge:} \quad \Psi(x) \longrightarrow \Psi'(x) = U(\vec{\alpha})\Psi(x) = e^{ig\alpha_a(x)T_a}\Psi(x) \tag{24.52}$$

where T_a $(a = 1, \ldots, N^2 - 1)$ are the generators of $SU(N)$ (see Appendix B.9), $\alpha_a = (\alpha_1, \ldots, \alpha_{N^2-1})$ and g represents the associated charge. A gauge transformation is characterized by the $N^2 - 1$ independent phase parameters α_a.

In order to describe QCD, we consider a gauge field theory based on the $SU(3)_{color}$ exact symmetry, so $N = 3$ and we have $N^2 - 1 = 8$ independent generators of the group and the charge is the color represented by g_s. We use the formalism defined in Chapter 18. Let us denote with q_f^i a quark field of color i $(i = 1, 2, 3)$ and flavor f. To simplify the equations, we use the vector notation in color space (see Eq. (18.4)). The free Lagrangian is given by Eq. (18.5):

$$\mathcal{L}_{Dirac,color} = \sum_f \bar{q}_f \left(i\gamma^\mu \partial_\mu - m_f \right) q_f \tag{24.53}$$

The Lagrangian is invariant under a global $SU(3)_C$ transformation in color space:

$$q_f^\alpha \longrightarrow (q_f^\alpha)' = U^\alpha_{\ \beta} q_f^\beta \qquad \text{where} \quad U(\alpha_a) \equiv e^{ig_s \alpha_a T_a} = e^{ig_s \frac{\lambda^a}{2} \alpha_a} \tag{24.54}$$

where $T_a = \lambda^a/2$ $(a = 1, 2, \ldots, 8)$ denote the Gell-Mann matrices (see Eq. (B.59)), and α_a are arbitrary parameters.

We require the Lagrangian to be invariant under *local* $SU(3)_C$ transformations, $\alpha_a = \alpha_a(x)$. To satisfy this requirement, we need to change the quark derivatives by covariant objects. Since we now have eight independent gauge parameters, eight different gauge bosons $G_a^\mu(x)$, the so-called *gluons*, are needed (cf. Eq. (24.33)):

$$D^\mu q_f \equiv= \left[\mathbb{1}\partial^\mu + ig_s T^a G_a^\mu(x) \right] q_f = \left[\mathbb{1}\partial^\mu + i\frac{g_s}{2} \lambda^a G_a^\mu(x) \right] q_f \tag{24.55}$$

Under an infinitesimal $SU(3)_C$ transformation, the gauge boson fields will transform as in the $SU(2)$ case (see Eq. (24.44)) but with the corresponding structure constants of the $SU(3)$ group. Hence:

$$(G_a^\mu)' = G_a^\mu - g_s f_{abc} \alpha_b G_c^\mu - \partial^\mu \alpha_a \tag{24.56}$$

As expected, the non-Abelian nature of the $SU(3)_C$ symmetry gives rise to the additional term involving the gluon fields themselves.

To build a gauge-invariant kinetic term for the gluon fields, we introduce the corresponding field strengths as in Eq. (24.48):

$$[D^\mu, D^\nu] q_f \equiv ig_s \frac{\lambda^a}{2} G_a^{\mu\nu} q_f \tag{24.57}$$

which can be shown to lead to the following gluons' field strength tensors $G_a^{\mu\nu}(x)$ as in Eq. (24.49), which is exactly what we had postulated in Eq. (18.19):

$$G_a^{\mu\nu}(x) = \partial^\mu G_a^\nu(x) - \partial^\nu G_a^\mu(x) - g_s f_{abc} G_b^\mu(x) G_c^\nu(x) \tag{24.58}$$

Taking the proper normalization for the gluon kinetic term, we finally have the $SU(3)_C$-invariant QCD Lagrangian:

$$\mathcal{L}_{QCD} \equiv \sum_f \bar{q}_f \left(i\gamma^\mu D_\mu - m_f \mathbb{1} \right) q_f - \frac{1}{4} \vec{G}^{\mu\nu} \vec{G}_{\mu\nu} \tag{24.59}$$

which is exactly Eq. (18.38), which was described in Chapter 18.

24.5   Spontaneous Symmetry Breaking

After seeing the importance of local gauge invariance, we will introduce **spontaneous symmetry breaking (SSB)**, which is another truly fundamental principle in quantum field theories. As was clear from the discussion in the previous sections, the principle of local gauge invariance enforces the gauge bosons to be massless. This poses a critical problem in the realization of a local gauge field theory of the weak interaction, since we have been considering a very massive charged intermediate vector boson $W^{\pm}$ as the carrier of the weak force. This assumption seems to be in sharp contradiction to the principle of local gauge invariance. Spontaneous symmetry breaking is a fundamental concept introduced to explain how gauge bosons can acquire a non-vanishing mass (effectively through a breaking of the symmetry). However, the spontaneous breaking of the symmetry is induced by the choice of the vacuum state, not by the fundamental properties of the Lagrangian. In this way, the Lagrangian will retain its fundamental symmetry, and the apparent breaking of the symmetry will result in the choice of one particular ground state among a set of possibilities, which could potentially be infinite.

We begin with a trivial case of SSB. We consider a classic real Klein–Gordon field $\phi(x)$ with the quartic interaction term $\phi^4(x)$ (see Section 9.14):

$$\mathcal{L}_\phi = \mathcal{L}_{kin} + \mathcal{L}_{mass} + \mathcal{L}_{int.} = \frac{1}{2}\left(\partial_\mu \phi\right)^2 - \frac{1}{2}\mu^2 \phi^2 - \frac{\lambda}{4}\phi^4 \tag{24.60}$$

There are two free parameters in the theory: μ and λ. We can rewrite the Lagrangian labeling the mass and interaction terms as the potential:

$$\mathcal{L}_\phi \equiv \frac{1}{2}\left(\partial_\mu \phi\right)^2 - V(\phi) \quad \text{where} \quad V(\phi) \equiv \frac{1}{2}\mu^2 \phi^2 + \frac{\lambda}{4}\phi^4 \tag{24.61}$$

We note that the so-constructed ϕ^4 theory possesses a discrete symmetry: **the Lagrangian remains unchanged under the transformation** $\phi \to -\phi$ (this is called the discrete Z_2 transformation where Z_2 is the cyclic group of order n).

The ground state of the theory (or vacuum) is the state with the lowest energy of the field and it corresponds to the minimum of the potential V. It can be found through a variation of the potential with respect to the field:

$$\frac{\partial V}{\partial \phi} = 0 = \frac{\partial}{\partial \phi}\left(\frac{1}{2}\mu^2 \phi^2 + \frac{\lambda}{4}\phi^4\right) = \phi\left(\mu^2 + \lambda \phi^2\right) \tag{24.62}$$

This equation admits the two different types of solutions:

$$\phi_0 = 0 \quad \text{or} \quad \phi_0^2 = -\frac{\mu^2}{\lambda} \tag{24.63}$$

One can assume $\lambda > 0$ to have the potential bounded from below and μ^2 is a priori arbitrary. We can now distinguish the two cases, depending on the chosen sign of μ^2.

- $\lambda > 0$ and $\mu^2 > 0$: the resulting potential is illustrated in Figure 24.1(left); there is only one solution $\phi_0 \equiv 0$, since the other solution is not real. The theory describes massive scalar particles of mass μ with four-point self-interactions proportional to $\lambda \phi^4$.

- $\lambda > 0$ and $\mu^2 < 0$: the resulting potential is illustrated in Figure 24.1(right); the parameter μ can no longer be associated with the rest mass of the particle. The lowest energy state does not occur at $\phi = 0$. There are two solutions for the minimum, which we can write as:

$$\phi_0 \equiv \pm v = \pm\sqrt{-\frac{\mu^2}{\lambda}} \quad \rightarrow \quad \mu^2 = -\lambda v^2 \tag{24.64}$$

and the field has acquired a non-zero vacuum expectation value. The actual vacuum state can be either solution, i.e., $\phi_0 = +v$ or $\phi_0 = -v$. The choice of the solution is known as the **spontaneous breaking of the symmetry**, since the two solutions are no longer symmetric (the choice is either one or the other).

The constant v is called the **vacuum expectation value (vev)**. Spontaneous symmetry breaking is often seen as a dynamical process. In the original theory, the symmetry is unbroken, and the system evolves into a configuration where the symmetry is broken. When the symmetry is unbroken, the minimum is at $\phi = 0$. As the symmetry breaks and the potential develops two minima, the state at $\phi = 0$ becomes unstable. The asymmetric potential enforces an asymmetric outcome, with the choice of one of the two possible distinct minima. This is graphically illustrated in Figure 24.1, where the state of the system is represented by the ball.

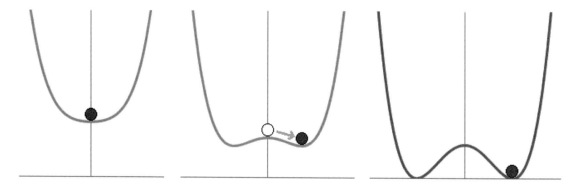

Figure 24.1 Illustration of spontaneous symmetry breaking: when the symmetry is unbroken, the minimum is at $\phi = 0$. As the symmetry breaks, the asymmetric potential enforces an asymmetric outcome, with two possible distinct minima.

Let us discuss the broken case $\lambda > 0$ and $\mu^2 < 0$ further. How can we physically interpret this solution in terms of particles? We consider the field in its minimum and its variation around this minimum. The field can be expanded, for example, around its $+v$ minimum in the following way:

$$\phi(x) \equiv v + h(x) \tag{24.65}$$

where the function $h(x)$ describes the variation of the field relative to its ground state given by v. Note that we could have expanded the field around the $-v$ minimum, so we have made a *choice* of the minimum. Once we have chosen the minimum, we lose the $\phi \to -\phi$ symmetry, since this latter transforms one minimum into the other! Hence, when we decide to expand the field around a chosen minimum, we spontaneously break the symmetry. Note that the symmetry is still underlying in the Lagrangian and this symmetry translates into the freedom of choice of the minimum.

We can rewrite the Lagrangian of the theory in the spontaneously broken condition:

$$\begin{aligned}
\mathcal{L}_\phi &= \frac{1}{2}\left(\partial_\mu \phi\right)^2 - \frac{1}{2}\mu^2\phi^2 - \frac{\lambda}{4}\phi^4 \\
&= \frac{1}{2}\left(\partial_\mu(v + h(x))\right)^2 - \frac{1}{2}\mu^2(v + h(x))^2 - \frac{\lambda}{4}(v + h(x))^4 \\
&= \frac{1}{2}\left(\partial_\mu h\right)^2 - \frac{1}{2}\mu^2 v^2 - \mu^2 vh - \frac{1}{2}\mu^2 h^2 - \frac{\lambda}{4}(v^4 + 4v^3 h + 6v^2 h^2 + 4vh^3 + h^4) \\
&= \frac{1}{2}\left(\partial_\mu h\right)^2 - \frac{1}{2}\mu^2 v^2 - \left(\mu^2 v + \lambda v^3\right)h - \left(\frac{1}{2}\mu^2 + \frac{3}{2}\lambda v^2\right)h^2 - \left(\frac{\lambda}{4}v^4 + \lambda vh^3 + \frac{\lambda}{4}h^4\right) \tag{24.66}
\end{aligned}$$

where we have regrouped the terms in powers of h. We note that the *vev* condition $v^2 = -\mu^2/\lambda$ (see Eq. (24.64)) implies:

$$\begin{cases} \mu^2 v + \lambda v^3 = 0 \\ \dfrac{1}{2}\mu^2 + \dfrac{3}{2}\lambda v^2 = \dfrac{1}{2}\mu^2 - \dfrac{3}{2}\lambda\dfrac{\mu^2}{\lambda} = -\mu^2 \end{cases} \tag{24.67}$$

The terms $(1/2)\mu^2 v^2$ and $\lambda v^4/4$ are constant and do not play any role in the dynamics. We can collect the remaining terms to construct the Lagrangian density for the field $h(x)$:

$$\mathcal{L}_h = \frac{1}{2}(\partial_\mu h)^2 + \mu^2 h^2 - \lambda v h^3 - \frac{\lambda}{4}h^4 = \frac{1}{2}(\partial_\mu h)^2 + \frac{1}{2}\left(\sqrt{2}\mu\right)^2 h^2 - V(h) \tag{24.68}$$

In the new variable h, the Lagrangian density is of the form of that of a massive scalar field of mass m_h in a potential V:

$$\mathcal{L}_h = \frac{1}{2}(\partial_\mu h)^2 - \frac{1}{2}m_h^2 h^2 - V(h) \quad \text{with} \quad V(h) = \lambda v h^3 + \frac{\lambda}{4}h^4 \tag{24.69}$$

and where the mass term is:

$$-\frac{1}{2}m_h^2 h^2 = \frac{1}{2}\left(\sqrt{2}\mu\right)^2 h^2 \tag{24.70}$$

consequently, the mass is *positive*:

$$m_h^2 = -\left(\sqrt{2}\mu\right)^2 > 0 \quad \text{since} \quad \mu^2 < 0 \tag{24.71}$$

Hence, the excitation of the field around its minimum creates massive particles. The field $h(x)$ describes a scalar particle of positive mass $m_h = \sqrt{-2\mu^2} = \sqrt{2\lambda}v$!

The potential V contains terms proportional to h^3 and h^4 which lead to **self-interactions of the massive scalar particle**. By labeling it h, these terms can be identified as the triple and quadruple self-interactions of the h particle, as illustrated in Figure 24.2.

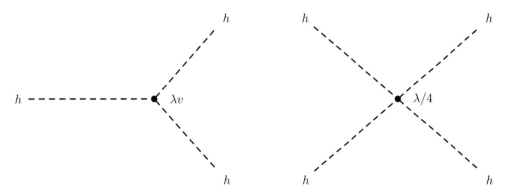

Figure 24.2 Self-interaction of the h particle: (left) triple coupling and (right) quadruple coupling.

It is important to keep in mind that the Lagrangian, Eq. (24.69), represents the same physics as the original Lagrangian, Eq. (24.60). However, it is now expressed in terms of excitations around the vacuum represented by the minimum at $\phi = +v$. The original symmetry is no longer apparent. Since we had to choose one of the two possible vacuum states, the h^3 term appeared in the potential, and the original symmetry $\phi \to -\phi$ is broken for h. In principle, the same predictions can be obtained from both forms. However, perturbation theory can only be applied to the small perturbations around the vacuum state.

24.6 Spontaneous Breaking of a Continuous Symmetry

The process of spontaneous symmetry breaking discussed above for a real field led to the breaking of the discrete $\phi \to -\phi$ symmetry. If we consider a complex field $\phi(x)$, we can then obtain the breaking of a continuous symmetry. We can always write a complex field as the following linear combination of a real and an imaginary part:

$$\phi(x) \equiv \frac{1}{\sqrt{2}} \left(\phi_1(x) + i\phi_2(x) \right) \tag{24.72}$$

The corresponding Lagrangian density for the complex field is (note the absence of the factors $1/2$ compared to that of the real field case Eq. (24.60)):

$$\begin{aligned}
\mathcal{L}_\phi &= \left(\partial_\mu \phi^* \right) \left(\partial^\mu \phi \right) - \mu^2 \phi^* \phi - \lambda (\phi^* \phi)^2 \\
&= \frac{1}{2} \left(\partial_\mu \phi_1 \right)^2 + \frac{1}{2} \left(\partial_\mu \phi_2 \right)^2 - \frac{1}{2} \mu^2 \left(\phi_1^2 + \phi_2^2 \right) - \frac{\lambda}{4} \left(\phi_1^2 + \phi_2^2 \right)^2
\end{aligned} \tag{24.73}$$

We note that the Lagrangian possesses the continuous global $U(1)$ symmetry $\phi(x) \to e^{i\alpha}\phi(x)$, where α is a real phase. This symmetry is best exposed going to polar coordinates in the field space. The complex field ϕ and its derivative are then expressed in terms of the two real functions $\varphi(x)$ and $\vartheta(x)$:

$$\phi(x) \equiv \varphi(x)e^{i\vartheta(x)} \quad \to \quad \partial^\mu \phi = \left(\partial^\mu \varphi + i\varphi \partial^\mu \vartheta \right) e^{i\vartheta} \tag{24.74}$$

Its complex conjugate is:

$$\phi^*(x) = \varphi(x)e^{-i\vartheta(x)} \quad \to \quad \partial_\mu \phi^* = \left(\partial_\mu \varphi - i\varphi \partial_\mu \vartheta \right) e^{-i\vartheta} \tag{24.75}$$

We express the Lagrangian in the corresponding polar coordinates and find:

$$\mathcal{L}_\phi = \left(\partial_\mu \varphi - i\varphi \partial_\mu \vartheta \right) \left(\partial^\mu \varphi + i\varphi \partial^\mu \vartheta \right) - \mu^2 \varphi^2 - \lambda \varphi^4 = \left(\partial_\mu \varphi \right)^2 + \varphi^2 \left(\partial_\mu \vartheta \right)^2 - \mu^2 \varphi^2 - \lambda \varphi^4 \tag{24.76}$$

As expected, the Lagrangian is invariant under the continuous rotation around the real phase ϑ.

As before, the potential has a minimum for $\lambda > 0$. The shape of the potential depends on the sign of μ^2, as illustrated in Figure 24.3. For $\mu^2 > 0$, the minimum of the potential occurs for $\phi_1 = \phi_2 = 0$, or equivalently $|\phi| = 0$. Considering the interesting non-trivial case where $\lambda > 0$ and $\mu^2 < 0$, we obtain a degenerate minimum given by the infinite set of points defined by:

$$\frac{\partial V}{\partial \varphi} = 0 = \frac{\partial}{\partial \varphi} \left(\mu^2 \varphi^2 + \lambda \varphi^4 \right) = 2\varphi \left(\mu^2 + 2\lambda \varphi^2 \right) \tag{24.77}$$

or (beware the factor 2 difference compared to that of the discrete case, Eq. (24.63)):

$$\phi_0 \equiv \varphi_0 e^{i\vartheta} \quad \text{where} \quad \varphi_0 = \sqrt{-\frac{\mu^2}{2\lambda}} \tag{24.78}$$

The minimum corresponds to a circle in the complex plane given by the condition, as indicated in Figure 24.3(right):

$$|\phi(x)|^2 = \phi(x)^* \phi(x) = \frac{1}{2} \left(\phi_1^2(x) + \phi_2^2(x) \right) = -\frac{\mu^2}{2\lambda} \equiv \frac{v^2}{2} \tag{24.79}$$

where we used the *vev* definition, Eq. (24.64).

The vacuum state corresponds to any particular point on the circle defined by the radius:

$$\sqrt{\phi_1^2(x) + \phi_2^2(x)} = v \tag{24.80}$$

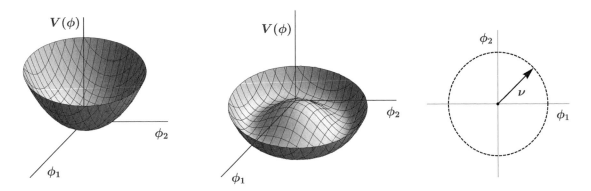

Figure 24.3 Illustration of spontaneous symmetry breaking for the case of a continuous symmetry: (left) unbroken; (middle) "Mexican hat" potential with the broken symmetry; (right) the minimum circle.

In order to spontaneously break the symmetry, we must choose a particular vacuum. Without loss of generality, we can decide to fix it at $(\phi_1, \phi_2) = (v, 0)$, hence the complex field $\phi(x)$ can be expanded around the vacuum with:

$$\begin{cases} \phi_1(x) = v + h(x) & (24.81) \\ \phi_2(x) = \pi(x) & (24.82) \end{cases}$$

or simply:

$$\phi(x) = \frac{1}{\sqrt{2}}(v + h(x) + i\pi(x)) \quad \rightarrow \quad \phi^*\phi = \frac{1}{2}\left((v+h)^2 + \pi^2\right) \tag{24.83}$$

The Lagrangian, Eq. (24.73), can then be written in terms of v, $h(x)$, and $\pi(x)$:

$$\mathcal{L}_\phi = \frac{1}{2}(\partial_\mu h)^2 + \frac{1}{2}(\partial_\mu \pi)^2 - \underbrace{\left[\frac{\mu^2}{2}\left((v+h)^2 + \pi^2\right) + \frac{\lambda}{4}\left((v+h)^2 + \pi^2\right)^2\right]}_{\equiv V(v,h,\pi)} \tag{24.84}$$

where, setting $\mu^2 = -\lambda v^2$ (see Eq. (24.64)), we get:

$$
\begin{aligned}
V(v, h, \pi) &= -\frac{\lambda v^2}{2}\left(v^2 + 2vh + h^2 + \pi^2\right) \\
&\quad + \frac{\lambda}{4}\left(h^4 + 4h^3 v + 2h^2\pi^2 + 6h^2 v^2 + 4h\pi^2 v + 4hv^3 + \pi^4 + 2\pi^2 v^2 + v^4\right) \\
&= -\frac{\lambda v^4}{4} + \underbrace{\lambda v^2 h^2}_{\text{mass}} + \underbrace{\lambda v h^3 + \lambda v h \pi^2}_{\text{triple}} + \underbrace{\frac{\lambda h^4}{4} + \frac{\lambda \pi^4}{4} + \frac{\lambda}{2}h^2\pi^2}_{\text{quadruple}}
\end{aligned}
\tag{24.85}
$$

Let us analyze this result. The first term proportional to λv^4 is constant and can be ignored in the dynamics. The next term describes the massive field $h(x)$ with a positive mass. The next terms proportional to three and four powers of the fields h and π correspond to the self-interaction diagrams shown in Figure 24.4. Therefore, the full Lagrangian can be expressed as:

$$\mathcal{L}_\phi = \frac{1}{2}(\partial_\mu h)^2 - \frac{1}{2}m_h^2 h^2 + \frac{1}{2}(\partial_\mu \pi)^2 - \mathcal{L}_3 - \mathcal{L}_4 \tag{24.86}$$

with

$$m_h = \sqrt{2\lambda}\, v, \qquad m_\pi = 0 \tag{24.87}$$

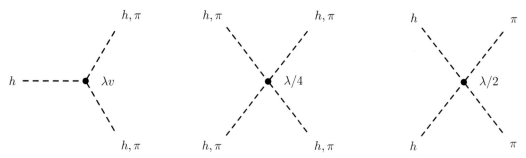

Figure 24.4 Self-interactions of the h and π scalar fields after symmetry breaking of the complex scalar field: (left) triple coupling hhh, $h\pi\pi$; (middle) quadruple coupling $hhhh$, $\pi\pi\pi\pi$; (right) quadruple coupling $hh\pi\pi$

and the triple and quadruple field interaction terms are equal to:

$$\mathcal{L}_3 = \lambda v h^3 + \lambda v h \pi^2, \quad \text{and} \quad \mathcal{L}_4 = \frac{\lambda h^4}{4} + \frac{\lambda \pi^4}{4} + \frac{\lambda}{2} h^2 \pi^2 \tag{24.88}$$

The massive scalar field h is associated with the excitations of the complex field in the direction where the potential varies quadratically. On the other hand, the degree of freedom represented by π in the direction where the potential does not change (around the circle in Figure 24.3) is associated with the so-called massless **Goldstone boson**[2] or **Nambu–Goldstone boson (NGB)**.[3] The massless nature of the Goldstone boson is easily understood by noting that along the direction of excitation of $\pi(x)$, the potential remains constant. Hence, no energy is associated with this mode of excitation. While in the direction of the $h(x)$ field, excitations are matched to change of potential, thus they represent massive particles.

In conclusion, the breaking of a continuous $U(1)$ symmetry yields *two* new particles: a massive scalar particle associated with the field $h(x)$ and a massless scalar particle called the Goldstone boson. First conjectured by Goldstone in 1961 [327], Goldstone, Salam, and Weinberg proved in 1962 this fundamental and general result: **massless Goldstone bosons are generated during the spontaneous breaking of a continuous symmetry** [328].

24.7 Spontaneously Broken Continuous $U(1)$ Theory

We now discuss the wonderful outcome when one mixes the principle of gauge invariance with spontaneous breaking of a continuous symmetry. Let us begin with the local Abelian $U(1)$ symmetry. We consider a complex field $\phi(x)$ whose Lagrangian density is given by Eq. (24.73). We can write this as:

$$\mathcal{L}_\phi = \left(\partial_\mu \phi^*\right)\left(\partial^\mu \phi\right) - V(\phi) = |\partial_\mu \phi|^2 - V(\phi) \tag{24.89}$$

where $|\ |^2 = ()^*()$. The potential $V(\phi)$ is given by:

$$V(\phi) = \mu^2 |\phi|^2 + \lambda |\phi|^4 \tag{24.90}$$

To impose local $U(1)$ invariance $\phi(x) \rightarrow e^{ie\alpha(x)}\phi(x)$, we replace the derivative by its covariant form, Eq. (24.13), applied to the ϕ field:

$$\partial_\mu \rightarrow D_\mu \equiv \partial_\mu + ieA_\mu \tag{24.91}$$

---

2   Jeffrey Goldstone (born 1933), British theoretical physicist.
3   Yoichiro Nambu (1921–2015), Japanese–American physicist.

where e is the coupling and A^μ is the field of the gauge boson, which transforms as (see Eq. (24.15)):

$$A'_\mu(x) \equiv A_\mu(x) - \partial_\mu \alpha(x) \tag{24.92}$$

We can add the common kinetic term of the gauge field to get the gauge-covariant Lagrangian density of the field ϕ:

$$\mathcal{L}_\phi = |D_\mu \phi|^2 - \mu^2 |\phi|^2 - \lambda |\phi|^4 - \frac{1}{4} F_{\mu\nu} F^{\mu\nu} \tag{24.93}$$

where $F_{\mu\nu} \equiv \partial_\mu A_\nu - \partial_\nu A_\mu$ (see Eq. (10.8)). **The gauge field A_μ is forced to be massless from the requirement of gauge invariance** (see Eq. (24.23)). Direct substitution of the covariant derivative gives:

$$\begin{aligned} |D_\mu \phi|^2 &= (\partial_\mu - ieA_\mu)\phi^* (\partial^\mu + ieA^\mu)\phi \\ &= |\partial_\mu \phi|^2 + ie(\partial_\mu \phi^*)A^\mu \phi - ieA_\mu \phi^* \partial^\mu \phi + e^2 A_\mu A^\mu |\phi|^2 \end{aligned} \tag{24.94}$$

By insertion into $\mathcal{L}_\phi$ and rearrangement of terms, we get the full Lagrangian of our theory:

$$\mathcal{L}_\phi = \underbrace{|\partial_\mu \phi|^2 - \mu^2 |\phi|^2 - \lambda |\phi|^4}_{\text{complex scalar field}} \underbrace{- \frac{1}{4} F_{\mu\nu} F^{\mu\nu}}_{\text{gauge field}} + \underbrace{ie(\partial_\mu \phi^*)A^\mu \phi - ieA_\mu \phi^* \partial^\mu \phi + e^2 A_\mu A^\mu |\phi|^2}_{\text{interactions}} \tag{24.95}$$

As before we take for the scalar field potential $\lambda > 0$ and for $\mu^2 < 0$, its vacuum state is degenerate with a vacuum expectation value given by (see Eq. (24.79)):

$$|\phi_0|^2 = -\frac{\mu^2}{2\lambda} = \frac{v^2}{2} \tag{24.96}$$

The choice of a particular vacuum state breaks the $U(1)$ symmetry. By expanding the field near its minimum chosen without loss of generality at $(\phi_1, \phi_2) = (v, 0)$, we obtain (see Eq. (24.83)):

$$\phi(x) = \frac{1}{\sqrt{2}}(v + h(x) + i\pi(x)) \tag{24.97}$$

By inserting this into the Lagrangian density, we recover the terms as in the case of the simple complex field. The new terms become the interactions between the scalar and the gauge fields. For instance:

$$\begin{aligned} ie(\partial_\mu \phi^*)A^\mu \phi - ieA_\mu \phi^* \partial^\mu \phi &= \frac{ie}{2}(\partial_\mu (h - i\pi))(v + h + i\pi)A^\mu - \frac{ie}{2}(v + h - i\pi)A^\mu (\partial_\mu (h + i\pi)) \\ &= \frac{ie}{2}A^\mu \left[(\partial_\mu (h - i\pi))(v + h + i\pi) - (v + h - i\pi)(\partial_\mu (h + i\pi))\right] \\ &= -eA^\mu \left[(\partial_\mu h)\pi - (\partial_\mu \pi)(v + h)\right] \\ &= evA^\mu (\partial_\mu \pi) - eA^\mu \left[(\partial_\mu h)\pi - (\partial_\mu \pi)h\right] \end{aligned} \tag{24.98}$$

and, see Eq. (24.83):

$$e^2 A_\mu A^\mu |\phi|^2 = \frac{e^2}{2} A_\mu A^\mu ((v + h)^2 + \pi^2) = \underbrace{\frac{e^2 v^2}{2} A_\mu A^\mu}_{\text{massive gauge field}} + \underbrace{e^2 v A_\mu A^\mu h}_{hAA} + \underbrace{\frac{e^2}{2} A_\mu A^\mu (h^2 + \pi^2)}_{hhAA, \pi\pi AA} \tag{24.99}$$

Inserting and grouping all the various terms, we get the full Lagrangian of our spontaneously broken theory:

$$\mathcal{L}_\phi = \underbrace{\frac{1}{2}(\partial_\mu h)^2 - \frac{1}{2}m_h^2 h^2}_{\text{massive } h} + \underbrace{\frac{1}{2}(\partial_\mu \pi)^2}_{\text{massless } \pi} \underbrace{-\mathcal{L}_3 - \mathcal{L}_4}_{h, \pi \text{ interactions}} \underbrace{-\frac{1}{4}F_{\mu\nu}F^{\mu\nu} + \frac{1}{2}m_A^2 A_\mu A^\mu + evA^\mu (\partial_\mu \pi)}_{\text{massive gauge field}} + \mathcal{L}_{int} \tag{24.100}$$

where $\mathcal{L}_3$ and $\mathcal{L}_4$ are given in Eq. (24.88) and

$$\mathcal{L}_{int} = e^2 v A_\mu A^\mu h + \frac{e^2}{2} A_\mu A^\mu \left(h^2 + \pi^2 \right) - e A^\mu \left[(\partial_\mu h)\pi - (\partial_\mu \pi)h \right] \qquad (24.101)$$

Finally, one obtains three interacting particles:

- A **massive scalar boson** h with mass $m_h = \sqrt{2\lambda}v$.

- A **massive gauge boson** B^μ with mass $m_A = ev$; the mass is given by the product of the coupling e and the field's vacuum expectation value (vev) v. This supports the origin of the mass being the consequence of the interaction between the gauge boson and the complex field.

- A **massless Goldstone boson**.

The vacuum expectation value has generated a term which corresponds to a mass for the gauge boson field. **Thus, the SSB mechanism can generate a massive gauge boson in an invariant local gauge theory, although in the unbroken state the gauge bosons are forced to be massless**. It's pure magic!

However, in Section 10.7, we have seen that gauge invariance imposed a massless spin-1 gauge boson (e.g., the photon) to have two polarization states, the longitudinal state being unphysical. On the other hand, a spin-1 massive particle possesses $2S + 1 = 3$ polarization states. What shall we conclude on the polarization states of our A^μ field, for which gauge invariance imposes a massless boson, but which acquired a mass during spontaneous symmetry breaking?

We note in this respect that the original Lagrangian had in total four degrees of freedom: two for the complex field ϕ and two for the massless gauge field B^μ. After the symmetry is broken, we expect the gauge boson to have acquired an extra degree of freedom, this latter corresponding to a longitudinal polarization state.

The broken Lagrangian does possess a new term $evA^\mu(\partial_\mu \pi)$ that directly couples (via a derivative coupling) the gauge boson to the Goldstone boson. It would appear that the spin-1 gauge boson is interacting with the spin-0 Goldstone boson. The solution to this puzzle is to notice that gauge freedom allows us to eliminate the Goldstone field π from the Lagrangian by absorbing it into the gauge boson B^μ. Gauge freedom allows us to write for the mass term of the gauge field (see Eq. (24.92)):

$$
\begin{aligned}
\frac{1}{2} m_A^2 \left(A'_\mu(x) A^{\mu\prime}(x) \right) &= \frac{1}{2} m_A^2 \left(A_\mu(x) - \partial_\mu \alpha(x) \right) \left(A^\mu(x) - \partial^\mu \alpha(x) \right) \\
&= \frac{1}{2} m_A^2 A_\mu A^\mu - e^2 v^2 A_\mu (\partial^\mu \alpha(x)) + \frac{1}{2} e^2 v^2 (\partial^\mu \alpha(x))^2 \qquad (24.102)
\end{aligned}
$$

With the association $\pi(x) \equiv ev\alpha(x)$, the mass term becomes just what we need:

$$\frac{1}{2} m_A^2 \left(A'_\mu(x) A^{\mu\prime}(x) \right) = \frac{1}{2} m_A^2 A_\mu A^\mu - ev A_\mu (\partial^\mu \pi(x)) + \frac{1}{2} (\partial^\mu \pi)^2 \qquad (24.103)$$

Hence, the full Lagrangian, Eq. (24.100), can now be expressed **without the Goldstone boson field**, having absorbed it into the massive gauge boson mass term (allowed due to the gauge freedom). Hence (using $m_h = \sqrt{2\lambda}v$ and $m_A = ev$) we find:

$$\mathcal{L}_\phi = \underbrace{\frac{1}{2} (\partial_\mu h)^2 - \lambda v^2 h^2}_{\text{massive } h} \underbrace{-\frac{1}{4} F_{\mu\nu} F^{\mu\nu} + \frac{1}{2} e^2 v^2 A'_\mu(x) A^{\mu\prime}(x)}_{\text{massive gauge field } B^\mu} - \mathcal{L}_3 - \mathcal{L}_4 + \mathcal{L}_{int} \qquad (24.104)$$

where $\mathcal{L}_3$ and $\mathcal{L}_4$ are given in Eq. (24.88) and $\mathcal{L}_{int}$ in Eq. (24.101). Since the new Lagrangian was obtained using the gauge freedom, the observables of the theory remain unchanged. The gauge in which the Goldstone boson $\pi(x)$ is removed from the Lagrangian is called the **unitary gauge**. It was introduced by Weinberg [329, 330] in the context of the electroweak theory (discussed in Chapter 25). It is said that the Goldstone boson has been "eaten" by the gauge field in order to increase this latter's polarization degrees of freedom and obtain a longitudinal state.

It is important to recall that the observables of the theory do not depend on the choice of gauge. However, in the unitary gauge the fields that appear in the Lagrangian are only the "physical" ones with their correct degrees of freedom (i.e., three for the massive gauge boson and one for the complex field). If we look back to our complex field expansion, Eq. (24.97), the unitary gauge corresponds to neglecting the $\pi(x)$ term and expanding the field simply as:

$$\phi(x) = \frac{1}{\sqrt{2}}\left(v + h(x)\right) \tag{24.105}$$

Repeating the derivations performed above under this definition gives the new Lagrangian in the unitary gauge where only physical fields and their interactions are expressed. This effectively absorbs also the interaction terms of the π field in $\mathcal{L}_3$ and $\mathcal{L}_4$ (see Eq. (24.88)), and $\mathcal{L}_{int}$ (see Eq. (24.101)), and leaves the following Lagrangian with the corresponding interactions between the massive gauge field A^μ and the scalar h field:

$$\mathcal{L}_\phi = \underbrace{\frac{1}{2}\left(\partial_\mu h\right)^2 - \lambda v^2 h^2}_{\text{massive } h} \underbrace{-\frac{1}{4}F_{\mu\nu}F^{\mu\nu} + \frac{1}{2}e^2 v^2 A_\mu(x)A^\mu(x)}_{\text{massive gauge field } A^\mu} \underbrace{-\lambda v h^3 - \frac{\lambda h^4}{4}}_{h \text{ self-interaction}} \underbrace{+ e^2 v A_\mu A^\mu h + \frac{e^2}{2}A_\mu A^\mu h^2}_{A,h \text{ interactions}} \tag{24.106}$$

The Lagrangian describes within a $U(1)$ local symmetry the dynamics of a massive gauge boson A^μ, a massive scalar field h, and their interactions. These latter are composed of interactions between the gauge boson and the scalar field, as well as self-interactions of the scalar field. The corresponding Feynman diagrams are illustrated in Figure 24.5.

The mass of the gauge boson is given by the product of the coupling strength and the vacuum expectation value of the scalar field:

$$m_A = ev \tag{24.107}$$

and is therefore fixed once e and v are known. The mass of the scalar boson h is given by:

$$m_h = \sqrt{2\lambda}v \tag{24.108}$$

so while the measurement of m_A and e could determine the value of v, the mass of the scalar boson requires the knowledge of both v and λ. Hence, there is no way to "predict" the mass m_h. Its determination fixes λ once v is known.

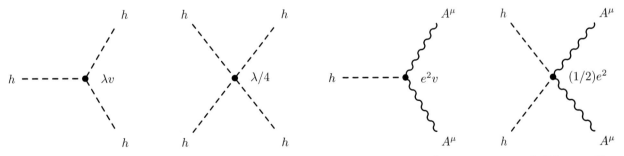

Figure 24.5 Self-interactions of the scalar field h and interactions between the massive gauge field boson A^μ and the scalar field h. Compare with Figure 24.2.

Finally, we note that the hAA coupling between the scalar field h and the gauge boson A^μ is predicted to be proportional to $e^2 v$, while we have seen that the mass of the gauge boson is $m_A = ev$. Hence, the coupling to the scalar field h depends on the mass of the gauge boson. The heavier the boson, the stronger the coupling to the h scalar (and vice versa).

24.8 Spontaneously Broken $SU(2)$ Yang–Mills Theory

We can now repeat the same considerations in the case of the non-Abelian Yang–Mills theory. Let us assume that the field is **an $SU(2)$ doublet of complex scalar fields**:

$$\phi(x) \equiv \begin{pmatrix} \phi_a(x) \\ \phi_b(x) \end{pmatrix} = \frac{1}{\sqrt{2}} \begin{pmatrix} \phi_1(x) + i\phi_2(x) \\ \phi_3(x) + i\phi_4(x) \end{pmatrix} \tag{24.109}$$

where $\phi_a(x)$ and $\phi_b(x)$ are two complex scalar fields. The Lagrangian of our theory can be expressed as (see Eq. (24.89)):

$$\mathcal{L}_\phi = \left(\partial_\mu \phi^\dagger\right)\left(\partial^\mu \phi\right) - V(\phi) = |\partial_\mu \phi|^2 - V(\phi) \quad \text{where} \quad |\ |^2 = (\ )^\dagger(\ ) \tag{24.110}$$

which is manifestly invariant under a global $SU(2)$ transformation of the type (see Eqs. (24.30) and (24.45)):

$$\Psi \to U(\vec{\alpha})\Psi = e^{i\vec{\alpha}\cdot\vec{\tau}/2}\Psi = e^{i\alpha_i t_i}\Psi \tag{24.111}$$

In order to achieve local $SU(2)$ invariance, we consider the $SU(2)$ covariant derivative defined in Eq. (24.33). The Lagrangian then describes a system of four scalar particles ϕ_i with mass terms determined by μ^2 interacting with three massless gauge bosons W_μ^i. For the case $\lambda > 0$ and $\mu^2 < 0$, the potential breaks the symmetry and the minimum of the potential is at the value:

$$|\phi|^2 \equiv \phi^\dagger \phi = \frac{1}{2}\left(\phi_1^2 + \phi_2^2 + \phi_3^2 + \phi_4^2\right) = -\frac{\mu^2}{2\lambda} \tag{24.112}$$

The set of points where $V(\phi)$ is minimized ($=$ a manifold) is invariant under $SU(2)$ transformations. Hence, **from the four degrees of freedom ϕ_i only one represents a massive scalar particle while the others give us three massless Goldstone bosons**. We choose a particular vacuum state among the points of this manifold to spontaneously break the symmetry. We can take for example:

$$\phi_1 = \phi_2 = \phi_4 = 0 \quad \text{and} \quad \phi_3^2 = -\frac{\mu^2}{\lambda} \equiv v^2 \tag{24.113}$$

The Goldstone bosons can be gauged away and the field can then be expressed as a real expansion around this vacuum state determined by the vacuum expectation value v and one field $h(x)$:

$$\phi(x) = \frac{1}{\sqrt{2}} \begin{pmatrix} 0 \\ v + h(x) \end{pmatrix} \tag{24.114}$$

The $SU(2)$ covariant derivative Eq. (24.33) leads to a term $|D_\mu\phi|^2$ which generates mass terms for the gauge bosons of the type:

$$\frac{g^2}{8}(0\ v)\tau_i\tau_j \begin{pmatrix} 0 \\ v \end{pmatrix} W_\mu^i W^{\mu,j} \tag{24.115}$$

which, making use of the property of the Pauli matrices, Eq. (C.99), can be simplified to:

$$\frac{g^2 v^2}{8} W_\mu^i W^{\mu,i} \equiv \frac{1}{2} M_W^2 W_\mu^i W^{\mu,i} \quad (i = 1, 2, 3) \tag{24.116}$$

This means that all three gauge bosons have a non-vanishing, degenerate mass of $M_W = gv/2$. Impressive! The Lagrangian describes three massive gauge bosons and one massive scalar field h and their interactions. The three Goldstone bosons have been "eaten" up in order for the gauge bosons to become massive. This fundamental result will be critical in building the electroweak theory, as discussed in the next chapter.

Problems

Ex 24.1 Goldstone bosons generation. The Lagrangian for three interacting real fields ϕ_1, ϕ_2, ϕ_3 is:

$$\mathcal{L} = \frac{1}{2}\left(\partial_\mu \phi_i\right)^2 - \frac{1}{2}\mu^2 \phi_i^2 - \frac{1}{4}\lambda(\phi_i^2)^2 \tag{24.117}$$

with $\mu^2 < 0$ and $\lambda > 0$, and where a summation of ϕ_i^2 over i is implied. Show that it describes a massive field of $\sqrt{-2\mu^2}$ and two massless Goldstone bosons.

Ex 24.2 Higgs mechanism in QED with a massive photon. Consider the hypothetical QED model with electrons and a massive photon. This model shall be described by the following Lagrangian density:

$$\mathcal{L} = \overline{\Psi}(i\gamma^\mu D_\mu - m_e)\Psi - \frac{1}{4}F^{\mu\nu}F_{\mu\nu} + \frac{1}{2}m_A^2 A^\mu A_\mu \tag{24.118}$$

(a) Show that the mass term of the photon violates the $U(1)$ gauge symmetry, while the rest of the Lagrangian does not.

(b) Introduce such a mass term via the Higgs mechanism with a scalar complex field ϕ.

(c) Show that a Yukawa interaction term of the type $f_e|\phi|\overline{\Psi}\Psi$ modifies the electron mass. Express the new electron mass in terms of the "bare" mass m_e, the coupling f_e, and the vacuum expectation value of the Higgs field v.

25 The Electroweak Theory

In contradistinction to quantum electrodynamics, the Fermi theory is not renormalizable. This difficulty could not be solved by smoothing the point-like interaction by a massive, and therefore short-range, charged vector particle exchange (the so-called W^+ and W^- bosons): theories with massive charged vector bosons are not renormalizable either. In the early 1960s, there seemed to be insuperable obstacles to formulating a consistent theory with short-range forces mediated by massive vectors. It is the notion of spontaneous broken symmetry as adapted to gauge theory that provided the clue to the solution.

Francois Englert [1]

25.1 Neutral Weak Currents

The concept of neutral weak currents can be traced back to the 1930s in papers from **Gamow** and **Teller** [331] as a "generalization of beta theory," from **Kemmer** [332, 333], and **Wentzel**[2] [334]. Most arguments are related to the observation that the weak interaction is insensitive to the electric charge. Wentzel wrote:

> *"In particular, the question arises whether in the β decay (in the Pauli–Fermi interpretation), electron and neutrino are not replaceable by other light particles, i.e., whether there are also elementary nuclear processes in which two charged or two uncharged particles are emitted. (...) It is likely that there are elementary β conversions besides the known β-conversions nuclear processes in which neutral pairs of light particles (electron + positron or neutrino + antineutrino) are emitted."* [3]

In 1958 **Bludman**[4] derived, from what he called *Fermi gauge invariance*, the weak interaction from charge-raising and -lowering chiral Fermi charge operators [335]. Considering for the moment just the first lepton family, the electron and the electron neutrino are the members of a **weak isospin** $I = 1/2$ **doublet**, in analogy to the strong isotopic spin. The third component of the weak isospin I_3 differentiates the electron from the neutrino. This is graphically represented in Figure 25.1. With the charge raising and lowering operators, one can change the weak isospin and effectively transform an electron into a neutrino or vice versa. Hence, anticipating the $SU(2)$ Yang–Mills gauge field structure of the weak isospin, we define the corresponding $SU(2)$

---

1 Reprinted from "A brief course in spontaneous symmetry breaking: II. The BEH mechanism," in *Corfu Summer Institute on Elementary Particle Physics (Corfu 2001)*, arXiv:hep-th/0203097.
2 Gregor Wentzel (1898–1978), German physicist.
3 *"Insbesondere stellt sich die Frage, ob nicht beim β-Zerfall (in der Pauli–Fermi'schen Deutung) Elektron und Neutrino durch andere leichte Teilchen ersetzbar sind, d.h. ob es nicht auch elementare Kernprozesse gibt, bei denen zwei geladene oder zwei ungeladene Teilchen emittiert werden. (...) Es ist wahrscheinlich, dass es ausser den bekannten β-Umwandlungen elementare Kernprozesse gibt, bei denen neutrale Paare leichter Teilchen (Elektron + Positron oder Neutrino + Antineutrino) emittiert werden."* Translated with permission from Springer Nature: G. Wentzel, "Zur Theorie der β-Umwandlung und der Kernkräfte. I," *Zeitschrift für Physik*, vol. 104, pp. 34–47, Jan 1937.
4 Sidney Arnold Bludman (born 1927), American theoretical physicist.

Dirac field iso-doublet (see Eq. (24.25) assigning $\Psi_1 = \nu_e$, $\Psi_2 = e$):

$$\Psi \equiv \begin{pmatrix} \nu_e \\ e \end{pmatrix} \quad \text{and} \quad \overline{\Psi} = (\overline{\nu_e} \ \ \bar{e}) \tag{25.1}$$

where e and ν_e represent, respectively, an electron and an electron-neutrino Dirac field. Note that $\overline{\nu_e}$ and $\bar{e}$ represent the adjoint of the Dirac fields. Also be aware that a spinor has indeed four degrees of freedom so it can represent a particle or an antiparticle, each with two helicity states.

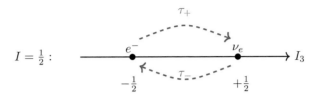

Figure 25.1 Representation of the electron and neutrino as a doublet in the weak isospin space.

A new ingredient that did not exist in the strong isospin is that we need to discriminate left-handed and right-handed chiral projections. This is necessary because the charged weak force only couples to the left-handed chiral states (i.e., $V - A$). In order to prepare for the chiral Lorentz structure of the weak interaction, we will need the gauge invariance principle to be applied on individual chiral projections. We can define the **weak isospin chiral left-handed L and right-handed R doublets** of the electron family:

$$\Psi_L \equiv \begin{pmatrix} \nu_e \\ e \end{pmatrix}_L = \begin{pmatrix} \nu_{eL} \\ e_L \end{pmatrix} \quad \text{and} \quad \Psi_R \equiv \begin{pmatrix} \nu_e \\ e \end{pmatrix}_R = \begin{pmatrix} \nu_{eR} \\ e_R \end{pmatrix} \tag{25.2}$$

where $e_{L,R} = \frac{1}{2}(1 \mp \gamma^5)e$ and similarly for the neutrino field. The subscript $L(R)$ means that one projects out the left-handed (right-handed) chiral components of the field. A general $SU(2)$ transformation on the left-handed doublet can be written in the form:

$$\Psi_L \xrightarrow{SU(2)_L} e^{\frac{i}{2}\vec{\tau} \cdot \vec{\alpha}} \Psi_L \tag{25.3}$$

where $\vec{\alpha}$ is a particular vector. The conjugate doublets are given by:

$$\overline{\Psi_L} = \begin{pmatrix} \overline{\nu_{eL}} & \overline{e_L} \end{pmatrix} \quad \text{and} \quad \overline{\Psi_R} = \begin{pmatrix} \overline{\nu_{eR}} & \overline{e_R} \end{pmatrix} \tag{25.4}$$

where, as before, $\overline{\nu_e}$ and $\bar{e}$ represent the adjoint of the Dirac fields. The charged-current weak interaction is expressed in terms of the I_3 raising and lowering operators τ^+ and τ^- (the factor $1/2$ is conventional):

$$\tau_\pm \equiv \frac{1}{2}(\tau_1 \pm i\tau_2) \tag{25.5}$$

Explicitly, using the Pauli matrices, we have:

$$\tau_\pm \equiv \frac{1}{2}(\tau_1 \pm i\tau_2) = \frac{1}{2}\left(\begin{pmatrix} 0 & 1 \\ 1 & 0 \end{pmatrix} \pm i \begin{pmatrix} 0 & -i \\ i & 0 \end{pmatrix} \right) = \begin{cases} \tau_+ = \begin{pmatrix} 0 & 1 \\ 0 & 0 \end{pmatrix} \\ \tau_- = \begin{pmatrix} 0 & 0 \\ 1 & 0 \end{pmatrix} \end{cases} \tag{25.6}$$

With the $\tau^\pm$ operators, the **left-handed weak charge-raising current** (ν, e) can be written as:

$$\overline{\Psi_L}\gamma^\mu \tau_+ \Psi_L = \begin{pmatrix} \overline{\nu_{eL}} & \overline{e_L} \end{pmatrix} \gamma^\mu \begin{pmatrix} 0 & 1 \\ 0 & 0 \end{pmatrix} \begin{pmatrix} \nu_{eL} \\ e_L \end{pmatrix} = \begin{pmatrix} \overline{\nu_{eL}} & \overline{e_L} \end{pmatrix} \gamma^\mu \begin{pmatrix} e_L \\ 0 \end{pmatrix} = \overline{\nu_{eL}}\gamma^\mu e_L \tag{25.7}$$

Type	Diagram	Current
$W^- \to e^- \bar{\nu}_e$:		$\overline{e}\gamma^\mu \left(\dfrac{1-\gamma^5}{2} \right) \nu_e = \overline{e_L}\gamma^\mu \nu_{eL}$
$W^+ \to e^+ \nu_e$:		$\overline{\nu_e}\gamma^\mu \left(\dfrac{1-\gamma^5}{2} \right) e = \overline{\nu_{eL}}\gamma^\mu e_L$
$W^0 \to e^+ e^-$:		$\dfrac{1}{2}\overline{e_L}\gamma^\mu e_L$
$W^0 \to \nu_e \bar{\nu}_e$:		$\dfrac{1}{2}\overline{\nu_{eL}}\gamma^\mu \nu_{eL}$

Table 25.1 Basic charged and neutral weak couplings between $W^\pm$, W^0, electrons/positrons, and neutrinos as predicted by the $SU(2)_L$ weak isospin symmetry.

which corresponds for example to the coupling of the charged boson $W^+ \to e^+ \nu_e$, as shown in Table 25.1. Similarly the **left-handed weak charge-lowering current** is given by:

$$\overline{\Psi_L}\gamma^\mu \tau_- \Psi_L = \begin{pmatrix} \overline{\nu_{eL}} & \overline{e_L} \end{pmatrix} \gamma^\mu \begin{pmatrix} 0 & 0 \\ 1 & 0 \end{pmatrix} \begin{pmatrix} \nu_{eL} \\ e_L \end{pmatrix} = \begin{pmatrix} \overline{\nu_{eL}} & \overline{e_L} \end{pmatrix} \gamma^\mu \begin{pmatrix} 0 \\ \nu_{eL} \end{pmatrix} = \overline{e_L}\gamma^\mu \nu_{eL} \tag{25.8}$$

which this time corresponds for example to the coupling of the charged boson $W^- \to e^- \bar{\nu}_e$.

In his 1958 paper Bludman states: *"These results, which may be subject to experimental check, are necessary consequences of our development. If τ^+ and τ^- have been introduced in order to obtain the ordinary charge-exchange interactions, then τ_3* **must occur if a group is to be defined**.*"* [5] In other words, the existence of an underlying $SU(2)_L$ symmetry in the weak currents enforces the existence of three weak currents given by τ^+, τ^-, and τ_3, where τ_3 defines a weak neutral current of the form:

$$\overline{\Psi_L}\gamma^\mu \frac{1}{2}\tau_3 \Psi_L = \overline{\Psi_L}\gamma^\mu \frac{1}{2} \begin{pmatrix} 1 & 0 \\ 0 & -1 \end{pmatrix} \Psi_L = \frac{1}{2}\overline{\nu_{eL}}\gamma^\mu \nu_{eL} - \frac{1}{2}\overline{e_L}\gamma^\mu e_L \tag{25.9}$$

The weak neutral currents are unavoidable and are mediated by **a third electrically neutral vector boson** W^0. The corresponding Feynman diagrams are shown in Table 25.1. Consequently, the $SU(2)_L$ generates a

5   Reprinted by permission of Springer Nature: S. A. Bludman, "On the universal Fermi interaction," *Nuovo Cim.*, vol. 9, pp. 433–445, 1958. Copyright 2007, Società Italiana di Fisica.

left-handed weak iso-triplet set of currents:

$$SU(2)_L: \quad \overline{\Psi_L}\gamma^\mu \frac{1}{2}\tau_i\Psi_L \quad (i=1,2,3) \tag{25.10}$$

with the corresponding three intermediate vector bosons W^0, W^1, W^2. In a similar way, a right-handed weak iso-triplet of currents can be generated by the $SU(2)_R$ group:

$$SU(2)_R: \quad \overline{\Psi_R}\gamma^\mu \frac{1}{2}\tau_i\Psi_R \quad (i=1,2,3) \tag{25.11}$$

with corresponding intermediate right-handed vector bosons W_R^0, W_R^1, W_R^2. **Such right-handed bosons were until now never found in experiments, and within the range of explored energies, the $SU(2)_R$ currents have never been observed**.

25.2   Experimental Discovery of Weak Neutral Currents (1973)

While the existence of weak neutral currents seemed to be theoretically well motivated, their experimental detection has been very difficult and somewhat strained by ambiguous results. Weak neutral currents were conclusively discovered in 1973 by studying neutrinos produced by the CERN PS neutrino beam and interacting in the Gargamelle bubble chamber [336–338]. Gargamelle was a heavy liquid bubble chamber detector with an active volume of 3 m^3, initiated by **André Lagarrigue**.[6] The chamber was in operation at CERN between 1970 and 1979. In the experiment that discovered neutral currents, the chamber was filled with liquid freon CF_3Br (density $1\,500$ kg/m^3) and so had a total mass of about 4 tons. The following **neutral-current (NC) neutrino–nucleon events** were searched for:

$$\text{NC:} \qquad \nu_\mu + \mathcal{N} \to \nu_\mu + \text{hadrons}, \qquad \overline{\nu}_\mu + \mathcal{N} \to \overline{\nu}_\mu + \text{hadrons}$$

among the **charged-current neutrino–nucleon (CC) events**, which are distinguishable by the presence of the muon (or in general electron or muon) in the final state:

$$\text{CC:} \qquad \nu_\mu + \mathcal{N} \to \mu^- + \text{hadrons}, \qquad \overline{\nu}_\mu + \mathcal{N} \to \mu^+ + \text{hadrons}$$

The search for NC events relied on a set of 83,000 pictures in neutrino mode and 207,000 in antineutrino mode. The bubble chamber resolution allowed one to identify the majority of the hadrons in a unique way, relying on their range, momentum and stopping power ($\mathrm{d}E/\mathrm{d}x$). In addition, a set of 15,000 pictures were taken without the beam to study cosmic-ray backgrounds. The analysis of all the pictures resulted in the following observations:

1. ν_μ data: 102 NC candidates, 428 CC candidates, and 15 associated events (AE).

2. $\overline{\nu}_\mu$ data: 64 NC candidates, 148 CC candidates, and 12 AE.

Typical NC candidates are shown in Figure 25.2. The associated events (AE) correspond to those where an identified CC event is accompanied by a "star" produced by a secondary interaction of a neutral hadron (such as neutrons, K_L^0s or Λs). Such interactions were used to estimate the number of neutrino interactions in the shielding that would produce high-energy neutral hadrons interacting in the bubble chamber. The number of observed events was corrected for the background induced by the neutral hadrons and the measured neutral-to-charged current ratio ($R \equiv NC/CC$) for neutrinos and antineutrinos was found to be:

$$R^\nu = \frac{\sigma^{\mathrm{NC}(\nu N)}}{\sigma^{\mathrm{CC}(\nu N)}} = 0.22 \pm 0.04 \quad \text{and} \quad R^{\overline{\nu}} = \frac{\sigma^{\mathrm{NC}(\overline{\nu}N)}}{\sigma^{\mathrm{CC}(\overline{\nu}N)}} = 0.43 \pm 0.12 \tag{25.12}$$

6   André Lagarrigue (1924–1975), French particle physicist.

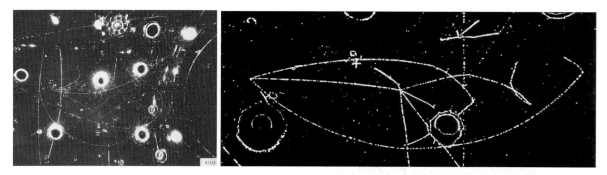

Figure 25.2 Neutral-current (NC) neutrino interaction candidates observed with the Gargamelle bubble chamber. Reprinted figures by permission from CERN. Copyright 1973–2021 CERN.

These results appeared in the 1973 Gargamelle paper [336], which was conservatively entitled "Observation of Neutrino Like Interactions Without Muon Or Electron in the Gargamelle Neutrino Experiment."

Soon after the observations of the candidates for the neutral-current neutrino–nucleon scattering, Gargamelle also found an event consistent with a purely leptonic neutral interaction [337]. The goal was to search for the purely leptonic scattering of muon neutrinos off electrons:

$$\nu_\mu + e^- \to \nu_\mu + e^-$$
$$\overline{\nu}_\mu + e^- \to \overline{\nu}_\mu + e^- \tag{25.13}$$

The final state is characterized by a single, very forward electron produced within the target volume, and which should not be accompanied by hadronic particles or debris indicating an interaction on a target nucleon, as well as no secondary photons coming from the vertex. The Gargamelle bubble chamber filled with freon (CF_3Br) was particularly adapted since the radiation length of the medium is 11 cm, leading to a clean identification of the electron. The electron is constrained by kinematics to be in the very forward direction relative to the incoming neutrino direction. The CERN PS neutrino beam had the maximum of its neutrino flux in an energy range between 1 and 2 GeV. In the search for leptonic neutral currents the threshold for the electron energy was set to 300 MeV. With this choice, all electrons produced in the leptonic neutral current would fall within $5°$ of the incoming beam direction.

In total, 375,000 pictures in neutrino mode and 360,000 pictures in antineutrino mode were collected! All pictures were manually scanned twice. After this gigantic work, only *one* event in the antineutrino mode survived! It is shown in Figure 25.3. The energy of the electron was reconstructed to be 385 ± 100 MeV and the scattering angle $1.4°\ {}^{+1.6°}_{-1.4°}$, consistent with zero. The estimated background was computed to be 0.3 ± 0.2 for the neutrino mode and 0.03 ± 0.02 for the antineutrino mode. With such low background the single event measured with the quality of the bubble chamber was sufficient to provide very strong evidence for the existence of purely leptonic weak neutral currents.

These discoveries had a very large impact, in particular because when it occurred the theoretical description of the weak interaction was already very advanced with the electroweak theory (described in the next chapter), so at the time experimenters were seen to lag. The success of the Gargamelle experiment relied in great part on its excellent imaging properties. The images of very high granularity and the information on the stopping power (dE/dx) provided the way to exclusively and uniquely identify most particles produced and hence fully reconstruct the interactions. Although limited in mass, and hence statistics, the bubble chambers provided a method to study interactions with very high quality, thereby compensating for the limited statistics. A few events were sufficient to claim discoveries!

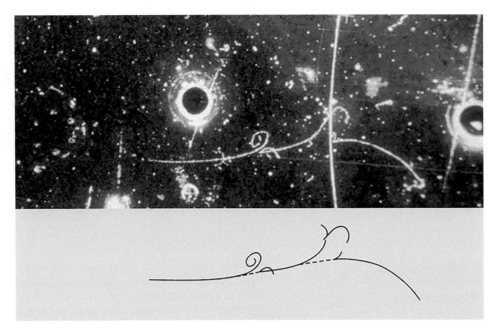

Figure 25.3 The first leptonic neutral-current event observed in Gargamelle. An antineutrino coming from the left knocks an electron forwards, creating a characteristic shower of electron–positron pairs. Reprinted from https://cds.cern.ch/record/39468, under CC BY license http://creativecommons.org/licenses/by/4.0/. Reprinted figure by permission from CERN. Copyright 1973–2021 CERN.

25.3   Interpretation of the Weak Neutral Currents

It is tempting to interpret the weak neutral currents as those due to the exchange of the neutral intermediate vector boson W^3 of the $SU(2)_L$ local gauge symmetry through the current $\overline{L}\gamma^\mu \frac{1}{2}\tau_3 L$. This would imply that the weak neutral current couples only to the left-handed chiral projection of fermions. **This is, however, in contradiction to experiments, which show that the physical electrically neutral weak boson (called the Z^0) couples to both left-handed and right-handed chiral states of electrically charged particles (although not equally)**. We will illustrate such an experiment in the next section. The solution to this apparent problem lies in the electroweak unified theory, as discussed in Section 25.5.

25.4   Polarized Electron Deep Inelastic Scattering (1978)

Is the weak neutral current, similar to the charged current, a pure $V - A$ parity-violating current? This question can be addressed with the study of ep scattering at high Q^2, where **the interference between the electromagnetic and neutral weak amplitudes** can be measured. In 1978 such an experiment has been performed at SLAC [339]. It showed that parity violation occurs in eN deep inelastic scattering, by studying the interaction of polarized electrons on deuterium:

$$e^-(\text{polarized}) + d \to e^- + X \tag{25.14}$$

where the incoming electron had energies between 16 and 22 GeV. The interaction occurs dominantly through the electromagnetic exchange of a photon (see Chapter 16). The existence of a neutral weak boson coupling to electrons and quarks (with couplings to be specified) implies that we should consider a second diagram where the photon is replaced by a spin-1 neutral weak Z^0 boson. Both diagrams contribute and interfere. At the kinematical region ($Q^2 \simeq 1 - 2$ GeV^2) considered by the SLAC experiment, the effect was, however, expected to be very small. A specific experiment had to be developed, as shown in Figure 25.4. The basic idea of the

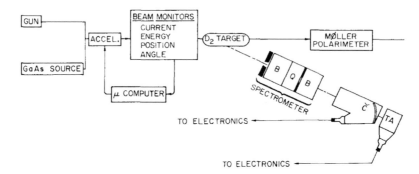

Figure 25.4 Schematic layout of the SLAC experiment that measured polarized electron deep inelastic scattering on an unpolarized target. Reprinted from C. Prescott *et al.*, "Parity nonconservation in inelastic electron scattering," *Phys. Lett. B*, vol. 77, pp. 347–352, 1978, with permission from Elsevier.

experiment is to measure the asymmetry of the cross-section with left-handed and right-handed longitudinally polarized electrons:

$$A_{eN \to eX} \equiv \frac{\sigma_R - \sigma_L}{\sigma_R + \sigma_L} \tag{25.15}$$

where σ_R and σ_L are the cross-sections measured with left and right longitudinal polarizations of the electrons. A crucial aspect is the development of an intense source of polarized electrons. In the experiment, the polarized electrons were produced by optically pumping electrons from the valence to the conduction bands of GaAs with circularly polarized photons. The photon source was a dye laser. Linearly polarized light from the laser was converted into circularly polarized light by a **Pockels[7] cell**. This latter is composed of a crystal with birefringence proportional to an applied electric field. The plane of polarization could be varied by rotating a linearly polarizing prism. Reversing the direction of the field reversed the helicity of the photons, which in turn reversed the helicity of the electrons.

The polarized electrons were accelerated and the sign and magnitude of the polarization were determined by the Møller polarimeter (see Figure 25.4), which measures the asymmetry in Møller scattering from a magnetized iron foil. This method was, for example, successfully used to determine the helicity of electrons in muon decays (see Section 22.7).

The target was filled with liquid deuterium. The electrons scattered at an angle of about $4°$ were measured in a spectrometer, composed of a dipole magnet followed by a single quadrupole and a second dipole. The electrons were identified by a Cherenkov counter and a lead-glass calorimeter. The beam intensity varied between $1 - 4 \times 10^{11}$ electrons per pulse. Approximately 1000 scattered electrons per pulse entered the counters. The experiment did not measure each particle individually but integrated the output of the detector during a pulse. In this way, the high statistics required to measure the small asymmetry was reached.

As mentioned above, the polarization of the electrons depends on the relative orientation of the linearly polarizing prism and on the sign of the voltage in the Pockels cell. This effect was tested, and the results are shown in Figure 25.5(left), which clearly shows the expected behavior. Although the helicity was well-defined at the exit of the Pockels cell, the helicity at the deuterium target depended on the beam energy, because the spin of the electrons precessed in the transport magnets which brought them from the source to the target. This effect is shown in Figure 25.5(right), where the measured asymmetry is plotted as a function of the beam energy E_0.

---

7   Friedrich Carl Alwin Pockels (1865–1913), German physicist.

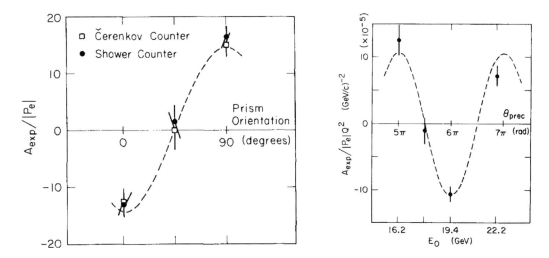

Figure 25.5 (left) The measured asymmetry as a function of the angle of rotation of the polarizing prism (see text). (right) The measured asymmetry as a function of the electron beam energy. Reprinted from C. Prescott *et al.*, "Parity nonconservation in inelastic electron scattering," *Phys. Lett. B*, vol. 77, pp. 347–352, 1978, with permission from Elsevier.

The above finding and several other cross-checks convinced the authors that the observed asymmetries were due to a dependence of the cross-section on the electron helicity (hence parity violation!). Their combined result is presented as an asymmetry per Q^2 given by:

$$A/Q^2 = (-9.5 \pm 1.6) \times 10^{-5} \; (\text{GeV}^{-2}) \tag{25.16}$$

It is clearly significantly different from zero. The power of this experiment is that it not only demonstrated the existence of parity violation but also allowed the testing of specific models which predicted the amount of parity violation in weak neutral currents (as mentioned previously, unlike the weak charged current which is consistent with pure $V - A$, the weak neutral current contains a mixture of V and A leading to a non-maximal parity violation). In particular, this measurement allowed one to predict the value of the weak mixing angle $\sin^2 \theta_W = 0.20 \pm 0.03$ within the electroweak theory to be discussed in the next section. Impressive!

25.5   The Electroweak $SU(2)_L \times U(1)_Y$ **Local Gauge Theory**

In 1957 **Schwinger** had first suggested the existence of an isospin "triplet" of vector fields (i.e., W^0, W^1, and W^2) to describe both the electromagnetic and weak interactions [340]. In 1960 **Glashow** pointed out that the simplest model reproducing the observed electromagnetic and weak interactions of leptons (note: this is before the NC discovery) required the existence of at least *four vector-boson fields* (including the photon) and suggested to extend the use of conserved quantities such as strangeness and isospin to the electromagnetic and weak interactions [341].

In 1967 **Weinberg** and independently **Salam** developed a model that unites electromagnetic and weak interactions. The model is based on two exact symmetries, spontaneously broken by the vacuum. The (unwanted) massless Goldstone bosons are absorbed into the spin degrees of freedom of the massive gauge bosons. One of the groups is $SU(2)_L$, which represents the **broken symmetry of the weak isospin** I. It is tempting to use the $U(1)$ group giving a scalar charge, however **this charge cannot be the electric charge** Q **since the symmetry is not broken** (the photon *is* massless!).

Consequently, in the electroweak theory, called the **Glashow, Salam, and Weinberg (GSW) model**, the $U(1)$ local gauge symmetry couples a gauge spin-1 boson field B^μ to a kind of charge called the **hypercharge** Y (in analogy to QED where the electric charge Q couples to the gauge field A^μ). A general $U(1)_Y$ transformation can be written in the form (see Eq. (24.11)):

$$\Psi \xrightarrow{U(1)_Y} e^{ig_1(Y/2)\alpha(x)}\Psi(x) \tag{25.17}$$

where g_1 **is the coupling constant associated with the** $U(1)_Y$ **local gauge symmetry**. The resulting interaction term is:

$$\text{Hypercharge:} \qquad \mathcal{L}_Y = -g_1\frac{Y}{2}B_\mu\overline{\Psi}\gamma^\mu\Psi \tag{25.18}$$

which is of the QED form where Q is replaced by $g_1Y/2$.

We then consider the second quantum number, i.e., the weak isospin with its $SU(2)_L$ symmetry. The general $SU(2)_L$ transformation on an $SU(2)$ doublet field Ψ_L is expressed as (see Eq. (24.31)):

$$\Psi_L(x) \xrightarrow{\text{Local } SU(2)} U(\vec{\alpha}(x))\Psi_L(x) = e^{ig_2\alpha_i(x)\tau_i/2}\Psi_L = e^{ig_2\vec{\alpha}(x)\cdot\vec{\tau}/2}\Psi_L \tag{25.19}$$

Local gauge invariance under the $SU(2)_L$ symmetry gives us the term:

$$\text{Weak isospin:} \qquad \mathcal{L}_{SU(2)_L} = -g_2\left(\overline{\Psi_L}\gamma^\mu\frac{\vec{\tau}}{2}\Psi_L\right)\cdot\vec{W}_\mu \tag{25.20}$$

where g_2 **is the coupling constant associated with the** $SU(2)_L$ **local gauge symmetry** and where the iso-triplet of spin-1 gauge bosons $\vec{W}_\mu$ is given by Eq. (24.34):

$$W_i^\mu \equiv (W_1^\mu, W_2^\mu, W_3^\mu) \tag{25.21}$$

For clarity, we adopt the convention that the Lorentz index and the field index are placed symmetrically, hence, we will write W_i^μ and W_μ^i synonymously.

To describe the charged weak interaction, we need to express it in terms of the I_3 raising and lowering operators τ^+ and τ^- (see Eq. (25.5)). This can be done by defining the **charged vector boson gauge fields** $W^\pm$:

$$W_\mu^\pm \equiv \frac{1}{\sqrt{2}}\left(W_\mu^1 \mp iW_\mu^2\right) \tag{25.22}$$

We note that:

$$\begin{aligned}
\tau_+W_\mu^+ + \tau_-W_\mu^- &= \frac{1}{2}(\tau_1 + i\tau_2)\frac{1}{\sqrt{2}}\left(W_\mu^1 - iW_\mu^2\right) + \frac{1}{2}(\tau_1 - i\tau_2)\frac{1}{\sqrt{2}}\left(W_\mu^1 + iW_\mu^2\right) \\
&= \frac{1}{2\sqrt{2}}\left[(\tau_1 + i\tau_2)\left(W_\mu^1 - iW_\mu^2\right) + (\tau_1 - i\tau_2)\left(W_\mu^1 + iW_\mu^2\right)\right] \\
&= 2\frac{1}{2\sqrt{2}}\left[\tau_1W_\mu^1 + \tau_2W_\mu^2\right] = \frac{1}{\sqrt{2}}\left[\tau_1W_\mu^1 + \tau_2W_\mu^2\right]
\end{aligned} \tag{25.23}$$

Consequently, the $SU(2)_L$ gauge interaction term, Eq. (25.20), can be expressed as the sum of charged and neutral weak currents:

$$\mathcal{L}_{SU(2)_L} = \underbrace{-\frac{g_2}{\sqrt{2}}\left(\overline{\Psi_L}\gamma^\mu\tau_+\Psi_L\right)W_\mu^+ - \frac{g_2}{\sqrt{2}}\left(\overline{\Psi_L}\gamma^\mu\tau_-\Psi_L\right)W_\mu^-}_{\text{charged current}} \underbrace{-g_2\left(\overline{\Psi_L}\gamma^\mu\frac{1}{2}\tau_3\Psi_L\right)W_\mu^3}_{\text{neutral current}} \tag{25.24}$$

Collecting all the pieces together, the $SU(2)_L \times U(1)_Y$ local gauge-invariant Lagrangian of the GSW model is just:

$$
\begin{aligned}
\mathcal{L}_{SU(2)_L \times U(1)_Y} &= -g_1 \frac{Y}{2} \left(\overline{\Psi} \gamma_\mu \Psi \right) B^\mu - g_2 \left(\overline{\Psi_L} \gamma_\mu \frac{\vec{\tau}}{2} \Psi_L \right) \cdot \vec{W}^\mu \\
&= \underbrace{-g_1 \frac{Y}{2} \left(\overline{\Psi_L} \gamma_\mu \Psi_L \right) B^\mu - g_1 \frac{Y}{2} \left(\overline{\Psi_R} \gamma_\mu \Psi_R \right) B^\mu}_{U(1)_Y} \underbrace{-g_2 \left(\overline{\Psi_L} \gamma_\mu \frac{\vec{\tau}}{2} \Psi_L \right) \cdot \vec{W}^\mu}_{SU(2)_L} \qquad (25.25)
\end{aligned}
$$

where in the second line we have separated the spinor Ψ into its left and right chiral projections L and R. We see that the hypercharge coupling treats L and R equally. However, here we note that the $SU(2)_L$ term forces us to take Ψ_L as an $SU(2)$ doublet. On the contrary, in the case of the $U(1)_Y$ there is no such restriction, and hence Ψ_R does not need to be an $SU(2)_L$ doublet. As a matter of fact, the Ψ_R are $SU(2)$ iso-singlets in the GSW model!

At this stage, the gauge fields B^μ and $\vec{W}^\mu$ are still somewhat arbitrary. The goal is to identify and unify the electromagnetic and weak forces into a single local gauge-invariant theory. **The electromagnetic photon interaction belongs to the class of neutral currents**. So, let us collect all the neutral current terms in the Lagrangian, Eq. (25.25). For instance, in the case of a left-handed electron and neutrino doublet, and an electron right-handed singlet, we find:

$$
L = \underbrace{\begin{pmatrix} \nu_e \\ e \end{pmatrix}_L}_{I=1/2 \; SU(2)_L \; \text{doublet}} \qquad \text{and} \qquad R = \underbrace{e_R}_{I=0 \; SU(2)_L \; \text{singlet}} \qquad (25.26)
$$

We obtain, replacing the hypercharge operator Y by the corresponding quantum numbers $Y_{\nu_{eL}}$, Y_{e_L}, and Y_{e_R}:

$$
\begin{aligned}
&-\frac{g_1}{2} \left(Y_{\nu_{eL}} \overline{\nu_L} \gamma_\mu \nu_L + Y_{e_L} \overline{e_L} \gamma_\mu e_L \right) B^\mu - \frac{g_1}{2} Y_{e_R} \left(\overline{e_R} \gamma_\mu e_R \right) B^\mu - \frac{g_2}{2} \left(\overline{\nu_L} \gamma_\mu \nu_L - \overline{e_L} \gamma_\mu e_L \right) W_3^\mu \\
=& -\frac{g_1}{2} \left(Y_L \overline{\nu_L} \gamma_\mu \nu_L + Y_L \overline{e_L} \gamma_\mu e_L \right) B^\mu - \frac{g_1}{2} Y_R \left(\overline{e_R} \gamma_\mu e_R \right) B^\mu - \frac{g_2}{2} \left(\overline{\nu_L} \gamma_\mu \nu_L - \overline{e_L} \gamma_\mu e_L \right) W_3^\mu \qquad (25.27)
\end{aligned}
$$

where in the second line we have used the fact that $Y_{\nu_{eL}} = Y_{e_L} \equiv Y_L$ since the hypercharges of particles in the same weak isospin doublet must be the same as a consequence of the symmetry, and set $Y_{e_R} \equiv Y_R$. We can rearrange the terms according to the interacting particles:

$$
\underbrace{-\left(\frac{g_1}{2} Y_L B^\mu + \frac{g_2}{2} W_3^\mu \right) \overline{\nu_L} \gamma_\mu \nu_L}_{\text{purely weak}} \underbrace{- \left(\frac{g_1}{2} Y_L B^\mu - \frac{g_2}{2} W_3^\mu \right) \overline{e_L} \gamma_\mu e_L + \left(\frac{g_1}{2} Y_R B^\mu \right) \overline{e_R} \gamma_\mu e_R}_{\text{electroweak}} \qquad (25.28)
$$

Using the same format, the **electromagnetic coupling of the electron to the photon** would be written as:

$$
\underbrace{Q_e e A^\mu \overline{e} \gamma_\mu e}_{\text{purely electromagnetic}} = (Q_e e A^\mu) \overline{e_L} \gamma_\mu e_L + (Q_e e A^\mu) \overline{e_R} \gamma_\mu e_R \qquad (25.29)
$$

For the GSW model to work, it must be able to reproduce this coupling. From the four gauge bosons of the $SU(2)_L \times U(1)$ symmetry, two are electrically charged $W^\pm$ and two are neutrals. One of them must be the photon and the other is called the Z^0. **These physical neutral gauge bosons** are linear combinations of the underlying gauge bosons B^μ and W_3^μ of the $SU(2)_L \times U(1)$ symmetry:

$$
\begin{pmatrix} A_\mu \\ Z_\mu \end{pmatrix} = \begin{pmatrix} \cos\theta_W & \sin\theta_W \\ -\sin\theta_W & \cos\theta_W \end{pmatrix} \begin{pmatrix} B_\mu \\ W_\mu^3 \end{pmatrix} \qquad (25.30)
$$

where θ_W is the **weak mixing angle** or **Weinberg angle**. When we observe the coupling to the neutrinos in Eq. (25.28), we note that the corresponding bracket is purely weak since the photon does not couple to neutrinos and therefore represents the Z^0 boson! We hence have $Z^\mu \propto g_1 Y_L B^\mu + g_2 W_3^\mu$, or properly normalizing the Z^0 gauge boson w.r.t. to the B^μ and W_3^μ bosons:

$$Z_\mu = \frac{g_1 Y_L B_\mu + g_2 W_\mu^3}{\sqrt{g_1^2 Y_L^2 + g_2^2}} \equiv -\sin\theta_W B_\mu + \cos\theta_W W_\mu^3 \tag{25.31}$$

The photon gauge field should be orthogonal to the Z^0, resulting in $A^\mu \propto g_2 B^\mu - g_1 Y_L W_3^\mu$. Consequently, normalizing the photon field, we get:

$$A_\mu = \frac{g_2 B_\mu - g_1 Y_L W_\mu^3}{\sqrt{g_1^2 Y_L^2 + g_2^2}} \equiv \cos\theta_W B_\mu + \sin\theta_W W_\mu^3 \tag{25.32}$$

We can easily invert these expressions to get back to the original vector gauge bosons as a function of the physical ones:

$$B^\mu = \frac{g_1 Y_L Z^\mu + g_2 A^\mu}{\sqrt{g_1^2 Y_L^2 + g_2^2}} \quad \text{and} \quad W_3^\mu = \frac{g_2 Z^\mu - g_1 Y_L A^\mu}{\sqrt{g_1^2 Y_L^2 + g_2^2}} \tag{25.33}$$

We inject these equalities for the gauge boson fields into the original interaction expression of Eq. (25.28) to find the couplings of the electron and the neutrino in terms of the physical fields:

$$\frac{1}{\sqrt{g_1^2 Y_L^2 + g_2^2}} \left(\frac{g_1}{2} Y_L (g_1 Y_L Z^\mu + g_2 A^\mu) + \frac{g_2}{2} (g_2 Z^\mu - g_1 Y_L A^\mu) \right) \overline{\nu_L}\gamma_\mu \nu_L$$
$$+ \frac{1}{\sqrt{g_1^2 Y_L^2 + g_2^2}} \left(\frac{g_1}{2} Y_L (g_1 Y_L Z^\mu + g_2 A^\mu) - \frac{g_2}{2} (g_2 Z^\mu - g_1 Y_L A^\mu) \right) \overline{e_L}\gamma_\mu e_L$$
$$+ \frac{1}{\sqrt{g_1^2 Y_L^2 + g_2^2}} \left(\frac{g_1}{2} Y_R (g_1 Y_L Z^\mu + g_2 A^\mu) \right) \overline{e_R}\gamma_\mu e_R \tag{25.34}$$

Collecting the A^μ terms, which should account for the electron coupling to the photon, we obtain (see Eq. (25.29)):

$$\frac{g_1 g_2}{\sqrt{g_1^2 Y_L^2 + g_2^2}} A^\mu \left(Y_L \overline{e_L}\gamma_\mu e_L + \frac{Y_R}{2}\overline{e_R}\gamma_\mu e_R \right) = \underbrace{(Q_e e A^\mu)\overline{e_L}\gamma_\mu e_L + (Q_e e A^\mu)\overline{e_R}\gamma_\mu e_R}_{\text{purely electromagnetic}} \tag{25.35}$$

In order to obtain a parity-conserving equal treatment of the left-handed and right-handed components of the electron, we must set:

$$Y_L = \frac{Y_R}{2} \tag{25.36}$$

We are further allowed to set $Y_L = -1$, since this would simply give a trivial redefinition of the g couplings, once we set $Y_L = Y_R/2$. In doing so, we have assigned the hypercharges of the electron and of the neutrino. We note that the left-handed and right-handed chiral projections of the electron acquire a different hypercharge:

$$Y_L = Y_{eL} = Y_{\nu_e L} = -1 \quad \text{and} \quad Y_R = Y_{eR} = -2 \tag{25.37}$$

• **Electromagnetic coupling.** Setting the electric charge $Q_e = -1$ and $Y_L^2 = 1$, the electromagnetic coupling of the electron with the photon, Eq. (25.35), becomes:

$$\frac{Y_L g_1 g_2}{\sqrt{g_1^2 + g_2^2}} A^\mu \left(\overline{e_L}\gamma_\mu e_L + \overline{e_R}\gamma_\mu e_R \right) = -\frac{g_1 g_2}{\sqrt{g_1^2 + g_2^2}} A^\mu \left(\overline{e_L}\gamma_\mu e_L + \overline{e_R}\gamma_\mu e_R \right) = -e A^\mu \left(\overline{e_L}\gamma_\mu e_L + \overline{e_R}\gamma_\mu e_R \right) \tag{25.38}$$

which allows us to define the electromagnetic coupling e in terms of the $SU(2)_L \times U(1)_Y$ coupling constants g_1 and g_2:

$$e \equiv \frac{g_1 g_2}{\sqrt{g_1^2 + g_2^2}} \tag{25.39}$$

• **Weinberg (or weak) angle.** A direct consequence of the definition of Eqs. (25.31) and (25.32) is that the Weinberg angle is tied to the coupling constants g_1 and g_2:

$$\sin\theta_W = \frac{-g_1 Y_L}{\sqrt{g_1^2 Y_L^2 + g_2^2}} = \frac{g_1}{\sqrt{g_1^2 + g_2^2}} \quad \text{and} \quad \cos\theta_W = \frac{g_2}{\sqrt{g_1^2 Y_L^2 + g_2^2}} = \frac{g_2}{\sqrt{g_1^2 + g_2^2}} \tag{25.40}$$

thus

$$\tan\theta_W = \frac{g_1}{g_2} \tag{25.41}$$

Combined with Eq. (25.39), these equations give us the direct relationship between the couplings g_1 and g_2 of our fundamental symmetry $SU(2)_L \times U(1)_Y$ as a function of the known physical electromagnetic coupling e and **the physical observable θ_W, to be determined experimentally**:

$$e = g_1 \underbrace{\cos\theta_W}_{\equiv c_W} = g_2 \underbrace{\sin\theta_W}_{\equiv s_W} \tag{25.42}$$

• **Useful relations.** There are several equivalent ways to express the relation between the gauge couplings and the physical observables e and θ_W which follow directly from the expressions of Eqs. (25.39) and (25.42). For example, setting as above $c_W \equiv \cos\theta_W$ and $s_W \equiv \sin\theta_W$:

$$g_1 = \frac{e}{c_W} \quad \text{and} \quad g_2 = \frac{e}{s_W} \tag{25.43}$$

Or also:

$$g_1^2 + g_2^2 = \frac{e^2}{c_W^2} + \frac{e^2}{s_W^2} = e^2 \left(\frac{s_W^2 + c_W^2}{c_W^2 s_W^2} \right) = \frac{e^2}{c_W^2 s_W^2} \tag{25.44}$$

and

$$g_1^2 - g_2^2 = e^2 \left(\frac{s_W^2 - c_W^2}{c_W^2 s_W^2} \right) = \frac{e^2(2s_W^2 - 1)}{c_W^2 s_W^2} \tag{25.45}$$

• **Weak neutral currents as a function of the physical parameters e and θ_W.** The coupling of neutrinos to the Z^0 is pure $V - A$ and can be extracted directly from Eq. (25.34) by setting $Y_L = -1$:

$$\frac{1}{\sqrt{g_1^2 + g_2^2}} \left(\frac{g_1^2}{2} + \frac{g_2^2}{2} \right) Z^\mu \overline{\nu_L}\gamma_\mu\nu_L = \frac{1}{2} \frac{g_1^2 + g_2^2}{\sqrt{g_1^2 + g_2^2}} \left(\overline{\nu_L}\gamma_\mu\nu_L \right) Z^\mu = \frac{e}{2 s_W c_W} \left(\overline{\nu_L}\gamma_\mu\nu_L \right) Z^\mu \tag{25.46}$$

The coupling of the electron to the Z^0 is also directly derivable from Eq. (25.34) and the expression is slightly more complicated since it involves both $\overline{e_L}\gamma_\mu e_L$ and $\overline{e_R}\gamma_\mu e_R$ terms:

$$\frac{1}{\sqrt{g_1^2 + g_2^2}} \left(\left(\frac{g_1^2}{2} - \frac{g_2^2}{2} \right) \overline{e_L}\gamma_\mu e_L + \frac{g_1^2}{2} Y_R Y_L \overline{e_R}\gamma_\mu e_R \right) Z^\mu = \frac{s_W c_W}{2e} \left((g_1^2 - g_2^2) \overline{e_L}\gamma_\mu e_L + g_1^2 Y_R Y_L \overline{e_R}\gamma_\mu e_R \right) Z^\mu$$

$$= \frac{e s_W c_W}{2} \left(\frac{(2s_W^2 - 1)}{c_W^2 s_W^2} \overline{e_L}\gamma_\mu e_L - \frac{Y_R}{c_W^2} \overline{e_R}\gamma_\mu e_R \right) Z^\mu$$

$$= \frac{e s_W c_W}{2 c_W^2 s_W^2} \left((2s_W^2 - 1)\overline{e_L}\gamma_\mu e_L + 2s_W^2 \overline{e_R}\gamma_\mu e_R \right) Z^\mu$$

$$= \frac{e}{s_W c_W} \left(\left(-\frac{1}{2} + s_W^2 \right) \overline{e_L}\gamma_\mu e_L + s_W^2 \overline{e_R}\gamma_\mu e_R \right) Z^\mu \tag{25.47}$$

• **Weak neutral universal couplings.** The last two expressions, Eqs. (25.46) and (25.47), can be shrunk into a single term for a particular fermion f depending on its weak isospin I_3^f and hypercharge Y^f, valid for both left-handed and right-handed projections, by noting that the universal Z^0 coupling is simply:

$$\text{Universal } Z^0 \text{ coupling}: \quad \frac{e}{s_W c_W} \left(I_3^f c_W^2 - \frac{Y^f}{2} s_W^2 \right) \equiv g_Z \left(I_3^f c_W^2 - \frac{Y^f}{2} s_W^2 \right) \tag{25.48}$$

Indeed:

- For ν_{eL}, we have $I_3^{\nu_{eL}} = +1/2$, $Y^{\nu_{eL}} = -1$ and find:

$$g_Z \left(+\frac{1}{2} c_W^2 + \frac{1}{2} s_W^2 \right) = \frac{e}{2 s_W c_W} \qquad \square \tag{25.49}$$

- For e_L, we have $I_3^{e_L} = -1/2$, $Y^{e_L} = -1$ and write:

$$g_Z \left(-\frac{1}{2} c_W^2 + \frac{1}{2} s_W^2 \right) = \frac{e}{s_W c_W} \left(-\frac{1}{2}(1 - s_W^2) + \frac{1}{2} s_W^2 \right) = \frac{e}{s_W c_W} \left(-\frac{1}{2} + s_W^2 \right) \qquad \square \tag{25.50}$$

- For e_R, we have $I_3^{e_R} = 0$, $Y^{e_R} = -2$ and write:

$$g_Z \left(-\frac{-2}{2} s_W^2 \right) = \frac{e}{s_W c_W} s_W^2 \qquad \square \tag{25.51}$$

- **Universal couplings as a function of the electric charge and the Weinberg angle.** The assignment of the hypercharge is constrained by the weak isospin and the electric charge since the theory must reproduce the electromagnetic coupling to the photon. The relation between hypercharge, weak isospin, and electric charge for a fermion of given chirality then simply reads:

$$Q^f = I_3^f + \frac{Y^f}{2} \qquad \Leftrightarrow \qquad Y^f = 2 \left(Q^f - I_3^f \right) \tag{25.52}$$

It can be checked directly for $f = \nu_{eL}$, e_L, and e_R. The relation can be extended to fractionally charged quarks and even right-handed singlet neutrinos ν_R. In the latter case, we would have $I = 0$, $I_3 = 0$, $Q = 0$, and hence $Y = 0$, which implies that it has no interactions. We say it is *sterile* within the theory. The assigned values for charged leptons and neutrinos are summarized in Table 25.2.

f	I^f	I_3^f	Y^f	Q^f
$\nu_{eL}, \nu_{\mu L}, \nu_{\tau L}$	1/2	+1/2	-1	0
e_L^-, μ_L^-, τ_L^-	1/2	$-1/2$	-1	-1
e_R^-, μ_R^-, τ_R^-	0	0	-2	-1
$\nu_{eR}, \nu_{\mu R}, \nu_{\tau R}$	0	0	0	0

Table 25.2 Quantum numbers of the charged and neutral leptons in the $SU(2)_L \times U(1)_Y$ classification.

Finally we can express the coupling to the Z^0 boson (Eq. (25.48)) with an expression valid for any fermion f and for both left-handed (L) and right-handed (R) projections as a function of its weak isospin I_3^f and its electric charge Q^f:

$$\begin{aligned}
g_Z \left(I_3^f c_W^2 - \frac{Y^f}{2} s_W^2 \right) &= g_Z \left(I_3^f - I_3^f s_W^2 - \frac{Y^f}{2} s_W^2 \right) = g_Z \left(I_3^f - \left(I_3^f + \frac{Y^f}{2} \right) s_W^2 \right) \\
&= \frac{e}{s_W c_W} \left(I_3^f - Q^f s_W^2 \right)
\end{aligned} \tag{25.53}$$

- **Universal couplings for all quarks and leptons.** We have up to this point considered only the electron and its neutrino. It is now straightforward to extend the model to the three lepton families by extending the following iso-doublets and singlets to the muon and tau:

$$L_e = \begin{pmatrix} \nu_{eL} \\ e_L \end{pmatrix}, \quad R_e = e_R, \quad L_\mu = \begin{pmatrix} \nu_{\mu L} \\ \mu_L \end{pmatrix}, \quad R_\mu = \mu_R, \quad L_\tau = \begin{pmatrix} \nu_{\tau L} \\ \tau_L \end{pmatrix}, \quad R_\tau = \tau_R \tag{25.54}$$

The Lagrangian of the theory for three lepton families can then be expressed as:

$$\mathcal{L}_{leptons} = \sum_{f=e,\mu,\tau} \overline{L_f} \left(-g_1 \frac{Y_L^f}{2} \gamma_\mu B^\mu - g_2 \gamma_\mu \frac{\vec{\tau}}{2} \cdot \vec{W}^\mu \right) L_f + \sum_{f=e,\mu,\tau} \overline{R_f} \left(-g_1 \frac{Y_R^f}{2} \gamma_\mu B^\mu \right) R_f \tag{25.55}$$

where we have explicitly introduced the hypercharges Y_L^f and Y_R^f for left-handed and right-handed chiral projections of the particular fermion f. Note that we could also have added the right-handed neutrinos, but this would not have any consequence on the interaction Lagrangian since they are sterile:

$$R_{\nu_e} = \nu_{eR}, \quad R_{\nu_\mu} = \nu_{\mu R}, \quad R_{\nu_\tau} = \nu_{\tau R} \tag{25.56}$$

Inclusion of quarks can be accomplished in a similar fashion by considering left-handed quark weak eigenstate doublets $Y_L^u = Y_L^c = Y_L^t = 2(+2/3 - (+1/2)) = +1/3$ and $Y_L^d = Y_L^s = Y_L^b = 2(-1/3 - (-1/2)) = +1/3$ (as already noticed, members of an iso-doublet have the same hypercharges, see Eq. (25.52))

$$L_u = \begin{pmatrix} u_L \\ d'_L \end{pmatrix}, \quad L_c = \begin{pmatrix} c_L \\ s'_L \end{pmatrix}, \quad L_t = \begin{pmatrix} t_L \\ b'_L \end{pmatrix} \tag{25.57}$$

together with the right-handed weak eigenstate singlets with $Y_R^{u,c,t} = 2(+2/3) = +4/3$:

$$R_u = u_R, \quad R_c = c_R, \quad R_t = t_R \tag{25.58}$$

and $Y_R^{d,s,b} = 2(-1/3) = -2/3$:

$$R_d = d'_R, \quad R_s = s'_R, \quad R_b = b'_R \tag{25.59}$$

then the Lagrangian for the quarks is:

$$\mathcal{L}_{quarks} = \sum_{f=u,c,t} \overline{L_f} \left(-g_1 \frac{Y_L^f}{2} \gamma_\mu B^\mu - g_2 \gamma_\mu \frac{\vec{\tau}}{2} \cdot \vec{W}^\mu \right) L_f + \sum_{f=u,d,s,c,b,t} \overline{R_f} \left(-g_1 \frac{Y_R^f}{2} \gamma_\mu B^\mu \right) R_f \tag{25.60}$$

The relevant quantum numbers for the electroweak theory of the six quarks and six leptons are summarized in Table 25.3.

f	I^f	I_3^f	Y^f	Q^f
u_L, c_L, t_L	1/2	+1/2	+1/3	+2/3
d'_L, s'_L, b'_L	1/2	-1/2	+1/3	-1/3
u_R, c_R, t_R	0	0	+4/3	+2/3
d'_R, s'_R, b'_R	0	0	-2/3	-1/3

Table 25.3 Quantum numbers of the quarks in the $SU(2)_L \times U(1)_Y$ classification.

• **Simple yet powerful.** The interactions are fully determined within the theory. The quantum numbers I^f, I_3^f, and Y^f uniquely fix the coupling of the fermion with the gauge bosons. The left-handed doublets have $I = 1/2$, $I_3 = \pm 1/2$, and $Y = -1$, while the right-handed singlets have $I = I_3 = 0$ and $Y = -2$. As a consequence and as noted previously, the right-handed and left-handed components have different couplings with the gauge bosons.

- **Gauge bosons.** In order to complete the theory, we now add the kinetic terms of the gauge fields B^μ and $\vec{W}^\mu$ to promote them to propagating fields. The pure gauge kinetic terms are then:

$$\mathcal{L}_G = -\frac{1}{4} F_{\mu\nu} F^{\mu\nu} - \frac{1}{4} \vec{W}_{\mu\nu} \cdot \vec{W}^{\mu\nu} \tag{25.61}$$

where $F_{\mu\nu} \equiv \partial_\mu B_\nu - \partial_\nu B_\mu$ (see Eq. (24.21)) and

$$\vec{W}_{\mu\nu} \equiv \partial_\mu \vec{W}_\nu - \partial_\nu \vec{W}_\mu - g_2 \vec{W}_\mu \times \vec{W}_\nu \tag{25.62}$$

(see Eq. (24.50)). Due to the gauge invariance, the gauge bosons are necessarily massless in the unbroken stage, since terms of the type:

$$\frac{1}{2} M_B^2 B_\mu B^\mu + \frac{1}{2} M_W^2 \vec{W}_\mu \cdot \vec{W}^\mu \tag{25.63}$$

would break the $SU(2)_L \times U(1)_Y$ invariance. This enforces $M_B^2 = M_W^2 = 0$. The gauge bosons will acquire mass when the $SU(2)_L \times U(1)_Y$ is spontaneously broken (see Section 25.6).

- **What about the fermion masses?** To obtain a complete Lagrangian, we also need to include the terms corresponding to the free Dirac fields. The point here is that the mass term of the Dirac fermions will be of the kind (see Eq. (10.160)):

$$m_f \overline{\Psi}\Psi = m_f \overline{(\Psi_L + \Psi_R)}(\Psi_L + \Psi_R) = m_f \left(\overline{\Psi_R}\Psi_R + \overline{\Psi_R}\Psi_L + \overline{\Psi_L}\Psi_R + \overline{\Psi_L}\Psi_L \right) \tag{25.64}$$

where we obtain four possible combinations RR, RL, LR, LL. We note that only two of the four are non-vanishing (cf. Eq. (10.162)). For example:

$$\overline{\Psi_L}\Psi_L = \Psi_L^\dagger \gamma^0 \Psi_L = \Psi^\dagger P_L^\dagger \gamma^0 \Psi_L = \Psi^\dagger P_L \gamma^0 \Psi_L = \Psi^\dagger \gamma^0 P_R \Psi_L = \overline{\Psi} P_R P_L \Psi = 0 \tag{25.65}$$

where we used $P_L \gamma^0 = \gamma^0 P_R$. Consequently, we find that for the mass term, both left-handed and right-handed chiral components are necessary. In fact, the mass term represents the coupling between the left-handed and right-handed chiral components of the spinor:

$$m_f \overline{\Psi}\Psi = m_f \left(\overline{\Psi_R}\Psi_L + \overline{\Psi_L}\Psi_R \right) \tag{25.66}$$

However, these terms would also break the $SU(2)_L \times U(1)_Y$ gauge symmetry, since the $SU(2)_L$ is directly broken in the sense that the left-handed and right-handed chiral projections belong to different representations of the group! So **all fermions of the unbroken electroweak theory must be massless not to violate the gauge symmetry**, and the Lagrangian term for the Dirac fields for the unbroken theory is then (assuming no right-handed neutrinos):

$$\mathcal{L}_f = \sum_{f=e,\mu,\tau,u,c,t} \overline{L_f}\left(i\partial_\mu \gamma^\mu \right) L_f + \sum_{f=e,\mu,\tau,u,d,c,s,t,b} \overline{R_f}\left(i\partial_\mu \gamma^\mu \right) R_f \tag{25.67}$$

We note that the left-handed and right-handed components are totally decoupled. This apparent conundrum will be solved by invoking the spontaneous symmetry breaking, also commonly referred to as the Brout–Englert–Higgs mechanism, to introduce the masses of the gauge bosons *and* those of the fermions.

Finally, collecting all the terms of Eqs. (25.55), (25.60), (25.61), and (25.67), the Lagrangian of the unbroken GSW electroweak theory can be expressed as:

$$\mathcal{L}_{GSW} = \mathcal{L}_f + \mathcal{L}_G + \mathcal{L}_{leptons} + \mathcal{L}_{quarks} \tag{25.68}$$

It's gorgeous!

25.6   The Brout–Englert–Higgs Mechanism

Using the concepts introduced in Chapter 24, the spontaneous symmetry breaking in the electroweak theory is realized via the so-called **Brout[8]–Englert[9]– Higgs[10] mechanism**. As a consequence of the spontaneous breaking, we want to obtain three massive and one massless gauge bosons:

$$\text{SSB } SU(2)_L \times U(1)_Y \Longrightarrow \begin{cases} 3 \text{ massive gauge bosons} \\ 1 \text{ massless gauge boson} \end{cases} \tag{25.69}$$

To break the symmetry, we need to consider a new scalar field called the **Higgs field** ϕ, which possesses specific transformation properties under our underlying symmetry. In the **minimal Higgs sector** (which is totally consistent with experimental data), the **Higgs field has the quantum number of weak isospin $I = 1/2$ and hypercharge $Y = 1$**. Consequently, one introduces an $SU(2)_L$ doublet complex scalar field:

$$\phi(x) \equiv \begin{pmatrix} \frac{1}{\sqrt{2}}(\phi_1(x) + i\phi_2(x)) \\ \frac{1}{\sqrt{2}}(\phi_3(x) + i\phi_4(x)) \end{pmatrix} = \begin{pmatrix} \phi^+(x) \\ \phi^0(x) \end{pmatrix} \tag{25.70}$$

where ϕ_i ($i = 1,2,3,4$) are four real scalar fields, and where, since $Q = I_3 + Y/2$, we note that the Higgs field can be expressed as two complex scalar fields, one being electrically charged ϕ^+ and the other electrically neutral ϕ^0. We note that the $SU(2)$ transformation of the Higgs field is by construction exactly the same as that of an $SU(2)_L$ doublet of fermion fields (see Eq. (25.19)):

$$\phi(x) \xrightarrow{\text{Local } SU(2)} U(\vec{\alpha}(x))\phi(x) = e^{ig_2\alpha_i(x)\tau_i/2}\phi = e^{ig_2\vec{\alpha}(x)\cdot\vec{\tau}/2}\phi \tag{25.71}$$

The Lagrangian density of the Higgs field ϕ is then written as (see Eq. (24.89)):

$$\mathcal{L}_\phi = (D_\mu\phi)^\dagger(D_\mu\phi) - \mu^2\phi^\dagger\phi - \lambda(\phi^\dagger\phi)^2 = |D_\mu\phi|^2 - \mu^2|\phi|^2 - \lambda|\phi|^4 \tag{25.72}$$

where the **covariant derivative** D_μ is determined by the $SU(2)_L \times U(1)_Y$ symmetry to be:

$$D_\mu \equiv \partial_\mu + ig_1\frac{Y}{2}B_\mu + ig_2\frac{\vec{\tau}}{2}\cdot\vec{W}_\mu \tag{25.73}$$

In order to spontaneously break the symmetry, we set $\lambda > 0$ and $\mu^2 < 0$ and find the minimum condition to be (see Eq. (24.112)):

$$|\phi|^2 \equiv \phi^\dagger\phi = \frac{1}{2}\left(\phi_1^2 + \phi_2^2 + \phi_3^2 + \phi_4^2\right) = -\frac{\mu^2}{2\lambda} \tag{25.74}$$

The symmetry is explicit using this notation and we can directly conclude that there are three Goldstone bosons. To break the symmetry, we fix the following vacuum (see Eq. (24.113)):

$$\phi_1 = \phi_2 = \phi_4 = 0 \quad \text{and} \quad \phi_3^2 = -\frac{\mu^2}{\lambda} \equiv v^2 \tag{25.75}$$

hence the Higgs doublet vacuum and its expansion around the minimum are just given by (see Eq. (24.114)):

$$\phi_0 \equiv \frac{1}{\sqrt{2}}\begin{pmatrix} 0 \\ v \end{pmatrix} \quad \text{and} \quad \phi(x) = \frac{1}{\sqrt{2}}\begin{pmatrix} 0 \\ v + h(x) \end{pmatrix} \tag{25.76}$$

8   Robert Brout (1928–2011), Belgian theoretical physicist.
9   François, Baron Englert (born 1932), Belgian theoretical physicist.
10   Peter Ware Higgs (born 1929), British theoretical physicist.

We insert this vacuum into the Lagrangian to compute the gauge boson fields:

$$
\begin{aligned}
|D_\mu \phi|^2 &= \left| \frac{1}{2\sqrt{2}} \begin{pmatrix} g_1 B_\mu + g_2 W_\mu^3 & g_2(W_\mu^1 - iW_\mu^2) \\ g_2(W_\mu^1 + iW_\mu^2) & g_1 B_\mu - g_2 W_\mu^3 \end{pmatrix} \begin{pmatrix} 0 \\ v \end{pmatrix} \right|^2 + \cdots \\
&= \left| \frac{v}{2\sqrt{2}} \begin{pmatrix} g_2(W_\mu^1 - iW_\mu^2) \\ g_1 B_\mu - g_2 W_\mu^3 \end{pmatrix} \right|^2 + \cdots \\
&= \frac{v^2}{8} \left(g_2^2 (W_\mu^1)^2 + g_2^2 (W_\mu^2)^2 + (g_1 B_\mu - g_2 W_\mu^3)^2 \right) + \cdots
\end{aligned}
\tag{25.77}
$$

The mass term of the charged bosons becomes (cf. Eq. (24.116)):

$$
M_W^2 W_\mu^+ W^{\mu -} = M_W^2 \frac{1}{2} \left(W_\mu^1 - iW_\mu^2 \right)\left(W^{1\mu} + iW^{2\mu} \right) = \frac{1}{2} M_W^2 \left((W_\mu^1)^2 + (W_\mu^2)^2 \right)
\tag{25.78}
$$

hence the mass of the charged weak boson is given by:

$$
M_{W^\pm} = g_2 \frac{v}{2}
\tag{25.79}
$$

The coupling of the $W^\pm$ boson to the Higgs field vacuum expectation value is illustrated in Figure 25.6. The

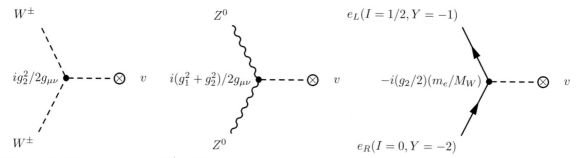

Figure 25.6 Coupling of the $W^\pm$, Z^0 bosons and of the electron to the Higgs field. The (minimal) Higgs field has weak isospin $I = 1/2$ and hypercharge $Y = 1$.

mass of the neutral weak boson Z^0 is found through its definition, Eq. (25.31), setting $Y_L = -1$:

$$
Z_\mu = \frac{-g_1 B^\mu + g_2 W_3^\mu}{\sqrt{g_1^2 + g_2^2}}
\tag{25.80}
$$

which when compared to the mass term:

$$
\frac{1}{2} M_Z^2 Z_\mu Z^\mu = \frac{1}{2} M_Z^2 \frac{(-g_1 B^\mu + g_2 W_3^\mu)^2}{g_1^2 + g_2^2} = \frac{v^2}{8} (-g_1 B^\mu + g_2 W_3^\mu)^2
\tag{25.81}
$$

yields

$$
M_Z = \sqrt{g_1^2 + g_2^2} \, \frac{v}{2}
\tag{25.82}
$$

The coupling of the Z^0 boson is also illustrated in Figure 25.6. Finally, there is no term proportional to the photon field (see Eq. (25.32)):

$$
A_\mu = \frac{g_2 B^\mu - g_1 Y_L W_3^\mu}{\sqrt{g_1^2 Y_L^2 + g_2^2}}
\tag{25.83}
$$

so the *photon remains massless* as desired after the spontaneous breaking:

$$M_A = 0 \qquad (25.84)$$

Because the Z^0 boson is a mixture between W_μ^3 and B_μ, it has a different mass than the $W^\pm$ bosons. The ratio of masses is simply given by:

$$\frac{M_W}{M_Z} = \frac{g_2 \frac{v}{2}}{\sqrt{g_1^2 + g_2^2}\, \frac{v}{2}} = \frac{e/s_W}{\sqrt{e^2/s_W^2 + e^2/c_W^2}} = \frac{e/s_W}{\frac{e/s_W}{c_W}\sqrt{c_W^2 + s_W^2}} = c_W \qquad (25.85)$$

or

$$M_W = M_Z \cos\theta_W \qquad (25.86)$$

Thus the Brout–Englert–Higgs mechanism has provided precise predictions of the masses for the vector bosons, which can be exactly calculated (at tree level) only as a function of the coupling constants g_1 and g_2, and the vacuum expectation value v.

A great feature of having introduced the Higgs field to generate the gauge boson masses is that the same field can be used for the fermions. In order to generate the fermion mass terms, which were prohibited in the unbroken theory, we add an $SU(2)_L \times U(1)_Y$ **gauge-invariant Yukawa term**, for example for the electron, of the form:

$$\mathcal{L}_e = -\lambda_e\left(\overline{L_e}\phi R_e + \overline{R_e}\phi^\dagger L_e\right) = -\lambda_e\left(\begin{pmatrix}\overline{\nu_{eL}} & \overline{e_L}\end{pmatrix}\begin{pmatrix}\phi^+ \\ \phi^0\end{pmatrix}e_R + \overline{e_R}\begin{pmatrix}(\phi^\dagger)^* & (\phi^0)^*\end{pmatrix}\begin{pmatrix}\nu_e \\ e\end{pmatrix}_L\right) \qquad (25.87)$$

where λ_e is a new coupling to be defined. We recall that the (minimal) Higgs field has a weak isospin $I = 1/2$ and hypercharge $Y = 1$, so it has exactly the required quantum numbers to couple the left-handed and right-handed fermion fields, see Figure 25.6. Gauge invariance can be noted by considering an infinitesimal gauge transformation on the iso-doublet field L_e and the Higgs field (see Eqs. (25.19) and (25.71)):

$$L_e \to L'_e \simeq (\mathbb{1} + ig_2\vec{\alpha}(x)\cdot\vec{\tau}/2)\, L_e \quad\Longrightarrow\quad \overline{L_e} \to \overline{L_e}' \simeq \overline{L_e}\,(\mathbb{1} - ig_2\vec{\alpha}(x)\cdot\vec{\tau}/2)$$

$$\phi \to \phi' \simeq (\mathbb{1} + ig_2\vec{\alpha}(x)\cdot\vec{\tau}/2)\,\phi \qquad (25.88)$$

and the iso-singlet R_e is by definition invariant under $SU(2)_L$. Hence the product $\overline{L_e}\phi R_e$ is invariant under $SU(2)_L$ transformation, and so is its Hermitian conjugate $\left(\overline{L_e}\phi R_e\right)^\dagger = \overline{R_e}\phi^\dagger L_e$. The Yukawa term is therefore $SU(2)_L$ and $U(1)_Y$ invariant and can be added to the Lagrangian of the GSW theory to generate the fermion masses.

With the chosen vacuum state, Eq. (25.76), we obtain

$$\mathcal{L}_e = -\lambda_e\frac{1}{\sqrt{2}}\left(\begin{pmatrix}\overline{\nu_{eL}} & \overline{e_L}\end{pmatrix}\begin{pmatrix}0 \\ v\end{pmatrix}e_R + \overline{e_R}\begin{pmatrix}0 & v\end{pmatrix}\begin{pmatrix}\nu_e \\ e\end{pmatrix}_L\right) = -\frac{\lambda_e}{\sqrt{2}}v\left(\overline{e_L}e_R + \overline{e_R}e_L\right) \qquad (25.89)$$

and similarly, for the muon and the tau lepton. These terms generate the mass terms for the charged leptons with rest masses:

$$m_e = \frac{\lambda_e v}{\sqrt{2}}, \quad m_\mu = \frac{\lambda_\mu v}{\sqrt{2}}, \quad \text{and} \quad m_\tau = \frac{\lambda_\tau v}{\sqrt{2}} \qquad (25.90)$$

where the **three Yukawa couplings λ_e, λ_μ, and λ_τ need to be determined experimentally**. We note that the Yukawa coupling of, for example, the electron is then:

$$\lambda_e \equiv \frac{\sqrt{2}m_e}{v} = \frac{\sqrt{2}g_2 m_e}{2M_{W^\pm}} = \frac{g_2}{\sqrt{2}}\frac{m_e}{M_{W^\pm}} \qquad (25.91)$$

Because the non-zero vacuum expectation value occurs in the lower component of the Higgs field doublet, the chosen Yukawa term $\overline{L}\phi R + \overline{R}\phi^\dagger L$ is only able to generate masses for the fermions with $I_3^f = -1/2$. This successfully generates the mass of the charged leptons and of the down-type quarks, but not of the up-type quarks or of the neutrinos. We can put aside the mass of the neutrinos, which will be discussed in Chapter 31.

In order to have a mechanism to give masses to the up-type quarks, we can consider a new Yukawa mass term involving the charge-conjugated doublet of the Higgs field ϕ^C of the type:

$$\mathcal{L}_u = -\lambda_u \left(\overline{L_u} \phi^C R_u + \overline{R_u} \phi^{C\dagger} L_u \right) \tag{25.92}$$

where

$$\phi^C = -i\sigma_2 \phi^* = \begin{pmatrix} \frac{1}{\sqrt{2}}(-\phi_3(x) + i\phi_4(x)) \\ \frac{1}{\sqrt{2}}(\phi_1(x) - i\phi_2(x)) \end{pmatrix} = \begin{pmatrix} -\phi^{0*}(x) \\ \phi^-(x) \end{pmatrix} \tag{25.93}$$

For example, for the u_L, d_L iso-doublet and the u_R iso-singlet, this term generates, after spontaneous symmetry breaking, the mass term:

$$\mathcal{L}_u = -\frac{\lambda_u}{\sqrt{2}} v \left(\overline{u_L} u_R + \overline{u_R} u_L \right) \tag{25.94}$$

Hence, mass terms for all leptons and quarks can be generated with gauge-invariant Yukawa terms of either $\overline{L_f}\phi R_f + \overline{R_f}\phi^\dagger L_f$ or $\overline{L_f}\phi^C R_f + \overline{R_f}\phi^{C\dagger} L_f$. The Yukawa couplings of the Higgs fields to the fermions then give rise to the mass of the fermions:

$$m_f = \frac{\lambda_f v}{\sqrt{2}} \tag{25.95}$$

Summarizing, the Brout–Englert–Higgs mechanism is able to generate the masses of the gauge bosons as well as those of the fermions. The masses of the gauge bosons are completely determined from the free parameters of the theory, i.e., the coupling constants of the theory g_1 and g_2, and the vacuum expectation value v. The masses of the fermions are, on the other hand, undetermined and need to be fixed to the experimentally observed value by setting the Yukawa couplings λ_f, which can be considered as additional free parameters of the theory.

25.7 Trilinear and Quadrilinear Gauge Boson Interactions

The non-Abelian nature of the theory leads to self-boson couplings. In order to derive these in terms of the physical bosons $W^\pm$, γ, and Z^0, we start from the Lagrangian term Eq. (25.61) and replace $F_{\mu\nu}$ and $W_{\mu\nu}$ with their definitions. We find:

$$\begin{aligned} \mathcal{L}_G = &-\frac{1}{4}\left(\partial_\mu B_\nu - \partial_\nu B_\mu\right)\left(\partial^\mu B^\nu - \partial^\nu B^\mu\right) \\ &-\frac{1}{4}\left(\partial_\mu \vec{W}_\nu - \partial_\nu \vec{W}_\mu - g_2 \vec{W}_\mu \times \vec{W}_\nu\right) \cdot \left(\partial^\mu \vec{W}^\nu - \partial^\nu \vec{W}^\mu - g_2 \vec{W}^\mu \times \vec{W}^\nu\right) \end{aligned} \tag{25.96}$$

Defining for the physical fields $A_{\mu\nu} = \partial_\mu A_\nu - \partial_\nu A_\mu$, $Z_{\mu\nu} = \partial_\mu Z_\nu - \partial_\nu Z_\mu$, and $W^\pm_{\mu\nu} = \partial_\mu W^\pm_\nu - \partial_\nu W^\pm_\mu$ (see Eq. (25.22)) leads, after some algebra, to:

$$\begin{aligned} \mathcal{L}_G = &-\frac{1}{4}A_{\mu\nu}A^{\mu\nu} - \frac{1}{4}W^{+\dagger}_{\mu\nu}W^{+\mu\nu} - \frac{1}{4}W^{-\dagger}_{\mu\nu}W^{-\mu\nu} - \frac{1}{4}Z_{\mu\nu}Z^{\mu\nu} \\ &\underbrace{+\frac{g_2}{4}\left[(\vec{W}_\mu \times \vec{W}_\nu)(\partial^\mu \vec{W}^\nu - \partial^\nu \vec{W}^\mu) + (\partial_\mu \vec{W}_\nu - \partial_\nu \vec{W}_\mu)(\vec{W}^\mu \times \vec{W}^\nu)\right]}_{\equiv \mathcal{L}_3} \\ &\underbrace{-\frac{g_2^2}{4}(\vec{W}_\mu \times \vec{W}_\nu) \cdot (\vec{W}^\mu \times \vec{W}^\nu)}_{\equiv \mathcal{L}_4} \end{aligned} \tag{25.97}$$

By replacing with the physical fields, the $\mathcal{L}_3$ term leads to the trilinear gauge boson interaction:

$$\begin{aligned}\mathcal{L}_3 \;=\;& ig_2\big[W_\mu^- W_\nu^+ (\cos\theta_W Z^{\mu\nu} + \sin\theta_W A^{\mu\nu})\\ &+W_\mu^+ W^{-\mu\nu}(\cos\theta_W Z_\nu + \sin\theta_W A_\nu) - W^{+\mu\nu} W_\mu^- (\cos\theta_W Z_\nu + \sin\theta_W A_\nu)\big]\end{aligned} \quad (25.98)$$

Similarly, the $\mathcal{L}_4$ term leads to the quadrilinear gauge boson interaction:

$$\begin{aligned}\mathcal{L}_4 \;=\;& -\frac{g_2^2}{4}\big[2W_\mu^+ W^{-\mu} W_\nu^+ W^{-\nu} - 2W_\mu^+ W^{+\mu} W_\nu^- W^{-\nu}\\ &+4W_\mu^+ W^{-\mu} W_{3\nu} W^{3\nu} - 4W_\mu^+ W_3^\mu W_\nu^- W^{3\nu}\big]\end{aligned} \quad (25.99)$$

These terms can be shown to lead to the interactions between the gauge bosons shown in Table 26.3.

Problems

Ex 25.1 Ratio of neutral-current to charged-current cross-sections. From the known formulae for the total cross-sections for neutrinos scattering off electrons, study the expected ratio of neutral-current (NC) events to charged-current (CC) events for neutrinos or antineutrinos scattering on electrons. Explicitly:

(a) Calculate $R_\nu = NC(\nu_e e)/CC(\nu_e e)$.

(b) Calculate $R_{\bar\nu} = NC(\bar\nu_e e)/CC(\bar\nu_e e)$.

(c) Do your results agree with the experimental results obtained in the Gargamelle neutrino experiment at CERN from 1973, discussed in Section 25.2?

(d) What were the main sources of background for the Gargamelle neutrino experiment?

Ex 25.2 Trilinear and quadrilinear gauge boson interactions. Derive Eqs. (25.97), (25.98), and (25.99).

26 Computations in the Electroweak Theory

> At this point I decided to develop a computer program that could do this work. (...) I developed a general-purpose symbolic manipulation program. Working furiously, I completed the first version of this program in about three months.
>
> *Martinus Veltman[1]*

26.1 Fermion Couplings to the Gauge Bosons

In the electroweak theory, we need to consider the couplings of the leptons and quarks to three different physical gauge bosons: the photon, the $W^\pm$, and the Z^0. We have seen that the electroweak theory is constructed to englobe QED, in which the coupling to the photon is simply given by:

$$-ie(\overline{\Psi}_f \gamma_\mu |Q_f| \Psi_f) A^\mu \equiv -ie(j_\mu^{em}) A^\mu \tag{26.1}$$

where Q_f is the electric charge of the fermion–antifermion $f\bar{f}$ pair: $|Q_e| = 1$, $|Q_{u,c,t}| = 2/3$, and $|Q_{d,s,b}| = 1/3$. The corresponding Feynman diagram vertex rule **[GSW1]** is shown in Table 26.1.

The coupling to the $W^\pm$ is given by the charged current term in the Lagrangian, Eq. (25.24). Generically defining it as:

$$-ig_W(j_\mu^\pm) W^{\pm\mu} \quad \text{where} \quad g_W \equiv g_2 = \frac{e}{\sin \theta_W} \tag{26.2}$$

We have:

$$j_\mu^\pm = \frac{1}{\sqrt{2}} \left(\overline{\Psi}_L \gamma^\mu \tau_\pm \Psi_L \right) = \frac{1}{\sqrt{2}} \overline{\Psi} \left(\gamma^\mu P_L \tau_\pm \right) \Psi = \frac{1}{\sqrt{2}} \overline{\Psi} \left(\gamma^\mu \frac{1 - \gamma^5}{2} \tau_\pm \right) \Psi \tag{26.3}$$

where Ψ_L represents one of the following $SU(2)_L$ iso-doublets of the weak eigenstates:

$$\Psi_L \equiv \begin{pmatrix} \nu_e \\ e \end{pmatrix}_L, \begin{pmatrix} \nu_\mu \\ \mu \end{pmatrix}_L, \begin{pmatrix} \nu_\tau \\ \tau \end{pmatrix}_L, \begin{pmatrix} u \\ d' \end{pmatrix}_L, \begin{pmatrix} c \\ s' \end{pmatrix}_L, \begin{pmatrix} t \\ b' \end{pmatrix}_L \tag{26.4}$$

Because of the CKM mixing (see Section 23.14), the weak charged current involving quarks is effectively given by (see Eq. (23.233)):

$$j_\mu^\pm = \frac{1}{\sqrt{2}} \overline{q}_1 V_{q_1, q_2} \left(\gamma^\mu \frac{1 - \gamma^5}{2} \tau_\pm \right) q_2 \tag{26.5}$$

where q_1 and q_2 represent the up-type and down-type quark fields and V_{q_1, q_2} is the corresponding CKM matrix element. The Feynman diagram vertex rules **[GSW2]** and **[GSW3]** are shown in Table 26.1.

---

1   Reprinted with permission from M. J. G. Veltman, "Nobel lecture: From weak interactions to gravitation," *Rev. Mod. Phys.*, vol. 72, pp. 341–349, Apr 2000. Copyright 2000 by the American Physical Society.

Type	Example	Vertex factor
• **[GSW1]** Photon vertex:		$-ie\lvert Q_f\rvert\gamma^\mu$
• **[GSW2]** Charged weak current (leptons):		$-i\dfrac{g_W}{\sqrt{2}}\gamma^\mu\left(\dfrac{1-\gamma^5}{2}\right)$
• **[GSW3]** Charged weak current (quarks):		$-i\dfrac{g_W}{\sqrt{2}}V_{q_1,q_2}\gamma^\mu\left(\dfrac{1-\gamma^5}{2}\right)$
• **[GSW4]** Neutral weak current:		$-i\dfrac{g_Z}{2}\gamma^\mu\left(c_V^f - c_A^f\gamma^5\right)$ $= -ig_Z\gamma^\mu\left(c_L^f\dfrac{(1-\gamma^5)}{2} + c_R^f\dfrac{(1+\gamma^5)}{2}\right)$

$$Q_e = -1,\ Q_{u,c,t} = 2/3,\ Q_{d,s,b} = -1/3,\ g_W = \frac{e}{\sin\theta_W},\ g_Z = \frac{g_W}{\cos\theta_W} = \frac{e}{\sin\theta_W\cos\theta_W}$$

$$c_V^f = I_{3L}^f - 2Q^f\sin^2\theta_W,\quad c_A^f = I_{3L}^f;\qquad c_L^f = I_{3L}^f - Q^f\sin^2\theta_W,\quad c_R^f = -Q^f\sin^2\theta_W$$

Table 26.1 Summary of Feynman rules for vertices in the electroweak theory.

The weak neutral current is the most complicated. Starting from Eq. (25.47) valid for the electron and combining it with Eq. (25.53), the coupling of any fermion pair $f\bar{f}$ to the Z^0 can be expressed as a function of its left-handed and right-handed couplings as:

$$\frac{e}{\sin\theta_W \cos\theta_W}\left(\left(I_{3L}^f - Q^f \sin^2\theta_W\right)\overline{f_L}\gamma_\mu f_L - Q^f \sin^2\theta_W \overline{f_R}\gamma_\mu f_R\right)Z^\mu$$

$$= \frac{e}{\sin\theta_W \cos\theta_W}\overline{\Psi}_f\left(\left(I_{3L}^f - Q^f \sin^2\theta_W\right)\gamma_\mu P_L - Q^f \sin^2\theta_W \gamma_\mu P_R\right)\Psi_f\, Z^\mu \tag{26.6}$$

where I_{3L}^f corresponds to the third component of the weak isospin of the left-handed projection of the fermion (by definition $I_{3R}^f = 0$). In analogy to the photon vertex of QED, Eq. (26.1), we can write:

$$-ig_Z(j_\mu^Z)Z^\mu \quad \text{where} \quad g_Z \equiv \frac{g_W}{\cos\theta_W} = \frac{e}{\sin\theta_W \cos\theta_W} \tag{26.7}$$

and the current contains both left-handed and right-handed components:

$$j_\mu^Z \equiv \overline{\Psi}_f \gamma_\mu \left(c_L^f P_L + c_R^f P_R\right)\Psi_f \tag{26.8}$$

with the left-handed and right-handed neutral weak couplings c_L^f and c_R^f given by:

$$c_L^f \equiv I_{3L}^f - Q^f \sin^2\theta_W \quad \text{and} \quad c_R^f \equiv -Q^f \sin^2\theta_W \tag{26.9}$$

Here **we have made explicit that the Z^0 couples to both left-handed and right-handed chiral components, but with different strengths given by the fermion-dependent c_L^f and c_R^f factors.** When we insert the chiral projection operators we can express the weak neutral current as a function of its vector (V) and axial-vector (A) couplings:

$$\begin{aligned} j_\mu^Z &= \overline{\Psi}_f \gamma_\mu \left(c_L^f \frac{(1-\gamma^5)}{2} + c_R^f \frac{(1+\gamma^5)}{2}\right)\Psi_f = \overline{\Psi}_f \gamma_\mu \frac{1}{2}\left((c_L^f + c_R^f) - (c_L^f - c_R^f)\gamma^5\right)\Psi_f \\ &\equiv \overline{\Psi}_f \gamma_\mu \frac{1}{2}\left(c_V^f - c_A^f \gamma^5\right)\Psi_f \end{aligned} \tag{26.10}$$

where the **vector and axial-vector neutral weak couplings c_V^f and c_A^f** are given by:

$$c_V^f = c_L^f + c_R^f = I_{3L}^f - 2Q^f \sin^2\theta_W \quad \text{and} \quad c_A^f = c_L^f - c_R^f = I_{3L}^f \tag{26.11}$$

from which we obtain $c_V^f + c_A^f = 2c_L^f$ and $c_V^f - c_A^f = 2c_R^f$, hence the **left-handed and right-handed couplings c_L^f and c_R^f** are given by:

$$c_L^f = \frac{1}{2}\left(c_V^f + c_A^f\right) \quad \text{and} \quad c_R^f = \frac{1}{2}\left(c_V^f - c_A^f\right) \tag{26.12}$$

Once the value of the Weinberg mixing angle is known, the couplings of the Z^0 boson to the leptons and quarks are predicted exactly. They are summarized in Table 26.2 for an assumed value $\sin^2\theta_W = 0.23$, a value consistent with the data.

The Feynman rule associated with the Z^0 to fermion vertex can be written in different ways:

$$-i\frac{g_Z}{2}\gamma^\mu\left(c_V^f - c_A^f\gamma^5\right) = -ig_Z\gamma^\mu\left(c_L^f \frac{(1-\gamma^5)}{2} + c_R^f \frac{(1+\gamma^5)}{2}\right) \tag{26.13}$$

where

$$g_Z = \frac{g_W}{\cos\theta_W} = \frac{e}{\sin\theta_W \cos\theta_W} \tag{26.14}$$

	Q_f	I_{3L}^f	$c_L^f = I_{3L}^f - Q_f s_W^2$	$c_R^f = -Q_f s_W^2$	$c_V^f = I_{3L}^f - 2Q_f s_W^2$	$c_A^f = I_{3L}$
ν_e, ν_μ, ν_τ	0	$+\frac{1}{2}$	$+\frac{1}{2}$	0	$+\frac{1}{2}$	$+\frac{1}{2}$
e^-, μ^-, τ^-	-1	$-\frac{1}{2}$	$-\frac{1}{2} + s_W^2 \simeq -0.27$	$s_W^2 \simeq +0.23$	$-\frac{1}{2} + 2s_W^2 \simeq -0.04$	$-\frac{1}{2}$
u, c, t	$+\frac{2}{3}$	$+\frac{1}{2}$	$+\frac{1}{2} - \frac{2}{3}s_W^2 \simeq +0.35$	$-\frac{2}{3}s_W^2 \simeq -0.15$	$+\frac{1}{2} - \frac{4}{3}s_W^2 \simeq +0.19$	$+\frac{1}{2}$
d', s', b'	$-\frac{1}{3}$	$-\frac{1}{2}$	$-\frac{1}{2} + \frac{1}{3}s_W^2 \simeq -0.42$	$\frac{1}{3}s_W^2 \simeq 0.08$	$-\frac{1}{2} + \frac{2}{3}s_W^2 \simeq -0.35$	$-\frac{1}{2}$

Table 26.2 Quantum number assignments and couplings of the fermions to the Z^0 and numerical values under the assumption $\sin^2 \theta_W = 0.23$.

The corresponding Feynman diagram vertex rule **[GSW4]** is shown in Table 26.1.

Because the weak neutral current contains both vector (V) and axial-vector (A) couplings, it will violate parity. This is a consequence of the current having contributions from the purely weak W_μ^3 interaction which couples only to the left-handed projection, and from the hypercharge interaction with B_μ, which treats left-handed and right-handed projections equally.

26.2   Trilinear and Quadrilinear Gauge Boson Vertices

The Feynman rules for the trilinear and quadrilinear gauge boson couplings can be derived directly from the Lagrangian, expressing the gauge boson fields as a function of the physical fields (see Section 25.7). This leads to the following vertices:

$$W^+W^-\gamma, W^+W^-Z^0, \gamma\gamma W^+W^-, \gamma Z^0 W^+W^-, Z^0 Z^0 W^+W^-, W^+W^-W^+W^- \tag{26.15}$$

They are summarized in Table 26.3.

26.3   Neutrino Scattering Off Electrons

We consider the scattering of neutrinos off target electrons. The following reactions involving neutrinos or antineutrinos are possible:

$$\nu_e + e^- \rightarrow \nu_e + e^- \qquad\qquad \overline{\nu}_e + e^- \rightarrow \overline{\nu}_e + e^- \tag{26.16}$$
$$\nu_x + e^- \rightarrow \nu_x + e^- \qquad\qquad \overline{\nu}_x + e^- \rightarrow \overline{\nu}_x + e^- \tag{26.17}$$

where $x = \mu, \tau$. We have written separately the reactions involving electron neutrinos and electron antineutrinos from those involving muon or tau neutrinos or antineutrinos. As can be seen from the tree-level Feynman diagrams illustrated in Figure 26.1, the reactions involving muon or tau flavors can only occur via neutral current (NC). In case of the electronic flavor, both charged-current (CC) and neutral-current (NC) interactions are possible. We first treat the NC interaction mediated by the Z^0 boson. The kinematical variables are defined as follows:

$$\nu_x(k^\mu) + e^-(p^\mu) \rightarrow \nu_x(k'^\mu) + e^-(p'^\mu), \tag{26.18}$$

Type	Diagram	Vertex factor
• **[GSW5]** Trilinear $WW\gamma$:		$= -ie \left[g_{\nu\rho}(p_1 - p_2)_\mu \right.$ $+ g_{\rho\mu}(p_2 - p_3)_\nu$ $\left. + g_{\mu\nu}(p_3 - p_1)_\rho \right]$
• **[GSW6]** Trilinear WWZ:		$= -ie \cot\theta_W \left[g_{\nu\rho}(p_1 - p_2)_\mu \right.$ $+ g_{\rho\mu}(p_2 - p_3)_\nu$ $\left. + g_{\mu\nu}(p_3 - p_1)_\rho \right]$
• **[GSW7]** Quadrilinear $\gamma\gamma WW$:		$= -ie^2 (2g_{\alpha\beta}g_{\mu\nu} - g_{\alpha\mu}g_{\beta\nu}$ $- g_{\alpha\nu}g_{\beta\mu})$
• **[GSW8]** Quadrilinear γZWW:		$= -ie^2 \cot\theta_W (2g_{\alpha\beta}g_{\mu\nu} - g_{\alpha\mu}g_{\beta\nu}$ $- g_{\alpha\nu}g_{\beta\mu})$
• **[GSW9]** Quadrilinear $ZZWW$:		$= -ie^2 \cot\theta_W (2g_{\alpha\beta}g_{\mu\nu} - g_{\alpha\mu}g_{\beta\nu}$ $- g_{\alpha\nu}g_{\beta\mu})$
• **[GSW10]** Quadrilinear $WWWW$:		$= i\dfrac{e^2}{\sin\theta_W}(2g_{\mu\alpha}g_{\nu\beta} - g_{\mu\beta}g_{\alpha\nu}$ $- g_{\mu\nu}g_{\alpha\beta})$

Table 26.3 Summary of Feynman rules for trilinear and quadrilinear boson vertices in the electroweak theory.

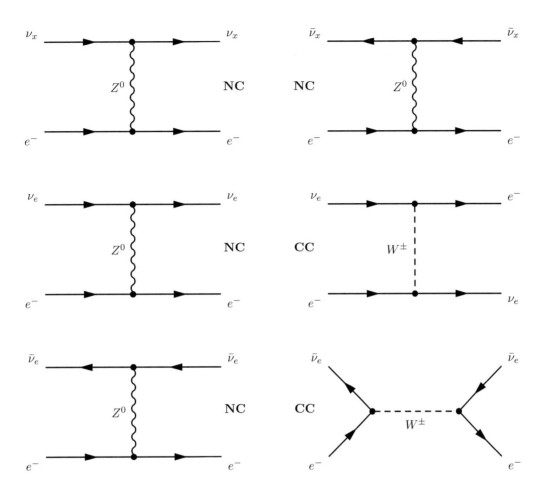

Figure 26.1 Tree-level Feynman diagrams for neutrino scattering off target electrons.

where $x = e, \mu, \tau$. The invariant amplitude is:

$$
\begin{aligned}
i\mathcal{M}^{\text{NC}} &= \left\{ -i\frac{g_Z}{2}\overline{\nu}(k')\gamma_\mu \left(\frac{1-\gamma^5}{2}\right)\nu(k) \right\} G_Z^{\mu\nu} \left\{ -i\frac{g_Z}{2}\overline{e}(p')\gamma_\nu \left(c_V^e - c_A^e\gamma^5\right)e(p) \right\} \\
&= \frac{ig_Z^2}{8M_Z^2} \left\{ \overline{\nu}(k')\gamma_\mu \left(1-\gamma^5\right)\nu(k) \right\} \left\{ \overline{e}(p')\gamma^\mu \left(c_V^e - c_A^e\gamma^5\right)e(p) \right\}
\end{aligned}
\tag{26.19}
$$

where $G_Z^{\mu\nu}$ is the Z^0 propagator (similar to Eq. (23.22) for the $W^\pm$ boson propagator). We note that $M_W = M_Z \cos\theta_W$ (Eq. (25.86)) yields:

$$
\frac{g_Z^2}{8M_Z^2} = \frac{g_Z^2 \cos^2\theta_W}{8M_W^2} = \frac{g_W^2}{8M_W^2}\frac{g_Z^2 \cos^2\theta_W}{g_W^2} \equiv \frac{G_F}{\sqrt{2}}\rho
\tag{26.20}
$$

where we used Eq. (23.25) and the (historical) parameter ρ takes into account the ratio of the charged and neutral couplings:

$$
\rho = \frac{g_Z^2 \cos^2\theta_W}{g_W^2} = \frac{g_Z^2}{M_Z^2}\bigg/\frac{g_W^2}{M_W^2}
\tag{26.21}
$$

with $\rho \equiv 1$ in the electroweak theory when assuming Eq. (26.7). The invariant amplitude, neglecting the structure of the boson propagator and using Fermi's approximation is then:

$$\mathcal{M}^{\mathrm{NC}} = \frac{\rho G_F}{\sqrt{2}} \left\{ \bar{\nu} \gamma^\mu \left(1 - \gamma^5\right) \nu \right\} \left\{ \bar{e} \gamma_\mu \left(c_V^e - c_A^e \gamma^5\right) e \right\} \tag{26.22}$$

We average over the electron spin in the initial state and sum over the final-state electron spin. Neglecting the electron rest mass $(m_e = 0)$, we obtain using standard trace techniques:

$$\langle |\mathcal{M}^{\mathrm{NC}}|^2 \rangle = \frac{1}{2} \sum_{\mathrm{spin}} |\mathcal{M}^{\mathrm{NC}}|^2 = \frac{\rho^2 G_F^2}{4} \mathrm{Tr} \left\{ \gamma^\mu \left(1 - \gamma^5\right) \not{k} \gamma^\nu \left(1 - \gamma^5\right) \not{k}' \right\}$$
$$\mathrm{Tr} \left\{ \gamma_\mu \left(c_V^e - c_A^e \gamma^5\right) \not{p} \gamma_\nu \left(c_V^e - c_A^e \gamma^5\right) \not{p}' \right\} \tag{26.23}$$

and finally:

$$\langle |\mathcal{M}^{\mathrm{NC}}|^2 \rangle = \frac{\rho^2 G_F^2}{4} \cdot 64 \left[\left(c_V^e + c_A^e\right)^2 (p \cdot k)(p' \cdot k') + \left(c_V^e - c_A^e\right)^2 (p \cdot k')(p' \cdot k) \right] \tag{26.24}$$

The channel corresponding to the NC is a t-channel exchange. We introduce the Mandelstam variables in order to simplify the expressions

$$s = (p + k)^2 = (p' + k')^2 \approx 2(p \cdot k) \approx 2(p' \cdot k') \quad \text{and} \quad u = (p - k')^2 = (p' - k)^2 \approx -2(p \cdot k') \approx -2(p' \cdot k)$$

We obtain:

$$\langle |\mathcal{M}^{\mathrm{NC}}|^2 \rangle = 16 \rho^2 G_F^2 \left[\left(c_V^e + c_A^e\right)^2 \frac{s^2}{4} + \left(c_V^e - c_A^e\right)^2 \frac{u^2}{4} \right] \tag{26.25}$$

In the center-of-mass reference system where $p = (E, \vec{p})$, $k = (E, -\vec{p})$, and $k' = (E, \vec{k}')$, we find that the differential cross-section is given by

$$\left(\frac{\mathrm{d}\sigma^{NC}(\nu_x e^-)}{\mathrm{d}\Omega} \right)_{\mathrm{CMS}} = \frac{1}{64\pi^2 s} \langle |\mathcal{M}^{\mathrm{NC}}|^2 \rangle = \frac{\rho^2 G_F^2}{4\pi^2 s} \frac{s^2}{4} \left[\left(c_V^e + c_A^e\right)^2 + \left(c_V^e - c_A^e\right)^2 \frac{u^2}{s^2} \right] \tag{26.26}$$

We define the dimensionless inelasticity variable y, as follows:

$$1 - y \equiv \frac{p \cdot k'}{p \cdot k} \approx -\frac{u}{s} = \frac{E^2 - \vec{p} \cdot \vec{k}'}{E^2 + \vec{p}^2} \approx \frac{E^2(1 + \cos\theta)}{2E^2} = \frac{1}{2}(1 + \cos\theta) \tag{26.27}$$

where θ is the scattering angle of the electron in the center-of-mass system. Then, we can use alternatively y or θ and the range of y is just given by the angular boundaries. We note that we have $y = 0$ for $\cos\theta = 1$ and $y = 1$ for $\cos\theta = -1$. Hence:

$$\mathrm{d}\Omega = 2\pi \left(\mathrm{d}\cos\theta\right) = 4\pi \mathrm{d}y \quad \text{and} \quad 0 \le y \le 1 \Leftrightarrow -1 \le \cos\vartheta \le 1 \tag{26.28}$$

We then find the differential cross-section for the neutral current process as a function of y:

$$\left(\frac{\mathrm{d}\sigma^{\mathrm{NC}}(\nu_x e^-)}{\mathrm{d}y} \right)_{\mathrm{CMS}} = \frac{\rho^2 G_F^2 s}{4\pi} \left[\left(c_V^e + c_A^e\right)^2 + \left(c_V^e - c_A^e\right)^2 (1 - y)^2 \right] \tag{26.29}$$

where $x = e, \mu, \tau$. In a similar way, we can now treat the CC interaction with the exchange of a $W^\pm$ boson. In this case, only electron neutrinos will participate on an electron target. The kinematical variables are defined as follows:

$$\nu_e(k^\mu) + e^-(p^\mu) \rightarrow \nu_e(k'^\mu) + e^-(p'^\mu)$$

The invariant amplitude, neglecting the structure of the boson propagator and using Fermi's approximation, of the diagram is:

$$\mathcal{M}^{\mathrm{CC}} = \frac{G_F}{\sqrt{2}} \left(\bar{u}(k')\gamma^\mu \left(1 - \gamma^5\right) u(p)\right) \left(\bar{u}(p')\gamma_\mu \left(1 - \gamma^5\right) u(k)\right) \tag{26.30}$$

We average over the electron spin in the initial state and sum over the final-state spin. Neglecting the electron rest mass ($m_e = 0$), we obtain using standard trace techniques:

$$\langle|\mathcal{M}^{\mathrm{CC}}|^2\rangle = \frac{1}{2}\sum_{\mathrm{spin}}|\mathcal{M}^{\mathrm{CC}}|^2 = \frac{G_F^2}{4}\,\mathrm{Tr}\left\{\gamma^\mu \left(1-\gamma^5\right)\not{p}\gamma^\nu \left(1-\gamma^5\right)\not{k}'\right\}\mathrm{Tr}\left\{\gamma_\mu \left(1-\gamma^5\right)\not{k}\gamma_\nu \left(1-\gamma^5\right)\not{p}'\right\} \tag{26.31}$$

Using the trace theorem we directly get:

$$\langle|\mathcal{M}^{\mathrm{CC}}|^2\rangle = \frac{G_F^2}{4}\cdot 256\,(p\cdot k)\,(p'\cdot k') = 64\,G_F^2\,(p\cdot k)\,(p'\cdot k') \tag{26.32}$$

so the matrix element simply becomes directly proportional to s^2:

$$\langle|\mathcal{M}^{\mathrm{CC}}|^2\rangle \approx 64\,G_F^2\left(\frac{s}{2}\right)\left(\frac{s}{2}\right) = 16\,G_F^2 s^2 \tag{26.33}$$

In this case, recalling $\mathrm{d}\Omega = 4\pi\mathrm{d}y$, we obtain:

$$\boxed{\left(\frac{\mathrm{d}\sigma^{\mathrm{CC}}\left(\nu_e e^-\right)}{\mathrm{d}y}\right)_{\mathrm{CMS}} = \frac{4\pi}{64\pi^2 s}\left\langle|\mathcal{M}^{\mathrm{CC}}|^2\right\rangle = \frac{G_F^2 s}{\pi}} \tag{26.34}$$

The differential cross-section is in this case isotropic in the center-of-mass system and is proportional to s. Since $s \approx 2m_e E_\nu^{\mathrm{Lab}}$, we note that the cross-section increases linearly with the incoming neutrino energy in fixed target mode:

$$\sigma \propto G_F^2 m_e E_\nu^{\mathrm{Lab}} \tag{26.35}$$

The isotropic angular dependence can be explained with the arguments of conservation of angular momentum and the $V - A$ structure. In the ultra-relativistic case, the helicity of the particles is only negative. Hence the incoming electron and the neutrino have opposite spins in the center-of-mass system and the total angular momentum will therefore vanish, as illustrated in Figure 26.2. Since the process has a total $J = 0$, the differential cross-section is expected to be isotropic.

For the scattering of the electron neutrino, we should add the charged current to the neutral current contribution:

$$\mathcal{M} = \mathcal{M}^{\mathrm{NC}} - \mathcal{M}^{\mathrm{CC}} \tag{26.36}$$

where the minus sign comes from the exchange of the outgoing fermions. Explicitly we have:

$$\mathcal{M}^{\mathrm{NC}} = \frac{\rho G_F}{\sqrt{2}}\left(\bar{\nu}_e\gamma^\mu \left(1-\gamma^5\right)\nu_e\right)\left(\bar{e}\gamma_\mu \left(c_V^e - c_A^e\gamma^5\right)e\right)$$

$$\mathcal{M}^{\mathrm{CC}} = \frac{G_F}{\sqrt{2}}\left(\bar{e}\gamma^\mu \left(1-\gamma^5\right)\nu_e\right)\left(\bar{\nu}_e\gamma_\mu \left(1-\gamma^5\right)e\right)$$

We can use a **Fierz transformation** (see Section 23.4) to rearrange the terms in the amplitudes so that the CC and NC amplitudes can be directly written as (see Eq. (23.48)):

$$\mathcal{M}^{\mathrm{CC}} = -\frac{G_F}{\sqrt{2}}\left(\bar{\nu}_e\gamma^\mu \left(1-\gamma^5\right)\nu_e\right)\left(\bar{e}\gamma_\mu \left(1-\gamma^5\right)e\right) \tag{26.37}$$

Setting $\rho = 1$ the total amplitude becomes:

$$\mathcal{M}^{\mathrm{CC+NC}} = \frac{G_F}{\sqrt{2}}\left(\bar{\nu}_e\gamma^\mu \left(1-\gamma^5\right)\nu_e\right)\left(\bar{e}\gamma_\mu \left(\left(c_V^e + 1\right) - \left(c_A^e + 1\right)\gamma^5\right)e\right) \tag{26.38}$$

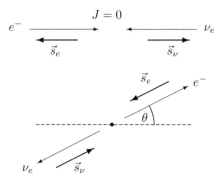

Figure 26.2 Spin configuration in $\nu - e$ scattering.

We note that **this result is exactly identical to the NC term under the simple replacement**:

$$c_V^e \to c_V^e + 1 \quad \text{and} \quad c_A^e \to c_A^e + 1 \tag{26.39}$$

We can therefore use the differential cross-section obtained for the NC and just replace the coupling constants to find the total differential cross-section with both CC, NC, and interference contributions:

$$\left(\frac{\mathrm{d}\sigma^{\mathrm{NC+CC}}\left(\nu_e e^-\right)}{\mathrm{d}y} \right) = \frac{G_F^2 s}{4\pi} \left[\left(c_V^e + c_A^e + 2\right)^2 + \left(c_V^e - c_A^e\right)^2 \left(1 - y\right)^2 \right] \tag{26.40}$$

We now summarize all the results obtained so far and extend them to the antineutrino case. The charged current electron-neutrino differential cross-section in the center-of-mass system Eq. (26.34) can be written as:

$$\left(\frac{\mathrm{d}\sigma^{\mathrm{CC}}\left(\nu_e e^-\right)}{\mathrm{d}y} \right) = \frac{G_F^2 s}{\pi} = \frac{G_F^2}{\pi s} \cdot s^2 \tag{26.41}$$

where we have made explicit that the matrix element $|\mathcal{M}|^2$ is proportional to s^2, while the phase-space factor introduces the s^{-1} factor. We can now apply crossing symmetry to consider the electron-antineutrino case, by replacing the incoming (outgoing) neutrino by an outgoing (incoming) antineutrino. Therefore, the differential cross-section for the antineutrino scattering off electrons is given by the replacement $s^2 \leftrightarrow u^2$. Consequently:

$$\left(\frac{\mathrm{d}\sigma^{\mathrm{CC}}\left(\overline{\nu}_e e^-\right)}{\mathrm{d}y} \right) = \frac{G_F^2}{\pi s} u^2 \approx \frac{G_F^2 s^2}{\pi s}(1 - y)^2 = \frac{G_F^2 s}{\pi}(1 - y)^2 \tag{26.42}$$

The corresponding behaviors of the differential cross-sections for neutrinos and antineutrinos are sketched in Figure 26.3.

The corresponding total CC cross-sections can be computed by direct integration over y:

$$\sigma^{\mathrm{CC}}(\nu_e e^-) = \int_0^1 \frac{\mathrm{d}\sigma}{\mathrm{d}y}\,\mathrm{d}y = \frac{G_F^2 s}{\pi} \tag{26.43}$$

$$\sigma^{\mathrm{CC}}(\overline{\nu}_e e^-) = \frac{G_F^2 s}{\pi} \int_0^1 (1 - y)^2\,\mathrm{d}y = \frac{G_F^2 s}{3\pi} \tag{26.44}$$

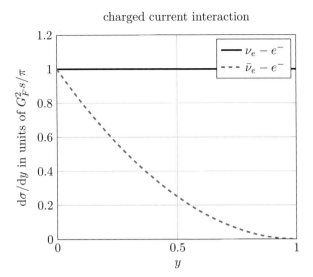

Figure 26.3 Differential cross-sections for the $\nu_e e^-$ and $\overline{\nu}_e e^-$ charged-current scattering.

The differential cross-section for the NC process of any neutrino is given by Eq. (26.29). For antineutrinos, we apply the crossing symmetry by swapping the factors proportional to 1 and $(1-y)^2$ and get:

$$\left(\frac{d\sigma^{\text{NC}}\left(\overline{\nu}_x e^-\right)}{dy}\right) = \frac{\rho^2 G_F^2 s}{4\pi}\left[(c_V^e - c_A^e)^2 + (c_V^e + c_A^e)^2 (1-y)^2\right] \tag{26.45}$$

where $x = e, \mu, \tau$. We can rewrite these results using the left-handed and right-handed couplings instead of the vector and axial ones. By insertion of Eq. (26.11) into the NC differential cross-sections for neutrinos (see Eq. (26.29)), we find:

$$\begin{aligned}\left(\frac{d\sigma^{\text{NC}}\left(\nu_x e^-\right)}{dy}\right) &= \frac{\rho^2 G_F^2 s}{4\pi}\left[(c_L^e + c_R^e + c_L^e - c_R^e)^2 + (c_L^e + c_R^e - c_L^e + c_R^e)^2 (1-y)^2\right] \\ &= \frac{\rho^2 G_F^2 s}{4\pi}\left[(2c_L^e)^2 + (2c_R^e)^2 (1-y)^2\right]\end{aligned} \tag{26.46}$$

And similarly for antineutrinos. So, finally:

$$\begin{aligned}\left(\frac{d\sigma^{\text{NC}}\left(\nu_x e^-\right)}{dy}\right) &= \sigma^0\left[(c_L^e)^2 + (c_R^e)^2 (1-y)^2\right] \\ \left(\frac{d\sigma^{\text{NC}}\left(\overline{\nu}_x e^-\right)}{dy}\right) &= \sigma^0\left[(c_R^e)^2 + (c_L^e)^2 (1-y)^2\right]\end{aligned} \tag{26.47}$$

where the overall normalization factor is (compare with Eq. (23.168)):

$$\sigma^0 \equiv \frac{\rho^2 G_F^2 s}{\pi} \simeq \frac{2\rho^2 G_F^2 m_e E_\nu}{\pi} \approx 1.720 \times 10^{-41} \rho^2 \left(\frac{E_\nu}{\text{GeV}}\right)\, \text{cm}^2 \tag{26.48}$$

having made use of the laboratory frame where $s \approx 2m_e E_\nu$. The dependence on $(1-y)^2$ can be understood in terms of angular momentum conservation, as illustrated in Figure 26.4. For the case involving antineutrinos, the terms proportional to 1 and $(1-y)^2$ must be swapped.

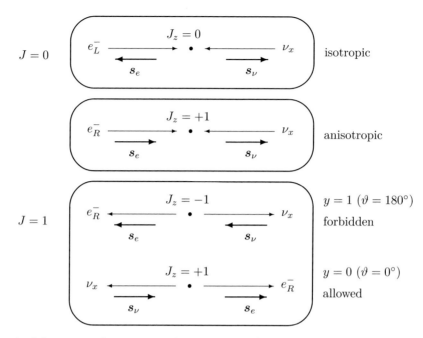

Figure 26.4 Sketch of the spin configurations and consequence of angular momentum conservation in neutrino–electron scattering.

The total cross-sections for the electron-neutrino and electron-antineutrino scattering, including both charged and neutral-current contributions, can be directly obtained by the replacements $c_V^e \longrightarrow c_V^e + 1$ and $c_A^e \longrightarrow c_A^e + 1$, or:

$$c_L^e \longrightarrow c_L^e + 1 \quad \text{and} \quad c_R^e \longrightarrow c_R^e \tag{26.49}$$

to get:

$$
\begin{aligned}
\left(\frac{d\sigma^{\text{NC+CC}} \left(\nu_e e^- \right)}{dy} \right) &= \sigma^0 \left[\left(c_L^e + 1 \right)^2 + \left(c_R^e \right)^2 \left(1 - y \right)^2 \right] \\
\left(\frac{d\sigma^{\text{NC+CC}} \left(\overline{\nu}_e e^- \right)}{dy} \right) &= \sigma^0 \left[\left(c_R^e \right)^2 + \left(c_L^e + 1 \right)^2 \left(1 - y \right)^2 \right]
\end{aligned}
\tag{26.50}
$$

The total cross-sections are computed by integrating each one over y:

$$\sigma^{\text{NC+CC}} \left(\nu_e e^- \right) = \sigma^0 \left[\left(c_L^e + 1 \right)^2 + \frac{1}{3} \left(c_R^e \right)^2 \right], \qquad \sigma^{\text{NC+CC}} \left(\overline{\nu}_e e^- \right) = \sigma^0 \left[\left(c_R^e \right)^2 + \frac{1}{3} \left(c_L^e + 1 \right)^2 \right] \tag{26.51}$$

and

$$\sigma^{\text{NC}} \left(\nu_x e^- \right) = \sigma^0 \left[\left(c_L^e \right)^2 + \frac{1}{3} \left(c_R^e \right)^2 \right], \qquad \sigma^{\text{NC}} \left(\overline{\nu}_x e^- \right) = \sigma^0 \left[\left(c_R^e \right)^2 + \frac{1}{3} \left(c_L^e \right)^2 \right] \tag{26.52}$$

Hence, the measurements of at least two of these cross-sections give a direct determination of the weak couplings c_R and c_L, and thereby on $\sin^2 \theta_W$, as discussed in Section 27.2. These cross-sections are plotted in Figure 28.13.

26.4   The $W^\pm$ Boson Decay

The $W^\pm$ bosons are very short lived due to their large mass and are experimentally detected by their decay products. We consider the leptonic mode $W \to e\nu_e$. The amplitude of the decay process:

$$W^-(P^\mu) \to e^-(p) + \bar\nu_e(k) \tag{26.53}$$

is given by

$$i\mathcal{M} = -i\frac{g_W}{\sqrt{2}}\epsilon_\mu^\lambda(P)\bar u(p)\gamma^\mu\frac{1}{2}(1-\gamma^5)v(k) \tag{26.54}$$

where $\epsilon_\mu^\lambda(P)$ is the polarization four-vector of the W boson. Averaging the matrix element squared $|\mathcal{M}|^2$ over the W polarizations using Eq. (10.153), and summing over the electron spin, we find in the case where the rest masses of the neutrino and electron are neglected:

$$
\begin{aligned}
\frac{1}{3}\langle|\mathcal{M}|^2\rangle = \frac{1}{3}\sum_{\lambda=1}^{3}\sum_{spins}|\mathcal{M}|^2 &= \frac{g_W^2}{6}\left(-g^{\mu\nu}+\frac{P^\mu P^\nu}{M_W^2}\right)\mathrm{Tr}\left(\not{p}\gamma_\mu\not{k}\gamma_\nu\left(\frac{1-\gamma^5}{2}\right)\right)\\
&= \frac{g_W^2}{3}\left(-g^{\mu\nu}+\frac{P^\mu P^\nu}{M_W^2}\right)(p_\mu k_\nu + k_\nu p_\mu - g_{\mu\nu}(k\cdot p))\\
&= \frac{g_W^2}{3}\left(p\cdot k + 2\frac{(P\cdot p)(P\cdot k)}{M_W^2}\right)
\end{aligned} \tag{26.55}
$$

Momentum conservation implies $P^\mu = p^\mu + k^\mu$, hence:

$$P^2 \equiv M_W^2 = (p+k)^2 = p^2 + k^2 + 2p\cdot k \implies p\cdot k \simeq \frac{M_W^2}{2} \tag{26.56}$$

therefore

$$
\begin{aligned}
\frac{1}{3}\langle|\mathcal{M}|^2\rangle &\simeq \frac{g_W^2}{3}\left(\frac{M_W^2}{2}+2\frac{((p+k)\cdot p)((p+k)\cdot k)}{M_W^2}\right) = \frac{g_W^2}{3}\left(\frac{M_W^2}{2}+\frac{2}{M_W^2}\left((p^2+k\cdot p)(p\cdot k+k^2)\right)\right)\\
&= \frac{g_W^2}{3}\left(\frac{M_W^2}{2}+\frac{2}{M_W^2}\left((p^2(p\cdot k)+(k\cdot p)(p\cdot k)+k^2(k\cdot p)+p^2 k^2)\right)\right)\\
&\simeq \frac{g_W^2}{3}\left(\frac{M_W^2}{2}+\frac{2}{M_W^2}\left(\frac{M_W^2}{2}\frac{M_W^2}{2}\right)\right) = \frac{1}{3}g_W^2 M_W^2
\end{aligned} \tag{26.57}
$$

Consequently, the unpolarized differential $W \to e\nu_e$ decay rate in the W boson rest frame is given by:

$$
\begin{aligned}
\mathrm{d}\Gamma(W^- \to e^-\bar\nu_e) &= \frac{1}{2M_W}\left(\frac{1}{3}g_W^2 M_W^2\right)\frac{\mathrm{d}^3\vec p}{(2\pi)^3 2E_e}\frac{\mathrm{d}^3\vec k}{(2\pi)^3 2E_{\bar\nu_e}}(2\pi)^4\delta^4(P-p-k)\\
&= \frac{g_W^2 M_W}{6(2\pi)^2}\frac{\mathrm{d}^3\vec p}{2E_e}\frac{\mathrm{d}^3\vec k}{2E_{\bar\nu_e}}\delta^4(P-p-k)
\end{aligned} \tag{26.58}
$$

The phase-space integration simply gives:

$$\int \frac{\mathrm{d}^3\vec p}{2E_e}\frac{\mathrm{d}^3\vec k}{2E_{\bar\nu_e}}\delta^4(P-p-k) = \frac{1}{2}\int\frac{p^2\mathrm{d}p\mathrm{d}\Omega}{4E^2}\delta(E-M_W/2) = \frac{1}{2}\frac{1}{4}(4\pi) = \frac{1}{2}\pi \tag{26.59}$$

where neglecting masses we have $E \equiv |\vec p| = E_e = E_{\bar\nu_e} = M_W/2$. Hence integrating over the phase space gives the total width:

$$\Gamma_W^e \equiv \Gamma(W^- \to e^-\bar\nu_e) = \frac{g_W^2 M_W}{24\pi^2}\frac{1}{2}\pi = \frac{g_W^2 M_W}{48\pi} \tag{26.60}$$

When we neglect the mass of the leptons, this result can readily be applied to the other charged leptons to give:

$$\Gamma_W^\ell \equiv \Gamma(W^- \to e^- \bar{\nu}_e) = \Gamma(W^- \to \mu^- \bar{\nu}_\mu) = \Gamma(W^- \to \tau^- \bar{\nu}_\tau) = \frac{g_W^2 M_W}{48\pi} \tag{26.61}$$

For the quarks, we compute the decay into a quark–antiquark pair $q'\bar{q}$ coupling an up-type and a down-type (anti)quark, and consider that the latter will always fragment into hadrons. The partial width into such quark–antiquark pairs can be translated from the width Γ_W^ℓ by taking into account the color factor for quarks and the Cabibbo–Kobayashi–Maskawa matrix element $V_{qq'}$ to give (again neglecting quark masses):

$$\Gamma_W^{qq'} = \Gamma(W^- \to q'\bar{q}) = N_C |V_{qq'}|^2 \Gamma_W^\ell \tag{26.62}$$

where N_C is equal to 3. We note that an "on-shell" real W boson cannot decay into the top quark since $m_t > m_W$, consequently we have the possible hadronic decay modes:

$$\begin{aligned}
\Gamma(W^- \to d\bar{u}) &= 3|V_{ud}|^2 \Gamma_W^\ell, & \Gamma(W^- \to d\bar{c}) &= 3|V_{cd}|^2 \Gamma_W^\ell, \\
\Gamma(W^- \to s\bar{u}) &= 3|V_{us}|^2 \Gamma_W^\ell, & \Gamma(W^- \to s\bar{c}) &- 3|V_{cs}|^2 \Gamma_W^\ell, \\
\Gamma(W^- \to b\bar{u}) &= 3|V_{ub}|^2 \Gamma_W^\ell, & \Gamma(W^- \to b\bar{c}) &= 3|V_{cb}|^2 \Gamma_W^\ell
\end{aligned} \tag{26.63}$$

hence, exploiting the unitarity of the Cabibbo–Kobayashi–Maskawa matrix, we can write:

$$\begin{aligned}
\Gamma(W^- \to d\bar{u}) + \Gamma(W^- \to s\bar{u}) + \Gamma(W^- \to b\bar{u}) &= 3\left(|V_{ud}|^2 + |V_{us}|^2 + |V_{ub}|^2\right)\Gamma_W^\ell = 3\Gamma_W^\ell \\
\Gamma(W^- \to d\bar{c}) + \Gamma(W^- \to s\bar{c}) + \Gamma(W^- \to b\bar{c}) &= 3\left(|V_{cd}|^2 + |V_{cs}|^2 + |V_{cb}|^2\right)\Gamma_W^\ell = 3\Gamma_W^\ell
\end{aligned} \tag{26.64}$$

Finally:

$$\Gamma(W^- \to \text{leptons}) = 3\Gamma_W^\ell, \quad \Gamma(W^- \to \text{hadrons}) = 6\Gamma_W^\ell \quad \text{and} \quad \Gamma(W \to \text{all}) = 9\Gamma_W^\ell \tag{26.65}$$

where the total width Γ_W is given by the sum of the decay to the leptons and hadrons. The branching ratios are then:

$$\begin{aligned}
Br(W^- \to e^- \bar{\nu}_e) = Br(W^- \to \mu^- \bar{\nu}_\mu) &= \frac{\Gamma_W^\ell}{\Gamma(W \to \text{all})} = \frac{1}{9} \\
Br(W^- \to \text{hadrons}) &= \frac{6}{9}
\end{aligned} \tag{26.66}$$

Since $g_W^2 = 8M_W^2 G_F/\sqrt{2}$, we can write

$$\Gamma_W^\ell = \frac{g_W^2 M_W}{48\pi} = \frac{G_F}{\sqrt{2}} \frac{M_W^3}{6\pi} \simeq 0.23 \text{ GeV} \tag{26.67}$$

where we used $M_W = 80.4$ GeV. The total width is approximately $\Gamma_W \simeq 2$ GeV, so the lifetime is only approximately 10^{-25} s. These results, computed at tree level, are very comparable to the experimentally measured values (values taken from the Particle Data Group's Review [5]):

$$\boxed{W^\pm}: \quad \begin{aligned}
M_W &= (80.385 \pm 0.015) \text{ GeV} \\
\Gamma_W &= (2.085 \pm 0.042) \text{ GeV} \\
Br(W^- \to e^- + \bar{\nu}_e) &= (10.71 \pm 0.16)\% \\
Br(W^- \to \mu^- + \bar{\nu}_\mu) &= (10.63 \pm 0.15)\% \\
Br(W^- \to \tau^- + \bar{\nu}_\tau) &= (11.38 \pm 0.21)\% \\
Br(W^- \to \text{hadrons}) &= (67.41 \pm 0.27)\%
\end{aligned} \tag{26.68}$$

26.5   The Z^0 Boson Decay

The calculation of the Z^0 boson width follows closely that of the $W^\pm$ boson. One main difference is that unlike the $W^\pm$ which couples universally to the left-handed chiral projections of the fermions, the Z^0 couples to both left-handed and right-handed projections with specific couplings which depend on the electric charge of the fermion. We can first consider the decay to a pair of leptons $Z^0 \to \ell^+\ell^-$, where $\ell = e, \mu$ or τ. The amplitude of the decay process:

$$Z^0(P^\mu) \to \ell^-(p) + \ell^+(k) \tag{26.69}$$

is

$$i\mathcal{M} = -i\frac{g_Z}{2}\epsilon_\mu^\lambda(P)\bar{u}(p)\gamma^\mu(c_V^\ell - c_A^\ell\gamma^5)v(k) \tag{26.70}$$

where $\epsilon_\mu^\lambda(P)$ is the polarization four-vector of the Z^0 boson. In the massless lepton approximation, a calculation similar to that of the W boson case gives:

$$\Gamma(Z^0 \to \ell^+\ell^-) = \frac{g_Z^2 M_Z}{48\pi}\left((c_V^\ell)^2 + (c_A^\ell)^2\right) \equiv \Gamma_Z^0\left((c_V^\ell)^2 + (c_A^\ell)^2\right) \tag{26.71}$$

where, using $g_Z^2 = 8M_Z^2 G_F/\sqrt{2}$ and $M_Z = 91.2$ GeV, we obtain:

$$\Gamma_Z^0 \equiv \frac{g_Z^2 M_Z}{48\pi} = \frac{G_F}{6\pi\sqrt{2}}M_Z^3 \simeq 0.33 \text{ GeV} \tag{26.72}$$

For the decay into a neutrino–antineutrino pair, we get the same result with the added simplification that:

$$\Gamma(Z^0 \to \nu_\ell\bar{\nu}_\ell) = \Gamma_Z^0\left((c_V^\nu)^2 + (c_A^\nu)^2\right) = \Gamma_Z^0\left(\frac{1}{4} + \frac{1}{4}\right) = \frac{1}{2}\Gamma_Z^0 \tag{26.73}$$

For the decay into a quark–antiquark pair, we need an additional factor to take into account the color:

$$\Gamma(Z^0 \to q\bar{q}) = 3\Gamma_Z^0\left((c_V^f)^2 + (c_A^f)^2\right) \tag{26.74}$$

where $f = u, d, s, c, \ldots$. In terms of the weak mixing angle, we have:

$$\begin{aligned}(c_V^f)^2 + (c_A^f)^2 &= (I_{3L}^f - 2Q^f\sin^2\theta_W)^2 + (I_{3L}^f)^2 = 2(I_{3L}^f)^2 - 4Q^f I_{3L}^f \sin^2\theta_W + 4(Q^f)^2\sin^4\theta_W \\ &= \frac{1}{2} - 4Q^f I_{3L}^f \sin^2\theta_W + 4(Q^f)^2\sin^4\theta_W\end{aligned} \tag{26.75}$$

We define (for the experimental value see Chapter 27):

$$x_W \equiv \sin^2\theta_W \simeq 0.23 \quad \to \quad \cos^2\theta_W = 1 - x_W \approx 0.77 \tag{26.76}$$

The partial widths at tree level are then:

$$\Gamma(Z^0 \to \nu_\ell\bar{\nu}_\ell) = \frac{1}{2}\Gamma_Z^0 \simeq 0.17 \text{ GeV}$$

$$\Gamma(Z^0 \to \ell^+\ell^-) = \Gamma_Z^0\left(\frac{1}{2} - 2x_W + 4x_W^2\right) \simeq 0.09 \text{ GeV}$$

$$\Gamma(Z^0 \to u\bar{u}) = \Gamma(Z^0 \to c\bar{c}) = 3\Gamma_Z^0\left(\frac{1}{2} - \frac{4}{3}x_W + \frac{16}{9}x_W^2\right) \simeq 0.30 \text{ GeV}$$

$$\Gamma(Z^0 \to d\bar{d}) = \Gamma(Z^0 \to s\bar{s}) = \Gamma(Z^0 \to b\bar{b}) = 3\Gamma_Z^0\left(\frac{1}{2} - \frac{2}{3}x_W + \frac{4}{9}x_W^2\right) \simeq 0.39 \text{ GeV} \tag{26.77}$$

and there is no decay into $t\bar{t}$ since $2m_t > m_Z$. The total width is found by adding all contributions:

$$
\begin{aligned}
\Gamma_Z &= 3\Gamma(Z^0 \to \nu_\ell \bar{\nu}_\ell) + 3\Gamma(Z^0 \to \ell^+ \ell^-) + 2\Gamma(Z^0 \to u\bar{u}) + 3\Gamma(Z^0 \to d\bar{d}) \\
&= \Gamma_Z^0 \left[\frac{3}{2} + 3\left(\frac{1}{2} - 2x_W + 4x_W^2\right) + 6\left(\frac{1}{2} - \frac{4}{3}x_W + \frac{16}{9}x_W^2\right) + 9\left(\frac{1}{2} - \frac{2}{3}x_W + \frac{4}{9}x_W^2\right) \right] \\
&= \Gamma_Z^0 \left[\frac{21}{2} - 20x_W + \frac{80}{3}x_W^2 \right] \simeq 7.31\,\Gamma_Z^0 \approx 2.4 \text{ GeV}
\end{aligned}
\tag{26.78}
$$

These results, computed at tree level, are very comparable to the experimentally measured values (values taken from the Particle Data Group's Review [5]):

$$
\boxed{Z^0}: \quad
\begin{aligned}
&M_{Z^0} = 91.1876 \pm 0.0021 \text{ GeV} \\
&\Gamma_Z = (2.4952 \pm 0.0023) \text{ GeV} \\
&Br(Z^0 \to e^+ + e^-) = (3.363 \pm 0.00)\% \\
&Br(Z^0 \to \mu^+ + \mu^-) = (3.366 \pm 0.007)\% \\
&Br(Z^0 \to \tau^+ + \tau^-) = (3.370 \pm 0.008)\% \\
&Br(Z^0 \to \text{hadrons}) = (69.91 \pm 0.06)\%
\end{aligned}
\tag{26.79}
$$

The precise measurement of the Z^0 boson properties will be described further in Chapter 27.

26.6 Heavy Lepton Pair Production

Within the electroweak theory, the cross-section for $e^+ e^- \to \ell^+ \ell^-$ includes the contribution from photon exchange, Z^0 exchange, and the interference between the two:

$$
\sigma(e^+ e^- \to V \to \ell^+ \ell^-) \propto |\mathcal{M}_\gamma + \mathcal{M}_{Z^0}|^2
\tag{26.80}
$$

Figure 26.5 illustrates the tree-level Feynman diagrams for the $e^+ e^- \to \gamma/Z^0 \to \ell^+ \ell^-$ process. The resulting

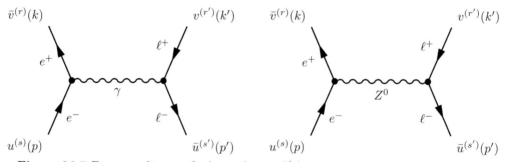

Figure 26.5 Feynman diagram for heavy lepton $\ell^+ \ell^-$ pair creation, where $\ell = \mu$ or τ.

cross-section depends strongly on the center-of-mass energy $\sqrt{s}$. For $\sqrt{s} < 40$ GeV, the cross-section is dominated by the photon exchange and the result is basically the QED prediction which we computed in Chapter 11. In the region $50 < \sqrt{s} < 80$ GeV, both photon and Z^0 processes are important and interfere. Close to $\sqrt{s} \simeq M_Z \approx 91$ GeV, the Z boson exchange dominates and is about three orders of magnitude greater than the pure QED contribution. This huge peak is the Z^0 **resonance**. For $\sqrt{s} \gg M_Z$ the contributions from the photon and Z^0 exchange are of the same order of magnitude, which is the region where we observe the unification of the electromagnetic and weak force, with $g_Z \simeq e$.

At the resonance, we cannot neglect the width of the Z^0 boson and its propagator is expressed with the use of the Breit–Wigner shape defined in Section 4.5. The cross-section in the region of the resonance becomes proportional to

$$\sigma \propto \left| \frac{1}{s - M_Z^2 + iM_Z\Gamma_Z} \right|^2 = \frac{1}{(s - M_Z^2)^2 + M_Z^2\Gamma_Z^2} \tag{26.81}$$

In the total cross-section for $e^-(p, s) + e^+(k, r) \rightarrow \ell^-(p', s') + \ell^+(k', r')$ near the Z^0 resonance, the QED contribution can be neglected. The amplitude can be obtained by using the massive boson propagator and the couplings to the Z^0:

$$i\mathcal{M} = -\frac{ig_Z^2}{s - M_Z^2 + iM_Z\Gamma_Z} g_{\mu\nu} \left[\bar{v}(k)\gamma^\mu \frac{1}{2} \left(c_V^e - c_A^e \gamma^5 \right) u(p) \right] \left[\bar{u}(p')\gamma^\nu \frac{1}{2} \left(c_V^\ell - c_A^\ell \gamma^5 \right) v(k') \right] \tag{26.82}$$

where c_V^e, c_A^e, c_V^ℓ, and c_A^ℓ are the vector (V) and axial-vector (A) couplings to the electron and lepton. This expression is very similar to Eq. (11.168), which we derived in QED to compute the *polarized* differential cross-section, and we can follow the same procedure and use results for the present case. The cross-section at high energy was subdivided into four combinations of spins, namely $e_R^+ e_L^- \rightarrow \ell_R^+ \ell_L^-$, $e_R^+ e_L^- \rightarrow \ell_L^+ \ell_R^-$, $e_L^+ e_R^- \rightarrow \ell_R^+ \ell_L^-$, and $e_L^+ e_R^- \rightarrow \ell_L^+ \ell_R^-$ (labeled LL, LR, RL, and RR following the handedness of the incoming electron and outgoing negative lepton).

Because of angular momentum conservation, each of these configurations has a well-defined angular differential cross-section in the center-of-mass system, namely:

$$\frac{d\sigma}{d\Omega} \left(e_R^+ e_L^- \rightarrow \ell_R^+ \ell_L^- \right) \propto (1 + \cos\theta)^2, \qquad \frac{d\sigma}{d\Omega} \left(e_L^+ e_R^- \rightarrow \ell_L^+ \ell_R^- \right) \propto (1 + \cos\theta)^2$$

$$\frac{d\sigma}{d\Omega} \left(e_L^+ e_R^- \rightarrow \ell_R^+ \ell_L^- \right) \propto (1 - \cos\theta)^2, \qquad \frac{d\sigma}{d\Omega} \left(e_R^+ e_L^- \rightarrow \ell_L^+ \ell_R^- \right) \propto (1 - \cos\theta)^2 \tag{26.83}$$

where θ is the scattering angle. In QED, parity conservation implied that the left and right contributions were identical. On the other hand, in the case of the weak Z^0 channel, we need to insert the corresponding coupling factors. Recalling that $c_V = c_L + c_R$ and $c_A = c_L - c_R$ (see Eq. (26.11)), we can write these factors in the corresponding combinations to find:

$$\frac{d\sigma}{d\Omega} \left(e_R^+ e_L^- \rightarrow \ell_R^+ \ell_L^- \right) \propto (c_L^e)^2(c_L^\ell)^2(1 + \cos\theta)^2, \qquad \frac{d\sigma}{d\Omega} \left(e_L^+ e_R^- \rightarrow \ell_L^+ \ell_R^- \right) \propto (c_R^e)^2(c_R^\ell)^2(1 + \cos\theta)^2$$

$$\frac{d\sigma}{d\Omega} \left(e_L^+ e_R^- \rightarrow \ell_R^+ \ell_L^- \right) \propto (c_R^e)^2(c_L^\ell)^2(1 - \cos\theta)^2, \qquad \frac{d\sigma}{d\Omega} \left(e_R^+ e_L^- \rightarrow \ell_L^+ \ell_R^- \right) \propto (c_L^e)^2(c_R^\ell)^2(1 - \cos\theta)^2 \tag{26.84}$$

For unpolarized incoming electron–positron beams, we should take the average over both incoming spin states to get:

$$\begin{aligned}
\frac{d\sigma}{d\Omega} \propto \ & \frac{1}{4} \left[\left((c_L^e)^2(c_L^\ell)^2 + (c_R^e)^2(c_R^\ell)^2 \right) (1 + \cos\theta)^2 + \left((c_R^e)^2(c_L^\ell)^2 + (c_L^e)^2(c_R^\ell)^2 \right) (1 - \cos\theta)^2 \right] \\
= \ & \frac{1}{4} \left[\left((c_L^e)^2(c_L^\ell)^2 + (c_R^e)^2(c_R^\ell)^2 + (c_R^e)^2(c_L^\ell)^2 + (c_L^e)^2(c_R^\ell)^2 \right) (1 + \cos^2\theta) + \right. \\
& \left. 2 \left((c_L^e)^2(c_L^\ell)^2 + (c_R^e)^2(c_R^\ell)^2 - (c_R^e)^2(c_L^\ell)^2 - (c_L^e)^2(c_R^\ell)^2 \right) \cos\theta \right] \\
= \ & \frac{1}{4} \left[\left((c_R^e)^2 + (c_L^e)^2 \right) \left((c_R^\ell)^2 + (c_L^\ell)^2 \right) (1 + \cos^2\theta) + \right. \\
& \left. 2 \left((c_R^e)^2 - (c_L^e)^2 \right) \left((c_R^\ell)^2 - (c_L^\ell)^2 \right) \cos\theta \right]
\end{aligned} \tag{26.85}$$

This result can be rewritten in terms of the original vector and axial-vector couplings of the electron and negative lepton to the Z^0 boson, noting that:

$$c_V^2 + c_A^2 = 2(c_L^2 + c_R^2) \quad \text{and} \quad c_V c_A = c_L^2 - c_R^2 \tag{26.86}$$

and getting:

$$\frac{d\sigma}{d\Omega} \propto \frac{1}{4}\left((c_V^e)^2 + (c_A^e)^2\right)\left((c_V^\ell)^2 + (c_A^\ell)^2\right)(1+\cos^2\theta) + 2c_V^e c_A^e c_V^\ell c_A^\ell \cos\theta \tag{26.87}$$

Finally, a rigorous calculation yields the unpolarized $e^+e^- \to \ell^+\ell^-$ annihilation differential cross-section in the region of the Z^0 resonance:

$$\frac{d\sigma}{d\Omega} = \frac{g_Z^4 s}{256\pi^2}D_Z(s)\left[\frac{1}{4}\left((c_V^e)^2 + (c_A^e)^2\right)\left((c_V^\ell)^2 + (c_A^\ell)^2\right)(1+\cos^2\theta) + 2c_V^e c_A^e c_V^\ell c_A^\ell \cos\theta\right] \tag{26.88}$$

where $D_Z(s)$ is the resonance term:

$$D_Z(s) \equiv \frac{1}{(s-M_Z^2)^2 + M_Z^2\Gamma_Z^2} \tag{26.89}$$

Integrating $(1+\cos^2\theta)$ over the solid angle $d\Omega$ gives:

$$\int(1+\cos^2\theta)d\Omega = (2\pi)\int_{-1}^{+1}dx(1+x^2) = (2\pi)\frac{8}{3} = \frac{16\pi}{3} \tag{26.90}$$

while over $\cos\theta$ it vanishes. The resulting total cross-section is:

$$\sigma(e^+e^+ \to Z^0 \to \ell^+\ell^-) = \frac{g_Z^4 s}{48\pi}D_Z(s)\frac{1}{4}\left((c_V^e)^2 + (c_A^e)^2\right)\left((c_V^\ell)^2 + (c_A^\ell)^2\right) \tag{26.91}$$

This result can be more elegantly written using the expression of Eq. (26.71) for the partial width of the Z^0 into leptons, which gives:

$$\left((c_V^\ell)^2 + (c_A^\ell)^2\right) = \frac{48\pi}{g_Z^2 M_Z}\Gamma(Z^0 \to \ell^+\ell^-) \equiv \frac{48\pi}{g_Z^2 M_Z}\Gamma_{\ell\ell} \tag{26.92}$$

then

$$\sigma(e^+e^+ \to Z^0 \to \ell^+\ell^-) = \frac{12\pi s}{M_Z^2}D_Z(s)\Gamma_{ee}\Gamma_{\ell\ell} \tag{26.93}$$

This result remains valid for any fermion–antifermion pair and can be expressed as:

$$\sigma(e^+e^+ \to Z^0 \to f\bar{f}) = \frac{12\pi s}{M_Z^2}\frac{\Gamma_{ee}\Gamma_{ff}}{(s-M_Z^2)^2 + M_Z^2\Gamma_Z^2} \tag{26.94}$$

At the "resonance peak" defined by a center-of-mass energy equal to M_Z or $s = M_Z^2$, the cross-section is the largest and is defined by σ_{ff}^0:

$$\sigma_{ff}^0 = 12\pi\frac{\Gamma_{ee}\Gamma_{ff}}{M_Z^2\Gamma_Z^2} \tag{26.95}$$

Hence, measuring the peak cross-section for a particular fermion σ_{ff}^0 can be used to measure the partial width Γ_{ff}.

26.7 The Forward–Backward Asymmetry

The forward–backward asymmetry A_{FB} in $e^+e^- \to \ell^+\ell^-$ collisions has been seen to be a powerful tool to explore the parity-violating nature of interactions. It was defined in Eq. (13.13):

$$A_{FB} \equiv \frac{\sigma_F - \sigma_B}{\sigma_F + \sigma_B} \tag{26.96}$$

where σ_F and σ_B are the cross-sections corresponding to the negative lepton being scattered in the forward direction ($\cos\theta > 0$) and in the backward direction ($\cos\theta < 0$). In the case of the $e^+e^- \to Z^0 \to \ell^+\ell^-$ diagram, the different couplings of the Z^0 boson to the left-handed and right-handed leptons will lead to differences in the differential cross-sections for the four helicity configurations, which are listed in Eq. (26.84).

The differential cross-section Eq. (26.88) is of the form (cf. Eq. (13.12)):

$$\frac{d\sigma}{d\Omega} \propto A(1 + \cos^2\theta) + B\cos\theta \tag{26.97}$$

The integration of the scattering angle setting $x = \cos\theta$ gives a term proportional to:

$$\int dx\, \left[A(1 + x^2) + Bx\right] = A\left(x + \frac{x^3}{3}\right) + B\frac{x^2}{2} \tag{26.98}$$

Then integrating for the forward ($0 \le x \le 1$) and backward ($-1 \le x \le 0$) cases, we get:

$$\sigma_F \propto \frac{4}{3}A + \frac{1}{2}B \quad \text{and} \quad \sigma_B \propto \frac{4}{3}A - \frac{1}{2}B \tag{26.99}$$

and then

$$A_{FB} = \frac{3}{8}\frac{B}{A} = \frac{3}{4}\frac{4c_V^e c_A^e c_V^\ell c_A^\ell}{\left((c_V^e)^2 + (c_A^e)^2\right)\left((c_V^\ell)^2 + (c_A^\ell)^2\right)} \tag{26.100}$$

So we can write:

$$A_{FB} \equiv \frac{3}{4}\mathcal{A}_e\mathcal{A}_\ell \tag{26.101}$$

where we have inserted the asymmetry parameter $\mathcal{A}_f$ for fermion f:

$$\mathcal{A}_f \equiv \frac{2c_V^f c_A^f}{(c_V^f)^2 + (c_A^f)^2} = \frac{2c_V^f/c_A^f}{1 + (c_V^f/c_A^f)^2} \tag{26.102}$$

where in the second term we have expressed the asymmetry as a function of the ratio of the vector and axial couplings. Within the electroweak theory, the ratio is directly related to the weak mixing angle since:

$$\frac{c_V^f}{c_A^f} = \frac{I_{3L}^f - 2Q_f \sin^2\theta_W}{I_{3L}^f} = 1 - \frac{2Q_f \sin^2\theta_W}{I_{3L}^f} \tag{26.103}$$

For the charged leptons, with $Q_f = -1$ and $I_{3L}^f = -1/2$, we directly have:

$$\frac{c_V^\ell}{c_A^\ell} = 1 - 4\sin^2\theta_W \simeq 0.08 \implies \mathcal{A}_e = \mathcal{A}_\ell \simeq 0.16 \tag{26.104}$$

where we used $\sin^2\theta_W \simeq 0.23$. Hence, the measurement of the forward–backward asymmetry in $e^+e^- \to Z^0 \to \ell^+\ell^-$ provides a powerful method to determine the weak mixing angle. For the up-type quarks, with $Q_f = +2/3$ and $I_{3L}^f = +1/2$, we get:

$$\frac{c_V^{u,c,t}}{c_A^{u,c,t}} = 1 - \frac{8}{3}\sin^2\theta_W \simeq 0.39 \implies \mathcal{A}_{u,c,t} \simeq 0.68 \tag{26.105}$$

and for the down-type quarks $Q_f = -1/3$ and $I_{3L}^f = -1/2$, so we find:

$$\frac{c_V^{d,s,b}}{c_A^{d,s,b}} = 1 - \frac{4}{3}\sin^2\theta_W \simeq 0.69 \implies \mathcal{A}_{d,s,b} \simeq 0.93 \tag{26.106}$$

These asymmetries are plotted as a function of the Weinberg mixing angle in Figure 26.6. They are expected to be largest for down-type quarks and smallest for charged leptons. However, the factors in front of the $\sin^2\theta_W$ make the charged leptons' asymmetry most sensitive to variations of $\sin^2\theta_W$, and hence provide the best method to determine the weak mixing angle.

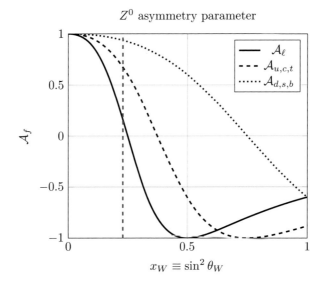

Figure 26.6 The asymmetry parameter as a function of the Weinberg mixing angle. The vertical dashed line indicates the value $x_W = 0.23$.

26.8   QED Radiative Corrections in e^+e^- Collisions

When the center-of-mass energy in e^+e^- collisions is in the region of the Z^0 pole, the cross-section is greatly enhanced due to the resonance which is described by the Breit–Wigner shape (see Eq. (26.81)). In practice, the observed shape has not quite the form of a Breit–Wigner because of QED radiative corrections. As illustrated in Figure 26.7, there are two diagrams where a photon is radiated from either the electron or positron in the initial state. This is called **initial-state radiation** or ISR. Let us assume that the ISR photon is emitted with an energy E_γ and, as is often the case, collinear to the incoming electron or positron. For a collision at a beam center-of-mass energy $\sqrt{s}$, the effect of ISR is to reduce the effective energy of the hard e^+e^- collision. Writing the four-momenta p_1, p_2 of the colliding particles, neglecting the rest masses at high energy, we have after the photon emission for example from the particle 1:

$$p_1 = (E - E_\gamma, 0, 0, E - E_\gamma), \qquad p_2 = (E, 0, 0, -E) \tag{26.107}$$

consequently the effective center-of-mass energy at the e^+e^- vertex is going to be:

$$s' = (p_1 + p_2)^2 = (2E - E_\gamma)^2 - E_\gamma^2 = 4E^2 \left(1 - \frac{E_\gamma}{E}\right) = s\left(1 - \frac{E_\gamma}{E}\right) \tag{26.108}$$

The impact of ISR is to reduce the center-of-mass energy of the collision to $\sqrt{s'}$. The probability for ISR can be precisely calculated within QED so it does not introduce significant errors and can simply be taken into account by writing the ISR-corrected cross-section as:

$$\sigma(s) = \int \sigma(s') p_{ISR}(s', s) \mathrm{d}s' \tag{26.109}$$

where $p_{ISR}(s', s)$ is the probability of an initial state of energy $\sqrt{s}$ being effectively at $\sqrt{s'}$ due to the ISR effect. Experimentally it is also possible to deconvolute the measured cross-section to reconstruct the ISR

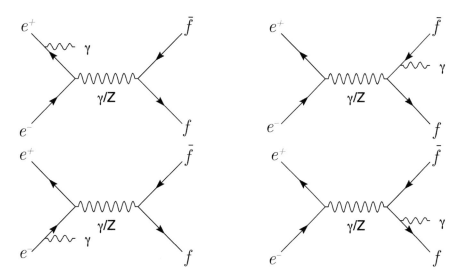

Figure 26.7 The QED radiative corrections with emission of a photon in the initial or final state in an $e^+e^- \to f\bar{f}$ process. When the photon is emitted in the initial state, we have an initial-state radiation (ISR).

unfolded line shape, which is going to be given by the Breit–Wigner shape of the Z^0. This effect is illustrated in Figure 26.8. The graph shows the measured cross-section of the process $e^+e^- \to q\bar{q} \to$ hadrons as a function of the beam center-of-mass energy. The smooth curve represents the Z^0 resonance with the ISR corrections. The dashed curve represents the shape of the Z^0 after the unfolding (deconvolution) of the expected ISR effects. Close to and below the peak of the resonance, the ISR reduces the cross-section because the effective center-of-mass energy is moved further away (to the left) from the peak of the resonance. Above the peak of the resonance, the cross-section is actually higher because of the "return to Z" phenomenon: although the beam center-of-mass energy is above resonance ($\sqrt{s} > M_Z$), the emission of ISR that returns the effective center-of-mass energy on the resonance is favored and increases the cross-section! After correcting for QED effects, the dashed curve on the plot is used to define the Z^0 parameters that are used to define the theory.

26.9 $W^\pm/Z^0$ **Production at Hadron Colliders**

The productions of $W^\pm$ and Z^0 bosons in proton–proton or proton–antiproton collisions are extremely important processes. First of all, the two bosons were discovered in $p\bar{p}$ collisions. Secondly, they can be calculated precisely in the electroweak theory and therefore provide an important "candle" at current hadron experiments (see Section 27.4). The cross-sections for $W^\pm$ and Z^0 production are computed in a similar way as the Drell–Yan cross-section (see Section 20.2) and in general are described as hard hadron–hadron collisions (see Section 19.10).

• **Single $W^\pm$ production.** We begin with the W^+ production process. At the partonic level, we first consider the fusion of the up-type and down-type quark $u\bar{d} \to W^+$:

$$u(p) + \bar{d}(k) \to W^+(P^\mu) \tag{26.110}$$

The amplitude is given by (see Eq. (26.54)):

$$i\mathcal{M} = -i\frac{g_W}{\sqrt{2}}|V_{ud}|\epsilon_\mu^{*\lambda}(P)\bar{v}(p)\gamma^\mu\frac{1}{2}(1-\gamma^5)u(k) \tag{26.111}$$

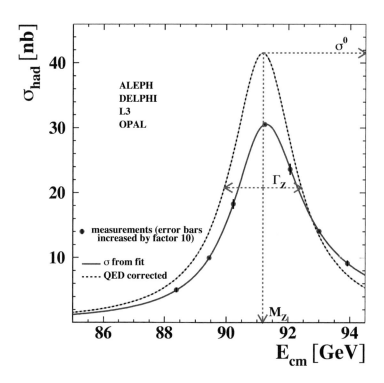

Figure 26.8 Illustration of the effect of QED radiative corrections on the shape of the Z^0 resonance as a function of the center-of-mass energy $\sqrt{s}$. Correcting for QED effects yields the dashed curve, which defines the Z^0 parameters. Reprinted from S. Schael *et al.*, "Precision electroweak measurements on the Z resonance," *Phys. Rept.*, vol. 427, pp. 257–454, 2006, with permission from Elsevier.

where, as before, $\epsilon_\mu^\lambda(P)$ is the polarization four-vector of the W boson. We should now average over the initial-state spins and sum over the final-state W^+ polarization states. We proceed as before (see Eq. (26.57)) and find:

$$\langle|\mathcal{M}(u\bar{d}\to W^+)|^2\rangle = \frac{1}{2}\frac{1}{2}|V_{ud}|^2 g_W^2 M_W^2 = \frac{1}{4}|V_{ud}|^2\frac{8G_F}{\sqrt{2}}M_W^4 \qquad (26.112)$$

Adding the kinematical factors for relativistic collisions leads to the partonic cross-section (see Eq. (5.117)):

$$\hat{\sigma}(u\bar{d}\to W^+) = \frac{1}{4}|V_{ud}|^2\frac{8G_F}{\sqrt{2}}\frac{M_W^4}{2\hat{s}}(2\pi)^4\int\frac{\mathrm{d}^3\vec{P}}{(2\pi)^3 2E_P}\delta^4(P-p-k) \qquad (26.113)$$

where, neglecting quark masses, we used the partonic center-of-mass energy squared $\hat{s}=(p+k)^2\simeq 2(p\cdot k)$ and the flux factor $F=4\sqrt{(p\cdot k)^2}=4\sqrt{(\hat{s}/2)^2}=2\hat{s}$ (see Eq. (5.121)). The phase-space integration is computed neglecting the W boson width and simply gives the constraint that the two quarks have the energy of the rest mass of the boson:

$$(2\pi)^4\int\frac{\mathrm{d}^3\vec{P}}{(2\pi)^3 2E_P}\delta^4(P-p-k) = (2\pi)^4\int\frac{\mathrm{d}^3\vec{P}}{(2\pi)^3 2E_P}\delta^3(\vec{P}-\vec{p}-\vec{k})\delta(E_P-\underbrace{(E_p+E_k)}_{\sqrt{\hat{s}}}) \qquad (26.114)$$

In the center-of-mass frame, we have $E_P=M_W$ and $\vec{P}=0$, hence the integration over $\mathrm{d}^3\vec{P}$ leaves only:

$$(2\pi)^4\frac{1}{(2\pi)^3 2M_W}\delta(M_W-\sqrt{\hat{s}}) = (2\pi)\delta(M_W^2-\hat{s}) \qquad (26.115)$$

Consequently, the Cabibbo enhanced partonic cross-section for the W^+ production becomes:

$$\hat{\sigma}(u\bar{d} \to W^+) = \frac{1}{4}|V_{ud}|^2 \frac{8G_F}{\sqrt{2}} \frac{M_W^4}{2\hat{s}}(2\pi)\delta(M_W^2 - \hat{s}) = (2\pi)|V_{ud}|^2 \frac{G_F}{\sqrt{2}} M_W^2 \delta(M_W^2 - \hat{s}) \qquad (26.116)$$

where in the second term we used $\hat{s} = M_W^2$. Similarly, one would obtain for the Cabibbo suppressed cross-section:

$$\hat{\sigma}(u\bar{s} \to W^+) = (2\pi)|V_{us}|^2 \frac{G_F}{\sqrt{2}} M_W^2 \delta(M_W^2 - \hat{s}) \qquad (26.117)$$

The total W^+ production cross-section in hard hadron–hadron collisions is then the convolution of the partonic cross-section with the parton distribution functions (see Eq. (19.45)):

$$\sigma(P_1 P_2 \to W^+)(s) = \underbrace{K}_{\text{QCD corr.}} \underbrace{\frac{1}{N_C}}_{\text{color factor}} \sum_{\{i,j\}} \int\int \mathrm{d}x_1 \mathrm{d}x_2 f_{q_i/P_1}(x_1) f_{q_j/P_2}(x_2) \hat{\sigma}_{ij \to W^+}(x_1 x_2 s) \qquad (26.118)$$

where we have included the color factor to produce a colorless W^+ (see Eq. (20.10)), and set $\hat{s} = x_1 x_2 s$ where x_1, x_2 are the fractions of momentum carried by the partons i and j relative to their parents P_1 and P_2 (see Eq. (19.44)). The sum over i, j runs over the pairs of compatible quarks to produce the W^+ boson, i.e., $u\bar{d}$, $u\bar{s}$, ... The parton distribution functions must be evaluated at the scale $Q^2 = \hat{s} = M_W^2$. The effect of QCD corrections to the total cross-section are conventionally taken into account by a multiplicative number called the K**-factor**.

In the case of proton–proton collisions, the cross-section becomes:

$$\begin{aligned}
\sigma(pp \to W^+)(s) &= \frac{2\pi K G_F}{3\sqrt{2}} \int\int \mathrm{d}x_1 \mathrm{d}x_2 M_W^2 \delta(M_W^2 - x_1 x_2 s) \\
&\quad \times \left[|V_{ud}|^2 \left(u(x_1)\bar{d}(x_2) + u(x_2)\bar{d}(x_1) \right) + |V_{us}|^2 \left(u(x_1)\bar{s}(x_2) + u(x_2)\bar{s}(x_1) \right) \right] \\
&\simeq \frac{2\pi K G_F}{3\sqrt{2}} \int\int \mathrm{d}x_1 \mathrm{d}x_2 M_W^2 \delta(M_W^2 - x_1 x_2 s) \left(u(x_1)\bar{d}(x_2) + u(x_2)\bar{d}(x_1) \right) \qquad (26.119)
\end{aligned}$$

where we assumed that the sea is flavor symmetric (i.e., $\bar{d} = \bar{s}$) and $|V_{ud}|^2 + |V_{us}|^2 \approx 1$. In Section 19.11, we have seen that the x_1 and x_2 needed to produce a particle of rest mass M_W are constrained by $\hat{s} = x_1 x_2 s = M_W^2$. We can compute them as a function of the rapidity y of the produced particle (see Eq. (19.53)):

$$x_{1W} = \frac{M_W}{\sqrt{s}} e^y \qquad \text{and} \qquad x_{2W} = \frac{M_W}{\sqrt{s}} e^{-y} \qquad (26.120)$$

and we have by construction since $x_{1W} \le 1$ and $x_{2W} \le 1$ that the allowed kinematical rapidity range is given by:

$$-\ln\frac{\sqrt{s}}{M_W} \le y \le \ln\frac{\sqrt{s}}{M_W} \qquad (26.121)$$

In addition, we find:

$$x_{1W} x_{2W} = \frac{M_W^2}{s} \qquad \text{and} \qquad \mathrm{d}x_{1W}\mathrm{d}x_{2W} = \frac{1}{s}\mathrm{d}\hat{s}\mathrm{d}y \qquad (26.122)$$

Consequently, we can write the differential cross-section for the production of the W^+ boson at rapidity y as:

$$\begin{aligned}
\frac{\mathrm{d}\sigma(pp \to W^+)}{\mathrm{d}y} &= \frac{\sqrt{2}\pi K G_F}{3} \int \frac{1}{s}\mathrm{d}\hat{s} M_W^2 \delta(M_W^2 - \hat{s}) \left(u(x_{1W})\bar{d}(x_{2W}) + u(x_{2W})\bar{d}(x_{1W}) \right) \\
&= \frac{\sqrt{2}\pi K G_F}{3} x_{1W} x_{2W} \left(u(x_{1W})\bar{d}(x_{2W}) + u(x_{2W})\bar{d}(x_{1W}) \right) \\
&\approx (8.68\text{ nb}) \times x_{1W} x_{2W} \left(u(x_{1W})\bar{d}(x_{2W}) + u(x_{2W})\bar{d}(x_{1W}) \right) \qquad (26.123)
\end{aligned}$$

where we used $G_F \approx 1.16 \times 10^{-5}$ GeV$^{-2} \approx 1.5$ nb (see Table 1.6) and a K-factor of 1.3 [342]. For the production of the W^- boson, we simply need to interchange the quark and antiquark parton distribution functions:

$$\frac{\mathrm{d}\sigma(pp \to W^-)}{\mathrm{d}y} = \frac{\sqrt{2}\pi K G_F}{3} x_{1W} x_{2W} \left(d(x_{1W}) \bar{u}(x_{2W}) + d(x_{2W}) \bar{u}(x_{1W}) \right)$$

$$\approx (8.68 \text{ nb}) \times x_{1W} x_{2W} \left(d(x_{1W}) \bar{u}(x_{2W}) + d(x_{2W}) \bar{u}(x_{1W}) \right) \quad (26.124)$$

At low center-of-mass energies (where valence quarks dominate), the cross-section for W^+ is about twice that of W^- because the proton is composed of *uud* valence quarks, therefore the term that depends on the parton distributions in brackets is larger for W^+ than for W^-.

For proton–antiproton collisions, we express the cross-section as a function of the *proton* parton distribution function but this time replacing the antiquark by its respective quark parton distribution function. Neglecting the contribution from antiquarks, one simply finds:

$$\frac{\mathrm{d}\sigma(p\bar{p} \to W^+)}{\mathrm{d}y} = \frac{\mathrm{d}\sigma(p\bar{p} \to W^-)}{\mathrm{d}y} = \frac{\sqrt{2}\pi K G_F}{3} x_{1W} x_{2W} \left(u(x_{1W}) d(x_{2W}) \right)$$

$$\approx (8.68 \text{ nb}) \times x_{1W} x_{2W} \left(u(x_{1W}) d(x_{2W}) \right) \quad (26.125)$$

The cross-sections for W^+ and W^- are necessarily equal to each other in $p\bar{p}$ collisions. The QCD correction that leads to the K-factor is illustrated in Figure 26.9, which shows the rapidity distribution for $p\bar{p} \to W^+ \to \ell^+\nu + X$ at the Tevatron Run II computed in LO (dotted), NLO (dashed), and NNLO (solid) QCD corrections. The plot takes into account the branching fraction of the W bosons to leptons $Br(W^\pm \to \ell^\pm\nu) \approx 11\%$ (see Eq. (26.68)). The ratio NLO/LO which defines the K-factor is approximately constant in the central region and its average value is about 1.3 (see Ref. [342]).

The total cross-section is found by integrating over the rapidity of the boson y within the allowed kinematical region defined in Eq. (26.121). For a fixed center-of-mass energy $\sqrt{s}$ each value of y corresponds to fixed values of x_{1W} and x_{2W}. The result therefore depends on the values of the parton distribution functions, as discussed in the following.

• **Parton luminosities.** In general, for the process $P_1 P_2 \to X$ it is convenient to introduce the scaling variables x and τ, where:

$$x_1 \equiv x \qquad \text{and} \qquad x_2 \equiv \frac{\tau}{x} \quad (26.126)$$

with the dimensionless $\tau \equiv \hat{s}/s$. For the $W^\pm$ boson production, we have $\hat{s} = M_W^2$ and we recover x_{1W} and x_{2W}. The general cross-section for X production in hadron–hadron collisions can then be expressed as [343] (cf. Eq. (26.118)):

$$\sigma(P_1 P_2 \to X)(s) = \frac{K}{N_C} \sum_{i,j} \frac{1}{1 + \delta_{ij}} \int \int \frac{\mathrm{d}x \mathrm{d}\tau}{x} \left[f_{q_i/P_1}(x) f_{q_j/P_2}\left(\frac{\tau}{x}\right) + f_{q_j/P_1}(x) f_{q_i/P_2}\left(\frac{\tau}{x}\right) \right] \hat{\sigma}_{ij \to X}(\hat{s})$$

$$(26.127)$$

where we assumed for simplicity that X is a colorless particle. This expression can be conveniently rearranged to look like:

$$\sigma(P_1 P_2 \to X)(s) = \frac{K}{N_C} \sum_{i,j} \int \frac{\mathrm{d}\tau}{\tau} \left(\frac{\tau}{\hat{s}} \frac{\mathrm{d}\mathcal{L}_{ij}}{\mathrm{d}\tau} \right) \left[\hat{s}\hat{\sigma}_{ij \to X}(\hat{s}) \right] \quad (26.128)$$

where the quantity $\hat{s}\hat{\sigma}_{ij \to X}(\hat{s})$ is dimensionless and where we introduced the **parton differential luminosity** [343]:

$$\frac{\tau}{\hat{s}} \frac{\mathrm{d}\mathcal{L}_{ij}}{\mathrm{d}\tau} \equiv \frac{1}{1 + \delta_{ij}} \frac{\tau}{\hat{s}} \int_\tau^1 \frac{\mathrm{d}x}{x} \left[f_{q_i/P_1}(x) f_{q_j/P_2}\left(\frac{\tau}{x}\right) + f_{q_j/P_1}(x) f_{q_i/P_2}\left(\frac{\tau}{x}\right) \right] \quad (26.129)$$

In this form, high-energy colliders collide in essence a broadband unseparated beam of partons (quarks, antiquarks, and gluons) of a given differential "luminosity." These luminosities can be calculated once for a given

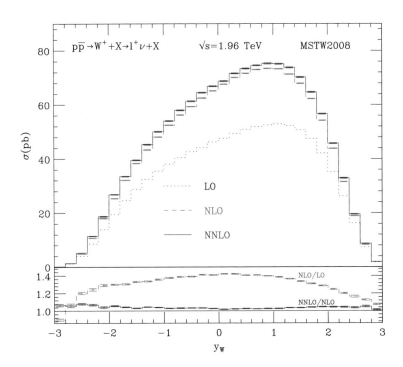

Figure 26.9 Rapidity distribution for $p\bar{p} \to W^+ \to \ell^+\nu + X$ at the Tevatron Run II in LO (dotted), NLO (dashed), and NNLO (solid) QCD. Reprinted by permission from Springer Nature: S. Catani, G. Ferrera, and M. Grazzini, "W boson production at hadron colliders: The lepton charge asymmetry in NNLO QCD," *JHEP*, vol. 05, p. 006, 2010. Copyright 2010 SISSA.

collider (proton–proton or proton–antiproton) of a given center-of-mass energy $\sqrt{s}$ for pairs of partons i, j. Then, partonic processes can be directly inserted and computed. In order to evaluate the parton luminosities, one must choose a set of parton distribution functions. As an example, the parton luminosities for $u\bar{d}$ and for two gluons gg in proton–proton collisions are shown in Figure 26.10 as a function of $\sqrt{\hat{s}}$ (TeV) for different center-of-mass energies $\sqrt{s} = 0.9, 2, 4, 6, 7, 10, 14$ TeV. The dashed curve corresponds to the proton–antiproton at the Tevatron energy. The parton luminosities increase rapidly with the center-of-mass energy $\sqrt{s}$ of the collider. Hence, for a given collider luminosity $\mathcal{L}$ they will be much more copiously produced as the center-of-mass energy $\sqrt{s}$ increases. We also note that gluon luminosities are higher than the quark–antiquark ones. Accordingly, we say that very high-energy proton–proton colliders are dominantly gluon–gluon colliders. In the case of the $W^{\pm}$ production, we look at the $u\bar{d}$ parton luminosities around the $\sqrt{\hat{s}} \approx 0.1$ TeV region.

The results are illustrated in Figure 19.22 as the line labeled σ_W. As already mentioned, the σ_W cross-section increases sharply with the center-of-mass energy of the collisions. However, it represents only an extremely small fraction ($\simeq 10^{-7}$) of the total cross-section labeled σ_{tot}.

• **Single Z^0 production.** The calculation for a single Z^0 boson follows a similar path as above. At the partonic level, one considers the annihilation of a quark–antiquark pair:

$$q(p) + \bar{q}(k) \to Z^0(P^\mu) \tag{26.130}$$

The amplitude is given by (see Eq. (26.70)):

$$i\mathcal{M} = -ig_Z \epsilon_\mu^{*\lambda}(P)\bar{u}(p)\gamma^\mu \frac{1}{2}(c_V^q - c_A^q \gamma^5)v(k) \tag{26.131}$$

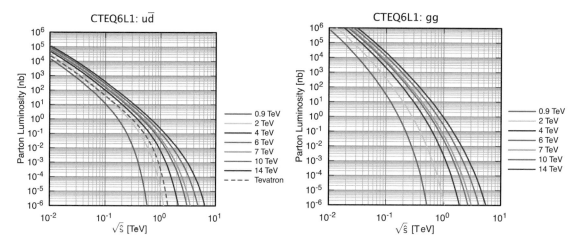

Figure 26.10 The parton luminosities in proton–proton collisions as a function of $\sqrt{\hat{s}}$ (TeV) for different center-of-mass energies $\sqrt{s} = 0.9, 2, 4, 6, 7, 10, 14$ TeV. The dashed curve corresponds to the proton–antiproton at the Tevatron energy. Reprinted with permission from E. Eichten, I. Hinchliffe, K. D. Lane, and C. Quigg, "Super collider physics," *Rev. Mod. Phys.*, vol. 56, pp. 579–707, 1984. [Addendum: *Rev. Mod. Phys.*, vol. 58, pp. 1065–1073 (1986)]. Copyright 1984 by the American Physical Society.

where $\epsilon_\mu^\lambda(P)$ is the polarization four-vector of the Z^0 boson. A calculation as in the single $W^\pm$ production case, leads to the spin-averaged cross-section:

$$\hat{\sigma}(q\bar{q} \to Z^0) = (8\pi)\left((c_V^q)^2 + (c_A^q)^2\right)\frac{G_F}{\sqrt{2}}M_Z^2\delta(M_Z^2 - \hat{s}) \tag{26.132}$$

The resulting color-averaged differential cross-section in hard collisions is then:

$$\frac{d\sigma(P_1 P_2 \to Z^0)}{dy} = K\frac{2\pi}{3}\frac{G_F}{\sqrt{2}}\sum_q \left((c_V^q)^2 + (c_A^q)^2\right) x_{1Z}x_{2Z}q(x_{1Z})\bar{q}(x_{2Z}) \tag{26.133}$$

where x_{1Z} and x_{2Z} are defined in a similar way as x_{1W} and x_{2W} in Eq. (26.120). In proton–proton collisions, this leads to the following expression:

$$\begin{aligned}
\frac{d\sigma(pp \to Z^0)}{dy} &= \frac{\sqrt{2}\pi K G_F}{3}x_{1Z}x_{2Z}\Big[\left((c_V^u)^2 + (c_A^u)^2\right)\left(u(x_{1Z})\bar{u}(x_{2Z}) + \bar{u}(x_{1Z})u(x_{2Z})\right) \\
&\quad + \left((c_V^d)^2 + (c_A^d)^2\right)\left(d(x_{1Z})\bar{d}(x_{2Z}) + \bar{d}(x_{1Z})d(x_{2Z})\right) \\
&\quad + \left((c_V^s)^2 + (c_A^s)^2\right)\left(s(x_{1Z})\bar{s}(x_{2Z}) + \bar{s}(x_{1Z})s(x_{2Z})\right)\Big]
\end{aligned} \tag{26.134}$$

Using the results of Eq. (26.77), we can write:

$$\begin{aligned}
\left((c_V^u)^2 + (c_A^u)^2\right) &= \left(\frac{1}{2} - \frac{4}{3}x_W + \frac{16}{9}x_W^2\right) \approx 0.28 \\
\left((c_V^d)^2 + (c_A^d)^2\right) &= \left((c_V^s)^2 + (c_A^s)^2\right) = \left(\frac{1}{2} - \frac{2}{3}x_W + \frac{4}{9}x_W^2\right) \approx 0.37
\end{aligned} \tag{26.135}$$

where we used $x_W \equiv \sin^2\theta_W \simeq 0.23$ (see Eq. (26.76)). This leads to the following result:

$$\begin{aligned}
\frac{d\sigma(pp \to Z^0)}{dy} &\approx (8.68 \text{ nb}) \times x_{1Z}x_{2Z}\Big[0.28(u(x_{1Z})\bar{u}(x_{2Z}) + \bar{u}(x_{1Z})u(x_{2Z})) \\
&\quad + 0.37\left(d(x_{1Z})\bar{d}(x_{2Z}) + \bar{d}(x_{1Z})d(x_{2Z})\right)\Big]
\end{aligned} \tag{26.136}$$

where we have neglected the strange quark contribution.

For proton–antiproton collisions, we simply replace the parton distribution functions of the antiquarks by those of quarks and get:

$$
\begin{aligned}
\frac{\mathrm{d}\sigma(p\bar{p} \to Z^0)}{\mathrm{d}y} &\approx (8.68 \text{ nb}) \times x_{1Z}x_{2Z}\Big[0.28(u(x_{1Z})u(x_{2Z}) + \bar{u}(x_{1Z})\bar{u}(x_{2Z})) \\
&\quad +0.37\left(d(x_{1Z})d(x_{2Z}) + \bar{d}(x_{1Z})\bar{d}(x_{2Z})\right)\Big]
\end{aligned}
\tag{26.137}
$$

Hence, at low center-of-mass energies where the valence quarks dominate, the cross-section for $p\bar{p} \to Z^0$ is higher than in $pp \to Z^0$. Overall, the couplings to the Z^0 make its production less abundant than that of the $W^\pm$ boson. Including also the branching fraction to leptons $Br(Z^0 \to \ell^+\ell^-) \approx 3\%$ (see Eq. (26.79)), the number of easily detectable Z^0 events at hadron colliders is about $1/10$ of that of $W^\pm$ events.

The total cross-section for single Z^0 production at hadron colliders is illustrated in Figure 19.22 as the line labeled σ_Z. As already mentioned, the σ_Z cross-section increases sharply with the center-of-mass energy of the collisions. It however represents only an extremely small fraction ($\simeq 10^{-7}$) of the total cross-section labeled σ_{tot}.

• **Transverse momentum of the bosons.** In the leading order (LO) fusion processes considered in the preceding paragraph, the W and Z bosons are produced without transverse momentum, since we assume the longitudinal approximation of the incoming partons. The higher-order QCD corrections were included as an overall normalization called the K-factor.

In practice, there are higher-order QCD processes which can produce bosons with large transverse momenta (P_T), where the P_T is balanced by a quark or a gluon (i.e., a hadronic jet in the detector). As an illustration, the Feynman diagrams for the $q + \bar{q} \to W + g$ (annihilation) and $g + q \to W + q$ (Compton-like) are shown in Figure 26.11. The corresponding cross-sections can be found in, e.g., Ref. [343]. Similar arguments hold for the Z^0. The transverse momentum is an important factor to take into account experimentally and becomes more important as the center-of-mass energy increases.

26.10 W^+W^- **Production in** e^+e^- **Collisions**

The process $e^+e^- \to W^+W^-$, which was observed at LEP-II, is an important channel in which several electroweak diagrams contribute at tree level. The γW^+W^- and $Z^0W^+W^-$ vertices involve the tri-linear couplings between the gauge bosons, and therefore provide a clear signature of the Yang–Mills (non-Abelian) nature of the electroweak interaction. This process is analogous to the QED pair annihilation $e^+e^- \to \gamma\gamma$ discussed in Section 11.15 and to the QCD pair annihilation into gluons $q\bar{q} \to gg$ discussed in Section 18.10. The kinematics is given by:

$$
e^-(p,s) + e^+(k,r) \to W^+(k_1,\epsilon_1) + W^-(k_2,\epsilon_2)
\tag{26.138}
$$

The corresponding Feynman diagrams at tree level are shown in Figure 26.12. We shall use the following kinematical variables:

$$
s = (p+k)^2 = (k_1+k_2)^2, \quad t = (p-k_1)^2, \quad \text{and} \quad u = (p-k_2)^2
\tag{26.139}
$$

We note that (see Eq. (11.88)):

$$
s + t + u = 2m_e^2 + 2M_W^2 \approx 2M_W^2 \quad \Longrightarrow \quad u \simeq 2M_W^2 - s - t
\tag{26.140}
$$

For a fixed s, the kinematical configuration of the process can then be determined by t. In the center-of-mass frame (see Section 11.12), we can define, neglecting the electron mass compared to that of the W boson:

$$
p^\mu = (E^\star, \vec{p}^{\,\star}) \qquad k^\mu = (E^\star, -\vec{p}^{\,\star}) \qquad k_1^\mu = (\omega^\star, \vec{k}^{\,\star}) \qquad k_2^\mu = (\omega^\star, -\vec{k}^{\,\star})
\tag{26.141}
$$

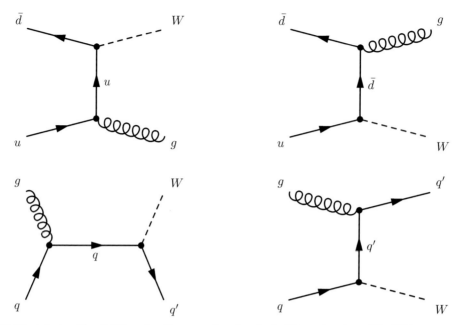

Figure 26.11 Next-to-leading QCD order diagrams for the single W production. The W boson recoils against a hadronic jet.

with $E^\star = |\vec{p}^\star|$ and $\omega^\star = \sqrt{\vec{k}^{\star 2} + M_W^2}$. With these definitions, we obtain:

$$s \simeq 2(p \cdot k) = (2E^\star)^2 = (2\omega^\star)^2 \implies E^\star = \omega^\star = \sqrt{s}/2$$

$$t = p^2 + k_1^2 - 2p \cdot k_1 \simeq M_W^2 - 2E^\star \omega^\star (1 - \beta \cos \theta^\star) = M_W^2 - \frac{s}{2}(1 - \beta \cos \theta^\star)$$

$$u = 2M_W^2 - s - M_W^2 + \frac{s}{2}(1 - \beta \cos \theta^\star) = M_W^2 - \frac{s}{2}(1 + \beta \cos \theta^\star) \tag{26.142}$$

where $\theta^\star$ is the scattering angle of the first boson and β is the velocity of the W bosons given by:

$$\beta \equiv \frac{|\vec{k}^\star|}{\omega^\star} = \frac{\sqrt{\omega^{\star 2} - M_W^2}}{\omega^\star} = \sqrt{1 - \frac{4M_W^2}{s}} \tag{26.143}$$

Hence, the kinematical threshold for creating two on-shell W bosons is $s \geq 4M_W^2$ or $\sqrt{s} \geq 2M_W$, as expected. Further, since t is negative, the upper value of t is found with $\theta^\star = 0$. In addition, $|\cos \theta^\star| \leq 1$, so the range of allowed values for t is bounded by:

$$t_\pm = M_W^2 - \frac{s}{2}(1 \mp \beta) = -\frac{1}{2}\left(s - 2M_W^2\right) \pm \frac{1}{2}s\sqrt{1 - \frac{4M_W^2}{s}} \tag{26.144}$$

or

$$-\frac{1}{2}\left(s - 2M_W^2 - \sqrt{s(s - 4M_W^2)}\right) \leq t \leq -\frac{1}{2}\left(s - 2M_W^2 + \sqrt{s(s - 4M_W^2)}\right) \tag{26.145}$$

We will use these results in the following.

• **Tree-level amplitudes.** We consider the tree-level diagrams in Figure 26.12 one-by-one. To write down the

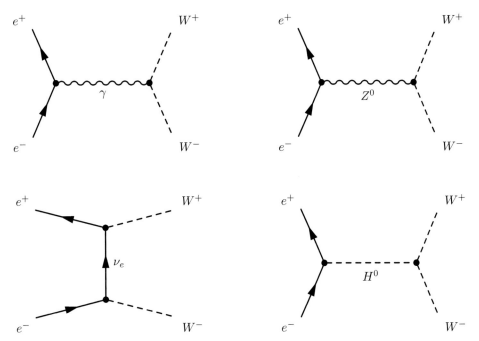

Figure 26.12 Tree-level Feynman diagrams for the $e^+e^- \to W^+W^-$ process.

first amplitude we use the trilinear $WW\gamma$ vertex rule **[GSW5]** and find:

$$\mathcal{M}_1 = \underbrace{\epsilon_2^{\mu*}(k_2)\epsilon_1^{\nu*}(k_1)}_{\text{final state bosons}} \underbrace{\bar{v}(k,r)\,(-ie\gamma_\delta)\,u(p,s)}_{ee\gamma \text{ vertex}} \underbrace{\left(\frac{-ig^{\delta\lambda}}{q^2}\right)}_{\text{photon propagator}}$$
$$\times \underbrace{\{-ie\,[g_{\mu\nu}(-k_2+k_1)_\lambda + g_{\nu\lambda}(-k_1-q)_\mu + g_{\lambda\mu}(q+k_2)_\nu]\}}_{\text{trilinear vertex}} \tag{26.146}$$

where $q = p + k = k_1 + k_2$. Hence, after simplification, one obtains:

$$\mathcal{M}_1 = i\epsilon_2^{\mu*}(k_2)\epsilon_1^{\nu*}(k_1)\bar{v}(k,r)\left(\frac{e^2\gamma^\lambda}{q^2}\right)u(p,s)$$
$$\times \{g_{\mu\nu}(-k_2+k_1)_\lambda + g_{\nu\lambda}(-k_1-q)_\mu + g_{\lambda\mu}(q+k_2)_\nu\} \tag{26.147}$$

At this stage, it is convenient to introduce a tensor that describes the WW vertex. Accordingly, we write:

$$\mathcal{M}_1 = i\epsilon_2^{\mu*}(k_2)\epsilon_1^{\nu*}(k_1)\bar{v}(k,r)W_{1\mu\nu}u(p,s) \tag{26.148}$$

where the 4×4 tensor matrix is equal to:

$$W_{1\mu\nu} \equiv \left(\frac{e^2\gamma^\lambda}{q^2}\right)\{g_{\mu\nu}(-k_2+k_1)_\lambda + g_{\nu\lambda}(-k_1-q)_\mu + g_{\lambda\mu}(q+k_2)_\nu\} \tag{26.149}$$

The expression can be further simplified by noting that $q = k_1 + k_2$ and $s = q^2$, thus:

$$W_{1\mu\nu} = \left(\frac{e^2\gamma^\lambda}{s}\right)\{g_{\mu\nu}(k_1-k_2)_\lambda - g_{\nu\lambda}(2k_1+k_2)_\mu + g_{\lambda\mu}(k_1+2k_2)_\nu\}$$
$$= \left(\frac{e^2}{s}\right)\{g_{\mu\nu}(\slashed{k}_1 - \slashed{k}_2)_\lambda - \gamma_\nu(2k_1+k_2)_\mu + \gamma_\mu(k_1+2k_2)_\nu\} \tag{26.150}$$

The second diagram with the Z^0 exchange can be derived promptly by replacing the photon with the Z^0 propagator and inserting the corresponding couplings. One finds:

$$\mathcal{M}_2 = i\epsilon_2^{\mu*}(k_2)\epsilon_1^{\nu*}(k_1)\bar{v}(k,r)W_{2\mu\nu}u(p,s) \tag{26.151}$$

with

$$
\begin{aligned}
W_{2\mu\nu} &= \left(\frac{g_Z\left(c_V^e + c_A^e\gamma^5\right)}{2(s - M_Z^2 + iM_Z\Gamma_Z)}\right)[e\cot\theta_W\{g_{\mu\nu}(\slashed{k}_1 - \slashed{k}_2)_\lambda - \gamma_\nu(2k_1 + k_2)_\mu + \gamma_\mu(k_1 + 2k_2)_\nu\}]\\
&= \left(\frac{e^2\left(c_V^e + c_A^e\gamma^5\right)}{2\sin^2\theta_W(s - M_Z^2 + iM_Z\Gamma_Z)}\right)\{g_{\mu\nu}(\slashed{k}_1 - \slashed{k}_2)_\lambda - \gamma_\nu(2k_1 + k_2)_\mu + \gamma_\mu(k_1 + 2k_2)_\nu\} \tag{26.152}
\end{aligned}
$$

where $c_V^e = -1/2 + 2\sin^2\theta_W$ and $c_A^e = -1/2$. The third diagram (t-channel neutrino exchange) can readily be written (cf. Eq. (11.190)):

$$
\begin{aligned}
\mathcal{M}_3 &= \epsilon_2^{\mu*}(k_2)\epsilon_1^{\nu*}(k_1)\bar{v}(k,r)\left(-i\frac{g_W}{\sqrt{2}}\gamma_\mu\left(\frac{1-\gamma^5}{2}\right)\right)\frac{i\slashed{q}}{q^2}\left(-i\frac{g_W}{\sqrt{2}}\gamma_\nu\left(\frac{1-\gamma^5}{2}\right)\right)u(p,s)\\
&= -i\frac{g_W^2}{2s}\epsilon_2^{\mu*}(k_2)\epsilon_1^{\nu*}(k_1)\bar{v}(k,r)\left(\gamma_\mu\left(\frac{1-\gamma^5}{2}\right)\right)\slashed{q}\left(\gamma_\nu\left(\frac{1-\gamma^5}{2}\right)\right)u(p,s)\\
&= -i\frac{g_W^2}{2s}\epsilon_2^{\mu*}(k_2)\epsilon_1^{\nu*}(k_1)\bar{v}(k,r)\left(\frac{1+\gamma^5}{2}\right)^2\gamma_\mu\slashed{q}\gamma_\nu u(p,s) \tag{26.153}
\end{aligned}
$$

Hence, as before:

$$\mathcal{M}_3 = i\epsilon_2^{\mu*}(k_2)\epsilon_1^{\nu*}(k_1)\bar{v}(k,r)W_{3\mu\nu}u(p,s) \tag{26.154}$$

with

$$W_{3\mu\nu} = -i\frac{e^2\left(1+\gamma^5\right)}{8\sin^2\theta_W s}\left(\gamma_\mu\slashed{q}\gamma_\nu\right) \tag{26.155}$$

Finally, the amplitude $\mathcal{M}_4$ with the Higgs boson vanishes if we neglect the incoming electron rest mass, so we do not consider it further. We can therefore ultimately write the total amplitude in compact form as:

$$\mathcal{M} = i\epsilon_2^{\mu*}(k_2)\epsilon_1^{\nu*}(k_1)\bar{v}(k,r)W_{\mu\nu}u(p,s) \tag{26.156}$$

with

$$
\begin{aligned}
W_{\mu\nu} &= e^2\left(\frac{1}{s} + \frac{(-1/2 + 2x_W - 1/2\gamma^5)}{2x_W(s - M_Z^2 + iM_Z\Gamma_Z)}\right)\{g_{\mu\nu}(\slashed{k}_1 - \slashed{k}_2)_\lambda - \gamma_\nu(2k_1 + k_2)_\mu + \gamma_\mu(k_1 + 2k_2)_\nu\}\\
&\quad - i\frac{e^2\left(1+\gamma^5\right)}{8x_W s}\left(\gamma_\mu\slashed{q}\gamma_\nu\right) \tag{26.157}
\end{aligned}
$$

where, as in the previous sections, $x_W \equiv \sin^2\theta_W \simeq 0.23$ (see Eq. (26.76)). We recognize three distinct terms corresponding to the photon trilinear vertex, the Z^0 trilinear vertex, and the t-channel neutrino exchange.

• **$W^\pm$ polarization and Goldstone boson equivalence theorem.** The $W^\pm$ bosons are massive spin-1 particles, hence they possess three independent states of polarization. As already pointed out in Section 10.14, the longitudinal polarization can pose problems of unitarity at high energy $E \gg M_W$. The three polarization vectors of a W can be written as (see Eq. (10.150)):

$$\underbrace{\epsilon_1^\mu = (0,1,0,0); \quad \epsilon_2^\mu = (0,0,1,0);}_{\text{transverse}} \quad \underbrace{\epsilon_3^\mu = \frac{1}{M_W}(|\vec{k}|,0,0,\omega)}_{\text{longitudinal}} \to \frac{E}{M_W}(1,0,0,1) \tag{26.158}$$

The longitudinal polarization vector becomes infinite since $E/M_W \to \infty$ as $E \to \infty$. Naively, for $M_W \simeq 80$ GeV and $g \simeq 0.1$, the unitarity bound is violated at $E \simeq 1$ TeV. One can say that the perturbation regime of the electroweak theory would break down at this energy scale. This would be unwelcome, but would happen *unless* there is cancellation of the diagrams in $W_{\mu\nu}$ (strictly speaking this occurs at the highest energies also due to the diagram containing the Higgs propagator). Although this cancellation is not immediately evident from Eq. (26.157), this is actually what happens! As a matter of fact, this is required by a result called the **Goldstone boson equivalence theorem**. It states that as the energy increases, the cross-section for producing the longitudinal W bosons should be equal to the cross-section for the production of the corresponding spin-0 scalar Goldstone bosons, which is well behaved as $1/s$. We can say that the gauge bosons "remember" the origin of their longitudinal polarization acquired during spontaneous symmetry breaking (see Section 24.8). Unbelievable! In the following we will perform the explicit computations.

- **Differential cross-sections for each diagram.** We shall now compute the amplitude squared of each diagram. The amplitude squared averaged over initial-state spins and summed over final-state polarizations of the W bosons is given by:

$$\langle |\mathcal{M}|^2 \rangle = \frac{1}{4} \sum_{s,r} \sum_{\lambda_1,\lambda_2} |\mathcal{M}|^2 \tag{26.159}$$

where the s, r indices define the spin of the incoming electron and positron, and the λ_1 and λ_2 define the three polarization states of the Ws. A lengthy calculation leads to:

$$\langle |\mathcal{M}_1|^2 \rangle = \frac{8\pi^2 \alpha^2}{M_W^4} \frac{1}{s^2} \big[2sM_W^2 \left(2s^2 + 3st + 2t^2\right) - M_W^4 \left(17s^2 + 20st + 12t^2\right) + 4M_W^6 (s + 6t) $$
$$ - 12 M_W^8 - s^2 t(s+t) \big] \tag{26.160}$$

Using Eq. (5.145), we find:

$$\left(\frac{d\sigma}{d\Omega} \right)_{CMS} = \left(\frac{1}{64\pi^2 s} \right) \langle |\mathcal{M}|^2 \rangle \implies \left(\frac{d\sigma}{dt} \right) = \left(\frac{1}{16\pi s^2} \right) \langle |\mathcal{M}|^2 \rangle \tag{26.161}$$

We conveniently label:

$$\left(\frac{d\sigma_{\gamma\gamma}}{dt} \right) = \left(\frac{d\sigma_1}{dt} \right), \quad \left(\frac{d\sigma_{ZZ}}{dt} \right) = \left(\frac{d\sigma_2}{dt} \right), \quad \text{and} \quad \left(\frac{d\sigma_{\nu\nu}}{dt} \right) = \left(\frac{d\sigma_3}{dt} \right) \tag{26.162}$$

Hence, the photon s-channel is given by:

$$\left(\frac{d\sigma_{\gamma\gamma}}{dt} \right) = \left(\frac{\pi \alpha^2}{2x_W^2 M_W^4 s^2} \right) \frac{x_W^2}{s^2} \left(\frac{d\sigma_A}{dt} \right) \tag{26.163}$$

where

$$\frac{d\sigma_A}{dt} = 2sM_W^2 \left(2s^2 + 3st + 2t^2\right) - M_W^4 \left(17s^2 + 20st + 12t^2\right) + 4M_W^6(s+6t) - 12M_W^8 - s^2 t(s+t)$$
$$ = 4s^3 M_W^2 + 6s^2 t M_W^2 - 17s^2 M_W^4 + 4st^2 M_W^2 - 20st M_W^4 + 4s M_W^6 - 12t^2 M_W^4 + 24t M_W^6 $$
$$ - 12 M_W^8 - s^3 t - s^2 t^2 \tag{26.164}$$

An equally lengthy calculation leads to the Z^0 s-channel:

$$\left(\frac{d\sigma_{ZZ}}{dt} \right) = \left(\frac{\pi \alpha^2 (1 - x_W)^2 \left(8x_W^2 - 4x_W + 1\right)}{16 M_W^4 x_W^2 s^2 \cos^2 \theta_W \left(s - M_Z^2 + iM_Z \Gamma_Z\right)^2} \right) \big[2s M_W^2 \left(2s^2 + 3st + 2t^2\right) $$
$$ - M_W^4 \left(17s^2 + 20st + 12t^2\right) + 4M_W^6(s+6t) - 12M_W^8 - s^2 t(s+t) \big] $$
$$ \equiv \left(\frac{\pi \alpha^2}{2x_W^2 M_W^4 s^2} \right) \frac{(1 - x_W)\left(x_W^2 - \frac{1}{2}x_W + \frac{1}{8}\right)}{(s - M_Z^2 + iM_Z\Gamma_Z)^2} \left(\frac{d\sigma_A}{dt} \right) \tag{26.165}$$

which is similar to the photon s-channel (check the terms in the square brackets) except for the factors in the front which determine the coupling and the Z^0 propagator. Another equally lengthy calculation leads to the ν_e-exchange t-channel:

$$
\begin{aligned}
\left(\frac{\mathrm{d}\sigma_{\nu\nu}}{\mathrm{d}t}\right) &= \frac{\pi\alpha^2}{16x_W^2 M_W^4}\frac{1}{s^2 t^2}\left[2t^2 M_W^2(2s+t) - tM_W^4(4s+5t) + 8tM_W^6 - 4M_W^8 - t^3(s+t)\right] \\
&\equiv \left(\frac{\pi\alpha^2}{2x_W^2 M_W^4 s^2}\right)\left(\frac{\mathrm{d}\sigma_B}{\mathrm{d}t}\right)
\end{aligned}
\tag{26.166}
$$

where

$$
\left(\frac{\mathrm{d}\sigma_B}{\mathrm{d}t}\right) = \frac{1}{8t^2}\left[4st^2 M_W^2 - 4stM_W^4 + 2t^3 M_W^2 - 5t^2 M_W^4 + 8tM_W^6 - 4M_W^8 - st^3 - t^4\right]
\tag{26.167}
$$

Yet another calculation yields the interference between the γ and Z-exchange diagrams:

$$
\begin{aligned}
\left(\frac{\mathrm{d}\sigma_{\gamma Z}}{\mathrm{d}t}\right) &= \left(\frac{1}{16\pi s^2}\right)\langle 2\Re e(\mathcal{M}_1\mathcal{M}_2^*)\rangle \\
&= \left(\frac{\pi\alpha^2}{4M_W^4 s^2}\right)\frac{1}{s}\frac{(4x_W-1)}{x_W\left(s-M_Z^2+iM_Z\Gamma_Z^2\right)}\left[M_W^4\left(17s^2+20st+12t^2\right)\right. \\
&\qquad \left. -2sM_W^2\left(2s^2+3st+2t^2\right) - 4M_W^6(s+6t) + 12M_W^8 + s^2 t(s+t)\right] \\
&= -\left(\frac{\pi\alpha^2}{4M_W^4 s^2}\right)\frac{1}{s}\frac{(4x_W-1)}{x_W\left(s-M_Z^2+iM_Z\Gamma_Z^2\right)}\left(\frac{\mathrm{d}\sigma_A}{\mathrm{d}t}\right)
\end{aligned}
\tag{26.168}
$$

And the interference between the γ and ν_e-exchange diagram is given by:

$$
\begin{aligned}
\left(\frac{\mathrm{d}\sigma_{\gamma\nu}}{\mathrm{d}t}\right) &= \left(\frac{1}{16\pi s^2}\right)\langle 2\Re e(\mathcal{M}_1\mathcal{M}_3^*)\rangle \\
&= -\left(\frac{\pi\alpha^2}{8M_W^4 x_W s^2}\right)\frac{1}{st}\left[2tM_W^2\left(2s^2+2st+t^2\right) - 5stM_W^4 - M_W^6(6t-8s) + 4M_W^8\right. \\
&\qquad \left. -st^2(s+t)\right] \\
&\equiv -\left(\frac{\pi\alpha^2}{8M_W^4 x_W s^2}\right)\frac{1}{st}\left(\frac{\mathrm{d}\sigma_C}{\mathrm{d}t}\right)
\end{aligned}
\tag{26.169}
$$

where $\mathrm{d}\sigma_C/\mathrm{d}t$ is the polynomial in the square brackets. Finally, the interference between the Z^0 and ν_e-exchange diagram is given by:

$$
\begin{aligned}
\left(\frac{\mathrm{d}\sigma_{Z\nu}}{\mathrm{d}t}\right) &= \left(\frac{1}{16\pi s^2}\right)\langle 2\Re e(\mathcal{M}_2\mathcal{M}_3^*)\rangle \\
&= -\left(\frac{\pi^2\alpha^2}{16\pi M_W^4 x_W^2 s^2}\right)\frac{(1-2x_W)}{t\left(s-M_Z^2+iM_Z\Gamma_Z\right)}\left[2tM_W^2\left(2s^2+2st+t^2\right) - 5stM_W^4\right. \\
&\qquad \left. -M_W^6(6t-8s) + 4M_W^8 - st^2(s+t)\right] \\
&= -\left(\frac{\pi^2\alpha^2}{16\pi M_W^4 x_W^2 s^2}\right)\frac{(1-2x_W)}{t\left(s-M_Z^2+iM_Z\Gamma_Z\right)}\left(\frac{\mathrm{d}\sigma_C}{\mathrm{d}t}\right)
\end{aligned}
\tag{26.170}
$$

We note that the interference terms between the diagrams containing trilinear gauge boson couplings and the ν_e-exchange are negative, and this is good news in the context of the Goldstone boson equivalence theorem!

• **Total cross-section.** To obtain the cross-sections, we have to integrate over the kinematically allowed values of t (see Eq. (26.145)). One gets for σ_A:

$$
\sigma_A = \int_{t_-}^{t_+} \frac{d\sigma_A}{dt} dt
$$

$$
= [4s^3 t M_W^2 + 3s^2 t^2 M_W^2 - 17s^2 t M_W^4 + \frac{4}{3}st^3 M_W^2 - 10st^2 M_W^4 + 4st M_W^6 - 4t^3 M_W^4 + 12t^2 M_W^6
$$

$$
-12t M_W^8 - \frac{s^3 t^2}{2} - \frac{s^2 t^3}{3}]\Big|_{t_-}^{t_+} \tag{26.171}
$$

where $t_\pm$ are given in Eq. (26.145). Consequently:

$$
\sigma_A = \frac{\beta s^2}{12} \left(4\left(\beta^2 + 6\right)s^2 M_W^2 - 12\left(\beta^2 + 9\right)s M_W^4 - \left(\beta^2 - 3\right)s^3 - 144 M_W^6\right)
$$

$$
= \frac{\beta s^2}{12} \left(4\beta^2 s^2 M_W^2 - 12\beta^2 s M_W^4 - \beta^2 s^3 + 24s^2 M_W^2 - 108 s M_W^4 - 144 M_W^6 + 3s^3\right)
$$

$$
= s M_W^4 \frac{\beta s^2}{12} \left(4\beta^2 s M_W^{-2} - 12\beta^2 - \beta^2 s^2 M_W^{-4} + 24 s M_W^{-2} - 108 - 144\frac{M_W^2}{s} + 3s^2 M_W^{-4}\right)
$$

$$
= s M_W^4 \frac{\beta s^2}{12} \left(4\beta^2 \zeta - 12\beta^2 - \beta^2 \zeta^2 + 24\zeta - 108 - 144\zeta^{-1} + 3\zeta^2\right)
$$

$$
= s M_W^4 \frac{\beta s^2}{12} \left((3 - \beta^2)\zeta^2 + (4\beta^2 + 24)\zeta - 12\beta^2 - 108 - 144\zeta^{-1}\right) \tag{26.172}
$$

where

$$
\zeta \equiv \frac{s}{M_W^2} \geq 4 \tag{26.173}
$$

We now introduce the pre-factors to obtain:

$$
\sigma_{\gamma\gamma} = \left(\frac{\pi\alpha^2}{2x_W^2 M_W^4 s^2}\right) \frac{x_W^2}{s^2} \left(s M_W^4 \frac{\beta s^2}{12}\left((3 - \beta^2)\zeta^2 + (4\beta^2 + 24)\zeta - 12\beta^2 - 108 - 144\zeta^{-1}\right)\right)
$$

$$
= \left(\frac{\pi\alpha^2}{2x_W^2}\right) \frac{x_W^2}{s^2} \left(\frac{\beta s}{12}\left((3 - \beta^2)\zeta^2 + (4\beta^2 + 24)\zeta - 12\beta^2 - 108 - 144\zeta^{-1}\right)\right)
$$

$$
= \left(\frac{\pi\alpha^2}{8x_W^2}\right) \frac{\beta}{s} x_W^2 \underbrace{\left((1 - \frac{\beta^2}{3})\zeta^2 + (\frac{4}{3}\beta^2 + 8)\zeta - 4\beta^2 - 36 - 48\zeta^{-1}\right)}_{\equiv A} \tag{26.174}
$$

Similarly:

$$
\sigma_{ZZ} = \left(\frac{\pi\alpha^2}{2x_W^2 M_W^4 s^2}\right) \frac{(1 - x_W)\left(x_W^2 - \frac{1}{2}x_W + \frac{1}{8}\right)}{(s - M_Z^2 + iM_Z\Gamma_Z)^2} \frac{\beta s^2}{12} [\cdots]
$$

$$
= \left(\frac{\pi\alpha^2}{2x_W^2 s^2}\right) \frac{(1 - x_W)\left(x_W^2 - \frac{1}{2}x_W + \frac{1}{8}\right)}{(s - M_Z^2 + iM_Z\Gamma_Z)^2} \frac{\beta s^2}{4} sA
$$

$$
= \left(\frac{\pi\alpha^2}{8x_W^2}\right) \frac{\beta}{s}\left(1 - x_W\right)\left(x_W^2 - \frac{1}{2}x_W + \frac{1}{8}\right) \frac{s^2}{(s - M_Z^2 + iM_Z\Gamma_Z)^2} A \tag{26.175}
$$

The indefinite integration of the ν_e-exchange channel gives:

$$
\int \frac{d\sigma_B}{dt} dt = \frac{1}{8}\left[t M_W^2(4s + t) + \frac{4M_W^8}{t} - \frac{1}{6}t^2(3s + 2t) - M_W^4(4s\ln(t) + 5t) + 8M_W^6 \ln(t)\right]
$$

$$
= \frac{1}{8}t M_W^2(4s + t) + \frac{M_W^8}{2t} - \frac{5t M_W^4}{8} - \frac{1}{48}t^2(3s + 2t) - \frac{1}{2}M_W^4 \ln(t)(s - 2M_W^2) \tag{26.176}
$$

The integral contains log-free terms and one logarithm term. We shall now compute the definite integral within the boundaries of t. For the log-free terms, we find:

$$
\begin{aligned}
\sigma_{\nu\nu}^{lnfree} &= \left(\frac{\pi\alpha^2}{2x_W^2 M_W^4 s^2}\right)\left[\frac{1}{8}tM_W^2(4s+t)+\frac{M_W^8}{2t}-\frac{5tM_W^4}{8}-\frac{1}{48}t^2(3s+2t)\right]\Bigg|_{t_-}^{t_+} \\
&= \left(\frac{\pi\alpha^2}{8x_W^2}\frac{\beta}{s}\right)\frac{\left(8\left(4\beta^2-3\right)\zeta-4\left(11\beta^2-45\right)-\left(\beta^4-4\beta^2+3\right)\zeta^2-336\zeta^{-1}+384\zeta^{-2}\right)}{24s\left(1+\beta-2\zeta^{-1}\right)\left(\beta-1+2\zeta^{-1}\right)} \\
&= \left(\frac{\pi\alpha^2}{96x_W^2 s^2}\right)\sqrt{s(s-4M_W^2)}\left(20\zeta-48+\zeta^2\right) \quad (26.177)
\end{aligned}
$$

For the term with the logarithm, we get:

$$
\begin{aligned}
\sigma_{\nu\nu}^{ln} &= \left(\frac{\pi\alpha^2}{2x_W^2 M_W^4 s^2}\right)\left[-\frac{1}{2}M_W^4\ln(t)(s-2M_W^2)\right]\Bigg|_{t_-}^{t_+} = \left(\frac{\pi\alpha^2}{4x_W^2 s}\right)\left(1-2M_W^2/s\right)\ln\left(\frac{-\beta-2M_W^2/s+1}{\beta-2M_W^2/s+1}\right) \\
&= \left(\frac{\pi\alpha^2}{8x_W^2}\frac{\beta}{s}\right)\frac{2\left(\zeta-1\right)L}{\beta} \quad (26.178)
\end{aligned}
$$

where

$$
L \equiv \ln\left(\frac{1+\beta-2\zeta^{-1}}{1-\beta-2\zeta^{-1}}\right) \geq 0 \quad (26.179)
$$

The log term is dependent on the velocity of the boson and vanishes for ultra-relativistic W's. We note that when $s \gg 4M_W^2 \to \infty$, $\beta \to 1$ and $\sigma_{\nu\nu}^{ln} \to 0$, hence:

$$
\sigma_{\nu\nu} \simeq \left(\frac{\pi\alpha^2}{96x_W^2 s^2}\right)s\zeta^2 = \frac{\pi\alpha^2 s}{96x_W^2 M_W^4} \qquad \text{when } s \to \infty \quad (26.180)
$$

The rise of the cross-section as a function of s is dramatic!

To obtain the total cross-section σ_{tot}, we should add all contributions together. The total cross-section is plotted as the continuous curve in Figure 26.13 as a function of the center-of-mass energy $\sqrt{s}$. The dotted curve in the figure includes the ν_e exchange and the γWW diagram. The dashed curve shows the behavior when one considers only the ν_e exchange, where we recognize the dramatic divergence at high energy. The various contributions are tabulated in Table 26.4. Summing the various terms of the cross-section, we find that the total cross-section is well behaved:

$$
\sigma(e^+e^- \to W^+W^-) \simeq \frac{\pi\alpha^2}{2x_W^2 s}\ln(\zeta) = \frac{\pi\alpha^2}{2x_W^2 s}\ln\left(\frac{s}{M_W^2}\right) \qquad \text{when } s \to \infty \quad (26.181)
$$

As mentioned previously, each diagram separately yields an increasing cross-section with s which violates unitarity at high energies (see Table 26.4). However, the sum of all diagrams with their interferences produces cancellations that cause the total cross-section to behave like $\ln(s)/s$, reaching a maximum at finite s and then tending to zero, so consistent with unitarity. Comparison with data is discussed in the next chapter.

26.11 WW and ZZ Production at Hadron Colliders

The production of two gauge bosons is an important process which is also experimentally accessible at hadron colliders, for example:

$$
p+p \to W^+W^- + X \quad \text{and} \quad p+p \to Z^0 Z^0 + X \quad (26.182)
$$

The expressions derived in the previous section can be used to compute the relevant partonic processes $q\bar{q} \to W^+W^-, Z^0 Z^0, \ldots$ for the production of two gauge bosons at hadron colliders. For example, the expression of

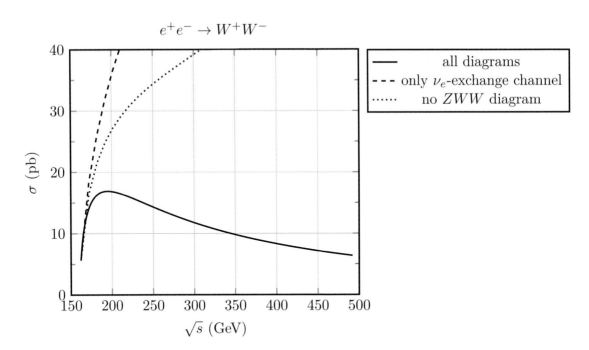

Figure 26.13 Total cross-section for $e^+e^- \to W^+W^-$ in picobarns as a function of the center-of-mass energy $\sqrt{s}$. The dashed curve shows the contribution from the ν_e-exchange t-channel. The dotted curve includes the ν_e exchange and the γWW diagram. The continuous line includes all contributions.

$\sqrt{s}$ **(GeV)**	β	L	σ_{tot} **(pb)**	$\sigma_{\gamma\gamma}$ **(pb)**	$\sigma_{\nu\nu}$ **(pb)**	$\sigma_{\nu\nu}^{lnfree}$ **(pb)**	$\sigma_{\nu\nu}^{ln}$ **(pb)**
165	0.2242	0.912	9.896	5.108	11.017	5.649	5.368
200	0.5946	2.739	16.856	14.018	35.688	21.550	14.138
250	0.7657	4.039	14.311	19.294	52.832	37.195	15.637
300	0.8442	4.943	11.755	23.500	64.935	50.589	14.347
350	0.8882	5.654	9.805	27.904	76.197	63.604	12.593
400	0.9156	6.245	8.316	32.815	88.026	77.081	10.945
450	0.9340	6.755	7.155	38.327	100.972	91.447	9.526
500	0.9469	7.203	6.231	44.473	115.267	106.933	8.334

Table 26.4 Cross-section for the process $e^+e^- \to W^+W^-$ as a function of $\sqrt{s}$. See text for the definitions of the variables.

Eq. (26.150) can be generalized to a pair of incoming fermion–antifermion $f\bar{f}$ (instead of electron–positron) by writing:

$$W_{1\mu\nu} = \left(\frac{Q_f e^2}{s}\right) \{g_{\mu\nu}(\slashed{k}_1 - \slashed{k}_2)_\lambda - \gamma_\nu(2k_1 + k_2)_\mu + \gamma_\mu(k_1 + 2k_2)_\nu\} \tag{26.183}$$

where Q_f is the charge of the fermion f (for quarks one should in addition include the color factors). The same can be done with the other diagrams. Finally, one can compute through a similar procedure the total cross-sections at the partonic level. These must then be embedded into the hadron–hadron process using standard techniques, as in Section 26.9.

The final results are illustrated in Figure 19.22 as the lines labeled σ_{WW} and σ_{ZZ}. As already mentioned, the cross-sections increase sharply with the center-of-mass energy of the collisions. They are suppressed by about four orders of magnitude compared to single gauge boson production. Consequently, they are only an extremely small fraction ($\simeq 10^{-11}$) of the total cross-section labeled σ_{tot} but represent very important signals to test the prediction of the electroweak theory.

Problems

Ex 26.1 Polarized W^- decay. We consider the three possible polarization states of the W^- boson: the longitudinal with $h = 0$ and the two transverse polarizations with $h = \pm 1$. Show that the decay distributions of the electron in the W^- rest frame are given by:

$$\frac{\mathrm{d}\Gamma_L}{\mathrm{d}\cos\theta} \propto \sin^2\theta \quad \text{and} \quad \frac{\mathrm{d}\Gamma_\pm}{\mathrm{d}\cos\theta} \propto (1 \mp \cos^2\theta)^2 \tag{26.184}$$

Ex 26.2 Complete production and decay of W bosons. We consider the complete production and decay reaction $u\bar{d} \to W^+ \to e^+\nu_e$.

(a) Draw the tree-level Feynman diagram and write the corresponding amplitude.

(b) Compute the matrix element squared averaged over initial spins and summed over final-state spins, neglecting the lepton and quark rest masses.

(c) Show that in the W rest frame, the differential cross-section can be expressed as:

$$\frac{\mathrm{d}\sigma(u\bar{d} \to e^+\nu_e)}{\mathrm{d}\cos\theta^\star} = \left(\frac{G_F|V_{ud}|^2 M_W^2}{8\pi\sqrt{2}}\right)^2 \left(\frac{\hat{s}}{(\hat{s} - M_W^2)^2 + (\Gamma_W M_W)^2}\right)(1 + \cos\theta^\star)^2 \tag{26.185}$$

Ex 26.3 Forward–backward asymmetry near the Z^0 resonance peak. The tree-level QED differential cross-section in the center of mass for the $e^+e^- \to \mu^+\mu^-$ reaction proceeding via photon exchange was found to be (see Eq. (11.153)):

$$\left(\frac{\mathrm{d}\sigma}{\mathrm{d}\Omega}\right)_\gamma = \left(\frac{\alpha^2}{4s}\right)(1 + \cos^2\theta) \tag{26.186}$$

(a) Show that the differential cross-section for this process proceeding via the Z^0 exchange given in Eq. (26.88) can be similarly written as:

$$\left(\frac{\mathrm{d}\sigma}{\mathrm{d}\Omega}\right)_{Z^0} = \left(\frac{\alpha^2}{4s}\right)\left(\frac{s^2}{(s - M_Z^2)^2 + M_Z^2\Gamma_Z^2}\right)\frac{1}{16\sin^4\theta_W \cos^4\theta_W}$$
$$\times \left[\left(c_V^2 + c_A^2\right)^2(1 + \cos^2\theta) + 8c_V^2 c_A^2 \cos\theta\right] \tag{26.187}$$

where we have set $c_V^e = c_V^\mu = c_V$ and $c_A^e = c_A^\mu = c_A$.

(b) Compute the interference term between the photon and the Z^0 diagram.

(c) Write the full differential cross-section including the photon, Z^0 and interference diagram in the form:

$$\left(\frac{\mathrm{d}\sigma}{\mathrm{d}\Omega}\right) = \left(\frac{\alpha^2}{4s}\right)\left(F_\gamma + F_{Z^0}\left(\frac{s^2}{(s - M_Z^2)^2 + M_Z^2\Gamma_Z^2}\right) + F_{\gamma Z^0}\left(\frac{s(s - M_Z^2)}{(s - M_Z^2)^2 + M_Z^2\Gamma_Z^2}\right)\right) \tag{26.188}$$

where F_γ, F_{Z^0}, and $F_{\gamma Z^0}$ are functions of $\cos\theta$ that need to be determined.

(d) Plot the differential cross-section $d\sigma/d\cos\theta$ as a function of $\cos\theta$ for a center-of-mass energy of (1) $M_Z - 5$ GeV, (2) M_Z, and (3) $M_Z + 5$ GeV. What can you say about the forward–backward asymmetry?

(e) Compute the actual asymmetry A_{FB} by integrating each term in the differential cross-section. Plot it as a function of the center-of-mass energy.

(f) In **Ex 19.1**, such collisions were generated using the PYTHIA Monte-Carlo generator. Modify that program to generate 10,000 $e^+e^- \to \mu^+\mu^-$ events at each of the following center-of-mass energies (1) $M_Z - 5$ GeV, (2) M_Z, and (3) $M_Z + 5$ GeV. Plot the $\cos\theta$ distribution of one of the outgoing muons and compute A_{FB}. Compare with the theoretical values.

Ex 26.4 The $e^+e^- \to W^+W^-$ process. Use FeynCalc (see Section 11.13) to compute and verify the matrix elements in Section 26.10.

27 Experimental Tests of the Electroweak Theory

> Well, those were great days (. . .) a time when experimentalists and theorists were really interested in what each other had to say, and made great discoveries through their mutual interchange.
>
> *Steven Weinberg[1]*

27.1 The Free Parameters of the Electroweak Theory

The electroweak theory is a very powerful theory which can be used to compute many processes. **The gauge sector is fully determined by only three independent parameters which completely fix the gauge boson masses and their coupling to fermions.** The basic set of parameters can be chosen to be those appearing in the Lagrangian:

$$g_1, g_2, v \tag{27.1}$$

However, it is customary to choose instead more convenient parameters which can be accurately measured experimentally. Therefore, in practice, the fine-structure constant α, the Fermi constant G_F (e.g., given by the muon lifetime) and the weak mixing angle $\sin^2 \theta_W$ are preferred and are commonly used:

$$\alpha, G_F, \sin^2 \theta_W \tag{27.2}$$

With this convention, the strength of the electromagnetic force and of the "effective" weak force are fixed, and the electroweak theory becomes essentially dependent on one parameter $\sin^2 \theta_W$, with the basic parameters of the Lagrangian derivable from the following relations (see Eq. (25.42)):

$$e = \sqrt{4\pi\alpha}, \quad g_1 = e/\cos\theta_W, \quad g_2 = g_W = e/\sin\theta_W \tag{27.3}$$

and (see Eq. (25.79))

$$v = \frac{2M_W}{g_2} = \left(\frac{1}{\sqrt{2}G_F}\right)^{1/2} \approx 246 \text{ GeV} \tag{27.4}$$

An important related constant precisely determined experimentally (see PDG [28]) is A_0 defined as (see Eq. (23.27)):

$$A_0 \equiv \sqrt{\frac{\pi\alpha}{\sqrt{2}G_F}} \simeq (37.28039 \pm 0.00001) \text{ GeV} \tag{27.5}$$

such that the weak boson masses are simply:

$$M_W = A_0/\sin\theta_W \quad \text{and} \quad M_Z = A_0/(\sin\theta_W \cos\theta_W) \tag{27.6}$$

---

1 Reprinted by permission from Springer Nature: S. Weinberg, "The making of the Standard Model," in *Symposium on Prestigious Discoveries at CERN : 1973 : Neutral currents. 1983 : W & Z Bosons*, CERN, Geneva, Switzerland, 16 Sep 2003 (EP-2003-073) and *Eur. Phys. J. C*, vol. 34, pp. 5–13, 2004. Copyright 2004.

Alternatively one can equally well use M_Z or M_W instead of $\sin^2 \theta_W$ as the third parameter. Indeed, in recent years, one prefers using the very precisely determined value of M_Z and relying on the three following parameters to fix the theory:

$$\alpha, \quad G_F, \quad \text{and} \quad M_Z = (91.1876 \pm 0.0021) \text{ GeV} \tag{27.7}$$

All electroweak observables can be determined from the three chosen parameters and the predicted values within the theory can be compared to direct measurements. In doing so, **one must take into account that all above relations are tree-level and are consequently affected by higher-order corrections**. For example, the W boson mass is related to the renormalized constants evaluated at $Q^2 = M_W^2$, while the weak mixing angle can be measured in, e.g., neutrino scattering at low Q^2 (as discussed in the next section). The relation between the weak Fermi constant and the weak mixing angle will be affected as well by radiative corrections. We write (cf. Eq. (23.25)):

$$\frac{G_F}{\sqrt{2}} \equiv \frac{g_2^2}{8M_W^2} \left(1 + \Delta r(m_t, m_H, \dots)\right) \tag{27.8}$$

Just as α and α_S, the weak mixing angle should therefore be considered as a "running coupling constant." However, there are different conventions in defining the weak mixing angle. In the so-called "**on-shell scheme**," the tree-level formula is promoted to a definition of the renormalized angle to all orders in perturbation theory. It is written as:

$$\text{on-shell scheme:} \quad x_W = s_W^2 \equiv 1 - M_W^2/M_Z^2 \tag{27.9}$$

then the gauge boson masses are:

$$M_W^2 = \frac{g_2^2 \sqrt{2}}{8G_F} \quad \longrightarrow \quad M_W = \frac{A_0}{s_W \sqrt{1 - \Delta r}}, \quad \text{and} \quad M_Z = \frac{M_W}{c_W} \tag{27.10}$$

and Δr includes the radiative corrections relating α, $\alpha(M_Z)$, G_F, M_W, and M_Z.

Another definition uses the vector and axial-vector couplings of the fermions. Recalling that at tree level we have:

$$\frac{c_V^f}{c_A^f} = 1 - \frac{2Q_f}{I_{3L}^f} \sin^2 \theta_W = 1 - 4|Q_f| \sin^2 \theta_W \tag{27.11}$$

one defines the "effective weak mixing angle" from a fermion pair s_f^2 as a function of the couplings of the given fermions (which depends on radiative corrections):

$$\text{effective mixing angle :} \quad s_f^2 \equiv \frac{1}{4|Q_f|} \left(1 - \frac{c_V^f}{c_A^f}\right) \tag{27.12}$$

Experimentally the couplings are most precisely measured with charged leptons, and one defines, assuming universality, the effective weak mixing angle based on charged leptons:

$$\text{leptonic effective mixing angle :} \quad s_\ell^2 \equiv \frac{1}{4} \left(1 - \frac{c_V^\ell}{c_A^\ell}\right) \tag{27.13}$$

In the absence of radiative corrections, we would of course have $s_W^2 = s_\ell^2$, but this is not the case. When testing the electroweak theory at very high energies, such as $\sqrt{s} \approx M_Z$ or above, a significant radiative effect already comes from the running of α itself. We have seen already that $\alpha(0) \simeq 1/137$, while $\alpha(M_Z) \simeq 1/127$ (see Section 12.6). In addition, electroweak radiative corrections depend in principle on all other parameters, such as the masses of the fermions entering in higher-order loops, etc. It turns out that the dominant contribution is from the top quark mass m_t. In addition, the Higgs boson itself enters in loops and hence radiative corrections will also depend on the Higgs mass m_H. So, the radiatively corrected electroweak theory will practically depend on five parameters, which can be chosen to be α, G_F, M_Z, m_t, and M_H. Finally, several observables will also depend on QCD radiative corrections, which implies that α_s will also play a role.

27.2 Measurement of Neutrinos Scattering Off Electrons

The importance of the purely leptonic interaction is that the electroweak theory makes clear predictions. Purely leptonic processes involve only fundamental fermions and only the weak interaction, without the need to worry about the nucleon structure or fragmentation. Because the cross-sections are small, one requires high-energy neutrinos and massive targets. High-energy neutrino beams at accelerators are predominantly of the muon type, with a fraction of a few percent of electron-type neutrinos. Massive neutrino detectors are often very large coarse calorimeters.

The scattering of muon neutrinos or muon antineutrinos off electrons can then be readily studied. The total cross-section $\nu_\mu - e$ scattering is given by (see Eq. (26.52)):

$$
\begin{aligned}
\sigma^{\mathrm{NC}}(\nu_\mu e^-) &= \sigma^0 \left\{ (c_L^e)^2 + \frac{1}{3}(c_R^e)^2 \right\} = \frac{\sigma^0}{4} \left\{ (c_V^e + c_A^e)^2 + \frac{1}{3}(c_V^e - c_A^e)^2 \right\} \\
&= \frac{\sigma^0}{4} \left\{ (c_V^e)^2 + (c_A^e)^2 + 2c_V^e c_A^e + \frac{1}{3}(c_V^e)^2 + \frac{1}{3}(c_A^e)^2 - \frac{2}{3}c_V^e c_A^e \right\} \\
&= \frac{\sigma^0}{3} \left\{ (c_V^e)^2 + (c_A^e)^2 + c_V^e c_A^e \right\}
\end{aligned}
\tag{27.14}
$$

where we used Eq. (26.12). In a similar way, one can show that for $\bar{\nu}_\mu - e$ scattering, we have:

$$
\sigma^{\mathrm{NC}}(\bar{\nu}_\mu e^-) = \frac{\sigma^0}{3} \left\{ (c_V^e)^2 + (c_A^e)^2 - c_V^e c_A^e \right\}
\tag{27.15}
$$

and $\sigma^0/E_\nu \approx 1.7 \times 10^{-41}$ cm^2 GeV^{-1} (see Eq. (26.48)). The cross-sections also depend on the electron weak couplings $c_V^e = -\frac{1}{2} + 2\sin^2\theta_W$ and $c_A^e = -1/2$, hence their experimental measurement leads to the determination of the weak mixing angle:

$$
\sigma^{\mathrm{NC}}(\nu_\mu e^-) = \sigma^0 \left[\frac{1}{4} - \sin^2\theta_W + \frac{4}{3}\sin^4\theta_W \right] \quad \text{and} \quad \sigma^{\mathrm{NC}}(\bar{\nu}_\mu e^-) = \frac{\sigma^0}{3} \left[\frac{1}{4} - \sin^2\theta_W + 4\sin^4\theta_W \right]
\tag{27.16}
$$

Comparing neutrinos and antineutrinos also allows us to fix the absolute signs of the couplings. Their ratio provides an experimentally aesthetic method to measure the weak mixing angle since several experimental errors cancel in the ratio:

$$
R \equiv \frac{\sigma^{\mathrm{NC}}(\nu_\mu e^-)/E_\nu}{\sigma^{\mathrm{NC}}(\bar{\nu}_\mu e^-)/E_\nu} = 3 \frac{1 - 4\sin^2\theta_W + \frac{16}{3}\sin^4\theta_W}{1 - 4\sin^2\theta_W + 16\sin^4\theta_W}
\tag{27.17}
$$

Since both cross-sections rise linearly with energy and both are measured in the same detector, the ratio R can be expressed as:

$$
R = \frac{N(\nu_\mu e^-) \int \phi_{\bar\nu} E_{\bar\nu} \mathrm{d}E_{\bar\nu}}{N(\bar{\nu}_\mu e^-) \int \phi_\nu E_\nu \mathrm{d}E_\nu}
\tag{27.18}
$$

where N is the number of events collected and the ϕs are the computed relative fluxes as a function of energy of the incident neutrinos in neutrino and antineutrino mode. This method is free of most systematic errors, since uncertainties in the knowledge of the absolute neutrino fluxes and of the experimental detection efficiencies cancel in the ratio. Only the relative normalization of fluxes in neutrino and antineutrino fluxes is relevant.

The measurement of neutrino scattering off electrons is experimentally challenging due to large potential backgrounds. Experimentally the signature for neutrino-electron scattering is an isolated electromagnetic shower initiated by a single electron which is kinematically constrained to a strongly forward peaked cone (see **Ex 27.1**). Most of the neutrino reactions consist of semi-leptonic neutrino reactions on nucleons: they outnumber purely leptonic scattering events by four orders of magnitude. The electron neutrino–nucleon scattering has a much higher cross-section and needs to be suppressed:

$$
\nu_e N \to e^- X \quad \text{or} \quad \bar{\nu}_e N \to e^+ X
$$

As mentioned, the electron-neutrino contamination is only at the few percent level and the presence of a hadronic system X helps identify and suppress this reaction. Another kind of background comes from the dominant ν_μ flux via the charged current nucleon scattering of the kind:

$$\nu_\mu N \to \mu^- (n\pi^0) X \qquad \text{or} \qquad \overline{\nu}_\mu N \to \mu^+ (n\pi^0) X \qquad (27.19)$$

where the muon escapes detection (e.g., when it has too low energy), or the neutral current reaction of the kind:

$$\nu_\mu N \to \nu_\mu (n\pi^0) X \qquad \text{or} \qquad \overline{\nu}_\mu N \to \overline{\nu}_\mu (n\pi^0) X \qquad (27.20)$$

There is also a source of background from the coherent production of neutral pions on the target nuclei:

$$\nu_\mu A \to \nu_\mu \pi^0 A \qquad \text{or} \qquad \overline{\nu}_\mu A \to \overline{\nu}_\mu \pi^0 A \qquad (27.21)$$

In all cases above the presence of neutral pions generates electromagnetic showers which can be misinterpreted as originating from electrons. The kinematics of the neutrino scattering leads to very forward peaked and energetic electrons. Often the hadronic system X is separated from the shower and can be identified to suppress these kinds of events. When a neutral pion mimics the characteristics of the very forward electron it is called a leading π^0, in the sense that it carries most of the energy of the hadronic system, leaving little energy for the other hadrons, which hence have a higher probability of going undetected. Thus, efficient discrimination between electron and hadron showers must be an essential feature of the detector design.

After the first measurements with Gargamelle in 1973, new experiments were performed to collect more events to increase statistical precision. In 1982 the world average was the following [344]:

$$\sigma(\nu_\mu e)/E_\nu = (1.49 \pm 0.24) \times 10^{-42} \text{ cm}^2 \text{ GeV}^{-1}$$
$$\sigma(\overline{\nu}_\mu e)/E_\nu = (1.69 \pm 0.33) \times 10^{-42} \text{ cm}^2 \text{ GeV}^{-1} \qquad (27.22)$$

with a precision of $\approx \pm 20\%$ due to limited statistics, and consistent with the electroweak predictions with a value $\sin^2 \theta_W \simeq 0.23$. In the late 1980s and early 1990s three new improved massive experiments aimed at the precise measurement of neutrino scattering off electrons were performed. They are listed below with their collected statistics:

1. The CHARM experiment with a mass of ≈ 100 t collected 83 $\nu_\mu e^-$ and 112 $\overline{\nu}_\mu e^-$ candidates with a neutrino beam with an average energy of 25 GeV produced by the 450 GeV CERN SPS protons.

2. The E734 experiment (≈ 170 t) collected 160 $\nu_\mu e^-$ and 98 $\overline{\nu}_\mu e^-$ candidates with an average energy of 1.5 GeV produced from 28 GeV protons in the AGS at BNL.

3. The CHARM-II experiment with a large mass of ≈ 700 t finally collected 2677 ± 82 $\nu_\mu e^-$ and 2752 ± 88 $\overline{\nu}_\mu e^-$ candidates also at the CERN SPS neutrino beam during the late 1980s and early 1990s.

The results are clearly statistically dominated by the CHARM-II experiment.

The **CHARM-II detector** was a modular detector, optimized to detect electrons in neutrino interactions. It was composed of ≈ 700 t of glass plates, which were interleaved with position counters (see Figure 27.1). The target contained 420 modules, each covering an area of 3.7×3.7 m^2. Each module contained a 48 mm thick (equivalent to 0.5 radiation lengths) glass plate, followed by 352 plastic streamer tubes. After each fifth module there was a scintillator plate for triggering purposes. The orientation of consecutive streamer tubes was perpendicularly arranged, so that both transverse coordinates of passing particles could be reconstructed. In total, energy and direction could be well determined.

The required separation of events induced by neutrinos scattering off electrons required a good energy and a precise direction (angular) determination for electromagnetic showers, induced by the leading electron in the event. This was achieved in a calorimeter like CHARM-II by a high-Z target complemented by detectors to provide an excellent granularity. Figure 27.2 shows an event collected with a test beam of 10 GeV electrons and

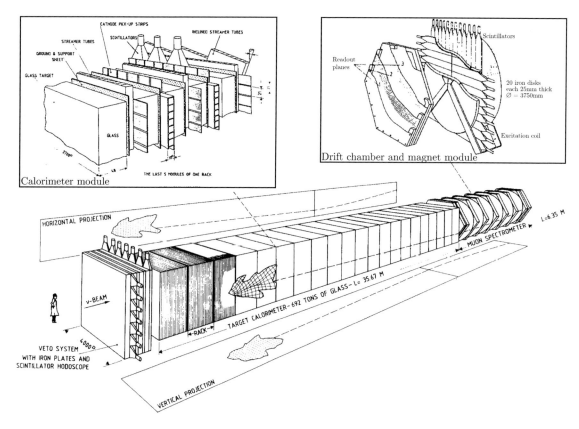

Figure 27.1 Schematic representation of the CHARM-II detector. Reprinted from K. De Winter *et al.*, "Experimental results obtained from a low Z, fine grained electromagnetic calorimeter," *Nucl. Instrum. Meth. A*, vol. 277, pp. 83–91, 1989, with permission from Elsevier.

pions. The black points represent the response of the streamer tubes, while the size of the squares represents the measured loss in that point. The magnetic measurement of the electric charge of the electron is not practical in such a large detector, because of the huge volume that would need to be magnetized with a high field.

The CHARM-II detector achieved an energy and angular resolution for electromagnetic shower at the level of:

$$\frac{\Delta E}{E} \approx \frac{0.23}{\sqrt{E/\text{GeV}}} + 0.05 \qquad \text{and} \qquad \Delta\theta \approx 17\,\frac{\text{mrad}}{\sqrt{E/\text{GeV}}} \qquad (27.23)$$

which provided the required energy and angular resolution for electrons. The events which passed the cuts for the identification of electromagnetic showers were used for further analysis. Electromagnetic showers between 3 and 24 GeV were kept and classified according to the so-called energy-angle-squared variable, which is kinematically constrained to be:

$$E_e\theta^2 \leq 2m_e \approx 1\text{ MeV} \qquad (27.24)$$

The observed distribution of this variable is shown in Figure 27.3, corresponding to the neutrino and antineutrino mode collected by CHARM-II [345]. The signal events are strongly forward peaked with a small $E_e\theta^2$. Backgrounds from electron neutrino–nucleon scattering have a much larger angle and so do the showers in misidentified leading π^0 events, since they originate from fragmentation which introduces transverse momentum, in addition to the scattering angle of the hadronic system already recoiling against the outgoing lepton.

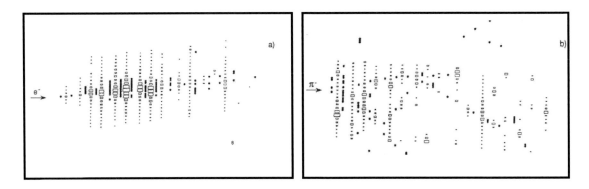

Figure 27.2 Test beam particles of 10 GeV: (a) electron, (b) pion. See text for explanation of the figure. Reprinted from D. Geiregat *et al.*, "Calibration and performance of the CHARM-II detector," *Nucl. Instrum. Meth. A*, vol. 325, pp. 92–108, 1993, with permission from Elsevier.

The data are fitted simultaneously in the two-dimensional plane $E_e\theta^2$ vs. E_e. The fit includes the signal

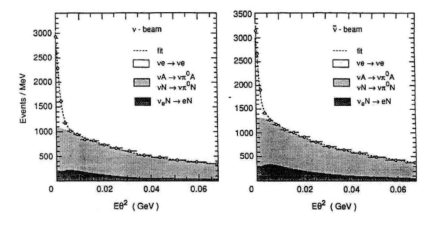

Figure 27.3 Distribution of $E_e\theta^2$ for the ν_μ and $\overline{\nu}_\mu$ scattering observed by CHARM-II. Reprinted from P. Vilain *et al.*, "Precision measurement of electroweak parameters from the scattering of muon neutrinos on electrons," *Phys. Lett.*, vol. B335, pp. 246–252, 1994, with permission from Elsevier.

labeled $\nu e \to \nu e$ and the backgrounds $\nu A \to \nu \pi^0 A$, $\nu N \to \nu \pi^0 N A$, and $\nu_e N \to eN$.

In the $c_V - c_A$ plane, a given absolute cross-section corresponds to an ellipse centered at the center $(0,0)$. The axes of the ellipses for neutrinos and antineutrinos are perpendicular and comparing both cross-sections leads to a fourfold ambiguity in the solutions for c_V and c_A. The ellipses are represented as dashed lines in Figure 27.4, and their overlap determines the four solutions.

The absolute normalization of neutrino fluxes and the presence of ν_e and $\overline{\nu}_e$ events detected in the same experiment was used to reduce the fourfold ambiguity of c_V and c_A to two solutions. The $\nu_e e^-$ scattering cannot be extracted on an event-by-event basis, but statistically they populate the higher-energy region of the E_e distribution because of the higher energy of the incoming $\nu_e(\overline{\nu}_e)$ present in the neutrino beam. Since the $\nu_e e^-$ contains both neutral and charged current reactions, we have seen in Section 26.3 that this corresponds to adding one to the couplings of the $\nu_\mu e^-$ cross-section. Hence, the absolute measurements of the $\nu_e e^-$ and

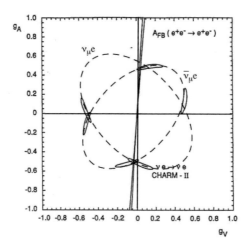

Figure 27.4 90% CL for the vector and axial-vector couplings of the electron from the absolute cross-section measurement of purely leptonic neutrino scattering off electrons in CHARM-II. Reprinted from P. Vilain *et al.*, "Precision measurement of electroweak parameters from the scattering of muon neutrinos on electrons," *Phys. Lett.*, vol. B335, pp. 246–252, 1994, with permission from Elsevier.

$\bar{\nu}_e e^-$ leads to two similar ellipses but this time with their center shifted to $(-1, -1)$. Hence, the ambiguity of the overlap can be reduced to two solutions. Finally, from the contours in the $c_V - c_A$ plane, CHARM-II extracted the values of the vector and axial-vector coupling constants of the electron:

$$\text{CHARM-II:} \quad c_V^e = (-35 \pm 17) \times 10^{-3} \quad \text{and} \quad c_A^e = (-503 \pm 17) \times 10^{-3} \tag{27.25}$$

which was consistent with the electroweak predictions and allowed the determination of the mixing angle with impressive percent level precision [345]:

$$\text{CHARM-II:} \quad \sin^2 \theta_W = 0.2324 \pm 0.0058 (\text{stat.}) \pm 0.0059 (\text{syst.}) \tag{27.26}$$

The vector couplings of the electron and of all other charged leptons turn out to be small because $\sin^2 \theta_W$ is almost equal to $1/4$.

27.3 Measurement of the Inverse Muon Decay at High Energy

The inverse muon decay (IMD) process:

$$\nu_\mu + e^- \rightarrow \mu^- + \nu_e \tag{27.27}$$

is a purely leptonic process which can be used to test weak interactions without hadronic uncertainties. We have computed its cross-section in Section 23.9 assuming a pure $V - A$ current. Measuring this process can therefore provide a direct test of these assumptions, and in particular any admixture of $V + A$ current and opposite helicity contributions from the neutrino can be tested. The rate of $\approx 0.1\%$ is small compared to the dominant neutrino–nucleon scattering process $\nu_\mu N \rightarrow \mu + X$, and the threshold is relatively high with $E_\nu \geq 11$ GeV (see Eq. (23.161) and Chapter 28).

The experimental signature is characterized by a single negatively charged high-energy muon, with no hadronic activity. Because of the small mass of the electron, the final-state muon is highly forward-boosted with a very small angle θ_μ in the laboratory, and the reconstructed Q^2 is low. In order to estimate background, one can search for oppositely charged muons (so called wrong-sign sample) since the following reaction is forbidden by charge conservation:

$$\bar{\nu}_\mu + e^- \nrightarrow \mu^+ + \nu_e \tag{27.28}$$

In the opposite charge sample, the events will be dominated by $\bar{\nu}_\mu N \to \mu^+ + X$ which can then be rescaled to predict the $\nu_\mu N \to \mu^- + X$ background in the right-sign sample.

We illustrate here the measurement of CCFR [346], which used the Tevatron Quadrupole Triplet neutrino beam (QTB). The beam was composed of neutrinos and antineutrinos with a ratio $\simeq 2:1$ and had a range $0 \leq E_\nu \leq 600$ GeV. The list of parameters of the CCFR detector is given in Table 28.1. The muon momentum resolution is 11% dominated by multiple scattering in the toroidal spectrometer (see Figure 28.16). The hadronic energy resolution can be parameterized as $\sigma_{E_{had}}/E_{had} \simeq 0.89/\sqrt{E_{had}\,(\text{GeV})}$.

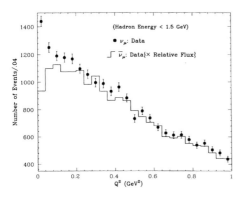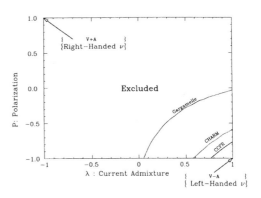

Figure 27.5 (left) Distribution of $Q^2 = E_\nu E_\mu \theta_\mu^2$ for events with hadronic energy $E_{had} \leq 1.5$ GeV. (right) Limits on the polarization and current admixture at the 90% CL. Reprinted with permission from S. Mishra *et al.*, "Measurement of inverse muon decay $\nu_\mu + e \to \mu^- + \nu_e$ at Fermilab Tevatron energies 15–600 GeV," *Phys. Rev. Lett.*, vol. 63, pp. 132–135, 1989. Copyright 1981 by the American Physical Society.

The distribution of the reconstructed $Q^2 = 4E_\nu E_{had} \sin^2(\theta_\mu/2)$ (cf. Eq. (28.21)) where $E_\nu = E_\mu + E_{had}$ in the data samples with a hadronic energy $E_{had} \leq 1.5$ GeV is plotted in Figure 27.5(left). The dots show the right-sign muons, while the histogram represents the wrong-sign muon sample rescaled by the expected ratio of the flux. One can clearly observe an excess of events at very low $Q^2 \lesssim 0.2$ GeV^2, which corresponds to the signal. The measured rate w.r.t to all events is $R = (0.125 \pm 0.009(\text{stat.}) \pm 0.003(\text{syst.})) \times 10^{-2}$, which translates into the following cross-section [346]:

$$\sigma(\nu_\mu + e^- \to \mu^- + \nu_e) = (17.00 \pm 1.22(\text{stat.}) \pm 0.41(\text{syst.})) \times 10^{-42} E_\nu(\text{GeV})\ \text{cm}^2 \tag{27.29}$$

perfectly consistent with the calculation of Eq. (23.168). Consequently, the measurements support the $V-A$ current structure, the left-handedness of the neutrino and the linearly rising cross-section with E_ν. This can be expressed more quantitatively by writing the differential cross-section for IMD for arbitrary couplings c_V and c_A and neutrino polarization [346]:

$$\left(\frac{d\sigma}{d\Omega}\right)_{CMS} = \frac{G_F^2 s}{64\pi^2} \left[(1+P)(1-\lambda)(1-\cos\theta)^2 + 4(1-P)(1+\lambda)\right] \tag{27.30}$$

where the scattering angle θ is given in the center-of-mass system, P denotes the neutrino polarization, and:

$$\lambda = -2\frac{\Re e(c_V^* c_A)}{c_V^2 + c_A^2} \tag{27.31}$$

The number of experimentally observed events can be used to set a constraint in the $P - \lambda$ plane. This is summarized in Figure 27.5(right). The observations are in clear agreement with a left-handed neutrino with a $V - A$ weak coupling. We have already discussed similar constraints coming from the μ decays (see Section 23.8). The measurement with neutrinos, however, is at much higher energies and hence complements the observations with muon decays.

27.4   The Discovery of the $W^{\pm}/Z^0$ Gauge Bosons

A crucial milestone for the establishment of the electroweak theory was the discovery and precise determination of the properties and couplings of the W and Z bosons.

• **Original idea.** In 1976 **Cline**,[2] **McIntyre**,[3] and **C. Rubbia**[4] proposed converting existing proton accelerators into proton–antiproton colliders [347], which could then produce the heavy gauge bosons predicted by the electroweak model. They wrote: *"The search for these massive bosons requires three separate elements to be successful: a reliable physical mechanism for production, very high center of mass energies, and an unambiguous experimental signature to observe them. In this note we outline a scheme which satisfies these requirements and that could be carried out with a relatively modest program at existing proton accelerators."* They pointed out that in proton–antiproton collisions the production mechanism $q + \bar{q} \rightarrow Z^0$ is similar to $e^+e^- \rightarrow Z^0$, for which, however, no sufficiently powerful electron–positron collider existed at the time. On the other hand, large proton synchrotrons had just been completed at CERN and at Fermilab, which could accelerate protons up to energies of several hundreds of giga-electronvolts. They estimated a cross-section of $\sigma(p\bar{p} \rightarrow Z^0 \rightarrow \mu^+\mu^-) \simeq 10^{-33}$–$10^{-32}$ cm^2 = 1–10 nb (see Eq. (1.25)).

To produce $R \simeq 1$ event per hour hence required a luminosity $\mathcal{L} \simeq R/\sigma \approx 3 \times 10^{28-29}$ cm^{-2} s^{-1}. They computed the luminosity for two bunches colliding head-on using the relation $\mathcal{L} = N_p N_{\bar{p}} \phi / a$, where N_p and $N_{\bar{p}}$ are the number of protons and antiprotons circulating in the machine, respectively, ϕ is the revolution frequency, and a is the effective area of interaction of the two beams (see Eq. (3.7)). They took $N_p = 10^{12}$ protons and estimated that for standard beam sizes, the necessary luminosity required $N_{\bar{p}} = 3 \times 10^{10}$ antiprotons. They also estimated that the half-life of the luminosity due to beam-gas scattering is about 24 h for an average residual pressure in the accelerator. So, the question was how to produce enough antiprotons per day, which could be injected into the same storage ring as protons (however circulating in the opposite direction)?

Antiprotons can be produced if protons of sufficient energy hit on a target (see Section 15.9). In the original proposal, the antiproton yield was assumed to be 4×10^{-6} antiprotons per proton. This is an apparently small number, but still raised questions among the sceptics. This yield would require 750 pulses of 10^{13} protons to accumulate the needed $N_{\bar{p}} = 3 \times 10^{10}$ antiprotons.

• **CERN** $Sp\bar{p}S$ **collider.** CERN saw the chances and supported the idea, where a first step was to demonstrate on a large scale that antiprotons could be produced and stored in a ring, ready for injection into the CERN 400 GeV SPS accelerator (see Figure 27.6(left) and also Figure 3.7). In case they should succeed, the **CERN SPS accelerator** would then become the $Sp\bar{p}S$ collider! The 26 GeV **CERN PS proton accelerator** (also illustrated in Figure 27.6) had sufficient energy, however, antiprotons are produced in a wide phase space, contrary to what can be accepted by a storage ring where the momentum, the transverse position, and the divergence should be small for particles to remain within the focusing capabilities of the magnets. Antiparticles outside the acceptance of the storage ring would simply be lost and not stored. A dedicated antiproton accumulator ring called **Antiproton Accumulator** (AA) was built to capture with relatively large acceptance the antiprotons of 3.5 GeV. Still the efficiency of capturing antiprotons from the PS target was in the range of 10^{-5}. Of course, the optics of the SPS accelerator could not be changed and an accelerator of that large acceptance was totally unpractical. The idea was then to "cool" the antiprotons stored in the AA such that they would be accepted by the SPS ring. An accelerator physicist **Van der Meer**[5] who had invented the method of "**stochastic**

---

2   David Bruce Cline (1933–2015), American particle physicist.
3   Peter M. McIntyre (born 1947), American physicist.
4   Carlo Rubbia (born 1934), Italian physicist.
5   Simon van der Meer (1925–2011), Dutch particle accelerator physicist.

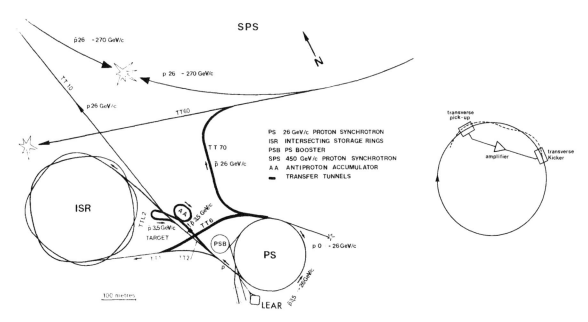

Figure 27.6 (left) Layout of the CERN proton–antiproton accelerator complex. Reprinted from Staff of the CERN proton–antiproton project, "First proton–antiproton collisions in the CERN SPS collider," *Phys. Lett. B*, vol. 107, no. 4, pp. 306–309, 1981, with permission from Elsevier. (right) Block diagram of the principle of stochastic cooling. Reprinted from D. Mohl, G. Petrucci, L. Thorndahl, and S. Van Der Meer, "Physics and technique of stochastic cooling," *Phys. Rept.*, vol. 58, pp. 73–119, 1980, with permission from Elsevier.

cooling" [348] played a key role in obtaining the required antiproton intensities.

In an electron storage ring, cooling happens as electrons lose some energy by synchrotron radiation (see Section 3.2). The lost energy being compensated only longitudinally by the accelerating cavities eventually leads to the reduction of the longitudinal and transverse spread of the electrons. This effect, however, is negligible in comparison for protons (and antiprotons) since the radiation lost per turn is proportional to the fourth power of the Lorentz factor γ (see Eq. (3.5)). In the stochastic cooling method, the spread of the antiprotons is continuously measured as they circulate in the ring and are actively corrected. Each antiproton in the beam cannot be corrected individually but the correction occurs on the ensemble of the particles, hence the name "stochastic cooling." The block diagram is shown in Figure 27.6(right). In a given location of the ring, the particles' transverse spread is measured (or "picked up") by a sensor, which measures the average beam position. At a later time, a "transverse kicker" magnet corrects the spread. The action of the magnet is indeed steered at each turn by the signal measured in the pick-up sensor. Since the particles travel nearly at the speed of light, the signal takes a short-cut with respect to the particles by proceeding via a straight line rather than around the arc of the circular ring. In this way, the pickup and the kicker effect could be synchronized to act on the correct antiproton bunch. After many passages through such systems, the antiprotons would be cooled enough so that they could be injected into the PS ring and accelerated up to 26 GeV, after which they would be transferred to the SPS and accelerated up to 270 GeV and be brought into collisions with the protons circulating in the opposite directions. The first $p\bar{p}$ collisions were observed in 1981 [349].

• **Leptons and the transverse mass.** Already in 1976 it was known that the production of heavy gauge bosons in hadron–hadron collisions would be orders of magnitude smaller than the dominant strong interactions leading to jets. For a precise estimate, see Section 26.9 and Figure 19.22. It would therefore be impossible to find the $W^{\pm} \to$ hadrons or $Z^0 \to$ hadrons decays within the background at hadronic colliders.

A very clear signature of the heavy Z^0 production is provided by its leptonic decays into electron–positron or muon–antimuon pairs, although the leptonic branching ratio is only 3.4% (see Eq. (26.79)). We have mentioned in Section 26.9 that while the longitudinal motion of the boson varies from event to event (see discussion on rapidity), the transverse motion is expected to be small and produced only by higher-order QCD corrections. Under this assumption, the two leptons from Z^0 decay have equal magnitude but opposite directed transverse momenta. If we label these $\vec{P}_T^{\ell^+}$ and $\vec{P}_T^{\ell^-}$, we consequently have:

$$p\bar{p} \to Z^0 + X \to \ell^+\ell^- + X : \quad \vec{P}_T^{\ell^+} + \vec{P}_T^{\ell^-} \approx 0 \tag{27.32}$$

This makes it relatively easy to detect the Z^0 by searching for pairs of oppositely charged electrons or muons with large momenta and in a back-to-back configuration in the transverse plane.

In the case of the $W^\pm$ boson, only one lepton can be observed, the other being a neutrino:

$$p\bar{p} \to W^\pm + X \to \ell^\pm + E_{miss}^T + X : \quad E_{miss}^T > 0 \tag{27.33}$$

where $\ell = e, \mu$. However, it is possible to also exploit information about the neutrino momentum. If one detects all hadrons produced in X precisely, then the overall imbalance of the visible particles should give an approximate measure of the undetected P_T^ν. So the signature for the W boson is given by a single lepton with large missing transverse energy. More formally, one can define the **transverse mass** m_T following the relation of the mass $m^2 = E^2 - p^2$ applied to the transverse plane:

$$m_T^2 = (E_\ell^T + E_{miss}^T)^2 - (\vec{p}_\ell^T + \vec{p}_{miss}^T)^2 \tag{27.34}$$

By construction $m_T^2 \leq m^2 \approx M_W^2$. Hence:

$$0 \leq m_T^2 \leq M_W \tag{27.35}$$

We note that the transverse mass is unaffected by the longitudinal motion (boost) of the W since it depends only on the transverse momenta. In the approximation that the transverse momentum of the W is zero, we in addition simply get $\vec{p}_\ell^T + \vec{p}_{miss}^T = 0$ and $E_\ell^T = E_{miss}^T$ (assuming no measurement errors) and:

$$m_T = 2E_\ell^T \tag{27.36}$$

One can show that the differential distribution of the transverse mass is given by (see **Ex 27.3**):

$$\frac{d\hat{s}}{dm_T^2} \propto \frac{2 - m_T^2/\hat{s}}{\sqrt{1 - m_T^2/\hat{s}}} \tag{27.37}$$

and it consequently exhibits a Jacobian peak. The shape of m_T close to its endpoint is sensitive to the mass M_W and the width Γ_W of the W boson. Experimentally it is affected by the missing energy resolution of the detector. One should also take care of the "background" produced by the reaction $p\bar{p} \to W^\pm + X \to \tau^\pm + \nu_\tau + X \to \ell^\pm \nu_\ell \nu_\tau \nu_\tau + X$, which produces single leptons with a softer spectrum due to the additional neutrinos in the final state. These properties were used to determine the $W^\pm$ boson rest mass, as will be illustrated in the following. In the case of the Z^0 boson, the mass can be directly determined by the invariant mass of the lepton–antilepton pair.

● **The CERN UA1 and UA2 experiments.** The two experiments UA1 and UA2 at the CERN $Sp\bar{p}S$ collider (see Figure 27.7) were optimized to detect the expected signatures of the leptonic decays of the $W^\pm$ and Z^0 bosons, as described above. The UA1 detector [350] was composed of a large drift chamber with 23,000 wires, surrounding the beam pipe, which was embedded in a transverse dipole field. An electromagnetic calorimeter surrounded the drift detector. The iron yoke of the magnet was segmented with scintillator planes so that it could act as a hadron calorimeter. The magnet was surrounded by muon chambers. Caution was taken to achieve **hermeticity**, in order to best measure the missing energy of the events. Only particles along the beam pipe with an angle smaller than 0.2° would escape detection. The UA2 detector [351] was more modest. It

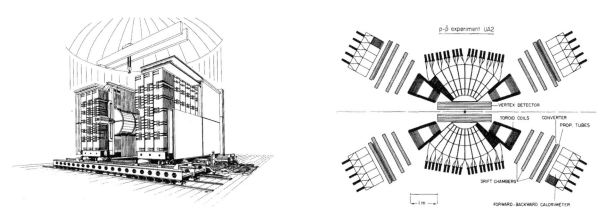

Figure 27.7 (left) Illustration of the UA1 detector at the CERN $Sp\bar{p}S$. Reprinted from M. Barranco-Luque *et al.*, "The construction of the central detector for an experiment at the CERN anti-p p collider," *Nucl. Instrum. Meth.*, vol. 176, p. 175, 1980, with permission from Elsevier. (right) Illustration of the UA2 detector at the CERN $Sp\bar{p}S$. Reprinted from M. Banner *et al.*, "Observation of very large transverse momentum jets at the CERN anti-p p Collider," *Phys. Lett. B*, vol. 118, pp. 203–210, 1982, with permission from Elsevier.

had a magnetic field only in a limited region in front of the forward and backward calorimeters covering the angles $20°$–$40°$ and $140°$–$160°$. UA2 primarily relied on calorimetry and used pattern recognition to distinguish electromagnetic and hadronic showers.

During the first run in 1981 at $\sqrt{s} = 540$ GeV, focus was on the observation of high-p_T jets produced by the strong interactions of quark–antiquark pairs. The transverse energy distribution of the collected events by UA2 is shown in Figure 27.8(left). The LEGO plot of the reconstructed energy in the $\theta - \phi$ plane for the highest-energy event with $E_T = 127$ GeV is displayed in Figure 27.8(right). Two beautiful jets emitted in opposite directions nearly perpendicular to the beam line are observable.

A clear signal for the production of $W^\pm$ was observed by both UA1 and UA2 in the 1982 run. UA1 observed six events with a high-p_T lepton balanced with a missing energy which could be attributed to the neutrino produced in the decay [292]. Similarly, UA2 found four. In 1983 more events were observed and the rapidity distribution of the events was shown to be compatible with expectations (see Figure 26.9). In addition, four events compatible with a Z^0 decaying into an electron–positron pair and one decay into a pair of oppositely charged muons were reported by UA1 [352]. Soon after, the UA2 experiment reported eight events compatible with $Z^0 \rightarrow e^+e^-$. Some of these beautiful events are shown in Figures 27.9–27.11. These experiments continued to take data until 1990, after which CERN turned to LEP, as discussed in Section 27.7.

27.5   Measurements of $W^\pm$ and Z^0 Bosons at Hadron Colliders

The mass and width of the gauge bosons are key parameters of the electroweak theory and should be measured as precisely as possible. The first observations at the CERN $Sp\bar{p}S$ collider allowed the estimation of their masses at the 1% level. Table 27.1 summarizes the $W^\pm$ mass measurements as a function of the calendar year.

Following the success at CERN, FNAL had decided to convert its very large proton synchrotron called the Tevatron into a $p\bar{p}$ collider at a center-of-mass energy $\sqrt{s} = 1$ TeV. In 1985 the first proton–antiproton collisions were observed at Fermilab by the CDF and DØ experiments. The Tevatron would later be upgraded to reach $\sqrt{s} = 1.96$ TeV. The increased energy allowed the production of a significant number of $W^\pm$ and Z^0 bosons, compared to the CERN program. The top quark was also discovered in 1995, as will be discussed in

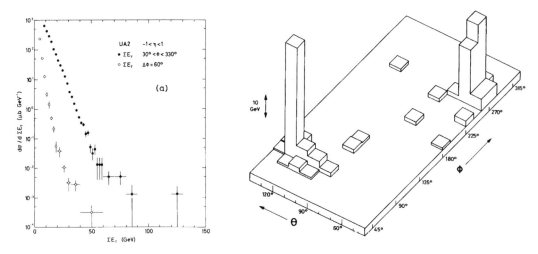

Figure 27.8 (left) The transverse energy distribution in proton–antiproton collisions at $\sqrt{s} = 540$ GeV. (right) The LEGO plot of the reconstructed energy in the $\theta - \phi$ plane for the highest-energy event with $E_T = 127$ GeV. Reprinted from M. Banner *et al.*, "Observation of very large transverse momentum jets at the CERN anti-p p collider," *Phys. Lett. B*, vol. 118, pp. 203–210, 1982, with permission from Elsevier.

Section 29.2.

With the ever-increasing statistics, the precision in determining the $W^\pm$ boson mass at hadron colliders has been significantly improved. Eventually, it became limited by systematic errors, as can be seen in Table 27.1. In the year 2000, competing measurements became available with the advent of the LEP e^+e^- collider, which provided 10^{-5} precision on the Z^0 mass and permil on the $W^\pm$ mass, as will be discussed in the next section.

● **Measurements of the W boson at the FNAL Tevatron.** The CDF and DØ detectors will be described in Section 29.4. Here we illustrate the measurement of the $W^\pm$ boson mass. The measured transverse mass in $W \rightarrow e\nu$ and $W \rightarrow \mu\nu$ candidate events collected by CDF in $p\bar{p}$ collisions at $\sqrt{s} = 1.96$ TeV is shown in Figure 27.12. The shaded histrogram represents the estimated background. A similar measurement was performed by DØ. In summary, they obtain:

$$\text{CDF} : M_W = 80.387 \pm 0.012(\text{stat.}) \pm 0.015(\text{syst.}) \quad \text{and} \quad \text{DØ} : M_W = 80.375 \pm 0.013(\text{stat.}) \pm 0.022(\text{syst.})$$
$$(27.38)$$

More recently, improved determinations of the $W^\pm$ mass were achieved at hadron colliders with the increased luminosity at the Tevatron and finally at the CERN LHC collider where very copious samples of more than 10 million $W^\pm$ candidates were collected. The value obtained by the ATLAS experiment is reported in Table 27.1.

Finally, as will be discussed in Section 27.11, the LEP collider at a center-of-mass energy sufficient to study the reaction $e^+e^- \rightarrow W^+W^-$ provided the possibility to precisely measure the $W^\pm$ boson mass and width. The results are reported in Table 27.1. Combining the values obtained at hadron and lepton colliders, yields the most precise determination of the parameters:

$$\text{World:} \quad M_W = 80.379 \pm 0.013 \text{ GeV}, \quad \Gamma_W = 2.085 \pm 0.042 \text{ GeV} \tag{27.39}$$

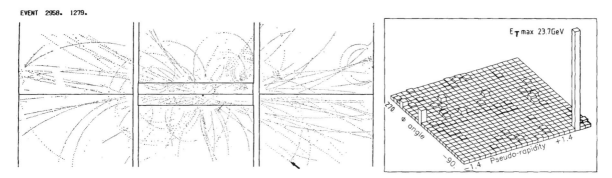

Figure 27.9 An event recorded by UA1 compatible with $p\bar{p} \to W \to e$ plus missing energy; (left) tracks in the central tracker (right) energy in the electromagnetic calorimeters. Reprinted from G. Arnison *et al.*, "Experimental observation of isolated large transverse energy electrons with associated missing energy at $\sqrt{s}$ = 540-GeV," *Phys. Lett.*, vol. B122, pp. 103–116, 1983, with permission from Elsevier.

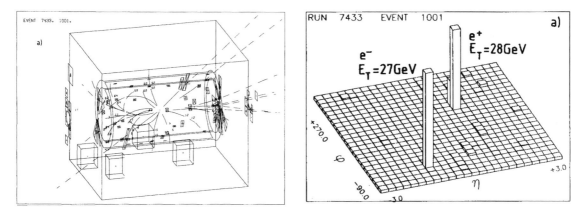

Figure 27.10 An event recorded by UA1 compatible with $p\bar{p} \to Z^0 \to e^+e^-$; (left) tracks in the central tracker, (right) energy in the electromagnetic calorimeters. Reprinted from G. Arnison *et al.*, "Experimental observation of lepton pairs of invariant mass around 95-GeV/c^2 at the CERN SPS collider," *Phys. Lett. B*, vol. 126, pp. 398–410, 1983, with permission from Elsevier.

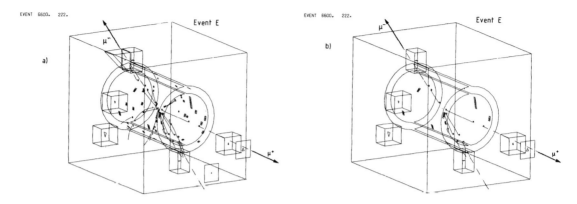

Figure 27.11 An event recorded by UA1 compatible with $p\bar{p} \to Z^0 \to \mu^+\mu^-$ (left) without cuts; (right) with $p_T > 1$ GeV threshold for tracks and $E_T > 0.5$ GeV for calorimeter hits. Reprinted from G. Arnison *et al.*, "Experimental Observation of Lepton Pairs of Invariant Mass Around 95-GeV/c^2 at the CERN SPS Collider," *Phys. Lett. B* 126, pp. 398–410, 1983, with permission from Elsevier.

Year	Reaction	$\sqrt{s}$ (GeV)	M_W (GeV)	N_{events}	Ref.
1983	$p\bar{p}$	546	81 ± 5	6	UA1 [292]
1983	$p\bar{p}$	546	80^{+10}_{-6}	4	UA2 [293]
1989	$p\bar{p}$	546–630	$82.7 \pm 1.0 \pm 2.7$	149	UA1 [353]
1989	$p\bar{p}$	546–630	$80.84 \pm 0.22 \pm 0.83$	2065	UA2 [354]
2001	$p\bar{p}$	1800	80.433 ± 0.079	53841	CDF [355]
2002	$p\bar{p}$	1800	80.483 ± 0.084	49247	DØ [356]
2000	e^+e^-	161–209	$80.440 \pm 0.043 \pm 0.027$	8692	ALEPH [357]
2000	e^+e^-	161–209	$80.270 \pm 0.046 \pm 0.031$	9909	L3 [358]
2000	e^+e^-	161–209	$80.336 \pm 0.055 \pm 0.039$	10300	DELPHI [359]
2001	e^+e^-	170–209	$80.415 \pm 0.042 \pm 0.031$	11830	OPAL [360]
2012	$p\bar{p}$	1960	80.387 ± 0.019	1095k	CDF [361]
2014	$p\bar{p}$	1960	80.375 ± 0.023	2177k	DØ [362]
2018	pp	7000	$80.370 \pm 0.007 \pm 0.017$	13.7M	ATLAS [363]

Table 27.1 A summary of the $W^\pm$ mass measurements.

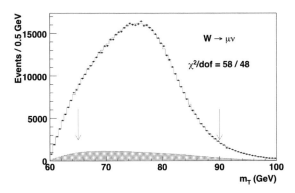 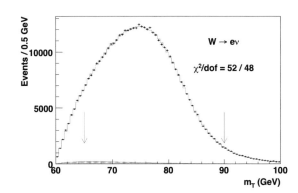

Figure 27.12 The measured transverse mass in $W \to \mu\nu$ and $W \to e\nu$ candidate events collected by CDF in $p\bar{p}$ collisions at $\sqrt{s} = 1.96$ TeV. Reprinted with permission from T. Aaltonen *et al.*, "Precise measurement of the W-boson mass with the CDF II detector," *Phys. Rev. Lett.*, vol. 108, p. 151803, 2012. Copyright 2012 by the American Physical Society.

27.6   The Detectors at LEP and SLC

The LEP and SLC e^+e^- high-energy colliders (see Section 3.4) were built with a primary goal to precisely study the properties of the Z^0 and W bosons. The cross-section into hadrons as a function of the center-of-mass energy over the range $10 < \sqrt{s} < 220$ GeV is shown in Figure 27.13. The high statistics coupled with optimal and clean experimental conditions provided by the pure e^+e^- initial state allowed a number of very important precision electroweak measurements, which were combined including data from all experiments in an over-constrained fit of the electroweak theory [364, 365].

The LEP center-of-mass energy was initially tuned to be on the Z^0 resonance mass where the cross-section $e^+e^- \to Z^0$ increases by more than two orders. This phase was called LEP1 or LEP-I. SLC could only operate in this regime. Over the years from 1989 to 1995, LEP produced 15 million Z^0 (see Table 27.2 for a summary), which were studied by the four experiments ALEPH, DELPHI, L3, and OPAL, and 600,000 Z^0 decays by the SLD experiment using a polarized beam at SLC. As such, they were dubbed "Z^0 factories." From 1996 to 2000, the energy of the LEP collider was increased to reach $\sqrt{s} > 2M_W$, as discussed in Section 27.11. This phase is often called LEP-II.

The designs of the LEP and SLC detectors were quite similar to each other, although the details varied significantly among them. The DELPHI, ALEPH, and OPAL detectors are shown in Figure 27.14. The L3 detector was described in detail in Section 13.7. All five detectors had almost complete solid angle coverage; the only holes being at polar angles below the coverage of the luminosity detectors. Thus, most events were fully contained in the active elements of the detectors, allowing straightforward identification. Starting radially from the interaction point, there was first a vertex detector, followed by a gas drift chamber to measure the parameters of charged particle tracks. Surrounding the tracking system was a calorimeter system, usually divided into two sections. The first section was designed to measure the position and energy of electromagnetic showers from photons, including those from π^0 decay, and electrons. The electromagnetic calorimeter was followed by a hadronic calorimeter to measure the energy of hadronic particles. Finally, an outer tracking system designed to measure the parameters of penetrating muons completed the system. The central part of the detector was immersed in a solenoidal magnetic field to allow the measurement of the momentum of charged particles. In addition, particle identification systems including dE/dx ionization loss measurements in the central chamber, time-of-flight, and ring-imaging Cherenkov detectors were used.

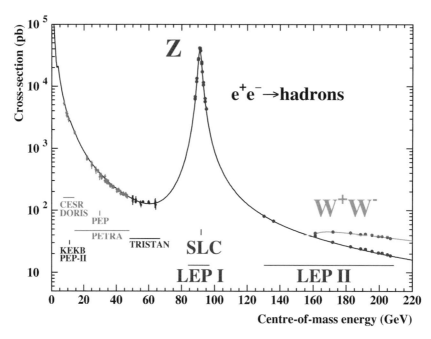

Figure 27.13 The cross-section into hadrons as a function of center-of-mass energy. The solid line is the theoretical prediction, and the points are the experimental measurements. Also indicated are the energy ranges of various e^+e^- accelerators. The cross-sections have been corrected for the effects of photon radiation. Reprinted from S. Schael *et al.*, "Precision electroweak measurements on the Z resonance," *Phys. Rept.*, vol. 427, pp. 257–454, 2006, with permission from Elsevier.

These measurements could be used to determine the velocity of particles; coupled with the momentum, they yielded the particle masses. The experiment also included multi-layer silicon devices as vertex detectors, which significantly improved the ability to measure impact parameters and to identify secondary vertices with a resolution of approximately 300 μm.

Images of $q\bar{q}$, e^+e^-, $\mu^+\mu^-$, and $\tau^+\tau^-$ final states from Z^0 decays, visualized with the event displays of the OPAL, DELPHI, L3, and ALEPH experiments, are illustrated in Figure 27.15. In all views, the electron–positron beam axis is perpendicular to the plane of the page. As can be seen, the events were extremely clean, with practically no detector activity unrelated to the products of the annihilation event, allowing high-efficiency and high-purity selections to be made. The electron–positron and muon–antimuon events are extremely clean and this allows these fundamental Z^0 decays to be directly observed. The low-multiplicity products of τ decays are confined to well-isolated cones and can also be readily recognized. Hadronic Z^0 decays result in higher-multiplicity jets of particles produced in the QCD cascades formed by the initial $q\bar{q}$ pair (see Section 19.2). Shown in Figure 27.16 is a side view of an SLD event interpreted as the decay of a Z^0 into $b\bar{b}$. The displaced vertex from the decay of a B hadron is clearly visible.

The clean e^+e^- environment of LEP also allowed the efficient separation and classification of the main decay modes of the Z^0 boson into four main categories: $Z^0 \to e^+e^-$, $Z^0 \to \mu^+\mu^-$, $Z^0 \to \tau^+\tau^-$, and $Z^0 \to$ hadrons (for some analyses the decays to heavy quarks, i.e., $Z^0 \to c\bar{c}$ and $Z^0 \to b\bar{b}$, were separated statistically within the $Z^0 \to$ hadrons sample).

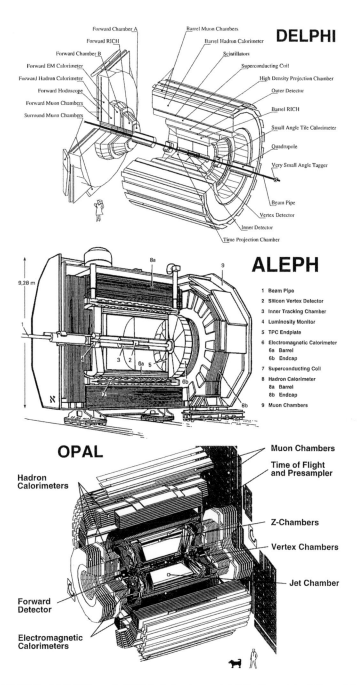

Figure 27.14 The DELPHI, ALEPH, and OPAL detectors at LEP. Reprinted from M. Grunewald, "Experimental tests of the electroweak standard model at high-energies," *Phys. Rept.*, vol. 322, pp. 125–346, 1999, with permission from Elsevier.

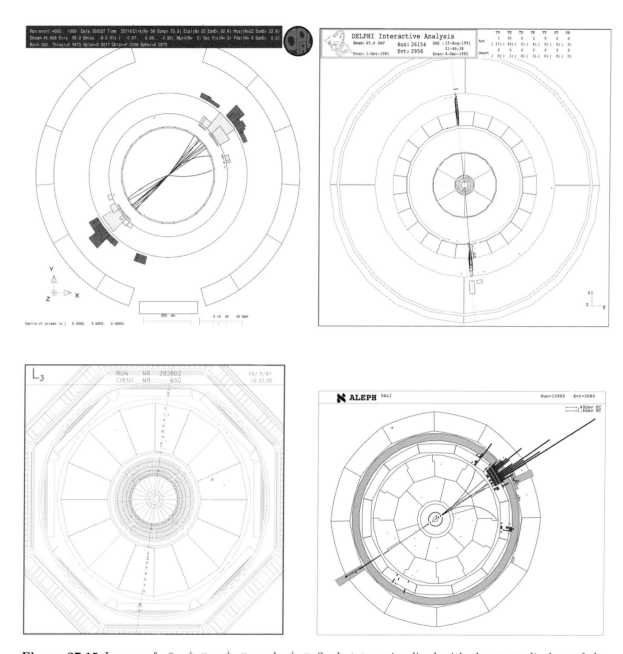

Figure 27.15 Images of $q\bar{q}$, e^+e^-, $\mu^+\mu^-$, and $\tau^+\tau^-$ final states, visualized with the event displays of the OPAL, DELPHI, L3, and ALEPH experiments, respectively. Reprinted from S. Schael *et al.*, "Precision electroweak measurements on the Z resonance," *Phys. Rept.*, vol. 427, pp. 257–454, 2006, with permission from Elsevier.

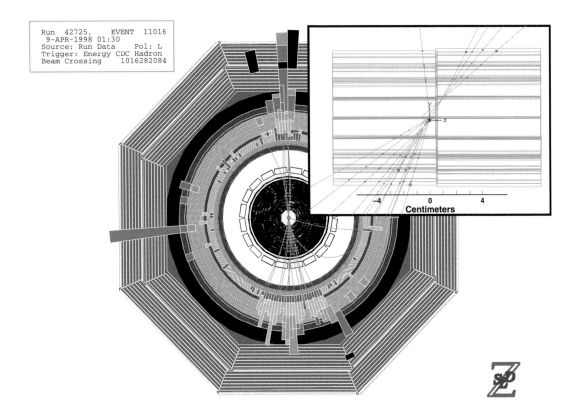

Figure 27.16 Front view of an event classified as $Z^0 \to b\bar{b}$. The displaced secondary vertex, reconstructed with the silicon vertex detector, is visible in the expanded side view (r–z view) of the beam interaction point. Reprinted from S. Schael *et al.*, "Precision electroweak measurements on the Z resonance," *Phys. Rept.*, vol. 427, pp. 257–454, 2006, with permission from Elsevier.

Year	$\sqrt{s}$	$\int \mathcal{L}$	Number of events									
			$Z^0 \to q\bar{q}$					$Z^0 \to \ell^+\ell^-$				
	(GeV)	(pb^{-1})	A	D	L	O	LEP	A	D	L	O	LEP
1990/91	88.2–94.2	27.5	433	357	416	454	1660	53	36	39	58	186
1992	91.3	28.6	633	697	678	733	2741	77	70	59	88	294
1993	89.4, 91.2, 93.0	40.0	630	682	646	649	2607	78	75	64	79	296
1994	91.2	64.5	1640	1310	1359	1601	5910	202	137	127	191	657
1995	89.4, 91.3, 93.0	39.8	735	659	526	659	2579	90	66	54	81	291
Total			4071	3705	3625	4096	15497	500	384	343	497	1724

Table 27.2 The $q\bar{q}$ and $\ell\ell$ event statistics, in units of 10^3, used for Z^0 analyses by the experiments ALEPH (A), DELPHI (D), L3 (L), and OPAL (O).

27.7 Z^0 **Resonance Precision Studies at LEP and SLC**

• Z^0 **lineshape measurement.** The measurement of the Z^0 **resonance lineshape** determines the peak and thereby the resonance M_Z, the width of the resonance Γ_Z, and the partial widths Γ_f for each of the charged leptons and the hadrons (all kinematically accessible quarks). In order to obtain the lineshape, one must collect data to measure the cross-sections to different final states as a function of the center-of-mass energy $\sqrt{s}$. As already mentioned in Section 3.4, a significant effort was dedicated to the determination of the energy of the LEP colliding beams. **A precision of about 2 MeV in the center-of-mass energy was achieved**, corresponding to a relative uncertainty of about 2×10^{-5} on the absolute energy scale. This level of accuracy was vital for the precision of the measurements of the mass and width of the Z^0. In particular the off-peak energies in the 1993 and 1995 scans were carefully calibrated employing the technique of resonant depolarization of the transversely polarized beams.

The precision measurements obtained by the four LEP experiments as published by the individual experiments are listed in Table 27.3 and compared to the results obtained at hadron colliders. The combined result for all LEP experiments for M_Z is impressively precise, at the level of 2×10^{-5} relative error [364]:

$$\text{LEP}: \quad M_Z = 91187.5 \pm 2.1 \text{ MeV} \tag{27.40}$$

A similar procedure is applied to the four LEP independent measurements of the width Γ_Z (see Figure 27.17) and their combined result is:

$$\text{LEP}: \quad \Gamma_Z = 2495.2 \pm 2.3 \text{ MeV} \tag{27.41}$$

The cross-section at the pole (corrected for ISR) was shown to be directly related to the partial width (Eq. (26.95)). In the case of the decay to hadrons, which is the most abundant with approximately 70% of the decays (Eq. (26.79)), we have:

$$\sigma^0_{had} = 12\pi \frac{\Gamma_{ee}\Gamma(Z^0 \to \text{hadrons})}{M_Z^2 \Gamma_Z^2} \equiv 12\pi \frac{\Gamma_{\ell\ell}\Gamma_{\text{had}}}{M_Z^2 \Gamma_Z^2} \tag{27.42}$$

where in the last line we have assumed lepton universality and set $\Gamma_{\ell\ell} = \Gamma_{ee} = \Gamma_{\mu\mu} = \Gamma_{\tau\tau}$ (see Section 23.12). This is advantageous and a very reasonable assumption that allows us to increase statistics of the samples which

Year	Reaction	$\sqrt{s}$ (GeV)	M_Z (GeV)	N_{events}	Ref.
1989	$p\bar{p}$	546–630	$93.1 \pm 1.0 \pm 3.0$	24	UA1 [353]
1989	$p\bar{p}$	1800	$90.9 \pm 0.3 \pm 0.2$	188	CDF [366]
2000	e^+e^-	88-94	91.1885 ± 0.0031	4.57M	ALEPH [357]
2000	e^+e^-	88-94	91.1898 ± 0.0031	3.96M	L3 [358]
2000	e^+e^-	88-94	91.1863 ± 0.0028	4.08M	DELPHI [359]
2001	e^+e^-	88-94	91.1852 ± 0.0030	4.57M	OPAL [360]

Table 27.3 A summary of the Z^0 mass measurements.

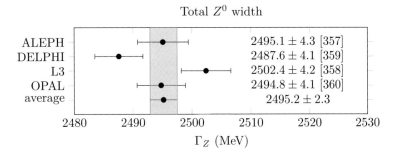

Figure 27.17 Precision measurement of the Z^0 boson width at LEP. The average value is from P. A. Zyla *et al.* (Particle Data Group), *Prog. Theor. Exp. Phys.* **2020**, 083C01 (2020).

otherwise individually are limited by the branching ratio to a given pair of charged leptons $Br(Z^0 \to \ell^+\ell^-)$ of approximately 3%. By combining all leptonic samples, the statistics is about 10% of the total. The naive prediction for the hadronic pole cross-section can be estimated from Eq. (26.95) to be:

$$
\begin{aligned}
\sigma^0_{had} &= 12\pi \frac{\Gamma(Z^0 \to \ell^+\ell^-)\left(2\Gamma(Z^0 \to u\bar{u}) + 3\Gamma(Z^0 \to d\bar{d})\right)}{M_Z^2 \Gamma_Z^2} \\
&\simeq 12\pi \frac{(0.09 \text{ GeV})(1.77 \text{ GeV})}{(91.2 \text{ GeV})^2(2.4 \text{ GeV})^2} \simeq 12\pi(3.3 \times 10^{-6} \text{ GeV}^{-2}) \simeq 49 \text{ nb}
\end{aligned}
\tag{27.43}
$$

The precision measurements obtained by the four LEP experiments and the combined result for the hadronic pole cross-section are shown in Figure 27.18. The combined result is:

$$
\text{LEP}: \quad \sigma^0_{had} = 41.541 \pm 0.037 \text{ nb}
\tag{27.44}
$$

This value is smaller than our naive expectation, although the ballpark is certainly correct. Indeed, comparison with data requires the introduction of radiative corrections and other effects, as discussed in Section 26.8. In particular, QED radiative corrections broaden the Z^0 lineshape and hence reduce the cross-section at the peak. See Figure 26.8.

The branching ratio into a particular fermion–antifermion pair is given by:

$$
Br(Z^0 \to f\bar{f}) = \frac{\Gamma_{ff}}{\Gamma_Z} = \frac{N_C \Gamma_Z^0 \left((c_V^f)^2 + (c_A^f)^2\right)}{\Gamma(Z^0 \to \text{all})}
\tag{27.45}
$$

where N_C is three for quarks and one otherwise. So the measurement of the branching ratios provides information on the couplings of fermions to the Z^0 bosons. Figure 27.18 also shows the measurements for the

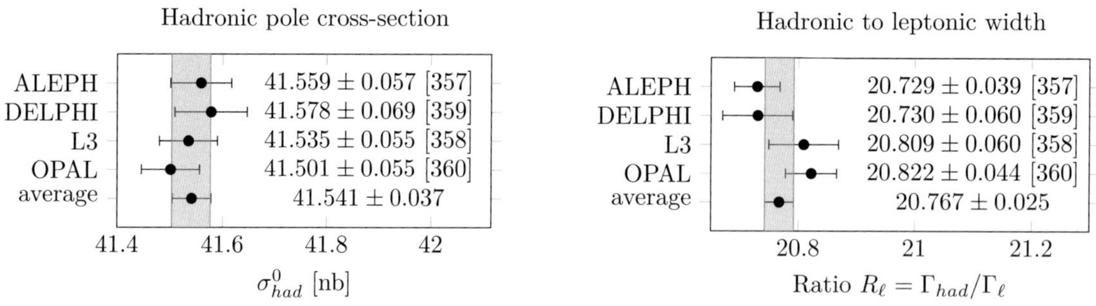

Hadronic pole cross-section

Hadronic to leptonic width

ALEPH		41.559 ± 0.057 [357]
DELPHI		41.578 ± 0.069 [359]
L3		41.535 ± 0.055 [358]
OPAL		41.501 ± 0.055 [360]
average		41.541 ± 0.037

41.4   41.6   41.8   42

σ_{had}^0 [nb]

ALEPH		20.729 ± 0.039 [357]
DELPHI		20.730 ± 0.060 [359]
L3		20.809 ± 0.060 [358]
OPAL		20.822 ± 0.044 [360]
average		20.767 ± 0.025

20.8   21   21.2

Ratio $R_\ell = \Gamma_{had}/\Gamma_\ell$

Figure 27.18 Precision measurement of the hadronic pole cross-section of the ratio of the hadronic to leptonic width at LEP. The average value is from P. A. Zyla *et al.* (Particle Data Group), *Prog. Theor. Exp. Phys.* **2020**, 083C01 (2020).

ratio of the hadronic to leptonic width. The electroweak theory predicts this ratio to be simply given by (see Eq. (26.78)):

$$R_\ell \equiv \frac{\Gamma_{had}}{\Gamma_\ell} = \frac{2\Gamma(Z^0 \to u\bar{u}) + 3\Gamma(Z^0 \to d\bar{d})}{\Gamma(Z^0 \to \ell^+\ell^-)} = \frac{6\left(\frac{1}{2} - \frac{4}{3}x_W + \frac{16}{9}x_W^2\right) + 9\left(\frac{1}{2} - \frac{2}{3}x_W + \frac{4}{9}x_W^2\right)}{\left(\frac{1}{2} - 2x_W + 4x_W^2\right)} \simeq 20$$

This result is valid at tree level and a precise comparison with data requires the introduction of radiative corrections and other effects. Nonetheless, it can be compared to the combined experimental value in Figure 27.18:

$$\text{LEP:} \quad R_\ell = 20.767 \pm 0.025 \tag{27.46}$$

A more precise comparison between data and electroweak predictions will be discussed in Section 29.12.

27.8   The Forward–Backward and Left–Right Asymmetries

• **Lepton forward–backward asymmetries.** The **forward–backward asymmetry** A_{FB} in $e^+e^- \to \ell^+\ell^-$ collisions defined in Eqs. (13.13) and (26.96) is a probe of the effects related to the helicities of the particles and hence determines the parity-violating nature of the interaction, differentiating the vector and axial-vector couplings. It consequently determines the effective weak mixing angle, as defined in Eq. (27.13). The tree-level prediction for leptons is estimated with Eqs. (26.101) and (26.102) to be:

$$A_{FB}^\ell \equiv \frac{3}{4}\mathcal{A}_e\mathcal{A}_\ell \simeq \frac{3}{4}(0.16)(0.16) \simeq 0.0192 \tag{27.47}$$

This is actually a small number, in the range of a couple of percent. There are a number of ways of measuring the asymmetry parameter. The angular distributions in the differential cross-sections $e^+e^- \to \ell^+\ell^-$ provide a direct measurement of the forward–backward asymmetries for each individual lepton family. The combined results at the Z^0 pole from the LEP experiments are:

$$A_{FB}^{0,e} = 0.0145 \pm 0.0025, \qquad A_{FB}^{0,\mu} = 0.0169 \pm 0.0013, \qquad A_{FB}^{0,\tau} = 0.0188 \pm 0.0017 \tag{27.48}$$

Assuming lepton universality, the experiments derive leptonic forward–backward asymmetries measured at the Z^0 pole. The combined LEP result is:

$$\text{LEP:} \quad A_{FB}^{0,\ell} = 0.0171 \pm 0.0010 \tag{27.49}$$

• **Tau polarization.** Another very elegant method to test the electroweak theory is the study of the tau polarization in the $e^+e^- \rightarrow Z^0 \rightarrow \tau^+\tau^-$ final states. The amount of longitudinal polarization is predicted by the theory to be [364] (see **Ex 27.5**):

$$\mathcal{P}_\tau(\cos\theta_{\tau^-}) = -\frac{\mathcal{A}_\tau(1+\cos^2\theta_{\tau^-})+2\mathcal{A}_e\cos\theta_{\tau^-}}{(1+\cos^2\theta_{\tau^-})+\frac{8}{3}A^\tau_{FB}\cos\theta_{\tau^-}} \tag{27.50}$$

where $\mathcal{A}_f$ is defined in Eq. (26.102). The average tau polarization is directly determined by $\mathcal{A}_\tau$:

$$\boxed{\mathcal{A}_\tau = -\langle\mathcal{P}_\tau\rangle} \tag{27.51}$$

Hence, the τ polarization measurements allow the determination of $\mathcal{A}_\tau$ and $\mathcal{A}_e$. The polarization measurements rely on the dependence of kinematical distributions of the observed τ decay products on the helicity of the parent τ lepton. Because the helicity of the parent cannot be determined on an event-by-event basis, the polarization measurement is performed by fitting the observed kinematical spectrum of a particular decay mode to a linear combination of the positive and negative helicity spectra associated with that mode. The LEP combined result assuming lepton universality is [364]:

$$\boxed{\text{LEP}: \quad \mathcal{A}_\ell(\mathcal{P}_\tau) = 0.1465 \pm 0.0033} \tag{27.52}$$

• **Left–right asymmetry.** The measurement of the left–right cross-section asymmetry, A_{LR}, provides a determination of the asymmetry parameter $\mathcal{A}_e$, which is the most precise single measurement of this quantity, with the smallest systematic error. At the Z^0 pole, one has [364]:

$$\boxed{A_{LR} \equiv \mathcal{A}_e} \tag{27.53}$$

where $\mathcal{A}_e$ is defined in Eq. (26.102). The analysis is straightforward in principle: one counts the numbers of Z^0 bosons produced by left and right longitudinally polarized electron bunches, N_L and N_R, forms the asymmetry, and then divides by the luminosity-weighted e^- beam polarization magnitude $\langle\mathcal{P}_e\rangle$, since the e^+ beam is not polarized (see Section 3.4):

$$A_{LR} = \frac{N_L - N_R}{N_L + N_R}\frac{1}{\langle\mathcal{P}_e\rangle}. \tag{27.54}$$

Assuming lepton universality and accounting for correlated uncertainties, the combined result is:

$$\boxed{\text{SLD}: \quad \mathcal{A}_\ell = 0.1513 \pm 0.0021} \tag{27.55}$$

where the total error includes the systematic error of ± 0.0011. The average electron polarization was in the range of 80% (see Section 3.4).

A_{LR}, along with the tau polarization measurements, is the observable that is the most sensitive to the effective weak mixing angle among the asymmetries, with $\delta A_{LR} \approx 8\delta\sin^2\theta^\ell_{eff}$. By itself, this measurement is equivalent to a determination of:

$$\sin^2\theta_{eff}(\mathcal{A}_\ell(\text{SLD})) = 0.23098 \pm 0.00026 \tag{27.56}$$

where the total error includes the systematic error of ± 0.00013.

• **Heavy quarks.** The tree-level predictions for quarks are estimated with Eq. (26.101) to be:

$$A^{u,c}_{FB} \equiv \frac{3}{4}\mathcal{A}_e\mathcal{A}_{u,c} \simeq \frac{3}{4}(0.16)(0.68) \simeq 0.0816 \quad \text{and} \quad A^{d,s,b}_{FB} \equiv \frac{3}{4}\mathcal{A}_e\mathcal{A}_{d,s,b} \simeq \frac{3}{4}(0.16)(0.93) \simeq 0.1116 \tag{27.57}$$

which are much bigger than those expected for leptons. However, hadronic final states are more difficult to handle than their leptonic counterparts, in particular when it comes to distinguishing their charge. Nonetheless, LEP provided ideal conditions to study the forward–backward asymmetry of heavy quark final states, *tagging* the presence of heavy flavors via detection of charmed or bottom particles in the hadronic jets. The exclusive tagging method allowed for the separation between quark and antiquark, in particular in the semi-leptonic decay of charged or bottom particles, and hence provided means to compute the forward–backward asymmetry. Independently, one could also study jets inclusively. These measurements relied on the observation that the average charge of all particles in a jet, or jet charge, retains some information on the original quark charge. It can then statistically be used to extract the forward–backward asymmetry. The experiments measured the partial widths of charm and bottom events, $R_c^0 \equiv \sigma_c/\sigma_{had}$ and $R_b^0 \equiv \sigma_b/\sigma_{had}$ at the Z^0 pole, and their forward–backward asymmetries $A_{FB}^{0,c}$ and $A_{FB}^{0,b}$ also at the pole. The LEP combined measured values are [364]:

$$\text{LEP} + \text{SLD}: \quad R_c^0 = 0.1721 \pm 0.0030, \quad R_b^0 = 0.21629 \pm 0.00066 \tag{27.58}$$

and

$$\text{LEP}: \quad A_{FB}^{0,c} = 0.0707 \pm 0.0035, \qquad A_{FB}^{0,b} = 0.0992 \pm 0.0016 \tag{27.59}$$

The inclusive jet charge method relies on many inputs and was performed at LEP in the context of the theory, and for this reason, the quoted result gives directly the effective mixing angle:

$$\sin^2 \theta_{eff}(Q_{fb}^{had}) = 0.2324 \pm 0.0012 \tag{27.60}$$

Due to the availability of electron polarization, the SLD experiment allows for a measurement of the left–right/forward–backward asymmetries in heavy quark final states defined as:

$$A_{\text{LRFB}}^{q\bar{q}} \equiv \frac{1}{|\mathcal{P}_e|} \frac{(\sigma_{\text{F}} - \sigma_{\text{B}})_{\text{L}} - (\sigma_{\text{F}} - \sigma_{\text{B}})_{\text{R}}}{(\sigma_{\text{F}} + \sigma_{\text{B}})_{\text{L}} + (\sigma_{\text{F}} + \sigma_{\text{B}})_{\text{R}}} \tag{27.61}$$

where L, R denote the cross-sections with left-handed and right-handed electron beam polarization. This asymmetry as a function of the scattering angle can be expressed as [364]:

$$A_{\text{LRFB}}^{q\bar{q}}(\cos\theta) = |\mathcal{P}_e|\, \mathcal{A}_q \frac{2\cos\theta}{1 + \cos^2\theta} \tag{27.62}$$

Consequently, the asymmetry $\mathcal{A}_q$ can be directly measured using this method! The results are based on several techniques: measurements of $\mathcal{A}_c$ and $\mathcal{A}_b$ using lepton tags [367], measurement of $\mathcal{A}_c$ using D-mesons [368]; measurement of $\mathcal{A}_b$ using jet charge [369]; measurement of $\mathcal{A}_b$ using vertex charge [370]; measurement of $\mathcal{A}_b$ using kaons [371]; measurement of $\mathcal{A}_c$ using vertex charge and kaons [370]. The combination of all these results yields the following most precise values [364]:

$$\text{SLD}: \quad \mathcal{A}_c = 0.670 \pm 0.027, \quad \mathcal{A}_b = 0.923 \pm 0.020 \tag{27.63}$$

• **Lepton asymmetry parameters.** The measurements of the leptonic forward–backward asymmetries can be combined to yield the most precise determinations of the lepton asymmetry parameters $\mathcal{A}_\ell$, defined in Eq. (26.102). Alternatively, each measurement for a pair of final state leptons can be interpreted as the measurement of the asymmetry parameter of the individual lepton flavor:

$$A_{FB}^e = \frac{3}{4}\mathcal{A}_e^2, \qquad A_{FB}^\mu = \frac{3}{4}\mathcal{A}_e\mathcal{A}_\mu, \qquad A_{FB}^\tau = \frac{3}{4}\mathcal{A}_e\mathcal{A}_\tau \tag{27.64}$$

such that the $e^+e^- \to e^+e^-$ uniquely determines $\mathcal{A}_e$ and combination with the muon and tau final states gives $\mathcal{A}_\mu$ and $\mathcal{A}_\tau$. The measurements of $\mathcal{A}_e$ and $\mathcal{A}_\tau$ performed by the various experiments are reported in

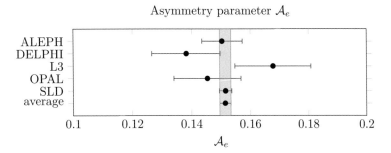

Figure 27.19 Precision measurements of the Z^0 asymmetry parameter $\mathcal{A}_e$. The average value is from P. A. Zyla *et al.* (Particle Data Group), *Prog. Theor. Exp. Phys.* **2020**, 083C01 (2020).

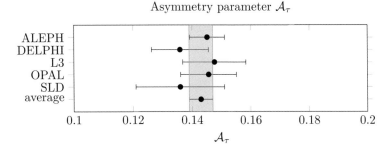

Figure 27.20 Precision measurements of the Z^0 asymmetry parameter $\mathcal{A}_\tau$. The average value is from P. A. Zyla *et al.* (Particle Data Group), *Prog. Theor. Exp. Phys.* **2020**, 083C01 (2020).

Figure 27.19 and 27.20. They are consistent among experiments. The average values for leptons are (values taken from the Particle Data Group's Review [5]):

$$\mathcal{A}_e = 0.1515 \pm 0.0019, \quad \mathcal{A}_\mu = 0.142 \pm 0.015, \quad \mathcal{A}_\tau = 0.143 \pm 0.004 \tag{27.65}$$

to be compared to the naive expectation $\mathcal{A}_\ell \simeq 0.16$ (see Eq. (26.104)).

• **Effective weak mixing angle.** All the effective electroweak mixing angles derived from LEP and SLD measurements of the lepton asymmetries only ($A_{fb}^{0,\ell}$, $A_\ell(\mathcal{P}_\tau)$, $A_\ell(\text{SLD})$) and from the quark couplings ($A_{fb}^{0,b}$, $A_{fb}^{0,c}$, Q_{fb}^{had}) are shown in Figure 27.21. The combined result on the effective weak mixing angle is:

$$\text{LEP/SLD}: \quad \sin^2\theta_{eff}^\ell = 0.23153 \pm 0.00016 \tag{27.66}$$

The achieved precision is indeed very impressive!

27.9   Test of the Electroweak Theory in the Drell–Yan Process

As discussed in Section 20.2, the Drell–Yan process $p\bar{p} \to \ell^+\ell^- + X$ and the production of quark pairs in high-energy e^+e^- collisions are analogous processes. In the electroweak theory, the process at leading order proceeds via the γ-mediated and the Z^0-mediated diagrams:

$$q\bar{q} \to \gamma^* \to \ell^+\ell^- \quad \text{and} \quad q\bar{q} \to Z^0 \to \ell^+\ell^- \tag{27.67}$$

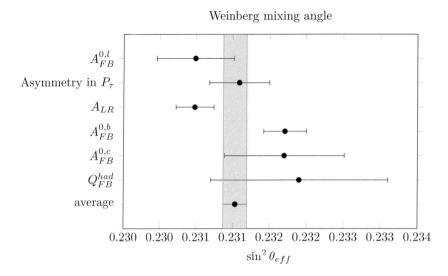

Figure 27.21 Precision measurements of the Weinberg angle. Data from S. Schael *et al.*, "Precision electroweak measurements on the *Z* resonance," *Phys. Rept.*, vol. 427, pp. 257–454, 2006.

When the center-of-mass energy of the $\ell^+\ell^-$ pair is close to the Z^0 boson mass, the latter becomes dominant and the diagrams interfere significantly. In this regime, the angular distributions of $\ell^+\ell^-$ pairs in the final state of the Drell–Yan process are sensitive to the vector and axial-vector couplings and consequently can be used to determine the Weinberg mixing angle, just as was the case for the $e^+e^- \to f\bar{f}$ reaction at high energy.

Experimentally both CDF and DØ have analyzed their data selecting events with high-p_T lepton pairs in the final state. CDF used approximately 485,000 electron pairs and 277,000 muon pairs in the measurements. DØ selected approximately 560,000 electron and 481,000 muon pairs in the final data sample. The angular kinematical properties of leptons from the Drell–Yan process were defined in the rest frame of the exchanged boson and were used to compute the asymmetries. The combined CDF and DØ result has been reported and yielded [372]:

$$\sin^2\theta_{\text{eff}}^\ell(\text{Tevatron}) = 0.23148 \pm 0.00033 \tag{27.68}$$

The result is consistent with, and approaches in precision, the best measurements from the LEP and SLC colliders.

27.10 The Number of Light Neutrinos

The coupling of neutrinos to the Z^0 and $W^\pm$ bosons is precisely predicted by the electroweak model. Under the assumption of the model, one can measure the total width of the boson decays and derive the number of light neutrinos, since each family of neutrino will contribute the same amount to the decay rate.

The indirect measurement of the number of light neutrinos was first performed at $p\bar{p}$ at the CERN SPS by the UA1 [373] and UA2 experiments [374]. They measured the ratio of the rates for the processes $W \to \ell\nu$ relative to $Z \to \ell\ell$. In order to compute the number of neutrino families the W boson decayed into, one can rescale for the branching ratio of the Z into leptons and for the ratio of the production cross-section of the bosons in $p\bar{p}$ collisions (and also assume that the top quark was heavier than the bosons). The results gave limits on the number of light neutrino families: for both experiments combined $N_\nu < 6$.

After these experiments, the number of neutrinos was measured in e^+e^- collisions via the reaction

$$e^+e^- \to \nu\bar{\nu}\gamma$$

The idea is to detect the radiative photon emitted by the initial-state leptons, otherwise the tree-level reaction $e^+e^- \to \nu\bar{\nu}$ is unobservable in a collider experiment! Some of the initial-state radiated photons are emitted at a large enough angle to be detected in the detector (most of them are collinear to the initial particles and tend to be lost around the accelerator beam pipe).

Before LEP and SLC, the experiments ASP, CELLO, MAC, MARK J, and VENUS searched at energies below the Z^0 resonance for such events. Their rate is however very small and they found only 3.9 events above background. This was enough to set the following limit on the number of light neutrinos:

$$N_\nu < 4.8 \quad (95\% \text{ CL}) \tag{27.69}$$

With the advent of LEP and SLC, the reaction could be studied with much higher statistics. In particular, for a center-of-mass energy above the Z^0 mass, the cross-section increases rapidly because of the "go back to resonance" phenomenon. The initial-state radiation leaves the e^+e^- pair with a lower center-of-mass energy due to energy conservation. The cases where the remaining energy is near the Z^0 pole are favored from the cross-section. Without the radiation, the process $e^+e^- \to Z^{0*}$ is off the resonance if $\sqrt{s} > M_Z$. With the initial radiation, the process $e^+e^- \to \gamma + Z^0$, where the Z^0 can basically be on-shell, becomes possible, although $\sqrt{s} > M_Z$. Consequently, above 90 GeV the radiative process $e^+e^- \to \nu\bar{\nu}\gamma$ becomes favored. It was observed by the ALEPH, DELPHI, L3, and OPAL experiments at LEP. As example, the measured cross-section as a function of the center-of-mass energy in L3 is shown in Figure 27.22. In total the LEP experiments have

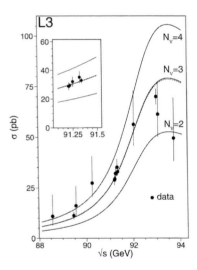

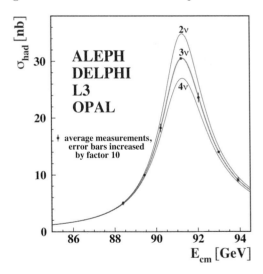

Figure 27.22 Measured cross-section for the process $e^+e^- \to \nu\bar{\nu}\gamma$ as a function of the center-of-mass energy by the L3 experiment. Reprinted from O. Adriani *et al.*, "Determination of the number of light neutrino species," *Phys. Lett. B*, vol. 292, pp. 463–471, 1992, with permission from Elsevier.

Figure 27.23 Z^0 lineshape: measured cross-section for the reaction $e^+e^- \to$ hadrons as a function of the center-of-mass energy by the LEP experiments, compared to the prediction for two, three, and four light neutrinos. Reprinted from S. Schael *et al.*, "Precision electroweak measurements on the Z resonance," *Phys. Rept.*, vol. 427, pp. 257–454, 2006, with permission from Elsevier.

collected several thousand such events and the combined result for the number of light neutrinos obtained with this direct method is:

$$N_\nu = 3.00 \pm 0.08 \tag{27.70}$$

in very strong support of the existence of only three light neutrino families!

The most precise determination of the number of light neutrinos was finally obtained at LEP with the precision measurement of the Z^0 lineshape. The total Z^0 width is given by (see Eq. (26.78)):

$$\Gamma_{\text{total}} \equiv \Gamma_{inv} + 3\Gamma_\ell + \Gamma_{hadrons} = N_\nu \Gamma_\nu + 3\Gamma_\ell + \Gamma_{hadrons} \tag{27.71}$$

where Γ_{inv} represents the "invisible" decays of the Z^0 and N_ν is the total number of light neutrino families. The total width can be measured in all decay channels. For example, the averaged measured cross-section for the reaction $e^+e^- \rightarrow$ hadrons as a function of the center-of-mass energy by the four LEP experiments is shown in Figure 27.23. The predicted lineshapes for respectively $N_\nu = 2, 3, 4$ are also plotted. The value $N_\nu = 3$ is clearly favored!

By precisely measuring the visible partial widths into hadrons and leptons as well as the total width, one can indirectly measure the invisible width:

$$\Gamma_{inv} = \Gamma_{\text{total}} - 3\Gamma_\ell - \Gamma_{hadrons} \tag{27.72}$$

The procedure actually uses all data to perform a "fit" to the electroweak theory (actually the Standard Model) to extract the Γ_{inv}. The number of light neutrinos is then defined as:

$$N_\nu = \left(\frac{\Gamma_{inv}}{\Gamma_\ell}\right) \left(\frac{\Gamma_\ell}{\Gamma_\nu}\right)_{\text{SM}} \tag{27.73}$$

The original combined LEP result yielded:

$$N_\nu = 2.984 \pm 0.008 \tag{27.74}$$

which is again a very impressive number! For a long time, however, this result remained a puzzle. It could indicate a 2σ deviation from the expected value of three. About 20 years later, in 2020, an updated and more accurate prediction of the Bhabha cross-section was reported which reduced the Bhabha cross-section at the relevant energies by about 0.048% and its uncertainty to $\pm0.037\%$ [375]. When accounted for, these changes modify the number of light neutrino species (and its accuracy) to

$$\text{LEP(corrected)}: \quad N_\nu = 2.9963 \pm 0.0074 \tag{27.75}$$

thereby removing the 20-year-old 2σ tension. Can't beat Nature!

27.11 W^+W^- Production at LEP

From 1996 to 2000, the energy of the LEP collider was increased to reach $\sqrt{s} > 2M_W$ above the production threshold to study the reaction:

$$\text{LEP} - \text{II}: \quad e^+e^- \rightarrow Z^0 \rightarrow W^+W^- \qquad \sqrt{s} > 2M_W \approx 160 \text{ GeV} \tag{27.76}$$

During the years 1996–2000, about 30,000 W^+W^- events were detected by the four experiments. So, LEP-II was effectively a "W-factory," which allowed several properties of the W boson to be determined precisely. As an example, the W mass measurements at LEP are shown in Figure 27.24 and can be compared to those at hadron colliders in Table 27.1. The combined result from the four LEP experiments yields the value:

$$\text{LEP}: \quad M_W = 80.376 \pm 0.025(\text{stat.}) \pm 0.022(\text{syst.}) \text{ GeV} \tag{27.77}$$

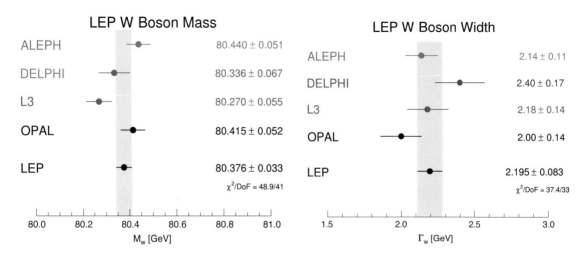

Figure 27.24 The measurements of the $W^\pm$ boson mass (left) and width (right) obtained by the four LEP collaborations (as published) together with the LEP combined result. Reprinted from S. Schael *et al.*, "Electroweak measurements in electron–positron collisions at W-boson-pair energies at LEP," *Phys. Rept.*, vol. 532, pp. 119–244, 2013, with permission from Elsevier.

Masses are consistent between lepton and hadron colliders. Currently, the most precise value is given by the result obtained at the LHC. However, consistency between the masses obtained from totally different methods represents a crucial consistency check.

The decay width of the W is also an important predictable and its measurement at LEP is summarized in Figure 27.24(right). The combined LEP result is:

$$\text{LEP}: \quad \Gamma_W = 2.195 \pm 0.063(\text{stat.}) \pm 0.055(\text{syst.}) \text{ GeV} \tag{27.78}$$

As a further important example, Figure 27.25 reports the cross-section measurement for the WW boson production as a function of the center-of-mass energy. By successive improvements, the center-of-mass energy could be increased up to 209 GeV, with the primary goal to search for the reaction $e^+e^- \to Z^{0*} \to Z^0 H^0$, but the energy turned out to be insufficient to produce the Higgs boson and the reaction was never observed.

The σ_{WW} cross-section is particularly relevant at high energies as it can directly test the existence of the self-coupling among the electroweak gauge bosons, as predicted by the theory (see Section 26.10). The data collected by the LEP experiment is compared to predictions computed with all diagrams. For comparison, the hypothetical cross-sections without ZWW vertex or with only neutrino exchange (see Figure 26.12) are also presented as dashed lines and are significantly incompatible with the data, which is very strong evidence for the existence of the self-coupling of gauge bosons, as predicted by the electroweak theory. Compare also with the theoretical curves presented in Figure 26.13.

Problems

Ex 27.1 Neutrino scattering off electrons. Compute the kinematics of the reaction $\nu_\mu e \to \nu_\mu e$ in the laboratory frame. Show that the outgoing electron is very forward peaked.

Ex 27.2 The CERN $p\bar{p}$ collider. The $W^\pm$ and Z^0 bosons were first discovered at the CERN $p\bar{p}$ collider. Explain why a center-of-mass energy of $270 + 270$ GeV was sufficient to produce them, although with reduced margin.

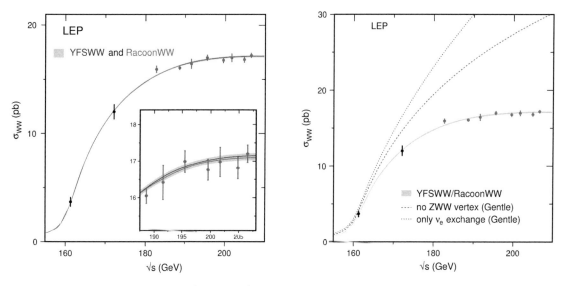

Figure 27.25 Measurements of the $e^+e^- \to W^+W^-$ cross-section at LEP as a function of the center-of-mass energy compared to the electroweak prediction (curve). Also shown are the predictions of individual diagrams (dashed, dotted). Reprinted from S. Schael *et al.*, "Electroweak measurements in electron–positron collisions at W-boson-pair energies at LEP," *Phys. Rept.*, vol. 532, pp. 119–244, 2013, with permission from Elsevier. Compare with Figure 26.13.

Ex 27.3 Transverse mass in $W^\pm$ decay. Using the results derived in **Ex 26.2**, show that the transverse momenta $\hat{p}_T$ of the electron and the neutrino exhibit a Jacobian peak given by:

$$\frac{\mathrm{d}\sigma}{\mathrm{d}\hat{p}_T^2} \propto \frac{1 - 2\hat{p}_T^2/\hat{s}}{\sqrt{1 - 4\hat{p}_T^2/\hat{s}}} \tag{27.79}$$

Hence derive the result of Eq. (27.37).

Ex 27.4 Transverse momentum of the W boson. By boosting the electron and the neutrino momenta in the transverse direction, show that the transverse mass m_T is uncharged to order β, where β is the transverse velocity of the decaying W.

Ex 27.5 Tau longitudinal polarization in $e^+e^- \to Z^0 \to \tau^+\tau^-$ final states. Show that the amount of longitudinal polarization is given by:

$$\mathcal{P}_\tau(\cos\theta_{\tau^-}) = -\frac{\mathcal{A}_\tau(1 + \cos^2\theta_{\tau^-}) + 2\mathcal{A}_e \cos\theta_{\tau^-}}{(1 + \cos^2\theta_{\tau^-}) + \frac{8}{3}A_{FB}^\tau \cos\theta_{\tau^-}} \tag{27.80}$$

where $\mathcal{A}_f$ is defined in Eq. (26.102) and A_{FB}^τ in Eq. (26.101).

28   Neutrino–Nucleon Interactions

The study of neutrino interaction physics played an important role in establishing the validity of the theory of weak interactions and electroweak unification. Today, however, the study of interactions of neutrinos takes a secondary role to studies of the properties of neutrinos, such as masses and mixings.

28.1   High-Energy Neutrino Beams

The general scheme for producing modern high-energy neutrinos follows from the one pioneered in the BNL-Columbia experiment (see Sections 21.10 and 21.11). At high-energy accelerators (typically with primary protons of tens of giga-electronvolts), neutrinos produced in decays of secondaries have some interesting properties:

- neutrino energies above the charged-current threshold for ν_μ or ν_τ are available;

- the flavor composition of the beam can be varied;

- directed beams can be produced;

- neutrino pulses with a well-defined time structure can be produced.

Several types of neutrino beams are therefore possible and there is a very large range of possible optimizations. In most cases, high-energy neutrino beams are produced by **bursts** of a few milliseconds of duration. This is a consequence of the acceleration scheme of the proton. Typically, protons are accelerated in a set of accelerators (e.g., at CERN from the booster, to the PS to the SPS – see Figure 3.7) up to a chosen maximum energy. At CERN this corresponded to a value in the range 400–450 GeV. The total number of protons was typically of the order of 10^{13} at each cycle. These protons were extracted from the ring with a process called "slow extraction." Protons were effectively put into betatron oscillations and scraped off the main ring, and then sent to the primary target. The duration of the extraction determined the length of the neutrino burst. Once all protons were extracted, a new cycle was started. This typically lasted 6–10 s (longer when the SPS was shared with the other CERN programs). We illustrate some more features of high-energy neutrino beams in the following sections.

28.2   Layout of Modern Neutrino Beams

The schematic layout of the so-called **wide-band** and **narrow-band neutrino beams** is illustrated in Figure 28.1. The distance L is always very large compared to R. The principle of operation of a neutrino beam is the following (the basics were already discussed in Section 21.11; here we describe the modern versions of their implementation that have developed after the BNL experiment): high-energy protons are sent onto a target; a hadronic shower develops as a result of the proton–nucleus interaction; the target is optimized so as to allow

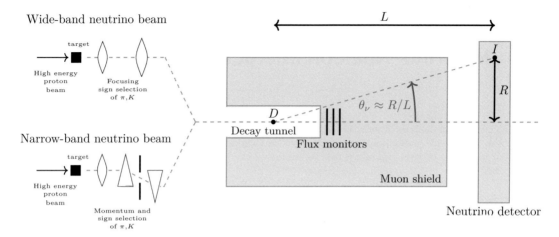

Figure 28.1 Schematic layout of a wide-band and narrow-band neutrino beam. D represents the meson decay point and I the neutrino interaction point. It is assumed that $L \gg R$.

the shower to leave the target; a major part of the shower leakage consists of long-lived mesons such as kaons and pions; mesons of the desired charge are focused by a possibly multi-stage magnetic focusing system. This is where narrow-band and wide-band differ:

- In a narrow-band beam (NBB), only secondary mesons of a given momentum with a narrow momentum "bite" (typically $\Delta p/p \approx \pm 5\%$) within a small solid angle and given charge are selected by a **system of magnetic lenses**. This leads to a relatively low-intensity beam since few of the totally produced mesons are selected. However, the beam properties are easier to compute and the sign selection is very efficient. Considering pions and kaons, this setup leads to a **dichromatic** energy spectrum, as shown later in this chapter.

- In a wide-band beam (WBB), the optimization is made to yield a maximal neutrino flux by accepting a broad range of mesons into the decay tunnel with a **horn focusing system**. The average energy of the mesons is small compared to the proton momentum where the secondary yield is highest. Wrong-sign contamination is higher than in the case of a NBB.

The produced meson beam is sent into a decay tunnel of appropriate length tuned to the mean energy of the mesons. A large fraction of the mesons decay in the tunnel via the following channels (see Eqs. (15.18), (15.61), and (15.62)):

- $\pi^+ \to \mu^+ + \nu_\mu$ ($Br \approx 100\%$)

- $K^+ \to \mu^+ + \nu_\mu$ ($Br \approx 63.5\%$)

- $K^+ \to \pi^0 + \mu^+ + \nu_\mu$ (so-called $K_{\mu 3}$, $Br \approx 3.2\%$)

- $K^+ \to \pi^0 + e^+ + \nu_e$ (so-called K_{e3}, $Br \approx 4.8\%$)

- $K_L^0 \to \pi^\mp + \mu^\pm + \nu_\mu(\bar{\nu}_\mu)$ ($Br \approx 27\%$)

- $K_L^0 \to \pi^\mp + e^\pm + \nu_e(\bar{\nu}_e)$ ($Br \approx 38.7\%$)

and their corresponding charge-conjugated channels. The decay tunnel is generally followed by a dense shielding (a hadron stop – not shown in Figure 28.1). Following the hadron stop, a certain number of muon detectors are used to measure the flux of muons. This can be used to monitor the neutrino beam, since muons and muon neutrinos are produced in pairs. Typically, the monitoring of the muon can be used to make sure that the beam is properly steered (e.g., proper alignment of the proton beam stop onto the target, and relative alignment of the focusing system, etc.). The decay tunnel is followed by the muon shield (e.g., of length about 400 m at CERN SPS) capable of ranging out all muons up to the maximum energy of 400 GeV. This is a critical component to ensure that the neutrino detector is not flooded by muons accompanying the neutrino burst.

• **Horn focusing system.** Different magnetic focusing systems have been designed, built, and operated. We mention here briefly the basic principle of the horn focusing system, shown schematically in Figure 28.2. A conducting horn, azimuthally symmetric, is built and centered around the beam axis and located at a certain distance (typically meters) from the target.

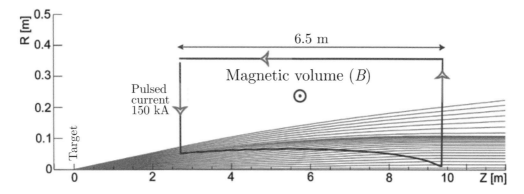

Figure 28.2 Magnetic kick provided by a horn focusing system. The lines represent the trajectories of charged particles emitted from the source at different angles. The shape is chosen such that most particles move forward after the horn.

A pulsed current which can typically reach values from 150 to 300 kA is injected into the horn, creating a large magnetic field within the volume of the horn. The pulsed current is synchronized with the sending of the protons onto the target (fast extraction approximately microseconds, slow extraction approximately milliseconds). Within the volume of the horn enclosed by the current, the field strength is:

$$B(r) = \frac{\mu_0 I}{2\pi r} \approx 2 \left(\frac{I}{100 \text{ kA}} \right) \left(\frac{1 \text{ cm}}{r} \right) \text{ T} \tag{28.1}$$

The lines represent the trajectories of charged particles emitted from the source at different angles. The shape of the inner conductor of the horn is determined in such a way that most particles, regardless of their initial transverse momentum, move forward with negligible transverse momentum after the horn.

The inner shape of the horn can be adjusted in order to more or less focalize the particles depending on their angle of emission. Typically, one would want to focalize all particles such that after the horn their transverse momentum vanishes. Assuming that the source of hadrons is point-like, one can show that parabolic or triangular shapes can be used to focalize particles of a given momentum or a given transverse momentum. Examples of conical and parabolic horns are shown in Figure 28.3.

The change of transverse momentum (called the $p_\perp$ "kick") is given by:

$$p_\perp^{horn} = p\Delta\theta = B \cdot x = \frac{\mu_0 I}{2\pi r} x \tag{28.2}$$

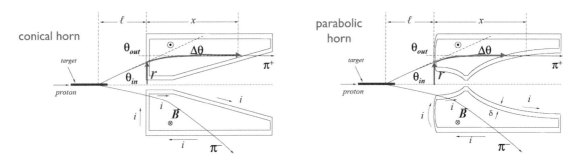

Figure 28.3 Magnetic kick provided by a horn focusing system. The lines represent the trajectories of charged particles emitted from the source at different angles. The shape is chosen such that most particles, regardless of their initial transverse momentum, move forward with negligible transverse momentum after the horn. Reprinted from S. E. Kopp, "Accelerator-based neutrino beams," *Phys. Rept.*, vol. 439, pp. 101–159, 2007, with permission from Elsevier.

where x is defined in Figure 28.3. The change of angle $\Delta\theta = \theta_{out} - \theta_{in} = \theta_{out} - r/\ell$. In useful units, we can write:

$$p_\perp^{horn}(\text{GeV}) = 0.3B(\text{T}) \cdot x(\text{m}) = 0.3 \cdot \frac{\mu_0 I(\text{A})}{2\pi r(\text{m})} \cdot x(\text{m}) \tag{28.3}$$

To give a kick $p_\perp^{horn} = 0.2$ GeV, the current must be:

$$I = \frac{p_\perp^{horn}(\text{GeV})}{0.3\mu_0 x(\text{m})} 2\pi r(\text{m}) = \frac{0.2 \cdot 2\pi}{0.3(2 \times 10^{-7})} \frac{r(\text{m})}{x(\text{m})} \quad \text{A} \tag{28.4}$$

With reasonable design parameters $x = 6.5$ m and $r = 0.05$ m, we get $I \simeq 160$ kA.

In order to improve the focalization efficiency, one can complement the horn by a so-called reflector.

• **Two-body meson decay kinematics.** We derive the meson decay kinematics in the simple case of the dominant two-body decays of the type $M(p^\mu) \to \mu(k^\mu)\nu_\mu(k'^\mu)$, where $M = \pi, K$. The decay kinematics can be expressed as $p^\mu \equiv k^\mu + k'^\mu$, so:

$$k^\mu = p^\mu - k'^\mu \implies m_\mu^2 = m_M^2 - 2(p \cdot k') \tag{28.5}$$

In the meson rest frame, we have:

$$p^\mu = (m_M, \vec{0}) \quad \text{and} \quad k'^\mu = (E_\nu^\star, E_\nu^\star \sin\theta^\star, 0, E_\nu^\star \cos\theta^\star) \tag{28.6}$$

Hence, the **neutrino is emitted monoenergetically in the rest frame of the two-body decay**:

$$m_\mu^2 = m_M^2 - 2m_M E_\nu^\star \implies E_\nu^\star = \frac{m_M^2 - m_\mu^2}{2m_M} = \begin{cases} 29.8 \text{ MeV for } M = \pi \\ 236.3 \text{ MeV for } M = K \end{cases} \tag{28.7}$$

In the laboratory system, the neutrino energy E_ν and decay angle θ_ν are given by (star quantities are defined in the meson M rest system):

$$E_\nu = \gamma E_\nu^\star (1 + \beta\cos\theta_\nu^\star) \quad \text{and} \quad \cos\theta_\nu = \frac{\cos\theta_\nu^\star + \beta}{1 + \beta\cos\theta_\nu^\star} \tag{28.8}$$

where $\beta = p_M/E_M$ and $\gamma = E_M/m_M$. The minimum and maximum neutrino energies in the laboratory correspond to the $\cos\theta^\star = \pm 1$ cases. Consequently:

$$E_{\nu,min} = \gamma E_\nu^\star (1 - \beta) \approx 0 \tag{28.9}$$

and

$$E_{\nu,max} = \gamma E_\nu^\star (1 + \beta) \approx \gamma \frac{m_M^2 - m_\mu^2}{m_M} = \left(1 - \frac{m_\mu^2}{m_M^2}\right) E_M = \begin{cases} 0.427 E_\pi & \text{for } M = \pi \\ 0.954 E_K & \text{for } M = K \end{cases} \qquad (28.10)$$

Since the decay is isotropic in the rest frame, the neutrino energy in the laboratory has a flat spectrum between the minimum and the maximum allowed values.

• **Energy-angle correlation.** Once the secondary meson yield and the efficiency of the focusing system are known, the neutrino flux in the neutrino detector can be computed with the help of the expressions derived in the preceding paragraphs. In fact, the difficult problem is to estimate the yield in momentum and angular distribution of hadrons in the very forward direction as produced by very high-energy protons impinging on a nuclear target, which in turn, changes the yield of particles that are focalized by the focusing system. After that, the problem is simply determined by kinematics. Considering again the decay kinematics, we note that the inverse transformations can be found by replacing β by $-\beta$ in Eq. (28.8). Therefore:

$$\cos \theta^\star = \frac{\cos \theta_\nu - \beta}{1 - \beta \cos \theta_\nu} \qquad (28.11)$$

Then

$$E_\nu = \gamma E_\nu^\star \left(1 + \beta \frac{\cos \theta_\nu - \beta}{1 - \beta \cos \theta_\nu}\right) = \gamma \left(\frac{m_M^2 - m_\mu^2}{2 m_M}\right)\left(\frac{1 - \beta^2}{1 - \beta \cos \theta_\nu}\right) = \frac{m_M^2 - m_\mu^2}{2 \gamma m_M (1 - \beta \cos \theta_\nu)} \qquad (28.12)$$

The neutrino energy in the laboratory is therefore completely correlated with its decay angle relative to the parent's flight direction:

$$E_\nu(\theta_\nu) = \frac{m_M^2 - m_\mu^2}{2 \gamma m_M (1 - \beta \cos \theta_\nu)} = \frac{E_{\nu,max}}{2 \gamma^2 (1 - \beta \cos \theta_\nu)} \qquad (28.13)$$

For small forward angles, we can expand $\cos \theta_\nu \approx 1 - \theta_\nu^2/2$. The neutrino energy is in this case just given by:

$$E_\nu(\theta_\nu) \approx \frac{E_{\nu,max}}{2 \gamma^2 (1 - \beta(1 - \theta_\nu^2/2))} = \frac{E_{\nu,max}}{2 \gamma^2 (1 - \beta + \frac{1}{2}\beta \theta_\nu^2))} = \frac{E_{\nu,max}}{2 \left(\frac{1-\beta}{1-\beta^2} + \frac{1}{2}\beta \gamma^2 \theta_\nu^2\right)} = \frac{E_{\nu,max}}{2 \left(\frac{1}{1+\beta} + \frac{1}{2}\beta \gamma^2 \theta_\nu^2\right)} \qquad (28.14)$$

For a relativistic parent with $\beta \to 1$, the expression simplifies further to:

$$E_\nu(\theta_\nu) \approx \frac{E_{\nu,max}}{1 + \gamma^2 \theta_\nu^2} \qquad (28.15)$$

Since $E_{\nu,max}$ (see Eq. (28.10)) is solely determined by the mass of the parent, there is a direct correlation between the neutrino energy in the laboratory and its emission angle for a given parent energy.

28.3   Narrow-Band Beam Fluxes

We have seen that in a narrow-band beam, the secondary mesons that are allowed to enter the decay tunnel have a narrow momentum "bite" (typically ±5%). Given the geometry of the decay tunnel, we can assume that only those mesons going straight will produce neutrinos (the others mostly hit the walls before they decay). Using the symbols of Figure 28.1, we define L as the distance between the decay point and the neutrino detector. Within the neutrino detector of a given surface, we consider an interaction located at a distance R from the center of the beam. The decay angle is therefore $\theta_\nu \approx R/L$, as shown in Figure 28.1.

As discussed above, the majority of the neutrinos are produced by the two-body decays $\pi \to \mu \nu_\mu$ and $K \to \mu \nu_\mu$. Consequently, we can apply Eq. (28.15). This leads to a flat flux distribution with a maximum

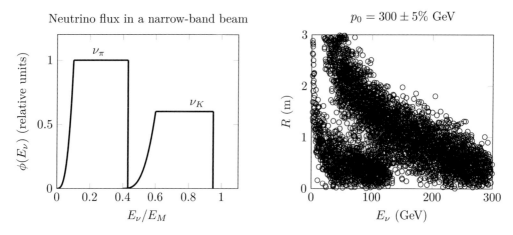

Figure 28.4 (left) Illustration of the neutrino flux in a narrow band beam. (right) Simulated correlation between the radial position and the neutrino energy.

determined by $E_{\nu,max}$ and a minimum determined by the angular acceptance of the detector (i.e., its transverse dimensions) smeared out by the length of the decay tunnel. This results in the fluxes shown in Figure 28.4(left) where ν_π and ν_K label the neutrino coming from the decays of pions and kaons, respectively. This correlation can be measured experimentally. The results of a simple beam simulation are shown in Figure 28.4(right). The distance between the target and the neutrino detector was assumed to be 800 m. The beam of mesons was distributed transversely according to a Gaussian with a standard deviation of 0.3 m. The length of the decay tunnel was set to 300 m, with its radius set to 1 m. Neutrinos hitting the detector plane within a radius of 3 m were recorded in the figure. The fraction of kaons with respect to pions was set to 30%. The two very distinct populations of events can easily be identified and also separated. Hence, a narrow-band beam is actually a dichromatic beam with events coming from pion and kaon decays easily separable in the R–E_ν plane! In reality, there are additional sources of events outside the two bands not included in the simulation: (a) those from $K_{\mu 3}$ decays and (b) "wide-band" background due to mesons decaying before muon momentum selection. These latter populate the graph mostly at low energy and all angles since the decay length increases with γ and they are not focalized by the momentum selection magnets.

• **The FNAL Sign Selected Quadrupole Train (SSQT).** A relatively high-intensity and very high-purity helicity-selectable beam was produced at FNAL with the Sign Selected Quadrupole Train (SSQT) [376]. Its layout is shown in Figure 28.5. The secondary mesons produced by 800 GeV protons on the target are focalized by a

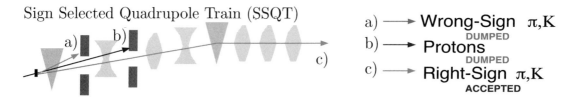

Figure 28.5 Schematic layout of the FNAL Sign Selected Quadrupole Train (SSQT) beam line. Reprinted from R. Bernstein *et al.*, "Sign selected quadrupole train," FERMILAB-TM-1884 (1994).

series of magnetic lenses. The wrong-sign mesons are unfocalized and are captured by beam dumps at the early stage of the beamline. Protons which have not interacted are also dumped into a second stage. Only right-sign mesons are accepted in the beam line, which bends them in the direction of the decay tunnel. The central momentum of the secondaries accepted by the focusing system is set at 250 GeV/c. The resulting numbers of neutrino interactions, for both polarities, are shown in Figure 28.6. The dominant flux has the recognizable and distinct shoulders produced by the pion and kaon decays. The contamination from opposite-helicity neutrinos is impressively small, at the level of permil, as well as that from the electron neutrinos, which is even smaller. This beam has been used by the CCFR experiment to perform the most precise measurement to date of the neutral-current to charged-current ratio in deep inelastic scattering, as described in Section 28.11.

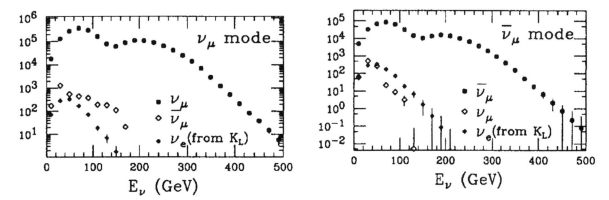

Figure 28.6 Predicted number of neutrino interactions for 10^{18} protons-on-target for both polarities (labeled neutrino and antineutrino modes). Reprinted from R. Bernstein *et al.*, "Sign selected quadrupole train," FERMILAB-TM-1884 (1994).

28.4   Wide-Band Beam Fluxes

We illustrate the neutrino flux from a wide-band beam for the case of CERN SPS (450 GeV). Figure 28.7 shows the predicted components of neutrinos at the CERN NOMAD experiment [377]. This kind of beam is much more intense compared to a narrow-band beam, as can readily be appreciated from the logarithmic scale of the graph. They also do not exhibit the dichromatic nature of the NBB, but rather are composed of two almost exponentially falling spectra, one at low energy from pions, and another very long tail up to the highest energies from kaon decays. As pointed out previously, WBB are optimized for flux and provide a very wide energy spectrum. Relative to the dominant composition of muon neutrinos, the relative abundance of muon antineutrinos is 6.78%, that of electron neutrinos is about 1%, and of electron antineutrinos is 0.3%. The graphs show the total flux as well as the contributions of individual mesons. One recognizes the features from the dominant decays of pions and kaons. In addition, the contributions from secondary muons and K_L^0s are also shown.

28.5   Off-Axis Long-Baseline Neutrino Beams

In order to study **neutrino flavor oscillations** (see Section 30.6), the distance between the source and the detector must be of the order of the oscillation wavelength. This distance can be several hundreds of kilometers. In the context of **long-baseline neutrino oscillation experiments**, one has considered the case of an "off-axis" neutrino beam. In order to discuss the effect on the neutrino energy spectrum, we consider the momentum sphere defined in Section 5.9. This latter is represented for neutrinos from pion decays in Figure 28.8(left). The transverse momentum is invariant under the longitudinal boost and is constrained within 30 MeV. The

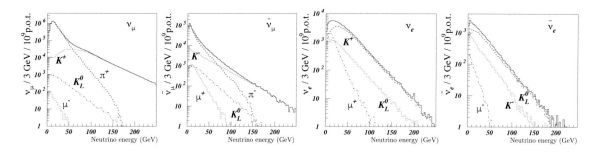

Figure 28.7 The prediction of the ν_μ, $\bar{\nu}_\mu$, ν_e, $\bar{\nu}_e$ neutrino fluxes at the CERN NOMAD experiment. The individual contributions from the various secondary particles are shown. Reprinted from P. Astier *et al.*, "Prediction of the neutrino fluxes in the NOMAD experiment," *Nucl. Instrum. Meth. A*, vol. 515, pp. 800–828, 2003, with permission from Elsevier.

longitudinal momentum is multiplied by the parent γ factor in the direction of flight. At a given angle θ_ν, there is an accumulation of neutrinos. The same effect can also be noted by plotting the correlation between

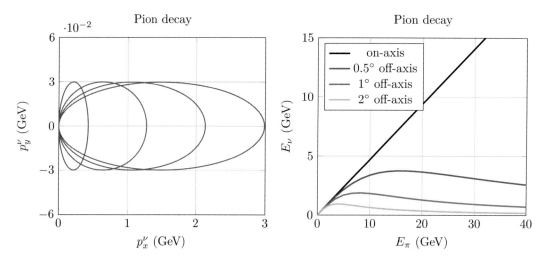

Figure 28.8 (left) Neutrino momentum sphere in the case of pions. (right) The correlation between the neutrino energy E_ν and the parent pion energy E_π at different scattering angles $\theta_\nu = 0$, 0.5°, 1°, and 2° off-axis.

the neutrino energy E_ν and the parent pion energy E_π. The correlation is shown for different scattering angles $\theta_\nu = 0, 0.5°, 1°$, and 2° in Figure 28.8(right). The "on-axis" corresponds to the $\theta_\nu = 0$ situation where we are measuring the neutrinos along the path of the decay channel of the beam. In the "off-axis" configuration, the neutrinos are detected around the axis of the main direction, at some distance from the ideal beam position. The angles considered lead to the a priori surprising effect that regardless of the parent energy E_π, neutrinos cluster around low energy.

The off-axis neutrino beam configuration has been adopted by the T2K experiment [378]. The spectra of the muon-neutrino flux at a distance of 295 km from the T2K neutrino beam source for two off-axis angles of 2° and 2.5° are plotted in Figure 28.9. The hypothetical spectrum at 0° (on-axis) is shown for comparison.

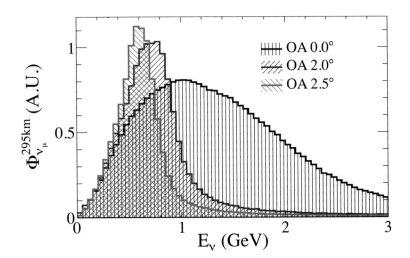

Figure 28.9 Spectrum of the muon-neutrino flux at a distance of 295 km from the T2K neutrino beam source for two off-axis angles of 2° and 2.5°. The hypothetical spectrum at 0° (on-axis) is shown for comparison. Reprinted with permission from K. Abe *et al.*, "T2K neutrino flux prediction," *Phys. Rev.*, vol. D87, no. 1, p. 012001, 2013. [Addendum: *Phys. Rev.* vol. D87, no.1, p.019902(2013)]. Copyright 2013 by the American Physical Society.

28.6   Neutrino Scattering on Free Nucleons

Neutrino scattering on point-like electrons was discussed in Section 26.3. Here we are interested in interactions of neutrinos on targets with internal structure, such as nucleons. The basis for the formalism has already been introduced in the context of electron–proton scattering in Chapter 16. The differences in going from e-p to ν-$\mathcal{N}$ scattering is that the parity-conserving electromagnetic force should be replaced by the parity-violating weak force. We will see that this has several consequences.

Neutrino–nucleon scattering processes can be categorized into four fundamental charged-current (CC) and neutral-current (NC) reactions:

$$\text{CC:} \quad \nu_\ell + \mathcal{N} \longrightarrow \ell^- + X \qquad \overline{\nu}_\ell + \mathcal{N} \longrightarrow \ell^+ + X$$

$$\text{NC:} \quad \nu_\ell + \mathcal{N} \longrightarrow \nu_\ell + X \qquad \overline{\nu}_\ell + \mathcal{N} \longrightarrow \overline{\nu}_\ell + X \,, \tag{28.16}$$

where $\mathcal{N}$ represents the target nucleon, a proton, or a neutron, and X is the hadronic system (which corresponds to the final state of the hit nucleon, which can be composed of one proton or a neutron, or of several hadronic particles resulting from the breaking of the hit nucleon). This latter occurs more often at higher energies, in particular in the deep inelastic regime, where jets are formed as a result of the quark fragmentation.

Lepton flavor conservation constrains type and charge of the outgoing charged lepton in CC interactions. Neutrino CC interactions produce negatively charged leptons, while antineutrino CC interactions produce positively charged antileptons.

The assumption that the target nucleon is free in neutrino scattering is only possible experimentally in the case of a liquid hydrogen target providing free protons, or in hydrogenated materials (such as water or scintillators) which contain a fraction of free protons. Quasi-free neutrons are accessible with deuterium (heavy hydrogen or fractionally in heavy water) as a target. Nucleons in nuclei are bound within the nuclear potential. If we consider neutrinos in the giga-electronvolt range, the nuclear structure (and its levels at the mega-electronvolt scale) can be factors and the nucleons can be considered as quasi-free within a potential well. In the deep inelastic regime, the interacting neutrino resolves the partons (quarks) which are constituting the nucleon within the nucleus. In general, calculations of neutrino interactions on nuclear targets are computed assuming quasi-free nucleons and subsequently correcting the computations for different "nuclear effects."

The nucleons inside the nucleus do not find themselves at rest and are subject to the so-called **Fermi motion**. This motion is a consequence of the exclusion principle which prevents all nucleons of the same type being in the ground state. Nucleons fill up layers (the analog to electrons around the nucleus). The Fermi momentum is typically 250 MeV/c and depends on the detailed nuclear physics of the particular element.

Let us consider the kinematics of the CC reactions (similar conclusions hold for the NC reactions). As in the case of the electron–proton scattering, the reaction is defined as (cf. Eq. (16.73)):

$$
\text{CC}: \quad
\begin{cases}
\nu_\ell\left(k^\mu\right) + \mathcal{N}\left(p^\mu\right) \rightarrow \ell^-\left(k'^\mu\right) + X\left(p'^\mu\right) \\
\bar{\nu}_\ell\left(k^\mu\right) + \mathcal{N}\left(p^\mu\right) \rightarrow \ell^+\left(k'^\mu\right) + X\left(p'^\mu\right)
\end{cases}
\tag{28.17}
$$

where k, k', p, and p' are the four-vectors of the incoming neutrino, scattered charged lepton, target nucleon, and hadronic system. The corresponding tree-level Feynman diagram is drawn in Figure 28.10(left). Similar definitions can be used for the case of antineutrino scattering or in the case of neutral currents. It is customary to neglect the mass difference between the proton and neutron and define the **nucleon mass as** $M = (M_p + M_n)/2$, assuming that the target nucleon is at rest in the laboratory frame. Without loss of generality, we can use the z-axis as the direction of the incoming neutrino. Then:

$$
k^\mu = \left(E_\nu, 0, 0, E_\nu\right), \qquad p^\mu = \left(M, \vec{0}\right), \qquad k'^\mu = \left(E' = \sqrt{\vec{k}'^2 + m_\ell^2}, |\vec{k}'|\sin\theta, 0, |\vec{k}'|\cos\theta\right)
\tag{28.18}
$$

where θ is the scattering angle, and m the mass of the scattered lepton. See Figure 28.10(right).

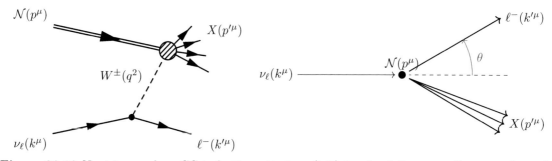

Figure 28.10 Neutrino–nucleon CC inelastic scattering: (left) tree-level Feynman diagram, where the blob illustrates the fact that the nucleon is not a point-like Dirac particle and breaks up and $q^2 = (k - k')^2$; (right) kinematics in the nucleon rest frame.

For the charged-current interactions, energy–momentum conservation requires a minimum energy for the incoming neutrino. Since the masses of the charged leptons are significantly different from one another, namely $m_e \approx 0.5$ MeV, $m_\mu \approx 106$ MeV, and $m_\tau \approx 1770$ MeV, the threshold energy will strongly depend on the neutrino flavor. For the electron antineutrino, the threshold was discussed for the inverse beta decay to be $E_{\bar{\nu}} > 1.806$ MeV (see Section 21.5). For muons and taus, the CC threshold energy is $\sqrt{s} \geq m_\ell + M$, so $s = (k + p)^2 = 2ME_\nu + M^2 \geq (m_\ell + M)^2 \Rightarrow 2ME_\nu \geq m_\ell^2 + 2Mm_\ell$ and finally:

$$
E_\nu \geq \frac{m_\ell^2 + 2Mm_\ell}{2M} =
\begin{cases}
\simeq 0 \text{ GeV for electrons} \\
0.110 \text{ GeV for muons} \\
3.5 \text{ GeV for taus}
\end{cases}
\tag{28.19}
$$

The kinematical variables commonly used for electron–proton scattering defined in Section 16.7 are directly applicable to the neutrino scattering case (see Table 16.2). We can express them as a function of the laboratory quantities defined above. We first compute all useful scalar products:

$$
\begin{cases}
(p \cdot q) = M\,(E_\nu - E') \\
(p \cdot k) = M E_\nu \\
(p \cdot k') = M E' \\
(q \cdot k) = \left(E_\nu (E_\nu - E') - E_\nu \left(E_\nu - |\vec{k}'| \cos\theta \right) \right) = -E_\nu \left(E' - |\vec{k}'| \cos\theta \right) \\
(q \cdot k') = |\vec{k}'|^2 + E_\nu E' - E'^2 - E_\nu |\vec{k}'| \cos\theta = -m_\ell^2 + E_\nu \left(E' - |\vec{k}'| \cos\theta \right)
\end{cases}
\tag{28.20}
$$

Then we introduce the conventional kinematical variables, in analogy to the electron–proton case described in Section 16.7.

- The four-momentum transfer $Q^2 \equiv -q^2$:

$$
Q^2 = -q \cdot (k - k') = q \cdot k' - q \cdot k = -m_\ell^2 + 2E_\nu \left(E' - |\vec{k}'| \cos\theta \right) \approx 2 E_\nu E' (1 - \cos\theta)
$$
$$
= 4 E_\nu E' \sin^2 \frac{\theta}{2}
\tag{28.21}
$$

- The energy transferred ν is equal to the energy transferred to the nucleon:

$$
\nu \equiv \frac{p \cdot q}{M} = E_\nu - E' = E_X - M
\tag{28.22}
$$

where E_X is the energy of the hadronic system.

- The invariant mass of the hadronic system W:

$$
W^2 \equiv p'^2 = (p + q)^2 = p^2 + q^2 + 2(p \cdot q) = M^2 - Q^2 + 2M\nu
\tag{28.23}
$$

- The dimensionless Bjorken-scaling variable x:

$$
x \equiv \frac{Q^2}{2\,(p \cdot q)} = \frac{Q^2}{2M\nu}
\tag{28.24}
$$

- The dimensionless inelasticity y:

$$
y = \frac{p \cdot q}{p \cdot k} = \frac{M(E_\nu - E')}{M E_\nu} = \frac{E_\nu - E'}{E_\nu} = \frac{\nu}{E_\nu}
\tag{28.25}
$$

Combining the kinematical bound $Q^2 \leq 2M\nu$ (see Eq. (16.88)) with x gives an upper bound on x. With the lower bound for x, one gets:

$$
\frac{m_\ell^2}{2M E_\nu} \leq x \leq 1 \Rightarrow 0 \leq x \leq 1 \quad \text{for} \quad m_\ell \ll E_\nu
\tag{28.26}
$$

The kinematical bounds for y are given by $A - B \leq y \leq A + B$, where:

$$
A = \frac{1}{2} \frac{1 - \dfrac{m_\ell^2}{2M E_\nu x} - \dfrac{m_\ell^2}{2 E_\nu^2}}{1 + x\dfrac{M}{2E_\nu}} \quad \text{and} \quad B = \frac{1}{2} \frac{1 - \dfrac{m_\ell^2}{2M E_\nu x} - \dfrac{m_\ell^2}{E_\nu^2}}{1 + x\dfrac{M}{2E_\nu}}
\tag{28.27}
$$

Neglecting the mass of the outgoing lepton, this reduces to:

$$0 \leq y \leq \left(1 + x\frac{M}{2E_\nu}\right)^{-1} \quad \text{for} \quad m_\ell \ll E_\nu \tag{28.28}$$

and subsequently neglecting also the target nucleon mass we simply obtain:

$$0 \leq y \leq 1 \quad \text{for} \quad m_\ell, M \ll E_\nu \tag{28.29}$$

The kinematics of an event, which for a fixed incoming neutrino energy, depends on two independent variables, can be analyzed for example in the Q^2 and ν plane. One could also choose the dimensionless scaling variables x and y. In practice, one considers the plane Q^2 vs. $2M\nu$, such that the range for both variables is the same $[0, 2ME_\nu]$. This is illustrated in Figure 28.11.

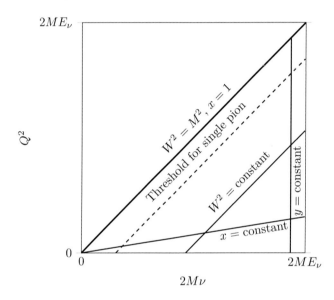

Figure 28.11 Graphical representation of the kinematical plane Q^2 vs. $2M\nu$.

It is straightforward to identify specific kinematical regions in this plane. On the diagonal of the plane defined by $Q^2 = 2M\nu$, we have by construction:

$$Q^2 = 2M\nu \quad \rightarrow x = 1 \tag{28.30}$$

in which case we have simply:

$$W^2 = M^2 \underbrace{-Q^2 + 2M\nu}_{= 0} = M^2 \tag{28.31}$$

This kinematical region corresponds to the *quasi-elastic* process, by which the nucleon in the initial state is not broken in the interaction and the final hadronic system is composed of just one nucleon. The process, comparable to the *elastic* $ep \rightarrow ep$ is labeled *quasi-elastic* since in charged current interactions the proton is exchanged into a neutron or vice versa.

The region above the $x = 1$ diagonal is kinematically forbidden by the condition $Q^2 \leq 2M\nu$. Lines parallel to the $x = 1$ diagonal correspond to constant invariant hadronic masses M', obeying:

$$W^2 = M^2 \underbrace{-Q^2 + 2M\nu}_{= \text{constant}} = M'^2 \tag{28.32}$$

Lines to the right of the $x = 1$ diagonal and parallel to it consequently correspond to constant hadronic systems of increasing invariant mass. Since $x = Q^2/2M\nu$, lines passing through the origin of the axes correspond to a constant x. Vertical lines correspond to constant y, since $2M\nu = 2MyE_\nu$.

Finally, it is customary to therefore define the physical kinematical plane into three distinct areas, and decompose the cross-section into three parts, where different types of underlying processes will dominate. For

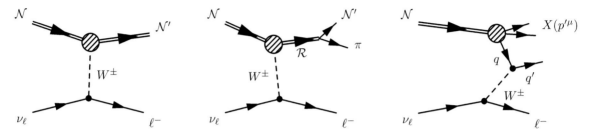

Figure 28.12 Neutrino–nucleon scattering: (left) quasi-elastic, (middle) resonance production, and (right) deep-inelastic. It is straightforward to derive corresponding diagrams for antineutrinos or neutral current interactions.

example, for the CC reactions one considers the following processes as illustrated in Figure 28.12, leading to the following channels:

- The **quasi-elastic domain** with $x = 1$, $W^2 = M^2$:

$$\text{QE} : \begin{cases} \nu_\ell + n & \to \ell^- + p \\ \overline{\nu}_\ell + p & \to \ell^+ + n \end{cases} \qquad \text{quasi-elastic}$$

- The **hadronic resonance production domain** which is actually separated from the quasi-elastic line by the threshold $W \simeq M + m_\pi \approx 1080$ MeV (as illustrated in Figure 28.11):

$$\text{RES} : \begin{cases} \nu_\ell + p & \to \ell^- + \mathcal{R}^{++} \to \ell^- + \mathcal{N} + n\pi \\ \nu_\ell + n & \to \ell^- + \mathcal{R}^+ \to \ell^- + \mathcal{N} + n\pi \\ \overline{\nu}_\ell + p & \to \ell^+ + \mathcal{R}^0 \to \ell^+ + \mathcal{N} + n\pi \\ \overline{\nu}_\ell + n & \to \ell^+ + \mathcal{R}^- \to \ell^+ + \mathcal{N} + n\pi \end{cases} \qquad \text{resonance production}$$

where $\mathcal{R}$ is a short-lived hadronic resonance. This can typically be seen in the $\Delta(1232)$, and $n \geq 1$.

- The **non-resonant, inelastic domain** where the target nucleon breaks into debris which form a recoiling jet of hadrons:

$$\text{DIS} : \begin{cases} \nu_\ell + \mathcal{N} & \to \ell^- + X \\ \overline{\nu}_\ell + \mathcal{N} & \to \ell^+ + X \end{cases} \qquad \text{inelastic process}$$

It is straightforward to derive corresponding diagrams for antineutrinos or neutral-current interactions. The **total neutrino cross-section** as a function of neutrino energy is then defined as (e.g., for CC):

$$\sigma_{CC}^{\nu,\overline{\nu}}(E_\nu) \equiv \sigma_{QE}^{\nu,\overline{\nu}}(E_\nu) + \sigma_{RES}^{\nu,\overline{\nu}}(E_\nu) + \sigma_{DIS}^{\nu,\overline{\nu}}(E_\nu) \tag{28.33}$$

and similarly for neutral-current reactions. The magnitudes of the cross-sections are illustrated in Figure 28.13. These domains are commonly treated separately, as described in the following sections.

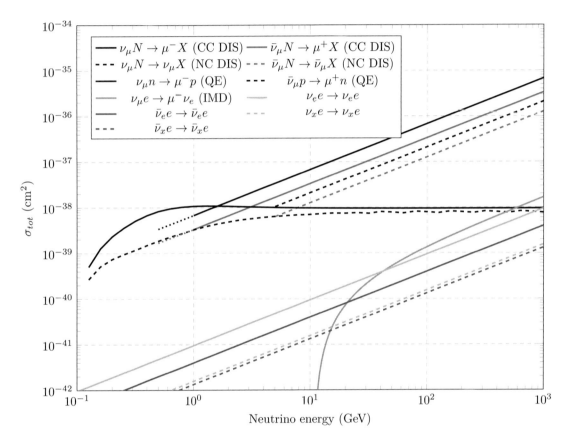

Figure 28.13 Total neutrino and antineutrino cross-sections for different types of reactions as a function of energy.

28.7   The Quasi-Elastic Neutrino–Nucleon Scattering

We discuss the calculation of the quasi-elastic processes $\nu_\ell + n \to \ell^- + p$ and $\overline{\nu}_\ell + p \to \ell^+ + n$. **While the elementary quark vertex is well known from the electroweak theory, this is not the case for composite particles. From the electron–proton scattering experiment and neutron decay, we know that the proton structure is to be taken into account with form factors to parametrize our ignorance on low-energy QCD bound states.** In the case of the electromagnetic interaction, we had written the general form of a conserving current with the Dirac and the Pauli form factor (see Eq. (16.34)). To describe the weak interaction of a neutrino and a nucleon we can write similarly as in the case of neutron decay (see Section 23.16):

$$\langle N(p')|J^\mu(q^2)|N(p)\rangle \equiv \overline{N}(p') \left[\gamma^\mu F_V(q^2) + \frac{i}{2M}\sigma^{\mu\nu}q_\nu F_M(q^2) \right] N(p) \tag{28.34}$$

where F_V **is the vector form factor** and F_M **the weak magnetic form factor**, two dimensionless functions of q^2 with the same form as the Dirac and Pauli terms for the electromagnetic case. However, these two terms are *not* sufficient to describe the parity-violating nature of the weak force. We need to include parity-violating axial-vector and pseudotensor terms. Consequently, the general form of the nucleon parity-violating current is:

$$\overline{N}(p') \left[F_V(q^2)\gamma^\mu + F_A(q^2)\gamma^\mu\gamma^5 + \frac{i}{2M}F_M(q^2)\sigma^{\mu\nu}q_\nu + \frac{i}{2M}F_T(q^2)\sigma^{\mu\nu}\gamma^5 q_\nu \right] N(p) \tag{28.35}$$

where F_A is the **axial-vector** and F_T the **pseudotensor form factor**. Two more terms depending on q^μ can still be added to obtain the most general form of current [379]:

$$\overline{N}(p') \left[F_V(q^2)\gamma^\mu + F_A(q^2)\gamma^\mu\gamma^5 + \frac{i}{2M}F_M(q^2)\sigma^{\mu\nu}q_\nu + \frac{i}{2M}F_T(q^2)\sigma^{\mu\nu}\gamma^5 q_\nu \right.$$
$$\left. + \frac{F_S(q^2)}{M}q^\nu + \frac{F_P(q^2)}{M}\gamma^5 q^\mu \right] N(p) \tag{28.36}$$

where F_S and F_P are the induced scalar and induced pseudoscalar form factors. The six form factors are dimensionless scalar functions of q^2 which must be determined experimentally. In 1958 Weinberg [380], inspired by G-parity conservation in strong interactions, notes that weak interactions can be subdivided into first-class and second-class currents. In the general form factor derived above, **the forms depending on F_V, F_M, F_A and F_P are called first-class currents, while the others are called second-class currents**. According to Weinberg, hadronic form factors should "be influenced" by fundamental symmetries of the strong interaction. In this case, the G-parity conservation implies the absence of second-class currents. Another way to express this is to say that second-class currents, if not zero, are to be suppressed compared to first-class currents, and will therefore be neglected. Consequently, only the first-class currents are retained, yielding:

$$\langle N(p')|J^\mu(q^2)|N(p)\rangle = \cos\theta_C \overline{N}(p') \left[\underbrace{F_V(q^2)\gamma^\mu + \frac{i}{2M}F_M(q^2)\sigma^{\mu\nu}q_\nu}_{\text{vector}} + \underbrace{F_A(q^2)\gamma^\mu\gamma^5}_{\text{axial}} + \underbrace{\frac{F_P(q^2)}{M}\gamma^5 q^\mu}_{\text{pseudoscalar}} \right] N(p)$$
$$\tag{28.37}$$

If we assume CP or T invariance, then all form factors are real [379].

Now comes the **hypothesis of conserved vector current (CVC)** – see Sections 21.9 and 23.16. This latter defines a tight relationship between the form factors F_V and F_M of the weak interaction and F_1 and F_2, those of the electromagnetic interaction Eq. (16.34). Basically, if we accept that the form factors characterize the structure of the target, then the *vector part* of both weak and electromagnetic currents should be the same. Hence, under the CVC hypothesis, the form factors F_V and F_M can be determined through e–$\mathcal{N}$ scattering data. We saw in Section 16.5 that in practice, linear combinations were used, which correspond to the **electric**

and magnetic Sachs form factors for the proton and the neutron (see Eq. (16.67)):

$$G_E^p(Q^2) \simeq \frac{G_M^p(Q^2)}{\mu_p} \simeq \frac{G_M^n(Q^2)}{\mu_n} \simeq G_D(Q^2) \tag{28.38}$$

with the static values $G_E^p(0) = 1$, $G_E^n(0) = 0$, $G_M^p(0) = \mu_p$, and $G_M^n(0) = \mu_n$, and a **vector dipole form factor** [193]:

$$G_D(Q^2) \equiv \frac{1}{(1 + Q^2/M_V^2)^2} = \frac{1}{(1 - q^2/M_V^2)^2} \tag{28.39}$$

where the fitted parameter is $M_V^2 = 0.71$ GeV^2 or $M_V = 0.84$ GeV.

In order to apply these results to the quasi-elastic neutrino scattering, we need to consider the current between a proton and a neutron. Under the CVC hypothesis, the vector part of the weak current is given by the electromagnetic current, however combining the proton and the neutron within the nucleon iso-doublet. This leads to the following vector and weak magnetic form factors:

$$F_V = F_1^p - F_1^n \quad \text{and} \quad F_M = F_2^p - F_2^n \tag{28.40}$$

which can be expressed as a function of the electric and magnetic Sachs form factors (see Section 16.5) as (note the negative signs between the proton and the neutron form factors):

$$F_V = \frac{G_E^p + \tau G_M^p}{1 + \tau} - \frac{G_E^n + \tau G_M^n}{1 + \tau} \quad \text{and} \quad F_M = \frac{G_M^p - G_E^p}{1 + \tau} - \frac{G_M^n - G_E^n}{1 + \tau} \tag{28.41}$$

where $\tau \equiv Q^2/(4M_p^2)$ (see Eqs. (16.62) and (16.63)).

Hence, under CVC, only the **axial-vector form factor** needs to be determined in neutrino interactions. It characterizes the parity-violating nature of the interaction, hence was absent in the electromagnetic process. The q^2 dependence of the axial-vector form factor is taken to be of the dipole form, just like the vector dipole factor (see Eq. (16.68)), however with a specific parameter, the **axial mass** M_A, to be determined by experiments:

$$F_A(Q^2) = \frac{F_A(Q^2 = 0)}{\left(1 + Q^2/M_A^2\right)^2} \tag{28.42}$$

The static value $F_A(Q^2 = 0) = -g_A = 1.2732 \pm 0.0023$ is extracted from the β decay of the neutron (see Section 23.16).

The differential cross-section for the quasi-elastic process can then be computed from the assumed form factors. The result is given in Ref. [379] (this equation corresponds to the Rosenbluth formula (see Section 16.3) for neutrino (antineutrino) quasi-elastic scattering!):

$$\frac{d\sigma_{QE}^{\nu,\bar{\nu}}}{dQ^2} = \frac{M^2 G_F^2 \cos^2 \theta_C}{8\pi E_\nu^2} \left(\frac{M_W^2}{Q^2 + M_W^2}\right)^2 \left[A(Q^2) \pm B(Q^2)\frac{s-u}{M^2} + C(Q^2)\left(\frac{s-u}{M^2}\right)^2\right] \tag{28.43}$$

where the effect of the $W^\pm$ boson propagator has been included in the second term, $s - u = 4ME_\nu - Q^2 - m_\ell^2$ with:

$$A(Q^2) = \frac{Q^2 + m_\ell^2}{M^2}\left[(1+\tau)F_A^2 - (1-\tau)F_V^2 + \tau(1-\tau)F_M^2 + 4\tau F_V F_M - D(Q^2)\right]$$

$$B(Q^2) = -4\tau F_A[F_V + F_M] \tag{28.44}$$

$$C(Q^2) = \frac{1}{4}\left[F_V^2 + \tau F_M^2 + F_A^2\right]$$

$$D(Q^2) = \frac{m_\ell^2}{4M^2}\left((F_V + F_M)^2 + (F_A + 2F_P)^2 - \left(\frac{Q^2}{M^2} + 4\right)F_P^2\right)$$

The term D proportional to the pseudoscalar F_P leads to a cross-section factor proportional to the mass of the outgoing lepton over the nucleon squared m_ℓ^2/M^2, thus it can be neglected for ν_e and ν_μ but becomes important for ν_τ. The difference between neutrinos and antineutrinos is visible in the sign in front of the B term, which is proportional to F_A. Hence, as expected, the axial form factor is the part that contributes to the asymmetry between neutrinos and antineutrinos. The extraction of the weak axial form factor from neutrino scattering data is sensitive to the assumptions that are used for the vector form factors from electron scattering data. Using the latest experiments, one obtains the value of the axial-vector mass [381]:

$$M_A \simeq 1.00 \text{ GeV} \tag{28.45}$$

The good agreement with the data is shown in Figure 28.14.

28.8   The Inclusive Inelastic Neutrino–Nucleon Scattering

In the inelastic regime, the target nucleon breaks because of the interaction and the final state is characterized by a hadronic system X composed of several hadrons. The inelastic ep scattering has been discussed in Section 16.6. Here we want to describe in general terms the inelastic neutrino nucleon scattering, Eq. (28.17). At first order, we can write the amplitude as the product of the leptonic and hadronic currents using the electroweak Feynman rules (see Table 26.1), including also the $W^\pm$ propagator, Eq. (23.8):

$$
\begin{aligned}
-i\mathcal{M} &= -i \left(\frac{g_W}{\sqrt{2}} \right)^2 \bar{u}(k') \gamma_\mu \left(\frac{1-\gamma^5}{2} \right) u(k) \frac{i\left(-g^{\mu\nu} + \frac{q^\mu q^\nu}{M_W^2} \right)}{q^2 - M_W^2} \langle X(p')|J_\nu(q^2)|N(p)\rangle \\
&\simeq 2 \left(\frac{g_W^2}{8M_W^2} \right) \bar{u}(k') \gamma^\mu \left(1-\gamma^5 \right) u(k) \langle X(p')|J_\mu(q^2)|N(p)\rangle \\
&= G_F^2 \, \bar{u}(k') \gamma^\mu \left(1-\gamma^5 \right) u(k) \langle X(p')|J_\mu(q^2)|N(p)\rangle
\end{aligned}
\tag{28.46}
$$

where we assumed $q^2 \ll M_W^2$ and used $G_F/\sqrt{2} \equiv g_W^2/(8M_W^2)$ (see Eq. (23.25)). Consequently, the spin-averaged matrix element will be the product of two tensors:

$$\langle |\mathcal{M}|^2 \rangle = \frac{G_F^2}{2} L^{\mu\nu} W_{\mu\nu} \tag{28.47}$$

where the leptonic tensor $L^{\mu\nu}$ is given by the exact $V-A$ structure of the weak current (including the outgoing lepton mass m_ℓ):

$$L^{\mu\nu} = \text{Tr}\left\{ (\slashed{k}' + m_\ell) \gamma^\mu \left(1-\gamma^5 \right) \slashed{k} \gamma^\nu \left(1-\gamma^5 \right) \right\} \tag{28.48}$$

For the hadronic tensor $W_{\mu\nu}$, we should consider the most general form which is consistent with a covariant Lorentz structure, which is now more complicated in weak interactions than in the electromagnetic electron–proton scattering because parity is no longer conserved. Just as in the case of the *polarized* electron–proton scattering, we should include both symmetric and antisymmetric terms (cf. Eq. (19.18)). This can be achieved with a sum of five terms with corresponding form factors $W_i = W_i(Q^2, \nu)$:

$$W_{\mu\nu} = -g_{\mu\nu} W_1 + \frac{p_\mu p_\nu}{M^2} W_2 - \frac{i\varepsilon_{\mu\nu\alpha\beta} p^\alpha q^\beta}{2M^2} W_3 + \frac{q_\mu q_\nu}{M^2} W_4 + \frac{p_\mu q_\nu + p_\nu q_\mu}{2M^2} W_5 \tag{28.49}$$

The form factors parameterize our ignorance on the structure of the target nucleon. We can simplify the leptonic tensor with the use of the trace theorems **[TS3]** and **[TS4]** in Appendix E.3, as follows:

$$
\begin{aligned}
L^{\mu\nu} &= \text{Tr}\left\{ \slashed{k}' \gamma^\mu \left(1-\gamma^5 \right) \slashed{k} \gamma^\nu \left(1-\gamma^5 \right) \right\} + \underbrace{m_\ell \, \text{Tr}\left\{ \gamma^\mu \left(1-\gamma^5 \right) \slashed{k} \gamma^\nu \left(1-\gamma^5 \right) \right\}}_{=0} \\
&= 8 \left[k^\mu k'^\nu + k^\nu k'^\mu - (k \cdot k') g^{\mu\nu} \right] + 8i\varepsilon^{\mu\alpha\nu\beta} k_\alpha k'_\beta
\end{aligned}
\tag{28.50}
$$

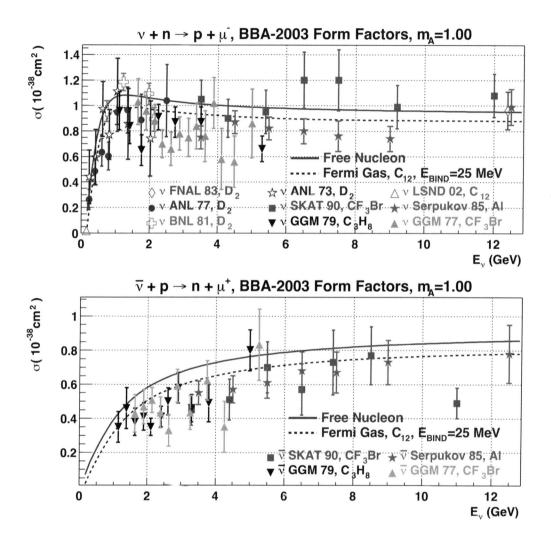

Figure 28.14 The quasi-elastic neutrino–nucleon scattering cross-section along with data from various experiments for (top) neutrinos and (bottom) antineutrinos. The calculation uses $M_A = 1.00$ GeV, $F_A(q^2 = 0) = 1.267$, $M_V^2 = 0.71$ GeV^2, and BBA-2003 form factors. Reprinted from H. S. Budd, A. Bodek, and J. Arrington, "Modeling quasielastic form-factors for electron and neutrino scattering," in *2nd International Workshop on Neutrino–Nucleus Interactions in the Few GeV Region (NuInt 02) Irvine, California, December 12–15, 2002,* 2003.

For antineutrinos, we have a similar term but for the signs in front of the chirality matrices γ^5 and the mass changes, resulting in the following expression:

$$L^{\mu\nu} = \text{Tr}\left\{ (\not{k}' - m_\ell)\gamma^\mu (1+\gamma^5)\not{k}\gamma^\nu (1+\gamma^5) \right\} = 8\left[k^\mu k'^\nu + k^\nu k'^\mu - (k\cdot k')g^{\mu\nu}\right] - 8i\varepsilon^{\mu\alpha\nu\beta}k_\alpha k'_\beta \quad (28.51)$$

Hence, the expression for the leptonic tensor valid for neutrinos $(+)$ and antineutrinos $(-)$ can be written as:

$$L^{\mu\nu} = 8\left[k^\mu k'^\nu + k^\nu k'^\mu - (k\cdot k')g^{\mu\nu}\right] \pm 8i\varepsilon^{\mu\alpha\nu\beta}k_\alpha k'_\beta \quad (28.52)$$

We should now multiply the leptonic and hadronic tensors, term by term. We consider each W_i separately:

$$
\begin{aligned}
-g_{\mu\nu}W_1 L^{\mu\nu} &= -8g_{\mu\nu}W_1\left[k^\mu k'^\nu + k^\nu k'^\mu - (k\cdot k')g^{\mu\nu}\right] \mp \underbrace{8iW_1 g_{\mu\nu}\varepsilon^{\mu\alpha\nu\beta}k_\alpha k'_\beta}_{=0} \\
&= -8W_1\left[(k\cdot k') + (k\cdot k') - (k\cdot k')\underbrace{g_{\mu\nu}g^{\mu\nu}}_{=4}\right] = 16(k\cdot k')W_1 \quad (28.53)
\end{aligned}
$$

and

$$
\begin{aligned}
\frac{p_\mu p_\nu}{M^2}W_2 L^{\mu\nu} &= \frac{8W_2}{M^2}\left[(p\cdot k)(p\cdot k') + (p\cdot k')(p\cdot k) - (k\cdot k')(p\cdot p) \pm \underbrace{i\varepsilon^{\mu\alpha\nu\beta}k_\alpha k'_\beta p_\mu p_\nu}_{=0}\right] \\
&= \frac{8W_2}{M^2}\left[2(p\cdot k)(p\cdot k') - (k\cdot k')(p\cdot p)\right] \quad (28.54)
\end{aligned}
$$

and

$$
\begin{aligned}
-\frac{i\varepsilon_{\mu\nu\alpha\beta}p^\alpha q^\beta}{2M^2}W_3 L^{\mu\nu} &= -\frac{4iW_3\varepsilon_{\mu\nu\alpha\beta}p^\alpha q^\beta}{M^2}\left[k^\mu k'^\nu + k^\nu k'^\mu - (k\cdot k')g^{\mu\nu} \pm i\varepsilon^{\mu\gamma\nu\delta}k_\gamma k'_\delta\right] \\
&= \mp\frac{4iW_3\varepsilon_{\mu\nu\alpha\beta}p^\alpha q^\beta}{M^2}\left[i\varepsilon^{\mu\gamma\nu\delta}k_\gamma k'_\delta\right] = \pm\frac{4W_3}{M^2}p^\alpha q^\beta k_\gamma k'_\delta\left[\varepsilon_{\mu\nu\alpha\beta}\varepsilon^{\mu\gamma\nu\delta}\right] \\
&= \pm\frac{8W_3}{M^2}p^\alpha q^\beta k_\gamma k'_\delta\left[\delta_\alpha^\gamma\delta_\beta^\delta - \delta_\beta^\gamma\delta_\alpha^\delta\right] = \pm\frac{8W_3}{M^2}\left[p^\alpha q^\beta k_\alpha k'_\beta - p^\alpha q^\beta k_\beta k'_\alpha\right] \\
&= \pm\frac{8W_3}{M^2}\left[(p\cdot k)(q\cdot k') - (p\cdot k')(q\cdot k)\right] \quad (28.55)
\end{aligned}
$$

and

$$\frac{q_\mu q_\nu}{M^2}W_4 L^{\mu\nu} = \frac{8W_4}{M^2}q_\mu q_\nu\left[k^\mu k'^\nu + k^\nu k'^\mu - (k\cdot k')g^{\mu\nu} \pm i\varepsilon^{\mu\gamma\nu\delta}k_\gamma k'_\delta\right] = \frac{8W_4}{M^2}\left[2(q\cdot k)(q\cdot k') - (k\cdot k')q^2 \pm 0\right] \quad (28.56)$$

and finally:

$$
\begin{aligned}
\frac{p_\mu q_\nu + p_\nu q_\mu}{2M^2}W_5 L^{\mu\nu} &= \frac{8W_5}{2M^2}(p_\mu q_\nu + p_\nu q_\mu)\left[k^\mu k'^\nu + k^\nu k'^\mu - (k\cdot k')g^{\mu\nu} \pm i\varepsilon^{\mu\gamma\nu\delta}k_\gamma k'_\delta\right] \\
&= \frac{8W_5}{M^2}\left[(p\cdot k)(q\cdot k') + (p\cdot k')(q\cdot k) - (k\cdot k')(p\cdot q)\right]
\end{aligned}
$$

In order to simplify the products of four-vectors, we note that:

$$
\begin{aligned}
q^2 &= (k-k')^2 = k^2 - 2(k\cdot k') + m_\ell^2 \Rightarrow (k\cdot k') = \frac{-q^2 + m_\ell^2}{2} = \frac{Q^2 + m_\ell^2}{2} \\
(q\cdot k) &= (k-k')\cdot k = k^2 - k'\cdot k \approx -(k\cdot k') = -\frac{Q^2 + m_\ell^2}{2} \\
(q\cdot k') &= (k-k')\cdot k' = k\cdot k' - k'^2 = (k\cdot k') - m_\ell^2 = \frac{Q^2 + m_\ell^2}{2} - m_\ell^2 = \frac{Q^2 - m_\ell^2}{2} \quad (28.57)
\end{aligned}
$$

Collecting all the terms and inserting the expressions above, we find:

$$L^{\mu\nu}W_{\mu\nu} = 16(k \cdot k')W_1 + \frac{8W_2}{M^2}\left[2(p \cdot k)(p \cdot k') - (k \cdot k')(p \cdot p)\right] \pm \frac{8W_3}{M^2}\left[(p \cdot k)(q \cdot k') - (p \cdot k')(q \cdot k)\right]$$

$$+ \frac{8W_4}{M^2}\left[2(q \cdot k)(q \cdot k') - (k \cdot k')q^2\right] + \frac{8W_5}{M^2}\left[(p \cdot k)(q \cdot k') + (p \cdot k')(q \cdot k) - (k \cdot k')(p \cdot q)\right]$$

$$= 16\left(\frac{Q^2 + m_\ell^2}{2}\right)W_1 + \frac{8W_2}{M^2}\left\{2ME_\nu ME' - \left(\frac{Q^2 + m_\ell^2}{2}\right)M^2\right\}$$

$$\pm \frac{8W_3}{M^2}\left\{ME_\nu \frac{Q^2 - m_\ell^2}{2} + ME'\frac{Q^2 + m_\ell^2}{2}\right\} + \frac{8W_4}{M^2}\left\{-2\left(\frac{Q^2 + m_\ell^2}{2}\right)\frac{Q^2 - m_\ell^2}{2} + \frac{Q^2 + m_\ell^2}{2}Q^2\right\}$$

$$+ \frac{8W_5}{M^2}\left\{ME_\nu \frac{Q^2 - m_\ell^2}{2} - ME'\frac{Q^2 + m_\ell^2}{2} - M(E_\nu - E')\frac{Q^2 + m_\ell^2}{2}\right\}$$

With some algebraic manipulation, the expression can be further simplified to:

$$L^{\mu\nu}W_{\mu\nu} = 4\left\{2\left(Q^2 + m_\ell^2\right)W_1 + \left[4E_\nu(E_\nu - \nu) - \left(Q^2 + m_\ell^2\right)\right]W_2\right.$$

$$\left. \pm \left[\frac{(2E_\nu - \nu)Q^2 - \nu m_\ell^2}{M}\right]W_3 + \frac{m_\ell^2}{M^2}\left[\left(Q^2 + m_\ell^2\right)W_4 - 2ME_\nu W_5\right]\right\}$$

The differential cross-section is found by adding the phase-space factors:

$$d\sigma = \frac{1}{4ME_\nu}\langle|\mathcal{M}|^2\rangle(2\pi)^4\delta^4(k + p - k' - p')\frac{d^3k'}{(2\pi)^3\,2E'}\frac{d^3p'}{(2\pi)^3\,2E_X} \tag{28.58}$$

By integration over the phase space and recalling that the inelastic differential cross-section depends on two variables, we can express the result as a function of Q^2 and ν valid for neutrinos $(+)$ and antineutrinos $(-)$:

$$\frac{d^2\sigma}{dQ^2\,d\nu} = \frac{G_F^2}{8\pi ME_\nu^2}\left\{2\left(Q^2 + m_\ell^2\right)W_1 + \left[4E_\nu(E_\nu - \nu) - (Q^2 + m_\ell^2)\right]W_2\right.$$

$$\left. \pm \left[\frac{(2E_\nu - \nu)Q^2 - \nu m_\ell^2}{M}\right]W_3 + \frac{m_\ell^2}{M^2}\left[\left(Q^2 + m_\ell^2\right)W_4 - 2ME_\nu W_5\right]\right\} \tag{28.59}$$

where the five form factors must be determined experimentally. If we neglect the mass of the outgoing lepton, then the two terms with W_4 and W_5 disappear, and the differential cross-section depends on only three functions.

We can also write this result as a function of the Lorentz-invariant kinematics and the Bjorken scaling variables. Using Eq. (16.103), we directly find:

$$\frac{d^2\sigma}{dx\,dy} = \frac{G_F^2}{8\pi ME_\nu^2}\cdot(2ME_\nu^2 y)\cdot\left\{2(2ME_\nu xy + m_\ell^2)W_1 + \left[4E_\nu(E_\nu - yE_\nu) - (2ME_\nu xy + m_\ell^2)\right]W_2\right.$$

$$\left. \pm \left[\frac{(2E_\nu - yE_\nu)2ME_\nu xy - yE_\nu m_\ell^2}{M}\right]W_3 + \frac{m_\ell^2}{M^2}\left[(2ME_\nu xy + m_\ell^2)W_4 - 2ME_\nu W_5\right]\right\}$$

or after rearranging:

$$
\begin{aligned}
\frac{d^2\sigma}{dx\,dy} = \frac{G_F^2 M E_\nu}{\pi} &\left\{ \left[xy^2 + \frac{m_\ell^2}{2ME_\nu}y \right] W_1 + \left[(1-y) - \frac{M}{2E_\nu}xy - \frac{m_\ell^2}{4E_\nu^2} \right] \frac{\nu}{M} W_2 \right. \\
&\left. \pm \left[\left(1 - \frac{y}{2}\right)xy - \frac{m_\ell^2}{4ME_\nu}y \right] \frac{\nu}{M} W_3 + \frac{m_\ell^2}{M^2}\left[\left(\frac{M}{2E_\nu}xy + \frac{m_\ell^2}{4E_\nu^2} \right)\frac{\nu}{M} W_4 - \frac{M}{2E_\nu}\frac{\nu}{M}W_5 \right] \right\}
\end{aligned}
\tag{28.60}
$$

Here all the factors in brackets are dimensionless.

The above general expression appears quite complex and depends on five a priori unknown functions $W_i(Q^2, \nu)$. It can be simplified at high neutrino energies with $m_\ell \ll E_\nu$ and $M \ll E_\nu$, where the mass of the outgoing lepton and target nucleon can be neglected:

$$
\frac{d^2\sigma}{dx\,dy} = \frac{G_F^2 M E_\nu}{\pi}\left\{ (xy^2)W_1 + (1-y)\frac{\nu}{M}W_2 \pm \left(1 - \frac{y}{2}\right)xy\frac{\nu}{M}W_3 \right\}
\tag{28.61}
$$

In this expression, the specific energy and angular dependence of the terms in front of each form factor can be used to extract the information on these latter from experimental data.

We can now proceed as in the case of the electron–proton scattering and replace the general form factors with Bjorken's scaling structure functions (see Eq. (16.109)), extending the concept to five scaling functions:

$$
W_1 = F_1 \quad \text{and} \quad \frac{\nu}{M}W_i = F_i \quad (i = 2, \dots, 5)
\tag{28.62}
$$

Experimental data shows that the F_4 contribution is small and can be neglected. Hence, one sets:

$$
F_4 = 0
\tag{28.63}
$$

With the structure functions F_i, the differential cross-section for neutrinos $(+)$ and antineutrinos $(-)$ can be written as:

$$
\begin{aligned}
\frac{d^2\sigma}{dx\,dy} = \frac{G_F^2 M E_\nu}{\pi} &\left\{ \left[xy^2 + \frac{m_\ell^2}{2ME_\nu}y \right] F_1 + \left[(1-y) - \frac{M}{2E_\nu}xy - \frac{m_\ell^2}{4E^2} \right] F_2 \right. \\
&\left. \pm \left[\left(1 - \frac{y}{2}\right)xy - \frac{m_\ell^2}{4ME_\nu}y \right] F_3 - \frac{m_\ell^2}{M^2}\left[\frac{M}{2E_\nu}F_5 \right] \right\}
\end{aligned}
\tag{28.64}
$$

The structure functions describe the nucleon target. In actual experiments, it is difficult to separate the interactions on protons and neutrons on an event-by-event basis. Except in the case of a hydrogen target, neutrino–nucleon interactions will always include a mixture of interactions on protons and neutrons. In general, one defines an ideal **isoscalar target** which is composed of an equal number of protons and neutrons. For example, helium, carbon, or oxygen are almost isoscalar. For other targets, such as iron for example, where $Z/A \approx 0.47$, one can correct experimental data on a statistical basis in order to derive data for an isoscalar target. In this way, results from different nuclear targets are more readily compared.

One defines accordingly isoscalar structure functions F_i^N based on differential cross-sections σ^N obtained on, or corrected for, the isoscalar case. We have:

$$
\begin{aligned}
\frac{d^2\sigma^N}{dx\,dy} = \frac{G_F^2 M E_\nu}{\pi} &\left\{ \left[xy^2 + \frac{m_\ell^2}{2ME_\nu}y \right] F_1^N + \left[(1-y) - \frac{M}{2E_\nu}xy - \frac{m_\ell^2}{4E_\nu^2} \right] F_2^N \right. \\
&\left. \pm \left[\left(1 - \frac{y}{2}\right)xy - \frac{m_\ell^2}{4ME_\nu}y \right] F_3^N - \frac{m_\ell^2}{M^2}\left[\frac{M}{2E_\nu}F_5^N \right] \right\}
\end{aligned}
\tag{28.65}
$$

In modern experiments, it is customary to neglect the mass of the outgoing lepton and express the differential cross-section as a function of the three structure functions F_2, xF_3, and R_L, taking into account also the effect of the W boson propagator, such that:

$$\frac{\mathrm{d}^2\sigma^{\nu(\overline{\nu})N}}{\mathrm{d}x\,\mathrm{d}y} = \frac{G_F^2 M E_\nu}{\pi\left(1 + Q^2/M_W^2\right)}\left\{\left[(1-y) - \frac{M}{2E_\nu}xy + \frac{y^2 + 4M^2x^2y^2/Q^2}{2 + 2R_L^{\nu(\overline{\nu})N}(x,Q^2)}\right]F_2^{\nu(\overline{\nu})N}(x,Q^2)\right.$$
$$\left. \pm \left[\left(1 - \frac{y}{2}\right)y\right]xF_3^{\nu(\overline{\nu})N}(x,Q^2)\right\} \tag{28.66}$$

where the longitudinal structure function $R_L^{\nu(\overline{\nu})N}(x,Q^2)$ is defined as:

$$R_L^{\nu(\overline{\nu})N}(x,Q^2) \equiv \frac{F_2^{\nu(\overline{\nu})N}(x,Q^2)(1 + 4M^2x^2/Q^2) - 2xF_1^{\nu(\overline{\nu})N}(x,Q^2)}{2xF_1^{\nu(\overline{\nu})N}(x,Q^2)} \tag{28.67}$$

The reason one prefers to use R_L instead of F_1 is that R_L can be shown to physically represent the ratio of the longitudinal to the transverse polarization of the W boson.

28.9 Neutrino Scattering Experiments and Results

During the 1970s, 1980s, and 1990s there were several neutrino experiments aimed at precisely studying neutrino–nucleon scattering at high energy. The main parameters of the most relevant experiments are summarized in Table 28.1. We have already mentioned the CHARM and CHARM-II experiments at CERN in the context of the studies of the purely leptonic neutrino–electron scattering (see Section 27.2).

	CDHS	**CHARM**	**CCFR/NuTeV**	**BEBC**
Technology	Calorimeter	Calorimeter	Calorimeter	Bubble chamber
Target	Fe	Low Z	Fe	Liquid H, D, Ne, Freon
Target mass	$\simeq 600$ t	$\simeq 100$ t	$\simeq 690$ t	$\simeq 1 - 10$ t
Muon identification	Excellent	Excellent	Excellent	External detectors
Muon charge	yes	yes	yes	yes
σ_{p_μ}/p_μ	11%	10–15% (toroids)	11%	4% for contained tracks
Total hadron energy resolution $\sigma_{E_{had}}/E_{had}$	$\simeq 0.60/\sqrt{E_{had}}$	$\simeq 0.53/\sqrt{E_{had}}$	$\simeq 0.89/\sqrt{E_{had}}$	Charged hadrons individually measured

Table 28.1 Key parameters of CDHS, CHARM, CCFR/NuTeV, and BEBC neutrino detectors. All energies are expressed in giga-electronvolts.

Another example is the CDHS experiment that took data at CERN from 1976 until 1984. CDHS was composed of 19 magnetized iron modules, interleaved with wire drift chambers that provided a precise measurement of the muon tracks produced in muon-neutrino CC interactions. The iron modules were built of several iron plates with plastic scintillator planes in between. The muon momentum was reconstructed from its radius of curvature in the magnetic field while the hadronic energy was estimated from the total energy deposited in the scintillator detectors. The total (target) mass of CDHS was 1250 (600) t, providing very large samples of ν_μ and $\bar{\nu}_\mu$ interactions. The side view of the detector is shown in Figure 28.15.

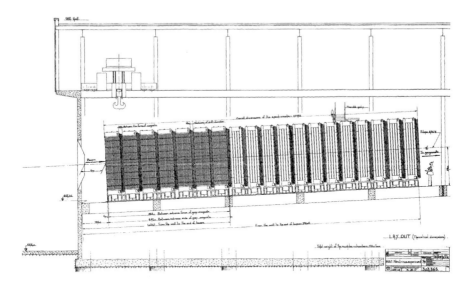

Figure 28.15 Side view of the CDHS experiment at CERN. Reprinted from P. Bergé *et al.*, "A measurement of differential cross-sections and nucleon structure functions in charged-current neutrino interactions on iron," CERN-EP-89-103, August 1989, under CC BY license http://creativecommons.org/licenses/by/4.0/.

Based on a similar principle, the CCFR/NuTeV experiment (shown in Figure 28.16) is an unmagnetized large calorimeter acting as a neutrino target followed by a magnetized muon spectrometer. The mass of the calorimeter is about 690 t. It is composed of 84 iron plates of size $3000 \times 3000 \times 100$ mm^3, interleaved with scintillators and drift chambers. The toroidal magnetic spectrometer had five sets of drift chambers. The experiment could reach very high energies due to the 800 GeV protons available at FNAL. Candidate events

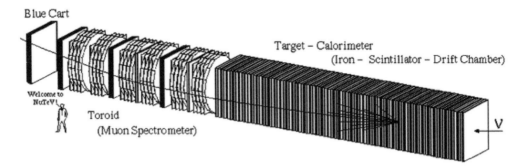

Figure 28.16 Schematic representation of the CCFR/NuTeV experiment at FNAL. Reprinted from D. Harris *et al.*, "Precision calibration of the NuTeV calorimeter," *Nucl. Instrum. Meth. A*, vol. 447, pp. 377–415, 2000, with permission from Elsevier.

for CC and NC neutrino interactions collected in the NuTeV detector are shown in Figure 28.17. A ν_μ CC event is identified by the presence of a muon, which is seen as a very long, penetrating track. The hadronic activity around the vertex corresponds to the energy of the remnant of the hit nucleon. For each event, one can determine the muon energy E_μ from the penetrating track, the hadronic energy E_X from the activity around the vertex, and the incoming neutrino energy and the inelasticity y with:

$$E_\nu \approx E_\mu + E_X, \qquad E_\mu = (1-y)E_\nu \quad \longrightarrow \quad y = \left(1 - \frac{E_\mu}{E_\nu}\right) \tag{28.68}$$

This method allows the determination of the differential cross-section $\mathrm{d}\sigma/\mathrm{d}y$. The muon scattering angle θ_μ can then be used to compute the full kinematics with $Q^2 \simeq 4E_\nu E_\mu \sin^2(\theta_\mu/2)$ (see Eq. (28.21)) from which $x = Q^2/(2MyE_\nu)$ (see Eq. (28.24)).

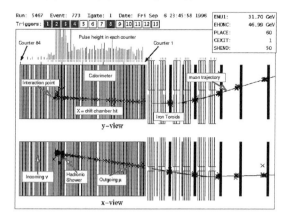 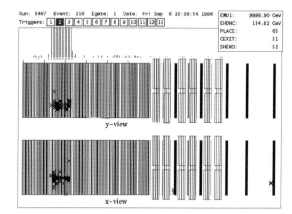

Figure 28.17 (left) Event candidate for CC and (right) NC interactions in the NuTeV detector. Figures accessed at //www-e815.fnal.gov/wma-lab-web/pix.html.

Having obtained x and y, one can measure $\mathrm{d}^2\sigma/\mathrm{d}x\mathrm{d}y$ and then extract the structure functions. The measurement of both F_2 and xF_3 by the CCFR experiment at Fermilab are shown in Figure 28.18. These precise data allow the study of the structure functions over a large kinematical domain. They are generally reinterpreted in the context of the parton model and QCD, as will be discussed in the next section.

BEBC stands for "'Big European Bubble Chamber." It had an active volume of 10 m^3 embedded in a 3 T magnetic field produced by a superconducting magnet. An external surface of about 150 m^2 was instrumented with muon detectors. The chamber was operated at CERN in the late 1970s and 1980s with both NBB and WBB beams with different target. The results with highest statistics were obtained with deuterium (CERN WA25) and neon (CERN WA59).

28.10 Deep Inelastic Neutrino Scattering (Quark–Parton Model)

The differential cross-section, Eq. (28.66), makes no assumption on the underlying structure of the target nucleon. In the deep inelastic regime, the interaction resolves the subconstituents of the nucleon, and the neutrino scatters off quarks inside the nucleon. We consequently derive the formalism of the scattering of neutrinos off point-like spin-1/2 quarks, factorizing as in the case of the electron–proton scattering, the nucleon structure with the parton distribution functions. This process is illustrated in Figure 28.12(right) within the quark–parton model.

We readily use the results derived in the context of the quark–parton model in Section 17.6 and recalling that the center-of-mass energy in the neutrino–parton system is given by $\hat{s} = (k+xp)^2 = k^2 + 2x(k\cdot p) + x^2p^2 \approx 2x(k\cdot p) \approx xs$, we consider for example the scattering process of a neutrino off the d quark (cf. Eq. (26.41)):

$$\nu_\ell d \to \ell^- u \qquad \leftrightarrow \qquad \frac{\mathrm{d}\sigma}{\mathrm{d}y} = \frac{G_F^2 \cos^2\theta_\mathrm{C}}{\pi}\hat{s} = \frac{G_F^2\cos^2\theta_\mathrm{C}}{\pi}xs \tag{28.69}$$

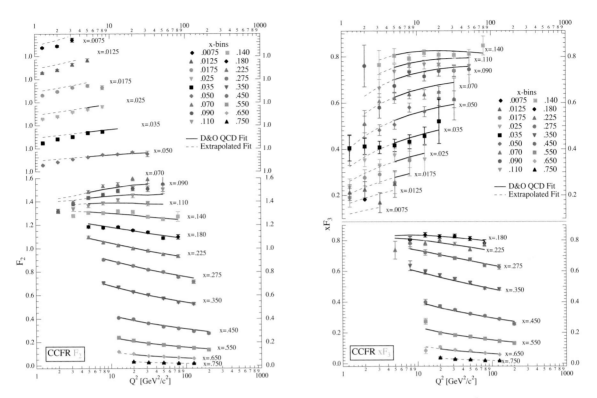

Figure 28.18 Measurement of the F_2 and xF_3 structure functions as a function of Q^2 for various x by the CCFR experiment at Fermilab. Reprinted with permission from W. G. Seligman *et al.*, "Improved determination of alpha(s) from neutrino nucleon scattering," *Phys. Rev. Lett.*, vol. 79, pp. 1213–1216, 1997. Copyright 1997 by the American Physical Society.

where we used the term $\cos^2 \theta_{\rm C}$ since the transition is Cabibbo favored. We note that the scattering of a neutrino off the u quark is prohibited by electric charge conservation: $\nu_\ell u \not\rightarrow \ell^- d$. On the other hand, we can consider the interaction off the u quark with an antineutrino (cf. Eq. (26.42)):

$$\bar{\nu}_\ell u \rightarrow \ell^+ d \qquad \leftrightarrow \qquad \frac{{\rm d}\sigma}{{\rm d}y} = \frac{G_F^2 \cos^2 \theta_{\rm C}}{\pi} xs(1-y)^2 \tag{28.70}$$

where this time the $(1-y)^2$ term comes from helicity conservation. Similar arguments can be used in the case of antiquarks and all possible combinations of Cabibbo favored and suppressed transitions. They can then be summarized in Table 28.2. The total cross-section shown in the last column is found by direct integration over y. All other combinations are prohibited by either electric charge or lepton charge conservation.

We now embed the parton-level processes into the *proton* by considering all possible quarks and adding their parton distribution functions accordingly (where we ignore the contribution from charmed quarks) in the case of the neutrino and the antineutrino:

$$\nu_\ell p \rightarrow \ell^- X : \qquad \frac{G_F^2 xs}{\pi} \left[d(x) + s(x) + \overline{u}(x)(1-y)^2 \right] \tag{28.72}$$

$$\overline{\nu}_\ell p \rightarrow \ell^+ Y : \qquad \frac{G_F^2 xs}{\pi} \left[u(x)(1-y)^2 + \overline{d}(x) + \overline{s}(x) \right] \tag{28.73}$$

The isospin symmetry allows us to write the neutron parton distribution functions as a function of those of

Reaction 1	Reaction 2	$d\sigma/dy$	σ
$\nu_\ell d \to \ell^- u$	$\overline{\nu}_\ell \overline{d} \to \ell^+ \overline{u}$	$\cos^2\theta_C \dfrac{G_F^2 xs}{\pi}$	$\cos^2\theta_C \dfrac{G_F^2 xs}{\pi}$
$\nu_\ell s \to \ell^- u$	$\overline{\nu}_\ell \overline{s} \to \ell^+ \overline{u}$	$\sin^2\theta_C \dfrac{G_F^2 xs}{\pi}$	$\sin^2\theta_C \dfrac{G_F^2 xs}{\pi}$
$\nu_\ell \overline{u} \to \ell^- \overline{d}$	$\overline{\nu}_\ell u \to \ell^+ d$	$\cos^2\theta_C \dfrac{G_F^2 xs}{\pi}(1-y)^2$	$\cos^2\theta_C \dfrac{G_F^2 xs}{3\pi}$
$\nu_\ell \overline{u} \to \ell^- \overline{s}$	$\overline{\nu}_\ell u \to \ell^+ s$	$\sin^2\theta_C \dfrac{G_F^2 xs}{\pi}(1-y)^2$	$\sin^2\theta_C \dfrac{G_F^2 xs}{3\pi}$

$$(28.71)$$

Table 28.2 Partonic-level interactions in neutrino nucleon deep inelastic scattering.

the proton, provided we swap the u and d quarks. Hence, for the neutrino or antineutrino scattering off the *neutron* we have:

$$\nu_\ell n \to \ell^- X' : \qquad \frac{G_F^2 xs}{\pi}\left[u(x) + s(x) + \overline{d}(x)(1-y)^2\right] \qquad (28.74)$$

$$\overline{\nu}_\ell n \to \ell^+ Y' : \qquad \frac{G_F^2 xs}{\pi}\left[d(x)(1-y)^2 + \overline{u}(x) + \overline{s}(x)\right] \qquad (28.75)$$

where we have used the parton distribution functions of the proton. In the case of an *isoscalar* target N, we must compute the average of the proton and neutron, to get:

$$\nu_\ell N \to \ell^- A : \qquad \frac{G_F^2 xs}{2\pi}\left[u(x) + d(x) + 2s(x) + (\overline{u}(x) + \overline{d}(x))(1-y)^2\right] \qquad (28.76)$$

$$\overline{\nu}_\ell N \to \ell^+ B : \qquad \frac{G_F^2 xs}{2\pi}\left[(u(x) + d(x))(1-y)^2 + \overline{u}(x) + \overline{d}(x) + 2\overline{s}(x)\right] \qquad (28.77)$$

These expressions are valid for energies above the threshold for charm production via the $s \to c$ and $d \to c$ transitions for neutrinos (and correspondingly for antineutrinos). This is the reason why the Cabibbo factors are absent in the expressions. For the case where the energy is below the charm threshold, one should include the corresponding Cabibbo factors.

We now compare these predictions from the naive quark–parton model to the general expression, Eq. (28.66). We recall that the **Callan–Gross relation**, which is a consequence of the quarks being spin-1/2 particles, states that $2xF_1 = F_2$ (see Eq. (17.100)). From this relation, it follows directly that the longitudinal R_L structure function is small, since:

$$R_L^{\nu(\overline{\nu})N}(\nu, Q^2) = \frac{F_2^{\nu(\overline{\nu})N}(\nu, Q^2)(1 + 4M^2x^2/Q^2) - F_2(\nu, Q^2)}{F_2^{\nu(\overline{\nu})N}(\nu, Q^2)} = 4\frac{M^2x^2}{Q^2} \approx 0 \qquad (28.78)$$

Therefore, the differential cross-section on an isoscalar target for neutrinos and antineutrinos can be written as:

$$\frac{d^2\sigma^{\nu(\bar{\nu})N}}{dx\,dy} = \frac{G_F^2 M E_\nu}{\pi} \left\{ \left[(1-y) - \frac{M}{2E_\nu}xy + \frac{y^2 + 4M^2x^2y^2/Q^2}{2} \right] F_2^{\nu(\bar{\nu})N} \pm \left[\left(1 - \frac{y}{2}\right) y \right] xF_3^{\nu(\bar{\nu})N} \right\}$$

$$\approx \frac{G_F^2 M E_\nu}{\pi} \left\{ \left[1 - y + \frac{y^2}{2} \right] F_2^{\nu(\bar{\nu})N} \pm \left[\left(y - \frac{y^2}{2} \right) \right] xF_3^{\nu(\bar{\nu})N} \right\}$$

$$= \frac{G_F^2 M E_\nu}{2\pi} \left\{ \left[1 + (1-y)^2 \right] F_2^{\nu(\bar{\nu})N} \pm \left[1 - (1-y)^2 \right] xF_3^{\nu(\bar{\nu})N} \right\}$$

$$= \frac{G_F^2 M E_\nu}{2\pi} \left\{ \left(F_2^{\nu(\bar{\nu})N} \pm xF_3^{\nu(\bar{\nu})N} \right) + \left(F_2^{\nu(\bar{\nu})N} \mp xF_3^{\nu(\bar{\nu})N} \right) (1-y)^2 \right\} \tag{28.79}$$

We compare this last expression with those predicted by the naive parton model and find the expressions for the structure functions F_2 and xF_3 as a function of the quark's parton distribution functions. For the proton:

$$F_2^{\nu p}(x) = 2x \left\{ d(x) + s(x) + \bar{u}(x) \right\}, \quad xF_3^{\nu p}(x) = 2x \left\{ d(x) + s(x) - \bar{u}(x) \right\} \tag{28.80}$$

and for the neutron as a function of the proton's parton distribution functions:

$$F_2^{\nu n}(x) = 2x \left\{ u(x) + s(x) + \bar{d}(x) \right\}, \quad xF_3^{\nu n}(x) = 2x \left\{ u(x) + s(x) - \bar{d}(x) \right\} \tag{28.81}$$

and finally for an isoscalar target as a function of the proton's parton distribution functions:

$$F_2^{\nu N}(x) = x \left\{ u(x) + d(x) + 2s(x) + \bar{u}(x) + \bar{d}(x) \right\}$$
$$xF_3^{\nu N}(x) = x \left\{ u(x) + d(x) + 2s(x) - \bar{u}(x) - \bar{d}(x) \right\} \tag{28.82}$$

One can also consider the cases for an incident antineutrino and finds for the proton:

$$F_2^{\bar{\nu} p}(x) = 2x \left\{ u(x) + \bar{d}(x) + \bar{s}(x) \right\}, \quad xF_3^{\bar{\nu} p}(x) = 2x \left\{ u(x) - \bar{d}(x) - \bar{s}(x) \right\} \tag{28.83}$$

and for the neutron:

$$F_2^{\bar{\nu} n}(x) = 2x \left\{ d(x) + \bar{u}(x) + \bar{s}(x) \right\}, \quad xF_3^{\bar{\nu} n}(x) = 2x \left\{ d(x) - \bar{u}(x) - \bar{s}(x) \right\} \tag{28.84}$$

and for the isoscalar target:

$$F_2^{\bar{\nu} N}(x) = x \left\{ u(x) + d(x) + 2\bar{s}(x) + \bar{u}(x) + \bar{d}(x) \right\}$$
$$xF_3^{\bar{\nu} N}(x) = x \left\{ u(x) + d(x) - 2\bar{s}(x) - \bar{u}(x) - \bar{d}(x) \right\} \tag{28.85}$$

These expressions can be further approximated by assuming a strangeness-symmetric sea, hence $s(x) = \bar{s}(x)$, so we finally have for an isoscalar target that $F_2^{\nu N}(x)$ for neutrinos and $F_2^{\bar{\nu} N}(x)$ for antineutrinos provide information on the total distribution of quarks:

$$F_2^{\nu N}(x) = F_2^{\bar{\nu} N}(x) = x \sum_{i=u,d,\dots} (q_i(x) + \bar{q}_i(x)) \tag{28.86}$$

while the average $x\bar{F}_3$ of xF_3 for neutrinos and antineutrinos is directly sensitive to valence quarks:

$$x\bar{F}_3 \equiv \frac{1}{2} \left(xF_3^{\nu N}(x) + xF_3^{\bar{\nu} N}(x) \right) = x \sum_{i=u,d,\dots} (q_i(x) - \bar{q}_i(x)) = x \sum_{i=u,d,\dots} q_i^V(x) \tag{28.87}$$

where V means valence quarks.

The double differential cross-section can be integrated over x (neglecting strangeness) to provide the y-dependence of the neutrino scattering cross-section. We have for the isoscalar target:

$$\frac{\mathrm{d}\sigma(\nu N)}{\mathrm{d}y} = \frac{G_F^2 s}{2\pi}\left[f_q + f_{\bar{q}}(1-y)^2\right], \qquad \frac{\mathrm{d}\sigma(\bar{\nu}N)}{\mathrm{d}y} = \frac{G_F^2 s}{2\pi}\left[f_q(1-y)^2 + f_{\bar{q}}\right] \qquad (28.88)$$

where f_q and $f_{\bar{q}}$ are the total momentum fractions carried by the quarks and antiquarks, respectively, within the proton:

$$f_q \equiv f_u + f_d = \int_0^1 x\left[u(x) + d(x)\right]\mathrm{d}x, \qquad f_{\bar{q}} \equiv f_{\bar{u}} + f_{\bar{d}} = \int_0^1 x\left[\bar{u}(x) + \bar{d}(x)\right]\mathrm{d}x; \qquad (28.89)$$

For incident neutrinos, the differential cross-section is proportional to the quark content while the antiquark content has a factor $(1-y)^2$ in front of it. For incident antineutrinos, the situation is reversed, and the factor $(1-y)^2$ appears in front of the quark content. This y dependence can therefore be used to provide a direct measurement of the quark and antiquark content of the nucleon by comparing neutrino and antineutrino scattering cross-sections. The predicted y-distributions for νN and $\bar{\nu}N$ are shown in Figures 28.19 and 28.20.

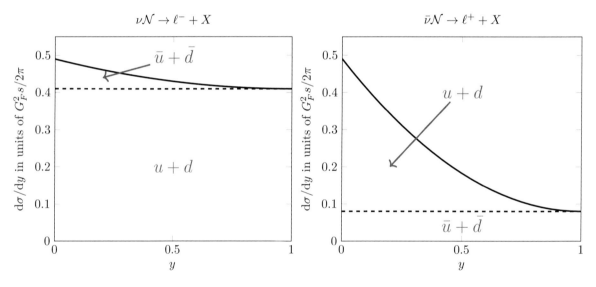

Figure 28.19 Predicted y-dependence of the differential cross-section of neutrino scattering.

Figure 28.20 Predicted y-dependence of the differential cross-section of antineutrino scattering.

For comparison, the y-distributions for neutrino energies between 30 and 200 GeV measured by the CDHS experiment [382] are shown in Figure 28.21. The neutrino data clearly shows a contribution from the νq $(q = u, d)$ scattering with a flat y-distribution with a smaller contribution from $\nu\bar{q}$ $(\bar{q} = \bar{u}, \bar{d})$ scattering with a $(1-y)^2$ distribution. On the other hand, the antineutrino data has contributions reversed with the $\bar{\nu}q$ exhibiting a $(1-y)^2$ distribution, and the $\bar{\nu}\bar{q}$ yielding a smaller flat y-distribution. Therefore, the observed shapes of the y-distribution of the events in neutrino and antineutrino modes are consistent with the nucleon being composed predominantly of quarks with a smaller antiquark component coming from the sea.

The total cross-sections can be obtained by integrating over y to find:

$$\sigma(\nu N) = \int_0^1 \frac{\mathrm{d}\sigma}{\mathrm{d}y}\,\mathrm{d}y = \frac{G_F^2 s}{2\pi}\left[f_q + \frac{1}{3}f_{\bar{q}}\right], \qquad \sigma(\bar{\nu}N) = \frac{G_F^2 s}{2\pi}\left[\frac{1}{3}f_q + f_{\bar{q}}\right] \qquad (28.90)$$

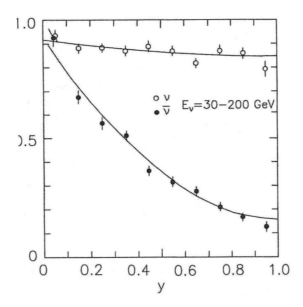

Figure 28.21 y-Dependence of the differential cross-section for neutrino–nucleon and antineutrino–nucleon scattering measured by the CDHS experiment. Reprinted with permission from Springer Nature: J. G. H. de Groot *et al.*, "Inclusive interactions of high-energy neutrinos and antineutrinos in iron," *Z. Phys.*, vol. C1, p. 143, 1979. Copyright 1979 Springer-Verlag.

We can express these results as a function of the neutrino (or antineutrino) energy in the laboratory frame where the target is at rest. We have in this case $s \to 2ME_\nu$, so:

$$\sigma(\nu N) = \sigma^0 \left[f_q + \frac{1}{3} f_{\bar{q}} \right], \qquad \sigma(\bar{\nu} N) = \sigma^0 \left[\frac{1}{3} f_q + f_{\bar{q}} \right] \tag{28.91}$$

where the overall normalization factor is (N.B. beware not to confuse with Eq. (26.48)!):

$$\sigma^0 \equiv \frac{G_F^2 s}{2\pi} \simeq \frac{G_F^2 M E_\nu}{\pi} \approx 1.583 \times 10^{-38} \frac{E_\nu}{\text{GeV}} \text{ cm}^2 \tag{28.92}$$

which shows that the total deep inelastic cross-section is proportional to the incoming neutrino energy E_ν in the laboratory frame. The ratio of the total cross-section for neutrino and antineutrinos is just:

$$R \equiv \frac{\sigma(\nu N)}{\sigma(\bar{\nu} N)} = \frac{3f_q + f_{\bar{q}}}{f_q + 3f_{\bar{q}}} \tag{28.93}$$

Figure 28.22 shows a summary of the experimental measurements of the total deep inelastic cross-sections for neutrinos $\sigma(\nu N)/E_\nu$ and antineutrinos $\sigma(\bar{\nu} N)/E$ normalized by the energy E_ν, for neutrinos in the range $25 < E_\nu < 250$ GeV. The closed symbols refer to neutrino and open symbols to antineutrino measurements. The data on the left are from CCFR (squares) and CDHS (triangles). The slopes are reported to the right and come from CCFR, CHARM, and the FNAL-15 foot, BEBC, and CITFR bubble chambers. The cross-sections are consistent with being proportional to energies (above 25 GeV). The fitted slopes give:

$$\sigma(\nu N)/E_\nu = (0.677 \pm 0.014) \times 10^{-38} \text{cm}^2 \text{ GeV}^{-1}, \ \sigma(\bar{\nu} N)/E_\nu = (0.334 \pm 0.008) \times 10^{-38} \text{cm}^2 \text{ GeV}^{-1} \tag{28.94}$$

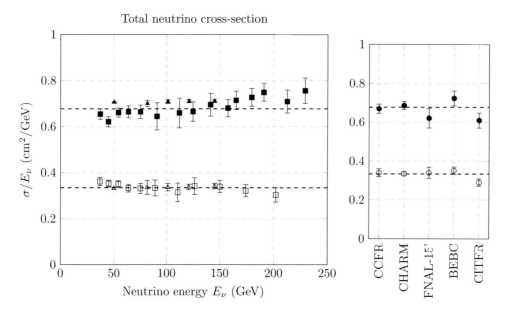

Figure 28.22 Deep inelastic cross-section for neutrinos and antineutrinos. (left) Measurements on iron for neutrinos (closed symbols) and antineutrinos (open symbols) from CCFR [383] (squares) and CDHS [384] (statistical error only – triangles). (right) Slope measurements from other experiments. The slope data are from CCFR [383], CHARM [385], FNAL-15' bubble chamber [386,387], BEBC [388], CITFR [389].

corresponding to the measured ratio:

$$R \equiv \frac{\sigma(\nu N)}{\sigma(\bar{\nu} N)} = 1.984 \pm 0.012 \tag{28.95}$$

The CCFR experiment has tested the linearity of the deep inelastic cross-section for energies between 30 and 350 GeV with a function of the form:

$$\frac{\sigma}{E_\nu} = \frac{G_F^2 s}{2\pi E_\nu} \left(1 + \varepsilon_{\nu(\bar{\nu})} F_\nu\right) = \frac{\sigma^0}{E_\nu} \left(1 + \varepsilon_{\nu(\bar{\nu})} E_\nu\right) \tag{28.96}$$

and found:

$$\varepsilon_\nu = \frac{(-22 \pm 7) \times 10^{-3}}{100 \text{ GeV}} \qquad \text{and} \qquad \varepsilon_{\bar{\nu}} = \frac{(-2 \pm 13) \times 10^{-3}}{100 \text{ GeV}} \tag{28.97}$$

The small apparent slope in the neutrino cross-section can be interpreted as a consequence of the QCD evolution of the structure functions at high Q^2.

The measured cross-sections can be used to determine the fractions of proton momentum carried by the quarks f_q and that carried by the antiquarks $f_{\bar{q}}$, since:

$$\frac{3\sigma(\nu N) - \sigma(\bar{\nu} N)}{\sigma^0} = \frac{8}{3} f_q \qquad \text{and} \qquad \frac{3\sigma(\bar{\nu} N) - \sigma(\nu N)}{\sigma^0} = \frac{8}{3} f_{\bar{q}} \tag{28.98}$$

Using these expressions, the experimental data is consistent with:

$$f_q \simeq 0.41 \qquad \text{and} \qquad f_{\bar{q}} \simeq 0.08 \tag{28.99}$$

The quarks and antiquarks carry together about 50% of the momentum of the protons. The valence quarks carry about 2/3 of that and the sea about 1/3. This result is direct evidence for the existence of antiquarks in the nucleon. If there were no antiquarks, the ratio would be $R = \sigma(\nu N)/\sigma(\bar{\nu} N) = 3$. The remaining 50% of momentum is carried by gluons which do not interact with the W boson.

There are practical sum rules that can be defined in the case of the deep inelastic neutrino scattering which can be used to test QCD predictions. The **Gross–Llewellyn Smith sum rule (GLS)** gives the total number of valence quarks in the nucleon and should be equal to three for protons, neutrons, or isoscalar targets, up to QCD radiative corrections:

$$
\begin{aligned}
\int_0^1 x\overline{F}_3(x,Q^2)\,\frac{\mathrm{d}x}{x} &= \frac{1}{2}\int_0^1 \left\{ F_3^{\nu N}(x,Q^2) + F_3^{\bar{\nu} N}(x,Q^2) \right\}\,\mathrm{d}x = \sum_{i=u,d,\dots}\int_0^1 \left\{ q_i(x) - \bar{q}_i(x) \right\}\,\mathrm{d}x \\
&= \sum_{i=u,d,\dots}\int_0^1 q_i^V(x)\,\mathrm{d}x = 3\left\{ 1 - \frac{\alpha_{\mathrm{s}}(Q^2)}{\pi} + \cdots \right\}
\end{aligned}
\tag{28.100}
$$

The GLS sum rule is well verified by experiments.

28.11 Deep Inelastic Neutral Current Scattering

The neutral current deep inelastic scattering reactions $\nu_\ell + \mathcal{N} \to \nu_\ell + X$ and $\bar{\nu}_\ell + \mathcal{N} \to \bar{\nu}_\ell + X$ (see Eq. (28.16)) can be derived within the quark–parton model from the four basic neutral current scatterings off quarks:

$$
\nu + q \to \nu + q, \quad \nu + \bar{q} \to \nu + \bar{q}, \quad \bar{\nu} + q \to \bar{\nu} + q, \quad \text{and} \quad \bar{\nu} + \bar{q} \to \bar{\nu} + \bar{q}
\tag{28.101}
$$

These reactions are formally very comparable to the neutrino scattering off electrons discussed in Section 26.3. We can directly use the results of Eqs. (26.29) and (26.47), and write the following expressions at the partonic level:

$$
\begin{aligned}
\left(\frac{\mathrm{d}\sigma^{\mathrm{NC}}(\nu q)}{\mathrm{d}y}\right) &= \left(\frac{\mathrm{d}\sigma^{\mathrm{NC}}(\bar{\nu}\bar{q})}{\mathrm{d}y}\right) = \frac{G_F^2 \hat{s}}{4\pi}\left[(c_V^q + c_A^q)^2 + (c_V^q - c_A^q)^2(1-y)^2 \right] \\
&= \frac{G_F^2 \hat{s}}{\pi}\left[(c_L^q)^2 + (c_R^q)^2(1-y)^2 \right]
\end{aligned}
\tag{28.102}
$$

Swapping $c_A^q \to -c_A^q$ or $g_L^q \leftrightarrow g_R^q$, we obtain the corresponding cross-sections:

$$
\begin{aligned}
\left(\frac{\mathrm{d}\sigma^{\mathrm{NC}}(\nu\bar{q})}{\mathrm{d}y}\right) &= \left(\frac{\mathrm{d}\sigma^{\mathrm{NC}}(\bar{\nu}q)}{\mathrm{d}y}\right) = \frac{G_F^2 \hat{s}}{4\pi}\left[(c_V^q - c_A^q)^2 + (c_V^q + c_A^q)^2(1-y)^2 \right] \\
&= \frac{G_F^2 \hat{s}}{\pi}\left[(c_R^q)^2 + (c_L^q)^2(1-y)^2 \right]
\end{aligned}
\tag{28.103}
$$

For up-type quarks (we will consider only u and c), we then define (see Table 26.2):

$$
g_L \equiv c_L^u = +\frac{1}{2} - \frac{2}{3}s_W^2 \quad \text{and} \quad g_R \equiv c_R^u = -\frac{2}{3}s_W^2
\tag{28.104}
$$

and for the down-type quarks (we will consider only d and s):

$$
g_L' \equiv c_L^d = -\frac{1}{2} + \frac{1}{3}s_W^2 \quad \text{and} \quad g_R' \equiv c_R^d = +\frac{1}{3}s_W^2
\tag{28.105}
$$

where $s_W^2 = \sin^2\theta_W$. As in the case of the charged current deep inelastic scattering described in the previous section, we can write the equivalent expression in the quark–parton model for the neutral current deep inelastic scattering off the proton:

$$\frac{\mathrm{d}\sigma^{\mathrm{NC}}(\overset{(-)}{\nu} p \to \overset{(-)}{\nu} X)}{\mathrm{d}x\mathrm{d}y} = \sum_q q(x)\frac{\mathrm{d}\sigma^{\mathrm{NC}}(\overset{(-)}{\nu} q)}{\mathrm{d}y} + \sum_{\bar{q}} \bar{q}(x)\frac{\mathrm{d}\sigma^{\mathrm{NC}}(\overset{(-)}{\nu} \bar{q})}{\mathrm{d}y} \tag{28.106}$$

where, as before, $q(x)$ and $\bar{q}(x)$ are the parton distribution functions of the relevant quarks. This naturally leads to the following expressions:

$$\begin{aligned}
\frac{\mathrm{d}\sigma^{\mathrm{NC}}(\overset{(-)}{\nu} p)}{\mathrm{d}x\mathrm{d}y} &= \frac{G_F^2 xs}{\pi}\left\{\left(g_{L,R}^2 + g_{R,L}^2(1-y)^2\right)[u+c] + \left(g_{L,R}^2 + g_{R,L}^2(1-y)^2\right)[d+s]\right. \\
&\quad\left. + \left(g_{L,R}^2(1-y)^2 + g_{R,L}^2\right)[\bar{u}+\bar{c}] + \left(g_{L,R}'^2(1-y)^2 + g_{R,L}'^2\right)[\bar{d}+\bar{s}]\right\} \\
&\simeq \frac{G_F^2 ME_\nu}{\pi}\cdot 2x\left\{\left(g_{L,R}^2 + g_{R,L}^2(1-y)^2\right)[u+c] + \left(g_{L,R}'^2 + g_{R,L}'^2(1-y)^2\right)[d+s]\right. \\
&\quad\left. + \left(g_{L,R}^2(1-y)^2 + g_{R,L}^2\right)[\bar{u}+\bar{c}] + \left(g_{L,R}'^2(1-y)^2 + g_{R,L}'^2\right)[\bar{d}+\bar{s}]\right\}
\end{aligned} \tag{28.107}$$

where the indices for g and g' correspond to the neutrino and antineutrino cases. In analogy to the charged current case, we define the structure functions $\tilde{F}_i(x)$ for the neutral current scattering as in Eq. (28.79):

$$\frac{\mathrm{d}\sigma^{\mathrm{NC}}(\overset{(-)}{\nu} p)}{\mathrm{d}x\mathrm{d}y} = \frac{G_F^2 ME_\nu}{\pi}\cdot\frac{1}{2}\left\{\left(\tilde{F}_2^{\nu p,\bar{\nu}p} \pm x\tilde{F}_3^{\nu p,\bar{\nu}p}\right) + \left(\tilde{F}_2^{\nu p,\bar{\nu}p} \mp x\tilde{F}_3^{\nu p,\bar{\nu}p}\right)(1-y)^2\right\} \tag{28.108}$$

where by direct comparison we obtain:

$$\begin{aligned}
\tilde{F}_2^{\nu p,\bar{\nu}p} &= 2x\left\{\left(g_L^2 + g_R^2\right)[u+c+\bar{u}+\bar{c}] + \left(g_L'^2 + g_R'^2\right)[d+s+\bar{d}+\bar{s}]\right\} \\
x\tilde{F}_3^{\nu p,\bar{\nu}p} &= 2x\left\{\left(g_L^2 - g_R^2\right)[u+c-\bar{u}-\bar{c}] + \left(g_L'^2 - g_R'^2\right)[d+s-\bar{d}-\bar{s}]\right\}
\end{aligned} \tag{28.109}$$

For the case of the neutron, we just need to swap u and d and get:

$$\begin{aligned}
\tilde{F}_2^{\nu n,\bar{\nu}n} &= 2x\left\{\left(g_L^2 + g_R^2\right)[d+c+\bar{d}+\bar{c}] + \left(g_L'^2 + g_R'^2\right)[u+s+\bar{u}+\bar{s}]\right\} \\
x\tilde{F}_3^{\nu n,\bar{\nu}n} &= 2x\left\{\left(g_L^2 - g_R^2\right)[d+c-\bar{d}-\bar{c}] + \left(g_L'^2 - g_R'^2\right)[u+s-\bar{u}-\bar{s}]\right\}
\end{aligned} \tag{28.110}$$

so that for an isoscalar target we simply find, averaging:

$$\begin{aligned}
\tilde{F}_2^{\nu N,\bar{\nu}N}(x) &= x\left\{\left(g_L^2 + g_R^2\right)[u+d+2c+\bar{u}+\bar{d}+2\bar{c}] + \left(g_L'^2 + g_R'^2\right)[u+d+2s+\bar{u}+\bar{d}+2\bar{s}]\right\} \\
x\tilde{F}_3^{\nu N,\bar{\nu}N}(x) &= x\left\{\left(g_L^2 - g_R^2\right)[u+d+2c-\bar{u}-\bar{d}-2\bar{c}] + \left(g_L'^2 - g_R'^2\right)[u+d+2s-\bar{u}-\bar{d}-2\bar{s}]\right\}
\end{aligned} \tag{28.111}$$

These expressions can be simplified if we neglect the c quarks and assume a symmetric $s(x) = \bar{s}(x)$ sea, to yield:

$$\begin{aligned}
\tilde{F}_2^{\nu N,\bar{\nu}N}(x) &= x\left\{\left(g_L^2 + g_R^2\right)[u+d+\bar{u}+\bar{d}] + \left(g_L'^2 + g_R'^2\right)[u+d+2s+\bar{u}+\bar{d}+2\bar{s}]\right\} \\
x\tilde{F}_3^{\nu N,\bar{\nu}N}(x) &= x\left\{\left(g_L^2 - g_R^2\right)[u+d-\bar{u}-\bar{d}] + \left(g_L'^2 - g_R'^2\right)[u+d-\bar{u}-\bar{d}]\right\}
\end{aligned} \tag{28.112}$$

which are directly comparable to the charged current case, Eqs. (28.82) and (28.85). As before, $x\tilde{F}_3$ is directly sensitive to the valence quark distributions. When we further neglect the strange sea quark contribution, we can easily relate the CC and NC cross-sections. Summarizing, we have in this case:

$$\begin{aligned}
\frac{\mathrm{d}\sigma^{\mathrm{CC}}(\nu N)}{\mathrm{d}x\mathrm{d}y} &= \sigma^0\cdot x\left\{(u+d) + (\bar{u}+\bar{d})(1-y)^2\right\} = \sigma^0\cdot x\left\{q + \bar{q}(1-y)^2\right\} \\
\frac{\mathrm{d}\sigma^{\mathrm{CC}}(\bar{\nu} N)}{\mathrm{d}x\mathrm{d}y} &= \sigma^0\cdot x\left\{(\bar{u}+\bar{d}) + (u+d)(1-y)^2\right\} = \sigma^0\cdot x\left\{\bar{q} + q(1-y)^2\right\}
\end{aligned} \tag{28.113}$$

where $q = u + d$ and $\bar{q} = \bar{u} + \bar{d}$, and σ^0 is given by Eq. (28.92). For neutral currents, we get similar expressions:

$$\frac{d\sigma^{\mathrm{NC}}(\nu N)}{dxdy} = \sigma^0 \cdot x \left\{ \left(g_L^2 + g_L'^2\right)\left(q + \bar{q}(1-y)^2\right) + \left(g_R^2 + g_R'^2\right)\left(\bar{q} + q(1-y)^2\right) \right\}$$

$$\frac{d\sigma^{\mathrm{NC}}(\bar{\nu} N)}{dxdy} = \sigma^0 \cdot x \left\{ \left(g_R^2 + g_R'^2\right)\left(q + \bar{q}(1-y)^2\right) + \left(g_L^2 + g_L'^2\right)\left(\bar{q} + q(1-y)^2\right) \right\} \qquad (28.114)$$

Therefore, we can see the strikingly straightforward result:

$$\frac{d\sigma^{\mathrm{NC}}(\nu N)}{dxdy} = \left(g_L^2 + g_L'^2\right)\frac{d\sigma^{\mathrm{CC}}(\nu N)}{dxdy} + \left(g_R^2 + g_R'^2\right)\frac{d\sigma^{\mathrm{CC}}(\bar{\nu} N)}{dxdy}$$

$$\frac{d\sigma^{\mathrm{NC}}(\bar{\nu} N)}{dxdy} = \left(g_R^2 + g_R'^2\right)\frac{d\sigma^{\mathrm{CC}}(\nu N)}{dxdy} + \left(g_L^2 + g_L'^2\right)\frac{d\sigma^{\mathrm{CC}}(\bar{\nu} N)}{dxdy} \qquad (28.115)$$

The above relations remain valid after phase-space integration over x and y. We can consequently introduce the ratios of CC to NC cross-sections:

$$R_\nu \equiv \frac{\sigma^{\mathrm{NC}}(\nu N)}{\sigma^{\mathrm{CC}}(\nu N)}, \quad R_{\bar{\nu}} \equiv \frac{\sigma^{\mathrm{NC}}(\bar{\nu} N)}{\sigma^{\mathrm{CC}}(\bar{\nu} N)}, \quad \text{and} \quad r \equiv \frac{\sigma^{\mathrm{CC}}(\bar{\nu} N)}{\sigma^{\mathrm{CC}}(\nu N)} = \frac{1}{R} \qquad (28.116)$$

where the measured value $R = 1.984 \pm 0.012$ (see Eq. (28.95)). Solving the system of equations, one infers the strengths of the left-handed and right-handed couplings:

$$g_L^2 + g_L'^2 = \frac{R_\nu - r^2 R_{\bar{\nu}}}{1 - r^2} \quad \text{and} \quad g_R^2 + g_R'^2 = \frac{r\left(R_{\bar{\nu}} - R_\nu\right)}{1 - r^2} \qquad (28.117)$$

Using the values predicted by the electroweak theory, this leads to a measurement of the Weinberg angle. Inserting Eqs. (28.104) and (28.105) yields:

$$R_\nu = \left(g_L^2 + g_L'^2\right) + r\left(g_R^2 + g_R'^2\right) = \frac{1}{2} - \sin^2\theta_W + (1+r)\frac{5}{9}\sin^4\theta_W$$

$$R_{\bar{\nu}} = \left(g_L^2 + g_L'^2\right) + \frac{1}{r}\left(g_R^2 + g_R'^2\right) = \frac{1}{2} - \sin^2\theta_W + \left(1 + \frac{1}{r}\right)\frac{5}{9}\sin^4\theta_W \qquad (28.118)$$

Thus, with the known value of r, the measurement of the weak mixing angle can be performed with either neutrinos or antineutrinos. However, experiments with antineutrinos have less statistics due to the smaller cross-sections and the generally lower flux. Experimentally charged and neutral current events at high energy can be differentiated quite easily, as was illustrated in Figure 28.17, by looking at the presence of a long track to be identified as the muon in a charged current. This track is obviously absent in the case of the neutral current since the final-state neutrino is not visible.

The ratio R_ν of neutral- to charged-current cross-sections has been measured to 1% accuracy by CHARM [390] and CDHS [391] at CERN. CCFR [392] at Fermilab has obtained an even more precise result. These measurements are summarized in Table 28.3. One of the advantages of this method is that many of the uncertainties from the strong interactions and neutrino spectra cancel in the ratio R^ν since the hadronic part is mostly insensitive to the charge of the probe. A theoretical uncertainty is associated with the threshold for producing the charm quark, which mainly affects σ^{CC} through the $s \to c$ (Cabibbo favored) and $d \to c$ (Cabibbo suppressed) partonic reactions. This introduces a dependence on the effective charmed quark mass as shown in the table. The value of m_c can be determined from dimuon production in neutrino interactions, where the charm quark decays semi-leptonically leading to the two opposite signed muons in the final states. The measured value is $m_c = 1.31 \pm 0.24$ GeV [394]. This leads to a dominant uncertainty on the estimate of $\sin^2\theta_W$ of ± 0.003. However, this uncertainty largely cancels in the Paschos–Wolfenstein ratio [395] (see **Ex 28.4**):

$$R^- = \frac{\sigma_{\nu N}^{\mathrm{NC}} - \sigma_{\bar{\nu} N}^{\mathrm{NC}}}{\sigma_{\nu N}^{\mathrm{CC}} - \sigma_{\bar{\nu} N}^{\mathrm{CC}}} = \left(g_L^2 + g_L'^2\right) - \left(g_R^2 + g_R'^2\right) = \frac{1}{2} - \sin^2\theta_W \qquad (28.119)$$

Experiment	Beam	$R_\nu (R_{\bar\nu})$	$\sin^2 \theta_W$
CHARM [390]	160 GeV NBB	0.3093 ± 0.0031	$0.236 \pm 0.005 \pm 0.003 + 0.012(m_c - 1.5)$
CDHS [391]	450 GeV NBB	0.3072 ± 0.0033 (0.382 ± 0.016)	$0.228 \pm 0.005 \pm 0.003 + 0.013(m_c - 1.5)$
CCFR [392]	800 GeV TQT	$-$	$0.2236 \pm 0.0028 \pm 0.0030 + 0.0111 \, (m_c - 1.31)$
NuTeV [393]	800 GeV SSQT	R^-	$0.2277 \pm 0.0013 \pm 0.0009$

Table 28.3 Measurements of the $R_\nu \equiv \sigma^{\rm NC}(\nu N)/\sigma^{\rm CC}(\nu N)$ in neutrino experiments. The first error on $\sin^2 \theta_W$ is experimental, the second is theoretical. The last term comes from the charm contribution and is given as a function of the charm effective mass m_c (GeV).

This, however, requires a high-intensity and high-energy antineutrino beam. R^- is more difficult to measure than R^ν, primarily because the neutral-current scatterings of neutrinos and antineutrinos yield identical observed final states which can be distinguished only through an a priori knowledge of the initial-state neutrino. It was measured for the first time in 2001 by Fermilab's NuTeV collaboration [393]. The high-purity ν and $\bar\nu$ beams were provided by the Sign Selected Quadrupole Train (SSQT) beam line at the Fermilab Tevatron (see Section 28.3) during the 1996–1997 fixed target run. Somewhat surprisingly, the result is:

$$\sin^2 \theta_W (\text{NuTeV}) = 0.2277 \pm 0.0013(\text{stat.}) \pm 0.0009(\text{syst.}) \qquad (28.120)$$

With such a small error, the value measured by NuTeV is higher and in strong tension with the measurements of $\sin^2 \theta_W$ at the Z^0 pole (see Eq. (27.66)). Some concerns have been raised about some of the assumptions used in the determination of $\sin^2 \theta_W$ by NuTeV. In absence of careful evaluation of these latter, the recommendation is to ignore this value in the overall determination of the Weinberg mixing angle (see Ref. [28] for further details).

Problems

Ex 28.1 Narrow-band beam. Write a program to simulate a narrow-band beam and reproduce the distributions shown in Figure 28.4.

Ex 28.2 Wide-band beam. Write a program to simulate a wide-band beam and reproduce the features of the muon-neutrino fluxes shown in Figure 28.7. Use the analytical formulae from **Bonesini–Marchionni–Pietropaolo–Tabarelli de Fatis** (BMPT) (see Ref. [396]) for the calculation of secondary particle yields in the primary proton interactions on the target.

Ex 28.3 Off-axis long-baseline beams. Use the results from the previous exercise to discuss the neutrino fluxes expected at the **T2K** experiment (see Figure 28.9 and Ref. [378]) and the **NOvA** experiment [397,398].

Ex 28.4 Paschos–Wolfenstein ratio. Derive Eq. (28.119).

29   Completing the Standard Model

(...) the Standard Model remains a work in progress. So there is room for big discoveries and big surprises in our field, there is room for new ideas, which makes our future thrilling (...).

Fabiola Gianotti[1]

29.1   Putting it All Together

We have now collected all the building blocks to write down the most up-to-date theory of particle physics. As we have seen in the previous chapters, it took several decades to develop it, basically in several stages during the second half of the 20th century. Its current form was finalized during the 1970s. The Standard Model (SM) describes the electromagnetic, weak, and strong interactions (but not gravity) among fundamental fermions of Nature. The interactions are described by the exchange of spin-1 gauge bosons, whose couplings are uniquely determined by the local gauge invariance principle. The underlying gauge symmetry group of the SM is $U(1)_Y \times SU(2)_L \times SU(3)_C$, so it includes the unified electroweak theory and QCD within a single model. The main building blocks are the 12 fundamental spin-1/2 fermions (six quarks and six leptons), which satisfy the Dirac equation, and their 12 corresponding antiparticles. The forces are mediated by the γ, $W^\pm$, Z^0 spin-1 gauge bosons, and the eight spin-1 gluons. Spontaneous symmetry breaking leaves one physical scalar spin-0 Higgs boson H^0. These particles are summarized in Table 29.1.

The precise predictions of the SM were confronted with several equally precise experimental measurements, in particular those discussed over the course of the previous chapters. The SM has consequently demonstrated huge successes in providing experimental predictions. However, it leaves some phenomena unexplained and falls short of being a complete theory of fundamental interactions. It does not fully explain the baryon asymmetry in the observable Universe, incorporate the full theory of gravitation as described by general relativity, or account for the accelerating expansion of the Universe as possibly described by dark energy. The model does not contain any viable dark matter particle that possesses all of the required properties deduced from observational cosmology. It also does not incorporate neutrino oscillations and their non-zero masses. These aspects will be discussed in the coming chapters.

In the following, we illustrate one of the very powerful aspects of the SM: its predictive aspect, but before this we address general considerations on the top quark.

---

[1]   Reprinted by permission from Springer Nature: F. Gianotti's contribution to the "Panel discussion on the future of particle physics," in *Symposium on Prestigious Discoveries at CERN : 1973 : Neutral currents. 1983 : W & Z Bosons*, CERN, Geneva, Switzerland, 16 Sep 2003, (EP-2003-073) and *Eur. Phys. J. C*, vol. 34, pp. 91–102, 2004. Copyright 2004.

12 fermions			5 bosons		
spin-1/2			spin-1		spin-0
1st	2nd	3rd			
u	c	t			H^0
2.2 MeV	1.7 GeV	173 GeV			125.1 GeV
d	s	b	g (gluon)		
4.7 MeV	96 MeV	4.8 GeV	0		
e	μ	τ	γ (photon)		
511 keV	105.7 MeV	1.777 GeV	0		
ν_1	ν_2	ν_3	$W^\pm$	Z^0	
≤ 2 eV	≤ 190 keV	≤ 18.2 MeV	80.4 GeV	91.2 GeV	

Table 29.1 Fundamental fermions and bosons of the SM. In the SM, the three neutrino masses are considered to be equal to zero.

29.2 The Top Quark

The **top quark** is by far the most massive of the quarks, with the current mass average (value taken from the Particle Data Group's Review [5]) equal to:

$$m_t = 173.1 \pm 0.6 \text{ GeV} \tag{29.1}$$

In fact, the top quark is the most massive particle in the SM, since $m_t > M_H > M_Z > M_W$. A heavy top quark with mass $m_t > M_W + m_q$ will decay into an on-shell (real) boson W and a lighter quark q, as illustrated in Figure 29.1. The W boson itself is unstable and will decay either hadronically into two jets or leptonically

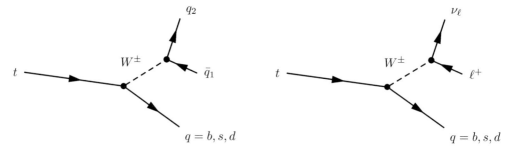

Figure 29.1 Tree-level diagram for top decay: (left) hadronic and (right) semi-leptonic.

into a charged lepton and a neutrino. The tree-level amplitude for the decay $t(Q^\mu) \to W(k^\mu) + q(q^\mu)$ is simply (see rule **[GSW3]** from Table 26.1):

$$-i\mathcal{M} = \frac{-ig_W}{\sqrt{2}} V_{tq} \bar{u}(q) \gamma^\mu \left(\frac{1 - \gamma_5}{2} \right) u(Q) \epsilon_\mu^*(k) \tag{29.2}$$

where ϵ_μ is the four-polarization of the $W^\pm$ boson. The spin-averaged matrix element squared summed over the polarizations of the $W^\pm$ boson is then given by (neglecting the mass of the light quark):

$$
\begin{aligned}
\frac{1}{2} \sum_{\text{spins}} |\mathcal{M}|^2 &= \frac{g_W^2}{16} |V_{tq}|^2 \, \text{Tr} \left\{ \not{q} \gamma^\mu (1 - \gamma_5)(\not{Q} + m_t) \gamma^\nu (1 - \gamma_5) \right\} \left(-g_{\mu\nu} + \frac{k_\mu k_\nu}{M_W^2} \right) \\
&= \frac{g_W^2}{16} |V_{tq}|^2 \, 8 \left(q^\mu Q^\nu + q^\nu Q^\mu - g^{\mu\nu} q \cdot Q + i \epsilon^{\mu\rho\nu\lambda} q_\lambda Q_\rho \right) \left(-g_{\mu\nu} + \frac{k_\mu k_\nu}{M_W^2} \right) \\
&= \frac{g_W^2}{2} |V_{tq}|^2 \left(q \cdot Q + \frac{2(q \cdot k)(Q \cdot k)}{M_W^2} \right)
\end{aligned}
\tag{29.3}
$$

where we used the trace theorems **[TS4]** and **[TS3]** from Appendix E.3. This expression can be simplified by noting that kinematically we have $Q = k + q$, hence:

$$
\begin{aligned}
Q^2 &= (k+q)^2 &\implies& \quad q \cdot k = \frac{1}{2}(Q^2 - q^2 - k^2) \simeq \frac{1}{2}(m_t^2 - M_W^2) \\
k^2 &= (Q-q)^2 &\implies& \quad q \cdot Q = \frac{1}{2}(Q^2 + q^2 - k^2) \simeq \frac{1}{2}(m_t^2 - M_W^2) \\
q^2 &= (Q-k)^2 &\implies& \quad Q \cdot k = \frac{1}{2}(Q^2 - q^2 + k^2) \simeq \frac{1}{2}(m_t^2 + M_W^2)
\end{aligned}
\tag{29.4}
$$

Consequently, we arrive at:

$$
\begin{aligned}
\langle |\mathcal{M}|^2 \rangle &= \frac{1}{2} \sum_{\text{spins}} |\mathcal{M}|^2 = \frac{g_W^2}{2} |V_{tq}|^2 \frac{1}{2} \left((m_t^2 - M_W^2) + \frac{(m_t^2 - M_W^2)(m_t^2 + M_W^2)}{M_W^2} \right) \\
&= \frac{g_W^2}{4 M_W^2} |V_{tq}|^2 \left(M_W^2 (m_t^2 - M_W^2) + m_t^4 - M_W^4 \right) \\
&= \frac{g_W^2 m_t^4}{4 M_W^2} |V_{tq}|^2 \left[1 + \frac{M_W^2}{m_t^2} - \frac{2 M_W^4}{m_t^4} \right] = \frac{g_W^2 m_t^4}{4 M_W^2} |V_{tq}|^2 \left(1 - \frac{M_W^2}{m_t^2} \right) \left(1 + \frac{2 M_W^2}{m_t^2} \right)
\end{aligned}
\tag{29.5}
$$

From this last expression, one can easily include the phase-space factor and integrate the two-body decay (see Section 5.17). Using Eqs. (5.163) and (5.164), we find:

$$
\Gamma(t \to Wq) = \int d\Gamma = \frac{\langle |\mathcal{M}|^2 \rangle}{32 \pi^2 m_t^2} \frac{4\pi \lambda^{1/2}(m_t^2, M_W^2, m_q^2)}{2 m_t} \simeq \frac{\langle |\mathcal{M}|^2 \rangle}{16 \pi m_t^3} \lambda^{1/2}(m_t^2, M_W^2, 0) = \frac{\langle |\mathcal{M}|^2 \rangle}{16 \pi m_t} \left(1 - \frac{M_W^2}{m_t^2} \right)
\tag{29.6}
$$

where we used the Källén function λ (see Eq. (5.132)). Then, finally the width of the decay $t \to Wq$ is given by:

$$
\Gamma(t \to Wq) = \frac{g_W^2 m_t^3}{64 \pi M_W^2} |V_{tq}|^2 \left(1 + \frac{2 M_W^2}{m_t^2} \right) \left(1 - \frac{M_W^2}{m_t^2} \right)^2
\tag{29.7}
$$

What is the scale of this width? It is tempting to collect the terms in the front using $G_F \equiv \sqrt{2} g^2 / 8 M_W^2$ (see Eq. (23.25)) to get:

$$
G_t \equiv \frac{\sqrt{2} g_W^2}{8 M_W^2} = \frac{e^2 \sqrt{2}}{8 \sin^2 \theta_W M_W^2} = \frac{\pi \alpha(m_t^2)}{\sqrt{2} x_W M_W^2} \approx 1.162 \times 10^{-5} \text{ GeV}^{-2}
\tag{29.8}
$$

where we have taken into account the running of $\alpha(m_t^2) \approx 1/127$, as described in Section 12.6 (see Table 12.1 and Figure 12.7) and $x_W(m_t^2) \approx 0.23$. This value is, as expected, of the weak scale $G_F \approx 1.166 \times 10^{-5} \text{ GeV}^{-2}$ (see Eq. (21.14)). Accordingly, we can write:

$$
\Gamma(t \to Wq) = \frac{G_t m_t^3}{8 \pi \sqrt{2}} |V_{tq}|^2 \left(1 + \frac{2 M_W^2}{m_t^2} \right) \left(1 - \frac{M_W^2}{m_t^2} \right)^2
\tag{29.9}
$$

Including QCD corrections, the lifetime becomes [399]:

$$\Gamma(t \to Wq) = \frac{G_t m_t^3}{8\pi\sqrt{2}} |V_{tq}|^2 \left(1 + 2\frac{M_W^2}{m_t^2}\right)\left(1 - \frac{M_W^2}{m_t^2}\right)^2 \left[1 - \frac{2\alpha_s}{3\pi}\left(\frac{2\pi^2}{3} - \frac{5}{2}\right)\right] \tag{29.10}$$

Perhaps surprisingly, the decay rate of the top quark is proportional to G_t and not G_t^2, therefore it is much larger than for example four-fermion processes that depend on G_F^2 (compare for example with the muon lifetime, Eq. (23.87)). This is because the W in top quark decays is on-shell (real) whereas in the four-fermion processes the boson enters as a virtual propagator with two weak vertices! Consequently, and given its very high mass, we expect **the lifetime of the top quark to be very short**. Numerically:

$$\Gamma_t \simeq 1.4 \text{ GeV} \quad \Longrightarrow \quad \tau \approx 5 \times 10^{-25} \text{ s} \tag{29.11}$$

The top quark does not have a chance to form "toponium" $t\bar{t}$ bound states, like in the case of the charmonium or bottonium. Its lifetime is also so short that the top mean free path is less than the typical scale for fragmentation (see Section 18.14). So, the top quark is assumed *not to hadronize before it decays*. Consequently, soon after it is formed, the top quark primarily decays via the production of a bottom quark and a W boson (with a probability of $\approx 96\%$), and the rest via the Cabibbo suppressed modes with strange or down quarks:

$$Br(t \to Wb) \gg Br(t \to Ws) > Br(t \to Wd) \tag{29.12}$$

the exact fractions being determined by the CKM matrix elements V_{tb}, V_{ts}, and V_{td} (see Section 30.2).

Below we list some of the measured properties of the top quark (values taken from the Particle Data Group's Review [5]):

$$\boxed{t}: \begin{array}{l} m_t = 173.1 \pm 0.6 \text{ GeV} \\ \Gamma_t = (1.41^{+0.19}_{-0.15}) \text{ GeV} \\ Br(t \to e\nu_e b) = (13.3 \pm 0.6)\% \\ Br(t \to \mu\nu_\mu b) = (13.4 \pm 0.6)\% \\ Br(t \to \tau\nu_\tau b) = (7.1 \pm 0.6)\% \\ Br(t \to jets) = (66.5 \pm 1.4)\% \end{array} \tag{29.13}$$

29.3 Higher-Order (Quantum Loop) Electroweak Corrections

In Section 27.1, we have already pointed out that the gauge sector of the electroweak theory is fully determined by only three parameters. When we include higher-order corrections, the number of parameters increases to five, which is still a small number. The three best measured quantities can be used for three of those parameters: (a) the fine-structure constant α measured in low-energy experiments, the Fermi constant G_F measured in muon decays (see Section 23.6), and the mass of the Z^0 boson measured at LEP and SLC (see Section 27.7).

These precise electroweak measurements can provide a "prediction" with a theoretical model for the massive particles entering through higher-level quantum loops. For instance, the W boson mass can be expressed in the following way (see Eq. (27.10) and Ref. [364]):

$$M_W^2 = \frac{(A^0)^2}{\sin^2\theta_W(1 - \Delta r)} = \frac{\pi\alpha}{\sqrt{2}G_F \sin^2\theta_W(1 - \Delta r)} \tag{29.14}$$

Hence, introducing the M_Z parameter, one gets:

$$M_W^2 = \frac{M_Z^2}{2}\left(1 + \sqrt{1 - 4\frac{\pi\alpha}{\sqrt{2}G_F M_Z^2(1 - \Delta r)}}\right) \tag{29.15}$$

where α, G_F and M_Z are precisely measured, and Δr contains radiative quantum loop corrections.

The power of the last expression is that it only contains the precisely measured constants α, G_F, and M_Z (renormalized – see Section 12.6). Contributions to Δr originate from the top quark by the one-loop diagrams shown in Figure 29.2, which contribute to the W and Z masses via:

$$(\Delta r)_{\text{top}} \simeq -\frac{3G_{\text{F}}}{8\sqrt{2}\pi^2 \tan^2\theta_W} m_{\text{t}}^2 \tag{29.16}$$

Also the Higgs boson contributes to Δr via the one-loop diagrams shown in Figure 29.3:

$$(\Delta r)_{\text{Higgs}} \simeq \frac{3G_{\text{F}}M_W^2}{8\sqrt{2}\pi^2} \left(\ln\frac{m_{\text{H}}^2}{M_Z^2} - \frac{5}{6} \right) \tag{29.17}$$

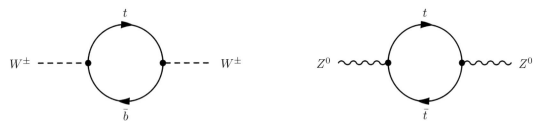

Figure 29.2 Virtual top quark loops contributing to the W and Z^0 boson masses.

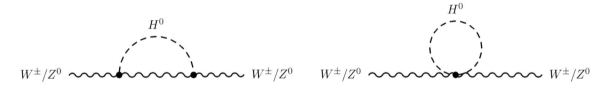

Figure 29.3 Virtual Higgs boson loops contributing to the W and Z^0 boson masses.

• **Fit of the top quark mass.** While the leading m_t dependence of Δr is quadratic (i.e., very strong), the leading m_{H} dependence is only logarithmic (i.e., rather weak). Therefore the inferred constraints on m_{H} are much weaker than those on m_t. The strong m_t dependence was used to successfully predict the top quark mass from the electroweak precision data before it was discovered by CDF and DØ in 1995, as discussed in Section 29.4.

The top quark mass can be *indirectly* fitted to the LEP/SLC data assuming that the loop corrections to the $W^\pm$ boson mass were correctly predicted by the SM. Neutral current weak interaction data, such as e^+e^- annihilation near the Z^0 pole, νN and eN deep inelastic scattering, νe elastic scattering, and atomic parity violation can also be used to constrain the top quark mass. Figure 29.4 shows the χ^2 of the SM electroweak fit to the precision data as a function of the assumed top quark mass for three different choices of the Higgs boson mass: $m_{\text{H}} = 50$ GeV was the lower limit of the Higgs boson mass from direct searches at LEP-I at the time, 1000 GeV is the theoretical upper limit of the Higgs boson mass, and 300 GeV was chosen to be a representative, central value as a logarithmic average between the two extremes. The minimum of the χ^2 curve indicates the best estimate of the top quark mass, the width of the curves gives an estimate of the uncertainty of this determination. The indirect measurements of the top quark mass using the Z^0 pole data together with the direct measurements of the W boson mass and total width and several other electroweak quantities yields [364]:

$$\text{LEP/SLD/M}_\text{W}/\Gamma_\text{W}: \quad m_{\text{top}} = 181.1^{+12.3}_{-9.5} \text{GeV} \tag{29.18}$$

The actual detection of the top quark (as discussed in the next sections) and its *direct* mass measurement **confirmed the validity of the SM theory.**

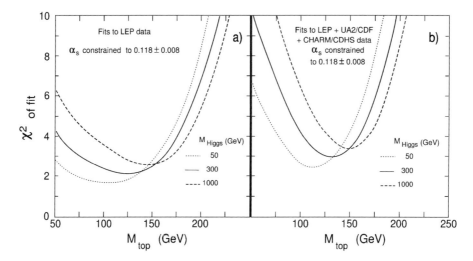

Figure 29.4 Graph of the χ^2 of the fit of the SM to the precision electroweak data as a function of the top quark mass. The mass of the Higgs boson was chosen to be 50, 300, and 1000 GeV to give an estimate of the uncertainty of this determination. Reprinted from G. Alexander *et al.*, "Electroweak parameters of the Z^0 resonance and the Standard Model: the LEP collaborations," *Phys. Lett. B*, vol. 276, pp. 247–253, 1992, with permission from Elsevier.

29.4   The Top Quark Discovery and Measurements

We have discussed the postulate and discovery of the bottom quark in Section 20.9. After the bottom quark discovery in 1977, the existence of its weak isospin partner was anticipated and the experimental search for the top quark began. The e^+e^- colliders started looking for the production and decays of the type $e^+e^- \to t\bar{t}$. Also at hadron colliders, the search began in particular at the CERN $Sp\bar{p}S$ and at the FNAL Tevatron. The top quark was too heavy to be produced in e^+e^- colliders even at LEP and was only discovered in 1995 in proton–antiproton interactions recorded at the Tevatron collider by the experiments CDF and DØ. Both experiments measured its mass directly, exploiting various decay chains. The quest for the top quark at colliders is summarized in Table 29.2.

Year	Collider	Beams	Limit on m_{t}
1979–1984	PETRA (DESY)	e^+e^-	> 23.3 GeV
1987–1990	TRISTAN (KEK)	e^+e^-	> 30.2 GeV
1989–1990	SLC (SLAC), LEP (CERN)	e^+e^-	> 45.8 GeV
1984	$Sp\bar{p}S$ (CERN)	$p\bar{p}$	> 45.0 GeV
1990	$Sp\bar{p}S$ (CERN)	$p\bar{p}$	> 69 GeV
1991	Tevatron (FNAL)	$p\bar{p}$	> 77 GeV
1992	Tevatron (FNAL)	$p\bar{p}$	> 91 GeV
1994	Tevatron (FNAL)	$p\bar{p}$	> 131 GeV
1995	Tevatron (FNAL)	$p\bar{p}$	$= 174 \pm 10^{+13}_{-12}$ GeV
			$= 199^{+19}_{-21} \pm 22$ GeV

Table 29.2 Summary of the search for the top quark at high-energy colliders.

• **Top pair production.** At hadron colliders, the favored mode to search for the top quark is via the pair production reaction:

$$pp(p\bar{p}) \to t\bar{t} + X \to (bW^+)(\bar{b}W^-) + X \tag{29.19}$$

The tree-level diagrams for the quark–antiquark annihilation into a $t\bar{t}$ pair are shown in Figure 29.5. They consist of: (a) gluon–gluon fusion, (b) t-channel, and (c) quark–antiquark annihilation diagrams. Following the techniques illustrated in Sections 19.10 and 19.11, the cross-section for the top quark pair production can be expressed as:

$$\sigma\left(P_1 P_2 \to t\bar{t}\right)(s) = \sum_{i,j=q,\bar{q},g} \int \mathrm{d}x_1 \mathrm{d}x_2 f_{i/P_1}\left(x_1, Q^2\right) f_{j/P_2}\left(x_2, Q^2\right) \hat{\sigma}_{ij \to t\bar{t}}(x_1 x_2 s) \tag{29.20}$$

Setting $x_1 \simeq x_2 \equiv x$ yields:

$$x \simeq \frac{2m_t}{\sqrt{s}} \approx \begin{cases} 0.18 \text{ at the Tevatron} \\ 0.025 \text{ at the LHC} \end{cases} \tag{29.21}$$

Hence, at the Tevatron, the production is dominated by valence quark processes (high x) while at the LHC, the dominant contribution is from gluon-initiated processes (small x) – see Section 19.9. The predicted cross-section as a function of $\sqrt{s}$ is illustrated in Figure 19.22. It increases rapidly with $\sqrt{s}$. Its value for $m_t = 175$ GeV at next-to-leading order is estimated to be [400, 401]:

$$\sigma(t\bar{t}) = \begin{cases} (5.24 \pm 6\%) \text{ pb at } \sqrt{s} = 1.96 \text{ TeV} \\ (0.833 \pm 15\%) \text{ nb at } \sqrt{s} = 14 \text{ TeV} \end{cases} \tag{29.22}$$

At the Tevatron, it is several orders of magnitude smaller than the W and Z production, and is rather at the scale of the WW or ZZ production. However, top quark pair production becomes really large at the LHC! The LHC is a "top-factory."

Once the top quark pairs have been produced, these latter very quickly subsequently decay into mostly Wb pairs (see Eq. (29.7)). There are different distinct final-state topologies that can be searched for:

$$
\begin{aligned}
&t\bar{t} \to (bq_1 q_2)(\bar{b}q_3 q_4) \to 6 \; jets \quad (\approx 46\%) \\
&t\bar{t} \to (b\ell^+ \nu_\ell)(\bar{b}q_3 q_4) \to 4 \; jets + 1 \text{ charged lepton} + \text{missing energy} \quad (\approx 44\%) \\
&t\bar{t} \to (b\ell^+ \nu_\ell)(\bar{b}\ell^- \bar{\nu}_\ell) \to 2 \; jets + 2 \text{ oppositely charged leptons} + \text{missing energy} \quad (\approx 10\%)
\end{aligned}
\tag{29.23}
$$

where ℓ^+ and ℓ^- can be either electrons or muons and the figures in brackets show the probability for the channel. Hence, the combinations ee, $e\mu$, and $\mu\mu$ are possible. In the case of the tau lepton, its semi-leptonic decay also leads to these final states. The dominant backgrounds for the $t\bar{t}$ searches in the channels with at least a lepton are the W+jets, Z+jets, and the WW, WZ, and ZZ processes. The all-jets search is dominated by QCD multijet events. In all cases, the presence of b quarks in the final state is an important experimental signature, as discussed later.

• **Single top production.** While the production of a pair of quarks was a strong (hard QCD) process, the production of a single top quark requires a W and is hence an electroweak process. It is therefore a rarer process than the top quark pair production, however it requires less energy since only one heavy top quark needs to be produced kinematically and this compensates the first effect. Overall, however, single top production remains a rarer process than the top pair production at the Tevatron and the LHC. The tree-level Feynman diagrams are illustrated in Figure 29.6. They consist of (a) an s or W^*-channel, which is similar to a Drell–Yan process; (b) a t or W–g fusion channel, and (c) a tW-channel. The cross-sections at NLO in QCD have been

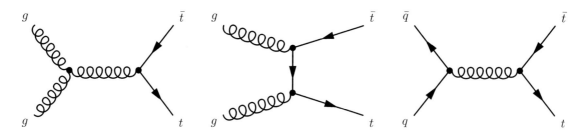

Figure 29.5 Tree-level partonic diagrams for top quark pair production at hadron colliders: (left) gluon–gluon fusion, (middle) t-channel, and (right) quark–antiquark annihilation.

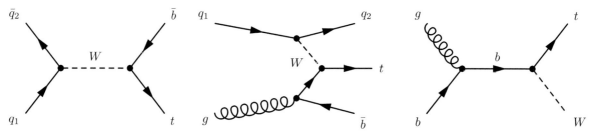

Figure 29.6 Tree-level partonic diagrams for the single top quark production at hadron colliders: (left) s or W^* channel; (middle) t or W–g fusion channel, and (right) tW-channel.

calculated in the different channels. For the value $m_t = 175$ GeV, they are estimated to be [402–404]:

$$\sigma(p\bar{p} \to t + X)(\sqrt{s} = 1.8 \text{ TeV}) = \begin{cases} 0.88 \pm 0.14 \text{ pb for the } s\text{-channel} \\ 1.98 \pm 0.30 \text{ pb for the } t\text{-channel} \\ \approx 0 \text{ for the } tW\text{-channel} \end{cases}$$

$$\sigma(pp \to t + X)(\sqrt{s} = 14 \text{ TeV}) = \begin{cases} 10.2 \pm 0.7 \text{ pb for the } s\text{-channel} \\ 245 \pm 27 \text{ pb for the } t\text{-channel} \\ 62.2^{+16.6}_{-3.7} \text{ pb for the } tW\text{-channel} \end{cases} \qquad (29.24)$$

- **Measurements at the FNAL Tevatron.** Before the start of the CERN LHC, the $p\bar{p}$ collider Tevatron at Fermilab (see Section 3.5) was the world's highest-energy particle accelerator, with a center-of-mass energy of $\sqrt{s} = 1.96$ TeV. During the data-taking period called Run-I (from 1992–1996), the CDF and DØ experiments each collected an integrated luminosity (see Section 3.7) of about 125 pb^{-1} of $p\bar{p}$ collisions data at a center-of-mass energy of 1.8 TeV. This significant amount of data led to the discovery of the top quark and a first direct measurement of its mass. A peak luminosity of 2×10^{31} cm^{-2} s^{-1} was achieved. After a long shutdown and several upgrades, the Run-II data-taking period started in March 2001 with the aim of increasing the luminosity by at least a factor 20 compared to Run-I. A record peak luminosity of 4×10^{32} cm^{-2} s^{-1} was indeed achieved at a center-of-mass energy of 1.96 TeV. The integrated luminosity delivered at 1.96 TeV approached 12 fb^{-1}, with approximately 10 fb^{-1} recorded by the CDF and DØ experiments. The Tevatron ceased operations in September 2011.

The **CDF detector** [405] was a general-purpose solenoidal detector which combined precision charged particle tracking with fast projective calorimetry and finely grained muon detection. The detector is shown in Figure 29.7. It consisted of three main functional sections going radially outwards from the beamline. The tracking system was used for particle charge and three-momentum measurements. It was contained in a superconducting solenoid, 1.5 m in radius and 4.8 m in length, which generated a 1.4 T magnetic field parallel to the beam axis. The solenoid was surrounded by the scintillator-based calorimeter system with separate electromagnetic and hadronic measurements, which covered the rapidity region $|\eta| \leq 3$. Outside the calorimeters, layers of steel absorbed the remaining hadrons leaving only muons, which were detected by external muon detectors.

An important component was the central tracking system (see Figure 29.8). It consisted of a silicon microstrip detector and an open-cell wire drift chamber that surrounded the silicon detector. The silicon microstrip detector consisted of seven layers in a barrel geometry that extended from a radius of 1.5 cm from the beam line to $r = 28$ cm, with a length from 90 cm to nearly 2 m. The readout pitch was 50 μm. The two outer layers comprised the intermediate silicon layer (ISL) system. The ISL consisted of two symmetric silicon layers in the forward and backward region ($|\eta| \geq 1.1$) located at radii of $R \simeq 20$ cm and $R \simeq 29$ cm, respectively, and one in the central region ($|\eta| < 1.1$) at $R \simeq 23$ cm. It provided one space point in the central region which improved the matching between the inner silicon layers (SVXII) tracks and outer tracking chamber (COT) tracks and its fine granularity helped to resolve ambiguities in dense track environments. This entire silicon tracking system allowed track reconstruction in three dimensions. The impact parameter resolution of the combination of SVXII and ISL was 40 μm including a 30 μm contribution from the transverse width of the beam-line. The position of the vertex along the beam pipe direction could be measured with the SVXII and ISL with a resolution of 70 μm. **The tracker was able to identify displaced vertices for b-quark tagging**.

The **DØ detector** [406], originally lacking a magnetic field, was composed of three major subsystems: (a) a central tracking detector, (b) a uranium/liquid argon calorimeter, and (c) an external muon spectrometer. The detector tracking system was significantly upgraded for Run-II (see Figure 29.9). The new system included a silicon microstrip tracker and a scintillating-fibre tracker located within a 2 T solenoidal magnet. As in the case of CDF, the silicon microstrip tracker was able to identify displaced vertices for b-quark tagging. The magnetic field enabled a measurement of the energy-to-momentum ratio for electron identification and calorimeter calibration, opened new capabilities for τ identification and hadron spectroscopy, and allowed for precision muon momentum measurements. Between the solenoidal magnet and the central calorimeter and in front of the forward calorimeters, preshower detectors have been added for improved electron identification. In the forward muon system, proportional drift chambers have been replaced by mini-drift tubes and trigger scintillation counters have been added for improved triggering. Also a forward proton detector was added for the study of diffractive physics.

The parameters of the CDF and DØ detectors are summarized and compared in Table 29.3.

• **Hermeticity.** Both CDF and DØ were designed to be very hermetic detectors, covering a large rapidity region around the vertex (as can be appreciated from Figures 29.7 and 29.9). As already discussed in Section 27.4, this capability is an important feature at hadron colliders to measure the missing transverse energy, leading to the reconstruction of the transverse mass.

• **b-tagging.** A key new feature of CDF and DØ compared to previous experiments at hadron colliders was the ability to classify the reconstructed jets according to their quark flavor, distinguishing "light jet flavors" (originating from the hadronization of the u-, d-, s-quark, or a gluon) from "heavy jet flavors" (originating from c- or b-quarks). Two techniques can be used to separate a heavy jet from a light one:

1. **Soft-lepton tagging (SLT).** The presence of a soft electron or a muon within the jet can indicate the semi-leptonic decay of a b or c hadron (average branching ratio of about 10%) – see Sections 20.8 and 20.9.

2. **Lifetime or secondary vertex tagging.** Given the relatively long lifetimes of the charmed and bottom hadrons and the Lorentz boost, they can decay significantly far away from the primary vertex. Hence, tracks emanating from their decays cross in what are called secondary vertices. The secondary vertex of a jet consequently relies on searching for such displaced vertices inside a jet. The efficiency to reconstruct such secondary vertices relies mainly on the very precise tracking capabilities of the vertex detectors (e.g., SVX in CDF). As an illustration, the b-tagging efficiency in CDF for b-jets coming from the top quark decays reaches about 50% in the central region. Similar considerations apply also for DØ.

Parameter	CDF	DØ	
Length	14 m	17 m	
Height	10 m	11 m	
Weight	5000 t	4600 t	
Solenoid field	1.4 T	2 T	
ECAL resolution $\left(\frac{\sigma_E}{E}\right)$	$\frac{13.5\%}{\sqrt{E}} \oplus 2\%$	$\frac{15\%}{\sqrt{E}} \oplus 0.4\%$	E (GeV)
HCAL resolution $\left(\frac{\sigma_E}{E}\right)$	$\frac{75\%}{\sqrt{E}} \oplus 3\%$	$\frac{50\%}{\sqrt{E}}\%$	E (GeV)
Calorimeter coverage η	≤ 3.64	≤ 4.0	
Tracking coverage η	≤ 2.0	≤ 3.0	
Nr. track points	$8 + 96$	$8 + 8$	
Track resolution $\sigma\left(1/p_{\mathrm{T}}\right)^{\eta=0}$	0.0017	$0.0018 \oplus \frac{0.015}{p_{\mathrm{T}}}$	p_T (GeV)
Muon coverage η	≤ 1.5	≤ 2.0	
Muon resolution ($p_{\mathrm{T}} < 0.1$ TeV)		10%–50%	

Table 29.3 The performance parameters of the CDF and DØ detectors. Compare with Table 3.7.

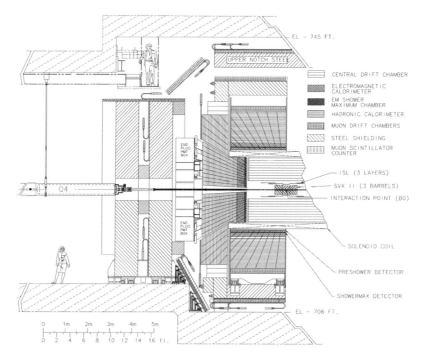

Figure 29.7 Layout of the CDF detector at the FNAL Tevatron. Reprinted from R. Blair *et al.*, "The CDF-II detector: Technical design report," FERMILAB-DESIGN-1996-01, FERMILAB-PUB-96-390-E.

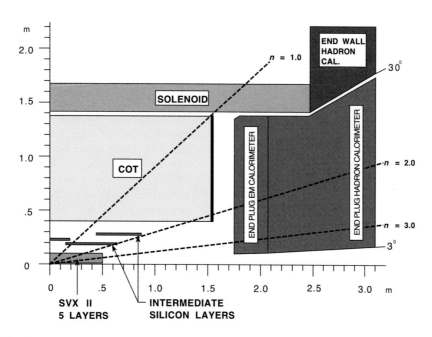

Figure 29.8 The layout of the CDF tracker at the FNAL Tevatron with the inner silicon layers (SVX), the intermediate silicon layer system (ISL), and the outer tracking chamber (COT). Reprinted from R. Blair *et al.*, "The CDF-II detector: Technical design report," FERMILAB-DESIGN-1996-01, FERMILAB-PUB-96-390-E.

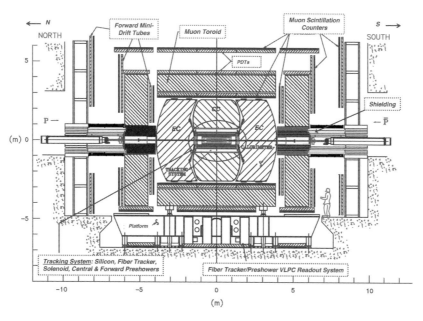

Figure 29.9 Layout of the DØ detector at the FNAL Tevatron. Reprinted from V. Abazov *et al.*, "The upgraded D0 detector," *Nucl. Instrum. Meth. A*, vol. 565, pp. 463–537, 2006, with permission from Elsevier.

• **Top quark discovery.** In 1995, CDF and DØ published evidence for the existence of the top quark [407,408]. CDF used a data sample of 67 pb^{-1}. The first searches focused on the dilepton (ee, $e\mu$, or $\mu\mu$) and the leptons+jets samples. The preselection required an isolated electron with $E_T^e > 20$ GeV or an isolated muon with $p_T^\mu > 20$ GeV. If there were two leptons, their invariant mass should be incompatible with that of the Z^0 boson. The possible existence of displaced vertices was searched for in jets according to the b-tagging algorithm. CDF found 37 b-tagged $W+$ 3-jet events that contained 27 SVX tags compared to 6.7 ± 2.1 expected from background and 23 SLT tags with an estimated background of 15.4 ± 2.0. Also, they found six dilepton events compared to 1.3 ± 0.3 events expected from background. Combining these excesses together, the probability of a fluctuation of the background was $\approx 10^{-6}$, which corresponds to a 4.8σ effect. A CDF event compatible with top pair production consisting of one high-p_T electron plus four jets of which two were b-tagged is shown in Figure 19.21. The inset to the bottom right shows a zoom on the vertex and indicates the position of the primary and two secondary vertices. DØ performed a similar analysis on a data sample of 50 pb^{-1} and observed 17 events over an expected background of 3.8 ± 0.6 events. The probability for an upward fluctuation of the background to produce the observed signal is in this case computed to be 2×10^{-6}, which is equivalent to a 4.6σ effect.

• **Single top measurement.** The first evidence for single top processes was published by the DØ collaboration in 2006 [409]. In 2009 the CDF and DØ collaborations released twin articles with the definitive observation of these processes [410, 411].

29.5 Top Quark Measurements at the LHC

By the end of 2011, the CERN LHC had achieved a luminosity almost 10 times higher than that of the Tevatron and a beam energy of 3.5 TeV each, surpassing the capabilities of the Tevatron. As mentioned above, the increase with energy of the cross-section for producing top quark pairs makes the LHC a real "top-factory." Many studies on the top quark mass and properties could be performed with much more abundant statistics than at the Tevatron. As an example, we can estimate the number of $t\bar{t}$ produced at the LHC at the highest energy of $\sqrt{s} = 13$ TeV. If we assume $\sigma_{t\bar{t}} \simeq 0.840$ nb and take the integrated luminosity of the LHC delivered for example to ATLAS over the years 2015–2018 (see Section 3.7), we find that:

$$N_{t\bar{t}} = \mathcal{L}_{integrated} \times \sigma_{t\bar{t}}(\sqrt{s} = 13 \text{ TeV}) \approx 156 \text{ fb}^{-1} \times 0.840 \times 10^6 \text{ fb} = 1.3 \times 10^8 \qquad (29.25)$$

A large number indeed! The high precision granted by statistics provided an important probe for testing the SM and possible deviations from it. First of all, higher-order QCD corrections could be tested by measuring the production cross-sections. In this context, the summary of the LHC and Tevatron measurements of the top pair production cross-section as a function of the center-of-mass energy compared to the NNLO QCD calculation complemented with NNLL resummation with $m_t = 172.5$ GeV is shown in Figure 29.10. The figure shows the Tevatron combined results at $\sqrt{s} = 1.96$ TeV as well as the measurements obtained at the LHC with ATLAS and CMS in various decay modes and increasing center-of-mass energies of $\sqrt{s} = 5, 7, 8$, and 13 TeV. Note the vertical logarithmic scale. The cross-section increases by about an order of magnitude between 5 and 13 TeV. These measurements can also help constrain the parton distribution functions, in particular those of the gluons since top quark production from gluon fusion corresponds to about 90% of the cross-section at the LHC. Direct reconstruction of the top final state leads to precise estimates on the top mass m_t and its width Γ_t. In addition to the mass and width, an important parameter of the SM is also the corresponding CKM matrix element $|V_{tb}|$. A measurement of this parameter was for instance reported in Ref. [412], based on an analysis of single top quark production. The measured value is:

$$|V_{tb}| = 1.02 \pm 0.04(\text{exp.}) \pm 0.02(\text{theory}) \qquad (29.26)$$

Overall, the agreement with the SM predictions is excellent. The top quark behaves exactly like the "very massive big brother" of the up and charm quarks! The measurements allowed the constraining of new physics beyond the SM, for instance in terms of potential anomalous couplings, or searches for $t\bar{t}$ resonances, and the existence of a heavy gauge boson W' decaying as $W' \to tb$.

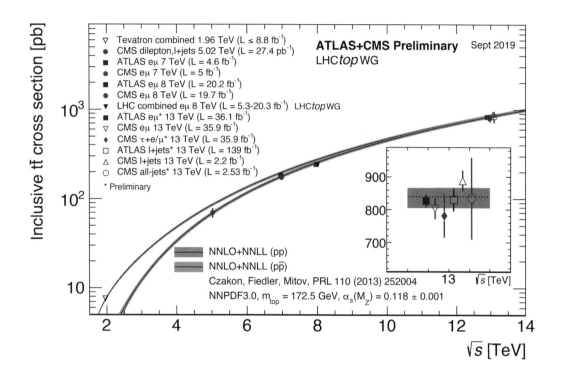

Figure 29.10 Summary of the LHC and Tevatron measurements of the top quark pair production cross-section as a function of the center-of-mass energy compared to the NNLO QCD calculation complemented with NNLL resummation with $m_t = 172.5$ GeV. Reprinted from the LHC Top Working Group https://lpcc.web.cern.ch/lhc-top-wg-wg-top-physics-lhc.

29.6   The Scalar Higgs Boson

In Section 25.6 we considered the Higgs scalar field as a constant with a non-vanishing vacuum expectation value v. We should now include the possible fluctuations or excitations of this field and write (cf. Eq. (25.76)):

$$\phi \equiv \frac{1}{\sqrt{2}} \begin{pmatrix} 0 \\ v + H(x) \end{pmatrix} \tag{29.27}$$

where the real scalar field $H(x)$ represents the neutral scalar (spin-0) Higgs particle H^0. Inserting this vacuum into the Lagrangian density, Eq. (25.72), and collecting only the terms that are relevant for the Higgs particle, we find that the **mass of the Higgs boson particle** is:

$$m_H = \sqrt{2\lambda}v \tag{29.28}$$

Hence, unlike the masses of the $W^\pm$ and Z^0 gauge bosons, the Higgs boson mass cannot be predicted from the theory since it depends on the coupling λ which is a priori unknown. From theoretical arguments related to radiative corrections and self-consistency of the theory, one can set an upper bound at about 1000 GeV, above which the Higgs boson would not be treatable perturbatively any longer. Therefore, from a naive theoretical point of view, the Higgs boson could possess any mass in the range $0 \leq m_H \leq 1000$ GeV.

Although the mass of the Higgs boson is not predicted, the SSB mechanism predicts uniquely the coupling between the Higgs boson and the other gauge bosons and fermions of the theory. Because of its nature, the coupling between the Higgs boson and a particle is proportional to the particle's rest mass. The corresponding Higgs coupling diagrams and the Feynman rules are listed in Table 29.4.

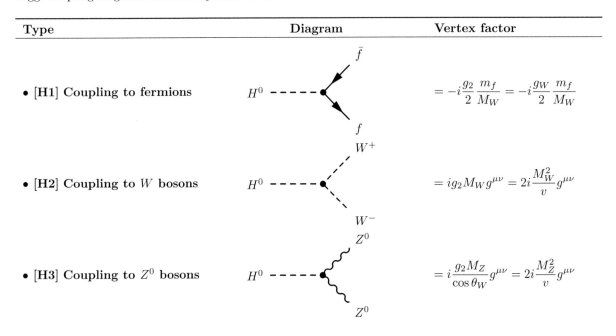

Type	Diagram	Vertex factor
• [H1] Coupling to fermions	H^0 f / $\bar{f}$	$= -i\dfrac{g_2}{2}\dfrac{m_f}{M_W} = -i\dfrac{g_W}{2}\dfrac{m_f}{M_W}$
• [H2] Coupling to W bosons	H^0 W^+ / W^-	$= ig_2 M_W g^{\mu\nu} = 2i\dfrac{M_W^2}{v}g^{\mu\nu}$
• [H3] Coupling to Z^0 bosons	H^0 Z^0 / Z^0	$= i\dfrac{g_2 M_Z}{\cos\theta_W}g^{\mu\nu} = 2i\dfrac{M_Z^2}{v}g^{\mu\nu}$

Table 29.4 Summary of Feynman rules for the Higgs boson couplings.

The Higgs boson couples to all SM particles and can, if kinematically allowed, decay into all these particles: $H^0 \to f\bar{f}$, $H^0 \to W^+W^-$, and $H^0 \to Z^0Z^0$. However, because its coupling is proportional to the mass of the particles involved, the largest branching ratios are decays to the heaviest particles. **The Higgs boson decays preferentially to the most massive particles that are kinematically accessible.**

Because the Higgs boson is a spin-0 scalar particle, it decays isotropically and the matrix element has no angular dependence. Furthermore, angular momentum conservation simply implies that the particles produced decay back-to-back in the center-of-mass of the Higgs and must have opposite spins, and hence they have the same helicities. Ignoring the mass of the fermions in the phase-space factors, the tree-level partial decay width into fermions can be derived in a straightforward way using the Feynman rule **[H1]**. The amplitude for the process $H^0 \to f(p_1)\bar{f}(p_2)$ is equal to:

$$\mathcal{M} = \frac{-ig_W}{2}\frac{m_f}{M_W}\bar{u}(p_1)v(p_2) = -i\left(2\frac{g_W^2}{8M_W^2}\right)^{1/2}m_f\bar{u}(p_1)v(p_2) = -i\left(\sqrt{2}G_F\right)^{1/2}m_f\bar{u}(p_1)v(p_2) \qquad (29.29)$$

After the calculation of the matrix element, and the inclusion of the phase space and of the color factor, the total width turns out to be:

$$\Gamma(H^0 \to f\bar{f}) = \frac{N_C}{8\pi}g_f^2 m_H \qquad (29.30)$$

where the factor $g_f \equiv (\sqrt{2}G_F)^{1/2}m_f$ is the coupling of the fermions to the Higgs boson, which is proportional to the fermion mass, as expected. The color factor N_C is equal to one for leptons and three for quarks.

Similarly, the diagrams **[H2]** and **[H3]** lead to the decay into two massive gauge bosons $H^0 \to W^+W^-$ and $H^0 \to Z^0Z^0$. These partial decay widths can be expressed as [413]:

$$\Gamma\left(H^0 \to VV\right) = \frac{G_F m_H^3}{16\sqrt{2}\pi}\delta_V\sqrt{1-4x}\left(1-4x+12x^2\right), \qquad x \equiv \frac{M_V^2}{m_H^2} \tag{29.31}$$

where $V = W$ or Z and $\delta_W = 2$, $\delta_Z = 1$. They become dominant for a Higgs boson mass above the WW and ZZ thresholds. Below these thresholds, the decay modes into off-shell gauge bosons are still important. In practice, it is sufficient to consider one off-shell boson, and the decays are consequently labeled as $H^0 \to WW^*$ and $H^0 \to ZZ^*$. The contribution of these decay modes will be illustrated in the next section.

Finally, it should be noted that the Higgs boson can also decay into massless particles through a loop of virtual fermions (predominantly top) and bosons (W), as shown in Figure 29.11. The particles in the loop are virtual but again because their coupling to the Higgs boson is so large, these loop-level diagrams compete with the tree-level branching ratios. The tree-level partial decay width for $H^0 \to \gamma\gamma$ is given by [414]:

$$\Gamma(H^0 \to \gamma\gamma) = \frac{\alpha^2}{8\sqrt{2}\pi^3}G_F m_H^3|I|^2 \tag{29.32}$$

where the unitless function I contains the amplitudes due to the "triangle" diagrams, as shown in Figure 29.11. One must include all possible diagrams, i.e., $I \equiv I_e + I_\mu + I_\tau + \sum_q I_q + I_W + \dots$. These are amplitudes and as a result the interference between the diagrams is taken into account. Their calculation is discussed in Ref. [415], which also presents the expected branching ratio as a function of the Higgs boson mass. This latter is very small, with the branching ratio being at the level of two permil. However, the Higgs boson decay into two photons, although rare, is experimentally clean at hadron colliders (see Section 29.9).

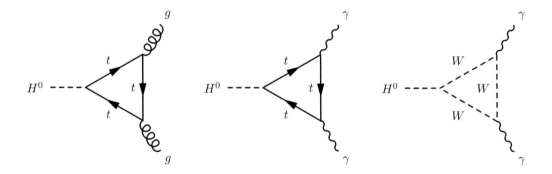

Figure 29.11 The loop diagrams for the Higgs boson decays into gg and $\gamma\gamma$.

The partial width to two gluons can be computed similarly at tree level. However, the QCD radiative corrections are large in the $H^0 \to gg$, nearly doubling the decay width compared to its tree level. The partial decay width, including NLO QCD corrections, can be expressed as [413]:

$$\Gamma\left(H^0 \to gg\right) \approx \frac{G_F \alpha_s^2 m_H^3}{36\sqrt{2}\pi^3}\left[1 + \underbrace{\frac{\alpha_s}{\pi}\left(\frac{95}{4} - \frac{7}{6}N_F\right)}_{\text{QCD correction}}\right] \tag{29.33}$$

where N_F is the number of quark–antiquark loops. The QCD correction term is equal to 0.7 for $N_F = 5$. This decay mode contributes to the inclusive hadronic decay mode of the Higgs boson, since the basic experimental picture after gluon fragmentation will be $H^0 \rightarrow \mathrm{jet} + \mathrm{jet}$.

29.7 Predictions on the Higgs Boson Before the LHC

With the e^+e^- precision electroweak data and the independent hadron collider data on M_W and m_t, constraints on the SM Higgs boson mass could be extracted. This "window" provided by higher-order corrections coupled to precision measurements did not *per se* provide a proof-of-existence of the Higgs boson, but stated that if a Higgs boson consistent with the SM prediction existed, then it must lie within a given mass range to be compatible with the experimental data. The precision electroweak measurements, performed at LEP and by SLD, CDF, and DØ, as a function of the Higgs boson mass, assuming the SM to be the correct theory of Nature preferred a light value for its mass.

We show the results computed with GFITTER (available at http://project-gfitter.web.cern.ch/project-gfitter/). This software package consists of tools for the handling of the data, the fitting, and statistical analyses such as toy Monte-Carlo sampling. Theoretical models are inserted as plug-in packages, which may be hierarchically organized. In the following, we use the "Standard Model" theory implemented in GFITTERv2.2, which included the latest theoretical predictions of electroweak observables. In particular, the mass of the W boson, the effective weak mixing angle, the partial and total widths of the Z^0, and the hadronic peak cross-section are calculated with the full two-loop corrections and the known beyond-two-loop corrections [416]. The main input parameters were chosen to be:

$$M_Z = 91.1875 \pm 0.0021\,\mathrm{GeV}, \quad M_W = 80.379 \pm 0.013\,\mathrm{GeV}, \quad m_t = 173.34 \pm 0.76\,\mathrm{GeV},$$
$$\alpha_s(M_Z^2) = 0.1196 \pm 0.0030 \tag{29.34}$$

The result of a 1D scan of the SM Higgs boson mass shows the impressively predictive power of the quantum loop corrections:

$$\text{SM fit:} \quad m_H = 87.1^{+28.4}_{-23.4}\,\mathrm{GeV} \quad m_H \leq 190\,\mathrm{GeV} \quad (3\sigma\ \mathrm{CL}) \tag{29.35}$$

The corresponding χ^2 distribution is plotted in Figure 29.12 as a function of the assumed Higgs boson mass parameter m_H. The $\Delta\chi^2 = \chi^2 - \min(\chi^2) = 1, 4, 9$ correspond to the $1, 2, 3\sigma$ errors. While this was not a proof that the SM Higgs boson actually existed, it did serve as a guideline on which mass range to look for it at the CERN LHC.

• **Decay modes of a light Higgs boson.** The partial width for decays to a fermion–antifermion pair is dominated by the direct coupling between the Higgs boson and the fermions. Including the phase-space factor for finite fermion masses, it becomes (cf. Eq. (29.30)):

$$\Gamma\left(H^0 \rightarrow f\bar{f}\right) = \frac{N_C}{8\pi} g_f^2 m_H \beta^3 = \frac{N_C G_F m_H}{4\sqrt{2}\pi} m_f^2 \left(1 - \frac{4m_f^2}{m_H^2}\right)^{3/2} \tag{29.36}$$

where β is the threshold phase-space factor introduced by the finite fermion mass. Since the width depends on the square of the fermion mass, the decay to light charged leptons and light quarks is really negligible compared to those of the tau and the heavy quarks, so we only need to consider τ, c, b, and t. Since the top quark is so heavy, it is kinematically inaccessible to a light Higgs boson with a mass less than $m_H \lesssim 2m_t$, so finally we have to consider $\Gamma(H^0 \rightarrow \tau^+\tau^-)$, $\Gamma(H^0 \rightarrow c\bar{c})$, and $\Gamma(H^0 \rightarrow b\bar{b})$. All of these fermions are in fact much lighter than the Higgs boson, so finally the β factor can be neglected. On the other hand, for quarks, QCD corrections are important. With massless quarks and including second-order QCD corrections, a more

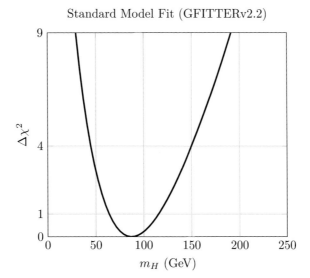

Figure 29.12 Fit of the Higgs boson mass within the SM computed with GFITTER.

precise expression for the width is [413]:

$$\Gamma\left(H^0 \to q\bar{q}\right) = \frac{3G_F m_H}{4\sqrt{2}\pi} m_q^2\left(m_H\right)\left[1 + 5.67\frac{\alpha_s}{\pi} + (35.94 - 1.36N_F)\frac{\alpha_s^2}{\pi^2}\right] \tag{29.37}$$

where N_F is as before. The strong coupling constant must be computed at the scale $q^2 = m_H^2$. The mass of the quark is also to be taken at $q^2 = m_H^2$, so they correspond to the running quark masses. This mass is computed for a given q^2 as a function of the pole quark mass m_q. Its evolution as a function of the scale can be found in Ref. [413].

• **Branching fractions of the Higgs boson.** For a Higgs boson of mass 125 GeV, the decay into $t\bar{t}$ is kinematically forbidden, and the decays to WW and ZZ are suppressed since one of the two bosons would have to be virtual. These latter decays are written as $H^0 \to WW^*$ and $H^0 \to ZZ^*$ to indicate that one of the two bosons is going to be off-mass-shell. Nonetheless, the large coupling to the Higgs boson due to the large gauge boson masses makes that compete with the other decay modes. The dominant decay mode is into $b\bar{b}$ fermions with a width of about 2 MeV. The other dominant fermions are $c\bar{c}$ and $\tau^+\tau^-$, their branching width given by:

$$\Gamma(H^0 \to b\bar{b}) : \Gamma(H^0 \to c\bar{c}) : \Gamma(H^0 \to \tau^+\tau^-) \simeq 3m_b^2 : 3m_c^2 : m_\tau^2 \tag{29.38}$$

The branching ratios predicted by the SM for a Higgs boson with a mass of 125 GeV are summarized in Table 29.5.

29.8   Discovery of the Higgs Boson at the CERN LHC

We have seen in the previous section that the precision measurements at e^+e^- colliders combined with the top quark measurements at the Tevatron provided a rather narrow window for the Higgs boson of the SM. This did not imply that such a boson existed, but that if the SM was the right theory and the Higgs boson existed, then it would be unlikely to have a mass above 152 GeV. At the same time, direct searches at LEP via the channel $e^+e^- \to Z^*H^0$ did not find evidence for the Higgs boson, and set a limit $m_H > 114$ GeV. Consequently, before

Decay mode	Branching ratio
$H \to b\bar{b}$	57.8%
$H \to WW^*$	21.6%
$H \to gg$	8.6%
$H \to \tau^+\tau^-$	6.4%
$H \to c\bar{c}$	2.9%
$H \to ZZ^*$	2.7%
$H \to \gamma\gamma$	0.2%
$H \to \mu^+\mu^-$	0.02%

Table 29.5 The predicted branching ratios of the SM Higgs boson of mass 125 GeV.

the start of the LHC, the SM Higgs boson had to be in the narrow range $114 < m_H < 152$ GeV or else, something else must have appeared in place of the SM Higgs boson!

In the year 2011, the CERN LHC pp collider operated at a center-of-mass energy of $\sqrt{s} = 7$ TeV, and in 2012 the energy was increased to $\sqrt{s} = 8$ TeV, reaching $\sqrt{s} = 13$ TeV in 2015. The LHC also provided high integrated luminosities, see Figure 3.15. With this excellent performance of the LHC collider, the two main experiments, ATLAS and CMS (see Figures 3.35 and 3.36), could already collect enough data to claim the discovery of a "Higgs-boson-like" particle with a mass around 125 GeV. The main mechanisms for the production of the SM Higgs boson are shown in Figure 29.13. The dominant mode is via gluon–gluon fusion, which can be expressed as:

$$\sigma_{ggH} \equiv \sigma(pp \to H^0 X) = \int dx_1 dx_2 g(x_1) g(x_2) \hat{\sigma}(gg \to H) \tag{29.39}$$

where $g(x)$ are the gluon parton distribution functions. As was evident from Figure 19.23, the production of the Higgs boson with a mass around 125 GeV is kinematically bound to the very low x region, where gluons dominated. In this sense, the LHC is really a gluon–gluon collider when it comes to Higgs boson production. As we discussed previously, the precise knowledge of the parton distribution function obtained (e.g., at HERA) was crucial in order to obtain the gluon parton distribution functions in the kinematical region of interest for the Higgs boson production at the LHC, which ultimately (after inclusion of QCD radiative corrections) led to a theoretical prediction for the Higgs boson cross-sections at the level of 10% or so. The second mechanism for the production is the WW boson fusion which is relevant because it tests directly the WWH coupling (see Figure 19.22):

$$\sigma_{VBF} \equiv \sigma(pp \to Hjj) = \int dx_1 dx_2 q_1(x_1) q_2(x_2) \hat{\sigma}(q_1 q_2 \to Hq_1' q_2') \tag{29.40}$$

Its tree-level diagram is shown in Figure 29.13(right). Since it involves two quarks in the initial state, it is less frequent than the gluon–gluon fusion process. However, the experimental signature is characterized by the Higgs boson accompanied by two forward jets (from the fragmentation of the hard q_1' and q_2' remnants). These cross-sections are plotted in Figure 19.22 as a function of the center-of-mass energy. We clearly see that $\sigma_{ggH} > \sigma_{VBF}$.

At present the study of the Higgs boson continues at the LHC. The combined LHC results indicate that the discovered particle at 125 GeV is consistent with being a spin-0 scalar and that its decay properties are consistent with those expected from the SM Higgs boson, as illustrated by the signal strength in a particular final state f defined by the cross-section times branching ratio in this channel normalized to the SM value

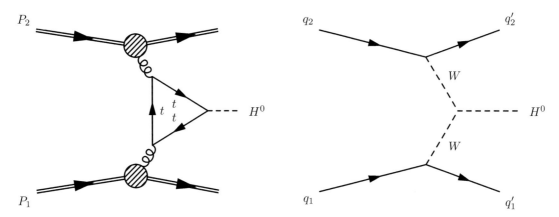

Figure 29.13 The tree-level diagrams for the Higgs boson production in hadron–hadron collisions: (left) gluon fusion; (right) WW fusion.

$\sigma Br(H^0 \to f)/(\sigma Br(H^0 \to f))_{SM}$. In particular, below we list the known properties (values taken from the Particle Data Group's Review [5]):

$$\boxed{H^0}: \quad \begin{aligned} & m_{H^0} = 125.09 \pm 0.24 \text{ GeV} \\ & J = 0 \\ & \Gamma_H < 0.013 \text{ GeV} \quad (95\%\,\text{CL}) \\ & \sigma Br(H^0 \to WW^*)/(\sigma Br(H^0 \to WW^*))_{SM} = 1.08^{+0.18}_{-0.16} \\ & \sigma Br(H^0 \to ZZ^*)/(\sigma Br(H^0 \to ZZ^*))_{SM} = 1.29^{+0.26}_{-0.23} \\ & \sigma Br(H^0 \to \gamma\gamma)/(\sigma Br(H^0 \to \gamma\gamma))_{SM} = 1.16 \pm 0.18 \\ & \sigma Br(H^0 \to all)/(\sigma Br(H^0 \to all))_{SM} = 1.10 \pm 0.11 \end{aligned}$$

$$(29.41)$$

The discovery of a scalar particle compatible with the SM Higgs boson is not the end of the story. In the coming decade, the LHC experiment will continue to study its properties with the hope of obtaining a detailed understanding of all properties of the Higgs boson. So far, all seems consistent with the SM predictions (see also Section 29.11), but it is too soon to know if this will stand the test of precision.

29.9   The Higgs Boson Search in Two Photon Events

The $H \to \gamma\gamma$ channel has been studied since the initial planning of the LHC as an important channel for the discovery of Higgs particles at masses beyond the upper reach of LEP and below about 150 GeV (see, e.g., Ref. [417]). The signature sought in the inclusive analysis is two high transverse energy (E_T) and isolated photons. The challenge for the discovery of a Higgs boson in this mode is the small branching fraction of about 0.002, since in this mass range the dominant decay mode of the Higgs boson is $b\bar{b}$ (see Table 29.5). The $\gamma\gamma$ decay mode can be well identified experimentally but the signal rate is small compared to the backgrounds coming both from two prompt photons (irreducible), and from those in which one or more of the photons are due to decay products or misidentified particles in jets (reducible). **It has long been understood that** $H \to \gamma\gamma$ **can be detected as a narrow mass peak above a large monotonically decreasing background in the distribution of the invariant mass of two reconstructed high-energy photons**. The background magnitude can be determined from the region outside the peak. The Higgs boson mass is measured from the position of the narrow resonant peak in the distribution of the invariant mass of the two photons.

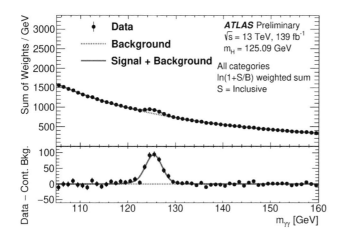
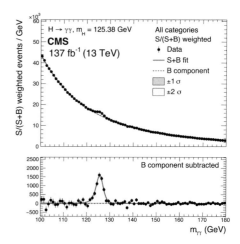

Figure 29.14 The observed and expected diphoton invariant mass distributions for the selected Higgs boson candidates at $\sqrt{s} = 13$ TeV from ATLAS and CMS. Reprinted with permission from "Measurement of the properties of Higgs boson production at $\sqrt{s} = 13$ TeV in the $H \to \gamma\gamma$ channel using 139 fb^{-1} of pp collision data with the ATLAS experiment," in *40th International Conference on High Energy Physics, Prague, Czech Republic*, 28 Jul–6 Aug 2020 and A. M. Sirunyan *et al.*, "Measurements of Higgs boson production cross sections and couplings in the diphoton decay channel at $\sqrt{s} = 13$ TeV," arXiv:2103.06956 (2021).

The idea of measuring the rate of background by using the mass regions adjoining the Higgs peak is extended to also measure the characteristics of the background, and using this information to help separate the background from the signal. The $H \to \gamma\gamma$ channel is particularly well suited to this technique because the signal is relatively small and can be confined to a narrow mass region due to the excellent photon energy and position resolution of the LHC detectors.

By using the photon isolation and kinematical information, significant additional discrimination between the signal and background can be achieved. The data analyses use this information to discriminate between the two by comparing data in mass side-bands. The backgrounds with two real prompt high E_T photons are called irreducible, although they can be somewhat reduced due to kinematical differences from signal processes in which high mass particles are produced. Prompt photons can be produced in higher-order QCD processes, but these latter generally lead to additional jet activity. Those backgrounds in which at least one final state jet is interpreted as a photon are called reducible.

The results from both ATLAS and CMS are shown in Figure 29.14. In both cases, the plot of the distribution of the invariant mass of two photons in the relevant mass range shows a smoothly distributed background with no structure, as expected from the QCD continuum, except in one place around 125 GeV. In that region, there is a clear excess of events! These are the ones coming from the Higgs boson. The continuous background can be fitted with a smooth curve and then statistically subtracted. This leads to the two plots also shown in the figure where the Higgs boson resonance clearly appears as a peak with a Gaussian shape above a flat background.

29.10 The Higgs Boson Search in Four Lepton Final States

One of the most promising roads towards a discovery of the Higgs boson at the LHC is via its single production followed by a cascade decay into charged leptons $H^0 \to ZZ^{(*)} \to \ell_1^+ \ell_1^- \ell_2^+ \ell_2^-$, where $\ell_i = e$ or μ. The branching ratio is sizeable for any Higgs boson mass above 130 GeV. For the value of 125 GeV, it is about 2.7% (see

Table 29.5). With this percentage, one must multiply the Z^0 branching fraction to leptons. Although the total branching ratio is small, it offers a very clean and simple multi-lepton final-state signature. In fact, this process is often called a golden channel. Ultimately, the channel can provide a precision determination of the Higgs boson mass and production cross-section. An irreducible background comes from the $pp \rightarrow ZZ^{(*)}$ production (see Section 26.11). Other high Q^2 reactions, such as $t\bar{t}$ production, $Z + $ jets, etc. are negligible relatively.

The results from the ATLAS and CMS are shown in Figure 29.15. To ensure that the final state could have come from a Higgs boson decay, it was required that the four leptons were electrons or muons, two were positively charged, and two were negatively charged, and that pairs of oppositely charged leptons were the same flavor. Figure 29.15 shows the four-lepton invariant mass distributions for the selected events in the relevant mass range. There is a clear resonance peak around 90 GeV which corresponds to the production of the Z^0 boson. The Z^0 boson self-coupling leads to a decay into two off-shell Z^0 bosons from which emerge four leptons. This background is irreducible. Above 180 GeV (not shown here), one expects the background to rise, due to the opening of the production threshold of a pair of Z^0 bosons, which peaks at $\approx 2M_Z$. In between, the background is expected to be relatively flat. However, at an invariant mass of around 125 GeV, there is a clear excess of events above the expected background. Both experiments saw this very significant peak above the background in the same mass region. This is again interpreted as due to the production of the Higgs boson!

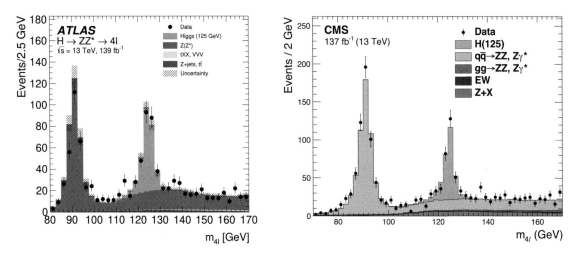

Figure 29.15 The observed and expected inclusive four-lepton invariant mass distributions for the selected Higgs boson candidates in ATLAS and CMS at $\sqrt{s} = 13$ TeV. Reprinted from G. Aad *et al.*, "Measurements of the Higgs boson inclusive and differential fiducial cross-sections in the 4ℓ decay channel at $\sqrt{s} = 13$ TeV," *Eur. Phys. J. C*, vol. 80, no. 10, p. 942, 2020, under CC BY license http://creativecommons.org/licenses/by/4.0/ and A. M. Sirunyan *et al.*, "Measurements of production cross sections of the Higgs boson in the four-lepton final state in proton–proton collisions at $\sqrt{s} = 13$ TeV," arXiv:2103.04956 (2021).

29.11   The Higgs Boson Properties

In the previous sections, we have seen the two "golden" channels which led to the discovery of the Higgs boson. Both a priori independent two-photon and four-lepton channels were consistent with its mass being around 125 GeV. This was very important as the two processes are subject to very different sources of backgrounds.

The fact that two independent channels lead to the same conclusion was a very important milestone. Both channels lead to very clean peaks in the invariant mass distributions. These can be used to fit the Higgs boson mass with high precision. As an example, the results from CMS are shown in Figure 29.16. The measurements were performed at different center-of-mass energies $\sqrt{s} = 7$, 8, and 13 TeV. The fits were performed in both $\gamma\gamma$ and 4ℓ channels and the resulting fitted Higgs boson masses are shown in the figure. They all consistently pointed to the existence of the Higgs boson with a mass in the region of 125 GeV. The consistency between measurements allowed the combination of all of them into a single most precise estimation of the Higgs boson mass with a precision at the level of 150 MeV. This is quite an impressive achievement!

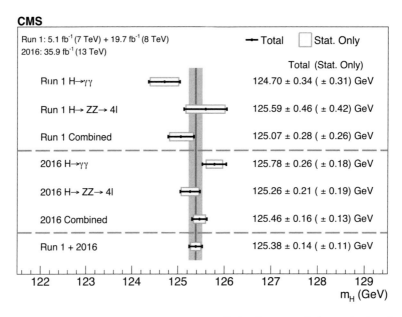

Figure 29.16 Fit of the Higgs boson mass within the SM. Reprinted from A. M. Sirunyan *et al.*, "A measurement of the Higgs boson mass in the diphoton decay channel," *Phys. Lett. B*, vol. 805, p. 135425, 2020., under CC BY license http://creativecommons.org/licenses/by/4.0/. Copyright CMS Collaboration.

Enabled by the increased statistics collected thanks to the excellent running of the LHC collider, both ATLAS and CMS were able to perform extended studies of the Higgs boson production cross-section and its properties. Further discovery channels were investigated. In particular, Table 29.5 shows the list of branching ratios of the Higgs boson that are expected. Many of these channels can be searched for. Overall, the comparison of the various channels can be used to test a fundamental property of the Higgs boson, namely that **its coupling to particles is proportional to the particle mass, a "smoking gun" for the boson being the one associated with the Higgs field!** An example of such an analysis performed by the CMS Collaboration is outlined below. Figure 29.17 shows the measured Higgs coupling as a function of the particle mass. The combination of various analyses allowed the independent determination of this coupling to muons, tau leptons, bottom quarks, to the two weak gauge bosons $W^{\pm}$ and Z^0, and even inferring the coupling to the top quark. When plotted versus the particle mass, the coupling exhibits a clear linear dependence. The points are clearly aligned along one line! The inset shows the ratio of the measured coupling to that expected within the SM. They are all consistent within errors to be equal to one. This kind of analysis strongly supports that the resonance found at 125 GeV indeed possesses the properties expected from the boson associated with the Higgs field! With the running of the LHC in the coming years, one expects to even further improve the knowledge of the Higgs boson, another truly magnificent test of the validity of the SM theory.

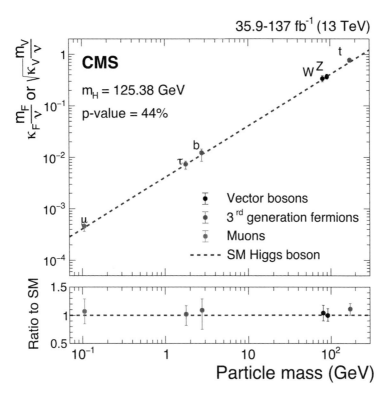

Figure 29.17 The measured Higgs boson coupling to other particles as a function of the particle mass. Reprinted from A. M. Sirunyan *et al.*, "Evidence for Higgs boson decay to a pair of muons," *JHEP*, vol. 01, p. 148, 2021, under CC BY license http://creativecommons.org/licenses/by/4.0/. Copyright CMS Collaboration.

29.12   Does it All Fit Together?

We have seen in the previous sections how the electroweak precision data from LEP, SLD, and Tevatron were used together with accurate theoretical higher-order calculations to predict parameters of the SM. A first impressive confirmation of the predictive power of global fits was described above: the discovery of the top quark at the Tevatron with a mass in agreement with the predictions from global fits! The knowledge of the top quark mass subsequently made it possible to constrain the mass of the Higgs boson (see Section 29.6).

With all the very precise experimental observables available, including also the recent direct measurements of the Higgs boson at the LHC, **the theory is over-constrained and it is worth asking if all those measurements do fit together**. Basically, we want to perform an accurate test of the internal consistency of the SM.

The input to the SM fit is summarized in Table 29.6. The mass of the Z^0 boson M_Z is the combined lineshape result from the LEP experiments (see Eq. (27.40)), and the same for its width Γ_Z (see Eq. (27.41)). The hadronic pole cross-section σ_{had}^0 is also coming from LEP, as shown in Eq. (27.44). The ratio of the Z^0 hadronic to leptonic branching ratios R_ℓ was also precisely measured at LEP (see Eq. (27.46)). The value of the lepton asymmetry A_ℓ is taken as the average of LEP ($A_\ell(P_\tau) = 0.1465 \pm 0.0033$, see Eq. (27.52)) and SLD ($A_\ell = 0.1513 \pm 0.0021$, see Eq. (27.55)) measurements. The forward–backward charge asymmetry for leptons $A_{FB}^{0,\ell}$ is taken from Eq. (27.49). A measurement of the effective Weinberg angle is taken from the inclusive jet charge method performed at LEP, whose result for $\sin^2\theta_{\text{eff}}(Q_{fb}^{had})$ is given in Eq. (27.60). Another independent measurement is the combined result from the analysis of Drell–Yan events at the Tevatron collider labeled $\sin^2\theta_{\text{eff}}^\ell(\text{Tevatron})$ (see Eq. (27.68)). The partial branching ratios into heavy quarks R_c^0 and R_b^0 are taken from the combined LEP/SLD result (see Eq. (27.58)). The forward–backward charge asymmetries for heavy quarks $A_{FB}^{0,c}$, $A_{FB}^{0,b}$ are taken from their LEP combined measurement (see Eq. (27.59)). The SLD result yields the heavy quark asymmetries A_c and A_b (see Eq. (27.63)). The $W^\pm$ boson mass and its width Γ_W are taken from the world averages (see Eq. (27.39)).

The result of the fit of the SM theory to precision experimental data is shown in the fourth column of Table 29.6. The last column of the table lists the difference between the measured and fitted values normalized to the expected error. This number is called the *pull*, and represents how well the variable predicted by the globally fitted parameters describes the given experimental value. All observables are within 2σ, as expected, of their fitted values, which gives strong support to the SM as the "correct description" of Nature. This is quite an incredible achievement that concludes several decades of theoretical developments and experimental investigations. The precision with which several observables have been measured and their overall consistency within a single theory certainly represents a point of culmination of the field. Nonetheless, as we will discuss in the next chapters, there are still open questions that will drive further progress in the field.

Parameter	Measured value	Free in fit	Fit result	Pull
M_Z [GeV]	91.1875 ± 0.0021	yes	91.1882 ± 0.0020	$+0.3$
Γ_Z [GeV]	2.4952 ± 0.0023	–	2.4947 ± 0.0014	-0.2
σ_{had}^0 [nb]	41.541 ± 0.037	–	41.484 ± 0.015	-1.5
R_ℓ	20.767 ± 0.025	–	20.742 ± 0.017	-1.0
A_ℓ (LEP+SLD)	0.1499 ± 0.0018	–	0.1470 ± 0.0005	-1.6
$\sin^2\theta_{\text{eff}}(Q_{fb}^{had})$	0.23240 ± 0.0012	–	0.23153 ± 0.00006	-0.7
$\sin^2\theta_{\text{eff}}^\ell(\text{Tevatron})$	0.23148 ± 0.00033	–	0.23153 ± 0.00006	$+0.2$
R_c^0 (LEP+SLD)	0.17210 ± 0.0030	–	0.17224 ± 0.00008	$+0.0$
R_b^0 (LEP+SLD)	0.21629 ± 0.00066	–	0.21582 ± 0.00011	-0.7
$A_{FB}^{0,\ell}$ (LEP)	0.0171 ± 0.0010	–	0.01620 ± 0.0001	-0.9
$A_{FB}^{0,c}$ (LEP)	0.0707 ± 0.0035	–	0.0736 ± 0.0003	$+0.8$
$A_{FB}^{0,b}$ (LEP)	0.0992 ± 0.0016	–	0.1030 ± 0.0003	$+2.4$
A_c (SLD)	0.670 ± 0.027	–	0.6679 ± 0.00021	-0.1
A_b (SLD)	0.923 ± 0.020	–	0.93475 ± 0.00004	$+0.6$
M_W [GeV]	80.379 ± 0.013	–	80.359 ± 0.006	-1.5
Γ_W [GeV]	2.085 ± 0.042	–	2.091 ± 0.001	$+0.1$
m_t [GeV]$^{(\triangledown)}$	172.47 ± 0.68	yes	172.83 ± 0.65	$+0.5$
m_H [GeV]	125.1 ± 0.2	yes	125.1 ± 0.2	0

$^{(\triangledown)}$Combination of experimental (0.46 GeV) and theory uncertainty (0.5 GeV).

Table 29.6 Measured values and fit results for the observables used in the global electroweak fit. The first and second columns list, respectively, the observables/parameters used in the fit and their experimental values or phenomenological estimates (see text for references). The third column indicates whether a parameter is floating in the fit. The fourth column gives the results of the fit including all experimental data. The fifth column is the *pull*, which represents how well the variable predicted by the globally fitted parameters describes the given experimental value. Table adapted from J. Haller, A. Hoecker, R. Kogler, K. Mönig, T. Peiffer, and J. Stelzer, "Update of the global electroweak fit and constraints on two-Higgs-doublet models," *Eur. Phys. J. C*, vol. 78, no. 8, p. 675, 2018, under CC BY license http://creativecommons.org/licenses/by/4.0/. Copyright the authors.

Problems

Ex 29.1 $H^0 \rightarrow f\bar{f}$ **decay width.** Compute the decay width of the Higgs boson to a pair of fermions, $H^0 \rightarrow f\bar{f}$, at tree level using FeynCalc (see Section 11.13).

Ex 29.2 **The measured width of the** H. Consider the decay of the Higgs boson into two photons, which is one of the channels that was used to discover the Higgs boson. Discuss the measured width of the Higgs. What are the experimental resolutions which enter in the measured width?

Ex 29.3 **Higgsstrahlung.** Future e^+e^- colliders might operate above the kinematical threshold $\sqrt{s} \geq m_H + M_Z$ where the Higgsstrahlung $e^+e^- \rightarrow Z^* \rightarrow ZH$ process becomes likely.

    (a) Show that the cross-section can be measured by studying the recoil mass distribution of the Z^0.

    (b) Show that the total width of the Higgs boson can be measured from the combination of two cross-section measurements.

Ex 29.4 **Muon collider.** A hypothetical muon–antimuon collider operating at a center-of-mass energy in the range of 125 GeV would offer the possibility of a direct s-channel production of the Higgs boson (sometimes called the muon–antimuon fusion into the Higgs boson).

    (a) Estimate the cross-section at the peak of the Higgs boson resonance.

    (b) According to you, what are the main technological challenges in building such a collider?

    (c) Motivate why this direct fusion measurement cannot be practically repeated with electron–positron colliders.

30   Flavor Oscillations and CP Violation

CP violation and more generally the flavour structure observed in elementary particle interactions still remains one of the unexplained mysteries in high-energy physics. In the framework of the standard model flavour mixing and CP violation is encoded in the CKM matrix for the quarks and in the PMNS matrix for the leptons. However, this is only a parametrization, in which CP violation appears through irreducible phases in these matrices and which is up to now completely consistent with the experimental facts, at least with what is found at accelerator experiments.

Thomas Mannel [1]

30.1   About Masses and Complex Mixing

The understanding of CP violation in particle physics is a key input ingredient to the cosmology of the early Universe. It is required to explain the observed dominance of matter over antimatter in the Universe (see Section 32.3). In 1967, **Sakharov**[2] proposed a set of three necessary conditions that a baryon-generating interaction must satisfy to produce matter and antimatter at different rates. They are [418]: (1) **baryon number B violation**; (2) **C symmetry and CP symmetry violation**; (3) **interactions out of thermal equilibrium**. Without CP violation an equal density of baryon and antibaryon would be predicted by the Big Bang theory. But this would be in contradiction with the observed matter-dominated Universe. Consequently, we expect CP violation during **baryogenesis**. An ideal place where CP violation can be naturally accommodated in the SM is in the weak interactions. To date CP violation has convincingly only been observed in interactions involving quarks; however, there is strong evidence that this also occurs in interactions involving leptons. While observing CP violation in both quark and lepton interactions would provide very strong support to the realization of the Sakharov conditions, actual models of baryogenesis (or alternatively **leptogenesis**) embed the existence of CP violation in a broader framework than the pure SM. In this context, the phenomenology of CP violation at low energies is considered as supporting evidence for CP violation at very high energy scales (e.g., GUT – see Section 31.5) directly relevant in the context of cosmology of the early Universe.

The masses and mixings of quarks have a common origin in the SM. The mass of the electron arises from the Yukawa interactions with the Higgs *vev* (see Section 25.6):

$$\mathcal{L}_e = -\lambda_e \left(\overline{L}\phi R + \overline{R}\phi^\dagger L \right) \tag{30.1}$$

where L was the electron-neutrino LH doublet and R the RH singlet **in the weak eigenstate basis**. This can be generalized to **several families by promoting the Yukawa couplings to a matrix**, for example for down-type quarks:

$$\mathcal{L}_d = -Y_{ij}^d \left(\overline{Q_{Li}}\phi d_{Rj} \right) + \text{h.c.} \tag{30.2}$$

---

1   Reprinted from T. Mannel, "Theory and phenomenology of CP violation," *Nucl. Phys. B Proc. Suppl.*, vol. 167, pp. 115–119, 2007.
2   Andrei Sakharov (1921–1989), eminent Soviet–Russian nuclear physicist.

where Q_{Li} are the left-handed quark doublets, and d_{Rj} are the right-handed down-type quark singlets **in the weak eigenstate basis**, and Y^d is the 3×3 Yukawa coupling matrix for the down quarks. For the up-type quarks, we need to consider a similar term:

$$\mathcal{L}_u = -Y^u_{ij} \left(\overline{Q_{Li}} \phi^C u_{Rj} \right) + \text{h.c.} \tag{30.3}$$

Q_{Li} are still the left-handed quark doublets, and u_{Rj} are the right-handed up-type quark singlets in the weak eigenstate basis.

The Yukawa matrices do not need to be diagonal in the weak basis! When the Higgs condensate acquires a *vev* and the symmetry is spontaneously broken, the quarks acquire a mass term which depends on the 3×3 matrices Y^d_{ij} and Y^u_{ij}. The mass eigenstates can then be found by diagonalizing the Y^d_{ij} and Y^u_{ij} matrices (cf. Eq. (25.95)):

$$M^d = V^d_L Y^d V^{d\dagger}_R \left(\frac{v}{\sqrt{2}} \right) = \begin{pmatrix} m_d & 0 & 0 \\ 0 & m_s & 0 \\ 0 & 0 & m_b \end{pmatrix}, \quad M^u = V^u_L Y^u V^{u\dagger}_R \left(\frac{v}{\sqrt{2}} \right) = \begin{pmatrix} m_u & 0 & 0 \\ 0 & m_c & 0 \\ 0 & 0 & m_t \end{pmatrix} \tag{30.4}$$

This is where the mixing between fermions is generated! At this stage, we note the striking observation that for the top quark, we have:

$$m_t = \lambda_t \frac{v}{\sqrt{2}} \simeq \lambda_t \frac{246 \text{ GeV}}{\sqrt{2}} = 173 \times \lambda_t \text{ GeV} \tag{30.5}$$

so with $m_t = 173.1 \pm 0.6$ GeV (see Eq. (29.13)), we have that the Yukawa coupling of the top quark is

$$\lambda_t \simeq 1.000 \pm 0.003 \tag{30.6}$$

Why is this value so incredibly close to one? **Nature is telling us something!** On the other side of the spectrum, we have:

$$m_e = \lambda_e \frac{v}{\sqrt{2}} \simeq 0.511 \text{ MeV} \implies \lambda_e \simeq 3 \times 10^{-6} \tag{30.7}$$

so $\lambda_e \ll \lambda_t \simeq 1$. For neutrinos, we only have upper bounds on the masses, so invoking the Higgs mechanism to generate their (Dirac) masses, we would get (using a conservative bound):

$$m_{\nu_e} = \lambda_{\nu_e} \frac{v}{\sqrt{2}} \lesssim 1 \text{ eV} \implies \lambda_{\nu_e} < 10^{-11} \quad ?? \tag{30.8}$$

We are generally used to thinking that the heaviness of the top quark is an oddity. However, from the above point of view, and under the assumption that couplings should be of order one, the top quark mass represents normality and the question is why are the other fundamental fermions so light? Why are the Yukawa couplings so small? In particular, in the case of neutrinos, a coupling of less than 10^{-11} is difficult to explain. **The origin of the Yukawa couplings is currently an open issue of particle physics.** It is called the **flavor problem.** In the case of neutrinos, it is not yet clear if they acquire their masses via a Dirac mass term or by some other mechanism.

Going back to the weak interaction, the result is that the charged current weak interaction couples to the physical mass quark eigenstates u_{Lj} and d_{Lk} with the current:

$$-\frac{g_W}{\sqrt{2}} (\bar{u} \ \bar{c} \ \bar{t})_L \gamma^\mu W^+_\mu V_{\text{CKM}} \begin{pmatrix} d \\ s \\ b \end{pmatrix}_L + \text{h.c.} \tag{30.9}$$

where

$$V_{\text{CKM}} \equiv V^u_L V^{d\dagger}_L = \begin{pmatrix} V_{ud} & V_{us} & V_{ub} \\ V_{cd} & V_{cs} & V_{cb} \\ V_{td} & V_{ts} & V_{tb} \end{pmatrix} \tag{30.10}$$

This is exactly the **Cabibbo–Kobayashi–Maskawa (CKM)** that we introduced in Section 23.14 as the generalization of the Cabibbo mechanism to three families.

30.2 Mixing and CP Violation in the Quark Sector

The CKM matrix can be simply defined as the matrix describing the mixing between the weak eigenstates and the mass eigenstates (see also Eq. (23.232)):

$$\begin{pmatrix} d' \\ s' \\ b' \end{pmatrix} = V_{\text{CKM}} \begin{pmatrix} d \\ s \\ b \end{pmatrix} = \begin{pmatrix} V_{ud} & V_{us} & V_{ub} \\ V_{cd} & V_{cs} & V_{cb} \\ V_{td} & V_{ts} & V_{tb} \end{pmatrix} \begin{pmatrix} d \\ s \\ b \end{pmatrix} \tag{30.11}$$

defined in such a way that the CKM vertex factor enters in the weak charged current when the down-type quark enters as a spinor (see Eq. (23.233)). For example, for V_{ud}:

$$-\frac{g_W}{\sqrt{2}} \bar{u} \gamma^\mu \left(\frac{1 - \gamma^5}{2} \right) d' \implies j^\mu = -\frac{g_W}{\sqrt{2}} V_{ud} \bar{u} \gamma^\mu \left(\frac{1 - \gamma^5}{2} \right) d \tag{30.12}$$

where d is the physical mass and d' the weak interaction eigenstate of the down quark. By taking the complex conjugate (i.e., if the down-type quark enters as an adjoint spinor) one must take the complex conjugate of the CKM element, since we should always assume that the matrix can be complex:

$$j^\mu = -\frac{g_W}{\sqrt{2}} V_{ud}^* \bar{d} \gamma^\mu \left(\frac{1 - \gamma^5}{2} \right) u \tag{30.13}$$

Hence, if the CKM matrix contains complex elements, then the weak current for a reaction and its conjugate will be different. For example, the coupling $u \to \bar{d}$ and $d \to \bar{u}$ will have different phases if $V_{ud} \neq V_{ud}^*$. In the SM, the CKM must be unitary (since there are no other families than the three known ones), and the condition $V^\dagger V = \mathbb{1}$ implies that:

$$\begin{cases} |V_{ud}|^2 + |V_{us}|^2 + |V_{ub}|^2 = 1 \\ |V_{cd}|^2 + |V_{cs}|^2 + |V_{cb}|^2 = 1 \\ |V_{td}|^2 + |V_{ts}|^2 + |V_{tb}|^2 = 1 \end{cases} \tag{30.14}$$

Since quarks do not propagate as free particles, the phenomenology of mixing appears in their bound states, i.e., the baryons and mesons. Consequently, the elements of the CKM matrix can be determined from observable quantities related to weak nuclear or weak hadronic processes. The most precise determination of $|V_{ud}|$ comes from the study of super-allowed $0^+ \to 0^+$ nuclear beta decays, which are pure vector transitions:

$$|V_{ud}| = 0.97417 \pm 0.00021 \tag{30.15}$$

Within the unitary constraint, the magnitude of the CKM elements are fitted to be (values taken from the Particle Data Group's Review [5]):

$$|V^{\text{CKM}}| = \begin{pmatrix} |V_{ud}| & |V_{us}| & |V_{ub}| \\ |V_{cd}| & |V_{cs}| & |V_{cb}| \\ |V_{td}| & |V_{ts}| & |V_{tb}| \end{pmatrix} \approx \begin{pmatrix} 0.9743 & 0.2250 & 0.0035 \\ 0.2249 & 0.9735 & 0.0411 \\ 0.0088 & 0.0403 & 0.9991 \end{pmatrix} \tag{30.16}$$

The data shows that the off-diagonal elements of the CKM matrix are relatively small. A unitary 3×3 matrix ($N = 3$) can be parameterized by $N(N-1)/2 = 3$ rotation angles and $(N-1)(N-2)/2 = 1$ complex phases:

$$V_{\text{CKM}} = \begin{pmatrix} V_{ud} & V_{us} & V_{ub} \\ V_{cd} & V_{cs} & V_{cb} \\ V_{td} & V_{ts} & V_{tb} \end{pmatrix} = \begin{pmatrix} 1 & 0 & 0 \\ 0 & c_{23} & s_{23} \\ 0 & -s_{23} & c_{23} \end{pmatrix} \times \begin{pmatrix} c_{13} & 0 & s_{13}e^{-i\delta} \\ 0 & 1 & 0 \\ -s_{13}e^{-i\delta} & 0 & c_{13} \end{pmatrix} \times \begin{pmatrix} c_{12} & s_{12} \\ -s_{12} & c_{12} & 0 \\ 0 & 0 & 1 \end{pmatrix} \tag{30.17}$$

where $s_{ij} = \sin\phi_{ij}$ and $c_{ij} = \cos\phi_{ij}$. Comparing this form to the data, one finds that the rotation angles between quark mass and weak eigenstates are rather small:

$$\phi_{12} = (13.04 \pm 0.05)°, \quad \phi_{23} = (2.38 \pm 0.06)°, \phi_{13} = (0.201 \pm 0.011)°, \delta = 1.20 \pm 0.08 \text{ rad} \tag{30.18}$$

We note that ϕ_{12} corresponds to the original angle θ_C introduced by Cabibbo in his original work. The weak interactions of quarks of different generations are suppressed relative to those of the same generation due to the smallness of the off-diagonal terms. In 1983 **Wolfenstein**[3] used this fact to express an approximation of the CKM matrix in terms of four real parameters λ, A, ρ, and η:

$$V_{\text{CKM}} = \begin{pmatrix} V_{ud} & V_{us} & V_{ub} \\ V_{cd} & V_{cs} & V_{cb} \\ V_{td} & V_{ts} & V_{tb} \end{pmatrix} = \begin{pmatrix} 1 - \lambda^2/2 & \lambda & A\lambda^3(\rho - i\eta) \\ -\lambda & 1 - \lambda^2/2 & A\lambda^2 \\ A\lambda^3(1 - \rho - i\eta) & -A\lambda^2 & 1 \end{pmatrix} + \mathcal{O}(\lambda^4) \tag{30.19}$$

The CKM matrix can always be redefined up to an overall complex phase. The relative phases have, however, physically observable consequences. In the Wolfenstein parameterization, this relative phase is defined by the η parameter, which appears in the two elements V_{td} and V_{ub}. If $\eta \neq 0$, then the CKM matrix contains a non-trivial complex phase.

Using the values of the CKM matrix, the best determination of the Wolfenstein parameters is:

$$\lambda = 0.2257^{+0.0009}_{-0.0010}, \quad A = 0.814^{+0.021}_{-0.022}, \quad \rho = 0.135^{+0.031}_{-0.016}, \text{ and } \eta = 0.349^{+0.015}_{-0.017} \tag{30.20}$$

where $\lambda \simeq \sin\phi_{12} = \sin\theta_C$, the Cabibbo angle.

30.3   Mixing of the Neutral Kaon System

$K^0(d\bar{s})$ and $\bar{K}^0(s\bar{d})$ are the lightest strange mesons. We will call these the strong eigenstates. They are antiparticles of each other, so they can be transformed into each other via charge conjugation (and hence they possess opposite strangeness). They can only decay via weak interactions, since strangeness is conserved in other types of interactions. Physically one observes two neutral kaons with very similar mass but very different lifetimes. These are called the K "short" K_S^0 and the K "long" K_L^0 (see Eq. (15.63)):

$$\tau(K_S^0) = (8.954 \pm 0.004) \times 10^{-11} \text{ s} \quad (c\tau \approx 2.68 \text{ cm}), \quad \tau(K_L^0) = (5.116 \pm 0.021) \times 10^{-8} \text{ s} \quad (c\tau \approx 1530 \text{ cm}) \tag{30.21}$$

How can this observation be explained in the SM? The solution is understood in terms of the weak box diagrams shown in Figure 30.1, which introduce the **mixing of the neutral kaon system**. If there were no weak interactions, the $K^0(d\bar{s})$ and $\bar{K}^0(s\bar{d})$ would be stable and equal in mass. The weak box diagrams allow transitions between the strong $K^0(d\bar{s})$ and $\bar{K}^0(s\bar{d})$ eigenstates, and consequently, the neutral kaon system propagates as a linear combination of $K^0(d\bar{s})$ and $\bar{K}^0(s\bar{d})$ strong and weak eigenstates! These mixtures correspond to the physical K_S^0 and K_L^0.

The spin and parity of the strong eigenstates $K^0(d\bar{s})$ and $\bar{K}^0(s\bar{d})$ is $J^P = 0^-$. Therefore, the parity operation will yield:

$$P|K^0\rangle = -|K^0\rangle \quad \text{and} \quad P|\bar{K}^0\rangle = -|\bar{K}^0\rangle \tag{30.22}$$

Considering the charge conjugation operation C, we note that the $K^0(d\bar{s})$ and $\bar{K}^0(s\bar{d})$ have opposite particle–antiparticle content so we can write one as a function of the other, allowing for an overall phase in the C operation:

$$C|K^0\rangle = \eta_C|\bar{K}^0\rangle \quad \text{and} \quad C|\bar{K}^0\rangle = \eta_C^*|K^0\rangle \tag{30.23}$$

---

3   Lincoln Wolfenstein (1923–2015), American particle physicist.

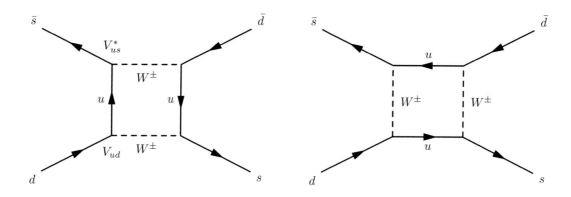

Figure 30.1 The box diagrams generating $K^0 - \bar{K}^0$ mixing.

with the phase η_C which is conventionally fixed to -1, so finally we have simply:

$$C\,|K^0\rangle = -\,|\bar{K}^0\rangle \quad \text{and} \quad C\,|\bar{K}^0\rangle = -\,|K^0\rangle \tag{30.24}$$

With this assignment, we can now consider the CP operation which turns out to be just:

$$CP\,|K^0\rangle = +\,|\bar{K}^0\rangle \quad \text{and} \quad CP\,|\bar{K}^0\rangle = +\,|K^0\rangle \tag{30.25}$$

In order to find CP eigenstates, we note that the following linear combinations fulfill exactly that:

$$|K_1\rangle = \frac{1}{\sqrt{2}}\left(|K^0\rangle + |\bar{K}^0\rangle\right) \quad \text{and} \quad |K_2\rangle = \frac{1}{\sqrt{2}}\left(|K^0\rangle - |\bar{K}^0\rangle\right) \tag{30.26}$$

since

$$CP\,|K_1\rangle = +\,|K_1\rangle \quad \text{and} \quad CP\,|K_2\rangle = -\,|K_2\rangle \tag{30.27}$$

Hence, the state $|K_1\rangle$ is a CP-even eigenstate and the state $|K_2\rangle$ is a CP-odd eigenstate.

Let us now analyze the decays of kaons. Neutral kaons can decay semi-leptonically or hadronically, typically in two or three pion final states (see Eq. (15.63)). We first consider the decays:

$$K^0 \to \pi^0\pi^0 \quad \text{and} \quad K^0 \to \pi^+\pi^- \tag{30.28}$$

The final-state pions have $J^P = 0^-$, like the initial-state neutral kaon. In order to conserve angular momentum, the relative angular momentum L between the final-state pions must vanish ($L = 0$). Hence, the parity of the final state is:

$$P(|\pi^0\pi^0\rangle) = (-1)(-1)(-1)^L = +1 \quad \text{and} \quad P(|\pi^+\pi^-\rangle) = +1 \tag{30.29}$$

where we have used the intrinsic parity of the pions $P(\pi) = -1$. Similarly, we note that charge conjugation does not change the final states:

$$C(|\pi^0\pi^0\rangle) = |\pi^0\pi^0\rangle \quad \text{and} \quad C(|\pi^+\pi^-\rangle) = |\pi^+\pi^-\rangle \tag{30.30}$$

Hence

$$CP(|\pi^0\pi^0\rangle) = (+1)\,|\pi^0\pi^0\rangle \quad \text{and} \quad CP(|\pi^+\pi^-\rangle) = (+1)\,|\pi^+\pi^-\rangle \tag{30.31}$$

So finally, the decay of a neutral kaon into two pions produces a CP-even eigenstate. Similar arguments can be repeated for the case of the neutral kaon decaying into three pions. This time we need to take into account two relative angular momenta: e.g. (1) the first between two pions, (2) the second between the third pion and the first two pions. In practice, it can be shown that phase space enforces these to be zero, i.e., $L_1 = L_2 = 0$. For this reason, one finds that the decay of a neutral kaon into three pions produces a CP-odd eigenstate:

$$CP(|\pi^0\pi^0\pi^0\rangle) = (-1)|\pi^0\pi^0\pi^0\rangle \quad \text{and} \quad CP(|\pi^+\pi^-\pi^0\rangle) = (-1)|\pi^+\pi^-\pi^0\rangle \tag{30.32}$$

From a kinematical point of view, we expect that the decay into three pions is more suppressed than the decay into two pions because of phase space. Hence, it is tempting to identify the K_S^0 with the $|K_1\rangle$ eigenstate and the K_L^0 with the $|K_2\rangle$ eigenstate:

$$K_S^0 \simeq K_1 = \frac{1}{\sqrt{2}}\left(|K^0\rangle + |\bar{K}^0\rangle\right) \quad \text{and} \quad K_L^0 \simeq K_2 = \frac{1}{\sqrt{2}}\left(|K^0\rangle - |\bar{K}^0\rangle\right) \tag{30.33}$$

However, for this to be *exactly* true, we must make the assumption that CP is conserved in the weak decays of kaons. But this turns out to be *not* the case!

30.4 *CP* Violation in the Neutral Kaon System

Let us consider the production of neutral kaons. Because $K^0(d\bar{s})$ and $\bar{K}^0(s\bar{d})$ have well-defined strangeness and strangeness is conserved in hadronic collisions, these are the states that are directly produced. For example, a common reaction is via proton–antiproton annihilation at low energy:

$$p\bar{p} \to K^-K^0\pi^+ \tag{30.34}$$

In this reaction, we assume that the neutral kaon is created at a time defined as $t = 0$ as the strong eigenstate:

$$\Psi(t = 0) \equiv |K^0\rangle = \frac{1}{\sqrt{2}}\left(|K_1\rangle + |K_2\rangle\right) \simeq \frac{1}{\sqrt{2}}\left(|K_S^0\rangle + |K_L^0\rangle\right) \tag{30.35}$$

Since the $|K_S^0\rangle$ and $|K_L^0\rangle$ are the physical states, their quantum-mechanical time evolution can be written in their rest frame, as:

$$\begin{cases} |K_S^0\rangle(t) = |K_S^0\rangle\, e^{-iEt}e^{-\Gamma_S t/2} = |K_S^0\rangle\, e^{-im_S t}e^{-\Gamma_S t/2} \\ |K_L^0\rangle(t) = |K_L^0\rangle\, e^{-iEt}e^{-\Gamma_L t/2} = |K_L^0\rangle\, e^{-im_L t}e^{-\Gamma_L t/2} \end{cases} \tag{30.36}$$

where m_S and m_L (resp. Γ_S and Γ_L) are the rest masses (resp. decay widths) of the $|K_S^0\rangle$ and $|K_L^0\rangle$ states. The second exponential is the correct term to describe the exponential decrease of the state normalization due to the decay of the particles with the lifetime $\tau = 1/\Gamma$, for example for $|K_S^0\rangle$:

$$\langle K_S^0(t)|K_S^0(t)\rangle = \langle K_S^0|K_S^0\rangle\, e^{-\Gamma_S t/2}e^{-\Gamma_S t/2} = e^{-\Gamma_S t} = e^{-t/\tau_s} \tag{30.37}$$

and equivalently for $\langle K_L^0(t)|K_L^0(t)\rangle$. Consequently, the quantum-mechanical time evolution of the state of the neutral kaon can be expressed as:

$$\begin{aligned} \Psi(t) &\simeq \frac{1}{\sqrt{2}}\left(|K_S^0\rangle\, e^{-(im_S + \Gamma_S/2)t} + |K_L^0\rangle\, e^{-(im_L + \Gamma_L/2)t}\right) \\ &= \frac{1}{\sqrt{2}}\left(|K_S^0\rangle\, e^{-i\lambda_S t} + |K_L^0\rangle\, e^{-i\lambda_L t}\right) \end{aligned} \tag{30.38}$$

where

$$\lambda_{S,L} \equiv m_{S,L} - i\Gamma_{S,L}/2 \tag{30.39}$$

If CP is conserved in weak decays (keep in mind that in Nature it is only slightly violated as we will see later), then we have seen that $K_S^0 \simeq K_1$ and $K_L^0 \simeq K_2$, and therefore the initial K^0 state will evolve as a linear combination of a short-lived component rapidly decaying into two pions (CP-even) and a long-lived component slowing decaying into three pions (CP-odd). The fraction $F(t)$ of K^0 particles after a time t is given by:

$$
\begin{aligned}
F(t) &= \left| \langle K^0 | \Psi(t) \rangle \right|^2 = \left| \frac{1}{\sqrt{2}} \left(\langle K_S^0 | + \langle K_L^0 | \right) \frac{1}{\sqrt{2}} \left(| K_S^0 \rangle e^{-i\lambda_S t} + | K_L^0 \rangle e^{-i\lambda_L t} \right) \right|^2 \\
&= \left| \frac{1}{2} \left(\langle K_S^0 | K_S^0 \rangle e^{-i\lambda_S t} + \langle K_L^0 | K_L^0 \rangle e^{-i\lambda_L t} \right) \right|^2 \\
&= \frac{1}{4} \left(e^{-\Gamma_S t} + e^{-\Gamma_L t} + 2\cos(\Delta m t) e^{-\bar{\Gamma} t} \right)
\end{aligned} \tag{30.40}
$$

where we used $|a + b|^2 = |a|^2 + |b|^2 + 2\mathcal{R}e(ab^*)$ and defined:

$$
\bar{\Gamma} \equiv (\Gamma_S + \Gamma_L)/2 \quad \text{and} \quad \Delta m \equiv |m_S - m_L| \tag{30.41}
$$

Similarly, the fraction $\bar{F}(t)$ of $\bar{K}^0$ particles after a time t is given by:

$$
\bar{F}(t) = \left| \langle \bar{K}^0 | \Psi(t) \rangle \right|^2 = \frac{1}{4} \left(e^{-\Gamma_S t} + e^{-\Gamma_L t} - 2\cos(\Delta m t) e^{-\bar{\Gamma} t} \right) \tag{30.42}
$$

where we used:

$$
|\bar{K}^0\rangle = \frac{1}{\sqrt{2}} \left(|K_1\rangle - |K_2\rangle \right) \simeq \frac{1}{\sqrt{2}} \left(|K_S^0\rangle - |K_L^0\rangle \right) \tag{30.43}
$$

The predicted fractions of K^0 and $\bar{K}^0$ as a function of the time t of an initially pure K^0 state are shown in Figure 30.2. The assumed values for the plot are $\Gamma_S = 1.12 \times 10^{10}$ s^{-1}, $\Gamma_L = 1.93 \times 10^7$ s^{-1}, and $\Delta m = 0.535 \times 10^{10}$ s^{-1}, which are close to the measured values. It can be seen that there is an oscillatory behavior, with a frequency proportional to the mass difference Δm, which is multiplied by an exponential factor from the K_S^0 decays. **After several K_S^0 lifetimes, the K^0 and $\bar{K}^0$ components become equal and there is no more oscillatory behavior. This corresponds to the situation where only K_L^0 particles are left because all of the K_S^0 have decayed.**

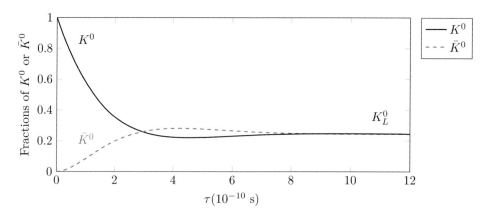

Figure 30.2 Predicted fractions of K^0 and $\bar{K}^0$ as a function of the time t of an initially pure K^0 state, for $\Gamma_S = 1.12 \times 10^{10}$ s^{-1}, $\Gamma_L = 1.93 \times 10^7$ s^{-1}, and $\Delta m = 0.535 \times 10^{10}$ s^{-1}.

If CP was conserved in weak interactions, then after several K_S^0 lifetimes, we would only expect to observe the $K_L^0 \simeq K_2$, i.e., the CP-odd eigenstate. Therefore, one should only observe decays into three pions.

In 1964 **Cronin**,[4] **Fitch**,[5] **Christenson, and Turlay** provided clear evidence from kaon decay that CP symmetry is broken. They observed 45 ± 10 events consistent with $K_L^0 \to \pi^+\pi^-$ out of a total of 22,700 K_L^0 decays at large distance from the production point [419]. These correspond to the branching ratio (see Eq. (15.63)):

$$Br(K_L^0 \to \pi^+\pi^-) = (1.967 \pm 0.010) \times 10^{-3} \tag{30.44}$$

Although a rare process, it provided clear evidence for the violation of the CP symmetry in weak decays! Given that the observed level of CP violation is small, we can rewrite the K_S^0 and K_L^0 as mixtures of CP-even and CP-odd eigenstates (cf. Eq. (30.33)):

$$|K_S^0\rangle \equiv \frac{1}{\sqrt{1+|\epsilon|^2}}\left(|K_1\rangle + \epsilon|K_2\rangle\right) \quad \text{and} \quad |K_L^0\rangle \equiv \frac{1}{\sqrt{1+|\epsilon|^2}}\left(\epsilon|K_1\rangle + |K_2\rangle\right) \tag{30.45}$$

For $\epsilon = 0$, we recover $K_S^0 = |K_1\rangle$ and $K_L^0 = |K_2\rangle$. For $\epsilon \neq 0$, we explain the observed rare process $K_L^0 \to \pi\pi$ from the fact that it contains a fraction of K_1:

$$|K_L^0\rangle = \frac{1}{\sqrt{1+|\epsilon|^2}}\left(\epsilon|K_1\rangle \quad + \quad |K_2\rangle\right) \tag{30.46}$$
$$\quad\quad\quad\quad\quad\quad\quad \hookrightarrow \pi\pi \quad\quad\quad \hookrightarrow \pi\pi\pi$$

So, the relative decay rates to two and three pions can determine the parameter ϵ. In this case, CP is conserved in weak decays but the physical states are mixtures of CP-even and CP-odd states. Additionally, there is also the possibility of **direct CP violation** in the weak decay of a CP eigenstate. This would mean that the K_2 state directly decayed into two and three pions. This can be written as:

$$|K_L^0\rangle = |K_2\rangle \tag{30.47}$$
$$\quad\quad\quad\quad \hookrightarrow \pi\pi$$
$$\quad\quad\quad\quad \hookrightarrow \pi\pi\pi$$

The amount of direct CP violation is commonly parameterized by ϵ', which is defined as:

$$\epsilon' \equiv \frac{\Gamma(K_2 \to \pi\pi)}{\Gamma(K_2 \to \pi\pi\pi)} \tag{30.48}$$

The results of the CERN NA48 experiment and those of the KTeV experiment at Fermilab demonstrate that direct CP violation is a very small effect (values taken from the Particle Data Group's Review [5]):

$$\mathcal{Re}\left(\frac{\epsilon'}{\epsilon}\right) = (1.66 \pm 0.23) \times 10^{-3} \tag{30.49}$$

Direct CP violation is expected in the SM. However, the numerical value cannot be reliably predicted because of theoretical uncertainties. Since ϵ' is small compared to an already small ϵ, the main contribution to CP violation in the neutral kaon system is from the $K^0 - \bar{K}^0$ mixing.

The physical states K_S^0 and K_L^0 of the neutral kaon defined in Eq. (30.45) can also be defined in terms of the flavor eigenstates as:

$$|K_S^0\rangle \equiv \frac{\left((1+\epsilon)|K^0\rangle + (1-\epsilon)|\bar{K}^0\rangle\right)}{\sqrt{2(1+|\epsilon|^2)}} \quad \text{and} \quad |K_L^0\rangle \equiv \frac{\left((1+\epsilon)|K^0\rangle - (1-\epsilon)|\bar{K}^0\rangle\right)}{\sqrt{2(1+|\epsilon|^2)}} \tag{30.50}$$

---

4   James Watson Cronin (1931–2016), American particle physicist.
5   Val Logsdon Fitch (1923–2015), American nuclear physicist.

Let us study again the experimental reaction $p\bar{p} \to K^- K^0 \pi^+$ (Eq. (30.34)) and its accompanying reaction (note the difference in the charges of the associated pion and kaon, which is enforced by charge and strangeness conservation in strong interactions):

$$\begin{cases} p\bar{p} \to K^- \pi^+ K^0 \\ p\bar{p} \to K^+ \pi^- \bar{K}^0 \end{cases} \tag{30.51}$$

If we neglect direct CP violation, then the two pion decays arise only from the CP-even K_1 component. Then, the branching ratio into two pions is proportional to:

$$\Gamma(K^0 \to \pi\pi) \propto |\langle K_1 | \Psi(t) \rangle|^2 \tag{30.52}$$

where the state $|\Psi(t)\rangle$ is given by Eq. (30.38):

$$\begin{aligned} \Psi(t) &= \frac{1}{\sqrt{2}} \left(|K_S^0\rangle e^{-i\lambda_S t} + |K_L^0\rangle e^{-i\lambda_L t} \right) \\ &= \frac{1}{\sqrt{2}} \frac{1}{\sqrt{1+|\epsilon|^2}} \left((|K_1\rangle + \epsilon |K_2\rangle) e^{-i\lambda_S t} + (\epsilon |K_1\rangle + |K_2\rangle) e^{-i\lambda_L t} \right) \end{aligned} \tag{30.53}$$

where we made use of Eq. (30.45). Hence:

$$\Gamma(K^0 \to \pi\pi) \propto \frac{1}{2} \frac{1}{1+|\epsilon|^2} \left| e^{-i\lambda_S t} + \epsilon e^{-i\lambda_L t} \right|^2 \tag{30.54}$$

The second term can be simplified to get:

$$\left| e^{-i\lambda_S t} + \epsilon e^{-i\lambda_L t} \right|^2 = e^{-\Gamma_S t} + |\epsilon|^2 e^{-\Gamma_L t} + 2|\epsilon| e^{-\bar{\Gamma} t} \cos(\Delta m t - \phi) \tag{30.55}$$

where $\epsilon = |\epsilon| e^{i\phi}$. The first term is the contribution from the K_S^0 decay, the second term comes from the K_L^0 and is small since $|\epsilon| \ll 1$, and the last term is the quantum interference between the two states K_S^0 and K_L^0! By repeating the same calculation for the initial state formed of $\bar{K}^0$, one gets a similar result but with the sign of the interference term swapped:

$$\begin{cases} \Gamma(K^0 \to \pi\pi) \propto e^{-\Gamma_S t} + |\epsilon|^2 e^{-\Gamma_L t} + 2|\epsilon| e^{-\bar{\Gamma} t} \cos(\Delta m t - \phi) \\ \Gamma(\bar{K}^0 \to \pi\pi) \propto e^{-\Gamma_S t} + |\epsilon|^2 e^{-\Gamma_L t} - 2|\epsilon| e^{-\bar{\Gamma} t} \cos(\Delta m t - \phi) \end{cases} \tag{30.56}$$

The above expressions are similar for $t \ll \tau_S$ or $t \gg \tau_L$, however, for intermediate times where the interference terms are important, introduce a significant difference of behavior between K^0 and $\bar{K}^0$. This difference of behavior between particle and antiparticle is a direct evidence for CP violation.

The **CPLEAR** experiment ran at CERN during the years 1990–1996. It used a low-energy 200 MeV antiproton beam impinging on a high-pressure H_2 target. See Figure 30.3. The cylindrical detector was placed inside a solenoid of 1 m radius and 3.6 m length, which provided a magnetic field of 0.44 T. The charged tracking system consisted of two proportional chambers, six drift chambers, and two layers of streamer tubes. The detector also had a microvertex chamber. Fast kaon identification was provided by a threshold Cerenkov counter sandwiched between two scintillators. An electromagnetic calorimeter made of 18 layers of Pb converters and streamer tubes was used for photon detection and electron identification. Consequently, the detector could reconstruct the sign and identify the final-state particles and thereby the strangeness of the neutral kaon at its creation (i.e., the formation of a K^0 or a $\bar{K}^0$ state at $t = 0$) could be tagged on an event-by-event basis. An example of event $p\bar{p} \to K^- \pi^+ K^0$ where the K^0 decays as $\bar{K}^0 \to \pi^+ e^- \bar{\nu}_e$ is shown in the figure at the right. Four charged tracks are visible. Two come from the vertex and are identified as K^- and π^+. The neutral kaon decay is visible as a displaced vertex (corresponding to the decay point due to its long lifetime), from which emerge an electron and a positive pion.

Due to the small branching ratio of the desired annihilation channels, CPLEAR ran at high beam intensities ($\approx 10^6 \bar{p}/\text{s}$), in order to collect very high statistics of about 7.0×10^7 events for the decay analysis [420]. The

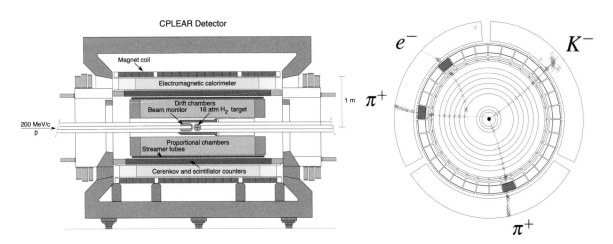

Figure 30.3 (left) Schematic view of the detector at the CERN CPLEAR experiment. Reprinted with permission from Springer Nature: A. Apostolakis *et al.*, "A detailed description of the analysis of the decay of neutral kaons to $\pi^+\pi^-$ in the CPLEAR experiment," *Eur. Phys. J.*, vol. C18, pp. 41–55, 2000. (right) An example of event $p\bar{p} \to K^-\pi^+K^0$ where the K^0 decays as $\bar{K}^0 \to \pi^+e^-\bar{\nu}_e$.

CPLEAR measurement of the decay time into two pions of the tagged K^0 and $\bar{K}^0$ events is shown in Figure 30.4. The decay rates for K^0 and $\bar{K}^0$ are shown separately and clearly show a difference in the interference region! This experimental result can be expressed with the asymmetry A^{+-} as a function of the decay proper time:

$$A^{+-} \equiv \frac{\Gamma(\bar{K}^0 \to \pi^+\pi^-) - \Gamma(K^0 \to \pi^+\pi^-)}{\Gamma(\bar{K}^0 \to \pi^+\pi^-) + \Gamma(K^0 \to \pi^+\pi^-)} \tag{30.57}$$

The asymmetry can be shown to depend upon the parameters as [420]:

$$A^{+-}(t) \simeq -2\frac{|\epsilon|e^{-\bar{\Gamma}t}\cos(\Delta mt - \phi)}{e^{-\Gamma_S t} + |\epsilon|^2 e^{-\Gamma_L t}} = -2\frac{|\epsilon|e^{(\Gamma_S - \Gamma_L)t/2}\cos(\Delta mt - \phi)}{1 + |\epsilon|^2 e^{(\Gamma_S - \Gamma_L)t}} \tag{30.58}$$

from which CPLEAR extracted the values:

$$|\epsilon| = (2.264 \pm 0.023(\text{stat.}) \pm 0.026(\text{syst.})) \times 10^{-3} \quad \text{and} \quad \phi = 43.19° \pm 0.53°(\text{stat.}) \pm 0.28°(\text{syst.}) \tag{30.59}$$

The non-zero value of $|\epsilon|$ represented clear evidence for CP violation in the weak decay of the neutral kaon system.

30.5 CP **Violation and the CKM Matrix**

In Figure 30.1 we have seen the box diagrams responsible in the SM for the mixing between the K^0 and $\bar{K}^0$ states. There are two sets of diagrams (as shown in the figure) and for each set one can consider three up-type quarks (u, c, and t) for each fermion propagator, hence, in total there are nine contributing diagrams. We note that at each vertex there is a corresponding CKM matrix element, such that, if we label q and q' the two quarks in each fermion propagator, the matrix element corresponding to one such diagram will be proportional to:

$$\mathcal{M}_{qq'}(K^0 \to \bar{K}^0) \propto V_{qs}^* V_{qd} V_{q's} V_{q'd}^* \tag{30.60}$$

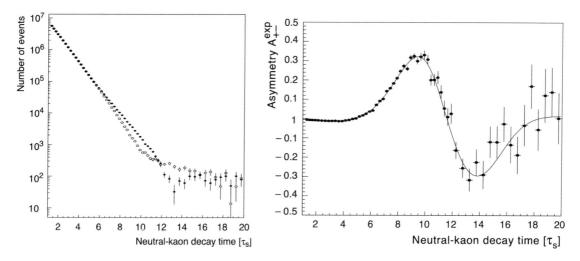

Figure 30.4 (left) The measured decay rates for K^0 (open circles) and $\bar{K}^0$ (filled circles) by the CPLEAR experiment at CERN. (right) The measured decay-rate asymmetry. Reprinted with permission from Springer Nature: A. Apostolakis *et al.*, "A detailed description of the analysis of the decay of neutral kaons to pi+ pi- in the CPLEAR experiment," *Eur. Phys. J.*, vol. C18, pp. 41–55, 2000.

For the time-reversed transition, we have:

$$\mathcal{M}_{qq'}(\bar{K}^0 \to K^0) = \mathcal{M}^*{}_{qq'}(K^0 \to \bar{K}^0) \tag{30.61}$$

Hence, the difference of amplitude is:

$$\mathcal{M}_{qq'} - \mathcal{M}^*{}_{qq'} = 2\mathcal{I}m\mathcal{M}_{qq'} \tag{30.62}$$

So, the rates can only be different if the CKM matrix has imaginary components. The CP **violation in the quark sector is related to the non-trivial imaginary phase of the CKM matrix**.

The elements of the CKM matrix are fundamental parameters of the SM, however, there is no theoretical prediction for them, and therefore their precise experimental measurement is important. The oscillations of neutral mesons are not confined to kaons. They have also been observed in the heavier charm and bottom neutral systems, namely $D^0 \leftrightarrow \bar{D}^0$, $B^0 \leftrightarrow \bar{B}^0$, and $B_s^0 \leftrightarrow \bar{B}_s^0$. The experimental study of these systems has provided crucial information on the CKM matrix as well as CP violation. In particular, the results from the studies by BaBar and Belle experiments (see Section 20.10) have been very important. The mathematical treatment of the oscillations of the heavy neutral systems follows that developed for the neutral kaons and will not be repeated here.

The CKM matrix must be unitary. Quarks can only transform into each other at a weak vertex. An important goal is to combine several experimental measurements to over-constrain the CKM elements to check if unitarity holds. Unitarity imposes the following relations:

$$\sum_i V_{ij}V_{ik}^* = \delta_{jk} \quad \text{and} \quad \sum_j V_{ij}V_{kj}^* = \delta_{ik} \tag{30.63}$$

We therefore obtain six sums which vanish. These latter can be represented as triangles in a complex plane. For example, one of these triangles can be expressed as:

$$V_{ud}V_{ub}^* + V_{cd}V_{cb}^* + V_{td}V_{tb}^* = 0 \tag{30.64}$$

In the Wolfenstein parameterization, Eq. (30.19), only $V_{ub} = A\lambda^3(\rho - i\eta)$ and $V_{td} = A\lambda^3(1 - \rho - i\eta)$ are complex, while the other elements are real and only dependent on A and λ. It is therefore customary to divide the unitarity relation by $V_{cd}V_{cb}^*$ to obtain the following complex expression:

$$\underbrace{\frac{V_{ud}}{V_{cd}V_{cb}^*}V_{ub}^*}_{\vec{a}} + \underbrace{(1+0i)}_{\vec{b}} + \underbrace{\frac{V_{tb}^*}{V_{cd}V_{cb}^*}V_{td}}_{\vec{c}} = 0 \qquad (30.65)$$

This equation can be interpreted as a vectorial equation in the (ρ, η) plane with the three vectors $\vec{a}$, $\vec{b}$, $\vec{c}$ forming the closed triangle shown in Figure 30.5. As can be seen from the figure, the angles of the unitarity

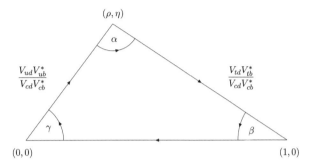

Figure 30.5 Sketch of the CKM unitarity triangle.

triangle are defined as follows:

$$\alpha \equiv \arg\left(-\frac{V_{td}V_{tb}^*}{V_{ud}V_{ub}^*}\right), \quad \beta \equiv \arg\left(-\frac{V_{cd}V_{cb}^*}{V_{td}V_{tb}^*}\right), \quad \text{and} \quad \gamma \equiv \arg\left(-\frac{V_{ud}V_{ub}^*}{V_{cd}V_{cb}^*}\right) \qquad (30.66)$$

The description of all the measurements performed to determine the elements of the CKM goes beyond the scope of this book. Here, we show the constraints from the experimental measurements in Figure 30.6. The constraints from the measurements are shown as shaded areas. One can recognize regions related to a class of direct measurements of the angles indicated as α, $\sin 2\beta$, and γ. They are combined with the other measurements labeled $|V_{ub}|$, Δm_d, Δm_s, and ϵ_K. Overall, all the experimental constraints are consistent with a common point in the (ρ, η) plane, as indicated by the small shaded ellipse at the apex of the triangle. Thus, the unitary relation can be said to be confirmed within the precision of the ellipse. This is quite an achievement! These results provide an important test of the SM predictions, since any deviation from unitarity could indicate new sources of CP violation beyond what is described by the SM. Since this is presently not the case, we can state that all the experimental measurements support the quark flavor phenomenology described by the SM, consistent with a CKM unitary matrix and where the observed CP violation is fully described by a single phase!

30.6   Neutrino Flavor Oscillations

Neutrino oscillations were first discussed by **Pontecorvo** [421, 422] and independently by **Maki**, **Nakagawa**, and **Sagata** [423] as a consequence of the lepton mixing (see Section 23.14). We assume the relation between the weak eigenstates ν_α ($\alpha = e, \mu, \tau$) and the mass eigenstates is given by the **Pontecorvo–Maki–Nakagawa–Sakata** **(PMNS)** mixing matrix (see Eq. (23.235)). In vacuum the time evolution of a neutrino mass eigenstate ν_i,

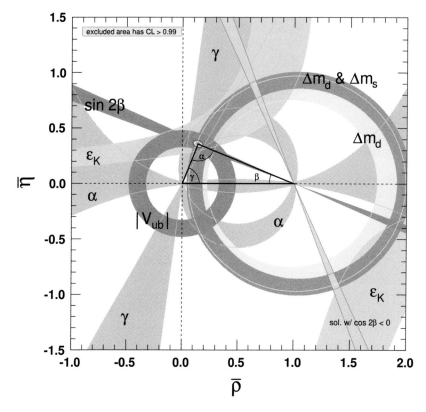

Figure 30.6 Experimental constraints on the CKM unitarity triangle. Reprinted with permission from P. A. Zyla *et al.* (Particle Data Group), *Prog. Theor. Exp. Phys.* **2020**, 083C01 (2020).

which is indeed a stationary solution of the free Hamiltonian, is just given by the phase factor:

$$e^{-iE_i t} \tag{30.67}$$

where E_i is the energy of the eigenstate. Neutrinos, on the other hand, are produced as eigenstates of the weak interaction. For example, let us assume that a neutrino is produced in a specific weak eigenstate at time $t = 0$:

$$|\nu(t=0)\rangle = |\nu_\alpha\rangle \qquad \alpha = e, \mu, \tau \tag{30.68}$$

We can compute the time evolution of this state by considering its linear combination of mass eigenstates ν_j $(j = 1, 2, 3)$:

$$|\nu(t=0)\rangle = |\nu_\alpha\rangle = \sum_j U_{\alpha j}^{PMNS} |\nu_j\rangle = \sum_j U_{\alpha j} |\nu_j\rangle \tag{30.69}$$

where in the last term we dropped the PMNS label for simplicity, setting $U \equiv U^{\mathrm{PMNS}}$. Since each mass eigenstate ν_j acquires a phase $e^{-iE_j t}$, the state's time evolution can be expressed as:

$$|\nu(t)\rangle = \sum_j U_{\alpha j} e^{-iE_j t} |\nu_j\rangle \tag{30.70}$$

We immediately notice that states with different energies will acquire different phases. The **probability** P_α that a state is measured with flavor α at a given time t is given following the rules of quantum mechanics by:

$$P_\alpha = P\left(\nu_\alpha \to \nu_\alpha\right) \equiv |\langle \nu_\alpha | \nu(t) \rangle|^2 \tag{30.71}$$

The formula $P\left(\nu_\alpha \to \nu_\alpha\right)$ expresses the probability that a neutrino produced at $t = 0$ in a weak eigenstate $|\nu_\alpha\rangle$ is detected with flavor α at time t.

It is useful to introduce the following vector notation in the basis of the weak eigenstates:

$$\vec{\nu} \equiv \begin{pmatrix} \nu_e \\ \nu_\mu \\ \nu_\tau \end{pmatrix} \tag{30.72}$$

The time evolution of the state vector with momentum p can then be readily written as:

$$i\frac{d\vec{\nu}}{dt} = U H U^\dagger \, \vec{\nu}, \qquad \text{with} \qquad H \equiv \begin{pmatrix} E_1 & 0 & 0 \\ 0 & E_2 & 0 \\ 0 & 0 & E_3 \end{pmatrix} \tag{30.73}$$

where H represents the (diagonal) Hamiltonian in the basis of the mass eigenstates, and where given the chosen momentum p the energies are simply:

$$E_i = \sqrt{p^2 + m_i^2} \approx p + \frac{m_i^2}{2p} \approx p + \frac{m_i^2}{2E} \tag{30.74}$$

In the second term we have assumed that the neutrinos are highly relativistic, so we write $E \simeq p$, where E becomes the neutrino energy. Since we are interested in the interference between the different phases, we can separate the energy matrix into two parts:

$$i\frac{d\vec{\nu}}{dt} \approx U \begin{pmatrix} p & 0 & 0 \\ 0 & p & 0 \\ 0 & 0 & p \end{pmatrix} U^\dagger \vec{\nu} + \frac{1}{2E} U \begin{pmatrix} m_1^2 & 0 & 0 \\ 0 & m_2^2 & 0 \\ 0 & 0 & m_3^2 \end{pmatrix} U^\dagger \vec{\nu} \tag{30.75}$$

The first term contributes an overall phase which can be ignored when we are interested in the interference phenomena. We only keep the second term, which describes the phase differences. In addition, we can normalize the phases by subtracting an overall phase proportional to m_1^2. We then get the following time-evolution equation:

$$i\frac{d\vec{\nu}}{dt} = \frac{1}{2E} U \begin{pmatrix} 0 & 0 & 0 \\ 0 & \Delta m_{21}^2 & 0 \\ 0 & 0 & \Delta m_{31}^2 \end{pmatrix} U^\dagger \vec{\nu} = \frac{1}{2E} U H_\Delta U^\dagger \vec{\nu} \tag{30.76}$$

where we have made explicit the very important parameters called the **mass squared differences** Δm_{ij}^2:

$$\Delta m_{ij}^2 \equiv m_i^2 - m_j^2 \tag{30.77}$$

Time integration can be formally written as:

$$|\vec{\nu}(t)\rangle = U e^{-iH_\Delta t/2E} U^\dagger |\vec{\nu}(t=0)\rangle \tag{30.78}$$

The **probability for a flavor oscillation of a neutrino of energy E from $\nu_\alpha \to \nu_\beta$** at time t is then given by:

$$P(\nu_\alpha \to \nu_\beta; E, t) \equiv |\langle \nu_\beta | S(t) | \nu_\alpha \rangle|^2 \tag{30.79}$$

where the time-evolution operator is:

$$S(t) = U e^{-iH_\Delta t/2E} U^\dagger \tag{30.80}$$

We can write it in the form of matrix elements between α and β:

$$
\begin{aligned}
|S_{\beta\alpha}(t)|^2 &= \left| \sum_{i,j} U_{\beta i} \left(e^{-iH_\Delta t/2E} \right)_{ij} U_{\alpha j}^* \right|^2 = \left| \sum_j U_{\beta j} \left(e^{-i\Delta m_{j1}^2 t/2E} \right)_{jj} U_{\alpha j}^* \right|^2 \\
&= \sum_{j,k} U_{\beta j} U_{\beta k}^* U_{\alpha j}^* U_{\alpha k} e^{-i\Delta m_{jk}^2 t/2E}
\end{aligned}
\tag{30.81}
$$

We define:

$$
|S_{\beta\alpha}(L)|^2 \equiv \sum_{j,k} J_{\alpha\beta jk} e^{-i\Delta_{jk}}
\tag{30.82}
$$

with

$$
\Delta_{jk} \equiv \frac{\left(m_j^2 - m_k^2 \right) t}{2E} \approx \frac{\Delta m_{jk}^2 L}{2E}
\tag{30.83}
$$

where we have introduced the **neutrino path length** $(L = vt \approx ct)$ between the source and the detector, and the **Jarlskog**[6] determinant J is given by:

$$
J_{\alpha\beta jk} \equiv U_{\beta j} U_{\beta k}^* U_{\alpha j}^* U_{\alpha k}
\tag{30.84}
$$

These last expressions show explicitly that the phase differences between the neutrino eigenstates arise from the mass squared differences Δm_{jk}^2. Neutrino flavor oscillations are consequently sensitive to the difference between the neutrino masses but are not sensitive to the absolute value of the neutrino masses. We note that with three neutrinos, there are only two independent mass squared differences, since:

$$
\Delta m_{12}^2 + \Delta m_{23}^2 = \Delta m_{13}^2
\tag{30.85}
$$

Are oscillations the same for neutrinos and antineutrinos? For neutrinos, we had defined $|\nu_\alpha\rangle = \sum_j U_{\alpha j} |\nu_j\rangle$, however for antineutrinos we should consider the complex conjugate of the mixing matrix elements:

$$
|\overline{\nu}_\alpha\rangle = \sum_j U_{\alpha j}^* |\overline{\nu}_j\rangle
\tag{30.86}
$$

The oscillation probabilities for antineutrinos can also be found from a derivation analogous to the previous one. We can define the function $\overline{S}$ for antineutrinos, which will be related to Eq. (30.82), by noting that:

$$
U \to U^* \qquad \text{implies} \qquad J_{\alpha\beta jk} \to J_{\alpha\beta jk}^* = J_{\beta\alpha jk}
\tag{30.87}
$$

Consequently, if the PMNS mixing matrix U is complex, then in general:

$$
\overline{S}_{\beta\alpha} = S_{\alpha\beta} \neq S_{\beta\alpha}
\tag{30.88}
$$

Therefore, in general, the oscillation probabilities for neutrinos and antineutrinos will be different. Explicitly, one can consider the following related probabilities:

$$
\begin{aligned}
P(\overline{\nu}_\alpha \to \overline{\nu}_\beta) &= P(\nu_\beta \to \nu_\alpha) & CPT\text{-invariance} \\
P(\nu_\alpha \to \nu_\beta) &\neq P(\nu_\beta \to \nu_\alpha) & T\text{-violation} \\
P(\nu_\alpha \to \nu_\beta) &\neq P(\overline{\nu}_\alpha \to \overline{\nu}_\beta) & CP\text{-violation}
\end{aligned}
\tag{30.89}
$$

---

6   Cecilia Jarlskog (born in 1941), Swedish theoretical physicist.

The probability of oscillation from flavor α to β can be further written as:

$$
\begin{aligned}
P(\nu_\alpha \to \nu_\beta) &= \sum_j U_{\beta j} U^*_{\beta j} U^*_{\alpha j} U_{\alpha j} + \sum_{j \neq k} U_{\beta j} U^*_{\beta k} U^*_{\alpha j} U_{\alpha k} e^{-i\Delta_{jk}} \\
&= \sum_j |U_{\beta j} U_{\alpha j}|^2 + \sum_{j>k} \left(U_{\beta j} U^*_{\beta k} U^*_{\alpha j} U_{\alpha k} e^{-i\Delta_{jk}} + U_{\beta k} U^*_{\beta j} U^*_{\alpha k} U_{\alpha j} e^{-i\Delta_{kj}} \right) \\
&= \sum_j |U_{\beta j} U_{\alpha j}|^2 + \sum_{j>k} \left(U_{\beta j} U^*_{\beta k} U^*_{\alpha j} U_{\alpha k} e^{-i\Delta_{jk}} + \left(U_{\beta j} U^*_{\beta k} U^*_{\alpha j} U_{\alpha k} e^{-i\Delta_{jk}} \right)^* \right) \\
&= \delta_{\alpha\beta} + 2\Re e \sum_{j>k} \left(J_{\alpha\beta jk} e^{-i\Delta_{jk}} \right)
\end{aligned}
\tag{30.90}
$$

where in the last line, we have used the unitarity of the U matrix. At the source (namely for a baseline $L = 0$), we do not expect any oscillation (!). We have in this case $e^{-i\Delta_{jk}} = 1$ and consequently as expected:

$$
P(\nu_\alpha \to \nu_\beta; L = 0) = \delta_{\alpha\beta} + 2\Re e \underbrace{\sum_{j>k} J_{\alpha\beta jk}}_{=0} = \delta_{\alpha\beta}
\tag{30.91}
$$

The vanishing of the sum is a direct consequence of the unitarity of U, whose columns form an orthonormal basis. If CP or T were not violated, then the U matrix would be real. In this case, the expression for the oscillation probability would reduce to:

$$
\begin{aligned}
P(\nu_\alpha \to \nu_\beta) &= P(\bar\nu_\alpha \to \bar\nu_\beta) = \delta_{\alpha\beta} + 2\sum_{j>k} \left(J_{\alpha\beta jk} \cos(\Delta_{jk}) \right) = \delta_{\alpha\beta} + 2\sum_{j>k} \left(J_{\alpha\beta jk} \left(1 - 2\sin^2\left(\frac{\Delta_{jk}}{2}\right) \right) \right) \\
&= \delta_{\alpha\beta} + 2\underbrace{\sum_{j>k} J_{\alpha\beta jk}}_{=0} - 4\sum_{j>k} J_{\alpha\beta jk} \sin^2\left(\frac{\Delta_{jk}}{2}\right) \\
&= \delta_{\alpha\beta} - 4\sum_{j>k} J_{\alpha\beta jk} \sin^2\left(\frac{\Delta_{jk}}{2}\right)
\end{aligned}
\tag{30.92}
$$

However, just like the CKM matrix in the quark sector, we expect that the PMNS matrix will also be complex. In this case, we find the general expression:

$$
\begin{aligned}
P(\overset{(-)}{\nu_\alpha} \to \overset{(-)}{\nu_\beta}) &= \delta_{\alpha\beta} + 2\Re e \sum_{j>k} \left(J^{(*)}_{\alpha\beta jk} \cos(\Delta_{jk}) \right) + 2\Re e \sum_{j>k} \left(J^{(*)}_{\alpha\beta jk} i \sin\Delta_{jk} \right) \\
&= \delta_{\alpha\beta} - 4\sum_{j>k} J_{\alpha\beta jk} \sin^2\left(\frac{\Delta_{jk}}{2}\right) \mp 2\Im m \sum_{j>k} J_{\alpha\beta jk} \sin\Delta_{jk}
\end{aligned}
\tag{30.93}
$$

where we used $\Re e(iz) = -\Im m(z)$. **Hence, we do expect U to be complex, and CP and T violation in the lepton sector to occur. The question is, by how much is it violated?** This is the subject of intense experimental investigations (see Section 30.10).

30.7   Neutrino Oscillations Between Only Two Flavors

If we consider the mixing only between two families, then the unitary mixing matrix is simply given by one angle and it is always real (so there is no CP/T violation):

$$
\begin{pmatrix} \nu_\alpha \\ \nu_\beta \end{pmatrix} = \begin{pmatrix} \cos\theta & \sin\theta \\ -\sin\theta & \cos\theta \end{pmatrix} \begin{pmatrix} \nu_1 \\ \nu_2 \end{pmatrix}
\tag{30.94}
$$

In this case, we have:

$$P(\nu_\alpha \to \nu_\beta) = \delta_{\alpha\beta} - 4(2\delta_{\alpha\beta}-1)\cos^2\theta\sin^2\theta\sin^2\left(\frac{\Delta_{12}}{2}\right) = \delta_{\alpha\beta} - (2\delta_{\alpha\beta}-1)(2\cos\theta\sin\theta)^2\sin^2\left(\frac{\Delta_{12}}{2}\right) \quad (30.95)$$

Consequently, the **oscillation survival probability** has a very simple form:

$$P(\nu_\alpha \to \nu_\alpha) = P(\nu_\beta \to \nu_\beta) = P(\overline{\nu}_\alpha \to \overline{\nu}_\alpha) = P(\overline{\nu}_\beta \to \overline{\nu}_\beta)$$
$$= 1 - \sin^2 2\theta \sin^2\frac{\Delta_{12}}{2} \quad (30.96)$$

with (cf. Eq. (30.83))

$$\Delta_{12} \equiv \frac{\Delta m_{12}^2 L}{2E} \quad (30.97)$$

where L is the neutrino path length between the source and the detector. This expression is given in natural units (see Table 1.6) and can be converted into practical units to find:

$$\Delta_{12} \equiv \frac{\Delta m_{12}^2 L}{2\hbar c E} \approx 2.534 \times \frac{\Delta m_{12}^2(\text{eV}^2)L(\text{km})}{E(\text{GeV})} \quad (30.98)$$

The probability is called survival because it is related to the probability that the neutrino created with flavor α is detected as flavor α at a distance L from the detector. Similarly, the **oscillation appearance probability** is given by:

$$P(\nu_\alpha \to \nu_\beta) = P(\nu_\beta \to \nu_\alpha) = P(\overline{\nu}_\alpha \to \overline{\nu}_\beta) = P(\overline{\nu}_\beta \to \overline{\nu}_\alpha)$$
$$= 1 - P(\nu_\alpha \to \nu_\alpha) = \sin^2 2\theta \sin^2\frac{\Delta_{12}}{2} \quad (30.99)$$

Both disappearance and appearance probabilities contain the product of two terms: (a) the mixing term which depends on the mixing angle θ and (b) the oscillation term that depends on Δ_{12}, so from the ratio L/E. The term $\sin^2 2\theta$ describes the maximal transition probability from the weak eigenstate α to the weak eigenstate β via all possible mass eigenstates. The second term contains the information from the quantum-mechanical phase, which leads to an L/E dependence. This dependence on the baseline (space) and inverse energy is expected as the oscillation occurs in the rest frame of the neutrino. For an external observer the neutrino clock slows down with the Lorentz boost, while in the neutrino rest frame the traveled distance is Lorentz contracted by the Lorentz factor, hence the L/E behavior.

The survival and appearance probabilities are illustrated in Figure 30.7 for the following chosen set of parameters $\Delta m^2 = 0.003$ eV^2 (where we removed the 12 index for clarity), $\sin^2 2\theta = 0.8$, and for a neutrino energy $E = 1$ GeV. The wavelength of the oscillation λ_{osc} for a given neutrino energy E is consequently given by:

$$\frac{\Delta_{12}(\lambda_{\text{osc}})}{2} = \pi \quad \Rightarrow \quad \lambda_{\text{osc}} = \frac{4\pi E}{\Delta m^2} \quad (30.100)$$

The wavelength can be converted into practical units to find:

$$\lambda_{\text{osc}} = 4\pi\hbar c \frac{E}{\Delta m^2} \approx 2.480 \times \frac{E(\text{GeV})}{\Delta m^2(\text{eV}^2)} \text{ km} \quad (30.101)$$

So, for example, for a value $\Delta m^2 = 0.003$ eV^2, we find a wavelength on the scale of 1000 km for energies of the giga-electronvolt scale:

$$\lambda_{\text{osc}}(\Delta m^2 = 0.003 \text{ eV}^2) \approx 2.480 \times \frac{E(\text{GeV})}{0.003} \text{ km} = 826.7 \times E(\text{GeV}) \text{ km} \quad (30.102)$$

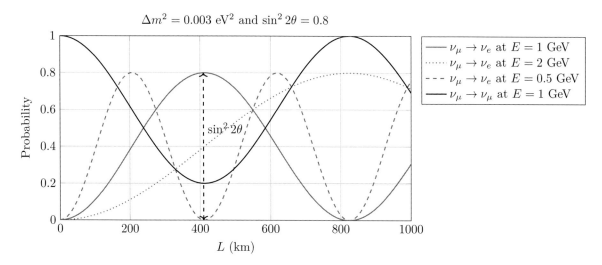

Figure 30.7 The two flavor oscillation probabilities for appearance $P(\nu_\mu \to \nu_e)$ and survival $P(\nu_\mu \to \nu_\mu)$ as a function of the distance L for a chosen set of parameters.

This is illustrated in Figure 30.7. The oscillatory term can also be written in practical units by noting:

$$\sin^2 \frac{\Delta_{12}}{2} = \sin^2 \left(\frac{\Delta m^2 L}{4\hbar c E} \right) \approx \sin^2 \left(\frac{1.267 \times \Delta m^2 (\text{eV}^2) L(\text{km})}{E(\text{GeV})} \right) \tag{30.103}$$

Experimentally, the distance between the source and the detector is generally fixed (actually we can set the detector at a distance L from the neutrino source of size d with $L \gg d$, such that the source appears point-like) and the oscillation probability is studied as a function of the neutrino energy for a given energy spectrum of the source (see, e.g., high-energy neutrino beams in Section 28.1). The oscillation probability $P(E)$ as a function of the neutrino energy E at a given distance L from the source is illustrated in Figure 30.8 for a given choice of parameters. The oscillation appearance probability exhibits maxima and minima, where the "maxima" are defined by ($n = 0, 1, ...$):

$$\frac{\Delta_{12}^{max}}{2} = \frac{(2n+1)\pi}{2} = \frac{\pi}{2}, \frac{3\pi}{2}, ... \implies E^{max} = \frac{\Delta m^2 L}{2(2n+1)\pi\hbar c} \approx \frac{2.534 \times \Delta m^2 (\text{eV}^2) L(\text{km})}{(2n+1)\pi} \text{ GeV} \tag{30.104}$$

and the "minima" by:

$$\frac{\Delta_{12}^{min}}{2} = n\pi = 0, \pi,, ... \implies E^{min} = \frac{\Delta m^2 L}{4n\pi\hbar c} \approx \frac{1.267 \times \Delta m^2 (\text{eV}^2) L(\text{km})}{n\pi} \text{ GeV} \tag{30.105}$$

For example, for our chosen parameters $\Delta m^2 = 0.003$ eV^2 and $L = 400$ km, the first maxima is at $E \approx 0.968$ GeV. At this energy, the oscillation probability is just given by $\sin^2 2\theta$. The first minimum is at $E \approx 0.484$ GeV. At high energy, the appearance probability falls off as $\sin^2(\Delta_{12}/2) \to 1/E^2$. As the energy decreases, the oscillations become more and more frequent. At some point, the oscillation becomes so fast that experimentally the oscillatory term averages out to $1/2$. Hence, we can distinguish three scenarios:

$$P(\nu_\alpha \to \nu_\beta) = \begin{cases} \sin^2 2\theta \left(\frac{\Delta m^2 L}{4E} \right)^2 \approx (\sin^2 2\theta)(\Delta m^2)^2 \left(\frac{L}{4E} \right)^2 & \text{for } L/E \ll \frac{1}{\Delta m^2} \\ \sin^2 2\theta \sin^2 \left(\frac{\Delta m^2 L}{4E} \right) & \text{for } L/E \simeq \frac{1}{\Delta m^2} \\ \frac{1}{2} \sin^2 2\theta & \text{for } L/E \gg \frac{1}{\Delta m^2} \end{cases} \tag{30.106}$$

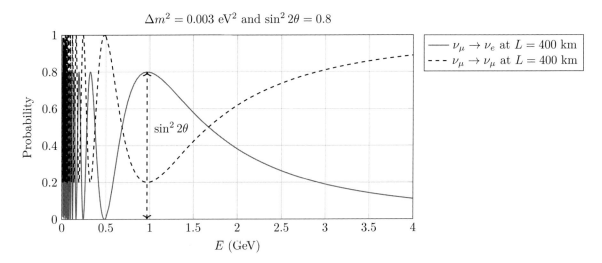

Figure 30.8 The two flavor oscillation probabilities for appearance $P(\nu_\mu \to \nu_e)$ and survival $P(\nu_\mu \to \nu_\mu)$ as a function of the energy E for a chosen set of parameters.

Hence, an oscillatory behavior is only visible when $L/E \simeq 1/\Delta m^2$. At very long baselines or small energies such that $L/E \gg 1/\Delta m^2$ the oscillation washes out and the oscillation appearance probability is $(1/2)\sin^2 2\theta$. For maximal mixing of $\theta = 45°$, it is just $1/2$. An initial beam of flavor α becomes fully mixed with just 50% ν_α's and 50% ν_β's. At short baselines or high energies such that $L/E \ll 1/\Delta m^2$, the probability behaves like κ/E^2 where $\kappa \propto \sin^2 2\theta(\Delta m^2)^2$. Hence, the oscillation appearance probability is determined by the product $\sin^2 2\theta(\Delta m^2)^2$. Consequently, it is impossible to distinguish a small mixing angle with "high" Δm^2 or a large mixing angle with "low" Δm^2, since oscillations with κ being constant are equiprobable.

Having introduced the phenomenology of neutrino oscillations, we can now turn to some experimental aspects. One generally starts with a neutrino source, whose flavor composition is known or measured near the source. Then, at a certain distance from the source, a detector is located and measures neutrino interactions. The detector must be able to distinguish the flavor of the interacting neutrinos. One usually distinguishes the two possible signatures for neutrino oscillations corresponding to the two types of oscillation probabilities derived in Eqs. 30.96 and 30.99. In **disappearance experiments**, neutrino oscillations can be observed as the disappearance of a flavor, translating for example into the observation of fewer than expected electrons in a pure beam initially composed of only electron neutrinos. Alternatively, in **appearance experiments**, neutrino oscillations can result in the appearance of a different flavor, translating for example into the observation of electrons or tau leptons produced in the neutrino interactions, from an initially pure muon-neutrino beam.

30.8   Three-Neutrino Mixing and One Mass Scale Approximation

As we did for the CKM matrix (see Section 30.2), the PMNS matrix can also be written as the product of three rotations plus a complex phase:

$$
U_{PMNS} = \begin{pmatrix} 1 & 0 & 0 \\ 0 & c_{23} & s_{23} \\ 0 & -s_{23} & c_{23} \end{pmatrix} \begin{pmatrix} c_{13} & 0 & s_{13}e^{-i\delta} \\ 0 & 1 & 0 \\ -s_{13}e^{i\delta} & 0 & c_{13} \end{pmatrix} \begin{pmatrix} c_{12} & s_{12} & 0 \\ -s_{12} & c_{12} & 0 \\ 0 & 0 & 1 \end{pmatrix}
$$

$$
= \begin{pmatrix}
c_{12}c_{13} & s_{12}c_{13} & s_{13}e^{-i\delta} \\
-s_{12}c_{23} - c_{12}s_{13}s_{23}e^{i\delta} & c_{12}c_{23} - s_{12}s_{13}s_{23}e^{i\delta} & c_{13}s_{23} \\
s_{12}s_{23} - c_{12}s_{13}c_{23}e^{i\delta} & -c_{12}s_{23} - s_{12}s_{13}c_{23}e^{i\delta} & c_{13}c_{23}
\end{pmatrix}
\tag{30.107}
$$

where $s_{ij} = \sin\theta_{ij}$ and $c_{ij} = \cos\theta_{ij}$. The magnitude of the CP violation in $\nu_\alpha \to \nu_\beta$ and $\bar{\nu}_\alpha \to \bar{\nu}_\beta$ oscillations is determined by the Jarkslog invariant (see Eq. (30.84)). **The sum of the invariant will vanish if any one of the mixing angles is zero!**, so CP or T violation will only be possible if all three neutrino flavors are involved in the process. The general expressions for the three flavor oscillations are rather lengthy [424]. Let us ignore for the moment the possibility of CP/T violation and take a real mixing matrix. Then, according to Section 30.6, the oscillation probability between a weak eigenstate α and a different weak eigenstate β is given by:

$$
P(\nu_\alpha \to \nu_\beta; E, L) = -4 \sum_{j>k} J_{\alpha\beta jk} \sin^2 \frac{\Delta m_{jk}^2 L}{4E} \qquad (\alpha \neq \beta)
\tag{30.108}
$$

This implies that there are three independent oscillations with the following oscillatory terms:

$$
\sin^2 \frac{\Delta m_{21}^2 L}{4E}, \qquad \sin^2 \frac{\Delta m_{31}^2 L}{4E}, \quad \text{and} \quad \sin^2 \frac{\Delta m_{32}^2 L}{4E}
\tag{30.109}
$$

In practice, one makes approximations appropriate for given experimental conditions. This is made even easier since the experimental results indicate that oscillations can in most cases be separated into two classes with very different wavelengths, because $\Delta m_{21}^2 \ll |\Delta m_{31}^2|$ (see Table 30.2). These correspond to the so-called **solar oscillations** and the **atmospheric oscillations** (these are historical names, see Section 30.11). In the **one mass scale approximation (OMS)**, we accordingly write:

$$
|\Delta m_{21}^2| \ll |\Delta m_{31}^2| \approx |\Delta m_{32}^2| \equiv |\Delta M^2|
\tag{30.110}
$$

Under this assumption, the phenomenology of the **three flavor oscillations** is simply determined by three parameters: ΔM^2 and the two angles θ_{13} and θ_{23}. Basically, we have ignored the oscillations with relatively very long wavelengths due to Δm_{21}^2 and θ_{12}. Consequently, experimentally the approximation will be valid when the energy is sufficiently high compared to the baseline, such that the following inequality is verified:

$$
\Delta m_{21}^2 \ll \frac{E}{L}
\tag{30.111}
$$

Finally, the oscillation probabilities under these conditions then simply become:

One mass scale approximation (OMS) :
$$
\begin{cases}
P(\nu_e \to \nu_e) & \simeq 1 - \sin^2 2\theta_{13} \sin^2 \frac{\Delta M^2 L}{4E} \\
P(\nu_e \to \nu_\mu) = P(\nu_\mu \to \nu_e) & \simeq \sin^2 2\theta_{13} \sin^2 \theta_{23} \sin^2 \frac{\Delta M^2 L}{4E} \\
P(\nu_e \to \nu_\tau) & \simeq \sin^2 2\theta_{13} \cos^2 \theta_{23} \sin^2 \frac{\Delta M^2 L}{4E} \\
P(\nu_\mu \to \nu_\tau) & \simeq \cos^4 \theta_{13} \sin^2 2\theta_{23} \sin^2 \frac{\Delta M^2 L}{4E}
\end{cases}
\tag{30.112}
$$

The expressions for antineutrinos are the same due to CP conservation (in our approximation). In the high-energy neutrino regime which is dominantly composed of muon neutrinos (or muon antineutrinos), we will therefore have the following two oscillations occurring with the same wavelength:

OMS :
$$
\begin{cases}
P(\nu_\mu \to \nu_e) & \simeq \sin^2 2\theta_{13} \sin^2 \theta_{23} \sin^2 \frac{\Delta M^2 L}{4E} \\
P(\nu_\mu \to \nu_\tau) & \simeq \cos^4 \theta_{13} \sin^2 2\theta_{23} \sin^2 \frac{\Delta M^2 L}{4E} \\
P(\nu_\mu \to \nu_\mu) & = 1 - P(\nu_\mu \to \nu_e) - P(\nu_\mu \to \nu_\tau)
\end{cases}
\tag{30.113}
$$

If we approximate the current experimental values $\sin^2 \theta_{13} = 0.02237$ and $\sin^2 \theta_{23} = 1/2$ (see Table 30.2), we have $\cos^4 \theta_{13} = \left(1 - \sin^2 \theta_{13}\right)^2 \approx 1$ and $\sin^2 2\theta_{23} = 1$, hence:

$$\text{OMS} + \text{data}: \quad \begin{cases} P(\nu_\mu \to \nu_e) & \simeq \frac{1}{2} \sin^2 2\theta_{13} \sin^2 \left(\frac{\Delta M^2 L}{4E}\right) \approx 0.04372 \sin^2 \left(\frac{\Delta M^2 L}{4E}\right) \\ P(\nu_\mu \to \nu_\tau) & \simeq \sin^2 \left(\frac{\Delta M^2 L}{4E}\right) \\ P(\nu_\mu \to \nu_\mu) & = 1 - P(\nu_\mu \to \nu_e) - P(\nu_\mu \to \nu_\tau) \end{cases} \tag{30.114}$$

Thus, we have a maximal appearance of the tau flavor corresponding to a maximal disappearance of muon flavor, with an approximate 4% conversion to electron neutrinos.

30.9 Neutrino Flavor Oscillations in Matter

When the neutrinos propagate in matter (e.g., when they cross the Sun or the Earth) rather than in vacuum, their flavor oscillations are influenced because of their coherent interactions with the electrons, protons, and neutrons of the matter they traverse. In this case, the oscillation probabilities can be significantly different than in vacuum. Neutrino oscillations in matter were first studied by **Wolfenstein** and then by **Mikheyev**[7] and **Smirnov**,[8] who investigated the effect of the so-called "MSW resonance" [425, 426]: the oscillation probability will grow significantly for a given neutrino energy and a given matter density, regardless of the vacuum mixing angle, in particular even if this latter is "tiny" (small mixing).

When neutrinos traverse a dense medium, the forward coherent elastic scatterings with protons and neutrons are occurring through neutral currents and are hence the same for all flavors. For the coherent scattering on electrons, the cross-section is different for ν_e than for ν_μ and ν_τ, since in addition to the neutral current interaction, the electron neutrinos feel also the charged current interaction. The time evolution of the neutrino state vector can then be written as in the flavor basis:

$$i \frac{d\vec{\nu}}{dt} = \frac{1}{2E} \left[U \begin{pmatrix} m_1^2 & 0 & 0 \\ 0 & m_2^2 & 0 \\ 0 & 0 & m_3^2 \end{pmatrix} U^\dagger + 2\sqrt{2} G_F N_e E \begin{pmatrix} 1 & 0 & 0 \\ 0 & 0 & 0 \\ 0 & 0 & 0 \end{pmatrix} \right] \vec{\nu} \tag{30.115}$$

where the second term describes the coherent scattering on electrons. In this expression, N_e is the density of electrons, i.e., the number of electrons per unit volume of target. It is given by:

$$2\sqrt{2} G_F N_e = 2\sqrt{2} G_F \frac{\rho Y_e}{m_N} \tag{30.116}$$

where $Y_e \approx 0.5$ is the fraction of electrons per nucleon. Numerically, we obtain:

$$2\sqrt{2} \times \left(1.16 \times 10^{-5} \text{ GeV}^{-2}\right) \frac{0.5\rho}{1.67 \times 10^{-27} \text{ kg}} (\hbar c)^3 = \left(9.82 \times 10^{21} \text{ GeV}^{-2} \text{ kg}^{-1}\right) \rho \left(197 \text{ MeV fm}\right)^3$$

$$= 75.6 \times 10^{-18} \text{ eV} \left(\frac{\rho}{\text{kg/m}^3}\right) \tag{30.117}$$

We continue the discussion in the case of two neutrinos (the general case of three neutrino oscillations can be found in Ref. [424]), where one can derive simple expressions for the resonance in matter. The time-evolution equation is:

$$i \frac{d\vec{\nu}}{dt} = \frac{1}{2E} \begin{pmatrix} A + m_1^2 \cos^2 \theta + m_2^2 \sin^2 \theta & \left(m_2^2 - m_1^2\right) \cos \theta \sin \theta \\ \left(m_2^2 - m_1^2\right) \cos \theta \sin \theta & m_2^2 \cos^2 \theta + m_1^2 \sin^2 \theta \end{pmatrix} \vec{\nu}, \quad \text{where} \quad \vec{\nu} = \begin{pmatrix} \nu_e \\ \nu_\mu \end{pmatrix} \tag{30.118}$$

---

7   Stanislav Pavlovich Mikheyev (1940–2011), Russian physicist.
8   Alexei Yuryevich Smirnov (born 1951), Russian physicist.

The effective potential is determined by A:

$$A \equiv 2\sqrt{2}G_F N_e E \approx 75.6 \times 10^{-9} \text{ eV}^2 \left(\frac{\rho}{\text{kg/m}^3} \right) \left(\frac{E}{\text{GeV}} \right) \tag{30.119}$$

These expressions are valid for neutrinos. For antineutrinos, the coherent interactions with the electrons of the matter have an opposite sign, and therefore for antineutrino oscillations in matter one must replace A by $-A$. The expression can be simplified by writing:

$$i\frac{d\vec{\nu}}{dt} = \frac{1}{2E} \begin{pmatrix} A + \frac{1}{2}\Sigma - \frac{1}{2}D\cos 2\theta & \frac{1}{2}D\sin 2\theta \\ \frac{1}{2}D\sin 2\theta & \frac{1}{2}\Sigma + \frac{1}{2}D\cos 2\theta \end{pmatrix} \vec{\nu} \tag{30.120}$$

with

$$\Sigma = m_1^2 + m_2^2 \quad \text{and} \quad D = m_2^2 - m_1^2 = \Delta m^2 \tag{30.121}$$

After diagonalization of the matrix, the mass eigenstates in matter defined as ν_{1m} and ν_{2m} have the following masses:

$$m_{1,2m}^2 = \frac{1}{2}\left\{ \Sigma + A \mp \sqrt{(A - D\cos 2\theta)^2 + D^2 \sin^2 2\theta} \right\} \tag{30.122}$$

We can then rewrite the oscillation probability in matter using the same expression for the probability of oscillation in vacuum while redefining the parameters such as the mixing angle and the mass squared difference with new parameters that are valid in matter. The mixing matrix can accordingly be expressed with the mixing angle in matter θ_m by:

$$\begin{pmatrix} \nu_e \\ \nu_\mu \end{pmatrix} = \begin{pmatrix} \cos\theta_m & \sin\theta_m \\ -\sin\theta_m & \cos\theta_m \end{pmatrix} \begin{pmatrix} \nu_{1m} \\ \nu_{2m} \end{pmatrix} \tag{30.123}$$

The oscillation appearance probability becomes:

$$P_m(\nu_e \to \nu_\mu) = \sin^2 2\theta_m \sin^2 \frac{\Delta M^2 L}{4E} \tag{30.124}$$

where for the effective angle in matter we have:

$$\sin^2 2\theta_m = \frac{\sin^2 2\theta}{\sin^2 2\theta + (x - \cos 2\theta)^2} \tag{30.125}$$

The parameter x is given by:

$$x = \frac{A}{D} \approx 38 \times 10^{-9} \frac{\dfrac{\rho}{\text{kg m}^{-3}} \dfrac{E}{\text{GeV}}}{\dfrac{\Delta m^2}{\text{eV}^2}} \tag{30.126}$$

The effective mass squared difference in matter is given by:

$$\Delta M^2 = m_{2m}^2 - m_{1m}^2 = \frac{1}{2}\left\{ 2\sqrt{(A - D\cos 2\theta)^2 + D^2 \sin^2 2\theta} \right\} = \sqrt{D^2\left[\left(\frac{A}{D} - \cos 2\theta \right)^2 + \sin^2 2\theta \right]} \tag{30.127}$$

hence

$$\Delta M^2 = |\Delta m^2|\sqrt{\sin^2 2\theta + (x - \cos 2\theta)^2} \tag{30.128}$$

Because of matter effects, the transition probabilities are different for neutrinos and antineutrinos, since $A \to -A$ and therefore $x \to -x$! We also note that for $x \to 0$, we recover the mixing angle and the mass squared difference in vacuum, as expected.

Oscillations in matter provide a way to determine the sign of Δm^2 (this was not the case for oscillations in vacuum). For neutrinos and $\Delta m^2 > 0$, there is a resonance condition for $x(E, \Delta m^2, \rho) \approx \cos 2\theta$. At this location, the oscillation amplitude is maximal:

$$A \approx \Delta m^2 \cos 2\theta \quad \Rightarrow \quad \sin^2 2\theta_\mathrm{m}(A) \approx 1 \tag{30.129}$$

Then, the mixing in matter is maximal and independent of the level of mixing in vacuum. This is the resonance effect. We also note that:

$$A \gg 2\Delta m^2 \cos 2\theta \quad \Rightarrow \quad \sin^2 2\theta_\mathrm{m}(A) \ll 1 \tag{30.130}$$

A suppression takes place when the mixing angle in matter is smaller than in vacuum. While the neutrino mixing probability is enhanced at the resonance, it is suppressed for antineutrinos, since one must replace A by $-A$. For $\Delta m^2 < 0$ (i.e., $D < 0$), the situation is reversed. The enhanced oscillations become suppressed, and vice versa. Hence, **matter effects in oscillations can be used to determine the sign of Δm^2.**

30.10   Long-Baseline Neutrino Oscillation Experiments

Neutrinos are detected via their interactions with matter,[9] as discussed in Chapter 28. The charged current interaction leads to a final-state charged lepton which can be used to tag the flavor of the incoming neutrino. It is therefore ideal to study flavor oscillations in experiments that combine disappearance and appearance signatures. The kinematical thresholds for producing charged currents is very low for electrons, about 110 MeV for muons and 3.5 GeV for taus, as calculated in Eq. (28.19). Hence, muon and tau appearance experiments require neutrino beams in the GeV range. This has an impact on the baselines needed to observe an appreciable effect. Indeed, as illustrated in Section 30.7, this will lead to wavelengths of several hundreds of kilometers for the atmospheric neutrino mass difference. For these reasons, the community has focused on what is called **long-baseline neutrino oscillation experiments**. In such experiments, one typically uses two neutrino detectors, one sufficiently close to the source to allow a measurement of the *unoscillated* neutrino beam (ideally as a function of the neutrino energy) and a second detector located several hundred kilometers away (i.e., far from the source) to measure the *oscillated* neutrino beam (again as a function of the neutrino energy). The comparison of the near and far measurements ensures that many uncertainties in the prediction of the beam and in the neutrino cross-section modelling cancel, and that a precise determination of the oscillation parameters taking full advantage of the statistics can be achieved. Given the baselines generally involved, the effects introduced by matter outlined in Section 30.9 will have to be included to compare the oscillation parameters with the observations.

• **The T2K experiment.** T2K is a long-baseline neutrino experiment that uses beams of muon neutrinos and antineutrinos, with energy spectra peaked at 0.6 GeV. The experiment measures interactions of neutrinos at a near detector facility 280 m from the beam production point which characterizes the beam and the interactions of the neutrinos before oscillations. The beam then propagates 295 km through the Earth to the T2K far detector, Super-Kamiokande (SK). The SK detector was described in Section 3.18. SK measures the oscillated beam, which is sensitive to the oscillation parameters.

The T2K neutrino beam layout is shown in Figure 30.10. If the horn current is run in one direction, positively charged secondaries are focused, which decay to produce a beam that is primarily composed of muon neutrinos. This is referred to as "forward horn current" (FHC) or neutrino mode. Alternatively, the horn current can be reversed to give a beam of primarily muon antineutrinos, which is referred to as "reversed horn current" (RHC) or antineutrino mode. In either case, secondary hadrons produced in the very forward

---

9   We focus here on the dominant neutrino–nucleon interactions, and neglect interactions with electrons from the target.

direction pass through the field-free necks of the horns and contribute to a wrong-sign flux that is of order 1% of the intended right-sign component in FHC mode, and order 10% in RHC mode.

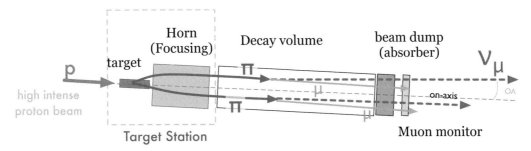

Figure 30.9 Layout of the off-axis neutrino beam of the T2K experiment. Courtesy T2K Collaboration.

T2K uses an off-axis beam, as discussed in Section 28.5. The beam axis is directed 2.5° away from the SK detector, taking advantage of the kinematics of the two-body pion decay to produce a narrow neutrino spectrum peaked at the expected energy of maximum oscillation probability. The spectrum of the muon neutrino flux at the chosen off-axis angle of 2.5° is plotted in Figure 28.9.

In order to estimate the oscillation phenomena at T2K (we neglect for the moment the potential difference between neutrino and antineutrino oscillation probability, which happens in the case of CP violation), we can use the one mass scale approximation, whereby T2K tests the atmospheric mass scale and the solar mass scale is neglected, see Eq. (30.113):

$$P(\nu_\mu \to \nu_e) \simeq \sin^2 2\theta_{13} \sin^2 \theta_{23} \sin^2 \frac{\Delta_{32}}{2} \quad \text{and} \quad P(\nu_\mu \to \nu_\tau) \simeq \cos^4 \theta_{13} \sin^2 2\theta_{23} \sin^2 \frac{\Delta_{32}}{2} \tag{30.131}$$

where Δ_{32} is given in Eq. (30.98). This yields, for the disappearance of muon neutrinos, the following probability:

$$P(\nu_\mu \to \nu_\mu) \simeq 1 - \left(\cos^4 \theta_{13} \sin^2 2\theta_{23} + \sin^2 2\theta_{13} \sin^2 \theta_{23}\right) \sin^2 \frac{\Delta_{32}}{2} \tag{30.132}$$

Because we expect θ_{13} to be relatively small compared to θ_{23}, the dominant term in the disappearance probability is the first term $\cos^4 \theta_{13} \sin^2 2\theta_{23} \simeq \sin^2 2\theta_{23}$. Moreover, for maximal mixing (i.e., $\theta_{23} \simeq \pi/2$) we anticipate that the muon neutrinos have totally oscillated away into tau neutrinos and electron neutrinos at the energy of the maximum of the oscillation given by Eq. (30.104), namely:

$$E^{max} \approx \frac{2.534 \times \Delta m^2 (\text{eV}^2) \times 295 \text{ km}}{\pi} \approx 0.6 \text{ GeV} \quad \text{for} \quad \Delta m^2 (\text{eV}^2) = 2.5 \times 10^{-3} \text{ eV}^2 \tag{30.133}$$

Given the threshold for tau charged current events, Eq. (28.19), the tau neutrinos only interact via neutral currents.

At neutrino energies of 0.6 GeV, the dominant interaction process is quasi-elastic scattering (see Section 28.6). The T2K near detector facility consists of two detectors both located 280 m downstream of the beam production target. The INGRID detector, located on the beam axis, monitors the direction and stability of the neutrino beam. The ND280 detector is located at the same angle away from the beam axis as SK, and characterizes the rate of neutrino interactions from the beam before oscillations have occurred. ND280 is magnetized so that charged leptons and antileptons bend in opposite directions as they traverse the detector, thereby allowing the neutrino and antineutrino interaction rates in each beam mode to be measured independently.

Neutrinos that interact in the far detector SK produce signals that can be analyzed in different event categories. At T2K's beam energy, the leptons produced in charged-current interactions will be either positive or negative muons, electrons, or positrons, so events are categorized as muon-like or electron-like based on the pattern of Cherenkov rings in the detector. SK has no magnetic field, so the sign of the horn current is used to categorize events as a proxy for neutrino/antineutrino identification, and the kinematics give some discrimination between signal events and backgrounds.

As at the time of writing, T2K has been collecting data for about a decade. A total of 1.5×10^{21} protons-on-target in FHC mode and 1.6×10^{21} protons-on-target in RHC mode have been collected. After event selection and categorization, the remaining sample is analyzed. For example, the reconstructed energy spectrum of the selected muon-like and electron-like events are shown in Figure 30.10, categorized according to the FHC and RHC beam modes. A total of 243 muon-like events were collected in FHC mode, and 140 in RHC mode. The number of collected electron-like events is 75 in FHC mode and 15 in RHC mode. The expectations are shown as histograms in the plots. They were computed for a reference set of oscillation parameters given by: $\sin^2 \theta_{23} = 0.528$, $\sin^2 \theta_{13} = 0.0212$, $\sin^2 \theta_{12} = 0.304$, $\delta = -1.601$, $\Delta m_{32}^2 = 2.509 \times 10^{-3}$ eV^2 (normal mass ordering), and $\Delta m_{21}^2 = 7.53 \times 10^{-5}$ eV^2. The T2K data clearly shows very strong evidence for muon disappearance and for electron appearance. It's very impressive!

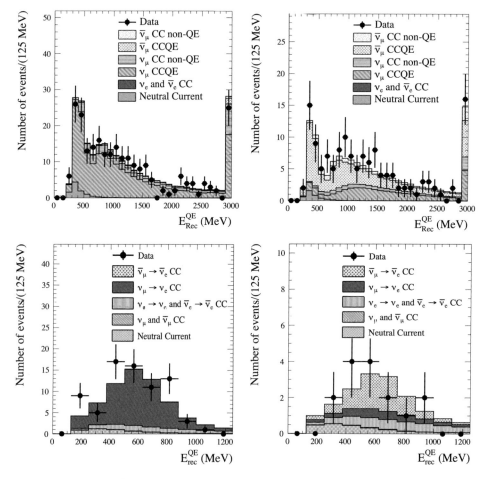

Figure 30.10 Reconstructed energy spectrum of the T2K events collected at SK: (upper-left) muon-like FHC; (upper-right) muon-like RHC; (lower-left) electron-like FHC; (lower-right) electron-like RHC. Reprinted figures from K. Abe *et al.*, "Improved constraints on neutrino mixing from the T2K experiment with 3.13×10^{21} protons on target," arXiv:2101.03779.

With the use of the FHC and RHC modes of the neutrino beam, one can enhance or suppress the flux of neutrinos vs. antineutrinos, as is clearly visible from Figure 30.10. The upper-left and lower-left plots are dominated by oscillations of neutrinos, while the upper-right and lower-right plots are dominated by oscillations of antineutrinos. This allows the independent study of the oscillations of neutrinos and antineutrinos. Are their oscillation probabilities identical or different, as one would expect in the case of CP violation? We note that the reference values taken for the histrograms in Figure 30.10 used a value $\delta = -1.601$, consequently, consistent with CP violation! Hence, at a glance, we could already say that T2K data is consistent with CP violation in the leptonic sector. However, statistics is limited, in particular in the electron-like RHC sample and the dependence of the oscillation probabilities on δ is subleading (i.e., small), and therefore a quantitative statistical analysis is needed in order to ascertain the range of δ values compatible with the data. We will illustrate this point in the following paragraph.

• **Search for CP violation in the leptonic sector.** Assuming CPT conservation, direct evidence for CP-violating effects must be present in flavor appearance measurements. At conventional neutrino beams, searches are favorably based on the electron appearance channels $\nu_\mu \to \nu_e$ and $\bar{\nu}_\mu \to \bar{\nu}_e$. Including higher-order terms and matter effects, the $\nu_\mu \to \nu_e$ oscillation probability can be approximated as [427]:

$$
\begin{aligned}
P(\nu_\mu \to \nu_e) &\approx \sin^2\theta_{23} \frac{\sin^2 2\theta_{13}}{(\hat{A}-1)^2} \sin^2((\hat{A}-1)\hat{\Delta}_{31}) \\
&- \alpha \frac{J_{CP}\sin\delta}{\hat{A}(1-\hat{A})} \sin(\hat{\Delta}_{31})\sin(\hat{A}\hat{\Delta}_{31})\sin((1-\hat{A})\hat{\Delta}_{31}) \\
&+ \alpha \frac{J_{CP}\cos\delta}{\hat{A}(1-\hat{A})} \cos(\hat{\Delta}_{31})\sin(\hat{A}\hat{\Delta}_{31})\sin((1-\hat{A})\hat{\Delta}_{31}) \\
&+ \alpha^2 \frac{\cos^2\theta_{23}\sin^2 2\theta_{12}}{\hat{A}^2} \sin^2(\hat{A}\hat{\Delta}_{31})
\end{aligned}
\tag{30.134}
$$

with $\Delta m_{21}^2 = \alpha \Delta m_{31}^2$ and $\hat{A} = A/\Delta m_{31}^2 = 2VE_\nu/\Delta m_{31}^2 \approx E_\nu(\text{GeV})/11$ for the Earth's crust. The potential V is proportional to the electron density. The parameter δ is the complex phase that violates CP symmetry. The corresponding probability for $\bar{\nu}_\mu \to \bar{\nu}_e$ transitions is obtained by replacing $\delta \to -\delta$ and $A \to -A$. The second term, containing $\sin\delta$, is the CP-odd violating term which flips the sign between ν and $\bar{\nu}$ and thus introduces CP asymmetry if $\sin\delta$ is non-zero. Matter effects are caused by coherent forward scattering in matter, and they also produce an asymmetry between neutrinos and antineutrinos. As seen from the definition of A, the amount of asymmetry due to the matter effect is proportional to the neutrino energy at a fixed value of L/E_ν. The last term in Eq. (30.134) is due to the *solar term*.

As can be seen from the oscillation probability in expression Eq. (30.134), the CP-violating effects of δ are modulated by those of all three mixing angles and their interplay, resulting in complicated dependencies and leading to an a priori eightfold parameter degeneracy [428]. In addition, the situation in all long baseline experiments is complicated by matter effects. Overall, it is known that the energy dependence of the probability can resolve several of these issues and allows in particular the disentangling of the CP-driven and the matter-driven effects, provided the baseline is long enough.

In this context, it is instructive to define two asymmetries between the probabilities of oscillations of neutrinos and antineutrinos, one related to the CP effect computed in vacuum $\mathcal{A}_{CP}^{vac}(\delta)$:

$$
\mathcal{A}_{CP}^{vac}(\delta) \equiv abs\left(\frac{P^{vac}(\nu) - P^{vac}(\bar{\nu})}{P^{vac}(\nu) + P^{vac}(\bar{\nu})} \right)
\tag{30.135}
$$

and the other to the matter effects $\mathcal{A}_{CP}(\rho)$ computed for a fixed value of δ:

$$
\mathcal{A}_{CP}(\rho) \equiv abs\left(\frac{P^{mat}(\nu) - P^{mat}(\bar{\nu})}{P^{mat}(\nu) + P^{mat}(\bar{\nu})} \right)
\tag{30.136}
$$

where ρ represents the traversed Earth matter density (in the constant density approximation). These two variables, plotted in the two-dimensional plane of neutrino energy E_ν vs. baseline L, are shown in Figure 30.11. In these graphs, the black regions correspond to combinations of neutrino energy and baseline at which the

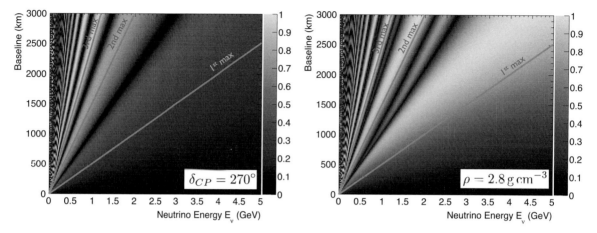

Figure 30.11 The two asymmetries, (left) $\mathcal{A}_{CP}^{vac}(\delta)$ and (right) $\mathcal{A}_{CP}(\rho)$, in the (E_ν, L) plane. The first, second, and third oscillation maxima are represented by straight lines at constant L/E_ν.

oscillation phenomenon is insensitive to the effect, while the light regions correspond to those where the effect is largest. We note:

- The CP asymmetries are largest at the second, third, etc. maxima.

- The matter asymmetry covers a broad region just below and dominates around the first maximum.

- Longer baselines and wide-band beams to cover several maxima are needed to resolve degeneracies.

- Assuming a reasonable energy threshold of 500 MeV, which is a realistic value taking into account realizable conventional neutrino beam fluxes, and the vanishing neutrino cross-sections at low energies (in particular for antineutrinos), the measurement of the second maximum requires a baseline greater than 1000 km: $E_\nu^{2nd\,max} \gtrsim 0.5\,\mathrm{GeV} \implies \mathrm{L} \gtrsim 1000$ km.

If the distance between the source and detector is fixed, the oscillatory behavior can easily be translated to that of the expected neutrino energy spectrum of the oscillated events. If the neutrino energy spectrum of the oscillated events can be reconstructed with sufficiently good resolution in order to distinguish first and second maxima, the spectral information obtained is invaluable for the unambiguous determination of the oscillation parameters.

• **Hint of CP violation in the T2K experiment.** As we have seen before, the beam flux of the T2K experiment is peaked on the first oscillation maximum and the baseline is 295 km. This configuration optimized for the discovery potential of the electron appearance signal is not able to measure the second oscillation maximum nor disentangle matter effects from CP violation effects, in the spirit of what is shown above. However, assuming a given mass hierarchy, matter effects in principle can be computed and taken into account, as shown in Eq. (30.134). High statistics and small systematic errors in the range of a few percent on the ratio of electron neutrino to electron antineutrino event rates can lead to a constraint on the oscillation parameters, including δ. The current constraint obtained by T2K is displayed in Figure 30.12. The plot shows the region in the δ–$\sin^2\theta_{13}$ plane, which is compatible with the data at the 90% CL. The parameter space is clearly constrained and provides very important information for the overall understanding of the PMNS matrix, as will be discussed in the next section. The T2K oscillation parameter measurements are, however, limited by statistics. T2K will collect more data in both neutrino and antineutrino beam operation mode in the coming years to help improve the constraints on the parameters. Longer-term next-generation experiments are being considered to ascertain the existence of CP violation in the leptonic sector.

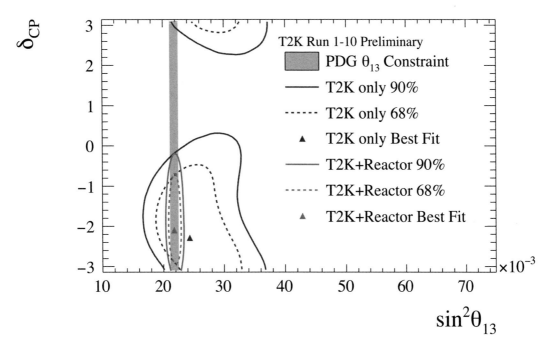

Figure 30.12 The 68% and 90% credible interval contours from the T2K data fit, both with and without reactor constraint and assuming normal hierarchy. Courtesy T2K Collaboration.

30.11   Global Fit of Neutrino Oscillation Parameters

The description of all experimental searches for neutrino oscillations is beyond the scope of this book. In the previous sections, it was only possible to describe the main features of these experiments and give a few notable examples. When results from all the experiments are taken together, one can over-constrain the parameters of the PMNS matrix, leading to an improved overall picture. Global analyses [429–432] of all neutrino oscillation data, including the recent highest precision measurements, indicate that **the "minimal" three-neutrino PMNS framework described in the previous sections, is sufficient to completely describe the observed oscillation phenomenology** (apart from some "anomalies" in terrestrial short baseline experiments).

The very long list of the main neutrino oscillation experiments that led to this conclusion is summarized in Table 30.1. The "solar sector," also called "1-2 sector" because driven by θ_{12} and Δm_{21}^2, has been measured by solar and reactor neutrino experiments. On the other hand, the "atmospheric sector" or "2-3 sector" driven by θ_{23} and $|\Delta m_{32}^2|$ (only its absolute value) has been measured by atmospheric neutrino, reactor, and accelerator neutrino experiments. As previously mentioned, these experimental data indicate that $\Delta m_{21}^2 \ll |\Delta m_{31}^2| \approx |\Delta m_{32}^2|$.

• **Neutrino mass hierarchy.** From the MSW effect described in Section 30.9, it follows from the observation of

Experiment	Type	Source	Typ. E	Typ. L	Measured
Homestake [433]	Solar	ν_e	0.1–10 MeV	10^{11} m	Rate
GALLEX & GNO [434]	Solar	ν_e	0.2–10 MeV	10^{11} m	Rate
SAGE [435]	Solar	ν_e	0.2–10 MeV	10^{11} m	Rate
Super-Kamiokande I-IV [436]	Solar	ν_e	4–10 MeV	10^{11} m	e-scattering
SNO [437]	Solar	ν_e	5–20 MeV	10^{11} m	CC, NC, e-scattering
Borexino [438–440]	Solar	ν_e	0.862 MeV	10^{11} m	e-scattering
Super-Kamiokande I–IV [441, 442]	Atm	ν_μ, $\bar{\nu}_\mu$, ν_e, $\bar{\nu}_e$	0.1–100 GeV	10–10^4 km	Rate, L/E
IceCube/DeepCore [443]	Atm	ν_μ, $\bar{\nu}_\mu$, ν_e, $\bar{\nu}_e$	0.1–100 GeV	10–10^4 km	Rate, L/E
Kamland [444–446]	Reactor	$\bar{\nu}_e$	1–10 MeV	100 km	Rate, L/E
Double-Chooz [447]	Reactor	$\bar{\nu}_e$	1–10 MeV	1 km	Rate, L/E
Daya-Bay [448]	Reactor	$\bar{\nu}_e$	1–10 MeV	1 km	Rate, L/E
Reno [449]	Reactor	$\bar{\nu}_e$	1–10 MeV	1 km	Rate, L/E
MINOS [450]	Accelerator	ν_μ, $\bar{\nu}_\mu$	2 GeV	732 km	Rate, L/E
T2K [451]	Accelerator	ν_μ, $\bar{\nu}_\mu$	0.6 GeV	295 km	Rate, L/E
NOνA [452, 453]	Accelerator	ν_μ, $\bar{\nu}_\mu$	1 GeV	810 km	Rate, L/E

Table 30.1 Summary of the neutrino experiments that most strongly constrain the oscillation parameters.

solar neutrinos that:

$$\Delta m_{12}^2 > 0 \tag{30.137}$$

On the other hand, the sign of Δm_{31}^2 is not yet known: whether m_3 is the heaviest mass eigenvalue (normal hierarchy, $\Delta m_{31}^2 > 0$) or m_3 is the lightest one (inverted hierarchy, $\Delta m_{31}^2 < 0$) remains to be experimentally determined in the future. The two hierarchy cases are called **normal ordering (NO)** and **inverted ordering (IO)**, as illustrated in Figure 30.13:

$$\text{NO:} \quad m_1 < m_2 < m_3, \qquad \text{IO:} \quad m_3 < m_1 < m_2 \tag{30.138}$$

The normal ordering can lead to a scenario where there is a strong **hierarchy of the neutrino masses**, i.e.

$$m_1 \ll m_2 \ll m_3 \qquad \text{Normal hierarchy} \tag{30.139}$$

On the other hand, if the mass m_1 is large compared to any of the mass-squared differences, then we find a degenerate neutrino mass scenario, in which:

$$m_1 \approx m_2 \approx m_3 \tag{30.140}$$

In the inverted ordering, the situation is similar except that m_3 plays the role of the lowest mass, hence the **inverted mass hierarchy** of the neutrino masses corresponds to the case:

$$m_3 \ll m_1 < m_2 \qquad \text{Inverted hierarchy} \tag{30.141}$$

An experimental determination is crucial for the understanding of the origin of neutrino masses, which is expected to relate to physics beyond the SM (see Chapter 31). In particular, an inverted hierarchy would be a strong sign that some totally unexpected physics is underlying the masses and flavor problem. Last but not least, the knowledge of the mass hierarchy is a necessary ingredient to resolve the CP/T violation problem.

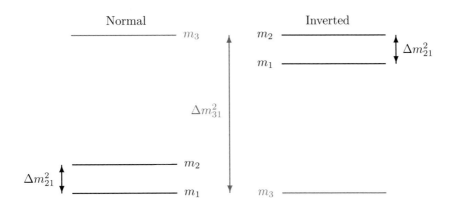

Figure 30.13 Normal $\Delta m_{31}^2 > 0$ (NH) and inverted hierarchy $\Delta m_{31}^2 < 0$ (IH).

• **Neutrino oscillation parameters.** The lack of knowledge on the sign of Δm_{32}^2 implies that numerical fits must presently be performed twice, once for each of the mass hierarchy hypotheses. The results from one such global analysis [432] are summarized in Table 30.2. The global current knowledge on the solar sector can be summarized to:

$$\left.\begin{array}{l} \sin^2 \theta_{12} = 0.310^{+0.013}_{-0.012} \\ \Delta m_{21}^2 = (7.39^{+0.21}_{-0.20}) \times 10^{-5} \text{ eV}^2 \end{array}\right\} \text{Solar sector} \qquad (30.142)$$

and the current knowledge on the atmospheric sector is (considering both signs of Δm_{31}^2 separately):

$$\left.\begin{array}{l} \Delta m_{31}^2 = \left\{ \begin{array}{l} 2.528^{+0.029}_{-0.031} \\ -(2.510^{+0.030}_{-0.031}) \end{array} \times 10^{-3} \text{ eV}^2 \right. \\ \sin^2 \theta_{23} = \left\{ \begin{array}{ll} 0.563^{+0.018}_{-0.024} & \text{for } \Delta m_{31}^2 > 0 \\ 0.565^{+0.017}_{-0.022} & \text{for } \Delta m_{31}^2 < 0 \end{array} \right. \end{array}\right\} \text{Atmospheric sector} \qquad (30.143)$$

Finally, the current knowledge on the 1-3 sector, also depending on the assumed sign of Δm_{31}^2, is:

$$\sin^2 \theta_{13} = \left.\left\{ \begin{array}{ll} 0.02237^{+0.00066}_{-0.00065} & \text{for } \Delta m_{31}^2 > 0 \\ 0.02259 \pm 0.00065 & \text{for } \Delta m_{31}^2 < 0 \end{array} \right\} \right. \text{1–3 sector} \qquad (30.144)$$

dominated by the precise measurements at the recent reactor experiments Double-CHOOZ, Daya Bay, and RENO.

The present level of understanding of the PMNS matrix represents an incredible experimental achievement, which was essentially initiated with the first atmospheric results of Super-Kamiokande in 1998 [441], complemented by the spectacular solar neutrino observations, and then completed by long baseline measurements at accelerators and reactors. During the last decade progress has been striking, yielding today's impressive accuracies on the parameters, opening a new window to explore physics beyond the SM.

When globally fitted to all available data, the oscillation parameters are constrained with high precision. Perhaps not surprisingly, the determination of the δ is an area where global fits give a large range. This is presently an area of intense experimental investigation! Is CP symmetry violated in the leptonic sector? This question will presumably be answered in a definite way by the year 2035.

	Normal Hierarchy (best fit)		**Inverted Hierarchy** $\left(\Delta\chi^2 = 10.4\right)$	
	bfp $\pm 1\sigma$	3σ range	bfp $\pm 1\sigma$	3σ range
$\frac{\Delta m_{21}^2}{10^{-5}\text{eV}^2}$	$7.39^{+0.21}_{-0.20}$	$6.79 \to 8.01$	$7.39^{+0.21}_{-0.20}$	$6.79 \to 8.01$
$\frac{\Delta m_{3\ell}^2}{10^{-3}\text{eV}^2}$	$+2.528^{+0.029}_{-0.031}$	$+2.436 \to +2.618$	$-2.510^{+0.030}_{-0.031}$	$-2.601 \to -2.419$
$\sin^2\theta_{12}$	$0.310^{+0.013}_{-0.012}$	$0.275 \to 0.350$	$0.310^{+0.013}_{-0.012}$	$0.275 \to 0.350$
$\theta_{12}/°$	$33.82^{+0.78}_{-0.76}$	$31.61 \to 36.27$	$33.82^{+0.78}_{-0.75}$	$31.61 \to 36.27$
$\sin^2\theta_{23}$	$0.563^{+0.018}_{-0.024}$	$0.433 \to 0.609$	$0.565^{+0.017}_{-0.022}$	$0.436 \to 0.610$
$\theta_{23}/°$	$48.6^{+1.0}_{-1.4}$	$41.1 \to 51.3$	$48.8^{+1.0}_{-1.2}$	$41.4 \to 51.3$
$\sin^2\theta_{13}$	$0.02237^{+0.00066}_{-0.00065}$	$0.02044 \to 0.02435$	$0.02259^{+0.00065}_{-0.00065}$	$0.02064 \to 0.02457$
$\theta_{13}/°$	$8.60^{+0.13}_{-0.13}$	$8.22 \to 8.98$	$8.64^{+0.12}_{-0.13}$	$8.26 \to 9.02$
$\delta/°$	221^{+39}_{-28}	$144 \to 357$	282^{+23}_{-25}	$205 \to 348$

Table 30.2 Three-flavor oscillation parameters from a global fit to all experimental data. "bfp" stands for best-fit-point.   Reprinted from I. Esteban, M. Gonzalez-Garcia, A. Hernandez-Cabezudo, M. Maltoni, and T. Schwetz, "Global analysis of three-flavour neutrino oscillations: Synergies and tensions in the determination of θ_{23}, δ, and the mass ordering," *JHEP*, vol. 01, p. 106, 2019, under CC BY license http://creativecommons.org/licenses/by/4.0/. Most up-to-date results can be found at www.nu-fit.org.

The 3σ CL ranges of the magnitudes of the elements of the three-flavor leptonic mixing matrix under the assumption that the matrix U^{PMNS} is unitary are given in the following:

$$\left|U^{PMNS}\right| = \begin{pmatrix} 0.797 \to 0.842 & 0.518 \to 0.585 & 0.143 \to 0.156 \\ 0.243 \to 0.490 & 0.473 \to 0.674 & 0.651 \to 0.772 \\ 0.295 \to 0.525 & 0.493 \to 0.688 & 0.618 \to 0.744 \end{pmatrix} \tag{30.145}$$

The structure of the PMNS consequently looks very different from that of the CKM (cf. Eq. (30.16)), which is very diagonal. The leptonic sector is characterized by a very large degree of mixing. As of today, the origin of this pattern remains an unsolved and profound puzzle!

Problems

Ex 30.1 The CKM unitarity triangle. Use the Wolfenstein parameterization of the CKM matrix to show that unitarity leads to the triangle shown in Figure 30.5.

Ex 30.2 The K_S^0 regeneration. Discuss quantitatively the phenomenon of the K_S^0 regeneration.

Ex 30.3 Atmospheric neutrinos. Atmospheric neutrinos are produced by the interactions of primary cosmic rays in the Earth's upper atmosphere. Such interactions produce pions which then decay into muons. Those muons, whose energies are not too high, also decay before reaching the ground. In this case, what is the expected ratio between muon neutrinos and electron neutrinos $(\nu_\mu + \bar{\nu}_\mu)/(\nu_e + \bar{\nu}_e)$ on the ground? How does it vary at higher energies? Discuss how this ratio is expected to change if one includes the known neutrino flavor oscillations.

Ex 30.4 *CP* **violation in future long-baseline experiments.** Use Eq. (30.134) to plot the expected oscillation probabilities for $\nu_\mu \to \nu_e$ and $\bar{\nu}_\mu \to \bar{\nu}_e$ over the full range of values of the δ parameter in the future Hyper-Kamiokande experiment (see Figure 129 of K. Abe *et al.*, *Hyper-Kamiokande Design Report*, arXiv:1805.04163 (2018)). Explain Figure 133 of the same reference. How does it compare with the case of the DUNE experiment described in B. Abi *et al.*, "Long-baseline neutrino oscillation physics potential of the DUNE experiment," *Eur. Phys. J. C*, vol. 80, no. 10, p. 978, 2020?

31   Beyond the Standard Model

> There is still a lot to learn and there is always great stuff out there. Even mistakes can be wonderful.
>
> *Robin Williams*

31.1   Why?

The SM is without doubt a great success of particle physics. It describes a very wide range of precise experimental measurements remarkably well. Nonetheless, it is still considered as a model based on perturbative quantum field theory constructed on fundamental gauge symmetries that are spontaneously broken. There is no "explanation" for the large number of input parameters:

- the three gauge couplings;

- the twelve Yukawa couplings of the fermions to the Higgs field;

- the eight independent parameters of the Cabibbo–Kobayashi–Maskawa and Pontecorvo–Maki–Nakagawa–Sakata mixing matrices.

- There is also an additional phase θ_{CP} in the QCD Lagrangian, which we have not spoken about, that could however lead to CP violation in the strong interaction. Experimental data is consistent with $\theta_{CP} \simeq 0$ and this is set "by hand" in the SM. This issue is dubbed the **strong CP problem**.

One additional caveat is that there is no explanation for the three generations (or families) of elementary fermions. The three generations (u, d, ν_e, e^-), (c, s, ν_μ, μ^-), and (t, b, ν_τ, τ^-) share the same properties except their rest masses. There seem to be no more than three generations. The distribution of their masses suggests a certain pattern, but the reasons remain a mystery. The hope would be to find a broader theoretical framework where all (or at least some) of the above parameters could be determined or constrained within the theory.

There are also open questions in particle physics and astrophysics that indicate that the SM cannot be the final theory of particle physics. Some of these are listed below.

- What are the properties of the scalar Higgs boson? At present they seem to be compatible with the SM predictions (see Eq. (29.41)) but will this stand the test of time as the precision of measurements improves with more experimental data?

- What is the origin of the flavor and of CP violation? This question is somewhat related to a more precise understanding of the spontaneous symmetry breaking in the SM, as the fields acquire their masses and mixing as a consequence of this transition.

- Is there a grand unification of the electroweak and strong force at very high energy (see Section 31.5)?

- Are neutrinos Dirac or Majorana particles (see Section 31.3)?

- Is supersymmetry realized in Nature (see Section 31.6)?

- What is the particle nature of dark matter? Is there a dark sector? What is dark energy? (See Section 32.5.)

- Are there surprises?

All these unanswered questions point to the existence of new physics beyond the SM. New and presently unresolved issues related to neutrino masses are illustrated in the following sections.

31.2   Neutrino Mass Terms

Neutrinos were assumed to be massless in the SM, where they remain without a mass term after spontaneous symmetry breaking. This choice is based on the following facts:

- The a priori absence of theoretical reasons for a right-handed neutrino. Experimentally the neutrino was found to be compatible with a 100% left-handed polarization. As a matter of fact, right-handed neutrinos possess no interactions in the SM theory and represent therefore a **sterile** particle.

- All experiments that attempt to directly measure non-vanishing neutrino masses have led to upper limits that are significantly smaller than the masses of the corresponding charged leptons.

- The experimental observations indicate that the lepton flavor and total lepton number are conserved in all known elementary interactions.

The observation of neutrino flavor oscillations is clear evidence for the existence of non-degenerate neutrino masses. Although neutrino oscillations do not probe the absolute neutrino masses, the non-vanishing observed mass squared differences allow us to write, assuming three neutrino families, the following relations:

$$m_1, \quad m_2^2 = m_1^2 + (m_2^2 - m_1^2) = m_1^2 + \Delta m_{21}^2, \quad \text{and} \quad m_3^2 = m_1^2 + \Delta m_{31}^2 = m_2^2 + \Delta m_{32}^2 \qquad (31.1)$$

where m_1 can be defined as the presently unknown **neutrino absolute mass scale**, and where the differences squared Δm_{ij}^2 can be determined by neutrino oscillation experiments. It is generally accepted in the community that experimental results point to the following indicative bounds:

$$0 \leq m_1 \lesssim 1 \text{ eV} \qquad (31.2)$$

So, certainly, neutrinos, although massive, are much lighter than the other known fermions. As of today, there is no clear response as to why this is so. A related issue is the question of the **nature of the neutrino**. For several decades, neutrinos were assumed to be massless and 100% polarized due to the left-handed weak current. Is this a *consequence* of the interaction or does it represent *an intrinsic property* of neutrinos?

A main difference between neutrinos and other fundamental fermions, such as the charged leptons and the quarks, is indeed that neutrinos are the only known fundamental fermions that are electrically neutral. Hence, unlike other fermions, the measurement of their electric charge *cannot* determine if the particle is a neutrino or an antineutrino. To distinguish a neutrino from an antineutrino we must introduce the separate global **lepton number**. There is, however, no a priori reason for this *global* quantum number to be *exactly* conserved in Nature. As a matter of fact, it is not known if the neutrino and antineutrino are indeed different particles or not!

The massiveness of neutrinos also opens up new possibilities: are neutrinos stable particles or can they decay? What are the neutrinos' electromagnetic properties, such as their anomalous magnetic moment?

Under these assumptions, we now discuss which modifications we can introduce to the fundamental Lagrangian of our theory to take into account these new effects. The a priori easiest solution to give mass to neutrinos is to add a **Dirac mass term**. It seems straightforward to include this, since all other fermions acquire mass in this way. If we consider for the moment only one flavor of neutrino then we have, in analogy to the electron, a Dirac neutrino mass term in the Lagrangian of the type:

$$\mathcal{L}_{D} = -m_{D}\bar{\nu}\nu \tag{31.3}$$

where m_{D} is the **Dirac neutrino mass**.

In order to describe the antineutrino case, we use the charge conjugation operator C defined in Eq. (8.225) and write:

$$\nu^{c} \equiv C\nu^{*} \tag{31.4}$$

where we considered both left-handed and right-handed chiral projections of the charge-conjugated state. We note that the definition of the charge conjugation operator $C = -i\gamma^{2}$ implies that:

$$C\left(\frac{1 \pm \gamma^{5}}{2}\right) = (-i\gamma^{2})\left(\frac{1 \pm \gamma^{5}}{2}\right) = -i\left(\frac{1 \mp \gamma^{5}}{2}\right)\gamma^{2} = \left(\frac{1 \mp \gamma^{5}}{2}\right)C \tag{31.5}$$

Consequently, we can for example consider the left-handed chiral projection of the charge-conjugated neutrino state:

$$(\nu^{c})_{L} = \left(\frac{1 - \gamma_{5}}{2}\right)C\nu^{*} = C\left(\frac{1 + \gamma_{5}}{2}\nu^{*}\right) = C\left(\frac{1 + \gamma_{5}}{2}\nu\right)^{*} = C(\nu_{R})^{*} = (\nu_{R})^{c} \tag{31.6}$$

and similarly for $(\nu^{c})_{R}$. We can summarize these two results in the following way:

$$(\nu^{c})_{L} = (\nu_{R})^{c} \quad \text{and} \quad (\nu^{c})_{R} = (\nu_{L})^{c} \tag{31.7}$$

It is customary to introduce a simpler notation where one removes the parentheses from the above expressions and to first consider the charge conjugation and then the chiral projection:

$$\nu_{L}^{c} \equiv (\nu^{c})_{L} = (\nu_{R})^{c} \quad \text{and} \quad \nu_{R}^{c} \equiv (\nu^{c})_{R} = (\nu_{L})^{c} \tag{31.8}$$

How shall we interpret this notation? The Dirac spinor ν has four components, and it can therefore represent any state: a left-handed or right-handed neutrino as well as a left-handed or right-handed antineutrino! Therefore, the spinor ν^{c} corresponds to **a notation with which we describe the charge-conjugated state** of any spinor ν.

We can rewrite the Dirac mass term using this notation. Using $\nu = \nu_{L} + \nu_{R}$, we find:

$$\mathcal{L}_{D} = -m_{D}\,\bar{\nu}\,\nu = -m_{D}\left(\overline{\nu_{L}}\,\nu_{R} + \overline{\nu_{R}}\,\nu_{L}\right) \tag{31.9}$$

However, we could as well have started from the charge-conjugated state ν^{c} and write:

$$\mathcal{L}_{D} = -m_{D}\,\overline{\nu^{c}}\,\nu^{c} = -m_{D}\left(\overline{\nu_{R}^{c}}\,\nu_{L}^{c} + \overline{\nu_{L}^{c}}\,\nu_{R}^{c}\right) \tag{31.10}$$

Both expressions lead to a valid Dirac mass term! They are equivalent. We note that both mass terms impose the inclusion of right-handed neutrinos ν_{R}, which, as already mentioned, have no interactions in the electroweak theory, hence are sterile. At this stage, the right-handed neutrinos enter only in our Lagrangian through the mass terms above. Although this looks like a trivial addition to the SM theory, it raises the following two issues: (1) Does the right-handed neutrino exist and if so, which role does it play in the Universe? (2) If the neutrino masses are to be generated like for other fermions by spontaneous symmetry breaking (Brout–Englert–Higgs mechanism), why are their masses so much lighter than those of the other fundamental charged fermions? These are unsolved questions.

Let us further consider the **consequences of the Dirac nature of the neutrino field** on the physical neutrinos ν and antineutrinos $\bar{\nu}$. The reader should pay attention that this time the bar over the symbol "ν" implies that we are dealing with an antineutrino. Since we are assuming a Dirac field, we have a priori four degrees of freedom, corresponding to the left-handed and right-handed chiral states with positive and negative energy solutions. In terms of helicity states, the four degrees of freedom correspond to two helicity states for each neutrino and antineutrino (so in total $2 \times 2 = 4$ degrees of freedom). Considering the neutrino case, we label ν_- for the negative helicity state and ν_+ for the positive helicity state. Similarly, the antineutrino can be in the $\bar{\nu}_-$ or $\bar{\nu}_+$ helicity states. So finally, we have our four cases:

$$\nu_-, \quad \nu_+, \quad \bar{\nu}_-, \quad \bar{\nu}_+ \tag{31.11}$$

These four states are graphically pictured in Figure 31.1. The momenta and the spins are also drawn.

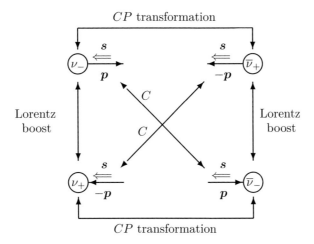

Figure 31.1 Transformation of Dirac neutrinos in a Lorentz boost which reverses momentum, and in the C and CP conjugations.

We note that for a massive particle, it is always possible to find a Lorentz boost that will reverse the momentum of the particles (since a massive particle must travel with a velocity $v < c$, it is always possible to consider a boosted inertial system with a velocity $|u| > v$ from which the momentum of the neutrino is reversed; however, the boost does not alter the spin direction). The two helicity states of a neutrino ($\nu_- \leftrightarrow \nu_+$) or of an antineutrino ($\bar{\nu}_- \leftrightarrow \bar{\nu}_+$) are therefore connected by such a Lorentz transformation, as illustrated in Figure 31.1.

What is the connection between the neutrino and the antineutrino? Let us consider the charge conjugation operator C (see Figure 31.1). The C operator does not change the spin nor the momentum, and hence not the helicity. This can be expressed as the C transformation changing a particle into its antiparticle while leaving its helicity unchanged. Accordingly, we can write:

$$(\nu_-)^c = \eta_c \bar{\nu}_- \quad \text{and} \quad (\nu_+)^c = \eta_c \bar{\nu}_+ \tag{31.12}$$

where η_c is an independent phase. At this stage, we should however recall neutrinos only participate in the weak interaction and with a pure $V - A$ coupling. This means that, from the four helicity neutrino states $\nu_-, \nu_+, \bar{\nu}_-, \bar{\nu}_+$, the ν_- and $\bar{\nu}_+$ are enhanced by their polarization factor m_ν/E while the two ν_+ and $\bar{\nu}_-$ are highly suppressed. Hence, the neutrino ν_- and antineutrino $\bar{\nu}_+$ that are enhanced in the weak interaction are connected CP-conjugates of one another:

$$(\nu_-)^{CP} = \eta_{CP} \bar{\nu}_+ \tag{31.13}$$

where η_{CP} is an independent phase. Similarly, the helicity states that are highly suppressed are evidently also CP-conjugates of one another:

$$(\nu_+)^{CP} = \eta_{CP} \bar{\nu}_- \tag{31.14}$$

Given the lightness of neutrinos (i.e., $m \lesssim 1$ eV), the helicity suppression is very strong for all practical cases. For example, a reactor neutrino with an energy $E \simeq 1$ MeV has a polarization of more than 99.9999%. The suppression of the wrong helicity is more than 10^6! Neutrinos are so strongly polarized that it is practically impossible to experimentally identify their two helicity states.

31.3   Majorana Neutrinos

Neutrinos differentiate themselves from other fundamental fermions in the fact that they are electrically neutral. As already mentioned, it is therefore *not* possible to distinguish the neutrino and the antineutrino from their charge. There must be another quantum number, called the lepton number L. On the other hand, this is just an assumption that needs to be verified experimentally. One can consider other scenarios: could the neutrino, like the photon, be an eigenstate of the charge conjugation? Starting from any Dirac spinor Ψ, it is always possible to build **a Majorana[1] spinor** $\Psi_{\rm M}$ by writing:

$$\Psi_{\rm M} \equiv \Psi + \eta \Psi^c \tag{31.15}$$

where η is a complex phase. If neutrinos were described by such a spinor, they would be called **Majorana neutrinos** instead of Dirac neutrinos and would possess only *two* independent states, as illustrated in Figure 31.2. In the case where neutrinos are Majorana particles, the neutrino and the antineutrino CP coupled states are

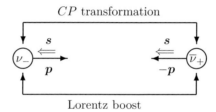

Figure 31.2 Predicted behavior of Majorana neutrinos under a Lorentz boost which reverses momentum and a CP transformation.

also directly connected by a Lorentz boost. We have lost degrees of freedom, and the four degrees of freedom of the Dirac neutrinos are reduced to two degrees of freedom of the Majorana neutrinos. This is equivalent to stating that **the known neutrinos in Nature exist only in a left-handed chiral state**.

What then is the mass term for a Majorana neutrino? A Dirac mass term necessarily requires a left-handed and a right-handed chiral state, since it is of the form $\overline{\Psi}\Psi$ or equivalently $\overline{\Psi^c}\Psi^c$. A Majorana mass term $\overline{\Psi}_M \Psi_M$ must therefore rely on only one chirality which is naturally given by factors of the type $\overline{\Psi}\Psi^c$ or $\overline{\Psi^c}\Psi$. We consider for example the **left-handed Majorana neutrino mass term** $\mathcal{L}_{\rm ML}$ given by:

$$\mathcal{L}_{\rm ML} = -\frac{1}{2}m_L\left(\overline{\nu_L}\,\nu_R^c + \overline{\nu_R^c}\,\nu_L\right) \qquad \Delta L = \pm 2 \tag{31.16}$$

where m_L is the left-handed Majorana neutrino mass. In this way, we have introduced a neutrino mass using only one chirality of the particle and its charge conjugate. This means that this mass term effectively needs only two degrees of freedom, unlike the Dirac mass term which uses four. The price to pay is the violation of the lepton number by two units, since the mass term couples a neutrino ($L = +1$) with an antineutrino

---

[1]   Ettore Majorana (1906–missing since 1938), Italian theoretical physicist.

$(L = -1)$. This affirmation can be verified by considering that a global gauge transformation of our spinor $\Psi \to e^{+i\alpha}\Psi$ implies:

$$\Psi^c = C\Psi^* \to Ce^{-i\alpha}\Psi^* = e^{-i\alpha}\Psi^c \tag{31.17}$$

and also $\overline{\Psi} \to e^{-i\alpha}\overline{\Psi}$ and $\overline{\Psi^c} \to e^{+i\alpha}\overline{\Psi^c}$. Hence, the Dirac mass terms transforms under gauge transformation as:

$$\overline{\Psi}\Psi \to e^{-i\alpha}\overline{\Psi}e^{+i\alpha}\Psi = \overline{\Psi}\Psi \quad \text{and} \quad \overline{\Psi^c}\Psi^c \to \overline{\Psi^c}\Psi^c \tag{31.18}$$

and is therefore invariant under our global gauge transformation. On the other hand, the Majorana mass terms lead to a gauge violation. For example:

$$\overline{\Psi}\Psi^c \to e^{-i\alpha}\overline{\Psi}e^{-i\alpha}\Psi^c \neq \overline{\Psi}\Psi^c \tag{31.19}$$

This is interpreted as the violation of the lepton number in our theory.

Consequently, a Majorana mass term violates the lepton number and is therefore customarily labeled $\Delta L = \pm 2$. On the other hand a Dirac mass term conserves the lepton number and is therefore by contrast a $\Delta L = 0$ process. Is this a problem? As we discussed in Section 1.6, the lepton number arises as the consequence of a $U(1)$ global symmetry (and not a *local* one), so there is no fundamental principle that allows us to state that this conservation should be exact. In the SM, violation of the lepton number is not allowed, and a lepton number violating Majorana mass term hence clearly represents physics beyond the SM.

31.4   General Neutrino Mass Matrix

Up until now we have considered the following two types of neutrino mass terms:

$$-\mathcal{L}_{\mathrm{D}} = m_{\mathrm{D}}\left(\overline{\nu_L}\,\nu_R + \overline{\nu_R}\,\nu_L\right) \quad \text{and} \quad -\mathcal{L}_{\mathrm{ML}} = \frac{1}{2}m_L\left(\overline{\nu_L}\,\nu_R^c + \overline{\nu_R^c}\,\nu_L\right)$$

We can in addition consider a "right-handed" Majorana mass term $\mathcal{L}_{\mathrm{MR}}$ by swapping the order of the particles and their charge-conjugated states in our original Majorana term $\mathcal{L}_{\mathrm{ML}}$:

$$-\mathcal{L}_{\mathrm{MR}} = \frac{1}{2}m_R\left(\overline{\nu_L^c}\,\nu_R + \overline{\nu_R}\,\nu_L^c\right) \tag{31.20}$$

where m_R is the right-handed Majorana neutrino mass. The Dirac mass term can conveniently be rewritten with the spinors and their charge conjugates by noting:

$$-\mathcal{L}_{\mathrm{D}} = m_{\mathrm{D}}\left(\overline{\nu_L}\,\nu_R + \overline{\nu_R}\,\nu_L\right) = \frac{1}{2}m_{\mathrm{D}}\left(\overline{\nu_L}\,\nu_R + \overline{\nu_R}\,\nu_L + \overline{\nu_L^c}\,\nu_R^c + \overline{\nu_R^c}\,\nu_L^c\right) \tag{31.21}$$

It then makes sense to express all three types of mass terms in matrix form, which then becomes the general **Dirac–Majorana neutrino mass term** $\mathcal{L}_{\mathrm{DM}}$:

$$
\begin{aligned}
-\mathcal{L}_{\mathrm{DM}} &= \frac{1}{2}m_{\mathrm{D}}\left(\overline{\nu_L}\,\nu_R + \overline{\nu_R}\,\nu_L + \overline{\nu_L^c}\,\nu_R^c + \overline{\nu_R^c}\,\nu_L^c\right) + \frac{1}{2}m_L\left(\overline{\nu_L}\,\nu_R^c + \overline{\nu_R^c}\,\nu_L\right) + \frac{1}{2}m_R\left(\overline{\nu_L^c}\,\nu_R + \overline{\nu_R}\,\nu_L^c\right) \\
&= \frac{1}{2}m_{\mathrm{D}}\left(\overline{\nu_L}\,\nu_R + \overline{\nu_L^c}\,\nu_R^c\right) + \frac{1}{2}m_L\left(\overline{\nu_L}\,\nu_R^c\right) + \frac{1}{2}m_R\left(\overline{\nu_L^c}\,\nu_R\right) + \text{h.c.} \\
&= \frac{1}{2}\begin{pmatrix} \overline{\nu_L} & \overline{\nu_L^c} \end{pmatrix}\begin{pmatrix} m_L & m_{\mathrm{D}} \\ m_{\mathrm{D}} & m_R \end{pmatrix}\begin{pmatrix} \nu_R^c \\ \nu_R \end{pmatrix} + \text{h.c.}
\end{aligned}
\tag{31.22}
$$

The expression $\mathcal{L}_{\mathrm{DM}}$ can be interpreted in different ways. It shows quite generally the richness and complexity of the neutrino mass theory compared to that of the other fermions, which we recall stems principally from the fact that the neutrinos are the only elementary fermions that are electrically neutral.

We can at this stage consider that the "right-handed" degrees of freedom do not play a role in the physical states of the light known neutrinos. In this case, it is customary to express these states with a different notation, hence getting:

$$-\mathcal{L}_{\text{DM}} = \frac{1}{2} \left(\overline{\nu_L} \ \overline{N_L^c} \right) \begin{pmatrix} m_L & m_D \\ m_D & m_R \end{pmatrix} \begin{pmatrix} \nu_R^c \\ N_R \end{pmatrix} + \text{h.c.} \tag{31.23}$$

where N_R (N_L^c) represent a right-handed (left-handed) neutrino field, which can be decoupled from the ordinary light neutrino fields ν_L (resp. ν_R^c). The parameters to be determined are the two Majorana masses m_L and m_R and the Dirac mass m_D.

Since the mass matrix written in this form is not diagonal, it can be interpreted as describing a Lagrangian mass term in a basis of fields, which does not coincide with the mass eigenstates. We should therefore diagonalize the mass matrix to find the corresponding mass eigenstates of given mass. Consequently, we should find the basis transformation T that satisfies:

$$T^\dagger \begin{pmatrix} m_L & m_D \\ m_D & m_R \end{pmatrix} T = \begin{pmatrix} m_1 & 0 \\ 0 & m_2 \end{pmatrix} \tag{31.24}$$

where $T^\dagger T = \mathbb{1}$.

In the two-dimensional case, and for real masses, we can simply parametrize the transformation by rotation of an angle θ:

$$T = \begin{pmatrix} \cos\theta & \sin\theta \\ -\sin\theta & \cos\theta \end{pmatrix} \equiv \begin{pmatrix} c & s \\ -s & c \end{pmatrix} \tag{31.25}$$

Therefore, we have:

$$\underbrace{\begin{pmatrix} m_L & m_D \\ m_D & m_R \end{pmatrix}}_{M} T = T \begin{pmatrix} m_1 & 0 \\ 0 & m_2 \end{pmatrix} \tag{31.26}$$

or

$$M \begin{pmatrix} c & s \\ -s & c \end{pmatrix} = \begin{pmatrix} c & s \\ -s & c \end{pmatrix} \begin{pmatrix} m_1 & 0 \\ 0 & m_2 \end{pmatrix} = \begin{pmatrix} cm_1 & sm_2 \\ -sm_1 & cm_2 \end{pmatrix} \tag{31.27}$$

We can separate this matrix equation into two vectors to obtain:

$$M \begin{pmatrix} c \\ -s \end{pmatrix} = m_1 \begin{pmatrix} c \\ -s \end{pmatrix} \quad \text{and} \quad M \begin{pmatrix} s \\ c \end{pmatrix} = m_2 \begin{pmatrix} s \\ c \end{pmatrix} \tag{31.28}$$

which tells us that $\vec{e}_1 = (c, \ s)$ and $\vec{e}_2 = (s, c)$ are the eigenvectors of M. We can rewrite this as two equations of the type:

$$M\vec{e}_i = m_i \vec{e}_i \quad \Rightarrow \quad (M - m_i)\vec{e}_i = 0 \qquad (i = 1, 2) \tag{31.29}$$

A solution of this problem requires that the determinant vanishes:

$$\det(M - m_i) = 0 \tag{31.30}$$

Hence:

$$\det \begin{pmatrix} m_L - m_i & m_D \\ m_D & m_R - m_i \end{pmatrix} = 0 \quad \Rightarrow (m_L - m_i)(m_R - m_i) - m_D^2 = 0$$

$$\Rightarrow m_i^2 - m_i(m_L + m_R) + (m_L m_R - m_D^2) = 0$$

This leads to the two solutions:

$$m_\pm = \frac{1}{2} \left\{ (m_L + m_R) \pm \sqrt{(m_L - m_R)^2 + 4m_D^2} \right\} \tag{31.31}$$

where we have replaced the indices $i = 1, 2$ by "$\pm$" corresponding to the signs in front of the square root. The rotation angle can be derived directly by noting that Eq. (31.29) can be expressed as:

$$\begin{pmatrix} m_L - m_1 & m_D \\ m_D & m_R - m_1 \end{pmatrix} \begin{pmatrix} c \\ -s \end{pmatrix} = 0 \quad \text{and} \quad \begin{pmatrix} m_L - m_2 & m_D \\ m_D & m_R - m_2 \end{pmatrix} \begin{pmatrix} s \\ c \end{pmatrix} = 0$$

from which one can get the relations:

$$\frac{s}{c} = \frac{m_L - m_1}{m_D} \quad \text{and} \quad \frac{c}{s} = \frac{m_R - m_1}{m_D}$$

or

$$\tan 2\theta = \frac{\sin 2\theta}{\cos 2\theta} = \frac{2sc}{c^2 - s^2} = \frac{2}{\dfrac{c}{s} - \dfrac{s}{c}} = \frac{2m_D}{m_R - m_L} \tag{31.32}$$

We can now insert the diagonalized mass matrix into the Lagrangian density. Recalling $TT^\dagger = \mathbb{1}$, we obtain:

$$\begin{aligned} -\mathcal{L}_{\text{DM}} &= \frac{1}{2} \begin{pmatrix} \overline{\nu_L} & \overline{N_L^c} \end{pmatrix} \begin{pmatrix} m_L & m_D \\ m_D & m_R \end{pmatrix} \begin{pmatrix} \nu_R^c \\ N_R \end{pmatrix} + \text{h.c.} \\ &= \frac{1}{2} \begin{pmatrix} \overline{\nu_L} & \overline{N_L^c} \end{pmatrix} TT^\dagger \begin{pmatrix} m_L & m_D \\ m_D & m_R \end{pmatrix} TT^\dagger \begin{pmatrix} \nu_R^c \\ N_R \end{pmatrix} + \text{h.c.} \\ &= \frac{1}{2} \begin{pmatrix} \overline{\nu_L} & \overline{N_L^c} \end{pmatrix} T \begin{pmatrix} m_1 & 0 \\ 0 & m_2 \end{pmatrix} T^\dagger \begin{pmatrix} \nu_R^c \\ N_R \end{pmatrix} + \text{h.c.} \end{aligned} \tag{31.33}$$

We can now mention the most important limiting cases, often considered in the literature:

- **Only Dirac.** One assumes $m_L = m_R = 0$.
  This means $\theta = 45°$ and $m_\pm = m_D$.
  A Dirac neutrino can be seen as a couple of degenerate Majorana neutrinos.

- **Pseudo-Dirac.** One assumes $m_D \gg m_L, m_R$.
  This means $\theta \approx 45°$ and $m_\pm \approx m_D$. This is still essentially a Dirac neutrino.

- **Only Majorana.** One assumes $m_D = 0$.
  This means $\theta = 0°$, $m_+ = m_L$, and $m_- = m_R$.
  This is a pure Majorana neutrino case, with two independent neutrinos of different masses, each with two degrees of freedom.

- **Seesaw mechanism.** One assumes $m_R \gg m_D$, and $m_L = 0$.
  Hence in this case, we simply have:

$$m_\pm = \frac{1}{2} \left\{ m_R \pm \sqrt{m_R^2 + 4m_D^2} \right\} = \frac{m_R}{2} \left\{ 1 \pm \sqrt{1 + \frac{4m_D^2}{m_R^2}} \right\} \approx \frac{m_R}{2} \left\{ 1 \pm 1 + \frac{2m_D^2}{m_R^2} \right\} \tag{31.34}$$

This means $\theta = m_D / m_R \ll 1$, and one obtains the peculiar rest mass values:

$$m_- \approx \frac{m_D^2}{m_R} \quad \text{and} \quad m_+ \approx m_R \left(1 + \frac{m_D^2}{m_R^2} \right) \approx m_R \tag{31.35}$$

The seesaw mechanism is the most interesting case from a theoretical point of view, which can be interpreted as a natural explanation for the lightness of the ordinary neutrinos compared to their charged lepton partners.

In this model, one has effectively assumed that each ordinary (Majorana) neutrino ν_L has a heavy right-handed partner N_R. It is assumed that the neutrino N_R has escaped experimental detection because its very high mass prohibits its direct kinematical production in experiments (e.g., at accelerators). The seesaw Lagrangian $\mathcal{L}_{\rm SS}$ can be rearranged in the following way to give directly the terms responsible for the light and heavy neutrinos (since $m_L = 0$):

$$
-\mathcal{L}_{\rm SS} = \frac{1}{2} \begin{pmatrix} \overline{\nu_L} & \overline{N_L^c} \end{pmatrix} \begin{pmatrix} 0 & m_D \\ m_D & m_R \end{pmatrix} \begin{pmatrix} \nu_R^c \\ N_R \end{pmatrix} + \text{h.c.} = \frac{1}{2} \begin{pmatrix} \overline{\nu_L} & \overline{N_L^c} \end{pmatrix} \begin{pmatrix} m_D N_R \\ m_D \nu_R^c + m_R N_R \end{pmatrix} + \text{h.c.}
$$
$$
= \frac{1}{2} m_D\, \overline{\nu_L} N_R + \frac{1}{2} m_D\, \overline{N_L^c} \nu_R^c + \frac{1}{2} m_R\, \overline{N_L^c} N_R + \text{h.c.}
\tag{31.36}
$$

so finally:

$$
-\mathcal{L}_{\rm SS} = m_D\, \overline{\nu_L} N_R + \frac{1}{2} m_R\, \overline{N_L^c} N_R + \text{h.c.}
\tag{31.37}
$$

One can go a step further by assuming that in models beyond the SM (such as, e.g., GUT models discussed in the next section), the Dirac mass m_D is taken at the scale of the charged leptons or the quarks. The right-handed Majorana neutrino mass becomes therefore very large, $m_R > 10^{10}$ GeV, in order to explain the lightness of the ordinary neutrinos (hence the name seesaw). The Majorana neutrino mass defines an energy scale for the new physics (NP) beyond the SM $\Lambda_{\rm NP} \approx m_R$. After diagonalization into the physical states, we therefore have the ordinary light neutrinos with masses given by m_ν and their heavy partners at $m_R \approx \Lambda_{\rm NP}$:

$$
m_\nu = \frac{m_f^2}{\Lambda_{\rm NP}} \ll m_f \quad \text{and} \quad m_R \approx \Lambda_{\rm NP}
\tag{31.38}
$$

where m_f is the scale of the mass of the charged fermions (leptons or quarks). The existence of a new scale $\Lambda_{\rm NP}$ motivates the lightness of the ordinary neutrinos. Conversely, we can interpret the **lightness of the ordinary neutrinos as a hint for new physics at the new, very high energy scale** $\Lambda_{\rm NP}$. As stated at the beginning of this chapter, the theory of neutrino masses can be seen as a potential door to new physics beyond the SM! We will come back to this point in Section 32.1.

31.5   Grand Unification Theories

The SM is a gauge field theory based on the symmetry $SU(3)_C \times SU(2)_L \times U(1)_Y$ (see Section 29.1). While strongly theoretically motivated, parts of its construction are mostly empirical and certainly leave unexplained the question of the different gauge couplings and the assignment of the fermion quantum numbers. Even within the electroweak model (see Section 25.5), the $SU(2)_L$ group, with the coupling g_2, and the $U(1)_Y$, with the coupling g_1, are separated, in the sense that the two couplings are not related by the theory. Their ratio is given by the tangent of the Weinberg angle (recall Eq. (25.41)):

$$
\tan\theta_W = \frac{g_1}{g_2}
\tag{31.39}
$$

which must be determined experimentally. **The idea of the grand unified theories (GUT) is that the $SU(3)$, $SU(2)$, and $U(1)$ are subgroups of a larger symmetry group G and that quarks and leptons belong to the same multiplets of G.** Symbolically, one can write:

$$
G \supset SU(3) \times SU(2) \times U(1)
\tag{31.40}
$$

The group G links the previously disconnected groups $SU(3)$, $SU(2)$, and $U(1)$. When the three groups are embedded in G, their coupling constants are related by a set of Clebsch–Gordan coefficients of the group

G. Rather than the historical g_1, g_2, and α_s, we can decide to generically label three coupling constants g_i ($i = 1, 2, 3$). Following the conventional choice of couplings, we define:

$$g_1 \equiv C g_1(U(1)), \qquad g_2 \equiv g_2(SU(2)), \quad \text{and} \quad g_3 \equiv \sqrt{4\pi\alpha_s} \tag{31.41}$$

where C is a Clebsch–Gordan coefficient depending on the actual group G, and $g_1(U(1))$ and $g_2(SU(2))$ refer to the terms in the electroweak Lagrangian defined in Chapter 25. This leads to:

$$\tan\theta_W \equiv \frac{1}{C}\frac{g_1}{g_2} \tag{31.42}$$

Under the GUT assumption, the electroweak and the strong interactions would be described by a grand unified gauge theory with a single coupling constant g_G, to which all other couplings g_i are related in a specific way determined by the gauge group G. One can further speculate that the higher symmetry G is unbroken above some very large mass scale called Λ_{GUT}. This scale was already introduced in Chapter 1. All the coupling constants g_i actually depend on the scale Q^2.

Above Λ_{GUT}, the gauge couplings g_i are all equal to the coupling constant g_G and they evolve as a single coupling constant:

$$g_1(Q^2) = g_2(Q^2) = g_3(Q^2) = g_G(Q^2) \qquad \text{for } Q^2 \geq \Lambda_{GUT}^2 \tag{31.43}$$

The strong force merges with the electroweak force and the sharp separation between colored quarks and colorless leptons disappears. This underlying symmetry between quarks and leptons restored at very high energies explains why we observe the same number of quarks and leptons! It also implies that there must be a transformation of the G group that converts quarks into leptons and vice versa. Hence, non-conservation of the baryonic and leptonic numbers becomes possible. Another amazing consequence is that if C is given by the group structure, then the Weinberg angle is determined at the GUT scale by Eq. (31.42) to be $\tan\theta_W(\Lambda_{GUT}^2) = 1/C$, since $g_1(\Lambda_{GUT}^2) = g_2(\Lambda_{GUT}^2)$. We recall that in the SM it was a free parameter to be determined experimentally.

Below the breaking scale Λ_{GUT}, individual coupling constants g_i evolve separately according to their respective subgroup. The value of $g_1(Q^2)$ associated with the $U(1)$ symmetry increases with Q^2 just like the conventional electric charge increases due to the charge screening of electromagnetism (see Section 12.5). On the other hand, $g_2(Q^2)$ and $g_3(Q^2)$ are asymptotically free due to the non-Abelian nature of the groups. Consequently, they decrease with Q^2 (see Figure 18.15). **A crucial test of the GUT is that the evolution of the coupling constants from their known values at low energy should meet at the common scale Λ_{GUT} (see Figure 1.2). Let us look at all this more quantitatively.**

Defining the structure coupling constants α_i in analogy to the electromagnetic case $e^2 = 4\pi\alpha$, we get:

$$\alpha_i \equiv \frac{g_i^2}{4\pi} \qquad (i = 1, 2, 3) \tag{31.44}$$

We use Eq. (25.42), which constrains $g_1^2 = e^2/\cos^2\theta_W$ in the electroweak theory to write in the case of the GUT:

$$g_1^2 = C^2 e^2/\cos^2\theta_W \quad \Longrightarrow \quad \frac{4\pi}{g_1^2} = \frac{4\pi}{C^2}\frac{\cos^2\theta_W}{e^2} = \frac{1}{C^2}\frac{(1 - \sin^2\theta_W)}{e^2/4\pi} \tag{31.45}$$

We then obtain using the value of the fine-structure constant α at M_Z from Eq. (12.65) and the experimentally determined value for $\sin^2\theta_W(M_Z^2)$ from Eq. (27.66):

$$\frac{1}{\alpha_1(M_Z^2)} = \frac{1}{C^2}\frac{(1 - \sin^2\theta_W(M_Z^2))}{\alpha(M_Z^2)} \approx \frac{3}{5}\frac{(1 - 0.232)}{1/129} \approx 60 \tag{31.46}$$

where we used the educated guess that $C = \sqrt{5/3}$ (this will be motivated later). Also, similarly, recalling Eq. (25.42) which constrains $g_2^2 = e^2/\sin^2\theta_W$:

$$\frac{1}{\alpha_2(M_Z^2)} = \frac{\sin^2\theta_W(M_Z^2)}{\alpha(M_Z^2)} \approx \frac{0.232}{1/129} \approx 30 \tag{31.47}$$

and finally using the value of the strong coupling constant Eq. (18.133):

$$\frac{1}{\alpha_3(M_Z^2)} = \frac{1}{\alpha_s(M_Z^2)} \approx \frac{1}{0.118} \approx 8.5 \tag{31.48}$$

The Q^2 dependence can be generically written as for the strong and electromagnetic case (cf. Eqs. (12.57) and (18.125)), and as well for the weak force as:

$$\alpha_i(Q^2) = \frac{\alpha_i(\mu^2)}{1 + \alpha_i(\mu^2)\frac{\beta_i}{4\pi}\ln\left(\frac{Q^2}{\mu^2}\right)} \tag{31.49}$$

where the coefficients β_i depend on the structure of the group. Rearranging the terms leads to the following important result:

$$\frac{1}{\alpha_i(Q^2)} = \frac{1}{\alpha_i(\mu^2)} + \frac{\beta_i}{4\pi}\ln\left(\frac{Q^2}{\mu^2}\right) \tag{31.50}$$

which can equally be expressed as a function of the g_i's:

$$\frac{1}{g_i^2(Q^2)} = \frac{1}{g_i^2(\mu^2)} + \frac{\beta_i}{(4\pi)^2}\ln\left(\frac{Q^2}{\mu^2}\right) \tag{31.51}$$

For the case of the strong force ($\alpha_3 \equiv \alpha_s$ and the $SU(3)$ group), we can extract β_3 from Eq. (18.125):

$$\beta_3 = \frac{11N_C - 2N_f}{3} = 11 - \frac{2}{3}N_f = 11 - \frac{4}{3}N_g \tag{31.52}$$

where N_C is the number of colors and N_f is the number of quark flavors. In the last term, we have replaced the number of quark flavors by the number of families (or generations), i.e., $N_f = 2N_g$. There is in fact a general result that states that for the group $SU(N)$, the corresponding β_N coefficient is given by [203]:

$$SU(N): \qquad \beta_N = \frac{11}{3}N - \frac{4}{3}N_g \tag{31.53}$$

This implies that for the electroweak force based on the $SU(2)$ group, we should have:

$$\beta_2 = \frac{11}{3}2 - \frac{4}{3}N_g = \frac{22}{3} - \frac{4}{3}N_g \tag{31.54}$$

For the Abelian $U(1)$ we use the result from QED, however, we take into account multiple families in the loops to find:

$$\beta_1 = -\frac{4}{3}N_g \tag{31.55}$$

Please be aware of the negative sign of β_1. As already mentioned, the corresponding interaction is not asymptotically free, in analogy to electromagnetism. Finally, we have:

$$\beta_1 = -\frac{4}{3}N_g; \quad \beta_2 = \frac{22}{3} + \beta_1; \quad \beta_3 = 11 + \beta_1 \tag{31.56}$$

We now use Eqs. (31.50) and (31.56) to compute the values of the inverses of the coupling constants $1/\alpha_i$ as a function of Q^2. We assume three families $N_g = 3$. These lead to the three straight lines with different slopes in a semi-logarithmic plot, as shown in Figure 31.3. In the figure we have taken the measured values of the coupling constants at low energy and extrapolated them to very high energies with the use of the computed β_i slopes, and have also taken the educated guess $C = \sqrt{5/3}$. These are orders of magnitude. It's a huge bold approach to consider such an extrapolation over more than 10 orders of magnitude! The good news is that, given their initial values and slopes, the coupling constants do seem to meet at some very high scale! From the experimental values, we find approximately:

$$\Lambda_{GUT} \approx 10^{15\pm1} \text{ GeV} \tag{31.57}$$

One can argue that this was bound to happen since we are effectively intersecting three straight lines in the graph. However, what is impressive is that they *all* seem to meet in the same range of energy given by Λ_{GUT}. We can certainly say that the unification scale Λ_{GUT} is very large compared to our everyday life. The scale is

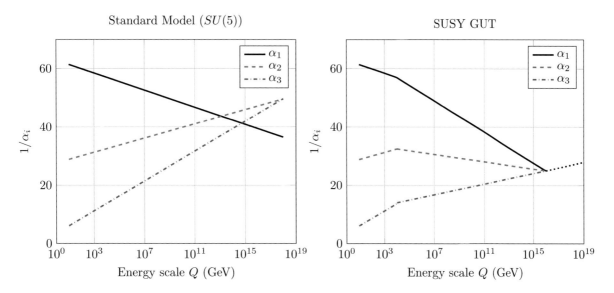

Figure 31.3 The variation of the coupling constants α_i as a function of the energy scale Q.

Figure 31.4 The variation of the coupling constants α_i as a function of the energy scale Q for SUSY GUT. The energy scale is shifted to higher values above 10^{15} GeV.

also much higher than the achievable center-of-mass energies by our currently operating or planned accelerators (see Section 3.1).

The extrapolation of the coupling constants relied on the β_i coefficients that were calculated assuming a given set of fermions and bosons. If for some reason, sets of new very heavy fermions or bosons do exist in Nature, they would not enter in our loops at low energy. However, at some given scale of the order of their masses, they would start playing a role and "curb" the slopes of the coupling constants. For instance, this situation is illustrated in Figure 31.4, where we have envisioned that around 10^4 GeV the slopes would change due to new physics, effectively producing "kinks" in the graphs of the coupling constants as a function of Q. One particular model where this happens is called SUSY GUT and will be discussed in Section 31.6. If this were the case, then it would be possible to obtain the scenario depicted in Figure 31.4, where all coupling constants do meet neatly in a single point. This is definitely very fascinating.

On the fundamental side, although large, the GUT scale is definitely smaller than the Planck scale defined in Eq. (1.5). Consequently, this would tend to point to the existence of new particle physics phenomena at very high energies, albeit where quantum gravitational effects can still be neglected. It is very intriguing to realize that within GUT, we are contemplating a new physics scale which is getting closer to, while different from, the ultimate Planck scale! In Section 31.4, we have also seen that the lightness of the ordinary neutrinos can be elucidated by the **seesaw mechanism**. The mechanism also requires a new physics scale Λ_{NP}, which could perhaps coincide with the Λ_{GUT} scale, but this latter is not a necessary requirement. What is interesting is the existence of two independent pieces of physics beyond the SM that might be explained with a new mass scale at very high energies.

• **The Weinberg angle.** Starting from Eq. (31.42), we can express the Weinberg angle in a way that is practical for the phenomenology (cf. Eq. (27.66)), namely:

$$\sin^2 \theta_W(Q^2) = \frac{g_1^2(Q^2)}{g_1^2(Q^2) + C^2 g_2^2(Q^2)} \tag{31.58}$$

We can use Eq. (31.50) to relate the individual coupling constant to their (common) value at the GUT scale:

$$\frac{1}{\alpha_i(Q^2)} = \underbrace{\frac{1}{\alpha_i(\Lambda_{GUT}^2)}}_{\equiv 1/\alpha_G} + \frac{\beta_i}{4\pi} \ln \left(\frac{Q^2}{\Lambda_{GUT}^2} \right) \tag{31.59}$$

Also, we have as before:

$$\begin{cases} g_1^2 = C^2 e^2 / \cos^2 \theta_W \longrightarrow \alpha_1 = \frac{g_1^2}{4\pi} = \frac{C^2 \alpha}{\cos^2 \theta_W} & (31.60) \\[2ex] g_2^2 = e^2 / \sin^2 \theta_W \longrightarrow \alpha_2 = \frac{g_2^2}{4\pi} = \frac{\alpha}{\sin^2 \theta_W} & (31.61) \\[2ex] g_3^2 = g_s^2 \longrightarrow \alpha_3 = \frac{g_3^2}{4\pi} = \alpha_s & (31.62) \end{cases}$$

Combining these we obtain:

$$\begin{cases} \dfrac{1}{\alpha_1} = \dfrac{\cos^2 \theta_W}{C^2 \alpha(Q^2)} = \dfrac{1}{\alpha_G} + \dfrac{\beta_1}{4\pi} \ln \left(\dfrac{Q^2}{\Lambda_{GUT}^2} \right) \\[2ex] \dfrac{1}{\alpha_2} = \dfrac{\sin^2 \theta_W}{\alpha(Q^2)} = \dfrac{1}{\alpha_G} + \dfrac{\beta_2}{4\pi} \ln \left(\dfrac{Q^2}{\Lambda_{GUT}^2} \right) \\[2ex] \dfrac{1}{\alpha_3} = \dfrac{1}{\alpha_s(Q^2)} = \dfrac{1}{\alpha_G} + \dfrac{\beta_3}{4\pi} \ln \left(\dfrac{Q^2}{\Lambda_{GUT}^2} \right) \end{cases} \tag{31.63}$$

We now use a trick and compute $1/\alpha_1 - 3/\alpha_2 + 2/\alpha_3$ to "magically" cancel the $1/\alpha_G$ terms to find:

$$\frac{\cos^2 \theta_W}{C^2 \alpha(Q^2)} - 3 \frac{\sin^2 \theta_W}{\alpha(Q^2)} + 2 \frac{1}{\alpha_s(Q^2)} = \frac{1}{4\pi} (\beta_1 - 3\beta_2 + 2\beta_3) \ln \left(\frac{Q^2}{\Lambda_{GUT}^2} \right) \tag{31.64}$$

A second "miracle" is that:

$$\beta_1 - 3\beta_2 + 2\beta_3 = \beta_1 - 3 \left(\frac{22}{3} + \beta_1 \right) + 2 (11 + \beta_1) = 0! \tag{31.65}$$

hence the right-hand side of the equation above vanishes. Consequently, we can rearrange our expression to find:

$$1 - \sin^2 \theta_W - 3C^2 \sin^2 \theta_W = -2C^2 \frac{\alpha(Q^2)}{\alpha_s(Q^2)} \tag{31.66}$$

or

$$\sin^2 \theta_W(Q^2) = \left(\frac{1}{1 + 3C^2} \right) \left(1 + 2C^2 \frac{\alpha(Q^2)}{\alpha_s(Q^2)} \right) = \frac{1}{6} + \frac{5}{9} \frac{\alpha(Q^2)}{\alpha_s(Q^2)} \tag{31.67}$$

where in the last term we introduced the educated value $C^2 = 5/3$. Similar gymnastics starting from $(8/3)/\alpha_3 - 1/\alpha_2 - (5/3)/\alpha_1$ give the following result:

$$\ln \left(\frac{Q^2}{\Lambda_{GUT}^2} \right) = \frac{4\pi}{22} \left(\frac{8}{3\alpha_s(Q^2)} - \frac{1}{\alpha(Q^2)} \left(\sin^2 \theta_W(Q^2) + \frac{5}{3C^2} \cos^2 \theta_W(Q^2) \right) \right) = \frac{4\pi}{22} \left(\frac{8}{3\alpha_s(Q^2)} - \frac{1}{\alpha(Q^2)} \right) \tag{31.68}$$

The evolution of the Weinberg mixing angle $\sin^2\theta_W(Q^2)$ as a function of the scale Q^2 can be found by eliminating α_s from the two previous equations to finally give:

$$\sin^2\theta_W(Q^2) = \frac{3}{8}\left[1 + \frac{55\alpha(Q^2)}{18\pi}\ln\left(\frac{Q^2}{\Lambda_{GUT}^2}\right)\right] \xrightarrow{Q^2\to\Lambda_{GUT}^2} \frac{3}{8} \approx 0.375 \qquad (31.69)$$

Therefore, the choice of C leads to $\sin^2\theta_W(\Lambda^2) = 3/8 \approx 0.375$ at the GUT scale. If one takes as before $\alpha(M_Z^2)/\alpha_s(M_Z^2) \approx (1/129)/0.118 \approx 1/15$, one gets $\sin^2\theta_W(M_Z^2) \approx 0.202$, to be compared with the experimental value of $\sin^2\theta_W \approx 0.232$ (see Eq. (27.66)). This result is in the ballpark but it is, however, not a perfect fit. So, it looks like our choice of C is not exactly optimal. The value of C was motivated by the choice of G to be $SU(5)$, as will be discussed below. Of course, different GUTs with larger groups than $SU(5)$ can be constructed.

• **The $SU(5)$ GUT.** In 1974 **Georgi**[2] and **Glashow** showed that the smallest group G to obey $G \supset SU(3) \times SU(2) \times U(1)$ is the $SU(5)$ group [454]. It was the first attempt to embed the SM in an underlying simple group. It represents the simplest GUT theory which is viable. Once the group is chosen, one should assign the quarks and leptons to the irreducible representations of the group.

The general properties of the $SU(N)$ group are summarized in Appendix B.9. The $SU(5)$ group has $5^2 - 1 = 24$ traceless generators, which we can label as T^a ($a, b = 1, \ldots, 24$) with the commutation rule $[T^a, T^b] = if^{abc}T^c$ defined by a specific set of structure constants f^{abc}. A five-dimensional representation of the 5×5 unitary matrices can be constructed with a set of 24 generalized Gell-Mann matrices λ_a with the usual convention $T_a \equiv \lambda_a/2$ (see Eq. (B.59)). By construction we have $\text{Tr}(\lambda_a) = 0$ (see Eq. (B.63) and the normalization is constrained by the condition $\text{Tr}(\lambda_a\lambda_b) = 2\delta_{ab}$ (see Eq. (B.65)). The five-dimensional generators of $SU(5)$ operate on five-dimensional column vectors. These matrices can be chosen in such a way that the generators for subgroups $SU(3)$ and $SU(2)$ are practical and evident to obtain a $SU(3) \times SU(2)$ content. To do so, it is customary to identify the first three indices of the $SU(5)$ indices with the color indices and the remaining two as the $SU(2)_L$ indices. This leads, for example, to the following λ_i matrices:

$$\lambda_1 = \begin{pmatrix} 0 & 0 & 0 & 0 & 0 \\ 0 & 0 & 0 & 0 & 0 \\ 0 & 0 & 0 & 0 & 0 \\ 0 & 0 & 0 & 0 & 1 \\ 0 & 0 & 0 & 1 & 0 \end{pmatrix}; \lambda_2 = \begin{pmatrix} 0 & 0 & 0 & 0 & 0 \\ 0 & 0 & 0 & 0 & 0 \\ 0 & 0 & 0 & 0 & 0 \\ 0 & 0 & 0 & 0 & -i \\ 0 & 0 & 0 & i & 0 \end{pmatrix}; \lambda_3 = \begin{pmatrix} 0 & 0 & 0 & 0 & 0 \\ 0 & 0 & 0 & 0 & 0 \\ 0 & 0 & 0 & 0 & 0 \\ 0 & 0 & 0 & 1 & 0 \\ 0 & 0 & 0 & 0 & -1 \end{pmatrix}; \ldots;$$

$$\lambda_{24} = \frac{1}{\sqrt{15}}\begin{pmatrix} -2 & 0 & 0 & 0 & 0 \\ 0 & -2 & 0 & 0 & 0 \\ 0 & 0 & -2 & 0 & 0 \\ 0 & 0 & 0 & 3 & 0 \\ 0 & 0 & 0 & 0 & 3 \end{pmatrix} \qquad (31.70)$$

Compare with Eq. (B.61). Two of the total of four diagonal matrices of $SU(5)$ are shown. They have been chosen to represent the diagonal generators of the $SU(2)$ and $U(1)$ subgroups of $SU(5)$.

The $SU(5)$ symmetry is assumed to be local and consequently to each of the generators corresponds a gauge boson. It will be convenient to choose the commuting generators to be the $SU(3)$ (color) isospin and hypercharge operators, and the corresponding $SU(2)$ (weak) isospin and hypercharge operators. So, four in total, which we can label as:

$$\underbrace{I_3^c, Y^c}_{SU(3)_C}, \underbrace{I_3^L, Y}_{SU(2)_L} \qquad (31.71)$$

We have already mentioned that the assumption of a GUT triggers a symmetry between quarks and leptons. We are tempted to regroup, for instance for the first family, the following fields (u, d, ν_e, e^-) (and

2   Howard Mason Georgi (born 1947), an American theoretical physicist

the same for their antiparticles). Because all the gauge couplings are vector and conserve helicity as in the construction of the SM, we should separate left-handed and right-handed chiral states. These latter cannot be members of the same representation. The way around this issue is to consider the charge-conjugated states which are left-handed and not right-handed states. This means that we include, for example, the positron in our representation in place of the right-handed electron. Let us then contemplate the left-handed fundamental fermions and take into account the quark colors. This gives us, again for the first family:

$$
\begin{pmatrix} u_R \\ d_R \end{pmatrix}_L, \quad \begin{pmatrix} u_G \\ d_G \end{pmatrix}_L, \quad \begin{pmatrix} u_B \\ d_B \end{pmatrix}_L, \quad u_R, u_G, u_B, d_R, d_G, d_B, \quad \begin{pmatrix} \nu_e \\ e^- \end{pmatrix}_L, e_L^+ \tag{31.72}
$$

We have left out the right-handed neutrino ν_R. There are thus 15 left-handed fundamental fermion fields for a given family. When the generators of the group are traceless, the sum of the eigenvalues taken over all the members of the representation must vanish. This can be used to ensure that the sum of the electric charges of the fermions assigned to a representation must vanish. In $SU(5)$, the fundamental left-handed fermion states and their charge-conjugated states can be accommodated into two different representations: a fundamental $\bar{5}$ and a 10 representation, as follows:

$$
\bar{5} = \underbrace{(\bar{3},1)}_{(d_R^c, d_G^c, d_B^c)_L} + \underbrace{(1,2)}_{(e^-, \nu_e)_L} \quad \text{and} \quad 10 = \underbrace{(\bar{3},1)}_{(u_R^c, u_G^c, u_B^c)_L} + \underbrace{(3,2)}_{(u,d)_L} + \underbrace{(1,1)}_{e_L^+} \tag{31.73}
$$

where we show the $(SU(3), SU(2))$ decomposition. For the right-handed partners, we have for example:

$$
5 = \underbrace{(3,1)}_{(d_R, d_G, d_B)_R} + \underbrace{(1,2)}_{(e^+, \bar{\nu}_e)_R} \tag{31.74}
$$

Note that it contains, as it should, the right-handed chiral state of the antineutrino. While in the SM the separation of left-handed and right-handed states was an ad-hoc construction, in $SU(5)$ the structure of the group naturally leads to this separation. Their quantum number assignments are shown in Table 31.1. Compare with Tables 25.2 and 25.3.

	$(d_R^c)_L$	$(d_G^c)_L$	$(d_B^c)_L$	e_L^-	ν_{eL}
I_3^c	$\frac{1}{2}$	$-\frac{1}{2}$	0	0	0
Y^c	$\frac{1}{2\sqrt{3}}$	$\frac{1}{2\sqrt{3}}$	$-\frac{1}{\sqrt{3}}$	0	0
I_3^L	0	0	0	$-\frac{1}{2}$	$+\frac{1}{2}$
Y	$\frac{2}{3}$	$\frac{2}{3}$	$\frac{2}{3}$	-1	-1

Table 31.1 Quantum numbers of the left-handed fundamental fermions in the $SU(5)$ classification.

The electric charges of the fermions are totally determined by their $SU(2)$ (weak) isospins and hypercharges, recall Eq. (25.52):

$$
Q = I_3^L + \frac{Y}{2} = T_3 + CT_{24} \tag{31.75}
$$

where T_3 and T_{24} are the diagonal generators of the $SU(2)_L$ and $U(1)_Y$ subgroup, represented by the λ_3 and λ_{24} matrices given in Eq. (31.70). At this stage, we can determine the C constant by noting that we should necessarily have:

$$
Q \begin{pmatrix} d_R \\ d_G \\ d_B \\ e^+ \\ \bar{\nu}_e \end{pmatrix}_R = \begin{pmatrix} -\frac{1}{3} \\ -\frac{1}{3} \\ -\frac{1}{3} \\ 1 \\ 0 \end{pmatrix} \tag{31.76}
$$

or equivalently:

$$\left(\frac{\lambda_3}{2} + C\frac{\lambda_{24}}{2}\right) = \begin{pmatrix} 0 & 0 & 0 & 0 & 0 \\ 0 & 0 & 0 & 0 & 0 \\ 0 & 0 & 0 & 0 & 0 \\ 0 & 0 & 0 & \frac{1}{2} & 0 \\ 0 & 0 & 0 & 0 & -\frac{1}{2} \end{pmatrix} + \frac{C}{\sqrt{15}}\begin{pmatrix} -1 & 0 & 0 & 0 & 0 \\ 0 & -1 & 0 & 0 & 0 \\ 0 & 0 & -1 & 0 & 0 \\ 0 & 0 & 0 & \frac{3}{2} & 0 \\ 0 & 0 & 0 & 0 & \frac{3}{2} \end{pmatrix} = \begin{pmatrix} -\frac{1}{3} & 0 & 0 & 0 & 0 \\ 0 & -\frac{1}{3} & 0 & 0 & 0 \\ 0 & 0 & -\frac{1}{3} & 0 & 0 \\ 0 & 0 & 0 & 1 & 0 \\ 0 & 0 & 0 & 0 & 0 \end{pmatrix}$$

$$(31.77)$$

This equation is satisfied by setting $C = \sqrt{\frac{5}{3}}$, which yields $C^2 = 5/3$. This explains our educated choice for C in the previous paragraphs!

Overall, **the multiplet assignment has defined the electric charges of the fields relative to that of the electron and we note that it necessarily must contain fractional charges**. For example, we see that the charges of the three d^c quarks must balance that of the electron and hence the down quarks have charge $-1/3$:

$$3Q_{\bar{d}} + Q_\nu + Q_{e^-} = 0 \implies Q_{\bar{d}} = -\frac{1}{3}Q_e \qquad (31.78)$$

Consequently, **GUT symmetry can explain the charge quantization of the leptons and quarks!** This is a truly fundamental result that has implications concerning the charge of the proton q_p being equal and opposite to that of the electron q_e. Imagine a Universe where q_p would not balance q_e. Atoms would certainly behave very differently than what we are used to. While this was an ad-hoc assignment in the SM, it can now be motivated by the existence of an underlying GUT symmetry.

The gauge bosons belong to the 24 representation:

$$24 = \underbrace{(3,2) + (\bar{3},2)}_{X,Y} + \underbrace{(8,1)}_{gluons} + \underbrace{(1,3) + (1,1)}_{W^\pm,\gamma,Z^0} \qquad (31.79)$$

There are 12 new gauge bosons labeled X and Y, with electric charges $Q_X = \pm 4/3$ and $Q_Y = \pm 1/3$. They belong to a triplet and antitriplet of $SU(3)$ and to a doublet of $SU(2)$.

• **Proton decay.** In GUT theories there lies a symmetry between quarks and leptons, thereby opening the possibility of transformation between them, and this has the fundamental consequence of making proton decay possible (violating the baryon number in the process). In $SU(5)$ it can proceed via the exchange of a virtual superheavy X gauge boson, mediating for example $u \leftrightarrow \bar{u}$ and $d \leftrightarrow e^+$ transitions, as illustrated in Figure 31.5. The process depicted here is $p \rightarrow e^+ + \text{meson(s)}$, such as $p \rightarrow e^+\pi^0$. The decay into a neutrino plus meson(s) is also possible, such as $p \rightarrow \bar{\nu}_e\pi^+$. The order of magnitude of the lifetime can be computed in analogy to

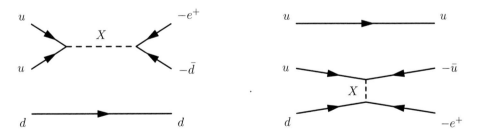

Figure 31.5 Feynman diagrams for quark–lepton transmutation in $SU(5)$ GUT leading to the proton decay process $p \rightarrow e^+\pi^0$.

the muon lifetime. Indeed, at low energies with $Q^2 \simeq M_p^2 \simeq 1 \text{ GeV}^2$, we certainly have $Q^2 \ll M_X^2$, hence the interaction can be seen in analogy to the Fermi weak interaction as a four-point interaction. An order of magnitude for the proton lifetime can then be estimated from the lifetime of the muon by the proper replacement (cf. Eq. (23.25)):

$$\frac{G_F}{\sqrt{2}} = \frac{g^2}{8M_W^2} \quad \Longrightarrow \quad \frac{G_G}{\sqrt{2}} = \frac{g_G^2}{8M_X^2} \tag{31.80}$$

Hence, using Eq. (23.87), the order of magnitude of the proton decay width is (ignoring the constant factors):

$$\Gamma_p \simeq G_G^2 M_p^5 = \frac{g_G^4 M_p^5}{M_X^4} \quad \Longrightarrow \quad \tau_p \simeq \frac{M_X^4}{g_G^4 M_p^5} \approx 10^{31} \text{ years} \quad \text{for } M_X = 5 \times 10^{15} \text{ GeV} \tag{31.81}$$

Despite significant experimental effort, proton decay has never been observed. The prediction disagrees for example with the experimental bound set by Superkamiokande on the $p \to e^+ + \pi^0$ decay rate [455]:

$$\tau(p \to e^+ + \pi^0) > 2.4 \times 10^{34} \text{ years} \qquad (90\% \text{ CL}) \tag{31.82}$$

• **To GUT or not to GUT?** Although the lack of observation of proton decay and the predicted value of $\sin^2 \theta_W$ now exclude the $SU(5)$ model, we should nevertheless look at it as the first concrete development of a GUT theory, with a relatively "near miss." The underlying idea of GUT is extremely attractive. It can explain many of the features that are put in "by hand" in the SM. For instance, charge quantization and the necessity of fractional charges are two examples. As already mentioned, the actual evolutions of the coupling constants depend on the assumed particle content in the loops, which we have assumed to be the same as at the low-energy range, containing three families of quarks and leptons and the usual gauge bosons, and we have extrapolated over the enormous range of energies, basically from M_Z up to Λ_{GUT}. This is sometimes dubbed as the **desert** between M_Z and M_X. It may well be that there exists an intermediate mass scale that introduces the effect of new particles and thereby alters the evolution of the couplings. We have illustrated this in Figure 31.4. We illustrate one such concrete theory example with an expanded group in the following section.

31.6 SUSY GUT

Supersymmetry (SUSY) is a new kind of symmetry which connects fermions and bosons. In its simplest form, the SUSY algebra has an operator that changes the total angular momentum by half a unit, thereby turning a boson field into a fermion field:

$$Q |boson\rangle = |fermion\rangle \quad \text{and} \quad Q |fermion\rangle = |boson\rangle \tag{31.83}$$

In a supersymmetric world the number of degrees of freedom of the fundamental fermions and bosons must match. For example, for each left-handed quark fermion q_L there must exist a supersymmetric boson partner, labeled $\tilde{q}_L$. The supersymmetric partner of a particle is called a "spartner." The commonly used spartners of the SM particles are listed in Table 31.2.

A new quantum number called the R-parity is introduced in order to distinguish ordinary particles from their supersymmetric partners. It can be expressed as:

$$R \equiv (-1)^{3B+L+2J} \tag{31.84}$$

where B is the baryon number, L is the lepton number, and J is the angular momentum (spin). We find that $R = +1$ for ordinary particles and $R = -1$ for the spartners. The supersymmetry must be broken, otherwise all particles and their spartners should have the same rest mass, and this was not observed experimentally. **As of the writing of this book, there has been no confirmed experimental evidence of SUSY particles of any kind.** It is generally assumed that the R-parity is conserved, hence spartners would always be produced in pairs. When they decay, the so-called lightest supersymmetric particle (LSP) would remain, which would be

Particle	Spin	Type	Sparticle	Spin	Type
lepton $\ell_{L,R}$	$\frac{1}{2}$	fermion	slepton $\tilde{\ell}_{L,R}$	0	boson
neutrino ν_L	$\frac{1}{2}$	fermion	sneutrino $\tilde{\nu}_L$	0	boson
quark $q_{L,R}$	$\frac{1}{2}$	fermion	squark $\tilde{q}_{L,R}$	0	boson
photon γ	1	boson	photino $\tilde{\gamma}$	$\frac{1}{2}$	fermion
gluon g	1	boson	gluino $\tilde{g}$	$\frac{1}{2}$	fermion
$W^\pm$	1	boson	wino $\tilde{W}$	$\frac{1}{2}$	fermion
Z^0	1	boson	zino $\tilde{Z}$	$\frac{1}{2}$	fermion
H^0	0	boson	higgsino $\tilde{H}$	$\frac{1}{2}$	fermion

Table 31.2 Quantum numbers of the fundamental fermions and bosons in the SUSY classification.

stable if the R-parity conservation were exact. There have been attempts to release this constraint but even in this case no conclusive evidence for SUSY was found.

• **SUSY GUT.** One generally also assumes that the supersymmetry is a spontaneously broken symmetry at some given scale between the electroweak and the GUT scale. The mass degeneracy between particles and spartners can then be lifted but the relations among the couplings remain intact.

In a supersymmetric $SU(5)$ GUT the evolution of the coupling constants is altered by the presence of new fermions and bosons in the loops. As a matter of fact, for each fermion loop there is by construction a boson loop and vice versa. This results in a "cancellation" of diagrams that reduces the slopes of the evolution of the coupling constants, as illustrated in the graph shown in Figure 31.4. The GUT energy scale is also shifted to a cleaner convergence at higher values above 10^{15} GeV.

31.7   Left–Right Symmetric Models

One of the open problems that beg for an answer is related to the parity and charge violation of the weak force, while all other known forces do obey those symmetries. This question is related to the puzzle of why left-handed fields are $SU(2)$ doublets while right-handed fields are singlets. Left–right symmetric models have been proposed in which the Lagrangian is both P and C-invariant and the observed parity violation is introduced as the consequence of spontaneous symmetry breaking [456, 457]. The basic underlying group is chosen to be $SU(2)_L \times SU(2)_R \times U(1)_Y$ and the interacting gauge-invariant part of the Lagrangian of the theory becomes (cf. Eq. (25.25)):

$$\mathcal{L}_{LR} = -g_1 \frac{Y}{2} \left(\overline{\Psi} \gamma_\mu \Psi \right) B^\mu - g_{2L} \left(\overline{\Psi_L} \gamma_\mu \frac{\vec{\tau}}{2} \Psi_L \right) \cdot \vec{W}_L^\mu - g_{2R} \left(\overline{\Psi_R} \gamma_\mu \frac{\vec{\tau}}{2} \Psi_R \right) \cdot \vec{W}_R^\mu \qquad (31.85)$$

with the three parameters g_1, g_{2L}, and g_{2R}. Both Ψ_L and Ψ_R are now doublets. Their quantum number assignments are shown in Table 31.3. They can be defined using the extended relation (cf. Eq. (25.52)):

$$Q = I_{3L} + I_{3R} + \frac{Y}{2} = I_{3L} + I_{3R} + \frac{B-L}{2} \qquad (31.86)$$

where, in the last equality, we noted that the hypercharge (unlike in the SM, where it lacked a direct physical meaning) has the attractive property that it is given by $B-L$, where B is the baryon number and L the lepton

f	I_L^f	I_{3L}^f	I_R^f	I_{3R}^f	Y^f	Q^f
$\nu_{eL}, \nu_{\mu L}, \nu_{\tau L}$	$+1/2$	$+1/2$	0	0	-1	0
e_L^-, μ_L^-, τ_L^-	$+1/2$	$-1/2$	0	0	-1	-1
$N_{eR}, N_{\mu R}, N_{\tau R}$	0	0	$+1/2$	$+1/2$	-1	0
e_R^-, μ_R^-, τ_R^-	0	0	$+1/2$	$-1/2$	-1	-1
u_L, c_L, t_L	$+1/2$	$+1/2$	0	0	$+1/3$	$+2/3$
d_L', s_L', b_L'	$+1/2$	$-1/2$	0	0	$+1/3$	$-1/3$
u_R, c_R, t_R	0	0	$+1/2$	$+1/2$	$+1/3$	$+2/3$
d_R', s_R', b_R'	0	0	$+1/2$	$-1/2$	$+1/3$	$-1/3$

Table 31.3 Quantum numbers of the fermions in the $SU(2)_L \times SU(2)_R \times U(1)_Y$ classification.

number [458]. Indeed, inspection of Table 31.3 indicates that all leptons independent of their chiral state have $Y = -1$ and all quarks, up and down types, have $Y = 1/3$. Therefore, the left–right symmetric models are often dubbed as being based on the $SU(2)_L \times SU(2)_R \times U(1)_{B-L}$ symmetry. The right-handed neutrinos have been written with a different label $N_{\ell R}$ from the ordinary neutrinos, to leave the possibility open that they represent two different sets of Majorana neutrinos.

Compared to the SM, there is a new triplet of right-handed W gauge bosons labeled W_R^μ. Clearly, we recover the phenomenology of the SM if one sets $g_{2R} \equiv 0$. However, an unbroken parity symmetry in the LR symmetric model imposes that $g_{2L} \equiv g_{2R}$ at very high energies. As mentioned above, this symmetry is assumed to be broken with a first spontaneous breaking at some high scale Λ_X. At the weak scale, a second symmetry breaking occurs similar to the one happening in the SM. This complicated set of symmetry breakings requires a Higgs sector that is more complex than in the SM, containing several Higgs multiplets. In practice, there are several ways of achieving this. A minimal Higgs sector with a bi-doublet, and left and right triplets is generally adopted [456, 457]. During the first stage of symmetry breaking, the charged right-handed gauge bosons labeled $W_R^\pm$ and a neutral gauge boson called Z' acquire a mass proportional to the *vev* of the right-handed Higgs sector and become heavy (in fact, much heavier than the ordinary $W^\pm$ and Z^0). The ordinary gauge bosons acquire their mass during the second symmetry breaking phase, bringing their masses to their experimentally measured values. One can define the Weinberg angle as:

$$\sin\theta_W = \frac{e}{g_{2L}} \tag{31.87}$$

and recover the SM phenomenology of the Z^0 and γ leading to the conventional neutral currents. In total there are two new gauge boson masses and two new mixing angles in both charged and neutral currents appearing in the LR model. It is clear that for $M_{W_L} \ll M_{W_R}$, the charged current weak interaction will appear nearly maximally parity violating at low energies, because the right-handed current is highly suppressed by the heaviness of the right-handed gauge boson compared to the left-handed one. Any deviation from the pure $V - A$ structure of charged weak currents could constitute evidence for the existence of the right-handed current and therefore for the underlying left–right symmetric structure of the weak interactions. The physical charged gauge bosons can actually be mixtures of the W_L and W_R and one accordingly defines:

$$W_1 = W_L \cos\zeta + W_R \sin\zeta \quad \text{and} \quad W_2 = -W_L \sin\zeta + W_R \cos\zeta \tag{31.88}$$

where ζ is a parameter to be determined. A similar situation occurs also in the weak neutral currents, under the assumption that $M_{Z'} \gg M_{Z^0}$. Again the physical neutral gauge bosons Z_1 and Z_2 can be mixtures of the Z^0 and Z' bosons, parameterized by the angle ξ.

• **Experimental constraints.** No direct evidence for the existence of the gauge bosons W_R or Z' has been found. Somewhat model-dependent limits can be derived from low-energy data, where one searches for deviations from the SM. The indirect limits on the Z' are best set by neutral current data with neutrinos and electrons. A review of low-energy experimental data leads to the following 90% CL limits (assuming possible mixing between gauge bosons) [459]:

$$M_{W_2} \geq 715 \text{ GeV}, |\zeta| \leq 0.013 \quad \text{and} \quad M_{Z_2} \geq 1205 \text{ GeV}, |\xi| \leq 0.0042 \tag{31.89}$$

The analysis of polarized muon decays yields $M_{W_2} \geq 592$ GeV [460], which is valid for the mass of the right-handed neutrinos $M_{N_R} \leq m_\mu$. Otherwise, the most stringent bound seems to arise from the analysis of the $K_L - K_S$ mass difference. A recent analysis including also the $B_d - \bar{B}_d$ and $B_s - \bar{B}_s$ mixing derives a lower limit $M_{W_R} \geq 2.5$ TeV [461]. The same authors find an improved lower bound $M_{W_R} \geq 4 - 8$ TeV including also data from CP-violating processes. Finally, it has been observed that if neutrinos are Dirac particles or Majorana particles with $m_{N_R} \lesssim 10$ MeV, then the observation of the neutrino signal from the SN1987A supernova provides the impressive constraints $M_{W_R} \geq 22$ TeV and $|\zeta| \lesssim 10^{-5}$ [462]. Of course, the most striking signature in favor of left–right symmetric models would be the direct detection of the new gauge bosons. This is an active search at the high-energy frontier, in particular at the LHC, but so far no evidence has been found.

Problems

Ex 31.1 Neutrino masses and lepton flavor violation. In the SM, the total lepton number and the individual lepton numbers are accidental global symmetries. If the PMNS matrix is non-diagonal, as implied by neutrino oscillations, individual lepton numbers will be violated. Consider the decays $\mu \to e\gamma$ and $\tau \to \mu\gamma$.

  (a) Draw a Feynman diagram that could explain such a decay.

  (b) Estimate the branching ratio for $\mu \to e\gamma$ as a function of the neutrino masses.

  (c) What is its upper limit if we assume $m_\nu \lesssim 1$ eV?

Ex 31.2 Experimental search for proton decay. An experiment is constructed to search for proton decays as predicted by the GUT theories. The experiment focuses on the "golden" channel $p \to e^+\pi^0$. The detector consists of a huge water tank of cubic shape. The potential proton decays are observed through the Cherenkov radiation produced by the decay products.

  (a) Estimate the size of the detector required to fully contain the electromagnetic cascades of the electron and $\pi^0 \to \gamma\gamma$.

  (b) Estimate the total number of photons produced in an event assuming that the Cherenkov photon yield in the relevant detection wavelength interval is 400 per centimeter of track.

  (c) Estimate the size of the detector needed to test a proton decay lifetime of 10^{35} years.

32   Outlook

> The possibility that the universe was generated from nothing is very interesting and should be further studied. A most perplexing question relating to the singularity is this: what preceded the genesis of the Universe? This question appears to be absolutely metaphysical but our experience with metaphysics tells us that metaphysical questions are sometimes given answers by physics.
>
> *Andrei Linde[1]*

32.1   The Standard Model as an Effective Theory

The **effective Lagrangian method** was developed by Weinberg [463] and independently by Wilczek and Zee [464]. It can be seen as a general, powerful method which allows us to quantitatively describe the effects of physics beyond the SM. The idea is that the SM is very effective at describing with high precision all experimental observations up to the tera-electronvolt scale, i.e., at "low energy." However, we can expect that at a high energy scale it will eventually break down. For instance, exact baryon or lepton number conservation can be seen as a consequence of a global symmetry, however, we have already stated in Section 1.6 that such symmetries could appear to be respected with a very high level of accuracy in our low-energy world, but that would not constitute a "proof" for their exactness up to the higher energy scales. In fact, their exactitude is questionable: for example, the excess of matter over antimatter in our Universe (see Section 32.3) provides a positive clue that some sort of physical process actually violated baryon number conservation in the early phases of our Universe!

The effective Lagrangian is constructed as a $SU_L(2) \times U_Y(1)$ invariant, non-renormalizable Lagrangian built from SM particle fields. One adds to the SM Lagrangian the following terms:

$$\mathcal{L} = \mathcal{L}_{SM} + \sum_{n=1,2,\ldots} \mathcal{L}_{4\,|\,n}^{\text{eff}} = \mathcal{L}_{SM} + \sum_{n=1,2,\ldots} \frac{O_{4+n}}{\Lambda^n} + \text{h.c.} \tag{32.1}$$

We note that $\mathcal{L}_{SM}$ has dimension E^4, i.e., $[\mathcal{L}_{SM}] = 4$ (see Section 6.3). Here O_{4+n} is a $SU_L(2) \times U_Y(1)$ invariant operator which has dimension E^{4+n} and Λ is a constant of the dimension of energy (or mass). **The operators of higher dimensionality of 4+n must be suppressed by powers of Λ^n in order to be dimensionally acceptable.** The constant Λ characterizes a scale of a new, beyond the SM, physics above which the new physics becomes of order unity. If Λ is large enough, their effect is highly suppressed at low energy. The higher the dimensionality n, the more the effect is suppressed and hence presumably more "difficult" to be observed at low energy. We can make a parallel with Fermi's theory of weak interactions here (see Section 21.3). The four-point interaction suppressed by G_F acts as an "effective" theory when the effects of the IVB propagation can be neglected.

The exercise is now to create an exhaustive list of all possible operators, starting with dimension five, then dimension six, and so on. In the following, we illustrate a fascinating example.

---

1   Reprinted from A. Linde's contribution to the proceedings of the Nuffield workshop "The Very Early Universe," Cambridge, June 21 to July 9, 1982.

• **Neutrino masses.** If the right-handed neutrinos ν_R do exist, then we can generate the mass of the neutrino with the Dirac term, for example for the electron neutrino (see Eq. (25.92)):

$$
\begin{aligned}
\mathcal{L}_{\nu_e} &= -\lambda_{\nu_e}\left(\overline{L_e}\phi^C R_\nu + \overline{R_\nu}\phi^{C\dagger}L_e\right) \overset{SSB}{=} -\lambda_{\nu_e}\frac{1}{\sqrt{2}}\left(\left(\overline{\nu_{eL}}\;\;\overline{e_L}\right)\begin{pmatrix} v \\ 0 \end{pmatrix}\nu_{eR} + \overline{\nu_{eR}}\left(v\;\;0\right)\begin{pmatrix} \nu_{eL} \\ e_L \end{pmatrix}\right) \\
&= -\frac{\lambda_{\nu_e}}{\sqrt{2}}v\left(\overline{\nu_{eL}}\,\nu_{eR} + \overline{\nu_{eR}}\,\nu_{eL}\right)
\end{aligned}
\tag{32.2}
$$

which, as it should, coincides exactly with the Dirac mass term, Eq. (31.9). This term is illustrated in Figure 32.1(left). We have treated the neutrino mass on an equal footing as that of other elementary fermions, but we are left with the fine-tuning problem of explaining why the Yukawa coupling for neutrinos is so small compared to that of other fermions (see Section 30.1). On the other hand, the effective Lagrangian prescription

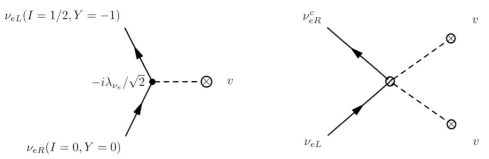

Figure 32.1 (left) Coupling of the Dirac neutrino to the Higgs field. (right) The dimension-five operator leading to the lepton number-violating Majorana neutrino mass. Its form reminds us of the four-fermion Fermi theory of weak interaction.

raises the question of the existence of operators of dimension at least five, which could do the job. Indeed, perhaps surprisingly, one can already construct at the first order $n = 1$ (hence suppressed by only one power of Λ) a Lagrangian term that leads to the generation of neutrino masses, even in the absence of the ν_R! Let us examine the following dimension-five lepton number-violating term involving the neutrino and the Higgs field:

$$
\begin{aligned}
\mathcal{L}_5 &= -\frac{\lambda_\nu}{\Lambda}\left(\overline{L_e}\phi^C\phi^{C\dagger}L_e^c + \text{h.c.}\right) \overset{SSB}{=} -\frac{\lambda_\nu}{2\Lambda}\left(\overline{\nu_{eL}}\;\;\overline{e_L}\right)\begin{pmatrix} v \\ 0 \end{pmatrix}\left(v\;\;0\right)\begin{pmatrix} \nu_{eL}^c \\ e_L^c \end{pmatrix} + \text{h.c.} \\
&= -\frac{\lambda_\nu v^2}{2\Lambda}\overline{\nu_{eL}}\nu_{eR}^c + \text{h.c.}
\end{aligned}
\tag{32.3}
$$

The dimension-five Lagrangian term represents a four-point contact term (see Figure 32.1(right)), which reminds us of Fermi's first attempt to describe weak interaction, without an IVB propagator (see Section 21.3). In Fermi's theory, the weak boson propagator is absent since its effect can be neglected in β decays. In the case of $\mathcal{L}_5$, we can interpret the exact physics happening inside the four-point vertex as being at a very high-energy scale. Hence, it only manifest itself through an effective four-point coupling at lower energies. The consequences of this term are spectacular, as this latter leads to an expression which is **exactly** identical to the lepton number-violating Majorana mass term, Eq. (31.16), with a prediction that the Majorana neutrino mass is given by:

$$
m_L = \frac{\lambda_\nu v^2}{\Lambda} = \frac{v}{\Lambda} \times \underbrace{(\lambda_\nu v)}_{\text{typ. Dirac mass}}
\tag{32.4}
$$

This result is so amazing that it needs to be repeated differently: **Perhaps not surprisingly, neutrino masses are the first evidence of deviation from the SM since the corresponding dimension-five operators are suppressed by only one power of Λ!** Clearly we are seeing neutrino masses as evidence for physics beyond the SM, which does not allow for lepton number violation. If we recall the discussion of the global symmetry in Section 1.6, we note that we did anticipate that *global* symmetries such as the lepton number would be expected to be approximate, as opposed to *local* symmetries, which are believed to be exact. So definitely we should not be afraid of this result. The mass term also clarifies why the neutrinos are so much lighter compared to the other Dirac fermions. Compared to an ordinary fermion of mass $\simeq \lambda v$, the dimension-five operator leads to a mass of the same scale but suppressed by a factor v/Λ, which is a small number since v is at the weak scale and Λ at a very high energy scale. Truly amazing! If we assume an upper bound for the neutrino masses, for example $m_L \lesssim 0.1$ eV, we can find a lower bound for Λ:

$$m_L = \frac{\lambda_\nu v^2}{\Lambda} \lesssim 0.1 \text{ eV} \quad \longrightarrow \quad \Lambda \gtrsim \frac{\lambda_\nu v^2}{0.1 \text{ eV}} \gtrsim 6 \times 10^{14} \text{ GeV} \tag{32.5}$$

where we set $\lambda_\nu \simeq 1$. Going back to the analogy with Fermi's weak theory, we can hope that if history were to repeat itself, the "blob" in the interaction will one day reveal new physics at a very high scale of the order of Λ, presumably in the form of the coupling to, and the propagation of, a new super-heavy particle with rest mass $M \simeq \Lambda$. Seen from this angle, this derivation can be seen as a **generalization of the seesaw mechanism** described in Section 31.4, where the super-heavy particles were actually identified with the right-handed neutrinos N_R.

32.2   Neutrinoless Double-Beta Decay

A major open question of particle physics is whether neutrinos are Dirac or Majorana particles. Neutrinos are the only fundamental fermions that can be Dirac or Majorana particles, since all other fundamental fermions are electrically charged. The most promising method to find out if neutrinos are Majorana particles is to search for so-called neutrinoless double-beta decays. Let us first consider the two-neutrino double-beta decay. Its existence was first suggested by **Goeppert-Mayer**[2] in 1935 [465]. It is a higher-order weak process, consisting of the simultaneous beta decay of two nucleons in the same nucleus (cf. Eq. (21.5)):

$$\begin{aligned} \beta\beta_{2\nu}^- : & \quad P(Z, A) \to D(Z+2, A) + 2e^- + 2\bar{\nu}_e \\ \beta\beta_{2\nu}^+ : & \quad P(Z, A) \to D(Z-2, A) + 2e^+ + 2\nu_e \end{aligned} \tag{32.6}$$

In 1939 **Furry**[3] proposed the neutrinoless double-beta decay processes $\beta\beta_{0\nu}$ or $0\nu\beta\beta$ [466]:

$$\begin{aligned} \beta\beta_{0\nu}^- : & \quad P(Z, A) \to D(Z+2, A) + 2e^- \\ \beta\beta_{0\nu}^+ : & \quad P(Z, A) \to D(Z-2, A) + 2e^+ \end{aligned} \tag{32.7}$$

which are characterized by the complete absence of neutrinos in the final state. These reactions are clearly prohibited in the SM due to lepton number conservation. So why consider them? Neutrinoless double-beta decays are possible if neutrinos are massive Majorana particles! In this case, a nucleus can decay through $\beta\beta_{2\nu}$ as well as $\beta\beta_{0\nu}$, although with a different probability.

The inverse of the $\beta\beta_{0\nu}$ half-life $T_{1/2}^{0\nu}$ of a nucleus $\mathcal{N}$ is given by [467]:

$$\frac{1}{T_{1/2}^{0\nu}} = G_{0\nu}^{\mathcal{N}} |\mathcal{M}_{0\nu}^{\mathcal{N}}|^2 \frac{|m_{\beta\beta}|^2}{m_e} \tag{32.8}$$

where $G_{0\nu}^{\mathcal{N}}$ is the phase-space factor and $\mathcal{M}_{0\nu}^{\mathcal{N}}$ the nuclear matrix element of the given nucleus. The phase-space factor can be computed with high accuracy. On the other hand, the nuclear matrix elements of the various decaying nuclei generally suffer from large uncertainties as their calculations require detailed and accurate

---

2   Maria Goeppert Mayer (1906–1972), German-born American theoretical physicist.
3   Wendell Hinkle Furry (1907–1984), American theoretical physics.

nuclear models of the different isotopes. The right-hand term is proportional to the square of the **effective Majorana mass** $m_{\beta\beta}$, whose value is given by (see, e.g., Ref. [468]):

$$m_{\beta\beta} \equiv \sum_{k=1}^{3} U_{ek}^2 m_k \tag{32.9}$$

Here, the mixing matrix U is related to the Pontecorvo–Maki–Nakagawa–Sakata mixing matrix (see Eqs. (23.235) and (30.107)) with the addition of two extra phases α_1 and α_2, which appear because the neutrinos are assumed to be Majorana particles:

$$
\begin{aligned}
U &\equiv U_{PMNS} \begin{pmatrix} e^{i\alpha_1/2} & & \\ & e^{i\alpha_2/2} & \\ & & 1 \end{pmatrix} \\
&= \begin{pmatrix} c_{12}c_{13} & s_{12}c_{13} & s_{13}e^{-i\delta} \\ -s_{12}c_{23}-c_{12}s_{13}s_{23}e^{i\delta} & c_{12}c_{23}-s_{12}s_{13}s_{23}e^{i\delta} & c_{13}s_{23} \\ s_{12}s_{23}-c_{12}s_{13}c_{23}e^{i\delta} & -c_{12}s_{23}-s_{12}s_{13}c_{23}e^{i\delta} & c_{13}c_{23} \end{pmatrix} \begin{pmatrix} e^{i\alpha_1/2} & & \\ & e^{i\alpha_2/2} & \\ & & 1 \end{pmatrix}
\end{aligned} \tag{32.10}
$$

Hence, one can see that the effective Majorana mass depends on the values of the Majorana neutrino masses as well as on the elements of the first row of the mixing matrix, which connect the electron neutrino with the mass eigenstates:

$$
\begin{aligned}
m_{\beta\beta} &= \sum_j m_j U_{ej}^2 = m_1 |U_{e1}|^2 + m_2 |U_{e2}|^2 e^{i(\alpha_2-\alpha_1)} + |U_{e3}|^2 e^{i(-\alpha_1-2\delta)} \\
&= c_{12}^2 c_{13}^2 m_1 + s_{12}^2 c_{13}^2 e^{i(\alpha_2-\alpha_1)} m_2 + s_{13}^2 e^{i(-\alpha_1-2\delta)} m_3
\end{aligned} \tag{32.11}
$$

where, as before, s_{ij} and c_{ij} stand for the sine and cosine of the angles θ_{ij} (the absolute phase is arbitrary so it is conventionally shifted by a factor $-i\alpha_1$ such that the term with m_1 has no apparent phase term).

• **Experimental results.** The experimental searches for neutrinoless double-beta decay are a top priority in the field, as the $\beta\beta_{0\nu}$ process provides the most sensitive test of lepton number violation. They represent a unique probe of the Majorana nature of neutrinos. However, experiments studying double-beta decays are extremely difficult and challenging. This is connected with the fact that the expected decay rates are extremely small. The matrix element is second order in the weak perturbation theory and in the case of $\beta\beta_{0\nu}$ proportional to the square of the effective neutrino mass. The smallness of the neutrino masses is an additional reason for the severe suppression of the probability for the process to occur. Last but not least, the $\beta\beta_{0\nu}$ process can *only* occur if the neutrino is a Majorana particle. Overall, it does require some faith to pursue such experiments! We summarize the most recent and relevant results in Table 32.1. Different experiments use different isotopes and different techniques and suffer from different sources of backgrounds and systematic errors. The exposure is defined as the mass of the isotope in the experiment multiplied by the observation time. The greater the better. **Up until now, neutrinoless double-beta decay was not observed.** In the absence of a signal, one can set a limit on the half-life of the process $T_{1/2}^{0\nu}$. For the most sensitive experiments, they are above 10^{25} years. We refer the interested reader to the given references in the table to explore the complexities and challenges of the experimental setups required to reach lifetime sensitivities beyond 10^{25} years! These limits can be converted into limits on the effective Majorana mass, as listed in the table. The range shown corresponds to the theoretical uncertainty on the nuclear matrix element, which changes from isotope to isotope. The order of magnitude constraint on the Majorana effective mass is around 0.1–0.2 eV! That's an impressive result when we, for instance, compare this to the rest mass of the lightest charged fermion, i.e., the electron. It is more than six orders of magnitude smaller. Neutrinos are definitely special (see also Section 21.1)!

• **The possible values of $m_{\beta\beta}$.** While the Dirac angles θ_{ij} and δ have been precisely measured in neutrino flavor oscillations (see Section 30.6), at the moment there is only a hint at the values that δ can possess.

Experiment	Isotope	Exposure (kg×yr)	$T_{1/2}^{0\nu}/10^{25}$ yr (90% CL)	Limits on $|m_{\beta\beta}|$ (eV)
GERDA [469]	^{76}Ge	127.2	18	0.07–0.16
KamLAND-Zen [470]	^{136}Xe	504	10.7	0.08–0.24
EXO-200 [471]	^{136}Xe	234.1	3.5	0.06–0.21
CUORE [472]	^{130}Te	372	3.2	0.08–0.35
CUPID-0 [473]	^{82}Se	5.29	0.35	0.31–0.64

Table 32.1 Lower limits of half-lives and upper limits for the effective Majorana mass, obtained in recent experiments in the search for $0\nu\beta\beta$-decay.

Moreover, the Majorana phases α_1 and α_2 cannot be measured in neutrino oscillations, so their values are totally unknown and we must assume $0 \leq \alpha_1, \alpha_2 < 2\pi$. Since in general we can have non-vanishing complex phases, a cancellation between terms in the sum becomes possible. Hence, the magnitude of $|m_{\beta\beta}|$ does not necessarily have to be greater than the magnitude of the individual mass eigenstates $|m_i|$.

We consider here the three ordinary neutrinos with mass eigenstates ν_i ($i = 1, 2, 3$) of rest masses m_i. Under the assumption of the three neutrino masses, we have *two* independent mass squared differences Δm_{ij}^2 (see Eq. (30.77)). From the global fit of the experimental data (see Table 30.2), it follows that one mass squared difference is about 30 times smaller than the other two. The small (large) mass squared difference is usually called solar (atmospheric) mass squared difference and is denoted by Δm_{sol}^2 (Δm_{atm}^2). As already discussed in Section 30.11, there are presently two open scenarios for the neutrino mass hierarchy. See Figure 30.13. The normal ordering (NO) and the inverted ordering (IO). Hence, we have (cf. Eq. (30.110)):

$$\Delta m_{sol}^2 \equiv \Delta m_{21}^2 \ll \Delta m_{atm}^2 = \begin{cases} |\Delta m_{32}^2| & \text{for NO} \\ |\Delta m_{31}^2| & \text{for IO} \end{cases} \tag{32.12}$$

In the case of the NO, we further find:

$$m_2 = \sqrt{m_1^2 + \Delta m_{sol}^2}, \qquad m_3 = \sqrt{m_1^2 + \Delta m_{sol}^2 + \Delta m_{atm}^2} \tag{32.13}$$

while in the case of IO, we obviously have:

$$m_1 = \sqrt{m_3^2 + \Delta m_{atm}^2}, \qquad m_2 = \sqrt{m_3^2 + \Delta m_{atm}^2 + \Delta m_{sol}^2} \tag{32.14}$$

In order to estimate the value of the effective Majorana mass, we first consider a normal hierarchy (see Eq. (30.139)). In this case, we have:

$$m_1 \ll \sqrt{m_{sol}^2}, \quad m_2 \simeq \sqrt{m_{sol}^2}, \quad m_3 \simeq \sqrt{m_{atm}^2} \tag{32.15}$$

If we neglect m_1, the effective Majorana mass is then given by:

$$m_{\beta\beta} \simeq s_{12}^2 c_{13}^2 e^{i(\alpha_2 - \alpha_1)} \sqrt{m_{sol}^2} + s_{13}^2 e^{i(-\alpha_1 - 2\delta)} \sqrt{m_{atm}^2} \approx (2.5 \times 10^{-3} \text{ eV}) e^{i(\alpha_2 - \alpha_1)} + (1.1 \times 10^{-3} \text{ eV}) e^{i(-\alpha_1 - 2\delta)} \tag{32.16}$$

Consequently, the effective Majorana neutrino mass is bounded to be $|m_{\beta\beta}| \lesssim 4 \times 10^{-3}$ eV (normal hierarchical model). The value of 4 meV is significantly smaller than the currently existing negative bounds on the order of 100 meV (see Table 32.1). This might explain the absence of experimental signal.

The situation looks different in the inverted hierarchical scenario, Eq. (30.140). In this case, we can write:

$$m_3 \ll \sqrt{m_{atm}^2}, \quad m_1 \simeq \sqrt{m_{atm}^2}, \quad m_2 \simeq \sqrt{m_{atm}^2} \left(1 + \frac{\sqrt{m_{sol}^2}}{2\sqrt{m_{atm}^2}}\right) \simeq \sqrt{m_{atm}^2} \tag{32.17}$$

We can therefore neglect the term proportional to m_3 in Eq. (32.11) and find:

$$m_{\beta\beta} \simeq c_{13}^2 \left(c_{12}^2 + s_{12}^2 e^{i(\alpha_2 - \alpha_1)}\right) \sqrt{m_{atm}^2} \implies |m_{\beta\beta}| \simeq c_{13}^2 \left(1 - \sin^2(2\theta_{12}) \sin^2(\alpha_2 - \alpha_1)\right) \sqrt{m_{atm}^2} \tag{32.18}$$

The only unknown parameter is the difference of the Majorana phases $\alpha_2 - \alpha_1$. This leads to the following lower and upper bounds:

$$c_{13}^2 \left(1 - \sin^2(2\theta_{12})\right) \sqrt{m_{atm}^2} \leq |m_{\beta\beta}| \leq c_{13}^2 \sqrt{m_{atm}^2} \tag{32.19}$$

or 10^{-2} eV$\lesssim |m_{\beta\beta}| \lesssim 5 \times 10^{-2}$ eV (inverted hierarchical). The upper bound is still below the current experimental results (see Table 32.1), however, it could be within the reach of next-generation experiments, as outlined below.

• **Future experiments.** Many new experiments on the search for the $0\nu\beta\beta$-decays are in preparation. They should improve the sensitivity of current experiments by one to two orders of magnitude. We mention a few of them. The KamLAND-Zen experiment $\left(^{136}\text{Xe}\right)$ should reach the sensitivity $T_{1/2}^{0\nu} > 2 \times 10^{27}$ yr after 5 years of running. The SNO+ $\left(^{130}\text{Te}\right)$, LEGEND $\left(^{76}\text{Ge}\right)$, n-EXO $\left(^{136}\text{Xe}\right)$, CUPID $\left(^{100}\text{Mo}\right)$, and NEXT-HD $\left(^{136}\text{Xe}\right)$ sensitivities should reach $T_{1/2}^{0\nu} > 1 \times 10^{27}$ yr, $> 2 \times 10^{28}$ yr, $> 5.7 \times 10^{27}$ yr, $> 1.1 \times 10^{27}$ yr and $> 3 \times 10^{27}$ yr after 10 years of running (see, e.g., Ref. [474] for more details). Hence, in these experiments the inverted hierarchy region and possibly parts of the normal hierarchy region will be probed. Only the future will tell what there is to be found! A positive result would have a dramatic impact. A negative result would only indicate that either the effective Majorana mass is too small to lead to a detectable signal, or that neutrinos are not Majorana particles, or more generally that no physics beyond the SM is driving such a decay at a rate observable by those experiments.

32.3   Matter–Antimatter Asymmetry in the Universe

It is logical to assume that the Big Bang created exactly the same amount of matter and antimatter. Astrophysical observations, on the other hand, indicate that our current Universe is **matter-dominated**. Basically, one does not find evidence that the excess of matter that characterizes our galaxy is a local phenomenon. There does not seem to be a "place" in the Universe where our local excess of matter would be compensated by a corresponding local antimatter domination. The boundaries between matter-dominated and antimatter-dominated regions would be a very perturbed place and no such evidence has been found. It is therefore believed that the matter dominance is a Universe-wide phenomenon and that some sort of mechanism has created this matter–antimatter asymmetry in the due course of the cosmic evolution.

We have already mentioned Sakharov's necessary conditions for **baryogenesis** (see Section 30.1). One of the key ingredients is the violation of CP symmetry. The CP violation observed is in the weak interactions of quarks, and it is too small to explain the matter–antimatter imbalance of the Universe. It has been shown that CP violation in the lepton sector could generate the matter–antimatter disparity through the process called **leptogenesis** [475, 476]. In this context, the definitive observation of CP violation in the leptonic sector would represent a hint that another source of violation in addition to that observed in quarks exists in Nature.

So far, we have only considered the weak interaction. What about the strong interaction? Can it be a source of CP violation? The answer is in principle yes, but it has not been observed. This is what is called the **strong CP problem.** At this point, we can state that our current understanding cannot quantitatively explain the observed imbalance of matter and antimatter in the Universe. And it is not clear what will be the solution to this problem.

32.4   Compositeness

Historically, many of the particles which were initially thought to be elementary, have revealed substructures when probed at larger energy scales, and this has been central to the progress in our understanding of the structure of matter. It is therefore natural to explore the possibility that some, or all, of the particles which we today call elementary, are in fact composite objects. For some time, and before the discovery of the Higgs boson, a theory without a fundamental scalar particle called Technicolor was developed. Technicolor was introduced to dynamically generate masses for the W and Z^0 bosons through new gauge interactions. This theory has now been superseded by our knowledge on the Higgs boson. Nonetheless, there remains the possibility that the quarks, the leptons, or the gauge bosons are composites of more fundamental particles called **preons**. The Higgs boson itself could be a composite particle. Preons could help explain the eminent regularities between the families of fundamental fermions. Theoretically, a unification of the quarks and the leptons in terms of a small number of common constituents is very attractive. On the other hand, it is worth mentioning the fact that the absence of compositeness would have the consequence to declare our fundamental fermions as the "ultimate" building blocks of Nature, an idea related to the existence of a minimal distance. Even if this latter exists, it is intellectually limiting to assume that we have already reached this limit. Hence, even in the absence of well-defined theoretical predictions, compositeness remains on the table of possible outcomes to be discovered in the future.

What kind of information can be derived from present experimental data? Since, with the currently available measurements, the fermions appear to be point-like, the scale of compositeness Λ_c must be large. Let us focus on the charged leptons for the sake of simplicity. Similar arguments can be extended to the other particles. If the electron, muon, or tau lepton are in fact composite objects, they must be simultaneously light in mass and small in spatial extension. They must then be thought of as being composed of very tightly bound sub-constituents of much larger masses. The most natural consequence of composite models is the existence of excited states of the leptons, labeled ℓ^* ($\ell = e, \mu, \tau$). For simplicity, the excited states are assumed to be fermions with spin-1/2, however, a higher spin assignment is possible and has been studied in the literature. Their effect is to alter the angular distributions of production and decays of the excited states but from a phenomenological point of view, the experimental signatures that are searched for are identical to those of spin-1/2 particles. Hence, we do not consider them here and limit ourselves to spin-1/2 particles.

Should the leptons have a spatial extension, then they could acquire additional anomalous moments like an electric dipole moment or the anomalous magnetic moment. The existence of excited leptons would also influence the anomalous magnetic moments of leptons via higher-order radiative corrections. This poses a problem for the electron and the muon, for which very stringent experimental constraints exist, as discussed in Chapter 14. To estimate the effect of excited leptons on the anomalous magnetic moment, one can generalize Eq. (13.20) to include vector and axial-vector contributions. The general form of the magnetic interaction $\ell^*\ell\gamma$ between excited leptons, ordinary leptons, and photons is then written as:

$$\mathcal{L}_{\ell^*\ell\gamma} = \frac{\lambda^\ell e}{2m_{\ell^*}} \overline{\Psi_{\ell^*}} \sigma^{\mu\nu}(a_\gamma - b_\gamma\gamma^5)\Psi_\ell F_{\mu\nu} + \text{h.c.} \tag{32.20}$$

where λ^ℓ is the unitless coupling constant, m_{ℓ^*} the mass of the excited lepton, and a_γ, b_γ are the vector and axial-vector couplings. The extra anomalous magnetic moment χ_l^B and the electric dipole moment χ_l^E of the ordinary leptons calculated with the inclusion of the $\ell\ell^*\gamma$ contributions are [477]:

$$\begin{aligned}
\chi_l^B &\simeq \frac{16\alpha}{\pi}(\lambda^l)^2 \left(|a_\gamma|^2 - |b_\gamma|^2\right) \frac{m_l}{m_{l^*}} + \frac{18\alpha}{\pi}(\lambda^l)^2 \left(|a_\gamma|^2 + |b_\gamma|^2\right) \frac{m_l^2}{m_{l^*}^2} \\
\chi_l^E &\simeq \frac{32\alpha}{\pi}(\lambda^l)^2 \Re e \left(a_\gamma b_\gamma^*\right) \left(\frac{m_l}{m_{l^*}}\right)
\end{aligned} \tag{32.21}$$

Comparing the measured values with the expected theoretical calculations of the anomalous magnetic moment, very stringent limits on deviations from QED are set. To interpret the results in terms of limits on the mass of the excited leptons, we note that the expression for χ_l^B is actually composed of two parts, one proportional to $\frac{m_L}{m_{l*}}$ and a second proportional to $(\frac{m_L}{m_{l*}})^2$. In the case where one assumes $|a_\gamma| \approx |b_\gamma|$, the leading-order term vanishes and only the contribution proportional to $\frac{m_L}{m_{l*}}$ is left. This case would exhibit a chiral symmetry, i.e., the left (right)-handed ordinary fermions couple to the left (right)-handed excited leptons similarly. The limits on the excited lepton masses are much weaker in this case. **The Lagrangian describing the transition between excited leptons and ordinary leptons should therefore respect the chiral symmetry to protect the light leptons from radiatively acquiring large anomalous magnetic moments that are incompatible with the stringent experimental bounds from the $g - 2$ experiments.** Unlike the case of the ordinary leptons where left-handed states form a weak iso-doublet and the right-handed states form a weak singlet, the existence of the chiral symmetry imposes that both the left-handed and the right-handed states of excited leptons are in weak iso-doublets:

$$l_L = \begin{pmatrix} \nu_l \\ l \end{pmatrix} ; \quad l_R \quad ; \quad L_L = \begin{pmatrix} \nu_l^* \\ l^* \end{pmatrix}_L ; \quad L_R = \begin{pmatrix} \nu_l^* \\ l^* \end{pmatrix}_R \tag{32.22}$$

These excited leptons are called "homo-doublets" since their left and right components have the same quantum numbers. In analogy to the electroweak interactions of ordinary leptons (see Eq. (25.25)), the $SU(2) \times U(1)$ interaction Lagrangian for the couplings of excited leptons to ordinary gauge bosons can be expressed as [478]:

$$\mathcal{L}_{\ell^* \bar{\ell}^*} = \overline{L}\gamma^\mu \left[g_1 \frac{Y}{2} B_\mu + g_2 \frac{\vec{\tau}}{2} \vec{W}_\mu \right] L + g_1 \frac{\kappa'}{2m_{l*}} \overline{L} \frac{Y}{2} \sigma^{\mu\nu} \partial_\mu B_\nu L + g_2 \frac{\kappa}{2m_{l*}} \overline{L} \frac{\vec{\tau}}{2} \sigma^{\mu\nu} \partial_\mu \vec{W}_\nu L \tag{32.23}$$

where $L = L_R + L_L$ and κ, κ' are the anomalous magnetic moments included because of the composite nature of the particles. This Lagrangian can then be rewritten as an effective Lagrangian that describes the interaction between the excited leptons and the physical gauge bosons with given couplings. See Ref. [478] for details.

In addition to these couplings, the form of the $\ell^* \ell V$ interaction given in Eq. (32.20) can be generalized within the SM framework and expressed in $SU(2) \times U(1)$ invariant form, as:

$$\mathcal{L}_{ff^*} = \frac{g_1 f'}{2\Lambda} \overline{L} \sigma^{\mu\nu} \frac{Y}{2} l_L \partial_\mu B_\nu + \frac{g_2 f}{2\Lambda} \overline{L} \sigma^{\mu\nu} \frac{\vec{\tau}}{2} l_L \partial_\mu \vec{W}_\nu + \text{h.c.} \tag{32.24}$$

where Λ is the interaction scale and the f, f' parameters allow for the possibility of different scales associated with the $SU(2)$ and $U(1)$ groups. In a similar way as before, the interaction can be rewritten in terms of the physical gauge bosons with well-defined relations between their couplings.

The experimental searches for excited leptons have been presented in Section 13.9. No evidence for such states has been found, setting very stringent lower limits on their masses in the range of several tera-electronvolts. Other searches have also focused on the possible compositeness of the intermediate gauge bosons, but no evidence for deviations from a point-like behavior has been gathered. Consequently, we will continue to consider the elementary particles of the SM as point-like. Perhaps, some day a deviation from a point-like behavior will be observed at very high energies, unless the quantization of space-time enters into play beforehand. Only further experimental investigations can give us an answer.

32.5   Dark Matter, Dark Energy, and Gravity

The effect of the gravitational force, determined by the Newtonian gravitational constant G, is so small on individual particles compared to other forces, that all particle physics experiments are insensitive to it and consequently any potential effects can be safely neglected. On a more profound level, attempts to reformulate **Einstein's theory of general relativity (GR)** as a quantum field theory in which the gravitational field is quantized, have been unsuccessful. Furthermore, one expects quantum fluctuations of gravity to become large only at the Planck scale $\Lambda_{Pl} \approx 10^{19}$ GeV/c^2 (see Eq. (1.5)) or at the corresponding Planck length of $l_{Pl} \approx 10^{-35}$ m. This domain is way beyond the range of sensitivity of our current experiments, which, as has been described in Chapter 31, does not even reach the GUT scale. In the SM, we can simply interpret the rest masses of the elementary particles, acquired after the spontaneous symmetry breaking, as the "inertial" masses of these latter. The rest masses entered in the kinematics of the problem (energy, momentum), however, do not engender any specific interactions.

Consequently, as was already mentioned in Chapter 1, a unified theory able to describe all forces is still missing. A model that was revived as a candidate for a unified theory including gravity is based on **strings**. The various theories discussed so far have always been formulated on the four-dimensional Minkowski space (this is true for both QFTs and Einstein's GR). Around 1920, **Kaluza**[4] and later **Klein** investigated the possibility that space contains additional dimensions. The extra dimensions, however, have a different topology than the other common four, basically, they are curled and periodic, or "compactified," into a very small radius. Graphically, one can think of a thin tube which when viewed from far away appears to be composed of a thin line. The aim of the work of Kaluza and Klein was to unify Einstein's GR and electromagnetism. Unlike QFT, their approach to electromagnetism was *geometrical*. Accordingly, through the extension of the dimensionality of space-time, gravity and electromagnetism become related. In **superstring theories**, the fundamental elementary particles are no longer to be seen as point particles but rather as a string and each particle species is one of its vibrational modes. For reasons that go beyond the scope of this book, the superstring theory becomes a quantized gauge theory that incorporates supersymmetry in it which extends GR in a 10-dimensional space-time [479]. In this case, the six extra dimensions are all "compactified." Another possible more recent interpretation is that our 4-dimensional "brane" floats in a 10-dimensional world. Such theoretical developments are really exciting and certainly represent a major breakthrough in the development of a fully unified theory. Many fascinating predictions come out of the theory, but unfortunately they remain untestable experimentally, essentially because the model is focused on phenomena at the Planck scale. Perhaps in the future some of the predictions of superstring theories will be experimentally accessible. In this context, an intriguing development is the possible existence of **Planck stars** [480] as opposed to black holes. In a Planck star, the black hole singularity is replaced by a tiny domain where matter is compressed all the way down to the Planck length at which point its fate would be governed by physics at the Planck scale. Even if this would create a rapid bounce of the collapsing star, this process would last billions of years for an observer outside of the star due to the extreme red shift! According to the authors, the hypothesis of Planck stars can be tested, and hence an indirect test of physics at the Planck scale can be performed, as Planck stars are predicted to emit radiation, which could be observable.

Recently, new problems have emerged that the SM and more generally particle physics are not able to address that are indirectly related to the problem of gravity. The first one is the existence of **dark matter** first suggested by **Zwicky**[5] in 1933. The issue was revived by **Rubin**[6] and others in the 1970s based on the observations of rotational curves of spiral galaxies. The motion of stars in spiral galaxies (including ours) showed that the rotational velocity stays constant as a function of the radius from the center of the galaxy after a certain distance, an observation which is incompatible with Newtonian mechanics and the observed visible mass distribution in the galaxies, which predicts that the velocity should decrease with distance. Hence, this observation unambiguously proved the existence of a new form of matter that does not interact strongly nor electromagnetically. It is hoped that apart from its gravitational effect, dark matter will be composed of elementary particles that possess weak interactions with ordinary matter. From cosmological observations, it can be concluded that dark matter must be primarily composed of non-relativistic particles (i.e., cold dark matter). It should be stressed that candidates for dark matter particles do not exist in the SM. A popular candidate for dark matter is the lightest neutral SUSY particle, known as the **neutralino**, which is stable by virtue of R-parity conservation (see Section 31.6). It is a mixture of the superpartners of the photon, the Z^0, and the Higgs particle. Another possible candidate is the **axion**, a very light scalar particle introduced to solve the so-called strong CP problem of QCD. Another option could be that as-of-yet-unknown sterile neutrinos with a mass in the range of keV contribute to the dark matter. Several experiments are underway to try to

4   Theodor Franz Eduard Kaluza (1885–1954), German mathematician and physicist.
5   Fritz Zwicky (1898–1974), Swiss astronomer.
6   Vera Florence Cooper Rubin (1928–2016), American astronomer.

detect evidence for dark matter, for instance by looking for the tiny recoil signal produced by the hypothetical collision of an ordinary matter nucleus with the flux of dark matter particles as the solar system, including the Earth, moves through the dark matter halo. There was also hope that new particles compatible with being dark matter candidates could be produced at the CERN LHC, although so far, this hasn't happened. Although less popular, another solution to the dark matter problem could be that our assumption on the validity of Newtonian mechanics at very large distances is incorrect. It has also been argued that the emission of the hypothetical Planck stars could solve the dark matter problem [481]. Only the future will tell which solution is the correct one.

Last but not least, another problem that is related to gravity, or more generally, to the equations of cosmic evolution, is the existence of a cosmological constant. The first direct evidence was obtained with the observation of distant supernovae, which demonstrated that the expansion of the Universe is actually accelerating. In this context, while particle physics is described by the SM, cosmology is discussed in the context of its own "Standard Model," which is called the **Concordance Model**. This latter has been established in the last decades with the advent of very high precision data on the **cosmic microwave background** (CMB), which was analyzed and provided a very consistent view of the dynamical history of the Universe. According to the model, most of the Universe is filled with **dark energy** ($\approx 70\%$) and dark matter ($\approx 25\%$), the remaining percents being composed by ordinary matter and radiation that are described by the SM of particle physics. This is indeed embarrassing for particle physicists! Another outcome of the precision data is that the curvature of the Universe on very large scales is consistent with zero, in other words, the Universe is perfectly flat. As of the writing of this book, the composition and the origin of dark energy is not fully understood. Qualitatively, we can say that while the presence of matter (energy) will curve space and slow its expansion (Newton's universal attractive force), the presence of dark energy acts as a negative pressure which generates a repulsive effect, thereby accelerating the expansion of the Universe. Is dark energy associated with a new field? We know from the Higgs mechanism that a scalar field that breaks a symmetry is able to create a non-zero vacuum energy. **Inflation**, which represents an exponential expansion phase in the very early Universe, is believed to have been produced by such a field. Is this the connection between cosmology and particle physics? Maybe. Quantitively the density of the dark energy is unfortunately orders of magnitude weaker than that of the Higgs field of particle physics, so the naive direct connection between the two fails. The dark energy could be a static vacuum energy or it could even change with time if it has a dynamical origin, something referred to as **quintessence**. We don't know the answer, but for certain its properties will determine the fate of the Universe. At this point, we are reaching the border of experimental science. From a particle physics point of view, dark matter, dark energy and in general gravitation, do remain extremely challenging problems, whose solution will lead to an improved SM or maybe even something totally new!

32.6   Final Words

Let us finish with some lines from the Hollywood movie *Shrek 2*[7]: – Donkey : "Are we there yet?" – Shrek : "Yes." – Donkey : "Really?" – Shrek : "No!"

In the coming years, the CERN LHC will continue to take data at the highest energies with the hope of elucidating some of the open questions in particle physics. In addition, a new generation of experiments at accelerators and without accelerators will be proposed, constructed and executed with the potential of new discoveries. Theoretical advances will never stop. The field will also be abundantly cross-fertilized by new discoveries and developments in astrophysics and astroparticle physics, and other disciplines. Only time will tell if new results will require perfecting the SM or whether something completely unexpected will be necessary to describe Nature. It would be truly exciting if it turned out to be the latter!

---

7   http://imdb.com/title/tt0298148

Appendix A   Mathematical and Calculus Tools

A.1   Series

A mathematical series is the sum of a list of numbers that are generated according to some pattern or rule. Useful results:

$$\frac{1}{1-x} = \sum_{k=0}^{\infty} x^k = 1 + x + x^2 + x^3 + \cdots \qquad \text{for } |x| < 1 \tag{A.1}$$

$$\ln(1+x) = \sum_{k=0}^{\infty} (-1)^{k+1} \frac{x^k}{k} = x - \frac{1}{2}x^2 + \frac{1}{3}x^3 - \frac{1}{4}x^4 + \cdots \tag{A.2}$$

$$\sin x = \sum_{k=0}^{\infty} \frac{(-1)^k x^{2k+1}}{(2k+1)!} = x - \frac{1}{6}x^3 + \frac{1}{120}x^5 - \frac{1}{5040}x^7 + \cdots \tag{A.3}$$

$$\cos x = \sum_{k=0}^{\infty} \frac{(-1)^k x^{2k}}{(2k)!} = 1 - \frac{1}{2}x^2 + \frac{1}{24}x^4 - \frac{1}{720}x^6 + \cdots \tag{A.4}$$

$$\sinh z = \sum_{k=0}^{\infty} \frac{z^{2k+1}}{(2k+1)!} = z + \frac{1}{6}z^3 + \frac{1}{120}z^5 + \frac{1}{5040}z^7 + \cdots \tag{A.5}$$

$$\cosh z = \sum_{k=0}^{\infty} \frac{z^{2k}}{(2k)!} = 1 + \frac{1}{2}z^2 + \frac{1}{24}z^4 + \frac{1}{720}z^6 \cdots \tag{A.6}$$

$$\tanh z = \sum_{k=1}^{\infty} \frac{(2^{2k}-1)2^{2k}B_{2k}z^{2k-1}}{2k!} = z - \frac{1}{3}z^3 + \frac{2}{15}z^5 - \frac{17}{315}z^7 + \cdots, \quad |z| < \pi/2 \tag{A.7}$$

where $B_n(x) = \sum_{k=0}^{n} \binom{n}{k} B_{n-k} x^k$ are the Bernoulli[1] polynomials. In addition:

$$e^x = \sum_{k=0}^{\infty} \frac{x^k}{k!} = 1 + x + \frac{1}{2}x^2 + \frac{1}{6}x^3 + \frac{1}{24}x^4 + \frac{1}{120}x^5 + \frac{1}{720}x^6 + \frac{1}{5040}x^7 + \cdots \tag{A.8}$$

$$e^{ix} = \sum_{k=0}^{\infty} \frac{(ix)^k}{k!} = 1 + ix - \frac{1}{2}x^2 - \frac{i}{6}x^3 + \frac{1}{24}x^4 + \frac{i}{120}x^5 - \frac{1}{720}x^6 - \frac{i}{5040}x^7 + \cdots \tag{A.9}$$

---

[1]   Jacob Bernoulli (1655–1705), mathematician (member of the Bernoulli family that generated many prominent mathematicians and physicists).

Now assuming x and y do not necessarily commute, i.e., $xy \neq yx$, if for example x, y are operators:

$$
\begin{aligned}
e^{ix}e^{iy} &= \left(1 + ix - \frac{1}{2}x^2 + \cdots\right)\left(1 + iy - \frac{1}{2}y^2 + \cdots\right) \\
&= 1 + i(x+y) - xy - \frac{1}{2}(x^2 + y^2) + \cdots \\
&= 1 + i(x+y) - xy - \frac{1}{2}(x^2 + xy + yx + y^2) + \frac{1}{2}(xy + yx) + \cdots \\
&= 1 + i(x+y) - \frac{1}{2}xy + \frac{1}{2}yx - \frac{1}{2}(x+y)^2 + \cdots \\
&= 1 + i(x+y) + \frac{i^2}{2}[x,y] - \frac{1}{2}(x+y)^2 + \cdots \\
&= 1 + i\left(x+y+\frac{i}{2}[x,y]\right) - \frac{1}{2}(x+y)^2 + \cdots \\
&\simeq e^{i(x+y)+\frac{1}{2}[x,y]}
\end{aligned}
\tag{A.10}
$$

where $[x,y] = xy - yx$. This result is a specific case of the **Baker–Campbell–Hausdorff formula**, which states if $e^z = e^x e^y$ then $z = x + y + \frac{1}{2}[x,y] + \frac{1}{12}[x,[x,y]] - \frac{1}{12}[y,[x,y]] + \cdots$.

A.2   Einstein Summation Notation

The summation convention is a notational convention that implies summation over a set of indexed terms in a formula, thus achieving brevity. For example:

$$
y = a_1 x^1 + a_2 x^2 + a_3 x^3 = \sum_{i=1}^{3} a_i x^i \equiv a_i x^i
\tag{A.11}
$$

where it is implied that we sum over the three coordinates $i = 1, 2, 3$.

A.3   The Totally Antisymmetric Levi-Civita and Kronecker Symbols

The **totally antisymmetric Levi-Civita**[2] **symbol** ϵ_{ijk} in three dimensions is defined as:

$$
\epsilon_{ijk} = \begin{cases} +1 & \text{even number of permutations of } 1,2,3 \\ -1 & \text{odd number of permutations of } 1,2,3 \\ 0 & \text{otherwise} \end{cases}
\tag{A.12}
$$

The symbol being antisymmetric, we get a minus sign under interchange of any pair of indices. This means for instance:

$$
\epsilon_{123} = \epsilon_{312} = \epsilon_{231} = 1 \text{ and } \epsilon_{321} = \epsilon_{132} = \epsilon_{213} = -1
\tag{A.13}
$$

Useful equalities:

$$
\begin{aligned}
\epsilon_{ijk}\epsilon_{lmn} &= \delta_{il}\delta_{jm}\delta_{kn} + \delta_{im}\delta_{jn}\delta_{kl} + \delta_{in}\delta_{jl}\delta_{km} - \delta_{il}\delta_{jn}\delta_{km} - \delta_{in}\delta_{jm}\delta_{kl} - \delta_{im}\delta_{jl}\delta_{kn} & \text{(A.14)} \\
\epsilon_{ijk}\epsilon_{lmk} &= \delta_{il}\delta_{jm} - \delta_{im}\delta_{jl} & \text{(A.15)} \\
\epsilon_{ijk}\epsilon_{ljk} &= 2\delta_{il} & \text{(A.16)} \\
\epsilon_{ijk}\epsilon_{ijk} &= 6 & \text{(A.17)}
\end{aligned}
$$

2   Tullio Levi-Civita (1873–1941), Italian mathematician.

where we used the **Kronecker**[3] symbol δ_{ij}:

$$\delta_{ij} \equiv \begin{cases} 1 \text{ if } i = j \\ 0 \text{ if } i \neq j \end{cases} \tag{A.18}$$

With the antisymmetric Levi-Civita symbol, we can easily write the ith component of the cross-product of two vectors (using the Einstein summation convention):

$$(\vec{a} \times \vec{b})_i = \epsilon_{ijk}a^j b^k \tag{A.19}$$

Also the triple product of three vectors (a number) is simply equal to:

$$\vec{a} \cdot (\vec{b} \times \vec{c}) = \epsilon_{ijk}a^i b^j c^k \tag{A.20}$$

A.4   The Heaviside Step Function

The Heaviside function $\Theta(x)$ represents a signal that switches on at a specified time, distance, or position and stays switched on indefinitely and is defined as:

$$\Theta(x) = \begin{cases} 0 & \text{if } x < 0 \\ 1 & \text{if } x > 0 \end{cases} \tag{A.21}$$

It is undefined at $x = 0$.

An integral representation of the Heaviside function is given by:

$$\Theta(t) = \lim_{\epsilon \to 0^+} \frac{-1}{2\pi i} \int_{-\infty}^{+\infty} \mathrm{d}z \frac{e^{-izt}}{z + i\epsilon} \tag{A.22}$$

where the "$i\epsilon$" prescription was used to regularize the pole at $z = 0$ in the denominator. By adding the $i\epsilon$ term, the integrand is analytic except at the pole in $z = -i\epsilon$, so it is now shifted in the lower-half complex plane. See Figure A.1. We must choose a contour to perform the integral. For e^{-izt} to remain finite, we must close the contour in the lower half complex plane when $t > 0$. In this case, $z \to -i\infty$ yields $e^{-izt} = e^{-|z|t} \to 0$. On the other hand, when $t < 0$, we must close the contour in the upper half complex plane, so that $e^{+iz|t|} \to 0$ when $z \to +i\infty$. With the **Cauchy's residue theorem**, we then note that the integral over the upper half complex plane is zero, while the integral over the lower half complex plane has one simple pole with residue $Res(z = -i\epsilon) = -1$ (the negative sign accounts for the fact that the contour must be counter-clockwise oriented), so finally:

$$\Theta(t) = \lim_{\epsilon \to 0^+} \frac{-1}{2\pi i} \int_{-\infty}^{+\infty} \mathrm{d}z \frac{e^{-izt}}{z + i\epsilon} = \begin{cases} \dfrac{-1}{2\pi i}(2\pi i Res(z = -i\epsilon)) = 1 & \text{if } t > 0 \\ 0 & \text{if } t < 0 \end{cases} \tag{A.23}$$

A.5   The Fourier Transform

A **Fourier series** is an expansion of a periodic function $f(x)$ in terms of an infinite sum of sines and cosines. Fourier series make use of the orthogonality relationships of the functions. The **Fourier transform** is a generalization of the complex Fourier series in the continuous limit, replacing discrete Fourier coefficients with a

3   Leopold Kronecker (1823–1891), German mathematician.

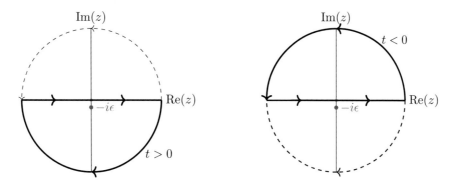

Figure A.1 Prescription for the contour integral.

function and changing a sum to an integral:

$$f(x) = \frac{1}{\sqrt{2\pi}} \int_{-\infty}^{\infty} A(k)e^{ikx}\mathrm{d}k \quad \text{where} \quad A(k) = \frac{1}{\sqrt{2\pi}} \int_{-\infty}^{\infty} f(x)e^{-ikx}\mathrm{d}x \tag{A.24}$$

We note that the $1/\sqrt{2\pi}$ in front of both equations above is conventional! Mathematically, the direct generalization of the Fourier series yields:

$$f(x) = \int_{-\infty}^{\infty} A(k)e^{2\pi ikx}\mathrm{d}k \quad \text{where} \quad A(k) = \int_{-\infty}^{\infty} f(x)e^{-2\pi ikx}\mathrm{d}x \tag{A.25}$$

From this we would, however, obtain the asymmetrical situation:

$$f(x) = \int_{-\infty}^{\infty} A(k)e^{ikx}\mathrm{d}k \quad \text{where} \quad A(k) = \frac{1}{2\pi} \int_{-\infty}^{\infty} f(x)e^{-ikx}\mathrm{d}x \tag{A.26}$$

so we prefer the symmetrical definition of Eq. (A.24). [Adapted from E. W. Weisstein, "Fourier Series," http://mathworld.wolfram.com/FourierSeries.html]

A.6   The Dirac δ Function

Dirac's δ function (mathematically it is not a function) is infinitely peaked at 0 with a unity integral. It can be represented as **the derivative of the Heaviside function** $\Theta(x)$. The δ function is mathematically defined by the following property:

$$\delta(x - y) = \begin{cases} 0 & \text{if } x \neq y \\ \infty & \text{if } x = y \end{cases} \quad \text{with} \quad \int_{t_1}^{t_2} \delta(t)\mathrm{d}t = 1 \text{ if } t_1 < 0 < t_2. \tag{A.27}$$

The important property of the δ function is the following relation:

$$\int_{-\infty}^{\infty} \mathrm{d}x f(x)\delta(x) = f(0) \tag{A.28}$$

where f is a continuous function. To show this, we assume that the function $f(x)$ vanishes at infinity, then we can use the integration by parts:

$$\int_{-\infty}^{\infty} \mathrm{d}x f(x)\delta(x) = [f(x)\Theta(x)]_{-\infty}^{\infty} - \int_{-\infty}^{\infty} \mathrm{d}x f'(x)\Theta(x) = -\int_{0}^{\infty} \mathrm{d}x f'(x) = f(0) \tag{A.29}$$

where we used the fact that the Dirac δ function is the derivative of the Heaviside step function. Since the δ function vanishes for $x \neq 0$, the integrand is non-zero only for $x = 0$, and thus the behavior of $f(x)$ at infinity does not affect the value of the integral. So, it is reasonable to state by a change of variable and regardless of the behavior of the function at infinity, that the important property of the δ function can be written as the following relation:

$$\int_{-\infty}^{\infty} dx f(x)\delta(x - y) = f(y) \tag{A.30}$$

where f is any continuous function. Calculations with δ functions can often be simplified with the following properties:

$$\delta(x) = \delta(-x), \qquad x\delta(x) = 0, \qquad \delta(ax) = \frac{1}{|a|}\delta(x), \qquad \delta(x^2 - a^2) = \frac{1}{2|a|}\left(\delta(x - a) + \delta(x + a)\right),$$

$$\delta(g(x)) = \sum_i \frac{\delta(x - x_i)}{\left|\frac{dg(x_i)}{dx}\right|} \quad \text{where} \quad g(x_i) = 0 \quad (x_i \text{ are the roots of } g). \tag{A.31}$$

• **Derivative of the Dirac δ function.** We can also consider the following integral with the derivative:

$$\int_{-\infty}^{\infty} dx \delta(x)\frac{d}{dx}(xf(x)) = \int_{-\infty}^{\infty} dx \delta(x)\left(f(x) + xf'(x)\right) = \int_{-\infty}^{\infty} dx \delta(x)f(x) + \underbrace{\int_{-\infty}^{\infty} dx \delta(x)xf'(x)}_{=0} \tag{A.32}$$

Also:

$$\int_{-\infty}^{\infty} dx \delta(x)\frac{d}{dx}(xf(x)) = \underbrace{\int_{-\infty}^{\infty} dx \frac{d}{dx}\left(\delta(x)xf(x)\right)}_{=0} - \int_{-\infty}^{\infty} dx \delta'(x)xf(x) = -\int_{-\infty}^{\infty} dx \delta'(x)xf(x) \tag{A.33}$$

Hence:

$$\delta(x) = -\delta'(x)x \quad \Longrightarrow \quad \delta(x - y) = -\frac{\partial \delta(x - y)}{\partial x}(x - y) = -\left(\frac{\partial}{\partial x}\delta(x - y)\right)(x - y) \tag{A.34}$$

and

$$\delta(x - y) = \left(\frac{\partial}{\partial y}\delta(x - y)\right)(x - y) \tag{A.35}$$

• **Fourier transform.** Another way to define the δ function is with the help of the Fourier transformation. We note that:

$$f(x) = \frac{1}{\sqrt{2\pi}}\int_{-\infty}^{\infty} dk e^{ikx}\left(\frac{1}{\sqrt{2\pi}}\int_{-\infty}^{\infty} dy f(y)e^{-iky}\right)$$

$$= \int_{-\infty}^{\infty} dy f(y)\int_{-\infty}^{\infty}\frac{dk}{2\pi}e^{+ik(x-y)} = \int_{-\infty}^{\infty} dy f(y)\delta(x - y) \tag{A.36}$$

where in the last term we wrote $f(x)$ using Eq. (A.30). Therefore the Dirac δ function can also be defined as:

$$\delta(x - y) = \int_{-\infty}^{\infty}\frac{dk}{2\pi}e^{+ik(x-y)} \qquad \text{or} \qquad 2\pi\delta(x - y) = \int_{-\infty}^{\infty} dk e^{+ik(x-y)} \tag{A.37}$$

• **Multi-dimensional δ function.** The Dirac δ^n function can be defined in the **multi-dimensional space** of dimension n:

$$\delta^n(\vec{x}) = 0 \quad \text{for } \vec{x} \neq 0 \quad \text{where } \vec{x} \in \mathbb{R}^n \quad \text{and} \quad \int_{\mathbb{R}^n} d^n\vec{x}\,\delta^n(\vec{x}) = 1 \tag{A.38}$$

The multi-dimensional δ^n function can be simply written as the product of one-dimensional δ functions:

$$\delta^n(\vec{x}) = \prod_{k=1}^{n} \delta(x_k) \quad \text{where } \vec{x} = (x_1, x_2, ..., x_n) \tag{A.39}$$

For example, in three dimensions with $\vec{x} = (x_1, x_2, x_3)$ and $\vec{y} = (y_1, y_2, y_3)$, we get:

$$\delta^3(\vec{x} - \vec{y}) = \delta(x_1 - y_1)\delta(x_2 - y_2)\delta(x_3 - y_3). \tag{A.40}$$

This result can be expressed with the following integral form:

$$(2\pi)^3 \delta^3(\vec{x} - \vec{y}) = \int d^3\vec{k}\, e^{i\vec{k}\cdot(\vec{x}-\vec{y})} \tag{A.41}$$

• **Relation to Green's function.** Let us assume a function (or field) $\phi(x)$ which satisfies the **differential equation** of the form:

$$D\phi(x) = \rho(x) \tag{A.42}$$

where D is some differential operator (e.g., Poisson's equation is $\nabla^2\phi(x) = \rho(x)$). The **Green's function** $G(x, y)$ is defined as the solution for a point source at $x = y$:

$$DG(x, y) = \delta(x - y) \tag{A.43}$$

Then

$$D \int G(x, y)\rho(y)\mathrm{d}y = \int \delta(x - y)\rho(y)\mathrm{d}y = \rho(x) \tag{A.44}$$

In other words, the function

$$\phi(x) = \int G(x, y)\rho(y)\mathrm{d}y \tag{A.45}$$

is a solution of the differential equation, Eq. (A.42). The Green's function lets us convert the problem of solving a differential equation to the problem of doing an integral.

A.7   Time-Ordering Dyson Operator

The **Dyson time-ordering operator** returns the time-ordered product of two time-dependent operators $A(t)$ and $B(t)$, putting the operator at the earlier time on the right and the operator at the later time on the left, and hence is defined as:

$$T[A(t_1)B(t_2)] = \begin{cases} A(t_1)B(t_2) & \text{if } t_1 > t_2 \\ B(t_2)A(t_1) & \text{if } t_1 < t_2 \end{cases} \tag{A.46}$$

The definition is equivalent to saying that operators are ordered from right to left with increasing time, or also that the latest operator is on the left. It can also be expressed with the Heaviside step function as:

$$T[A(t_1)B(t_2)] = \Theta(t_1 - t_2)A(t_1)B(t_2) + \Theta(t_2 - t_1)B(t_2)A(t_1) \tag{A.47}$$

A.8   Jacobi Determinant

Often we need to change variables in an integral. In order to change variables in an integral one uses the **Jacobian**[4] of the variable transformation. For example, to change the double integral of the function f over x, y to u, v, we write:

$$\int \int f(x, y)\mathrm{d}x\mathrm{d}y = \int \int f(g(u, v), h(u, v)) \left| \frac{\partial(x, y)}{\partial(u, v)} \right| \mathrm{d}u\mathrm{d}v \tag{A.48}$$

where the Jacobian of the transformation $x = g(u, v)$ and $y = h(u, v)$ is defined as:

$$\frac{\partial(x, y)}{\partial(u, v)} \equiv \begin{vmatrix} \dfrac{\partial x}{\partial u} & \dfrac{\partial x}{\partial v} \\ \dfrac{\partial y}{\partial u} & \dfrac{\partial y}{\partial v} \end{vmatrix} = \frac{\partial x}{\partial u}\frac{\partial y}{\partial v} - \frac{\partial x}{\partial v}\frac{\partial y}{\partial u} \tag{A.49}$$

This result can be generalized to a function $f : \Re^n \rightarrow \Re^m$ with the determinant of the $m \times n$ matrix J:

$$J \equiv \frac{\partial(f_1, \ldots, f_m)}{\partial(x_1, \ldots, x_n)} \tag{A.50}$$

whose individual elements are defined by ($i = 1, \ldots, m$; $j = 1, \ldots, n$):

$$J_{ij} = \frac{\partial f_i}{\partial x_j} \tag{A.51}$$

A.9   The Gradient (or Three-Gradient)

The **gradient of a scalar function** f is denoted ∇f, where ∇ (the nabla symbol) denotes the vector differential operator. The gradient of f is defined as the unique vector field whose dot product with any vector $\vec{v}$ at each point $\vec{x}(x, y, z)$ is the directional derivative of f along $\vec{v}$. In the three-dimensional Cartesian coordinates with a Euclidean metric, the gradient is given by the following expression in different equivalent notations:

$$\nabla f \equiv \frac{\partial}{\partial \vec{x}} f = \left(\frac{\partial}{\partial x}, \frac{\partial}{\partial y}, \frac{\partial}{\partial z} \right) f = (\partial x, \partial y, \partial z)f = \left(\frac{\partial f}{\partial x}, \frac{\partial f}{\partial y}, \frac{\partial f}{\partial z} \right) \tag{A.52}$$

A.10   The Laplacian

The **Laplace operator**, or **Laplacian**, is a differential operator given by the divergence ($\nabla \cdot$) of the gradient of a function (∇f):

$$\nabla^2 f = \nabla \cdot \nabla f \tag{A.53}$$

In the three-dimensional Cartesian coordinates with a Euclidean metric, the Laplacian is given by:

$$\nabla^2 f = \frac{\partial^2 f}{\partial x^2} + \frac{\partial^2 f}{\partial y^2} + \frac{\partial^2 f}{\partial z^2} = \left(\frac{\partial^2}{\partial x^2} + \frac{\partial^2}{\partial y^2} + \frac{\partial^2}{\partial z^2} \right) f \tag{A.54}$$

For example:

4   Carl Gustav Jacob Jacobi (1804–1851), German mathematician.

- The Laplacian on $1/|\vec{x} - \vec{x'}|$:

$$\nabla^2 \left(\frac{1}{|\vec{x} - \vec{x'}|} \right) = -4\pi\delta(|\vec{x} - \vec{x'}|) \tag{A.55}$$

Proof: $\nabla^2 \left(\frac{1}{r} \right) = -4\pi\delta(r)$; $\vec{r} = (x, y, z)$; $r = \sqrt{x^2 + y^2 + z^2}$.

(a) For $r \neq 0$: $\nabla \frac{1}{r} = \left(\frac{\partial}{\partial x}, \frac{\partial}{\partial y}, \frac{\partial}{\partial z} \right) \frac{1}{r}$:

$$\frac{\partial}{\partial x} \frac{1}{\sqrt{x^2 + y^2 + z^2}} = -\frac{1}{2} \frac{2x}{(x^2 + y^2 + z^2)^{3/2}} = -\frac{x}{r^3} \tag{A.56}$$

Same for $\partial/\partial y$ and $\partial/\partial z$. Hence:

$$\nabla \frac{1}{r} = -\frac{\vec{r}}{r^3} \tag{A.57}$$

Then

$$\nabla^2 \frac{1}{r} = \nabla \cdot \nabla \frac{1}{r} = -\nabla \cdot \frac{\vec{r}}{r^3} \tag{A.58}$$

It follows:

$$\frac{\partial}{\partial x} \frac{x}{r^3} = \frac{r^3 - 3x^2 r}{r^6} \tag{A.59}$$

where we have used $(\partial/\partial x)r^3 = 3xr$, and similarly for y and z. Finally, adding all three components, we have:

$$-\nabla \cdot \frac{\vec{r}}{r^3} = -\frac{3r^3 - 3r(x^2 + y^2 + z^2)}{r^6} = 0 \tag{A.60}$$

(b) For $r = 0$: we define a vector field $\vec{F} = \nabla \frac{1}{r} = -\frac{\vec{r}}{r^3}$. With Gauss's theorem for any vector field $\vec{F}$, we find:

$$\int_V (\nabla^2 \frac{1}{r}) \mathrm{d}^3 \vec{x} = \int_V (\nabla \cdot \vec{F}) \mathrm{d}^3 \vec{x} = \int_{A = \partial V} \vec{F} \cdot \mathrm{d}\vec{A} = -\int_A \frac{\vec{r}}{r^3} \cdot \mathrm{d}\vec{A} \tag{A.61}$$

We can take a sphere centered at zero for the volume V. Then we have:

$$\int_A \frac{\vec{r}}{r^3} \cdot \mathrm{d}\vec{A} = \oint_{sphere\ surface} \frac{r\mathrm{d}A}{r^3} = \oint_{sphere\ surface} \frac{\mathrm{d}A}{r^2} = \oint_{sphere\ surface} \mathrm{d}\Omega = 4\pi \tag{A.62}$$

Hence:

$$\int_V (\nabla^2 \frac{1}{r}) \mathrm{d}^3 \vec{x} = -\oint \mathrm{d}\Omega = -4\pi \quad \square \tag{A.63}$$

A.11   Gauss's Theorem

Gauss's theorem, or the divergence theorem, relates the flux (that is, flow) of a vector field through a surface to the behavior of the vector field in the volume surrounded by the surface:

$$\int_{A = \partial V} \vec{F} \cdot \mathrm{d}\vec{A} = \int_V (\nabla \cdot \vec{F}) \mathrm{d}^3 \vec{x} \tag{A.64}$$

A.12   Stokes's Theorem

In its common form, **Stokes's theorem** relates the line integral of a vector field to the behavior of the vector field on the surface surrounded by the line:

$$\int_{L=\partial A} \vec{F} \cdot \mathrm{d}\vec{l} = \int_A (\nabla \times \vec{F}) \cdot \mathrm{d}\vec{A} \tag{A.65}$$

In general differential geometry, the theorem (also called the generalized Stokes theorem or the Stokes–Cartan theorem) is a statement about the integration of differential forms on manifolds, stating that the integral of a differential form $\mathcal{F}$ over the boundary $\partial\Omega$ of some orientable manifold Ω is equal to the integral of its exterior derivative ∂ over the entirety of Ω:

$$\int_{\partial\Omega} \mathcal{F} \cdot \mathrm{d}^{n-1}S = \int_{\Omega} \mathrm{d}^n x \, \partial\mathcal{F} \tag{A.66}$$

A.13   Green's Identities

We consider Stokes's theorem for the specific choice $\vec{F} = \phi\nabla\psi$. For this case, the theorem says:

$$\int_V \nabla \cdot (\phi\nabla\psi) \mathrm{d}^3\vec{x} = \int_{A=\partial V} \phi\nabla\psi \cdot \mathrm{d}\vec{A} \tag{A.67}$$

Using the identity $\nabla \cdot (\phi\vec{F}) = \nabla\phi \cdot \vec{F} + \phi\nabla \cdot \vec{F}$, we find **Green's first identity**:

$$\int_V \nabla\phi \cdot \nabla\psi \mathrm{d}^3\vec{x} + \int_V \phi\nabla^2\psi \mathrm{d}^3\vec{x} = \int_{A=\partial V} \phi\nabla\psi \cdot \mathrm{d}\vec{A} \tag{A.68}$$

We can reverse the roles of ϕ and ψ to find:

$$\int_V \nabla\psi \cdot \nabla\phi \mathrm{d}^3\vec{x} + \int_V \psi\nabla^2\phi \mathrm{d}^3\vec{x} = \int_{A=\partial V} \psi\nabla\phi \cdot \mathrm{d}\vec{A} \tag{A.69}$$

By taking the difference of the last two results, we find **Green's second identity**:

$$\int_V \left(\phi\nabla^2\psi - \psi\nabla^2\phi\right) \mathrm{d}^3\vec{x} = \int_{A=\partial V} (\phi\nabla\psi - \psi\nabla\phi) \cdot \mathrm{d}\vec{A} \tag{A.70}$$

Appendix B   Linear Algebra Tools

B.1   The Metric Tensor

In differential geometry, a *metric tensor* takes as input a pair of tangent vectors at a point of a surface of a manifold and produces a real number scalar in a way that generalizes the scalar product of vectors in Euclidean space. In the same way as a dot product, metric tensors are used to define the length of, and angle between, tangent vectors. Through integration, the metric tensor allows one to define and compute the length of curves on the manifold.

B.2   Covariance and Contravariance of Vectors

Covariance and contravariance describe how the quantitative description of certain geometric or physical entities changes under a transformation. Assume that the coordinates of these entities are given relative to a particular basis. If we change the units of the reference axes, e.g., from meters to centimeters (divide by 100), the components of a *measured velocity vector* would multiply by 100. Vectors exhibit this behavior of changing scale inversely to changes in scale to the reference axes: they are **contravariant**. In contrast, **covariant** vectors have inverse units in the sense that their components change in the same way as changes to scale of the reference axes. An example of a covector is the gradient, which has units of a spatial derivative, or inverse distance.

B.3   Some Basic Notions of Group Theory

A **group** G **is a set of related elements**, which satisfy the following rules:

1. The product $a \cdot b \in G$ for all $a, b \in G$.

2. The product is associative $(a \cdot b) \cdot c = a \cdot (b \cdot c)$.

3. One element is the identity $I \in G$, for which $a \cdot I = I \cdot a = a$ for all $a \in G$.

4. The inverse $a^{-1} \in G$ where $a \cdot a^{-1} = a^{-1} \cdot a = I$ for all $a \in G$ (hence $I^{-1} = I$).

A group can contain a finite or an infinite number of elements. A **continuous group** is a group with an infinite number of elements, which can be defined by a continuous parameter. For example, the set of finite rotations in space form such a group.

- An **Abelian group** is a group for which, in addition to the above rules, the product also commutes:

$$a \cdot b = b \cdot a \quad \text{for all } a, b \in G \tag{B.1}$$

If the group is not Abelian then it is **non-Abelian**.

● **Relation between groups.** Consider two groups with their corresponding products: $(G, \cdot)$ and $(g, *)$. A **homomorphism** is a function that connects the two groups $h : G \to g$ such that:

$$h(u \cdot v) = h(u) * h(v) \quad \text{for all } u, v \in G \tag{B.2}$$

A homomorphism preserves the algebraic structure. A group **isomorphism** is a group homomorphism that is bijective, that is there is a one-to-one correspondence between the elements of the groups in a way that respects the given group operations.

B.4   Matrices

A **matrix is defined as a rectangular array of real or complex numbers**. These numbers are organized with two subscripts, the first indicating the row (horizontal) and the second the column (vertical). For example, we define a matrix M with m rows and n columns:

$$M = \begin{pmatrix} M_{11} & M_{12} & \dots & M_{1n} \\ M_{21} & M_{22} & \dots & M_{2n} \\ & & \vdots & \\ M_{m1} & M_{m2} & \dots & M_{mn} \end{pmatrix} \in \text{Mat}_{m \times n}(\mathbb{R}) \quad \text{or} \quad \text{Mat}_{m \times n}(\mathbb{C}) \tag{B.3}$$

The set of $m \times n$ matrices and their addition form an Abelian group. A square matrix M is a $n \times n$ matrix ($M \in \text{Mat}_n(\mathbb{R})$). The identity matrix $\mathbb{1}$ is defined as $M_{ij} = \delta^i_j$. The multiplication (inner product) $A = MN$ is defined by the elements $A_{ij} = M_{ik}N_{kj}$ (with summation convention). Matrix multiplication is associative, but not necessarily commutative ($MN \neq NM$), and is distributive with respect to the addition. The inverse of the matrix M, if it exists, is defined as M^{-1}, where $MM^{-1} = M^{-1}M = \mathbb{1}$. The transpose M^T is found by interchanging the columns with the respective rows. A matrix is orthogonal if and only if $M^T = M^{-1}$.

For complex matrices, the adjoint $M^\dagger$ is formed by complex conjugating its transpose $M^\dagger = M^{T*}$. A Hermitian matrix satisfies $M^\dagger = M$. A unitary matrix satisfies $M^\dagger = M^{-1}$ (generalization of the orthogonality). The similarity transformation is given by

$$M' = UMU^{-1} \overset{unitary}{=} UMU^\dagger \tag{B.4}$$

where the last term is true if U is unitary.

B.5   Lie Groups, Lie Algebras, and Representations

Lie groups and algebras and their representations play a major role in particle physics with the transformation playing the role of a symmetry of a physical system. An **algebra** consists of a vector space $\mathcal{V}$ with a product $\times$ such that

$$\times : (A, B) \to A \times B \quad (A, B \in \mathcal{V}) \tag{B.5}$$

which satisfies distributivity:

$$A \times (aB + bC) = aA \times B + bA \times C, \quad (aA + bB) \times C = aA \times C + bB \times C \tag{B.6}$$

where a, b are scalars and A, B, C vectors.

   A **Lie[1] group** G **is an algebraic group whose elements are specified by one or more continuous parameters, which vary smoothly.** An infinitesimal transformation is a limiting form of small transformations. For example one may talk about an infinitesimal rotation around a fixed axis in space. The group is characterized by a single parameter: the angle of rotation around the chosen axis. Two consecutive rotations are represented by the products of the individual rotations, e.g., $R = R_1(\theta_1)R_2(\theta_2) = R(\theta_1 + \theta_2)$.

   A Lie group is composed of transformations which can be obtained as a product of infinitesimal transformations. Consider for simplicity the case of a group defined by a single parameter θ, a small transformation $T \in G$ is close to the identity and can therefore be expanded linearly as a function of an infinitesimal parameter ϵ:

$$T(\epsilon) = \mathbb{1} + \epsilon \left.\frac{dT}{d\epsilon}\right|_0 + \mathcal{O}(\epsilon^2) \tag{B.7}$$

The **generator of the Lie group** $\mathcal{G}$ is defined as follows, where the factor $1/i = -i$ is conventional to obtain Hermitian generators:

$$T(\epsilon) = \mathbb{1} + i\epsilon\mathcal{G} \quad \text{where} \quad \mathcal{G} \equiv \frac{1}{i}\left.\frac{dT}{d\epsilon}\right|_{\epsilon=0} \tag{B.8}$$

A finite transformation $T(\theta) \in G$ with finite parameter θ can be constructed as a product of an (infinite) number n of infinitesimal transformations with infinitesimal parameter $\epsilon = \theta/n$ $(n \to \infty)$. Then, we have:

$$T(\theta) = \left[T\left(\frac{\theta}{n}\right)\right]^n = [T(\epsilon)]^n = [\mathbb{1} + i\epsilon\mathcal{G}]^n = \left[\mathbb{1} + \frac{i\theta\mathcal{G}}{n}\right]^n \quad (n \to \infty) \tag{B.9}$$

Let us compute:

$$\ln \lim_{n\to\infty}\left(1 + \frac{\theta}{n}\right)^n = \lim_{n\to\infty} n\ln\left(1 + \frac{\theta}{n}\right) = \lim_{n\to\infty}\frac{\ln(1+\theta/n)}{1/n} \tag{B.10}$$

With L'Hôpital's rule, we find after derivation w.r.t. n:

$$\lim_{n\to\infty}\frac{\ln(1+\theta/n)}{1/n} = \lim_{n\to\infty}\frac{\frac{-\theta/n^2}{(1+\theta/n)}}{-1/n^2} = \lim_{n\to\infty}\frac{\theta}{(1+\theta/n)} = \theta \tag{B.11}$$

Hence (**binomial theorem**):

$$\lim_{n\to\infty}\left(1 + \frac{\theta}{n}\right)^n = e^\theta \tag{B.12}$$

A finite transformation $T(\theta)$ can therefore be expressed in terms of its generator $\mathcal{G}$ by **exponentiation of the generator**:

$$T(\theta) = e^{i\theta\mathcal{G}} \tag{B.13}$$

For small transformations $\theta_1, \theta_2 \ll 1$, the structure of the group near the identity transformation is given by (see Eq. (A.10)):

$$T(\theta_1)T(\theta_2) = e^{i\theta_1\mathcal{G}}e^{i\theta_2\mathcal{G}} \simeq e^{i(\theta_1+\theta_2)\mathcal{G}+\frac{1}{2}[\theta_1\mathcal{G},\theta_2\mathcal{G}]} = e^{i(\theta_1+\theta_2)\mathcal{G}} = T(\theta_1 + \theta_2) \tag{B.14}$$

where we used the fact that there is only one generator $\mathcal{G}$ which commutes with itself. However, this last relation takes on importance when the group has more than one generator which do not necessarily commute. Hence, the statement $T(\theta_1)T(\theta_2) = T(\theta_1 + \theta_2)$ is only true for commuting generators. Therefore, the commutation rules of the generators are key properties of the group.

• **Lie Algebra.** The Lie algebra is the mathematical construct that accompanies the Lie group and mimics the Lie group structure. It is a linear algebra related to the generators of the group but not a group itself. The

1   Marius Sophus Lie (1842–1899), Norwegian mathematician.

idea of the Lie algebra is to summarize and store in itself all of the information about the structure of the Lie group! Once the generators are defined by the Lie algebra, the entire list of Lie group elements can be recovered by exponentiation – see Eq. (B.13). This will best be illustrated with examples in the next sections. Mathematically, a **real Lie algebra** $\mathcal{A}$ of dimension N is a real vector space of dimension N with a commutator ("Lie product") $[a, b]$ defined for each $a, b \in \mathcal{A}$, which satisfies for all $a, b, c \in \mathcal{A}$:

1. $[a, b] \in \mathcal{A}$, $[b, a] \in \mathcal{A}$ and $[a, b] = -[b, a]$.

2. $[\alpha a + \beta b, c] = \alpha[a, c] + \beta[b, c]$ for α, β real numbers.

3. The Jacobi identity holds $[a, [b, c]] + [b, [c, a]] + [c, [a, b]] = 0$.

• **Representation of a group.** Up until now we have never worried (and we did not have to) about the number dimensions of space. For this, we need a **representation of the group**. For example, if we consider the group for rotations we have two, three, or more dimensions, acting on a particular set of axes. There are many ways to represent transformations and one often uses matrices. There are generally several matrix representations of a given group. Representations having the same dimensionality can be equivalent to one another in that their matrices have a one-to-one relationship with each other and each pair of matrices is related through the same similarity transformation.

Mathematically, the representation ρ of a group G is a *homomorphism* of all the elements of the group into the group of *non-singular linear transformations of a vector space of dimension N*, namely the set of the $N \times N$ invertible complex matrices:

$$\rho : G \to GL_N(\mathbb{C}) \tag{B.15}$$

In addition, the group multiplication is preserved, hence the representation of the product of two elements is equal to the products of their matrix representations.

• **Representation of an algebra.** Analogously, a representation of an algebra $\mathcal{A}$ of dimension N is a *homomorphism* of the algebra into $N \times N$ matrices $\in \mathrm{Mat}_N(\mathbb{C})$.

• **Irreducible representation.** A **reducible representation** can be expressed by a set of representations of lower dimensions ($n < N$), denoted with the direct sum:

$$R(g) = R_1(g) \oplus \cdots \oplus R_n(g) \tag{B.16}$$

with the dimensions of the $R_i(g)$ ($i = 1, \ldots, n$) adding up to the total dimension of $R(g)$. An **irreducible representation** cannot be written as a direct sum of other representations.

• **Dual representation of a group.** The dual representation ρ^* of a group G is defined on the dual-vector space such that for all elements of the group g, the dual representation of g is the transpose of the representation of the inverse element:

$$\rho^*(g) = \rho(g^{-1})^T \quad \text{for all } g \in G \tag{B.17}$$

B.6   The $SO(n)$ Group

A general spatial rotation in three dimensions in a system with three coordinates (x, y, z) is of the form:

$$\begin{pmatrix} x' \\ y' \\ z' \end{pmatrix} = (R) \begin{pmatrix} x \\ y \\ z \end{pmatrix} \tag{B.18}$$

where R is a 3×3 matrix. Since rotations preserve the distance from the origin, we have $x'^2 + y'^2 + z'^2 = x^2 + y^2 + z^2$ and it can be shown that the matrix R must be orthogonal. The rotations form a group called

$SO(3)$, where "S" is for special (a.k.a. determinant equals 1), "O" for orthogonal, and "3" is for three dimensions. In n dimensions the group of such matrices with their multiplication is called $SO(n)$:

$$SO(n) = \left\{ R \in \mathrm{Mat}_n(\mathbb{R}) : M^T M = MM^T = \mathbb{1}, \det(M) = 1 \right\} \tag{B.19}$$

We may denote three special rotations $R_x(\phi)$, $R_y(\psi)$, and $R_z(\theta)$:

$$R_x(\phi) = \begin{pmatrix} 1 & 0 & 0 \\ 0 & \cos\phi & \sin\phi \\ 0 & -\sin\phi & \cos\phi \end{pmatrix}; \quad R_y(\psi) = \begin{pmatrix} \cos\psi & 0 & -\sin\psi \\ 0 & 1 & 0 \\ \sin\psi & 0 & \cos\psi \end{pmatrix}; \quad R_z(\theta) = \begin{pmatrix} \cos\theta & \sin\theta & 0 \\ -\sin\theta & \cos\theta & 0 \\ 0 & 0 & 1 \end{pmatrix} \tag{B.20}$$

Note that in general the rotations do **not** commute, e.g.:

$$R_x(\phi) R_z(\theta) \neq R_z(\theta) R_x(\phi) \tag{B.21}$$

Hence, the **rotation group $SO(n)$ is a non-Abelian Lie group**.

It is known that a general rotation can be written as a function of three consecutive (Euler) rotations. This means that three parameters are necessary and sufficient to define a given general rotation. Since the 3×3 matrix R has *nine* elements, we should recall that the orthogonality conditions give *six* conditions reducing the problem to the desired *three* parameters. To the three parameters correspond **three generators of the $SO(3)$ group**. Using Eq. (B.8), they can be, for example, defined as J_x, J_y, and J_z, where:

$$J_x = \frac{1}{i} \frac{\mathrm{d}R_x(\phi)}{\mathrm{d}\phi}\bigg|_{\phi=0} = \begin{pmatrix} 0 & 0 & 0 \\ 0 & 0 & -i \\ 0 & i & 0 \end{pmatrix}; \qquad J_y = \frac{1}{i} \frac{\mathrm{d}R_y(\psi)}{\mathrm{d}\psi}\bigg|_{\psi=0} = \begin{pmatrix} 0 & 0 & i \\ 0 & 0 & 0 \\ -i & 0 & 0 \end{pmatrix} \tag{B.22}$$

and

$$J_z = \frac{1}{i} \frac{\mathrm{d}R_z(\theta)}{\mathrm{d}\theta}\bigg|_{\theta=0} = \begin{pmatrix} 0 & -i & 0 \\ i & 0 & 0 \\ 0 & 0 & 0 \end{pmatrix} \tag{B.23}$$

We note that they are Hermitian. Infinitesimal rotations are directly found using the generators, for example:

$$R_x(\epsilon_x) = \mathbb{1} + i\epsilon_x J_x; \quad R_y(\epsilon_y) = \mathbb{1} + i\epsilon_y J_y; \quad R_z(\epsilon_z) = \mathbb{1} + i\epsilon_z J_z. \tag{B.24}$$

Finite rotations are found by exponentiation, as in Eq. (B.13):

$$R_x(\phi) = e^{i\phi J_x}; \quad R_y(\psi) = e^{i\psi J_y}; \quad R_z(\theta) = e^{i\theta J_z}. \tag{B.25}$$

For example, we can verify:

$$
\begin{aligned}
R_x(\phi) &= e^{i\phi J_x} = \mathbb{1} + iJ_x\phi - \frac{\phi^2}{2!}J_x^2 + \frac{\phi^3}{3!}J_x^3 + \cdots \\
&= \begin{pmatrix} 1 & 0 & 0 \\ 0 & 1 & 0 \\ 0 & 0 & 1 \end{pmatrix} + \phi\begin{pmatrix} 0 & 0 & 0 \\ 0 & 0 & 1 \\ 0 & -1 & 0 \end{pmatrix} - \frac{\phi^2}{2!}\begin{pmatrix} 0 & 0 & 0 \\ 0 & 1 & 0 \\ 0 & 0 & 1 \end{pmatrix} + \cdots \\
&= \begin{pmatrix} 1 & 0 & 0 \\ 0 & \cos\phi & \sin\phi \\ 0 & -\sin\phi & \cos\phi \end{pmatrix} \qquad \square
\end{aligned}
\tag{B.26}
$$

where we used the series of Eqs. (A.9), (A.3), and (A.4). In general, a rotation by a finite angle α around a unity (normalized) axis $\hat{n} = (n_1, n_2, n_3)$ is given by:

$$R_n(\alpha) = e^{i\alpha(n_x J_x + n_y J_y + n_z J_z)} = e^{i\alpha \hat{n} \cdot \vec{J}} \tag{B.27}$$

By considering the commutators of the generators, we can find the **Lie algebra of the** $SO(3)$ **group**. For example:

$$[J_x, J_y] = J_x J_y - J_y J_x = \begin{pmatrix} 0 & 0 & 0 \\ 0 & 0 & -i \\ 0 & i & 0 \end{pmatrix} \begin{pmatrix} 0 & 0 & i \\ 0 & 0 & 0 \\ -i & 0 & 0 \end{pmatrix} - \begin{pmatrix} 0 & 0 & i \\ 0 & 0 & 0 \\ -i & 0 & 0 \end{pmatrix} \begin{pmatrix} 0 & 0 & 0 \\ 0 & 0 & -i \\ 0 & i & 0 \end{pmatrix}$$

$$= \begin{pmatrix} 0 & 0 & 0 \\ -1 & 0 & 0 \\ 0 & i & 0 \end{pmatrix} - \begin{pmatrix} 0 & -1 & 0 \\ 0 & 0 & 0 \\ 0 & 0 & 0 \end{pmatrix} = \begin{pmatrix} 0 & 1 & 0 \\ -1 & 0 & 0 \\ 0 & 0 & 0 \end{pmatrix} = i \begin{pmatrix} 0 & -i & 0 \\ i & 0 & 0 \\ 0 & 0 & 0 \end{pmatrix} = iJ_z \quad \text{(B.28)}$$

Finally, repeating also for the other commutators, we get the **Lie algebra** of the $SO(3)$ group:

$$[J_x, J_y] = iJ_z, \quad [J_y, J_z] = iJ_x, \quad \text{and} \quad [J_z, J_x] = iJ_y \tag{B.29}$$

This property can be written in compact form with the antisymmetric Levi-Civita tensor ϵ_{ijk} (see Eq. (A.12))

$$[J_i, J_j] = i\epsilon_{ijk} J_k \equiv ig_{ijk} J_k \tag{B.30}$$

where $g_{ijk} \equiv \epsilon_{ijk}$ are called the **structure constants of the group**. These constants, and hence the cyclic commutation rules, completely define the group of rotation. Although we started from the three-dimensional case, this algebra is valid for any representation of any dimensionality of the $SO(n)$ group. The fact that the generators do not commute is a consequence of the non-Abelian nature of the $SO(n)$ group.

B.7   The $U(n)$ Group

$U(n)$ **is the Lie group of all unitary** $n \times n$ **matrices with the operation of matrix multiplication**:

$$U(n) = \left\{ U \in \text{Mat}_n(\mathbb{C}) : U^\dagger U = UU^\dagger = \mathbb{1} \right\} \tag{B.31}$$

A complex matrix in n dimensions is determined by $2n^2$ real numbers. The unitarity condition means that there are n^2 relations among these numbers, so that a member of $U(n)$ is characterized by n^2 numbers.

In the case $n = 1$, we find the group $U(1)$, which consists of all complex numbers with absolute value equal to 1. In mathematics, the $U(1)$ group is called the **circle group** since it can be parametrized by an angle α of rotation by

$$\alpha \quad \rightarrow \quad z = e^{i\alpha} = \cos\alpha + i\sin\alpha \tag{B.32}$$

B.8   The $SU(2)$ Group

$SU(2)$ is the Lie group of two-dimensional complex matrices, which are unitary and special. We can define U in general and find its inverse:

$$U = \begin{pmatrix} a & b \\ c & d \end{pmatrix}; \qquad U^{-1} = \frac{1}{ad - bc} \begin{pmatrix} d & -b \\ -c & a \end{pmatrix} \tag{B.33}$$

where we used $\det U = ad - bc$. The unitary condition can be expressed as $U^\dagger = U^{-1}$, and since $\det U = 1$, we get:

$$\begin{pmatrix} a^* & c^* \\ b^* & d^* \end{pmatrix} = \begin{pmatrix} d & -b \\ -c & a \end{pmatrix} \tag{B.34}$$

Finally, $a^* = d$, $b^* = -c$, and $\det U = |a|^2 + |b|^2$. The representation of an $SU(2)$ matrix is given by two complex numbers a and b:

$$U(a,b) = \begin{pmatrix} a & b \\ -b^* & a^* \end{pmatrix} \qquad \text{where } |a|^2 + |b|^2 = 1 \tag{B.35}$$

The elements of the $SU(2)$ group can therefore be defined as:

$$SU(2) = \left\{ \begin{pmatrix} a & b \\ -b^* & a^* \end{pmatrix} \in \mathrm{Mat}_2(\mathbb{C}) : |a|^2 + |b|^2 = 1 \right\} \tag{B.36}$$

The identity is given by $U(1,0)$. The two complex numbers with the condition $|a^2| + |b^2| = 1$ can be defined by three real parameters, like $SO(3)$. We can enforce without loss of generality that a is purely imaginary, and b is complex. Then $a = i\alpha$ and $b = \beta + i\delta$:

$$U(\alpha, \beta, \delta) = \begin{pmatrix} i\alpha & \beta + i\delta \\ -\beta + i\delta & -i\alpha \end{pmatrix} \tag{B.37}$$

The generators of $SU(2)$ can be computed by differentiation:

$$G_1 = \frac{1}{i}\frac{dU}{d\delta}\bigg|_{\delta=0} = \frac{1}{i}\begin{pmatrix} 0 & i \\ i & 0 \end{pmatrix} = \begin{pmatrix} 0 & 1 \\ 1 & 0 \end{pmatrix}; \quad G_2 = \frac{1}{i}\frac{dU}{d\beta}\bigg|_{\beta=0} = \frac{1}{i}\begin{pmatrix} 0 & 1 \\ -1 & 0 \end{pmatrix} = \begin{pmatrix} 0 & -i \\ i & 0 \end{pmatrix} \tag{B.38}$$

and

$$G_3 = \frac{1}{i}\frac{dU}{d\alpha}\bigg|_{\alpha=0} = \frac{1}{i}\begin{pmatrix} i & 0 \\ 0 & -i \end{pmatrix} = \begin{pmatrix} 1 & 0 \\ 0 & -1 \end{pmatrix} \tag{B.39}$$

These look familiar! We have found that the Pauli matrices (see Appendix C.12) are the generators of the $SU(2)$ group. They are traceless and their determinants are equal to -1. The generators follow the $SU(2)$ **Lie algebra**:

$$[G_1, G_2] = 2iG_3, \quad [G_2, G_3] = 2iG_1, \quad \text{and} \quad [G_3, G_1] = 2iG_2 \tag{B.40}$$

For example:

$$\begin{aligned} [G_1, G_2] &= G_1G_2 - G_2G_1 = \begin{pmatrix} 0 & 1 \\ 1 & 0 \end{pmatrix}\begin{pmatrix} 0 & -i \\ i & 0 \end{pmatrix} - \begin{pmatrix} 0 & -i \\ i & 0 \end{pmatrix}\begin{pmatrix} 0 & 1 \\ 1 & 0 \end{pmatrix} \\ &= \begin{pmatrix} i & 0 \\ 0 & -i \end{pmatrix} - \begin{pmatrix} -i & 0 \\ 0 & i \end{pmatrix} = \begin{pmatrix} 2i & 0 \\ 0 & -2i \end{pmatrix} = 2i\begin{pmatrix} 1 & 0 \\ 0 & -1 \end{pmatrix} = 2iG_3 \end{aligned} \tag{B.41}$$

These results can be summarized with the totally antisymmetric Levi-Civita tensor ϵ_{ijk} (see Eq. (A.12)) as:

$$[G_i, G_j] = 2i\epsilon_{ijk}G_k \equiv ig_{ijk}G_k \tag{B.42}$$

where $g_{ijk} \equiv 2\epsilon_{ijk}$ are the structure constants of the group. We note that the constant 2 can be absorbed in the definition of the generators, so if we define new $SU(2)$ **generators** t_i ($i = 1, 2, 3$) with an added factor $(1/2)$, then the Lie algebra becomes:

$$[t_i, t_j] = i\epsilon_{ijk}t_k \equiv if_{ijk}t_k \tag{B.43}$$

where now $f_{ijk} \equiv \epsilon_{ijk}$ are the **structure constants of the t generators of the group**. In this way, the t_i operators satisfy exactly the Lie algebra of the $SO(3)$ group, Eq. (B.30).

• **Vectors of $SU(2)$.** The two-dimensional generators of $SU(2)$ operate on two-dimensional column vectors. A vector χ and its adjoint $\chi^\dagger$ can be written as:

$$\chi = \begin{pmatrix} \chi_1, \\ \chi_2 \end{pmatrix} = c_1u_1 + c_2u_2 \quad \text{and} \quad \chi^\dagger = \begin{pmatrix} \chi_1^* & \chi_2^* \end{pmatrix} = c_1^*u_1^\dagger + c_2^*u_2^\dagger \tag{B.44}$$

with the two linearly independent basis vectors:

$$u_1 = \begin{pmatrix} 1 \\ 0 \end{pmatrix}, \qquad u_2 = \begin{pmatrix} 0 \\ 1 \end{pmatrix} \tag{B.45}$$

By construction, the vector χ and its adjoint $\chi^\dagger$ transform differently under a transformation of the group:

$$\chi \to U\chi \quad \text{and} \quad \chi^\dagger \to \chi^\dagger U^\dagger \tag{B.46}$$

• **Homomorphism between $SO(3)$ and $SU(2)$.** Let us take a three-dimensional vector $\vec{x} = (x, y, z)$ and for each $\vec{x}$ construct the following matrix with the use of the Pauli matrices:

$$U(x, y, z) \equiv \vec{\sigma} \cdot \vec{x} = x\sigma_1 + y\sigma_2 + z\sigma_3 = \begin{pmatrix} z & x - iy \\ x + iy & -z \end{pmatrix} \tag{B.47}$$

By construction, the determinant is $\det(U) = x^2 + y^2 + z^2 = |\vec{x}|^2$ and it will be preserved under rotations. Here z is real, but consider the following diagonal $SU(2)$ matrix defined as:

$$U_z = \begin{pmatrix} e^{i\alpha} & 0 \\ 0 & e^{-i\alpha} \end{pmatrix} \implies U_z\sigma_1 U_z^\dagger = \begin{pmatrix} e^{i\alpha} & 0 \\ 0 & e^{-i\alpha} \end{pmatrix} \begin{pmatrix} 0 & 1 \\ 1 & 0 \end{pmatrix} \begin{pmatrix} e^{-i\alpha} & 0 \\ 0 & e^{i\alpha} \end{pmatrix} = \begin{pmatrix} 0 & e^{2i\alpha} \\ e^{-2i\alpha} & 0 \end{pmatrix} \tag{B.48}$$

Therefore

$$U_z\sigma_1 U_z^\dagger = \begin{pmatrix} 0 & \cos(2\alpha) + i\sin(2\alpha) \\ \cos(2\alpha) - i\sin(2\alpha) & 0 \end{pmatrix} = \cos(2\alpha)\sigma_1 - \sin(2\alpha)\sigma_2 \tag{B.49}$$

Similarly

$$U_z\sigma_2 U_z^\dagger = \sin(2\alpha)\sigma_1 + \cos(2\alpha)\sigma_2 \quad \text{and} \quad U_z\sigma_3 U_z^\dagger = \sigma_3 \tag{B.50}$$

Hence, there exists a **half-angle correspondence** between the 2×2 unitary transformation U_z and the spatial rotation R_z! See Eq. (B.20). This can be expressed as:

$$U_z(\theta/2) = \begin{pmatrix} e^{i\theta/2} & 0 \\ 0 & e^{-i\theta/2} \end{pmatrix} \quad \Leftrightarrow \quad R_z(\theta) = \begin{pmatrix} \cos\theta & \sin\theta & 0 \\ -\sin\theta & \cos\theta & 0 \\ 0 & 0 & 1 \end{pmatrix} \tag{B.51}$$

The correspondence is not one-to-one, since $R_z(\theta + 2\pi) = R_z(\theta)$ while $U_z((\theta + 2\pi)/2) = -U_z(\theta/2)$. So both $U_z(\theta/2)$ and $U_z(\theta/2 + \pi)$ correspond to $R_z(\theta)$. The same arguments can be repeated for x and y to find:

$$U_x(\phi/2) = \begin{pmatrix} \cos(\phi/2) & i\sin(\phi/2) \\ i\sin(\phi/2) & \cos(\phi/2) \end{pmatrix} \quad \Leftrightarrow \quad R_x(\phi) = \begin{pmatrix} 1 & 0 & 0 \\ 0 & \cos\phi & \sin\phi \\ 0 & -\sin\phi & \cos\phi \end{pmatrix} \tag{B.52}$$

and

$$U_y(\psi)/2) = \begin{pmatrix} \cos(\psi/2) & \sin(\psi/2) \\ -\sin(\psi/2) & \cos(\psi/2) \end{pmatrix} \quad \Leftrightarrow \quad R_y(\psi) = \begin{pmatrix} \cos\psi & 0 & -\sin\psi \\ 0 & 1 & 0 \\ \sin\psi & 0 & \cos\psi \end{pmatrix} \tag{B.53}$$

This two-to-one correspondence represents the homomorphism between the $SO(3)$ and the $SU(2)$ groups. This implies that the representations of $SO(3)$ are also representations of $SU(2)$.

• **Alternate representation.** Another way to represent $SU(2)$ with the three real parameters θ, ϕ, ψ with $|a|^2 + |b|^2 = 1$ is to choose:

$$a = \cos(\theta/2)e^{i(\psi+\phi)/2} \implies b = \sin(\theta/2)e^{i(\psi-\phi)/2} \tag{B.54}$$

with $0 \leq \theta < \pi$, $0 \leq \psi < 4\pi$, and $0 \leq \phi < 2\pi$. The representation of an $SU(2)$ matrix is then given by:

$$U(\theta, \phi, \psi) = \begin{pmatrix} \cos(\theta/2)e^{i(\psi+\phi)/2} & \sin(\theta/2)e^{i(\psi-\phi)/2} \\ -\sin(\theta/2)e^{-i(\psi-\phi)/2} & \cos(\theta/2)e^{-i(\psi+\phi)/2} \end{pmatrix} \tag{B.55}$$

which corresponds exactly to the combined unitary transformation obtained using the three Euler angles:

$$U(\theta, \phi, \psi) = U_z(\psi/2)U_y(\theta/2)U_z(\phi/2) = \begin{pmatrix} e^{i\psi/2} & 0 \\ 0 & e^{-i\psi/2} \end{pmatrix} \begin{pmatrix} \cos(\theta/2) & \sin(\theta/2) \\ -\sin(\theta/2) & \cos(\theta/2) \end{pmatrix} \begin{pmatrix} e^{i\phi/2} & 0 \\ 0 & e^{-i\phi/2} \end{pmatrix} \tag{B.56}$$

B.9   The $SU(N)$ Group

The **special unitary group** $SU(N)$ is composed of the $N \times N$-dimensional unitary matrices that have a determinant equal to unity ($\det U = 1$). Consequently, the group members are defined by $M^2 - 1$ numbers. The generators of the $SU(N)$ algebra, T^a ($a = 1, 2, \ldots, N^2 - 1$), are hermitian, traceless matrices satisfying the commutation relations:

$$[T^a, T^b] = if^{abc}T^c \tag{B.57}$$

the **structure constants** f^{abc} being real and totally antisymmetric. For $SU(3)$, the only non-zero (up to permutations) f^{abc} constants are:

$$\frac{1}{2}f^{123} = f^{147} = -f^{156} = f^{246} = f^{257} = f^{345} = -f^{367} = \frac{1}{\sqrt{3}}f^{458} = \frac{1}{\sqrt{3}}f^{678} = \frac{1}{2} \tag{B.58}$$

We can construct N-dimensional representations of the algebra of $SU(N)$ defining a set of matrices λ_i with:

$$T^a \equiv \frac{\lambda_a}{2} \tag{B.59}$$

For $N = 2$, we can choose these matrices to be the familiar Pauli spin matrices (see Section C.12):

$$\sigma_1 = \begin{pmatrix} 0 & 1 \\ 1 & 0 \end{pmatrix}, \quad \sigma_2 = \begin{pmatrix} 0 & -i \\ i & 0 \end{pmatrix}, \quad \sigma_3 = \begin{pmatrix} 1 & 0 \\ 0 & -1 \end{pmatrix} \tag{B.60}$$

For $N = 3$ the generalization of the Pauli matrices are the **Gell-Mann matrices** λ_i ($i = 1, \ldots, 8$):

$$\lambda_1 = \begin{pmatrix} 0 & 1 & 0 \\ 1 & 0 & 0 \\ 0 & 0 & 0 \end{pmatrix}; \quad \lambda_2 = \begin{pmatrix} 0 & -i & 0 \\ i & 0 & 0 \\ 0 & 0 & 0 \end{pmatrix}; \quad \lambda_3 = \begin{pmatrix} 1 & 0 & 0 \\ 0 & -1 & 0 \\ 0 & 0 & 0 \end{pmatrix}$$

$$\lambda_4 = \begin{pmatrix} 0 & 0 & 1 \\ 0 & 0 & 0 \\ 1 & 0 & 0 \end{pmatrix}; \quad \lambda_5 = \begin{pmatrix} 0 & 0 & -i \\ 0 & 0 & 0 \\ i & 0 & 0 \end{pmatrix}$$

$$\lambda_6 = \begin{pmatrix} 0 & 0 & 0 \\ 0 & 0 & 1 \\ 0 & 1 & 0 \end{pmatrix}; \quad \lambda_7 = \begin{pmatrix} 0 & 0 & 0 \\ 0 & 0 & -i \\ 0 & i & 0 \end{pmatrix}; \quad \lambda_8 = \frac{1}{\sqrt{3}}\begin{pmatrix} 1 & 0 & 0 \\ 0 & 1 & 0 \\ 0 & 0 & -2 \end{pmatrix} \tag{B.61}$$

• **Multiplets of the group.** The N-dimensional generators of $SU(N)$ operate on N-dimensional column vectors. Clearly, there are N linearly independent vectors, which we may denote by u_a ($a = 1, 2, \ldots, N$). In $SU(3)$ we have for instance:

$$u_1 = \begin{pmatrix} 1 \\ 0 \\ 0 \end{pmatrix}, \quad u_2 = \begin{pmatrix} 0 \\ 1 \\ 0 \end{pmatrix}, \quad u_3 = \begin{pmatrix} 0 \\ 0 \\ 1 \end{pmatrix} \tag{B.62}$$

- **Useful traces.** By construction, the Gell-Mann matrices are traceless:

$$\text{Tr}(\lambda_a) = 0 \tag{B.63}$$

Very often we will be interested in the traces of products of Gell-Mann matrices. We can directly compute:

$$\lambda_1\lambda_1 = \lambda_2\lambda_2 = \lambda_3\lambda_3 = \begin{pmatrix} 1 & 0 & 0 \\ 0 & 1 & 0 \\ 0 & 0 & 0 \end{pmatrix} ; \quad \lambda_4\lambda_4 = \lambda_5\lambda_5 = \begin{pmatrix} 1 & 0 & 0 \\ 0 & 0 & 0 \\ 0 & 0 & 1 \end{pmatrix} ; \quad \lambda_6\lambda_6 = \lambda_7\lambda_7 = \begin{pmatrix} 0 & 0 & 0 \\ 0 & 1 & 0 \\ 0 & 0 & 1 \end{pmatrix}$$

$$\lambda_8\lambda_8 = \begin{pmatrix} \frac{1}{3} & 0 & 0 \\ 0 & \frac{1}{3} & 0 \\ 0 & 0 & \frac{4}{3} \end{pmatrix} ; \quad \lambda_1\lambda_2 = \begin{pmatrix} i & 0 & 0 \\ 0 & -i & 0 \\ 0 & 0 & 0 \end{pmatrix} ; \quad \lambda_1\lambda_3 = \begin{pmatrix} 0 & -1 & 0 \\ 1 & 0 & 0 \\ 0 & 0 & 0 \end{pmatrix} ; \quad \lambda_1\lambda_4 = \begin{pmatrix} 0 & 0 & 0 \\ 0 & 0 & 1 \\ 0 & 0 & 0 \end{pmatrix}$$

$$\lambda_1\lambda_5 = \begin{pmatrix} 0 & 0 & 0 \\ 0 & 0 & -i \\ 0 & 0 & 0 \end{pmatrix} ; \quad \lambda_1\lambda_6 = \begin{pmatrix} 0 & 0 & 1 \\ 0 & 0 & 0 \\ 0 & 0 & 0 \end{pmatrix} ; \quad \lambda_1\lambda_7 = \begin{pmatrix} 0 & 0 & -i \\ 0 & 0 & 0 \\ 0 & 0 & 0 \end{pmatrix} ; \quad \lambda_1\lambda_8 = \begin{pmatrix} 0 & \frac{1}{\sqrt{3}} & 0 \\ \frac{1}{\sqrt{3}} & 0 & 0 \\ 0 & 0 & 0 \end{pmatrix}$$

$$\lambda_2\lambda_3 = \begin{pmatrix} 0 & i & 0 \\ i & 0 & 0 \\ 0 & 0 & 0 \end{pmatrix} ; \quad \lambda_2\lambda_4 = \begin{pmatrix} 0 & 0 & 0 \\ 0 & 0 & i \\ 0 & 0 & 0 \end{pmatrix} \tag{B.64}$$

and so on. From these products, we can deduce that the traces of a product of two different Gell-Mann matrices is zero, whereas the trace of the product of two identical matrices is always equal to 2. We can summarize these results as:

$$\text{Tr}(\lambda_a\lambda_b) = 2\delta_{ab} \tag{B.65}$$

Sometimes we need to sum over all eight matrices (or colors), so it is useful to note that:

$$\sum_a \text{Tr}(\lambda_a\lambda_a) = 2\sum_a \delta_{aa} = 16 \tag{B.66}$$

Finally, we will need the trace of the product of four Gell-Mann matrices of the type $\lambda_a\lambda_b\lambda_a\lambda_c$. We can for instance consider:

$$\sum_{a=1}^{8} \lambda_a\lambda_1\lambda_a\lambda_1 = \sum_{a=1}^{8} \lambda_a\lambda_2\lambda_a\lambda_2 = \sum_{a=1}^{8} \lambda_a\lambda_3\lambda_a\lambda_3 = \begin{pmatrix} -\frac{2}{3} & 0 & 0 \\ 0 & -\frac{2}{3} & 0 \\ 0 & 0 & 0 \end{pmatrix}$$

$$\sum_{a=1}^{8} \lambda_a\lambda_4\lambda_a\lambda_4 = \sum_{a=1}^{8} \lambda_a\lambda_5\lambda_a\lambda_5 = \begin{pmatrix} -\frac{2}{3} & 0 & 0 \\ 0 & 0 & 0 \\ 0 & 0 & -\frac{2}{3} \end{pmatrix}$$

$$\sum_{a=1}^{8} \lambda_a\lambda_6\lambda_a\lambda_6 = \sum_{a=1}^{8} \lambda_a\lambda_7\lambda_a\lambda_7 = \begin{pmatrix} 0 & 0 & 0 \\ 0 & -\frac{2}{3} & 0 \\ 0 & 0 & -\frac{2}{3} \end{pmatrix}$$

$$\sum_{a=1}^{8} \lambda_a\lambda_8\lambda_a\lambda_8 = \begin{pmatrix} -\frac{2}{9} & 0 & 0 \\ 0 & -\frac{2}{9} & 0 \\ 0 & 0 & -\frac{8}{9} \end{pmatrix} \tag{B.67}$$

We note that the traces are always equal to $-4/3$. On the other hand, the trace will always vanish for products of the type:

$$\sum_{a=1}^{8} \lambda_a\lambda_1\lambda_a\lambda_2 = \begin{pmatrix} -\frac{2i}{3} & 0 & 0 \\ 0 & \frac{2i}{3} & 0 \\ 0 & 0 & 0 \end{pmatrix} ; \quad \sum_{a=1}^{8} \lambda_a\lambda_1\lambda_a\lambda_3 = \begin{pmatrix} 0 & \frac{2}{3} & 0 \\ -\frac{2}{3} & 0 & 0 \\ 0 & 0 & 0 \end{pmatrix} ; \quad \dots \tag{B.68}$$

Hence, we can summarize these results with the following expression:

$$\sum_a \mathrm{Tr}(\lambda_a \lambda_b \lambda_a \lambda_c) = -\frac{4}{3}\delta_{bc} \tag{B.69}$$

Last but not least, we compute the traces of the following type of products $\lambda_a \lambda_a \lambda_b \lambda_c$:

$$\sum_{a=1}^{8} \lambda_a \lambda_a \lambda_1 \lambda_1 = \sum_{a=1}^{8} \lambda_a \lambda_a \lambda_2 \lambda_2 = \sum_{a=1}^{8} \lambda_a \lambda_a \lambda_3 \lambda_3 = \begin{pmatrix} \frac{16}{3} & 0 & 0 \\ 0 & \frac{16}{3} & 0 \\ 0 & 0 & 0 \end{pmatrix}$$

$$\sum_{a=1}^{8} \lambda_a \lambda_a \lambda_4 \lambda_4 = \sum_{a=1}^{8} \lambda_a \lambda_a \lambda_5 \lambda_5 = \begin{pmatrix} \frac{16}{3} & 0 & 0 \\ 0 & 0 & 0 \\ 0 & 0 & \frac{16}{3} \end{pmatrix}$$

$$\sum_{a=1}^{8} \lambda_a \lambda_a \lambda_6 \lambda_6 = \sum_{a=1}^{8} \lambda_a \lambda_a \lambda_7 \lambda_7 = \begin{pmatrix} 0 & 0 & 0 \\ 0 & \frac{16}{3} & 0 \\ 0 & 0 & \frac{16}{3} \end{pmatrix}$$

$$\sum_{a=1}^{8} \lambda_a \lambda_a \lambda_8 \lambda_8 = \begin{pmatrix} \frac{16}{9} & 0 & 0 \\ 0 & \frac{16}{9} & 0 \\ 0 & 0 & \frac{64}{9} \end{pmatrix} \quad \ldots \tag{B.70}$$

while the trace will vanish in other cases. Consequently, we can summarize these results with the following expression:

$$\sum_a \mathrm{Tr}(\lambda_a \lambda_a \lambda_b \lambda_c) = \frac{32}{3}\delta_{bc} \tag{B.71}$$

B.10   The Lorentz $SO(3,1)$ Group

The **Lorentz $SO(3,1)$ Lie group** contains all the homogeneous transformations of special relativity with rotations and boosts or more generally the transformations in four dimensions. We work in flat Minkowski space (see Appendix D) with the metric tensor $g_{\mu\nu} = diag(1,-1,-1,-1)$. Consequently, the group has a signature "1–3" (one time coordinate, three space coordinates) and is called the $SO(3,1)$ group. The Lorentzian **inner product** is not positive definite (contrary to the Euclidian case) and is given by:

$$x \cdot y \equiv x^T g y \tag{B.72}$$

or in components:

$$x \cdot y = g_{\mu\nu} x^\mu y^\nu = x^\mu y_\mu = x_\mu y^\mu \tag{B.73}$$

Lorentz transformations are those transformations $x' = \Lambda x$ that leave the inner product invariant:

$$(\Lambda x)\cdot(\Lambda y) = x \cdot y \quad \Longrightarrow \quad x^T \Lambda^T g \Lambda y = x^T g y \tag{B.74}$$

This leads to the **relation of orthogonality** (or **Lorentz condition**):

$$\Lambda^T g \Lambda = g \tag{B.75}$$

Written in components, it reads:

$$g_{\mu\nu} \Lambda^\mu{}_\alpha \Lambda^\nu{}_\beta = g_{\alpha\beta} \tag{B.76}$$

Since the metric tensor is symmetric, this gives 10 constraints; the Lorentz transformation Λ is a 4×4 matrix, so it depends on $16 - 10 = 6$ independent parameters.

Lorentz transformations preserve the norm $a^2 = a \cdot a$ in Minkowski space, which is positive for time-like four-vectors, negative for space-like four-vectors, or zero for light-like four-vectors (see Section D.2). Therefore, they are transformations along the hypersurfaces of constant norm, and cannot change the likeness (time-like, space-like, or light-like) of a four-vector.

The Lorentz group can be subdivided into four classes according to the determinant and the orthochronicity of the Λ matrix. For a given transformation, we have:

$$\det(\Lambda^T g \Lambda) = \det(g \Lambda \Lambda) = \det(g)(\det(\Lambda))^2 = \det(g) \implies (\det \Lambda)^2 = 1 \tag{B.77}$$

and from $g_{\mu\nu}\Lambda^\mu_{\ 0}\Lambda^\nu_{\ 0} = g_{00} = 1$:

$$\Lambda^0_{\ 0}\Lambda^0_{\ 0} - \sum_i \Lambda^i_{\ 0}\Lambda^i_{\ 0} = (\Lambda^0_{\ 0})^2 - \sum_i (\Lambda^i_{\ 0})^2 = 1 \implies (\Lambda^0_{\ 0})^2 = 1 + \sum_i (\Lambda^i_{\ 0})^2. \tag{B.78}$$

These two conditions define the four classes:

$$
\begin{array}{cccc}
\det \Lambda = +1 & \Lambda^0_{\ 0} \geq 1 & \text{proper orthochronous} & (1) \\
\det \Lambda = +1 & \Lambda^0_{\ 0} \leq -1 & \text{proper not orthochronous} & (PT) \\
\det \Lambda = -1 & \Lambda^0_{\ 0} \geq 1 & \text{improper orthochronous} & (P) \\
\det \Lambda = -1 & \Lambda^0_{\ 0} \leq -1 & \text{improper not orthochronous} & (T)
\end{array} \tag{B.79}
$$

Depending on the signs of $\det \Lambda$ and $\Lambda^0_{\ 0}$, the Lorentz group has four disconnected components. The four classes are disconnected, because the determinant and $\Lambda^0_{\ 0}$ cannot be varied in a continuous way from values smaller or larger than 1. In each class, we find a "characteristic transformation":

1. The identity ($\mathbb{1}$)

$$\mathbb{1} = \begin{pmatrix} 1 & 0 & 0 & 0 \\ 0 & 1 & 0 & 0 \\ 0 & 0 & 1 & 0 \\ 0 & 0 & 0 & 1 \end{pmatrix} \tag{B.80}$$

2. The parity P

$$P = \begin{pmatrix} 1 & 0 & 0 & 0 \\ 0 & -1 & 0 & 0 \\ 0 & 0 & -1 & 0 \\ 0 & 0 & 0 & -1 \end{pmatrix} \tag{B.81}$$

3. The time-reversal T

$$T = \begin{pmatrix} -1 & 0 & 0 & 0 \\ 0 & 1 & 0 & 0 \\ 0 & 0 & 1 & 0 \\ 0 & 0 & 0 & 1 \end{pmatrix} \tag{B.82}$$

4. The strong reflection (PT) – see Section 1.7.

The first class generates a subgroup. It contains the transformations that just act on the spatial coordinates which are the purely **spatial rotations** since these preserve the lengths, and those combining the time with a spatial coordinate, forming the **boosts**. Other subgroups can be generated by combining the other classes (P, T, or PT) with the first one. Each transformation in each subgroup can be transformed in a continuous way into another transformation of the same subgroup.

The general space transformation of the Lorentz group is defined as (see Appendix D.3):

$$x'^\mu = \Lambda^\mu_{\ \nu} x^\nu \tag{B.83}$$

For example, for a Lorentz boost along x, y, and z, we have:

$$\Lambda^\mu_{\ \nu}(\beta_x) = \begin{pmatrix} \gamma & -\beta_x\gamma & 0 & 0 \\ -\beta_x\gamma & \gamma & 0 & 0 \\ 0 & 0 & 1 & 0 \\ 0 & 0 & 0 & 1 \end{pmatrix} ; \quad \Lambda^\mu_{\ \nu}(\beta_y) = \begin{pmatrix} \gamma & 0 & -\beta_y\gamma & 0 \\ 0 & 1 & 0 & 0 \\ -\beta_y\gamma & 0 & \gamma & 0 \\ 0 & 0 & 0 & 1 \end{pmatrix} \tag{B.84}$$

and

$$\Lambda^\mu_{\ \nu}(\beta_z) = \begin{pmatrix} \gamma & 0 & 0 & -\beta_z\gamma \\ 0 & 1 & 0 & 0 \\ 0 & 0 & 1 & 0 \\ -\beta_z\gamma & 0 & 0 & \gamma \end{pmatrix} \tag{B.85}$$

where $\beta = v/c$ and $\gamma^2 = 1/(1-\beta^2)$. For the spatial rotations, we have (cf. Eq. (B.20)):

$$\Lambda^\mu_{\ \nu}(\phi) = \begin{pmatrix} 1 & 0 & 0 & 0 \\ 0 & 1 & 0 & 0 \\ 0 & 0 & \cos\phi & \sin\phi \\ 0 & 0 & -\sin\phi & \cos\phi \end{pmatrix} ; \quad \Lambda^\mu_{\ \nu}(\psi) = \begin{pmatrix} 1 & 0 & 0 & 0 \\ 0 & \cos\psi & 0 & -\sin\psi \\ 0 & 0 & 1 & 0 \\ 0 & \sin\psi & 0 & \cos\psi \end{pmatrix} \tag{B.86}$$

and

$$\Lambda^\mu_{\ \nu}(\theta) = \begin{pmatrix} 1 & 0 & 0 & 0 \\ 0 & \cos\theta & \sin\theta & 0 \\ 0 & -\sin\theta & \cos\theta & 0 \\ 0 & 0 & 0 & 1 \end{pmatrix} \tag{B.87}$$

Since such rotations clearly form a group themselves, the spatial rotations form a subgroup of the Lorentz group. However, the set of pure boosts does not form a subgroup! In fact, the composition of two non-collinear Lorentz boosts results in a Lorentz transformation that is not a pure boost but is the composition of a boost and a rotation. This rotation is called the Thomas rotation, **Thomas–Wigner rotation**, or Wigner rotation!

If we write the identity transformation as $\Lambda^\mu_{\ \nu} = \delta^\mu_\nu$, then the **infinitesimal Lorentz transformation** derives from it in the following way:

$$x'^\mu = \Lambda^\mu_{\ \nu}(\omega)x^\nu = (\delta^\mu_\nu + \omega^\mu_{\ \nu})x^\nu \tag{B.88}$$

where the $\omega^\mu_{\ \nu}$ are infinitesimal. A direct consequence of the **Lorentz orthogonality condition**, Eq. (5.10), is:

$$\begin{aligned} g_{\alpha\beta} = g_{\mu\nu}\Lambda^\mu_{\ \alpha}\Lambda^\nu_{\ \beta} &= g_{\mu\nu}(\delta^\mu_\alpha + \omega^\mu_{\ \alpha})(\delta^\nu_\beta + \omega^\nu_{\ \beta}) = g_{\mu\nu}\delta^\mu_\alpha\delta^\nu_\beta + g_{\mu\nu}\delta^\mu_\alpha\omega^\nu_{\ \beta} + g_{\mu\nu}\omega^\mu_{\ \alpha}\delta^\nu_\beta + \cdots \\ &\simeq g_{\alpha\beta} + g_{\alpha\nu}\omega^\nu_{\ \beta} + g_{\mu\beta}\omega^\mu_{\ \alpha} \\ &= g_{\alpha\beta} + \omega_{\alpha\beta} + \omega_{\beta\alpha} \end{aligned} \tag{B.89}$$

therefore $\omega_{\alpha\beta}$ is an antisymmetric tensor (be careful, $\omega_{\alpha\beta} \neq \omega^\beta_\alpha$!), so it has six independent components with zeros on the diagonal:

$$\omega_{\alpha\beta} = -\omega_{\beta\alpha} \tag{B.90}$$

For the infinitesimal $\Lambda^\mu_{\ \nu}(\beta_x)$ we find:

$$\Lambda^\mu_{\ \nu}(\beta_x) \simeq \begin{pmatrix} 1 & -\beta_x & 0 & 0 \\ -\beta_x & 1 & 0 & 0 \\ 0 & 0 & 1 & 0 \\ 0 & 0 & 0 & 1 \end{pmatrix} = \mathbb{1} + \begin{pmatrix} 0 & -\beta_x & 0 & 0 \\ -\beta_x & 0 & 0 & 0 \\ 0 & 0 & 0 & 0 \\ 0 & 0 & 0 & 0 \end{pmatrix} \tag{B.91}$$

We can write $\omega_{10} \equiv \beta_x$ and define a 4×4 matrix M^{10}, such that:

$$\Lambda^\mu_{\ \nu}(\beta_x) = \mathbb{1} + \omega_{10}\begin{pmatrix} 0 & -1 & 0 & 0 \\ -1 & 0 & 0 & 0 \\ 0 & 0 & 0 & 0 \\ 0 & 0 & 0 & 0 \end{pmatrix} \equiv \mathbb{1} + \omega_{10}M^{10} \tag{B.92}$$

Since $\omega_{\alpha\beta} = -\omega_{\beta\alpha}$ from the Lorentz orthogonality condition, we can define a new 4×4 matrix $M^{01} \equiv -M^{10}$ and write:

$$\omega_{10} M^{10} = \frac{1}{2} \left(\omega_{10} M^{10} + \omega_{01} M^{01} \right) \tag{B.93}$$

With

$$M^{10} = \begin{pmatrix} 0 & -1 & 0 & 0 \\ -1 & 0 & 0 & 0 \\ 0 & 0 & 0 & 0 \\ 0 & 0 & 0 & 0 \end{pmatrix} ; \quad M^{20} = \begin{pmatrix} 0 & 0 & -1 & 0 \\ 0 & 0 & 0 & 0 \\ -1 & 0 & 0 & 0 \\ 0 & 0 & 0 & 0 \end{pmatrix} ; \quad M^{30} = \begin{pmatrix} 0 & 0 & 0 & -1 \\ 0 & 0 & 0 & 0 \\ 0 & 0 & 0 & 0 \\ -1 & 0 & 0 & 0 \end{pmatrix} \tag{B.94}$$

the general boost generator matrix is given with $\omega_{10} = \beta_x$, $\omega_{20} = \beta_y$, $\omega_{30} = \beta_z$ by (sum over $i = 1, 2, 3$):

$$\frac{1}{2} \left(\omega_{i0} M^{i0} + \omega_{0i} M^{0i} \right) \tag{B.95}$$

Similarly, for infinitesimal spatial rotations, we use M^{23} for the rotation about x, M^{31} for the rotation about y, and M^{12} for the rotation about z, where:

$$M^{23} = \begin{pmatrix} 0 & 0 & 0 & 0 \\ 0 & 0 & 0 & 0 \\ 0 & 0 & 0 & -1 \\ 0 & 0 & 1 & 0 \end{pmatrix} ; \quad M^{31} = \begin{pmatrix} 0 & 0 & 0 & 0 \\ 0 & 0 & 0 & 1 \\ 0 & 0 & 0 & 0 \\ 0 & -1 & 0 & 0 \end{pmatrix} ; \quad M^{12} = \begin{pmatrix} 0 & 0 & 0 & 0 \\ 0 & 0 & -1 & 0 \\ 0 & 1 & 0 & 0 \\ 0 & 0 & 0 & 0 \end{pmatrix} \tag{B.96}$$

and the spatial rotation generator matrix is (sum over $i, j = 1, 2, 3$):

$$\frac{1}{2} \left(\omega_{ij} M^{ij} + \omega_{ji} M^{ji} \right) \tag{B.97}$$

with ω_{23}, ω_{31}, ω_{12} representing the three Euler angles. With in addition $\omega_{\alpha\alpha} = 0$ ($\alpha = 0, \dots, 3$), the matrix representing a finite Lorentz transformation (including both boosts and rotations) is obtained by exponentiating the infinitesimal results, so:

$$\Lambda(\vec{\beta}, \vec{\theta}) = e^{\frac{1}{2} \omega_{\mu\nu} M^{\mu\nu}} \quad \text{where} \quad \omega_{\mu\nu} = \begin{pmatrix} 0 & \beta_x & \beta_y & \beta_z \\ -\beta_x & 0 & \theta_z & \theta_y \\ -\beta_y & -\theta_z & 0 & \theta_x \\ -\beta_z & -\theta_y & -\theta_x & 0 \end{pmatrix} \tag{B.98}$$

We now look for the explicit space-time **representation of the generators of the Lorentz group**. The generators of the spatial rotations can be trivially extracted in analogy to those of the $SO(3)$ group, which can be found in Section B.6, or just matching the M^{ij} matrices above. We define:

$$\vec{J} = i(M^{23}, M^{31}, M^{12}) \tag{B.99}$$

and write **the 4×4 matrix representation of the generators of the spatial rotations** J_x, J_y, and J_z:

$$J_x = \frac{1}{i} \begin{pmatrix} 0 & 0 & 0 & 0 \\ 0 & 0 & 0 & 0 \\ 0 & 0 & 0 & 1 \\ 0 & 0 & -1 & 0 \end{pmatrix} ; \quad J_y = \frac{1}{i} \begin{pmatrix} 0 & 0 & 0 & 0 \\ 0 & 0 & 0 & -1 \\ 0 & 0 & 0 & 0 \\ 0 & 1 & 0 & 0 \end{pmatrix} ; \quad J_z = \frac{1}{i} \begin{pmatrix} 0 & 0 & 0 & 0 \\ 0 & 0 & 1 & 0 \\ 0 & -1 & 0 & 0 \\ 0 & 0 & 0 & 0 \end{pmatrix} \tag{B.100}$$

These generators obey, as they should, the **commutation rules of the $SO(3)$ Lie algebra** (see Eq. (B.30)). When going from the Ms to Js, we have used the convention with the i factor for an operator acting on a Hilbert space (see Section C.2). Similarly, for the Lorentz boosts, we define:

$$\vec{K} = i(M^{10}, M^{20}, M^{30}) \tag{B.101}$$

We can also directly derive them from the boost matrices. We note that it is convenient to relabel the boost parameter by substituting $\tanh \phi = \beta$ (see Eq. (D.27)):

$$\cosh \phi = \gamma \quad \text{and} \quad \sinh \phi = \beta\gamma \tag{B.102}$$

With this definition, the boost along the x-axis is for example expressed as (cf. Eq. (D.28)):

$$B_x(\phi_1) = \Lambda^\mu{}_\nu(\tanh \phi_1) = \begin{pmatrix} \cosh \phi_1 & -\sinh \phi_1 & 0 & 0 \\ -\sinh \phi_1 & \cosh \phi_1 & 0 & 0 \\ 0 & 0 & 1 & 0 \\ 0 & 0 & 0 & 1 \end{pmatrix} \tag{B.103}$$

The most general transformation is composed of boosts in three directions $B_x(\phi_1)$, $B_y(\phi_2)$, and $B_z(\phi_3)$, which we define in analogy with B_x. To the three parameters correspond **three generators of the Lorentz boost group**. They can, for example, be defined as K_x, K_y, and K_z where:

$$K_x = \frac{1}{i}\frac{dB_x(\phi_1)}{d\phi_1}\bigg|_{\phi_1=0} = \begin{pmatrix} 0 & -i & 0 & 0 \\ -i & 0 & 0 & 0 \\ 0 & 0 & 0 & 0 \\ 0 & 0 & 0 & 0 \end{pmatrix}; \quad K_y = \frac{1}{i}\frac{dB_y(\phi_2)}{d\phi_2}\bigg|_{\phi_2=0} = \begin{pmatrix} 0 & 0 & -i & 0 \\ 0 & 0 & 0 & 0 \\ -i & 0 & 0 & 0 \\ 0 & 0 & 0 & 0 \end{pmatrix} \tag{B.104}$$

and

$$K_z = \frac{1}{i}\frac{dB_z(\phi_3)}{d\phi_3}\bigg|_{\phi_3=0} = \begin{pmatrix} 0 & 0 & 0 & -i \\ 0 & 0 & 0 & 0 \\ 0 & 0 & 0 & 0 \\ -i & 0 & 0 & 0 \end{pmatrix} \tag{B.105}$$

That's all – the most general Lorentz transformation is composed of three boosts in three directions, and three rotations about three axes, and we have defined the corresponding six generators J_x, J_y, J_z, K_x, K_y, and K_z above. A finite Lorentz transformation can then be expressed as:

$$\Lambda(\vec{\theta}, \vec{\phi}) = e^{\frac{i}{2}(\vec{\theta}\cdot\vec{J}+\vec{\phi}\cdot\vec{K})} \tag{B.106}$$

where $\vec{\theta} = (\theta_x, \theta_y, \theta_z)$ and $\vec{\phi} = (\phi_1, \phi_2, \phi_3)$. The **Lorentz Lie algebra** is defined by the following set of commutation rules. Those of the $\vec{J}$ operators can be derived directly from the $SO(3)$ ones. They can easily be verified to satisfy:

$$[J_i, J_j] = i\epsilon_{ijk}J_k \tag{B.107}$$

Similarly, we can consider:

$$
[K_x, K_y] = \begin{pmatrix} 0 & -i & 0 & 0 \\ -i & 0 & 0 & 0 \\ 0 & 0 & 0 & 0 \\ 0 & 0 & 0 & 0 \end{pmatrix}\begin{pmatrix} 0 & 0 & -i & 0 \\ 0 & 0 & 0 & 0 \\ -i & 0 & 0 & 0 \\ 0 & 0 & 0 & 0 \end{pmatrix} - \begin{pmatrix} 0 & 0 & -i & 0 \\ 0 & 0 & 0 & 0 \\ -i & 0 & 0 & 0 \\ 0 & 0 & 0 & 0 \end{pmatrix}\begin{pmatrix} 0 & -i & 0 & 0 \\ -i & 0 & 0 & 0 \\ 0 & 0 & 0 & 0 \\ 0 & 0 & 0 & 0 \end{pmatrix}
$$

$$
= \begin{pmatrix} 0 & 0 & 0 & 0 \\ 0 & 0 & -1 & 0 \\ 0 & 0 & 0 & 0 \\ 0 & 0 & 0 & 0 \end{pmatrix} - \begin{pmatrix} 0 & 0 & 0 & 0 \\ 0 & 0 & 0 & 0 \\ 0 & -1 & 0 & 0 \\ 0 & 0 & 0 & 0 \end{pmatrix} = \begin{pmatrix} 0 & 0 & 0 & 0 \\ 0 & 0 & -1 & 0 \\ 0 & 1 & 0 & 0 \\ 0 & 0 & 0 & 0 \end{pmatrix}
$$

$$
= -iJ_z \tag{B.108}
$$

or we can compute:

$$
\begin{aligned}
[K_x, J_y] &= \frac{1}{i}\begin{pmatrix} 0 & -i & 0 & 0 \\ -i & 0 & 0 & 0 \\ 0 & 0 & 0 & 0 \\ 0 & 0 & 0 & 0 \end{pmatrix}\begin{pmatrix} 0 & 0 & 0 & 0 \\ 0 & 0 & 0 & -1 \\ 0 & 0 & 0 & 0 \\ 0 & 1 & 0 & 0 \end{pmatrix} - \frac{1}{i}\begin{pmatrix} 0 & 0 & 0 & 0 \\ 0 & 0 & 0 & -1 \\ 0 & 0 & 0 & 0 \\ 0 & 1 & 0 & 0 \end{pmatrix}\begin{pmatrix} 0 & -i & 0 & 0 \\ -i & 0 & 0 & 0 \\ 0 & 0 & 0 & 0 \\ 0 & 0 & 0 & 0 \end{pmatrix} \\
&= \frac{1}{i}\begin{pmatrix} 0 & 0 & 0 & i \\ 0 & 0 & 0 & 0 \\ 0 & 0 & 0 & 0 \\ 0 & 0 & 0 & 0 \end{pmatrix} - \frac{1}{i}\begin{pmatrix} 0 & 0 & 0 & 0 \\ 0 & 0 & 0 & 0 \\ 0 & 0 & 0 & 0 \\ -i & 0 & 0 & 0 \end{pmatrix} = -\frac{1}{i}\begin{pmatrix} 0 & 0 & 0 & -i \\ 0 & 0 & 0 & 0 \\ 0 & 0 & 0 & 0 \\ -i & 0 & 0 & 0 \end{pmatrix} \\
&= -\frac{1}{i}K_z = +iK_z
\end{aligned}
\tag{B.109}
$$

By computing the other commutators, and including also the algebra for $\vec{J}$, we then obtain:

$$
[J_i, J_j] = i\epsilon_{ijk}J_k, \quad [K_i, K_j] = -i\epsilon_{ijk}J_k, \quad \text{and} \quad [K_i, J_j] = i\epsilon_{ijk}K_k
\tag{B.110}
$$

The first commutator is the familiar expression for the swap of two rotations. The last rule indicates that a boost before or after a rotation will result in a different boost, which is intuitively understandable. The commutation of two boosts is perhaps surprising. **A swapping of two boosts in different directions results in a rotation rather than another boost, and that explains the commutators mixing Ks and Js.** So, as mentioned, the subgroup of rotations is closed, however, the boosts do not form a subgroup of the Lorentz group. Swapping the order of boosts introduces rotations. This property is crucial when transporting, for example, the polarization of particles from different inertial frames.

• **Complexification of the Lorentz group and transformation of complex fields.** We note that the generators of the Lorentz group do contain complex numbers, while the Minkowski space is real valued. This means that the representation of the Lorentz group (boosts/rotations) on the Minkowski space are 4×4 real-valued matrices, which when applied to four-vectors in Minkowski space preserve their norm. In general, we must consider the Lorentz transformation of complex relativistic tensor fields (scalar, vector, tensor, spinor, ...), see Chapters 6–8. Starting from the generators $\vec{J}$ and $\vec{K}$ defined above, one considers a new set of generators $\vec{L}$ and $\vec{R}$ defined as the following linear combinations:

$$
\vec{L} = \frac{1}{2}(\vec{J} + i\vec{K}) \quad \text{and} \quad \vec{R} = \frac{1}{2}(\vec{J} - i\vec{K})
\tag{B.111}
$$

Their properties are defined by their commutation rules. Let us consider:

$$
\begin{aligned}
[L_i, L_j] &= \left[\frac{1}{2}(J_i + iK_i), \frac{1}{2}(J_j + iK_j)\right] = \frac{1}{4}[J_i, J_j] + \underbrace{\frac{i}{4}[J_i, K_j]}_{[J_i,K_j]=-[K_j,J_i]=-i\epsilon_{jik}K_k} + \frac{i}{4}[K_i, J_j] - \frac{1}{4}[K_i, K_j] \\
&= \frac{1}{4}i\epsilon_{ijk}J_k - \frac{i}{4}i\epsilon_{jik}K_k + \frac{i}{4}i\epsilon_{ijk}K_k + \frac{1}{4}i\epsilon_{ijk}J_k = i\epsilon_{ijk}\left\{\frac{1}{4}J_k + \frac{i}{4}K_k + \frac{i}{4}K_k + \frac{1}{4}J_k\right\} \\
&= i\epsilon_{ijk}\left\{\frac{1}{2}J_k + \frac{i}{2}K_k\right\} = i\epsilon_{ijk}L_k
\end{aligned}
\tag{B.112}
$$

and similarly for the other combinations, one finally finds:

$$
[L_i, L_j] = i\epsilon_{ijk}L_k, \quad [R_i, R_j] = i\epsilon_{ijk}R_k, \quad \text{and} \quad [L_i, R_j] = 0
\tag{B.113}
$$

Therefore, **the generators of $SO(3,1)$ can be split into two subsets which commute with each other. Each satisfy the same commutation relations as $SO(n)$, Eq. (B.30)** or by extension as the spin if we consider the application on an internal degree of freedom. Therefore, we obtain in a natural way the representation of scalar (spin-0), spinor (spin-1/2), and vector (spin-1) particles as a consequence of group theory! This result is of enormous relevance. In particular, in the case of the Dirac equation where we use spinors, these generators map directly to the left-handed and right-handed chiral or Weyl representation (see Section 8.19).

B.11 The Poincaré Group

The **Poincaré group**, also sometimes called the **inhomogeneous Lorentz group**, is the **Lie group** that contains all the symmetries of the Lorentz group adding translations to rotations and boosts. The components of a four-vector transform accordingly under the Poincaré group:

$$x'^\mu = \Lambda^\mu_{\ \nu} x^\nu + a^\mu \tag{B.114}$$

Since the Lorentz group is defined by six parameters, a Poincaré transformation is defined by $6 + 4 = 10$ parameters. Its representation on the Hilbert space can be expressed as:

$$U(\Lambda, a) = U(a, \mathbb{1}) U(0, \Lambda) \tag{B.115}$$

The rotations and boosts can be taken directly from the exponentiation of Eq. (B.97). Adding also the translation, we find:

$$U(\Lambda, a) = U(a, \mathbb{1}) U(0, \Lambda) = e^{-ia^\alpha P_\alpha} e^{\frac{i}{2} \omega_{\mu\nu} M^{\mu\nu}} \tag{B.116}$$

where the convention of the i factor has been adopted. The identity of the Poincaré group is trivially given by $\Lambda^\mu_{\ \nu} = \delta^\mu_\nu$ and $a = (0,0,0,0)$, and an infinitesimal transformation is found by:

$$\Lambda^\mu_{\ \nu} = \delta^\mu_\nu + \omega^\mu_{\ \nu} \quad \text{and} \quad a^\mu = \epsilon^\mu \tag{B.117}$$

where ω and ϵ are infinitesimal. The representation of the operators on the Hilbert space in the infinitesimal case is:

$$U(\omega, \epsilon) = 1 - i\epsilon_\alpha P^\alpha + \frac{i}{2} \omega_{\beta\delta} M^{\beta\delta} + \cdots \tag{B.118}$$

Because the translation is Abelian, the commutators of P are vanishing:

$$[P^\mu, P^\nu] = 0 \tag{B.119}$$

From the Ms, we can define the $\vec{J}$ and $\vec{K}$, and their Lie algebra has been given in Eqs. (B.107) and (B.110).

B.12 Young Tableaux

A **Young**[2] tableau is an elegant mathematical tool that can describe the symmetries of a collection of identical objects. It is composed of boxes arranged from left to right, conventionally with rows, where **any row is not longer than the row on top of it**. It can be used to study the symmetries of a collection of m identical particles, each of which can be in one given state, from several possible states. Then, every Young tableau represents an irreducible representation of the group.

Let us begin with an $SU(2)$ symmetry. The basis vectors of its fundamental doublet are u_1 and u_2 (see Eq. (B.45)). These states are represented with a single box by means of a Young tableau:

$$u_1 \equiv \boxed{1}, \qquad u_2 \equiv \boxed{2} \tag{B.120}$$

2   Alfred Young (1873–1940), British mathematician.

Now we assume that we have $m = 2$ such particles, each with a possible state u_1 or u_2. With two particles, we can create the symmetric state $\Psi_s \equiv (u_1 u_2 + u_2 u_1)/\sqrt{2}$ and the antisymmetric state $\Psi_a \equiv (u_1 u_2 - u_2 u_1)/\sqrt{2}$. In the form of Young tableaux, these would look like:

$$\Psi_s = \boxed{1\ 2}, \qquad \Psi_a = \begin{array}{c}\boxed{1}\\\boxed{2}\end{array} \tag{B.121}$$

and the symmetric states $u_1 u_1$ and $u_2 u_2$ would obviously be:

$$\boxed{1\ 1}, \qquad \boxed{2\ 2} \tag{B.122}$$

Note that the arrangement $\boxed{2\ 1}$ is the same as $\boxed{1\ 2}$ and does not need to be considered.

A box without a label stands for all members of the multiplet. In our case, an empty box would represent both possible states of the doublet. Consequently, the **three** symmetric states Ψ_s and the **single** antisymmetric state Ψ_a can be jointly expressed as:

$$\Psi_s - \boxed{\ \ }, \qquad \Psi_u = \boxed{\ } \tag{B.123}$$

Young tableaux with only one row describe a symmetric representation. A column is associated with an antisymmetric representation.

- **Tensor product.** The product of two states can be conveniently built using the Young tableaux product, which corresponds to the tensor product of the irreducible representations:

$$\square \otimes \square \tag{B.124}$$

which stands for the four states:

$$u_1 u_1, \; u_1 u_2, \; u_2 u_1, \; u_2 u_2 \tag{B.125}$$

The rule to build the product tableau is to take the linear combinations that correspond to the symmetric and the antisymmetric Young tableaux! Hence we can add the second box to the first, either to make a row or a column with two boxes:

$$\square \otimes \square = \square\square \oplus \begin{array}{c}\square\\\square\end{array} \tag{B.126}$$

which translates directly in terms of **multiplicities** (and respectively the dimensions of the irreducible representations) to the result:

$$SU(2): \qquad 2 \otimes 2 = 3 \oplus 1 \tag{B.127}$$

Now we consider the combination of three states in $SU(2)$. We need to add another box to the two-box row and column found in Eq. (B.126). Under the constraint that any row is not longer than the row on top of it, we have the following possible topologies:

$$\square\square \otimes \square = \square\square\square \oplus \begin{array}{c}\square\square\\\square\end{array} \tag{B.128}$$

and

$$\begin{array}{c}\square\\\square\end{array} \otimes \square = \begin{array}{c}\square\\\square\\\square\end{array} \oplus \begin{array}{c}\square\square\\\square\end{array} \tag{B.129}$$

Here we need to consider an additional rule: **An $SU(N)$ tableau that has more than N boxes in any column has dimension zero and does not need to be considered.** Hence, in the case of $SU(2)$ we note that the tableau with a column with three vertical boxes disappears. Finally, we obtain:

$$\square \otimes \square \otimes \square = \square\square\square \oplus \begin{array}{c}\square\square\\\square\end{array} \oplus \begin{array}{c}\square\square\\\square\end{array} \tag{B.130}$$

which translates into the multiplicities:

$$SU(2): \qquad (2 \otimes 2) \otimes 2 = (4 \oplus 2) \oplus 2 \tag{B.131}$$

So far we have only considered the $SU(2)$ group. We now want to extend these concepts to the more general $SU(N)$ case and easily build the irreducible representations. For the $N = 2$ case, we concluded that a box represented two states and hence had a dimension of two. In $SU(N)$ a box is associated with the fundamental representation of dimension N:

$$SU(N): \square = \text{dimension } N \tag{B.132}$$

The **conjugate (dual) representation** $\bar{N}$ is associated with a column of $N - 1$ boxes:

$$SU(N): \quad \begin{smallmatrix} \boxed{1} \\ \boxed{2} \\ \boxed{3} \\ \vdots \\ \square \end{smallmatrix} \quad = N - 1 \,\text{boxes} \tag{B.133}$$

So far we considered $SU(2)$, so in this case the box representations for 2 (the fundamental representation of $SU(2)$) and $\bar{2}$ (its conjugate) are identical (isomorph). But in $SU(3)$ we find:

$$SU(3): \qquad \square = 3, \qquad \begin{smallmatrix}\square\\\square\end{smallmatrix} = \bar{3} \tag{B.134}$$

and so the fundamental and the conjugate representations are the same for $SU(2)$ but differ for $SU(N \geq 3)$. Young tableaux make the construction of irreducible representations rapid. We can readily use the results of Eq. (B.126) for the $SU(3)$ case to find:

$$
\begin{aligned}
&\qquad\qquad \square \otimes \square \;=\; \square\square \;\oplus\; \begin{smallmatrix}\square\\\square\end{smallmatrix} \\
SU(2): &\qquad 2 \otimes 2 \;=\; 3 \oplus 1 \\
SU(3): &\qquad 3 \otimes 3 \;=\; 6 \oplus \bar{3}
\end{aligned} \tag{B.135}
$$

So incredibly easy! A big effort can be saved noting that there are rules to compute the dimension (or state multiplicities). These rules are summarized for typical topologies in Table B.1. The results in the table can

Tableau	Dimension	$SU(2)$	$SU(3)$
⬛⬛	$\dfrac{N(N+1)}{2}$	3	6
⬛/⬛	$\dfrac{N(N-1)}{2}$	1	3
⬛⬛⬛	$\dfrac{N(N+1)(N+2)}{6}$	4	10
⬛/⬛/⬛	$\dfrac{N(N-1)(N-2)}{6}$	–	1
⬛⬛/⬛	$\dfrac{N(N-1)(N+1)}{3}$	2	8

Table B.1 Commonly used Young tableaux and their dimensions in $SU(2)$ and $SU(3)$.

readily be used to construct the various (infinitely many) irreducible representations combining two or three objects. For example, the result of Eq. (B.130) can be extended to:

$$\square \otimes \square \otimes \square \;=\; \square\square\square \;\oplus\; \begin{array}{c}\square\square\\\square\end{array} \;\oplus\; \begin{array}{c}\square\square\\\square\end{array} \;\oplus\; \begin{array}{c}\square\\\square\\\square\end{array}$$

$$SU(2): \qquad (2 \otimes 2) \otimes 2 \;=\; (4 \oplus 2) \oplus 2$$

$$SU(3): \qquad (3 \otimes 3) \otimes 3 \;=\; (10 \oplus 8) \oplus (8 \oplus 1) \tag{B.136}$$

These results are of great physical interest for hadron building (see Chapter 17). Finally, we can consider the product of the fundamental representation with its conjugate (or dual), to find for example for $SU(3)$:

$$\square \otimes \begin{array}{c}\square\\\square\end{array} \;=\; \begin{array}{c}\square\square\\\square\end{array} \;\oplus\; \begin{array}{c}\square\\\square\\\square\end{array}$$

$$SU(3): \qquad 3 \otimes \bar{3} \;=\; 8 \oplus 1 \tag{B.137}$$

Physically this result is used, for example, to combine a quark of flavor u, d, s with an antiquark of flavor $\bar{u}, \bar{d}, \bar{s}$, as discussed in Chapter 17. For those with great appetite, one can also easily find other products of physical interest for hadron building, from the $SU(6)$ symmetry. In this case, the combination of three quarks among six flavors u, d, s, c, b, t gives the following irreducible representation:

$$SU(6): \qquad (6 \otimes 6) \otimes 6 = (56 \oplus 70) \oplus (70 \oplus 20) \tag{B.138}$$

Enjoy!

Appendix C Notions of Non-relativistic Quantum Mechanics

C.1 Bohr Atomic Model

The classical description of the nuclear atom is based on the Coulomb attraction between the positively charged nucleus and the negative electrons orbiting the nucleus. To simplify the discussion, we focus on the hydrogen atom, with a single proton and electron. Furthermore, we consider only circular orbits of radius r with constant tangential velocity v. The total force on the electron is thus:

$$F = \frac{e^2}{4\pi\epsilon_0}\frac{1}{r^2} = \frac{\alpha\hbar c}{r^2} = \frac{m_e v^2}{r} \tag{C.1}$$

where the **fine-structure constant** α is:

$$\alpha \equiv \frac{e^2}{4\pi\epsilon_0\hbar c} \tag{C.2}$$

The sum of the kinetic and the potential energy is:

$$E = T + V = \frac{1}{2}m_e v^2 - \frac{\alpha\hbar c}{r} = \frac{1}{2}\frac{\alpha\hbar c}{r} - \frac{\alpha\hbar c}{r} = -\frac{1}{2}\frac{\alpha\hbar c}{r} \tag{C.3}$$

Bohr proposed that the orbiting electron could only exist in certain special states of motion – called stationary states (in which no electromagnetic radiation is emitted by the orbiting electron). In these stationary states, the electron angular momentum can take on values $\hbar$, $2\hbar$, $3\hbar$, ..., but never non-integer values:

$$L = m_e v r = n\hbar \tag{C.4}$$

with $n = 1, 2, 3, \ldots$. This is equivalent to stating that the wave orbits are an integer number of de Broglie wavelengths:

$$m_e v r = n\hbar = n\frac{h}{2\pi} \quad \rightarrow \quad 2\pi r = n\frac{h}{mv_e} = n\lambda_e \tag{C.5}$$

where $\lambda_e = h/m_e v$ is the de Broglie wavelength of the electron. Hence:

$$\frac{\alpha\hbar c}{r} = m_e v^2 \quad \Longrightarrow \quad v^2 r^2 = \frac{\alpha\hbar c}{m_e}r \tag{C.6}$$

and

$$m_e v r = n\hbar \quad \Longrightarrow \quad v^2 r^2 = \left(\frac{n\hbar}{m_e}\right)^2 \tag{C.7}$$

So

$$r = \left(\frac{n\hbar}{m_e}\right)^2\frac{m_e}{\alpha\hbar c} = \left(\frac{\hbar}{\alpha m_e c}\right)n^2 \equiv a_0 n^2 \tag{C.8}$$

where a_0 is the **Bohr radius**:

$$a_0 = \frac{\hbar}{\alpha m_e c} = \frac{4\pi\varepsilon_0 \hbar^2}{m_e e^2} \tag{C.9}$$

It is known with very high experimental precision (CODATA [10]):

$$a_0 = (0.52917721067 \pm 0.00000000012) \times 10^{-10} \text{ m} \tag{C.10}$$

The allowed **Bohr energy levels** are then given by:

$$E_n = -\frac{1}{2}\frac{\alpha\hbar c}{a_0 n^2} = -\frac{1}{2}\frac{\alpha^2 m_e c^2}{n^2} = -\frac{1}{2}\frac{e^4 m_e}{(4\pi\epsilon_0)^2\hbar^2 c}\frac{1}{n^2} \equiv -\frac{Ry}{n^2} \tag{C.11}$$

with $n = 1, 2, 3, \ldots$, and where Ry is the **Rydberg unit of energy**, which is related to the **Rydberg constant** R_∞:

$$Ry \equiv \frac{\alpha^2 m_e c^2}{2} = \frac{1}{2}\frac{e^4 m_e}{(4\pi\epsilon_0)^2\hbar^2 c}, \qquad R_\infty \equiv \frac{Ry}{hc} = \frac{\alpha^2 m_e c}{2h} = \frac{m_e e^4}{8\epsilon_0^2 h^3 c} \tag{C.12}$$

The CODATA recommended value is [10]:

$$Ry = \frac{R_\infty}{hc} = (13.605693122994 \pm 0.000000000026) \text{ eV} \tag{C.13}$$

C.2   The Hilbert Space

The mathematical concept of a Hilbert space generalizes the notion of Euclidean space. It extends the methods of vector algebra and calculus from the two-dimensional Euclidean plane and three-dimensional space to spaces with any finite or infinite number of dimensions. In quantum mechanics, the possible states of a system are represented by unit vectors (called state vectors) residing in a complex separable Hilbert space, known as the state space, well defined up to a complex number of norm 1 (the phase factor). A quantum-mechanical state can be completely described by a state vector in the Hilbert space. The abstract Hilbert space $\mathcal{H}$ is given by a set of elements represented by kets of the type $|\psi\rangle$, for which addition, multiplication by a complex number, and a scalar product are defined. The space $\mathcal{H}$ is complete, which means that every ket can be expressed as a linear combination of an orthonormal basis according to:

$$|\psi\rangle = \sum_{n=1}^{\infty} a_n |\phi_n\rangle \quad \text{and} \quad \langle\psi| \equiv |\psi\rangle^* \tag{C.14}$$

where the scalar product gives:

$$\langle\phi_m|\psi\rangle = \sum_{n=1}^{\infty} a_n \langle\phi_m|\phi_n\rangle = \sum_{n=1}^{\infty} a_n \delta_n^m = a_m \tag{C.15}$$

from which follows that each state can be represented as:

$$|\psi\rangle = \sum_{n=1}^{\infty} |\phi_n\rangle \langle\phi_n|\psi\rangle \quad \text{and} \quad \sum_{n=1}^{\infty} |\phi_n\rangle \langle\phi_n| = 1 \tag{C.16}$$

where the states $|\phi_n\rangle$ form a complete set.

A relation

$$A|\psi\rangle = |A\psi\rangle = |\psi'\rangle \tag{C.17}$$

is a linear operator if $A(a\,|\psi\rangle + b\,|\phi\rangle) = aA(|\psi\rangle) + bA(|\phi\rangle)$. The adjoint operator $A^\dagger$ is defined by:

$$\langle\phi|A|\psi\rangle = \langle A^\dagger\phi|\psi\rangle \tag{C.18}$$

A **Hermitian operator** satisfies:

$$\text{Hermitian}: \quad A^\dagger = A \tag{C.19}$$

The expectation value of an **observable** A in the state ψ is given by:

$$\langle A\rangle_\psi = \frac{\langle\psi|A|\psi\rangle}{\langle\psi|\psi\rangle} = \langle\psi|A|\psi\rangle \tag{C.20}$$

if the state is normalized $\langle\psi|\psi\rangle = 1$. The expectation value of a Hermitian operator is real:

$$\langle A\rangle_\psi = \langle\psi|A|\psi\rangle = \langle A^\dagger\psi|\psi\rangle = \langle A\psi|\psi\rangle = \langle\psi|A\psi\rangle^* = \langle\psi|A|\psi\rangle^* = \left[\langle A\rangle_\psi\right]^* \tag{C.21}$$

An **unitary operator** is defined by $U^\dagger = U^{-1}$ or:

$$U^\dagger U = UU^\dagger = \mathbb{1} \tag{C.22}$$

The operators can be represented in matrix form by noting that the completeness relation allows us to define a state as a column vector:

$$|\psi\rangle = \begin{pmatrix} \langle\phi_1|\psi\rangle \\ \langle\phi_2|\psi\rangle \\ \vdots \\ \langle\phi_n|\psi\rangle \\ \vdots \end{pmatrix} = \begin{pmatrix} a_1 \\ a_2 \\ \vdots \\ a_n \\ \vdots \end{pmatrix} \tag{C.23}$$

An operator is represented by the matrix of the linear map:

$$A = [\langle\phi_m|A|\phi_n\rangle] = [A_{mn}] \tag{C.24}$$

Once in matrix and column form, all rules of linear algebra can be applied.

The eigenvectors of an operator do not in general form a complete orthonormal basis. If, for a given operator, they do, then one can use these eigenvectors to form a complete basis of the system. In this basis, the operator will have a diagonal matrix representation. This can be stated by saying that an operator is diagonal in its own representation. This is especially true for all Hermitian operators A with non-degenerate eigenvalues, since in this case:

$$\langle\phi_m|A\phi_n\rangle = \lambda_n\langle\phi_m|\phi_n\rangle = \langle\phi_n|A\phi_m\rangle^* = \lambda_m^*\langle\phi_n|\phi_m\rangle^* = \lambda_m\langle\phi_m|\phi_n\rangle \tag{C.25}$$

where we used the fact that the eigenvalues of a Hermitian operator are real. Then we have:

$$(\lambda_n - \lambda_m)\langle\phi_m|\phi_n\rangle = 0 \tag{C.26}$$

and consequently for $\lambda_n \neq \lambda_m$, we must have $\langle\phi_m|\phi_n\rangle = 0$, so the eigenstates are orthogonal (or orthonormal after the proper normalization).

C.3 Commuting and Non-commuting Operators

We consider two observables represented by two Hermitian operators A and B. If the physical quantum system yields with certainty the results a and b for the expectations of A and B, then both operators are said to *commute*. The physical meaning is that the two operators possess a common set of eigenstates:

$$A\,|\Psi_i\rangle = a\,|\Psi_i\rangle\,, \qquad B\,|\Psi_i\rangle = b\,|\Psi_i\rangle \tag{C.27}$$

where the $|\Psi_i\rangle$ ($n = 1, \ldots, n$) are the common eigenstates. An obvious and equivalent consequence is that:

$$(AB - BA)\,|\Psi_i\rangle = (ab - ba)\,|\Psi_i\rangle = 0 \tag{C.28}$$

Hence, the two operators must commute:

$$[A, B] = AB - BA = 0 \tag{C.29}$$

Therefore, if two observables have commuting operators, they possess a complete set of common (orthonormal) eigenstates. These two variables may be simultaneously precisely measured. They are *compatible*. The commutation of operators is often dealt with by commutator algebra. With A, B, and C being three independent operators, we have:

$$\begin{aligned}
[A, B] &= -[B, A] \\
[A, B + C] &= [A, B] + [A, C] \\
[A, BC] &= [A, B]\,C + B\,[A, C] \\
0 &= [A, [B, C]] + [B, [C, A]] + [C, [A, B]] \quad \text{(Jacobi identity)}
\end{aligned} \tag{C.30}$$

As an example, let us consider the common operators of *position* x and *momentum* p (we assume here for simplicity only one dimension). The space-position wave function is given by:

$$\Psi(x) \equiv \langle x|\Psi\rangle \tag{C.31}$$

The probability of finding the particle within positions x and $x + \mathrm{d}x$ is given by $|\langle x|\Psi\rangle|^2\mathrm{d}x = |\Psi(x)|^2\mathrm{d}x$. The $|x\rangle$ and $|p\rangle$ are not aligned bases, the state of the quantum system cannot be an eigenvector of position and momentum simultaneously. In addition, since the measurements change the state of the system, the order of measurements is important. We have that:

$$x\,p\,|\Psi\rangle \neq p\,x\,|\Psi\rangle \tag{C.32}$$

The following **commutation relation** between *position* x and *momentum* p is postulated:

$$\boxed{[x, p] \equiv x\,p - p\,x = i} \tag{C.33}$$

As a consequence, we can determine the representation of the operators on the wave function. We first note that:

$$\langle x|[x, p]|y\rangle = \langle x|xp|y\rangle - \langle x|px|y\rangle = (x - y)\,\langle x|p|y\rangle \tag{C.34}$$

where we used $x\,|x\rangle = x$ and $x\,|y\rangle = y$ (x is the space operator acting on the $|y\rangle$ state). Using the commutation relation, we can also write:

$$\langle x|[x, p]|y\rangle = i\,\langle x|y\rangle = i\delta(x - y) \tag{C.35}$$

Consequently, using Eq. (A.35), we find:

$$\langle x|p|y\rangle = i\frac{\delta(x - y)}{x - y} = i\frac{\partial}{\partial y}\delta(x - y) \tag{C.36}$$

Using the property of an orthonormal complete basis, we can write:

$$\langle x|p|\Psi\rangle = \int \langle x|p|y\rangle \langle y|\Psi\rangle \, \mathrm{d}y = i \int \frac{\partial}{\partial y}\left(\delta(x-y)\right)\langle y|\Psi\rangle \, \mathrm{d}y = -i\frac{\partial}{\partial x}\langle x|\Psi\rangle \tag{C.37}$$

Hence, the suitable representation of the (one-dimensional) momentum operator is given by:

$$p = -i\frac{\partial}{\partial x} \tag{C.38}$$

We can further consider the projection of a momentum ket on a space bra and, solving the differential equation, we get:

$$\langle x|p|p\rangle = p\langle x|p\rangle = -i\frac{\partial}{\partial x}\langle x|p\rangle \quad \Longrightarrow \quad \langle x|p\rangle = \frac{1}{\sqrt{2\pi}}e^{ipx} \tag{C.39}$$

We can then note that the space and momentum bases are related by the Fourier transformation (see Eq. (A.24)):

$$|p\rangle = \frac{1}{\sqrt{2\pi}}\int e^{ipx}|x\rangle \, \mathrm{d}x, \qquad |x\rangle = \frac{1}{\sqrt{2\pi}}\int e^{-ipx}|p\rangle \, \mathrm{d}p \tag{C.40}$$

These equations translate into the **uncertainty principle** of Heisenberg.

C.4   The Hamilton Operator or Hamiltonian

Let us consider as an operator the Hamiltonian (Hamilton operator), which represents the energy of a system. For a single particle, it can be written as $H = T + V$, where T is the kinetic energy and V the classical potential. The Schrödinger equation and its complex conjugate yield:

$$i\frac{\partial\Psi}{\partial t} = H\Psi, \qquad -i\frac{\partial\Psi^*}{\partial t} = (H\Psi)^* \tag{C.41}$$

If we take any operator A, its time dependence is given by:

$$\frac{\mathrm{d}\langle A(t)\rangle_\psi}{\mathrm{d}t} = \frac{\mathrm{d}}{\mathrm{d}t}\langle\psi|A|\psi\rangle = \langle\frac{\partial}{\partial t}\psi|A|\psi\rangle + \langle\psi|A|\frac{\partial}{\partial t}\psi\rangle + \langle\psi|\frac{\partial}{\partial t}A|\psi\rangle \tag{C.42}$$

This last equation can be rewritten with the help of the Schrödinger equation and its conjugate:

$$\begin{aligned}
\frac{\mathrm{d}\langle A(t)\rangle_\psi}{\mathrm{d}t} &= -\frac{1}{i}\langle H\psi|A|\psi\rangle + \frac{1}{i}\langle\psi|A|H\psi\rangle + \langle\psi|\frac{\partial}{\partial t}A|\psi\rangle = \frac{1}{i}\langle\psi|AH - HA|\psi\rangle + \langle\psi|\frac{\partial}{\partial t}A|\psi\rangle \\
&= i\langle\psi|[H,A]|\psi\rangle + \langle\psi|\frac{\partial A}{\partial t}|\psi\rangle
\end{aligned} \tag{C.43}$$

Hence, the time evolution of the expectation value of an operator without explicit time dependence can be written through the commutation with the Hamiltonian as:

$$\frac{\mathrm{d}A}{\mathrm{d}t} = i[H,A] \tag{C.44}$$

Accordingly, an observable is called a *constant of motion* if A commutes with H, that is:

$$[H,A] = 0 \quad \leftrightarrow \quad \frac{\mathrm{d}A}{\mathrm{d}t} = 0 \tag{C.45}$$

Its expectation value is constant.

C.5 Stationary States

We consider the case where the potential is time independent, $V = V(\vec{x})$. By the method of separation of variables, we can assume the following ansatz for the solution to the Schrödinger equation:

$$\psi(\vec{x}, t) = \psi(\vec{x})\xi(t) \tag{C.46}$$

Insertion in Eq. (C.41) yields:

$$i\psi(\vec{x})\frac{\partial \xi}{\partial t} = H\psi(\vec{x})\xi(t) = \frac{\vec{p}^2}{2m}\psi(\vec{x})\xi(t) + V(\vec{x})\psi(\vec{x})\xi(t) \tag{C.47}$$

Hence, dividing by $\psi(\vec{x})\xi(t)$ on both sides, we find:

$$i\frac{1}{\xi(t)}\frac{\partial \xi}{\partial t} = -\frac{1}{2m}\frac{1}{\psi(\vec{x})}\nabla^2\psi + V(\vec{x}) \tag{C.48}$$

The left-hand side depends only on t and the right-hand side only on $\vec{x}$. Consequently, both sides must be equal to a constant which is labeled E. Both sides can be solved independently. The left-hand side gives:

$$i\frac{1}{\xi(t)}\frac{d\xi}{dt} = E \implies \xi(t) = Ae^{-iEt} \tag{C.49}$$

The factor A can be absorbed in the spatial part of the wave function. The right-hand side yields:

$$-\frac{1}{2m}\nabla^2\psi + V(\vec{x})\psi(\vec{x}) \equiv H\psi(\vec{x}) = E\psi(\vec{x}) \tag{C.50}$$

This equation is called the **time-independent Schrödinger equation** with the time-independent Hamiltonian H. The solutions represent the **stationary solutions** and their wave functions are expressed as:

$$\psi(\vec{x}, t) = \psi(\vec{x})e^{-iEt} \tag{C.51}$$

C.6 The Schrödinger and Heisenberg Pictures

In the Schrödinger equation, the time evolution of the wave function is defined from the Hamiltonian operator. We can write:

$$i\frac{\partial \Psi}{\partial t} = H\Psi \implies \langle x|i\frac{\partial}{\partial t}|\Psi\rangle = \langle x|H|\Psi\rangle \tag{C.52}$$

The time-evolution operator $U(t, t_0)$ is defined as:

$$|\Psi(t)\rangle = U(t, t_0)|\Psi(t_0)\rangle \implies \langle x|i\frac{\partial}{\partial t}|U(t, t_0)\Psi(t_0)\rangle = \langle x|H|U(t, t_0)\Psi(t_0)\rangle \tag{C.53}$$

and consequently the time-evolution operator can be defined formally as:

$$i\frac{\partial}{\partial t}U(t, t_0) = HU(t, t_0) \implies U(t, t_0) = e^{-i(t-t_0)H} \tag{C.54}$$

where we assumed in the last step that the operator H is constant.

In the **Heisenberg picture**, the **state vector remains fixed** and the **operators change with time**. The Heisenberg state vector is simply defined at time t_0 and the time-dependent operators are given by implementing the time evolution operator:

$$O_H(t) = U^\dagger(t, t_0)O_S U(t, t_0) \tag{C.55}$$

where the H subscript is for the Heisenberg and the S subscript for the Schrödinger picture. Naturally, any expectation value is independent of the chosen picture, so we have:

$$\langle \phi|_H O_H(t)|\Psi\rangle_H = \langle \phi(t)|_S U(t, t_0)U^\dagger(t, t_0)O_S U(t, t_0)U^\dagger(t, t_0)|\Psi\rangle_S \tag{C.56}$$

C.7   The Simple Linear Harmonic Oscillator

The Hamiltonian of the linear harmonic oscillator (where for simplicity the mass is set to $m = 1$) is:

$$H_{HO} = \frac{1}{2}p^2 + \frac{1}{2}\omega^2 x^2 \tag{C.57}$$

where ω is the angular frequency and $[x, p] = i$. Dirac introduced the **ladder operators** a and $a^\dagger$:

$$x = \frac{1}{\sqrt{2\omega}}(a + a^\dagger) \quad \text{and} \quad p = -i\sqrt{\frac{\omega}{2}}(a - a^\dagger) \tag{C.58}$$

or equivalently:

$$a = \sqrt{\frac{\omega}{2}}x + \frac{i}{\sqrt{2\omega}}p \quad \text{and} \quad a^\dagger = \sqrt{\frac{\omega}{2}}x - \frac{i}{\sqrt{2\omega}}p \tag{C.59}$$

from which follow the **commutation rule** $[a, a^\dagger] = aa^\dagger - a^\dagger a = 1$, since:

$$
\begin{aligned}
aa^\dagger &= \left(\sqrt{\frac{\omega}{2}}x + \frac{i}{\sqrt{2\omega}}p\right)\left(\sqrt{\frac{\omega}{2}}x - \frac{i}{\sqrt{2\omega}}p\right) = \frac{\omega}{2}x^2 - \frac{i}{2}xp + \frac{i}{2}px + \frac{p^2}{2\omega} = \frac{\omega}{2}x^2 + \frac{1}{2\omega}p^2 - \frac{i}{2}\underbrace{(xp - px)}_{i} \\
&= \frac{\omega}{2}x^2 + \frac{p^2}{2\omega} + \frac{1}{2}
\end{aligned}
\tag{C.60}
$$

and similarly $a^\dagger a = \frac{\omega}{2}x^2 + \frac{p^2}{2\omega} - \frac{1}{2}$. So finally:

$$[a, a^\dagger] = aa^\dagger - a^\dagger a = 1 \tag{C.61}$$

In this case, we have:

$$aa^\dagger = a^\dagger a + 1 = \frac{\omega}{2}x^2 + \frac{p^2}{2\omega} + \frac{1}{2} = \frac{1}{\omega}H_{HO} + \frac{1}{2} \tag{C.62}$$

and the Hamiltonian becomes:

$$H_{HO} = \omega\left(a^\dagger a + \frac{1}{2}\right) = \omega\left(N + \frac{1}{2}\right)$$

where $N \equiv a^\dagger a$ is the **number operator**. From this, we see that the eigenstate of N is also the eigenstate of energy. We define and write such eigenstates as $|n\rangle$:

$$N|n\rangle = n|n\rangle \quad \text{with} \quad E_n = \omega\left(n + \frac{1}{2}\right) \tag{C.63}$$

The **ground state** $|0\rangle$ is defined as:

$$a|0\rangle = 0 \quad \text{and} \quad E_0 = \frac{1}{2}\omega \text{ (zero point energy)} \tag{C.64}$$

The $a^\dagger$ plays then the role of "ladder" or **creation operator**, since it transforms a state $|n\rangle$ with increased number operator by one unit:

$$Na^\dagger|n\rangle = a^\dagger aa^\dagger|n\rangle = a^\dagger(a^\dagger a + 1)|n\rangle = a^\dagger(N + 1)|n\rangle = (n + 1)a^\dagger|n\rangle \tag{C.65}$$

The nth excitation of the system can therefore be obtained by successive applications of the ladder operator:[1]

$$|n\rangle \equiv (a^\dagger)^n|0\rangle \tag{C.66}$$

---

[1]   The normalization is neglected here. The normalized eigenstates are: $|n\rangle \equiv \frac{(a^\dagger)^n}{\sqrt{n!}}|0\rangle$.

The a-operator plays the role of **"annihilation" operator**:

$$
\begin{aligned}
Na\,|n\rangle &= a^\dagger aa\,|n\rangle = (aa^\dagger - 1)a\,|n\rangle = a(a^\dagger a - 1)\,|n\rangle \\
&= a(N-1)\,|n\rangle = (n-1)a\,|n\rangle
\end{aligned}
\tag{C.67}
$$

Finally, the eigenvalues of N must be positive, since:

$$
\langle n|N|n\rangle = n\,\langle n|n\rangle = \langle n|a^\dagger a|n\rangle = \langle na^\dagger|an\rangle = |\,|an\rangle\,|^2 \geq 0 \quad \implies \quad n \geq 0
\tag{C.68}
$$

and with $a\,|0\rangle = 0$, we have $n \in \mathbb{N}$ (n must be an integer).

C.8 Anticommutation Rules for Ladder-Like Operators

The **commutation rule** $[a, a^\dagger] = 1$ leads to a system with a ground state $|0\rangle$ and states $|n\rangle$ of increased level of excitation. In quantum field theory, this is an adequate description for bosons.

On the other hand, an **anticommutation rule** is adequate for describing a system of fermions. We define a ladder operator b which anticommutes as follows:

$$
\{b, b^\dagger\} = bb^\dagger + b^\dagger b = 1
\tag{C.69}
$$

The ground state (vacuum) is given by $|0\rangle$, where $b(|0\rangle) = 0$. The first excited state is found as $|1\rangle = b^\dagger(|0\rangle)$. The number operator is $N = b^\dagger b$. We note that:

$$
Nb^\dagger\,|1\rangle = b^\dagger bb^\dagger\,|1\rangle = b^\dagger(1 - b^\dagger b)\,|1\rangle = b^\dagger(1 - N)\,|1\rangle \overset{!}{=} 0
\tag{C.70}
$$

which implies:

$$
b^\dagger\,|1\rangle = 0 \quad \text{or} \quad b^\dagger b^\dagger\,|0\rangle = 0
\tag{C.71}
$$

For such a system, there are only two possible independent states: the ground state $|0\rangle$ and the first excited state $|1\rangle$.

C.9 Angular Momentum

The *canonical* angular momentum operator $\vec{L}$ is defined with the position and momentum operators as:

$$
\boxed{\vec{L} \equiv \vec{x} \times \vec{p} = -i\vec{x} \times \nabla}
\tag{C.72}
$$

In Cartesian coordinates, we have for each component:

$$
L_x = yp_z - zp_y = -i\left(y\frac{\partial}{\partial z} - z\frac{\partial}{\partial y}\right)
\tag{C.73}
$$

$$
L_y = zp_x - xp_z = -i\left(z\frac{\partial}{\partial x} - x\frac{\partial}{\partial z}\right)
\tag{C.74}
$$

$$
L_z = xp_y - yp_x = -i\left(x\frac{\partial}{\partial y} - y\frac{\partial}{\partial x}\right)
\tag{C.75}
$$

Starting from the commutation rules for the space coordinates and the momentum $[x, p_x] = [y, p_y] = [z, p_z] = i$, one finds the commutation rules for the components of the angular momentum:

$$
[L_x, L_y] = iL_z, \; [L_y, L_z] = iL_x, \; \text{and} \quad [L_z, L_x] = iL_y
\tag{C.76}
$$

Writing the indices x, y, z as $1, 2, 3$ we can use the Levi-Civita tensor (see Eq. (A.12)) to define the **Lie algebra of the angular momentum operator**:

$$[L_i, L_j] = i\epsilon_{ijk}L_k, \text{ where } \vec{L} = (L_1, L_2, L_3) \tag{C.77}$$

The operator $\vec{L}^2$, interpreted as the square of the total momentum, is a Casimir operator. It is diagonal and commutes with each component:

$$\vec{L}^2 = L_1^2 + L_2^2 + L_3^2 \quad \rightarrow \quad \left[\vec{L}^2, L_i\right] = 0 \tag{C.78}$$

Conventionally, $\vec{L}^2$ and L_3 are chosen to define a common diagonal representation, with states denoted $|\ell, m\rangle$, where ℓ is the angular momentum quantum number and m is the angular momentum projection onto the z-axis. They satisfy the eigenvalue equations:

$$\begin{cases} \vec{L}^2 |\ell, m\rangle = \ell(\ell + 1) |\ell, m\rangle \\ L_3 |\ell, m\rangle = m |\ell, m\rangle \end{cases} \tag{C.79}$$

where $m = -\ell, -\ell + 1, \ldots, \ell$. In this basis, the raising and lowering operators $L_\pm = L_1 \pm iL_2$ can be used to alter the value of m:

$$L_\pm |\ell, m\rangle = \sqrt{(\ell \mp m)(\ell \pm m + 1)} |\ell, m \pm 1\rangle \tag{C.80}$$

The **spherical harmonics** $Y_{\ell m}$ (see https://dlmf.nist.gov/14.30.i) are the spatial eigenfunctions of the angular momentum operator. In spherical coordinates, they are determined by the two quantum numbers $\ell = 0, 1, 2, \ldots$ and $m = -\ell, -\ell + 1, \ldots, \ell$:

$$Y_{\ell, m}(\theta, \phi) = \sqrt{\frac{(2\ell + 1)(l - m)!}{4\pi(l + m)!}} P_{l,m}(\cos\theta)e^{im\phi} \tag{C.81}$$

where the associated **Legendre polynomial** (see https://dlmf.nist.gov/18.3) is equal to:

$$P_{l,m}(x) = \frac{(-1)^m}{2^\ell} \frac{1}{\ell!}(1 - x^2)^{m/2} \frac{d^{l+m}}{dx^{l+m}}(x^2 - 1)^\ell \tag{C.82}$$

We list the spherical harmonics for $\ell = 0, 1$ and the possible values of m:

$$\begin{aligned} Y_{0,0}(\theta, \phi) &= \frac{1}{\sqrt{4\pi}} \\ Y_{1,\pm 1}(\theta, \phi) &= \sqrt{\frac{3}{4\pi}} \sin\theta \frac{e^{\pm i\phi}}{\sqrt{2}}, \qquad Y_{1,0}(\theta, \phi) = \sqrt{\frac{3}{4\pi}} \cos\theta \end{aligned} \tag{C.83}$$

The parity transformation in spherical coordinates can be expressed as $\theta \to \pi - \theta$ and $\phi \to \phi + \pi$. Hence:

$$\begin{aligned} P_{l,m}(\cos\theta) &\to P_{l,m}(\cos(\pi - \theta)) = (-1)^{l+m}P_{l,m}(\cos\theta) \\ e^{im\phi} &\to e^{im(\phi+\pi)} = (-1)^m e^{im\phi} \end{aligned} \tag{C.84}$$

This implies that the parity of the spherical harmonics is $(-1)^\ell$, since:

$$PY_{l,m}(\theta, \phi) = (-1)^\ell Y_{\ell, m}(\theta, \phi) \tag{C.85}$$

C.10   The Hydrogen Atom

The Hamiltonian of the hydrogen atom is the radial kinetic energy operator and Coulomb attraction force between the positive proton and negative electron. Using the time-independent Schrödinger equation, ignoring all spin-coupling interactions and using the reduced mass $\mu = m_e M/(m_e + M)$, the equation is written as:

$$\left(-\frac{\hbar^2}{2\mu}\nabla^2 - \frac{e^2}{4\pi\epsilon_0 r} \right) \psi(r,\theta,\phi) = E\psi(r,\theta,\phi) \tag{C.86}$$

where

$$\nabla^2 = \frac{1}{r^2}\frac{\partial}{\partial r}\left(r^2 \frac{\partial}{\partial r} \right) + \frac{1}{r^2 \sin\theta}\frac{\partial}{\partial \theta}\left(\sin\theta \frac{\partial}{\partial \theta} \right) + \frac{1}{r^2 \sin^2\theta}\frac{\partial^2}{\partial \phi^2} \tag{C.87}$$

This partial differential equation is separable. It can be solved in terms of special functions. The normalized position wave functions are found to be:

$$\psi_{n\ell m}(r,\theta,\phi) = \sqrt{\left(\frac{2}{na_0^*}\right)^3 \frac{(n-\ell-1)!}{2n(n+\ell)!}}\, e^{-\rho/2}\rho^\ell L_{n-\ell-1}^{2\ell+1}(\rho) Y_{\ell,m}(\theta,\phi) \tag{C.88}$$

where $n = 1, 2, 3, \ldots$ is the principal quantum number, $\ell = 0, 1, 2, \ldots, n-1$ is the orbital or azimuthal quantum number, and $m = -\ell, \ldots, \ell$ is the magnetic quantum number. The spectroscopic notation uses letters for ℓ: s (sharp) for $\ell = 0$, p (principal) for $\ell = 1$, d (diffuse) for $\ell = 2$, and f (fundamental) for $\ell = 3$. The variable ρ is given by:

$$\rho = \frac{2r}{na_0^*} = \frac{\mu e^2}{4\pi\epsilon_0\hbar^2}\frac{2r}{n} \tag{C.89}$$

where a_0^* is the reduced Bohr radius (see Eq. (C.9)). The $L_{n-\ell-1}^{2\ell+1}(\rho)$ functions are the **generalized Laguerre**[2] **polynomials** of degree $n - \ell - 1$ (see also https://dlmf.nist.gov/18.3), and the $Y_{\ell m}$ are the spherical harmonics given in Eq. (C.81). They are tabulated for $n = 1$ and 2 in Table C.1. The eigenvalues of the Hamiltonian coincide with the Bohr energy levels (see Eq. (C.11)).

1s level	$\psi_{n=1,\ell=0,m=0} = \frac{1}{\sqrt{\pi a_0^3}} e^{-r/a_0}$
2s level	$\psi_{n=2,\ell=0,m=0} = \frac{1}{\sqrt{8\pi a_0^3}}\left(1 - \frac{r}{2a_0}\right) e^{-r/2a_0}$
2p level	$\psi_{n=2,\ell=1,m=1} = -\frac{1}{8\sqrt{\pi a_0^3}}\frac{r}{a_0} e^{-r/2a_0}\sin\theta e^{i\phi}$
	$\psi_{n=2,\ell=1,m=0} = \frac{1}{4\sqrt{2\pi a_0^3}}\frac{r}{a_0} e^{-r/2a_0}\cos\theta$
	$\psi_{n=2,\ell=1,m=-1} = \frac{1}{8\sqrt{\pi a_0^3}}\frac{r}{a_0} e^{-r/2a_0}\sin\theta e^{-i\phi}$

Table C.1 The wave-function solutions of the hydrogen atom for $n = 1$ and 2.

C.11   Addition of Angular Momenta and Clebsch–Gordan Coefficients

Very often it is necessary to add angular momenta, be it orbital and spin $\vec{J} = \vec{L} + \vec{S}$ or just orbital $\vec{J} = \vec{L}_1 + \vec{L}_2$. Both problems are solved in similar manners. Let us consider in general the addition of $\vec{J} = \vec{J}_1 + \vec{J}_2$, where

---

2   Edmond Nicolas Laguerre (1834–1886), French mathematician.

$\vec{J}_1$ and $\vec{J}_2$ are each obeying angular momentum commutation rules and are also distinct, i.e., $\left[\vec{J}_1, \vec{J}_2\right] \neq 0$. The total angular momentum operator $\vec{J}$ obeys the same commutation rules as the individual operators $\vec{J}_i$ ($i = 1, 2$).

For each angular momentum operator, we have a set of eigenstates $|j_1, m_1\rangle$ and $|j_2, m_2\rangle$ where $m_i = -j_i, \ldots, j_i$, which provide a basis for their independent Hilbert spaces with independent eigenvalues $\vec{J}_i^2$ and third components $J_{i,z}$. Each set of states $|j_i, m_i\rangle$ spans a space of $(2j_i + 1)$ dimensions. The combined Hilbert space can be expressed using the **direct product** describing the state of the coupled momenta:

$$|j_1, m_1, j_2, m_2\rangle \equiv |j_1, m_1\rangle \otimes |j_2, m_2\rangle \qquad (C.90)$$

These states are also eigenstates of the third component of the total momentum J_z with eigenvalue $m_1 + m_2$, but not of its magnitude squared $\vec{J}^2$. We hence seek a basis in which the total angular momentum squared $\vec{J}^2$ is also diagonal. We want simultaneous eigenstates of the mutually commuting operators $\vec{J}^2$, J_z, $\vec{J}_1^2$, and $\vec{J}_2^2$. Such states are labeled with four quantum numbers corresponding to each operator, namely $|J, M, j_1, j_2\rangle$. They satisfy the eigenvalue equations:

$$\begin{cases} \vec{J}^2 |J, M, j_1, j_2\rangle = J(J + 1) |J, M, j_1, j_2\rangle \\ J_z |J, M, j_1, j_2\rangle = M |J, M, j_1, j_2\rangle \end{cases} \qquad (C.91)$$

where $M = -J, -J + 1, \ldots, J$. Note that $|J, M, j_1, j_2\rangle$ is often simply written as $|J, M\rangle$. The relation between the two individual and combined bases is expressed as:

$$|J, M, j_1, j_2\rangle = \sum_{m_1, m_2} |j_1, m_1, j_2, m_2\rangle \langle j_1, m_1, j_2, m_2 | J, M, j_1, j_2\rangle \qquad (C.92)$$

where the elements $\langle j_1, m_1, j_2, m_2 | J, M, j_1, j_2\rangle$ are called the **Clebsch**[3]–**Gordan**[4] **coefficients**. The determination of these coefficients from first principles is a somewhat lengthy exercise and for practical purposes one refers to tables found in the literature. For example, they can be found at http://pdg.lbl.gov/2018/reviews/rpp2018-rev-clebsch-gordan-coefs.pdf.

C.12   The Pauli Matrices

The **Pauli matrices** are complex 2×2 matrices, defined as follows:

$$\sigma_1 = \begin{pmatrix} 0 & 1 \\ 1 & 0 \end{pmatrix}, \quad \sigma_2 = \begin{pmatrix} 0 & -i \\ i & 0 \end{pmatrix}, \quad \sigma_3 = \begin{pmatrix} 1 & 0 \\ 0 & -1 \end{pmatrix} \qquad (C.93)$$

The **eigenvalues** of each matrix are ± 1. The matrices are **involutory** and **hermitian**:

$$\sigma_i^2 = \mathbb{1}, \qquad \sigma_i^+ = \sigma_i \qquad (C.94)$$

The matrices are **traceless** and the **determinants** are:

$$\mathrm{Tr}\, \sigma_i = 0, \qquad \det \sigma_i = -1 \qquad (C.95)$$

They obey the following **commutation relations**, which represent their $SU(2)$ Lie algebra (see Eq. (B.42)):

$$[\sigma_i, \sigma_j] = 2i\epsilon_{ijk}\sigma_k \qquad (C.96)$$

---

3   Rudolf Friedrich Alfred Clebsch (1833–1872), German mathematician.
4   Paul Albert Gordan (1837–1912), German mathematician.

The **Pauli matrices vector** is defined as $\vec{\sigma} = (\sigma_1, \sigma_2, \sigma_3)$. Useful identities are:

$$\sigma_i \sigma_j = \mathbb{1} \delta_{ij} + i \epsilon_{ijk} \sigma_k \quad \text{or} \quad \sigma_i \sigma_j = \delta_j^i + i \epsilon_{ijk} \sigma_k \tag{C.97}$$

and hence $(\vec{\sigma} \cdot \vec{a})(\vec{\sigma} \cdot \vec{b}) = \sigma_i \sigma_j a_i b_j$, or:

$$(\vec{\sigma} \cdot \vec{a})(\vec{\sigma} \cdot \vec{b}) = (\vec{a} \cdot \vec{b}) \mathbb{1} + i \vec{\sigma} \cdot (\vec{a} \times \vec{b}) \tag{C.98}$$

where $\vec{a}$ and $\vec{b}$ are any two vectors, and $\mathbb{1}$ the 2×2 identity matrix. For $\vec{a} = \vec{b}$ we obtain the special case:

$$(\vec{\sigma} \cdot \vec{a})^2 = (\vec{a})^2 \mathbb{1} \tag{C.99}$$

Also useful for any vector $\vec{p}$:

$$\vec{\sigma} \cdot \vec{p} = p_x \begin{pmatrix} 0 & 1 \\ 1 & 0 \end{pmatrix} + p_y \begin{pmatrix} 0 & -i \\ i & 0 \end{pmatrix} + p_z \begin{pmatrix} 1 & 0 \\ 0 & -1 \end{pmatrix} = \begin{pmatrix} p_z & p_x - i p_y \\ p_x + i p_y & -p_z \end{pmatrix} \tag{C.100}$$

Using polar coordinates $\vec{p} = |\vec{p}|(\sin\theta \cos\phi, \sin\theta \sin\phi, \cos\theta)$, this last result can be expressed as:

$$\vec{\sigma} \cdot \vec{p} = \begin{pmatrix} p_z & p_x - i p_y \\ p_x + i p_y & -p_z \end{pmatrix} = |\vec{p}| \begin{pmatrix} \cos\theta & \sin\theta e^{-i\phi} \\ \sin\theta e^{+i\phi} & -\cos\theta \end{pmatrix} \tag{C.101}$$

Appendix D   Lorentz Transformations and 4D Mathematical Tools

> The views of space and time which I wish to lay before you have sprung from the soil of experimental physics, and therein lies their strength. They are radical. Henceforth space by itself, and time by itself, are doomed to fade away into mere shadows, and only a kind of union of the two will preserve an independent reality.[1]
>
> *Hermann Minkowski (1909)*

D.1   Minkowski Space

Let us first consider the ideas behind the four-dimensional Minkowski space. The combination of three-dimensional Euclidean space and one-dimensional time forms the four-dimensional space-time. One of the simplest combinations is called **Minkowski space**.

In three-dimensional Euclidean space, rotations are a particularly important class of transformations because they change the direction of a three-vector but preserve its length. Such "rotations" in four-dimensional Minkowski space are termed **Lorentz transformations**. They generalize the idea of abstract rotations in the four-dimensional Minkowski space. These consists of the well-known spatial rotations in the three-dimensional space-like subspace as well as boosts which correspond to rotations in subspaces that mix space and time dimensions.

We seek a set of transformations that leave the "length" of a four-vector invariant in Minkowski space. We begin with the Euclidean rotation in three dimensions. The three-dimensional rotation can be represented by a 3×3 matrix acting on the three coordinates of an arbitrary vector $\vec{x}$:

$$\vec{x}\,' = R\vec{x} \tag{D.1}$$

For the length of the vector $\vec{x}$ to be unchanged by this transformation means that:

$$\vec{x}\,' \cdot \vec{x}\,' = (R\vec{x}) \cdot (R\vec{x}) = \vec{x} \cdot \vec{x} \tag{D.2}$$

The scalar product can be expressed as the multiplication of a row vector with a column vector:

$$\vec{x} \cdot \vec{y} = x^T y = x^T \mathbb{1} y \tag{D.3}$$

where we have introduced the identity matrix $\mathbb{1} = diag(1,1,1)$. Hence:

$$\vec{x}\,' \cdot \vec{x}\,' = (R\vec{x})^T \mathbb{1} (R\vec{x}) = (\vec{x})^T R^T \mathbb{1} R\vec{x} = (\vec{x})^T \mathbb{1}\vec{x} \tag{D.4}$$

---

[1]   Translated from *"Die Anschauungen über Raum und Zeit, die ich Ihnen entwickeln möchte, sind auf experimentell-physikalischem Boden erwachsen. Darin liegt ihre Stärke. Ihre Tendenz ist eine radikale. Von Stund' an sollen Raum für sich und Zeit für sich völlig zu Schatten herabsinken und nur noch eine Art Union der beiden soll Selbständigkeit bewahren."*

where the last equality is satisfied if R is **orthogonal**:

$$R^T \mathbb{1} R = \mathbb{1} \tag{D.5}$$

Hence, rotations in Euclidean space are implemented by orthogonal matrices.

Note that the use of the identity matrix is not truly necessary but will be helpful in view of its generalization to Minkowski space. Indeed, the requirement of orthogonality is valid generally, not just for Euclidean spaces. Therefore, we can use it as guidance for constructing generalized rotations in Minkowski space.

In Minkowski space, a four-dimensional vector is termed a **four-vector** x^μ (with the **Lorentz index** $\mu = 0, 1, 2, 3$). We adopt the convention that the first component corresponds to the time axis. The four-vector x^μ defines an **event in space-time** and is denoted:

$$x^\mu \equiv (x^0, x^1, x^2, x^3) = (x^0, \vec{x}) \tag{D.6}$$

where the three-vector $\vec{x}$ defines a point in space. The **inner product of any two four-vectors** is given in matrix form by:

$$x \cdot y \equiv x^T g y \tag{D.7}$$

where g is the **Minkowski metric**. For flat space and using our convention for x^μ, the metric is simply given, now using the index notation, by:

$$g_{\mu\nu} = g^{\mu\nu} = \begin{pmatrix} 1 & 0 & 0 & 0 \\ 0 & -1 & 0 & 0 \\ 0 & 0 & -1 & 0 \\ 0 & 0 & 0 & -1 \end{pmatrix} \tag{D.8}$$

(both $g_{\mu\nu}$ and $g^{\mu\nu}$ here are the same for Cartesian coordinates, but this does not hold in general for curved space). In our case, the metric will be used to convert and keep track of covariant and contravariant vectors, and will therefore have a sort of "passive" role. In general relativity, the metric contains all the information about the geometry of space-time and depends on its local energy (or matter) density, which might lead to non-diagonal elements. On the contrary, we can say that in QFT, space-time is *fixed*, it has no dynamics of its own. With this flat space metric, we obtain:

$$x \cdot y = x^0 y^0 - \vec{x} \cdot \vec{y} \tag{D.9}$$

The metric is sometimes called a pseudo-Euclidean metric to emphasize that it is Euclidean-like except for the difference in sign between the time and space terms. However, this definition already marks a clear distinction between the Minkowski space-time and a four-dimensional Euclidean space, which would have an identity metric $\mathbb{1} = diag(1, 1, 1, 1)$. **The opposite sign between time and space coordinates is the fundamental distinction between the Euclidean and Minkowski spaces!**

A four-dimensional transformation in Minkowski space acts on a four-vector. Generalizing the rotation concept, Lorentz transformations are those transformations of type (matrix notation):

$$x' = \Lambda x \tag{D.10}$$

that leave the inner product invariant:

$$x' \cdot y' = (\Lambda x) \cdot (\Lambda y) = x \cdot y \quad \Longrightarrow \quad x^T \Lambda^T g \Lambda y = x^T g y \tag{D.11}$$

This leads to the **relation of orthogonality** (or **Lorentz condition**):

$$\Lambda^T g \Lambda = g \tag{D.12}$$

Compare with Eq. (D.5).

D.2 The Light Cone of Minkowski Space

The light cone is a three-dimensional surface in the four-dimensional space-time. Events in space-time may be characterized according to whether they are inside of, outside of, or on the light cone. Consider two events in space-time x^μ and y^μ. Let's use the **space-time interval** Δs^μ defined as:

$$\Delta s^\mu \equiv x^\mu - y^\mu = (\Delta x^0, \Delta x^1, \Delta x^2, \Delta x^3) = (\Delta x^0, \Delta \vec{x}). \tag{D.13}$$

The notion of distance between two points is generalized to the interval between two points in space-time. According to Eq. (D.11), the inner product is **invariant**. Hence:

$$(\Delta s)^2 = (\Delta x^0)^2 - (\Delta \vec{x})^2 = (c\Delta t)^2 - (\Delta \vec{x})^2 \tag{D.14}$$

is invariant, where we reintroduced t, x, y, z, and c for clarity. The light-cone classification illustrates another distinction between Minkowski space-time and four-dimensional Euclidean space, in that **two points in the Minkowski space-time may be separated by a distance whose square $(\Delta s)^2$ could be (1) positive $(\Delta s)^2 > 0$, (2) negative $(\Delta s)^2 < 0$, or (3) zero $(\Delta s)^2 = 0$.** This is impossible for a Euclidean space! The Minkowski light cone is illustrated in Figure D.1 for the case of two space (x, y) and one time (ct) dimensions. Time flows vertically from past to future. For the case $(\Delta s)^2 > 0$, we have:

$$(c\Delta t)^2 - (\Delta \vec{x})^2 > 0 \quad \longrightarrow \quad c^2 > \left(\frac{\Delta \vec{x}}{\Delta t}\right)^2 \quad \longrightarrow \quad |\vec{v}| < c \tag{D.15}$$

By similar arguments, $(\Delta s)^2 = 0$ implies $|\vec{v}| = c$ and $(\Delta s)^2 < 0$ implies $|\vec{v}| > c$.

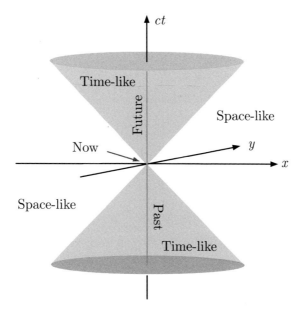

Figure D.1 Illustration of the light cone for two space and one time dimensions.

Hence, events in Minkowski space separated by a null space-time interval $(\Delta s = 0)$ are connected by signals moving at light velocity $v = c$. Since the time (ct) and space axes have the same scales in Figure D.1, this means that the **worldline** of a freely propagating massless particle moving at the speed of light always make

$\pm 45°$ angles in the light-cone diagram. Such space-time intervals are called **light-like intervals**. Events with $(\Delta s)^2 > 0$ have a so-called **time-like separation** (inside the light cone) and are connected by signals with $v < c$. Those with $(\Delta s)^2 < 0$ are termed **space-like separations** (outside the light cone) and could be connected only by signals with $v > c$, which would violate causality.

To summarize, the causal properties of Minkowski space-time are encoded in its light-cone structure, which requires that $v \leq c$ for all signals. Each point in space-time may be viewed as lying at the apex of a light cone ("now"). An event at the origin of a light cone may influence any event in its forward light cone (the "future"). The event at the origin of the light cone may be influenced by events in its backward light cone (the "past"). Events at space-like separations are causally disconnected from the event at the origin. Events on the light cone are connected by signals that travel exactly at c. In conclusion, one can say that the light cone is a surface separating the *knowable* from the *unknowable* for an observer at the apex of the light cone.

D.3   Proper Lorentz Transformations

We want to generalize the concept of rotations to the four-dimensional Minkowski space. These rotations can mix space and time. Properties that can appear surprising at first follow from the Minkowski metric g since the invariant space-time interval that is being conserved is not the length of vectors in space nor the length of time intervals separately. Rather, it is the specific mixture of time and space intervals implied by the indefinite metric $g = diag(1, -1, -1, -1)$.

• **Space rotations.** We begin with *pure space rotations*. For rotations about the $x^3 = z$ axis, the transformation may be written in matrix notation as:

$$
\begin{pmatrix} ct' \\ x^{1}{}' \\ x^{2}{}' \\ x^{3}{}' \end{pmatrix} = \Lambda(R) \begin{pmatrix} ct \\ x^1 \\ x^2 \\ x^3 \end{pmatrix} = \begin{pmatrix} 1 & 0 & 0 & 0 \\ 0 & a & b & 0 \\ 0 & c & d & 0 \\ 0 & 0 & 0 & 1 \end{pmatrix} \begin{pmatrix} ct \\ x^1 \\ x^2 \\ x^3 \end{pmatrix}
\tag{D.16}
$$

To define the rotation in space, we use the Euclidian metric and the orthogonality relation of Eq. (D.5):

$$
R^T \mathbb{1} R = \begin{pmatrix} 1 & 0 & 0 & 0 \\ 0 & a & c & 0 \\ 0 & b & d & 0 \\ 0 & 0 & 0 & 1 \end{pmatrix} \begin{pmatrix} 1 & 0 & 0 & 0 \\ 0 & 1 & 0 & 0 \\ 0 & 0 & 1 & 0 \\ 0 & 0 & 0 & 1 \end{pmatrix} \begin{pmatrix} 1 & 0 & 0 & 0 \\ 0 & a & b & 0 \\ 0 & c & d & 0 \\ 0 & 0 & 0 & 1 \end{pmatrix} = \begin{pmatrix} 1 & 0 & 0 & 0 \\ 0 & a^2 + c^2 & ab + cd & 0 \\ 0 & ab + cd & b^2 + d^2 & 0 \\ 0 & 0 & 0 & 1 \end{pmatrix}
\tag{D.17}
$$

Comparison of the equation with $R^T \mathbb{1} R = \mathbb{1}$ implies that:

$$
a^2 + c^2 = 1; \quad b^2 + d^2 = 1; \quad ab + cd = 0
\tag{D.18}
$$

These requirements are satisfied by the choices of one parameter θ and the following relations:

$$
a = \cos\theta; \quad b = \sin\theta; \quad c = -\sin\theta; \quad d = \cos\theta
\tag{D.19}
$$

Hence, we obtain the expected result for an ordinary rotation:

$$
\Lambda^{\mu}{}_{\nu}(R(\theta)) = \begin{pmatrix} 1 & 0 & 0 & 0 \\ 0 & \cos\theta & \sin\theta & 0 \\ 0 & -\sin\theta & \cos\theta & 0 \\ 0 & 0 & 0 & 1 \end{pmatrix}
\tag{D.20}
$$

• **Boosts.** Now we apply the same technique to determine the elements of a Lorentz boost transformation with the use of the Minkowski metric. We consider two inertial systems that move relative to one another with a constant velocity v. At time $t = t' = 0$ the origins of the two systems overlap and their coordinate axes are always parallel. See Figure D.2.

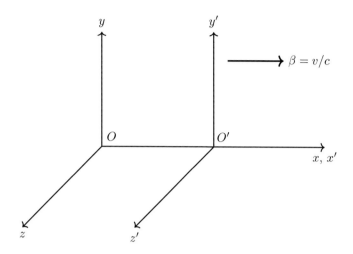

Figure D.2 Coordinate axes for the boost along the x-axis. The reference system O' moves relative to O with a velocity v in the x direction.

We find for the boost along the x-axis:

$$
\begin{pmatrix} ct' \\ x^{1\,\prime} \\ x^{2\,\prime} \\ x^{3\,\prime} \end{pmatrix} = \Lambda(B) \begin{pmatrix} ct \\ x^1 \\ x^2 \\ x^3 \end{pmatrix} = \begin{pmatrix} a & b & 0 & 0 \\ c & d & 0 & 0 \\ 0 & 0 & 1 & 0 \\ 0 & 0 & 0 & 1 \end{pmatrix} \begin{pmatrix} ct \\ x^1 \\ x^2 \\ x^3 \end{pmatrix}
\tag{D.21}
$$

and hence:

$$
\Lambda^T g \Lambda = \begin{pmatrix} a & c & 0 & 0 \\ b & d & 0 & 0 \\ 0 & 0 & 1 & 0 \\ 0 & 0 & 0 & 1 \end{pmatrix} \begin{pmatrix} 1 & 0 & 0 & 0 \\ 0 & -1 & 0 & 0 \\ 0 & 0 & -1 & 0 \\ 0 & 0 & 0 & -1 \end{pmatrix} \begin{pmatrix} a & b & 0 & 0 \\ c & d & 0 & 0 \\ 0 & 0 & 1 & 0 \\ 0 & 0 & 0 & 1 \end{pmatrix} = \begin{pmatrix} a^2 - c^2 & ab - cd & 0 & 0 \\ ab - cd & b^2 - d^2 & 0 & 0 \\ 0 & 0 & -1 & 0 \\ 0 & 0 & 0 & -1 \end{pmatrix}
\tag{D.22}
$$

With the Lorentz condition $\Lambda^T g \Lambda = g$ (Eq. (D.12)), we find the following relations:

$$
a^2 - c^2 = 1; \quad b^2 - d^2 = -1; \quad ab - cd = 0
\tag{D.23}
$$

These requirements are satisfied by the choices of one parameter ϕ and the following relations:

$$
a = \cosh \phi; \quad b = -\sinh \phi; \quad c = -\sinh \phi; \quad d = \cosh \phi
\tag{D.24}
$$

Hence, we obtain the result for a boost along the x-axis:

$$\Lambda^\mu_{\ \nu}(B(\phi)) = \begin{pmatrix} \cosh\phi & -\sinh\phi & 0 & 0 \\ -\sinh\phi & \cosh\phi & 0 & 0 \\ 0 & 0 & 1 & 0 \\ 0 & 0 & 0 & 1 \end{pmatrix} \tag{D.25}$$

The derivation makes it clear that the appearance of hyperbolic functions in the boost, rather than trigonometric functions as in spatial rotations, is a consequence of the indefinite metric. We can relate the angle ϕ to the velocity of the boosted system v. The origin of the boosted primed system relative to the non-primed one is found by solving for $x^{1\,\prime} = 0$:

$$x^{1\,\prime} = (-\sinh\phi)ct + (\cosh\phi)x^1 = 0 \tag{D.26}$$

Therefore, we conclude that:

$$\beta \equiv \frac{v}{c} = \frac{x^1}{ct} = \frac{\sinh\phi}{\cosh\phi} = \tanh\phi \quad \Longrightarrow \quad \cosh\phi = \gamma, \quad \sinh\phi = \beta\gamma \tag{D.27}$$

with the Lorentz factor $\gamma = 1/\sqrt{1-\beta^2}$. Inserting these results in the Lorentz boost matrix gives the common representation of the boost matrix as a function of β and γ:

$$\Lambda^\mu_{\ \nu}(\beta_x) = \begin{pmatrix} \gamma & -\beta_x\gamma & 0 & 0 \\ -\beta_x\gamma & \gamma & 0 & 0 \\ 0 & 0 & 1 & 0 \\ 0 & 0 & 0 & 1 \end{pmatrix} \tag{D.28}$$

• **Rapidity.** If we perform two Lorentz boosts of the type of Eq. (D.28) with parameters β_1 and β_2, the result is given by:

$$\Lambda(\beta_1)\Lambda(\beta_2) = \begin{pmatrix} \gamma_1 & -\beta_1\gamma_1 & 0 & 0 \\ -\beta_1\gamma_1 & \gamma_1 & 0 & 0 \\ 0 & 0 & 1 & 0 \\ 0 & 0 & 0 & 1 \end{pmatrix} \begin{pmatrix} \gamma_2 & -\beta_2\gamma_2 & 0 & 0 \\ -\beta_2\gamma_2 & \gamma_2 & 0 & 0 \\ 0 & 0 & 1 & 0 \\ 0 & 0 & 0 & 1 \end{pmatrix}$$

$$= \begin{pmatrix} \gamma_1\gamma_2(1+\beta_1\beta_2) & \gamma_1\gamma_2(-\beta_1-\beta_2) & 0 & 0 \\ \gamma_1\gamma_2(-\beta_1-\beta_2) & \gamma_1\gamma_2(1+\beta_1\beta_2) & 0 & 0 \\ 0 & 0 & 1 & 0 \\ 0 & 0 & 0 & 1 \end{pmatrix} \equiv \begin{pmatrix} \gamma_3 & -\beta_3\gamma_3 & 0 & 0 \\ -\beta_3\gamma_3 & \gamma_3 & 0 & 0 \\ 0 & 0 & 1 & 0 \\ 0 & 0 & 0 & 1 \end{pmatrix} \tag{D.29}$$

with

$$\beta_3 = (\beta_1+\beta_2)/(1+\beta_1\beta_2) \quad \text{and} \quad \gamma_3 = \gamma_1\gamma_2(1+\beta_1\beta_2) \tag{D.30}$$

Consequently, the two parallel boosts are equivalent to a single boost with the parameter β_3 given above. To see it more clearly, we can revert to the ϕ parameter of Eq. (D.27) and define the **rapidity** (usually labeled η) as:

$$\beta = \tanh\eta, \quad \gamma = \cosh\eta, \quad \text{and} \quad \sinh\eta = \beta_\gamma \tag{D.31}$$

We then have:

$$\beta_3 = \tanh \eta_3 = \frac{(\tanh \eta_1 + \tanh \eta_2)}{(1 + \tanh \eta_1 \tanh \eta_2)} = \tanh (\eta_1 + \eta_2) \tag{D.32}$$

or $\eta_3 = \eta_1 + \eta_2$. Thus, we found that **the rapidity η is additive under two consecutive parallel Lorentz transformations**.

• **General boost direction.** We can use the results above to derive the boost in a general direction $\vec{\beta} = (\beta_x, \beta_y, \beta_z)$. We consider the components orthogonal and parallel to the direction of the boost direction. The transformation of the spatial coordinates of a four-vector $x^\mu = (x^0, \vec{x} = \vec{x}_\parallel + \vec{x}_\perp)$ can then be expressed as:

$$\vec{x}_\parallel' = \gamma \left(\vec{x}_\parallel - \vec{\beta} x^0 \right) \quad \text{and} \quad \vec{x}_\perp' = \vec{x}_\perp \tag{D.33}$$

where we used the fact that components parallel to the boost change as in Eq. (D.28) while the perpendicular components remain unchanged. By construction, $\vec{\beta}$ and $\vec{x}_\parallel \equiv \left(\vec{x} \cdot \vec{\beta}/|\beta| \right) (\vec{\beta}/|\beta|)$ are parallel, so the vectorial equations of Eq. (D.33) hold! We can rewrite:

$$
\begin{aligned}
\vec{x}' &= \vec{x}_\parallel' + \vec{x}_\perp' = \gamma \left(\vec{x}_\parallel - \vec{\beta} x^0 \right) + \underbrace{\vec{x}_\perp}_{=\vec{x}-\vec{x}_\parallel} = \vec{x} + (\gamma - 1)\vec{x}_\parallel - \gamma\vec{\beta} x^0 \\
&= \vec{x} + (\gamma - 1)\frac{\left(\vec{x} \cdot \vec{\beta} \right)}{|\beta|} \left(\frac{\vec{\beta}}{|\beta|} \right) - \gamma\vec{\beta} x^0 \\
&= \vec{x} + (\gamma - 1)\frac{\left(\vec{x} \cdot \vec{\beta} \right)}{\beta^2} \vec{\beta} - \gamma\vec{\beta} x^0
\end{aligned}
\tag{D.34}
$$

Finally, by extrapolation the time coordinate must change as:

$$x^{0\prime} = \gamma \left(x^0 - \vec{\beta} \cdot \vec{x} \right) = \gamma \left(x^0 - \beta_x x_1 - \beta_y x_2 - \beta_z x_3 \right) \tag{D.35}$$

Collecting all the terms into the transformation matrix and defining $B \equiv (\gamma - 1)/\beta^2$, we find:

$$
\Lambda^\mu{}_\nu(\vec{\beta}) = \begin{pmatrix}
\gamma & -\beta_x \gamma & -\beta_y \gamma & -\beta_z \gamma \\
-\beta_x \gamma & 1 + B\beta_x^2 & B\beta_x\beta_y & B\beta_x\beta_z \\
-\beta_y \gamma & B\beta_x\beta_y & 1 + B\beta_y^2 & B\beta_y\beta_z \\
-\beta_z \gamma & B\beta_x\beta_z & B\beta_y\beta_z & 1 + B\beta_z^2
\end{pmatrix}
\tag{D.36}
$$

D.4   Einstein Summation in Four Dimensions

In four-dimensional Minkowski space, we use Greek indices to imply summation over the four components $0, 1, 2, 3$, with the inner product defined with a covariant and contravariant four-vector to be:

$$x \cdot y = x^\mu y_\mu = x_\mu y^\mu = g_{\mu\nu} x^\mu y^\nu \tag{D.37}$$

where $g_{\mu\nu}$ is the flat-space Minkowski metric defined in Eq. (D.8).

D.5   The Four-Gradient

We generalize the three-gradient (see Section A.9) to the **four-dimensional space-time four-gradient**:

$$\frac{\partial}{\partial x^\mu} \equiv \left(\frac{\partial}{\partial x^0}, \frac{\partial}{\partial x^1}, \frac{\partial}{\partial x^2}, \frac{\partial}{\partial x^3} \right) \tag{D.38}$$

We compute the properties of the four-gradient under a Lorentz transformation:

$$\begin{aligned}
\frac{\partial}{\partial x^\mu} &= \left(\frac{\partial x'^\nu}{\partial x^\mu} \right) \left(\frac{\partial}{\partial x'^\nu} \right) = \left(\frac{\partial}{\partial x^\mu} \Lambda^\nu{}_\alpha x^\alpha \right) \left(\frac{\partial}{\partial x'^\nu} \right) = \Lambda^\nu{}_\alpha \left(\frac{\partial x^\alpha}{\partial x^\mu} \right) \left(\frac{\partial}{\partial x'^\nu} \right) \\
&= \Lambda^\nu{}_\alpha \delta^\alpha_\mu \left(\frac{\partial}{\partial x'^\nu} \right) = \Lambda^\nu{}_\mu \left(\frac{\partial}{\partial x'^\nu} \right)
\end{aligned} \tag{D.39}$$

By noting the position of the indices in the Λ matrix, we see that the four-gradient $\left(\frac{\partial}{\partial x^\mu} \right)$ is a covariant four-vector like $x_\mu = \Lambda^\nu{}_\mu (x_\nu)'$ (see Eq. (5.16)). We introduce the following notation:

$$\partial_\mu \equiv \left(\frac{\partial}{\partial x^\mu} \right) = \left(\frac{\partial}{\partial t}, \vec{\nabla} \right) \qquad \text{covariant} \tag{D.40}$$

$$\partial^\mu \equiv \left(\frac{\partial}{\partial x_\mu} \right) = \left(\frac{\partial}{\partial t}, -\vec{\nabla} \right) \qquad \text{contravariant} \tag{D.41}$$

D.6   The D'Alembert Operator

The **D'Alembert[2] operator**, also called the d'Alembertian, wave operator, or box operator, is the Laplace operator of Minkowski space-time. It is sometimes represented by a box, $\Box$, but we prefer the form $\Box^2$ to recall its scalar nature:

$$\Box^2 \equiv \partial^2 = \partial^\mu \partial_\mu = g_{\mu\nu} \partial^\mu \partial^\nu = \frac{\partial^2}{\partial t^2} - \vec{\nabla}^2 \tag{D.42}$$

The symbol $\Box$ is called the **quabla**. Lorentz transformations leave the operator invariant, so the d'Alembertian yields a Lorentz scalar, hence it really makes sense to write it as $\Box^2$.

D.7   Fourier Transform in Minkowski Space

One can define a Fourier transform with respect to the four space-time variables, which will take functions of x^μ to functions of the Fourier transform variables p^μ. Unlike the convention used in Eq. (A.24) and for the Dirac function definition, the convention here is to put all factors of 2π with the p integrals. Hence, we have:

$$f(x) = \int \frac{\mathrm{d}^4 p}{(2\pi)^4} e^{-ip \cdot x} \tilde{f}(p) \tag{D.43}$$

where

$$\tilde{f}(p) = \int \mathrm{d}^4 x\, e^{ip \cdot x} f(x) \tag{D.44}$$

2   Jean-Baptiste le Rond d'Alembert (1717–1783), French mathematician, mechanical engineer, physicist, philosopher, and music theorist.

D.8   Dirac δ Function

The Dirac δ function in Minkowski space is directly extrapolated from its definition in other dimensions, Eq. (A.41):

$$(2\pi)^4 \delta^4(x-y) = \int d^4k\, e^{ik\cdot(x-y)} \tag{D.45}$$

where x and y are four-vectors.

D.9   Levi-Civita and Kronecker Tensors

The Levi-Civita and Kronecker symbols were defined in Appendix A.3. In four dimensions, we use them as ordinary tensors. For the Levi-Civita tensor, we start from $\epsilon_{1234} = +1$ in ascending order, and any even permutation gives $+1$ and any odd permutation -1, and 0 otherwise. It's pretty straightforward and boring! For example, $\epsilon_{2143} = -\epsilon_{2134} = \epsilon_{1234} = +1$. The useful equalities can be extended to the four-dimensional Minkowski space to give:

$$\epsilon^{\alpha\beta\mu\nu}\epsilon_{\alpha\beta\rho\sigma} = -2\left(\delta^\mu_{\ \rho}\delta^\nu_{\ \sigma} - \delta^\mu_{\ \sigma}\delta^\nu_{\ \rho}\right) \tag{D.46}$$

$$\epsilon^{\alpha\beta\gamma\mu}\epsilon_{\alpha\beta\gamma\nu} = -6\delta^\mu_{\ \nu} \tag{D.47}$$

$$\epsilon^{\alpha\beta\gamma\delta}\epsilon_{\alpha\beta\gamma\delta} = -24 \tag{D.48}$$

The Kronecker symbol can be generalized to the **Kronecker tensor** written δ^μ_ν with a covariant index μ and contravariant index ν:

$$\delta^\mu_\nu \equiv \begin{cases} 1 \text{ if } \mu = \nu \\ 0 \text{ if } \mu \neq \nu \end{cases} \tag{D.49}$$

We note that:

$$\delta^{\mu\nu} = g^{\mu\alpha}\delta^\nu_\alpha = g^{\mu\nu} = \begin{cases} 1 \text{ if } \mu = \nu = 0 \\ -1 \text{ if } \mu = \nu = 1, 2, 3 \\ 0 \text{ if } \mu \neq \nu \end{cases} \tag{D.50}$$

Appendix E   Dirac Matrices and Trace Theorems

E.1   Dirac γ Matrices

The Dirac or gamma matrices $\gamma^\mu \equiv \gamma^0, \ldots, \gamma^3$ are a set of conventional matrices with specific anticommutation relations that ensure they generate a 4×4 matrix representation of the Clifford algebra. Higher-dimensional representations are also possible. When interpreted as the matrices of the action of a set of orthogonal basis vectors for contravariant vectors in Minkowski space, the column vectors on which the matrices act become a space of spinors, on which the Clifford algebra of space-time acts. The **Clifford algebra** is, with $\mu, \nu = 0, 1, 2, 3$:

$$\{\gamma^\mu, \gamma^\nu\} = \gamma^\mu \gamma^\nu + \gamma^\nu \gamma^\mu = 2g^{\mu\nu}\mathbb{1} \tag{E.1}$$

which yields:

$$(\gamma^0)^2 = \mathbb{1}, \quad (\gamma^k)^2 = -\mathbb{1} \quad \text{and} \quad \gamma^\mu \gamma^\nu = -\gamma^\nu \gamma^\mu \ (\mu \neq \nu) \tag{E.2}$$

Another consequence is that the commutation of two γ matrices is just:

$$[\gamma^\mu, \gamma^\nu] = \gamma^\mu \gamma^\nu - \gamma^\nu \gamma^\mu = 2\gamma^\mu \gamma^\nu - 2g^{\mu\nu}\mathbb{1} \tag{E.3}$$

The properties $(\gamma^0)^\dagger = \gamma^0$ and $(\gamma^k)^\dagger = -\gamma^k$ can be combined in:

$$(\gamma^\mu)^\dagger = \gamma^0 \gamma^\mu \gamma^0 \tag{E.4}$$

Useful products of γ^μ matrices are listed below:

- **[D1]**: $\gamma^\mu \gamma_\mu = 4 \cdot \mathbb{1}$
- **[D2]**: $\gamma^\mu \gamma^\nu \gamma_\mu = (\gamma^\mu \gamma^\nu + \gamma^\nu \gamma^\mu - \gamma^\nu \gamma^\mu)\gamma_\mu = (2g^{\mu\nu} - \gamma^\nu \gamma^\mu)\gamma_\mu = 2\gamma^\nu - 4\gamma^\nu = -2\gamma^\nu$
- **[D3]**: $\gamma^\mu \gamma^\nu \gamma^\rho \gamma_\mu = 4g^{\nu\rho}\mathbb{1}$
- **[D4]**: $\gamma^\mu \gamma^\nu \gamma^\rho \gamma^\sigma \gamma_\mu = -2\gamma^\sigma \gamma^\rho \gamma^\nu$

From these inequalities, we have the following for any slashed momenta:

- **[S1]**: $\gamma^\mu \slashed{a} \gamma_\mu = -2\slashed{a}$
- **[S2]**: $\gamma^\mu \slashed{a}\slashed{b} \gamma_\mu = 4(a \cdot b)\mathbb{1}$
- **[S3]**: $\gamma^\mu \slashed{a}\slashed{b}\slashed{c} \gamma_\mu = -2\slashed{c}\slashed{b}\slashed{a}$

Specific representations are described in Chapter 8. One also defines γ^5, sometimes called the "fourth" gamma matrix, as:

$$\gamma^5 \equiv i\gamma^0 \gamma^1 \gamma^2 \gamma^3 \tag{E.5}$$

And it follows that γ^5 is Hermitian and has eigenvalues ± 1:

- **[G51]**: $(\gamma^5)^\dagger = (i\gamma^0\gamma^1\gamma^2\gamma^3)^\dagger = -i\gamma^{3\dagger}\gamma^{2\dagger}\gamma^{1\dagger}\gamma^{0\dagger} = i\gamma^3\gamma^2\gamma^1\gamma^0 = -i\gamma^0\gamma^3\gamma^2\gamma^1 = -i\gamma^0\gamma^1\gamma^3\gamma^2$
  $= i\gamma^0\gamma^1\gamma^2\gamma^3 = \gamma^5$

- **[G52]**: $(\gamma^5)^2 = -\gamma^0\gamma^1\gamma^2\gamma^3\gamma^0\gamma^1\gamma^2\gamma^3 = \gamma^0\gamma^0\gamma^1\gamma^2\gamma^3\gamma^1\gamma^2\gamma^3 = \gamma^1\gamma^1\gamma^2\gamma^3\gamma^2\gamma^3 = -\gamma^2\gamma^3\gamma^2\gamma^3$
  $= \gamma^2\gamma^2\gamma^3\gamma^3 = \mathbb{1}$

- **[G53]**: $\{\gamma^5, \gamma^\mu\} = i\left\{\gamma^0\gamma^1\gamma^2\gamma^3, \gamma^\mu\right\} = \begin{cases} i(\gamma^0\gamma^0\gamma^1\gamma^2\gamma^3 + \underbrace{\gamma^0\gamma^1\gamma^2\gamma^3\gamma^0}_{-\gamma^0\gamma^0\gamma^1\gamma^2\gamma^3}) = 0 \\ i(\underbrace{\gamma^1\gamma^0\gamma^1\gamma^2\gamma^3}_{-\gamma^0\gamma^1\gamma^1\gamma^2\gamma^3} + \underbrace{\gamma^0\gamma^1\gamma^2\gamma^3\gamma^1}_{\gamma^0\gamma^1\gamma^1\gamma^2\gamma^3}) = 0 \\ i(\underbrace{\gamma^2\gamma^0\gamma^1\gamma^2\gamma^3}_{\gamma^0\gamma^1\gamma^2\gamma^2\gamma^3} + \underbrace{\gamma^0\gamma^1\gamma^2\gamma^3\gamma^2}_{-\gamma^0\gamma^1\gamma^2\gamma^2\gamma^3}) = 0 \\ i(\underbrace{\gamma^3\gamma^0\gamma^1\gamma^2\gamma^3}_{-\gamma^0\gamma^1\gamma^2\gamma^3\gamma^3} + \gamma^0\gamma^1\gamma^2\gamma^3\gamma^3) = 0 \end{cases} \implies \gamma^5\gamma^\mu = -\gamma^\mu\gamma^5$

The totally antisymmetric tensors $S^{\mu\nu}$ and $\sigma^{\mu\nu}$ are defined as:

$$S^{\mu\nu} \equiv \frac{i}{4}[\gamma^\mu, \gamma^\nu] \quad \text{and} \quad \sigma^{\mu\nu} \equiv \frac{i}{2}[\gamma^\mu, \gamma^\nu] = 2S^{\mu\nu} \tag{E.6}$$

- **[G54]**: $[S^{\mu\nu}, \gamma^5] = 0$

E.2   Trace Theorems

Dirac traces do not depend on the specific representation of the γ matrices but are completely determined by the Clifford algebra.

In order to prove the properties of the traces of products of γ matrices, one uses the following mathematical properties of traces valid for any matrix A, B, and C: $\mathrm{Tr}(A+B) = \mathrm{Tr}(A) + \mathrm{Tr}(B)$, $\mathrm{Tr}(aA) = a \cdot \mathrm{Tr}(A)$ and $\mathrm{Tr}(ABC) = \mathrm{Tr}(CAB) = \mathrm{Tr}(BCA)$ (cyclic).

- **[T1]**: $\mathrm{Tr}(\mathbb{1}) = 4$

- **[T2]**: $\mathrm{Tr}(\gamma^\mu) = 0$

$$\mathrm{Tr}\,\gamma^\mu = \mathrm{Tr}\,\gamma^5\gamma^5\gamma^\mu = -\mathrm{Tr}\,\gamma^5\gamma^\mu\gamma^5 \overset{cyclic}{=} -\mathrm{Tr}\,\gamma^5\gamma^5\gamma^\mu = -\mathrm{Tr}\,\gamma^\mu = 0 \qquad \square$$

- **[T3]**: The trace of any product of an *odd number* of γ^μ is zero.

$$\mathrm{Tr}(\underbrace{\gamma^\mu \dots \gamma^\nu}_{odd\#}) = \mathrm{Tr}(\gamma^5\gamma^5\gamma^\mu \dots \gamma^\nu) = -\mathrm{Tr}(\gamma^5\gamma^\mu \dots \gamma^\nu\gamma^5) \overset{cyclic}{=} -\mathrm{Tr}(\gamma^5\gamma^5\gamma^\mu \dots \gamma^\nu) = 0 \qquad \square$$

- **[T4]**: $\mathrm{Tr}(\gamma^\mu\gamma^\nu) = 4g^{\mu\nu}$

$$\mathrm{Tr}(\gamma^\mu\gamma^\nu) = \mathrm{Tr}(2g^{\mu\nu} - \gamma^\nu\gamma^\mu) = 2g^{\mu\nu}\,\mathrm{Tr}(1) - \mathrm{Tr}(\gamma^\nu\gamma^\mu) = 8g^{\mu\nu} - \mathrm{Tr}(\gamma^\nu\gamma^\mu) \qquad \square$$

- **[T5]**: $\mathrm{Tr}(\gamma^\mu\gamma^\nu\gamma^\rho\gamma^\sigma) = 4(g^{\mu\nu}g^{\rho\sigma} - g^{\mu\rho}g^{\nu\sigma} + g^{\mu\sigma}g^{\nu\rho})$

$$\begin{aligned} \mathrm{Tr}(\gamma^\mu\gamma^\nu\gamma^\rho\gamma^\sigma) &= \mathrm{Tr}(2g^{\mu\nu}\gamma^\rho\gamma^\sigma - \gamma^\nu\gamma^\mu\gamma^\rho\gamma^\sigma) = \mathrm{Tr}(2g^{\mu\nu}\gamma^\rho\gamma^\sigma - \gamma^\nu 2g^{\mu\rho}\gamma^\sigma + \gamma^\nu\gamma^\rho 2g^{\mu\sigma} - \gamma^\nu\gamma^\rho\gamma^\sigma\gamma^\mu) \\ \implies \quad \mathrm{Tr}(\gamma^\mu\gamma^\nu\gamma^\rho\gamma^\sigma) &= g^{\mu\nu}\,\mathrm{Tr}(\gamma^\rho\gamma^\sigma) - g^{\mu\rho}\,\mathrm{Tr}(\gamma^\nu\gamma^\sigma) + g^{\mu\sigma}\,\mathrm{Tr}(\gamma^\nu\gamma^\rho) \\ &= 4(g^{\mu\nu}g^{\rho\sigma} - g^{\mu\rho}g^{\nu\sigma} + g^{\mu\sigma}g^{\nu\rho}) \qquad \square \end{aligned}$$

- **[T6]**: $\mathrm{Tr}\,\gamma^5 = 0$

$$\mathrm{Tr}\,\gamma^5 = \mathrm{Tr}(\gamma^0\gamma^0\gamma^5) = -\mathrm{Tr}(\gamma^0\gamma^5\gamma^0) \overset{cyclic}{=} -\mathrm{Tr}(\gamma^0\gamma^0\gamma^5) = 0 \qquad \square$$

- **[T7]**: $\mathrm{Tr}(\gamma^\mu\gamma^\nu\gamma^5) = 0$

$$\mathrm{Tr}(\gamma^\mu\gamma^\nu\gamma^5) = \mathrm{Tr}(\gamma^\alpha\gamma^\alpha\gamma^\mu\gamma^\nu\gamma^5) = -\mathrm{Tr}(\gamma^\alpha\gamma^\mu\gamma^\nu\gamma^5\gamma^\alpha) \overset{cyclic}{=} -\mathrm{Tr}(\gamma^\alpha\gamma^\alpha\gamma^\mu\gamma^\nu\gamma^5) = 0 \qquad \square$$

- **[T8]**: The trace of γ^5 times a product of an *odd number* of γ^μ is zero.

$$\mathrm{Tr}(\gamma^5\underbrace{\gamma^\mu\ldots\gamma^\nu}_{odd\#}) = -\mathrm{Tr}(\gamma^\mu\ldots\gamma^\nu\gamma^5) \overset{cyclic}{=} -\mathrm{Tr}(\gamma^5\gamma^\mu\ldots\gamma^\nu) = 0 \qquad \square$$

- **[T9]**: $\mathrm{Tr}(\gamma^\mu\gamma^\nu\gamma^\rho\gamma^\sigma\gamma^5) = -4i\epsilon^{\mu\nu\rho\sigma}$

  Interchanging any two γ matrices will flip the sign, so $\mathrm{Tr} \propto \epsilon^{\mu\nu\rho\sigma}$ and $\mathrm{Tr}(\gamma^0\gamma^1\gamma^2\gamma^3\gamma^5) = -i\,\mathrm{Tr}(\gamma^5\gamma^5) = -4i$ $\quad\square$.

- **[T10]**: This combination appears in the calculation of weak interactions:

$$
\begin{aligned}
T &= \mathrm{Tr}\left[\gamma^\alpha\gamma^\mu\gamma^\beta\gamma^\gamma\left(C_1 - C_2\gamma^5\right)\right]\mathrm{Tr}\left[\gamma_\theta\gamma_\mu\gamma_\phi\gamma_\gamma\left(C_3 - C_4\gamma^3\right)\right] \\
&= 16\left[C_1\left(g^{\alpha\mu}g^{\beta\nu} + g^{\alpha\nu}g^{\mu\beta} - g^{\alpha\beta}g^{\mu\nu}\right) - iC_2\varepsilon^{\alpha\mu\beta\nu}\right]\left[C_3\left(g_{\theta\mu}g_{\phi\nu} + g_{\theta\nu}g_{\mu\phi} - g_{\theta\phi}g_{\mu\nu}\right) - iC_4\varepsilon_{\theta\mu\phi\nu}\right] \\
&= 32\left[C_1C_3\left(\delta^\alpha_\theta\delta^\beta_\phi + \delta^\alpha_\phi\delta^\beta_\theta\right) + C_2C_4\left(\delta^\alpha_\theta\delta^\beta_\phi - \delta^\alpha_\phi\delta^\beta_\theta\right)\right] \\
&= 32\left[(C_1C_3 + C_2C_4)\left(\delta^\alpha_\theta\delta^\beta_\phi\right) + (C_1C_3 - C_2C_4)\left(\delta^\alpha_\phi\delta^\beta_\theta\right)\right]
\end{aligned}
$$

  For pure $V - A$, we have $C_1 = C_2 = C_3 = C_4 = 1$. The term proportional to $\delta^\alpha_\phi\delta^\beta_\theta$ vanishes, and we are left with $T = 64\delta^\alpha_\theta\delta^\beta_\phi$.

Also recall $g^{\mu\nu}g_{\mu\nu} = 4$.

E.3 Trace Theorems with Slashes

After application of Casimir's trick, we get traces that include four-vectors multiplied by a gamma matrix, which correspond to a trace with slashes matrix. These are easily computed using the previous theorems. For example, using **[T4]**, we directly have:

- **[TS1]**: $\mathrm{Tr}(\slashed{a}\slashed{b}) = 4(a \cdot b)$

Using **[T5]**, we also get:

- **[TS2]**: $\mathrm{Tr}(\slashed{a}\slashed{b}\slashed{c}\slashed{d}) = 4[(a \cdot b)(c \cdot d) - (a \cdot c)(b \cdot d) + (a \cdot d)(b \cdot c)]$

and

- **[TS3]**: $\mathrm{Tr}(\gamma^\mu\slashed{a}\gamma^\nu\slashed{b}) = 4\left[a^\mu b^\nu + a^\nu b^\mu - (a \cdot b)g^{\mu\nu}\right]$

We can also consider the case with a γ^5 matrix which is often encountered in $V - A$ calculations:

- **[TS4]**: $\mathrm{Tr}\left[\gamma^\mu\left(1 - \gamma^5\right)\slashed{a}\gamma^\nu\left(1 - \gamma^5\right)\slashed{b}\right] = 2\,\mathrm{Tr}(\gamma^\mu\slashed{a}\gamma^\nu\slashed{b}) + 8i\epsilon^{\mu\alpha\nu\beta}a_\alpha b_\beta$

One often also encounters products of traces which can be conveniently calculated beforehand. For instance, taking the products of two traces with two slashes of the type [**TS3**]:

- **[TS6]**: $\mathrm{Tr}(\gamma^\mu \slashed{a}\gamma^\nu \slashed{b})\,\mathrm{Tr}(\gamma_\mu \slashed{c}\gamma_\nu \slashed{d}) = 32\,[(a\cdot c)(b\cdot d) + (a\cdot d)(b\cdot c)]$ and

- **[TS7]**: $\mathrm{Tr}(\gamma^\mu \slashed{a}\gamma^\nu \gamma^5 \slashed{b})\,\mathrm{Tr}(\gamma_\mu \slashed{c}\gamma_\nu \gamma^5 \slashed{d}) = 32\,[(a\cdot c)(b\cdot d) - (a\cdot d)(b\cdot c)]$

And finally:

- **[TS8]**: $\mathrm{Tr}(\gamma^\mu (1-\gamma^5)\slashed{a}\gamma^\nu (1-\gamma^5)\slashed{b})\,\mathrm{Tr}(\gamma_\mu (1-\gamma^5)\slashed{c}\gamma_\nu (1-\gamma^5)\slashed{d}) = 256\,[(a\cdot c)(b\cdot d)]$

We also note that:

- **[TS9]**: $\mathrm{Tr}\left(\gamma^5 \slashed{a}\slashed{b}\right) = 0$

and

- **[TS10]**: $\mathrm{Tr}\left(\gamma^5 \slashed{a}\slashed{b}\slashed{c}\slashed{d}\right) = 4i\epsilon_{\mu\nu\alpha\beta}a^\mu b^\nu c^\alpha d^\beta$

E.4　　　More Traces and Fierz Identity

A 4×4 Γ matrix has 16 independent components, hence it can be expressed in a basis of 16 independent 4×4 matrices. A convenient basis is given by the following 16 matrices ($C = 1,\ldots,16$):

$$\Gamma_C: \quad \mathbb{1}, \gamma^0, \gamma^1, \gamma^2, \gamma^3, i\sigma^{01}, i\sigma^{02}, i\sigma^{03}, i\sigma^{12}, i\sigma^{13}, i\sigma^{23}, i\gamma^0\gamma^5, i\gamma^1\gamma^5, i\gamma^2\gamma^5, i\gamma^3\gamma^5, \gamma^5 \tag{E.7}$$

A property that follows from their algebra, is that the trace of the product of any two of the 16 matrices obeys ($A, B = 1,\ldots,16$):

$$\mathrm{Tr}\left(\Gamma_A \Gamma_B\right) = 4\delta_{AB} \tag{E.8}$$

For example:

$$\mathrm{Tr}\left(\Gamma_0 \Gamma_0\right) = \mathrm{Tr}\left(\mathbb{1}\right) = 4, \quad \mathrm{Tr}\left(\Gamma_0 \Gamma_B\right) = \mathrm{Tr}\left(\mathbb{1}\Gamma_B\right) = \mathrm{Tr}\left(\Gamma_B\right) = 0\,(\text{if } B \neq 0) \tag{E.9}$$

etc.

The **Fierz identity** [294] states further that:

$$\sum_{C=1}^{16} \left(\Gamma_C\right)_{ad}\left(\Gamma_C\right)_{cb} = 4\delta_{ab}\delta_{cd} \tag{E.10}$$

Appendix F — Some Tools to Compute Higher-Order Diagrams

F.1 Gamma Function

The integral form of the Gamma function for $z \in \mathbb{C}$ is given by (https://dlmf.nist.gov/5.2.i):

$$\Gamma(z) = \int_0^\infty \mathrm{d}x \, x^{z-1} e^{-x} \tag{F.1}$$

Graphical representations for real and complex arguments can be found at https://dlmf.nist.gov/5.3.i and https://dlmf.nist.gov/5.3.ii. For $n \in \mathbb{N}$:

$$\Gamma(n+1) = n! \quad \text{and} \quad \Gamma(n + \tfrac{1}{2}) = \frac{(2n-1)(2n-3)\cdots 3 \cdot 1}{2^n} \sqrt{\pi} \tag{F.2}$$

The function is tabulated in Table F.1.

x	$\Gamma(x)$	x	$\Gamma(x)$	x	$\Gamma(x)$
0	∞			1/2	$\sqrt{\pi}$
1	1	-1	∞	1 + 1/2	$(1/2)\sqrt{\pi}$
2	1	-2	∞	2 + 1/2	$(3/4)\sqrt{\pi}$
3	2	-3	∞	3 + 1/2	$(15/8)\sqrt{\pi}$
4	6	-4	∞	4 + 1/2	$(105/16)\sqrt{\pi}$

Table F.1 Commonly used values of $\Gamma(x)$.

For $n \in \mathbb{N}$ and x near zero, one has:

$$\Gamma(-n+x) \approx \frac{(-1)^n}{n!}\left(\frac{1}{x} - \gamma_E + \sum_{k=1}^{n}\frac{1}{k}\right) \quad \text{and} \quad \Gamma(x) = \frac{1}{x} - \gamma_E + \mathcal{O}(x) \tag{F.3}$$

where γ_E is the **Euler–Mascheroni constant**. The constant is defined as (https://dlmf.nist.gov/5.2.ii):

$$\gamma_E = \lim_{n\to\infty}\left(-\ln(n) + \sum_{k=1}^{n} n\frac{1}{k}\right) = -\Gamma'(1) \approx 0.57721\,56649\,01532\,86060\ldots \tag{F.4}$$

F.2   Exponential Integrals

First we solve the following integral, which is needed in order to solve the photon self-energy in Section 12.3. The integral can be solved by substitution:

$$(-i) \int_0^\infty dz e^{iz(t+i\epsilon)} = \lim_{z' \to \infty} \frac{-i}{i(t+i\epsilon)} \int_0^{iz'(t+i\epsilon)} du\, e^u = \lim_{z' \to \infty} \frac{-1}{(t+i\epsilon)} e^u \Big|_0^{iz'(t+i\epsilon)}$$

$$= \frac{-1}{(t+i\epsilon)} \left(\underbrace{\lim_{z' \to \infty} e^{iz'(t+i\epsilon)}}_{=0 \text{ when } \epsilon > 0} - 1 \right) = \frac{1}{t+i\epsilon} \qquad (\epsilon > 0) \qquad \square \qquad (F.5)$$

where we substituted $u = iz(t+i\epsilon)$, and hence, $du = i(t+i\epsilon)dz$.

• **Exponential integral (of Legendre).** For a real number x, the exponential integral is defined as (https://dlmf.nist.gov/6.2.E5):

$$E_i(x) = \int_{-\infty}^x dt \frac{e^t}{t} \qquad (F.6)$$

The exponential integral for a real number x can be represented by the following series (https://encyclopediaofmath.org/index.php?title=Integral_exponential_function):

$$E_i(x) = \sum_{k=1}^\infty \frac{x^k}{kk!} + \ln(-x) + \gamma_E \qquad (F.7)$$

where γ_E is the Euler–Mascheroni constant defined in Eq. (F.4).

The function is extended to the complex plane by analytic continuation of the function on the interval $(0, \infty)$. For real or complex arguments off the negative real axis, instead of E_i, the following infinite series is used:

$$E_1(z) = -\sum_{k=1}^\infty \frac{(-z)^k}{kk!} - \ln(z) - \gamma_E \qquad (F.8)$$

where we note that for positive real x, one has $E_1(x) = -E_i(-x)$. The integral form of $E_1(z)$ then becomes (https://dlmf.nist.gov/6.2.i):

$$E_1(z) = \int_z^\infty dt \frac{e^{-t}}{t} \qquad (F.9)$$

Graphical representations for real and complex arguments can be found at https://dlmf.nist.gov/6.3.i.

F.3   Wick Rotation

The Wick rotation is a method to transform an integral in Minkowski space into the related problem in Euclidean space. Consider the generic space-time integral over the four-vector ℓ^μ of the type:

$$\lim_{\epsilon \to 0^+} \int \frac{d^4\ell}{(2\pi)^4} \frac{1}{(\ell^2 - \Delta + i\epsilon)^2} = \lim_{\epsilon \to 0^+} \int \frac{d^4\ell}{(2\pi)^4} \frac{1}{((\ell^0)^2 - \vec{\ell}^2 - \Delta + i\epsilon)^2} \qquad (F.10)$$

with $\Delta > 0$. The time-coordinate integration over ℓ^0 is equivalent to:

$$\int_{-\infty}^\infty \frac{d\ell^0}{(2\pi)} \frac{1}{\left(\ell^0 + \sqrt{\vec{\ell}^2 + \Delta} - i\epsilon \right)^2 \left(\ell^0 - \sqrt{\vec{\ell}^2 + \Delta} + i\epsilon \right)^2} \qquad (F.11)$$

Extending the integral to the complex plane, where the two poles are shifted by the $i\epsilon$ prescription, we can "distort" the contour along the real axis in Figure F.1. We can assume that the integrals over the arcs on dashed lines are negligible if the integrand falls off sufficiently rapidly as $|\ell^0| \to \infty$. Therefore, we are left with the integration along the imaginary axis. The Wick rotation has been performed when we pass from the integral on the real axis to that on the imaginary axis. Formally, the Wick rotation is the following analytic

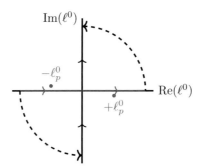

Figure F.1 The complex plane of the time coordinate ℓ^0. The integral on the real axis can be distorted to that on the imaginary axis.

continuation of the time coordinate, keeping the space coordinate unchanged:

$$\ell^0 = i\ell_E^0 \quad \text{and} \quad \vec{\ell} = \vec{\ell}_E \quad \Longrightarrow \quad d^4\ell = id^4\ell_E \quad \text{and} \quad \ell^2 = -\ell_E^2 \tag{F.12}$$

where the E index stands for the four-dimensional Euclidian space and $-\ell_E^2 = -(\ell_E^0)^2 - \vec{\ell}_E^2 = (i\ell_E^0)^2 - \vec{\ell}^2$. We then have:

$$\int \frac{d^4\ell}{(2\pi)^4} \frac{1}{(\ell^2 - \Delta)^2} = i \int \frac{d^4\ell_E}{(2\pi)^4} \frac{1}{(\ell_E^2 + \Delta)^2} = i \int \frac{d\ell_E \ell_E^3 d\Omega_4}{(2\pi)^4} \frac{1}{(\ell_E^2 + \Delta)^2} \tag{F.13}$$

where we have introduced the four-dimensional solid angle Ω_4. The integration over the solid angle gives the surface area of a four-dimensional unit sphere. The volume V_n of a sphere of radius R in n-dimensional space and the surface S_n in $n+1$-dimensional space are given by:

$$V_n(R) = \frac{\pi^{n/2}}{\Gamma(\frac{n}{2}+1)} R^n \quad \text{and} \quad S_n(R) = \frac{2\pi^{(n+1)/2}}{\Gamma(\frac{n+1}{2})} R^n \tag{F.14}$$

Consequently (see Eq. (F.2)):

$$\int d\Omega_4 = S_3 = \frac{2\pi^2}{\Gamma(2)} = 2\pi^2 \tag{F.15}$$

Our original integral is then equal to:

$$\int \frac{d^4\ell}{(2\pi)^4} \frac{1}{(\ell^2 - \Delta)^2} = 2\pi^2 i \int_0^\infty \frac{d\ell_E \ell_E^3}{(2\pi)^4} \frac{1}{(\ell_E^2 + \Delta)^2} = \frac{i}{8\pi^2} \int_0^\infty d\ell_E \frac{\ell_E^3}{(\ell_E^2 + \Delta)^2} \tag{F.16}$$

The integral is divergent! We can replace the upper bound by a cutoff Λ to find:

$$\int \frac{d^4\ell}{(2\pi)^4} \frac{1}{(\ell^2 - \Delta)^2} = \frac{i}{8\pi^2} \int_0^\Lambda d\ell_E \frac{\ell_E^3}{(\ell_E^2 + \Delta)^2} = \frac{i}{16\pi^2} \left(-\frac{\Lambda^2}{\Lambda^2 + \Delta} + \ln\left(\Lambda^2 + \Delta\right) - \ln(\Delta) \right)$$
$$\approx \frac{i}{16\pi^2} \left(-1 + \ln\left(\frac{\Lambda^2}{\Delta}\right) \right) \tag{F.17}$$

We can consider a similar integral by changing the exponent $m > 2$ at the denominator, to find simply:

$$\int \frac{d^4\ell}{(2\pi)^4} \frac{1}{(\ell^2 - \Delta)^m} \approx \frac{i(-1)^m}{16\pi^2(m-2)(m-1)\Delta^m} \left(\Delta^2 - \frac{\Delta^m(m-1)}{\Lambda^{2(m-2)}} \right) \approx \frac{i(-1)^m}{16\pi^2(m-2)(m-1)\Delta^{m-2}} \tag{F.18}$$

F.4   Feynman Integrals

In order to compute higher-order loops in a Feynman diagram where the four-momentum in the loop must be integrated upon, there are clever tricks that were invented in order to solve them. The idea is to find ways to regroup the product of propagators in the denominator into a single term, however, at the cost of additional integrations. We list a set of useful formulas. We begin by noting that:

$$\int_a^b \frac{\mathrm{d}z}{z^2} = -\frac{1}{z}\Big|_a^b = \frac{1}{a} - \frac{1}{b} \quad \Longrightarrow \quad \frac{1}{b-a}\int_a^b \frac{\mathrm{d}z}{z^2} = \frac{\frac{1}{a}-\frac{1}{b}}{b-a} = \frac{1}{ab} \tag{F.19}$$

Setting $z = ax + b(1-x)$ such that $z = a$ for $x = 1$ and $z = b$ for $x = 0$ and $\mathrm{d}z = (a-b)\mathrm{d}x$ leads to the identity:

$$\int_0^1 \mathrm{d}x \frac{1}{(ax+b(1-x))^2} = \frac{1}{ab} \tag{F.20}$$

and by differentiation with respect to a one immediately finds:

$$\int_0^1 \mathrm{d}x \frac{2x}{(ax+b(1-x))^3} = \frac{1}{a^2 b} \tag{F.21}$$

The generalized version of Eq. (F.20) goes as follows:

$$\int_0^1 \mathrm{d}x_1 \mathrm{d}x_2 \ldots \mathrm{d}x_n \delta(x_1 + x_2 + \cdots + x_n - 1)\frac{(n-1)!}{(x_1 a + x_2 b + \cdots)^n} = \frac{1}{abc\ldots} \tag{F.22}$$

(which one can prove when one has 3 minutes of spare time). This general expression can, for instance, be used to find:

$$\int_0^1 \mathrm{d}x_1 \mathrm{d}x_2 \mathrm{d}x_3 \delta(x_1 + x_2 + x_3 - 1)\frac{2}{(x_1 a + x_2 b + x_3 c)^3} = \frac{1}{abc} \tag{F.23}$$

F.5   Four-Dimensional Integrals

Once the denominators of products of propagators have been combined into a single term at a given power by the so-called **Feynman parameter technique**, one can perform the space-time integration first and then integrate over the parameters. To perform the space-time integration, we can find a set of useful integrals that will be forthcoming.

- **[FI1]**: Starting from Eq. (F.18), we can write:

$$\int \frac{\mathrm{d}^4\ell}{(2\pi)^4}\frac{1}{(\ell^2 - \Delta + i\epsilon)^3} = \frac{-i}{32\pi^2 \Delta} \tag{F.24}$$

- **[FI2]**: Now we consider the vector variant. Since we are integrating over the full phase space, symmetry will ensure that the following integral vanishes:

$$\int \frac{\mathrm{d}^4\ell}{(2\pi)^4}\frac{\ell^\mu}{(\ell^2 - \Delta + i\epsilon)^3} = 0 \tag{F.25}$$

- **[FI3]**: By shifting $\ell \to k - p$ and $\Delta \to -s + p^2$, we have:

$$\int \frac{\mathrm{d}^4 k}{(2\pi)^4}\frac{1}{(k^2 - 2k\cdot p + s + i\epsilon)^3} = \frac{i}{32\pi^2(s-p^2)} \tag{F.26}$$

- **[FI4]**: With $k^\mu = \ell^\mu + p^\mu$, and using **[FI2]**:

$$\int \frac{\mathrm{d}^4 k}{(2\pi)^4} \frac{k^\mu}{(k^2 - 2k \cdot p + s + i\epsilon)^3} = \frac{-ip^\mu}{32\pi^2(-s + p^2)} = \frac{ip^\mu}{32\pi^2(s - p^2)} \tag{F.27}$$

- **[FI5]**: We consider a related integral with an exponential term. For the scalar form, we obtain using Wick's rotation with $d^4\ell = id^4\ell_E$ and $\ell^2 = -\ell_E^2$:

$$
\begin{aligned}
\int \frac{\mathrm{d}^4 \ell}{(2\pi)^4} e^{i\ell^2 z} &= i \int \frac{\mathrm{d}^4 \ell_E}{(2\pi)^4} e^{-i\ell_E^2 z} = i \int \frac{\mathrm{d}\ell_E \ell_E^3 \mathrm{d}\Omega_4}{(2\pi)^4} e^{-i\ell_E^2 z} = i \frac{2\pi^2}{(2\pi)^4} \int_0^\infty \mathrm{d}\ell_E \ell_E^3 e^{-i\ell_E^2 z} = i \frac{2\pi^2}{(2\pi)^4} \frac{1}{2i^2 z^2} \\
&= \frac{1}{16\pi^2 i z^2}
\end{aligned}
\tag{F.28}
$$

- **[FI6]**: For the vector form, we note that it will vanish by symmetry (see **[FI2]**):

$$\int \frac{\mathrm{d}^4 \ell}{(2\pi)^4} \ell^\mu e^{i\ell^2 z} = 0 \tag{F.29}$$

- **[FI7]**: To solve the tensor variant, we should first realize that the integral will vanish by symmetry with $\mu \neq \nu$, since the indices can be interchanged. For the cases where $\mu = \nu$, Lorentz covariance implies that the result must be proportional to $g^{\mu\nu}$. It can be written as:

$$\int \frac{\mathrm{d}^4 \ell}{(2\pi)^4} \ell^\mu \ell^\nu e^{i\ell^2 z} = C g^{\mu\nu} = \frac{1}{32\pi^2 i z^2} \frac{ig^{\mu\nu}}{z} \tag{F.30}$$

where to find C one considers the equation multiplied by $g_{\mu\nu}$:

$$g_{\mu\nu} \int \frac{\mathrm{d}^4 \ell}{(2\pi)^4} \ell^\mu \ell^\nu e^{i\ell^2 z} = C g_{\mu\nu} g^{\mu\nu} \tag{F.31}$$

hence, using a Wick rotation as before:

$$C = \frac{1}{4} \int \frac{\mathrm{d}^4 \ell}{(2\pi)^4} \ell^2 e^{i\ell^2 z} = -\frac{1}{4} \frac{i2\pi^2}{(2\pi)^4} \int_0^\infty \mathrm{d}\ell_E \ell_E^5 e^{-i\ell_E^2 z} = -\frac{1}{4} \frac{i2\pi^2}{(2\pi)^4} \frac{1}{i^3 z^3} = \frac{1}{32\pi^2} \frac{i}{iz^3} \tag{F.32}$$

- **[FI8]**: In a similar manner, we also compute:

$$\int \frac{\mathrm{d}^4 \ell}{(2\pi)^4} \ell^2 e^{i\ell^2 z} = -i \int \frac{\mathrm{d}^4 \ell_E}{(2\pi)^4} (\ell_E^2) e^{-i\ell_E^2 z} = -i \int \frac{\mathrm{d}\ell_E \ell_E^3 \ell_E^2 \mathrm{d}\Omega_4}{(2\pi)^4} e^{-i\ell_E^2 z} = \frac{-i2\pi^2}{16\pi^4} \frac{1}{i^3 z^3} = \frac{i}{8\pi^2} \frac{1}{iz^3} \tag{F.33}$$

F.6 D-Dimensional Integrals

Dimensional regularization is a very elegant technique to deal with the divergences of the space-time integrals encountered in higher-order calculations. Following the technique valid in four dimensions, we can first simplify the denominator of a product of propagators by introducing Feynman parameters. Then, combine the denominators into a single term of the form $(\ell^2 - \Delta)^m$ by shifting the integration variable to ℓ^μ, where m is the power that depends on the number of propagators we started with. The numerator will then be rewritten in terms of the new shifted momentum ℓ^μ. Once the integral has been put in the generic form of integration over a momentum $d^D\ell$ in D dimensions with the mth power of a typical propagator at the denominator, we can use the result [96]:

$$I_m(\Delta) = \int \frac{\mathrm{d}^D \ell}{(2\pi)^D} \frac{1}{(\ell^2 - \Delta)^m} = i \frac{(-1)^m}{2^D \pi^{D/2}} \frac{\Gamma(m - D/2)}{\Gamma(m)} \frac{1}{\Delta^{m-D/2}} \tag{F.34}$$

where we can see that we recover Eq. (F.24) for $D = 4$ and $m = 3$. When the integral converges, one can use directly $D = 4$. Otherwise, setting $D = 4 - 2\epsilon$, we can study the behavior near $D = 4$ by expanding in ϵ. An important relation for $m = 2$ is:

$$\frac{\Gamma(2 - D/2)}{2^D \pi^{D/2}} \frac{1}{\Delta^{2-D/2}} \approx \frac{1}{4\pi^2} \left(\frac{1}{\epsilon} - \ln(\Delta) - \gamma_E + \ln(4\pi) \right) \tag{F.35}$$

Similarly, we note (cf. Eq. (F.25)):

$$I_m^\mu(\Delta) = \int \frac{\mathrm{d}^D \ell}{(2\pi)^D} \frac{\ell^\mu}{(\ell^2 - \Delta)^m} = 0 \tag{F.36}$$

In general, the terms in the numerator with an odd number of powers of ℓ^μ will vanish in the integration. And one also has [96]:

$$I_m^{\mu\nu}(\Delta) = \int \frac{\mathrm{d}^D \ell}{(2\pi)^D} \frac{\ell^\mu \ell^\nu}{(\ell^2 - \Delta)^m} = i \frac{(-1)^{m-1}}{2^D \pi^{D/2}} \frac{g^{\mu\nu}}{2} \frac{\Gamma(m - D/2 - 1)}{\Gamma(m)} \frac{1}{\Delta^{m-D/2-1}} \tag{F.37}$$

which implies, multiplying by $g_{\mu\nu}$, that:

$$I_m^2(\Delta) = \int \frac{\mathrm{d}^D \ell}{(2\pi)^D} \frac{\ell^2}{(\ell^2 - \Delta)^m} = i \frac{(-1)^{m-1}}{2^D \pi^{D/2}} \frac{D}{2} \frac{\Gamma(m - D/2 - 1)}{\Gamma(m)} \frac{1}{\Delta^{m-D/2-1}} \tag{F.38}$$

Appendix G   Statistics

G.1   The Poisson Distribution

Let us consider a process that has some constant probability of occurring per unit time. The average number of occurring processes (or events) per unit time is defined as μ. We divide the time interval into a (large) arbitrary number M of equal time slices. The probability of the process occurring in any one of these M slices is μ/M and accordingly the probability of the process *not* occurring in the slice is $1 - \mu/M$. Therefore, the probability of the process occurring in k and only k slices is given by the Bernoulli distribution:

$$P_0(k; \mu, M) = \left(\frac{\mu}{M} \right)^k \left(1 - \frac{\mu}{M} \right)^{M-k} \tag{G.1}$$

We should now consider all possible combinations in which k events can occur in the time interval. This is equivalent to the number of combinations taking k objects at a time out of a total of M objects:

$$\frac{M(M-1)(M-2)\ldots(M-k+1)}{k!} = \frac{M!}{k!(M-k)!} \tag{G.2}$$

By multiplying the above two results, we obtain the binomial distribution:

$$P_M(k; \mu) = \frac{M!}{k!(M-k)!} \left(\frac{\mu}{M} \right)^k \left(1 - \frac{\mu}{M} \right)^{M-k} \tag{G.3}$$

which is conventionally expressed using $p = \mu/M$ and $q = 1 - p$:

$$P_M(k) = \frac{M!}{k!(M-k)!} p^k q^{M-k} \tag{G.4}$$

This result is interpreted in the following more general way: a random process can have two outcomes A or B which are mutually exclusive. The probability of A occurring in a trial is given by p and the probability of B occurring is then $1 - p = q$. The probability in M trials of obtaining k outcomes A and $M - k$ outcomes B is then given by $P_M(k)$.

    We now note that the binomial expansion of the polynomial $(p + q)^M$ is just:

$$(p+q)^M = \sum_{k=0}^{M} \frac{M!}{k!(M-k)!} p^k q^{M-k} = (p + (1-p))^M = 1 \tag{G.5}$$

hence the probabilities are properly normalized:

$$\sum_{k=0}^{M} P_M(k) = 1 \tag{G.6}$$

We now take the limit for $M \to \infty$ and use the mathematical formula:

$$\lim_{M \to \infty} \frac{M!}{k!(M-k)!} = \frac{M^k}{k!} \quad \text{and} \quad \lim_{M \to \infty} \left(1 - \frac{\mu}{M}\right)^{M-\mu} = e^{-\mu} \tag{G.7}$$

to find the **Poisson**[1] **formula**:

$$P(k; \mu) = \frac{\mu^k}{k!} e^{-\mu} \tag{G.8}$$

We note that the average number of events can also be written as (going back to the discrete case for a moment):

$$\mu = \sum_{k=0}^{M} k P_M(k) = \sum_{k=0}^{M} k \frac{M!}{k!(M-k)!} p^k q^{M-k} \tag{G.9}$$

And on the other hand, we can use this trick:

$$\frac{\partial}{\partial p}(p+q)^M = \sum_{k=0}^{M} k \frac{M!}{k!(M-k)!} p^{k-1} q^{M-k} \implies \mu = p \frac{\partial}{\partial p}(p+q)^M = pM(p+q)^{M-1} = pM \tag{G.10}$$

since $p + q = 1$. We now estimate the **variance**, which is defined as the average value of $(k - \mu)^2$ or $\sigma^2 \equiv \langle (k - \mu)^2 \rangle$, where σ is the standard deviation. We note:

$$\sigma^2 = \langle (k - \mu)^2 \rangle = \langle (k - pM)^2 \rangle = \langle k^2 \rangle - 2pM \underbrace{\langle k \rangle}_{=pM} + p^2 M^2 = \langle k^2 \rangle - p^2 M^2 \tag{G.11}$$

or simply $\sigma^2 = \langle k^2 \rangle - (\langle k \rangle)^2$. Then

$$
\begin{aligned}
\langle k^2 \rangle - p^2 M^2 &= \sum_{k=0}^{M} k^2 \frac{M!}{k!(M-k)!} p^k q^{M-k} - p^2 M^2 = \left(p^2 \frac{\partial^2}{\partial p^2} + p \frac{\partial}{\partial p}\right) \sum_{k=0}^{M} \frac{M!}{k!(M-k)!} p^k q^{M-k} - p^2 M^2 \\
&= \left(p^2 \frac{\partial^2}{\partial p^2} + p \frac{\partial}{\partial p}\right)(p+q)^M - p^2 M^2 = M(M-1)p^2 + Mp - p^2 M^2 \\
&= -Mp^2 + Mp = Mp(-p+1) = Mpq
\end{aligned} \tag{G.12}
$$

so

$$\sigma = \sqrt{Mpq} = \sqrt{M(\mu/M)(1 - \mu/M)} = \sqrt{\mu(1 - \mu/M)} \tag{G.13}$$

In the limit $M \to \infty$ we obtain that the variance is equal to the mean:

$$\sigma^2 = \mu \tag{G.14}$$

---

[1]   Siméon Denis Poisson (1781–1840), French mathematician, engineer, and physicist.

Appendix H   Monte-Carlo Techniques

H.1   Basic Notions

Monte-Carlo (MC) techniques offer a way of calculating complex integrals that may not be tractable analytically or by ordinary numerical interpolation methods (see, e.g., Ref. [482]). The MC method is primarily used when the number of integration variables becomes large. As an example, by standard integration methods, the d-dimensional integration with 10 interpolation points per independent variable of the integrand requires 10^d evaluations. In contrast, the n-body phase space routinely used in kinematics (see Chapter 5) has $(3n - 4)$ dimensions and therefore one quickly approaches the practical limit when $n \geq 4$. Other cases involve folding complicated factors, such as complex integration boundaries, representing for example particle detector effects like acceptances or efficiencies.

MC techniques rely on **random sampling**. An integral over the range (a, b) can be written as:

$$I \equiv \int_a^b f(x)\mathrm{d}x \tag{H.1}$$

Without loss of generality, we can set $a = 0$ and $b = 1$. The reason for this is that most computer languages have a built-in function which returns a **pseudo-random number** uniformly distributed in the interval $[0, 1)$. For example, in Python/Numpy, one can generate $N = 10$ samples with:

```
>>> import numpy as np
>>> np.random.rand(10)
array([ 0.66587209,  0.91315361,  0.23806812,  0.97277416,  0.55758549,
        0.38613788,  0.83831719,  0.36918088,  0.10571988,  0.6919927 ])
```

The MC techniques rely on the fact that the integral over a given range can be approximated by sampling the integrand N times and taking the average. Consequently:

$$\int_0^1 f(x)\mathrm{d}x = \langle f \rangle \approx \frac{1}{N} \sum_{i=1}^N f(x(i)) \tag{H.2}$$

where $\langle f \rangle$ is the mean value of the function in the given interval, and $x(i)$ is the ith sample of a random variable x uniformly distributed between $[0, 1)$. The first term is just the mathematical definition of the mean of the function, while the second term is its statistical estimation. See Figure H.1.

This result can be generalized to an n-fold integration of the n-dimensional function $f = f(x_1, x_2, \ldots, x_n)$ by taking a set of variables $x_1, \ldots, x_n$ randomly distributed between $[0, 1)$:

$$\underbrace{\int_0^1 f(x_1, x_2, \ldots, x_n)\mathrm{d}x_1\mathrm{d}x_2 \ldots \mathrm{d}x_n}_{=\langle f \rangle} \approx \underbrace{\frac{1}{N} \sum_{i=1}^N f(x_1(i), \ldots x_n(i))}_{=f_N} \tag{H.3}$$

For example, the code to generate $N = 10$ samples with $n = 3$ variables would give for the array of $(x_1(i), x_2(i), x_3(i))$ $(i = 1, 10)$:

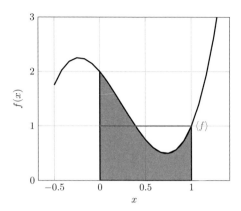

Figure H.1 Illustration of the mean value of $f(x)$. Here we set $f(x) = 4x^3 - 3x^2 - 2x + 2$, so $F(x) = \int f(x)\mathrm{d}x = x^4 - x^3 - x^2 + 2x + C$ and $F(1) - F(0) = 1$. The mean value of $f(x)$ over the interval is therefore $\langle f \rangle = 1$, as illustrated by the straight line on the graph.

```
>>> import numpy as np
>>> a=np.random.rand(3,10)
>>> a           # dump x_1(i),x_2(i), x_3(i)
array([[ 0.83821063,  0.81616484,  0.56310384,  0.48539242,  0.52897497,
         0.60054432,  0.86523752,  0.31800811,  0.94706172,  0.02790464],
       [ 0.00358234,  0.61912253,  0.35881049,  0.95247522,  0.09212247,
         0.23318401,  0.1600166 ,  0.55794396,  0.09443015,  0.28902217],
       [ 0.18324684,  0.31287035,  0.60206872,  0.46079945,  0.79117349,
         0.90706786,  0.51824877,  0.85446816,  0.88821168,  0.18442265]])
>>> a[0]        # dump x_1(i)
array([ 0.83821063,  0.81616484,  0.56310384,  0.48539242,  0.52897497,
        0.60054432,  0.86523752,  0.31800811,  0.94706172,  0.02790464])
>>> a[1]        # dump x_2(i)
array([ 0.00358234,  0.61912253,  0.35881049,  0.95247522,  0.09212247,
        0.23318401,  0.1600166 ,  0.55794396,  0.09443015,  0.28902217])
>>> a[2]        # dump x_3(i)
array([ 0.18324684,  0.31287035,  0.60206872,  0.46079945,  0.79117349,
        0.90706786,  0.51824877,  0.85446816,  0.88821168,  0.18442265])
```

By implementing the **Central Limit Theorem**, one expects f_N to tend to the exact value $\langle f \rangle$ as $N \to \infty$. Thus, for a reasonably large N, f_N is a Gaussian statistical estimator for the mean value with a mean $\mu = \langle f \rangle$ and a variance $\sigma^2 = \sigma_1^2/N$. Therefore, the error in estimating $\langle f \rangle$ by MC scales like $1/\sqrt{N}$. For example, if one wishes to estimate the integral of the inverse cosine (arccos):

$$\int_0^1 \arccos(x)\mathrm{d}x = \left(x \arccos(x) - \sqrt{1 - x^2} \right)\Big|_0^1 = 1 \tag{H.4}$$

the convergence can be estimated by increasing the number of samples, e.g.:

```
>>> import numpy as np
>>> a=np.random.rand(10)
>>> numpy.average(numpy.arccos(a))
1.0859304851526712           # exact result is 1.0
>>> a=np.random.rand(1000)
```

```
>>> numpy.average(numpy.arccos(a))
1.0036057671058796              # exact result is 1.0
>>> a=np.random.rand(100000)
>>> numpy.average(numpy.arccos(a))
0.99960971019345379             # exact result is 1.0
```

Although this convergence rate might seem modest, the idea of the MC calculation is that N can be effectively very large since the only computation needed is the value of $f = f(x_1, x_2, \ldots, x_n)$ for each set of randomly distributed points. In practice, it can easily be taken as $N = 10^5, 10^6, \ldots$, as shown above. The main advantage of MC integration is the fact that the error estimate is independent of the dimension d of the integral.

Intuitively it is easy to realize that the MC approximation is exact for $f(x) =$ constant. This illustrates the general principle that **the approximation is better when the function is "flatter" or "smoother."** On the contrary, a very "wiggly" function is more difficult to integrate since the "random" x_i might miss some peaks or valleys. Hence, MC techniques perform badly for singular or near-singular integrands (even if integrable) as they often occur in our practical cases (e.g., differential cross-sections diverge in a particular area of phase space). For the same reason, discontinuities of $f(x)$ create problems and so do Dirac δ-functions! These latter must be integrated beforehand and the other problems can be tackled by optimizing the MC, as discussed in the following.

We can estimate the accuracy of the MC calculation achieved with a given number N of trials by comparing independent "runs." This means, for example, computing the result with $N = 10^5$ and computing it again with the same number $N = 10^5$ but for a *different* set of $x_j(i)$.

```
>>> a=np.random.rand(100000)
>>> numpy.average(numpy.arccos(a))
1.0012119311266925
>>> a=np.random.rand(100000)
>>> numpy.average(numpy.arccos(a))
1.0006693701921081
```

The difference between the two samples here is at the level of 5×10^{-4}, affected by the random fluctuations in the samples.

One can also directly calculate the **variance of the estimator** by accumulating the statistical averages of the function $\langle f \rangle$ and its square $\langle f^2 \rangle$. After N trials, the variance of a single trial is estimated by:

$$\sigma_1^2 \approx \frac{1}{N}\sum_{i=1}^{N}(f(x_i) - \langle f \rangle)^2 = \frac{1}{N}\sum_{i=1}^{N}f(x_i)^2 - 2\langle f \rangle \underbrace{\frac{1}{N}\sum_{i=1}^{N}f(x_i)}_{=\langle f \rangle} + \underbrace{\frac{1}{N}\sum_{i=1}^{N}(\langle f \rangle)^2}_{=N(\langle f \rangle)^2}$$

$$= \langle f^2 \rangle - 2(\langle f \rangle)^2 + (\langle f \rangle)^2 = \langle f^2 \rangle - (\langle f \rangle)^2 \tag{H.5}$$

and the accuracy of the MC integration with N samples is then given by:

$$\sigma_N \approx \frac{\sigma_1}{\sqrt{N}} \tag{H.6}$$

These values are themselves statistical estimators with their own errors. Using our example above, we get:

```
>>> nsamples=100000
>>> a=np.random.rand(nsamples)
>>> sigma1=(numpy.average(numpy.arccos(a)*numpy.arccos(a)))
                        -numpy.power(numpy.average(numpy.arccos(a)),2)
>>> sigma1
0.14282591866369243
>>> sigmaN=sigma1/math.sqrt(float(nsamples))
>>> sigmaN
0.00045165521188322052
```

This result coincides with our previous estimate of the uncertainty of the integration at the level of 5×10^{-4}.

The above arguments point to possible improvements or optimizations, which generally lead to faster convergence and greater accuracy: the basic idea is the *reduction of the variance* of the function over the chosen interval.

- **Stratified sampling.** For the one-dimensional case, we understand intuitively that the random fluctuations in the MC integration are reduced if the integrand is more uniform. One can achieve this by subdividing the full range into equal sub-ranges (k cells) and choosing an equal number of samples randomly within each cell. For the n-dimensional case, it is still possible but more complicated as one has to deal with k^n hypercubes, so in practice k cannot be too large. A yet-better way is to subdivide the range into unequally sized hypercubes, such that, e.g., the variance is equal in each hypercube. In this case, however, one has to take care of the book-keeping such that the hypercubes are sampled according to their sizes.

- **Importance sampling.** Another intuitive solution is to sample more often the regions where the integrand is largest, as these are the ones which we expect to contribute most to the integral. One can also alternatively change variables of integration such that the integrand plotted as a function of the new variables is much flatter than the original function.

H.2   Event Generation

A Monte-Carlo integration can be seen as a direct simulation of what happens in Nature. If we consider a cross-section of a scattering process or a decay width (see Chapter 5), then we will be confronted with the multi-dimensional integration of the phase space. Mathematically the result is an integration of a weight function $f = f(x_1, x_2, \ldots, x_n)$ where the variables $x_1, x_2, \ldots, x_n$ parametrize the physical configurations (e.g., the momenta of the final-state particles in a given reaction).

The MC simulation calculates the integral (i.e., the total cross-section or the decay width within the allowed phase space) by generating random configurations ("events") and averaging the integrand. One can see the set of events as a set of "real events." At the end of the simulation, this set can be compared directly to a set of events collected in an actual experiment. Any interesting observable quantity (e.g., total visible energy, etc.) can be computed equally for both sets (real and simulated) and compared. The simulation is thus akin to collecting data in a Gedanken experiment. Of course, assumptions can be changed to study their effect on observables. Furthermore, sets of simulations for given input parameters ("templates") can be compared to real data. In this case, the input parameters can be extracted from, or fitted to, the data.

- **Weighted and unweighted events.** There is a priori one subtle difference between the simulation and real data. The MC simulation generates events that are uniformly random in the $(x_1, x_2, \ldots, x_n)$ coordinates, and for each configuration the contribution is given by its "weight," which is just $f(x_1, x_2, \ldots, x_n)$. These are called weighted events as they do not really occur with the same frequency as events in the real data. Since physicists are lazy and do not want to carry weights (pun!), modern MC event generators simulate unweighted events, which means that the frequency of events in their generated sample is what occurs in real data (in other words, each event in the output has a weight of 1).

- **Hit-or-miss (acceptance–rejection) method.** The simplest way to convert weighted events to unweighted events is to generate many more weighted events and reject a fraction of them based on their weight (namely, less probable events will be rejected more often than more probable ones). This is usually implemented with the hit-or-miss algorithm. For simplicity, in one dimension:

  1. first find the maximum weight f_{max} over the range;

  2. during event generation, pick a random configuration x_i;

  3. generate a second random number y in the range $[0, 1)$;

4. if $f(x_i)/f_{max} > y$ then accept the event, otherwise reject it and start again.

This is illustrated in Figure H.2.

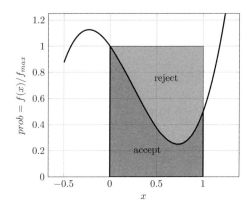

Figure H.2 Illustration of a hit-or-miss algorithm: the probability of accepting an unweighted event is given by $f(x)/f_{max}$.

Textbooks

- *Collider Physics*, Vernon D. Barger and Roger J. N. Phillips, "Frontiers in physics," Addison-Wesley Publishing Company; 1st edition (1996). ISBN 978-0201149456.

- *Relativistic Quantum Mechanics*, J. D. Bjorken and S. D. Drell, "International series in pure and applied physics," MCGraw-Hill Book Company; 1st edition (1964).

- *The Principles of Quantum Mechanics*, P. A. M. Dirac (International Series of Monographs on Physics), Clarendon Press; 4th edition (1982). ISBN 978-0198520115.

- *Elementary Particle Physics*, S. Gasiorowicz, John Wiley & Sons Inc. (1966). ISBN 978-0471292876.

- *Introduction to Elementary Particles*, D. Griffiths, Wiley VCH; 2nd revised edition (2008). ISBN 978-3527406012.

- *Quarks and Leptons: Introductory Course in Modern Particle Physics*, F. Halzen, A. D. Martin, John Wiley & Sons; International edition (1984). ISBN 978-0471811879.

- *Introductory Nuclear Physics*, K. S. Krane, John Wiley & Sons (1988). ISBN 978-0471805533.

- *Techniques for Nuclear and Particle Physics Experiments*, William R. Leo, Springer-Verlag (1994). ISBN 978-3540572800.

- *Gauge Theories of Strong, Weak, and Electromagnetic Interactions*, C. Quigg, Advanced Book Classics, Westview Press (1997). ISBN 978-0201328325.

- Class notes *Quantum Field Theory*, V. Kaplunovsky (UTA), http://bolvan.ph.utexas.edu/ vadim/class_pages.html

- *Leptons and Quarks*, L. B. Okun, North Holland (1982). ISBN 978-0444869241.

- *Introduction to High Energy Physics*, D. H. Perkins, Cambridge University Press; 4th edition (2000). ISBN 978-0521621960.

- *An Introduction to Quantum Field Theory*, M. E. Peskin and D. V. Schroeder, Perseus Books Publishing, 1st edition (1995). ISBN 978-0201503975.

- *Electroweak Interactions: An Introduction to the Physics of Quarks and Leptons*, P. Renton, Cambridge University Press; 1st edition (1990). ISBN 978-0521266031.

- *Quantum Field Theory*, L. H. Ryder, Cambridge University Press; 2nd edition (2008). ISBN 978-0521749091.

- *Advanced Quantum Mechanics*, J. J. Sakurai, Pearson; 1st edition (1967). ISBN 978-0201067101.

- *Neutrinophysik*, N. Schmitz, Teubner Studienbücher Physik, Vieweg+Teubner Verlag (1997). ISBN 978-3519032366.

- *Modern Particle Physics*, M. Thomson, Cambridge University Press (2013). ISBN 978-1107034266.

- *The Quantum Theory of Fields*, S. Weinberg, Cambridge University Press, 3 volumes (2005). ISBN 978-0521670562.

- *Quantum Field Theory*, J.-B. Zuber and C. Itzykson, Dover Books on Physics (2006). ISBN 978-0486445687.

References

[1] E. P. Wigner, "On unitary representations of the inhomogeneous Lorentz group," *Annals Math.*, vol. 40, pp. 149–204, 1939.

[2] A. Einstein, "Zur Elektrodynamik bewegter Körper," *Ann. Phys.*, vol. 322, no. 10, pp. 891–921, 1905.

[3] A. Einstein, "Ist die Trägheit eines Körpers von seinem Energieinhalt abhängig?," *Ann. Phys.*, vol. 323, no. 13, pp. 639–641, 1905.

[4] D. B. Newell *et al.*, "The CODATA 2017 values of h, e, k, and n_a for the revision of the SI," *Metrologia*, vol. 55, no. 1, pp. L13–L16, 2018.

[5] P. Zyla *et al.*, "Review of particle physics," *PTEP*, vol. 2020, no. 8, p. 083C01, 2020.

[6] W. Pauli, "The connection between spin and statistics," *Phys. Rev.*, vol. 58, pp. 716–722, 1940.

[7] De Broglie, L., "Recherches sur la théorie des Quanta," *Ann. Phys.*, vol. 10, no. 3, pp. 22–128, 1925.

[8] E. Noether, "Invariante Variationsprobleme," *Nachrichten von der Gesellschaft der Wissenschaften zu Göttingen, Mathematisch-Physikalische Klasse*, vol. 1918, pp. 235–257, 1918.

[9] G. Lüders, "Proof of the TCP theorem," *Ann. Phys.*, vol. 2, no. 1, pp. 1–15, 1957.

[10] P. J. Mohr, D. B. Newell, and B. N. Taylor, "CODATA recommended values of the fundamental physical constants: 2014," *Rev. Mod. Phys.*, vol. 88, no. 3, p. 035009, 2016.

[11] H. Cavendish, "XIX. Three papers, containing experiments on factitious air," *Philos. Trans.*, vol. 56, pp. 141–184, 1766.

[12] J. Balmer, "Notiz über die Spektrallinien des Wasserstoffs," *Ann. Phys. Chem.*, vol. 25, pp. 80–85, 1885.

[13] L. J. Boya, "The thermal radiation formula of Planck (1900)," *arXiv e-prints*, p. physics/0402064, Feb. 2004.

[14] W. C. Röntgen, "Über eine neue Art von Strahlen," *Ann. Phys.*, vol. 300, no. 1, pp. 12–17, 1898.

[15] H. Geiger and E. Marsden, "On a diffuse reflection of the α-particles," *Proc. Roy. Soc. Lond., Ser. A*, vol. 82, no. 557, pp. 495–500, 1909.

[16] H. Geiger and E. Marsden, "The laws of deflexion of α particles through large angles," *Philos. Mag.*, vol. 25, no. 148, pp. 604–623, 1913.

[17] J. D. Jackson, *Classical Electrodynamics*. Wiley, 2 ed., 1975.

[18] A. H. Compton, "A quantum theory of the scattering of x-rays by light elements," *Phys. Rev.*, vol. 21, pp. 483–502, 1923.

[19] A. H. Compton, "The spectrum of scattered x-rays," *Phys. Rev.*, vol. 22, pp. 409–413, 1923.

[20] W. Bothe and H. Becker, "Künstliche Erregung von Kern-γ-Strahlen," *Z. Phys.*, vol. 66, no. 5, pp. 289–306, 1930.

[21] I. Curie and F. Joliot, "Effet dabsorption de rayons γ de très haute fréquence par projection de noyaux légers," *C. R. Acad. Sci. Paris*, vol. 174, p. 708, 1932.

[22] J. Chadwick, "Possible existence of a neutron," *Nature*, vol. 129, p. 312, 1932.

[23] D. Craig and T. Thirunamachandran, *Molecular Quantum Electrodynamics: An Introduction to Radiation-molecule Interactions*. Dover Books on Chemistry Series, Dover Publications, 1998.

[24] H. Bethe, "Zur Theorie des Durchgangs schneller Korpuskularstrahlen durch Materie," *Ann. Phys.*, vol. 397, no. 3, pp. 325–400.

[25] H. Bethe, "Bremsformel für Elektronen relativistischer Geschwindigkeit," *Z. Phys.*, vol. 76, no. 5, pp. 293–299, 1932.

[26] F. Bloch, "Zur Bremsung rasch bewegter Teilchen beim Durchgang durch Materie," *Ann. Phys.*, vol. 408, no. 3, pp. 285–320.

[27] C. Møller, "Zur Theorie des Durchgangs schneller Elektronen durch Materie," *Ann. Phys.*, vol. 406, no. 5, pp. 531–585.

[28] C. Patrignani *et al.*, "Review of particle physics," *Chin. Phys.*, vol. C40, no. 10, p. 100001, 2016.

[29] H. Bethe, "Moliere's theory of multiple scattering," *Phys. Rev.*, vol. 89, pp. 1256–1266, 1953.

[30] G. R. Lynch and O. I. Dahl, "Approximations to multiple Coulomb scattering," *Nucl. Instrum. Meth. B*, vol. 58, pp. 6–10, 1991.

[31] M. Chadwick *et al.*, "Endf/b-vii.1 nuclear data for science and technology: Cross sections, covariances, fission product yields and decay data," *Nuclear Data Sheets*, vol. 112, no. 12, pp. 2887–2996, 2011. Special Issue on ENDF/B-VII.1 Library.

[32] R. Brun, R. Hagelberg, M. Hansroul, and J. C. Lassalle, *Simulation Program for Particle Physics Experiments, GEANT: User Guide and Reference Manual.* Geneva: CERN, 1978.

[33] S. Agostinelli *et al.*, "GEANT4 – a simulation toolkit," *Nucl. Instrum. Meth. A*, vol. 506, pp. 250–303, 2003.

[34] G. Ising, "Prinzip einer Methode zur Herstellung von Kanalstrahlen hoher Voltzahl," *Arkiv för Matematik, Astronomi och Fysik*, vol. 18, no. 30, pp. 1–4, 1924.

[35] *LEP Design Report.* Geneva: CERN, 1984. Copies shelved as reports in LEP, PS, and SPS libraries.

[36] I. Levine *et al.*, "Measurement of the electromagnetic coupling at large momentum transfer," *Phys. Rev. Lett.*, vol. 78, pp. 424–427, 1997.

[37] R. Assmann *et al.*, "Calibration of center-of-mass energies at LEP-1 for precise measurements of Z properties," *Eur. Phys. J.*, vol. C6, pp. 187–223, 1999.

[38] M. Woods, "The polarized electron beam for the SLAC linear collider," in *12th International Symposium on high-energy Spin Physics (SPIN 96)*, pp. 623–627, 1996.

[39] L. R. Evans and P. Bryant, "LHC Machine," *J. Instrum.*, vol. 3, no. 08, pp. S08001–S08001, 2008. This report is an abridged version of the LHC Design Report (CERN-2004-003).

[40] W. Herr and B. Muratori, "Concept of luminosity," *CERN*, 2006.

[41] C. T. R. Wilson, "On an expansion apparatus for making visible the tracks of ionising particles in gases and some results obtained by its use," *Proc. Roy. Soc. Lond., Ser. A*, vol. 87, no. 595, pp. 277–292, 1912.

[42] W. R. Leo, *Techniques for Nuclear and Particle Physics Experiments.* Springer-Verlag, 1994.

[43] B. Hyams *et al.*, "A silicon counter telescope to study shortlived particles in high-energy hadronic interactions," *Nucl. Instrum. Meth.*, vol. 205, pp. 99–105, 1983.

[44] C. Broennimann *et al.*, "Development of an Indium bump bond process for silicon pixel detectors at PSI," *Nucl. Instrum. Meth. A*, vol. 565, no. 1, pp. 303–308, 2006. Proceedings of the International Workshop on Semiconductor Pixel Detectors for Particles and Imaging.

[45] G. Aad *et al.*, "The ATLAS experiment at the CERN Large Hadron Collider," *J. Instrum.*, vol. 3, no. 08, p. S08003, 2008.

[46] S. Chatrchyan *et al.*, "The CMS experiment at the CERN LHC," *J. Instrum.*, vol. 3, no. 08, p. S08004, 2008.

[47] A. A. Alves, Jr. *et al.*, "The LHCb detector at the LHC," *J. Instrum.*, vol. 3, p. S08005, 2008.

[48] K. Aamodt *et al.*, "The ALICE experiment at the CERN LHC," *J. Instrum.*, vol. 3, p. S08002, 2008.

[49] A. Aab *et al.*, "The Pierre Auger Cosmic Ray Observatory," *Nucl. Instrum. Meth. A*, vol. 798, pp. 172–213, 2015.

[50] M. G. Aartsen *et al.*, "The IceCube Neutrino Observatory: Instrumentation and online systems," *J. Instrum.*, vol. 12, no. 03, p. P03012, 2017.

[51] R. Agnese *et al.*, "Silicon detector dark matter results from the final exposure of CDMS II," *Phys. Rev. Lett.*, vol. 111, no. 25, p. 251301, 2013.

[52] M. Aguilar *et al.*, "The Alpha Magnetic Spectrometer (AMS) on the international space station: Part I results from the test flight on the space shuttle," *Phys. Rep.*, vol. 366, no. 6, pp. 331–405, 2002.

[53] A. Bettini, "The world deep underground laboratories," *Eur. Phys. J. Plus*, vol. 127, no. 9, p. 114, 2012.

[54] Y. Fukuda *et al.*, "The Super-Kamiokande detector," *Nucl. Instrum. Meth. A*, vol. 501, pp. 418–462, 2003.

[55] C. J. Davisson and L. H. Germer, "Reflection of electrons by a crystal of nickel," *Proc. Nat. Acad. Sci.*, vol. 14, no. 4, pp. 317–322, 1928.

[56] C. Jönsson, "Elektroneninterferenzen an mehreren künstlich hergestellten Feinspalten," *Z. Phys.*, vol. 161, no. 4, pp. 454–474, 1961.

[57] W. Heisenberg, "Über quantentheoretische Umdeutung kinematischer und mechanischer Beziehungen," *Z. Phys.*, vol. 33, no. 1, pp. 879–893, 1925.

[58] M. Born, W. Heisenberg, and P. Jordan, "Zur Quantenmechanik. II.," *Z. Phys.*, vol. 35, no. 8–9, pp. 557–615, 1926.

[59] J. Mehra and H. Rechenberg, *The Historical Development of Quantum Theory – Volume 5*. Springer-Verlag, 1987.

[60] W. Gerlach and O. Stern, "Der experimentelle Nachweis der Richtungsquantelung im Magnetfeld," *Z. Phys.*, vol. 9, no. 1, pp. 349–352, 1922.

[61] W. Pauli, "Zur Quantenmechanik des magnetischen Elektrons," *Z. Phys.*, vol. 43, no. 9, pp. 601–623, 1927.

[62] R. J. Alder and R. A. Martin, "The electron g-factor and factorization of the Pauli equation," *Am. J. Phys.*, vol. 60, no. 9, pp. 837–839, 1992.

[63] A. Messiah, *Quantum Mechanics (Vols 1-2)*. North-Holland, 1961.

[64] W. Ehrenberg and R. E. Siday, "The refractive index in electron optics and the principles of dynamics," *Proc. Phys. Soc. B*, vol. 62, no. 1, p. 8, 1949.

[65] Y. Aharonov and D. Bohm, "Significance of electromagnetic potentials in the quantum theory," *Phys. Rev.*, vol. 115, pp. 485–491, 1959.

[66] R. G. Chambers, "Shift of an electron interference pattern by enclosed magnetic flux," *Phys. Rev. Lett.*, vol. 5, pp. 3–5, 1960.

[67] B. A. Lippmann and J. Schwinger, "Variational principles for scattering processes. I," *Phys. Rev.*, vol. 79, pp. 469–480, 1950.

[68] T. Sjöstrand, S. Mrenna, and P. Skands, "A brief introduction to PYTHIA 8.1," *Comput. Phys. Commum.*, vol. 178, no. 11, pp. 852–867, 2008.

[69] R. H. Dalitz, "On the analysis of tau-meson data and the nature of the tau-meson," *Phil. Mag. Ser. 7*, vol. 44, pp. 1068–1080, 1953.

[70] F. E. James, "Monte Carlo phase space," CERN, Geneva, 1 May 1968.

[71] R. Kleiss, W. Stirling, and S. Ellis, "A new Monte Carlo treatment of multiparticle phase space at high energies," *Comput. Phys. Commum.*, vol. 40, no. 2, pp. 359–373, 1986.

[72] A. Puig and J. Eschle, "Phasespace: n-body phase space generation in Python," *Journal of Open Source Software*, 2019.

[73] A. A. Alves Junior, "MCBooster: a tool for MC generation for massively parallel platforms." Oct 2016.

[74] P. A. M. Dirac, "The quantum theory of the electron," *Proc. Roy. Soc. Lond., Ser. A*, vol. 117, pp. 610–624, 1928.

[75] C. D. Anderson, "The positive electron," *Phys. Rev.*, vol. 43, pp. 491–494, 1933.

[76] P. Blackett and G. Occhialini, "Some photographs of the tracks of penetrating radiation," *Proc. Roy. Soc. Lond., Ser. A*, vol. 139, no. 839, pp. 699–720, 1933.

[77] E. C. G. Stueckelberg, "Relativistisch invariante Störungstheorie des Diracschen Elektrons I. Teil: Streustrahlung und Bremsstrahlung," *Ann. Phys.*, vol. 413, no. 4, pp. 367–389, 1934.

[78] R. P. Feynman, "The Theory of Positrons," *Phys. Rev.*, vol. 76, pp. 749–759, 1949.

[79] E. P. Wigner, "Normal form of antiunitary operators," *Journal of Mathematical Physics*, vol. 1, no. 5, pp. 409–413, 1960.

[80] G. C. Wick, "The evaluation of the collision matrix," *Phys. Rev.*, vol. 80, pp. 268–272, 1950.

[81] P. A. M. Dirac, *The Principles of Quantum Mechanics*. Comparative Pathobiology – Studies in the Postmodern Theory of Education, Clarendon Press, 1981.

[82] C. V. Raman and S. Bhagavantam, "Experimental proof of the spin of the photon," *Nature*, vol. 129, pp. 22–23, 1932.

[83] N. F. Mott, "The scattering of fast electrons by atomic nuclei," *Proc. Roy. Soc. Lond., Ser. A*, vol. 124, no. 794, pp. 425–442, 1929.

[84] R. Mertig, M. Bohm, and A. Denner, "FEYNCALC: Computer algebraic calculation of Feynman amplitudes," *Comput. Phys. Commun.*, vol. 64, pp. 345–359, 1991.

[85] V. Shtabovenko, R. Mertig, and F. Orellana, "New developments in FeynCalc 9.0," *Comput. Phys. Commun.*, vol. 207, pp. 432–444, 2016.

[86] T. Hahn, "Generating Feynman diagrams and amplitudes with FeynArts 3," *Comput. Phys. Commun.*, vol. 140, pp. 418–431, 2001.

[87] O. Klein and Y. Nishina, "Über die Streuung von Strahlung durch freie Elektronen nach der neuen relativistischen Quantendynamik von Dirac," *Z. Phys.*, vol. 52, no. 11, pp. 853–868, 1929.

[88] W. Pauli and F. Villars, "On the invariant regularization in relativistic quantum theory," *Rev. Mod. Phys.*, vol. 21, pp. 434–444, 1949.

[89] M. Kobel, "Direct measurements of the electromagnetic coupling constant at large Q^2," *Freiburg preprint Freiburg-EHEP*, pp. 97–12, 1997.

[90] I. Levine *et al.*, "Measurement of the electromagnetic coupling at large momentum transfer," *Phys. Rev. Lett.*, vol. 78, pp. 424–427, 1997.

[91] G. Abbiendi *et al.*, "Tests of the standard model and constraints on new physics from measurements of fermion pair production at 189-GeV to 209-GeV at LEP," *Eur. Phys. J.*, vol. C33, pp. 173–212, 2004.

[92] J. H. Kuhn and M. Steinhauser, "A theory driven analysis of the effective QED coupling at M(Z)," *Phys. Lett.*, vol. B437, pp. 425–431, 1998.

[93] A. O. G. Kallen and A. Sabry, "Fourth order vacuum polarization," *Kong. Dan. Vid. Sel. Mat. Fys. Med.*, vol. 29, no. 17, pp. 1–20, 1955.

[94] K. G. Chetyrkin, J. H. Kuhn, and M. Steinhauser, "Heavy quark vacuum polarization to three loops," *Phys. Lett.*, vol. B371, pp. 93–98, 1996.

[95] V. F. Weisskopf, "On the self-energy and the electromagnetic field of the electron," *Phys. Rev.*, vol. 56, pp. 72–85, 1939.

[96] M. E. Peskin and D. V. Schroeder, *An Introduction to Quantum Field Theory*. Reading, MA: Addison-Wesley, 1995.

[97] J. Schwinger, "On quantum-electrodynamics and the magnetic moment of the electron," *Phys. Rev.*, vol. 73, pp. 416–417, 1948.

[98] G. 't Hooft and M. J. G. Veltman, "Regularization and renormalization of gauge fields," *Nucl. Phys.*, vol. B44, pp. 189–213, 1972.

[99] B. Adeva *et al.*, "Electroweak studies in e+ e- collisions: 12 < s**(1/2) < 46.78 GeV," *Phys. Rev. D*, vol. 38, pp. 2665–2678, 1988.

[100] A. Bacala *et al.*, "Measurements of cross-sections and charge asymmetries for $e^+e^- \to \tau^+\tau^-$ and $e^+e^- \to \mu^+\mu^-$ for $\sqrt{s}$ from 52 GeV to 57-GeV," *Phys. Lett. B*, vol. 218, pp. 112–118, 1989.

[101] C. Velissaris *et al.*, "Measurements of cross-section and charge asymmetry for e+ e- —> mu+ mu- and e+ e- —> tau+ tau- at s**(1/2) = 57.8-GeV," *Phys. Lett. B*, vol. 331, pp. 227–235, 1994.

[102] H. U. Martyn, "Test of QED by high-energy electron–positron collisions," *Adv. Ser. Direct. High Energy Phys.*, vol. 7, pp. 92–161, 1990.

[103] F. E. Low, "Heavy electrons and muons," *Phys. Rev. Lett.*, vol. 14, pp. 238–239, 1965.

[104] W. Bartel *et al.*, "Measurement of the processes $e^+e^- \to e^+e^-$ and $e^+e^- \to \gamma\gamma$ at {PETRA}," *Z. Phys. C*, vol. 19, p. 197, 1983.

[105] F. A. Berends and R. Kleiss, "Distributions for electron-positron annihilation into two and three photons," *Nucl. Phys. B*, vol. 186, pp. 22–34, 1981.

[106] F. Mandl and T. H. R. Skyrme, "The theory of the double Compton effect," *Proc. Roy. Soc. Lond., Ser. A*, vol. 215, pp. 497–507, 1952.

[107] F. Berends, P. De Causmaeceker, R. Gastmans, R. Kleiss, W. Troost, and T. T. Wu, "Multiple Bremsstrahlung in gauge theories at high energies: (IV). The process $e^+e^- \to \gamma\gamma\gamma\gamma$," *Nucl. Phys. B*, vol. 239, no. 2, pp. 395–409, 1984.

[108] M. Acciarri *et al.*, "Hard-photon production and tests of QED at LEP," *Phys. Lett. B*, vol. 475, pp. 198–205, 2000.

[109] F. Boudjema, A. Djouadi, and J. L. Kneur, "Excited fermions at e+ e- and e P colliders," *Z. Phys. C*, vol. 57, pp. 425–450, 1993.

[110] M. Nowakowski, E. A. Paschos, and J. M. Rodriguez, "All electromagnetic form factors," *Eur. J. Phys.*, vol. 26, no. 4, pp. 545–560, 2005.

[111] J. E. Nafe, E. B. Nelson, and I. I. Rabi, "The hyperfine structure of atomic hydrogen and deuterium," *Phys. Rev.*, vol. 71, pp. 914–915, 1947.

[112] V. Bargmann, L. Michel, and V. L. Telegdi, "Precession of the polarization of particles moving in a homogeneous electromagnetic field," *Phys. Rev. Lett.*, vol. 2, pp. 435–436, 1959.

[113] G. Breit, "Does the electron have an intrinsic magnetic moment?," *Phys. Rev.*, vol. 72, pp. 984–984, 1947.

[114] P. Kusch and H. M. Foley, "The magnetic moment of the electron," *Phys. Rev.*, vol. 74, pp. 250–263, 1948.

[115] F. J. Dyson, "The radiation theories of Tomonaga, Schwinger, and Feynman," *Phys. Rev.*, vol. 75, pp. 486–502, 1949.

[116] D. T. Wilkinson and H. R. Crane, "Precision measurement of the g factor of the free electron," *Phys. Rev.*, vol. 130, pp. 852–863, 1963.

[117] A. Rich and H. R. Crane, "Direct measurement of the g factor of the free positron," *Phys. Rev. Lett.*, vol. 17, pp. 271–275, 1966.

[118] G. Graff, F. G. Major, R. W. H. Roeder, and G. Werth, "Method for measuring the cyclotron and spin resonance of free electrons," *Phys. Rev. Lett.*, vol. 21, pp. 340–342, 1968.

[119] G. Gräff, E. Klempt, and G. Werth, "Method for measuring the anomalous magnetic moment of free electrons," *Z. Phys. A*, vol. 222, no. 3, pp. 201–207, 1969.

[120] D. Hanneke, S. Fogwell, and G. Gabrielse, "New measurement of the electron magnetic moment and the fine structure constant," *Phys. Rev. Lett.*, vol. 100, p. 120801, 2008.

[121] T. Aoyama, M. Hayakawa, T. Kinoshita, and M. Nio, "Tenth-order electron anomalous magnetic moment: Contribution of diagrams without closed lepton loops," *Phys. Rev. D*, vol. 91, p. 033006, 2015.

[122] R. H. Parker, C. Yu, W. Zhong, B. Estey, and H. Müller, "Measurement of the fine-structure constant as a test of the standard model," *Science*, vol. 360, no. 6385, pp. 191–195, 2018.

[123] T. Aoyama, T. Kinoshita, and M. Nio, "Theory of the anomalous magnetic moment of the electron," *Atoms*, vol. 7, no. 1, p. 28, 2019.

[124] H. Suura and E. H. Wichmann, "Magnetic moment of the mu meson," *Phys. Rev.*, vol. 105, pp. 1930–1931, 1957.

[125] A. Petermann, "Fourth order magnetic moment of electron," *Helv. Phys. Acta*, vol. 30, p. 407, 1957.

[126] T. Coffin, R. L. Garwin, S. Penman, L. M. Lederman, and A. M. Sachs, "Magnetic moment of the free muon," *Phys. Rev.*, vol. 109, pp. 973–979, 1958.

[127] G. Charpak, F. J. M. Farley, R. L. Garwin, T. Muller, J. C. Sens, V. L. Telegdi, and A. Zichichi, "Measurement of the anomalous magnetic moment of the muon," *Phys. Rev. Lett.*, vol. 6, pp. 128–132, 1961.

[128] F. J. M. Farley *et al.*, "The anomalous magnetic moment of the negative muon," *Il Nuovo Cimento A (1965–1970)*, vol. 45, no. 1, pp. 281–286, 1966.

[129] J. Bailey, W. Bartl, G. Von Bochmann, R. Brown, F. Farley, H. Joestlein, E. Picasso, and R. Williams, "Precision measurement of the anomalous magnetic moment of the muon," *Phys. Lett. B*, vol. 28, pp. 287–290, 1968.

[130] R. L. Garwin, L. M. Lederman, and M. Weinrich, "Observations of the failure of conservation of parity and charge conjugation in meson decays: the magnetic moment of the free muon," *Phys. Rev.*, vol. 105, pp. 1415–1417, 1957.

[131] R. L. Garwin, D. P. Hutchinson, S. Penman, and G. Shapiro, "Accurate determination of the μ^+ magnetic moment," *Phys. Rev.*, vol. 118, pp. 271–283, 1960.

[132] G. Charpak, F. Farley, R. Garwin, T. Muller, J. Sens, V. Telegdi, and A. Zichichi, "Measurement of the anomalous magnetic moment of the muon," *Phys. Rev. Lett.*, vol. 6, pp. 128–132, 1961.

[133] G. Charpak, F. Farley, and R. Garwin, "A new measurement of the anomalous magnetic moment of the muon," *Phys. Lett.*, vol. 1, p. 16, 1962.

[134] J. Bailey *et al.*, "New measurement of (g-2) of the muon," *Phys. Lett. B*, vol. 55, pp. 420–424, 1975.

[135] J. Bailey *et al.*, "Final report on the CERN muon storage ring including the anomalous magnetic moment and the electric dipole moment of the muon, and a direct test of relativistic time dilation," *Nucl. Phys. B*, vol. 150, pp. 1–75, 1979.

[136] H. Brown *et al.*, "Improved measurement of the positive muon anomalous magnetic moment," *Phys. Rev. D*, vol. 62, p. 091101, 2000.

[137] H. Brown *et al.*, "Precise measurement of the positive muon anomalous magnetic moment," *Phys. Rev. Lett.*, vol. 86, pp. 2227–2231, 2001.

[138] G. Bennett *et al.*, "Measurement of the positive muon anomalous magnetic moment to 0.7 ppm," *Phys. Rev. Lett.*, vol. 89, p. 101804, 2002. [Erratum: *Phys.Rev.Lett.* vol. 89, 129903 (2002)].

[139] G. Bennett *et al.*, "Measurement of the negative muon anomalous magnetic moment to 0.7 ppm," *Phys. Rev. Lett.*, vol. 92, p. 161802, 2004.

[140] G. Bennett *et al.*, "Final report of the muon E821 anomalous magnetic moment measurement at BNL," *Phys. Rev. D*, vol. 73, p. 072003, 2006.

[141] B. Lautrup, A. Peterman, and E. de Rafael, "Recent developments in the comparison between theory and experiments in quantum electrodynamics," *Phys. Rep.*, vol. 3, no. 4, pp. 193–259, 1972.

[142] A. Antognini *et al.*, "The Lamb shift in muonic hydrogen and the proton radius," *Phys. Proc.*, vol. 17, pp. 10–19, 2011. 2nd International Workshop on the Physics of fundamental Symmetries and Interactions - PSI2010.

[143] R. C. Williams, "The fine structures of Hα and Dα under varying discharge conditions," *Phys. Rev.*, vol. 54, pp. 558–567, 1938.

[144] W. E. Lamb and R. C. Retherford, "Fine structure of the hydrogen atom by a microwave method," *Phys. Rev.*, vol. 72, pp. 241–243, 1947.

[145] S. Mohorovii, "Möglichkeit neuer Elemente und ihre Bedeutung für die Astrophysik," *Astro. Nachrichten*, vol. 253, no. 4, pp. 93–108, 1934.

[146] A. E. Ruark, "Positronium," *Phys. Rev.*, vol. 68, pp. 278–278, 1945.

[147] M. Deutsch, "Evidence for the formation of positronium in gases," *Phys. Rev.*, vol. 82, pp. 455–456, 1951.

[148] J. A. Wheeler, "Polyelectrons," *Ann. N.Y. Acad. Sci.*, vol. 48, no. 3, pp. 219–238, 1946.

[149] J. Pirenne *Arch. Sci. Phys. Nat.*, vol. 29, p. 265, 1947.

[150] A. Ore and J. L. Powell, "Three-photon annihilation of an electron–positron pair," *Phys. Rev.*, vol. 75, pp. 1696–1699, 1949.

[151] S. N. Gninenko, N. V. Krasnikov, V. A. Matveev, and A. Rubbia, "Some aspects of positronium physics," *Phys. Part. Nucl.*, vol. 37, no. 3, pp. 321–346, 2006.

[152] A. H. Al-Ramadhan and D. W. Gidley, "New precision measurement of the decay rate of singlet positronium," *Phys. Rev. Lett.*, vol. 72, pp. 1632–1635, 1994.

[153] D. Sillou, "Status of ortho-positronium decay rate measurements," *Int. J. Mod. Phys A*, vol. 19, no. 23, pp. 3919–3925, 2004.

[154] O. Jinnouchi, S. Asai, and T. Kobayashi, "Precision measurement of orthopositronium decay rate using SiO(2) powder," *Phys. Lett.*, vol. B572, pp. 117–126, 2003.

[155] R. S. Vallery, P. W. Zitzewitz, and D. W. Gidley, "Resolution of the orthopositronium-lifetime puzzle," *Phys. Rev. Lett.*, vol. 90, p. 203402, 2003.

[156] C. I. Westbrook, D. W. Gidley, R. S. Conti, and A. Rich, "Precision measurement of the orthopositronium vacuum decay rate using the gas technique," *Phys. Rev. A*, vol. 40, pp. 5489–5499, 1989.

[157] J. S. Nico, D. W. Gidley, A. Rich, and P. W. Zitzewitz, "Precision measurement of the orthopositronium decay rate using the vacuum technique," *Phys. Rev. Lett.*, vol. 65, p. 1344, 1990.

[158] S. Asai, S. Orito, and N. Shinohara, "New measurement of orthopositronium decay rate," *Phys. Lett. B*, vol. 357, pp. 475–480, 1995.

[159] H. Yukawa, "On the interaction of elementary particles. I," *Proc. Phys.-Math. Soc. Jpn., Ser. 3*, vol. 17, pp. 48–57, 1935.

[160] S. Tomonaga and G. Araki, "Effect of the nuclear coulomb field on the capture of slow mesons," *Phys. Rev.*, vol. 58, pp. 90–91, 1940.

[161] H. Primakoff, "Theory of muon capture," *Rev. Mod. Phys.*, vol. 31, pp. 802–822, 1959.

[162] D. F. Measday, "The nuclear physics of muon capture," *Phys. Rep.*, vol. 354, pp. 243–409, 2001.

[163] B. Rossi and D. B. Hall, "Variation of the rate of decay of mesotrons with momentum," *Phys. Rev.*, vol. 59, pp. 223–228, 1941.

[164] M. Conversi, E. Pancini, and O. Piccioni, "On the disintegration of negative mesons," *Phys. Rev.*, vol. 71, pp. 209–210, 1947.

[165] D. H. Perkins, "Nuclear disintegration by meson capture," *Nature*, vol. 159, pp. 126–127, 1947.

[166] C. M. G. Lattes, H. Muirhead, G. P. S. Occhialini, and C. F. Powell, "Processes involving charged mesons," *Nature*, vol. 159, pp. 694–697, 1947.

[167] M. Garçon and J. W. Van Orden, *The Deuteron: Structure and Form Factors*, pp. 293–378. Springer-Verlag, 2001.

[168] W. B. Cheston, "On the reactions $\pi^+ + d \leftrightarrows p + p$," *Phys. Rev.*, vol. 83, pp. 1118–1122, 1951.

[169] W. F. Cartwright, C. Richman, M. N. Whitehead, and H. A. Wilcox, "The production of positive pions by 341-MeV protons on protons," *Phys. Rev.*, vol. 91, no. 3, p. 677, 1953.

[170] R. Durbin, H. Loar, and J. Steinberger, "The spin of the pion via the reaction $\pi^+ + d \leftrightarrows p + p$," *Phys. Rev.*, vol. 83, pp. 646–648, 1951.

[171] W. Chinowsky and J. Steinberger, "Absorption of negative pions in deuterium: Parity of the pion," *Phys. Rev.*, vol. 95, pp. 1561–1564, 1954.

[172] C.-N. Yang, "Selection rules for the dematerialization of a particle into two photons," *Phys. Rev.*, vol. 77, pp. 242–245, 1950.

[173] N. M. Kroll and W. Wada, "Internal pair production associated with the emission of high-energy gamma rays," *Phys. Rev.*, vol. 98, pp. 1355–1359, 1955.

[174] E. Abouzaid *et al.*, "Determination of the parity of the neutral pion via the four-electron decay," *Phys. Rev. Lett.*, vol. 100, p. 182001, 2008.

[175] L. Leprince-Ringuet and M. Lhéritier, "Existence probable d'une particule de masse (990 ± 12%) m_0 dans le rayonnement cosmique," *J. Phys. Radium*, vol. 7, no. 3, pp. 65–69, 1946.

[176] B. Degrange, G. Fontaine, and P. Fleury, "Tracking Louis Leprince-Ringuets contributions to cosmic-ray physics," *Physics Today*, vol. 66, 2013.

[177] R. Brown, U. Camerini, P. H. Fowler, H. Muirhead, C. F. Powell, and D. M. Ritson, "Observations with electron sensitive plates exposed to cosmic radiation," *Nature*, vol. 163, p. 82, 1949.

[178] R. Armenteros, K. H. Barker, C. C. Butler, A. Cachon, and A. H. Chapman, "Decay of V-particles," *Nature*, vol. 167, p. 501, 1951.

[179] A. Pais, "Some remarks on the V-particles," *Phys. Rev.*, vol. 86, pp. 663–672, 1952.

[180] T. Nakano and K. Nishijima, "Charge independence for V-particles," *Progr. Theoret. Phys.*, vol. 10, no. 5, pp. 581–582, 1953.

[181] K. Nishijima, "Charge independence theory of V particles," *Progr. Theoret. Phys.*, vol. 13, no. 3, pp. 285–304, 1955.

[182] M. Gell-Mann, "The interpretation of the new particles as displaced charge multiplets," *Il Nuovo Cimento (1955–1965)*, vol. 4, no. 2, pp. 848–866, 1956.

[183] O. Chamberlain, E. Segrè, C. Wiegand, and T. Ypsilantis, "Observation of antiprotons," *Phys. Rev.*, vol. 100, pp. 947–950, 1955.

[184] B. Cork, G. R. Lambertson, O. Piccioni, and W. A. Wenzel, "Antineutrons produced from antiprotons in charge-exchange collisions," *Phys. Rev.*, vol. 104, pp. 1193–1197, 1956.

[185] D. J. Prowse and M. Baldo-Ceolin, "Anti-lambda hyperon," *Phys. Rev. Lett.*, vol. 1, pp. 179–180, 1958.

[186] H. L. Anderson, E. Fermi, E. A. Long, and D. E. Nagle, "Total cross sections of positive pions in hydrogen," *Phys. Rev.*, vol. 85, pp. 936–936, 1952.

[187] K. A. Brueckner, "Meson-nucleon scattering and nucleon isobars," *Phys. Rev.*, vol. 86, pp. 106–109, 1952.

[188] C. Alff, D. Berley, D. Colley, N. Gelfand, U. Nauenberg, D. Miller, J. Schultz, J. Steinberger, T. H. Tan, H. Brugger, P. Kramer, and R. Plano, "Production of pion resonances in $\pi^+ p$ interactions," *Phys. Rev. Lett.*, vol. 9, pp. 322–324, 1962.

[189] M. N. Rosenbluth, "High energy elastic scattering of electrons on protons," *Phys. Rev.*, vol. 79, pp. 615–619, 1950.

[190] R. W. McAllister and R. Hofstadter, "Elastic scattering of 188-MeV electrons from the proton and the alpha particle," *Phys. Rev.*, vol. 102, pp. 851–856, 1956.

[191] R. Hofstadter, "Electron scattering and nuclear structure," *Rev. Mod. Phys.*, vol. 28, pp. 214–254, 1956.

[192] D. R. Yennie, M. M. Lévy, and D. G. Ravenhall, "Electromagnetic structure of nucleons," *Rev. Mod. Phys.*, vol. 29, pp. 144–157, 1957.

[193] T. Janssens, R. Hofstadter, E. B. Hughes, and M. R. Yearian, "Proton form factors from elastic electron-proton scattering," *Phys. Rev.*, vol. 142, pp. 922–931, 1966.

[194] V. Punjabi, C. F. Perdrisat, M. K. Jones, E. J. Brash, and C. E. Carlson, "The structure of the nucleon: Elastic electromagnetic form factors," *Eur. Phys. J. A*, vol. 51, no. 7, p. 79, 2015.

[195] E. D. Bloom *et al.*, "High-energy inelastic $e - p$ scattering at 6 and 10 deg," *Phys. Rev. Lett.*, vol. 23, pp. 930–934, 1969.

[196] E. D. Bloom *et al.*, "Recent results in inelastic electron scattering," in *15th International Conference on High Energy Physics*, 9, 1970.

[197] S. Drell and J. Walecka, "Electrodynamic processes with nuclear targets," *Ann. Phys.*, vol. 28, no. 1, pp. 18–33, 1964.

[198] V. E. Barnes *et al.*, "Observation of a hyperon with strangeness minus three," *Phys. Rev. Lett.*, vol. 12, pp. 204–206, 1964.

[199] M. Gell-Mann, "A schematic model of baryons and mesons," *Phys. Lett.*, vol. 8, pp. 214–215, 1964.

[200] M. Gell-Mann, "Quarks," *Acta Phys. Austriaca Suppl.*, vol. 9, pp. 733–761, 1972.

[201] K. Gottfried, "Sum rule for high-energy electron–proton scattering," *Phys. Rev. Lett.*, vol. 18, pp. 1174–1177, 1967.

[202] L. Faddeev and V. Popov, "Feynman diagrams for the Yang–Mills field," *Phys. Lett. B*, vol. 25, no. 1, pp. 29–30, 1967.

[203] P. Langacker, "Grand unified theories and proton decay," *Phys. Rep.*, vol. 72, p. 185, 1981.

[204] J. D. Bjorken and S. J. Brodsky, "Statistical model for electron–positron annihilation into hadrons," *Phys. Rev. D*, vol. 1, pp. 1416–1420, 1970.

[205] S. D. Drell, D. J. Levy, and T.-M. Yan, "Theory of deep-inelastic lepton-nucleon scattering and lepton-pair annihilation processes. I," *Phys. Rev.*, vol. 187, pp. 2159–2171, 1969.

[206] N. Cabibbo, G. Parisi, and M. Testa, "Hadron production in e^+e^- collisions," *Lett. Nuovo Cimento (1969–1970)*, vol. 4, no. 1, pp. 35–39, 1970.

[207] G. Hanson *et al.*, "Evidence for jet structure in hadron production by e^+e^- annihilation," *Phys. Rev. Lett.*, vol. 35, pp. 1609–1612, 1975.

[208] H.-J. Behrend *et al.*, "Determination of α_s and $\sin^2\theta_W$ from measurements of the total hadronic cross section in e^+e^- annihilation," *Phys. Lett. B*, vol. 183, no. 3, pp. 400–411, 1987.

[209] S. L. Wu and G. Zobernig, "A method of three-jet analysis in e^+e^- annihilation," *Z. Phys. C*, vol. 2, pp. 107–110, Jun 1979.

[210] R. Brandelik *et al.*, "Evidence for planar events in e^+e^- annihilation at high-energies," *Phys. Lett.*, vol. 86B, pp. 243–249, 1979.

[211] D. P. Barber *et al.*, "Discovery of three-jet events and a test of quantum chromodynamics at PETRA," *Phys. Rev. Lett.*, vol. 43, pp. 830–833, 1979.

[212] C. Berger *et al.*, "Evidence for gluon Bremsstrahlung in e^+e^- annihilations at high-energies," *Phys. Lett.*, vol. 86B, pp. 418–425, 1979.

[213] W. Bartel *et al.*, "Observation of planar three jet events in e^+e^- annihilation and evidence for gluon Bremsstrahlung," *Phys. Lett.*, vol. 91B, pp. 142–147, 1980.

[214] S. Moretti, L. Lonnblad, and T. Sjostrand, "New and old jet clustering algorithms for electron–positron events," *J. High Energy Phys.*, vol. 08, p. 001, 1998.

[215] V. W. Hughes and J. Kuti, "Internal spin structure of the nucleon," *Ann. Rev. Nucl. Part. Sci.*, vol. 33, no. 1, pp. 611–644, 1983.

[216] J. Ellis and R. Jaffe, "Sum rule for deep-inelastic electroproduction from polarized protons," *Phys. Rev. D*, vol. 9, pp. 1444–1446, 1974.

[217] V. Alexakhin *et al.*, "The deuteron spin-dependent structure function g1d and its first moment," *Phys. Lett. B*, vol. 647, no. 1, pp. 8–17, 2007.

[218] I. Abt *et al.*, "The H1 detector at HERA," *Nucl. Instrum. Meth.*, vol. A386, pp. 310–347, 1997.

[219] H. Abramowicz *et al.*, "Combination of measurements of inclusive deep inelastic $e^\pm p$ scattering cross sections and QCD analysis of HERA data," *Eur. Phys. J. C*, vol. 75, no. 12, p. 580, 2015.

[220] A. Benvenuti *et al.*, "A high statistics measurement of the proton structure functions $F_2(x, Q^2)$ and R from deep inelastic muon scattering at high Q^2," *Phys. Lett. B*, vol. 223, pp. 485–489, 1989.

[221] M. Arneodo *et al.*, "Measurement of the proton and deuteron structure functions, $F_2(p)$ and $F_2(d)$, and of the ratio σ_L/σ_T," *Nucl. Phys. B*, vol. 483, pp. 3–43, 1997.

[222] V. Bertone, S. Carrazza, and J. Rojo, "APFEL: A PDF evolution library with QED corrections," *Comput. Phys. Commun.*, vol. 185, pp. 1647–1668, 2014.

[223] S. Carrazza, A. Ferrara, D. Palazzo, and J. Rojo, "APFEL Web," *J. Phys.*, vol. G42, no. 5, p. 057001, 2015.

[224] S. D. Drell and T.-M. Yan, "Partons and their applications at high energies," *Ann. Phys.*, vol. 66, no. 2, pp. 578–623, 1971.

[225] B. Combridge, J. Kripfganz, and J. Ranft, "Hadron production at large transverse momentum and QCD," *Phys. Lett. B*, vol. 70, no. 2, pp. 234–238, 1977.

[226] M. H. Seymour and M. Marx, *Monte Carlo Event Generators*, pp. 287–319. Springer International, 2015.

[227] T. Gleisberg, S. Höche, F. Krauss, M. Schönherr, S. Schumann, F. Siegert, and J. Winter, "Event generation with SHERPA 1.1," *J. High Energy Phys.*, vol. 2009, no. 02, p. 007, 2009.

[228] M. Bähr *et al.*, "Herwig++ physics and manual," *Eur. Phys. J. C*, vol. 58, no. 4, pp. 639–707, 2008.

[229] V. Khachatryan *et al.*, "Jet energy scale and resolution in the CMS experiment in pp collisions at 8 TeV," *J. Instrum.*, vol. 12, no. 02, p. P02014, 2017.

[230] B. Bjorken and S. Glashow, "Elementary particles and $SU(4)$," *Phys. Lett.*, vol. 11, no. 3, pp. 255–257, 1964.

[231] J. H. Christenson, G. S. Hicks, L. M. Lederman, P. J. Limon, B. G. Pope, and E. Zavattini, "Observation of massive muon pairs in hadron collisions," *Phys. Rev. Lett.*, vol. 25, pp. 1523–1526, 1970.

[232] J. H. Christenson, G. S. Hicks, L. M. Lederman, P. J. Limon, B. G. Pope, and E. Zavattini, "Observation of muon pairs in high-energy hadron collisions," *Phys. Rev.*, vol. D8, pp. 2016–2034, 1972.

[233] S. D. Drell and T.-M. Yan, "Massive lepton-pair production in hadron–hadron collisions at high energies," *Phys. Rev. Lett.*, vol. 25, pp. 316–320, 1970.

[234] Y. Li and F. Petriello, "Combining QCD and electroweak corrections to dilepton production in the framework of the FEWZ simulation code," *Phys. Rev. D*, vol. 86, p. 094034, 2012.

[235] G. Altarelli, R. Ellis, and G. Martinelli, "Large perturbative corrections to the Drell-Yan process in QCD," *Nucl. Phys. B*, vol. 157, pp. 461–497, 1979.

[236] J. Badier *et al.*, "Drell-Yan events from 400-GeV/c protons: Determination of the K factor in a large kinematical domain," *Z. Phys. C*, vol. 26, p. 489, 1985.

[237] G. Moreno *et al.*, "Dimuon production in proton–copper collisions at $\sqrt{s} = 38.8$-GeV," *Phys. Rev. D*, vol. 43, pp. 2815–2836, 1991.

[238] P. L. McGaughey, J. M. Moss, D. M. Alde, H. W. Baer, T. A. Carey, G. T. Garvey, A. Klein, C. Lee, M. J. Leitch, J. Lillberg, C. S. Mishra, J. C. Peng, C. N. Brown, W. E. Cooper, Y. B. Hsiung, M. R. Adams, R. Guo, D. M. Kaplan, R. L. McCarthy, G. Danner, M. Wang, M. Barlett, and G. Hoffmann, "Cross sections for the production of high-mass muon pairs from 800 GeV proton bombardment of ^{2}H," *Phys. Rev. D*, vol. 50, pp. 3038–3045, 1994. [Erratum: *Phys. Rev. D*, vol. 60, p 119903].

[239] M. K. Gaillard and B. W. Lee, "Rare decay modes of the K mesons in gauge theories," *Phys. Rev. D*, vol. 10, pp. 897–916, 1974.

[240] R. Van Royen and V. Weisskopf, "Hadron decay processes and the quark model," *Nuovo Cim. A*, vol. 50, pp. 617–645, 1967. [Erratum: *Nuovo Cim. A* 51, 583 (1967)].

[241] M. Kobayashi and T. Maskawa, "Cp violation in the renormalizable theory of weak interaction," *Prog. Theor. Phys.*, vol. 49, pp. 652–657, 1973.

[242] B. Aubert *et al.*, "The BaBar detector," *Nucl. Instrum. Meth. A*, vol. 479, pp. 1–116, 2002.

[243] A. Abashian *et al.*, "The Belle detector," *Nucl. Instrum. Meth. A*, vol. 479, pp. 117–232, 2002.

[244] M. Davier, "Low-energy e^+e^- hadronic annihilation cross sections," *Ann. Rev. Nucl. Part. Sci.*, vol. 63, pp. 407–434, 2013.

[245] J. Kühn and P. Zerwas, "The toponium scenario," *Phys. Rep.*, vol. 167, no. 6, pp. 321–403, 1988.

[246] H. Abramowicz, "Diffraction and the pomeron," *Int. J. Mod. Phys.*, vol. A15S1, pp. 495–520, 2000.

[247] F. E. Low, "Model of the bare pomeron," *Phys. Rev. D*, vol. 12, pp. 163–173, 1975.

[248] S. Nussinov, "Colored-quark version of some hadronic puzzles," *Phys. Rev. Lett.*, vol. 34, pp. 1286–1289, 1975.

[249] E. Fermi, "Tentativo di una teoria dei raggi β," *Il Nuovo Cimento (1924–1942)*, vol. 11, no. 1, p. 1, 2008.

[250] E. Fermi, "Versuch einer Theorie der β-Strahlen. I," *Z. Phys.*, vol. 88, pp. 161–177, 1934.

[251] F. L. Wilson, "Fermi's theory of beta decay," *Am. J. Phys.*, vol. 36, no. 12, pp. 1150–1160, 1968.

[252] L. M. Langer and R. J. D. Moffat, "The beta-spectrum of tritium and the mass of the neutrino," *Phys. Rev.*, vol. 88, pp. 689–694, 1952.

[253] A. Strumia and F. Vissani, "Precise quasielastic neutrino/nucleon cross-section," *Phys. Lett.*, vol. B564, pp. 42–54, 2003.

[254] M. James, "Energy released in fission," *J. Nucl. Energy*, vol. 23, no. 9, pp. 517–536, 1969.

[255] T. A. Mueller *et al.*, "Improved predictions of reactor antineutrino spectra," *Phys. Rev. C*, vol. 83, p. 054615, 2011.

[256] L. Zhan, "Recent results from Daya Bay," *PoS*, vol. NEUTEL2015, p. 017, 2015.

[257] S.-H. Seo, "New results from RENO and the 5 MeV excess," *AIP Conf. Proc.*, vol. 1666, p. 080002, 2015.

[258] Y. Abe *et al.*, "Improved measurements of the neutrino mixing angle θ_{13} with the Double Chooz detector," *J. High Energy Phys.*, vol. 10, p. 086, 2014. [Erratum: JHEP02 074 (2015)].

[259] C. L. Cowan, F. Reines, F. B. Harrison, E. C. Anderson, and F. N. Hayes, "Large liquid scintillation detectors," *Phys. Rev.*, vol. 90, pp. 493–494, 1953.

[260] F. Reines and C. L. Cowan, "A proposed experiment to detect the free neutrino," *Phys. Rev.*, vol. 90, pp. 492–493, 1953.

[261] F. Reines and C. L. Cowan, "Detection of the free neutrino," *Phys. Rev.*, vol. 92, pp. 830–831, 1953.

[262] C. L. Cowan, F. Reines, F. B. Harrison, H. W. Kruse, and A. D. McGuire, "Detection of the free neutrino: A confirmation," *Science*, vol. 124, pp. 103–104, 1956.

[263] F. Reines and C. L. Cowan, "Free anti-neutrino absorption cross-section. 1: Measurement of the free anti-neutrino absorption cross-section by protons," *Phys. Rev.*, vol. 113, pp. 273–279, 1959.

[264] J. Steinberger, "On the range of the electrons in meson decay," *Phys. Rev.*, vol. 74, pp. 500–501, 1948.

[265] J. Tiomno and J. A. Wheeler, "Energy spectrum of electrons from meson decay," *Rev. Mod. Phys.*, vol. 21, pp. 144–152, 1949.

[266] R. Davis, "Attempt to detect the antineutrinos from a nuclear reactor by the $Cl^{37}(\bar{\nu}, e^-)Ar^{37}$ reaction," *Phys. Rev.*, vol. 97, pp. 766–769, 1955.

[267] G. Feinberg, "Decays of the μ meson in the intermediate-meson theory," *Phys. Rev.*, vol. 110, pp. 1482–1483, 1958.

[268] J. Steinberger and H. B. Wolfe, "Electrons from muon capture," *Phys. Rev.*, vol. 100, pp. 1490–1493, 1955.

[269] T. D. Lee and C. N. Yang, "Theoretical discussions on possible high-energy neutrino experiments," *Phys. Rev. Lett.*, vol. 4, pp. 307–311, 1960.

[270] B. Pontecorvo, "Electron and muon neutrinos," *Sov. Phys. JETP*, vol. 10, pp. 1236–1240, 1960. [*Zh. Eksp. Teor. Fiz.* 37, 1751 (1959)].

[271] R. P. Feynman and M. Gell-Mann, "Theory of the fermi interaction," *Phys. Rev.*, vol. 109, pp. 193–198, 1958.

[272] M. Schwartz, "Feasibility of using high-energy neutrinos to study the weak interactions," *Phys. Rev. Lett.*, vol. 4, pp. 306–307, 1960.

[273] G. Danby, J.-M. Gaillard, K. Goulianos, L. M. Lederman, N. Mistry, M. Schwartz, and J. Steinberger, "Observation of high-energy neutrino reactions and the existence of two kinds of neutrinos," *Phys. Rev. Lett.*, vol. 9, pp. 36–44, 1962.

[274] M. L. Perl *et al.*, "Evidence for anomalous lepton production in $e^+ - e^-$ annihilation," *Phys. Rev. Lett.*, vol. 35, pp. 1489–1492, 1975.

[275] K. Belous *et al.*, "Measurement of the τ-lepton lifetime at Belle," *Phys. Rev. Lett.*, vol. 112, no. 3, p. 031801, 2014.

[276] K. Kodama *et al.*, "Observation of tau neutrino interactions," *Phys. Lett.*, vol. B504, pp. 218–224, 2001.

[277] K. Kodama *et al.*, "Final tau-neutrino results from the DONuT experiment," *Phys. Rev.*, vol. D78, p. 052002, 2008.

[278] T. D. Lee and C. N. Yang, "Question of parity conservation in weak interactions," *Phys. Rev.*, vol. 104, pp. 254–258, 1956.

[279] C. S. Wu, E. Ambler, R. W. Hayward, D. D. Hoppes, and R. P. Hudson, "Experimental test of parity conservation in beta decay," *Phys. Rev.*, vol. 105, pp. 1413–1415, 1957.

[280] E. C. G. Sudarshan and R. e. Marshak, "Chirality invariance and the universal Fermi interaction," *Phys. Rev.*, vol. 109, pp. 1860–1860, 1958.

[281] M. Goldhaber, L. Grodzins, and A. W. Sunyar, "Helicity of neutrinos," *Phys. Rev.*, vol. 109, pp. 1015–1017, 1958.

[282] P. C. Macq, K. M. Crowe, and R. P. Haddock, "Helicity of the electron and positron in muon decay," *Phys. Rev.*, vol. 112, pp. 2061–2071, 1958.

[283] D. Schwartz, "Helicity of electrons from μ^- decay," *Phys. Rev.*, vol. 162, p. 1306, 1967.

[284] A. Kresnin and L. Rosentsveig, "Polarization effects in the scattering of electrons and positrons by electrons," *J. Exp. Theoret. Phys.*, vol. 5, p. 288, 1957.

[285] A. Jodidio *et al.*, "Search for right-handed currents in muon decay," *Phys. Rev.*, vol. D34, p. 1967, 1986. [Erratum: *Phys. Rev.D* 37, 237 (1988)].

[286] B. Balke *et al.*, "Precise measurement of the asymmetry parameter delta in muon decay," *Phys. Rev.*, vol. D37, pp. 587–617, 1988.

[287] W. Fetscher, "Helicity of the muon-neutrino in π^+ decay: a comment on the measurement of $P(\mu)\xi\delta/\rho$ in muon decay," *Phys. Lett.*, vol. 140B, pp. 117–118, 1984.

[288] N. Cabibbo, "Unitary symmetry and leptonic decays," *Phys. Rev. Lett.*, vol. 10, pp. 531–533, 1963.

[289] A. Salam and J. C. Ward, "Electromagnetic and weak interactions," *Phys. Lett.*, vol. 13, pp. 168–171, 1964.

[290] A. Salam, "Weak and electromagnetic interactions," *Conf. Proc.*, vol. C680519, pp. 367–377, 1968.

[291] S. Weinberg, "A model of leptons," *Phys. Rev. Lett.*, vol. 19, pp. 1264–1266, 1967.

[292] G. Arnison *et al.*, "Experimental observation of isolated large transverse energy electrons with associated missing energy at $\sqrt{s} = 540$-GeV," *Phys. Lett.*, vol. B122, pp. 103–116, 1983.

[293] M. Banner *et al.*, "Observation of single isolated electrons of high transverse momentum in events with missing transverse energy at the CERN anti-p p collider," *Phys. Lett.*, vol. B122, pp. 476–485, 1983.

[294] M. Fierz, "Zur Fermischen Theorie des β-Zerfalls," *Z. Phys.*, vol. 104, pp. 553–565, Jul 1937.

[295] T. van Ritbergen and R. G. Stuart, "On the precise determination of the Fermi coupling constant from the muon lifetime," *Nucl. Phys.*, vol. B564, pp. 343–390, 2000.

[296] T. Kinoshita and A. Sirlin, "Radiative corrections to Fermi interactions," *Phys. Rev.*, vol. 113, pp. 1652–1660, 1959.

[297] S. M. Berman, "Radiative corrections to muon and neutron decay," *Phys. Rev.*, vol. 112, pp. 267–270, 1958.

[298] T. van Ritbergen and R. G. Stuart, "Complete two loop quantum electrodynamic contributions to the muon lifetime in the Fermi model," *Phys. Rev. Lett.*, vol. 82, pp. 488–491, 1999.

[299] A. Pak and A. Czarnecki, "Mass effects in muon and semileptonic $b \to c$ decays," *Phys. Rev. Lett.*, vol. 100, p. 241807, 2008.

[300] V. Tishchenko *et al.*, "Detailed report of the MuLan measurement of the positive muon lifetime and determination of the Fermi constant," *Phys. Rev.*, vol. D87, no. 5, p. 052003, 2013.

[301] J. Duclos, A. Magnon, and J. Picard, "A new measurement of the muon lifetime," *Phys. Lett.*, vol. 47B, pp. 491–493, 1973.

[302] M. P. Balandin, V. M. Grebenyuk, V. G. Zinov, A. D. Konin, and A. N. Ponomarev, "Measurement of the lifetime of the positive muon," *Sov. Phys. JETP*, vol. 40, pp. 811–814, 1975.

[303] K. L. Giovanetti *et al.*, "Mean life of the positive muon," *Phys. Rev.*, vol. D29, pp. 343–348, 1984.

[304] G. Bardin, J. Duclos, A. Magnon, J. Martino, E. Zavattini, A. Bertin, M. Capponi, M. Piccinini, and A. Vitale, "A new measurement of the positive muon lifetime," *Phys. Lett.*, vol. 137B, pp. 135–140, 1984.

[305] D. B. Chitwood *et al.*, "Improved measurement of the positive muon lifetime and determination of the Fermi constant," *Phys. Rev. Lett.*, vol. 99, p. 032001, 2007.

[306] A. Barczyk *et al.*, "Measurement of the Fermi constant by FAST," *Phys. Lett.*, vol. B663, pp. 172–180, 2008.

[307] M. Bardon, P. Norton, J. Peoples, A. M. Sachs, and J. Lee-Franzini, "Measurement of the momentum spectrum of positrons from muon decay," *Phys. Rev. Lett.*, vol. 14, pp. 449–453, 1965.

[308] L. Michel, "Interaction between four half spin particles and the decay of the μ meson," *Proc. Phys. Soc.*, vol. A63, pp. 514–531, 1950.

[309] C. Bouchiat and L. Michel, "Theory of μ-meson decay with the hypothesis of nonconservation of parity," *Phys. Rev.*, vol. 106, pp. 170–172, 1957.

[310] T. Kinoshita and A. Sirlin, "Polarization of electrons in muon decay with general parity-nonconserving interactions," *Phys. Rev.*, vol. 108, pp. 844–850, 1957.

[311] W. Fetscher and H. J. Gerber, "Muon decay parameters: in *Review of Particle Physics* (RPP 1998)," *Eur. Phys. J.*, vol. C3, pp. 282–284, 1998.

[312] W. J. Marciano and A. Sirlin, "Radiative corrections to $\pi_{\ell 2}$ decays," *Phys. Rev. Lett.*, vol. 71, pp. 3629–3632, 1993.

[313] S. Jadach, Z. Was, R. Decker, and J. H. Kuhn, "The tau decay library TAUOLA: Version 2.4," *Comput. Phys. Commun.*, vol. 76, pp. 361–380, 1993.

[314] P. Golonka, B. Kersevan, T. Pierzchala, E. Richter-Was, Z. Was, and M. Worek, "The Tauola-photos-F environment for the TAUOLA and PHOTOS packages: Release. 2.," *Comput. Phys. Commun.*, vol. 174, pp. 818–835, 2006.

[315] P. Ilten, "Tau decays in Pythia 8," *Nucl. Phys. Proc. Suppl.*, vol. 253–255, pp. 77–80, 2014.

[316] J. H. Kuhn and A. Santamaria, "Tau decays to pions," *Z. Phys.*, vol. C48, pp. 445–452, 1990.

[317] M. Finkemeier and E. Mirkes, "The scalar contribution to $\tau \to K\pi\nu_\tau$," *Z. Phys.*, vol. C72, pp. 619–626, 1996.

[318] D. M. Asner *et al.*, "Hadronic structure in the decay $\tau^- \to \nu_\tau \pi^- \pi^0 \pi^0$ and the sign of the tau-neutrino helicity," *Phys. Rev.*, vol. D61, p. 012002, 2000.

[319] M. Finkemeier and E. Mirkes, "Tau decays into kaons," *Z. Phys.*, vol. C69, pp. 243–252, 1996.

[320] R. Decker, E. Mirkes, R. Sauer, and Z. Was, "Tau decays into three pseudoscalar mesons," *Z. Phys.*, vol. C58, pp. 445–452, 1993.

[321] A. E. Bondar, S. I. Eidelman, A. I. Milstein, T. Pierzchala, N. I. Root, Z. Was, and M. Worek, "Novosibirsk hadronic currents for $\tau \to 4\pi$ channels of tau decay library TAUOLA," *Comput. Phys. Commun.*, vol. 146, pp. 139–153, 2002.

[322] J. H. Kuhn and Z. Was, "Tau decays to five mesons in TAUOLA," *Acta Phys. Polon.*, vol. B39, pp. 147–158, 2008.

[323] S. L. Glashow, J. Iliopoulos, and L. Maiani, "Weak interactions with lepton–hadron symmetry," *Phys. Rev. D*, vol. 2, pp. 1285–1292, 1970.

[324] D. H. Wilkinson, "Analysis of neutron beta decay," *Nucl. Phys.*, vol. A377, pp. 474–504, 1982.

[325] M. A. P. Brown *et al.*, "New result for the neutron β-asymmetry parameter A_0 from UCNA," *Phys. Rev.*, vol. C97, no. 3, p. 035505, 2018.

[326] C. N. Yang and R. L. Mills, "Conservation of isotopic spin and isotopic gauge invariance," *Phys. Rev.*, vol. 96, pp. 191–195, 1954.

[327] J. Goldstone, "Field theories with superconductor solutions," *Nuovo Cim.*, vol. 19, pp. 154–164, 1961.

[328] J. Goldstone, A. Salam, and S. Weinberg, "Broken symmetries," *Phys. Rev.*, vol. 127, pp. 965–970, 1962.

[329] S. Weinberg, "Physical processes in a convergent theory of the weak and electromagnetic interactions," *Phys. Rev. Lett.*, vol. 27, pp. 1688–1691, 1971.

[330] S. Weinberg, "General theory of broken local symmetries," *Phys. Rev.*, vol. D7, pp. 1068–1082, 1973.

[331] G. Gamow and E. Teller, "Some generalizations of the beta transformation theory," *Phys. Rev.*, vol. 51, pp. 289–289, 1937.

[332] N. Kemmer, "Field theory of nuclear interaction," *Phys. Rev.*, vol. 52, pp. 906–910, 1937.

[333] G. Wentzel, "Zur Frage der β-Wechselwirkung," *Phys. Acta*, vol. 10, p. 108, 1937.

[334] G. Wentzel, "Zur Theorie der β-Umwandlung und der Kernkräfte. I," *Z. Phys.*, vol. 104, pp. 34–47, Jan 1937.

[335] S. A. Bludman, "On the universal Fermi interaction," *Nuovo Cim.*, vol. 9, pp. 433–445, 1958.

[336] F. J. Hasert *et al.*, "Observation of neutrino like interactions without muon or electron in the Gargamelle neutrino experiment," *Phys. Lett.*, vol. B46, pp. 138–140, 1973.

[337] F. J. Hasert *et al.*, "Search for elastic ν_μ electron scattering," *Phys. Lett.*, vol. B46, pp. 121–124, 1973.

[338] F. J. Hasert *et al.*, "Observation of neutrino like interactions without muon or electron in the Gargamelle neutrino experiment," *Nucl. Phys.*, vol. B73, pp. 1–22, 1974.

[339] C. Prescott *et al.*, "Parity nonconservation in inelastic electron scattering," *Phys. Lett. B*, vol. 77, pp. 347–352, 1978.

[340] J. S. Schwinger, "A theory of the fundamental interactions," *Ann. Phys.*, vol. 2, pp. 407–434, 1957.

[341] S. L. Glashow, "Partial symmetries of weak interactions," *Nucl. Phys.*, vol. 22, pp. 579–588, 1961.

[342] S. Catani, G. Ferrera, and M. Grazzini, "W boson production at hadron colliders: The lepton charge asymmetry in NNLO QCD," *J. High Energy Phys.*, vol. 05, p. 006, 2010.

[343] E. Eichten, I. Hinchliffe, K. D. Lane, and C. Quigg, "Super collider physics," *Rev. Mod. Phys.*, vol. 56, pp. 579–707, 1984. [Addendum: *Rev.Mod.Phys.* 58, 1065–1073 (1986)].

[344] G. Myatt, "Experimental verification of the Weinberg–Salam theory," *Rept. Prog. Phys.*, vol. 45, pp. 1–46, 1982.

[345] P. Vilain *et al.*, "Precision measurement of electroweak parameters from the scattering of muon-neutrinos on electrons," *Phys. Lett.*, vol. B335, pp. 246–252, 1994.

[346] S. Mishra *et al.*, "Measurement of inverse muon decay $\nu_\mu + e \to \mu^- + \nu_e$ at Fermilab Tevatron energies 15-600 GeV," *Phys. Rev. Lett.*, vol. 63, pp. 132–135, 1989.

[347] C. Rubbia, P. McIntyre, and D. Cline, "Producing massive neutral intermediate vector bosons with existing accelerators," in *30 years of weak neutral currents*, pp. 683–687, 1976.

[348] D. Mohl, G. Petrucci, L. Thorndahl, and S. Van Der Meer, "Physics and technique of stochastic cooling," *Phys. Rept.*, vol. 58, pp. 73–119, 1980.

[349] Staff of the CERN Proton–Antiproton Project, "First proton–antiproton collisions in the CERN SPS collider," *Phys. Lett. B*, vol. 107, no. 4, pp. 306–309, 1981.

[350] A. Astbury *et al.*, "A 4π solid angle detector for the SPS used as a proton–antiproton collider at a centre of mass energy of 540 GeV," Tech. Rep. CERN-SPSC-78-6. CERN-SPSC-78-06. SPSC-P-92, CERN, Geneva, Jan 1978.

[351] M. Banner *et al.*, "Observation of very large transverse momentum jets at the CERN anti-p p collider," *Phys. Lett. B*, vol. 118, pp. 203–210, 1982.

[352] G. Arnison *et al.*, "Experimental observation of lepton pairs of invariant mass around 95-GeV/c^2 at the CERN SPS collider," *Phys. Lett. B*, vol. 126, pp. 398–410, 1983.

[353] C. Albajar *et al.*, "Studies of intermediate vector boson production and decay in UA1 at the CERN proton–antiproton collider," *Z. Phys. C*, vol. 44, pp. 15–61, 1989.

[354] J. Alitti *et al.*, "An improved determination of the ratio of W and Z masses at the CERN $\bar{p}p$ collider," *Phys. Lett. B*, vol. 276, pp. 354–364, 1992.

[355] T. Affolder *et al.*, "Measurement of the W boson mass with the collider detector at Fermilab," *Phys. Rev. D*, vol. 64, p. 052001, 2001.

[356] V. Abazov *et al.*, "Improved W boson mass measurement with the D0 detector," *Phys. Rev. D*, vol. 66, p. 012001, 2002.

[357] R. Barate *et al.*, "Measurement of the Z resonance parameters at LEP," *Eur. Phys. J. C*, vol. 14, pp. 1–50, 2000.

[358] M. Acciarri *et al.*, "Measurements of cross-sections and forward backward asymmetries at the Z resonance and determination of electroweak parameters," *Eur. Phys. J. C*, vol. 16, pp. 1–40, 2000.

[359] P. Abreu *et al.*, "Cross-sections and leptonic forward backward asymmetries from the Z0 running of LEP," *Eur. Phys. J. C*, vol. 16, pp. 371–405, 2000.

[360] G. Abbiendi *et al.*, "Precise determination of the Z resonance parameters at LEP: 'Zedometry'," *Eur. Phys. J. C*, vol. 19, pp. 587–651, 2001.

[361] T. Aaltonen *et al.*, "Precise measurement of the W-boson mass with the CDF II detector," *Phys. Rev. Lett.*, vol. 108, p. 151803, 2012.

[362] V. M. Abazov *et al.*, "Measurement of the W boson mass with the D0 detector," *Phys. Rev. D*, vol. 89, no. 1, p. 012005, 2014.

[363] M. Aaboud *et al.*, "Measurement of the W-boson mass in pp collisions at $\sqrt{s} = 7$ TeV with the ATLAS detector," *Eur. Phys. J. C*, vol. 78, no. 2, p. 110, 2018. [Erratum: *Eur. Phys. J. C* 78, 898 (2018)].

[364] S. Schael *et al.*, "Precision electroweak measurements on the Z resonance," *Phys. Rep.*, vol. 427, pp. 257–454, 2006.

[365] S. Schael *et al.*, "Electroweak measurements in electron–positron collisions at W-boson-pair energies at LEP," *Phys. Rep.*, vol. 532, pp. 119–244, 2013.

[366] F. Abe *et al.*, "Measurement of the mass and width of the Z^0 boson at the Fermilab Tevatron," *Phys. Rev. Lett.*, vol. 63, p. 720, 1989.

[367] K. Abe *et al.*, "Improved direct measurement of A(b) and A(c) at the Z0 pole using a lepton tag," *Phys. Rev. Lett.*, vol. 88, p. 151801, 2002.

[368] K. Abe *et al.*, "Measurement of A(c) with charmed mesons at SLD," *Phys. Rev. D*, vol. 63, p. 032005, 2001.

[369] K. Abe *et al.*, "Improved direct measurement of the parity violation parameter A(b) using a mass tag and momentum weighted track charge," *Phys. Rev. Lett.*, vol. 90, p. 141804, 2003.

[370] K. Abe *et al.*, "Direct measurements of A(b) and A(c) using vertex/kaon charge tags at SLD," *Phys. Rev. Lett.*, vol. 94, p. 091801, 2005.

[371] K. Abe *et al.*, "Direct measurement of A(b) in Z0 decays using charged kaon tagging," *Phys. Rev. Lett.*, vol. 83, pp. 1902–1907, 1999.

[372] T. A. Aaltonen *et al.*, "Tevatron Run II combination of the effective leptonic electroweak mixing angle," *Phys. Rev. D*, vol. 97, no. 11, p. 112007, 2018.

[373] C. Albajar *et al.*, "Events with large missing transverse energy at the CERN collider. 3. mass limits on super-symmetric particles," *Phys. Lett.*, vol. B198, pp. 261–270, 1987.

[374] R. Ansari *et al.*, "Measurement of the Standard Model parameters from a study of W and Z bosons," *Phys. Lett.*, vol. B186, pp. 440–451, 1987.

[375] P. Janot and S. a. Jadach, "Improved Bhabha cross section at LEP and the number of light neutrino species," *Phys. Lett. B*, vol. 803, p. 135319, 2020.

[376] R. Bernstein, "Sign selected quadrupole train," tech. rep., 4 1994. FERMILAB-TM-1884.

[377] P. Astier *et al.*, "Prediction of neutrino fluxes in the NOMAD experiment," *Nucl. Instrum. Meth. A*, vol. 515, pp. 800–828, 2003.

[378] K. Abe *et al.*, "T2K neutrino flux prediction," *Phys. Rev.*, vol. D87, no. 1, p. 012001, 2013. [Addendum: *Phys. Rev.D* 87, no.1, 019902 (2013)].

[379] C. H. Llewellyn Smith, "Neutrino reactions at accelerator energies," *Phys. Rept.*, vol. 3, pp. 261–379, 1972.

[380] S. Weinberg, "Charge symmetry of weak interactions," *Phys. Rev.*, vol. 112, pp. 1375–1379, 1958.

[381] H. S. Budd, A. Bodek, and J. Arrington, "Modeling quasielastic form-factors for electron and neutrino scattering," in *2nd International Workshop on Neutrino–Nucleus Interactions in the Few GeV Region (NuInt 02) Irvine, California, December 12–15, 2002*, 2003.

[382] J. G. H. de Groot *et al.*, "Inclusive interactions of high-energy neutrinos and anti-neutrinos in iron," *Z. Phys.*, vol. C1, p. 143, 1979.

[383] R. Blair *et al.*, "Measurement of the rate of increase of neutrino cross-sections with energy," *Phys. Rev. Lett.*, vol. 51, pp. 343–346, 1983.

[384] P. Bergé *et al.*, "Total neutrino and anti-neutrino charged current cross-section measurements in 100-GeV, 160-GeV and 200-GeV Narrow Band Beams," Tech. Rep. CERN-EP/87-09, 1987.

[385] J. Allaby *et al.*, "Total cross-sections of charged current neutrino and anti-neutrino interactions on isoscalar nuclei," *Z. Phys. C*, vol. 38, pp. 403–410, 1988.

[386] N. Baker *et al.*, "Measurement of the muon-neutrino charged current cross-section," *Phys. Rev. Lett.*, vol. 51, pp. 735–738, 1983.

[387] G. Taylor *et al.*, "Anti-muon-neutrino nucleon charged current total cross-section for 5-GeV to 250-GeV," *Phys. Rev. Lett.*, vol. 51, pp. 739–742, 1983.

[388] P. Bosetti *et al.*, "Total cross-sections for ν_μ and $\bar\nu_\mu$ charged current interactions between 20-GeV and 200-GeV," *Phys. Lett. B*, vol. 110, pp. 167–172, 1982.

[389] B. Barish *et al.*, "Measurements of $\nu_\mu n$ and $\bar\nu_\mu n$ charged current total cross-sections," *Phys. Rev. Lett.*, vol. 39, p. 1595, 1977.

[390] J. Allaby *et al.*, "A precise determination of the electroweak mixing angle from semileptonic neutrino scattering," *Z. Phys. C*, vol. 36, p. 611, 1987.

[391] A. Blondel *et al.*, "Electroweak parameters from a high statistics neutrino nucleon scattering experiment," *Z. Phys. C*, vol. 45, pp. 361–379, 1990.

[392] K. S. McFarland *et al.*, "A precision measurement of electroweak parameters in neutrino–nucleon scattering," *Eur. Phys. J. C*, vol. 1, pp. 509–513, 1998.

[393] G. Zeller *et al.*, "A precise determination of electroweak parameters in neutrino nucleon scattering," *Phys. Rev. Lett.*, vol. 88, p. 091802, 2002. [Erratum: *Phys. Rev. Lett.* 90, 239902 (2003)].

[394] S. Rabinowitz *et al.*, "Measurement of the strange sea distribution using neutrino charm production," *Phys. Rev. Lett.*, vol. 70, pp. 134–137, 1993.

[395] E. Paschos and L. Wolfenstein, "Tests for neutral currents in neutrino reactions," *Phys. Rev. D*, vol. 7, pp. 91–95, 1973.

[396] M. Bonesini, A. Marchionni, F. Pietropaolo, and T. Tabarelli de Fatis, "On particle production for high-energy neutrino beams," *Eur. Phys. J. C*, vol. 20, pp. 13–27, 2001.

[397] P. Adamson *et al.*, "The NuMI neutrino beam," *Nucl. Instrum. Meth. A*, vol. 806, pp. 279–306, 2016.

[398] D. S. Ayres *et al.*, "NOvA: Proposal to build a 30 kiloton off-axis detector to study $\nu_\mu \to \nu_e$ oscillations in the NuMI beamline," 3 2004.

[399] M. Jezabek and J. H. Kuhn, "QCD corrections to semileptonic decays of heavy quarks," *Nucl. Phys. B*, vol. 314, pp. 1–6, 1989.

[400] R. Bonciani, S. Catani, M. L. Mangano, and P. Nason, "NLL resummation of the heavy quark hadroproduction cross-section," *Nucl. Phys. B*, vol. 529, pp. 424–450, 1998. [Erratum: *Nucl. Phys. B* 803, 234 (2008)].

[401] N. Kidonakis, "High order corrections and subleading logarithms for top quark production," *Phys. Rev. D*, vol. 64, p. 014009, 2001.

[402] Z. Sullivan, "Understanding single-top-quark production and jets at hadron colliders," *Phys. Rev. D*, vol. 70, p. 114012, 2004.

[403] B. Harris, E. Laenen, L. Phaf, Z. Sullivan, and S. Weinzierl, "The fully differential single top quark cross-section in next to leading order QCD," *Phys. Rev. D*, vol. 66, p. 054024, 2002.

[404] T. Stelzer, Z. Sullivan, and S. Willenbrock, "Single top quark production via W-gluon fusion at next-to-leading order," *Phys. Rev. D*, vol. 56, pp. 5919–5927, 1997.

[405] R. Blair *et al.*, "The CDF-II detector: Technical design report," 11 1996. FERMILAB-DESIGN-1996-01, FERMILAB-PUB-96-390-E.

[406] V. Abazov *et al.*, "The upgraded D0 detector," *Nucl. Instrum. Meth. A*, vol. 565, pp. 463–537, 2006.

[407] F. Abe *et al.*, "Observation of top quark production in $\bar{p}p$ collisions," *Phys. Rev. Lett.*, vol. 74, pp. 2626–2631, 1995.

[408] S. Abachi *et al.*, "Observation of the top quark," *Phys. Rev. Lett.*, vol. 74, pp. 2632–2637, 1995.

[409] V. Abazov *et al.*, "Evidence for production of single top quarks and first direct measurement of |Vtb|," *Phys. Rev. Lett.*, vol. 98, p. 181802, 2007.

[410] T. Aaltonen *et al.*, "First observation of electroweak single top quark production," *Phys. Rev. Lett.*, vol. 103, p. 092002, 2009.

[411] V. Abazov *et al.*, "Observation of single top quark production," *Phys. Rev. Lett.*, vol. 103, p. 092001, 2009.

[412] M. Aaboud *et al.*, "Combinations of single-top-quark production cross-section measurements and $|f_{LV} V_{tb}|$ determinations at $\sqrt{s} = 7$ and 8 TeV with the ATLAS and CMS experiments," *J. High Energy Phys.*, vol. 05, p. 088, 2019.

[413] A. Djouadi, "Decays of the Higgs bosons," in *International Workshop on Quantum Effects in the Minimal Supersymmetric Standard Model*, pp. 197–222, 1997.

[414] L. Resnick, M. Sundaresan, and P. Watson, "Is there a light scalar boson?," *Phys. Rev. D*, vol. 8, pp. 172–178, 1973.

[415] J. Ellis, M. K. Gaillard, and D. Nanopoulos, "A phenomenological profile of the Higgs boson," *Nucl. Phys. B*, vol. 106, pp. 292–340, 1976.

[416] J. Haller, A. Hoecker, R. Kogler, K. Mönig, T. Peiffer, and J. Stelzer, "Update of the global electroweak fit and constraints on two-Higgs-doublet models," *Eur. Phys. J. C*, vol. 78, no. 8, p. 675, 2018.

[417] *CMS, the Compact Muon Solenoid: Technical proposal.* LHC technical proposal CERN-LHCC-94-38, CERN-LHCC-P-1, Geneva: CERN, 1994.

[418] A. D. Sakharov, "Violation of CP invariance, C asymmetry, and baryon asymmetry of the universe," *Pisma Zh. Eksp. Teor. Fiz.*, vol. 5, pp. 32–35, 1967.

[419] J. H. Christenson, J. W. Cronin, V. L. Fitch, and R. Turlay, "Evidence for the 2π decay of the K_2^0 meson," *Phys. Rev. Lett.*, vol. 13, pp. 138–140, 1964.

[420] A. Apostolakis *et al.*, "A detailed description of the analysis of the decay of neutral kaons to pi+ pi- in the CPLEAR experiment," *Eur. Phys. J.*, vol. C18, pp. 41–55, 2000.

[421] B. Pontecorvo, "Neutrino experiments and the problem of conservation of leptonic charge," *Sov. Phys. JETP*, vol. 26, pp. 984–988, 1968.

[422] V. Gribov and B. Pontecorvo, "Neutrino astronomy and lepton charge," *Phys. Lett. B*, vol. 28, p. 493, 1969.

[423] Z. Maki, M. Nakagawa, and S. Sakata, "Remarks on the unified model of elementary particles," *Prog. Theor. Phys.*, vol. 28, pp. 870–880, 1962.

[424] A. Bueno, M. Campanelli, and A. Rubbia, "Physics potential at a neutrino factory: Can we benefit from more than just detecting muons?," *Nucl. Phys.*, vol. B589, pp. 577–608, 2000.

[425] L. Wolfenstein, "Neutrino oscillations in matter," *Phys. Rev.*, vol. D17, pp. 2369–2374, 1978.

[426] S. Mikheev and A. Smirnov, "Resonance enhancement of oscillations in matter and solar neutrino spectroscopy," *Sov. J. Nucl. Phys.*, vol. 42, 1985.

[427] M. Freund, "Analytic approximations for three neutrino oscillation parameters and probabilities in matter," *Phys. Rev. D*, vol. 64, p. 053003, 2001.

[428] V. Barger, D. Marfatia, and K. Whisnant, "Breaking eight fold degeneracies in neutrino CP violation, mixing, and mass hierarchy," *Phys. Rev. D*, vol. 65, p. 073023, 2002.

[429] P. Machado, H. Minakata, H. Nunokawa, and R. Zukanovich Funchal, "Combining accelerator and reactor measurements of θ_{13}: The first result," *J. High Energy Phys.*, vol. 05, p. 023, 2012.

[430] G. Fogli, E. Lisi, A. Marrone, D. Montanino, A. Palazzo, and A. Rotunno, "Global analysis of neutrino masses, mixings and phases: entering the era of leptonic CP violation searches," *Phys. Rev. D*, vol. 86, p. 013012, 2012.

[431] D. Forero, M. Tortola, and J. Valle, "Global status of neutrino oscillation parameters after Neutrino-2012," *Phys. Rev. D*, vol. 86, p. 073012, 2012.

[432] I. Esteban, M. Gonzalez-Garcia, A. Hernandez-Cabezudo, M. Maltoni, and T. Schwetz, "Global analysis of three-flavour neutrino oscillations: synergies and tensions in the determination of θ_{23}, δ_{CP}, and the mass ordering," *J. High Energy Phys.*, vol. 01, p. 106, 2019.

[433] B. Cleveland, T. Daily, J. Davis, Raymond, J. R. Distel, K. Lande, C. Lee, P. S. Wildenhain, and J. Ullman, "Measurement of the solar electron neutrino flux with the Homestake chlorine detector," *Astrophys. J.*, vol. 496, pp. 505–526, 1998.

[434] F. Kaether, W. Hampel, G. Heusser, J. Kiko, and T. Kirsten, "Reanalysis of the GALLEX solar neutrino flux and source experiments," *Phys. Lett. B*, vol. 685, pp. 47–54, 2010.

[435] J. Abdurashitov *et al.*, "Measurement of the solar neutrino capture rate with gallium metal. III: Results for the 2002–2007 data-taking period," *Phys. Rev. C*, vol. 80, p. 015807, 2009.

[436] J. Hosaka *et al.*, "Solar neutrino measurements in super-Kamiokande-I," *Phys. Rev. D*, vol. 73, p. 112001, 2006.

[437] B. Aharmim *et al.*, "Combined analysis of all three phases of solar neutrino data from the Sudbury Neutrino Observatory," *Phys. Rev. C*, vol. 88, p. 025501, 2013.

[438] G. Bellini *et al.*, "Precision measurement of the 7Be solar neutrino interaction rate in Borexino," *Phys. Rev. Lett.*, vol. 107, p. 141302, 2011.

[439] G. Bellini *et al.*, "Measurement of the solar 8B neutrino rate with a liquid scintillator target and 3 MeV energy threshold in the Borexino detector," *Phys. Rev. D*, vol. 82, p. 033006, 2010.

[440] G. Bellini *et al.*, "Neutrinos from the primary proton–proton fusion process in the Sun," *Nature*, vol. 512, no. 7515, pp. 383–386, 2014.

[441] Y. Fukuda *et al.*, "Evidence for oscillation of atmospheric neutrinos," *Phys. Rev. Lett.*, vol. 81, pp. 1562–1567, 1998.

[442] K. Abe *et al.*, "Atmospheric neutrino oscillation analysis with external constraints in Super-Kamiokande I-IV," *Phys. Rev. D*, vol. 97, no. 7, p. 072001, 2018.

[443] M. Aartsen *et al.*, "Measurement of atmospheric neutrino oscillations at 6–56 GeV with IceCube DeepCore," *Phys. Rev. Lett.*, vol. 120, no. 7, p. 071801, 2018.

[444] K. Eguchi *et al.*, "First results from KamLAND: Evidence for reactor anti-neutrino disappearance," *Phys. Rev. Lett.*, vol. 90, p. 021802, 2003.

[445] S. Abe *et al.*, "Precision measurement of neutrino oscillation parameters with KamLAND," *Phys. Rev. Lett.*, vol. 100, p. 221803, 2008.

[446] A. Gando *et al.*, "Reactor on–off antineutrino measurement with KamLAND," *Phys. Rev. D*, vol. 88, no. 3, p. 033001, 2013.

[447] Y. Abe *et al.*, "Indication of reactor $\bar{\nu}_e$ disappearance in the Double Chooz experiment," *Phys. Rev. Lett.*, vol. 108, p. 131801, 2012.

[448] F. An *et al.*, "Observation of electron-antineutrino disappearance at Daya Bay," *Phys. Rev. Lett.*, vol. 108, p. 171803, 2012.

[449] J. Ahn *et al.*, "Observation of Reactor Electron Antineutrino Disappearance in the RENO Experiment," *Phys. Rev. Lett.*, vol. 108, p. 191802, 2012.

[450] P. Adamson *et al.*, "Measurement of neutrino and antineutrino oscillations using beam and atmospheric data in MINOS," *Phys. Rev. Lett.*, vol. 110, no. 25, p. 251801, 2013.

[451] K. Abe *et al.*, "Indication of electron neutrino appearance from an accelerator-produced off-axis muon neutrino beam," *Phys. Rev. Lett.*, vol. 107, p. 041801, 2011.

[452] P. Adamson *et al.*, "First measurement of electron neutrino appearance in NOvA," *Phys. Rev. Lett.*, vol. 116, no. 15, p. 151806, 2016.

[453] P. Adamson *et al.*, "First measurement of muon-neutrino disappearance in NOvA," *Phys. Rev. D*, vol. 93, no. 5, p. 051104, 2016.

[454] H. Georgi and S. Glashow, "Unity of all elementary particle forces," *Phys. Rev. Lett.*, vol. 32, pp. 438–441, 1974.

[455] A. Takenaka *et al.*, "Search for proton decay via $p \to e^+\pi^0$ and $p \to \mu^+\pi^0$ with an enlarged fiducial volume in Super-Kamiokande I–IV," 10 2020.

[456] J. C. Pati and A. Salam, "Lepton number as the fourth color," *Phys. Rev. D*, vol. 10, pp. 275–289, 1974. [Erratum: *Phys. Rev. D* 11, 703–703 (1975)].

[457] R. N. Mohapatra and J. C. Pati, "Left-right gauge symmetry and an isoconjugate model of CP violation," *Phys. Rev. D*, vol. 11, pp. 566–571, 1975.

[458] R. Marshak and R. N. Mohapatra, "Quark–lepton symmetry and $B - L$ as the $U(1)$ generator of the electroweak symmetry group," *Phys. Lett. B*, vol. 91, pp. 222–224, 1980.

[459] M. Czakon, J. Gluza, and M. Zralek, "Low-energy physics and left-right symmetry: Bounds on the model parameters," *Phys. Lett. B*, vol. 458, pp. 355–360, 1999.

[460] J. F. Bueno *et al.*, "Precise measurement of parity violation in polarized muon decay," *Phys. Rev.*, vol. D84, p. 032005, 2011.

[461] Y. Zhang, H. An, X. Ji, and R. N. Mohapatra, "General CP violation in minimal left-right symmetric model and constraints on the right-handed scale," *Nucl. Phys. B*, vol. 802, pp. 247–279, 2008.

[462] R. Barbieri and R. N. Mohapatra, "Limits on right-handed interactions from SN1987A observations," *Phys. Rev. D*, vol. 39, p. 1229, 1989.

[463] S. Weinberg, "Baryon and lepton nonconserving processes," *Phys. Rev. Lett.*, vol. 43, pp. 1566–1570, 1979.

[464] F. Wilczek and A. Zee, "Operator analysis of nucleon decay," *Phys. Rev. Lett.*, vol. 43, pp. 1571–1573, 1979.

[465] M. Goeppert-Mayer, "Double beta-disintegration," *Phys. Rev.*, vol. 48, pp. 512–516, 1935.

[466] W. Furry, "On transition probabilities in double beta-disintegration," *Phys. Rev.*, vol. 56, pp. 1184–1193, 1939.

[467] W. Haxton and G. Stephenson, "Double beta decay," *Prog. Part. Nucl. Phys.*, vol. 12, pp. 409–479, 1984.

[468] S. M. Bilenky, "Neutrinos: Majorana or Dirac?," *Universe*, vol. 6, no. 9, p. 134, 2020.

[469] M. Agostini *et al.*, "Final results of GERDA on the search for neutrinoless double-β decay," *Phys. Rev. Lett.*, vol. 125, p. 252502, 2020.

[470] A. Gando *et al.*, "Search for Majorana neutrinos near the inverted mass hierarchy region with KamLAND-Zen," *Phys. Rev. Lett.*, vol. 117, no. 8, p. 082503, 2016. [Addendum: *Phys. Rev. Lett.* 117, 109903 (2016)].

[471] G. Anton *et al.*, "Search for neutrinoless double-β decay with the complete EXO-200 dataset," *Phys. Rev. Lett.*, vol. 123, no. 16, p. 161802, 2019.

[472] D. Adams *et al.*, "Improved limit on neutrinoless double-beta decay in ^{130}Te with CUORE," *Phys. Rev. Lett.*, vol. 124, no. 12, p. 122501, 2020.

[473] O. Azzolini *et al.*, "Final result of CUPID-0 phase-I in the search for the ^{82}Se Neutrinoless Double-β Decay," *Phys. Rev. Lett.*, vol. 123, no. 3, p. 032501, 2019.

[474] A. Giuliani, J. Gomez Cadenas, S. Pascoli, E. Previtali, R. Saakyan, K. Schäffner, and S. Schönert, "Double Beta Decay APPEC Committee Report," 10 2019.

[475] M. Fukugita and T. Yanagida, "Baryogenesis without grand unification," *Phys. Lett. B*, vol. 174, pp. 45–47, 1986.

[476] S. Davidson, E. Nardi, and Y. Nir, "Leptogenesis," *Phys. Rept.*, vol. 466, pp. 105–177, 2008.

[477] F. M. Renard, "Limits on masses and couplings of excited electrons and muons," *Phys. Lett. B*, vol. 116, pp. 264–268, 1982.

[478] K. Hagiwara, D. Zeppenfeld, and S. Komamiya, "Excited lepton production at LEP and HERA," *Z. Phys. C*, vol. 29, p. 115, 1985.

[479] M. B. Green, J. H. Schwarz, and E. Witten, *Superstring Theory Vol. 1: 25th Anniversary Edition.* Cambridge Monographs on Mathematical Physics, Cambridge University Press, 11 2012.

[480] C. Rovelli and F. Vidotto, "Planck stars," *Int. J. Mod. Phys. D*, vol. 23, no. 12, p. 1442026, 2014.

[481] I. Nikitin, "On dark stars, Planck cores and the nature of dark matter," *Bled Workshops Phys.*, vol. 21, pp. 221–246, 2020.

[482] F. James, "Monte-Carlo phase space (CERN-68-15)," tech. rep., 1968.

Index